BASIC ELECTRICAL ENGINEERING

For B.E./B.Tech. and Other Engineering Examinations

V.K. MEHTA

ROHIT MEHTA

S. CHAND
PUBLISHING

S Chand And Company Limited
(ISO 9001 Certified Company)

S Chand And Company Limited

(ISO 9001 Certified Company)

Head Office: D-92, Sector–2, Noida – 201301, U.P. (India), Ph. 91-120-4682700

Registered Office: A-27, 2nd Floor, Mohan Co-operative Industrial Estate, New Delhi – 110 044, Phone: 011-49731800

www.**schandpublishing**.com; e-mail: info@**schandpublishing**.com

Marketing Offices:

Chennai	:	Ph: 23632120; chennai@schandpublishing.com
Guwahati	:	Ph: 2738811, 2735640; guwahati@schandpublishing.com
Hyderabad	:	Ph: 40186018; hyderabad@schandpublishing.com
Jalandhar	:	Ph: 4645630; jalandhar@schandpublishing.com
Kolkata	:	Ph: 23357458, 23353914; kolkata@schandpublishing.com
Lucknow	:	Ph: 4003633; lucknow@schandpublishing.com
Mumbai	:	Ph: 25000297; mumbai@schandpublishing.com
Patna	:	Ph: 2260011; patna@schandpublishing.com

S. CHAND'S *Seal of Trust*

In our endeavour to protect you against counterfeit/fake books, we have pasted a hologram over the cover of this book. The hologram displays full visible effect, emboss effect, relief effect, mirror lens effect, pearl effect, motion effect, animated text, kinetic effect, concealed effect, micro structure, multicolour small text 'S.CHAND', nanotext '50 micron' 'ORIGINAL' 'S.CHAND', mirror strip '6.5 mm', mirror lens 3 mm with text 'SC', microtext 'OK', scratch strip '7 mm', color sparkling effect under scratch and QR code size 10 mm x 10 mm, etc.

A fake hologram does not display ALL these effects.

First Edition 1988
Subsequent Editions and Reprints 1991, 95, 97, 98, 2001, 2006, 2007, 2008 (Thrice), 2009 (Twice), 2010 (Twice); 2011
Revised Edition 2012
Reprints 2016 (Thrice), 2018, 2019, 2020 (Twice), 2021 (Thrice), 2022

Reprint 2023

ISBN: 978-81-219-0871-9   **Product Code:** H3BEE52BEEL10ENAF12O

PRINTED IN INDIA

By Vikas Publishing House Private Limited, Plot 20/4, Site-IV, Industrial Area Sahibabad, Ghaziabad – 201 010 and Published by S Chand And Company Limited, A-27, 2nd Floor, Mohan Co-operative Industrial Estate, New Delhi – 110 044.

Preface to Sixth Edition

The general response to the Fifth Edition of the book was very encouraging. Authors feel that their work has been amply rewarded and wish to express their deep sense of gratitude to the large number of readers who have used it and in particular to those of them who have sent helpful suggestions from time to time for the improvement of the book.

The popularity of the book is judged from the fact that authors frequently receive feedback from many quarters including teachers, students and serving engineers. This feedback helps the authors to make the book up-to-date. In the present edition, many new topics/numericals/illustrations have been added to make the book more useful.

Authors lay no claim to the original research in preparing the book. Liberal use of materials available in the works of eminent authors has been made. What they may claim, in all modesty, is that they have tried to fashion the vast amount of material available from primary and secondary sources into coherent body of description and analysis.

The authors wish to thank their colleagues and friends who have contributed many valuable suggestions regarding the scope and content sequence of the book. Authors are also indebted to S. Chand & Company Ltd., New Delhi for bringing out this revised edition in a short time and pricing the book moderately inspite of heavy cost of paper and printing.

Errors might have crept in despite utmost care to avoid them. Authors shall be grateful if these are pointed out along with other suggestions for the improvement of the book.

V.K. MEHTA

ROHIT MEHTA

Contents

Superposition Principle—Electric Field—Properties of Electric Lines of Force—Electric Intensity or Field Strength (E)—Electric Flux (ψ)—Electric Flux Density (D)—Gauss's Theorem—Proof of Gauss's Law—Electric Potential Energy—Electric Potential—Electric Potential Difference—Potential at a Point Due to a Point Charge—Potential at a Point Due to Group of Point Charges—Behaviour of Metallic Conductors in Electric Field—Potential of a Charged Conducting Sphere—Potential Gradient—Breakdown Voltage or Dielectric Strength—Uses of Dielectrics—Refraction of Electric Flux—Equipotential Surface—Motion of a Charged Particle in Uniform Electric Field—Objective Questions.

Magnetic Energy Stored Per Unit Volume—Lifting Power of a Magnet—Closing and Breaking an Inductive Circuit—Rise of Current in an Inductive Circuit—Time Constant—Decay of Current in an Inductive Circuit—Eddy Current Loss—Formula for Eddy Current Power Loss—Objective Questions.

1

Basic Concepts

Introduction

Everybody is familiar with the functions that electricity can perform. It can be used for lighting, heating, traction and countless other purposes. The question always arises, "What is electricity" ? Several theories about electricity were developed through experiments and by observation of its behaviour. The only theory that has survived over the years to explain the nature of electricity is the *Modern Electron theory of matter*. This theory has been the result of research work conducted by scientists like Sir William Crooks, J.J. Thomson, Robert A. Millikan, Sir Earnest Rutherford and Neils Bohr. In this chapter, we shall deal with some basic concepts concerning electricity.

1.1. Nature of Electricity

We know that matter is electrical in nature *i.e.* it contains particles of electricity *viz.* protons and electrons. The positive charge on a proton is equal to the negative charge on an electron. Whether a given body exhibits electricity (*i.e.* charge) or not depends upon the relative number of these particles of electricity.

(*i*) If the number of protons is equal to the number of electrons in a body, the resultant charge is zero and the body will be electrically neutral. Thus, the paper of this book is electrically neutral (*i.e.* paper exhibits no charge) because it has the same number of protons and electrons.

(*ii*) If from a neutral body, some *electrons are removed, there occurs a deficit of electrons in the body. Consequently, the body attains a *positive charge*.

(*iii*) If a neutral body is supplied with electrons, there occurs an excess of electrons. Consequently, the body attains a *negative charge*.

1.2. Unit of Charge

The charge on an electron is so small that it is not convenient to select it as the unit of charge. In practice, *coulomb* is used as the unit of charge *i.e.* SI unit of charge is coulomb abbreviated as C. *One coulomb of charge is equal to the charge on 625×10^{16} electrons, i.e.*

$$1 \text{ coulomb} = \text{Charge on } 625 \times 10^{16} \text{ electrons}$$

Thus when we say that a body has a positive charge of one coulomb (*i.e.* +1 C), it means that the body has a deficit of 625×10^{16} electrons from normal due share. The charge on one electron is given by ;

$$\text{Charge on electron} = -\frac{1}{625 \times 10^{16}} = -1.6 \times 10^{-19} \text{ C}$$

1.3. The Electron

Since electrical engineering generally deals with tiny particles called electrons, these small particles require detailed study. We know that an electron is a negatively charged particle having negligible mass. Some of the important properties of an electron are :

(*i*) Charge on an electron, $e = 1.602 \times 10^{-19}$ coulomb

(*ii*) Mass of an electron, $m = 9.0 \times 10^{-31}$ kg

(*iii*) Radius of an electron, $r = 1.9 \times 10^{-15}$ metre

---

*   Electrons have very small mass and, therefore, are much more mobile than protons. On the other hand, protons are powerfully held in the nucleus and cannot be removed or detached.

The ratio e/m of an electron is 1.77×10^{11} coulombs/kg. This means that mass of an electron is very small as compared to its charge. It is due to this property of an electron that it is very mobile and is greatly influenced by electric or magnetic fields.

1.4. Energy of an Electron

An electron moving around the nucleus possesses two types of energies *viz.* kinetic energy due to its motion and potential energy due to the charge on the nucleus. The total energy of the electron is the sum of these two energies. The energy of an electron increases as its distance from the nucleus increases. Thus, an electron in the second orbit possesses more energy than the electron in the first orbit ; electron in the third orbit has higher energy than in the second orbit. It is clear that electrons in the last orbit possess very high energy as compared to the electrons in the inner orbits. These last orbit electrons play an important role in determining the physical, chemical and electrical properties of a material.

1.5. Valence Electrons

The electrons in the outermost orbit of an atom are known as **valence electrons**.

The outermost orbit can have a maximum of 8 electrons *i.e.* the maximum number of valence electrons can be 8. The valence electrons determine the physical and chemical properties of a material. These electrons determine whether or not the material is chemically active; metal or non-metal or, a gas or solid. These electrons also determine the electrical properties of a material.

On the basis of electrical conductivity, materials are generally classified into *conductors, insulators* and *semi-conductors*. As a rough rule, one can determine the electrical behaviour of a material from the number of valence electrons as under :

(*i*) *When the number of valence electrons of an atom is less than 4* (*i.e.* half of the maximum eight electrons), the material is usually *a metal and a conductor*. Examples are sodium, magnesium and aluminium which have 1, 2 and 3 valence electrons respectively.

(*ii*) *When the number of valence electrons of an atom is more than 4*, the material is usually *a non-metal and an insulator*. Examples are nitrogen, sulphur and neon which have 5, 6 and 8 valence electrons respectively.

(*iii*) *When the number of valence electrons of an atom is 4* (*i.e.* exactly one-half of the maximum 8 electrons), the material has both metal and non-metal properties and is usually a *semi-conductor*. Examples are carbon, silicon and germanium.

1.6. Free Electrons

We know that electrons move around the nucleus of an atom in different orbits. The electrons in the inner orbits (*i.e.*, orbits close to the nucleus) are tightly bound to the nucleus. As we move away from the nucleus, this binding goes on decreasing so that electrons in the last orbit (called valence electrons) are quite loosely bound to the nucleus. In certain substances, especially metals (*e.g.* copper, aluminium etc.), the valence electrons are so weakly attached to their nuclei that they can be easily removed or detached. Such electrons are called free electrons.

Those valence electrons which are very loosely attached to the nucleus of an atom are called **free electrons.**

The free electrons move at random from one atom to another in the material. Infact, they are so loosely attached that they do not know the atom to which they belong. It may be noted here that all valence electrons in a metal are not free electrons. It has been found that one atom of a metal can provide at the most one free electron. Since a small piece of metal has billions of atoms, one can expect a very large number of free electrons in metals. For instance, one cubic centimetre of copper has about 8.5×10^{22} free electrons at room temperature.

(*i*) A substance which has a large number of free electrons at room temperature is called a **conductor** of electricity *e.g.* all metals. If a voltage source (*e.g.* a cell) is applied across the wire of a conductor material, free electrons readily flow through the wire, thus constituting electric current. The best conductors are silver, copper and gold in that order. Since copper is the least expensive out of these materials, it is widely used in electrical and electronic industries.

(*ii*) A substance which has very few free electrons is called an **insulator** of electricity. If a voltage source is applied across the wire of insulator material, practically no current flows through the wire. Most substances including plastics, ceramics, rubber, paper and most liquids and gases fall in this category. Of course, there are many practical uses for insulators in the electrical and electronic industries including wire coatings, safety enclosures and power-line insulators.

(*iii*) There is a third class of substances, called **semi-conductors**. As their name implies, they are neither conductors nor insulators. These substances have crystalline structure and contain very few free electrons at room temperature. Therefore, at room temperature, a semiconductor practically behaves as an insulator. However, if suitable controlled impurity is imparted to a semi-conductor, it is possible to provide controlled conductivity. Most common semi-conductors are silicon, germanium, carbon etc. However, *silicon* is the principal material and is widely used in the manufacture of electronic devices (*e.g.* crystal diodes, transistors etc.) and integrated circuits.

1.7. Electric Current

The directed flow of free electrons (or charge) is called **electric current**. The flow of electric current can be beautifully explained by referring to Fig. 1.1. The copper strip has a large number of free electrons. When electric pressure or voltage is applied, then free electrons, being negatively charged, will start moving towards the positive terminal around the circuit as shown in Fig. 1.1. This directed flow of electrons is called electric current.

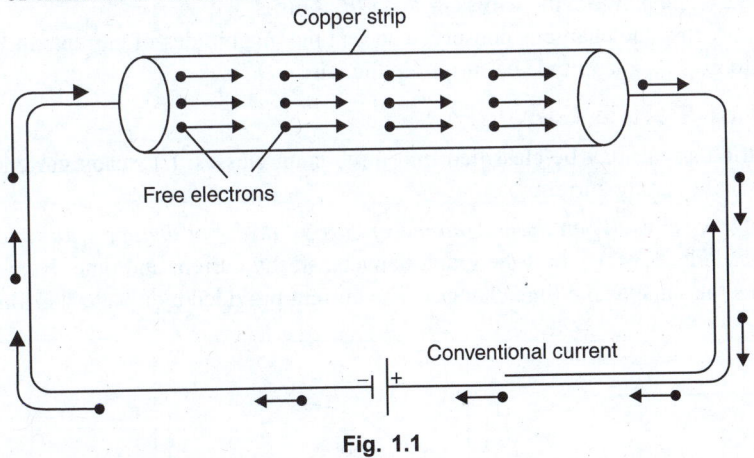

Fig. 1.1

The reader may note the following points :

(*i*) Current is flow of electrons and electrons are the constituents of matter. Therefore, electric current is matter (*i.e.* free electrons) in motion.

(*ii*) The actual direction of current (*i.e.* flow of electrons) is from negative terminal to the positive terminal through that part of the circuit external to the cell. However, prior to Electron theory, it was assumed that current flowed from positive terminal to the negative terminal of the cell *via* the circuit. This convention is so firmly established that it is still in use. This assumed direction of current is now called *conventional current*.

Unit of Current. The strength of electric current I is the rate of flow of electrons *i.e.* charge flowing per second.

$$\therefore \qquad \text{Current, } I = \frac{Q}{t}$$

The charge Q is measured in coulombs and time t in seconds. Therefore, the unit of electric current will be *coulombs/sec or ampere*. If $Q = 1$ coulomb, $t = 1$ sec, then $I = 1/1 = 1$ ampere.

One ampere *of current is said to flow through a wire if at any cross-section one coulomb of charge flows in one second.*

Thus, if 5 amperes current is flowing through a wire, it means that 5 coulombs per second flow past any cross-section of the wire.

Note. 1 C $=$ charge on 625×10^{16} electrons. Thus when we say that current through a wire is 1 A, it means that 625×10^{16} electrons per second flow past any cross-section of the wire.

$$\therefore \qquad I = \frac{Q}{t} = \frac{ne}{t} \qquad \text{where } e = -1.6 \times 10^{-19}\ \text{C}\ ; \ n = \text{number of electrons}$$

1.8. Electric Current is a Scalar Quantity

(*i*)   Electric current, $I = \dfrac{Q}{t}$

As both charge and time are scalars, electric current is a scalar quantity.

(*ii*)   We show electric current in a wire by an arrow to indicate the direction of flow of positive charge. But such arrows are not vectors because they do not obey the laws of vector algebra. This point can be explained by referring to Fig. 1.2. The wires OA and OB carry currents of 3 A and 4 A respectively. The total current in the wire CO is $3 + 4 = 7$ A irrespective of the angle between the wires OA and OB. This is

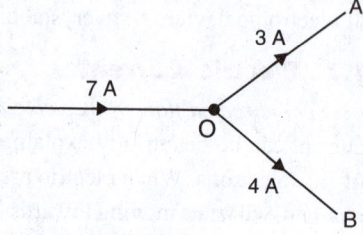

Fig. 1.2

not surprising because the charge is conserved so that the magnitudes of currents in wires OA and OB must add to give the magnitude of current in the wire CO.

1.9. Types of Electric Current

The electric current may be classified into three main classes: (*i*) steady current (*ii*) varying current and (*iii*) alternating current.

(*i*) **Steady current.** *When the magnitude of current does not change with time, it is called a steady current.* Fig. 1.3 (*i*) shows the graph between steady current and time. Note that value of current remains the same as the time changes. The current provided by a battery is almost a steady current (*d.c.*).

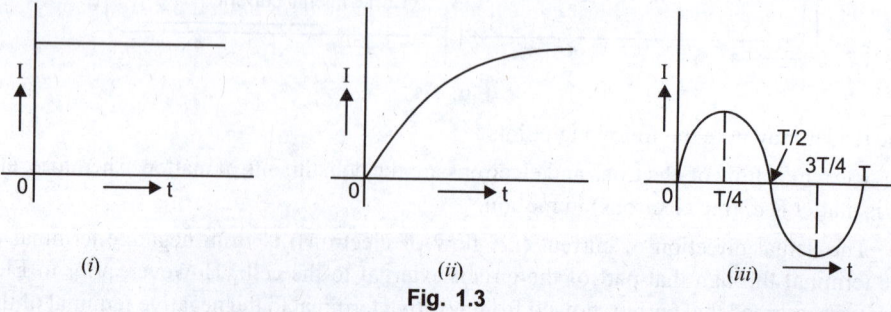

(*i*)   (*ii*)   (*iii*)

Fig. 1.3

(ii) **Varying current.** *When the magnitude of current changes with time, it is called a varying current.* Fig. 1.3 (*ii*) shows the graph between varying current and time. Note that value of current varies with time.

(iii) **Alternating current.** *An alternating current is one whose magnitude changes continuously with time and direction changes periodically.* Due to technical and economical reasons, we produce alternating currents that have sine waveform (or cosine waveform) as shown in Fig. 1.3 (*iii*). It is called *alternating current* because current flows in alternate directions in the circuit, *i.e.,* from 0 to $T/2$ second (T is the time period of the wave) in one direction and from $T/2$ to T second in the opposite direction. The current provided by an a.c. generator is alternating current that has sine (or cosine) waveform.

1.10. Mechanism of Current Conduction in Metals

Every metal has a large number of free electrons which wander randomly within the body of the conductor somewhat like the molecules in a gas. The average speed of free electrons is sufficiently high ($\simeq 10^5$ ms^{-1}) at room temperature. During random motion, the free electrons collide with positive ions (positive atoms of metal) again and again and after each collision, their direction of motion changes. When we consider all the free electrons, their random motions average to zero. In other words, there is no net flow of charge (electrons) in any particular direction. Consequently, no current is established in the conductor.

When potential difference is applied across the ends of a conductor (say copper wire) as shown in Fig. 1.4, electric field is applied at every point of the copper wire. The electric field exerts force on the free electrons which start accelerating towards the positive terminal (*i.e.,* opposite to the direction of the field). As the free electrons move, they *collide again and again with positive ions of the metal. Each collision destroys the extra velocity gained by the free electrons.

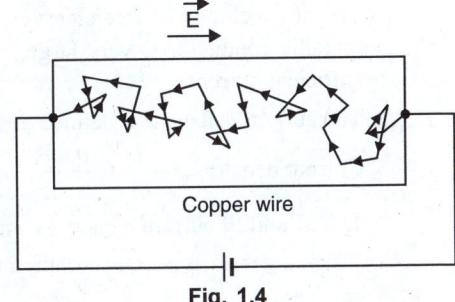

Copper wire

Fig. 1.4

The average time that an electron spends between two collisions is called the **relaxation time** (τ). Its value is of the order of 10^{-14} second.

Although the free electrons are continuously accelerated by the electric field, collisions prevent their velocity from becoming large. The result is that electric field provides a small constant velocity towards positive terminal which is superimposed on the random motion of the electrons. This constant velocity is called the drift velocity.

The average velocity with which free electrons get drifted in a metallic conductor under the influence of electric field is called **drift velocity** $(\vec{v}_d)$. The drift velocity of free electrons is of the order of 10^{-5} ms^{-1}.

Thus when a metallic conductor is subjected to electric field (or potential difference), free electrons move towards the positive terminal of the source with drift velocity $\vec{v}_d$. Small though it is, the drift velocity is entirely responsible for electric current in the metal.

Note. The reader may wonder that if electrons drift so slowly, how room light turns on quickly when switch is closed ? The answer is that propagation of electric field takes place with the speed of light. When we apply electric field (*i.e.,* potential difference) to a wire, the free electrons everywhere in the wire begin drifting almost at once.

---

*   What happens to an electron after collision with an ion ? It moves off in some new and quite random direction. However, it still experiences the applied electric field, so it continues to accelerate again, gaining a velocity in the direction of the positive terminal. It again encounters an ion and loses its directed motion. This situation is repeated again and again for every free electron in a metal.

1.11. Relation Between Current and Drift Velocity

Consider a portion of a copper wire through which current I is flowing as shown in Fig. 1.5. Clearly, copper wire is under the influence of electric field.

Let A = area of X-section of the wire

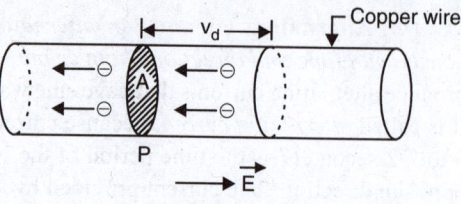

   n = electron density, *i.e.*, number of free electrons per unit volume

   e = charge on each electron

   v_d = drift velocity of free electrons

In one second, all those free electrons within a distance v_d to the right of cross-section at P (*i.e.*, in a volume Av_d) will flow through the cross-section at P as shown in Fig. 1.5. This volume contains n Av_d electrons and, hence, a charge $(nAv_d)e$. Therefore, a charge of $neAv_d$ per second passes the cross-section at P.

Fig. 1.5

∴ $I = n e A v_d$

Since A, n and e are constant, $I \propto v_d$.

Hence, current flowing through a conductor is directly proportional to the drift velocity of free electrons.

(*i*) The drift velocity of free electrons is very small. Since the number of free electrons in a metallic conductor is very large, even small drift velocity of free electrons gives rise to sufficient current.

(*ii*) The current density J is defined as current per unit area and is given by ;

Current density, $J = \dfrac{I}{A} = \dfrac{n e Av_d}{A} = ne\, v_d$

The SI unit of current density is amperes/m^2.

Note. Current density is a vector quantity and is denoted by the symbol $\vec{J}$. Therefore, in vector notation, the relation between I and $\vec{J}$ is $I = \vec{J}.\vec{A}$

where $\vec{A}$ = Area vector

Example 1.1. *A 60 W light bulb has a current of 0.5 A flowing through it. Calculate (i) the number of electrons passing through a cross-section of the filament (ii) the number of electrons that pass the cross-section in one hour.*

Solution. (*i*) $I = \dfrac{Q}{t} = \dfrac{ne}{t}$

∴ $n = \dfrac{It}{e} = \dfrac{0.5 \times 1}{1.6 \times 10^{-19}} =$ **3.1 × 10^{18} electrons/s**

(*ii*) Charge passing the cross-section in one hour is

$Q = It = (0.5) \times (60 \times 60) = 1800$ C

Now, $Q = ne$

∴ $n = \dfrac{Q}{e} = \dfrac{1800}{1.6 \times 10^{-19}} =$ **1.1 × 10^{22} electrons/hour**

Example 1.2. *A copper wire of area of X-section 4 mm^2 is 4 m long and carries a current of 10 A. The number density of free electrons is 8 × 10^{28} m^{-3}. How much time is required by an electron to travel the length of wire ?*

Solution. $I = n A e v_d$

Here $I = 10$ A ; $A = 4$ mm$^2 = 4 \times 10^{-6}$ m^2 ; $e = 1.6 \times 10^{-19}$ C ; $n = 8 \times 10^{28}$ m^{-3}

$\therefore$ Drift velocity, $v_d = \dfrac{I}{nAe} = \dfrac{10}{8 \times 10^{28} \times (4 \times 10^{-6}) \times 1.6 \times 10^{-19}} = 1.95 \times 10^{-4}$ ms^{-1}

$\therefore$ Time taken by the electron to travel the length of the wire is

$$t = \frac{l}{v_d} = \frac{4}{1.95 \times 10^{-4}} = 2.05 \times 10^4 \text{s} = \textbf{5.7 hours}$$

Example 1.3. *The area of X-section of copper wire is 3×10^{-6} m^2. It carries a current of 4.2 A. Calculate (i) current density in the wire and (ii) the drift velocity of electrons. The number density of conduction electrons is 8.4×10^{28} m^{-3}.*

Solution. (*i*) Current density, $J = \dfrac{I}{A} = \dfrac{4.2}{3 \times 10^{-6}} = \textbf{1.4} \times \textbf{10}^6 \textbf{A/m}^2$

(*ii*) $I = neAv_d$

$\therefore$ Drift velocity, $v_d = \dfrac{I}{nAe} = \dfrac{4.2}{(8.4 \times 10^{28}) \times (1.6 \times 10^{-19}) \times 3 \times 10^{-6}} = \textbf{1.04} \times \textbf{10}^{-4} \textbf{ms}^{-1}$

$\boxed{\text{\textbf{Tutorial Problems}}}$

1. How much current is flowing in a circuit where 1.27×10^{15} electrons move past a given point in 100 ms ?   **[2.03 A]**

2. The current in a certain conductor is 40 mA.

    (*i*) Find the total charge in coulombs that passes through the conductor in 1.5 s.

    (*ii*) Find the total number of electrons that pass through the conductor in that time.

                                                                 [(*i*) 60 mC (*ii*) 3.745 $\times$ 10^{17} electrons]

3. The density of conduction electrons in a wire is 10^{22} m^{-3}. If the radius of the wire is 0.6 mm and it is carrying a current of 2 A, what will be the average drift velocity ?   **[1.1 $\times$ 10^{-3} ms^{-1}]**

4. Find the velocity of charge leading to 1 A current which flows in a copper conductor of cross-section 1 cm^2 and length 10 km. Free electron density of copper = 8.5×10^{28} per m^3. How long will it take the electric charge to travel from one end of the conductor to the other ?   **[0.735 μm/s; 431 years]**

1.12. Electric Potential

When a body is charged, work is done in charging it. This work done is stored in the body in the form of potential energy. The charged body has the capacity to do work by moving other charges either by attraction or repulsion. The ability of the charged body to do work is called electric potential.

The capacity of a charged body to do work is called its **electric potential.**

The greater the capacity of a charged body to do work, the greater is its electric potential. Obviously, the work done to charge a body to 1 coulomb will be a measure of its electric potential *i.e.*

Electric potential, $V = \dfrac{\text{Work done}}{\text{Charge}} = \dfrac{W}{Q}$

The work done is measured in joules and charge in coulombs. Therefore, the unit of electric potential will be *joules/coulomb* or *volt*. If $W = 1$ joule, $Q = 1$ coulomb, then $V = 1/1 = 1$ volt.

Hence a body is said to have an electric potential of **1 volt** *if 1 joule of work is done to give it a charge of 1 coulomb.*

Thus, when we say that a body has an electric potential of 5 volts, it means that 5 joules of work has been done to charge the body to 1 coulomb. In other words, every coulomb of charge possesses an energy of 5 joules. The greater the joules/coulomb on a charged body, the greater is its electric potential.

1.13. Potential Difference

The difference in the potentials of two charged bodies is called **potential difference.**

If two bodies have different electric potentials, a potential difference exists between the bodies. Consider two bodies *A* and *B* having potentials of 5 volts and 3 volts respectively as shown in Fig. 1.6 (*i*). Each coulomb of charge on body *A* has an energy of 5 joules while each coulomb of charge on body *B* has an energy of 3 joules. Clearly, body *A* is at higher potential than the body *B*.

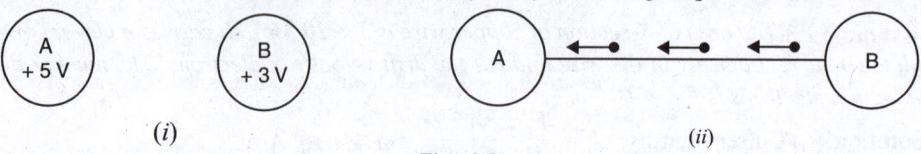

| (*i*) | (*ii*) |

Fig. 1.6

If the two bodies are joined through a conductor [See Fig. 1.6 (*ii*)], then electrons will *flow from body *B* to body *A*. When the two bodies attain the same potential, the flow of current stops. Therefore, we arrive at a very important conclusion that current will flow in a circuit if potential difference exists. No potential difference, no current flow. It may be noted that potential difference is sometimes called voltage.

Unit. Since the unit of electric potential is volt, one can expect that unit of potential difference will also be *volt*. It is defined as under :

The potential difference between two points is **1 volt** *if one joule of work is* **done or *released in transferring 1 coulomb of charge from one point to the other.*

1.14. Maintaining Potential Difference

A device that maintains potential difference between two points is said to develop electromotive force (e.m.f.). A simple example is that of a cell. Fig. 1.7 shows the familiar voltaic cell. It consists of a copper plate (called anode) and a zinc rod (called cathode) immersed in dilute H_2SO_4.

The chemical action taking place in the cell removes electrons from copper plate and transfers them to the zinc rod. This transference of electrons takes place through the agency of dil. H_2SO_4 (called electrolyte). Consequently, the copper plate attains a positive charge of $+Q$ coulombs and zinc rod a charge of $-Q$ coulombs. The chemical action of the cell has done a certain amount of work (say W joules) to do so. Clearly, the potential difference between the two plates will be W/Q volts. If the two plates are joined through a wire, some electrons from zinc rod will be attracted through the wire to copper plate. The chemical action of the cell now transfers an equal amount of electrons from copper plate to zinc rod internally through the cell to maintain original potential difference (*i.e.* W/Q). This process continues so long as the

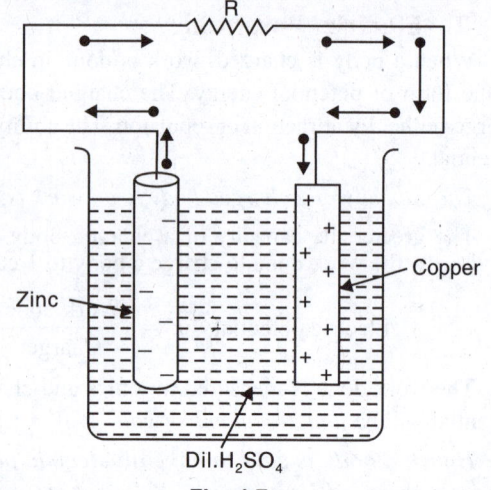

Fig. 1.7

---

*   The conventional current flow will be in the opposite direction *i.e.* from body *A* to body *B*.

**   1 joule of work will be done in the case if 1 coulomb is transferred from point of lower potential to that of higher potential. However, 1 joule of work will be released (as heat) if 1 coulomb of charge moves from a point of higher potential to a point of lower potential.

circuit is complete or so long as there is chemical energy. The flow of electrons through the external wire from zinc rod to copper plate is the electric current.

Thus potential difference causes current to flow while an *e.m.f.* maintains the potential difference. Although both *e.m.f.* and *p.d.* are measured in volts, they do not mean exactly the same thing.

1.15. Concept of E.M.F. and Potential Difference

There is a distinct difference between *e.m.f.* and potential difference. The *e.m.f.* of a device, say a battery, is a measure of the energy the battery gives to each coulomb of charge. Thus if a battery supplies 4 joules of energy per coulomb, we say that it has an *e.m.f.* of 4 volts. The energy given to each coulomb in a battery is due to the chemical action.

The potential difference between two points, say *A* and *B*, is a measure of the energy used by one coulomb in moving from *A* to *B*. Thus if potential difference between points *A* and *B* is 2 volts, it means that each coulomb will give up an energy of 2 joules in moving from *A* to *B*.

Illustration. The difference between e.m.f. and p.d. can be made more illustrative by referring to Fig. 1.8. Here battery has an e.m.f. of 4 volts. It means battery supplies 4 joules of energy to each coulomb continuously. As each coulomb travels from the positive terminal of the battery, it gives up its most of energy to resistances (2 Ω and 2 Ω in this case) and remaining to connecting wires. When it returns to the negative terminal, it has lost all its energy originally supplied by the battery. The battery now supplies fresh energy to each coulomb (4 joules in the present case) to start the journey once again.

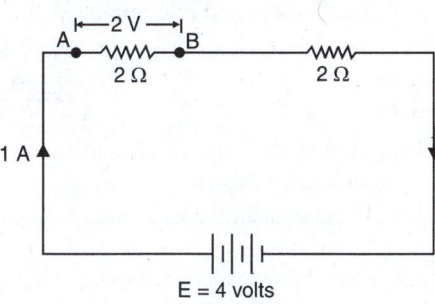

Fig. 1.8

The *p.d.* between any two points in the circuit is the energy used by one coulomb in moving from one point to another. Thus in Fig. 1.8, p.d. between *A* and *B* is 2 volts. It means that 1 coulomb will give up an energy of 2 joules in moving from *A* to *B*. This energy will be released as heat from the part *AB* of the circuit.

The following points may be noted carefully :

(i) The name e.m.f. at first sight implies that it is a force that causes current to flow. This is not correct because it is not a force but energy supplied to charge by some active device such as a battery.

(ii) Electromotive force (e.m.f.) maintains potential difference while p.d. causes current to flow.

1.16. Potential Rise and Potential Drop

Fig. 1.9 shows a circuit with a cell and a resistor. The cell provides a potential difference of 1.5 V. Since it is an energy source, there is a *rise* in potential associated with a cell. The cell's potential difference represents an e.m.f. so that symbol *E* could be used. The resistor is also associated with a potential difference. Since it is a consumer (converter) of energy, there is a *drop* in potential across the resistor. We can combine the idea of potential rise or drop with the popular term "voltage". It is customary to refer to the potential difference across the cell as a *voltage rise* and to the potential difference across the resistor as a *voltage drop*.

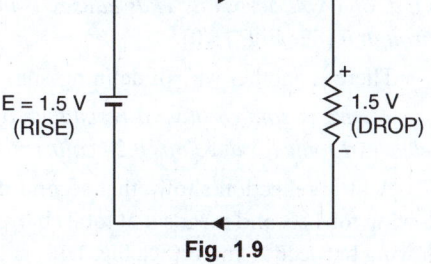

Fig. 1.9

Note. The term voltage refers to a potential difference across two points. There is no such thing as a voltage at one point. In cases where a single point is specified, some reference must be used as the other point. Unless stated otherwise, the ground or common point in any circuit is the reference when specifying a voltage at some other point.

Example 1.4. *A charge of 4 coulombs is flowing between points A and B of a circuit. If the potential difference between A and B is 2 volts, how many joules will be released by part AB of the circuit ?*

Solution. The p.d. of 2 volts between points *A* and *B* means that each coulomb of charge will give up an energy of 2 joules in moving from *A* to *B*. As the charge flowing is 4 coulombs, therefore, total energy released by part *AB* of the circuit is = 4 × 2 = **8 joules.**

Example 1.5. *How much work will be done by an electric energy source with a potential difference of 3 kV that delivers a current of 1 A for 1 minute ?*

Solution. We know that 1 A of current represents a charge transfer rate of 1 C/s. Therefore, the total charge for a period of 1 minute is $Q = It = 1 \times 60 = 60$ C.

$$\text{Total work done, } W = Q \times V = 60 \times (3 \times 10^3) = 180 \times 10^3 \text{ J} = \textbf{180 kJ}$$

Tutorial Problems

1. Calculate the potential difference of an energy source that provides 6.8 J for every milli-coulomb of charge that it delivers.   **[6.8 kV]**

2. The potential difference across a battery is 9 V. How much charge must it deliver to do 50 J of work ?   **[5.56 C]**

3. A 300 V energy source delivers 500 mA for 1 hour. How much energy does this represent ?   **[540 kJ]**

1.17. Resistance

The opposition offered by a substance to the flow of electric current is called its **resistance.**

Since current is the flow of free electrons, resistance is the opposition offered by the substance to the flow of free electrons. This opposition occurs because atoms and molecules of the substance obstruct the flow of these electrons. Certain substances (*e.g.* metals such as silver, copper, aluminium etc.) offer very little opposition to the flow of electric current and are called conductors. On the other hand, those substances which offer high opposition to the flow of electric current (*i.e.* flow of free electrons) are called insulators *e.g.* glass, rubber, mica, dry wood etc.

It may be noted here that resistance is the electric friction offered by the substance and causes production of heat with the flow of electric current. The moving electrons collide with atoms or molecules of the substance ; each collision resulting in the liberation of minute quantity of heat.

Unit of resistance. The practical unit of resistance is ohm and is represented by the symbol Ω. It is defined as under :

A wire is said to have a resistance of **1 ohm** *if a p.d. of 1 volt across its ends causes 1 ampere to flow through it* (See Fig. 1.10).

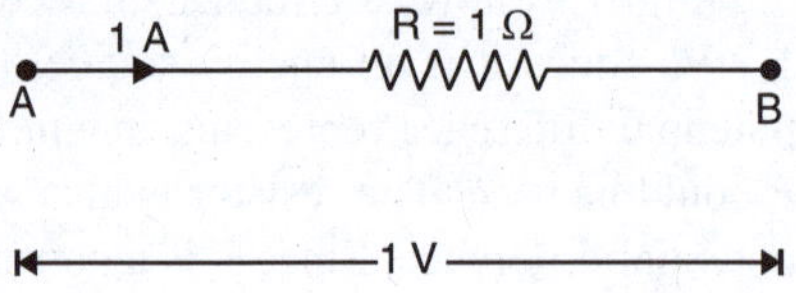

There is another way of defining ohm.

Fig. 1.10

A wire is said to have a resistance of **1 ohm** *if it releases 1 joule (or develops 0.24 calorie of heat) when a current of 1 A flows through it for 1 second.*

A little reflection shows that second definition leads to the first definition. Thus 1 A current flowing for 1 second means that total charge flowing is $Q = I \times t = 1 \times 1 = 1$ coulomb. Now the charge flowing between *A* and *B* (See Fig. 1.10) is 1 coulomb and energy released is 1 joule (or 0.24 calorie). Obviously, by definition, p.d. between *A* and *B* should be 1 volt.

1.18. Factors Upon Which Resistance Depends

The resistance R of a conductor

(**i**) is directly proportional to its length *i.e.*
$$R \propto l$$

(**ii**) is inversely proportional to its area of X-section *i.e.*
$$R \propto \frac{1}{a}$$

(**iii**) depends upon the nature of material.

(**iv**) depends upon temperature.

From the first three points (leaving temperature for the time being), we have,
$$R \propto \frac{l}{a} \quad \text{or} \quad R = \rho \frac{l}{a}$$

where ρ (Greek letter 'Rho') is a constant and is known as *resistivity* or *specific resistance* of the material. Its value depends upon the nature of the material.

1.19. Specific Resistance or Resistivity

We have seen above that $\quad R = \rho \dfrac{l}{a}$

If $\ l = 1$ m, $a = 1$ m^2, then, $\ R = \rho$

Hence **specific resistance** *of a material is the resistance offered by 1 m length of wire of material having an area of cross-section of 1 m^2* [See Fig. 1.11 (*i*)].

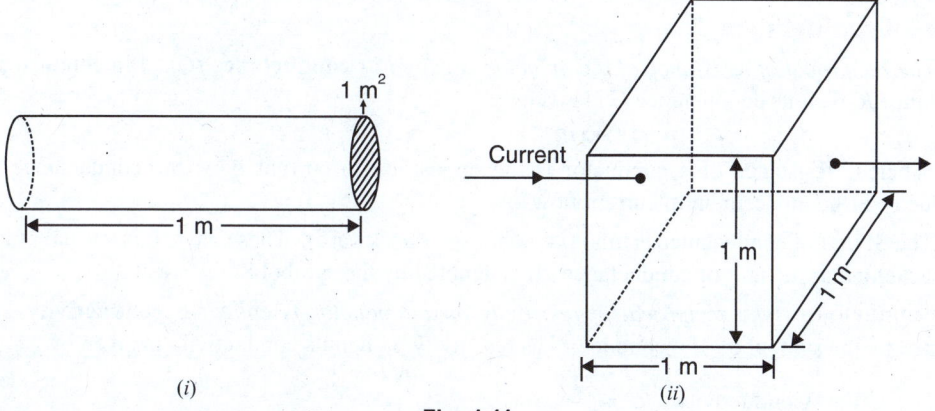

1 m^2

Current

1 m

1 m

1 m

1 m

(*i*)   (*ii*)

Fig. 1.11

Specific resistance can also be defined in another way. Take a cube of the material having each side 1 m. Considering any two opposite faces, the area of cross-section is 1 m^2 and length is 1 m [See Fig. 1.11 (*ii*)] *i.e.* $l = 1$ m, $a = 1$ m^2.

Hence **specific resistance** *of a material may be defined as the resistance between the opposite faces of a metre cube of the material.*

Unit of resistivity. We know $\quad R = \dfrac{\rho l}{a} \quad$ or $\quad \rho = \dfrac{R\,a}{l}$

Hence the unit of resistivity will depend upon the units of area of cross-section (a) and length (l).

(**i**)   If the length is measured in metres and area of cross-section in m^2, then unit of resistivity will be ohm-metre (Ω m).
$$\rho = \frac{\text{ohm} \times \text{m}^2}{\text{m}} = \text{ohm-m}$$

(*ii*) If length is measured in cm and area of cross-section in cm^2, then unit of resistivity will be ohm-cm (Ω cm).

$$\rho = \frac{\text{ohm} \times \text{cm}^2}{\text{cm}} = \text{ohm-cm}$$

The resistivity of substances varies over a wide range. To give an idea to the reader, the following table may be referred :

S.No.	Material	Nature	Resistivity (Ω-m) at room temperature
1	Copper	metal	1.7×10^{-8}
2	Iron	metal	9.68×10^{-8}
3	Manganin	alloy	48×10^{-8}
4	Nichrome	alloy	100×10^{-8}
5	Pure silicon	semiconductor	2.5×10^3
6	Pure germanium	semiconductor	0.6
7	Glass	insulator	10^{10} to 10^{14}
8	Mica	insulator	10^{11} to 10^{15}

The reader may note that resistivity of metals and alloys is very small. Therefore, these materials are good conductors of electric current. On the other hand, resistivity of insulators is extremely large. As a result, these materials hardly conduct any current. There is also an intermediate class of semiconductors. The resistivity of these substances lies between conductors and insulators.

1.20. Conductance

The reciprocal of resistance of a conductor is called its **conductance** *(G).* If a conductor has resistance R, then its conductance G is given by ;

$$G = 1/R$$

Whereas resistance of a conductor is the opposition to current flow, the conductance of a conductor is the inducement to current flow.

The SI unit of conductance is mho (*i.e.*, ohm spelt backward). These days, it is a usual practice to use **siemen** as the unit of conductance. It is denoted by the symbol S.

Conductivity. *The reciprocal of resistivity of a conductor is called its* **conductivity.** It is denoted by the symbol σ. If a conductor has resistivity ρ, then its conductivity is given by ;

$$\text{Conductivity,} \quad \sigma = \frac{1}{\rho}$$

We know that $G = \dfrac{1}{R} = \dfrac{a}{\rho l} = \sigma \dfrac{a}{l}$. Clearly, the SI unit of conductivity is *Siemen metre*$^{-1}$ (S m^{-1}).

Example 1.6. *A coil consists of 2000 turns of copper wire having a cross-sectional area of 0.8 mm^2. The mean length per turn is 80 cm and the resistivity of copper is 0.02 $\mu\Omega$ m. Find the resistance of the coil and power absorbed by the coil when connected across 110 V d.c. supply.*

Solution. Length of coil, $l = 0.8 \times 2000 = 1600$ m; cross-sectional area of coil, $a = 0.8$ mm^2 $= 0.8 \times 10^{-6}$ m^2; Resistivity of copper, $\rho = 0.02 \times 10^{-6}$ Ωm

$$\therefore \quad \text{Resistance of coil,} \quad R = \rho\frac{l}{a} = 0.02 \times 10^{-6}\,\frac{1600}{0.8 \times 10^{-6}} = \textbf{40 } \boldsymbol{\Omega}$$

$$\text{Power absorbed,} \quad P = \frac{V^2}{R} = \frac{(110)^2}{40} = \textbf{302.5 W}$$

Example 1.7. *Find the resistance of 1000 metres of a copper wire 25 sq. mm in cross-section. The resistance of copper is 1/58 ohm per metre length and 1 sq. mm cross-section. What will be the resistance of another wire of the same material, three times as long and one-half area of cross-section ?*

Solution. For the first case, $R_1 = ?$; $a_1 = 25$ mm^2 ; $l_1 = 1000$ m

For the second case, $R_2 = 1/58\ \Omega$; $a_2 = 1$ mm^2 ; $l_2 = 1$ m

$$R_1 = \rho\,(l_1/a_1)\ ;\quad R_2 = \rho\,(l_2/a_2)$$

$\therefore$
$$\frac{R_1}{R_2} = \frac{l_1}{l_2} \times \frac{a_2}{a_1} = \left(\frac{1000}{1}\right)\left(\frac{1}{25}\right) = 40$$

or
$$R_1 = 40\,R_2 = 40 \times \frac{1}{58} = \frac{20}{29}\,\Omega$$

For the third case, $R_3 = ?$; $a_3 = a_1/2$; $l_3 = 3l_1$

$\therefore$
$$\frac{R_3}{R_1} = \left(\frac{l_3}{l_1}\right) \times \left(\frac{a_1}{a_3}\right) = (3) \times (2) = 6$$

or
$$R_3 = 6R_1 = 6 \times \frac{20}{29} = \frac{120}{29}\,\Omega$$

Example 1.8. *A copper wire of diameter 1 cm had a resistance of 0.15 Ω. It was drawn under pressure so that its diameter was reduced to 50%. What is the new resistance of the wire ?*

Solution. Area of wire before drawing, $a_1 = \dfrac{\pi}{4}\,(1)^2 = 0.785$ cm^2

Area of wire after drawing, $a_2 = \dfrac{\pi}{4}\,(0.5)^2 = 0.196$ cm^2

As the volume of wire remains the same before and after drawing,

$\therefore$ $\qquad\qquad a_1 l_1 = a_2 l_2$

or $\qquad\qquad l_2/l_1 = a_1/a_2 = 0.785/0.196 = 4$

For the first case, $\qquad R_1 = 0.15\ \Omega$; $a_1 = 0.785$ cm^2 ; $l_1 = l$

For the second case, $\qquad R_2 = ?$; $a_2 = 0.196$ cm^2 ; $l_2 = 4l$

Now $\qquad\qquad R_1 = \rho\dfrac{l_1}{a_1};\ R_2 = \rho\dfrac{l_2}{a_2}$

$\therefore$
$$\frac{R_2}{R_1} = \left(\frac{l_2}{l_1}\right) \times \left(\frac{a_1}{a_2}\right) = (4) \times (4) = 16$$

or $\qquad\qquad R_2 = 16R_1 = 16 \times 0.15 = \textbf{2.4}\ \boldsymbol{\Omega}$

Example 1.9. *Two wires of aluminium and copper have the same resistance and same length. Which of the two is lighter? Density of copper is 8.9×10^3 kg/m^3 and that of aluminium is 2.7×10^3 kg/m^3. The resistivity of copper is $1.72 \times 10^{-8}\ \Omega$ m and that of aluminium is $2.6 \times 10^{-8}\ \Omega$ m.*

Solution. That wire will be lighter which has less mass. Let suffix 1 represent aluminium and suffix 2 represent copper.

$$R_1 = R_2 \quad \text{or} \quad \rho_1\frac{l_1}{A_1} = \rho_2\frac{l_2}{A_2}$$

or
$$\frac{\rho_1}{A_1} = \frac{\rho_2}{A_2} \qquad\qquad (\because l_1 = l_2)$$

or
$$\frac{A_1}{A_2} = \frac{\rho_1}{\rho_2} = \frac{2.6 \times 10^{-8}}{1.72 \times 10^{-8}} = 1.5$$

Now $$\frac{m_1}{m_2} = \frac{(A_1 l_1)d_1}{(A_2 l_2)d_2} = \frac{A_1 d_1}{A_2 d_2} \qquad (\because l_1 = l_2)$$

or $$\frac{m_1}{m_2} = \left(\frac{A_1}{A_2}\right) \times \left(\frac{d_1}{d_2}\right) = 1.5 \times \frac{2.7 \times 10^3}{8.9 \times 10^3} = 0.46$$

or $$m_1/m_2 = 0.46$$

It is clear that for the same length and same resistance, **aluminium wire is lighter than copper wire**. For this reason, aluminium wires are used for overhead power transmission lines.

Example 1.10. *A rectangular metal strip has the dimensions x = 10 cm, y = 0.5 cm and z = 0.2 cm. Determine the ratio of the resistances R_x, R_y and R_z between the respective pairs of opposite faces.*

Solution. $$R_x : R_y : R_z = \frac{\rho x}{yz} : \frac{\rho y}{zx} : \frac{\rho z}{xy} = \frac{10}{0.5 \times 0.2} : \frac{0.5}{0.2 \times 10} : \frac{0.2}{10 \times 0.5}$$

$$= \frac{10}{0.1} : \frac{1}{4} : 0.04 = \mathbf{2500 : 6.25 : 1}$$

Example 1.11. *Calculate the resistance of a copper tube 0.5 cm thick and 2 m long. The external diameter is 10 cm. Given that resistance of copper wire 1 m long and 1 mm^2 in cross-section is 1/58 Ω.*

Solution. External diameter, $D = 10$ cm

Internal diameter, $d = 10 - 2 \times 0.5 = 9$ cm

Area of cross-section, $$a = \frac{\pi}{4}(D^2 - d^2) = \frac{\pi}{4}\left[(10)^2 - (9)^2\right] \text{cm}^2$$

$$= \frac{\pi}{4}\left[(10)^2 - (9)^2\right] \times 100 \text{ mm}^2$$

$\therefore$ Resistance of copper tube $$= \frac{\rho l}{a} = \frac{1}{58} \times \frac{\text{length in metres}}{\text{area of X-section in mm}^2}$$

$$= \frac{1}{58} \times \frac{2}{\frac{\pi}{4}\left[(10)^2 - (9)^2\right] \times 100} = 23.14 \times 10^{-6}\ \Omega = \mathbf{23.14\ \mu\Omega}$$

Example 1.12. *A copper wire is stretched so that its length is increased by 0.1%. What is the percentage change in its resistance ?*

Solution. $$R = \rho\frac{l}{a}; \ R' = \rho\frac{l'}{a'}$$

Now $$l' = l + \frac{0.1}{100} \times l = 1.001\ l$$

As the volume remains the same, $al = a'l'$.

$\therefore$ $$a' = a\frac{l}{l'} = \frac{a}{1.001}$$

$\therefore$ $$\frac{R'}{R} = \left(\frac{l'}{l}\right) \times \left(\frac{a}{a'}\right) = (1.001) \times (1.001) = 1.002$$

or $$\frac{R' - R}{R} = 0.002$$

$\therefore$ Percentage increase $$= \frac{R' - R}{R} \times 100 = 0.002 \times 100 = \mathbf{0.2\%}$$

Example 1.13. *A lead wire and an iron wire are connected in parallel. Their respective specific resistances are in the ratio 49 : 24. The former carries 80% more current than the latter and the latter 47% longer than the former. Determine the ratio of their cross-sectional areas.*

Solution. Let us represent lead and iron by suffixes 1 and 2 respectively. Then as per the conditions of the problem, we have,

$$\frac{\rho_1}{\rho_2} = \frac{49}{24} \; ; \; I_1 = 1.8\, I_2 \; ; \; l_2 = 1.47\, l_1$$

Now

$$R_1 = \rho_1 \frac{l_1}{a_1} \; ; \; R_2 = \rho_2 \frac{l_2}{a_2}$$

$$I_1 = \frac{V}{R_1} \quad \text{and} \quad I_2 = \frac{V}{R_2}$$

∴

$$\frac{I_2}{I_1} = \frac{R_1}{R_2} = \frac{\rho_1 l_1}{a_1} \times \frac{a_2}{\rho_2 l_2} = \left(\frac{\rho_1}{\rho_2}\right) \times \left(\frac{l_1}{l_2}\right) \times \left(\frac{a_2}{a_1}\right)$$

or

$$\frac{1}{1.8} = \frac{49}{24} \times \frac{1}{1.47} \times \frac{a_2}{a_1}$$

∴

$$\frac{a_2}{a_1} = \frac{1}{1.8} \times \frac{24}{49} \times 1.47 = \mathbf{0.4}$$

Example 1.14. *An aluminium wire 7.5 m long is connected in parallel with a copper wire 6 m long. When a current of 5 A is passed through the combination, it is found that the current in the aluminium wire is 3 A. The diameter of the aluminium wire is 1 mm. Determine the diameter of the copper wire. Resistivity of copper is 0.017 μΩm ; that of the aluminium is 0.028 μΩ m.*

Solution. Let us assign subscripts a and c to aluminium and copper respectively.

Current through *Al* wire, $I_a = 3$ A

∴ Current through Cu wire, $I_c = 5 - 3 = 2$ A

Since R_a and R_c are in parallel, the voltage across them is the same [See Fig. 1.12] *i.e.*

$$I_a R_a = I_c R_c \quad \text{or} \quad \frac{R_a}{R_c} = \frac{I_c}{I_a} = \frac{2}{3}$$

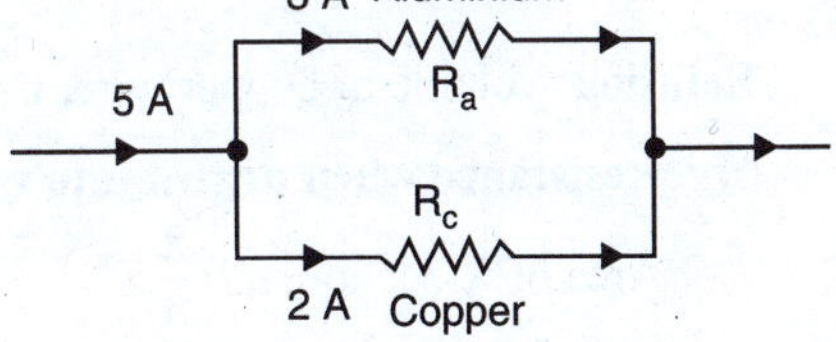

Now

$$R_a = \frac{\rho_a l_a}{a_a} \; ; \; R_c = \frac{\rho_c l_c}{a_c}$$

∴

$$\frac{R_c}{R_a} = \frac{\rho_c}{\rho_a} \times \frac{l_c}{l_a} \times \frac{a_a}{a_c}$$

Here

$$\frac{R_c}{R_a} = \frac{3}{2} \; ; \; \frac{\rho_c}{\rho_a} = \frac{0.017}{0.028} \; ; \; \frac{l_c}{l_a} = \frac{6}{7.5} \; ;$$

$$a_a = \frac{\pi}{4}d^2 = \frac{\pi \times (1)^2}{4} = \frac{\pi}{4} \text{ mm}^2$$

∴

$$\frac{3}{2} = \frac{0.017}{0.028} \times \frac{6}{7.5} \times \frac{\pi/4}{a_c}$$

or

$$a_c = \frac{2}{3} \times \frac{0.017}{0.028} \times \frac{6}{7.5} \times \frac{\pi}{4} = 0.2544 \text{ mm}^2$$

or

$$\frac{\pi}{4}d_c^2 = 0.2544 \quad \therefore \quad d_c = \mathbf{0.569 \text{ mm}}$$

3 A Aluminium

R_a

5 A

R_c

2 A Copper

Fig. 1.12

Example 1.15. *A transmission line cable consists of 19 strands of identical copper conductors, each 1.5 mm in diameter. The length of the cable is 2 km but because of the twist of the strands, the actual length of each conductor is increased by 5 percent. What is resistance of the cable ? Take the resistivity of the copper to be 1.78×10^{-8} Ω m.*

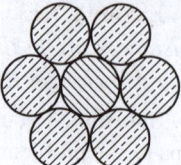

Solution. Fig. 1.13 shows the general shape of a stranded conductor. Allowing for twist, the length of the strands is

Fig. 1.13

$$l = 2000 \text{ m} + 5\% \text{ of } 2000 \text{ m} = 2100 \text{ m}$$

Area of X-section of 19 strands, $a = (19)\left(\dfrac{\pi}{4}\right) \times (1.5 \times 10^{-3})^2 = 33.576 \times 10^{-6} \text{ m}^2$

$\therefore$ Resistance of line, $R = \rho\dfrac{l}{a} = 1.72 \times 10^{-8} \times \dfrac{2100}{33.576 \times 10^{-6}} = \textbf{1.076 } \boldsymbol{\Omega}$

Example 1.16. *The resistance of the wire used for telephone is 35 Ω per kilometre when the weight of the wire is 5 kg per kilometre. If the specific resistance of the material is 1.95×10^{-8} Ω m, what is the cross-sectional area of the wire ? What will be the resistance of a loop to a subscriber 8 km from the exchange if wire of the same material but weighing 20 kg per kilometre is used?*

Solution. For the first case, $R = 35 \, \Omega$; $l = 1000$ m ; $\rho = 1.95 \times 10^{-8}$ Ω m

Now $R = \rho\dfrac{l}{a}$ $\therefore$ $a = \dfrac{\rho l}{R} = \dfrac{1.95 \times 10^{-8} \times 1000}{35} = \textbf{55.7} \times \textbf{10}^{-8} \textbf{ m}^2$

Since weight of conductor is directly proportional to the area of cross-section, for the second case, we have,

$$a = \dfrac{20}{5} \times 55.7 \times 10^{-8} = 222.8 \times 10^{-8} \text{ m}^2 \; ; \; l = 2 \times 8 = 16 \text{ km} = 16000 \text{ m}$$

$\therefore$ $R = \rho\dfrac{l}{a} = 1.95 \times 10^{-8} \times \dfrac{16000}{222.8 \times 10^{-8}} = \textbf{140.1 } \boldsymbol{\Omega}$

Example 1.17. *Find the resistance of a cubic centimetre of copper (i) when it is drawn into a wire of diameter 0.32 mm and (ii) when it is hammered into a flat sheet of 1.2 mm thickness, the current flowing through the sheet from one face to another, specific resistance of copper is 1.6×10^{-8} Ω-m.*

Solution. Volume of copper wire, $v = 1 \text{ cm}^3 = 1 \times 10^{-6} \text{ m}^3$

(i) **Resistance when drawn into wire.**

Area of X-section, $a = \dfrac{\pi}{4}d^2 = \dfrac{\pi}{4}(0.32 \times 10^{-3})^2 = 0.804 \times 10^{-7} \text{ m}^2$

Length of wire, $l = \dfrac{v}{a} = \dfrac{1 \times 10^{-6}}{0.804 \times 10^{-7}} = 12.43 \text{ m}$

$\therefore$ Resistance of wire, $R = \rho\dfrac{l}{a} = 1.6 \times 10^{-8} \dfrac{12.43}{0.804 \times 10^{-7}} = \textbf{2.473 } \boldsymbol{\Omega}$

(ii) **Resistance when hammered into flat sheet.**

Length of flat sheet, $l = 1.2 \times 10^{-3}$ m ; Area of cross-section of flat sheet is

$$a = \dfrac{v}{l} = \dfrac{1 \times 10^{-6}}{1.2 \times 10^{-3}} = \dfrac{10^{-3}}{1.2} \text{ m}^2$$

$\therefore$ Resistance of copper flat sheet is $R = \rho\dfrac{l}{a} = 1.6 \times 10^{-8} \dfrac{1.2 \times 10^{-3}}{10^{-3}/1.2} = \textbf{2.3} \times \textbf{10}^{-8} \boldsymbol{\Omega}$

Tutorial Problems

1. Calculate the resistance of 915 metres length of a wire having a uniform cross-sectional area of 0.77 cm^2 if the wire is made of copper having a resistivity of 1.7×10^{-6} Ω cm. **[0.08 Ω]**

2. A wire of length 1 m has a resistance of 2 ohms. What is the resistance of second wire, whose specific resistance is double the first, if the length of wire is 3 metres and the diameter is double of the first? **[3 Ω]**

3. A rectangular copper strip is 20 cm long, 0.1 cm wide and 0.4 cm thick. Determine the resistance between (i) opposite ends (ii) opposite sides. The resistivity of copper is 1.7×10^{-6} Ω cm.
   [(i) 0.85×10^{-4} Ω (ii) 0.212×10^{-6} Ω]

4. A cube of a material of side 1 cm has a resistance of 0.001 Ω between its opposite faces. If the same material has a length of 9 cm and a uniform cross-sectional area 1 cm^2, what will be the resistance of this length ? **[0.009 Ω]**

5. An aluminium wire 10 metres long and 2 mm in diameter is connected in parallel with a copper wire 6 metres long. A total current of 2 A is passed through the combination and it is found that current through the aluminium wire is 1.25 A. Calculate the diameter of copper wire. Specific resistance of copper is 1.6×10^{-6} Ω cm and that of aluminium is 2.6×10^{-6} Ω cm. **[0.94 mm]**

6. A copper wire is stretched so that its length is increased by 0.1%. What is the percentage change in its resistance ? **[0.2%]**

1.21. Types of Resistors

A component whose function in a circuit is to provide a specified value of resistance is called a **resistor**. The principal applications of resistors are to limit current, divide voltage and in certain cases, generate heat. Although there are a variety of different types of resistors, the following are the commonly used resistors in electrical and electronic circuits :

    (i) Carbon composition types   (ii) Film resistors

    (iii) Wire-wound resistors   (iv) Cermet resistors

  (i) Carbon composition type. This type of resistor is made with a mixture of finely ground carbon, insulating filler and a resin binder. The ratio of carbon and insulating filler decides the resistance value [See Fig. 1.14]. The mixture is formed into a rod and lead connections are made. The entire resistor is then enclosed in a plastic case to prevent the entry of moisture and other harmful elements from outside.

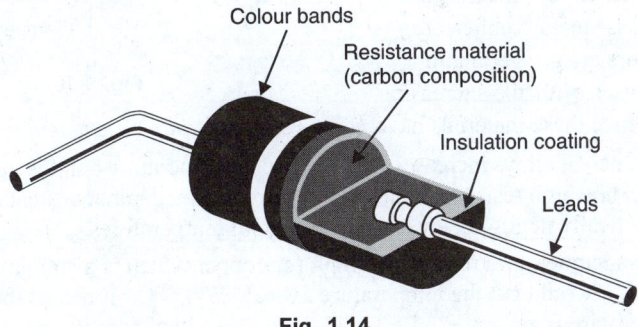

Fig. 1.14

Carbon resistors are relatively inexpensive to build. However, they are highly sensitive to temperature variations. The carbon resistors are available in power ratings ranging from 1/8 to 2 W.

  (ii) Film resistors. In a film resistor, a resistive material is deposited uniformly onto a high-grade ceramic rod. The resistive film may be carbon (carbon film resistor) or nickel-chromium (metal film resistor). In these types of resistors, the desired resistance value is obtained by removing a part of the resistive material in a helical pattern along the rod as shown in Fig. 1.15.

Metal film resistors have better characteristics as compared to carbon film resistors.

(*iii*) **Wire-wound resistors.** A wire-wound resistor is constructed by winding a resistive wire of some alloy around an insulating rod. It is then enclosed in an insulating cover. Generally, nickle-chromium alloy is used because of its very small temperature coefficient of resistance. Wire-wound resistors can safely operate at higher temperatures than carbon types. These resistors have high power ratings ranging from 12 to 225 W.

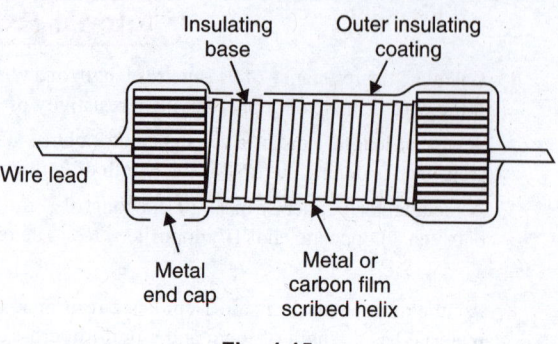

Fig. 1.15

(*iv*) **Cermet resistors.** A cermet resistor is made by depositing a thin film of metal such as nichrome or chromium cobalt on a ceramic substrate. They are cermet which is a contraction for ceramic and metal. These resistors have very accurate values.

1.22. Effect of Temperature on Resistance

In general, the resistance of a material changes with the change in temperature. The effect of temperature upon resistance varies according to the type of material as discussed below :

(*i*) The resistance of pure metals (*e.g.* copper, aluminium) increases with the increase of temperature. The change in resistance is fairly regular for normal range of temperatures so that temperature/resistance graph is a straight line as shown in Fig. 1.16 (for copper). Since the resistance of metals increases with the rise in temperature, they have *positive temperature co-efficient of resistance.*

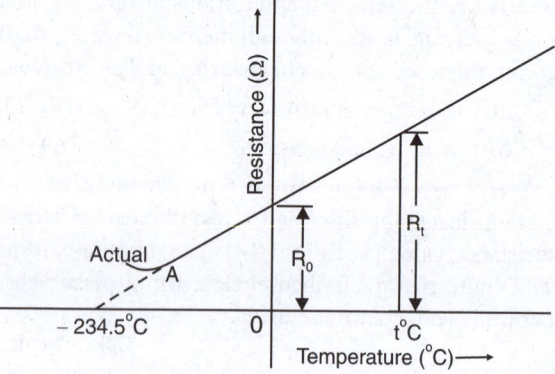

Fig. 1.16

(*ii*) The resistance of electrolytes, insulators (*e.g.* glass, mica, rubber etc.) and semiconductors (*e.g.* germanium, silicon etc.) decreases with the increase in temperature. Hence these materials have *negative temperature co-efficient of resistance.*

(*iii*) The resistance of alloys increases with the rise in temperature but this increase is very small and irregular. For some high resistance alloys (*e.g.* Eureka, manganin, constantan etc.), the change in resistance is practically negligible over a wide range of temperatures.

Fig. 1.16 shows temperature/resistance graph for copper which is a straight line. If this line is extended backward, it would cut the temperature axis at −234.5°C. It means that theoretically, the resistance of copper wire is zero at −234.5°C. However, in actual practice, the curve departs (point *A*) from the straight line path at very low temperatures.

1.23. Temperature Co-efficient of Resistance

Consider a conductor having resistance R_0 at 0°C and R_t at t °C. It has been found that in the normal range of temperatures, the increase in resistance (*i.e.* $R_t - R_0$)

(*i*) is directly proportional to the initial resistance *i.e.*

$$R_t - R_0 \propto R_0$$

(*ii*) is directly proportional to the rise in temperature *i.e.*

$$R_t - R_0 \propto t$$

(*iii*) depends upon the nature of material.

Combining the first two, we get,

$$R_t - R_0 \propto R_0 \, t$$

or

$$R_t - R_0 = {}^*\alpha_0 \, R_0 \, t \qquad \qquad ...(i)$$

where α_0 is a constant and is called temperature co-efficient of resistance at 0°C. Its value depends upon the nature of material and temperature.

Rearranging eq. (*i*), we get,

$$R_t = R_0 \, (1 + \alpha_0 \, t) \qquad \qquad ...(ii)$$

Definition of α_0. From eq. (*i*), we get,

$$\alpha_0 = \frac{R_t - R_0}{R_0 \times t}$$

= Increase in resistance/ohm original resistance/°C rise in temperature

Hence **temperature co-efficient of resistance** *of a conductor is the increase in resistance per ohm original resistance per °C rise in temperature.*

A little reflection shows that unit of α will be ohm/ohm°C *i.e.*/°C. Thus, copper has a temperature co-efficient of resistance of 0.00426/°C. It means that if a copper wire has a resistance of 1 Ω at 0°C, then it will increase by 0.00426 Ω for 1°C rise in temperature *i.e.* it will become 1.00426 Ω at 1°C. Similarly, if temperature is raised to 10°C, then resistance will become 1 + 10 × 0.00426 = 1.0426 ohms.

The following points may be noted carefully :

(*i*)   Those substances (*e.g.* pure metals) whose resistance increases with rise in temperature are said to have *positive* temperature co-efficient of resistance. On the other hand, those substances whose resistance decreases with increase in temperature are said to have *negative* temperature co-efficient of resistance.

(*ii*)   If a conductor has a resistance R_0, R_1 and R_2 at 0°C, t_1°C and t_2°C respectively, then,

$$R_1 = R_0 \, (1 + \alpha_0 \, t_1)$$

$$R_2 = R_0 \, (1 + \alpha_0 \, t_2)$$

$$\therefore \quad \frac{R_2}{R_1} = \frac{1 + \alpha_0 \, t_2}{1 + \alpha_0 t_1} \qquad \qquad ...(iii)$$

This relation is often utilised in determining the rise of temperature of the winding of an electrical machine. The resistance of the winding is measured both before and after the test run. Let R_1 and t_1 be the resistance and temperature before the commencement of the test. After the operation of the machine for a given period, let these values be R_2 and t_2. Since R_1 and R_2 can be measured and t_1 (ambient temperature) and α_0 are known, the value of t_2 can be calculated from eq. (*iii*). The average rise in temperature of the winding will be $(t_2 - t_1)$°C.

Note. The life expectancy of electrical apparatus is limited by the temperature of its insulation; the higher the temperature, the shorter the life. The useful life of electrical apparatus reduces approximately by half every time the temperature increases by 10°C. This means that if a motor has a normal life expectancy of eight years

---

*   It will be shown in Art. 1.25 that value of α depends upon temperature. Therefore, it is referred to the original temperature *i.e.* 0°C in this case. Hence the symbol α_0.

at a temperature of 100°C, it will have a life expectancy of only four years at a temperature of 110°C, of two years at a temperature of 120°C and of only one year at 130°C.

1.24. Graphical Determination of α

The value of temperature co-efficient of resistance can also be determined graphically from temperature/resistance graph of the material. Fig. 1.17 shows the temperature/resistance graph for a conductor. The graph is a straight line AX as is the case with all conductors. The resistance of the conductor is R_0 (represented by OA) at 0°C and it becomes R_t at t°C. By definition,

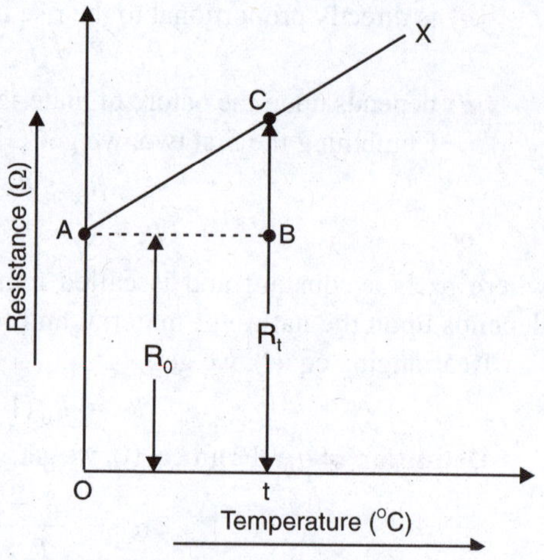

$$\alpha_0 = \frac{R_t - R_0}{R_0 \times t}$$

But

$$R_t - R_0 = BC$$

and

$$t = \text{rise in temperature} = AB$$

$\therefore$

$$\alpha_0 = \frac{BC}{R_0 \times AB}$$

Fig. 1.17

But BC/AB is the slope of temperature/resistance graph.

$\therefore$

$$\alpha_0 = \frac{\text{Slope of temp./resistance graph}}{\text{Original resistance}} \qquad \ldots(i)$$

*Hence, **temperature co-efficient** of resistance of a conductor at 0°C is the slope of temp./resistance graph divided by resistance at 0°C (i.e. R_0).*

The following points may be particularly noted :

(*i*) The value of α depends upon temperature. At any temperature, α can be calculated by using eq. (*i*).

Thus,

$$\alpha_0 = \frac{\text{Slope* of temperature/resistance graph}}{R_0}$$

and

$$\alpha_t = \frac{\text{Slope of temperature/resistance graph}}{R_t}$$

(*ii*) The value of α_0 is maximum and it decreases as the temperature is increased. This is clear from the fact that the slope of temperature/resistance graph is constant and R_0 has the minimum value.

1.25. Temperature Co-efficient at Various Temperatures

Consider a conductor having resistances R_0 and R_1 at temperatures 0°C and t_1°C respectively. Let α_0 and α_1 be the temperature co-efficients of resistance of the conductor at 0°C and t_1°C respectively. It is desired to establish the relationship between α_1 and α_0. Fig. 1.18 shows the temperature/resistance graph of the conductor. As proved in Art. 1.24,

$$\alpha_0 = \frac{\text{Slope of graph}}{R_0}$$

$\therefore \quad$ Slope of graph $= \alpha_0 R_0$

---

* The slope of temp./resistance graph of a conductor is always constant (being a straight line).

Similarly, $\qquad \alpha_1 = \dfrac{\text{Slope of graph}}{R_1}$

or $\qquad$ Slope of graph $= \alpha_1 R_1$

Since the slope of temperature/resistance graph is constant,

$\therefore \qquad\qquad \alpha_0 R_0 = \alpha_1 R_1$

or $\qquad \alpha_1 = \dfrac{\alpha_0 R_0}{R_1} = \dfrac{\alpha_0 R_0}{R_0(1+\alpha_0 t_1)}$

$$[\because R_1 = R_0 (1 + \alpha_0 t_1)]$$

$\therefore \qquad\qquad \alpha_1 = \dfrac{\alpha_0}{1+\alpha_0 t_1} \qquad ...(i)$

Similarly,* $\quad \alpha_2 = \dfrac{\alpha_0}{1+\alpha_0 t_2} \qquad ...(ii)$

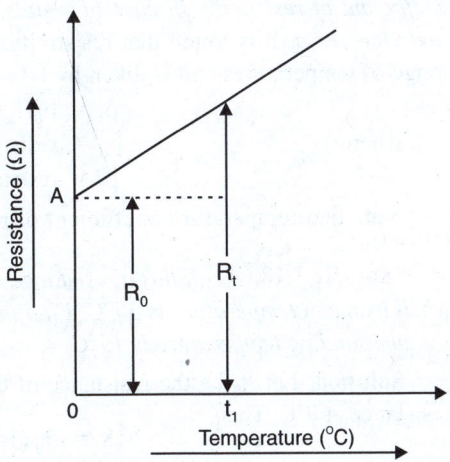

Fig. 1.18

Subtracting the reciprocal of eq. (i) from the reciprocal of eq. (ii),

$$\frac{1}{\alpha_2} - \frac{1}{\alpha_1} = \frac{1+\alpha_0 t_2}{\alpha_0} - \frac{1+\alpha_0 t_1}{\alpha_0} = t_2 - t_1$$

$\therefore \qquad\qquad \alpha_2 = \dfrac{1}{\dfrac{1}{\alpha_1} + (t_2 - t_1)} \qquad\qquad\qquad ...(iii)$

Eq. (i) gives the relation between α_1 and α_0 while Eq. (iii) gives the relation between α_2 and α_1.

1.26. Summary of Temperature Co-efficient Relations

(i) If R_0 and α_0 are the resistance and temperature co-efficient of resistance of a conductor at $0°C$, then its resistance R_t at $t°C$ is given by ;

$$R_t = R_0(1 + \alpha_0 t)$$

(ii) If α_0, α_1 and α_2 are the temperature co-efficients of resistance at $0°C$, $t_1°C$ and $t_2°C$ respectively, then,

$$\alpha_1 = \frac{\alpha_0}{1+\alpha_0 t_1} \;;\; \alpha_2 = \frac{\alpha_0}{1+\alpha_0 t_2} \;;\; \alpha_2 = \frac{1}{\dfrac{1}{\alpha_1} + (t_2 - t_1)}$$

(iii) Suppose R_1 and R_2 are the resistances of a conductor at $t_1°C$ and $t_2°C$ respectively. If α_1 is the temperature co-efficient of resistance at $t_1°C$, then,

$$R_2** = R_1[1 + \alpha_1(t_2 - t_1)]$$

1.27. Variation of Resistivity With Temperature

Not only resistance but resistivity or specific resistance of a material also changes with temperature. The change in resistivity per °C change in temperature is called *temperature*

---

* $\qquad\qquad\qquad\qquad \alpha_0 R_0 = \alpha_2 R_2$ where R_2 is the resistance at $t_2°C$

or $\qquad\qquad\qquad \alpha_2 = \dfrac{\alpha_0 R_0}{R_2} = \dfrac{\alpha_0 R_0}{R_0(1+\alpha_0 t_2)} = \dfrac{\alpha_0}{1+\alpha_0 t_2}$

** $\qquad$ Slope of graph, $\tan\theta = R_0 \alpha_0 = R_1 \alpha_1 = R_2 \alpha_2$

Increase in resistance as temperature is raised from $t_1°C$ to $t_2°C$

$$= \tan\theta(t_2 - t_1) = R_1\alpha_1(t_2 - t_1)$$

$\therefore \qquad$ Resistance at $t_2°C$, $R_2 = R_1 + R_1\alpha_1(t_2 - t_1) = R_1[1 + \alpha_1(t_2 - t_1)]$

coefficient of resistivity. In case of metals, the resistivity increases with increase in temperature and vice-versa. It is found that resistivity of a metallic conductor increases linearly over a wide range of temperatures and is given by ;

$$\rho_t = \rho_0(1 + \alpha_0 t)$$

where ρ_0 = resistivity of metallic conductor at 0°C

               ρ_t = resistivity of metallic conductor at temperature t°C

Note that temperature coefficient of resistivity is equal to temperature coefficient of resistance α_0.

Example 1.18. *A coil has a resistance of 18 Ω when its mean temperature is 20°C and of 20 Ω when its mean temperature is 50°C. Find its mean temperature rise when its resistance is 21Ω and the surrounding temperature is 15°C.*

Solution. Let R_0 be the resistance of the coil at 0°C and α_0 be its temperature coefficient of resistance at 0°C. Then,

$$18 = R_0(1 + \alpha_0 \times 20) \quad \text{and} \quad 20 = R_0(1 + \alpha_0 \times 50)$$

$$\therefore \quad \frac{20}{18} = \frac{1 + 50\alpha_0}{1 + 20\alpha_0} \quad \text{or} \quad \alpha_0 = \frac{1}{250} = 0.004/°C$$

If t°C is the temperature of the coil when its resistance is 21Ω, then,

$$21 = R_0(1 + 0.004\,t)$$

$$\therefore \quad \frac{21}{18} = \frac{R_0(1 + 0.004\,t)}{R_0(1 + 0.004 \times 20)} \quad \text{or} \quad t = 65°C$$

$$\therefore \quad \text{Temperature rise} = t - 15 = 65° - 15° = \mathbf{50°C}$$

Example 1.19. *The resistance of the field coils of a dynamo is 173 Ω at 16°C. After working for 6 hours on full-load, the resistance of the coils increases to 212 Ω. Calculate (i) the temperature of the coils (ii) mean rise of temperature of the coils. Assume temperature co-efficient of resistance of copper is 0.00426/°C at 0°C.*

Solution. (*i*) Let t°C be the final temperature.

$$\frac{R_{16}}{R_t} = \frac{R_0(1 + \alpha_0 \times 16)}{R_0(1 + \alpha_0 \times t)}$$

or $$\frac{173}{212} = \frac{1 + 0.00426 \times 16}{1 + 0.00426 \times t}$$

or $$0.816 = \frac{1.068}{1 + 0.00426t} \quad \therefore \quad t = \mathbf{72.5°C}$$

(*ii*) Rise in temperature $= t - 16 = 72.5 - 16 = \mathbf{56.5°C}$

Example 1.20. *The resistance of a transformer winding is 460 Ω at room temperature of 25°C. When the transformer is running and the final temperature is reached, the resistance of the winding increases to 520 Ω. Find the average temperature rise of winding, assuming that $\alpha_{20} = 1/250$ per °C.*

Solution. $$\alpha_{25} = \frac{1}{1/\alpha_{20} + (25 - 20)} = \frac{1}{250 + 5} = \frac{1}{255}/°C$$

Let t°C be the final temperature of the winding. Then, the rise in temperature is $t - 25$.

Now, $R_{25} = 460\ \Omega\ ;\ R_t = 520\ \Omega$

$$R_t = R_{25}[1 + \alpha_{25}(t - 25)]$$

or $$t - 25 = \frac{1}{\alpha_{25}}\left(\frac{R_t}{R_{25}} - 1\right) = 255(520/460 - 1) = 33.26°C$$

$$\therefore \quad \text{Temperature rise} = t - 25 = \mathbf{32.26°C}$$

Example 1.21. *The filament of a 60 watt, 230 V lamp has a normal working temperature of 2000°C. Find the current flowing in the filament at the instant of switching, when the lamp is cold. Assume the temperature of cold lamp to be 15°C and $\alpha_{15} = 0.005/°C$.*

Solution. Resistance of lamp at 2000°C is

$$R_{2000} = V^2/P = (230)^2/60 = 881.67 \ \Omega$$
$$R_{2000} = R_{15}[1 + \alpha_{15}(2000 - 15)]$$
$$\therefore \qquad R_{15} = \frac{R_{2000}}{1 + 0.005(1985)} = \frac{881.67}{10.925} = 80.7 \ \Omega$$

∴ Current taken by cold lamp (*i.e.* at the time of switching) is

$$I = V/R_{15} = 230/80.7 = \textbf{2.85 A}$$

Example 1.22. *Two coils connected in series have resistances of 600 Ω and 300 Ω and temperature coefficients of 0.1% and 0.4% per °C at 20°C respectively. Find the resistance of combination at a temperature of 50°C. What is the effective temperature coefficient of the combination at 50°C ?*

Solution. Resistance of 600 Ω coil at 50°C

$$= 600 \ [1 + 0.001(50 - 20)] = 618 \ \Omega$$

Resistance of 300 Ω coil at 50°C

$$= 300 \ [1 + 0.004 \ (50 - 20)] = 336 \ \Omega$$

Resistance of series combination at 50°C is

$$R_{50} = 618 + 336 = \textbf{954} \ \Omega$$

Resistance of series combination at 20°C is

$$R_{20} = 600 + 300 = 900 \ \Omega$$

Now $$R_{50} = R_{20} \ [1 + \alpha_{20} \ (t_2 - t_1)]$$

$$\therefore \qquad \alpha_{20} = \frac{\dfrac{R_{50}}{R_{20}} - 1}{t_2 - t_1} = \frac{\dfrac{954}{900} - 1}{50 - 20} = 0.002$$

Now $$\alpha_{50} = \frac{1}{1/\alpha_{20} + (t_2 - t_1)} = \frac{1}{1/0.002 + (50 - 20)} = \frac{1}{\textbf{530}} /°C$$

Example 1.23. *The coil of a relay takes a current of 0.12 A when it is at the room temperature of 15°C and connected across a 60 V supply. If the minimum operating current of the relay is 0.1 A, calculate the temperature above which the relay will fail to operate when connected to the same supply. Resistance temperature coefficient of the coil material is 0.0043 per °C at 0°C.*

Solution. Resistance of relay coil at 15°C, $R_{15} = 60/0.12 = 500 \ \Omega$

If the temperature increases, the resistance of relay coil increases and current in relay coil decreases. Let $t°C$ be the temperature at which the current in relay coil becomes 0.1 A (= the minimum relay coil current for its operation). Clearly, $R_t = 60/0.1 = 600 \ \Omega$.

Now, $$R_{15} = R_0 \ (1 + 15 \ \alpha_0) = R_0 \ (1 + 15 \times 0.0043)$$
$$R_t = R_0 \ (1 + \alpha_0 t) = R_0 \ (1 + 0.0043 \ t)$$
$$\therefore \qquad \frac{R_t}{R_{15}} = \frac{1 + 0.0043 \ t}{1.0645}$$
or $$\frac{600}{500} = \frac{1 + 0.0043 \ t}{1.0645}$$

On solving, $t = \textbf{64.5°C}$

If the temperature of relay coil increases above 64.5°C, the resistance of relay coil will increase and the relay coil current will be less than 0.1 A. As a result, the relay will fail to operate.

Example 1.24. *Two materials, A and B, have resistance temperature coefficients of 0.004 and 0.0004 respectively at a given temperature. In what proportion must A and B be joined in series to produce a circuit having a temperature coefficient of 0.001 ?*

Solution. Let the resistance of A be 1 Ω and that of B be x Ω i.e. $R_A = 1$ Ω and $R_B = x$ Ω.

Resistance of series combination $= R_A + R_B = (1 + x)$ Ω

Suppose the temperature rises by $t°C$.

Resistance of series combination at the raised temperature $= (1 + x)(1 + 0.001\ t)$ ...(i)

Resistance of A at the raised temperature $= 1(1 + 0.004\ t)$ ...(ii)

Resistance of B at the raised temperature $= x(1 + 0.0004\ t)$ ...(iii)

As per the conditions of the problem, we have, $(ii) + (iii) = (i)$

or $1(1 + 0.004\ t) + x(1 + 0.0004\ t) = (1 + x)(1 + 0.001\ t)$

or $0.004\ t + 0.0004\ t\ x = (1 + x) \times 0.001\ t$

Dividing by t and multiplying throughout by 10^4, we have,

$$40 + 4x = 10(1 + x)   \therefore x = 5$$

$\therefore$ $R_A : R_B = $ **1 : 5**   *i.e.* R_B should be 5 times R_A.

Example 1.25. *A resistor of 80 Ω resistance, having a temperature coefficient of 0.0021/°C is to be constructed. Wires of two materials of suitable cross-sectional areas are available. For material A, the resistance is 80 Ω per 100 m and the temperature coefficient is 0.003/°C. For material B, the corresponding figures are 60 Ω per 100 m and 0.0015/°C. Calculate suitable lengths of wires of materials A and B to be connected in series to construct the required resistor. All data are referred to the same temperature.*

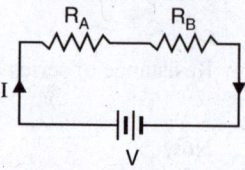

Fig. 1.19

Solution. Let R_A and R_B be the required resistances of materials A and B respectively which when joined in series have a combined temperature coefficient of 0.0021 [See Fig. 1.19].

Resistance of series combination $= R_A + R_B$

Resistance of series combination at raised temperature $= (R_A + R_B)(1 + 0.0021\ t)$ ...(i)

Resistance of A at raised temperature $= R_A(1 + 0.003\ t)$ ...(ii)

Resistance of B at raised temperature $= R_B(1 + 0.0015\ t)$ ...(iii)

As per conditions of the problem, $(ii) + (iii) = (i)$.

$\therefore$ $R_A(1 + 0.003\ t) + R_B(1 + 0.0015\ t) = (R_A + R_B)(1 + 0.0021\ t)$

On solving, $\dfrac{R_B}{R_A} = \dfrac{3}{2}$ ...(iv)

Now, $R_A + R_B = 80$ ...(v)

From eqs. (iv) and (v), $R_A = 32$ Ω and $R_B = 48$ Ω

$\therefore$ Length of wire A, $L_A = (100/80) \times 32 = $ **40 m**

          Length of wire B, $L_B = (100/60) \times 48 = $ **80 m**

Example 1.26. *Two wires A and B are connected in series at 0°C and resistance of B is 3.5 times that of A. The resistance temperature coefficient of A is 0.4% and that of combination is 0.1%. Find the resistance temperature coefficient of B.*

Solution. Let the temperature coefficient of resistance of wire B be α_B. If R is the resistance of wire A, then,

$$R_A = R \; ; \; R_B = 3.5\,R$$

Total resistance of two wires at 0°C $= R_A + R_B = R + 3.5\,R = 4.5\,R$

Increase in resistance of wire A per °C rise $= \alpha_A\,R = 0.004\,R$

Increase in resistance of wire B per °C rise $= \alpha_B \times 3.5\,R = 3.5\,R\,\alpha_B$

Total increase in the resistance of combination per °C rise $= 0.004\,R + 3.5\,R\,\alpha_B$ ... (i)

Also, total increase in the resistance of combination per °C rise $= \alpha_C \times$ Total resistance of combination $= 0.001 \times 4.5\,R = 0.0045\,R$ ... (ii)

From eqs. (i) and (ii), $0.004\,R + 3.5\,R\alpha_B = 0.0045\,R$

$$\therefore \qquad \alpha_B = \frac{0.0045\,R - 0.004\,R}{3.5\,R} = \mathbf{0.000143/°C} \text{ or } \mathbf{0.0143\%}$$

Example 1.27. *Two conductors, one of copper and the other of iron, are connected in parallel and carry equal currents at 25°C. What proportion of current will pass through each if the temperature is raised to 100°C? The temperature co-efficients of resistance at 0°C are 0.0043/°C and 0.0063/°C for copper and iron respectively.*

Solution. Since copper and iron conductors carry equal currents at 25°C, their resistances are the same at this temperature. Let their resistance be R ohms at 25°C. If R_1 and R_2 are the resistances of copper and iron conductors respectively at 100°C, then,

$$R_1 = R\,[1 + 0.0043\,(100 - 25)] = 1.3225\,R$$
$$R_2 = R\,[1 + 0.0063\,(100 - 25)] = 1.4725\,R$$

If I is the total current at 100°C, then,

Current in copper conductor $= I \times \dfrac{R_2}{R_1 + R_2} = I \times \dfrac{1.4725\,R}{1.3225\,R + 1.4725\,R} = 0.5268\,I$

Current in iron conductor $= I \times \dfrac{R_1}{R_1 + R_2} = I \times \dfrac{1.3225\,R}{1.3225\,R + 1.4725\,R} = 0.4732\,I$

Therefore, at 100°C, the copper conductor will carry **52.68%** of total current and the remaining **47.32%** will be carried by iron conductor.

Example 1.28. *A semi-circular ring of copper has an inner radius 6 cm, radial thickness 3 cm and an axial thickness 4 cm. Find the resistance of the ring at 50°C between its two end-faces. Assume specific resistance of copper at 20°C = 1.724 × 10^{-6} Ω-cm and resistance temperature coefficient of copper at 0°C = 0.0043/°C.*

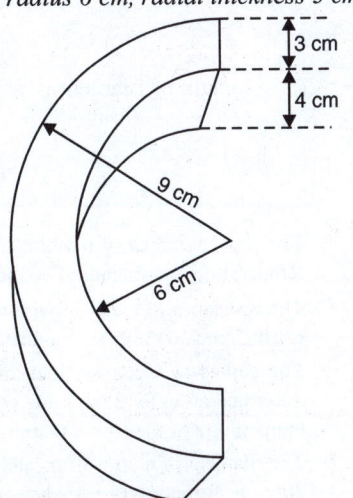

Solution. Fig. 1.20 shows the semi-circular ring.

Mean radius of the ring, $r_m = (6 + 9)/2 = 7.5$ cm

Mean length between end faces is

$$l_m = \pi r_m = \pi \times 7.5 = 23.56 \text{ cm}$$

Cross-sectional area of the ring is

$$a = 3 \times 4 = 12 \text{ cm}^2$$

Now $$\alpha_{20} = \frac{\alpha_0}{1 + \alpha_0\,t} = \frac{0.0043}{1 + 0.0043 \times 20}$$

$$= 0.00396/°C$$

Fig. 1.20

Also
$$\rho_{50} = \rho_{20}[1 + \alpha_{20}(t-20)]$$
$$= 1.724 \times 10^{-6}[1 + 0.00396 \times (50-20)]$$
$$= 1.93 \times 10^{-6} \ \Omega \ cm$$

∴
$$R_{50} = \frac{\rho_{50} \, l_m}{a} = \frac{1.93 \times 10^{-6} \times 23.56}{12} = \mathbf{3.79 \times 10^{-6} \ \Omega}$$

This example shows that resistivity of a conductor increases with the increase in temperature and vice-versa.

Example 1.29. *A copper conductor has its specific resistance of 1.6 ×10⁻⁶ Ω cm at 0°C and a resistance temperature coefficient of 1/254.5 per °C at 20°C. Find (i) specific resistance and (ii) the resistance temperature coefficient at 60°C.*

Solution.
$$\alpha_{20} = \frac{\alpha_0}{1 + \alpha_0 \times 20} \quad or \quad \frac{1}{254.5} = \frac{\alpha_0}{1 + \alpha_0 \times 20} \quad \therefore \ \alpha_0 = \frac{1}{234.5} / °C$$

(i)
$$\rho_{60} = \rho_0(1 + \alpha_0 \times 60) = 1.6 \times 10^{-6}(1 + 60/234.5) = \mathbf{2.01 \times 10^{-6} \ \Omega \ cm}$$

(ii)
$$\alpha_{60} = \frac{1}{\dfrac{1}{\alpha_{20}} + (t_2 - t_1)} = \frac{1}{254.5 + (60-20)} = \frac{1}{\mathbf{294.5}} / °C$$

Example 1.30. *The filament of a 240 V metal-filament lamp is to be constructed from a wire having a diameter of 0.02 mm and a resistivity at 20° C of 4.3 μΩ cm. If α₂₀ = 0.005/°C, what length of filament is necessary if the lamp is to dissipate 60 W at a filament temperature of 2420°C ?*

Solution. Power to be dissipated by the lamp at 2420°C is
$$\frac{V^2}{R_{2420}} = 60 \quad \therefore \ R_{2420} = \frac{V^2}{60} = \frac{(240)^2}{60} = 960\Omega$$

Now
$$R_{2420} = R_{20}[1 + \alpha_{20}(2420-20)]$$

or
$$960 = R_{20}[1 + 0.005(2420-20)]$$

∴
$$R_{20} = 960/13 \ \Omega$$

Now
$$\rho_{20} = 4.3 \times 10^{-6} \ \Omega \ cm \ ; \quad a = \frac{\pi}{4}d^2 = \frac{\pi}{4}(0.02 \times 10^{-1})^2 \ cm^2$$

∴   Length of filament is
$$l = \frac{a \times R_{20}}{\rho_{20}} = \frac{\pi}{4} \times \frac{(0.02 \times 10^{-1})^2 \times 960}{4.3 \times 10^{-6} \times 13} = \mathbf{54 \ cm}$$

Tutorial Problems

1. The shunt winding of a motor has a resistance of 35.1 Ω at 20°C. Find its resistance at 32.6°C. The temperature co-efficient of copper is 0.00427/°C at 0°C.   **[39.6 Ω]**

2. The resistance of a coil of wire increases from 40 Ω at 10°C to 48.25 Ω at 60°C. Find the temperature coefficient at 0°C of the conductor material.   **[0.0043/°C]**

3. The coil of an electromagnet, made of copper wire, has resistance of 4 Ω at a temperature of 22°C. After operating for 2 days, the coil current is 42 A at a terminal voltage of 210 V. Calculate the average temperature of the coil at that time.   **[86.1°C]**

4. The filament of a 60 watt incandescent lamp possesses a cold resistance of 17.6 Ω at 20°C. The lamp draws a current of 0.25 A when connected to a 240 V source. Calculate the temperature of hot filament. Take temperature co-efficient at 0°C as 0.0055/°C.   **[2571°C]**

5. A nichrome heater is operated at 1500°C. What is the percentage increase in its resistance over that at room temperature (20°C) ? Temperature co-efficient of nichrome is 0.00016/°C at 0°C.   **[23.6%]**

6. Two wires A and B are connected in series at 0°C and resistance of B is 3.5 times that of A. The resistance temperature coefficient of A is 0.4% and that of the combination is 0.1%. Find the resistance temperature coefficient of B.   **[0.0143%]**

7. A d.c. shunt motor after running for several hours on constant voltage mains of 400 V takes a field current of 1.6 A. If the temperature rise is known to be 40°C, what value of extra circuit resistance is required to adjust the field current to 1.6 A when starting from cold at 20°C ? Temperature coefficient = 0.0043/°C at 20°C.   **[36.69 Ω]**

8. A potential difference of 250 V is applied to a copper field coil at a temperature of 15°C and the current is 5 A. What will be the mean temperature of the coil when the current has fallen to 3.91 A, the applied voltage being the same as before ?   **[85°C]**

9. An insulating material has an insulation resistance of 100% at 0°C. For each rise in temperature of 5°C its resistance is reduced by 10%. At what temperature is the insulation resistance halved ?   **[33°C]**

10. A carbon electrode has a resistance of 0.125 Ω at 20°C. The temperature coefficient of carbon is −0.0005 at 20°C. What will the resistance of the electrode be at 85°C ?   **[0.121 Ω]**

1.28. Ohm's Law

The relationship between voltage (V), the current (I) and resistance (R) in a d.c. circuit was first discovered by German scientist George Simon *Ohm. This relationship is called Ohm's law and may be stated as under :

The ratio of potential difference (V) between the ends of a conductor to the current (I) flowing between them is constant, provided the physical conditions (e.g. temperature etc.) do not change i.e.

$$\frac{V}{I} = \text{Constant} = R$$

where R is the resistance of the conductor between the two points considered.

For example, if in Fig. 1.21 (*i*), the voltage between points A and B is V volts and current flowing is I amperes, then V/I will be constant and equal to R, the resistance between points A and B. If the voltage is doubled up, the current will also be doubled up so that the ratio V/I remains constant. If we draw a graph between V and I, it will be a straight line passing through the origin as shown in Fig. 1.21 (*ii*). The resistance R between points A and B is given by slope of the graph *i.e.*

$$R = \tan \theta = V/I = \text{Constant}$$

Ohm's law can be expressed in three forms *viz.*

$$I = V/R \; ; \; V = IR \; ; \; R = V/I$$

These formulae can be applied to any part of a d.c. circuit or to a complete circuit. It may be noted that if voltage is measured in volts and current in amperes, then resistance will be in ohms.

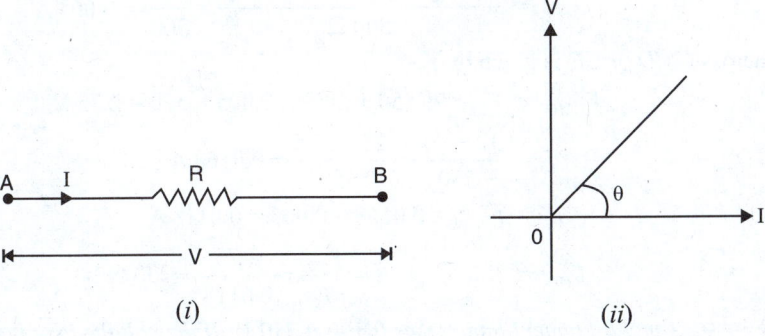

(*i*)   (*ii*)

Fig. 1.21

* The unit of resistance (*i.e.* ohm) was named in his honour.

1.29. Non-ohmic Conductors

Those conductors which do not obey Ohm's law (1 ∝ V) are called non-ohmic conductors e.g., vacuum tubes, transistors, electrolytes, etc. A non-ohmic conductor may have one or more of the following properties :

   (*i*) The *V-I* graph is non-linear *i.e. V/I* is variable.

   (*ii*) The *V-I* graph may not pass through the origin as in case of an ohmic conductor.

   (*iii*) A non-ohmic conductor may conduct poorly or not at all when the p.d. is reversed.

   The non-linear circuit problems are generally solved by graphical methods.

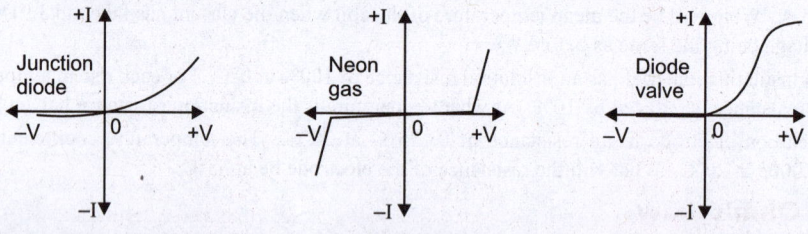

Fig. 1.22

Fig. 1.22 illustrates the graphs of non-ohmic conductors. Note that *V-I* graphs for these non-ohmic conductors are not a straight line.

Example 1.31. *What is the value of the unknown resistor R in Fig. 1.23 (i) if the voltage drop across the 500 Ω resistor is 2.5 volts ? All resistances are in ohm.*

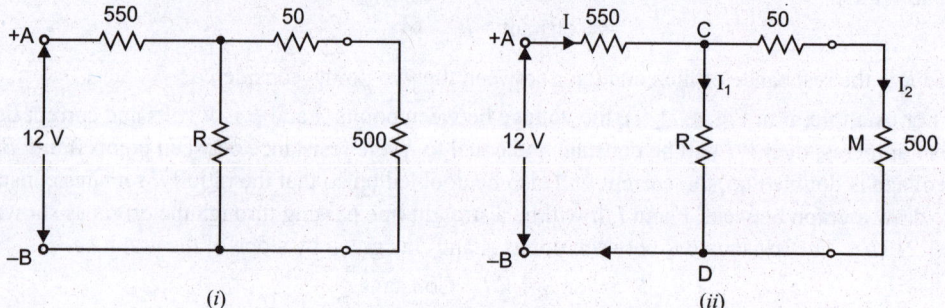

Fig. 1.23

Solution. Fig. 1.23 (*ii*) shows the various currents in the circuit.

$$I_2 = \frac{\text{Voltage drop across 500 }\Omega}{500\,\Omega} = \frac{2.5}{500} = 0.005 \text{ A}$$

Voltage across *CMD* or *CD* is given by ;

$$V_{CMD} = V_{CD} = I_2(50 + 500) = 0.005 \times 550 = 2.75 \text{ V}$$

Now

$$I = \frac{12 - V_{CD}}{550} = \frac{12 - 2.75}{550} = 0.0168 \text{ A}$$

∴

$$I_1 = I - I_2 = 0.0168 - 0.005 = 0.0118 \text{ A}$$

Now

$$V_{CD} = I_1 R \quad \therefore \quad R = \frac{V_{CD}}{I_1} = \frac{2.75}{0.0118} = \textbf{233 }\Omega$$

Example 1.32. *A metal filament lamp takes 0.3 A at 230 V. If the voltage is reduced to 115 V, will the current be halved ? Explain your answer.*

Solution. No. It is because Ohm's law is applicable only if the resistance of the circuit does not change. In the present case, when voltage is reduced from 230 V to 115 V, the temperature of the lamp will decrease too much, resulting in an enormous decrease of lamp resistance. Consequently, Ohm's law ($I = V/R$) cannot be applied. To give an idea to the reader, the hot resistance (*i.e.* at normal operating temperature) of an incandescent lamp is more than 10 times its cold resistance.

Example 1.33. *A coil of copper wire has resistance of 90 Ω at 20°C and is connected to a 230 V supply. By how much must the voltage be increased in order to maintain the current constant if the temperature of the coil rises to 60°C? Take α_0 for copper = 0.00428/°C.*

Solution.
$$R_{20} = R_0(1 + \alpha_0 \times 20) \quad ; \quad R_{60} = R_0(1 + \alpha_0 \times 60)$$

$$\therefore \quad \frac{R_{60}}{R_{20}} = \frac{1 + 0.00428 \times 60}{1 + 0.00428 \times 20} = \frac{1.2568}{1.0856}$$

or
$$R_{60} = R_{20} \times \frac{1.2568}{1.0856} = 90 \times \frac{1.2568}{1.0856} = 104.2\,\Omega$$

Now, current at 20°C $= \dfrac{230}{90} = \dfrac{23}{9}$ A

The wire resistance has become 104.2 Ω at 60°C. Therefore, in order to keep the current constant at the previous value, the new voltage required = (23/9) × 104.2 = 266.3 V.

∴ Required voltage increase = 266.3 – 230 = **36.3 V**

Tutorial Problems

1. A battery has an e.m.f. of 12.8 V and supplies a current of 3.2 A. What is the resistance of the circuit ? How many coulombs leave the battery in 5 minutes ?   **[4 Ω ; 960 C]**

2. In a discharge tube, the number of hydrogen ions (*i.e.* protons) drifting across a cross-section per second is 1.2×10^{18} while the number of electrons drifting in the opposite direction is 2.8×10^{18} per second. If the supply voltage is 220 V, what is the effective resistance of the tube ?   **[344 Ω]**

3. An electromagnet of resistance 12.4 Ω requires a current of 1.5 A to operate it. Find the required voltage.   **[18.6 V]**

4. The cold resistance of a certain gas-filled tungsten lamp is 18.2 Ω and its hot resistance at the operating voltage of 220 V is 202 Ω. Find the current (*i*) at the instant of switching (*ii*) under normal operating conditions.   **[(*i*) 12.08 A (*ii*) 1.09 A]**

1.30. Electric Power

The rate at which work is done in an electric circuit is called its **electric power** *i.e.*

$$\text{Electric power} = \frac{\text{Work done in electric circuit}}{\text{Time}}$$

When voltage is applied to a circuit, it causes current (*i.e.* electrons) to flow through it. Clearly, work is being done in moving the electrons in the circuit. This work done in moving the electrons in a unit time is called the electric power. Thus referring to the part *AB* of the circuit (See Fig. 1.24),

V = P.D. across *AB* in volts

I = Current in amperes

R = Resistance of *AB* in Ω

t = Time in sec. for which current flows

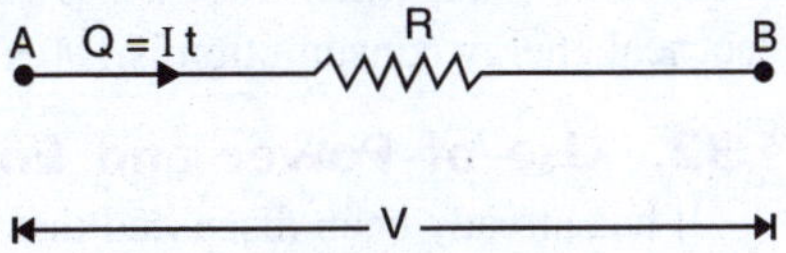

Fig. 1.24

The total charge that flows in *t* seconds is $Q = I \times t$ coulombs and by definition (See Art. 1.12),

$$V = \frac{\text{Work}}{Q}$$

or   Work $= VQ = VIt$ $(\because Q = It)$

$\therefore$ Electric power, $P = \dfrac{\text{Work}}{t} = \dfrac{VIt}{t} = VI$ joules/sec or watts

$\therefore$ $P = VI = I^2R = \dfrac{V^2}{R}$ $[\because V = IR \text{ and } I = V/R]$

The above three formulae are equally valid for calculation of electric power in a d.c. circuit. Which one is to be used depends simply on which quantities are known or most easily determined.

Unit of electric power. The basic unit of electric power is *joules/sec* or *watt*. The power consumed in a circuit is 1 watt if a p.d. of 1 V causes 1 A current to flow through the circuit.

Power in watts = Voltage in volts × Current in amperes

The bigger units of electric power are kilowatts (kW) and megawatts (MW).

1 kW $= 1000$ watts   ;   1 MW $= 10^6$ watts or 10^3 kW

1.31. Electrical Energy

The total work done in an electric circuit is called **electrical energy** *i.e.*

Electrical energy = Electrical power × Time

$$= VIt = I^2Rt = \dfrac{V^2}{R}t$$

The reader may note that formulae for electrical energy can be readily derived by multiplying the electric power by 't', the time for which the current flows. The unit of electrical energy will depend upon the units of electric power and time.

(*i*) If power is taken in watts and time in seconds, then the unit of electrical energy will be *watt-sec. i.e.*

Energy in watt-sec. = Power in watts × Time in sec.

(*ii*) If power is expressed in watts and time in hours, then unit of electrical energy will be *watt-hour i.e.*

Energy in watt-hours = Power in watts × Time in hours

(*iii*) If power is expressed in kilowatts and time in hours, then unit of electrical energy will be *kilowatt-hour* (kWh) *i.e.*

Energy in kWh = Power in kW × Time in hours

It may be pointed out here that in practice, electrical energy is measured in kilowatt-hours (kWh). Therefore, it is profitable to define it.

One kilowatt-hour (kWh) *of electrical energy is expended in a circuit if 1 kW (1000 watts) of power is supplied for 1 hour.*

The electricity bills are made on the basis of total electrical energy consumed by the consumer. The unit for charge of electricity is 1 kWh. One kWh is also called Board of Trade (B.O.T.) unit or simply unit. Thus when we say that a consumer has consumed 100 units of electricity, it means that electrical energy consumption is 100 kWh.

1.32. Use of Power and Energy Formulas

It has already been discussed that electric power as well as electrical energy consumed can be expressed by three formulas. While using these formulas, the following points may be kept in mind:

(*i*)   Electric power, $P = I^2R = \dfrac{V^2}{R}$ watts

Electrical energy consumed, $W = I^2Rt = \dfrac{V^2}{R}t$ joules

The above formulas apply *only* to resistors and to devices (*e.g.* electric bulb, heater, electric kettle etc) where all electrical energy consumed is converted into heat.

(*ii*)   Electric power, $P = VI$ watts

                Electrical energy consumed, $W = VIt$ joules

These formulas apply to any type of load including the one mentioned in point (*i*).

Example 1.34. *A 100 V lamp has a hot resistance of 250 Ω. Find the current taken by the lamp and its power rating in watts. Calculate also the energy it will consume in 24 hours.*

Solution. Current taken by lamp, $I = V/R = 100/250 =$ **0.4 A**

        Power rating of lamp, $P = VI = 100 \times 0.4 =$ **40 W**

Energy consumption in 24 hrs. $=$ Power $\times$ time $= 40 \times 24 =$ **960 watt-hours**

Example 1.35. *A heating element supplies 300 kilojoules in 50 minutes. Find the p.d. across the element when current is 2 amperes.*

Solution.   Total charge, $Q = I \times t = 2 \times 50 \times 60 = 6000$ C

$$P.D., V = \frac{\text{Work}}{\text{Charge}} = \frac{300 \times 10^3}{6000} = \textbf{50 V}$$

Example 1.36. *A 10 watt resistor has a value of 120 Ω. What is the rated current through the resistor ?*

Solution.   Rated power, $P = I^2 R$

$\therefore$ Rated current, $I = \sqrt{\dfrac{P}{R}} = \sqrt{\dfrac{10}{120}} =$ **0.2887 A**

If current through the resistor exceeds this value, the resistor will be burnt due to excessive heat.

Note. Every electrical equipment has power and current ratings marked on its body. While the equipment is in operation, care should be taken that neither of these limits is exceeded, otherwise the equipment may be damaged/burnt due to excessive heat.

Example 1.37. *The following are the details of load on a circuit connected through a supply metre :*

   (*i*) *Six lamps of 40 watts each working for 4 hours per day*

  (*ii*) *Two flourescent tubes 125 watts each working for 2 hours per day*

 (*iii*) *One 1000 watt heater working for 3 hours per day*

If each unit of energy costs 70 P, what will be the electricity bill for the month of June ?

Solution. Total wattage of lamps $= 40 \times 6 = 240$ watts

              Total wattage of tubes $= 125 \times 2 = 250$ watts

              Wattage of heater $= 1000$ watts

Energy consumed by the appliances per day

$$= (240 \times 4) + (250 \times 2) + (1000 \times 3)$$
$$= 4460 \text{ watt-hours} = 4.46 \text{ kWh}$$

Total energy consumed in the month of June (*i.e.* in 30 days)

$$= 4.46 \times 30 = 133.8 \text{ kWh}$$

Bill for the month of June $=$ Rs. $0.7 \times 133.8 =$ **Rs. 93.66**

(Tutorial Problems)

1. A resistor of 50 Ω has a p.d. of 100 volts d.c. across it for 1 hour. Calculate (*i*) power and (*ii*) energy.

                                                       [(*i*) 200 watts (*ii*) 7.2×10^5 J]

2. A current of 10 A flows through a resistor for 10 minutes and the power dissipated by the resistor is 100 watts. Find the p.d. across the resistor and the energy supplied to the circuit.   **[10 V ; 6×10^4 J]**

3. A factory is supplied with power at 210 volts through a pair of feeders of total resistance 0.0225 Ω. The load consists of 354, 250 V, 60 watt lamps and 4 motors each taking 40 amperes. Find :

    (*i*) total current required

    (*ii*) voltage at the station end of feeders

    (*iii*) power wasted in feeders.   [(*i*) 231.4 A (*ii*) 215.78 V (*iii*) 1.4 kW]

4. How many kilowatts will be required to light a factory in which 250 lamps each taking 1.3 A at 230 V are used ?   [74.75 kW]

1.33. Power Rating of a Resistor

The ability of a resistor to dissipate power as heat without destructive temperature build-up is called **power rating** *of the resistor.*

Power rating of resistor $= I^2R$ or V^2/R [See Fig. 1.25]

Suppose the power rating of a resistor is 2 W. It means that I^2R or V^2/R should not exceed 2 W. Suppose the quantity I^2R (or V^2/R) for this resistor becomes 4 W. The resistor is able to dissipate 2 W as heat and the remaining 2 W will start building up the temperature. In a matter of seconds, the resistor will burn out.

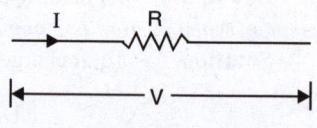

Fig. 1.25

The physical size of a resistor is not necessarily related to its resistance value but rather to its *power rating*. A large resistor is able to dissipate (throw off) more heat because of its large physical size. In general, the greater the physical size of a resistor, the greater is its power rating and vice-versa.

Example 1.38. *A 0.1 Ω resistor has a power rating of 5 W. Is this resistor safe when conducting a current of 10 A ?*

Solution. Power developed in the resistor is
$$P = I^2R = (10)^2 \times 0.1 = 10 \text{ W}$$
The resistor is **not safe** since the power developed in the resistor exceeds its dissipation rating.

Example 1.39. *What is the maximum safe current flow in a 47 Ω, 2 W resistor ?*

Solution.   Power rating $= I^2R$

or $2 = I^2 \times 47$ ∴   Maximum safe current, $I = \sqrt{\dfrac{2}{47}} = $ **0.21 A**

Example 1.40. *What is the maximum voltage that can be applied across a 100 Ω, 10 W resistor in order to keep within the resistor's power rating ?*

Solution.   Power rating $= V^2/R$

or $10 = V^2/100$ ∴   Max. safe voltage, $V = \sqrt{10 \times 100} = $ **31.6 volts**

Tutorial Problems

1. A 200 Ω resistor has a 2 W power rating. What is the maximum current that can flow in the resistor without exceeding the power rating ?   [100 mA]

2. A 6.8 kΩ, 0.25 W resistor shows a potential difference of 40 V. Is the resistor safe ?   [Yes]

3. A 1.5 kΩ resistor has 1 W power rating. What maximum voltage can be applied across the resistor without exceeding the power rating ?   [38.73 V]

1.34. Nonlinear Resistors

A device or circuit element whose V/I characteristic is not a straight line is said to exhibit **nonlinear resistance.**

The examples of nonlinear resistors are thermistors, varistors, diodes, filaments of incandescent lamps etc.

1. Thermistors. *A* **thermistor** *is a heat sensitive device usually made of a semiconductor material whose resistance changes very rapidly with change of temperature.* A thermistor has the following important properties :

(*i*) The resistance of a thermistor changes very rapidly with change of temperature.

(*ii*) The temperature coefficient of a thermistor is very high.

(*iii*) The temperature co-efficient of a thermistor can be both positive and negative.

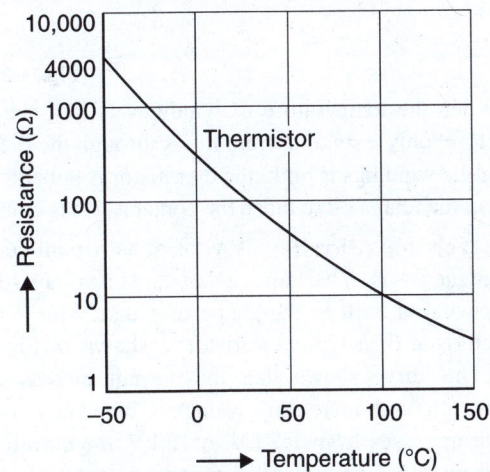

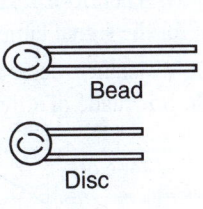

Fig. 1.26

Fig. 1.27

Construction. Thermistors are made from semiconductor oxides of iron, nickel and cobalt. They are generally in the form of beads, discs or rods (See Fig. 1.26). A pair of platinum leads are attached at the two ends for electrical connections. The arrangement is enclosed in a very small glass bulb and sealed.

Fig. 1.27 shows the resistance/temperature characteristic of a typical thermistor with negative temperature coefficient. The resistance decreases progressively from 4000 Ω to 3 Ω as its temperature varies from − 50°C to +150°C.

Applications

(*a*) A thermistor with negative temperature coefficient of resistance may be used to safeguard against current surges in a circuit where this could be harmful *e.g.* in a circuit where the heaters of the radio valves are in series (See Fig. 1.28).

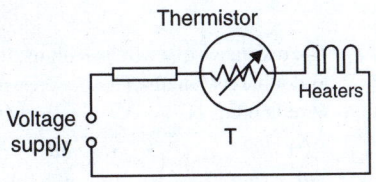

Fig. 1.28

A thermistor *T* is included in the circuit. When the supply voltage is switched on, the thermistor has a high resistance at first because it is cold. It thus limits the current to a moderate value. As it warms up, the thermistor resistance drops appreciably and an increased current then flows through the heaters.

(*b*) A thermistor with a negative temperature coefficient can be used to issue an alarm for excessive temperature of winding of motors, transformers and generators [See Fig. 1.29].

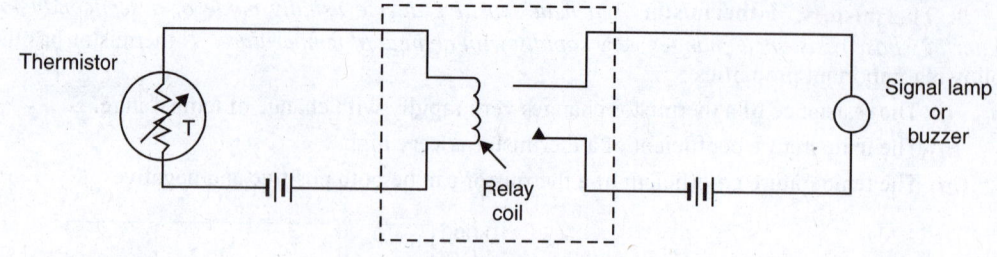

Fig. 1.29

When the temperature of windings is low, the thermistor is cool and its resistance is high. Therefore, only a small current flows through the thermistor and the relay coil. When the temperature of the windings is high, the thermistor is hot and its resistance is low. Therefore, a large current flows in the relay coil to close the contacts. This completes the circuit for the signal lamp or buzzer.

2. Varistor (Thyrite). A varistor is a nonlinear resistor whose resistance decreases as the voltage increases. Therefore, a varistor is a voltage-dependent resistor. It is made of silicon-carbide powder and is built in the shape of a disc. The *V-I* characteristic of a typical varistor is shown in Fig. 1.30. The curve shows that the current increases dramatically with increasing voltage. Thus when the voltage increases from 1.5 kV to 10 kV, the current rises from 1 mA to 100 A. Varistors are placed in parallel with critical components which might be damaged by high transient voltages. Under normal conditions, the varistor remains in high-resistance state and draws very little current. On the application of surge, the varistor is driven to its low-resistance state. The varistor then conducts a relatively large amount of current and dissipates much of the surge as heat. Thus the component is saved from damage.

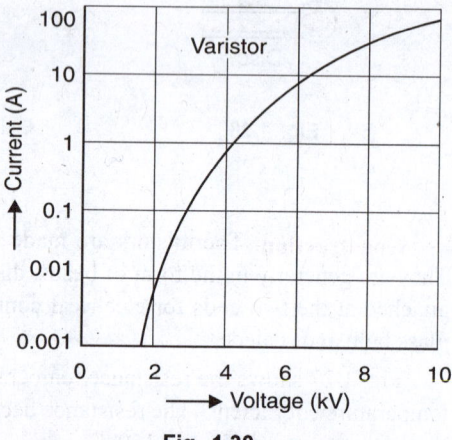

Fig. 1.30

OBJECTIVE QUESTIONS

1. The resistance of a wire is R ohms. It is stretched to double its length. The new resistance of the wire in ohms is

    (*i*) $R/2$ (*ii*) $2R$

    (*iii*) $4R$ (*iv*) $R/4$

2. The example of non-ohmic resistance is

    (*i*) copper wire

    (*ii*) carbon resistance

    (*iii*) tungsten wire

    (*iv*) diode

3. In which of the following substances, the resistance decreases with the increase of temperature

    (*i*) carbon   (*ii*) constantan

    (*iii*) copper   (*iv*) silver

4. The resistance of a wire of uniform diameter d and length l is R. The resistance of another wire of the same material but diameter $2d$ and length $4l$ will be

    (*i*) $2R$ (*ii*) R

    (*iii*) $R/2$ (*iv*) $R/4$

5. The temperature coefficient of resistance of a wire is 0.00125 °C^{-1}. At 300 K, its resistance is one ohms. The resistance of the wire will be 2 ohms at

    (*i*) 1154 K   (*ii*) 1100 K

    (*iii*) 1400 K   (*iv*) 1127 K

6. The resistance of 20 cm long wire is 5 ohms. The wire is stretched to a uniform wire of 40 cm length. The resistance now will be (in ohms)

    (*i*) 5   (*ii*) 10

    (*iii*) 20   (*iv*) 200

7. A current of 4.8 A is flowing in a conductor. The number of electrons flowing per second through the X-section of conductor will be

(i) 3×10^{19} electrons

(ii) 76.8×10^{20} electrons

(iii) 7.68×10^{20} electrons

(iv) 3×10^{20} electrons

8. A carbon resistor has coloured strips as brown, green, orange and silver respectively. The resistance is

(i) $15\,k\Omega \pm 10\%$ (ii) $10\,k\Omega \pm 10\%$

(iii) $15\,k\Omega \pm 5\%$ (iv) $10\,k\Omega \pm 5\%$

9. A wire has a resistance of $10\,\Omega$. It is stretched by one-tenth of its original length. Then its resistance will be

(i) $10\,\Omega$ (ii) $12.1\,\Omega$

(iii) $9\,\Omega$ (iv) $11\,\Omega$

10. A 10 m long wire of resistance $20\,\Omega$ is connected in series with a battery of e.m.f. 3 V (negligible internal resistance) and a resistance of $10\,\Omega$. The potential gradient along the wire in volt per metre is

(i) 0.02   (ii) 0.1

(iii) 0.2   (iv) 1.2

11. The diameter of an atom is about

(i) 10^{-10} m   (ii) 10^{-8} m

(iii) 10^{-2} m   (iv) 10^{-15} m

12. 1 cm^3 of copper at room temperature has about

(i) 200 free electrons

(ii) 20×10^{10} free electrons

(iii) 8.5×10^{22} free electrons

(iv) 3×10^5 free electrons

13. The electric current is due to the flow of

(i) positive charges only

(ii) negative charges only

(iii) both positive and negative charges

(iv) neutral particles only

14. The quantity of charge that will be transferred by a current flow of 10 A over 1 hour period is

(i) 10 C   (ii) 3.6×10^4 C

(iii) 2.4×10^3 C   (iv) 1.6×10^2 C

15. The drift velocity of electrons is of the order of

(i) $1\ ms^{-1}$ (ii) $10^{-3}\ ms^{-1}$

(iii) $10^6\ ms^{-1}$ (iv) $3 \times 10^8\ ms^{-1}$

16. Insulators have .... temperature co-efficient of resistance.

(i) zero   (ii) positive

(iii) negative   (iv) none of the above

17. Eureka has .... temperature co-efficient of resistance.

(i) almost zero   (ii) negative

(iii) positive   (iv) none of the above

18. Constantan wire is used for making standard resistances because it has ....

(i) low specific resistance

(ii) high specific resistance

(iii) negligibly small temperature co-efficient of resistance

(iv) high melting point

19. Two resistors A and B have resistances R_A and R_B respectively with $R_A < R_B$. The resistivities of their materials are ρ_A and ρ_B. Then,

(i) $\rho_A > \rho_B$ (ii) $\rho_A = \rho_B$

(iii) $\rho_A < \rho_B$

(iv) Information insufficient

20. In case of liquids, Ohm's law is ....

(i) fully obeyed

(ii) partially obeyed

(iii) there is no relation between current and p.d.

(iv) none of the above.

ANSWERS

1. (iii)	2. (ii) and (iii)	3. (i)	4. (ii)	5. (ii)
6. (iii)	7. (i)	8. (i)	9. (ii)	10. (iii)
11. (i)	12. (iii)	13. (iii)	14. (ii)	15. (ii)
16. (iii)	17. (i)	18. (iii)	19. (iv)	20. (i)

2

D.C. Circuits

Introduction

It is well known that electric current flows in a closed path. The closed path followed by electric current is called an electric circuit. The essential parts of an electric circuit are (*i*) the source of power (*e.g.* battery, generator etc.), (*ii*) the conductors used to carry current and (*iii*) the load* (*e.g.* lamp, heater, motor etc.). The source supplies electrical energy to the load which converts it into heat or other forms of energy. Thus, conversion of electrical energy into other forms of energy is possible only with suitable circuits. For instance, conversion of electrical energy into mechanical energy is achieved by devising a suitable motor circuit. In fact, the innumerable uses of electricity have been possible only due to the proper use and application of electric circuits. In this chapter, we shall confine our discussion to d.c. circuits only *i.e.* circuits carrying direct current.

2.1. D.C. Circuit

The closed path followed by direct current (d.c.) is called a **d.c. circuit.**

A d.c. circuit essentially consists of a source of d.c. power (*e.g.* battery, d.c. generator etc.), the conductors used to carry current and the load. Fig. 2.1 shows a torch bulb connected to a battery through conducting wires. The direct current **starts from the positive terminal of the battery and comes back to the starting point *via* the load. The direct current follows the path *ABCDA* and *ABCDA* is a d.c. circuit. The load for a d.c. circuit is usually a *** resistance. In a d.c. circuit, loads (*i.e.* resistances) may be connected in series or parallel or series-parallel. Accordingly, d.c. circuits can be classified as :

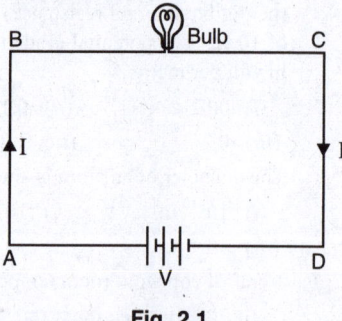

Fig. 2.1

    (*i*) Series circuits

   (*ii*) Parallel circuits

  (*iii*) Series-parallel circuits.

2.2. D.C. Series Circuit

The d.c. circuit in which resistances are connected end to end so that there is only one path for current to flow is called a **d.c. series circuit.**

Consider three resistances R_1, R_2 and R_3 ohms connected in series across a battery of V volts as shown in Fig. 2.2 (*i*). Obviously, there is only one path for current I *i.e.* current is same throughout the circuit. By Ohm's law, voltage across the various resistances is

$$V_1 = I R_1 \; ; \; V_2 = I R_2 \; ; \; V_3 = I R_3$$

Now
$$V = V_1 + V_2 + V_3$$
$$= I R_1 + I R_2 + I R_3$$

---

*   The device which utilises electrical energy is called load. For instance, heater converts electrical energy supplied to it into heat. Therefore, heater is the load.

**   This is the direction of conventional current. However, the electron flow will be in the opposite direction.

*** Other passive elements *viz.* inductance and capacitance are relevant only in a.c. circuits.

$$= I(R_1 + R_2 + R_3)$$

or
$$\frac{V}{I} = R_1 + R_2 + R_3$$

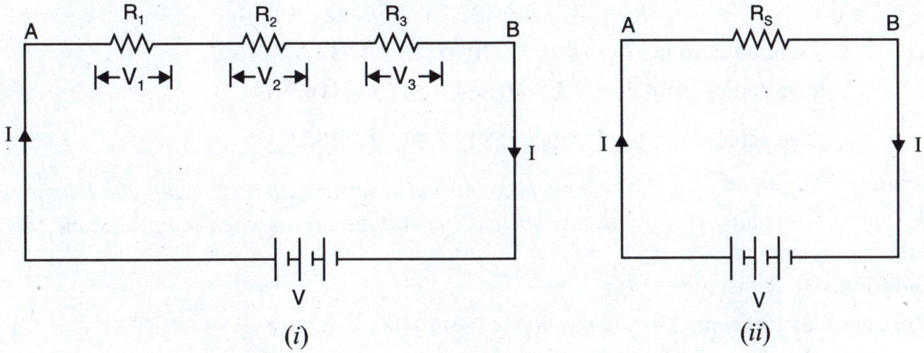

Fig. 2.2

But V/I is the total resistance R_S between points A and B. Note that R_S is called the *total or equivalent resistance of the three resistances.

$\therefore$
$$R_S = R_1 + R_2 + R_3$$

Hence when a number of resistances are connected in series, the total resistance is equal to the sum of the individual resistances.

The total conductance G_S of the circuit is given by ;

$$G_S = \frac{1}{R_S} = \frac{1}{R_1 + R_2 + R_3}$$

Also
$$\frac{1}{G_S} = \frac{1}{G_1} + \frac{1}{G_2} + \frac{1}{G_3}$$

The main characteristics of a series circuit are :

(*i*) The current in each resistor is the same.

(*ii*) The total resistance in the circuit is equal to the sum of individual resistances.

(*iii*) The total power dissipated in the circuit is equal to the sum of powers dissipated in individual resistances. Thus referring to Fig. 2.2 (*i*),

$$R_S = R_1 + R_2 + R_3$$

or
$$I^2 R_S = I^2 R_1 + I^2 R_2 + I^2 R_3$$

or
$$P_S = P_1 + P_2 + P_3$$

Thus total power dissipated in a series circuit is equal to the sum of powers dissipated in individual resistances. As we shall see, this is also true for parallel and series-parallel d.c. circuits.

Note. A series resistor circuit [See Fig. 2.2 (*i*)] can be considered to be a *voltage divider circuit* because the potential difference across any one resistor is a fraction of the total voltage applied across the series combination; the fraction being determined by the values of the resistances.

Example 2.1. *Two filament lamps A and B take 0.8 A and 0.9 A respectively when connected across 110 V supply. Calculate the value of current when they are connected in series across a 220-V supply, assuming the filament resistances to remain unaltered. Also find the voltage across each lamp.*

---

*   Total or equivalent resistance is the single resistance, which if substituted for the series resistances, would provide the same current in the circuit.

Solution.   For lamp A, $R_A = 110/0.8 = 137.5\ \Omega$

For lamp B, $R_B = 110/0.9 = 122.2\ \Omega$

When the lamps are connected in series, total resistance is

$$R_S = 137.5 + 122.2 = 259.7\ \Omega$$

$\therefore$ Circuit current, $I = V/R_S = 220/259.7 = \textbf{0.847 A}$

Voltage across lamp $A = I\,R_A = 0.847 \times 137.5 = \textbf{116.5 V}$

Voltage across lamp $B = I\,R_B = 0.847 \times 122.2 = \textbf{103.5 V}$

Example 2.2. *A 100 watt, 250 V lamp is connected in series with a 100 watt, 200 V lamp across 250 V supply. Calculate (i) circuit current and (ii) voltage across each lamp. Assume the lamp resistances to remain unaltered.*

Solution. (*i*) Resistance, $R = \dfrac{V^2}{P}$

Resistance of 100 watt, 250 V lamp, $R_1 = (250)^2/100 = 625\ \Omega$

Resistance of 100 watt, 200 V lamp, $R_2 = (200)^2/100 = 400\ \Omega$

When the lamps are connected in series, total resistance is

$$R_S = 625 + 400 = 1025\ \Omega$$

$\therefore$ Circuit current, $I = V/R_S = 250/1025 = \textbf{0.244 A}$

(*ii*) Voltage across 100 W, 250 V lamp $= I\,R_1 = 0.244 \times 625 = \textbf{152.5 V}$

Voltage across 100 W, 200 V lamp $= I\,R_2 = 0.244 \times 400 = \textbf{97.6 V}$

Example 2.3. *The element of 500 watt electric iron is designed for use on a 200 V supply. What value of resistance is needed to be connected in series in order that the iron can be operated from 240 V supply?*

Solution. Current rating of iron, $I = \dfrac{\text{Wattage}}{\text{Voltage}} = \dfrac{500}{200} = 2.5\ \text{A}$

If R ohms is the required value of resistance to be connected in series, then voltage to be dropped across this resistance $= 240 - 200 = 40$ V.

$\therefore$ $R = 40/2.5 = \textbf{16 }\boldsymbol{\Omega}$

Example 2.4. *Determine the resistance and the power dissipation of a resistor that must be placed in series with a 75 - ohm resistor across 120 V source in order to limit the power dissipation in the 75 - ohm resistor to 90 watts.*

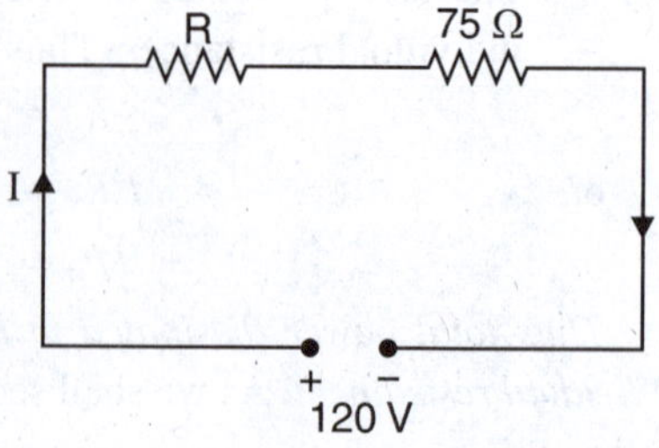

Solution. Fig. 2.3 represents the conditions of the problem.

$$I^2 \times 75 = 90$$

$\therefore$ $I = \sqrt{90/75} = 1.095\ \text{A}$

Now, $I = \dfrac{120}{R + 75}$

or $1.095 = \dfrac{120}{R + 75}$

$\therefore$ $R = \textbf{34.6 }\boldsymbol{\Omega}$

Fig. 2.3

Power dissipation in $R = I^2 R = (1.095)^2 \times 34.6 = \textbf{41.5 watts}$

Example 2.5. *A generator of e.m.f. E volts and internal resistance r ohms supplies current to a water heater. Calculate the resistance R of the heater so that three-quarter of the total energy developed by the generator is absorbed by the water.*

Solution. Current supplied by generator, $I = \dfrac{E}{R+r}$

Power developed by generator $= E\,I = \dfrac{E^2}{R+r}$

Power dissipated by heater $= I^2 R = R \times \dfrac{E^2}{(R+r)^2} = \dfrac{E^2 R}{(R+r)^2}$

As per the conditions of the problem, we have,

$$\dfrac{E^2 R}{(R+r)^2} = \dfrac{3}{4} \times \dfrac{E^2}{R+r} \quad \text{or} \quad \dfrac{R}{R+r} = \dfrac{4}{3} \quad \therefore \quad R = \mathbf{3\,r}$$

Example 2.6. *A direct current arc has a voltage/current relation expressed as :*

$$V = 44 + \dfrac{30}{I} \ volts$$

It is connected in series with a resistor across 100 V supply. If voltages across the arc and resistor are equal, find the ohmic value of the resistor.

Solution. Let R ohms be resistance of the resistor. The voltage across the arc as well as resistor $= 50$ volts.

Now

$$50 = 44 + \dfrac{30}{I} \quad \therefore \quad I = 5\,A$$

$\therefore$

$$R = \dfrac{V}{I} = \dfrac{50}{5} = \mathbf{10\,\Omega}$$

Tutorial Problems

1. If the resistance of a circuit having 12 V source is increased by 4 Ω, the current drops by 0.5 A. What is the original resistance of the circuit ? **[8 Ω]**

2. A searchlight takes 100 A at 80 V. It is to be operated from a 220 V supply. Find the value of the resistor to be connected in series. **[1.4 Ω]**

3. The maximum resistance of a rheostat is 4.8 Ω and the minimum resistance is 0.5 Ω. Find for each condition the voltage across the rheostat when current is 1.2 A. **[5.76V ; 0.6V]**

4. What is the drop across the 150 Ω resistor in Fig. 2.4 ? **[5.33 V]**

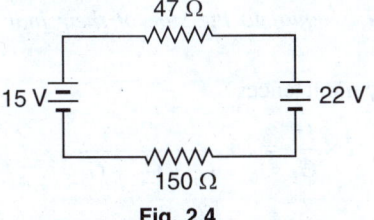

Fig. 2.4   Fig. 2.5

5. Calculate the current flow for Fig. 2.5. **[3.51 mA]**

2.3. D.C. Parallel Circuit

When one end of each resistance is joined to a common point and the other end of each resistance is joined to another common point so that there are as many paths for current flow as the number of resistances, it is called a **parallel circuit.**

Consider three resistances R_1, R_2 and R_3 ohms connected in parallel across a battery of V volts as shown in Fig. 2.6 (*i*). The total current I divides into three parts : I_1 flowing through R_1, I_2 flowing through R_2 and I_3 flowing through R_3. Obviously, the voltage across each resistance is the same (*i.e. V volts in this case*) and there are as many current paths as the number of resistances. By Ohm's law, current through each resistance is

$$I_1 = V/R_1 \; ; \; I_2 = V/R_2 \; ; \; I_3 = V/R_3$$

Now,
$$I = I_1 + I_2 + I_3$$

$$= \frac{V}{R_1} + \frac{V}{R_2} + \frac{V}{R_3}$$

$$= V\left(\frac{1}{R_1} + \frac{1}{R_2} + \frac{1}{R_3}\right)$$

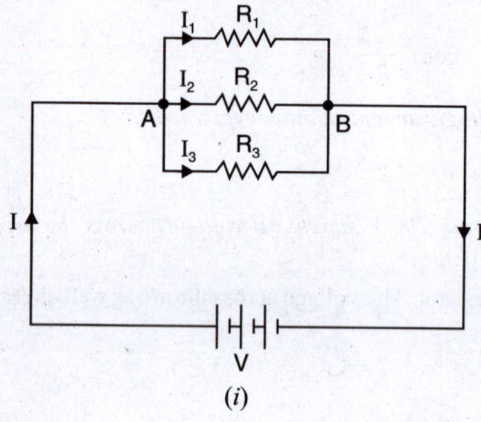 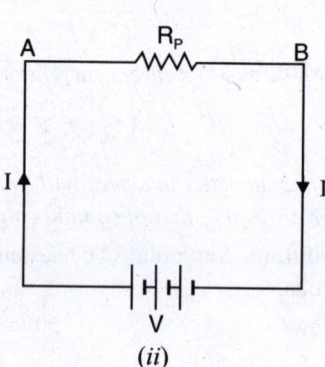

(i)   *(ii)*

Fig. 2.6

or
$$\frac{I}{V} = \frac{1}{R_1} + \frac{1}{R_2} + \frac{1}{R_3}$$

But V/I is equivalent resistance R_P of the parallel resistances [See Fig. 2.6 *(ii)*] so that $I/V = 1/R_P$.

$\therefore$
$$\frac{1}{R_P} = \frac{1}{R_1} + \frac{1}{R_2} + \frac{1}{R_3}$$

Hence when a number of resistances are connected in parallel, the reciprocal of total resistance is equal to the sum of the reciprocals of the individual resistances.

Also
$$G_P = G_1 + G_2 + G_3$$

Hence total conductance G_P of resistors in parallel is equal to the sum of their individual conductances.

We can also express currents I_1, I_2 and I_3 in terms of conductances.

$$I_1 = \frac{V}{R_1} = VG_1 = \frac{I}{G_P} G_1 = I \times \frac{G_1}{G_P} = I \times \frac{G_1}{G_1 + G_2 + G_3}$$

Similarly,
$$I_2 = I \times \frac{G_2}{G_1 + G_2 + G_3} \; ; \quad I_3 = I \times \frac{G_3}{G_1 + G_2 + G_3}$$

2.4. Main Features of Parallel Circuits

The following are the characteristics of a parallel circuit :

(i) The voltage across each resistor is the same.

(ii) The current through any resistor is inversely proportional to its resistance.

(iii) The total current in the circuit is equal to the sum of currents in its parallel branches.

(iv) The reciprocal of the total resistance is equal to the sum of the reciprocals of the individual resistances.

(v) As the number of parallel branches is increased, the total resistance of the circuit is decreased.

(vi) The total resistance of the circuit is always less than the smallest of the resistances.

(vii) If n resistors, each of resistance R, are connected in parallel, then total resistance $R_P = R/n$.

(viii) The conductances are additive.

(ix) The total power dissipated in the circuit is equal to the sum of powers dissipated in the individual resistances. Thus referring to Fig. 2.6 *(i)*,

$$\frac{1}{R_P} = \frac{1}{R_1} + \frac{1}{R_2} + \frac{1}{R_3}$$

or

$$\frac{V^2}{R_P} = \frac{V^2}{R_1} + \frac{V^2}{R_2} + \frac{V^2}{R_3}$$

or

$$P_P = P_1 + P_2 + P_3$$

Like a series circuit, *the total power dissipated in a parallel circuit is equal to the sum of powers dissipated in the individual resistances.*

Note. A parallel resistor circuit [See Fig. 2.6 *(i)*] can be considered to be a *current divider circuit* because the current through any one resistor is a fraction of the total circuit current; the fraction depending on the values of the resistors.

2.5. Two Resistances in Parallel

A frequent special case of parallel resistors is a circuit that contains two resistances in parallel. Fig. 2.7 shows two resistances R_1 and R_2 connected in parallel across a battery of V volts. The total current I divides into two parts; I_1 flowing through R_1 and I_2 flowing through R_2.

(i) **Total resistance R_P.** $\dfrac{1}{R_P} = \dfrac{1}{R_1} + \dfrac{1}{R_2} = \dfrac{R_2 + R_1}{R_1 R_2}$

$\therefore$ $R_P = \dfrac{R_1 R_2}{R_1 + R_2} = \dfrac{\text{Product}}{\text{Sum}}$

Hence the total value of two resistors connected in parallel is equal to the product divided by the sum of the two resistors.

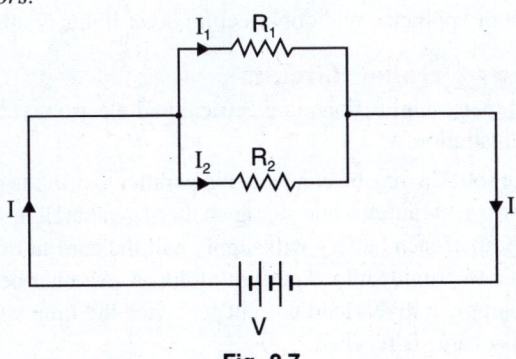

Fig. 2.7

(ii) **Branch Currents.** $R_P = \dfrac{R_1 R_2}{R_1 + R_2}$

$$V = I R_P = I \frac{R_1 R_2}{R_1 + R_2}$$

Current through R_1, $I_1 = \dfrac{V}{R_1} = I \dfrac{R_2}{R_1 + R_2}$ $\left[\text{Putting } V = I \dfrac{R_1 R_2}{R_1 + R_2} \right]$

Current through R_2, $I_2 = \dfrac{V}{R_2} = I \dfrac{R_1}{R_1 + R_2}$

Hence in a parallel circuit of two resistors, the current in one resistor is the line current (i.e. total current) times the opposite resistor divided by the sum of the two resistors.

We can also express currents in terms of conductances.

$$G_P = G_1 + G_2$$

$$I_1 = \frac{V}{R_1} = VG_1 = \frac{I}{G_P} \times G_1 = I \times \frac{G_1}{G_P} = I \times \frac{G_1}{G_1 + G_2}$$

$$I_2 = \frac{V}{R_2} = VG_2 = \frac{I}{G_P} \times G_2 = I \times \frac{G_2}{G_P} = I \times \frac{G_2}{G_1 + G_2}$$

Note. When two resistances are connected in parallel and one resistance is much greater than the other, then the total resistance of the combination is very nearly equal to the smaller of the two resistances. For example, if $R_1 = 10 \ \Omega$ and $R_2 = 10 \ k\Omega$ and they are connected in parallel, then total resistance R_P of the combination is given by ;

$$R_P = \frac{R_1 R_2}{R_1 + R_2} = \frac{10 \times 10^4}{10 + 10^4} = \frac{10^5}{10,010} = 9.99 \ \Omega \eqsim R_1$$

In general, if R_2 is 10 times (or more) greater than R_1, then their combined resistance in parallel is nearly equal to R_1.

2.6. Advantages of Parallel Circuits

The most useful property of a parallel circuit is the fact that potential difference has the same value between the terminals of each branch of parallel circuit. This feature of the parallel circuit offers the following advantages :

(*i*) The appliances rated for the same voltage but different powers can be connected in parallel without disturbing each other's performance. Thus a 230 V, 230 W TV receiver can be operated independently in parallel with a 230 V, 40 W lamp.

(*ii*) If a break occurs in any one of the branch circuits, it will have no effect on other branch circuits.

Due to above advantages, electrical appliances in homes are connected in parallel. We can switch on or off any light or appliance without affecting other lights or appliances.

2.7. Applications of Parallel Circuits

Parallel circuits find many applications in electrical and electronic circuits. We shall give two applications by way of illustration.

(*i*) Identical voltage sources may be connected in parallel to provide a greater current capacity. Fig. 2.8 shows two 12 V automobile storage batteries in parallel. If the starter motor draws 400 A at starting, then each battery will supply half the current *i.e.* 200 A. A single battery might not be able to provide a load current of 400 A. Another benefit is that two batteries in parallel will supply a given load current for twice the time when compared to a single battery before discharge is reached.

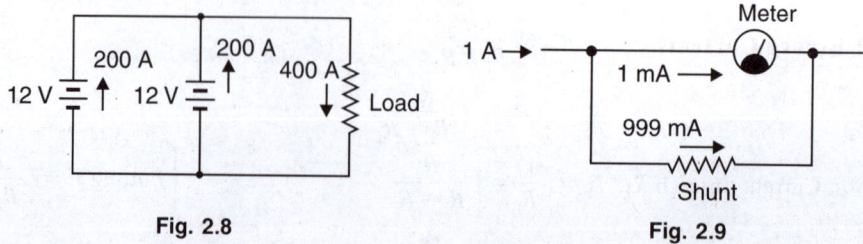

Fig. 2.8   **Fig. 2.9**

(*ii*) Fig. 2.9 shows another application for parallel connection. A low resistor, called a *shunt*, is connected in parallel with an ammeter to increase the current range of the meter. If shunt is not used, the ammeter is able to measure currents up to 1 mA. However, the use of shunt permits to measure currents up to 1 A. Thus shunt increases the range of the ammeter.

Example 2.7. *Two coils connected in series have a resistance of 18 Ω and when connected in parallel have a resistance of 4 Ω. Find the value of resistances.*

Solution. Let R_1 and R_2 be the resistances of the coils. When resistances are connected in series, $R_S = 18\ \Omega$.

$$\therefore \qquad\qquad R_1 + R_2 = 18 \qquad\qquad\qquad ...(i)$$

When resistances are connected in parallel, $R_P = 4\ \Omega$.

$$\therefore \qquad\qquad 4 = \frac{R_1 R_2}{R_1 + R_2} \qquad\qquad\qquad ....(ii)$$

Multiplying Eqns. (*i*) and (*ii*), we get, $R_1 R_2 = 18 \times 4 = 72$

Now $\qquad\qquad R_1 - R_2 = \sqrt{(R_1 + R_2)^2 - 4R_1 R_2} = \sqrt{(18)^2 - 4 \times 72}$

$$\therefore \qquad\qquad R_1 - R_2 = \pm 6 \qquad\qquad\qquad ...(iii)$$

Solving Eqns. (*i*) and (*iii*), we get, $R_1 = $ **12 Ω** or **6 Ω** ; $R_2 = $ **6 Ω** or **12 Ω**

Example 2.8. *A 100 watt, 250 V lamp is connected in parallel with an unknown resistance R across a 250 V supply. The total power dissipated in the circuit is 1100 watts. Find the value of unknown resistance. Assume the resistance of lamp remains unaltered.*

Solution. The total power dissipated in the circuit is equal to the sum of the powers consumed by the lamp and unknown resistance R.

$$\therefore \qquad \text{Power consumed by } R = 1100 - 100 = 1000 \text{ watts}$$

$$\therefore \qquad \text{Value of resistance, } R = \frac{V^2}{\text{Power consumed}} = \frac{(250)^2}{1000} = \textbf{62.5 Ω}$$

Example 2.9. *A coil has a resistance of 5.2 ohms; the resistance has to be reduced to 5 Ω by connecting a shunt across the coil. If this shunt is made of manganin wire of diameter 0.025 cm, find the length of wire required. Specific resistance for manganin is 47×10^{-8} Ω m.*

Solution. Let R ohms be the required resistance of the shunt.

$$R_P = \frac{R \times 5.2}{R + 5.2} \quad \text{or} \quad 5 = \frac{5.2R}{R + 5.2} \qquad \therefore \quad R = 130\ \Omega$$

$$a = \frac{\pi}{4}\left(0.025 \times 10^{-2}\right)^2 = 490 \times 10^{-10}\ \text{m}^2 \;;\; \rho = 47 \times 10^{-8}\ \Omega\text{-}m$$

Now $\qquad\qquad R = \rho\dfrac{l}{a}$

$$\therefore \qquad l = \frac{Ra}{\rho} = \frac{130 \times (490 \times 10^{-10})}{47 \times 10^{-8}} = \textbf{13.55 m}$$

Example 2.10. *Three equal resistors are connected as shown in Fig 2.10. Find the equivalent resistance between points A and B.*

Solution. The reader may observe that one end of each resistor is connected to point A and the other end of each resistor is connected to point B. Hence the three resistors are in parallel.

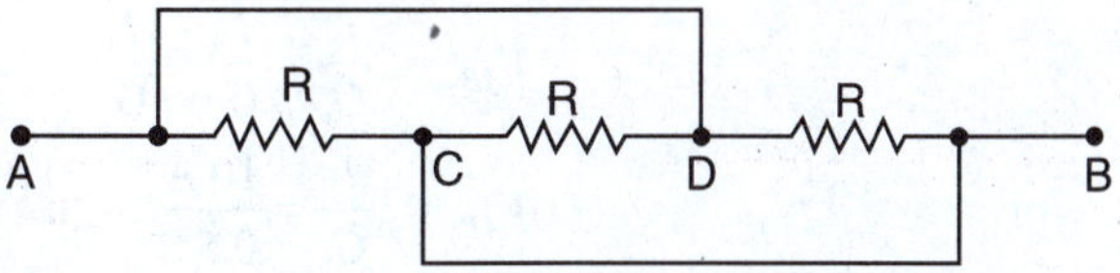

Fig. 2.10

$$\therefore \qquad \frac{1}{R_{AB}} = \frac{1}{R} + \frac{1}{R} + \frac{1}{R} = \frac{3}{R} \quad \text{or} \quad R_{AB} = \frac{R}{3}$$

Example 2.11. *Find the branch currents for Fig. 2.11 using the current divider rule for parallel conductances.*

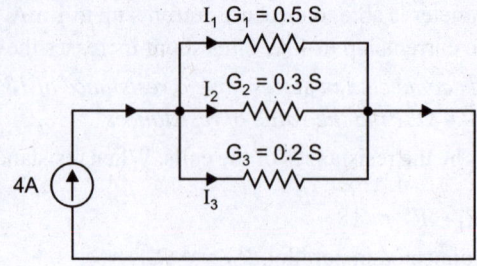

Fig. 2.11

Solution.
$$G_P = G_1 + G_2 + G_3 = 0.5 + 0.3 + 0.2 = 1 \text{ S}$$

∴
$$I_1 = I\frac{G_1}{G_P} = 4 \times \frac{0.5}{1} = \textbf{2 A}$$

$$I_2 = I\frac{G_2}{G_P} = 4 \times \frac{0.3}{1} = \textbf{1.2 A}$$

$$I_3 = I\frac{G_3}{G_P} = 4 \times \frac{0.2}{1} = \textbf{0.8 A}$$

Example 2.12. *Find the three branch currents in the circuit shown in Fig. 2.12. What is the potential difference between points A and B?*

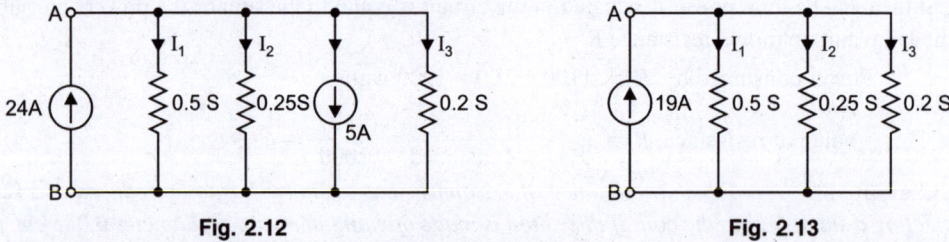

Fig. 2.12   **Fig. 2.13**

Solution. Current sources in parallel add *algebraically*. Therefore, the two current sources can be combined to give the resultant current source of current $I = 24 - 5 = 19$ A as shown in Fig. 2.13. Referring to Fig. 2.13,

$$G_P = G_1 + G_2 + G_3 = 0.5 + 0.25 + 0.2 = 0.95 \text{ S}$$

∴
$$I_1 = I \times \frac{G_1}{G_P} = 19 \times \frac{0.5}{0.95} = \textbf{10 A}$$

$$I_2 = I \times \frac{G_2}{G_P} = 19 \times \frac{0.25}{0.95} = \textbf{5 A}$$

$$I_3 = I \times \frac{G_3}{G_P} = 19 \times \frac{0.2}{0.95} = \textbf{4 A}$$

The voltage across each conductance is the same.

∴
$$V_{AB} = \frac{I_1}{G_1} = \frac{I_2}{G_2} = \frac{I_3}{G_3}$$

or
$$V_{AB} = \frac{I_1}{G_1} = \frac{10 \text{ A}}{0.5 \text{ S}} = \textbf{20 V}$$

Example 2.13. *A current of 90 A is shared by three resistances in parallel. The wires are of the same material and have their lengths in the ratio 2 : 3 : 4 and their cross-sectional areas in the ratio 1 : 2 : 3. Determine current in each resistance.*

Solution. Conductance, $G = \sigma \dfrac{a}{l}$ so that $G \propto \dfrac{a}{l}$ ($\because \sigma$ is same)

$\therefore \qquad G_1 : G_2 : G_3 :: \dfrac{a_1}{l_1} : \dfrac{a_2}{l_2} : \dfrac{a_3}{l_3}$

or $\qquad G_1 : G_2 : G_3 :: \dfrac{1}{2} : \dfrac{2}{3} : \dfrac{3}{4}$

or $\qquad G_1 : G_2 : G_3 :: 6 : 8 : 9$

$\therefore \qquad I_1 = I \times \dfrac{G_1}{G_1 + G_2 + G_3} = 90 \times \dfrac{6}{6+8+9} = \mathbf{23.48\ A}$

$\qquad I_2 = I \times \dfrac{G_2}{G_1 + G_2 + G_3} = 90 \times \dfrac{8}{6+8+9} = \mathbf{31.30\ A}$

$\qquad I_3 = I \times \dfrac{G_3}{G_1 + G_2 + G_3} = 90 \times \dfrac{9}{6+8+9} = \mathbf{35.22\ A}$

Example 2.14. *An aluminium wire 7.5 m long is connected in parallel with a copper wire 6 m long. When a current of 5 A is passed through the combination, it is found that current in the aluminium wire is 3 A. The diameter of aluminium wire is 1 mm. Determine the diameter of copper wire. Resistivity of copper is 0.017 $\mu\Omega$ m and that of aluminium is 0.028 $\mu\Omega$ m.*

Solution. Assign suffix A to aluminium and C to copper. Then,

$$I_A = 3\ A \quad \text{and} \quad I_C = 5 - I_A = 5 - 3 = 2\ A$$

In a parallel circuit, the current in any branch is directly proportional to conductance (G) of that branch ($\because I = VG$).

$\therefore \qquad I_A \propto G_A \quad \text{and} \quad I_C \propto G_C$

$\therefore \qquad \dfrac{G_C}{G_A} = \dfrac{I_C}{I_A} = \dfrac{2}{3}$

Now, $\qquad G_C = \dfrac{a_C}{\rho_C l_C} \quad \text{and} \quad G_A = \dfrac{a_A}{\rho_A l_A}$

$\therefore \qquad \dfrac{G_C}{G_A} = \dfrac{a_C}{\rho_C l_C} \times \dfrac{\rho_A l_A}{a_A}$

or $\qquad \dfrac{2}{3} = \dfrac{a_C}{0.017 \times 6} \times \dfrac{0.028 \times 7.5}{a_A}$

or $\qquad \dfrac{a_C}{a_A} = \dfrac{2}{3} \times \dfrac{0.017 \times 6}{0.028 \times 7.5} = 0.3238$

$\therefore \qquad a_C = 0.3238 \times a_A = 0.3238 \times \dfrac{\pi}{4}(1\ \text{mm})^2$

or $\qquad \dfrac{\pi}{4}(d_C)^2 = 0.3238 \times \dfrac{\pi}{4}$

$\therefore \qquad d_C = \sqrt{0.3238} = \mathbf{0.57\ mm}$

Example 2.15. *A voltage of 200 V is applied to a tapped resistor of 500 Ω. Find the resistance between two tapping points connected to a circuit needing 0.1 A at 25 V. Calculate the total power consumed.*

Solution. Fig. 2.14 shows the conditions of the problem.

Current in $AB = 0.1 + \dfrac{25}{R}$

Also current in $AB = \dfrac{200 - 25}{500 - R} = \dfrac{175}{500 - R}$

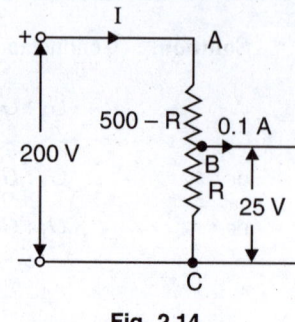

Fig. 2.14

$$\therefore \qquad 0.1 + \frac{25}{R} = \frac{175}{500 - R}$$

or
$$\frac{0.1R + 25}{R} = \frac{175}{500 - R}$$

or $\quad (500 - R)(0.1R + 25) = 175\,R$

or $\quad 0.1\,R^2 + 150\,R - 12500 = 0$

On solving and taking the positive value, $R = \mathbf{79\ \Omega}$.

$$\text{Total current, } I = \text{Current in } AB$$
$$= 0.1 + \frac{25}{79} = 0.4165\ \text{A}$$

$$\therefore \qquad \text{Total power} = 200 \times I = 200 \times 0.4165 = \mathbf{83.5\ W}$$

Example 2.16. *A heater has two similar elements controlled by a 3-heat switch. Draw a connection diagram of each position of the switch. What is the ratio of heat developed for each position of the switch?*

Solution. Fig. 2.15 shows the connections of 3-heat switch controlling two similar elements. Suppose the supply voltage is V.

With points 1 and 3 linked and supply connected across 1 and 3, the two elements will be in parallel.

$$\therefore \qquad \text{Power dissipated, } P_1 = \frac{V^2}{R/2} = \frac{2V^2}{R}$$

With voltage across 1 and 2 or 2 and 3, only one element is in the circuit.

$$\therefore \qquad \text{Power dissipated, } P_2 = \frac{V^2}{R}$$

With voltage across 1 and 3, the two elements are in series.

$$\therefore \qquad \text{Power dissipated, } P_3 = \frac{V^2}{2R}$$

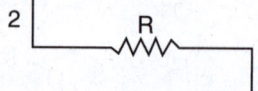

Fig. 2.15

$$\therefore \qquad P_1 : P_2 : P_3 = \frac{2V^2}{R} : \frac{V^2}{R} : \frac{V^2}{2R} = 2 : 1 : \frac{1}{2} = \mathbf{4 : 2 : 1}$$

Example 2.17. *The frame of an electric motor is connected to three earthing plates having resistance to earth of 10 Ω, 20 Ω and 30 Ω respectively. Due to a fault, the frame becomes live. What proportion of total fault energy is dissipated at each earth connection ?*

Solution. The three resistances are in parallel. During the fault, suppose voltage to ground is V. Then ratios of energy dissipated are :

$$\frac{V^2}{10} : \frac{V^2}{20} : \frac{V^2}{30} = \frac{1}{10} : \frac{1}{20} : \frac{1}{30} = 6 : 3 : 2$$

% of fault energy dissipated in 10 Ω $= \dfrac{6}{6 + 3 + 2} \times 100 = \mathbf{54.5\%}$

% of fault energy dissipated in 20 Ω $= \dfrac{3}{6 + 3 + 2} \times 100 = \mathbf{27.3\%}$

% of fault energy dissipated in 30 Ω $= \dfrac{2}{6 + 3 + 2} \times 100 = \mathbf{18.2\%}$

Example 2.18. *A 50 Ω resistor is in parallel with 100 Ω resistor. Current in 50 Ω resistor is 7.2 A. How will you add a third resistor and what will be its value if the line current is to be 12.1 A?*

Solution. Source voltage $= 50 \times 7.2 = 360\ \text{V}$

$$\therefore \quad \text{Current in 100 Ω resistor} = \frac{360}{100} = 3.6\ \text{A}$$

Total current drawn by 50 Ω and 100 Ω resistors = 7.2 + 3.6 = 10.8 A

In order that line current is 12.1 A, **some resistance R must be added in parallel.** The current in R is to be = 12.1 – 10.8 = 1.3 A.

$$\therefore \qquad \text{Value of } R = \frac{360}{1.3} = \mathbf{277\ \Omega}$$

Tutorial Problems

1. Two resistors of 4 Ω and 6 Ω are connected in parallel. If the total current is 30 A, find the current through each resistor. **[18 A ; 12 A]**

2. Four resistors are in parallel. The currents in the first three resistors are 4 mA, 5 mA and 6 mA respectively. The voltage drop across the fourth resistor is 200 volts. The total power dissipated is 5 watts. Determine the values of the resistances of the branches and the total resistance.

   [50 k Ω, 40 k Ω, 33.33 k Ω, 8 k Ω, 5 k Ω]

3. Four resistors of 2 Ω, 3 Ω, 4 Ω and 5 Ω respectively are connected in parallel. What potential difference must be applied to the group in order that total power of 100 watts may be absorbed ? **[8.826 volts]**

4. Three resistors are in parallel. The current in the first resistor is 0.1 A. The power dissipated in the second is 3 watts. The voltage drop across the third is 100 volts. Determine the ohmic values of resistors and the total resistance if total current is 0.2 A. **[1000 Ω, 3333.3 Ω, 1428.5 Ω, 500 Ω]**

5. Two coils each of 250 Ω resistance are connected in series across a constant voltage mains. Calculate the value of resistance to be connected in parallel with one of the coils to reduce the p.d. across its terminals by 1%. **[12,375 Ω]**

6. When a resistor is placed across a 230 volt supply, the current is 12 A. What is the value of resistor that must be placed in parallel to increase the load to16 A ? **[57.5 Ω]**

7. A 50-ohm resistor is in parallel with a 100-ohm resistor. The current in 50 Ω resistor is 7.2 A. What is the value of third resistance to be added in parallel to make the line current 12.1 A ? **[276.9 Ω]**

8. Five equal resistors each of 2 Ω are connected in a network as shown in Fig. 2.16. Find the equivalent resistance between points A and B. **[2 Ω]**

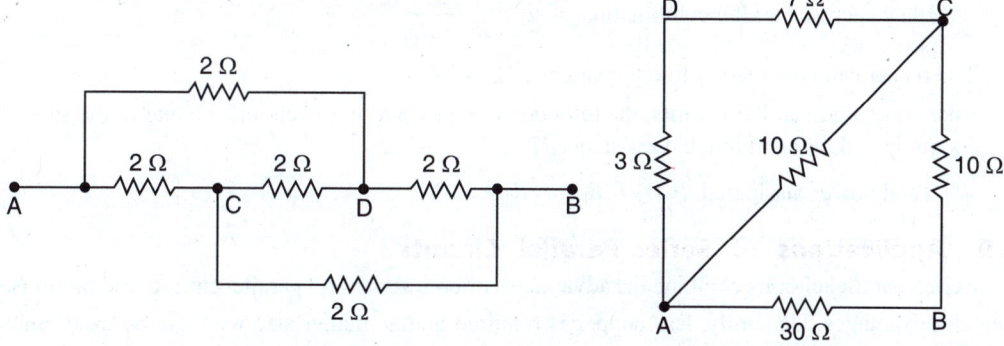

Fig. 2.16   **Fig. 2.17**

9. Find the equivalent resistance between points A and B in the circuit shown in Fig. 2.17. **[10 Ω]**

10. Fig. 2.18 shows a 50 V source connected to three resistances : $R_1 = 5$ kΩ; $R_2 = 25$ kΩ and $R_3 = 10$ kΩ. Calculate (i) branch currents (ii) total current for the given figure.

    [(i) $I_1 = 10$ mA ; $I_2 = 2$ mA; $I_3 = 5$ mA (ii) $I = 17$ mA]

11. A parallel circuit consists of four parallel-connected 480 Ω resistors in parallel with six 360 Ω resistors. What is the total resistance and total conductance of the circuit? **[40 Ω ; 0.025 S]**

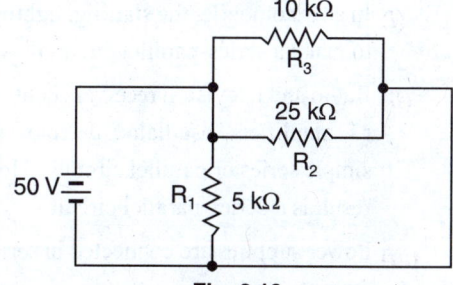

Fig. 2.18

2.8. D.C. Series-Parallel Circuit

As the name suggests, this circuit is a combination of series and parallel circuits. A simple example of such a circuit is illustrated in Fig. 2.19. Note that R_2 and R_3 are connected in parallel with each other and that both together are connected in series with R_1. One simple rule to solve such circuits is to first reduce the parallel branches to an equivalent series branch and then solve the circuit as a simple series circuit.

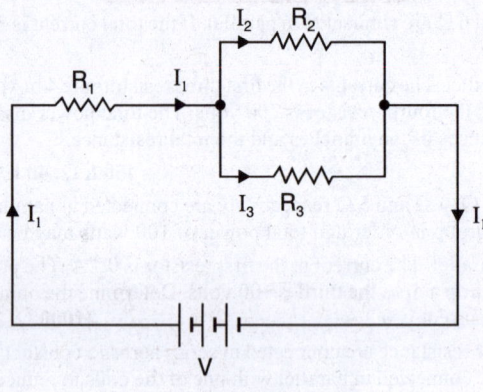

Fig. 2.19

Referring to the series-parallel circuit shown in Fig. 2.19,

$$R_P \text{ for parallel combination } = \frac{R_2 R_3}{R_2 + R_3}$$

$$\text{Total circuit resistance } = R_1 + \frac{R_2 R_3}{R_2 + R_3}$$

$$\text{Voltage across parallel combination } = I_1 \times \frac{R_2 R_3}{R_2 + R_3}$$

The reader can now readily find the values of I_1, I_2, I_3.

Like series and parallel circuits, the total power dissipated in the circuit is equal to the sum of powers dissipated in the individual resistances *i.e.*,

$$\text{Total power dissipated, } P = I_1^2 R_1 + I_2^2 R_2 + I_3^2 R_3$$

2.9. Applications of Series-Parallel Circuits

Series-parallel circuits combine the advantages of both series and parallel circuits and minimise their disadvantages. Generally, less copper is required and a smaller size wire can be used. Such circuits are used whenever various types of circuits must be fed from the same power supply. A few common applications of series-parallel circuits are given below :

(i) In an automobile, the starting, lighting and ignition circuits are all individual circuits joined to make a series-parallel circuit drawing its power from one battery.

(ii) Radio and television receivers contain a number of separate circuits such as tuning circuits, r.f. amplifiers, oscillator, detector and picture tube circuits. Individually, they may be simple series or parallel circuits. However, when the receiver is considered as a whole, the result is a series-parallel circuit.

(iii) Power supplies are connected in series to get a higher voltage and in parallel to get a higher current.

2.10. Internal Resistance of a Supply

All supplies (*e.g.* a cell) must have some internal resistance, however small. This is shown as a series resistor external to the supply. Fig 2.20 shows a cell of *e.m.f.* E volts and internal resistance r. When the cell is delivering no current (*i.e.* on no load), the p.d. across the terminals will be equal to *e.m.f.* E of the cell as shown in Fig. 2.20 (*i*).

When some load resistance R is connected across the terminals of the cell, the current I starts flowing in the circuit. This current causes a voltage drop across internal resistance r of the cell so that terminal voltage V available will be less than E. The relationship between E and V can be easily established [See Fig. 2.20 (*ii*)].

$$I = \frac{E}{R+r}$$

or $$IR = E - Ir$$

But $$IR = V, \text{ the terminal voltage of the cell.}$$

$$\therefore \qquad V = E - Ir$$

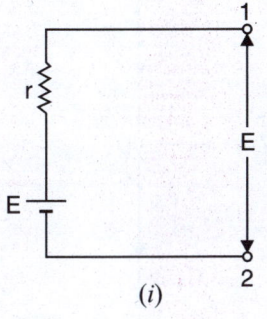 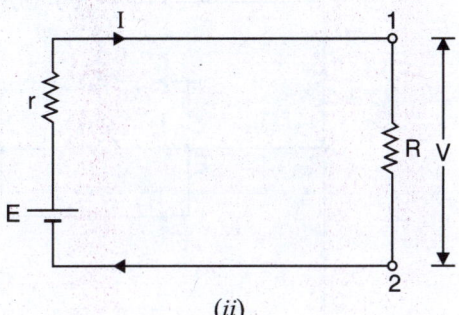

(*i*)   (*ii*)

Fig. 2.20

Internal resistance of cell, $r = \dfrac{E-V}{I} = \dfrac{(E-V)}{V}R$ $\left(\because I = \dfrac{V}{R} \right)$

2.11. Equivalent Resistance

Sometimes we come across a complicated circuit consisting of many resistances. The resistance between the two desired points (or terminals) of such a circuit can be replaced by a single resistance between these points using laws of series and parallel resistances. Then this single resistance is called equivalent resistance of the circuit between these points.

*The **equivalent resistance** of a circuit or network between its any two points (or terminals) is that single resistance which can replace the entire circuit between these points (or terminals).*

Once equivalent resistance is found, we can use Ohm's law to solve the circuit. It is important to note that resistance between two points of a circuit is different for different point-pairs. This is illustrated in Fig. 2.21.

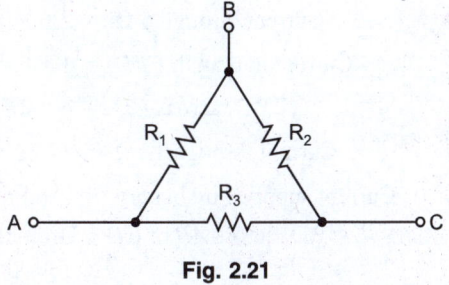

Fig. 2.21

(*i*) Between points A and B, R_1 is in parallel with the series combination of R_2 and R_3 *i.e.*

$$R_{AB} = R_1 \| (R_2 + R_3) = \frac{R_1(R_2 + R_3)}{R_1 + R_2 + R_3}$$

(*ii*) Between points A and C, R_3 is in parallel with the series combination of R_1 and R_2 i.e.

$$R_{AC} = R_3 \| (R_1 + R_2) = \frac{R_3(R_1 + R_2)}{R_1 + R_2 + R_3}$$

(*iii*) Between points B and C, R_2 is in parallel with the series combination of R_1 and R_3 i.e.

$$R_{BC} = R_2 \| (R_1 + R_3) = \frac{R_2(R_1 + R_3)}{R_1 + R_2 + R_3}$$

Example 2.19. *A battery having an e.m.f. of E volts and internal resistance 0.1 Ω is connected across terminals A and B of the circuit shown in Fig. 2.22. Calculate the value of E in order that power dissipated in 2 Ω resistor shall be 2 W.*

Solution. Resistance between E and F is given by ;

$$\frac{1}{R_{EF}} = \frac{1}{3} + \frac{1}{2} + \frac{1}{6} = \frac{6}{6}$$

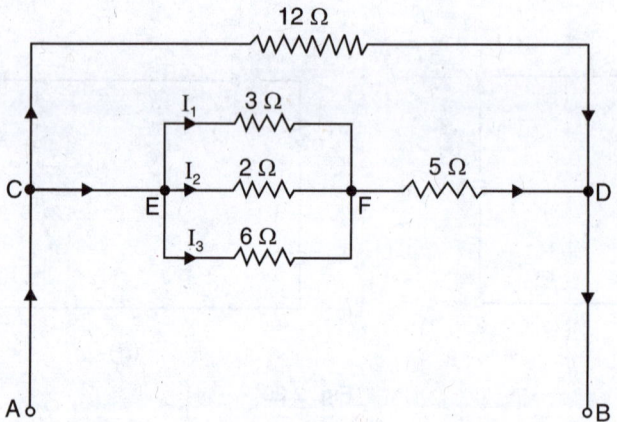

Fig. 2.22

$\therefore$ $R_{EF} = 6/6 = 1\ \Omega$

Resistance of branch $CEFD = 1 + 5 = 6\ \Omega$

$$\text{Current through 2 }\Omega = \sqrt{\frac{\text{Power loss}}{\text{Resistance}}} = \sqrt{\frac{2}{2}} = 1\text{A}$$

P.D. across $EF = 1 \times 2 = 2\text{V}$

Current through 3 $\Omega = 2/3 = 0.67$ A

Current through 6 $\Omega = 2/6 = 0.33$ A

Current in branch $CED = 1 + 0.67 + 0.33 = 2$ A

P.D. across $CD = 6 \times 2 = 12$ V

Current through 12 $\Omega = 12/12 = 1$ A

Current supplied by battery $= 2 + 1 = 3$ A

$\therefore$ E = P.D. across AB or CD + Drop in battery resistance

$$= 12 + 0.1 \times 3 = \mathbf{12.3\ V}$$

Example 2.20. *Calculate the values of various currents in the circuit shown in Fig. 2.23. What is total circuit conductance and total resistance?*

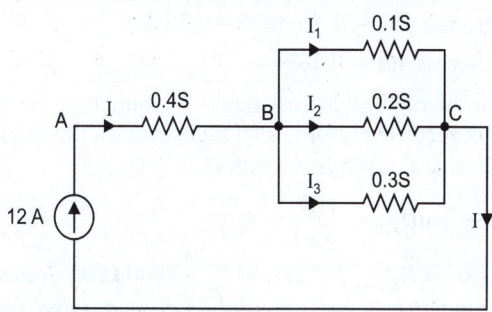

Fig. 2.23

Solution. $I = 12$ A ; $G_{BC} = 0.1 + 0.2 + 0.3 = 0.6$ S

$$\therefore \quad I_1 = I \times \frac{0.1}{G_{BC}} = 12 \times \frac{0.1}{0.6} = \mathbf{2\ A}\ ;\ I_2 = I \times \frac{0.2}{0.6} = 12 \times \frac{0.2}{0.6} = \mathbf{4\ A};$$

$$I_3 = I \times \frac{0.3}{0.6} = \mathbf{6\ A}\ ;\ I = \mathbf{12\ A}$$

Now, $G_{AB} = 0.4$ S and $G_{BC} = 0.6$ S are in series.

$$\therefore \qquad \frac{1}{G_{AC}} = \frac{1}{G_{AB}} + \frac{1}{G_{BC}} = \frac{1}{0.4} + \frac{1}{0.6} = \frac{25}{6} \quad \therefore \quad G_{AC} = \mathbf{\frac{6}{25}}\ \mathbf{S}$$

Total circuit resistance, $R_{AC} = \frac{1}{G_{AC}} = \frac{1}{6/25} = \mathbf{\frac{25}{6}}\ \mathbf{\Omega}$

Example 2.21. *Six resistors are connected as shown in Fig. 2.24. If a battery having an e.m.f. of 24 volts and internal resistance of 1 Ω is connected to the terminals A and B, find (i) the current from the battery, (ii) p.d. across 8 Ω and 4 Ω resistors and (iii) the current taken from the battery if a conductor of negligible resistance is connected in parallel with 8 Ω resistor.*

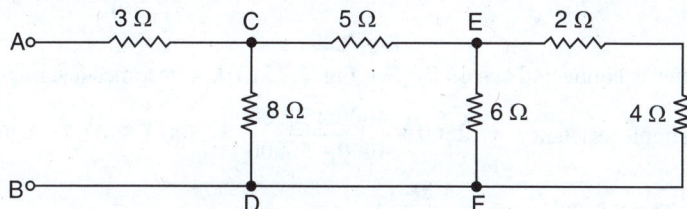

Fig. 2.24

Solution.

Resistance between E and F, $R_{EF} = \dfrac{(4+2) \times 6}{(4+2)+6} = 3\ \Omega$

Resistance between C and D, $R_{CD} = \dfrac{(5+3) \times 8}{(5+3)+8} = 4\ \Omega$

Resistance between A and B, $R_{AB} = 3 + 4 = 7\ \Omega$

Total circuit resistance, $R_T = R_{AB} +$ Supply resistance $= 7 + 1 = 8\ \Omega$

(i)   Current from battery, $I = E/R_T = 24/8 = \mathbf{3\ A}$

(ii)   P.D. across $8\ \Omega = E - I(3+1) = 24 - 3(4) = \mathbf{12\ V}$

Current through $8\ \Omega = 12/8 = 1.5$ A

Current through $5\ \Omega = 3 - 1.5 = 1.5$A

P.D. across $EF = 12 - 1.5 \times 5 = 4.5$ V

Current through $6\Omega = 4.5/6 = 0.75$A

$\therefore$ Current through 4 Ω = 1.5 − 0.75 = 0.75 A

$\therefore$ Voltage across 4Ω = 0.75 × 4 = **3 V**

(iii) When a conductor of negligible resistance is connected across 8 Ω, then resistance between C and D is zero. Therefore, total resistance in the circuit is now 3 Ω resistor in series with 1 Ω internal resistance of battery.

$\therefore$ Current from battery = $\dfrac{24}{3+1}$ = **6 A**

Example 2.22. *Two resistors R_1 = 2500 Ω and R_2 = 4000 Ω are joined in series and connected to a 100 V supply. The voltage drops across R_1 and R_2 are measured successively by a voltmeter having a resistance of 50000 Ω. Find the sum of two readings.*

Solution. When voltmeter is connected across resistor R_1 [See Fig. 2.25 (*i*)], it becomes a series-parallel circuit and total circuit resistance decreases.

Total circuit resistance = $4000 + \dfrac{2500 \times 50000}{2500 + 50000} = 4000 + 2381 = 6381\ \Omega$

Circuit current, $I = \dfrac{100}{6381}$A

Voltmeter reading, $V_1 = I \times 2381 = \dfrac{100}{6381} \times 2381 = 37.3$ V

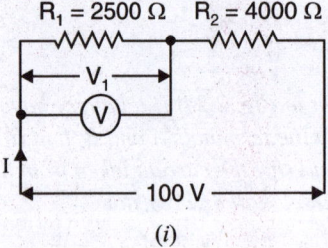 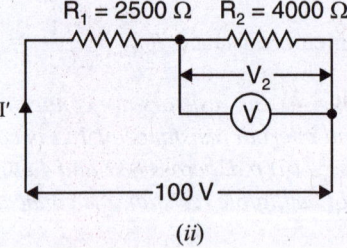

Fig. 2.25

When voltmeter is connected across R_2 [See Fig. 2.25 (*ii*)], it becomes a series-parallel circuit.

Total circuit resistance = $2500 + \dfrac{4000 \times 50000}{4000 + 50000} = 2500 + 3703.7 = 6203.7\ \Omega$

Circuit current, $I' = \dfrac{100}{6203.7}$A

Voltmeter reading, $V_2 = I' \times 3703.7 = \dfrac{100}{6203.7} \times 3703.7 = 59.7$ V

$\therefore$ Sum of two readings = $V_1 + V_2$ = 37.3 + 59.7 = **97 V**

Example 2.23. *A battery of unknown e.m.f. is connected across resistances as shown in Fig. 2.26. The voltage drop across the 8 Ω resistor is 20 V. What will be the current reading in the ammeter? What is the e.m.f. of the battery?*

Solution. The current through 8 Ω resistance is I = 20/8 = 2.5 A. At point A in Fig. 2.26, the current I is divided into two paths viz I_2 flowing in path ABC of 15 + 13 = 28 Ω resistance and current I_1 flowing in path AC of 11 Ω resistor. By current divider rule, the value of I_2 is given by ;

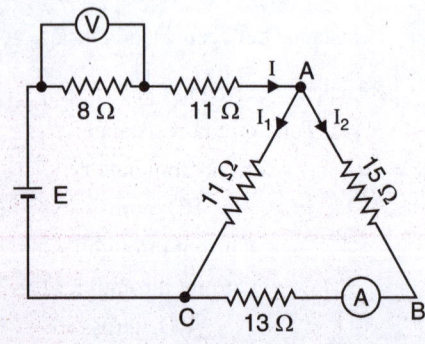

Fig. 2.26

$$I_2 = I \times \frac{11}{11+28} = 2.5 \times \frac{11}{39} = 0.7 \text{ A}$$

Therefore, ammeter reads **0.7 A**.

Resistance between A and C = $(28 \times 11)/39 = 308/39 \ \Omega$

Total circuit resistance, $R_T = 8 + 11 + (308/39) = 1049/39 \ \Omega$

$\therefore$ $E = I \times R_T = 2.5 \times (1049/39) = $ **67.3 V**

Example 2.24. *Find the voltage V_{AB} in the circuit shown in Fig. 2.27.*

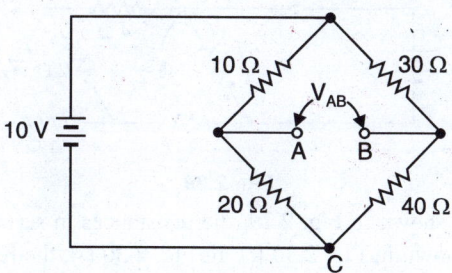

Fig. 2.27

Solution. The resistors $10 \ \Omega$ and $20 \ \Omega$ are in series and voltage across this combination is 10 V.

$\therefore$ $$V_{AC} = \frac{20}{10+20} \times 10 = 6.667 \text{ V}$$

The resistors $30 \ \Omega$ and $40 \ \Omega$ are in series and voltage across this combination is 10 V.

$\therefore$ $$V_{BC} = \frac{40}{30+40} \times 10 = 5.714 \text{ V}$$

The point A is positive *w.r.t.* point B.

$\therefore$ $$V_{AB} = V_{AC} - V_{BC} = 6.667 - 5.714 = \textbf{0.953 V}$$

Example 2.25. *A circuit consists of four 100 W lamps connected in parallel across a 230 V supply. Inadvertently, a voltmeter has been connected in series with the lamps. The resistance of the voltmeter is 1500 Ω and that of the lamps under the conditions stated is six times their value then burning normally. What will be the reading of the voltmeter ?*

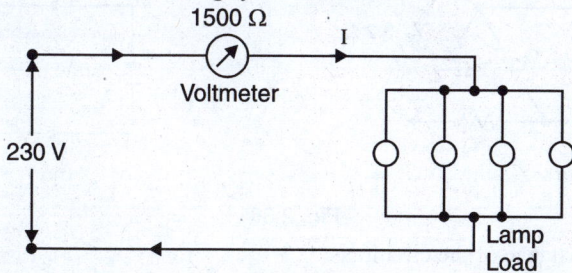

Fig. 2.28

Solution. Fig. 2.28 shows the conditions of the problem. When burning normally, the resistance of each lamp is $R = V^2/P = (230)^2/100 = 529 \ \Omega$. Under the conditions shown in Fig. 2.28, resistance of each lamp = $6 \times 529 = 3174 \ \Omega$.

$\therefore$ Equivalent resistance of 4 lamps under stated conditions is $R_P = 3174/4 = 793 \ \Omega$

Total circuit resistance = $1500 + R_P$

$= 1500 + 793.5 = 2293.5 \ \Omega$

$\therefore$ Circuit current, $I = \dfrac{230}{2293.5}$ A

$\therefore$ Voltage drop across voltmeter $= I \times 1500 = \dfrac{230}{2293.5} \times 1500 \simeq$ **150 V**

Example 2.26. *Find the current supplied by the d.c. source in the circuit shown in Fig. 2.29.*

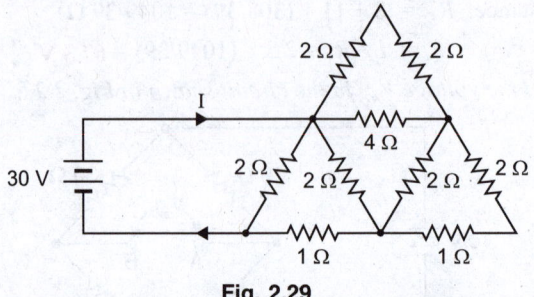

Fig. 2.29

Solution. In the circuit shown in Fig. 2.29, the resistances in series can be combined and the circuit reduces to the one shown in Fig. 2.30 (*i*). In Fig. 2.30 (*i*), the resistances in parallel can be combined using the formula product divided by sum and the circuit reduces to the one shown in Fig. 2.30 (*ii*).

In Fig. 2.30 (*ii*), the resistances in series can be combined and the circuit reduces to the one shown in Fig. 2.30 (*iii*). In Fig. 2.30 (*iii*), 3.2 Ω and 2 Ω are in parallel and their combined resistance is 16/13 Ω. Now 16/13 Ω and 1 Ω are in series and this series combination is in parallel with 2 Ω.

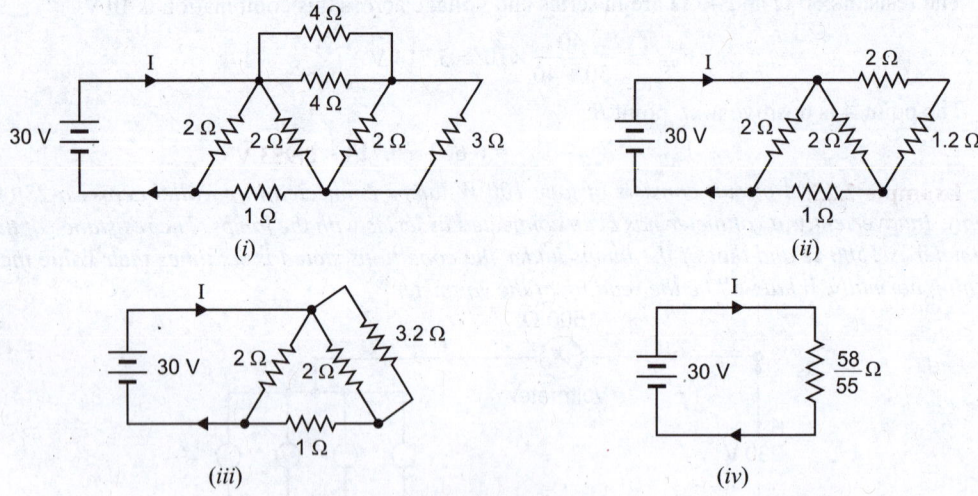

Fig. 2.30

$\therefore$ Effective resistance of the circuit is

$$R_{eff} = \left(\dfrac{16}{13}+1\right)\Omega \,\|\, 2\,\Omega = \dfrac{58}{55}\,\Omega \qquad\qquad \text{[See Fig. 2.30 (\textit{iv})]}$$

$\therefore$ Current supplied by source $= \dfrac{30}{R_{eff}} = \dfrac{30}{58/55} =$ **28.45 A**

Example 2.27. *Determine the current drawn by a 12 V battery with internal resistance 0.5 Ω by the following infinite network (See Fig. 2.31).*

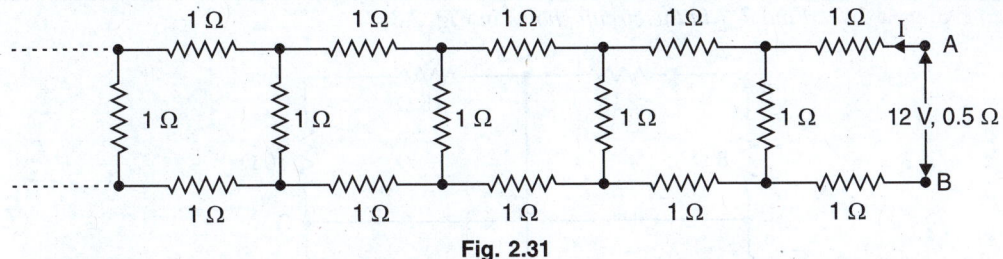

Fig. 2.31

Solution. Let x be the equivalent resistance of the network. Since the network is infinite, the addition of one set of three resistances, each of 1 Ω, will not change the total resistance, *i.e.*, it will remain x. The network would then become as shown in Fig. 2.32. The resistances x and 1 Ω are in parallel and their total resistance is R_P given by ;

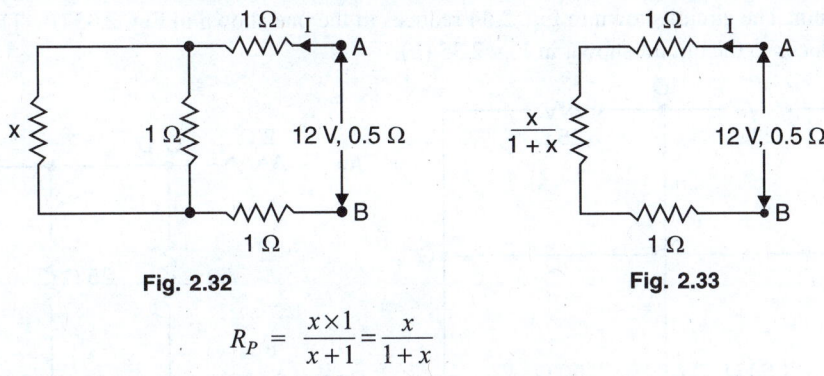

Fig. 2.32   **Fig. 2.33**

$$R_P = \frac{x \times 1}{x+1} = \frac{x}{1+x}$$

The circuit then reduces to the one shown in Fig. 2.33. Referring to Fig. 2.33,

Total resistance of the network $= 1 + 1 + \dfrac{x}{1+x} = 2 + \dfrac{x}{1+x}$

But total resistance of the network is x as mentioned above.

$\therefore \qquad\qquad\qquad\qquad x = 2 + \dfrac{x}{1+x}$

or $\qquad\qquad\qquad x + x^2 = 2 + 2x + x$

or $\qquad\qquad\qquad x^2 - 2x - 2 = 0$

$\therefore \qquad\qquad\qquad\qquad x = \dfrac{2 \pm \sqrt{4+8}}{2} = \dfrac{2 \pm \sqrt{12}}{2} = \dfrac{2 \pm 2\sqrt{3}}{2}$

or $\qquad\qquad\qquad\qquad x = 1 \pm \sqrt{3}$

As the value of the resistance cannot be negative,

$\therefore \qquad\qquad\qquad\qquad x = 1 + \sqrt{3} = 1 + 1.732 = 2.732 \ \Omega$

Total circuit resistance, $R_T = x +$ internal resistance of the supply

$$= 2.732 + 0.5 = 3.232 \ \Omega$$

$\therefore$ Current drawn by the network is

$$I = \frac{E}{R_T} = \frac{12}{3.232} = \textbf{3.71 A}$$

Example 2.28. *Find R_{AB} in the circuit shown in Fig. 2.34.*

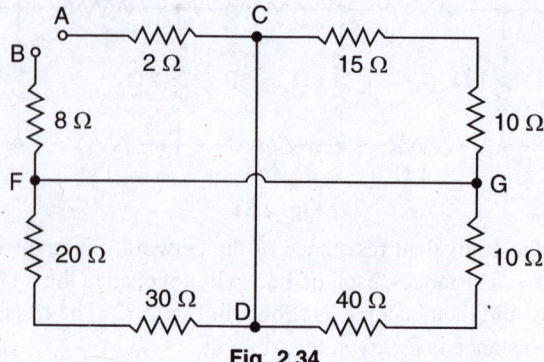

Fig. 2.34

Solution. The circuit shown in Fig. 2.34 reduces to the one shown in Fig. 2.35 (*i*). This circuit further reduces to the circuit shown in Fig. 2.35 (*ii*).

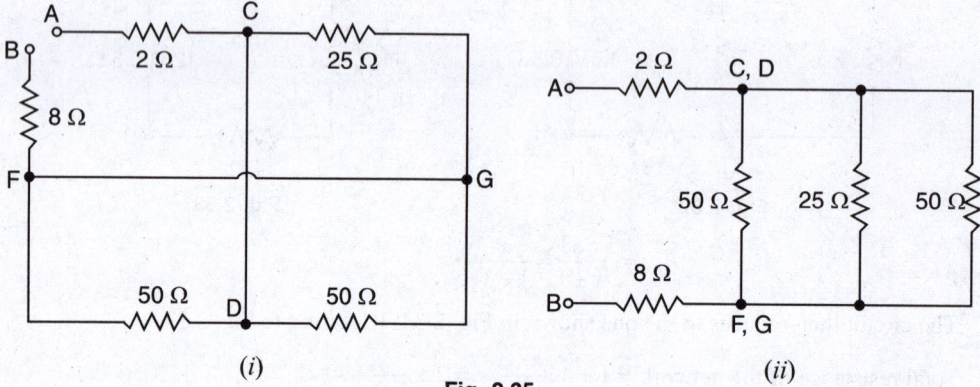

       (*i*)   **Fig. 2.35**   (*ii*)

Referring to Fig. 2.35 (*ii*), we have,

$$R_{AB} = 2 + (50 \parallel 25 \parallel 50) + 8$$
$$= 2 + (25 \parallel 25) + 8$$
$$= 2 + 12.5 + 8 = \mathbf{22.5 \ \Omega}$$

Example 2.29. *What is the equivalent resistance between the terminals A and B in Fig. 2.36?*

Solution. The network shown in Fig. 2.36 can be redrawn as shown in Fig. 2.37 (*i*). It is a balanced Wheatstone bridge. Therefore, points C and D are at the same potential. Since no current flows in the branch CD, this branch is ineffective in determining the equivalent resistance between terminals A and B and can be removed. The circuit then reduces to that shown in Fig. 2.37 (*ii*).

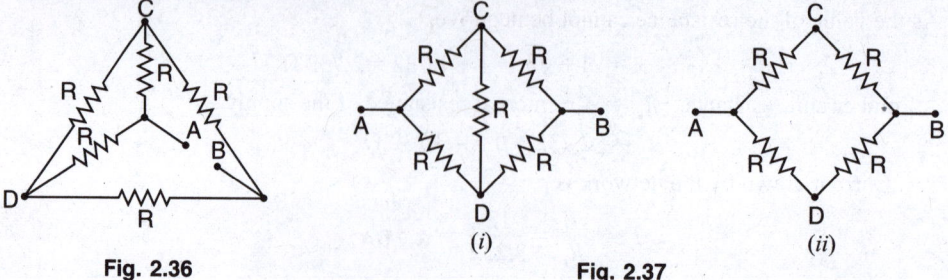

     Fig. 2.36   (*i*)   **Fig. 2.37**   (*ii*)

The branch ACB (= $R + R = 2R$) is in parallel with branch ADB (= $R + R = 2R$).

$$\therefore \qquad R_{AB} = \frac{(2R) \times (2R)}{2R + 2R} = \mathbf{R}$$

Example 2.30. *An electrical network is arranged as shown in Fig. 2.38. Find the value of current in the branch AF.*

Solution. Resistance between E and C,

$$R_{EC} = \frac{(5+9) \times 14}{(5+9) + 14} = 7\,\Omega$$

Resistance between B and E,

$$R_{BE} = \frac{(11+7) \times 18}{(11+7) + 18} = 9\,\Omega$$

Resistance between A and E,

$$R_{AE} = \frac{(13+9) \times 22}{(13+9) + 22} = 11\,\Omega$$

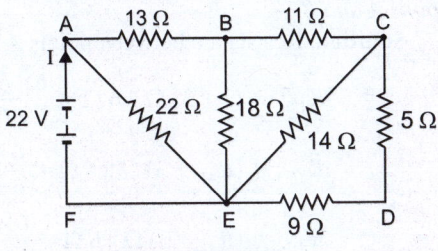

Fig. 2.38

i.e., Total circuit resistance, $R_T = 11\,\Omega$

∴   Current in branch AF, $I = V/R_T = 22/11 = $ **2 A**

Example 2.31. *A resistor of 5 Ω is connected in series with a parallel combination of a number of resistors each of 5 Ω. If the total resistance of the combination is 6 Ω, how many resistors are in parallel?*

Solution. Let n be the required number of 5 Ω resistors to be connected in parallel. The equivalent resistance of this parallel combination is

$$\frac{1}{R_P} = \frac{1}{5} + \frac{1}{5} + \frac{1}{5} \ldots n \text{ times} = \frac{n}{5}$$

Therefore, $R_P = 5/n$

Now $R_P (= 5/n)$ in series with 5 Ω is equal to 6 Ω *i.e.*,

$$\frac{5}{n} + 5 = 6 \quad \therefore \quad n = \mathbf{5}$$

Example 2.32. *A letter A consists of a uniform wire of resistance 1 Ω per cm. The sides of the letter are each 20 cm long and the cross-piece in the middle is 10 cm long while the apex angle is 60°. Find the resistance of the letter between the two ends of the legs.*

Solution. Fig. 2.39 shows the conditions of the problem. Point B is the mid-point of AC, point D is the mid-point of EC and $BD = 10$ cm.

∴ $AB = BC = CD = DE = BD = 10$ cm

or $R_1 = R_2 = R_3 = R_4 = R_5 = 10\,\Omega$ (∵ 1 cm = 1 Ω)

Now R_2 and R_3 are in series and their total resistance = $10 + 10 = 20\,\Omega$. This 20 Ω resistance is in parallel with R_5.

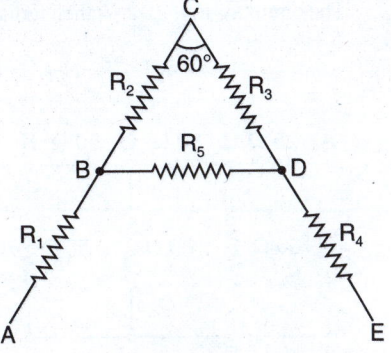

Fig. 2.39

∴ $R_{BD} = 20\,\Omega \parallel R_5 = 20\,\Omega \parallel 10\,\Omega$

$$= \frac{20 \times 10}{20 + 10} = \frac{20}{3}\,\Omega$$

Now, R_1, R_{BD} and R_4 are in series so that :

$$R_{AE} = R_1 + R_{BD} + R_4$$

$$= 10 + \frac{20}{3} + 10 = \mathbf{26.67\,\Omega}$$

Example 2.33. *All the resistances in Fig. 2.40 are in ohms. Find the effective resistance between the points A and B.*

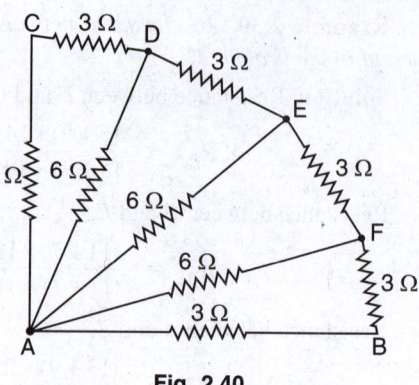

Solution. Resistance between points A and D is

$$R_{AD} = (3+3)\,\Omega \,\|\, 6\,\Omega = \frac{6\times 6}{6+6} = 3\,\Omega$$

$$R_{AE} = (R_{AD}+3)\,\Omega \,\|\, 6\,\Omega = \frac{6\times 6}{6+6} = 3\,\Omega$$

$$R_{AF} = (R_{AE}+3)\,\Omega \,\|\, 6\,\Omega = \frac{6\times 6}{6+6} = 3\,\Omega$$

∴   Resistance between points A and B is

$$R_{AB} = (R_{AF}+3)\,\Omega \,\|\, 3\,\Omega$$

$$= \frac{6\times 3}{6+3} = \frac{18}{9} = \mathbf{2\,\Omega}$$

Fig. 2.40

Example 2.34. *What is the equivalent resistance of the ladder network shown in Fig. 2.41?*

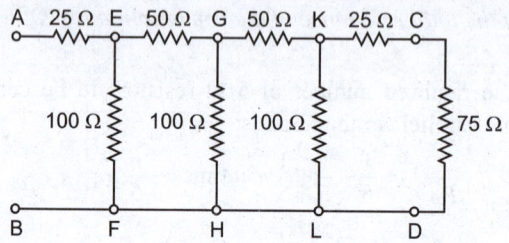

Fig. 2.41

Solution. Referring to Fig. 2.41, the resistance between points K and L is

$$R_{KL} = (25+75)\,\Omega \,\|\, 100\,\Omega = \frac{100\times 100}{100+100} = 50\,\Omega$$

The circuit of Fig. 2.41 then reduces to the one shown in Fig. 2.42 (*i*). Referring to Fig. 2.42 (*i*),

$$R_{GH} = (50+50)\,\Omega \,\|\, 100\,\Omega = \frac{100\times 100}{100+100} = 50\,\Omega$$

The circuit of Fig. 2.42 (*i*) then reduces to the one shown in Fig. 2.42 (*ii*). Referring to Fig. 2.42 (*ii*),

$$R_{EF} = (50+50)\,\Omega \,\|\, 100\,\Omega = \frac{100\times 100}{100+100} = 50\,\Omega$$

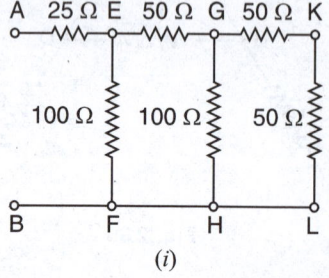   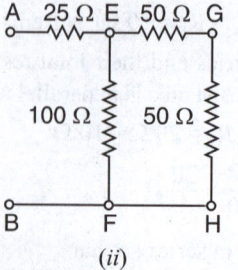   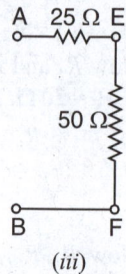

    (*i*)   (*ii*)   (*iii*)

Fig. 2.42

The circuit of Fig. 2.42 (*ii*) then reduces to the one shown in Fig. 2.42 (*iii*). Referring to Fig. 2.42 (*iii*),

Equivalent resistance of the ladder network

$$= 25 + 50 = \mathbf{75\,\Omega}$$

Tutorial Problems

1. A resistor of 3.6 Ω is connected in series with another of 4.56 Ω. What resistance must be placed across 3.6 Ω so that the total resistance of the circuit shall be 6 Ω ? **[2.4 Ω]**

2. A circuit consists of three resistors of 3 Ω, 4 Ω and 6 Ω in parallel and a fourth resistor of 4 Ω in series. A battery of e.m.f. 12 V and internal resistance 6 Ω is connected across the circuit. Find the total current in the circuit and terminal voltage across the battery. **[1.059 A, 5.65 V]**

3. A resistance R is connected in series with a parallel circuit comprising two resistors of 12 Ω and 8 Ω respectively. The total power dissipated in the circuit is 70 W when the applied voltage is 22 volts. Calculate the value of R. **[0.91 Ω]**

4. Two resistors R_1 and R_2 of 12 Ω and 6 Ω are connected in parallel and this combination is connected in series with a 6.25 Ω resistance R_3 and a battery which has an internal resistance of 0.25 Ω. Determine the e.m.f. of the battery. **[12.6 V]**

5. Find the voltage across and current through 4 kΩ resistor in the circuit shown in Fig. 2.43. **[4 V ; 1 mA]**

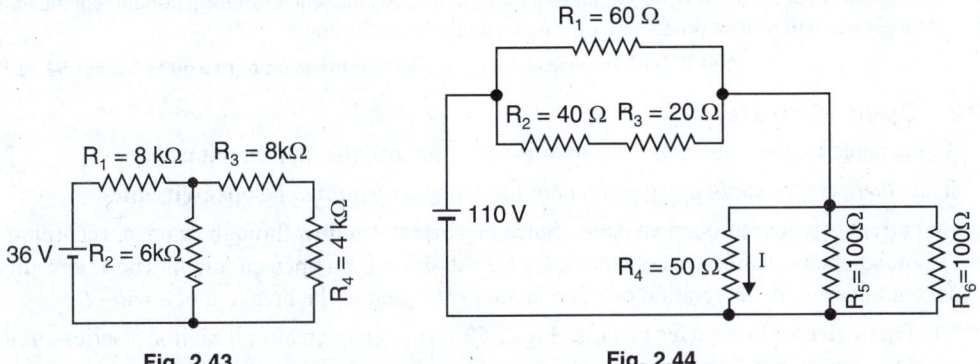

Fig. 2.43   Fig. 2.44

6. Find the current I in the 50 Ω resistor in the circuit shown in Fig. 2.44. **[1 A]**

7. Find the current in the 1 kΩ resistor in Fig. 2.45. **[6.72 mA]**

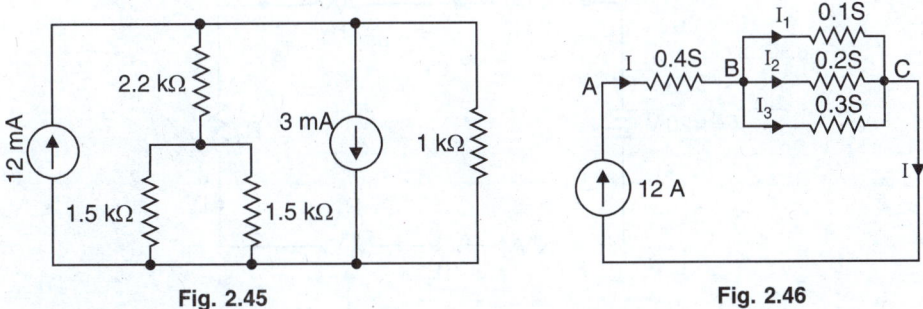

Fig. 2.45   Fig. 2.46

8. Calculate the value of different currents for the circuit shown in Fig. 2.46. What is the total circuit conductance and resistance ? **[I = 12 A ; I_1 = 2 A ; I_2 = 4 A ; I_3 = 6 A ; G_{AC} = 6/25 S ; R_{AC} = 25/6 Ω]**

9. For the parallel circuit of Fig. 2.47, calculate (i) V (ii) I_1 (iii) I_2. **[(i) 20 V (ii) 5 A (iii) –5 A]**

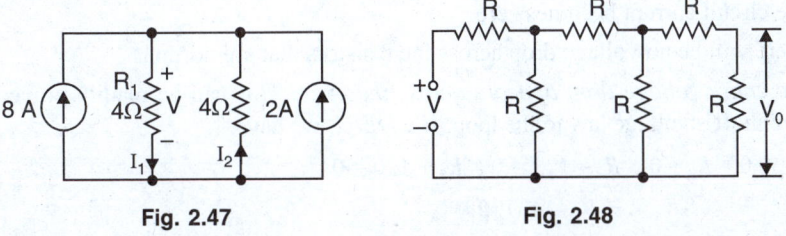

Fig. 2.47   Fig. 2.48

10. Prove that output voltage V_0 in the circuit of Fig. 2.48 is $V/13$.

11. Find the current I supplied by the 50 V source in Fig. 2.49.   **[$I = 13.7$ A]**

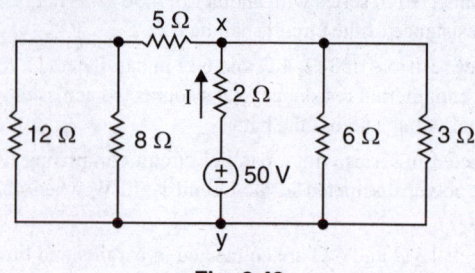

Fig. 2.49

12. An electric heating pad rated at 110 V and 55 W is to be used at a 220 V source. It is proposed to connect the heating pad in series with a series-parallel combination of light bulbs, each rated at 100 V ; bulbs are having ratings of 25 W, 60 W, 75 W and 100 W. Obtain a possible scheme of the pad-bulbs combination. At what rate will heat be produced by the pad with this modification ?

[100 W bulb in series with a parallel combination of two 60 W bulbs ; 54.54 W]

2.12. Open Circuits

As the name implies, an *open* is a gap or break or interruption in a circuit path.

When there is a break in any part of a circuit, that part is said to be **open-circuited.**

No current can flow through an open. Since no current can flow through an open, according to Ohm's law, an open has infinite resistance ($R = V/I = V/0 = \infty$). An open circuit may be as a result of component failure or disintegration of a conducting path such as the breaking of a wire.

1. Open circuit in a series circuit. Fig. 2.50 shows an open circuit fault in a series circuit. Here resistor R_4 is burnt out and an open develops. Because of the open, no current can flow in the circuit.

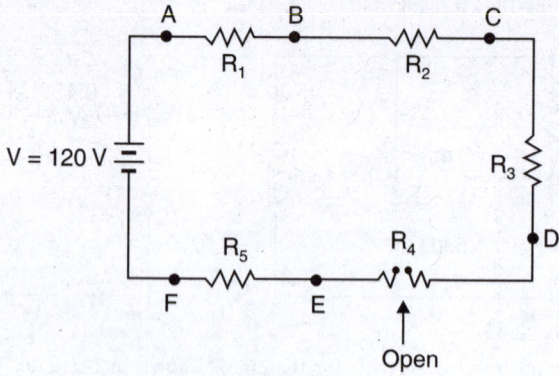

Fig. 2.50

When an open occurs in a series circuit, the following symptoms can be observed :

(i) The circuit current becomes zero.

(ii) There will be no voltage drop across the resistors that are normal.

(iii) *The entire voltage drop appears across the open.* This can be readily proved. Applying Kirchhoff's voltage law to the loop *ABCDEFA*, we have,

$$-0 \times R_1 - 0 \times R_2 - 0 \times R_3 - V_{DE} - 0 \times R_5 + 120 = 0$$

$$\therefore \qquad\qquad V_{DE} = 120 \text{ V}$$

(*iv*) Since the circuit current is zero, there is no voltage drop in the internal resistance of the source. Therefore, terminal voltage may appear higher than the normal.

2. Open circuit in a parallel circuit. One or more branches of a parallel circuit may develop an open. Fig. 2.51 shows a parallel circuit with an open. Here resistor R_3 is burnt out and now has infinite resistance.

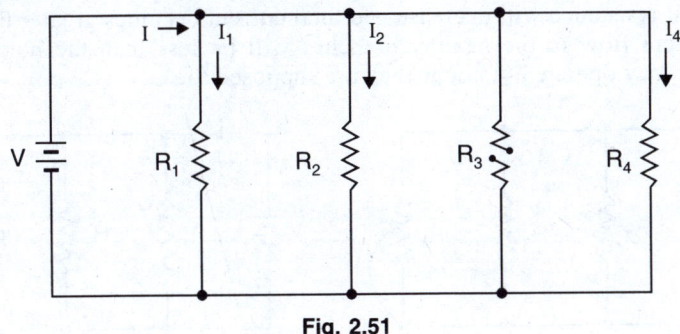

Fig. 2.51

The following symptoms can be observed :

  (*i*) Branch current I_3 will be zero because R_3 is open.

 (*ii*) The total current I will be less than the normal.

(*iii*) The operation of the branches without opens will be normal.

(*iv*) The open device will not operate. If R_3 is a lamp, it will be out. If it is a motor, it will not run.

2.13. Short Circuits

A short circuit or short is a path of low resistance. *A* **short circuit** *is an unwanted path of low resistance.* When a short circuit occurs, the resistance of the circuit becomes low. As a result, current greater than the normal flows which can cause damage to circuit components. The short circuit may be due to insulation failure, components get shorted etc.

1. Partial short in a series circuit. Fig. 2.52 (*i*) shows a **series circuit** with a **partial short.** An unwanted path has connected R_1 to R_3 and has eliminated R_2 from the circuit. Therefore, the circuit resistance decreases and the circuit current becomes greater than normal. The voltage drop across components that are not shorted will be higher than normal. Since current is increased, the power dissipation in the components that are not shorted will be greater than the normal. A partial short may cause healthy component to burn out due to abnormally high dissipation.

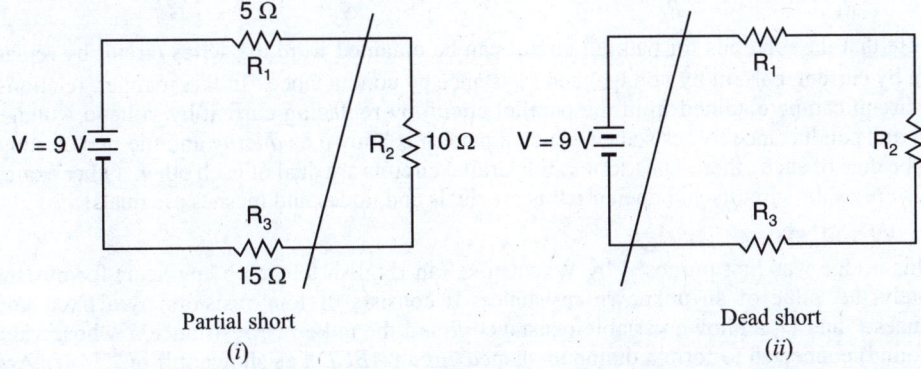

Partial short   Dead short
   (*i*)   (*ii*)

Fig. 2.52

2. Dead short in a series circuit. Fig. 2.52 (*ii*) shows a **series circuit** with a **dead short.** Here all the loads (*i.e.* resistors in this case) have been removed by the unwanted path. Therefore, the circuit resistance is almost zero and the circuit current becomes extremely high. If there are no protective devices (fuse, circuit breaker etc.) in the circuit, drastic results (smoke, fire, explosion etc.) may occur.

3. Partial short in a parallel circuit. Fig. 2.53 (*i*) shows a **parallel circuit** with a **partial short.** The circuit resistance will decrease and total current becomes greater than the normal. Further, the current flow in the healthy branches will be less than the normal. Therefore, healthy branches may operate but not as they are supposed to.

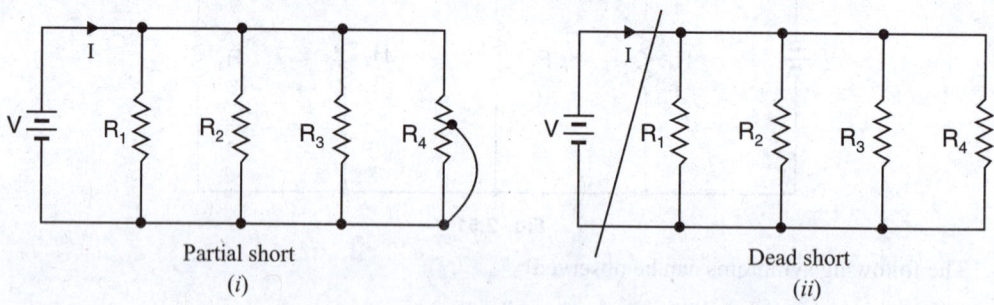

Partial short   Dead short
(*i*)   (*ii*)

Fig. 2.53

4. Dead short in a parallel circuit. Fig. 2.53 (*ii*) shows a **parallel circuit** with a **dead short.** Note that all the loads are eliminated by the short circuit so that the circuit resistance is almost zero. As a result, the circuit current becomes abnormally high and may cause extensive damage unless it has protective devices (*e.g.* fuse, circuit breaker etc.).

2.14. Duality Between Series and Parallel Circuits

Two physical systems or circuits are called dual if they are described by equations of the same mathematical form.

This peculiar pattern of relationship exists between series and parallel circuits. For example, consider the following table for d.c. series circuit and d.c. parallel circuit.

D.C. series circuit	**D.C. parallel circuit**
$I_1 = I_2 = I_3 = ...$	$V_1 = V_2 = V_3 = ...$
$V = V_1 + V_2 + V_3 + ...$	$I = I_1 + I_2 + I_3 + ...$
$R_S = R_1 + R_2 + R_3 + ...$	$G_P = G_1 + G_2 + G_3 + ...$
$I = \dfrac{V_1}{R_1} = \dfrac{V_2}{R_2} = \dfrac{V_3}{R_3} = ...$	$V = \dfrac{I_1}{G_1} = \dfrac{I_2}{G_2} = \dfrac{I_3}{G_3} = ...$
$V_1 = V\dfrac{R_1}{R_S} \quad ; \quad V_2 = V\dfrac{R_2}{R_S}$	$I_1 = I\dfrac{G_1}{G_P} \quad ; \quad I_2 = I\dfrac{G_2}{G_P}$

Note that the relations for parallel circuit can be obtained from the series circuit by replacing voltage by current, current by voltage and resistance by conductance. In like manner, relations for series circuit can be obtained from the parallel circuit by replacing current by voltage, voltage by current and conductance by resistance. Such a pattern is known as *duality* and the two circuits are said to be dual of each other. Thus series and parallel circuits are dual of each other. Other examples of duals are : short circuits and open circuits are duals and nodes and meshes are duals.

2.15. Wheatstone Bridge

This bridge was first proposed by Wheatstone (an English telegraph engineer) for measuring accurately the value of an unknown resistance. It consists of four resistors (two fixed known resistances P and Q, a known variable resistance R and the unknown resistance X whose value is to be found) connected to form a diamond-shaped circuit *ABCDA* as shown in Fig.2.54 (*i*). Across one pair of opposite junctions (*A* and *C*), battery is connected and across the other opposite pair of

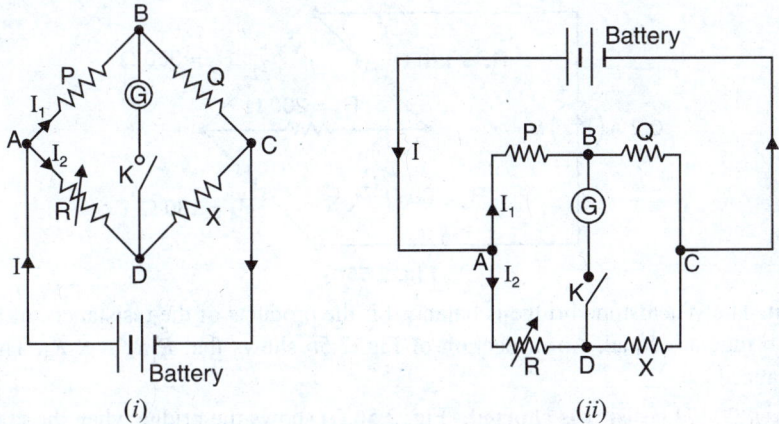

Fig. 2.54

junctions (*B* and *D*), a galvanometer is connected through the key *K*. The circuit is called a bridge because galvanometer bridges the opposite junctions *B* and *D*. Fig. 2.54 (*ii*) shows another* way of drawing the Wheatstone bridge.

Working. The values of *P* and *Q* are properly fixed. The value of *R* is varied such that on closing the key *K*, there is no current through the galvanometer. Under such conditions, the bridge is said to be *balanced*. The point at which the bridge is balanced is called the *null point*. Let I_1 and I_2 be the currents through *P* and *R* respectively when the bridge is balanced. Since there is no current through the galvanometer, the currents in *Q* and *X* are also I_1 and I_2 respectively. As the galvanometer reads zero, points *B* and *D* are at the same potential. This means that voltage drops from *A* to *B* and *A* to *D* must be equal. Also voltage drops from *B* to *C* and *D* to *C* must be equal. Hence,

$$I_1 P = I_2 R \qquad \qquad ...(i)$$

and
$$I_1 Q = I_2 X \qquad \qquad ...(ii)$$

Dividing exp. (*i*) by (*ii*), we get,

$$P/Q = R/X$$

or
$$PX = QR$$

i.e.   *Product of opposite arms* = *Product of opposite arms*

Unknown resistance, $X = \dfrac{Q}{P} \times R$ ...(*iii*)

Since the **values of *Q*, *P* and *R* are known, the value of unknown resistance *X* can be calculated. It should be noted that exp. (*iii*) is true only under the balanced conditions of Wheatstone bridge.

Note. When the bridge is balanced, $V_B = V_D$ so the voltage across galvanometer is zero *i.e.* $V_{BD} = V_B - V_D = 0$. When there is zero voltage across the galvanometer, there is also zero current though the galvanometer. Consequently, **in a balanced Wheatstone bridge, galvanometer can be replaced by either a short circuit or an open circuit without affecting the voltages and currents anywhere else in the circuit.**

Example 2.35. *Verify that the Wheatstone bridge shown in Fig. 2.55 is balanced. Then find the voltage V_T across the 0.2 A current source by (i) replacing the 200 Ω resistor with a short. (ii) replacing the 200 Ω resistor with an open.*

---

*   Note the four points *A,B, C* and *D*, each lying at the junction between two resistors. A galvanometer should bridge a pair of opposite points such as *B* and *D* and the battery the other pair *A* and *C*.

**   Resistances *P* and *Q* are called the ratio arms of bridge and are usually made equal to definite ratio such as 1 to 1, 10 to 1 or 100 to 1. The resistance *R* is ca]lled the rheostat arm and is made continuously variable from 1 to 1000 ohms or from 1 to 10,000 ohms.

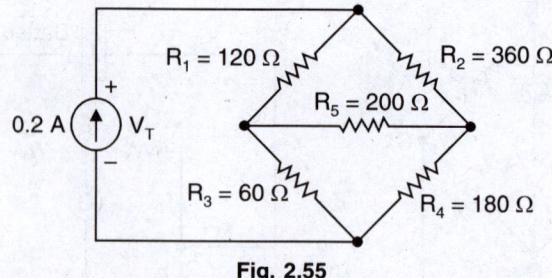

Fig. 2.55

Solution. The Wheatstone bridge is balanced if the products of the resistances of the opposite arms of the bridge are equal. An inspection of Fig. 2.55 shows that $R_1 R_4 = R_2 R_3$. Therefore, the bridge is balanced.

(i) **When 200 Ω resistor is shorted.** Fig. 2.56 *(i)* shows the bridge when the 200 Ω resistor (R_5) is replaced by a short. In this case, the circuit is equivalent to a series-parallel circuit as shown in Fig. 2.56 *(ii)*. Referring to Fig. 2.56 *(ii)*, the circuit is equivalent to parallel combination of R_1 and R_2 in series with the parallel combination of R_3 and R_4.

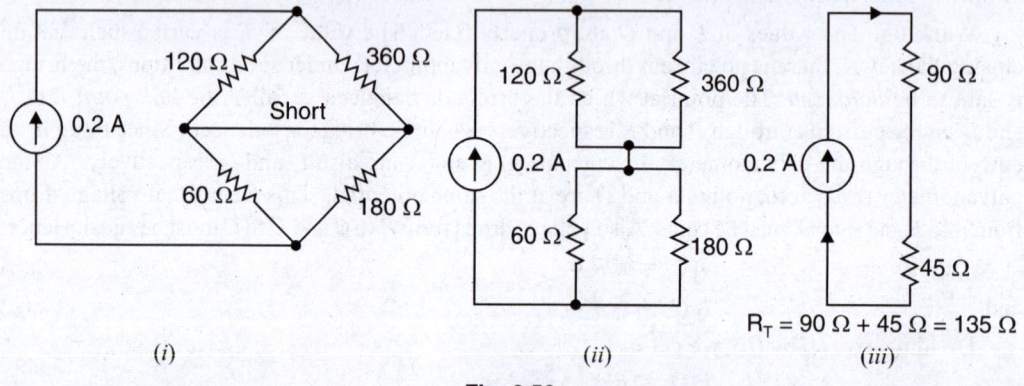

| *(i)* | *(ii)* | *(iii)* |

Fig. 2.56

The circuit shown in Fig. 2.56 *(ii)* further reduces to the one shown in Fig. 2.57 *(iii)*. Therefore, total circuit resistance, $R_T = 90 + 45 = 135\ \Omega$.

∴   Voltage across 0.2 A current source is

$$V_T = I R_T = 0.2 \times 135 = \textbf{27 V}$$

(ii) **When 200 Ω resistor is open-circuited.** Fig. 2.57 *(i)* shows the bridge when 200 Ω resistor is replaced by an open. In this case, the circuit is equivalent to a series-parallel circuit in which series combination of R_1 and R_3 is in parallel with the series combination of R_2 and R_4. This is shown in Fig. 2.57 *(ii)*.

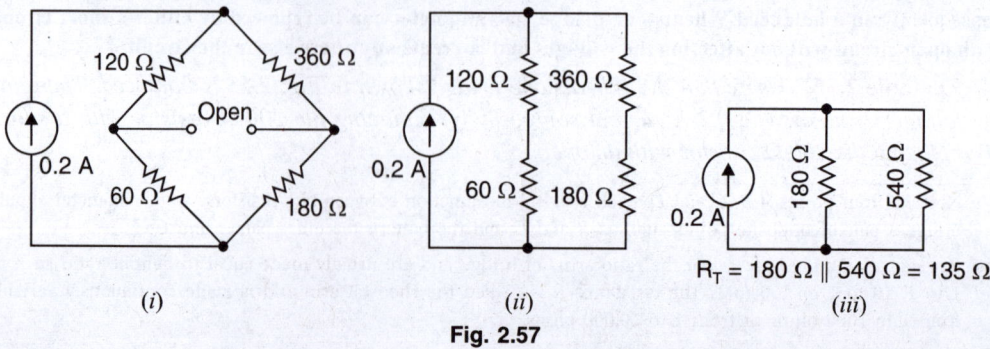

| *(i)* | *(ii)* | *(iii)* |

Fig. 2.57

The circuit shown in Fig. 2.57 (*ii*) further reduces to the one shown in Fig. 2.57 (*iii*). Referring to Fig. 2.57 (*iii*), the total circuit resistance R_T is given by ;

$$R_T = \frac{180 \times 540}{180 + 540} = 135\ \Omega$$

∴   Voltage across 0.2 A current source, $V_T = I\,R_T = 0.2 \times 135 = \textbf{27 V}$

Note that the voltage across current source is unaffected whether 200 Ω resistor is replaced by a short or an open.

2.16. Complex Circuits

Sometimes we encounter circuits where simplification by series and parallel combinations is impossible. Consequently, Ohm's law cannot be applied to solve such circuits. This happens when there is more than one e.m.f. in the circuit or when resistors are connected in a complicated manner. Such circuits are called *complex circuits*. We shall discuss two such circuits by way of illustration.

(*i*) Fig. 2.58 shows a circuit containing two sources of *e.m.f.* E_1 and E_2 and three resistors. This circuit cannot be solved by series-parallel combinations. Are resistors R_1 and R_3 in parallel? Not quite, because there is an *e.m.f.* source E_1 between them. Are they in series? Not quite, because same current does not flow between them.

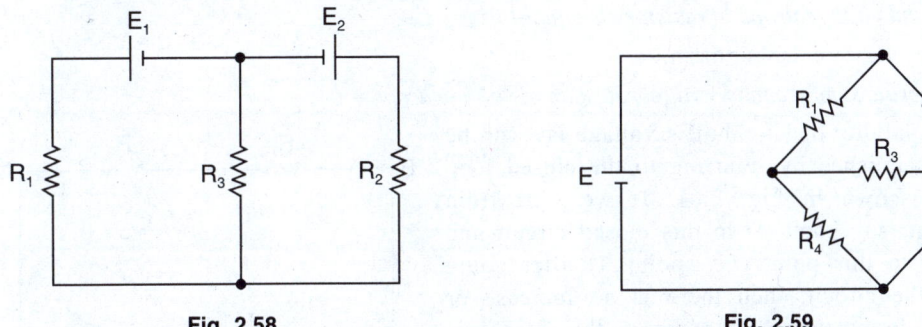

Fig. 2.58   Fig. 2.59

(*ii*) Fig. 2.59 shows another circuit where we cannot solve the circuit by series-parallel combinations. Though this circuit has one source of *e.m.f.* (*E*), it cannot be solved by using series and parallel combinations. Thus resistors R_1 and R_2 are neither in series nor in parallel; the same is true for other pair of resistors.

In order to solve such complex circuits, Gustav Kirchhoff gave two laws, known as Kirchhoff's laws.

2.17. Kirchhoff's Laws

Kirchhoff gave two laws to solve complex circuits, namely ;

1. Kirchhoff's Current Law (KCL)   2. Kirchhoff's Voltage Law (KVL)

1.   KIRCHHOFF'S CURRENT LAW (KCL)

This law relates to the currents at the *junctions of an electric circuit and may be stated as under :

The algebraic sum of the currents meeting at a junction in an electrical circuit is zero.

An algebraic sum is one in which the sign of the quantity is taken into account. For example, consider four conductors carrying currents I_1, I_2, I_3 and I_4 and meeting at point O as shown in Fig. 2.60.

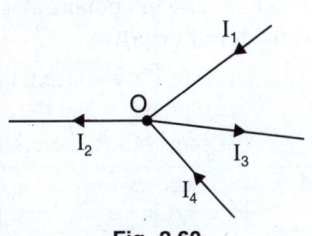

Fig. 2.60

*   A junction is that point in an electrical circuit where *three or more* circuit elements meet.

If we take the signs of currents flowing towards point O as positive, then currents flowing away from point O will be assigned negative sign. Thus, applying Kirchhoff's current law to the junction O in Fig. 2.60, we have,

$$(I_1) + (I_4) + (-I_2) + (-I_3) = 0$$

or $$I_1 + I_4 = I_2 + I_3$$

i.e., Sum of incoming currents = Sum of outgoing currents

Hence, Kirchhoff's current law may also be stated as under :

The sum of currents flowing towards any junction in an electrical circuit is equal to the sum of currents flowing away from that junction. Kirchhoff's current law is also called junction rule.

Kirchhoff's current law is true because electric current is merely the flow of free electrons and they cannot accumulate at any point in the circuit. This is in accordance with the law of conservation of charge. Hence, Kirchhoff's current law is based on the law of conservation of charge.

2. KIRCHHOFF'S VOLTAGE LAW (KVL)

This law relates to *e.m.fs* and voltage drops in a closed circuit or loop and may be stated as under :

In any closed electrical circuit or mesh, the algebraic sum of all the electromotive forces (e.m.fs) and voltage drops in resistors is equal to zero, i.e.,

In any closed circuit or mesh,

Algebraic sum of e.m.fs + Algebraic sum of voltage drops = 0

The validity of Kirchhoff's voltage law can be easily established by referring to the closed loop *ABCDA* shown in Fig. 2.61. If we start from any point (say point *A*) in this closed circuit and go back to this point (*i.e.,* point *A*) after going around the circuit, then there is no increase or decrease in potential. This means that algebraic sum of the *e.m.fs* of all the sources (here only one *e.m.f.* source is considered) met on the way *plus* the algebraic sum of the voltage drops in the resistances

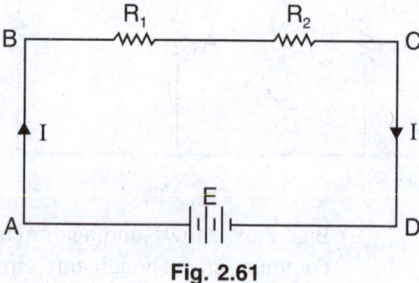

Fig. 2.61

must be zero. Kirchhoff's voltage law is based on the law of *conservation of energy, i.e.,* net change in the energy of a charge after completing the closed path is zero.

Note. Kirchhoff's voltage law is also called *loop rule.*

2.18. Sign Convention

While applying Kirchhoff's voltage law to a closed circuit, algebraic sums are considered. Therefore, it is very important to assign proper signs to *e.m.fs* and voltage drops in the closed circuit. The following convention may be followed :

A **rise in potential should be considered positive and fall in potential should be considered negative.

    (i) Thus if we go from the positive terminal of the battery to the negative terminal, there is fall of potential and the *e.m.f.* should be assigned negative sign. Thus in Fig. 2.62 (*i*), as we go from *A* to *B*, there is a fall in potential and the *e.m.f.* of the cell will be assigned negative

---

*   As a charge traverses a loop and returns to the starting point, the sum of rises of potential energy associated with *e.m.fs* in the loop must be equal to the sum of the drops of potential energy associated with resistors.

**   The reverse convention is equally valid *i.e.* rise in potential may be considered negative and fall in potential as positive.

sign. On the other hand, if we go from the negative terminal to the positive terminal of the battery or source, there is a rise in potential and the e.m.f should be

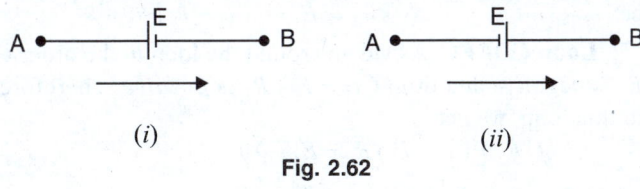

Fig. 2.62

assigned positive sign. Thus in Fig. 2.62 (*ii*) as we go from *A* to *B*, there is a rise in potential and the *e.m.f.* of the cell will be assigned positive sign. *It may be noted that the sign of e.m.f. is independent of the direction of current through the branch under consideration.*

(*ii*) When current flows through a resistor, there is a voltage drop across it. If we go through the resistor in the same direction as the current, there is a fall in potential because current flows from higher potential to lower potential. Hence this voltage drop should be assigned negative sign. In Fig. 2.63 (*i*), as we go from *A* to *B*, there is a fall in potential and the voltage drop across the resistor will be assigned negative sign.

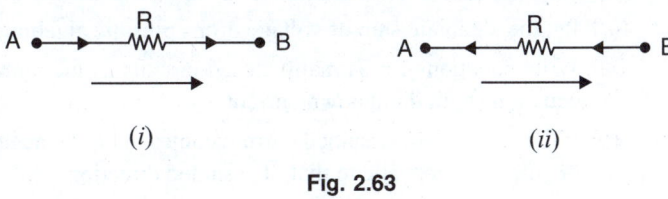

Fig. 2.63

On the other hand, if we go through the resistor against the current flow, there is a rise in potential and the voltage drop should be given positive sign. Thus referring to Fig. 2.63 (*ii*), as we go from *A* to *B*, there is a rise in potential and this voltage drop will be given positive sign. *It may be noted that sign of voltage drop depends on the direction of current and is independent of the polarity of the e.m.f. of source in the circuit under consideration.*

2.19. Illustration of Kirchhoff's Laws

Kirchhoff's Laws can be beautifully explained by referring to Fig. 2.64. Mark the directions of currents as indicated. The direction in which currents are assumed to flow is unimportant, since if wrong direction is chosen, it will be indicated by a negative sign in the result.

(*i*) The magnitude of current in any branch of the circuit can be found by applying Kirchhoff's current law. Thus at junction *C* in Fig. 2.64, the incoming currents to the junction are I_1 and I_2. Obviously, the current in branch *CF* will be $I_1 + I_2$.

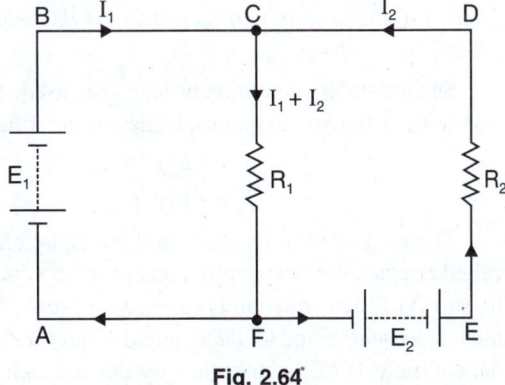

Fig. 2.64

(*ii*) There are three closed circuits in Fig 2.64 *viz. ABCFA, CDEFC* and *ABCDEFA.* Kirchhoff's voltage law can be applied to these closed circuits to get the desired equations.

Loop ABCFA. In this loop, *e.m.f.* E_1 will be given *positive* sign. It is because as we consider the loop in the order *ABCFA*, we go from −ve terminal to the positive terminal of the battery in the branch *AB* and hence there is a rise in potential. The voltage drop in branch *CF* is $(I_1 + I_2) R_1$ and shall bear *negative* sign. It is because as we consider the loop in the order *ABCFA*, we go with current in branch *CF* and there is a fall in potential. Applying Kirchhoff's voltage law to the loop *ABCFA*,

$$-(I_1 + I_2) R_1 + E_1 = 0$$

or $$E_1 = (I_1 + I_2)R_1 \qquad \qquad ...(i)$$

Loop CDEFC. As we go around the loop in the order *CDEFC*, drop $I_2 R_2$ is *positive,* e.m.f. E_2 is *negative* and drop $(I_1 + I_2)R_1$ is *positive*. Therefore, applying Kirchhoff's voltage law to this loop, we get,

$$I_2 R_2 + (I_1 + I_2)R_1 - E_2 = 0$$

or $$I_2 R_2 + (I_1 + I_2)R_1 = E_2 \qquad \qquad ...(ii)$$

Since E_1, E_2, R_1 and R_2 are known, we can find the values of I_1 and I_2 from the above two equations. Hence currents in all branches can be determined.

2.20. Method to Solve Circuits by Kirchhoff's Laws

(*i*) Assume unknown currents in the given circuit and show their direction by arrows.

(*ii*) Choose any closed circuit and find the algebraic sum of voltage drops *plus* the algebraic sum of e.m.fs in that loop.

(*iii*) Put the algebraic sum of voltage drops plus the algebraic sum of e.m.fs equal to zero.

(*iv*) Write equations for as many closed circuits as the number of unknown quantities. Solve equations to find unknown currents.

(*v*) If the value of the assumed current comes out to be negative, it means that actual direction of current is opposite to that of assumed direction.

Note. It may be noted that Kirchhoff's laws are also applicable to a.c. circuits. The only thing to be done is that **I**, **V** and **Z** are substituted for *I*, *V* and *R*. Here **I**, **V** and **Z** are phasor quantities.

2.21. Matrix Algebra

The solution of two or three simultaneous equations can be achieved by a method that uses *determinants*. A determinant is a numerical value assigned to a square arrangement of numbers called a *matrix*. The advantage of determinant method is that it is less difficult for three unknowns and there is less chance of error. The theory behind this method is not presented here but is available in any number of mathematics books.

Second-order determinant. A 2×2 matrix has four numbers arranged in two rows and two columns. The value of such a matrix is called a *second-order determinant* and is *equal to the product of the principal diagonal minus the product of the other diagonal*. For example, value of the matrix $= ad - cb$.

Second-order determinant can be used to solve simultaneous equations with two unknowns. Consider the following equations :

$$a_1 x + b_1 y = c_1$$
$$a_2 x + b_2 y = c_2$$

The unknowns are x and y in these equations. The numbers associated with the unknowns are called *coefficients*. The coefficients in these equations are a_1, a_2, b_1 and b_2. The right hand number (c_1 or c_2) of each equation is called a *constant*. The coefficients and constants can be arranged as a *numerator matrix* and as a *denominator matrix*. The matrix for the numerator is formed by replacing the coefficients of the unknown by the constants. The denominator matrix is called *characteristic matrix* and is the same for each fraction. It is formed by the coefficients of the simultaneous equations.

$$x = \frac{\begin{vmatrix} c_1 & b_1 \\ c_2 & b_2 \end{vmatrix}}{\begin{vmatrix} a_1 & b_1 \\ a_2 & b_2 \end{vmatrix}} \quad ; \quad y = \frac{\begin{vmatrix} a_1 & c_1 \\ a_2 & c_2 \end{vmatrix}}{\begin{vmatrix} a_1 & b_1 \\ a_2 & b_2 \end{vmatrix}}$$

Note that the characteristic determinant (denominator) is the same in both cases and needs to be evaluated only once. Also note that the coefficients for x are replaced by the constants when solving for x and that the coefficients for y are replaced by the constants when solving for y.

Third-order determinant. A third-order determinant has 9 numbers arranged in 3 rows and 3 columns. Simultaneous equations with three unknowns can be solved with third-order determinants. Consider the following equations :

$$a_1x + b_1y + c_1z = d_1$$
$$a_2x + b_2y + c_2z = d_2$$
$$a_3x + b_3y + c_3z = d_3$$

The characteristic matrix forms the denominator and is the same for each fraction. It is formed by the coefficients of the simultaneous equations.

$$\text{Denominator} = \begin{vmatrix} a_1 & b_1 & c_1 \\ a_2 & b_2 & c_2 \\ a_3 & b_3 & c_3 \end{vmatrix}$$

The matrix for each numerator is formed by replacing the coefficient of the unknown with the constant.

$$x = \frac{\begin{vmatrix} d_1 & b_1 & c_1 \\ d_2 & b_2 & c_2 \\ d_3 & b_3 & c_3 \end{vmatrix}}{\text{Denominator}} \quad ; \quad y = \frac{\begin{vmatrix} a_1 & d_1 & c_1 \\ a_2 & d_2 & c_2 \\ a_3 & d_3 & c_3 \end{vmatrix}}{\text{Denominator}} \quad ; \quad z = \frac{\begin{vmatrix} a_1 & b_1 & d_1 \\ a_2 & b_2 & d_2 \\ a_3 & b_3 & d_3 \end{vmatrix}}{\text{Denominator}}$$

Example 2.36. *In the network shown in Fig. 2.65, the different currents and voltages are as under :*

$i_2 = 5e^{-2t} \; ; \; i_4 = 3 \sin t \; ; \; v_3 = 4e^{-2t}$

Using KCL, find voltage v_1.

Solution. Current through capacitor is

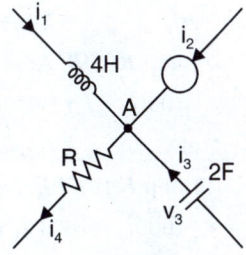

$$i_3 = C\frac{dv_3}{dt} = C\frac{d}{dt}(v_3) = \frac{2d}{dt}(4e^{-2t})$$

$$= -16e^{-2t}$$

Applying *KCL* to junction *A* in Fig. 2.65,

$$i_1 + i_2 + i_3 + (-i_4) = 0$$

or $i_1 + 5e^{-2t} - 16e^{-2t} - 3 \sin t = 0$

Fig. 2.65

or . $\qquad\qquad i_1 = 3 \sin t + 11e^{-2t}$

∴ Voltage developed across 4H coil is

$$v_1 = L\frac{di_1}{dt} = L\frac{d}{dt}(i_1) = 4\frac{d}{dt}(3 \sin t + 11e^{-2t})$$

$$= 4(3 \cos t - 22e^{-2t}) = \mathbf{12 \cos t - 88e^{-2t}}$$

Example 2.37. *For the circuit shown in Fig. 2.66, find the currents flowing in all branches.*

Solution. Mark the currents in various branches as shown in Fig. 2.66. Since there are two unknown quantities I_1 and I_2, two loops will be considered.

Loop *ABCFA*. Applying *KVL*,

$$30 - 2 I_1 - 10 + 5 I_2 = 0$$

or $\qquad 2 I_1 - 5 I_2 = 20 \qquad ...(i)$

Loop *FCDEF*. Applying *KVL*,

$$-5 I_2 + 10 - 3 (I_1 + I_2) - 5 - 4 (I_1 + I_2) = 0$$

or $\qquad 7 I_1 + 12 I_2 = 5 \qquad ...(ii)$

Multiplying eq. *(i)* by 7 and eq. *(ii)* by 2, we get,

$$14 I_1 - 35 I_2 = 140 \qquad ...(iii)$$
$$14 I_1 + 24 I_2 = 10 \qquad ...(iv)$$

Subtracting eq. *(iv)* from eq. *(iii)*, we get,

$$-59 I_2 = 130$$

∴ $\qquad I_2 = -130/59 = -2.2 A = $ **2.2 A** *from C to F*

Substituting the value of $I_2 = -2.2$ A in eq. *(i)*, we get, $I_1 = $ **4.5 A**

Current in branch $CDEF = I_1 + I_2 = (4.5) + (-2.2) = $ **2.3 A**

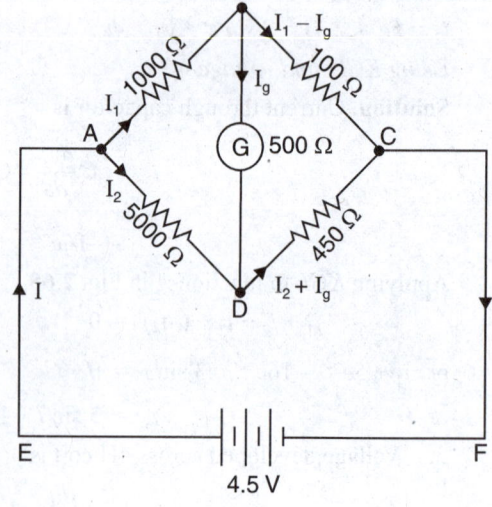

Fig. 2.66

Example 2.38. *A Wheatstone bridge ABCD has the following details ; AB = 1000 Ω ; BC = 100 Ω; CD = 450 Ω ; DA = 5000 Ω.*

A galvanometer of resistance 500 Ω is connected between B and D. A 4.5-volt battery of negligible resistance is connected between A and C with A positive. Find the magnitude and direction of galvanometer current.

Solution. Fig. 2.67 shows the Wheatstone bridge *ABCD*. Mark the currents in the various sections as shown. Since there are three unknown quantities (*viz.* I_1, I_2 and I_g), three loops will be considered.

Loop *ABDA*. Applying *KVL*,

$$-1000 I_1 - 500 I_g + 5000 I_2 = 0$$

or $\qquad 2 I_1 + I_g - 10 I_2 = 0 \qquad ...(i)$

Loop *BCDB*. Applying *KVL*,

$$-100(I_1 - I_g) + 450(I_2 + I_g) + 500 I_g = 0$$

or $\qquad 2 I_1 - 21 I_g - 9 I_2 = 0 \qquad ...(ii)$

Loop *EABCFE*. Applying *KVL*,

$$-1000 I_1 - 100 (I_1 - I_g) + 4.5 = 0$$

or $\qquad 1100 I_1 - 100 I_g = 4.5 \qquad ...(iii)$

Subtracting eq. *(ii)* from eq. *(i)*, we get,

$$22 I_g - I_2 = 0 \qquad ...(iv)$$

Multiplying eq. *(i)* by 550 and subtracting eq. *(iii)* from it, we get,

$$650 I_g - 5500 I_2 = -4.5 ...(v)$$

Multiplying eq. *(iv)* by 5500 and subtracting eq. *(v)* from it, we get,

$$120350 I_g = 4.5$$

∴ $\qquad I_g = \dfrac{4.5}{120350} = 37.4 \times 10^{-6} A = $ **37.4 μA** *from B to D*

Example 2.39. *A Wheatstone bridge ABCD is arranged as follows : AB = 1 Ω ; BC = 2 Ω ; CD = 3 Ω ; DA = 4 Ω. A resistance of 5 Ω is connected between B and D. A 4-volt battery of internal resistance 1 Ω is connected between A and C. Calculate (i) the magnitude and direction of current in 5 Ω resistor and (ii) the resistance between A and C.*

Solution. (*i*) Fig. 2.68 shows the Wheatstone bridge *ABCD*. Mark the currents in the various branches as shown. Since there are three unknown quantities (*viz.* I_1, I_2 and I_3), three loops will be considered.

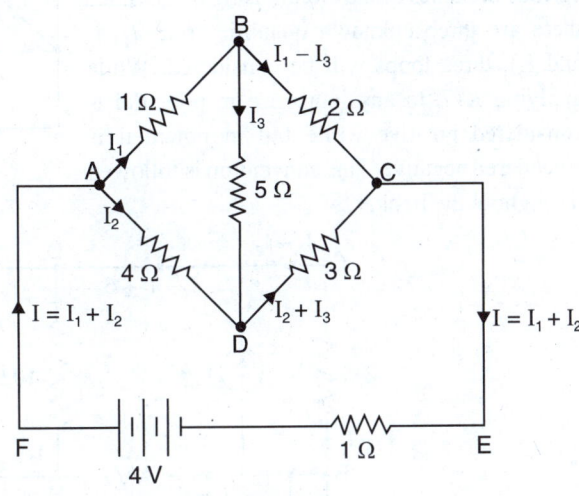

Fig. 2.68

Loop ABDA. Applying *KVL*,

$$-1 \times I_1 - 5 I_3 + 4 I_2 = 0$$

or $\qquad I_1 + 5 I_3 - 4 I_2 = 0 \quad ...(i)$

Loop BCDB. Applying *KVL*,

$$-2 (I_1 - I_3) + 3 (I_2 + I_3) + 5I_3 = 0$$

or $\qquad 2 I_1 - 10 I_3 - 3 I_2 = 0 \quad ...(ii)$

Loop FABCEF. Applying *KVL*,

$$-I_1 \times 1 - 2 (I_1 - I_3) - 1 (I_1 + I_2) + 4 = 0$$

or $\qquad 4 I_1 - 2 I_3 + I_2 = 4 \quad ...(iii)$

Multiplying eq.(*i*) by 2 and subtracting *eq.* (*ii*) from it, we get,

$$20 I_3 - 5 I_2 = 0 \qquad\qquad ...(iv)$$

Multiplying *eq.* (*i*) by 4 and subtracting *eq.* (*iii*) from it, we get,

$$22 I_3 - 17 I_2 = -4 \qquad\qquad ...(v)$$

Multiplying *eq.* (*iv*) by 17 and eq. (*v*) by 5, we get,

$$340 I_3 - 85 I_2 = 0 \qquad\qquad ...(vi)$$

$$110 I_3 - 85 I_2 = -20 \qquad\qquad ...(vii)$$

Subtracting *eq.* (*vii*) from *eq.* (*vi*), we get,

$$230 I_3 = 20$$

∴ $\qquad\qquad I_3 = 20/230 = 0.087$ A

i.e $\qquad$ Current in 5 Ω, $I_3 =$ **0.087 A** *from B to D*

(*ii*) Substituting the value of $I_3 = 0.087$ A in eq. (*iv*), we get, $I_2 = 0.348$ A.

Substituting values of $I_3 = 0.087$ A and $I_2 = 0.348$ A in *eq.* (*ii*), $I_1 = 0.957$ A.

Current supplied by battery, $I = I_1 + I_2 = 0.957 + 0.348 = 1.305$ A

P.D. between *A* and *C* = E.M.F. of battery − Drop in battery = 4 − 1.305 × 1 = 2.695 V

∴ $\qquad$ Resistance between *A* and *C* = $\dfrac{\text{P.D. across } AC}{\text{Battery current}} = \dfrac{2.695}{1.305} =$ **2.065 Ω**

Example 2.40. *Determine the current in 4 Ω resistance of the circuit shown in Fig. 2.69.*

Solution. The given circuit is redrawn as shown in Fig. 2.70. Mark the currents in the various branches of the circuit using *KCL*. Since there are three unknown quantities (*viz.* I_1, I_2 and I_3), three loops will be considered. While applying *KVL* to any loop, rise in potential is considered positive while fall in potential is considered negative. This convention is followed throughout the book.

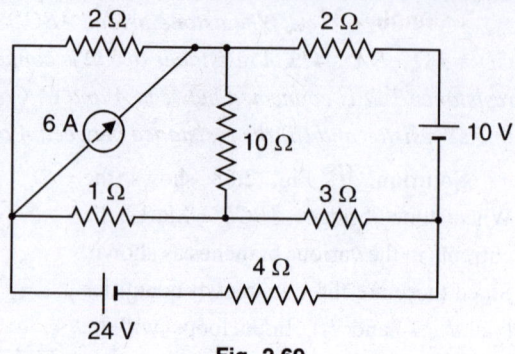

Fig. 2.69

Fig. 2.70

Loop BCDHB. Applying *KVL*, we have,

$$-2(I_1 - I_2) + 10I_3 + 1 \times (I_2 - 6) = 0$$

or
$$2I_1 - 3I_2 - 10I_3 = -6 \qquad \qquad ...(i)$$

Loop DEFHD. Applying K*VL*, we have,

$$-2(I_1 - I_2 + 6 + I_3) - 10 + 3(I_2 - 6 - I_3) - 10I_3 = 0$$

or
$$2I_1 - 5I_2 + 15I_3 = -40 \qquad \qquad ...(ii)$$

Loop BHFGAB. Applying *KVL*, we have,

$$-1(I_2 - 6) - 3(I_2 - 6 - I_3) - 4I_1 + 24 = 0$$

or
$$4I_1 + 4I_2 - 3I_3 = 48 \qquad \qquad ...(iii)$$

Solving eqs. (*i*), (*ii*) and (*iii*), we get, $I_1 = 4.1$ A.

∴   Current in 4 Ω resistance = I_1 = **4.1 A**

Example 2.41. *Two batteries E_1 and E_2 having e.m.fs of 6V and 2V respectively and internal resistances of 2 Ω and 3 Ω respectively are connected in parallel across a 5 Ω resistor. Calculate (i) current through each battery and (ii) terminal voltage.*

Solution. Fig. 2.71 shows the conditions of the problem. Mark the currents in the various branches. Since there are two unknown quantities I_1 and I_2, two loops will be considered.

(i) Loop HBCDEFH. Applying Kirchhoff's voltage law to loop *HBCDEFH*, we get,

$$2I_1 - 6 + 2 - 3I_2 = 0$$

or
$$2I_1 - 3I_2 = 4 \qquad \qquad ...(i)$$

Loop ABHFEGA. Applying Kirchhoff's voltage law to loop *ABHFEGA*, we get,

$$3I_2 - 2 + 5(I_1 + I_2) = 0$$

or
$$5I_1 + 8I_2 = 2 \qquad \qquad ...(ii)$$

Multiplying eq. (*i*) by 8 and eq. (*ii*) by 3 and then adding them, we get,

$$31\ I_1 = 38$$

or $\qquad I_1 = \dfrac{38}{31} = \mathbf{1.23\ A}$

i.e. battery E_1 is being discharged at 1.23 A. Substituting $I_1 = 1.23$ A in eq. (*i*), we get, $I_2 = -\ \mathbf{0.52A}$ *i.e.* battery E_2 is being charged.

(*ii*) Terminal voltage $= (\ I_1 + I_2\ )\ 5$

$= (1.23 - 0.52)\ 5 = \mathbf{3.55\ V}$

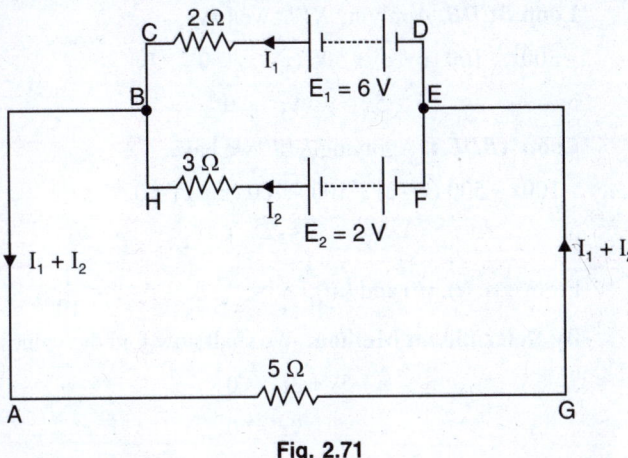

Fig. 2.71

Example 2.42. *Twelve wires, each of resistance r, are connected to form a skeleton cube. Find the equivalent resistance between the two diagonally opposite corners of the cube.*

Solution. Let *ABCDEFGH* be the skeleton cube formed by joining 12 wires, each of resistance *r* as shown in Fig. 2.72. Suppose a current of 6*I* enters the cube at the corner *A*. Since the resistance of each wire is the same, the current at corner *A* is divided into three equal parts: 2*I* flowing in *AE*, 2*I* flowing in *AB* and 2*I* flowing in *AD*. At points *B*, *D* and *E*, these currents are divided into equal parts, each part being equal to *I*. Applying Kirchhoff's current law, 2*I* current flows in each of the wires *CG*, *HG* and *FG*. These three currents add up at the corner *G* so that current flowing out of this corner is 6*I*.

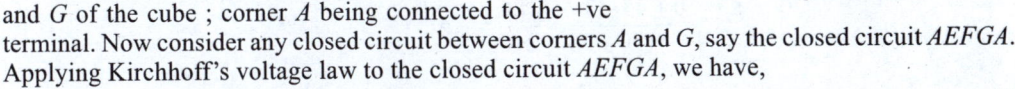

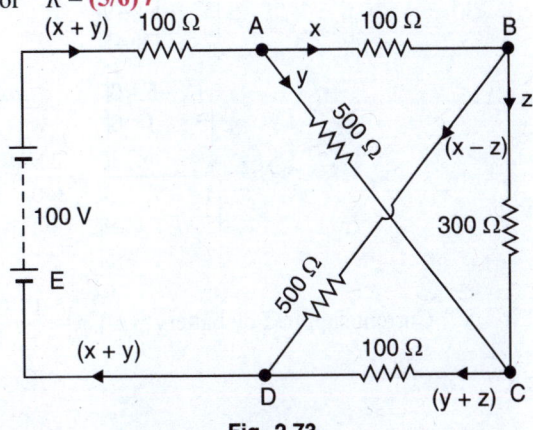

Fig. 2.72

Let *E* = e.m.f. of the battery connected to corners *A* and *G* of the cube ; corner *A* being connected to the +ve terminal. Now consider any closed circuit between corners *A* and *G*, say the closed circuit *AEFGA*. Applying Kirchhoff's voltage law to the closed circuit *AEFGA*, we have,

$$-2\ I\ r - I\ r - 2\ I\ r\ =\ -E \quad \text{or} \quad 5\ I\ r = E \qquad \qquad ...(i)$$

Let *R* be the equivalent resistance between the diagonally opposite corners *A* and *G*.

Then, $\qquad\qquad\qquad E\ =\ 6\ I\ R \qquad\qquad\qquad ...(ii)$

From eqs. (*i*) and (*ii*), we get, $6IR = 5I \cdot r$ or $\quad R = \mathbf{(5/6)\ r}$

Example 2.43. *Determine the current supplied by the battery in the circuit shown in Fig. 2.73.*

Solution. Mark the currents in the various branches as shown in Fig. 2.73. Since there are three unknown quantities *x*, *y* and *z*, three equations must be formed by considering three loops.

Loop *ABCA*. Applying *KVL*, we have,

$$-100x - 300z + 500y\ =\ 0$$

or $\qquad\qquad x - 5y + 3z\ =\ 0 \qquad\qquad\qquad ...(i)$

Fig. 2.73

Loop BCDB. Applying *KVL*, we have,

$$-300x - 100(y + z) + 500(x - z) = 0$$

or

$$5x - y - 9z = 0 \qquad \qquad ...(ii)$$

Loop ABDEA. Applying *KVL*, we have,

$$-100x - 500(x - z) + 100 - 100(x + y) = 0$$

or

$$7x + y - 5z = 1 \qquad \qquad ...(iii)$$

From eqs. (*i*), (*ii*) and (*iii*), $x = \dfrac{1}{5}\text{A}$; $y = \dfrac{1}{10}\text{A}$; $z = \dfrac{1}{10}\text{A}$

By Determinant Method. We shall now find the values of x, y and z by determinant method.

$$x - 5y + 3z = 0 \qquad \qquad ...(i)$$

$$5x - y - 9z = 0 \qquad \qquad ...(ii)$$

$$7x + y - 5z = 1 \qquad \qquad ...(iii)$$

$$\begin{bmatrix} 1 & -5 & 3 \\ 5 & -1 & -9 \\ 7 & 1 & -5 \end{bmatrix}\begin{bmatrix} x \\ y \\ z \end{bmatrix} = \begin{bmatrix} 0 \\ 0 \\ 1 \end{bmatrix}$$

$$\therefore \quad x = \frac{\begin{vmatrix} 0 & -5 & 3 \\ 0 & -1 & -9 \\ 1 & 1 & -5 \end{vmatrix}}{\begin{vmatrix} 1 & -5 & 3 \\ 5 & -1 & -9 \\ 7 & 1 & -5 \end{vmatrix}} = \frac{0\begin{vmatrix} -1 & -9 \\ 1 & -5 \end{vmatrix} + 5\begin{vmatrix} 0 & -9 \\ 1 & -5 \end{vmatrix} + 3\begin{vmatrix} 0 & -1 \\ 1 & 1 \end{vmatrix}}{1\begin{vmatrix} -1 & -9 \\ 1 & -5 \end{vmatrix} + 5\begin{vmatrix} 5 & -9 \\ 7 & -5 \end{vmatrix} + 3\begin{vmatrix} 5 & -1 \\ 7 & 1 \end{vmatrix}}$$

$$= \frac{0\,[(-1\times-5)-(1\times-9)]+5\,[(0\times-5)-(1\times-9)]+3\,[(0\times1)-(1\times-1)]}{1\,[(-1\times-5)-(1\times-9)]+5\,[(5\times-5)-(7\times-9)]+3\,[(5\times1)-(7\times-1)]}$$

$$= \frac{0+45+3}{14+190+36} = \frac{48}{240} = \frac{1}{5}\text{A}$$

$$y = \frac{\begin{vmatrix} 1 & 0 & 3 \\ 5 & 0 & -9 \\ 7 & 1 & -5 \end{vmatrix}}{\begin{vmatrix} 1 & -5 & 3 \\ 5 & -1 & -9 \\ 7 & 1 & -5 \end{vmatrix}} = \frac{24}{240} = \frac{1}{10}\text{A}$$

$$z = \frac{\begin{vmatrix} 1 & -5 & 0 \\ 5 & -1 & 0 \\ 7 & 1 & 1 \end{vmatrix}}{\begin{vmatrix} 1 & -5 & 3 \\ 5 & -1 & -9 \\ 7 & 1 & -5 \end{vmatrix}} = \frac{24}{240} = \frac{1}{10}\text{A}$$

$$\therefore \quad \text{Current supplied by battery} = x + y = \frac{1}{5} + \frac{1}{10} = \frac{3}{10}\text{A}$$

Example 2.44. *Use Kirchhoff's voltage law to find the voltage V_{ab} in Fig. 2.74.*

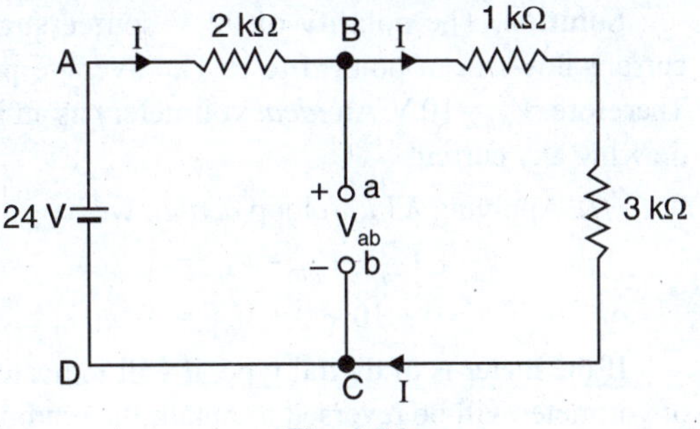

Solution. We shall use Kirchhoff's voltage law to solve this problem, although other methods can be used.

Total circuit resistance, $R_T = 2 + 1 + 3 = 6\ k\Omega$

Circuit current, $I = \dfrac{V}{R_T} = \dfrac{24\ V}{6\ k\Omega} = 4\ mA$

Applying Kirchhoff's voltage law to loop *ABCDA*, we have,

$$24 - 4\ mA \times 2\ k\Omega - {}^*V_{ab} = 0$$

or $\qquad 24 - 8 - V_{ab} = 0 \qquad \therefore \quad V_{ab} = 24 - 8 = \mathbf{16\ V}$

Fig. 2.74

Example 2.45. *For the ladder network shown in Fig. 2.75, find the source voltage V_s which results in a current of 7.5 mA in the 3 Ω resistor.*

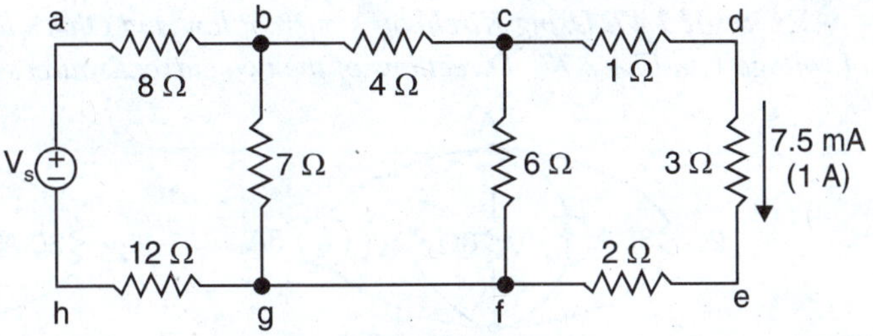

Solution. Let us assume that current in branch *de* is 1 A.

Since the circuit is linear, the voltage necessary to produce 1 A is in the same ratio to 1 A as V_s to 7.5 mA.

Fig. 2.75

Voltage between *c* and *f*, $V_{cf} = 1\,(1 + 3 + 2) = 6\ V$

$\therefore \qquad$ Current in branch *cf*, $I_{cf} = 6/6 = 1\ A$

Applying *KCL* at junction *c*,

$$I_{bc} = 1 + 1 = 2\ A$$

Applying *KVL* to loop *bcfgb*, we have,

$$-4 \times 2 - 6 \times 1 + V_{bg} = 0 \quad \therefore \quad V_{bg} = 8 + 6 = 14\ V$$

$\therefore \qquad$ Current in branch *bg*, $I_{bg} = \dfrac{V_{bg}}{7} = \dfrac{14}{7} = 2\ A$

Applying *KCL* to junction *b*, we have, $I_{ab} = 2 + 2 = 4\ A$

Applying *KVL* to loop *abgha*, we have,

$$-8 \times 4 - 7 \times 2 - 12 \times 4 + V_{ah} = 0 \quad \therefore \quad V_{ah} = 94\ V$$

Now $\qquad \dfrac{V_{ah}}{1\ A} = \dfrac{V_s}{7.5\ mA} \quad$ or $\quad \dfrac{94}{1\ A} = \dfrac{V_s}{7.5 \times 10^{-3}\ A} \qquad \therefore \quad V_s = \mathbf{0.705\ V}$

Example 2.46. *Determine the readings of an ideal voltmeter connected in Fig. 2.76 to (i) terminals a and b, (ii) terminals c and g. The average power dissipated in the 5 Ω resistor is equal to 20 W.*

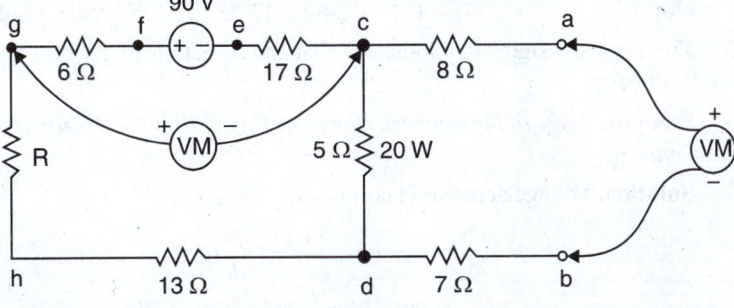

Fig. 2.76

---

*   Note that point *a* is positive w.r.t. point *b*.

Solution. The polarity of 90 V source suggests that point d is positive w.r.t. c. Therefore, current flows from point d to c. The average power in 5 Ω resistor is 20 W so that $V_{dc}^2/5 = 20$. Therefore, $V_{dc} = 10$ V. An *ideal* voltmeter has an infinite resistance and indicates the voltage without drawing any current.

(*i*) Applying *KVL* to loop *acdba*, we have,

$$V_{ac} + V_{cd} + V_{db} + V_{ba} = 0$$

or $\qquad 0 + 10 + 0 + V_{ba} = 0 \qquad \therefore \qquad V_{ba} = -\textbf{10 V}$

If the meter is of digital type, it will indicate -10 V. For moving-coil galvanometer, the leads of voltmeter will be reversed to obtain the reading.

(*ii*) Applying *KVL* to loop *cefgc*, we have,

$$-V_{ce} + V_{ef} - V_{fg} - V_{gc} = 0$$

or $-17 \times 2 + 90 - 6 \times 2 - V_{gc} = 0 \qquad \therefore \qquad V_{gc} = \textbf{44 V}$

Example 2.47. *Using Kirchhoff's current law and Ohm's law, find the magnitude and polarity of voltage V in Fig. 2.77. Directions of the two current sources are shown.*

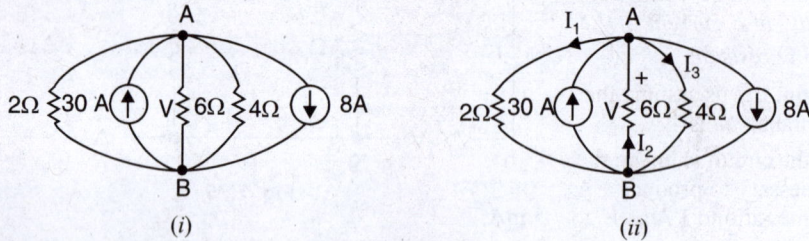

Fig. 2.77

Solution. Let us assign the directions of I_1, I_2 and I_3 and polarity of V as shown in Fig. 2.77 (*ii*). We shall see in the final result whether our assumptions are correct or not. Referring to Fig. 2.77 (*ii*) and applying *KCL* to junction A, we have,

$$\text{Incoming currents } = \text{ Outgoing currents}$$

or $\qquad I_2 + 30 = I_1 + I_3 + 8$

$\therefore \qquad I_1 - I_2 + I_3 = 22 \qquad\qquad\qquad ...(i)$

Applying Ohm's law to Fig. 2.77 (*ii*), we have,

$$I_1 = \frac{V}{2} \quad ; \quad I_3 = \frac{V}{4} \quad ; \quad I_2 = -\frac{V}{6}$$

Putting these values of I_1, I_2 and I_3 in eq. (*i*), we have,

$$\frac{V}{2} - \left(-\frac{V}{6}\right) + \frac{V}{4} = 22 \quad \text{or} \quad V = \textbf{24 V}$$

Now $\qquad I_1 = V/2 = 24/2 = 12$ A $\quad ; \quad I_2 = -24/6 = -4$ A $\quad ; \quad I_3 = 24/4 = 6$ A

The negative sign of I_2 indicates that the direction of its flow is opposite to that shown in Fig. 2.77 (*ii*).

Example 2.48. *In the network shown in Fig. 2.78, $v_1 = 4$ volts ; $v_4 = 4 \cos 2t$ and $i_3 = 2e^{-t/3}$. Determine i_2.*

Solution. Voltage across 6 H coil is

$$v_3 = L\frac{di_3}{dt} = L\frac{d}{dt}(i_3)$$

$$= 6\frac{d}{dt}(2e^{-t/3}) = -4e^{-t/3}$$

Applying *KVL* to loop *ABCDA*, we have,

$$-v_1 - v_2 + v_3 + v_4 = 0$$

or $\quad -4 - v_2 - 4e^{-t/3} + 4\cos 2t = 0$

$$\therefore \qquad\qquad v_2 = 4\cos 2t - 4e^{-t/3} - 4$$

Current through 8 F capacitor is

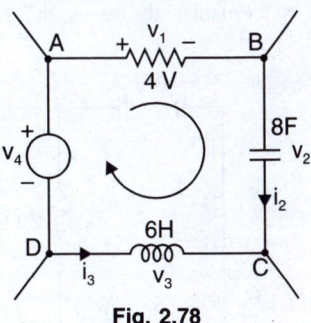

Fig. 2.78

$$i_2 = C\frac{dv_2}{dt} = C\frac{d}{dt}(v_2)$$

$$= 8\frac{d}{dt}(4\cos 2t - 4e^{-t/3} - 4)$$

$$= 8\left(-8\sin 2t + \frac{4}{3}e^{-t/3}\right)$$

$$= -64\sin 2t + \frac{32}{3}e^{-t/3}$$

Tutorial Problems

1. Using Kirchhoff's laws, find the current in various resistors in the circuit shown in Fig. 2.79.

   [6.574 A, 3.611 A ,10.185 A]

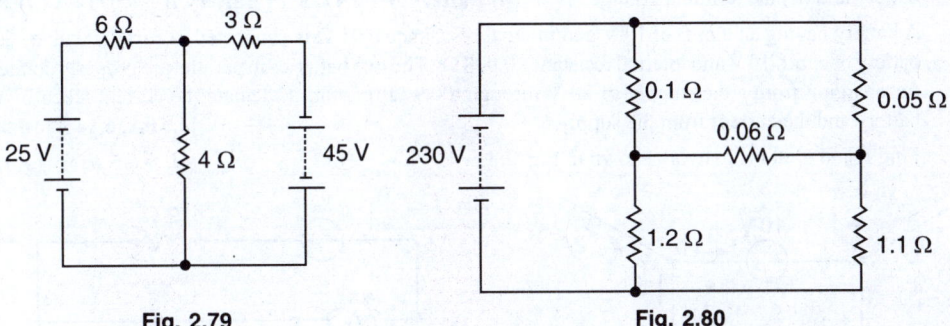

Fig. 2.79   **Fig. 2.80**

2. For the circuit shown in Fig. 2.80, determine the branch currents using Kirchhoff's laws.

   [151.35A, 224.55A, 27.7A , 179.05 A, 196.84 A]

3. Two batteries *A* and *B* having e.m.fs. 12 V and 8 V respectively and internal resistances of 2 Ω and 1 Ω respectively, are connected in parallel across 10 Ω resistor. Calculate (*i*) the current in each of the batteries and the external resistor and (*ii*) p.d. across external resistor.

   [(*i*) I_A = 1.625 A *discharge* ; I_B = 0.75 A *charge*; 0.875 A (*ii*) 8.75 V]

4. A Wheatstone bridge *ABCD* is arranged as follows : *AB* = 20 Ω, *BC* = 5 Ω, *CD* = 4 Ω and *DA* = 10 Ω. A galvanometer of resistance 6Ω is connected between *B* and *D*. A 100-volt supply of negligible resistance is connected between *A* and *C* with *A* positive. Find the magnitude and direction of galvanometer current.

   [0.667 A *from D to B*]

5. A network *ABCD* consists of the following resistors : *AB* = 5 kΩ, *BC* = 10 kΩ, *CD* = 15 kΩ and *DA* = 20 kΩ. A fifth resistor of 10 kΩ is connected between *A* and *C*. A dry battery of e.m.f. 120 V and internal resistance 500Ω is connected across the resistor *AD*. Calculate (*i*) the total current supplied by the battery, (*ii*) the p.d. across points *C* and *D* and (*iii*) the magnitude and direction of current through branch *AC*.

   [(*i*) 11.17 mA (*ii*) 81.72 V (*iii*) 3.27 mA *from A to C*]

6. A Wheatstone bridge *ABCD* is arranged as follows : *AB* = 10 Ω, *BC* = 30 Ω, *CD* = 15Ω and *DA* = 20Ω. A 2 volt battery of internal resistance 2Ω is connected between *A* and *C* with *A* positive. A galvanometer of resistance 40Ω is connected between *B* and *D*. Find the magnitude and direction of galvanometer current.

   [11.5 mA *from B to D*]

7. Two batteries E_1 and E_2 having *e.m.fs* 6 V and 2 V respectively and internal resistances of 2 Ω and 3 Ω respectively are connected in parallel across a 5 Ω resistor. Calculate (*i*) current through each battery and (*ii*) terminal voltage.

   [(*i*) 1.23A; –0.52A (*ii*) 3.55V]

8. Calculate the current in 20 Ω resistor in Fig. 2.81.   **[26.67 mA]**

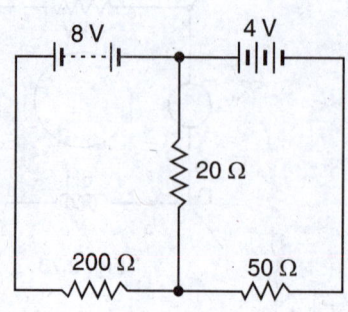

Fig. 2.81   **Fig. 2.82**

9. In the circuit shown in Fig. 2.82, find the current in each branch and the current in the battery. What is the p.d. between A and C ?

[Branch ABC = 0.581 A ; Branch ADC = 0.258 A ; Branch AC = 0.839 A ; V_{AC} = 2.32 V]

10. Two batteries A and B having e.m.f.s of 20 V and 21 V respectively and internal resistances of 0.8 Ω and 0.2 Ω respectively, are connected in parallel across 50 Ω resistor. Calculate (*i*) the current through each battery and (*ii*) the terminal voltage.   **[(*i*) Battery A = 0.4725 A ; Battery B = 0.0714 A (*ii*) 20 V]**

11. A battery having an e.m.f. of 10 V and internal resistance 0.01 Ω is connected in parallel with a second battery of e.m.f. 10 V and internal resistance 0.008 Ω. The two batteries in parallel are properly connected for charging from a d.c. supply of 20 V through a 0.9 Ω resistor. Calculate the current taken by each battery and the current from the supply.   **[4.91 A, 6.14 A, 10.05 A]**

12. Find i_x and v_x in the network shown in Fig. 2.83.   **[i_x = – 5 A; v_x = – 15 V]**

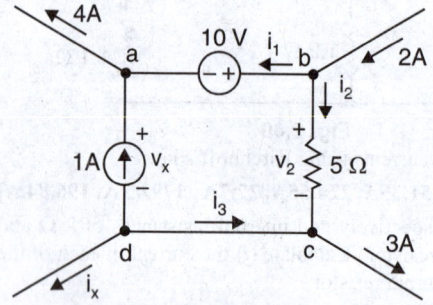

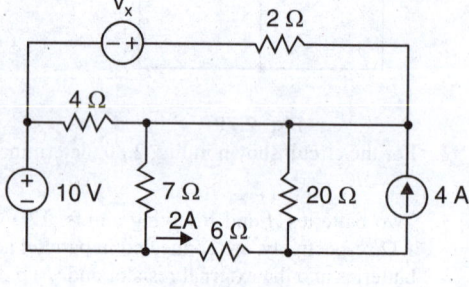

Fig. 2.83   **Fig. 2.84**

13. Find v_x for the network shown in Fig. 2.84.   **[31 V]**

14. Find i and v_{ab} for the network shown in Fig. 2.85.   **[3 A ; 19 V]**

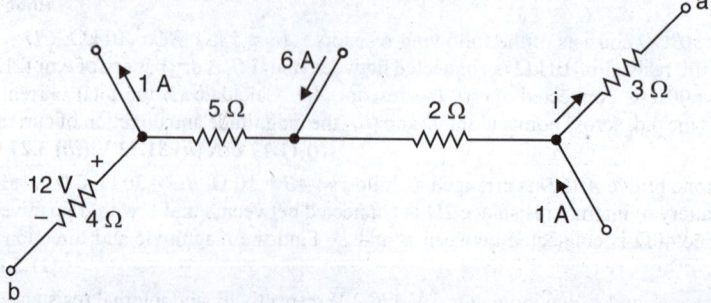

Fig. 2.85

2.22. Voltage and Current Sources

The term *voltage source* is used to describe a source of energy which establishes a potential difference across its terminals. Most of the sources encountered in everyday life are voltage sources *e.g.*, batteries, d.c. generators, alternators etc. The term *current source* is used to describe a source of energy that provides a current *e.g.*, collector circuits of transistors. Voltage and current sources are called active elements because they provide electrical energy to a circuit.

Unlike a voltage source, which we can imagine as two oppositely charged electrodes, it is difficult to visualise the structure of a current source. However, as we will learn in later sections, a real current source can always be converted into a real voltage source. In other words, we can regard a current source as a convenient fiction that aids in solving circuit problems, yet we feel secure in the knowledge that the current source can be replaced by the equivalent voltage source, if so desired.

2.23. Ideal Voltage Source or Constant-Voltage Source

*An **ideal voltage source** (also called constant-voltage source) is one that maintains a constant terminal voltage, no matter how much current is drawn from it.*

An ideal voltage source has zero internal resistance. Therefore, it would provide constant terminal voltage regardless of the value of load connected across its terminals. For example, an ideal 12V source would maintain 12V across its terminals when a 1 MΩ resistor is connected (so $I = 12$ V/1 MΩ = 12A) as well as when a 1 kΩ resistor is connected ($I = 12$ mA) or when a 1 Ω resistor is connected ($I = 12$A). This is illustrated in Fig. 2.86.

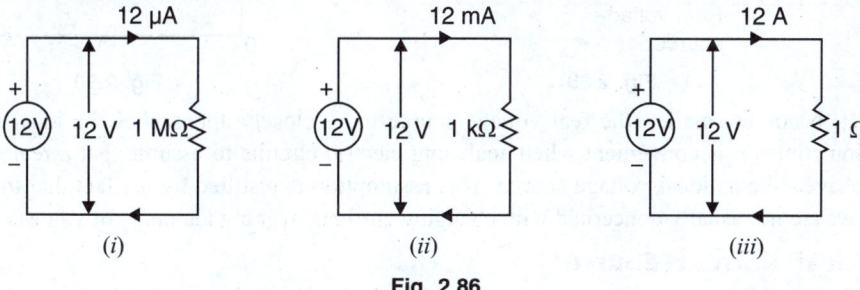

Fig. 2.86

It is not possible to construct an ideal voltage source because every voltage source has some internal resistance that causes the terminal voltage to fall due to the flow of current. *However, if the internal resistance of voltage source is very small, it can be considered as a constant voltage source.* This is illustrated in Fig. 2.87. It has a d.c. source of 6 V with an internal resistance R_i = 0.005 Ω. If the load current varies over a wide

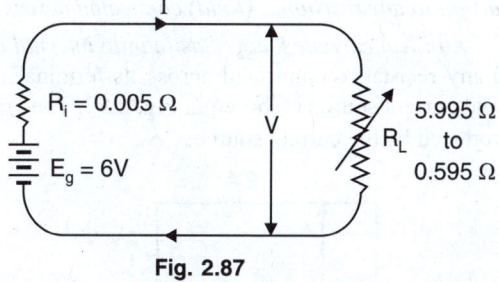

Fig. 2.87

range of 1 to 10 A, for any of these values, the internal drop across R_i (= 0.005 Ω) is less than 0.05 volt. Therefore, the voltage output of the source is between 5.995 and 5.95 volts. This can be considered constant voltage compared with wide variations in load current. The practical example of a constant voltage source is the lead-acid cell. The internal resistance of lead-acid cell is very small (about 0.01 Ω) so that it can be regarded as a constant voltage source for all practical purposes. A constant voltage source is represented by the symbol shown in Fig. 2.88.

Fig. 2.88

2.24. Real Voltage Source

A real or non-ideal voltage source has low but finite internal resistance (R_{int}) that causes its terminal voltage to decrease when load current is increased and *vice-versa*. *A **real voltage source** can be represented as an ideal voltage source in series with a resistance equal to its internal resistance (R_{int}) as shown in Fig. 2.89.*

When load R_L is connected across the terminals of a real voltage source, a load current I_L flows through the circuit so that output voltage V_o is given by ;

$$V_o = E - I_L R_{int}$$

Here E is the voltage of the ideal voltage source *i.e.*, it is the potential difference between the terminals of the source when no current (*i.e.*, $I_L = 0$) is drawn. Fig. 2.90 shows the graph of output voltage V_o versus load current I_L of a real or non-ideal voltage source.

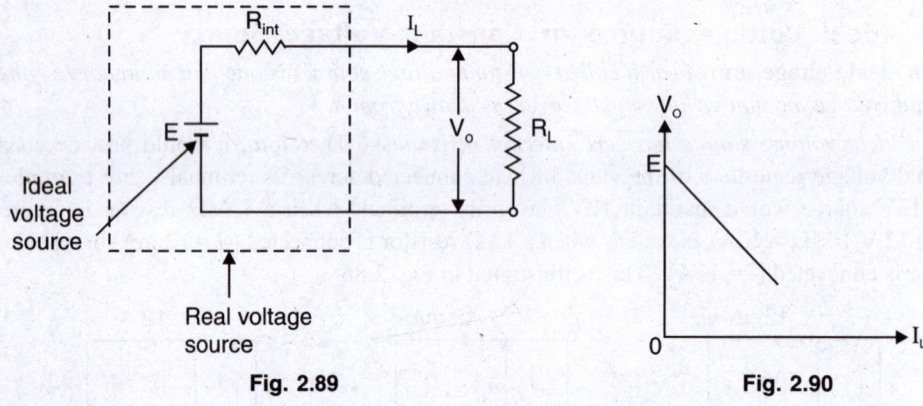

Fig. 2.89   Fig. 2.90

As R_{int} becomes smaller, the real voltage source more closely approaches the ideal voltage source. Sometimes it is convenient when analysing electric circuits to assume that a real voltage source behaves like an ideal voltage source. This assumption is justified by the fact that in circuit analysis, we are not usually concerned with changing currents over a wide range of values.

2.25. Ideal Current Source

*An **ideal current source** or **constant current source** is one which will supply the same current to any resistance (load) connected across its terminals.*

An ideal current source has infinite internal resistance. Therefore, it supplies the same current to any resistance connected across its terminals. This is illustrated in Fig. 2.91. The symbol for ideal current source is shown in Fig. 2.92. The arrow shows the direction of current (conventional) produced by the current source.

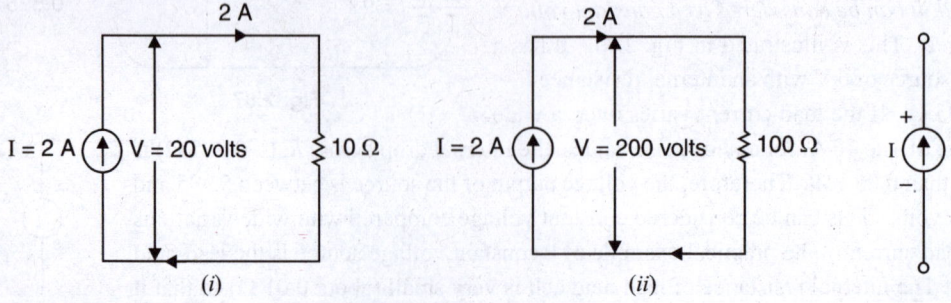

Fig. 2.91   Fig. 2.92

Since an ideal current source supplies the same current regardless of the value of resistance connected across its terminals, it is clear that the terminal voltage V of the current source will depend on the value of load resistance. For example, if a 2 A current source has 10 Ω across its terminals, then terminal voltage of the source is $V = 2\ A \times 10\ \Omega = 20$ volts. If load resistance is changed to 100 Ω, then terminal voltage of the current source becomes $V = 2\ A \times 100\ \Omega = 200$ volts. This is illustrated in Fig. 2.91.

2.26. Real Current Source

A real or non-ideal current source has high but finite internal resistance (R_{int}). Therefore, the load current (I_L) will change as the value of load resistance (R_L) changes. *A real current source can be represented by an ideal current source (I) in parallel with its internal resistance (R_{int}) as shown in Fig. 2.93.* When load resistance R_L is connected across the terminals of the real current source, the load current I_L is equal to the current I from the ideal current source *minus* that part of the current that passes through the parallel internal resistance (R_{int}) *i.e.*,

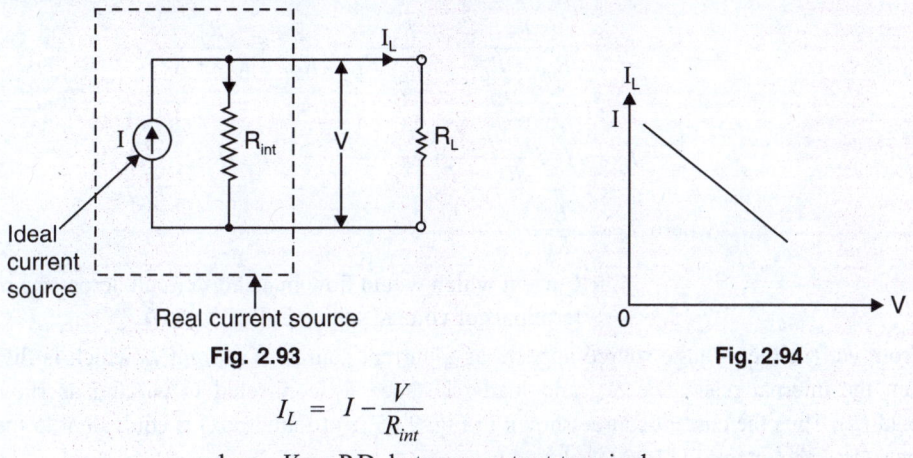

Ideal current source

Real current source

Fig. 2.93

Fig. 2.94

$$I_L = I - \frac{V}{R_{int}}$$

where V = P.D. between output terminals

Fig. 2.94 shows the graph of load current I_L versus output voltage V of a real current source.

Note that load current I_L is less than it would be if the source were ideal. As the internal resistance of real current source becomes greater, the current source more closely approaches the ideal current source.

Note. Current sources in parallel add *algebraically*. If two current sources are supplying currents in the same direction, their equivalent current source supplies current equal to the sum of the individual currents. If two current sources are supplying currents in the opposite directions, their equivalent current source supplies a current equal to the difference of the currents of the two sources.

2.27. Source Conversion

A real voltage source can be converted to an *equivalent* real current source and *vice-versa*. When the conversion is made, the sources are equivalent in every sense of the word; it is impossible to make any measurement or perform any test at the external terminals that would reveal whether the source is a voltage source or its equivalent current source.

(i) **Voltage to current source conversion.** Let us see how a real voltage source can be converted to an equivalent current source. We know that a real voltage source can be represented by constant voltage E in series with its internal resistance R_{int} as shown in Fig. 2.95 (*i*).

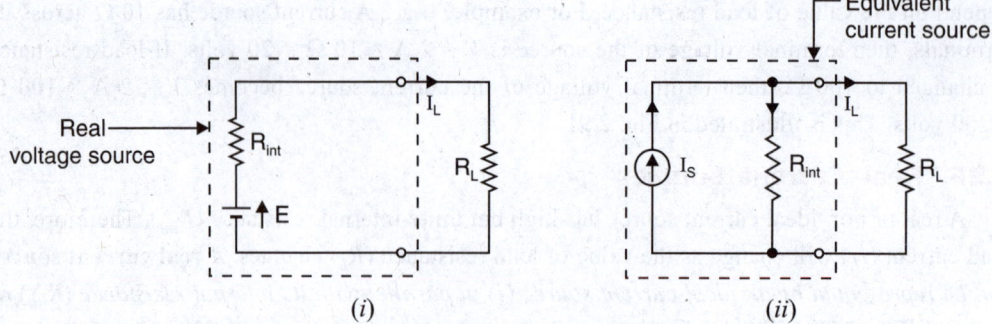

Fig. 2.95

It is clear from Fig. 2.95 (*i*) that load current I_L is given by ;

$$I_L = \frac{E}{R_{int} + R_L} = \frac{\dfrac{E}{R_{int}}}{\dfrac{R_{int} + R_L}{R_{int}}} = \frac{E}{R_{int}} \times \frac{R_{int}}{R_{int} + R_L}$$

$$\therefore \qquad I_L = I_S \times \frac{R_{int}}{R_{int} + R_L} \qquad\qquad\qquad ...(i)$$

$$\text{where } I_S = \frac{E^*}{R_{int}}$$

= Current which would flow in a short circuit across the output terminals of voltage source in Fig. 2.95 (*i*)

From eq. (*i*), the voltage source appears as a current source of current I_S which is dividing between the internal resistance R_{int} and load resistance R_L connected in parallel as shown in Fig. 2.95 (*ii*). Thus the current source shown in Fig. 2.95 (*ii*) (dotted box) is equivalent to the real voltage source shown in Fig. 2.95 (*i*) (dotted box).

Thus a real voltage source of constant voltage E and internal resistance R_{int} is equivalent to a current source of current $I_S = E/R_{int}$ and R_{int} in parallel with current source.

Note that internal resistance of the equivalent current source has the same value as the internal resistance of the original voltage source but is in parallel with current source. The two circuits shown in Fig. 2.95 are equivalent and either can be used for circuit analysis.

(*ii*) Current to voltage source conversion. Fig. 2.96 (*i*) shows a real current source whereas Fig. 2.96 (*ii*) shows its equivalent voltage source. Note that series resistance R_{int} of the voltage source

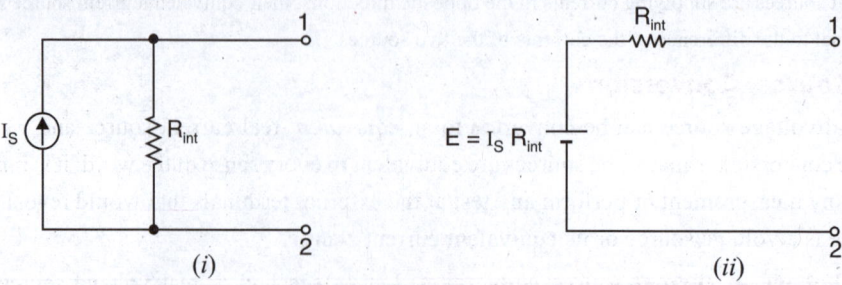

Fig. 2.96

---

* The source voltage is E and its internal resistance is R_{int}. Therefore, E/R_{int} is the current that would flow when source terminals in Fig. 2.95 (*i*) are shorted.

has the same value as the parallel resistance of the original current source. The value of voltage of the equivalent voltage source is $E = I_S R_{int}$ where I_S is the magnitude of current of the current source.

Note that the two circuits shown in Fig. 2.96 are equivalent and either can be used for circuit analysis.

Note. The source conversion (voltage source into equivalent current source and vice-versa) often simplifies the analysis of many circuits. Any resistance that is in series with a voltage source, whether it be internal or external resistance, can be included in its conversion to an equivalent current source. Similarly, any resistance in parallel with current source can be included when it is converted to an equivalent voltage source.

Example 2.49. *Show that the equivalent sources shown in Fig. 2.97 have exactly the same terminal voltage and produce exactly the same external current when the terminals (i) are shorted, (ii) are open and (iii) have a 500 Ω load connected.*

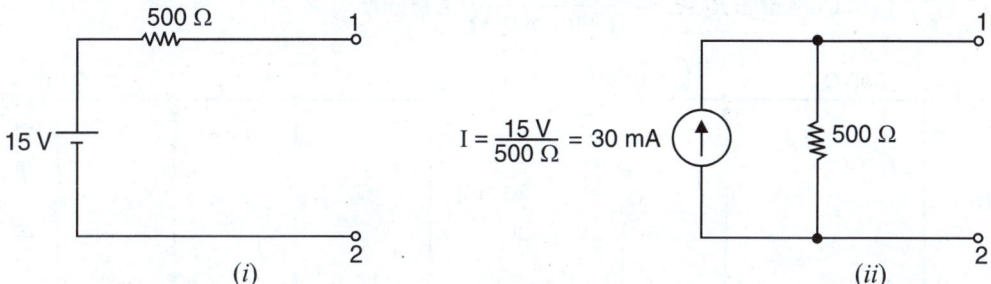

Fig. 2.97

Solution. Fig 2.97 (*i*) shows a voltage source whereas Fig. 2.97 (*ii*) shows its equivalent current source.

(*i*) **When terminals are shorted.** Referring to Fig. 2.98, the terminal voltage is 0 V in both circuits because the terminals are shorted.

$$I_L = \frac{15 \text{ V}}{500 \ \Omega} = 30 \text{ mA} \ ... \text{ voltage source}$$

$$I_L = 30 \text{ mA} \ ... \text{current source}$$

Note that in case of current source, 30 mA flows in the shorted terminals because the short diverts all of the source current around the 500 Ω resistor.

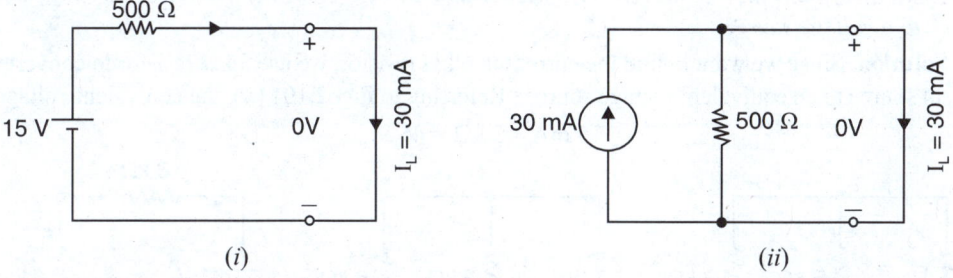

Fig. 2.98

(*ii*) **When the terminals are open.** Referring to Fig. 2.99 (*i*), the voltage across the open terminals of voltage source is 15 V because no current flows and there is no voltage drop across 500 Ω resistor. Referring to Fig. 2.99 (*ii*), the voltage across the open terminals of the current source is also 15 V ; $V = 30$ mA × 500 Ω = 15 V. The current flowing from one terminal into the other is zero in both cases because the terminals are open.

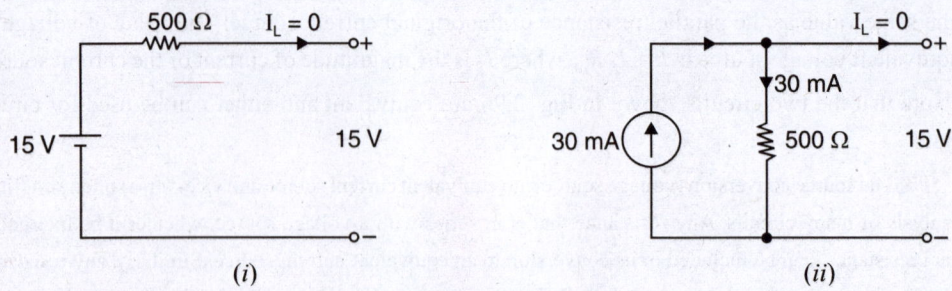

Fig. 2.99

(iii) Terminals have a 500 Ω load connected.

(a) Voltage source. Referring to Fig. 2.100 (i),

$$\text{Current in } R_L, \ I_L = \frac{15 \text{ V}}{(500 + 500) \ \Omega} = 15 \text{ mA}$$

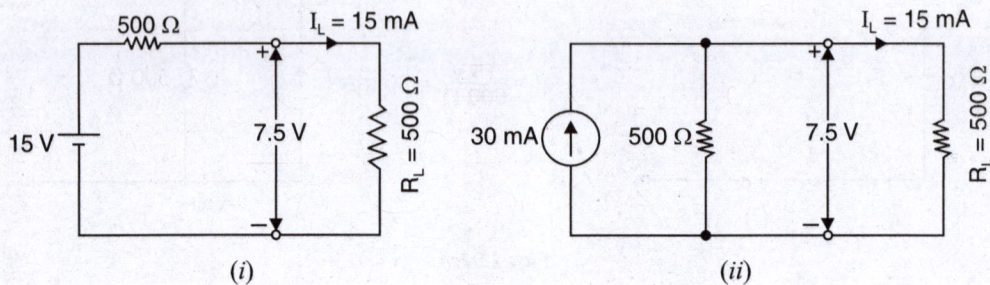

Fig. 2.100

Terminal voltage of source, $V = I_L R_L = 15 \text{ mA} \times 500 \ \Omega = 7.5 \text{ V}$

(b) Current source. Referring to Fig. 2.100 (ii),

$$\text{Current in } R_L, \ I_L = 30 \times \frac{500}{500 + 500} = 15 \text{ mA}$$

Terminal voltage of source $= I_L R_L = 15 \text{ mA} \times 500 \ \Omega = 7.5 \text{ V}$

We conclude that equivalent sources produce exactly the same voltages and currents at their external terminals, no matter what the load and that they are therefore indistinguishable.

Example 2.50. *Find the current in 6 kΩ resistor in Fig. 2.101 (i) by converting the current source to a voltage source.*

Solution. Since we want to find the current in 6 kΩ resistor, we use 3 kΩ resistor to convert the current source to an equivalent voltage source. Referring to Fig. 2.101 (ii), the equivalent voltage is

$$E = 15 \text{ mA} \times 3 \text{ k}\Omega = 45 \text{ V}$$

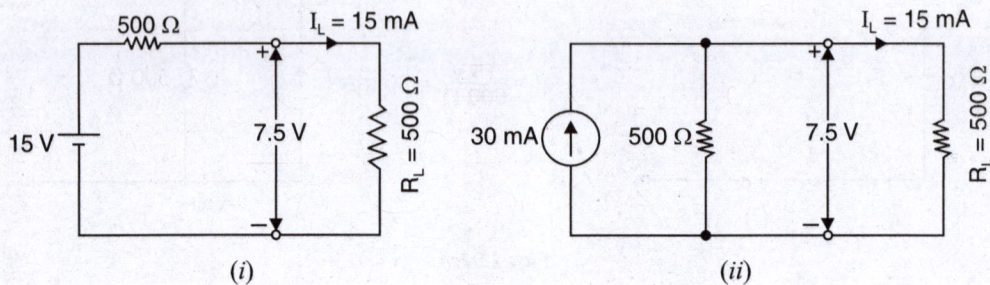

Fig. 2.101

The circuit then becomes as shown in Fig. 2.101 (*iii*). Note that polarity of the equivalent voltage source is such that it produces current in the same direction as the original current source.

Referring to Fig. 2.101 (*iii*), the current in 6 kΩ is

$$I = \frac{45 \text{ V}}{(3+6)\text{ k}\Omega} = \textbf{5 mA}$$

In the series circuit shown in Fig. 2.101 (*iii*), it would appear that current in 3 kΩ resistor is also 5 mA. However, 3 kΩ resistor was involved in source conversion, so we *cannot* conclude that there is 5 mA in the 3 kΩ resistor of the original circuit [See Fig. 2.101 (*i*)]. Verify that the current in the 3 kΩ resistor in that circuit is, in fact, 10 mA.

Example 2.51. *Find the current in the 3 kΩ resistor in Fig. 2.101 (i) above by converting the current source to a voltage source.*

Solution. The circuit shown in Fig. 2.101 (*i*) is redrawn in Fig. 2.102 (*i*). Since we want to find the current in 3 kΩ resistor, we use 6 kΩ resistor to convert the current source to an equivalent voltage source. Referring to Fig. 2.102 (*i*), the equivalent voltage is

$$E = 15 \text{ mA} \times 6 \text{ k}\Omega = 90 \text{ V}$$

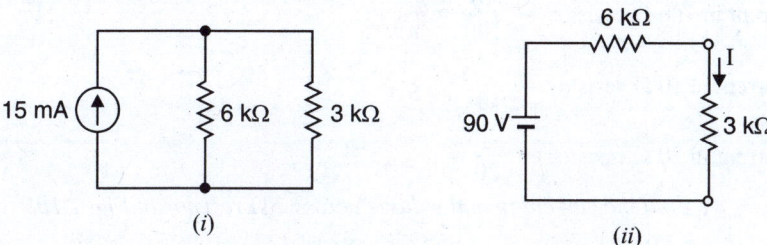

Fig. 2.102

The circuit then reduces to that shown in Fig. 2.102 (*ii*). The current in 3 kΩ resistor is

$$I = \frac{90 \text{ V}}{(6+3)\text{ k}\Omega} = \frac{90 \text{ V}}{9 \text{ k}\Omega} = \textbf{10 mA}$$

Example 2.52. *Find the current in various resistors in the circuit shown in Fig. 2.103 (i) by converting voltage sources into current sources.*

Solution. Referring to Fig. 2.103 (*i*), the 100 Ω resistor can be considered as the internal resistance of 15 V battery. The equivalent current is

$$I = \frac{15 \text{ V}}{100 \text{ }\Omega} = 0.15 \text{ A}$$

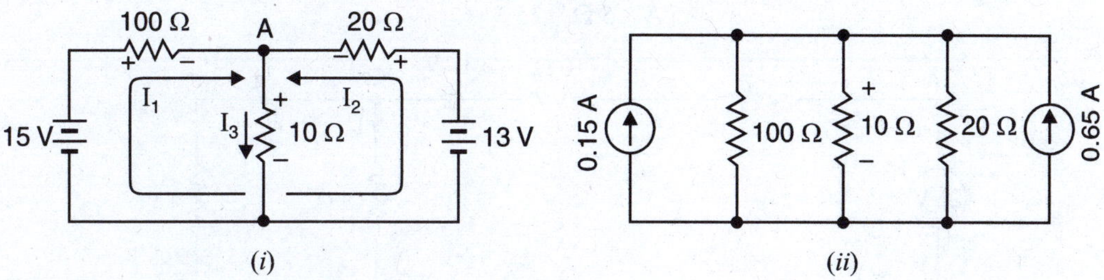

Fig. 2.103

Similarly, 20 Ω resistor can be considered as the internal resistance of 13 V battery. The equivalent current is

$$I = \frac{13\,V}{20\,\Omega} = 0.65\,A$$

Replacing the voltage sources with current sources, the circuit becomes as shown in Fig. 2.103 (*ii*). The current sources are parallel-aiding for a total flow = 0.15 + 0.65 = 0.8 A. The parallel resistors can be combined.

$$100\,\Omega \parallel 10\,\Omega \parallel 20\,\Omega = 6.25\,\Omega$$

The total current flowing through this resistance produces the drop :

$$0.8\,A \times 6.25\,\Omega = 5\,V$$

This 5 V drop can now be "transported" back to the original circuit. It appears across 10 Ω resistor [See Fig. 2.104]. Its polarity is negative at the bottom and positive at the top. Applying Kirchhoff's voltage law (KVL), the voltage drop across 100 Ω resistor = 15 – 5 = 10 V and drop across 20 Ω resistor = 13 – 5 = 8 V.

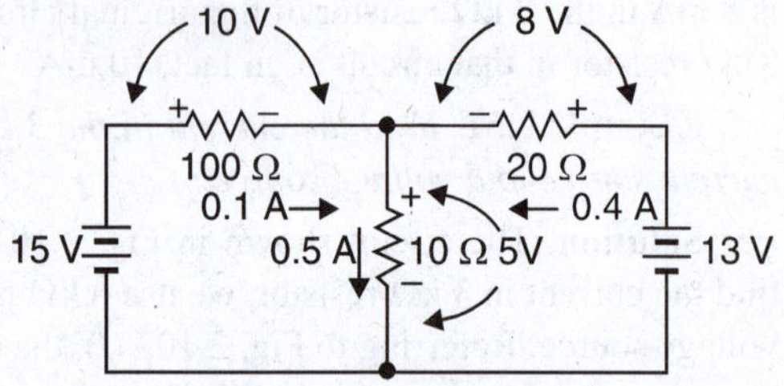

∴ Current in 100 Ω resistor = $\frac{10}{100}$ = **0.1 A**

Current in 10 Ω resistor = $\frac{5}{10}$ = **0.5 A**

Current in 20 Ω resistor = $\frac{8}{20}$ = **0.4 A**

Fig. 2.104

Example 2.53. *Find the current in and voltage across 2 Ω resistor in Fig. 2.105.*

Fig. 2.105

Solution. We use 5 Ω resistor to convert the current source to an equivalent voltage source. The equivalent voltage is

$$E = 5\,A \times 5\,\Omega = 25\,V$$

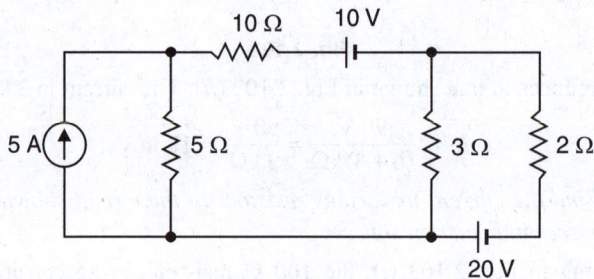

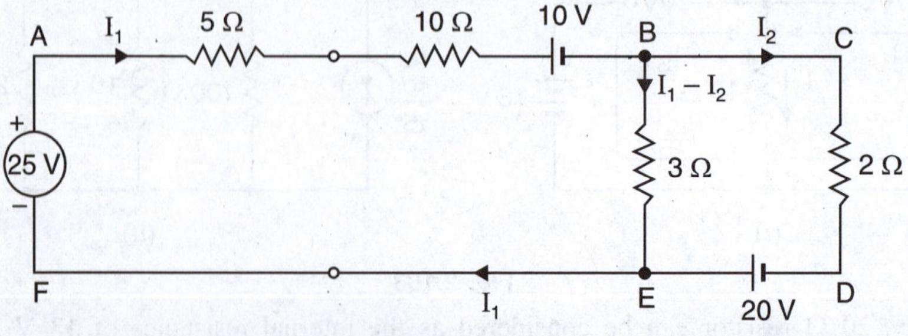

Fig. 2.106

The circuit shown in Fig. 2.105 then becomes as shown in Fig. 2.106.

Loop *ABEFA*. Applying Kirchhoff's voltage law to loop *ABEFA*, we have,

$$- 5\,I_1 - 10\,I_1 - 10 - 3\,(I_1 - I_2) + 25 = 0$$

or
$$- 18\,I_1 + 3\,I_2 = - 15 \qquad\qquad ...(i)$$

Loop *BCDEB*. Applying Kirchhoff's voltage law to loop *BCDEB*, we have,

$$- 2\,I_2 + 20 + 3\,(I_1 - I_2) = 0$$

or
$$3\,I_1 - 5\,I_2 = - 20 \qquad\qquad ...(ii)$$

Solving equations (*i*) and (*ii*), we get, $I_2 = 5$ A.

∴ Current through 2 Ω resistor $= I_2 = $ **5 A**

Voltage across 2 Ω resistor $= I_2 \times 2 = 5 \times 2 = $ **10 V**

Example 2.54. *Find the current in 28 Ω resistor in the circuit shown in Fig. 2.107.*

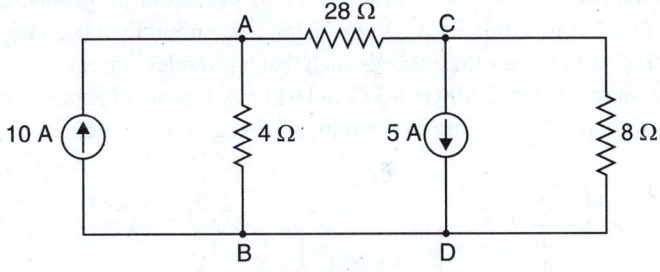

Fig. 2.107

Solution. The two current sources cannot be combined together because 28 Ω resistor is present between points *A* and *C*. However, this difficulty is overcome by converting current sources into equivalent voltage sources. Now 10 A current source in parallel with 4 Ω resistor can be converted into equivalent voltage source of voltage $= 10$ A $\times 4$ Ω $= 40$ V in series with 4 Ω resistor as shown in Fig. 2.108 (*i*). Note that polarity of the equivalent voltage source is such that it provides current in the same direction as the original current source.

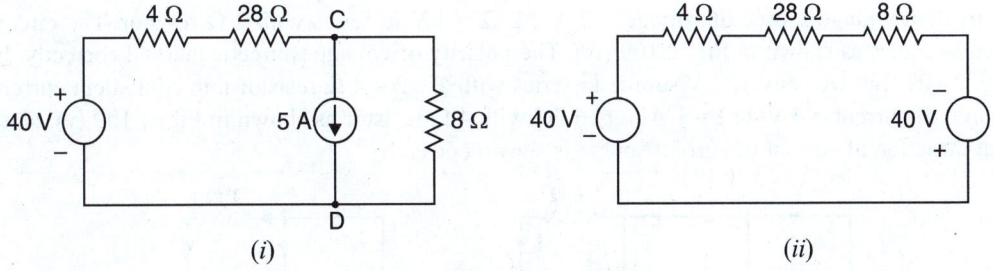

(*i*)   (*ii*)

Fig. 2.108

Similarly, 5 A current source in parallel with 8 Ω resistor can be converted into equivalent voltage source of voltage $= 5$ A $\times 8$ Ω $= 40$ V in series with 8 Ω resistor. The circuit then becomes as shown in Fig. 2.108 (*ii*). Note that polarity of the voltage source is such that it provides current in the same direction as the original current source. Referring to Fig. 2.108 (*ii*),

Total circuit resistance $= 4 + 28 + 8 = 40$ Ω

Total voltage $= 40 + 40 = 80$ V

∴ Current in 28 Ω resistor $= \dfrac{80}{40} = $ **2 A**

Example 2.55. *Using source conversion technique, find the load current I_L in the circuit shown in Fig. 2.109 (i).*

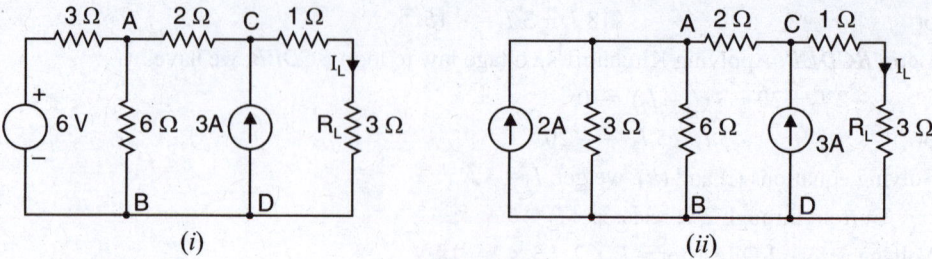

Fig. 2.109

Solution. We first convert 6 V source in series with 3 Ω resistor into equivalent current source of current = 6 V/3 Ω = 2 A in parallel with 3 Ω resistor. The circuit then becomes as shown in Fig. 2.109 (ii). Note that polarity of current source is such that it provides current in the same direction as the original voltage source. In Fig. 2.109 (ii), 3 Ω and 6 Ω resistors are in parallel and their equivalent resistance = (3 × 6)/3 + 6 = 2 Ω. Therefore, circuit of Fig. 2.109 (ii) reduces to the one shown in Fig. 2.109 (iii).

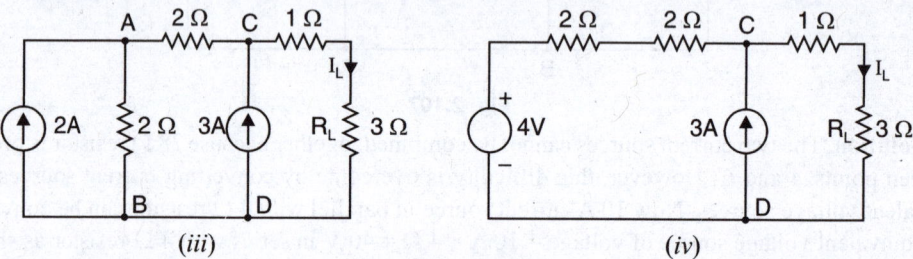

Fig. 2.109

In Fig. 2.109 (iii), we now convert 2 A current source in parallel with 2 Ω resistor into equivalent voltage source of voltage = 2 A × 2 Ω = 4 V in series with 2 Ω resistor. The circuit then becomes as shown in Fig. 2.109 (iv). The polarity of voltage source is marked correctly. In Fig. 2.109 (iv), we convert 4 V source in series with 2 + 2 = 4 Ω resistor into equivalent current source of current = 4 V/4 Ω = 1 A in parallel with 4 Ω resistor as shown in Fig. 2.109 (v). Note that direction of current of current source is shown correctly.

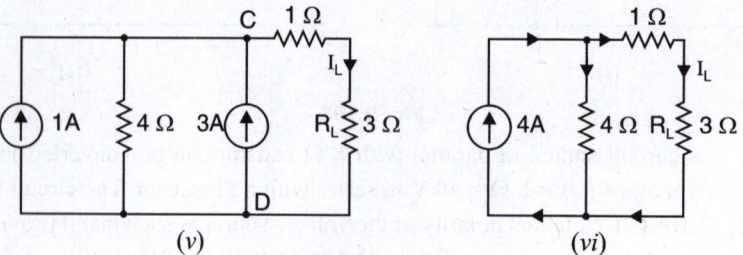

Fig. 2.109

In Fig. 2.109 (v), the two current sources can be combined together to give resultant current source of 3 + 1 = 4 A. The circuit then becomes as shown in Fig. 2.109 (vi). Referring to Fig. 2.109 (vi) and applying current-divider rule,

$$\text{Load current, } I_L = 4 \times \frac{4}{(3+1)+4} = \mathbf{2\,A}$$

Tutorial Problems

1. By performing an appropriate source conversion, find the voltage across 120 Ω resistor in the circuit shown in Fig. 2.110.   **[20 V]**

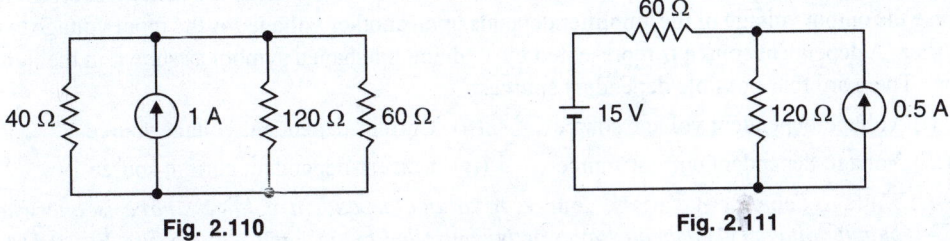

Fig. 2.110   **Fig. 2.111**

2. By performing an appropriate source conversion, find the voltage across 120 Ω resistor in the circuit shown in Fig. 2.111.   **[30 V]**

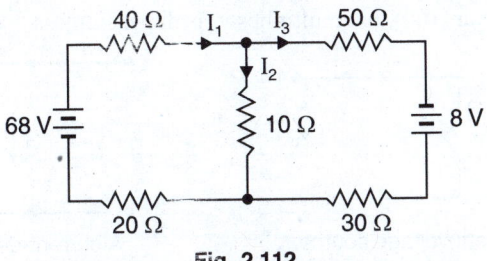

Fig. 2.112

3. By performing an appropriate source conversion, find the currents I_1, I_2 and I_3 in the circuit shown in Fig. 2.112.   **[$I_1 = 1$ A; $I_2 = 0.2$ A; $I_3 = 0.8$ A]**

2.28. Independent Voltage and Current Sources

So far we have been dealing with independent voltage and current sources. We now give brief description about these two active elements.

(i) Independent voltage source. *An independent voltage source is a two-terminal element (e.g. a battery, a generator etc.) that maintains a specified voltage between its terminals.*

An independent voltage source provides a voltage independent of any other voltage or current. The symbol for independent voltage source having v volts across its terminals is shown in Fig. 2.113. (*i*). As shown, the terminal a is v volts above terminal b. If v is greater than zero, then terminal a is at a higher potential than terminal b. In Fig. 2.113 (*i*), the voltage v may be time varying or it may be constant in which case we label it V.

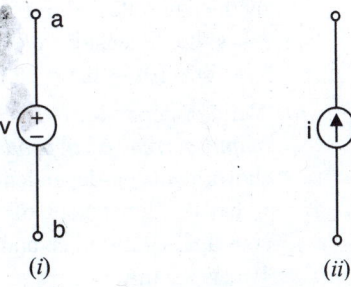

Fig. 2.113

(ii) Independent current source. *An independent current source is a two-terminal element through which a specified current flows.*

An independent current source provides a current that is completely independent of the voltage across the source. The symbol for an independent current source is shown in Fig. 2.113 (*ii*) where i is the specified current. The direction of the current is indicated by the arrow. In Fig. 2.113 (*ii*), the current i may be time varying or it may be constant in which case we label it I.

2.29. Dependent Voltage and Current Sources

A dependent source provides a voltage or current between its output terminals which depends upon another variable such as voltage or current.

For example, a voltage amplifier can be considered to be a dependent voltage source. It is because the output voltage of the amplifier depends upon another voltage *i.e.* the input voltage to the amplifier. A dependent source is represented by a *diamond-shaped symbol as shown in the figures below. There are four possible dependent sources :

    (*i*) Voltage-dependent voltage source   (*ii*) Current-dependent voltage source

    (*iii*) Voltage-dependent current source   (*iv*) Current-dependent current source

    (*i*) **Voltage-dependent voltage source.** *A voltage-dependent voltage source is one whose output voltage* (v_0) *depends upon or is controlled by an input voltage* (v_1). Fig. 2.114 (*i*) shows a voltage-dependent voltage source. Thus if in Fig. 2.114 (*i*), $v_1 = 20$ mV, then $v_0 = 60 \times 20$ mV $= 1.2$ V. If v_1 changes to 30 mV, then v_0 changes to 60×30 mV $= 1.8$ V. Note that the constant (60) that multiplies v_1 is dimensionless.

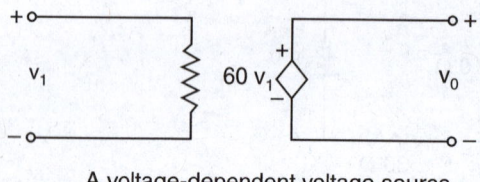 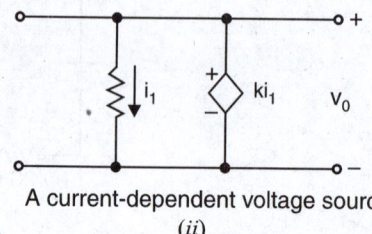

    A voltage-dependent voltage source   A current-dependent voltage source
             (*i*)   (*ii*)

Fig. 2.114

    (*ii*) **Current-dependent voltage source.** *A current-dependent voltage source is one whose output voltage* (v_0) *depends on or is controlled by an input current* (i_1). Fig. 2.114 (*ii*) shows a current-dependent voltage source. Note that the controlling current i_1 is in the same circuit as the controlled source itself. The constant that multiplies the value of voltage produced by the controlled source is sometimes designated by a letter k or β. Note that the constant k has the dimensions of V/A or ohm. Thus if $i_1 = 50$ μA and constant k is 0.5 V/A, then $v_0 = 50 \times 10^{-6} \times 0.5 = 25$ μV.

    (*iii*) **Voltage-dependent current source.** *A voltage-dependent current source is one whose output current* (*i*) *depends upon or is controlled by an input voltage* (v_1). Fig. 2.115 (*i*) shows a voltage-dependent current source. The constant that multiplies the value of voltage v_1 has the dimensions of *A/V i.e.* mho or siemen. For example, in Fig. 2.115. (*i*), if the constant is 0.2 siemen and if input voltage v_1 is 10 mV, then the output current $i = 0.2$ S $\times 10$ mV $= 2$ mA.

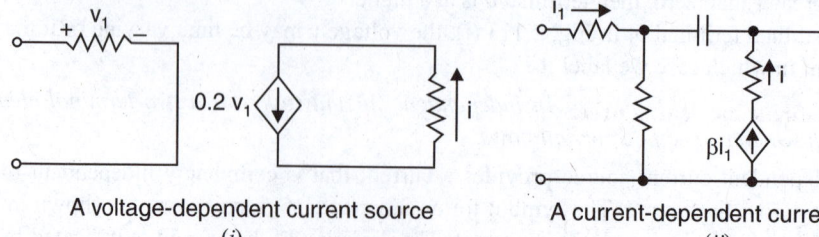

    A voltage-dependent current source   A current-dependent current source
             (*i*)   (*ii*)

Fig. 2.115

---

*   So as not to confuse with the symbol of independent source.

(*iv*) **Current-dependent current source.** *A current-dependent current source is one whose output current* (*i*) *depends upon or is controlled by an input current* (*i*₁). Fig. 2.115 (*ii*) shows a current-dependent current source. Note that controlling current i_1 is in the same circuit as the controlled source itself. The constant (β) that multiplies the value of current produced by the controlled source is dimensionless. Thus in Fig. 2.115 (*ii*), if $i_1 = 50 \ \mu A$ and if constant β equals 100, then the current produced by the controlled current source is $i = 100 \times 50 \ \mu A = 5 \ mA$. If i_1 changes to 20 μA, then i changes to $i = 100 \times 20 \ \mu A = 2 \ mA$.

2.30. Circuits With Dependent-Sources

Fig. 2.116 shows the circuit that has an independent source, a dependent-source and two resistors. The dependent-source is a voltage source controlled by the current i_1. The constant for the dependent-source is 0.5 V/A. Dependent sources are essential components in amplifier circuits. Circuits containing dependent-sources are analysed in the same manner as those without dependent-sources. That is, Ohm's law for resistors and Kirchhoff's voltage and current laws apply, as well as the concepts of equivalent resistance and voltage and current division. We shall solve a few examples by way of illustration.

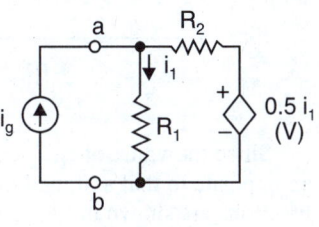

Fig. 2.116

Example 2.56. *Find the value of v in the circuit shown in Fig. 2.117. What is the value of dependent-current source ?*

Solution. By applying *KCL* to node* *A* in Fig. 2.117, we get,

$$4 - i_1 + 2i_1 = \frac{v}{2} \qquad ...(i)$$

Fig. 2.117

By Ohm's law, $\qquad i_1 = \dfrac{v}{6}$

Putting $i_1 = v/6$ in eq. (*i*), we get,

$$4 - \frac{v}{6} + \frac{2v}{6} = \frac{v}{2} \quad \therefore \quad v = \textbf{12 V}$$

Value of dependent-current source $= 2 i_1 = \dfrac{2v}{6} = \dfrac{2 \times 12}{6} = \textbf{4 A}$

Example 2.57. *Find the values of v, i_1 and i_2 in the circuit shown in Fig. 2.118 (i) which contains a voltage-dependent current source. Resistance values are in ohms.*

Solution. Applying *KCL* to node *A* in Fig. 2.118 (*i*), we get,

$$2 - i_1 + 4v = i_2 \qquad\qquad ...(i)$$

Now $\qquad$ By Ohm's law, $i_1 = \dfrac{v}{3}$ and $i_2 = \dfrac{v}{6}$

Putting $\quad i_1 = \dfrac{v}{3}$ and $i_2 = \dfrac{v}{6}$ in eq. (*i*), we get,

$$2 - \frac{v}{3} + 4v = \frac{v}{6} \quad \therefore \quad v = \frac{-4}{7} \ \textbf{V}$$

$$\therefore \qquad\qquad i_1 = \frac{v}{3} = \frac{1}{3} \times v = \frac{1}{3} \times \frac{-4}{7} = \frac{-4}{21} \ \textbf{A}$$

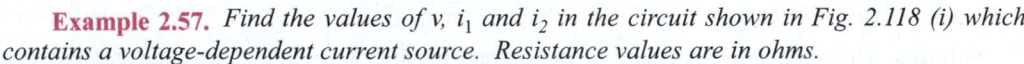

*   A node of a network is an equipotential surface at which *two or more* circuit elements are joined.

$$\therefore \qquad i_2 = \frac{v}{6} = \frac{1}{6} \times v = \frac{1}{6} \times \frac{-4}{7} = \frac{-2}{21} \text{ A}$$

Value of dependent current source $= 4v = 4 \times \dfrac{-4}{7} = \dfrac{-16}{7}$ A

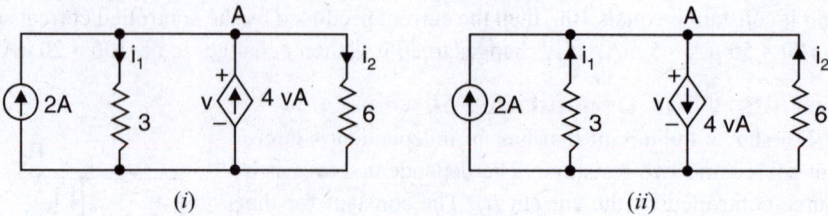

(i) (ii)

Fig. 2.118

Since the value of i_1, i_2 comes out to be negative, it means that directions of flow of currents are opposite to that assigned in Fig. 2.118. (i). The same is the case for current source. The actual directions are shown in Fig. 2.118 (ii).

Example 2.58. *Find the value of i in the circuit shown in Fig. 2.119 if R = 10 Ω.*

Solution. Applying *KVL* to the loop *ABEFA*, we have,

$$5 - 10\,i_1 + 5\,i_1 = 0 \qquad \therefore \quad i_1 = 1 \text{ A}$$

Applying *KVL* to the loop *BCDEB*, we have,

$$10\,i - 25 - 5\,i_1 = 0$$

or $\qquad\qquad 10\,i - 25 - 5 = 0 \qquad \therefore \quad i = \mathbf{3\,A}$

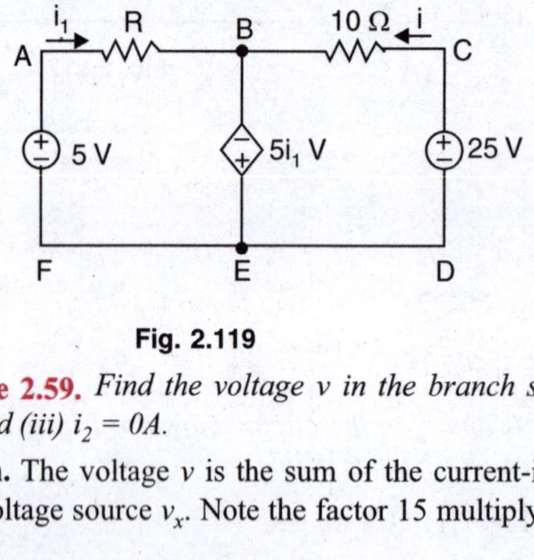

Fig. 2.119 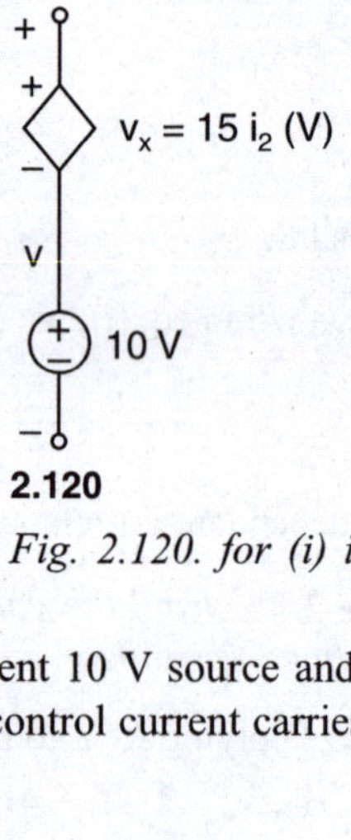 **Fig. 2.120**

Example 2.59. *Find the voltage v in the branch shown in Fig. 2.120. for (i) i_2 = 1 A, (ii) i_2 = − 2 A and (iii) i_2 = 0A.*

Solution. The voltage v is the sum of the current-independent 10 V source and the current-dependent voltage source v_x. Note the factor 15 multiplying the control current carries the units of ohm.

(i) $\quad v = 10 + v_x = 10 + 15\,(1) = \mathbf{25\,V}$

(ii) $\quad v = 10 + v_x = 10 + 15\,(-2) = \mathbf{-20\,V}$

(iii) $\quad v = 10 + v_x = 10 + 15\,(0) = \mathbf{10\,V}$

Example 2.60. *Find the values of current i and voltage drops v_1 and v_2 in the circuit of Fig. 2.121 which contains a current-dependent voltage source. What is the voltage of the dependent-source? All resistance values are in ohms.*

Solution. Note that the factor 4 multiplying the control current carries the units of ohms. Applying *KVL* to the loop *ABCDA* in Fig. 2.121, we have,

$$-v_1 + 4\,i - v_2 + 6 = 0$$

or $\qquad\qquad v_1 - 4\,i + v_2 = 6 \qquad\qquad\qquad ...(i)$

By Ohm's law, $v_1 = 2\,i$ and $v_2 = 4\,i$.

Putting the values of $v_1 = 2\,i$ and $v_2 = 4\,i$ in eq. (*i*), we have,

$$2\,i - 4\,i + 4\,i = 6 \qquad \therefore \quad i = 3\,\text{A}$$

$$\therefore \qquad v_1 = 2i = 2 \times 3 = 6\,\text{V} \; ; \; v_2 = 4\,i = 4 \times 3 = 12\,\text{V}$$

Voltage of the dependent source $= 4\,i = 4 \times 3 = 12\,\text{V}$

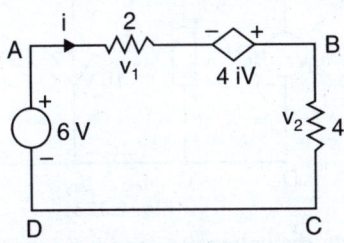

Fig. 2.121

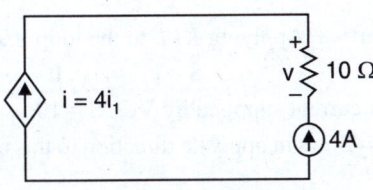

Fig. 2.122

Example 2.61. *Find the voltage v across the 10 Ω resistor in Fig. 2.122, if the control current i_1 in the dependent current-source is (i) 2A (ii) – 1A.*

Solution.

(*i*) $\quad v = (i - 4)10 = [4\,(2) - 4]10 = \textbf{40 V}$

(*ii*) $\quad v = (i - 4)10 = [4(-1) - 4]\,10 = \textbf{– 80 V}$

Example 2.62. *Calculate the power delivered by the dependent-source in Fig. 2.123.*

Solution. Applying *KVL* to the loop *ABCDA*, we have,

$$-2\,I - 4\,I - 3\,I + 10 = 0$$

$$\therefore \qquad I = 10/9 = 1.11\,\text{A}$$

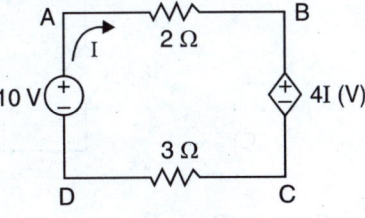

Fig. 2.123

The current *I* enters the positive terminal of dependent-source. Therefore, power absorbed $= 1.11 \times 4\,(1.11) = 4.93$ watts. Hence power delivered is **– 4.93 W.**

Example 2.63. *In the circuit of Fig. 2.124, find the values of i and v. All resistances are in ohms.*

Solution. Referring to Fig. 2.124, it is clear that $v_a = 12 + v$.

Therefore, $\qquad v = v_a - 12$

Voltage drop across left 2 Ω resistor $= 0 - v_a$

Voltage drop across top 2 Ω resistor $= v_a - 12$

Applying *KCL* to the node *a*, we have,

$$\frac{0 - v_a}{2} + \frac{v}{4} - \frac{v_a - 12}{2} = 0 \quad \text{or} \quad v_a = 4\,\text{V}$$

$$\therefore \qquad v = v_a - 12 = 4 - 12 = \textbf{– 8V}$$

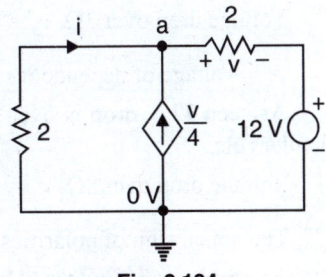

Fig. 2.124

The negative sign shows that the polarity of *v* is opposite to that shown in Fig. 2.124. The current that flows from point *a* to ground $= 4/2 = 2$ A. Hence $i = \textbf{– 2 A.}$

Example 2.64. *In Fig. 2.125, both independent and dependent-current sources drive current through resistor R. Is the value of R uniquely determined ?*

Solution. By definition of an independent source, the current *I* must be 10 A.

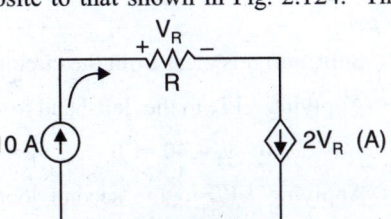

Fig. 2.125

$$\therefore \qquad I = 10\,A = 2\,V_R$$

or $\qquad V_R = 10/2 = 5\,V$

Now $\qquad 5\,V = (10)\,(R) \quad \therefore R = 5/10 = \mathbf{0.5\,\Omega}$

No other value of R is possible.

Example 2.65. *Find the value of current i_2 supplied by the voltage-controlled current source (VCCS) shown in Fig. 2.126.*

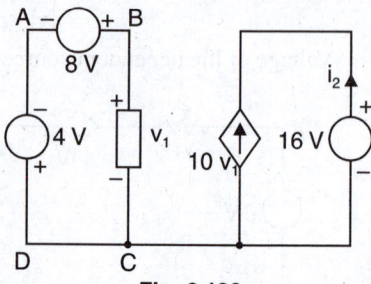

Solution. Applying *KVL* to the loop *ABCDA*, we have,

$$8 - v_1 - 4 = 0 \quad \therefore v_1 = 4V$$

The current supplied by VCCS $= 10\,v_1 = 10 \times 4 = 40A$

As i_2 flows in opposite direction to this current, therefore, $i_2 = -\,\mathbf{40A.}$

Fig. 2.126

Example 2.66. *By using voltage divider rule, calculate the voltages v_x and v_y in the circuit shown in Fig. 2.127.*

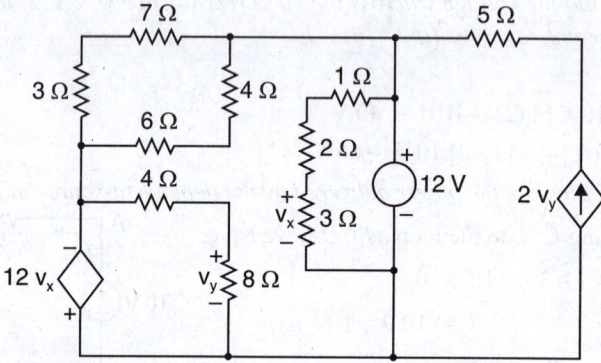

Fig. 2.127

Solution. As can be seen from Fig. 2.127, 12 V drop is over the series combination of 1Ω, 2Ω and 3Ω resistors. Therefore, by voltage divider rule,

Voltage drop over 3Ω, $v_x = 12 \times \dfrac{3}{1+2+3} = \mathbf{6V}$

$\therefore$ Voltage of dependent source $= 12v_x = 12 \times 6 = 72\,V$

As seen 72 V drop is over series combination of 4Ω and 8Ω resistors. Therefore, by voltage divider rule,

Voltage drop over 8Ω, $v_y = 72 \times \dfrac{8}{4+8} = 48\,V$

The actual sign of polarities of v_y is opposite to that shown in Fig. 2.127. Hence $v_y = -\,\mathbf{48\,V.}$

Example 2.67. *Find the values of i_1, v_1, v_x and v_{ab} in the network shown in Fig. 2.128 with its terminals a and b open.*

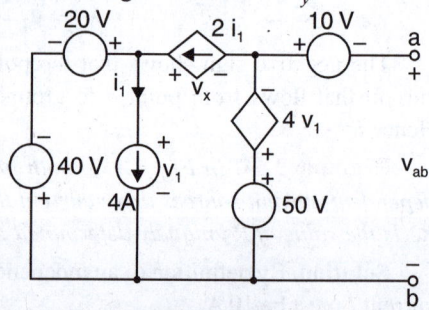

Solution. It is clear from the circuit that $i_1 = 4A.$

Applying *KVL* to the left-hand loop, we have,

$$20 - v_1 - 40 = 0 \quad \therefore v_1 = -\,\mathbf{20\,V}$$

Applying *KVL* to the second loop from left, we have,

Fig. 2.128

$$-v_x + 4v_1 - 50 + v_1 = 0$$

$$\therefore \qquad v_x = 5v_1 - 50 = 5(-20) - 50 = -\textbf{150 V}$$

Applying *KVL* to the third loop containing v_{ab}, we have,

$$-10 - v_{ab} + 50 - 4v_1 = 0$$

$$\therefore \qquad v_{ab} = -10 + 50 - 4v_1 = -10 + 50 - 4(-20) = \textbf{120 V}$$

(Tutorial Problems)

1. The circuit of Fig. 2.129 contains a voltage-dependent voltage source. Find the current supplied by the battery and power supplied by the voltage source. **[8A; 1920 W]**

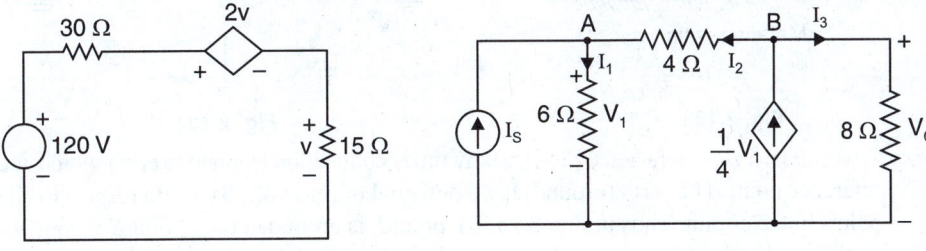

Fig. 2.129   **Fig. 2.130**

2. Applying Kirchhoff's current law, determine current I_S in the electric circuit of Fig. 2.130. Take $V_0 = 16V$. **[1A]**

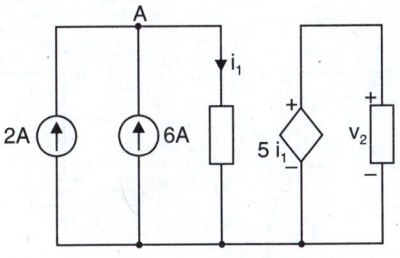

Fig. 2.131

3. Find the voltage drop v_2 across the current-controlled voltage source shown in Fig. 2.131. **[40 V]**

2.31. Ground

Voltage is relative. That is, the voltage at one point in a circuit is always measured relative to another point in the circuit. For example, if we say that voltage at a point in a circuit is + 100V, we mean that the point is 100V more positive than some reference point in the circuit. This reference point in a circuit is usually called the *ground point.* Thus ground is used as reference point for specifying voltages. The ground may be used as common connection (*common ground*) or as a zero reference point (*earth ground*). There are different symbols for chassis ground, common ground and earth ground as shown in Fig. 2.132. *However, earth ground symbol is often used in place of chassis ground or common ground.*

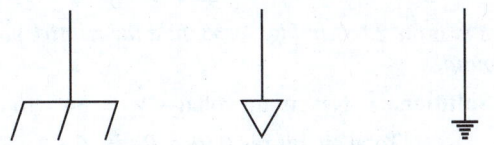

Chassis ground   Common ground   Earth ground

Fig. 2.132

(*i*) **Ground as a common connection.** It is a usual practice to mount the electronic and electrical components on a metal base called *chassis* (See Fig. 2.133). Since chassis is good conductor, it provides a conducting return path as shown in Fig. 2.134. It may be seen that all points connected to chassis are shown as grounded and represent the same potential. The adoption of this scheme (*i.e.* showing points of same potential as grounded) often simplifies the electrical and electronic circuits.

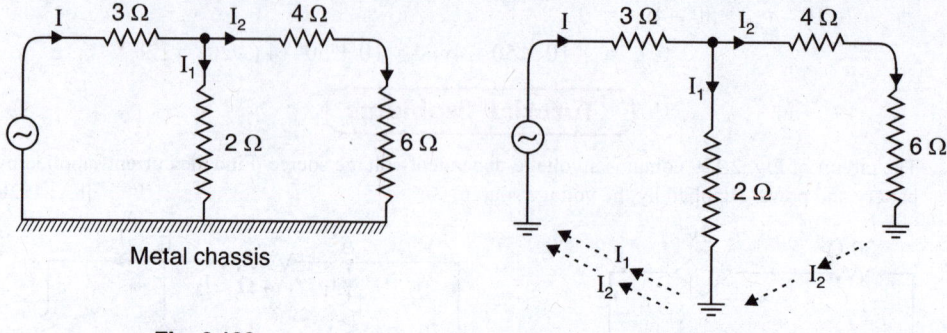

Fig. 2.133   Fig. 2.134

(*ii*) **Ground as a zero reference point.** Many times connection is made to earth which acts as a reference point. The earth (ground) has a potential of zero volt (0V) with respect to all other points in the circuit. Thus in Fig. 2.135(*i*), point E is grounded (*i.e.*, point E is connected to earth) and has zero potential. The voltage across each resistor is 25 volts. The voltages of the various points with respect to ground or earth (*i.e.*, point E) are :

$$V_E = 0V \; ; \; V_D = +25 \text{ V} \; ; \; V_C = +50 \text{ V} \; ; \; V_B = +75 \text{ V} \; ; \; V_A = +100V$$

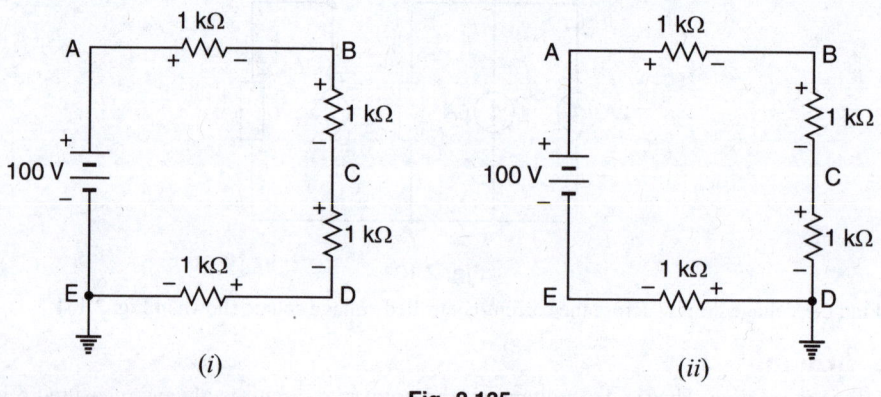

(*i*)   (*ii*)

Fig. 2.135

If instead of point E, the point D is grounded as shown in Fig. 2.135 (*ii*), then potentials of various points with respect to ground (*i.e.*, point D) will be :

$$V_E = -25 \text{ V} \; ; \; V_D = 0 \text{ V} \; ; \; V_C = +25 \text{ V} \; ; \; V_B = +50 \text{ V} \; ; \; V_A = +75 \text{ V}$$

Example 2.68. *In Fig. 2.136, find the relative potentials of points A, B, C, D and E when point A is grounded.*

Solution.   Net circuit voltage, $V = 34 - 10 = 24$ V

Total circuit resistance, $R_T = 6 + 4 + 2 = 12 \, \Omega$

Circuit current, $I = V/R_T = 24/12 = 2$ A

Drop across 2 Ω resistor $= 2 \times 2 = 4$ V

Drop across 4 Ω resistor $= 2 \times 4 = 8$ V

Drop across 6 Ω resistor $= 2 \times 6 = 12$ V

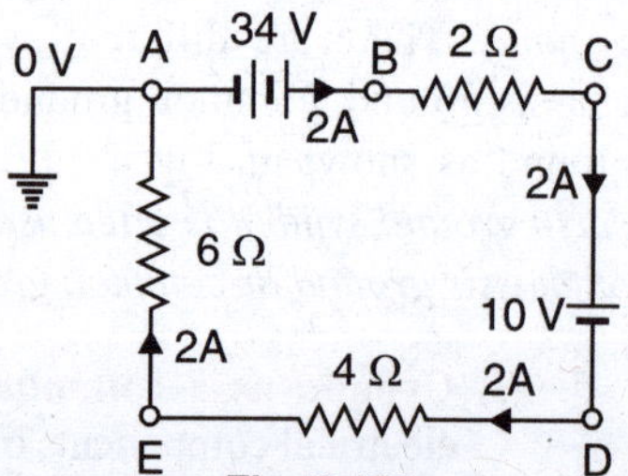

Fig. 2.136

∴   Potential at point B, V_B = 34 – 0 = **34 V**

Potential at point C, V_C = 34 – drop in 2 Ω

= 34 – 2 × 2 = **30 V**

Potential at point D, V_D = V_C – 10 = 30 – 10 = **20 V**

Potential at point E, V_E = V_D – drop in 4Ω = 20 – 2 × 4 = **12 V**

Potential at point A, V_A = V_E – drop in 6 Ω

= 12 – 6 × 2 = **0 V**

Example 2.69. *Fig. 2.137 shows the circuit with common ground symbols. Find the total current I drawn from the 25 V source.*

Solution. The circuit shown in Fig. 2.137 is redrawn by eliminating the common ground symbols. The equivalent circuit then becomes as shown in Fig. 2.138. (*i*). We see that 8 kΩ and 12 kΩ resistors are in parallel as are the 9 kΩ and 4.5 kΩ resistors. Fig. 2.138 (*ii*) shows the circuit when these parallel combinations are replaced by their equivalent resistances :

$$\frac{8 \times 12}{8 + 12} = 4.8 \text{ k}\Omega \quad \text{and} \quad \frac{9 \times 4.5}{9 + 4.5} = 3 \text{ k}\Omega$$

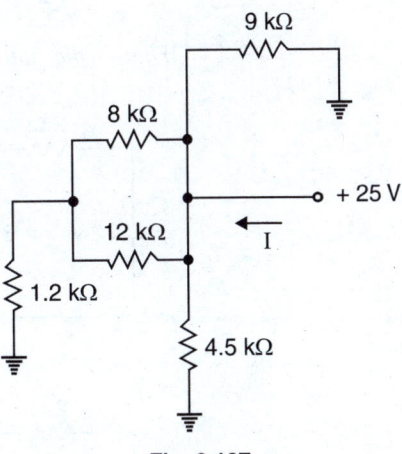

Fig. 2.137

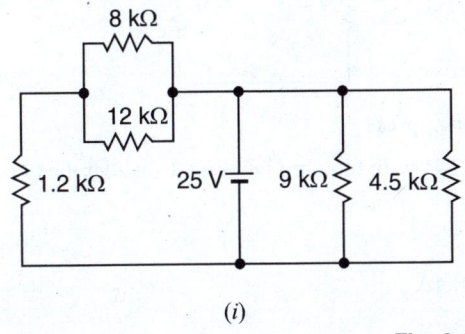

(*i*)

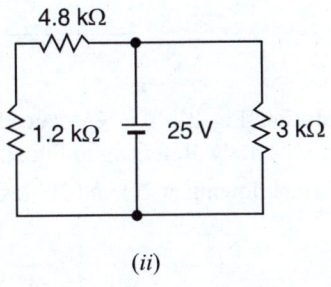

(*ii*)

Fig. 2.138

Referring to Fig. 2.138 (*ii*), it is clear that 4.8 kΩ resistance is in series with 1.2 kΩ resistance, giving an equivalent resistance of 4.8 + 1.2 = 6 kΩ.

The circuit then becomes as shown in Fig. 2.139 (*i*).

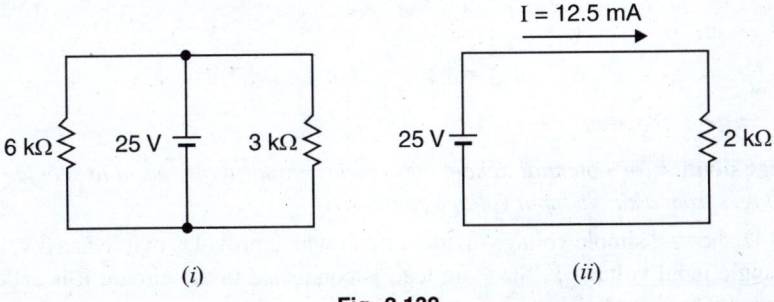

(*i*)   (*ii*)

Fig. 2.139

Referring to Fig. 2.139 (*i*), 6 kΩ is in parallel with 3 kΩ giving the total resistance R_T as :

$$R_T = \frac{6 \times 3}{6 + 3} = 2 \text{ k}\Omega$$

The circuit then reduces to the one shown in Fig. 2.139 (*ii*).

∴   Total current *I* drawn from 25 V source is

$$I = \frac{25 \text{ V}}{R_T} = \frac{25 \text{ V}}{2 \text{ k}\Omega} = 12.5 \text{ mA}$$

Example 2.70. *What is the potential difference between X and Y in the network shown in Fig. 2.140 ?*

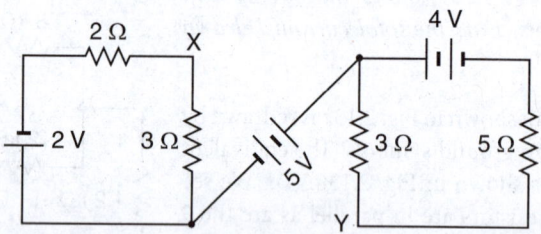

Fig. 2.140

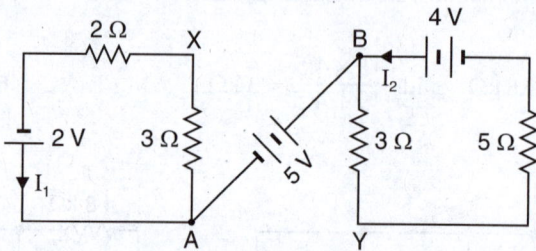

Fig. 2.141

Solution. Fig. 2.140 is reproduced as Fig 2.141 with required labeling. Consider the two battery circuits separately. Referring to Fig. 2.141,

Current flowing in 2Ω and 3Ω resistors is

$$I_1 = \frac{2}{2 + 3} = 0.4 \text{A}$$

Current flowing in 3Ω and 5Ω resistors is

$$I_2 = \frac{4}{3 + 5} = 0.5 \text{ A}$$

∴   Potential difference between *X* and *Y* is

$$V_{XY} = V_{XA} + V_{AB} - V_{BY} \qquad\qquad\qquad \text{[See Fig. 2.141]}$$
$$= 3I_1 + 5 - 3I_2$$
$$= 3 \times 0.4 + 5 - 3 \times 0.5 = \textbf{4.7 V}$$

2.32. Voltage Divider Circuit

A **voltage divider** *(or potential divider) is a series circuit that is used to provide two or more reduced voltages from a single input voltage source.*

Fig. 2.142 shows a simple voltage divider circuit which provides two reduced voltages V_1 and V_2 from a single input voltage *V*. Since no load is connected to the circuit, it is called **unloaded voltage divider**. The values of V_1 and V_2 can be found as under :

Circuit current, $I = \dfrac{V}{R_1 + R_2} = \dfrac{V}{R_T}$

where $\qquad\qquad R_T$ = Total resistance of the voltage divider

$\therefore\qquad\qquad V_1 = IR_1 = V \times \dfrac{R_1}{R_T}$

and $\qquad\qquad V_2 = IR_2 = V \times \dfrac{R_2}{R_T}$

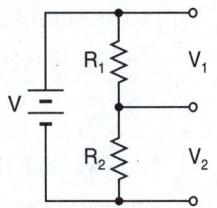

Fig. 2.142

Therefore, voltage drop across any resistor in an unloaded voltage divider is equal to the ratio of that resistance value to the total resistance multiplied by the source voltage.

Loaded voltage divider. When load R_L is connected to the output terminals of the voltage divider as shown in Fig. 2.143, the output voltage (V_2) is reduced by an amount depending on the value of R_L. It is because load resistor R_L is in parallel with R_2 and reduces the resistance from point A to point B. As a result, the output voltage is reduced. The larger the value of R_L, the less the output voltage is reduced from the unloaded value. Loading a voltage divider has the following effects :

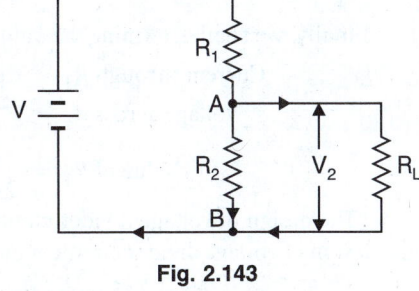

Fig. 2.143

  (i) The output voltage is reduced depending upon the value of load resistance R_L.

  (ii) The current drawn from the source is increased because total resistance of the circuit is reduced. The decrease in total resistance is due to the fact that loaded voltage divider becomes series-parallel circuit.

Example 2.71. *Design a voltage divider circuit that will operate the following loads from a 20 V source :*

5 V at 5 mA ; 12 V at 10 mA ; 15 V at 5 mA

The bleeder current is 4 mA.

Solution. A voltage divider that produces a *bleeder current requires $N+1$ resistors where N is the number of loads. In this example, the number of loads is three. Therefore, four resistors are required for this voltage divider. The required circuit is shown in Fig. 2.144. Here R_1 is the bleeder resistor. The loads are arranged in ascending order of their voltage requirements, starting at the bottom of the divider network.

Voltage across bleeder resistor $R_1 = 5$ V ; Current through R_1, $I_B = 4$ mA .

$\therefore$ Value of $R_1 = \dfrac{5\text{V}}{4\text{mA}} = \mathbf{1.25\ k\Omega}$

Next we shall find the value of resistor R_2. For this purpose, we find the current through R_2 and voltage across R_2.

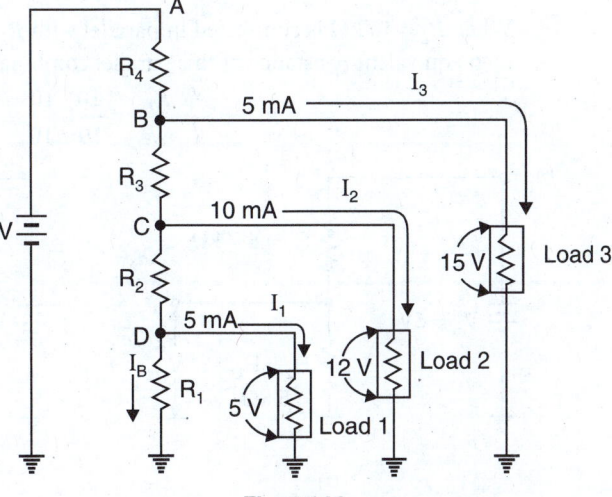

Fig. 2.144

---

* The current drawn continuously from a power supply by the resistive voltage divider circuit is called bleeder current. Without a bleeder current, the voltage divider outputs go up to full value of supply voltage if all the loads are disconnected.

Current through $R_2 = I_B + 5\,\text{mA} = 4\,\text{mA} + 5\,\text{mA} = 9\,\text{mA}$

Voltage across $R_2 = V_C - V_D = 12 - 5 = 7\,\text{V}$

$$\therefore \qquad \text{Value of } R_2 = \frac{7\,\text{V}}{9\,\text{mA}} = 778\,\Omega$$

Now we shall find the value of resistor R_3.

Current through R_3 = Current in $R_2 + 10\,\text{mA} = 9\,\text{mA} + 10\,\text{mA} = 19\,\text{mA}$

Voltage across $R_3 = V_B - V_C = 15 - 12 = 3\,\text{V}$

$$\therefore \qquad \text{Value of } R_3 = \frac{3\,\text{V}}{19\,\text{mA}} = 158\,\Omega$$

Finally, we shall determine the value of resistor R_4.

Current through R_4 = Current through $R_3 + 5\,\text{mA} = 19\,\text{mA} + 5\,\text{mA} = 24\,\text{mA}$

Voltage across $R_4 = V_A - V_B = 20 - 15 = 5\,\text{V}$

$$\therefore \qquad \text{Value of } R_4 = \frac{5\,\text{V}}{24\,\text{mA}} = 208\,\Omega$$

The design of voltage divider circuit means finding the values of R_1, R_2, R_3 and R_4. Therefore, the design of voltage divider circuit stands completed.

Example 2.72. *Fig. 2.145 shows the voltage divider circuit. Find (i) the unloaded output voltage, (ii) the loaded output voltage for $R_L = 10\,\text{k}\Omega$ and $R_L = 100\,\text{k}\Omega$.*

Solution. (*i*) When load R_L is removed, the voltage across R_2 is the unloaded output voltage of the voltage divider.

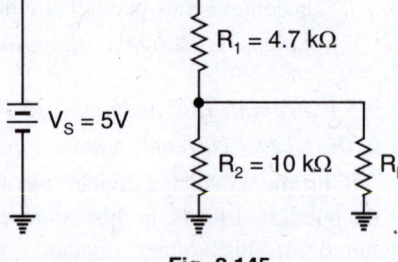

$$\therefore \quad \text{Unloaded output voltage} = \frac{R_2}{R_1 + R_2} \times V_S$$

$$= \frac{10}{4.7 + 10} \times 5$$

$$= \textbf{3.4 V}$$

Fig. 2.145

(*ii*) When $R_L = 10\,\text{k}\Omega$ is connected in parallel with R_2, then equivalent resistance of this parallel combination is

$$R_T = \frac{R_2 R_L}{R_2 + R_L} = \frac{10 \times 10}{10 + 10} = 5\,\text{k}\Omega$$

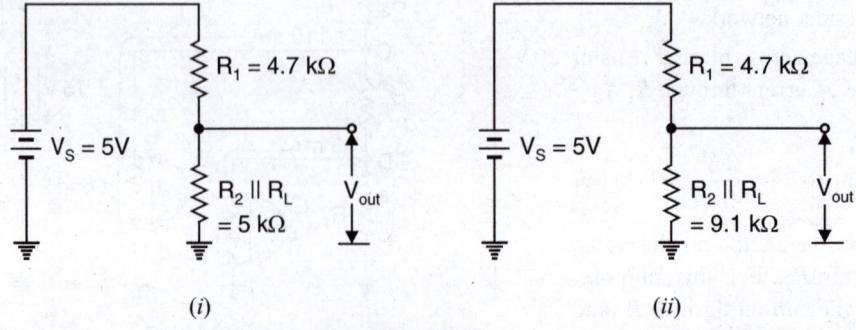

(*i*)   (*ii*)

Fig. 2.146

The circuit then becomes as shown in Fig. 2.146 (*i*).

$$\therefore \quad \text{Loaded output voltage} = \frac{R_T}{R_1 + R_T} \times V_S = \frac{5}{4.7 + 5} \times 5 = \textbf{2.58 V}$$

When $R_L = 100$ kΩ is connected in parallel with R_2, then equivalent resistance of this parallel combination is given by ; $R'_T = \dfrac{R_2 R_L}{R_2 + R_L} = \dfrac{10 \times 100}{10 + 100} = 9.1$ kΩ

The circuit then becomes as shown in Fig. 2.146 (*ii*).

∴   Loaded output voltage $= \dfrac{R'_T}{R_1 + R'_T} \times V_S = \dfrac{9.1}{4.7 + 9.1} \times 5 = $ **3.3 V**

Example 2.73. *Find the values of different voltages that can be obtained from 25V source with the help of voltage divider circuit of Fig. 2.147.*

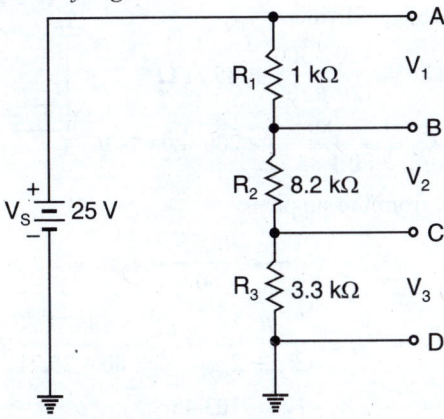

Fig. 2.147

Solution. Total circuit resistance, $R_T = R_1 + R_2 + R_3 = 1 + 8.2 + 3.3 = 12.5$ kΩ

Voltage drop across R_1, $V_1 = \dfrac{R_1}{R_T} \times V_S = \dfrac{1}{12.5} \times 25 = 2$ V

∴   Voltage at point B, $V_B = 25 - 2 = 23$ V

Voltage drop across R_2, $V_2 = \dfrac{R_2}{R_T} \times V_S = \dfrac{8.2}{12.5} \times 25 = 16.4$ V

∴   Voltage at point C, $V_C = V_B - V_2 = 23 - 16.4 = 6.6$ V

The different available load voltages are :

$V_{AB} = V_A - V_B = 25 - 23 = $ **2 V** ; $V_{AC} = V_A - V_C = 25 - 6.6 = $ **18.4 V**

$V_{BC} = V_B - V_C = 23 - 6.6 = $ **16.4 V** ; $V_{AD} = $ **25 V**; $V_{CD} = V_C - V_D = 6.6 - 0 = $ **6.6 V**

$V_{BD} = V_B - V_D = 23 - 0 = $ **23 V**

Example 2.74. *Fig. 2.148 shows a 10 kΩ potentiometer connected in a series circuit as an adjustable voltage divider. What total range of voltage V_1 can be obtained by adjusting the potentiometer through its entire range ?*

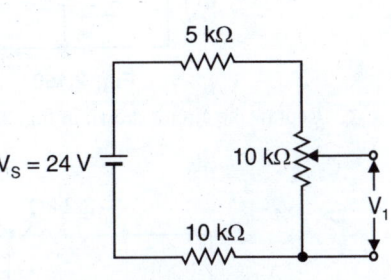

Fig. 2.148

Solution. Total circuit resistance is

$R_T = 5 + 10 + 10 = 25$ kΩ

The total voltage E that appears across the end terminals of potentiometer is

$E = \dfrac{10}{R_T} \times V_S = \dfrac{10}{25} \times 24 = 9.6$ V

When the wiper arm is at the top of the potentiometer,

$V_1 = \dfrac{10}{10} \times E = \dfrac{10}{10} \times 9.6 = 9.6$ V

When the wiper arm is at the bottom of the potentiometer,

$$V_1 = \frac{0}{10} \times E = \frac{0}{10} \times 9.6 = 0 \text{ V}$$

Therefore, V_1 can be adjusted between **0 and 9.6 V.**

Example 2.75. *Fig. 2.149 shows the voltage divider circuit. Find (i) the current drawn from the supply, (ii) voltage across the load R_L, (iii) the current fed to R_L and (iv) the current in the tapped portion of the divider.*

Solution. It is a loaded voltage divider.

(*i*) $R_{BC} = 120 \ \Omega \parallel 300 \ \Omega = \dfrac{120 \times 300}{120 + 300} = 85.71 \ \Omega$

$V_{AB} = \dfrac{R_{AB}}{R_{AB} + R_{BC}} \times V_S = \dfrac{80}{80 + 85.71} \times 200 = 96.55 \text{ V}$

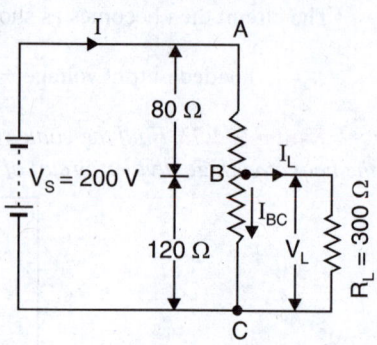

Fig. 2.149

$\therefore$ The current I drawn from the supply is

$$I = \frac{V_{AB}}{R_{AB}} = \frac{96.55}{80} = \textbf{1.21 A}$$

(*ii*) $\qquad V_{BC} = \dfrac{R_{BC}}{R_{AB} + R_{BC}} \times V_S = \dfrac{85.71}{80 + 85.71} \times 200 = \textbf{103.45 V}$

(*iii*) $\therefore \qquad$ Current fed to load, $I_L = \dfrac{V_{BC}}{R_L} = \dfrac{103.45}{300} = \textbf{0.35 A}$

(*iv*) Current in the tapped portion of the divider is

$$I_{BC} = I - I_L = 1.21 - 0.35 = \textbf{0.86A}$$

Tutorial Problems

1. Redraw the circuit shown in Fig. 2.150 using the common ground symbol.

   Ans.

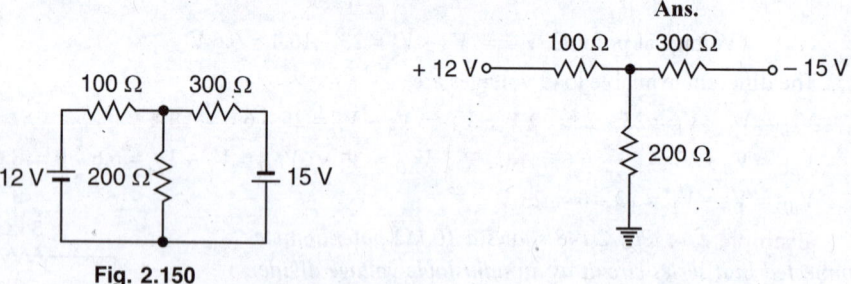

   Fig. 2.150

2. Redraw the circuit shown in Fig. 2.151 using the common ground symbol.

   Ans.

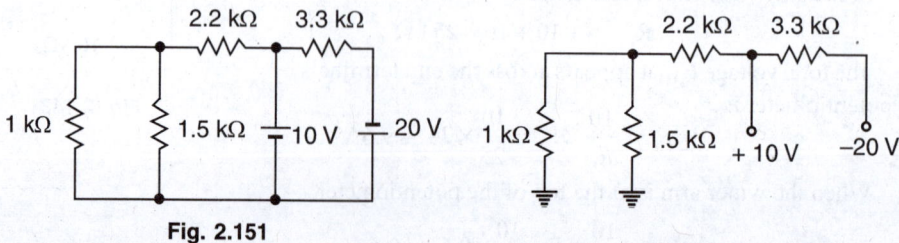

   Fig. 2.151

3. Draw the circuit shown in Fig. 2.152 by eliminating the common ground symbols.

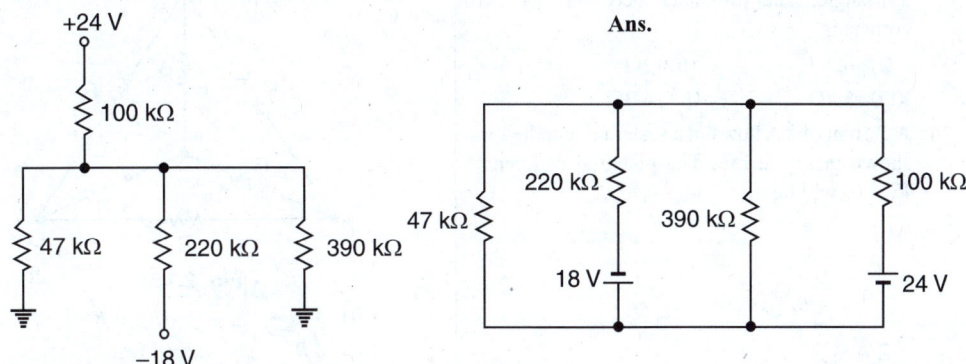

Fig. 2.152

4. A voltage of 200 V is applied to a tapped resistor of 500 Ω. Find the resistance between the tapped points connected to a circuit reading 0.1 A at 25 V. Also calculate the total power consumed. **[79Ω ; 83.3W]**

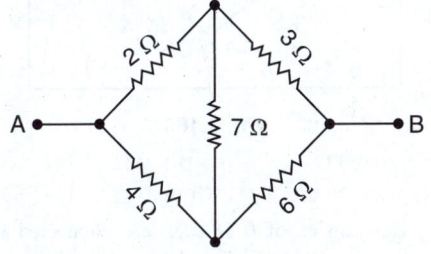

Objective Questions

1. Two resistances are joined in parallel whose resultant resistance is 6/5 ohms. One of the resistance wire is broken and the effective resistance becomes 2 ohms. Then the resistance of the wire that got broken is

    (*i*) 6/5 ohms   (*ii*) 3 ohms

    (*iii*) 2 ohms   (*iv*) 3/5 ohms

2. The smallest resistance obtained by connecting 50 resistances of 1/4 ohm each is

    (*i*) 50/4 Ω   (*ii*) 4/50 Ω

    (*iii*) 200 Ω   (*iv*) 1/200 Ω

3. Five resistances are connected as shown in Fig. 2.153. The effective resistance between points *A* and *B* is

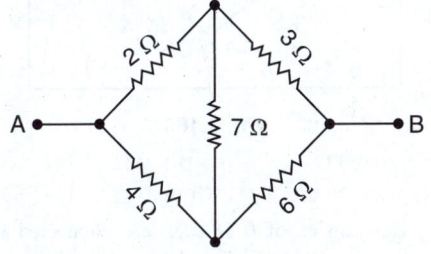

Fig. 2.153

    (*i*) 10/3 Ω   (*ii*) 20/3 Ω

    (*iii*) 15 Ω   (*iv*) 6 Ω

4. A 200 W and a 100 W bulb both meant for operation at 220 V are connected in series. When connected to a 220 V supply, the power consumed by them will be

    (*i*) 33 W   (*ii*) 100 W

    (*iii*) 66 W   (*iv*) 300 W

5. A wire has a resistance of 12 ohms. It is bent in the form of a circle. The effective resistance between two points on any diameter is

    (*i*) 6 Ω   (*ii*) 24 Ω

    (*iii*) 16 Ω   (*iv*) 3 Ω

6. A primary cell has an e.m.f. of 1.5 V. When short-circuited, it gives a current of 3 A. The internal resistance of the cell is

    (*i*) 4.5 Ω   (*ii*) 2 Ω

    (*iii*) 0.5 Ω   (*iv*) 1/4.5 Ω

7. Fig. 2.154 shows a part of a closed electrical circuit. Then $V_A - V_B$ is

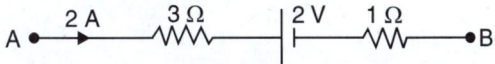

Fig. 2.154

    (*i*) − 8 V   (*ii*) 6 V

    (*iii*) 10 V   (*iv*) 3 V

8. The current *I* in the electric circuit shown in Fig. 2.155 is

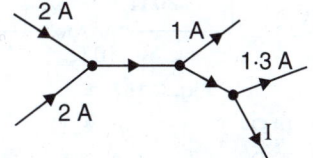

Fig. 2.155

    (*i*) 1.3 A   (*ii*) 3.7 A

    (*iii*) 1A   (*iv*) 1.7 A

9. Three 2 ohm resistors are connected to form a triangle. The resistance between any two corners is

   (*i*) 6Ω   (*ii*) 2Ω

   (*iii*) 3/4Ω   (*iv*) 4/3Ω

10. A current of 2 A flows in a system of conductors shown in Fig. 2.156. The potential difference $V_A - V_B$ will be

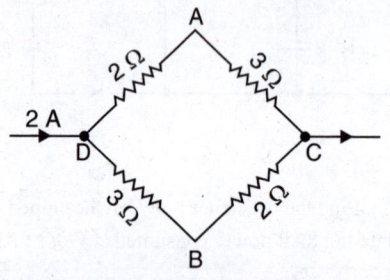

Fig. 2.156

   (*i*) +2 V   (*ii*) +1 V

   (*iii*) −1 V   (*iv*) −2 V

11. A uniform wire of resistance R is divided into 10 equal parts and all of them are connected in parallel. The equivalent resistance will be

   (*i*) 0.01 R (*ii*) 0.1 R

   (*iii*) 10 R (*iv*) 100 R

12. A cell of negligible resistance and e.m.f. 2 volts is connected to series combination of 2, 3 and 5 ohms. The potential difference in volts between the terminals of 3-ohm resistance will be

   (*i*) 0.6 V   (*ii*) $\dfrac{2}{3}$ V

   (*iii*) 3 V   (*iv*) 6 V

13. The equivalent resistance between points X and Y in Fig. 2.157 is

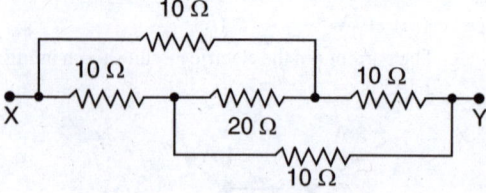

Fig. 2.157

   (*i*) 10 Ω   (*ii*) 22 Ω

   (*iii*) 20 Ω   (*iv*) 50 Ω

14. If each resistance in the network shown in Fig. 2.158 is R, what is the equivalent resistance between terminals A and B ?

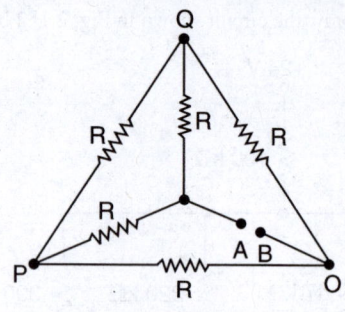

Fig. 2.158

   (*i*) 5 R (*ii*) 3 R

   (*iii*) 6 R (*iv*) R

15. Fig. 2.159 represents a part of a closed circuit. The potential difference between A and B (*i.e.* $V_A - V_B$) is

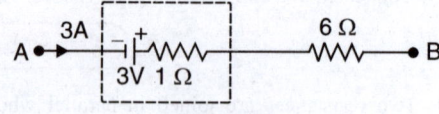

Fig. 2.159

   (*i*) 24 V   (*ii*) 0 V

   (*iii*) 18 V   (*iv*) 6 V

16. In the arrangement shown in Fig. 2.160, the potential difference between B and D will be zero if the unknown resistance X is

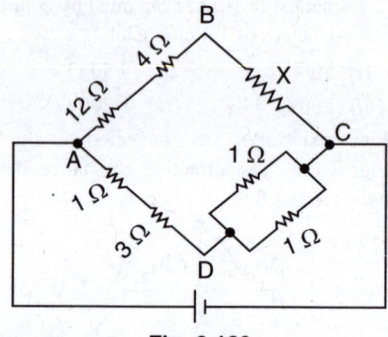

Fig. 2.160

   (*i*) 4 Ω   (*ii*) 2 Ω

   (*iii*) 20 Ω   (*iv*) 3 Ω

17. Resistances of 6 Ω each are connected in a manner shown in Fig. 2.161. With the current 0.5A as shown in the figure, the potential difference $V_P - V_Q$ is

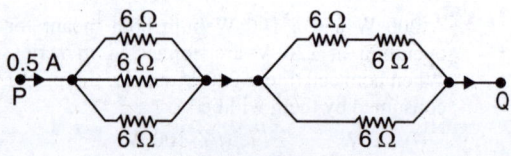

Fig. 2.161

　(*i*)　3.6 V　　　　　(*ii*)　6 V

　(*iii*)　3 V　　　　　(*iv*)　7.2 V

18. An electric fan and a heater are marked 100 W, 220 V and 1000 W, 220 V respectively. The resistance of the heater is

　(*i*)　zero

　(*ii*)　greater than that of fan

　(*iii*)　less than that of fan

　(*iv*)　equal to that of fan

19. In the circuit shown in Fig. 2.162, the final voltage drop across the capacitor C is

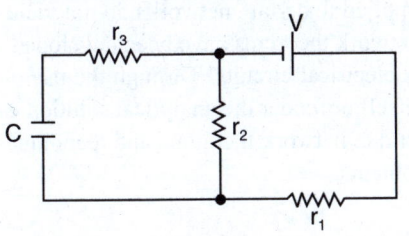

Fig. 2.162

(*i*) $\dfrac{V r_1}{r_1 + r_2}$　　　　　(*ii*) $\dfrac{V r_2}{r_1 + r_2}$

(*iii*) $\dfrac{V(r_1 + r_2)}{r_2}$　　　　(*iv*) $\dfrac{V(r_2 + r_1)}{r_1 + r_2 + r_3}$

20. A primary cell has an e.m.f. of 1.5 V. When short circuited, it gives a current of 3 A. The internal resistance of the cell is

　(*i*)　4.5 Ω　　　　　(*ii*)　2 Ω

　(*iii*)　0.5 Ω　　　　(*iv*)　(1/4.5) Ω

Answers

1.	(*ii*)	**2.**	(*iv*)	**3.**	(*i*)	**4.**	(*iii*)	**5.**	(*iv*)
6.	(*iii*)	**7.**	(*iii*)	**8.**	(*iv*)	**9.**	(*iv*)	**10.**	(*ii*)
11.	(*i*)	**12.**	(*i*)	**13.**	(*i*)	**14.**	(*iv*)	**15**	(*iii*)
16.	(*ii*)	**17.**	(*iii*)	**18.**	(*iii*)	**19.**	(*ii*)	**20.**	(*iii*)

3

D.C. Network Theorems

Introduction

Any arrangement of electrical energy sources, resistances and other circuit elements is called an electrical network. The terms *circuit* and *network* are used synonymously in electrical literature. In the text so far, we employed two network laws *viz* Ohm's law and Kirchhoff's laws to solve network problems. Occasions arise when these laws applied to certain networks do not yield quick and easy solution. To overcome this difficulty, some network theorems have been developed which are very useful in analysing both simple and complex electrical circuits. Through the use of these theorems, it is possible either to simplify the network itself or render the analytical solution easy. In this chapter, we shall focus our attention on important d.c. network theorems and techniques with special reference to their utility in solving network problems.

3.1. Network Terminology

While discussing network theorems and techniques, one often comes across the following terms:

(*i*) **Linear circuit.** A linear circuit is one whose parameters (*e.g.* resistances) are constant *i.e.* they do not change with current or voltage.

(*ii*) **Non-linear circuit.** A non-linear circuit is one whose parameters (*e.g.* resistances) change with voltage or current.

(*iii*) **Bilateral circuit.** A bilateral circuit is one whose properties are the same in either direction. For example, transmission line is a bilateral circuit because it can be made to perform its function equally well in either direction.

(*iv*) **Active element.** An active element is one which supplies electrical energy to the circuit. Thus in Fig. 3.1, E_1 and E_2 are the active elements because they supply energy to the circuit.

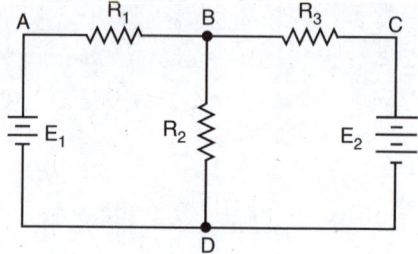

(*v*) **Passive element.** A passive element is one which receives electrical energy and then either converts it into heat (resistance) or stores in an electric field (capacitance) or magnetic field (inductance). In Fig. 3.1, there are three passive

Fig. 3.1

elements, namely R_1, R_2 and R_3. These passive elements (*i.e.* resistances in this case) receive energy from the active elements (*i.e.* E_1 and E_2) and convert it into heat.

(*vi*) **Node.** A node of a network is an equipotential surface at which *two or more* circuit elements are joined. Thus in Fig. 3.1, circuit elements R_1 and E_1 are joined at A and hence A is the node. Similarly, B, C and D are nodes.

(*vii*) **Junction.** A junction is that point in a network where *three or more* circuit elements are joined. In Fig. 3.1, there are only two junction points *viz.* B and D. That B is a junction is clear from the fact that three circuit elements R_1, R_2 and R_3 are joined at it. Similarly, point D is a junction because it joins three circuit elements R_2, E_1 and E_2.

(*viii*) **Branch.** A branch is that part of a network which lies between two junction points. Thus referring to Fig. 3.1, there are a total of three branches *viz. BAD, BCD* and *BD*. The branch

BAD consists of R_1 and E_1 ; the branch *BCD* consists of R_3 and E_2 and branch *BD* merely consists of R_2.

(*ix*) **Loop.** A loop is any closed path of a network. Thus in Fig. 3.1, *ABDA*, *BCDB* and *ABCDA* are the loops.

(*x*) **Mesh.** A mesh is the most elementary form of a loop and cannot be further divided into other loops. In Fig. 3.1, both loops *ABDA* and *BCDB* qualify as meshes because they cannot be further divided into other loops. However, the loop *ABCDA* cannot be called a mesh because it encloses two loops *ABDA* and *BCDB*.

(*xi*) **Network and circuit.** Strictly speaking, the term network is used for a circuit containing passive elements only while the term circuit implies the presence of both active and passive elements. However, there is no hard and fast rule for making these distinctions and the **terms "network" and "circuit" are often used interchangeably.**

(*xii*) **Parameters.** The various elements of an electric circuit like resistance (R), inductance (L) and capacitance (C) are called parameters of the circuit. These parameters may be lumped or distributed.

(*xiii*) **Unilateral circuit.** A unilateral circuit is one whose properties change with the direction of its operation. For example, a diode rectifier circuit is a unilateral circuit. It is because a diode rectifier cannot perform rectification in both directions.

(*xiv*) **Active and passive networks.** An active network is that which contains active elements as well as passive elements. On the other hand, a passive network is that which contains passive elements only.

3.2. Network Theorems and Techniques

Having acquainted himself with network terminology, the reader is set to study the various network theorems and techniques. In this chapter, we shall discuss the following network theorems and techniques :

(*i*) Maxwell's mesh current method   (*ii*) Nodal analysis

(*iii*) Superposition theorem   (*iv*) Thevenin's theorem

(*v*) Norton's theorem   (*vi*) Maximum power transfer theorem

(*vii*) Reciprocity theorem   (*viii*) Millman's theorem

(*ix*) Compensation theorem   (*x*) Delta/star or star/delta transformation

(*xi*) Tellegen's theorem

3.3. Important Points About Network Analysis

While analysing network problems by using network theorems and techniques, the following points may be noted :

(*i*) There are two general approaches to network analysis *viz.* (*a*) **direct method** (*b*) **network reduction method.** In direct method, the network is left in its original form and different voltages and currents in the circuit are determined. This method is used for simple circuits. Examples of direct method are Kirchhoff's laws, Mesh current method, nodal analysis, superposition theorem etc. In network reduction method, the original network is reduced to a simpler equivalent circuit. This method is used for complex circuits and gives a better insight into the performance of the circuit. Examples of network reduction method are Thevenin's theorem, Norton's theorem, star/delta or delta/star transformation etc.

(*ii*) *The above theorems and techniques are applicable only to networks that have linear, bilateral circuit elements.*

(*iii*) The network theorem or technique to be used will depend upon the network arrangement. The general rule is this. *Use that theorem or technique which requires a smaller number of independent equations to obtain the solution or which can yield easy solution.*

(*iv*) *Analysis of a circuit usually means to determine all the currents and voltages in the circuit.*

3.4. Maxwell's Mesh Current Method

In this method, Kirchhoff's voltage law is applied to a network to write mesh equations in terms of **mesh currents** *instead of branch currents.* Each mesh is assigned a separate mesh current. This mesh current is assumed to flow *clockwise* around the perimeter of the mesh without splitting at a junction into branch currents. Kirchhoff's voltage law is then applied to write equations in terms of unknown mesh currents. The branch currents are then found by taking the algebraic sum of the mesh currents which are common to that branch.

Explanation. Maxwell's mesh current method consists of following steps :

(*i*) Each mesh is assigned a separate mesh current. For convenience, all mesh currents are assumed to flow *clockwise direction. For example, in Fig. 3.2, meshes *ABDA* and *BCDB* have been assigned mesh currents I_1 and I_2 respectively. The mesh currents take on the appearance of a mesh fence and hence the name mesh currents.

(*ii*) If two mesh currents are flowing through a circuit element, the actual current in the circuit element is the algebraic sum of the two. Thus in Fig. 3.2, there are two mesh currents I_1 and I_2 flowing in R_2. If we go from *B* to *D*, current is $I_1 - I_2$ and if we go in the other direction (*i.e.* from *D* to *B*), current is $I_2 - I_1$.

(*iii*) **Kirchhoff's voltage law is applied to write equation for each mesh in terms of mesh currents. Remember, while writing mesh equations, rise in potential is assigned positive sign and fall in potential negative sign.

(*iv*) If the value of any mesh current comes out to be negative in the solution, it means that true direction of that mesh current is anticlockwise *i.e.* opposite to the assumed clockwise direction.

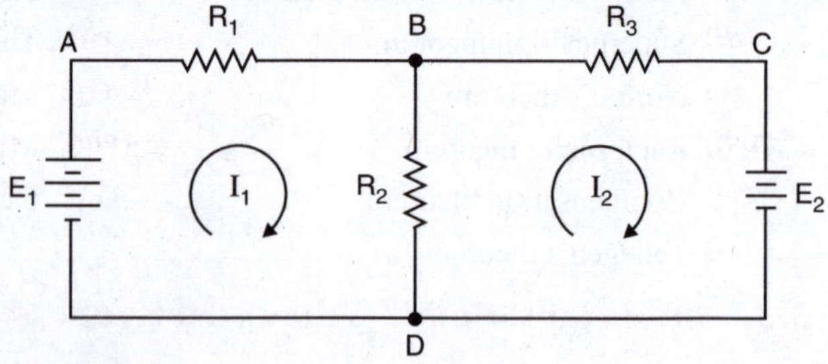

Fig. 3.2

Applying Kirchhoff's voltage law to Fig. 3.2, we have,

Mesh ABDA.

$$-I_1R_1 - (I_1 - I_2) R_2 + E_1 = 0$$

or $\qquad I_1 (R_1 + R_2) - I_2R_2 = E_1$ $\qquad\qquad\qquad\qquad$ ...(*i*)

---

*   It is convenient to consider all mesh currents in one direction (clockwise or anticlockwise). The same result will be obtained if mesh currents are given arbitrary directions.

**   Since the circuit unknowns are currents, the describing equations are obtained by applying *KVL* to the meshes.

Mesh BCDB.

$$-I_2R_3 - E_2 - (I_2 - I_1)R_2 = 0$$

or $\quad -I_1R_2 + (R_2 + R_3)I_2 = -E_2 \qquad\qquad\qquad ...(ii)$

Solving eq. (*i*) and eq. (*ii*) simultaneously, mesh currents I_1 and I_2 can be found out. Once the mesh currents are known, the branch currents can be readily obtained. *The advantage of this method is that it usually reduces the number of equations to solve a network problem.*

Note. Branch currents are the real currents because they actually flow in the branches and can be measured. However, mesh currents are fictitious quantities and cannot be measured except in those instances where they happen to be identical with branch currents. Thus in branch *DAB*, branch current is the same as mesh current and both can be measured. But in branch *BD*, mesh currents (I_1 and I_2) cannot be measured. Hence mesh current is a concept rather than a reality. However, it is a useful concept to solve network problems as it leads to the reduction of number of mesh equations.

3.5. Shortcut Procedure for Network Analysis by Mesh Currents

We have seen above that Maxwell mesh current method involves lengthy mesh equations. Here is a shortcut method to write mesh equations simply by inspection of the circuit. Consider the circuit shown in Fig. 3.3. The circuit contains resistances and independent voltage sources and has three meshes. Let the three mesh currents be I_1, I_2 and I_3 flowing in the clockwise direction.

Loop 1. Applying *KVL* to this loop, we have,

$$100 - 20 = I_1(60 + 30 + 50) - I_2 \times 50 - I_3 \times 30$$

or $\quad 80 = 140I_1 - 50I_2 - 30I_3 \qquad\qquad\qquad ...(i)$

We can write eq. (*i*) in a shortcut form as :

$$E_1 = I_1R_{11} - I_2R_{12} - I_3R_{13}$$

Here $\qquad E_1 =$ Algebraic sum of e.m.f.s in Loop (1) in the direction of I_1

$\qquad\qquad = 100 - 20 = 80 \text{ V}$

$\qquad R_{11} =$ Sum of resistances in Loop (1)

$\qquad\qquad =$ Self*-resistance of Loop (1)

$\qquad\qquad = 60 + 30 + 50 = 140 \ \Omega$

$\qquad R_{12} =$ Total resistance common to Loops (1) and (2)

$\qquad\qquad =$ Common resistance between Loops (1) and (2) $= 50 \ \Omega$

$\qquad R_{13} =$ Total resistance common to Loops (1) and (3) $= 30 \ \Omega$

It may be seen that the sign of the term involving self-resistances is positive while the sign of common resistances is negative. It is because the positive directions for mesh currents were all chosen clockwise. Although mesh currents are abstract currents, yet mesh current analysis offers the advantage that resistor polarities do not have to be considered when writing mesh equations.

Loop 2. We can use shortcut method to find the mesh equation for Loop (2) as under :

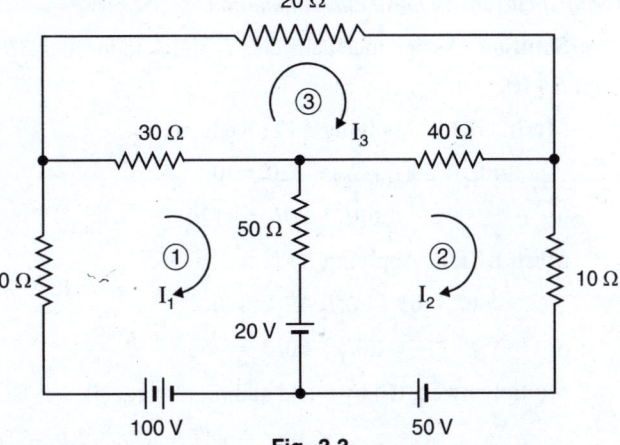

Fig. 3.3

---

* The sum of all resistances in a loop is called self-resistance of that loop. Thus in Fig. 3.3, self-resistance of Loop (1) = 60 + 30 + 50 = 140 Ω.

$$E_2 = -I_1 R_{21} + I_2 R_{22} - I_3 R_{23}$$

or $\qquad 50 + 20 = -50I_1 + 100I_2 - 40I_3$ ...(ii)

Here, $\qquad E_2 = $ Algebraic sum of e.m.f.s in Loop (2) in the direction of I_2

$\qquad\qquad = 50 + 20 = 70$ V

$\qquad R_{21} = $ Total resistance common to Loops (2) and (1) $= 50\ \Omega$

$\qquad R_{22} = $ Sum of resistances in Loop (2) $= 50 + 40 + 10 = 100\ \Omega$

$\qquad R_{23} = $ Total resistance common to Loops (2) and (3) $= 40\ \Omega$

Again the sign of self-resistance of Loop (2) (R_{22}) is positive while the sign of the terms of common resistances (R_{21}, R_{23}) is negative.

Loop 3. We can again use shortcut method to find the mesh equation for Loop (3) as under :

$$E_3 = -I_1 R_{31} - I_2 R_{32} + I_3 R_{33}$$

or $\qquad 0 = -30I_1 - 40I_2 + 90I_3$ ...(iii)

Again the sign of self-resistance of Loop (3) (R_{33}) is positive while the sign of the terms of common resistances (R_{31}, R_{32}) is negative.

Mesh analysis using matrix form. The three mesh equations are rewritten below :

$$E_1 = I_1 R_{11} - I_2 R_{12} - I_3 R_{13}$$
$$E_2 = -I_1 R_{21} + I_2 R_{22} - I_3 R_{23}$$
$$E_3 = -I_1 R_{31} - I_2 R_{32} + I_3 R_{33}$$

The matrix equivalent of above given equations is :

$$\begin{bmatrix} R_{11} & R_{12} & R_{13} \\ R_{21} & R_{22} & R_{23} \\ R_{31} & R_{32} & R_{33} \end{bmatrix} \begin{bmatrix} I_1 \\ I_2 \\ I_3 \end{bmatrix} = \begin{bmatrix} E_1 \\ E_2 \\ E_3 \end{bmatrix}$$

It is reminded again that (i) all self-resistances are positive (ii) all common resistances are negative and (iii) by their definition, $R_{12} = R_{21}$; $R_{23} = R_{32}$ and $R_{13} = R_{31}$.

Example 3.1. *In the network shown in Fig. 3.4 (i), find the magnitude and direction of each branch current by mesh current method.*

Solution. Assign mesh currents I_1 and I_2 to meshes *ABDA* and *BCDB* respectively as shown in Fig. 3.4 (i).

Mesh ABDA. Applying *KVL*, we have,

$$-40I_1 - 20(I_1 - I_2) + 120 = 0$$

or $\qquad\qquad 60I_1 - 20I_2 = 120$ ...(i)

Mesh BCDB. Applying *KVL*, we have,

$$-60I_2 - 65 - 20(I_2 - I_1) = 0$$

or $\qquad\qquad -20I_1 + 80I_2 = -65$ ...(ii)

Multiplying eq. (ii) by 3 and adding it to eq. (i), we get,

$$220I_2 = -75 \qquad \therefore \quad I_2 = -75/220 = -0.341 \text{ A}$$

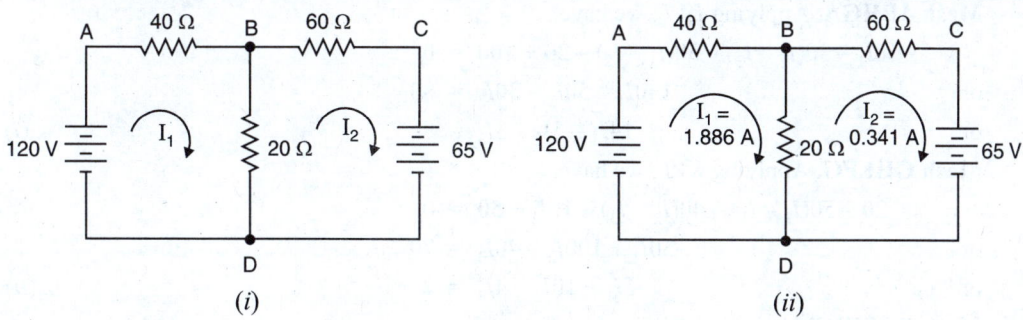

Fig. 3.4

The minus sign shows that true direction of I_2 is anticlockwise. Substituting $I_2 = -0.341$A in eq. (*i*), we get, $I_1 = 1.886$ A. The actual direction of flow of currents is shown in Fig. 3.4 (*ii*).

By determinant method

$$60I_1 - 20I_2 = 120$$

$$-20I_1 + 80I_2 = -65$$

$$\therefore \quad I_1 = \frac{\begin{vmatrix} 120 & -20 \\ -65 & 80 \end{vmatrix}}{\begin{vmatrix} 60 & -20 \\ -20 & 80 \end{vmatrix}} = \frac{(120 \times 80) - (-65 \times -20)}{(60 \times 80) - (-20 \times -20)} = \frac{8300}{4400} = 1.886 \text{ A}$$

$$I_2 = \frac{\begin{vmatrix} 60 & 120 \\ -20 & -65 \end{vmatrix}}{\text{Denominator}} = \frac{(60 \times -65) - (-20 \times 120)}{4400} = \frac{-1500}{4400} = -0.341 \text{A}$$

Referring to Fig. 3.4 (*ii*), we have,

Current in branch $DAB = I_1 = $ **1·886 A** ; Current in branch $DCB = I_2 = $ **0·341 A**

Current in branch $BD = I_1 + I_2 = 1.886 + 0.341 = $ **2·227 A**

Example 3.2. *Calculate the current in each branch of the circuit shown in Fig. 3.5.*

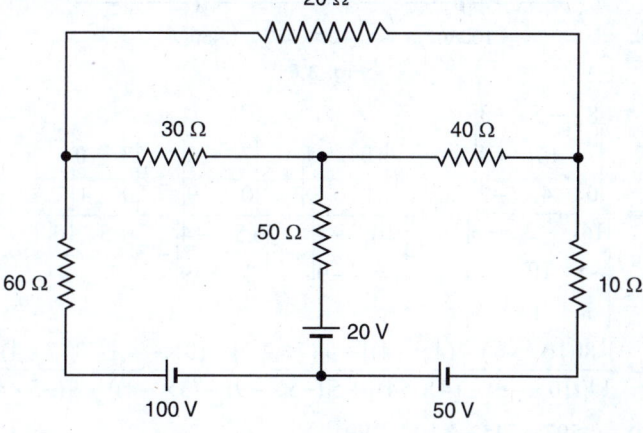

Fig. 3.5

Solution. Assign mesh currents I_1, I_2 and I_3 to meshes *ABHGA*, *HEFGH* and *BCDEHB* respectively as shown in Fig. 3.6.

Mesh ABHGA. Applying *KVL*, we have,

$$-60I_1 - 30(I_1 - I_3) - 50(I_1 - I_2) - 20 + 100 = 0$$

or $$140I_1 - 50I_2 - 30I_3 = 80$$

or $$14I_1 - 5I_2 - 3I_3 = 8 \qquad \qquad ...(i)$$

Mesh GHEFG. Applying *KVL*, we have,

$$20 - 50(I_2 - I_1) - 40(I_2 - I_3) - 10I_2 + 50 = 0$$

or $$-50I_1 + 100I_2 - 40I_3 = 70$$

or $$-5I_1 + 10I_2 - 4I_3 = 7 \qquad \qquad ...(ii)$$

Mesh BCDEHB. Applying *KVL*, we have,

$$-20I_3 - 40(I_3 - I_2) - 30(I_3 - I_1) = 0$$

or $$30I_1 + 40I_2 - 90I_3 = 0$$

or $$3I_1 + 4I_2 - 9I_3 = 0 \qquad \qquad ...(iii)$$

Solving for equations (*i*), (*ii*) and (*iii*), we get, $I_1 = 1 \cdot 65$ A ; $I_2 = 2 \cdot 12$ A ; $I_3 = 1 \cdot 5$ A

By determinant method

$$14I_1 - 5I_2 - 3I_3 = 8$$
$$-5I_1 + 10I_2 - 4I_3 = 7$$
$$3I_1 + 4I_2 - 9I_3 = 0$$

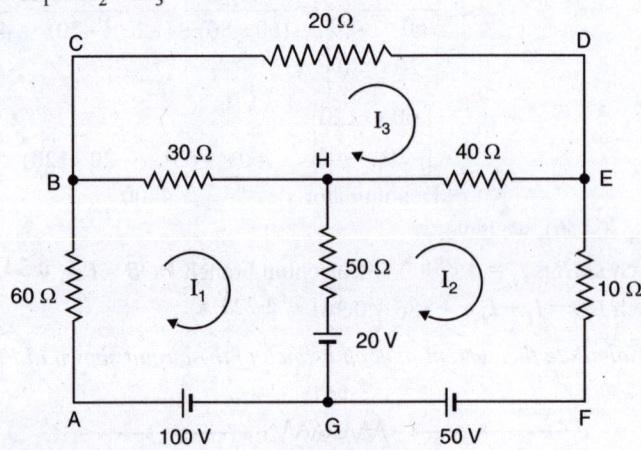

Fig. 3.6

$$\therefore \quad I_1 = \frac{\begin{vmatrix} 8 & -5 & -3 \\ 7 & 10 & -4 \\ 0 & 4 & -9 \end{vmatrix}}{\begin{vmatrix} 14 & -5 & -3 \\ -5 & 10 & -4 \\ 3 & 4 & -9 \end{vmatrix}} = \frac{8\begin{vmatrix} 10 & -4 \\ 4 & -9 \end{vmatrix} + 5\begin{vmatrix} 7 & -4 \\ 0 & -9 \end{vmatrix} - 3\begin{vmatrix} 7 & 10 \\ 0 & 4 \end{vmatrix}}{14\begin{vmatrix} 10 & -4 \\ 4 & -9 \end{vmatrix} + 5\begin{vmatrix} -5 & -4 \\ 3 & -9 \end{vmatrix} - 3\begin{vmatrix} -5 & 10 \\ 3 & 4 \end{vmatrix}}$$

$$= \frac{8[(10 \times -9) - (4 \times -4)] + 5[(7 \times -9) - (0 \times -4)] - 3[(7 \times 4) - (0 \times 10)]}{14[(10 \times -9) - (4 \times -4)] + 5[(-5 \times -9) - (3 \times -4)] - 3[(-5 \times 4) - (3 \times 10)]}$$

$$= \frac{-592 - 315 - 84}{-1036 + 285 + 150} = \frac{-991}{-601} = 1 \cdot 65 \text{ A}$$

$$I_2 = \frac{\begin{vmatrix} 14 & 8 & -3 \\ -5 & 7 & -4 \\ 3 & 0 & -9 \end{vmatrix}}{\text{Denominator}} = \frac{14[(-63)-(0)]-8[(45)-(-12)]-3[(0)-(21)]}{-601}$$

$$= \frac{-882-456+63}{-601} = \frac{-1275}{-601} = 2.12 \text{ A}$$

$$I_3 = \frac{\begin{vmatrix} 14 & -5 & 8 \\ -5 & 10 & 7 \\ 3 & 4 & 0 \end{vmatrix}}{\text{Denominator}} = \frac{14[(0)-(28)]+5[(0)-(21)]+8[(-20)-(30)]}{-601}$$

$$= \frac{-392-105-400}{-601} = \frac{-897}{-601} = 1.5 \text{ A}$$

∴   Current in 60 Ω $= I_1 =$ **1·65 A from A to B**

Current in 30 Ω $= I_1 - I_3 = 1.65 - 1.5 =$ **0·15 A from B to H**

Current in 50 Ω $= I_2 - I_1 = 2.12 - 1.65 =$ **0·47 A from G to H**

Current in 40 Ω $= I_2 - I_3 = 2.12 - 1.5 =$ **0·62 A from H to E**

Current in 10 Ω $= I_2 =$ **2·12 A from E to F**

Current in 20 Ω $= I_3 =$ **1·5 A from C to D**

Example 3.3. *By using mesh resistance matrix, determine the current supplied by each battery in the circuit shown in Fig. 3.7.*

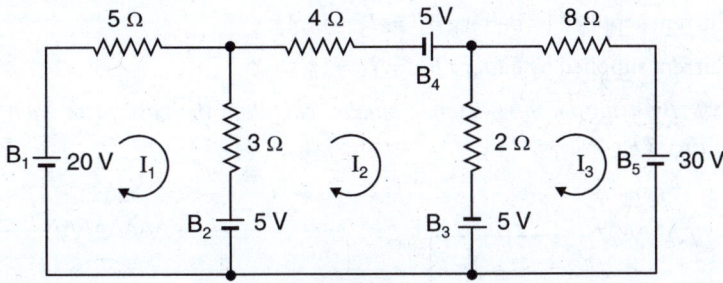

Fig. 3.7

Solution. Since there are three meshes, let the three mesh currents be I_1, I_2 and I_3, all assumed to be flowing in the clockwise direction. The different quantities of the mesh-resistance matrix are :

$$R_{11} = 5+3 = 8 \,\Omega \quad ; \quad R_{22} = 4+2+3 = 9 \,\Omega \quad ; \quad R_{33} = 8+2 = 10 \,\Omega$$
$$R_{12} = R_{21} = -3 \,\Omega \quad ; \quad R_{13} = R_{31} = 0 \quad ; \quad R_{23} = R_{32} = -2 \,\Omega$$
$$E_1 = 20-5 = 15 \text{ V} \quad ; \quad E_2 = 5+5+5 = 15 \text{ V} \quad ; \quad E_3 = -30-5 = -35 \text{ V}$$

Therefore, the mesh equations in the matrix form are :

$$\begin{bmatrix} R_{11} & R_{12} & R_{13} \\ R_{21} & R_{22} & R_{23} \\ R_{31} & R_{32} & R_{33} \end{bmatrix} \begin{bmatrix} I_1 \\ I_2 \\ I_3 \end{bmatrix} = \begin{bmatrix} E_1 \\ E_2 \\ E_3 \end{bmatrix}$$

or $$\begin{bmatrix} 8 & -3 & 0 \\ -3 & 9 & -2 \\ 0 & -2 & 10 \end{bmatrix} \begin{bmatrix} I_1 \\ I_2 \\ I_3 \end{bmatrix} = \begin{bmatrix} 15 \\ 15 \\ -35 \end{bmatrix}$$

By determinant method, we have,

$$I_1 = \dfrac{\begin{vmatrix} 15 & -3 & 0 \\ 15 & 9 & -2 \\ -35 & -2 & 10 \end{vmatrix}}{\begin{vmatrix} 8 & -3 & 0 \\ -3 & 9 & -2 \\ 0 & -2 & 10 \end{vmatrix}} = \dfrac{1530}{598} = 2\!\cdot\!56 \text{ A}$$

$$I_2 = \dfrac{\begin{vmatrix} 8 & 15 & 0 \\ -3 & 15 & -2 \\ 0 & -35 & 10 \end{vmatrix}}{\text{Denominator}} = \dfrac{1090}{598} = 1\!\cdot\!82 \text{ A}$$

$$I_3 = \dfrac{\begin{vmatrix} 8 & -3 & 15 \\ -3 & 9 & 15 \\ 0 & -2 & -35 \end{vmatrix}}{\text{Denominator}} = \dfrac{-1875}{598} = -3\!\cdot\!13 \text{ A}$$

The negative sign with I_3 indicates that actual direction of I_3 is opposite to that assumed in Fig. 3.7. Note that batteries B_1, B_3, B_4 and B_5 are discharging while battery B_2 is charging.

∴ Current supplied by battery B_1 = I_1 = **2·56 A**

Current supplied to battery B_2 = $I_1 - I_2$ = 2·56 – 1·82 = **0·74 A**

Current supplied by battery B_3 = $I_2 + I_3$ = 1·82 + 3·13 = **4·95 A**

Current supplied by battery B_4 = I_2 = **1·82 A**

Current supplied by battery B_5 = I_3 = **3·13 A**

Example 3.4. *By using mesh resistance matrix, calculate the current in each branch of the circuit shown in Fig. 3.8.*

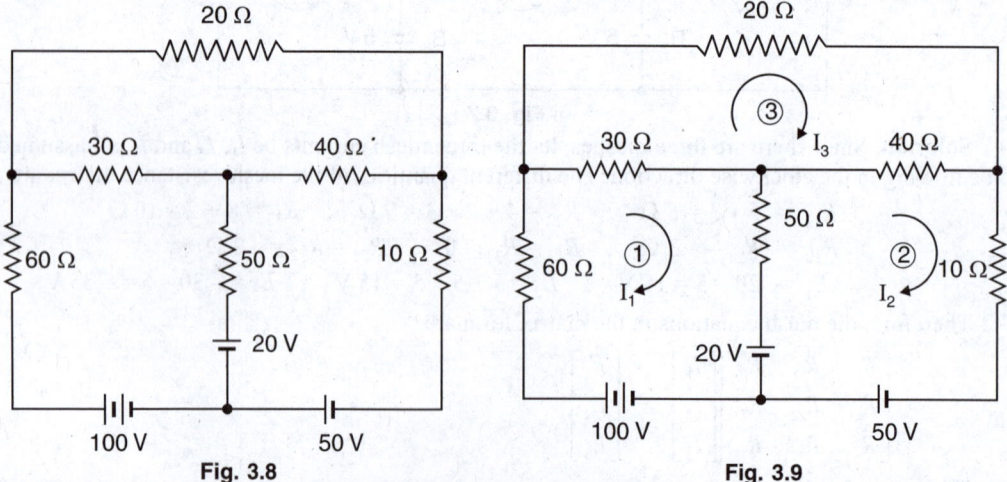

Fig. 3.8   Fig. 3.9

Solution. Since there are three meshes, let the three mesh currents be I_1, I_2 and I_3, all assumed to be flowing in the clockwise direction as shown in Fig. 3.9. The different quantities of the mesh resistance-matrix are :

$$R_{11} = 60 + 30 + 50 = 140 \ \Omega \ ; \ R_{22} = 50 + 40 + 10 = 100 \ \Omega \ ; \ R_{33} = 30 + 20 + 40 = 90 \ \Omega$$

$$R_{12} = R_{21} = -50 \ \Omega \ ; \quad R_{13} = R_{31} = -30 \ \Omega \ ; \quad R_{23} = R_{32} = -40 \ \Omega$$

$$E_1 = 100 - 20 = 80 \ \text{V} \ ; \quad E_2 = 50 + 20 = 70 \ \text{V} \ ; \quad E_3 = 0 \ \text{V}$$

Therefore, the mesh equations in the matrix form are :

$$\begin{bmatrix} R_{11} & R_{12} & R_{13} \\ R_{21} & R_{22} & R_{23} \\ R_{31} & R_{32} & R_{33} \end{bmatrix} \begin{bmatrix} I_1 \\ I_2 \\ I_3 \end{bmatrix} = \begin{bmatrix} E_1 \\ E_2 \\ E_3 \end{bmatrix}$$

or $$\begin{bmatrix} 140 & -50 & -30 \\ -50 & 100 & -40 \\ -30 & -40 & 90 \end{bmatrix} \begin{bmatrix} I_1 \\ I_2 \\ I_3 \end{bmatrix} = \begin{bmatrix} 80 \\ 70 \\ 0 \end{bmatrix}$$

By determinant method, we have,

$$I_1 = \frac{\begin{vmatrix} 80 & -50 & -30 \\ 70 & 100 & -40 \\ 0 & -40 & 90 \end{vmatrix}}{\begin{vmatrix} 140 & -50 & -30 \\ -50 & 100 & -40 \\ -30 & -40 & 90 \end{vmatrix}} = \frac{991000}{601000} = 1 \cdot 65 \ \text{A}$$

$$I_2 = \frac{\begin{vmatrix} 140 & 80 & -30 \\ -50 & 70 & -40 \\ -30 & 0 & 90 \end{vmatrix}}{\text{Denominator}} = \frac{1275000}{601000} = 2 \cdot 12 \ \text{A}$$

$$I_3 = \frac{\begin{vmatrix} 140 & -50 & 80 \\ -50 & 100 & 70 \\ -30 & -40 & 0 \end{vmatrix}}{\text{Denominator}} = \frac{897000}{601000} = 1 \cdot 5 \ \text{A}$$

∴   Current in 60 Ω = I_1 = **1·65 A in the direction of I_1**
　　Current in 30 Ω = $I_1 - I_3$ = **0·15 A in the direction of I_1**
　　Current in 50 Ω = $I_2 - I_1$ = **0·47 A in the direction of I_2**
　　Current in 40 Ω = $I_2 - I_3$ = **0·62 A in the direction of I_2**
　　Current in 10 Ω = I_2 = **2·12 A in the direction of I_2**
　　Current in 20 Ω = I_3 = **1·5 A in the direction of I_3**

Example 3.5. *Find mesh currents i_1 and i_2 in the electric circuit shown in Fig. 3.10.*

Solution. We shall use mesh current method for the solution. Mesh analysis requires that all the sources in a circuit be voltage sources. If a circuit contains any current source, convert it into equivalent voltage source.

Outer mesh. Applying *KVL* to this mesh, we have,

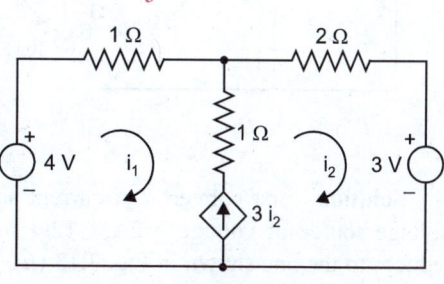

Fig. 3.10

$$-i_1 \times 1 - 2i_2 - 3 + 4 = 0 \quad \text{or} \quad i_1 + 2i_2 = 1 \qquad \qquad ...(i)$$

First mesh. Applying *KVL* to this mesh, we have,

$$-i_1 \times 1 - (i_1 - i_2) \times 1 - 3i_2 + 4 = 0 \quad \text{or} \quad i_1 + i_2 = 2 \qquad \qquad ...(ii)$$

From eqs. (*i*) and (*ii*), we have $i_1 = \mathbf{3A}$; $i_2 = \mathbf{-1\,A}$

Example 3.6. *Using mesh current method, determine current I_x in the circuit shown in Fig. 3.11.*

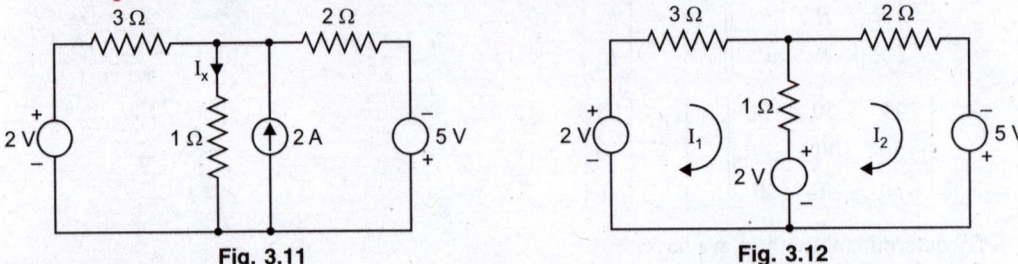

Fig. 3.11   Fig. 3.12

Solution. First convert 2A current source in parallel with 1Ω resistance into equivalent voltage source of voltage 2A × 1Ω = 2V in series with 1Ω resistance. The circuit then reduces to that shown in Fig. 3.12. Assign mesh currents I_1 and I_2 to meshes 1 and 2 in Fig. 3.12.

Mesh 1. Applying *KVL* to this mesh, we have,

$$-3I_1 - 1 \times (I_1 - I_2) - 2 + 2 = 0 \quad \text{or} \quad I_2 = 4I_1$$

Mesh 2. Applying *KVL* to this mesh, we have,

$$-2I_2 + 5 + 2 - (I_2 - I_1) \times 1 = 0$$

or $\qquad -2\,(4I_1) + 7 - (4I_1 - I_1) = 0 \quad (\because I_2 = 4I_1)$

$\therefore \quad I_1 = \dfrac{7}{11}\text{A} \text{ and } I_2 = 4I_1 = 4 \times \dfrac{7}{11} = \dfrac{28}{11}\text{A}$

$\therefore \quad$ Current in 3Ω resistance, $I_1 = \dfrac{7}{11}\text{A}$; Current in 2Ω resistance, $I_2 = \dfrac{28}{11}\text{A}$

Referring to the original Fig. 3.11, we have,

$$I_x = I_1 + (2 - I_2) = \frac{7}{11} + \left(2 - \frac{28}{11}\right) = \frac{1}{11}\mathbf{A}$$

Example 3.7. *Using mesh current method, find the currents in resistances R_3, R_4, R_5 and R_6 of the circuit shown in Fig. 3.13 (i).*

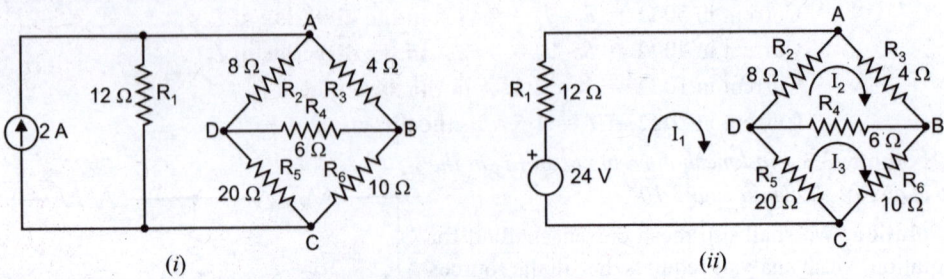

(*i*)   (*ii*)

Fig. 3.13

Solution. First convert 2 A current source in parallel with 12Ω resistance into equivalent voltage source of voltage = 2A × 12Ω = 24V in series with 12Ω resistance. The circuit then reduces to the one shown in Fig. 3.13 (*ii*). Assign the mesh currents I_1, I_2 and I_3 to three meshes 1, 2 and 3 shown in Fig. 3.13 (*ii*).

Mesh 1. Applying *KVL* to this mesh, we have,

$$-12I_1 - 8 \times (I_1 - I_2) - 20 \times (I_1 - I_3) + 24 = 0$$

or

$$10I_1 - 2I_2 - 5I_3 = 6 \qquad ...(i)$$

Mesh 2. Applying *KVL* to this mesh, we have,

$$- 4I_2 - 6 \times (I_2 - I_3) - 8(I_2 - I_1) = 0$$

or

$$- 4I_1 + 9I_2 - 3I_3 = 0 \qquad ...(ii)$$

Mesh 3. Applying *KVL* to this mesh, we have,

$$-10I_3 - 20 \times (I_3 - I_1) - 6 \times (I_3 - I_2) = 0$$

or

$$- 10I_1 - 3I_2 + 18I_3 = 0 \qquad ...(iii)$$

From eqs. (*i*), (*ii*) and (*iii*), $I_1 = 1.125$ A ; $I_2 = 0.75$ A ; $I_3 = 0.75$ A

∴ Current in R_3 (= 4Ω) = $I_2 = \mathbf{0.75\ A}$ *from* **A to B**

Current in R_4 (= 6Ω) = $I_2 - I_3 = 0.75 - 0.75 = \mathbf{0A}$

Current in R_5 (= 20Ω) = $I_1 - I_3 = 1.125 - 0.75 = \mathbf{0.375A}$ *from* **D to C**

Current in R_6 (= 10Ω) = $I_3 = \mathbf{0.75A}$ *from* **B to C**

Example 3.8. *Use mesh current method to determine currents through each of the components in the circuit shown in Fig. 3.14 (i).*

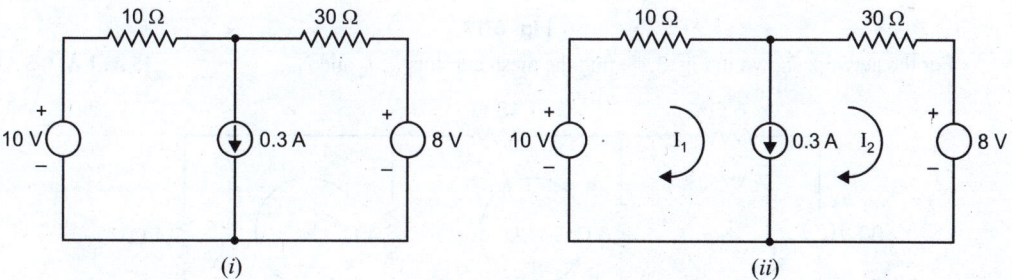

Fig. 3.14

Solution. Suppose voltage across current source is *v*. Assign mesh currents I_1 and I_2 in the meshes 1 and 2 respectively as shown in Fig. 3.14 (*ii*).

Mesh 1. Applying *KVL* to this mesh, we have,

$$10 - 10I_1 + v = 0 \qquad ...(i)$$

Mesh 2. Applying *KVL* to this mesh, we have,

$$- 30I_2 - 8 - v = 0 \qquad ...(ii)$$

Adding eqs. (*i*) and (*ii*), $2 - 10I_1 - 30I_2 = 0 \qquad ...(iii)$

Also current in the branch containing current source is

$$I_1 - I_2 = 0.3 \qquad ...(iv)$$

From eqs. (*iii*) and (*iv*), $I_1 = 0.275$ A ; $I_2 = -0.025A$

∴ Current in 10Ω = $I_1 = \mathbf{0.275A}$

Current in 30Ω = $I_2 = \mathbf{-0.025\ A}$

Current in current source = $I_1 - I_2 = 0.275 - (-0.025) = \mathbf{0.3A}$

Note that negative sign means current is in the opposite direction to that assumed in the circuit.

Tutorial Problems

1. Use mesh analysis to find the current in each resistor in Fig. 3.15.

   · [in 100 Ω = 0·1 A from *L* to *R* ; in 20 Ω = 0·4 A from *R* to *L* ; in 10 Ω = 0·5 A downward]

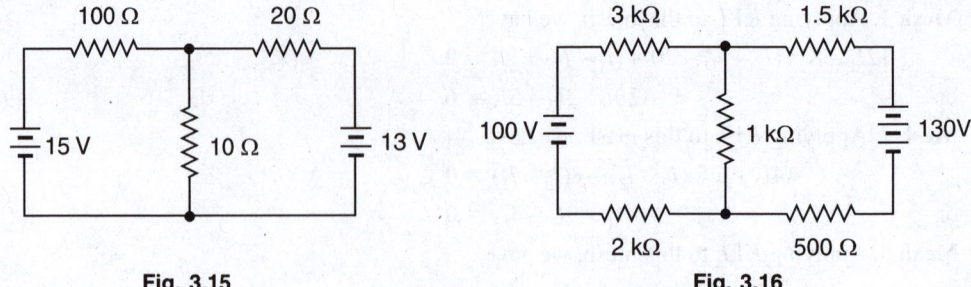

Fig. 3.15   Fig. 3.16

2. Using mesh analysis, find the voltage drop across the 1 kΩ resistor in Fig. 3.16.   [50 V]

3. Using mesh analysis, find the currents in 50 Ω, 250 Ω and 100 Ω resistors in the circuit shown in Fig. 3.17. $[I(50\,\Omega) = 0.171\,\text{A} \rightarrow ;\ I(250\,\Omega) = 0.237\,\text{A} \leftarrow ;\ I(100\,\Omega) = 0.408\,\text{A}\downarrow]$

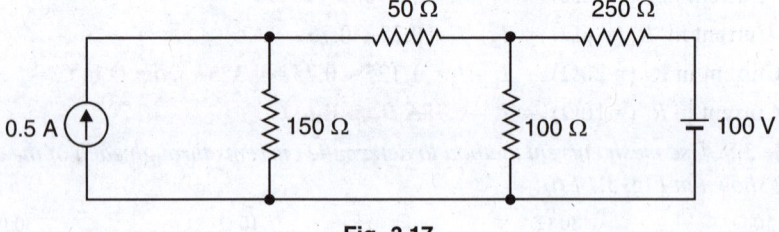

Fig. 3.17

4. For the network shown in Fig. 3.18, find the mesh currents I_1, I_2 and I_3.   [5 A, 1 A, 0.5 A]

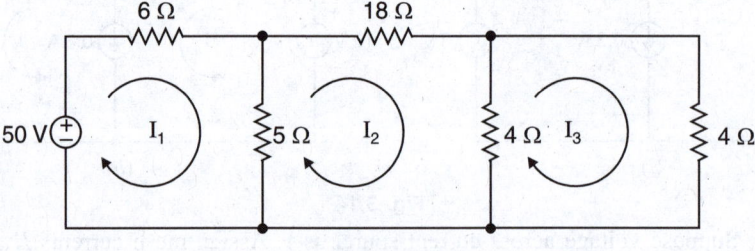

Fig. 3.18

5. In the network shown in Fig. 3.19, find the magnitude and direction of current in the various branches by mesh current method. $[FAB = 4\,\text{A} ;\ BF = 3\,\text{A} ;\ BC = 1\,\text{A} ;\ EC = 2\,\text{A} ;\ CDE = 3\,\text{A}]$

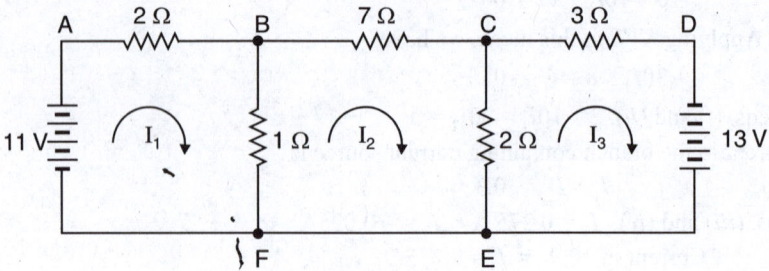

Fig. 3.19

3.6. Nodal Analysis

Consider the circuit shown in Fig. 3.20. The branch currents in the circuit can be found by Kirchhoff's laws or Maxwell's mesh current method. There is another method, called *nodal analysis* for determining branch currents in a circuit. In this method, one of the nodes (Remember a node is a point in a network where two or more circuit elements meet) is taken as the *reference node*. The potentials of all the points in the circuit are measured w.r.t. this reference node. In Fig. 3.20, A, B, C and D are four nodes and the node D has been taken as the *reference node. The fixed-voltage nodes are called *dependent nodes*. Thus in Fig. 3.20, A and C are fixed nodes because $V_A = E_1 = 120$ V and $V_C = 65$ V. The voltage from D to B is V_B and its magnitude depends upon the parameters of circuit elements and the currents through these elements. Therefore, node B is called *independent node*. Once we calculate the potential at the independent node (or nodes), each branch current can be determined because the voltage across each resistor will then be known.

Hence **nodal analysis** *essentially aims at choosing a reference node in the network and then finding the unknown voltages at the independent nodes w.r.t. reference node*. For a circuit containing N nodes, there will be $N-1$ node voltages, some of which may be known if voltage sources are present.

Circuit analysis. The circuit shown in Fig. 3.20 has only one independent node B. Therefore, if we find the voltage V_B at the independent node B, we can determine all branch currents in the circuit. We can express each current in terms of e.m.f.s, resistances (or conductances) and the voltage V_B at node B. Note that we have taken point D as the reference node.

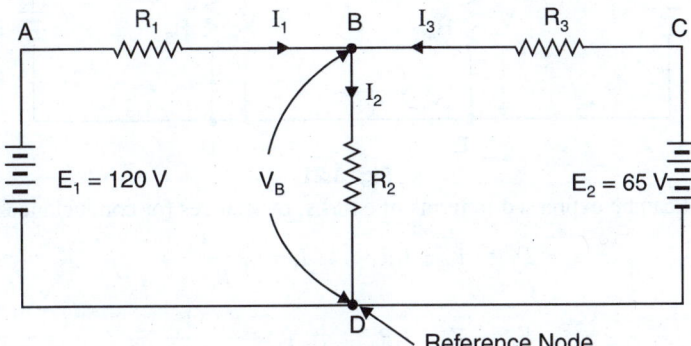

Fig. 3.20

The voltage V_B can be found by applying **Kirchhoff's current law at node B.

$$I_1 + I_3 = I_2 \qquad \qquad \text{...(i)}$$

In mesh $ABDA$, the voltage drop across R_1 is $E_1 - V_B$.

$$\therefore \qquad I_1 = \frac{E_1 - V_B}{R_1}$$

In mesh $CBDC$, the voltage drop across R_3 is $E_2 - V_B$.

$$\therefore \qquad I_3 = \frac{E_2 - V_B}{R_3}$$

Also $$I_2 = \frac{V_B}{R_2}$$

Putting the values of I_1, I_2 and I_3 in eq. (i), we get,

$$\frac{E_1 - V_B}{R_1} + \frac{E_2 - V_B}{R_3} = \frac{V_B}{R_2} \qquad \qquad \text{...(ii)}$$

All quantities except V_B are known. Hence V_B can be found out. Once V_B is known, all branch currents can be calculated. It may be seen that nodal analysis requires only one equation [eq. (ii)] for determining the branch currents in this circuit. However, Kirchhoff's or Maxwell's solution would have needed two equations.

---

* An obvious choice would be ground or common, if such a point exists.

** Since the circuit unknowns are voltages, the describing equations are obtained by applying *KCL* at the nodes.

Notes.

(*i*) We can mark the directions of currents at will. If the value of any current comes out to be negative in the solution, it means that actual direction of current is opposite to that of assumed.

(*ii*) We can also express the currents in terms of conductances.

$$I_1 = \frac{E_1 - V_B}{R_1} = (E_1 - V_B)G_1 \; ; \; I_2 = \frac{V_B}{R_2} = V_B G_2 \; ; \; I_3 = \frac{E_2 - V_B}{R_3} = (E_2 - V_B)G_3$$

3.7. Nodal Analysis with Two Independent Nodes

Fig. 3.21 shows a network with two independent nodes B and C. We take node D (or E) as the reference node. We shall use Kirchhoff's current law for nodes B and C to find V_B and V_C. Once the values of V_B and V_C are known, we can find all the branch currents in the network.

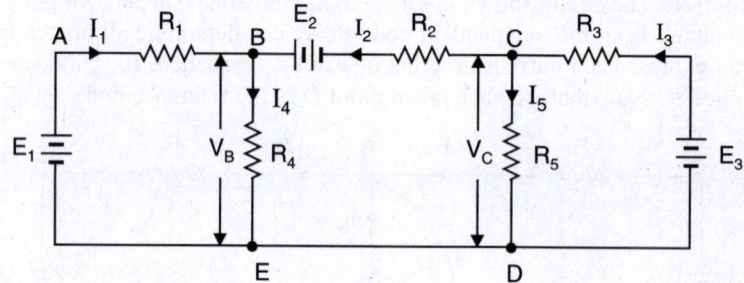

Fig. 3.21

Each current can be expressed in terms of e.m.f.s, resistances (or conductances), V_B and V_C.

$$E_1 = V_B + I_1 R_1 \quad \therefore I_1 = \frac{E_1 - V_B}{R_1}$$

$$E_3 = V_C + I_3 R_3 \quad \therefore I_3 = \frac{E_3 - V_C}{R_3}$$

$$E_2^* = V_B - V_C + I_2 R_2 \quad \therefore I_2 = \frac{E_2 - V_B + V_C}{R_2}$$

Similarly,
$$I_4 = \frac{V_B}{R_4} \; ; I_5 = \frac{V_C}{R_5}$$

At node B.
$$I_1 + I_2 = I_4$$

or
$$\frac{E_1 - V_B}{R_1} + \frac{E_2 - V_B + V_C}{R_2} = \frac{V_B}{R_4} \qquad \ldots(i)$$

At node C.
$$I_2 + I_5 = I_3$$

or
$$\frac{E_2 - V_B + V_C}{R_2} + \frac{V_C}{R_5} = \frac{E_3 - V_C}{R_3} \qquad \ldots(ii)$$

From eqs. (*i*) and (*ii*), we can find V_B and V_C since all other quantities are known. Once we know the values of V_B and V_C, we can find all the branch currents in the network.

Note.. We can also express currents in terms of conductances as under :

$$I_1 = (E_1 - V_B)\, G_1 \; ; \; I_2 = (E_2 - V_B + V_C)\, G_2$$

$$I_3 = (E_3 - V_C)\, G_3 \; ; \; I_4 = V_B\, G_4 \; ; \; I_5 = V_C\, G_5$$

---

* As we go from C to B, we have,

$$V_C - I_2 R_2 + E_2 = V_B$$

$\therefore$
$$E_2 = V_B - V_C + I_2 R_2$$

Example 3.9. *Find the currents in the various branches of the circuit shown in Fig. 3.22 by nodal analysis.*

Solution. Mark the currents in the various branches as shown in Fig. 3.22. If the value of any current comes out to be negative in the solution, it means that actual direction of current is opposite to that of assumed. Take point E (or F) as the reference node. We shall find the voltages at nodes B and C.

At node B. $\qquad I_2 + I_3 = I_1$

or $\qquad \dfrac{V_B}{10} + \dfrac{*V_B - V_C}{15} = \dfrac{100 - V_B}{20}$

or $\qquad 13V_B - 4V_C = 300$ $\qquad\qquad$ ...(i)

At node C. $\qquad I_4 + I_5 = I_3$

or $\qquad \dfrac{V_C}{10} + \dfrac{V_C + 80}{10} = \dfrac{V_B - V_C}{15}$

or $\qquad V_B - 4V_C = 120$ $\qquad\qquad$ ...(ii)

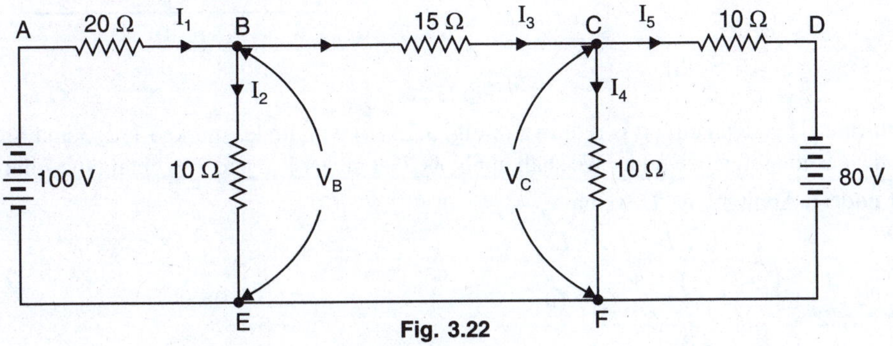

Fig. 3.22

Subtracting eq. (*ii*) from eq. (*i*), we get, $12V_B = 180$ $\quad\therefore\quad V_B = 180/12 = 15$ V

Putting $V_B = 15$ volts in eq. (*i*), we get, $V_C = -26.25$ volts.

By determinant method

$$13V_B - 4V_C = 300$$
$$V_B - 4V_C = 120$$

$$\therefore\qquad V_B = \frac{\begin{vmatrix} 300 & -4 \\ 120 & -4 \end{vmatrix}}{\begin{vmatrix} 13 & -4 \\ 1 & -4 \end{vmatrix}} = \frac{(300 \times -4) - (120 \times -4)}{(13 \times -4) - (1 \times -4)} = \frac{-720}{-48} = 15 \text{ V}$$

and $$V_C = \frac{\begin{vmatrix} 13 & 300 \\ 1 & 120 \end{vmatrix}}{\text{Denominator}} = \frac{(13 \times 120) - (1 \times 300)}{-48} = \frac{1260}{-48} = -26.25 \text{ V}$$

$\therefore\qquad$ Current $I_1 = \dfrac{100 - V_B}{20} = \dfrac{100 - 15}{20} = $ **4·25 A**

$\qquad$ Current $I_2 = V_B/10 = 15/10 = $ **1·5 A**

$\qquad$ Current $I_3 = \dfrac{V_B - V_C}{15} = \dfrac{15 - (-26.25)}{15} = $ **2·75 A**

---

* Note that the current I_3 is assumed to flow from B to C. Therefore, with this assumption, $V_B > V_C$.

Current $I_4 = V_C/10 = -26 \cdot 25/10 = -2 \cdot 625$ A

Current $I_5 = \dfrac{V_C + 80}{10} = \dfrac{-26.25 + 80}{10} = 5 \cdot 375$ A

The negative sign for I_4 shows that actual current flow is opposite to that of assumed.

Example 3.10. *Use nodal analysis to find the currents in various resistors of the circuit shown in Fig. 3.23 (i).*

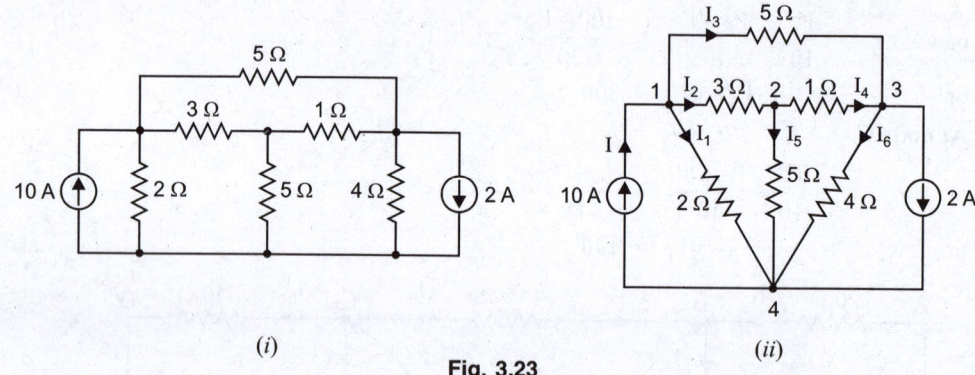

(i)   (ii)

Fig. 3.23

Solution. The given circuit is redrawn in Fig. 3.23 (*ii*) with nodes marked 1, 2, 3 and 4. Let us take node 4 as the reference node. We shall apply *KCL* at nodes 1, 2 and 3 to obtain the solution.

At node 1. Applying *KCL*, we have,

$$I_1 + I_2 + I_3 = I$$

or $\qquad \dfrac{V_1}{2} + \dfrac{V_1 - V_2}{3} + \dfrac{V_1 - V_3}{5} = 10$

or $\qquad 31V_1 - 10V_2 - 6V_3 = 300$ $\qquad\qquad$ ...(*i*)

At node 2. Applying *KCL*, we have,

$$I_2 = I_4 + I_5$$

or $\qquad \dfrac{V_1 - V_2}{3} = \dfrac{V_2 - V_3}{1} + \dfrac{V_2}{5}$

or $\qquad 5V_1 - 23V_2 + 15V_3 = 0$ $\qquad\qquad$ ...(*ii*)

At node 3. Applying *KCL*, we have,

$$I_3 + I_4 = I_6 + 2$$

or $\qquad \dfrac{V_1 - V_3}{5} + \dfrac{V_2 - V_3}{1} = \dfrac{V_3}{4} + 2$

or $\qquad 4V_1 + 20V_2 - 29V_3 = 40$ $\qquad\qquad$ ...(*iii*)

From eqs. (*i*), (*ii*) and (*iii*), $V_1 = \dfrac{6572}{545}$ V ; $V_2 = \dfrac{556}{109}$ V ; $V_3 = \dfrac{2072}{545}$ V

$\therefore \qquad$ Current $I_1 = \dfrac{V_1}{2} = \dfrac{6572}{545} \times \dfrac{1}{2} =$ **6.03 A**

$\qquad\qquad$ Current $I_2 = \dfrac{V_1 - V_2}{3} = \dfrac{1}{3}\left[\dfrac{6572}{545} - \dfrac{556}{109}\right] =$ **2.32A**

$\qquad\qquad$ Current $I_3 = \dfrac{V_1 - V_3}{5} = \dfrac{1}{5}\left[\dfrac{6572}{545} - \dfrac{2072}{545}\right] =$ **1.65 A**

$$\text{Current } I_4 = \frac{V_2 - V_3}{1} = \frac{556}{109} - \frac{2072}{545} = \textbf{1.3A}$$

$$\text{Current } I_5 = \frac{V_2}{5} = \frac{556}{109} \times \frac{1}{5} = \textbf{1.02A}$$

$$\text{Current } I_6 = \frac{V_3}{4} = \frac{2072}{545} \times \frac{1}{4} = \textbf{0.95A}$$

Example 3.11. *Find the total power consumed in the circuit shown in Fig. 3.24.*

Solution. Mark the direction of currents in the various branches as shown in Fig. 3.24. Take D as the reference node. If voltages V_B and V_C at nodes B and C respectively are known, then all the currents can be calculated.

At node B. $\qquad\qquad I_1 + I_3 = I_2$

or $\qquad \dfrac{15 - V_B}{1} + \dfrac{V_C - V_B}{0.5} = \dfrac{V_B}{1}$

or $\qquad 15 - V_B + 2(V_C - V_B) - V_B = 0$

or $\qquad 4V_B - 2V_C = 15$ $\qquad\qquad\qquad\qquad\qquad$ ...(i)

At node C. $\qquad\qquad I_3 + I_4 = I_5$

or $\qquad \dfrac{V_C - V_B}{0.5} + \dfrac{V_C}{2} = \dfrac{20 - V_C}{1}$

or $\quad 2(V_C - V_B) + 0.5V_C - (20 - V_C) = 0$

or $\qquad 3.5V_C - 2V_B = 20$

or $\qquad 4V_B - 7V_C = -40$ $\qquad\qquad\qquad\qquad\qquad$ ...(ii)

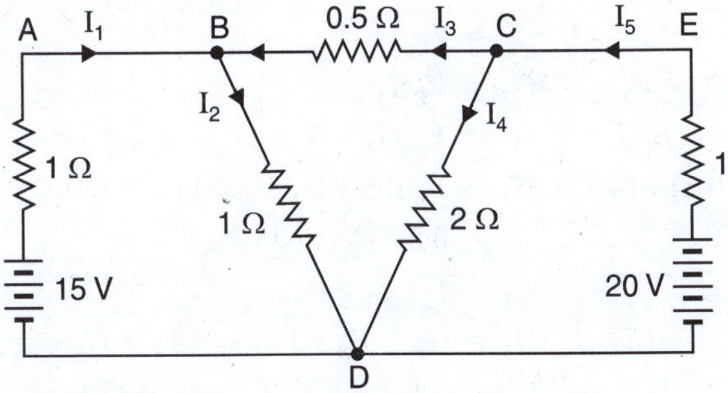

Fig. 3.24

Subtracting eq. (*ii*) from eq. (*i*), we get, $5V_C = 55$

$\therefore \qquad\qquad V_C = 55/5 = 11$ volts

Putting $V_C = 11$ V in eq. (*i*), we get, $V_B = 9.25$ V

$\therefore \qquad \text{Current } I_1 = \dfrac{15 - V_B}{1} = \dfrac{15 - 9.25}{1} = 5.75 \text{ A}$

$\text{Current } I_2 = V_B/1 = 9.25/1 = 9.25 \text{ A}$

$\text{Current } I_3 = \dfrac{V_C - V_B}{0.5} = \dfrac{11 - 9.25}{0.5} = 3.5 \text{ A}$

$\text{Current } I_4 = V_C/2 = 11/2 = 5.5 \text{ A}$

$\text{Current } I_5 = \dfrac{20 - V_C}{1} = \dfrac{20 - 11}{1} = 9 \text{ A}$

$\therefore$ Power loss in the circuit $= I_1^2 \times 1 + I_2^2 \times 1 + I_3^2 \times 0.5 + I_4^2 \times 2 + I_5^2 \times 1$

$$= (5.75)^2 \times 1 + (9.25)^2 \times 1 + (3.5)^2 \times 0.5 + (5.5)^2 \times 2 + (9)^2 \times 1$$

$$= \mathbf{266.25\ W}$$

Example 3.12. *Using nodal analysis, find node-pair voltages V_B and V_C and branch currents in the circuit shown in Fig. 3.25. Use conductance method.*

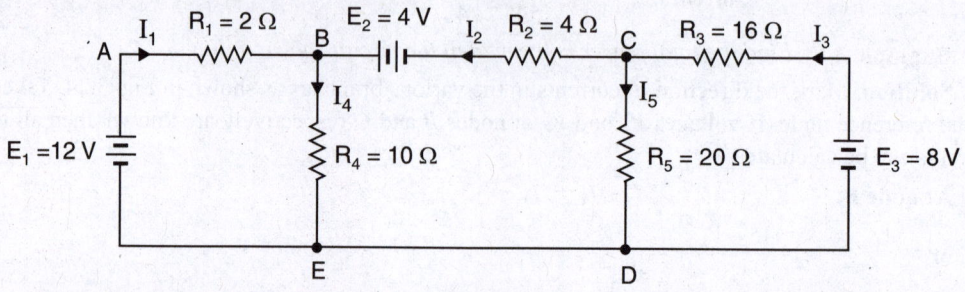

Fig. 3.25

Solution. Mark the currents in the various branches as shown in Fig. 3.25. If the value of any current comes out to be negative in the solution, it means that actual direction of current is opposite to that of assumed. Take point D (or E) as the reference node. We shall find the voltages at nodes B and C and hence the branch currents.

$$G_1 = \frac{1}{R_1} = \frac{1}{2} = 0.5\ \text{S}\ ;\ G_2 = \frac{1}{R_2} = \frac{1}{4} = 0.25\ \text{S}\ ;\ G_3 = \frac{1}{R_3} = \frac{1}{16} = 0.0625\ \text{S}\ ;$$

$$G_4 = \frac{1}{R_4} = \frac{1}{10} = 0.1\ \text{S}\ ;\quad G_5 = \frac{1}{R_5} = \frac{1}{20} = 0.05\ \text{S}$$

At node B. $I_1 + I_2 = I_4$

or $(E_1 - V_B)G_1 + (E_2 - V_B + V_C)G_2 = V_B G_4$

or $E_1 G_1 + E_2 G_2 = V_B(G_1 + G_2 + G_4) - V_C G_2$

or $(12 \times 0.5) + (4 \times 0.25) = V_B(0.5 + 0.25 + 0.1) - V_C \times 0.25$

or $7 = 0.85\ V_B - 0.25\ V_C$ ...(i)

At node C. $I_3 = I_2 + I_5$

or $(E_3 - V_C)G_3 = (E_2 - V_B + V_C)G_2 + V_C \times G_5$

or   . $E_3 G_3 - E_2 G_2 = -V_B G_2 + V_C(G_2 + G_3 + G_5)$

or $(8 \times 0.0625) - (4 \times 0.25) = -V_B(0.25) + V_C(0.25 + 0.0625 + 0.05)$

or $-0.5 = -0.25\ V_B + 0.362\ V_C$ ...(ii)

From equations (i) and (ii), we get, $V_B = \mathbf{9.82\ V}$; $V_C = \mathbf{5.4V}$

$\therefore$ $I_1 = (E_1 - V_B)G_1 = (12 - 9.82) \times 0.5 = \mathbf{1.09\ A}$

$I_2 = (E_2 - V_B + V_C)G_2 = (4 - 9.82 + 5.4) \times 0.25 = \mathbf{-0.105A}$

$I_3 = (E_3 - V_C)G_3 = (8 - 5.4) \times 0.0625 = \mathbf{0.162A}$

$I_4 = V_B G_4 = 9.82 \times 0.1 = \mathbf{0.982A}$

$I_5 = V_C G_5 = 5.4 \times 0.05 = \mathbf{0.27A}$

The negative sign for I_2 means that the actual direction of this current is opposite to that shown in Fig. 3.25.

Example 3.13. *Using nodal analysis, find the different branch currents in the circuit shown in* Fig. 3.26 *(i).*

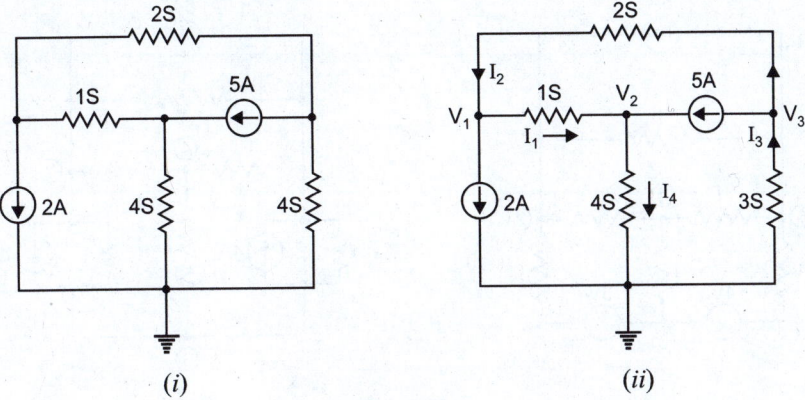

Fig. 3.26

Solution. Mark the currents in the various branches as shown in Fig. 3.26 (*ii*). Take ground as the reference node. We shall find the voltages at the other three nodes.

At first node. Applying *KCL* to the first node from left,

$$I_2 = I_1 + 2$$

or $\qquad (V_3 - V_1)2 = (V_1 - V_2)1 + 2$

or $\qquad 3V_1 - V_2 - 2V_3 = -2$ $\qquad\qquad\qquad$ ...(*i*)

At second node. Applying *KCL* to the second node from left,

$$I_1 + 5 = I_4$$

or $\qquad (V_1 - V_2)1 + 5 = V_2 \times 4$

or $\qquad V_1 - 5V_2 = -5$ $\qquad\qquad\qquad\qquad$ ...(*ii*)

At third node. Applying *KCL* to the third node from left,

$$I_3 = 5 + I_2$$

or $\qquad -V_3 \times 3 = 5 + (V_3 - V_1)2$

or $\qquad 2V_1 - 5V_3 = 5$ $\qquad\qquad\qquad\qquad$ ...(*iii*)

Solving eqs. (*i*), (*ii*) and (*iii*), we have, $V_1 = -\dfrac{3}{2} V$; $V_2 = \dfrac{7}{10} V$ and $V_3 = \dfrac{-8}{5} V$

$\therefore \qquad I_1 = (V_1 - V_2)1 = \left(-\dfrac{3}{2} - \dfrac{7}{10}\right)1 = \textbf{-2.2A}$

$$I_2 = (V_3 - V_1)2 = \left(-\dfrac{8}{5} + \dfrac{3}{2}\right)2 = \textbf{-0.2A}$$

$$I_3 = -V_3 \times 3 = \dfrac{8}{5} \times 3 = \textbf{4.8 A}$$

$$I_4 = V_2 \times 4 = \dfrac{7}{10} \times 4 = \textbf{2.8A}$$

The negative value of any current means that actual direction of current is opposite to that originally assumed.

Example 3.14. *Find the current I in Fig. 3.27 (i) by changing the two voltage sources into their equivalent current sources and then using nodal method. All resistances are in ohms.*

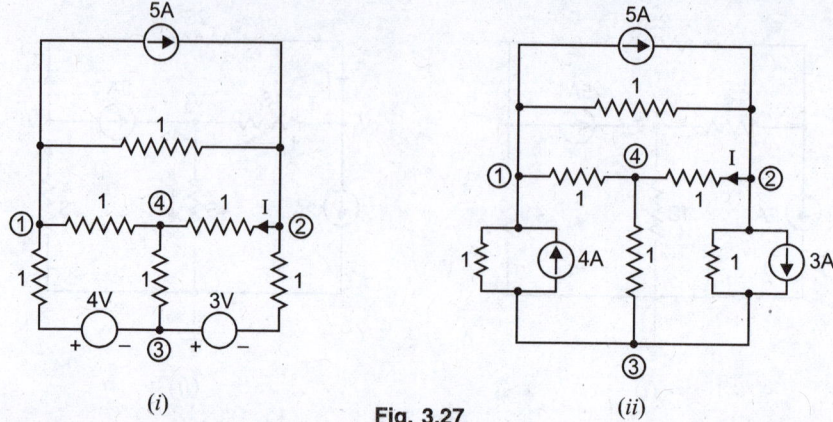

Fig. 3.27

(i)   (ii)

Solution. Since we are to find *I*, it would be convenient to take node 4 as the reference node. The two voltage sources are converted into their equivalent current sources as shown in Fig. 3.27. (*ii*). We shall apply *KCL* at nodes 1, 2 and 3 in Fig. 3.27 (*ii*) to obtain the required solution.

At node 1. Applying *KCL*, we have,

$$\frac{V_3 - V_1}{1} + 4 = \frac{V_1}{1} + \frac{V_1 - V_2}{1} + 5$$

or $3V_1 - V_2 - V_3 = -1$ ...(*i*)

At node 2. Applying *KCL*, we have,

$$5 + \frac{V_1 - V_2}{1} = \frac{V_2}{1} + \frac{V_2 - V_3}{1} + 3$$

or $V_1 - 3V_2 + V_3 = -2$ ...(*ii*)

At node 3. Applying *KCL*, we have,

$$\frac{V_2 - V_3}{1} + 3 - \frac{V_3}{1} = \frac{V_3 - V_1}{1} + 4$$

or $V_1 + V_2 - 3V_3 = 1$ ...(*iii*)

From eqs. (*i*), (*ii*) and (*iii*), we get, $V_2 = 0.5$ V.

∴   Current $I = \dfrac{V_2 - 0}{1} = \dfrac{0.5 - 0}{1} = $ **0.5A**

Example 3.15. *Use nodal analysis to find the voltage across and current through 4 Ω resistor in Fig. 3.28 (i).*

Solution. We must first convert the 2V voltage source to an equivalent current source. The value of the equivalent current source is $I = 2V/2\Omega = 1$ A. The circuit then becomes as shown in Fig. 3.28 (*ii*).

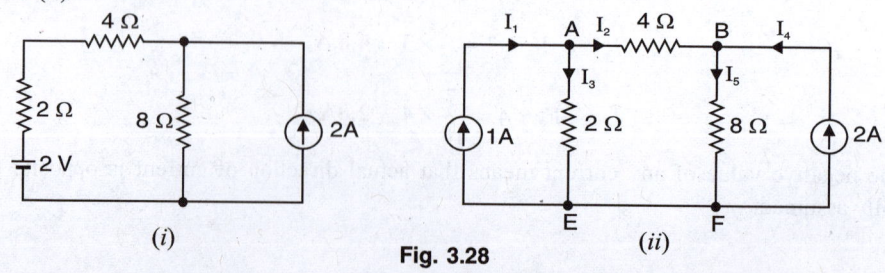

(i)   (ii)

Fig. 3.28

Mark the currents in the various branches as shown in Fig. 3.28 (*ii*). Take point E (or F) as the reference node. We shall calculate the voltages at nodes A and B.

At node A. $I_1 = I_2 + I_3$

or $1 = \dfrac{*V_A - V_B}{4} + \dfrac{V_A}{2}$

or $3V_A - V_B = 4$ ...(*i*)

At node B. $I_2 + I_4 = I_5$

or $\dfrac{V_A - V_B}{4} + 2 = \dfrac{V_B}{8}$

or $2V_A - 3V_B = -16$ ...(*ii*)

Solving equations (*i*) and (*ii*), we find $V_A = 4$V and $V_B = 8$V. Note that $V_B > V_A$, contrary to our initial assumption. Therefore, actual direction of current is from node B to node A.

By determinant method

$$3V_A - V_B = 4$$
$$2V_A - 3V_B = -16$$

$$\therefore \quad V_A = \frac{\begin{vmatrix} 4 & -1 \\ -16 & -3 \end{vmatrix}}{\begin{vmatrix} 3 & -1 \\ 2 & -3 \end{vmatrix}} = \frac{(-12) - (16)}{(-9) - (-2)} = \frac{-28}{-7} = 4V$$

$$V_B = \frac{\begin{vmatrix} 3 & 4 \\ 2 & -16 \end{vmatrix}}{\text{Denominator}} = \frac{(-48) - (8)}{-7} = \frac{-56}{-7} = 8V$$

Voltage across 4Ω resistor = $V_B - V_A = 8 - 4 = $ **4V**

Current through 4Ω resistor = $\dfrac{4V}{4Ω} = $ **1A**

We can also find the currents in other resistors.

$$I_3 = \frac{V_A}{2} = \frac{4}{2} = 2A$$

$$I_5 = \frac{V_B}{8} = \frac{8}{8} = 1A$$

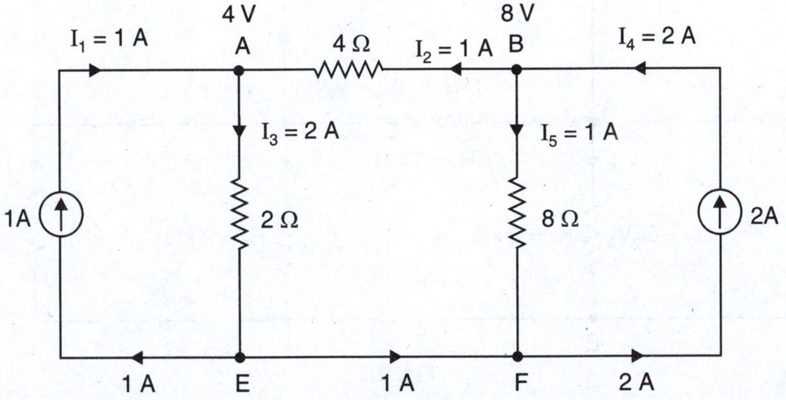

Fig. 3.29

---

*   We assume that $V_A > V_B$. On solving the circuit, we shall see whether this assumption is correct or not.

Fig. 3.29 shows the various currents in the circuit. You can verify Kirchhoff's current law at each node.

Example 3.16. *Use nodal analysis to find current in the 4 kΩ resistor shown in Fig. 3.30.*

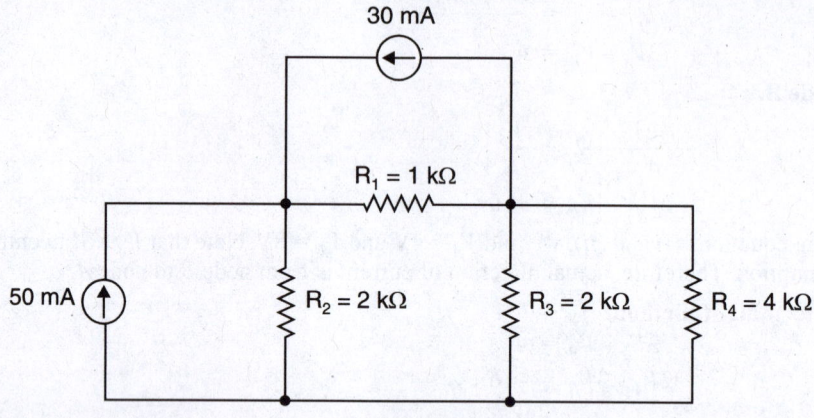

Fig. 3.30

Solution. We shall solve this example by expressing node currents in terms of conductance than expressing them in terms of resistance. The conductance of each resistor is

$$G_1 = \frac{1}{R_1} = \frac{1}{1 \times 10^3} = 10^{-3} \text{ S} \; ; \; G_2 = \frac{1}{R_2} = \frac{1}{2 \times 10^3} = 0.5 \times 10^{-3} \text{ S}$$

$$G_3 = \frac{1}{R_3} = \frac{1}{2 \times 10^3} = 0.5 \times 10^{-3} \text{ S} \; ; \; G_4 = \frac{1}{R_4} = \frac{1}{4 \times 10^3} = 0.25 \times 10^{-3} \text{ S}$$

Mark the currents in the various branches as shown in Fig. 3.31. Take point E (or F) as the reference node. We shall find voltages at nodes A and B.

At node A. $I_5 + I_6 = I_1 + I_2$

or $50 \times 10^{-3} + 30 \times 10^{-3} = G_1(V_A - V_B) + G_2 V_A$

or $80 \times 10^{-3} = 10^{-3}(V_A - V_B) + 0.5 \times 10^{-3} V_A$

or $1.5 V_A - V_B = 80$ ...(i)

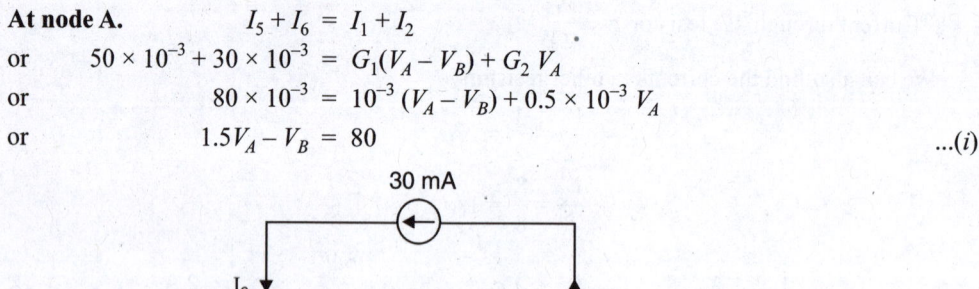

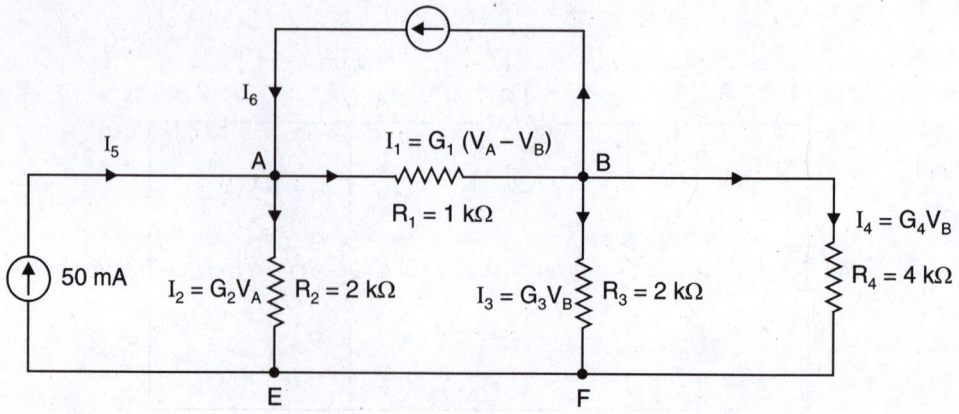

Fig. 3.31

At node B. $I_1 = I_6 + I_3 + I_4$

or $G_1(V_A - V_B) = 30 \times 10^{-3} + G_3 V_B + G_4 V_B$

or $\qquad 10^{-3}\,(V_A - V_B) = 30 \times 10^{-3} + 0.5 \times 10^{-3}\,V_B + 0.25 \times 10^{-3}\,V_B$

or $\qquad V_A - 1.75\,V_B = 30 \qquad\qquad\qquad\qquad\qquad\qquad ...(ii)$

Solving equations (*i*) and (*ii*), we get, $V_B = 21.54$ V.

By determinant method

$$1.5\,V_A - V_B = 80$$

$$V_A - 1.75\,V_B = 30$$

$\therefore \qquad V_B = \dfrac{\begin{vmatrix} 1.5 & 80 \\ 1 & 30 \end{vmatrix}}{\begin{vmatrix} 1.5 & -1 \\ 1 & -1.75 \end{vmatrix}} = \dfrac{(45)-(80)}{(-2.625)-(-1)} = \dfrac{-35}{-1.625} = 21.54$ V

$\therefore$ Current in 4 kΩ resistor, $I_4 = G_4 V_B = 0.25 \times 10^{-3} \times 21.54 = 5.39 \times 10^{-3}$ A = **5·39 mA**

Example 3.17. *For the circuit shown in Fig. 3.32 (i), find (i) voltage v and (ii) current through 2Ω resistor using nodal method.*

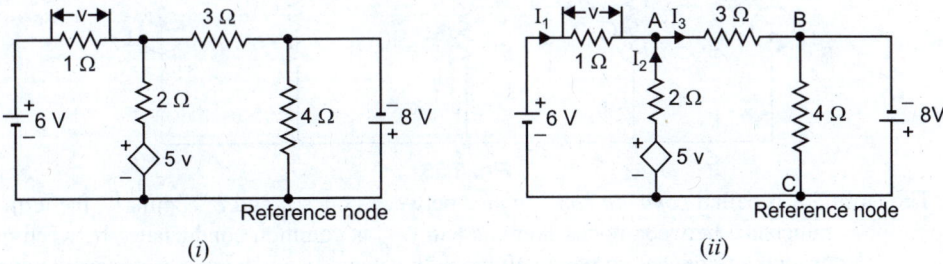

(i)   (ii)

Fig. 3.32

Solution. Mark the direction of currents in the various branches as shown in Fig. 3.32 (*ii*). Let us take node C as the reference node. It is clear from Fig. 3.32 (*ii*) that $V_B = -8$V ($\because V_C = 0$V). Also, $v = 6 - V_A$.

Applying *KCL* to node A, we have,

$$I_1 + I_2 = I_3$$

or $\qquad \dfrac{6 - V_A}{1} + \dfrac{5v - V_A}{2} = \dfrac{V_A - V_B}{3}$

or $\qquad \dfrac{6 - V_A}{1} + \dfrac{5(6 - V_A) - V_A}{2} = \dfrac{V_A - (-8)}{3}$

On solving, we get, $V_A = \dfrac{55}{13}$ V

(*i*) Voltage $v = 6 - V_A = 6 - \dfrac{55}{13} = \dfrac{23}{13}$ V

(*ii*) Current through 2Ω, $I_2 = \dfrac{5v - V_A}{2} = \dfrac{5(23/13) - (55/13)}{2} = \dfrac{30}{13}$ A

3.8. Shortcut Method for Nodal Analysis

There is a shortcut method for writing node equations similar to the form for mesh equations. Consider the circuit with three independent nodes A, B and C as shown in Fig. 3.33.

The node equations in shortcut form for nodes A, B and C can be written as under :

$$V_A G_{AA} + V_B G_{AB} + V_C G_{AC} = I_A$$
$$V_A G_{BA} + V_B G_{BB} + V_C G_{BC} = I_B$$
$$V_A G_{CA} + V_B G_{CB} + V_C G_{CC} = I_C$$

Let us discuss the various terms in these equations.

$$G_{AA} = \text{Sum of all conductances connected to node } A$$
$$= G_1 + G_2 \text{ in Fig. 3.33.}$$

The term G_{AA} is called *self-conductance* at node A. Similarly, G_{BB} and G_{CC} are self-conductances at nodes B and C respectively. *Note that product of node voltage at a node and self-conductance at that node is always a **positive** quantity.* Thus $V_A G_{AA}$, $V_B G_{BB}$ and $V_C G_{CC}$ are all positive.

$$G_{AB} = \text{Sum of all conductances } directly \text{ connected}$$
$$\text{between nodes } A \text{ and } B$$
$$= G_2 \text{ in Fig. 3.33}$$

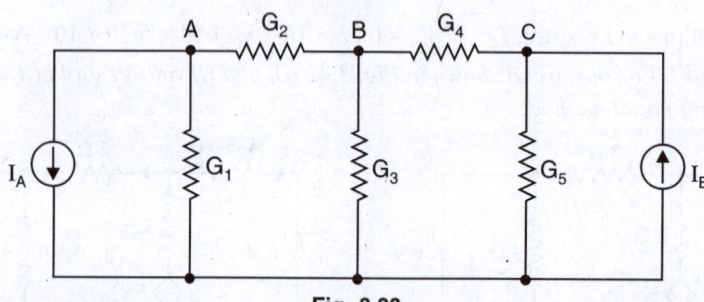

Fig. 3.33

The term G_{AB} is called *common conductance* between nodes A and B. Similarly, the term G_{BC} is common conductance between nodes B and C and G_{CA} is common conductance between nodes C and A. *The product of connecting node voltage with common conductance is always a **negative** quantity.* Thus $V_B G_{AB}$ is a negative quantity. Here connecting node voltage is V_B and common conductance is G_{AB}. Note that $G_{AB} = G_{BA}$, $G_{AC} = G_{CA}$ and so on.

Note the direction of current provided by current source connected to the node. *A current leaving the node is shown as negative and a current entering a node is positive. If a node has no current source connected to it, set the term equal to zero.*

Node A. Refer to Fig. 3.33. At node A, $G_{AA} = G_1 + G_2$ and is a positive quantity. The product $V_B G_{AB}$ is a negative quantity. The current I_A is leaving the node A and will be assigned a negative sign. Therefore, node equation at node A is

$$V_A G_{AA} - V_B G_{AB} = -I_A$$
or $$V_A(G_1 + G_2) - V_B(G_2) = -I_A$$

Similarly, for **nodes B and C,** the node equations are :

$$V_B(G_2 + G_3 + G_4) - V_A(G_2) - V_C(G_4) = 0$$
$$V_C(G_4 + G_5) - V_B(G_4) = I_B$$

Example 3.18. *Solve the circuit shown in Fig. 3.34 using nodal analysis.*

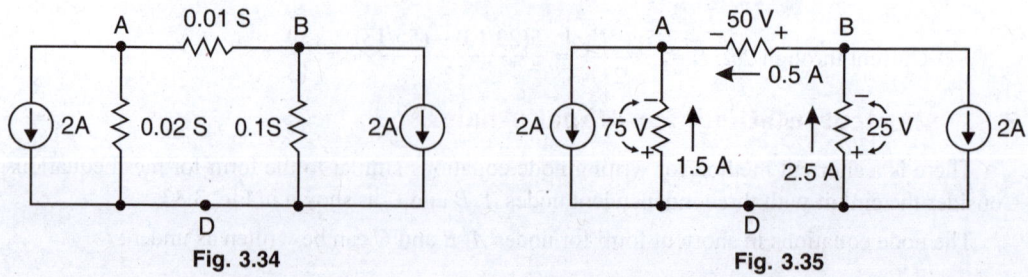

Fig. 3.34   Fig. 3.35

Solution. Here point D is chosen as the reference node and A and B are the independent nodes.

Node A. $V_A(0.02 + 0.01) - V_B(0.01) = -2$

or $\qquad\qquad 0.03\, V_A - 0.01\, V_B = -2$ ...(i)

Node B. $V_B(0.01 + 0.1) - V_A(0.01) = -2$

or $\qquad\qquad -0.01\, V_A + 0.11\, V_B = -2$ ...(ii)

From equations (i) and (ii), we have, $V_A = -75\text{V}$ and $V_B = -25\text{V}$

Fig. 3.35 shows the circuit redrawn with solved voltages.

$$\text{Current in } 0.02\ \text{S} = VG = 75 \times 0.02 = \textbf{1·5A}$$
$$\text{Current in } 0.1\ \text{S} = VG = 25 \times 0.1 = \textbf{2·5A}$$
$$\text{Current in } 0.01\ \text{S} = VG = 50 \times 0.01 = \textbf{0·5 A}$$

The directions of currents will be as shown in Fig. 3.35.

Example 3.19. *Solve the circuit shown in Fig. 3.36 using nodal analysis.*

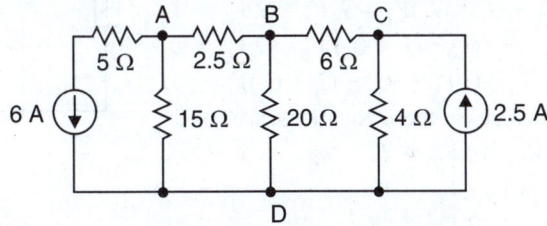

Fig. 3.36

Solution. Here A, B and C are the independent nodes and D is the reference node.

Node A. $\qquad\qquad V_A^*\left(\dfrac{1}{15} + \dfrac{1}{2.5}\right) - V_B\left(\dfrac{1}{2.5}\right) = -6$

or $\qquad\qquad 0.467\, V_A - 0.4\, V_B = -6$ ...(i)

Node B. $\quad V_B\left(\dfrac{1}{2.5} + \dfrac{1}{20} + \dfrac{1}{6}\right) - V_A\left(\dfrac{1}{2.5}\right) - V_C\left(\dfrac{1}{6}\right) = 0$

or $\qquad\qquad -0.4\, V_A + 0.617\, V_B - 0.167\, V_C = 0$ ...(ii)

Node C. $\qquad\qquad V_C\left(\dfrac{1}{6} + \dfrac{1}{4}\right) - V_B\left(\dfrac{1}{6}\right) = 2.5$

or $\qquad\qquad -0.167\, V_B + 0.417\, V_C = 2.5$ ...(iii)

From equations (i), (ii) and (iii), $V_A = -30\ \text{V}$; $V_B = -20\ \text{V}$; $V_C = -2\ \text{V}$

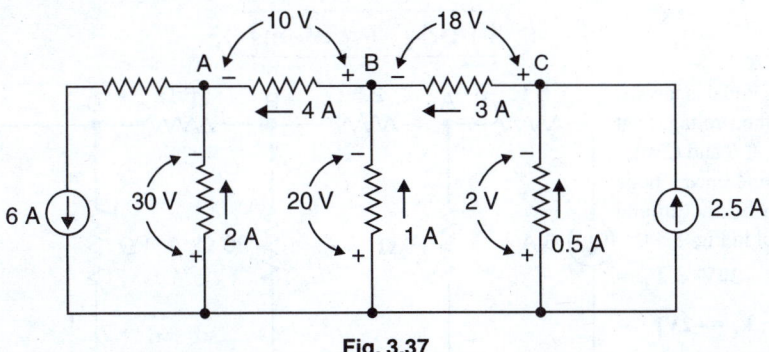

Fig. 3.37

---

* Note that 5Ω is omitted from the equation for node A because it is in series with the current source.

Fig. 3.37 shows the circuit redrawn with solved voltages.

$$\text{Current in } 15\ \Omega = 30/15 = \mathbf{2\ A}$$
$$\text{Current in } 20\ \Omega = 20/20 = \mathbf{1\ A}$$
$$\text{Current in } 4\ \Omega = 2/4 = \mathbf{0.5\ A}$$
$$\text{Current in } 6\ \Omega = 18/6 = \mathbf{3\ A}$$
$$\text{Current in } 2.5\ \Omega = 10/2.5 = \mathbf{4\ A}$$
$$\text{Current in } 5\ \Omega = 4 + 2 = \mathbf{6\ A}$$

The directions of currents will be as shown in Fig. 3.37.

Example 3.20. *Find the value of I_x in the circuit shown in Fig. 3.38 using nodal analysis. The various values are :*

$G_u = 10\ S\ ;\ G_v = 1S\ ;\ G_w = 2S\ ;$
$G_x = 1S\ ;\ G_y = 1S\ ;\ G_z = 1S$ and $I = 100\ A.$

Solution.

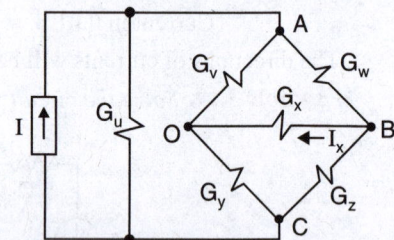

Fig. 3.38

Node A. $(G_u + G_v + G_w)V_A - G_wV_B - G_uV_C = I$
Node B. $- G_wV_A + (G_w + G_x + G_z)\,V_B - G_zV_C = 0$
Node C. $- G_uV_A - G_zV_B + (G_u + G_y + G_z)V_C = -I$

Putting the various values in these equations, we have,

$$13\,V_A - 2\,V_B - 10\,V_C = I$$
$$-2\,V_A + 4\,V_B - V_C = 0$$
$$-10\,V_A - V_B + 12\,V_C = -I$$

Now V_B can be calculated as the ratio of two determinants N_B/D where

$$D = \begin{vmatrix} 13 & -2 & -10 \\ -2 & 4 & -1 \\ -10 & -1 & 12 \end{vmatrix} = 624 - 20 - 20 - (400 + 48 + 13) = 123$$

and

$$N_B = \begin{vmatrix} 13 & I & -10 \\ -2 & 0 & -1 \\ -10 & -I & 12 \end{vmatrix} = 10I - 20I - (13I - 24I) = I$$

∴

$$V_B = \frac{N_B}{D} = \frac{I}{123}$$

$$\text{Current } I_x = G_xV_B = 1 \times \frac{I}{123} = 1 \times \frac{100}{123} = \mathbf{0.813A}$$

Tutorial Problems

1. Using nodal analysis, find the voltages at nodes A, B and C w.r.t. the reference node shown by the ground symbol in Fig. 3.39.

   $[V_A = -30V\ ;\ V_B = -20V\ ;\ V_C = -2V]$

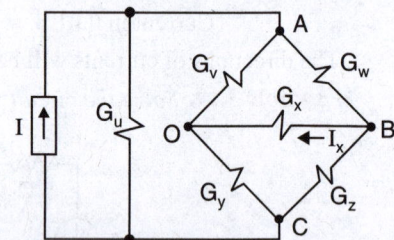

Fig. 3.39

2. Using nodal analysis, find the current through 0.05 S conductance in Fig. 3.40. **[0.264 A]**

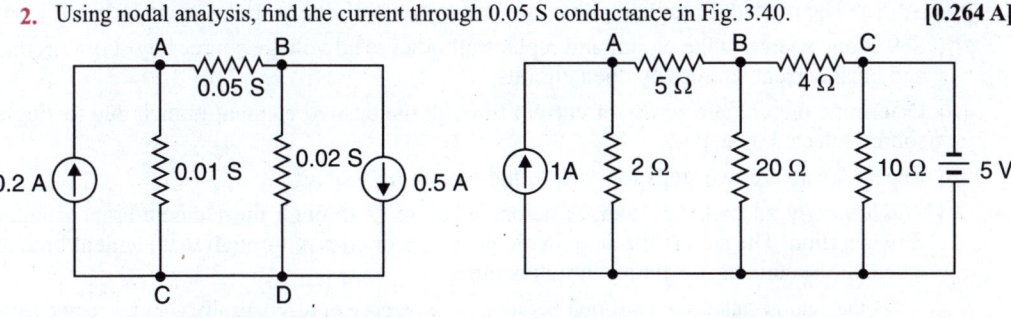

Fig. 3.40   **Fig. 3.41**

3. Using nodal analysis, find the current flowing in the battery in Fig. 3.41. **[1.21 A]**

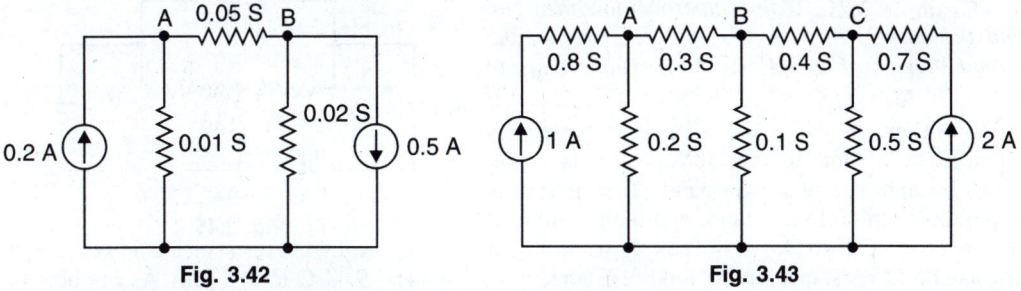

Fig. 3.42   **Fig. 3.43**

4. In Fig. 3.42, find the node voltages. $[V_A = -6.47 \text{ V}; V_B = -11.8\text{V}]$

5. In Fig. 3.42, find current through 0.05 S conductance. Use nodal analysis. **[264 mA]**

6. In Fig. 3.43, find the node voltages. $[V_A = 4.02 \text{ V}; V_B = 3.37 \text{ V}; V_C = 3.72 \text{ V}]$

7. By using nodal analysis, find current in 0.3 S in Fig. 3.43. **[196 mA]**

8. Using nodal analysis, find current in 0.4 S conductance in Fig. 3.43. **[141 mA]**

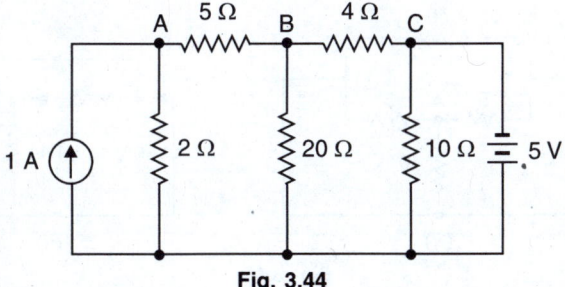

Fig. 3.44

9. Find node voltages in Fig. 3.44. $[V_A = 0.806 \text{ V}; V_B = -2.18 \text{ V}; V_C = -5 \text{ V}]$

10. Using nodal analysis, find current through the battery in Fig. 3.44. **[1. 21A]**

3.9. Superposition Theorem

Superposition is a general principle that allows us to determine the effect of several energy sources (voltage and current sources) acting simultaneously in a circuit by considering the effect of each source acting alone, and then combining (superposing) these effects. This theorem as applied to d.c. circuits may be stated as under :

In a linear, bilateral d.c. network containing more than one energy source, the resultant potential difference across or current through any element is equal to the algebraic sum of potential differences or currents for that element produced by each source acting alone with all other independent ideal voltage sources replaced by short circuits and all other independent ideal current sources replaced by open circuits (non-ideal sources are replaced by their internal resistances).

Procedure. The procedure for using this theorem to solve d.c. networks is as under :

(*i*) Select one source in the circuit and replace all other ideal voltage sources by short circuits and ideal current sources by open circuits.

(*ii*) Determine the voltage across or current through the desired element/branch due to single source selected in step (*i*).

(*iii*) Repeat the above two steps for each of the remaining sources.

(*iv*) *Algebraically* add all the voltages across or currents through the element/branch under consideration. The sum is the actual voltage across or current through that element/branch when all the sources are acting simultaneously.

Note. This theorem is called *superposition* because we superpose or algebraically add the components (currents or voltages) due to each independent source acting alone to obtain the total current in or voltage across a circuit element.

Example 3.21. *Using superposition theorem, find the current through the 40 Ω resistor in the circuit shown in Fig. 3.45 (i). All resistances are in ohms.*

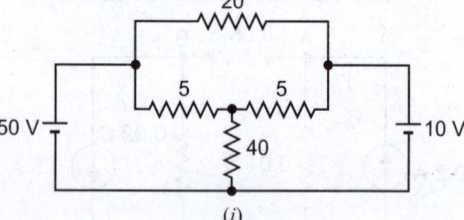

(*i*)

Fig. 3.45

Solution. In Fig. 3.45 (*ii*), 10V battery is replaced by a short so that 50V battery is acting alone. It can be seen that right-hand 5 Ω resistance is in parallel with 40 Ω resistance and their combined resistance = 5 Ω || 40 Ω = 4.44 Ω as shown in Fig. 3.45 (*iii*). The 4.44 Ω resistance is in series with left-hand 5 Ω resistance giving total resistance of (5 + 4.44) = 9.44 Ω to this path. As can be seen from Fig. 3.45 (*iii*), there are two parallel branches of resistances 20 Ω and 9.44 Ω across the 50 V battery. Therefore, current through 9.44 Ω branch is $I = 50/9.44 = 5.296$ A. Thus in Fig. 3.45 (*ii*), the current $I (= 5.296$ A) at point A divides between 5 Ω resistance and 40 Ω resistance. By current-divider rule, current I_1 in 40 Ω resistance is

$$I_1 = I \times \frac{5}{5+40} = 5.296 \times \frac{5}{45} = 0.589 \text{ A downward}$$

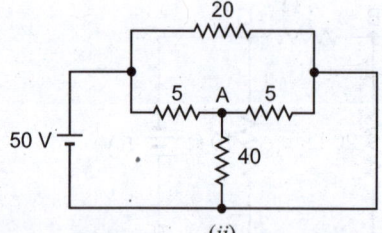

(*ii*)

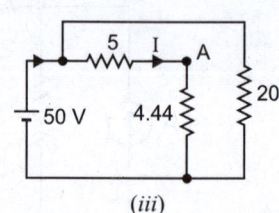

(*iii*)

Fig. 3.45

In Fig. 3.45 (*iv*), the 50 V battery is replaced by a short so that 10 V battery is acting alone. Again, there are two parallel branches of resistances 20 Ω and 9.44 Ω across the 10V battery [See Fig. 3.45 (*v*)]. Therefore, current through 9.44 Ω branch is $I = 10/9.44 = 1.059$ A.

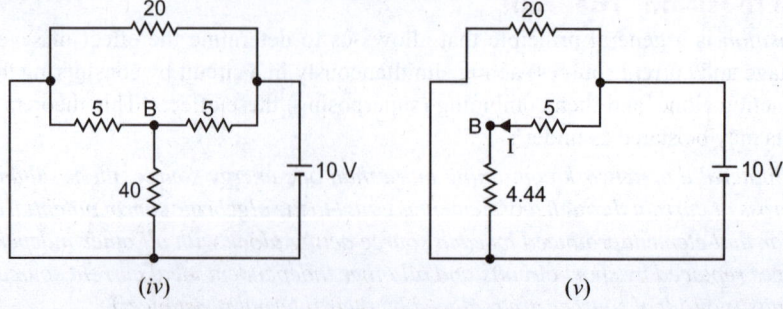

(*iv*)   (*v*)

Fig. 3.45

Thus in Fig. 3.45 (*iv*), the current I (= 1.059 A) at point B divides between 5 Ω resistance and 40 Ω resistance. By current-divider rule, current in 40 Ω resistance is

$$I_2 = 1.059 \times \frac{5}{5+40} = 0.118 \text{ A downward}$$

∴ By superposition theorem, the total current in 40 Ω

$$= I_1 + I_2 = 0.589 + 0.118 = \textbf{0.707 A downward}$$

Example 3.22. *In the circuit shown in Fig. 3.46 (i), the internal resistances of the batteries are* 0·12 Ω *and* 0·08 Ω. *Calculate (i) current in load (ii) current supplied by each battery.*

Solution. In Fig. 3.46 (*ii*), the right-hand 12 V source is replaced by its internal resistance so that left-hand battery of 12 V is acting alone. The various branch currents due to left-hand battery of 12 V alone [See Fig. 3.46 (*ii*)] are :

$$\text{Total circuit resistance} = 0.12 + \frac{0.08 \times 0.5}{0.08 + 0.5} = 0 \cdot 189 \text{ Ω}$$

$$\text{Total circuit current, } I_1' = 12/0 \cdot 189 = 63 \cdot 5 \text{ A}$$

$$\text{Current in } 0 \cdot 08 \text{ Ω, } I_2' = 63.5 \times \frac{0.5}{0.08 + 0.5} = 54 \cdot 74 \text{ A}$$

$$\text{Current in } 0 \cdot 5 \text{ Ω, } I_3' = 63.5 \times \frac{0.08}{0.08 + 0.5} = 8 \cdot 76 \text{ A}$$

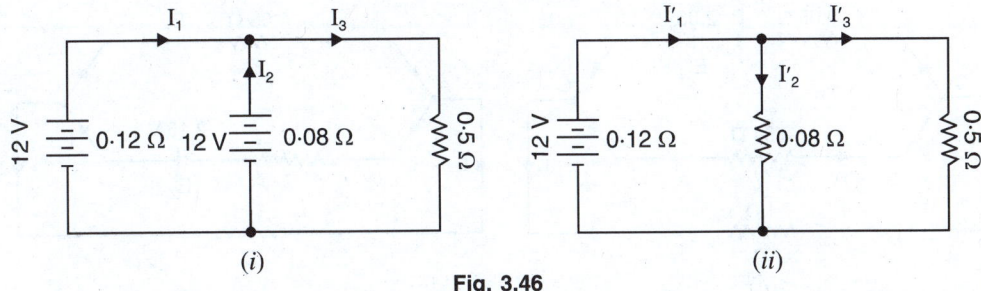

(*i*) (*ii*)

Fig. 3.46

In Fig. 3.46 (*iii*), left-hand 12 V source is replaced by its internal resistance so that now right-hand 12 V source is acting alone.

$$\text{Total circuit resistance} = 0.08 + \frac{0.12 \times 0.5}{0.12 + 0.5}$$

$$= 0 \cdot 177 \text{ Ω}$$

$$\text{Total circuit current, } I_2'' = 12/0 \cdot 177 = 67 \cdot 8 \text{ A}$$

$$\text{Current in } 0 \cdot 12 \text{ Ω, } I_1'' = 67.8 \times \frac{0.5}{0.12 + 0.5}$$

$$= 54 \cdot 6 \text{ A}$$

$$\text{Current in } 0 \cdot 5 \text{ Ω, } I_3'' = 67.8 \times \frac{0.12}{0.12 + 0.5} = 13 \cdot 12 \text{ A}$$

(*iii*)

Fig. 3.46

The actual current values of I_1 (current in first battery), I_2 (current in second battery) and I_3 (load current) can be found by algebraically adding the component values.

$$I_1 = I_1' - I_1'' = 63 \cdot 5 - 54 \cdot 6 = \textbf{8·9 A}$$

$$I_2 = I_2'' - I_2' = 67 \cdot 8 - 54 \cdot 74 = \textbf{13·06 A}$$

$$I_3 = I_3' + I_3'' = 8 \cdot 76 + 13 \cdot 12 = \textbf{21·88 A}$$

Example 3.23. *By superposition theorem, find the current in resistance R in Fig. 3.47 (i).*

Solution. In Fig. 3.47 (*ii*), battery E_2 is replaced by a short so that battery E_1 is acting alone. It is clear that resistances of 1Ω (= R) and 0.04Ω are in parallel across points A and C.

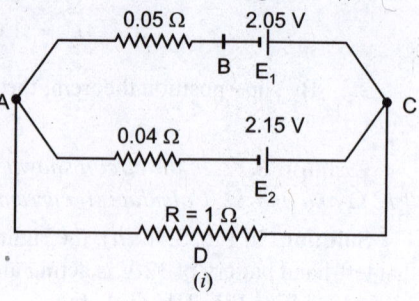

$$\therefore \quad R_{AC} = 1\Omega \parallel 0.04\Omega = \frac{1 \times 0.04}{1 + 0.04} = 0.038\ \Omega$$

This resistance (*i.e.*, R_{AC}) is in series with $0.05\ \Omega$.

Total resistance to battery $E_1 = 0.038 + 0.05 = 0.088\Omega$

$\therefore$ Current supplied by battery E_1 is

$$I = \frac{E_1}{0.088} = \frac{2.05}{0.088} = 23.2A$$

Fig. 3.47

The current I(= 23.2A) is divided between the parallel resistances of 1Ω (= R) and 0.04Ω.

$\therefore$ Current in 1Ω (= R) resistance is

$$I_1 = 23.2 \times \frac{0.04}{1 + 0.04} = 0.892\ \text{A from } C \text{ to } A$$

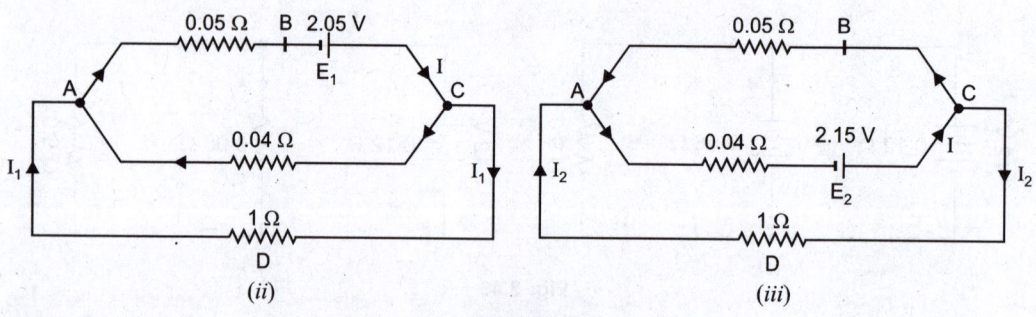

Fig. 3.47

In Fig. 3.47 (*iii*), battery E_1 is replaced by a short so that battery E_2 is acting alone.

Total resistance offered to battery E_2

$$= (1\Omega \parallel 0.05\Omega) + 0.04\Omega$$

$$= \frac{1 \times 0.05}{1 + 0.05} + 0.04 = 0.088\Omega$$

$\therefore$ Current supplied by battery E_2 is

$$I = \frac{2.15}{0.088} = 24.4A$$

The current I(= 24.4A) is divided between two parallel resistances of 1Ω (= R) and 0.05Ω.

$\therefore$ Current in 1Ω (= R) resistance is

$$I_2 = 24.4 \times \frac{0.05}{1 + 0.05} = 1.16A \text{ from } C \text{ to } A$$

$\therefore$ Current through 1Ω resistance when both batteries are present

$$= I_1 + I_2 = 0.892 + 1.16 = \textbf{2.052A}$$

Example 3.24. *Using the superposition principle, find the voltage across 1kΩ resistor in Fig. 3·48. Assume the sources to be ideal.*

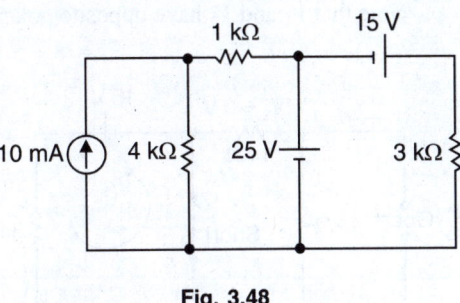

Fig. 3.48

Solution. (*i*) The voltage across 1kΩ resistor due to **current source acting alone** is found by replacing 25-V and 15-V sources by short circuit as shown in Fig. 3.49 (*i*). Since 3 kΩ resistor is shorted out, the current in 1 kΩ resistor is, by current divider rule,

$$I_{1 k\Omega} = \left(\frac{4}{1+4}\right)10 = 8 \text{ mA}$$

∴   Voltage V_1 across 1 kΩ resistor is

$$V_1 = (8 \text{ mA})(1 \text{ k}\Omega) = {}^+8V^-$$

The + and − symbols indicate the polarity of the voltage due to current source acting alone as shown in Fig. 3.49 (*i*).

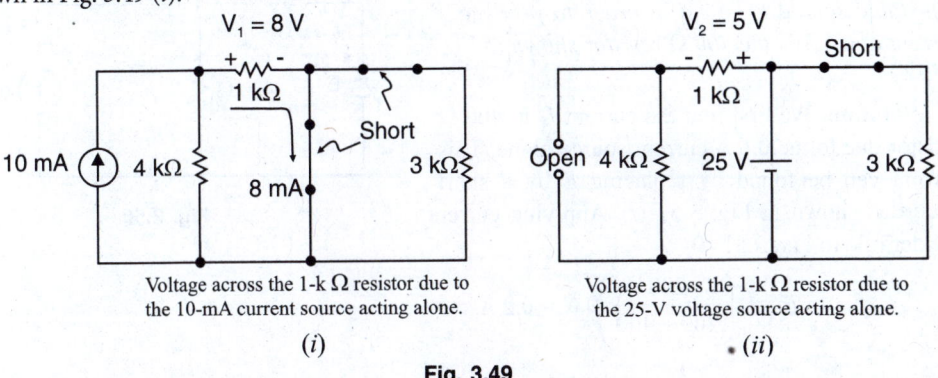

Voltage across the 1-k Ω resistor due to the 10-mA current source acting alone.

(*i*)

Voltage across the 1-k Ω resistor due to the 25-V voltage source acting alone.

(*ii*)

Fig. 3.49

(*ii*) The voltage across the 1 k Ω resistor due to **25 V source acting alone** is found by replacing the 10 mA current source by an open circuit and 15 V source by a short circuit as shown in Fig. 3.49 (*ii*). Since the 25 V source is across the series combination of the 1 kΩ and 4 k Ω resistors, the voltage V_2 across 1 kΩ resistor can be found by the voltage divider rule.

∴

$$V_2 = \left(\frac{1}{4+1}\right)25 = {}^-5V^+$$

Note that 3 kΩ resistor has no effect on this computation.

(*iii*) The voltage V_3 across 1 kΩ resistor due to **15 V source acting alone** is found by replacing the 25 V source by a short circuit and the 10 mA current source by an open circuit as shown in Fig. 3.49 (*iii*). The short circuit prevents any current from flowing in the 1 k Ω resistor.

∴ $$V_3 = 0$$

(*iv*) Applying superposition principle, the voltage across the 1kΩ resistor due to all the three sources acting simultaneously [See Fig. 3.49 (*iv*)] is

$$V_{1 k\Omega} = V_1 + V_2 + V_3$$
$$= {}^+8 \text{ V}^- + {}^-5 \text{ V}^+ + 0 \text{ V}$$
$$= {}^+\mathbf{3} \textbf{ V}^-$$

Note that V_1 and V_2 have opposite polarities so that the sum (net) voltage is actually

$$= 8 - 5 = 3 \text{ V}$$

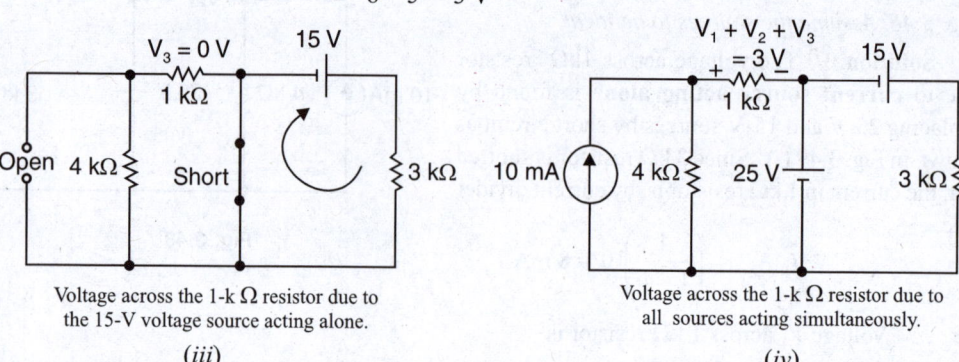

Voltage across the 1-k Ω resistor due to
the 15-V voltage source acting alone.

(iii)

Voltage across the 1-k Ω resistor due to
all sources acting simultaneously.

(iv)

Fig. 3.49

Example 3.25. *To what voltage should adjustable source E be set in order to produce a current of 0.3 A in the 400 Ω resistor shown in Fig. 3.50?*

Solution. We first find the current I_1 in 400 Ω resistor due to the 0.6 A current source alone. This current can be found by replacing E by a short circuit as shown in Fig. 3.51 (*i*). Applying current divider rule to Fig. 3.51 (*i*),

$$I_1 = \left(\frac{200}{200 + 400} \right) 0.6 = 0.2 \text{ A}$$

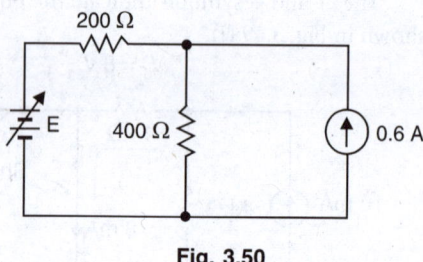

Fig. 3.50

(i)

(ii)

Fig. 3.51

In order that current in the 400 Ω resistor is equal to 0.3 A, the current produced in the resistor by the voltage source acting alone must be $= 0.3 - 0.2 = 0.1$ A. The current in the 400 Ω resistor due to voltage source alone can be calculated by open-circuiting the current source as shown in Fig 3.51 (*ii*). Referring to Fig. 3.51 (*ii*) and applying Ohm's law, we have,

$$I = \frac{E}{200 + 400} = \frac{E}{600}$$

or

$$0.1 = \frac{E}{600} \qquad \therefore E = 600 \times 0.1 = \textbf{60 V}$$

BCA. Applying *KVL*, we have,

$$-0.1 I_1 + (20 - I_1) \times 0.05 + 0.1 I_2 = 0$$
$$0.15 I_1 - 0.1 I_2 = 1 \qquad \qquad ...(ii)$$

quations (*i*) and (*ii*), we get, $I_2 = 40/7$A.

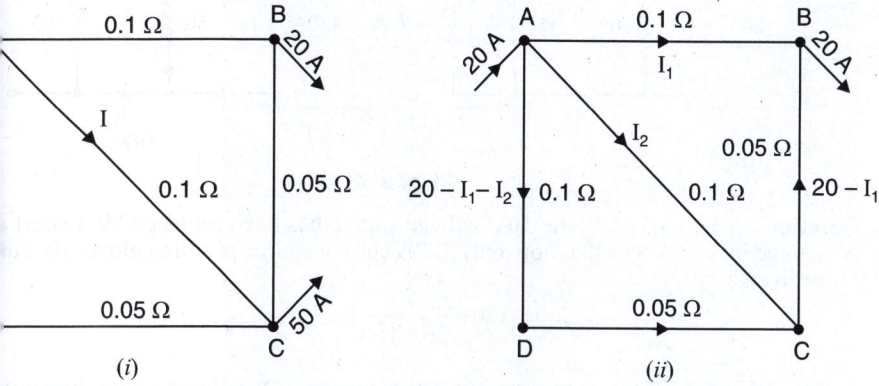

Fig. 3.53

er now 50 A load acting alone

and I_2' be the currents through *AB* and *AC* respectively. Then the current distribution will
 in Fig. 3.54 (*i*).

BCA. Applying *KVL*, we have,

$$-0.15 I_1' + 0.1 I_2' = 0$$
$$0.15 I_1' - 0.1 I_2' = 0 \qquad \qquad ...(iii)$$

DCA. Applying *KVL*, we have,

$$(50 - I_1' - I_2') \times 0.15 + 0.1 I_2' = 0$$
$$0.15 I_1' + 0.25 I_2' = 7.5 \qquad \qquad ...(iv)$$

quations (*iii*) and (*iv*), we get, $I_2' = 150/7$A.

er now 30A load acting alone

currents circulate as shown in Fig. 3.54 (*ii*). It is required to find I_2''.

BCA. Applying *KVL*, we have,

$$-0.15 I_1'' + 0.1 I_2'' = 0$$
$$0.15 I_1'' - 0.1 I_2'' = 0 \qquad \qquad ...(v)$$

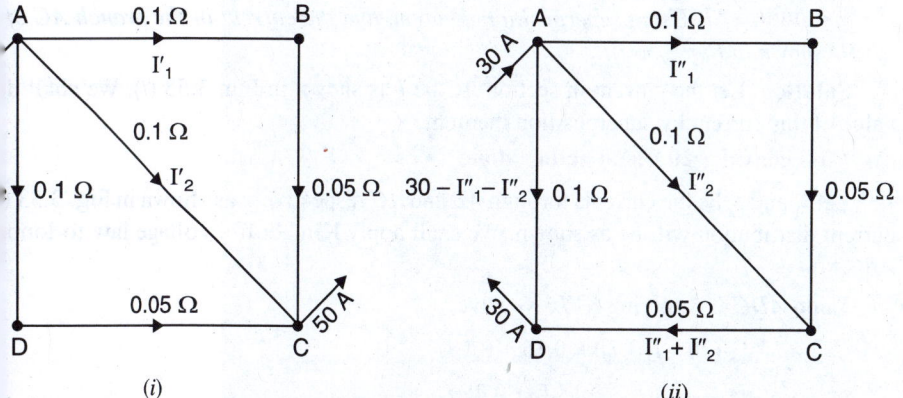

Fig. 3.54

Example 3.26. *Use superposition theorem to find current I in*
All resistances are in ohms.

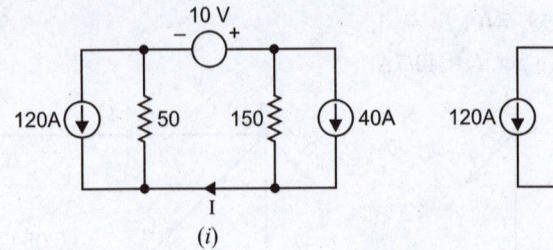

10 V

120A   50   150   40A   120A

I

(i)

Fig. 3.52

Solution. In Fig. 3.52 (*ii*), the 10V voltage source has been
current source by an open so that now only 120A current source is
rule, I_1 is given by ;

$$I_1 = 120 \times \frac{50}{50 + 150} = 30 \text{ A}$$

In Fig. 3.52 (*iii*), 40A current source is acting alone; 10 V v
short and 120A current source by an open. By current-divider rule

$$I_2 = 40 \times \frac{150}{50 + 150} = 30 \text{A}$$

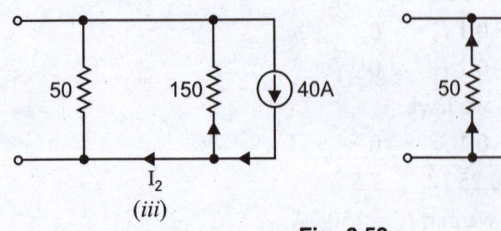

50   150   40A   50

I_2

(*iii*)

Fig. 3.52

In Fig. 3.52 (*iv*), 10V voltage source is acting alone. By Ohm

$$I_3 = \frac{10}{50 + 150} = 0.05 \text{A}$$

Currents I_1 and I_2, being equal and opposite, cancel out so tha

$$I = I_3 = \mathbf{0.05 \text{ A}}$$

Example 3.27. *Using superposition theorem, find the curren*
ABCD shown in Fig. 3.53 (i).

Solution. Let the current in section *AC* be *I* as shown in Fig
value of this current by superposition theorem.

First consider 20A load acting alone

Let I_1 and I_2 be the currents through *AB* and *AC* respectively a
current distribution will be as shown. We shall apply Kirchhoff's
ABCA.

Loop *ADCA*. Applying *KVL*, we have,

$$-(20 - I_1 - I_2) \times 0.15 + 0.1 I_2 = 0$$

or $0.15 I_1 + 0.25 I_2 = 3$

Loop

or

From

100 A

0.1 Ω

30 A

Cons

Let I_1
be as shov

Loop

or

Loop

or

From

Consi

Let th

Loop

or

50 A

$50 - I'_1 -$

Loop *ADCA*. Applying *KVL*, we have,

$$-(30 - I_1'' - I_2'') \times 0.1 + 0.05 (I_1'' + I_2'') + 0.1I_2'' = 0$$

or $\qquad\qquad\qquad\qquad 0.15 I_1'' + 0.25 I_2'' = 3$ \qquad\qquad\qquad ...(vi)

From equations (v) and (vi), we get, $I_2'' = 60/7$A.

According to superposition theorem, the total current in *AC* is equal to the algebraic sum of the component values.

$$I = I_2 + I_2' + I_2''$$
$$= 40/7 + 150/7 + 60/7$$
$$= 250/7 = \mathbf{35.7A}$$

Example 3.28. *Using superposition theorem, find the current in the each branch of the network shown in Fig. 3.55 (i).*

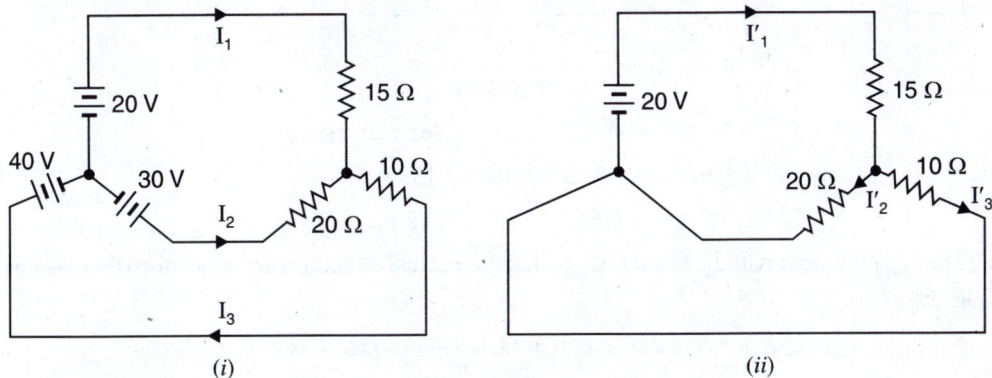

Fig. 3.55

Solution. Since there are three sources of e.m.f., three circuits [Fig. 3.55 (ii), Fig. 3.56 (i) and (ii)] are required for analysis by superposition theorem.

In Fig. 3.55 (ii), it is shown that only 20 V source is acting.

$$\text{Total resistance across source} = 15 + \frac{20 \times 10}{20 + 10} = 21.67\Omega$$

∴ $\qquad\qquad$ Total circuit current, $I_1' = 20/21.67 = 0.923$ A

$\qquad\qquad$ Current in 20 Ω, $I_2' = 0.923 \times 10/30 = 0.307$A

$\qquad\qquad$ Current in 10 Ω, $I_3' = 0.923 \times 20/30 = 0.616$ A

In Fig. 3.56 (i), only 40V source is acting in the circuit.

$$\text{Total resistance across source} = 10 + \frac{20 \times 15}{20 + 15} = 18.57\Omega$$

$\qquad\qquad$ Total circuit current, $I_3'' = 40/18.57 = 2.15$A

$\qquad\qquad$ Current in 20 Ω, $I_2'' = 2.15 \times 15/35 = 0.92$ A

$\qquad\qquad$ Current in 15 Ω, $I_1'' = 2.15 \times 20/35 = 1.23$ A

In Fig. 3.56 (ii), only 30 *V* source is acting in the circuit.

$\qquad\qquad$ Total resistance across source $= 20 + 10 \times 15/(10 + 15) = 26\ \Omega$

$\qquad\qquad$ Total circuit current, $I_2''' = 30/26 = 1.153$ A

$\qquad\qquad$ Current in 15 Ω, $I_1''' = 1.153 \times 10/25 = 0.461$ A

$\qquad\qquad$ Current in 10 Ω, $I_3''' = 1.153 \times 15/25 = 0.692$ A

The actual values of currents I_1, I_2 and I_3 shown in Fig. 3.55 (*i*) can be found by algebraically adding the component values.

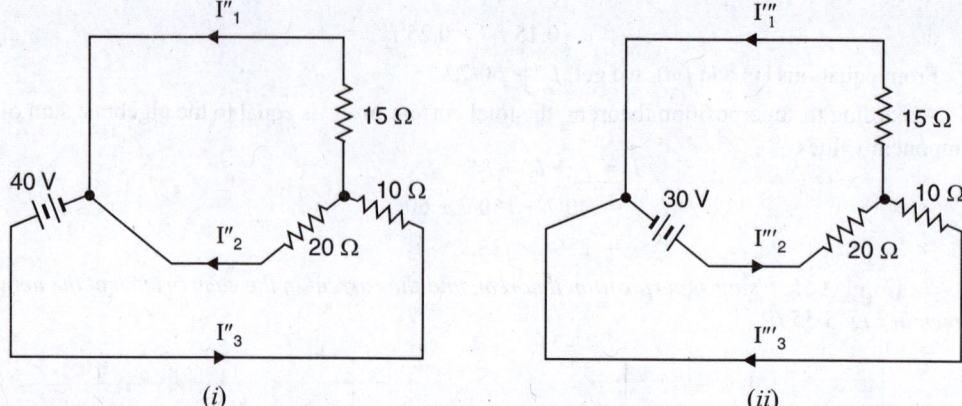

Fig. 3.56

$$I_1 = I_1' - I_1'' - I_1''' = 0.923 - 1.23 - 0.461 = \mathbf{-0.768\ A}$$

$$I_2 = -I_2' - I_2'' + I_2''' = -0.307 - 0.92 + 1.153 = \mathbf{-0.074\ A}$$

$$I_3 = I_3' - I_3'' + I_3''' = 0.616 - 2.15 + 0.692 = \mathbf{-0.842\ A}$$

The negative signs with I_1, I_2 and I_3 show that their actual directions are opposite to that assumed in Fig. 3.55 (*i*).

Example 3.29. *Use superposition theorem to find the voltage V in Fig. 3.57 (i).*

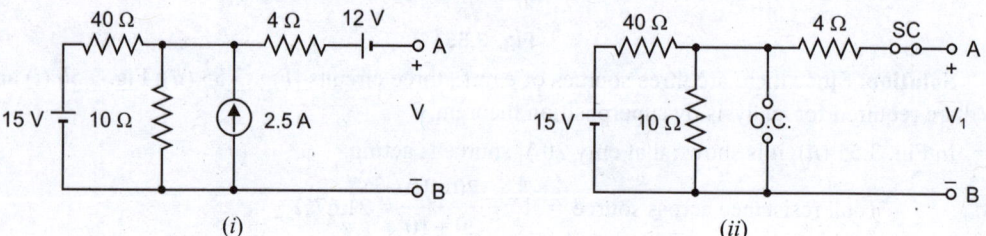

Fig. 3.57

Solution. In Fig. 3.57 (*ii*), 12 V battery is replaced by a short and 2.5A current source by an open so that 15V battery is acting alone. Therefore, voltage V_1 across open terminals A and B is

$$V_1 = \text{Voltage across } 10\Omega \text{ resistor}$$

By voltage-divider rule, V_1 is given by ;

$$V_1 = 15 \times \frac{10}{40 + 10} = 3V$$

In Fig. 3.57 (*iii*), 15 V and 12 V batteries are replaced by shorts so that 2.5A current source is acting alone. Therefore, voltage V_2 across open terminals A and B is

$$V_2 = \text{Voltage across } 10\ \Omega \text{ resistor}$$

By current-divider rule, current in $10\ \Omega = 2.5 \times \dfrac{40}{50} = 2A$

$$\therefore \qquad V_2 = 2 \times 10 = 20V$$

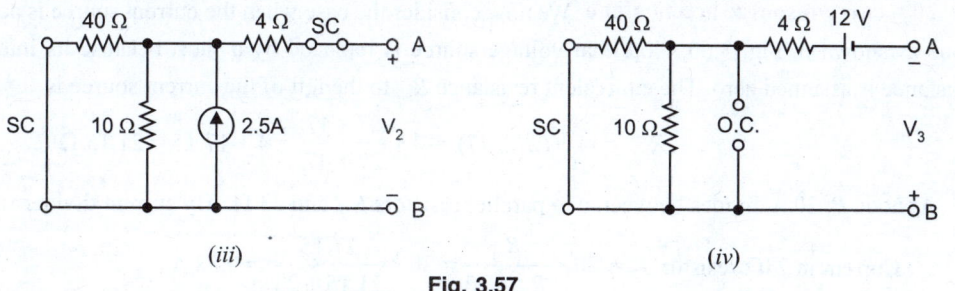

Fig. 3.57

In Fig. 3.57 (*iv*), 15 V battery is replaced by a short and 2.5 A current source by an open so that 12V battery is acting alone. Therefore, voltage V_3 across open terminals *A* and *B* is

$$V_3 = -*12V$$

The minus sign is given because the negative terminal of the battery is connected to point *A* and positive terminal to point *B*.

∴   Voltage across open terminals *AB* when all sources are present is

$$V = V_1 + V_2 + (-V_3) = 3 + 20 - 12 = \textbf{11V}$$

Example 3.30. *Using superposition theorem, find the current in 23 Ω resistor in the circuit shown in Fig. 3.58.*

Solution.

200 V source acting alone. We first consider the case when 200 V voltage source is acting alone as shown in Fig. 3.59. Note that current source is replaced by an open. The total resistance R_T presented to the voltage source is 47 Ω in series with the parallel combination of 27 Ω and (23 + 4) Ω. Therefore, the value of R_T is given by ;

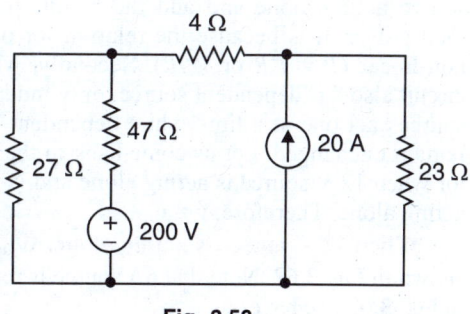

Fig. 3.58

$$R_T = 47 + [27 \parallel (23 + 4)] = 47 + \frac{27 \times 27}{27 + 27} = 47 + 13.5 = 60.5 \ \Omega$$

∴   Current supplied by 200 V source is given by ;

$$I_T = \frac{V}{R_T} = \frac{200}{60.5} = 3.31 \ A$$

At the node *A*, $I_T (= 3.31 \ A)$ divides between the parallel resistors of 27 Ω and (23 + 4) Ω.

∴   Current through 23 Ω, $I_1 = 3.31 \times \frac{27}{27 + 27} = 1.65 \ A$ downward

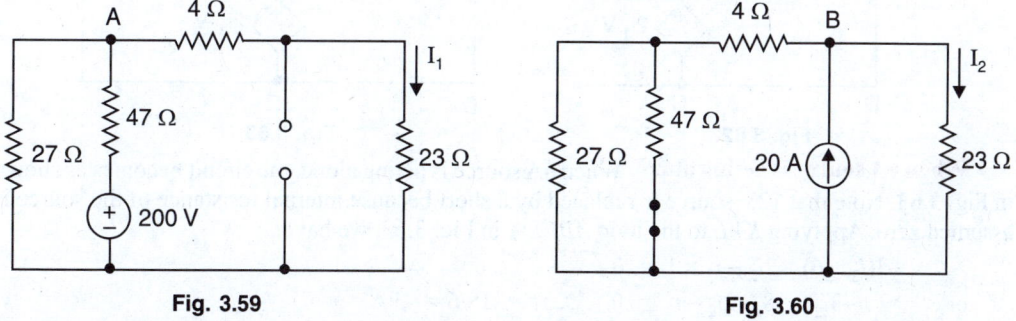

Fig. 3.59   **Fig. 3.60**

---

*   The total circuit resistance at terminals *AB* = 4 + (40‖10) = 12Ω. The circuit behaves as a 12V battery having internal resistance of 12Ω with terminals *A* and *B* open.

20 A current source acting alone. We now consider the case when the current source is acting alone as shown in Fig. 3.60. Note that voltage source is replaced by a short because its internal resistance is assumed zero. The equivalent resistance R_{eq} to the left of the current source is

$$R_{eq} = 4 + (27 \parallel 47) = 4 + \frac{27 \times 47}{27 + 47} = 4 + 17.15 = 21.15 \ \Omega$$

At node B, 20 A divides between two parallel resistors R_{eq} and 23 Ω. By current divider rule,

$$\text{Current in } 23\Omega \text{ resistor, } I_2 = 20 \times \frac{R_{eq}}{R_{eq} + 23} = 20 \times \frac{21.15}{21.15 + 23} = 9.58 \text{ A}$$

Note that I_2 in 23Ω resistor is downward.

$$\therefore \quad \text{Total current in } 23 \ \Omega = I_1 + I_2 = 1.65 + 9.58 = \textbf{11.23 A}$$

Example 3.31. *Fig. 3.61 shows the circuit with two independent sources and one dependent source. Find the power delivered to the 3 Ω resistor.*

Solution. While applying superposition theorem, two points must be noted carefully. First, we *cannot* find the power due to each independent source acting alone and add the results to obtain total power. It is because the relation for power is non-linear ($P = I^2R$ or V^2/R). Secondly, when the circuit also has dependent source, only independent sources act one at a time while dependent sources

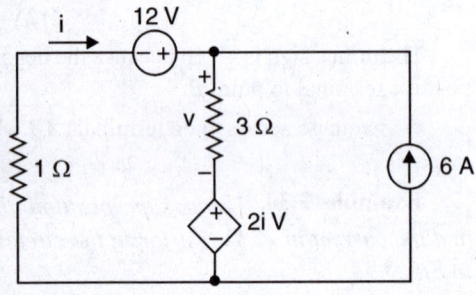

Fig. 3.61

remain unchanged. Let us come back to the problem. Suppose v_1 is the voltage across 3 Ω resistor when 12 V source is acting alone and v_2 is the voltage across 3 Ω resistor when 6 A source is acting alone. Therefore, $v = v_1 + v_2$.

When 12 V source is acting alone. When 12 V source is acting alone, the circuit becomes as shown in Fig. 3.62. Note that 6A source is replaced by an open. Applying *KVL* to the loop *ABCDA* in Fig. 3.62, we have,

$$12 - v_1 - 2i_1 - i_1 \times 1 = 0$$

or
$$12 - 3i_1 - 2i_1 - i_1 = 0 \quad \therefore \quad i_1 = 12/6 = 2 \text{ A}$$

$$\therefore \qquad v_1 = 3i_1 = 3 \times 2 = 6\text{V}$$

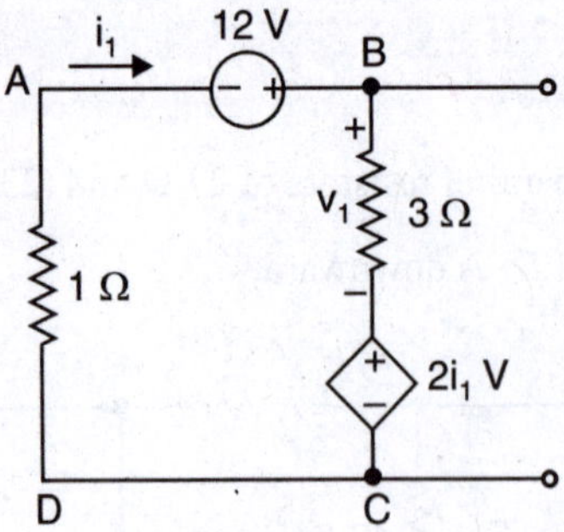

Fig. 3.62 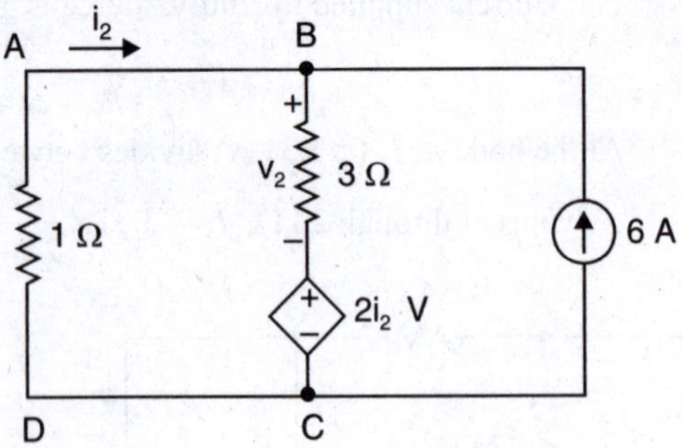   **Fig. 3.63**

When 6A source is acting alone. When 6A source is acting alone, the circuit becomes as shown in Fig. 3.63. Note that 12V source is replaced by a short because internal resistance of the source is assumed zero. Applying *KVL* to the loop *ABCDA* in Fig. 3.63, we have,

$$-3(i_2 + 6) - 2i_2 - i_2 \times 1 = 0$$

or
$$-3i_2 - 18 - 2i_2 - i_2 = 0 \quad \therefore \quad i_2 = -18/6 = -3 \text{ A}$$

$$\therefore \qquad v_2 = 3(i_2 + 6) = 3 \times 3 = 9 \text{ V}$$

$$\therefore \qquad v = v_1 + v_2 = 6 + 9 = 15 \text{ V}$$

$\therefore$ Power delivered to 3Ω, $P = \dfrac{v^2}{3} = \dfrac{(15)^2}{3} = $ **75 W**

Example 3.32. *Using superposition principle, find the current through G_C conductance in the circuit shown in Fig. 3.64. Given that $G_A = 0.3$ S ; $G_B = 0.4$ S and $G_C = 0.1$ S.*

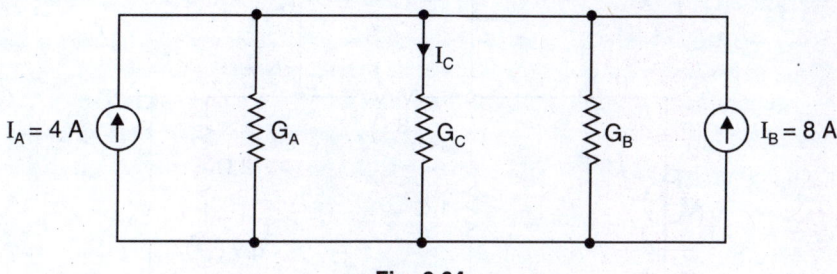

Fig. 3.64

Solution.

Current source I_A acting alone. We first consider the case when current source I_A is acting alone as shown in Fig. 3.65. Note that current source I_B is replaced by an open.

Total conductance, $G_T = G_A + G_C + G_B = 0.3 + 0.1 + 0.4 = 0.8$S

Voltage across G_C, $\quad V' = \dfrac{I_A}{G_T} = \dfrac{4}{0.8} = 5$ V

$\therefore$ Current through G_C, $\quad I'_C = V'G_C = 5 \times 0.1 = 0.5$ A

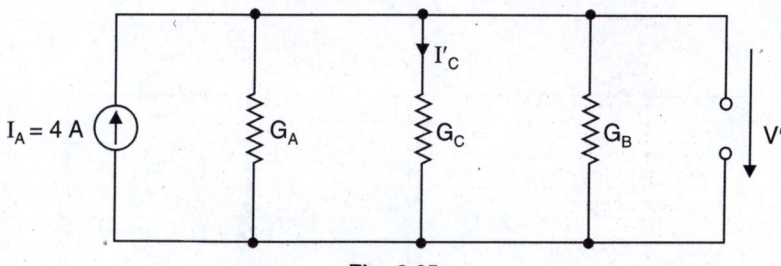

Fig. 3.65

Current source I_B acting alone. We now consider the case when current source I_B acts alone as shown in Fig. 3.66. Note that current source I_A is replaced by an open.

Voltage across G_C, $V'' = \dfrac{I_B}{G_T} = \dfrac{8}{0.8} = 10$ V

Current through G_C, $I''_C = V''G_C = 10 \times 0.1 = 1$ A

$\therefore$ Total current through G_C, $I_C = I'_C + I''_C = 0.5 + 1 = $ **1·5 A**

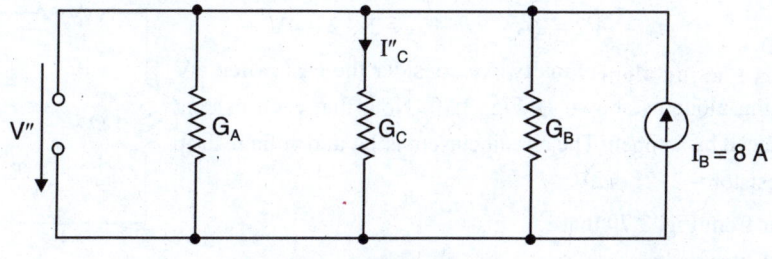

Fig. 3.66

Note. It is important to note that superposition theorem applies to currents and voltages; it does not mean that powers from two sources can be superimposed. It is because power varies as the *square* of the voltage or the current and this relationship is nonlinear.

Example 3.33. *Using superposition theorem, find the value of output voltage V_0 in the circuit shown in Fig. 3.67.*

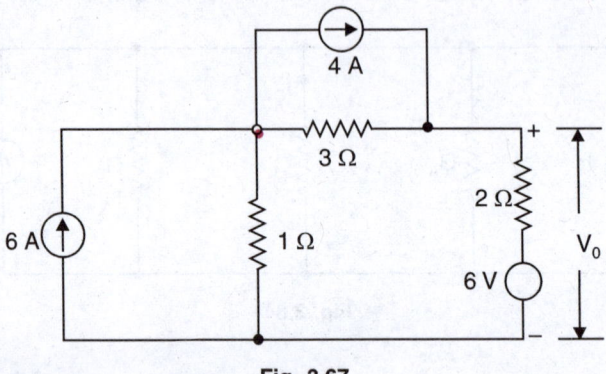

Fig. 3.67

Solution. The problem will be divided into three parts using one source at a time.

6A source acting alone. We first consider the case when 6 A source is acting alone as shown in Fig. 3.68. Note that voltage source is replaced by a short and the current source of 4 A is replaced by an open. According to current-divider rule, current i_1 through 2 Ω resistor is

$$i_1 = 6 \times \frac{1}{1+2+3} = 1A \quad \therefore \quad V_{01} = 1 \times 2 = 2V$$

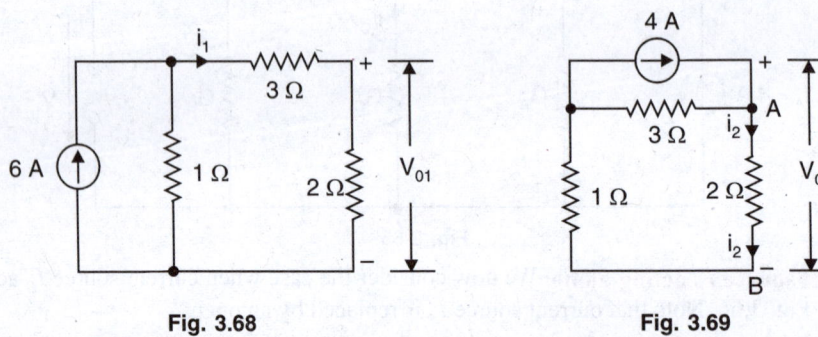

Fig. 3.68   **Fig. 3.69**

4A source acting alone. We now consider the case when 4A source is acting alone as shown in Fig. 3.69. Note that voltage source is replaced by a short and current source of 6A is replaced by an open. At point A, the current 4A finds two parallel paths; one of resistance 3 Ω and the other of resistance = 2 + 1 = 3 Ω. Therefore, current i_2 through 2 Ω resistor is

$$i_2 = 4/2 = 2A \quad \therefore \quad V_{02} = 2 \times 2 = 4V$$

6 V source acting alone. Finally, we consider the case when 6 V source is acting alone as shown in Fig. 3.70. Note that each current source is replaced by an open. The circuit current is 1A and voltage drop across 2 Ω resistor = 2 × 1 = 2V.

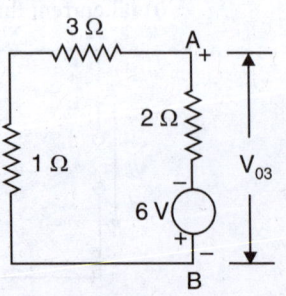

It is clear from Fig. 3.70 that :

$$V_A - 2V + 6V = V_B \quad \therefore \quad V_A - V_B = V_{03} = -4V$$

Fig. 3.70

According to superposition theorem, we have,

$$V_0 = V_{01} + V_{02} + V_{03} = 2 + 4 - 4 = \textbf{2V}$$

Example 3.34. *Using superposition theorem, find voltage across 4Ω resistance in Fig. 3.71 (i).*

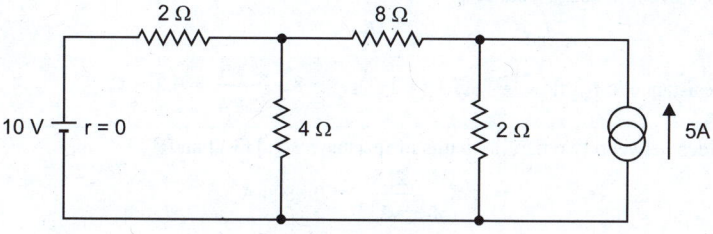

Fig. 3.71 *(i)*

Solution. In Fig. 3.71 (*ii*), the 5A current source is replaced by an open so that 10V source is acting alone. Referring to Fig. 3.71 (*ii*), the total circuit resistance R_T offered to 10V source is

$$R_T = 2\Omega + [4\Omega \parallel (2+8)\Omega] = 2 + \frac{4 \times 10}{4 + 10} = 4.857\Omega$$

∴ Current *I* supplied by 10 V source is given by ;

$$I = \frac{10V}{R_T} = \frac{10V}{4.857\Omega} = 2.059 \text{ A}$$

At point *A* in Fig. 3.71 (*ii*), the current 2.059 A divides into two parallel paths consisting of 4Ω resistance and (8 + 2) = 10Ω resistance.

∴ By current-divider rule, current I_1 in 4Ω due to 10 V alone is

$$I_1 = 2.059 \times \frac{10}{4+10} = 1.471 \text{ A in downward direction}$$

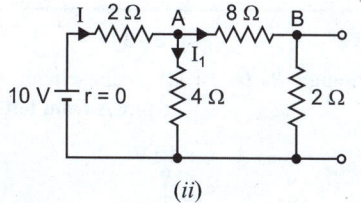 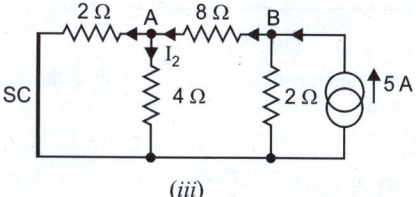

(ii)   *(iii)*

Fig. 3.71

In Fig. 3.71 (*iii*), the 10V battery is replaced by a short so that 5A current source is acting alone. At point *B* in Fig. 3.71 (*iii*), current 5A divides into two parallel paths consisting of 2Ω resistance and 8Ω + (2Ω∥4Ω) = 8 + (2 × 4)/(2 + 4) = 9.333Ω.

∴ By current-divider rule, current in 8Ω resistance is

$$I_{8\Omega} = 5 \times \frac{2}{2 + 9.333} = 0.8824 \text{ A}$$

At point A in Fig. 3.71 (*iii*), current 0.8824A divides into two parallel paths consisting of 2Ω resistance and 4Ω resistance.

∴ By current-divider rule, current I_2 in 4Ω due to 5A alone is

$$I_2 = 0.8824 \times \frac{2}{2 + 4} = 0.294 \text{ A in downward direction}$$

By superposition theorem, total current in 4 Ω

$$= I_1 + I_2 = 1.471 + 0.294 = 1.765A \text{ in downward direction}$$

∴   Voltage across $4\Omega = 1.765 \times 4 = $ **7.06V**

Note. We can also find I_2 in another way. Current in left-hand side 2Ω resistance will be $2I_2$ because $2\Omega \parallel 4\Omega$. By *KCL*, current in 8Ω resistance is

$$I_{8\Omega} = I_2 + 2I_2 = 3I_2$$

$$\text{Resistance to } I_{8\Omega} \text{ flow} = 8\Omega + (4\Omega \parallel 2\Omega) = 8 + \frac{2 \times 4}{2 + 4} = 9.333 \ \Omega$$

Now 5A divides between two parallel paths of resistances $9.333 \ \Omega$ and $2 \ \Omega$.

∴ $$I_{8\Omega} = 5 \times \frac{2}{2 + 9.333} = 0.8824 \text{ A}$$

or $$3I_2 = 0.8824 \quad \therefore I_2 = \frac{0.8824}{3} = 0.294 \text{ A}$$

Tutorial Problems

1. Use the superposition theorem to find the current in R_1 ($= 60 \ \Omega$) in the circuit shown in Fig. 3.72.
   [0.125 A from left to right]

2. Use the superposition theorem to find the current through R_1 ($= 1k \ \Omega$) in the circuit shown in Fig 3.73.
   [2 mA from right to left]

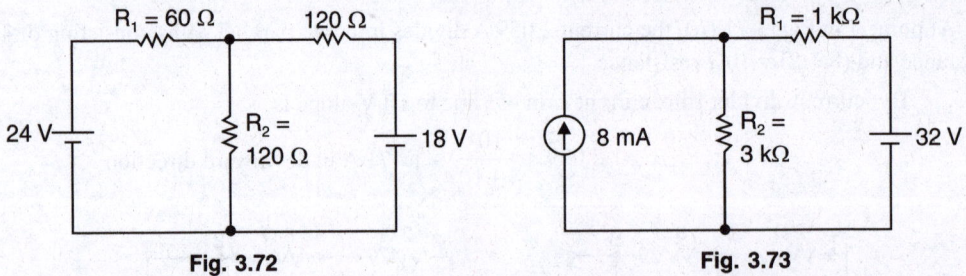

            Fig. 3.72   **Fig. 3.73**

3. Use the superposition theorem to find the current through R_1 ($= 10 \ \Omega$) in the circuit shown in Fig. 3.74.
   [4.6 A from left to right]

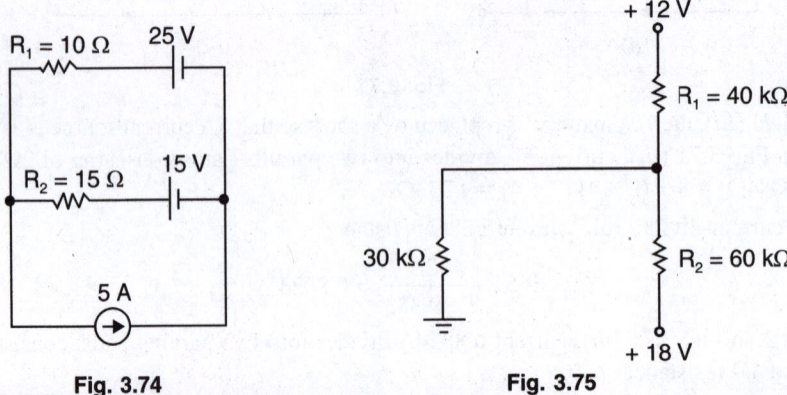

            Fig. 3.74   **Fig. 3.75**

4. Use superposition principle to find the current through resistance R_1 ($= 40 \ k\Omega$) in the circuit shown in Fig. 3.75.
   [1 mA downward]

5. Use superposition principle to find the voltage across R_1 ($= 1 \ k \ \Omega$) in the circuit shown in Fig. 3.76. Be sure to indicate the polarity of the voltage.
   [− (11 V) +]

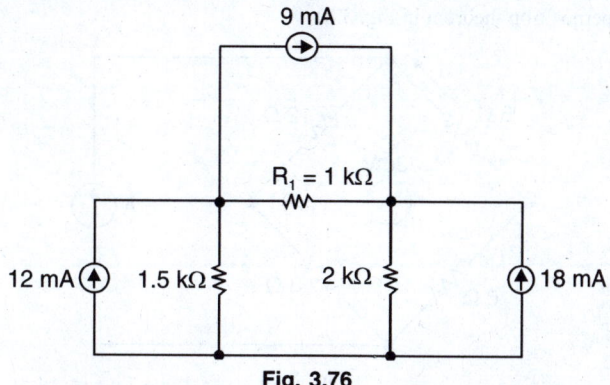

Fig. 3.76

6. Using superposition principle, find the current through 10 Ω resistor in Fig. 3.77. **[0.5 A↓]**

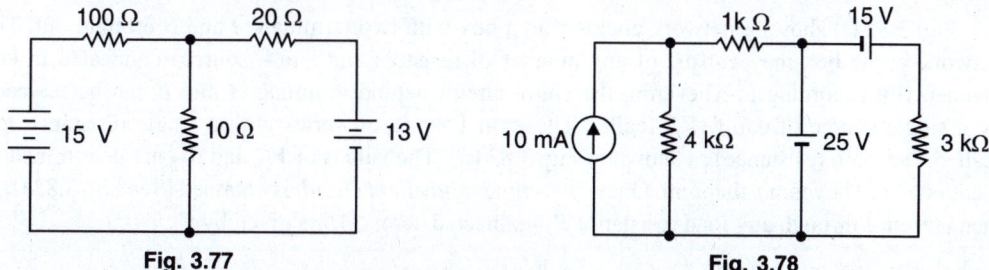

Fig. 3.77　　　　　　　　　　　　　**Fig. 3.78**

7. Using superposition principle, find the voltage across 4 kΩ resistor in Fig. 3.78. **[28 V$_-^+$]**

8. Referring to Fig. 3.79, the internal resistance R_S of the current source is 100 Ω. The internal resistance R_S of the voltage source is 10 Ω. Use superposition principle to find the power dissipated in 50 Ω resistor. **[8.26 W]**

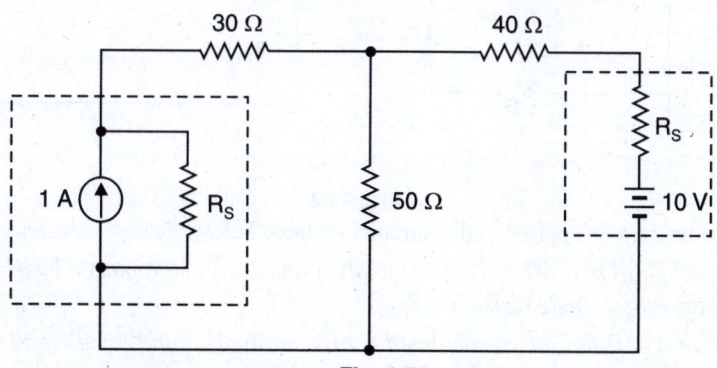

Fig. 3.79

9. Find v using superposition principle if $R = 2Ω$ in Fig. 3.80. **[8 V]**

10. State whether true or false.

　(i) Superposition theorem is applicable to multiple source circuits.

　(ii) Superposition theorem is restricted to linear circuits. **[(i) True (ii) True]**

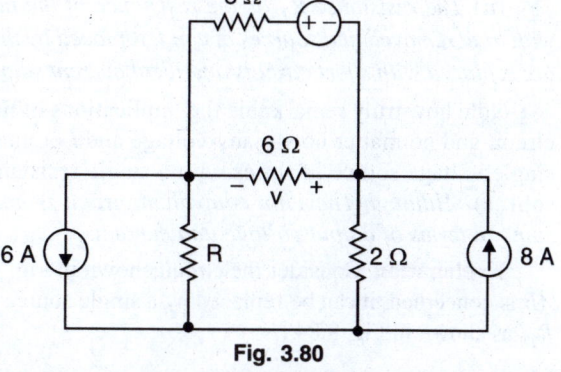

Fig. 3.80

11. Find i using superposition theorem in Fig. 3.81.   **[–6 A]**

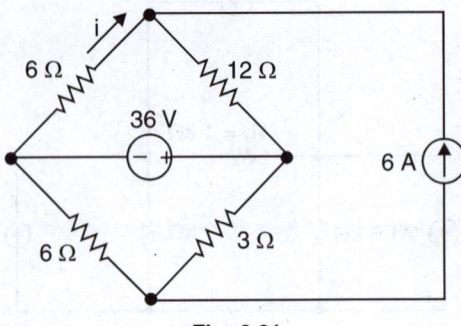

Fig. 3.81

3.10. Thevenin's Theorem

Fig. 3.82 (i) shows a network enclosed in a box with two terminals A and B brought out. The network in the box may consist of any number of resistors and e.m.f. sources connected in any manner. But according to Thevenin, the entire circuit behind terminals A and B can be replaced by a single source of e.m.f. V_{Th} (called Thevenin voltage) in series with a single resistance R_{Th} (called Thevenin resistance) as shown in Fig. 3.82 (ii). The values of V_{Th} and R_{Th} are determined as mentioned in Thevenin's theorem. Once *Thevenin's equivalent circuit* is obtained [See Fig. 3.82 (ii)], then current I through any load resistance R_L connected across AB is given by ;

$$I = \frac{V_{TH}}{R_{TH} + R_L}$$

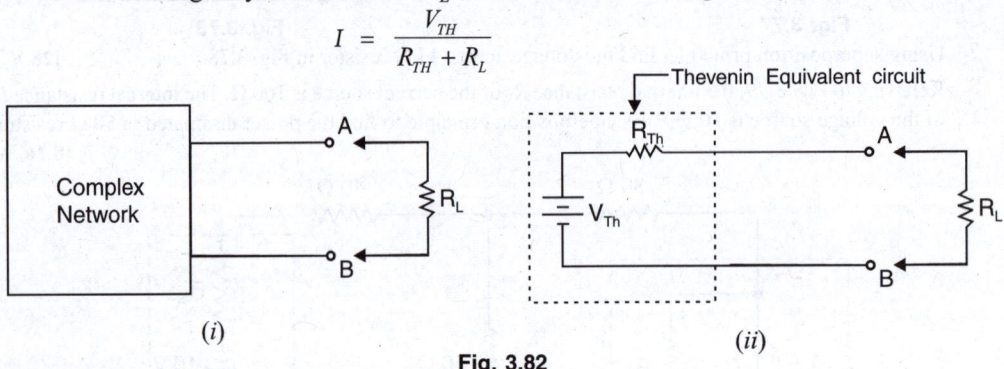

Fig. 3.82

Thevenin's theorem as applied to d.c. circuits is stated below :

Any linear, bilateral network having terminals A and B can be replaced by a single source of e.m.f. V_{Th} in series with a single resistance R_{Th}.

(i) *The e.m.f. V_{Th} is the voltage obtained across terminals A and B with load, if any removed i.e. it is open-circuited voltage between terminals A and B.*

(ii) *The resistance R_{Th} is the resistance of the network measured between terminals A and B with load removed and sources of e.m.f. replaced by their internal resistances. Ideal voltage sources are replaced with short circuits and ideal current sources are replaced with open circuits.*

Note how truly remarkable the implications of this theorem are. No matter how complex the circuit and no matter how many voltage and / or current sources it contains, it is equivalent to a single voltage source in series with a single resistance (*i.e.* **equivalent to a single real voltage source).** *Although Thevenin equivalent circuit is not the same as its original circuit, it acts the same in terms of output voltage and current.*

Explanation. Consider the circuit shown in Fig. 3.83 (i). As far as the circuit behind terminals AB is concerned, it can be replaced by a single source of e.m.f. V_{Th} in series with a single resistance R_{Th} as shown in Fig. 3.84 (ii).

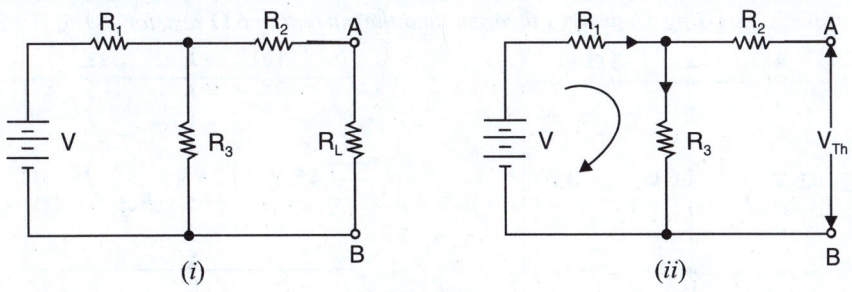

Fig. 3.83

(*i*) **Finding V$_{Th}$.** The e.m.f. V_{Th} is the voltage across terminals *AB* with load (*i.e.* R_L) removed as shown in Fig. 3.83 (*ii*). With R_L disconnected, there is no current in R_2 and V_{Th} is the voltage appearing across R_3.

∴ V_{Th} = Voltage across R_3 = $\dfrac{V}{R_1 + R_3} \times R_3$

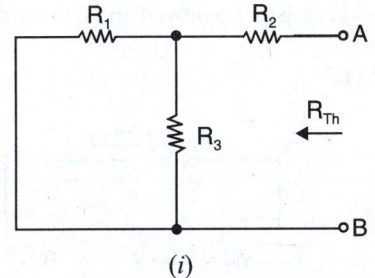
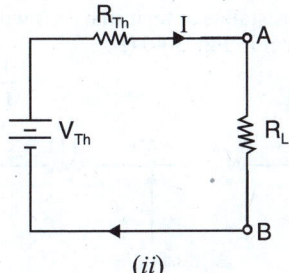

Fig. 3.84

(*ii*) **Finding R$_{Th}$.** To find R_{Th}, remove the load R_L and replace the battery by a short-circuit because its internal resistance is assumed zero. Then resistance between terminals *A* and *B* is equal to R_{Th} as shown in Fig. 3.84 (*i*). Obviously, at the terminals *AB* in Fig. 3.84 (*i*), R_1 and R_3 are in parallel and this parallel combination is in series with R_2.

∴ R_{Th} = $R_2 + \dfrac{R_1 R_3}{R_1 + R_3}$

When load R_L is connected between terminals *A* and *B* [See Fig. 3.84 (*ii*)], then current in R_L is given by ;

$$I = \dfrac{V_{Th}}{R_{Th} + R_L}$$

3.11. Procedure for Finding Thevenin Equivalent Circuit

(*i*) Open the two terminals (*i.e.*, remove any load) between which you want to find Thevenin equivalent circuit.

(*ii*) Find the open-circuit voltage between the two open terminals. It is called Thevenin voltage V_{Th}.

(*iii*) Determine the resistance between the two open terminals with all ideal voltage sources shorted and all ideal current sources opened (a non-ideal source is replaced by its internal resistance). It is called Thevenin resistance R_{Th}.

(*iv*) Connect V_{Th} and R_{Th} in series to produce Thevenin equivalent circuit between the two terminals under consideration.

(*v*) Place the load resistor removed in step (*i*) across the terminals of the Thevenin equivalent circuit. The load current can now be calculated using only Ohm's law and it has the same value as the load current in the original circuit.

Note. Thevenin's theorem is sometimes called *Helmholtz's theorem.*

Example 3.35. *Using Thevenin's theorem, find the current in 6 Ω resistor in Fig. 3·85 (i).*

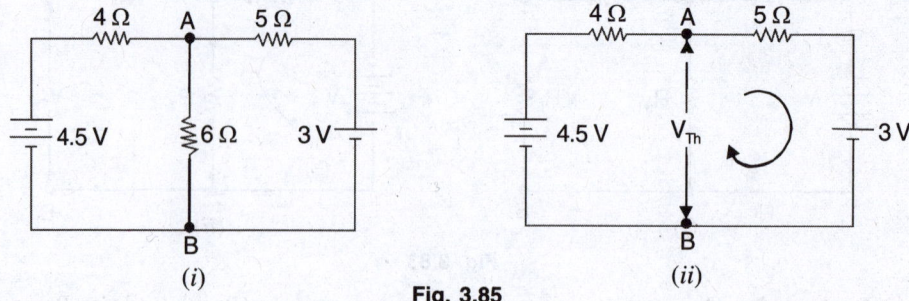

Fig. 3.85

Solution. Since internal resistances of batteries are not given, it will be assumed that they are zero. We shall find Thevenin's equivalent circuit at terminals *AB* in Fig. 3.85 (*i*).

V_{Th} = Voltage across terminals *AB* with load (*i.e.* 6 Ω resistor) removed as shown in Fig. 3·85 (*ii*).

$$= *4·5 - 0.167 \times 4 = 3·83 \text{ V}$$

R_{Th} = Resistance at terminals *AB* with load (*i.e.* 6 Ω resistor) removed and battery replaced by a short as shown in Fig. 3·86 (*i*).

$$= \frac{4 \times 5}{4 + 5} = 2.22 \ \Omega$$

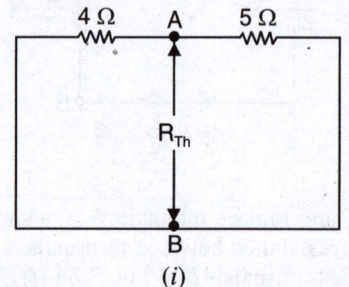
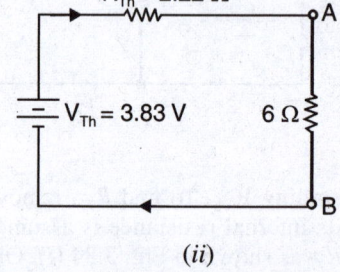

Fig. 3.86

Thevenin's equivalent circuit at terminals *AB* is V_{Th} (= 3·83 *V*) in series with R_{Th} (= 2·22 Ω). When load (*i.e.* 6 Ω resistor) is connected between terminals *A* and *B*, the circuit becomes as shown in Fig. 3·86 (*ii*).

∴   Current in 6 Ω resistor $= \dfrac{V_{Th}}{R_{Th} + 6} = \dfrac{3.83}{2.22 + 6} = \textbf{0.466A}$

Example 3.36. *Using Thevenin's theorem, find p.d. across terminals AB in Fig. 3·87 (i).*

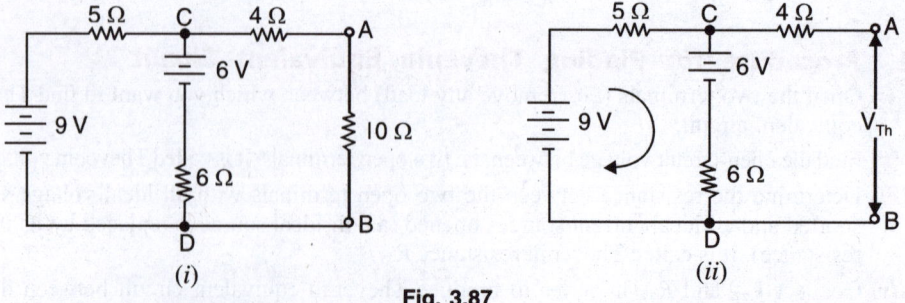

Fig. 3.87

* Net e.m.f. in the circuit shown in Fig. 3·85 (*ii*) is 4·5 − 3 = 1·5 V and total circuit resistance is 9 Ω.

∴   Circuit current = 1·5/9 = 0·167 A

The voltage across *AB* is equal to 4·5 V *less* drop in 4 Ω resistor.

∴ V_{Th} = 4·5 − 0·167 × 4 = 3·83 V

Solution. We shall find Thevenin's equivalent circuit at terminals *AB* in Fig. 3.87 (*i*).

V_{Th} = Voltage across terminals *AB* with load (*i.e.* 10 Ω resistor) removed as shown in Fig. 3.87 (*ii*).

    = Voltage across terminals *CD*

    = 9 – drop in 5 Ω resistor

    = 9* – 5 × 0·27 = 7·65 V

R_{Th} = Resistance at terminals *AB* with load (*i.e.* 10 Ω resistor) removed and batteries replaced by a short as shown in Fig. 3.88 (*i*).

$$= 4 + \frac{5 \times 6}{5 + 6} = 6.72\Omega$$

Thevenin's equivalent circuit to the left of terminals *AB* is V_{Th} (= 7·65 *V*) in series with R_{Th} (= 6·72 Ω). When load (*i.e.* 10 Ω resistor) is connected between terminals *A* and *B*, the circuit becomes as shown in Fig. 3.88 (*ii*).

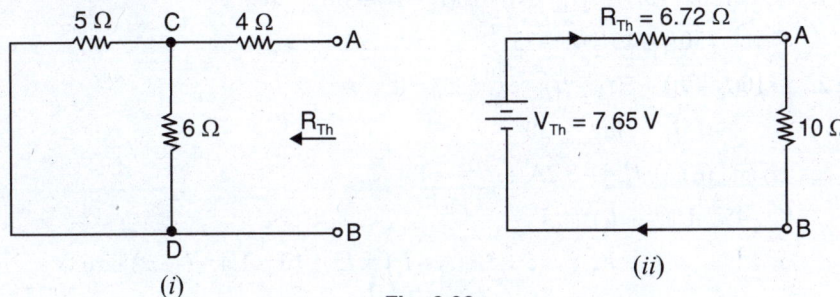

(*i*)   (*ii*)

Fig. 3.88

∴   Current in 10 Ω resistor $= \dfrac{V_{Th}}{R_{Th} + 10} = \dfrac{7.65}{6.72 + 10} = 0.457$ A

    P.D. across 10 Ω resistor = 0·457 × 10 = **4·57 V**

Example 3.37. *Using Thevenin's theorem, find the current through resistance R connected between points a and b in Fig. 3.89 (i).*

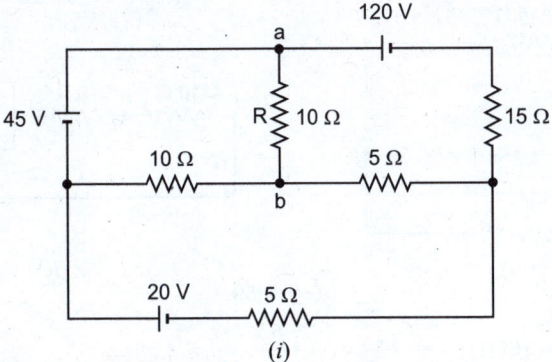

(*i*)

Fig. 3.89

Solution. (*i*) **Finding** V_{Th}**.** Thevenin voltage V_{Th} is the voltage across terminals *ab* with

---

*   The net e.m.f. in the loop of circuit shown in Fig. 3·87 (*ii*) is 9 – 6 = 3V and total resistance is 5 + 6 = 11 Ω.

∴   Circuit current = 3/11 = 0·27 A

resistance R (= 10Ω) removed as shown in Fig. 3.89 (*ii*). It can be found by Maxwell's mesh current method.

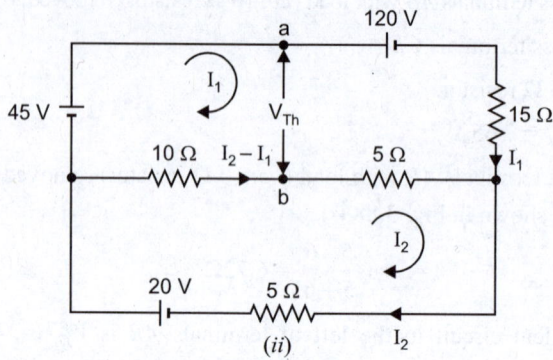

Fig. 3.89

Mesh 1. $45 - 120 - 15I_1 - 5(I_1 - I_2) - 10(I_1 - I_2) = 0$

or $30I_1 - 15I_2 = -75$ ...(*i*)

Mesh 2. $-10(I_2 - I_1) - 5(I_2 - I_1) - 5I_2 + 20 = 0$

or $-15I_1 + 20I_2 = 20$ ...(*ii*)

From eqs. (*i*) and (*ii*), $I_1 = -3.2$ A ; $I_2 = -1.4$ A

Now, $V_a - 45 - 10(I_2 - I_1) = V_b$

or $V_a - V_b = 45 + 10(I_2 - I_1) = 45 + 10[-1.4 - (-3.2)] = 63$V

∴ $V_{Th} = V_{ab} = V_a - V_b = 63$V

   (*ii*) **Finding R_{Th}.** Thevenin resistance R_{Th} is the resistance at terminals *ab* with resistance R (= 10Ω) removed and batteries replaced by a short as shown in Fig. 3.89 (*iii*). Using laws of series and parallel resistances, the circuit is reduced to the one shown in Fig. 3.89 (*iv*).

   ∴ R_{Th} = Resistance at terminals *ab* in Fig 3.89 (*iv*).

$$= 10Ω \,||\, [5Ω + (15Ω \,||\, 5Ω)] = 10Ω \,||\, (5Ω + 3.75Ω) = \frac{14}{3}Ω$$

(*iii*)

(*iv*)

Fig. 3.89

∴ Current in R (= 10Ω) = $\dfrac{V_{Th}}{R_{Th} + R} = \dfrac{63}{(14/3) + 10} = $ **4.295A**

   Example 3.38. *A Wheatstone bridge ABCD has the following details : AB = 10 Ω, BC = 30 Ω, CD = 15 Ω and DA = 20 Ω. A battery of e.m.f. 2 V and negligible resistance is connected between A and C with A positive. A galvanometer of 40 Ω resistance is connected between B and D. Using Thevenin's theorem, determine the magnitude and direction of current in the galvanometer.*

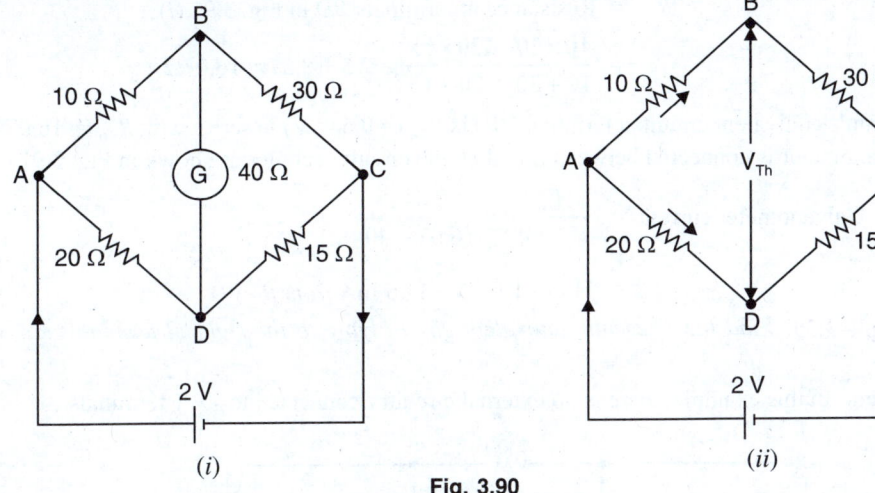

Fig. 3.90

Solution. We shall find Thevenin's equivalent circuit at terminals BD in Fig. 3.90 (*i*).

(*i*) Finding V_{Th}. To find V_{Th} at terminals BD, remove the load (*i.e.* 40 Ω galvanometer) as shown in Fig. 3.90 (*ii*). The voltage between terminals B and D is equal to V_{Th}.

$$\text{Current in branch } ABC = \frac{2}{10+30} = 0.05 \text{ A}$$

$$\text{P.D. between } A \text{ and } B, V_{AB} = 10 \times 0.05 = 0.5 \text{ V}$$

$$\text{Current in branch } ADC = \frac{2}{20+15} = 0.0571 \text{A}$$

$$\text{P.D. between } A \text{ and } D, V_{AD} = 0.0571 \times 20 = 1.142 \text{ V}$$

$\therefore$ P.D. between B and D, $V_{Th} = V_{AD} - V_{AB} = 1.142 - 0.5 = 0.642$ V

Obviously, point B* is positive w.r.t. point D *i.e.* current in the galvanometer, when connected between B and D, will flow from B to D.

(*ii*) Finding R_{Th}. In order to find R_{Th}, remove the load (*i.e.* 40 Ω galvanometer) and replace the battery by a short (as its internal resistance is assumed zero) as shown in Fig. 3.91 (*i*). Then resistance measured between terminals B and D is equal to R_{Th}.

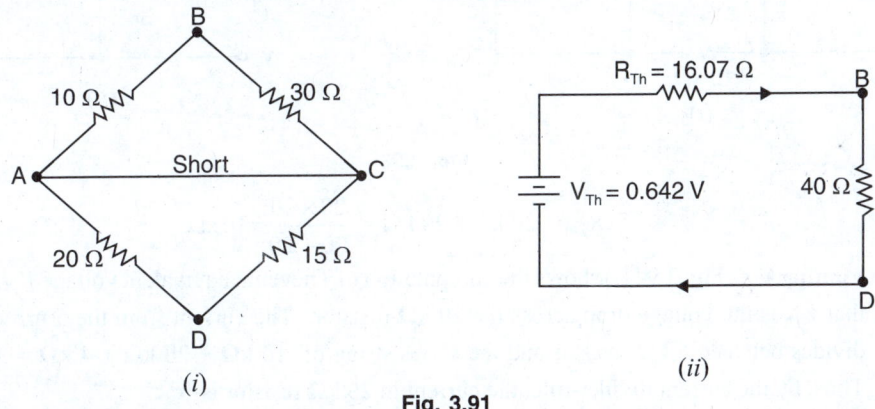

Fig. 3.91

---

*   The potential at point D is 1·142 V lower than at A. Also potential of point B is 0·5 V lower than A. Hence point B is at higher potential than point D.

$$R_{Th} = \text{Resistance at terminals } BD \text{ in Fig. 3.91 } (i).$$

$$= \frac{10 \times 30}{10 + 30} + \frac{20 \times 15}{20 + 15} = 7.5 + 8.57 = 16.07\Omega$$

Thevenin's equivalent circuit at terminals BD is V_{Th} ($= 0.642$ V) in series with R_{Th} ($= 16 \cdot 07$ Ω). When galvanometer is connected between B and D, the circuit becomes as shown in Fig. 3.91 (ii).

$$\therefore \quad \text{Galvanometer current} = \frac{V_{Th}}{R_{Th} + 40} = \frac{0.642}{16.07 + 40}$$

$$= 11.5 \times 10^{-3} \text{ A} = \textbf{11.5 mA } \textit{from B to D}$$

Example 3·39. *Find the Thevenin equivalent circuit lying to the right of terminals $x - y$ in Fig. 3.92.*

Solution. In this example, there is no external circuitry connected to $x - y$ terminals.

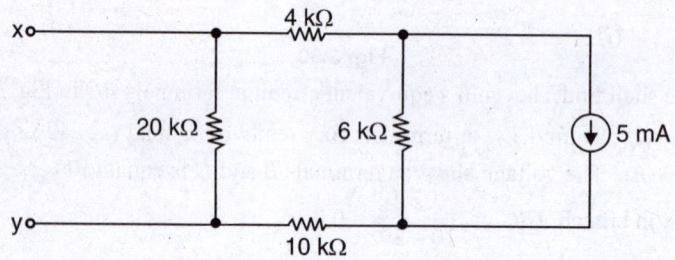

Fig. 3.92

(*i*) **Finding R_{Th}.** To find Thevenin equivalent resistance R_{Th}, we open-circuit the current source as shown in Fig. 3.93 (*i*). Note that 4 kΩ, 6 kΩ and 10 kΩ resistors are then in series and have a total resistance of 20 kΩ. Thus R_{Th} is the parallel combination of that 20 kΩ resistance and the other 20 kΩ resistor as shown in Fig. 3·93 (*ii*).

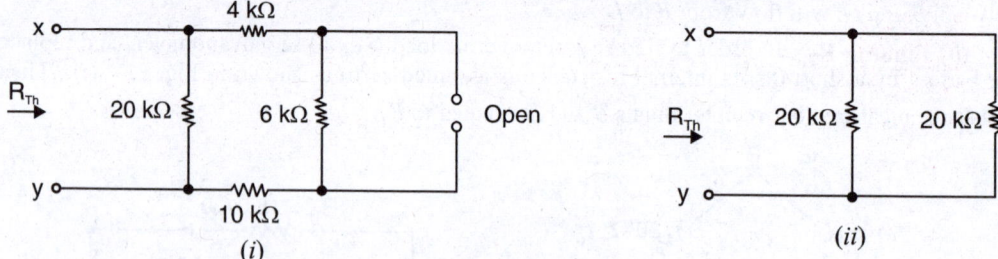

Fig. 3.93

$$\therefore \qquad R_{Th} = 20 \text{ k}\Omega \parallel 20 \text{ k}\Omega = \frac{20 \times 20}{20 + 20} = 10 \text{ k}\Omega$$

(*ii*) **Finding V_{Th}.** Fig. 3.94 (*i*) shows the computation of Thevenin equivalent voltage V_{Th}. Note that V_{Th} is the voltage drop across the 20 kΩ resistor. The current from the 5 mA source divides between 6 kΩ resistor and the series string of 10 kΩ + 20 kΩ + 4 kΩ = 34 kΩ. Thus, by the current divider-rule, the current in 20 kΩ resistor is

$$I_{20 \text{ k}\Omega} = \left(\frac{6}{34 + 6} \right) \times 5 = 0.75 \text{ mA}$$

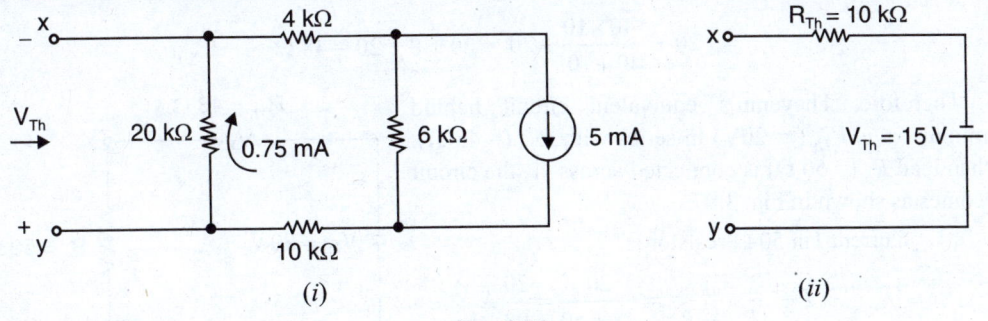

Fig. 3.94

Voltage across 20 kΩ resistor is given by ;

$$V_{Th} = (0.75 \text{ mA}) (20 \text{ k}\Omega) = 15 \text{ V}$$

Notice that terminal y is positive with respect to terminal x. **Fig. 3.94 (ii) shows the Thevenin equivalent circuit.** The polarity of V_{Th} is such that terminal y is positive with respect to terminal x, as required.

Example 3.40. *Calculate the power which would be dissipated in a 50 Ω resistor connected across xy in the network shown in Fig. 3.95.*

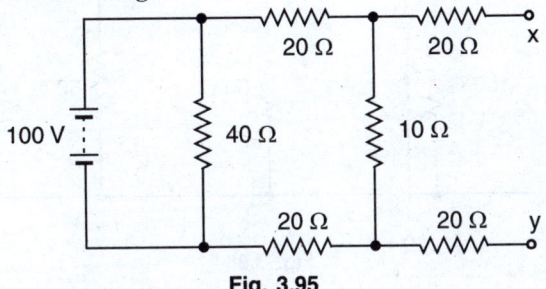

Fig. 3.95

Solution. We shall find Thevenin equivalent circuit to the left of terminals xy. With xy terminals open, the current in 10 Ω resistor is given by ;

$$*I = \frac{100}{20 + 10 + 20} = 2\text{A}$$

∴   Open circuit voltage across xy is given by ;

$$V_{Th} = I \times 10 = 2 \times 10 = 20\text{V}$$

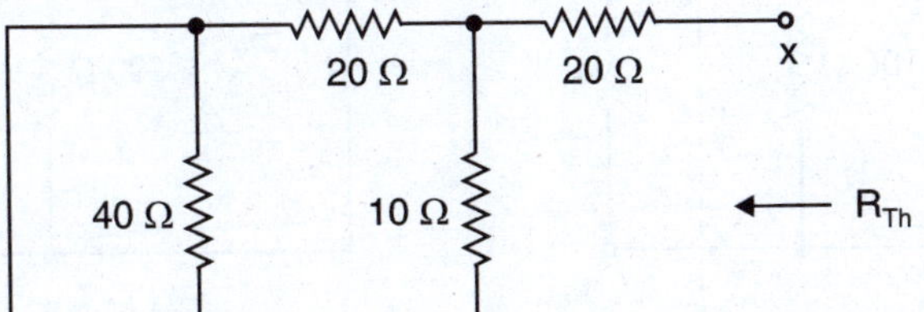

Fig. 3.96

In order to find R_{Th} replace the battery by a short since its internal resistance is assumed to be zero [See Fig. 3.96].

$$R_{Th} = \text{Resistance looking into the terminals } xy \text{ in Fig. 3.96.}$$
$$= 20 + [(20 + 20) \| 10] + 20$$

---

*   It is clear that $(20 + 10 + 20)$ Ω is in parallel with 40 Ω resistor across 100 V source.

$$= 20 + \frac{*40 \times 10}{40 + 10} + 20 = 20 + 8 + 20 = 48 \ \Omega$$

Therefore, Thevenin's equivalent circuit behind terminals xy is V_{Th} ($= 20V$) in series with R_{Th} ($= 48 \ \Omega$). When load R_L ($= 50 \ \Omega$) is connected across xy, the circuit becomes as shown in Fig. 3.97.

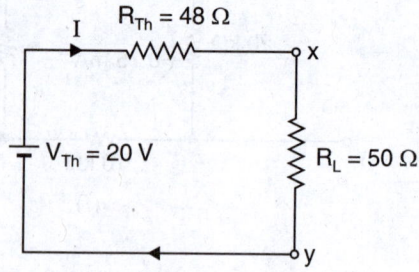

∴   Current I in 50 Ω resistor is

$$I = \frac{V_{Th}}{R_{Th} + R_L} = \frac{20}{48 + 50} = \frac{20}{98} A$$

∴   Power dissipated in 50 Ω resistor is

$$P = I^2 R_L = \left(\frac{20}{98}\right)^2 \times 50 = \textbf{2·08 W}$$

Fig. 3.97

Example 3.41. *Calculate the current in the 50 Ω resistor in the network shown in Fig. 3.98.*

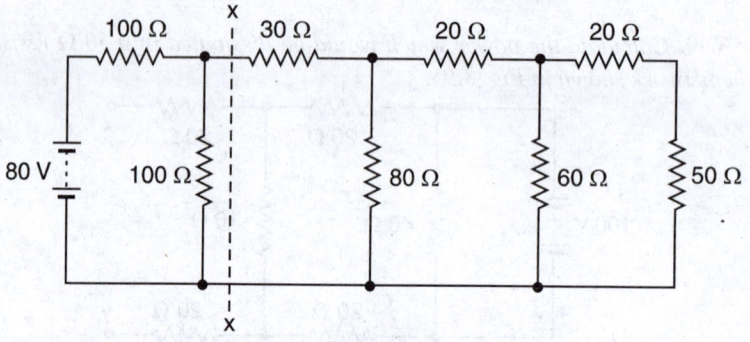

Fig. 3.98

Solution. We shall simplify the circuit shown in Fig. 3.98 by the repeated use of Thevenin's theorem. We first find Thevenin's equivalent circuit to the left of ******xx.

---

*   Note that 40 Ω resistor is shorted and may be considered as removed in the circuit shown in Fig. 3.96.

**

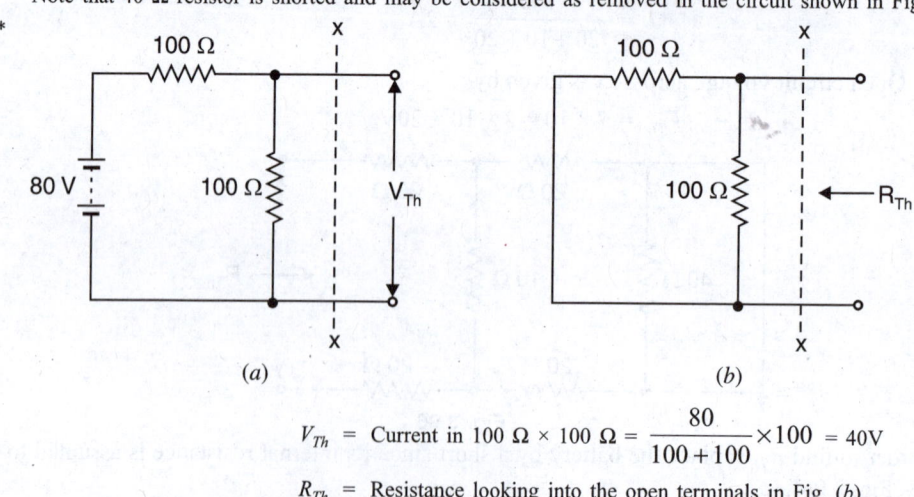

(a)   (b)

$$V_{Th} = \text{Current in } 100 \ \Omega \times 100 \ \Omega = \frac{80}{100 + 100} \times 100 = 40V$$

$$R_{Th} = \text{Resistance looking into the open terminals in Fig. } (b)$$

$$= 100 \parallel 100 = \frac{100 \times 100}{100 + 100} = 50 \Omega$$

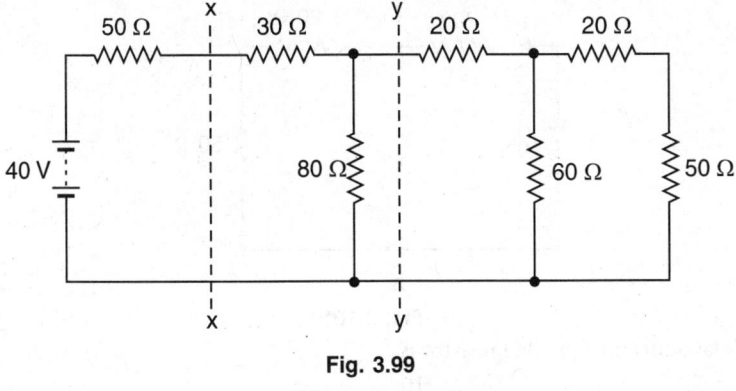

Fig. 3.99

$$V_{Th} = \frac{80}{100+100} \times 100 = 40V$$

$$R_{Th} = 100 \parallel 100 = \frac{100 \times 100}{100+100} = 50\Omega$$

Therefore, we can replace the circuit to the left of *xx* in Fig. 3.98 by its Thevenin's equivalent circuit *viz.* V_{Th} (= 40V) in series with R_{Th} (= 50 Ω). The original circuit of Fig. 3.98 then reduces to the one shown in Fig. 3.99.

We shall now find Thevenin's equivalent circuit to the left of *yy* in Fig. 3.99.

$$V'_{Th} = \frac{40}{50+30+80} \times 80 = 20\ V$$

$$R'_{Th} = (50+30) \parallel 80 = \frac{80 \times 80}{80+80} = 40\ \Omega$$

We can again replace the circuit to the left of *yy* in Fig. 3.99 by its Thevenin's equivalent circuit. Therefore, the original circuit reduces to that shown in Fig. 3.100.

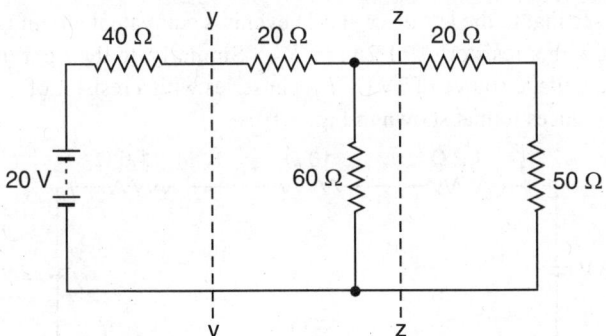

Fig. 3.100

Using the same procedure to the left of *zz*, we have,

$$V''_{Th} = \frac{20}{40+20+60} \times 60 = 10\ V$$

$$R''_{Th} = (40+20) \parallel 60 = \frac{60 \times 60}{60+60} = 30\Omega$$

The original circuit then reduces to that shown in Fig. 3.101.

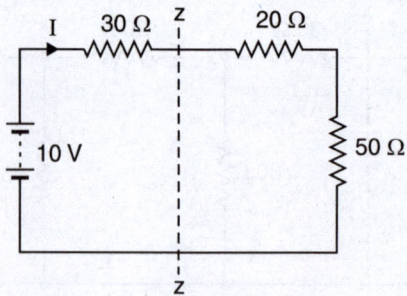

Fig. 3.101

By Ohm's law, current I in 50 Ω resistor is

$$I = \frac{10}{30 + 20 + 50} = 0.1 \text{ A}$$

Example 3.42. *Calculate the current in the 10 Ω resistor in the network shown in Fig. 3·102.*

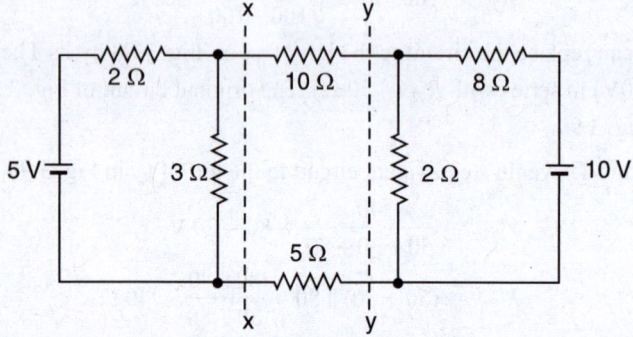

Fig. 3.102

Solution. We can replace circuits to the left of *xx* and right of *yy* by the Thevenin's equivalent circuits. It is easy to see that to the left of *xx*, the Thevenin's equivalent circuit is a voltage source of 3V (= V_{Th}) in series with a resistor of *1·2 Ω (= R_{Th}). Similarly, to the right of *yy*, the Thevenin's equivalent circuit is a voltage source of 2V (= V_{Th}) in series with a resistor of **1·6 Ω (= R_{Th}). The original circuit then reduces to that shown in Fig. 3.103.

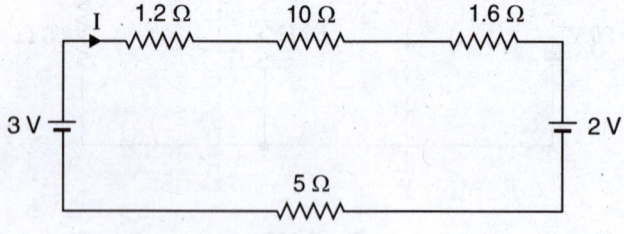

Fig. 3.103

$\therefore$ Current through 10 Ω resistor is given by ;

$$I = \frac{\text{Net voltage}}{\text{Total resistance}} = \frac{3-2}{1.2 + 10 + 1.6 + 5} = 56 \cdot 2 \times 10^{-3} \text{A} = \textbf{56·2 mA}$$

---

* $R_{Th} = 2 \parallel 3 = \dfrac{2 \times 3}{2+3} = 1.2\Omega$

** $R_{Th} = 2 \parallel 8 = \dfrac{2 \times 8}{2+8} = 1.6\Omega$

Example 3.43. *Calculate the values of V_{Th} and R_{Th} between terminals A and B in Fig. 3.104 (i).
All resistances are in ohms.*

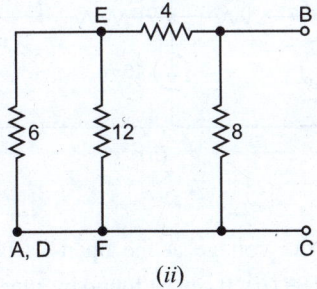

Fig. 3.104

Solution. *(i)* **Finding V_{Th}.** Between points E and F [See Fig. 3.104 *(i)*], $12\Omega \parallel (4+8)\Omega$.

$$\therefore \qquad R_{EF} = 12\Omega \parallel (4+8)\Omega = 12\Omega \parallel 12\Omega = 6\Omega$$

By voltage-divider rule, we have,

$$V_{DE} = 48 \times \frac{6}{6+6} = 24V \; ; \; V_{EF} = 48 \times \frac{6}{6+6} = 24V$$

Now $V_{EF} (= 24V)$ is divided between 4Ω and 8Ω resistances in series.

$$\therefore \qquad V_{EG} = 24 \times \frac{4}{4+8} = 8V$$

In going from A to B via D, E and G, there is fall in potential from D to E, fall in potential from E to G and rise in potential from B to A. Therefore, by *KVL*,

$$V_{BA} - V_{DE} - V_{EG} = 0 \quad \text{or} \quad V_{BA} = V_{DE} + V_{EG} = 24 + 8 = 32V$$

$$\therefore \qquad V_{Th} = V_{BA} = \textbf{32V} \; ; \; A \text{ positive } w.r.t \; B.$$

(ii) **Finding R_{Th}.** R_{Th} is the resistance between open terminals AB with voltage source replaced by a short as shown in Fig. 3.104 *(ii)*. Shorting voltage source brings points A, D and F together. Now combined resistance of parallel combination of 6Ω and $12\Omega = 6\Omega \parallel 12\Omega = 4\Omega$ and the circuit reduces to the one shown in Fig. 3.104 *(iii)*.

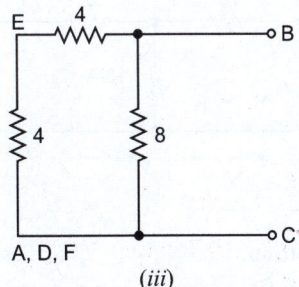

Fig. 3.104

$$\therefore \qquad R_{Th} = R_{AB} \text{ in Fig. 3.104 } (iii) = 8\Omega \parallel (4+4)\Omega = \textbf{4}\Omega$$

Example 3.44. *The circuit shown in Fig. 3.105
consists of a current source $I = 10\,A$ paralleled by $G
= 0 \cdot 1S$ and a voltage source $E = 200\,V$ with a $10\,\Omega$
series resistance. Find Thevenin equivalent circuit
to the left of terminals AB.*

Solution. With terminals A and B open-
circuited, the current source will send a current
through conductance G as shown in Fig. 3.106.

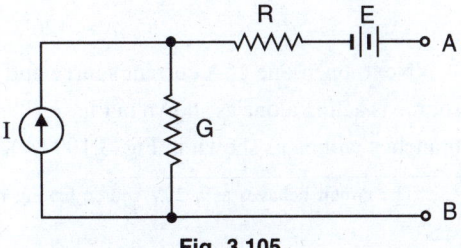

Fig. 3.105

$\therefore$ Voltage across G, $V_G = \dfrac{I}{G} = \dfrac{10}{0.1} = 100$ V

Thevenin voltage, V_{Th} = Open-circuited voltage at terminals AB in Fig. 3.106.

$$= E + V_G = 200 + 100 = 300 \text{ V}$$

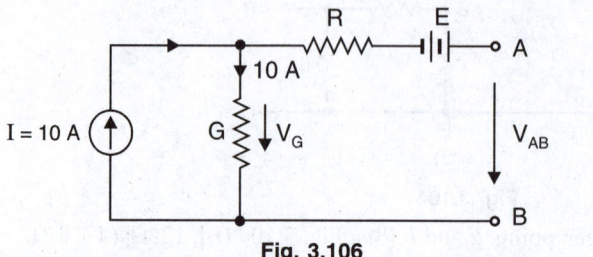

Fig. 3.106

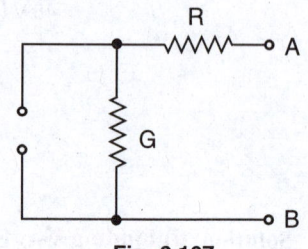

Fig. 3.107

In order to find Thevenin resistance R_{Th}, replace the voltage source by a short and current source by an open. The circuit then becomes as shown in Fig. 3.107.

R_{Th} = Resistance looking into terminals AB in Fig. 3.107.

$$= R + \dfrac{1}{G} = 10 + \dfrac{1}{0.1} = 10 + 10 = 20\Omega$$

Fig. 3.108

Therefore, Thevenin equivalent circuit consists of **300V voltage source in series with a resistance of 20 Ω** as shown in Fig. 3.108.

Example 3.45. *Using Thevenin's theorem, find the voltage across 3Ω resistor in Fig. 3.109 (i).*

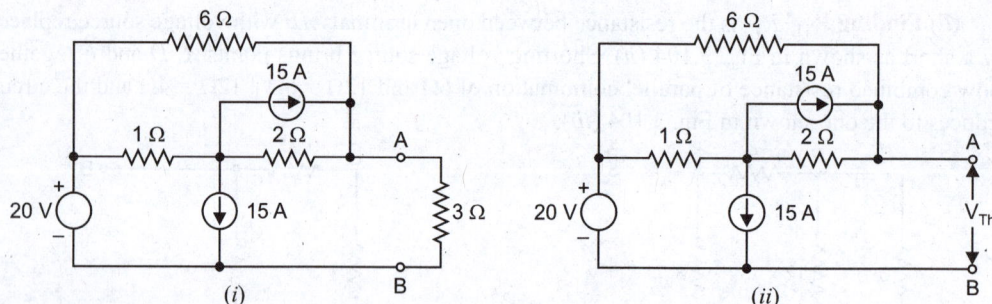

Fig. 3.109

Solution. (*i*) **Finding V_{Th}.** Thevenin voltage V_{Th} is the voltage at the open-circuited load terminals AB (*i.e.,* when 3Ω is removed) as shown in Fig. 3.109 (*ii*). It can be found by superposition theorem. First, open circuit both 15A current sources so that 20V voltage source is acting alone as shown in Fig. 3.109 (*iii*). It is clear that :

$$V_{AB1} = {}^{*}20\text{V}$$

Next, open one 15A current source and replace 20V source by a short so that the second 15A source is acting alone as shown in Fig. 3.109 (*iv*). By current-divider rule, the currents in the various branches will be as shown in Fig. 3.109 (*iv*).

---

*   The circuit behaves as a 20V source having internal resistance of $(1 + 2)\Omega \parallel 6\Omega$ with terminals AB open.

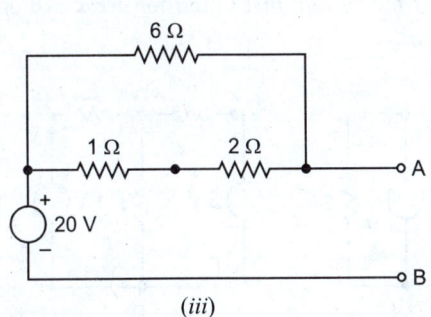

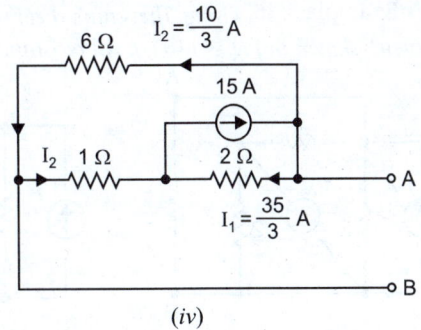

(iii)   *(iv)*

Fig. 3.109

Referring to Fig. 3.109 (*iv*), we have,

$$V_A - I_1 \times 2 + I_2 \times 1 = V_B$$

$$\therefore \quad V_A - V_B = I_1 \times 2 - I_2 \times 1 = \frac{35}{3} \times 2 - \frac{10}{3} \times 1 = 20 \text{ V}$$

$$\therefore \quad V_{AB2} = V_A - V_B = 20 \text{V}$$

Finally, open the second 15A source and replace the 20V source by a short as shown in Fig. 3.109 (*v*). By current-divider rule, the currents in the various branches will be as shown in Fig. 3.109 (*v*).

Now, $\quad V_A - I_3 \times 2 + I_4 \times 1 = V_B$

$$\therefore \quad V_A - V_B = I_3 \times 2 - I_4 \times 1 = \frac{5}{3} \times 2 - \frac{40}{3} \times 1 = -10 \text{V}$$

$$\therefore \quad V_{AB3} = V_A - V_B = -10 \text{V}$$

By superposition theorem, the open-circuited voltage at terminals *AB* (*i.e.*, V_{Th}) with all sources present is

$$V_{Th} = V_{AB1} + V_{AB2} + V_{AB3} = 20 + 20 - 10 = 30 \text{V}$$

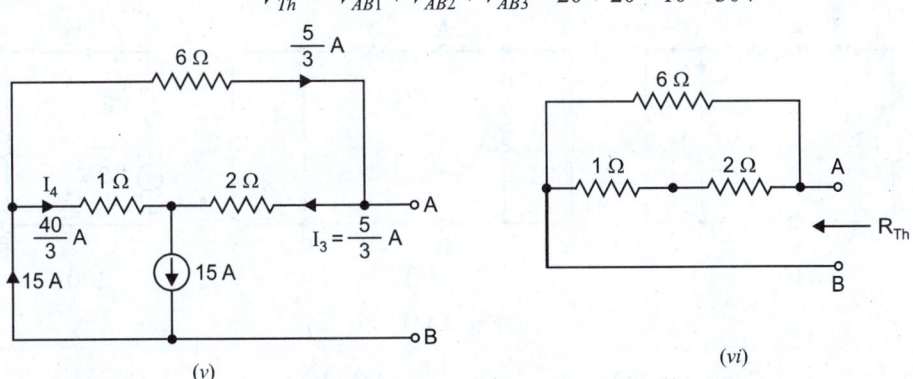

(v)   *(vi)*

Fig. 3.109

(ii) Finding R_{Th}. Thevenin resistance R_{TH} is the resistance at terminals *AB* when 3Ω is removed and current sources replaced by open and voltage source replaced by short as shown in Fig. 3.109 (*vi*).

$$\therefore \quad R_{Th} = (1\Omega + 2\Omega) \,\|\, 6\Omega = 2\Omega$$

$$\therefore \quad \text{Current in 3Ω, } I = \frac{V_{Th}}{R_{Th} + 3} = \frac{30}{2 + 3} = 6\text{A}$$

$$\therefore \quad \text{Voltage across 3Ω} = I \times 3 = 6 \times 3 = \textbf{18V}$$

Example 3.46. *Using Thevenin's theorem, determine the current in 1 Ω resistor across AB of the network shown in Fig. 3.110 (i). All resistances are in ohms.*

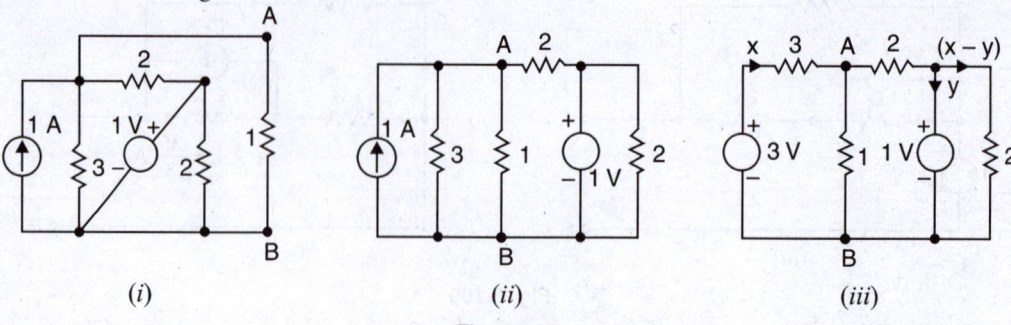

Fig. 3.110

Solution. The circuit shown in Fig. 3.110 (i) can be redrawn as shown in Fig. 3.110 (ii). If we convert the current source into equivalent voltage source, the circuit becomes as shown in Fig. 3.110 (iii). In order to find V_{Th}, remove 1 Ω resistor from the terminals AB. Then voltage at terminals AB is equal to V_{Th} (See Fig. 3.111 (i)). Applying *KVL* to the first loop in Fig. 3.111 (i), we have,

$$3 - (3 + 2)\, x - 1 = 0 \qquad \therefore \qquad x = 0.4 \text{ A}$$

$$\therefore \qquad V_{Th} = V_{AB} = 3 - 3x = 3 - 3 \times 0.4 = 1.8 \text{ V}$$

In order to find R_{Th}, replace the voltage sources by short circuits and current sources by open circuits in Fig. 3.110 (ii). The circuit then becomes as shown in Fig. 3.111 (ii). Then resistance at terminals AB is equal to R_{Th}.

Clearly, $$R_{Th} = 2 \,\|\, 3 = \frac{2 \times 3}{2 + 3} = 1.2 \ \Omega$$

Thevenin's equivalent circuit is 1.8 V voltage source in series with 1.2 Ω resistor. When 1 Ω resistor is connected across the terminals AB of the Thevenin's equivalent circuit, the circuit becomes as shown in Fig. 3.111 (iii).

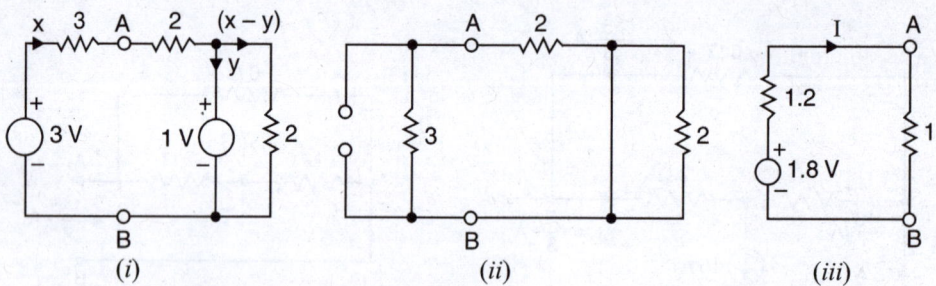

Fig. 3.111

$$\therefore \qquad \text{Current in 1 }\Omega = \frac{V_{Th}}{R_{Th} + 1} = \frac{1.8}{1.2 + 1} = \mathbf{0.82 \text{ A}}$$

Example 3.47. *At no-load, the terminal voltage of a d.c. generator is 120 V. When delivering its rated current of 40 A, its terminal voltage drops to 112 V. Represent the generator by its Thevenin equivalent.*

Solution. If R is the internal resistance of the generator, then,

$$E = V + IR \qquad \text{or} \qquad R = \frac{E - V}{I} = \frac{120 - 112}{40} = 0.2\Omega$$

Therefore, V_{Th} = No-load voltage = 120 V and $R_{Th} = R = 0.2$ Ω.

Hence Thevenin equivalent circuit of the generator is **120 V source in series with 0.2 Ω resistor.**

Example 3.48. *Calculate V_{Th} and R_{Th} between the open terminals A and B of the circuit shown in Fig. 3.112 (i). All resistance values are in ohms.*

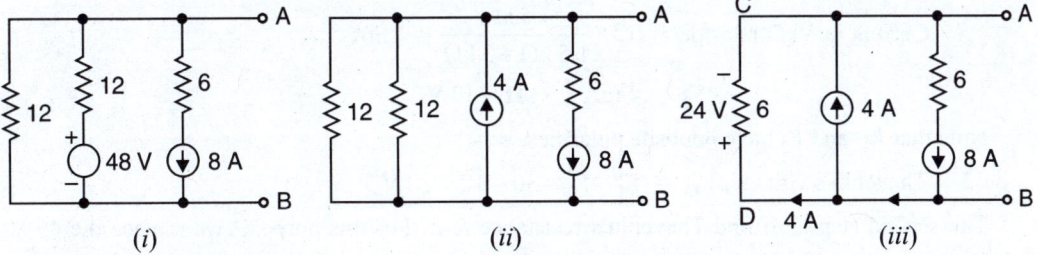

Fig. 3.112

Solution. If we replace the 48 V voltage source into equivalent current source, the circuit becomes as shown in Fig. 3.112 (*ii*). The two 12 Ω resistors are in parallel and can be replaced by 6 Ω resistor. The circuit then reduces to the one shown in Fig. 3.112 (*iii*). It is clear that 4 A current flows through 6 Ω resistor.

$$\therefore \quad V_{Th} = \text{Voltage across terminals } AB \text{ in Fig. 3.112 (iii)}$$
$$= \text{Voltage across 6 Ω resistor} = 4 \times 6 = 24 \text{ V}$$

Note that terminal *A* is negative w.r.t. *B*. Therefore, $V_{Th} = -\textbf{24 V.}$

$$R_{Th} = \text{Resistance between terminals } AB \text{ in Fig. 3.112 (i) with 48V}$$
$$\text{source replaced by a short and 8 A source replaced by an open}$$
$$= 12 \parallel 12 = \textbf{6 Ω}$$

Example 3.49. *Find the voltage across R_L in Fig. 3.113 when (i) $R_L = 1 \text{ k}\Omega$ (ii) $R_L = 2 \text{ k}\Omega$ (iii) $R_L = 9 \text{ k}\Omega$. Use Thevenin's theorem to solve the problem.*

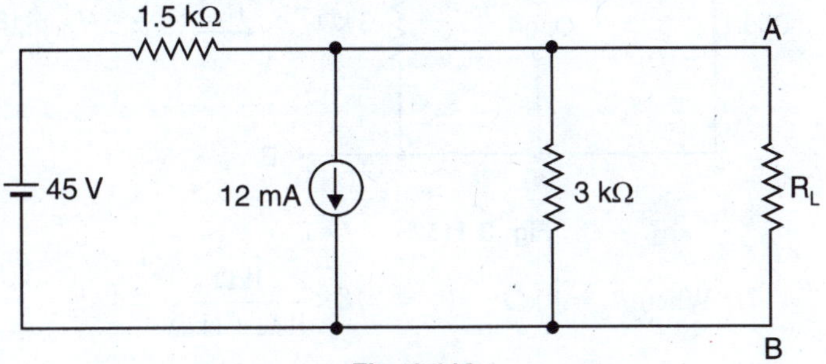

Solution. It is required to find the voltage across R_L when R_L has three different values. We shall find Thevenin's equivalent circuit to the left of the terminals *AB*. The solution involves two steps.

The first step is to find the open-circuited voltage V_{Th} at terminals *AB*. For this purpose, we shall use the superposition principle. With the current source removed (opened), we find voltage V_1 due to the 45 V source acting alone as shown in Fig. 3.114 (*i*). Since V_1 is the voltage across the 3 kΩ resistor, we have by voltage-divider rule :

Fig. 3.113

$$V_1 = 45 \times \frac{3 \text{ k}\Omega}{1.5 \text{ k}\Omega + 3 \text{ k}\Omega} = 30 \text{V}^+_-$$

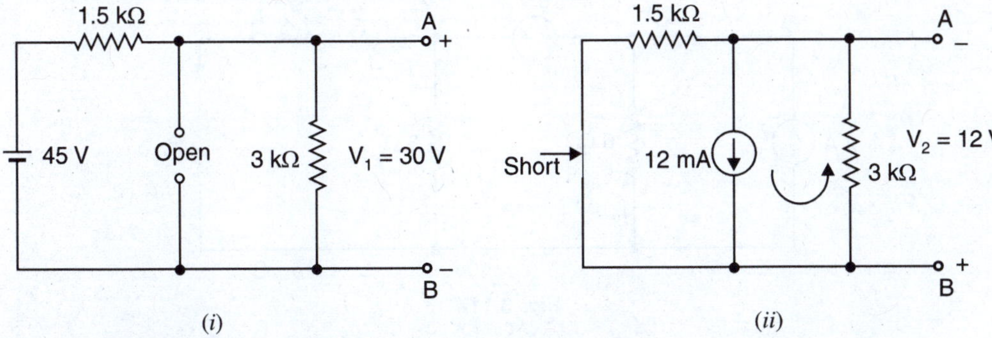

Fig. 3.114

The voltage V_2 due to the current source acting alone is found by shorting 45 V voltage source as shown in Fig. 3.114 (*ii*). By current-divider rule,

$$\text{Current in 3 k}\Omega \text{ resistor} = 12 \times \frac{1.5 \text{ k}\Omega}{1.5 \text{ k}\Omega + 3 \text{ k}\Omega} = 4 \text{ mA}$$

$$\therefore \qquad V_2 = 4 \text{ mA} \times 3 \text{ k}\Omega = 12 \text{ V}_+^-$$

Note that V_1 and V_2 have opposite polarities.

$$\therefore \qquad \text{Thevenin's voltage, } V_{Th} = V_1 - V_2 = 30 - 12 = 18 \text{ V}_-^+$$

The second step is to find Thevenin's resistance R_{Th}. For this purpose, we replace the 45 V voltage source by a short circuit and the 12 mA current source by an open circuit as shown in Fig. 3.115. As can be seen in the figure, R_{Th} is equal to parallel equivalent resistance of 1·5 kΩ and 3 kΩ resistors.

$$\therefore \qquad R_{Th} = 1\cdot5 \text{ k}\Omega \parallel 3 \text{ k}\Omega = 1 \text{ k}\Omega$$

Fig. 3.116 shows Thevenin's equivalent circuit.

$$\text{Voltage across } R_L, \quad V_L = 18 \times \frac{R_L}{1 \text{ k}\Omega + R_L}$$

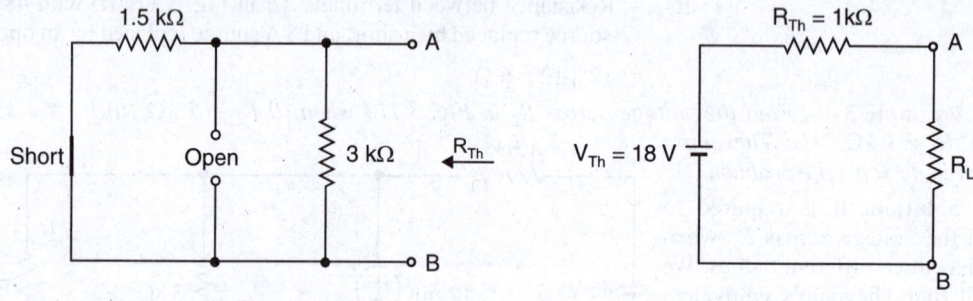

| Fig. 3.115 | Fig. 3.116 |

(*i*) When $R_L = 1 \text{ k}\Omega$; $V_L = 18 \times \dfrac{1\text{k}\Omega}{1\text{k}\Omega + 1\text{k}\Omega} = \textbf{9 V}$

(*ii*) When $R_L = 2 \text{ k}\Omega$; $V_L = 18 \times \dfrac{2\text{k}\Omega}{1\text{k}\Omega + 2\text{k}\Omega} = \textbf{12 V}$

(*iii*) When $R_L = 9 \text{ k}\Omega$; $V_L = 18 \times \dfrac{9\text{k}\Omega}{1\text{k}\Omega + 9\text{k}\Omega} = \textbf{16·2 V}$

Example 3.50. *Find Thevenin's equivalent circuit to the left of terminals AB in Fig. 3.117.*

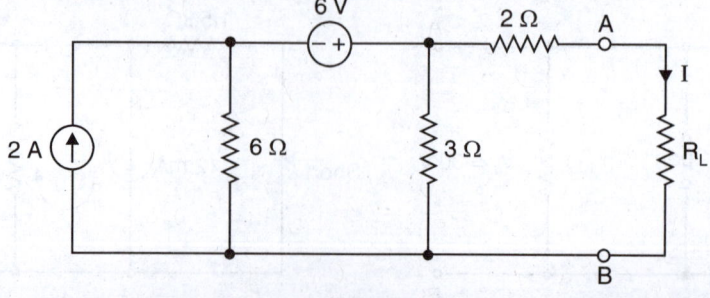

Fig. 3.117

Solution. To find V_{Th}, remove R_L from terminals AB. The circuit then becomes as shown in Fig. 3.118 (*i*).

$$\therefore \quad V_{Th} = \text{Voltage across terminals } AB \text{ in Fig. 3.118 (*i*)}$$

$$= \text{Voltage across 3 } \Omega \text{ resistor in Fig. 3.118 (*i*)}$$

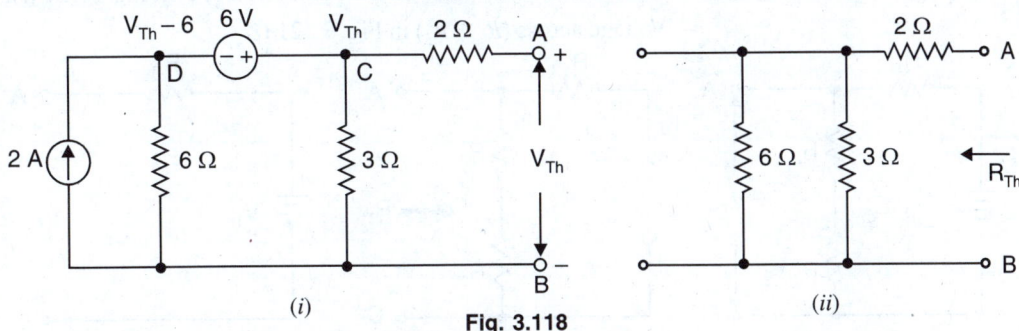

(*i*)   (*ii*)

Fig. 3.118

Note that voltage at point C is V_{Th} and voltage at point D is $V_{Th} - 6$. Therefore, nodal equation becomes :

$$\frac{V_{Th} - 6}{6} + \frac{V_{Th}}{3} = 2 \quad \text{or} \quad V_{Th} = 6 \text{ V}$$

In order to find R_{Th}, remove R_L and replace voltage source by a short and current source by an open in Fig. 3.117. The circuit then becomes as shown in Fig. 3.118 (*ii*).

$$\therefore \quad R_{Th} = \text{Resistance looking into terminals } AB \text{ in Fig. 3.118 (*ii*).}$$

$$= 2 + (3 || 6) = 2 + \frac{3 \times 6}{3 + 6} = 4 \, \Omega$$

Therefore, Thevenin equivalent circuit to the left of terminals AB is a **voltage source of 6 V (= V_{Th}) in series with a resistor of 4 Ω (= R_{Th}).** When load R_L is connected across the output terminals of Thevenin equivalent circuit, the circuit becomes as shown in Fig. 3.119. We can use Ohm's law to find current in the load R_L.

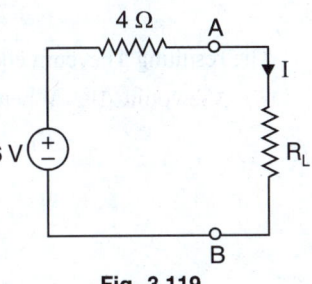

$$\therefore \quad \text{Current in } R_L, I = \frac{V_{Th}}{R_{Th} + R_L} = \frac{6}{4 + R_L}$$

Fig. 3.119

Example 3.51. *Find Thevenin's equivalent circuit in Fig. 3.120 when we view from (i) between points A and C (ii) between points B and C.*

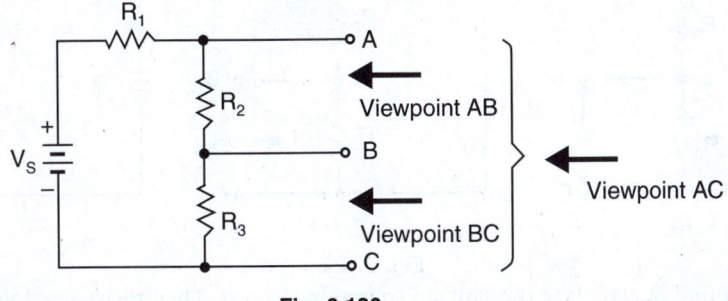

Fig. 3.120

Solution. The Thevenin equivalent for any circuit depends on the location of the two points from between which circuit is "viewed". Any given circuit can have more than one Thevenin equivalent, depending on how the viewpoints are designated. For example, if we view the circuit in Fig. 3.120

from between points A and C, we obtain a completely different result than if we view it from between points A and B or from between points B and C.

 (i) Viewpoint AC. When the circuit is viewed from between points A and C,

$$V_{Th} = \text{Voltage between open-circuited points } A \text{ and } C \text{ in Fig. 3.121 } (i).$$
$$= \text{Voltage across } (R_2 + R_3) \text{ in Fig. 3.121 } (i)$$

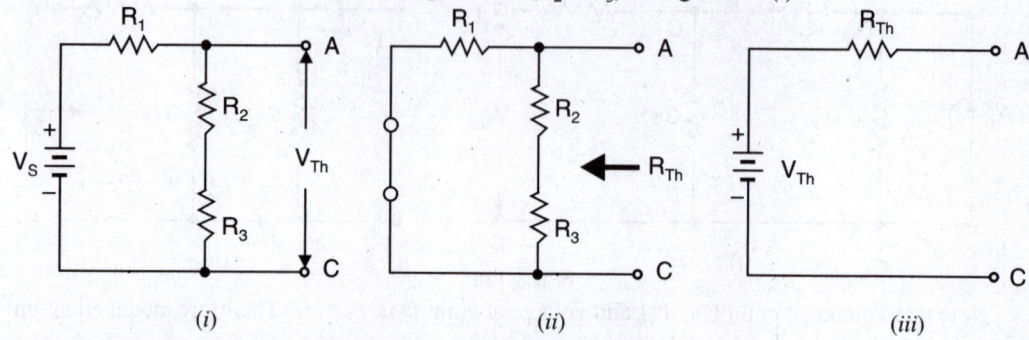

(i)   *(ii)*   *(iii)*

Fig. 3.121

$$= \frac{V_S}{R_1 + R_2 + R_3} \times (R_2 + R_3) = \left(\frac{R_2 + R_3}{R_1 + R_2 + R_3} \right) V_S$$

In order to find R_{Th}, replace the voltage source by a short. Then resistance looking into the open-circuited terminals A and C [See Fig. 3.121 *(ii)*] is equal to R_{Th}.

$$\therefore \qquad R_{Th} = R_1 \| (R_2 + R_3) = \frac{R_1 (R_2 + R_3)}{R_1 + R_2 + R_3}$$

The resulting Thevenin equivalent circuit is shown in Fig. 3.121 *(iii)*.

 (ii) Viewpoint BC. When the circuit is viewed from between points B and C,

$$V_{Th} = \text{Voltage between open-circuited points } B \text{ and } C \text{ in Fig.3.122 } (i).$$
$$= \text{Voltage across } R_3$$
$$= \frac{V_S}{R_1 + R_2 + R_3} \times R_3 = \left(\frac{R_3}{R_1 + R_2 + R_3} \right) V_S$$

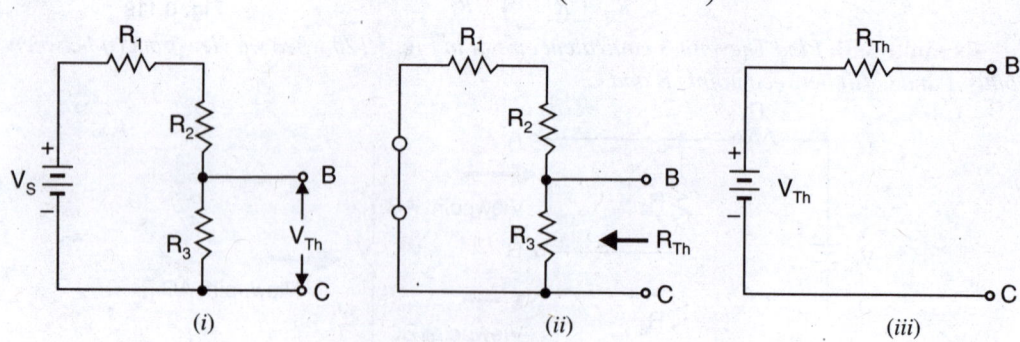

(i)   *(ii)*   *(iii)*

Fig. 3.122

In order to find R_{Th}, replace the voltage source by a short. Then resistance looking into the open-circuited terminals B and C [See Fig. 3.122 *(ii)*] is equal to R_{Th}.

$$\therefore \qquad R_{Th} = (R_1 + R_2) \| R_3 = \frac{R_3 (R_1 + R_2)}{R_1 + R_2 + R_3}$$

The resulting Thevenin equivalent circuit is shown in Fig. 3.122 *(iii)*.

Example 3.52. *Calculate (i) V_{Th} and (ii) R_{Th} between the open terminals A and B in the circuit shown in Fig. 3.123 (i). All resistance values are in ohms.*

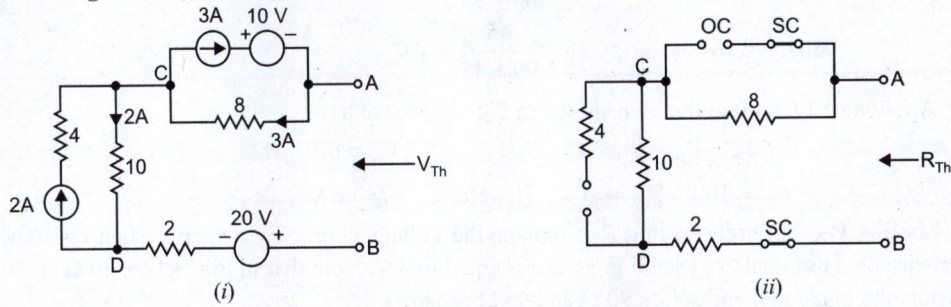

Fig. 3.123

Solution. Since terminals A and B are open, it is clear from the circuit that 10V and 20V voltage sources are ineffective in producing current in the circuit. However, current sources will circulate currents in their respective loops. Therefore, 2A current circulating in its loop will produce a voltage drop across 10 Ω resistance = 2A × 10 Ω = 20 V. Similarly, 3A current will produce a voltage drop across 8 Ω resistance = 3A × 8Ω = 24V. Tracing the circuit from A to B via points C and D [See Fig. 3.123 (*i*)], we have,

$$V_A - 24 - 20 + 20 = V_B$$

or
$$V_A - V_B = 24 + 20 - 20 = 24V$$

∴
$$V_{Th} = V_{AB} = V_A - V_B = \textbf{24V}$$

In order to find R_{Th}, open circuit the current sources and replace the voltage sources by a short as shown in Fig. 3.123 (*ii*). The resistance at the open-circuited terminals AB is R_{Th}.

∴
$$R_{Th} = \text{Resistance at terminals } AB \text{ in Fig. 3.123 } (ii)$$
$$= 8\Omega + 10\Omega + 2\Omega = \textbf{20}\boldsymbol{\Omega}$$

Example 3.53. *Find the current in the 25 Ω resistor in Fig. 3.124 (i) when E = 3 V.*

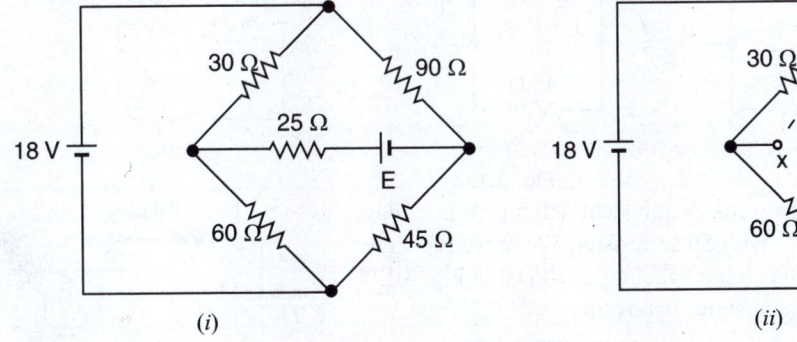

Fig. 3.124

Solution. Finding V_{Th}. Remove the voltage source E and the 25 Ω resistor, leaving the terminals $x - y$ open-circuited as shown in Fig. 3.124 (*ii*). The circuit shown in Fig. 3.124 (*ii*) can be redrawn as shown in Fig. 3.125. The voltage between terminals xy in Fig. 3.125 is equal to V_{Th}. We can use voltage-divider rule to find voltage drops across 60 Ω and 45 Ω resistors.

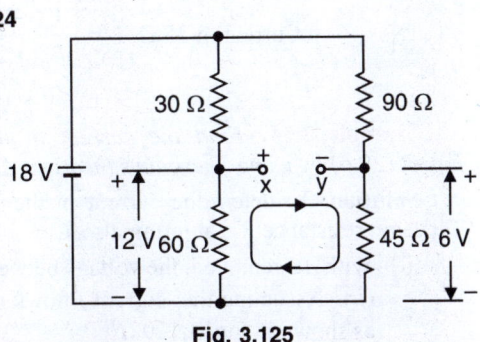

Fig. 3.125

$$\text{Voltage across } 60\ \Omega = 18 \times \frac{60}{60+30} = 12\text{V}$$

$$\text{Voltage across } 45\ \Omega = 18 \times \frac{45}{90+45} = 6\text{V}$$

Applying *KVL* around the loop shown in Fig. 3.125, we have,

$$12 - V_{xy} - 6 = 0 \qquad \therefore \qquad V_{xy} = 6\text{ V}$$

But $V_{xy} = V_{Th}$. Therefore, $V_{Th} = 6$ V.

Finding R$_{Th}$. In order to find R_{Th}, replace the voltage source by a short. Then resistance at open-circuited terminals xy (See Fig. 3.126) is equal to R_{Th}. Note that in Fig. 3.126, 30 Ω and 60 Ω resistors are in parallel and so are 90 Ω and 45 Ω resistors.

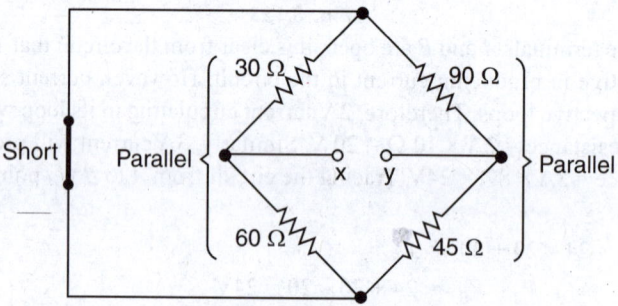

Fig. 3.126

The circuit shown in Fig. 3.126 can be redrawn as shown in Fig. 3.127 (*i*). This further reduces to the circuit shown in Fig. 3.127 (*ii*).

$$\therefore \qquad R_{Th} = 20 + 30 = 50\ \Omega$$

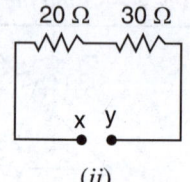

(*i*)

Fig. 3.127

(*ii*)

Therefore, the Thevenin equivalent circuit is a voltage source of 6 V in series with 50 Ω resistor. When we reconnect E and 25 Ω resistor, the circuit becomes as shown in Fig. 3.128. Note that V_{Th} and E are in series opposition.

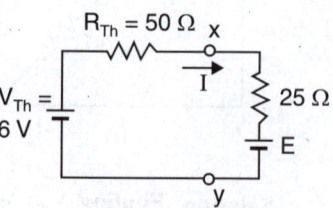

Fig. 3.128

$$\therefore \qquad \text{Current in 25 } \Omega, I = \frac{V_{Th} - E}{R_{Th} + 25} = \frac{6-3}{50+25}$$

$$= 40 \times 10^{-3}\text{ A} = \textbf{40 mA}$$

Example 3.54. *Find the current in the feeder AC of the distribution circuit shown in Fig. 3.129 (i) by using Thevenin's theorem. Also determine the currents in other branches.*

Solution. To determine current in the feeder *AC*, we shall find Thevenin voltage V_{Th} and Thevenin resistance R_{Th} at terminals *AC*.

(*i*) With *AC* removed, the voltage between *A* and *C* will be equal to V_{Th} as shown in Fig. 3.129 (*ii*). Assuming that current *I* flows in *AB*, then current distribution in the network will be as shown in Fig. 3.129 (*ii*).

Voltage drop along ADC = Voltage drop across ABC

or $\qquad 0.05\,(100 - I) + 0.05\,(80 - I) = 0.1\,I + 0.1\,(I - 30)$

or $\qquad\qquad\qquad\qquad\qquad 0.3\,I = 12 \quad \therefore \quad I = 12/0.3 = 40\,\text{A}$

$\therefore \qquad$ P.D. between A and C, V_{Th} = Voltage drop from A to C

$\qquad\qquad\qquad\qquad\qquad\qquad\qquad = 0.05\,(100 - 40) + 0.05\,(80 - 40) = 5\,\text{V}$

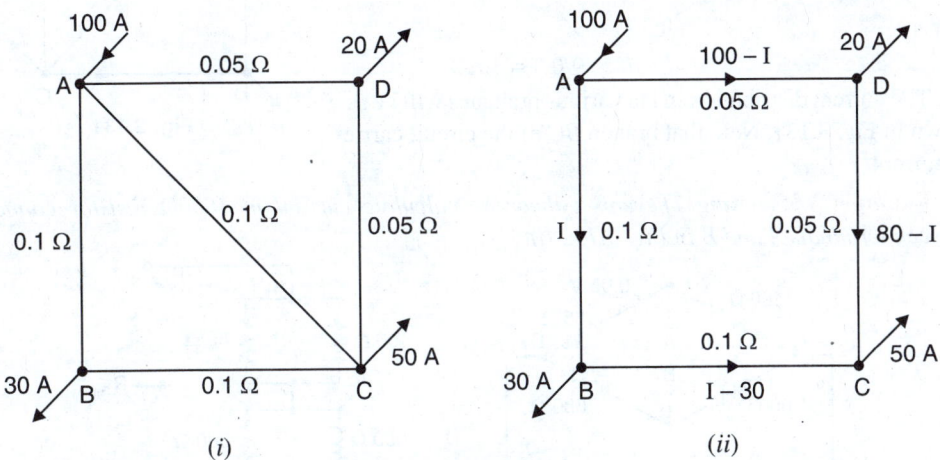

Fig. 3.129

(ii) With AC removed, the resistance between terminals A and C is equal to R_{Th}. Referring to Fig. 3.129 (ii), there are two parallel paths viz ADC (= 0.05 + 0.05 = 0.1 Ω) and ABC (= 0.1 + 0.1 = 0.2 Ω) between terminals A and C.

$\therefore \qquad\qquad\qquad R_{Th} = \dfrac{0.2 \times 0.1}{0.2 + 0.1} = 0.067\,\Omega$

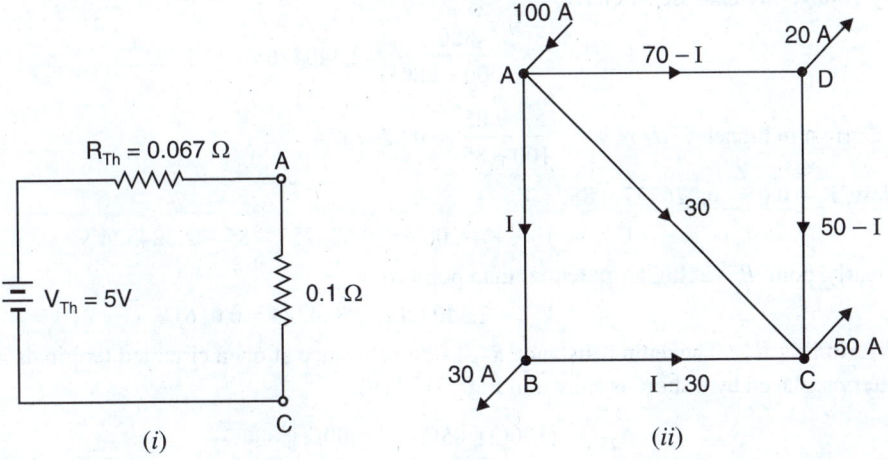

Fig. 3.130

The Thevenin equivalent circuit at terminals AC will be V_{Th} (= 5 V) in series with R_{Th} (= 0.067 Ω). When feeder AC (= 0.1 Ω) is connected between A and C, the circuit becomes as shown in Fig. 3.130 (i).

$\therefore \qquad\qquad$ Current in $AC = \dfrac{V_{Th}}{R_{Th} + 0.1} = \dfrac{5}{0.067 + 0.1} = \mathbf{30\,A}$

To find currents in other branches, refer to Fig. 3.130 (*ii*). With current in *AC* calculated (*i.e.* 30A) and current in *AB* assumed to be *I*, the current distribution will be as shown in Fig. 3.130 (*ii*). It is clear that voltage drop along the path *ADC* is equal to the voltage drop along the path *ABC* i.e.

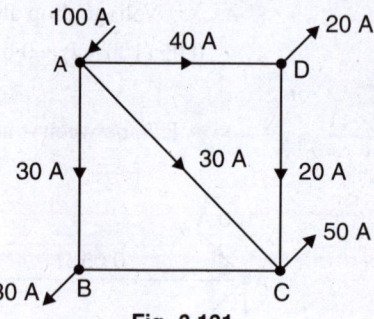

$$0.05 (70 - I) + 0.05 (50 - I) = 0.1\, I + 0.1\, (I - 30)$$

or $$0.3\, I = 9$$

∴ $$I = 9/0.3 = \textbf{30 A}$$

The current distribution in the various branches will be as shown in Fig. 3.131. Note that branch *BC* of the circuit carries no current.

Fig. 3.131

Example 3.55. *Using Thevenin's theorem, calculate current in 1000Ω resistor connected between terminals A and B in Fig. 3.132 (i).*

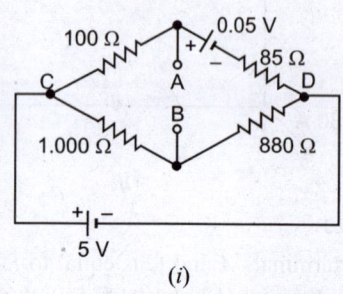

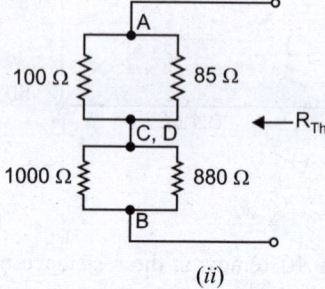

(*i*)   (*ii*)

Fig. 3.132

Solution. (*i*) **Finding V_{Th}.** Thevenin voltage V_{Th} is the voltage across open circuited terminals *AB* in Fig. 3.132 (*i*). Refer to Fig. 3.132 (*i*).

By voltage-divider rule, we have,

$$V_{BD} = 5 \times \frac{880}{1000 + 880} = 2.340426 \text{V}$$

Current in branch *CAD* is $I = \dfrac{5 - 0.05}{100 + 85} = 0.026757 \text{A}$

Now, $V_A - 0.05 - 0.026757 \times 85 = V_D$

∴ $$V_{AD} = V_A - V_D = 0.05 + 0.026757 \times 85 = 2.324324 \text{ V}$$

Clearly, point *B* is at higher potential than point *A*.

∴ $$V_{Th} = V_{BA} = 2.340426 - 2.324324 = 0.0161 \text{V}$$

(*ii*) **Finding R_{Th}.** Thevenin resistance R_{Th} is the resistance at open circuited terminals *AB* with 5V battery replaced by a short as shown in Fig. 3.132 (*ii*).

∴ $$R_{Th} = (100\Omega \parallel 85\Omega) + (1000\Omega \parallel 880\Omega)$$

$$= \frac{100 \times 85}{100 + 85} + \frac{1000 \times 880}{1000 + 880} = 514\Omega$$

∴   Current in 1000Ω connected between terminals *A* and *B*

$$= \frac{V_{Th}}{R_{Th} + 1000} = \frac{0.0161}{514 + 1000} = 10.634 \times 10^{-6} \text{A}$$

$$= \textbf{10.634 } \boldsymbol{\mu}\textbf{A } \textit{from B to A}$$

Example 3.56. *Calculate the values of* V_{Th} *and* R_{Th} *between the open terminals A and B of the circuit shown in Fig. 3.133 (i). All resistance values are in ohms.*

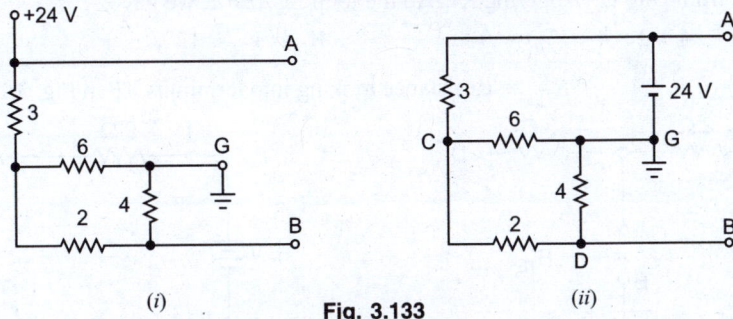

(i)   **Fig. 3.133**   (ii)

Solution. If we eliminate the ground symbols in the circuit shown in Fig. 3.133 (*i*), we get the circuit shown in Fig. 3.133 (*ii*). Referring to Fig. 3.133 (*ii*),

Total resistance offered to 24V battery

$$= 3\Omega + (6\Omega \parallel 6\Omega) = 3\Omega + 3\Omega = 6\Omega$$

Current delivered by 24V battery = 24/6 = 4A

The distribution of currents in the various branches of the circuit is shown in Fig. 3.133 (*iii*).

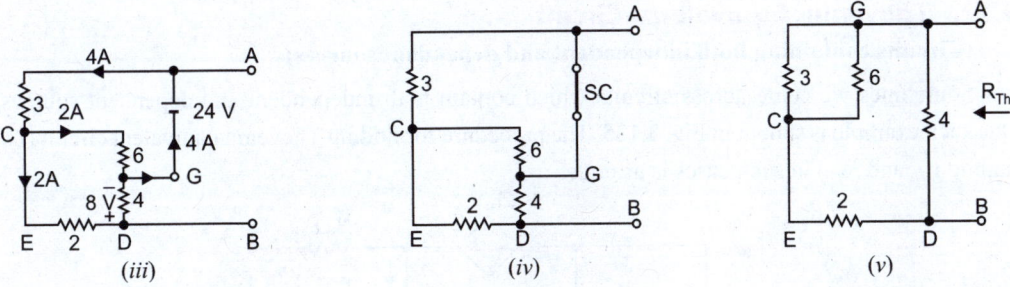

(iii)   (iv)   (v)

Fig. 3.133

Referring to Fig. 3.133 (*iii*) and tracing the circuit from point *A* to point *B via* points *C* and *D*, we have,

$$V_A - 3 \times 4 - 2 \times 6 + 4 \times 2 = V_B \quad \therefore \quad V_A - V_B = 3 \times 4 + 2 \times 6 - 4 \times 2 = 16V$$

$$\therefore \qquad V_{Th} = V_{AB} = V_A - V_B = \mathbf{16V}$$

In order to find R_{Th}, we replace the 24V source by a short and the circuit becomes as shown in Fig. 3.133 (*iv*). This circuit further reduces to the one shown in Fig. 3.133 (*v*).

$$\therefore \qquad R_{Th} = R_{AB} = [(3\Omega \parallel 6\Omega) + 2\Omega] \parallel 4\Omega = [2\Omega + 2\Omega] \parallel 4\Omega = \mathbf{2\Omega}$$

Example 3.57. *Using Thevenin theorem, find current in 1 Ω resistor in the circuit shown in Fig. 3.134 (i).*

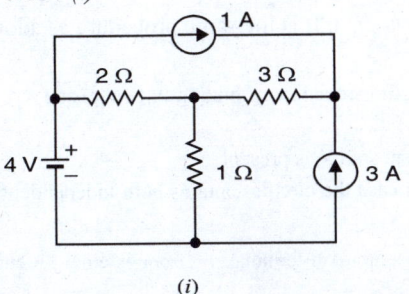

(i)

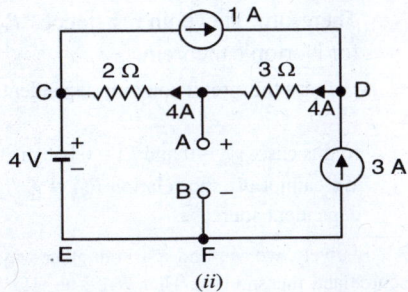

(ii)

Fig. 3.134

Solution. In order to find V_{Th}, remove the load as shown in Fig. 3.134 (ii). Then voltage between the open-circuited terminals A and B is equal to V_{Th}. It is clear from Fig. 3.134 (ii) that 4 A ($= 3 + 1$) flows from D to C. Applying *KVL* to the loop *ECABFE*, we have,

$$4 + 2 \times 4 - V_{AB} = 0 \qquad \therefore \qquad V_{AB} = V_{Th} = 12 \text{ V}$$

$$R_{Th} = \text{Resistance looking into terminals } AB \text{ in Fig. 3.134 (iii)} = 2 \, \Omega$$

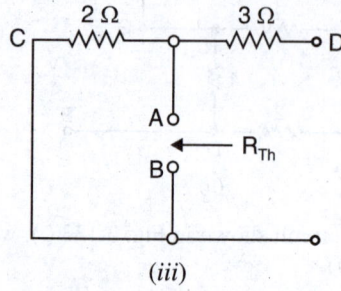   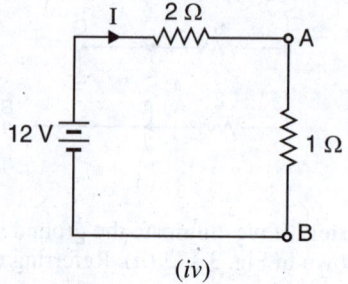

(iii)   (iv)

Fig. 3.134

When load (*i.e.* 1 Ω resistor) is reconnected, circuit becomes as shown in Fig. 3.134 (iv).

$$\therefore \qquad \text{Current in } 1 \, \Omega = \frac{12}{2+1} = 4 \text{ A}$$

3.12. Thevenin Equivalent Circuit

(Circuits containing both independent and dependent sources)

Sometimes we come across circuits which contain both independent and dependent sources. One such example is shown in Fig. 3.135. The procedure for finding Thevenin equivalent circuit (*i.e.* finding v_{Th} and R_{Th}) in such cases is as under :

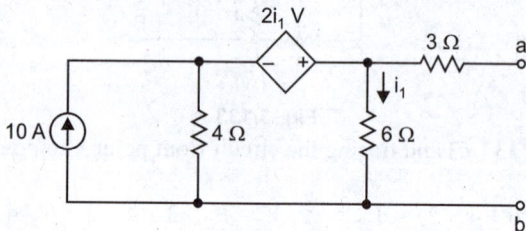

Fig. 3.135

(*i*) The open-circuit voltage v_{oc} ($= v_{Th}$) at terminals *ab* is determined as usual with sources present.

(*ii*) We cannot find R_{Th} at terminals *ab* simply by calculating equivalent resistance because of the presence of the dependent source. Instead, we place a short circuit across the terminals *ab* and find the value of short-circuit current i_{sc} at terminals *ab*.

(*iii*) Therefore, Thevenin resistance *$R_{Th} = v_{oc}/i_{sc} (= v_{Th}/i_{sc})$. It is the same procedure as adopted for Norton's theorem.

Note. In case the circuit contains dependent sources *only*, the procedure of finding $v_{oc} (= v_{Th})$ and R_{Th} is as under :

(*a*) In this case, $v_{oc} = 0$ and $i_{sc} = 0$ because no independent source is present.

(*b*) We cannot use the relation $R_{Th} = v_{oc}/i_{sc}$ as we do in case the circuit contains both independent and dependent sources.

---

*   Alternatively, we can find R_{Th} in another way. We excite the circuit at terminals *ab* from external 1A current source and measure v_{ab}. Then $R_{Th} = v_{ab}/1\Omega$.

(c) In order to find R_{Th}, we excite the circuit at terminals *ab* by connecting 1A source to the terminals *a* and *b* and calculate the value v_{ab}. Then $R_{Th} = v_{ab}/1\Omega$.

Example 3.58. *Find the values of v_{Th} and R_{Th} at terminals ab for the circuit shown in Fig. 3.136 (i).*

Solution. We first put a short circuit across terminals *a* and *b* and find short-circuit current i_{sc} at terminals *ab* as shown in Fig. 3.136 *(ii)*. Applying *KCL* at node *C*,

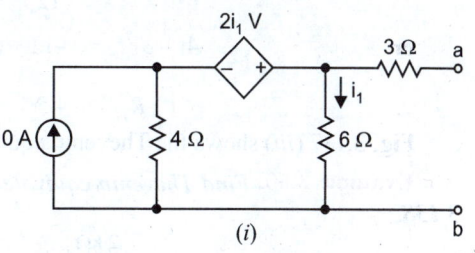

$$10 = i_1 + i_2 + i_{sc}$$

or $$i_2 = 10 - i_1 - i_{sc}$$

Applying *KVL* to loops 1 and 2, we have,

$$-4i_2 + 6i_1 - 2i_1 = 0 \qquad \text{... Loop 1}$$

or $$-4(10 - i_1 - i_{sc}) + 4i_1 = 0 \qquad ...(i)$$

Also $$-6i_1 + 3i_{sc} = 0 \qquad ...(ii) ... \text{Loop 2}$$

From eqs. *(i)* and *(ii)*, $i_{sc} = 5A$.

(i)

Fig. 3.136

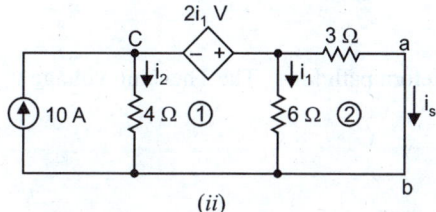

(ii)

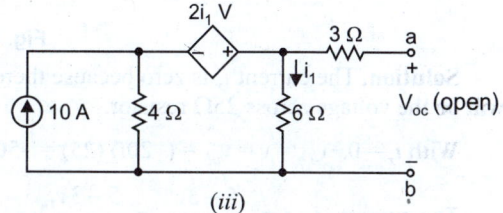

(iii)

Fig. 3.136

In order to find $v_{oc}\ (= v_{Th})$, we refer to Fig. 3.136 *(iii)* where we have,

$$v_{oc} = 6i_1 \qquad ...(iii)$$

Applying *KVL* to the central loop in Fig. 3.136 *(iii)*,

$$-4(10 - i_1) + 6i_1 - 2i_1 = 0 \qquad ...(iv)$$

From eqs. *(iii)* and *(iv)*, we have, $v_{oc} = v_{Th} = $ **30V**

Also $$R_{Th} = \frac{v_{oc}}{i_{sc}} = \frac{30}{5} = 6\Omega$$

Example 3.59. *Find Thevenin equivalent circuit for the network shown in Fig. 3.137 (i) which contains only a dependent source.*

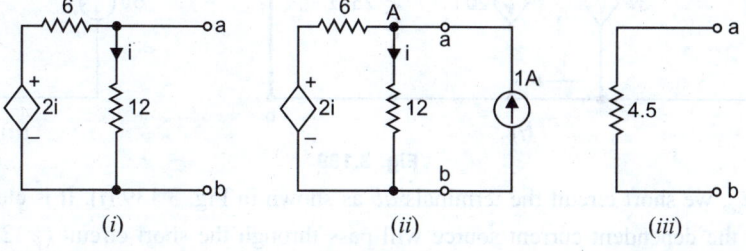

(i)   **(ii)**   **(iii)**

Fig. 3.137

Solution. In order to find R_{Th}, we connect 1A current source to terminals *a* and *b* as shown in Fig. 3.137 *(ii)*. Then by finding the value of v_{ab}, we can determine the value of $R_{Th} = v_{ab}/1\Omega$. It may be seen that potential at point *A* is the same as that at *a*.

$\therefore$ $$v_{ab} = \text{Voltage across } 12\Omega \text{ resistor}$$

Applying *KCL* to point *A*, we have,

$$\frac{2i - v_{ab}}{6} + 1 = \frac{v_{ab}}{12}$$

or $\qquad 4i - 3v_{ab} = -12 \quad$ or $\quad 4\left(\frac{v_{ab}}{12}\right) - 3v_{ab} = -2 \quad \therefore v_{ab} = 4.5V$

$\therefore \qquad\qquad R_{Th} = 4.5/1 = \mathbf{4.5\Omega}$

Fig. 3.137 (*iii*) shows the Thevenin equivalent circuit.

Example 3.60. *Find Thevenin equivalent circuit at terminals ab for the circuit shown in Fig. 3.138.*

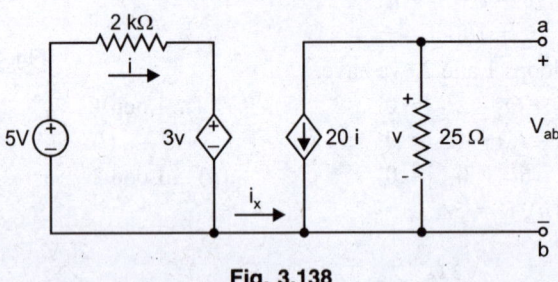

Fig. 3.138

Solution. The current i_x is zero because there is no return path for i_x. The Thevenin voltage v_{Th} will be the voltage across 25Ω resistor.

With $i_x = 0$, $v_{Th} = v = v_{ab} = (-20i)(25) = -500i$

The current i is, $i = \dfrac{5 - 3v}{2 \times 1000} = \dfrac{5 - 3v_{Th}}{2000} \qquad (\because v = v_{Th})$

$\therefore \qquad\qquad v_{Th} = -500\left(\dfrac{5 - 3v_{Th}}{2000}\right) \quad$ or $\quad v_{Th} = -5V$

In order to find Thevenin resistance R_{Th}, we find the short-circuit current i_{sc} at terminals *ab*. Then,

$$R_{Th} = \frac{v_{Th}}{i_{sc}}$$

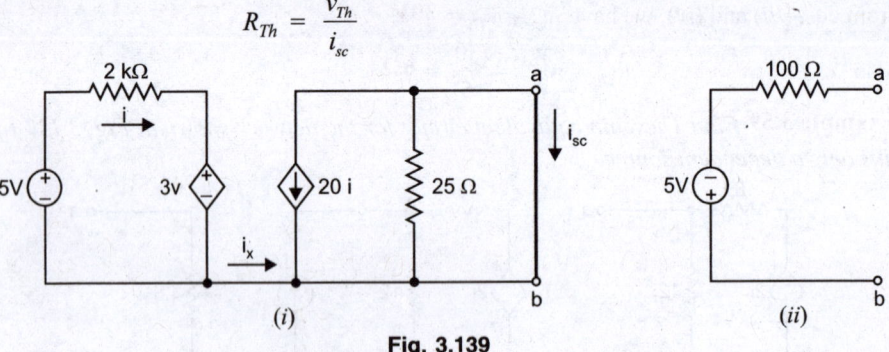

(*i*)   (*ii*)

Fig. 3.139

To find i_{sc}, we short circuit the terminals *ab* as shown in Fig. 3.139 (*i*). It is clear that all the current from the dependent current source will pass through the short circuit ($\because$ 25Ω resistor is shunted by the short circuit).

$\therefore \qquad\qquad i_{sc} = -20i$

Now, $i = \dfrac{5}{2000} = 2.5$ mA so that $i_{sc} = -20 \times 2.5 = -50$ mA

$\therefore \qquad\qquad R_{Th} = \dfrac{v_{Th}}{i_{sc}} = \dfrac{-5}{-50 \times 10^{-3}} = 100\ \Omega$

Fig. 3.139 (*ii*) shows the **Thevenin equivalent circuit at terminals *ab*.**

Tutorial Problems

1. Using Thevenin's theorem, find the current in 10 Ω resistor in the circuit shown in Fig. 3.140. **[0.481 A]**

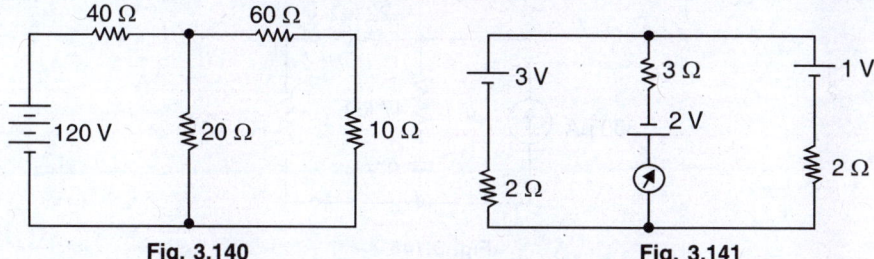

Fig. 3.140   **Fig. 3.141**

2. Using Thevenin's theorem, find current in the ammeter shown in Fig. 3.141.   **[1 A]**
3. Using Thevenin's theorem, find p.d. across branch *AB* of the network shown in Fig. 3.142.   **[4.16 V]**

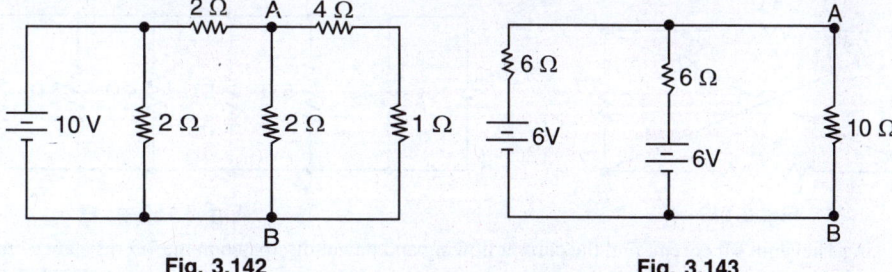

Fig. 3.142   **Fig. 3.143**

4. Determine Thevenin's equivalent circuit to the left of *AB* in Fig. 3.143.

[A 6 V source in series with 3 Ω]

5. A Wheatstone bridge *ABCD* is arranged as follows : *AB* = 100 Ω, *BC* = 99 Ω, *CD* = 1000 Ω and *DA* = 1000 Ω. A battery of e.m.f. 10 V and negligible resistance is connected between *A* and *C* with *A* positive. A galvanometer of resistance 100 Ω is connected between *B* and *D*. Using Thevenin's theorem, determine the galvanometer current. **[38.6 μA]**

6. Find the Thevenin equivalent circuit of the circuitry, excluding R_1, connected to the terminals $x - y$ in Fig. 3.144. **[10 V in series with 9Ω ; *x* positive w.r.t. *y*]**

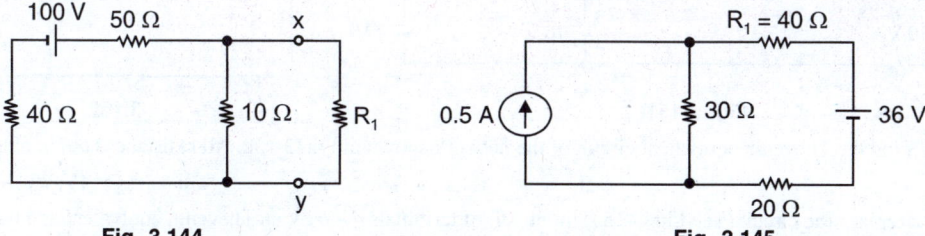

Fig. 3.144   **Fig. 3.145**

7. Find the voltage across R_1 in Fig. 3.145 by constructing Thevenin equivalent circuit at the R_1 terminals. Be sure to indicate the polarity of the voltage. **[– (9.33V) +]**

8. By using Thevenin Theorem, find current *I* in the circuit shown in Fig. 3.146. **[2·5 A]**

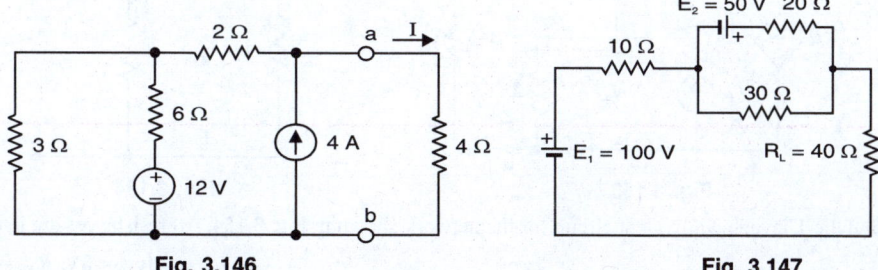

Fig. 3.146   **Fig. 3.147**

9. Find Thevenin equivalent circuit in Fig. 3.147. $[V_{Th} = 130 \text{ V}; \ R_{Th} = 22 \ \Omega]$

10. Find the Thevenin equivalent circuit of the circuitry, excluding R_1, connected to terminals $x - y$ in Fig. 3.148. $[V_{Th} = 23 \cdot 1 \text{ V}; \ R_{Th} = 69 \text{ k}\Omega]$

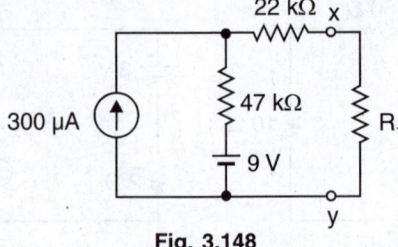

Fig. 3.148

11. Using Thevenin's theorem, find the magnitude and direction of current in 2Ω resistor in the circuit shown in Fig. 3.149.   [0.25A from D to B]

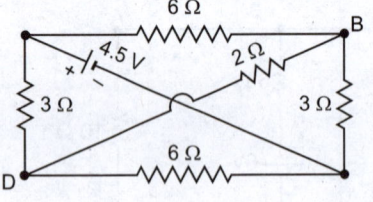

Fig. 3.149

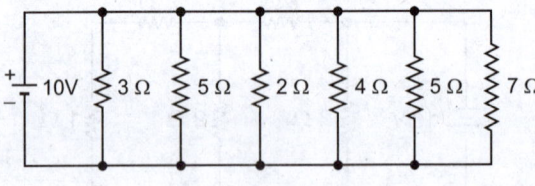

Fig. 3.150

12. Using Thevenin's theorem, find the current flowing and power dissipated in the 7Ω resistance branch in the circuit shown in Fig. 3.150.   [1.43A; 14.3W]

13. Find Thevenin's equivalent circuit at terminals BC of Fig. 3.151. Hence determine current through the resistor $R = 1\Omega$. $[V_{Th} = 76/7 \text{ V}; \ R_{Th} = 32/7\Omega; \ 76/39\text{A}]$

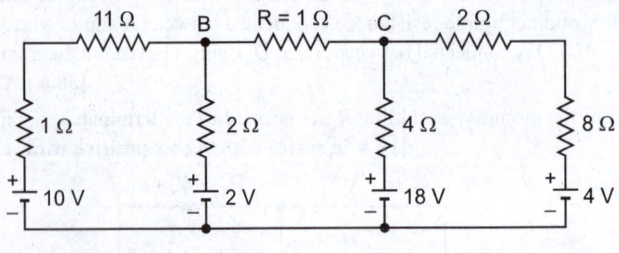

Fig. 3.151

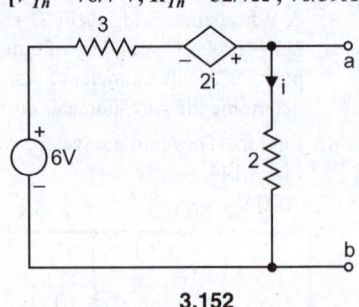

3.152

14. Find the Thevenin equivalent circuit of the network shown in Fig. 3.152. All resistances are in ohms.
                          $[v_{Th} = 4\text{V}; \ R_{Th} = 8\Omega]$

15. Replace the circuit (See Fig. 3.153) to the left of terminals $a - b$ by its Thevenin equivalent and use the result to find v. $[v_{Th} = 12\text{V}; \ R_{Th} = 8\Omega; \ v = 4\text{V}]$

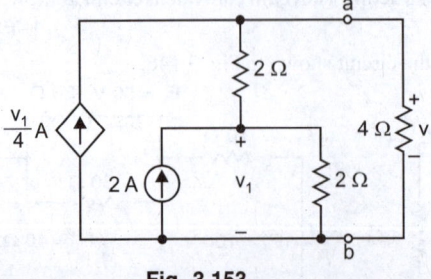

Fig. 3.153

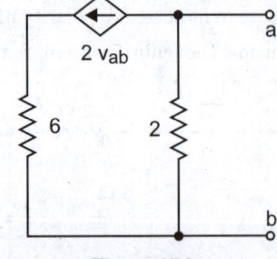

Fig. 3.154

16. Find the Thevenin equivalent circuit for the network shown in Fig. 3.154. All resistances are in ohms.
                          $[v_{Th} = 0\text{V}; \ R_{Th} = 2/5\Omega]$

3.13. Advantages of Thevenin's Theorem

The Thevenin equivalent circuit is *always* an equivalent voltage source (V_{Th}) in series with an equivalent resistance (R_{Th}) regardless of the original circuit that it replaces. Although the Thevenin equivalent is not the same as its original circuit, it acts the same in terms of output voltage and current. It is worthwhile to give the advantages of Thevenin's theorem.

(*i*) It reduces a complex circuit to a simple circuit *viz.* a single source of e.m.f. V_{Th} in series with a single resistance R_{Th}.

(*ii*) It greatly simplifies the portion of the circuit of lesser interest and enables us to view the action of the output part directly.

(*iii*) This theorem is particularly useful to find current in a particular branch of a network as the resistance of that branch is varied while all other resistances and sources remain constant.

(*iv*) Thevenin's theorem can be applied in successive steps. Any two points in a circuit can be chosen and all the components to one side of these points can be reduced to Thevenin's equivalent circuit.

3.14. Norton's Theorem

Fig. 3.155 (*i*) shows a network enclosed in a box with two terminals A and B brought out. The network in the box may contain any number of resistors and e.m.f. sources connected in any manner. But according to Norton, the entire circuit behind AB can be replaced by a current source I_N in parallel with a resistance R_N as shown in Fig. 3.155 (*ii*). The resistance R_N is the same as Thevenin resistance R_{Th}. The value of I_N is determined as mentioned in Norton's theorem. Once *Norton's equivalent circuit* is determined [See Fig. 3.155 (*ii*)], then current in any load R_L connected across AB can be readily obtained.

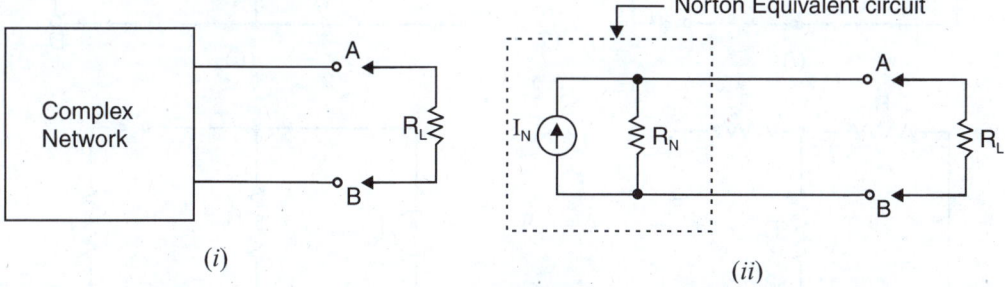

Fig. 3.155

Hence Norton's theorem as applied to d.c. circuits may be stated as under :

Any linear, bilateral network having two terminals A and B can be replaced by a current source of current output I_N in parallel with a resistance R_N.

(*i*) *The output I_N of the current source is equal to the current that would flow through AB when A and B are short-circuited.*

(*ii*) *The resistance R_N is the resistance of the network measured between A and B with load removed and the sources of e.m.f. replaced by their internal resistances. Ideal voltage sources are replaced with short circuits and ideal current sources are replaced with open circuits.*

Norton's Theorem is *converse* of Thevenin's theorem in that Norton equivalent circuit uses a current generator instead of voltage generator and the resistance R_N (which is the same as R_{Th}) in parallel with the generator instead of being in series with it. *Thus the use of either of these theorems enables us to replace the entire circuit seen at a pair of terminals by an equivalent circuit made up of a single source and a single resistor.*

Illustration. Fig. 3.156 illustrates the application of Norton's theorem. As far as the circuit behind terminals AB is concerned [See Fig. 3.156 (i)], it can be replaced by a current source I_N in parallel with a resistance R_N as shown in Fig. 3.156 (iv). The output I_N of the current generator is equal to the current that would flow through AB when terminals A and B are short-circuited as shown in Fig. 3.156 (ii). The load on the source when terminals AB are short-circuited is given by ;

$$R' = R_1 + \frac{R_2 R_3}{R_2 + R_3} = \frac{R_1 R_2 + R_1 R_3 + R_2 R_3}{R_2 + R_3}$$

Source current, $I' = \dfrac{V}{R'} = \dfrac{V(R_2 + R_3)}{R_1 R_2 + R_1 R_3 + R_2 R_3}$

Short-circuit current, I_N = Current in R_2 in Fig. 3.156 (ii)

$$= I' \times \frac{R_3}{R_2 + R_3} = \frac{V R_3}{R_1 R_2 + R_1 R_3 + R_2 R_3}$$

To find R_N, remove the load R_L and replace battery by a short because its internal resistance is assumed zero [See Fig. 3.156 (iii)].

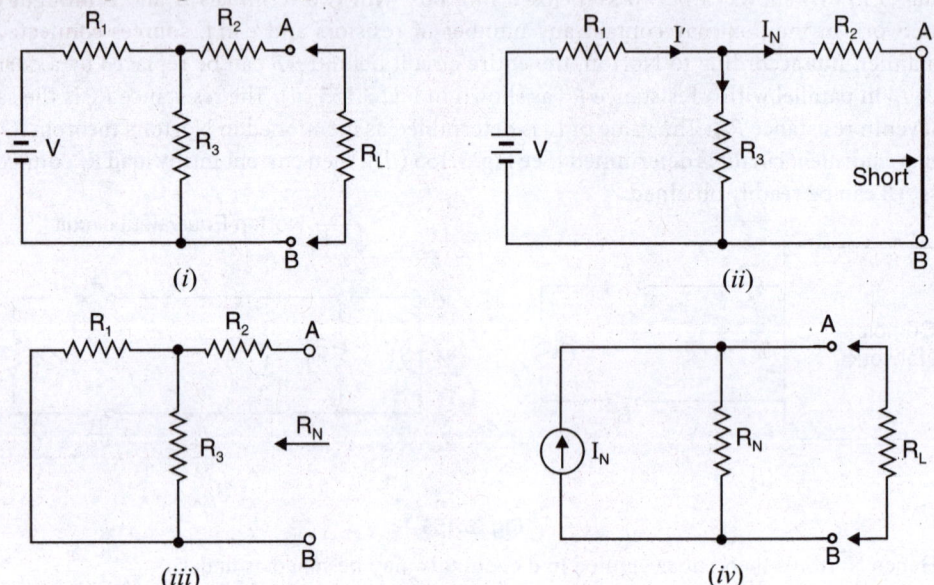

Fig. 3.156

$\therefore$ $\qquad R_N$ = Resistance at terminals AB in Fig. 3.156 (iii).

$$= R_2 + \frac{R_1 R_3}{R_1 + R_3}$$

Thus the values of I_N and R_N are known. The Norton equivalent circuit will be as shown in Fig. 3.156 (iv).

3.15. Procedure for Finding Norton Equivalent Circuit

(i) Open the two terminals (*i.e.* remove any load) between which we want to find Norton equivalent circuit.

(ii) Put a short-circuit across the terminals under consideration. Find the short-circuit current flowing in the short circuit. It is called Norton current I_N.

(*iii*) Determine the resistance between the two open terminals with all ideal voltage sources shorted and all ideal current sources opened (a non-ideal source is replaced by its internal resistance). It is called Norton's resistance R_N. It is easy to see that $R_N = R_{Th}$.

(*iv*) Connect I_N and R_N in parallel to produce Norton equivalent circuit between the two terminals under consideration.

(*v*) Place the load resistor removed in step (*i*) across the terminals of the Norton equivalent circuit. The load current can now be calculated by using current-divider rule. This load current will be the same as the load current in the original circuit.

Example 3.61. *Show that when Thevenin's equivalent circuit of a network is converted into Norton's equivalent circuit, $I_N = V_{Th}/R_{Th}$ and $R_N = R_{Th}$. Here V_{Th} and R_{Th} are Thevenin voltage and Thevenin resistance respectively.*

Solution. Fig. 3.157 (*i*) shows a network enclosed in a box with two terminals A and B brought out. Thevenin's equivalent circuit of this network will be as shown in Fig. 3.157 (*ii*). To find Norton's equivalent circuit, we are to find I_N and R_N. Referring to Fig. 3.157 (*ii*),

$$I_N = \text{Current flowing through short-circuited } AB \text{ in Fig. 3.157 (ii)}$$
$$= V_{Th}/R_{Th}$$
$$R_N = \text{Resistance at terminals } AB \text{ in Fig. 3.157 (ii)}$$
$$= R_{Th}$$

Fig. 3.157 (*iii*) shows Norton's equivalent circuit. Hence we arrive at the following two important conclusions :

(*i*) To convert Thevenin's equivalent circuit into Norton's equivalent circuit,

$$I_N = V_{Th}/R_{Th} \;;\; R_N = R_{Th}$$

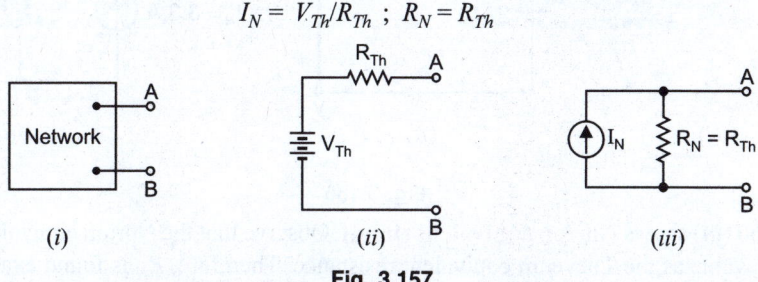

(i) (ii) (iii)

Fig. 3.157

(*ii*) To convert Norton's equivalent circuit into Thevenin's equivalent circuit,

$$V_{Th} = I_N R_N \;;\; R_{Th} = R_N$$

Example 3.62. *Find the Norton equivalent circuit at terminals $x - y$ in Fig. 3.158.*

Solution. We shall first find the Thevenin equivalent circuit and then convert it to an equivalent current source. This will then be Norton equivalent circuit.

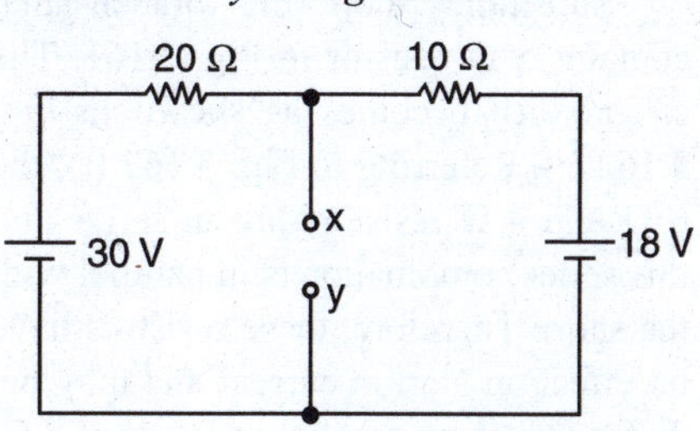

Finding Thevenin equivalent circuit. To find V_{Th}, refer to Fig. 3.159 (*i*). Since 30 V and 18 V sources are in opposition, the circuit current I is given by ;

$$I = \frac{30 - 18}{20 + 10} = \frac{12}{30} = 0.4 \text{ A}$$

Fig. 3.158

Applying Kirchhoff's voltage law to loop $ABCDA$, we have,

$$30 - 20 \times 0.4 - V_{Th} = 0 \quad \therefore \quad V_{Th} = 30 - 8 = 22 \text{ V}$$

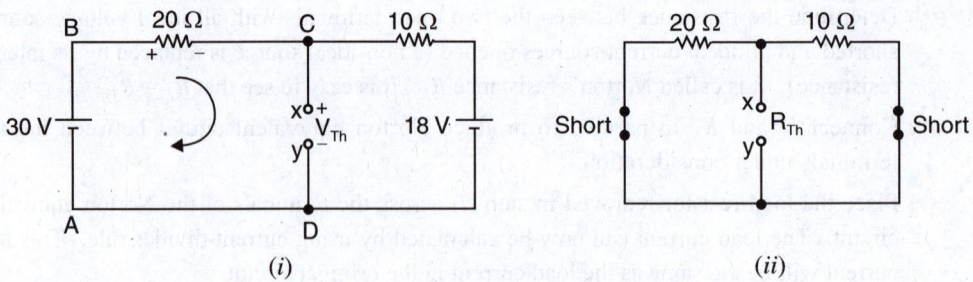

Fig. 3.159

To find R_{Th}, we short both voltage sources as shown in Fig. 3.159 (ii). Notice that 10 Ω and 20 Ω resistors are then in parallel.

$$\therefore \qquad R_{Th} = 10\ \Omega\ ||\ 20\ \Omega = \frac{10 \times 20}{10 + 20} = 6.67\ \Omega$$

Therefore, Thevenin equivalent circuit will be as shown in Fig. 3.160 (i). Now it is quite easy to convert it into equivalent current source.

$$I_N = \frac{V_{Th}}{R_{Th}} = \frac{22}{6.67} = 3.3\text{A} \qquad\qquad [\text{See Fig. 3.160 (ii)}]$$

$$R_N = R_{Th} = 6.67\ \Omega$$

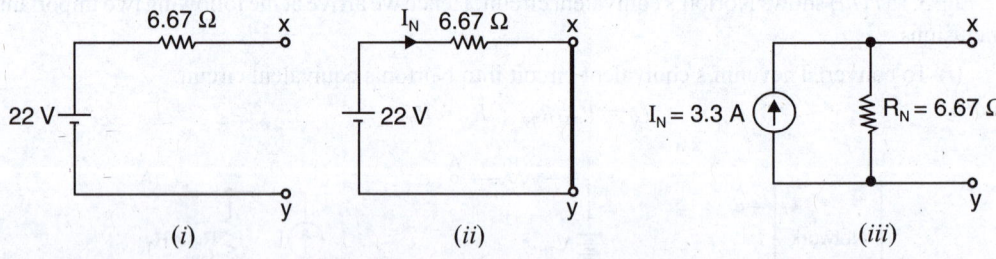

Fig. 3.160

Fig. 3.160 (iii) shows **Norton equivalent circuit.** Observe that the Norton equivalent resistance has the same value as the Thevenin equivalent resistance. Therefore, R_N is found exactly the same way as R_{Th}.

Example 3.63. *Using Norton's theorem, calculate the current in the 5 Ω resistor in the circuit shown in Fig. 3.161.*

Solution. Short the branch that contains 5 Ω resistor in Fig. 3.161. The circuit then becomes as shown in Fig. 3.162 (i). Referring to Fig. 3.162 (i), the 6 Ω and 4 Ω resistors are in series and this series combination is in parallel with the short. Therefore, these resistors have

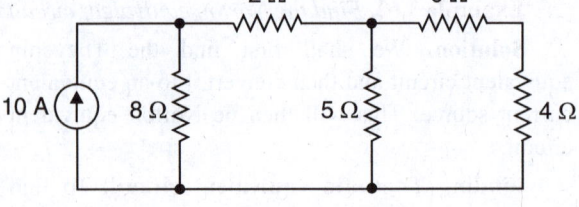

Fig. 3.161

no effect on Norton current and may be considered as removed from the circuit. As a result, 10 A divides between parallel resistors of 8 Ω and 2 Ω.

$$\therefore \qquad \text{Norton current, } I_N = \text{Current in 2 Ω resistor}$$

$$= 10 \times \frac{8}{8 + 2} = 8\text{ A} \qquad\qquad \text{... Current-divider rule}$$

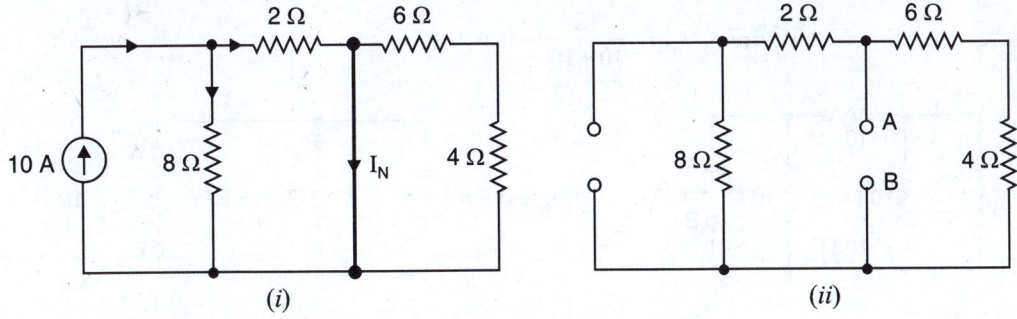

Fig. 3.162

In order to find Norton resistance $R_N (= R_{Th})$, open circuit the branch containing the 5 Ω resistor and replace the current source by an open in Fig. 3.161. The circuit then becomes as shown in Fig. 3.162 (*ii*).

$$\text{Norton resistance, } R_N = \text{Resistance at terminals } AB \text{ in Fig. 3.162 (}ii\text{).}$$

$$= (2 + 8) \| (4 + 6) = 10 \| 10 = \frac{10 \times 10}{10 + 10} = 5\,\Omega$$

Therefore, Norton equivalent circuit consists of a current source of 8 A $(= I_N)$ in parallel with a resistance of 5 Ω $(= R_N)$ as shown in Fig. 3.163 (*i*). When the branch containing 5 Ω resistor is connected across the output terminals of Norton's equivalent circuit, the circuit becomes as shown in Fig. 3.163 (*ii*).

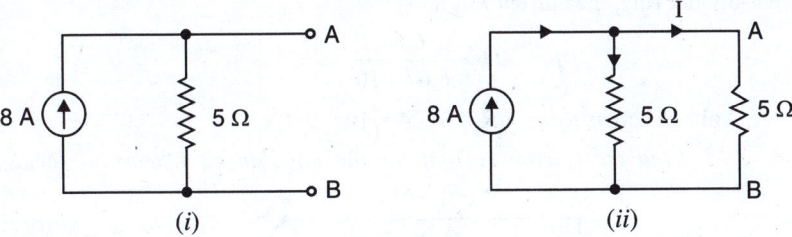

Fig. 3.163

By current-divider rule, the current I in 5 Ω resistor is

$$I = 8 \times \frac{5}{5 + 5} = \textbf{4 A}$$

Example 3.64. *Find Norton equivalent circuit for Fig. 3.164 (i). Also solve for load current and load voltage.*

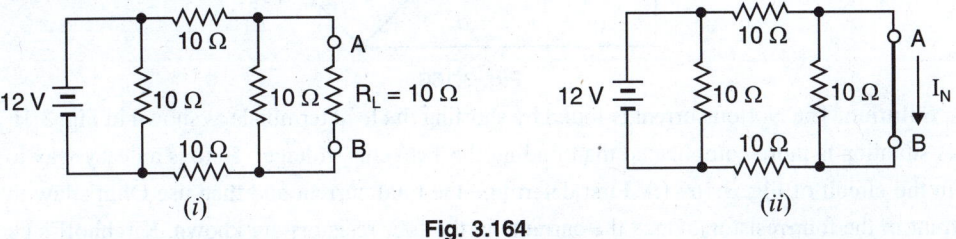

Fig. 3.164

Solution. Short the branch that contains R_L (= 10 Ω) in Fig. 3.164 (*i*). The circuit then becomes as shown in Fig. 3.164 (*ii*). The resistor that is in parallel with the battery has no effect on the Norton current (I_N). The resistor in parallel with the short also has no effect. Therefore, these resistors may be considered as removed from the circuit shown in Fig. 3.164 (*ii*). The circuit then contains two 10 Ω resistors in series.

$$\therefore \quad \text{Norton current, } I_N = \frac{12}{10+10} = 0.6 \text{ A}$$

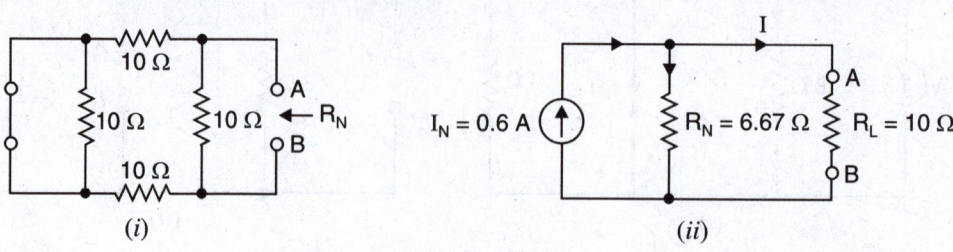

(i)   (ii)

Fig. 3.165

In order to find Norton resistance R_N (= R_{Th}), open circuit the branch containing R_L and replace the voltage source by a short ($\because$ internal resistance of the voltage source is assumed zero) in Fig. 3.164 (i). The circuit then becomes as shown in Fig. 3.165 (i).

Norton resistance, *R_N = Resistance at terminals AB in Fig. 3.165 (i)

$$= (10+10) \parallel 10 = \frac{20 \times 10}{20+10} = 6.67 \text{ } \Omega$$

Therefore, Norton equivalent circuit consists of a **current source of 0·6 A (= I_N) in parallel with a resistance of 6·67 Ω (=R_N).** When the branch containing R_L (= 10 Ω) is connected across the output terminals of Norton equivalent circuit, the circuit becomes as shown in Fig. 3.165 (ii).

By current-divider rule, the current I in R_L is

$$I = 0.6 \times \frac{6.67}{6.67+10} = \textbf{0·24 A}$$

Voltage across $R_L = I R_L = 0.24 \times 10 = \textbf{2·4 V}$

Example 3.65. *Find the Norton current for the unbalanced Wheatstone bridge shown in Fig. 3.166.*

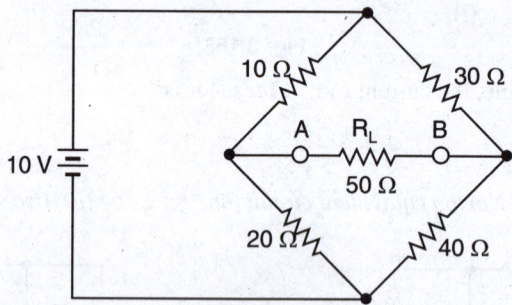

Fig. 3.166

Solution. The Norton current is found by shorting the load terminals as shown in Fig. 3.167 (i). This situation is more complicated than finding the Thevenin voltage. Here is an easy way to find I_N in the circuit of Fig. 3.167 (i). First determine the total current and then use Ohm's law to find current in the four resistors. Once the currents in the four resistors are known, Kirchhoff's current law can be used to determine Norton current I_N.

---

* The resistor 10 Ω that is in parallel with short is ineffective and may be considered as removed from the circuit of Fig. 3.165 (i). Therefore, two 10 Ω resistors are in series and this series combination is in parallel with 10 Ω resistor.

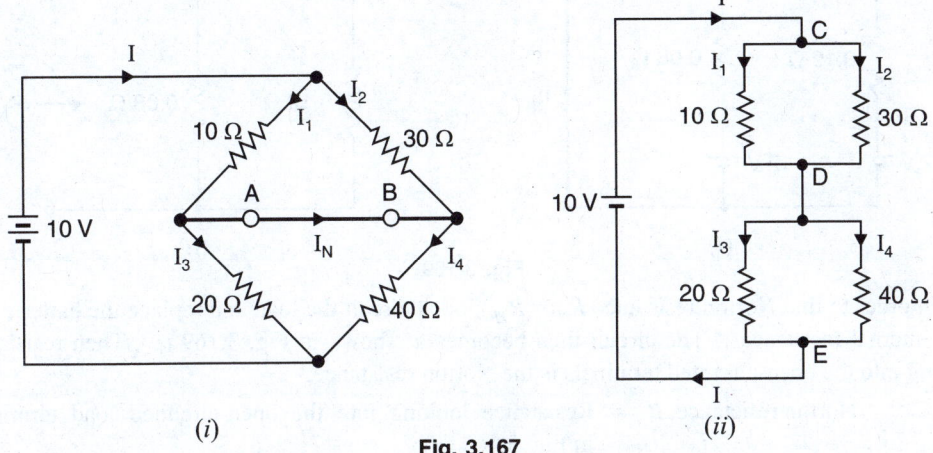

(i)

Fig. 3.167

(ii)

Fig. 3.167 (*ii*) shows the equivalent circuit of Fig. 3.167 (*i*). The total circuit resistance R_T to 10 V source is

$$R_T = \frac{10 \times 30}{10 + 30} + \frac{20 \times 40}{20 + 40} = 7 \cdot 5 + 13 \cdot 33 = 20 \cdot 83 \ \Omega$$

Total circuit current, $I = \dfrac{10}{20.83} = 0 \cdot 48 \ A$

Referring to Fig. 3.167 (*ii*), we have,

$$V_{CD} = I \times R_{CD} = 0 \cdot 48 \times 7 \cdot 5 = 3 \cdot 6 \ V$$

$$V_{DE} = I \times R_{DE} = 0 \cdot 48 \times 13 \cdot 33 = 6 \cdot 4 \ V$$

$$\therefore \quad I_1 = \frac{V_{CD}}{10} = \frac{3.6}{10} = 0 \cdot 36 \ A \ ; \quad I_2 = \frac{V_{CD}}{30} = \frac{3.6}{30} = 0 \cdot 12 \ A$$

$$I_3 = \frac{V_{DE}}{20} = \frac{6.4}{20} = 0 \cdot 32 \ A \ ; \quad I_4 = \frac{V_{DE}}{40} = \frac{6.4}{40} = 0 \cdot 16 \ A$$

Referring to Fig. 3.167 (*i*), it is now clear that $I_1 (= 0 \cdot 36 \ A)$ is greater than $I_3 (= 0 \cdot 32 \ A)$. Therefore, current I_N will flow from *A* to *B* and its value is

$$I_N = I_1 - I_3 = 0 \cdot 36 - 0 \cdot 32 = \mathbf{0 \cdot 04 \ A}$$

Example 3.66. *Two batteries, each of e.m.f. 12 V, are connected in parallel to supply a resistive load of 0·5 Ω. The internal resistances of the batteries are 0·12 Ω and 0·08 Ω. Calculate the current in the load and the current supplied by each battery.*

Solution. Fig. 3.168 shows the conditions of the problem. If a short circuit is placed across the load, the circuit becomes as shown in Fig. 3.169 (*i*). The total short circuit current is given by ;

$$I_N = \frac{12}{0.12} + \frac{12}{0.08}$$

$$= \ 100 + 150 = 250 \ A$$

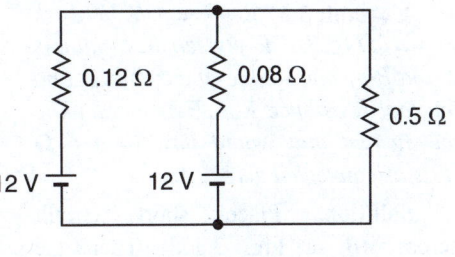

Fig. 3.168

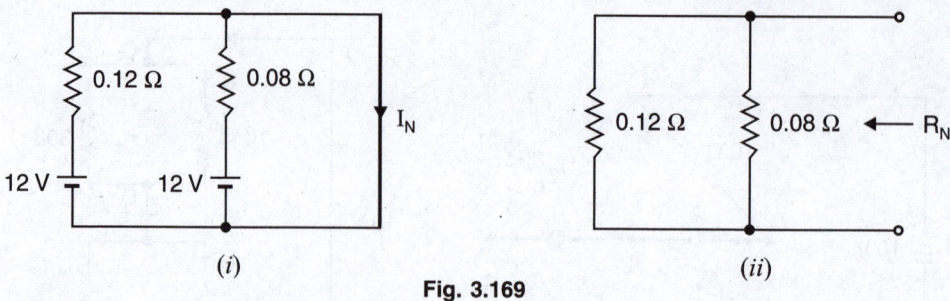

(i)   (ii)

Fig. 3.169

In order to find Norton resistance $R_N (= R_{Th})$, open circuit the load and replace the batteries by their internal resistances. The circuit then becomes as shown in Fig. 3.169 (ii). Then resistance looking into the open-circuited terminals is the Norton resistance.

∴   Norton resistance, R_N = Resistance looking into the open-circuited load terminals in Fig. 3.169 (ii)

$$= 0.12 \parallel 0.08 = \frac{0.12 \times 0.08}{0.12 + 0.08} = 0.048 \; \Omega$$

Therefore, Norton equivalent circuit consists of a current source of 250 A $(= I_N)$ in parallel with a resistance of 0.048 Ω $(= R_N)$. When load $(= 0.5$ $\Omega)$ is connected across the output terminals of Norton equivalent circuit, the circuit becomes as shown in Fig. 3.170. By current-divider rule, the current I in load $(= 0.5$ $\Omega)$ is given by ;

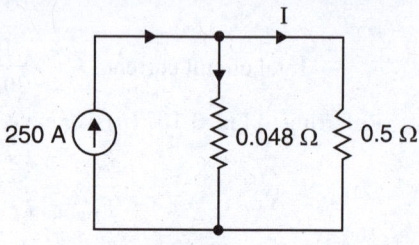

$$I = 250 \times \frac{0.048}{0.048 + 0.5} = \textbf{21·9 A}$$

Fig. 3.170

Battery terminal voltage = $I R_L = 21.9 \times 0.5$

$$= 10.95 \; V$$

Current in first battery = $\dfrac{12 - 10.95}{0.12} = \textbf{8·8 A}$

Current in second battery = $\dfrac{12 - 10.95}{0.08} = \textbf{13·1 A}$

Example 3.67. *Represent the network shown in Fig. 3.171 between the terminals A and B by one source of current I_N and internal resistance R_N. Hence calculate the current that would flow in a 6 Ω resistor connected across AB.*

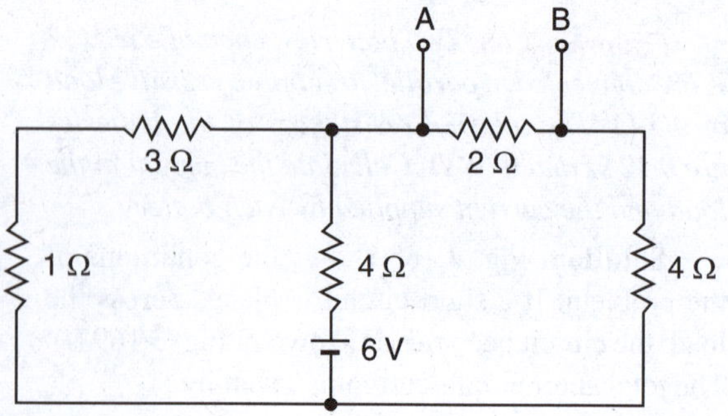

Solution. Place short circuit across *AB* in Fig. 3.171. Then the circuit becomes as shown in Fig. 3.172 (*i*). Note that 2 Ω resistor is shorted and may be considered as removed

Fig. 3.171

from the circuit. The total resistance R_T presented to the 6 V source is a parallel combination of $(3 + 1)$ Ω and 4 Ω in series with 4 Ω. Therefore, the value of R_T is given by ;

$$R_T = [(3 + 1) \parallel 4] + 4 = \frac{4 \times 4}{4 + 4} + 4 = 2 + 4 = 6 \; \Omega$$

∴ Current supplied by 6 V source, $I = 6/6 = 1$ A

At node D, 1 A current divides between two parallel resistors of $(3 + 1)$ Ω and 4 Ω.

∴ Norton current, $I_N = 1 \times \dfrac{4}{4+4} = 0.5$ A

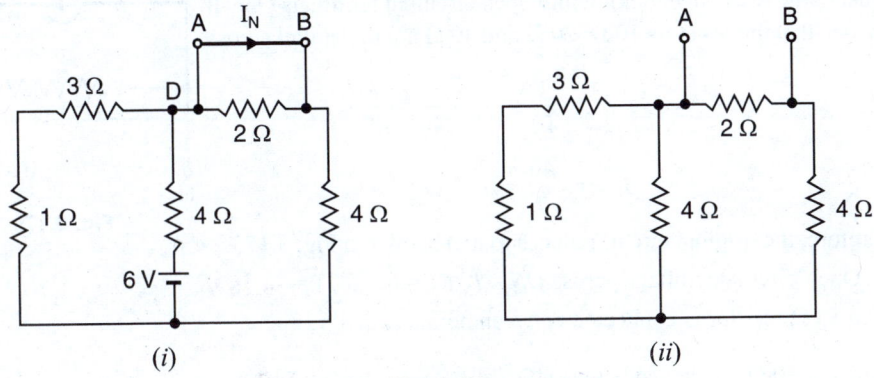

(i)   *(ii)*

Fig. 3.172

Now Norton resistance $R_N (= R_{Th})$ is the resistance between open-circuited terminals AB with voltage source replaced by a short as shown in Fig. 3.172 *(ii)*. Referring to Fig. 3.172 *(ii)*, $(3 + 1)$ Ω resistance is in parallel with 4 Ω, giving equivalent resistance of 2 Ω. Now $(2 + 4)$ Ω resistance is in parallel with 2 Ω.

∴
$$R_N = (2 + 4) \| 2 = 6 \| 2$$
$$= \frac{6 \times 2}{6 + 2} = \frac{12}{8} = 1.5 \ \Omega$$

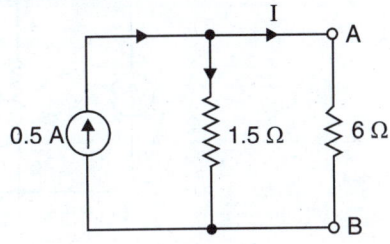

Therefore, Norton equivalent circuit is a **current source of 0·5 A in parallel with resistance of 1·5 Ω.** When a 6 Ω resistor is connected across AB, the circuit becomes as shown in Fig. 3.173. By current-divider rule, current in 6 Ω,

Fig. 3.173

$$I = 0.5 \times \frac{1.5}{1.5 + 6} = 0.5 \times \frac{1.5}{7.5} = \mathbf{0.1 \ A}$$

Example 3.68. *For the circuit shown in Fig. 3.174, calculate the potential difference between the points O and N and what current would flow in a 50 Ω resistor connected between these points?*

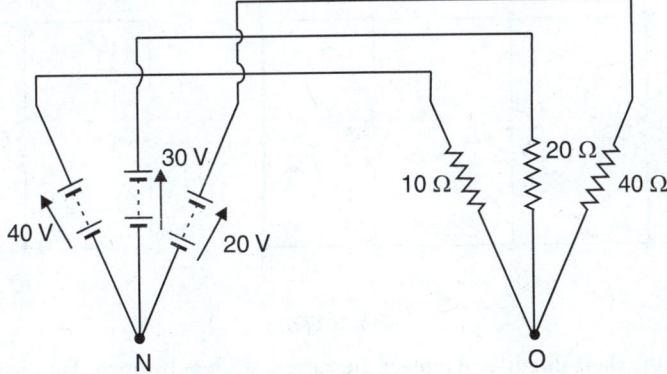

Fig. 3.174

Solution. Place a short circuit across ON in Fig. 3.174. Then total short circuit current in ON is

$$I_N = \frac{40}{10} + \frac{30}{20} + \frac{20}{40} = 4 + 1\cdot5 + 0\cdot5 = 6\,\text{A}$$

In order to find $R_N (= R_{Th})$, replace the voltage sources by short. Then R_N is equal to the resistance looking into open circuited terminals ON. It is easy to see that the resistors $10\,\Omega$, $20\,\Omega$ and $40\,\Omega$ are in parallel across ON.

$$\therefore \qquad \frac{1}{R_N} = \frac{1}{10} + \frac{1}{20} + \frac{1}{40} = \frac{7}{40}$$

or $$R_N = \frac{40}{7} = 5\cdot71\,\Omega$$

Therefore, the original circuit reduces to that shown in Fig. 3.175.

$\therefore$ Open-circuited voltage across $ON = I_N\,R_N = 6 \times 5\cdot71 = \textbf{34·26 V}$

When $50\,\Omega$ resistor is connected between points O and N,

Current in $50\,\Omega$ connected between $ON = 6 \times \dfrac{5.71}{5.71 + 50} = \textbf{0·62 A}$

Example 3.69. *Find Norton equivalent circuit to the left of terminals AB in the circuit shown in Fig. 3.176. The current sources are $I_1 = 10\,A$ and $I_2 = 15\,A$. The conductances are $G_1 = 0\cdot2\,S$, $G_2 = 0\cdot3\,S$ and G_3 is variable.*

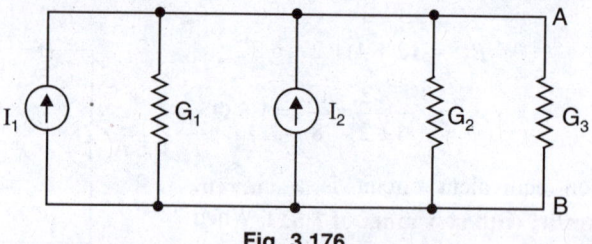

Fig. 3.176

Solution. First, disconnect branch G_3 and short circuit the terminals AB as shown in Fig. 3.177 (*i*). Since the short circuit has infinite conductance, the total current of 25 A ($= I_1 + I_2$) supplied by the two sources would pass through the short-circuited terminals *i.e.*

Norton current, $I_N = I_1 + I_2 = 10 + 15 = 25\,\text{A}$

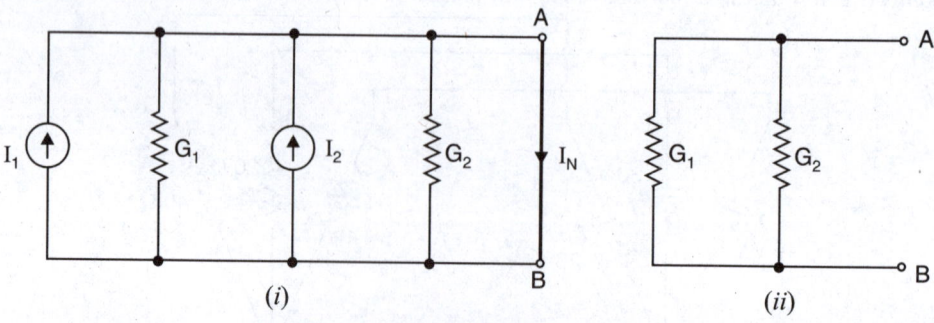

(*i*)   (*ii*)

Fig. 3.177

Next, remove the short-circuit and replace the current sources by open. The circuit then becomes as shown in Fig. 3.177 (*ii*).

Norton conductance, $G_N =$ Conductance at terminals AB in Fig. 3.177 (*ii*).

$$= G_1 + G_2 = 0.2 + 0.3 = 0.5 \text{ S}$$

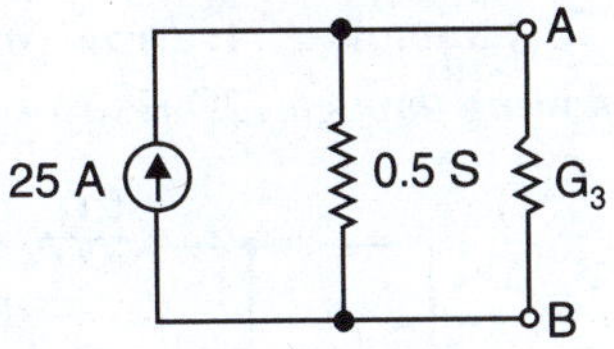

Therefore, Norton equivalent circuit consists of a **25 A current source in parallel with a conductance of 0·5 S.** When conductance G_3 is connected across terminals AB, the circuit becomes as shown in Fig. 3·178. Although Norton equivalent circuit is not the same as its original circuit, it acts the same in terms of output voltage and current.

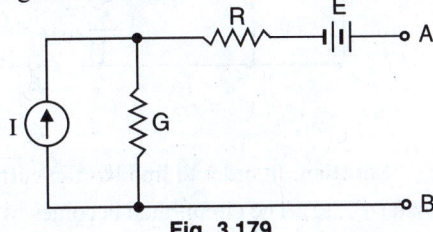

Fig. 3.178

Example 3.70. *The circuit shown in Fig. 3.179 consists of a current source I = 10 A paralleled by G = 0·1 S and a voltage source E = 200 V with 10 Ω series resistance. Find Norton equivalent circuit to the left of terminals AB.*

Fig. 3.179

Solution. We are to find Norton current and Norton resistance. In order to find Norton current I_N, short-circuit the terminals AB as shown in Fig. 3.180 (*i*). Then current that flows in AB is I_N. It is easy to see that current which flows in conductance G is *$I_G = 5$ A (upward).

∴   Norton current, $I_N = I + I_G = 10 + 5 = 15$ A

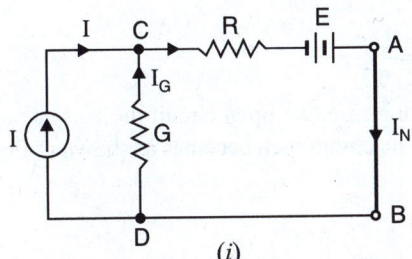   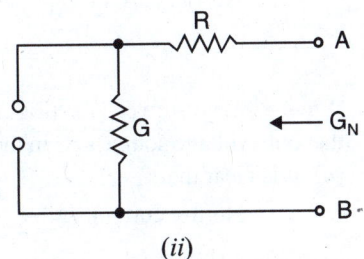

                  (*i*)   (*ii*)

Fig. 3.180

In order to find Norton resistance, remove the short circuit and replace the voltage source by a short and current source by an open. The circuit then becomes as shown in Fig. 3.180 (*ii*).

R_N = Resistance looking into terminals AB in Fig. 3.180 (*ii*).

$$= R + \frac{1}{G} = 10 + \frac{1}{0.1} = 10 + 10 = 20 \ \Omega$$

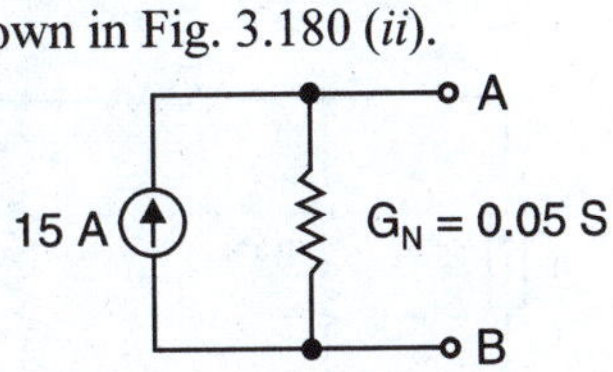

∴   Norton conductance, $G_N = \dfrac{1}{R_N} = \dfrac{1}{20} = 0.05$ S

Fig. 3.181

Therefore, Norton equivalent circuit consists of a **15 A current source paralleled with 0·05 S conductance G_N** as shown in Fig. 3·181.

---

*   Applying *KVL* to loop *CABDC*, we have,

$$- (I + I_G) R + 200 - \frac{I_G}{G} = 0$$

or $- (10 + I_G) 10 + 200 - \dfrac{I_G}{0.1} = 0$

or $- 100 - 10 I_G + 200 - 10 I_G = 0$

∴ $I_G = 100/20 = 5$ A

Example 3.71. *Draw Norton's equivalent circuit at terminals AB and determine the current flowing through 12Ω resistor for the network shown in Fig. 3.182 (i).*

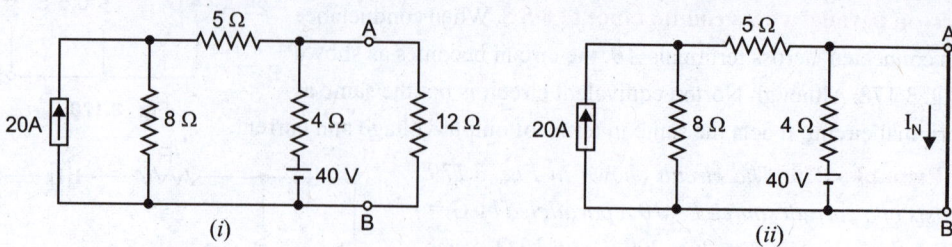

Fig. 3.182

Solution. In order to find Norton current I_N, short circuit terminals A and B after removing the load ($= 12Ω$). The circuit then becomes as shown in Fig. 3.182 (*ii*). The current flowing in the short circuit is the Norton current I_N. It can be found by using superposition theorem.

   (*i*) When current source is acting alone. In this case, we short circuit the voltage source so that only current source acts in the circuit. The circuit then becomes as shown in Fig. 3.183 (*i*). It is clear that :

$$\text{Norton current, } I_{N1} = \text{*Current in 5Ω resistor}$$

$$= 20 \times \frac{8}{8+5} = \frac{160}{13}\text{A}$$

   (*ii*) When voltage source is acting alone. In this case, we open circuit the current source so that only voltage source acts in the circuit. The circuit then becomes as shown in Fig. 3.183 (*ii*). It is clear that :

$$\text{Norton current, } I_{N2} = \frac{40}{4} = 10\text{A}$$

Therefore, when both voltage and current sources are present in the circuit, we have,

$$\text{Norton current, } I_N = I_{N1} + I_{N2} = \frac{160}{13} + 10 = \frac{290}{13}\text{A}$$

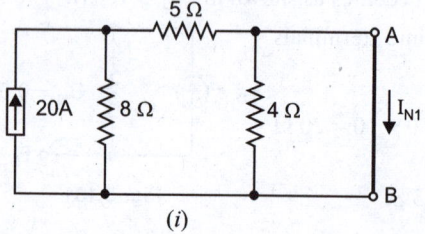 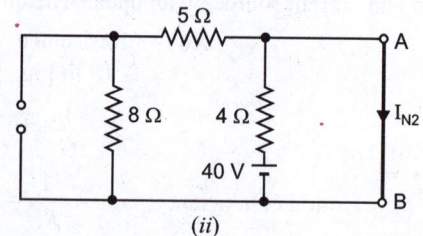

Fig. 3.183

In order to find R_N, open circuit 12Ω resistor and replace current source by open circuit and voltage source by short circuit. Then circuit becomes as shown in Fig. 3.184 (*i*).

∴ R_N = Resistance at terminals *AB* in Fig. 3.184 (*i*)

$$= 4 \parallel (5+8) = 4 \parallel 13 = \frac{4 \times 13}{4+13} = \frac{52}{17}Ω$$

---

*   No current flows in 4Ω resistor because it is short circuited at terminals *A* and *B*. Therefore, 20A divides between 8Ω and 5Ω connected in parallel.

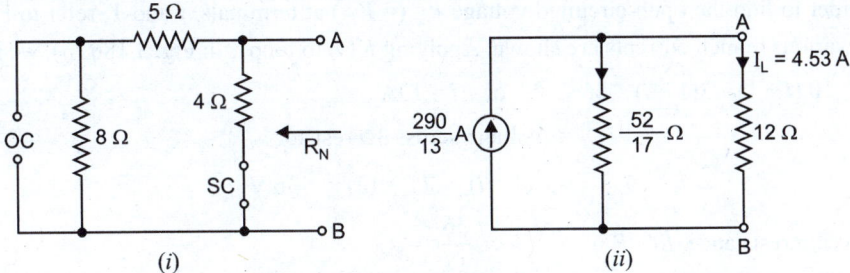

Fig. 3.184

Thus Norton equivalent circuit at terminals *AB* is a current source of current 290/13 A in parallel with 52/17Ω resistance. When load resistor of 12Ω is connected across Norton's equivalent circuit, the circuit becomes as shown in Fig. 3.184 (*ii*).

∴   Load current, $I_L = I_N \times \dfrac{R_N}{R_N + R_L} = \dfrac{290}{13} \times \dfrac{52/17}{52/17 + 12} =$ **4.53 A**

Example 3.72. *Determine the values of I and R in the circuit shown in Fig. 3.185.*

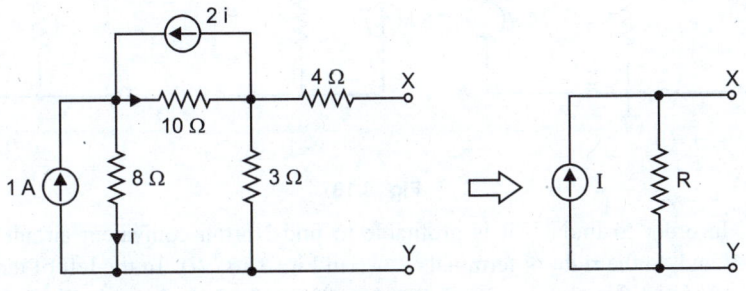

Fig. 3.185

Solution. Short the terminals *XY* in Fig. 3.185 and we get the circuit shown in Fig. 3.186 (*i*). The currents in the various branches will be as shown. In order to find the short-circuit current $I_{sc} (= I = I_N)$, we apply *KVL* to loops 1 and 2 in Fig. 3.186 (*i*).

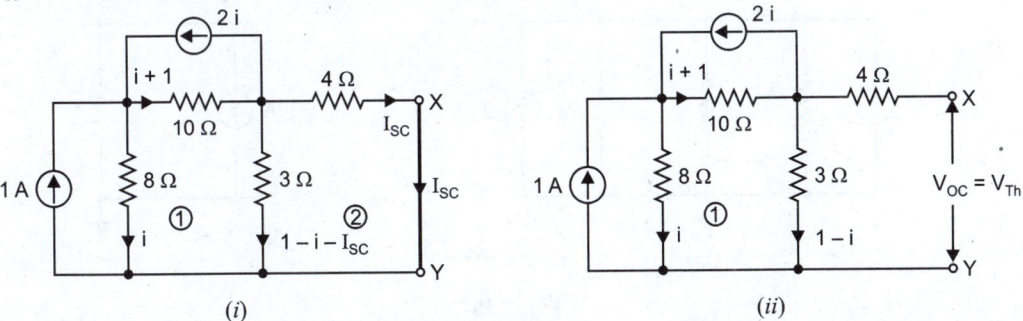

Fig. 3.186

Loop 1. $-10(i + 1) - 3(1 - i - I_{sc}) + 8i = 0$

or $i + 3I_{sc} = 13$ ...(*i*)

Loop 2. $-4I_{sc} + 3(1 - i - I_{sc}) = 0$

or $3i + 7I_{sc} = 3$ ...(*ii*)

From eqs. (*i*) and (*ii*), we have, $I_{sc} = 18$A.

In order to find the open-circuited voltage V_{oc} ($= V_{Th}$) at terminals X and Y, refer to Fig 3.186 (ii). The various branch currents are shown. Applying KVL to loop 1 in Fig. 3.186 (ii), we have,

$$-10\,(i+1) - 3(1-i) + 8i = 0 \quad \text{or} \quad i = 13A$$

$$\therefore \qquad V_{oc} = \text{Voltage across } 3\Omega \text{ resistor}$$

$$= 3(1-i) = 3(1-13) = -36 \text{ V}$$

$$\text{Thevenin resistance, } R(=R_N) = \frac{V_{oc}}{I_{sc}} = \frac{36}{18} = 2\Omega$$

$$\text{Current } I = I_N = -18A$$

Note the polarity of current source I ($= I_N$).

Example 3.73. *With the help of Norton's theorem, find V_o in the circuit shown in Fig. 3.187 (i). All resistances are in ohms.*

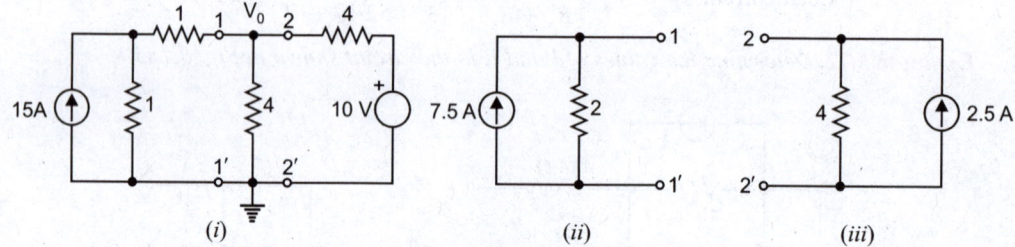

(i)   (ii)   (iii)

Fig. 3.187

Solution. In order to find V_o, it is profitable to find Norton equivalent circuit to the left of terminals $1-1'$ and to the right of terminals $2-2'$ in Fig. 3.187 (i). To the left of terminals $1-1'$, $V_{oc} = 15 \times 1 = 15$ V and $R_N = 1 + 1 = 2\Omega$ so that $I_N = 15/2 = 7.5$A as shown in Fig. 3.187 (ii). To the right of terminals $2-2'$, $V_{oc} = 10$ V and $R_{Th} = R_N = 4\Omega$ so that $I_N = 10/4 = 2.5$A as shown in Fig. 3.187 (iii). The two Norton equivalent circuits are put back at terminals $1-1'$, and $2-2'$ as shown in Fig. 3.187 (iv).

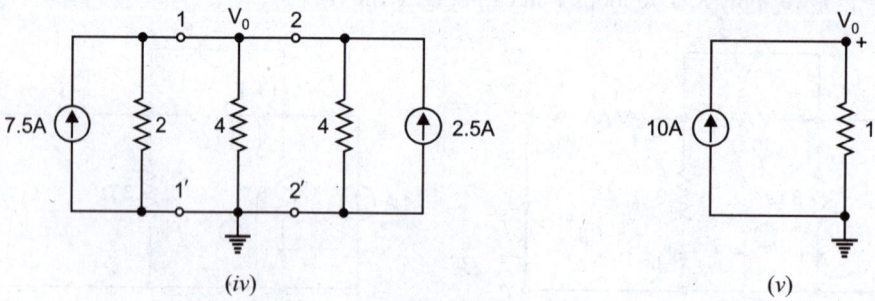

(iv)   (v)

Fig. 3.187

In Fig. 3.187 (iv), the two current sources, being parallel and carrying currents in the same direction, can be combined into a single current source of $7.5 + 2.5 = 10$A. The three resistances are in parallel and can be combined to give a single resistance $= 2\Omega \,\|\, 4\Omega \,\|\, 4\Omega = 1\Omega$. Therefore, the circuit of Fig. 3.187 (iv) reduces to the circuit shown in Fig. 3.187 (v).

$$\therefore \qquad V_o = 10A \times 1\Omega = 10V$$

Example 3.74. *Find current in the 4 ohm resistor by any three methods for the circuit shown in Fig. 3.188(i).*

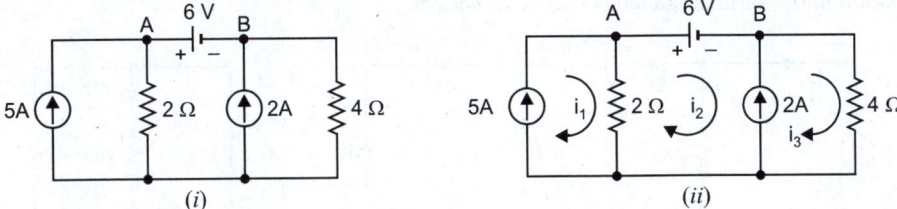

Fig. 3.188

Solution. **Method 1.** We shall find current in 4Ω resistor by **mesh current method.** Mark three mesh currents i_1, i_2 and i_3 in the three loops as shown in Fig. 3.188 (*ii*). The describing circuit equations are :

$$i_1 = 5A \text{ due to the current source of 5A}$$
$$V_A - V_B = 6V \text{ due to voltage source of 6V}$$
$$i_3 - i_2 = 2A \text{ due to current source of 2A}$$
$$V_A = (i_1 - i_2)2 \ ; \ V_B = i_3 \times 4$$

Now, $-6 - 4i_3 - 2(i_2 - i_1) = 0$ ... Applying *KVL*

or $-6 - 4(2 + i_2) - 2(i_2 - 5) = 0$

or $-6i_2 = 4$

∴ $i_2 = -\dfrac{4}{6} = -\dfrac{2}{3}A$ and $i_3 = i_2 + 2 = -\dfrac{2}{3} + 2 = \dfrac{4}{3}A$

∴   Current in 4Ω resistor $= i_3 = \dfrac{4}{3}\mathbf{A}$

Method 2. We now find current in 4Ω resistor by **Thevenin's theorem.** Remove 4Ω resistor (*i.e.* load) and the circuit becomes as shown in Fig. 3.188 (*iii*).

Current in 2Ω resistor $= 5 + 2 = 7A$

It is because 6V source is ineffective in producing any current.

In going from point X to point Y via B and A, we have,

$$V_X + 6 - 7 \times 2 = V_Y$$
or $V_X - V_Y = 7 \times 2 - 6 = 8V$
∴ $V_{Th} = V_{XY} = V_X - V_Y = 8V$

In order to find R_{Th}, short circuit the voltage source and open-circuit the current sources in Fig. 3.188 (*iii*). Then circuit becomes as shown in Fig. 3.188 (*iv*). The resistance at the open-circuited terminals XY in Fig. 3.188 (*iv*) is R_{Th}.

∴ $R_{Th} = 2\Omega$

∴   Current in 4Ω resistor $= \dfrac{V_{Th}}{R_{Th} + 4} = \dfrac{8}{2 + 4} = \dfrac{4}{3}\mathbf{A}$

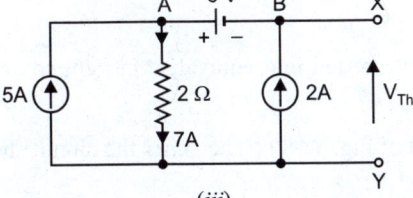

Fig. 3.188

Method 3. Finally, we find current in 4Ω resistor by **Norton's theorem.** To find I_N, short-circuit 4Ω resistor in Fig. 3.188 (*i*). The circuit then becomes as shown in Fig. 3.188 (*v*). The current distribution in the various branches will be as shown.

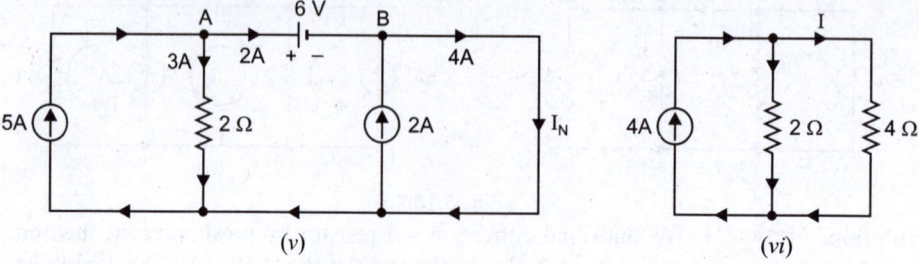

Fig. 3.188

It is clear from Fig. 3.188 (*v*) that :

$$I_N = 2 + 2 = 4\text{A}$$

$$R_N = R_{Th} = 2\Omega \qquad \qquad \text{...as calculated above}$$

When 4Ω resistor is connected to Norton equivalent circuit, it becomes as shown in Fig. 3.188 (*vi*).

∴ Current in 4Ω resistor is given by (current-divider rule) ;

$$I = 4 \times \frac{2}{2+4} = \frac{8}{6} = \frac{4}{3}\text{A}$$

Example 3.75. *Using Norton's theorem, find current through 1Ω resistor in Fig. 3.189 (i). All resistances are in ohms,*

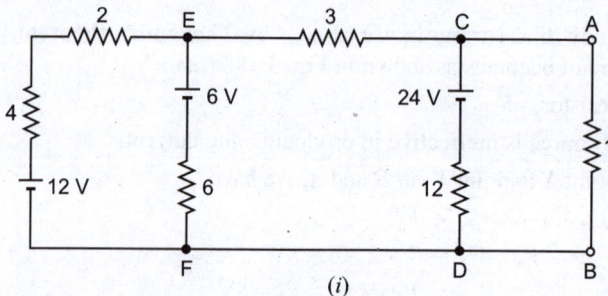

(*i*)

Fig. 3.189

Solution. To find the answers, we convert the three voltage sources into their equivalent current sources.

(*a*) 12 V source in series with (4 + 2) = 6Ω resistance is converted into equivalent current source of 12V/6Ω = 2A in parallel with 6Ω resistance.

(*b*) 6V source in series with 6Ω resistance is converted into equivalent current source of 6V/6Ω = 1A in parallel with 6Ω resistance.

(*c*) 24V source in series with 12Ω resistance is converted into equivalent current source of 24V/12Ω = 2A in parallel with 12Ω resistance.

After the above source conversions, the circuit of Fig. 3.189 (*i*) becomes the circuit shown in Fig. 3.189 (*ii*).

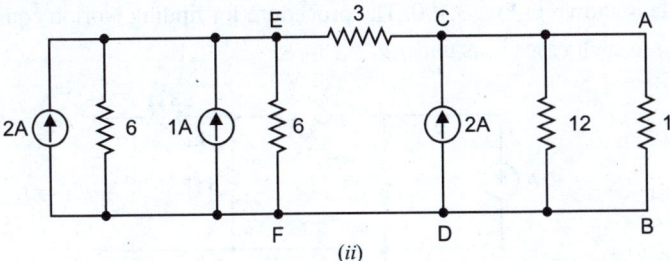

Fig. 3.189

Referring to Fig. 3.189 (*ii*), we can combine the two current sources to the left of *EF* but cannot combine 2A source across *CD* with them because 3Ω resistance is between *E* and *C*. Therefore, combining the two current sources to the left of *EF*, we have a single current source of 2 + 1 = 3A and a single resistance of 6Ω || 6Ω = 3Ω in parallel with it. As a result, Fig. 3.189 (*ii*) reduces to the circuit shown in Fig. 3.189 (*iii*).

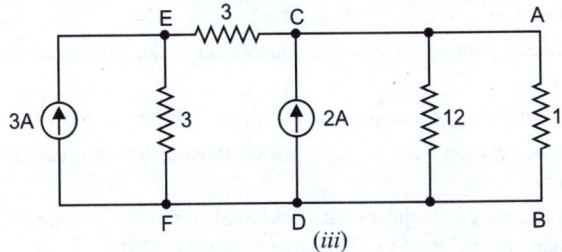

Fig. 3.189

We now convert the circuit to the left of *CD* in Fig. 3.189 (*iii*) into Norton equivalent circuit. Fig. 3.189 (*iv*) shows this circuit to the left of *CD*. Its Norton equivalent circuit values are :

$$I_N = 3 \times \frac{3}{3+3} = 1.5A \; ; \; R_N = 3\Omega + 3\Omega = 6\Omega$$

Therefore, replacing the circuit to the left of *CD* in Fig. 3.189 (*iii*) by its Norton equivalent circuit, we get the circuit shown in Fig. 3.189 (*v*).

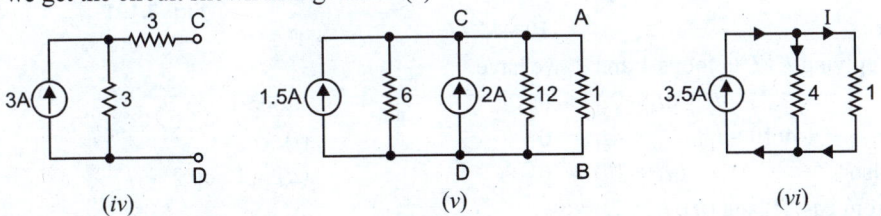

Fig. 3.189

Referring to Fig. 3.189 (*v*), we can combine the two current sources into a single current source of 1.5 + 2 = 3.5 A and a single resistance of 6Ω || 12Ω = 4Ω in parallel with it. The circuit then reduces to the one shown in Fig. 3.189 (*vi*). By current-divider rule [See Fig. 3.189 (*vi*)],

$$\text{Current in 1}\Omega \text{ resistor, } I = 3.5 \times \frac{4}{4+1} = \textbf{2.8 A}$$

3.16. Norton Equivalent Circuit

(Circuits containing both independent and dependent sources)

Sometimes we come across circuits which contain both independent and dependent sources.

One such example is shown in Fig. 3.190. The procedure for finding Norton equivalent circuit (*i.e.* finding i_N and R_N) in such cases is as under :

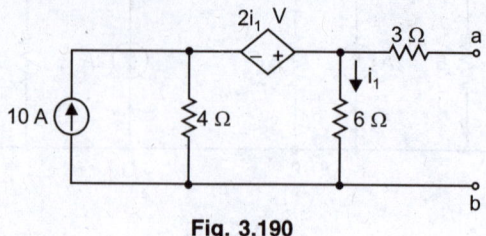

Fig. 3.190

(*i*) The open-circuited voltage $v_{oc}(= v_{Th})$ at terminals *ab* is determined as usual with sources present.

(*ii*) We cannot find R_N (= R_{Th}) at terminals *ab* simply by calculating equivalent resistance because of the presence of the dependent source. Instead, we place a short circuit across the terminals *ab* and find the value of short-circuit current i_{sc} (= i_N) at terminals *ab*.

(*iii*) Norton resistance, $R_N = v_{oc}/i_{sc}$ (= v_{Th}/i_{sc}).

Note. In case the circuit contains dependent sources *only*, the procedure for finding v_{oc} (= v_{Th}) and $R_N (= R_{Th})$ is as under :

(*a*) In this case, $v_{oc} = 0$ and $i_{sc} = 0$ because no independent source is present.

(*b*) We cannot use the relation $R_N = v_{oc}/i_{sc}$ as we do in case the circuit contains both independent and dependent sources.

(*c*) In order to find R_N, we excite the circuit at terminals *ab* by connecting 1A source to the terminals *a* and *b* and calculate the value of v_{ab}. Then $R_N (= R_{Th}) = v_{ab}/1\Omega$.

Example 3.76. *Find the values of i_N and R_N at terminals ab for the circuit shown in Fig. 3.191 (i).*

Solution. We first put a short circuit across terminals *a* and *b* to find short-circuit current i_{sc} (= i_N) at terminals *ab* as shown in Fig. 3.191 (*ii*). Applying *KCL* at node *c*, we have,

$$10 = i_1 + i_2 + i_{sc}$$

or $i_2 = 10 - i_1 - i_{sc}$

Applying *KVL* to loops 1 and 2, we have,

$$-4i_2 + 6i_1 - 2i_1 = 0 \qquad \text{... Loop 1}$$

or $-4(10 - i_1 - i_{sc}) + 4i_1 = 0 \qquad ...(i)$

Also $-6i_1 + 3i_{sc} = 0 \qquad ...(ii) \text{ ... Loop 2}$

From eqs. (*i*) and (*ii*), $i_{sc} = i_N = $ **5A.**

(i)

Fig. 3.191

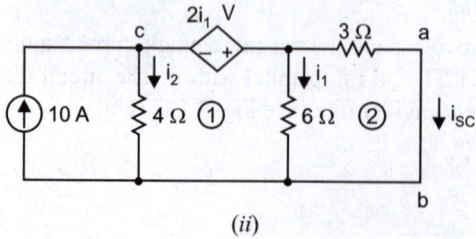

(ii)

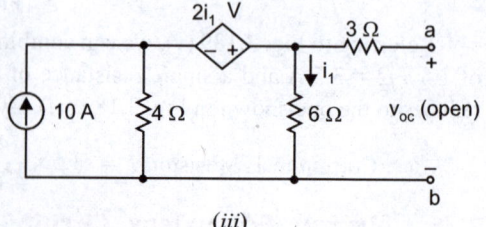

(iii)

Fig. 3.191

In order to find v_{oc} (= v_{Th}), we refer to Fig. 3.191 (*iii*) where we have,

$$v_{oc} = 6i_1 \qquad ...(iii)$$

Applying *KVL* to the central loop in Fig. 3.191 (*iii*),

$$-4(10 - i_1) + 6i_1 - 2i_1 = 0 \qquad \qquad ...(iv)$$

From eqs. (*iii*) and (*iv*), we have, $v_{oc} = v_{Th} = 30$V.

Also $$R_N (= R_{Th}) = \frac{v_{oc}}{i_{sc}} = \frac{30}{5} = 6\Omega$$

Tutorial Problems

1. Using Norton's theorem, find the current in 8 Ω resistor of the network shown in Fig. 3.192.   **[1.55 A]**

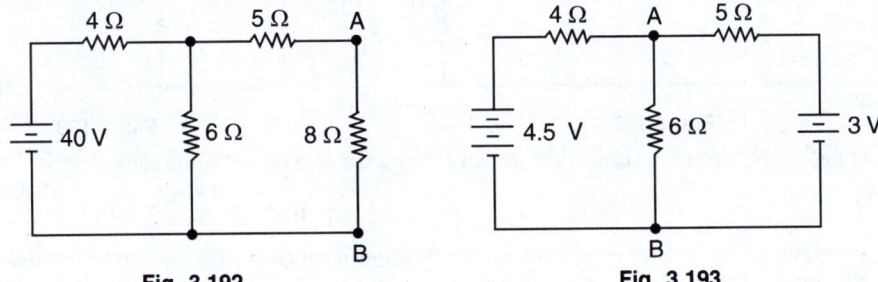

Fig. 3.192   **Fig. 3.193**

2. Using Norton's theorem, find the current in the branch *AB* containing 6 Ω resistor of the network shown in Fig. 3.193.   **[0.466 A]**

3. Show that when Thevenin's equivalent circuit of a network is converted into Norton equivalent circuit, $I_N = V_{TH}/R_{Th}$ and $R_N = R_{Th}$.

4. Find the voltage between points *A* and *B* in the network shown in Fig. 3.194 using Norton's theorem.   **[2·56 V]**

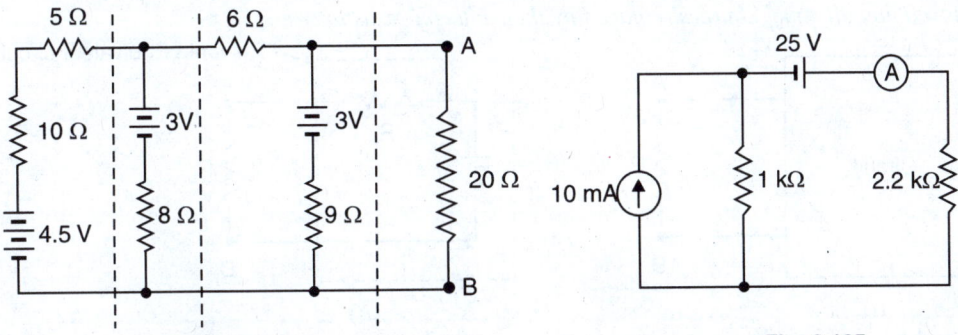

Fig. 3.194   **Fig. 3.195**

5. The ammeter labelled A in Fig. 3.195 reads 35 mA. Is the 2·2 kΩ resistor shorted? Assume that ammeter has zero resistance.   **[Shorted]**

6. Find Norton equivalent circuit to the left of terminals *a – b* in Fig. 3.196.   **[$I_N = 1·5$ A; $R_N = 4$ Ω]**

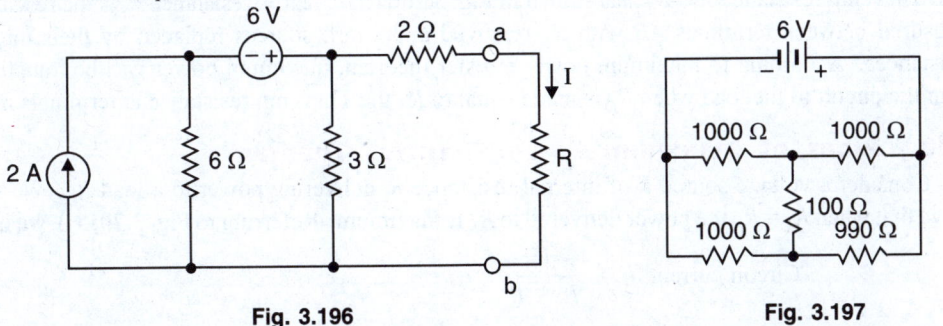

Fig. 3.196   **Fig. 3.197**

7. What is the current in the 100 Ω resistor in Fig. 3.197 if the 990 Ω resistor is changed to 1010 Ω ? Use Norton theorem to obtain the result.   **[13·45 μA]**

8. Determine the Norton equivalent circuit and the load current in R_L in Fig. 3.198. The various circuit values are :

$$E' = 64 \text{ V} ; \quad R_1 = 230 \text{ Ω} ; \quad R_2 = 450 \text{ Ω} ;$$
$$R_3 = 260 \text{ Ω} ; \quad R_4 = 550 \text{ Ω} ; \quad R_5 = 440 \text{ Ω} ; \quad R_L = 360 \text{ Ω}$$

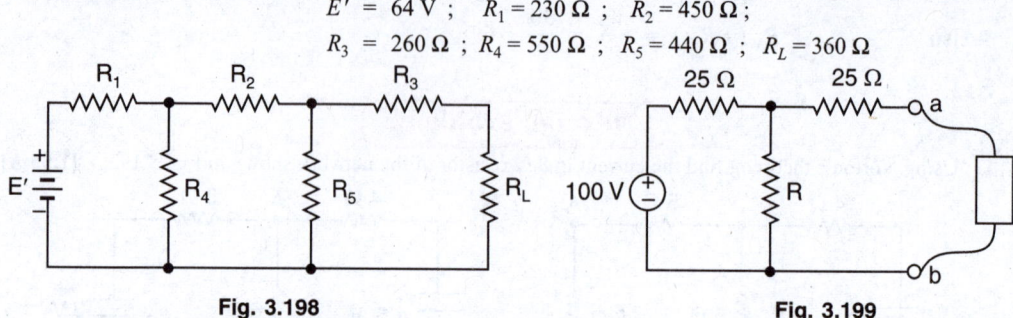

Fig. 3.198   **Fig. 3.199**

9. In Fig. 3.199, replace the network to the left of terminals *ab* with its Norton equivalent.

$$[I_N = \frac{2R}{R+12.5} \text{A} ; R_N = \frac{50R+625}{R+25} \text{Ω}]$$

10. When any source (voltage or current) is delivering maximum power to a load, prove that overall circuit efficiency is 50%.

3.17. Maximum Power Transfer Theorem

This theorem deals with transfer of maximum power from a source to load and may be stated as under :

In d.c. circuits, maximum power is transferred from a source to load when the load resistance is made equal to the internal resistance of the source as viewed from the load terminals with load removed and all e.m.f. sources replaced by their internal resistances.

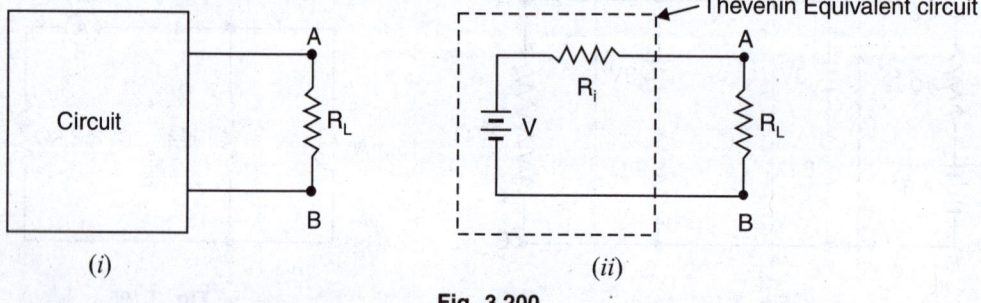

Fig. 3.200

Fig. 3.200 (*i*) shows a circuit supplying power to a load R_L. The circuit enclosed in the box can be replaced by Thevenin's equivalent circuit consisting of Thevenin voltage $V = V_{Th}$ in series with Thevenin resistance $R_i(=R_{Th})$ as shown in Fig. 3.200 (*ii*). Clearly, resistance R_i is the resistance measured between terminals *AB* with R_L removed and e.m.f. sources replaced by their internal resistances. According to maximum power transfer theorem, maximum power will be transferred from the circuit to the load when R_L is made equal to R_i, the Thevenin resistance at terminals *AB*.

3.18. Proof of Maximum Power Transfer Theorem

Consider a voltage source V of internal resistance R_i delivering power to a load R_L. We shall prove that when $R_L = R_i$, the power delivered to R_L is maximum. Referring to Fig. 3.201 (*i*), we have,

Circuit current, $I = \dfrac{V}{R_L + R_i}$

Power delivered to load, $P = I^2 R_L$

$$= \left(\frac{V}{R_L + R_i}\right)^2 R_L \qquad \qquad ...(i)$$

For a given source, generated voltage V and internal resistance R_i are constant. Therefore, power delivered to the load depends upon R_L. In order to find the value of R_L for which the value of P is maximum, differentiate eq. (i) w.r.t. R_L and set the result equal to zero.

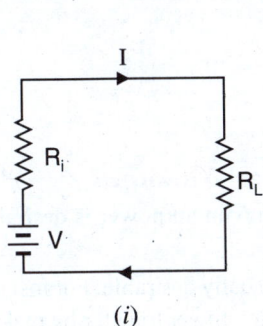

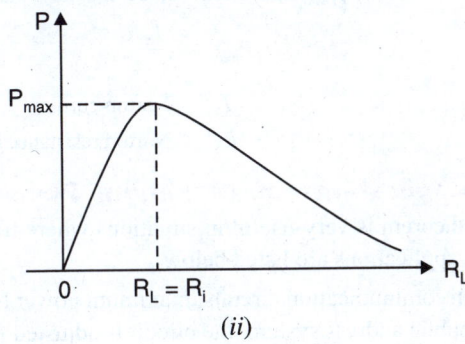

(i) (ii)

Fig. 3.201

Thus,
$$\frac{dP}{dR_L} = V^2 \left[\frac{(R_L + R_i)^2 - 2R_L(R_L + R_i)}{(R_L + R_i)^4}\right] = 0$$

or $(R_L + R_i)^2 - 2R_L(R_L + R_i) = 0$

or $(R_L + R_i)(R_L + R_i - 2R_L) = 0$

or $(R_L + R_i)(R_i - R_L) = 0$

Since $R_L + R_i$ cannot be zero,

$\therefore$ $R_i - R_L = 0$

or $R_L = R_i$

or   **Load resistance = Internal resistance of the source**

Thus, for maximum power transfer, load resistance R_L must be equal to the internal resistance R_i of the source. Fig. 3.201 (ii) shows the graph between power delivered (P) and R_L. We may extend the maximum power transfer theorem to a linear circuit rather than a single source by means of Thevenin's theorem as under :

The maximum power is obtained from a linear circuit at a given pair of terminals when terminals are loaded by Thevenin's resistance (R_{Th}) of the circuit.

The above statement is obviously true because by Thevenin's theorem, the circuit is equivalent to a voltage source in series with internal resistance (R_{Th}) of the circuit.

Important Points. The following points are worth noting about maximum power transfer theorem :

(i) *The circuit efficiency at maximum power transfer is only 50% as one-half of the total power generated is dissipated in the internal resistance R_i of the source.*

$$\text{Efficiency} = \frac{\text{Output power}}{\text{Input power}} = \frac{I^2 R_L}{I^2(R_L + R_i)}$$

$$= \frac{R_L}{2R_L} = \frac{1}{2} = 50\% \qquad \qquad (\because R_L = R_i)$$

(ii) *Under the conditions of maximum power transfer, the load voltage is one-half of the open-circuited voltage at the load terminals.*

$$\text{Load voltage} = I\,R_L = \left(\frac{V}{R_L + R_i}\right)R_L = \frac{V\,R_L}{2\,R_L} = \frac{V}{2}$$

(*iii*) Max. power transferred $= \left(\dfrac{V}{R_L + R_i}\right)^2 R_L = \left(\dfrac{V}{2\,R_L}\right)^2 R_L = \dfrac{V^2}{4\,R_L}$

Note. In case of a practical current source, the maximum power delivered is given by ;

$$P_{max} = \frac{I_N^2 R_N}{4}$$

where $\qquad\qquad I_N$ = Norton current

$\qquad\qquad\qquad R_N$ = Norton resistance $(= R_{Th} = R_i)$

3.19. Applications of Maximum Power Transfer Theorem

This theorem is very useful in situations where transfer of maximum power is desirable. Two important applications are listed below :

(*i*) In communication circuits, maximum power transfer is usually desirable. For instance, in a public address system, the circuit is adjusted for maximum power transfer by making load (*i.e.* speaker) resistance equal to source (*i.e.* amplifier) resistance. When source and load have the same resistance, they are said to be *matched.*

In most practical situations, the internal resistance of the source is fixed. Also, the device that acts as a load has fixed resistance. In order to make $R_L = R_i$, we use a transformer. We can use the reflected-resistance characteristic of the transformer to make the load resistance appear to have the same value as the source resistance, thereby "fooling" the source into "thinking" that there is a match (*i.e.* $R_L = R_i$). This technique is called **impedance matching.**

(*ii*) Another example of maximum power transfer is found in starting of a car engine. The power delivered to the starter motor of the car will depend upon the effective resistance of the motor and internal resistance of the battery. If the two resistances are equal (as is the case when battery is fully charged), maximum power will be transferred to the motor to turn on the engine. This is particularly desirable in winter when every watt that can be extracted from the battery is needed by the starter motor to turn on the cold engine. If the battery is weak, its internal resistance is high and the car does not start.

Note. Electric power systems are never operated for maximum power transfer because the efficiency under this condition is only 50%. This means that 50% of the generated power will be lost in the power lines. This situation cannot be tolerated because power lines must operate at much higher than 50% efficiency.

Example 3.77. *Two identical cells connected in series deliver a maximum power of 1W to a resistance of 4 Ω. What is the internal resistance and e.m.f. of each cell ?*

Solution. Let E and r be the e.m.f. and internal resistance of each cell. The total internal resistance of the battery is $2r$. For maximum power transfer,

$$2\,r = R_L = 4 \qquad \therefore \quad r = R_L/2 = 4/2 = 2\,\Omega$$

$$\text{Maximum power} = \frac{{}^*(2E)^2}{4\,R_L}$$

or $\qquad\qquad\qquad 1 = \dfrac{4E^2}{4\,R_L} \qquad \therefore \quad E = \sqrt{R_L} = \sqrt{4} = 2\text{ V}$

---

*   Here total voltage $= 2E$.

Example 3.78. *Find the value of resistance R to have maximum power transfer in the circuit shown in Fig. 3.202 (i). Also obtain the amount of maximum power. All resistances are in ohms.*

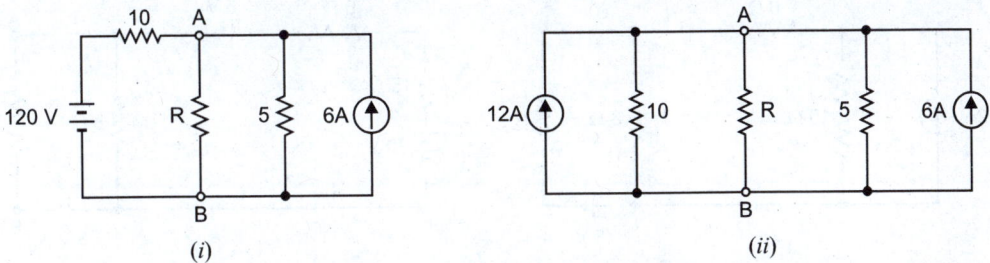

Fig. 3.202

Solution. To find the desired answers, we should find V_{Th} and R_{Th} at the load (*i.e.* R) terminals. For this purpose, we first convert 120V voltage source in series with 10Ω resistance into equivalent current source of 120/10 = 12A in parallel with 10Ω resistance. The circuit then becomes as shown in Fig. 3.202. (*ii*).

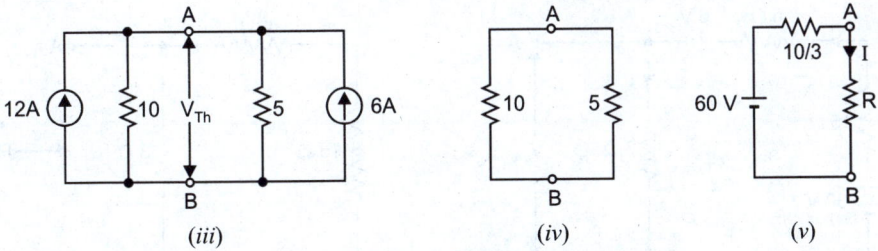

Fig. 3.202

To find V_{Th}, remove R (*i.e.* load) from the circuit in Fig. 3.202 (*ii*), and the circuit becomes as shown in Fig. 3.202 (*iii*). Then voltage across the open-circuited terminals AB is V_{Th}. Referring to Fig. 3.202 (*iii*) and applying *KCL*, we have,

$$\frac{V_{Th}}{10} + \frac{V_{Th}}{5} = 12 + 6 \quad \text{or} \quad V_{Th} = 60\text{V}$$

In order to find R_{Th}, remove R and replace the current sources by open in Fig. 3.202 (*ii*). Then circuit becomes as shown in Fig. 3.202 (*iv*). Then resistance at the open-circuited terminals AB is R_{Th}.

∴ $$R_{Th} = 10Ω \,\|\, 5Ω = \frac{10 \times 5}{10 + 5} = \frac{10}{3}Ω$$

When R is connected to the terminals of Thevenin equivalent circuit, the circuit becomes as shown in Fig. 3.202 (*v*).

For maximum power transfer, the condition is

$$R = R_{Th} = \frac{10}{3}Ω$$

Max. power transferred, $$P_{max} = \frac{V_{Th}^2}{4R_L} = \frac{V_{Th}^2}{4R} = \frac{(60)^2}{4 \times (10/3)} = \textbf{270 W}$$

Example 3.79. *Calculate the value of R which will absorb maximum power from the circuit of Fig. 3.203 (i). Also find the value of maximum power.*

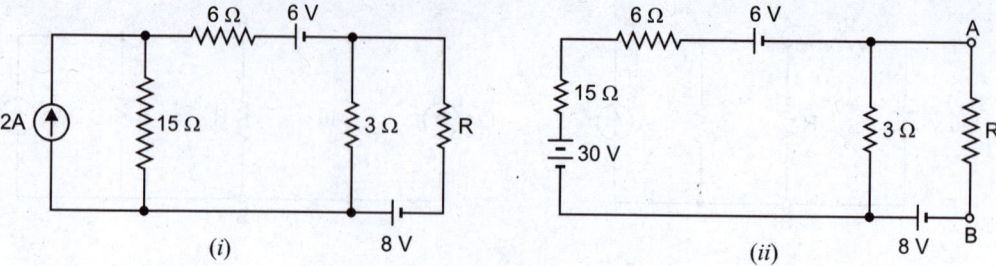

Fig. 3.203

Solution. To find the desired answers, we should find V_{Th} and R_{Th} at the load (*i.e. R*) terminals. For this purpose, we first convert 2A current source in parallel with 15Ω resistance into equivalent voltage source of 2A × 15Ω = 30 V in series with 15Ω resistance. The circuit then becomes as shown in Fig. 3.203 (*ii*).

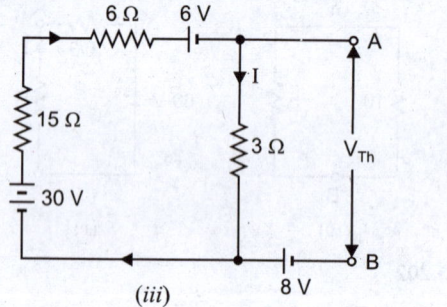
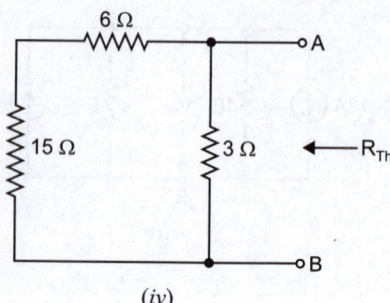

Fig. 3.203

To find V_{Th}, remove R (*i.e.* load) from the circuit in Fig. 3.203 (*ii*) and the circuit becomes as shown in Fig. 3.203 (*iii*). Then voltage across the open-circuited terminals *AB* is V_{Th}. Referring to Fig. 3.203 (*iii*),

Current in 3Ω resistor, $I = \dfrac{30-6}{15+6+3} = 1A$

In Fig. 3.203 (*iii*), as we go from point *A* to point *B* via 3Ω resistor, we have,

$$V_A - I \times 3 - 8 = V_B$$

or $$V_A - V_B = I \times 3 + 8 = 1 \times 3 + 8 = 11V$$

∴ $$V_{Th} = V_{AB} = V_A - V_B = 11V$$

In order to find R_{Th}, remove R and replace the voltage sources by short in Fig. 3.203 (*ii*). Then circuit becomes as shown in Fig. 3.203 (*iv*). Then resistance at open-circuited terminals *AB* is R_{Th}.

∴ $$R_{Th} = (15+6)\Omega \parallel 3\Omega = \frac{21 \times 3}{21+3} = \frac{21}{8}\Omega$$

For maximum power transfer, the condition is

$$R = R_{Th} = \frac{\mathbf{21}}{\mathbf{8}}\Omega$$

Max. power transferred, $P_{max} = \dfrac{V_{Th}^2}{4R_L} = \dfrac{V_{Th}^2}{4R} = \dfrac{(11)^2}{4 \times (21/8)} = \mathbf{11.524\ W}$

Example 3.80. *Determine the value of R_L in Fig. 3.204 (i) for maximum power transfer and evaluate this power.*

Solution. The three current sources in Fig. 3.204 (i) are in parallel and supply current in the same direction. Therefore, they can be replaced by a single current source supplying $0.8 + 1 + 0.9 = 2.7$ A as shown in Fig. 3.204 (ii). The circuit to the left of R_L in Fig. 3.204 (ii) can be replaced by Thevenin's equivalent circuit as under :

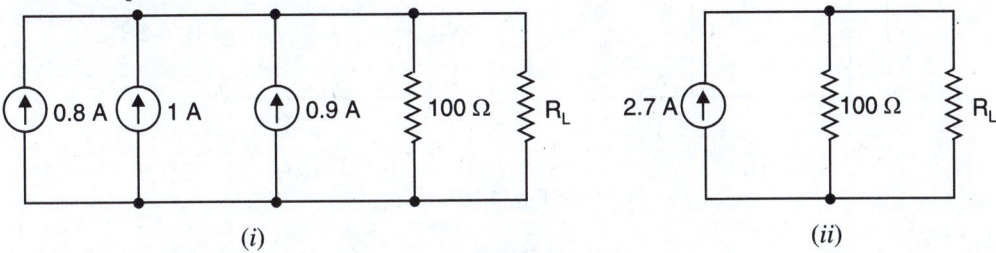

(i)   (ii)

Fig. 3.204

$$V_{Th} = I_N R_N = 2.7 \times 100 = 270 \text{ V}$$

$$R_i = R_N = 100 \ \Omega$$

The Thevenin's equivalent circuit to the left of R_L is $V_{Th}(= 270 \text{ V})$ in series with R_i (= 100 Ω). When load R_L is connected, the circuit becomes as shown in Fig. 3.205. It is clear that maximum power will be transferred when

$$R_L = R_i = \mathbf{100 \ \Omega}$$

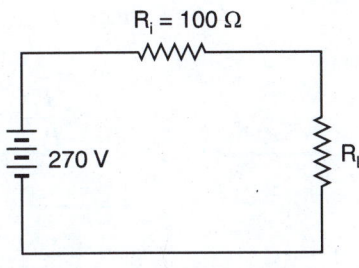

$$\text{Max. power} = \frac{V_{Th}^2}{4 R_L} = \frac{(270)^2}{4 \times 100}$$

$$= \mathbf{182 \cdot 25 \text{ watts}}$$

Fig. 3.205

Example 3.81. *Determine the maximum power that can be delivered by the circuit shown in Fig. 3.206 (i).*

Solution. Fig. 3.206 (ii) shows the Norton's equivalent circuit. Maximum power transfer occurs when $R_L = R_N = 300 \ \Omega$.

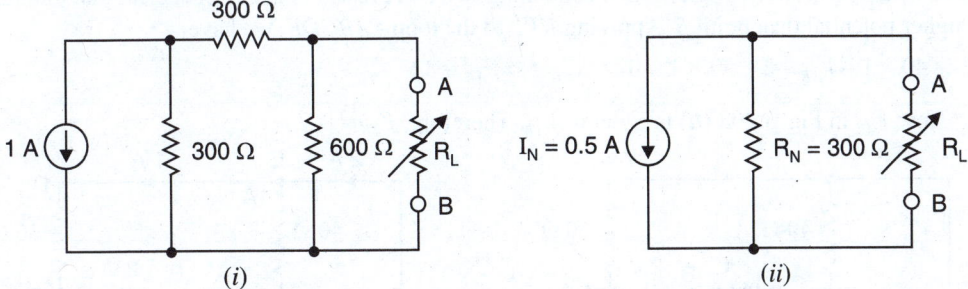

(i)   (ii)

Fig. 3.206

Referring to Fig. 3.206 (ii), current in R_L (= 300 Ω) $= I_N/2 = 0.5/2 = 0.25$ A

∴   Max. power transferred $= (0.25)^2 \times R_L = (0.25)^2 \times 300 = \mathbf{18 \cdot 8 \text{ W}}$

* **Example 3.82.** *What percentage of maximum possible power is delivered to R_L in Fig. 3.207 (i) when $R_L = 2 R_{Th}$?*

Solution. Fig. 3.207 (ii) shows the circuit when $R_L = 2 R_{Th}$.

$$\text{Circuit current} = \frac{V_{Th}}{R_{Th} + 2 R_{Th}} = \frac{V_{Th}}{3 R_{Th}}$$

Voltage across load, $V_L = \dfrac{V_{Th}}{3\,R_{Th}} \times 2\,R_{Th} = \dfrac{2}{3}\,V_{Th}$

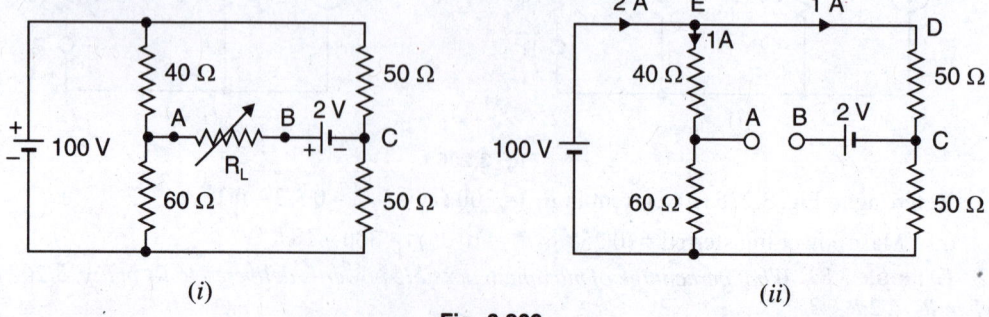

Fig. 3.207

Power delivered to load, $P_L = \dfrac{V_L^2}{R_L} = \dfrac{\left(\dfrac{2}{3}V_{Th}\right)^2}{2\,R_{Th}} = \dfrac{4V_{Th}^2}{18\,R_{Th}}$

Since $P_{max} = V_{Th}^2/4\,R_{Th}$, the ratio of P_L/P_{max} is

$$\dfrac{P_L}{P_{max}} = \dfrac{\dfrac{4V_{Th}^2}{18\,R_{Th}}}{\dfrac{V_{Th}^2}{4\,R_{Th}}} = \dfrac{16}{18}$$

∴ $P_L = \dfrac{16}{18}P_{max} \times 100 = $ **88·89% of P_{max}**

Example 3.83. *Find the maximum power in R_L which is variable in the circuit shown in Fig. 3.208 (i).*

Solution. We shall use Thevenin theorem to obtain the result. For this purpose, remove the load R_L as shown in Fig. 3.208 (*ii*). The open-circuited voltage at terminals AB in Fig. 3.208 (*ii*) is equal to V_{Th}. It is clear from Fig. 3.208 (*ii*) that current in the branch containing 40 Ω and 60 Ω resistors is 1 A. Similarly, current in the branch containing two 50 Ω resistors is 1 A. It is clear that point A is at higher potential than point B. Applying *KVL* to the loop *EABCDE*, we have,

$$- 40 \times 1 - V_{AB} - 2 + 50 \times 1 = 0 \qquad \therefore \qquad V_{AB} = 8\ V$$

Now V_{AB} in Fig. 3.208 (*ii*) is equal to V_{Th}. Therefore, $V_{Th} = 8$ V.

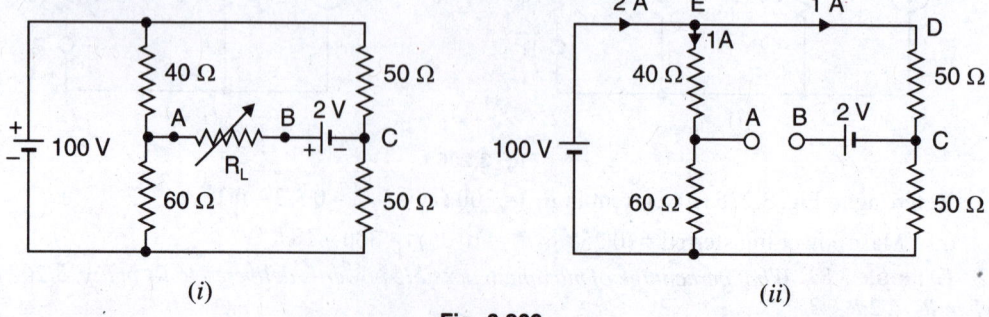

Fig. 3.208

In order to find Thevenin's resistance R_{Th}, replace 100V and 2V batteries by a short in Fig. 3.208 (*ii*). Then resistance at terminals AB is the R_{Th}. It is clear that 40 Ω and 60 Ω resistors are in parallel and so the two 50 Ω resistors.

$$\therefore \quad R_{Th} = (40 \parallel 60) + (50 \parallel 50) = \frac{40 \times 60}{40 + 60} + \frac{50 \times 50}{50 + 50} = 24 + 25 = 49 \ \Omega$$

Therefore, for maximum power, R_L should be 49 Ω. The Thevenin equivalent circuit is a voltage source of 8 V in series with a resistance of 49 Ω. When load R_L is connected across the terminals of Thevenin equivalent circuit, the total circuit resistance = 49 + 49 = 98 Ω.

$$\therefore \qquad \text{Circuit current, } I = \frac{V_{Th}}{R_{Th} + R_L} = \frac{8}{49 + 49} = \frac{8}{98} = 0\cdot08163 \ A$$

$$\therefore \qquad P_{max} = I^2 R_L = (0\cdot08163)^2 \times 49 = \mathbf{0\cdot3265 \ W}$$

Example 3.84. *For the circuit shown in Fig. 3.209 (i), find the value of R that will receieve maximum power. Determine this power.*

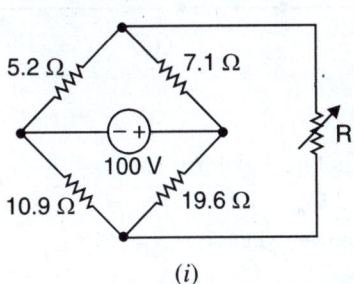

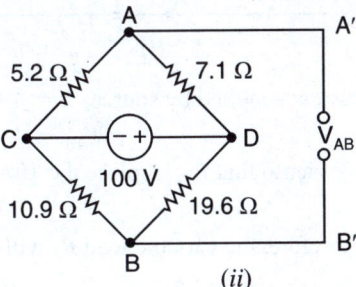

(i)   (ii)

Fig. 3.209

Solution. We will use Thevenin's theorem to obtain the results. In order to find V_{Th}, remove the variable load R as shown in Fig. 3.209 (ii). Then open-circuited voltage across terminals AB is equal to V_{Th}.

$$\text{Current in branch } DAC = \frac{100}{7.1 + 5.2} = 8\cdot13 \ A$$

$$\text{Current in branch } DBC = \frac{100}{19.6 + 10.9} = 3\cdot28 \ A$$

It is clear from Fig. 3.209 (ii) that point A is at higher* potential than point B. Applying *KVL* to the loop $A'ACBB'A'$, we have,

$$-5\cdot2 \times 8\cdot13 + 10\cdot9 \times 3\cdot28 + V_{AB} = 0$$

$$\therefore \qquad\qquad V_{AB} = 6\cdot52 \ V$$

Now V_{AB} in Fig. 3.209 (ii) is equal to V_{Th} so that $V_{Th} = 6\cdot52$ V.

In order to find R_{Th}, replace the 100 V source in Fig. 3.209 (ii) by a short. The circuit becomes as shown in Fig. 3.209 (iii). The resistance across terminals AB is the Thevenin resistance. Referring to Fig. 3.209 (iii),

$$R_{AB} = R_{Th} = (5\cdot2 \parallel 7\cdot1) + (10\cdot9 \parallel 19\cdot6)$$
$$= 3 + 7 = 10 \ \Omega$$

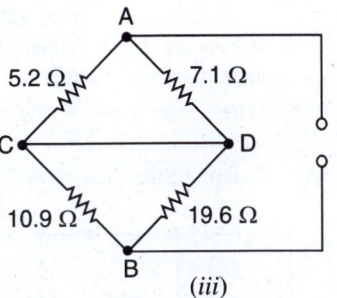

Therefore, for maximum power transfer, $R = R_{Th} = \mathbf{10 \ \Omega}$.

$$P_{max} = \frac{(V_{Th})^2}{4R} = \frac{(6.52)^2}{4 \times 10} = \mathbf{1\cdot06 \ W}$$

Fig. 3.209

---

*   The fall in potential along *DA* is less than the fall in potential along *DB*. Since point *D* is common, point *A* will be at higher potential than point *B*.

Example 3.85. *For the circuit shown in Fig. 3.210 (i), what will be the value of R_L to get maximum power? Also find this power.*

Solution. We shall use Thevenin's theorem to obtain the results. In order to find V_{Th}, remove the load R_L as shown in Fig. 3.210 (ii). Then voltage at the open-circuited terminals AB is equal to V_{Th} i.e. $V_{AB} = V_{Th}$. The total load on 10 V source is

$$R_T = (90 \,||\, 60 \,||\, 180) + 20 = 30 + 20 = 50 \ \Omega$$

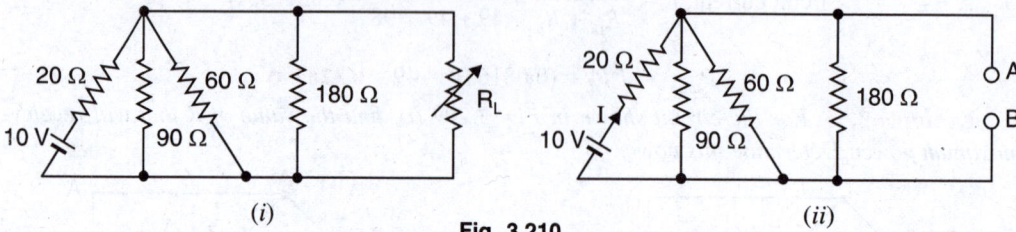

(i)   (ii)

Fig. 3.210

Current supplied by source, $I = 10/50 = 0.2$ A

$\therefore \qquad\qquad V_{AB} = V_{Th} = 10 - 20 \times 0.2 = 6\text{V}$

In order to find R_{Th}, replace the 10 V source by a short in Fig. 3.210 (ii). Then,

$$R_{Th} = 20 \,||\, 90 \,||\, 60 \,||\, 180 = 12 \ \Omega$$

Therefore, the variable load R_L will receive maximum power when $R_L = R_{Th} = \textbf{12 } \pmb{\Omega}$.

$$\therefore \qquad\qquad P_{max} = \frac{(V_{Th})^2}{4 R_L} = \frac{(6)^2}{4 \times 12} = \textbf{0.75 W}$$

Tutorial Problems

1. Find the value of R_L in Fig. 3.211 necessary to obtain maximum power in R_L. Also find the maximum power in R_L.
   [150Ω ; 1.042 W]

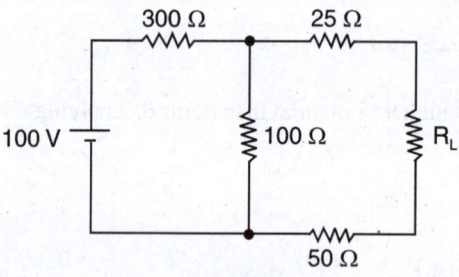
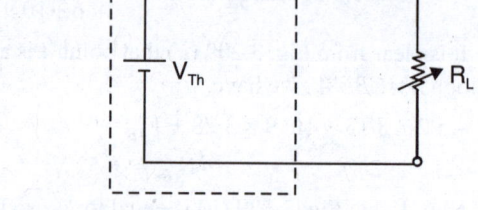

Fig. 3.211   **Fig. 3.212**

2. If R_L in Fig. 3.211 is fixed at 100 Ω, what alternation (s) can be made in the rest of the circuit to obtain maximum power in R_L?
   [Short out 50 Ω resistor]

3. What percentage of the maximum possible power is delivered to R_L in Fig. 3.212, when $R_L = R_{Th}/2$?
   [88.9%]

4. Determine the value of R_L for maximum power transfer in Fig. 3.213 and evaluate this power.
   [100 Ω; 182·25 W]

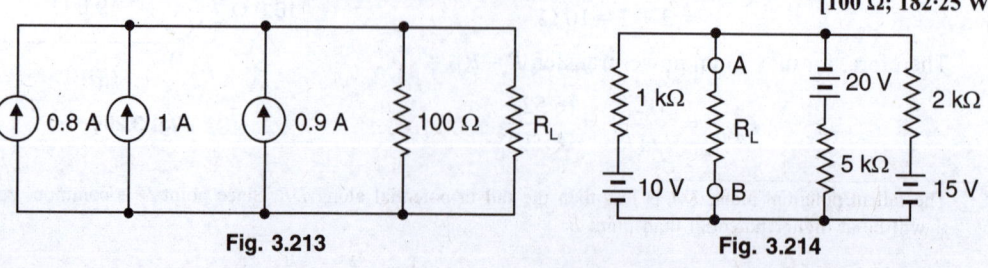

Fig. 3.213   **Fig. 3.214**

5. What value should R_L be in Fig. 3.214 to achieve maximum power transfer to the load? [588 Ω]

6. For the circuit shown in Fig. 3.215, find the value of R_L for which power transferred is maximum. Also calculate this power. [50 Ω; 0·72 W]

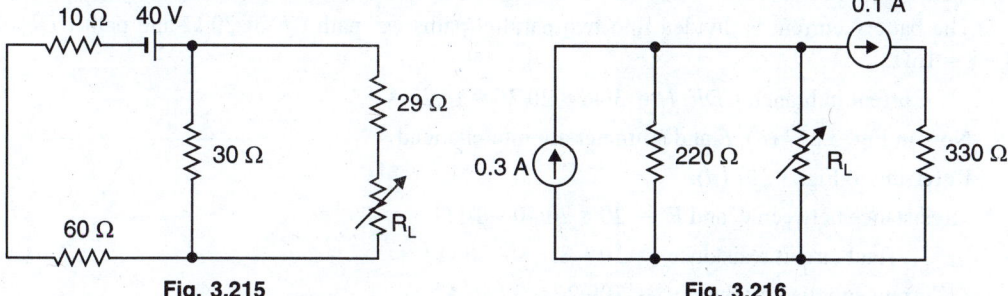

Fig. 3.215   Fig. 3.216

7. Calculate the value of R_L for transference of maximum power in Fig. 3.216. Evaluate this power. [220 Ω; 2·2 W]

3.20. Reciprocity Theorem

This theorem permits us to transfer source from one position in the circuit to another and may be stated as under :

In any linear, bilateral network, if an e.m.f. E acting in a branch X causes a current I in branch Y, then the same e.m.f. E located in branch Y will cause a current I in branch X. However, currents in other parts of the network will not remain the same.

Explanation. Consider the circuit shown in Fig. 3.217 (*i*). The e.m.f. E (=100 V) acting in the branch *FAC* produces a current I amperes in branch *CDF* and is indicated by the ammeter. According to reciprocity theorem, if the e.m.f. E and ammeter are interchanged* as shown in Fig. 3.217 (*ii*), then the ammeter reading does not change *i.e.* the ammeter now connected in branch *FAC* will read I amperes. In fact, the essence of this theorem is that E and I are interchangeable. The ratio E/I is constant and is called *transfer resistance* (or impedance in case of a.c. system).

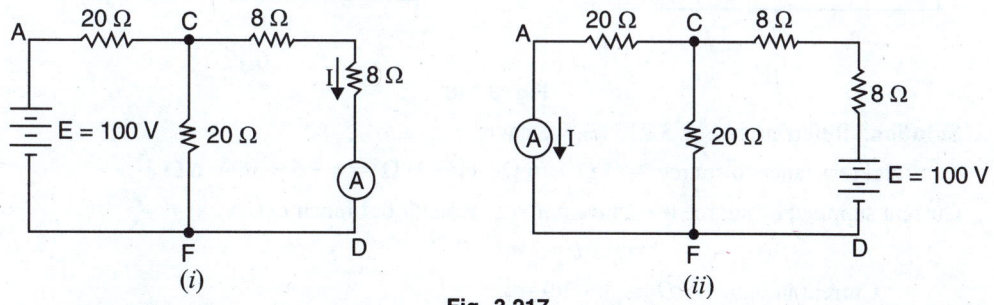

Fig. 3.217

Note. Suppose an ideal current source is connected across points *ab* of a network and this causes a voltage *v* to appear across points *cd* of the network. The reciprocity theorem states that if the current source is now connected across *cd*, the same amount of voltage *v* will appear across *ab*. This is sometimes stated as follows: An ideal current source and an ideal voltmeter can be interchanged without changing the reading of the voltmeter. However, voltages in other parts of the network will not remain the same.

Example 3.86. *Verify the reciprocity theorem for the network shown in Fig. 3.217 (i). Also find the transfer resistance.*

Solution. In Fig. 3.217 (*i*), e.m.f. E (= 100V) is in branch *FAC* and ammeter is in branch *CDF*. Referring to Fig. 3.217 (*i*),

---

* If the source of e.m.f in the original circuit has an internal resistance, this resistance must remain in the original branch and cannot be transferred to the new location of the e.m.f.

Resistance between C and F = 20 Ω || $(8 + 8)$ Ω = 20 × 16/36 = 8·89 Ω

Total circuit resistance = 20 + 8·89 = 28·89 Ω

$\therefore$ Current supplied by battery = 100/28·89 = 3·46 A

The battery current is divided into two parallel paths *viz.* path CF of 20 Ω and path CDF of $8 + 8 = 16\Omega$.

Current in branch CDF, I = 3·46 × 20/36 = 1·923 A

Now in Fig. 3.217 (*ii*), E and ammeter are interchanged.

Referring to Fig. 3.217 (*ii*),

Resistance between C and F = 20 × 20/40 = 10 Ω

Total circuit resistance = 10 + 8 + 8 = 26 Ω

Current supplied by battery = 100/26 = 3·846 A

The battery current is divided into two parallel paths of 20 Ω each.

$\therefore$ Current in branch CAF = 3·846/2 = 1·923A

Hence, ammeter reading in both cases is the same. This verifies the reciprocity theorem.

Transfer resistance = E/I = 100/1·923 = **52 Ω**

Example 3.87. *Find the currents in the various branches of the circuit shown in Fig. 3.218 (i). If a battery of 9V is added in branch BCD, find current in 4 Ω resistor using reciprocity theorem and superposition theorem.*

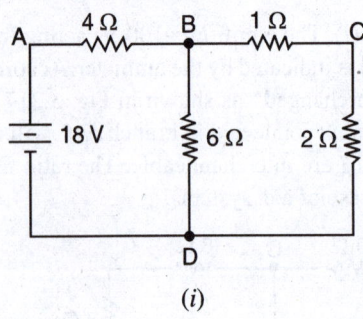
(*i*)

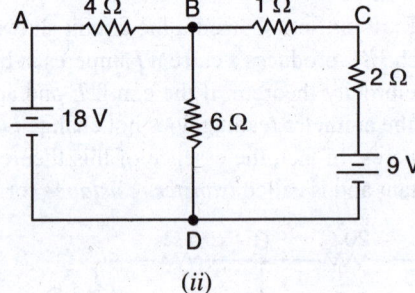
(*ii*)

Fig. 3.218

Solution. Referring to Fig. 3.218 (*i*), we have,

Total resistance to source = 4 Ω + [6 Ω || (1 + 2) Ω] = 4 + 6 × 3/9 = 6 Ω

Current supplied by source (*i.e.* current in 4 Ω resistor or branch DAB)

= 18/6 = **3 A**

Current in branch BD = 3 × 3/9 = **1 A**

Current in branch BCD = 3 × 6/9 = **2 A**

In Fig. 3.218 (*i*), the current in branch BCD due to 18 V source acting alone is 2 A. If the 18V source is placed in branch BCD, then according to reciprocity theorem, the current in 4 Ω will be 2 A flowing from B to A. If a battery of 9 V is placed in branch BCD, then current in 4 Ω resistor due to it alone would be 2 × 9/18 = 1 A (By proportion).

Now referring to Fig. 3.218 (*ii*), the current in 4 Ω due to 18 V battery alone is 3 A flowing from A to B. The current in 4 Ω resistor due to 9 V acting alone in branch BCD is 1 A flowing from B to A. By superposition theorem, the current in 4 Ω is the algebraic sum of the two currents *i.e.*

Current in 4 Ω = 3 − 1 = **2 A** *from A to B*

Example 3.88. *Prove the reciprocity theorem.*

Solution. We now prove the reciprocity theorem for the circuit shown in Fig. 3.219. In Fig. 3.219 (*i*), the e.m.f. E is acting in the branch *FAC* and the current in the branch *CDF* is I_2. If the same e.m.f. E now acts in branch *CDF* [See Fig. 3.219 (*ii*)], then current I_b in the branch *FAC* will be equal to I_2. We now show that $I_b = I_2$. Referring to Fig. 3.219 (*i*), we have,

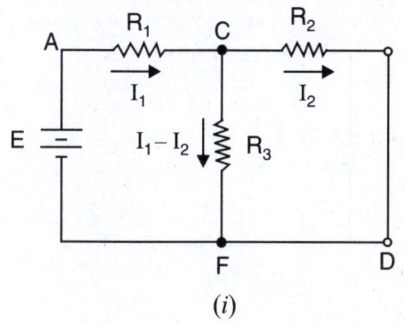 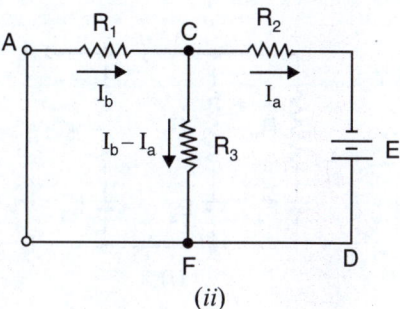

| (*i*) | (*ii*) |

Fig. 3.219

$$E = I_1 R_T$$

where

$$R_T = R_1 + (R_2 \| R_3) = \left[R_1 + \frac{R_2 R_3}{R_2 + R_3} \right]$$

$\therefore$

$$E = I_1 \left[R_1 + \frac{R_2 R_3}{R_2 + R_3} \right]$$

$$= I_1 \left[\frac{R_1 R_2 + R_2 R_3 + R_3 R_1}{R_2 + R_3} \right] \qquad ...(i)$$

Also in Fig. 3.219 (*i*),

$$0 = -(I_1 - I_2) R_3 + I_2 R_2$$

or

$$I_2 = I_1 \left[\frac{R_3}{R_2 + R_3} \right] \qquad ...(ii)$$

Dividing eq. (*i*) by eq. (*ii*), we have,

$$\frac{E}{I_2} = \frac{R_1 R_2 + R_2 R_3 + R_3 R_1}{R_3} \qquad ...(iii)$$

Similarly, it can be shown that in Fig. 3.219 (*ii*), we have,

$$\frac{E}{I_b} = \frac{R_1 R_2 + R_2 R_3 + R_3 R_1}{R_3} \qquad ...(iv)$$

From eqs. (*iii*) and (*iv*), $\quad I_b = I_2$

Therefore, reciprocity theorem stands proved.

3.21. Millman's Theorem

Millman's theorem is a combination of Thevenin's and Norton's theorems. It is used to reduce any number of parallel voltage/current sources to an equivalent circuit containing only one source. It has the advantage of being easier to apply to some networks than mesh analysis, nodal analysis or superposition. This theorem can be stated in terms of voltage sources or current sources or both.

1. Parallel voltage sources. Millman's theorem provides a method of calculating the common voltage across different parallel-connected voltage sources and may be stated as under :

The voltage sources that are directly connected in parallel can be replaced by a single equivalent voltage source.

Obviously, the above statement is true by virtue of Thevenin's theorem. Fig 3.220 (*i*) shows three parallel-connected voltage sources E_1, E_2 and E_3. Then common terminal voltage V_{AB} of these parallel voltage sources is given by ;

$$V_{AB} = \frac{E_1/R_1 + E_2/R_2 + E_3/R_3}{1/R_1 + 1/R_2 + 1/R_3} = \frac{I_1 + I_2 + I_3}{G_1 + G_2 + G_3} = \frac{\Sigma I}{\Sigma G} \qquad ...(i)$$

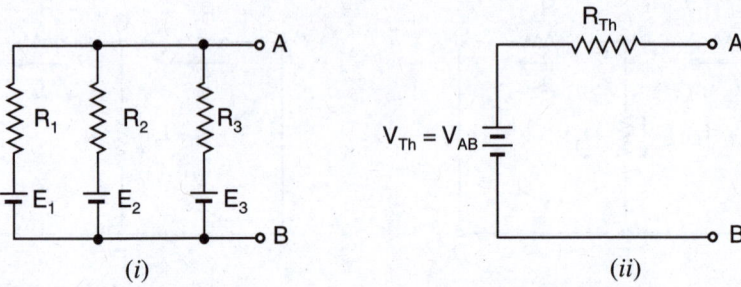

(*i*)   (*ii*)

Fig. 3.220

This voltage represents the Thevenin's voltage V_{Th}. The denominator represents Thevenin's resistance R_{Th} *i.e.*

$$R_{Th} = \frac{1}{1/R_1 + 1/R_2 + 1/R_3}$$

Therefore, parallel-connected voltage sources in Fig 3.220 (*i*) can be replaced by a single voltage source as shown in Fig 3.220 (*ii*). If load R_L is connected across terminals *AB*, then load current I_L is given by ;

$$I_L = \frac{V_{Th}}{R_{Th} + R_L}$$

Note. If a branch does not contain any voltage source, the same procedure is used except that current in that branch will be zero. This is illustrated in example 3.89.

2. Parallel current sources. The Millman's theorem states as under :

The current sources that are directly connected in parallel can be replaced by a single equivalent current source. The current of this single current source is the algebraic sum of the individual source currents. The internal resistance of the single current source is equal to the combined resistance of the parallel combination of the source resistances.

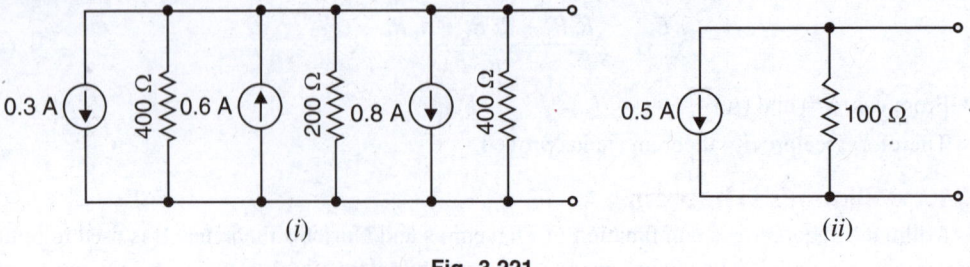

(*i*)   (*ii*)

Fig. 3.221

Fig. 3.221 (*i*) shows three parallel connected current sources. The resultant current of the three sources is

$$0\cdot3\downarrow + 0\cdot6\uparrow + 0\cdot8\downarrow = 0\cdot5\,A\downarrow$$

The internal resistance of the single current source is equal to the equivalent resistance of three parallel resistors.

$$400 \parallel 200 \parallel 400 = 100 \,\Omega$$

Thus the single equivalent current source has value 0·5 A and internal resistance 100 Ω in parallel as shown in Fig. 3.221 (ii).

3. Voltage sources and current sources in parallel. The Millman's theorem is also applicable if the circuit has a mixture of parallel voltage and current sources. Each parallel-connected voltage source is converted to an equivalent current source. The result is a set of parallel-connected current sources and we can replace them by a single equivalent current source. Alternatively, each parallel-connected current source can be converted to an equivalent voltage source and the set of parallel-connected voltage sources can be replaced by an equivalent voltage source.

Example 3.89. *Using Millman's theorem, determine the common voltage* V_{xy} *and the load current in the circuit shown in Fig. 3.222 (i).*

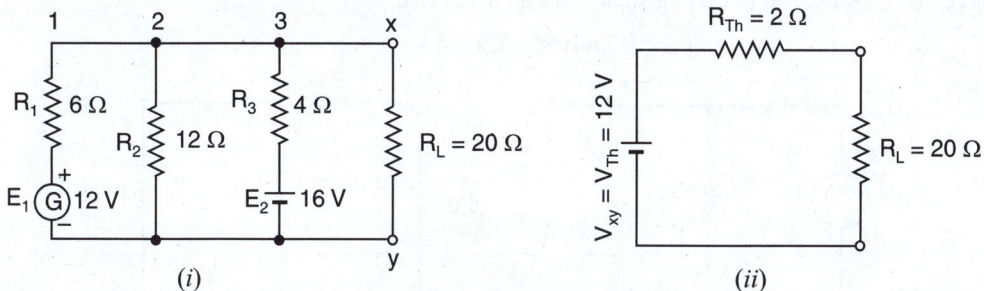

Fig. 3.222

Solution.
$$V_{xy} = V_{Th} = \frac{E_1/R_1 + E_2/R_2 + E_3/R_3}{1/R_1 + 1/R_2 + 1/R_3}$$

$$= \frac{12/6 + 0/2 + 16/4}{1/6 + 1/12 + 1/4} = \frac{2 + 0 + 4}{0.167 + 0.083 + 0.25} = \frac{6}{0.5} = \textbf{12V}$$

$$R_{Th} = \frac{1}{1/6 + 1/12 + 1/4} = 2\Omega$$

Therefore, the circuit shown in Fig. 3.222 (*i*) can be replaced by the one shown in Fig. 3.222 (*ii*).

$$\text{Load current} = \frac{V_{Th}}{R_{Th} + R_L} = \frac{12}{2 + 20} = \textbf{0.545 A}$$

Example 3.90. *Find the current in the 1 k Ω resistor in Fig. 3.223 by finding Millman equivalent voltage source with respect to terminals x – y.*

Solution. As shown Fig. 3.224 (*i*), each of the three voltage sources is converted to an equivalent current source. For example, the 36 V source in series with 18 kΩ resistor becomes a 36 V/18 kΩ = 2 mA current source in parallel with 18 kΩ. Note that the polarity of each current source is such that it produces current in the same direction as the voltage source it replaces.

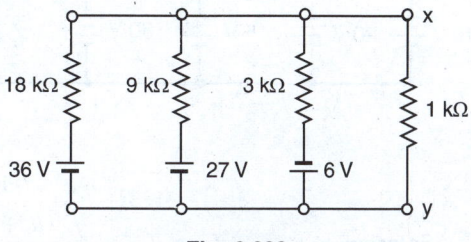

Fig. 3.223

The resultant current of the three current sources

$$= 2 \text{ mA} \uparrow + 3 \text{ mA} \uparrow + 2 \text{ mA} \downarrow = 3 \text{ mA} \uparrow$$

The parallel equivalent resistance of three resistors

$$= 18 \text{ k}\Omega \parallel 9 \text{ k}\Omega \parallel 3 \text{ k}\Omega = 2 \text{ k}\Omega$$

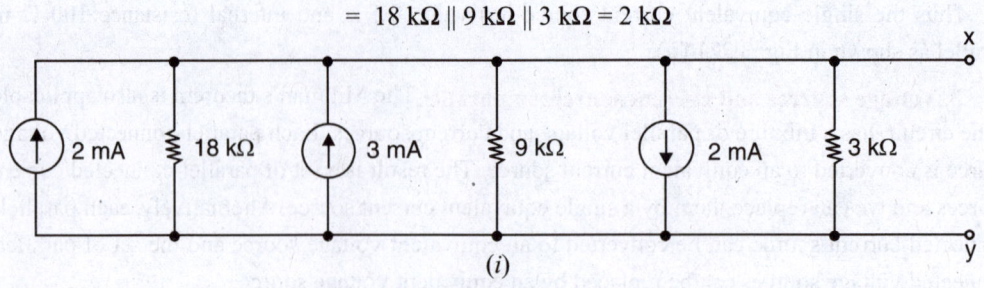

(i)

Fig. 3.224

Fig. 3.224 (ii) shows the single equivalent current source. Fig. 3.224 (iii) shows the voltage source that is equivalent to current source in Fig. 3.224 (ii).

$$V_{Th} = 3 \text{ mA} \times 2 \text{ k}\Omega = 6 \text{ V}$$

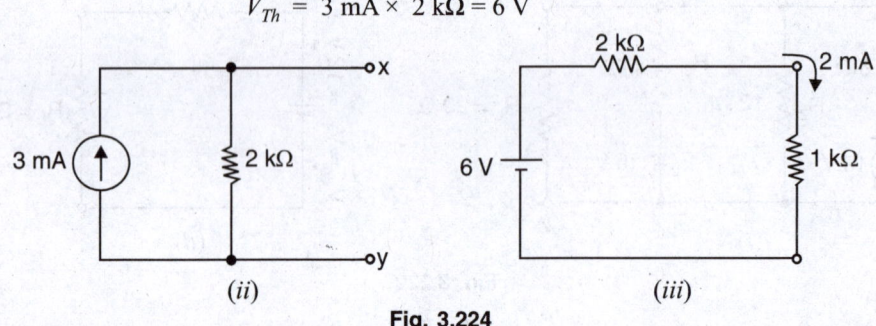

(ii)   (iii)

Fig. 3.224

When the 1 kΩ resistor is connected across the $x - y$ terminals, the current is

$$I = \frac{6\text{V}}{3\text{k}\Omega} = \textbf{2 mA}$$

Example 3.91. *Find an equivalent voltage source for the circuit shown in Fig. 3.225 (i). What is the load current?*

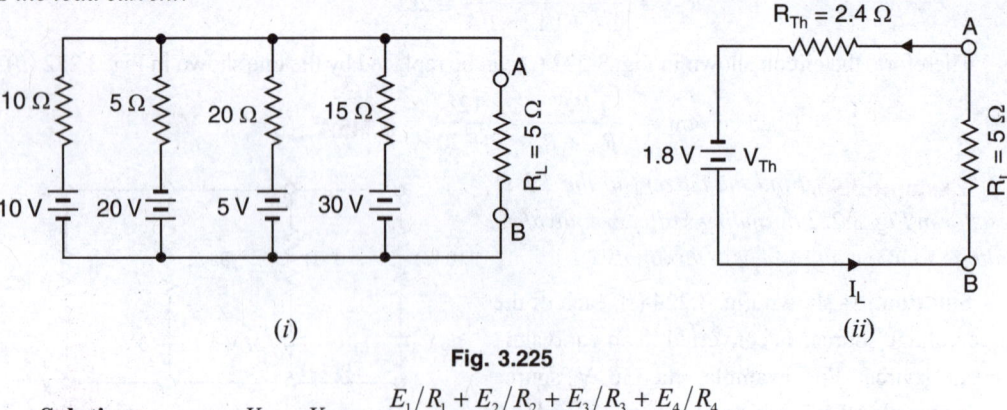

(i)   (ii)

Fig. 3.225

Solution.
$$V_{AB} = V_{Th} = \frac{E_1/R_1 + E_2/R_2 + E_3/R_3 + E_4/R_4}{1/R_1 + 1/R_2 + 1/R_3 + 1/R_4}$$

$$= \frac{10/10 - *20/5 + 5/20 + 30/15}{1/10 + 1/5 + 1/20 + 1/15} = \frac{-0.75}{0.417} = -1{\cdot}8 \text{ V}$$

Negative sign shows that terminal *A* is negative *w.r.t.* terminal *B*.

* Note that polarity is opposite as compared to other sources.

$$R_{Th} = \frac{1}{1/R_1 + 1/R_2 + 1/R_3 + 1/R_4}$$

$$= \frac{1}{1/10 + 1/5 + 1/20 + 1/15} = 2.4\Omega$$

Therefore, equivalent voltage source consists of **1·8 V source in series with 2·4 Ω resistor** as shown in Fig. 3.225 (*ii*).

$$\therefore \qquad \text{Load current, } I_L = \frac{V_{Th}}{R_{Th} + R_L} = \frac{1.8}{2.4 + 5} = \mathbf{0.24A}$$

Example 3.92. *For the circuit shown in Fig. 3.225 (i) above, find the equivalent current source. Also find load current.*

Solution. Convert the voltage sources to current sources as shown in Fig. 3.226 (*i*). The arrow for each current source corresponds to the polarity of each voltage source in the original circuit.

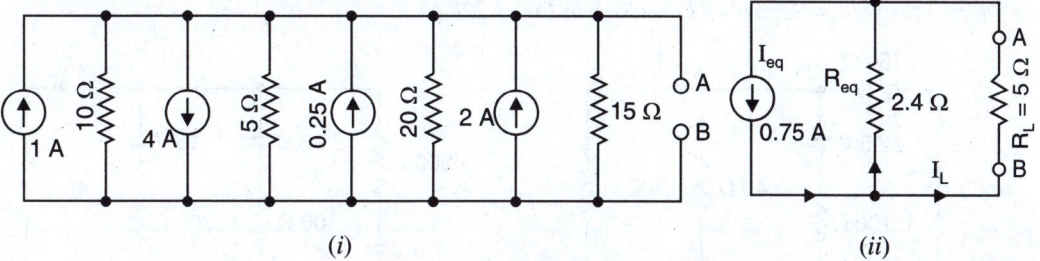

$$(i) \hspace{10em} (ii)$$

Fig. 3.226

The equivalent current source is found by algebraically adding the currents of individual sources.

$$I_{eq} = 1\,A\uparrow + 4\,A\downarrow + 0.25\,A\uparrow + 2\,A\uparrow = 0.75\,A\downarrow$$

The downward arrow for I_{eq} shows that terminal *A* is negative *w.r.t.* terminal *B*.

$$R_{eq} = 10\,\Omega \parallel 5\,\Omega \parallel 20\,\Omega \parallel 15\,\Omega = 2.4\,\Omega$$

Therefore, the equivalent current source consists of **0·75 A current source in parallel with 2·4 Ω resistor** as shown in Fig. 3.226 (*ii*). By current-divider rule, the load current I_L is

$$I_L = 0.75 \times \frac{2.4}{2.4 + 5} = \mathbf{0.243A}$$

Example 3.93. *Find the load current for Fig. 3.227 (i) using the dual of Millman's theorem.*

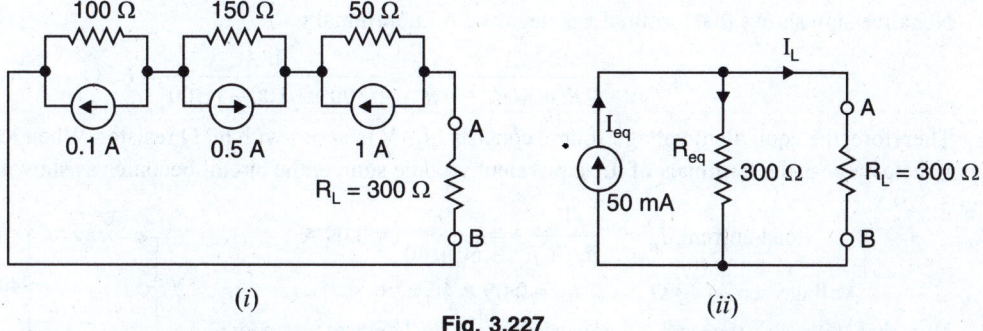

$$(i) \hspace{10em} (ii)$$

Fig. 3.227

Solution. There is a dual for Millman's theorem and it is useful for solving circuits with series current sources [See Fig. 3.227 (*i*)]. In such a case, the following equations are used to find the current and resistance of the equivalent circuit.

$$I_{eq} = \frac{I_1 R_1 + I_2 R_2 + I_3 R_3}{R_1 + R_2 + R_3}$$

$$R_{eq} = R_1 + R_2 + R_3$$

Thus referring to Fig. 3.227 (*i*), we have,

$$I_{eq} = \frac{-0.1 \times 100 + 0.5 \times 150 - 1 \times 50}{100 + 150 + 50} = \frac{15}{300} A = 50 \text{ mA}$$

$$R_{eq} = 100 + 150 + 50 = 300 \ \Omega$$

The equivalent circuit is shown in Fig. 3.227 (*ii*). By current-divider rule, the load current I_L is

$$I_L = 50 \times \frac{300}{300 + 300} = \mathbf{25 \ mA}$$

Example 3.94. *By constructing a Millman equivalent voltage source with respect to terminals* $x - y$, *find the voltage across 40 Ω resistor in Fig. 3.228 (i).*

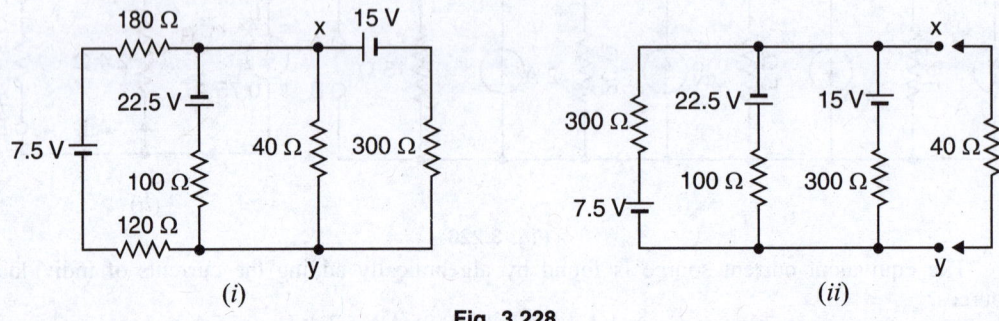

Fig. 3.228

Solution. Note that 120 Ω and 180 Ω resistors are in a series path and can therefore be combined into an equivalent resistance of 300 Ω. The circuit is *redrawn as shown in Fig. 3.228 (*ii*). It is clear that redrawn circuit has three parallel-connected voltage sources. Referring to Fig. 3.228 (*ii*), we have,

$$V_{xy} = V_{Th} = \frac{E_1/R_1 - E_2/R_2 + E_3/R_3}{1/R_1 + 1/R_2 + 1/R_3}$$

$$= \frac{7.5/300 - 22.5/100 + 15/300}{1/300 + 1/100 + 1/300} = \frac{-0.15}{0.0167} = -9 \text{V}$$

Negative sign shows that terminal x is negative w.r.t. terminal y.

$$R_{Th} = \frac{1}{1/R_1 + 1/R_2 + 1/R_3} = \frac{1}{1/300 + 1/100 + 1/300} = 60 \ \Omega$$

Therefore, the equivalent voltage source consists of 9 V in series with 60 Ω resistor. When load is connected across the terminals of the equivalent voltage source, the circuit becomes as shown in Fig. 3.229.

Load current, $I_L = \dfrac{V_{Th}}{R_{Th} + R_L} = \dfrac{9}{60 + 40} = 0.09$ A

Voltage across 40 Ω = $I_L R_L = 0{\cdot}09 \times 40 = \mathbf{3{\cdot}6 \ V}$

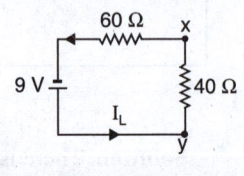

Fig. 3.229

Note that Millman's theorem is a powerful tool in the hands of engineers to solve many problems which cannot be solved easily by the usual methods of circuit analysis.

---

*   It makes no difference on which side of each voltage source its series resistance is drawn.

Tutorial Problems

1. Find the single equivalent current source for the circuit shown in Fig. 3.230.
   $\left[3\text{ mA} \bigoplus \quad \lessgtr 10\ \Omega \right]$

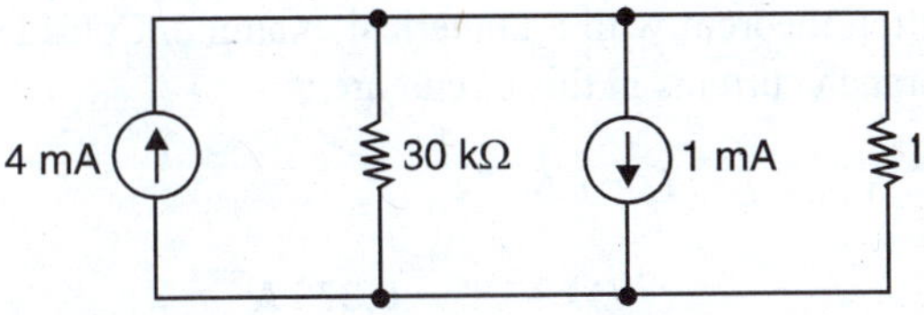

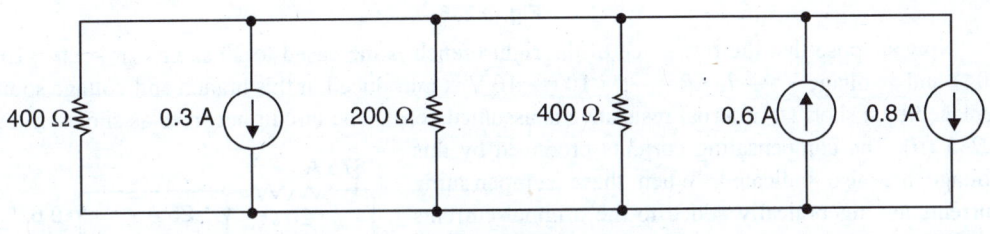

Fig. 3.230 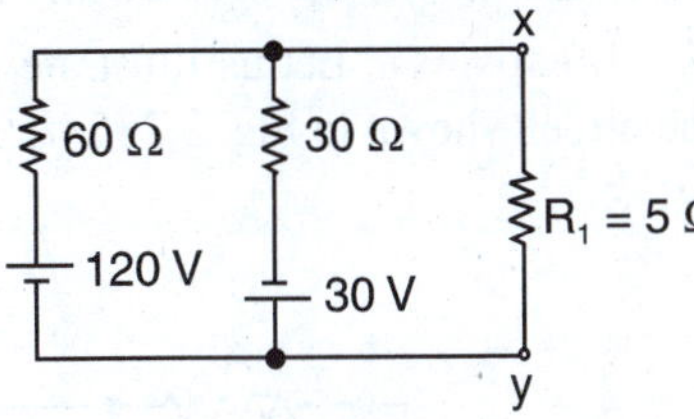   **Fig. 3.231**

2. By constructing a Millman equivalent voltage source at terminals $x - y$, find the voltage across R_1 (= 5 Ω) in the circuit shown in Fig. 3.231.   **[4 V ±]**

3. Find the single equivalent current source for the circuit shown in Fig. 3.232.
   $\left[0.5\text{ A} \bigoplus \quad \lessgtr 100\ \Omega \right]$

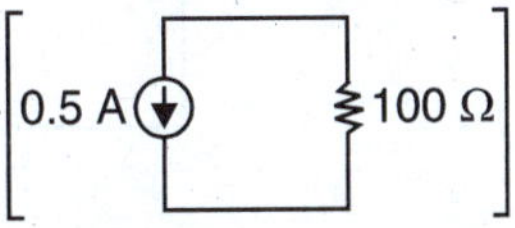

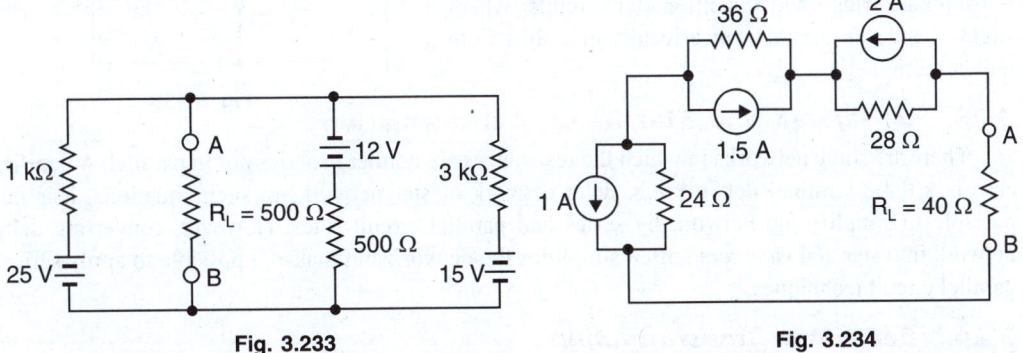

Fig. 3.232

4. What is the current flowing in the load resistor in Fig. 3.233 ?   **[2·25 mA]**

Fig. 3.233   **Fig. 3.234**

5. What is the drop and polarity of the load in Fig. 3.234 ?   **[8·13V and terminal A is negative]**

3.22. Compensation Theorem

It is sometimes necessary to know, when making a change in one branch of a network, what effect this change will have on the various currents and voltages throughout the network. The compensation theorem deals with this situation and may be stated for d.c. circuits as under :

The compensation theorem states that any resistance R in a branch of a network in which current I is flowing can be replaced, for the purpose of calculations, by a voltage equal to − IR. It follows from Kirchhoff's voltage law that the current I is unaltered if an e.m.f. − IR is substituted for the voltage drop IR.

Or

If the resistance of any branch of a network is changed from R to (R + ΔR) where the current was originally I, then the change of current at any point in the network may be calculated by assuming than an e.m.f. – IΔR has been introduced into the modified branch while all other sources have their e.m.f.s. suppressed and are represented by their internal resistances only.

Illustration. Let us illustrate the compensation theorem with a numerical example. Consider the circuit shown in Fig. 3.235 (*i*). The various branch currents in this circuit are :

$$I_1 = \frac{50}{20+5} = 2 \text{ A} \quad ; \quad I_2 = I_3 = 1 \text{ A}$$

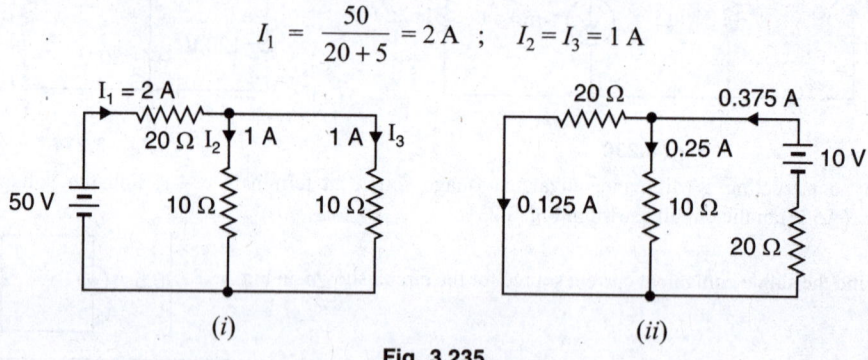

(*i*)   (*ii*)

Fig. 3.235

Now suppose that the resistance of the right branch is increased to 20 Ω *i.e.* ΔR = 20 – 10 = 10 Ω and a voltage V = – I_3 ΔR = – 1 × 10 = – 10 V is introduced in this branch and voltage source replaced by a short (∵ internal resistance is assumed zero). The circuit becomes as shown in Fig. 3.235 (*ii*). The compensating currents produced by this voltage are also indicated. When these compensating currents are algebraically added to the original currents in their respective branches, the new branch currents will be as shown in Fig. 3.236. The compensation theorem is useful in bridge and potentiometer circuits, where a slight change in one resistance results in a shift from a null condition.

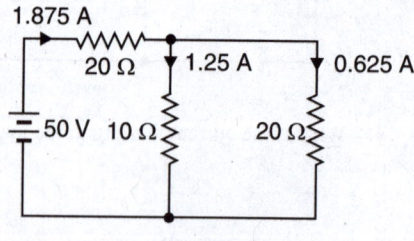

Fig. 3.236

3.23. Delta/Star and Star/Delta Transformation

There are some networks in which the resistances are neither in series nor in parallel. A familiar case is a three terminal network *e.g.* delta network or star network. In such situations, it is not possible to simplify the network by series and parallel circuit rules. However, converting delta network into star and *vice-versa* often simplifies the network and makes it possible to apply series-parallel circuit techniques.

3.24. Delta/Star Transformation

Consider three resistors R_{AB}, R_{BC} and R_{CA} connected in delta to three terminals *A*, *B* and *C* as shown in Fig. 3.237 (*i*). Let the equivalent star-connected network have resistances R_A, R_B and R_C. Since the two arrangements are electrically equivalent, the resistance between any two terminals of one network is equal to the resistance between the corresponding terminals of the other network.

Let us consider the terminals *A* and *B* of the two networks.

Resistance between *A* and *B* for star = Resistance between *A* and *B* for delta

or $R_A + R_B = R_{AB} \| (R_{BC} + R_{CA})$

or $R_A + R_B = \dfrac{R_{AB}(R_{BC} + R_{CA})}{(R_{AB} + R_{BC} + R_{CA})}$ ...(*i*)

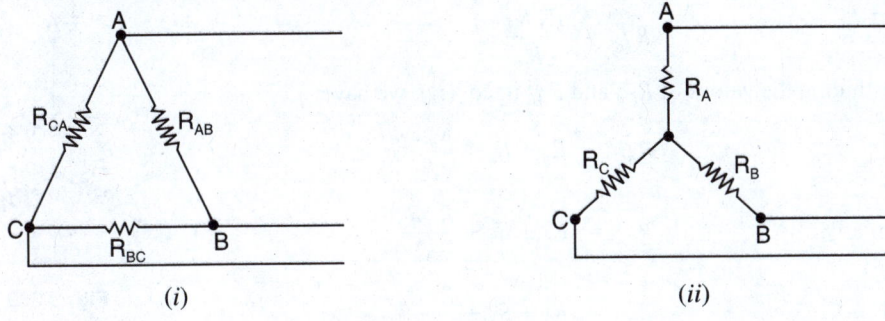

Fig. 3.237

Similarly,
$$R_B + R_C = \frac{R_{BC}(R_{CA} + R_{AB})}{R_{AB} + R_{BC} + R_{CA}} \qquad \text{...(ii)}$$

and
$$R_C + R_A = \frac{R_{CA}(R_{AB} + R_{BC})}{R_{AB} + R_{BC} + R_{CA}} \qquad \text{...(iii)}$$

Subtracting eq. (*ii*) from eq. (*i*) and adding the result to eq. (*iii*), we have,

$$R_A = \frac{R_{AB}\,R_{CA}}{R_{AB} + R_{BC} + R_{CA}} \qquad \text{...(iv)}$$

Similarly,
$$R_B = \frac{R_{BC}\,R_{AB}}{R_{AB} + R_{BC} + R_{CA}} \qquad \text{...(v)}$$

and
$$R_C = \frac{R_{CA}\,R_{BC}}{R_{AB} + R_{BC} + R_{CA}} \qquad \text{...(vi)}$$

How to remember ? There is an easy way to remember these relations. Referring to Fig. 3.238, star-connected resistances R_A, R_B and R_C are electrically equivalent to delta-connected resistances R_{AB}, R_{BC} and R_{CA}. We have seen above that :

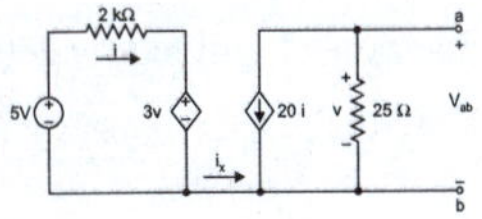

$$R_A = \frac{R_{AB}\,R_{CA}}{R_{AB} + R_{BC} + R_{CA}}$$

i.e. Any arm of star-connection $= \dfrac{\text{Product of two adjacent arms of } \Delta}{\text{Sum of arms of } \Delta}$

Fig. 3.238

Thus to find the star resistance that connects to terminal *A*, *divide the product of the two delta resistors connected to A by the sum of the delta resistors*. Same is true for terminals *B* and *C*.

3.25. Star/Delta Transformation

Now let us consider how to replace the star-connected network of Fig. 3.237 (*ii*) by the equivalent delta-connected network of Fig. 3.237 (*i*).

Dividing eq. (*iv*) by (*v*), we have,

$$R_A/R_B = R_{CA}/R_{BC}$$

∴
$$R_{CA} = \frac{R_A\,R_{BC}}{R_B}$$

Dividing eq. (*iv*) by (*vi*), we have,

$$R_A/R_C = R_{AB}/R_{BC}$$

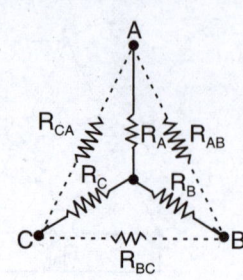

$$\therefore \qquad R_{AB} = \frac{R_A R_{BC}}{R_C}$$

Substituting the values of R_{CA} and R_{AB} in eq. (*iv*), we have,

$$R_{BC} = R_B + R_C + \frac{R_B R_C}{R_A}$$

Similarly,

$$R_{CA} = R_C + R_A + \frac{R_C R_A}{R_B}$$

and

$$R_{AB} = R_A + R_B + \frac{R_A R_B}{R_C}$$

Fig. 3.239

How to remember ? There is an easy way to remember these relations.

Referring to Fig. 3.239, star-connected resistances R_A, R_B and R_C are electrically equivalent to delta-connected resistances R_{AB}, R_{BC} and R_{CA}. We have seen above that :

$$R_{AB} = R_A + R_B + \frac{R_A R_B}{R_C}$$

i.e. Resistance between two = Sum of star resistances connected to those terminals *plus* product of
terminals of delta   same two resistances divided by the third star resistance

Note. Figs. 3.240 (*i*) to (*iii*) show three ways that a wye (Y) arrangement might appear in a circuit. Because the wye-connected components may appear in the equivalent form shown in Fig. 3.240 (*ii*), the arrangement is also called a *tee* (*T*) arrangement. Figs. 3.240 (*iv*) to (*vi*) show equivalent delta forms. Because the delta (Δ) arrangement may appear in the equivalent form shown in Fig. 3.240 (*vi*), it is also called a *pi* (π) arrangement. The figures show only a few of the ways the wye (*Y*) and delta (Δ) networks might be drawn in a schematic diagram. Many equivalent forms can.be drawn by rotating these basic arrangements through various angles. Note that each network has three terminals.

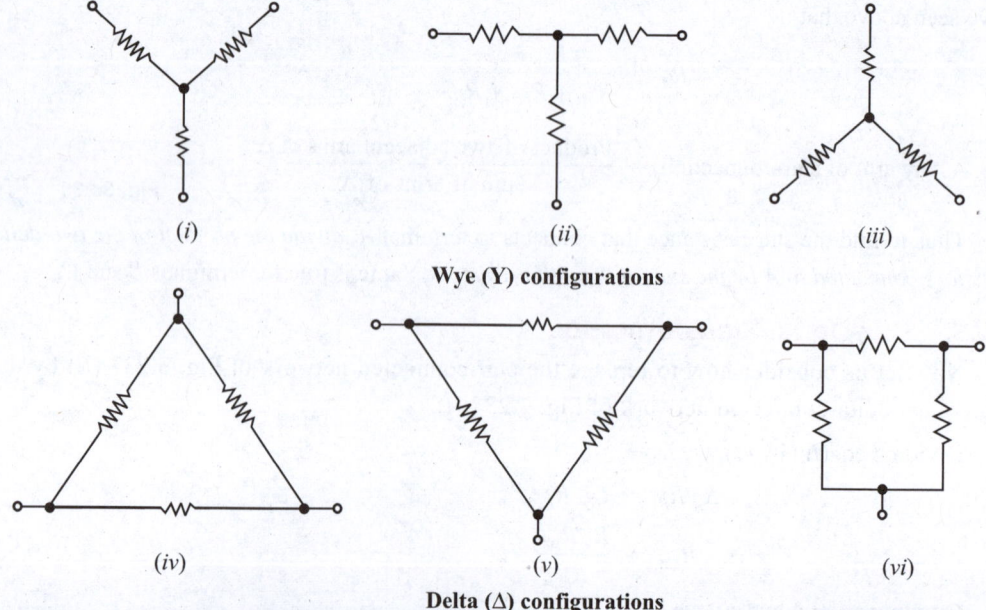

(*i*)   (*ii*)   (*iii*)

Wye (Y) configurations

(*iv*)   (*v*)   (*vi*)

Delta (Δ) configurations

Fig. 3.240

Example 3.95. *Using delta/star transformation, find the galvanometer current in the Wheatstone bridge shown in Fig. 3.241 (i).*

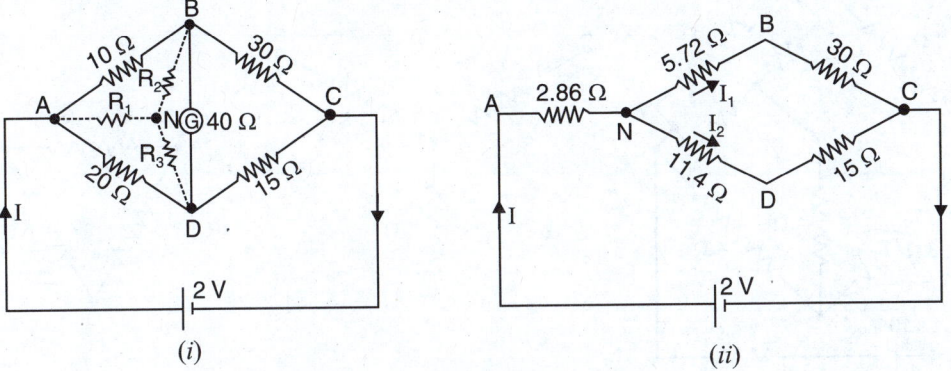

Fig. 3.241

Solution. The network *ABDA* in Fig. 3.241 (*i*) forms a delta. These delta-connected resistances can be replaced by equivalent star-connected resistances R_1, R_2 and R_3 as shown in Fig. 3.241 (*i*).

$$R_1 = \frac{R_{AB}\,R_{DA}}{R_{AB} + R_{BD} + R_{DA}} = \frac{10 \times 20}{10 + 40 + 20} = 2.86\ \Omega$$

$$R_2 = \frac{R_{AB}\,R_{BD}}{R_{AB} + R_{BD} + R_{DA}} = \frac{10 \times 40}{10 + 40 + 20} = 5.72\ \Omega$$

$$R_3 = \frac{R_{DA}\,R_{BD}}{R_{AB} + R_{BD} + R_{DA}} = \frac{20 \times 40}{10 + 40 + 20} = 11.4\ \Omega$$

Thus the network shown in Fig. 3.241 (*i*) reduces to the network shown in Fig. 3.241 (*ii*).

$$R_{AC} = 2.86 + \frac{(30 + 5.72)\,(15 + 11.4)}{(30 + 5.72) + (15 + 11.4)} = 18.04\ \Omega$$

Battery current, $I = 2/18 \cdot 04 = 0 \cdot 11\ A$

The battery current divides at *N* into two parallel paths.

∴ Current in branch *NBC*, $I_1 = 0.11 \times \dfrac{26.4}{26.4 + 35.72} = 0 \cdot 047\ A$

Current in branch *NDC*, $I_2 = 0.11 \times \dfrac{35.72}{26.4 + 35.72} = 0 \cdot 063\ A$

Potential of *B w.r.t. C* $= 30 \times 0 \cdot 047 = 1 \cdot 41\ V$

Potential of *D w.r.t. C* $= 15 \times 0 \cdot 063 = 0 \cdot 945\ V$

Clearly, point *B* is at higher potential than point *D* by

$$1 \cdot 41 - 0 \cdot 945 = 0 \cdot 465\ V$$

∴ Galvanometer current $= \dfrac{\text{P.D. between } B \text{ and } D}{\text{Galvanometer resistance}}$

$$= 0 \cdot 465 \,/\, 40 = 11 \cdot 6 \times 10^{-3}\ A = \textbf{11·6 mA } \textit{from B to D}$$

Example 3.96. *With the help of star/delta transformation, obtain the value of current supplied by the battery in the circuit shown in Fig. 3.242 (i).*

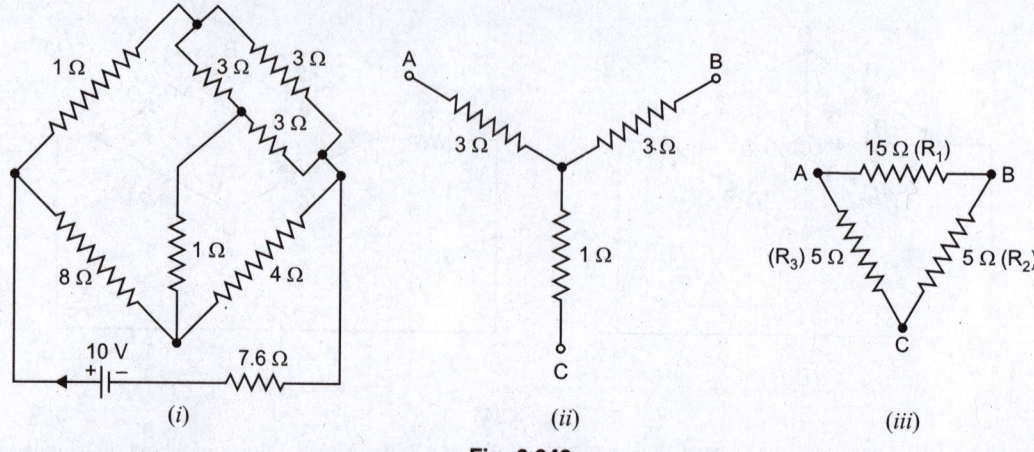

Fig. 3.242

Solution. The star-connected resistances $3\,\Omega$, $3\,\Omega$ and $1\,\Omega$ in Fig. 3.242 (*i*), are shown separately in Fig. 3.242 (*ii*). These star-connected resistances are converted into equivalent delta-connected resistances R_1, R_2 and R_3 as shown in Fig. 3.242 (*iii*).

$$R_1 = 3+3+\frac{3\times 3}{1} = 15\,\Omega$$

$$R_2 = 3+1+\frac{3\times 1}{3} = 5\,\Omega$$

$$R_3 = 1+3+\frac{1\times 3}{3} = 5\,\Omega$$

After above star-delta conversion, the circuit reduces to the one shown in Fig. 3.242 (*iv*). This circuit can be further simplified by combining parallel resistances and the circuit becomes as shown in Fig. 3.242 (*v*).

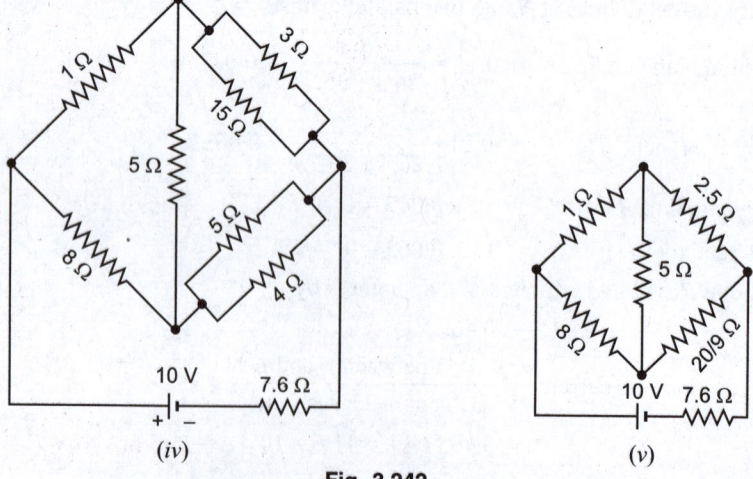

Fig. 3.242

The three delta-connected resistances $1\,\Omega$, $5\,\Omega$ and $8\,\Omega$ in Fig. 3.242 (*v*) are shown separately in Fig. 3.242 (*vi*). These delta-connected resistances can be converted into equivalent star-connected resistances R'_1, R'_2 and R'_3 as shown in Fig. 3.242 (*vii*).

$$R'_1 = \frac{1 \times 8}{1 + 5 + 8} = \frac{4}{7}\Omega$$

$$R'_2 = \frac{5 \times 1}{1 + 5 + 8} = \frac{5}{14}\Omega$$

$$R'_3 = \frac{8 \times 5}{1 + 5 + 8} = \frac{20}{7}\Omega$$

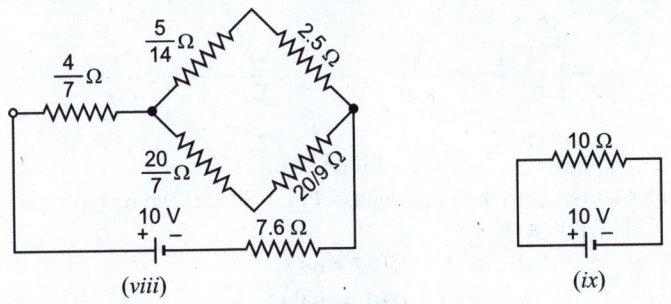

(vi)   (vii)

Fig. 3.242

After above delta-star conversion, the circuit reduces to the one shown in Fig. 3.242 (*viii*).

(viii)   (ix)

Fig. 3.242

Total resistance offered by the circuit to the battery is

$$R_T = \frac{4}{7} + \left[\left(\frac{5}{14} + 2.5\right) \| \left(\frac{20}{7} + \frac{20}{9}\right)\right] + 7.6$$

$$= \frac{4}{7} + \left(\frac{20}{7} \| \frac{320}{63}\right) + 7.6 = 10\,\Omega$$

∴   Current supplied by the battery [See Fig. 3.242 (*ix*)] is

$$I = \frac{V}{R_T} = \frac{10}{10} = \mathbf{1\,A}$$

Example 3.97. *A network of resistors is shown in Fig. 3.243 (i). Find the resistance (i) between terminals A and B (ii) B and C and (iii) C and A.*

Solution. The star-connected resistances 6 Ω, 3 Ω and 4 Ω in Fig. 3.243 (*i*) are shown separately in Fig. 3.243 (*ii*). These star-connected resistances can be converted into equivalent delta-connected resistances R_1, R_2 and R_3 as shown in Fig. 3.243 (*ii*).

$$R_1 = 4 + 6 + (4 \times 6/3) = 18\,\Omega$$
$$R_2 = 6 + 3 + (6 \times 3/4) = 13\cdot5\,\Omega$$
$$R_3 = 4 + 3 + (4 \times 3/6) = 9\,\Omega$$

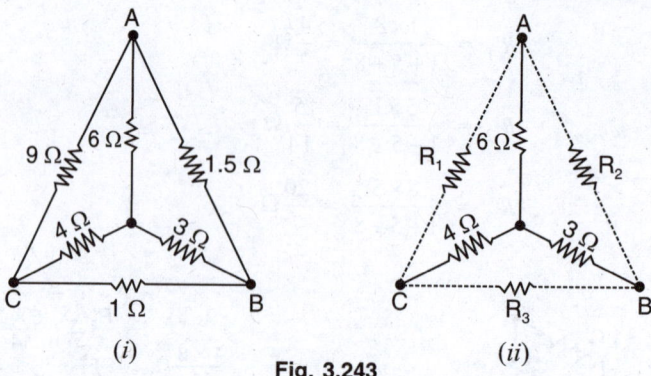

Fig. 3.243

These delta-connected resistances R_1, R_2 and R_3 come in parallel with the original delta-connected resistances. The circuit shown in Fig. 3.243 (*i*) reduces to the circuit shown in Fig. 3.244(*i*).

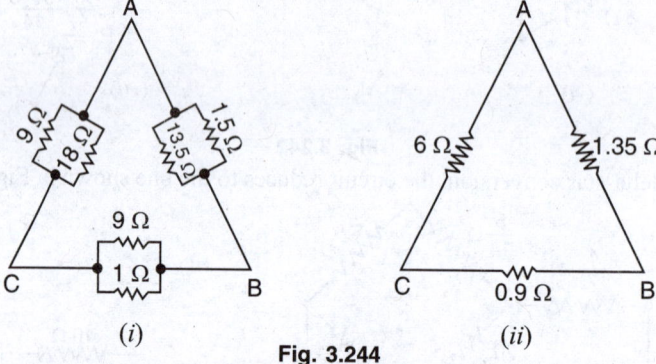

Fig. 3.244

The parallel resistances in each leg of delta in Fig. 3.244 (*i*) can be replaced by a single resistor as shown in Fig. 3.244 (*ii*) where

$$R_{AC} = 9 \times 18/27 = 6\,\Omega$$
$$R_{BC} = 9 \times 1/10 = 0{\cdot}9\,\Omega$$
$$R_{AB} = 1{\cdot}5 \times 13{\cdot}5/15 = 1{\cdot}35\,\Omega$$

(*i*) Resistance between A and B = $1{\cdot}35\,\Omega \,\|\, (6+0{\cdot}9)\,\Omega$ = $1{\cdot}35 \times 6{\cdot}9/8{\cdot}25 = \textbf{1·13}\,\Omega$

(*ii*) Resistance between B and C = $0{\cdot}9\,\Omega \,\|\, (6+1{\cdot}35)\,\Omega$ = $0{\cdot}9 \times 7{\cdot}35/8{\cdot}25 = \textbf{0·8}\,\Omega$

(*iii*) Resistance between A and C = $6\,\Omega \,\|\, (1{\cdot}35+0{\cdot}9)\,\Omega$ = $6 \times 2{\cdot}25/8{\cdot}25 = \textbf{1·636}\,\Omega$

Example 3.98. *Determine the load current in branch EF in the circuit shown in Fig. 3.245 (i).*

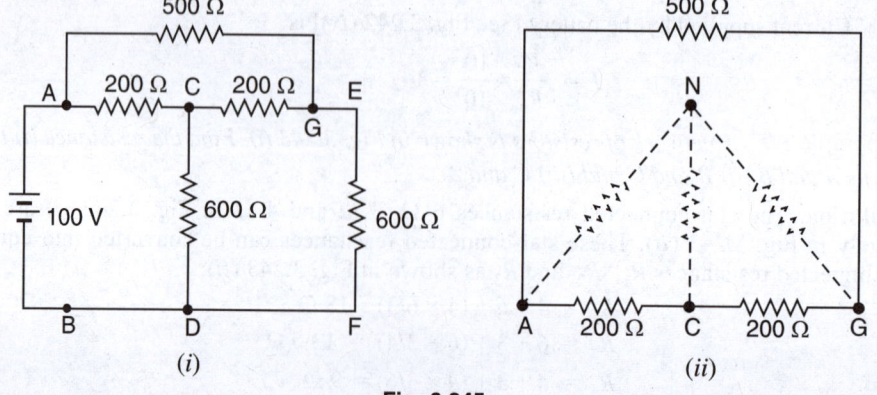

Fig. 3.245

Solution. The circuit $ACGA$ forms delta and is shown separately in Fig. 3.245 (*ii*) for clarity. Changing this delta connection into equivalent star connection [See Fig. 3.245 (*ii*)], we have,

$$R_{AN} = \frac{500 \times 200}{500 + 200 + 200} = 111.11\ \Omega\ ;\quad R_{CN} = \frac{200 \times 200}{500 + 200 + 200} = 44.44\ \Omega\ ;$$

$$R_{GN} = \frac{500 \times 200}{500 + 200 + 200} = 111.11\ \Omega$$

Thus the circuit shown in Fig. 3.245 (*i*) reduces to the circuit shown in Fig. 3.246 (*i*). The branch *NEF* ($= 111{\cdot}11 + 600 = 711{\cdot}11\ \Omega$) is in parallel with branch *NCD* ($= 44{\cdot}44 + 600 = 644{\cdot}44\ \Omega$) and the equivalent resistance of this parallel combination is

$$= \frac{711.11 \times 644.44}{711.11 + 644.44} = 338\ \Omega$$

The circuit shown in Fig. 3.246 (*i*) reduces to the circuit shown in Fig. 3.246 (*ii*).

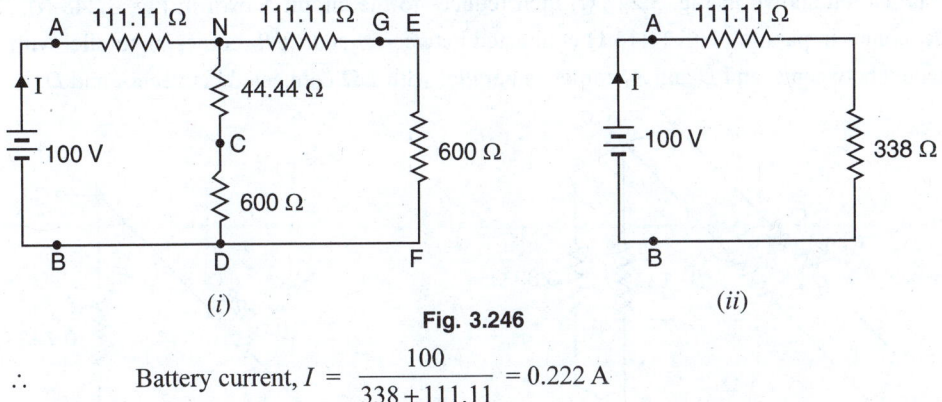

Fig. 3.246

$$\therefore\qquad \text{Battery current,}\ I\ =\ \frac{100}{338 + 111.11} = 0.222\ \text{A}$$

This battery current divides into two parallel paths [See Fig. 3.246 (*i*)] *viz.* branch *NEF* and branch *NCD*.

$\therefore$ Current in branch *NEF i.e.* in branch *EF*

$$=\ 0.222 \times \frac{644.44}{711.11 + 644.44} = \textbf{0.1055 A}$$

Example 3.99. *A square and its diagonals are made of a uniform covered wire. The resistance of each side is 1 Ω and that of each diagonal is 1·414 Ω. Determine the resistance between two opposite corners of the square.*

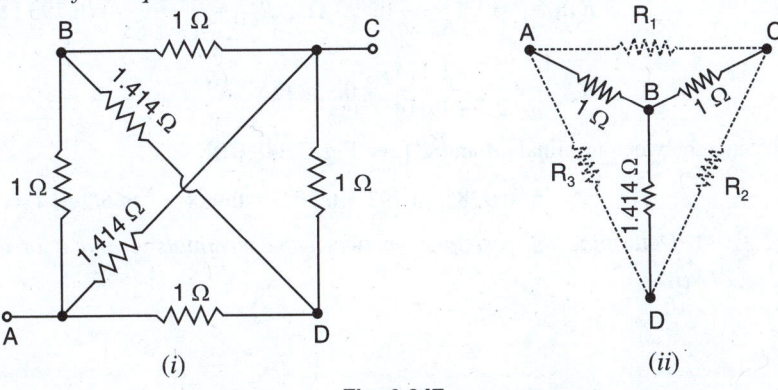

Fig. 3.247

Solution. Fig. 3.247 (*i*) shows the given square. It is desired to find the resistance between terminals *A* and *C*. The star-connected resistances 1 Ω, 1 Ω and 1·414 Ω (with star point at *B*) are shown separately in Fig. 3.247 (*ii*). These star-connected resistances can be converted into equivalent delta connected resistances R_1, R_2 and R_3 as shown in Fig. 3.247 (*ii*) where

$$R_1 = R_{AB} + R_{BC} + \frac{R_{AB} \cdot R_{BC}}{R_{BD}}$$

$$= 1 + 1 + \frac{1 \times 1}{1.414} = 2.7 \ \Omega$$

$$R_2 = 1 + 1.414 + \frac{1 \times 1.414}{1} = 3\cdot83 \ \Omega$$

$$R_3 = 1 + 1.414 + \frac{1 \times 1.414}{1} = 3\cdot83 \ \Omega$$

The circuit shown in Fig. 3.247 (*i*) then reduces to the circuit shown in Fig. 3.248 (*i*). Note that R_1 comes in parallel with 1·414 Ω connected between *A* and *C*; R_2 comes in parallel with 1 Ω connected between *C* and *D* and R_3 comes in parallel with 1 Ω connected between *A* and *D*.

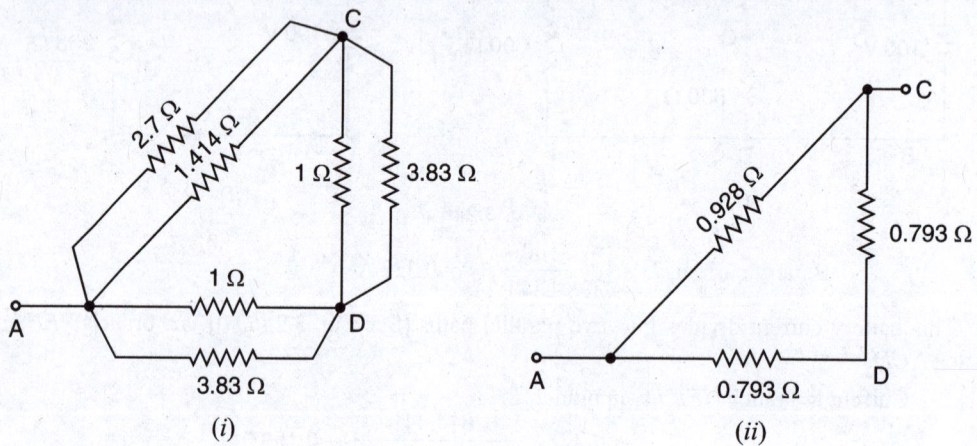

Fig. 3.248

In Fig. 3.248 (*i*), branch *AD* has 1 Ω and 3·83 Ω resistances in parallel.

$$\therefore \qquad R_{AD} = \frac{1 \times 3.83}{1 + 3.83} = 0.793 \ \Omega \ ; \ R_{CD} = \frac{1 \times 3.83}{1 + 3.83} = 0.793 \ \Omega \ ;$$

$$R_{AC} = \frac{2.7 \times 1.414}{2.7 + 1.414} = 0.928 \ \Omega$$

∴   Resistance between terminals *A* and *C* [See Fig. 3.248 (*ii*)]

$$= 0\cdot928 \ \| \ (0\cdot793 + 0\cdot793) = 0\cdot928 \times 1\cdot586/2\cdot514 = \mathbf{0\cdot585 \ \Omega}$$

Example 3.100. *Determine the resistance between the terminals A and B of the network shown in Fig. 3.249 (i).*

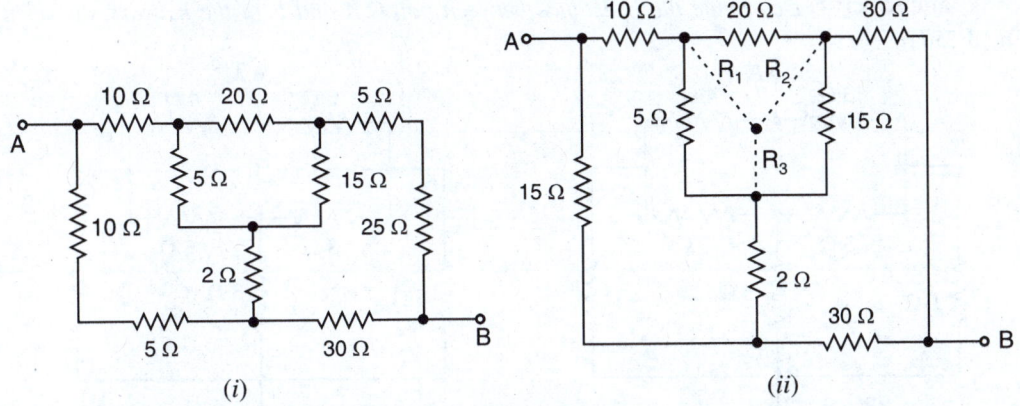

Fig. 3.249

Solution. We can combine series resistances on the right and left of Fig. 3.249 (*i*). The circuit then reduces to the one shown in Fig. 3.249 (*ii*). The resistances 5 Ω, 20 Ω and 15 Ω form a delta circuit and can be replaced by a star network where

$$R_1 = \frac{\text{Product of two adjacent arms of delta}}{\text{Sum of arms of delta}} = \frac{20 \times 5}{5 + 20 + 15} = \frac{100}{40} = 2.5 \ \Omega \ ;$$

$$R_2 = \frac{20 \times 15}{40} = 7.5 \ \Omega \ ; \qquad R_3 = \frac{5 \times 15}{40} = 1.875 \ \Omega$$

Referring to Fig. 3.249 (*ii*), R_1 is in series with 10 Ω resistor and their total resistance is $10 + R_1 = 10 + 2.5 = 12.5 \ \Omega$. Similarly, we have $30 + R_2 = 30 + 7.5 = 37.5 \ \Omega$ and $2 + R_3 = 2 + 1.875 = 3.875 \ \Omega$. The circuit then reduces to the one shown in Fig. 3.249 (*iii*).

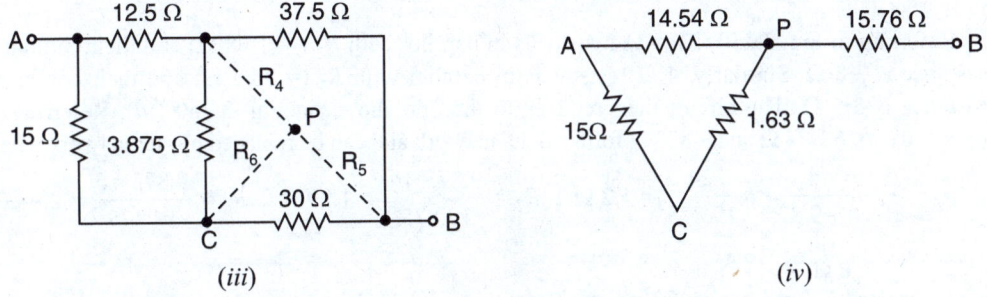

(*iii*)   (*iv*)

Fig. 3.249

Referring to Fig. 3.249 (*iii*), 3·875 Ω, 37·5 Ω and 30 Ω form a delta network and can be reduced to star network where

$$R_4 = \frac{3.875 \times 37.5}{3.875 + 37.5 + 30} = \frac{3.875 \times 37.5}{71.375} = 2.04 \ \Omega \ ;$$

$$R_5 = \frac{37.5 \times 30}{71.375} = 15.76 \ \Omega \ ; \ R_6 = \frac{3.875 \times 30}{71.375} = 1.63 \ \Omega$$

Referring to Fig. 3.249 (*iii*), R_4 is in series with 12.5 Ω resistor and their combined resistance = $R_4 + 12.5 = 2.04 + 12.5 = 14.54 \ \Omega$. The circuit then reduces to the one shown in Fig. 3.249 (*iv*). The resistance between terminals A and B is given by ;

$$R_{AB} = 15.76 + [14.54 \ || \ (15 + 1.63)] = 15.76 + \frac{14.54 \times 16.63}{31.17} = \mathbf{23 \cdot 5 \ \Omega}$$

Example 3.101. *Determine the resistance between points A and B in the network shown in Fig. 3.250 (i).*

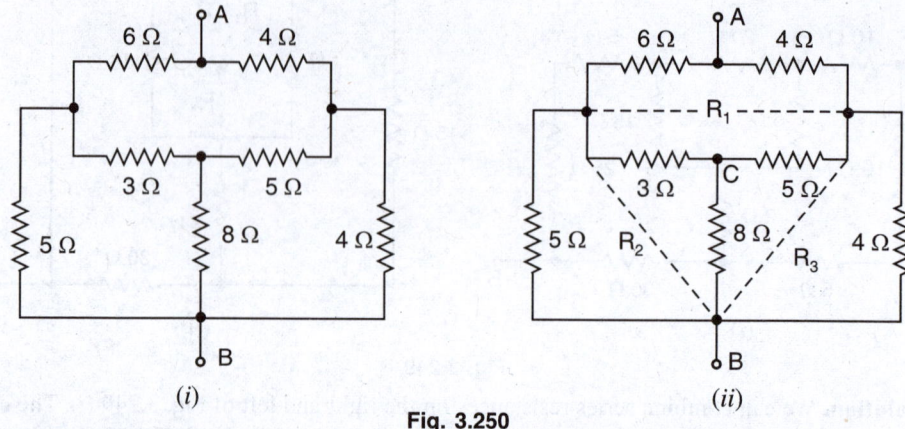

Fig. 3.250

Solution. The 3 Ω, 5 Ω and 8 Ω form star network and can be replaced by delta network where

Resistance between two terminals of delta = Sum of star resistances connected to those terminals *plus* product of same two resistances divided by the third star resistance.

∴

$$R_1 = 3 + 5 + \frac{3 \times 5}{8} = 9.875 \ \Omega$$

$$R_2 = 3 + 8 + \frac{3 \times 8}{5} = 15.8 \ \Omega$$

$$R_3 = 5 + 8 + \frac{5 \times 8}{3} = 26.3 \ \Omega$$

Referring to Fig. 3.250 (*ii*), 5 Ω resistor is in parallel with R_2 (= 15.8 Ω) and their combined resistance is 3.8 Ω. Similarly, 4 Ω resistor is in parallel with R_3 (= 26.3 Ω) and their combined resistance is 3.5 Ω. The circuit then reduces to the one shown in Fig. 3.250 (*iii*). Referring to Fig. 3.250 (*iii*), 6 Ω, 4 Ω and 9.875 Ω form a delta network and can be replaced by star network where

$$R_6 = \frac{6 \times 4}{6 + 4 + 9.875} = \frac{24}{19.875} = 1.2 \ \Omega \ ; \ R_7 = \frac{9.875 \times 4}{19.875} = 1.99 \ \Omega \ ; \ R_8 = \frac{9.875 \times 6}{19.875} = 2.98 \ \Omega$$

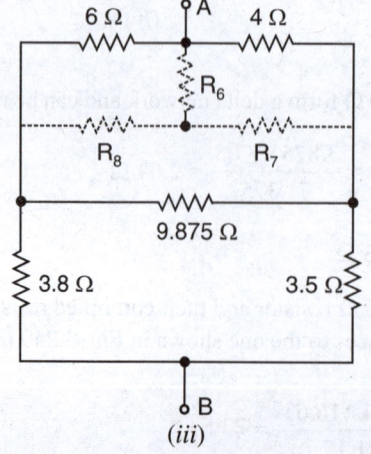

(*iii*)

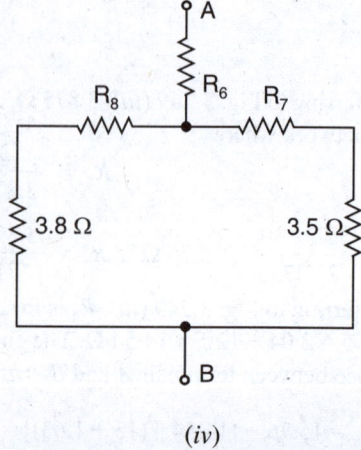

(*iv*)

Fig. 3.250

Therefore, the circuit shown in Fig. 3.250 (*iii*) reduces to the one shown in Fig. 3.250 (*iv*). It is clear that :

$$R_{AB} = (3 \cdot 8 + R_8) \parallel (R_7 + 3 \cdot 5) + R_6 = (3 \cdot 8 + 2 \cdot 98) \parallel (1 \cdot 99 + 3 \cdot 5) + 1 \cdot 2$$
$$= (6 \cdot 78 \parallel 5 \cdot 49) + 1 \cdot 2 = \mathbf{4 \cdot 23 \ \Omega}$$

Example 3.102. *A π network is to be constructed as shown in Fig. 3.251 (i) so that the resistance R_{XZ} looking into the X – Z terminals (with Y – Z open) equals the resistance R_{YZ} looking into the Y – Z terminals (with X – Z open). If that resistance must equal 1 kΩ, find the value of R_Δ that should be used in the π network.*

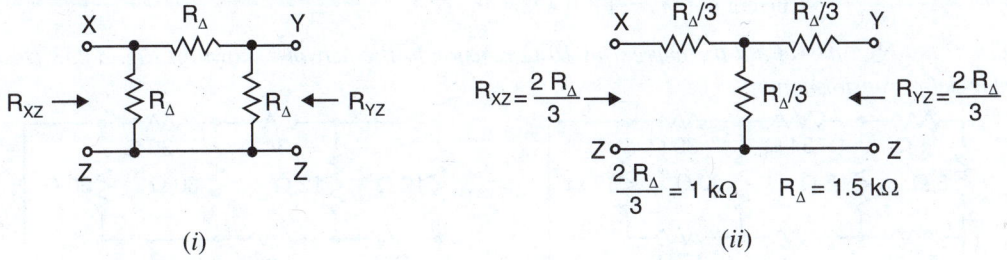

Fig. 3.251

Solution. The delta network shown in Fig. 3.251 (*i*) can be converted into star network as shown in Fig. 3.251 (*ii*). Note that the star network has equal-valued resistors $R_\Delta/3$. It is clear from this figure that :

$$R_{XZ} = R_{YZ} = \frac{R_\Delta}{3} + \frac{R_\Delta}{3} = \frac{2R_\Delta}{3}$$

or $\qquad 1 \text{ k}\Omega = \dfrac{2R_\Delta}{3} \quad$ or $\quad R_\Delta = \mathbf{1.5 \ k\Omega}$

Therefore, the π network must have three 1.5 kΩ resistors as shown in Fig 3.251 (*iii*).

Fig. 3.251

Example 3.103. *Find the current distribution in the network shown in Fig. 3.252 (i).*

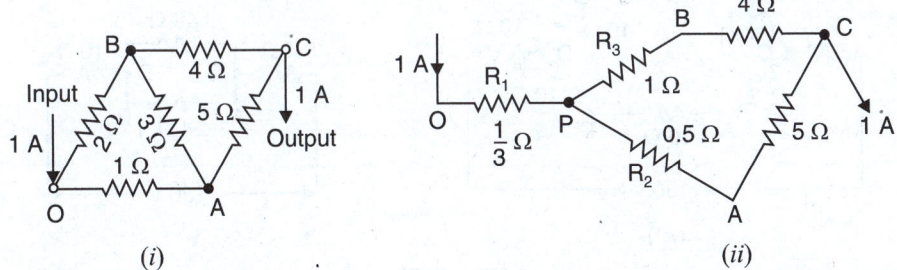

Fig. 3.252

Solution. The network *OAB* forms a delta and can be replaced by star where :

$$R_1 = \frac{1 \times 2}{6} = \frac{1}{3}\Omega \quad ; \quad R_2 = \frac{1 \times 3}{6} = 0.5 \ \Omega \quad ; \quad R_3 = \frac{2 \times 3}{6} = 1 \ \Omega$$

The network then reduces to the one shown in Fig. 3.252 (*ii*). The current through *OP* is 1 A and divides between two parallel paths at point *P*. By current-divider rule :

$$\text{Current in } PA = \text{Current in } AC = 1 \times \frac{5}{1 + 4 + 0.5 + 5} = 1 \times \frac{5}{10.5} = \mathbf{0.477 \ A}$$

$$\text{Current in } PB = \text{Current in } BC = 1 - 0.477 = \mathbf{0.523A}$$

$$\text{Voltage drop in } PB = 1 \times 0{\cdot}523 = 0.523 \text{ V}$$

$$\text{Voltage drop in } PA = 0{\cdot}5 \times 0{\cdot}477 = 0.238 \text{ V}$$

$$\therefore \qquad V_{AB} = 0{\cdot}523 - 0{\cdot}238 = 0.285 \text{ V}$$

$$\therefore \qquad I_{AB} = 0{\cdot}285/3 = \mathbf{0.095 \text{ A}}$$

$$\text{Current in } OB = \text{Current in } BC - \text{Current in } AB$$

$$= 0.523 - 0.095 = \mathbf{0.428 \text{ A}}$$

$$\text{Current in } OA = 1 - 0.428 = \mathbf{0.572 \text{ A}}$$

Example 3.104. *Find the current in 10 Ω resistor in the network shown in Fig. 3.253 (i) by star-delta transformation.*

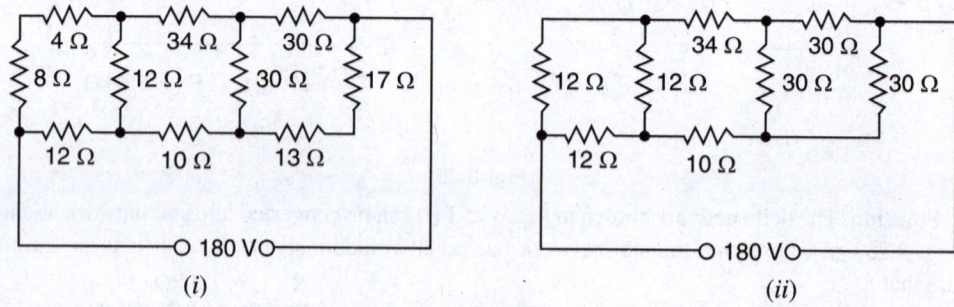

Fig. 3.253

Solution. In Fig. 3.253 (*i*), the 4 Ω and 8 Ω resistors are in series and their total resistance is 8 + 4 = 12 Ω. Similarly, at the right end of figure, 17 Ω and 13 Ω are in series so that their total resistance becomes 17 + 13 = 30 Ω. The circuit then reduces to the one shown in Fig. 3.253 (*ii*). Replacing the two deltas at the left end and right end in Fig. 3.253 (*ii*) by their equivalent star, we get the circuit shown in Fig. 3.253 (*iii*).

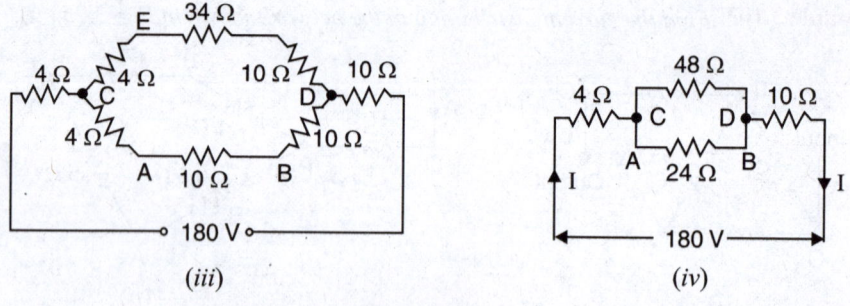

Fig. 3.253

Referring to Fig. 3.253 (*iii*), the path *CED* has resistance = 4 + 34 + 10 = 48 Ω and path *CABD* has resistance = 4 + 10 + 10 = 24 Ω. The circuit then reduces to the one shown in Fig. 3.253 (*iv*). The total resistance R_T presented to 180V source is

$$R_T = 4 + (48 \parallel 24) + 10 = 30 \ \Omega$$

$$\therefore \qquad \text{Circuit current, } I = 180/30 = 6 \text{ A}$$

$$\therefore \qquad \text{Voltage across parallel combination} = I \times (48 \parallel 24) = 6 \times 16 = 96 \text{ V}$$

$$\therefore \qquad \text{Current in 10 Ω resistor [part of 24 Ω in Fig. 3.253 (iv)]} = 96/24 = \mathbf{4 \text{ A}}$$

Example 3.105. *Using Norton's theorem, find the current through the 8 Ω resistor shown in Fig 3.254 (i). All resistance values are in ohms.*

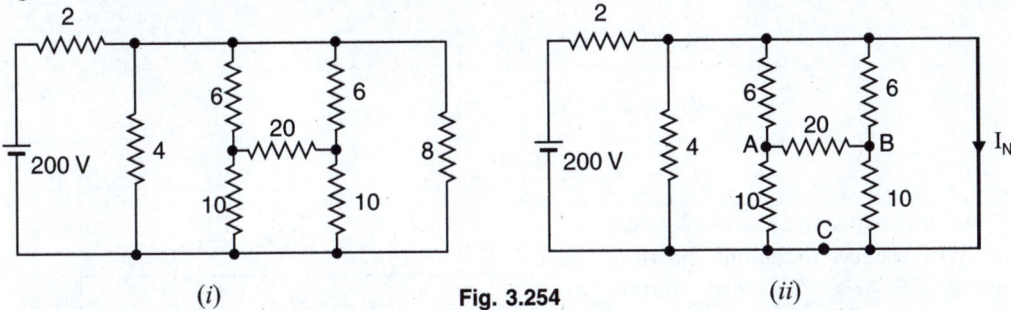

(*i*)   **Fig. 3.254**   (*ii*)

Solution. In order to find Norton current I_N, place short circuit across the load of 8 Ω resistor. The circuit then becomes as shown in Fig. 3.254 (*ii*). The short circuit bypasses all the resistors except 2 Ω resistor. Therefore, $I_{SC} = I_N = 200/2 = 100$ A. In order to find R_N, replace 200 V source by a short. Then R_N is the resistance looking into open-circuited terminals A and B in Fig. 3.254 (*iii*).

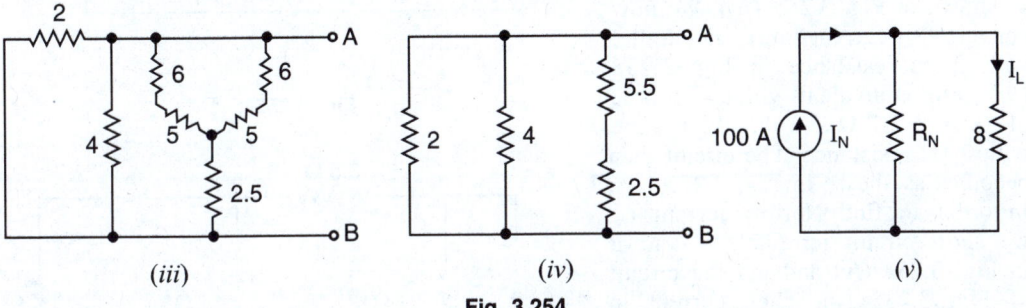

(*iii*)   (*iv*)   (*v*)

Fig. 3.254

In Fig. 3.254 (*ii*), ABC network forms a delta and can be replaced by equivalent star network as shown in Fig. 3.254 (*iii*). This circuit reduces to the one shown in Fig. 3.254 (*iv*).

Norton's resistance, R_N = Resistance at the open-circuited terminals in Fig. 3.254 (*iv*)

$$= 2 \,\|\, 4 \,\|\, (5 \cdot 5 + 2 \cdot 5) = 8/7 \ \Omega$$

Therefore, Norton equivalent circuit consists of 100A current source in parallel with a resistance of 8/7 Ω. When load R_L (= 8 Ω) is connected at the output terminals of Norton's equivalent circuit, the circuit becomes as shown in Fig 3.254 (*v*). By current-divider rule, the load current I_L through R_L (= 8 Ω) is given by ;

$$I_L = 100 \times \frac{8/7}{8 + (8/7)} = \textbf{12·5 A}$$

Example 3.106. *In the network shown in Fig. 3.255 (i), find (i) Norton equivalent circuit at terminals AB (ii) the maximum power that can be provided to a resistor R connected between terminals A and B.*

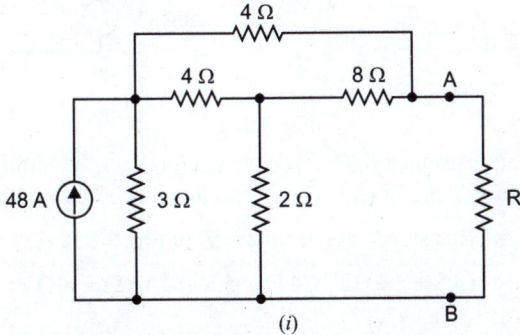

(*i*)

Fig. 3.255

Solution. (*i*) The star- connected resistances 4 Ω, 8 Ω and 2 Ω in Fig. 3.255 (*i*) can be converted into equivalent delta-connected resistances R_{ab}, R_{bc} and R_{ca} as shown in Fig. 3.255 (*ii*).

$$R_{ab} = 4+8+\frac{4\times8}{2} = 28\ \Omega$$

$$R_{bc} = 8+2+\frac{8\times2}{4} = 14\ \Omega$$

$$R_{ca} = 2+4+\frac{2\times4}{8} = 7\ \Omega$$

After above star-delta conversion, the circuit reduces to the one shown in Fig. 3.255 (*ii*). We can further simplify the circuit in Fig. 3.255 (*ii*) by combining the parallel resistances (4 Ω || 28 Ω = 3.5 Ω and 3 Ω || 7 Ω = 2.1 Ω). The circuit then becomes as shown in Fig. 3.255 (*iii*). We now convert 48A current source in parallel with 2.1Ω resistance in Fig. 3.255 (*iii*) into equivalent voltage source of 48 A × 2.1 Ω = 100.8 V in series with 2.1Ω resistance. The circuit then becomes as shown in Fig. 3.255 (*iv*). In order to find Norton current I_N, we short circuit terminals A and B in Fig. 3.255 (*iv*) and get the circuit of Fig. 3.255 (*v*). Then current in the short-circuit is I_N. Referring to Fig. 3.255 (*v*) and applying Ohm's law, the value of I_N is given by ;

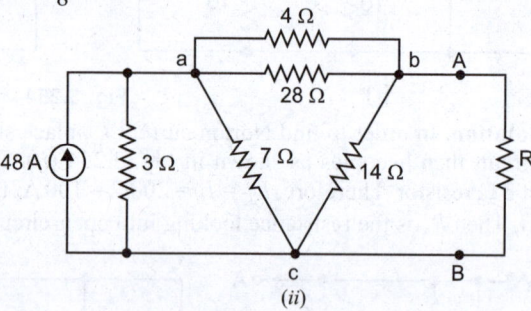

(*ii*)

Fig. 3.255

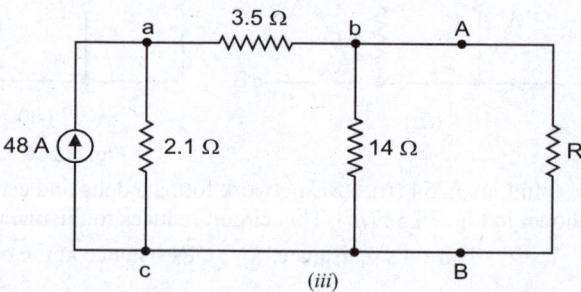

(*iii*)

Fig. 3.255

$$I_N = \frac{100.8}{2.1+3.5} = 18A$$

Note that no current will pass through 14 Ω resistor in Fig. 3.255 (*v*). It is because there is a short across this resistor and the entire current (= I_N) will pass through the short.

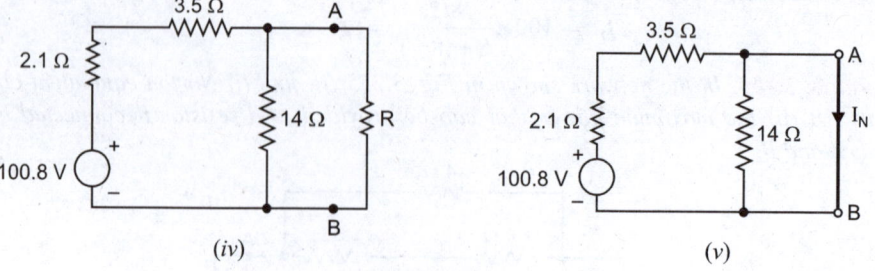

(*iv*)   (*v*)

Fig. 3.255

In order to find Norton resistance $R_N (= R_{Th})$, we open circuit the terminals AB and replace the voltage source by a short in Fig. 3.255 (*iv*). The circuit then becomes as shown in Fig. 3.255 (*vi*).

∴ R_N = Resistance at terminals AB in Fig. 3.255 (*vi*)

= (3.5 + 2.1) Ω || 14 Ω = 5.6 Ω || 14 Ω = 4 Ω

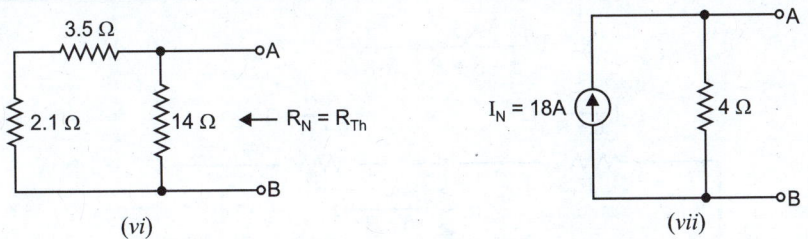

Fig. 3.255

The Norton equivalent circuit at terminals AB is shown in Fig. 3.255 (*vii*).

(*ii*) Maximum power will be provided to resistance R connected between terminals A and B when resistance R is equal to Norton resistance R_N i.e.

$$R = R_N = 4\,\Omega$$

When $R(=4\,\Omega)$ is connected across terminals A and B in Fig. 3.255 (*vii*), then by current-divider rule,

$$\text{Current in } R\ (=4\,\Omega),\quad I = 18 \times \frac{4}{4+4} = 9A$$

∴   Maximum power (P_{max}) provided to R is

$$P_{max} = I^2 R = (9)^2 \times 4 = \textbf{324 W}$$

Remember that under the condition of maximum power transfer, the circuit efficiency is *only* 50% and the remaining 50% is dissipated in the circuit.

Example 3.107. *Determine a non-negative value of R such that the power consumed by the 2 Ω resistor in Fig. 3.256 (i) is maximum.*

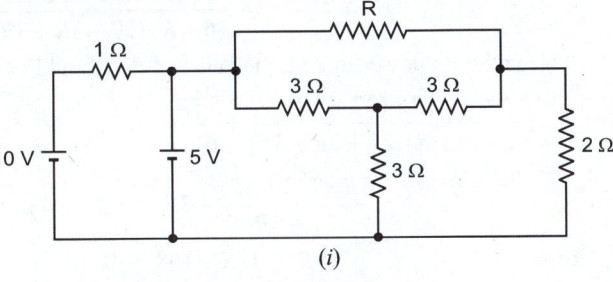

(*i*)

Fig. 3.256

Solution. In order to find maximum power consumed in 2 Ω resistor (*i.e.* load), we should find Thevenin resistance R_{Th} at 2 Ω terminals. For this purpose, we open circuit the load terminals (*i.e.* remove 2 Ω resistor) and short circuit the voltage sources as shown in Fig. 3.256 (*ii*). The resistance at the open-circuited load (*i.e.* 2Ω) terminals XY is the R_{Th}.

R_{Th} = Resistance at terminals XY in Fig. 3.256 (*ii*).

In order to facilitate the determination of R_{Th}, we convert delta-connected resistances $R\,\Omega$, 3 Ω and 3 Ω in Fig. 3.256 (*ii*) into equivalent star-connected resistances R_1, R_2 and R_3 as shown in Fig. 3.256 (*iii*). The values of R_1, R_2 and R_3 are given by ;

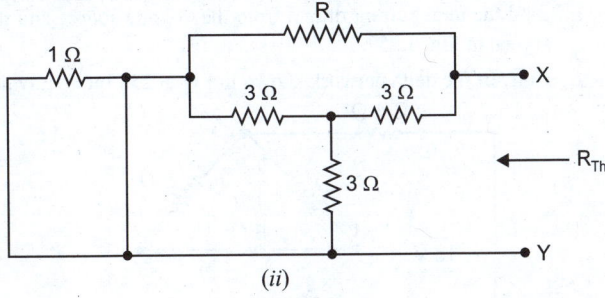

(*ii*)

Fig. 3.256

$$R_1 = \frac{3 \times R}{3+3+R} = \frac{3R}{6+R}$$

$$R_2 = \frac{3 \times R}{3+3+R} = \frac{3R}{6+R}$$

$$R_3 = \frac{3 \times 3}{3 + 3 + R} = \frac{9}{6 + R}$$

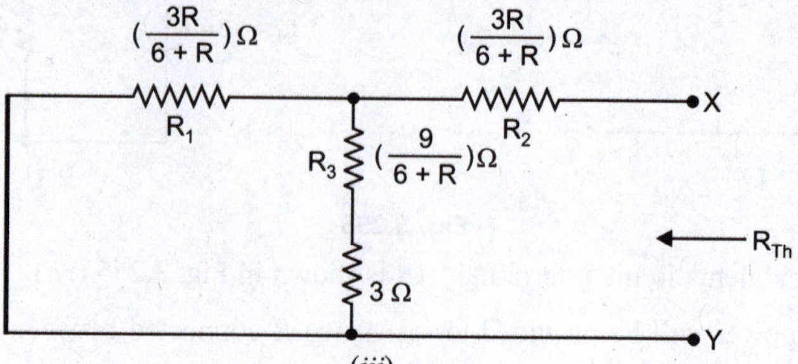

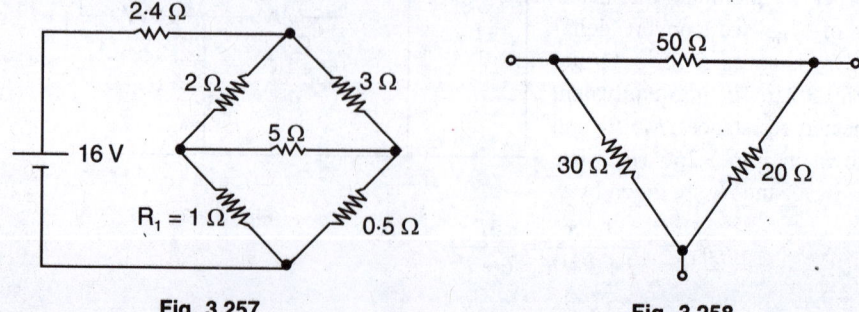

Fig. 3.256

After above delta-star conversion, the circuit becomes as shown in Fig. 3.256 (*iii*). Then resistance at open-circuited terminals *XY* is R_{Th}.

Referring to Fig. 3.256 (*iii*),

$$R_{Th} = \left[\left(\frac{3R}{6 + R} \right) \| \left(\frac{9}{6 + R} + 3 \right) \right] + \frac{3R}{6 + R}$$

$$= \left[\frac{3R}{6 + R} \| \frac{27 + 3R}{6 + R} \right] + \frac{3R}{6 + R}$$

$$= \frac{3R \times (27 + 3R)}{(6 + R)(27 + 3R + 3R)} + \frac{3R}{6 + R}$$

For maximum power in 2 Ω, the value of R_{Th} should be equal to 2 Ω.

$$\therefore \qquad \frac{3R \times (27 + 3R)}{(6 + R)(27 + 3R + 3R)} + \frac{3R}{6 + R} = 2$$

or $$\qquad \frac{3R \times (27 + 3R)}{27 + 6R} + 3R = 2(6 + R)$$

or $$\qquad 5R^2 + 12R - 108 = 0 \qquad\qquad\qquad \text{...after simplification}$$

$$\therefore \qquad R = +3.6\,\Omega \quad \text{or} \quad -6\,\Omega$$

Accepting the positive value, **R = 3.6 Ω**.

Tutorial Problems

1. Find the total current drawn from the voltage source and the current through R_1 (= 1 Ω) in the circuit shown in Fig. 3.257.   [4 A ; 2 A]

2. Convert the delta network shown in Fig. 3.258 into equivalent wye network.

Fig. 3.257   Fig. 3.258

3. Convert the wye network shown in Fig. 3.259 into equivalent delta network.

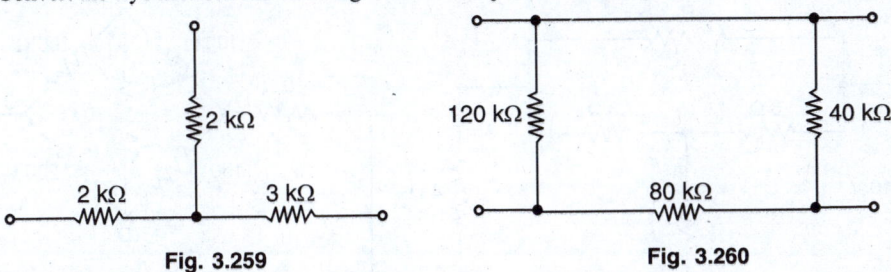

Fig. 3.259   Fig. 3.260

4. Convert the delta network shown in Fig. 3.260 into the equivalent wye network.

5. In the network shown in Fig. 3.261, find the resistance between terminals B and C using star/delta transformation.   [17/12 Ω]

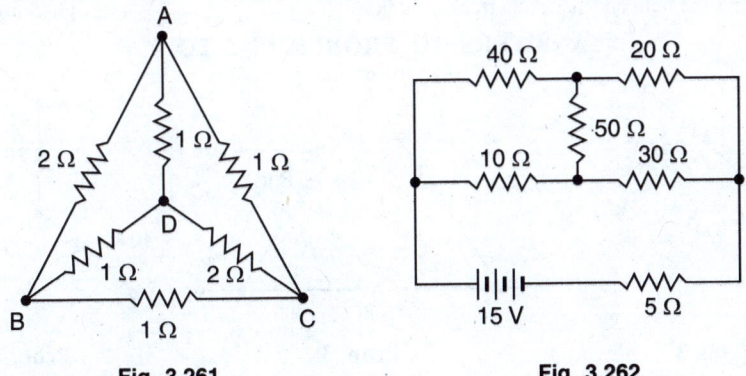

Fig. 3.261   Fig. 3.262

6. In the network shown in Fig. 3.262, find the current supplied by the battery using star/delta transformation.   [0·452 A]

7. What is the resistance between terminals A and B of the network shown in Fig. 3.263 ?   [274·2 Ω]

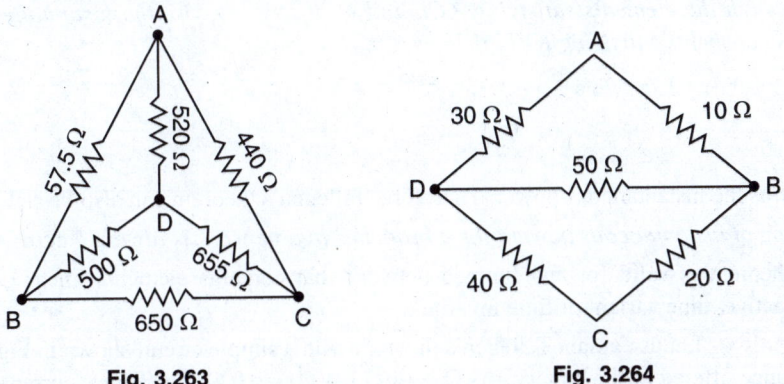

Fig. 3.263   Fig. 3.264

8. Using delta/star transformation, find the resistance between terminals A and C of the network shown in Fig. 3.264.

9. Using star/delta transformation, determine the value of R for the network shown in Fig. 3.265 such that 4Ω resistor consumes the maximum power.   [$R = 36\Omega$]

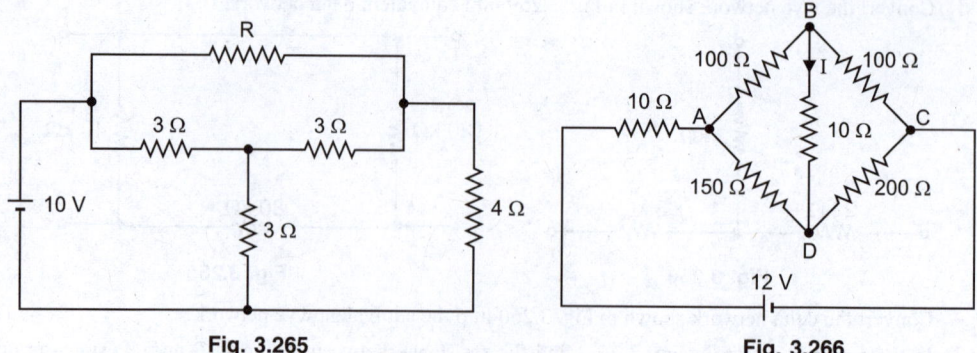

Fig. 3.265   **Fig. 3.266**

10. Calculate the current I flowing through the 10 Ω resistor in the circuit shown in Fig. 3.266. Apply Thevenin's theorem and star/delta transformation.   **[5.45 mA from D to B]**

ANSWERS TO PROBLEMS 2 TO 4

Prob. 2	Prob. 3	Prob. 4
15 Ω 10 Ω 6 Ω	5.333 kΩ 8 kΩ 8 kΩ	20 kΩ 40 kΩ 13.33 kΩ

Prob. 2   **Prob. 3**   **Prob. 4**

3.26. Tellegen's Theorem

This theorem has wide applications in electric networks and may be stated as under :

For a network consisting of n elements if i_1, i_2, i_3 .... i_n are the instantaneous currents flowing through the elements satisfying KCL and v_1, v_2, v_3 ... v_n are the instantaneous voltages across these elements satisfying KVL, then,

$$v_1 i_1 + v_2 i_2 + v_3 i_3 + ... + v_n i_n = 0$$

or

$$\sum_{n=1}^{n} v_n i_n = 0$$

Now vi is the instantaneous power. Therefore, Tellegen's theorem can also be stated as under :

*The sum of instantaneous powers for n branches in a network is always * zero.*

This theorem is valid for any lumped network that contains elements linear or non-linear, passive or active, time variant or time invariant.

Explanation. Let us explain Tellegen's theorem with a simple circuit shown in Fig. 3.267. The total resistance offered to the battery = 8 Ω + (4 Ω ‖ 4 Ω) = 10 Ω. Therefore, current supplied by battery is $I = 100/10 = 10$ A. This current divides equally at point A.

Voltage drop across 8 Ω = – (10 × 8) = – 80 V

Voltage drop across 4 Ω = – (5 × 4) = – 20 V

Voltage drop across 1 Ω = – (5 × 1) = – 5 V

Voltage drop across 3 Ω = – (5 × 3) = – 15 V

---

*   This is in accordance with the law of conservation of energy because power delivered by the battery is consumed in the circuit elements.

According to Tellegen's theorem,

Sum of instantaneous powers = 0

or $\quad v_1i_1 + v_2i_2 + v_3i_3 + v_4i_4 + v_5i_5 = 0$

or $\quad (100 \times 10) + (-80 \times 10) + (-20 \times 5) + (-5 \times 5)$
$$+ (-15 \times 5) = 0$$

or $\quad 1000 - 800 - 100 - 25 - 75 = 0$

or $\quad 0 = 0$ which is true

Thus Tellegen's theorem stands proved.

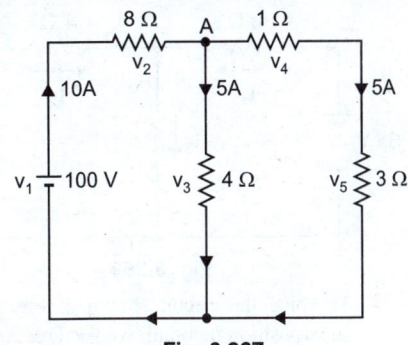

Fig. 3.267

Objective Questions

1. An active element in a circuit is one which ....

   (*i*) receives energy

   (*ii*) supplies energy

   (*iii*) both receives and supplies energy

   (*iv*) none of the above

2. A passive element in a circuit is one which ....

   (*i*) supplies energy

   (*ii*) receives energy

   (*iii*) both supplies and receives energy

   (*iv*) none of the above

3. An electric circuit contains ....

   (*i*) active elements only

   (*ii*) passive elements only

   (*iii*) both active and passive elements

   (*iv*) none of the above

4. A linear circuit is one whose parameters (*e.g.* resistances etc.) ....

   (*i*) change with change in current

   (*ii*) change with change in voltage

   (*iii*) do not change with voltage and current

   (*iv*) none of the above

5. In the circuit shown in Fig. 3.268, the number of nodes is ....

   (*i*) one   (*ii*) two

   (*iii*) three   (*iv*) four

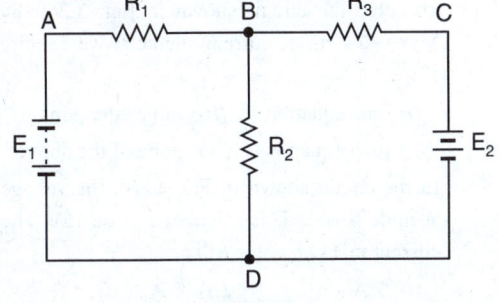

Fig. 3.268

6. In the circuit shown in Fig. 3.268, there are .... junctions.

   (*i*) three   (*ii*) four

   (*iii*) two   (*iv*) none of the above

7. The circuit shown in Fig. 3.268 has .... branches.

   (*i*) two   (*ii*) four

   (*iii*) three   (*iv*) none of these

8. The circuit shown in Fig. 3.268 has .... loops.

   (*i*) two   (*ii*) four

   (*iii*) three   (*iv*) none of the above

9. In the circuit shown in Fig. 3.268, there are .... meshes.

   (*i*) two   (*ii*) three

   (*iii*) four   (*iv*) five

10. To solve the circuit shown in Fig. 3.268 by Kirchhoff's laws, we require ....

   (*i*) one equation   (*ii*) two equations

   (*iii*) three equations (*iv*) none of the above

11. To solve the circuit shown in Fig. 3.268 by nodal analysis, we require ....

   (*i*) one equation   (*ii*) two equations

   (*iii*) three equations (*iv*) none of the above

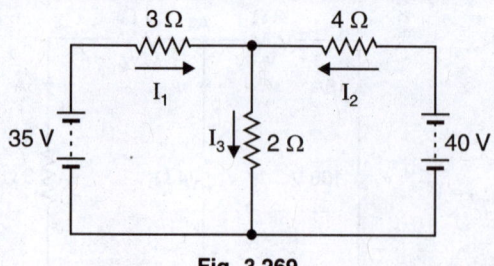

Fig. 3.269

12. To solve the circuit shown in Fig. 3.269 by superposition theorem, we require ....

    (*i*) one circuit   (*ii*) two circuits

   (*iii*) three circuits   (*iv*) none of the above

13. To solve the circuit shown in Fig. 3.269 by Maxwell's mesh current method, we require ....

    (*i*) one equation   (*ii*) three equations

   (*iii*) two equations (*iv*) none of the above

14. In the circuit shown in Fig. 3.270, the voltage at node *B w.r.t. D* is calculated to be 15V. The current in 3 Ω resistor will be ....

    (*i*) 2 A   (*ii*) 5 A

   (*iii*) 2·5 A   (*iv*) none of the above

15. The current in 2 Ω horizontal resistor in Fig. 3.270 is ....

    (*i*) 10 A   (*ii*) 5 A

   (*iii*) 2 A   (*iv*) 2·5 A

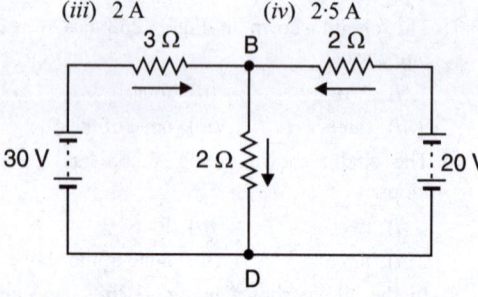

Fig. 3.270

16. In order to solve the circuit shown in Fig. 3.270 by nodal analysis, we require ....

    (*i*) one equation   (*ii*) two equations

   (*iii*) three equations (*iv*) none of the above

17. The superposition theorem is used when the circuit contains ....

    (*i*) a single voltage source

   (*ii*) a number of voltage sources

   (*iii*) passive elements only

   (*iv*) none of the above

18. Fig. 3.271 (*ii*) shows Thevenin's equivalent circuit of Fig. 3.271 (*i*). The value of Thevenin's voltage V_{Th} is ....

    (*i*) 20 V   (*ii*) 24 V

   (*iii*) 12 V   (*iv*) 36 V

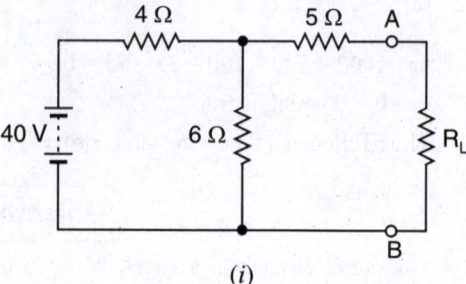

(*i*)

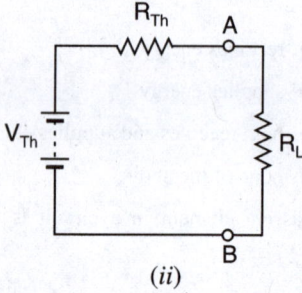

(*ii*)

Fig. 3.271

19. The value of R_{Th} in Fig. 3.271 (*ii*) is ....

    (*i*) 15 Ω (*ii*) 3·5 Ω

   (*iii*) 6·4 Ω (*iv*) 7·4 Ω

20. The open-circuited voltage at terminals *AB* in Fig. 3.271 (*i*) is ....

    (*i*) 12 V   (*ii*) 20 V

   (*iii*) 24 V   (*iv*) 40 V

21. Find the value of R_L in Fig. 3.272 to obtain maximum power in R_L.

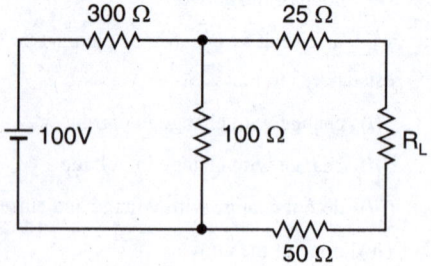

Fig. 3.272

    (*i*) 100 Ω (*ii*) 75 Ω

   (*iii*) 250 Ω (*iv*) 150 Ω

22. In Fig. 3.272, find the maximum power in R_L.

    (*i*) 2 W   (*ii*) 1·042 W

   (*iii*) 2·34 W   (*iv*) 4·52 W

23. What percent of the maximum power is delivered to R_L in Fig. 3.273 when $R_L = 2R_{Th}$?

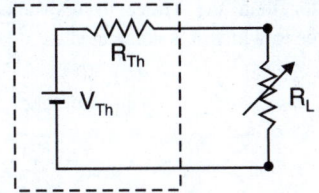

Fig. 3.273

   (i) 79 % of P_L (max)

   (ii) 65 % of P_L (max)

   (iii) 88·89 % of P_L (max)

   (iv) none of above

24. What percent of the maximum power is delivered to R_L in Fig. 3.273 when $R_L = R_{Th}/2$?

   (i) 65 %   (ii) 70 %

   (iii) 88·89 %   (iv) none of above

25. Find Millman's equivalent circuit *w.r.t.* terminals $x - y$ in Fig. 3.274.

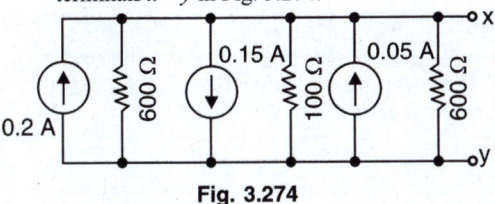

Fig. 3.274

   (i) Single current source of 0·1A and resistance 75 Ω

   (ii) Single current source of 2 A and resistance 50 Ω

   (iii) Single current source of 1 A and resistance 25 Ω

   (iv) none of above

26. Use superposition principle to find current through R_1 in Fig. 3.275.

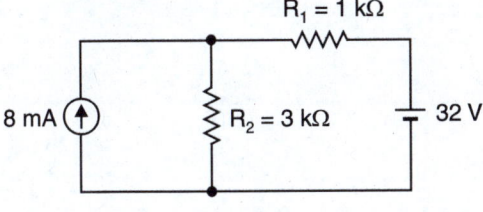

Fig. 3.275

   (i) 1 mA ←   (ii) 2 mA ←

   (iii) 1·5 mA →   (iv) 2·5 A ←

27. Use superposition principle to find current through R_1 in the circuit shown in Fig. 3.276.

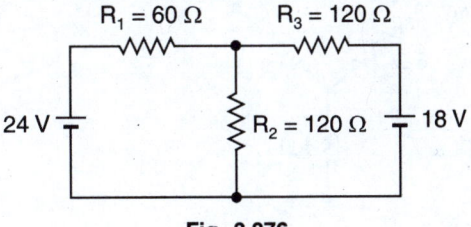

Fig. 3.276

   (i) 0·2 A ←   (ii) 0·25 A →

   (iii) 0·125 A →   (iv) 0·5 A →

28. Find Thevenin equivalent circuit to the left of terminals $x - y$ in Fig. 3.277.

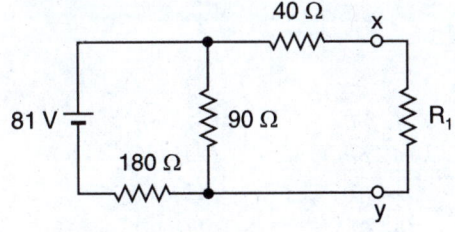

Fig. 3.277

   (i) $V_{Th} = 5$ V ; $R_{Th} = 4·5$ Ω

   (ii) $V_{Th} = 6$ V ; $R_{Th} = 5$ Ω

   (iii) $V_{Th} = 4·5$ V ; $R_{Th} = 10$ Ω

   (iv) $V_{Th} = 10$ V ; $R_{Th} = 9$ Ω

29. Convert delta network shown in Fig. 3.278 to equivalent Wye network.

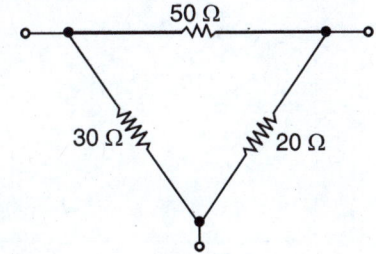

Fig. 3.278

(i)

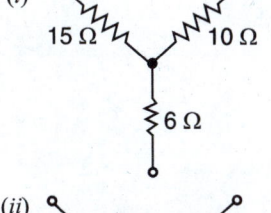

(ii)

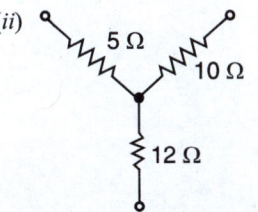

(iii)

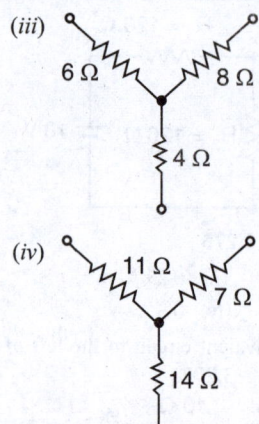

(iv)

30. What percentage of the maximum power is delivered to a load if load resistance is 10 times greater than the Thevenin resistance of the source to which it is connected ?

    (i) 25 %   (ii) 40 %

    (iii) 35 %   (iv) 33·06 %

Answers

1. *(ii)*	**2.** *(ii)*	**3.** *(iii)*	**4.** *(iii)*	**5.** *(iv)*
6. *(iii)*	**7.** *(iii)*	**8.** *(iii)*	**9.** *(i)*	**10.** *(ii)*
11. *(i)*	**12.** *(ii)*	**13.** *(iii)*	**14.** *(ii)*	**15.** *(iv)*
16. *(i)*	**17.** *(ii)*	**18.** *(ii)*	**19.** *(iv)*	**20.** *(iii)*
21. *(iv)*	**22.** *(ii)*	**23.** *(iii)*	**24.** *(iii)*	**25.** *(i)*
26. *(ii)*	**27.** *(iii)*	**28.** *(iv)*	**29.** *(i)*	**30.** *(iv)*

4

Units—Work, Power and Energy

Introduction

Engineering is an applied science dealing with a very large number of *physical quantities like distance, time, speed, temperature, force, voltage, resistance *etc*. Although it is possible to assign a standard unit for each quantity, it is rarely necessary to do so because many of the quantities are functionally related through experiment, derivation or definition. In the study of mechanics, for example, the units of only three quantities *viz. mass, length* and *time* need to be selected. All other quantities (*e.g.* area, volume, velocity, force *etc*.) can be expressed in terms of the units of these three quantities by means of experimental, derived and defined **relationship between the physical quantities. The units selected for these three quantities are called *fundamental units*. In order to cover the entire subject of engineering, three more fundamental quantities have been selected *viz.* †*electric current, temperature* and *luminous intensity. Thus there are in all six fundamental quantities (viz, mass, length, time, current, temperature and luminous intensity) which need to be assigned proper and standard units*. The units of all other physical quantities can be derived from the units of these six fundamental quantities. In this chapter, we shall focus our attention on the mechanical, electrical and thermal units of work, power and energy.

4.1. International System of Units

Although several systems were evolved to assign units to the above mentioned six fundamental quantities, only international system of units (abbreviated as SI) has been universally accepted. The units assigned to these six fundamental quantities in this system are given below.

Quantity	Symbol	Unit name	Unit symbol
Length	l, L	metre	m
Mass	m	kilogram	kg
Time	t	second	s
Electric Current	I	ampere	A
Temperature	T	degree kelvin	K
Luminous Intensity	I	candela	Cd

It may be noted that the units of all other physical quantities in science and engineering (*i.e.* other than six fundamental or basic quantities above) can be derived from the above basic units and are called *derived units*. Thus unit of velocity (= 1 m/s) results when the unit of length (= 1 m) is divided by the unit of time (= 1 s). Similarly, the unit of force (= 1 newton) results when unit of mass (= 1 kg) is multiplied by the unit of acceleration (= 1 m/s²). Therefore, units of velocity and force are the derived units.

---

* A physical quantity is one which can be measured.

** For example, by definition, speed is the distance travelled per second. Therefore, speed is related to distance (*i.e.* length) and time.

† For practical reasons, electric current and *not* charge has been taken as the fundamental quantity, though one is derivable from the other. The important consideration which led to the selection of current as the fundamental quantity is that it serves as the link between electric, magnetic and mechanical quantities and can be readily measured.

4.2. Important Physical Quantities

It is profitable to give a brief description of the following physical quantities much used in science and engineering :

(*i*) **Mass.** It is the quantity of matter possessed by a body. The SI unit of mass is kilogram (kg). The mass of a body is a constant quantity and is independent of place and position of the body. Thus the mass of a body is the same whether it is on Earth's surface, the Moon's surface, on the top of a mountain or down a deep well.

$$1 \text{ quintal} = 100 \text{ kg} \; ; \quad 1 \text{ tonne} = 10 \text{ quintals} = 1000 \text{ kg}$$

(*ii*) **Force.** It is the product of mass (kg) and acceleration (m/s²). The unit of force is newton (N) ; being the force required to accelerate a mass of 1 kg through an acceleration of 1 m/s².

$$\therefore \qquad F = m\,a \text{ newtons}$$

where $\qquad m$ = mass of the body in kg

$$a = \text{acceleration in m/s}^2$$

(*iii*) **Weight.** The force with which a body is attracted towards the centre of Earth is called the weight of the body. Now, force = mass × acceleration. If m is the mass of a body in kg and g is the acceleration due to gravity in m/s², then,

$$\text{Weight, } W = m\,g \text{ newtons}$$

As the value of g* varies from place to place on earth's surface, therefore, the weight of the body varies accordingly. However, for practical purposes, we take $g = 9.81$ m/s² so that weight of the body = $9.81\,m$ newtons. Thus if a mass of 1 kg rests on a table, the downward force on the table *i.e.*, weight of the body is $W = 9.81 \times 1 = 9.81$ newtons.

The following points may be noted carefully :

(*a*) *The mass of a body is a constant quantity whereas its weight depends upon the place or position of the body.* However, it is reasonably accurate to express weight $W = 9.81\,m$ newtons where m is the mass of the body in kg.

(*b*) Sometimes weight is given in kg. wt. units. One kg-wt means weight of mass of 1 kg *i.e.* $9.81 \times 1 = 9.81$ newtons.

$$\therefore \qquad 1 \text{ kg. wt.} = 9.81 \text{ newtons}$$

Thus, when we say that a body has a weight of 100 kg, it means that it has a mass of 100 kg and that it exerts a downward force of 100×9.81 newtons.

4.3. Units of Work or Energy

Work is said to be done on a body when a force acts on it and the body moves through some distance. This work done is stored in the body in the form of energy. Therefore, work and energy are measured in the same units. The SI unit of work or energy is *joule* and is defined as under :

The work done on a body is **one joule** *if a force of one newton moves the body through 1 m in the direction of force.*

It may be noted that work done or energy possessed in an electrical circuit or mechanical system or thermal system is measured in the same units viz. joules. This is expected because mechanical, electrical and thermal energies are interchangeable. For example, when mechanical work is transferred into heat or heat into work, the quantity of work in joules is equal to the quantity of **heat in joules.

---

*   The value of g is about 9.81 m/s² at sea level whereas at equator, it is about 9.78 m/s² and at each pole it is about 9.832 m/s².

**   Although heat energy was assigned a separate unit *viz.* calorie but the reader remembers that 1 calorie = 4.186 joules. In fact, the thermal unit calorie is obsolete and now-a-days heat is expressed in joules.

Note. To gain some appreciation for the magnitude of a joule of heat energy, it would require about 90,000 J to heat a cup of water from room temperature to boiling.

4.4. Some Cases of Mechanical Work or Energy

It may be helpful to give a few important cases of work done or energy possessed in a mechanical system :

(*i*) When a force of F newtons is exerted on a body through a distance 'd' metres in the direction of force, then,

Work done = $F \times d$ joules or Nm

(*ii*) Suppose a force of F newtons in maintained tangentially at a radius r metres from O as shown in Fig. 4.1. In one revolution, the point of application of force travels through a distance of $2\pi r$ metres.

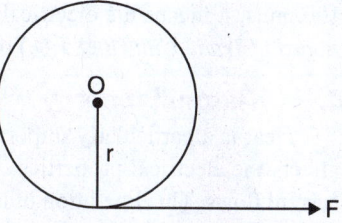

Fig. 4.1

∴ Work done in one revolution

$$= \text{Force} \times \text{Distance moved in 1 revolution}$$
$$= F \times 2\pi r$$
$$= 2\pi \times T \text{ joules or Nm}$$

where $T = F r$ is the torque. Clearly, the SI unit of torque will be joules or Nm. If the body makes N revolutions per minute, then,

$$\text{Work done/minute} = 2\pi N T \text{ joules}$$

(*iii*) If a body of mass m kg is moving with a speed of v m/s, then kinetic energy possessed by the body is given by ;

$$\text{K.E. of the body} = \frac{1}{2}mv^2 \text{ joules}$$

(*iv*) If a body having a mass of m kg is lifted vertically through a height of h metres and g is acceleration due to gravity in m/s^2, then,

$$\text{Potential energy of body} = \text{Work done in lifting the body} = \text{Force required} \times \text{height}$$
$$= \text{Weight of body} \times \text{height} = m g \times h$$
$$= m g h \text{ joules}$$

4.5. Electrical Energy

The SI unit of electrical work done or electrical energy expended in a circuit is also joule—exactly the same as for mechanical energy. It is defined as under :

One joule *of energy is expended electrically when one coulomb is moved through a p.d. of 1 volt.*

Suppose a charge of Q coulomb moves through a p.d. of V volts in time t in part AB of a circuit as shown in Fig. 4.2. Then electrical energy expended is given by ;

Electrical energy expended

$$= V Q \text{ joules}$$
$$= V I t \text{ joules} \qquad (\because Q = I t)$$
$$= I^2 R t \text{ joules} \qquad (\because V = I R)$$
$$= \frac{V^2 t}{R} \text{ joules} \qquad \left(\because I = \frac{V}{R} \right)$$

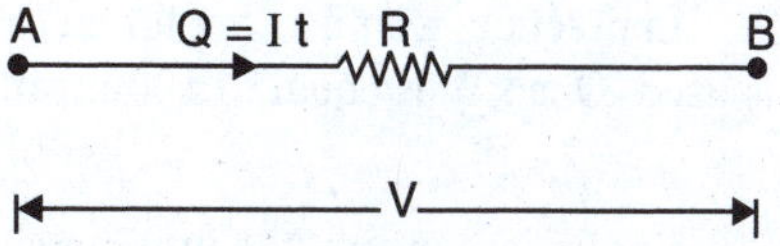

Fig. 4.2

It may be mentioned here that joule is also known as watt-second *i.e.* 1 joule = 1 watt-sec. When we are dealing with large amount of electrical energy, it is often convenient to express it in *kilowatt hours* (kWh).

$$1 \text{ kWh} = 1000 \text{ watt-hours} = 1000 \times 3600 \text{ watt-sec or joules}$$

$$\therefore \qquad 1 \text{ kWh} = 36 \times 10^5 \text{ joules or watt-sec}$$

Although practical unit of electrical energy is kWh, yet it is easy to see that this unit is readily convertible to joules with the help of above relation.

The electricity bills are made on the basis of total electrical energy consumed by the consumer. *The unit for billing of electrical energy is 1 kWh.* Thus when we say that a consumer has consumed 100 units, it means the electrical energy consumption is 100 kWh. Note that 1 kWh is also called *Board of Trade Unit* (*B.O.T.U.*) or unit of electricity.

4.6. Thermal Energy

Heat is a particularly important form of energy in the study of electricity, not only because it affects the electrical properties of the materials but also because it is *liberated* whenever electric current flows. This liberation of heat is infact the conversion of electrical energy to heat energy.

The thermal energy was originally assigned the unit 'calorie'. One calorie is the amount of heat required to raise the temperature of 1 gm of water through 1°C. If S is the specific heat of a body, then amount of heat required to raise the temperature of m gm of body through θ°C is given by ;

$$\text{Heat gained} = (m\,S\,\theta) \text{ calories}$$

It has been found experimentally that 1 calorie = 4·186 joules so that heat energy in calories can be expressed in joules as under :

$$\text{Heat gained} = (m\,S\,\theta) \times 4\cdot186 \text{ joules}$$

The reader may note that SI unit of heat is also joule. In fact, the thermal unit calorie is obsolete and unit joule is preferred these days.

4.7. Units of Power

Power is the *rate* at which energy is expended or the rate at which work is performed. Since energy and work both have the units of joules, it follows that power, being rate, has the units joule/second. Now Joule/second is also called **watt.** In general,

$$\text{Power} = \frac{W}{t} \text{ watts}$$

where W is the total number of joules of work performed or total joules of energy expended in t seconds.

Suppose a charge of Q coulomb moves through a p.d. of V volts in time t in part AB of a circuit as shown in Fig. 4.2. Then,

$$\text{Electrical energy expended} = VQ = VIt = I^2Rt = \frac{V^2 t}{R}$$

$$\therefore \qquad \text{Power of circuit, } P = \frac{VIt}{t} = \frac{I^2 Rt}{t} = \frac{V^2 t}{Rt}$$

$$\text{or} \qquad P = VI = I^2R = \frac{V^2}{R}$$

In practice, watt is often found to be inconveniently small, consequently the unit kilowatt (kW) is used. One kW is equal to 1000 watts *i.e.*

$$1 \text{ kW} = 1000 \text{ watts}$$

For larger powers, the unit megawatt (MW) is used. One megawatt is equal to 1000 kW *i.e.*

$$1 \text{ MW} = 1000 \text{ kW} = 1000 \times 1000 \text{ watts}$$

$$\therefore \qquad 1 \text{ MW} = 10^6 \text{ watts}$$

It may be noted that power of an electrical system or mechanical system or thermal system is measured in the same units viz joules/sec. or watts.

Important points. The following points are worth noting :

(*i*) Sometimes power is measured in *horse power (h.p.).

$$1 \text{ h.p.} = 746 \text{ watts}$$

(*ii*) If a body makes N r.p.m. and the torque acting is T newton-metre, then,

$$\text{Work done/minute} = 2\pi N T \text{ joules} \qquad [\text{See Art. 4·4}]$$

$$\text{Work done/sec} = \frac{2\pi NT}{60} \text{ joules/sec or watts}$$

i.e.,

$$\text{Power} = \frac{2\pi NT}{60} \text{ watts}$$

Since 746 watts = 1 h.p., we have,

$$\text{Power} = \frac{2\pi NT}{60 \times 746} \text{ h.p.}$$

where T is in newton-m and N is in r.p.m.

(*iii*) Power can also be expressed in terms of force and velocity.

$$\text{Power} = \text{Work done/sec} = \text{Force} \times \text{distance/sec}$$

∴ $$\text{Power} = \text{Force} \times \text{velocity}$$

4.8. Efficiency of Electric Device

The efficiency of an electric device is the ratio of useful output power to the input power, i.e.

$$\text{Efficiency, } \eta = \frac{\text{Useful output power}}{\text{Input power}}$$

$$= \frac{\text{Useful output Energy}}{\text{Input Energy}}$$

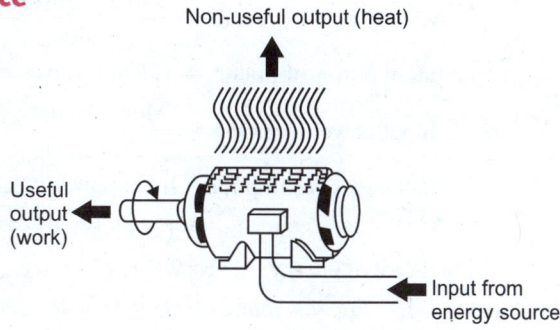

Fig. 4.3

The law of conservation of energy states that "energy cannot be created or destroyed but can be converted from one form to another." Some of the input energy to an electric device may be converted into a form that is not useful. For example, consider an electric motor shown in Fig. 4.3. The purpose of the motor is to convert electric energy into mechanical energy. It does this but it also converts a part of input energy into heat. The heat produced is not useful. Therefore, the useful output energy is less than the input energy. In other words, the efficiency of motor is less than 100%. While selecting an electric device, its efficiency is an important consideration because the operating cost of the device depends upon this factor.

Some electric devices are nearly 100% efficient. An electric heater is an example. In a heater, the heat is useful output energy and practically all the input electric energy is converted into heat energy.

4.9. Harmful Effects of Poor Efficiency

The poor (or low) efficiency of a device or of a circuit has the following harmful effects :

(*i*) Poor efficiency means waste of energy on non-useful output.

---

* This unit for power was conceived by James Watt, a Scottish scientist who invented the steam engine. In his experiments, he compared the output of his engine with the power a horse could put out. He found that an "average" horse could do work at the rate of 746 joules/sec. Although power can be expressed in watts or kW, the unit h.p. is still used.

(*ii*) Non-useful output of a device or circuit usually appears in the form of heat. Therefore, poor efficiency means a significant temperature rise. High temperature is one of the major limiting factors in producing reliable electric and electronic devices. Circuits and devices that run hot are more likely to fail.

(*iii*) The heat produced as a result of poor efficiency has to be dissipated *i.e.*, heat has to be transferred to the atmosphere or some other mass. Heat removal can become quite difficult in high power circuits and adds to the cost and size of the equipment.

Example 4.1. *An electrically driven pump lifts 80 m^3 of water per minute through a height of 12 m. Allowing an overall efficiency of 70% for the motor and pump, calculate the input power to motor. If the pump is in operation for an average of 2 hours per day for 30 days, calculate the energy consumption in kWh and the cost of energy at the rate of Rs 2 per kWh. Assume 1 m^3 of water has a mass of 1000 kg and g = 9·81 m/s^2.*

Solution. Mass of 80 m^3 of water, $m = 80 \times 1000 = 8 \times 10^4$ kg

Weight of water lifted, $W = m\,g = 8 \times 10^4 \times 9·81$ N

Height through which water lifted, $h = 12\ m$

$$\text{W.D. by motor/minute} = m\,g\,h = 8 \times 10^4 \times 9·81 \times 12 \text{ joules}$$

$$\text{W.D. by motor/second} = \frac{8 \times 10^4 \times 9.81 \times 12}{60} = 156960 \text{ watts}$$

∴   Output power of motor = 156960 watts

$$\text{Input power to motor} = \frac{\text{Motor output}}{\text{Efficiency}} = \frac{156960}{0.7} = 2,24,228 \text{ W} = \textbf{224·228 kW}$$

Total energy consumption = Input power × Time of operation

$$= (224·228) \times (2 \times 30) \text{ kWh} = \textbf{13453 kWh}$$

Total cost of energy   =Rs 2 × 13453 = **Rs. 26906**

Example 4.2. *Fig. 4.4 shows an electric motor driving an electric generator. The 2 h.p. motor draws 14·6 A from a 120 V source and the generator supplies 56 A at 24 V.*

(*i*)   *Find the motor efficiency and generator efficiency*

(*ii*)   *Find the overall efficiency.*

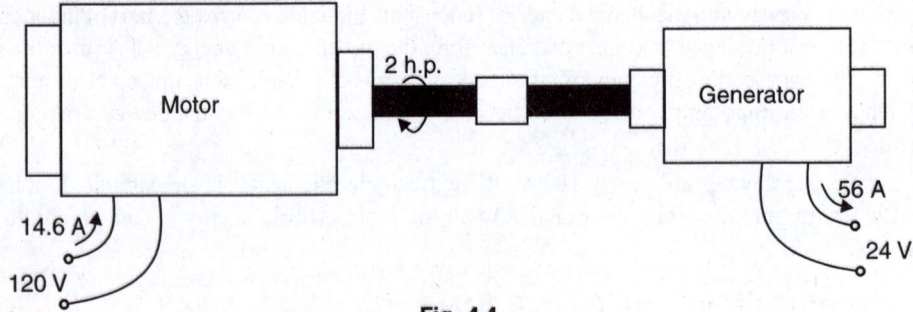

Fig. 4.4

Solution. Efficiency of a machine is output power (P_o) divided by input power (P_i).

(*i*) P_i (motor) = 120 × 14·6 = 1752 W

P_o (motor) = 2 h.p. = 2 × 746 = 1492 W

∴ η (motor) = $\dfrac{1492}{1752}$ = **0.8516 or 85·16%**

$$P_i \text{ (generator)} = 2 \text{ h.p.} = 1492 \text{ W}$$

$$P_o \text{ (generator)} = 24 \times 56 = 1344 \text{ W}$$

$$\therefore \qquad \eta \text{ (generator)} = \frac{1344}{1492} = 0\text{·}90 \quad \text{or} \quad \mathbf{90\%}$$

(ii) $\qquad \eta \text{ (overall)} = \dfrac{P_o \text{(generator)}}{P_i \text{(motor)}} = \dfrac{1344}{1752} = \mathbf{0\text{·}767} \text{ or } \mathbf{76\text{·}7\%}$

Note that overall η is the product of efficiencies of the individual machines.

$$\eta \text{ (overall)} = \eta \text{ (motor)} \times \eta \text{ (generator)} = 0\text{·}8516 \times 0\text{·}90 = 0\text{·}767.$$

Example 4.3. *Neglecting losses, at what horse power rate could energy be obtained from Bhakra dam which has an average height of 225 m and water flows at a rate of 500,000 kg/minute ? If the overall efficiency of conversion were 25%, how many 100 watt light bulbs could Bhakra dam supply ?*

Solution. Wt. of water flowing/minute

$$= m\,g = 500,000 \times 9\text{·}81 \text{ N}$$

$$\text{Work done/minute} = m\,g\,h = 500,000 \times 9\text{·}81 \times 225 \text{ joules}$$

$$\text{Work done/second} = \frac{500,000 \times 9.81 \times 225}{60} = 18394 \times 10^3 \text{ watts}$$

$$\therefore \qquad \text{Gross power obtained} = 18394 \times 10^3 \text{ watts} = 18394 \text{ kW}$$

$$\text{Useful output power} = 18394 \times 0\text{·}25 = 4598\text{·}5 \text{ kW}$$

$$= \frac{4598.5 \times 10^3}{746} \text{ h.p.} = \mathbf{6164 \text{ h.p.}}$$

No. of 100-watt bulbs that could be lighted

$$= \frac{4598.5 \times 10^3}{100} = \mathbf{45985}$$

Example 4.4. *A 100 MW hydro-electric station is supplying full-load for 10 hours a day. Calculate the volume of water which has been used. Assume effective head of station as 200 m and overall efficiency of the station as 80%.*

Solution. Energy supplied by the station in 10 hours

$$= (100 \times 10^3) \times 10 = 10^6 \text{ kWh}$$

$$= 36 \times 10^5 \times 10^6 = 36 \times 10^{11} \text{ joules}$$

$$\text{Energy input of station} = 36 \times 10^{11}/0\text{·}8 = 45 \times 10^{11} \text{ joules}$$

Suppose m kg is the mass of water used in 10 hours.

Then, $\qquad\qquad m\,g\,h = 45 \times 10^{11}$

or $\qquad\qquad m = \dfrac{45 \times 10^{11}}{9.81 \times 200} = 22.93 \times 10^8 \text{ kg}$

Since 1 m^3 of water has a mass of 1000 kg,

$\therefore \quad$ Volume of water used $= 22\text{·}93 \times 10^8/10^3 = \mathbf{22\text{·}93 \times 10^5 \text{ m}^3}$

Example 4.5. *Two coils are connected in parallel and a voltage of 200 V is applied to the terminals. The total current taken is 15A and the power dissipated in one of the coils is 1500 W. What is the resistance of each coil ?*

Solution. Let R_1 and R_2 be the resistances of the coils and I_1 and I_2 be the current drawn from the supply. Since the coils are connected in parallel, voltage across each coil is the same *i.e.* 200 V.

$$V I_1 = W \quad \text{or} \quad I_1 = W/V = 1500/200 = 7 \cdot 5 \text{A}$$

$$\therefore \qquad R_1 = V/I_1 = 200/7 \cdot 5 = \textbf{26.7} \ \Omega$$

$$I_1 + I_2 = 15 \quad ..... \text{ given} \quad \therefore \quad I_2 = 15 - I_1 = 15 - 7 \cdot 5 = 7 \cdot 5 \text{A}$$

$$\therefore \qquad R_2 = \frac{V}{I_2} = \frac{200}{7.5} = \textbf{26.7} \ \Omega$$

Although not technically correct usage, it is convenient to say that resistance "dissipates power", meaning that it dissipates (liberates) heat at a certain rate.

Example 4.6. *A motor is being self-started against a resisting torque of 60 N-m and at each start, the engine is cranked at 75 r.p.m. for 8 seconds. For each start, energy is drawn from a lead-acid battery. If the battery has the capacity of 100 Wh, calculate the number of starts that can be made with such a battery. Assume an overall efficiency of the motor and gears as 25%.*

Solution. Angular speed, $\omega = 2\pi N/60 \text{ rad/s} = 2\pi \times 75/60 = 7 \cdot 85 \text{ rad/s}$

$$\text{Power required per start, } P = \frac{\text{Torque} \times \text{Angular speed}}{\text{Efficiency of motor}} = \frac{60 \times 7.85}{0.25} = 1884 \text{ W}$$

$$\text{Energy required/start} = P \times \text{Time for start}$$

$$= 1884 \times 8 = 15072 \text{ Ws} = 15072 \text{ J}$$

$$= 15072/3600 = 4 \cdot 187 \text{ Wh}$$

$\therefore$ No. of starts with a fully-charged battery

$$= 100/4 \cdot 187 \simeq \textbf{24}$$

Example 4.7. *A hydro-electric power station has a reservoir of area 2·4 square kilometres and capacity $5 \times 10^6 \ m^3$. The effective head of water is 100 m. The penstock, turbine and generator efficiencies are 95%, 90% and 85% respectively.*

(i) *Calculate the total energy in kWh which can be generated from the power station.*

(ii) *If a load of 15,000 kW has been supplied for 3 hours, find the fall in reservoir level.*

Solution.

(*i*) Wt. of water available, $W =$ Volume of reservoir $\times 1000 \times 9.81$ N

$$= (5 \times 10^6) \times (1000) \times (9 \cdot 81) = 49 \cdot 05 \times 10^9 \text{ N}$$

Overall efficiency, $\eta_{overall} = 0 \cdot 95 \times 0 \cdot 90 \times 0 \cdot 85 = 0 \cdot 726$

Electrical energy that can be generated from the station

$$= W \times \text{Effective head} \times \eta_{overall}$$

$$= (49 \cdot 05 \times 10^9) \times (100) \times (0 \cdot 726) = 35 \cdot 61 \times 10^{11} \text{ watt-sec.}$$

$$= \frac{35.61 \times 10^{11}}{1000 \times 3600} \text{ kWh} = \textbf{9,89,116 kWh}$$

(*ii*) Level of reservoir $= \dfrac{\text{Volume of reservoir}}{\text{Area of reservoir}} = \dfrac{5 \times 10^6}{2.4 \times 10^6} = 2 \cdot 083 \text{ m}$

kWh generated in 3 hrs $= 15000 \times 3 = 45,000 \text{ kWh}$

Using unitary method, we get,

$$\text{Fall in reservoir level} = \frac{2.083}{9,89,166} \times 45,000 = 0 \cdot 0947 \text{ m} = \textbf{9·47 cm}$$

Example 4.8. *A large hydel power station has a head of 324 m and an average flow of 1370 m³/sec. The reservoir is a lake covering an area of 6400 sq. km. Assuming an efficiency of 90% for the turbine and 95% for the generator, calculate (i) the available electric power and (ii) the number of days this power could be supplied for a drop in water level by 1 metre.*

Solution. Water discharge = 1370 m³/sec ; Water head, $h = 324$ m ; $\eta_{overall} = 0.9 \times 0.95$

(*i*) As mass of 1 m³ of water is 1000 kg,

$\therefore$ Mass of water flowing/sec, $m = 1370 \times 1000$ kg $= 137 \times 10^4$ kg

Weight of water flowing/sec, $W = mg = 137 \times 10^4 \times 9.81$ N

Energy or work available per second (*i.e.* power) is

$$\text{Power available, } P = Wh \times \eta_{overall}$$
$$= (137 \times 10^4 \times 9.81) \times 324 \times (0.9 \times 0.95)$$
$$= 3723 \times 10^6 \text{ W} = \textbf{3723 MW}$$

(*ii*)   Area of reservoir, $A = 6400$ km² $= 6400 \times 10^6$ m²

Rate of water discharge, $Q = 1370$ m³/sec

Fall of reservoir level, $h' = 1$ m

Volume of water used $= A \times h'$

$\therefore$ Required time, $t = \dfrac{A \times h'}{Q} = \dfrac{6400 \times 10^6 \times 1}{1370}$

$$= 4.67 \times 10^6 \text{ sec.} = \textbf{54.07 days}$$

Example 4.9. *Calculate the current required by a 500 V d.c. locomotive when drawing 100 tonne load at 25 km/hr with a tractive resistance of 7 kg/tonne along (i) level road and (ii) a gradient 1 in 100. Given that the efficiency of motor and gearing is 70%.*

Solution. Weight of locomotive, $W = 100$ tonne $= 100,000$ kg

Tractive resistance, $F = 7 \times 100 = 700$ kg-wt $= 700 \times 9\cdot81 = 6867$ N

(*i*) **Level Track.** In this case, the force required is equal to the tractive resistance F [See Fig. 4.5 (*i*)].

$$\text{Distance travelled/sec} = \frac{25 \times 1000}{3600} = 6\cdot94 \text{ m}$$

Work done/sec $=$ Force $\times$ Distance/sec

or   Motor output $= 6867 \times 6\cdot94 = 47,657$ watts

Motor input $= 47,657/0\cdot7 = 68,081$ watts

$\therefore$ Current drawn $= 68,081/500 = \textbf{136·16A}$

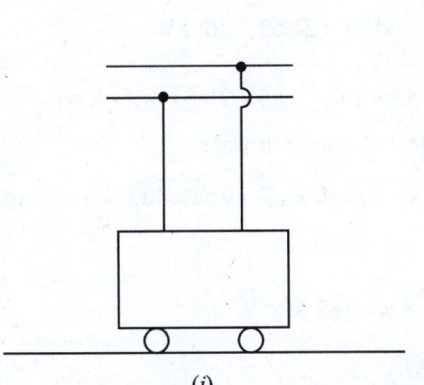

(*i*)

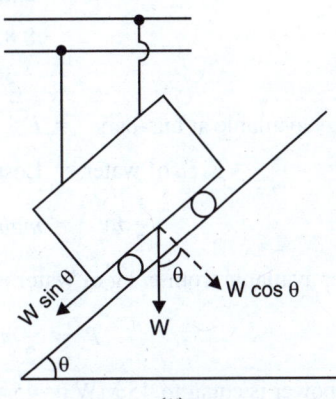

(*ii*)

Fig. 4.5

(ii) **Inclined plane.** In this case, the total force required is the sum of tractive resistance F and component $W \sin\theta$ of locomotive weight [See Fig. 4.5 (ii)]. Clearly, $\sin\theta = 1/100 = 0 \cdot 01$.

∴   Force required $= W \sin\theta + F$

$$= (100,000 \times 0 \cdot 01 + 700)\, 9 \cdot 81 \text{ N} = 16,677 \text{ N}$$

Work done/sec $=$ Force $\times$ distance travelled/sec

$$= 16,677 \times 6 \cdot 94 = 1,15,738 \text{ watts}$$

∴   Motor output $= 1,15,738$ watts

Motor input $= 1,15,738/0 \cdot 7 = 1,65,340$ watts

∴   Current drawn $= 1,65,340/500 = \mathbf{330 \cdot 68A}$

Example 4.10. *A diesel-electric generator set supplies an output of 25 kW. The calorific value of the fuel oil used is 12,500 kcal/kg. If the overall efficiency of the unit is 35%, calculate (i) the mass of oil required per hour (ii) the electric energy generated per tonne of the fuel.*

Solution. Output power of set $= 25 \text{ kW}$; $\eta_{overall} = 35\% = 0.35$

∴   Input power to set $= 25/0.35 = 71.4 \text{ kW}$

(*i*) Input energy/hour $= 71.4 \text{ kW} \times 1\text{h} = 71.4 \text{ kWh} = 71.4 \times 860 \text{ kcal}$

As 1 kg of fuel oil produces 12,500 kcal,

∴   Mass of fuel oil required/hour $= \dfrac{71.4 \times 860}{12,500} = \mathbf{4.91 \text{ kg}}$

(*ii*) Heat content in 1 tonne fuel oil ($= 1000 \text{ kg}$) $= 1000 \times 12,500 = 12.5 \times 10^6 \text{ kcal}$

$$= \dfrac{12.5 \times 10^6}{860} \text{ kWh} = 14,534 \text{ kWh}$$

∴   Energy generated/tonne $= 14,534 \times 0.35 = \mathbf{5087 \text{ kWh}}$

Example 4.11. *The reservoir for a hydro-electric station is 230 m above the turbine house. The annual replenishment of the reservoir is 45×10^{10} kg. What is the energy available at the generating station bus-bars if the loss of head in the hydraulic system is 30 m and the overall efficiency of the station is 85% ? Also, calculate the diameter of the steel pipes needed if a maximum demand of 45 MW is to be supplied using two pipes.*

Solution. Actual available head, $h = 230 - 30 = 200 \text{ m}$

Energy available at turbine house is given by ;

$$E = mgh = 45 \times 10^{10} \times 9 \cdot 81 \times 200 = 8 \cdot 829 \times 10^{14} \text{ J}$$

$$= \dfrac{8.829 \times 10^{14}}{36 \times 10^5} \text{ kWh} = 24 \cdot 52 \times 10^7 \text{ kWh}$$

Energy available at bus-bars $= E \times \eta = 24 \cdot 52 \times 10^7 \times 0.85 = \mathbf{20 \cdot 84 \times 10^7 \text{ kWh}}$

K.E. of water $=$ Loss of potential energy of water

or $\dfrac{1}{2}mv^2 = mgh$ ∴ $v = \sqrt{2gh} = \sqrt{2 \times 9.81 \times 200} = 62 \cdot 65 \text{ m/s}$

Power available from m kg of water is

$$P = \dfrac{1}{2}mv^2 = \dfrac{1}{2} \times m \times (62.65)^2 \text{ W}$$

This power is equal to 45 MW ($= 45 \times 10^6$ W).

$$\therefore \qquad P = 45 \times 10^6 \text{ W}$$

$$\text{or} \qquad \frac{1}{2} \times m \times (62.65)^2 = 45 \times 10^6 \quad \therefore \quad m = 22930 \text{ kg/s}$$

If A is the total area of two pipes in m^2, then flow of water is Av m³/s.

$\therefore$ Mass of water flowing/second $= Av \times 10^3$ kg ($\because$ 1 m³ of water $= 1000$ kg)

$$\therefore \qquad Av \times 10^3 = 22930 \quad \text{or} \quad A = \frac{22930}{62.65 \times 10^3} = 0 \cdot 366 \text{ m}^2$$

$$\text{Area of each pipe } = 0 \cdot 366/2 = 0 \cdot 183 \text{ m}^2$$

If d is the diameter of each pipe, then,

$$\frac{\pi}{4} d^2 = 0.183 \quad \text{or} \quad d = \sqrt{\frac{0.183 \times 4}{\pi}} = \mathbf{0 \cdot 4826 \text{ m}}$$

Example 4.12. *A proposed hydro-electric station has an available head of 30 m, catchment area of 50×10^6 m², the rainfall for which is 120 cm per annum. If 70% of the total rainfall can be collected, calculate the power that could be generated. Assume the following efficiencies: Penstock 95%, Turbine 80% and Generator 85%.*

Solution. Available head, $h = 30$ m ; $\eta_{overall} = 0.95 \times 0.8 \times 0.85 = 0.646$

Volume of water *available/annum $= 0.7(50 \times 10^6 \times 1.2) = 4.2 \times 10^7$ m³

Mass of water available/annum $= 4.2 \times 10^7 \times 1000 = 4.2 \times 10^{10}$ kg

$$\text{Mass of water available/sec; } m = \frac{4.2 \times 10^{10}}{365 \times 24 \times 3600} = 1.33 \times 10^3 \text{ kg}$$

Potential energy available/sec $= mgh = 1.33 \times 10^3 \times 9.8 \times 30 = 391 \times 10^3$ J/s

$\therefore$ Power that could be generated $= \eta_{overall} \times 391 \times 10^3$ W

$$= 0.646 \times 391 \times 10^3 = 253 \times 10^3 \text{ W} = \mathbf{253 \text{ kW}}$$

Example 4·13. *A current of 20A flows for one hour in a resistance across which there is a voltage of 8V. Determine the velocity in metres per second with which a weight of one tonne must move in order that kinetic energy shall be equal in amount to the energy dissipated in the resistance.*

Solution. Energy dissipated in resistance

$$= V It = 8 \times 20 \times 3600 = 576 \times 10^3 \text{ J}$$

Mass of body, $m = 1$ tonne $= 1000$ kg

Let v m/s be the required velocity of the weight.

$$\text{Kinetic energy } = \frac{1}{2} mv^2 \text{ joules}$$

In order that K.E. of weight is equal to energy dissipated in resistance,

$$\frac{1}{2} mv^2 = 576 \times 10^3 \quad \therefore v = \sqrt{\frac{2 \times 576 \times 10^3}{1000}} = \mathbf{33 \cdot 9 \text{ m/s}}$$

Example 4·14. *What must be the horse-power of an engine to drive by means of a belt a generator supplying 7000 lamps each taking 0·5 A at 250 V ? The line drop is 5V and the efficiency of the generator is 95%. There is a 2·5% loss in the belt drive.*

Solution. Total current supplied by generator, $I = 0 \cdot 5 \times 7000 = 3500$ A

$$\text{Generated voltage, } E = \text{Load voltage} + \text{Line drop} = 250 + 5 = 255 \text{ V}$$

$$\text{Generator output } = EI = 255 \times 3500 \text{ W}$$

---

* 0.7 × (Catchment area in m² × Rainfall in m)

$\therefore$ Engine output $= \dfrac{255 \times 3500}{0.95 \times 0.975} = 963562$ W $= \dfrac{963562}{746}$ h.p. $\doteq$ **1292 h.p.**

Example 4.15. *Find the head in metres of a hydroelectric generating station in which the reservoir of area 4000 m^2 falls by 30 cm when 75 kWh is developed in the turbine. The efficiency of the turbine is 70%.*

Solution. Hydroelectric generating stations are generally built in hilly areas.

$$\text{Volume of water used, } V = 4000 \times 0.3 = 1200 \text{ m}^3$$

$$\text{Mass of water used, } m = 1200 \times 10^3 = 1.2 \times 10^6 \text{ kg}$$

$$\text{Useful energy developed in turbine} = mgh \times \eta = 1.2 \times 10^6 \times 9.81 \times h \times 0.7$$

$$\text{But useful energy developed in turbine} = 75 \text{ kWh} = 75 \times 3.6 \times 10^6 \text{ J}$$

$\therefore$ $1.2 \times 10^6 \times 9.81 \times h \times 0.7 = 75 \times 3.6 \times 10^6$

or $h =$ **32·76 m**

Example 4.16. *A room measures 3m × 4m × 4·75m and air in it has to be always kept 10°C higher than that of the incoming air. The air inside has to be renewed every 30 minutes. Neglecting radiation losses, find the necessary rating of electric heater for this purpose. Take specific heat of air as 0·24 and density as 1·28 kg/m^3.*

Solution. It is desired to find the power of the electric heater.

$$\text{Volume of air to be changed/second} = \dfrac{3 \times 4 \times 4.75}{30 \times 60} = 0.032 \text{ m}^3$$

$$\text{Mass of air to be changed/second} = 0.032 \times 1.28 = 0.041 \text{ kg}$$

$$\text{Heat required/second} = \text{Mass/second} \times \text{Specific heat} \times \text{Rise in temp.}$$

$$= 0.041 \times 0.24 \times 10 \text{ kcal}$$

$$= 0.041 \times 0.24 \times 10 \times 4186 \text{ W} = \textbf{411 W}$$

Here, we have neglected radiation losses. However, in practice, radiation losses do occur so that heater power required would be greater than the calculated value. $\left(\because \dfrac{1 \text{ kcal}}{\text{sec.}} = 4186 \text{ W} \right)$

Example 4.17. *An electric lift is required to raise a load of 5 tonne through a height of 30 m. One quarter of electrical energy supplied to the lift is lost in the motor and gearing. Calculate the energy in kWh supplied. If the time required to raise the load is 27 minutes, find the kW rating of the motor and the current taken by the motor, the supply voltage being 230V d.c. Assume the efficiency of the motor at 90%.*

Solution. Work done by lift $= mgh = (5 \times 10^3) \times 9.8 \times 30 = 1.47 \times 10^6$ J

$$\text{Input energy to lift} = \dfrac{1.47 \times 10^6}{\eta_{lift}{}^*} = \dfrac{1.47 \times 10^6}{0.75} = 1.96 \times 10^6 \text{ J}$$

$$= \dfrac{1.96 \times 10^6}{36 \times 10^5} \text{ kWh} = \textbf{0.545 kWh}$$

$$\text{Motor energy output} = \text{Input energy to lift} = 1.96 \times 10^6 \text{ J}$$

$$\text{Motor energy input} = \dfrac{1.96 \times 10^6}{\eta_{motor}} = \dfrac{1.96 \times 10^6}{0.9} = 2.18 \times 10^6 \text{ J}$$

---

* Since 25% energy is *wasted* in the motor and gearing, the efficiency of the lift is 75%.

$$\text{Power rating of motor} = \frac{\text{Work done}}{\text{Time taken}} = \frac{2.18 \times 10^6}{27 \times 60} = \textbf{1346 W}$$

$$\text{Current taken by motor} = \frac{1346}{230} = \textbf{5.85 A}$$

Example 4.18. *An electric hoist makes 10 double journeys per hour. In each journey, a load of 6000 kg is raised to a height of 60 m in 90 seconds and the hoist returns empty in 75 seconds. The hoist cage weighs 500 kg and has a balance weight of 3000 kg. The efficiency of the hoist is 80% and that of the driving motor 88%. Calculate (i) the electrical energy absorbed per double journey (ii) the hourly consumption in kWh (iii) the horse-power of the motor (iv) the cost of electric energy if hoist works for 4 hours per day for 30 days. Cost per kWh is Rs 4.50.*

Solution. When the hoist cage goes up, the balance weight goes down and when the cage goes down, the balance weight goes up.

$$\text{Total mass lifted on upward journey} = \text{Load} + \text{mass of cage} - \text{mass of balance weight}$$
$$= 6000 + 500 - 3000 = 3500 \text{ kg}$$
$$\text{Work done during upward journey} = mgh = 3500 \times 9 \cdot 8 \times 60 \text{ J}$$
$$\text{Total mass moved on downward journey} = \text{Mass of balance wt.} - \text{Mass of cage}$$
$$= 3000 - 500 = 2500 \text{ kg}$$
$$\text{Work done during downward journey} = mgh = 250 \times 9 \cdot 8 \times 60 \text{ J}$$
$$\text{Work done during each double journey} = 9 \cdot 8 \times 60 \, (3500 + 2500) \text{J} = 353 \times 10^4 \text{ J}$$
$$\text{Overall } \eta = 0 \cdot 8 \times 0 \cdot 88 = 0 \cdot 704$$

(i)
$$\text{Input energy per double journey} = 353 \times 10^4 / 0 \cdot 704 = 501 \times 10^4 \text{ J}$$
$$= \frac{501 \times 10^4}{3.6 \times 10^6} \text{ kWh} = \textbf{1·4 kWh}$$

(ii)
$$\text{Hourly consumption} = 1 \cdot 4 \times \text{No. of double journeys/hr}$$
$$= 1 \cdot 4 \times 10 = \textbf{14 kWh}$$

(iii) The maximum rate of working is during upward journey.

$$\therefore \quad \text{h.p. rating of motor} = \frac{\text{Work done in upward journey}}{\text{Hoist efficiency} \times \text{time for up journey} \times 746}$$

$$= \frac{3500 \times 9.8 \times 60}{0.8 \times 90 \times 746} = \textbf{38·4 h.p.}$$

(iv) Energy consumption for 30 days = Hourly consumption $\times 4 \times 30 = 14 \times 4 \times 30 = 1680 \text{ kWh}$

Total cost of energy = Rs. $1680 \times 4 \cdot 5 = \textbf{Rs. 7560}$

Example 4.19. *A generator supplies power to a factory through cables of total resistance 20 ohms. The potential difference at the generator is 5000 V and power output is 50 kW. Calculate (i) power supplied by the generator, (ii) potential difference at the factory.*

Solution. Fig. 4.6 shows the conditions of the problem.

Output power of generator is given by ;

$$P = 50 \text{ kW} = 50 \times 10^3 \text{ W}$$

P.D. at the generator, $E = 5000 \text{ V}$

$\therefore$ Current in cables is given by ;

$$I = \frac{P}{E} = \frac{50 \times 10^3}{5000} = 10 \text{ A}$$

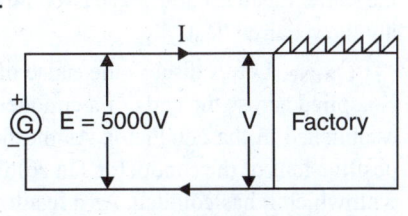

Fig. 4.6

   (*i*)   Power loss in cables $= I^2 R = (10)^2 \times 20 = 2000$ W

∴   Power supplied at the factory $= 50 \times 10^3 - 2000 =$ **48,000 W**

   (*ii*)   Voltage drop in cables $= I R = 10 \times 20 = 200$ V

∴   P.D. at the factory, $V = E - I R = 5000 - 200 =$ **4800 V**

Tutorial Problems

1. The power required to drive a certain machine at 350 r.p.m. is 600 kW. Calculate the driving torque.

   [16370 Nm]

2. An electrically driven pump lifts 1500 litres of water per minute through a height of 25 m. Allowing an overall efficiency of 75%, calculate the input power to the motor. If the pump is in operation for an average of 8 hours per day for 30 days, calculate the energy consumed in kWh and the cost of energy at the rate of 50 P/kWh. Assume 1 litre of water has a mass of 1000 kg and $g = 9\cdot81$ m/s^2.

   [8·167 kW, 1960 kWh, Rs. 980]

3. A 440-volt motor is used to drive an irrigation pump. The efficiency of motor is 85% and the efficiency of pump is 66%. The pump is required to lift 240 tonne of water per hour to a height of 30 metres. Calculate the current taken by the motor. **[79·48 A]**

4. A hydro-electric generating plant is supplied from a reservoir of capacity 2×10^7 m^3 with a head of 200 m. The hydraulic efficiency of the plant is $0\cdot8$ and electric efficiency is $0\cdot9$. What is the total available energy ? **[7·85 × 10^9 watt-hours]**

5. A 460-V d.c. motor drives a hoist which raises a load of 100 kg with a velocity of 15 m/s. Calculate :

   (*i*) The power output of the motor assuming the hoist gearing to have an efficiency of $0\cdot8$.

   (*ii*) The motor current, assuming the motor efficiency to be $0\cdot75$. **[(*i*) 18·4 kW (*ii*) 53·2 A]**

6. When a certain electric motor is operated for 30 minutes, it consumes $0\cdot75$ kWh of energy. During that time, its total energy loss is 3×10^5 J.

   (*i*) What is the efficiency of the motor ?

   (*ii*) How many joules of work does it perform in 30 minutes ? **[(*i*) 88·8% (*ii*) 2·4 × 10^{6}J]**

7. The total power supplied to an engine that drives an electric generator is 40·25 kW. If the generator delivers 15A to a 100 Ω load, what is the efficiency of the system ? **[55·9%]**

8. A certain system consists of three identical devices in cascade, each having efficiency $0\cdot85$. The first device draws 3A from a 20V source. How much current does the third device deliver to a 50Ω load ?

   [0·027 A]

4.10. Heating Effect of Electric Current

When electric current is passed through a conductor, heat is produced in the conductor. This effect is called **heating effect of electric current.**

It is a matter of common experience that when electric current is passed through the element of an electric heater, the element becomes red hot. It is because electrical energy is converted into heat energy. This is called heating effect of electric current and is utilised in the manufacture of many heating appliances, *e.g.,* electric iron, electric kettle, etc. The basic principle of all these devices is the same. Electric current is passed through a high resistance (called *heating element*), thus producing the required heat.

Cause. Let us discuss the cause of heating effect of electric current. When potential difference is applied across the ends of a conductor, the free electrons move with drift velocity and current is established in the conductor. As the free electrons move through the conductor, they collide with positive ions of the conductor. On collision, the kinetic energy of an electron is transferred to the ion with which it has collided. As a result, the kinetic energy of vibration of the positive ion increases, *i.e.,* temperature of the conductor increases. Therefore, as current flows through a conductor, the free electrons lose energy which is converted into heat. Since the source of e.m.f. (*e.g.,* a battery) is maintaining current in the conductor, it is clear that electrical energy supplied by the battery is converted into heat in the conductor.

Applications. The heating effect of electric current is utilised in the manufacture of many heating appliances such as electric heater, electric toaster, electric kettle, soldering iron etc. The basic principle of all these appliances is the same. Electric current is passed through a high resistance (called heating element), thus producing the required heat. There are a number of substances used for making a heating element. One that is commonly used is an alloy of nickel and chromium, called **nichrome.** This alloy has a resistance more than 50 times that of copper. The heating element may be either nichrome wire or ribbon wound on some insulating material that is able to withstand heat.

4.11.   Heat Produced in a Conductor by Electric Current

On the basis of his experimental results, Joule found that the amount of heat produced (H) when current I amperes flows through a conductor of resistance R ohms for time t seconds is $H = I^2Rt$ joules. This equation is known as Joule's law of heating.

Suppose a battery maintains a potential difference of V volts across the ends of a conductor AB of resistance R ohms as shown in Fig. 4.7. Let the steady current that passes from A to B be I amperes. If this current flows for t seconds, then charge transferred from A to B in t seconds is

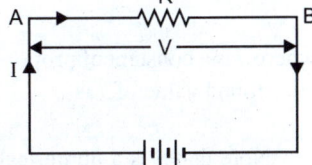

Fig. 4.7

$$q = It$$

The electric potential energy lost (W) by the charge q as it moves from A to B is given by ;

$$W = \text{Charge} \times \text{P.D. between } A \text{ and } B$$

$$= qV = (It)\,V = I^2Rt \quad (\because V = IR)$$

or

$$W = I^2Rt$$

This loss of electric potential energy of charge is converted into heat (H) because the conductor AB has resistance only.

∴ $$H = W = I^2Rt \text{ joules} = \frac{I^2Rt}{4.18} \text{ calories} \qquad \qquad ...(i)$$

It is found experimentally that 1 cal = 4.18 J.

Eq. (*i*) is known as **Joule's law of heating.** It is because Joule was the first scientist who studied the heating effect of electric current through a resistor. Thus according to Joule, heat produced in a conductor is directly proportional to

   (*i*) square of current through the conductor

   (*ii*) resistance of the conductor

   (*iii*) time for which current is passed through the conductor.

Note. $$H = VIt = I^2Rt = \frac{V^2}{R}t \text{ joules}$$

$$= \frac{VIt}{4.18} = \frac{I^2Rt}{4.18} = \frac{V^2t}{R \times 4.18} \text{ calories}$$

Important points. While dealing with problems on heating effect of electric current, the following points may be kept in mind :

   (*i*) The electrical energy in kWh can be converted into joules by the following relation :

$$1 \text{ kWh} = 36 \times 10^5 \text{ joules}$$

(*ii*) The heat energy in calories can be converted into joules by the following relation :

$$1 \text{ calorie} = 4 \cdot 186 \text{ joules}$$
$$1 \text{ kcal} = 4186 \text{ joules}$$

(*iii*) The electrical energy in kWh can be converted into calories (or kilocalories) by the following relation :

$$1 \text{ kWh} = 36 \times 10^5 \text{ joules} = \frac{36 \times 10^5}{4.186} \text{ calories} = 860 \times 10^3 \text{ calories}$$

$$\therefore \qquad 1 \text{ kWh} = 860 \text{ kcal}$$

(*iv*) The electrical energy supplied to the heating appliance forms the *input energy*. The heat obtained from the device is the *output energy*. The difference between the two, if any, represents the loss of energy during conversion from electrical into heat energy.

4.12. Mechanical Equivalent of Heat (J)

Joule performed a series of experiments to establish the relationship between the mechanical work done and heat produced. He found that heat produced (*H*) is directly proportional to the amount of mechanical work done (*W*) *i.e.*,

$$H \propto W \quad \text{or} \quad W = JH$$

where *J* is a constant of proportionality and is called *mechanical equivalent of heat*. The experimentally found value of *J* is

$$J = 4 \cdot 2 \text{ J/cal}$$

Note that *J* is a numerical factor relating mechanical units to heat units. Let us interpret the meaning of *J*. It takes 4·2 J of mechanical work to raise the temperature of 1g of water by 1°C. In other words, 4·2J of mechanical energy is equivalent to 1 calorie of heat energy.

Example 4.20. *In Fig. 4.8, the heat produced in 5 Ω resistor due to current flowing through it is 10 calories per second. Calculate the heat generated in 4 Ω resistor.*

Solution. Let I_1 and I_2 be the currents in the two parallel branches as shown in Fig. 4.8. The p.d. across the parallel branches is the same *i.e.*

$$I_1 (4 + 6) = 5 I_2 \quad \therefore \quad I_2 = 2 I_1$$

Heat produced per second in 5Ω resistor is

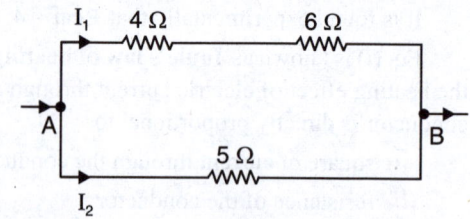

$$H_1 = \frac{I_2^2 \times 5}{4.2}$$

or

$$10 = \frac{(2I_1)^2 \times 5}{4.2}$$

$$\therefore \qquad I_1^2 = 2 \cdot 1$$

Fig. 4.8

Heat produced in 4Ω resistor per second

$$= \frac{I_1^2 \times 4}{4.2} = \frac{2.1 \times 4}{4.2} = \textbf{2 cal/sec}$$

Example 4.21. *An electric heater contains 4 litres of water initially at a mean temperature of 15°C. 0·25 kWh is supplied to the water by the heater. Assuming no heat losses, what is the final temperature of the water ?*

Solution. Let *t*°C be the final temperature of water.

Heat received by water (*i.e.* output energy)

$$= \text{mass} \times \text{sp. heat} \times \text{rise in temp.} = 4 \times 1 \times (t - 15) \text{ kcal}$$

Electrical energy supplied to heater (*i.e.* input energy)

$$= 0{\cdot}25 \text{ kWh} = 0{\cdot}25 \times 860 \text{ kcal} \qquad\qquad (\because 1 \text{ kWh} = 860 \text{ kcal})$$

As there are no losses, output energy is equal to the input energy *i.e.*

$$4 \times 1 \times (t - 15) = 0{\cdot}25 \times 860 \quad \text{or} \quad t = \textbf{68·8°C}$$

Example 4.22. *An immersion heater takes 1 hour to heat 50 kg of water from 20°C to boiling point. Calculate the power rating of the heater, assuming the heating equipment to have an efficiency of 90%.*

Solution. Heat received by water (*i.e.* output energy)

$$= \text{mass} \times \text{specific heat} \times \text{rise in temperature}$$

$$= 50 \times 1 \times 80 = 4000 \text{ kcal} = 4000/860 = 4{\cdot}65 \text{ kWh}$$

Electrical energy supplied to heater (*i.e.* input energy)

$$= 4{\cdot}65/0{\cdot}9 = 5{\cdot}167 \text{ kWh}$$

$$\therefore \qquad \text{Power rating} = \frac{\text{Energy}}{\text{Time}} = \frac{5.167}{1 \text{ hour}} = \textbf{5.167 kW}$$

Example 4.23. *The cost of boiling 2 kg of water in an electric kettle is 12 paise. The kettle takes 6 minutes to boil water from an ambient temperature of 20°C. Calculate (i) the efficiency of kettle and (ii) the wattage of kettle if cost of 1 kWh is 40 paise.*

Solution. (*i*) Heat received by water (*i.e.* output energy)

$$= 2 \times 1 \times 80 = 160 \text{ kcal}$$

Electrical energy supplied (*i.e.* input energy)

$$= 12/40 \text{ kWh} = 860 \times 12/40 = 258 \text{ kcal}$$

$$\therefore \qquad \text{Kettle efficiency} = \frac{160}{258} \times 100 = \textbf{62\%}$$

(*ii*) Let W kilowatt be the power rating of the kettle.

$$\text{Input energy} = W \times \text{time in hours}$$

or $$12/40 = W \times 6/60$$

$$\therefore \qquad \text{Wattage of kettle, } W = \frac{12}{40} \times \frac{60}{6} = \textbf{3 kW}$$

Example 4.24. *How long will it take to raise the temperature of 880 gm of water from 16°C to boiling point ? The heater takes 2 amperes at 220 V and its efficiency is 90%.*

Solution. Heat received by water (*i.e.* output energy)

$$= 0{\cdot}88 \times 1 \times (100 - 16) = 73{\cdot}92 \text{ kcal} = 73{\cdot}92/860 = 0{\cdot}086 \text{ kWh}$$

Electrical energy supplied to the heater (*i.e.* input energy)

$$= 0{\cdot}086/0{\cdot}9 = 0{\cdot}096 \text{ kWh}$$

The heater is supplying a power of $220 \times 2 = 440$ watts $= 0{\cdot}44$ kW. Let t hours be the required time.

$$\text{Input energy} = \text{wattage} \times \text{time} \quad \text{or} \quad 0{\cdot}096 = 0{\cdot}44 \times t$$

$$\therefore \qquad t = 0{\cdot}096/0{\cdot}44 = 0{\cdot}218 \text{ hours} = 0{\cdot}218 \times 60 = \textbf{13·08 minutes}$$

Example 4.25. *An electric kettle is required to raise the temperature of 2 kg of water from 20°C to 100°C in 15 minutes. Calculate the resistance of the heating element if the kettle is to be used on a 240 volts supply. Assume the efficiency of the kettle to be 80%.*

Solution. Heat received by water (*i.e.* output energy)

$$= 2 \times 1 \times (100 - 20) = 160 \text{ kcal} = 160/860 = 0.186 \text{ kWh}$$

Electrical energy supplied to the kettle

$$= 0.186/0.8 = 0.232 \text{ kWh}$$

The electrical energy of 0·232 kWh is supplied in 15/60 = 0·25 hours.

∴   Power rating of kettle = 0·232/0·25 = 0·928 kW = 928 watts

Let *R* ohms be the resistance of the heating element.

∴ $V^2/R = 928$ or $R = \dfrac{240 \times 240}{928} = 62\ \Omega$

Example 4.26. *The heater element of an electric kettle has a constant resistance of 100 Ω and the applied voltage is 250 V. Calculate the time taken to raise the temperature of one litre of water from 15°C to 90°C assuming that 85% of the power input to the kettle is usefully employed. If the water equivalent of the kettle is 100g, find how long will it take to raise a second litre of water through the same temperature range immediately after the first.*

Solution. Mass of water, m = 1 litre = 1 kg ; $\theta = 90 - 15 = 75°C$; $S = 1$

Heat taken by water = $mS\theta = 1 \times 1 \times 75 = 75$ kcal

Heat taken by kettle = water equivalent of kettle $\times \theta = 0.1 \times 75 = 7.5$ kcal

Heat taken by both = 75 + 7.5 = 82.5 kcal

Now, $I = \dfrac{250}{100} = 2.5$ A ; $J = 4200$ J/kcal

Heat produced electrically = $\dfrac{I^2 R t}{J}$ kcal ... t in seconds

Heat available for heating = $0.85 \times \dfrac{I^2 R t}{J}$ kcal

or $0.85 \times \dfrac{I^2 R t}{J} = 82.5$

or $0.85 \times \dfrac{(2.5)^2 \times 100 \times t}{4200} = 82.5$

∴ $t = 652$ s = **10 min. 52 seconds**

In the second case, heat would be required to heat water only because kettle would be already hot.

∴ $\dfrac{0.85 \times (2.5)^2 \times 100 \times t}{4200} = 75$ or $t = $ **9 min. 53 seconds**

As expected, the time required for heating in the second case is less than the first case.

Example 4.27. *The heaters A and B are in parallel across the supply voltage V. Heater A produces 500 kcal in 20 minutes and B produces 1000 kcal in 10 minutes. The resistance of heater A is 10 Ω. What is the resistance of heater B ? If the same heaters are connected in series, how much heat will be produced in 5 minutes ?*

Solution.   Heat produced = $\dfrac{V^2 t}{R \times J}$ kcal

For heater *A*, 500 = $\dfrac{V^2 \times (20 \times 60)}{10 \times J}$ ...(*i*)

For heater *B*, 1000 = $\dfrac{V^2 \times (10 \times 60)}{R \times J}$ ....(*ii*)

Dividing eq. (*i*) by eq. (*ii*), we get,

$$\frac{500}{1000} = \frac{20 \times 60}{10 \times 60} \times \frac{R}{10} \qquad \therefore \quad R = \textbf{2·5 } \Omega$$

When the heaters are connected in series, the total resistance becomes $R_T = 10 + 2·5 = 12·5\ \Omega$.

∴   Heat produced in 5 minutes

$$= \frac{V^2 t}{R_T \times J} = \frac{V^2}{J} \times \frac{t}{R_T}$$

$$= \frac{5,000}{20 \times 60} \times \frac{5 \times 60}{12.5} = \textbf{100 kcal} \qquad \left[\begin{array}{l} \text{From eq. } (i) \\ \dfrac{V^2}{J} = \dfrac{5000}{20 \times 60} \end{array} \right]$$

Example 4.28. *A soldering iron is rated at 50 watts when connected to a 250 V supply. If the soldering iron takes 5 minutes to heat to a working temperature of 190°C from 20°C, find its mass, assuming it to be made of copper. Given specific heat capacity of copper is 390 J/kg°C.*

Solution. Let m kg be the mass of soldering iron.

Heat gained by the soldering iron $= mS\theta = m \times 390 \times (190 - 20) = 66,300\ m$ joules

Heat released by the heating element $=$ power $\times$ time $= (50) \times (5 \times 60) = 15,000$ joules

Assuming all the heat released by the element is absorbed by the copper *i.e.* soldering iron is 100% efficient,

$$15,000 = 66,300\ m \quad \therefore \quad m = 15,000/66,300 = \textbf{0.226 kg}$$

Example 4.29. *A cubic water tank has surface area of 6 m² and is filled to 90% capacity 6 times daily. The water is heated from 20°C to 65°C. The losses per square metre of tank surface per 1°C temperature difference are 6.3 W. Find the loading in kW and the efficiency of the tank. Assume specific heat of water = 4200 J/kg/°C and 1 kWh = 3.6 MJ.*

Solution. Rise in temp, $\theta = 65 - 20 = 45°C$; $S = 4200$ J/kg/°C. If l metres is one side of the tank, then surface area of the tank is $6l^2$.

∴ $6l^2 = 6m^2$ or $l = 1m$

Volume of tank $= l^3 = (1)^3 = 1m^3$

Volume of water to be heated daily $= 6 \times 0.9 = 5.4\ m^3$. As the mass of 1 m³ of water is 1000 kg,

∴   Mass of water to be heated daily, $m = 5.4 \times 1000 = 5400$ kg

Heat required to heat water to the desired temperature is

$$H_1 = mS\theta = 5400 \times 4200 \times 45 = 1020.6 \times 10^6\ J$$

$$= \frac{1020.6 \times 10^6}{36 \times 10^5}\ kWh = 283.5\ kWh$$

Heat losses, $H_2 = \dfrac{6.3 \times 6 \times \theta \times 24}{1000}$ kWh

$$= \frac{6.3 \times 6 \times 45 \times 24}{1000} = 40.82\ kWh$$

Total energy supplied, $H = H_1 + H_2 = 283.5 + 40.82 = 324.32$ kWh

Loading in kW $= \dfrac{H}{24\ hr} = \dfrac{324.32\ kWh}{24\ hr} = \textbf{13.5 kW}$

Efficiency of tank $= \dfrac{H_1}{H} \times 100 = \dfrac{283.5}{324.32} \times 100 = \textbf{87.4\%}$

Example 4.30. *An electric furnace is being used to melt 10 kg of aluminium. The initial temperature of aluminium is 20°C. Assume the melting point of aluminium to be 660°C, its specific heat capacity to be 950 J/kg°C and its specific latent heat of fusion to be 387000 J/kg. Calculate the power required to accomplish the conversion in 20 minutes, assuming the efficiency of conversion to be 75%. What is the cost of energy consumed if tariff is 50 paise per kWh ?*

Solution. Heat used to melt aluminium (*i.e.* output energy)

$$= 10 \times 950 \times (660 - 20) + 10 \times 387000 = 995 \times 10^4 \text{ joules}$$

$$= \frac{995 \times 10^4}{36 \times 10^5} = 2 \cdot 76 \text{ kWh}$$

Electrical energy supplied to the heating element

$$= 2 \cdot 76 / 0 \cdot 75 = 3 \cdot 68 \text{ kWh}$$

This much energy (*i.e.* 3·68 kWh) is to be supplied in 20/60 = 1/3 hour.

$$\therefore \qquad \text{Power required} = \frac{3.68}{1/3} = 3 \cdot 68 \times 3 = \textbf{11·04 kW}$$

$$\text{Cost of energy} = \text{Rs. } 0 \cdot 5 \times 3 \cdot 68 = \textbf{Rs. 1·84}$$

Example 4.31. *A transmitting valve is cooled by water circulating through its hollow electrodes. The water enters the valve at 25°C and leaves it at 85°C. Calculate the rate of flow in kg/second needed per kW of cooling. The temperature of 1 kg of water is raised to 1°C by 4178 joules.*

Solution. Heat to be taken away/sec = 1 kW × 1 sec = 1000 × 1 = 1000 joules. Let the required flow of water be *m* kg per second.

$$\text{Heat produced/sec} = \text{mass} \times \text{Sp. heat} \times \text{rise in temp.}$$

$$= m \times 4178 \times (85 - 25) = 250,680 \, m \text{ joules}$$

$$\therefore \qquad 250,680 \, m = 1000 \quad \text{or} \quad m = \frac{1000}{250,680} = \textbf{0·004 kg/sec}$$

Tutorial Problems

1. An electric kettle marked 1 kW, 230 V, takes 7·5 minutes to bring 1 kg of water at 15°C to boiling point (100°C). Find the efficiency of the kettle. **[79·07%]**

2. An electric kettle contains 1·5 kg of water at 15°C. It takes 2·5 hours to raise the temperature to 90°C. Assuming the heat losses due to radiation and heating the kettle to be 15 kcal, find (*i*) wattage of the kettle and (*ii*) current taken if supply voltage is 230 V. **[(*i*) 59·2 W (*ii*) 0·257 A]**

3. A soldering iron is rated at 50 watts when connected to a 250 *V* supply. If the soldering iron takes 5 minutes to heat to a working temperature of 190°C from 20°C, find its mass, assuming it to be made of copper. Given specific heat capacity of copper is 390 J/kg°C. **[0·226 kg]**

4. Find the amount of electrical energy expended in raising the temperature of 45 litres of water by 75°C. To what height could a weight of 5 tonnes be raised with the expenditure of the same energy ? Assume efficiencies of heating and lifting equipment to be 90% and 70% respectively **[4·36 kWh, 224 m]**

5. Calculate the time taken for a 25 kW furnace, having an overall efficiency of 80% to melt 20 kg of aluminium. Take the specific heat capacity, melting point and latent heat of fusion of aluminium as 896 J/kg°C, 657°C and 402 kJ/kg respectively. **[16 min 13 sec]**

6. An electric boiler has two heating elements each of 230 V, 3·5 kW rating and containing 8 litres of water at 30°C. Assuming 10% loss of heat from the boiler, find how long after switching on the heater circuit will the water boil at atmospheric pressure
   (*i*) if the two elements are in parallel
   (*ii*) if the two elements are in series ? The supply voltage is 230 V. **[(*i*) 373·3 s (*ii*) 1493·2 s]**

7. A coil of resistance 100 Ω is immersed in a vessel containing 0·5 kg of water at 16°C and is connected to a 220 V electric supply. Calculate the time required to boil away all the water. Given *J* = 4200 J/kcal; latent heat of steam = 536 kcal/kg. **[44 min 50 sec]**

Objective Questions

1. A 25W, 220 V bulb and a 100 W, 220 V bulb are joined in parallel and connected to 220 V supply. Which bulb will glow more brightly ?
   - (i) 25 W bulb
   - (ii) 100 W bulb
   - (iii) both will glow with same brightness
   - (iv) neither bulb will glow

2. A 25 W, 220 V bulb and a 100 W, 220 V bulb are joined in series and connected to 220 V supply. Which bulb will glow brighter ?
   - (i) 25 W bulb
   - (ii) 100 W bulb
   - (iii) both will glow with same brightness
   - (iv) neither bulb will glow

3. You are given three bulbs of 25 W, 40 W and 60 W. Which of them has the lowest resistance ?
   - (i) 25 W bulb
   - (ii) 40 W bulb
   - (iii) 60 W bulb
   - (iv) information incomplete

4. You have the following electric appliances :
   - (a) 1 kW, 250 V electric heater
   - (b) 1 kW, 250 V electric kettle
   - (c) 1 kW, 250 V electric bulb

   Which of these has the highest resistance ?
   - (i) heater
   - (ii) kettle
   - (iii) all have equal resistances
   - (iv) electric bulb

5. The time required for 1 kW electric heater to raise the temperature of 10 litres of water through 10°C is
   - (i) 210 sec
   - (ii) 420 sec
   - (iii) 42 sec
   - (iv) 840 sec

6. Two electric bulbs rated at P_1 watt, V volt and P_2 watt, V volt are connected in series across V volt. The total power consumed is
   - (i) $P_1 + P_2$
   - (ii) $\sqrt{P_1 P_2}$
   - (iii) $\dfrac{P_1 + P_2}{2}$
   - (iv) $\dfrac{P_1 P_2}{P_1 + P_2}$

7. A tap supplies water at 22°C. A man takes 1 litre of water per minute at 37°C from the geyser. The power of geyser is
   - (i) 1050 W
   - (ii) 1575 W
   - (iii) 525 W
   - (iv) 2100 W

8. A 3°C rise in temperature is observed in a conductor by passing a certain amount of current. When the current is doubled, the rise in temperature is
   - (i) 15°C
   - (ii) 12°C
   - (iii) 9°C
   - (iv) 3°C

9. How much electrical energy in kWh is consumed in operating ten 50 W bulbs for 10 hours in a day in a month of 30 days ?
   - (i) 500
   - (ii) 15000
   - (iii) 150
   - (iv) 15

10. Two heater wires of equal length are first connected in series and then in parallel. The ratio of heat produced in the two cases will be
    - (i) 2 : 1
    - (ii) 1 : 2
    - (iii) 4 : 1
    - (iv) 1 : 4

11. Two identical heaters each marked 1000 W, 250 V are placed in series and connected to 250 V supply. Their combined rate of heating is
    - (i) 500 W
    - (ii) 2000 W
    - (iii) 1000 W
    - (iv) 250 W

12. A constant voltage is applied between the ends of a uniform metallic wire. Some heat is developed in it. If both length and radius of the wire are halved, the heat developed during the same duration will become
    - (i) half
    - (ii) twice
    - (iii) one fourth
    - (iv) same

13. What is immaterial for a fuse ?
    - (i) its specific resistance
    - (ii) its radius
    - (iii) its length
    - (iv) current flowing through it

14. If the current in an electric bulb drops by 2%, then power decreases by
    - (i) 1%
    - (ii) 2%
    - (iii) 4 %
    - (iv) 16%

15. The fuse wire is made of
    - (i) tin-lead alloy
    - (ii) copper
    - (iii) tungsten
    - (iv) nichrome

Answers

1. *(ii)*	2. *(i)*	3. *(iii)*	4. *(iii)*	5. *(ii)*
6. *(iv)*	7. *(i)*	8. *(ii)*	9. *(iii)*	10. *(iv)*
11. *(i)*	12. *(i)*	13. *(iii)*	14. *(iii)*	15. *(i)*

5

Electrostatics

Introduction

So far we have discussed that if two oppositely charged bodies are connected through a conductor, electrons will flow from the negative charge (excess of electrons) to the positive charge (deficiency of electrons). This directed flow of electrons is called electric current. The electric current will continue to flow so long as the 'excess' and 'deficiency' of electrons exist in the bodies. In other words, electric current will continue to flow so long as we maintain the potential difference between the bodies. The branch of engineering which deals with the flow of electrons (*i.e.* electric current) is called **current electricity** and is important in many ways. For example, it is the electric current by means of which electrical energy can be transferred from one point to another for utilisation.

There can be another situation where charges (*i.e.* electrons) do not move but remain static or stationary on the bodies. Such a situation will arise when the charged bodies are separated by some insulating medium, disallowing the movement of electrons. This is called **static electricity** and the branch of engineering which deals with static electricity is called **electrostatics**. Although current electricity is of greater practical use, yet the importance of static electricity cannot be ignored. Many of the advancements made in the field of electricity owe their developments to the knowledge scientists obtained from electrostatics. *The most useful outcomes of static electricity are the development of* **lightning rod** *and the* **capacitor.** In this chapter, we shall confine our attention to the behaviour and applications of static electricity.

5.1. Electrostatics

The branch of engineering which deals with charges at rest is called **electrostatics**.

When a glass rod is rubbed with silk and then separated, the former becomes positively charged and the latter attains equal negative charge. It is because during rubbing, some electrons are transferred from glass to silk. Since glass rod and silk are separated by an insulating medium (*i.e.*, air), they retain the charges. In other words, the charges on them are static or stationary. Note that the word 'electrostatic' means electricity at rest.

5.2. Importance of Electrostatics

During the past century, there was considerable increase in the practical importance of electrostatics. A few important applications of electrostatics are given below :

(*i*) Electrostatic generators can produce voltages as high as 10^6 volts. Such high voltages are required for X-ray work and nuclear bombardment.

(*ii*) We use principles of electrostatics for spray of paints, powder, etc.

(*iii*) The principles of electrostatics are used to prevent pollution.

(*iv*) The problems of preventing sparks and breakdown of insulators in high voltage engineering are essentially electrostatic.

(*v*) *The development of lightning rod and capacitor are the outcomes of electrostatics.*

5.3. Methods of Charging a Conductor

An uncharged conductor can be charged by the following two methods :

(*i*) By conduction   (*ii*) By induction

(*i*) **By conduction.** *In this method, a charged body is brought in contact with the uncharged conductor.* Fig. 5.1 (*i*) shows the uncharged conductor *B* kept on an insulating stand. When the positively charged conductor *A* provided with insulating handle is touched with uncharged conductor *B* [See Fig. 5.1 (*ii*)], free electrons from conductor *B* move to conductor *A*. As a result, there occurs a deficit of electrons in conductor *B* and it becomes positively charged. Similarly, if the conductor *A* is negatively charged, the conductor *B* will also get negatively charged.

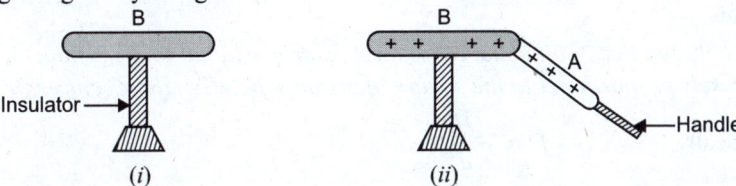

Fig. 5.1

It may be noted that conductor *A* is provided with an insulting handle so that its charge does not escape to the ground through our body. For the same reason, the conductor *B* is kept on the insulating stand.

(*ii*) **By Induction.** *In this method, a charged body is brought close to the uncharged conductor but does not touch it.* Fig. 5.2 (*i*) shows a negatively charged plastic rod (provided with insulating handle) kept near an uncharged metal sphere. The free electrons of the sphere near the rod are repelled to the farther end. As a result, the region of the sphere near the rod becomes positively charged and the farthest end of sphere becomes equally negatively charged. If now the sphere is connected to the ground through a wire as shown in Fig. 5.2 (*ii*), its free electrons at the farther end flow to the ground. On removing the wire to the ground [See Fig. 5.2 (*iii*)], the positive charge at the near end of sphere remains held there due to the attractive force of external negative charge. Finally, when the plastic rod is removed [See Fig. 5.2 (*iv*)], the positive charge spreads uniformly on the sphere. Thus, the sphere is positively charged by induction. Note that in the process, the negatively charged plastic rod loses none of its negative charge. Similarly, the metal sphere can be negatively charged by bringing a positively charged rod near it.

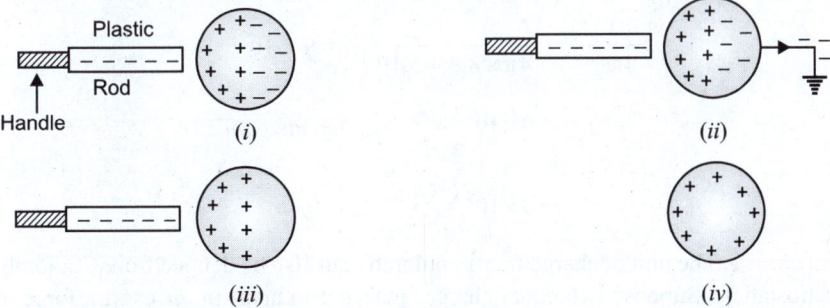

Fig. 5.2

Note that charging a body by induction requires no contact with the body inducing the charge. This is in contrast to charging a body by conduction which does require contact between the two bodies.

5.4. Coulomb's Laws of Electrostatics

Charles Coulomb, a French scientist, observed that when two charges are placed near each other, they experience a force. He performed a number of experiments to study the nature and magnitude of the force between the charged bodies. He summed up his conclusions into two laws, known as Coulomb's laws of electrostatics.

First law. This law relates to the nature of force between two charged bodies and may be stated as under :

Like charges repel each other while unlike charges attract each other.

In other words, if two charges are of the same nature (*i.e.* both positive or both negative), the force between them is repulsion. On the other hand, if one charge is positive and the other negative, the force between them is an attraction.

Second law. This law tells about the magnitude of force between two charged bodies and may be stated as under :

The force between two *point charges is directly proportional to the product of their magnitudes and inversely proportional to the square of distance between their centres.

Mathematically, $\qquad F \propto \dfrac{Q_1 Q_2}{d^2}$

or $\qquad F = k\dfrac{Q_1 Q_2}{d^2}$ ...(*i*)

where k is a constant whose value depends upon the medium in which the charges are placed and the system of units employed. In SI units, force is measured in newtons, charge in coulombs, distance in metres and the value of k is given by ;

$$k = \dfrac{1}{4\pi\varepsilon_0 \varepsilon_r}$$

Fig. 5.3

where $\qquad \varepsilon_0 =$ Absolute permittivity of vacuum or air.

$\varepsilon_r =$ Relative permittivity of the medium in which the charges are placed. For vacuum or air, its value is 1.

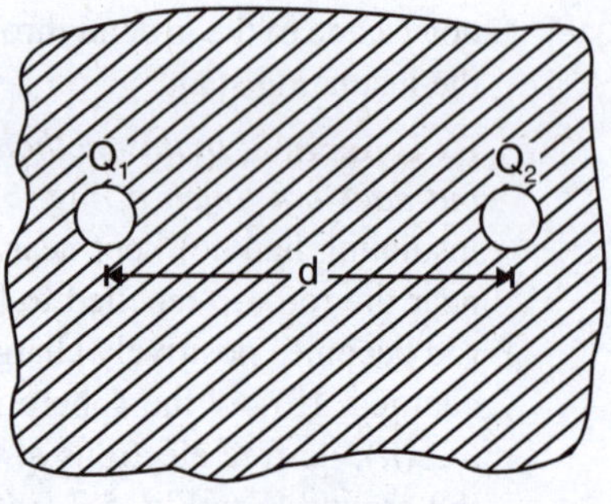

The value of $\varepsilon_0 = 8\cdot854 \times 10^{-12}$ F/m and the value of ε_r is different for different media.

∴ $\qquad F = \dfrac{Q_1 Q_2}{4\pi\varepsilon_0 \varepsilon_r d^2}$ ...(*ii*)

Now $\qquad \dfrac{1}{4\pi\varepsilon_0} = \dfrac{1}{4\pi \times 8.854 \times 10^{-12}} = 9 \times 10^9$

∴ $\qquad F = 9 \times 10^9 \dfrac{Q_1 Q_2}{\varepsilon_r d^2}$ ...*in a medium*

$\qquad\qquad = 9 \times 10^9 \dfrac{Q_1 Q_2}{d^2}$ ...*in air*

Unit of charge. The unit of charge (*i.e.* 1 coulomb) can also be defined from Coulomb's second law of electrostatics. Suppose two equal charges placed 1 m apart in *air* exert a force of 9×10^9 newtons *i.e.*

$$Q_1 = Q_2 = Q \ ; \ d = 1m \ ; \ F = 9 \times 10^9 \text{ N}$$

∴ $\qquad F = 9 \times 10^9 \dfrac{Q_1 Q_2}{d^2}$

or $\qquad 9 \times 10^9 = 9 \times 10^9 \dfrac{Q^2}{(1)^2}$

*   Charged bodies approximate to point charges if they are small compared to the distance between them.

or $\quad\quad\quad\quad\quad\quad\quad\quad Q^2 = 1$

or $\quad\quad\quad\quad\quad\quad\quad\quad Q = \pm 1 = 1\,\text{coulomb}$

Hence **one coulomb** *is that charge which when placed in air at a distance of one metre from an equal and similar charge repels it with a force of* 9×10^9 *N.*

Note that coulomb is very large unit of charge in the study of electrostatics. In practice, charges produced experimentally range between pico-coulomb (pC) and micro-coulomb (μC).

$$1\text{pC} = 10^{-12}\text{C} \; ; \quad 1\mu\text{C} = 10^{-6}\text{C}$$

Note. One disadvantage of SI units is that coulomb is an inconveniently large unit. This is clear from the fact that the force exerted by a charge of 1C on another equal charge at a distance of 1m is 9×10^9N. Could you hold two one-coulomb charges a metre apart ?

5.5. Absolute and Relative Permittivity

Permittivity is the property of a medium and affects the magnitude of force between two point charges. The greater the permittivity of a medium, the lesser the force between the charged bodies placed in it and *vice-versa*. Air or vacuum has a minimum value of permittivity. The absolute (or actual) permittivity ε_0 (Greek letter 'epsilon') of air or vacuum is $8 \cdot 854 \times 10^{-12}$ F/m. The absolute (or actual) permittivity ε of all other insulating materials is greater than ε_0. The ratio $\varepsilon/\varepsilon_0$ is called the *relative permittivity of the material and is denoted by ε_r i.e.

$$\varepsilon_r = \frac{\varepsilon}{\varepsilon_0}$$

where $\quad\quad\quad\quad \varepsilon =$ absolute (or actual) permittivity of the material

$\quad\quad\quad\quad\quad\quad \varepsilon_0 =$ absolute (actual) permittivity of air or vacuum ($8 \cdot 854 \times 10^{-12}$ F/m)

$\quad\quad\quad\quad\quad\quad \varepsilon_r =$ relative permittivity of the material.

Obviously, ε_r for air would be $\varepsilon_0/\varepsilon_0 = 1$.

Permittivity of a medium plays an important role in electrostatics. For instance, the relative permittivity of insulating oil is 3. It means that for the same charges (Q_1 and Q_2) and distance (d), the force between the two charges in insulating oil will be one-third of that in air [See eq. (*ii*) in Art.5·4].

5.6. Coulomb's Law in Vector Form

Consider two like point charges Q_1 and Q_2 separated by distance d in vacuum. Clearly, charges will repel each other [See Fig. 5.4].

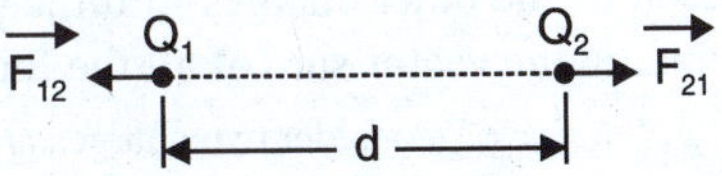

Let $\quad\quad\quad\quad \vec{F}_{21} =$ force on Q_2 due to Q_1

$\quad\quad\quad\quad\quad \vec{F}_{12} =$ force on Q_1 due to Q_2

$\quad\quad\quad\quad\quad \hat{d}_{12} =$ unit vector pointing from Q_1 to Q_2

$\quad\quad\quad\quad\quad \hat{d}_{21} =$ unit vector pointing from Q_2 to Q_1

Fig. 5.4

According to Coulomb's law,

$$\vec{F}_{21} = k\frac{Q_1 Q_2}{d^2}\hat{d}_{12}$$

or $\quad\quad\quad\quad \vec{F}_{21} = \frac{1}{4\pi\varepsilon_0}\frac{Q_1 Q_2}{d^2}\hat{d}_{12} \quad\quad\quad\quad ...(i)$

Similarly, $\quad\quad\quad \vec{F}_{12} = \frac{1}{4\pi\varepsilon_0}\frac{Q_1 Q_2}{d^2}\hat{d}_{21} \quad\quad\quad\quad ...(ii)$

Eqs. (*i*) and (*ii*) express Coulomb's law in vector form.

---

* Thus when we say that relative permittivity of a material is 10, it means that its absolute or acutal permittivity $\varepsilon = \varepsilon_0\,\varepsilon_r = 8.854 \times 10^{-12} \times 10 = 8.854 \times 10^{-11}$ F/m.

Importance of vector form. The reader may wonder about the utility of Coulomb's law in vector form over the scalar form. The answer will be readily available from the following discussion :

(*i*) The vector form shows at a glance that forces $\vec{F}_{21}$ and $\vec{F}_{12}$ are equal and opposite.

$$\vec{F}_{21} = \frac{1}{4\pi\varepsilon_0}\frac{Q_1Q_2}{d^2}\hat{d}_{12}$$

$$\vec{F}_{12} = \frac{1}{4\pi\varepsilon_0}\frac{Q_1Q_2}{d^2}\hat{d}_{21}$$

As $\quad\quad\quad\quad \hat{d}_{12} = -\hat{d}_{21}$

$\therefore \quad\quad\quad\quad \vec{F}_{21} = -\vec{F}_{12}$

That is $\vec{F}_{21}$ is equal in magnitude to $\vec{F}_{12}$ but opposite in direction. The scalar form does not show this fact. This is a distinct advantage over the scalar form.

(*ii*) $\quad\quad\quad\quad \vec{F}_{21} = -\vec{F}_{12}$

This means that $\vec{F}_{21}$ and $\vec{F}_{12}$ act along the same line *i.e.* along the line joining charges Q_1 and Q_2. In other words, the electrostatic force between two charges is a central force *i.e.* it acts along the line joining the centres of the two charges. However, scalar form does not show such a nature of electrostatic force between two charges.

5.7. The Superposition Principle

If we are given two charges, the electrostatic force between them can be found by using Coulomb's laws. However, if a number of charges are present, the force on any charge due to the other charges can be found by superposition principle stated below :

When a number of charges are present, the total force on a given charge is equal to the vector sum of the forces due to the remaining other charges on the given charge.

This simply means that we first find the force on the given charge (by Coulomb's laws) due to each of the other charges in turn. We then determine the total or net force on the given charge by finding the vector sum of all the forces.

Notes. (*i*) Consider two charges Q_1 and Q_2 located in air. If a third charge Q_3 is brought nearby, it has been found experimentally that presence of the third charge (Q_3) has no effect on the force between Q_1 and Q_2. This fact permits us to use superposition principle for electric forces.

(*ii*) The superposition principle holds good for electric forces and electric fields. This fact has made the mathematical description of electrostatic phenomena simpler than it otherwise would be.

(*iii*) We can use superposition principle to find (*a*) net force (*b*) net field (*c*) net flux (*d*) net potential and (*e*) net potential energy due to a number of charges.

Example 5.1. *A small sphere is given a charge of $+ 20\mu C$ and a second sphere of equal diameter is given a charge of $-5\ \mu C$. The two spheres are allowed to touch each other and are then spaced 10 cm apart. What force exists between them ? Assume air as the medium.*

Solution. When the two spheres touch each other, the resultant charge = $(20) + (-5) = 15\ \mu C$. When the spheres are separated, charge on each sphere, $Q_1 = Q_2 = 15/2 = 7\cdot5\ \mu C$.

$\therefore \quad\quad$ Force, $F = 9 \times 10^9 \times \dfrac{Q_1Q_2}{d^2}$

$$= 9 \times 10^9 \times \frac{(7.5 \times 10^{-6})\ (7.5 \times 10^{-6})}{(0.1)^2} = \textbf{50.62 N } \textbf{\textit{repulsive}}$$

Example 5.2. *A charge q is divided into two parts in such a way that they repel each other with a maximum force when held at a certain distance apart. Find the distribution of the charge.*

Solution. Let the two parts be q' and $(q - q')$. Therefore, force F between them is

$$F = \frac{1}{4\pi\varepsilon_0} \frac{q'(q - q')}{d^2} = \frac{1}{4\pi\varepsilon_0} \frac{qq' - q'^2}{d^2}$$

For maximum value of F, $\frac{dF}{dq'} = 0$ $\therefore$ $\frac{dF}{dq'} = \frac{1}{4\pi\varepsilon_0 d^2}(q - 2q') = 0$

or $q - 2q' = 0$ $\therefore$ $q' = \dfrac{q}{2}$

Hence in order to have maximum force, q should be divided into two equal parts.

Example 5.3. *Three point charges of $+ 5\mu C$, $+ 5\mu C$ and $+ 5\mu C$ are placed at the vertices of an equilateral triangle which has sides 10 cm long. Find the force on each charge.*

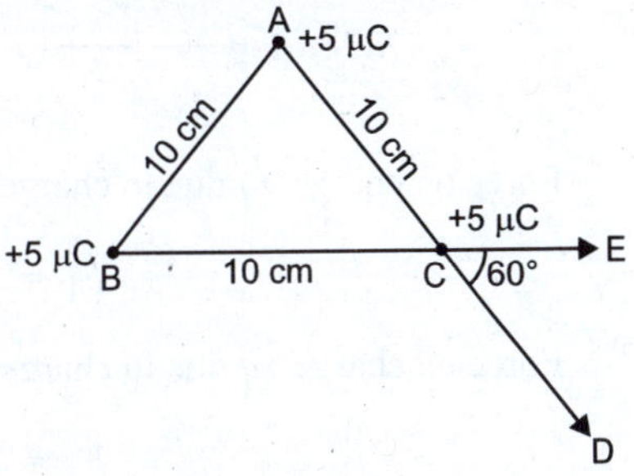

Solution. The conditions of the problem are represented in Fig. 5.5. Consider $+ 5\mu C$ placed at the corner C. It is being repelled by the charges at A and B along ACD and BCE respectively. These two forces are equal, each being given by ;

$$F = 9\times10^9 \frac{(5\times10^{-6})(5\times10^{-6})}{(0.1)^2} = 22.5 \text{ N}$$

Fig. 5.5

Resultant force at $C = 2F \cos 30° = 2 \times 22.5 \times \dfrac{\sqrt{3}}{2} = \textbf{38.97 N}$

The forces acting on the charges placed at A and B will also be the same (*i.e.*, 38.97 N)

Example 5.4. *Two small spheres, each having a mass of 0·1g are suspended from a point by threads 20 cm long. They are equally charged and they repel each other to a distance of 24cm. What is the charge on each sphere ?*

Solution. Fig. 5.6 shows the conditions of the problem. Let B and C be the spheres, each carrying a charge q. The force of repulsion between the spheres is given by ;

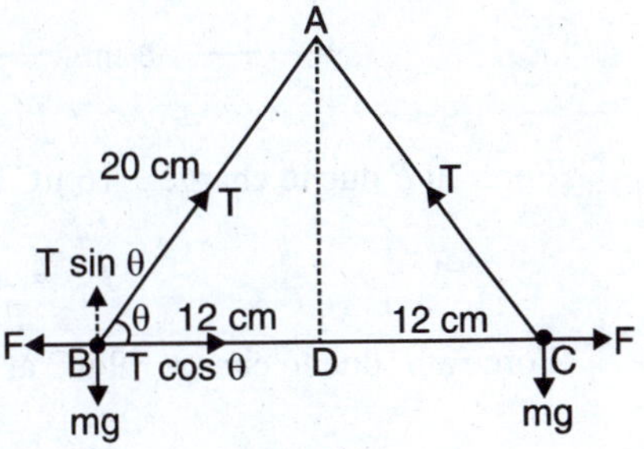

$$F = 9\times10^9 \frac{q^2}{(0.24)^2}$$
$$= 156.25 \times 10^9 \, q^2$$

Each sphere is under the action of three forces :

(*i*) weight $m\,g$ acting vertically downward, (*ii*) tension T, and (*iii*) electrostatic force F. Considering the sphere B and resolving T into rectangular components, we have,

$$m\,g = T\sin\theta \; ; \; F = T\cos\theta$$

$\therefore$ $\tan\theta = mg/F$

Now, $AD = \sqrt{AB^2 - BD^2} = \sqrt{(20)^2 - (12)^2} = 16$ cm

Fig. 5.6

$\therefore$ $\tan\theta = \dfrac{AD}{BD} = \dfrac{16}{12}$ $\therefore$ $\dfrac{16}{12} = \dfrac{mg}{F}$

or $\qquad F = \dfrac{12}{16} mg = 0.75\, mg = 0.75 \times 10^{-4} \times 9.8 = 7.4 \times 10^{-4}\,\text{N}$

But $\qquad F = 156 \cdot 25 \times 10^9\, q^2$

$\therefore \qquad 156 \cdot 25 \times 10^9\, q^2 = 7 \cdot 4 \times 10^{-4} \quad$ or $\quad q^2 = \dfrac{7.4 \times 10^{-4}}{156.25 \times 10^9} = 4.8 \times 10^{-15}$

$\therefore \qquad\qquad q = 6 \cdot 9 \times 10^{-8}\,\text{C}$

Example 5.5. *Two point charges* $+Q$ *and* $+4Q$ *are placed at a distance 'a' apart on a horizontal plane. Where should the third charge be placed for it to be in equilibrium ?*

Solution. Let the point charge $+q$ be placed at a distance x from the charge $+4Q$ [See Fig. 5.7].

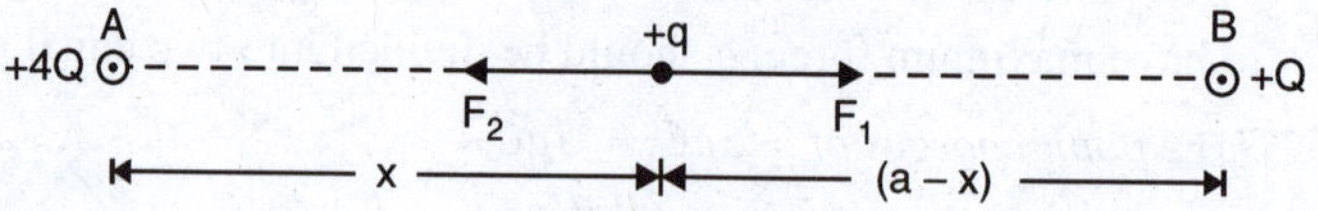

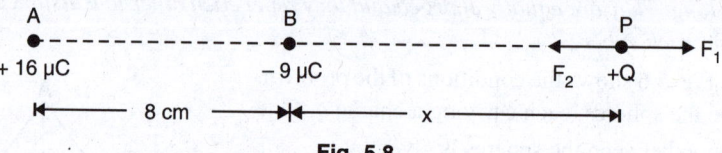

Fig. 5.7

Force on charge $+q$ due to charge $+4Q$ is

$$F_1 = \dfrac{q(4Q)}{4\pi\varepsilon_0 x^2} \quad \text{from } A \text{ to } B$$

Force on charge $+q$ due to charge $+Q$ is

$$F_2 = \dfrac{q(Q)}{4\pi\varepsilon_0 (a - x)^2} \quad \text{from } B \text{ to } A$$

In order that charge $+q$ is in equilibrium, $F_1 = F_2$.

$\therefore \qquad \dfrac{q(4Q)}{4\pi\varepsilon_0 x^2} = \dfrac{q(Q)}{4\pi\varepsilon_0 (a - x)^2} \quad$ or $\quad x = \boldsymbol{2a/3}$

Example 5.6. *Two point charges of* $+16\,\mu C$ *and* $-9\,\mu C$ *are 8 cm apart in air. Where can a third charge be located so that no net electrostatic force acts on it ?*

Solution. Let the third charge $+Q$ be located at P at a distance x from the charge $-9\mu C$ as shown in Fig. 5.8.

Fig. 5.8

Force at P due to charge $+16\,\mu C$ at A is

$$F_1 = k\dfrac{16 \times 10^{-6} \times Q}{(x + 0.08)^2} \quad \text{along } AP$$

Force at P due to charge $-9\,\mu C$ at B is

$$F_2 = k\dfrac{9 \times 10^{-6} \times Q}{x^2} \quad \text{along } PB$$

For zero electrostatic force at P, $F_1 = F_2$.

$\therefore \qquad k\dfrac{16 \times 10^{-6} \times Q}{(x + 0.08)^2} = k\dfrac{9 \times 10^{-6} \times Q}{x^2}$

or $\qquad \dfrac{16}{(x + 0.08)^2} = \dfrac{9}{x^2} \quad$ or $\quad \dfrac{4}{x + 0.08} = \dfrac{3}{x}$

$\therefore \qquad\qquad x = 0.24\,\text{m} = \boldsymbol{24\ cm}$

Example 5.7. *Two small balls are having equal charge Q (coulomb). The balls are suspended by two insulating strings of equal length L (metre) from a hook fixed to a stand. The whole set up is taken in a satellite into space where there is no gravity.*

(i) *What is the angle between the two strings ?*

(ii) *What is the tension in each string ?*

Solution. (*i*) In the absence of gravity, the tension in the strings is only due to Coulomb's repulsive force. Therefore, the strings become horizontal due to the electric force between the charges. Consequently, the angle between the strings is **180°**.

(ii) $$F = 9 \times 10^9 \times \frac{Q_1 Q_2}{d^2}$$

Here $$Q_1 = Q_2 = Q ; \quad d = 2L$$

∴ $$F = 9 \times 10^9 \frac{Q^2}{4L^2}$$

Example 5.8. *Two identical charged spheres are suspended by strings of equal length. The strings make an angle of 30° with each other. When suspended in a liquid of density 800 kg m^{-3}, the angle remains the same. What is the dielectric constant of the liquid ? The density of the material of the sphere is 1600 kg m^{-3}.*

Solution. Fig. 5.9 shows the conditions of the problem. Suppose the mass of each sphere is m kg, the charge on each q coulomb and in equilibrium, the distance between them is r. Each sphere is in equilibrium under the action of three forces as shown. Considering the sphere A,

$$F = \frac{1}{4\pi\varepsilon_0} \times \frac{q^2}{r^2}$$

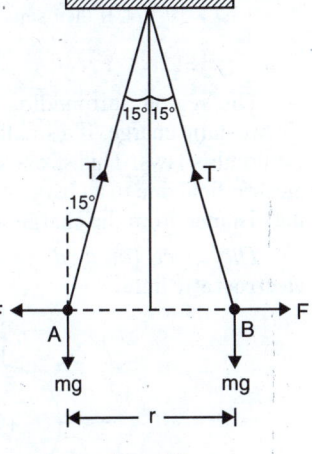

Now $$T \cos 15° = mg ; T \sin 15° = F = \frac{1}{4\pi\varepsilon_0} \frac{q^2}{r^2}$$

∴ $$\tan 15° = \frac{1}{4\pi\varepsilon_0} \frac{q^2}{mg\, r^2} \qquad \qquad ...(i)$$

When the spheres are immersed in the liquid, the effective weight of each sphere and the force of repulsion both decrease. Consequently, tension also decreases.

Weight of sphere in liquid $$= mg * \left(1 - \frac{800}{1600}\right) = \frac{mg}{2}$$

Electric force in liquid, $$F' = \frac{1}{4\pi\varepsilon_0 K} \times \frac{q^2}{r^2}$$

Here K is the dielectric constant of the liquid. If the reduced tension is T', then for the equilibrium of sphere A, we have,

$$T' \cos 15° = \frac{mg}{2} \quad \text{and} \quad T' \sin 15° = \frac{1}{4\pi\varepsilon_0 K} \times \frac{q^2}{r^2}$$

∴ $$\tan 15° = \frac{1}{4\pi\varepsilon_0 K} \frac{2q^2}{mg\, r^2} \qquad \qquad ...(ii)$$

From eqs. (*i*) and (*ii*), we have,

Fig. 5.9

---

* Weight of sphere in liquid, W' = Weight in air – Weight of liquid displaced.

Now, Weight in air = mg

Also, weight of liquid displaced $$= m\left(\frac{\sigma}{\rho}\right)g = mg\left(\frac{\sigma}{\rho}\right) = mg\left(\frac{800}{1600}\right)$$

∴ $$W' = mg - mg\left(\frac{800}{1600}\right) = mg\left(1 - \frac{800}{1600}\right) = \frac{mg}{2}$$

$$\frac{1}{4\pi\varepsilon_0} \frac{q^2}{mg\,r^2} = \frac{1}{4\pi\varepsilon_0 K} \frac{2q^2}{mg\,r^2} \qquad \therefore K = 2$$

Tutorial Problems

1. Two copper spheres A and B have their centres separated by 50 cm. If charge on each sphere is $6 \cdot 5 \times 10^{-7}$ C, what is the mutual force of repulsion between them ? The radii of the spheres are negligible compared to the distance of separation. What will be the magnitude of force if the two spheres are placed in water ? (Dielectric constant of water = 80). **[1·52 × 10⁻² N; 1·9 × 10⁻⁴ N]**

2. Charges q_1 and q_2 lie on the x-axis at points $x = -4$ cm and $x = +4$ cm respectively. How must q_1 and q_2 be related so that net electrostatic force on a charge placed at $x = +2$ cm is zero ? **[$q_1 = 9q_2$]**

3. Two small spheres of equal size are 10 cm apart in air and carry charges $+1$ μC and -3 μC. Where should a third charge be located so that no net electrostatic force acts on it ? **[24 cm from −3 μC]**

4. Two identical spheres, having unequal and opposite charges are placed at a distance of 90 cm apart. After touching them mutually, they are again separated by same distance. Now they repel each other with a force of 0·025N. Find the final charge on each of them. **[1·5 μC on each]**

5. Two small spheres, each of mass 0·05 g are suspended by silk threads from the same point. When given equal charges, they separate the threads making an angle of 10° with each other. What is the force of repulsion acting on each sphere ? **[4·3 × 10⁻⁵ N]**

6. Point charges of 2×10^{-9} C lie at each of the three corners of a square of side 20cm. Find the magnitude of force on a charge of -1×10^{-9} C placed at the centre of square. **[9 × 10⁷ N]**

7. The electrostatic force of repulsion between two positively charged ions carrying equal charge is $3 \cdot 7 \times 10^{-9}$ N. If their separation is 5 Å, how many electrons are missing from each ion ? **[2]**

5.8. Electric Field

The region surrounding a charged body is always under stress and strain because of the electrostatic charge. If a small charge is placed in this region, it will experience a force according to Coulomb's laws. This stressed region around a charged body is called electric field. Theoretically, electric field due to a charge extends upto infinity but its effect practically dies away very quickly as the distance from the charge increases.

The space (or field) in which a charge experiences a force is called an **electric field** *or* **electrostatic field.**

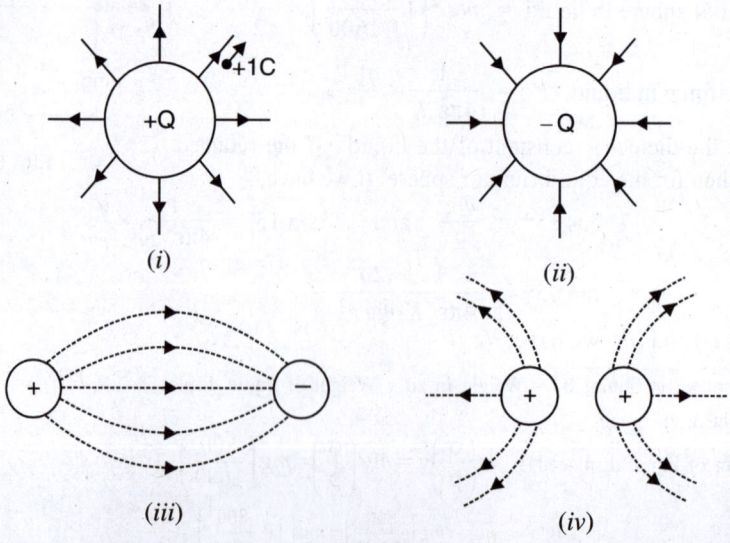

Fig. 5.10

The electric field around a charged body is represented by imaginary lines, called *electric lines of *force.* By convention, the direction of these lines of force at any point is the direction along which a unit positive charge (*i.e.,* positive charge of 1C) placed at that point would move or tend to move. The unit positive charge is sometimes called a *test charge* because it is used as an indicator to find the direction of electric field. Following this convention, it is clear that electric lines of force would always originate from a positive charge and end on a negative charge. The electric lines of force leave or enter the charged surface **normally.

Fig. 5.10 shows typical field distribution. Fig. 5.10 (*i*) shows electric field due to an isolated positively charged sphere. A unit positive charge placed near it will experience a force directed radially away from the sphere. Therefore, the direction of electric field will be radially outward as shown in Fig. 5.10 (*i*). For the negatively charged sphere [See Fig. 5.10 (*ii*)], the force acting on the unit positive charge would be directed radially towards the sphere. Fig. 5.10 (*iii*) shows the electric field between a positive charge and a negative charge while Fig. 5.10 (*iv*) shows electric field between two similarly charged (*i.e.* + vely charged) bodies.

5.9. Properties of Electric Lines of Force

(*i*) The electric field lines are directed away from a positive charge and towards a negative charge so that at any point, the tangent to a field line gives the direction of electric field at that point.

(*ii*) Electric lines of force start from a positive charge and end on a negative charge.

(*iii*) Electric lines of force leave or enter the charged surface normally.

(*iv*) Electric lines of force cannot pass through a ***conductor. This means that electric field inside a conductor is zero.

(*v*) Electric lines of force can never intersect each other. In case the two electric lines of force intersect each other at a point, then two tangents can be drawn at that point. This would mean two directions of electric field at that point which is impossible.

(*vi*) Electric lines of force have the tendency to contract in length. This explains attraction between oppositely charged bodies.

(*vii*) Electric lines of force have the tendency to expand laterally *i.e.* they tend to separate from each other in the direction perpendicular to their lengths. This explains repulsion between two like charges.

5.10. Electric Intensity or Field Strength (E)

To describe an electric field, we must specify its intensity or strength. The intensity of electric field at any point is determined by the force acting on a unit positive charge placed at that point.

Electric intensity (*or* **field strength***) at a point in an electric field is the force acting on a unit positive charge placed at that point. Its direction is the direction along which the force acts.*

Electric intensity at a point, $E = \dfrac{F}{+Q}$ N/C

where Q = Charge in coulombs placed at that point

F = Force in newtons acting on Q coulombs

---

*   So called because forces are experienced by charges in this region.

**   If a line of force is at an angle other than 90°, it will have a tangential component. This tangential component would cause redistribution (*i.e.* movement) of charge. By definition, electrostatic charge is static and hence tangential component cannot exist.

*** However, electric lines of force can pass through an insulator.

Thus, if a charge of 2 coulombs placed at a point in an electric field experiences a force of 10N, then electric intensity at that point will be $10/2 = 5$N/C. The following points may be noted carefully:

(i) Since electric intensity is a force, it is a vector quantity possessing both magnitude and direction.

(ii) Electric intensity can also be *described in terms of lines of force. Where the lines of force are close together, the intensity is high and where the lines of force are widely separated, intensity will be low.

(iii) Electric intensity can also be expressed in V/m.

$$1 \text{ V/m} = 1 \text{ N/C (See foot note on page 284)}$$

Electric intensity due to a point charge. The value of electric intensity at any point in an electric field due to a point charge can be calculated by Coulomb's laws. Suppose it is required to find the electric intensity at point P situated at a distance d metres from a charge of $+ Q$ coulomb (See Fig. 5.11). Imagine a unit positive charge (i.e. $+1$C) is placed at point P. Then, by definition, electric intensity at P is the force acting on $+1$C placed at P i.e.

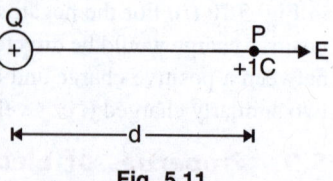

Fig. 5.11

Electric intensity at P, E = Force on $+ 1$C placed at P

$$= 9 \times 10^9 \frac{Q \times 1}{\varepsilon_r d^2}$$

$$\therefore \quad E = 9 \times 10^9 \frac{Q}{\varepsilon_r d^2} \quad \textit{...in a medium}$$

$$= 9 \times 10^9 \frac{Q}{d^2} \quad \textit{...in air}$$

Note the direction of electric intensity. It is acting radially away from $+ Q$. For a negative charge (i.e. $- Q$), its direction would have been radially towards the charge.

The electric field intensity in vector form is given as :

$$\vec{E} = 9 \times 10^9 \frac{Q}{d^2} \hat{d} \quad \textit{... in air}$$

$$= 9 \times 10^9 \frac{Q}{\varepsilon_r d^2} \hat{d} \quad \textit{... in a medium}$$

where $\hat{d}$ is a unit vector directed from $+ Q$ to $+ 1$C.

Electric field intensity due to a group of point charges. The resultant (or net) electric field intensity at a point due to a group of point charges can be found by applying **superposition principle. Thus electric field intensity at a point P due to n point charges ($q_1, q_2, q_3 \dots q_n$) is equal to the vector sum of electric field intensities due to $q_1, q_2, q_3 \dots q_n$ at point P i.e.

$$\vec{E} = \vec{E_1} + \vec{E_2} + \vec{E_3} + \dots + \vec{E_n}$$

where
$$\vec{E} = \text{Net or resultant electric field intensity at } P$$

$$\vec{E_1} = \text{Electric field intensity at } P \text{ due to } q_1$$

$$\vec{E_2} = \text{Electric field intensity at } P \text{ due to } q_2$$
$$\text{and so on.}$$

---

*   It may be noted that electric lines of force do not actually exist. It is only a way of representing an electric field. However, it is a useful method of representation. It is a usual practice to indicate high field strength by drawing lines of force close together and low field strength by widely spaced lines.

** Since the electric force obeys the superposition principle, so does the electric field intensity—the force per unit charge.

Example 5.9. *Two equal and opposite charges of magnitude* 2×10^{-7} *C are placed 15cm apart. (i) What is the magnitude and direction of electric intensity (E) at a point mid-way between the charges? (ii) What force would act on a proton (charge = + 1·6 × 10^{-19} C) placed there ?*

Solution. Fig. 5.12 shows two equal and opposite charges separated by a distance of 15cm *i.e.* 0·15 m. Let *M* be the mid-point *i.e. AM* = *MB* = 0·075 m.

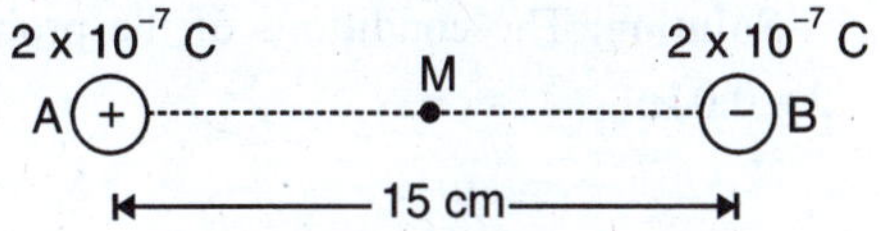

2×10^{-7} C   M 2×10^{-7} C

A (+) - - - - - - - - - - - - - • - - - - - - - - - - - - (−) B

|← —————— 15 cm—————— →|

Fig. 5.12

   (*i*)   Imagine a charge of + 1 C placed at *M*.

   ∴   Electric intensity at *M* due to charge + 2 × 10^{-7} C is

$$E_1 = 9 \times 10^9 \times \frac{2 \times 10^{-7}}{(0.075)^2} = 0.32 \times 10^6 \text{ N/C } along AM$$

Electric intensity at *M* due to charge −2 × 10^{-7}C is

$$E_2 = 9 \times 10^9 \times \frac{2 \times 10^{-7}}{(0.075)^2} = 0.32 \times 10^6 \text{ N/C } along MB$$

Since electric intensities are acting in the same direction, the resultant intensity *E* is the sum of E_1 and E_2.

   ∴   Resultant intensity at point *M* is

$$E = 0·32 \times 10^6 + 0·32 \times 10^6 = \mathbf{0·64 \times 10^6} \text{ N/C } along AB$$

   (*ii*)   Electric intensity *E* at *M* is 0·64 × 10^6 N/C. Therefore, force *F* acting on a proton (charge, $Q = +1·6 \times 10^{-19}$ C) placed at *M* is

$$F = EQ = (0·64 \times 10^6) \times (1·6 \times 10^{-19}) = \mathbf{1·024 \times 10^{-13}} \text{ N } along AB$$

Example 5.10. *A charged oil drop remains stationary when situated between two parallel plates 25mm apart. A p.d. of 1000V is applied to the plates. If the mass of the drop is 5 ×10^{-15} kg, find the charge on the drop (take g = 10ms^{-2}).*

Solution. Let *Q* coulomb be the charge on the oil drop. Since the drop is stationary,

Upward force on drop = Weight of drop [See Fig. 5.13]

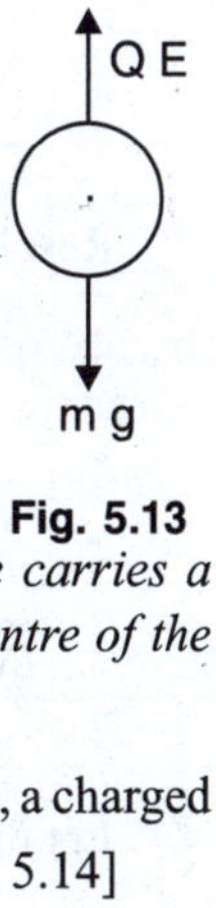

or $QE = mg$

Here $E = \dfrac{V}{d} = \dfrac{1000}{25 \times 10^{-3}} = 4 \times 10^4$ V/m

∴ $Q = \dfrac{mg}{E} = \dfrac{(5 \times 10^{-15}) \times 10}{4 \times 10^4} = \mathbf{1.25 \times 10^{-18}}$ C

↑ QE

⊙

↓ mg

Fig. 5.13

Example 5.11. *The diameter of a hollow metallic sphere is 60cm and the sphere carries a charge of 500μC. Find the electric field intensity (i) at a distance of 100cm from the centre of the sphere and (ii) at the surface of sphere.*

Solution. The electric field due to a charged sphere has spherical symmetry. Therefore, a charged sphere behaves for external points as if the whole charge is placed at its centre. [See Fig. 5.14]

   (*i*)   *d* = *OP* = 100cm = 1m ; *Q* = 500 μC = 500 × 10^{-6}C

   ∴ $E = 9 \times 10^9 \dfrac{Q}{d^2} = 9 \times 10^9 \times \dfrac{500 \times 10^{-6}}{1} = \mathbf{4.5 \times 10^6}$ N/C

   (*ii*)   *d* = *OP'* =30 cm = 0·3 m ; *Q* = 500μC = 500 × 10^{-6}C

   ∴ $E = 9 \times 10^9 \dfrac{Q}{d^2} = 9 \times 10^9 \times \dfrac{500 \times 10^{-6}}{(0.3)^2} = \mathbf{5 \times 10^7}$ N/C

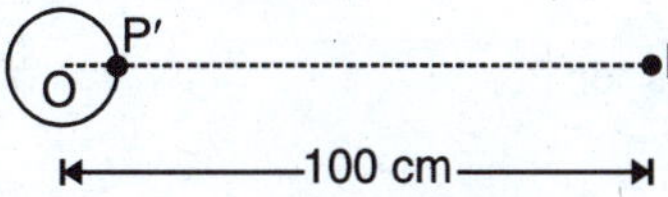

O ◯ P' - - - - - - - - - - - - - - - • P

|← —————— 100 cm —————— →|

Fig. 5.14

Example 5.12. *Three point charges of +8 × 10⁻⁹ C, +32 × 10⁻⁹ C and +24 × 10⁻⁹ C are placed at the corners A, B and C of a square ABCD having each side 4 cm. Find the electric field intensity at the corner D. Assume that the medium is air.*

Solution. The conditions of the problem are represented in Fig. 5.15. It is clear that $BD = \sqrt{2} \times 0.04\,\text{m}$.

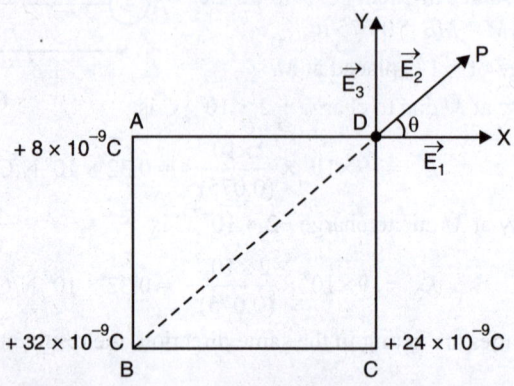

Fig. 5.15

Magnitude of electric field intensity at D due to charge $+8 \times 10^{-9}$ C is

$$E_1 = 9 \times 10^9 \times \frac{8 \times 10^{-9}}{(0.04)^2} = 4 \cdot 5 \times 10^4 \text{ N/C} \quad \text{along } DX$$

Magnitude of electric field intensity at D due to charge $+32 \times 10^{-9}$ C is

$$E_2 = 9 \times 10^9 \times \frac{32 \times 10^{-9}}{\left(\sqrt{2} \times 0.04\right)^2} = 9 \times 10^4 \text{ N/C along } DP$$

Magnitude of electric field intensity at D due to charge $+24 \times 10^{-9}$ C is

$$E_3 = 9 \times 10^9 \times \frac{24 \times 10^{-9}}{(0.04)^2} = 13 \cdot 5 \times 10^4 \text{ N/C along } DY$$

It is easy to see that $\theta = 45°$.

Resolving electric field intensities along X-axis and Y-axis, we have,

Total X-component $= E_1 + E_2 \cos \theta + 0$

$\qquad\qquad\qquad = 4 \cdot 5 \times 10^4 + 9 \times 10^4 \times \cos 45° = 10 \cdot 86 \times 10^4$ N/C

Total Y-component $= 0 + E_2 \sin 45° + E_3$

$\qquad\qquad\qquad = 0 + 9 \times 10^4 \sin 45° + 13 \cdot 5 \times 10^4 = 19 \cdot 86 \times 10^4$ N/C

∴   Magnitude of resultant electric intensity at D

$$= \sqrt{(10.86 \times 10^4)^2 + (19.86 \times 10^4)^2} = \mathbf{22 \cdot 63 \times 10^4 \text{ N/C}}$$

Let the resultant intensity make an angle ϕ with DX.

∴ $\qquad\qquad\qquad \tan\phi = \dfrac{Y - \text{component}}{X - \text{component}} = \dfrac{19.86 \times 10^4}{10.86 \times 10^4} = 1 \cdot 828$

or $\qquad\qquad\qquad \phi = \tan^{-1} 1 \cdot 828 = \mathbf{61 \cdot 32°}$

Tutorial   Problems

1. What is the magnitude of a point charge chosen so that electric field 20 cm away from it has a magnitude of 18×10^6 N/C ?   [80μC]

2. Two point charge s of 0.12 μC and -0.06 μC are situated 3m apart in air. Calculate the electric field strength at a point midway between them on the line joining their centres.

[720 N/C towards −ve charge]

3. An oil drop of 12 excess electrons is held stationary in a uniform electric field of 2.55×10^4 N/C. If the density of oil is 12600 kg/m³, find (*i*) mass of the drop (*ii*) radius of the drop.

[(*i*) 1.5×10^{-15} kg (*ii*) 9.8×10^{-7} m]

4. A point charge of 0.33×10^{-8} C is placed in a medium of relative permittivity of 5. Calculate electric field intensity at a point 10cm from the charge. **[525 N/C]**

5. Three point charges of $+0.33 \times 10^{-8}$ C, $+0.33 \times 10^{-8}$ C and 0.165×10^{-8} C are at the points *A*, *B* and *C* respectively of a square *ABCD*. Find the electric field intensity at the corner *D*. **[1.63×10^4 N/C]**

5.11. Electric Flux (ψ)

Fig. 5.16 shows electric field between two equal and oppositely charged parallel plates. The electric field is considered to be filled with electric flux and each unit of charge is assumed to give rise to one unit of electric flux. The symbol for electric flux is the Greek letter ψ(psi) and it is measured in coulombs. Thus in Fig. 5.16, the charge on each plate is *Q* coulombs so that electric flux between the plates is

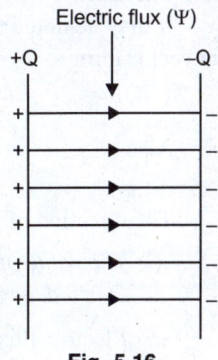

Electric flux (Ψ)

Fig. 5.16

Electric flux, ψ = *Q* coulombs

Electric flux is a measure of electric lines of force. The greater the electric flux passing through an area, the greater is the number of electric lines of force passing through that area and *vice-versa*. Suppose there is a charge of *Q* coulombs in a medium of absolute permittivity ε ($= ε_0 ε_r$) where $ε_r$ is the relative permittivity of the medium. Then number of electric lines of force *N* produced by this charge is

$$N = \frac{Q}{ε} = \frac{Q}{ε_0 ε_r}$$

(*i*) The electric flux through a surface area has maximum value when the surface is perpendicular to the electric field.

(*ii*) The electric flux through the surface is zero when the surface is parallel to the electric field.

5.12. Electric Flux Density (D)

The **electric flux density** *at any section in an electric field is the electric flux crossing normally per unit area of that section i.e.*

Electric flux density, $D = \dfrac{ψ}{A}$

The SI unit of electric flux density is *C/m².

For example, when we say that electric flux density in an electric field is 4C/m², .it means that 4C of electric flux passes normally through an area of 1m². Electric flux density is a vector quantity; possessing both magnitude and direction. Its direction is the same as the direction of electric intensity.

Relation between D and E. Consider a charge of $+Q$ coulombs placed in a medium of relative permittivity $ε_r$ as shown in Fig. 5.17. The electric flux density at *P* at a distance *d* metres from the charge can be found as follows. With centre at the charge and radius *d* metres, an imaginary sphere can be considered. The electric flux of *Q* coulombs will pass normally through this imaginary sphere. Now area of sphere $= 4π\,d^2$.

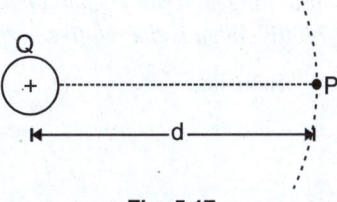

Fig. 5.17

---

* $D = ε_0 ε_r E = [C^2\,N^{-1}\,m^{-2}]\,[N/C] = Cm^{-2} = C/m^2$

$$\therefore \quad \text{Flux density at } P, D = \frac{\text{Flux}}{\text{Area}} = \frac{Q}{4\pi d^2}$$

Also, Electric intensity at P, $E = \dfrac{Q}{4\pi\varepsilon_0\varepsilon_r d^2} = \dfrac{Q}{4\pi d^2} \times \dfrac{1}{\varepsilon_0\varepsilon_r}$

$$= \frac{D}{\varepsilon_r\varepsilon_0} \qquad\qquad \left[\because D = \frac{Q}{4\pi d^2}\right]$$

$$\therefore \qquad\qquad D = \varepsilon_0\,\varepsilon_r\,E$$

Hence flux density at any point in an electric field is $\varepsilon_0\,\varepsilon_r$ times the electric intensity at that point.

The electric flux density (D) is also called **electric displacement.**

It may be noted that D and E are vector quantities having magnitude and direction. Therefore, in vector form,

$$\vec{D} = \varepsilon_0\varepsilon_r\,\vec{E}$$

Also
$$\vec{D} = \frac{Q}{4\pi d^2}\hat{d}$$

The direction of $\vec{D}$ at every point is the same as that of $\vec{E}$ but its magnitude is $D = \varepsilon_0\varepsilon_r E$.

(i) *The value of E depends upon the permittivity $\varepsilon(= \varepsilon_0\varepsilon_r)$ of the surrounding medium, that of D is independent of it.*

(ii) Electric flux density (D) is directly related to electric field intensity (E); permittivity $\varepsilon(= \varepsilon_0\varepsilon_r)$ of the medium being the factor by which one quantity differs from the other.

(iii) The importance of relation $D = \varepsilon_0\varepsilon_r E$ lies in the fact that it relates density concept to intensity concept.

(iv) Electric intensity at a point is also defined as equal to the electric lines of force passing normally through a unit cross-sectional area at that point. If Q coulombs is the charge, then number of electric lines of force produced by it is Q/ε. If these lines fall normally on area A m^2 surrounding the point, then electric intensity E at the point is

$$E = \frac{Q/\varepsilon}{A} = \frac{Q}{\varepsilon A}$$

But $\dfrac{Q}{A} = D = $ Electric flux density over the area.

$$\therefore \qquad\qquad E = \frac{D}{\varepsilon} = \frac{D}{\varepsilon_0\varepsilon_r} \qquad \text{... in a medium}$$

$$= \frac{D}{\varepsilon_0} \qquad \text{... in air}$$

Example 5.13. *Calculate the dielectric flux between two parallel flat metal plates each 35 cm square with an air gap of $1\cdot5$ mm between; the potential difference being 3000 V. A sheet of insulating material $1\cdot5$ mm thick is inserted between the plates and the potential difference raised to 7400V. What is the relative permittivity of this material if the charge is now 32 μC ?*

Solution. $\qquad\qquad E = V/d \; ; \;\; D = \varepsilon_0\varepsilon_r E = \dfrac{\varepsilon_0\varepsilon_r V}{d} \; ; \;\; \psi = DA$

$$\therefore \qquad\qquad \psi = \left(\frac{\varepsilon_0\varepsilon_r V}{d}\right)\times A$$

When medium is air $(\varepsilon_r = 1)$

$$\psi = \frac{\varepsilon_0 V}{d}\times A = \frac{(8.85\times10^{-12})\times 3000\times(35\times35\times10^{-4})}{1.5\times10^{-3}}$$

$$= 21 \cdot 6 \times 10^{-7} C = \textbf{2·16 μC}$$

When medium is insulating material

$$\psi = \frac{\varepsilon_0 \varepsilon_r V}{d} \times A$$

Here $\psi = Q = 32\ \mu C = 32 \times 10^{-6}$ C ; $V = 7400$ volts ; $d = 1 \cdot 5 \times 10^{-3}$ m

$$\therefore \quad \varepsilon_r = \frac{\psi \times d}{\varepsilon_0 VA} = \frac{32 \times 10^{-6} \times 1.5 \times 10^{-3}}{8.85 \times 10^{-12} \times 7400 \times (35)^2 \times 10^{-4}} = \textbf{6}$$

Tutorial Problems

1. What is the total flux passing through a 10 cm × 6 cm surface in a region where the electric flux density is 2700 μC/m^2 ?   **[1·62 × 10^{-5} C]**

2. At a certain point in a material, the flux density is 0·09 C/m^2 and electric field intensity is $1 \cdot 2 \times 10^9$ V/m. What is the absolute permittivity of the material ?   **[7·5 × 10^{-11} C^2 N^{-1} m^{-2}]**

5.13. Gauss's Theorem

This theorem was first expressed by a German scientist Karl Fredrich Gauss (1777–1855) and may be stated as under :

The electric flux passing through a closed surface surrounding a number of charges is equal to the algebraic sum of the charges inside the closed surface.

To illustrate Gauss's theorem, consider Fig. 5.18 where charges Q_1, Q_2, Q_3 and $-Q_4$ coulombs are placed inside a closed surface. According to Gauss, the total electric flux ψ passing through this closed surface is given by the algebraic sum of the charges inside the closed surface *i.e.*

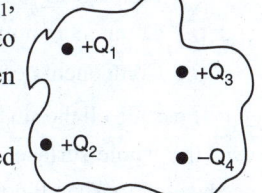

ψ = Algebraic sum of the charges inside the closed surface

$$= (Q_1) + (Q_2) + (Q_3) + (-Q_4)$$
$$= Q_1 + Q_2 + Q_3 - Q_4 \text{ coulombs}$$

Fig. 5.18

The following points may be noted :

(a) The location of charge/charges inside the closed surface does not matter.

(b) The shape of the surface does not matter provided it is a closed surface enclosing the charge/charges.

Explanation. *(i)* Consider a charge of $+Q$ coulomb placed at the centre of sphere of radius r as shown in Fig. 5.19 *(i)*. Since the charge is at the centre of the sphere, electric flux density (D) is uniform over all the surface and perpendicular to the surface at every point.

$$D = \frac{\text{Charge}}{\text{Area of sphere}} = \frac{Q}{4\pi r^2}$$

Therefore, the electric flux ψ passing outward through the sphere is

$$\psi = D \times \text{Area} = \frac{Q}{4\pi r^2} \times 4\pi r^2 = Q \text{ coulomb}$$

The number of electric lines of force passing through the closed surface normally is Q/ε_0.

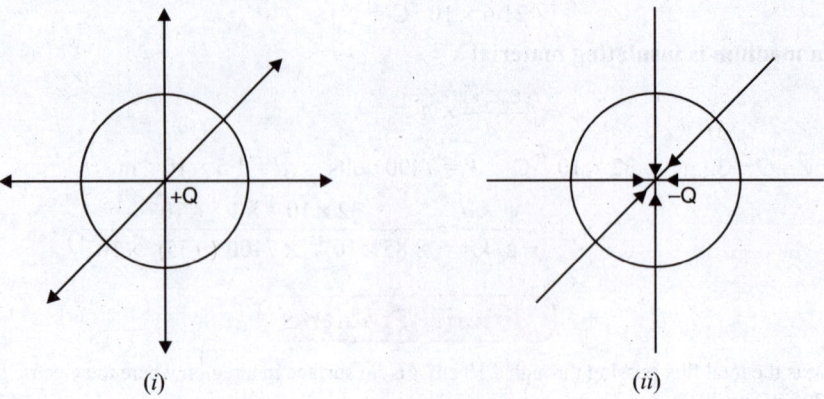

(i)   (ii)

Fig. 5.19

Thus the electric flux passing through the surface of sphere is equal to Q, the charge enclosed in the sphere. This establishes Gauss's theorem.

If the sphere were enclosing a charge $-Q$ placed at the centre [See Fig. 5.19 (ii)], then electric flux $\psi = Q$ coulomb would pass inward through the surface and terminate at the charge.

(**ii**) Now consider that the charge $+Q$ coulomb is placed at any other point (other than centre O) inside the sphere as shown in Fig. 5.20. The electric lines of force flow outward but not normal to the surface. However, at any point on the sphere (such as point P), electric flux can be resolved into two rectangular components viz

(**a**) Component normal to the surface i.e., $\cos \theta$ component.

(**b**) Component perpendicular to the normal to the surface i.e. $\sin \theta$ component.

If we add all the $\sin \theta$ components of electric flux over the whole surface, the result will be zero. It is because various $\sin \theta$ components cancel each other. However, all $\cos \theta$ components of flux are normal to the sphere surface and meet at the centre if produced backward. Hence the resultant of all $\cos \theta$ components over the surface of sphere is equal to Q coulomb i.e.

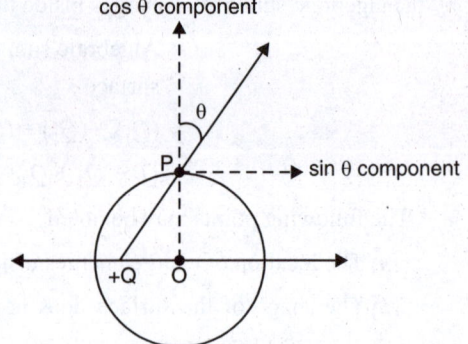

$$\psi = Q \text{ coulomb}$$

The number of electric lines of force passing through the closed surface normally is Q/ε_0.

Fig. 5.20

Thus irrespective of the position of charge Q within the sphere, the flux passing through the sphere surface is Q coulomb. This establishes Gauss's theorem. Similarly, it can be shown that if a surface encloses a number of charges, the electric flux passing through the surface is equal to the algebraic sum of charges inside the closed surface.

Gauss's law can also be expressed *mathematically*.

We know that : $\psi = \oint \vec{E} \cdot \vec{dS}$

where $\oint \vec{E} \cdot \vec{dS}$ is the surface integral of electric field $(\vec{E})$ over the entire closed surface enclosing the charge Q.

$\therefore \qquad\qquad\qquad\qquad \psi = \oint \vec{E} \cdot \vec{dS} = \dfrac{Q}{\varepsilon_0}$

Hence, Gauss's law may be stated as under :

If a closed surface encloses a net charge (Q), then surface integral of electric field $(\vec{E})$ over the closed surface is equal to $1/\varepsilon_0$ times the charge enclosed.

5.14. Proof of Gauss's Law

Consider a positive charge $+Q$ located at point O as shown in Fig. 5.21. We draw a sphere of radius r with charge $+Q$ as its centre. We now show that total electric flux (*i.e.* total number of electric lines of force) passing through the closed surface is Q/ε_0. The magnitude of electric field at any point on the spherical surface is given by ;

$$E = \frac{Q}{4\pi\varepsilon_0 r^2}$$

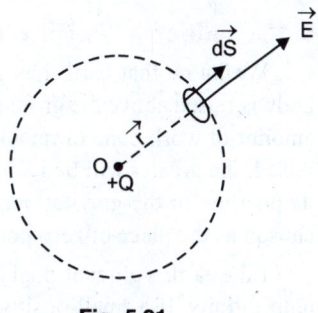

Fig. 5.21

The electric field is directed radially outward from $+Q$. The spherical surface is only imaginary and is called *Gaussian surface*.

Consider a small elementary area $\vec{dS}$ on the surface of sphere as shown in Fig. 5.21. It is clear that $\vec{E}$ is * parallel to $\vec{dS}$ *i.e.* angle between $\vec{E}$ and $\vec{dS}$ is zero. Therefore, electric flux through the entire closed spherical surface is

$$\psi = \oint \vec{E} \cdot \vec{dS} = \oint E \, dS \cos 0° = \oint E \, dS$$

Since E (magnitude of $\vec{E}$) is constant over the considered closed surface, it can be taken out of integral.

$$\therefore \qquad\qquad \psi = E \oint dS$$

Now $\qquad\qquad E = \dfrac{Q}{4\pi\varepsilon_0 r^2}$ and $\oint dS$ = Surface area of sphere = $4\pi r^2$

$$\therefore \qquad\qquad \psi = \frac{Q}{4\pi\varepsilon_0 r^2} \times 4\pi r^2 = \frac{Q}{\varepsilon_0}$$

Hence, $\qquad\qquad \psi = \oint \vec{E}.\vec{dS} = \dfrac{Q}{\varepsilon_0}$

Note. We know : $\qquad \psi = \oint \vec{E}.\vec{dS} = \dfrac{Q}{\varepsilon_0}$

$$= \oint \varepsilon_0 \vec{E}.\vec{dS} = Q$$

$$\therefore \qquad\qquad \psi = \oint \vec{D}.\vec{dS} = Q \qquad\qquad (\because \ \varepsilon_0 \vec{E} = \vec{D})$$

Note that ψ can be expressed in Q or Q/ε_0.

Hence Gauss's law may be stated in terms of flux density $(\vec{D})$ as under :

If a closed surface encloses a net charge (Q), then surface integral of $\vec{D}$ (electric flux density) over the closed surface is equal to the charge enclosed by the closed surface.

Example 5.14. *A spherical surface 50 cm in diameter is penetrated by an inward flux uniformly distributed over the surface, the electric flux density being $2\cdot5 \times 10^{-7}$ C/m^2. What is the magnitude and sign of the charge enclosed by this surface ?*

Solution. Area of spherical surface is

$$A = 4\pi r^2 = 4\pi \times (25 \times 10^{-2})^2 = 0\cdot785 \text{ m}^2$$

---

*   This is true for every elementary area on the surface.

$$\text{Electric flux, } \psi = D \times A = (2 \cdot 5 \times 10^{-7}) \times (0 \cdot 785) = 0 \cdot 1962 \times 10^{-6} \text{ C}$$

$$\therefore \qquad \text{Charge enclosed} = 0 \cdot 1962 \times 10^{-6} \text{ C} = \mathbf{0 \cdot 1962 \ \mu C}$$

Since the electric flux is passing inward through the sphere, the charge enclosed is **negative**.

5.15. Electric Potential Energy

We know that earth has gravitational field which attracts the bodies towards earth. When a body is raised above the ground level, it possesses mechanical potential energy which is equal to the amount of work done in raising the body to that point. The greater the height to which the body is raised, the greater will be its potential energy. Thus, the potential energy of the body depends upon its position in the gravitational field; being zero on earth's surface. Strictly speaking, sea level is chosen as the place of zero potential energy.

Like earth's gravitational field, every charge ($+ Q$) has electric field which theoretically extends upto infinity. If a small positive test charge $+ q_0$ is placed in this electric field, the test charge will experience a force of repulsion. If test charge $+ q_0$ is moved towards $+ Q$, work will have to be done against the force of repulsion. This work done is stored in $+ q_0$ in the form of potential energy. We say the charge $+ q_0$ has electric potential energy. The electric potential energy of $+ q_0$ depends upon its position in the electric field ; being zero if q_0 is situated at infinity.

From the above discussion, it follows that just as a mass has mechanical potential energy in the gravitational field, similarly a charge has electric potential energy in the electric field. The electric potential energy of a charge is positive or negative depending upon the kind of charge.

5.16. Electric Potential

Just as we define electric field intensity as the force per unit charge, similarly *electric potential is defined as the electric potential energy per unit charge.*

Consider an isolated charge $+Q$ fixed in space as shown in Fig. 5.22. If a unit positive charge (*i.e.* $+1$C) is placed at infinity, the force on it due to charge $+Q$ is *zero. If the unit positive charge at infinity is moved towards

Fig. 5.22

$+Q$, a force of repulsion acts on it (like charges repel) and hence work is required to be done to bring it to a point like A. Hence when the unit positive charge is at A, it has some amount of electric potential energy which is a measure of electric potential. The closer the point to the charge, the higher will be the electric potential energy and hence the electric potential at that point. Therefore, electric potential at a point due to a charge depends upon the position of the point; being zero if the point is situated at infinity. Obviously, in electric field, infinity is chosen as the point of **zero potential.

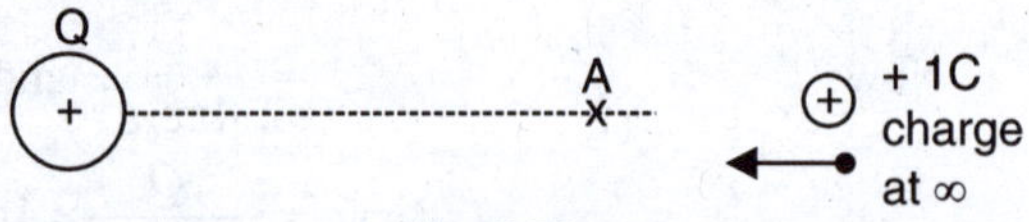

Hence **electric potential** *at a point in an electric field is the amount of work done in bringing a unit positive charge (i.e. $+1$ C) from infinity to that point i.e.*

$$\text{Electric potential} = \frac{\text{Work}}{\text{Charge}} = \frac{W}{Q}$$

where W is the work done to bring a charge of Q coulombs from infinity to the point under consideration.

---

* $F = 9 \times 10^9 \times \dfrac{Q \times 1}{d^2}$; As $d \to \infty$, $F \to 0$

**   In practice, earth is chosen to be at zero electric potential. It is because earth is such a huge conductor that its electric potential practically remains constant.

Unit. The *SI* unit of electric potential is *volt and may be defined as under :

The electric potential at a point in an electric field is **1 volt** *if 1 joule of work is done in bringing a unit positive charge (i.e. + 1 C) from infinity to that point **against the electric field.*

Thus when we say that potential at a point in an electric field is +5V, it simply means that 5 joules of work has been done in bringing a unit positive charge from infinity to that point.

5.17. Electric Potential Difference

In practice, we are more concerned with potential difference between two points rather than their †absolute potentials. The potential difference (p.d.) between two points may be defined as under :

The potential difference between two points is the amount of work done in moving a unit positive charge (i.e. + 1C) from the point of lower potential to the point of higher potential.

Consider two points A and B in the electric field of a charge $+Q$ as shown in Fig. 5.23. Let V_2 and V_1 be the absolute potentials at A and B respectively. Clearly, $V_2 > V_1$. The potential V_1 at B means that V_1 joules of work has been done in bringing a unit positive charge from infinity to point B. Let the extra work done to bring the unit positive charge from B to A be W joules.

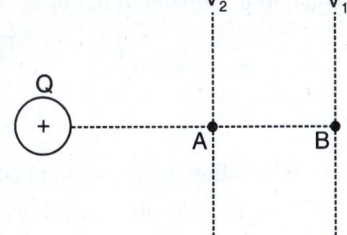

∴ Potential at $A = V_1 + W$

∴ P.D. between A and $B = (V_1 + W) - V_1$

or $V_2 - V_1 = W = W.D.$ to move + 1C from B to A **Fig. 5.23**

The SI unit of potential difference is volt and may be defined as under :

The p.d. between two points is **1 V** *if 1 joule of work is done in bringing a unit positive charge (i.e. + 1 C) from the point of lower potential to the point of higher potential.*

Thus when we say that p.d. between two points is 5 volts, it simply means that 5 joules of work will have to be done to bring +1C of charge from the point of lower potential to the point of higher potential. Conversely, 5 joules of work or energy will be released if + 1 C charge moves from the point of higher potential to the point of lower potential.

5.18. Potential at a Point Due to a Point Charge

Consider an isolated positive charge of Q coulombs placed in a medium of relative permittivity ε_r. It is desired to find the electric potential at point P due to this charge. Let P be at a distance d metres from the charge. Imagine a unit positive charge (*i.e.* + 1 C) placed at A and situated x metres from the charge. Then the force acting on this unit charge (*i.e.* electric intensity) is given by [See Fig. 5.24] ;

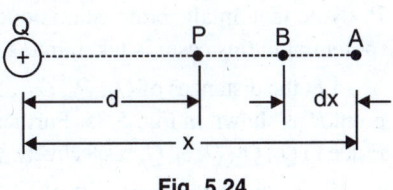

Fig. 5.24

$$F = E = \frac{Q}{4\pi\varepsilon_0\varepsilon_r x^2}$$

If this unit positive charge at A is moved through a small distance dx towards the charge $+Q$, then work done is given by ;

$$dW = \frac{Q}{4\pi\varepsilon_0\varepsilon_r x^2} \times (-\text{††} \, dx) = -\frac{Q}{4\pi\varepsilon_0\varepsilon_r x^2} dx$$

---

* Electric potential = W/Q = joules/coulomb. Now joule/coulomb has been given a special name *viz* volt.

** Note if the field is due to a positive charge (as is in this case), work will be done against the electric field. However, if the field is due to a negative charge, work is done by the electric field.

† The potential at a point with infinity as reference is termed as absolute potential.

†† The negative sign is taken because dx is considered in the negative direction of distance (x).

Total work done in bringing a unit positive charge from infinity to point P is

$$\text{Total work done, } W = \int_{\infty}^{d} -\frac{Q}{4\pi\varepsilon_0\varepsilon_r x^2}\,dx = -\frac{Q}{4\pi\varepsilon_0\varepsilon_r}\int_{\infty}^{d}\frac{1}{x^2}\,dx$$

$$= -\frac{Q}{4\pi\varepsilon_0\varepsilon_r}\left[-\frac{1}{x}\right]_{\infty}^{d} = \frac{-Q}{4\pi\varepsilon_0\varepsilon_r}\left[-\frac{1}{d}-\left(-\frac{1}{\infty}\right)\right]$$

$$= \frac{Q}{4\pi\varepsilon_0\varepsilon_r d}$$

$$= 9\times10^9\,\frac{Q}{\varepsilon_r d}\text{ joules} \qquad\qquad \left[\because \frac{1}{4\pi\varepsilon_0}=9\times10^9\right]$$

By definition, the work done in joules to bring a unit positive charge from infinity to point P is equal to potential at P in volts.

$$\therefore \qquad V_P = 9\times10^9\,\frac{Q}{\varepsilon_r d}\text{ volts} \qquad\qquad ...in\ a\ medium$$

$$= 9\times10^9\,\frac{Q}{d}\text{ volts} \qquad\qquad ...in\ air$$

The following points may be noted carefully :

(*i*) The potential varies inversely with the distance d from the point charge Q. If the distance is increased three times, the potential is reduced one-third of its value and so on.

(*ii*) Electric potential is a scalar quantity.

(*iii*) At $d = \infty$ in air/vacuum, $V_P = 9\times10^9\,\dfrac{q}{\infty} = 0$.

(*iv*) If Q is positive, then potential at P is *positive. On the other hand, if Q is negative, then potential at P is negative.

5.19. Potential at a Point Due to Group of Point Charges

Electric potential obeys superposition principle. Therefore, electric potential at any point P due to a group of point charges $Q_1, Q_2, Q_3 .... Q_n$ is equal to the algebraic sum of potentials due to $Q_1, Q_2, Q_3 ... Q_n$ at point P. Note that an algebraic sum is one in which sign of the physical quantity (potential in this case) is taken into account.

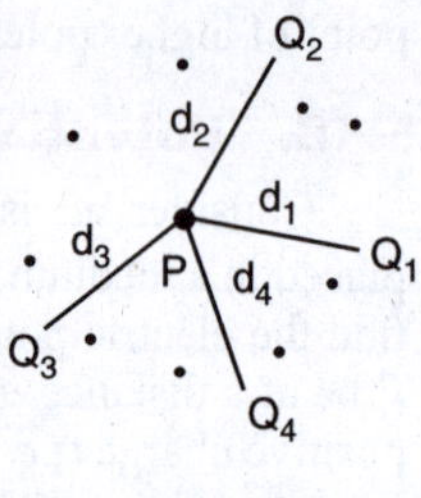

Let the distances of $Q_1, Q_2, Q_3, ... Q_n$ be $d_1, d_2, d_3 ... d_n$ respectively from point P as shown in Fig. 5.25. Further, let $V_1, V_2, V_3 ... V_n$ be the potentials at P due to $Q_1, Q_2, Q_3 .... Q_n$ respectively. Assuming the medium to be free space/air,

Fig. 5.25

$$\text{Total potential at } P,\ V_P = V_1 + V_2 + V_3 + ... + V_n$$

$$= \frac{1}{4\pi\varepsilon_0}\frac{Q_1}{d_1} + \frac{1}{4\pi\varepsilon_0}\frac{Q_2}{d_2} + \frac{1}{4\pi\varepsilon_0}\frac{Q_3}{d_3} + ... + \frac{1}{4\pi\varepsilon_0}\frac{Q_n}{d_n}$$

$$= \frac{1}{4\pi\varepsilon_0}\left[\frac{Q_1}{d_1} + \frac{Q_2}{d_2} + \frac{Q_3}{d_3} + ... + \frac{Q_n}{d_n}\right]$$

$$\therefore \qquad V_P = 9\times10^9\left[\frac{Q_1}{d_1} + \frac{Q_2}{d_2} + \frac{Q_3}{d_3} + ... + \frac{Q_n}{d_n}\right]$$

---

*   The potential near an isolated positive charge is positive because work is done by an external agency to push a test charge (positive) from infinity to that point. The potential near an isolated negative charge is negative because outside agent must exert a restraining force as test charge comes in from infinity.

If the system of charges is placed in a medium of relative permittivity ε_r, then,

$$V_P = \frac{9 \times 10^9}{\varepsilon_r}\left[\frac{Q_1}{d_1} + \frac{Q_2}{d_2} + \frac{Q_3}{d_3} + \dots + \frac{Q_n}{d_n}\right]$$

5.20. Behaviour of Metallic Conductors in Electric Field

When a metallic conductor (solid or hollow) is placed in an electric field, there is a momentary flow of charges (*i.e.*, free electrons). Once the flow of charges ceases, the conductor is said to be in *electrostatic equilibrium*. It has been seen experimentally that under the conditions of electrostatic equilibrium, a conductor (solid or hollow) shows the following properties [See Fig. 5.26] :

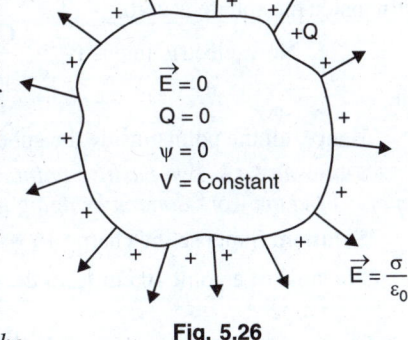

Fig. 5.26

(*i*) *The net electric field inside a charged conductor is zero i.e., no electric lines of force exist inside the conductor.*

(*ii*) *The net charge inside a charged conductor is zero.*

(*iii*) *The electric field (i.e., electric lines of force) on the surface of a charged conductor is perpendicular to the surface of the conductor at every point.*

(*iv*) *The magnitude of electric field just outside a charged conductor is σ/ε_0 where σ is the surface charge density.*

(*v*) *The electric potential is the same (i.e., constant) at the surface and inside a charged conductor.*

Inside a charged conductor, $E = 0$

Now $$E = -\frac{dV}{dS} \quad \text{or} \quad 0 = -\frac{dV}{dS}$$

This means that V is constant.

5.21. Potential of a Charged Conducting Sphere

Consider an isolated conducting sphere of radius r metres placed in air and charged uniformly with Q coulombs. The field has spherical symmetry *i.e.* lines of force spread out normally from the surface and meet at the centre of the sphere if produced backward. *Outside the sphere*, the field is exactly the same as though the charge Q on sphere were concentrated at its centre.

(*i*) **Potential at the sphere surface.** Due to spherical symmetry of the field, we can imagine the charge Q on the sphere as concentrated at its centre O [See Fig. 5.27 (*i*)]. The problem then reduces to find the potential at a point r metres from a charge Q.

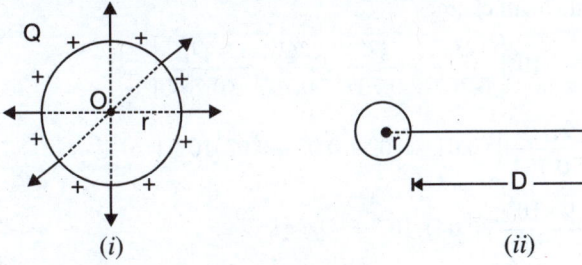

(*i*)   (*ii*)

Fig. 5.27

∴   Potential at the surface of sphere

$$= \frac{Q}{4\pi\varepsilon_0 r} \text{ volts} \qquad\qquad \text{[See Art. 5·18]}$$

$$= 9 \times 10^9 \frac{Q*}{r} \text{ volts}$$

(ii) **Potential outside the sphere.** Consider a point P outside the sphere as shown in Fig. 5.27 *(ii)*. Let this point be at a distance of D metres from the surface of sphere.

$$\text{Then potential at } P = 9 \times 10^9 \frac{Q}{(D+r)} \text{ volts}$$

(iii) **Potential inside the sphere.** Since there is no electric flux inside the sphere, electric intensity inside the sphere is zero.

$$\text{Now, electric intensity} = \frac{\text{Change in potential}}{r}$$

or $$0 = \text{Change in potential}$$

Hence, all the points inside the sphere are at the same potential as the points on the surface.

Example 5.15. *Two positive point charges of 16×10^{-10} C and 12×10^{-10} C are placed 10 cm apart. Find the work done in bringing the two charges 4 cm closer.*

Solution. Suppose the charge 16×10^{-10} C to be fixed.

Potential of a point 10 cm from the charge 16×10^{-10} C

$$= 9 \times 10^9 \frac{16 \times 10^{-10}}{0.1} = 144 \text{ V}$$

Potential of a point 6 cm from the charge 16×10^{-10} C

$$= 9 \times 10^9 \frac{16 \times 10^{-10}}{0.06} = 240 \text{ V}$$

$$\therefore \qquad \text{Potential difference} = 240 - 144 = 96 \text{ V}$$

$$\text{Work done} = \text{Charge} \times \text{p.d.} = 12 \times 10^{-10} \times 96 = \mathbf{11 \cdot 52 \times 10^{-8} \text{ joules}}$$

Example 5.16. *A square ABCD has each side of 1 m. Four point charges of $+0.01$ μC, -0.02 μC, $+0.03$ μC and $+0.02$ μC are placed at A, B, C and D respectively. Find the potential at the centre of the square.*

Solution. Fig. 5.28 shows the square $ABCD$ with charges placed at its corners. The diagonals of the square intersect at point P. Clearly, point P is the centre of the square. The distance of each charge from point P (*i.e.* centre of square) is

$$= \frac{1}{2}\sqrt{1^2 + 1^2} = 0.707 \text{ m}$$

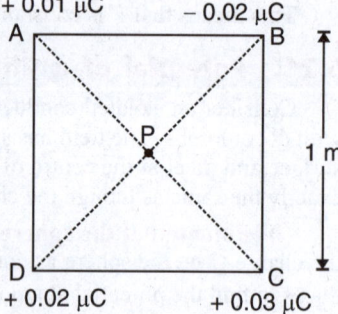

Fig. 5.28

The potential at point P due to all charges is equal to the algebraic sum of potentials due to each charge.

$$\therefore \quad \text{Potential at } P \text{ due to all charges}$$

$$= 9 \times 10^9 \left[\frac{Q_1}{0.707} + \frac{Q_2}{0.707} + \frac{Q_3}{0.707} + \frac{Q_4}{0.707} \right]$$

$$= \frac{9 \times 10^9}{0.707} \left[(0.01 - 0.02 + 0.03 + 0.02) 10^{-6} \right]$$

$$= \frac{9 \times 10^9}{0.707} \times 0.04 \times 10^{-6} = \mathbf{509.2V}$$

$*$ If the sphere is placed in a medium (ε_r), then potential is

$$= 9 \times 10^9 \frac{Q}{\varepsilon_r r}$$

Example 5.17. *A hollow sphere is charged to 12μC. Find the potential (i) at its surface (ii) inside the sphere (iii) at a distance of 0·3m from the surface. The radius of the sphere is 0·1m.*

Solution. (*i*) The potential at the surface of the sphere in air is

$$V = \frac{Q}{4\pi\varepsilon_0 d} = 9\times10^9 \times \frac{Q}{d}$$

Here $Q = 12\ \mu C = 12\times10^{-6}\ C$; $d = 0\cdot1m$

∴ $V = 9\times10^9 \times \dfrac{12\times10^{-6}}{0.1} = \textbf{108}\times\textbf{10}^\textbf{4}\ \textbf{volts}$

(*ii*) Potential inside the sphere is the same as at the surface *i.e.* **108 × 10⁴ volts.**

(*iii*) Distance of the point from the centre = $0\cdot3 + 0\cdot1 = 0\cdot4m$

∴   Potential = $9\times10^9 \times \dfrac{12\times10^{-6}}{0.4} = \textbf{27}\times\textbf{10}^\textbf{4}\ \textbf{volts}$

Example 5.18. *If 300 J of work is done in carrying a charge of 3 C from a place where the potential is −10 V to another place where potential is V, calculate the value of V.*

Solution. $V_B - V_A = \dfrac{W}{Q}$

Here $V_B = V$; $V_A = -10V$; $W = 300\ J$; $Q = 3\ C$

∴ $V - (-10) = 300/3$ or $V + 10 = 100$

∴ $V = 100 - 10 = \textbf{90 volts}$

Example 5.19. *The electric field at a point due to a point charge is 30 N/C and the electric potential at that point is 15 J/C. Calculate the distance of the point from the charge and magnitude of charge.*

Solution. Suppose q coulomb is the magnitude of charge and its distance from the point is r metres.

Now, $E = \dfrac{k\,q}{r^2} = 30$; $V = \dfrac{k\,q}{r} = 15$

∴ $\dfrac{E}{V} = \dfrac{1}{r}$ or $r = \dfrac{V}{E} = \dfrac{15}{30} = \textbf{0·5 m}$

Now $kq = 15\,r = 15\times0\cdot5 = 7\cdot5$

∴ $q = \dfrac{7.5}{k} = \dfrac{7.5}{9\times10^9} = \textbf{0·83}\times\textbf{10}^{-\textbf{9}}\ \textbf{C}$

Example 5.20. *Two point charges of +4 μC and −6 μC are separated by a distance of 20 cm in air. At what point on the line joining the two charges is the electric potential zero ?*

Solution. Fig. 5.29 shows the conditions of the problem. Suppose C is the point of zero potential. Potential at point C is given by ;

$$V = \frac{1}{4\pi\varepsilon_0}\left[\frac{4\times10^{-6}}{d_1} - \frac{6\times10^{-6}}{d_2}\right]$$

or $0 = \dfrac{10^{-6}}{4\pi\varepsilon_0}\left[\dfrac{4}{d_1} - \dfrac{6}{d_2}\right]$

or $\dfrac{4}{d_1} - \dfrac{6}{d_2} = 0$ or $d_1 = \dfrac{2}{3}d_2$ ...(*i*)

Also $d_1 + d_2 = 20$ cm   ...(*ii*)

Solving eqs. (*i*) and (*ii*), we get, $d_1 = 8$ cm ; $d_2 = 12$ cm.

Therefore, the point of zero potential lies **8 cm from the charge of +4 μC** or at 12 cm from the charge of −6 μC.

Tutorial Problems

1. A charge of -4.5×10^{-7} C is carried from a distant point upto a charged metal sphere. What is the electrical potential of the body if the work done is 1.8×10^{-3} joule ?   [4×10^{3} V]

2. The difference of potentials between two points in an electric field is 6 volts. How much work is required to move a charge of 300 μC between these points ?   [1.8×10^{-3} joule]

3. A force of 0.032 N is required to move a charge of 42 μC in an electric field between two points 25 cm apart. What potential difference exists between the two points ?   [1.9×10^{2} V]

4. What is the magnitude of an isolated positive charge to give an electric potential of 100V at 10 cm from the charge ?   [1.11×10^{-9} C]

5. A square ABCD has each side of 1m. Four charges of +0.02 μC, +0.04 μC, +0.06 μC and +0.02 μC are placed at A, B, C and D respectively. Find the potential at the centre of the square.   [1000V]

6. A sphere of radius 0.1 m has a charge of 5×10^{-8} C. Determine the potential (i) at the surface of sphere, (ii) inside the sphere and (iii) at a distance of 1m from the surface of the sphere. Assume air as the medium.   [(i) **4500 V** (ii) **4500 V** (iii) **409 V**]

5.22. Potential Gradient

The change of potential per unit distance is called **potential gradient** *i.e.*

$$\text{Potential gradient} = \frac{V_2 - V_1}{S}$$

where $V_2 - V_1$ is the change in potential (or p.d.) between two points S metres apart. Obviously, the unit of potential gradient will be volts/m.

Consider a charge $+Q$ and let there be two points A and B situated S metres apart in its electric field as shown in Fig. 5.30. Clearly, potential at point A is more than the potential at point B. If distance S is small, then the electric intensity will be approximately the same in this small distance. Let it be E newtons/coulomb. It means that a force of E newtons will act on a unit positive charge (*i.e.* + 1C) placed anywhere between A and B. If a unit positive charge is moved from B to A, then work done to do so is given by ;

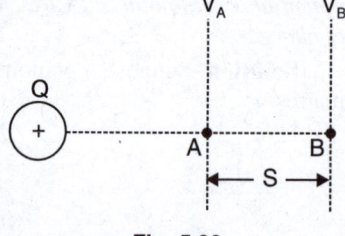

Fig. 5.30

$$\text{Work done} = E \times S \text{ joules}$$

But work done in bringing a unit positive charge from B to A is the potential difference ($V_A - V_B$) between A and B.

∴ $$E \times S = V_A - V_B$$

or $$E = \frac{V_A - V_B}{S} = \text{Potential gradient}$$

In differential form, $$E = -\frac{*dV}{dS}$$

Hence electric intensity at a point is numerically equal to the potential gradient at that point.

Since electric intensity is numerically equal to potential gradient at any point, both must be measured in the same units. Clearly, electric intensity can also be measured in V/m. For example, when we say that potential gradient at a point is 1000 V/m, it means that electric intensity at that point is also **1000 V/m or 1000 N/C.

---

*   Since work done in moving +1C from B to A is against electric field, a negative sign must be used to make the equation technically correct.

**   It can be shown that 1 V/m = 1 N/C.

$$1 \text{ V/m} = \frac{\text{joule/coulomb}}{\text{metre}} = \frac{\text{newton} \times \text{metre}}{\text{metre} \times \text{coulomb}} = 1 \text{ N/C}$$

5.23. Breakdown Voltage or Dielectric Strength

In an insulator or dielectric, the valence electrons are tightly bound so that no free electrons are available for current conduction. However, when voltage applied to a dielectric is gradually increased, a point is reached when these electrons are torn away, a large current (much larger than the usual leakage current) flows through the dielectric and the material loses its insulating properties. Usually, a *spark or arc occurs which burns up the material. The minimum voltage required to break down a dielectric is called breakdown voltage or dielectric strength.

The maximum voltage which a unit thickness of a dielectric can withstand without being punctured by a spark discharge is called ***dielectric strength** of the material.*

The dielectric strength (or breakdown voltage) is generally measured in kV/cm or kV/mm. For example, air has a dielectric strength of 30kV/cm. It means that maximum p.d. which 1 cm thickness of air can withstand across it without breaking down is 30kV. If p.d. exceeds this value, the breakdown of air insulation will occur; allowing a large current to flow through it. Below is given the table showing dielectric constant and dielectric strength of some common insulators or dielectrics :

S.No.	Dielectric	Dielectric Constant (ε_r)	Dielectric strength (kV/cm)
1	Air	1	30
2	Paper (oiled)	2	400
3	Paraffin	2·25	350
4	Mica	6	500
5	Glass	8	1000

The following points may be noted :

(i) The value of dielectric strength of an insulator (or dielectric) depends upon temperature, moisture content, shape *etc.*

(ii) The electric intensity, potential gradient and dielectric strength are numerically equal *i.e.*

Electric intensity = Potential gradient = Dielectric strength

(iii) The breakdown of solid insulating material (dielectric) usually renders it unfit for further use by puncturing, burning, cracking or otherwise damaging it. Gaseous and liquid dielectrics are self-healing and may be used repeatedly following breakdown.

(iv) *For reasons of safety, electric field applied to a dielectric is only 10% of the dielectric strength of the dielectric material.*

Note. To avoid electric breakdown of dielectric, capacitors are rated according to their *working voltage,* meaning the maximum safe voltage that can be applied to the capacitor.

5.24. Uses of Dielectrics

The insulating materials (or dielectrics) are widely used to provide electrical insulation to electrical and electronics apparatus. The choice of a dielectric for a particular situation will depend upon service requirements. A few cases are given below by way of illustration :

(i) If the dielectric is to be subjected to a great heat, as in soldering irons or toasters; mica should be used.

(ii) If space, flexibility and a fair dielectric strength are the deciding factors, as in the dielectric for small fixed capacitors, cellulose and animal tissue materials are used.

---

*   This spark may burn a path through such dielectrics as paper, cloth, wood or mica. Hard materials such as porcelain or glass will crack or allow a small path to be melted through them.

**   Dielectric strength should not be confused with dielectric constant (relative permittivity).

(*iii*) If a high dielectric strength is desired, as in case of high voltage transformers, glass and porcelain should be used.

(*iv*) If the insulation must remain liquid, like that used in large switches and circuit breakers to quench the arc when the circuit is opened, then various oils are used.

Example 5.21. *A parallel plate capacitor has plates 1 mm apart and a dielectric with relative permittivity of 3·39. Find (i) electric intensity and (ii) the voltage between plates if the surface charge density is 3×10^{-4} C/m².*

Solution. (*i*) The surface charge density is equal to electric flux density D.

Now, $D = \varepsilon_0 \varepsilon_r E$

∴   Electric intensity, $E = \dfrac{D}{\varepsilon_0 \varepsilon_r} = \dfrac{3 \times 10^{-4}}{8.854 \times 10^{-12} \times 3.39} = \mathbf{10^7\ V/m}$

(*ii*)   P.D. between plates, $V = E \times dx = 10^7 \times (1 \times 10^{-3}) = \mathbf{10^4 V}$

Example 5.22. *The electric potential difference between the parallel deflection plates in an oscilloscope is 300V. If the potential drops uniformly when going from one plate to the other and if distance between the plates is 0·75 cm, what is the magnitude of the electric field between them and in which direction does it point?*

Solution. Let us choose the positive direction of ΔS to be in the direction of increasing potential.

∴ $E = -\dfrac{\Delta V}{\Delta S}$

Here $\Delta V = +300\text{V}; \quad \Delta S = +0.75\text{ cm} = 0.75 \times 10^{-2}\text{ m}$

∴ $E = -\dfrac{300}{0.75 \times 10^{-2}} = \mathbf{-40{,}000\ V/m}$

The negative value of E tells us that E is directed opposite to ΔS. Thus E is directed from the higher-voltage plate towards the lower-voltage one.

Example 5.23. *A uniform electric field is acting from left to right. If a + 2C charge moves from a to b, a distance of 4m, [See Fig. 5.31], find (i) electric field strength and (ii) potential energy of charge at b w.r.t. a. Given that p.d. between a and b is 50 volts.*

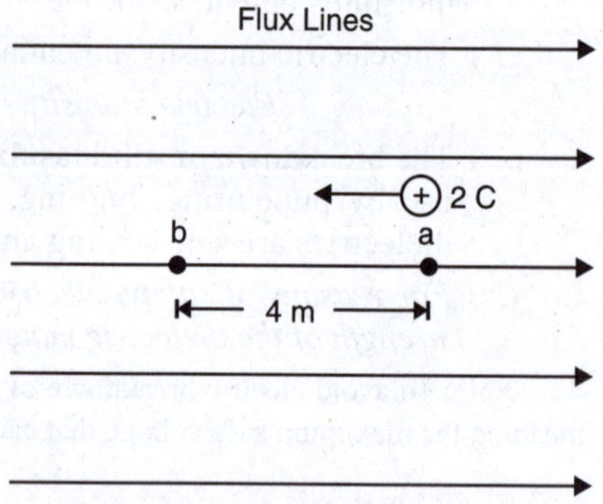

Solution. Referring to Fig. 5.31, we have,

(*i*) Electric intenstity = Potential gradient = 50/4

= **12·5 V/m**

(*ii*) Potential energy of charge (*i.e.*, +2C) at *b* w.r.t. *a*

= Work per unit charge × Charge

= Voltage between *a* and *b* × Charge

= 50 joules/C × (2C) = **100 joules**

Flux Lines

Fig. 5.31

Example 5.24. *A sheet of glass 1·5 cm thick and of relative permittivty 7 is introduced between two parallel brass plates 2 cm apart. The remainder of the space between the plates is occupied by air. If a p.d. of 10,000 V is applied between the plates, calculate (i) electric intensity in air film between glass and plate and (ii) in the glass sheet.*

Solution. Fig. 5.32 shows the arrangement. Let V_1 and V_2 be the p.d. across air and glass respectively and E_1 and E_2 the corresponding electric intensities.

Now, $V_1 = E_1 x_1 = E_1 \times (0.5 \times 10^{-2})$

and $\qquad V_2 = E_2 x_2 = E_2 \times (1\cdot5 \times 10^{-2})$

Now $\qquad V = V_1 + V_2$

or $\qquad 10{,}000 = (0\cdot5\,E_1 + 1\cdot5\,E_2)10^{-2}$

or $\qquad E_1 + 3E_2 = 2 \times 10^6$ $\qquad$ ...(i)

Now electric flux density $D\,(= \varepsilon_0\,\varepsilon_r\,E)$ is the same in the two media because it is independent of the surrounding medium.

$\therefore \qquad \varepsilon_0\,\varepsilon_{r2}\,E_1 = \varepsilon_0\,\varepsilon_{r2}\,E_2$

or $\qquad E_1 = 7\,E_2$ $\qquad$ ...(ii)

From exps. (i) and (ii), we get,

(i) $\quad$ Electric intensity in air $= \mathbf{1\cdot4 \times 10^6\ V/m}$

(ii) Electric intensity in glass $= \mathbf{0\cdot2 \times 10^6\ V/m}$

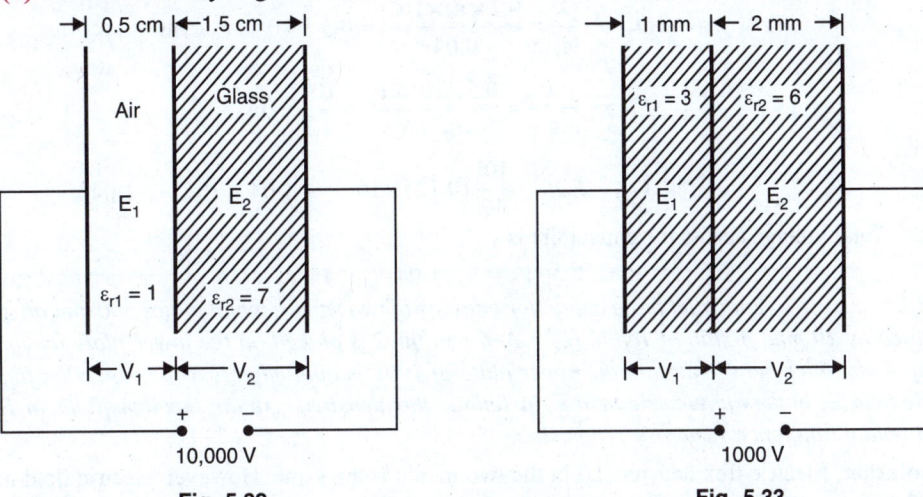

Fig. 5.32 $\qquad\qquad\qquad\qquad\qquad\qquad\qquad$ Fig. 5.33

Example 5.25. *A capacitor has two dielectrics 1 mm and 2 mm thick. The relative permittivities of these dielectrics are 3 and 6 respectively. Calculate the potential gradient along the dielectrics if a p.d. of 1000 V is applied between the plates.*

Solution. Fig. 5.33 shows the arrangement. Finding the potential gradient means to find the electric intensity (or electric stress).

$$V_1 = E_1\,x_1 = E_1 \times (1 \times 10^{-3})$$
$$V_2 = E_2\,x_2 = E_2 \times (2 \times 10^{-3})$$

Now $\qquad V = V_1 + V_2$

or $\qquad 1000 = (E_1 + 2E_2)\,10^{-3}$

or $\qquad E_1 + 2E_2 = 10^6$ $\qquad$ ...(i)

Since flux density $D\,(= \varepsilon_0\,\varepsilon_r\,E)$ is the same in the two media,

$\therefore \qquad \varepsilon_0\,\varepsilon_{r1}\,E_1 = \varepsilon_0\,\varepsilon_{r2}\,E_2$

or $\qquad 3\,E_1 = 6\,E_2$ $\qquad$ ...(ii)

From exps. (i) and (ii), we get, $E_1 = \mathbf{0\cdot5 \times 10^6\ V/m}$; $\quad E_2 = \mathbf{0\cdot25 \times 10^6\ V/m}$

Example 5.26. *Two series connected parallel plate capacitors have plate areas of 0·2 m² and 0·04 m², plate separation of 0·5 mm and 0·125 mm and relative permittivities of 1 and 6 respectively. Calculate the total voltage across the capacitors that will produce a potential gradient of 100 kV/cm between the plates of first capacitor.*

Solution. We shall use suffix 1 for the first capacitor and suffix 2 for the second capacitor. Suppose for a potential gradient of 100 kV/cm between the plates of first capacitor, the voltages across first and second capacitors are V_1 and V_2 respectively. Then,

Total voltage V across capacitors is

$$V = V_1 + V_2$$

For the first capacitor $E_1 = 100 \text{ kV/cm} = 100 \times 10^3 \times 10^2 = 10^7 \text{ V/m}$

$\therefore$ $V_1 = E_1 d_1 = (10^7) \times (0.5 \times 10^{-3}) = 5 \times 10^3 \text{ V} = 5\text{kV}$

$D_1 = \varepsilon_0 \varepsilon_{r1} E_1 = \varepsilon_0 \times 10^7$ $(\because \varepsilon_{r1} = 1)$

$Q_1 = A_1 D_1 = (0.2) \times \varepsilon_0 \times 10^7 \text{ C}$

For the second capacitor. Since the capacitors are connected in series, the charge on them is the same *i.e.*

$$Q_1 = Q_2 = 0.2 \times \varepsilon_0 \times 10^7 \text{ C}$$

$\therefore$ $D_2 = \dfrac{Q_2}{A_2} = \dfrac{0.2 \times \varepsilon_0 \times 10^7}{0.04} = 0.5 \times 10^8 \times \varepsilon_0 \text{ C/m}^2$

$\therefore$ $E_2 = \dfrac{D_2}{\varepsilon_0 \varepsilon_{r2}} = \dfrac{0.5 \times 10^8 \times \varepsilon_0}{\varepsilon_0 \times 6} = \dfrac{10^8}{12} \text{V/m}$ $(\because \varepsilon_{r2} = 6)$

$\therefore$ $V_2 = E_2 d_2 = \dfrac{10^8}{12}(0.125 \times 10^{-3}) = 1.04 \times 10^3 \text{ V} = 1.04 \text{ kV}$

$\therefore$ Total voltage across the capacitors is

$$V = V_1 + V_2 = 5 + 1.04 = \textbf{6.04 kV}$$

Example 5.27. *A parallel plate capacitor consists of two square metal plates 500 mm on a side separated by 10 mm. A slab of Teflon (ε_r = 2) 6 mm thick is placed on the lower plate leaving an air gap 4 mm thick between it and the upper plate. If 100V is applied across the capacitor, find the electric field E_a in the air, electric field E_t in Teflon, flux density D_a in air, flux density D_t in Teflon and potential difference V_t across Teflon slab.*

Solution. Electric flux density (*D*) in the two media is the same. However, electric field intensity (*E*) is inversely proportional to the relative permittivity (ε_r) of the medium. If E_a is the electric intensity in air, then electric intensity in Teflon is $E_t = E_a/2$ ($\because$ relative permittivity of Teflon = 2).

Thickness of air, t_a = 4 mm ;   Thickness of Teflon, t_t = 6 mm

Voltage between two plates, $V = E_a t_a + E_t t_t$

or $100 = E_a \times 4 + \dfrac{E_a}{2} \times 6$ $\left[\because E_t = \dfrac{E_a}{2} \right]$

$\therefore$ $E_a = \dfrac{100}{7}$ volts/mm = **14.286 kV/m**

Electric field in Teflon, $E_t = \dfrac{14.286}{2} = \textbf{7.143 kV/m}$

As electric flux density is the same in the two media,

$\therefore$ $D_a = D_t = \varepsilon_0 \varepsilon_r E_a = 8.854 \times 10^{-12} \times 1 \times 14.286 \times 1000$

$$= \textbf{1.265} \times \textbf{10}^{-7} \textbf{ C/m}^2$$

P.D. across Teflon slab, $V_t = E_t \times t_t = 7.143 \times 1000 \times 6 \times 10^{-3} = \textbf{42.86 V}$

Tutorial Problems

1. An electron (charge = 1.6×10^{-19} C; mass = 9.1×10^{-31} kg) is released in a vacuum between two flat, parallel metal plates that are 10cm apart and are maintained at a constant electric potential difference of 750V. If the electron is released at the negative plate, what is the speed just before it strikes the positive plate ?   [**1.6×10^7 ms^{-1}**]

2. To move a charged particle through an electric potential difference of 10^{-3} V requires 2×10^{-6} J of work. What is the magnitude of charge ? $[2 \times 10^{-3}$ C]

3. A proton of mass $1 \cdot 67 \times 10^{-27}$ kg and charge $= 1 \cdot 6 \times 10^{-19}$ C is accelerated from rest through an electric potential of 400 kV. What is its final speed ? $[8 \cdot 8 \times 10^6$ ms$^{-1}]$

5.25. Refraction of Electric Flux

When electric flux passes from one uniform dielectric medium to another of different permittivities, the electric flux gets refracted at the boundary of the two dielectric media. Under this condition, the following two conditions exist at the boundary (called **boundary conditions**) :

(*i*) The normal components of electric flux density are equal *i.e.*

$$D_{1n} = D_{2n}$$

(*ii*) The tangential components of electric field intensities are equal *i.e.*

$$E_{1t} = E_{2t}$$

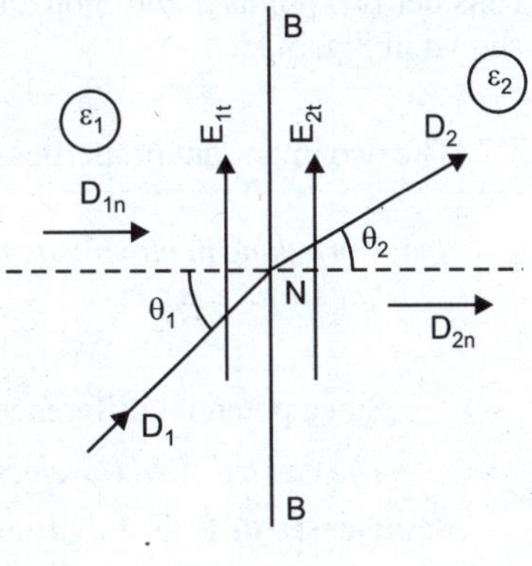

Fig. 5.34

Fig. 5.34 shows the refraction of electric flux at the boundary *BB* of two dielectric media of permittivities ε_1 and ε_2. As shown, the electric flux in the first medium (ε_1) approaches the boundary *BB* at an angle θ_1 and leaves it at θ_2. D_{1n} and D_{2n} are the normal components of D_1 and D_2 while E_{1t} and E_{2t} are the tangential components of E_1 and E_2. Referring to Fig. 5.34,

$$D_{1n} = D_1 \cos \theta_1 \text{ and } D_{2n} = D_2 \cos \theta_2$$

Also

$$E_1 = D_1/\varepsilon_1 \text{ and } E_{1t} = D_1 \sin\theta_1/\varepsilon_1$$

Similarly,

$$E_2 = D_2/\varepsilon_2 \text{ and } E_{2t} = D_2 \sin \theta_2/\varepsilon_2$$

$$\therefore \quad \frac{D_{1n}}{E_{1t}} = \frac{\varepsilon_1}{\tan \theta_1} \text{ and } \frac{D_{2n}}{E_{2t}} = \frac{\varepsilon_2}{\tan \theta_2}$$

Since $D_{1n} = D_{2n}$ and $E_{1t} = E_{2t}$,

$$\therefore \quad \frac{\tan \theta_1}{\tan \theta_2} = \frac{\varepsilon_1}{\varepsilon_2} \qquad \qquad \dots (i)$$

Eq. (*i*) gives the law of refraction of electric flux at the boundary of two dielectric media whose permittivities are different.

It is clear that if $\varepsilon_2 > \varepsilon_1$, then $\theta_2 > \theta_1$.

Note. When electric flux passes from one of the commonly used dielectrics (ε being 2 to 8) into another or air, there is hardly any refraction of electric flux.

Example 5.28. *An electric field in a medium with relative permittivity 7 passes into a medium of relative permittivity 2. If E makes an angle of 60° with the normal to the boundary in the first dielectric, what angle does the field make with the normal in the second dielectric ?*

Solution. As proved in Art 5.25,

$$\frac{\tan \theta_1}{\tan \theta_2} = \frac{\varepsilon_1}{\varepsilon_2}$$

Here $\theta_1 = 60°$; $\varepsilon_1 = 7$; $\varepsilon_2 = 2$; $\theta_2 = ?$

$$\therefore \quad \frac{\tan 60°}{\tan \theta_2} = \frac{7}{2} \quad \text{or} \quad \tan \theta_2 = \sqrt{3} \times \frac{2}{7} = 0.495$$

$$\therefore \quad \theta_2 = \tan^{-1} 0.495 = \mathbf{26.33°}$$

5.26. Equipotential Surface

Any surface over which the potential is constant is called an **equipotential surface.**

In other words, the potential difference between any two points on an equipotential surface is zero. For example, consider two points A and B on an equipotential surface as shown in Fig. 5.35.

$$V_B - V_A = 0 \quad \therefore \; V_B = V_A$$

The two important properties of equipotential surfaces are :

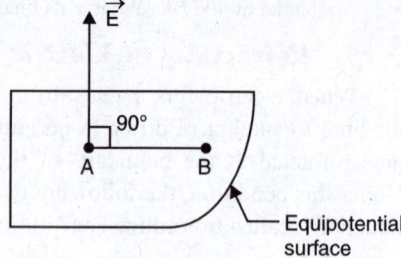

    Fig. 5.35

   (*a*) *Work done in moving a charge over an equipotential surface is zero.*

$$\text{Work done} = \text{Charge} \times \text{P.D.}$$

Since potential difference (P.D.) over an equipotential surface is zero, work done is zero.

   (*b*) *The electric field (or electric lines of force) are *perpendicular to an equipotential surface.*

Some cases of Equipotential surfaces. The fact that the electric field lines and equipotential surfaces are mutually perpendicular helps us to locate the equipotential surfaces when the electric field lines are known.

   (*i*) **Isolated point charge.** The potential at a point P at a distance r from a point charge $+q$ is given by ;

$$V_P = k\frac{q}{r} \quad \text{where} \quad k = \frac{1}{4\pi\varepsilon_0}$$

It is clear that potential at various points equidistant from the point charge is the same. Hence, in case of an isolated point charge, the spheres concentric with the charge will be the equipotential surfaces as shown in Fig. 5.36. Note that in drawing the equipotential surfaces, the potential difference is kept the same, *i.e.,* 10 V in this case. It may be seen that distance between charge and equipotential surface I is small so that $E \;(= dV/dr = 10/dr)$ is high. However, the distance between charge and equipotential surfaces II and III is large so that $E \;(= dV/dr = 10/dr)$ is small. It follows, therefore, that equipotential surfaces near the charge are crowded (*i.e.,* more E) and become widely spaced as we move away from the charge.

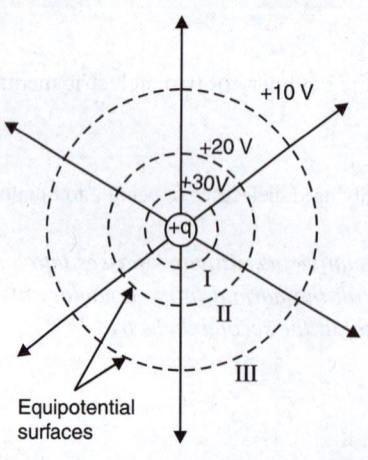

          Fig. 5.36

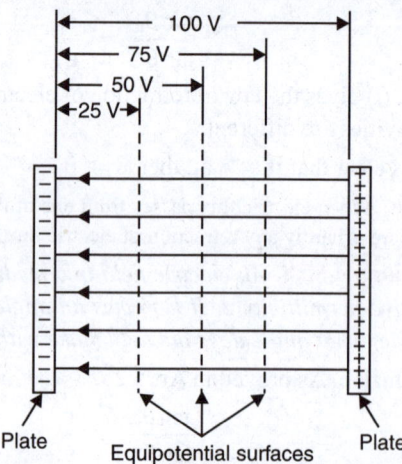

          Fig. 5.37

---

*   If this were not so that is if there were a component of $\vec{E}$ parallel to the surface — it would require work to move the charge along the surface against this component of $\vec{E}$; and this would contradict that it is an equipotential surface.

(*ii*) **Uniform electric field.** In case of uniform electric field (*e.g.*, electric field between the plates of a charged parallel-plate capacitor), the field lines are straight and equally spaced. Therefore, equipotential surfaces will be parallel planes at right angles to the field lines as shown in Fig. 5.37.

5.27. Motion of a Charged Particle in Uniform Electric Field

Consider that a charged particle of charge +q and mass m enters at right angles to a uniform electric field of strength E with velocity v along OX-axis as shown in Fig. 5.38. The electric field is along OY-axis and acts over a horizontal distance x.

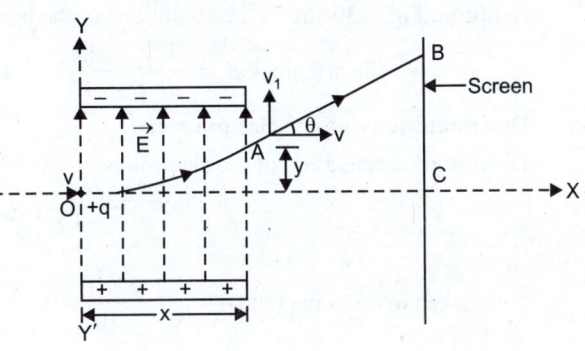

Fig. 5.38

Since the electric field is along OY-axis, no horizontal force acts on the charged particle entering the field. Therefore, the horizontal velocity v of the charged particle remains the same throughout the journey. *The electric field accelerates the charged particle along OY-axis only.*

Force on the charged particle, $F = qE$ ... along OY

Acceleration of the charged particle, $a = \dfrac{qE}{m}$ ...along OY

Time taken to traverse the field, $t = \dfrac{x}{v}$

If y is the displacement of the charged particle along OY direction in the electric field during the time t, then,

$$y = *(0)t + \frac{1}{2}at^2$$

or

$$y = \frac{1}{2}at^2 = \frac{1}{2}\left(\frac{qE}{m}\right)\left(\frac{x}{v}\right)^2$$

or

$$y = \frac{qE}{2mv^2}x^2$$

or

$$y = kx^2 \qquad\qquad \left(\because \frac{qE}{2mv^2} = \text{Constant} = k\right)$$

This is the equation of a parabola. *Therefore, inside the electric field, the charged particle follows a parabolic path OA.* As the charged particle leaves the electric field at A, it follows a straight line path AB tangent to path OA at A.

Note. When an electron (or a charged particle) at rest is accelerated through a potential difference (P.D.) of V volts, then,

Energy imparted to electron = Charge × P.D. = $e \times V$

K.E. gained by electron = $\dfrac{1}{2}mv^2$

∴

$$\frac{1}{2}mv^2 = eV \quad \text{or} \quad v = \sqrt{\frac{2eV}{m}}$$

Here e is the charge on electron and m is the mass of electron. The velocity acquired by the electron is v.

---

*   At the time the charged particle enters the electric field, its velocity along OY-axis is zero.

Example 5.29. *An electron moving with a velocity of 10^7 ms^{-1} enters mid-way between two horizontal plates P, Q in a direction parallel to the plates as shown in Fig. 5.39. The length of the plates is 5 cm and their separation is 2 cm. If a p.d. of 90 V is applied between the plates, calculate the transverse deflection produced by the electric field when the electron just passes the field. Assume e/m = 1.8 × 10^{11} C kg^{-1}.*

Solution. Fig. 5.39 shows the conditions of the problem.

$$\text{Electric field, } E = \frac{V}{d} = \frac{90}{2 \times 10^{-2}} = 45 \times 10^2 \text{ Vm}^{-1}$$

Downward force on the electron = eE

Downward acceleration of the electron is

$$a = \frac{eE}{m} = (1.8 \times 10^{11}) \times (45 \times 10^2) = 81 \times 10^{13} \text{ ms}^{-2}$$

$$\text{Time taken to cross the field, } t = \frac{x}{v} = \frac{5 \times 10^{-2}}{10^7} = 5 \times 10^{-9} \text{ s}$$

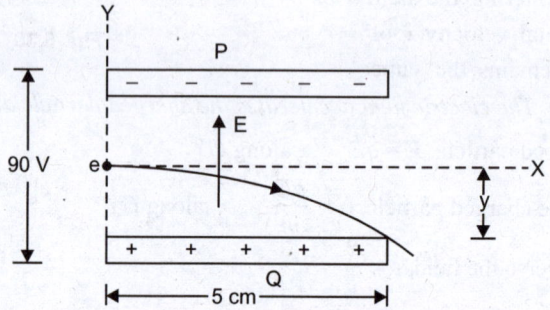

Fig. 5.39

∴   Transverse deflection, $y = \frac{1}{2}at^2 = \frac{1}{2}(81 \times 10^{13}) \times (5 \times 10^{-9})^2 = 0.01 \text{ m} = \mathbf{1cm}$

Example 5.30. *A potential gradient of 3 × 10^6 V/m is maintained between two horizontal parallel plates 1 cm apart. An electron starts from rest at the negative plate, travels under the influence of potential gradient to the positive plate. Given the mass of electron = 9.1 × 10^{-31} kg and the charge on electron = 1.603 × 10^{-19} C. Calculate (i) the force acting on the electron (ii) the ratio of electric force to gravitational force (iii) acceleration (iv) time taken to reach the positive plate.*

Solution. $E = 3 \times 10^6$ V/m ; $e = 1.603 \times 10^{-19}$ C ; $m = 9.1 \times 10^{-31}$ kg ; $S = 1 \times 10^{-2}$ m

(i) Force on electron, $F = eE = 1.603 \times 10^{-19} \times 3 \times 10^6 = \mathbf{4.81 \times 10^{-13}}$ **N**

(ii) Ratio of electric force to gravitational force

$$= \frac{F}{mg} = \frac{4.81 \times 10^{-13}}{9.1 \times 10^{-31} \times 9.81} = \mathbf{5.39 \times 10^{16}}$$

Note that electric force is very large compared to the gravitational force.

(iii) Acceleration of electron, $a = \frac{F}{m} = \frac{4.81 \times 10^{-13}}{9.1 \times 10^{-31}} = \mathbf{51.66 \times 10^{16}}$ **m/s²**

(iv) Distance travelled, $S = \frac{1}{2}at^2$

∴   Time taken to reach + ve plate, $t = \sqrt{\frac{2S}{a}} = \sqrt{\frac{2 \times 1 \times 10^{-2}}{51.66 \times 10^{16}}} = \mathbf{1.968 \times 10^{-10}}$ **s**

Example 5.31. *An electron of charge 1.6 × 10⁻¹⁹ C can move freely for a distance of 2 cm in a field of 1000 V/cm. The mass of the electron is 9.1 × 10⁻²⁸ g. If the electron starts with an initial velocity of zero, what velocity will it attain, what will be the time taken and what will be its kinetic energy?*

Solution. $e = 1.6 \times 10^{-19}$ C ; $m = 9.1 \times 10^{-31}$ kg ; $E = 1000$ V/cm $= 10^5$ V/m

Distance of free movement, $d = 2$ cm $= 0.02$ m

∴ Potential difference applied, $V = E \times d = 10^5 \times 0.02$ volts

Energy imparted to electron = Charge × P.D. = $e \times V$

$$= 1.6 \times 10^{-19} \times 10^5 \times 0.02 = 3.2 \times 10^{-16} \text{ J}$$

Now, Energy imparted = K.E. of electron = **3.2 × 10⁻¹⁶ J**

Also $$v = \sqrt{\frac{2eV}{m}} = \sqrt{\frac{2 \times 3.2 \times 10^{-16}}{9.1 \times 10^{-31}}} = \textbf{2.652} \times \textbf{10}^7 \textbf{ m/s}$$

Force on electron, $F = eE = 1.6 \times 10^{-19} \times 10^5 = 1.6 \times 10^{-14}$ N

Acceleration of electron, $$a = \frac{F}{m} = \frac{1.6 \times 10^{-14}}{9.1 \times 10^{-31}} = 1.758 \times 10^{16} \text{ m/s}^2$$

Distance travelled, $$d = \frac{1}{2}at^2$$

∴ Time taken, $$t = \sqrt{\frac{2d}{a}} = \sqrt{\frac{2 \times 0.02}{1.758 \times 10^{16}}} = \textbf{1.51} \times \textbf{10}^{-9} \textbf{ s}$$

Objective Questions

1. The force between two electrons separated by a distance r varies as
   - (i) r^2
   - (ii) r
   - (iii) r^{-1}
   - (iv) r^{-2}

2. Two charges are placed at a certain distance apart. A brass sheet is placed between them. The force between them will
   - (i) increase
   - (ii) decrease
   - (iii) remain unchanged
   - (iv) none of the above

3. Which of the following appliance will be studied under electrostatics ?
   - (i) incandescent lamp
   - (ii) electric iron
   - (iii) lightning rod
   - (iv) electric motor

4. The relative permittivity of air is
   - (i) 0
   - (ii) 1
   - (iii) 8.854×10^{-12}
   - (iv) none of the above

5. The relative permittivity of a material is 10. Its absolute permittivity will be
   - (i) 8.854×10^{-11} F/m
   - (ii) 9×10^8 F/m
   - (iii) 5×10^{-5} F/m
   - (iv) 9×10^5 F/m

6. Another name for relative permittivity is
   - (i) dielectric constant
   - (ii) dielectric strength
   - (iii) potential gradient
   - (iv) none of the above

7. The relative permittivity of most materials lies between
   - (i) 20 and 100
   - (ii) 10 and 20
   - (iii) 100 and 200
   - (iv) 1 and 10

8. When the relative permittivity of the medium is increased, the force between two charges placed at a given distance apart
   - (i) increases
   - (ii) decreases
   - (iii) remains the same
   - (iv) none of the above

9. Two charges are placed at a distance apart. If a glass slab is placed between them, the force between the charges will
   - (i) be zero
   - (ii) increase
   - (iii) decrease
   - (iv) remain the same

10. There are two charges of $+1$ µC and $+5$ µC. The ratio of the forces acting on them will be

(i) 1 : 1   (ii) 1 : 5

(iii) 5 : 1   (iv) 1 : 25

11. A soap bubble is given a negative charge. Its radius

(i) decreases   (ii) increases

(iii) remains unchanged

(iv) information is incomplete to say anything

12. The ratio of force between two small spheres with constant charge in air and in a medium of relative permittivity K is

(i) $K^2 : 1$ (ii) $1 : K$

(iii) $1 : K^2$ (iv) $K : 1$

13. An electric field can deflect

(i) x-rays   (ii) neutrons

(iii) α-particles   (iv) γ-rays

14. Electric lines of force enter or leave a charged surface at an angle

(i) of 90°   (ii) of 30°

(iii) of 60°

(iv) depending upon surface conditions

15. The relation between absolute permittivity of vacuum (ε_0), absolute permeability of vacuum (μ_0) and velocity of light (c) in vacuum is

(i) $\mu_0 \varepsilon_0 = c^2$ (ii) $\mu_0/\varepsilon_0 = c$

(iii) $\varepsilon_0/\mu_0 = c$ (iv) $\dfrac{1}{\mu_0 \varepsilon_0} = c^2$

16. As one penetrates a uniformly charged sphere, the electric field strength E

(i) increases   (ii) decreases

(iii) is zero at all points

(iv) remains the same as at the surface

17. If the relative permittivity of the medium increases, the electric intensity at a point due to a given charge

(i) decreases   (ii) increases

(iii) remains the same

(iv) none of the above

18. Electric lines of force about a negative point charge are

(i) circular, anticlockwise

(ii) circular, clockwise

(iii) radial, inward   (iv) radial, outward

19. A hollow sphere of charge does not produce an electric field at any

(i) outer point   (ii) interior point

(iii) beyond 2 m   (iv) beyond 10 m

20. Two charged spheres of radii 10 cm and 15 cm are connected by a thin wire. No current will flow if they have

(i) the same charge (ii) the same energy

(iii) the same field on their surface

(iv) the same potential

21. Electric potential is a

(i) scalar quantity (ii) vector quantity

(iii) dimensionless

(iv) nothing can be said

22. A charge Q_1 exerts some force on a second charge Q_2. A third charge Q_3 is brought near. The force of Q_1 exerted on Q_2

(i) decreases   (ii) increases

(iii) remains unchanged

(iv) increases if Q_3 is of the same sign as Q_1 and decreases if Q_3 is of opposite sign

23. The potential at a point due to a charge is 9 V. If the distance is increased three times, the potential at that point will be

(i) 27 V   (ii) 3 V

(iii) 12 V   (iv) 18 V

24. A hollow metal sphere of radius 5 cm is charged such that the potential on its surface is 10 V. The potential at the centre of the sphere is

(i) 10 V   (ii) 0 V

(iii) same as at point 5 cm away from the surface

(iv) same as at point 25 cm away from the surface

25. If a unit charge is taken from one point to another over an equipotential surface, then,

(i) work is done on the charge

(ii) work is done by the charge

(iii) work on the charge is constant

(iv) no work is done

Answers

1. (iv)	2. (ii)	3. (iii)	4. (ii)	5. (i)
6. (i)	7. (iv)	8. (ii)	9. (iii)	10. (i)
11. (ii)	12. (iv)	13. (iii)	14. (i)	15. (iv)
16. (iii)	17. (i)	18. (iii)	19. (ii)	20. (iv)
21. (i)	22. (iii)	23. (ii)	24. (i)	25. (iv)

Capacitance and Capacitors

Introduction

It is well known that different bodies hold different charge when given the same potential. This charge holding property of a body is called *capacitance* or *capacity* of the body. In order to store sufficient charge, a device called capacitor is purposely constructed. A capacitor essentially consists of two conducting surfaces (say metal plates) separated by an insulating material (*e.g.*, air, mica, paper etc.). It has the property to store electrical energy in the form of electrostatic charge. The capacitor can be connected in a circuit so that this stored energy can be made to flow in a desired circuit to perform a useful function. Capacitance plays an important role in d.c. as well as a.c. circuits. In many circuits (*e.g.*, radio and television circuits), capacitors are intentionally inserted to introduce the desired capacitance. In this chapter, we shall confine our attention to the role of capacitance in d.c. circuits only.

6.1. Capacitor

Any two conducting surfaces separated by an insulating material is called a ***capacitor** *or* **condenser.** Its purpose is to store charge in a small space.

The conducting surfaces are called the *plates* of the capacitor and the insulating material is called the ***dielectric.* The most commonly used dielectrics are air, mica, waxed paper, ceramics *etc.* The following points may be noted carefully :

(*i*) The ability of a capacitor to store charge (*i.e.* its capacitance) depends upon the area of plates, distance between plates and the nature of insulating material (or dielectric).

(*ii*) A capacitor is generally named after the dielectric used *e.g.* air capacitor, paper capacitor, mica capacitor *etc.*

(*iii*) The capacitor may be in the form of parallel plates, concentric cylinder or other arrangement.

6.2. How does a Capacitor Store Charge ?

Fig. 6.1 shows how a capacitor stores charge when connected to a d.c. supply. The parallel plate capacitor having plates A and B is connected across a battery of V volts as shown in Fig. 6.1 (*i*). When the switch S is open as shown in Fig. 6.1 (*i*), the capacitor plates are neutral *i.e.* there is no charge on the plates. When the switch is closed as shown in Fig. 6.1 (*ii*), the electrons from plate A will be attracted by the +ve terminal of the battery and these electrons start ***accumulating on plate B. The result is that plate A attains more and more positive charge and plate B gets more and more negative charge. This action is referred to as charging a capacitor because the capacitor plates are becoming charged. This process of electron flow or charging (*i.e.* detaching electrons from plate A and accumulating on B) continues till p.d. across capacitor plates becomes equal to battery voltage V. When the capacitor is charged to battery voltage V, the current flow ceases as shown in Fig. 6.1

---

*   The name is derived from the fact that this arrangement has the capacity to store charge. The name condenser is given to the device due to the fact that when p.d. is applied across it, the electric lines of force are condensed in the small space between the plates.

**   A steady current cannot pass through an insulator but an electric field can. For this reason, an insulator is often referred to as a dielectric.

*** The electrons cannot flow from plate B to A as there is insulating material between the plates. Hence electrons detached from plate A start piling up on plate B.

(*iii*). If now the switch is opened as shown in Fig. 6.1 (*iv*), the capacitor plates will retain the charges. Thus the capacitor plates which were neutral to start with now have charges on them. This shows that a capacitor stores charge. The following points may be noted about the action of a capacitor :

(*i*) When a d.c. potential difference is applied across a capacitor, a charging current will flow until the capacitor is fully charged when the current will cease. This whole charging process takes place in a very short time, a fraction of a second. *Thus a capacitor once charged, prevents the flow of direct current.*

(*ii*) *The current does not flow through the capacitor i.e. between the plates.* There is only transference of electrons from one plate to the other.

(*iii*) When a capacitor is charged, the two plates carry equal and opposite charges (say + Q and –Q). This is expected because one plate loses as many electrons as the other plate gains. *Thus charge on a capacitor means charge on either plate.*

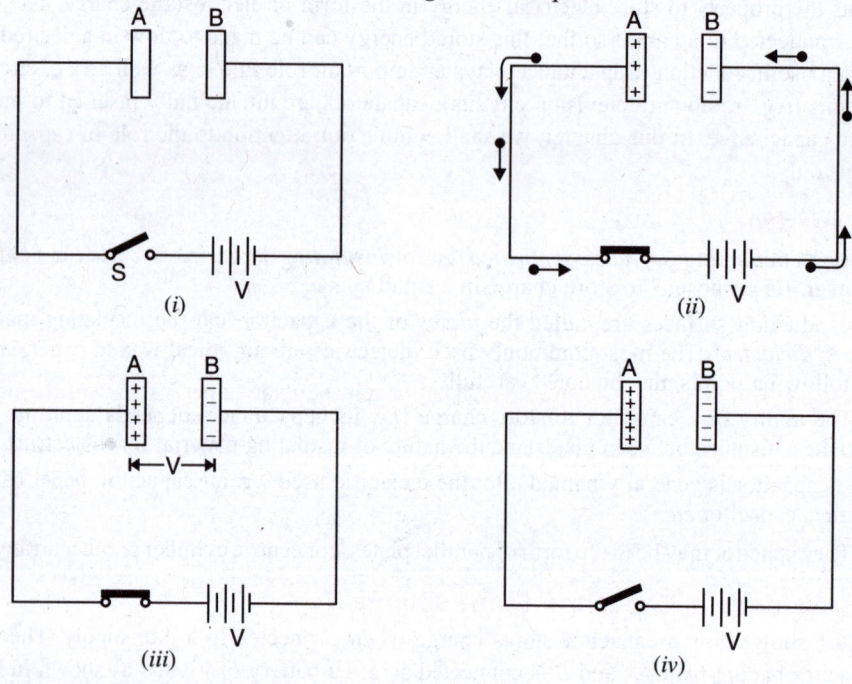

Fig. 6.1

(*iv*) The energy required to charge the capacitor (*i.e.* transfer of electrons from one plate to the other) is supplied by the battery.

6.3. Capacitance

The ability of a capacitor to store charge is known as its capacitance. It has been found experimentally that charge Q stored in a capacitor is directly proportional to the p.d. V across it *i.e.*

$$Q \propto V$$

or

$$\frac{Q}{V} = \text{Constant} = C$$

The constant C is called the capacitance of the capacitor. Hence capacitance of a capacitor can be defined as under :

The ratio of charge on capacitor plates to the p.d. across the plates is called **capacitance** *of the capacitor.*

Unit of capacitance

We know that : $C = Q/V$

The SI unit of charge is 1 coulomb and that of voltage is 1 volt. Therefore, the SI unit of capacitance is one coulomb/volt which is also called *farad* (Symbol F) in honour of Michael Faraday.

$$1 \text{ farad } = 1 \text{ coulomb/volt}$$

A capacitor is said to have a capacitance of **1 farad** *if a charge of 1 coulomb accumulates on each plate when a p.d. of 1 volt is applied across the plates.*

Thus if a charge of 0·1C accumulates on each plate of a capacitor when a p.d. of 10V is applied across its plates, then capacitance of the capacitor = 0·1/10 = 0·01 F. The farad is an extremely large unit of capacitance. Practical capacitors have capacitances of the order of microfarad (μF) and micro-microfarad (μμF) or picofarad (pF).

$$1 \text{ μF } = 10^{-6}\text{F} \quad ; \quad 1\text{μμF (or 1 pF)} = 10^{-12} \text{ F}$$

6.4. Factors Affecting Capacitance

The ability of a capacitor to store charge (*i.e.* its capacitance) depends upon the following factors :

(*i*) **Area of plate.** The greater the area of capacitor plates, the larger is the capacitance of the capacitor and *vice-versa*. It is because larger the plates, the greater the charge they can hold for a given p.d. and hence greater will be the capacitance.

(*ii*) **Thickness of dielectric.** The capacitance of a capacitor is inversely proportional to the thickness (*i.e.* distance between plates) of the dielectric. The smaller the thickness of dielectric, the greater the capacitance and *vice-versa*. When the plates are brought closer, the electrostatic field is intensified and hence capacitance increases.

(*iii*) **Relative permittivity of dielectric.** The greater the relative permittivity of the insulating material (*i.e.*, dielectric), the greater will be the capacitance of the capacitor and *vice-versa*. It is because the nature of dielectric affects the electrostatic field between the plates and hence the charge that accumulates on the plates.

6.5. Dielectric Constant or Relative Permittivity

The insulating material between the plates of a capacitor is called dielectric. When the capacitor is charged, the electrostatic field extends across the dielectric. The presence of dielectric* increases the concentration of electric lines of force between the plates and hence the charge on each plate. The degree of concentration of electric lines of force between the plates depends upon the nature of dielectric.

The ability of a dielectric material to concentrate electric lines of force between the plates of a capacitor is called **dielectric constant** *or* **relative permittivity** *of the material.*

Air has been assigned a reference value of dielectric constant (or relative permittivity) as 1. The dielectric constant of all other insulating materials is greater than unity. The dielectric constants of materials commonly used in capacitors range from 1 to 10. For example, dielectric constant of mica is 6. It means that if mica is used as a dielectric between the plates of a capacitor, the charge on each plate will be 6 times the value when air is used; other things remaining equal. In other words, with mica as dielectric, the capacitance of the capacitor becomes 6 times as great as when air is used.

---

*   Normally, the electrons of the atoms of the dielectric revolve around their nuclei in their regular orbits. When the capacitor is charged, electrostatic field causes distortion of the orbits of the electrons of the dielectric. This distortion of orbits causes an additional electrostatic field within the dielectric which causes more electrons to be transferred from one plate to the other. Hence, the presence of dielectric increases the charge on the capacitor plates and hence the capacitance.

Let $\qquad$ V = Potential difference between capacitor plates

Q = Charge on capacitor when air is dielectric

Then, $\qquad C_{air} = Q/V$

When mica is used as a dielectric in the capacitor and the same p.d. is applied, the capacitor will now hold a charge of $6Q$.

$$\therefore \qquad C_{mica} = \frac{6Q}{V} = 6\frac{Q}{V} = 6\,C_{air}$$

or $\qquad \dfrac{C_{mica}}{C_{air}} = 6$ = Dielectric constant of mica

Hence **dielectric constant** (or **relative permittivity**) *of a dielectric material is the ratio of capacitance of a capacitor with that material as a dielectric to the capacitance of the same capacitor with air as dielectric.*

6.6. Capacitance of an Isolated Conducting Sphere

We can find the capacitance of an isolated spherical conductor by assuming that "missing" plate is earth (zero potential). Suppose an isolated conducting sphere of radius r is placed in a medium of relative permittivity ε_r as shown in Fig. 6.2. Let charge $+Q$ be given to this spherical conductor. The charge is spread *uniformly over the surface of the sphere. Therefore, in order to find the potential at any point on the surface of sphere (or outside the sphere), we can assume that entire charge $+Q$ is concentrated at the centre O of the sphere.

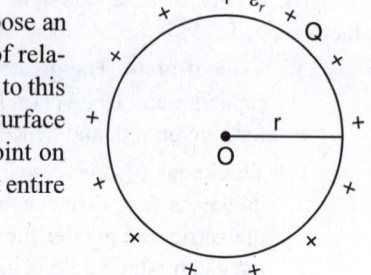

Fig. 6.2

Potential at the surface of the sphere, $V = \dfrac{Q}{4\pi\varepsilon_0\varepsilon_r r}$

$\therefore$ Capacitance of isolated sphere, $C = \dfrac{Q}{V} = 4\pi\varepsilon_0\varepsilon_r r$

$\therefore \qquad **C = 4\pi\,\varepsilon_0\,\varepsilon_r\,r \qquad$ ... *in a medium*

$\qquad\qquad = 4\pi\,\varepsilon_0\,r \qquad$ ... *in air*

The following points may be noted :

(*i*) *The capacitance of an isolated spherical conductor is directly proportional to its radius.* Therefore, for a given potential, a large spherical conductor (more r) will hold more charge $Q\,(= CV)$ than the smaller one.

(*ii*) Unit of $\varepsilon_0 = C/4\pi r$ = F/m. Hence, the SI unit of ε_0 is F/m.

Example 6.1. *Twenty seven spherical drops, each of radius 3 mm and carrying 10^{-12} C of charge are combined to form a single drop. Find the capacitance and potential of the bigger drop.*

Solution. Let r and R be the radii of smaller and bigger drops respectively.

Volume of bigger drop = 27 × Volume of smaller drop

or $\qquad \dfrac{4}{3}\pi R^3 = 27 \times \dfrac{4}{3}\pi r^3$

or $\qquad R = 3r = 3 \times 3 = 9$ mm $= 9 \times 10^{-3}$ m

Capacitance of bigger drop, $C = 4\pi\,\varepsilon_0\,R = \dfrac{1}{9\times10^9} \times 9 \times 10^{-3} = 10^{-12}$ F $= \mathbf{1\ pF}$

---

* Note that a charged conductor is an equipotential surface. Therefore, electric lines of force emerging from the sphere are everywhere normal to the sphere.

** Note that values of Q and V do not occur in the expression for capacitance. This again reminds us that capacitance is a property of physical construction of a capacitor.

Since charge is conserved, the charge on the bigger drop is 27×10^{-12} C.

$\therefore$ Potential of bigger drop, $V = \dfrac{Q}{C} = \dfrac{27 \times 10^{-12}}{10^{-12}} = $ **27 V**

6.7. Capacitance of Spherical Capacitor

We shall discuss two cases.

(*i*) **When outer sphere is earthed.** A spherical capacitor consists of two concentric hollow metallic spheres *A* and *B* which do not touch each other as shown in Fig. 6.3. The outer sphere *B* is earthed while charge is given to the inner sphere *A*. Suppose the medium between the two spheres has relative permittivity ε_r.

Let r_A = radius of inner sphere *A*

r_B = radius of outer sphere *B*

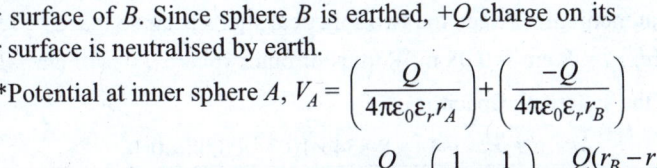

When a charge $+Q$ is given to the inner sphere *A*, it induces a charge $-Q$ on the inner surface of outer sphere *B* and $+Q$ on the outer surface of *B*. Since sphere *B* is earthed, $+Q$ charge on its outer surface is neutralised by earth.

*Potential at inner sphere A, $V_A = \left(\dfrac{Q}{4\pi\varepsilon_0\varepsilon_r r_A} \right) + \left(\dfrac{-Q}{4\pi\varepsilon_0\varepsilon_r r_B} \right)$

Fig. 6.3

$$= \dfrac{Q}{4\pi\,\varepsilon_0\varepsilon_r}\left(\dfrac{1}{r_A} - \dfrac{1}{r_B} \right) = \dfrac{Q(r_B - r_A)}{4\pi\,\varepsilon_0\varepsilon_r\,r_A\,r_B}$$

Since sphere *B* is earthed, its potential is zero (*i.e.*, $V_B = 0$).

$\therefore$ P.D. between *A* and *B*, $V_{AB} = V_A - V_B = V_A - 0 = V_A$

$\therefore$ Capacitance of spherical capacitor, $C = \dfrac{Q}{V_A} = \dfrac{4\pi\,\varepsilon_0\varepsilon_r\,r_A\,r_B}{(r_B - r_A)}$

$\therefore$
$$C = \dfrac{4\pi\,\varepsilon_0\varepsilon_r\,r_A\,r_B}{(r_B - r_A)} \qquad \textit{... in a medium}$$

$$= \dfrac{4\pi\,\varepsilon_0\,r_A\,r_B}{(r_B - r_A)} \qquad \textit{... in air}$$

(*ii*) **When inner sphere is earthed.** Fig. 6.4 shows the situation. The system constitutes two capacitors in parallel.

(*a*) One capacitor (C_{BA}) consists of the inner surface of *B* and outer surface of *A*. Its capacitance as found above is

$$C_{BA} = \dfrac{4\pi\,\varepsilon_0\,\varepsilon_r\,r_A\,r_B}{r_B - r_A}$$

(*b*) The second capacitor (C_{BG}) consists of outer surface of *B* and earth. Its capacitance is that of an isolated sphere.

$\therefore$ $C_{BG} = 4\pi\,\varepsilon_0\,r_B$ ... if surrounding medium is air

$\therefore$ Total capacitance $= C_{BA} + C_{BG}$

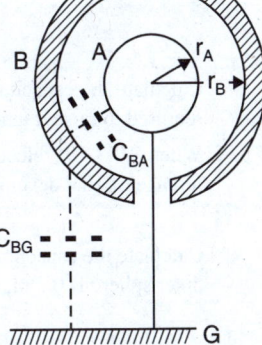

Fig. 6.4

Note. Unless stated otherwise, the outer sphere of a spherical capacitor is assumed to be earthed.

---

* Potential on sphere *A* = (Potential on sphere *A* due to its own charge $+Q$) + (Potential on sphere *A* due to charge $-Q$ on sphere *B*) = $\left(\dfrac{Q}{4\pi\,\varepsilon_0\varepsilon_r\,r_A} \right) + \left(\dfrac{-Q}{4\pi\,\varepsilon_0\varepsilon_r\,r_B} \right)$

Example 6.2. *The thickness of air layer between two coatings of a spherical capacitor is 2 cm. The capacitor has the same capacitance as the capacitance of sphere of 1.2 m diameter. Find the radii of its surfaces.*

Solution. Given : $\dfrac{4\pi\,\varepsilon_0\,r_A\,r_B}{r_B-r_A} = 4\pi\,\varepsilon_0\,R \quad \therefore \quad \dfrac{r_A\,r_B}{r_B-r_A} = R$

Here, $\qquad r_B - r_A = 2$ cm and $R = 1\cdot2/2 = 0\cdot6$ m $= 60$ cm

$\therefore \qquad \dfrac{r_A\,r_B}{2} = 60 \quad$ or $\quad r_A\,r_B = 120$

Now $(r_B + r_A)^2 = (r_B - r_A)^2 + 4r_A\,r_B = (2)^2 + 4 \times 120 = 484$

$\therefore \qquad r_B + r_A = \sqrt{484} = 22$ cm

Since $r_B - r_A = 2$ cm and $r_B + r_A = 22$ cm, $r_B = $ **12 cm** ; $r_A = $ **10 cm**

Example 6.3. *A capacitor has two concentric thin spherical shells of radii 8 cm and 10 cm. The outer shell is earthed and a charge is given to the inner shell. Calculate (i) the capacitance of this capacitor and (ii) the final potential acquired by the inner shell if the outer shell is removed after the inner shell has acquired a potential of 200 V.*

Solution. It is assumed that medium between the two spherical shells is air so that $\varepsilon_r = 1$.

(*i*) Radius of inner sphere, $r_A = 8$ cm $= 0.08$ m; Radius of outer sphere, $r_B = 10$ cm $= 0.1$ m

The capacitance C of the spherical capacitor is

$$C = \frac{4\pi\varepsilon_0\varepsilon_r\,r_A r_B}{r_B - r_A} = \frac{4\pi \times 8.854 \times 10^{-12} \times 0.08 \times 0.1}{0.1 - 0.08} = \mathbf{44.44 \times 10^{-12}\,F}$$

(*ii*) Charge on the capacitor when the inner sphere acquires a potential of 200 V is

$$Q = CV = 44.44 \times 10^{-12} \times 200 = 8888 \times 10^{-12}\,C$$

When the outer shell is removed, the capacitance C' of the resulting isolated sphere is

$$C' = 4\pi\varepsilon_0\varepsilon_r\,r_A = \frac{1}{9 \times 10^9} \times 1 \times 0.08 = 8.88 \times 10^{-12}\,F$$

$\therefore$ Potential V' acquired by the inner shell when outer shell is removed is

$$V' = \frac{Q}{C'} = \frac{8888 \times 10^{-12}}{8.88 \times 10^{-12}} = \mathbf{1000\,V}$$

Tutorial Problems

1. Calculate the capacitance of a conducting sphere of radius 10 cm situated in air. How much charge is required to raise it to a potential of 1000 V? **[11 pF; 1.1×10^{-8} C]**

2. When 1.0×10^{12} electrons are transferred from one conductor to another of a capacitor, a potential difference of 10V develops between the two conductors. Calculate the capacitance of the capacitor.

   [1.6×10^{-8} F]

3. Calculate the capacitance of a spherical capacitor if the diameter of inner sphere is 0.2 m and that of the outer sphere is 0.3 m, the space between them being filled with a liquid having dielectric constant 12.

   [4×10^{-10} F]

4. The stratosphere acts as a conducting layer for the earth. If the stratosphere extends beyond 50 km from the surface of the earth, then calculate the capacitance of the spherical capacitor formed between stratosphere and earth's surface. Take radius of earth as 6400 km. **[0.092 F]**

5. A spherical capacitor has an outer sphere of radius 0.15 m and the inner sphere of radius 0.1m. The outer sphere is earthed and inner sphere is given a charge of 6μC. The space between the concentric spheres is filled with a material of dielectric constant 18. Calculate the capacitance and potential of the inner sphere. **[6×10^{-10} F; 10^4 V]**

6.8. Capacitance of Parallel-Plate Capacitor with Uniform Medium

We have seen that the capacitance of a capacitor can be determined from its electrical properties using the relation $C = Q/V$. However, it is often desirable to determine the capacitance of a capacitor in terms of its dimensions and relative permittivity of the dielectric. Although there are many forms of capacitors, the most important arrangement is the parallel-plate capacitor.

Consider a parallel plate capacitor consisting of two plates, each of area A square metres and separated by a *uniform dielectric* of thickness d metres and relative permittivity ε_r as shown in Fig. 6.5. Let a p.d. of V volts applied between the plates place a charge of $+Q$ and $-Q$ on the plates. With reasonable accuracy, it can be assumed that electric field between the plates is uniform.

Electric flux density between plates is

$$D = Q/A \text{ coulomb/m}^2$$

Electric intensity between plates is

$$E = V/d$$

But

$$D = \varepsilon_0 \varepsilon_r E \qquad \text{...See Art. 5.12}$$

or

$$\frac{Q}{A} = \varepsilon_0 \varepsilon_r \frac{V}{d}$$

or

$$\frac{Q}{V} = \frac{\varepsilon_0 \varepsilon_r A}{d}$$

The ratio Q/V is the capacitance C of the capacitor.

$$\therefore \qquad C = \frac{\varepsilon_0 \varepsilon_r A}{d} \qquad \text{...in a medium}$$

$$= \frac{\varepsilon_0 A}{d} \qquad \text{...in air}$$

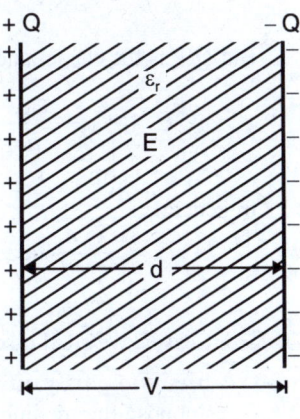

Fig. 6.5

The following points may be noted carefully :

(*i*) Capacitance is directly proportional to ε_r and A and inversely proportional to d.

(*ii*) $\dfrac{C_{med}}{C_{air}} = \varepsilon_r = $ Relative permittivity of medium

(*iii*) Re-arranging the relation for C in air

$$\varepsilon_0 = \frac{Cd}{A} = \frac{\text{farad} \times \text{m}}{\text{m}^2} = \text{F/m}$$

Obviously, permittivity can also be measured in F/m.

6.9. Parallel-Plate Capacitor with Composite Medium

Suppose the space between the plates is occupied by three dielectrics of thicknesses d_1, d_2 and d_3 metres and relative permittivities ε_{r1}, ε_{r2} and ε_{r3} respectively as shown in Fig. 6.6. *The electric flux density D in the dielectrics remains the *same and is equal to Q/A.* However, the electric intensities in the three dielectrics will be different and are given by ;

$$E_1 = \frac{D}{\varepsilon_0 \varepsilon_{r1}} \quad ; \quad E_2 = \frac{D}{\varepsilon_0 \varepsilon_{r2}} \quad ; \quad E_3 = \frac{D}{\varepsilon_0 \varepsilon_{r3}}$$

If V is the total p.d. across the capacitor and V_1, V_2 and V_3 the p.d.s. across the three dielectrics respectively, then,

$$V = V_1 + V_2 + V_3$$
$$= E_1 d_1 + E_2 d_2 + E_3 d_3$$

---

*   The total charge on each plate is Q. Hence Q coulombs is also the total electric flux through each dielectric.

$$= \frac{D}{\varepsilon_0 \, \varepsilon_{r1}} d_1 + \frac{D}{\varepsilon_0 \, \varepsilon_{r2}} d_2 + \frac{D}{\varepsilon_0 \, \varepsilon_{r3}} d_3$$

$$= \frac{D}{\varepsilon_0} \left[\frac{d_1}{\varepsilon_{r1}} + \frac{d_2}{\varepsilon_{r2}} + \frac{d_3}{\varepsilon_{r3}} \right]$$

$$= \frac{Q}{\varepsilon_0 \, A} \left[\frac{d_1}{\varepsilon_{r1}} + \frac{d_2}{\varepsilon_{r2}} + \frac{d_3}{\varepsilon_{r3}} \right] \quad \left(\because D = \frac{Q}{A} \right)$$

or $\quad \dfrac{Q}{V} = \dfrac{\varepsilon_0 \, A}{\left(\dfrac{d_1}{\varepsilon_{r1}} + \dfrac{d_2}{\varepsilon_{r2}} + \dfrac{d_3}{\varepsilon_{r3}} \right)}$

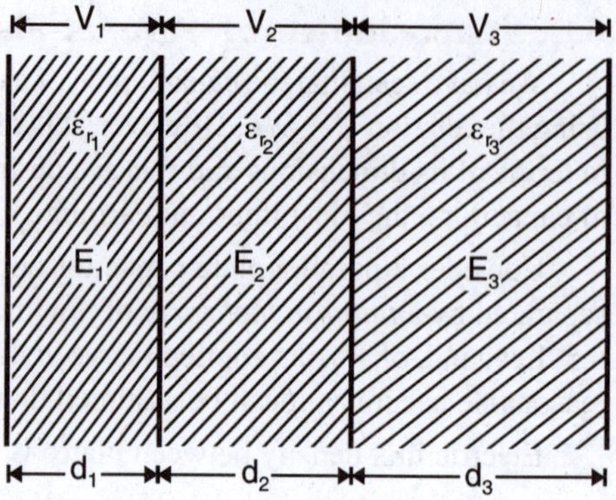

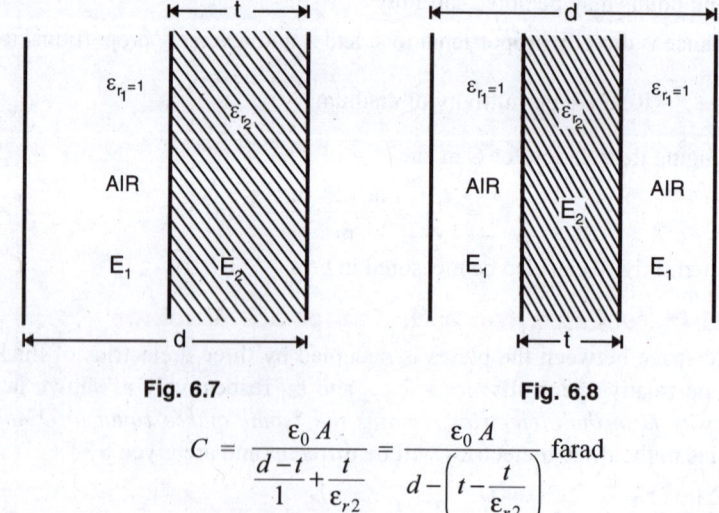

Fig. 6.6

But Q/V is the capacitance C of the capacitor.

$$\therefore \quad C = \frac{\varepsilon_0 \, A}{\left(\dfrac{d_1}{\varepsilon_{r1}} + \dfrac{d_2}{\varepsilon_{r2}} + \dfrac{d_3}{\varepsilon_{r3}} \right)} \text{ farad}$$

In general,

$$C = \frac{\varepsilon_0 A}{\sum \dfrac{d}{\varepsilon_r}} \text{ farad} \qquad \qquad \dots(i)$$

Different cases. We shall discuss the following two cases :

(i) Medium partly air. Fig. 6.7 shows a parallel plate capacitor having plates d metres apart. Suppose the medium between the plates consists partly of air and partly of dielectric of thickness t metres and relative permittivity ε_{r2}. Then thickness of air is $d - t$. Using the relation (i) above, we have,

Fig. 6.7   **Fig. 6.8**

$$C = \frac{\varepsilon_0 \, A}{\dfrac{d-t}{1} + \dfrac{t}{\varepsilon_{r2}}} = \frac{\varepsilon_0 \, A}{d - \left(t - \dfrac{t}{\varepsilon_{r2}} \right)} \text{ farad}$$

(ii) When dielectric slab introduced. Fig. 6.8 shows a parallel-plate air capacitor having plates d metres apart. Suppose a dielectric slab of thickness t metres and relative permittivity ε_{r2} is introduced between the plates of the capacitor.

Using the relation (i) above, we have,

$$C = \frac{\varepsilon_0 \, A}{\dfrac{d-t}{1} + \dfrac{t}{\varepsilon_{r2}}} = \frac{\varepsilon_0 \, A}{d - \left(t - \dfrac{t}{\varepsilon_{r2}} \right)} \text{ farad}$$

6.10. Special Cases of Parallel-Plate Capacitor

We have seen that capacitance of a capacitor depends upon plate area, thickness of dielectric and value of relative permittivity of the dielectric.

We consider two cases by way of illustration.

(i) Fig. 6.9 shows that dielectric thickness is d but plate area is divided into two parts; area A_1 having air as the dielectric and area A_2 having dielectric of relative permittivity ε_r. The arrangement is equivalent to two capacitors in parallel. Their capacitances are :

$$C_1 = \frac{\varepsilon_0 A_1}{d} \quad ; \quad C_2 = \frac{\varepsilon_0 \varepsilon_r A_2}{d}$$

The total capacitance C of this parallel-plate capacitor is

$$C = C_1 + C_2$$

(ii) Fig. 6.10 shows that plate area is divided into two parts ; area A_1 has dielectric (air) of thickness d and area A_2 has a dielectric (ε_r) of thickness t and the remaining thickness is occupied by air. The arrangement is equivalent to two capacitors connected in parallel. Their capacitances are :

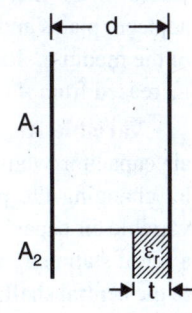

Fig. 6.9   **Fig. 6.10**

$$C_1 = \frac{\varepsilon_0 A_1}{d} \quad ; \quad C_2 = \frac{\varepsilon_0 A_2}{[d - (t - t/\varepsilon_r)]}$$

The total capacitance C of this parallel plate capacitor is

$$C = C_1 + C_2$$

6.11. Multiplate Capacitor

The most *convenient way of achieving large capacitance is by using large plate area. Increasing the plate area may increase the physical size of the capacitor enormously. In order to obtain a large area of plate surface without using too bulky a capacitor, multiplate construction is employed. In this construction, the capacitor is built up of alternate sheets of metal foil (i.e. plates) and thin sheets of dielectric. The odd-numbered metal sheets are connected together to form one terminal T_1 and even-numbered metal sheets are connected together to form the second terminal T_2.

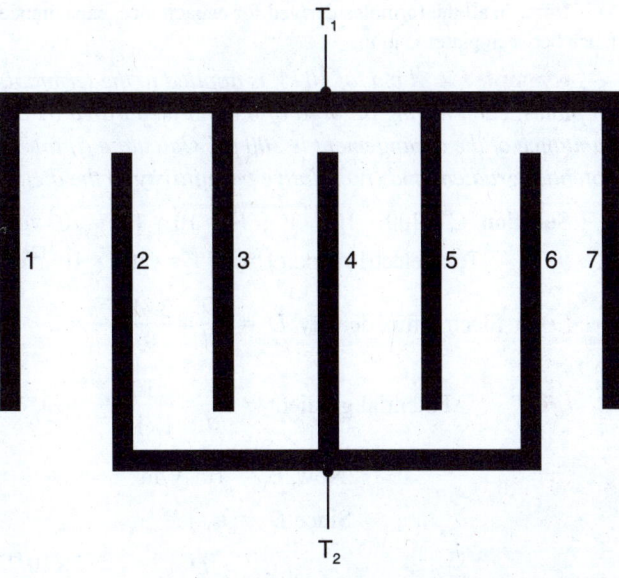

Fig. 6.11

Fig. 6.11 shows a multiplate capacitor with seven plates. A little

---

* The capacitance of a capacitor can also be increased by (i) using a dielectric of high ε_r and (ii) decreasing the distance between plates. High cost limits the choice of dielectric and dielectric strength of the insulating material limits the reduction in spacing between the plates..

reflection shows that this arrangement is equivalent to 6 capacitors in parallel. The total capacitance will, therefore, be 6 times the capacitance of a single capacitor (formed by say plates 1 and 2). If there are n plates, each of area A, then $(n-1)$ capacitors will be in parallel.

∴   Capacitance of n plate capacitor is

$$C = (n-1)\frac{\varepsilon_0\,\varepsilon_r\,A}{d}$$

where d is the distance between any two adjacent plates and ε_r is the relative permittivity of the medium. It may be seen that plate area is increased from A to $A\,(n-1)$.

Variable Air capacitor. It is a multiplate air capacitor whose capacitance can be varied by changing the plate area. Fig. 6.12 shows a

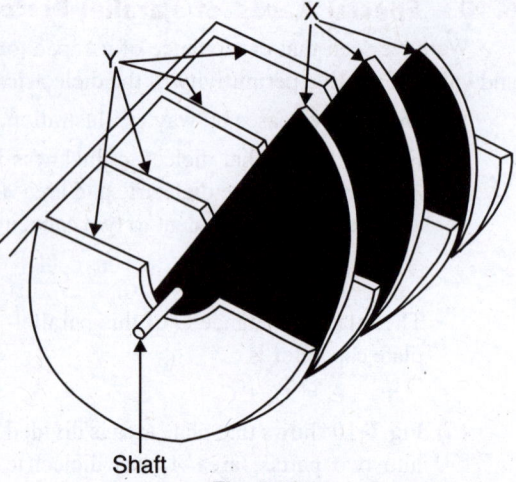

Shaft

Fig. 6.12

variable air capacitor commonly used to "tune in" radio stations in the radio receiver. It consists of a set of stationary metal plates Y fixed to the frame and another set of movable metal plates X fixed to the central shaft. The two sets of plates are electrically insulated from each other. Rotation of the shaft moves the plates X into the spaces between plates Y, thus changing the *common (or effective) plate area and hence the capacitance. The capacitance of such a capacitor is given by ;

$$C = (n-1)\frac{\varepsilon_0\,A}{d} \qquad\qquad (\because \varepsilon_r = 1)$$

When the movable plates X are completely rotated in (*i.e.* the two sets of plates completely over-lap each other), the common plate area 'A' is maximum and so is the capacitance of the capacitor. Minimum capacitance is obtained when the movable plates X are completely rotated out of stationary plates Y. The capacitance of such variable capacitors is from zero to about 4000 pF.

Note. In all the formulas derived for capacitance, capacitance will be in farad if area is in m^2 and the distance between plates is in m.

Example 6.4. *A p.d. of 10 kV is applied to the terminals of a capacitor consisting of two parallel plates, each having an area of 0·01 m^2 separated by a dielectric 1 mm thick. The resulting capacitance of the arrangement is 300 pF. Calculate (i) total electric flux (ii) electric flux density (iii) potential gradient and (iv) relative permittivity of the dielectric.*

Solution. $C = 300 \times 10^{-12}$ F ; $V = 10 \times 10^3 = 10^4$ volts

(*i*)   Total electric flux, $Q = CV = (300 \times 10^{-12}) \times 10^4 = 3 \times 10^{-6}$ C $= \mathbf{3\mu C}$

(*ii*)   Electric flux density, $D = \dfrac{Q}{A} = \dfrac{3 \times 10^{-6}}{0\cdot 01} = \mathbf{3 \times 10^{-4}\ C/m^2}$

(*iii*)   Potential gradient $= \dfrac{V}{d} = \dfrac{10^4}{1 \times 10^{-3}} = \mathbf{10^7\ V/m}$

(*iv*)   Now, $E = 10^7$ V/m

                              Since $D = \varepsilon_0 \varepsilon_r\, E$

∴ $\varepsilon_r = \dfrac{D}{\varepsilon_0 E} = \dfrac{3 \times 10^{-4}}{(8.854 \times 10^{-12}) \times 10^7} = \mathbf{3\cdot 39}$

---

*   Remember in the formula for capacitance, A is the common plate area *i.e.* plate area facing the opposite polarity plate area.

Example. 6.5. *A capacitor is composed of two plates separated by 3mm of dielectric of permittivity 4. An additional piece of insulation 5mm thick is now inserted between the plates. If the capacitor now has capacitance one-third of its original capacitance, find the relative permittivity of the additional dielectric.*

Solution. Figs. 6.13 (*i*) and 6.13 (*ii*) respectively show the two cases.

For the first case,
$$C = \frac{\varepsilon_0 \varepsilon_{r1} A}{d} = \frac{\varepsilon_0 \times 4 \times A}{3 \times 10^{-3}} \qquad \text{...(i)}$$

For the second case,
$$\frac{C}{3} = \frac{\varepsilon_0 A}{\dfrac{d_1}{\varepsilon_{r1}} + \dfrac{d_2}{\varepsilon_{r2}}}$$

$$= \frac{\varepsilon_0 A}{\dfrac{3 \times 10^{-3}}{4} + \dfrac{5 \times 10^{-3}}{\varepsilon_{r2}}} \qquad \text{...(ii)}$$

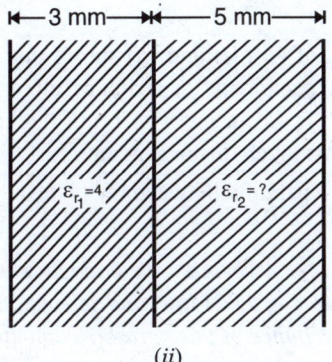

Fig. 6.13

Dividing eq. (*i*) by eq. (*ii*), we get,

$$3 = \frac{4}{3}\left(\frac{3}{4} + \frac{5}{\varepsilon_{r2}}\right)$$

or
$$9 = 3 + 20/\varepsilon_{r2} \quad \therefore \ \varepsilon_{r2} = 20/6 = \textbf{3·33}$$

Example 6.6. *Determine the dielectric flux in microcoulombs between two parallel plates each 0.35 metre square with an air gap of 1.5 mm between them, the p. d. being 3000 V. A sheet of insulating material 1 mm thick is inserted between the plates, the relative permittivity of the insulating material being 6. Find out the potential gradient in the insulating material and also in air if the voltage across the plates is raised to 7500 V.*

Solution. $A = 0.35 \times 0.35 = 0.1225 \text{ m}^2$; $d = 1.5 \text{ mm} = 1.5 \times 10^{-3} \text{ m}$; $\varepsilon_r = 1(\text{air})$.

Capacitance C of the parallel-plate air capacitor is

$$C = \frac{\varepsilon_0 \varepsilon_r A}{d} = \frac{8.854 \times 10^{-12} \times 1 \times 0.1225}{1.5 \times 10^{-3}} = 723 \times 10^{-12} \text{ F}$$

Dielectric flux, $\psi = Q = CV = 723 \times 10^{-12} \times 3000 = 2.17 \times 10^{-6} \text{ C} = \textbf{2.17 μC}$

Suppose the potential gradient in air is g_a. Then potential gradient in the insulating material is $g_i = g_a / \varepsilon_r = g_a / 6$.

Thickness of air ; $t_a = 1.5 - 1 = 0.5 \text{ mm} = 0.5 \times 10^{-3} \text{ m}$; Thickness of insulating material, $t_i = 1\text{mm} = 10^{-3} \text{ m}$.

$\therefore$ Applied voltage, $V = g_a t_a + g_i t_i$

or
$$7500 = g_a \times 0.5 \times 10^{-3} + \frac{g_a}{6} \times 10^{-3}$$

$\therefore \qquad g_a = \textbf{11.25} \times \textbf{10}^6 \textbf{ V/m}$

and
$$g_i = \frac{g_a}{6} = \frac{11.25 \times 10^6}{6} = \textbf{1.875} \times \textbf{10}^6 \textbf{ V/m}$$

Example 6.7. *An air capacitor has two parallel plates of 1500 cm² in area and 5 mm apart. If a dielectric slab of area 1500 cm², thickness 2 mm and relative permittivity 3 is now introduced between the plates, what must be the new separation between the plates to bring the capacitance to the original value?*

Solution. This is a case of introduction of dielectric slab into an air capacitor. As proved in Art. 6.9, the capacitance under this condition becomes :

$$C = \frac{\varepsilon_0 A}{d - (t - t/\varepsilon_r)} \qquad \qquad ...(i)$$

If the medium were totally air, capacitance would have been

$$C_{air} = \frac{\varepsilon_0 A}{d} \qquad \qquad ...(ii)$$

Inspection of eqs. (*i*) and (*ii*) shows that with the introduction of dielectric slab between the plates of air capacitor, its capacitance increases. The distance between the plates is effectively reduced by $t - (t/\varepsilon_r)$. In order to bring the capacitance to the original value, the plates must be separated by this much distance in air.

$\therefore$ New separation between the plates
$$= d + (t - t/\varepsilon_r) = 5 + (2 - 2/3) = \textbf{6·33 mm}$$

Example 6.8. *A variable air capacitor has 11 movable plates and 12 stationary plates. The area of each plate is 0·0015 m² and separation between opposite plates is 0·001 m. Determine the maximum capacitance of this variable capacitor.*

Solution. The capacitance will be maximum when the movable plates are completely rotated in *i.e.* when the two sets of plates completely overlap each other. Under this condition, the common (or effective) area is equal to the physical area of each plate.

$$C = (n-1)\frac{\varepsilon_0 \varepsilon_r A}{d}$$

Here
$$n = 11 + 12 = 23 ; \quad \varepsilon_r = 1 ; \quad A = 0{\cdot}0015 \text{ m}^2 ; \quad d = 0{\cdot}001 \text{ m}$$

$\therefore \qquad C = (23-1) \times \dfrac{8.854 \times 10^{-12} \times 1 \times 0.0015}{0.001} = 292 \times 10^{-12}\text{F} = \textbf{292 pF}$

Example 6.9. *The capacitance of a variable radio capacitor can be changed from 50 pF to 950 pF by turning the dial from 0° to 180°. With dial set at 180°, the capacitor is connected to 400 V battery. After charging, the capacitor is disconnected from the battery and the dial is tuned at 0°. What is the potential difference across the capacitor when the dial reads 0° ?*

Solution. With dial at 0°, the capacitance of the capacitor is
$$C_1 = 50 \text{ pF} = 50 \times 10^{-12} \text{ F}$$

With dial at 180°, the capacitance of the capacitor is
$$C_2 = 950 \text{ pF} = 950 \times 10^{-12} \text{ F}$$

P.D. across $C_2, V_2 = 400$ V

$\therefore \qquad$ Charge on $C_2, Q = C_2V_2 = (950 \times 10^{-12}) \times 400 = 380 \times 10^{-9}$ C

When the battery is disconnected, charge Q remains the same. Suppose V_1 is the potential difference across the capacitor when the dial reads 0°.

$\therefore \qquad Q = C_1V_1$

or $\qquad V_1 = \dfrac{Q}{C_1} = \dfrac{380 \times 10^{-9}}{50 \times 10^{-12}} = 7600 \text{ V}$

Example 6.10. *A parallel plate capacitor has plates of area 2 m^2 spaced by three layers of different dielectric materials. The relative permittivities are 2, 4, 6 and thicknesses are 0.5, 1.5 and 0.3 mm respectively. Calculate the combined capacitance and the electric stress (potential gradient) in each material when applied voltage is 1000 V.*

Solution.   Capacitance, $C = \dfrac{\varepsilon_0 A}{\dfrac{d_1}{\varepsilon_{r1}} + \dfrac{d_2}{\varepsilon_{r2}} + \dfrac{d_3}{\varepsilon_{r3}}}$

$$= \dfrac{8 \cdot 854 \times 10^{-12} \times 2}{\dfrac{0 \cdot 5 \times 10^{-3}}{2} + \dfrac{1 \cdot 5 \times 10^{-3}}{4} + \dfrac{0 \cdot 3 \times 10^{-3}}{6}} = 0 \cdot 0262 \times 10^{-6} \text{ F}$$

Charge on each plate, $Q = CV = (0 \cdot 0262 \times 10^{-6}) \times 1000 = 26 \cdot 2 \times 10^{-6} \text{ C}$

Electric flux density, $D = \dfrac{Q}{A} = \dfrac{26 \cdot 2 \times 10^{-6}}{2} = 13 \cdot 1 \times 10^{-6} \text{ C/m}^2$

Electric stress in the material with $\varepsilon_{r1} = 2$ is

$$E_1 = \dfrac{D}{\varepsilon_0 \varepsilon_{r1}} = \dfrac{13 \cdot 1 \times 10^{-6}}{8 \cdot 854 \times 10^{-12} \times 2} = 74 \times 10^4 \text{ V/m}$$

Electric stress in the material with $\varepsilon_{r2} = 4$ is

$$E_2 = \dfrac{13 \cdot 1 \times 10^{-6}}{8.854 \times 10^{-12} \times 4} = 37 \times 10^4 \text{ V/m}$$

Electric stress in the material with $\varepsilon_{r3} = 6$ is

$$E_3 = \dfrac{13 \cdot 1 \times 10^{-6}}{8854 \times 10^{-12} \times 6} = 24.67 \times 10^4 \text{ V/m}$$

It is clear from the above example that electric stress is greatest in the material having the least relative permittivity. Since air has the lowest relative permittivity, efforts should be made to avoid air pockets in the dielectric materials.

Example 6.11. *A parallel plate capacitor is maintained at a certain potential difference. When a 3 mm slab is introduced between the plates in order to maintain the same potential difference, the distance between the plates is increased by 2·4 mm. Find the dielectric constant of the slab.*

Solution. The capacitance of parallel-plate capacitor in air is

$$C = \dfrac{\varepsilon_0 A}{d} \qquad\qquad\qquad ...(i)$$

With the introduction of slab of thickness t, the new capacitance is

$$C' = \dfrac{\varepsilon_0 A}{d' - t(1 - 1/\varepsilon_r)} \qquad\qquad\qquad ...(ii)$$

Now the charge ($Q = CV$) remains the same in the two cases.

$\therefore \qquad \dfrac{\varepsilon_0 A}{d} = \dfrac{\varepsilon_0 A}{d' - t(1 - 1/\varepsilon_r)}$

or $\qquad\qquad d = d' - t(1 - 1/\varepsilon_r)$

Here, $d' = d + 2 \cdot 4 \times 10^{-3}$ m ; $t = 3$ mm $= 3 \times 10^{-3}$ m

$\therefore \qquad\qquad d = d + 2 \cdot 4 \times 10^{-3} - 3 \times 10^{-3}\left(1 - \dfrac{1}{\varepsilon_r}\right)$

or $\qquad 2 \cdot 4 \times 10^{-3} = 3 \times 10^{-3}\left(1 - \dfrac{1}{\varepsilon_r}\right)$

$\therefore \qquad\qquad \varepsilon_r = 5$

Example 6.12. *A parallel plate capacitor has three similar parallel plates. Find the ratio of capacitance when the inner plate is mid-way between the outers to the capacitance when inner plate is three times as near one plate as the other.*

Solution. Fig. 6.14 (*i*) shows the condition when the inner plate is mid-way between the outer plates. This arrangement is equivalent to two capacitors in parallel.

$$\therefore \quad \text{Capacitance of the capacitor } C_1 \;=\; \frac{\varepsilon_0\,\varepsilon_r\,A}{d/2} + \frac{\varepsilon_0\,\varepsilon_r\,A}{d/2} = \frac{4\varepsilon_0\,\varepsilon_r\,A}{d}$$

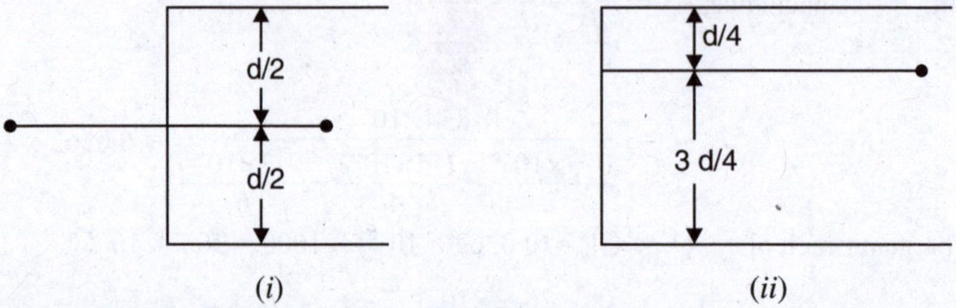

d/2

d/2

d/4

3 d/4

(*i*)   (*ii*)

Fig. 6.14

Fig. 6.14 (*ii*) shows the condition when inner plate is three times as near as one plate as the other.

$$\therefore \text{ Capacitance of the capacitor } C_2 = \frac{\varepsilon_0\,\varepsilon_r\,A}{d/4} + \frac{\varepsilon_0\,\varepsilon_r\,A}{3d/4} = \frac{16\varepsilon_0\,\varepsilon_r\,A}{3d}$$

$$\therefore \qquad\qquad C_1/C_2 \;=\; \mathbf{0.75}$$

Example 6.13. *The permittivity of the dielectric material between the plates of a parallel-plate capacitor varies uniformly from ε_1 at one plate to ε_2 at other plate. Show that the capacitance is given by ;*

$$C \;=\; \frac{A}{d}\,\frac{\varepsilon_2 - \varepsilon_1}{\log_e \varepsilon_2/\varepsilon_1}$$

where A and d are the area of each plate and separation between the plates respectively.

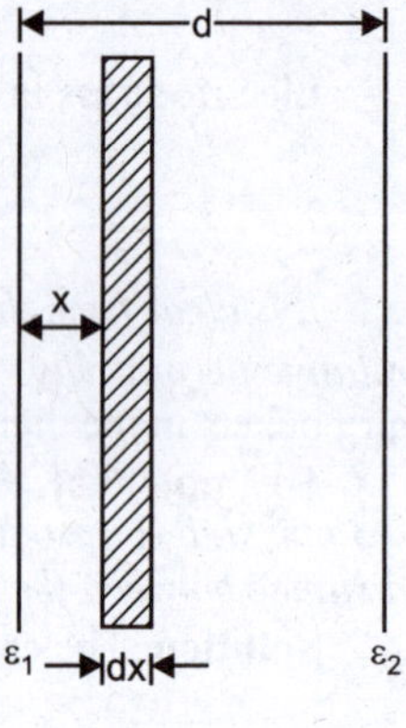

Solution. Fig. 6.15 shows the conditions of the problem. The permittivity of the dielectric material at a distance x from the left plate is

$$\varepsilon_x \;=\; \varepsilon_1 + \frac{x}{d}(\varepsilon_2 - \varepsilon_1)$$

Consider an elementary strip of width dx at a distance x from the left plate. The capacitance C of this strip is

Fig. 6.15

$$C \;=\; \frac{\varepsilon_x A}{dx}$$

or

$$\frac{1}{C} \;=\; \frac{dx}{\varepsilon_x A} = \frac{dx}{A\left[\varepsilon_1 + \dfrac{x}{d}(\varepsilon_2 - \varepsilon_1)\right]} = \frac{d}{A}\,\frac{dx}{\varepsilon_1 d + x(\varepsilon_2 - \varepsilon_1)}$$

$\therefore$ Total capacitance C_T between the plates is

$$\frac{*1}{C_T} \;=\; \int_{x=0}^{x=d}\frac{1}{C} = \frac{d}{A}\int_0^d \frac{dx}{\varepsilon_1 d + x(\varepsilon_2 - \varepsilon_1)}$$

$$=\; \frac{d}{A} \,{**}\left[\frac{\log_e\{\varepsilon_1 d + (\varepsilon_2 - \varepsilon_1)x\}}{\varepsilon_2 - \varepsilon_1}\right]_0^d$$

---

*   The arrangement constitutes capacitors in series.

** $\displaystyle\int \frac{dx}{a + bx} = \frac{\log_e(a+bx)}{b}$

$$= \frac{d}{A(\varepsilon_2 - \varepsilon_1)}[\log_e(\varepsilon_1 d + \varepsilon_2 d - \varepsilon_1 d) - \log_e \varepsilon_1 d]$$

$$= \frac{d}{A(\varepsilon_2 - \varepsilon_1)}\log_e \frac{\varepsilon_2 d}{\varepsilon_1 d} = \frac{d}{A(\varepsilon_2 - \varepsilon_1)}\log_e \frac{\varepsilon_2}{\varepsilon_1}$$

$$\therefore \qquad C_T = \frac{A}{d} \frac{\varepsilon_2 - \varepsilon_1}{\log_e \frac{\varepsilon_2}{\varepsilon_1}}$$

Tutorial Problems

1. A capacitor consisting of two parallel plates 0·5 mm apart in air and each of effective area 500 cm^2 is connected to a 100V battery. Calculate (i) the capacitance and (ii) the charge. **[(i) 885 pF (ii) 0·0885 μC]**

2. A capacitor consisting of two parallel plates in air, each of effective area 50 cm^2 and 1 mm apart, carries a charge of 1770×10^{-12} C. Calculate the p.d. between the plates. If the distance between the plates is increased to 5mm, what will be the electrical effect ? **[40 V ; p.d. across plates is increased to 200 V]**

3. Two insulated parallel plates each of 600 cm^2 effective area and 5 mm apart in air are charged to a p.d. of 1000 V. Calculate (i) the capacitance and (ii) the charge on each plate.

   The source of supply is now disconnected, the plates remaining insulated. Calculate (iii) the p.d. between the plates when their spacing is increased to 10 mm and (iv) the p.d. when the plates, still 10 mm apart, are immersed in oil of relative permittivity 5. **[(i) 106·2 pF (ii) 106·2 $\times$ 10^{-12} C (iii) 2000 V (iv) 400 V]**

4. A p.d. of 500 V is applied across a parallel plate capacitor with a plate area of 0·025 m^2. The plates are separated by a dielectric of relative permittivity 2·5. If the capacitance of the capacitor is 500 μF, find (i) the electric flux (ii) electric flux density and (iii) the electric intensity.
   [(i) 0·25 μC (ii) 0·01 mC/m^2 (iii) 45.3 $\times$ 10^6 V/m]

5. A capacitor consists of two parallel metal plates, each of area 2000 cm^2 and 5 mm apart. The space between the plates is filled with a layer of paper 2 mm thick and a sheet of glass 3 mm thick. The relative permittivities of paper and glass are 2 and 8 respectively. A p.d. of 5 kV is applied across the plates. Calculate (i) the capacitance of the capacitor and (ii) the potential gradient in each dielectric.
   [(i) 1290 pF (ii) 1820 V/mm (paper); 453 V/mm (glass)]

6. A parallel plate capacitor has a plate area of 20 cm^2 and the plates are separated by three dielectric layers each 1 mm thick and of relative permittivity 2, 4 and 5 respectively. Find the capacitance of the capacitor and the electric stress in each dielectric if applied voltage is 1000 V.
   [18·6 pF ; 5·26 $\times$ 10^5 V/m; 2·63 $\times$ 10^5 V/m; 2·11 $\times$ 10^5 V/m]

7. A 1μF parallel plate capacitor that can just withstand a p.d. of 6000 V uses a dielectric having a relative permittivity 5, which breaks down if the electric intensity exceeds 30×10^6 V/m. Find (i) the thickness of dielectric required and (ii) the effective area of each plate. **[(i) 0·2 mm (ii) 4·5 m^2]**

8. An air capacitor has two parallel plates 10 cm^2 in area and 5 mm apart. When a dielectric slab of area 10 cm^2 and thickness 5 mm was inserted between the plates, one of the plates has to be moved by 0·4 cm to restore the capacitance. What is the dielectric constant of the slab ? **[5]**

9. A multiplate parallel capacitor has 6 fixed plates connected in parallel, interleaved with 5 similar plates; each plate has effective area of 120 cm^2. The gap between the adjacent plates is 1 mm. The capacitor is immersed in oil of relative permittivity 5. Calculate the capacitance. **[5·31 pF]**

10. Calculate the number of sheets of tin foil and mica for a capacitor of 0·33 μF capacitance if area of each sheet of tin foil is 82 cm^2, the mica sheets are 0·2 mm thick and have relative permittivity 5.
    [182 sheets of mica; 183 sheets of tin foil]

6.12. Cylindrical Capacitor

A cylindrical capacitor consists of two co-axial cylinders separated by an insulating medium. This is an important practical case since *a single core cable is in effect a capacitor of this kind.* The conductor (or core) of the cable is the inner cylinder while the outer cylinder is represented by lead sheath which is at earth potential. The two co-axial cylinders have insulation between them.

Consider a single core cable with conductor diameter d metres and inner sheath diameter D metres (See Fig. 6.16). Let the charge per metre axial length of the cable be Q coulombs and ε_r be the relative permittivity of the insulating material. Consider a cylinder of radius x metres. According to Gauss's theorem, electric flux passing through this cylinder is Q coulombs. The surface area of this cylinder is

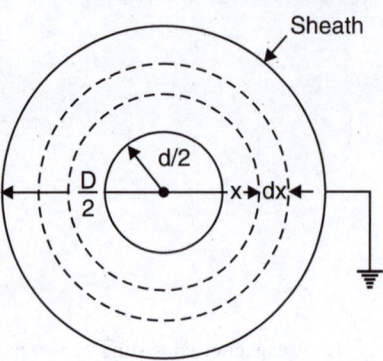

$$= 2\pi x \times 1 = 2\pi x \text{ m}^2$$

∴ Electric flux density at any point P on the considered cylinder is given by ;

$$D_x = \frac{Q}{2\pi x} \text{ C/m}^2$$

Fig. 6.16

Electric intensity at point P is given by;

$$E_x = \frac{D_x}{\varepsilon_0\,\varepsilon_r} = \frac{Q}{2\pi x \varepsilon_0\,\varepsilon_r} \text{ V/m}$$

The work done in moving a unit positive charge from point P through a distance dx in the direction of electric field is $E_x\,dx$. Hence the work done in moving a unit positive charge from conductor to sheath, which is the p.d. V between the conductor and sheath, is given by ;

$$V = \int_{d/2}^{D/2} E_x\,dx = \int_{d/2}^{D/2} \frac{Q}{2\pi x \varepsilon_0\,\varepsilon_r}\,dx = \frac{Q}{2\pi \varepsilon_0\,\varepsilon_r}\log_e\frac{D}{d}$$

∴ Capacitance of cable, $C = \dfrac{Q}{V} = \dfrac{Q}{\dfrac{Q}{2\pi\varepsilon_0\,\varepsilon_r}\log_e\dfrac{D}{d}} \text{F/m} = \dfrac{2\pi\varepsilon_0\,\varepsilon_r}{\log_e(D/d)}\text{F/m}$

$$= \frac{2\pi\times 8\cdot 854\times 10^{-12}\times\varepsilon_r}{2\cdot 303\log_{10}(D/d)}\text{F/m} = \frac{\varepsilon_r}{41\cdot 4\log_{10}(D/d)}\times 10^{-9}\text{ F/m}$$

If the cable has a length of l metres, then capacitance of the cable is

$$= \frac{\varepsilon_r l}{41\cdot 4\log_{10}(D/d)}\times 10^{-9}\text{ F} = \frac{24\varepsilon_r l}{\log_{10}(D/d)}\text{pF}$$

Example 6.14. *In a concentric cable 20 cm long, the diameter of inner and outer cylinders are 15 cm and 15·4 cm respectively. The relative permittivity of the insulation is 5. If a p.d. of 5000 V is maintained between the two cylinders, calculate :*

(i) capacitance of cylindrical capacitor

(ii) the charge

(iii) the electric flux density and electric intensity in the dielectric.

Solution. *(i)* Capacitance of the cylindrical capacitor is

$$C = \frac{\varepsilon_r l}{41\cdot 4\log_{10}(D/d)}\times 10^{-9} = \frac{5\times 0\cdot 2}{41\cdot 4\log_{10}(15\cdot 4/15)}\times 10^{-9}\text{ F} = \mathbf{2.11\times 10^{-9}\ F}$$

(ii) Charge on capacitor, $Q = CV = (2\cdot 11\times 10^{-9})\times 5000 = 10\cdot 55\times 10^{-6}\text{ C} = \mathbf{10{\cdot}55\ \mu C}$

(iii) To determine D and E in the dielectric, we shall consider the average radius of dielectric, *i.e.,*

Average radius of dielectric, $x = \dfrac{1}{2}\left[\dfrac{15}{2} + \dfrac{15\cdot 4}{2}\right] = 7\cdot 6\text{ cm} = 0\cdot 076\text{ m}$

Flux density in dielectric, $D = \dfrac{Q}{2\pi x l}$ C/m^2 $= \dfrac{10 \cdot 55 \times 10^{-6}}{2\pi \times 0 \cdot 076 \times 0 \cdot 2} = \mathbf{110 \cdot 47 \times 10^{-6}}$ **C/m^2**

Electric intensity in dielectric, $E = \dfrac{D}{\varepsilon_0 \varepsilon_r} = \dfrac{110 \cdot 47 \times 10^{-6}}{8 \cdot 854 \times 10^{-12} \times 5} = \mathbf{2.5 \times 10^6}$ **V/m**

Example 6.15. *A 33 kV, 50 Hz, 3-phase underground cable, 4 km long uses three single core cables. Each of the conductor has a diameter of 2·5 cm and the radial thickness of insulation is 0·5 cm. Determine (i) capacitance of the cable/phase (ii) charging current/phase (iii) total charging kVAR. The relative permittivity of insulation is 3.*

Solution. *(i)* Capacitance of cable/phase, $C = \dfrac{\varepsilon_r \, l}{41 \cdot 4 \, \log_{10}(D/d)} \times 10^{-9}$ F

Here $\qquad \varepsilon_r = 3$; $l = 4$ km $= 4000$ m

$\qquad\qquad\qquad d = 2 \cdot 5$ cm ; $D = 2.5 + 2 \times 0.5 = 3.5$ cm

Putting these values in the above expression, we get,

$$C = \dfrac{3 \times 4000 \times 10^{-9}}{41 \cdot 4 \times \log_{10}(3 \cdot 5 / 2 \cdot 5)} = \mathbf{1984 \times 10^{-9}}\ \textbf{F}$$

(ii) $\qquad$ Voltage/phase, $V_{ph} = \dfrac{33 \times 10^3}{\sqrt{3}} = 19 \cdot 05 \times 10^3$ V

Charging current/phase, $I_C = \dfrac{V_{ph}}{X_C} = 2\pi f C\, V_{ph}$

$\qquad\qquad\qquad = 2\pi \times 50 \times 1984 \times 10^{-9} \times 19 \cdot 05 \times 10^3 = \mathbf{11 \cdot 87}$ **A**

(iii) $\qquad$ Total charging kVAR $= 3 V_{ph} I_C = 3 \times 19 \cdot 05 \times 10^3 \times 11 \cdot 87 = \mathbf{678 \cdot 5 \times 10^3}$ **kVAR**

6.13. Potential Gradient in a Cylindrical Capacitor

Under operating conditions, the insulation of a cable is subjected to electrostatic forces. This is known as dielectric stress. The dielectric stress at any point in a cable is infact the potential gradient (or *electric intensity) at that point.

Consider a single core cable with core diameter d and internal sheath diameter D. As proved in Art. 6.12, the electric intensity at a point x metres from the centre of the cable is

$$E_x = \dfrac{Q}{2\pi \varepsilon_0 \varepsilon_r x} \ \text{volts/m}$$

By definition, electric intensity is equal to potential gradient. Therefore, potential gradient g at a point x metres from the centre of the cable is

$$g = E_x$$

or $\qquad\qquad g = \dfrac{Q}{2\pi \varepsilon_0 \varepsilon_r x}$ volts/m $\qquad$ ...(i)

As proved in Art. 6.12, potential difference V between conductor and sheath is

$$V = \dfrac{Q}{2\pi \varepsilon_0 \varepsilon_r} \log_e \dfrac{D}{d} \ \text{volts}$$

or $\qquad\qquad Q = \dfrac{2\pi \varepsilon_0 \varepsilon_r V}{\log_e \dfrac{D}{d}}$ $\qquad$ ...(ii)

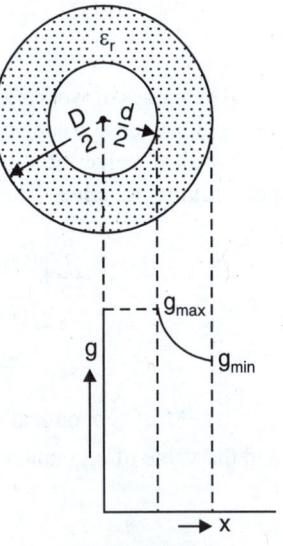

Fig. 6.17

* It may be recalled that potential gradient at any point is equal to the electric intensity at that point.

Substituting the value of Q from exp. (*ii*) in exp. (*i*), we get,

$$g = \frac{2\pi\varepsilon_0\varepsilon_r V}{\log_e D/d} = \frac{V}{x\log_e \dfrac{D}{d}} \text{ volts/m} \qquad ...(iii)$$

It is clear from exp. (*iii*) that potential gradient varies inversely as the distance x. Therefore, potential gradient will be maximum when x is minimum *i.e.*, when $x = d/2$ or at the surface of the conductor. On the other hand, potential gradient will be minimum at $x = D/2$ or at sheath surface.

$\therefore$ Maximum potential gradient, $g_{max} = \dfrac{2V}{d\log_e \dfrac{D}{d}}$ volts/m $\qquad$ [Putting $x = d/2$ in exp. (*iii*)]

Minimum potential gradient, $g_{min} = \dfrac{2V}{D\log_e \dfrac{D}{d}}$ volts/m $\qquad$ [Putting $x = D/2$ in exp. (*iii*)]

$\therefore \qquad \dfrac{g_{max}}{g_{min}} = \dfrac{\dfrac{2V}{d\log_e D/d}}{\dfrac{2V}{D\log_e D/d}} = \dfrac{D}{d}$

The variation of stress in the dielectric is shown in Fig. 6.17. *It is clear that dielectric stress is maximum at the conductor surface and its value goes on decreasing as we move away from the conductor*. It may be noted that maximum stress is an important consideration in the design of a cable. For instance, if a cable is to be operated at such a voltage that *maximum stress is 5 kV/mm, then the insulation used must have a dielectric strength of atleast 5 kV/mm, otherwise breakdown of the cable will become inevitable.

6.14. Most Economical Conductor Size in a Cable

It has already been shown that maximum stress in a cable occurs at the surface of the conductor. For safe working of the cable, dielectric strength of the insulation should be more than the maximum stress. Rewriting the expression for maximum stress, we get,

$$g_{max} = \frac{2V}{d\log_e \dfrac{D}{d}} \text{ volts/m} \qquad ...(i)$$

The values of working voltage V and internal sheath diameter D have to be kept fixed at certain values due to design considerations. This leaves conductor diameter d to be the only variable in exp. (*i*). For given values of V and D, the most economical conductor diameter will be one for which g_{max} has a minimum value. The value of g_{max} will be minimum when $d\log_e D/d$ is maximum *i.e.*

$$\frac{d}{dd}\left[d\log_e \frac{D}{d}\right] = 0 \quad \text{or} \quad \log_e \frac{D}{d} + d.\frac{d}{D}.\frac{-D}{d^2} = 0$$

$\therefore \qquad \log_e (D/d) - 1 = 0$

or $\qquad\qquad \log_e (D/d) = 1 \quad \text{or} \quad (D/d) = e = 2.718$

$\therefore$ Most economical conductor diameter, $d = \dfrac{D}{2.718}$

and the value of g_{max} under this condition is

$$g_{max} = \frac{2V}{d} \text{ volts/m} \qquad\qquad \text{[Putting } \log_e D/d = 1 \text{ in exp. (i)]}$$

---

* Of course, it will occur at the conductor surface.

For low and medium voltage cables, the value of conductor diameter arrived at by this method (*i.e., $d = 2V/g_{max}$*) is often too small from the point of view of current density. Therefore, the conductor diameter of such cables is determined from the consideration of safe current density. For high voltage cables, designs based on this theory give a very high value of *d*, much too large from the point of view of current carrying capacity and it is, therefore, advantageous to increase the conductor diameter to this value. There are three ways of doing this without using excessive copper :

(*i*) Using aluminium instead of copper because for the same current, diameter of aluminium will be more than that of copper.

(*ii*) Using copper wires stranded around a central core of hemp.

(*iii*) Using a central lead tube instead of hemp.

Example 6.16. *The maximum and minimum stresses in the dielectric of a single core cable are 40 kV/cm (r.m.s.) and 10 kV/cm (r.m.s.) respectively. If the conductor diameter is 2 cm, find :*

(*i*) *thickness of insulation (ii) operating voltage*

Solution. Here, g_{max} = 40 kV/cm ; g_{min} = 10 kV/cm ; $d = 2$ cm ; $D = ?$

(*i*)   As proved in Art. 6.13,

$$\frac{g_{max}}{g_{min}} = \frac{D}{d} \quad \text{or} \quad D = \frac{g_{max}}{g_{min}} \times d = \frac{40}{10} \times 2 = 8 \text{ cm}$$

∴   Insulation thickness $= \dfrac{D-d}{2} = \dfrac{8-2}{2} = \textbf{3 cm}$

(*ii*) $g_{max} = \dfrac{2V}{d\log_e \dfrac{D}{d}}$

∴ $V = \dfrac{g_{max}\, d\log_e \dfrac{D}{d}}{2} = \dfrac{40 \times 2\log_e 4}{2} \text{ kV} = \textbf{55.45 kV} \textit{ r.m.s.}$

Example 6.17. *A single core cable for use on 11 kV, 50 Hz system has conductor area of 0·645 cm^2 and internal diameter of sheath is 2·18 cm. The permittivity of the dielectric used in the cable is 3·5. Find (i) the maximum electrostatic stress in the cable (ii) minimum electrostatic stress in the cable (iii) capacitance of the cable per km length (iv) charging current.*

Solution. Area of cross-section of conductor, $a = 0.645$ cm^2

Diameter of the conductor, $d = \sqrt{\dfrac{4a}{\pi}} = \sqrt{\dfrac{4 \times 0 \cdot 645}{\pi}} = 0 \cdot 906$ cm

Internal diameter of sheath, $D = 2 \cdot 18$ cm

(*i*) Maximum electrostatic stress in the cable is

$$g_{max} = \frac{2V}{d\log_e \dfrac{D}{d}} = \frac{2 \times 11}{0.906 \log_e \dfrac{2.18}{0.906}} \text{ kV/cm} = \textbf{27.65 kV/cm} \textit{ r.m.s.}$$

(*ii*) Minimum electrostatic stress in the cable is

$$g_{min} = \frac{2V}{D\log_e \dfrac{D}{d}} = \frac{2 \times 11}{2.18 \log_e \dfrac{2.18}{0.906}} \text{ kV/cm} = \textbf{11.5 kV/cm} \textit{ r.m.s.}$$

(*iii*) Capacitance of cable, $C = \dfrac{\varepsilon_r\, l}{41.4 \log_{10} \dfrac{D}{d}} \times 10^{-9}$ F

Here $\varepsilon_r = 3.5$; $l = 1$ km $= 1000$ m

$$\therefore \qquad C = \frac{3 \cdot 5 \times 1000}{41 \cdot 4 \log_{10} \frac{2 \cdot 18}{0 \cdot 906}} \times 10^{-9} = \mathbf{0 \cdot 22 \times 10^{-6} \ F}$$

(iv)   Charging current, $I_C = \dfrac{V}{X_C} = 2\pi f C V = 2\pi \times 50 \times 0 \cdot 22 \times 10^{-6} \times 11000 = \mathbf{0 \cdot 76 \ A}$

Example 6.18. *Find the most economical size of a single-core cable working on a 132 kV, 3-phase system, if a dielectric stress of 60 kV/cm can be allowed.*

Solution. Phase voltage of cable $= 132/\sqrt{3} = 76 \cdot 21$ kV

Peak value of phase voltage, $V = 76 \cdot 21 \times \sqrt{2} = 107 \cdot 78$ kV

Max. permissible stress, $g_{max} = 60$ kV/cm

$\therefore$ Most economical conductor diameter is

$$d = \frac{2V}{g_{max}} = \frac{2 \times 107 \cdot 78}{60} = \mathbf{3 \cdot 6 \ cm}$$

Internal diameter of sheath, $D = 2 \cdot 718 \ d = 2 \cdot 718 \times 3 \cdot 6 = \mathbf{9 \cdot 78 \ cm}$

Therefore, the cable should have a conductor diameter of 3.6 cm and internal sheath diameter of 9·78 cm.

Example 6.19. *The radius of the copper core of a single-core rubber-insulated cable is 2.25 mm. Calculate the radius of the lead sheath which covers the rubber insulation and the cable capacitance per metre. A voltage of 10 kV may be applied between the core and the lead sheath with a safety factor of 3. The rubber insulation has a relative permittivity of 4 and breakdown field strength of 18×10^6 V/m.*

Solution. As proved in Art 6.13,

$$g_{max} = \frac{2V}{d \log_e \dfrac{D}{d}}$$

Here, $g_{max} = E_{max} = 18 \times 10^6$ V/m ; $V =$ Breakdown voltage $\times$ Safety factor

$= 10^4 \times 3 = 30{,}000$ volts ; $d = 2.25 \times 2 = 4.5$ mm

$$\therefore \qquad 18 \times 10^6 = \frac{2 \times 30{,}000}{4.5 \times 10^{-3} \times \log_e \dfrac{D}{d}}$$

or $\dfrac{D}{d} = 2.1 \ \therefore D = 2.1 \times d = 2.1 \times 4.5 = 9.45$ mm

$\therefore$ Radius of sheath $= \dfrac{D}{2} = \dfrac{9.45}{2} = \mathbf{4.72 \ mm}$

Capacitance, $C = \dfrac{\varepsilon_r l}{41.4 \log_{10} \dfrac{D}{d}} \times 10^{-9} \text{F} = \dfrac{4 \times 1}{41.4 \log_{10} \dfrac{9.45}{4.5}} \times 10^{-9} = \mathbf{0.3 \times 10^{-9} \ F}$

6.15. Capacitance Between Parallel Wires

This case is of practical importance in overhead transmission lines. The simplest system for power transmission is 2-wire d.c. or a.c. system. Consider 2-wire transmission line consisting of two parallel conductors A and B spaced d metres apart in air. Suppose that radius of each conductor is r metres. Let their respective charges be $+ Q$ and $- Q$ coulombs per metre length [See Fig. 6.18].

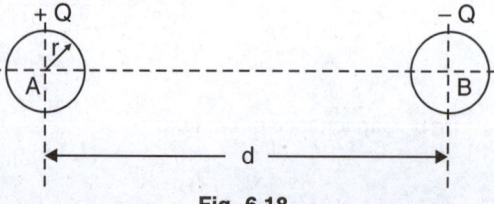

Fig. 6.18

The total p.d. between conductor A and neutral "infinite" plane is

$$V_A^* = \int_r^\infty \frac{Q}{2\pi\,x\varepsilon_0}\,dx + \int_d^\infty \frac{-Q}{2\pi\,x\,\varepsilon_0}\,dx$$

$$= \frac{Q}{2\pi\,\varepsilon_0}\left[\log_e\frac{\infty}{r} - \log_e\frac{\infty}{d}\right]\text{volts} = \frac{Q}{2\pi\,\varepsilon_0}\log_e\frac{d}{r}\;\text{volts}$$

Similarly, p.d. between conductor B and neutral "infinite" plane is

$$V_B = \int_r^\infty \frac{-Q}{2\pi x\varepsilon_0}\,dx + \int_d^\infty \frac{Q}{2\pi x\varepsilon_0}\,dx$$

$$= \frac{-Q}{2\pi\,\varepsilon_0}\left[\log_e\frac{\infty}{r} - \log_e\frac{\infty}{d}\right] = \frac{-Q}{2\pi\,\varepsilon_0}\log_e\frac{d}{r}\;\text{volts}$$

Both these potentials are *w.r.t.* the same neutral plane. Since the unlike charges attract each other, the potential difference between the conductors is

$$V_{AB} = 2V_A = \frac{2Q}{2\pi\,\varepsilon_0}\log_e\frac{d}{r}\;\text{volts}$$

$\therefore$ Capacitance, $C_{AB} = Q/V_{AB} = \dfrac{Q}{\dfrac{2Q}{2\pi\,\varepsilon_0}\log_e\dfrac{d}{r}}$ F/m

$\therefore$

$$C_{AB} = \frac{\pi\,\varepsilon_0}{\log_e\dfrac{d}{r}}\;\text{F/m} \qquad\qquad \text{...(i)}$$

The capacitance for a length l is given by ;

$$C_{AB} = \frac{\pi\varepsilon_0\,l}{\log_e\dfrac{d}{r}}\text{F} \qquad\qquad \textit{... in air}$$

$$= \frac{\pi\varepsilon_0\,\varepsilon_r\,l}{\log_e\dfrac{d}{r}}\text{F} \qquad\qquad \textit{... in a medium}$$

Example 6.20. *A 3-phase overhead transmission line has its conductors arranged at the corners of an equilateral triangle of 2 m side. Calculate the capacitance of each line conductor per km. Given that diameter of each conductor is 1·25 cm.*

Solution. Conductor radius, $r = 1\cdot25/2 = 0\cdot625$ cm ; Spacing of conductors, $d = 2$ m $= 200$ cm

Capacitance of each line conductor $= \dfrac{2\pi\,\varepsilon_0}{\log_e d/r}$ F/m $= \dfrac{2\pi\times8\cdot854\times10^{-12}}{\log_e 200/0.625}$ F/m

$$= 0\cdot0096\times10^{-9}\;\text{F/m} = 0\cdot0096\times10^{-6}\;\text{F/km} = \mathbf{0\cdot0096\;\mu F/km}$$

---

* The electric intensity E at a distance x from the centre of the conductor in air is given by ;

$$E = \frac{Q}{2\pi x\,\varepsilon_0}\;\text{volts/m}$$

Here, Q = charge per metre length ; ε_0 = permittivity of air

As x approaches infinity, the value of E approaches zero. Therefore, the potential difference between the conductors and infinity distant neutral plane is

$$V_A = \int_r^\infty \frac{Q}{2\pi x\,\varepsilon_0}\,dx$$

6.16. Insulation Resistance of a Cable Capacitor

The cable conductor is provided with a suitable thickness of insulating material in order to prevent leakage current. The path for leakage current is radial through the insulation. The opposition offered by insulation to leakage current is known as insulation resistance of the cable. For satisfactory operation, the insulation resistance of the cable should be very high.

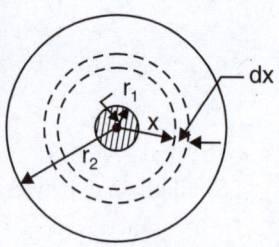

Consider a single-core cable of conductor radius r_1 and internal sheath radius r_2 as shown in Fig. 6.19. Let l be the length of the cable and ρ be the resistivity of the insulation.

Fig. 6.19

Consider a very small layer of insulation of thickness dx at a radius x. The length through which leakage current tends to flow is dx and the area of X-section offered to this flow is $2\pi x\, l$.

∴   Insulation resistance of considered layer

$$= \rho \, \frac{dx}{2\pi\, x\, l}$$

Insulation resistance of the whole cable is

$$R = \int_{r_1}^{r_2} \rho \, \frac{dx}{2\pi\, x\, l} = \frac{\rho}{2\pi l} \int_{r_1}^{r_2} \frac{1}{x} dx$$

∴

$$R = \frac{\rho}{2\pi l} \log_e \frac{r_2}{r_1}$$

This shows that insulation resistance of a cable is inversely proportional to its length. In other words, if the cable length increases, its insulation resistance decreases and *vice-versa*.

Example 6.21. *Two underground cables having conductor resistances of 0.7Ω and 0.5Ω and insulation resistances of 300 MΩ and 600 MΩ respectively are joined (i) in series (ii) in parallel. Find the resultant conductor and insulation resistance.*

Solution. *(i)* **Series connection.** In this case, conductor resistances are added like resistances in series. However, insulation resistances are given by reciprocal relation.

∴   Total conductor resistance = 0.7 + 0.5 = **1.2Ω**

The total insulation resistance R is given by ;

$$\frac{1}{R} = \frac{1}{300} + \frac{1}{600} \quad \therefore R = \textbf{200 M}\Omega$$

(ii) **Parallel connection.** In this case, conductor resistances are governed by reciprocal relation while insulation resistances are added.

∴   Total conductor resistance $= \dfrac{0.7 \times 0.5}{0.7 + 0.5} = \textbf{0.3 }\Omega$

Total insulation resistance = 300 + 600 = **900 MΩ**

Example 6.22. *The insulation resistance of a single-core cable is 495 MΩ per km. If the core diameter is 2·5 cm and resistivity of insulation is $4·5 \times 10^{14}$ Ω-cm, find the insulation thickness.*

Solution.   Length of cable, l = 1 km = 1000 m

Cable insulation resistance, R = 495 MΩ = 495 × 10⁶Ω

Conductor radius, r_1 = 2·5/2 = 1·25 cm

Resistivity of insulation, ρ = $4·5 \times 10^{14}$ Ω-cm = $4·5 \times 10^{12}$ Ωm

Let r_2 cm be the internal sheath radius.

Now, $R = \dfrac{\rho}{2\pi l} \log_e \dfrac{r_2}{r_1}$

or $\qquad \log_e \dfrac{r_2}{r_1} = \dfrac{2\pi l R}{\rho} = \dfrac{2\pi \times 1000 \times 495 \times 10^6}{4 \cdot 5 \times 10^{12}} = 0 \cdot 69$

or $\qquad 2 \cdot 3 \log_{10} r_2/r_1 = 0 \cdot 69$

or $\qquad r_2/r_1 = \text{Antilog } 0 \cdot 69/2 \cdot 3 = 2$

or $\qquad r_2 = 2\, r_1 = 2 \times 1 \cdot 25 = 2 \cdot 5$ cm

∴ $\qquad$ Insulation thickness $= r_2 - r_1 = 2 \cdot 5 - 1 \cdot 25 = $ **1·25 cm**

Example 6.23. *The insulation resistance of a kilometre of the cable having a conductor diameter of 1.5 cm and an insulation thickness of 1.5 cm is 500 MΩ. What would be the insulation resistance if the thickness of the insulation were increased to 2.5 cm?*

Solution. $R_1 = 500$ MΩ ; $l = 100$ m ; $R_2 = ?$

For first case : $\qquad R_1 = \dfrac{\rho}{2\pi l} \log_e \dfrac{r_2}{r_1}$

For second case: $\qquad R_2 = \dfrac{\rho}{2\pi l} \log_e \dfrac{r_2'}{r_1'}$

∴ $\qquad \dfrac{R_2}{R_1} = \dfrac{\log_e (r_2'/r_1')}{\log_e (r_2/r_1)}$

Now, $r_1 = 1.5/2 = 0.75$ cm ; $r_2 = 0.75 + 1.5 = 2.25$ cm ∴ $r_2/r_1 = 3$

$r_1' = 0.75$ cm ; $r_2' = 0.75 + 2.5 = 3.25$ cm ; ∴ $r_2'/r_1' = 4.333$

∴ $\qquad \dfrac{R_2}{500} = \dfrac{\log_e (4.333)}{\log_e (3)} = 1.334$

or $\qquad R_2 = 500 \times 1.334 = $ **667.3 MΩ**

── **Tutorial Problems** ──

1. A single-core cable has a conductor diameter of 2·5 cm and insulation thickness of 1·2 cm. If the specific resistance of insulation is $4 \cdot 5 \times 10^{14}$ Ω cm, calculate the insulation resistance per kilometre length of the cable. **[305·5 MΩ]**

2. A single core cable 3 km long has an insulation resistance of 1820 MΩ. If the conductor diameter is 1·5 cm and sheath diameter is 5 cm, calculate the resistivity of the dielectric in the cable.

   [28·57 × 10¹² Ωm]

3. Determine the insulation resistance of a single-core cable of length 3 km and having conductor radius 12·5 mm, insulation thickness 10 mm and specific resistance of insulation of 5×10^{12} Ωm. **[156 MΩ]**

6.17. Leakage Resistance of a Capacitor

The resistance of the dielectric of the capacitor is called **leakage resistance**. The dielectric in an *ideal* capacitor is a perfect insulator (*i.e.,* it has infinite resistance) and zero current flows through it when a voltage is applied across its terminals. The dielectric in a *real* capacitor has a large but finite resistance so a very small current flows between the capacitor plates when a voltage is applied.

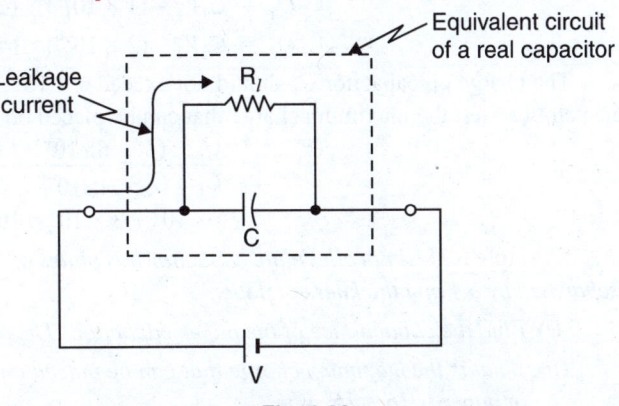

Fig. 6.20

Fig. 6.20 shows the equivalent circuit of a real capacitor consisting of an ideal capacitor in parallel with leakage resistance R_l. Typical values of leakage resistance may range from about 1 MΩ (considered a very "leaky" capacitor) to greater than 100,000 MΩ. A well designed capacitor has very high leakage resistance ($> 10^4$ MΩ) so that very little power is dissipated even when high voltage is applied across it.

6.18. Voltage Rating of a Capacitor

The maximum voltage that may be safely applied to a capacitor is usually expressed in terms of its d.c. working voltage.

The maximum d.c. voltage that can be applied to a capacitor without breakdown of its dielectric is called **voltage rating** *of the capacitor.*

If the voltage rating of a capacitor is exceeded, the dielectric may break down and conduct current, causing permanent damage to the capacitor. Both capacitance and voltage rating must be taken into consideration before a capacitor is used in a circuit application.

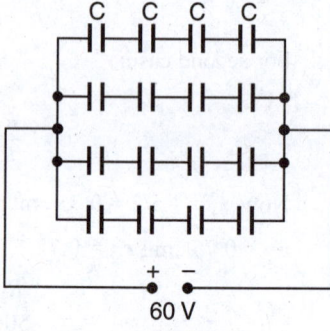

Example 6.24. *Given some capacitors of 0·1 μF capable of withstanding 15 V. Calculate the number of capacitors needed if it is desired to obtain a capacitance of 0·1 μF for use in a circuit involving 60 V.*

Fig. 6.21

Solution. Fig. 6.21 shows the conditions of the problem.

Capacitance of each capacitor, C = 0.1 μF

Voltage rating of each capacitor, V_C = 15 V

Supply voltage, V = 60 V

Since each capacitor can withstand 15 V, the number of capacitors to be connected in series = 60/15 = 4.

Capacitance of 4 series-connected capacitors, $C_T = C/4 = 0·1/4 = 0·025$ μF. Since it is desired to have a total capacitance of 0·1 μF, number of such rows in parallel = $C/C_T = 0·1/0·025 = 4$.

∴　　Total number of capacitors = 4 × 4 = **16**

Fig. 6.21 shows the arrangement of capacitors.

Example 6.25. *A capacitor of capacitance C_1 = 1 μF withstands the maximum voltage V_1 = 6 kV while another capacitance C_2 = 2 μF withstands the maximum voltage V_2 = 4 kV. What maximum voltage will the system of these two capacitors withstand if they are connected in series ?*

Solution. The maximum charges Q_1 and Q_2 that can be placed on C_1 and C_2 are :

$$Q_1 = C_1 V_1 = (1 \times 10^{-6}) \times (6 \times 10^3) = 6 \times 10^{-3} \text{ C}$$
$$Q_2 = C_2 V_2 = (2 \times 10^{-6}) \times (4 \times 10^3) = 8 \times 10^{-3} \text{ C}$$

The charge on capacitor C_1 should not exceed 6×10^{-3} C. Therefore, when capacitors are connected in series, the maximum charge that can be placed on the capacitors is 6×10^{-3} C ($= Q_1$).

∴　　　　$$V_{max} = \frac{Q_1}{C_1} + \frac{Q_1}{C_2} = \frac{6 \times 10^{-3}}{1 \times 10^{-6}} + \frac{6 \times 10^{-3}}{2 \times 10^{-6}}$$

$$= 6 \times 10^3 + 3 \times 10^3 = 10^3 (6 + 3) = 9 \times 10^3 \text{ V} = \textbf{9 kV}$$

Example 6.26. *A parallel plate capacitor has plates of dimensions 2 cm × 3 cm. The plates are separated by a 1 mm thickness of paper.*

(i) *Find the capacitance of the paper capacitor. The dielectric constant of paper is 3·7.*

(ii) *What is the maximum charge that can be placed on the capacitor ? The dielectric strength of paper is 16×16^6 V/m.*

Solution. (*i*)
$$C = \frac{\varepsilon_0 \varepsilon_r A}{d}$$

Here
$$\varepsilon_0 = 8\cdot85 \times 10^{-12} \text{ F/m} \; ; \; \varepsilon_r = 3\cdot7; \; A = 6 \times 10^{-4} \text{ m}^2; \; d = 1 \times 10^{-3} \text{ m}$$

∴
$$C = \frac{(8\cdot85\times10^{-12})\times(3\cdot7)\times(6\times10^{-4})}{1\times10^{-3}} = 19\cdot6 \times 10^{-12} \text{ F}$$

(*ii*) Since the thickness of the paper is 1 mm, the maximum voltage that can be applied before breakdown occurs is

$$V_{max} = E_{max} \times d$$

Here
$$E_{max} = 16 \times 10^6 \text{ V/m} \; ; \; d = 1 \text{ mm} = 1 \times 10^{-3} \text{ m}$$

∴
$$V_{max} = (16 \times 10^6) \times (1 \times 10^{-3}) = 16 \times 10^3 \text{ V}$$

∴ Maximum charge that can be placed on capacitor is

$$Q_{max} = CV_{max} = (19\cdot6 \times 10^{-12}) \times (16 \times 10^3) = 0.31 \times 10^{-6} \text{ C} = 0\cdot31 \text{ μC}$$

6.19. Capacitors in Series

Consider three capacitors, having capacitances C_1, C_2 and C_3 farad respectively, connected in series across a p.d. of V volts [See Fig. 6.22 (*i*)]. In series connection, charge on each capacitor is the *same (*i.e.* $+Q$ on one plate and $-Q$ on the other) but p.d. across each is different.

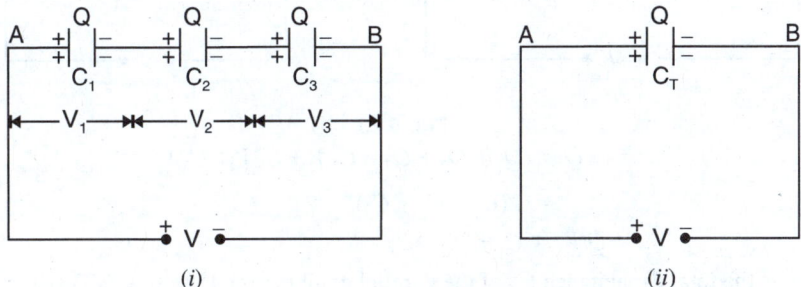

Fig. 6.22

Now,
$$V = V_1 + V_2 + V_3 = \frac{Q}{C_1} + \frac{Q}{C_2} + \frac{Q}{C_3}$$

$$= Q\left(\frac{1}{C_1} + \frac{1}{C_2} + \frac{1}{C_3}\right)$$

or
$$\frac{V}{Q} = \frac{1}{C_1} + \frac{1}{C_2} + \frac{1}{C_3}$$

But Q/V is the **total capacitance C_T between points A and B so that $V/Q = 1/C_T$ [See Fig. 6.22 (*ii*)].

∴
$$\frac{1}{C_T} = \frac{1}{C_1} + \frac{1}{C_2} + \frac{1}{C_3}$$

Thus capacitors in series are treated in the same manner as are resistors in parallel.

Special Case. Frequently we come across two capacitors in series. The total capacitance in such a case is given by ;

$$\frac{1}{C_T} = \frac{1}{C_1} + \frac{1}{C_2} = \frac{C_1 + C_2}{C_1 C_2}$$

---

* When voltage V is applied, a similar electron movement occurs on each plate. Hence the same charge is stored by each capacitor. Alternatively, current (charging) in a series circuit is the same. Since $Q = It$ and both I and t are the same for each capacitor, the charge on each capacitor is the same.

** Total or equivalent capacitance is the single capacitance which if substituted for the series capacitances, would provide the same charge for the same applied voltage.

or $\qquad C_T = \dfrac{C_1 C_2}{C_1 + C_2}$ *i.e.* $\dfrac{\text{Product}}{\text{Sum}}$

Note. The capacitors are connected in series when the circuit voltage exceeds the voltage rating of individual units. In using the series connection, it is important to keep in mind that the voltages across capacitors in series are not the same unless the capacitances are equal. The greater voltage will be across the smaller capacitance which may result in its failure if the capacitances differ very much.

6.20. Capacitors in Parallel

Consider three capacitors, having capacitances C_1, C_2 and C_3 farad respectively, connected in parallel across a p.d. of V volts [See Fig. 6.23 (*i*)]. In parallel connection, p.d. across each capacitor is the same but charge on each is different.

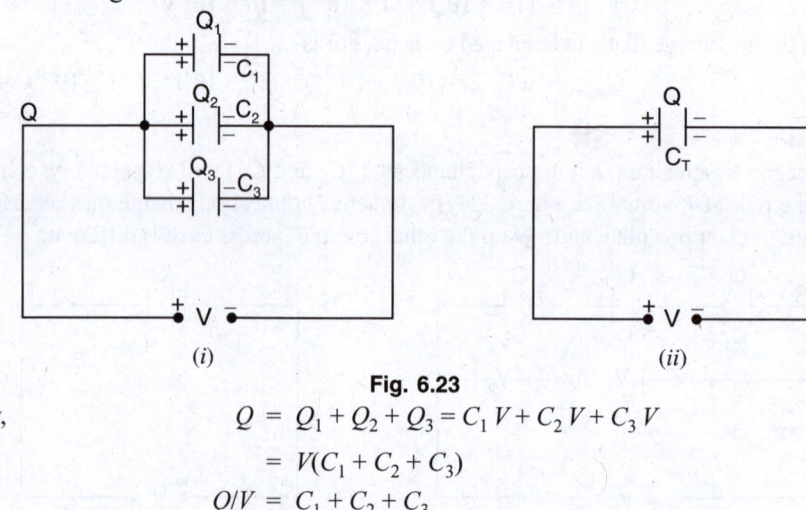

Fig. 6.23

Now, $\qquad Q = Q_1 + Q_2 + Q_3 = C_1 V + C_2 V + C_3 V$

$\qquad\qquad = V(C_1 + C_2 + C_3)$

or $\qquad Q/V = C_1 + C_2 + C_3$

But Q/V is the total capacitance C_T of the parallel combination [See Fig. 6.23 (*ii*)].

$\therefore \qquad\qquad C_T = C_1 + C_2 + C_3$

Thus capacitors in parallel are treated in the same manner as are resistors in series.

Note. Capacitors may be connected in parallel to obtain larger values of capacitance than are available from individual units.

Example 6.27. *In the circuit shown in Fig. 6.24, the total charge is 750 μC. Determine the values of V_1, V and C_2.*

Solution. $\qquad V_1 = \dfrac{Q}{C_1} = \dfrac{750 \times 10^{-6}}{15 \times 10^{-6}} = \textbf{50 V}$

$\qquad\qquad V = V_1 + V_2 = 50 + 20 = \textbf{70 V}$

$\qquad$ Charge on $C_3 = C_3 \times V_2$

$\qquad\qquad\qquad = (8 \times 10^{-6}) \times 20$

$\qquad\qquad\qquad = 160 \times 10^{-6} \, C = 160 \, \mu C$

$\therefore \qquad$ Charge on $C_2 = 750 - 160 = 590 \, \mu C$

$\therefore$ Capacitance of $C_2 = \dfrac{590 \times 10^{-6}}{20}$

$\qquad\qquad\qquad = 29 \cdot 5 \times 10^{-6} \, F = \textbf{29·5 μF}$

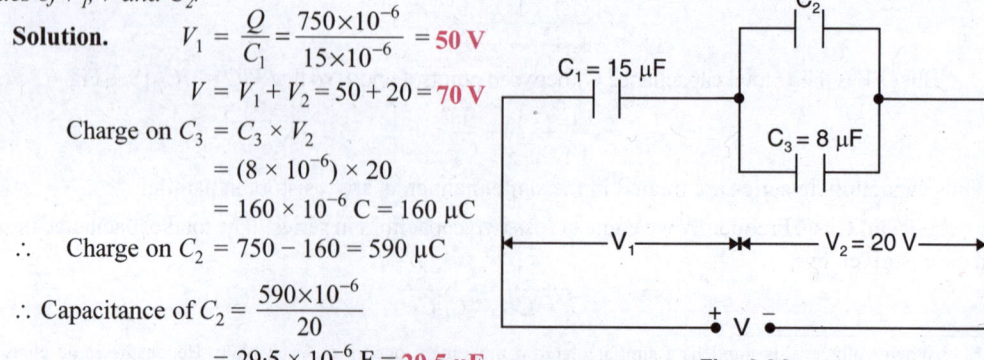

Fig. 6.24

Example 6.28. *Two capacitors A and B are connected in series across a 200 V d.c. supply. The p.d. across A is 120 V. This p.d. is increased to 140 V when a 3 μF capacitor is connected in parallel with B. Calculate the capacitances of A and B.*

Solution. Let C_1 and C_2 μF be the capacitances of capacitors A and B respectively. When the capacitors are connected in series [See Fig. 6.25 (i)], charge on each capacitor is the same.

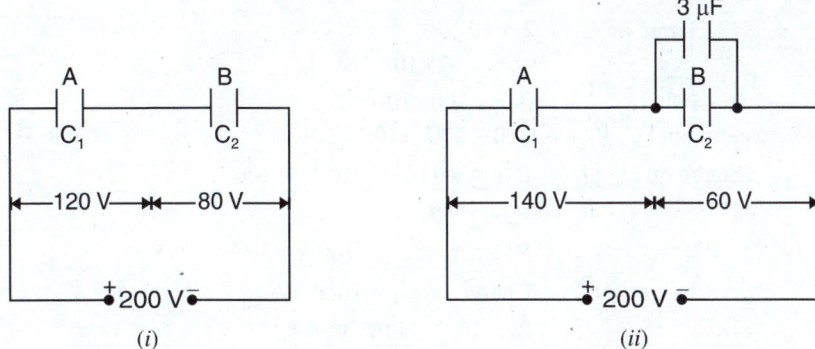

Fig. 6.25

∴ $C_1 \times 120 = C_2 \times 80$ or $C_2 = 1\cdot5\ C_1$ ...(i)

When a 3μF capacitor is connected in parallel with B [See Fig. 6.25 (ii)], the combined capacitance of this parallel branch is $(C_2 + 3)$. Thus the circuit shown in Fig. 6.25 (ii) can be thought as a series circuit consisting of capacitances C_1 and $(C_2 + 3)$ connected in series.

∴ $C_1 \times 140 = (C_2 + 3)\ 60$

or $7C_1 - 3\ C_2 = 9$ ...(ii)

Solving eqs. (i) and (ii), we get, $C_1 = \textbf{3.6 μF}$; $C_2 = \textbf{5·4 μF}$

Example 6.29. *Obtain the equivalent capacitance for the network shown in Fig. 6.26. For 300 V d.c. supply, determine the charge and voltage across each capacitor.*

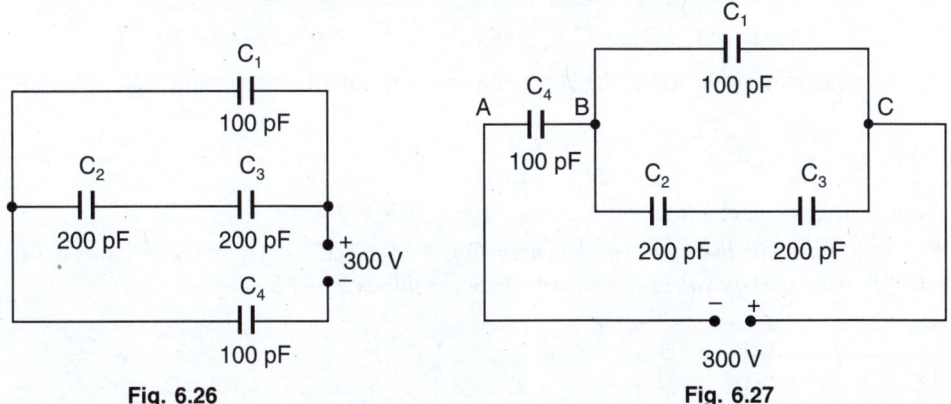

Fig. 6.26   **Fig. 6.27**

Solution. **Equivalent Capacitance.** The above network can be redrawn as shown in Fig. 6.27. The equivalent capacitance C' of series-connected capacitors C_2 and C_3 is

$$C' = \frac{C_2 \times C_3}{C_2 + C_3} = \frac{200 \times 200}{200 + 200} = 100 \text{ pF}$$

The equivalent capacitance of parallel combination C' (= 100 pF) and C_1 is

$$C_{BC} = C' + C_1 = 100 + 100 = 200 \text{ pF}$$

The entire circuit now reduces to two capacitors C_4 and C_{BC} (= 200 pF) in series.

∴   Equivalent capacitance of the network is

$$C = \frac{C_4 \times C_{BC}}{C_4 + C_{BC}} = \frac{100 \times 200}{100 + 200} = \frac{200}{3} \text{pF}$$

Charges and p.d. on various capacitors

$$\text{Total charge, } Q = CV = \left(\frac{200}{3} \times 10^{-12}\right) \times 300 = 2 \times 10^{-8} \text{ C}$$

$\therefore$ Charge on C_4 = **2×10^{-8} C**

$\therefore$ P.D. across C_4, V_4 = $\dfrac{Q}{C_4} = \dfrac{2 \times 10^{-8}}{100 \times 10^{-12}} =$ **200 V**

P.D. between B and C, V_{BC} = $300 - 200 = 100$ V

Charge on C_1, Q_1 = $C_1 V_{BC} = (100 \times 10^{-12}) \times 100 =$ **10^{-8} C**

P.D. across C_1, V_1 = $V_{BC} =$ **100 V**

P.D. across C_2 = P.D. across $C_3 = 100/2 =$ **50 V**

Charge on C_2 = Charge on C_3 = Total charge – Charge on C_1

= $(2 \times 10^{-8}) - (10^{-8}) =$ **10^{-8} C**

Example 6.30. *Two perfect insulated capacitors are connected in series. One is an air capacitor with a plate area of 0.01 m^2, the plates being 1 mm apart, the other has a plate area of 0.001 m^2, the plates separated by a solid dielectric of 0.1 mm thickness with a dielectric constant of 5. Determine the voltage across the combination if the potential gradient in the air capacitor is 200 V/mm.*

Solution. Capacitance C_1 of air capacitor is

$$C_1 = \frac{\varepsilon_0 \varepsilon_{r1} A_1}{t_1} = \frac{8.854 \times 10^{-12} \times 1 \times 0.01}{1 \times 10^{-3}} = 88.54 \times 10^{-12} \text{ F}$$

Capacitance C_2 of the capacitor with dielectric of $\varepsilon_{r_2} = 5$ is

$$C_2 = \frac{\varepsilon_0 \varepsilon_{r2} A_2}{t_2} = \frac{8.854 \times 10^{-12} \times 5 \times 0.001}{0.1 \times 10^{-3}} = 442.7 \times 10^{-12} \text{ F}$$

Voltage across C_1, V_1 = $g_1 \times t_1 = 200\text{V/mm} \times 1 \text{ mm} = 200$ V

Charge on C_1, Q_1 = $C_1 V_1 = 88.54 \times 10^{-12} \times 200 = 177.08 \times 10^{-10}$ C

As the capacitors are in series, the charge on each capacitor is the same *i.e.* $Q_2 = Q_1 = 177.08 \times 10^{-10}$ C.

$\therefore$ Voltage across C_2, $V_2 = \dfrac{Q_2}{C_2} = \dfrac{177.08 \times 10^{-10}}{442.7 \times 10^{-12}} = 40$ V

$\therefore$ Voltage across combination, $V = V_1 + V_2 = 200 + 40 =$ **240 volts**

Example 6.31. *In the network shown in Fig. 6.28 (i), $C_1 = C_2 = C_3 = C_4 = 8$ μF and $C_5 = 10$ μF. Find the equivalent capacitance between points A and B.*

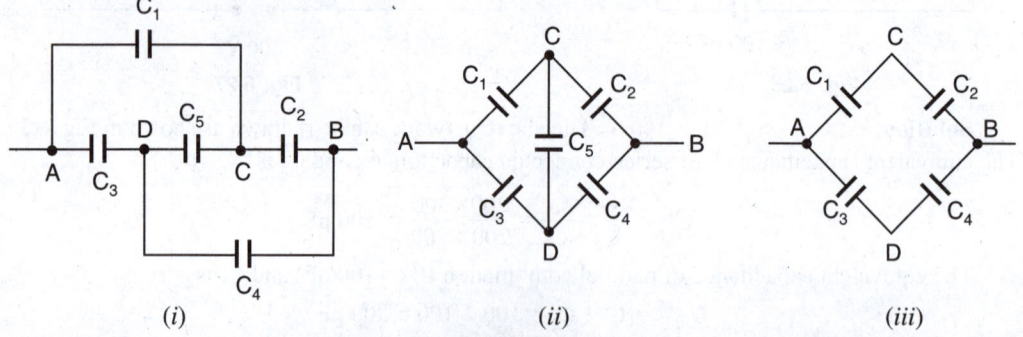

Fig. 6.28

Solution. A little reflection shows that circuit of Fig. 6.28 (*i*) can be redrawn as shown in Fig. 6.28 (*ii*). We find that the circuit is a Wheatstone bridge. Since the product of opposite arms of the bridge are equal ($C_1C_4 = C_2C_3$ because $C_1 = C_2 = C_3 = C_4$), the bridge is balanced. It means that points C and D are at the same potential. Therefore, there will be no charge on capacitor C_5. Hence, this capacitor is ineffective and can be removed from the circuit as shown in Fig. 6.28 (*iii*). Referring to Fig. 6.28 (*iii*), the equivalent capacitance C' of the series connected capacitors C_1 and C_2 is

$$C' = \frac{C_1C_2}{C_1+C_2} = \frac{8\times8}{8+8} = 4 \; \mu F$$

The equivalent capacitance C'' of series connected capacitors C_3 and C_4 [See Fig. 6.28 (*iii*)] is

$$C'' = \frac{C_3C_4}{C_3+C_4} = \frac{8\times8}{8+8} = 4 \; \mu F$$

Now $\qquad C_{AB} = C' \parallel C'' = 4 \parallel 4 = 4 + 4 = 8 \; \mu F$

Example 6.32. *Find the charge on 5 μF capacitor in the circuit shown in Fig. 6.29.*

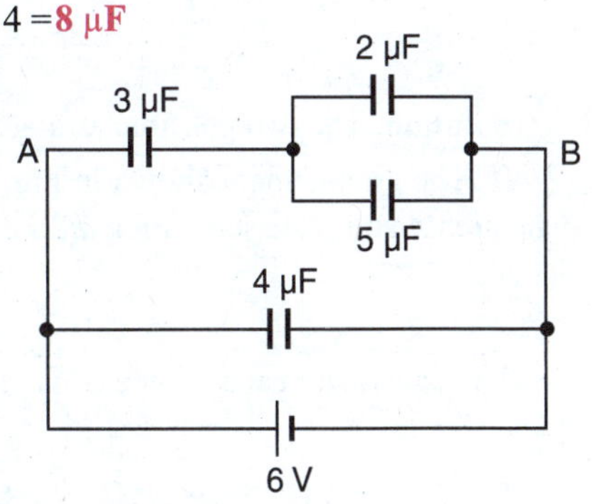

Solution. The p.d. between A and B is 6 V. Considering the branch AB, the capacitors 2 μF and 5 μF are in parallel and their equivalent capacitance = $2 + 5 = 7$ μF. The branch AB then has 7 μF and 3 μF in series. Therefore, the effective capacitance of branch AB is

$$C_{AB} = \frac{7\times3}{7+3} = \frac{21}{10} \; \mu F$$

Total charge in branch AB is

$$Q = C_{AB}V = \frac{21}{10} \times 6 = \frac{63}{5} \; \mu C$$

P.D. across 3 μF capacitor $= \dfrac{Q}{3} = \dfrac{63}{5} \times \dfrac{1}{3} = \dfrac{21}{5}$ volts

$\therefore$ P.D. across parallel combination $= 6 - \dfrac{21}{5} = \dfrac{9}{5}$ volts

Charge on 5 μF capacitor $= (5 \times 10^{-6}) \times \dfrac{9}{5} = 9 \times 10^{-6} \, C = 9 \; \mu C$

Fig. 6.29

Example 6.33. *Two parallel plate capacitors A and B having capacitances of 1 μF and 5 μF are charged separately to the same potential of 100 V. Now positive plate of A is connected to the negative plate of B and the negative plate of A is connected to the positive plate of B. Find the final charge on each capacitor.*

Solution. Initial charge on A, $Q_1 = C_1V = (1 \times 10^{-6}) \times 100 = 100 \; \mu C$

Initial charge on B, $Q_2 = C_2V = (5 \times 10^{-6}) \times 100 = 500 \; \mu C$

When the oppositely charged plates of A and B are connected together, the net charge is

$$Q = Q_2 - Q_1 = 500 - 100 = 400 \; \mu C$$

Final potential difference $= \dfrac{\text{Net charge}}{\text{Net capacitance}} = \dfrac{400\times10^{-6}}{(1+5)10^{-6}} = \dfrac{200}{3} \; V$

Final charge on A $= C_1 \times \dfrac{200}{3} = (1 \times 10^{-6}) \times \dfrac{200}{3} = \dfrac{200}{3} \mu C$

Final charge on B $= C_2 \times \dfrac{200}{3} = (5 \times 10^{-6}) \times \dfrac{200}{3} = \dfrac{1000}{3} \mu C$

Example 6.34. *A capacitor is filled with two dielectrics of the same dimensions but of dielectric constants K_1 and K_2 respectively. Find the capacitances in two possible arrangements.*

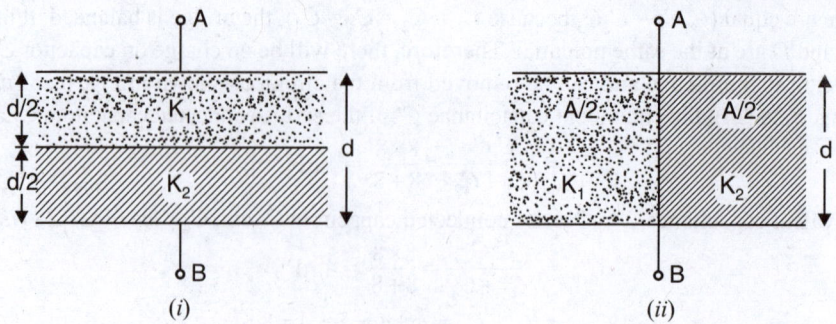

Fig. 6.30

Solution. The two possible arrangements are shown in Fig. 6.30.

(*i*) The arrangement shown in Fig. 6.30 (*i*) is equivalent to two capacitors in series, each with plate area A and plate separation $d/2$ *i.e.*,

$$C_1 = \frac{K_1 \varepsilon_0 A}{d/2} = \frac{2K_1 \varepsilon_0 A}{d} \quad ; \quad C_2 = \frac{K_2 \varepsilon_0 A}{d/2} = \frac{2K_2 \varepsilon_0 A}{d}$$

The equivalent capacitance C' is given by ;

$$\frac{1}{C'} = \frac{1}{C_1} + \frac{1}{C_2} = \frac{d}{2K_1 \varepsilon_0 A} + \frac{d}{2K_2 \varepsilon_0 A} = \frac{d}{2\varepsilon_0 A}\left(\frac{1}{K_1} + \frac{1}{K_2}\right)$$

$$= \frac{d}{2\varepsilon_0 A}\left(\frac{K_1 + K_2}{K_1 K_2}\right)$$

$$\therefore \qquad C' = \frac{2\varepsilon_0 A}{d}\left(\frac{K_1 K_2}{K_1 + K_2}\right)$$

(*ii*) The arrangement shown in Fig. 6.30 (*ii*) is equivalent to two capacitors in parallel, each with plate area $A/2$ and plate separation d *i.e.*,

$$C_1 = \frac{K_1 \varepsilon_0 (A/2)}{d} = \frac{K_1 \varepsilon_0 A}{2d} \quad ; \quad C_2 = \frac{K_2 \varepsilon_0 (A/2)}{d} = \frac{K_2 \varepsilon_0 A}{2d}$$

The equivalent capacitance C'' is given by ;

$$C'' = C_1 + C_2 = \frac{K_1 \varepsilon_0 A}{2d} + \frac{K_2 \varepsilon_0 A}{2d} = \frac{\varepsilon_0 A}{2d}(K_1 + K_2)$$

$$\therefore \qquad C'' = \frac{\varepsilon_0 A}{2d}(K_1 + K_2)$$

Example 6.35. *Determine the capacitance between terminals A and B of the network shown in Fig. 6.31. The values shown are capacitances in μF.*

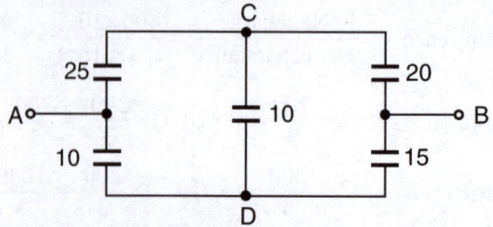

Fig. 6.31

Solution. The circuit shown in Fig. 6.31 is equivalent to the circuit shown in Fig. 6.32.

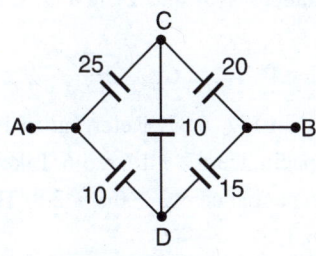

Fig. 6.32

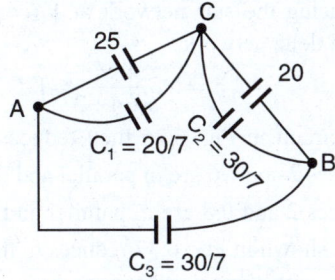

Fig. 6.33

Replacing the star network at D (consisting of capacitances 10, 10 and 15) by equivalent delta, we have,

$$C_1 = \frac{10\times10}{10+10+15} = \frac{20}{7} \qquad \text{(between } A \text{ and } C\text{)}$$

$$C_2 = \frac{10\times15}{10+10+15} = \frac{30}{7} \qquad \text{(between } B \text{ and } C\text{)}$$

$$C_3 = \frac{10\times15}{10+10+15} = \frac{30}{7} \qquad \text{(between } A \text{ and } B\text{)}$$

The circuit then reduces to the circuit shown in Fig. 6.33. Referring to Fig. 6.33,

$$C_{AC} = 25+\frac{20}{7} = \frac{195}{7} = 27{\cdot}86 \; ; \; C_{BC} = 20+\frac{30}{7} = \frac{170}{7} = 24{\cdot}29$$

The circuit then reduces to the circuit shown in Fig. 6.34.

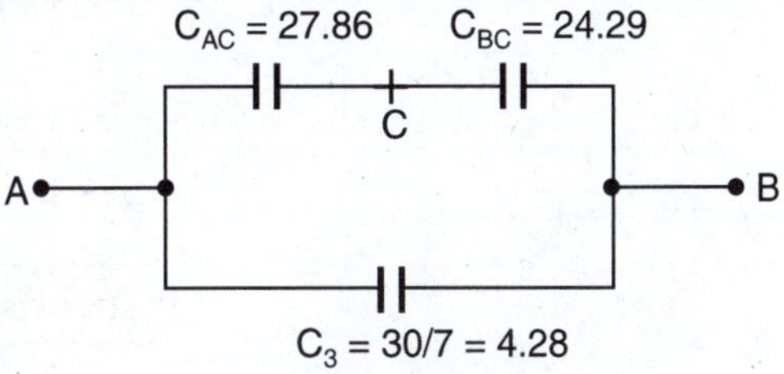

∴

$$C_{AB} = \frac{C_{AC}\times C_{BC}}{C_{AC}+C_{BC}} + C_3$$

$$= \frac{27{\cdot}86\times24{\cdot}29}{27{\cdot}86+24{\cdot}29} + 4{\cdot}28$$

$$= 12{\cdot}98 + 4{\cdot}28 = \mathbf{17{\cdot}3 \; \mu F}$$

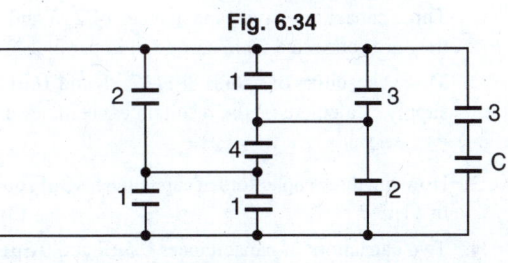

Fig. 6.34

Example 6.36. *In the network shown in Fig. 6.35, the capacitances are in μF. If the capacitance between terminals P and Q is 5 μF, find the value of C.*

Solution. The capacitances 1 and 1 are in parallel and their equivalent capacitance = $1 + 1 = 2$. Likewise, the capacitances 1 and 3 are in parallel and their equivalent capacitance = $1 + 3 = 4$. Therefore, the original circuit reduces to the circuit shown in Fig. 6.36.

Fig. 6.35

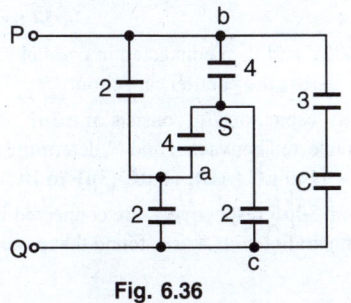

Fig. 6.36

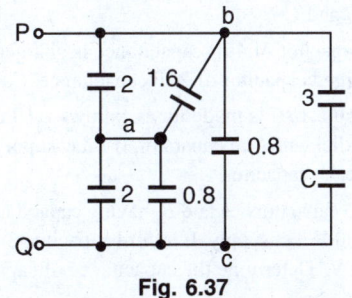

Fig. 6.37

Replacing the star network at S (consisting of capacitances 4, 4 and 2) in Fig. 6.36 by its equivalent delta network,

$$C_{ab} = \frac{4\times 4}{4+4+2} = 1\cdot 6 \; ; \quad C_{bc} = \frac{4\times 2}{4+4+2} = 0\cdot 8 \; ; \quad C_{ca} = \frac{4\times 2}{4+4+2} = 0\cdot 8$$

The circuit in Fig. 6.36 then reduces to the one shown in Fig. 6.37. Referring to Fig. 6.37, capacitances 2 and 1·6 are in parallel and their equivalent capacitance = 2 + 1·6 = 3·6. Likewise, the capacitances 2 and 0·8 are in parallel and their equivalent capacitance = 2 + 0·8 = 2·8. Therefore, the circuit shown in Fig. 6.37 reduces to that shown in Fig. 6.38.

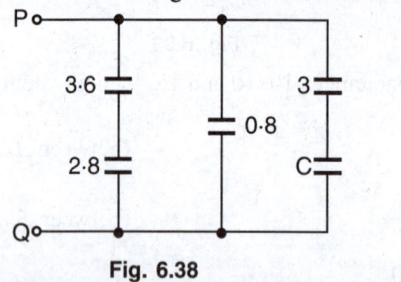

Fig. 6.38

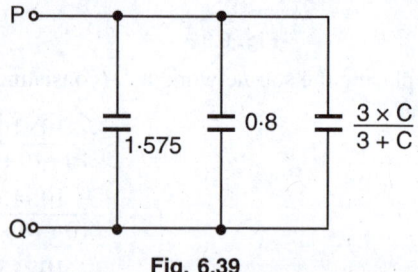

Fig. 6.39

Referring to Fig. 6.38, capacitances 3·6 and 2·8 are in series and their equivalent capacitance = 3·6 × 2·8/(3·6 + 2·8) = 1·575. Likewise, capacitances 3 and C are in series and their equivalent capacitance = 3 × C/(3 + C). The circuit shown in Fig. 6.38 reduces to that shown in Fig. 6.39. Referring to Fig. 6.39,

$$C_{PQ} = 1\cdot 575 + 0\cdot 8 + \frac{3\times C}{3+C}$$

or
$$5 = 1\cdot 575 + 0\cdot 8 + \frac{3C}{3+C} \qquad \text{[Given } C_{PQ} = 5 \text{ μF]}$$

$$\therefore \qquad C = \mathbf{21\ \mu F}$$

Tutorial Problems

1. Three capacitors have capacitances of 2, 3 and 4μF respectively. Calculate the total capacitance when they are connected (*i*) in series (*ii*) in parallel. **[(*i*) 0·923μF (*ii*) 9μF]**

2. Three capacitors of values 8μF, 12 μF and 16μF respectively are connected in series across a 240 V d.c. supply. Calculate (*i*) the resultant capacitance and (*ii*) p.d. across each capacitor.

   [(*i*) 3·7μF (*ii*) V₁ = 111V, V₂ = 74 V, V₃ = 55 V]

3. How can three capacitors of capacitances 3μF, 6μF and 9μF respectively be arranged to give a capacitance of 11μF ? **[3μF and 6μF in series, with 9μF in parallel with both]**

4. Two capacitors of capacitances 0·5μF and 0·3μF are joined in series. What value of capacitance joined in parallel with this combination would give a capacitance of 0·5μF ? **[0·31μF]**

5. Three capacitors A, B and C are connected in series across a 200 V d.c. supply. The p.d.s. across the capacitors are 40 V, 70V and 90V respectively. If the capacitance of A is 8μF, what are the capacitances of B and C ? **[4·57 μF, 3·56 μF]**

6. A capacitor of 4μF capacitance is charged to a p.d. of 400V and then connected in parallel with an uncharged capacitor of 2μF capacitance. Calculate the p.d. across the parallel capacitors. **[267 V]**

7. Circuit ABC is made up as follows : AB consists of a 3μF capacitor, BC consists of a 3μF capacitor in parallel with 5μF capacitor. If a d.c. supply of 100 V is connected between A and C, determine the charge on each capacitor. **[160 μC (AB); 60 μC (3μF in BC); 100 μC]**

8. Two capacitors, A and B, having capacitances of 20μF and 30μF respectively, are connected in series to a 600 V d.c. supply. If a third capacitor C is connected in parallel with A, it is found that p.d. across B is 400 V. Determine the capacitance of capacitor C. **[40μF]**

6.21. Joining Two Charged Capacitors

Consider two charged capacitors of capacitances C_1 and C_2 charged to potentials V_1 and V_2 respectively as shown in Fig. 6.40. With switch S open,

$$Q_1 = C_1 V_1 \quad \text{and} \quad Q_2 = C_2 V_2$$

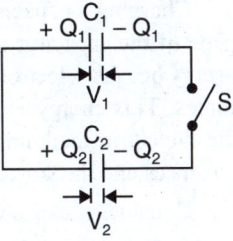

When switch S is closed, positive charge will flow from the capacitor of higher potential to the capacitor of lower potential. This flow of charge will continue till p.d. across each capacitor is the same. This is called *common potential* (V).

$$\text{Common potential, } V = \frac{\text{Total charge}}{\text{Total capacitance}} = \frac{Q_1 + Q_2}{C_1 + C_2}$$

Fig. 6.40

$$\therefore \qquad V = \frac{C_1 V_1 + C_2 V_2}{C_1 + C_2} \qquad \qquad ...(i)$$

The following points may be noted :

(i) Although there is a redistribution of charge on connecting the capacitors (*i.e.,* closing switch S), the total charge before and after the connection remains the same (Remember charge is a conserved quantity). This means that charge lost by one capacitor is *equal to the charge gained by the other capacitor.

(ii) When switch S is closed, the capacitors are in parallel.

(iii) Since the two capacitors acquire the same common potential V,

$$V = \frac{Q_1}{C_1} = \frac{Q_2}{C_2} \qquad \therefore \quad \frac{Q_1}{Q_2} = \frac{C_1}{C_2}$$

Therefore, the charges acquired by the capacitors are in the ratio of their capacitances.

(iv) In this process of charge sharing, the total stored energy of the capacitors decreases. It is because energy is dissipated as heat in the connecting wires when charge flows from one capacitor to the other.

Example 6.37. *Two capacitors of capacitances 4 μF and 6 μF respectively are connected in series across a p.d. of 250 V. The capacitors are disconnected from the supply and are reconnected in parallel with each other. Calculate the new p.d. and charge on each capacitor.*

Solution. In series-connected capacitors, p.d.s across the capacitors are in the inverse ratio of their capacitances.

$$\therefore \text{ P.D. across 4 μF capacitor} = 250 \times \frac{6}{4+6} = 150 \text{ V}$$

Charge on 4 μF capacitor $= (4 \times 10^{-6}) \times 150 = 0.0006$ C

Since the capacitors are connected in series, charge on each capacitor is the same.

$$\therefore \text{ Charge on both capacitors} = 2 \times 0.0006 = 0.0012 \text{ C}$$

Parallel connection. When the capacitors are connected in parallel, the total capacitance $C_T = 4 + 6 = 10$ μF. The total charge 0·0012 C is distributed between the capacitors to have a common p.d.

$$\therefore \qquad \text{P.D. across capacitors} = \frac{\text{Total charge}}{C_T} = \frac{0.0012}{10 \times 10^{-6}} = \textbf{120 V}$$

Charge on 4 μF capacitor $= (4 \times 10^{-6}) \times 120 = 480 \times 10^{-6}$ C $= \textbf{480 μC}$

Charge on 6 μF capacitor $= (6 \times 10^{-6}) \times 120 = 720 \times 10^{-6}$ C $= \textbf{720 μC}$

---

* Thus referring to exp. (i), $V(C_1 + C_2) = C_1 V_1 + C_2 V_2$ or $C_1 V_1 - C_1 V = C_2 V - C_2 V_2$

∴ Charge lost by one = Charge gained by the other

6.22. Energy Stored in a Capacitor

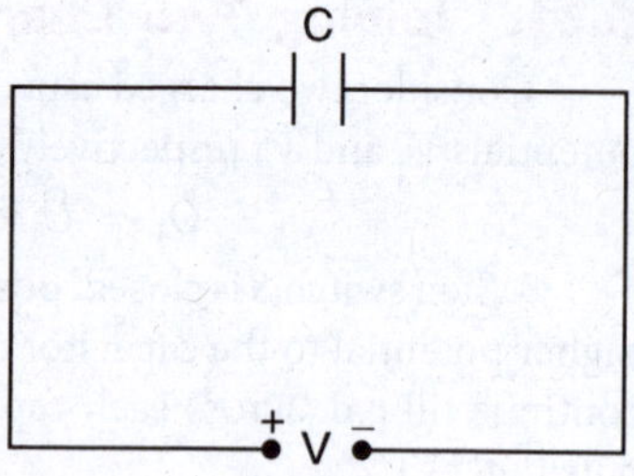

Charging a capacitor means transferring electrons from one plate of the capacitor to the other. This involves expenditure of energy because electrons have to be moved against the *opposing forces. This energy is stored in the electrostatic field set up in the dielectric medium. On discharging the capacitor, the field collapses and the stored energy is released.

Consider a capacitor of C farad being charged from a d.c. source of V volts as shown in Fig. 6.41. Suppose at any stage of

Fig. 6.41

charging, the charge on the capacitor is q coulomb and p.d. across the plates is v volts.

Then,
$$C = \frac{q}{v}$$

At this instant, v joules (by definition of v) of work will be done in transferring 1 C of charge from one plate to the other. If further small charge dq is transferred, then work done is

$$dW = v \, dq$$
$$= C v \, dv \qquad \left[\begin{array}{l} \because q = C v \\ \therefore dq = C \, dv \end{array} \right]$$

∴ Total work done in raising the potential of uncharged capacitor to V volts is

$$W = \int_0^V C v \, dv = C \left[\frac{v^2}{2} \right]_0^V$$

or
$$W = \frac{1}{2} C V^2 \text{ joules}$$

This work done is stored in the electrostatic field set up in the dielectric.

∴ Energy stored in the capacitor is

$$E = \frac{1}{2} C V^2 = \frac{1}{2} **Q V = \frac{†Q^2}{2 C} \text{ joules}$$

Note that an ideal (or pure) capacitor does not dissipate or consume energy; instead, it *stores* energy.

6.23. Energy Density of Electric Field

The energy stored per unit volume of the electric field is called **energy density** *of the electric field*

∴   Energy density, $u = \dfrac{\text{Total energy stored} (U)}{\text{Volume of electric field}}$

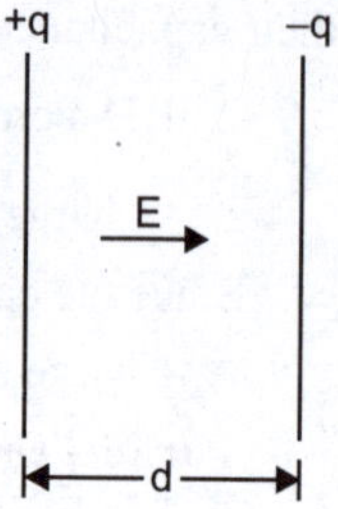

We have seen that energy is stored in the electric field of a capacitor. In fact, wherever electric field exists, there is stored energy. While dealing with electric fields, we are generally interested in energy density (u) *i.e.* energy stored per unit volume. Consider a charged parallel plate capacitor of plate area A and plate separation d as shown in Fig. 6.42.

$$\text{Energy stored} = \frac{1}{2} C V^2$$

Volume of space between plates $= A \, d$

Fig. 6.42

∴   Energy density, $u = \dfrac{\text{Energy stored}}{\text{Volume}} = \dfrac{C V^2}{2 A d}$

---

* Electrons are being pushed to the negative plate which tends to repel them. Similarly, electrons are removed from the positive plate which tends to attract them. In either case, forces oppose the transfer of electrons from one plate to the other. This opposition increases as the charge on the plates increases.

** Putting $C = Q/V$ in the exp., $E = \frac{1}{2} Q V$

† Putting $V = Q/C$ in the exp., $E = Q^2/2C$

We know that capacitance of a parallel plate capacitor is $C = \varepsilon_0 A/d$.

$$\therefore \qquad u = \frac{\varepsilon_0 A}{d} \times \frac{V^2}{2Ad} = \frac{1}{2}\varepsilon_0 \left(\frac{V}{d}\right)^2$$

But V/d is the electric field intensity (E) between the plates.

$$\therefore \qquad \text{Energy density, } u = \frac{1}{2}\varepsilon_0 E^2 \qquad\qquad \text{... in air} \qquad\qquad ...(i)$$

$$= \frac{1}{2}\varepsilon_0 \varepsilon_r E^2 \qquad\qquad \text{... in a medium} \qquad\qquad ...(ii)$$

Obviously, the unit of energy density will be joules/m^3.

Therefore, energy density (i.e., electric field energy stored per unit volume) in any region of space is directly proportional to the square of the electric field intensity in that region.

Note that we derived exps. (i) and (ii) for the special case of a parallel plate capacitor. But it can be shown to be true for any region of space where electric field exists.

Note. We can also express energy density of electric field in terms of electric flux density $D \ (= \varepsilon_0\varepsilon_r E)$.

$$u = \frac{1}{2}DE = \frac{D^2}{2\varepsilon_0\varepsilon_r}$$

Example 6.38. *A 16 μF capacitor is charged to 100 V. After being disconnected, it is immediately connected in parallel with an uncharged capacitor of capacitance 4μF. Determine (i) the p.d. across the combination, (ii) the electrostatic energies before and after the capacitors are connected in parallel and (iii) loss of energy.*

Solution. $\qquad\qquad C_1 = 16 \ \mu\text{F} \ ; \quad C_2 = 4 \ \mu\text{F}$

Before joining

Charge on 16 μF capacitor, $Q = C_1 V_1 = (16 \times 10^{-6}) \times 100 = 1.6 \times 10^{-3} \text{ C}$

$$\text{Energy stored, } E_1 = \frac{1}{2} C_1 V_1^2 = \frac{1}{2}(16 \times 10^{-6}) \times 100^2 = \textbf{0.08 J}$$

After joining. When the capacitors are connected in parallel, the total capacitance $C_T = C_1 + C_2$ $= 16 + 4 = 20 \ \mu\text{F}$. The charge 1.6×10^{-3} C distributes between the two capacitors to have a common p.d. of V volts.

$$\text{P.D. across parallel combination, } V = \frac{Q}{C_T} = \frac{1.6 \times 10^{-3}}{20 \times 10^{-6}} = \textbf{80 V}$$

$$\text{Energy stored, } E_2 = \frac{1}{2}C_T V^2 = \frac{1}{2}(20 \times 10^{-6}) \times (80)^2 = \textbf{0.064 J}$$

$$\text{Loss of energy} = E_1 - E_2 = 0.08 - 0.064 = \textbf{0.016 J}$$

It may be noted that there is a loss of energy. This is due to the heat dissipated in the conductor connecting the capacitors.

Example 6.39. *A capacitor-type stored-energy welder is to deliver the same heat to a single weld as a conventional weld that draws 20 kVA at 0.8 p.f. for 0.0625 second/weld. If C = 2000 μF, find the voltage to which it is charged.*

Solution. The energy supplied per weld in a conventional welder is

$$W = VA \times \cos \phi \times \text{time} = (20 \times 10^3) \times (0.8) \times 0.0625 = 1000 \text{ J}$$

The stored energy in the capacitor should be 1000 J.

$$\therefore \qquad\qquad 1000 = \frac{1}{2}CV^2$$

$$\text{or} \qquad\qquad V = \sqrt{\frac{2 \times 1000}{C}} = \sqrt{\frac{2 \times 1000}{2000 \times 10^{-6}}} = \textbf{1000 V}$$

Example 6.40. *A parallel plate 100 μF capacitor is charged to 500 V. If the distance between the plates is halved, what will be the new potential difference between the plates and what will be the new stored energy ?*

Solution. $C = 100 \ \mu F = 100 \times 10^{-6} \ F = 10^{-4} \ F$; $V = 500$ volts

When plate separation is decreased to half, the new capacitance C' becomes twice *i.e.*, $C' = 2C$. Since the capacitor is not connected to the battery, the charge on the capacitor remains the same. The potential difference between the plates must decrease to maintain the same charge.

$\therefore$ $Q = CV = C'V'$ or $V' = \dfrac{CV}{C'} = \dfrac{CV}{2C} = \dfrac{V}{2} = \dfrac{500}{2}$ **=250 volts**

$$\text{New stored energy} = \frac{1}{2}C'V'^2 = \frac{1}{2}(2C)\left(\frac{V}{2}\right)^2$$

$$= \frac{1}{2}\frac{CV^2}{2} = \frac{1}{2}\left(\frac{1}{2}CV^2\right)$$

$$= \frac{1}{2}\left[\frac{1}{2}\times 10^{-4}\times(500)^2\right] \text{ =6·25 J}$$

Example 6.41. *A parallel-plate capacitor is charged with a battery to a charge q_0 as shown in Fig. 6.43 (i). The battery is then removed and the space between the plates is filled with a dielectric of dielectric constant K. Find the energy stored in the capacitor before and after the dielectric is inserted.*

Solution. Energy stored in the capacitor in the absence of dielectric is

$$*E_0 = \frac{1}{2}C_0 V_0^2$$

Since $V_0 = q_0/C_0$, this can be expressed as :

$$E_0 = \frac{q_0^2}{2C_0} \qquad\qquad\qquad ...(i)$$

Eq. (*i*) gives the energy stored in the capacitor in the absence of dielectric.

After the battery is removed and the dielectric is inserted between the plates, *charge on the capacitor remains the same.* But the capacitance of the capacitor is increased K times *i.e.,* new capacitance is $C' = K\,C_0$ [See Fig. 6.43 (*ii*)].

$\therefore$ Energy stored in the capacitor after insertion of dielectric is

$$E = \frac{q_0^2}{2C'} = \frac{q_0^2}{2K\,C_0} = \frac{E_0}{K}$$

or $E = \dfrac{E_0}{K}$ ...(ii)

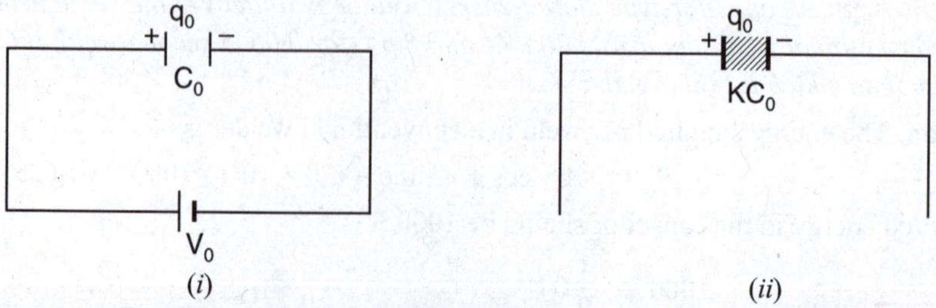

Fig. 6.43

---

* The subscript 0 indicates the conditions when the medium is air.

Since $K > 1$, we find that final energy is *less* than the initial energy by the factor $1/K$. How will you account for "missing energy" ? When the dielectric is inserted into the capacitor, it gets pulled into the device. The external agent must do negative work to keep the dielectric from accelerating. This work is simply $= E_0 - E$. Alternately, the positive work done by the system $= E_0 - E$.

Example 6.42. *Suppose in the above problem, the capacitor is kept connected with the battery and then dielectric is inserted between the plates. What will be the change in charge, the capacitance, the potential difference, the electric field and the stored energy ?*

Solution. Since the battery remains connected, the potential difference V_0 will **remain unchanged.**

As a result, electric field $(= V_0/d)$ will also **remain unchanged.**

The capacitance C_0 will increase to $C = K\,C_0.$

The charge will also increase to $q = K\,q_0$ as explained below.

$$q_0 = C_0\,V_0 \ ; \quad q = CV_0 = KC_0\,V_0 = K\,q_0$$

Initial stored energy, $E_0 = \dfrac{1}{2}C_0 V_0^2$

Final stored energy, $E = \dfrac{1}{2}CV_0^2 = \dfrac{1}{2}K\,C_0 V_0^2 = KE_0$

$$\therefore \qquad E = KE_0$$

Note that stored energy is increased K times. Will any work be done in inserting the dielectric? The answer is yes. In this case, the work will be done by the battery. The battery not only gives the increased energy to the capacitor but also provides the necessary energy for inserting the dielectric.

Example 6.43. *An air-capacitor of capacitance $0.005\ \mu F$ connected to a direct voltage of 500 V is disconnected and then immersed in oil with a relative permittivity of 2.5. Find the energy stored in the capacitor before and after immersion.*

Solution. Energy before immersion, $E_1 = \dfrac{1}{2}CV^2 = \dfrac{1}{2}\times 0\cdot005\times10^{-6}\times(500)^2 = \mathbf{625 \times 10^{-6}\ J}$

When the capacitor is immersed in oil, its capacitance becomes $C' = \varepsilon_r C = 2.5 \times 0.005 = 0.0125\ \mu F.$ Since charge remains the same $(V = Q/C)$, new voltage is decreased and becomes $V' = V/\varepsilon_r = 500/2.5 = 200\ V.$

$$\therefore \quad \text{Energy after immersion, } E_2 = \dfrac{1}{2}C'V'^2 = \dfrac{1}{2}\times 0\cdot0125\times10^{-6}\times(200)^2 = \mathbf{250 \times 10^{-6}\ J}$$

Example 6.44. *In the circuit shown in Fig. 6.44, the battery e.m.f. is 100 V and the capacitor has a capacitance of 1 μF. The switch is operated 100 times every second. Calculate (i) the average current through the switch between switching operations and (ii) the average power dissipated in the resistor. It may be assumed that the capacitor is ideal and that the capacitor is fully charged or discharged before the subsequent switching.*

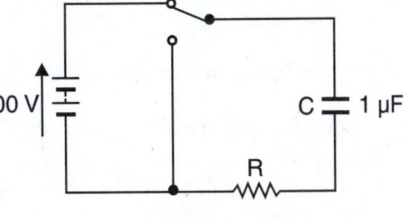

Fig. 6.44

Solution. (*i*) Maximum charge on capacitor, $Q = CV = (1 \times 10^{-6}) \times (100) = 10^{-4}\ C$

The time taken to acquire this charge (or to lose it) is

$$T = \dfrac{1}{f} = \dfrac{1}{100} = 0.01\ \text{s}$$

$$\therefore \quad \text{Average current, } I_{av} = \dfrac{\Delta Q}{\Delta T} = \dfrac{10^{-4}}{0\cdot01} = 0\cdot01\ \text{A} = \mathbf{10\ mA}$$

(*ii*) The maximum energy stored during charging is

$$E_m = \frac{1}{2}CV^2 = \frac{1}{2} \times 10^{-6} \times (100)^2 = 0.005 \text{ J}$$

During the charging period, a similar quantity of energy must be dissipated in the resistor. In the subsequent discharging period, the stored energy in the capacitor is dissipated in the resistor. Hence for every switching action, 0.005 J is dissipated in the resistor. For 100 switching operations, the energy E dissipated is

$$E = 100 \times 0.005 = 0.5 \text{ J}$$

$$\text{Average power taken} = \frac{\Delta E}{\Delta T} = \frac{0.5}{1} = \textbf{0.5 W}$$

Note that amount of energy stored in a capacitor is very small because the value of C is very small.

6.24. Force on Charged Plates

Consider two parallel conducting plates x metres apart and carrying constant charges of $+Q$ and $-Q$ coulombs respectively as shown in Fig. 6.45. Let the force of attraction between the two plates be F newtons. If one of the plates is moved away from the other by a small distance dx, then work done is

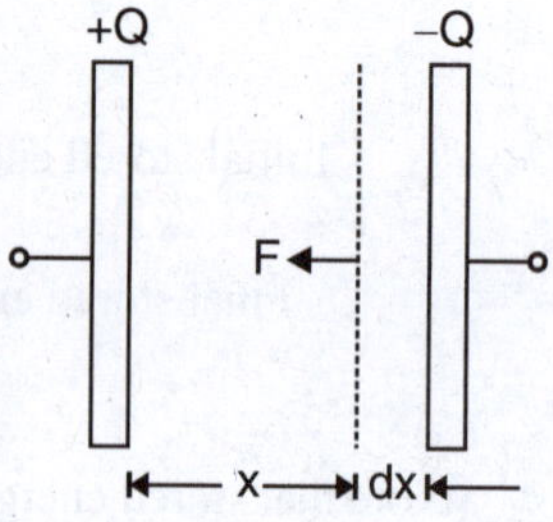

$$\text{Work done} = F \times dx \text{ joules} \qquad ...(i)$$

Since the charges on the plates remain constant, no electrical energy can enter or leave the system during the movement dx.

∴ Work done = Change in stored energy

$$\text{Initial stored energy} = \frac{1}{2}\frac{Q^2}{C} \text{ joules}$$

Since the separation of the plates has increased, the capacitance will decrease by dC. The final capacitance is, therefore, $(C - dC)$.

$$\text{Final stored energy} = \frac{1}{2}\frac{Q^2}{(C-dC)} = \frac{Q^2(C+dC)*}{2[C^2 - (dC)^2]}$$

Since dC is small compared to C, $(dC)^2$ can be neglected compared to C^2.

∴ $$\text{Final stored energy} = \frac{Q^2(C+dC)}{2C^2} = \frac{Q^2}{2C} + \frac{Q^2}{2C^2}dC$$

∴ $$\text{Change in stored enegry} = \left(\frac{Q^2}{2C} + \frac{Q^2}{2C^2}dC\right) - \frac{Q^2}{2C} = \frac{Q^2}{2C^2}dC \qquad ...(ii)$$

Equating eqs. (*i*) and (*ii*), we get,

$$F \times dx = \frac{Q^2}{2C^2}dC$$

or $$F = \frac{Q^2}{2C^2}\frac{dC}{dx}$$

$$= \frac{1}{2}V^2\frac{dC}{dx} \qquad ...(iii) \quad (\because V = Q/C)$$

Now $$C = \frac{\varepsilon_0 \varepsilon_r A}{x}$$

Fig. 6.45

---

* Note this exp. Multiply the numerator and denominator by $(C + dC)$.

$$\therefore \qquad \frac{dC}{dx} = -\frac{\varepsilon_0 \, \varepsilon_r \, A}{x^2}$$

$\therefore$ Substituting the value of dC/dx in eq. (*iii*), we get,

$$F = -\frac{1}{2}V^2 \frac{\varepsilon_0 \, \varepsilon_r \, A}{x^2} = -\frac{1}{2}\varepsilon_0 \, \varepsilon_r A\left(\frac{V}{x}\right)^2$$

$$= -\frac{1}{2}\varepsilon_0 \, \varepsilon_r \, A \, E^2 \qquad ...in \ a \ medium$$

$$= -\frac{1}{2}\varepsilon_0 \, A \, E^2 \qquad ...in \ air$$

This represents the force between the plates of a parallel-plate capacitor charged to a p.d. of V volts. The negative sign shows that it is a force of attraction.

Note. The force of attraction between charged plates may be utilised as a means of measuring potential difference. An instrument of this kind is known as an **electrostatic voltmeter**.

Example 6.45. *A parallel plate capacitor has its plates separated by 0·5 mm of air. The area of plates is 2 m^2 and they are charged to a p.d. of 100 V. The plates are pulled apart until they are separated by 1 mm of air. Assuming the p.d. to remain unchanged, what is the mechanical force experienced in separating the plates ?*

Solution. Here, $A = 2m^2$; $d = 0.5$ mm $= 0.5 \times 10^{-3}$ m ; $V = 100$ volts

$$\text{Initial capacitance, } C_1 = \frac{\varepsilon_0 \, A}{d} = \frac{8 \cdot 85 \times 10^{-12} \times 2}{0 \cdot 5 \times 10^{-3}} = 35 \cdot 4 \times 10^{-9} \text{F}$$

$$\text{Initial stored energy, } E_1 = \frac{1}{2}C_1 V^2 = \frac{1}{2} \times (35 \cdot 4 \times 10^{-9}) \times 100^2 = 17 \cdot 7 \times 10^{-5} \text{ J}$$

$$\text{Final capacitance, } C_2 = \frac{1}{2}C_1 = \frac{1}{2}(35.4 \times 10^{-9}) = 17.7 \times 10^{-9} \text{ F}$$

$$\text{Final stored energy, } E_2 = \frac{1}{2}C_2 V^2 = \frac{1}{2}(17 \cdot 7 \times 10^{-9}) \times 100^2 = 8 \cdot 85 \times 10^{-5} \text{ J}$$

Change in stored energy $= (17 \cdot 7 - 8 \cdot 85) \times 10^{-5} = 8 \cdot 85 \times 10^{-5}$ J

Suppose F newtons is the average mechanical force between the plates. The plates are separated by a distance $dx = 1 - 0 \cdot 5 = 0 \cdot 5$ mm.

$$\therefore \qquad F \times dx = \text{Change in stored energy}$$

or

$$F = \frac{8 \cdot 85 \times 10^{-5}}{0 \cdot 5 \times 10^{-3}} = 17.7 \times 10^{-2} \text{ N}$$

Note that small low-voltage capacitors store microjoules of energy.

6.25. Behaviour of Capacitor in a D.C. Circuit

When d.c. voltage is applied to an uncharged capacitor, there is transfer of electrons from one plate (connected to +ve terminal of source) to the other plate (connected to –ve terminal of source). This is called *charging current* because the capacitor is being charged. The capacitor is *quickly* charged to the applied voltage and charging current becomes zero. Under this condition, the capacitor is said to be fully charged. When a wire is connected across the charged capacitor, the excess electrons on the negative plate move through connecting wire to the positive plate. The energy stored in the capacitor is dissipated in the resistance of the wire. The charge is neutralised when the number of free electrons on both plates are again equal. At this time, the voltage across the capacitor is zero and the capacitor is fully discharged. The behaviour of a capacitor in a d.c. circuit is summed up below :

(*i*) When d.c. voltage is applied to an uncharged capacitor, the capacitor is quickly (*not instantaneously*) charged to the applied voltage.

Charging current, $i = \dfrac{dQ}{dt} = \dfrac{d}{dt}(CV) = C\dfrac{dV}{dt}$

When the capacitor is fully charged, capacitor voltage becomes constant and is equal to the applied voltage. Therefore, $dV/dt = 0$ and so is the charging current. Note that dV/dt is the slope of v–t graph of a capacitor.

(*ii*) A capacitor can have voltage across it even when there is no current flowing.

(*iii*) The voltage across a capacitor ($Q = CV$) is proportional to *charge* and not the current.

(*iv*) *There is no current through the dielectric of the capacitor during charging or discharging because the dielectric is an insulating material.* There is merely transfer of electrons from one plate to the other through the connecting wires.

(*v*) When the capacitor is fully charged, there is no circuit current. *Therefore, a fully charged capacitor appears as an open to d.c.*

(*vi*) *An uncharged capacitor is equivalent to a *short circuit as far as d.c. voltage is concerned.* Therefore, a capacitor must be charged or discharged by connecting a resistance in series with it to limit the charging or discharging current.

(*vii*) When the circuit containing capacitor is disconnected from the supply, the capacitor remains charged for a long period. *If the capacitor is charged to a high value, it can be dangerous to someone working on the circuit.*

Example 6.46. *A certain voltage source causes the current to an initially discharged 1000 μF capacitor to increase at a constant rate of 0·06 A/s. Find the voltage across the capacitor after t = 10 s.*

Solution. Charging current, $i_C = 0\cdot06t$

∴ Voltage across the capacitor after $t = 10$ s is

$$**v_C = \frac{1}{C}\int_0^{10} i_C\, dt = \frac{1}{1000\times10^{-6}}\int_0^{10} 0\cdot06t$$

$$= 10^3\times0\cdot06\int_0^{10} t\, dt = 60\left[\frac{t^2}{2}\right]_0^{10}$$

$$= 60\times\frac{10^2}{2} = \textbf{3000 V}$$

Example 6.47. *A voltage across a 100 μF capacitor varies as follows : (i) uniform increase from 0 V to 700 V in 10 sec (ii) a uniform decrease from 700 V to 400 V in 2 sec (iii) a steady value of 400 V (iv) an instantaneous drop from 400 V to zero. Find the circuit current during each period.*

Solution. $i = C\dfrac{dv}{dt} = 100\times10^{-6}\dfrac{dv}{dt} = 10^{-4}\dfrac{dv}{dt}$ A

(*i*) $dv = 700$ V ; $dt = 10$ sec

∴ $i = 10^{-4}\times\dfrac{700}{10} = 7\times10^{-3}$ A $= \textbf{7 mA}$

(*ii*) $dv = 700 - 400 = 300$ V ; $dt = 2$ sec

∴ $i = 10^{-4}\times\dfrac{300}{2} = 15\times10^{-3}$ A $= \textbf{15 mA}$

---

* When d.c. voltage is applied to an uncharged capacitor, the charging current is limited only by the small resistance of source and any wiring resistance present. The surge current that flows when no resistor is present may be great enough to damage the capacitor, the source or both.

** $i = C\dfrac{dv}{dt}$ or $\dfrac{dv}{dt} = \dfrac{i}{C}$ ∴ Integrating, $v = \dfrac{1}{C}\int_0^t i\,dt$

(*iii*) $dv/dt = 0$. Therefore, current is **zero**.

(*iv*) $dv = 400 - 0 = 400$ V ; $dt = 0$

∴ $i = 10^{-4} \times \dfrac{400}{0} = $ **infinite**

Note that in this period, the current is extremely high.

6.26. Charging of a Capacitor

Consider an uncharged capacitor of capacitance C farad connected in series with a resistor R to a d.c. supply of V volts as shown in Fig. 6.46. When the switch is closed, the capacitor starts charging up and charging current flows in the circuit. The charging current is maximum at the instant of switching and decreases gradually as the voltage across the capacitor increases. When the capacitor is charged to applied voltage V, the charging current becomes zero.

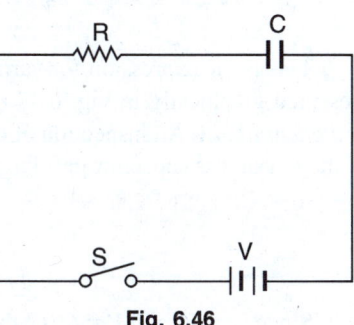

Fig. 6.46

1. **At switching instant.** At the instant the switch is closed, the voltage across capacitor is zero since we started with an uncharged capacitor. The entire voltage V is dropped across resistance R and charging current is maximum (call it I_m).

∴   Initial charging current, $I_m = V/R$

Voltage across capacitor $= 0$

Charge on capacitor $= 0$

2. **At any instant.** After having closed the switch, the charging current starts decreasing and the voltage across capacitor gradually increases. Let at any time t during charging :

$i = $ Charging current

$v = $ P.D. across C

$q = $ Charge on capacitor $= C v$

(*i*) Voltage across capacitor

According to Kirchhoff's voltage law, the applied voltage V is equal to the sum of voltage drops across resistor and capacitor.

∴ $V = v + iR$ ...(*i*)

or $V = v + CR* \dfrac{dv}{dt}$

or $-\dfrac{dv}{V-v} = -\dfrac{dt}{RC}$

Integrating both sides, we get,

$$\int -\dfrac{dv}{V-v} = \int -\dfrac{dt}{RC}$$

or $\log_e (V-v) = -\dfrac{t}{RC} + K$ ...(*ii*)

where K is a constant whose value can be determined from the initial conditions. At the instant of closing the switch S, $t = 0$ and $v = 0$.

Substituting these values in eq. (*ii*), we get, $\log_e V = K$.

Putting the value of $K = \log_e V$ in eq. (*ii*), we get,

$$\log_e (V-v) = -\dfrac{t}{RC} + \log_e V$$

---

* $i = \dfrac{dq}{dt} = \dfrac{d}{dt}(q) = \dfrac{d}{dt}(Cv) = C\dfrac{dv}{dt}$

or
$$\log_e \frac{V-v}{V} = -\frac{t}{RC}$$

or
$$\frac{V-v}{V} = e^{-t/RC}$$

$$\therefore \qquad v = V[1 - e^{-t/RC}] \qquad\qquad\qquad ...(iii)$$

This is the expression for variation of voltage across the capacitor (v) w.r.t. time (t) and is represented graphically in Fig. 6.47 (i). *Note that growth of voltage across the capacitor follows an exponential law.* An inspection of eq. (iii) reveals that as t increases, the term $e^{-t/RC}$ gets smaller and voltage v across capacitor gets larger.

(ii) Charge on Capacitor

$$q = \text{Charge at any time } t$$
$$Q = \text{Final charge}$$

Since $v = q/C$ and $V = Q/C$, the exp. (iii) becomes :

$$\frac{q}{C} = \frac{Q}{C}[1 - e^{-t/RC}]$$

or
$$q = Q(1 - e^{-t/RC}) \qquad\qquad\qquad ...(iv)$$

Again the increase of charge on capacitor plates follows exponential law.

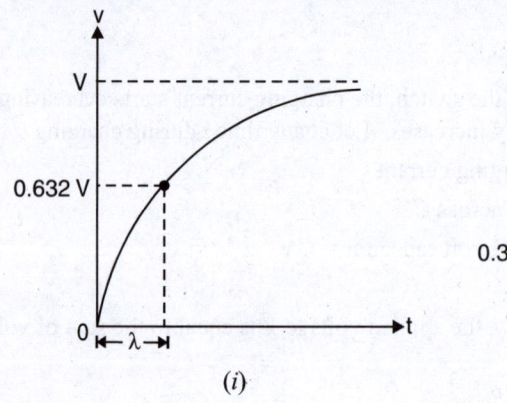
(i)

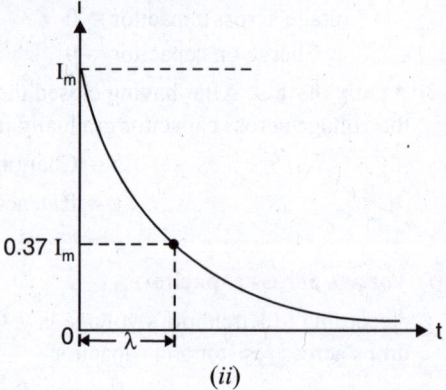
(ii)

Fig. 6.47

(iii) Charging current

From exp. (i), $V - v = iR$

From exp. (iii), $V - v = Ve^{-t/RC}$

$\therefore$ $iR = Ve^{-t/RC}$

or
$$i = \frac{V}{R}e^{-t/RC}$$

$\therefore$
$$i = I_m e^{-t/RC}$$

where I_m (= V/R) is the initial charging current. *Again the charging current decreases following exponential law.* This is also represented graphically in Fig. 6.47 (ii).

(iv) Rate of rise of voltage across capacitor

We have seen above that :
$$V = v + CR\frac{dv}{dt}$$

At the instant the switch is closed, $v = 0$.

$\therefore$
$$V = CR\frac{dv}{dt}$$

or   Initial rate of rise of voltage across capacitor is given by ;

$$\frac{dv}{dt} = \frac{V}{CR} \text{ volts/sec} \qquad \ldots(iv)$$

Note. *The capacitor is almost fully charged in a time equal to 5 RC i.e., 5 time constants.*

6.27. Time Constant

Consider the eq. (*iii*) above showing the rise of voltage across the capacitor :

$$v = V(1 - e^{-t/RC})$$

The exponent of *e* is *t/RC*. The quantity *RC* has the *dimensions of time so that exponent of *e* is a number. The quantity *RC* in called the *time constant* of the circuit and affects the charging (or discharging) time. It is represented by λ (or T or τ).

∴   Time constant, $\lambda = RC$ seconds

Time constant may be defined in one of the following ways :

(*i*) At the instant of closing the switch, p.d. across capacitor is zero. Therefore, putting $v = 0$ in the expression $V = v + CR\dfrac{dv}{dt}$, we have,

$$V = CR\frac{dv}{dt}$$

or

$$\frac{dv}{dt} = \frac{V}{CR}$$

If this rate of rise of voltage could continue, the capacitor voltage will reach the final value *V* in time $= V \div V/CR = RC$ seconds $=$ time constant λ.

Hence **time constant** *may be defined as the time required for the capacitor voltage to rise to its final steady value V if it continued rising at its initial rate (i.e., V/CR).*

(*ii*) If the time interval $t = \lambda$ (or *RC*), then,

$$v = V(1 - e^{-t/t}) = V(1 - e^{-1}) = 0.632\ V$$

Hence **time constant** *can also be defined as the time required for the capacitor voltage to reach 0·632 of its final steady value V.*

(*iii*) If the time interval $t = \lambda$ (or *RC*), then,

$$i = I_m e^{-t/t} = I_m e^{-1} = 0.37\ I_m$$

Hence **time constant** *can also be defined as the time required for the charging current to fall to 0·37 of its initial maximum value I_m.*

Fig. 6.48 as well as adjoining table shows the percentage of final voltage (*V*) after each time constant interval during voltage buildup (*v*) across the capacitor. An uncharged capacitor charges to about 63% of its fully charged voltage (*V*) in first time constant. A 5 time-constant interval is accepted as the time to fully charge (or discharge) a capacitor and is called the *transient time.*

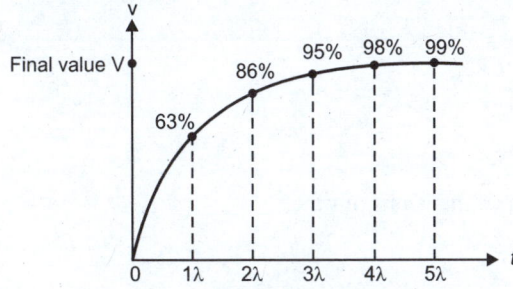

Number of time constants	% of final value
1	63
2	86
3	95
4	98
5	99 considered 100 %

Fig. 6.48

* $RC = \left(\dfrac{\text{Volt}}{\text{Ampere}}\right) \times \left(\dfrac{\text{Coulomb}}{\text{Volt}}\right) = \dfrac{\text{Volt}}{(\text{Coulomb/sec})} \times \left(\dfrac{\text{Coulomb}}{\text{Volt}}\right) = \text{seconds}$

6.28. Discharging of a Capacitor

Consider a capacitor of C farad charged to a p.d. of V volts and connected in series with a resistance R through a switch S as shown in Fig. 6.49 (*i*). When the switch is open, the voltage across the capacitor is V volts. When the switch is closed, the voltage across capacitor starts decreasing. The discharge current rises instantaneously to a value of $V/R \, (= I_m)$ and then decays gradually to zero.

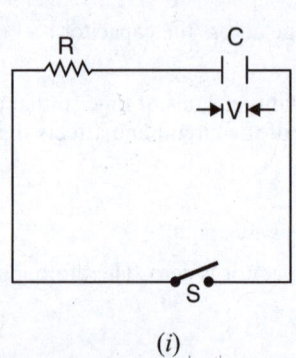

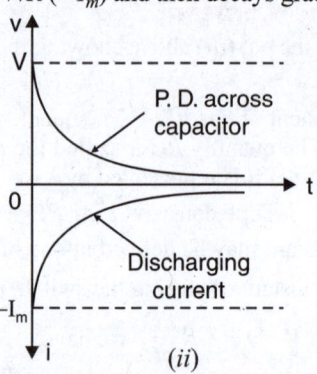

(*i*)   (*ii*)

Fig. 6.49

Let at any time t during discharging,

$$v = \text{p.d. across the capacitor}$$
$$i = \text{discharging current}$$
$$q = \text{charge on capacitor}$$

By Kirchhoff's voltage law, we have,

$$0 = v + RC \frac{dv}{dt}$$

or

$$\frac{dv}{v} = -\frac{dt}{RC}$$

Integrating both sides, we get,

$$\int \frac{dv}{v} = -\frac{1}{RC} \int dt$$

$\therefore$

$$\log_e v = -\frac{t}{RC} + K \qquad \qquad ...(i)$$

At the instant of closing the switch, $t = 0$ and $v = V$. Putting these values in eq. (*i*), we get,

$$\log_e V = K$$

$\therefore$ Equation (*i*) becomes : $\log_e v = (-t/RC) + \log_e V$

or

$$\log_e \frac{v}{V} = -\frac{t}{RC}$$

or

$$\frac{v}{V} = e^{-t/RC}$$

$\therefore$

$$v = V e^{-t/\lambda} \qquad \qquad ...(ii)$$

Again $\lambda \, (= RC)$ is the time constant and has the dimensions of time.

Similarly,

$$q = Q e^{-t/RC}$$

and

$$i = -I_m e^{-t/RC}$$

Note that negative sign is attached to I_m. This is because the discharging current flows in the opposite direction to that in which the charging current flows.

Fig. 6.50 as well as adjoining table shows the percentage of initial voltage (V) after each time constant interval during discharging of capacitor. A fully charged capacitor discharges to about 37% of its initial fully charged value in first time constant. The capacitor is fully discharged in a 5 time-constant interval.

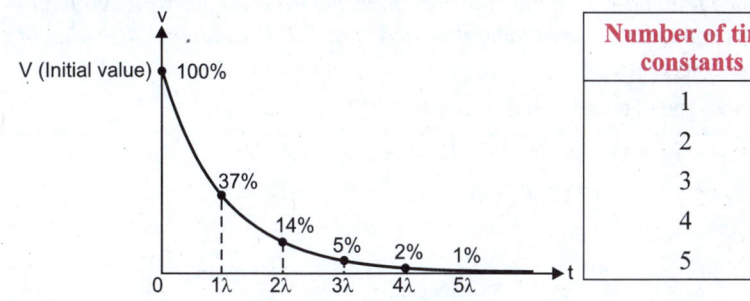

Number of time constants	% of Initial value
1	37
2	14
3	5
4	2
5	1 considered 0

Fig. 6.50

Example 6.48. *A 2 µF capacitor is connected, by closing a switch, to a supply of 100 volts through a 1 MΩ series resistance. Calculate (i) the time constant (ii) initial charging current (iii) the initial rate of rise of p.d. across capacitor (iv) voltage across the capacitor 6 seconds after the switch has been closed and (v) the time taken for the capacitor to be fully charged.*

Solution. (*i*)Time constant, $\lambda = RC = (10^6) \times (2 \times 10^{-6}) =$ **2 seconds**

(*ii*) Initial charging current, $I_m = \dfrac{V}{R} = \dfrac{100}{10^6} \times 10^6 =$ **100 µA**

(*iii*) Initial rate of rise of voltage across capacitor is

$$\frac{dv}{dt} = \frac{V}{CR} = \frac{100}{(2 \times 10^{-6}) \times 10^6} = \textbf{50 V/s}$$

(*iv*) $$v = V(1 - e^{-t/RC})$$

Here $V = 100$ volts ; $t = 6$ seconds ; $RC = 2$ seconds

∴ $v = 100 (1 - e^{-6/2}) = 100 (1 - e^{-3}) =$ **95.1 V**

(*v*) Time taken for the capacitor to be fully charged

$$= 5 RC = 5 \times 2 = \textbf{10 seconds}$$

Example 6.49. *A capacitor of 8 µF capacitance is connected to a d.c. source through a resistance of 1 megaohm. Calculate the time taken by the capacitor to receive 95% of its final charge. How long will it take the capacitor to be fully charged ?*

Solution. $$q = Q(1 - e^{-t/RC})$$

Here $$RC = (10)^6 \times 8 \times 10^{-6} = 8 \text{ seconds} ; \quad q/Q = 0.95$$

∴ $$0.95 = 1 - e^{-t/8} \quad \text{or} \quad e^{-t/8} = 0.05$$

∴ $$e^{t/8} = 1/0.05 = 20$$

or $$(t/8) \log_e e = \log_e 20$$

∴ $$t = 8 \log_e 20 = \textbf{23.96 seconds}$$

Time taken for the capacitor to be fully charged

$$= 5 RC = 5 \times 8 = \textbf{40 seconds}$$

Alternatively. $$t = \lambda \log_e \frac{V - V_0}{V - v_C} \quad \text{... See Art. 6.30}$$

or $$t = \lambda \log_e \frac{Q - q_0}{Q - q}$$

Here, $\lambda = 8s$; $q_0 = 0$; $q = 95\%$ of $Q = 0.95\ Q$

$$\therefore \qquad t = 8 \times \log_e \frac{Q-0}{Q-0.95Q} = 8 \times \log_e \frac{Q}{0.05Q} = \textbf{23.96 seconds}$$

Example 6.50. *A resistance R and a 4 μF capacitor are connected in series across a 200 V d.c. supply. Across the capacitor is connected a neon lamp that strikes at 120 V. Calculate the value of R to make the lamp strike after 5 seconds.*

Solution. The voltage across the neon lamp has to rise to 120 V in 5 seconds.

Now, $\qquad\qquad v = V(1-e^{-t/\lambda})$ or $120 = 200\,(1-e^{-5/\lambda})$

or $\qquad\qquad e^{-5/\lambda} = 1-(120/200) = 0 \cdot 4$ or $e^{5/\lambda} = 1/0 \cdot 4 = 2 \cdot 5$

$\therefore \qquad\qquad (5/\lambda) \log_e e = \log_e 2 \cdot 5$

or $\qquad\qquad \lambda = \dfrac{5}{\log_e 2 \cdot 5} = 5 \cdot 457$ seconds

or $\qquad\qquad RC = 5 \cdot 457$ $\therefore$ $R = \dfrac{5 \cdot 457}{4 \times 10^{-6}} = 1 \cdot 364 \times 10^6\ \Omega = \textbf{1·364 MΩ}$

Alternatively. $\qquad\qquad t = \lambda \log_e \dfrac{V-V_0}{V-v_C}$

Here, $t = 5s$; $V = 200$ volts ; $V_0 = 0$; $v_C = 120$ volts

Putting these values in the above expression, we get, $\lambda = 5.457$s.

Now $\lambda = RC$ or $R = \dfrac{\lambda}{C} = \dfrac{5.457}{4 \times 10^{-6}} = 1.364 \times 10^6 \Omega = \textbf{1.364 MΩ}$

Example 6.51. *A capacitor of 1 μF and resistance 82 kΩ are connected in series with an e.m.f. of 100 V. Calculate the magnitude of energy and the time in which energy stored in the capacitor will reach half of its equilibrium value.*

Solution.   Equilibrium value of energy $= \dfrac{1}{2}CV^2$

$\therefore \qquad\qquad$ Energy stored $\propto V^2$

Half energy of the equilibrium value will be stored when voltage across capacitor is $v = 100/\sqrt{2}$ $= 70 \cdot 7$ volts.

$\therefore \qquad\qquad$ Energy stored $= \dfrac{1}{2}Cv^2 = \dfrac{1}{2}(1 \times 10^{-6}) \times (70 \cdot 7) = \textbf{0·0025 J}$

Now, $\qquad\qquad v = V(1-e^{-t/RC})$

Here, $RC = (82 \times 10^3) \times (1 \times 10^{-6}) = 0.082$ s ; $v = 70 \cdot 7$ V ; $V = 100$ V

$\therefore \qquad\qquad 70 \cdot 7 = 100\,(1-e^{-t/0 \cdot 082})$ or $e^{-t/0 \cdot 082} = 1-(70 \cdot 7/100) = 0 \cdot 293$

$\therefore \qquad\qquad e^{t/0 \cdot 082} = 1/0 \cdot 293 = 3 \cdot 413$

or $\qquad\qquad (t/0 \cdot 082) \log_e e = \log_e 3 \cdot 413$

$\therefore \qquad\qquad t = 0 \cdot 082 \times \log_e 3 \cdot 413 = \textbf{0·1 second}$

Example 6.52. *When a capacitor C charges through a resistor R from a d.c. source voltage E, determine the energy appearing as heat.*

Solution. When $R - C$ series circuit is switched on to d.c. source of voltage E, the charging current i decreases at exponential rate given by ;

$$i = I e^{-t/\lambda}$$

where $I = E/R$; $\lambda = RC$

Energy appearing as heat in small time Δt is

$$\Delta W_R = i^2 R\, \Delta t$$

Total energy appearing as heat in the entire process of charging is

$$W_R = \int_0^\infty i^2 R\, dt = R\int_0^\infty (I\, e^{-t/\lambda})^2\, dt = R\int_0^\infty I^2 e^{-2t/\lambda} dt$$

$$= R \times I^2 \int_0^\infty e^{-2t/\lambda} dt = RI^2 \left[\frac{e^{-2t/\lambda}}{-2/\lambda}\right]_0^\infty$$

$$= R \times (E/R)^2 \left(\frac{-\lambda}{2}\right)[e^{-\infty} - e^0] = \frac{E^2}{R} \times \left(\frac{-RC}{2}\right) \times (-1)$$

$$\therefore \qquad W_R = \frac{1}{2}CE^2$$

Although energy stored in a capacitor is very small, it can provide a large current (and hence large power) for a short period of time.

Note. Energy stored in the capacitor at the end of charging process is $CE^2/2$. Also energy appearing as heat in the entire process of charging the capacitor is $CE^2/2$.

$$\therefore \quad \text{Total energy received from the source} = \frac{1}{2}CE^2 + \frac{1}{2}CE^2 = CE^2$$

Thus during charging of capacitor, the total energy received from the source is CE^2; half is converted into heat and the rest half stored in the capacitor.

Example 6.53. *Referring to the circuit shown in Fig. 6.51,*

(i) Write the mathematical expression for charging current i and voltage v across capacitor when the switch is placed in position 1.

(ii) Write the mathematical expression for the discharging current and voltage across capacitor when switch is placed in position 2 after having been in position 1 for 1 s.

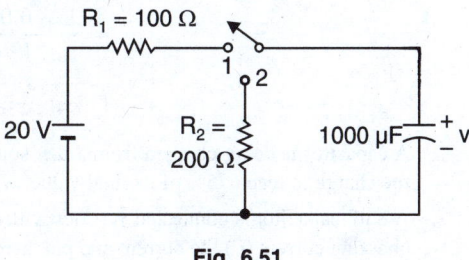

Fig. 6.51

Solution.

(i) When the switch is placed in position 1, the capacitor charges through R_1 only. Therefore, time constant during charging is

Time constant, $\lambda = R_1 C = (100) \times (1000 \times 10^{-6}) = 0.1$ s

Initial charging current, $I_m = V/R_1 = 20/100 = 0.2$ A

The charging current at any time t is given by ;

$$i = I_m\, e^{-t/\lambda} \quad \text{or} \quad i = 0.2\, e^{-t/0.1} \text{ A}$$

The voltage v across the capacitor at any time t is given by ;

$$v = V(1 - e^{-t/\lambda}) \quad \text{or} \quad v = 20\,(1 - e^{-t/0.1}) \text{ volts}$$

(ii) Since the switch remains in position 1 for 1 s or 10 time constants, the capacitor charges fully to 20 V. When the switch is placed in position 2, the capacitor discharges through R_2 only. Therefore, time constant during discharge is

Time constant, $\lambda = R_2 C = (200) \times (1000 \times 10^{-6}) = 0.2$ s

Initial discharging current, $I_m = V/R_2 = 20/200 = 0.1$ A

The discharging current at any time t is given by ;

$$i = -I_m\, e^{-t/\lambda} \quad \text{or} \quad i = -0.1\, e^{-t/0.2} \text{ A}$$

The voltage v across the capacitor at any time t is given by ;

$$v = V e^{-t/\lambda} \quad \text{or} \quad v = 20\, e^{-t/0.2} \text{ volts}$$

Example 6.54. *A cable 10 km long and of capacitance 2.5μF discharges through its insulation resistance of 50 MΩ. By what percentage the voltage would have fallen 1, 2 and 5 minutes respectively after disconnection from its bus-bars?*

Solution. Capacitance of cable capacitor, $C = 2.5 \times 10^{-6}$ F; Insulation resistance of cable, $R = 50$ MΩ $= 50 \times 10^{6}$ Ω

Time constant, $\lambda = RC = (50 \times 10^{6}) \times (2.5 \times 10^{-6}) = 125$ seconds

During discharging, decreasing voltage v across the capacitor is given by ;

$$v = Ve^{-t/\lambda} = Ve^{-t/125}$$

At $t = 1$ min. $= 60$ seconds, $v_1 = Ve^{-60/125} = 0.618$ V

At $t = 2$ min. $= 120$ seconds, $v_2 = Ve^{-120/125} = 0.383$ V

At $t = 5$ min. $= 300$ seconds, $v_3 = Ve^{-300/125} = 0.09$ V

∴ At $t = 1$ min, the % age fall in voltage across capacitor

$$= \frac{V - 0.618V}{V} \times 100 = \mathbf{38.2\%}$$

At $t = 2$ min; the % age fall in voltage across capacitor

$$= \frac{V - 0.383V}{V} \times 100 = \mathbf{61.7\%}$$

At $t = 5$ min; the % age fall in voltage across capacitor

$$= \frac{V - 0.09V}{V} \times 100 = \mathbf{91\%}$$

Tutorial Problems

1. A capacitor is being charged from a d.c. source through a resistance of 2MΩ. If it takes 0·2 second for the charge to reach 75% of its final value, what is the capacitance of the capacitor ?   [18×10^{-4} F]

2. A 8 μF capacitor is connected is series with 0·5 MΩ resistance across 200 V supply. Calculate (*i*) initial charging current (*ii*) the current and p.d. across capacitor 4 seconds after it is connected to the supply.
    [(*i*) **400 μA** (*ii*) **147 μA; 126·4 V**]

3. What resistance connected in series with a capacitance of 4μF will give the circuit a time constant of 2 seconds ?   [**500 kΩ**]

4. A series *RC* circuit is to have an initial charging current of 4 mA and a time constant of 3·6 seconds when connected to 120 V d.c. supply. Calculate the values of *R* and *C*. What will be the energy stored in the capacitor ?   [**30 kΩ ; 120 μF ; 0·864 J**]

5. A 20μF capacitor initially charged to a p.d. of 500V is discharged through an unknown resistance. After one minute, the p.d. at the terminals of the capacitor is 200 V. What is the value of the resistance ?
    [**3·274 MΩ**]

6.29. Transients in D.C. Circuits

When a circuit goes from one steady-state condition to another steady-state condition, it passes through a transient state which is of short duration. The word transient means temporary or short-lived. When a d.c. voltage source is first connected to a series *RC* network, the charging current flows only until the capacitor is fully charged. This charging current is called a **transient current**. In connection with d.c. circuits, a transient is a voltage or current that *changes* with time for a short duration of time and remains constant thereafter. As a capacitor charges, its voltage builds up (*i.e.,* changes) until the capacitor is fully charged and its voltage equals the source voltage. After that time, there is no further change in capacitor voltage. Thus the voltage across a capacitor during the time it is being charged is an example of a **transient voltage**.

6.30. Transient Relations During Charging/Discharging of Capacitor

When a capacitor is charging or discharging, it goes from one steady-state condition (called *initial condition*) to another steady-state condition (called *final condition*). During this change, the voltage across and current through the capacitor change continuously. These are called *transient conditions* and exist for a short duration. *It can be shown mathematically that voltage v_C across the capacitor at any time t during charging or discharging* is given by ;

$$v_C = V - (V - V_0)e^{-t/\lambda} \qquad \text{...(i)}$$

where
v_C = voltage across capacitor at any time t
V = Source voltage during charging
V_0 = Voltage across capacitor at $t = 0$
λ = Time constant ($= RC$)

Note that for discharging of capacitor, $V = 0$ because there is no source voltage.

1. Transient conditions during charging. When we charge an uncharged capacitor, $V_0 = 0$ so that eq. (*i*) becomes :

$$v_C = V - (V - 0)e^{-t/\lambda} = V - Ve^{-t/\lambda}$$

$$\therefore \qquad v_C = V(1 - e^{-t/\lambda}) \qquad \text{...(ii)}$$

This is the same equation that we derived in Art. 6.26 for charging of a capacitor.

From eq. (*ii*), $V - v_C = Ve^{-t/\lambda}$

But $V - v_C = iR$, where i is the charging current at time t.

$$\therefore \qquad iR = Ve^{-t/\lambda} \text{ or } i = \frac{V}{R}e^{-t/\lambda}$$

$$\therefore \qquad i = Ie^{-t/\lambda} \qquad \text{...(iii)}$$

where $I (= V/R)$ is the initial charging current.

Note that eq. (*iii*) is the same that we derived in Art. 6.26 for charging of a capacitor. Fig. 6.52 shows capacitor voltage (v_C) and charging current (i) waveforms for a charging capacitor. It may be seen that voltage across the capacitor is building up at an exponential rate while the charging current is decreasing at an exponential rate.

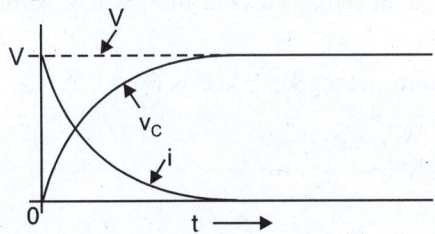
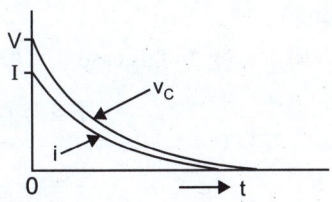

| Fig. 6.52 | Fig. 6.53 |

2. Transient conditions during discharging. For discharging of a capacitor, $V = 0$ because there is no source voltage. Therefore, eq. (*i*) becomes :

$$v_C = 0 - (0 - V_0)e^{-t/\lambda}$$

or
$$v_C = V_0 e^{-t/\lambda} \qquad \text{...(iv)}$$

Here V_0 is, of course, the voltage to which the capacitor was originally charged. Note that this is the same expression which derived in Art. 6.28 for discharging of a capacitor.

Now,
$$\frac{v_C}{C} = \frac{V_0}{C}e^{-t/\lambda}$$

or
$$q = Q_0 e^{-t/\lambda}$$

---
*   The word transient means temporary or short-lived.

where Q_0 is the initial charge on the capacitor and q is the charge on the capacitor at time t.

Similarly, $i = I_0 e^{-t/\lambda}$

where I is the initial discharging current and i is the discharging current at time t.

Fig. 6.53 shows the capacitor voltage and discharging current waveforms. Both decrease at exponential rate and reach zero value at the same time.

Time for charge or discharge. Sometimes it is desirable to determine how long will it take the capacitor in RC series circuit to charge or discharge to a specified voltage. This can be found as follows : From eq. (*i*),

$$v_C = V - (V - V_0)\, e^{-t/\lambda}$$

or

$$V - v_C = (V - V_0)e^{-t/\lambda}$$

or

$$\frac{V - v_C}{V - V_0} = e^{-t/\lambda}$$

or

$$\frac{V - V_0}{V - v_C} = e^{t/\lambda}$$

Taking the natural log, we have,

$$\frac{t}{\lambda}\log_e e = \log_e \frac{V - V_0}{V - v_C}$$

$\therefore$

$$t = \lambda \log_e \frac{V - V_0}{V - v_C} \qquad \qquad ...(v)$$

Exp. (v) is applicable for charging as well as discharging of a capacitor.

For charging. When C is charging from 0V (*i.e.* capacitor is uncharged), $V_0 = 0$. Therefore, putting $V_0 = 0$ in exp. (*v*), we have,

$$t = \lambda \log_e \frac{V - 0}{V - v_C} = \lambda \log_e \frac{V}{V - v_C}.$$

$\therefore$

$$t = \lambda \log_e \frac{V}{V - v_C}$$

If the capacitor has some initial charge instead of zero, then value of V_0 will be corresponding to that charge.

For discharging. In this case, $V = 0$. Therefore, putting $V = 0$ in exp. (*v*), we have,

$$t = \lambda \log_e \frac{0 - V_0}{0 - v_C} = \lambda \log_e \frac{V_0}{v_C}$$

$\therefore$

$$t = \lambda \log_e \frac{V_0}{v_C}$$

Example 6.55. *The uncharged capacitor in Fig. 6.54 is initially switched to position 1 of the switch for two seconds and then switched to position 2 for the next two seconds. What will be the voltage on the capacitor at the end of this period?*

Solution. When uncharged capacitor is switched to position 1, it will be instantaneously charged to 100 V because there is no resistance in the charging circuit. Therefore, after 2 seconds, the capacitor will be at 100 V. Now when switch is put to position 2, the time of discharge t is given by ;

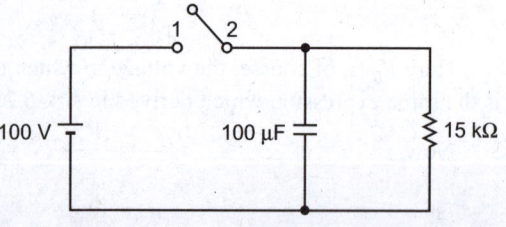

Fig. 6.54

$$t = \lambda \log_e \frac{V - V_0}{V - v_c}$$

Here $t = 2$s ; $\lambda = RC = 15000 \times 100 \times 10^{-6} = 1.5$s ; $V = 0$; $V_0 = 100$ volts

∴ $$2 = 1.5 \log_e \frac{0 - 100}{0 - v_c} = 1.5 \log_e \frac{100}{v_c}$$

On solving, $v_c = \textbf{26.36 V}$

Example 6.56. *A 50μF capacitor and a 20 kΩ resistor are connected in series across a battery of 100 V at the instant t = 0. At instant t = 0.5s, the applied voltage is suddenly increased to 150V. Find the charge on the capacitor at t = 0.75 s.*

Solution. Time constant, $\lambda = RC = 20,000 \times 50 \times 10^{-6} = 1$ sec.

For first case. $$t = \lambda \log_e \frac{V - V_0}{V - v_c}$$

Here, $t = 0.5$s ; $\lambda = 1$s ; $V = 100$ volts ; $V_0 = 0$; $v_C = ?$

∴ $$0.5 = 1 \times \log_e \frac{100 - 0}{100 - v_C} = \log_e \frac{100}{100 - v_C}$$

On solving, $v_C = 39.4$ volts

For second case. After 0.5 sec., the source voltage is increased to 150 V.

Now $$t = \lambda \log_e \frac{V - V_0}{V - v_C}$$

Here, $t = 0.75 - 0.5 = 0.25$s ; $\lambda = 1$s ; $V = 150$ volts ; $V_0 = 39.4$ volts ; $v'_C = ?$

∴ $$0.25 = 1 \times \log_e \frac{150 - 39.4}{150 - v'_C} = \log_e \frac{110.6}{150 - v'_C}$$

On solving, $v'_C = 63.6$ volts

∴ Charge on capacitor $= C \times v'_C = 50 \times 10^{-6} \times 63.6 = \textbf{3.18} \times \textbf{10}^{-3}$ **C**

Example 6.57. *Find how long it takes after the switch S is closed before the total current from the supply reaches 25 mA when V = 10 V, $R_1 = 500Ω$, $R_2 = 700Ω$ and C = 100μF.*

Solution. When switch S is closed, the current in $R_1 = 500Ω$ is set up instantaneously and its value is $= 10/R_1 = 10/500 = 0.02$A $= 20$ mA. In order to draw 25 mA current from the supply, current in capacitor circuit is $= 25 - 20 = 5$mA. Now when switch S is closed, the current in capacitor circuit is maximum and its value is $I = 10/R_2 = 10/700 = 0.0143$A $= 14.3$ mA and decreases at exponential rate. Our problem is to find the time t in which charging current in capacitor circuit decreases from 14.3 mA to 5 mA.

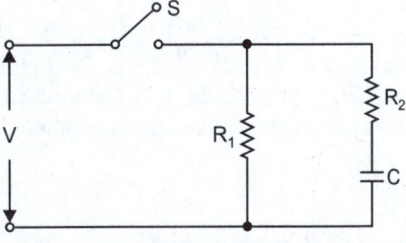

Fig. 6.55

Now, $$i = Ie^{-t/\lambda}$$

Here $i = 5$ mA ; $I = 14.3$ mA ; $\lambda = R_2C = 700 \times 100 \times 10^{-6} = 0.07$ s

∴ $$5 = 14.3\, e^{-t/0.07}$$

On solving, $t = \textbf{0.0735 s}$

Example 6.58. *In an RC series circuit, R = 2MΩ, C = 5μF and applied voltage V = 100 volts. Calculate (i) initial rate of change of capacitor voltage (ii) initial rate of change of capacitor current (iii) initial rate of change of voltage across 2MΩ resistor.*

Solution. Time constant, $\lambda = RC = 2 \times 10^6 \times 5 \times 10^{-6} = 10$ seconds

(i) $$v_C = V(1 - e^{-t/\lambda})$$

$$\therefore \quad \frac{dv_C}{dt} = 0 - Ve^{-t/\lambda}\left(-\frac{1}{\lambda}\right) = \frac{V}{\lambda}e^{-t/\lambda}$$

$$\text{At } t = 0, \ \frac{dv_C}{dt} = \frac{V}{\lambda}e^{-0/\lambda} = \frac{V}{\lambda} = \frac{100}{10} = \textbf{10 V/s}$$

(ii) $$i = Ie^{-t/\lambda}$$

$$\therefore \quad \frac{di}{dt} = Ie^{-t/\lambda}\left(-\frac{1}{\lambda}\right) = -\frac{I}{\lambda}e^{-t/\lambda}$$

$$\text{At } t = 0, \ \frac{di}{dt} = -\frac{I}{\lambda}e^{-0/\lambda} = -\frac{I}{\lambda} = -\frac{V/R}{\lambda} = -\frac{100/2 \times 10^6}{10} = \textbf{-5 } \boldsymbol{\mu}\textbf{A/s}$$

(iii) $$v_R = iR = (Ie^{-t/\lambda})R = \left(\frac{V}{R}e^{-t/\lambda}\right)R = Ve^{-t/\lambda}$$

$$\therefore \quad \frac{dv_R}{dt} = Ve^{-t/\lambda}\left(-\frac{1}{\lambda}\right) = -\frac{V}{\lambda}e^{-t/\lambda}$$

$$\text{At } t = 0, \ \frac{dv_R}{dt} = -\frac{V}{\lambda}e^{-0/\lambda} = -\frac{V}{\lambda} = -\frac{100}{10} = \textbf{-10 V/s}$$

Example 6.59. *Calculate the values of* i_2, i_3, v_2, v_3, v_C *and* v_L *in the network shown in Fig. 6.56 at the following times :*

   (*i*) *At time,* $t = 0$ *immediately after the switch S is closed.*

   (*ii*) *At time,* $t \to \infty$ *i.e. in the steady state. All resistances are in ohms.*

Solution. (*i*) At the instant of closing the switch (*i.e.* at $t = 0$), the inductance ($= 1\ H$) behaves as an open circuit so that no current flows in the coil.

$$\therefore \ i_2 = \textbf{0 A} \ ; \ v_2 = \textbf{0 V} \ ; \ v_L = \textbf{20 V}$$

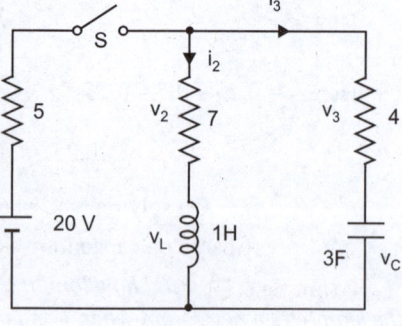

Fig. 6.56

At the instant of closing the switch, the capacitor behaves as a short circuit.

$$\therefore \ i_3 = \frac{20}{5+4} = \frac{20}{9}\textbf{A} \ ; \ v_3 = 4 \times \frac{20}{9} = \frac{80}{9}\textbf{V} \ ; \ v_C = \textbf{0 V}$$

(*ii*) Under steady state conditions (*i.e.* when the capacitor is fully charged), the capacitor behaves as an open circuit and the inductance ($= 1H$) as short.

$$\therefore \ i_2 = \frac{20}{5+7} = \frac{5}{3}\textbf{A} \ ; \ v_2 = 7 \times \frac{5}{3} = \frac{35}{3}\textbf{V} \ ; \ v_L = \textbf{0 V} \ ; \ i_3 = \textbf{0A} \ ;$$

$$v_3 = \textbf{0 V} \ ; \ v_C = \textbf{20 V}$$

Example 6.60. *In Fig. 6.57, the capacitor C is uncharged. Determine the final voltage on the capacitor after the switch has been in position 2 for 3s and then in position 3 for 5s.*

Solution. When the switch is in position 2, the voltage v_C across the capacitor is

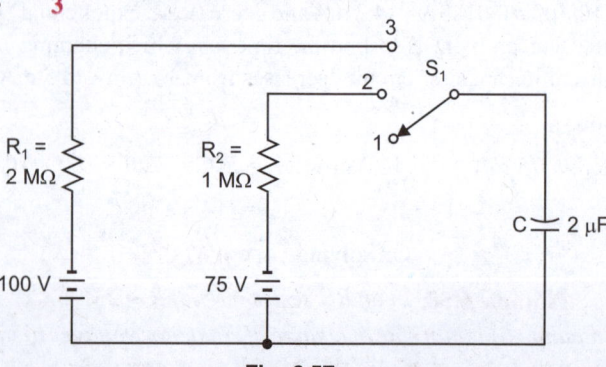

Fig. 6.57

$$v_C = V(1 - e^{-t/\lambda})$$

Here, $V = 75$ volts ; $t = 3$s ; $\lambda = R_2C = (1 \times 10^6) \times 2 \times 10^{-6} = 2$s

$\therefore$ $v_C = 75(1 - e^{-3/2}) = 75(1 - 0.223) = 58.3$ V

Therefore, after 2s, voltage across capacitor is 58.3 V.

When switch is in position 3, voltage v'_C across capacitor is

$$v'_C = V - (V - v_C)e^{-t/\lambda}$$

Here, $V = 100$ volts ; $t = 5$s ; $v_C = 58.3$ volts ; $\lambda = R_1C = 2 \times 10^6 \times 2 \times 10^{-6} = 4$s

$\therefore$ $v'_C = 100 - (100 - 58.3)e^{-5/4}$

$$= 100 - (100 - 58.3) \times 0.287 = \mathbf{88.0\ V}$$

Therefore, final voltage across the capacitor is 88.0 V.

Tutorial Problems

1. A capacitor of capacitance 12μF is allowed to discharge through its own leakage resistance and a fall of p.d. from 120 V to 100 V is recorded in a time interval of 300 seconds by an electrostatic voltmeter connected in parallel. Calculate the leakage resistance of the capacitor.   **[137 MΩ]**

2. When a capacitor charged to a p.d. of 400 V is connected to a voltmeter having a resistance of 25 MΩ, the voltmeter reading is observed to have fallen to 50 V at the end of an interval of 2 minutes. Find the capacitance of the capacitor.   **[2.31 μF]**

3. An 8μF capacitor is connected through a 1.5 MΩ resistance to a direct current source. After being on charge for 24 seconds, the capacitor is disconnected and discharged through a resistor. Determine what %age of the energy input from the supply is dissipated in the resistor.   **[43.2%]**

4. An 8μF capacitor is connected in series with a 0.5 MΩ resistor across a 200V d.c. supply. Calculate (*i*) the time constant (*ii*) the initial charging current (*iii*) the time taken for the p.d. across the capacitor to grow to 160 V and (*iv*) the current and the p.d. across the capacitor in 4 seconds after it is connected to the supply.   **[(*i*) 4s (*ii*) 0.4 mA (*iii*) 6.4s (*iv*) 0.14 mA ; 126.4 V]**

Objective Questions

1. The capacitance of a capacitor is .... relative permittivity.
    (*i*) directly proportional to
    (*ii*) inversely proportional to
    (*iii*) independent of
    (*iv*) directly proportional to square of

2. An air capacitor has the same dimensions as that of a mica capacitor. If the capacitance of mica capacitor is 6 times that of air capacitor, then relative permittivity of mica is
    (*i*) 36   (*ii*) 12
    (*iii*) 3   (*iv*) 6

3. The most convenient way of achieving large capacitance is by using
    (*i*) multiplate construction
    (*ii*) decreased distance between plates
    (*iii*) air as dielectric
    (*iv*) dielectric of low permittivity

4. Another name for relative permittivity is
    (*i*) dielectric strength
    (*ii*) breakdown voltage
    (*iii*) specific inductive capacity

   (*iv*) potential gradient

5. A capacitor opposes
    (*i*) change in current
    (*ii*) change in voltage
    (*iii*) both change in current and voltage
    (*iv*) none of the above

6. If a multiplate capacitor has 7 plates each of area 6 cm^2, then,
    (*i*) 6 capacitors will be in parallel
    (*ii*) 7 capacitors will be in parallel
    (*iii*) 7 capacitors will be in series
    (*iv*) 6 capacitors will be in series

7. The capacitance of three-plate capacitor [See Fig. 6.58 (*ii*)] is .... that of 2-plate capacitor.

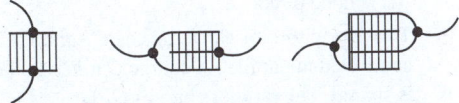

2-plate capacitor   3-plate capacitor   4-plate capacitor
    (*i*)   (*ii*)   (*iii*)

Fig. 6.58

   (*i*) 3 times   (*ii*) 6 times

   (*iii*) 4 times   (*iv*) 2 times

8. The capacitance of a 4-plate capacitor [See Fig. 6.58 (*iii*)] is .... that of 2-plate capacitor.

   (*i*) 2 times   (*ii*) 4 times

   (*iii*) 3 times   (*iv*) 6 times

9. Two capacitors of capacitances 3 μF and 6 μF in series will have a total capacitance of

   (*i*) 9 μF   (*ii*) 2 μF

   (*iii*) 18 μF   (*iv*) 24 μF

10. The capacitance of a parallel-plate capacitor does not depend upon

   (*i*) area of plates

   (*ii*) medium between plates

   (*iii*) separation between plates

   (*iv*) metal of plates

11. In order to increase the capacitance of a parallel-plate capacitor, one should introduce between the plates a sheet of

   (*i*) mica   (*ii*) tin

   (*iii*) copper   (*iv*) stainless steel

12. The capacitance of a parallel-plate capacitor depends upon

   (*i*) the type of metals used

   (*ii*) separation between plates

   (*iii*) thickness of plates

   (*iv*) potential difference between plates

13. The force between the plates of a parallel plate capacitor of capacitance C and distance of separation of plates d with a potential difference V between the plates is

   (*i*) $\dfrac{CV^2}{2d}$ (*ii*) $\dfrac{C^2V^2}{2d^2}$

   (*iii*) $\dfrac{C^2V^2}{d^2}$ (*iv*) $\dfrac{V^2d}{C}$

14. A parallel-plate air capacitor is immersed in oil of dielectric constant 2. The electric field between the plates is

   (*i*) increased 2 times

   (*ii*) increased 4 times

   (*iii*) decreased 2 times

   (*iv*) none of above

15. Two capacitors of capacitances C_1 and C_2 are connected in parallel. A charge Q given to them is shared. The ratio of charges Q_1/Q_2 is

   (*i*) C_2/C_1 (*ii*) C_1/C_2

   (*iii*) $C_1C_2/1$ (*iv*) $1/C_1C_2$

16. The dimensional formula of capacitance is

   (*i*) $M^{-1}L^{-2}T^{-4}A^2$ (*ii*) $M^{-1}L^2T^4A^2$

   (*iii*) $ML^2T^{-4}A$ (*iv*) $M^{-1}L^{-2}T^4A^2$

17. Four capacitors are connected as shown in Fig. 6.59. What is the equivalent capacitance between A and B ?

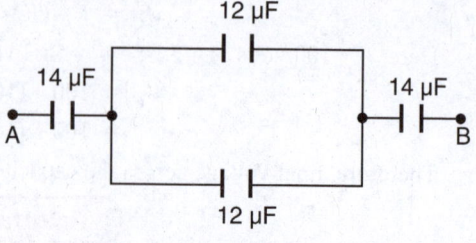

Fig. 6.59

   (*i*) 36 μF   (*ii*) 5·4 μF

   (*iii*) 52 μF   (*iv*) 11·5 μF

18. The empty space between the plates of a capacitor is filled with a liquid of dielectric constant K. The capacitance of capacitor

   (*i*) increases by a factor K

   (*ii*) decreases by a factor K

   (*iii*) increases by a factor K^2

   (*iv*) decreases by a factor K^2

19. A parallel plate capacitor is made by stacking n equally spaced plates connected alternately. If the capacitance between any two plates is C, then the resulting capacitance is

   (*i*) C (*ii*) nC

   (*iii*) $(n-1)C$ (*iv*) $(n+1)C$

20. 64 drops of radius r combine to form a bigger drop of radius R. The ratio of capacitances of bigger to smaller drop is

   (*i*) 1 : 4   (*ii*) 2 : 1

   (*iii*) 1 : 2   (*iv*) 4 : 1

21. Two capacitors have capacitances 25 μF when in parallel and 6 μF when in series. Their individual capacitances are

   (*i*) 12 μF and 13 μF

   (*ii*) 15 μF and 10 μF

   (*iii*) 10 μF and 8 μF

   (*iv*) none of above

22. A capacitor of 20 μF charged to 500 V is connected in parallel with another capacitor of 10 μF capacitance and charged to 200 V. The common potential is

   (*i*) 200 V   (*ii*) 250 V

   (*iii*) 400 V   (*iv*) 300 V

23. Which of the following does not change when a glass slab is introduced between the plates of a charged parallel plate capacitor?

    (*i*) electric charge   (*ii*) electric energy

(*iii*) capacitance

(*iv*) electric field intensity

24. A capacitor of 1 μF is charged to a potential of 50 V. It is now connected to an uncharged capacitor of capacitance 4 μF. The common potential is

    (*i*) 50 V   (*ii*) 20 V

(*iii*) 15 V   (*iv*) 10 V

25. Three parallel plates each of area A with separation d_1 between first and second and d_2 between second and third are arranged to form a capacitor. If the dielectric constants are K_1 and K_2, the capacitance of this capacitor is

(*i*) $\dfrac{\varepsilon_0 K_1 K_2}{A(d_1 + d_2)}$ (*ii*) $\dfrac{\varepsilon_0}{A\left(\dfrac{d_1}{K_1} + \dfrac{d_2}{K_2}\right)}$

(*iii*) $\dfrac{\varepsilon_0 A K_1 K_2}{d_1 + d_2}$ (*iv*) $\dfrac{\varepsilon_0 A}{\dfrac{d_1}{K_1} + \dfrac{d_2}{K_2}}$

Answers

1. (*i*)	**2.** (*iv*)	**3.** (*i*)	**4.** (*iii*)	**5.** (*ii*)
6. (*i*)	**7.** (*iv*)	**8.** (*iii*)	**9.** (*ii*)	**10.** (*iv*)
11. (*i*)	**12.** (*ii*)	**13.** (*i*)	**14.** (*iii*)	**15.** (*ii*)
16. (*iv*)	**17.** (*ii*)	**18.** (*i*)	**19.** (*iii*)	**20.** (*iv*)
21. (*ii*)	**22.** (*iii*)	**23.** (*i*)	**24.** (*iv*)	**25.** (*iv*)

7

Magnetism and Electromagnetism

Introduction

In the ancient times people believed that the invisible force of magnetism was purely a magical quality and hence they showed little practical interest. However, with steadily increasing scientific knowledge over the passing centuries, magnetism assumed a larger and larger role. Today magnetism has attained a place of pride in electrical engineering. Without the aid of magnetism, it is impossible to operate such devices as electric generators, electric motors, transformers, electrical instruments etc. Without the use of magnetism, we should be deprived of such valuable assets as the radio, television, telephone, telegraph and the ignition systems of our cars, airplanes, trucks etc. In fact, electrical engineering is so much dependent on magnetism that without it a very few of our modern devices would be possible. The purpose of this chapter is to present the salient features of magnetism.

7.1. Poles of a Magnet

If we take a bar magnet and dip it into iron filings, it will be observed that the iron filings cluster about the ends of the bar magnet. The ends of the bar magnet are apparently points of maximum magnetic effect and for convenience we call them the *poles of the magnet. A magnet has two poles *viz* north pole and south pole. In order to determine the polarity of a magnet, suspend or pivot it at the centre. The magnet will then come to rest in north-south direction. The end of the magnet pointing north is called *north pole* of the magnet while the end pointing south is called the *south pole*. The following points may be noted about the poles of a magnet :

- (*i*) The poles of a magnet cannot be separated. If a bar magnet is broken into two parts, each part will be complete magnet with poles at its ends. No matter how many times a magnet is broken, each piece will contain N-pole at one end and S-pole at the other.

- (*ii*) The two poles of a magnet are of equal strength. The pole strength is represented by m.

- (*iii*) Like poles repel each other and unlike poles attract each other.

7.2. Laws of Magnetic Force

Charles Coulomb, a French scientist observed that when two **isolated poles are placed near each other, they experience a force. He performed a number of experiments to study the nature and magnitude of force between the magnetic poles. He summed up his conclusions into two laws, known as Coulomb's laws of magnetic force. These laws give us the magnitude and nature of magnetic force between two magnetic poles.

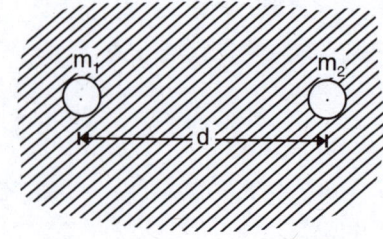

Fig. 7.1

---

*   Magnetic poles have no physical reality, but the concept enables us to appreciate magnetic effects more easily.

**   It is not possible to get an isolated pole because magnetic poles exist in pairs. However, if we take thin and long steel rods (about 50 cm long) with a small steel ball on either end and then magnetise them, N and S poles become concentrated in the steel balls. Such poles may be assumed point poles for all practical purposes.

(*i*) *Like poles repel each other while unlike poles attract each other.*

(*ii*) *The force between two magnetic poles is directly proportional to the product of their pole strengths and inversely proportional to the square of distance between their centres.*

Consider two poles of magnetic strength m_1 and m_2 placed at a distance d apart in a medium as shown in Fig. 7.1. According to Coulomb's laws, the force between the two poles is given by ;

$$F \propto \frac{m_1 m_2}{d^2}$$

$$= K \frac{m_1 m_2}{d^2}$$

where K is a constant whose value depends upon the surrounding medium and the system of units employed. In SI units, force is measured in newtons, pole strength in *weber, distance in metres and the value of K is given by ;

$$K = \frac{1}{4\pi\,\mu_0\,\mu_r}$$

where μ_0 = Absolute permeability of vacuum or air

μ_r = Relative permeability of the surrounding medium. For vacuum or air, its value is 1.

The value of $\mu_0 = 4\pi \times 10^{-7}$ H/m and the value of μ_r is different for different media.

$\therefore$ $$F = \frac{m_1 m_2}{4\pi\,\mu_0\,\mu_r\,d^2} \text{ newtons} \qquad\qquad ...in\ a\ medium$$

$$= \frac{m_1 m_2}{4\pi\mu_0 d^2} \text{ newtons} \qquad\qquad ...in\ air$$

Unit of pole strength. By unit pole strength we mean 1 weber. It can be defined from Coulomb's laws of magnetic force. Suppose two equal point poles placed 1 m apart in *air* exert a force of 62800 newtons *i.e.*

$$m_1 = m_2 = m ; \quad d = 1\text{ m} ; F = 62800\text{ N}$$

$\therefore$ $$F = \frac{m_1 m_2}{4\pi\mu_0 d^2} \qquad (\because\ \text{For air, } \mu_r = 1)$$

or $$62800 = \frac{m^2}{4\pi \times 4\pi \times 10^{-7} \times (1)^2}$$

or $$m^2 = (62800) \times (4\pi \times 4\pi \times 10^{-7} \times 1) = 1$$

$\therefore$ $$m = \pm 1\text{ Wb}$$

Hence a **pole of unit strength** *(i.e. 1 Wb) is that pole which when placed in air 1 m from an identical pole, repels it with a force of 62800 newtons.*

In vector form : $$\overrightarrow{F} = \frac{m_1 m_2}{4\pi\mu_0\,\mu_r d^2}\hat{d}$$

where $\hat{d}$ is a unit vector to indicate the direction of d.

Example 7.1. *Two magnetic S poles are located 5 cm apart in air. If each pole has a strength of 5 mWb, find the force of repulsion between them.*

Solution. $$F = \frac{m_1 m_2}{4\pi\mu_0 d^2} \qquad (\because\ \text{For air, } \mu_r = 1)$$

---

*   The unit of magnetic flux is named after Wilhelm Weber (1804–1890), the founder of electrical system of measurements.

Here $m_1 = m_2 = 5$ mWb $= 5 \times 10^{-3}$ Wb ; $d = 5$ cm $= 0.05$ m

$$\therefore \quad F = \frac{(5 \times 10^{-3}) \times (5 \times 10^{-3})}{4\pi \times 4\pi \times 10^{-7} \times (0.05)^2} = \mathbf{633\ N}$$

7.3. Magnetic Field

Just as electric field exists near a charged object, similarly magnetic field exists around a magnet. If an isolated magnetic pole is brought near a magnet, it experiences a force according to Coulomb's laws. The region near the magnet where forces act on magnetic poles is called a magnetic field. The magnetic field is strongest near the pole and goes on decreasing in strength as we move away from the magnet.

The space (or field) in which a magnetic pole experiences a force is called a **magnetic field.**

The magnetic field around a magnet is represented by imaginary lines called *magnetic lines of force.* By convention, the direction of these lines of force at any point is the direction along which an *isolated unit *N*-pole (*i.e.* *N*-pole of 1 Wb) placed at that point would move or tends to move. Following this convention, it is clear that magnetic lines of force would emerge from *N*-pole of the magnet, pass through the surrounding medium and re-enter the *S*-pole. Inside the magnet, each line of force passes from *S*-pole to

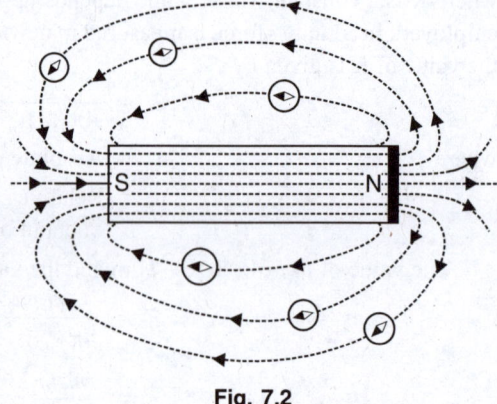

Fig. 7.2

N-pole (See Fig. 7.2), thus forming a closed loop or magnetic circuit. Although magnetic lines of force have no real existence and are purely imaginary, yet they are a useful concept to describe the various magnetic effects.

Properties of magnetic lines of force. The important properties of magnetic lines of force are :

   (i) *Each magnetic line of force forms a closed loop i.e.* outside the magnet, the direction of a magnetic line of force is from north pole to south pole and it continues through the body of the magnet to form a closed loop (See Fig. 7.2).

   (ii) *No two magnetic lines of force intersect each other.* If two magnetic lines of force intersect, there would be two directions of magnetic field at that point which is not possible.

   (iii) *Where the magnetic lines of force are close together, the magnetic field is strong and where they are well spaced out, the field is weak.*

   (iv) *Magnetic lines of force contract longitudinally and widen laterally.*

   (v) *Magnetic lines of force are always ready to pass through magnetic materials like iron in preference to pass through non-magnetic materials like air.*

It may be noted that in practice, magnetic fields are produced by (a) current carrying conductor or coil or (b) a permanent magnet. Both these means of producing magnetic fields are widely used in electrical engineering.

7.4. Magnetic Flux

The number of magnetic lines of force in a magnetic field determines the value of magnetic flux. The more the magnetic lines of force, the greater the magnetic flux and the stronger the magnetic field.

---

* Theoretically, it is not possible to get an isolated *N*-pole. However, a small compass needle well approximates to an isolated *N*-pole. The marked end (*N*-pole) of the compass needle indicates the direction of magnetic lines of force as shown in Fig. 7.2.

The total number of magnetic lines of force produced by a magnetic source is called **magnetic flux**. It is denoted by Greek letter ϕ (phi).

A unit *N*-pole is supposed to radiate out a flux of one weber. Therefore, the magnetic flux coming out of *N*-pole of *m* weber is

$$\phi = m \text{ Wb}$$

Now $\qquad\qquad 1 \text{Wb} = 10^8 \text{ lines of force}$

Sometimes we have to use smaller unit of magnetic flux *viz* microweber (μWb).

$$1 \ \mu\text{Wb} = 10^{-6} \text{ Wb} = 10^{-6} \times 10^8 \text{ lines} = 100 \text{ lines}$$

7.5. Magnetic Flux Density

The **magnetic flux density** *is defined as the magnetic flux passing normally per unit area i.e.*

Magnetic flux density, $B = \dfrac{\phi}{A} \text{ Wb/m}^2$

where $\quad \phi$ = flux in Wb

A = area in m^2 normal to flux

The SI unit of magnetic flux density is Wb/m^2 or *tesla. Flux density is a measure of field concentration *i.e.* amount of flux in each square metre of the field. In practice, it is much more important than the total amount of flux. Magnetic flux density is a *vector quantity*.

(*i*) When the plane of the coil is perpendicular to the flux direction [See Fig. 7.3], maximum flux will pass through the coil *i.e.*

Maximum flux, $\phi_m = B A \text{ Wb}$

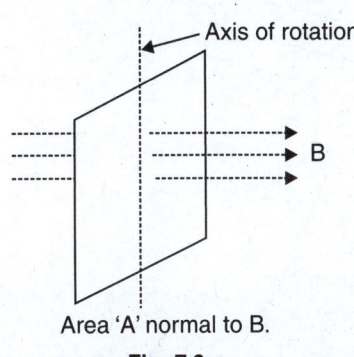

Area 'A' normal to B.

Fig. 7.3

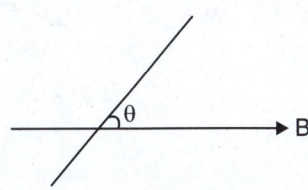

Fig. 7.4

(*ii*) When the plane of the coil is inclined at an angle θ to the flux direction [See Fig. 7.4], then flux ϕ through the coil is

$$\phi = B A \sin \theta \text{ Wb}$$

(*iii*) When the plane of the coil is parallel to the flux direction, $\theta = 0°$ so that no flux will pass through the coil ($\phi = BA \sin 0° = 0$).

Example 7.2. *A circular coil of 100 turns and diameter 3·18 cm is mounted on an axle through a diameter and placed in a uniform magnetic field, where the flux density is 0·01 Wb/m^2, in such a manner that axle is normal to the field direction. Calculate :*

(*i*) *the maximum flux through the coil and the coil position at which it occurs.*

(*ii*) *the minimum flux and the coil position at which it occurs.*

(*iii*) *the flux through the coil when its plane is inclined at 60° to the flux direction.*

---

*   Named in honour of Nikola Tesla (1857–1943), an American electrician and inventor.

Solution. Fig. 7.5 shows the conditions of the problem.

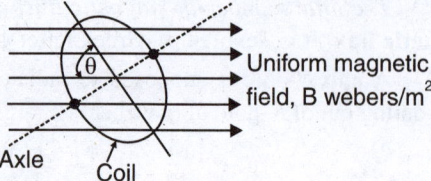

Fig. 7.5

(*i*) The maximum flux will pass through the coil when the plane of the coil is perpendicular to the flux direction.

∴ Maximum flux, ϕ_m = B × Total coil area

$$= (0.01) \times \pi\, r^2$$

$$= 0.01 \times \pi \times \left(\frac{3.18}{2}\right)^2 \times 10^{-4} = \mathbf{0.795 \times 10^{-5}\ Wb}$$

(*ii*) When the plane of the coil is parallel to the flux direction, no flux will pass through the coil. This is the minimum flux coil position and the minimum flux is **zero.**

(*iii*) When the plane of the coil is inclined at an angle θ to the flux direction, the flux φ through the coil is

$$\phi = B\,A \sin\theta = (B\,A)\sin\theta = (0.795 \times 10^{-5}) \times \sin 60°$$

$$= \mathbf{0.69 \times 10^{-5}\ Wb}$$

Example 7.3. *The total flux emitted from the pole of a bar magnet is 2×10^{-4} Wb (See Fig. 7.6).*

(i) If the magnet has a cross-sectional area of 1 cm^2, determine the flux density within the magnet.

(ii) If the flux spreads out so that a certain distance from the pole, it is distributed over an area of 2 cm by 2 cm, find the flux density at that point.

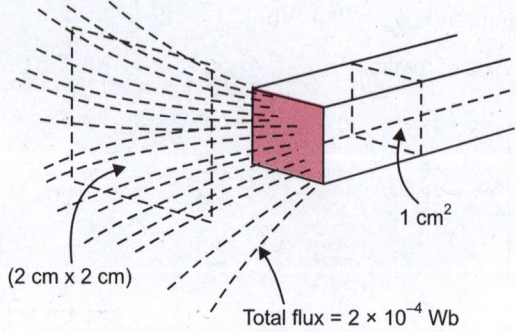

1 cm^2

(2 cm x 2 cm)

Total flux = 2×10^{-4} Wb

Fig. 7.6

Solution. (*i*) *Flux density within magnet.* $\phi = 2 \times 10^{-4}$ Wb ; $A = 1\ cm^2 = 1 \times 10^{-4}\ m^2$

∴   Flux density, B = $\dfrac{\phi}{A} = \dfrac{2 \times 10^{-4}}{1 \times 10^{-4}} = \mathbf{2\ Wb/m^2}$

(*ii*) *Flux density away from the pole.*

$$\phi = 2 \times 10^{-4}\ Wb\ ;\ A = 2 \times 2 = 4\ cm^2 = 4 \times 10^{-4}\ m^2$$

∴   Flux density, B = $\dfrac{\phi}{A} = \dfrac{2 \times 10^{-4}}{4 \times 10^{-4}} = \mathbf{0.5\ Wb/m^2}$

Example 7.4. *Flux density in the air gap between N and S poles is 2.5 Wb/m^2. The poles are circular with a diameter of 5.6 cm. Calculate the total flux crossing the air gap.*

Solution. $B = 2.5\ Wb/m^2$; Area of each pole, $A = \pi r^2 = \pi \times (5.6/2)^2 = 24.63\ cm^2 = 24.63 \times 10^{-4}\ m^2$

∴ Flux crossing the air gap is given by ;

$$\phi = B \times A = 2.5 \times 24.63 \times 10^{-4} = 6.16 \times 10^{-3}\ Wb = \mathbf{6.16\ mWb}$$

7.6. Magnetic Intensity or Magnetising Force (H)

Magnetic intensity (or field strength) at a point in a magnetic field is the force acting on a unit *N*-pole (*i.e.*, *N*-pole of 1 Wb) placed at that point. Clearly, the unit of *H* will be N/Wb.

Suppose it is desired to find the magnetic intensity at a point *P* situated at a distance *d* metres from a pole of strength *m* webers (See Fig. 7.7). Imagine a unit north pole (*i.e.* *N*-pole of 1 Wb) is placed at *P*. Then, by definition, magnetic intensity at *P* is the force acting on the unit *N*-pole placed at *P i.e.*

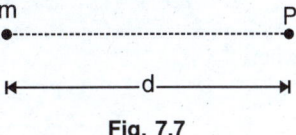

Fig. 7.7

Magnetic intensity at *P*, *H* = Force on unit *N*-pole placed at *P*

or $$H = \frac{m \times 1}{4\pi\mu_0 d^2} \text{N/Wb} \qquad [\because \ \mu_r = 1 \text{ for air}]$$

or $$H = \frac{m}{4\pi\mu_0 d^2} \text{N/Wb}$$

The reader may note the following points carefully :

(*i*) Magnetic intensity is a vector quantity, possessing both magnitude and direction. In vector form, it is given by ;

$$\vec{H} = \frac{m}{4\pi\mu_0 d^2}\hat{d}$$

(*ii*) If a pole of *m* Wb is placed in a uniform magnetic field of strength *H* newtons/Wb, then force acting on the pole, *F* = *m H* newtons.

7.7. Magnetic Potential

The **magnetic potential** *at any point in the magnetic field is measured by the work done in moving a unit N-pole (i.e. 1 Wb strength) from infinity to that point against the magnetic force.*

Consider a magnetic pole of strength *m* webers placed in a medium of relative permeability μ_r. At a point at a distance *x* metres from it, the force on unit *N*-pole is

$$F = \frac{m}{4\pi\mu_0\mu_r x^2}$$

If the unit *N*–pole is moved towards *m* through a small distance *dx*, then work done is

$$dW = \frac{m}{4\pi\mu_0\mu_r x^2} \times (-dx)$$

The negative sign is taken because *dx* is considered in the negative direction of *x*.

Therefore, the total work done (*W*) in bringing a unit *N*-pole from infinity to any point which is *d* metres from *m* is

$$W = \int_{x=\infty}^{x=d} -\frac{m}{4\pi\mu_0\mu_r x^2}\,dx = \frac{m}{4\pi\mu_0\mu_r d}\text{J/Wb}$$

By definition, *W* = Magnetic potential *V* at that point.

$\therefore$ Magnetic potential, $V = \dfrac{m}{4\pi\mu_0\mu_r d}\text{J/Wb}$

Note that magnetic potential is a scalar quantity.

7.8. Absolute and Relative Permeability

Permeability of a material means its conductivity for magnetic flux. The greater the permeability of a material, the greater is its conductivity for magnetic flux and *vice-versa*. Air or vacuum is the poorest conductor of magnetic flux. The absolute (or actual) permeability **μ_0* (Greek letter "*mu*")

---

*   The absolute (or actual) permeability of all non-magnetic materials is also $4\pi \times 10^{-7}$ H/m.

of air or vacuum is $4\pi \times 10^{-7}$ H/m. The absolute (or actual) permeability μ of magnetic materials is much greater than μ_0. The ratio μ/μ_0 is called the relative permeability of the material and is denoted by μ_r, *i.e.*

$$\mu_r = \frac{\mu}{\mu_0}$$

where μ = absolute (or actual) permeability of the material

μ_0 = absolute permeability of air or vacuum

μ_r = relative permeability of the material

Obviously, the relative permeability for air or vacuum would be $\mu_0/\mu_0 = 1$. The value of μ_r for all non-magnetic materials is also 1. However, relative permeability of magnetic materials is very high. For example, soft iron (*i.e.* pure iron) has a relative permeability of 8,000 whereas its value for permalloy (an alloy containing 22% iron and 78% nickel) is as high as 50,000.

Concept of relative permeability. The relative permeability of a material is a measure of the relative ease with which that material conducts magnetic flux compared with the conduction of flux in air. Fig. 7.8 illustrates the concept of relative permeability. In Fig. 7.8 (*i*), the magnetic flux passes between the poles of a magnet in air. Consider a soft iron ring (μ_r = 8,000) placed between the same poles as shown in Fig. 7.8 (*ii*). Since soft iron is a very good conductor of magnetic flux, the flux follows a path entirely within the soft iron itself. The flux density in the soft iron is much greater than it is in air. In fact, flux density in soft iron will be 8,000 times (*i.e.* μ_r times) the flux density in air.

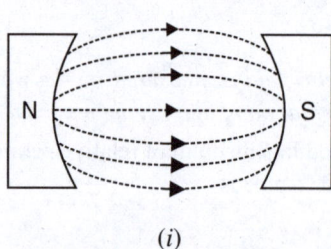 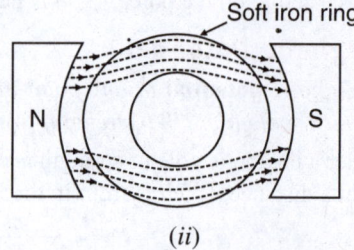

Soft iron ring

(*i*)   (*ii*)

Fig. 7.8

Due to high relative permeability of magnetic materials (*e.g.* iron, steel and other magnetic alloys), they are widely used for the cores of all electromagnetic equipment.

7.9. Relation Between B and H

The flux density B produced in a material is directly proportional to the applied magnetising force H. In other words, the greater the magnetising force, the greater is the flux density and *vice-versa i.e.*

$$B \propto H$$

or $$\frac{B}{H} = \text{Constant} = \mu$$

The ratio B/H in a material is always constant and is equal to the absolute permeability μ ($= \mu_0 \mu_r$) of the material. This relation gives yet another definition of absolute permeability of a material.

Obviously, $B = \mu_0 \mu_r H$ ...*in a medium*

$= \mu_0 H$ ...*in air*

Suppose a magnetising force H produces a flux density B_0 in air. Clearly, $B_0 = \mu_0 H$. If air is replaced by some other material (relative permeability μ_r) and the same magnetising force H is applied, then flux density in the material will be $B_{mat} = \mu_0 \mu_r H$.

$$\therefore \quad \frac{B_{mat}}{B_0} = \frac{\mu_0 \mu_r H}{\mu_0 H} = \mu_r$$

*Hence **relative permeability** of a material is equal to the ratio of flux density produced in that material to the flux density produced in air by the same magnetising force.*

Thus when we say that μ_r of soft iron is 8000, it means that for the same magnetising force, flux density in soft iron will be 8000 times its value in air. In other words, for the same cross-sectional area and H, the magnetic lines of force will be 8000 times greater in soft iron than in air.

7.10. Important Terms

(*i*) Intensity of magnetisation (I). When a magnetic material is subjected to a magnetising force, the material is magnetised. Intensity of magnetisation is a measure of the extent to which the material is magnetised and depends upon the nature of the material. It is defined as under :

The intensity of magnetisation of a magnetic material is defined as the magnetic moment developed per unit volume of the material.

$\therefore$ Intensity of magnetisation, $I = \dfrac{M}{V}$

where M = magnetic moment developed in the material

V = volume of the material

If m is the pole strength developed, a is the area of X-section of the material and $2l$ is the magnetic length, then,

$$I = \frac{m \times 2l}{a \times 2l} = \frac{m}{a}$$

Hence intensity of magnetisation of a material may be defined as the pole strength developed per unit area of cross-section of the material.

$$I = \frac{\text{magnetic moment}}{\text{volume}} = \frac{\text{Amp. (metre)}^2}{\text{(metre)}^3} = \text{A m}^{-1}$$

$\therefore$ SI units of I are A m^{-1}.

(*ii*) Magnetic susceptibility (χ_m). The magnetic susceptibility of a material indicates how easily the material can be magnetised. It is defined as under :

The magnetic susceptibility of a material is defined as the ratio of intensity of magnetisation (I) developed in the material to the applied magnetising force (H). It is represented by χ_m (Greek alphabet Chi).

$\therefore$ Magnetic susceptibility, $\chi_m = \dfrac{I}{H}$

The unit of I is the same as that of H so that χ_m is a number. Since I is magnetic moment per unit volume, χ_m is also called *volume susceptibility* of the material.

7.11. Relation Between μ_r and χ_m

Consider a current carrying toroid having core material of relative permeability μ_r. The total magnetic flux density in the material is given by ;

$$B = B_0 + B_M$$

where B_0 = magnetic flux density due to current in the coils.

B_M = magnetic flux density due to the magnetisation of the material.

Now $B_0 = \mu_0 H$ and $B_M = \mu_0 I*$

---

*   We can imagine that B_M is produced by a fictitious current I_M in the coils.

$\therefore \quad B_M = \mu_0\, n\, I_M = \mu_0 \dfrac{N}{l} I_M = \mu_0 \dfrac{N I_M\, A}{Al} = \mu_0\, I$

where $N\, I_M A$ = magnetic dipole moment developed and $A\, l$ is the volume of the specimen.

$$\therefore \qquad\qquad B = \mu_0 H + \mu_0 I = \mu_0 (H + I)$$

or $\qquad\qquad B = \mu_0 (H + I)$

Now $\qquad\qquad \chi_m = \dfrac{I}{H}$ so that $I = \chi_m H$

$\therefore \qquad\qquad B = \mu_0 (H + \chi_m H) = \mu_0 H (1 + \chi_m)$

But $\qquad\qquad B = \mu H = \mu_0 \mu_r H$

$\therefore \qquad\qquad \mu_0 \mu_r H = \mu_0 H (1 + \chi_m)$

or $\qquad\qquad \mu_r = 1 + \chi_m$

Example 7.5. *The magnetic moment of a magnet (10 cm × 2 cm × 1 cm) is 1 Am². What is the intensity of magnetisation ?*

Solution. Volume of the magnet, $V = 10 \times 2 \times 1 = 20 \text{ cm}^3 = 20 \times 10^{-6} \text{ m}^3$

Magnetic moment of magnet, $M = 1 \text{ Am}^2$

$\therefore \qquad$ Intensity of magnetisation, $I = \dfrac{M}{V} = \dfrac{1}{20 \times 10^{-6}} = \textbf{5} \times \textbf{10}^4 \textbf{ A/m}$

Example 7.6. *A specimen of iron is uniformly magnetised by a magnetising field of 500 A/m. If the magnetic induction in the specimen is 0·2 Wb/m², find the relative permeability and susceptibility.*

Solution. $\qquad\qquad B = \mu H = \mu_0 \mu_r H$

$\therefore \qquad$ Relative permeability of the specimen is

$$\mu_r = \dfrac{B}{\mu_0 H} = \dfrac{0 \cdot 2}{4\pi \times 10^{-7} \times 500} = \textbf{318·5}$$

Now $\qquad\qquad \mu_r = 1 + \chi_m$

$\therefore \qquad$ Susceptibility, $\chi_m = \mu_r - 1 = 318 \cdot 5 - 1 = \textbf{317·5}$

7.12. Refraction of Magnetic Flux

When magnetic flux passes from one medium to another of different permeabilities, the magnetic flux gets refracted at the boundary of the two media [See Fig. 7.9]. Under this condition, the following two conditions exist at the boundary (called **boundary conditions**) :

    (*i*) The normal components of magnetic flux density are equal *i.e.*

$$B_{1n} = B_{2n}$$

    (*ii*) The tangential components of magnetic field intensities are equal *i.e.*

$$H_{1t} = H_{2t}$$

As proved in Art. 5.25, in a similar way, it can be proved that :

$$\dfrac{\tan \theta_1}{\tan \theta_2} = \dfrac{\mu_1}{\mu_2}$$

This relation is called law of magnetic flux refraction.

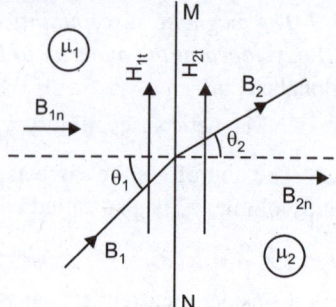

Fig. 7.9

7.13. Molecular Theory of Magnetism

The molecular theory of magnetism was proposed by Weber in 1852 and modified by Ewing in 1890. *According to this theory, every molecule of a magnetic substance (whether magnetised or not) is a complete magnet in itself having a north pole and a south pole of equal strength.*

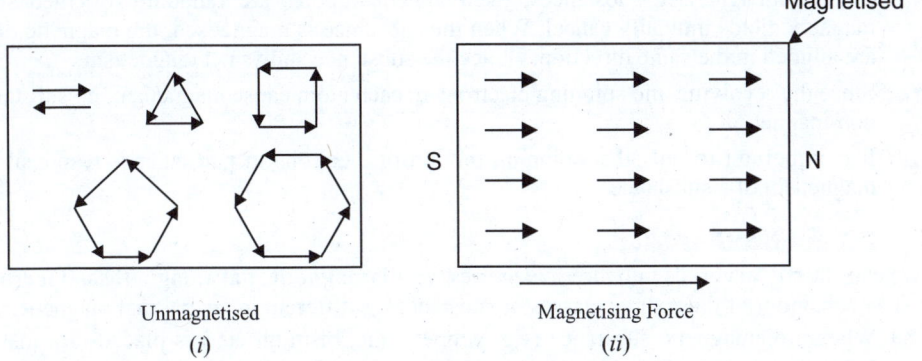

Fig. 7.10

(*i*) In an unmagnetised substance, the molecular magnets are randomly oriented and form closed chains as shown in Fig. 7.10 (*i*). The north pole of one molecular magnet cancels the effect of the south pole of the other so that the substance does not show any net magnetism.

(*ii*) When a magnetising force is applied to the substance (*e.g.* by rubbing a magnet or by passing electric current through a wire wound over it), the molecular magnets are turned and tend to align in the same direction with *N*-pole of one molecular magnet facing the *S*-pole of other as shown in Fig. 7.10 (*ii*). The result is that magnetic fields of the molecular magnets aid each other and two definite *N* and *S* poles are developed near the ends of the specimen ; the strength of the two poles being equal. Hence the substance gets magnetised.

(*iii*) The extent of magnetisation of the substance depends upon the extent of alignment of molecular magnets. When all the molecular magnets are fully aligned, the substance is said to be *saturated* with magnetism.

(*iv*) When a magnetised substance (or a magnet) is heated, the molecular magnets acquire kinetic energy and some of them go back to the closed chain arrangement. For this reason, a magnet loses some magnetism on heating.

Curie temperature. The magnetisation of a magnetised substance decreases with the increase in temperature. It is because when a magnetised substance is heated, random thermal motions tend to destroy the alignment of molecular magnets. As a results, the magnetisation of the substance decreases. At sufficiently high temperature, the magnetic property of the substance suddenly disappears and the substance loses magnetism.

The temperature at which a magnetised substance loses its magnetism is called **Curie temperature** *or* **Curie point** *of the substance.*

For example, the curie temperature of iron is 770°C. Therefore, if the temperature of the magnetised iron piece becomes 770°C, it will loose its magnetism. Similarly, the curie temperatures of nickel and cobalt are 358°C and 1121°C respectively.

7.14. Modern View about Magnetism

According to modern view, the magnetic properties of a substance are attributed to the motions of electrons (orbital and spin) in the atoms. We know that an atom consists of central nucleus with electrons revolving around the nucleus in different orbits. This motion of electrons is called *orbital motion* [See Fig. 7.11 (*i*)]. The electrons also rotate around their own axis. This motion of electrons is called *spin motion* [See Fig. 7.11 (*ii*)]. Due to these two motions, each atom is equivalent to a current loop *i.e.* each atom behaves as a magnetic dipole.

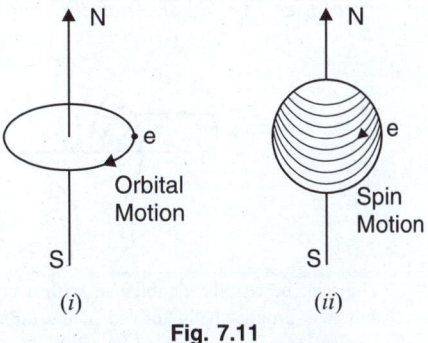

Fig. 7.11

(*i*) In the unmagnetised substances, the magnetic dipoles are randomly oriented so that magnetic fields mutually cancel. When the substance is magnetised, the magnetic dipoles are aligned in the same direction. Hence the substance shows net magnetism.

(*ii*) Since the revolving and spinning electrons in each atom cause magnetism, no substance is non-magnetic.

(*iii*) It is important to note that spinning motion of electrons in particular is responsible for magnetism of a substance.

7.15. Magnetic Materials

We can classify materials into three categories *viz.* **diamagnetic, paramagnetic** and **ferromagnetic**. The behaviour of these three classes of substances is different in an external magnetic field.

(*i*) When a diamagnetic substance (*e.g.* copper, zinc, bismuth etc.) is placed in a magnetic field, the substance if *feebly* magnetised in a direction opposite to that of the applied field. Therefore, a diamagnetic substance is feebly repelled by a strong magnet.

(*ii*) When a paramagnetic substance (*e.g.* aluminium, antimony etc.) is placed in a magnetic field, the substance is *feebly* magnetised in the direction of the applied field. Therefore, a paramagnetic substance is feebly attracted by a strong magnet.

(*iii*) When a ferromagnetic substance (*e.g.* iron, nickel, cobalt etc.) is placed in a magnetic field, the substance is *strongly* magnetised in the direction of the applied field. Therefore, a ferromagnetic substance is strongly attracted by a magnet.

Note that diamagnetism and paramagnetism are weak forms of magnetism. However, ferromagnetic substances exhibit very strong magnetic effects.

7.16. Electromagnetism

The first discovery of any connection between electricity and magnetism was made by Hans Christian Oersted, a Danish physicist in 1819. On one occasion at the end of his lecture, he inadvertently placed a wire carrying current parallel to a compass needle. To his surprise, needle was deflected. Upon reversing the current in the wire, the needle deflected in the opposite direction.

Oersted found that the compass deflection was due to a magnetic field established around the current carrying conductor. This accidental discovery was the first evidence of a long suspected link between electricity and magnetism. The production of magnetism from electricity (which we call electromagnetism) has opened a new era. The operation of all electrical machinery is due to the applications of magnetic effects of electric current in one form or the other.

7.17. Magnetic Effect of Electric Current

When an electric current flows through a conductor, magnetic field is set up all along the length of the conductor. Fig. 7.12 shows the magnetic field produced by the current flowing in a straight wire. The magnetic lines of force are in the form of *concentric circles around the conductor.

The direction of lines of force depends upon the direction of current and may be determined by **right-hand rule.** *Hold the conductor in the right-hand with the thumb pointing in the direction of current (See Fig. 7.12). Then the fingers will point in the direction of magnetic field around the*

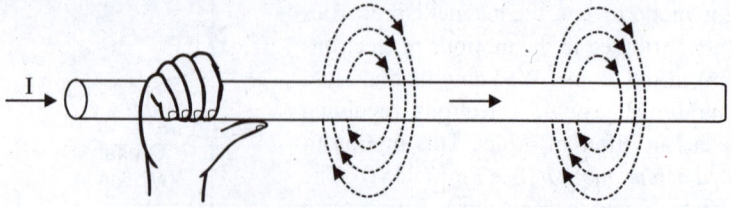

Fig. 7.12

---

* This can be readily established with a compass needle. If a compass needle is placed near the conductor and it is progressively moved in the direction of its north pole, it will be seen that the paths of magnetic lines of force are concentric circles.

conductor. Applying this rule to Fig. 7.12, it is clear that when viewed from left-hand side, the direction of magnetic lines of force will be clockwise.

The following points may be noted about the magnetic effect of electric current :

(*i*) The greater the current through the conductor, the stronger the magnetic field and *vice-versa.*

(*ii*) The magnetic field neart the conductor is stronger and becomes weaker and weaker as we move away from the conductor.

(*iii*) The magnetic lines of force around the conductor will be either clockwise or anticlockwise, depending upon the direction of current. One may use *right-hand rule* to determine the direction of magnetic field around the conductor.

(*iv*) The shape of the magnetic field depends upon the shape of the conductor.

7.18. Typical Electromagnetic Fields

The current carrying conductor may be in the form of a straight wire, a loop of one turn, a coil of several turns. The shape of the magnetic field would eventually depend upon the shape of conductor. By way of illustration, we shall discuss magnetic fields produced by some current carrying conductor arrangements.

(*i*) **Long straight conductor.** If a straight long conductor is carrying current, the magnetic lines of force will be concentric circles around the conductor as shown in Fig. 7.13. In Fig. 7.13 (*i*), the conductor is carrying current into the plane of paper (usually represented by a cross inside the *X*-section of the conductor). Applying right-hand rule, it is clear that direction of magnetic lines of force will be clockwise. In Fig. 7.13 (*ii*), the conductor is carrying current out of the plane of paper (usually represented by a dot inside the *X*-section of the conductor). Clearly, the direction of magnetic lines of force will be anticlockwise.

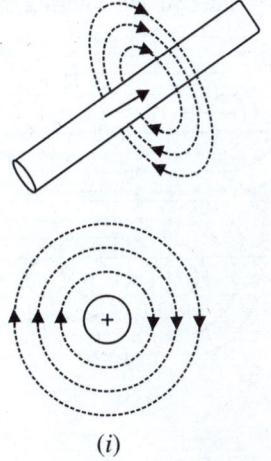 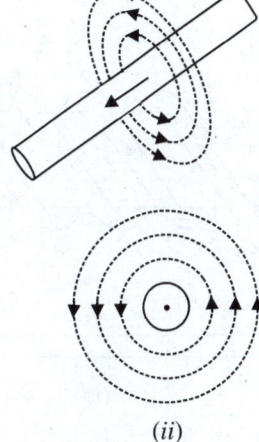

(*i*)   (*ii*)

Fig. 7.13

(*ii*) **Parallel conductors.** Consider two parallel conductors *A* and *B* placed close together and carrying current into the plane of the paper as shown in Fig. 7.14 (*i*). The magnetic lines of force will be clockwise around each conductor. In the space between *A* and *B*, the lines of force due to the conductors are in the opposite direction and hence they cancel out each other. This results in a field that entirely surrounds the conductors as shown in Fig. 7.14 (*ii*).

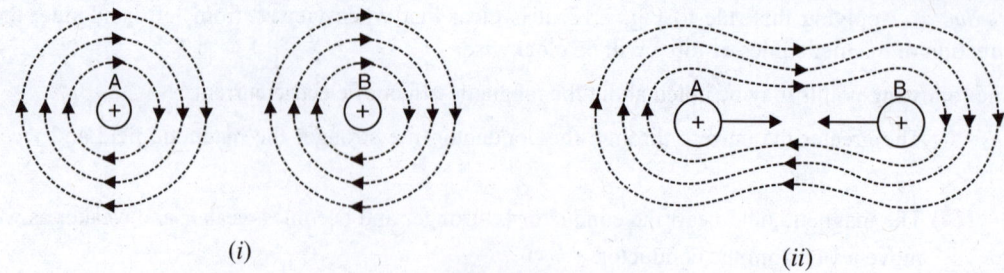

(i)   *(ii)*

Fig. 7.14

If there are several parallel conductors placed close together and carrying current into the plane of the paper as shown in Fig. 7.15 (*i*), the magnetic field envelops the conductors. If the direction of current is reversed, the direction of field is also reversed as shown in Fig. 7.15 (*ii*).

(i)   *(ii)*

Fig. 7.15

(iii) **Coil of several turns.** Consider a coil of several truns wound on a hollow tube or iron bar as shown in Fig. 7.16 (*i*). Such an arrangement is called a *solenoid. Suppose current flows through the coil in the direction shown. In the upper part of each turn (at points 1, 2, 3, 4 and 5), the current is flowing into the plane of the paper and in the lower part of each turn (at points 6, 7, 8 and 9), current is flowing out of the plane of paper. This is shown in the cross-sectional view of the coil in Fig. 7.16 (*ii*). It is clear that a clockwise field entirely surrounds the conductors 1, 2, 3, 4 and 5 while anticlockwise field completely envelops the conductors 6, 7, 8 and 9. As a result, the field becomes similar to that of a bar magnet with flux emerging from one end of the coil and entering the other.

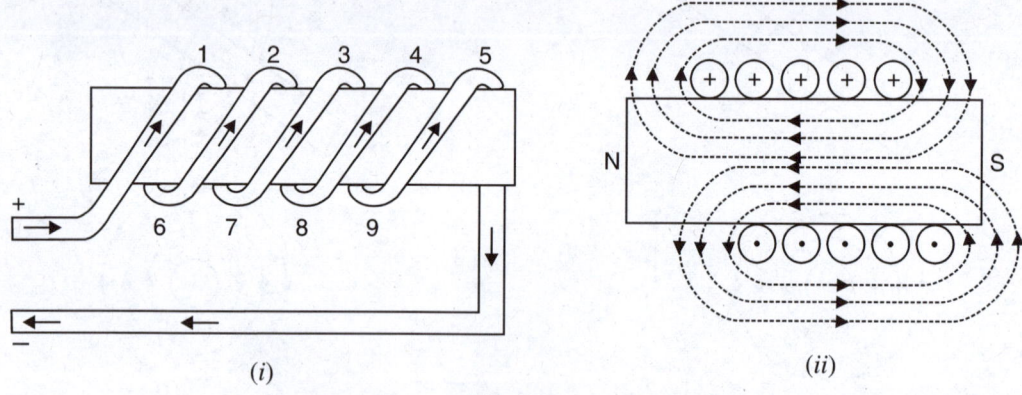

(i)   *(ii)*

Fig. 7.16

It is clear that left-hand face of the coil [See Fig. 7.16 (*ii*)] becomes a *N*-pole and right-hand face *S*-pole. The magnetic polarity of the coil can also be determined by the **right-hand rule for coil.** *Grasp the whole coil with right-hand so that the fingers are curled in the direction of current. Then thumb stretched parallel to the axis of the coil will point towards the N-pole end of the coil (See*

---

*   Solenoid is Greek word meaning "tube-like."

Fig. 7.17). It may be noted that both right-hand rules (for a conductor and for a coil) discussed so far can be applied in reverse. If we know the direction of magnetic field encircling a conductor or the magnetic polarity of a coil, we can determine the directiron of current by applying appropriate right-hand rule.

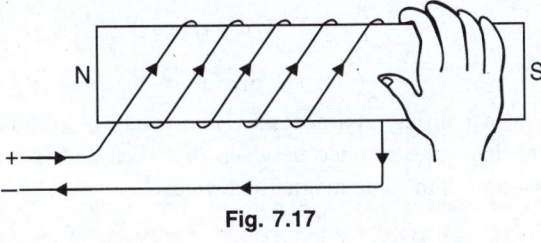

Fig. 7.17

7.19. Magnetising Force (H) Produced by Electric Current

The magnetic flux (ϕ) can be produced by (*i*) current-carrying conductor or coil or (*ii*) a permanent magnet. We generally use current-carrying conductor or coil to produce magnetic flux. Experiments show that magnetic flux (ϕ) produced by a current-carrying coil is directly proportional to the product of number of turns (N) of the coil and electric current (I) which the coil carries. The quantity NI is called *magnetomotive force (m.m.f)* and is measured in *ampere-turns* (AT) or **amperes* (A)

$$\therefore \qquad \text{m.m.f.} = NI \text{ Ampere-turns (AT)}$$

Just as e.m.f. (electromotive force) is required to produce electric current in an electric circuit, similarly, m.m.f. is required to produce magnetic flux in a **magnetic circuit. The greater the m.m.f., the greater is the magnetic flux produced in the magnetic circuit and *vice-versa*.

The **magnetising force** (*H*) *produced by an electric current is defined as the m.m.f. set up per unit length of the magnetic circuit i.e.*

$$\text{Magnetising force, } H = \frac{NI}{l} \text{ AT/m}$$

where $\qquad\qquad\qquad NI = \text{m.m.f. (AT)}$

$$l = \text{length of magnetic circuit in m}$$

Different current-carrying conductor arrangements produce different magnetising force. Magnetising force (*H*) is known by different names such as *magnetic field strength, magnetic intensity* and *magnetic potential gradient.*

Example 7.7. *A toroidal coil has a magnetic path length of 33 cm and a magnetic field strength of 650 A/m. The coil current is 250 mA. Determine the number of coil turns.*

Solution. $\qquad\qquad\qquad H = \dfrac{NI}{l}$

Here, $H = 650$ A/m ; $I = 250$ mA $= 0.25$A ; $l = 33$ cm $= 0.33$ m

$$\therefore \qquad\qquad 650 = \frac{N \times 0.25}{0.33} \quad \text{or} \quad N = \frac{650 \times 0.33}{0.25} = \textbf{858 turns}$$

Example 7.8. *Determine the m.m.f. required to generate a total flux of $100 \mu Wb$ in an air gap 0.2 cm long. The cross-sectional area of the air gap is 25 cm².*

Solution. $\phi = 100 \ \mu Wb = 100 \times 10^{-6}$ Wb ; $l = 0.2 \times 10^{-2}$ m ; $A = 25 \times 10^{-4}$ m²

$$\text{Flux density, } B = \frac{\phi}{A} = \frac{100 \times 10^{-6}}{25 \times 10^{-4}} = 4 \times 10^{-2} \text{ Wb/m}^2$$

$$\text{Magnetising force, } H = \frac{B}{\mu_0} = \frac{4 \times 10^{-2}}{4\pi \times 10^{-7}} = 3.18 \times 10^4 \text{ AT/m}$$

---

*   Since number of turns is dimensionless, ampere turns and amperes are the same as for as dimensions are concerned.

**   The closed path followed by magnetic flux is called a magnetic circuit; just as the closed path followed by electric current is called an electric circuit.

Now, $$H = \frac{\text{m.m.f.}}{l}$$

∴   m.m.f. $= H \times l = 3.18 \times 10^4 \times 0.2 \times 10^{-2} = $ **63.7 AT**

An air gap is a necessity in a rotating machine such as a motor or a generator. It provides mechanical clearance between the fixed and moving parts. Air gaps are also used to prevent saturation in some magnetic devices.

7.20. Force on Current-carrying Conductor Placed in a Magnetic Field

When a current-carrying conductor is placed at right angles to a magnetic field, it is found that the conductor experiences a force which acts in a direction perpendicular to the direction of both the field and the current. Consider a straight current-carrying conductor placed in a uniform magnetic field as shown in Fig. 7.18.

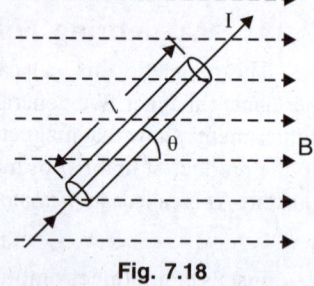

Let B = magnetic flux density in Wb/m^2

I = current through the conductor in amperes

l = effective length of the conductor in metres
i.e. the length of the conductor lying in the magnetic field

Fig. 7.18

θ = angle which the conductor makes with the direction of the magnetic field

It has been found experimentally that the magnitude of force (F) acting on the conductor is directly proportional to the magnitudes of flux density (B), current (I), length (l) and sin θ *i.e.*

$$F \propto Bil \sin \theta \text{ newtons}$$

or $$F = k\,Bil \sin \theta$$

where k is a constant of proportionality. Now SI unit of B is so defined that value of k becomes unity.

∴ $$F = Bil \sin \theta$$

By experiment, it is found that the direction of the force is always perpendicular to the plane containing the conductor and the magnetic field.

Both magnitude and direction of the force will be given by the following vector equation :

$$\vec{F} = I\left(\vec{l} \times \vec{B}\right)$$

The direction of this force is perpendicular to the plane containing $\vec{l}$ and $\vec{B}$. It can be found by using right-hand rule for cross product.

Special Cases. $F = Bil \sin \theta$

(*i*) When $\theta = 0°$ or $180°$; sin $\theta = 0$

∴ $$F = Bil \times 0 = 0$$

Therefore, if a current-carrying conductor is placed parallel to the direction of magnetic field, the conductor will experience no force.

(*ii*) When $\theta = 90°$; sin $\theta = 1$

∴ $$F = Bil \quad \text{...maximum value}$$

Therefore, a current-carrying conductor will experience a maximum force when it is placed at right angles to the direction of the magnetic field.

Direction of force. The direction of force $\vec{F}$ is always perpendicular to the plane containing $\vec{l}$ and $\vec{B}$ and can be determined by *right-hand rule for cross product* stated below :

Orient your right hand so that your outstretched fingers point along the direction of the conventional current; the orientation should be such that when you bend your fingers, they must point along the direction of the magnetic field ($\vec{B}$). *Then your extended thumb will point in the direction of the force on the conductor.*

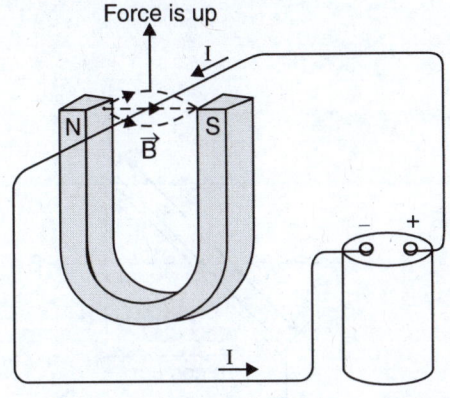

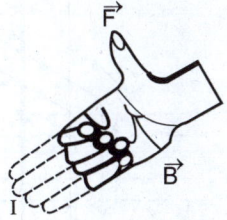

Right-hand rule

Fig. 7.19

Thus applying right-hand rule for cross product to Fig. 7.19, it is clear that magnetic force on the conductor is vertically upward.

Note. If the current-carrying conductor is at right angles to the magnetic field, the direction of force can also be found by Fleming's Left-hand rule stated below :

Fleming's Left-hand Rule. Stretch out the First finger, seCond finger and thuMb of your left hand so that they are at right angles to one another. If the first finger points in the direction of magnetic field (North to South) and second finger (*i.e.* middle finger) points towards the direction of current, then the thumb will point in the direction of motion of the conductor.

Example 7.9. *A conductor of length 100 cm and carrying 100 A is situated in and at right angles to a uniform magnetic field produced by the pole core of an electrical machine. If the pole core has a circular cross-section of 120 mm diameter and the total flux in the core is 16 mWb, find (i) the mechanical force on the conductor and (ii) power required to move the conductor at a speed of 10 m/s in a plane at right angles to the magnetic field.*

Solution. In this case, mechanical force acts on the conductor.

X-sectional area of pole core $= (\pi/4) \times (0 \cdot 12)^2 = 0 \cdot 0113$ m^2

$$\text{Flux density of field, } B = \frac{\text{Flux}}{\text{Pole core area}} = \frac{16 \times 10^{-3}}{0.0113} = 1 \cdot 416 \text{ Wb/m}^2$$

(*i*) Force on the conductor is given by ;

$$F = BIl = 1 \cdot 416 \times 100 \times 1 = \mathbf{141 \cdot 6 \ N}$$

(*ii*)   Power required $=$ Force $\times$ distance/second

$$= 141 \cdot 6 \times 10 = \mathbf{1416 \ watts}$$

Example 7.10. *The plane of a rectangular coil makes an angle of 60° with the direction of a uniform magnetic field of flux density 4×10^{-2} Wb/m^2. The coil is of 20 turns, measuring 20 cm by 10 cm, and carries a current of 0·5 A. Calculate the torque acting on the coil.*

Solution. Consider a rectangular coil, measuring b by l, of N turns carrying a current of I amperes and placed in a uniform magnetic field of B Wb/m^2. The coil is pivoted about the mid points of the sides b and is free to rotate about an axis in its own plane ; this axis being at right angles to the field density B [See Fig. 7.20 (*i*)]. When current is passed through the coil, forces acting on the coil sides are :

(*i*) The forces developed on each half of coil sides b are equal and produce torques of opposing sense. They, therefore, cancel each other.

(*ii*) The coil sides *l* always remain at right angles to the field as the coil rotates. The force *F* acting on each of the coil sides *l* gives rise to a torque as shown in Fig. 7.20 (*ii*).

Force on each coil side *l*, $F = B\,I\,l\,N$ newtons

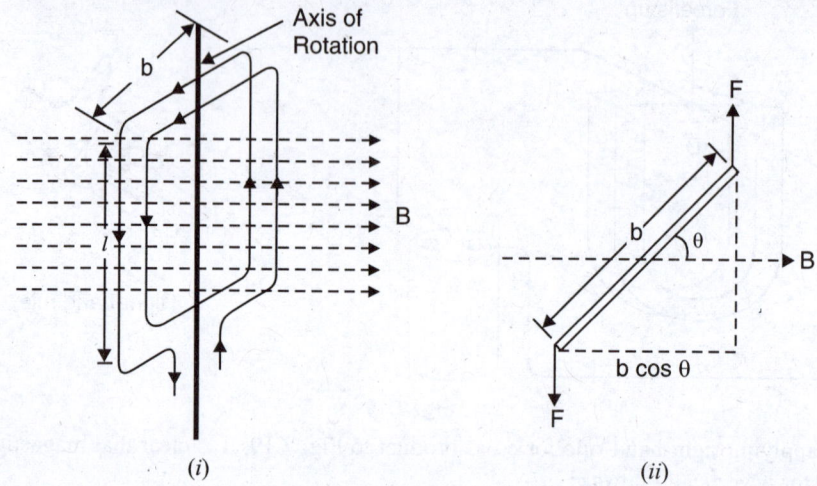

Fig. 7.20

The perpendicular distance between the lines of action of the two forces is $b \cos \theta$.

∴   Torque, $T = F \times b \cos \theta = (B\,I\,l\,N)\,b \cos \theta$

or $T = BINA \cos \theta$ newton-metre

where $A\,(= l \times b)$ is the area of the coil. By an extension of this reasoning, the expression may be proved quite generally for a coil of area *A* and of any shape.

In the given problem, the data is

$$B = 4 \times 10^{-2} \text{ Wb/m}^2 \text{ ; } A = 20 \times 10 = 200 \text{ cm}^2 = 2 \times 10^{-2} \text{ m}^2 \text{ ; } I = 0 \cdot 5 \text{ A ; } \theta = 60° \text{ ; } N = 20$$

∴   Torque, $T = (4 \times 10^{-2}) \times (0 \cdot 5) \times (20) \times (2 \times 10^{-2}) \times \cos 60° = \mathbf{4 \times 10^{-3} \text{ Nm}}$

Tutorial Problems

1. A straight conductor 0·4m long carries a current of 12 A and lies at right angles to a uniform field of 2·5 Wb/m². Find the mechanical force on the conductor when (*i*) it lies in the given position (*ii*) it lies in a position such that it is inclined at an angle of 30° to the direction of field.   **[(*i*) 12 N (*ii*) 6 N]**

2. A conductor of length 100 cm and carrying 100 A is situated in and at right angles to a uniform magnetic field of strength 1 Wb/m². Calculate the force and power required to move the conductor at a speed of 100 m/s in a plane at right angles to the magnetic field.   **[100 N ; 1000 watts]**

3. A d.c. motor consists of an armature winding of 400 turns (equivalent to 800 conductors). The effective lengths of conductor in the field is 160 mm and the conductors are situated at a radius of 100 mm from the centre of the motor shaft. The magnetic flux density is 0·6 Wb/m² and a current of 25 A flows through the winding. Calculate the torque available at the motor shaft.   **[192 Nm]**

4. A d.c. motor is to provide a torque of 540 Nm. The armature winding consists of 600 turns (equivalent to 1200 conductors). The effective length of a conductor in the field is 250 mm and the conductors are situated at a radius of 150 mm from the centre of the motor shaft. Each conductor carries a current of 10 A. Calculate the flux density which must be provided by the radial field in which the conductors lie.   **[1·2 Wb/m²]**

7.21. Ampere's Work Law or Ampere's Circuital Law

The magnetising force (*H*) at any point in an electromagnetic field is the force experienced by a unit *N*-pole placed at that point. If the unit *N*-pole is made to move in a complete path around *N* current-carrying conductors, then work is done provided the unit *N*-pole is moved in opposition to the lines of force. Conversely, if the unit N-pole moves in the direction of magnetic field, then work will be done by the magnetic force on whatever force is restraining the movement of the pole. In either case, unit N-pole makes one complete loop around the N conductors. The work done is given by Ampere's work law stated below :

The work done on or by a unit N-pole in moving once around any complete path is equal to the product of current and number of turns enclosed by that path i.e.

$$*\oint \vec{H_r} \cdot \vec{dr} = NI$$

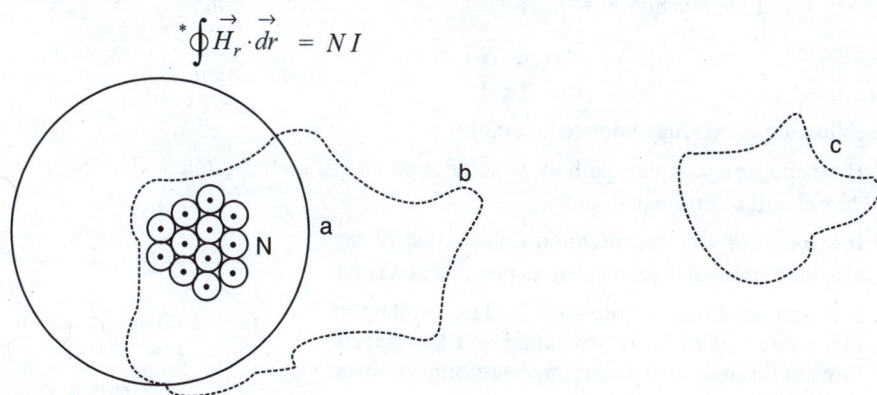

Fig. 7.21

where $\vec{H_r}$ is the magnetising force at a distance r. The circle around the integral sign indicates that the integral is around a complete path.

The work law is applicable regardless of the shape of complete path. Thus in Fig. 7.21, paths 'a' and 'b' completely enclose N conductors. If a unit N-pole is moved once around any of these complete paths, the work done in each case will be equal to NI. Although path 'c' is a complete path, it fails to enclose any current carrying conductor. Hence, no work is done in moving a unit N-pole around such a path.

Note. The work law is applicable for all magnetic fields, irrespective of the shape of the field or of the materials which may be present.

7.22. Applications of Ampere's Work Law

Ampere's work law can be used to find magnetising force (H) in simple conductor arrangements. We shall discuss two cases by way of illustration.

1. Magnetising force around a long straight conductor. Consider the case of a long straight conductor carrying a current of I amperes as shown in Fig. 7.22. The conductor will set up magnetic lines of force which encircle it. Consider a circular path of radius r metres. By symmetry, the field intensity H on all the points of this circular path will be the same. If a unit N-pole is moved once around this circular path, then work done is $= 2\pi rH$. By work law, this must be equal to the product of current and number of turns enclosed by this circular path.

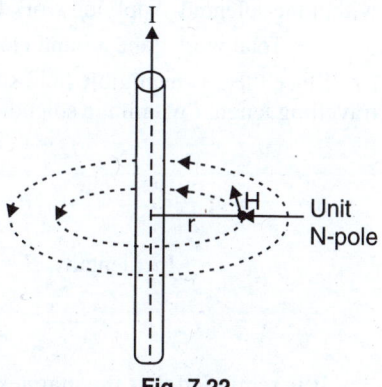

Fig. 7.22

$\therefore$ $2\pi r H = I$ $(\because N = 1)$

or $H = \dfrac{I}{2\pi r}$

---

*   This law can also be stated as *the closed line integral of magnetic field intensity (H) is equal to the encloed ampere-turns that produce the magnetic field.*

Note that magnetic lines of force encircle the conductor like concentric circles. The direction of magnetic lines of force can be determined by right-hand rule.

If there had been N turns enclosed by the path, then,

$$H = \frac{NI}{2\pi r}$$

Flux density, $B = \mu_0 H = \dfrac{\mu_0 NI}{2\pi r}$ *... in air*

$$= \frac{\mu_0 \mu_r NI}{2\pi r}$$ *... in a medium*

The following points may be noted carefully :

(*i*) If we choose a complete path for which r is smaller, H on that circle will be large. However, $2\pi r H$ will be still equal to NI.

(*ii*) Inspection of above expression reveals that H can also be expressed in ampere-turns per metre (AT/m).

(*iii*) It is reminded that the quantity NI (*i.e.* product of the number of turns in a winding and the current flowing through it) is called **magnetomotive force** (m.m.f.).

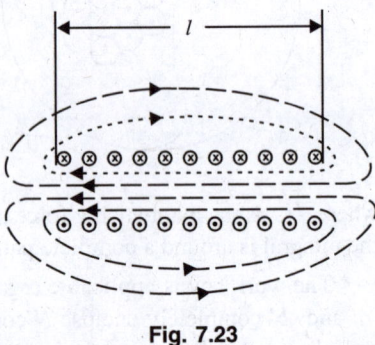

Fig. 7.23

m.m.f. $= NI$ ampere-turns

2. Magnetising force due to long solenoid. Consider a long solenoid of length l and wound uniformly with N turns (See Fig. 7.23). The length of the solenoid is much greater than the breadth, say 10 times greater. The following assumptions are permissible :

(*i*) The field strength external to the solenoid is effectively zero.

(*ii*) The field strength inside the solenoid is uniform.

Suppose the current I flowing through the solenoid produces uniform magnetic field strength H within the solenoid. Applying work law to any closed path say dotted one shown in Fig. 7.23,

Total work done around closed path = Ampere turns linked

Since there is negligible field strength (H) outside the solenoid, the only work done will be in travelling length l within the solenoid.

$\therefore$ $H \times l = NI$

or $H = \dfrac{NI}{l}$ AT/m or A/m

Incidentally, $B = \mu_0 H = \dfrac{\mu_0 NI}{l}$ Wb/m^2 *...in air*

$= \mu_0 \mu_r H = \dfrac{\mu_0 \mu_r NI}{l}$ Wb/m^2 *...in a medium*

It is reminded that the magnetic field strength (H) is a vector quantity since it has magnitude and direction.

Example 7.11. *An air-cored toroidal coil shown in Fig. 7.24 has 3000 turns and carries a current of 0.1A. The cross-sectional area of the coil is 4 cm^2 and the length of the magnetic circuit is 15 cm. Determine the magnetic field strength, the flux density and the total flux within the coil.*

Solution. $N = 3000$ turns ; $I = 0.1$ A ; $A = 4 \times 10^{-4}$ m^2 ; $l = 15 \times 10^{-2}$ m

Magnetic field strength, $H = \dfrac{NI}{l} = \dfrac{3000 \times 0.1}{15 \times 10^{-2}}$

$$= \mathbf{2000\ AT/m}$$

Flux density, $B = \mu_0 H = 4\pi \times 10^{-7} \times 2000$

$$= 2.5 \times 10^{-3} \text{ Wb/m}^2$$

Total flux, $\phi = B \times A = 2.5 \times 10^{-3} \times 4 \times 10^{-4}$

$$= 1 \times 10^{-6} \text{ Wb} = 1 \text{ } \mu\text{Wb}$$

Example 7.12. *An air-cored solenoid has length of 15 cm and inside diameter of 1.5 cm. If the coil has 900 turns, determine the total flux within the solenoid when the coil current is 100 mA.*

Solution. For a solenoid, the length of the magnetic circuit, l = coil length = 15×10^{-2} m.

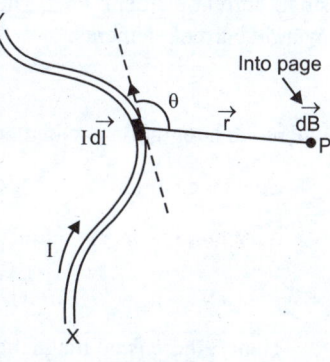

Fig. 7.24

$$D = 1.5 \times 10^{-2} \text{ m} \text{ ; } N = 900 \text{ turns ; } I = 100 \times 10^{-3} \text{ A}$$

$\therefore$ m.m.f. $= NI = 900 \times 100 \times 10^{-3} = 90$ AT

Magnetising force, $H = \dfrac{\text{m.m.f.}}{l} = \dfrac{90}{15 \times 10^{-2}} = 600$ AT/m

Magnetic flux density, $B = \mu_0 H = 4\pi \times 10^{-7} \times 600 = 24\pi \times 10^{-5}$ Wb/m^2

$\therefore$ Total flux, $\phi = BA = 24\pi \times 10^{-5} \times \pi \dfrac{D^2}{4}$

$$= 24\pi \times 10^{-5} \times \pi \times \dfrac{(1.5 \times 10^{-2})^2}{4} = \mathbf{1.33 \times 10^{-7} \text{ Wb}}$$

If the solenoid were iron-cored, the magnitude of the magnetic flux within the solenoid would have been much greater than the calculated value because of very high relative permeability of iron.

7.23. Biot-Savart Law

A conductor carrying current I produces a magnetic field around it. We can consider the current carrying conductor to be consisting of infinitesimally small *current elements $I \overrightarrow{dl}$* ; each current element contributing to magnetic field. *Biot-Savart law gives us expression for the magnetic field at a point due to a current element.*

Consider a current element $I \overrightarrow{dl}$ of a conductor XY carrying current I [See Fig. 7.25]. Let P be the point where the magnetic field $\overrightarrow{dB}$ due to the current element is to be found. Suppose $\overrightarrow{r}$ is the position vector of point P from the current element $I \overrightarrow{dl}$ and θ is the angle between $\overrightarrow{dl}$ and $\overrightarrow{r}$.

Fig. 7.25

According to Biot-Savart law, the magnitude dB of magnetic field at point P due to the current element depends upon the following factors :

(i) $dB \propto I$ *(ii)* $dB \propto dl$ *(iii)* $dB \propto 1/r^2$ *(iv)* $dB \propto \sin\theta$

Combining all these four factors, we get,

$$dB \propto \dfrac{I \, dl \sin\theta}{r^2}$$

or

$$dB = K \dfrac{I \, dl \sin\theta}{r^2}$$

---

* The current element $I \overrightarrow{dl}$ is a vector. Its direction is tangent to the element and acts in the direction of flow of current in the conductor.

where K is a constant of proportionality. Its value depends on the medium in which the conductor is situated and the system of units adopted.

For free space and SI units, $K = \dfrac{\mu_0}{4\pi} = 10^{-7}\, \text{Tm A}^{-1}$

where $\mu_0 = $ Absolute permeability of free space $= 4\pi \times 10^{-7}\, \text{Tm A}^{-1}$

$$\therefore \qquad dB = \frac{\mu_0}{4\pi}\cdot\frac{Idl\sin\theta}{r^2} \qquad\qquad\qquad ...(i)$$

Eq. (i) is known as *Biot-Savart law* and gives the magnitude of the magnetic field at a point due to small current element $I\,\vec{dl}$. **Note that Biot-Savart law holds strictly for steady currents.**

In vector form. $\qquad\qquad \vec{dB} = \dfrac{\mu_0}{4\pi}\dfrac{I(\vec{dl}\times\vec{r})}{r^3} \qquad\qquad\qquad ...(ii)$

The Biot-Savart law is analogous to Coulomb's law. Just as the charge q is the source of electrostatic field, similarly, the source of magnetic field is the current element $I\,\vec{dl}$.

Direction of $\vec{B}$. $\qquad\qquad \vec{dB} = \dfrac{\mu_0}{4\pi}\dfrac{I(\vec{dl}\times\vec{r})}{r^3}$

The direction of $\vec{dB}$ is perpendicular to the plane containing $\vec{dl}$ and $\vec{r}$. By right-hand rule for the cross product, the field is directed *inward*.

Magnetic field due to whole conductor. Eq. (ii) gives the magnetic field at point P due to a small current element $I\,\vec{dl}$. The total magnetic field at point P is found by summing (integrating) over all current elements.

$$\vec{B} = \int\vec{dB} = \int\frac{\mu_0}{4\pi}\frac{I(\vec{dl}\times\vec{r})}{r^3}$$

where the integration is taken over the entire conductor in which current I flows.

Special cases. $\qquad\qquad dB = \dfrac{\mu_0}{4\pi}\cdot\dfrac{I\,dl\sin\theta}{r^2}$

(*i*) **When $\theta = 0°$** *i.e.,* point P lies on the axis of the conductor, then,

$$dB = \frac{\mu_0}{4\pi}\cdot\frac{I\,dl\sin 0°}{r^2} = 0$$

Hence, there is no magnetic field at any point on the thin linear current carrying conductor.

(*ii*) **When $\theta = 90°$** *i.e.,* point P lies at a perpendicular position *w.r.t.* current element, then,

$$dB = \frac{\mu_0}{4\pi}\cdot\frac{I\,dl\sin 90°}{r^2} = \frac{\mu_0}{4\pi}\cdot\frac{I\,dl}{r^2} \qquad ...\textit{Maximum value}$$

Hence magnetic field due to a current element is maximum in a plane passing through the element and perpendicular to its axis.

(*iii*) **When $\theta = 0°$ or $180°$, $dB = 0$** $\qquad\qquad ...\textit{Minimum value}$

Important points about Biot-Savart law. This law has the following salient features :

   (*i*) Biot-Savart law is valid for symmetrical current distributions.

  (*ii*) Biot-Savart law cannot be proved experimentally because it is not possible to have a current carrying conductor of length dl.

 (*iii*) Like Coulomb's law in electrostatics, Biot-Savart law also obeys inverse square law.

 (*iv*) The direction of $\vec{dB}$ is perpendicular to the plane containing $I\,\vec{dl}$ and $\vec{r}$.

7.24. Applications of Biot-Savart Law

Biot-Savart law is very useful in determining magnetic flux density B and hence magnetising force H $(= B/\mu_0)$ due to current-carrying conductor arrangements. We shall discuss the following cases by way of illustration.

(*i*) Magnetic flux density at the centre of current-carrying circular coil.

(*ii*) Magnetic flux density due to straight conductor carrying current.

(*iii*) Magnetic flux density on the axis of circular coil carrying current.

7.25. Magnetic Field at the Centre of Current-Carrying Circular Coil

This is a practical case because the operation of many devices depends upon the magnetic field produced by the current-carrying circular coil. Consider a circular coil of radius r and carrying current I in the direction shown in Fig. 7.26. Suppose the loop lies in the plane of paper. It is desired to find the magnetic field at the centre O of the coil. Suppose the entire circular coil is divided into a large number of current elements, each of length dl. According to Biot-Savart law, the magnetic field $\overrightarrow{dB}$ at the centre O of the coil due to current element $I\overrightarrow{dl}$ is given by ;

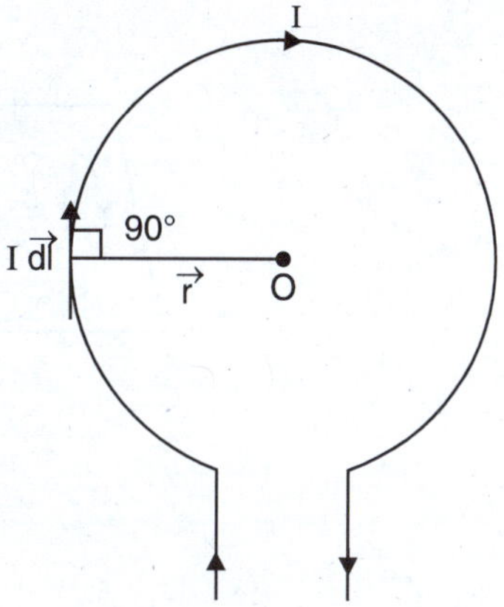

$$\overrightarrow{dB} = \frac{\mu_0}{4\pi} \frac{I\,(\overrightarrow{dl} \times \overrightarrow{r})}{r^3}$$

where $\overrightarrow{r}$ is the position vector of point O from the current element.

The magnitude of $\overrightarrow{dB}$ at the centre O is

$$dB = \frac{\mu_0}{4\pi} \frac{I\,dl\,r\sin\theta}{r^3} = \frac{\mu_0}{4\pi} \frac{I\,dl\sin\theta}{r^2} \qquad \text{...(i)}$$

Fig. 7.26

The direction of $\overrightarrow{dB}$ is perpendicular to the plane of the coil and is directed inwards. Since each current element contributes to the magnetic field in the same direction, the total magnetic field B at the centre O can be found by integrating eq. (*i*) around the loop *i.e.*

$$B = \int dB = \int \frac{\mu_0}{4\pi} \frac{I\,dl\sin\theta}{r^2}$$

For each current element, angle between $\overrightarrow{dl}$ and $\overrightarrow{r}$ is 90°. Also distance of each current element from the centre O is r.

$$\therefore \qquad B = \frac{\mu_0}{4\pi} \frac{I\sin 90°}{r^2} \int dl$$

Now, $$\int dl = \text{Total length of the coil} = 2\pi r$$

$$\therefore \qquad B = \frac{\mu_0}{4\pi} \frac{I}{r^2} (2\pi r)$$

or $$B = \frac{\mu_0 I}{2r}$$

Also, $$H = \frac{B}{\mu_0} = \frac{1}{\mu_0} \times \frac{\mu_0 I}{2r} = \frac{I}{2r}$$

If the coil has N turns, each carrying current in the same direction, then contributions of all the turns are added up. Therefore, the magnetic field at the centre of the coil is greatly increased and is given by ;

$$B = \frac{\mu_0 \, N \, I}{2r}$$

Also, $$H = \frac{B}{\mu_0} = \frac{NI}{2r}$$

Direction of $\vec{B}$. The direction of magnetic field $\vec{B}$ is perpendicular to the plane of the coil and for Fig. 7.27, magnetic field inside the coil is directed inwards. The magnetic lines of force are circular near the wire but practically straight near the centre of the coil. In the middle *M* of the coil, the magnetic field is uniform for a short distance on either side. The direction of magnetic field at the centre of a current-carrying circular coil can be determined by *right-hand palm rule*.

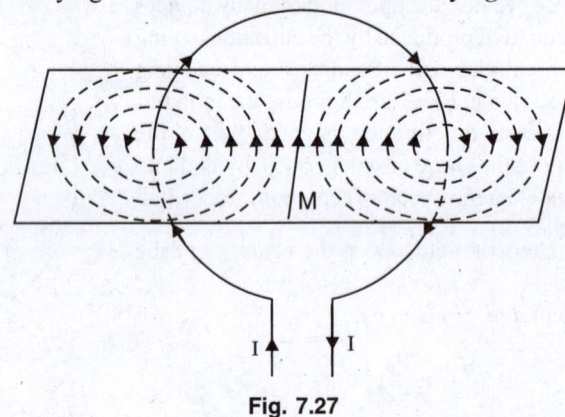

Fig. 7.27

Right-hand palm rule. *Orient the thumb of your right hand perpendicular to the grip of the fingers such that curvature of the fingers points in the direction of current in the circular coil. Then thumb will point in the direction of the magnetic field near the centre of the circular coil.*

7.26. Magnetic Field Due to Straight Conductor Carrying Current

Consider a straight conductor *XY* carrying current *I* in the direction shown in Fig. 7.28. It is desired to find the magnetic field at point *P* located at a perpendicular distance *a* from the conductor (*i.e. PQ = a*). Consider a small current element of length *dl*. Let $\vec{r}$ be the position vector of point *P* from the current element and θ be the angle between $\vec{dl}$ and $\vec{r}$ (*i.e.,* ∠*POQ* = θ). Let us further assume that *QO = l*.

According to Biot-Savart law, the magnitude of magnetic field $\vec{dB}$ at point *P* due to the considered current element is given by ;

$$dB = \frac{\mu_0}{4\pi} \frac{I \, dl \sin \theta}{r^2} \qquad ...(i)$$

To get the total magnetic field *B*, we must integrate eq. (*i*) over the whole conductor. As we move along the conductor, the quantities *dl*, θ and *r* change. The integration becomes much easier if we express everything in terms of angle φ shown in Fig. 7.28.

In the right angled triangle *PQO*, θ = 90° − φ.

∴   sin θ = sin (90° − φ) = cos φ

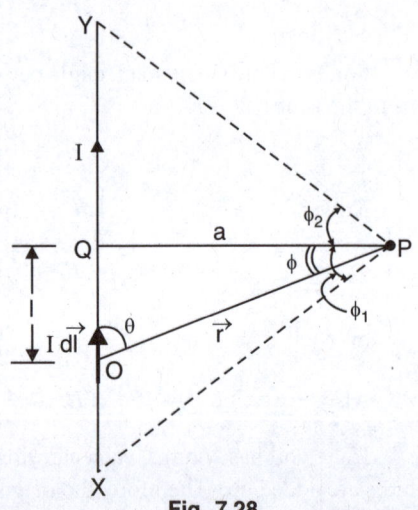

Fig. 7.28

Also, $\qquad \cos\phi = \dfrac{a}{r} \quad$ or $\quad r = \dfrac{a}{\cos\phi}$

Further, $\quad \tan\phi = \dfrac{l}{a} \quad$ or $\quad l = a\tan\phi$

or $\qquad dl = a\sec^2\phi \, d\phi$

Putting the values of $\sin\theta$, dl and r in eq. (i), we have,

$$dB = \frac{\mu_0}{4\pi} \frac{I\,(a\sec^2\phi\,d\phi)\cos\phi}{(a/\cos\phi)^2}$$

or $\qquad dB = \dfrac{\mu_0}{4\pi} \dfrac{I\cos\phi\,d\phi}{a} \qquad\qquad$ ...(ii)

The direction of $\overrightarrow{dB}$ is perpendicular to the plane of the conductor and is directed inwards (Right-hand grip rule, See section 7.17). Since each current element contributes to the magnetic field in the same direction, the total magnetic field B at point P can be found by integrating eq. (ii) over the length XY i.e.

$$B = \int_{-\phi_1}^{\phi_2} dB = \frac{\mu_0}{4\pi}\frac{I}{a}\int_{-\phi_1}^{\phi_2}\cos\phi\,d\phi$$

$$= \frac{\mu_0 I}{4\pi a}[\sin\phi]_{-\phi_1}^{\phi_2} = \frac{\mu_0 I}{4\pi a}(\sin\phi_2 + \sin\phi_1)$$

$\therefore \qquad B = \dfrac{\mu_0 I}{4\pi a}(\sin\phi_2 + \sin\phi_1) \qquad\qquad$ ...(iii)

Also, $\qquad H = \dfrac{B}{\mu_0} = \dfrac{I}{4\pi a}(\sin\phi_2 + \sin\phi_1)$

Eq. (iii) gives the value of B at point P due to a conductor of *finite length*.

Special cases. We shall discuss a few important cases.

(i) *When the conductor XY is of infinite length and point P lies at the centre of the conductor.* In this case, $\phi_1 = \phi_2 = 90° = \pi/2$.

$\therefore \qquad B = \dfrac{\mu_0}{4\pi}\dfrac{I}{a}(\sin\pi/2 + \sin\pi/2)$

or $\qquad B = \dfrac{\mu_0}{4\pi}\dfrac{2I}{a}$

Also, $\qquad H = \dfrac{B}{\mu_0} = \dfrac{1}{4\pi}\cdot\dfrac{2I}{a} = \dfrac{I}{2\pi a}$

(ii) *When conductor XY is of infinite length but point P lies near one end Y (or X).* In this case, $\phi_1 = 90°$ and $\phi_2 = 0°$.

$\therefore \qquad B = \dfrac{\mu_0}{4\pi}\dfrac{I}{a}(\sin 90° + \sin 0°)$

or $\qquad B = \dfrac{\mu_0 I}{4\pi a}$

Note that it is half of that for case (i).

Also, $\qquad H = \dfrac{B}{\mu_0} = \dfrac{I}{4\pi a}$

(iii) *If the length of the conductor is finite (say l) and point P lies on the right bisector of the conductor.* In this case, $\phi_1 = \phi_2 = \phi$.

Now,
$$\sin\phi = \frac{l/2}{\sqrt{a^2 + (l/2)^2}} = \frac{l}{\sqrt{4a^2 + l^2}}$$

$\therefore$
$$B = \frac{\mu_0}{4\pi}\frac{I}{a}(\sin\phi + \sin\phi) = \frac{\mu_0}{4\pi}\frac{2I}{a}\sin\phi$$

or
$$B = \frac{\mu_0}{4\pi}\frac{2I}{a}\frac{l}{\sqrt{4a^2 + l^2}}$$

Also
$$H = \frac{B}{\mu_0} = \frac{1}{4\pi}\frac{2I}{a}\frac{l}{\sqrt{4a^2 + l^2}}$$

Direction of $\vec{B}$. For a long straight conductor carrying current, the magnetic lines of force are concentric circles with conductor as the centre ; the direction of magnetic lines of force can be found by *right-hand grip rule*. The direction of $\vec{B}$ at any point is along the tangent to field line at that point as shown in Fig. 7.29.

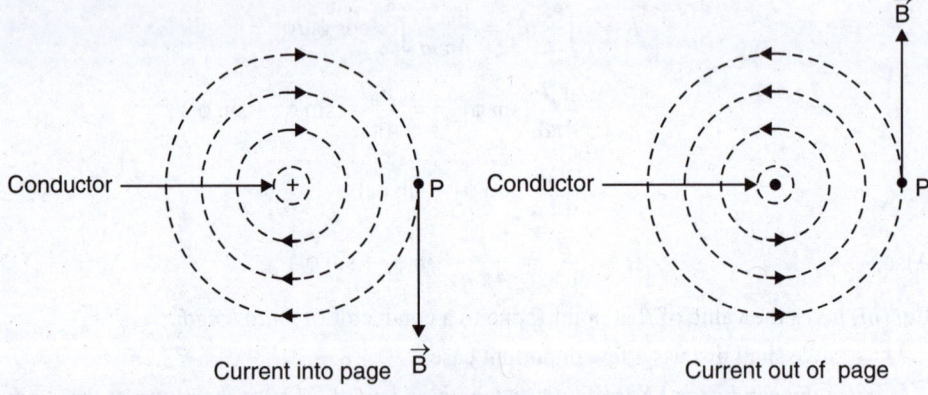

Fig. 7.29

Note. For a given current, $B \propto 1/a$ so that graph between B and a is a hyperbola.

7.27. Magnetic Field on the Axis of Circular Coil Carrying Current

Consider a circular coil of radius a, centre O and carrying a current I in the direction shown in Fig. 7.30. Let the plane of the coil be perpendicular to the plane of the paper. It is desired to find the magnetic field at a point P on the axis of the coil such that $OP = x$.

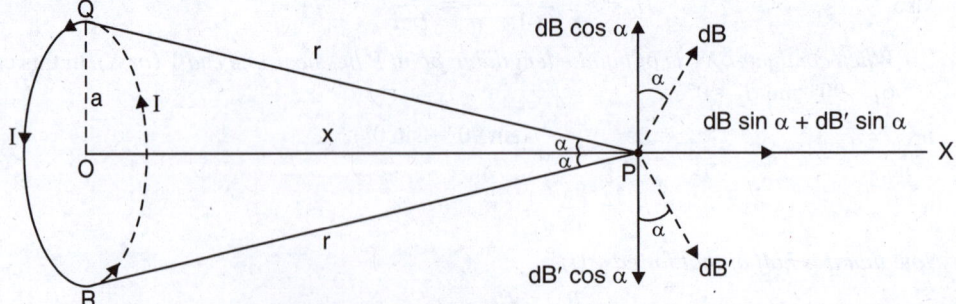

Fig. 7.30

Consider two small current elements, each of length dl, located diametrically opposite to each other at Q and R. Suppose the distance of Q or R from P is r i.e. $PQ = PR = r$.

$\therefore$ $$r = \sqrt{a^2 + x^2}$$

According to Biot-Savart law, the magnitude of magnetic field at P due to current element at Q is given by ;

$$dB = \frac{\mu_0}{4\pi} \frac{I \, dl \sin 90°}{r^2} \qquad\qquad (\because \theta* = 90°)$$

or $$dB = \frac{\mu_0}{4\pi} \frac{I \, dl}{(a^2 + x^2)} \qquad\qquad ...(i)$$

The magnetic field at P due to current element at Q is in the plane of paper and at right angles to $\overrightarrow{r}$ and in the direction shown.

Similarly, magnitude of magnetic field at point P due to current element at R is given by ;

$$dB' = \frac{\mu_0}{4\pi} \frac{I \, dl}{(a^2 + x^2)} \qquad\qquad ...(ii)$$

It also acts in the plane of paper and at right angles to $\overrightarrow{r}$ but in opposite direction to dB.

From eqs. (i) and (ii), $dB = dB'$.

It is clear that vertical components ($dB \cos \alpha$ and $dB' \cos \alpha$) will be equal and opposite and thus cancel each other. However, components along the axis of the coil ($dB \sin \alpha$ and $dB' \sin \alpha$) are added and act in the direction PX. This is true for all the diametrically opposite elements of the circular coil. Therefore, when we sum up the contributions of all the current elements of the coil, the perpendicular components will cancel. Hence the resultant magnetic field at point P is the vector sum of all the components $dB \sin \alpha$ over the entire coil.

$\therefore$ $$B = \int dB \sin \alpha = \int \frac{\mu_0 \, I \, dl \sin \alpha}{4\pi (a^2 + x^2)} = \frac{\mu_0 \, I \sin \alpha}{4\pi (a^2 + x^2)} \int dl$$

Now $$\sin \alpha = \frac{a}{\sqrt{a^2 + x^2}} \quad \text{and} \quad \int dl = 2\pi a$$

$\therefore$ $$B = \frac{\mu_0 \, I \, a^2}{2(a^2 + x^2)^{3/2}} \quad \text{along } PX \qquad\qquad ...(iii)$$

Also, $$H = \frac{B}{\mu_0} = \frac{Ia^2}{2(a^2 + x^2)^{3/2}}$$

If the circular coil has N turns, then,

$$B = \frac{\mu_0 \, NI \, a^2}{2(a^2 + x^2)^{3/2}} \quad \text{along } PX \qquad\qquad ...(iv)$$

Also, $$H = \frac{B}{\mu_0} = \frac{NIa^2}{2(a^2 + x^2)^{3/2}}$$

Different Cases. Let us discuss some special cases.

(i) *When point P is at the centre of the coil.* In this case, $x = 0$ and eq. (iv) becomes :

$$B = \frac{\mu_0 NIa^2}{2a^3} = \frac{\mu_0 NI}{2a}$$

This is the expression for the magnetic field at the centre of a current-carrying circular coil already derived in section 7.25. Note that the value of magnetic field is maximum at the centre of the coil.

---

*   The radius vector QP of each current element is perpendicular to it so that $\theta = 90°$ in each case.

Also, $$H = \frac{B}{\mu_0} = \frac{NI}{2a}$$

(ii) When point P is far away from the centre of coil. In this case, $x \gg a$ so that $a^2 + x^2 \simeq x^2$.

∴ $$B = \frac{\mu_0 \, NI \, a^2}{2x^3}$$

Also, $$H = \frac{B}{\mu_0} = \frac{NIa^2}{2x^3}$$

The magnetic field is directed along the axis of the coil and falls off as the cube of the distance from the coil.

Direction of $\vec{B}$. The magnetic field at the centre of a coil carrying current is along the axis of the coil as shown in Fig. 7.31. The direction of magnetic field can be determined by using **right-hand fist rule.** *Hold the axis of the coil in the right-hand fist in such a way that fingers point in the direction of current in the coil. Then outstretched thumb gives the direction of the magnetic field.* Applying this rule to Fig. 7.31, it is clear that direction of magnetic lines of force is along the axis of the coil as shown.

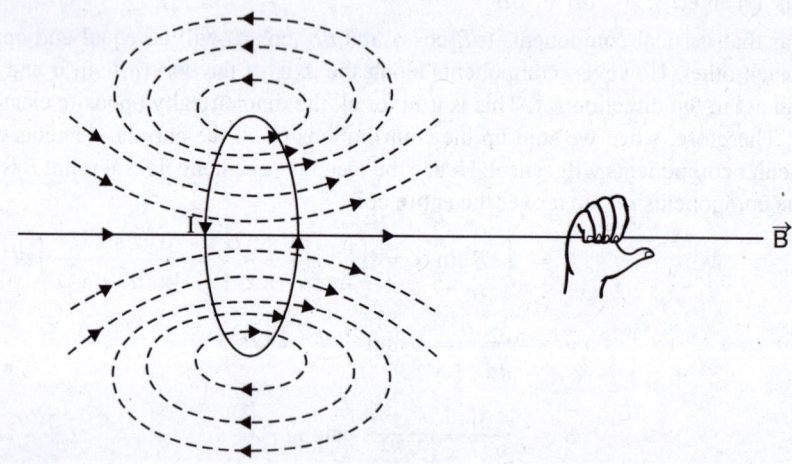

Fig. 7.31

Example 7.13. *How far from a compass should a wire carrying 1 A current be located if its magnetic field at the compass is not to exceed 1 percent of the *earth's magnetic field* $(3 \times 10^{-5} \, Wb/m^2)$?*

Solution. Let r metre be the desired distance.

Required flux density at the compass is

$$B = 1\% \text{ of Earth's flux density}$$

$$= 0{\cdot}01 \times 3 \times 10^{-5} = 3 \times 10^{-7} \, \text{Wb/m}^2$$

Required magnetising force at the compass is

$$H = \frac{B}{\mu_0} = \frac{3 \times 10^{-7}}{4\pi \times 10^{-7}} = 0{\cdot}239 \, \text{AT/m}$$

Now, $$H = \frac{I}{2\pi r} \qquad \therefore \quad r = \frac{I}{2\pi H} = \frac{1}{2\pi \times 0{\cdot}239} = \mathbf{0{\cdot}67 \, m}$$

Example 7.14. *A horizontal overhead power line carries a current of 50 A in west to east direction. What is the magnitude and direction of the magnetic field 1·5 m below the line ?*

_____

* **Earth's magnetic field.** The earth itself has a weak magnetic field. This is believed to be caused by electric currents circulating within its core. The currents are probably generated by convection in the liquid core maintained by radioactive heating of the earth's interior.

Solution. Figure 7.32 shows the conditions of the problem. The magnitude of magnetic field at point P, 1·5 m below the wire is given by ;

$$B = \frac{\mu_0\, I}{2\pi\, a}$$

Here,

$$\mu_0 = 4\pi \times 10^{-7}\ \text{H/m}\ ;\ I = 50\ \text{A}\ ;\quad a = 1\cdot5\ \text{m}$$

∴

$$B = \frac{4\pi\times10^{-7}}{2\pi}\times\frac{50}{1\cdot5} = \textbf{6.7} \times \textbf{10}^{-6}\ \textbf{T}$$

According to right-hand grip rule, the direction of magnetic field below the wire is from south to north.

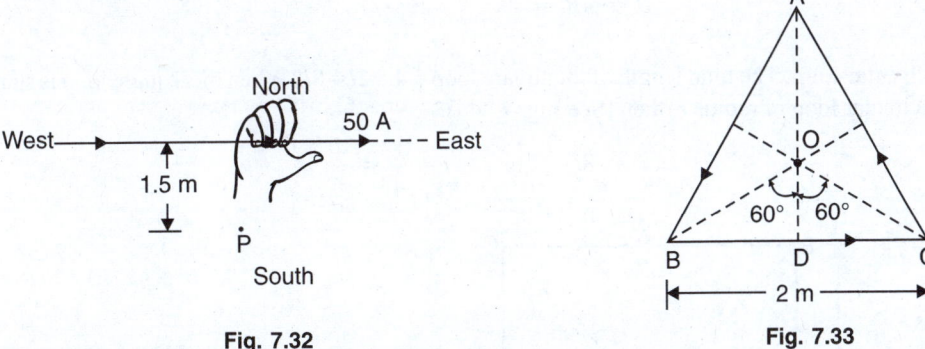

Fig. 7.32   Fig. 7.33

Example 7.15. *A current of 1 A is flowing in the sides of an equilateral triangle of side 2 m. Find the magnetic field at the centroid of the triangle.*

Solution. It is clear that all the three sides of the triangle will produce magnetic field at the centroid O in the same direction. Therefore, total magnetic field at O is = 3 × magnetic field due to one side.

Magnetic field at O due to side BC [See Fig. 7.33] is

$$B_1 = \frac{\mu_0\, I}{4\pi\, a}\ (\sin \phi_1 + \sin \phi_2)$$

Here, $I = 1\ \text{A}\ ;\ \ \phi_1 = \phi_2 = 60°\ ;\ \ \mu_0 = 4\pi \times 10^{-7}\ \text{H/m}$

$$a = OD = \frac{BD}{\tan 60°} = \frac{BC/2}{\tan 60°} = \frac{2/2}{\sqrt{3}} = \frac{1}{\sqrt{3}}$$

∴

$$B_1 = \frac{4\pi\times10^{-7}}{4\pi}\times\frac{1}{1/\sqrt{3}}\ (\sin 60° + \sin 60°)$$

$$= 10^{-7}\times\sqrt{3}\left(\frac{\sqrt{3}}{2}+\frac{\sqrt{3}}{2}\right) = 3 \times 10^{-7}\ \text{T}$$

∴   Magnetic field at O due to the whole triangle is

$$B = 3B_1 = 3(3 \times 10^{-7}) = \textbf{9} \times \textbf{10}^{-7}\ \textbf{T}$$

Example 7.16. *A square loop of wire of side 2l carries a current I. What is the magnetic field at the centre of the square ? If the square wire is reshaped into a circle, would the magnetic field increase or decrease at the centre ?*

Solution. Square loop. Figure 7.34 (*i*) shows the conditions of the problem. It is clear that each side of the square produces magnetic field at the centre O of the square in the same direction. Therefore, total magnetic field at O = 4 × magnetic field due to one side.

Magnetic field at O due to side AB is given by ;

$$B_1 = \frac{\mu_0}{4\pi} \frac{I}{a} (\sin\phi_1 + \sin\phi_2)$$

Here

$$\mu_0 = 4\pi \times 10^{-7} \text{ H/m} ; \phi_1 = \phi_2 = 45° ; a = OM = AB/2 = l$$

∴

$$B_1 = \frac{4\pi \times 10^{-7}}{4\pi} \times \frac{I}{l} (\sin 45° + \sin 45°)$$

$$= 10^{-7} \times \frac{I}{l}\left(\frac{2}{\sqrt{2}}\right) = \sqrt{2}\frac{I}{l} \times 10^{-7} \text{ T}$$

Magnetic field at O due to the whole square is

$$B = 4B_1 = 4\sqrt{2}\frac{I}{l} \times 10^{-7} T \qquad \qquad ...(i)$$

Circular loop. The total length of the square loop $= 4 \times 2l = 8l$. When this square loop is shaped into a circular loop of radius r, then [See Fig. 7.34 (ii)],

$$2\pi r = 8l \qquad \text{or} \qquad r = \frac{8l}{2\pi} = \frac{4l}{\pi}$$

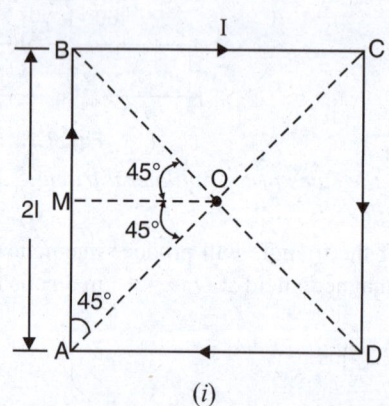

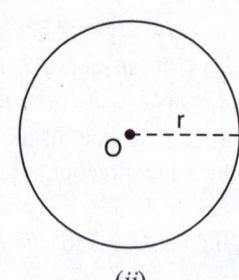

$$(i) \qquad \qquad \qquad \qquad (ii)$$

Fig. 7.34

Magnetic field at the centre of the circular loop is

$$B = \frac{\mu_0 I}{2r} = \frac{4\pi \times 10^{-7} \times I}{2(4l/\pi)} = \frac{\pi^2}{2} \times \frac{I}{l} \times 10^{-7}$$

∴

$$B = 4.93 \times \frac{I}{l} \times 10^{-7} T \qquad \qquad ...(ii)$$

Comments. Inspection of eqs. (i) and (ii) reveals that magnetic field in case of square loop will be more.

Example 7.17. *A current of 15A is passing along a straight wire. Calculate the force on a unit N-pole placed 0.15 metre from the wire. If the wire is bent to form into a loop, calculate the diameter of the loop so as to produce the same force at the centre of the coil upon a unit N-pole when carrying a current of 15A.*

Solution. By definition, the force on the unit N-pole is the magnetising force H. Therefore, force on a unit N-pole placed at a point 0.15 m (*i.e.* $a = 0.15$m) from a long straight wire carrying current $I(= 15$A) is given by ;

$$H = \frac{I}{2\pi a} = \frac{15}{2\pi \times 0.15} = \frac{50}{\pi} \text{AT/m or N/Wb}$$

Force on a unit N-pole placed at the centre of a loop of radius r when the loop carries a current $I(= 15$ A) is

$$H' = \frac{I}{2r} = \frac{15}{2r} \text{ AT/m}$$

As per the statement of the problem, $H' = H$.

$$\therefore \quad \frac{15}{2r} = \frac{50}{\pi} \quad \text{or} \quad r = \frac{15\pi}{2 \times 50} = 0.4713 \text{ m}$$

$\therefore$ Diameter of loop, $D = 2r = 2 \times 0.4713 = 0.9426$ m = **94.26 cm**

Tutorial Problems

1. A horizontal overhead power line carries a current of 90 A in east to west direction. What is the magnitude and direction of magnetic field due to the current 1·5 m below the wire ? **[1·2 × 10⁻⁵ T *towards south*]**

2. A long straight wire is turned into a loop of radius 10 cm as shown in Fig. 7.35. If a current of 8 A is passed, then find the value of magnetic field at the centre O of the loop.

   [3·4 × 10⁻⁵ T *perpendicular to plane of paper pointing upward*]

   [**Hint :** The magnetic field at O due to straight wire is perpendicular to the plane of paper and is directed downward. However, field due to circular loop is directed in opposite direction.]

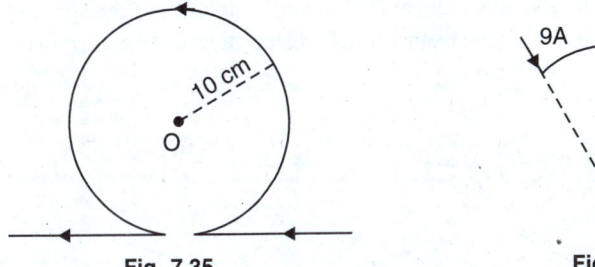

| Fig. 7.35 | Fig. 7.36 |

3. A circular segment of radius 10 cm subtends an angle of 60° at its centre. A current of 9 A is flowing through it. Find the magnitude and direction of magnetic field produced at the centre [See Fig. 7.36].

   [9·42 × 10⁻⁶ T *perpendicular to the plane of paper pointing downward*]

   [**Hint :** The magnetic field at the centre of a single turn circular coil is

   $$B = \frac{\mu_0 I}{2a} \qquad \dots a \text{ is the radius of coil.}$$

   For the given arc, $\quad B = \frac{60°}{360°}\left(\frac{\mu_0 I}{2a}\right)$

4. A long wire having a semicircular loop of radius a carries a current I amperes as shown in Fig. 7.37. Find the magnetic field at the centre of the semicircular arc. $\left[\dfrac{\mu_0 I}{4a}\right]$

   [**Hint :** The straight portions AB and DE do not contribute to any magnetic field at O. Therefore, magnetic field at O is only due to semicircular loop.]

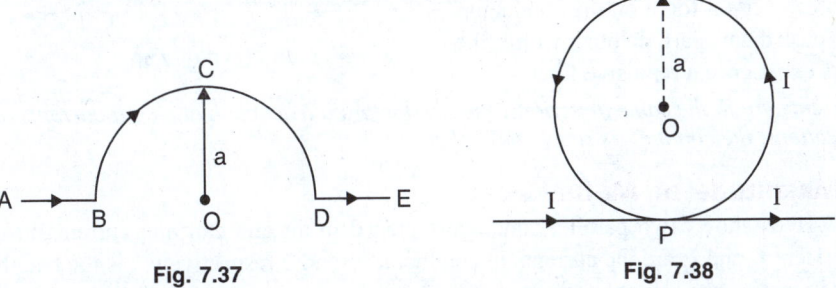

| Fig. 7.37 | Fig. 7.38 |

5. The wire shown in Fig. 7.38 carries a current I. What will be the magnitude and direction of magnetic field at the centre O ? Assume that various portions of wire do not touch each other at P.

$$\left[\frac{\mu_0 I}{2a}\left(1+\frac{1}{\pi}\right) \textit{perpendicular to the plane of paper directed upward} \right]$$

[**Hint :** The magnetic field due to straight conductor and that due to circular part aid each other at O.]

7.28. Force Between Current-Carrying Parallel Conductors

When two current-carrying conductors are parallel to each other, a mechanical force acts on each of the conductors. This force is the result of each conductor being acted upon by the magnetic field produced by the other. *If the currents are in the same direction, the forces are attractive ; if currents are in opposite direction, the forces are repulsive.* This can be beautifully illustrated by drawing the magnetic field produced by each conductor.

(*i*) **Currents in the same direction.** Consider two parallel conductors A and B carrying currents in the same direction (*i.e.* into the plane of paper) as shown in Fig. 7.39 (*i*). Each conductor will set up its own magnetic field as shown. It is clear that in the space between A and B, the two fields are in opposition and hence they tend to cancel each other. However, in the space outside A and B, the two fields assist each other. Hence the resultant field distribution will be as shown in Fig. 7.39 (*ii*).

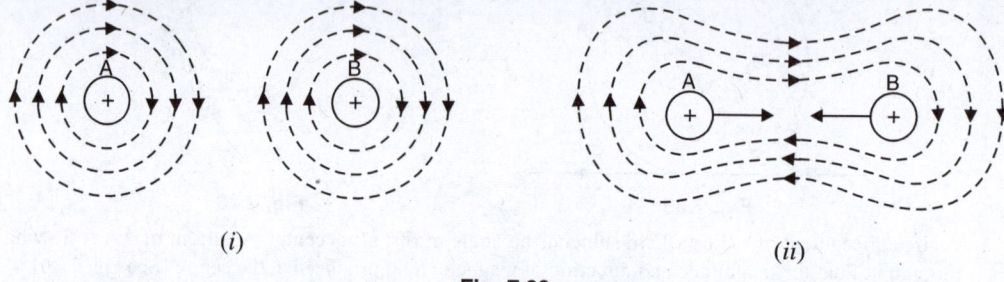

(*i*)   (*ii*)

Fig. 7.39

Since magnetic lines of force behave as stretched elastic cords, the two conductors are attracted towards each other. Alternatively, the conductors can be viewed as moving away from the relatively strong field (in the space outside A and B) into the weaker field between the conductors.

(*ii*) **Currents in opposite direction.** Consider two parallel conductors A and B carrying currents in the opposite direction as shown in Fig. 7.40. Each conductor will set up its own field as shown. It is clear that in the space outside A and B, the two fields are in opposition and hence they tend to cancel each other. However, in the space between A and B, the two fields assist each other. The lateral pressure between lines of force exerts a force on the conductors tending to push them apart. In other words, the conductors experience a repulsive force.

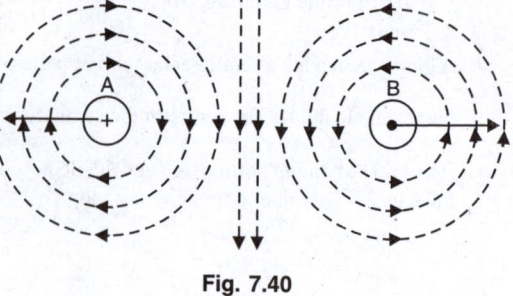

Fig. 7.40

If currents are in the same directions, the conductors attract each other ; if currents are in opposite directions, the conductors repel each other.

7.29. Magnitude of Mutual Force

Fig. 7.41 (*i*) shows two parallel conductors placed in air and carrying currents in the same direction. Here I_1 and I_2 are the currents in conductors 1 and 2 respectively, l is the length of each conductor in metres and d is the distance between conductors in metres. It is clear that each of the two parallel conductors lies in the magnetic field of the other conductor.

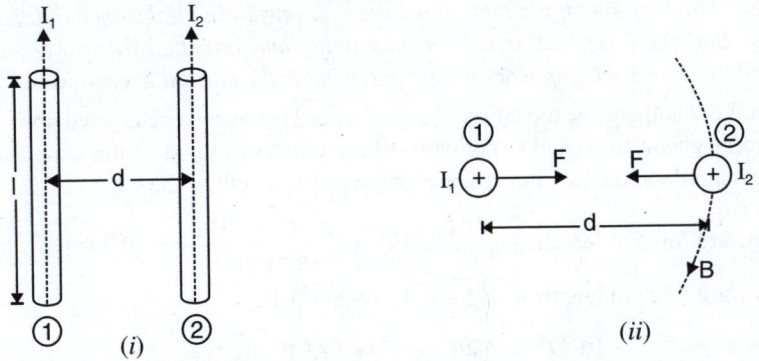

Fig. 7.41

In order to determine the magnitude of force, we can consider conductor 2 placed in the magnetic field produced by conductor 1 as shown in Fig. 7.41 (*ii*). Now field intensity H due to current I_1 in conductor 1 at the centre of conductor 2 is given by ;

$$H = \frac{I_1}{2\pi d}$$

But $B = \mu_0 \mu_r H = \mu_0 H = \dfrac{\mu_0 I_1}{2\pi d}$ [For air, $\mu_r = 1$]

Force acting on conductor 2 is given by ;

$$F = B I_2 l = \left(\frac{\mu_0 I_1}{2\pi d} \right) I_2 l$$

$$= \frac{4\pi \times 10^{-7} \, I_1 \, I_2 \, l}{2\pi \, d} = \frac{2 \, I_1 \, I_2 \, l}{d} \times 10^{-7} \text{ newtons}$$

$$\therefore \quad F = \frac{2 \, I_1 \, I_2 \, l}{d} \times 10^{-7} \text{ N}$$

It can be easily shown that conductor 1 will experience an equal force in the opposite direction [See Fig. 7.41 (*ii*)].

Force per metre run of the conductor is given by ;

$$F' = \frac{2 I_1 I_2}{d} \times 10^{-7} \text{ N/m}$$

According to Fleming's left-hand rule, the two conductors will attract each other.

7.30. Definition of Ampere

The force acting between two parallel conductors has led to the modern definition of an ampere. We have seen above that force between two parallel current-carrying conductors is

$$F = \frac{2 I_1 I_2 l}{d} \times 10^{-7} \text{ newtons}$$

If $I_1 = I_2 = 1\text{A}$; $l = 1$ m ; $d = 1$ m, then,

$$F = \frac{2 \times 1 \times 1 \times 1}{1} \times 10^{-7} = 2 \times 10^{-7} \text{ N}$$

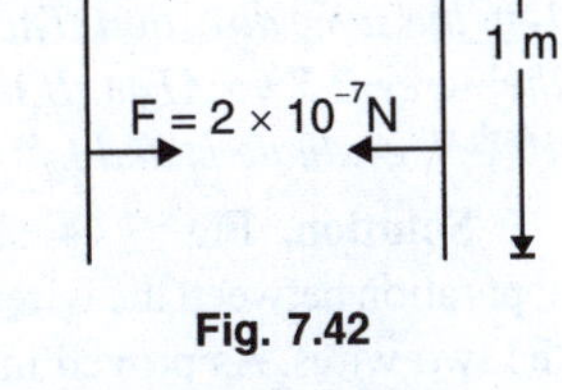

*Hence **one ampere** is that current which, if maintained in two long parallel conductors, and placed 1 m apart in vacuum, would produce between these conductors a force equal to 2×10^{-7} newton per metre of length (See Fig. 7.42).*

Fig. 7.42

Historically, the ampere was fixed originally in a very different way. The constant 2×10^{-7} that appears in the modern definition was chosen so as to keep the magnitude of ampere the same as formerly.

Example 7.18. *Two long horizontal wires are kept parallel at a distance of 0·2 cm apart in a vertical plane. Both the wires have equal currents in the same direction. The lower wire has a mass of 0·05 kg/m. If the lower wire appears weightless, what is the current in each wire ?*

Solution. Let I amperes be the current in each wire. The lower wire is acted upon by two forces *viz* (*i*) upward magnetic force and (*ii*) downward force due to weight of the wire. Since the lower wire appears weightless, the two forces are equal over 1m length of the wire.

$$\text{Upward force/m length} = \frac{2I_1 I_2}{d} \times 10^{-7} = \frac{2 \times I \times I \times 10^{-7}}{0.2 \times 10^{-2}} = 10^{-4} I^2 \text{ N}$$

$$\text{Downward force/m length} = mg = 0.05 \times 9.8 = 0.49 \text{ N}$$

∴
$$10^{-4} I^2 = 0.49 \quad \text{or} \quad I = \sqrt{0.49 \times 10^4} = \textbf{70 A}$$

Example 7.19. *A rectangular loop ABCD carrying a current of 16A in clockwise direction is placed with its longer side parallel to a straight conductor 4 cm apart and carrying a current of 20A as shown in Fig. 7.43. The sides of the loop are 15 cm and 6 cm. What is the net force on the loop ? What will be the difference in force if the direction of current in the loop is reversed ?*

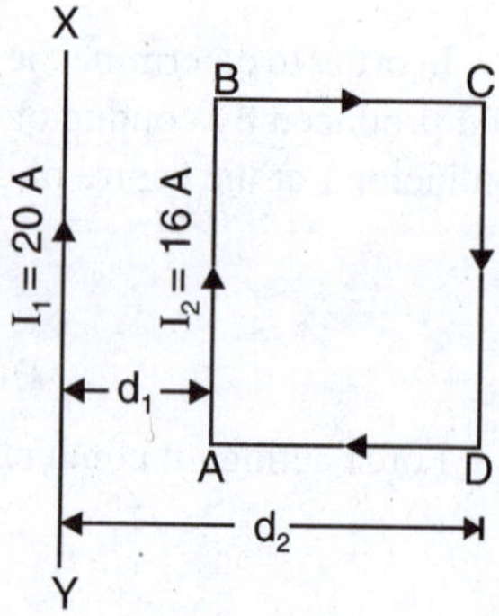

Solution. Fig. 7.43 shows the arrangement. The long straight conductor XY will exert an attractive force on arm AB of the loop while arm CD will experience a repulsive force. The forces on the arms BC and AD will be equal and opposite and hence cancel out. Referring to Fig. 7.43,

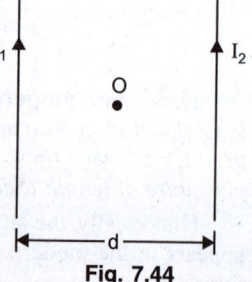

Fig. 7.43

$$d_1 = 4 \text{ cm} = 0.04 \text{ m} \quad ; \quad d_2 = 4 + 6 = 10 \text{ cm} = 0.1 \text{ m}$$

$$\text{Force on arm } AB, \ F_1 = \frac{2 I_1 I_2}{d_1} \times 10^{-7} \times \text{Length } AB \qquad \textit{... towards XY}$$

$$= \frac{2 \times 20 \times 16}{0.04} \times 10^{-7} \times 0.15 = 2.4 \times 10^{-4} \text{ N}$$

$$\text{Force on arm } CD, \ F_2 = \frac{2 I_1 I_2}{d_2} \times 10^{-7} \times \text{Length } CD \qquad \textit{... away from XY}$$

$$= \frac{2 \times 20 \times 16}{0.1} \times 10^{-7} \times 0.15 = 0.96 \times 10^{-4} \text{ N}$$

Net force on the loop is $F = F_1 - F_2 = 10^{-4} (2.4 - 0.96) = \textbf{1·44} \times \textbf{10}^{-4} \textbf{ N}$

Therefore, the net force on the loop is directed *towards* the current-carrying straight conductor XY. If the direction of current in the loop is reversed, the magnitude of net force on the loop remains the same (*i.e.* $F = 1.44 \times 10^{-4}$ N) but its direction will be away from the current-carrying straight conductor XY.

Example 7.20. *Two long straight parallel wires, standing in air 2m apart, carry currents I_1 and I_2 in the same direction. The magnetic intensity at a point midway between the wires is 7.95 AT/m. If the force on each wire per unit length is 2.4 × 10^{-4} N, evaluate I_1 and I_2.*

Solution. Fig. 7.44 shows the conditions of the problem. Here, separation between the wires is $d = 2$ m and O is the point midway between the two wires. As proved in Art. 7.26, the magnetic intensity H at a point distant a from a long straight current-carrying wire is

$$H = \frac{I}{2\pi a}$$

Fig. 7.44

Since the two wires are carrying currents in the same direction, the net magnetic intensity H at O is the difference of the magnetic intensities at O due to two currents *i.e.*

$$H = H_1 - H_2$$

or

$$7.95 = \frac{I_1}{2\pi \times 1} - \frac{I_2}{2\pi \times 1} \qquad (\because \text{ point } O \text{ is 1m from each wire})$$

$\therefore$

$$I_1 - I_2 = 50 \qquad \qquad ...(i)$$

As proved in Art. 7.29, force per unit length of the conductors is

$$F = \frac{2I_1 I_2}{d} \times 10^{-7}$$

or

$$2.4 \times 10^{-4} = \frac{2I_1 I_2}{2} \times 10^{-7}$$

$\therefore$

$$I_1 I_2 = 2400$$

Now,

$$(I_1 + I_2)^2 = (I_1 - I_2)^2 + 4 I_1 I_2 = (50)^2 + 4 \times 2400 = 12100$$

$\therefore$

$$I_1 + I_2 = 110 \qquad \qquad ...(ii)$$

From eqs. (*i*) and (*ii*), $I_1 = \mathbf{80A}$; $I_2 = \mathbf{30A}$

Example 7.21. *A horizontal straight wire 5 cm long of mass 1·2 g/m is placed perpendicular to a uniform magnetic field of 0·6 T. If resistance of the wire is 3·8 Ω m^{-1}, calculate the p.d. that has to be applied between the ends of the wire to make it just self-supporting.*

Solution. The current (I) in the wire is to be in such a direction that magnetic force acts on it vertically upward. To make the wire self-supporting, its weight should be equal to the upward magnetic force *i.e.*

$$B I l = m g \qquad (\because \theta = 90°)$$

or

$$I = \frac{mg}{Bl}$$

Here,

$$m = 1·2 \times 10^{-3} l \; ; \; B = 0·6 \text{ T} \; ; \; g = 9·8 \text{ ms}^{-2}$$

$\therefore$

$$I = \frac{(1·2 \times 10^{-3} l) \times 9·8}{0·6 \times l} = 19·6 \times 10^{-3} \text{ A}$$

Resistance of the wire, $R = 0·05 \times 3·8 = 0·19 \; \Omega$

$\therefore$

Required P.D., $V = I R = (19·6 \times 10^{-3}) \, 0·19 = \mathbf{3·7 \times 10^{-3} \text{ V}}$

Tutorial Problems

1. A pair of rising mains has a spacing of 200 mm between centres. If each conductor carries 500 A, determine the force between the conductors for each 10m length of run. **[2·5 N** *repulsive***]**

2. Two busbars, each 20 m long, feed a circuit and are spaced at a distance of 80 mm inbetween centres. If a short-circuit current of 20,000 A flows through the conductors, calculate the force per metre between the bars. **[1000 N]**

3. Two long straight parallel conductors carry the same current I in the same direction. The conductors are placed 20 cm apart in air. The magnetic flux density between the conductors 5 cm from one of them is $1·33 \times 10^{-5}$ Wb/m². If the force on each conductor per metre length is 25×10^{-6} N, find the current in each conductor. **[5 A]**

4. The wires that supply current to a 120 V, 2kW electric heater are 2 mm apart. What is the force per metre between the wires ? **[0·028 N/m]**

5. The busbars 10 cm apart are supported by insulators every metre along their length. The busbars each carry a current of 15 kA. What is the force acting on each insulator ? **[450 N]**

Objective Questions

1. When a magnet is heated,

   (i) it gains magnetism

   (ii) it loses magnetism

   (iii) it neither loses nor gains magnetism

   (iv) none of the above

2. The magnetic material used in permanent magnets is

   (i) iron   (ii) soft steel

   (iii) nickel   (iv) hardened steel

3. The magnetic material used in temporary magnets is

   (i) hardened steel   (ii) cobalt steel

   (iii) soft iron   (iv) tungsten steel

4. Magnetic flux density is a

   (i) vector quantity   (ii) scalar quantity

   (iii) phasor   (iv) none of the above

5. The relative permeability of a ferromagnetic material is 1000. Its absolute permeability will be

   (i) 10^6 H/m   (ii) $4\pi \times 10^{-3}$ H/m

   (iii) $4\pi \times 10^{-11}$ H/m (iv) none of the above

6. The main advantage of temporary magnets is that we can

   (i) change the magnetic flux

   (ii) use any magnetic material

   (iii) decrease the hysteresis loss

   (iv) none of the above

7. One weber is equal to

   (i) 10^6 lines   (ii) $4\pi \times 10^{-7}$ lines

   (iii) 10^{12} lines   (iv) 10^8 lines

8. Magnetic field intensity is a

   (i) scalar quantity   (ii) vector quantity

   (iii) phasor   (iv) none of the above

9. The absolute permeability of a material having a flux density of 1 Wb/m^2 is 10^{-3} H/m. The value of magnetising force is

   (i) 10^{-3} AT/m   (ii) $4\pi \times 10^{-3}$ AT/m

   (iii) 1000 AT/m   (iv) $4\pi \times 10^3$ AT/m

10. When the relative permeability of a material is slightly less than 1, it is called a

    (i) diamagnetic material

    (ii) paramagnetic material

    (iii) ferromagnetic material

    (iv) none of the above

11. The greater percentage of substances are

    (i) diamagnetic   (ii) paramagnetic

    (iii) ferromagnetic   (iv) none of the above

12. When the relative permeability of material is much greater than 1, it is called

    (i) diamagnetic material

    (ii) paramagnetic material

    (iii) ferromagnetic material

    (iv) none of the above

13. The magnetic flux density in an air-cored coil is 10^{-2} Wb/m^2. With a cast iron core of relative permeability 100 inserted, the flux density will become

    (i) 10^{-4} Wb/m^2 (ii) 10^4 Wb/m^2

    (iii) 10^{-2} Wb/m^2 (iv) 1 Wb/m^2

14. Which of the following is more suitable for the core of an electromagnet ?

    (i) soft iron   (ii) air

    (iii) steel   (iv) tungsten steel

15. The source of a magnetic field is

    (i) an isolated magnetic pole

    (ii) static electric charge

    (iii) magnetic substances

    (iv) current loop

16. A magnetic needle is kept in a uniform magnetic field. It experiences

    (i) a force and a torque

    (ii) a force but not a torque

    (iii) a torque but not a force

    (iv) neither a torque nor a force

17. The unit of pole strength is

    (i) A/m^2 (ii) Am

    (iii) Am^2 (iv) Wb/m^2

18. When the relative permeability of a material is slightly more than 1, it is called a

    (i) diamagnetic material

    (ii) paramagnetic material

    (iii) ferromagnetic material

    (iv) none of the above

19. AT/m is the unit of

    (i) m.m.f.

    (ii) reluctance

    (iii) magnetising force

    (iv) magnetic flux density

20. A magnetic needle is kept in a non-uniform magnetic field. It experiences

    (*i*) a force and a torque

    (*ii*) a force but not a torque

    (*iii*) a torque but not a force

    (*iv*) neither a force nor a torque

21. Magnetic flux passes more readily through

    (*i*) air   (*ii*) wood

    (*iii*) vacuum   (*iv*) iron

22. Iron is ferromagnetic

    (*i*) above 770°C

    (*ii*) below 770°C

    (*iii*) at all temperatures

    (*iv*) none of the above

23. The relative permeability of a material is 0·9998. It is

    (*i*) diamagnetic   (*ii*) paramagnetic

    (*iii*) ferromagnetic   (*iv*) none of the above

24. Magnetic lines of force

    (*i*) intersect at infinity

    (*ii*) intersect within the magnet

    (*iii*) cannot intersect at all

    (*iv*) none of the above

25. Demagnetising of magnets can be done by

    (*i*) rough handling (*ii*) heating

    (*iii*) magnetising in opposite direction

    (*iv*) all of the above

Answers

1. (*ii*)	**2.** (*iv*)	**3.** (*iii*)	**4.** (*i*)	**5.** (*ii*)
6. (*i*)	**7.** (*iv*)	**8.** (*ii*)	**9.** (*iii*)	**10.** (*i*)
11. (*ii*)	**12.** (*iii*)	**13.** (*iv*)	**14.** (*i*)	**15.** (*iv*)
16. (*iii*)	**17.** (*ii*)	**18.** (*ii*)	**19.** (*iii*)	**20.** (*i*)
21. (*iv*)	**22.** (*ii*)	**23.** (*i*)	**24.** (*iii*)	**25.** (*iv*)

8

Magnetic Circuits

Introduction

We have seen that magnetic lines of force form closed loops around and through the magnetic material. The closed path followed by magnetic flux is called a magnetic circuit just as the closed path followed by current is called an electric circuit. Many electrical devices (*e.g.* generator, motor, transformer etc.) depend upon magnetism for their operation. Therefore, such devices have magnetic circuits *i.e.* closed flux paths. In order that these devices function efficiently, their magnetic circuits must be properly designed to obtain the required magnetic conditions. In this chapter, we shall focus our attention on the basic principles of magnetic circuits and methods to obtain their solution.

8.1. Magnetic Circuit

The closed path followed by magnetic flux is called a **magnetic circuit.**

In a magnetic circuit, the magnetic flux leaves the N-pole, passes through the entire circuit, and returns to the starting point. A magnetic circuit usually consists of materials having high permeability *e.g.* iron, soft steel *etc.* It is because these materials offer very small opposition to the 'flow' of magnetic flux. The most usual way of producing magnetic flux is by passing electric current through a wire of number of turns wound over a magnetic material. This helps in exercising excellent control over the magnitude, density and direction of magnetic flux.

Consider a coil of N turns wound on an iron core as shown in Fig. 8.1. When current I is passed through the coil, magnetic flux ϕ is set up in the core. The flux follows the closed path *ABCDA* and hence *ABCDA* is the magnetic circuit. The following points may be noted carefully :

(*i*) The amount of magnetic flux set up in the core depends upon current (I) and number of turns (N). If we increase the current or number of turns, the amount of magnetic flux also increases and *vice-versa*. The product *NI is called the **magnetomotive force (m.m.f.)** and determines the amount of flux set up in the magnetic circuit.

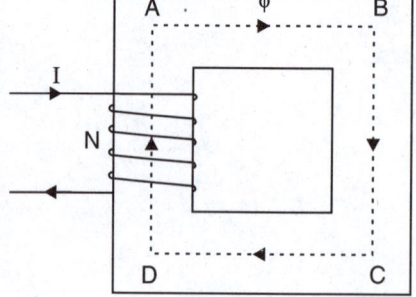

m.m.f. $= NI$ ampere-turns

It can just be compared to electromotive force (e.m.f.) which sends current in an electric circuit.

(*ii*) The opposition that the magnetic circuit offers to the magnetic flux is called **reluctance.** It depends upon length of magnetic circuit (*i.e.* length *ABCDA* in this case), area of X-section of the circuit and the nature of material that makes up the magnetic circuit.

Fig. 8.1

8.2. Analysis of Magnetic Circuit

Consider the magnetic circuit shown in Fig. 8.1. Suppose the mean length of the magnetic circuit (*i.e.* length *ABCDA*) is l metres, cross-sectional area of the **core is 'a' m^2 and relative

---

* Coiling a conductor into two or more turns has the effect of using the same current for more than once. For example, 5-turn coil carrying a current of 10A produces the same magnetic flux in a given magnetic circuit as a 1-turn coil carrying a current of 50A. Hence m.m.f. is equal to the product of N and I.

** The arrangement of magnetic materials to form a magnetic circuit is generally called a *core.*

permeability of core material is μ_r. When current I is passed through the coil, it will set up flux ϕ in the material.

Flux density in the material, $B = \dfrac{\phi}{a} \, \text{Wb/m}^2$

Magnetising force in the material, $H = \dfrac{B}{\mu_0 \mu_r} = \dfrac{\phi}{a\mu_0\mu_r} \, \text{AT/m}$

According to work law, the work done in moving a unit magnetic pole once around the magnetic circuit (*i.e.* path *ABCDA* in this case) is equal to the ampere-turns enclosed by the magnetic circuit.

$\therefore$ $\qquad\qquad\qquad *H \times l = NI$

or $\qquad\qquad\qquad \dfrac{\phi}{a\mu_0\mu_r} \times l = NI$

or $\qquad\qquad\qquad \phi = \dfrac{NI}{(l/a\mu_0\mu_r)}$

The quantity NI which produces the magnetic flux is called the magnetomotive force (m.m.f.) and is measured in ampere-turns. The quantity $l/a\,\mu_0\,\mu_r$ is called the reluctance of the magnetic circuit. Reluctance is the opposition that the magnetic circuit offers to magnetic flux.

$\therefore$ $\qquad\qquad$ Flux, $\phi = \dfrac{\text{m.m.f.}}{\text{reluctance}}$ $\qquad\qquad\qquad\qquad$ ...(*i*)

Note that the relationship expressed in eq. (*i*) has a strong resemblance to Ohm's law of electric circuit ($I = E/R$). The m.m.f. is analogous to e.m.f. in the electric circuit, reluctance is analogous to resistance and flux is analogous to current. Because of this similarity, eq. (*i*) is sometimes referred to as **Ohm's law of magnetic circuit.**

8.3. Important Terms

In the study of magnetic circuits, we generally come across the following terms :

(*i*) **Magnetomotive force (m.m.f.).** It is a magnetic pressure which sets up or tends to set up flux in a magnetic circuit and may be defined as under :

The work done in moving a unit magnetic pole once around the magnetic circuit is called the magnetomotive force (m.m.f.). It is equal to the product of current and number of turns of the coil *i.e.*

$$\text{m.m.f.} = NI \text{ ampere-turns (or AT)}$$

Magnetomotive force in a magnetic circuit corresponds to e.m.f. in an electric circuit. The only change in the definition is the substitution of unit magnetic pole in place of unit charge.

(*ii*) **Reluctance.** *The opposition that the magnetic circuit offers to magnetic flux is called reluctance.* The reluctance of a magnetic circuit depends upon its length, area of *X*-section and permeability of the material that makes up the magnetic circuit. Its unit is [†]AT/Wb.

$$\text{Reluctance, } S = \dfrac{l}{a\,\mu_0\,\mu_r}$$

Reluctance in a magnetic circuit corresponds to resistance ($R = \rho\,l/a$) in an electric circuit. Both of them vary as length ÷ area and are dependent upon the nature of material of the circuit. Magnetic materials (*e.g.* iron, steel *etc.*) have a low reluctance because the value of μ_r is very large in their case. On the other hand, non-magnetic materials (*e.g.* air, wood, copper, brass *etc.*) have a high reluctance because they possess least value of μ_r ; being 1 in case of all non-magnetic materials.

---

*   You may recall that H means force acting on a unit magnetic pole. If the unit pole is moved once around the magnetic circuit (*i.e.* distance covered is l), then work done = $H \times l$.

†   Reluctance $= \dfrac{\text{m.m.f.}}{\text{flux}} = \dfrac{\text{AT}}{\text{Wb}} = \text{AT/Wb}$

The reciprocal of permeability $\mu(= \mu_0\mu_r)$ corresponds to resistivity ρ of the electrical circuit and is called *reluctivity. It may be noted that magnetic permeability (μ) is the analog of electrical conductivity.*

(*iii*) **Permeance.** *It is the reciprocal of reluctance* and is a measure of the ease with which flux can pass through the material. Its unit is Wb/AT.

$$\text{Permeance} = \frac{1}{\text{Reluctance}} = \frac{a\mu_0\mu_r}{l}$$

Permeance of a magnetic circuit corresponds to conductance (reciprocal of resistance) in an electric circuit.

8.4. Comparison Between Magnetic and Electric Circuits

There are many points of similarity between magnetic and electric circuits. However, the two circuits are not anologous in all respects. A comparison of the two circuits is given below in the tabular form.

Magnetic Circuit   **Electric Circuit**

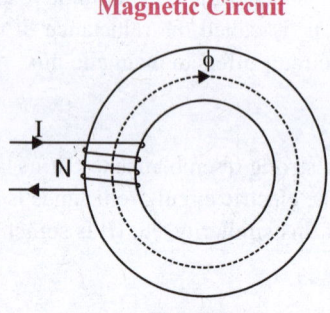   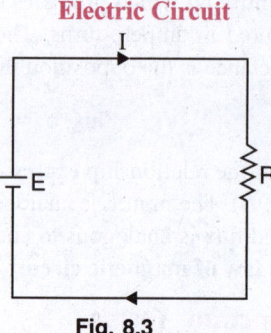

Fig. 8.2   Fig. 8.3

Similarities

1.	The closed path for magnetic flux is called a magnetic circuit.	1.	The closed path for electric current is called an electric circuit.
2.	Flux, $\phi = \dfrac{\text{m.m.f.}}{\text{reluctance}}$	2.	Current, $I = \dfrac{\text{e.m.f.}}{\text{resistance}}$
3.	m.m.f. (ampere-turns)	3.	e.m.f. (volts)
4.	Reluctance, $S = \dfrac{l}{a\mu_0\mu_r}$	4.	Resistance, $R = \rho\dfrac{l}{a}$
5.	Flux density, $B = \dfrac{\phi}{a} \text{Wb/m}^2$	5.	Current density, $J = \dfrac{I}{a} \text{A/m}^2$
6.	m.m.f. drop $= \phi S$	6.	Voltage drop $= I R$
7.	Magnetic intensity, $H = N I / l$	7.	Electric intensity, $E = V / d$
8.	Permeance	8.	Conductance.
9.	Permeability	9.	Conductivity

Dissimilarities

1.	Truly speaking, magnetic flux does not flow.	1.	The electric current acutally flows in an electric circuit.
2.	There is no magnetic insulator. For example, flux can be set up even in air (the best known magnetic insulator) with reasonable m.m.f.	2.	There are a number of electric insulators. For instance, air is a very good insulator and current cannot pass through it.

3. The value of μ_r is not constant for a given magnetic material. It varies considerably with flux density (B) in the material. This implies that reluctance of a magnetic circuit is not constant rather it depends upon B.	3. The value of resistivity (ρ) varies very slightly with temperature. Therefore, the resistance of an electric circuit is practically constant. This salient feature calls for different approach to the solution of magnetic and electric circuits.
4. No energy is expended in a magnetic circuit. In other words, energy is required in creating the flux, and not in maintaining it.	4. When current flows through an electric circuit, energy is expended so long as the current flows. The expended energy is dissipated in the form of heat.

8.5. Calculation of Ampere-Turns

In any magnetic circuit, flux produced is given by ;

$$\text{Flux, } \phi = \frac{\text{m.m.f.}}{\text{reluctance}} = \frac{AT}{(l/a\mu_0\mu_r)}$$

$\therefore$

$$AT \text{ required} = \phi \times \frac{l}{a\mu_0\mu_r} = \frac{\phi}{a} \times \frac{l}{\mu_0\mu_r}$$

$$= \frac{B}{\mu_0\mu_r} \times l \qquad\qquad \left(\because \ B = \frac{\phi}{a} \right)$$

$$= H \times l \qquad\qquad\qquad (\because \ H = B/\mu_0\mu_r)$$

i.e. **AT required for any part = Field strength H in that part × length of that part of magnetic circuit**

8.6. Series Magnetic Circuits

In a series magnetic circuit, the same flux ϕ flows through each part of the circuit. It can just be compared to a series electric circuit which carries the same current throughout.

Consider a *composite series magnetic circuit consisting of three different magnetic materials of different relative permeabilities along with an air gap as shown in Fig. 8.4. Each part of this series magnetic circuit will offer reluctance to the magnetic flux ϕ. The reluctance offered by each part will depend upon dimensions and μ_r of that part. Since the different parts of the circuit are in series, the total reluctance is equal to the sum of reluctances of individual parts, *i.e.*

$$\text{Total reluctance} = \frac{l_1}{a_1\,\mu_0\,\mu_{r1}} + \frac{l_2}{a_2\,\mu_0\,\mu_{r2}} + \frac{l_3}{a_3\,\mu_0\,\mu_{r3}} + \frac{l_g}{a_g\,\mu_0}{}^{**}$$

$$\text{Total m.m.f.} = \text{Flux} \times \text{Total reluctance}$$

$$= \phi \left[\frac{l_1}{a_1\,\mu_0\,\mu_{r1}} + \frac{l_2}{a_2\,\mu_0\,\mu_{r2}} + \frac{l_3}{a_3\,\mu_0\,\mu_{r3}} + \frac{l_g}{a_g\,\mu_0} \right]$$

$$= \frac{\phi}{a_1\,\mu_0\,\mu_{r1}} \times l_1 + \frac{\phi}{a_2\,\mu_0\,\mu_{r2}} \times l_2 + \frac{\phi}{a_3\,\mu_0\,\mu_{r3}} \times l_3 + \frac{\phi}{a_g\,\mu_0} \times l_g$$

$$= \frac{B_1}{\mu_0\,\mu_{r1}} \times l_1 + \frac{B_2}{\mu_0\,\mu_{r2}} \times l_2 + \frac{B_3}{\mu_0\,\mu_{r3}} \times l_3 + \frac{B_g}{\mu_0} \times l_g$$

$$= H_1\,l_1 + H_2\,l_2 + H_3\,l_3 + H_g\,l_g \qquad\qquad (\because \ H = B/\mu_0\,\mu_r)$$

---

* A series magnetic circuit that has parts of different dimensions and materials is called a composite series circuit.

** For air, $\mu_r = 1$.

Hence the total ampere-turns required for a series magnetic circuit can be found as under :

   (*i*) Find H for each part of the series magnetic circuit. For air, $H = B/\mu_0$ whereas for magnetic material, $H = B/\mu_0\mu_r$.

   (*ii*) Find the mean length (*l*) of magnetic path for each part of the circuit.

   (*iii*) Find AT required for each part of the magnetic circuit using the relation, $AT = H \times l$.

   (*iv*) The total AT required for the entire series circuit is equal to the sum of AT for various parts.

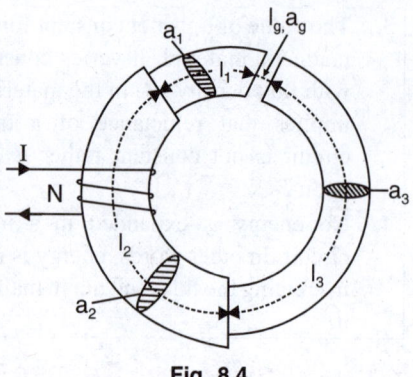

Fig. 8.4

8.7. Air Gaps in Magnetic Circuits

   In many practical magnetic circuits, air gap is indispensable. For example, in electromechanical conversion devices like electric motors and generators, the magnetic flux must pass through stator as well as rotor. This necessitates to have a small air gap between the stator and rotor to permit mechanical clearance.

   The magnitude of AT required for air gap is much greater than that required for iron part of the magnetic circuit. It is because reluctance of air is very large compared to that offered by iron. Consider a magnetic

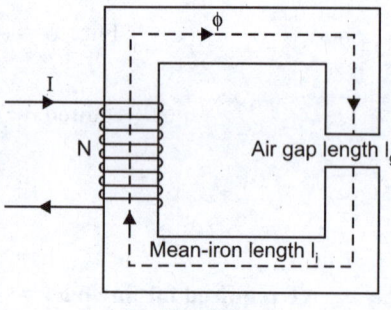

Fig. 8.5

circuit of uniform cross-sectional area *a* with an air gap as shown in Fig. 8.5. The length of the air gap is l_g and the mean length of iron part is l_i. The flux density $B(= \phi/a)$ is constant in the magnetic circuit.

$$\therefore \quad \text{Reluctance of air gap} = \frac{l_g}{a\mu_0}$$

$$\text{Reluctance of iron part} = \frac{l_i}{a\mu_0\mu_r}$$

   Now relative permeability μ_r of iron is very high (> 6000) so that reluctance of iron part is very small as compared to that of air gap inspite of the fact that $l_i > l_g$. In fact, most of ampere-turns (AT) are required in a magnetic circuit to force the flux through the air gap than through the iron part. In some magnetic circuits, we neglect reluctance of iron part compared to the air gap/gaps. This assumption leads to reasonable accuracy.

8.8. Parallel Magnetic Circuits

   A magnetic circuit which has more than one path for flux is called a parallel magnetic circuit. It can just be compared to a parallel electric circuit which has more than one path for electric current.

   The concept of parallel magnetic circuit is illustrated in Fig. 8.6. Here a coil of *N* turns wounded on limb *AF* carries a current of *I* amperes. The flux ϕ_1 set up by the coil divides at *B* into two paths, namely ;

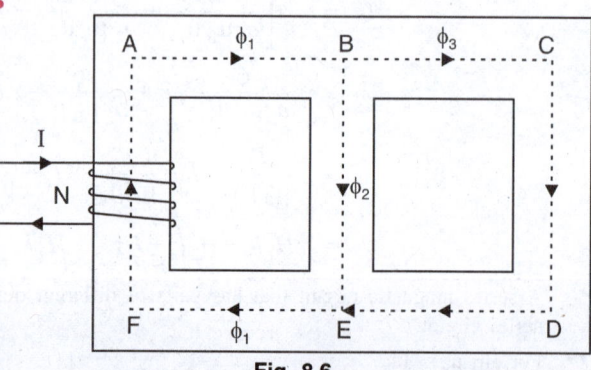

Fig. 8.6

(*i*) flux ϕ_2 passes along the path *BE*

(*ii*) flux ϕ_3 follows the path *BCDE*

Clearly, $\phi_1 = \phi_2 + \phi_3$

The magnetic paths *BE* and *BCDE* are in parallel and form a parallel magnetic circuit. The *AT* required for this parallel circuit is equal to *AT* required for any *one of the paths.

Let S_1 = reluctance of path *EFAB*

S_2 = reluctance of path *BE*

S_3 = reluctance of path *BCDE*

∴ Total m.m.f. required = m.m.f. for path *EFAB* + m.m.f. for path *BE* or path *BCDE*

or $NI = \phi_1 S_1 + \phi_2 S_2$

$= \phi_1 S_1 + \phi_3 S_3$

The reluctances S_1, S_2 and S_3 must be determined from a calculation of $l/a\mu_0\mu_r$ for those paths of the magnetic circuit in which ϕ_1, ϕ_2 and ϕ_3 exist respectively.

8.9. Magnetic Leakage and Fringing

The flux that does not follow the desired path in a magnetic circuit is called a **leakage flux.**

In most of practical magnetic circuits, a large part of flux path is through a magnetic material and the remainder part of flux path is through air. The flux in the air gap is known as *useful flux* because it can be utilised for various useful purposes. Fig. 8.7 shows an iron ring wound with a coil and having a narrow air gap. The total flux produced by the coil does not pass through the air gap as some of it **leaks through the air (path at '*a*') surrounding the iron. These flux lines as at '*a*' are called leakage flux.

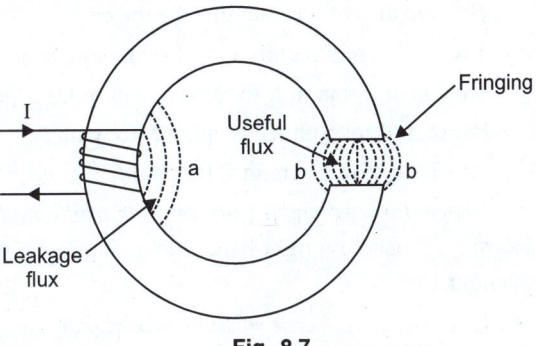

Fig. 8.7

Let ϕ_i = total flux produced *i.e.*, flux in the ***iron ring

ϕ_g = useful flux across the air gap

∴ Leakage flux, $\phi_{leak} = \phi_i - \phi_g$

Leakage coefficient, $\lambda = \dfrac{\text{Total flux}}{\text{Useful flux}} = \dfrac{\phi_i}{\phi_g}$

The value of leakage coefficients for electrical machines is usually about 1.15 to 1.25.

Magnetic leakage is undesirable in electrical machines because it increases the weight as well as cost of the machine. Magnetic leakage can be greatly reduced by placing source of m.m.f. close to the air gap.

Fringing. When crossing an air gap, magnetic lines of force tend to bulge out such as lines of force at *bb* in Fig. 8.7. It is because lines of force repel each other when passing through non-

---

* This means that we may consider either path, say path *BE*, and calculate *AT* required for it. The same *AT* will also send the flux (ϕ_3 in this case) through the other parallel path *BCDE*. The situation is similar to that of two resistors R_1 and R_2 in parallel in an electric circuit. The voltage *V* required to send currents (say I_1 and I_2) in the resistors is equal to that appearing across either resistor *i.e.* $V = I_1 R_1 = I_2 R_2$.

** Air is not a good magnetic insulator. Therefore, leakage of flux from iron to air takes place easily.

*** The flux ϕ_i is not constant all around the ring. However, for reasonable accuracy, it is assumed that the iron carries the whole of the flux produced by the coil.

magnetic material such as air. This effect is known as *fringing*. The result of bulging or fringing is to increase the effective area of air gap and thus decrease the flux density in the gap. The longer the air gap, the greater is the fringing and *vice-versa*.

Note. In a short air gap with large cross-sectional area, the fringing may be insignificant. In other situations, 10% is added to the air gap's cross-sectional area to allow for fringing.

8.10. Solenoid

*A long coil of wire consisting of closely packed loops is called a **solenoid.***

The word solenoid comes from Greek word meaning 'tube-like'. By a long solenoid we mean that length of the solenoid is very large as compared to its diameter. When current is passed through the coil of air-cored solenoid, magnetic field is set up as shown in Fig. 8.8. The path of the magnetic flux is made up of two components :

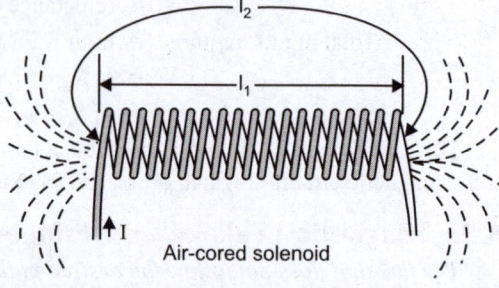

Air-cored solenoid

Fig. 8.8

   (*i*) length l_1 of the path within the coil

   (*ii*) length l_2 of the path outside the coil.

The total m.m.f. required for the solenoid is the sum of m.m.f.s required for these two paths *i.e.*

Total m.m.f. = m.m.f. for path l_1 + m.m.f. for path l_2

But m.m.f. for path l_1 *>> m.m.f. for path l_2

∴   Total m.m.f. = m.m.f. for path l_1

Hence, for a solenoid (air-cored or iron-cored), the length of the magnetic circuit is the coil length l_1. We can use right-hand rule to determine the direction of magnetic field in the core of the solenoid.

Example 8.1. *A cast steel electromagnet has an air gap length of 3 mm and an iron path of length 40 cm. Find the number of ampere-turns necessary to produce a flux density of 0.7 Wb/m² in the gap. Neglect leakage and fringing. Assume ampere-turns required for air gap to be 70% of the total ampere-turns.*

Solution.   Air-gap length, l_g = 3 mm = 3×10^{-3} m

   Flux density in air gap, B_g = 0.7 Wb/m²

∴   Magnetising force, $H_g = \dfrac{B_g}{\mu_0 \mu_r} = \dfrac{0.7}{4\pi \times 10^{-7} \times 1} = 5.57 \times 10^5$ AT/m

AT required for air gap, $AT_g = H_g \times l_g = 5.57 \times 10^5 \times 3 \times 10^{-3} = 1671$ AT

   It is given that : AT_g = 70% of total AT

∴   Total $AT = \dfrac{AT_g}{0.7} = \dfrac{1671}{0.7} = $ **2387 AT**

Example 8.2. *An iron ring has a cross-sectional area of 400 mm² and a mean diameter of 25 cm. It is wound with 500 turns. If the value of relative permeability is 250, find the total flux set up in the ring. The coil resistance is 474 Ω and the supply voltage is 240 V.*

---

*   The lengths l_2 and l_1 do not differ very much. However, the cross-sectional area of path l_2 is very large as compared to that of path l_1. Therefore, reluctance of path l_2 is very small as compared to that of path l_1.

   Now, m.m.f. = flux × reluctance

   Since reluctance of path l_2 is very small, the m.m.f. required for this path is negligible compared to that for path l_1.

Solution. The conditions of the problem are represented in Fig. 8.9.

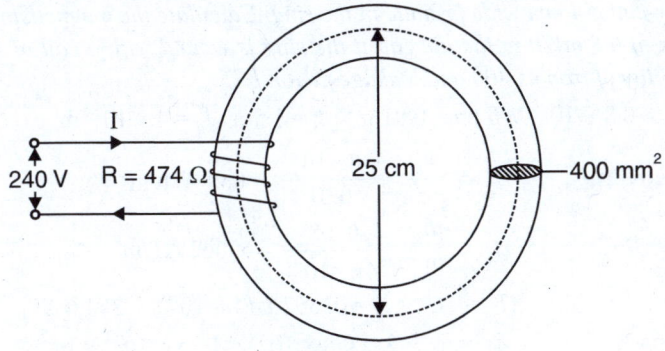

Fig. 8.9

Current through the coil, $I = V/R = 240/474 = 0.506$ A

Mean length of magnetic circuit is given by ;

$$l = \pi \times (25 \times 10^{-2}) = 0.7854 \text{ m}$$

Magnetising force, $H = \dfrac{Nl}{l} = \dfrac{500 \times 0.506}{0.7854} = 322.13$ AT/m

Flux density, $B = \mu_0 \mu_r H = (4\pi \times 10^{-7}) \times 250 \times 322 \cdot 13 = 0.1012$ Wb/m^2

∴   Flux in the ring, $\phi = B \times a = 0.1012 \times (400 \times 10^{-6}) =$ **$40 \cdot 48 \times 10^{-6}$ Wb**

Example 8.3. *An iron ring of cross-sectional area 6 cm^2 is wound with a wire of 100 turns and has a saw cut of 2 mm. Calculate the magnetising current required to produce a flux of 0·1 mWb if mean length of magnetic path is 30 cm and relative permeability of iron is 470.*

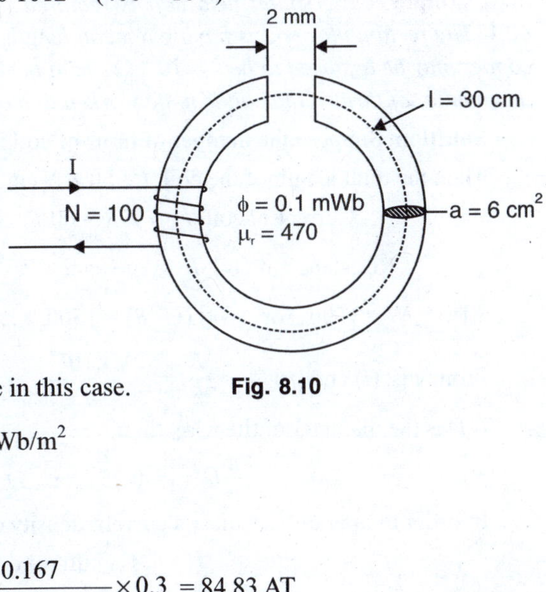

Solution. The conditions of the problem are represented in Fig. 8.10. It will be assumed that flux density in the air gap is equal to the flux density in the core *i.e.* fringing is neglected. This assumption is quite reasonable in this case.

Fig. 8.10

Flux density, $B = \dfrac{\phi}{a} = \dfrac{0.1 \times 10^{-3}}{6 \times 10^{-4}} = 0.167$ Wb/m^2

Ampere-turns required for iron will be :

$$AT_i = H_i \times l_i$$

$$= \dfrac{B}{\mu_0 \mu_r} \times l_i = \dfrac{0.167}{4\pi \times 10^{-7} \times 470} \times 0.3 = 84.83 \text{ AT}$$

Ampere-turns required for air will be :

$$AT_g = \dfrac{B}{\mu_0} \times l_g = \dfrac{0 \cdot 167}{4\pi \times 10^{-7}} \times (2 \times 10^{-3}) = 265 \cdot 8 \text{ AT}$$

∴   Total $AT = 265 \cdot 8 + 84 \cdot 83 = 350 \cdot 63$ AT

∴   Magnetising current, $I = 350 \cdot 63/N = 350 \cdot 63/100 =$ **3·51 A**

It may be seen that many more ampere-turns are required to produce the magnetic flux through 2 mm of air gap than through the iron part. This is expected because reluctance of air is much more than that of iron.

Example 8.4. *A circular iron ring has a mean circumference of 1·5 m and a cross-sectional area of 0·01 m^2. A saw-cut of 4 mm wide is made in the ring. Calculate the magnetising current required to produce a flux of 0·8 mWb in the air gap if the ring is wound with a coil of 175 turns. Assume relative permeability of iron as 400 and leakage factor 1·25.*

Solution. $\phi_g = 0.8 \times 10^{-3}$ Wb ; $a = 0.01$ m^2 ; $l_i = 1.5$ m ; $l_g = 4 \times 10^{-3}$ m

AT for air gap

$$B_g = \frac{\phi_g}{a} = \frac{0.8 \times 10^{-3}}{0.01} = 0.08 \text{ Wb/m}^2$$

$$H_g = \frac{B_g}{\mu_0} = \frac{0.08}{4\pi \times 10^{-7}} = 63662 \text{ AT/m}$$

$\therefore \qquad AT_g = H_g \times l_g = 63662 \times (4 \times 10^{-3}) = 254.6 \text{ AT}$

AT for iron path $\phi_i = \phi_g \times \lambda = 0.8 \times 10^{-3} \times 1.25 = 10^{-3}$ Wb

$B_i = \phi_i/a = 10^{-3}/0.01 = 0.1$ Wb/m^2

$$H_i = \frac{B_i}{\mu_0 \mu_r} = \frac{0.1}{4\pi \times 10^{-7} \times 400} = 199 \text{ AT/m}$$

$\therefore \qquad AT_i = H_i \times l_i = 199 \times 1.5 = 298.5 \text{ AT}$

$\therefore \qquad$ Total $AT = 254.6 + 298.5 = 553.1$ AT

$\therefore \qquad$ Magnetising current, $I = 553.1/N = 553.1/175 = $**3·16 A**

Example 8.5. *A shunt field coil is required to develop 1500 AT with an applied voltage of 60 V. The rectangular coil is having a mean length of 50 cm. Calculate the wire size. Resistivity of copper may be assumed to be 2×10^{-6} Ω-cm at the operating temperature of the coil. Estimate also the number of turns if the coil is to be worked at a current density of 3 A/mm^2.*

Solution. Suppose the number of turns of coil is N.

Then the total length of the coil, $l = 50 \times N$ cm

Current in coil, $I = V/R = 60/R$

Resistance of coil, $R = \rho \dfrac{l}{A} = 2 \times 10^{-6} \times \dfrac{50 \times N}{A} = \dfrac{N \times 10^{-4}}{A}$ ...(i)

Also $NI = 1500$ or $N \times (60/R) = 1500$ $\therefore$ $R = N/25$ ...(ii)

From eqs. (i) and (ii), $\dfrac{N}{25} = \dfrac{N \times 10^{-4}}{A}$ or $A = 25 \times 10^{-4}$ cm$^2 = 0.25$ mm^2

If D is the diameter of the wire, then,

$$\frac{\pi}{4} D^2 = 0.25 \quad \text{or} \quad D = \textbf{0·568 mm}$$

In order to operate the coil at a current density of 3 A/mm^2, the current in the coil is

$$I' = A \times \text{current density} = 0.25 \times 3 = 0.75 \text{ A}$$

$\therefore \qquad N'I' = 1500$ or $N' = 1500/I' = 1500/0.75 = $**2000**

Example 8.6. *An iron ring has a mean diameter of 15 cm, a cross-section of 20 cm^2 and a radial gap of 0·5 mm cut in it. It is uniformly wound with 1500 turns of insulated wire and a magnetising current of 1 A produces a flux of 1 mWb. Neglecting the effect of magnetic leakage and fringing, calculate (i) reluctance of the magnetic circuit, (ii) relative permeability of iron and (iii) inductance of the winding.*

Solution.(ii) $a = 20 \times 10^{-4}$ m^2 ; $l_i = \pi \times 0.15 = 0.471$ m ; $l_g = 0.5 \times 10^{-3}$ m

Flux density in air gap, $B = \dfrac{\phi}{a} = \dfrac{1 \times 10^{-3}}{20 \times 10^{-4}} = 0.5$ Wb/m^2

Magnetising force in air gap, $H_g = B/\mu_0 = 0.5/4\pi \times 10^{-7} = 398 \times 10^3$ AT/m

Ampere-turns for air gap, $AT_g = H_g \times l_g = (398 \times 10^3) \times 0.5 \times 10^{-3} = 199$ AT

Total AT provided $= NI = 1500 \times 1 = 1500$ AT

$\therefore$ AT available for iron part, $AT_i = 1500 - 199 = 1301$ AT

Magnetising force in iron, $H_i = \dfrac{AT_i}{l_i} = \dfrac{1301}{0.471} = 2762$ AT/m

Now, $B = \mu_0 \mu_r H_i$

$\therefore$ $\mu_r = \dfrac{B}{\mu_0 H_i} = \dfrac{0.5}{4\pi \times 10^{-7} \times 2762} \doteqdot \mathbf{144}$

(i)   Reluctance of air gap $= \dfrac{l_g}{a\mu_0} = \dfrac{0.5 \times 10^{-3}}{(20 \times 10^{-4}) \times 4\pi \times 10^{-7}} = 1.99 \times 10^5$ AT/Wb

Reluctance of iron part $= \dfrac{l_i}{a\mu_0\mu_r} = \dfrac{0.471}{(20 \times 10^{-4}) \times 4\pi \times 10^{-7} \times 144} = 13.01 \times 10^5$ AT/Wb

$\therefore$ Total circuit reluctance $= 10^5 (1.99 + 13.01) = \mathbf{15 \times 10^5 \ AT/Wb}$

(iii) Inductance of winding $= \dfrac{N\phi}{I} = \dfrac{(1500) \times (1 \times 10^{-3})}{1} = \mathbf{1.5 \ H}$

Example 8.7. *A magnetic circuit is constructed as shown in Fig. 8.11. Both sections A and B are of 20 mm by 20 mm square cross-section and the mean dimensions are 100 mm by 80 mm. The relative permeability of section A is 250 and of section B is 500. Find the reluctance of each section and the total circuit reluctance.*

If the joints between sections A and B have an air gap of 0·5 mm at each joint, find the total reluctance of the circuit.

Solution. The conditions of the problem are represented in Fig. 8.11. The area of X-section of the core, $a = 20 \times 20 = 400$ mm$^2 = 4 \times 10^{-4}$ m^2.

Section A

Length of magnetic path, $l_A = 80 + 10 + 10 = 100$ mm $= 0.1$ m

$\therefore$ Reluctance of section $A = \dfrac{l_A}{a\mu_0\mu_r} = \dfrac{0.1}{(4 \times 10^{-4}) \times 4\pi \times 10^{-7} \times 250} = \mathbf{0.796 \times 10^6 \ AT/Wb}$

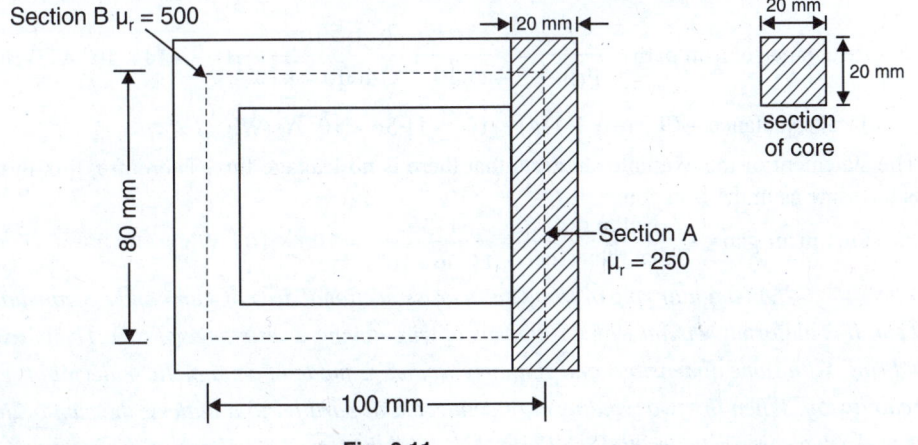

Fig. 8.11

Section B

Length of magnetic path, $l_B = 80 + 90 + 90 = 260$ mm $= 0.26$ m

$\therefore$ Reluctance of section $B = \dfrac{l_B}{a\mu_0\mu_r} = \dfrac{0.26}{(4\times10^{-4})\times4\pi\times10^{-7}\times500} = \mathbf{1.035 \times 10^6\ AT/Wb}$

$\therefore$ Total circuit reluctance $= 10^6 (0.796 + 1.035) = \mathbf{1.831 \times 10^6\ AT/Wb}$

Regarding the second part of the problem, the total length of air gaps is $l_g = 2 \times 0.5 = 1$ mm $= 0.001$ m.

$\therefore$ Reluctance of air gaps $= \dfrac{l_g}{a\mu_0} = \dfrac{0.1}{(4\times10^{-4})\times4\pi\times10^{-7}} = 1.99 \times 10^6\ AT/Wb$

$\therefore$ Total circuit reluctance $= 10^6 (1.831 + 1.99) = \mathbf{3.821 \times 10^6\ AT/Wb}$

The reader may note that the reluctance of even small air gaps is very large. It is very important, therefore, that the joints of magnetic circuits — for example, the core of a transformer — should be tightly bolted together.

Note. The air gap is very small. Therefore, the magnetic length of iron part is the same in the two cases.

Example 8.8. *A rectangular iron core is shown in Fig. 8.12. It has a mean length of magnetic path of 100 cm, cross-section of 2 cm × 2 cm, relative permeability of 1400 and an air gap of 5 mm cut in the core. The three coils carried by the core have number of turns $N_a = 335$, $N_b = 600$ and $N_c = 600$ and the respective currents are 1·6 A, 4 A and 3 A. The directions of the currents are as shown in Fig. 8.12. Find the flux in the air gap.*

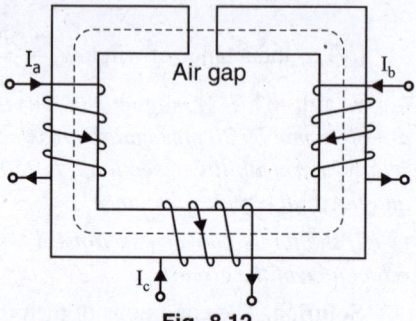

Fig. 8.12

Solution. By applying right-hand rule for the coil, it is easy to see that fluxes produced by currents I_a and I_b are in the clockwise direction through the iron core while the flux produced by current I_c is in the anticlockwise direction through the core.

$\therefore$ Net m.m.f. $= N_a I_a + N_b I_b - N_c I_c = 335 \times 1.6 + 600 \times 4 - 600 \times 3 = 1136$ AT

Reluctance of air gap $= \dfrac{l_g}{\mu_0 a} = \dfrac{5\times10^{-3}}{4\pi\times10^{-7}\times4\times10^{-4}} = 9.946 \times 10^6\ AT/Wb$

Reluctance of iron path $= \dfrac{l_i}{\mu_0\mu_r a} = \dfrac{(100 - 0.5)\times10^{-2}}{4\pi\times10^{-7}\times1400\times4\times10^{-4}} = 1.414 \times 10^6\ AT/Wb$

$\therefore$ Total reluctance $= (9.946 + 1.414) \times 10^6 = 11.36 \times 10^6\ AT/Wb$

The statement of the example suggests that there is no leakage flux. Therefore, flux in the air gap is the same as in the iron core.

$\therefore$ Flux in air gap $= \dfrac{\text{Net m.m.f.}}{\text{Total reluctance}} = \dfrac{1136}{11.36\times10^6} = 100 \times 10^{-6}\ Wb = \mathbf{100\ \mu Wb}$

Example 8.9. *An angular ring of wood has a cross-sectional area of 4 cm^2 and a mean diameter of 30 cm. It is uniformly wound with 1200 turns of wire having a resistance of 6 Ω. The core of the second ring, with same dimensions and similarly wound, is made of a magnetic material of relative permeability 50. When the two windings are connected in parallel to a battery, the sum of the two fluxes in the two cores is 0·2 mWb [See Fig. 8.13]. Find the terminal voltage of the battery.*

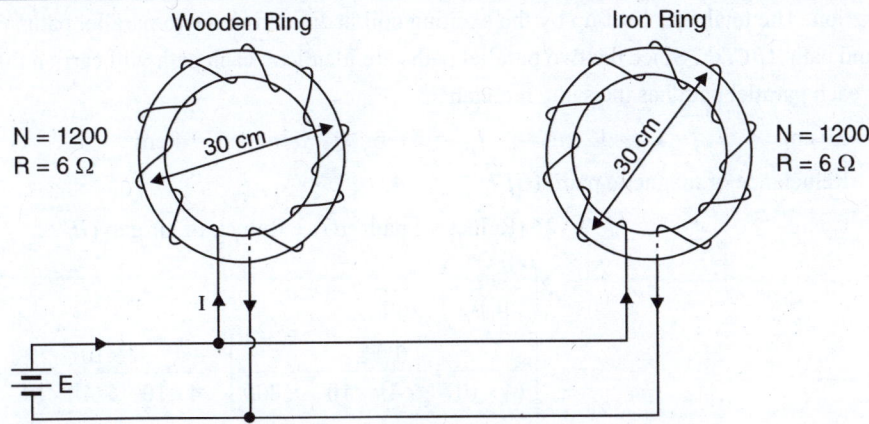

Fig. 8.13

Solution. The windings will carry the same current I as their resistances are equal. Moreover, each ring has the same mean magnetic length $l = \pi \times 0.3 = 0.942$ m.

Wooden ring. Reluctance $= \dfrac{l}{a\mu_0\mu_r} = \dfrac{0.942}{(4\times10^{-4})\times4\pi\times10^{-7}\times1} = 18.74 \times 10^8$ AT/Wb

Now, m.m.f. = flux × reluctance

∴ Flux in wooden ring, $\phi_1 = \dfrac{\text{m.m.f.}}{\text{reluctance}} = \dfrac{1200I}{18.74\times10^8} = 6.4 \times 10^{-7}\,I$ Wb

Iron ring. Reluctance $= \dfrac{l}{a\mu_0\mu_r} = \dfrac{0.942}{(4\times10^{-4})\times4\pi\times10^{-7}\times50} = 0.375 \times 10^8$ AT/Wb

∴ Flux in the iron ring, $\phi_2 = \dfrac{1200I}{0.375\times10^8} = 320 \times 10^{-7}\,I$ Wb

∴ Total flux in the two rings $= (6.4 + 320)\,10^{-7}\,I = 326.4 \times 10^{-7}\,I$ Wb

But the sum of two fluxes in the rings is given to be 0.2×10^{-3} Wb.

∴ $326.4 \times 10^{-7}\,I = 0.2 \times 10^{-3}$ or $I = \dfrac{0.2\times10^{-3}}{326.4\times10^{-7}} = 6.13$ A

∴ Battery terminal voltage $= IR = 6.13 \times 6 =$ **36·78 V**

Example 8.10. *In the magnetic circuit shown in Fig. 8.14, find (i) the total reluctance of the magnetic circuit and (ii) value of flux linking the coil. Assume that the relative permeability of the magnetic material is 800. The exciting coil has 1000 turns and carries a current of 1·25 A.*

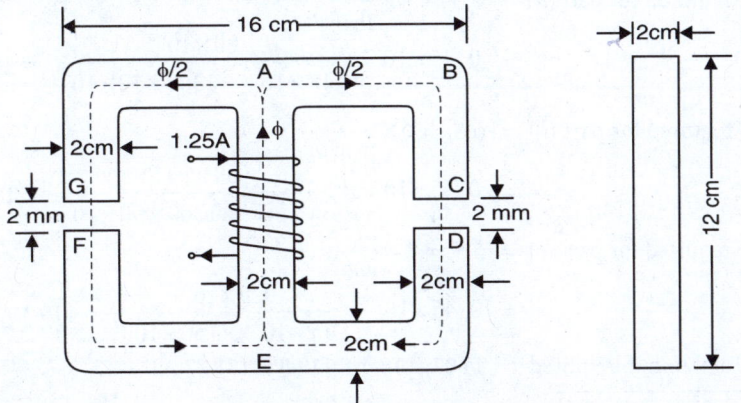

Fig. 8.14

Solution. The total flux ϕ set up by the exciting coil is divided into two parallel paths *viz.* path *AGFE* and path *ABCDE*. Since the two parallel paths are identical, each path will carry a flux = $\phi/2$ and that each parallel path has the same reluctance.

$$l_{AE} = 10 \text{ cm} \quad ; \quad l_{AG} = l_{FE} = 12 \text{ cm} \quad ; \quad l_{GF} = 2 \text{ mm} \quad ; \quad a = 2 \times 2 = 4 \text{ cm}^2$$

(*i*)   Reluctance of magnetic path *AGFE*

$$= 2^* \text{ (Reluct. of path } AG) + \text{Reluct. of air gap } GF$$

$$= 2\left(\frac{l_{AG}}{a\mu_0\mu_r}\right) + \frac{l_{GF}}{a\mu_0}$$

$$= 2\left[\frac{0 \cdot 12}{(4 \times 10^{-4}) \times 4\pi \times 10^{-7} \times 800}\right] + \frac{2 \times 10^{-3}}{4 \times 10^{-4} \times 4\pi \times 10^{-7}}$$

$$= 5 \cdot 968 \times 10^5 + 39 \cdot 788 \times 10^5 = 45 \cdot 756 \times 10^5 \text{ AT/Wb}$$

Reluctance of magnetic path *AE* will be

$$= \frac{l_{AE}}{a\mu_0\mu_r} = \frac{0 \cdot 1}{(4 \times 10^{-4}) \times 4\pi \times 10^{-7} \times 800} = 2 \cdot 486 \times 10^5 \text{ AT/Wb}$$

Total reluctance of magnetic circuit will be

$$= 45 \cdot 75 \times 10^5 + 2 \cdot 486 \times 10^5 = \mathbf{48 \cdot 242 \times 10^5 \ AT/Wb}$$

(*ii*)   m.m.f. = flux × reluctance

or $1000 \times 1 \cdot 25 = \phi \times (48.242 \times 10^5)$

$\therefore$ $\phi = \dfrac{1000 \times 1 \cdot 25}{48 \cdot 242 \times 10^5} = \mathbf{25 \cdot 9 \times 10^{-5} \ Wb}$

Example 8.11. *A magnetic circuit consists of three parts in series, each of uniform cross-sectional area. They are :*

(a) a length of 80 mm and cross-sectional area 50 mm^2

(b) a length of 60 mm and cross-sectional area 90 mm^2

(c) an air gap of length 0·5 mm and cross-sectional area 150 mm^2.

A coil of 4000 turns is wound on part (b) and the flux density in the air gap is 0·3 Wb/m^2. Assuming that all the flux passes through the given circuit, and that relative permeability μ_r is 1300, estimate the coil current to produce such a flux density.

Solution. Flux in the circuit, $\phi = B_g \times a_g = 0 \cdot 3 \times 1 \cdot 5 \times 10^{-4} = 0 \cdot 45 \times 10^{-4} \text{ Wb/m}^2$

m.m.f. required for part (*a*) $= \phi S_a = \phi \times \dfrac{l_a}{\mu_0 \mu_r a_a}$

$$= 0 \cdot 45 \times 10^{-4} \times \frac{80 \times 10^{-3}}{4\pi \times 10^{-7} \times 1300 \times 50 \times 10^{-6}} = 44 \cdot 07 \text{ AT}$$

m.m.f. required for part (*b*) $= \phi S_b = \phi \times \dfrac{l_b}{\mu_0 \mu_r a_b}$

$$= 0 \cdot 45 \times 10^{-4} \times \frac{60 \times 10^{-3}}{4\pi \times 10^{-7} \times 1300 \times 90 \times 10^{-6}} = 18 \cdot 4 \text{ AT}$$

m.m.f. required for part (*c*) $= \phi S_c = \phi \times \dfrac{l_c}{\mu_0 a_c}$

$$= 0 \cdot 45 \times 10^{-4} \times \frac{0 \cdot 5 \times 10^{-3}}{4\pi \times 10^{-7} \times 150 \times 10^{-6}} = 119 \cdot 3 \text{ AT}$$

Total m.m.f. required $= 44 \cdot 07 + 18 \cdot 4 + 119 \cdot 3 = 181 \cdot 77 \text{ AT}$

---

*   Reluctance of path *AG* = Reluctance of path *FE*

$\therefore$ $NI = 181{\cdot}77$ or $I = 181{\cdot}77/N = 181.77/4000 = 45{\cdot}4 \times 10^{-3}\,A = \mathbf{45{\cdot}4\ mA}$

Since the absolute permeability of air (μ_0) is much smaller than that of a ferromagnetic material, the value of reluctance of air gap ($= l_g/a_g\mu_0$) is much greater than the reluctance of adjacent magnetic material ($= l_i/a_i\mu_0\mu_r$). Therefore, the m.m.f. required to force flux through the air gap can be quite large.

Example 8.12. *A laminated soft-iron ring has a mean circumference of 600 mm, cross-sectional area 500 mm^2 and has a radial air gap of 1 mm cut through it. It is wound with a coil of 1000 turns. Estimate the current in the coil to produce a flux of 0·5 mWb in the air gap assuming :*

> *(i) the relative permeability of the soft iron is 1000, (ii) the leakage factor is 1·2, (iii) fringing is negligible, (iv) the space factor is 0·9.*

Solution. **AT for air-gap**

$$\phi_g = 0{\cdot}5\ \text{mWb} = 5 \times 10^{-4}\ \text{Wb}\ \ ;\ \ l_g = 1 \times 10^{-3}\ \text{m}\ \ ;\ \ a_g = 500 \times 10^{-6}\ \text{m}^2$$

$$\text{m.m.f. for air gap} = \phi_g S_g = \phi_g \times \frac{l_g}{\mu_0 a_g}$$

$$= 5 \times 10^{-4} \times \frac{1 \times 10^{-3}}{4\pi \times 10^{-7} \times 500 \times 10^{-6}} = 795{\cdot}7\ \text{AT}$$

AT for iron part

$$\phi_i = \phi_g \times 1{\cdot}2^* = 5 \times 10^{-4} \times 1{\cdot}2\ \text{Wb}\ ;\ l_i = 600 \times 10^{-3}\ \text{m}\ ;\ a_i = 500 \times 10^{-6} \times 0{\cdot}9^{**}\text{m}^2$$

$\therefore$ $\text{m.m.f. for iron part} = \phi_i S_i = \phi_i \times \dfrac{l_i}{\mu_0 \mu_r a_i}$

$$= 5 \times 10^{-4} \times 1{\cdot}2 \times \frac{600 \times 10^{-3}}{4\pi \times 10^{-7} \times 1000 \times 500 \times 10^{-6} \times 0{\cdot}9}$$

$$= 636{\cdot}6\ \text{AT}$$

$\therefore$ Total m.m.f. required $= 795{\cdot}7 + 636{\cdot}6 = 1432{\cdot}3\ \text{AT}$

Now $NI = 1432{\cdot}3$ $\therefore$ $I = 1432{\cdot}3/N = 1432.3/1000 = \mathbf{1{\cdot}432\ A}$

Note that *AT* for air-gap are comparable to that for iron part. It is because length of air gap is very small.

Example 8.13. *The ring-shaped core shown in Fig. 8.15 is made of material having relative permeability 1000. The flux density in the thicker section is 1·5 T. If the current through the coil is not to exceed 0·5 A, find the number of turns of the coil.*

Solution. The statement of the problem suggests that flux in the thicker as well as in thin section is the same *i.e.* it is a series magnetic circuit.

Flux in the magnetic circuit is

$$\phi = 1{\cdot}5 \times 6 \times 10^{-4}$$

$$= 9 \times 10^{-4}\ \text{Wb}$$

AT for thick section

$$H_1 = \frac{B_1}{\mu_0 \mu_r} = \frac{1{\cdot}5}{4\pi \times 10^{-7} \times 1000} = 1194\ \text{AT/m}$$

m.m.f. for thick section $= H_1 l_1 = (1194) \times (10 \times 10^{-2})$

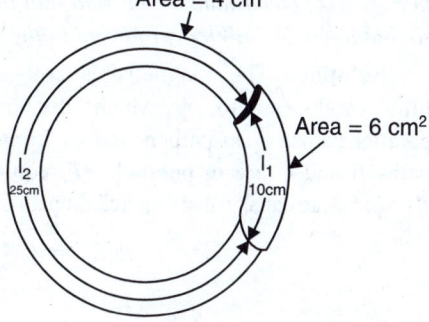

Area = 4 cm^2

Area = 6 cm^2

l_2
25cm

l_1
10cm

Fig. 8.15

---

* The leakage factor refers to the flux leakage in the iron part of the magnetic circuit.

$$\text{Leakage factor} = \frac{\text{Total flux}}{\text{Useful flux}}$$

** $$\text{Space factor} = \frac{\text{Useful area}}{\text{Total area}}$$

$$= 119{\cdot}4 \text{ AT}$$

AT for thin section $\quad B_2 = \dfrac{\phi}{a} = \dfrac{9\times10^{-4}}{4\times10^{-4}} = 2{\cdot}25 \text{ T}$

$$H_2 = \frac{B_2}{\mu_0\mu_r} = \frac{2{\cdot}25}{4\pi\times10^{-7}\times1000} = 1790 \text{ AT/m}$$

m.m.f. for thin section $= H_2\, l_2 = (1790)\times(25\times10^{-2}) = 448 \text{ AT}$

∴   Total m.m.f. required $= 119{\cdot}4 + 448 = 567{\cdot}4 \text{ AT}$

Now $NI = 567{\cdot}4 \quad \text{or} \quad N = 567{\cdot}4/I = 567{\cdot}4/0{\cdot}5 = \mathbf{1135}$

Example 8.14. *A steel ring 30 cm mean diameter and of circular section 2 cm in diameter has an air gap 1 mm long. It is wound uniformly with 600 turns of wire carrying current of 2·5 A. Find (i) total m.m.f., (ii) total reluctance and (iii) flux. Neglect magnetic leakage. The iron path takes 40% of the total m.m.f.*

Solution. (*i*)   Total m.m.f. $= NI = 600 \times 2{\cdot}5 = \mathbf{1500\ AT}$

(*ii*) Let M_1 and M_2 be the m.m.f. for iron part and air gap respectively and S_1 and S_2 their corresponding reluctances.

$$M_1 = 40\% \text{ of } 1500 = (40/100)\times1500 = 600 \text{ AT}$$
$$M_2 = 1500 - 600 = 900 \text{ AT}$$

Now, $M_1 = \phi S_1$ and $M_2 = \phi S_2$

∴ $\dfrac{S_1}{S_2} = \dfrac{M_1}{M_2} = \dfrac{600}{900} = 0{\cdot}67$

$$S_2 = \frac{l_g}{a\mu_0} = \frac{1\times10^{-3}}{\pi(1\times10^{-2})^2\times4\pi\times10^{-7}} = 2{\cdot}5\times10^6 \text{ AT/Wb}$$

∴ $S_1 = 0{\cdot}67 S_2 = 0{\cdot}67\times(2{\cdot}5\times10^6) = 1{\cdot}675\times10^6 \text{ AT/Wb}$

Total reluctance $= S_1 + S_2 = (1{\cdot}675 + 2{\cdot}5)\times10^6 = \mathbf{4{\cdot}175\times10^6\ AT/Wb}$

(*iii*)   Flux $= \dfrac{\text{Total m.m.f.}}{\text{Total reluctance}} = \dfrac{1500}{4{\cdot}175\times10^6}$

$$= 0{\cdot}36\times10^{-3} \text{ Wb} = \mathbf{0{\cdot}36\ mWb}$$

Example 8.15. *A cast steel magnetic structure made of a bar of section 2 cm × 2 cm is shown in Fig. 8.16. Determine the current that the 500 turn magnetising coil on the left limb should carry so that a flux of 2mWb is produced in the right limb. Take $\mu_r = 600$ and neglect leakage.*

Solution. The magnetising coil on the left limb produces flux ϕ which divides into two parallel paths; ϕ_1 in path B and ϕ_2 in path C. Since paths B and C are in parallel, AT required for path $B\,(=\phi_1 S_B)$ are equal to that required for path $C\,(=\phi_2 S_c)$ i.e.

$$\phi_1 S_B = \phi_2 S_c$$

or $\phi_1 \times \dfrac{l_B}{\mu_0\mu_r a} = \phi_2 \times \dfrac{l_C}{\mu_0\mu_r a}$

∴ $\phi_1 = \phi_2\times\dfrac{l_C}{l_B} = 2\times\dfrac{25}{15} = \dfrac{10}{3} \text{mWb}$ $(\because \phi_2 = 2\text{mWb})$

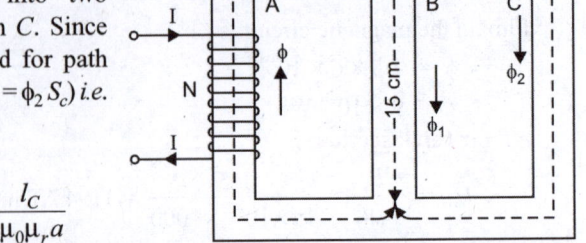

Fig. 8.16

Total flux in path A, $\phi = \phi_1 + \phi_2 = \dfrac{10}{3} + 2 = \dfrac{16}{3} \text{mWb}$

Total AT required for the whole magnetic circuit are equal to the sum of (i) AT required for path A and (ii) AT required for one of the parallel paths B or C.

$$\text{Flux density in path } A, B_A = \frac{\phi}{a} = \frac{(16/3)\times10^{-3}}{4\times10^{-4}} = \frac{40}{3}\,\text{Wb/m}^2$$

$$AT \text{ required for path } A = \frac{B_A}{\mu_0\mu_r}\times l_A \times = \frac{(40/3)}{4\pi\times10^{-7}\times600}\times0.25 = 4420\,\text{AT}$$

$$\text{Flux density in path } B, B_B = \frac{\phi_1}{a} = \frac{(10/3)\times10^{-3}}{4\times10^{-4}} = 8.33\,\text{Wb/m}^2$$

$$AT \text{ required for path } B = \frac{B_B}{\mu_0\mu_r}\times l_B = \frac{8.33}{4\pi\times10^{-7}\times600}\times0.15 = 1658\,\text{AT}$$

$$\therefore \quad \text{Total } AT \text{ required} = 4420 + 1658 = 6078\,\text{AT}$$

Now, $$NI = 6078 \quad \therefore I = \frac{6078}{N} = \frac{6078}{500} = \textbf{12.16 A}$$

Example 8.16. *A magnetic core made of annealed sheet steel has the dimensions as shown in Fig. 8.17. The X-section is 25 cm² everywhere. The flux in branches A and B is 3500 µWb but that in the branch C is zero. Find the required ampere-turns for coil A and for coil C. Relative permeability of sheet steel is 1000.*

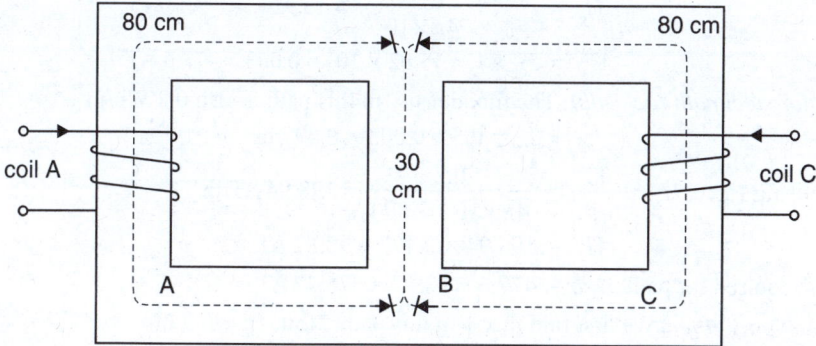

80 cm 80 cm

coil A 30 cm coil C

A B C

Fig. 8.17

Solution. AT for coil A. Flux paths B and C are in parallel. Therefore, AT required for coil A is equal to AT for path A plus AT for path B or path C.

$$AT \text{ for path } A = \text{ flux} \times \text{reluctance} = (3500\times10^{-6})\times\frac{0.8}{(25\times10^{-4})\times4\pi\times10^{-7}\times1000} = 891.3\,\text{AT}$$

$$AT \text{ for path } B = \text{ flux} \times \text{reluctance} = (3500\times10^{-6})\times\frac{0.3}{(25\times10^{-4})\times4\pi\times10^{-7}\times1000} = 334.2\,\text{AT}$$

Total AT for coil A = 891·3 + 334·2 = **1225·5 AT**

AT for coil C. The coil C produces flux $\phi_C\,\mu$Wb in the opposite direction to that produced by coil A.

$$\text{m.m.f. of path } B = \text{m.m.f. of path } C$$
$$\phi_B S_B = \phi_C S_C$$

or $$(3500\times10^{-6})\times\frac{l_B}{a\mu_0\mu_r} = \phi_C\times\frac{l_C}{a\mu_0\mu_r}$$

$$\therefore \quad \phi_C = (3500\times10^{-6})\times l_B/l_C = (3500\times10^{-6})\times0.3/0.8 = 1312.5\,\mu\text{Wb}$$

Total AT for coil C = $\phi_C \times$ reluctance

$$= (1312.5\times10^{-6})\times\frac{0.8}{(25\times10^{-4})\times4\pi\times10^{-7}\times1000} = \textbf{334.22 AT}$$

Example 8.17. *A magnetic circuit is shown in Fig. 8.18. It is made of cast steel 0.05 m thick. The length of air gap is 0.003 m. Find the m.m.f. to establish a flux of 5×10^{-4} Wb in the air gap. The relative permeability for the material is 800.*

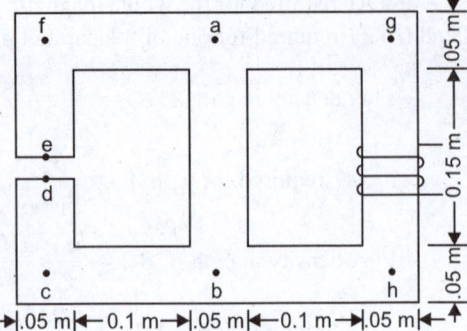

Fig. 8.18

Solution. The flux ϕ set up by the current-carrying coil in the path *bhga* divides into two parallel paths *viz* path *ab* and path *aedb*. Therefore, total m.m.f. required is equal to *AT* required for path *bhga* plus *AT* required for one of the parallel paths (*i.e.* path *aedb* or path *ab*) *i.e.*

Total m.m.f = *AT* for path *aedb* + *AT* for path *bhga*

1. AT for path aedb. The m.m.f. required for this path is equal to *AT* required for air gap *ed* *plus* *AT* required for steel path (*ae* + *db*)

(i) AT for air gap. $\phi_g = 5 \times 10^{-4}$ Wb ; $a_g = 0.05 \times 0.05 = 0.0025 \ m^2$; $l_g = 0.003$ m

$$\therefore \qquad B_g = \frac{\phi_g}{a_g} = \frac{5 \times 10^{-4}}{0.0025} = 0.2 \ Wb/m^2$$

Now,
$$H_g = \frac{B_g}{\mu_0} = \frac{0.2}{4\pi \times 10^{-7}} = 15.92 \times 10^4 \ AT/m$$

$$\therefore \qquad AT_g = H_g \times l_g = 15.92 \times 10^4 \times 0.003 = 477.6 \ AT$$

(ii) AT for steel path (ae + db). The flux density in this path is also 0.2 Wb/m².

$$l_{ae} + l_{bd} = 0.5 - 0.003 = 0.497 \ m$$

Magnetising force, $H_s = \dfrac{0.2}{\mu_0 \mu_r} = \dfrac{0.2}{4\pi \times 10^{-7} \times 800} = 198.94 \ AT/m$

$$\therefore \qquad AT_s = 198.94 \times 0.497 = 98.87 \ AT$$

$\therefore$ *AT* required for path *aedb* = 477.6 + 98.87 = 576.47 AT = AT_{ab}

2. AT for path bhga. We first find flux ϕ in this path. Now, $l_{ab} = 0.2$ m.

Also,
$$AT_{ab} = 576.47 \ AT \ ... \ Calculated \ above$$

$$\text{Flux density, } B_{ab} = \frac{AT_{ab} \times \mu_0 \mu_r}{l_{ab}} = \frac{576.47 \times 4\pi \times 10^{-7} \times 800}{0.2} = 2.898 \ Wb/m^2$$

$$\text{Flux, } \phi_{ab} = B_{ab} \times a = 2.898 \times 0.0025 = 0.007245 \ Wb$$

$$\therefore \qquad \phi = \phi_g + \phi_{ab} = 5 \times 10^{-4} + 0.007245 = 0.007745 \ Wb$$

$$\text{Flux density in path } bhga = \frac{\phi}{a} = \frac{0.007745}{0.0025} = 3.098 \ Wb/m^2$$

$$\text{Magnetising force, } H = \frac{3.098}{\mu_0 \mu_r} = \frac{3.098}{4\pi \times 10^{-7} \times 800} = 3081.63 \ AT/m$$

Length of path *bhga*, $l = 0.5$ m

AT for path *bhga* = $H \times l = 3081.63 \times 0.5 = 1540.815$ AT

$\therefore$ Total m.m.f. required = 576.47 + 1540.815 = **2117.285 AT**

Example 8.18. *The magnetic core shown in Fig. 8.19 has the following dimensions :*

$l_1 = 10 \ cm$; $l_2 = l_3 = 18 \ cm$; *cross-sectional area of l_1 path* $= 6.25 \times 10^{-4} \ m^2$; *cross-sectional areas of l_2 and l_3 paths* $= 3 \times 10^{-4} \ m^2$; *length of air gap, $l_4 = 2mm$.*

Determine the current that must be passed through the 600-turn coil to produce a total flux of 100 μWb in the air gap. Assume that the metal has relative permeability of 800.

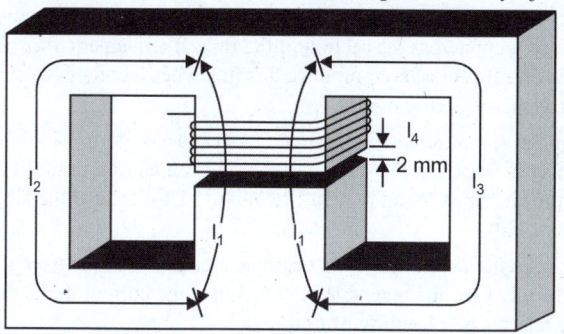

Fig. 8.19

Solution. $\phi_g = 100 \; \mu Wb = 100 \times 10^{-6}$ Wb ; $a_g = 6.25 \times 10^{-4} \; m^2$

AT for air gap.
$$B_g = \frac{\phi_g}{a_g} = \frac{100 \times 10^{-6}}{6.25 \times 10^{-4}} = 0.16 \; Wb/m^2$$

Now,
$$H_g = \frac{B_g}{\mu_0} = \frac{0.16}{4\pi \times 10^{-7}} = 1.27 \times 10^5 \; AT/m$$

$\therefore$
$$AT_g = H_g \times l_g = 1.27 \times 10^5 \times 2 \times 10^{-3} = 254 \; AT$$

AT for path l_1.
$$B_1 = 0.16 \; Wb/m^2 \; ; \; l_1 = 10 \times 10^{-2} \; m$$

Now,
$$H_1 = \frac{B_1}{\mu_0 \mu_r} = \frac{0.16}{4\pi \times 10^{-7} \times 800} = 159 \; AT/m$$

$\therefore$
$$AT_1 = H_1 \times l_1 = 159 \times 10 \times 10^{-2} = 15.9 \; AT$$

Here, we neglect l_g, being very small, compared to iron path. Paths l_2 and l_3 are similar so that total flux (= 100×10^{-6} Wb) divides equally between these two paths. Since paths l_2 and l_3 are in parallel, it is necessary to consider m.m.f. for only one of them. Let us find AT for path l_2.

AT for path l_2.
$$\phi_2 = 50 \times 10^{-6} \; Wb \; ; \; \mu_r = 800 \; ; \; l_2 = 18 \times 10^{-2} \; m$$

$\therefore$
$$B_2 = \frac{\phi_2}{a} = \frac{50 \times 10^{-6}}{3 \times 10^{-4}} = 0.167 \; Wb/m^2$$

Now,
$$H_2 = \frac{B_2}{\mu_0 \mu_r} = \frac{0.167}{4\pi \times 10^{-7} \times 800} = 166 \; AT/m$$

$\therefore$
$$AT_2 = H_2 \times l_2 = 166 \times 18 \times 10^{-2} = 29.9 \; AT$$

$\therefore$
$$\text{Total } AT = 254 + 15.9 + 29.9 = 300 \; AT$$

Now,
$$NI = 300 \quad \text{or} \quad I = \frac{300}{N} = \frac{300}{600} = 0.5 \; A = \textbf{500 mA}$$

Tutorial Problems

1. It is required to produce a flux density of $0.6 \; Wb/m^2$ in an air gap having a length of 8 mm. Calculate the m.m.f. required. **[480 × 10³ AT/m]**

2. A coil of 200 turns is wound uniformly over a wooden ring having a mean circumference of 60 cm and a uniform cross-sectional area of 5 cm². If the current through the coil is 4A, calculate (*i*) the magnetising force (*ii*) the flux density and (*iii*) the total flux. **[(*i*) 1333 AT/m (*ii*) 1675 μWb/m² (*iii*) 0·8375 μWb]**

3. A core forms a closed magnetic loop of path length 32 cm. Half of this path has a cross-sectional area of 2 cm² and relative permeability 800. The other half has a cross-sectional area of 4 cm² and relative

permeability 400. Find the current needed to produce a flux of 0·4 Wb in the core if it is wound with 1000 turns of insulated wire. Ignore leakage and fringing effects.   **[636·8 A]**

4. An iron ring has a cross-sectional area of 400 mm^2 and a mean diameter of 250 mm. An air gap of 1 mm has been made by a saw-cut across the section of the ring. If a magnetic flux of 0·3 mWb is required in the air gap, find the current necessary to produce this flux when a coil of 400 turns is wound on the ring. The iron has a relative permeability of 500.   **[3·84 A]**

5. An iron ring has a mean circumferential length of 60 cm and a uniform winding of 300 turns. An air gap has been made by a saw-cut across the section of the ring. When a current of 1 A flows through the coil, the flux density in the air gap is found to be 0·126 Wb/m^2. How long is the air gap ? Assume iron has a relative permeability of 300.   **[1 mm]**

6. An iron magnetic circuit has a uniform cross-sectional area of 5 cm^2 and a length of 25 cm. A coil of 120 turns is wound uniformly over the magnetic circuit. When the current in the coil is 1·5 A, the total flux is 0·3 Wb. Find the relative permeability of iron.   **[663]**

7. The uneven ring-shaped core shown in Fig. 8.20 has $\mu_r = 1000$ and the flux density in the thicker section is to be 0.75 T. If the current through a coil wound on the core is to be 500 mA, determine number of coil turns required.   **[567]**

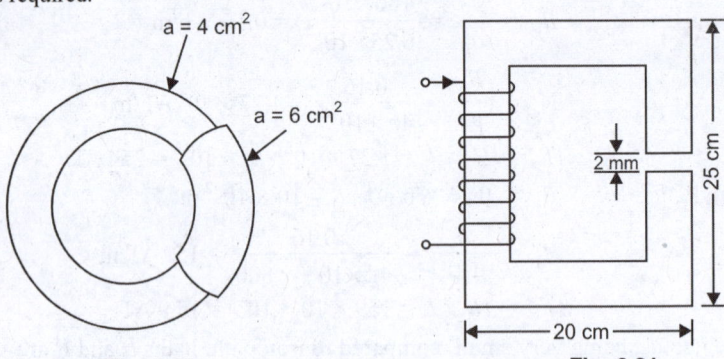

Fig. 8.20   **Fig. 8.21**

8. A rectangular magnetic core shown in Fig. 8.21 has square cross section of area 16 cm^2. An air gap of 2 mm is cut across one of its limbs. Find the exciting current needed in the coil having 1,000 turns wound on the core to create an air-gap flux of 4 mWb. The relative permeability of the core is 2000. **[4.713 A]**

9. The magnetic circuit of Fig. 8.22 is energised by a current of 3A. If the coil has 1500 turns, find the flux produced in the air gap. The relative permeability of the core material is 3000.   **[65.25 × 10^{-4} Wb]**

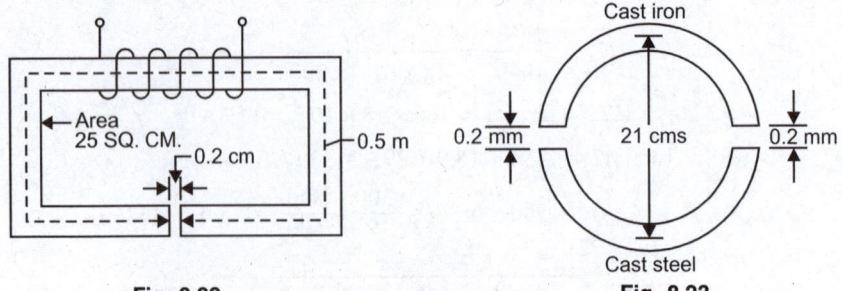

Fig. 8.22   **Fig. 8.23**

10. A ring [See Fig. 8.23] has a diameter of 21 cm and a cross-sectional area of 10 cm^2. The ring is made up of semicircular sections of cast iron and cast steel with each joint having a reluctance equal to an air gap of 0.2 mm. Find the ampere turns required to produce a flux of 8 × 10^{-4} Wb. The relative permeabilities of cast steel and cast iron are 800 and 166 respectively. Neglect leakage and fringing effects. **[1783 AT]**

8.11. B-H Curve

The *B-H* curve (or magnetisation curve) indicates the manner in which the flux density (*B*) varies with the magnetising force (*H*).

(i) **For non-magnetic materials.** For non-magnetic materials (*e.g.* air, copper, rubber, wood *etc.*), the relation between B and H is given by ;

$$B = \mu_0 H$$

Since $\mu_0 \,(= 4\pi \times 10^{-7} \text{H/m})$ is constant,

$$\therefore \qquad B \propto H$$

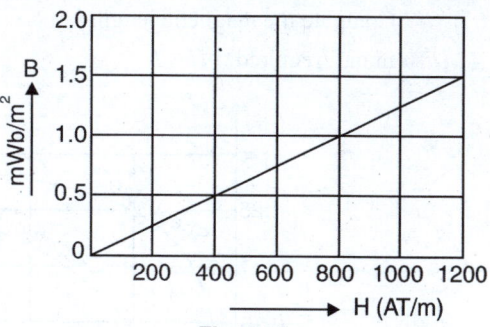

Fig. 8.24

Hence, the *B-H* curve of a non-magnetic material is a straight line passing through the origin as shown in Fig. 8.24. Two things are worth noting. First, the curve never saturates no matter how great the flux density may be. Secondly, a large m.m.f. is required to produce a given flux in the non-magnetic material *e.g.* air.

(ii) **For magnetic materials.** For magnetic materials (*e.g.* iron, steel *etc.*), the relation between B and H is given by ;

$$B = \mu_0 \,\mu_r \, H$$

Unfortunately, μ_r is not constant but varies with the flux density. Consequently, the *B-H* curve of a magnetic material is not linear. Fig. 8.25 *(i)* shows the general *shape of *B-H* curve of a magnetic material. The non-linearity of the curve indicates that relative permeability $\mu_r \,(= B/\mu_0 H)$ of a material is not constant but depends upon the flux density. Fig. 8.25 *(ii)* shows how relative permeability μ_r of a magnetic material (cast steel) varies with flux density.

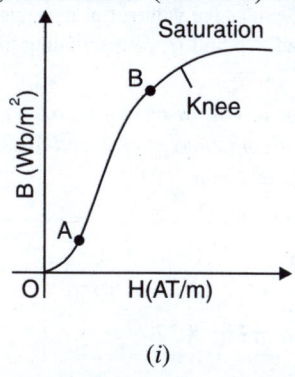

(i)

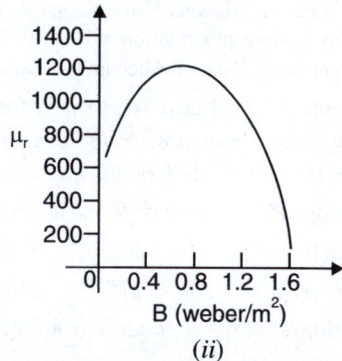

(ii)

Fig. 8.25

While carrying out magnetic calculations, it should be ensured that the values of μ_r and H are taken at the working flux density. For this purpose, the *B-H* curve of the material in question may be very helpful. In fact, the use of *B-H* curves permits the calculations of magnetic circuits with a fair degree of ease.

8.12. Magnetic Calculations From B-H Curves

The solution of magnetic circuits can be easily obtained by the use of *B-H* curves. The procedure is as under :

(i) Corresponding to the flux density B in the material, find the magnetising force H from the *B-H* curve of the material.

---

*   Note the shape of the curve. It is slightly concave up for 'low' flux densities (portion *OA*) and exhibits a straight line character (portion *AB*) for 'medium' flux densities. In the portion *AB* of the curve, the μ_r of the material is almost constant. For higher flux 'densities', the curve concaves down (called the *knee* of the curve). After knee of the curve, any further increase in H does not increase B. From now onwards, the curve is almost flat and the material is said to be *saturated*. In terms of molecular theory, saturation can be explained as the point at which all the molecular magnets are oriented in the direction of applied H.

(*ii*) Compute the magnetic length *l*.

(*iii*) m.m.f. required = $H \times l$

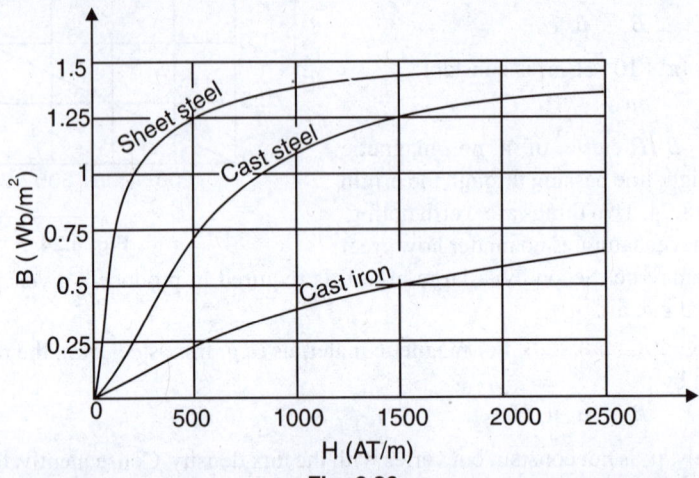

Fig. 8.26

The reader may note that the use of *B-H* curves for magnetic calculations saves a lot of time. Fig. 8.26 shows the *B-H* curves for sheet steel, cast steel and cast iron.

Note. We do not use *B-H* curve to find m.m.f. for air gap. We can find H_g directly from B_g/μ_0 and hence the m.m.f. = $H_g \times l_g$. However, in a magnetic material, $*H_i = B_i/\mu_0 \mu_r$. Since the value of μ_r depends upon the working flux density, this realtion will not yield correct result. Instead, we find H_i corresponding to B_i in the material from the *B-H* curve. Then m.m.f. required for iron path = $H_i \times l_i$.

Example 8.19. *A cast steel ring of mean diameter 30 cm having a circular cross-section of 5 cm² is uniformly wound with 500 turns. Determine the magnetising current required to establish a flux of 5×10^{-4} Wb (i) with no air gap (ii) with a radial air gap of 1 mm.*

The magnetisation curve for cast steel is given by the following :

$B(Wb/m^2)$	0·2	0·4	0·6	0·8	1	1·2
$H(AT/m)$	175	300	400	600	850	1250

Solution. Plot the *B-H* curve from the given data as shown in Fig. 8.27.

(*i*) **With no air gap**

$$B_i = \frac{\phi}{a} = \frac{5 \times 10^{-4}}{5 \times 10^{-4}} = 1 \text{ Wb/m}^2$$

From the *B-H* curve, we find that for a flux density of 1 Wb/m², the value of

$$H_i = 850 \text{ AT/m}$$

Now,

$$l_i = \pi D = \pi \times 30 \times 10^{-2}$$

$$= 0·942 \text{ m}$$

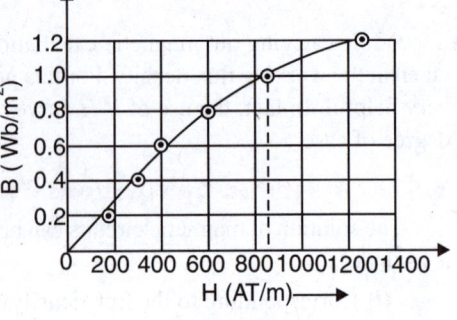

Fig. 8.27

∴ Total *AT* required = $H_i \times l_i$

$$= 850 \times 0·942 = 800·7 \text{ AT}$$

∴   Magnetising current, $I = 800·7/500 = $**1·6 A**

(*ii*) **With air gap of 1 mm**

Flux density in air gap, $B_g = 1 \text{ Wb/m}^2$ (same as in steel)

---

*   The suffix *i* denotes iron part while suffix *g* denotes air gap.

Magnetising force required, $^*H_g$ $= \dfrac{B}{\mu_0} = \dfrac{1}{4\pi \times 10^{-7}} = 7 \cdot 96 \times 10^5$ AT/m

$\qquad$ AT required for air gap $= H_g \times l_g = (7 \cdot 96 \times 10^5) \times (1 \times 10^{-3}) = 796$ AT

$\qquad\qquad$ Total AT required $= 800 \cdot 7 + 796 = 1596 \cdot 7$ AT

$\therefore$ $\quad$ Magnetising current, $I = 1596 \cdot 7/500 =$ **3·19 A**

Example 8.20. *A magnetic circuit made of wrought iron is arranged as shown in Fig. 8.28. The central limb has a cross-sectional area of 8 cm² and each of the side limbs has a cross-sectional area of 5 cm². Calculate the ampere-turns required to produce a flux of 1 mWb in the central limb, assuming the magnetic leakage is negligible. Given that for wrought iron (from B-H curve), H = 500 AT/m at B = 1·25 Wb/m² and H = 200 AT/m at B = 1 Wb/m².*

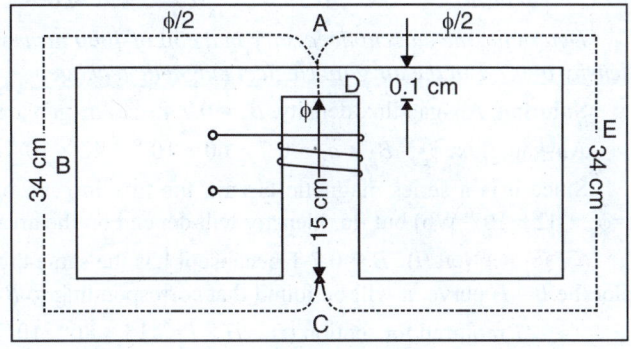

Fig. 8.28

Solution. The flux ϕ set up in the central limb divides equally into two identical parallel paths *viz.* path *ABC* and path *AEC*. The toal m.m.f. required for the entire circuit is the sum of the following three m.m.fs' :

$\quad$ (*i*) that required for path *CD*

$\quad$ (*ii*) that required for air gap *DA*

$\quad$ (*iii*) that required for either of parallel paths (*i.e.* path *ABC* or path *AEC*).

(*i*) AT for path CD

$$B = \frac{\phi}{a} = \frac{1 \times 10^{-3}}{8 \times 10^{-4}} = 1.25 \text{ Wb/m}^2$$

$\qquad$ Now H at 1·25 Wb/m² $= 500$ AT/m (*given*)

$\therefore$ $\quad$ AT required for path $CD = 500 \times 0 \cdot 15 = 75$ AT

(*ii*) AT for air gap DA

$$H \text{ in air gap} = \frac{B}{\mu_0} = \frac{1 \cdot 25}{4\pi \times 10^{-7}} = 994 \cdot 7 \times 10^3 \text{ AT/m}$$

$\therefore$ $\quad$ AT required for air gap $= (994 \cdot 7 \times 10^3) \times (0 \cdot 1 \times 10^{-2}) = 994 \cdot 7$ AT

(*iii*) AT for path ABC

$\qquad$ Flux in path $ABC = \phi/2 = 1/2 = 0 \cdot 5$ mWb

$$\text{Flux density in path } ABC = \frac{0.5 \times 10^{-3}}{5 \times 10^{-4}} = 1 \text{ Wb/m}^2$$

$\qquad$ Now H at 1 Wb/m² $= 200$ AT/m $\quad$ (*given*)

$\therefore$ $\quad$ AT required for path $ABC = 200 \times 0 \cdot 34 = 68$ AT

$\therefore$ $\qquad$ Total AT required $= 75 + 994 \cdot 7 + 68 =$ **1137·7 AT**

The reader may note that air gap "grabs" 87 per cent of the applied ampere-turns.

---

* $\quad$ We do not use *B-H* curve to find *AT* for air gap. It is because μ_r for air (in fact for all non-magnetic materials) is constant, being equal to 1, and *AT* can be calculated directly.

Example 8.21. *A series magnetic circuit comprises three sections (i) length of 80 mm with cross-sectional area 60 mm^2, (ii) length of 70 mm with cross-sectional area 80 mm^2 and (iii) air gap of length 0.5 mm with cross-sectional area 60 mm^2. Sections (i) and (ii) are of a material having magnetic characteristics given by the following table.*

H(AT/m)	100	210	340	500	800	1500
B(Tesla)	0.2	0.4	0.6	0.8	1.0	1.2

Determine the current necessary in a coil of 4000 turns wound on section (ii) to produce a flux density of 0.7 T in the air gap. Neglect magnetic leakage.

Solution. Air-gap flux density, $B_g = 0.7$ T ; Air-gap area, $a_g = 60 \times 10^{-6}$ m^2

Air-gap, flux, $\phi_g = B_g \times a_g = 0.7 \times 60 \times 10^{-6} = 42 \times 10^{-6}$ Wb

Since it is a series magnetic circuit, the flux in each of the three sections will be the same $(=\phi_g = 42 \times 10^{-6}$ Wb) but flux density will depend on the area of X-section of the section.

AT for section (i). $B = 0.7$ T because it has the same cross-sectional area as the air gap. If we plot the $B - H$ curve, it will be found that corresponding to $B = 0.7$ T, $H = 415$ AT/m.

∴   AT required for section (i) $= H \times l = 415 \times 80 \times 10^{-3} = 33.2$ AT

AT for section (ii). $B = \dfrac{\phi_g}{a} = \dfrac{42 \times 10^{-6}}{80 \times 10^{-6}} = 0.525$ T

From B–H curve, corresponding to $B = 0.525$ T, $H = 285$ AT/m.

∴   AT required for section (ii) $= H \times l = 285 \times 70 \times 10^{-3} = 19.95$ AT

AT for section (iii). This section is air gap.

$$B_g = 0.7 \text{ T and } H_g = \frac{B_g}{\mu_0} = \frac{0.7}{4\pi \times 10^{-7}} = 0.557 \times 10^6 \text{ AT/m}$$

∴   AT required for air gap $= H_g \times l_g = 0.557 \times 10^6 \times 0.5 \times 10^{-3} = 278.5$ AT

Total AT required $= 33.2 + 19.95 + 278.5 = 331.6$ AT

Now, $NI = 331.6$ or $I = \dfrac{331.6}{N} = \dfrac{331.6}{4000} = \mathbf{0.083 \ A}$

Example 8.22. *A magnetic circuit is made of mild steel arranged as shown in Fig. 8.29. The central limb is wound with 500 turns and has a cross-sectional area of 8cm^2. Each of the outer limbs has a cross-sectional area of 5 cm^2. The air gap has a length of 1 mm. Calculate the current required to set up a flux of 1·3 mWb in the central limb, assuming no magnetic leakage and fringing. The mean lengths of the various magnetic paths are shown in the diagram. Given that for mild steel (from B-H curve) H = 3800 AT/m at B = 1·625 T and H = 850 AT/m at B = 1·3 T.*

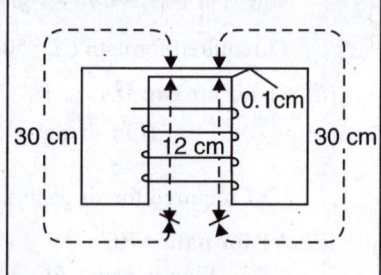

Fig. 8.29

Solution. Flux density in the central limb

$$= \frac{\text{Flux}}{\text{cross-sectional area}} = \frac{1.3 \times 10^{-3}}{8 \times 10^{-4}} = 1.625 \text{ T}$$

Given that $H = 3800$ AT/m at $B = 1·625$ T

∴   m.m.f. for central limb $= H_1 l_1 = 3800 \times 0·12 = 4·56$ AT

Since half the flux returns through one outer limb and half through the other, the two outer limbs are magnetically equivalent to a single limb having a cross-sectional area of 10 cm^2 and a length of 30 cm.

∴   Flux density in outer limbs $= \dfrac{1.3 \times 10^{-3}}{10 \times 10^{-4}} = 1.3$ T

Given that $H = 850$ AT/m at $B = 1.3$ T

$\therefore$ m.m.f. for outer limbs $= H_2 l_2 = 850 \times 0.3 = 255$ AT

Flux density in airgap, $B = 1.625$ T

Magnetising force for air gap is given by ;

$$H_3 = \frac{B}{\mu_0} = \frac{1.625}{4\pi \times 10^{-7}} = 1.294 \times 10^6 \text{ AT/m}$$

m.m.f. for air gap $= H_3 l_3 = (1.294 \times 10^6) \times (1 \times 10^{-3}) = 1294$ AT

Total m.m.f. $= 456 + 255 + 1294 = 2005$ AT

$\therefore$ Magnetising current, $I = \dfrac{\text{Total m.m.f.}}{\text{Turns}} = \dfrac{2005}{500} \simeq$ **4A**

Example 8.23. *Fig. 8.30 shows the cross-section of a simple relay. Calculate the ampere-turns required on the coil for a flux density of 0·1 Wb/m² in the air gaps from the following data :*

Cross-sectional area of yoke	$= 2 \; cm^2$
Magnetic length of yoke	$= 25 \; cm$
Cross-sectional area of armature	$= 3 \; cm^2$
Magnetic length of armature	$= 12 \; cm$
Air gap area	$= 6 \; cm^2$
Each air gap length	$= 5 \; mm$
Leakage coefficient	$= 1·33$

The yoke and armature material have the following magnetic characteristics :

H (AT/m)	100	210	340	500	800	1500
B (Wb/m²)	0·2	0·4	0·6	0·8	1·0	1·2

Solution. Plot the *B-H* curve from the given data as shown in Fig. 8.31.

Flux in air gap, $\phi_g = 6 \times 10^{-4} \times 0.1 = 6 \times 10^{-5}$ Wb $=$ Flux in armature

Flux in yoke, $\phi_y = \lambda \, \phi_g = 1.33 \times 6 \times 10^{-6} = 7.98 \times 10^{-5}$ Wb

AT for armature

Flux density in armature $= \dfrac{6 \times 10^{-5}}{3 \times 10^{-4}} = 0.2$ Wb/m²

Corresponding to $B = 0.2$ Wb/m² (See *B-H* curve), $H = 100$ AT/m.

$\therefore$ AT required for armature $= 100 \times 0.12 = 12$ AT

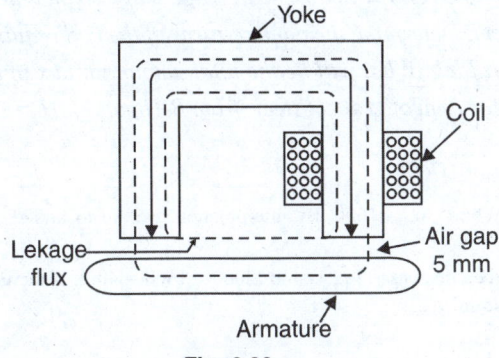

Fig. 8.30

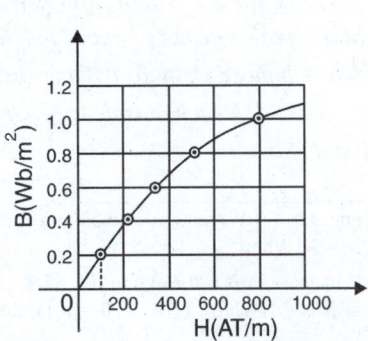

Fig. 8.31

AT for yoke

Flux density in the yoke $= \dfrac{7 \cdot 98 \times 10^{-5}}{2 \times 10^{-4}} = 0 \cdot 4 \text{ Wb/m}^2$

Corresponding to $B = 0 \cdot 4 \text{ Wb/m}^2, \quad H = 210 \text{ AT/m}.$

∴   AT required for yoke $= 210 \times 0 \cdot 25 = 52 \cdot 5 \text{ AT}$

AT for air gaps

Magnetising force in air gaps $= \dfrac{0 \cdot 1}{\mu_0} = \dfrac{0 \cdot 1}{4\pi \times 10^{-7}} = 7 \cdot 96 \times 10^4 \text{ AT/m}$

AT for two air gaps $= (7 \cdot 96 \times 10^4) \times (10 \times 10^{-3}) = 796 \text{ AT}$

Total AT required $= 12 + 52 \cdot 5 + 796 = \textbf{860·5 AT}$

Example 8.24. *An iron ring of mean diameter 19.1 cm and having cross-sectional area of 4 cm^2 is required to produce a flux of 0.44 mWb. Find the coil m.m.f. required. If a saw-cut 1 mm wide is made in the ring, how many extra ampere-turns are required to maintain the same flux ? $B - \mu_r$ curve is as follows :*

$B(\text{Wb/m}^2)$	0.8	1.0	1.2	1.4
μ_r	2300	2000	1600	1100

Solution. $D_m = 0.191 \text{ m} ; a = 4 \times 10^{-4} \text{ m}^2; \phi = 0.44 \times 10^{-3} \text{ Wb}$

Length of mean path, $l_m = \pi D_m = \pi \times 0.191 = 0.6 \text{ m}$

Flux density in ring, $B_i = \dfrac{\phi}{a} = \dfrac{0.44 \times 10^{-3}}{4 \times 10^{-4}} = 1.1 \text{ Wb/m}^2$

By *interpolation, for flux density of 1.1 Wb/m^2, $\mu_r = 1800$.

∴   Magnetising force, $H_i = \dfrac{B_i}{\mu_0 \mu_r} = \dfrac{1.1}{4\pi \times 10^{-7} \times 1800} = 486.5 \text{ AT/m}$

∴   m.m.f. required $= H_i \times l_m = 486.5 \times 0.6 = \textbf{292 AT}$

If a saw-cut of 1 mm wide is made in the ring, we require extra *AT* to maintain the same flux ($= 0.44 \times 10^{-3} \text{Wb}$).

Now $H_g = \dfrac{B_g}{\mu_0} = \dfrac{1.1}{4\pi \times 10^{-7}} = 875352 \text{ AT/m} ; l_g = 1 \times 10^{-3} \text{ m}$

∴   Extra m.m.f. required $= H_g \times l_g = 875352 \times 1 \times 10^{-3} = \textbf{875 AT}$

Example 8.25. *A transformer core made of annealed steel sheet has the form and dimensions shown in Fig. 8.32. A coil of N turns is wound on the central limb. The average length of magnetic circuit (i.e. path ABCDA or path EFGHE) is 30 cm. Determine the ampere-turns of the coil required to produce a flux density of 1 Wb/m^2 in the central leg. What will be the total amount of flux in the central leg and in each outside leg ? Given that for annealed sheet steel (from B-H curve), H = 200 AT/m at 1 Wb/m^2.*

---

*   For $B = 1.0 \text{ Wb/m}^2$, $\mu_r = 2000$ and for $B = 1.2 \text{ Wb/m}^2$, $\mu_r = 1600$. By interpolation, we are to find μ_r for $B = 1.1 \text{Wb/m}^2$.

If increase in B is 0.2 Wb/m^2 ($= 1.2 - 1.0 = 0.2$), then decrease in μ_r is 400 ($2000 - 1600 = 400$). If increase in B is 0.1/Wb/m^2 ($1.1 - 1.0 = 0.1$), then decrease in μ_r

$$= \dfrac{400}{0.2} \times 0.1 = 200$$

∴ μ_r at 1.1 Wb/m^2 = 2000 − 200 = 1800

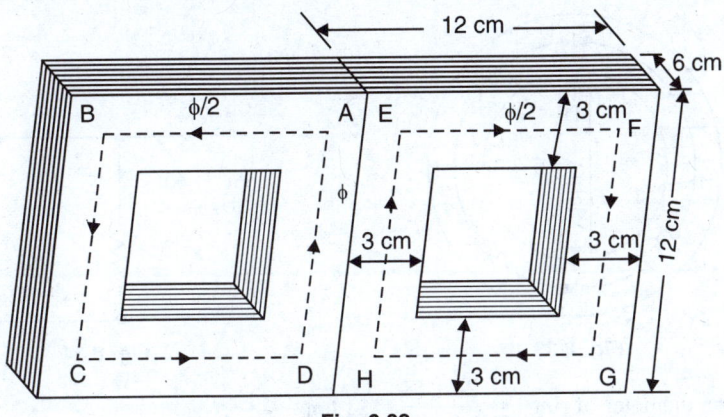

Fig. 8.32

Solution. It is a case of parallel magnetic circuit. It is clear from Fig. 8.32 that central leg has twice the area of an outside leg. The flux ϕ set up in the central limb divides equally into two parallel identical paths *viz.* path *ABCD* and path *EFGH*. It may be noted very carefully that flux density is the *same in the central leg, each outside leg and other parts.

Mean length of magnetic path (*i.e.* path *ABCDA* and *EFGHE*)

$$= 30 \text{ cm} = 0.3 \text{ m}$$

∴ AT required $= 200 \times 0.3 = \textbf{60 AT}$

Area of central leg $= 0.06 \times 0.06 = 0.0036 \text{ m}^2$

Flux in central leg $=$ Flux density $\times$ Area $= 1 \times 0.0036 = \textbf{0.0036 Wb}$

Area of each outside leg $= 0.03 \times 0.06 = 0.0018 \text{ m}^2$

Flux in each outside leg $= 1 \times 0.0018 = \textbf{0.0018 Wb}$

Alternatively, flux in each outside leg will be half that in the central leg *i.e.* $0.0036/2 = \textbf{0.0018 Wb.}$

Example 8.26. *A ring of cast steel has an external diameter of 24 cm and a square cross-section of 3 cm side. Inside and cross the ring, an ordinary steel bar 18 cm $\times$ 3 cm $\times$ 0.4 cm is fitted with negligible gap. Calculate the number of ampere-turns required to be applied to one half of the ring to produce a flux density of 1.0 weber per metre^2 in the other half. Neglect leakage. The B-H characteristics are as below :*

	For Cast Steel			For Ordinary Plate			
B in Wb/m^2	1.0	1.1	1.2	B in Wb/m^2	1.2	1.4	1.45
Amp-turn/m	900	1020	1220	Amp-turn/m	590	1200	1650

Solution. The conditions of the problem lead to the magnetic circuit shown in Fig. 8.33. The equivalent electrical circuit is shown in Fig. 8.34. Note that m.m.f. is shown as a battery and reluctances as resistances. Referring to Fig. 8.33, the flux paths *D* and *C* are in parallel. Therefore, total AT required is equal to AT for path *A* plus AT for path *C* or path *D*.

---

* The area of central leg is '*a*' and flux is ϕ so that $B = \phi/a$. The area of each outside and other part of flux path is $a/2$ and flux is $\phi/2$ so that B is again $= \phi/a$.

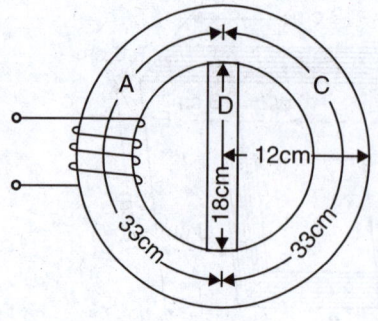

Fig. 8.33

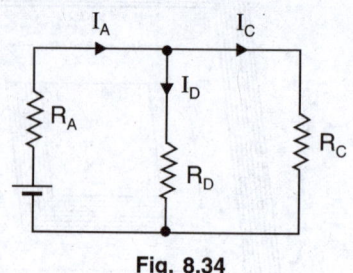

Fig. 8.34

$$\text{Mean diameter of ring} = \frac{24+18}{2} = 21 \text{ cm}$$

$$\text{Mean circumference} = \pi \times 21 = 66 \text{ cm}$$

$$\text{Length of path } A \text{ or } C = 66/2 = 33 \text{ cm} = 0.33 \text{ m}$$

AT for path C. We shall first determine AT required for path C because flux density in this path is known (1 Wb/m^2). From the B-H characteristic, H corresponding to 1 Wb/m^2 is 900 AT/m.

$$\therefore \quad \text{AT required for path } C = H \times \text{Length of path } C$$

$$= 900 \times 0.33 = 297 \text{ AT}$$

AT for path D. Since paths C and D are in parallel, AT required for path D = 297 AT and H = 297/0.18 = 1650 AT/m. From the B-H characteristic, B corresponding to 1650 AT/m is 1.45 Wb/m^2.

$$\text{Flux through } C, \phi_C = B \times A = 1 \times 9 \times 10^{-4} = 9 \times 10^{-4} \text{ Wb}$$

$$\text{Flux through } D, \phi_D = (1.45) \times (3 \times 0.4 \times 10^{-4}) = 1.74 \times 10^{-4} \text{ Wb}$$

$$\therefore \quad \text{Flux through } A, \phi_A = \phi_C + \phi_D = (9 + 1.74) \times 10^{-4} = 10.74 \times 10^{-4} \text{ Wb}$$

$$\text{Flux density in } A = \frac{10.74 \times 10^{-4}}{9 \times 10^{-4}} = 1.193 \text{ Wb/m}^2$$

From the B-H characteristics, H corresponding to 1.193 Wb/m^2 is 1200 AT/m (approx.).

$$\therefore \quad \text{AT for path } A = 1200 \times 0.33 = 396 \text{ AT}$$

$$\therefore \quad \text{Total AT required} = \text{AT for path } C + \text{AT for path } A$$

$$= 297 + 396 = \textbf{693 AT}$$

Tutorial Problems

1. A cast iron-cored toroidal coil has 3000 turns and carries a current of 0.1A . The length of the magnetic circuit is 15 cm and cross-sectional area of the coil is 4 cm^2. Find H, B and total flux. Use the following B–H curve for cast iron :

H(AT/m) :	200	400	1000	2000	3000
B(T) :	0.1	0.19	0.375	0.57	0.625

    [2000 AT/m ; 0.57 T; 2.28 × 10^{-4} Wb]

2. A series magnetic circuit has an iron path of length 50 cm and an air gap of length 1 mm. The cross-sectional area of the iron is 6 cm^2 and the exciting coil has 400 turns. Determine the current required to produce a flux of 0.9 mWb in the circuit. The following points are taken from the magnetisation characteristic :

B(Wb/m^2) :	1.2	1.35	1.45	1.55
H(AT/m) :	500	1000	2000	4500

    [6.35 A]

3. A cast-steel ring of mean circumference 50 cm has a cross-section of 0.52 cm^2. It has a saw-cut of 1 mm at one place. Given the following data :

$B(\text{Wb/m}^2)$:	1.0	1.25	1.46	1.60
μ_r :	714	520	360	247

Calculate how many ampere-turns are required to produce a flux of 0.052 mWb if leakage factor is 1.2.

[1647 AT]

4. A magnetic circuit with a uniform cross-sectional area of 6 cm^2 consists of a cast steel ring with a mean magnetic length of 80 cm and an air gap of 2 mm. The magnetising winding has 540 ampere-turns. Estimate the magnetic flux produced in the gap. The relevant points on the magnetisation curve of cast steel are :

$B(\text{Wb/m}^2)$:	0.12	0.14	0.16	0.18	0.20
$H(\text{AT/m})$:	200	230	260	290	320 **[0.1128 mWb]**

8.13. Determination of B/H or Magnetisation Curve

The variation of permeability $\mu \, (= \mu_0 \mu_r)$ with flux density creates a design problem. Permeability must be known in order to find the flux density ($B = \mu H$) but permeability changes with flux density. This necessitates a graphical approach to magnetic circuit design. We plot B-H curves or magnetisation curves for various magnetic materials. The value of permeability is determined from the B-H curve of the material. The B-H curve can be determined by the following two methods provided the material is in the form of a ring : (*i*) By means of ballistic galvanometer, (*ii*) By means of fluxmeter.

8.14. B-H Curve by Ballistic Galvanometer

A ballistic galvanometer is similar in principle to the permanent moving coil instrument. It has a moving coil suspended between the poles of a permanent magnet. The coil is wound on a *non-metallic* former so that there is very little damping. The first deflection or 'throw' is proportional to the charge passed through the galvanometer if the duration of the charge passed is short compared with the time of one oscillation.

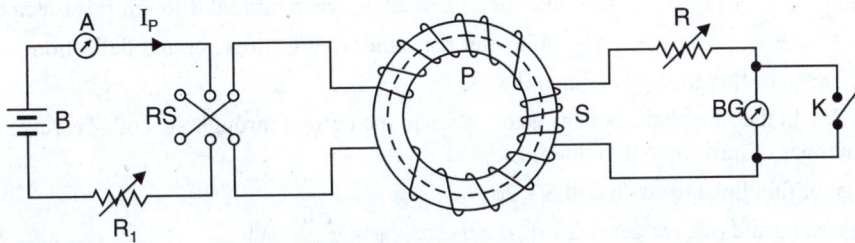

Fig. 8.35

Fig. 8.35 shows the circuit arrangement for the determination of B-H curve of a magnetic material by ballistic galvanometer. The specimen ring of uniform cross-section is wound uniformly with a coil P, thereby eliminating magnetic leakage. The primary coil P is connected to a battery through a reversing switch RS, an ammeter A and a variable resistor R_1. Another secondary coil S (called *search coil*) is wound over a small portion of the ring and is connected through a resistance R to the ballistic galvanometer BG.

Theory. We shall use subscript P for primary and subscript S for secondary.

Let $\quad \theta =$ first deflection or 'throw' of the galvanometer when primary current I_P is reversed

$\quad\quad k =$ ballistic constant of the galvanometer *i.e.* charge per unit deflection

$\therefore \quad$ Charge passing through $BG = k \, \theta$ coulombs $\hspace{4cm}$ ...(*i*)

If ϕ is the flux produced in the ring by I_P amperes through primary P and t the time in seconds of *reversal of flux, then,

---

* $\quad$ The flux changes from ϕ to $-\phi$ by changing reversing switch RS. Therefore, change in flux is 2ϕ Wb.

Rate of change of flux $= \dfrac{2\phi}{t}$ Wb/s

If N_S is the number of turns in the secondary or search coil, then,

Average e.m.f. induced in $S = N_S \times \dfrac{2\phi}{t}$ volts

If R_S is the total resistance in the secondary circuit, then,

Current through secondary or BG, $I_S = \dfrac{2 N_S \phi}{R_S t}$ amperes

$\therefore$ Charge through $BG = I_S \times t = \dfrac{2 N_S \phi}{R_S t} \times t = \dfrac{2 N_S \phi}{R_S}$ coulombs   ...(ii)

From eqs. (i) and (ii), we get, $k\theta = \dfrac{2 N_S \phi}{R_S}$ $\therefore$ $\phi = \dfrac{k \theta R_S}{2 N_S}$ Wb

If A is the area of cross-section of the ring in m^2, then,

Flux density in the ring, $B = \dfrac{\phi}{A} = \dfrac{k\theta R_S}{2 N_S A}$ Wb/m^2

If N_P is the number of turns on coil P, l the mean circumference of the ring and I_P is the current through coil P, then,

Magnetising force, $H = \dfrac{N_P I_P}{l}$

The above experiment is repeated with different values of primary current and from the data obtained, the B-H curve can be plotted.

8.15. B-H Curve by Fluxmeter

In this method, the BG is replaced by the fluxmeter which is a special type of ballistic galvanometer. Its operation is based on the change in flux linkages.

Theory.   Let $\theta =$ fluxmeter deflection when current through P is reversed

$c =$ fluxmeter constant *i.e.* Wb-turns per unit deflection

$\therefore$ Change of flux linkages with coil $S = c\,\theta$ ...(i)

If the flux in the ring changes from ϕ to $-\phi$ when the current through the coil P is reversed and N_S is the number of turns on coil S, then,

Change of flux linkages with coil $S = 2\phi\, N_S$ ...(ii)

From eqs. (i) and (ii), we get, $2\phi\, N_S = c\,\theta$ $\therefore$ $\phi = \dfrac{c\,\theta}{2 N_S}$ Wb

If A is the cross-sectional area of the ring in m^2, then,

Flux density, $B = \dfrac{\phi}{A} = \dfrac{c\,\theta}{2 N_S A}$ Wb/m^2

Also, $H = \dfrac{N_P I_P}{l}$

where $l =$ mean circumference of the ring in metres

Thus we can plot the B-H curve.

Example 8.27. *A fluxmeter is connected to a search coil having 600 turns and mean area of 4 cm². The search coil is placed at the centre of an air-cored solenoid 1 m long and wound with 1000 turns. When a current of 4A is reversed, there is a deflection of 20 scale divisions on the fluxmeter. Calculate the calibration in Wb-turns per scale division.*

Solution. Here, $N_P = 1000$ turns ; $I_P = 4$A ; $l = 1$m ; $N_S = 600$ turns ; $A = 4 \times 10^{-4}$ m^2.

Since the length of the solenoid is large compared to its diameter, the magnetising force inside the solenoid is uniform. Therefore, magnetising force H at the centre of the solenoid is

$$H = \frac{N_P I_P}{l} = \frac{1000 \times 4}{1} = 4000 \text{ AT/m}$$

$\therefore$ Flux density, $B = \mu_0 H = 4\pi \times 10^{-7} \times 4000 = 16\pi \times 10^{-4}$ Wb/m^2

Flux linked with search coil, $\phi = BA = 16\pi \times 10^{-4} \times 4 \times 10^{-4} = 64\pi \times 10^{-8}$ Wb

When current in the solenoid is reversed, the change in flux linkages with search coil

$$= 2 N_S \phi = 2 \times 600 \times 64\pi \times 10^{-8} = 7.68\pi \times 10^{-4} \text{ Wb-turns}$$

It c is the fluxmeter constant, then, value of c is given by ;

$$c = \frac{\text{Change in flux linkages}}{\text{Deflection produced}}$$

$$= \frac{7.68\pi \times 10^{-4}}{20} = \mathbf{1.206 \times 10^{-4} \text{ Wb-turns/division}}$$

Example 8.28. *A solenoid 1·2 m long is uniformly wound with a coil of 800 turns. A short coil of 50 turns, having a mean diameter of 30 mm, is placed at the centre of the solenoid and is connected to a ballistic galvanometer. The total resistance of the galvanometer circuit is 2000 Ω. When a current of 5 A through the solenoid primary winding is reversed, the initial deflection of the ballistic galvanometer is 85 divisions. Determine the ballistic constant.*

Solution. Within the solenoid, we have,

$$H = \frac{N_P I_P}{l} ; \quad B = \mu_0 H = \frac{\mu_0 N_P I_P}{l}$$

$\therefore$ Flux passing through the secondary or search coil of area A is

$$\phi = B \times A = \frac{\mu_0 N_P I_P \, A}{l}$$

Here, $N_P = 800$; $I_P = 5$ A ; $A = \pi \times (15)^2 \times 10^{-6}$ m^2 ; $l = 1\cdot2$ m

$\therefore$

$$\phi = \frac{4\pi \times 10^{-7} \times 800 \times 5 \times (\pi \times 15^2 \times 10^{-6})}{1\cdot2} = 2\cdot96 \times 10^{-6} \text{ Wb}$$

Ballistic constant, $k = \dfrac{2 N_S \phi}{R_S \, \theta} = \dfrac{2 \times 50 \times 2\cdot96 \times 10^{-6}}{2000 \times 85}$

$$= 1\cdot74 \times 10^{-9} \text{ C/div} = \mathbf{1740 \text{ pC/div.}}$$

Example 8.29. *A steel ring, 400 mm^2 cross-sectional area with a mean length 800 mm, is wound with a magnetising winding of 1000 turns. A secondary coil with 200 turns of wire is connected to a ballistic galvanometer having a constant of 1 μC/div. The total resistance of the secondary circuit is 2 kΩ. On reversing a current of 1 A in the magnetising coil, the galvanometer gives a throw of 100 scale divisions. Calculate :*

(i) The flux density in the specimen.

(ii) The relative permeability at this flux density.

Solution.

(i) As proved in Art. 8·14, the flux density B within the ring is given by ;

$$B = \frac{k \theta R_S}{2 N_S \, A}$$

Here, $k = 1 \, \mu\text{C/div} = 1 \times 10^{-6}$ C/div ; $\theta = 100$ divisions ;

$R_S = 2 \text{ k}\Omega = 2000 \, \Omega$; $N_S = 200$; $A = 400 \text{ mm}^2 = 400 \times 10^{-6}$ m^2

$\therefore$

$$B = \frac{(1 \times 10^{-6}) \times (100) \times (2000)}{2 \times 200 \times 400 \times 10^{-6}} = \mathbf{1\cdot25 \text{ T}}$$

(ii)
$$H = \frac{N_P I_P}{l} = \frac{1000 \times 1}{800 \times 10^{-3}} = 1 \cdot 25 \times 10^3 \text{ AT/m}$$

Now
$$B = \mu_0 \, \mu_r \, H$$

∴ Relative permeability, $\mu_r = \dfrac{B}{\mu_0 \, H} = \dfrac{1.25}{4\pi \times 10^{-7} \times 1.25 \times 10^3} = \textbf{796}$

Example 8.30. *An iron ring has a mean diameter of 0·1 m and a cross-section of 33·5 × 10⁻⁶ m². It is wound with a magnetising winding of 320 turns and the secondary winding of 220 turns. On reversing a current of 10 A in the magnetising winding, a ballistic galvanometer gives a throw of 272 scale divisions, while a Hilbert magnetic standard with 10 turns and a flux of 2·5 × 10⁻⁴ Wb gives a reading of 102 scale divisions, other conditions remaining the same. Find the relative permeability of the specimen.*

Solution. Within the iron ring, we have,

Length of magnetic path, $l = \pi D = 0 \cdot 1\pi$ m

$$H = \frac{N_P I_P}{l} = \frac{320 \times 10}{0.1\,\pi} = 10186 \text{ AT/m}$$

$$B = \mu_0 \, \mu_r \, H = 4\pi \times 10^{-7} \times \mu_r \times 10186 = 0.0128 \, \mu_r \qquad \ldots(i)$$

From Hilbert's magnetic standard, we have,

$$2.5 \times 10^{-4} \times 10 = k \times 102 \quad \therefore \quad k = 2.45 \times 10^{-5} \text{ Wb-turn/div.}$$

On reversing a current of 10 A in the primary coil, change in terms of Wb-turn is

$$2\phi \, N_S = k\theta \quad \text{or} \quad 2 \times \phi \times 220 = 2.45 \times 10^{-5} \times 272$$

∴
$$\phi = \frac{2.45 \times 10^{-5} \times 272}{2 \times 220} = 1.51 \times 10^{-5} \text{ Wb}$$

$$B = \frac{\phi}{A} = \frac{1.51 \times 10^{-5}}{33.5 \times 10^{-6}} = 0.45 \text{ Wb/m}^2$$

But $B = 0 \cdot 0128 \, \mu_r$ as is evident from eq. (*i*).

∴
$$0.45 = 0.0128 \, \mu_r \quad \text{or} \quad \mu_r = 0.45/0.0128 = \textbf{35.1}$$

Example 8.31. *A coil of 120 turns is wound uniformly over a steel ring having a mean circumference of 1 m and a cross-sectional area of 500 mm². A search coil of 15 turns, wound on the ring, is connected to a fluxmeter having a constant of 300 μWbt/div. When a current of 6 A through the 120-turn coil is reversed, the fluxmeter deflection is 64 divisions. Calculate :*

(i) *The flux density in the ring.*

(ii) *The corresponding value of relative permeability.*

Solution. (*i*) Fluxmeter constant, $c = \dfrac{2N_S \, \phi}{\theta}$

Here $c = 300 \times 10^{-6}$ Wbt/div. ; $N_S = 15$; $\theta = 64$ div.

∴
$$\phi = \frac{c\theta}{2N_S} = \frac{300 \times 10^{-6} \times 64}{2 \times 15} = 0 \cdot 64 \times 10^{-3} \text{ Wb}$$

Note that ϕ is the flux passing through the search coil.

∴   Flux density, $B = \dfrac{\phi}{A} = \dfrac{0.64 \times 10^{-3}}{500 \times 10^{-6}} = \textbf{1·28 Wb/m}^2$

(ii)   Within the ring, $H = \dfrac{N_P I_P}{l} = \dfrac{120 \times 6}{1} = 720 \text{ AT/m}$

Now,
$$B = \mu_0 \, \mu_r \, H$$

$$\therefore \qquad \mu_r = \frac{B}{\mu_0\,H} = \frac{1\cdot 28}{4\pi\times10^{-7}\times720} = 1400$$

Tutorial Problems

1. A moving coil ballistic galvanometer of 150Ω resistance gives a throw of 75 divisions when the flux through a search coil to which it is connected is reversed. Find the flux density given that the galvanometer constant is 110 μC per scale division and the search coil has 1400 turns, a mean area of 50 cm^2 and a resistance of 20Ω. **[0.1T]**

2. A fluxmeter is connected to a search coil having 500 turns and mean area of 5 cm^2. The search coil is placed at the centre of a solenoid one metre long wound with 800 turns. When a current of 5A is reversed, there is a deflection of 25 scale divisions on the fluxmeter. Calculate the fluxmeter constant.

   [10^{-4} Wb-turn/division]

3. A ballistic galvanometer connected to a search coil for measuring flux density in a core gives a throw of 100 scale divisions on reversal of flux. The galvanometer coil has a resistance of 180Ω. The galvanometer constant is 100μC per scale division. The search coil has an area of 50 cm^2 wound with 1000 turns having a resistance of 20Ω. Calculate the flux density in the core. **[0.2 T]**

8.16. Magnetic Hysteresis

When a magnetic material is subjected to a cycle of magnetisation (*i.e.* it is magnetised first in one direction and then in the other), it is found that flux density B in the material lags behind the applied magnetising force H. This phenomenon is known as hysteresis.

The phenomenon of lagging of flux density (B) behind the magnetising force (H) in a magnetic material subjected to cycles of magnetisation is known as **magnetic hysteresis.**

The term 'hysteresis' is derived from the Greek word *hysterein* meaning to lag behind. If a piece of magnetic material is subjected to *one cycle of magnetisation, the resultant *B-H* curve is a closed loop *abcdefa* called *hysteresis loop* [See Fig. 8.36 (*ii*)]. Note that B always lags behind H. Thus at point '*b*', H is zero but flux density B has a positive finite value *ob*. Similarly at point '*e*', H is zero, but flux density has a finite negative value *oe*. This tendency of flux density B to lag behind magnetising force H is known as magnetic hysteresis.

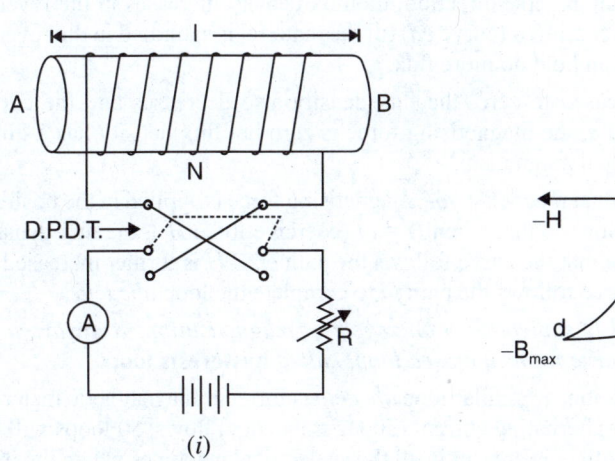

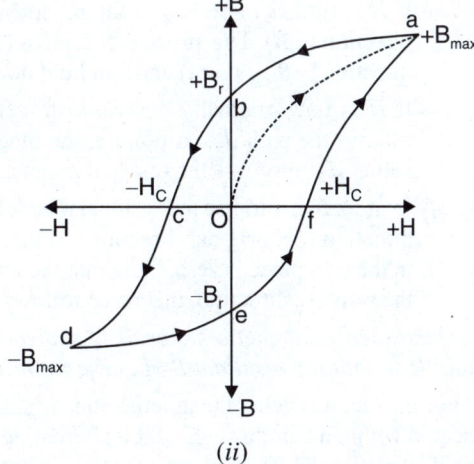

(i)   (ii)

Fig. 8.36

---

* If we start with unmagnetised iron piece, then magnetise it in one direction and then in the other direction and finally demagnetise it (*i.e.* obtain the original condition we started with), the piece is said to go through one cycle of magnetisation. Compare it with one cycle of alternating current or voltage.

Hysteresis Loop. Consider an unmagnetised iron bar *AB* wound with *N* turns as shown in Fig. 8.36 (*i*). The magnetising force H (= NI/l) produced by this solenoid can be changed by varying the current through the coil. The double-pole, double-throw switch (DPDT) is used to reverse the direction of current through the coil. We shall see that *when the iron piece is subjected to a cycle of magnetisation, the resultant B-H curve traces a loop abcdefa called hysteresis loop.*

(*i*) We start with unmagnetised solenoid *AB*. When the current in the solenoid is zero, $H = 0$ and hence B in the iron piece is 0. As H is increased (by increasing solenoid current), the flux density (+ B) also increases until the point of maximum flux density (+ B_{max}) is reached. The material is saturated and beyond this point, the flux density will not increase regardless of any increase in current or magnetising force. Note that *B-H* curve of the iron follows the path *oa*.

(*ii*) If now H is gradually reduced (by reducing solenoid current), it is found that the flux density B does not decrease along the same line by which it had increased but follows the path *ab*. At point *b*, the magnetising force H is zero but flux density in the material has a finite value + B_r (= *ob*) called **residual flux density.** It means that after the removal of *H*, the iron piece still retains some magnetism (*i.e.* + B_r). In other words, B lags behind H. The greater the lag, the greater is the residual magnetism (*i.e.* ordinate *ob*) retained by the iron piece. The power of retaining residual magnetism is called **retentivity** of the material.

The hysteresis effect (*i.e.* lagging of B behind H) in a magnetic material is due to the opposition offered by the magnetic domains (or molecular magnets) to the turning effect of magnetising force. Once arranged in an orderly position by the magnetising force, the magnetic domains do not return exactly to the original positions. In other words, the material retains some magnetism even after the removal of magnetising force. This results in the lagging of B behind H.

(*iii*) To demagnetise the iron piece (*i.e.* to remove the residual magnetism *ob*), the magnetising force H is reversed by reversing the current through the coil. When H is graudally increased in the reverse direction, the *B-H* curve follows the path *bc* so that when $H = oc$, the residual magnetism is zero. The value of H (= *oc*) required to wipe out residual magnetism is known as **coercive force** (H_c).

(*iv*) If H is further increased in the reverse direction, the flux density increases in the reverse direction (− B). This process continues (curve *cd*) till the material is saturated in the reverse direction (−B_{max} point) and can hold no more flux.

(*v*) If H is now gradually decreased to zero, the flux density also decreases and the curve follows the path *de*. At point *e*, the magnetising force is zero but flux density has a finite value −B_r (= *oe*) — the residual magnetism.

(*vi*) In order to neutralise the residual magnetism *oe*, magnetising force is applied in the positive direction (*i.e.* original direction) so that when $H = of$ (coercive force H_c), the flux density in the iron piece is zero. Note that the curve follows the path *ef*. If H is further increased in the positive direction, the curve follows the path *fa* to complete the loop *abcdefa*.

Thus when a magnetic material is subjected to one cycle of magnetisation, B always lags behind H so that the resultant B-H curve forms a closed loop, called **hysteresis loop.**

For the second cycle of magnetisation, a *similar loop *abcdefa* is formed. If a magnetic material is located within a coil through which alternating current (50 Hz frequency) flows, 50 loops will be formed every second. This hysteresis effect is present in all those electrical machines where the iron parts are subjected to cycles of magnetisation *e.g.* armature of a d.c. machine rotating in a stationary magnetic field, transformer core subjected to alternating flux etc.

---

*   Owing to the nature of magnetic material, a second or even third cycle of H would not exactly lie on the tops of the first one. After a relatively few cycles, the successive loops would follow a fixed path.

8.17. Hysteresis Loss

When a magnetic material is subjected to a cycle of magnetisation (*i.e.* it is magnetised first in one direction and then in the other), an energy loss takes place due to the *molecular friction in the material. That is, the domains (or molecular magnets) of the material resist being turned first in one direction and then in the other. Energy is thus expended in the material in overcoming this opposition. This loss is in the form of heat and is called *hysteresis loss*. Hysteresis loss is present in all those electrical machines whose iron parts are subjected to cycles of magnetisation. The obvious effect of hysteresis loss is the rise of temperature of the machine.

(*i*) Transformers and most electric motors operate on alternating current. In such devices, the flux in the iron changes continuously, both in value and direction. Hence hysteresis loss occurs in such machines.

(*ii*) Hysteresis loss also occurs when an iron part rotates in a constant magnetic field *e.g.* d.c. machines.

8.18. Calculation of Hysteresis Loss

We will now show that area of hysteresis loop represents the [†]energy loss/m³/cycle.

Let

l = length of the iron bar

A = area of X-section of bar

N = No. of turns of wire of solenoid

Suppose at any instant the current in the solenoid is i. Then,

$$H = \frac{Ni}{l} \quad \text{or} \quad i = \frac{Hl}{N}$$

Suppose the current increases by di in a small time dt. This will cause the flux density to increase by dB [See Fig. 8.37] and hence an increase in flux $d\phi \ (= AdB)$. This causes an e.m.f. e to be induced in the solenoid.

∴

$$e = N\frac{d\phi}{dt} = NA\frac{dB}{dt}$$

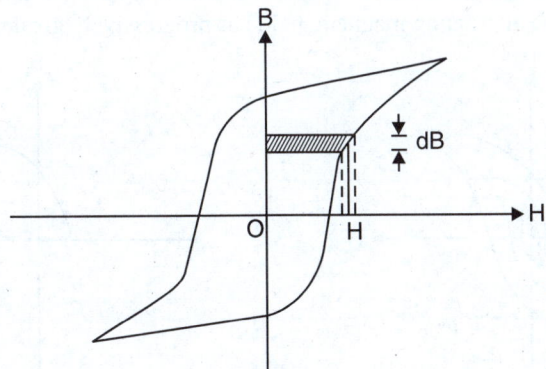

Fig. 8.37

By Lenz's law, this e.m.f. opposes the current i so that energy dW is spent in overcoming this opposing e.m.f.

---

* The opposition offered by the magnetic domains (or molecular magnets) to the turning effect of magnetising force is sometimes referred to as the molecular friction.

† In order to set up magnetic field, certain amount of energy has to be supplied which is stored in the field. If the field is in free space, the stored energy is returned to the circuit when the field collapses. If the field is in a magnetic material, not all the energy supplied can be returned ; part of it having been converted into heat due to hysteresis effect.

$$\therefore \qquad dW = ei \, dt \text{ joules}$$

$$= \left(NA\frac{dB}{dt} \right) \times \left(\frac{Hl}{N} \right) \times dt \text{ joules}$$

$$= Al \times H \times dB \text{ joules}$$

$$= V \times (H \times dB) \text{ joules}$$

$$\text{where} \quad Al = V = \text{volume of iron bar}$$

Now $H \times dB$ is the area of the shaded strip (See Fig. 8.37). For one cycle of magnetisation, the area $H \times dB$ will be equal to the area of hysteresis loop.

$\therefore$ Hysteresis energy loss/cycle, $W_h = V \times$ (area of loop) joules

If f is the frequency of reversal of magnetisation, then,

Hysteresis power loss, $P_h = W_h \times f = V \times$ (area of loop) $\times f$

Note. While calculating the area of hysteresis loop, proper scale factors of B and H must be considered.

For example, if the scales are :   1 cm = x AT/m   ...for H

$$1 \text{ cm} = y \text{ Wb/m}^2 \qquad \text{...for } B$$

Then, $W_h = xy \times$ (area of loop in cm^2) $\times V$ joules

where x and y are the scale factors.

8.19. Factors Affecting the Shape and Size of Hysteresis Loop

There are three factors that affect the shape and size of hysteresis loop.

(i) **The material.** The shape and size of the hysteresis loop largely depends upon the nature of the material. If the material is easily magnetised, the loop will be narrow. On the other hand, if the material does not get magnetised easily, the loop will be wide. Further, different materials will saturate at different values of magnetic flux density thus affecting the height of the loop.

(ii) **The maximum flux density.** The loop area also depends upon the maximum flux density that is established in the material. This is illustrated in Fig. 8.38. It is clear that the loop area increases as the alternating magnetic field has progressively greater peak values.

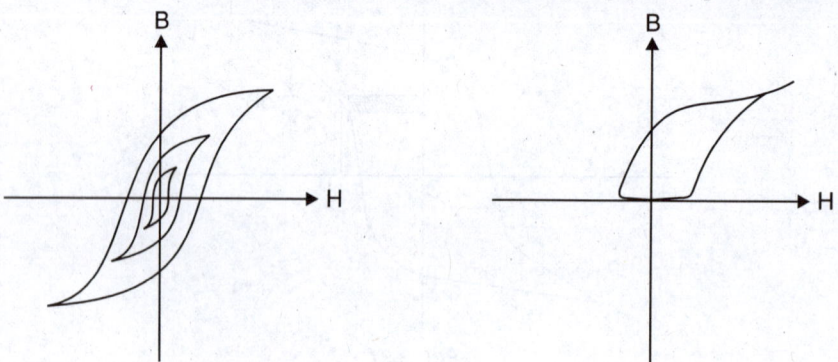

Variation of peak flux density
Fig. 8.38   **Fig. 8.39**

(iii) **The initial state of the specimen.** The shape and size of the hysteresis loop also depends upon the initial state of the specimen. To illustrate this point, refer to Fig. 8.39. It is clear that the specimen is already saturated to start with. The magnetic flux density is then reduced to zero and finally the specimen is returned to the saturated condition.

8.20. Importance of Hysteresis Loop

The shape and size of the hysteresis loop *largely depends upon the nature of the material. The choice of a magnetic material for a particular application often depends upon the shape and size of the hysteresis loop. A few cases are discussed below by way of illustration.

(*i*) *The smaller the hysteresis loop area of a magnetic material, the less is the hysteresis loss.* The hysteresis loop for silicon steel has a very small area [See Fig. 8.40 (*i*)]. For this reason, silicon steel is widely used for making transformer cores and rotating machines which are subjected to rapid reversals of magnetisation.

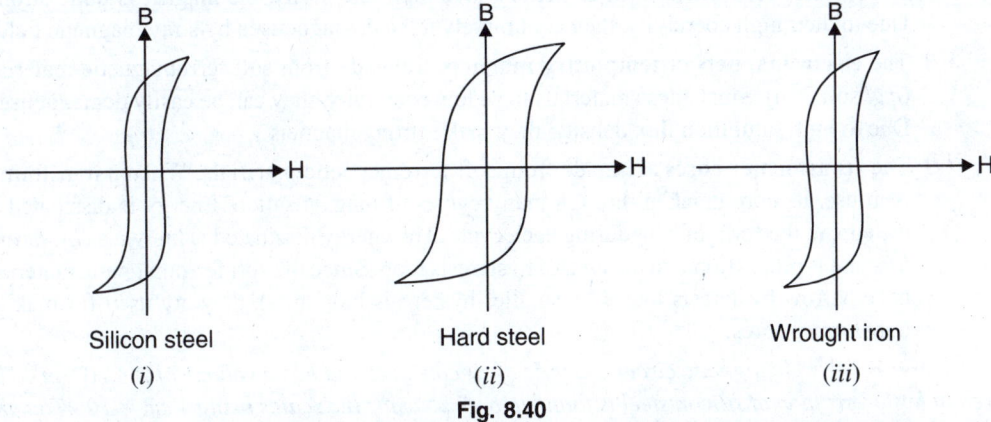

Silicon steel	Hard steel	Wrought iron
(*i*)	(*ii*)	(*iii*)

Fig. 8.40

(*ii*) The hysteresis loop for hard steel [See Fig. 8.40 (*ii*)] indicates that this material has high retentivity and coercivity. Therefore, hard steel is quite suitable for making permanent magnets. But due to the large area of the loop, there is greater hysteresis loss. For this reason, hard steel is not suitable for the construction of electrical machines.

(*iii*) The hysteresis loop for wrought iron [See Fig. 8.40 (*iii*)] shows that this material has fairly good residual magnetism and coercivity. Hence, it is suitable for making cores of electromagnets.

8.21. Applications of Ferromagnetic Materials

Ferromagnetic materials (*e.g.* iron, steel, nickel, cobalt etc.) are widely used in a number of applications. The choice of a ferromagnetic material for a particular application depends upon its magnetic properties such as retentivity, coercivity and area of the hysteresis loop. Ferromagnetic materials are classified as being either **soft** (soft iron) and **hard** (steel). Fig. 8.41 shows the hysteresis loop for soft and hard ferromagnetic materials. The table below gives the magnetic properties of hard and soft ferromagnetic materials.

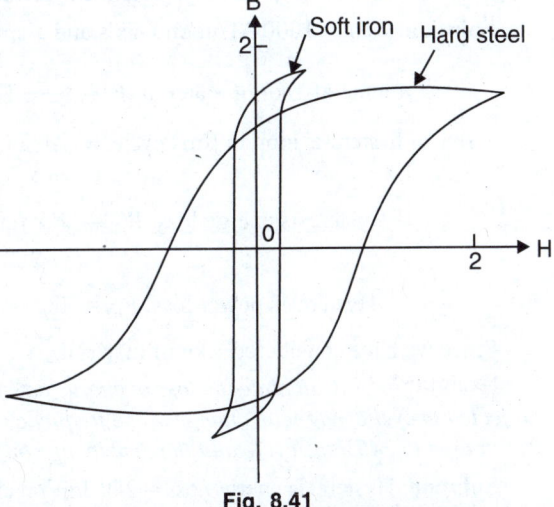

Fig. 8.41

---

*   It also depends upon (*i*) the maximum value of flux density established and (*ii*) the initial magnetic state of the material.

Magnetic property	Soft Iron	Hard Steel
Hysteresis loop	narrow	large area
Retentivity	high	high
Coercivity	low	high
Saturation flux density	high	good

(*i*) The **permanent magnets** are made from hard ferromagnetic materials (steel, cobalt steel, carbon steel etc). Since these materials have high retentivity, the magnet is quite strong. Due to their high coercivity, they are unlikely to be demagnetised by stray magnetic fields.

(*ii*) The **electromagnets** or **temporary magnets** are made from soft ferromagnetic materials (*e.g.* soft iron). Since these materials have low coercivity, they can be easily demagnetised. Due to high saturation flux density, they make strong magnets.

(*iii*) The **transformer cores** are made from soft ferromagnetic materials. When a transformer is in use, its core is taken through many cycles of magnetisation. Energy is dissipated in the core in the form of heat during each cycle. The energy dissipated is known as *hysteresis loss* and is proportional to the area of hysteresis loop. Since the soft ferromagnetic materials have narrow hysteresis loop (*i.e.* smaller hysteresis loop area), they are used for making transformer cores.

Example 8.32. *A magnetic circuit is made of silicon steel and has a volume of* $2 \times 10^{-3} m^3$. *The area of hysteresis loop of silicon steel is found to be* $7 \cdot 25$ cm^2 ; *the scales being 1 cm = 10 AT/m and 1 cm = 4 Wb/m^2. Calculate the hysteresis power loss when the flux is alternating at 50 Hz.*

Solution. 1 cm = 10 AT/m on x-axis and 1 cm = 4 Wb/m^2 on y-axis.

Area of hysteresis loop in J/m^3/cycle = (Area in cm^2) × (Scale factors) = $(7 \cdot 25) \times (xy)$

$$= (7 \cdot 25) \times (10 \times 4) = 290 \text{ J/m}^3/\text{cycle}$$

∴   Hysteresis power loss, P_h = Volume × area of loop × frequency

$$= (2 \times 10^{-3}) \times (290) \times (50) \text{ W} = \mathbf{29 \text{ W}}$$

Example 8.33. *The area of hysteresis loop obtained with a certain magnetic material was* $9 \cdot 3$ cm^2. *The co-ordinates were such that 1 cm = 1000 AT/m and 1 cm = 0·2 Wb/m^2. If the density of the given material is* $7 \cdot 8$ g/cm^3, *calculate the hysteresis loss in watts/kg at 50 Hz.*

Solution. 1 cm = 1000 AT/m on x-axis and 1 cm = 0.2 Wb/m^2 on y-axis.

Volume of 1 kg of material, $V = \dfrac{10^3}{7 \cdot 8} 10^{-6} = 1 \cdot 282 \times 10^{-4} \text{ m}^3$

Area of hysteresis loop in J/m^3/cycle = Area in cm^2 × scales factors

$$= (9 \cdot 3) \times (1000 \times 0 \cdot 2) = 1860 \text{ J/m}^3/\text{cycle}$$

Hysteresis energy loss, $W_h = V \times$ (area of loop in J/m^3/cycle)

$$= (1 \cdot 282 \times 10^{-4}) \times 1860 = 0 \cdot 238 \text{ J/cycle}$$

Hysteresis power loss, $P_h = W_h \times f = 0 \cdot 238 \times 50 = 11 \cdot 9 \text{ W}$

Since we have considered 1 kg of material, ∴ Hysteresis power loss, $P_h = \mathbf{11 \cdot 9 \text{ W/kg}}$

Example 8.34. *Calculate the loss of energy caused by hysteresis in 1 hour in 50 kg of iron when subjected to cyclic magnetic changes. The frequency is 25 Hz, the area of hysteresis loop is equivalent in area to 240 J/m^3/cycle and the density of iron is* $7 \cdot 8$ g/cm^3.

Solution. Hysteresis energy loss = 240 J/m^3/cycle

Volume of iron = $\dfrac{\text{mass}}{\text{density}} = \dfrac{50 \times 10^3}{7 \cdot 8} 10^{-6} = 6 \cdot 41 \times 10^{-3} \text{ m}^3$

No. of cycles/hour $= 25 \times 60 \times 60 = 9 \times 10^4$

$\therefore$ Energy loss/hour $=$ volume $\times$ (area of loop in J/m^3/cycle) $\times$ cycles/hour

$= (6\cdot41 \times 10^{-3}) \times (240) \times (9 \times 10^4) = \mathbf{138456\ J}$

Example 8.35. *The armature of a 4-pole d.c. generator has a volume of 12×10^{-3} m^3. During rotation, the armature is taken through a hysteresis loop whose area is 20 cm^2 when plotted to a scale of 1 cm = 100 AT/m, 1 cm = 0·1 Wb/m^2. Determine the hysteresis loss in watts when the armature rotates at a speed of 900 r.p.m.*

Solution. 1 cm = 100 AT/m on x–axis and 1 cm = 0.1 Wb/m^2 on y-axis. Since it is a 4-pole machine, two hysteresis loops will be formed in one revolution of the armature.

$\therefore$ No. of loops generated/second, $f = 2 \times 900/60 = 30$

Hysteresis energy loss/cycle $=$ Area of loop in cm^2 $\times$ scale factors

$= 20 \times (100 \times 0\cdot1) = 200$ J/m^3/cycle

Total hysteresis energy loss/second $=$ volume $\times$ (area of loop in J/m^3/cycle) $\times f$

$= (12 \times 10^{-3}) \times 200 \times 30 = 72$ W

i.e. Hysteresis power loss $= \mathbf{72\ W}$

Example 8.36. *A magnetic circuit core is made of silicon steel and has a volume of 1000000 mm^3. Using the hysteresis loop shown in Fig. 8.42, calculate the hysteresis power loss when the flux is alternating at 50 Hz.*

Solution. Hysteresis power loss, $P_h = V \times f \times$ (area of loop in J/m^3/cycle)

Volume of material, $V = 1000000$ mm$^3 = 1000000 \times 10^{-9}$ m^3

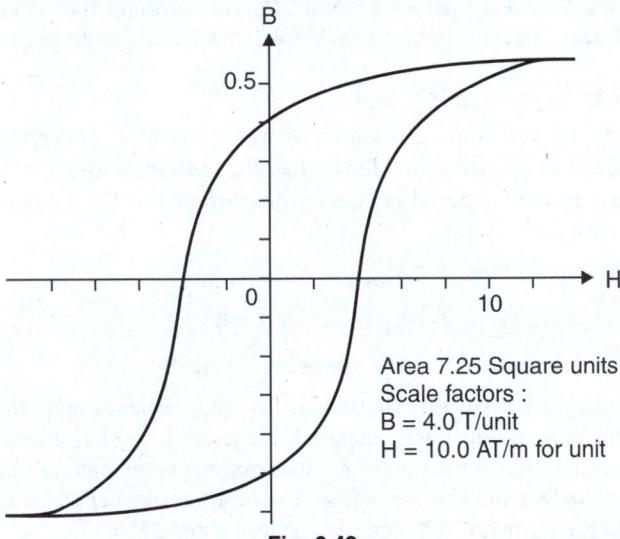

Fig. 8.42

Area of loop in J/m^3/cycle $=$ Area in square units $\times$ scale factors

$= 7\cdot25 \times 4 \times 10 = 290$ J/m^3/cycle

$\therefore$ $P_h = (1000000 \times 10^{-9}) \times 50 \times 290 = \mathbf{14\cdot5\ W}$

Example 8.37. *A hysteresis loop is plotted with horizontal axis scale of 1 cm = 1000 AT/m and vertical axis scale of 5 cm = 1T. The area of the loop is 9 cm^2 and overall height is 14 cm. Find (i) hysteresis loss in J/m^3/cycle (ii) B_m and (iii) hysteresis loss in W/kg if density is 7800 kg/m^3. The frequency is 50 Hz.*

Solution. **(i)** 1 cm = 1000 AT/m on x-axis and 1 cm = 0.2T on y-axis.

Area of hysteresis loop in J/m^3/cycle $=$ (Area of loop in cm^2) $\times$ scale factors

$$= (9) \times (1000 \times 0.2) = 1800 \text{ J/m}^3/\text{cycle}$$

i.e. Hysteresis energy loss = **1800 J/m³/cycle**

(*ii*) In a hysteresis loop, flux density varies from $+B_m$ to $-B_m$. The scale for B is 5 cm = 1T and the overall height of the loop is 14 cm.

$$\therefore \qquad 2\,B_m = \frac{14}{5} = 2.8 \text{ T or } B_m = \frac{2.8}{2} = 1.4 \text{ T}$$

(*iii*)   Volume of 1 kg of material, $V = \dfrac{\text{Mass}}{\text{Density}} = \dfrac{1}{7800} \text{m}^3$

$\therefore$ Hysteresis power loss, P_h = Energy loss/m³/cycle $\times V \times f$

$$= 1800 \times \frac{1}{7800} \times 50 = 11.538 \text{ W}$$

Since we have considered 1 kg of material, $\therefore P_h = \textbf{11.538 W/kg}$

<hr>

Tutorial Problems

1. The hysteresis loop for a specimen of mass 12 kg is equivalent to 30 W/mm³. Find the loss of energy in kWh in one hour at 50 Hz. The density of the specimen is 7·8 g/cm³. **[0·024 kWh]**

2. A transformer is made of 200 kg of steel plate with a specific gravity of 7·5. It may be assumed that the maximum operating flux density is 1·1 Wb/m² for all parts of the steel. When a specimen of the steel was tested, it was found to have a hysteresis loop of area 100 cm² for a maximum flux density of 1·1 Wb/m². If the scales of the hysteresis loop graph were 1 cm = 50 AT/m and 1 cm = 0·1 Wb/m², calculate the hysteresis power loss when the transformer is operated on 50 Hz mains. **[667 W]**

3. A magnetic core is made from sheet steel, the hysteresis loop of which has an area of 2·1 cm²; the scales being 1 cm = 400 AT/m and 1 cm = 0·4 Wb/m². The core measures 100 cm long and has an average cross-sectional area of 10 cm². The hysteresis loss is 16·8 W. Calculate the frequency of alternating flux.

[50 Hz]

8.22. Steinmetz Hysteresis Law

To eliminate the need of finding the area of hysteresis loop for computing the hysteresis loss, Steinmetz devised an empirical law for finding the hysteresis loss. He found that the area of hysteresis loop of a magnetic material is directly proportional to 1·6 the power of the maximum flux density established *i.e.*

Area of hysteresis loop $\propto$ *$B_{max}^{1.6}$

or   Hysteresis energy loss $\propto B_{max}^{1.6}$ joules/m³/cycle

or   Hysteresis energy loss = $\eta\, B_{max}^{1.6}$ joules/m³/cycle

where η is a constant called **hysteresis coefficient**. Its value depends upon the nature of material. The smaller the value of η of a magnetic material, the lesser is the hysteresis loss. The armatures of electrical machines and transformer cores are made of magnetic materials having low hysteresis coefficient in order to reduce the hysteresis loss. The best transformer steels have η values around 130, for cast steel they are around 2500 and for cast iron about 3750.

If V is the volume of the material in m³ and f is the frequency of reversal of magnetisation, then,

Hysteresis power loss; $P_h = \eta\, f\, B_{max}^{1.6}\, V$ J/s or watts

Example 8.38. *The volume of a transformer core built up of sheet steel laminations is 5000 cm³ and the gross cross-sectional area is 240 cm². Because of the insulation between the plates, the net cross-sectional area is 90% of the gross. The maximum value of flux is 22 mWb and the frequency is 50 Hz. Find (i) the hysteresis loss/m³/cycle and (ii) power loss in watts. Take hysteresis coefficient as 250.*

<hr>

*   The index 1·6 is called **Steinmetz index**. In fact, the value of this index depends upon the nature of material and may vary from 1·6 to 2·5. However, reasonable accuracy is obtained if it is taken as 1·6.

Solution. $a = 0.9 \times 240 = 216 \text{ cm}^2$; $B_{max} = \dfrac{22 \times 10^{-3}}{216 \times 10^{-4}} = 1.019 \text{ Wb/m}^2$

(i) Hysteresis energy loss $= \eta \, B^{1.6}_{max} = 250 \times (1.019)^{1.6} = \textbf{257.6 J/m}^3\textbf{/cycle}$

(ii) Hysteresis power loss, $P_h = \eta f B^{1.6}_{max} \times V = (257.6) \times (50) \times (5000 \times 10^{-6}) = \textbf{64.4 W}$

Example 8.39. *The area of hysteresis loop obtained with a certain specimen of iron was 9.3 cm². The co-ordinates were such that 1 cm = 1000 AT/m and 1 cm = 0.2 Wb/m². Calculate (i) the hysteresis loss in J/m³/cycle (ii) hysteresis loss in W/m³ at a frequency of 50 Hz. (iii) If the maximum flux density was 1.5 Wb/m², calculate the hysteresis loss/m³ for a maximum flux density of 1.2 Wb/m², and a frequency of 30 Hz, assuming the loss to be proportional to $B^{1.8}_{max}$.*

Solution. 1 cm = 1000 AT/m on x-axis and 1 cm = 0.2 Wb/m² on y-axis.

(i) Hysteresis energy loss $= (xy) \times (\text{area of loop}) \text{ J/m}^3\text{/cycle}$

$\qquad\qquad\qquad\qquad\qquad = (1000 \times 0.2) \times 9.3 = \textbf{1860 J/m}^3\textbf{/cycle}$

(ii) Hysteresis power loss $= 1860 \times 50 = \textbf{93,000 W/m}^3$

(iii) Hysteresis power loss/m³ $= \eta f (B_{max})^{1.8}$

or $\qquad\qquad\qquad 93000 = \eta \times 50 \times (1.5)^{1.8}$

$\therefore \qquad\qquad\qquad \eta = \dfrac{93000}{50 \times (1.5)^{1.8}} = 896.5$

For $B_{max} = 1.2 \text{ Wb/m}^2$ and $f = 30$ Hz,

Hysteresis loss/m³ $= \eta f (B_{max})^{1.8} \text{ W} = 896.5 \times 30 \times (1.2)^{1.8} = \textbf{37342 W}$

Example 8.40. *A cylinder of iron of volume $8 \times 10^{-3} \text{ m}^3$ revolves for 20 min at a speed of 3000 r.p.m. in a two-pole field of flux density 0.8 Wb/m². If the hysteresis coefficient of iron is 753.6 J/m³, specific heat of iron is 0.11, the loss due to eddy current is equal to that due to hysteresis and 25% of heat produced is lost by radiation, find the temperature rise of iron. Take density of iron as $7.8 \times 10^3 \text{ kg/m}^3$.*

Solution. When an armature revolves in a multipolar field, one magnetic reversal occurs after it passes a pair of poles. If P is the number of poles, the number of magnetic reversals in one revolution is $P/2$. If the speed of the armature is N r.p.m., then number of revolutions/second = $N/60$.

$\therefore$ No. of magnetic reversals/second = Reversal in one sec. × No. of revolutions/sec.

or Frequency of magnetic reversals $= \dfrac{P}{2} \times \dfrac{N}{60} = \dfrac{2}{2} \times \dfrac{3000}{60} = 50 \text{ cycles/sec}$

According to Steinmetz hysteresis law,

Hysteresis power loss, $P_h = \eta f B^{1.6}_{max} V \text{ joules/sec.}$

$\qquad\qquad\qquad = 753.6 \times 50 \times (0.8)^{1.6} \times 8 \times 10^{-3} = 211 \text{ J/s}$

$\therefore \qquad$ Energy loss in 20 min. $= 211 \times (20 \times 60) = 253.2 \times 10^3 \text{ J}$

$\qquad\qquad$ Eddy current loss $= 253.2 \times 10^3 \text{ J ... } given$

$\therefore \qquad\qquad$ Total energy loss $= 2 \times 253.2 \times 10^3 = 506.4 \times 10^3 \text{ J}$

$\qquad\qquad$ Heat produced $= \dfrac{506.4 \times 10^3}{J} = \dfrac{506.4 \times 10^3}{4200} = 120.57 \text{ kcal}$

It is given that 25% of heat produced is lost due to radiation.

$\therefore$ Heat used to heat iron cylinder = 0.75 × 120.57 = 90.43 kcal

Now, mass of iron cylinder, m = volume × density = $8 \times 10^{-3} \times 7.8 \times 10^3 = 62.4$ kg; specific heat, $S = 0.11$.

If θ°C is the rise of temperature of iron cylinder, then,

$$mS\theta = 90.43 \quad \text{or} \quad \theta = \dfrac{90.43}{62.4 \times 0.11} = \textbf{13.17°C}$$

Example 8.41. *In a certain transformer, the hysteresis loss was found to be 160 watts when the maximum flux density was 1·1 Wb/m² and the frequency 60 Hz. What will be the loss when the maximum flux density is reduced to 0·9 Wb/m² and frequency to 50 Hz ?*

Solution. According to Steinmetz hysteresis law,

$$\text{Hysteresis loss, } P_h \propto f(B_{max})^{1·6}$$

For the first case, $P_1 \propto 60 \times (1·1)^{1·6}$

For the second case, $P_2 \propto 50 \times (0·9)^{1·6}$

$$\therefore \qquad \frac{P_2}{P_1} = \frac{50 \times (0·9)^{1·6}}{60 \times (1·1)^{1·6}} = 0·604$$

$$\therefore \qquad P_2 = 0·604 \, P_1 = 0·604 \times 160 = \mathbf{96·64 \ W}$$

Example 8.42. *Calculate the loss of energy caused by hysteresis in one hour in 11·25 kg of iron if maximum flux density reached is 1·3 Wb/m² and frequency is 50 Hz. Assume Steinmetz coefficient as 500 J/m³/cycle and density of iron as 7·5 g/cm³.*

What will be the area of B/H curve (i.e. hysteresis loop) of this specimen if 1 cm = 50 AT/m and 1 cm = 0·1 Wb/m² ?

Solution.   Volume of iron, $V = \dfrac{11·25}{7·5 \times 10^3} = 1·5 \times 10^{-3} \text{ m}^3$

Hysteresis power loss, $P_h = \eta f (B_{max})^{1·6} \, V$ watts

$$= 500 \times 50 \times (1·3)^{1·6} \times (1·5 \times 10^{-3}) = 57·06 \text{ W}$$

$\therefore$ Hysteresis energy loss in 1 hour

$$= 57·06 \times 3600 = \mathbf{205416 \ J}$$

According to Steinmetz hysteresis law,

$$\text{Hysteresis energy loss} = \eta \, (B_{max})^{1·6} \text{ J/m}^3\text{/cycle}$$

1 cm = 50 AT/m on x-axis and 1 cm = 0.1 Wb/m² on y-axis.

$$\text{Hysteresis energy loss} = xy \times (\text{area of loop}) \text{ J/m}^3\text{/cycle}$$

Equating the two, we get,

$$500 \times (1·3)^{1·6} = (50 \times 0·1) \times \text{Area of loop}$$

$$\therefore \qquad \text{Area of loop} = \frac{500 \times (1·3)^{1·6}}{50 \times 0·1} = \mathbf{152·16 \ cm}^2$$

Tutorial Problems

1. The hysteresis loss in an iron specimen is given by the expression; Hysteresis loss is J/m³/cycle = $\eta B_{max}^{1.7}$ where B_{max} is the maximum flux density. If loss is 5.215 W/kg at a frequency of 50 Hz and a maximum flux density is 1.1 Wb/m², find the constant η if density of iron is 7600 kg/m³. Also find the hysteresis loss at 60 Hz if B_{max} = 1.7 Wb/m². **[674.11; 13.117 W/kg]**

2. A sample of silicon steel has a hysteresis coefficient of 100 and a corresponding Steinmetz index of 1.6. Calculate the hysteresis power loss in 10^6 mm³ when the flux is alternating at 50 Hz, such that the maximum flux density is 2T. **[15.2 W]**

3. The hysteresis loss in an iron specimen is proportional to $(B_{max})^{1.7}$. At B_{max} = 1.1T, the hyteresis loss is 320W at 50 Hz. Find hysteresis loss at 60 Hz if B_{max} = 1.6 T. **[726.05 W]**

8.23. Comparison of Electrostatics and Electromagnetic Terms

It may be worthwhile to compare the terms and symbols used in electrostatics with the corresponding terms and symbols used in electromagnetism. (See table on page 427).

Electrostatics		Electromagnetism	
Term	**Symbol**	**Term**	**Symbol**
Electric flux	ψ	Magnetic flux	ϕ
Electric flux density	D	Magnetic flux density	B
Electric field strength	E	Magnetic field strength	H
Electromotive force	E	Magnetomotive force	—
Electric potential difference	V	Magnetic potential difference	—
Permittivity of free space	ε_0	Permeability of free space	μ_0
Relative permittivity	ε_r	Relative permeability	μ_r
Absolute permittivity		Absolute permeability	
$= \dfrac{\text{electric flux density}}{\text{electric field strength}}$		$= \dfrac{\text{magnetic flux density}}{\text{magnetic field strength}}$	
i.e. $\varepsilon_0\varepsilon_r = \varepsilon = D/E$		*i.e.* $\mu_0\mu_r = \mu = B/H$	

Objective Questions

1. In Fig. 8.43, the magnetic circuit is the path

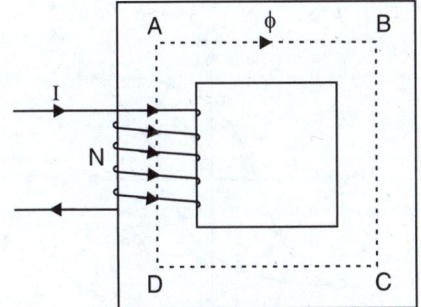

Fig. 8.43

   (*i*) *DAB*   (*ii*) *ABCDA*

   (*iii*) *ABC*   (*iv*) *ABCD*

2. If *l* is the magnetic path in Fig. 8.43, then magnetising force is

   (*i*) *NI*   (*ii*) *NI × l*

   (*iii*) *l/NI*   (*iv*) *NI/l*

3. The reluctance of the magnetic circuit shown in Fig. 8.43 is

   (*i*) *NI/l*   (*ii*) ϕ*/NI*

   (*iii*) *NI/*ϕ (*iv*) ϕ*/l*

4. The SI unit of reluctance is

   (*i*) AT/Wb   (*ii*) AT/m

   (*iii*) AT   (*iv*) N/Wb

5. A magnetic circuit has m.m.f. of 400 AT and reluctance of 2×10^5 AT/Wb. The magnetic flux in the magnetic circuit is

   (*i*) 3×10^{-5} Wb   (*ii*) 2×10^{-3} Wb

   (*iii*) $1 \cdot 5 \times 10^{-2}$ Wb   (*iv*) $2 \cdot 5 \times 10^{-4}$ Wb

6. A 2 cm long coil has 10 turns and carries a current of 750 mA. The magnetising force of the coil is

   (*i*) 225 AT/m   (*ii*) 675 AT/m

   (*iii*) 450 AT/m   (*iv*) 375 AT/m

7. A magnetic device has a core with cross-section of 1 inch^2. If the flux in the core is 1 mWb, then flux density (1 inch = 2.54 cm) is

   (*i*) 2.5 T   (*ii*) 1.3 T

   (*iii*) 1.55 T   (*iv*) 0.25 T

8. The reluctance of a magnetic circuit varies as ....

   (*i*) length × area   (*ii*) length ÷ area

   (*iii*) area ÷ length   (*iv*) (length)^2 + area

9. The reluctance of a magnetic circuit is ..... relative permeability of the material comprising the circuit.

   (*i*) directly proportional to

   (*ii*) inversely proportional to

   (*iii*) independent of

   (*iv*) none of the above

10. M.M.F. in a magnetic circuit corresponds to ...... in an electric circuit.

   (*i*) voltage drop   (*ii*) potential difference

   (*iii*) electric intensity (*iv*) e.m.f.

11. Permeance of a magnetic circuit is .... area of *x*-section of the circuit.

   (*i*) inversely proportional to

   (*ii*) directly proportional to

(iii) independent of

(iv) none of the above.

12. The magnitude of AT required for air gap is much greater than that required for iron part of a magnetic circuit because ....

(i) air is a gas

(ii) air has the lowest relative permeability

(iii) air is a conductor of magnetic flux

(iv) none of the above

13. In electro-mechanical conversion devices (*e.g.* motors and generators), a small air gap is left between the rotor and stator in order to .....

(i) complete the magnetic path

(ii) decrease the reluctance of magnetic path

(iii) permit mechanical clearance

(iv) increase flux density in air gap

14. A magnetic circuit carries a flux ϕ_i in the iron part and a flux ϕ_g in the air gap. Then leakage coefficient is ......

(i) ϕ_i/ϕ_g (ii) ϕ_g/ϕ_i

(iii) $\phi_g \times \phi_i$ (iv) none of the above

15. The value of leakage coefficient for electrical machines is usually about.......

(i) 0·5 to 1   (ii) 4 to 10

(iii) above 10   (iv) 1·15 to 1·25

16. The reluctance of a magnetic circuit depends upon ....

(i) current in the coil

(ii) no. of turns of coil

(iii) flux density in the circuit

(iv) none of the above

17. The *B-H* curve for .... will be a straight line passing through the origin.

(i) air   (ii) soft iron

(iii) hardened steel   (iv) silicon steel

18. Whatever may be the flux density in ...., the material will never saturate.

(i) soft iron   (ii) cobalt steel

(iii) air   (iv) silicon steel

19. The *B-H* curve of .... will not be a straight line.

(i) air   (ii) copper

(iii) wood   (iv) soft iron

20. The *B-H* curve is used to find the m.m.f. of .... in a magnetic circuit.

(i) air gap   (ii) iron part

(iii) both air gap and iron part

(iv) none of the above

21. A magnetising force of 800 AT/m will produce a flux density of .... in air.

(i) 1 mWb/m^2 (ii) 1 Wb/m^2

(iii) 10 mWb/m^2 (iv) $0·5 \text{ Wb/m}^2$

22. The saturation flux density for most magnetic materials is about ....

(i) $0·5 \text{ Wb/m}^2$ (ii) 10 Wb/m^2

(iii) 2 Wb/m^2 (iv) 1 Wb/m^2

23. Hysteresis is the phenomenon of .... in a magnetic circuit.

(i) lagging of *B* behind *H*

(ii) lagging of *H* behind *B*

(iii) setting up constant flux

(iv) none of the above

24. In Fig. 8.44, the point .... represents the saturation condition.

(i) *b*   (ii) *c*

(iii) *a*   (iv) *e*

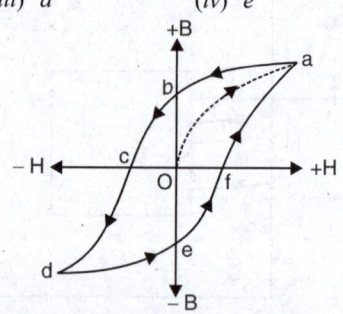

Fig. 8.44

25. In Fig. 8.44, ... represents the residual magnetism.

(i) *of*   (ii) *oc*

(iii) *ob*   (iv) none of the above

26. In Fig. 8.44, oc represents the ....

(i) residual magnetism

(ii) coercive force

(iii) retentivity   (iv) none of the above

27. If a magnetic material is located within a coil through which alternating current (50 Hz frequency) flows, then .... hysteresis loops will be formed every second.

(i) 50   (ii) 25

(iii) 100   (iv) 150

28. Out of the following materials, the area of hysteresis loop will be least for ....

(i) wrought iron   (ii) hard steel

(iii) silicon steel   (iv) soft iron

29. The materials used for the core of a good relay should have .... hysteresis loop.

   (*i*) large   (*ii*) very large

   (*iii*) narrow   (*iv*) none of the above

30. The magnetic material used for .... should have a large hysteresis loop.

   (*i*) transformers   (*ii*) d.c. generators

   (*iii*) a.c. motors   (*iv*) permanent magnets

Answers

1. (*ii*)	**2.** (*iv*)	**3.** (*iii*)	**4.** (*i*)	**5.** (*ii*)
6. (*iv*)	**7.** (*iii*)	**8.** (*ii*)	**9.** (*ii*)	**10.** (*iv*)
11. (*ii*)	**12.** (*ii*)	**13.** (*iii*)	**14.** (*i*)	**15.** (*iv*)
16. (*iii*)	**17.** (*i*)	**18.** (*iii*)	**19.** (*iv*)	**20.** (*ii*)
21. (*i*)	**22.** (*iii*)	**23.** (*i*)	**24.** (*iii*)	**25.** (*iii*)
26. (*ii*)	**27.** (*i*)	**28.** (*iii*)	**29.** (*iii*)	**30.** (*iv*)

9

Electromagnetic Induction

Introduction

In the beginning of nineteenth century, Oersted discovered that a magnetic field exists around a current-carrying conductor. In other words, magnetism can be created by means of an electric current. Can a magnetic field create an electric current in a conductor ? In 1831, Michael Faraday, the famous English scientist, discovered that this could be done. He demonstrated that when the magnetic flux linking a conductor changes, an e.m.f. is induced in the conductor. This phenomenon is known as *electromagnetic induction*. The great discovery of electromagnetic induction by Faraday through a series of brilliant experiments has brought a revolution in the engineering world. Most of the electrical devices (*e.g.* electric generator, transformer, telephones *etc.*) are based on this principle. In this chapter, we shall confine our attention to the various aspects of eletromagnetic induction.

9.1. Electromagnetic Induction

When the magnetic flux *linking a conductor changes, an e.m.f. is induced in the conductor. If the conductor forms a complete loop or circuit, a current will flow in it. This phenomenon is known as **electromagnetic induction.

The phenomenon of production of e.m.f. and hence current in a conductor or coil when the magnetic flux linking the conductor or coil changes is called **electromagnetic induction.**

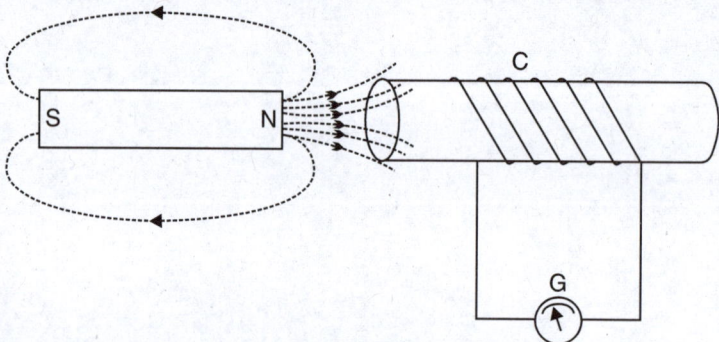

Fig. 9.1

To demonstrate the phenomenon of electromagnetic induction, consider a coil *C* of several turns connected to a centre zero galvanometer *G* as shown in Fig. 9.1. If a permanent magnet is moved towards the coil, it will be observed that the galvanometer shows deflection in one direction. If the magnet is moved away from the coil, the galvanometer again shows deflection but in the opposite direction. In either case, *the deflection will persist so long as the magnet is in motion.* The production of e.m.f. and hence current in the coil *C* is due to the fact that when the magnet is in motion (towards or away from the coil), the amount of flux linking the coil changes—the basic requirement for inducing e.m.f. in the coil. If the movement of the magnet is stopped, though the flux is linking the

---

*   Magnetic lines of force form closed loops. Flux linking the conductor means that the flux embraces it *i.e.* it encircles the conductor.

**   So called because electricity is produced from magnetism (*i.e. electromagnetic*) and that there is no physical connection (*induction*) between the magnetic field and the conductor.

coil, there is *no change in flux* and hence no e.m.f. is induced in the coil. Consequently, the deflection of the galvanometer reduces to zero.

The following points may be noted carefully :

(*i*) *The basic requirement for inducing e.m.f. in a coil is not the magnetic flux linking the coil but the change in flux linking the coil. No change in flux, no e.m.f. induced in the coil.*

(*ii*) The change in flux linking the coil can be brought about in two ways. First, the conductors (or coils) are moved through a stationary magnetic field as is the case with d.c. generators. Secondly, the conductors are stationary and the magnetic field is moving as is the case with a.c. generators. In either case, the basic principle is the same *i.e.* the amount of flux linking the conductors (or coils) is changed.

(*iii*) *The e.m.f. and hence current in the conductors (or coils) will persist so long as the magnetic flux linking them is changing.*

Note. We have seen that when magnetic flux linking a conductor changes, an e.m.f. is induced in it. An equivalent statement is like this : *When a conductor cuts magnetic field lines , an e.m.f. is induced in it.* If the conductor moves parallel to the magnetic field lines, no e.m.f. is induced. This terminology is very helpful in visualising the concept of production of e.m.f.

9.2. Flux Linkages

The product of number of turns (N) of the coil and the magnetic flux (ϕ) linking the coil is called **flux linkages** *i.e.*

Flux linkages $= N\phi$

Experiments show that the magnitude of e.m.f. induced in a coil is directly proportional to the rate of change of flux linkages. If N is the number of turns of the coil and the magnetic flux linking the coil changes (say increases) from ϕ_1 to ϕ_2 in t seconds, then,

Induced e.m.f., $e \propto$ Rate of change of flux linkages

or $$e \propto \frac{N\phi_2 - N\phi_1}{t}$$

9.3. Faraday's Laws of Electromagnetic Induction

Faraday performed a series of experiments to demonstrate the phenomenon of electromagnetic induction. He summed up his conclusions into two laws, known as Faraday's laws of electromagnetic induction.

First Law. It tells us about the condition under which an e.m.f. is induced in a conductor or coil and may be stated as under :

When the magnetic flux linking a conductor or coil changes, an e.m.f. is induced in it.

It does not matter how the change in magnetic flux is brought about. The essence of the first law is that the induced e.m.f. appears in a circuit subjected to a changing magnetic field.

Second Law. It gives the magnitude of the induced e.m.f. in a conductor or coil and may be stated as under :

The magnitude of the e.m.f. induced in a conductor or coil is directly proportional to the rate of change of flux linkages i.e.

Induced e.m.f., $e \propto \dfrac{N\phi_2 - N\phi_1}{t}$

or $$e = k\frac{N\phi_2 - N\phi_1}{t}$$

where the value of k is *unity in SI units.

$$\therefore \qquad e = \frac{N\phi_2 - N\phi_1}{t}$$

In differential form, we have, $e = N\dfrac{d\phi}{dt}$

The direction of induced e.m.f. (and hence of induced current if the circuit is closed) is given by **Lenz's law.** The magnitude and direction of induced e.m.f. should be written as :

$$e = -N\frac{d\phi}{dt} \qquad\qquad ...(i)$$

The minus sign on the R.H.S. represents Lenz's law mathematically. In SI units, e is measured in volts, ϕ in webers and t in seconds.

9.4. Direction of Induced E.M.F. and Current

The direction of induced e.m.f. and hence current (if the circuit is closed) can be determined by one of the following two methods :

    (*i*) Lenz's Law   (*ii*) Fleming's right-hand rule

(*i*) Lenz's law. Emil Lenz, a German scientist, gave the following simple rule (known as Lenz's law) to find the direction of the induced current :

The induced current will flow in such a direction so as to oppose the cause that produces it i.e. the induced current will set up magnetic flux to oppose the change in flux.

Note that Lenz's law is reflected mathematically in the minus sign on the R.H.S. of Faraday's second law viz. $e = -\,N\,d\phi/dt.$

The negative sign simply reminds us that the induced current *opposes* the changing magnetic field that caused the induced current. The negative sign has no other meaning.

Let us apply Lenz's law to Fig. 9.2. Here the N-pole of the magnet is approaching a coil of several turns. As the N-pole of the magnet moves towards the coil, the magnetic flux linking the coil increases. Therefore an e.m.f. and hence current is induced in the coil according to Faraday's laws of electromagnetic induction. According to Lenz's law, the direction of the induced current will be such so as to oppose the cause that produces it. In the present case, the cause of the induced current is the increasing magnetic flux linking the coil. Therefore, the induced current will set up magnetic flux that opposes the increase in flux through the coil. This is possible only if the left hand face of the coil becomes N-pole. Once we know the magnetic polarity of the coil face, the direction of the induced current can be easily determined by applying right-hand rule for the coil. If the magnet is moved away from the coil, then by Lenz's law, the left hand face of the coil will become S-pole. Therefore, by right-hand rule for the coil, the direction of induced current in the coil will be opposite to that in the first case.

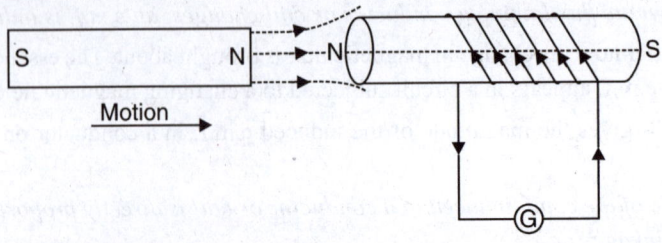

Fig. 9.2

---

   *   One volt (SI unit of e.m.f.) has been so defined that the value of k becomes unity. Thus 1V is said to be induced in a coil if the flux linkages change by 1 Wb-turn in 1 second.

     Here, $N\phi_2 - N\phi_1 = 1$ Wb-turn, $t = 1$ s and $e = 1$ volt $\quad\therefore\quad 1 = k\times\dfrac{1}{1} \quad$ or $\quad k = 1.$

It may be noted here that Lenz's law directly follows from the law of conservation of energy *i.e.* in order to set up induced current, some energy must be expended. In the above case, for example, when the *N*-pole of the magnet is approaching the coil, the induced current will flow in the coil in such a direction that the left-hand face of the coil becomes *N*-pole. The result is that the motion of the magnet is opposed. The mechanical energy spent in overcoming this opposition is converted into electrical energy which appears in the coil. Thus Lenz's law is consistent with the law of conservation of energy.

(*ii*) **Fleming's Right-Hand Rule.** This law is particularly suitable to find the direction of the induced e.m.f. and hence current when the conductor moves at right angles to a stationary magnetic field. It may be stated as under :

Stretch out the **forefinger, middle finger** *and* **thumb** *of your right hand so that they are at right angles to one another. If the forefinger points in the direction of magnetic field, thumb in the direction of motion of the conductor, then the middle finger will point in the direction of induced current.*

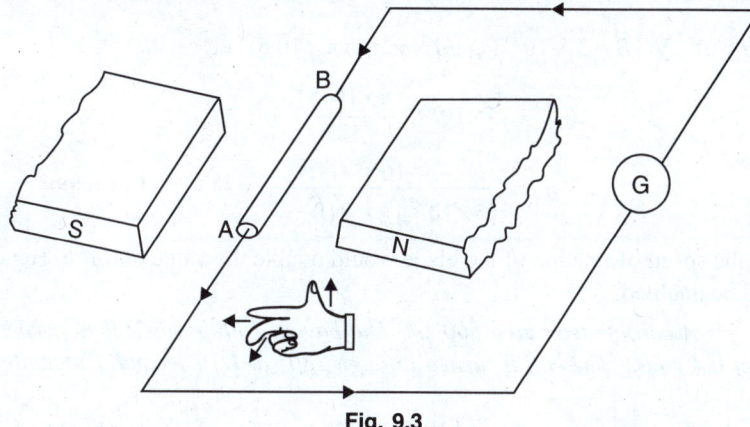

Fig. 9.3

Consider a conductor *AB* moving upwards at right angles to a uniform magnetic field as shown in Fig. 9.3. Applying Fleming's right-hand rule, it is clear that the direction of induced current is from *B* to *A*. If the motion of the conductor is downward, keeping the direction of magnetic field unchanged, then the direction of induced current will be from *A* to *B*.

Example 9.1. *A coil of 200 turns of wire is wound on a magnetic circuit of reluctance 2000 AT/Wb. If a current of 1A flowing in the coil is reversed in 10 ms, find the average e.m.f. induced in the coil.*

Solution.   Flux in the coil $= \dfrac{m.m.f.}{reluctance} = \dfrac{200 \times 1}{2000} = 0 \cdot 1$ Wb

When the current (*i.e.* 1A) in the coil is reversed, flux through the coil is also reversed.

$$e = N\frac{d\phi}{dt}$$

Here, $N = 200$; $d\phi = 0 \cdot 1 - (-0 \cdot 1) = 0 \cdot 2$ mWb ; $dt = 10 \times 10^{-3}$s

∴ $e = 200 \times \dfrac{0 \cdot 2 \times 10^{-3}}{10 \times 10^{-3}} = $ **4 V**

Example 9.2. *The field winding of a 4-pole d.c. generator consists of 4 coils connected in series, each coil being wound with 1200 turns. When the field is excited, there is a magnetic flux of 0·04 Wb/pole. If the field switch is opened at such a speed that the flux falls to the residual value of 0·004 Wb/pole in 0·1 second, calculate the average value of e.m.f. induced across the field winding terminals.*

Solution. Total no. of turns, $N = 1200 \times 4 = 4800$

Total initial flux $= 4 \times 0 \cdot 04 = 0 \cdot 16$ Wb

$$\text{Total residual flux} = 4 \times 0.004 = 0.016 \text{ Wb}$$

$$\text{Change in flux, } d\phi = 0.16 - 0.016 = 0.144 \text{ Wb}$$

$$\text{Time taken, } dt = 0.1 \text{ second}$$

$$\therefore \quad \text{Induced e.m.f., } e = N\frac{d\phi}{dt} = 4800 \times \frac{0.144}{0.1} = \mathbf{6912 \ V}$$

Example 9.3. *A fan blade of length 0·5m rotates perpendicular to a magnetic field of 5×10^{-5}T. If the e.m.f. induced between the centre and end of the blade is 10^{-2}V, find the rate of rotation of the blade.*

Solution. Let n be the required number of rotations in one second. The magnitude of induced e.m.f. is given by ;

$$e = N\frac{d\phi}{dt} = N\frac{d}{dt}(BA) = B\frac{dA}{dt} \qquad (\because N = 1)$$

Here dA is the area swept by the blade in one revolution and dt is the time taken to complete one revolution.

Now $e = 10^{-2}$ V ; $B = 5 \times 10^{-5}$T ; $dA = \pi r^2 = \pi \times (0.5)^2$ m^2 ; $dt = \dfrac{1}{n}$ s

$$\therefore \quad 10^{-2} = 5 \times 10^{-5} \times \frac{\pi \times (0.5)^2}{1/n}$$

or $$n = \frac{10^{-2}}{\left(5 \times 10^{-5}\right) \times \pi \times (0.5)^2} = \mathbf{254.7 \ rev \ / \ second}$$

Doubling the speed of rotation of the blade would double the value of dA/dt. Hence, the e.m.f. induced would be doubled.

Example 9·4. *A coil of mean area 500 cm^2 and having 1000 turns is held perpendicular to a uniform field of 0.4 gauss. The coil is turned through 180° in 1/10 second. Calculate the average induced e.m.f.*

Solution. $\phi = NBA \cos \theta$

When the plane of the coil is perpendicular to the field, $\theta = 0°$. When the coil is turned through 180°, $\theta = 180°$. Therefore, initial flux linked with the coil is

$$\phi_1 = NBA \cos 0° = NBA$$

Flux linked with coil when turned through 180° is

$$\phi_2 = NBA \cos 180° = -NBA$$

Change in flux linking the coil is

$$\Delta\phi = \phi_2 - \phi_1 = (-NBA) - (NBA) = -2 \ NBA$$

$$\therefore \quad \text{Average induced e.m.f., } e = -\frac{\Delta\phi}{\Delta t} = \frac{2NBA}{\Delta t}$$

Here $N = 1000$; $B = 0.4$ gauss $= 0.4 \times 10^{-4}$ T ; $A = 500 \times 10^{-4}$ m^2 ; $\Delta t = 0.1$ s

$$\therefore \quad e = \frac{2 \times 1000 \times (0.4 \times 10^{-4}) \times 500 \times 10^{-4}}{0.1} = \mathbf{0.04 \ V}$$

Example 9.5. *The magnetic flux passing perpendicular to the plane of the coil and directed into the paper (See Fig. 9.4) is varying according to the relation :*

$$\phi_B = 6t^2 + 7t + 1$$

where ϕ_B is in mWb and t in seconds.

(i) *What is the magnitude of induced e.m.f. in the loop when t = 2 seconds?*

(ii) *What is the direction of current through the resistor R ?*

Solution. $\phi_B = (6t^2 + 7t + 1) \text{ mWb} = (6t^2 + 7t + 1) \times 10^{-3}$ Wb

(i) Magnitude of induced e.m.f. is

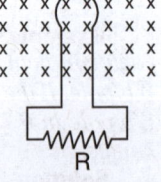

Fig. 9.4

$$e = \frac{d\phi_B}{dt} = \frac{d}{dt} (6t^2 + 7t + 1) \times 10^{-3} = (12t + 7) \times 10^{-3} \text{ V}$$

At $t = 2$ sec, $e = (12 \times 2 + 7) \times 10^{-3} = 31 \times 10^{-3}$ V = **31 mV**

(*ii*) According to Lenz's law, the direction of induced current will be such so as to oppose the change in flux. This means that direction of current in the loop will be such as to produce magnetic field opposite to the given field. For this (*i.e.*, upward field), the current induced in the loop will be anticlockwise. Therefore, **current in resistor R will be from left to right.**

Tutorial Problems

1. A square coil of side 5 cm contains 100 loops and is positioned perpendicular to a uniform magnetic field of 0.6 T. It is quickly removed from the field (moving perpendicular to the field) to a region where magnetic field is zero. It takes 0.1 s for the whole coil to reach field-free region. If resistance of the coil is 100 Ω, how much energy is dissipated in the coil ? **[2·3×10⁻³J]**

2. A flat search coil containing 50 turns each of area 2×10^{-4} m² is connected to a galvanometer; the total resistance of the circuit is 100Ω. The coil is placed so that its plane is normal to a magnetic field of flux density 0·25 T.
    (*i*) What is the change in magnetic flux linking the circuit when the coil is moved to a region of negligible magnetic field ?
    (*ii*) What charge passes through the galvanometer ? **[(i) 2·5×10⁻³ Wb (ii) 25 μC]**

3. The magnetic flux passing perpendicular to the plane of a coil and directed into the plane of the paper is varying according to the following equation :

    $$\phi = 5t^2 + 6t + 2$$

    where ϕ is in mWb and t in seconds. Find the e.m.f. induced in the coil at $t = 1$ s. **[16mV]**

4. A coil has an area of 0.04 m² and has 1000 turns. It is suspended in a magnetic field of 5×10^{-5} Wb/m² perpendicular to the field. The coil is rotated through 90° in 0·2s. Calculate the average e.m.f. induced in the coil due to rotation. **[0·01V]**

5. A gramophone disc of brass of diameter 30 cm rotates horizontally at the rate of 100/3 revolutions per minute. If the vertical component of earth's field is 0·01 T, calculate the e.m.f. induced between the centre and the rim of the disc. **[3·9 × 10⁻⁴ V]**

9.5. Induced E.M.F.

When the magnetic flux linking a conductor (or coil) changes, an e.m.f. is induced in it. This change in flux linkages can be brought about in the following two ways :

(*i*) The conductor is moved in a stationary magnetic field in such a way that the flux linking it changes in magnitude. The e.m.f. induced in this way is called **dynamically induced e.m.f.** (as in a d.c. generator). It is so called because e.m.f. is induced in the conductor which is in motion.

(*ii*) The conductor is stationary and the magnetic field is moving or changing. The e.m.f. induced in this way is called **statically induced e.m.f.** (as in a transformer). It is so called because the e.m.f. is induced in a conductor which is stationary.

It may be noted that in either case, the magnitude of induced e.m.f. is given by Ndφ/dt or derivable from this relation.

9.6. Dynamically Induced E.M.F.

Consider a single conductor of length *l* metres moving at *right angles to a uniform magnetic field of *B* Wb/m² with a velocity of *v* m/s [See Fig. 9.5 (*i*)]. Suppose the conductor moves through a small distance *dx* in *dt* seconds. Then area swept by the conductor is $= l \times dx$.

---

* If the conductor is moved parallel to the magnetic field, there would be no change in flux and hence no e.m.f. would be induced.

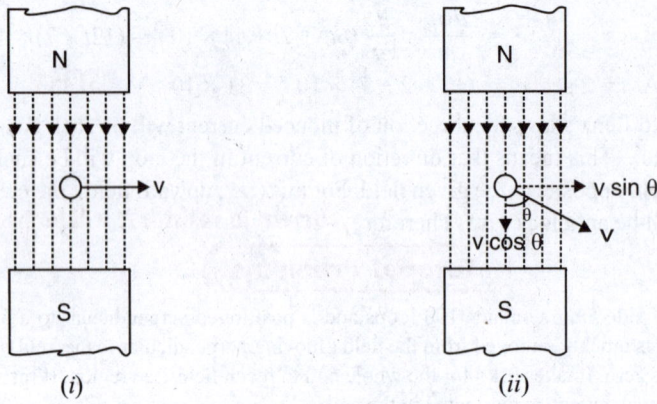

Fig. 9.5

∴   Flux cut, $d\phi$ = Flux density × Area swept = $B\,l\,dx$ Wb

According to Faraday's laws of electromagnetic induction, the magnitude of e.m.f. e induced in the conductor is given by ;

$$e = N\frac{d\phi}{dt} = \frac{B\,l\,dx}{dt} \qquad\qquad (\because N=1)$$

∴ $e = B\,l\,v$ volts $(\because dx\,/\,dt = v)$

Special case. If the conductor moves at angle θ to the magnetic field [See Fig. 9.5 (ii)], then the velocity at which the conductor moves across the field is *$v \sin \theta$.

∴ $e = B\,l\,v \sin \theta$

The direction of the induced e.m.f. can be determined by Fleming's right-hand rule.

Example 9.6. *An aircraft has a wing span of 56 m. It is flying horizontally at a speed of 810 km/hr and the vertical component of earth's magnetic field is 4×10^{-4} Wb/m². Calculate the potential difference between the wing tips of the aircraft.*

Solution. Induced e.m.f. = $B\,lv$

Here $B = 4 \times 10^{-4}$ Wb/m² ; $l = 56$ m ; $v = \dfrac{810 \times 1000}{3600} = 225$ m/s

∴   Induced e.m.f. = $(4 \times 10^{-4}) \times 56 \times (225) = 5 \cdot 04$ V

or   Potential difference = **5·04 V**

Example 9.7. *A d.c. generator consists of conductors lying in a radius of 10 cm and the effective length of a conductor in a constant radial field of strength 0·9 Wb/m² is 12 cm. The armature rotates at 1400 r.p.m. Given that the generator has 152 conductors in series, calculate the voltage being generated.*

Solution. Since the magnetic field is radial, the conductors cut the magnetic lines of force at right angles.

$$\text{Velocity, } v = \omega \times r = \frac{2\pi N}{60} \times r = \frac{2\pi \times 1400}{60} \times 0 \cdot 1 = 14 \cdot 66 \text{ m/s}$$

Voltage generated in each conductor = $B\,lv = 0 \cdot 9 \times 0 \cdot 12 \times 14 \cdot 66 = 1 \cdot 583$ V

Voltage generated in 152 conductors in series

$$= 1 \cdot 583 \times 152 = \mathbf{240 \cdot 6 \ V}$$

Note that effective length (l) is that portion of the conductor which takes part in the actual cutting of magnetic flux lines.

---

* The component $v \cos \theta$ is parallel to magnetic field and hence no e.m.f. is induced in the conductor due to this component.

Example 9.8. *A square metal wire loop of side 10 cm and resistance 1 Ω is moved with a constant velocity v_0 in a uniform magnetic field of induction B = 2 Wb/m² as shown in Fig. 9.6. The magnetic field lines are perpendicular to the plane of the loop directed into the paper. The loop is connected to a network of resistors each of value 3 Ω. The resistances of lead wires OS and PQ are negligible. What should be the speed v_0 of the loop so as to have a steady current of 1 mA in the loop ? Also indicate the direction of current in the loop.*

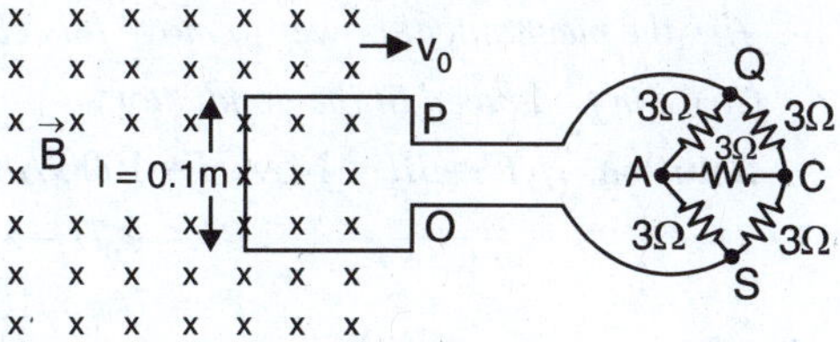

Fig. 9.6

Solution. We shall first find the equivalent resistance of the network. It is clear that network is a balanced Wheatstone bridge. Therefore, the resistance in the branch *AC* is ineffective. The equivalent resistance R' of the network is given by ;

$$\frac{1}{R'} = \frac{1}{6} + \frac{1}{6} = \frac{1}{3} \quad \text{or} \quad R' = 3 \, \Omega$$

The resistance of the loop is 1 Ω.

∴ Effective resistance of the circuit, $R = R' + 1 = 3 + 1 = 4 \, \Omega$

E.M.F. induced in the loop, $e = Bl \, v_0$

Current in the loop, $i = \frac{e}{R} = \frac{Blv_0}{R}$ ∴ Speed of the loop, $v_0 = \frac{iR}{Bl}$

Here $i = 1 \, \text{mA} = 10^{-3} \, \text{A}$; $R = 4 \, \Omega$; $B = 2 \, \text{Wb/m}^2$; $l = 0.1 \, \text{m}$

∴ $$v_0 = \frac{10^{-3} \times 4}{2 \times 0.1} = 2 \times 10^{-2} \, \text{ms}^{-1} = \textbf{2 cm/second}$$

According to Fleming's right-hand rule, direction of induced current is **clockwise from O to P**

Example 9.9. *A wheel with 10 metal spokes each 0.5 m long is rotated with a speed of 120 r.p.m. in a plane normal to earth's magnetic field at a place. If the magnitude of the field is 0.4 G, what is the magnitude of induced e.m.f. between the axle and rim of the wheel ?*

Solution. Length of spoke, l = radius $r = 0.5$ m

Frequency of rotation, $n = 120$ r.p.m. = 2 r.p.s.

Magnetic flux density, $B = 0.4 \, G = 0.4 \times 10^{-4} \, \text{T}$

Angular frequency, $\omega = 2\pi n = 2\pi \times 2 = 4\pi \, \text{rad s}^{-1}$

As the wheel rotates, the linear velocity of spoke end at the rim $= \omega r$ and linear velocity of spoke end at the axle = 0.

∴ Average linear velocity, $v = \frac{0 + \omega r}{2} = \frac{1}{2} \omega r$

Induced e.m.f. across the ends of each spoke is

$$e = Bl \, v = (B) \, (r) \left(\frac{1}{2} \omega r \right) = \frac{1}{2} B r^2 \, \omega$$

or $$e = \frac{1}{2} B r^2 \, \omega = \frac{1}{2} (0.4 \times 10^{-4}) \times (0.5)^2 \times 4\pi = \textbf{6.28} \times \textbf{10}^{-5} \, \textbf{V}$$

One end of all 10 spokes is connected to the rim and the other end to the axle. Therefore, the spokes are connected in parallel. As a result, e.m.f. between rim and axle is equal to the e.m.f. across the ends of each spoke.

Example 9.10. *A conductor 10 cm long and carrying a current of 50 A lies perpendicular to a field of strength 1000 A/m. Calculate :*

(i) the force acting on the conductor.

(ii) the mechanical power to move this conductor against the force with a speed of 1 m/s.

(iii) e.m.f. induced in the conductor.

Solution. *(i)* $F = BIl$.   Now $H = 1000$ A/m

$$\therefore \qquad B = \mu_0 H = 4\pi \times 10^{-7} \times 1000 = 4\pi \times 10^{-4} \text{ Wb/m}^2$$

$$\therefore \qquad F = (4\pi \times 10^{-4}) \times 50 \times 0.1 = \mathbf{6.28 \times 10^{-3} \text{ N}}$$

(ii) Mechanical power required is given by ;

$$P = F \times v = 6.28 \times 10^{-3} \times 1 = \mathbf{6.28 \times 10^{-3} \text{ W}}$$

(iii) E.M.F. induced in the conductor is given by ;

$$e = Blv = (4\pi \times 10^{-4}) \times 0.1 \times 1 = \mathbf{4\pi \times 10^{-5} \text{ V}}$$

Note that electric power developed $= eI = (4\pi \times 10^{-5}) \times 50 = 6.28 \times 10^{-3}$ W. This is equal to the mechanical input power. Therefore, law of conservation of energy is obeyed.

Tutorial Problems

1. A copper disc 40 cm in diameter is rotated at 3000 r.p.m. on a horizontal axis perpendicular to and through the centre of the disc, the axis lying in the magnetic meridian. Two brushes make contact with the disc, one at the edge and the other at the centre. If the horizontal component of earth's field be 0.02 m Wb/m^2, calculate the e.m.f. induced between the brushes. **[0.12 mV]**

2. A meter driving motor consists of a horizontal disc of aluminium 20 cm in diameter, pivoted on a vertical spindle and lying in a permanent magnetic field of density 0.3 Wb/m^2. The current flow is radial from the spindle to the circumference of the disc. The circuit resistance is 0.225 Ω and a p.d. of 2.3 V is required to pass a current of 10 A through the motor. Calculate the rotational speed of the disc and the power lost in friction. **[319 r.p.m. ; 0.5 W]**

3. If the vertical component of earth's magnetic field be 4×10^{-5} Wb/m^2, then what will be the induced potential difference produced between the rails of a metre-gauge when a train is running on them with a speed of 36 km/hr ? **[4 × 10⁻⁴ V]**

9.7. Statically Induced E.M.F.

When the conductor is stationary and the field is moving or changing, the e.m.f. induced in the conductor is called statically induced e.m.f. A statically induced e.m.f. can be further sub-divided into :

1. Self-induced e.m.f.   2. Mutually induced e.m.f.

1. Self-induced e.m.f. *The e.m.f. induced in a coil due to the change of its own flux linked with it is called* **self-induced e.m.f.**

When a coil is carrying current (See Fig. 9.7), a magnetic field is established through the coil. If current in the coil changes, then the flux linking the coil also changes. Hence an e.m.f. $(= N \, d\phi/dt)$ is induced in the coil. This is known as self-induced e.m.f. The direction of this e.m.f. (by Lenz's law) is such so as to oppose the cause producing it, namely the change of current (and hence field) in the coil. The self-induced e.m.f. will persist so long as the current in the coil is changing. The following points are worth noting :

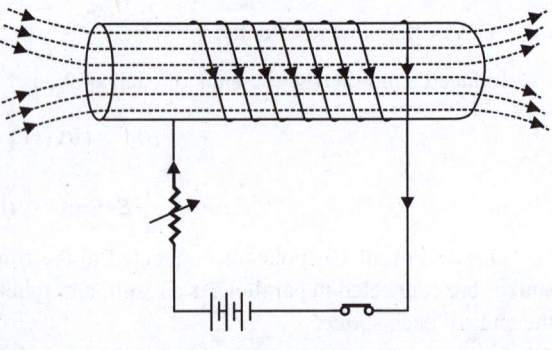

Fig. 9.7

(i) When current in a coil changes, the self-induced e.m.f. opposes the change of current in the coil. This property of the coil is known as its *self-inductance* or *inductance*.

(*ii*) *The self-induced e.m.f. (and hence inductance) does not prevent the current from changing ; it serves only to delay the change.* Thus after the switch is closed (See Fig. 9.7), the current will rise from zero ampere to its final steady value in some time (a fraction of a second). This delay is due to the self-induced e.m.f. of the coil.

2. Mutually induced e.m.f. *The e.m.f. induced in a coil due to the changing current in the neighbouring coil is called* **mutually induced e.m.f.**

Consider two coils *A* and *B* placed adjacent to each other as shown in Fig. 9.8. A part of the magnetic flux produced by coil *A* passes through or links with coil *B*. This flux which is common to both the coils *A* and *B* is called *mutual flux* (ϕ_m). If current in coil *A* is varied, the mutual flux also varies and hence e.m.f. is induced in both the coils. The e.m.f. induced in coil *A* is called self-induced e.m.f. as already discussed. The e.m.f. induced in coil *B* is known as *mutually induced e.m.f.*

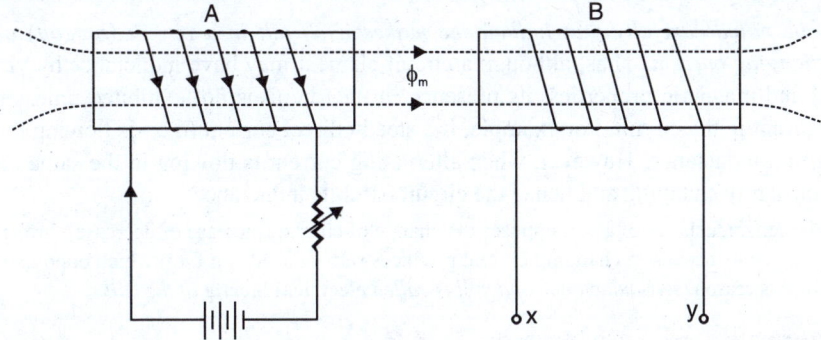

Fig. 9.8

The magnitude of mutually induced e.m.f. is given by Faraday's laws *i.e.* $e_M = N_B \, d\phi_m/dt$ where N_B is the number of turns of coil *B* and $d\phi_m/dt$ is the rate of change of mutual flux *i.e.* flux common to both the coils. The direction of mutually induced e.m.f. (by Lenz's law) is always such so as to oppose the very cause producing it. The cause producing the mutually induced e.m.f. in coil *B* is the changing mutual flux produced by coil *A*. Hence the direction of induced current (when the circuit is completed) in coil *B* will be such that the flux set up by it will oppose the changing mutual flux produced by coil *A*.

The following points may be noted carefully :

(*i*) The mutually induced e.m.f. in coil *B* persists so long as the current in coil *A* is changing. If current in coil *A* becomes steady, the mutual flux also becomes steady and mutually induced e.m.f. drops to zero.

(*ii*) The property of two neighbouring coils to induce voltatge in one coil due to the change of current in the other is called *mutual inductance*.

9.8. Self-inductance (L)

The property of a coil that opposes any change in the amount of current flowing through it is called its **self-inductance** *or* **inductance.**

This property (*i.e. inductance*) is due to the self-induced e.m.f. in the coil itself by the changing current. If the current in the coil is increasing, the self-induced e.m.f. is set up in such a direction so as to oppose the rise of current *i.e.* direction of self-induced e.m.f. is opposite to that of the applied voltage. Similarly, if the current in the coil is decreasing, self-induced voltage will be such so as to oppose the decrease in current *i.e.* self-induced e.m.f. will be in the same direction as the applied voltage. It may be noted that self-inductance does not prevent the current from changing ; it serves only to delay the change.

Factors affecting inductance. The greater the self-induced voltage, the greater the self-inductance of the coil and hence larger is the opposition to the changing current. According to Faraday's laws of electromagnetic induction, induced voltage in a coil depends upon the number of turns (N) and the rate of change of flux ($d\phi/dt$) linking the coil. Hence, the inductance of a coil depends upon these factors, *viz* :

(*i*) Shape and number of turns.

(*ii*) Relative permeability of the material surrounding the coil.

(*iii*) The speed with which the magnetic field changes.

In fact, anything that affects magnetic field also affects the inductance of the coil. Thus, increasing the number of turns of a coil increases its inductance. Similarly, substituting an iron core for air core increases its inductance.

It may be noted carefully that inductance makes itself felt in a circuit (or coil) only when there is a changing current. Thus, although a circuit element may have inductance by virtue of its geometrical and magnetic properties, its presence in the circuit is not exhibited unless there is a change of current in the circuit. For example, if a steady direct current (d.c.) is flowing in a circuit, there will be no inductance. However, when alternating current is flowing in the same circuit, the current is constantly changing and hence the circuit exhibits inductance.

Note. The self-inductance of a coil opposes the change of current (increase or decrease) through the coil. This opposition occurs because a changing current produces self-induced e.m.f. (*e*) which opposes the change of current. For this reason, *self-inductance of a coil is called* **electrical inertia** *of the coil.*

9.9. Magnitude of Self-induced E.M.F.

Consider a coil of N turns carrying a current of I amperes. If current in the coil changes, the flux linkages of the coil will also change. This will set up a self-induced e.m.f. e in the coil given by ;

$$e = N\frac{d\phi}{dt} = \frac{d}{dt}(N\phi)$$

Since flux is due to current in the coil, it follows that flux linkages ($= N\phi$) will be proportional to I.

$$\therefore \qquad e = \frac{d}{dt}(N\phi) \propto \frac{dI}{dt}$$

$$\therefore \qquad e = \text{Constant} \times \frac{dI}{dt}$$

or $\qquad\qquad\qquad e = L\frac{dI}{dt}$ (*in magnitude*) ...(*i*)

where L is a constant called **self-inductance** or **inductance** of the coil. The unit of inductance is henry (H). If in eq. (*i*) above, $e = 1$ volt, $dI/dt = 1$ A/second, then $L = 1$ H.

Hence a coil (or circuit) has an inductance of **1 henry** *if an e.m.f. of 1 volt is induced in it when current through it changes at the rate of* 1 *ampere per second.*

Note. The magnitude of self-induced e.m.f. is $e = LdI/dt$. However, the magnitude and direction of self-induced e.m.f. should be written as :

$$e = -L\frac{dI}{dt}$$

The minus sign is because the self-induced e.m.f. tends to send current in the coil in such a direction so as to produce magnetic flux which opposes the change in flux produced by the change in current in the coil. In fact, minus sign represents Lenz's law mathematically.

9.10. Expressions for Self-inductance

The self-inductance (L) of a circuit or coil can be determined by one of the following three ways :

(i) First Method. If the magnitude of self-induced e.m.f. (e) and the rate of change of current (dI/dt) are known, then inductance can be determined from the following relation :

$$e = L\frac{dI}{dt}$$

∴ $$L = \frac{e}{(dI/dt)}$$ ...(i)

(ii) Second Method. If the flux linkages of the coil and current are known, then inductance can be determined as under :

$$e = L\frac{dI}{dt} = \frac{d}{dt}(LI)$$

Also $$e = N\frac{d\phi}{dt} = \frac{d}{dt}(N\phi)$$

From the two expressions, we have,

$$LI = N\phi$$

∴ $$L = \frac{N\phi}{I}$$ ...(ii)

Thus, inductance is the flux linkages of the coil per ampere. If $N\phi = 1$ Wb-turn and $I = 1$ A, then $L = 1$H.

Hence a coil has an inductance of **1 henry** *if a current of 1 A in the coil sets up flux linkages of 1 Wb-turn.*

Note. Relation (ii) above reveals that inductance depends upon the ratio ϕ/I. Therefore, inductance is constant only when the flux changes uniformly with current. This condition is met only when the flux path is entirely composed of non-magnetic material *e.g.* air. But when the flux path is through a magnetic material (*e.g.* coil wound over iron bar), inductance of the coil will be constant only over the linear portion of the magnetisation curve.

(iii) Third Method. The inductance of a magnetic circuit can be found in terms of its physical dimensions. Consider an iron-cored *solenoid of dimensions as shown in Fig. 9.9. Inductance of the solenoid is given by [from exp. (ii) above] ;

$$L = N\frac{d\phi}{dI}$$

Now $$\phi = \frac{m.m.f.}{reluctance} = \frac{NI}{l/a\mu_0\mu_r}$$

Differentiating ϕ w.r.t. I, we get,

$$\frac{d\phi}{dI} = \frac{Na\mu_0\mu_r}{l}$$

∴ $$L = N\frac{(Na\mu_0\mu_r)}{l}$$

or $$L = \frac{N^2 a\mu_0\mu_r}{l}$$ ...(iii)

$$= \frac{N^2}{l/a\mu_0\mu_r} = \frac{N^2}{Reluctance\,(S)}$$ ...(iv)

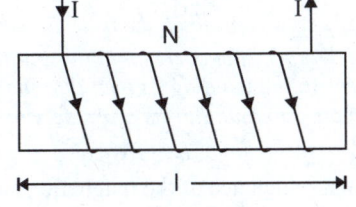

Fig. 9.9

Thus, inductance can be determined by using the relation (iii) or (iv). *It is important to note [See relation (iv)] that inductance is directly proportional to turns squared and inversely proportional*

---

*   Solenoid is an important winding arrangement, being simple to manufacture, it is found in relays, inductors, small transformers in the form considered.

to the reluctance of the magnetic path. The smaller the reluctance of the magnetic path, the larger the inductance and *vice-versa.* For this reason, an iron-cored coil has more inductance than the equivalent air-cored coil.

Example 9.11. *A coil wound on an iron core of permeability 400 has 150 turns and a cross-sectional area of 5 cm^2. Calculate the inductance of the coil. Given that a steady current of 3 mA produces a magnetic field of 10 lines/cm^2 when air is present as the medium.*

Solution.
$$\mu_i = \frac{\text{Flux density in iron}}{\text{Flux density in air}} = \frac{B_i}{10}$$

$$\therefore \quad B_i = 10 \times \mu_i = 10 \times 400 = 4000 \text{ lines/cm}^2$$

Flux produced by 3 mA current in the iron core is

$$\phi = B_i \times a = 4000 \times 5 = 20{,}000 \text{ lines} = 2 \times 10^{-4} \text{ Wb}$$

$$\therefore \quad L = \frac{N\phi}{I} = \frac{150 \times 2 \times 10^{-4}}{3 \times 10^{-3}} = \textbf{10 H}$$

Example 9.12. *A solenoid with 900 turns has a total flux of 1.33 × 10^{-7} Wb through its air core when the coil current is 100 mA. If the flux takes 75 ms to grow from zero to its maximum level, calculate the inductance of the coil. Also, calculate the induced e.m.f. in the coil during the flux growth.*

Solution. The magnitude of induced e.m.f. is given by the following two expressions :

$$e = L\frac{dI}{dt} \; ; e = N\frac{d\phi}{dt}$$

$$\therefore \quad L\frac{dI}{dt} = N\frac{d\phi}{dt} \quad \text{or} \quad L = N\frac{d\phi}{dI}$$

Here $\quad N = 900 \; ; d\phi = 1.33 \times 10^{-7} \text{ Wb} \; ; dt = 75 \text{ ms} = 75 \times 10^{-3} \text{ s} \; ;$

$$dI = 100 \text{ mA} = 100 \times 10^{-3} \text{A}$$

$$\therefore \quad L = 900 \times \frac{1.33 \times 10^{-7}}{100 \times 10^{-3}} = 1.2 \times 10^{-3} \text{ H} = \textbf{1.2 mH}$$

Induced e.m.f., $e = N\dfrac{d\phi}{dt} = 900 \times \dfrac{1.33 \times 10^{-7}}{75 \times 10^{-3}} = 1.6 \times 10^{-3} \text{ V} = \textbf{1.6 mV}$

Example 9.13. *An air-cored choke is designed to have an inductance of 20H when operating at a flux density of 1 Wb/m^2 ; the corresponding relative permeability of iron core is 4000. Determine the number of turns in the winding ; given that the flux path has a mean length of 22 cm in the iron core and 1 mm in air gap and that its cross-section is 10 cm^2.*

Solution. $\quad L = N^2/S_T$

where S_T is the total reluctance of the magnetic path.

$$S_T = S_{iron} + S_{air} = \frac{l_{iron}}{a\,\mu_0\,\mu_r} + \frac{l_{air}}{a\,\mu_0\,\mu_r}$$

$$= \frac{0 \cdot 22}{\left(10 \times 10^{-4}\right) \times 4\pi \times 10^{-7} \times 4000} + \frac{0 \cdot 001}{\left(10 \times 10^{-4}\right) \times 4\pi \times 10^{-7} \times 1}$$

$$= 43767 + 795774 = 839541 \text{ AT/Wb}$$

Now $\quad L = N^2/S_T$

$$\therefore \quad N = \sqrt{L\,S_T} = \sqrt{20 \times 839541} = \textbf{4097 turns}$$

Example 9.14. *An iron rod, 1 cm diameter and 50 cm long is formed into a closed ring and uniformly wound with 400 turns of wire. A direct current of 0·5 A is passed through the winding and produces a flux density of 0·75 Wb/m^2. If all the flux links with every turn of the winding, calculate*

(i) the relative permeability of iron (ii) the inductance of the coil (iii) the average value of e.m.f. induced when the interruption of current causes the flux in the iron to decay to 20% of its original value in 0·01 second.

Solution. *(i)*

$$H = \frac{NI}{l} = \frac{400 \times 0 \cdot 5}{0 \cdot 5} = 400 \text{ AT/m}$$

$$\mu_r = \frac{B}{\mu_0 H} = \frac{0.75}{4\pi \times 10^{-7} \times 400} = \mathbf{1492}$$

(ii)

$$\phi = B \times a = 0 \cdot 75 \times \frac{\pi}{4} (1 \times 10^{-2})^2 = 0 \cdot 589 \times 10^{-4} \text{ Wb}$$

$\therefore$

$$L = \frac{N\phi}{I} = \frac{(400) \times 0 \cdot 589 \times 10^{-4}}{0 \cdot 5} = \mathbf{0.0471 \ H}$$

(iii) Change in flux, $d\phi = 80\%$ of original flux $= 0 \cdot 8 \times 0 \cdot 589 \times 10^{-4} = 0 \cdot 47 \times 10^{-4}$ Wb

$\therefore$

$$e = N\frac{d\phi}{dt} = 400 \times \frac{0.47 \times 10^{-4}}{0.01} = \mathbf{1.88 \ V}$$

Example 9.15. *A circuit has 1000 turns enclosing a magnetic circuit 20 cm^2 in section. With 4A, the flux density is 1 Wb/m^2 and with 9A, it is 1.4 Wb/m^2. Find the mean value of the inductance between these current limits and the induced e.m.f. if the current falls from 9A to 4A in 0·05 seconds.*

Solution.

$$L = N\frac{d\phi}{dI} = N\frac{d}{dI}(BA) = NA\frac{dB}{dI}$$

Here $N = 1000$; $dB = 1 \cdot 4 - 1 = 0 \cdot 4$ Wb/m² ; $dI = 9 - 4 = 5$A

$\therefore$

$$L = (1000) \times (20 \times 10^{-4}) \times \frac{0.4}{5} = \mathbf{0.16 \ H}$$

Also

$$e = L\frac{dI}{dt} = 0 \cdot 16 \times \frac{5}{0 \cdot 05} = \mathbf{16 \ V}$$

Example 9.16. *A single element has the current and voltage functions graphed in Fig. 9.10 (i) and (ii). Determine the element.*

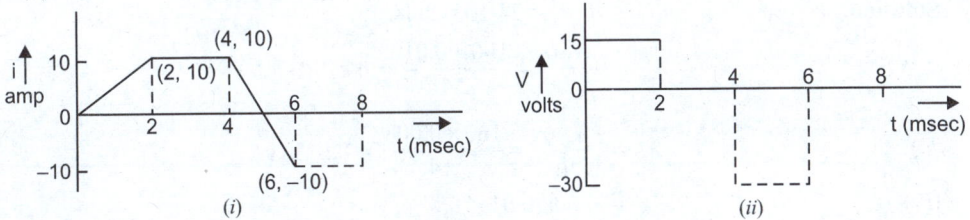

Fig. 9.10

Solution. From $i - t$ and $V - t$ graph of the element, we observe that :

Between 0 – 2 ms ; $di = 10$A ; $dt = 2$ ms ; $V = 15$ volts

$\therefore$

$$\frac{di}{dt} = \frac{10 \text{ A}}{2 \times 10^{-3} \text{ s}} = 5000 \text{ A/s. Now, } L = \frac{V}{di/dt} = \frac{15}{5000} = 3 \times 10^{-3} \text{ H} = 3 \text{ mH}$$

Between 4 – 6 ms ; $di = -20$A ; $dt = 2$ ms ; $V = -30$ volts

$\therefore$

$$\frac{di}{dt} = \frac{-20 \text{ A}}{2 \times 10^{-3} \text{ s}} = -10,000 \text{ A/s. Now, } L = \frac{V}{di/dt} = \frac{-30}{-10,000} = 3 \times 10^{-3} \text{ H} = 3 \text{ mH}$$

Note that when current is constant, $di/dt = 0$ so that voltage across L is zero. **Hence, the element is 3 mH inductor.**

Example 9.17. *A 300-turn coil has a resistance of 6 Ω and an inductance of 0·5 H. Determine the new resistance and new inductance if one-third of the turns are removed. Assume all the turns have the same circumference.*

Solution. As the resistance of a coil is directly proportional to its length,

$$\therefore \qquad R_1/R_2 = N_1/N_2 \quad \text{or} \quad 6/R_2 = 300/200$$

$$\therefore \qquad R_2 = 6 \times \frac{200}{300} = 4\,\Omega$$

Also

$$\frac{L_1}{L_2} = \frac{N_1^2/S}{N_2^2/S} \quad \text{or} \quad \frac{0.5}{L_2} = \frac{(300)^2}{(200)^2}$$

$$\therefore \qquad L_2 = 0.5 \times \frac{(200)^2}{(300)^2} = \textbf{0·22 H}$$

Example 9.18. *A battery of 24 V is connected to the primary (coil 1) of a two-winding transformer as shown in Fig. 9.11 and the secondary (coil 2) is open-circuited. The coil parameters are :*

$$R_1 = 10\,\Omega \qquad\qquad R_2 = 30\,\Omega$$
$$N_1 = 100 \text{ turns} \qquad N_2 = 160 \text{ turns}$$
$$\phi_1 = 0·01 \text{ Wb} \qquad \phi_2 = 0·008 \text{ Wb}$$

Calculate (i) the self-inductance of coil 1 (ii) the mutual inductance (iii) the coefficient of coupling and (iv) the self-inductance of coil 2.

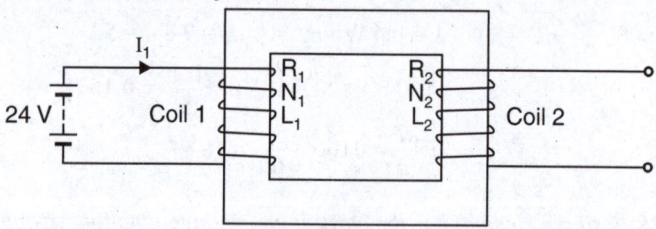

Fig. 9.11

Solution. *(i)*

$$I_1 = V/R_1 = 24/10 = 2·4 \text{A}$$

$$\therefore \qquad L_1 = \frac{N_1\,\phi_1}{I_1} = \frac{100 \times 0.01}{2·4} = \textbf{0.417 H}$$

(ii)

$$M = \frac{N_2\,\phi_2}{I_1} = \frac{160 \times 0.008}{2·4} = \textbf{0.533 H}$$

(iii)

$$k = 0·008/0·01 = \textbf{0·8}$$

(iv)

$$M = k\sqrt{L_1 L_2} \quad \text{or} \quad 0·533 = 0·8\sqrt{0·417 \times L_2} \quad \therefore \quad L_2 = \textbf{1·064 H}$$

Example 9.19. *A coil of 1000 turns is wound on a laminated core of steel having a cross-section of 5 cm². The core has an air gap of 2 mm cut at right angle. What value of current is required to have an air gap flux density of 0.5 T? Permeability of steel may be taken as infinity. Determine the coil inductance.*

Solution. $B_g = 0.5$ T ; $a = 5 \times 10^{-4}$ m² ; $N = 1000$ turns ; $l_g = 2 \times 10^{-3}$ m ; $\mu_r = \infty$

$$\text{Total } AT \text{ required} = H_i l_i + H_g l_g = \frac{B_g}{\mu_0 \mu_r} l_i + \frac{B_g}{\mu_0} l_g$$

$$= 0 + \frac{B_g}{\mu_0} l_g = 0 + \frac{0.5}{4\pi \times 10^{-7}} \times 2 \times 10^{-3} = 796 \text{ AT} \qquad (\because \mu_r = \infty)$$

Now $NI = 796 \quad \therefore I = \dfrac{796}{N} = \dfrac{796}{1000} = \textbf{0.796 A}$

Inductance of coil, $L = \dfrac{N\phi}{I} = \dfrac{N \times (B_g \times a)}{I} = \dfrac{1000 \times (0.5 \times 5 \times 10^{-4})}{0.796} = \textbf{0.314 H}$

Tutorial Problems

1. A current of 2·5 A flows through a 1000-turn coil that is air-cored. The coil inductance is 0·6 H. What magnetic flux is set up ? **[1·5 m Wb]**

2. A 2000-turn coil is uniformly wound on an ebonite ring of mean diameter 320 mm and cross-sectional area 400 mm^2. Calculate the inductance of the toroid so formed. **[2 mH]**

3. A coil has self-inductance of 10 H. If a current of 200 mA is reduced to zero in a time of 1 ms, find the average value of induced e.m.f. across the terminals of the coil. **[2000 V]**

4. A coil consists of 750 turns and a current of 10 A in the coil gives rise to a magnetic flux of 1200 μWb. Calculate the inductance of the coil and determine the average e.m.f. induced in the coil when this current is reversed in 0·01 second. **[0·09 H ; 180 V]**

5. Calculate the inductance of a solenoid of 2000 turns wound uniformly over a length of 50 cm on a cylindrical paper tube 4 cm in diameter. The medium is air. **[12·62 mH]**

6. A circular iron ring of mean diameter 100 mm and cross-sectional area 500 mm^2 has 200 turns of wire uniformly wound around the circumference. If the relative permeability of iron is assumed to be 1200, find the self-inductance of the coil. **[96 mH]**

7. A certain 40-turn coil has an inductance of 6 H. Determine the new inductance if 10 turns are added to the coil. **[9·38 H]**

8. The e.m.f. induced in a coil is 100V when current through it changes from 1A to 10 A in 0·1s. Calculate the inductance of the coil. **[1·11 H]**

9. A 6-pole, 500 V d.c. generator has a flux/pole of 50 mWb produced by a field current of 10 A. Each pole is wound with 600 turns. The resistance of entire field circuit is 50 Ω. If the field circuit is broken in 0·02s, calculate (*i*) the inductance of the field coils (*ii*) the induced e.m.f. and (*iii*) the value of discharge resistance so that the induced e.m.f. should not exceed 1000V. **[(*i*) 18 H (*ii*) 1500 V (*iii*) 50 Ω]**

10. What is the inductance of a single layer 10-turn air-cored coil that is 1 cm long and 0·5 cm in diameter ? **[0·214 μH]**

9.11. Magnitude of Mutually Induced E.M.F.

Consider two coils A and B placed adjacent to each other as shown in Fig. 9.12. If a current I_1 flows in the coil A, a flux is set up and a part ϕ_{12} (*mutual flux*) of this flux links the coil B. If current in coil A is varied, the mutual flux also varies and hence an e.m.f. is induced in the coil B. The e.m.f. induced in coil B is termed as mutually induced e.m.f. Note that coil B is not electrically connected to coil A ; the two coils being magnetically linked.

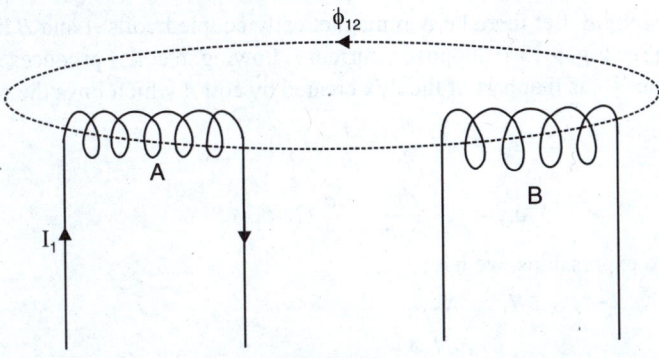

Fig. 9.12

The larger the rate of change of current in coil A, the greater is the e.m.f. induced in coil B. In other words, mutually induced e.m.f. in coil B is directly proportional to the rate of change of current in coil A i.e.,

Mutually induced e.m.f. in coil $B \propto$ Rate of change of current in coil A

or $$e_M \propto \frac{dI_1}{dt}$$

or $$e_M = M\frac{dI_1}{dt} \qquad \text{(in magnitude)} \qquad ...(i)$$

where M is a constant called **mutual inductance** between the two coils. The unit of mutual inductance is henry (H). If in exp. (i), $e_M = 1$ volt, $dI_1/dt = 1$ A/sec, then, $M = 1$ H.

Hence mutual inductance between two coils is **1 henry** *if current changing at the rate of 1 A/sec in one coil induces an e.m.f. of 1 V in the other coil.*

Mutual inductance comes into picture when two coils are placed close together in such a way that flux produced by one links the other. We say then that the two coils are coupled. Each coil has its own inductance but in addition, there is further inductance due to the induced voltage produced by coupling between the coils. We call this further inductance as mutual inductance. We say the two coils are coupled together by mutual inductance. The terms **magnetic** or **inductive coupling** are sometimes used.

Note. The magnitude of mutually induced e.m.f. in coil B (secondary) is $e_M = M\,dI_1/dt$ where dI_1 is the change of current in coil A (primary). However, the magnitude and direction of mutually induced e.m.f. in coil B should be written as :

$$e_M = -M\frac{dI_1}{dt}$$

The minus sign is because the mutually induced e.m.f. sends current in coil B in such a direction so as to produce magnetic flux which opposes the change in flux produced by change in current in coil A. In fact, minus sign represents Lenz's law mathematically.

9.12. Expressions for Mutual Inductance

The mutual inductance between two coils can be determined by one of the following three methods :

(*i*) **First Method.** If the magnitude of mutually induced e.m.f. (e_M) in one coil for the given rate of change of current in the other is known, then M between the two coils can be determined from the following relation :

$$e_M = M\frac{dI_1}{dt}$$

or $$M = \frac{e_M}{dI_1/dt} \qquad\qquad ...(i)$$

(*ii*) **Second Method.** Let there be two magnetically coupled coils A and B having N_1 and N_2 turns respectively (See Fig. 9.13). Suppose a current I_1 flowing in coil A produces a mutual flux ϕ_{12}. Note that mutual flux ϕ_{12} is that part of the flux created by coil A which links the coil B.

$$e_M = M\frac{dI_1}{dt} = \frac{d}{dt}(M\,I_1)$$

Also $$e_M = N_2\frac{d\phi_{12}}{dt} = \frac{d}{dt}(N_2\phi_{12})$$

From these two expressions, we have,

$$MI_1 = N_2\,\phi_{12}$$

or $$M = \frac{N_2\phi_{12}}{I_1} \qquad\qquad ...(ii)$$

Thus, mutual inductance between two coils is equal to the flux linkages of one coil ($N_2\phi_{12}$) due to one ampere in the other coil. If $N_2\,\phi_{12} = 1$ Wb-turn and $I_1 = 1$ A, then, $M = 1$ H.

Hence mutual inductance between two coils is **1 henry** *if a current of 1 A flowing in one coil produces flux linkages of 1 Wb-turn in the other.*

(*iii*) **Third Method.** The mutual inductance between the two coils can be determined in terms of physical dimensions of the magnetic circuit. Fig. 9.13 shows two magnetically coupled coils A and B having N_1 and N_2 turns respectively. Suppose l and 'a' are the length and area of cross-section of the magnetic circuit respectively. Let μ_r be the relative permeability of the material of which the magnetic circuit is composed.

$$\text{Mutual flux, } \phi_{12} = \frac{\text{m.m.f.}}{\text{reluctance}}$$

$$= \frac{N_1 I_1}{l\,/\,a\,\mu_0\,\mu_r}$$

or

$$\frac{\phi_{12}}{I_1} = \frac{N_1\,a\,\mu_0\,\mu_r}{l}$$

Now

$$M = \frac{N_2\,\phi_{12}}{I_1}$$

∴

$$M = \frac{N_1 N_2\,a\,\mu_0\,\mu_r}{l} \quad \text{...(iii)}$$

$$= \frac{N_1 N_2}{l\,/\,a\,\mu_0\,\mu_r}$$

$$= \frac{N_1\,N_2}{\text{Reluctance}\,(S)} \quad \text{...(iv)}$$

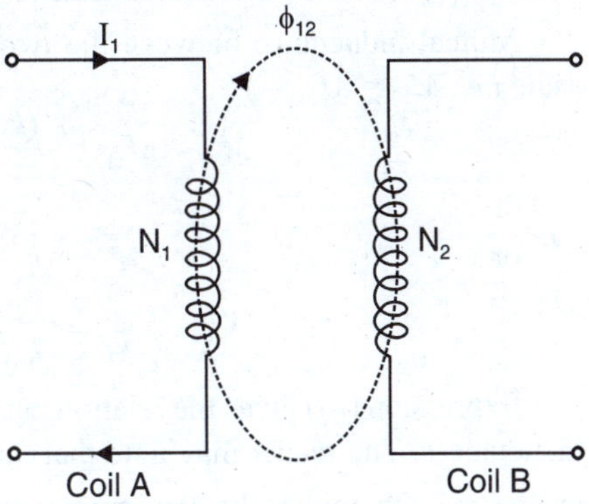

Fig. 9.13

The mutual inductance can be found by using relation (*iii*) or (*iv*). Note that mutual inductance is inversely proportional to the reluctance of the magnetic circuit. The smaller the reluctance of the magnetic circuit, the greater is the mutual inductance and *vice-versa*.

9.13. Coefficient of Coupling

The **coefficient of coupling (k)** *between two coils is defined as the fraction of magnetic flux produced by the current in one coil that links the other.*

When the entire flux of one coil links the other, coefficient of coupling is 1 (*i.e.*, 100%). If only half the flux set up in one coil links the other, then coefficient of coupling is 0·5 (or 50%). If two coils have self-inductances L_1 and L_2, then mutual inductance M between them is given by ;

$$M = k\sqrt{L_1 L_2}$$

where k = coefficient of coupling. Clearly, the mutual inductance M between the coils will be maximum when $k = 1$. If flux of one coil does not at all link with the other coil, then $k = 0$. Under such condition, mutual inductance (M) between the coils will be zero.

Proof. Consider two magnetically coupled coils 1 and 2 having N_1 and N_2 turns respectively (See Fig. 9.14). The current I_1 flowing in coil 1 produces a magnetic flux ϕ_1. Suppose the coefficient of coupling between the two coils is k. It means that flux $k\phi_1$ links with coil 2. Then, by definition,

$$L_1 = \frac{N_1\phi_1}{I_1}$$

and

$$M_{12} = \frac{k\phi_1 N_2}{I_1} \quad \text{...(i)}$$

where M_{12} represents mutual inductance of coil 1 to coil 2.

The current I_2 flowing in coil 2 will produce flux ϕ_2. Since the coefficient of coupling between the coils is k, it means that flux $k\phi_2$ will link with coil 1. Then,

$$L_2 = \frac{\phi_2 N_2}{I_2}$$

and

$$M_{21} = \frac{k\phi_2 N_1}{I_2} \qquad ...(ii)$$

where M_{21} represents mutual inductance of coil 2 to coil 1.

Mutual inductance between the two coils is exactly the same *i.e.*, $M_{12} = M_{21} = M$.

$\therefore$

$$M_{12} \times M_{21} = \frac{(k\phi_1 N_2)}{I_1} \times \frac{(k\phi_2 N_1)}{I_2}$$

Fig. 9.14

or

$$M^2 = k^2 \frac{\phi_1 N_1}{I_1} \times \frac{\phi_2 N_2}{I_2} = k^2 L_1 L_2$$

$\therefore$

$$M = k\sqrt{L_1 L_2} \qquad ...(iii)$$

Expression (*iii*) gives the relation between the mutual inductance of the two coils and their self-inductances. The reader may note that mutual inductance between the two coils will be maximum when $k = 1$. Obviously, the maximum value of mutual inductance between the two coils is $= \sqrt{L_1 L_2}$.

$\therefore$

$$k = \frac{M}{\sqrt{L_1 L_2}} = \frac{\text{Actual mutual inductance}}{\text{Max. possible mutual inductance}}$$

Hence, coefficient of coupling can also be defined *as the ratio of the actual mutual inductance* (*M*) *between the two coils to the maximum possible value* $(\sqrt{L_1 L_2})$.

When two coils are wound on a single ferromagnetic core as shown in Fig. 9.15 (*i*), effectively all of the magnetic flux produced by one coil links with the other. The coils are then said to be **tightly coupled.** Another way to ensure tight coupling is shown in Fig. 9.15 (*ii*) where each turn of the secondary winding is side by side with one turn of primary winding. Coils wound in this fashion are said to be bifilar and it is called **bifilar winding.**

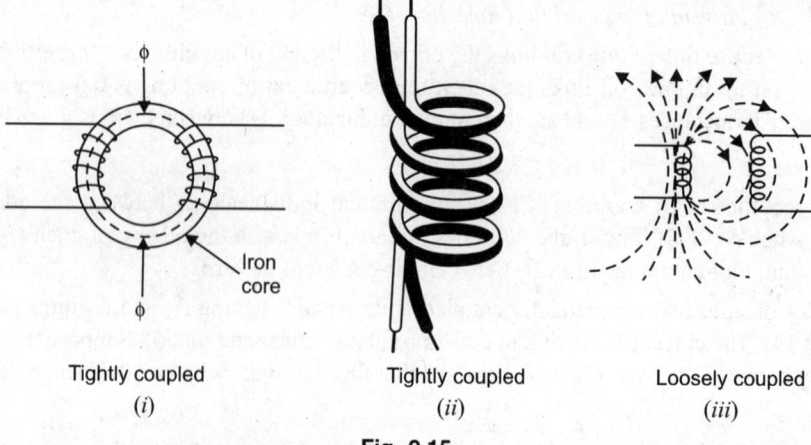

Tightly coupled
(*i*)

Tightly coupled
(*ii*)

Loosely coupled
(*iii*)

Fig. 9.15

When the two coils are air-cored as shown in Fig. 9.15 (*iii*), then only a fraction of magnetic flux produced by one coil may link with the other coil. The coils are then said to be **loosely coupled.**

Example 9.20. *Two identical coils A and B of 1000 turns each lie in parallel planes such that 80% of flux produced by one coil links with the other. A current of 5 A flowing in coil A produces a flux of 0·05 mWb in it. If the current in coil A changes from + 12A to −12A in 0·02 second, calculate (i) the mutual inductance and (ii) the e.m.f. induced in coil B.*

Solution. (*i*) $$M = \frac{N_2 \phi_{12}}{I_1}$$

Here $N_2 = 1000$; $I_1 = 5$ A ; *$\phi_{12} = 0.8 \times 0.05 \times 10^{-3} = 0.4 \times 10^{-4}$ Wb

∴ $$M = \frac{1000 \times 0.4 \times 10^{-4}}{5} = \textbf{0.008 H}$$

(*ii*)   E.M.F. in coil B, $e_B = M \dfrac{dI_1}{dt}$

Here $M = 0.008$ H ; $dI_1 = 12 - (-12) = 24$ A ; $dt = 0.02$ s

∴ $$e_B = 0.008 \times \frac{24}{0.02} = \textbf{9.6 V}$$

Example 9.21. *Coils A and B in a magnetic circuit have 600 and 500 turns respectively. A current of 8 A in coil A produces a flux of 0·04 Wb. If the coefficient of coupling is 0·2, calculate :*

(*i*) *Self-inductance of coil A, with B open-circuited.*

(*ii*) *Flux linking with coil B.*

(*iii*) *The average e.m.f. induced in coil B when the flux with it changes from zero to full value in 0·02 second.*

(*iv*) *Mutual inductance.*

(*v*) *Average e.m.f. in B when current in A changes from 0 to 8 A in 0·05 second.*

Solution. (*i*) Inductance of coil A, $L_A = \dfrac{N_A \phi_A}{I_A} = \dfrac{600 \times 0.04}{8} = \textbf{3 H}$

(*ii*)   Flux linking coil B, $\phi_B = k \times \phi_A = 0.2 \times 0.04 = \textbf{0·008 Wb}$

(*iii*)   e.m.f. in coil B, $e_B = N_B \dfrac{\phi_B - 0}{t} = 500 \dfrac{0.008}{0.02} = \textbf{200 V}$

(*iv*)   Mutual inductance, $M = \dfrac{k \phi_A N_B}{I_A} = \dfrac{0.2 \times 0.04 \times 500}{8} = \textbf{0.5 H}$

(*v*)   e.m.f. in coil B $= M \dfrac{dI_A}{dt} = 0.5 \times \dfrac{8 - 0}{0.05} = \textbf{80 V}$

Example 9.22. *Two identical coils are wound on a ring-shaped iron core that has a relative permeability of 500. Each coil has 100 turns and the core dimensions are : area, a = 3 cm² and magnetic path length, l = 20 cm. Calculate the inductance of each coil and the mutual inductance between the coils.*

Solution. $N = 100$ turns ; $\mu_r = 500$; $a = 3 \times 10^{-4}$ m² ; $l = 20 \times 10^{-2}$ m

The statement of the problem suggests that each coil has the same inductance.

∴ $$L_1 = L_2 = \mu_0 \mu_r N^2 \frac{a}{l}$$

$$= 4\pi \times 10^{-7} \times 500 \times (100)^2 \times \frac{3 \times 10^{-4}}{20 \times 10^{-2}} = 9.42 \times 10^{-3} \text{ H} = \textbf{9.42mH}$$

Since the coils are wound on the same iron core, coefficient of coupling $k = 1$.

∴ $$M = k\sqrt{L_1 L_2} = 1\sqrt{9.42 \times 9.42} = \textbf{9.42 mH}$$

---

*   Note that 80% of flux produced in coil *A* links with coil *B*. Therefore, mutual flux (ϕ_{12}) is 80% of 0·05 mWb.

Example 9.23. *Two identical 750-turn coils A and B lie in parallel planes. A current changing at the rate of 1500 A/s in coil A induces an e.m.f. of 11·25 V in coil B. Calculate the mutual inductance of the arrangement. If the self-inductance of each coil is 15 mH, calculate the flux produced in coil A per ampere and the percentage of this flux which links the turns of coil B.*

Solution. Induced e.m.f. in coil B, $e_B = M \dfrac{dI_A}{dt}$

or $\qquad\qquad 11\cdot25 = M \times 1500 \quad\therefore\quad M = 7\cdot5 \times 10^{-3} \text{ H} = \mathbf{7\cdot5 \text{ mH}}$

Now $\qquad\qquad L_1 = \dfrac{N_1 \phi_1}{I_1} \quad\therefore\quad \dfrac{\phi_1}{I_1} = \dfrac{L_1}{N_1} = \dfrac{15 \times 10^{-3}}{750} = \mathbf{2 \times 10^{-5} \text{ Wb/A}}$

Coefficient of coupling, $k = \dfrac{M}{\sqrt{L_1 L_2}} = \dfrac{M}{\sqrt{L^2}} = \dfrac{7\cdot5 \times 10^{-3}}{15 \times 10^{-3}} = 0\cdot5 \text{ or } \mathbf{50\%}$

Example 9.24. *Two coils A and B of 500 and 750 turns respectively are connected in series on the same magnetic circuit of reluctance $1\cdot55 \times 10^6$ AT/Wb. Assuming that there is no flux leakage, calculate (i) self-inductance of each coil and (ii) mutual inductance between coils.*

Solution. *(i)* $\qquad\qquad L_A = \dfrac{N_A{}^2}{\text{Reluctance}} = \dfrac{(500)^2}{1\cdot55 \times 10^6} = \mathbf{0.16 \text{ H}}$

$\qquad\qquad\qquad L_B = \dfrac{N_B{}^2}{\text{Reluctance}} = \dfrac{(750)^2}{1\cdot55 \times 10^6} = \mathbf{0.36 \text{ H}}$

(ii) $\qquad\qquad M = \dfrac{N_A N_B}{\text{Reluctance}} = \dfrac{500 \times 750}{1\cdot55 \times 10^6} = \mathbf{0.24 \text{ H}}$

Alternatively. $\qquad M = k\sqrt{L_1 L_2} = 1\sqrt{0\cdot16 \times 0\cdot36} = \mathbf{0.24 \text{ H}}$

Example 9.25. *Two coils A and B are wound side by side on a paper tube former. An e.m.f. of 0·25 V is induced in coil A when the flux linking it changes at the rate of 10^{-3} Wb/s. A current of 2 A in coil B causes a flux of 10^{-5} Wb to link coil A. What is the mutual inductance between the coils?*

Solution. Induced e.m.f. in coil $A = N_1 \dfrac{d\phi}{dt} \qquad$ or $\qquad 0\cdot25 = N_1 \times 10^{-3}$

$\therefore \qquad\qquad N_1 = 0\cdot25 / 10^{-3} = 250 \text{ turns}$

Flux linkages in coil A due to 2 A in coil $B = 250 \times 10^{-5}$ Wb-turns.

$\therefore \qquad\qquad M = \dfrac{\text{Flux linkages in coil } A}{\text{Current in coil } B}$

$\qquad\qquad\qquad = 250 \times 10^{-5}/2 = \mathbf{1\cdot25 \times 10^{-3} \text{ H}}$

Example 9.26. *The coefficient of coupling between two coils is 0·6 or 60%. The excited coil produces 0·1 Wb of magnetic flux. How much flux is coupled to the other coil ? What is the value of the leakage flux ?*

Solution. The coefficient of coupling is given by ;

$$k = \dfrac{\phi_m}{\phi_t}$$

where $\quad\phi_t$ = flux of the coil receiving current ; $\quad\phi_m$ = flux that links with the other coil

$\therefore \qquad\qquad 0\cdot6 = \phi_m/\phi_t$

or $\qquad\qquad \phi_m = 0\cdot6 \times \phi_t = 0\cdot6 \times 0\cdot1 = \mathbf{0\cdot06 \text{ Wb}}$

The difference between ϕ_t and ϕ_m is the leakage flux.

$\therefore$ Leakage flux, $\phi_l = \phi_t - \phi_m = 0.1 - 0.06 = $ **0.04 Wb**

Example 9.27. *Two coils, A of 12,500 turns and B of 16,000 turns, lie in parallel planes so that 60% of flux produced in A links coil B. It is found that a current of 5A in A produces a flux of 0.6 mWb while the same current in B produces 0.8 mWb. Determine (i) mutual inductance and (ii) coupling coefficient.*

Solution. (*i*) Mutual inductance, $M = \dfrac{k\phi_A N_B}{I_A}$

Here $k = 0.6$; $\phi_A = 0.6\,\text{mWb} = 0.6 \times 10^{-3}\,\text{Wb}$; $N_B = 16000$; $I_A = 5\text{A}$

$\therefore$ $M = \dfrac{0.6 \times 0.6 \times 10^{-3} \times 16000}{5} = $ **1.15 H**

(*ii*) Now $L_A = \dfrac{N_A \phi_A}{I_A} = \dfrac{12500 \times 0.6 \times 10^{-3}}{5} = 1500 \times 10^{-3}\,\text{H} = 1500\,\text{mH}$

and $L_B = \dfrac{N_B \phi_B}{I_B} = \dfrac{16000 \times 0.8 \times 10^{-3}}{5} = 2560 \times 10^{-3}\,\text{H} = 2560\,\text{mH}$

$\therefore$ Coefficient of coupling, $k = \dfrac{M}{\sqrt{L_A L_B}} = \dfrac{1.15 \times 10^3}{\sqrt{1500 \times 2560}} = $ **0.586**

The coefficient of coupling is a measure of how tightly the two coils are coupled. It is a pure number (no units) that varies from 0 to 1. The only way to closely approach $k = 1$ is to wind both coils on the same high-permeability core. This couples them tightly.

Example 9.28. *The coefficient of coupling between two coils is 0.85. Coil 1 has 250 turns. When the current in coil 1 is 2A, the total flux of this coil is 3×10^{-4} Wb. When I_1 is changed from 2A to zer linearly in 2 ms, the voltage induced in the coil 2 is 63.75 V. Find L_1, L_2, M and N_2.*

Solution. Inductance of coil 1, $L_1 = \dfrac{N_1 \phi_1}{I_1} = \dfrac{250 \times 3 \times 10^{-4}}{2} = 37.5 \times 10^{-3}\,\text{H}$

e.m.f. induced in coil 2, $e_2 = M \dfrac{dI_1}{dt}$

Here, $e_2 = 63.75\,\text{V}$; $dI_1 = 2 - 0 = 2\text{A}$; $dt = 2\text{ms} = 2 \times 10^{-3}\,\text{s}$

$\therefore$ $63.75 = M \times \dfrac{2}{2 \times 10^{-3}}$ or $M = $ **63.75 $\times 10^{-3}$ H**

Now, $M = k\sqrt{L_1 L_2}$

Here, $M = 63.75 \times 10^{-3}\,\text{H}$; $k = 0.85$; $L_1 = 37.5 \times 10^{-3}\,\text{H}$

$\therefore$ $63.75 \times 10^{-3} = 0.85 \times \sqrt{37.5 \times 10^{-3} \times L_2}$ or $L_2 = $ **150 $\times 10^{-3}$ H**

Now $\dfrac{L_1}{L_2} = \dfrac{N_1^2}{N_2^2}$ or $\dfrac{37.5 \times 10^{-3}}{150 \times 10^{-3}} = \dfrac{(250)^2}{N_2^2}$

$\therefore$ $0.25 = \dfrac{(250)^2}{N_2^2}$ or $N_2 = $ **500**

Example 9.29. *The dimensions of the magnetic core shown in Fig. 9.16 are :*

Cross-sectional area, $a = 3\,cm^2$; magnetic path length, $l = 10\,cm$ and the relative permeability is 250.

The primary coil has $N_P = 100$ turns and the secondary coil has $N_S = 75$ turns. If the current is increased from 0 to 5A in 0.1s, determine the e.m.f. induced in the secondary.

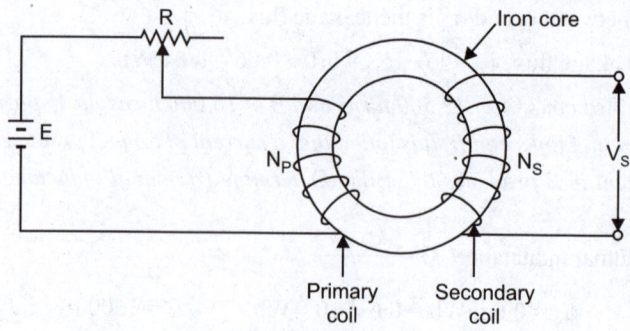

Fig. 9.16

Solution.

$$\text{m.m.f.} = N_P I = 100 \times 5 = 500 \text{ AT}$$

$$\text{Magnetising force, } H = \frac{\text{m.m.f.}}{l} = \frac{500}{10 \times 10^{-2}} = 5000 \text{ AT/m}$$

$$\text{Flux density in core, } B = \mu_0 \mu_r H = 4\pi \times 10^{-7} \times 250 \times 5000 = 1.57 \text{ Wb/m}^2$$

$$\text{Total flux in core, } \phi = B \times a = 1.57 \times 3 \times 10^{-4} = 471 \times 10^{-6} \text{ Wb}$$

∴ Induced e.m.f. in the secondary is given by ;

$$e_S = N_S \frac{d\phi}{dt} = 75 \times \frac{471 \times 10^{-6}}{0.1} = \mathbf{0.35 \text{ V}}$$

Example 9.30. *A long single layer solenoid has an effective diameter of 10 cm and is wound with 2500 T/m. There is a small concentrated coil having its plane lying in the centre cross-sectional plane of the solenoid. Calculate the mutual inductance between the two coils if the concentrated coil has 120 turns on an effective diameter of (i) 8 cm and (ii) 12 cm.*

Solution. Let I_1 be the current flowing through the solenoid.

(*i*) Fig. 9.17 (*i*) shows the conditions of the problem when the effective diameter of concentrated search coil is 8 cm (less than that of the solenoid).

Magnetising force H inside the solenoid is

$$H = \frac{NI_1}{l} = \frac{N}{l} I_1 = 2500 \, I_1 \qquad \qquad \left(\because \frac{N}{l} = 2500 \right)$$

∴   Flux density at the centre of the solenoid is

$$B = \mu_0 H = 2500 \, \mu_0 I_1 \text{ Wb/m}^2$$

Area of search coil, $a_S = \dfrac{\pi}{4} d^2 = \dfrac{\pi}{4}(0.08)^2 = 0.005 \text{ m}^2$

Flux linking with search coil is given by ;

$$\phi_2 = B \, a_S = 2500 \, \mu_0 I_1 \times 0.005 = 15.79 \times 10^{-6} I_1 \text{ Wb}$$

∴

$$M = \frac{N_2 \phi_2}{I_1} = \frac{120 \times 15.79 \times 10^{-6} I_1}{I_1} = \mathbf{1.895 \times 10^{-3} \text{ H}}$$

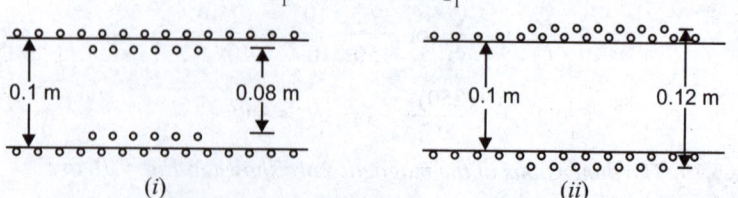

0.1 m   0.08 m   0.1 m   0.12 m

(*i*)   (*ii*)

Fig. 9.17

(*ii*) Fig. 9.17 (*ii*) shows the conditions of the problem when the effective diameter of concentrated search coil is 12 cm (*i.e.* more than that of the solenoid). Since the field strength outside the solenoid is negligible, the effective area of search coil will be equal to the area of solenoid *i.e.*

$$a'_S = \frac{\pi}{4}(0.1)^2$$

∴ Flux linking with the search coil is given by ;

$$\phi'_2 = B\, a'_S = 2500\ \mu_0\, I_1 \times \frac{\pi}{4} \times (0.1)^2$$

∴

$$M = \frac{N_2 \phi'_2}{I_1} = 120 \times \frac{2500 \mu_0 I_1 \times (\pi/4) \times (0.1)^2}{I_1} = 2.962 \times 10^{-3}\ \text{H}$$

Tutorial Problems

1. A solenoid 70 cm in length and of 2100 turns has a radius of 4·5 cm. A second coil of 750 turns is wound upon the middle part of the solenoid. Find the mutual inductance between the two coils. **[18·2 mH]**

2. Two coils having 150 and 200 turns respectively are wound side by side on a closed iron circuit of section 150 cm^2 and mean length of 300 cm. Determine the mutual inductance between the coils and e.m.f. induced in the second coil if current changes from zero to 10A in the first coil in 0·02 second. Relative permeability of iron = 2000. **[0·377 H; 188·5 V]**

3. The self-inductance of a coil of 500 turns is 0·25H. If 60% of the flux is linked with a second coil of 10,000 turns, calculate the mutual inductance between the two coils. **[3 H]**

4. The windings of a transformer has an inductance of $L_1 = 6$H; $L_2 = 0.06$ H and a coefficient of coupling $k = 0·9$. Find the e.m.f. in both the windings when current in primary increases at the rate of 1000 A/s.

    [6000 V; 540 V]

5. An air-cored solenoid with length 30 cm, area of X-section 25 cm^2 and number of turns 500 carries a current of 2·5 A. The current is suddenly switched off in a brief time of 10^{-4} second. How much average e.m.f. is induced across the ends of the open switch in the circuit ? Ignore the variation of magnetic field near the ends of the solenoid. **[6·5 V]**

9.14. Inductors in Series

Consider two coils connected in series as shown in Fig. 9.18.

Let $\quad\quad\quad\quad\quad\quad\quad\quad L_1 = $ inductance of first coil

$\quad\quad\quad\quad\quad\quad\quad\quad\quad L_2 = $ inductance of second coil

$\quad\quad\quad\quad\quad\quad\quad\quad\quad M = $ mutual inductance between the two coils

(i) **Series-aiding.** This is the case when the coils are so arranged that their fluxes *aid each other *i.e.* in the same direction as shown in Fig. 9.18 (*i*). Suppose the current is changing at the rate *di/dt*. The total induced e.m.f. in the circuit will be equal to the sum of e.m.f.s induced in L_1 and L_2 *plus* the mutually induced e.m.f.s, *i.e.*

$$e = L_1 \frac{di}{dt} + L_2 \frac{di}{dt} + M \frac{di}{dt} + M \frac{di}{dt} \quad\quad\quad \textit{... in magnitude}$$

$$= (L_1 + L_2 + 2M)\ di/dt$$

If L_T is the total inductance of the circuit, then,

$$e = L_T \frac{di}{dt}$$

∴

$$L_T = L_1 + L_2 + 2M \quad\quad\quad\quad\quad \textit{... fluxes additive}$$

---

*   **Dot notation.** It is generally not possible to state from the figure whether the fluxes of the two coils are additive or in opposition. Dot notation removes this confusion. The end of the coil through which the current enters is indicated by placing a dot behind it. If the current after leaving the dotted end of coil L_1 enters the dotted end of coil L_2, it means the fluxes of the two coils are additive otherwise in opposition.

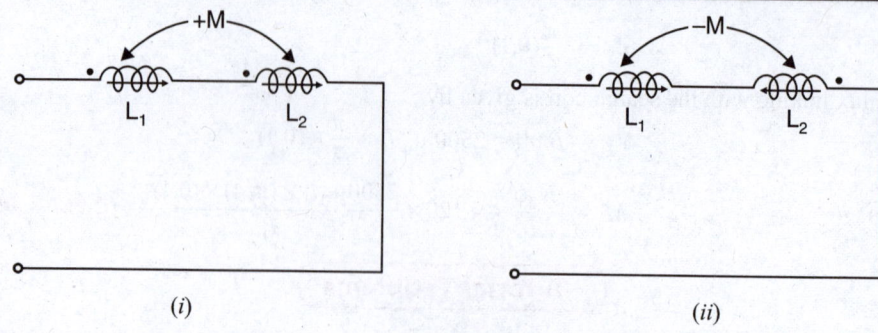

Fig. 9.18

(ii) Series-opposing. Fig. 9.18 (*ii*) shows the series-opposing connection *i.e.* the fluxes of the two coils oppose each other. Suppose the current is changing at the rate *di/dt*. The total induced e.m.f. in the circuit will be equal to sum of e.m.f.s induced in L_1 and L_2 *minus* the mutually induced e.m.f.s.

$$e = L_1\frac{di}{dt} + L_2\frac{di}{dt} - M\frac{di}{dt} - M\frac{di}{dt} = (L_1 + L_2 - 2M)\frac{di}{dt}$$

If L_T is the total inductance of the circuit, then,

$$e = L_T\frac{di}{dt}$$

$\therefore$ $\qquad L_T = L_1 + L_2 - 2M$ $\qquad\qquad\qquad\qquad$ ...*fluxes subtractive*

In general, $\quad L_T = L_1 + L_2 \pm 2M$

Use + sign if fluxes are additive and –ve sign if fluxes are subtractive.

If the two coils are so positioned that *$M = 0$, then, $L_T = L_1 + L_2$.

9.15. Inductors in Parallel with no Mutual Inductance

Consider three inductances L_1, L_2 and L_3 in parallel as shown in Fig. 9.19. Assume that mutual inductance between the coils is zero. Referring to Fig. 9.19, we have,

$$i_T = i_1 + i_2 + i_3$$

or $\qquad \dfrac{di_T}{dt} = \dfrac{di_1}{dt} + \dfrac{di_2}{dt} + \dfrac{di_3}{dt}$

But $\qquad e = L\dfrac{di}{dt}$ or $\dfrac{di}{dt} = \dfrac{e}{L}$

$\therefore \qquad \dfrac{e}{L_T} = \dfrac{e}{L_1} + \dfrac{e}{L_2} + \dfrac{e}{L_3}$

or $\qquad \dfrac{1}{L_T} = \dfrac{1}{L_1} + \dfrac{1}{L_2} + \dfrac{1}{L_3}$ ...(*i*)

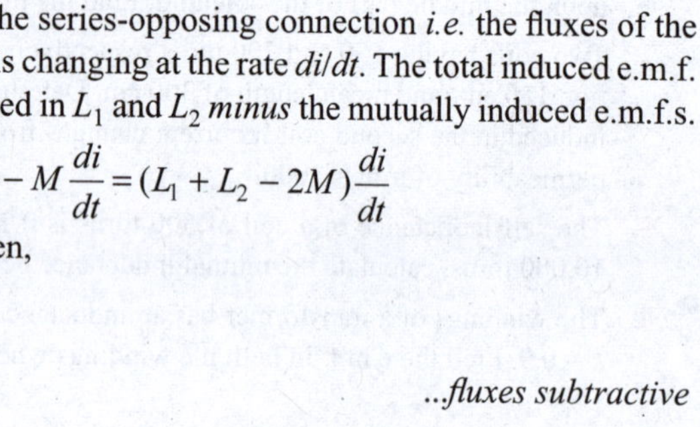

Fig. 9.19

If only two inductors L_1 and L_2 are in parallel, then,

$$\frac{1}{L_T} = \frac{1}{L_1} + \frac{1}{L_2} = \frac{L_1 + L_2}{L_1 L_2}$$

or $\qquad L_T = \dfrac{L_1 L_2}{L_1 + L_2}$ *i.e.* $\dfrac{\text{Product}}{\text{Sum}}$

---

* If the coils are so placed that fluxes produced by them are at right angles to each other, then mutual flux will be zero and hence $M = 0$.

9.16. Inductors in Parallel with Mutual Inductance

Consider two coils A and B of inductances L_1 and L_2 connected in parallel as shown in Fig. 9.20. Let the mutual inductance between the two coils be M. The supply current i divides into two branch currents i_1 and i_2.

$$\text{By } KCL, \quad i = i_1 + i_2$$

$$\therefore \quad \frac{di}{dt} = \frac{di_1}{dt} + \frac{di_2}{dt} \qquad \qquad ...(i)$$

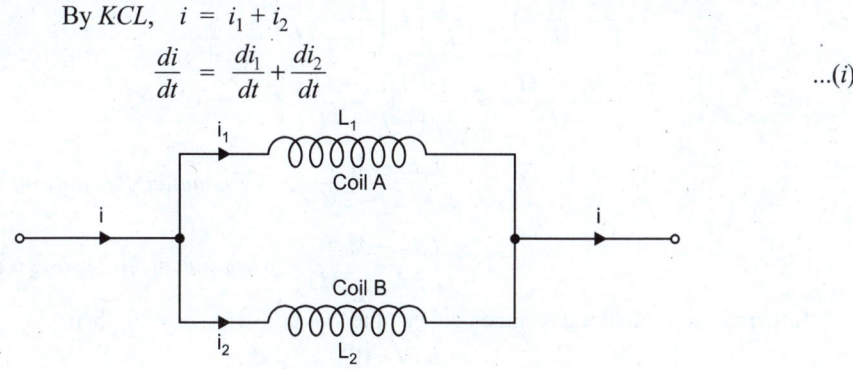

Fig. 9.20

$$\text{Self-induced e.m.f. in coil } A = -L_1\frac{di_1}{dt}$$

$$\text{Mutually induced e.m.f. in coil } A = -M\frac{di_2}{dt}$$

$$\text{Total e.m.f. induced in coil } A = -\left(L_1\frac{di_1}{dt} + M\frac{di_2}{dt}\right)$$

$$\text{Similarly, total e.m.f. induced in coil } B = -\left(L_2\frac{di_2}{dt} + M\frac{di_1}{dt}\right)$$

Since the two coils are in parallel, these e.m.f.s are equal *i.e.*

$$L_1\frac{di_1}{dt} + M\frac{di_2}{dt} = L_2\frac{di_2}{dt} + M\frac{di_1}{dt}$$

or
$$\frac{di_1}{dt}(L_1 - M) = \frac{di_2}{dt}(L_2 - M)$$

$$\therefore \quad \frac{di_1}{dt} = \left(\frac{L_2 - M}{L_1 - M}\right)\frac{di_2}{dt} \qquad \qquad ...(ii)$$

Putting this value of di_1/dt in eq. (i), we have,

$$\frac{di}{dt} = \left[\left(\frac{L_2 - M}{L_1 - M}\right) + 1\right]\frac{di_2}{dt} \qquad \qquad ...(iii)$$

If L_T is the equivalent inductance of the parallel combination, then,

$$\text{Induced e.m.f.} = -L_T\frac{di}{dt}$$

Since induced e.m.f. in the parallel combination is equal to the e.m.f. induced in any one coil (say coil A),

$$\therefore \quad L_T\frac{di}{dt} = L_1\frac{di_1}{dt} + M\frac{di_2}{dt}$$

or
$$\frac{di}{dt} = \frac{1}{L_T}\left(L_1\frac{di_1}{dt} + M\frac{di_2}{dt}\right)$$

Putting the value of di_1/dt from eq. (ii), we have,

$$\frac{di}{dt} = \frac{1}{L_T}\left[L_1\left(\frac{L_2 - M}{L_1 - M}\right) + M\right]\frac{di_2}{dt} \qquad \qquad ...(iv)$$

From eqs. (*iii*) and (*iv*), we have,

$$\frac{L_2 - M}{L_1 - M} + 1 = \frac{1}{L_T}\left[L_1\left(\frac{L_2 - M}{L_1 - M}\right) + M\right]$$

or $$\frac{L_1 + L_2 - 2M}{L_1 - M} = \frac{1}{L_T}\left(\frac{L_1 L_2 - M^2}{L_1 - M}\right)$$

$$\therefore \qquad \qquad L_T = \frac{L_1 L_2 - M^2}{L_1 + L_2 - 2M} \quad ... \textit{when mutual flux aids the individual fluxes}$$

$$= \frac{L_1 L_2 - M^2}{L_1 + L_2 + 2M} \quad ... \textit{when mutual flux opposes the individual fluxes}$$

If there is no mutual inductance between the two coils (*i.e.* $M = 0$), then,

$$L_T = \frac{L_1 L_2 - (0)^2}{L_1 + L_2 \pm 2(0)} = \frac{L_1 L_2}{L_1 + L_2}$$

Example 9.31. *When two coils are connected in series, their effective inductance is found to be 10 H. When the connections of one coil are reversed, the effective inductance is 6 H. If the co-efficient of coupling is 0·6, calculate the self-inductance of each coil and the mutual inductance.*

Solution. $\qquad \qquad \qquad 10 = L_1 + L_2 + 2M \qquad \qquad \qquad \qquad ...(i)$

$$6 = L_1 + L_2 - 2M \qquad \qquad \qquad \qquad ...(ii)$$

Subtracting (*ii*) from (*i*), we get, $4 = 4M$ or $M = 1$ **H**

Putting $M = 1$ H in eq. (*i*), we have, $L_1 + L_2 = 8 \qquad \qquad \qquad ...(iii)$

Also $\qquad \qquad L_1 L_2 = \frac{M^2}{k^2} = \frac{(1)^2}{(0.6)^2} = 2·78 \qquad \qquad \qquad ...(iv)$

Now $\qquad (L_1 - L_2)^2 = (L_1 + L_2)^2 - 4L_1 L_2 = (8)^2 - 4 \times 2·78 = 52·88$

$$\therefore \qquad \qquad L_1 - L_2 = \sqrt{52.88} = 7·27 \qquad \qquad \qquad ...(v)$$

Solving eqs. (*iii*) and (*v*), $L_1 = $ **7·635 H**   and $L_2 = $ **0·365 H**

Example 9.32. *The total inductance of two coils, A and B, when connected in series, is 0.5 H or 0.2H, depending upon the relative direction of the currents in the coils. Coil A, when isolated from coil B, has a self-inductance of 0.2 H. Calculate (i) the mutual inductance between the two coils, (ii) the self-inductance of coil B, (iii) the coupling factor between the coils, and (iv) the two possible values of the induced e.m.f. in coil A when the current is decreasing at 1000 A/s in the series circuit.*

Solution. (*i*) Combined inductance of two coils, $L = L_1 + L_2 \pm 2M$

For series-aiding : $L_1 + L_2 + 2M = 0.5 \qquad \qquad \qquad ...(i)$

For series-opposing : $L_1 + L_2 - 2M = 0.2 \qquad \qquad \qquad ...(ii)$

Subtracting eq. (*ii*) from eq. (*i*), we have,

$$4M = 0.3 \quad \therefore \; M = \textbf{0.075 H}$$

(*ii*) Adding eq. (*i*) and eq. (*ii*), we have,

$$2(L_1 + L_2) = 0.7 \; \text{ or } \; 2(0.2 + L_2) = 0.7 \; \therefore \; L_2 = \textbf{0.15 H}$$

(*iii*) Coefficient of coupling is given by ;

$$k = \frac{M}{\sqrt{L_1 L_2}} = \frac{0.075}{\sqrt{0.2 \times 0.15}} = \textbf{0.433 or 43.3\%}$$

(*iv*) $\qquad \qquad \qquad e_1 = L_1 \frac{di}{dt} \pm M \frac{di}{dt}$

$$\therefore \qquad e_1 = L_1 \frac{di}{dt} + M \frac{di}{dt} = 0.2 \times 1000 + 0.075 \times 1000 = \mathbf{275\ V}$$

or $\qquad\qquad e_1 = L_1 \frac{di}{dt} - M \frac{di}{dt} = 0.2 \times 1000 - 0.075 \times 1000 = \mathbf{125\ V}$

Example 9.33. *Two mutually coupled coils, A and B, are connected in series to a 360 V d.c. supply. Coil A has a resistance of 6 Ω and inductance 4 H. Coil B has resistance of 11 Ω and inductance 9 H. At a certain instant after the circuit is energised, the current is 10 A and is increasing at the rate of 10 A/s. Calculate (i) the mutual inductance between the coils and (ii) the coefficient of coupling.*

Solution. Fig. 9.21 shows the conditions of the problem.

(*i*) Total circuit resistance, $R_T = R_A + R_B$

$$= 6 + 11 = 17\ \Omega$$

Total circuit inductance, $L_T = L_A + L_B + 2M$

$$= 4 + 9 + 2M = 13 + 2M$$

Now $\qquad\qquad V = iR_T + L_T \frac{di}{dt}$

or $\qquad\qquad 360 = 10 \times 17 + (13 + 2M)\ 10 \quad \therefore \quad M = \mathbf{3\ H}$

(*ii*) Coefficient of coupling, $k = \dfrac{M}{\sqrt{L_A L_B}} = \dfrac{3}{\sqrt{4 \times 9}} = \mathbf{0.5}$

Example 9.34. *Two identical coils with terminals, $T_1 T_2$ and $T_3 T_4$ respectively are placed side by side. The inductances measured under different sets of connections are as follows :*

When T_2 is connected to T_3 and inductance measured between T_1 and T_4, it is 4H.

When T_2 is connected to T_4 and inductance measured between T_1 and T_3, it is 0.8 H.

Determine the self inductance of each coil, the mutual inductance between the coils and the coefficient of coupling.

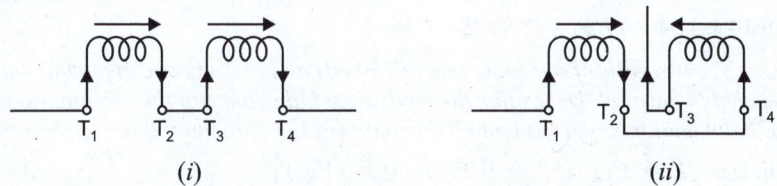

Fig. 9.22

Solution. Since the two coils are identical, each has inductance L (say).

When T_2 is connected to T_3 as shown in Fig. 9.22 (*i*), it is a series-aiding connection so that :

$$L + L + 2M = 4 \quad \text{or} \quad L + M = 2 \qquad\qquad\text{...(}i\text{)}$$

When T_2 is connected to T_4 as shown in Fig. 9.22 (*ii*), it is a series-opposing connection so that:

$$L + L - 2M = 0.8 \quad \text{or} \quad L - M = 0.4 \qquad\qquad\text{...(}ii\text{)}$$

From eqs. (*i*) and (*ii*), $L = \mathbf{1.2\ H}$; $M = \mathbf{0.8\ H}$

Coefficient of coupling, $k = \dfrac{M}{\sqrt{L_1 L_2}} = \dfrac{0.8}{\sqrt{1.2 \times 1.2}} = \mathbf{0.667}$ or $\mathbf{66.7\ \%}$

Example 9.35. *Find the total inductance of the circuit shown in Fig. 9.23.*

$$L_1 = 10\,H \qquad M_{12} = 5\,H$$
$$L_2 = 15\,H \qquad M_{23} = 3\,H$$
$$L_3 = 12\,H \qquad M_{13} = 1\,H$$

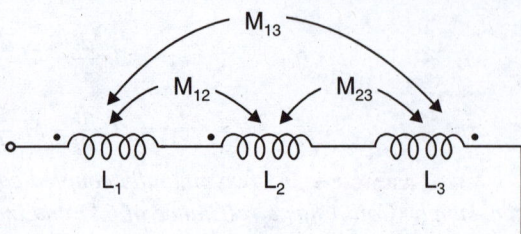

Solution. The fluxes of L_1 and L_2 add to each other and hence M_{12} is positive. The fluxes of L_1 and L_3 are in opposition so M_{13} is negative. Similarly, it can be seen that M_{23} is negative.

Fig. 9.23

$$\therefore \quad L_T = (L_1 + M_{12} - M_{13}) + (L_2 - M_{23} + M_{12}) + (L_3 - M_{23} - M_{13})$$
$$= (10 + 5 - 1) + (15 - 3 + 5) + (12 - 3 - 1)$$
$$= 14 + 17 + 8 = \textbf{39 H}$$

Example 9.36. *Fig. 9.24 shows three inductances in series. Find the total inductance of the circuit from the following data :*

$$L_1 = 12\,H \qquad k_1 = 0\cdot33$$
$$L_2 = 14\,H \qquad k_2 = 0\cdot37$$
$$L_3 = 14\,H \qquad k_3 = 0\cdot65$$

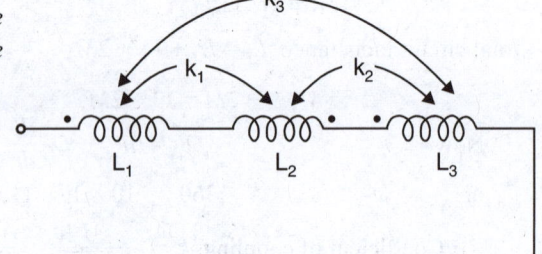

Solution.

$$M_{12} = k_1\sqrt{L_1 L_2} = 0.33\sqrt{12\times14} = 4\cdot28\,H$$

$$M_{23} = k_2\sqrt{L_2 L_3} = 037\sqrt{14\times14} = 5\cdot18\,H$$

Fig. 9.24

$$M_{13} = k_3\sqrt{L_1 L_3} = 065\sqrt{12\times14} = 8\cdot42\,H$$

$$\therefore \quad L_T = (L_1 - M_{12} + M_{13}) + (L_2 - M_{12} - M_{23}) + (L_3 + M_{13} - M_{23})$$
$$= (12 - 4.28 + 8.42) + (14 - 4.28 - 5.18) + (14 + 8.42 - 5.18)$$
$$= 16\cdot14 + 4\cdot54 + 17\cdot24 = \textbf{37·92 H}$$

Example 9.37. *Two coils of self-inductances 150 mH and 250 mH and of mutual inductance 120 mH are connected in parallel. Determine the equivalent inductance of the combination if (i) mutual flux helps the individual flux and (ii) mutual flux opposes the individual flux.*

Solution. Here, $L_1 = 0.15\,H$; $L_2 = 0.25\,H$; $M = 0.12\,H$

(*i*) Equivalent inductance L_T of the parallel combination when mutual flux helps the individual flux is

$$L_T = \frac{L_1 L_2 - M^2}{L_1 + L_2 - 2M} = \frac{0.15\times0.25 - (0.12)^2}{0.15 + 0.25 - 2\times0.12} = \textbf{0.144 H}$$

(*ii*) Equivalent inductance L_T of the parallel combination when the mutual flux opposes the individual flux is

$$L_T = \frac{L_1 L_2 - M^2}{L_1 + L_2 + 2M} = \frac{0.15\times0.25 - (0.12)^2}{0.15 + 0.25 + 2\times0.12} = \textbf{0.036 H}$$

Example 9.38. *Two coils of inductances 0.3 H and 0.8 H are connected in parallel. If the coefficient of coupling is 0.7, calculate the equivalent inductance of the combination if mutual inductance assists the self-inductance.*

Solution. Here, $L_1 = 0.3$ H ; $L_2 = 0.8$ H ; $k = 0.7$

Mutual inductance M between the two coils is

$$M = k\sqrt{L_1 L_2} = 0.7\sqrt{0.3 \times 0.8} = 0.343 \text{ H}$$

∴. Equivalent inductance L_T of the combination when mutual inductance assists the self-inductance is

$$L_T = \frac{L_1 L_2 - M^2}{L_1 + L_2 - 2M} = \frac{0.3 \times 0.8 - (0.343)^2}{0.3 + 0.8 - 2 \times 0.343} = \textbf{0.2955 H}$$

Example 9.39. *Find the equivalent inductance* L_{AB} *in Fig. 9.25.*

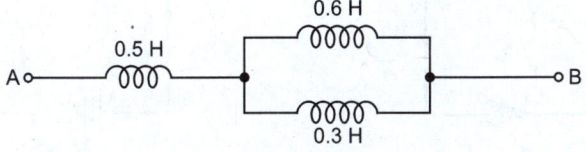

Fig. 9.25

Solution. It is understood that there is no mutual coupling between the coils because it is not given in the problem.

Here, $L_1 = 0.5$ H ; $L_2 = 0.6$H ; $L_3 = 0.3$ H

∴

$$L_{AB} = L_1 + \frac{L_2 L_3}{L_2 + L_3} = 0.5 + \frac{0.6 \times 0.3}{0.6 + 0.3} = \textbf{0.7 H}$$

Tutorial Problems

1. The mutual inductance between two coils in a radio receiver is 100 mH. One coil has 100 mH of self-inductance. What is the self-inductance of the other if coefficient of coupling between the coils is 0·5 ?

   [400 mH]

2. The self-inductances of two coils are $L_1 = 150$ mH, $L_2 = 250$ mH. When they are connected in series with their fluxes aiding, their total inductance is 620 mH. When the connection to one of the coils is reversed (they are still in series), their total inductance is 180 mH. How much mutual inductance exists between them ?

   [110 mH]

3. Two coils of self-inductances 5 H and 8 H are connected in series with their fluxes aiding. If the co-efficient of coupling between the coils is 0·45, find the total inductance of the circuit. **[18·06 H]**

4. The self-inductances of three coils are $L_A = 20$ H, $L_B = 30$ H and $L_C = 40$ H. The coils are connected in series in such a way that fluxes of L_A and L_B add, fluxes of L_A and L_C are in opposition and fluxes of L_B and L_C are in opposition. If $M_{AB} = 8$ H, $M_{BC} = 12$ H and $M_{AC} = 10$ H, find the total inductance of the circuit. **[62 H]**

9.17. Energy Stored in a Magnetic Field

In order to establish a magnetic field around a coil, energy is *required, though no energy is needed to **maintain it. This energy is stored in the magnetic field and is not used up. When the current is decreased, the flux surrounding the coil is decreased, causing the stored energy to be returned to the circuit. Consider an inductor connected to a d.c. source as shown in Fig. 9.26 (*i*). The inductor is equivalent to inductance L in series with a small resistance R as shown in Fig. 9.26 (*ii*). The energy supplied to the circuit is spent in two ways :

---

* When the coil is connected to supply, current increases from zero gradually and reaches the final value $I (= V/R)$ after some time. During this change of current, an e.m.f. is induced in L due to the change in flux linkages. This induced e.m.f. opposes the rise of current. Electrical energy must be supplied to meet this opposition. This supplied energy is stored in the magnetic field.

** To impart a kinetic energy of $\frac{1}{2}mv^2$ to a body, energy is required but no energy is required to maintain it at that energy level.

(*i*) A part of supplied energy is spent to meet I^2R losses and cannot be recovered.

(*ii*) The remaining part is spent to create flux around the coil (or inductor) and is stored in the magnetic field. When the field collapses, the stored energy is returned to the circuit.

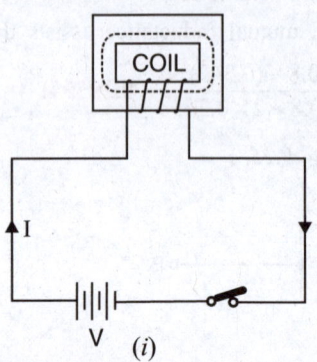
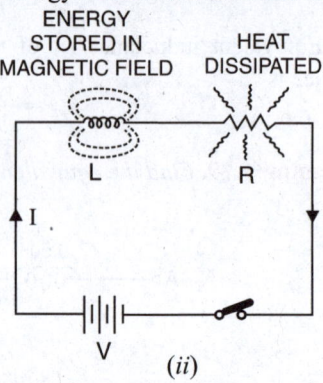

Fig. 9.26

Mathematical Expression. Suppose at any instant the current in the coil is *i* and is increasing at the rate of *di/dt*. The e.m.f. *e* across *L* is given by ;

$$e = L\frac{di}{dt}$$

∴   Instantaneous power, $p = ei = Li\frac{di}{dt}$

During a short interval of time *dt*, the energy *dw* put into the magnetic field is equal to power multiplied by time *i.e.*

$$dw = p.dt = \left(Li\frac{di}{dt} \right) dt = Li\,di$$

The total energy put into the magnetic field from the time current is zero until it has attained the final steady value *I* is given by ;

$$W = \int_0^I Li\,di = \frac{1}{2}LI^2$$

∴   Energy stored in magnetic field, $E = \frac{1}{2}LI^2$ joules

It is clear that energy stored in an inductor depends upon inductance and current through the inductor. For a given inductor, the amount of energy stored is determined by the maximum current through the inductor. Note that energy stored will be in joules if inductance (*L*) and current (*I*) are in henry and amperes respectively.

Note. If current in an inductor varies, the stored energy rises and falls in step with the current. Thus, whenever current increases, the coil absorbs energy and whenever current falls, energy is returned to the circuit.

Alternate method. In order to determine the amount of energy an inductor stores, we need to determine inductor's current and voltage during the time it is storing energy. Since the inductor stores energy only during the time the current is increasing, we must determine the average current during the time the current is rising. This can be done by referring to Fig. 9.27 which shows the current in an inductor increasing at a constant rate until it reaches the maximum value I_m. Since the current rises linearly from 0 to I_m, the average value of current is

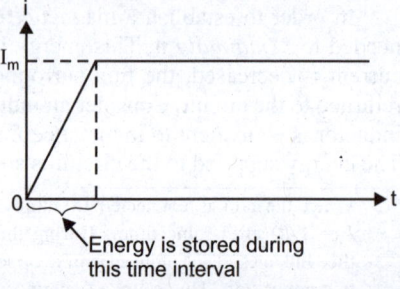

Energy is stored during
this time interval

Fig. 9.27

$$I_{av} = \frac{0 + I_m}{2} = 0.5\,I_m$$

The voltage V_L across the inductor during the time it is storing energy is

$$V_L = L\frac{dI}{dt}$$

Since current rises from 0 to I_m linearly, dI/dt remains constant. Therefore, V_L remains constant during the time the current in inductor is increasing. As a result, expression for V_L reduces to :

$$V_L = \frac{LI_m}{t} \qquad\qquad (\because dI = I_m \text{ and } dt = t)$$

∴   Energy stored in the inductor during time t is

$$E = V_L I_{av}\, t = \frac{LI_m}{t} \times 0.5 I_m \times t = 0.5\, I_m^2 L$$

or

$$E = \frac{1}{2}LI_m^2$$

The subscript m is usually dropped so that :

$$E = \frac{1}{2}LI^2$$

Note that I is the final steady current through the inductor. It may be kept in mind that an inductor stores energy in its magnetic field when the current is rising and returns energy to the circuit when the current is falling.

Note. In case of inductors connected in series, the energy stored is given by ;

$$E = \frac{1}{2}(L_1 + L_2 + 2M)I^2 \qquad \text{... series-aiding}$$

$$E = \frac{1}{2}(L_1 + L_2 - 2M)I^2 \qquad \text{... series - opposing}$$

Example 9.40. *A current of 20 mA is passed through a coil of self-inductance 500 mH. Find the magnetic energy stored. If the current is halved, find the new value of energy stored and the energy released back to the electrical circuit.*

Solution. Magnetic energy stored when current is 20 mA is

$$E_1 = \frac{1}{2}LI^2 = \frac{1}{2}(500\times10^{-3})\times(20\times10^{-3})^2 = \mathbf{100 \times 10^{-6}\ J}$$

Magnetic energy stored when current becomes 10 mA is

$$E_2 = \frac{1}{2}LI^2 = \frac{1}{2}(500\times10^{-3})(10\times10^{-3})^2 = \mathbf{25 \times 10^{-6}\ J}$$

Magnetic energy released back to the circuit

$$= E_1 - E_2 = (100 - 25) \times 10^{-6} = \mathbf{75 \times 10^{-6}\ J}$$

Example 9.41. *The field winding of a machine consists of 8 coils in series, each containing 1200 turns. When the current is 3A, flux linked with each coil is 20 mWb. Calculate (i) the inductance of the circuit, (ii) the energy stored in the circuit and (iii) the average value of induced e.m.f. if the circuit is broken in 0·1 s.*

Solution.

(*i*)   Inductance of each coil, $L = \dfrac{N\phi}{I} = \dfrac{1200\times20\times10^{-3}}{3} = 8\text{H}$

∴   Total inductance, $L_T = 8\,L = 8 \times 8 = \mathbf{64\ H}$

(*ii*)   Magnetic energy stored $= \dfrac{1}{2}L_T\, I^2 = \dfrac{1}{2}\times64\times3^2 = \mathbf{288\ J}$

(*iii*) Average e.m.f. induced, $e = L_T\dfrac{di}{dt} = 64\times\dfrac{3-0}{0\cdot1} = \mathbf{1920\ V}$

Example 9.42. *A coil of inductance 5 H and resistance 100 Ω carries a steady current of 2 A. Calculate the initial rate of fall of current in the coil after a short-circuiting switch connected across its terminals has been suddenly closed. What was the energy stored in the coil and in what form is it dissipated ?*

Solution. The conditions of the problem are represented in Fig. 9.28.

$$V = iR + L\frac{di}{dt}$$

or

$$0 = 2 \times 100 + 5\frac{di}{dt}$$

$$\therefore \quad \frac{di}{dt} = \frac{-200}{5} = -40 \text{ A/s}$$

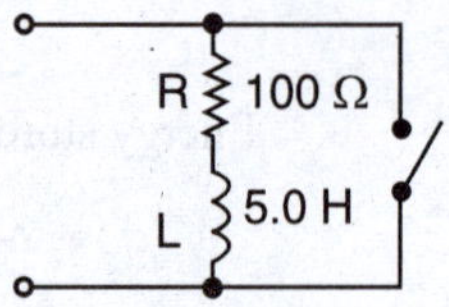

Fig. 9.28

Magnetic energy stored in coil $= \frac{1}{2}LI^2 = \frac{1}{2} \times 5 \times (2)^2 = \textbf{10 J}$

The stored magnetic energy is dissipated in the form of **heat.**

Example 9.43. *(a) A coil of 100 turns is wound on a toroidal magnetic core having a reluctance of 10^4 AT/Wb. When the coil current is 5A and is increasing at the rate of 200 A/s, determine (i) energy stored in the magnetic circuit and (ii) voltage applied across the coil. Assume coil resistance as zero.*

(b) How are your answers affected if the coil resistance is 2Ω ?

Solution. $N = 100$ turns ; Reluctance of core, $S = 10^4$ AT/Wb

(a) Inductance of coil, $L = \dfrac{N^2}{S} = \dfrac{(100)^2}{10^4} = 1 \text{ H}$

(i) Energy stored $= \dfrac{1}{2}LI^2 = \dfrac{1}{2} \times 1 \times (5)^2 = \textbf{12.5 J}$

(ii) Voltage applied across coil = Self-induced e.m.f. in the coil

$$= L\frac{dI}{dt} = 1 \times 200 = \textbf{200 V}$$

(b) If the coil resistance is 2Ω, the energy stored will remain the same *i.e.,* **12.5 J.**

Voltage across coil $= IR + L\dfrac{dI}{dt} = 5 \times 2 + 1 \times 200 = \textbf{210 V}$

However, there will be a loss of energy $= I^2R = (5)^2 \times 2 = 50\text{W}$

Example 9.44. *An iron ring 15 cm in diameter and 10 cm^2 in cross-section is wound with 200 turns of wire. For a flux density of 1 Wb/m^2 and a relative permeability of 500, find the exciting current, the inductance and the stored energy. Find the corresponding quantities when there is a 2 mm air gap.*

Solution.   Magnetic flux, $\phi = B \times a = 1 \times (10 \times 10^{-4}) = 10^{-3}$ Wb

Magnetic length, $l = 0.15 \times \pi$ m

Now   Flux density, $B = \mu_0 \, \mu_r \, H$

$\therefore$ Magnetising force, $H = \dfrac{B}{\mu_0 \mu_r} = \dfrac{1}{(4\pi \times 10^{-7}) \times 500} = 1590$ AT/m

Total ampere-turns $= H \times l = 1590 \times (0.15 \times \pi)$ AT

$\therefore$ Exciting current, $I = \dfrac{\text{Total AT}}{N} = \dfrac{1590 \times (0.15 \times \pi)}{200} = \textbf{3.75 A}$

Inductance, $L = \dfrac{N\phi}{I} = \dfrac{200 \times 10^{-3}}{3.75} = 53.4 \times 10^{-3}$ H $= \textbf{53.4 mH}$

Magnetic energy stored $= \dfrac{1}{2}LI^2 = \dfrac{1}{2} \times 534 \times 10^{-3} \times (3.75)^2 = \mathbf{0.375\ J}$

With 2 mm air gap. The length of air gap, $l_g = 2$ mm $= 2 \times 10^{-3}$ m

$$\text{Air gap } AT = H \times l_g = \dfrac{B}{\mu_0} \times l_g = \dfrac{1}{4\pi \times 10^{-7}} \times 2 \times 10^{-3} = 1590 \text{ AT}$$

Additional current required $= 1590/200 = 7.95$ A

$\therefore$ Total exciting current, $I_T = 3.75 + 7.95 = \mathbf{11.7\ A}$

$$\text{Inductance, } L = \dfrac{N\phi}{I_T} = \dfrac{200 \times 10^{-3}}{11.7} = 17.1 \times 10^{-3}\ H = \mathbf{17.1\ mH}$$

$$\text{Magnetic energy stored} = \dfrac{1}{2}LI_T^2 = \dfrac{1}{2} \times (17.1 \times 10^{-3}) \times (11.7)^2 = \mathbf{1.17\ J}$$

Example 9.45. *An inductor with 10 Ω resistance and 200 mH inductance is connected across 24 V d.c. source. Calculate (i) energy stored in the inductance, (ii) power dissipated by the resistance and (iii) power dissipated by the inductance.*

Solution. $V = 24$ volts ; $R = 10\ \Omega$; $L = 200$ mH $= 0.2$ H

(*i*) Final current in inductor, $I = \dfrac{V}{R} = \dfrac{24}{10} = 2.4$ A

Energy stored in inductance $= \dfrac{1}{2}LI^2 = \dfrac{1}{2} \times 0.2 \times (2.4)^2 = \mathbf{0.576\ J}$

(*ii*) Power dissipated by resistor $= I^2 R = (2.4)^2 \times 10 = \mathbf{57.6\ W}$

(*iii*) Power dissipated by inductance $= \mathbf{0\ W}$

Example 9.46. *A coil of inductance 0.25 H and negligible resistance is connected to a source of supply represented by v = 4 t volts. If the voltage is applied at t = 0 and switched off at t = 5 sec., find (i) the maximum value of current, (ii) r.m.s. value of current and (iii) the energy stored during this period.*

Solution. (*i*)
$$v = 4t \quad \text{or} \quad L\dfrac{di}{dt} = 4t \quad \text{or} \quad 0.25\dfrac{di}{dt} = 4t$$

$\therefore$
$$0.25 \int_0^I di = \int_0^5 4t\ dt$$

or
$$0.25\ I = \left|\dfrac{4t^2}{2}\right|_0^5 = 50$$

$\therefore$ Max. value of current, $I = 50/0.25 = \mathbf{200\ A}$

(*ii*) Suppose i is the current at any time t. Then,

$$0.25\ i = \int_0^t 4t\ dt = 2t^2$$

$\therefore$
$$i = 8t^2$$

The sum of squares of current from 0 to 5 sec.

$$= \int_0^5 64t^4 dt = \dfrac{64 \times 5^5}{5} = 64 \times 5^4$$

$\therefore$ Mean square value $= \dfrac{64 \times 5^4}{5} = 64 \times 5^3$

$\therefore$ $\qquad$ R.M.S. value $= \sqrt{64 \times 5^3} = 89.5\,A$

(iii) $\qquad$ Energy stored $= \int\limits_0^5 vi\,dt = \int\limits_0^5 (4t \times 8t^2)\,dt$

$$= \left. \frac{32t^4}{4} \right|_0^5 = \frac{32 \times 5^4}{4} = 5000\,J$$

Example 9.47. *A direct current of 1 A is passed through a coil of 5000 turns and produces a flux of 0·1 mWb. Assuming that whole of this flux threads all the turns, what is the inductance of the coil ? What would be the voltage developed across the coil if the current were interrupted in 10^{-3} second ? What would be the maximum voltage developed across the coil if a capacitor of 10 μF were connected across the switch breaking the d.c. supply ?*

Solution. Inductance of coil, $L = \dfrac{N\phi}{I} = \dfrac{5000 \times 0 \cdot 1 \times 10^{-3}}{1} = \mathbf{0 \cdot 5\,H}$

$\qquad$ E.M.F. induced in coil, $e = L\dfrac{dI}{dt} = 0 \cdot 5 \times \dfrac{1-0}{10^{-3}} = \mathbf{500\,V}$

When capacitor is connected, the voltage developed will be equal to the p.d. developed across the capacitor plates due to the energy stored in the coil. If V is the value of voltage developed, then,

$$\frac{1}{2}CV^2 = \frac{1}{2}LI^2$$

or $\qquad$ $\dfrac{1}{2} \times (10 \times 10^{-6})\,V^2 = \dfrac{1}{2} \times 0 \cdot 5 \times (1)^2$

$\therefore$ $\qquad$ $V = \mathbf{2 \cdot 24\,volts}$

$$\boxed{\text{Tutorial\ \ Problems}}$$

1. The field winding of a d.c. electromagnet is wound with 960 turns and has resistance of 50 Ω. The exciting voltage is 230 V and the magnetic flux linking the coil is 5 mWb. Find (*i*) self-inductance of the coil and (*ii*) the energy stored in the magnetic field. $\qquad$ [(*i*) 1.043H (*ii*) 11.04 J]

2. An iron ring of 20 cm mean diameter having a cross-section of 100 cm² is wound with 400 turns of wire. Calculate the exciting current required to establish a flux density of 1 Wb/m² if the relative permeability of iron is 1000. What is the value of energy stored? $\qquad$ [1.25 A ; 2.5 J]

3. The inductance of a coil is 0.15H. The coil has 100 turns. Find (*i*) total magnetic flux through the coil when the current is 4A (*ii*) energy stored in the magnetic field (*iii*) voltage induced in the coil when current is reduced to zero in 0.01 second. $\qquad$ [(*i*) 6 mWb (*ii*) 1.2 J (*iii*) 60 V]

4. An air-cored solenoid has a length of 50 cm and a diameter of 2 cm. Calculate its inductance if it has 1000 turns and also find the energy stored in it if the current rises from zero to 5A. $\qquad$ [0.7 mH ; 8.7 mJ]

9.18. Magnetic Energy Stored Per Unit Volume

Consider a coil of N turns wound over a magnetic circuit of length l metres and uniform cross-sectional area of '*a*' m².

$$\text{Magnetic energy stored} = \frac{1}{2}LI^2 = \frac{1}{2}\left(\frac{N^2 a\,\mu_0\,\mu_r}{l}\right)I^2 = \frac{1}{2}(a\,\mu_0\,\mu_r)l\left(\frac{NI}{l}\right)^2$$

$$= \frac{1}{2}(\mu_0\,\mu_r)(a\,l)\,H^2 \qquad\qquad (\because H = NI/l)$$

Now, $\qquad\qquad$ $al = $ volume of magnetic field in m³

$\therefore$ Magnetic energy stored/m^3 $= \dfrac{1}{2}\mu_0\mu_r H^2$

$$= \dfrac{1}{2}\mu_0\mu_r\left(\dfrac{B}{\mu_0\mu_r}\right)^2 \qquad\qquad \left[\because H = \dfrac{B}{\mu_0\mu_r}\right]$$

$$= \dfrac{B^2}{2\mu_0\mu_r} \text{ joules} \qquad\qquad\qquad\qquad \textit{...in a medium}$$

$$= \dfrac{B^2}{2\mu_0} \text{ joules} \qquad\qquad\qquad\qquad \textit{...in air}$$

Note that magnetic energy stored will be in joules if the flux density B is in Wb/m^2.

9.19.   Lifting Power of a Magnet

When two opposite polarity magnetic poles are separated by a short distance in air, there is a force of attraction tending to pull the two poles together. The magnitude of this force can be calculated in terms of flux density in the air gap and cross-sectional area of the pole.

Consider two poles, north and south, each of area 'a' square metres separated by a short distance in air as shown in Fig. 9.29. Let P newtons be the force of attraction between the two poles. If one of the poles, say S-pole, is pulled apart through a small distance dx, then work will have to be done against the force of attraction.

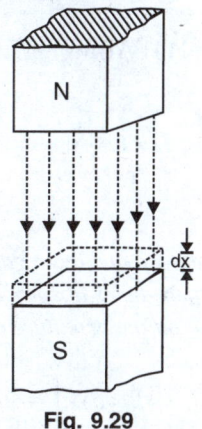

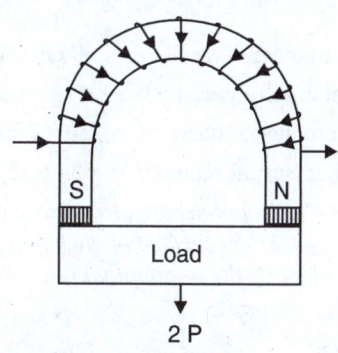

Fig. 9.29	**Fig. 9.30**

Work done $= P \times dx$ joules

This work done is stored in the additional volume of the magnetic field created.

Additional volume of magnetic field created

$$= a \times dx \text{ m}^3$$

$\therefore$ Increase in stored energy $= \dfrac{B^2}{2\mu_0}\times a\,dx$

But increase in stored energy $=$ Work done

or $\dfrac{B^2}{2\mu_0}\times a\,dx = P \times dx$

$\therefore$ $\qquad\qquad P = \dfrac{B^2 a}{2\mu_0}$ newtons

It may be noted that P will be in newtons if B is in Wb/m^2 and 'a' in m^2.

Note that P is the force of attraction at each pole. In a practical magnet, there are two poles (See Fig. 9.30) so that total force of attraction is $2P$. Electromagnets are widely used for commercial lifting jobs such as loading scrap iron into a truck or raising an armature to a higher position.

Example 9.48. *A lifting magnet of inverted U-shape is formed out of an iron bar 60 cm long and 10 cm^2 in cross-sectional area [See Fig. 9.31]. Exciting coils of 750 turns each are wound on the two side limbs and are connected in series. Calculate the exciting current necessary for the magnet to lift a load of 60 kg, assuming that the load has negligible reluctance and makes close contact with the magnet. Relative permeability of the material of magnet is 600.*

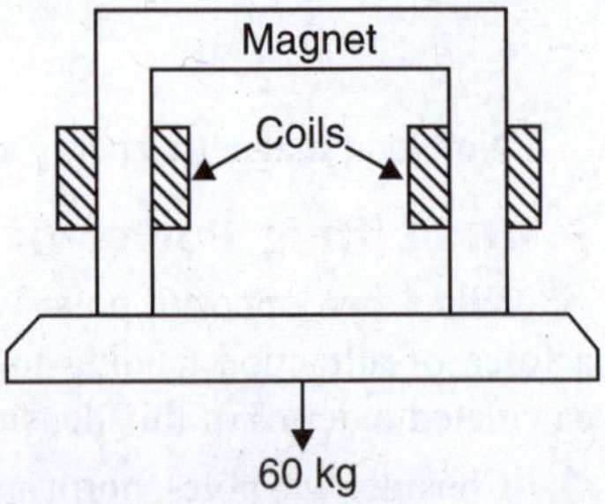

Solution. Attractive force at each pole is

$$P = \frac{60 \times 9.81}{2} = 294.3 \text{ N}$$

Now,
$$P = \frac{B^2 a}{2\mu_0}$$

or
$$B^2 = \frac{2\mu_0 P}{a}$$

$$= \frac{2 \times (4\pi \times 10^{-7}) \times 294.3}{10 \times 10^{-4}} = 0.74$$

∴
$$B = \sqrt{0.74} = 0.86 \text{ Wb/m}^2$$

Magnet

Coils

60 kg

Fig. 9.31

Magnetising force, $H = \dfrac{B}{\mu_0 \mu_r} = \dfrac{0.86}{4\pi \times 10^{-7} \times 600} = 1141 \text{ AT/m}$

Length of magnetic path, l = 60 cm = 0·6 m

Total AT required = 0·6 × 1141 = 684·6 AT

Total number of turns = 2 × 750 = 1500

∴ Exciting current required = 684·6/1500 = **0.456 A**

Example 9.49. *A smooth core armature working in a 4-pole field magnet has a gap (iron to iron) of 0·5 cm. The area of the surface of each pole is 0·1 m^2. The ampere-turns absorbed by each pole are 3000. Calculate (i) the mechanical force exerted by each pole on the armature and (ii) energy stored in the four air gaps.*

Solution. *(i)* AT per gap = Flux × Reluctance of air gap = $(B \times a) \times \left(\dfrac{l_g}{a\mu_0} \right) = \dfrac{B\, l_g}{\mu_0}$

or
$$\frac{B}{\mu_0} = \frac{AT \text{ per gap}}{l_g} = \frac{3000}{0\cdot5 \times 10^{-2}} = 6 \times 10^5$$

or
$$B = \mu_0 \times 6 \times 10^5 = (4\pi \times 10^{-7}) \times 6 \times 10^5 = 0\cdot75 \text{ Wb/m}^2$$

Mechanical force exerted by each pole is

$$P = \frac{B^2 a}{2\mu_0} = \frac{(0.75)^2 \times 0.1}{2 \times 4\pi \times 10^{-7}} = \textbf{22381 N}$$

(ii)   Volume of 4 air gaps = $4\, a\, l_g = 4 \times 0\cdot1 \times 0\cdot5 \times 10^{-2} = 0\cdot002 \text{ m}^3$

Energy stored in air gaps = $\dfrac{B^2}{2\mu_0}$ × Volume of 4 airgaps

$$= \frac{(0.75)^2}{2 \times 4\pi \times 10^{-7}} \times 0.002 = \textbf{448 J}$$

Example 9.50. *An iron ring having a mean circumference of 30 cm and a cross-sectional area of 1 cm² has two radial saw cuts at diameterically opposite points. A brass plate is inserted in each gap (thickness of each gap being 0·1 mm). If the ring is wound with 200 turns, calculate the magnetising current to exert a pull of 5 kg between the two halves. Assume the magnetic data for the iron to be :*

B (Wb/m²)	0·79	1·0	1·3
H (AT/m)	250	350	520

Solution. The total force of attraction at the two separations is $= 5 \times 9·80 = 49$ N. Therefore, force of attraction at each separation, $P = 49/2 = 24·5$ N.

Now,
$$P = \frac{B^2 a}{2\mu_0} \quad \therefore \quad B = \sqrt{\frac{2 \times 4\pi \times 10^{-7} \times 245}{1 \times 10^{-4}}} = 0·79 \text{ Wb/m}^2$$

Corresponding to $B = 0·79$ Wb/m², we have, $H = 250$ AT/m.

$$\text{Length of iron path} = 30 \text{ cm} = 0·3 \text{ m}$$
$$AT \text{ for iron path} = 250 \times 0·3 = 75 \text{ AT}$$
$$H \text{ for brass} = B/\mu_0 = 0·79/4\pi \times 10^{-7} = 628662 \text{ AT/m}$$
$$\text{Thickness of brass plates} = 0·1 + 0·1 = 0·2 \text{ mm} = 0·2 \times 10^{-3} \text{ m}$$
$$AT \text{ for brass paths} = 628662 \times 0·2 \times 10^{-3} = 125·73 \text{ AT}$$
$$\text{Total } AT \text{ required} = 75 + 125·73 = 200·73 \text{ AT}$$

$\therefore$ Magnetising current required $= 200·73/200 = $ **1 A**

Example 9.51. *The arm of a starter is held in the "ON" position by means of an electromagnet. The torque exerted by the spring is 5 Nm and the effective radius at which the force is exerted is 10 cm. Area of each pole face is 2·5 cm² and each air gap is 0·4 mm. Find the minimum number of AT required to keep the arm in the "ON" position.*

Solution. Fig. 9.32 shows the whole arrangement. Let F newtons be the force exerted by the electromagnet.

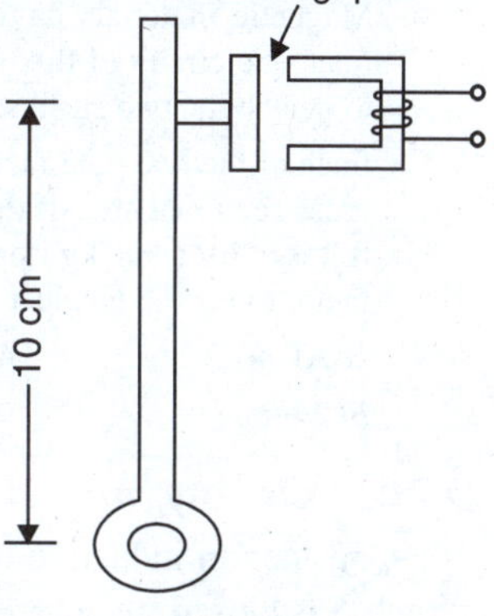

$$\text{Torque} = \text{Force} \times \text{radius}$$

or,
$$5 = F \times 0.1 \quad \therefore \quad F = 5/0.1 = 50 \text{ N}$$

The force exerted at each pole of the magnet, $P = 50/2 = 25$ N

Now
$$P = \frac{B^2 a}{2\mu_0}$$

$\therefore$
$$B = \sqrt{\frac{25 \times 2 \times 4\pi \times 10^{-7}}{25 \times 10^{-4}}} = 0.5 \text{ Wb/m}^2$$

The AT for iron path may be neglected ; being very small.

$$H \text{ in air gap} = B/\mu_0 = 0.5/4\pi \times 10^{-7} = 397887 \text{ AT/m}$$
$$\text{Total air gap length} = 2 \times 0.4 \times 10^{-3} = 0.8 \times 10^{-3} \text{ m}$$
$$AT \text{ required} = 397887 \times 0.8 \times 10^{-3} = \textbf{318.3 AT}$$

Air gap

Fig. 9.32

Example 9.52. *The electromagnet shown in Fig. 9.33 has pole pieces each having a cross-sectional area of 25 cm². The total flux crossing each pole is 250 μWb. Determine the maximum weight of iron plate that can be lifted by the magnet. Neglect magnetic leakage and fringing.*

Solution. Flux density in the air gap is

$$B = \frac{\phi}{a} = \frac{250 \times 10^{-6}}{25 \times 10^{-4}} = 0.1 \text{ Wb/m}^2$$

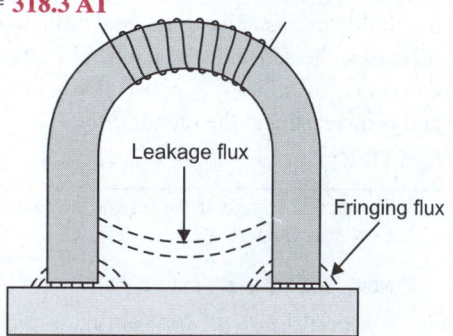

Leakage flux

Fringing flux

Fig. 9.33

Attractive force at each pole is

$$P = \frac{B^2 a}{2\mu_0} = \frac{(0.1)^2 \times 25 \times 10^{-4}}{2 \times 4\pi \times 10^{-7}} = 9.95 \text{ N}$$

Total force due to two poles $= 2P = 2 \times 9.95 = 19.9$ N

Let m be the maximum mass of the plate that can be lifted.

$$\therefore \qquad\qquad m \times g = 19.9$$

or $\qquad\qquad m = \dfrac{19.9}{g} = \dfrac{19.9}{9.8} = 2.03$ kg

Therefore, maximum weight of the plate that can be lifted is **2.03 kg.**

(**Tutorial Problems**)

1. Find the pull exerted on the plunger of an electromagnet when the total flux uniformly distributed is 500 µWb. Diameter of the plunger is 2·54 cm. **[196·3 N]**

2. A horse shoe magnet has two poles, each of area 5 cm². Find the pull between the poles and the keeper when the flux density at the contact surface is 1 Wb/m². **[398 N]**

3. The core material for use in an electromagnet should not have a flux density more than 1·5 Wb/m². How much area each of the two poles should have if the magnet is to lift 200 kg ? **[10·95 cm²]**

4. A circular crane magnet has an iron cross-section of 200 cm² and a mean magnetic path of 80 cm. Assuming the total length of each air gap to be 1·5 mm, calculate (i) the AT to produce a gap flux of 0·025 Wb (ii) the force to separate the contact surface, assuming no leakage or fringing.

B(Wb/m²)	1·0	1·2	1·4
H (AT/m)	900	1230	2100

   [(i) 4080 AT (ii) 2500 kg (force)]

5. In a telephone receiver, the size of each pole of the electromagnet is 1.2 cm × 0.2 cm and flux between each pole and diaphragm is 4×10^{-6} Wb. With what force is the diaphragm attracted towards the poles? **[0.532 N]**

6. Magnetic materials having a surface area of 100 cm² are in contact with each other. They are in a magnetic circuit of flux 0.01 Wb uniformly distributed across the surface. Calculate the force required to detach the two surfaces. **[3978 N]**

7. Each of the two pole faces of a lifting magnet has an area of 150 cm² and this may also be taken as the cross-sectional area of the 40 cm long flux path in the magnet. Determine the AT needed on the magnet if it is to lift a 900 kg iron block separated by 0·5 mm from the pole faces. Assume the magnetic leakage factor to be 1·2. Neglect fringing of the gap flux and reluctance of the flux path in the iron block.

H(AT/m)	400	600	800	1200	1600
B(Wb/m²)	0·81	0·98	1·1	1·24	1·35

   [955 AT]

9.20. Closing and Breaking an Inductive Circuit

Consider an inductive circuit shown in Fig. 9.34. When switch S is closed, the current increases gradually and takes some time to reach the final value. The reason the current does not build up *instantly to its final value is that as the current increases, the self-induced e.m.f. in L opposes the change in current (Lenz's Law). Suppose at any instant, the current is i and is increasing at the rate of di/dt.

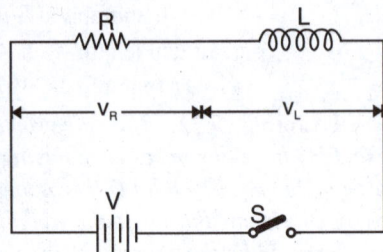

Then, $\qquad\qquad V = v_R + v_L$

Fig. 9.34

---

* The current is zero at the instant the switch is closed because it must start from zero.

Now, self-induced e.m.f., $v_L = L\dfrac{di}{dt}$

If current change (i.e.di) is instant, it means $di/dt = 0$. This means that L is infinite which is impossible. So it is not possible for current in inductance to change from one value to the other in zero time.

$$= iR + L\frac{di}{dt}$$

As the current increases, v_R (= iR) increases and v_L decreases since V is constant. The decrease in v_L (= $L\,di/dt$) means that di/dt decreases because L is constant. The result is that after some time, di/dt becomes zero and so does the self- induced e.m.f. v_L (= $L\,di/dt$). At this stage, the current attains the final fixed value I given by ;

$$V = IR + 0 \quad \text{or} \quad I = \frac{V}{R}$$

Thus, when a d.c. circuit containing inductance is switched on, the current takes some time to reach the final value I (= V/R). Note that the role of inductance is to delay the change; it cannot prevent the current from attaining the final value. Similarly, when an inductive circuit is opened, the current does not jump to zero, but falls gradually. In either case, the delay in change depends upon the values of L and R as explained in the next article.

9.21. Rise of Current in an Inductive Circuit

Consider an inductive circuit shown in Fig. 9.34. When switch S is closed, the current rises from zero to the final value I (= V/R) in a small time t. Suppose at any instant, the current is i and is increasing at the rate of di/dt. Then,

$$V = iR + L\frac{di}{dt} \quad \text{or} \quad V - iR = L\frac{di}{dt}$$

or
$$\frac{di}{V - iR} = \frac{dt}{L}$$

*Multiplying both sides by –R, we get,

$$\frac{-R\,di}{V - iR} = \frac{-R}{L}dt$$

Integrating both sides, we get,

$$\int -\frac{R\,di}{V - iR} = -\frac{R}{L}\int dt$$

or
$$\log_e (V - iR) = -\frac{R}{L}t + K \qquad \qquad ...(i)$$

where K is a constant whose value can be determined from the initial conditions. At $t = 0$, $i = 0$. Putting these values in exp. (i), we have, $\log_e V = K$.

∴   Equation (i) becomes :

$$\log_e (V - iR) = -\frac{R}{L}t + \log_e V$$

or
$$\log_e \frac{V - iR}{V} = -\frac{R}{L}t$$

or
$$\frac{V - iR}{V} = e^{-Rt/L}$$

or
$$V - iR = V e^{-Rt/L}$$

or
$$i = \frac{V}{R}(1 - e^{-Rt/L})$$

But $V/R = I$, the final value of current attained by the circuit.

∴
$$i = I(1 - e^{-Rt/L}) \qquad \qquad ...(ii)$$

_____
*   This step makes the numerator on the L.H.S. a differential of the denominator.

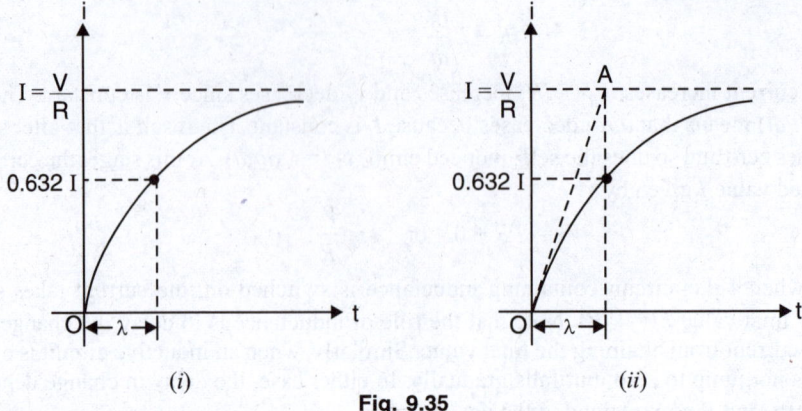

Fig. 9.35

Eq. (*ii*) shows that rise of current follows an exponential law (See Fig. 9.35). As *t* increases, the term $e^{-Rt/L}$ gets smaller and current *i* in the circuit gets larger. Theoretically, the current will reach its final value $I (= V/R)$ in an infinite time. However, practically it reaches this value in a short time.

Note. $V = iR + L\,di/dt$

At the instant the switch is closed, $i = 0$. $\qquad \therefore \quad V = L\,di/dt$

Initial rate of rise of current, $\dfrac{di}{dt} = \dfrac{V}{L}$ A/sec.

The initial rate of rise of current in an inductive circuit helps us in defining the time constant of the circuit.

9.22. Time Constant

Consider the eq. (*ii*) above showing the rise of current w.r.t. time *t*.

$$i = I\,(1 - e^{-Rt/L})$$

The exponent of *e* is Rt/L. The quantity L/R has the dimensions of time so that exponent of *e* (*i.e. Rt/L*) is a number. The quantity L/R is called the *time constant* of the circuit and affects the rise of current in the circuit. It is represented by λ.

$\therefore \qquad\qquad$ Time constant, $\lambda = L/R$ seconds

$\therefore \qquad\qquad\qquad i = I(1 - e^{-t/\lambda})$

Time constant of an inductive circuit can be defined in the following ways :

(*i*) Consider the graph showing the rise of current w.r.t. time *t* [See Fig. 9.35 (*ii*)]. The initial rate of rise of current (*i.e.* at $t = 0$) in the circuit is

$$\frac{di}{dt} = \frac{V}{L}$$

If this rate of rise of current were maintained, the graph would be linear [*i.e. OA* in Fig. 9.35 (*ii*)] instead of exponential. If this rate of rise could continue, the circuit current will reach the final value $I (= V/R)$ in time

$$= \frac{V}{R} \div \frac{V}{L} = \frac{L}{R} = \text{Time constant } \lambda$$

Hence **time constant** *may be defined as the time required for the current to rise to its final steady value if it continued rising at its initial rate (i.e. V/L).*

(*ii*) $\qquad\qquad$ If time interval, $t = \lambda$ (or L/R), then,

$$i = I\,(1 - e^{-Rt/L}) = I\,(1 - e^{-1}) = 0.632\,I$$

Hence **time constant** *can also be defined as the time required for the current to reach 0.632 of its final steady value while rising.*

Fig 9.36 as well as adjoining table shows the percentage of final current (I) after each time constant interval during current buildup (i) in the inductor. The current will increase to about 63% of its full value (I) in first time constant. A 5 time-constant time interval is accepted as the time for the current to attain its final value I.

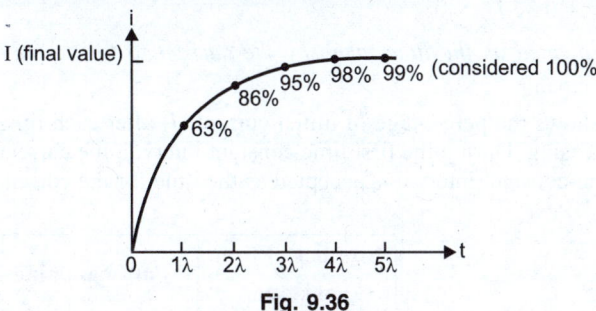

Fig. 9.36

Number of time constants	% of final value
1	63
2	86
3	95
4	98
5	99 (considered 100%)

9.23. Decay of Current in an Inductive Circuit

Consider an inductive circuit shown in Fig. 9.37. When switch S is thrown to position 2, the current in the circuit starts rising and attains the final value I (= V/R) after some time as explained above. If now switch is thrown to position 1, it is found that current in the $R-L$ circuit does not cease immediately but gradually reduces to zero. Suppose at any instant, the current is i and is decreasing at the rate of di/dt. Then,

$$0 = iR + L\frac{di}{dt}$$

or

$$\frac{di}{i} = -\frac{R}{L}dt$$

Integrating both sides, we get, $\log_e i = -\dfrac{R}{L}t + K$ ...(i)

where K is a constant whose value can be determined from the initial conditions. When $t = 0$, then $i = I$ (= V/R).

Putting these values in eq. (i), we have, $\log_e I = 0 + K$ or $K = \log_e I$

∴ Equation (i) becomes : $\log_e i = -\dfrac{R}{L}t + \log_e I$

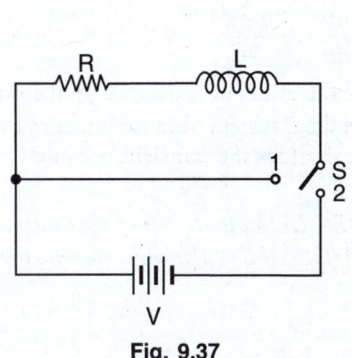

Fig. 9.37

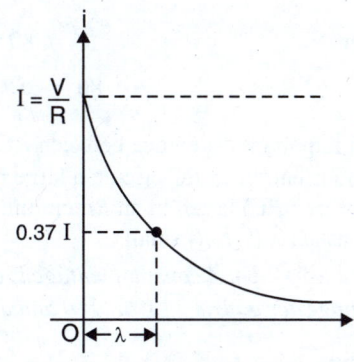

Fig. 9.38

or $\log_e\dfrac{i}{I} = -\dfrac{R}{L}t$ or $\dfrac{i}{I} = e^{-Rt/L}$

∴ $i = Ie^{-Rt/L}$ or $i = Ie^{-t/\lambda}$ ...(ii)

Eq. (*ii*) gives the decay of current in an $R - L$ series circuit with time t and is represented graphically in Fig. 9.38. Note that decay of current follows the exponential law.

Time constant. The quantity L/R in eq. (*ii*) is known as time constant of the circuit. When $t = \lambda \, (= L/R)$,

$$i = I e^{-1} = 0.37 \, I$$

Hence, **time constant** *may also be defined as the time taken by the current to fall to 0.37 of its final steady value I (= V/R) while decaying.*

Fig. 9.39 as well as adjoining table shows the percentage of initial current (I) after each time constant interval while the current is decreasing. During the first time constant interval, the current decreases 37% of its initial value. A 5 time-constant interval is accepted as the time for the current to reduce to zero value.

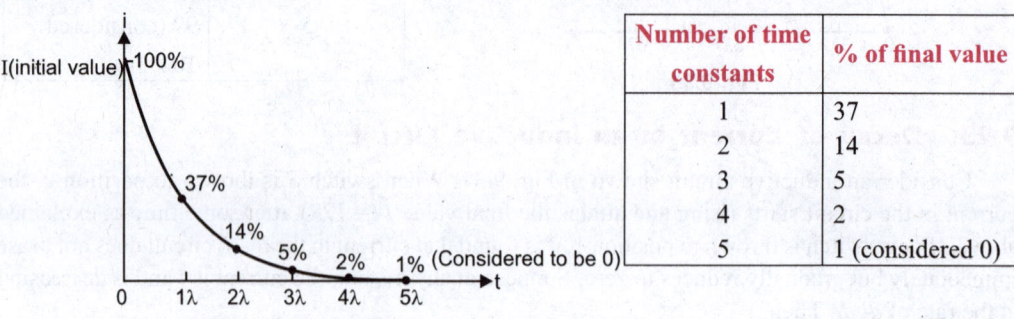

Number of time constants	% of final value
1	37
2	14
3	5
4	2
5	1 (considered 0)

Fig. 9.39

Example 9.53. *The resistance and inductance of a series circuit are 5 Ω and 20 H respectively. At the instant of closing the supply, the current increases at the rate of 4 A/s. Calculate (i) the applied voltage and (ii) the rate of growth of current when the current is 5 A.*

Solution. (*i*) The voltage equation of *R-L* series circuit is

$$V = iR + L\frac{di}{dt}$$

At the instant the switch is closed, $i = 0$.

∴ $$V = L\frac{di}{dt} = 20 \times 4 = \textbf{80 V}$$

(*ii*) $$V = iR + L\frac{di}{dt}$$

Here $$V = 80 \text{ volts} ; \quad i = 5 \text{ A} ; L = 20 \text{ H} ; \quad R = 5 \, \Omega$$

∴ $$80 = 5 \times 5 + 20\frac{di}{dt} \quad \text{or} \quad \frac{di}{dt} = \frac{80 - 25}{20} = \textbf{2.75 A/s}$$

An important difference between *RC* and *RL* circuits is the effect of resistance on the duration of the transient. In an *RC* circuit, a large resistance prolongs the transient because it makes the time constant $\lambda \, (= RC)$ large. In an *RL* circuit, a large resistance shortens the transient because it makes time constant $\lambda \, (= L/R)$ small.

Example 9.54. *A constant voltage is applied to a series $R - L$ circuit at t = 0 by closing a switch. The voltage across L is 25 V at t = 0 and drops to 5 V at t = 0.025s. If L = 2H, what must be the value of R ?*

Solution. Applied voltage, $V = iR + L\dfrac{di}{dt}$

At $t = 0$, $i = 0$ and $L\dfrac{di}{dt} = 25$ volts (*given*)

∴ $$V = 0 + 25 = 25 \text{ volts}$$

At $t = 0.025$ second, $L\dfrac{di}{dt} = 5V$ so that $iR = 25 - 5 = 20$ V.

Now,
$$i = I(1 - e^{-t/\lambda}) = \frac{V}{R}(1 - e^{-t/\lambda})$$

or
$$iR = \frac{V}{R} \times R(1 - e^{-t/\lambda})$$

$$\therefore \qquad iR = V(1 - e^{-t/\lambda})$$

At $t = 0.025$ second, $iR = 20$ V and $V = 25$ volts.

$$\therefore \qquad 20 = 25(1 - e^{-0.025/\lambda})$$

or
$$1 - e^{-0.025/\lambda} = 0.8 \quad \text{or} \quad e^{+0.025/\lambda} = 5$$

$$\therefore \qquad \frac{0.025}{\lambda} \log_e e = \log_e 5 \quad \text{or} \quad \lambda = \frac{0.025}{\log_e 5} = 0.0155$$

Now,
$$\lambda = \frac{L}{R} \quad \text{or} \quad R = \frac{L}{\lambda} = \frac{2}{0.0155} = \mathbf{129.03\Omega}$$

Example 9.55. *The steady current flowing in an inductor is 250 mA ; the current flowing 0.1 sec. after connecting the supply voltage is 120 mA. Calculate (i) time constant of the circuit and (ii) the time from closing the circuit at which circuit current has reached 200 mA.*

Solution. (i) $\qquad\qquad i = I(1 - e^{-t/\lambda})$

Here $\qquad\qquad\qquad i = 120$ mA ; $\quad I = 250$ mA ; $\quad t = 0.1$ sec.

$$\therefore \qquad 120 = 250(1 - e^{-0.1/\lambda}) \quad \text{or} \quad e^{-0.1/\lambda} = 1 - (120/250) = 0.52$$

$$\therefore \qquad e^{0.1/\lambda} = 1/0.52 = 1.923$$

or
$$(0.1/\lambda)\log_e e = \log_e 1.923$$

$$\therefore \qquad \text{Time constant, } \lambda = \frac{0.1}{\log_e 1.923} = \mathbf{0.153 \ s}$$

(ii) $\qquad\qquad\qquad i = I(1 - e^{-t/\lambda})$

Here $\qquad\qquad\qquad i = 200$ mA ; $I = 250$ mA ; $\lambda = 0.153$ sec.

$$\therefore \qquad 200 = 250(1 - e^{-t/0.153}) \quad \text{or} \quad e^{-t/0.153} = 1 - (200/250) = 0.2$$

$$\therefore \qquad e^{t/0.153} = 1/0.2 = 5$$

or
$$(t/0.153)\log_e e = \log_e 5$$

$$\therefore \qquad t = 0.153 \log_e 5 = \mathbf{0.25 \ s}$$

Example 9.56. *A coil having $L = 2.4$ H and $R = 4 \ \Omega$ is connected to a constant 100 V supply source. How long does it take the voltage across the resistance to reach 50 V ?*

Solution. $\qquad\qquad i = I(1 - e^{-t/\lambda}) = \frac{V}{R}(1 - e^{-t/\lambda})$

or
$$iR = V(1 - e^{-t/\lambda})$$

Here
$$iR = 50 \text{ volts} ; V = 100 \text{ volts} ; \lambda = L/R = 2.4/4 = 0.6 \text{ s}$$

$$\therefore \qquad 50 = 100(1 - e^{-t/0.6}) \quad \text{or} \quad e^{-t/0.6} = 1 - (50/100) = 0.5$$

$$\therefore \qquad e^{t/0.6} = 1/0.5 = 2$$

or
$$(t/0.6)\log_e e = \log_e 2$$

$$\therefore \qquad t = 0.6 \log_e 2 = \mathbf{0.416 \ s}$$

Example 9.57. *The time constant of a certain inductive coil was found to be 2.5 ms. With a resistance of 80 Ω added in series, a new time constant of 0.5ms was obtained. Find the inductance and resistance of the coil.*

Solution. Time constant, $\lambda = L/R$

For the first case, $L/R = 2.5$; For the second case, $L/(R + 80) = 0.5$

$$\therefore \qquad \frac{R+80}{R} = \frac{2\cdot 5}{0\cdot 5} = 5 \quad \text{or} \quad R = 20\,\Omega$$

Now $\qquad\qquad L/R = 2\cdot 5 \quad \therefore \quad L = 2\cdot 5\,R = 2\cdot 5 \times 20 = 50\,H$

Example 9.58. *A coil having an effective resistance of 25 Ω and an inductance of 5 H is suddenly connected across a 50 V d.c. supply. What is the rate at which energy is stored in the field of the coil when current is (i) 0·5 A, (ii) 1A and (iii) steady ? Also find the induced EMF in the coil under the above conditions.*

Solution.

(i) When current is 0·5 A

Power input $= 50 \times 0\cdot 5 = 25\,W$;   Power wasted as heat $= i^2 R = (0\cdot 5)^2 \times 25 = 6\cdot 25\,W$

$\therefore$ Rate of energy storage in the field of the coil $= 25 - 6\cdot 25 = \mathbf{18\cdot 75\,W}$

(ii) When current is 1 A

Power input $= 50 \times 1 = 50\,W$;   Power wasted as heat $= (1)^2 \times 25 = 25\,W$

$\therefore \qquad$ Rate of energy stored $= 50 - 25 = \mathbf{25\,W}$

(iii) When current is steady $= V/R = 50/25 = 2A$

Power input $= 50 \times 2 = 100\,W$;   Power wasted as heat $= (2)^2 \times 25 = 100\,W$

$\therefore \qquad$ Rate of energy stored $= 100 - 100 = \mathbf{0\,W}$

Induced e.m.f.

$\qquad$ Voltage across coil, $e_L = V - iR$

(*i*) $\qquad\qquad\qquad$ When $i = 0\cdot 5\,A$; $e_L = 50 - 0\cdot 5 \times 25 = \mathbf{37\cdot 5\,V}$

(*ii*) $\qquad\qquad\qquad$ When $i = 1\,A$; $e_L = 50 - 1 \times 25 = \mathbf{25\,V}$

(*iii*) $\qquad\qquad\qquad$ When $i = 2\,A$; $e_L = 50 - 2 \times 25 = \mathbf{0\,V}$

Example 9.59. *A circuit of resistance R ohms and inductance L henries has a direct voltage of 230 V applied to it. 0.3 second after switching on, the current in the circuit was found to be 5A. After the current had reached its final steady value, the circuit was suddenly short-circuited. The current was again found to be 5A at 0.3 second after short-circuiting the coil. Find the values of R and L.*

Solution. This is a case of growth and decay of current in $R - L$ series circuit. In both cases, $i = 5A$ and $t = 0.3$ s.

$\qquad$ *For growth :* $\qquad\qquad\qquad i = I(1 - e^{-t/\lambda})$

or $\qquad\qquad\qquad\qquad\qquad 5 = I(1 - e^{-0.3/\lambda})$ $\qquad\qquad\qquad\qquad\qquad$ ...(*i*)

$\qquad$ *For decay :* $\qquad\qquad\qquad i = I\,e^{-t/\lambda}$

or $\qquad\qquad\qquad\qquad\qquad 5 = I\,e^{-0.3/\lambda}$ $\qquad\qquad\qquad\qquad\qquad$ ...(*ii*)

From eqs. (*i*) and (*ii*), $I\,e^{-0.3/\lambda} = I(1 - e^{-0.3/\lambda})$

or $\qquad\qquad\qquad\qquad 2\,e^{-0.3/\lambda} = 1$

$\therefore \qquad\qquad\qquad\qquad e^{-0.3/\lambda} = 0.5 \quad \text{or} \quad \lambda = 0.4328$

Putting $\lambda = 0.4328$ in eq. (*ii*), we get,

$$5 = I\,e^{-0.3/0.4328} \quad \text{or} \quad I = 5\,e^{+0.3/0.4328} = 5 \times 2 = 10\,A$$

Now, $\qquad\qquad\qquad\qquad I = \frac{V}{R} \quad \therefore R = \frac{V}{I} = \frac{230}{10} = \mathbf{23\Omega}$

Also, $\qquad\qquad\qquad\qquad \lambda = \frac{L}{R} \quad \text{or} \quad L = R\lambda = 23 \times 0.4328 = \mathbf{9.95\,H}$

Example 9.60. *Two mutually coupled coils, A and B, are connected in series to a 400 V d.c. supply. Coil A has a resistance of 14 Ω and inductance of 4 H. Coil B has a resistance of 20 Ω and inductance of 9 H. At a certain instant after the circuit is energised, the current is 5 A and is increasing at the rate of 10 A/s. Calculate (i) the mutual inductance between the coils, and (ii) the coefficient of coupling.*

Solution. Fig. 9.40 shows the conditions of the problem. When not mentioned in the problem, it is understood that the mutual fluxes of the two coils aid each other.

(*i*)
$$V = i(R_A + R_B) + L_T \frac{di}{dt}$$

where L_T is the total inductance of the circuit.

or
$$400 = 5(14 + 20) + 10 L_T \quad \therefore \quad L_T = 23 \text{ H}$$

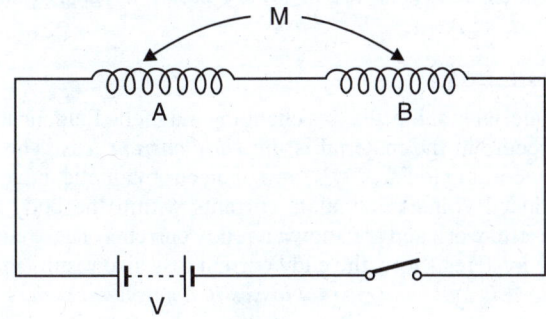

Fig. 9.40

Now,
$$L_T = L_A + L_B + 2M \quad \text{or} \quad 23 = 4 + 9 + 2M \quad \therefore \quad M = 5 \text{ H}$$

(*ii*) Coefficient of coupling, $k = \dfrac{M}{\sqrt{L_A L_B}} = \dfrac{5}{\sqrt{4 \times 9}} = 0.83$

Example 9.61. *The two circuits of Fig. 9.41 have the same time constant of 0·005 second. With the same d.c. voltage applied to the two circuits, it is found that steady state current of circuit (i) is 2000 times the initial current of circuit (ii). Find R_1, L and C.*

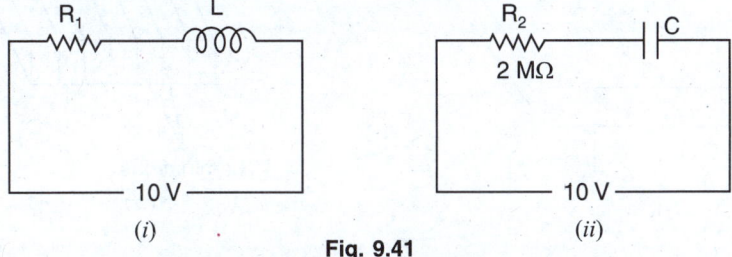

Fig. 9.41

Solution. The time constant for both the circuits is 0·005 s.

$\therefore$
$$R_2 C = 0.005 \quad \text{or} \quad C = \frac{0.005}{R_2}$$

$\therefore$
$$C = \frac{0.005}{2 \times 10^6} = 0.0025 \times 10^{-6} \text{ F} = 0.0025 \, \mu\text{F}$$

Steady state current in Fig. 9.41 (*i*) $= V/R_1 = 10/R_1$

Initial current in Fig. 9.41 (*ii*) $= V/R_2 = 10/2 \times 10^6 = 5 \times 10^{-6}$ A

As per statement of the problem, we have,
$$10/R_1 = 2000 \times (5 \times 10^{-6}) \quad \therefore \quad R_1 = 1000 \, \Omega$$

Now
$$L/R_1 = 0.005 \quad \therefore \quad L = 1000 \times 0.005 = 5 \text{ H}$$

Tutorial Problems

1. A 12 V battery is connected in series with 30 Ω resistor and a 220 mH inductor. How long will it take the current to reach half its maximum possible value ? At this instant, at what rate is energy being delivered by the battery ?   [5ms ; 2·4 W]

2. A p.d. of 100 V is applied to a circuit consisting of a resistance of 50 Ω and an inductance of 5 H. Determine the current in the circuit 0·1 second after the application of the voltage. **[1·264 A]**

3. How many time constants one should wait for the current in an *RL* circuit to grow within 0·1% of its steady value ? **[6·9 time constants]**

4. Calculate the back e.m.f. of a 1 H, 10 Ω coil 0·1 s after 100 V d.c. supply is connected to it. **[36·8 V]**

5. The resistance and inductance of a series circuit are 50 Ω and 20 H respectively. At the instant of closing the supply, the current increases at the rate of 4 A/s. Calculate (*i*) supply voltage (*ii*) the rate of growth of current when current is 5 A. **[(*i*) 80 V (*ii*) 2·75 A/s]**

9.24. Eddy Current Loss

When a magnetic material is subjected to a changing magnetic field, in addition to the hysteresis loss, another loss that occurs in the material is the *eddy current loss*. The changing flux induces voltages in the material according to Faraday's laws of electromagnetic induction. Since the material is conducting, these induced voltages circulate currents within the body of the material. These induced currents do no useful work and are known as eddy currents. These eddy currents develop i^2R loss in the material. Like hysteresis loss, the eddy current loss also results in the rise of temperature of the material. *The hysteresis and eddy current losses in a magnetic material are sometimes called* **core losses** or **iron losses.**

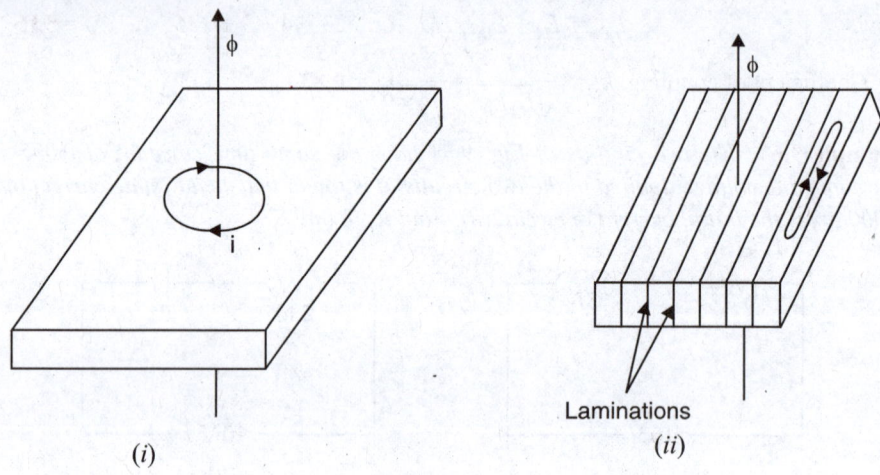

(*i*)   (*ii*)

Fig. 9.42

Fig. 9.42 (*i*) shows a solid block of iron subjected to a changing magnetic field. The eddy current power loss in the block will be i^2R where *i* is the eddy current and *R* is the resistance to the eddy current path. Since the block is a continuous iron piece of large *X*-section, the magnitude of *i* will be very *large and hence greater eddy current loss will result. The obvious method of reducing this loss is to reduce the magnitude of eddy current. This can be achieved by splitting the solid block into thin sheets (called **laminations**) in planes parallel to the magnetic flux as shown in Fig.9·42 (*ii*). Each lamination is insulated from the other by a layer of varnish. This arrangement reduces the area of each section and hence the induced e.m.f. It also increases the resistance of eddy current paths since the area through which the currents can pass is smaller. Both effects combine to reduce the eddy current and hence eddy current loss. Further, reduction in this loss can be obtained by using a magnetic material of high resistivity (*e.g.* silicon steel).

The only drawback of laminated core is that the total cross-sectional area of the magnetic material is reduced by the total thickness of the insulation. This is generally taken into account by allowing about 10% reduction in the thickness of core when making the magnetic calculations.

---

*   The large area of the block will have greater e.m.f. induced in it. Larger *X*-section also means smaller resistance to eddy current path. Both these effects increase the magnitude of eddy current to a great extent.

9.25. Formula for Eddy Current Power Loss

It is difficult to determine the eddy current power loss because the current and resistance values cannot be determined directly. Experiments have shown that eddy current power loss P_e in a magnetic material can be expressed as :

$$P_e = k_e B_m^2 t^2 f^2 V \text{ watts}$$

where
k_e = eddy current coefficient and its value depends upon the nature of the material.

B_m = maximum flux density in Wb/m^2

t = thickness of lamination in m

f = frequency of flux in Hz

V = volume of material in m^3

Example 9.62. *The flux in a magnetic core is alternating sinusoidally at 50 Hz. The maximum flux density is 1·5 Wb/m^2. The eddy current loss then amounts to 140 watts. Find the eddy current loss in the core when the frequency is 75 Hz and the flux density is 1·2 Wb/m^2.*

Solution. Eddy current power loss, $P_e \propto B_m^2 f^2$

For the first case, $P_{e1} \propto (1·5)^2 \times (50)^2$;   For the second case, $P_{e2} \propto (1·2)^2 \times (75)^2$

$$\therefore \qquad \frac{P_{e2}}{P_{e1}} = \left(\frac{1·2}{1·5}\right)^2 \times \left(\frac{75}{50}\right)^2 = 1.44$$

$$\therefore \qquad P_{e2} = 1·44\, P_{e1} = 1·44 \times 140 = \textbf{201·6 W}$$

Example 9.63. *Find the eddy current power loss in a 50 Hz transformer with a maximum flux density of 1 Wb/m^2. The core is of section 8 cm × 6 cm and total effective length is 50 cm constructed of laminations of thickness 0·4 mm. The eddy current coefficient is $6·58 \times 10^6$. Assume a space factor of 0·9.*

Solution.   Total core area = $8 \times 6 = 48$ cm^2 = 48×10^{-4} m^2

*Useful core area = $0·9 \times 48 \times 10^{-4} = 43·2 \times 10^{-4}$ m^2

Volume of iron in core, $V = 43·2 \times 10^{-4} \times 0·5 = 21·6 \times 10^{-4}$ m^3

Thickness of lamination, $t = 0·4$ mm $= 0·4 \times 10^{-3}$ m

$$\therefore \qquad P_e = k_e B_m^2 t^2 f^2 V \text{ watts}$$

$$= (6·58 \times 10^6) \times (1)^2 \times (0·4 \times 10^{-3})^2 \times (50)^2 \times 21·6 \times 10^{-4} = \textbf{5·68 W}$$

Example 9.64. *A transformer connected to 25 Hz supply has a core loss of 1500 watts of which 1000 watts are due to hysteresis and 500 watts due to eddy currents. If the flux density is kept constant and frequency is increased to 50 Hz, find the new value of the core loss.*

Solution.

Hysteresis power loss, $P_h \propto B_m^{1·6} f$

Eddy current power loss, $P_e \propto B_m^2 f^2$

Hysteresis loss

$$\frac{P_{h2}}{P_{h1}} = \left(\frac{B_{m2}}{B_{m1}}\right)^{1·6} \times \frac{f_2}{f_1} = (1)^{1·6} \times 50/25 = 2 \qquad (\because B_{m2} = B_{m1})$$

$$\therefore \qquad P_{h2} = 2 \times 1000 = 2000 \text{ W}$$

---

*   The core of the transformer is laminated to reduce the eddy current loss. The cross-sectional area of iron is now less than the apparent area due to the area taken up by the insulation.

Space factor = $\dfrac{\text{Useful area}}{\text{Total area}}$

Eddy current loss

$$\frac{P_{e2}}{P_{e1}} = \left(\frac{B_{m2}}{B_{m1}}\right)^2 \times \left(\frac{f_2}{f_1}\right)^2 = (1)^2 \times \left(\frac{50}{25}\right)^2 = 4$$

$$\therefore \qquad P_{e2} = 4\,P_{e1} = 4 \times 500 = 2000 \text{ W}$$

$$\therefore \qquad \text{New core loss} = P_{h2} + P_{e2} = 2000 + 2000 = \mathbf{4000 \text{ W}}$$

Example 9.65. *The core loss in a given specimen is found to be 65 W at a frequency of 30 Hz and a flux density of 1 Wb/m² and 190 W at 60 Hz and the same flux density. What are the hysteresis loss and the eddy current loss at each frequency ?*

Solution. Since the flux density, the volume of specimen and the thickness of laminations remain constant, the iron or core loss (= hysteresis loss + eddy current loss) can be written as :

$$\text{Core loss, } P_c = k_h' f + k_e' f^2 \qquad \qquad ...(i)$$

where $k_h' = k_h\,B_m^{1.6}\,V$ and $k_e' = k_e\,B_m^2\,t^2\,V$

Putting the given values in eq. (*i*), we have,

$$65 = k_h' \times 30 + k_e' \times (30)^2 \qquad \qquad ...(ii)$$
$$190 = k_h' \times 60 + k_e' \times (60)^2 \qquad \qquad ...(iii)$$

Solving eqs. (*ii*) and (*iii*), $k_h' = 1\cdot167$; $k_e' = 0\cdot0333$

At 30 Hz. At 30 Hz, these losses are :

$$P_h = k_h' \times 30 = 1\cdot167 \times 30 = \mathbf{35 \text{ W}}$$

$$P_e = k_e' \times (30)^2 = 0\cdot0333 \times (30)^2 = \mathbf{30 \text{ W}}$$

At 60 Hz. At 60 Hz, these losses are :

$$P_h = k_h' \times 60 = 1\cdot167 \times 60 = \mathbf{70 \text{ W}}$$

$$P_e = k_e' \times (60)^2 = 0\cdot0333 \times (60)^2 = \mathbf{120 \text{ W}}$$

Objective Questions

1. The basic requirement for inducing e.m.f. in a coil is that ....
   - (*i*) flux should link the coil
   - (*ii*) there should be change in flux linking the coil
   - (*iii*) coil should form a closed loop
   - (*iv*) none of the above

2. The e.m.f. induced in a coil is .... the rate of change in flux linkages.
   - (*i*) directly proportional to
   - (*ii*) inversely proportional to
   - (*iii*) independent of
   - (*iv*) none of the above

3. The e.m.f. induced in a coil of *N* turns is given by ....
   - (*i*) $d\phi/dt$
   - (*ii*) $N\,d\phi/dt$
   - (*iii*) $-N\,d\phi/dt$
   - (*iv*) $N\,dt/d\phi$

4. The direction of induced e.m.f. in a conductor (or coil) can be determined by ....
   - (*i*) work law
   - (*ii*) Ampere's law

   - (*iii*) Fleming's right-hand rule
   - (*iv*) Fleming's left-hand rule

5. In Fig. 9.43, the conductor is moving upward. The direction of induced e.m.f. is ....

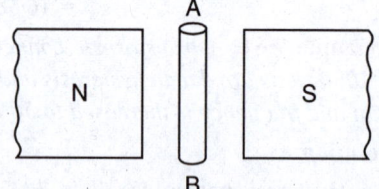

Fig. 9.43

   - (*i*) from *A* to *B*
   - (*ii*) from *B* to *A*
   - (*iii*) none of the above

6. In Fig. 9.44, the direction of induced e.m.f. in the conductor *A* is ....

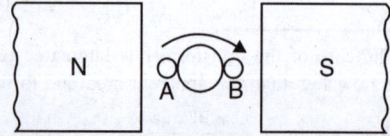

Fig. 9.44

     (*i*) into the plane of paper

     (*ii*) out of plane of paper

     (*iii*) none of the above

7. In Fig. 9.44, the rate of change of flux linkages of conductors *A* and *B* is ....

     (*i*) minimum   (*ii*) maximum

     (*iii*) mid-way between (*a*) and (*b*)

     (*iv*) none of the above

8. The e.m.f. induced in a .... is the statically induced e.m.f.

     (*i*) d.c. generator   (*ii*) transformer

     (*iii*) d.c. motor   (*iv*) none of the above

9. The e.m.f. induced in a .... is dynamically induced e.m.f.

     (*i*) alternator   (*ii*) transformer

     (*iii*) d.c. generator   (*iv*) none of the above

10. In Fig. 9.45, 1 single conductor of length *l* metres moves at right angles to a uniform field of *B* Wb/m^2 with a velocity of *v* m/s. The e.m.f. induced is ....

     (*i*) $B\,l/v$ (*ii*) Bv/l

     (*iii*) Blv (*iv*) lv/B

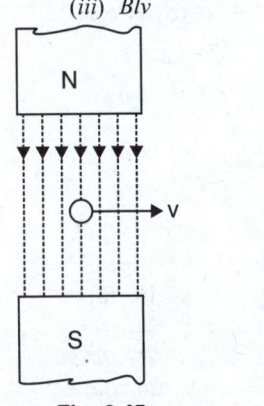
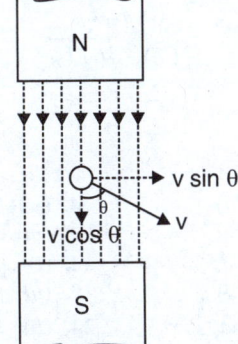

Fig. 9.45   **Fig. 9.46**

11. In Fig. 9.46, the component of velocity that does not induce any e.m.f. in the conductor is ....

     (*i*) $v \sin \theta$ (*ii*) $v \cos \theta$

     (*iii*) $v \tan \theta$ (*iv*) none of the above

12. Inductance opposes .... in current in a circuit.

     (*i*) only increase   (*ii*) only decrease

     (*iii*) change   (*iv*) none of the above

13. If the number of turns of a coil is increased, its inductance ....

     (*i*) remains the same   (*ii*) is increased

     (*iii*) is decreased   (*iv*) none of the above

14. If the relative permeability of the material surrounding the coil is increased, the inductance of the coil ....

     (*i*) is increased   (*ii*) is decreased

     (*iii*) remains unchanged

     (*iv*) none of the above

15. Inductance in a circuit ....

     (*i*) prevents the current from changing

     (*ii*) delays the change in current

     (*iii*) causes power loss

     (*iv*) causes the current to lead the voltage

16. The inductance of a coil is .... the reluctance of magnetic path.

     (*i*) independent of

     (*ii*) directly proportional to

     (*iii*) inversely proportional to

     (*iv*) none of the above

17. If the number of turns of a coil is increased two times, its inductance is ....

     (*i*) increased two times

     (*ii*) decreased two times

     (*iii*) decreased four times

     (*iv*) increased four times

18. A circuit has inductance of 2H. If the circuit current changes at the rate of 10 A/second, then self-induced e.m.f. is ....

     (*i*) 5 V   (*ii*) 0·2 V

     (*iii*) 20 V   (*iv*) 10 V

19. A current of 2 A through a coil sets up flux linkages of 4 Wb-turn. The inductance of the coil is ....

     (*i*) 8 H   (*ii*) 0·5 H

     (*iii*) 2 H   (*iv*) 1 H

20. An air-cored choke is used for .... applications.

     (*i*) radio frequency   (*ii*) audio frequency

     (*iii*) power frequency   (*iv*) none of the above

21. If a 10-turn coil has a second layer of 10 turns wound over the first, then total inductance will be about .... the original inductance.

     (*i*) two times   (*ii*) four times

     (*iii*) six times   (*iv*) three times

22. An iron-cored coil of 10 turns has reluctance of 100 AT/Wb. The inductance of the coil is ....

     (*i*) 1 H   (*ii*) 10 H

     (*iii*) 0·1 H   (*iv*) 5 H

23. An iron-cored coil has an inductance of 2 H. If the reluctance of the magnetic path is 200 AT/Wb, the number of turns on the coil is ....

    (i) 100   (ii) 400

    (iii) 50   (iv) 20

24. The mutual inductance between two coils is .... reluctance of magnetic path.

    (i) directly proportional to

    (ii) inversely proportional to

    (iii) independent of   (iv) none of the above

25. Mutual inductance between two coils can be decreased by ....

    (i) increasing the number of turns of either coil

    (ii) by moving the coils closer

    (iii) by moving the coils apart

    (iv) none of the above

26. Mutual inductance between two coils is 4H. If current in one coil changes at the rate of 2 A/second, then e.m.f. induced in the other coil is ....

    (i) 8 V   (ii) 2 V

    (iii) 0·5 V   (iv) none of the above

27. If in Fig. 9.47, ϕ_{12} = 2 Wb, N_2 = 20 and I_2 = 20 A, then mutual inductance between the coils is ....

    (i) 200 H   (ii) 20 H

    (iii) 4 H   (iv) 2 H

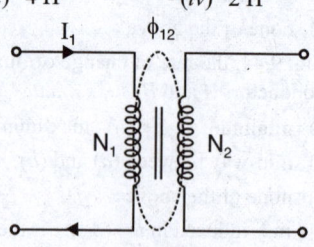

Fig. 9.47

28. If in Fig. 9.47, N_1 = 100, N_2 = 1000 and mutual inductance between the coils is 2H, the reluctance of magnetic circuit is ....

    (i) 5×10^4 AT/Wb   (ii) 10^5 AT/Wb

    (iii) 20 AT/Wb   (iv) 5 AT/Wb

29. If the coefficient of coupling between two coils is increased, mutual inductance between the coils ....

    (i) is increased   (ii) is decreased

    (iii) remains unchanged

    (iv) none of the above

30. The maximum mutual inductance between the coils shown in Fig. 9.47 is given by ....

    (i) $L_A L_B$ (ii) L_A / L_B

    (iii) $\sqrt{L_A L_B}$ (iv) $(L_A L_B)^2$

Answers

1. (ii)	2. (i)	3. (iii)	4. (iv)	5. (ii)
6. (ii)	7. (ii)	8. (ii)	9. (iii)	10. (iii)
11. (ii)	12. (iii)	13. (ii)	14. (i)	15. (ii)
16. (iii)	17. (iv)	18. (iii)	19. (iii)	20. (i)
21. (ii)	22. (i)	23. (iv)	24. (ii)	25. (iii)
26. (i)	27. (iv)	28. (i)	29. (i)	30. (iii)

10

Chemical Effects of Electric Current

Introduction

The reader is well acquainted with the passage of electric current through metallic conductors, *e.g.*, copper, aluminium etc. In such conductors, current conduction is due to the movement of free electrons and there is no chemical or physical change except the rise in temperature. However, conduction of current through *certain salt solutions is quite different. Such liquids provide a large number of oppositely charged atoms (called ions) and are known as *electrolytes e.g.*, acids (H_2SO_4, HCl etc.), solutions of inorganic compounds (NaCl, $CuSO_4$, $AgNO_3$ etc.), hydroxides of metals (KOH, NaOH etc). In an electrolyte, conduction is due to the movement of ions (an electrolyte has no free electrons) and chemical changes occur so long as the conduction takes place. Thus passage of electric current through an electrolyte causes chemical changes i.e. electrical energy is converted into chemical energy. The converse of this is also true *i.e.*, we can produce electrical energy from chemical energy. In this chapter, we shall study about the close relationship between electrical energy and chemical energy.

10.1. Electric Behaviour of Liquids

Some liquids conduct current while others do not permit the passage of current through them. On the basis of electrical conductivity, the liquids may be divided into three classes *viz.*

(*i*) Those liquids which do not conduct current are called **insulators** *e.g.*, mineral oils, distilled water etc.

(*ii*) Those liquids which conduct current due to drifting of free electrons are called **conductors** *e.g.*, mercury.

(*iii*) Those liquids which conduct current due to drifting of †ions are known as **electolytes** *e.g.*, solutions of $CuSO_4$, $AgNO_3$ etc. This is the most important class of liquid conductors.

10.2. Electrolytes

A liquid which conducts electric current due to the drifting of ions is called an **electrolyte**.

Salts like silver nitrate ($AgNO_3$), sodium chloride (NaCl), copper sulphate ($CuSO_4$), etc. when dissolved in water dissociate into ions. Their ionic dissociation can be represented as under :

$$AgNO_3 \longrightarrow Ag^+ + NO_3^-$$
$$NaCl \longrightarrow Na^+ + Cl^-$$
$$CuSO_4 \longrightarrow Cu^{++} + SO_4^{--}$$

The atom or group of atoms having positive charge is called a *positive ion*. On the other hand, the atom or group of atoms having negative charge is called a *negative ion*. For example, when NaCl

---

* Conduction of current is possible only in those liquids which break up into oppositely charged atoms called *ions*. Such liquids are called *electrolytes*. There are, however, many substances (*e.g.*, sugar) which dissolve without splitting into ions. Solutions of these substances do not conduct current and are called non-electrolytes.

† An electrolyte has no free electrons.

is dissolved in water, it splits into positve ions (Na^+) and negative ions (Cl^-). The conduction of current through an electrolyte is due to the drifting of negative and positive ions within the liquid.

10.3. Mechanism of Ionisation

The splitting up of an ionic compound in solution into ions is known as **ionisation** *or* **ionic dissociation**. Let us take the example of sodium chloride (NaCl). The structure of this solid crystalline salt is made up of Na^+ and Cl^- ions. When in solid state, there is a very strong force of attraction between Na^+ and Cl^- ions which holds them together as a molecule of NaCl. However, when sodium chloride is dissolved in water, the force of attraction between the ions (Na^+ and Cl^-) of sodium chloride molecule is tremendously *reduced due to high permittivity of water ($K = 81$). In fact, the force of attraction between ions reduces 81 times. The result is that sodium ion (Na^+) and Cl^- ion get separated. This process is called ionisation. It may be noted that as soon as sodium chloride is dissolved in water, ions are formed. In other words, ions are present in an electrolytic solution even before it conducts electric current.

10.4. Electrolysis

The condution of electric current through the solution of an electrolyte together with the resulting chemical changes is called **electrolysis.**

Fig. 10.1 shows the process of electrolysis in a copper voltameter. When copper sulphate ($CuSO_4$) is dissolved in water, it splits up into its components viz. the positive copper ions (Cu^{++}) and negative sulphate ions (SO_4^{--}). This process is called **ionisation. When d.c. voltage is applied across the electrodes, the negative sulphate ions (SO_4^{--}) move towards the anode (+ve electrode) and positive copper ions move towards the cathode (–ve electrode) causing the following chemical changes :

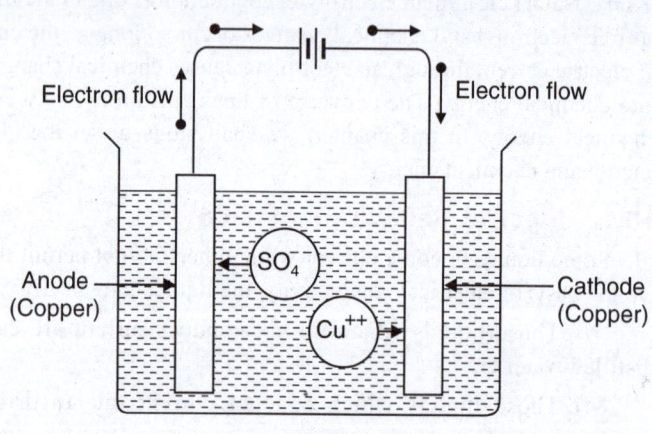

Fig. 10.1

At anode. A sulphate ion (SO_4^{--}) on reaching the anode gives its two extra electrons to it and becomes sulphate radical. These given up electrons continue their journey towards the cathode *via* the external circuit. Now the sulphate radical cannot exist and, therefore, it acts chemically on the anode material to form copper sulphate according to the following reaction :

$$Cu + SO_4 \longrightarrow CuSO_4$$

Thus copper from anode continuously dissolves into the solution so long as this action takes place.

At cathode. At the same time, a copper ion (Cu^{++}) on reaching the cathode takes two electrons from it (these are the same electrons given by the sulphate ion at the anode and have come to cathode

---

* $F = \dfrac{1}{4\pi\varepsilon_0 K} \dfrac{q_1 q_2}{r^2}$. For air, $K = 1$ and for water at room temperature, $K = 81$.

**   The reader may recall that copper sulphate is an ionic compound *i.e.,*, each molecule of $CuSO_4$ is formed due to the attraction between oppositely charged atoms viz Cu^{++} and SO_4^{--}. When dissolved in water, the force of attraction between them is tremendously reduced due to high relative permittivity of water. The result is that Cu^{++} and SO_4^{--} get separated. These charged atoms (Cu^{++} and SO_4^{--}) are called *ions*.

via the external circuit). The copper ion (Cu^{++}) combines with these two electrons to become copper atom and gets deposited on the cathode.

$$Cu^{++} + 2e \longrightarrow Cu \text{ atom}$$

Thus copper from the solution ($CuSO_4$) gets deposited on the cathode.

The following points may be carefully noted :

(*i*) *Electrolysis is possible only if d.c. potential is applied to the electrodes.* It is because we are to attract ions of only one kind to each electrode.

(*ii*) *During electrolysis, either anode material gets deposited over the cathode or gases are liberated at the two electrodes.*

(*iii*) The resulting chemical changes during electrolysis take place so long as the current flows through the electrolyte. When the current through the electrolyte ceases, chemical action also ceases.

10.5. Back e.m.f. or Polarisation Potential

The process of electrolysis is carried out in an apparatus called *voltameter* or *electrolytic cell*. When external d.c. voltage is applied across the electrodes, an e.m.f. is set up between each electrode and the electrolyte which opposes the external d.c. voltage. This opposing e.m.f. is called *back e.m.f.* (E_b) of the electrolyte and is produced due to the coating of electrodes by the products of electrolysis. This effect is called polarisation and for this reason, back e.m.f. is also called *polarisation potential*.

The e.m.f. set up in the voltameter which opposes the external d.c. voltage is called **back e.m.f.** *of the electrolyte.*

The value of back e.m.f. is different for different electrolytes. For acids and alkalies which evolve hydrogen and oxygen, its value is about 1.7 V. For other electrolytes, the value of back e.m.f. depends on the particular salt and generally lies between 0.5 V and 2 V for normal solutions.

Voltage equation for electrolysis. For electrolysis, the applied external d.c. voltage V must overcome the back e.m.f. (E_b) and voltage drop (IR_e) in the electrolyte *i.e.*

$$V = E_b + IR_e \qquad \qquad \text{... (i)}$$

$$\text{where} \quad V = \text{External d.c. voltage}$$

$$E_b = \text{Back e.m.f. of electrolyte}$$

$$I = \text{Circuit current}$$

$$R_e = \text{Resistance of electrolyte}$$

Therefore, in order to carry out electrolysis at an appreciable rate, the external d.c. voltage V must be atleast equal to $E_b + IR_e$. If the external d.c. voltage is less than this value, electrolysis will not take place.

10.6. Faraday's Laws of Electrolysis

Faraday performed a series of experiments to determine the factors which govern the mass of an element deposited or liberated during electrolysis. He summed up his conclusions into two laws, known as Faraday's laws of electrolysis.

First law. *The mass of an element deposited or liberated at an electrode is directly proportional to the quantity of electricity that passes through the electrolyte.*

If m is the mass of an element deposited or liberated due to the passage of I amperes for t seconds, then according to first law,

$$m \propto Q$$

or

$$m \propto It \qquad \qquad (\because Q = It)$$

or

$$m = ZIt \text{ or } m = ZQ$$

where Z is a constant known as *electro-chemical equivalent* (*E.C. E.*) of the element. It has the same value for one element but different for other elements.

If $Q = 1$ coulomb, then, $m = Z$.

Hence **electro-chemical equivalent** (*E.C.E.*) *of an element is equal to the mass of element deposited or liberated by the passage of 1 coulomb of electricity through the electrolyte.* Its unit is gm/C or kg/C.

For example, E.C.E. of copper is $0{\cdot}000304$ gm/C. It means that if 1 coulomb of electricity is passed through a solution of $CuSO_4$, then mass of copper deposited on the cathode will be $0{\cdot}000304$ gm.

The validity of first law is explained by the fact that current inside the electrolyte is carried by the ions themselves. Hence the masses of the chemical substances reaching the anode and cathode are proportional to the quantity of electricity carried by the ions *i.e.*, mass of an ion liberated at any electrode is proportional to the quantity of electricity passed through the electrolyte.

Second law. *The mass of an element deposited or liberated during electrolysis is directly proportional to the chemical equivalent weight of that element i.e.*

$$m \propto \text{Chemical equivalent weight of the element } (E)$$

Faraday's second law is illustrated in Fig. 10.2 where silver and copper voltameters are connected in series. When the same current is passed for the same time through the two voltameters, it will be seen that the masses of silver (Ag) and copper (Cu) deposited on the respective cathodes are in the ratio of 108 : 32. These values of 108 and 32 are respectively the equivalent weights of silver and copper.

$$\frac{\text{Mass of silver deposited}}{\text{Mass of copper deposited}} = \frac{\text{Eq. wt. of Ag}}{\text{Eq. wt. of Cu}} = \frac{108}{32}$$

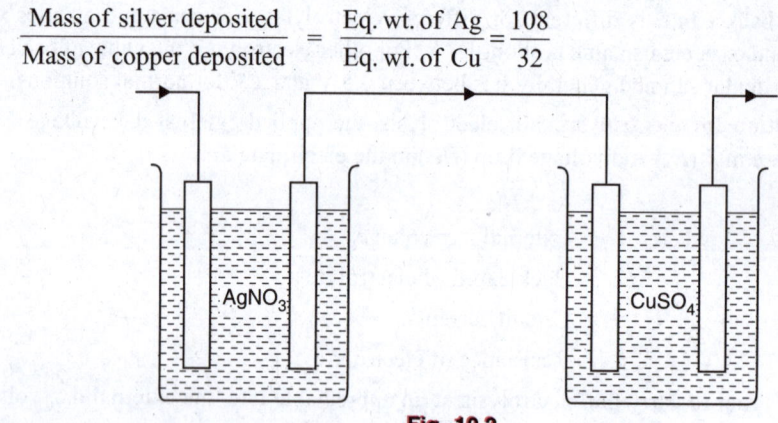

Fig. 10.2

Faraday's second law can be explained as follows. The negative ions (i.e. NO_3^- and SO_4^{--}) from the solutions give up their respective extra electrons to the anodes. These electrons come to cathodes *via* the external circuit and are taken up by the positive ions (Ag^+ and Cu^{++}) to become metallic atoms and get deposited on the respective cathodes. Suppose 10 electrons are flowing in the external circuit. Since silver is monovalent (*i.e.*, its valency is 1), 10 silver ions must be liberated at the cathode of silver voltameter. Again copper is bivalent (*i.e.*, its valency is 2) and hence 5 copper ions must be liberated at the cathode of copper voltameter. This means that mass of an element (silver or copper) liberated is directly proportional to the atomic weight and inversely proportional to the valency of that element *i.e.*

$$\text{Mass liberated, } m \propto \frac{\text{Atomic weight}}{\text{Valency}}$$

$$\propto \text{Chemical equivalent wt. of the element} \cdot$$

i.e. $$m \propto E$$

10.7. Relation Between E and Z

Suppose the same amount of charge (Q) is passed through the solutions of two electrolytes. If m_1 and m_2 are the masses of the substances liberated/deposited and Z_1 and Z_2 are their electro-chemical equivalents, then,

$$m_1 = Z_1Q \; ; \; m_2 = Z_2Q$$

$$\therefore \quad \frac{m_1}{m_2} = \frac{Z_1}{Z_2} \quad \text{But} \quad \frac{m_1}{m_2} = \frac{E_1}{E_2} \qquad \text{...Faraday's second law}$$

$$\therefore \quad \frac{E_1}{E_2} = \frac{Z_1}{Z_2}$$

or

$$\frac{E}{Z} = \text{Constant}$$

Thus the ratio E/Z is the same for all substances. This constant is called Faraday constant $F(=E/Z)$.

Faraday constant (F). The value of Faraday constant is found to be 96500 C *i.e.* $F = 96500$ C.

Hence **Faraday constant** *is the quantity of charge (i.e., 96500 C) required to liberate/deposit one gram equivalent (chemical equivalent in gram) of the substance during electrolysis.*

For example, chemical equivalent of silver is 108. When a charge of 96500 C is passed through a silver voltameter, then mass of silver deposited on the cathode will be 108g. Again chemical equivalent of copper is 31·75. If a charge of 96500 C is passed through a copper voltameter, then mass of copper deposited on the cathode will be 31.75 g.

Finding the value of F. According to Faraday's first law of electrolysis,

$$m = ZQ$$

Suppose M is the mass of one mole of the substance. If, during electrolysis, the mass of the substance to be deposited is M and p is the valency of the depositing atom, then $N_A (= 6.023 \times 10^{23})$ atoms will deposit on the electrode.

Now $m = M$ and $Q = N_A p e$

where e = Charge on electron

$$\therefore \quad M = Z N_A p e \qquad \text{...Faraday's first Law}$$

or

$$Z = \frac{1}{N_A e} \cdot \frac{M}{p}$$

But $\dfrac{M}{p} = \dfrac{\text{Mass of one mole}}{\text{Valency}}$ = Chemical Equivalent of the substance

$$\therefore \quad Z = \frac{1}{N_A e} E$$

Now, $N_A e$ is a constant, called Faraday constant F.

$$\therefore \quad Z = \frac{E}{F}$$

Now, $1F = N_A e = (6.023 \times 10^{23}) \times 1.602 \times 10^{-19} = 96485 \text{ C} \approx 96500 \text{ C}$

10.8. Deduction of Faraday's Laws of Electrolysis

Suppose a charge Q is passed through an electrolyte during electrolysis and the mass liberated/deposited on the cathode is m.

$$\therefore \quad m = ZQ$$

or

$$m = \frac{E}{F}Q \qquad \left(\because Z = \frac{E}{F} \right) \qquad \text{...(i)}$$

Equation (*i*) is the **fundamental equation of electrolysis** and contains Faraday's two laws of electrolysis.

(*i*) It is clear from equation (*i*) that :

$$m \propto Q \qquad \text{...Faraday's first law}$$

(*ii*) If the same charge is passed (*i.e.* Q is constant) through different electrolytes during electrolysis, then,

$$m \propto E \qquad \text{...Faraday's second law}$$

Example 10.1. *A current passes through two voltameters in series, one having silver plates and a solution of $AgNO_3$, and the other copper plates and a solution of $CuSO_4$. After the current has ceased to flow, 3·6 gm of silver have been deposited. How much copper will have deposited in the other voltameter ? Take E.C.E. of silver as 0·001118 gm/C and that of copper as $328·86 \times 10^{-6}$ gm/C.*

Solution. For silver voltameter, we have,

$$m_1 = Z_1 It$$

or
$$It = \frac{m_1}{Z_1} = \frac{3.6}{0.001118}$$

For copper voltameter, $m_2 = Z_2 It = (328.86 \times 10^{-6}) \times \dfrac{3.6}{0.001118} = \textbf{1.06 gm}$

Example 10.2. *If 16 amperes deposit 12 gm of silver in 9 minutes, how much copper would 10 amperes deposit in 15 minutes? At. wt. of silver = 108 and At. wt. of copper = 63.5.*

Solution. 16A in 9 minutes deposit silver = 12 gm

10A in 15 minutes deposit silver = $12 \times \dfrac{10}{16} \times \dfrac{15}{9} = 12.5$ gm

Eq. wt. of silver = At. wt./Valency = 108/1 = 108 ; Eq. wt. of copper = 63.5/2 = 31.75

Let *m* gm be the mass of copper deposited by 10 A in 15 minutes. Then by Faraday's second law of electrolysis,

$$\frac{\text{mass of Cu deposited}}{\text{mass of Ag deposited}} = \frac{\text{Eq. wt. of Cu}}{\text{Eq. wt. of Ag}}$$

or
$$\frac{m}{12.5} = \frac{31.75}{108} \qquad \therefore \quad m = \frac{31.75}{108} \times 12.5 = \textbf{3.67 gm}$$

Example 10.3. *A coating of nickel 1 mm thick is to be deposited on a cylinder 2 cm in diameter and 30 cm in length. Calculate the time taken if the current used is 100 A. The following data may be taken. Specific gravity of nickel = 8·9, At. wt. of nickel = 58·7 (divalent), E.C.E. of silver = 1·12 mg/C, At.wt. of silver = 108.*

Solution. Area of curved surface of cylinder = $\pi D \times l = \pi \times 2 \times 30 = 188.5$ cm^2

Volume of Ni to be deposited = Area of curved surface × thickness of Ni

$$= 188.5 \times 0.1 = 18.85 \text{ cm}^3$$

Mass of Ni to be deposited, $m = 18.85 \times 8.9 = 167.7$ gm

Eq. wt. of Ni = 58.7/2 = 29.35 ; Eq. wt. of Ag = 108/1 = 108

$$\frac{\text{E.C.E. of Ni}}{\text{E.C.E of Ag}} = \frac{\text{Eq. wt. of Ni}}{\text{Eq. wt. of Ag}} \qquad \text{or} \qquad \frac{\text{E.C.E of Ni}}{1.12} = \frac{29.35}{108}$$

$$\therefore \quad \text{E.C.E. of Ni} = \frac{29.35}{108} \times 1.12 = 0.304 \text{ mg/C}$$

Now,
$$m = ZIt$$

$$\therefore \quad t = \frac{m}{ZI} = \frac{167.7}{0.304 \times 10^{-3} \times 100} = 5516 \text{ seconds} = \textbf{91.93 minutes}$$

Example 10.4. *Find the thickness of copper deposited on a plate area of 0.00025 m^2 during electrolysis if a current of 1 A is passed for 100 minutes. Density of copper = 8900 kg/m^3 and E.C.E. of copper = 32.95 × 10^{-8} kg/C.*

Solution. According to Faraday's law of electrolysis, mass (m) of copper deposited is

$$m = ZIt$$

Here, $Z = 32.95 \times 10^{-8}$ kg/C ; $I = 1$ A ; $t = 100$ min $= 100 \times 60$ sec.

∴ $m = 32.95 \times 10^{-8} \times 1 \times 100 \times 60 = 0.001977$ kg

Volume of Cu deposited, $v = \dfrac{\text{Mass}}{\text{Density}} = \dfrac{0.001977}{8900} = 0.222 \times 10^{-6}$ m^3

∴ Thickness of Cu deposited $= \dfrac{v}{\text{Plate area}} = \dfrac{0.222 \times 10^{-6}}{0.00025} = 0.888 \times 10^{-3}$ m $=$ **0.888 mm**

Example 10.5. *An ammeter is being calibrated with the aid of copper voltameter. The ammeter continually reads 2 A when a current is passed through the voltameter for 1 hour. During this time, 2·34 gm of copper was liberated. Taking the electro-chemical equivalent of copper to be 330 × 10^{-9} kg/C, determine the magnitude of error of the ammeter.*

Solution. Let I amperes be the actual current.

Now, $m = ZIt$

Here, $m = 2.34 \times 10^{-3}$ kg ; $Z = 330 \times 10^{-9}$ kg/C ; $t = 1$ hr $= 3600$ s

∴ $I = \dfrac{m}{Zt} = \dfrac{2.34 \times 10^{-3}}{330 \times 10^{-9} \times 3600} = 1.97$A

∴ Error $= 2 - 1.97 =$ **0.03 A ; *ammeter reads more***

This example shows that the phenomenon of electrolysis can also be used to measure the magnitude of current.

Example 10.6. *Find the mass of zinc which has been dissolved in a simple zinc-copper voltaic cell when 2200 J of energy has been supplied. Assume that electromotive force (e.m.f.) is constant at 1.1 V and that electro-chemical equivalent of zinc is 0·34 × 10^{-6} kg/C.*

Solution. A cell is a device which converts chemical energy into electrical energy. Faraday's law $m = ZQ$ is applicable to cells also but in this case, the mass m refers to the mass dissolved instead of mass liberated.   Energy $=$ Volts × Coulombs

or $2200 = 1.1 Q$ ∴ $Q = 2200/1.1 = 2000$ C

Now, $m = ZQ = (0.34 \times 10^{-6}) \times (2000) = 0.68 \times 10^{-3}$ kg $= 0.68$ g

∴ Mass of zinc dissolved $=$ **0.68 g**

Example 10.7. *A steady direct current of 100A flows for 5 minutes through fused sodium chloride. How much sodium will be drawn off and how much chlorine will be evolved ? The atomic masses of sodium and chlorine are 23 and 35·5 respectively.*

Solution. $m = E\dfrac{Q}{F}$

Sodium $E_{Na} = \dfrac{23}{1} = 23$; $\dfrac{Q}{F} = \dfrac{100 \times 5 \times 60}{96500} = 0.311$

∴ $m = 23 \times 0.311 =$ **7.15 g**

Chlorine $E_{Cl} = 35.5/1 = 35.5$; $Q/F = 0.311$ (same as before)

∴ $m = 35.5 \times 0.311 =$ **11.04 g**

Example 10.8. *A steady current of 10.0 A is passed through a water voltameter for 300 s. Estimate the volume of hydrogen evolved at standard temperature and pressure. Use the known value of Faraday constant. Relative molecular mass of H_2 is 2.016 and molar volume = 22.4 litres (volume of 1 mole of an ideal gas at S.T.P.).*

Solution. $m = ZIt$

Now, $Z = \dfrac{E}{F} = \dfrac{M}{pF}$ or $m = \dfrac{MIt}{pF}$

Here, $M = 2.016$; $p = 2$; $F = 96500$ C ; $I = 10.0$ A ; $t = 300$ s

∴ $m = \dfrac{2.016 \times 10 \times 300}{2 \times 96500} = 0.0313$ g

∴ Volume of 0·0313 g of H_2 at STP $= \dfrac{22.4 \times 0.0313}{2.016} =$ **0.35 litres**

Example 10.9. *The potential difference across the terminals of a battery of e.m.f. 12 V and internal resistance 2 Ω drops to 10 V when it is connected to a silver voltameter. Calculate the silver deposited at the cathode in half an hour. Atomic weight of silver is 107.9 g.mol⁻¹.*

Solution. E.M.F. of battery, $E = 12$ volts ; Terminal p.d. of battery, $V = 10$ volts ; Internal resistance of battery, $r = 2\,\Omega$; Resistance of voltameter $= R$.

∴ $r = \dfrac{E-V}{V} \times R$ or $2 = \dfrac{12-10}{10} \times R$ ∴ $R = 10\,\Omega$

∴   Circuit current, $I = \dfrac{E}{R+r} = \dfrac{12}{10+2} = 1A$

Now, $E_{Ag} = \dfrac{\text{Atomic weight}}{\text{Valency}} = \dfrac{107.9}{1} = 107.9$ g mole⁻¹

Electrochemical equivalent, $Z = \dfrac{E_{Ag}}{F} = \dfrac{107.9}{96500}$ g C⁻¹

∴ Mass of silver deposited in half hour ($t = 30 \times 60$ s) is

           $m = ZIt = \dfrac{107.9}{96500} \times 1 \times 30 \times 60 =$ **2.01 g**

Example 10.10. *A silver and copper voltameters are connected across a 6 V battery of negligible resistance. In half hour, 1 g of copper and 2 g of silver are deposited. Calculate the rate at which energy is supplied by the battery. Given that E.C.E. of Cu is 3294×10^{-7} g/C and that of silver is 1118×10^{-6} g/C.*

Solution. We know that : $I = \dfrac{m}{Zt}$

For copper voltameter, $I_1 = \dfrac{m_1}{Z_1 t} = \dfrac{1}{3294 \times 10^{-7} \times 1800} = 1\cdot687$ A

For silver voltameter, $I_2 = \dfrac{m_2}{Z_2 t} = \dfrac{2}{1118 \times 10^{-6} \times 1800} = 0\cdot994$ A

Total current I drawn from the battery is

                $I = I_1 + I_2 = 1\cdot687 + 0\cdot994 = 2\cdot681$ A

Rate at which energy is supplied by the battery is

                $P = VI = 6 \times 2.681 =$ **16.1 W**

Example 10.11. *A refining plant employs 1000 electrolytic cells for copper refining. A current of 5000 A is used and the voltage per cell is 0.25 V. If the plant works for 100 hours/week, determine the annual output of refined copper and the energy consumed in kWh/tonne. The E.C.E. of copper = 1.1844 kg/1000 Ah.*

Solution. Since the voltage drop across each electrolytic cell is less than 1V, a number of cells are connected in series so that the generator can supply current at reasonable voltage.

$$\text{Supply voltage, } V = 0.25 \times 1000 = 250 \text{ volts}$$

$$\text{Circuit current, } I = 5000 \text{ A}$$

$$\text{Plant working time/year, } t = 100 \times 52 = 5200 \text{ hours/year}$$

$$\text{E.C.E. of copper, } Z = 1.1844 \text{ kg/1000 Ah}$$

$$= \frac{1.1844}{1000 \times 3600 As} = 0.329 \times 10^{-6} \text{ kg/C} \qquad (\because As = C)$$

The amount (*m*) of refined copper per year is

$$m = ZIt = 0.329 \times 10^{-6} \times 5000 \times 5200 \times 3600 = 30794 \text{ kg}$$

$$= \frac{30794}{1000} \text{tonne} = 30.794 \text{ tonne}$$

$$\text{Energy consumption/year} = VIt = 250 \times 5000 \times 5200 \text{ Wh} = 6500 \times 10^6 \text{ Wh}$$

$$= \frac{6500 \times 10^6}{1000} \text{kWh} = 6500 \times 10^3 \text{ kWh}$$

Since this energy consumption is for refining 30.794 tonne of copper,

$$\therefore \text{ Energy consumption/tonne} = \frac{6500 \times 10^3}{30.794} = \textbf{211.08} \times \textbf{10}^3 \textbf{ kWh/tonne}$$

Tutorial Problems

1. A current of 5 A flows for 40 minutes through an electrolyte which is a solution of a salt of chromium in water. Calculate the mass of chromium liberated. The electro-chemical equivalent of chromium is 90×10^{-9} kg/C. **[1.08 gm]**

2. How long will it take to deposit, from a copper sulphate solution, a coating of copper 0.05 mm thick on an area of 118 cm^2 if the supply p.d. is 4.5 volts and the total resistance of the circuit is 2.3 Ω. Specific gravity of copper is 8·93 and E.C.E. of copper = 0.329 mg/C. **[2.269 hr]**

3. A metal plate having a surface area of 115 cm^2 is to be silver plated. If a current of 1.5 A is passed for 1 hour and 30 minutes, what thickness of copper will be deposited ? Specific gravity of silver = 10.5 and E.C.E. of silver = 1.118 mg/C. **[0.075 mm]**

4. A worn shaft is to be reconditioned by depositing chromium on its curved surface to a thickness of 0·1 mm. The shaft has a diameter of 3.5 cm and a length of 80 cm. If a current of 4.4 A is passed, calculate how long the plate will take. Density of chromium = 6600 kg/m^3 and E.C.E. of chromium = 90×10^{-9} kg/C. **[41 hours, 44 minutes]**

5. Due to an error, a car battery is overcharged with a current of 5 A for 10 hours. Given that the electro-chemical equivalents of hydrogen and oxygen are 10.4×10^{-9} kg/C and 83.2×10^{-9} kg/C respectively, calculate the volume of distilled water which must be added to compensate for the loss. **[168 c.c.]**

10.9. Practical Applications of Electrolysis

The phenomenon of electrolysis has many industrial and commercial applications. A few of them are discussed below by way of illustration.

(*i*) **Electroplating.** The process of depositing a thin layer of superior metal (*e.g.*, gold, silver, nickel, etc.) over an inferior metal (*e.g.,* iron) is known as *electroplating*. The aim of electroplating is to provide good appearance and protect the object against corrosion. Fig. 10.3 shows a simple arrangement of silver plating. The object to be plated (*e.g.* a small iron key) is used as the cathode. The metal to be deposited (silver in this case) is made the anode. The solution of compound of the metal to be deposited — in this case $AgNO_3$ — is taken as the electrolyte. When d.c. supply is applied to the electrodes, the silver rod (*i.e.* anode) continuously dissolves into the solution and gets deposited on the cathode (*i.e.* the object).

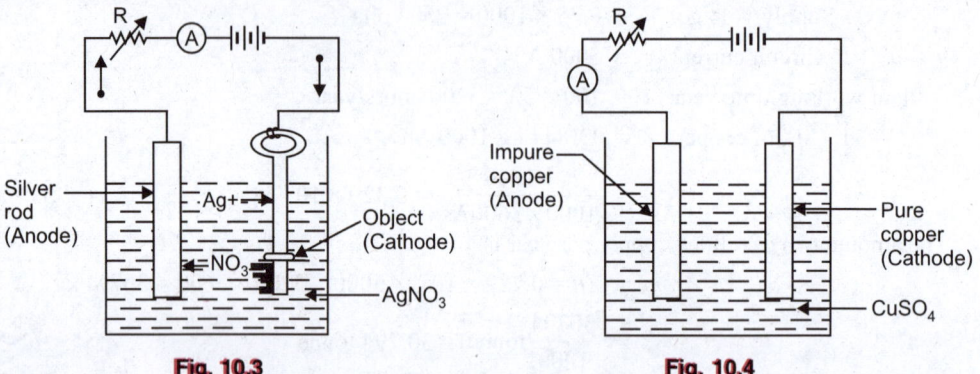

Fig. 10.3   **Fig. 10.4**

(*ii*) **Refining of metals.** Many metals are purified by the process of electrolysis. The impure metal is used as the anode and a small piece of pure metal is used as the cathode. The salt solution of the metal to be purified is used as the electrolyte. Fig. 10.4 shows a simple arrangement for electrolytic refining of *copper. When d.c. supply is applied to the electrodes, pure copper from anode goes into the solution and gets deposited on the cathode. The impurities sink to the bottom of the container and from an 'anode mud'. Similarly, electrolytic process can be employed for refining other metals such as silver, gold, platinum and nickel.

(*iii*) **Production of chemicals.** The process of electrolysis is being extensively used for the commercial production of chemicals like sodium carbonate, sodium bicarbonate, caustic soda etc.

(*iv*) **Extraction of metals from ores.** Many metals such as Al, Mg, Zn, Cu, etc. are extracted from their ores by electrolysis.

(*v*) **Production of oxygen and hydrogen.** By the electrolysis of acidic water, oxygen and hydrogen can be produced.

(*vi*) **Electrolytic capacitor.** An electrolytic capacitor consists of two aluminium foils in which a very thin dielectric is sandwitched by the process of electrolysis. Two foils of aluminium (which act as electrodes) are dipped in a solution of boric acid, glycerine and ammonia water. When current is passed between anode and cathode, a very thin film of aluminium oxide (Al_2O_3) is formed on the anode surface. This oxide film acts as a dielectric. Since the oxide film is very thin, the electrolytic capacitor has a very large capacitance.

(*vii*) **Anodising.** It is the process of coating aluminium oxide on the aluminium itself by electrolysis to protect it against corrosion. The aluminium article is made the anode and dil. H_2SO_4 is the electrolyte.

(*viii*) **Electro-typing.** This is the process in which we obtain an impression of a page of the book in **copper. The ordinary type is arranged to form a page of the book. Then a special kind of wax is poured over it. When the wax hardens and is peeled off, it bears an impression of that page. The *wax-mould* is uniformly coated with a very thin layer of powdered graphite to make it a conductor. The mould is made the cathode and a pure copper plate as the anode. The two electrodes are immersed in

---

*   Pure copper is a good conductor of electricity. But if it has even a small amount of impurity, its resistance
      is greatly increased.

**   We know that printing is accomplished by the use of type. The metal employed for type generally is an
      alloy of lead, antimony and tin. Such a type can give only a few thousand impressions because the alloy is
      quite soft. If several hundred thousands copies are to be printed, the type metal would soon wear out.

a bath containing copper sulphate solution ($CuSO_4$) as the electrolyte. When d.c. supply is applied to the electrodes, copper from anode goes into the solution and gets deposited over the cathode *i.e.* wax-mould. When sufficient thick layer (as thick as ordinary visiting card) of copper is deposited on the cathode (*i.e.* wax-mould), the circuit is disconnected. The wax is removed from the cathode and the copper plate obtained would have an impression of the page in copper. Several hundred thousands impressions of the page can be taken from this plate.

10.10. Cell

We have seen that when a current passes through the solution of an electrolyte, chemical changes take place i.e., electrical energy is converted into chemical energy. The converse of this is also true i.e., we can convert chemical energy into electrical energy. The device which accomplishes this job is called a cell.

A **cell** *is a source of e.m.f. in which chemical energy is converted into electrical energy.*

A cell essentially consists of two metal plates of different materials immersed in a suitable solution. The plates are called electrodes and the solution is called electrolyte. The magnitude of e.m.f. of a cell depends upon (*i*) the nature of material of electrodes used and (*ii*) the nature of electrolyte.

10.11. Types of Cells

Using various metals and methods of construction, a large variety of cells has been developed. However, the cells can be divided into two main classes *viz.*

    1. Primary cells   2. Secondary cells

1. Primary cells. *A cell in which chemical action is not reversible is called a* **primary cell** *e.g.,* Voltaic cell, Daniel cell, Lachlanche cell, dry cell etc. As a primary cell delivers current, the active materials are used up. When the active materials are nearly consumed, the cell stops delivering current. In order to renew the cell, fresh active materials are provided. Another drawback of a primary cell is that it cannot provide large and steady current for a longer period. This fact makes the primary cell rather an expensive source of electrical energy. Due to these drawbacks, the use of primary cells is limited to torch batteries and for experimental purposes in the laboratories.

2. Secondary cells. *A cell in which chemical action is reversible is called a* **secondary cell** or **storage cell.** A secondary cell operates on the same principle as a primary cell but differs in the method in which it may be renewed. In a secondary cell, there is no actual consumption of any plate and that the chemical process is reversible. When the cell is delivering current (*i.e., discharging*), the chemical action changes the composition of plates. When the cell is exhausted, the chemical action can be reversed (*i.e.,* plates can be restored to the original condition) by passing current through the cell in the reverse direction to that in which the cell provided current. This process is called *charging.* In other words, charging process reverses the chemical action and enables the plates to acquire original compositon. There are several types of secondary or storage cells; the more common ones being :

    (*i*) Lead-acid cell

    (*ii*) Nickel-iron-alkaline cell (or Edison cell)

    (*iii*) Nickel-cadmium-alkaline cell.

Note. When a secondary cell is charged, electrical energy is converted into chemical energy which is stored in the cell. When the cell discharges, the stored chemical energy starts converting into electrical energy. For this reason, a secondary cell is sometimes called a **storage cell.**

10.12. Lead-Acid Cell

The most inexpensive secondary cell is the lead–acid cell and is widely used for commercial purposes. A lead-acid cell when ready for *use contains two plates immersed in a dilute sulphuric acid (H_2SO_4) of specific gravity about 1·28. The positive plate (*anode*) is of **lead-peroxide** (PbO_2) which has chocolate brown colour and the negative plate (*cathode*) is **lead** (Pb) which is of grey colour.

(*i*) When the cell supplies current to a load (*i.e.* discharging), the chemical action that takes place forms lead sulphate ($PbSO_4$) on both the plates with water being formed in the electrolyte. After a certain amount of energy has been withdrawn from the cell, both plates are transformed into the same material (*i.e.* $PbSO_4$) and the specific gravity of the electrolyte (H_2SO_4) is lowered. The cell is then said to be discharged. There are several methods (See Art. 10.19) to ascertain whether the cell is discharged or not.

(*ii*) *To charge the cell, direct current is passed through the cell in the reverse direction to that in which the cell provided current*. This reverses the chemical process and again forms a lead peroxide (PbO_2) positive plate and a pure lead (Pb) negative plate. At the same time, H_2SO_4 is formed at the expense of water, restoring the electrolyte (H_2SO_4) to its original condition.

The chemical changes that occur during discharging and recharging of a lead-acid cell are discussed in the articles that follow.

10.13. Chemical Changes During Discharging

By discharging of a cell, we mean that it is delivering current to the external circuit. Consider a charged lead acid cell with anode of PbO_2 and cathode of Pb; the electrolyte being dilute H_2SO_4 [See Fig. 10.5]. Sulphuric acid splits up into hydrogen ions (H^+H^+) and sulphate ions (SO_4^{--}). The sulphate ions move towards the cathode and hydrogen ions move towards the anode causing the following chemical reactions :

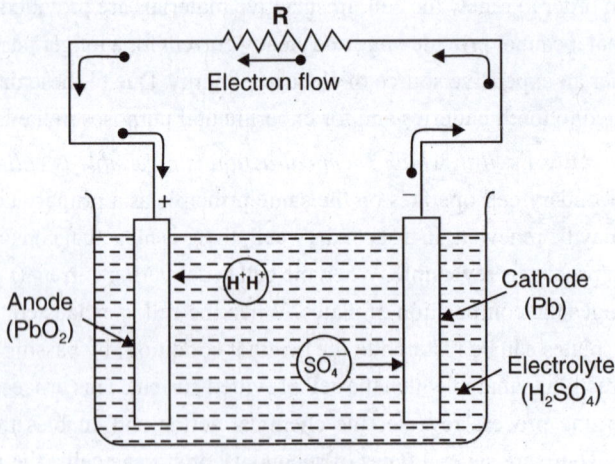

Fig. 10.5

At cathode. On reaching the cathode, a sulphate ion (SO_4^{--}) gives up its two extra electrons to become sulphate radical. These electrons given up at the cathode move through the external circuit to the anode where they are available to neutralise the positive ions (H^+H^+) arriving there. Since sulphate radical cannot exist, it enters into chemical reaction with cathode material (Pb) to form lead sulphate ($PbSO_4$).

---

*   Before the cell is put on the market, the manufacturer does the process of forming the plates i.e. making the positive plate PbO_2 and negative plate Pb.

$$SO_4^{--} - 2e \longrightarrow SO_4 \text{ (radical)}$$
$$Pb + SO_4 \longrightarrow PbSO_4$$

(Grey colour)   *(Whitish in colour)*

At anode. On reaching the anode, each hydrogen ion takes one electron from it to become hydrogen gas. This electron is given by the sulphate ion at the cathode and has come to the anode *via* the external circuit.

$$H^+H^+ + 2e \longrightarrow 2H$$

The hydrogen gas liberated at the anode acts chemically on the anode material (PbO_2) and reduces it to lead oxide (PbO).

$$PbO_2 + 2H \longrightarrow PbO + H_2O$$

Sulphuric acid reacts with PbO to form $PbSO_4$.

$$PbO + H_2SO_4 \longrightarrow PbSO_4 + H_2O$$

As the cell delivers current, both the plates start getting converted into lead sulphate ($PbSO_4$). The water produced in the chemical reactions above dilutes the electrolyte (H_2SO_4) and lowers its specific gravity. When the specific gravity (which can be measured by a *hydrometer*) of the electrolyte falls to 1.18, the cell is fully discharged. The chemical changes that take place during discharging of a lead-acid cell can be summed up as under :

   (*i*) *Both the plates are converted into lead sulphate (PbSO₄) which is whitish in colour.*

   (*ii*) *Water is formed which lowers the specific gravity of the electrolyte (H₂SO₄). When the cell is fully discharged, the specific gravity of H₂SO₄ falls to about 1.18.*

   (*iii*) *The e.m.f. of the cell falls.* The lead-acid cell should not be discharged beyond the point where its e.m.f. falls to about 1.8 volts.

   (*iv*) *The chemical energy stored in the cell is converted into electrical energy.*

It is important to note that e.m.f. of the cell provides little indication to the state of discharge of the cell since it remains close to 2 V for 90% of the discharge period. In practice, specific gravity of the electrolyte (H_2SO_4) is used to know the state of discharge. The cell should be recharged when specific gravity of H_2SO_4 falls to about 1.18.

10.14. Chemical Changes During Recharging

Consider a discharged lead-acid cell having both the plates converted to lead sulphate ($PbSO_4$). *In order to recharge the cell, direct current is passed through the cell in the reverse direction to that in which the cell provided current.* To do so, the anode is connected to the positive terminal of d.c. source and cathode to the negative terminal of the source as shown in Fig. 10.6. The electrolyte (H_2SO_4) breaks up into hydrogen ions (H^+H^+) and sulphate ions (SO_4^{--}). Hydrogen ions move towards cathode and sulphate ions move towards anode causing the following chemical reactions :

At anode. On reaching the anode, a sulphate ion (SO_4^{--}) gives up its two extra electrons to become sulphate radical. These electrons given up at the anode move through the external circuit to the cathode where they are available to neutralise the positive ions (H^+H^+) arriving there. Since sulphate radical cannot exist, it enters into chemical reaction with water as under :

$$SO_4^{--} - 2e \longrightarrow SO_4 \text{ (radical)}$$
$$SO_4 + H_2O \longrightarrow H_2SO_4 + O$$

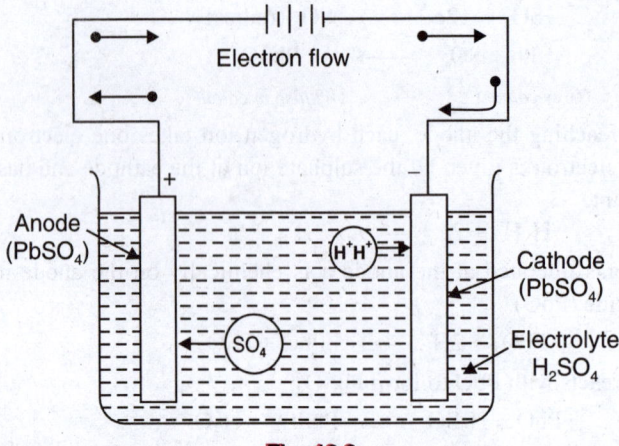

Fig. 10.6

The oxygen in the atomic state (*i.e.* O) is very active and reacts chemically with anode material ($PbSO_4$) to produce the following chemical change :

$$PbSO_4 + O + H_2O \longrightarrow PbO_2 + H_2SO_4$$

At cathode. On reaching the cathode, each hydrogen ion (H^+) takes one electron from it to become hydrogen gas. This electron is given up by the sulphate ion at anode and has come to the cathode *via* the external circuit.

$$H^+H^+ + 2e \longrightarrow 2H$$

The hydrogen gas liberated at the cathode reacts with cathode material ($PbSO_4$) to reduce it to lead (Pb) as under :

$$PbSO_4 + 2H \longrightarrow Pb + H_2SO_4$$

As the charging process goes on, the anode is converted into PbO_2 and cathode into Pb. The H_2SO_4 produced in the chemical reactions above increases the specific gravity of the electrolyte. The chemical changes that occur during recharging can be summed up as under :

(*i*) *The positive plate (anode) is converted into PbO_2 and the negative plate (cathode) into Pb.*

(*ii*) *H_2SO_4 is formed in the reactions. Therefore, the specific gravity of the electrolyte (H_2SO_4) is raised.* When the cell is fully charged, the specific gravity of H_2SO_4 rises to about 1.28.

(*iii*) *The e.m.f. of cell rises.* The e.m.f. of a fully charged lead-acid cell is about 2 volts.

(*iv*) *Electrical energy supplied is converted into chemical energy which is stored in the cell.*

Note. The charging and discharging of a lead-acid cell can be represented by a single reversible equation given below :

$$\underset{\substack{\text{Positive}\\\text{plate}}}{PbO_2} + \underset{\substack{\text{Negative}\\\text{plate}}}{2H_2SO_4 + Pb} \underset{\text{Charge}}{\overset{\text{Discharge}}{\rightleftharpoons}} \underset{\substack{\text{Positive}\\\text{plate}}}{PbSO_4} + \underset{\substack{\text{Negative}\\\text{plate}}}{2H_2O + PbSO_4}$$

Reading from left to right, the equation represents the chemical action during discharge while reading from right to left, the equation gives the chemical action during charge.

10.15. Formation of Plates of Lead-acid Cells

There are two methods of forming the plates of a lead-acid cell *i.e.*, making the positive plate PbO_2 and negative plate Pb. These are known after the names of their inventors *viz* (*i*) Plante method (*ii*) Faure' method.

(*i*) **Plante method.** In this method, we start with two plates of pure lead immersed in dilute sulphuric acid (H_2SO_4). When direct current is passed through this cell, the anode (*i.e.*, plate connected to +ve terminal of d.c. source) is *covered with an exceedingly thin layer of PbO_2. The cathode, however, remains unaffected. The cell is now discharged by connecting it to a load e.g., galvanometer. As the cell delivers current, both the plates are covered with a coating of **lead sulphate ($PbSO_4$). After some time, the cell is discharged and no more current can be drawn from it. If the cell is charged ***again, the anode is again covered with a layer (this time somewhat thicker than the initial charging) of PbO_2 and the cathode gets transformed to lead (Pb). If this *charge-discharge cycle* is repeated again and again, the chemical reactions make deeper and deeper penetration into the plates. The result is that a sufficient thick layer of PbO_2 is formed on the anode while the cathode gets transformed to spongy lead; spongy lead being lead in a somewhat porous condition. Plates produced in this way are called **formed plates.**

Plante' plates have a longer life and are less liable to disintegration. However, they suffer from two important drawbacks. First, the process of forming such plates is very time-consuming and expensive. Secondly, these plates are very heavy and cannot be used in portable batteries.

(*ii*) **Faure method.** In this method, the plates are constructed in the form of open metallic grids made of lead-antimony alloy. The openings in the positive grid are filled with a paste of red lead (Pb_3O_4) and that of a negative grid with litharge (PbO). A single "forming" charge is sufficient to cause the necessary chemical changes. The paste on the positive plate is converted into PbO_2 and that on the negative plate converted into spongy lead (Pb). It may be noted that the grid serves to hold the active materials in place and also to distribute the current evenly over the surface of the plates. The plates formed in this way are called **pasted plates** (See Fig. 10.7).

Bars at back of grid

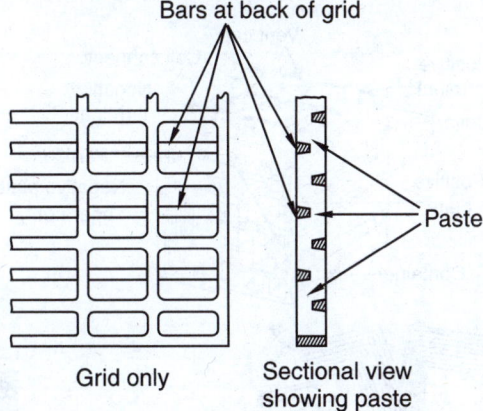

Grid only   Sectional view
showing paste

Fig. 10.7

Many advantages are claimed for pasted plates. First, such plates can be formed quickly. Secondly, pasted plates are much lighter than formed plates. For a given capacity, the weight of a pasted plate is only about one-third of that of a formed plate.

---

* H_2SO_4 breaks up into hydrogen ions (H^+ H^+) and sulphate ions (SO_4^{--}).

Hydrogen ions move towards the cathode and sulphate ions (SO_4^{--}) move towards anode.

At anode. $SO_4 + H_2O \longrightarrow H_2SO_4 + O$

$2Pb + 2O_2 \longrightarrow 2PbO_2$

At cathode. $2H^+ + 2e \longrightarrow H_2$

Since hydrogen can form no compound with Pb, cathode remains unaffected.

**   See the chemical changes during discharging in Art. 10.13.

***   See the chemical changes during recharging in Art. 10.14

10.16. Construction of a Lead-acid Battery

A lead-acid battery is a collection of a number of lead-acid cells connected in series; the most common ones being 6-volt type and 12-volt type. In case of 6-volt type, three cells are connected in series whereas for 12-volt type, six cells are series-connected. A commercial lead-acid cell incorporates many refinements in its construction that make it possible to increase its capacity and efficiency. Fig. 10.8 shows the cut-away view of a commercial lead-acid battery. The various parts are :

(i) **Container.** The container houses the plates and the electrolyte. The container of some batteries is made of glass or transparent synthetic plastic material. The main advantage of using these materials is that a visual examination can be made of the state of plates. However, if the battery is used in an automobile, a more robust material (*e.g.* hard rubber or a strong synthetic plastic material) is required to withstand the mechanical shocks. The container is sealed off at the top to prevent the spilling of electrolyte. A large space is left at the bottom of the container so that any sediment that drops from the plates may collect there without causing the short-circuit between positive and negative plates.

(ii) **Plates.** The capacity of a lead-acid cell *depends upon the plate area. To increase the effective area of plates without increasing the size of the cell, we use a large number of thin plates in the cell instead of two plates. The alternate positive and negative plates of the cell are sandwitched together with insulators (called *separators*). The negative plates are connected together as are the positive plates. A commercial cell always contains an odd number of plates such as 11,13, 15 or 17. The number of negative plates is always one more than the number of positive plates ; the outside plates being negative. This arrangement not only allows both sides of the positive plate to be used actively but also the tendency for one side of the plate to distort is balanced by a similar effect on the other.

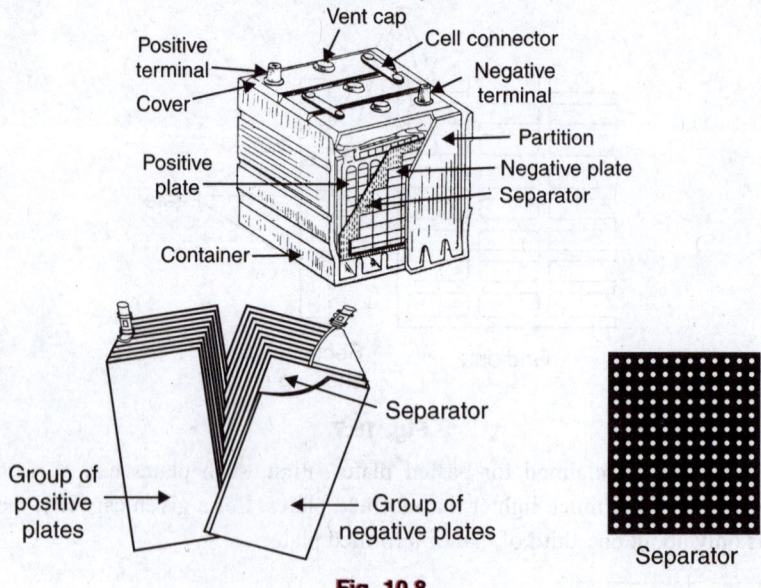

Fig. 10.8

A separate compartment is provided for each cell and each compartment has large space at the bottom so that any sediment from the plates may collect there. In a 6-V battery, three cells are assembled, each in an acid proof compartment, the cells are then connected in series. A final point about the "formation" of plates. The type of plate construction (*i.e.* Plante plate or Faure plate) to be used will depend upon the service requirements.

---

*   The e.m.f. of a secondary cell does not depend upon the size of its plates. However, the larger the plates, the more the chemical energy that can be stored in the cell and hence greater the current that can be drawn from it.

(*iii*) **Separators.** In order to save space and to reduce the internal resistance of the cell, the plates are placed close together. To prevent the plates from touching each other if they wrap or buckle, they are separated by non-conducting materials (*e.g.* wood, rubber *etc*) called *separators*. Separators are grooved vertically on one side and are smooth on the other side. The grooved side is placed against the positive plate to permit free circulation of electrolyte around the positive plate where *greater chemical action takes place.

(*iv*) **Electrolyte.** The electrolyte is dilute sulphuric acid (H_2SO_4) solution mixed in such a proportion so that with a fully charged battery, its specific gravity is about 1.28. During the normal life of a battery that is properly cared for, the electrolyte loses none of the acid. It is, therefore, necessary only occasionally to replace the water that has evaporated. To ensure normal battery life, only pure water should be added.

(*v*) **Cell cover.** Each cell compartment has a cover, usually made of moulded hard rubber. Openings are provided in these covers for the two terminal posts and for a vent. The joints between covers and container are sealed with an acid-resistant material.

(*vi*) **Vent caps.** Each cell cover has a hole into which is fitted the vent cap. This cap has a vent hole to allow the free exit of the gas formed in the cell. The vent caps can be easily removed for adding water or taking hydrometer reading.

(*vii*) **Inter-cell connector.** It is a lead alloy link that joins the cells in series. The positive terminal of the cell is marked by a large + sign or with a red colour.

(*viii*) **Cell terminals.** Each cell has two terminals. The terminals are generally made of lead as it does not corrode due to the electrolyte.

10.17. Characteristics of a Lead-acid Cell

A lead-acid cell is the most popular type of secondary cell. Therefore, it is profitable to study its important characteristics.

1. E.M.F. *The e.m.f. of a fully charged lead-acid cell is about 2.2 volts.* As the cell delivers current, its e.m.f. also decreases, though by a very small amount. The magnitude of e.m.f. depends upon :

(*i*) length of time since it was last charged

(*ii*) specific gravity of the electrolyte

(*iii*) temperature

When a cell has been recently charged, its e.m.f. is high but it gradually decreases even if left on open-circuit. The e.m.f. of the cell rises with the increase in specific gravity of the electrolyte. The surrounding temperature also affects the e.m.f. ; there being a slight increase in its value with the rise in temperature.

2. Terminal voltage. The open-circuit terminal voltage (*i.e.* e.m.f.) of a fully charged cell is about 2.2 V. When the cell delivers current, the terminal voltage is less than its e.m.f. due to voltage drop in the internal resistance of the cell. If the discharge is continued, the terminal voltage falls rapidly for a short time, then slowly for some time and again rapidly towards the end of discharge.

It may be noted that when the terminal voltage has fallen to about 1.8 volts, the discharge should be stopped. If it is continued, too much lead sulphate ($PbSO_4$) is formed, clogging the pores of the plates, possibly **damaging them and making the recharging increasingly difficult. Also excess lead

---

*   The positive active material (*i.e.* PbO_2) undergoes a large change in volume during charge and discharge as compared with negative active material (*i.e.* Pb). Therefore, the tendency of PbO_2 to disintegrate and fall to the bottom of the cell is much greater than for the Pb. The life of a positive plate is roughly half of the negative plate.

**   If a cell is discharged excessively (*i.e.* terminal voltage falls below 1.8 volts), crystalline lead sulphate will be formed. Since this lead sulphate occupies a larger volume than the material from which it is formed, the plates may buckle, the separators may be damaged and the active material may be dislodged.

sulphate in the plates raises the internal resistance of cell, thus lowering the terminal voltage. The Specific Gravity of the electrolyte too is reduced by the water that is formed and this lowers the cell's voltage even more.

3. Internal resistance. *The opposition offered to current within the cell is called the* **internal resistance** *of the cell.* It is made up of resistances of the plates, the active material and the electrolyte. The internal resistance of a lead-acid cell is very small (typical value being 0.01 Ω) and depends upon the following factors :

(*i*)   Area of plates—decreases with the increase in plate area

(*ii*) Spacing between plates—decreases with the decrease in spacing

(*iii*) Sp. Gravity of H_2SO_4 – decreases with the increase in sp. gravity.

The internal resistance of a lead-acid cell should be minimum in order to reduce the internal drop. This is achieved by using multiplate construction in a cell. As explained in Art. 10.16, the negative plates of a cell are connected together as are the positive plates. The effect of this arrangement is as if we have connected a number of cells in parallel. At the same time, the length of electrolyte between the plates is reduced. The result is that the internal resistance of the cell is lowered.

4. Capacity. *The* **capacity** *of a cell is the quantity of electricity which it can give out during single discharge until its terminal voltage falls to 1.8 volts.* It is measured by the product of current in amperes and the time in hours *i.e.*

$$\text{Capacity of cell} = I_d \times T_d \text{ ampere-hours (or Ah)}$$

where I_d = Steady discharging current in amperes

T_d = Time in hours for which the cell can supply current until its p.d. falls to 1.8 volts

The capacity of a lead-acid cell is taken up to a point till it discharges down to 1.8 volts. It is because a lead-acid cell is not permitted to discharge beyond this point to avoid permanent damage to the cell. Consider a cell of capacity 120 Ah. This means that theoretically the cell can deliver 1 A for 120 hours, 2 A for 60 hours or any combination of amperes and hours that, when multiplied together, gives 120. After that much time, the terminal voltage of the cell falls to 1.8 volts, requiring recharging. The capacity of a cell depends upon the following factors :

(*i*)   Rate of discharge   — Higher the rate of discharge, less the capacity

(*ii*) Temperature   — Increases with temperature

(*iii*) Area of plates   — Increases with plate area

(*iv*) Sp. gravity of electrolyte — Increases with Sp. gravity

5. Efficiency. There are two ways of expressing the efficiency of a secondary cell *viz*

(*i*) Ampere hour efficiency   (*ii*) Watt-hour efficiency.

(*i*) **Ampere-hour efficiency.** *It is the ratio of output ampere-hours to the input ampere-hours of the cell i.e.*

$$\text{Ampere-hour efficiency, } \eta_{Ah} = \frac{\text{Ampere} - \text{hours provided on discharge}}{\text{Ampere} - \text{hours of charge}} \times 100$$

$$= \frac{I_d \times T_d}{I_c \times T_c} \times 100$$

Note that efficiency is considered for a complete charge and complete discharge down to the state in which the cell was at the commencement.

Since ampere-hours is the quantity of electricity, this efficiency is sometimes called *quantity efficiency.* The ampere-hour efficiency of a lead-acid cell is about 90%. It means that if a lead-acid cell provides 90 Ah on complete discharge, then 100 Ah must be put back into the cell to restore it to its original condition.

(ii) Watt-hour efficiency. *It is the ratio of output energy to the input energy of the cell i.e.*

$$\text{Watt-hour efficiency, } \eta_{wh} = \frac{\text{Energy given on discharge}}{\text{Energy input of charge}} \times 100$$

$$= \frac{V_d \times I_d \times T_d}{V_c \times I_c \times T_c} \times 100 = \frac{I_d \times T_d}{I_c \times T_c} \times 100 \times \frac{V_d}{V_c}$$

$$= \eta_{Ah} \times \frac{V_d}{V_c}$$

Again, efficiency is considered for a complete charge and complete discharge down to the state in which the cell was at the commencement.

Since average p.d. during discharge (*i.e.* V_d) is lower than the average p.d. during charge (*i.e.* V_c) due to the effects of internal resistance and polarisation, η_{wh} is always less than η_{Ah}. The watt-hour (or energy) efficiency of a lead-acid cell is about 75%. This means that if a lead-acid cell provides 75 Wh on complete discharge, then 100 Wh must be put back into the cell to restore it to its original condition.

The efficiency of a lead-acid cell depends upon the following factors :

(*a*) Rate of charge and discharge

(*b*) Internal resistance and polarisation

(*c*) Time interval between the end of discharge and the commencement of recharge.

(*d*) Temperature.

10.18. Characteristic Curves of a Lead-acid Cell

The behaviour of a lead-acid cell can be easily studied from its characteristic curves.

(*i*) **Charging and discharging curves.** Fig. 10.9 shows the variations in the terminal P.D. of a cell with time on charge and discharge. When the cell is charged, its terminal voltage rises from 1.8 V (See charging curve) to 2.2 V during the first 2 hours and then increases slowly. Finally, after the charging period, the terminal voltage is about 2.7 V. When the cell is disconnected from the charging source, its terminal voltage rapidly falls to about 2.2 V without any discharge.

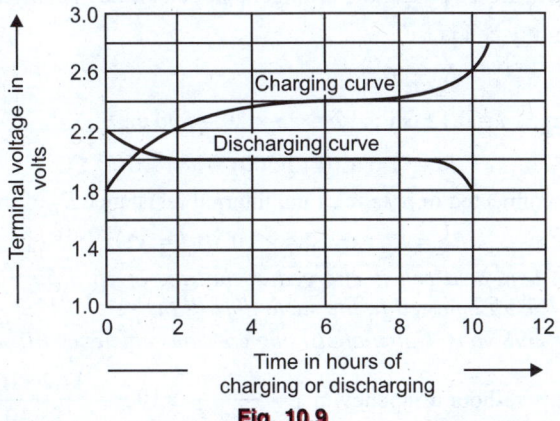

Fig. 10.9

On discharge, the terminal voltage falls to 2.0 V in the beginning, remains constant for sufficient time and finally drops to 1.8 V at which the cell should be recharged. If the cell is allowed to discharge below 1.8 V repeatedly, or it is left in an uncharged condition for a considerable time, a hard insoluble lead sulphate ($PbSO_4$) is formed on the plates of the cell. This increases the internal resistance of the cell considerably and finally ruins the cell.

(*ii*) **Capacity and rate of discharge.** Fig. 10.10 shows the effect of rate of discharge on the capacity of the cell. The ampere-hour capacity of the cell decreases with the increase in the rate of discharge. Rapid rate of discharge means heavier load current. As a result, the voltage of the cell falls more rapidly due to the increased voltage drop in the internal resistance of the cell. Moreover, the rapid discharge weakens the acid in the pores of the plates and chemical action becomes brisk. These effects reduce the capacity of the cell.

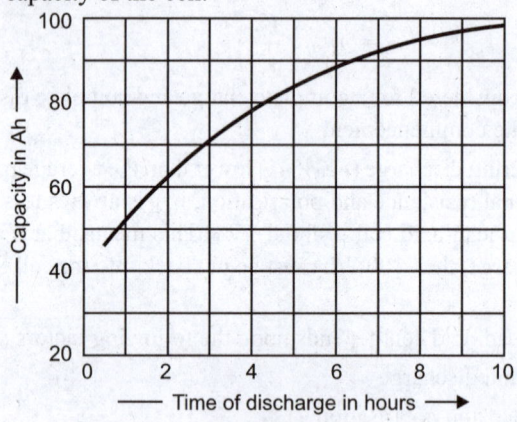

Fig. 10.10

The capacity of a cell is given in ampere-hours for a discharge of 8 hours or 10 hours. A rate of discharge greater than this value decreases the ampere-hour capacity of the cell while a slower rate increases its capacity.

(*iii*) **Capacity and temperature.** The ampere-hour capacity of a cell increases with the increase in temperature and *vice-versa*. It is because higher temperatures permit more diffusion of acid and increase the velocities of the ions taking part in the chemical changes.

Example 10.12. *A lead-acid cell has 13 plates, each 25 cm × 20 cm. The clearance between the neighbouring plates is 1·2 mm. If the resistivity of the acid is 1·6 Ω-cm, find the internal resistance of the cell.*

Solution. Since there are 13 plates, the arrangement constitutes 12 tiny cells in parallel.

Resistance of each tiny cell is

$$R = \rho \frac{l}{a}$$

Here, $\rho = 1.6 \ \Omega$ cm ; $l = 0.12$ cm ; $a = 25 \times 20 = 500$ cm^2

∴ $R = 1.6 \times (0.12/500) = 384 \times 10^{-6} \ \Omega$

Since the cells are connected in parallel, total internal resistance R_T of the cell is

$$R_T = R/12 = 384 \times 10^{-6}/12 = \mathbf{32 \times 10^{-6} \ \Omega}$$

Example 10.13. *A lead-acid cell is charged at the rate of 18 A for 10 hours at an average voltage of 2·26 volts. It is discharged in the same time at the rate of 17·2 A; the average voltage during discharge being 1·98 volts. Calculate (i) ampere-hour efficiency (ii) watt-hour efficiency.*

Solution. (*i*) Ampere-hour efficiency, $\eta_{Ah} = \dfrac{I_d \times T_d}{I_c \times T_c} \times 100 = \dfrac{17.2 \times 10}{18 \times 10} \times 100 = \mathbf{95.55\%}$

(*ii*) Watt-hour efficiency, $\eta_{wh} = \eta_{Ah} \times \dfrac{V_d}{V_c} = 95.55 \times \dfrac{1.98}{2.26} = \mathbf{83.71\%}$

Example 10.14. *A discharged battery is charged at 8A for 2 hours after which it is discharged through a resistor of R Ω. If discharge period is 6 hours and the terminal voltage remains fixed at 12V, find the value of R assuming the Ah efficiency of the battery as 80%.*

Solution. Ah efficiency of battery, $\eta_{Ah} = 80\% = 0.8$

$\qquad$ Input Ah to battery $= 8 \times 2 = 16$ Ah

$\qquad$ Output Ah of battery $= 16 \times \eta_{Ah} = 16 \times 0.8 = 12.8$ Ah

$\qquad$ Discharging current $= \dfrac{\text{Output Ah}}{\text{Hours of discharge}} = \dfrac{12.8}{6} = 2.133$ A

$\therefore \qquad$ Value of $R = \dfrac{12}{2.133} = \mathbf{5.625\ \Omega}$

Example 10.15. *Determine the mass of lead and lead peroxide required to produce 1 Ah in a lead-acid cell. The gram equivalent of lead is 103.6 g and Faraday's constant is 96500 C.*

Solution. Gram equivalent of lead $= 103.6$ g

$\qquad$ Faraday's constant $= 96500$ C

Now, $\qquad\qquad$ 1 Ah $= 1 \times (1 \times 3600) = 3600$ As $= 3600$ C

It means that in order to deliver 96500 C, the mass of lead required is 103.6 g. Therefore, in order to deliver 3600 C (*i.e.* 1 ampere-hour), the mass of lead required

$$= \dfrac{103.6}{96500} \times 3600 = \mathbf{3.865\ \textit{g}}$$

$\qquad$ Atomic wt. of lead peroxide $=$ Atomic wt. of lead $+ 2 \times$ Atomic wt. of oxygen

$$= 207.2 + 2 \times 16 = 239.2\ g$$

Gram equivalent of lead peroxide

$$= \dfrac{\text{Atomic wt. in g}}{\text{Valency}} = \dfrac{239.2\ g}{2} = 119.6\ g$$

$\therefore$ Mass of lead peroxide required to deliver 1 Ah (*i.e.* 3600 C)

$$= \dfrac{119.6}{96500} \times 3600 = \mathbf{4.46\ g}$$

$\boxed{\textbf{Tutorial Problems}}$

1. A lead-acid cell maintains a constant current of 1.5 A for 20 hours before its terminal voltage falls to 1.8 volts. What is the capacity of the cell? **[30 Ah]**

2. A discharged battery is put on charge at 5A for 3.5 hours at a mean charging voltage of 13.5 V. It is then discharged in 6 hours at constant terminal voltage of 12 V through a resistance of R ohms. Calculate the value of R for an Ah efficiency of 85%. Also calculate the watt-hour efficiency. **[4.84 Ω ; 75.56%]**

3. Determine the ampere-hour and watt-hour efficiencies of an accumulator which is charged for 12 hours at 25A at an average voltage of 2.5 V and is discharged in 10 hours at a load of 20 A at an average voltage of 2.25 V. **[66.67%; 60%]**

10.19. Indications of a Fully Charged Lead-acid Cell

During the charging process, it is very essential that the battery is taken out from the charging circuit as soon as it is fully charged. Overcharging as well as undercharging are undesirable and should always be avoided. The indications of a fully charged cell (or battery) are :

$\quad$ (*i*) Voltage $\qquad\qquad\qquad$ (*ii*) Specific gravity of electrolyte

$\quad$ (*iii*) Gassing $\qquad\qquad\qquad$ (*iv*) Colour of plates

$\quad$ (*i*) **Voltage.** During charging, the terminal potential of a cell increases and provides an indication to the state of charge. A fully charged lead-acid cell has a terminal voltage of about 2.1 volts.

(*ii*) **Specific gravity.** During the charging process, the specific gravity of the electrolyte (H_2SO_4) increases and provides an important indication to the state of charge of the cell. The specific gravity of the electrolyte of a fully charged lead-acid cell is about 1.28. This can be measured by means of a **hydrometer** (See Fig. 10.11). It consists of a long glass tube fitted with a thin rubber hose at its lower end and a rubber bulb at the upper end. Inside the tube is sealed glass float weighted at one end to make it float upright. By inserting the hose into the cell and then operating the bulb, a quantity of electrolyte (H_2SO_4) may be drawn into the glass tube. The depth to which the float sinks in the drawn liquid indicates the specific gravity of the electrolyte. If the float sinks low, the specific gravity of the liquid (*i.e.* electrolyte H_2SO_4), is low. If it floats high, the specific gravity is high. Hydrometer float is marked with a scale starting with 1 and extending to 1.3. When the cell is fully charged, the hydrometer reading approaches 1.28 mark.

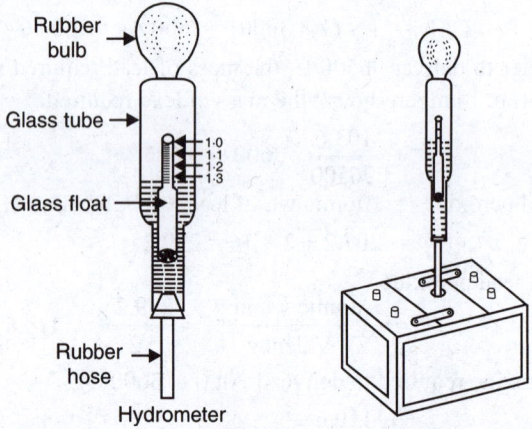

Rubber bulb
Glass tube
1·0
1·1
1·2
1·3
Glass float
Rubber hose
Hydrometer
Fig. 10.11

In practice, the state of charge of a lead-acid cell (or battery) is determined from the specific gravity of the electrolyte (H_2SO_4). The following table may be useful in this regard.

Sp. Gr. of H_2SO_4	State of charge
1.13	Discharged
1.19	25% charged
1.22	50% charged
1.25	75% charged
1.28	100% charged

(*iii*) **Gassing.** When the cell is fully charged, the charging current starts electrolysis of water. The result is that hydrogen is given off at the cathode and oxygen at the anode; the process being known as **gassing*. Gassing at both the plates indicates that the charging current is doing no useful work and hence should be stopped.

(*iv*) **Colour of plates.** The visual examination of colour of the plates of a lead-acid cell provides yet another important indication to its state of charge. When the cell is fully charged, the positive plate gets converted into PbO_2 (chocolate brown) and the negative plate to spongy lead (grey).

---

*   Gassing is harmless so long as it is not violent enough to dislodge the active material from the plates. However, it leads to waste of energy and the usual practice is to gradually reduce the charging current towards the finish of charge in order to reduce gassing and to avoid overheating of battery.

10.20. Load Characteristic of a Lead-acid Cell

The curve between terminal voltage of a lead-acid cell or battery of such cells and the load current is called load characteristic of the cell or battery. Fig. 10.12 shows the load characteristic of a lead-acid battery (or secondary battery). Note that the value of current is positive for discharge and negative for charge.

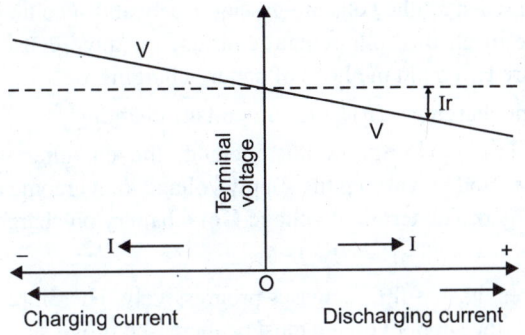

Fig. 10.12

Let V = Terminal voltage of the battery

E = E.M.F. of the battery

I = Discharging current

r = Internal resistance of the battery

∴   Terminal voltage V of the battery is given by ;

$$V = E - Ir \qquad \text{... for discharge}$$

For charging, I is negative so that :

$$V = E + Ir \qquad \text{...for charge}$$

The load characteristic shown in Fig. 10.12 is linear but it will be so if E and r remain constant. However, these conditions are not realised in practice because the values of E and r depend upon the state of charge of the battery. For example, the e.m.f. of a fully charged battery is about 2.2 volts/cell and its value decreases as the battery discharges ; being 1.8 volts/cell for a fully discharged battery.

10.21. Sulphation of Plates

If a lead-acid cell remains in a partially discharged condition for a long period or is not fully charged periodically, then the lead sulphate ($PbSO_4$) formed during discharge is not fully converted back to PbO_2 and Pb. Some of the unconverted lead sulphate gets deposited on the plates. This phenomenon is called *sulphation of plates and the plates are said to be sulphated. The lead sulphate is white in colour, hard, insoluble in acid and is an insulator. Sulphation reduces the active surface of the plates which raises the internal resistance of the cell. The result is that capacity as well as efficiency of the cell is reduced.

Cure. Sulphation of plates is undesirable and must be avoided. If the sulphation is slight, we can eliminate it by over-charging the battery using a low current for several days. Badly sulphated batteries can be given extra useful life by charging them with a current of one or two amperes for 50-100 hours.

---

*   Sulphation may be caused to a limited extent due to higher battery temperature (>40°C) and too much concentration of H_2SO_4 in the electrolyte.

10.22. Methods of Charging Batteries

For charging batteries, direct current is essential and it may be obtained from a d.c. generator. In case the available supply is a.c., it is converted into d.c. by such means as motor-generator set, rotary convertor set or rectifier. In order to charge a new or a discharged battery, direct current is passed through the battery in a direction opposite to that in which it supplied current. The battery is considered fully charged when all the cells are gassing freely and specific gravity of the electrolyte and the terminal voltage of all the cells remain constant for three consecutive readings taken at hourly intervals. There are two main methods of battery charging *viz* .

(*i*) Constant current charging   (*ii*) Constant voltage charging

(*i*) **Constant current charging.** In this method, the charging current is kept constant throughout the charging period by varying the supply voltage to overcome the increased back e.m.f. of the cells on being charged. The terminal voltage V of a battery on charge is given by ;

$$V = E + Ir$$

As the battery charges, its e.m.f. E increases progressively. Therefore, in order to maintain the charging current constant, the supply voltage must be increased progressively to compensate for the increased e.m.f. In case the d.c. shunt generator is used to charge the battrey, the necessary voltage change can be obtained by varying the field current of the generator. The drawback of this method is that it fails to take into account the state of charge of the battery because a high charging rate is required in the beginning when the battery is fully discharged and a low rate of charge is desirable as the battery approaches full state of charge. Further, this method requires longer time for complete charging.

(*ii*) **Constant voltage charging.** In this method, the charging voltage is kept constant throughout the charging process. A variable resistance is connected in series with the circuit as shown in Fig. 10.13. As the battery charges, its e.m.f. increases progressively and series resistance is reduced progressively to maintain the correct charging current. This method is the most common method of charging lead-acid batteries. It is because this method reduces the charging time by about 50% and increases cell capacity by about 20%.

10.23. Battery Charging Circuit

Fig. 10.13 shows the battery charging circuit. A d.c. source of suitable magnitude is connected in series with a rheostat R, ammeter and the battery to be charged. Ensure that polarity is correct *i.e.* positive terminal of d.c. source should be connected to the positive terminal of the battery. The charging current is adjusted to the required value with the help of rheostat. As the charging process proceeds, the terminal voltage of the battery rises but the charging current is kept constant by adjusting the value of rheostat R. The terminal voltage of the battery and specific gravity of electrolyte are checked at regular intervals of time. When the terminal voltage ceases to rise, the specific gravity of electrolyte reaches the value 1.28 and there is enough gassing at the plates, the battery is fully charged. It is then taken out of the charging circuit. The entire charging process may take several hours.

Calculations. When the battery is being charged, its e.m.f. acts in opposition to the applied voltage. The applied voltage V sends a charging current I against the back e.m.f. E_b of the battery. The input power is VI but the power being supplied to the battery is $E_b I$. The power $E_b I$ is converted into chemical energy which is stored in the battery.

Charging current, $I = \dfrac{V - E_b}{R + r}$

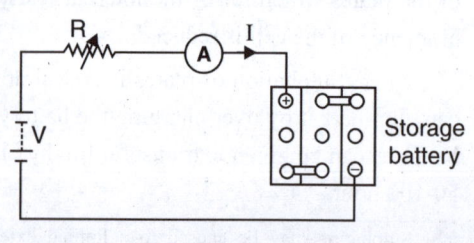

Fig. 10.13

where R = resistance of rheostat in the circuit

r = internal resistance of the battery

The charging current is kept constant throughout (by adjusting R) except towards the finish of charge.

The following points may be kept in mind during charging :

(*i*) When the battery is being charged, the vents must be open so that the gases (H_2 and O_2) may be able to escape. Otherwise the case may be cracked.

(*ii*) *The mixture of hydrogen and oxygen is explosive.* Therefore, care must be taken not to carry an open flame or lighted cigarette near a battery being charged.

(*iii*) The charging current should be such that battery temperature does not exceed 40°C and that violent gassing does not take place. Instead of a constant charging current, the usual practice is to charge the battery at a tapered rate *i.e.* at high rate at first but at a gradually reduced rate as the battery becomes fully charged.

(*iv*) After charging, water should be added to compensate for the loss of water by gassing and evaporation. The level of the electrolyte should be 1 cm above the tops of the plates. If water is not added, the excessive concentration of H_2SO_4 may char the separators, causing permanent damage to the battery.

10.24. Important Points About Charging of Lead-Acid Batteries

In order to ensure normal life for a battery, it should be maintained in the charged condition. While charging a battery, the following points may be kept in view :

(*i*) The charging source must be d.c. one. If a.c. supply is available, it must be converted to d.c. before being applied to the battery.

(*ii*) During charging, it should be ensured that polarity is correct. The polarity is correct when the positive terminal of d.c. charging source is connected to the positive terminal of the battery.

(*iii*) The charging voltage must be more than the e.m.f. of the battery to be charged. Approximately 2.5 volts per cell should be applied. For example, if a battery to be charged has 6 cells, then applied direct voltage should be about 15 volts.

(*iv*) The charging current should be set at proper value.

Charging rate. What should be the value of charging current? The answer is that it may be as high as the battery can take without excessive 'gassing' or heat. Too great a charging current may produce excessive heat. This will cause plates to buckle and short circuit may occur. Also the water of electrolyte will be lost due to electrolysis and evaporation, raising the concentration of sulphuric acid left behind. The strong acid then may char the separators, causing permanent damage to the battery. The temperature of the electrolyte should not be permitted to rise above 40°C. Efforts should be made to use the charging current value recommended by the manufacturer. In case it is not available, one of the following thumb rules may be applied :

(*i*) *The charging current should be 1 A for every positive plate of a single cell.* Thus, if the cell contains 13 plates, six of them will be positive. The charging current for the battery should, therefore, be 6A.

(*ii*) *The charging rate should be such that full charge can be obtained in 8 hours.* Thus 100 Ah battery should be charged at the rate of 100/8 = 12.5 A. This will ensure the maximum life of the battery.

10.25. Effects of Overcharging

Overcharging a lead-acid battery is very harmful due to the following reasons :

(*i*) It results in the loss of water.

    (*ii*) It results in shedding of active meterial due to excessive gassing.

    (*iii*) It produces heat which adversely affects the life of the various components such as positive plates and separators.

    (*iv*) The plates can buckle.

    (*v*) The life of the battery is reduced.

Example 10.16. *A battery of 40 cells is to be charged from a 180 V supply. The internal resistance of each cell is 0.05 Ω and the charging current is to be 4 A. If the average e.m.f. of each cell during charge is 2.5 V, what should be the value of series resistor? If the cost of energy is Rs. 2/kWh, what will it cost to charge the battery for 8 hours and what percent of energy supplied will be used in the form of heat?*

    Solution. Applied voltage, $V = 180$ volts

$$\text{Charging current, } I = 4\text{A}$$

$$\text{Back e.m.f. of battery, } E_b = 40 \times 2.5 = 100 \text{ volts}$$

$$\text{Internal resistance of battery, } r = 40 \times 0.05 = 2 \,\Omega$$

Let R ohms be the required value of series resistor.

$$\text{Charging current, } I = \frac{V - E_b}{R + r}$$

or $\qquad\qquad\qquad 4 = \dfrac{180 - 100}{R + 2} \qquad \therefore \quad R = \mathbf{18\,\Omega}$

$$\therefore \qquad \text{Input energy for 8 hours} = 180 \times 4 \times 8 = 5760 \text{ Wh} = 5.76 \text{ kWh}$$

$$\text{Cost of energy} = \text{Rs } 2 \times 5.76 = \mathbf{Rs\ 11.52}$$

$$\text{Total resistance} = (R + r) = (18 + 2) = 20 \,\Omega$$

$$\text{Power wasted as heat} = I^2(R + r) = 4^2 \times 20 = 320 \text{ W}$$

$$\text{Input power} = 180 \times 4 = 720 \text{ W}$$

$$\therefore \qquad \text{Percentage waste} = \frac{\text{Power wasted}}{\text{Power supplied}} \times 100 = \frac{320}{720} \times 100 = \mathbf{44.45\%}$$

Example 10.17. *It is desired to charge a 12-V car battery at 6A from a 230-V d.c. source. The d.c. source and battery are connected in series with a group of 60-watt, 220-V bulbs in parallel. How many lamps are required for the purpose?*

    Solution. Charging current, $I = \dfrac{V - E_b}{R}$ (Neglecting internal resistance)

or $\qquad\qquad\qquad 6 = \dfrac{230 - 12}{R}$

$\therefore$ Required circuit resistance, $R = 218/6 = 36.33 \,\Omega$

$$\text{Resistance of each bulb} = (220)^2/60 = 806.67 \,\Omega$$

Let N be the required number of bulbs in parallel. It means that N bulbs in parallel should give an equivalent resistance of $36.33 \,\Omega$.

$\therefore$ $\qquad\qquad\qquad\qquad 36.33 = 806.67/N \quad$ or $\quad N = 806.67/36.33 = \mathbf{22\ bulbs}$

Example 10.18. *A series battery of 12 lead accumulators, each of e.m.f. 2 V and internal resistance (1/24) Ω, is to be charged from 240 V d.c. mains. If the charging current is not to exceed 3A, what percentage of energy taken from the mains would be wasted?*

    Solution. In order that circuit current is 3 A, a series resistance R is required in the circuit.

Internal resistance of battery, $r_B = 12 \times r = 12 \times \dfrac{1}{24} = 0.5 \,\Omega$

Total circuit resistance, $R_T = R + r_B = (R + 0.5)\,\Omega$

Net voltage in the circuit $= 240 - 12 \times 2 = 216$ V

$\therefore$ Circuit current $= \dfrac{\text{Net voltage}}{(R + 0.5)}$ or $3 = \dfrac{216}{R + 0.5}$ $\therefore$ $R = 71.5\,\Omega$

Energy taken from mains $= VIt \ldots t$ is the time

$\qquad = 240 \times 3 \times t = 720\,t$

Energy wasted $= I^2 Rt + I^2 r_B t = I^2 (R + r_B)\,t = (3)^2\,t\,(71.5 + 0.5) = 648\,t$

$\therefore$ Percentage wasted $= \dfrac{648\,t}{720\,t} \times 100 = \mathbf{90\%}$

Example 10.19. *A battery of e.m.f. 12 V and internal resistance 0.5 Ω is connected across a resistor in series with a low resistance ammeter. When the circuit is switched on, the steady reading of the ammeter is 1 A. Calculate: (i) rate of consumption of chemical energy of the battery, (ii) rate of dissipation of energy inside the battery, (iii) rate of dissipation of energy in the resistor, (iv) power output of the source.*

Solution. Here, $E = 12$ V ; $I = 1$ A ; $r = 0.5\,\Omega$

(*i*) Rate of consumption of chemical energy, $P_1 = EI = 12 \times 1 = \mathbf{12\ W}$

(*ii*) Rate of energy dissipation inside the battery, $P_2 = I^2 r = (1)^2 \times 0.5 = \mathbf{0.5\ W}$

(*iii*) Rate of energy dissipation in the resistor $= P_1 - P_2 = 12 - 0.5 = \mathbf{11.5\ W}$

(*iv*) Power output of the source $= P_1 - P_2 = \mathbf{11.5\ W}$

Example 10.20. *Thirty-five lead-acid cells, each of discharging capacity 100 Ah at the 10-hour rate are to be fully charged at constant current for 8 hours. The d.c. supply is 120 V, the ampere-hour efficiency is 80% and the e.m.f. of each cell at the beginning and end of charge is 1·9 V and 2·6 V respectively. Calculate the maximum and minimum value of the necessary resistance. Ignore internal resistance of the cells.*

Solution. The output of each cell is 100 Ah. Since Ah efficiency is 80%, the input ampere hours will be $= 100/0.8 = 125$. As the number of hours of charge is 8,

$\therefore$ Charging current, $I = 125/8 = 15.62$ A

Beginning of charge. Back e.m.f. of battery, $E_b = 35 \times 1.9 = 66.5$ volts

Charging current, $I = (V - E_b)/R$

$\therefore$ External resistance, $R = \dfrac{V - E_b}{I} = \dfrac{120 - 66.5}{15.62} = \mathbf{3.42\ \Omega}$

The value of external resistance will be maximum at the beginning of charge because then back e.m.f. of the battery is minimum.

End of charging. Back e.m.f. of battery, $E_b = 35 \times 2.6 = 91$ volts

$\therefore$ External resistance, $R = \dfrac{V - E_b}{I} = \dfrac{120 - 91}{15.62} = \mathbf{1.86\ \Omega}$

Example 10.21. *A shunt generator is used to charge a battery of 100 cells in series. If each cell has a terminal p.d. of 2.5 V at the completion of charge and is of internal resistance 0·001 Ω, calculate the e.m.f. to be generated to give a charging current of 15 A at the end of charge. Assume that the armature and field resistances are 0.2 Ω and 200 Ω respectively and the cables connecting the generator to the battery have a resistance of 0.6 Ω.*

Solution. Fig. 10.14 shows the whole arrangement. The charging current $I = 15$ A.

Back e.m.f. of battery, $E_b = 100 \times 2.5 = 250$ volts

Battery terminal voltage, $V = 250 + $ Drop in battery

$\qquad = 250 + (15 \times 0.001 \times 100) = 251.5$ volts

Voltage at generator terminals = 251.5 + Drop in connecting cables
= 251.5 + 15 × 0.6 = 260.5 volts

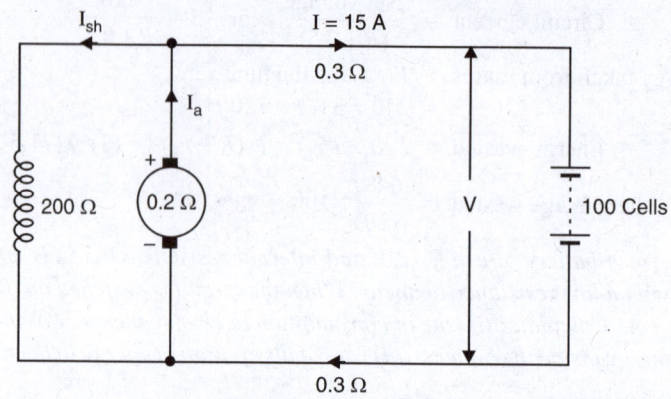

Fig. 10.14

Shunt current, I_{sh} = 260.5/200 = 1.3 A

Armature current, I_a = 15 + 1.3 = 16.3 A

∴   E.M.F. of generator = 260.5 + 16.3 × 0.2 = **263.76 volts**

Example 10.22. *A storage battery consists of 55 cells in series. Each has an e.m.f. of 2V and an internal resistance of 0.001 Ω. The full load current per cell is 0.04 A/cm^2 of positive plate surface and each cell has 10 positive plates each measuring 12 cm × 12 cm. Find the full-load terminal voltage and the power wasted in the battery if the connectors have a total resistance of 0.02 Ω.*

Solution. A commercial storage cell has odd number of plates. The number of negative plates is always one more than the number of positive plates; the outside plates being negative. Therefore, both sides of a positive plate are utilised and the areas of both sides of positive plates are taken into consideration.

∴   Total area (both sides) of 10 positive plates

$$= 2 × 10 × (12 × 12) = 2880 \text{ cm}^2$$

Full-load current, I = 0.04 × 2880 = 115.2 A

Battery e.m.f., E_b = 55 × 2 = 110 volts

Total resistance, R_T = 0.001 × 55 + 0.02 = 0.075 Ω

Total voltage drop in battery and across connectors

$$= IR_T = 115.2 × 0.075 = 8.64 \text{ volts}$$

Battery terminal voltage, V = $E_b - IR_T$ = 110 − 8.64 = **101.36 volts**

Power wasted in battery = $I^2R_T = (115.2)^2 × 0.075 =$ **995 W**

Example 10.23. *The relation between the discharge current I and the time of discharge t of a battery is given by $I^n t = K$ where n and K are constants. If a battery gives 20A for 8 hrs and 40 A for 2.6 hrs, find for how long it will give 30A.*

Solution. $I^n t = K$. Let t hours be the required time.

Now, $20^n × 8 = K$ (i) ; $40^n × 2.6 = K$ (ii) ; $30^n × t = K$ (iii)

From eqs. (i) and (ii), $40^n × 2.6 = 20^n × 8$

or $2^n = \dfrac{8}{2.6} = 3.077$

or $\qquad n \log_e 2 = \log_e 3.077 \qquad \therefore \quad n = \dfrac{\log_e 3.077}{\log_e 2} = 1.622$

From eqs. (ii) and (iii), $\quad 30^n \times t = 40^n \times 2.6$

$$\therefore \qquad t = \left(\frac{40}{30}\right)^n \times 2.6 = \left(\frac{40}{30}\right)^{1.622} \times 2.6 = \textbf{4.15 hr.}$$

i.e. the battery will give 30 A for 4.15 hr.

(**Tutorial Problems**)

1. A battery of accumulators of e.m.f. 50 volts and internal resistance $2\,\Omega$ is charged on 100 V direct mains. What series resistance will be required to give a charging current of 2 A? If the cost of energy is 70 paise per kWh, what will it cost to charge the battery for 8 hours and what percentage of energy supplied will be wasted in the form of heat? **[23 Ω ; Rs. 1.12 ; 50%]**

2. A battery of 4 lead-acid cells in series is to be charged and the only available source is 210 V d.c. supply. The desired charging current is 0.5 A and each cell on charge has an e.m.f. of 2.5 volts. If an electric lamp is used as the controlling resistance for the circuit, what voltage and size of the lamp will be necessary? **[200 V ; 100 W]**

3. A lead-acid battery of 50 cells is to be charged at 25 A from a 200 V d.c. supply using a series resistor. If the e.m.f. of each cell is 1.95 volts at the commencement of charge and 2.55 volts at the end of the charge, find the range of series resistance value required. The internal resistance of each cell may be assumed to remain constant at $0.002\,\Omega$ and the resistance of the connecting leads may be neglected. **[4 Ω to 2.8 Ω]**

4. Determine the variation in the applied voltage required to charge a battery of 100 lead-acid cells if charged from an e.m.f. of 1.8 volts to 2.6 volts per cell with a constant charging current of 10 A. The internal resistance of each cell is $0.01\,\Omega$.

    [Initial voltage = 190 V; Final voltage = 270 V ; Variation = 80 V]

10.26. Care of Lead-acid Batteries

The average life of a lead-acid battery is two to four years, depending upon its quality and the kind of care exercised in its use. To get the maximum life out of a lead-acid battery, the following procedures are recommended :

(*i*) **Keep the top of battery clean and dry** at all times to prevent corrosion and leakage of current. If corrosion has started, the battery may be cleaned by the use of a stiff brush and then wiped with a rag moistened in a solution of household ammonia.

(*ii*) **Keep the electrolyte at the proper level.** In the normal operation of a lead-acid battery, a certain amount of water is lost from the electrolyte (H_2SO_4) due to evaporation and gassing. The level of the electrolyte should be checked at regular intervals of one or two weeks and when found to be too low, it should be brought to the proper level by adding water free from impurities. It should always be 1 cm above the tops of plates and separators as this prevents decay of separators. As only water evaporates, it is never necessary to add acid during normal operation of a battery. Be sure to replace and tighten the vent caps.

(*iii*) **Take frequent hydrometer reading** to ascertain the state of charge of the battery. It is advisable to take these readings at the same time the level of electrolyte is being checked. When the specific gravity of the electrolyte drops to about 1.18, the battery should be recharged. Allowing the battery to remain in a low charged condition will produce an excess of lead sulphate ($PbSO_4$) on the plates. This may reduce the life of the battery.

(*iv*) **Do not charge the battery at a high rate.** Charging at too high a rate causes excessive heat and gassing and also a permanent loss in some of the active material of the plates. The battery should be charged at such a rate (See Art. 10.24) that battery temperature does not exceed 40°C and that excessive gassing does not take place.

(*v*) **Do not leave the battery in a discharged condition for a long period.** If left in a discharged condition for a long period, a coating of hard lead sulphate may form on the plates. This sulphate coating cannot be converted to active materials and the battery is permanently damaged. This action is commonly called **sulphation** and may also be caused due to excessive heat and too much concentration of sulphuric acid in the electrolyte.

(*vi*) **If a lead-acid battery is to be stored for several months,** the level of the electrolyte should be periodically checked. After intervals of 4 to 6 weeks, the battery should be given a freshening charge so that it will be kept fully charged at all times.

(*vii*) **Do not short circuit the battery.** Since the internal resistance of lead-acid battery is very small (typical value being 0·01 Ω), a short-circuit will give a damaging current of several hundred amperes. This may cause the plates to buckle, thus ruining the battery.

10.27. Applications of Lead-acid Batteries

The chief application of lead-acid batteries is at places where a d.c. generator cannot be conveniently installed. Such batteries are employed whenever low-voltage, high current d.c. source is required. Some of the important applications of lead-acid batteries are given below :

(*i*) *Lead-acid batteries are extensively used in automobiles.* A d.c. generator, battery and load are connected in parallel. Since the internal resistance of a lead-acid battery is very low, it provides a large output current required for *starting the engine of the car. In addition, the battery furnishes the power for lights, radio *etc.* when the engine is stopped. When the engine is running, it operates the d.c. generator which takes over the duties of the battery. The generator also charges up the battery so that it will be ready when needed.

(*ii*) Railway train lighting system is similar to the above mentioned system. When the train is in motion, axle-driven d.c. generators furnish power for lighting, fans *etc.* When the train runs at slow speeds or is stopped, lead-acid batteries take over the duties of the generators.

(*iii*) In a.c. generating plants, lead-acid batteries are used to energise the control apparatus *e.g.* switchgear. During shut down of generators, these batteries are used to supply emergency lights.

(*iv*) Hospitals and other places (*e.g.* theatres, banks *etc.*) where continuous supply is absolutely essential often use batteries as an emergency supply.

(*v*) These are used for lighting purposes in remote rural areas where there are no power lines. The batteries are kept in charge by means of a d.c. generator. When the generator fails, the batteries supply the load.

10.28. Voltage Control Methods

A lead-acid battery is often required to give approximately constant voltage to a circuit as in the case of a generating station stand-by battery. We know that a lead-acid battery has 2.2 V/cell when fully charged and as the battery discharges, the voltage per cell decreases. Therefore, it is necessary to use some means to keep the output voltage of the battery within permissible limits. We discuss three methods of controlling the output voltage of the battery.

(*i*) **Rheostatic control.** In this method, a variable resistance *R* is inserted in series with the battery as shown in Fig. 10.15. There is a certain amount of resistance in the circuit when

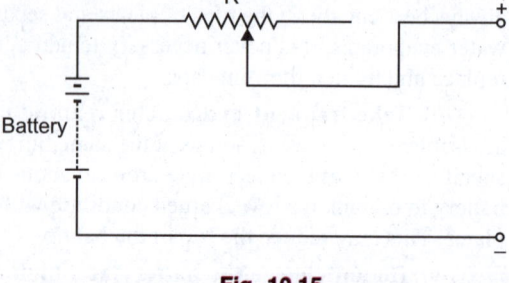

Fig. 10.15

---

*   The starting motor of an automobile may require more than 300 A at start. Because of its low internal resistance, a lead-acid battery can furnish currents of such large magnitude but only for a few seconds at a time.

the battery is fully charged. As the battery discharges, the resistance is gradually cut out of the circuit. The use of rheostat for controlling the battery voltage is objectionable particularly in large capacity installations where I^2R losses would be considerable. Therefore, this method is used for batteries of small capacity and where the cost of energy is low so that energy losses in the rheostat are not objectionable.

(*ii*) **End-cell control.** In this method, the battery voltage is controlled by changing the number of cells. Here, group of cells selected for this control are situated at one end of the battery and are called **end cells**. By suitable switching arrangement, the end cells are cut in or out of the battery circuit to control the battery voltage. Fig. 10.16 shows a simple circuit for end-cell control. The sliding end switch is made of two parts A and B separated by insulating material through the protective resistance R. This arrangement permits the end switch to operate without opening the circuit or shorting the cells during its passage from one cell to another. The resistance R limits the current in the short circuited cell at the instant the end switch passes from one cell to the next. By moving the end switch, the desired number of end cells can be cut in or out of the battery circuit to control the battery voltage.

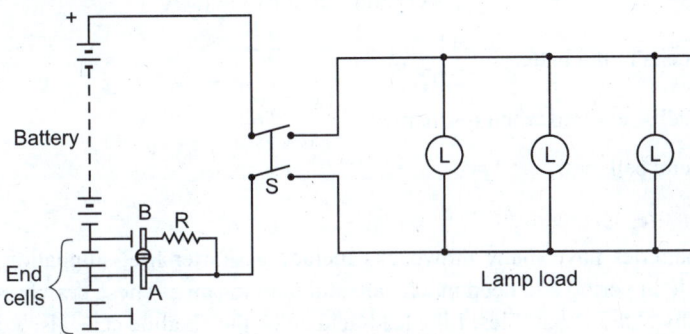

Fig. 10.16

Number of end cells. Suppose it is desired to get a constant voltage of 100 volts from a lead-acid battery. We know that a fully charged battery has a terminal voltage of about 2.1 V/cell while a fully discharged battery has a terminal voltage of about 1.8 V/cell.

∴   Number of cells required when the battery is fully charged

$$= \frac{100}{2.1} \simeq 48 \text{ cells}$$

Number of cells required when the battery is fully discharged

$$= \frac{100}{1.8} \simeq 56$$

∴   Number of end cells required = 56 − 48 = 8

Therefore, in this case, we need 8 end cells. When the battery discharges, its terminal voltage decreases. However, we can add one or more end cells in series with the battery to maintain the battery voltage at 100 volts.

(*iii*) **Booster control.** A booster is a low voltage generator inserted in series in a circuit to change its voltage. A battery booster is a small d.c. generator connected in series with the battery. The function of the battery booster is to add or subtract the voltage in the battery circuit. The amount of voltage produced by the booster can be regulated by the rheostat in its field circuit. By reversing the field of the booster (*i.e.* d.c. generator), it may be used to discharge the battery.

Example 10.24. *Two hundred and twenty lamps of 100 W each are to be run on a battery supply at 110 V. The cells of the battery when fully charged have an e.m.f. of 2.1 V each and when discharged 1.83 V each. If the internal resistance per cell is 0.00015 Ω, find (i) the number of cells in the battery and (ii) the number of end cells. Take the resistance of connecting wires as 0.005 Ω.*

Solution. Total power of lamps is given by ;

$$P = 220 \times 100 = 22 \times 10^3 \text{ W}$$

Current drawn by lamps, $I = \dfrac{P}{110} = \dfrac{22 \times 10^3}{110} = 200$ A

Voltage drop in connecting wires $= I \times 0.005 = 200 \times 0.005 = 1$V

$\therefore$ Battery supply voltage $= 110 + 1 = 111$ V

Terminal voltage/cell when fully charged and supplying load

$$= 2.1 - 200 \times 0.00015 = 2.07 \text{ V}$$

Terminal voltage/cell when discharged and supplying load

$$= 1.83 - 200 \times 0.00015 = 1.8 \text{ V}$$

(*i*) No. of cells in the battery $= \dfrac{111}{2.07} \simeq \mathbf{54}$

(*ii*) No. of cells required when discharged $= \dfrac{111}{1.8} = 62$

$\therefore$ No. of end cells $= 62 - 54 = \mathbf{8}$

10.29. Alkaline Batteries

Lead-acid batteries have many drawbacks including shorter life, sulphation of plates if left discharged for a long period and need much care and maintenance. These drawbacks are overcome to a large extent by alkaline batteries. Like lead-acid cells, the alkaline cells also consist of positive and negative plates immersed in an electrolyte (alkali). The plates and the electrolyte are placed in a suitable container. The two types of alkaline cells in general use are :

(*i*) Nickel-iron cell or Edison cell

(*ii*) Nickel-cadmium cell

10.30. Nickel-Iron Cell or Edison Cell

Nickel-iron cell was developed by American scientist Thomas A. Edison in 1909. It has lesser weight and longer life than that of a lead-acid cell. As a result, these cells are very suitable for portable work. The e.m.f. of this cell is about 1.36 V.

Construction. When nickel-iron cell is in the charged condition, the active material on positive plates is $Ni(OH)_4$ and that on the negative plates is iron (Fe). The positive and negative plates are held in a nickel-plated steal container; the plates being insulated from each other by hard rubber strips. The container contains 21 percent solution of KOH (electrolyte) to which is added a small amount of lithium hydrate (LiOH) for increasing the capacity of the cell.

(*i*) The positive plates are in the form of perforated nickel-plated steel tubes filled with $Ni(OH)_4$ and flakes of metallic nickel; the addition of flakes of nickel reduces the internal resistance of the cell.

(*ii*) The negative plates are also in the form of perforated nickel-plated steel tubes filled with powdered iron oxide and a little mercuric oxide. The purpose of mercuric oxide is to decrease the internal resistance of the cell.

Chemical changes. The molecules of electrolyte (KOH) dissociate into K^+ and OH^- ions.

$$KOH \longrightarrow K^+ + OH^-$$

(*i*) During **discharging**, the K^+ ions move towards the positive plate (anode) and reduce $Ni(OH)_4$ to $Ni(OH)_2$. The OH^- ions travel towards the negative plate (cathode) and oxidise iron. The chemical changes during discharging can be represented by the following equations :

Positive plate : $Ni(OH)_4 + 2K \longrightarrow Ni(OH)_2 + 2KOH$

Negative plate : $Fe + 2OH \longrightarrow Fe(OH)_2$

(*ii*) During **recharging**, the K^+ ions move towards negative plate (cathode) and OH^- ions go to positive plate (anode) causing the following chemical changes :

Positive plate : $Ni(OH)_2 + 2OH \longrightarrow Ni(OH)_4$

Negative plate : $Fe(OH)_2 + 2K \longrightarrow Fe + 2KOH$

The chemical reactions during discharging and recharging can be summed up in a single reversible equation as under :

$$\underset{\substack{\text{Positive} \\ \text{plate}}}{Ni(OH)_4} + KOH + \underset{\substack{\text{Negative} \\ \text{plate}}}{Fe} \underset{\text{Recharge}}{\overset{\text{Discharge}}{\rightleftharpoons}} \underset{\substack{\text{Positive} \\ \text{plate}}}{Ni(OH)_2} + KOH + \underset{\substack{\text{Negative} \\ \text{plate}}}{Fe(OH)_2}$$

It may be observed from the above equation that no water is formed in the reaction. Consequently, the specific gravity of the electrolyte (KOH) remains unchanged during charging or discharging. For this reason, a nickel-iron cell is not damaged if left in a fully discharged condition for a considerable period of time.

Since the electrolyte (KOH) does not undergo any change in specific gravity during charge or discharge, the state of charge of this cell cannot be determined by the specific gravity of the electrolyte. Instead, a voltmeter is employed to ascertain whether the cell is charged up to its rated voltage.

Advantages

(*i*)   Since the tspecific gravity of the electrolyte (KOH) does not change, a nickel-iron cell is not damaged by being left in a fully discharged condition for a considerable period of time.

(*ii*)   It is more rugged and can stand more mechanical and electrical abuse than a lead-acid cell.

(*iii*)   It weighs only about half as much as a lead-acid cell of equivalent capacity.

(*iv*)   It does not produce acid fumes that corrode nearby objects.

(*v*)   Its life is much longer than that of a lead-acid cell.

(*vi*)   It can be discharged at a high rate for long periods without danger to the battery and can be recharged safely at a rapid rate.

(*vii*)   It can withstand high temperatures better than a lead-acid cell.

Disadvantages

(*i*)   It is costlier than a lead-acid cell.

(*ii*)   The e.m.f. of a nickel-iron cell is about 1·2 V against 2 V of the lead-acid cell. A supply of 12 V needs a battery of 10 nickel-iron cells in series but only 6 lead-acid cells in series.

(*iii*)   It has higher internal resistance than the lead-acid cell and cannot provide a large current. For this reason, it is unsuitable for automobile starting service.

(*iv*)   The efficiencies of a nickel-iron cell are lower than that of a lead-acid cell. The ampere-hour and watt-hour efficiencies of such a cell are about 80% and 65% respectively.

Applications. Due to their high mechanical strength, lightness, ability to withstand arduous conditions and free from acid fumes, nickel-iron batteries have been found to be more satisfactory for traction purposes, mine locomotives, *submarines etc. They are particularly suitable in cases where exceptional vibrations are encountered as on motor cycles etc.

---

*   When a submarine is submerged, it cannot use those engines which consume air *e.g.* petrol engine or diesel engine. Hence it runs on electric motors operated by nickel-iron batteries. The lead-acid battery is not suitable because the acid fumes may corrode the equipment. On the surface, the submarine is run by diesel engines. These engines also operate a d.c. generator which charges the batteries.

10.31. Electrical Characteristics of Nickel-Iron Cell

Fig. 10.17 shows the charging and discharging curves of a nickel-iron cell. Note that shape of the curves is similar to those for the lead-acid cell.

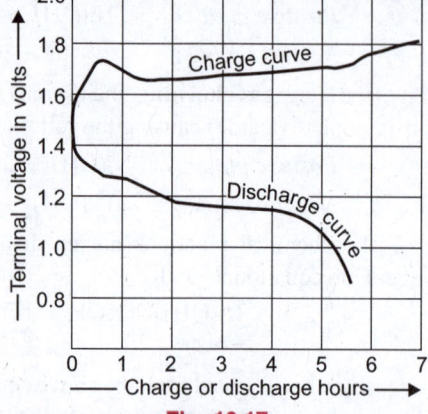

Fig. 10.17

(*i*) The e.m.f. of a nickel-iron cell is about 1.2 V against about 2V for lead-acid cell.

(*ii*) Since the internal resistance of a nickel-iron cell is greater than that of a lead-acid cell, the efficiencies (η_{Ah} and η_{Wh}) of nickel iron-cell are lower than those of lead-acid cell.

(*iii*) The e.m.f. of a nickel-iron cell increases slightly with increase in temperature and *vice-versa*.

(*iv*) The ampere-hour capacity of a nickel-iron cell increases appreciably with the increase in temperature and *vice-versa*. Therefore, Ah capacity of the cell decreases by an appreciable amount with the decrease in temperature and at about 4°C, the capacity of the cell practically becomes zero, though the cell is fully charged. As a result, nickel-iron cells pose problems in cold weather.

(*v*) The rated capacity of nickel-iron cells generally refers to either 5-hour or 8-hour discharge rate.

(*vi*) The internal resistance of a nickel iron cell is about 5 times that of a lead-acid cell. For this reason, the cell delivers much smaller current on short-circuit or to a low resistance load. For the same reason, the difference between the charge and discharge voltages of the cell is greater than that of a lead-acid cell as shown in Fig. 10.17.

10·32. Nickel-Cadmium Cell

Nickel-Cadmium cell was developed by Waldermar Junger, a Swede, in 1899. When the nickel-cadmium cell is in the charged condition, the active material on the positive plate is $Ni(OH)_3$ and that on the negative plate is cadmium (Cd). The electrolyte is the same as in the nickel-iron cell *i.e.* potassium hydroxide (KOH). When the cell is discharged, the positive plate is converted to $Ni(OH)_2$ and the negative plate to $Cd(OH)_2$. When the cell is recharged, the chemical process is reversed *i.e.* positive plate is converted to $Ni(OH)_3$ and negative plate to Cd. The equation of chemical reaction is :

$$\underset{\substack{\text{Positive} \\ \text{plate}}}{2\,Ni(OH)_3} + \underset{\substack{\text{Negative} \\ \text{plate}}}{Cd} + KOH \underset{\text{charge}}{\overset{\text{discharge}}{\rightleftharpoons}} \underset{\substack{\text{Positive} \\ \text{plate}}}{2\,Ni(OH)_2} + \underset{\substack{\text{Negative} \\ \text{plate}}}{Cd(OH)_2} + KOH$$

The equation is read from left to right for discharging and from right to left when charging. The following points may be noted in these electro-chemical reactions :

(*i*) There is merely transfer of hydroxyl ions ($2OH^-$) from one plate to the other. Thus during discharging, hydroxyl ions from the positive plate are transferred to the negative plate to form $Cd(OH)_2$. The reverse happens during charging.

(*ii*) The specific gravity of the electrolyte (KOH) remains unchanged during charge or discharge.

The e.m.f. of a nickel-cadmium cell is about 1·2 V when charged and falls to 1·1 V when discharged. Since the specific gravity of the electrolyte (KOH) remains unchanged during charge or discharge, it cannot be used as an indication to the state of charge of the cell. Instead, a voltmeter is employed to determine whether the cell is charged upto its rated voltage.

Advantages. The nickel-cadmium cell combines the best features of both the lead-acid and Edison types.

(*i*) It is rugged and has a very long active life. Periods upto 20 years of service can be expected from a nickel-cadmium battery.

(*ii*) It can be stored indefinitely in either a discharged or charged state without suffering any ill effects.

(*iii*) A nickel-cadmium (Ni-Cd) battery can be charged in a short time (say 1 hour).

(*iv*) Its internal resistance is very low which means that it can deliver very large current at constant terminal voltage.

(*v*) It can stand peak rates of discharge and charge upto 20 times the normal operating rate.

The chief disadvantage of a nickel-cadmium cell is that it is expensive.

Applications. Since Ni-Cd cells have high discharge rate, they are used in commercial airlines, military aeroplanes and helicopters for starting main engine or auxiliary turbines. Furthermore, because these cells can be charged and discharged hundreds of times, they are very useful in stationary emergency-power installations.

Small nickel-cadmium cells. The obvious advantages of nickel cadmium cells prompted the scientists to develop them in small sizes. Today small-sized nickel-cadmium cells are *used in cordless electric appliances such as electric shavers, hearing aids, photography equipment and in space exploration. A popular size is the 1·2 V, 4 Ah battery which has a 1-hour discharge rate.

The ingredients of the small nickel-cadmium cell are the same as for the larger type. The plates are formed of a woven nickel screen and a paste consisting of the active materials is pressed into the spaces within the screen. A separator is placed between the positive and negative plates and the whole rolled to form a cylinder which is sealed into a small can for protection. The metal can forms the negative terminal of the cell and an insulated metal button at the top is the positive terminal. The main disadvantage of a nickel-cadmium cell, compared to carbon-zinc primary cell (*i.e.* dry cell), is its relatively higher initial cost. Since nickel-cadmium cell is rechargeable, it offers a lower total cost because it does not have to be discarded after it has been exhausted.

10.33. Comparison of Lead-acid Cell and Edison Cell

Particulars	Lead-acid cell	Edison cell
1. Positive plate	PbO_2, lead-peroxide	Nickel hydroxide $Ni(OH)_4$ or NiO_2
2. Negative plate	Spongy lead	Iron oxide
3. Electrolyte	diluted H_2SO_4	KOH
4. Average e.m.f.	2.0 V/cell	1.2 V/cell
5. Internal resistance	Comparatively low	Comparatively higher
6. Efficiency :		
amp-hour	90–95%	nearly 80%
watt-hour	72–80%	about 60%
7. Cost	Comparatively less than alkali cell	Almost twice that of Pb-acid cell. Easy maintenance.
8. Life	Gives nearly 1250 charges and discharges	Five years atleast
9. Strength	Needs much care and maintenance. Sulphation occurs often due to incomplete charge or discharge.	Due to all-steel construction, they are robust, mechanically strong, can withstand vibration, are light, unlimited rates of charge and discharge, can be left discharged, free from corrosive liquids and fumes.

---

*   Previously, primary cells (e.g. dry cells) were used for this purpose.

10.34. Silver-Zinc Batteries

The active material of the positive plate is silver oxide which is pressed into plate and then subjected to heat treatment. The active material for the negative plate is a mixture of zinc oxide and zinc powder and this plate is prepared in the same manner as the positive plate. The charging and discharging of a silver-zinc cell can be represented by the following single reversible equation :

$$\underset{\substack{\text{Positive}\\\text{plate}}}{Ag_2O} + \underset{\substack{\text{Negative}\\\text{plate}}}{Zn} \underset{\text{Charge}}{\overset{\text{Discharge}}{\rightleftharpoons}} \underset{\substack{\text{Positive}\\\text{plate}}}{2Ag} + \underset{\substack{\text{Negative}\\\text{plate}}}{ZnO}$$

For discharge, the equation should be read from left to right and for charge from right to left.

Characteristics. The important characteristics of a silver-zinc cell or battery are :

(*i*) Its specific capacity (*i.e.* capacity per unit weight) is 4 to 5 times greater than that of other types of cells.

(*ii*) The efficiencies (η_{Ah} and η_{Wh}) of a silver-zinc cell are very high.

(*iii*) Silver-zinc batteries can withstand much heavier discharge currents compared to other types of batteries.

(*iv*) Silver-zinc batteries can be operated over a wide range of temperatures ($-20°C$ to $+60°C$).

(*v*) Silver-zinc batteries have small self-discharge.

(*vi*) The cost of a silver-zinc battery or cell is very high.

(*vii*) The energy density of silver-zinc batteries is as high as 150 Wh/kg.

Applications. Silver-zinc batteries are mainly used in military services *viz* communication equipment, portable radar sets and night-vision equipment. They are also used in heavy-weight torpedoes and for submarine propulsion.

10·35. Solar Cells

A solar cell is a device that converts light energy (e.g. sunlight) directly to electrical energy. Fig. 10.18 shows the construction of a simple solar cell. A wafer consisting of a pure silicon (semiconductor) is "*doped" with a specific amount of arsenic (donor impurity). This makes it *N*-type semiconductor. As a result of this, the wafer contains an excess of free electrons. This wafer is coated at its top with a very thin layer of silicon doped with appropriate amount of boron (acceptor impurity). This

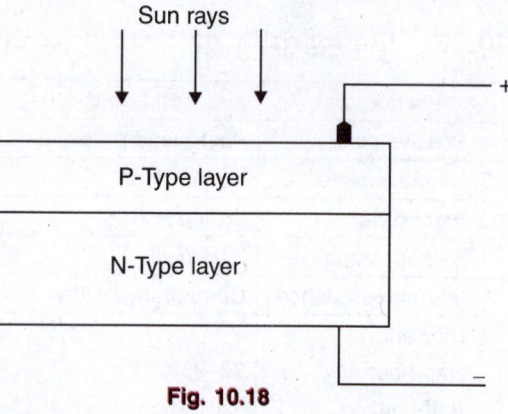

Fig. 10.18

makes the top layer a *P*-type semiconductor and the contact surface becomes a *P-N* junction. When sunlight shines on top *P*-type layer and penetrates into the *N*-type material just below it, the free electrons in *N*-type material receive energy and move across the *P-N* junction into *P*-type material. This movement of charge carriers (*i.e.* holes from *P*-type and free electrons from *N*-type) constitutes electric current. The bottom of the wafer (*N*-type material) and a spot on the *P*-type layer are tinned for the connection leads.

The operating voltage of one solar cell is about 0·39 V and the current may be between 30 and 40 mA. The output power of the cell depends upon the exposed area of the cell and the intensity of light falling upon it. The maximum output with the sun directly overhead on a clear day is 8 or

---

*   The process of adding a suitable impurity to a pure semiconductor is called "doping".

9 mW/cm^2. The *area of the cell cannot be made large because it is difficult to have large silicon crystals. The operating efficiency is about 10%. The life of a solar cell is estimated to be thousands of years. They do not deteriorate when not in use. The chief uses of solar cells are : to provide power for transistor portable radios, to charge nickel-cadmium batteries in satellites, to provide power for clocks and other devices such as aperture control for movie cameras, microwave relay stations etc.

10.36. Fuel Cells

Whenever a fuel such as wood, coal or natural gas burns, it combines with oxygen to create a new substance. The chemical reaction is called *oxidation* and is accompanied by the release of a large amount of energy in the form of heat. The heat may be used to produce steam, which in turn can drive a turbogenerator set. Unfortunately, the efficiency is very low (about 20%) when heat is converted into electrical energy in this way. Can we permit oxidation to take place without actually burning the fuel? This is what is done in a fuel cell. The energy of oxidation appears in the form of electrical energy at the electrodes of the fuel cell. The efficiency of a fuel cell is more than 40%. Clearly, this is an improvement compared to the efficiency obtained by burning the fuel.

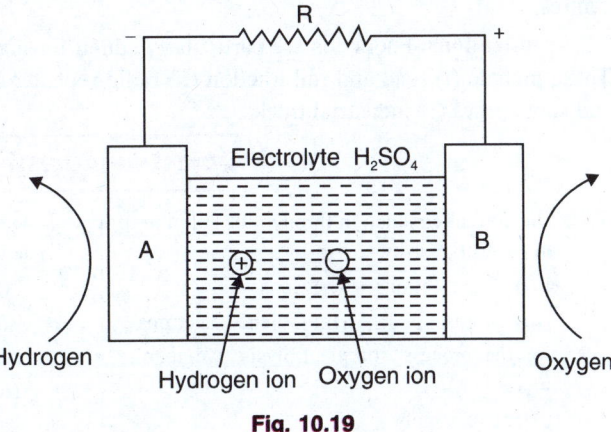

Fig. 10.19

Hydrogen-oxygen fuel cell. When hydrogen burns, it always combines in definite proportions with oxygen. For complete combustion, 1 kg of hydrogen consumes 8 kg of oxygen. In the process about 120×10^6 J of heat is released and the resulting product is water. In hydrogen-oxygen fuel cell, oxidation is permitted to take place without actually burning hydrogen. The energy of oxidation is converted into electrical energy.

Fig. 10.19 shows the various parts of a hydrogen-oxygen fuel cell. It consists of two platinum electrodes A and B immersed in a solution of H_2SO_4 (electrolyte). The hydrogen gas is continuously supplied to **electrode A and oxygen gas to electrode B. An electrical load is connected across the terminals.

Working. When hydrogen gas touches the electrode A, a special reaction takes place, causing every molecule of hydrogen to break up into 2 electrons and 2 positive hydrogen ions. The hydrogen ions move into the electrolyte and slowly travel towards electrode B. The electrons, on the other hand, are captured by electrode A and flow through the load and on towards electrode B. The oxygen molecules that touch electrode B capture electrons coming *via* the load, thereby becoming negative oxygen ions. These oxygen ions move into the electrolyte and combine with hydrogen ions to form water. Water is periodically removed to prevent contaminating the fuel cell. It may be seen that oxidation takes place inside the electrolyte where hydrogen and oxygen ions combine. However, no energy is released during the process; all the energy of oxidation appears as electrical energy at the electrodes. In fact, the electrons released at the electrode A (*i.e.* fuel terminal) are recaptured at the electrode B (*i.e.* oxygen terminal) with the result that a steady current flows through the load. Ideally, the electric power supplied to the load is equal to the thermal power that would be released if the fuel were burnt. There are, however, some losses but the efficiency of even a small fuel cell is more than 40%.

---

*   However, for higher voltages and currents, these cells can be connected in series-parallel.

**   The electrode in contact with fuel (*i.e.* hydrogen in this case) is always negative while that in contact with oxidising agent (*i.e.* oxygen in this case) is positive.

The voltage of a hydrogen-oxygen fuel cell is about 0·8 V. However, the current supplied by such a cell is very high—of the order of kA. Hence fuel cells, like batteries are basically high-current, low voltage d.c. devices. Higher voltages are obtained by connecting cells in series.

Advantages. It may be seen that fuel cell is basically a primary cell in which the reacting agents are fed continuously into the enclosure while the unwanted products (*e.g.* water in hydrogen-oxygen cell) are removed.

(*i*) It is not a thermal machine like a steam generator. Therefore, it has much higher efficiency.

(*ii*) It can be built in modular form and units can be added as the need arises.

(*iii*) It is noise-free and pollution-free. Consequently, it can be installed in the heart of urban centres.

Applications. Fuel cells are particularly suited for low voltage and high current applications. These include (*i*) road and rail traction (*ii*) radio repeater stations (*iii*) spaceships (*iv*) naval crafts and submarines (*v*) industrial trucks.

Objective Questions

1. The current conduction through the solution of an electrolyte is by
   (*i*) free electrons   (*ii*) ions
   (*iii*) atoms   (*iv*) valence electrons

2. For the process of electrolysis, we require ....
   (*i*) d. c. supply
   (*ii*) a.c. supply
   (*iii*) varying voltage
   (*iv*) both d.c. and a.c. supply

3. A positively charged atom is sometimes called ....
   (*i*) donor atom   (*ii*) acceptor atom
   (*iii*) cation   (*iv*) anion

4. The mass of an element deposited or liberated at an electrode during electrolysis is .... the quantity of electricity passed through the electrolyte.
   (*i*) directly proportional to
   (*ii*) inversely proportional to
   (*iii*) independent of
   (*iv*) none of the above

5. The E.C.E. of the copper is 304×10^{-6} gm/C. If during electrolysis, 2C of electricity is passed through the solution of $CuSO_4$, then, mass of copper deposited on cathode is ....
   (*i*) 2 gm   (*ii*) 152×10^{-6} gm
   (*iii*) 608×10^{-6} gm   (*iv*) 304×10^{-6} gm

6. Out of the following, E.C.E. of .... is largest.
   (*i*) copper   (*ii*) iron
   (*iii*) nickel   (*iv*) silver

7. During electrolysis, mass of an element liberated at the electrode is .... of the element.
   (*i*) directly proportional to the valency
   (*ii*) inversely proportional to atomic weight
   (*iii*) directly proportional to chemical eq. wt.
   (*iv*) none of the above

8. The relation between E.C.E. (*Z*) and chemical equivalent weight (*E*) is that ....
   (*i*) E/Z is constant for all elements
   (*ii*) E/Z is different for different elements
   (*iii*) E^2/Z is constant for all elements
   (*iv*) E/Z^2 is constant for all elements

9. When the same quantity of electricity is passed through silver and copper voltameters in series, the masses of silver (Ag) and copper (Cu) deposited on the respective cathodes will be in the ratio of their ....
   (*i*) atomic weights (*ii*) valencies
   (*iii*) chemical equivalent weights
   (*iv*) none of the above

10. The e.m.f. of a cell does not depend upon ....
    (*i*) nature of electrolyte
    (*ii*) nature of material of electrodes
    (*iii*) concentration of electrolyte
    (*iv*) size and spacing of electrodes

11. If two different elements in the electrochemical series are used as the electrodes in a cell, the one higher in the series ....
    (*i*) will be positive plate
    (*ii*) will be negative plate
    (*iii*) can be positive or negative plate
    (*iv*) none of the above

12. The major drawback of a primary cell is that
    ....
    (i)   chemical action is not reversible
    (ii) chemical action is reversible
    (iii) the electrolyte used is very costly
    (iv) it is not portable

13. The efficiency of a primary cell is about ....
    (i)   25 %   (ii) 15 %
    (iii) 70 %   (iv) 35 %

14. The most commonly used cell is ....
    (i)   lead-acid cell   (ii) nickel-iron cell
    (iii) nickel-cadmium cell
    (iv) fuel cell

15. In practice, the state of discharge of a lead-acid
    cell is determined by ....
    (i)   e.m.f. of the cell
    (ii) specific gravity of the electrolyte
    (iii) colour of plates of the cell
    (iv) none of the above

16. When a lead-acid cell is fully charged, ...
    (i)   anode is converted into Pb
    (ii) cathode is converted into PbO_2
    (iii) specific gravity of H_2SO_4 rises to about
          1·28
    (iv) both plates become red

17. A multiplate construction is used in a lead-acid
    cell to ....
    (i)   increase the capacity of the cell
    (ii) increase the e.m.f. of the cell
    (iii) increase the internal resistance of the cell
    (iv) facilitate cell connections

18. The commercial lead-acid cell has 15 plates.
    The number of negative plates will be ....
    (i)   7   (ii) 9
    (iii) 10   (iv) 8

19. The life of the positive plates of a lead-acid cell
    is roughly .... that of negative plates.
    (i)   the same as   (ii) half
    (iii) twice   (iv) thrice

20. If the specific gravity of electrolyte (H_2SO_4) in
    a lead-acid cell increases, the internal resistance
    of the cell ....
    (i)   remains unchanged
    (ii) is increased
    (iii) is decreased   (iv) none of the above

21. The material of positive plates of a lead-acid
    cell is .... than those of negative plates.

    (i)   mechanically stronger
    (ii) poorer in conductivity
    (iii) better in conductivity
    (iv) less chemically active

22. The specific gravity of the electrolyte (H_2SO_4)
    of a lead-acid cell is about 1·25. The cell is
    about ....
    (i)   25% charged   (ii) 50% charged
    (iii) 75% charged   (iv) 100% charged

23. The internal resistance of a lead-acid cell is
    mainly due to ....
    (i)   positive plates   (ii) negative plates
    (iii) both positive and negative plates
    (iv) electrolyte

24. A lead-acid cell has 13 plates. In the absence
    of manufacturer's data, the charging current
    should be ....
    (i)   13 A   (ii) 2 A
    (iii) 6 A   (iv) 3 A

25. The ampere hour efficiency of a lead-acid cell is
    about ....
    (i)   25%   (ii) 50%
    (iii) 75%   (iv) 90%

26. The state of charge of a nickel-iron cell is
    determined by ....
    (i)   the specific gravity of electrolyte (KOH)
    (ii) e.m.f. of the cell
    (iii) colour of plates (iv) none of the above

27. Nickel-iron batteries are used in ....
    (i)   starting of car engine
    (ii) hospitals as an emergency supply
    (iii) submarines
    (iv) railway train lighting system

28. The state of charge of nickel-cadmium cell is
    determined by ....
    (i)   e.m.f. of the cell
    (ii) specific gravity of the electrolyte (KOH)
    (iii) colour of plates (iv) none of the above

29. The chief disadvantage of a nickel-cadmium
    cell is that it ....
    (i)   takes a long time for charging
    (ii) has high internal resistance
    (iii) has a short life
    (iv) is very expensive

30. The Faraday constant F is equal to (N is
    Avogadro number and $e = 1.6 \times 10^{-19}$ C)
    (i) $F = Ne$ (ii) $F = N/e$
    (iii) $F = N^2 e$ (iv) $F = Ne^2$

Answers

1. *(ii)*	**2.** *(i)*	**3.** *(iii)*	**4.** *(i)*	**5.** *(iii)*
6. *(iv)*	**7.** *(iii)*	**8.** *(i)*	**9.** *(iii)*	**10.** *(iv)*
11. *(ii)*	**12.** *(i)*	**13.** *(iii)*	**14.** *(i)*	**15.** *(ii)*
16. *(iii)*	**17.** *(i)*	**18.** *(iv)*	**19.** *(ii)*	**20.** *(iii)*
21. *(ii)*	**22.** *(iii)*	**23.** *(iv)*	**24.** *(iii)*	**25.** *(iv)*
26. *(ii)*	**27.** *(iii)*	**28.** *(i)*	**29.** *(iv)*	**30.** *(i)*

11

A.C. Fundamentals

Introduction

We have dealt so far with cases in which the currents are steady and in one direction; this is called direct current (d.c.). The use of direct currents is limited to a few applications e.g. charging of batteries, electroplating, electric traction etc. For large scale power distribution there are, however, many advantages in using alternating current (a.c.). In an a.c. system, the voltage acting in the circuit changes polarity at regular intervals of time and the resulting current (called alternating current) changes direction accordingly. The a.c. system has offered so many advantages that at present electrical energy is universally generated, transmitted and used in the form of alternating current. Even when d.c. energy is necessary, it is a common practice to convert a.c. into d.c. by means of rotary converters or rectifiers.

Three principal advantages are claimed for a.c. system over the d.c. system. First, alternating voltages can be stepped up or stepped down efficiently by means of a transformer. This permits the transmission of electric power at high voltages to achieve economy and distribute the power at utilisation voltages. Secondly, a.c. motors (induction motors) are cheaper and simpler in construction than d.c. motors. Thirdly, the switchgear (*e.g.* switches, circuit breakers etc.) for a.c. system is simpler than the d.c. system. In this chapter, we shall confine our attention to the fundamentals of alternating currents.

11.1. Alternating Voltage and Current

A voltage which changes its polarity at regular intervals of time is called an **alternating voltage**. When an alternating voltage is applied in a circuit, the current flows first in one direction and then in the opposite direction; the direction of current at any instant depends upon the polarity of the voltage. Fig. 11.1 shows an alternating voltage source connected to a resistor R. In Fig. 11.1 (*i*), the upper terminal of alternating voltage source is positive and lower terminal negative so that current flows in the circuit as shown in Fig. 11.1 (*i*). After some time (a fraction of a second), the polarities of the voltage source are reversed [See Fig. 11.1 (*ii*)] so that current now flows in the opposite direction. This is called alternating current because the current flows in alternate directions in the circuit.

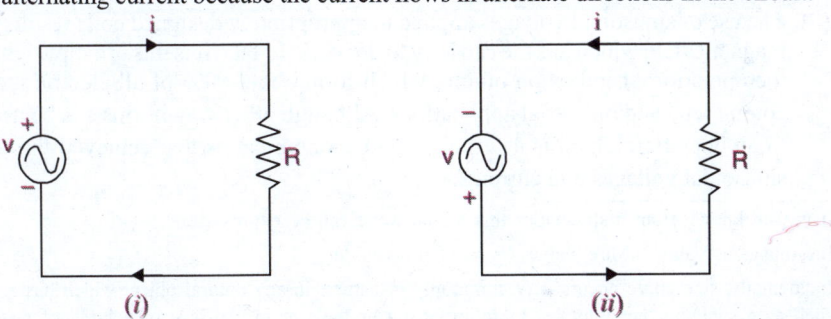

Fig. 11.1

11.2. Sinusoidal Alternating Voltage and Current

Commercial alternators produce sinusoidal alternating voltage *i.e.* alternating voltage is a sine wave. A sinusoidal alternating voltage can be produced (For proof, refer to Art. 11.5) by rotating a coil with a constant angular velocity (say ω rad/sec) in a uniform magnetic field. The sinusoidal alternating voltage can be *expressed by the equation :

$$v = V_m \sin \omega t$$

where v = Instantaneous value of alternating voltage

V_m = Max. value of alternating voltage

ω = Angular velocity of the coil

Sinusoidal voltages always produce sinusoidal currents, unless the circuit is non-linear. Therefore, a sinusoidal current can be expressed in the same way as voltage *i.e.* $i = I_m \sin \omega t$. Fig. 11.2 (*i*) shows the waveform of sinusoidal voltage whereas Fig. 11.2 (*ii*) shows the waveform of sinusoidal current. Note that sinusoidal voltage or current not only changes direction at regular intervals but the magnitude is also changing continuously.

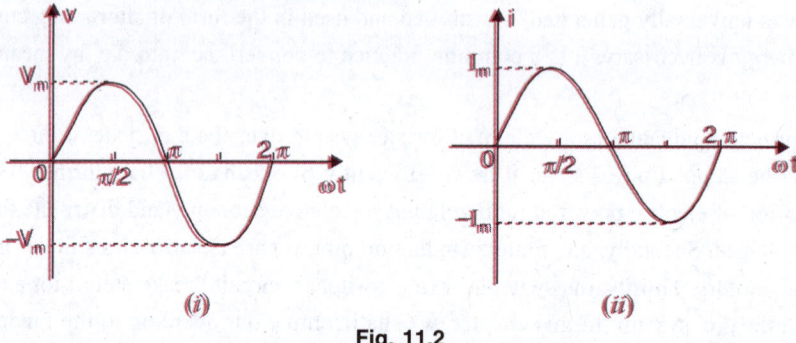

Fig. 11.2

Note. An alternating current can also be represented as a cosine function of time viz. $i = I_m \cos \omega t$. Similarly, alternating voltage can be represented as $v = V_m \cos \omega t$.

11.3. Why Sine Waveform?

Although it is possible to produce alternating voltages and currents with an endless variety of waveforms (*e.g.*, square waves, triangular waves, rectangular waves etc), yet the engineers choose to adopt **sine waveform. The following are the technical and economical advantages of producing sinusoidal alternating voltages and currents :

(*i*) *The sine waveform produces the least disturbance in the electrical circuit and is the smoothest and efficient waveform.* For example, when current in a capacitor, in an inductor or in a transformer is sinusoidal, the voltage across the element is also sinusoidal. This is not true of any other waveform.

(*ii*) The use of sinusoidal voltages applied to appropriately designed coils results in a revolving magnetic field which has the capacity to do work. In fact, it is this principle which underlines the operation of induction motors which form about 90% of all electric motors found in commercial and industrial applications. Although other waveforms can be used, none leads to an operation which is as efficient and economical as that achieved through the use of sinusoidal voltages and currents.

---

*   It is well known from mathematics that a sine wave can be expressed as :

Instantaneous value = Max. value × sine of time angle

**   Incidentally, sine curve occurs very commonly in nature. In any natural object which has a periodic motion such as a swinging pendulum, a vibrating string or the rippling surface of a body of water, we find this form of wave. *The sine curve is apparently nature's standard.*

(*iii*) The mathematical computations, connected with alternating current work, are much simpler with this waveform.

(*iv*) By means of Fourier series analysis, it is possible to represent any periodic function of whatever waveform in terms of sinusoids. This is a notable advantage in the mathematical sense because non-sinusoidal waves can be analysed in terms of sinusoids.

Due to above advantages, electric supply companies all over the world generate sinusoidal alternating voltages and currents. **It may be noted that alternating voltage and current mean sinusoidal alternating voltage and current unless stated otherwise.**

11.4. Generation of Alternating Voltages and Currents

An alternating voltage may be generated :

(*i*) by rotating a coil at constant angular velocity in a uniform magnetic field as shown in Fig. 11.3.

or

(*ii*) by rotating a magnetic field at a constant angular velocity within a stationary coil as shown in Fig. 11.4.

In either case, the generated voltage will be of sinusoidal waveform. The magnitude of generated voltage will depend upon the number of turns of coil, the strength of magnetic field and the speed of rotation. The first method is used for small a.c. generators while the second method is employed for large a.c. generators.

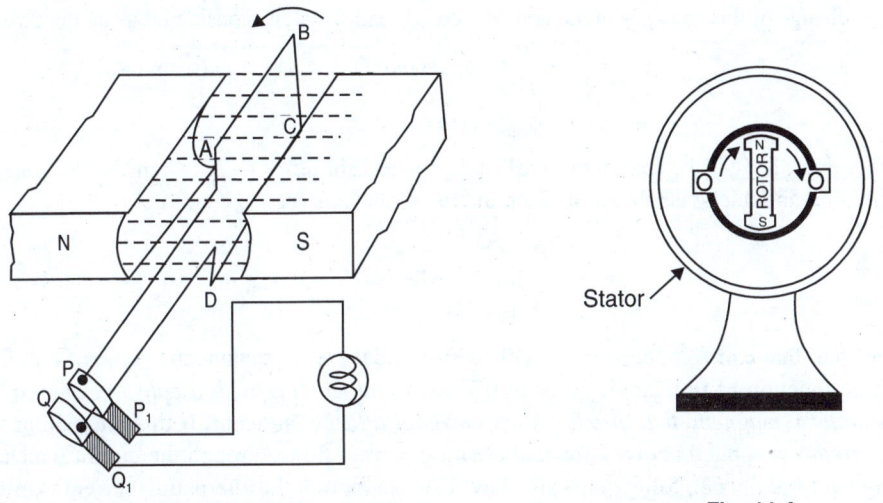

Fig. 11.3   Fig. 11.4

11.5. Equation of Alternating Voltage and Current

Consider a rectangular coil of n turns rotating in anticlockwise direction with an angular velocity of ω rad/sec in a uniform magnetic field as shown in Fig. 11.5. The e.m.f. induced in the coil will be sinusoidal. This can be readily established.

Let the time be measured from the instant the plane of the coil coincides with OX-axis. In this position of the coil [See Fig. 11.5 (*i*)], the flux linking with the coil has its maximum value ϕ_{max}. Let the coil turn through an angle θ (= ωt) in anticlockwise direction in t seconds and assumes the position shown in Fig. 11.5 (*ii*). In this position, the maximum flux ϕ_{max} acting vertically downward can be resolved into two perpendicular components *viz*.

(*i*) Component $\phi_{max} \sin \omega t$ parallel to the plane of the coil. This component induces *no e.m.f. in the coil.

---
*   A coil moving parallel to the flux has no change of flux linkage with it or there is no "cutting" of flux.

(*ii*) Component $\phi_{max} \cos \omega t$ perpendicular to the plane of the coil. This component induces e.m.f. in the coil.

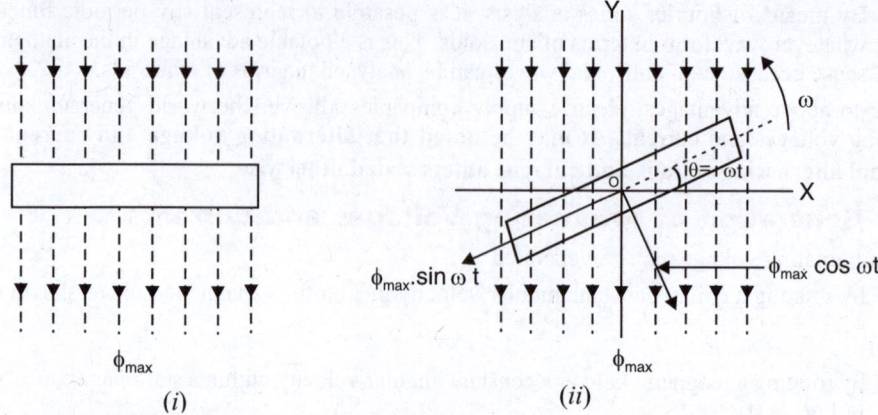

$$\text{(i)} \qquad\qquad\qquad\qquad \text{(ii)}$$

$$\textbf{Fig. 11.5}$$

∴ Flux linkages of the coil at the considered instant (*i.e.* at θ^0)

$$= \text{No. of turns} \times \text{Flux linking} = n\, \phi_{max} \cos \omega t$$

According to Faraday's laws of electromagnetic induction, the e.m.f. induced in a coil is equal to the rate of change of flux linkages of the coil. Hence, the e.m.f. v at the considered instant is given by ;

$$v = -\frac{d}{dt}(n\, \phi_{max} \cos \omega t) = -n\, \phi_{max}\, \omega\, (-\sin \omega t)$$

∴
$$v = n\, \phi_{max}\, \omega \sin \omega t$$

The value of v will be maximum (call it V_m) when $\sin \omega t = 1$ *i.e.*, when the coil has turned through 90° in anticlockwise direction from the reference axis (*i.e.*, OX-axis).

∴
$$V_m = n\, \phi_{max}\, \omega$$

∴
$$v = V_m \sin \omega t \quad \text{where} \quad V_m = n\, \phi_{max}\, \omega$$

or
$$v = V_m \sin \theta$$

It is clear that e.m.f. induced in the coil is sinusoidal *i.e.*, instantaneous value of e.m.f. varies as the sine function of time angle (θ or ωt). *Thus a coil rotating with a constant angular velocity in a uniform magnetic field produces a sinusoidal alternating e.m.f.* If this alternating voltage ($v = V_m \sin \omega t$) is applied across a load, alternating current flows through the circuit which would also vary sinusoidally *i.e.*, following a sine law. The equation of the alternating current is given by ;

$$i = I_m \sin \omega t \quad \text{provided the load is *resistive.}$$

Note. If at $t = 0$, θ is measured from the position of the coil when its plane is perpendicular to the direction of magnetic field, then, $v = V_m \sin \theta$. If θ (*i.e.*, at $t = 0$) is measured from the position of the coil when its plane makes an angle ϕ with the **normal** to the direction of field, then, $v = V_m \sin (\theta + \phi)$. If $\phi = 90°$ (*i.e.*, at $t = 0$, the plane of the coil is parallel to the field), then $v = V_m \sin (\theta + 90°) = V_m \cos \theta$ *i.e.*, $v = V_m \cos \omega t$.

Example 11.1. *A square coil of 10 cm side and with 100 turns is rotated at a uniform speed of 500 r.p.m. about an axis at right angles to a uniform field of 0·5 T. Calculate the maximum e.m.f. produced in the coil. What is the instantaneous value of e.m.f. when the plane of the coil makes an angle of 30° with the magnetic field ?*

Solution. The instantaneous value of e.m.f. is

$$v = V_m \sin \omega t = V_m \sin \theta$$

---

* It will be shown later that if the load is inductive or capacitive, the current equation is changed in time angle.

where time is measured from the position of the coil when its plane is perpendicular to the direction of the magnetic field.

Maximum induced e.m.f., $V_m = n\phi_{max}\omega = nBA\omega$

Here $n = 100$; $A = 10 \times 10 = 100$ cm$^2 = 100 \times 10^{-4}$ m$^2 = 10^{-2}$ m^2 ;

$B = 0.5$ T ; $\omega = 2\pi f = 2\pi \times (500/60) = 50\pi/3$ rad s^{-1}

$\therefore \qquad\qquad V_m = 100 \times 0.5 \times 10^{-2} \times 50\,\pi/3 = $ **26.18 V**

The angle between the plane of the coil and magnetic field is 30°. Therefore, angle between normal to the coil and the field is $\theta = 90° - 30° = 60°$.

$\therefore \qquad\qquad v = V_m \sin 60° = 26.18 \times 0.866 = $ **22.6 V**

Example 11.2. *An a.c. generator consists of a coil of 50 turns and area 2·5 m² rotating at an angular speed of 60 rad s⁻¹ in a uniform magnetic field B = 0·3 T. The resistance of the circuit including that of the coil is 500 Ω.*

(i) What is the maximum current drawn from the generator?

(ii) What is the flux through the coil when the current is zero? What is the flux when the current is maximum?

(iii) Would the generator work if the coil were stationary and the magnetic field (i.e., poles) rotated with the same speed as above?

Solution. (*i*) The current drawn from the generator will be maximum when induced e.m.f. is maximum.

Maximum induced e.m.f., $V_m = nAB\omega = 50 \times 2.5 \times 0.3 \times 60 = 2250$ V

$\therefore \qquad$ Maximum current, $I_m = V_m/R = 2250/500 = $ **4.5 A**

(*ii*) When induced current is zero, the magnetic flux through the coil is **maximum.** On the other hand, when induced current is maximum, the flux through the coil is **zero**.

(*iii*) **Yes,** the generator will work. It is because the basic condition for inducing an e.m.f. is that there should be relative motion between the coil and the magnetic field.

Example 11.3. *An a.c. generator consists of a coil of 100 turns and cross-sectional area of 3 m², rotating at a constant angular speed of 60 rad s⁻¹ in a uniform magnetic field of 0·04 T. The resistance of the coil is 500 Ω. Calculate maximum power dissipated in the coil.*

Solution. Maximum e.m.f. induced in the coil is

$$V_m = nAB\,\omega = 100 \times 3 \times 0.04 \times 60 = 720 \text{ V}$$

Maximum current supplied by the generator is

$$I_m = V_m/R = 720/500 = 1.44 \text{ A}$$

Maximum power dissipated in the coil is

$$P_{max} = V_m I_m = 720 \times 1.44 = \textbf{1036.8 W}$$

$$\boxed{\textbf{Tutorial Problems}}$$

1. A coil of wire has 50 turns and its area is 500 cm². It is rotating at the rate of 50 rounds per second at right angles to a magnetic field of 0.5T. Calculate the maximum e.m.f. induced in the coil . **[392.5V]**

2. A coil is wound with 300 turns on a square former having sides 5 cm in length. Calculate the maximum value of e.m.f. generated when it is rotated at 2000 revolutions per minute in a uniform magnetic field of 0.8T. What is the frequency of this e.m.f.? **[125.6V; 33.3Hz]**

3. A coil of 100 turns is rotated at 1500 revolutions per minute in a magnetic field having a uniform density of 0.05T, the axis of rotation being at right angles to the direction of magnetic lines. The mean area per turn is 40cm². Calculate (*i*) the frequency (*ii*) time period (*iii*) maximum value of the generated e.m.f. and (*iv*) the value of generated e.m.f. when the coil has rotated through 30° from the position of zero e.m.f. **[(*i*) 25Hz (*ii*) 0.04 s (*iii*) 3.14V (*iv*) 1.57 V]**

4. A circular coil of radius 8.0 cm and 20 turns rotates about its vertical diameter with an angular speed of 50 rad s^{-1} in a uniform horizontal magnetic field of magnitude 3.0×10^{-2} T. Find the maximum and average e.m.f. induced in the coil. If the coil forms a closed loop of resistance 10Ω, how much power is dissipated as heat? What is the source of this power? **[0.6032V ; zero ; 0.0182W]**

11.6. Important A.C. Terminology

An alternating voltage or current changes continuously in magnitude and alternates in direction at regular intervals of time. It rises from zero to maximum positive value, falls to zero, increases to a maximum in the reverse direction and falls back to zero again (See Fig. 11.6). From this point on indefinitely, the voltage or current repeats the procedure. The important a.c. terminology is defined below :

(*i*) **Waveform.** The shape of the curve obtained by plotting the instantaneous values of voltage or current as ordinate against *time as abcissa is called its *waveform* or *waveshape*. Fig. 11.6 shows the waveform of an alternating voltage varying sinusoidally.

(*ii*) **Instantaneous value.** The value of an alternating quantity at any instant is called instantaneous value. The instantaneous values of alternating voltage and current are represented by v and i respectively. As an example, the instantaneous values of voltage (See Fig. 11.6) at $0°$, $90°$ and $270°$ are 0, $+ V_m$, $-V_m$ respectively.

(*iii*) **Cycle.** One complete set of positive and negative values of an alternating quantity is known as a cycle. Fig. 11.6 shows one cycle of an alternating voltage.

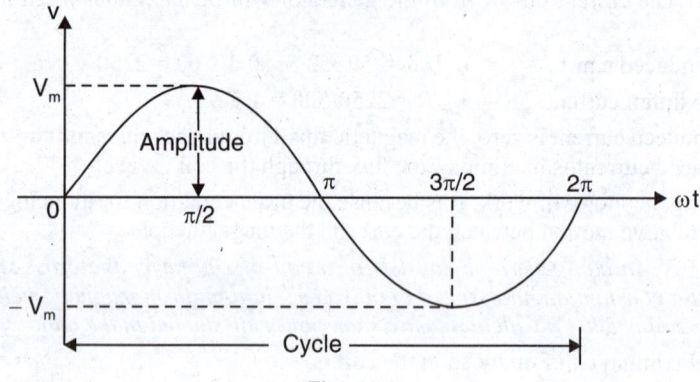

Fig. 11.6

A cycle can also be defined in terms of angular measure. One cycle corresponds to $360°$ **electrical or 2π radians. The voltage or current generated in a conductor will span $360°$ electrical (or complete one cycle) when the conductor moves past successive north and south poles.

(*iv*) **Alternation.** One-half cycle of an alternating quantity is called an alternation. An alternation spans $180°$ electrical. Thus in Fig. 11.6, the positive or negative half of alternating voltage is the alternation.

(*v*) **Time period.** The time taken in seconds to complete one cycle of an alternating quantity is called its time period. It is generally represented by T.

---

*   We know $\theta = \omega t$. Since ω is constant, $\theta \propto t$.

    Hence, we may take time t or θ or ωt along X-axis.

**   The time required to generate one cycle is divided into $360°$ divisions called electrical degrees. They differ from mechanical degrees. In one revolution, the coil will always traverse $360°$ mechanical. However, the electrical degrees spanned will depend upon the number of poles. For a 2-pole generator, electrical degrees spanned in one revolution of the coil are $360°$ *i.e.* one cycle is generated. For a 4-pole generator, 2 cycles (*i.e.* $720°$) will be generated for one revolution of the coil.

(*vi*) **Frequency.** The number of cycles that occur in one second is called the frequency (*f*) of the alternating quantity. It is measured in cycles/sec (C/s) or Hertz (Hz). One Hertz is equal to 1C/s.

The frequency of power system is low; the most common being 50 C/s or 50 Hz. It means that alternating voltage or current completes 50 cycles in one second. The 50 Hz frequency is the most popular because it gives the best results when used for operating both lights and machinery.

(*vii*) **Amplitude.** The maximum value (positive or negative) attained by an alternating quantity is called its amplitude or peak value. The amplitude of an alternating voltage or current is designated by V_m (or E_m) or I_m.

11.7. Important Relations

Having become familiar with a.c. terminology, we shall now establish some important relations.

(*i*) **Time period and frequency.** Consider an alternating quantity having a frequency of *f* C/s and time period *T* second.

Time taken to complete *f* cycles = 1 second   (*By definition*)

Time taken to complete 1 cycle = 1/*f* second

But the time taken to complete one cycle is the time period *T* (by definition).

$$\therefore \qquad\qquad T = \frac{1}{f} \quad \text{or} \quad f = \frac{1}{T}$$

(*ii*) **Angular velocity and frequency.** Referring to Fig. 11.5 (*ii*), the coil is rotating with an angular velocity of ω rad/sec in a uniform magnetic field. In one revolution of the coil, the angle turned is 2π radians and the voltage wave completes 1 cycle. The time taken to complete one cycle is the time period *T* of the alternating voltage.

$$\therefore \qquad \text{Angular velocity, } \omega = \frac{\text{Angle turned}}{\text{Time taken}} = \frac{2\pi}{T}$$

or $\qquad\qquad\qquad \omega = 2\,\pi f \qquad\qquad (\because f = 1/T)$

(*iii*) **Frequency and speed.** Consider a coil rotating at a speed of *N* r.p.m. in the field of *P* poles. As the coil moves past successive north and south poles, one complete cycle is generated. Obviously, in one revolution of the coil, *P*/2 cycles will be generated.

Now, $\qquad\qquad$ Frequency, *f* = No. of cycles/sec

$$= \text{(No. of cycles/revolution)} \times \text{(No. of revolutions/sec)}$$

$$= \left(\frac{P}{2}\right) \times \left(\frac{N}{60}\right) = \frac{P\,N}{120}$$

$$\therefore \qquad\qquad f = \frac{P\,N}{120}$$

For example, an a.c. generator having 10 poles and running at 600 r.p.m. will generate alternating voltage and current whose frequency is :

$$f = \frac{P\,N}{120} = \frac{10 \times 600}{120} = 50 \text{ Hz}$$

Example 11.4. *The maximum current in a sinusoidal a.c. circuit is 10A. What is the instantaneous current at 45° ?*

Solution. $\qquad\qquad\qquad i = I_m \sin \theta$

Here, $\quad I_m = 10\text{A} ; \theta = 45°$

$$\therefore \qquad\qquad\qquad i = 10 \times \sin 45° = \textbf{7·07 A}$$

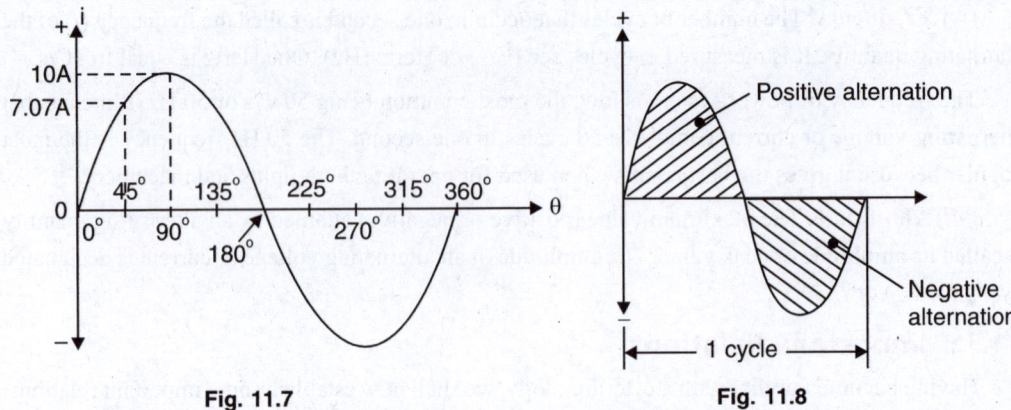

Fig. 11.7   **Fig. 11.8**

Fig. 11.7 shows one cycle of the alternating current. One cycle has 360 electrical degrees. The cycle begins at 0° where the magnitude of current is zero. At 45°, the instantaneous current is $i = I_m \sin \theta$ $= 10 \times \sin 45° = 7·07$ A. It achieves a maximum value I_m (= 10 A) at 90° and then begins to decrease until it reaches zero again at 180°. At 225°, it is $- 7·07$ A. It reaches its maximum negative value $- I_m$ (= $- 10$ A) at 270° and then returns to zero at 360°. One sine cycle has two **alternations** : one positive and the other negative as shown in Fig. 11.8.

    Example 11.5. *Write the mathematical expression for a 50 Hz sinusoidal voltage of peak value 80 V. Sketch the waveform versus time t.*

    Solution.   Here, $f = 50$ Hz ; $V_m = 80$ V ; $\omega = 2\pi f = 2\pi \times 50 = 314$ rad/s

$\therefore$ $v = V_m \sin \omega t = 80 \sin 314\, t$

or   **$v = 80 \sin 314\, t$**

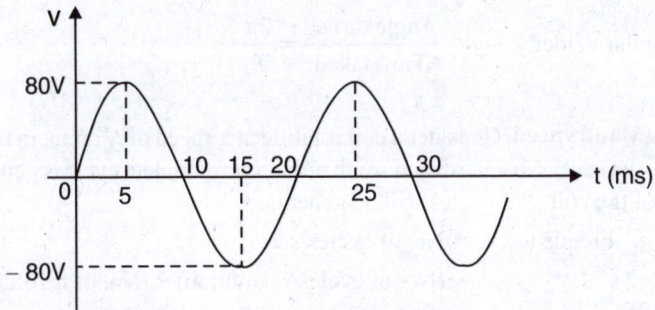

Fig. 11.9

In order to sketch the waveform versus time t, we first find the time period T of the wave.

Now, $T = \dfrac{1}{f} = \dfrac{1}{50} = 20 \times 10^{-3}$ s $= 20$ ms

Since time period corresponds to 360° and the waveform reaches its peak value at 90°, the instant of time at which the peak occurs = $(90/360) \times (20) = 5$ ms. Similarly, the waveform passes through zero at $t = (180/360) \times 20 = 10$ ms and it reaches its negative peak value at $t = (270/360) \times 20 = 15$ ms. The waveform is shown in Fig. 11.9.

11.8. Different Forms of Alternating Voltage

    The standard form of an alternating voltage is given by ;

$$v = V_m \sin \theta$$
$$= V_m \sin \omega t$$

$$= V_m \sin 2\pi f t \qquad\qquad (\because \omega = 2\pi f)$$

$$= V_m \sin \frac{2\pi}{T} t \qquad\qquad (\because f = 1/T)$$

Which of the above form of equations is to be used will depend upon the given data.

The following points may be noted carefully :

(*i*) *The maximum value of alternating voltage is given by the co-efficient of sine of the time angle i.e.*

Maximum value of voltage, V_m = Co-efficient of sine of time angle

(*ii*) *The frequency f of alternating voltage is given by dividing the co-efficient of time in the angle by 2π i.e.*

$$\text{Frequency, } f = \frac{\text{Co-efficient of time in the angle}}{2\pi}$$

For example, suppose the equation of an alternating voltage is given by $v = 100 \sin 314\, t$. Then the maximum value of voltage, $V_m = 100$ V ; frequency, $f = 314/2\pi = 50$ Hz and time period, $T = 1/f = 1/50 = 0\cdot02$ second.

Following similar procedure, maximum value, frequency, time period *etc.* can be found out from the various forms of current equation.

Example 11.6. *An alternating current i is given by ;*

$$i = 141\cdot4 \sin 314\, t$$

Find (i) the maximum value (ii) frequency (iii) time period and (iv) the instantaneous value when t is 3 ms.

Solution. Comparing the given equation of alternating current with the standard form $i = I_m \sin\omega t$, we have,

(*i*)　　　　　　　　Maximum value, I_m = **141·4 A**

(*ii*)　　　　　　　　Frequency, $f = \omega/2\pi = 314/2\pi =$ **50 Hz**

(*iii*)　　　　　　Time period, $T = 1/f = 1/50 =$ **0·02 s**

(*iv*)　　　　　　　　　$i = 141\cdot4 \sin 314\, t$

　　　　　　When $t = 3$ m s $= 3 \times 10^{-3}$ s,

　　　　　$i = 141\cdot4 \sin 314 \times 3 \times 10^{-3} =$ **114· 35 A**

Example 11.7. *An alternating current of frequency 60 Hz has a maximum value of 120 A.*

(*i*) *Write down the equation for the instantaneous value.*

(*ii*) *Reckoning time from the instant the current is zero and becoming positive, find the instantaneous value after 1/360 second.*

(*iii*) *Time taken to reach 96 A for the first time.*

Solution. Max. value of current, $I_m = 120$ A ; Frequency, $f = 60$ Hz

(*i*) The instantaneous value of current is given by ;

$$i = I_m \sin \omega\, t = I_m \sin 2\pi f t = 120 \sin 2\pi \times 60 \times t$$

$\therefore$　　　　　　　　**$i = 120 \sin 120\, \pi\, t$**

(*ii*) Fig. 11.10 shows the wavefrom of the given alternating current. Since point O has been taken as the reference, the current equation is :

$$i = 120 \sin 120\, \pi\, t$$

When　　$t = 1/360$ second,

$$i = 120 \sin 120\pi \times 1/360$$
$$= 120 \sin (\pi/3)$$
$$= \textbf{103·92 A}$$

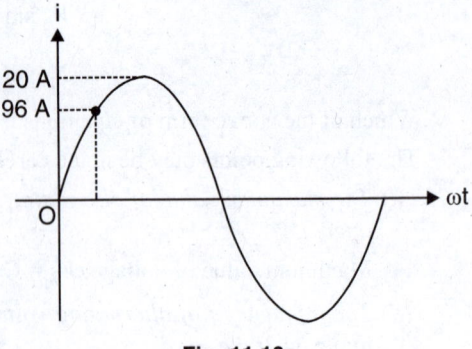

(*iii*) Suppose the current becomes 96A for the first time after t second as shown in Fig. 11.10.

Then, $\quad 96 = 120 \sin 120 \pi t$

or $\quad \sin 120 \pi t = 96/120 = 0·8$

or $\quad 120 \pi t = \sin^{-1} 0·8 = 0·927$ rad

∴ $\quad t = \dfrac{0.927}{120 \times \pi} = \textbf{0.00246 s}$

Fig. 11.10

Example 11.8. *An alternating current is given by ;*

$$i = 10 \sin 942\, t$$

Determine the time taken from t = 0 for the current to reach a value of + 6 A for a first and second time.

Solution. Fig. 11.11 shows the waveform of the given alternating current. Let the current become + 6A for the first time after t second. Then,

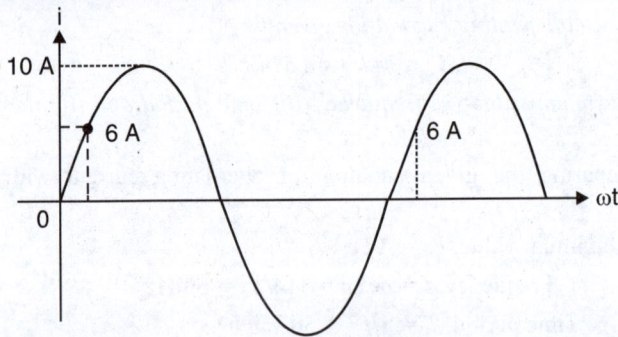

Fig. 11.11

$$6 = 10 \sin 942\, t \quad \text{or} \quad \sin 942\, t = 6/10 = 0·6$$

∴ $\quad 942\, t = \sin^{-1} 0·6 = 0·643$ rad

or $\quad t = \dfrac{0.643}{942} = 0·68 \times 10^{-3} \text{ s} = \textbf{0·68 ms}$

The frequency of the current is $f = 942/2\pi = 150$ Hz and the time period of the wave is $T = 1/150$ $= 6·66 \times 10^{-3}$ s $= 6·66$ ms.

∴   Time taken to reach + 6 A for the second time (See Fig. 11.11)

$$= 0·68 + \text{time period} = 0·68 + 6·66 = \textbf{7·34 ms}$$

Example 11.9. *A sinusoidal voltage of 50 Hz has a maximum value of $200\sqrt{2}$ volts. At what time measured from a positive maxmium value will the instantaneous voltage be 141·4 volts ?*

Solution. Fig. 11.12 shows the voltage waveform. The equation of this wave with point O as the reference is given by ;

$$v = V_m \sin \omega t$$
$$= V_m \sin 2\pi\, ft$$
$$= 200\sqrt{2}\, \sin 2\pi \times 50 \times t$$

∴ $\quad v = 282·8 \sin 314\, t$

This equation is valid when time is measured from the instant the voltage is zero *i.e.* point O. Since the time is measured from the positive maximum value (point A in Fig. 11.12), the above equation is modified to :

$$v = 282 \cdot 8 \sin (314\, t + \pi/2)$$
$$= 282 \cdot 8 \cos 314\, t$$

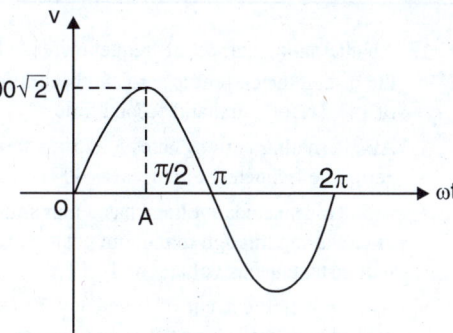

Fig. 11.12

Let the value of voltage become 141·4 volts t second after passing through the maximum positive value. Then,

$$141 \cdot 4 = 282 \cdot 8 \cos 314\, t$$

or $\quad \cos 314\, t = 141 \cdot 4/282 \cdot 8 = 0 \cdot 5$

or $\quad 314\, t = \cos^{-1} 0 \cdot 5 = 1 \cdot 047$ rad

$\therefore \qquad t = \dfrac{1.047}{314} = 3.33 \times 10^{-3}$ s $= \textbf{3.33 ms}$

Example 11.10. *An alternating current of frequency 50Hz has a maximum value of 100 A. Calculate (i) its value 1/600 second after the instant the current is zero and its value is decreasing there afterwards (ii) how many seconds after the current is zero and then increasing will attain the value of 86·6 A ?*

Solution. Fig. 11.13 shows the current waveform. With O as the origin, the equation of the current is : $\qquad i = I_m \sin \omega t = 100 \sin 2\pi \times 50 \times t$

$\therefore \qquad i = 100 \sin 100 \pi\, t$

(i) Since the current is measured from the instant the current is zero and is decreasing there afterwards (point A in Fig. 11.13), equation of the current with point A as the origin becomes :

$$i = 100 \sin (100 \pi\, t + \pi)$$
$$= -100 \sin 100 \pi t$$

When $t = 1/600$ second, then,
$$i = -100 \sin 100 \times 180°$$
$$\times (1/600) = \textbf{-50 A}$$

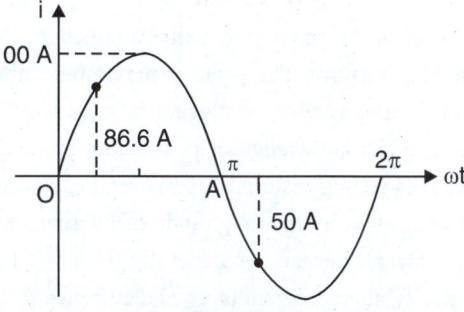

Fig. 11.13

(ii) When the current is measured from the instant the current is zero and is increasing thereafter (*i.e.*, point O in Fig. 11.13), the current equation is given by ;

$$i = 100 \sin 100 \pi\, t \quad \text{or} \quad 86 \cdot 6 = 100 \sin 100 \pi\, t$$

$\therefore \qquad \sin 100 \pi\, t = 86 \cdot 6/100 = 0 \cdot 866$

or $\qquad 100 \pi\, t = \sin^{-1} 0 \cdot 866 = 1 \cdot 047$ rad

$\therefore \qquad t = \dfrac{1.047}{100\pi} = 3 \cdot 33 \times 10^{-3}$ s $= \textbf{3·33 ms}$

Tutorial Problems

1. An alternating voltage is represented by :

$$v = 141 \cdot 4 \sin 377\, t$$

Find (*i*) the maximum value (*ii*) frequency (*iii*) time period and (*iv*) the instantaneous value of voltage when t is 3 ms.   **[(*i*) 141·4V (*ii*) 60 Hz (*iii*) 16·67 ms (*iv*) 127·8 V]**

2. An alternating current of frequency 50Hz has a maximum value of $200\sqrt{2}A$ Reckoning the time from the instant the current is zero and becoming positive, find the time taken by the current to reach a value of 141·4A for a first and second time.   **[1·67 ms ; 21·67 ms]**

3. An alternating current takes 3·375 ms to reach 15 A for the first time after becoming instantaneously zero. The frequency of current is 40 Hz. Find the maximum value of alternating current.   **[20A]**

4. A 50 Hz sinusoidal voltage has a maximum value of 56·56 V. Find the value of voltage 0·0025 second after passing through maximum positive value. At what time measured from a positive maximum value will instantaneous voltage be 14·14 V ?   **[40V ; 4·2 ms]**

5. An alternating current of frequency 50 Hz has a maximum value of 100A. Calculate its value 1/300 second after the instant the current is zero and its value is decreasing thereafter.   **[−86·6 A]**

11.9. Values of Alternating Voltage and Current

In a d.c. system, the voltage and current are constant so that there is no problem of specifying their magnitudes. However, an alternating voltage or current varies from instant to instant. A natural question arises how to express the magnitude of an alternating voltage or current. There are four ways of expressing it, namely ;

  (*i*) Peak value   (*ii*) Average value or mean value

  (*iii*) R.M.S. value or effective value   (*iv*) Peak-to-peak value

Although peak, average and peak-to-peak values may be important in some engineering applications, it is the r.m.s. or effective value which is used to express the magnitude of an alternating voltage or current.

11.10. Peak Value

It is the maximum value attained by an alternating quantity. The peak or maximum value of an alternating voltage or current is represented by V_m or I_m. The knowledge of peak value is important in case of testing materials. However, peak value is not used to specify the magnitude of alternating voltage or current. Instead, we generally use r.m.s. values to specify alternating voltages and currents.

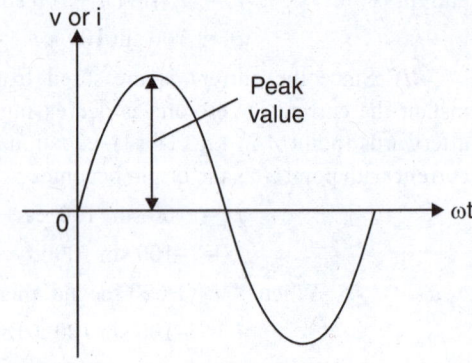

Fig. 11.14

11.11. Average Value

The average value of a waveform is the average of all its values over a period of time. In performing such a computation, we regard the area above the time axis as positive area and area below the time axis as negative area. The algebraic signs of the areas must be taken into account when computing the total (net) area. The time interval over which the net area is computed is the period T of the waveform.

∴   Average value $= \dfrac{\text{Total (net) area under curve for time } T}{\text{Time } T}$

(*i*) In case of *symmetrical waves (*e.g.* sinusoidal voltage or current), the average value over one cycle is zero. It is because positive half is exactly equal to the negative half so that net area is zero. However, the average value of positive or negative half is not zero. Hence in case of symmetrical waves, average value means the average value of half-cycle or one alternation.

Average value of a symmetrical wave $= \dfrac{\text{Area of one alternation}}{\text{Base length of one alternation}}$

---

* A symmetrical wave is one which has positive half-cycle exactly equal to the negative half-cycle.

$$= \frac{\text{Sum of *mid-ordinates over one alternation}}{\text{No. of mid-ordinates}}$$

$$= \frac{i_1 + i_2 + i_3 + ... + i_n}{n}$$

(*ii*) *In case of unsymmetrical waves (e.g. half-wave rectified voltage etc.), the average value is taken over the full cycle.*

$$\text{Average value of an unsymmetrical wave} = \frac{\text{Area over one cycle}}{\text{Base length of 1 cycle}}$$

The average value of a waveform is also called its *d.c. value.* In fact, when a waveform is measured with a d.c. instrument (d.c. ammeter or d.c. voltmeter), it is the average value of the waveform that is indicated by the instrument.

11.12. Average Value of Sinusoidal Current

The average value of alternating current (or voltage) over one cycle is zero. It is because the waveform is symmetrical about time axis and positive area exactly cancels the negative area. However, the average value over a half-cycle (positive or negative) is not zero. *Therefore, average value of alternating current (or voltage) means half-cycle average value unless stated otherwise.*

The **half-cycle average value** *of a.c. is that value of steady current (d.c.) which would send the same amount of charge through a circuit for half the time period of a.c. as is sent by the a.c. through the same circuit in the same time. It is represented by* I_{av}*.* This can be obtained by integrating the instantaneous value of current over one half cycle (*i.e.* area over half-cycle) and dividing the result by base length of half-cycle ($= \pi$).

The equation of an alternating current varying sinusoidally is given by ;

$$i = I_m \sin \theta$$

Consider an elementary strip of thickness $d\theta$ in the first half-cycle of current wave as shown in Fig. 11.15. Let i be the mid-ordinate of this strip.

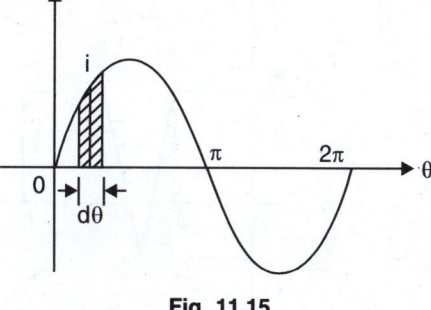

Fig. 11.15

$$\text{Area of strip} = i \, d\theta$$

$$\text{Area of half-cycle} = \int_0^{\pi} i \, d\theta$$

$$= \int_0^{\pi} I_m \sin \theta \, d\theta$$

$$= I_m \left[-\cos \theta \right]_0^{\pi} = 2I_m$$

$$\therefore \quad \text{Average value, } I_{av} = \frac{\text{Area of half-cycle}}{\text{Base length of half-cycle}} = \frac{2I_m}{\pi}$$

or $$I_{av} = 0\cdot637 \, I_m$$

Hence, the half-cycle average value of a.c. is 0·637 times the peak value of a.c.

For positive half-cycle, $I_{av} = + 0\cdot637 \, I_m$

For negative half-cycle, $I_{av} = - 0\cdot637 \, I_m$

Clearly, average value of a.c. over a complete cycle is zero. Similarly, it can be proved that for alternating voltage varying sinusoidally, $V_{av} = 0\cdot637 \, V_m$.

---

* Suppose positive half-cycle is divided into n equal parts. The middle value of each part is the mid-ordinate. The average value will be the sum of mid-ordinates divided by the number of mid-ordinates (*i.e. n* in this case).

Example. 11.11. *Find the average value of ac voltage whose waveform is shown in Fig. 11.16.*

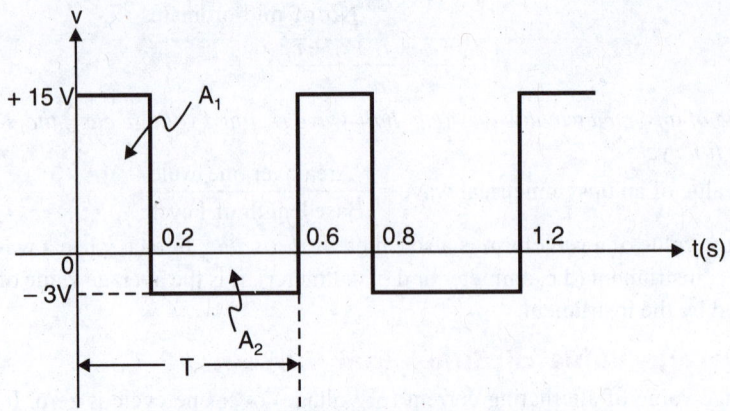

Fig. 11.16

Solution. One cycle of waveform extends from $t = 0$ to $t = 0.6$ s so that the time period $T = 0.6$s.

$$V_{av} = \frac{\text{Area under one cycle}}{\text{Time period } (T)}$$

$$\text{Area under one cycle} = \int_0^{0.2} 15 \, dt + \int_{0.2}^{0.6} -3 \, dt$$

$$= 15 \times 0.2 + (-3)(0.6 - 0.2) = 3 - 1.2 = 1.8 \text{ volt-sec}$$

$$\therefore \qquad V_{av} = 1.8/T = 1.8/0.6 = \textbf{3V}$$

Example 11.12. *Find the average value of the waveform shown in Fig. 11.17.*

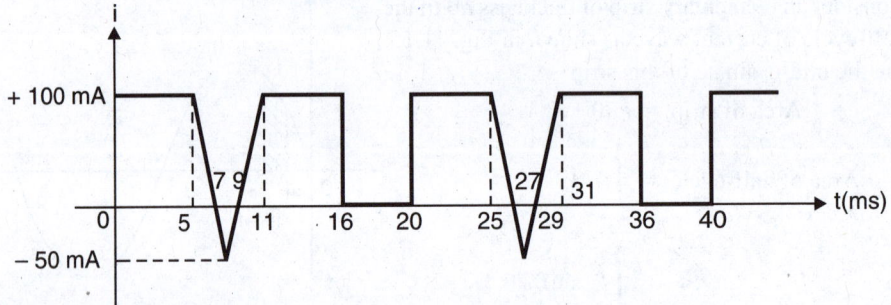

Fig. 11.17

Solution. The reader may note carefully the waveform from 0 to 20 ms. What the pattern of the wave is from 0 to 20 ms, the same pattern is from 20 to 40 ms. Therefore, time period of the wave is $T = 20$ ms.

$$I_{av} = \frac{\text{Area under one cycle}}{\text{Time period } (T)}$$

$$\text{Area under one cycle} = \text{Area of waveform from 0 to 20 ms}$$

$$= 100 \times 5 + \frac{1}{2} \times 2 \times 100 - \frac{1}{2} \times 50 \times 2 + \frac{1}{2} \times 2 \times 100 + 5 \times 100$$

$$= 500 + 100 - 50 + 100 + 500 = 1150 \text{ mA-ms}$$

$$\therefore \qquad I_{av} = 1150/20 = \textbf{57.5 mA}$$

Example 11.13. *A delayed full-wave rectified sinusoidal current has an average value equal to half its maximum value. Find the delay angle* θ.

Solution. The current wave as per the statement of the problem is shown in Fig. 11.18. The instantaneous value of current varying sinusoidally is

$$i = I_m \sin \theta$$

∴ Average value I_{av} of current wave in Fig. 11.18 is

$$I_{av} = \frac{1}{\pi}\int_\theta^\pi i\,d\theta = \frac{1}{\pi}\int_\theta^\pi I_m \sin\theta\,d\theta$$

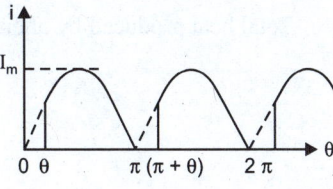

∴ $$I_{av} = \frac{I_m}{\pi}(-\cos\pi + \cos\theta)$$

Fig. 11.18

It is given that $I_{av} = I_m/2$.

∴ $$\frac{I_m}{\pi}(-\cos\pi + \cos\theta) = \frac{I_m}{2}$$

or $$(1 + \cos\theta) = \frac{\pi}{2} \quad \text{or} \quad \cos\theta = 0.57 \quad \therefore \theta = \textbf{55.25°}$$

Tutorial Problems

1. Find the average value of the waveform shown in Fig. 11.19.   **[12 mV]**

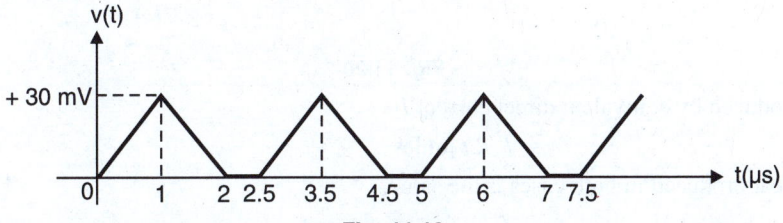

Fig. 11.19

2. Find the average value of $v(t) = 6 + 2\sin(2\pi \times 100t)$V.   **[6V]**

3. A sinusoidal current has a maximum value of 500 mA. What is its average value?   **[318 mA]**

11·13. R.M.S. or Effective Value

The average value cannot be used to specify a sinusoidal voltage or current. It is because its value over one-cycle is zero and cannot be used for power calculations. Therefore, we must search for a more suitable criterion to measure the effectiveness of an alternating current (or voltage). The obvious choice would be to measure it in terms of direct current that would do work (or produce heat) at the same average rate under similar conditions. This equivalent direct current is called the root-mean-square (r.m.s.) or effective value of alternating current.

*The **effective or r.m.s. value** of an alternating current is that steady current (d.c.) which when flowing through a given resistance for a given time produces the same amount of heat as produced by the alternating current when flowing through the same resistance for the same time.*

For example, when we say that the r.m.s. or effective value of an alternating current is 5A, it means that the alternating current will do work (or produce heat) at the same rate as 5A direct current under similar conditions.

Illustration. The r.m.s. or effective value of alternating current (or voltage) can be determined as follows. Consider the half-cycle of a non-sinusoidal alternating current i [See Fig. 11.20 (*i*)] flowing through a resistance $R\Omega$ for t seconds. Divide the time t in n equal intervals of time, each of duration t/n second. Let the mid-ordinates be $i_1, i_2, i_3, ...i_n$. Each current $i_1, i_2, i_3, ...i_n$ will produce heating effect when passed through the resistance R as shown in Fig. 11.20 (*ii*). Suppose the heating effect produced by current i in R is the same as produced by some direct current I flowing through the resistance R for the same time t. Then direct current I is the r.m.s. or effective value of alternating current i.

The heating effect of various components of alternating current will be i_1^2Rt/n......i_n^2Rt/n joules. Since the alternating current is varying, the heating effect will also vary.

Total heat produced by alternating current i

$$= (i_1^2R + i_2^2R + i_3^2R +........+ i_n^2R)\, t/n$$

$$= \frac{(i_1^2R + i_2^2R + i_3^2R +...+ i_n^2R)}{n}t \text{ joules}$$

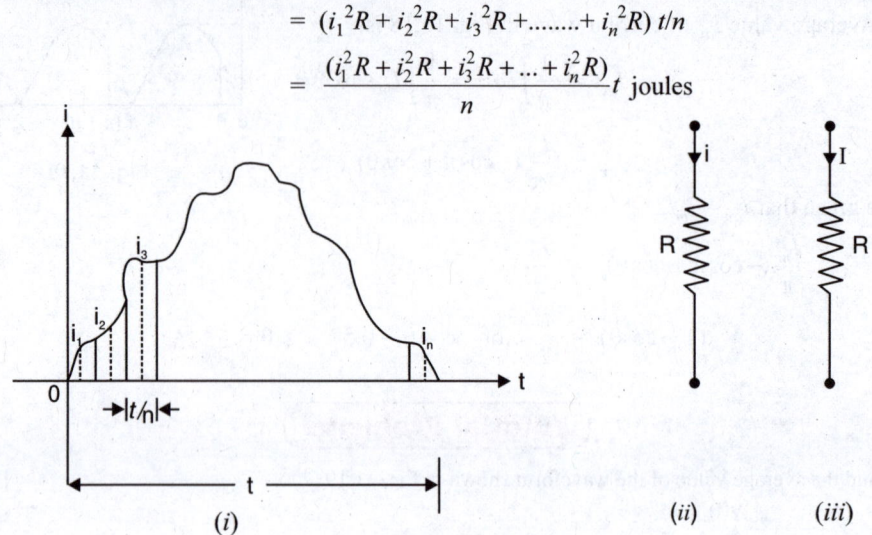

Fig. 11.20

Heat produced by equivalent direct current I

$$= I^2 R\, t \text{ joules}$$

Since heat produced in both cases is the same,

$$\therefore \quad I^2Rt = \left(\frac{i_1^2R + i_2^2R + i_3^2R +...+ i_n^2R}{n} \right)t$$

or

$$I = \sqrt{\frac{i_1^2 + i_2^2 + i_3^2 +...+ i_n^2}{n}} \qquad ...(i)$$

$$= \sqrt{*\text{mean value of } i^2}$$

$$= \text{Square } \textbf{root} \text{ of the } \textbf{mean} \text{ of the } \textbf{squares} \text{ of the current}$$

$$= \textbf{root-mean-square (r.m.s.)} \text{ value}$$

It hardly needs any explanation as to why the equivalent direct current (generally written as $I_{r.m.s.}$) is called the root-mean-square value. It is also called **effective value** because it is this value which tells the energy transfer capability of a.c. source. The following points may be noted :

(*i*) For symmetrical waves, the r.m.s. or effective value can be found by considering half-cycle or full-cycle. It is because second half is the negative of the first half and the r.m.s. value depends upon the squares of the instantaneous values. However, for unsymmetrical waves, full-cycle should be considered.

(*ii*) The r.m.s. value of symmetrical wave can also be expressed as :

$$**\text{R.M.S. value} = \sqrt{\frac{\text{Area of half-cycle of squared wave}}{\text{Half-cycle base}}}$$

---

*   The r.m.s. value is found by taking the square root of the average value of the sum of the squared instantaneous values.

**   This result is readily obtained if the numerator and denominator of exp. (*i*) under the root is multiplied by t/n.

(*iii*) The r.m.s. or effective value of an alternating voltage can similarly be expressed as :

$$V_{r.m.s.} = \sqrt{\frac{v_1^2 + v_2^2 + v_3^2 + ... + v_n^2}{n}}$$

Note. In case of an unsymmetrical wave, full-cycle should be considered to find the r.m.s. value.

11.14. R.M.S. Value of Sinusoidal Current

The equation of the alternating current varying sinusoidally is given by ;

$$i = I_m \sin \theta$$

Consider an elementary strip of thickness $d\theta$ in first half-cycle of the squared current wave (shown dotted in Fig. 11.21). Let i^2 be the mid-ordinate of this strip.

Area of strip $= i^2 d\theta$

Area of half-cycle of the squared wave

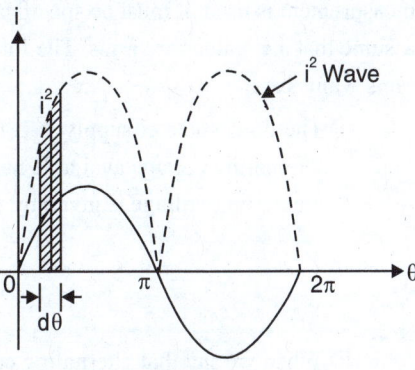

Fig. 11.21

$$= \int_0^\pi i^2 d\theta$$

$$= \int_0^\pi I_m^2 \sin^2 \theta d\theta$$

$$= I_m^2 \int_0^\pi \sin^2 \theta d\theta = \frac{{}^*\pi I_m^2}{2}$$

$$\therefore \qquad I_{r.m.s.} = \sqrt{\frac{\text{Area of half-cycle squared wave}}{\text{Half-cycle base}}}$$

$$= \sqrt{\frac{\pi I_m^2/2}{\pi}} = \frac{I_m}{\sqrt{2}} = 0.707 \, I_m$$

$$\therefore \qquad I_{r.m.s.} = \textbf{0·707} \, I_m$$

Similarly, it can be proved that for alternating voltage varying sinusoidally, $V_{r.m.s.} = \textbf{0·707} \, V_m$.

Alternate method. $\qquad i = I_m \sin \omega t$

If this current is passed through a resistance R, then power delivered at any instant is

$$p = i^2 R = (I_m \sin \omega t)^2 R$$

$$= I_m^2 R \sin^2 \omega t$$

or $\qquad p = I_m^2 R \sin^2 \omega t$

Because the current is squared, power is always positive. Since the value of $\sin^2 \omega t$ varies between 0 and 1, its average value is 1/2 [See Fig. 11.22].

$\therefore$ Average power delivered, $P = \dfrac{1}{2} I_m^2 R \qquad ...(i)$

If $I_{r.m.s.}$ is the r.m.s. (or effective) value of alternating current, then by definition,

Power delivered, $P = I_{r.m.s.}^2 R \qquad ...(ii)$

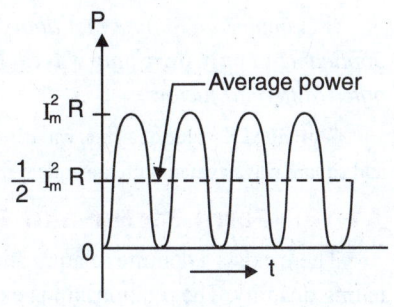

Fig. 11.22

$${}^* \int_0^\pi \sin^2 \theta d\theta = \int_0^\pi \frac{1 - \cos 2\theta}{2} = \frac{1}{2}\left[\theta - \frac{\sin 2\theta}{2}\right]_0^\pi = \frac{\pi}{2}$$

From eqs. (*i*) and (*ii*), we have,

$$I^2_{r.m.s.} R = \frac{1}{2} I_m^2 R$$

or $$I_{r.m.s.} = \frac{I_m}{\sqrt{2}} = 0.707\ I_m$$

∴ $$I_{r.m.s.} = 0{\cdot}707\ I_m$$

11.15. Importance of R.M.S. Values

An alternating voltage or current is always specified in terms of r.m.s. values. For example, common household appliances are rated at 230 V a.c. This is an r.m.s. value. If some other method of measurement is used, it must be specifically stated. Lacking any information to the contrary, always assume that a.c. values are r.m.s. The following points will give the reader a clear concept about the r.m.s. values :

(*i*) The domestic a.c. supply is 230 V, 50 Hz. It is the r.m.s. or effective value. It means that alternating voltage available has the same heating effect as 230 V d.c. The equation of this alternating voltage is given by ;

$$v = V_m \sin \omega\, t$$
$$= 230 \times \sqrt{2}\ \sin 2\pi \times 50 \times t \qquad (V_m = \sqrt{2}\ V_{r.m.s.})$$
∴ $$v = 230\ \sqrt{2}\ \sin 314\, t$$

(*ii*) When we say that alternating current in a circuit is 5 A, we are specifying the r.m.s. value. It means that the alternating current flowing in the circuit has the same heating effect as 5 A d.c.

(*iii*) A.C. ammeters and voltmeters record r.m.s. values of alternating current and voltage respectively.

Fig. 11.23 summarises the various ways to measure sinusoidal voltages and the conversion constants. The relationships apply for currents as well as voltages.

$$V_{p\text{-}p} = 2 \times V_m$$
$$V_{av} = 0.637 \times V_m$$
$$V_{r.m.s.} = 0.707 \times V_m$$
$$V_m = 1.414 \times V_{r.m.s.}$$

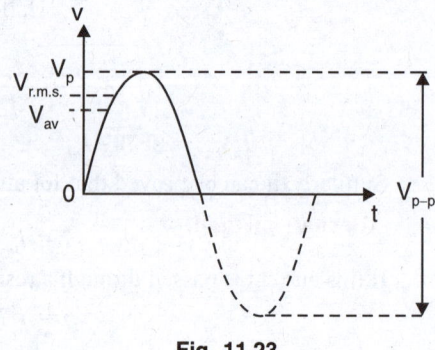

Fig. 11.23

*It is important to note that above relationship between peak, average and r.m.s. quantities are applicable to **only pure sine waves**. In the case of other waveforms, these quantities are related by other (different) factors.*

Note.R.M.S. value of an a.c. wave is always greater than the average value **except** in the case of rectangular and square waves when both are equal.

11.16. Form Factor and Peak Factor

There exists a definite relation among the peak value, average value and r.m.s. value of an alternating quantity. The relationship is expressed by two factors, namely ; form factor and peak factor.

(*i*) **Form factor.** *The ratio of r.m.s. value to the average value of an alternating quantity is known as* **form factor** *i.e.*

$$\text{Form factor} = \frac{\text{R.M.S. value}}{\text{Average value}}$$

The value of form factor depends upon the waveform of the alternating quantity. Its least value is 1 (*e.g.* for square wave, rectangular wave) and may be as high as 5 for other waveforms. The form factor for an alternating voltage or current varying sinusoidally is 1.11. *i.e.*

For a sinusoidal voltage or current,

$$\text{Form factor} = \frac{0.707 \times \text{Max. value}}{0.637 \times \text{Max. value}} = 1.11$$

The form factor gives a measure of the "peakiness" of the waveform. The peakier the wave, the greater is its form factor and *vice-versa*. For instance, a sine wave is peakier than a square wave. Hence the former has a greater form factor (1·11) than the latter. Similarly, a triangular wave is more peaky than a sine wave and has a form factor of 1·15. The form factor is useful in rectifier service.

(*ii*) **Peak factor.** *The ratio of maximum value to the r.m.s. value of an alternating quantity is known as* **peak factor** *i.e.*

$$\text{Peak factor} = \frac{\text{Max. value}}{\text{R.M.S. value}}$$

The value of peak factor also depends upon the waveform of the alternating quantity. For an alternating voltage or current varying sinusoidally, its value is 1·414 *i.e.*

For a sinusoidal voltage or current,

$$\text{Peak factor} = \frac{\text{Max. value}}{0.707 \times \text{Max. value}} = 1.414$$

The peak factor is of much greater importance because it indicates the maximum voltage being applied to the various parts of the apparatus. For instance, when an alternating voltage is applied across a cable or capacitor, the breakdown of insulation will depend upon the maximum voltage. The insulation must be able to withstand the maximum rather than the r.m.s. value of voltage.

Note. Peak factor is also called *crest factor* or *amplitude factor*.

Example 11.14. *An alternating current, when passed through a resistor immersed in water for 5 minutes, just raised the temperature of water to boiling point. When a direct current of 4 A was passed through the same resistor under identical conditions, it took 8 minutes to boil the water. Find the r.m.s. value of the alternating current.*

Solution. Let $I_{r.m.s.}$ amperes be the r.m.s. value of the alternating current and R ohms be the value of the resistor.

Heat produced by the alternating current

$$= I^2_{r.m.s.} \times R \times t = I^2_{r.m.s.} \times R \times 5 \times 60 \text{ joules} \qquad \qquad ...(i)$$

Heat produced by the direct current

$$= 4^2 \times R \times 8 \times 60 \text{ joules} \qquad \qquad ...(ii)$$

As the heat produced in the two cases is the same,

$$\therefore \qquad I^2_{r.m.s.} \times R \times 5 \times 60 = 4^2 \times R \times 8 \times 60$$

or $$I_{r.m.s.} = \sqrt{\frac{16 \times 8}{5}} = \textbf{5.06 A}$$

Example 11.15. *Find the average value, r.m.s. value, form factor and peak factor for (i) half-wave rectified alternating current and (ii) full-wave rectified alternating current.*

Solution. (*i*) **Half-wave rectified a.c.** Fig. 11.24 shows half-wave rectified a.c. in which one half-cycle is suppressed *i.e.* current flows for half the time during complete cycle.

$$I_{av} = \frac{\text{Area of one cycle}}{\text{Full-cycle base}} = \frac{2I_m + 0}{2\pi} = \frac{I_m}{\pi}$$

$$I_{r.m.s.} = \left[\frac{\text{Area of squared wave over one cycle}}{\text{Full-cycle base}}\right]^{1/2}$$

$$= \left[\frac{\pi(I_m^2/2)+0}{2\pi}\right]^{1/2} = \frac{I_m}{2}$$

Form factor $= \dfrac{I_{r.m.s.}}{I_{av}} = \dfrac{I_m/2}{I_m/\pi} = 1.57$

Peak factor $= \dfrac{I_{max}}{I_{r.m.s.}} = \dfrac{I_m}{I_m/2} = 2$

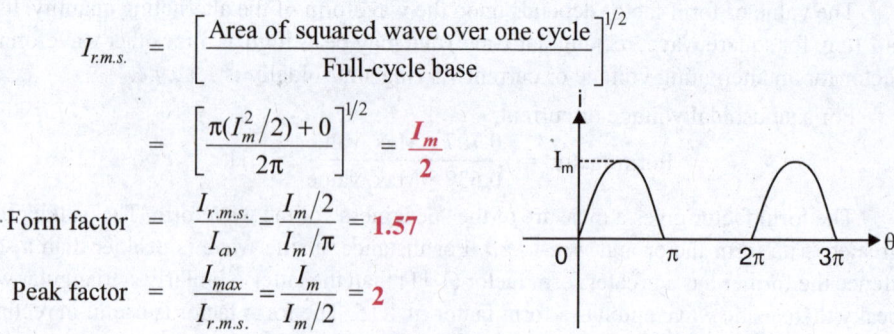

Fig. 11.24

(*ii*) **Full-wave rectified a.c.** Fig. 11.25 shows full-wave rectified a.c. in which both half-cycles appear in the output *i.e.* current flows in the same direction for both half-cycles. Since the wave is symmetrical, half-cycle may be considered for various computations.

$$I_{av} = \frac{\text{Area of half-cycle}}{\text{Half-cycle base}} = \frac{2I_m}{\pi}$$

$$I_{r.m.s.} = \left[\frac{\text{Area of squared wave over half-cycle}}{\text{Half-cycle base}}\right]^{1/2}$$

$$= \left[\frac{\pi I_m^2/2}{\pi}\right]^{1/2} = \frac{I_m}{\sqrt{2}}$$

Form factor $= \dfrac{I_{r.m.s}}{I_{av}} = \dfrac{I_m/\sqrt{2}}{(2/\pi)I_m} = 1.11$

Peak factor $= \dfrac{I_m}{I_{r.m.s.}} = \dfrac{I_m}{I_m/\sqrt{2}} = 1.414$

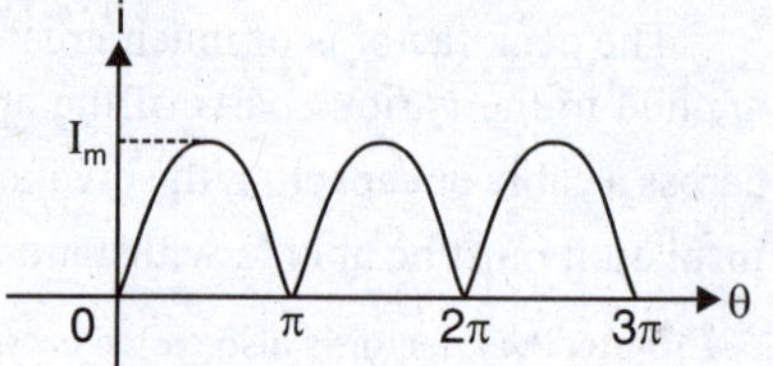

Fig. 11.25

Example 11.16. *An alternating voltage v = 200 sin 314t is applied to a device which offers an ohmic resistance of 20 Ω to the flow of current in one direction while entirely preventing the flow of current in the opposite direction. Calculate the r.m.s. value, average value and form factor.*

Solution. It is clear that the device is doing half-wave rectification. The maximum value of the rectified current is given by ;

$$I_m = V_m/R = 200/20 = 10 \text{ A}$$

For a half-wave rectified a.c. (See ex. 11.15),

$$I_{r.m.s.} = I_m/2 = 10/2 = 5 \text{ A}$$

$$I_{av} = I_m/\pi = 10/\pi = 3 \cdot 18 \text{ A}$$

Form factor $= I_{r.m.s.}/I_{av} = 5/3 \cdot 18 = 1 \cdot 57$

Example 11.17. *A current has the following steady values in amperes for equal intervals of time changing instantaneously from one value to the next [See Fig. 11.26].*

0, 10, 20, 30, 20, 10, 0, –10, –20, –30, –20, –10, 0, etc. Calculate (i) average value (ii) r.m.s. value (iii) form factor and (iv) peak factor.

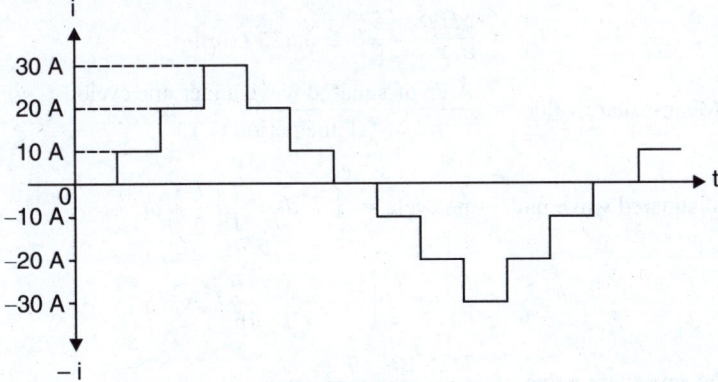

Fig. 11.26

Solution. (*i*) $I_{av} = \dfrac{i_1 + i_2 + i_3 + i_4 + i_5 + i_6}{6}$

$$= \frac{0 + 10 + 20 + 30 + 20 + 10}{6} = \frac{90}{6} = \textbf{15A}$$

(*ii*) $I^2_{r.m.s.} = \dfrac{i_1^2 + i_2^2 + i_3^2 + i_4^2 + i_5^2 + i_6^2}{6}$

$$= \frac{0^2 + 10^2 + 20^2 + 30^2 + 20^2 + 10^2}{6} = \frac{1900}{6} = 316.67$$

or $I_{r.m.s.} = \sqrt{316.67} = \textbf{17.8A}$

(*iii*)   Form factor $= \dfrac{I_{r.m.s}}{I_{av}} = \dfrac{17.8}{15} = \textbf{1.19}$

(*iv*)   Peak factor $= \dfrac{I_m}{I_{r.m.s}} = \dfrac{30}{17.8} = \textbf{1.68}$

Example 11.18. *A resistor eventually reaches a temperature of 65°C when it conducts a direct current of 75 mA. What peak value of sinusoidal alternating current will cause the resistor to reach the same temperature ?*

Solution. The heating effect of r.m.s. alternating current is the same as for direct current under similar conditions. Therefore, the r.m.s. value of the given alternating current must be 75 mA.

∴   Peak value of alternating current is

$$I_m = 1.414 \times I_{r.m.s.} = 1.414 \times 75 = \textbf{106 mA}$$

Example 11.19. *Determine (i) the average value and (ii) r.m.s. value of the current wave shown in Fig. 11.27.*

Solution. (*i*) Refer to Fig. 11.27. From $t = 0$ to $t = T/4$, we have, $i = I$. Then from $T/4$ to T, we have, $i = I/2$.

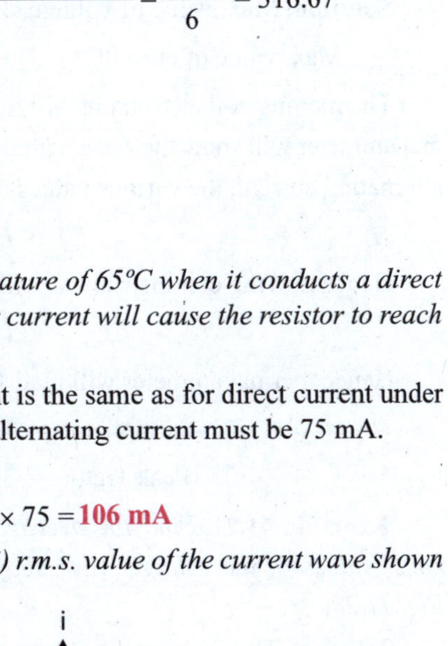

$$I_{av} = \frac{\text{Area under one cycle}}{\text{Time period } (T)}$$

Area under one cycle $= \displaystyle\int_{0}^{T/4} I\, dt + \int_{T/4}^{T} \frac{I}{2} dt$

$$= \frac{IT}{4} + \frac{1}{2}(T - T/4) = \frac{5IT}{8} \text{ amp-sec}$$

Fig. 11.27

$$\therefore \qquad I_{av} = \frac{5IT/8}{T} = \frac{5}{8}I = \textbf{0.625 I amp}$$

(ii) Mean-square value $= \dfrac{\text{Area of squared wave under one cycle}}{\text{Time period }(T)}$

$$\text{Area of squared wave under one cycle} = \int_0^{T/4} I^2 dt + \int_{T/4}^{T} \left(\frac{I}{2}\right)^2 dt$$

$$= \frac{I^2 T}{4} + \frac{I^2}{4}\left(T - \frac{T}{4}\right) = \frac{7T}{16}I^2 \text{A}^2\text{s}$$

$$\therefore \qquad \text{Mean square value} = \frac{7TI^2/16}{T} = \frac{7}{16}I^2 \text{ amp}^2$$

$$\text{R.M.S. value, } I_{r.m.s} = \sqrt{\frac{7}{16}I^2} = \textbf{0·662 I amp}$$

Example 11.20. *A moving coil ammeter, a thermal ammeter and a half-wave rectifier are connected in series with a resistor across 110 V a.c. supply. The circuit offers a resistance of 50Ω in one direction and an infinite resistance in the reverse direction. Calculate (i) the readings on the ammeters and (ii) the form factor and peak factor of current wave. Assume the supply voltage to be sinusoidal.*

Solution. Max. value of voltage, $V_m = 110/0.707 = 155.59$ V

Max. value of current, $I_m = 155.59/50 = 3.11$ A

The moving coil instrument will indicate the average value over the whole cycle while the thermal ammeter will show the r.m.s. value over the whole cycle. Since it is a case of half-wave rectified alternating current, the various values are (See ex. 11.15) :

(i) $\qquad\qquad I_{av} = I_m/\pi = 3.11/\pi = \textbf{0.99 A}$

Hence moving coil instrument will read 0.99 A.

$$I_{r.m.s.} = I_m/2 = 3.11/2 = \textbf{1.555 A}$$

Hence thermal ammeter will read 1.555 A.

(ii) $\qquad\qquad$ Form factor $= 1.555/0·99 = \textbf{1.57}$

$\qquad\qquad\qquad$ Peak factor $= 3.11/1.555 = \textbf{2}$

Example 11.21. *Find the average and r.m.s. values of the voltage wave shown in Fig. 11.28.*

Solution. The wave can be represented mathematically as :

$v = t$ for $0 < t < 1$; $v = 1$ for $1 < t < 2$

Average value, $V_{av} = \dfrac{1}{2}\left[\displaystyle\int_0^1 t\,dt + \int_1^2 dt\right]$

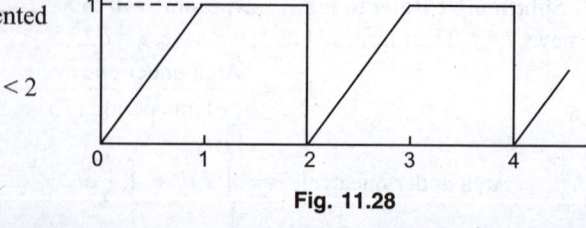

Fig. 11.28

$$= \frac{1}{2}\left[\frac{t^2}{2}\right]_0^1 + \frac{1}{2}[t]_1^2 = 0.25 + 0.5 = \textbf{0.75V}$$

Now,
$$V^2_{r.m.s.} = \frac{1}{2}\int_0^1 t^2 \, dt + \frac{1}{2}\int_1^2 dt$$

$$= \frac{1}{2}\left[\frac{t^3}{3}\right]_0^1 + \frac{1}{2}[t]_1^2 = \frac{1}{6} + \frac{1}{2} = \frac{4}{6}$$

$$\therefore \qquad V_{r.m.s.} = \sqrt{\frac{4}{6}} = \textbf{0.8165 V}$$

Example 11.22. *Determine the form factor of the sawtooth wave shown in Fig. 11.29.*

Solution. The equation of the voltage wave is given by $e = kt$ where k is the slope (*i.e.,* $\tan \theta$) of the curve. Referring to Fig. 11.29, the slope of the curve is $E_m/2$.

$$\therefore \qquad e = \frac{E_m}{2}t$$

$$E^2_{r.m.s.} = \frac{1}{2}\int_0^2 e^2 \, dt = \frac{1}{2}\int_0^2 \frac{E_m^2 t^2}{4} \, dt$$

$$= \frac{E_m^2}{8}\left[\frac{t^3}{3}\right]_0^2 = \frac{E_m^2}{3}$$

$$\therefore \qquad E_{r.m.s.} = E_m/\sqrt{3} = 50/\sqrt{3} = 28.87 \text{ V}$$

$$E_{av.} = \frac{\text{Area under curve}}{\text{Length of base}} = \frac{(1/2)\times 50 \times 2}{2} = 25 \text{ V}$$

$$\therefore \qquad \text{Form factor} = 28.87/25 = \textbf{1.155}$$

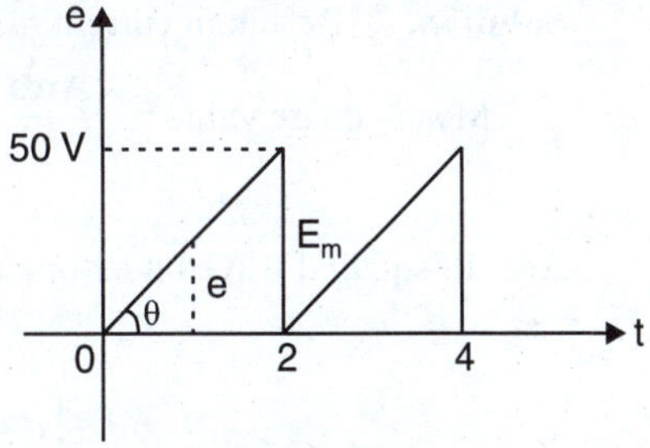

Fig. 11.29

Example 11.23. *A periodic sequence of rectangular pulses is shown in Fig. 11.30 with a peak value of 100 volts and time period of 0·1s. Find the form factor of the wave.*

Solution. The equation of the voltage wave is given by ;

$$e = 100 \text{ volts}$$

$$E^2_{r.m.s.} = \frac{1}{0.1}\int_{0.05}^{0.1} e^2 \, dt = \frac{1}{0.1}\int_{0.05}^{0.1} (100)^2 \, dt$$

$$= \frac{1}{0.1}[10^4 t]_{0.05}^{0.1}$$

$$= \frac{1}{0.1}[1000 - 500] = 5000$$

$$\therefore \qquad E_{r.m.s.} = \sqrt{5000} = 70.7 \text{ V}$$

$$E_{av} = \frac{\text{Area under curve}}{\text{Length of base}} = \frac{100 \times 0.05}{0.1} = 50 \text{ V}$$

Fig. 11.30

$$\therefore \qquad \text{Form factor} = 70·7/50 = \textbf{1·414}$$

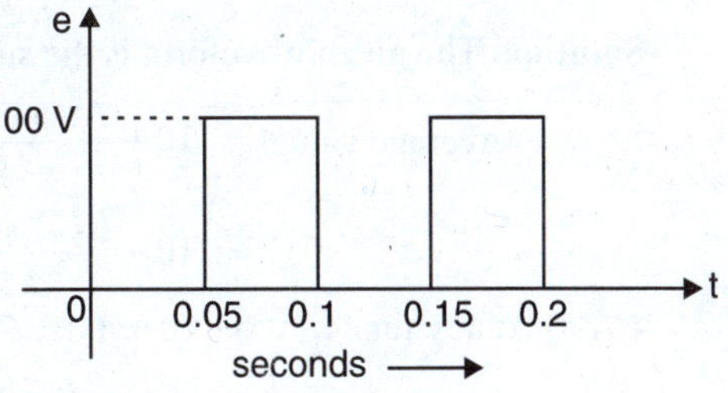

Example 11.24. *A current wave is represented as $i = 100e^{-200t}$. The time period of the wave is 0.05 second. Find the average and r.m.s. values of the wave.*

Solution. Average value, $I_{av} = \dfrac{1}{0.05}\displaystyle\int_0^{0.05} 100e^{-200t} \, dt = \dfrac{100}{0.05}\displaystyle\int_0^{0.05} e^{-200t} \, dt$

$$= \frac{100}{0.05(-200)} \left[e^{-200t} \right]_0^{0.05} = \mathbf{10A}$$

$$I_{r.m.s.}^2 = \frac{1}{0.05} \int_0^{0.05} (100 e^{-200t})^2 dt = \frac{10000}{0.05} \int_0^{0.05} e^{-400t} dt$$

$$= \frac{10000}{0.05 \times (-400)} \left[e^{-400t} \right]_0^{0.05} = 500$$

$$\therefore \quad I_{r.m.s.} = \sqrt{500} = \mathbf{22.36\ A}$$

Example 11.25. *A current is made up of two components viz. 3A d.c. component and an a.c. component given by i = 4 sin ωt amperes. Find (i) an expression for the resultant current (ii) r.m.s. value.*

Solution. (*i*) Resultant current $= I_{dc} + i = (3 + 4 \sin \omega t)$ A

(*ii*) Mean-square value $= \dfrac{\text{Area of square wave under one cycle}}{\text{Time period } (T)}$

Area of squared wave under one cycle $= \int_0^T (3 + 4 \sin \omega t)^2 dt = \int_0^T (9 + 16 \sin^2 \omega t + 24 \sin \omega t) dt$

$$= 9 \int_0^T dt + 16 \int_0^T \sin^2 \omega t\, dt + 24 \int_0^T \sin \omega t\, dt$$

$\therefore$ Mean square value $= 9 \left[\dfrac{1}{T} \int_0^T dt \right] + 16 \left[\dfrac{1}{T} \int_0^T \sin^2 \omega t\, dt \right] + 24 \left[\dfrac{1}{T} \int_0^T \sin \omega t\, dt \right]$

Now, $\dfrac{1}{T} \int_0^T \sin^2 \omega t\, dt = \dfrac{1}{2}$; $\dfrac{1}{T} \int_0^T \sin \omega t\, dt = 0$; $\dfrac{1}{T} \int_0^T dt = 1$

$\therefore$ Mean square value $= 9 \times 1 + 16 \times \dfrac{1}{2} + 24 \times 0 = 17$ $\quad \therefore I_{r.m.s.} = \sqrt{17}\,\mathbf{A}$

Example 11.26. *Find (i) the average value and (ii) r.m.s. value of the waveform shown in Fig. 11.31.*

Solution. The given waveform is the sum of constant component of 10 and sawtooth wave.

(*i*) Average value $= 10 + \dfrac{\text{Area of } \triangle ABC}{T}$

$$= 10 + \dfrac{(1/2) \times 10 \times T}{T} = 10 + 5 = \mathbf{15}$$

(*ii*) At any time t, $y = 10 + *(10/T)t$

$\therefore$ Mean square value $= \dfrac{1}{T} \int_0^T y^2 dt$

$$= \dfrac{1}{T} \int_0^T \{10 + (10/T)t\}^2 dt$$

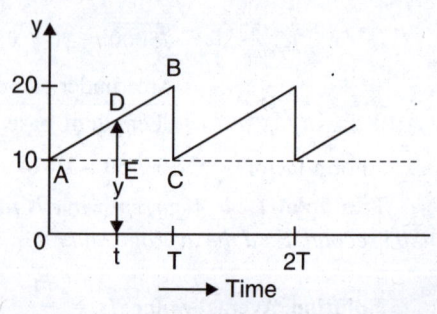

Fig. 11.31

---

* At any time t, $DE = mx = \dfrac{BC}{AC} \times t = \dfrac{10}{T} \times t$

$$= \frac{1}{T}\int_0^T \left(100 + \frac{100}{T^2}t^2 + \frac{200}{T}t \right) dt$$

$$= \frac{1}{T}\left| 100t + \frac{100t^3}{3T^2} + \frac{100t^2}{T} \right|_0^T = 700/3$$

∴   R.M.S. value $= \sqrt{700/3} = 15.2$

Example 11.27. *Determine (i) the average value (ii) r.m.s. value of a symmetrical square wave. Also find the form factor and peak factor.*

Solution. Consider the symmetrical square current wave of maximum value I_m [See Fig. 11.32]. The instantaneous value of current is given by ;

$$i = I_m \text{ for } 0 < \theta < \pi$$

(*i*) **Average value.** The average value of the square current wave is given by ;

$$I_{av} = \frac{1}{\pi}\int_0^\pi id\theta = \frac{1}{\pi}\int_0^\pi I_m d\theta \qquad (\because i = I_m)$$

$$= I_m\frac{(\pi - 0)}{\pi} = I_m$$

∴ $I_{av} = I_m$

Note that this is the half-cycle average value of the square wave. However, its average value over one cycle is zero.

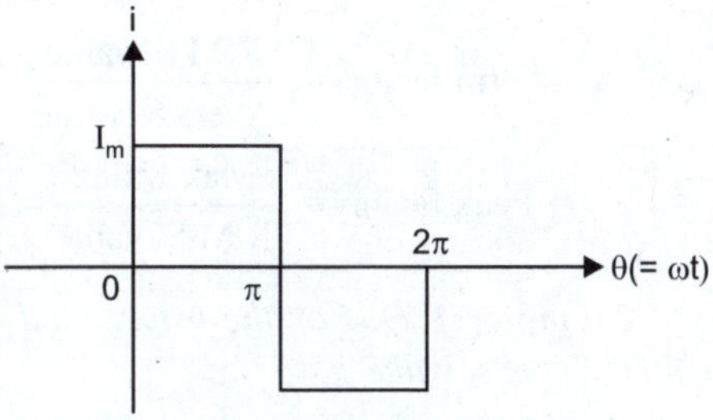

(*ii*) **R.M.S. value.** The r.m.s. value of square current wave is given by ;

Fig. 11.32

$$I_{r.m.s.} = \sqrt{\frac{1}{\pi}\int_0^\pi i^2 d\theta} = \sqrt{\frac{1}{\pi}\int_0^\pi I_m^2 d\theta} = \sqrt{\frac{1}{\pi}I_m^2(\pi - 0)} = I_m$$

∴ $I_{r.m.s.} = I_m$

$$\text{Form factor} = \frac{\text{R.M.S. value}}{\text{Average value}} = \frac{I_m}{I_m} = 1$$

$$\text{Peak factor} = \frac{\text{Max. value}}{\text{R.M.S. value}} = \frac{I_m}{I_m} = 1$$

Example 11.28. *Determine (i) the average value (ii) r.m.s. value of triangular wave. Also find the form factor and peak factor.*

Solution. Consider a triangular current wave of maximum value I_m as shown in Fig. 11.33. The instantaneous value of current is given by ;

$$*i = \frac{I_m}{\pi}\theta \text{ for } 0 < \theta < \pi$$

(*i*) **Average value.** The average value of triangular current wave is given by ;

---

*   The equation of the straight line is $y = mx$.

Here, $y = i$, $x = \theta$ and slope $m = \dfrac{I_m}{\pi}$

∴ $i = \dfrac{I_m}{\pi}\theta$

$$I_{av} = \frac{1}{\pi}\int_0^\pi i\, d\theta = \frac{1}{\pi}\int_0^\pi \frac{I_m}{\pi}\theta\, d\theta \qquad\qquad \left(\because i = \frac{I_m}{\pi}\theta\right)$$

$$= \frac{I_m}{\pi^2}\int_0^\pi \theta\, d\theta = \frac{I_m}{\pi^2}\left[\frac{\theta^2}{2}\right]_0^\pi = \frac{I_m}{2}$$

$$\therefore\qquad I_{av} = \frac{I_m}{2}$$

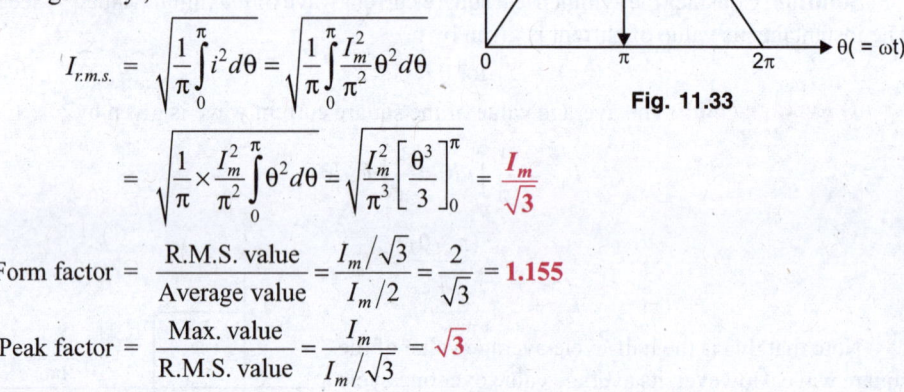

Fig. 11.33

(ii) R.M.S. value. The r.m.s. value of triangular current wave is given by ;

$$I_{r.m.s.} = \sqrt{\frac{1}{\pi}\int_0^\pi i^2\, d\theta} = \sqrt{\frac{1}{\pi}\int_0^\pi \frac{I_m^2}{\pi^2}\theta^2\, d\theta}$$

$$= \sqrt{\frac{1}{\pi}\times\frac{I_m^2}{\pi^2}\int_0^\pi \theta^2\, d\theta} = \sqrt{\frac{I_m^2}{\pi^3}\left[\frac{\theta^3}{3}\right]_0^\pi} = \frac{I_m}{\sqrt{3}}$$

$$\text{Form factor} = \frac{\text{R.M.S. value}}{\text{Average value}} = \frac{I_m/\sqrt{3}}{I_m/2} = \frac{2}{\sqrt{3}} = 1.155$$

$$\text{Peak factor} = \frac{\text{Max. value}}{\text{R.M.S. value}} = \frac{I_m}{I_m/\sqrt{3}} = \sqrt{3}$$

Example 11.29. *For the trapezoidal current waveform of Fig. 11.34, determine (i) average value (ii) r.m.s. value.*

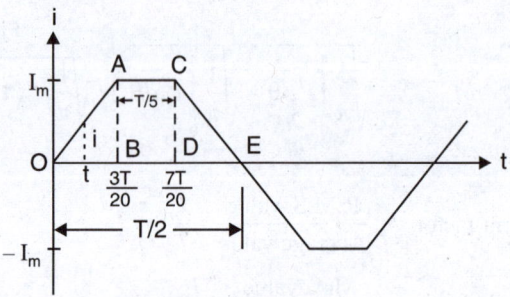

Fig. 11.34

Solution. The time period of the wave is *T*.

(i) Average value, $I_{av} = \dfrac{\text{Half-cycle area of wave}}{T/2} = \dfrac{\frac{1}{2}\left(\frac{T}{5} + \frac{T}{2}\right)I_m}{T/2} = \dfrac{7}{10}I_m$

(ii) Note that area of the squared wave for the portion *OA* of the wave is equal to the area of squared wave for the portion *CE* of the wave. The current equations are :

For portion *OA*, $i = mx = \dfrac{AB}{OB}\times t = \dfrac{I_m}{3T/20}\times t = \dfrac{20 I_m}{3T}t$

For portion *AC*, $i = I_m$

$\therefore$ Mean square value $= \dfrac{1}{T/2}\left[2\int_0^{3T/20} i^2\, dt + \int_{3T/20}^{7T/20} I_m^2\, dt\right]$

$$= \frac{2}{T}\left[2\left(\frac{20I_m}{3T}\right)^2 \int_0^{3T/20} t^2 \, dt + I_m^2 \int_{3T/20}^{7T/20} dt \right] = \frac{3}{5}I_m^2$$

$\therefore$ R.M.S. value, $I_{r.m.s.} = \sqrt{\frac{3}{5}I_m^2} = 0.775\, I_m$

Example 11.30. *Find the average and effective values of the sinusoidal waveform shown in Fig. 11.35. The maximum value is 100 V.*

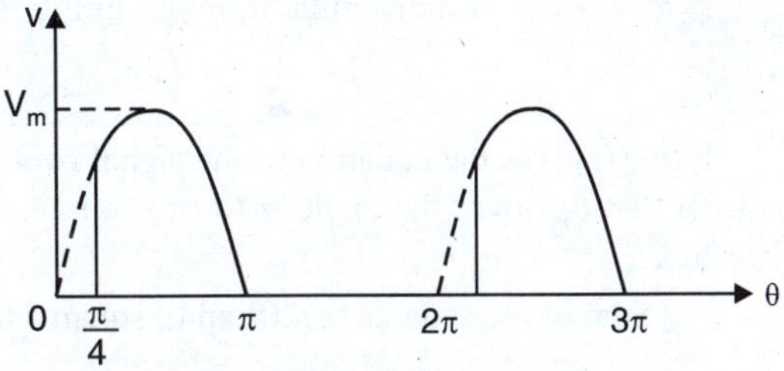

Solution. The time period of the wave is 2π. Since the wave is unsymmetrical, average value over the whole cycle is to be taken. The equation of the wave is $v = 100 \sin \theta$.

Average value, $V_{av} = \dfrac{1}{2\pi} \displaystyle\int_{\pi/4}^{\pi} 100 \sin \theta \, d\theta = \dfrac{100}{2\pi}\left|-\cos\theta\right|_{\pi/4}^{\pi} = 27.2\text{V}$

Mean square value, $V_{r.m.s.}^2 = \dfrac{1}{2\pi} \displaystyle\int_{\pi/4}^{\pi} (100)^2 \sin^2\theta d\theta = \dfrac{10^4}{2\pi} \displaystyle\int_{\pi/4}^{\pi} \dfrac{1-\cos 2\theta}{2} d\theta$

$$= \frac{10^4}{4\pi}\left|\theta - \frac{\sin 2\theta}{2}\right|_{\pi/4}^{\pi} \qquad \therefore \quad V_{r.m.s.} = 47.7 \text{ V}$$

Example 11.31. *Determine the r.m.s. value of semicircular current wave which has a maximum value of a.*

Solution. Fig. 11.36 shows the conditions of the problem. The equation of a semicircular wave is

$$x^2 + y^2 = a^2 \quad \text{or} \quad y^2 = a^2 - x^2$$

$\therefore$ Mean square value, $I_{r.m.s.}^2 = \dfrac{1}{2a}\displaystyle\int_{-a}^{+a}(a^2 - x^2)dx = \dfrac{1}{2a}\displaystyle\int_{-a}^{+a} a^2 dx - x^2 dx$

$$= \frac{1}{2a}\left|a^2 x - \frac{x^3}{3}\right|_{-a}^{+a} = \frac{2a^2}{3}$$

$\therefore \qquad I_{r.m.s.} = \sqrt{\dfrac{2a^2}{3}} = 0.816\, a$

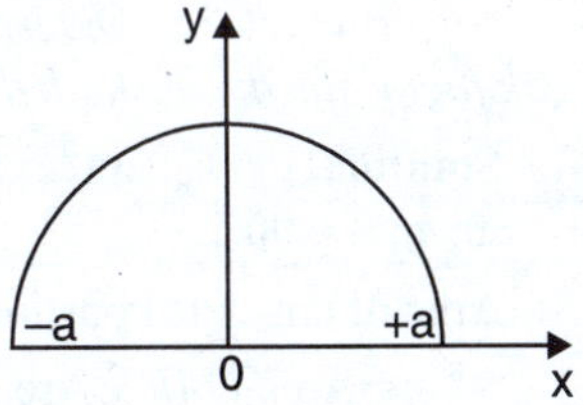

Fig. 11.35

Fig. 11.36

Example 11.32. *The half-cycle of an alternating signal is as follows :*

It increases uniformly from zero at $0°$ to F_m at $\alpha°$, remains constant from $\alpha°$ to $(180 - \alpha)°$, decreases uniformly from F_m at $(180 - \alpha)°$ to zero at $180°$. Calculate the average and effective values of the signal.

Solution. The half-cycle of the given alternating signal is shown in Fig. 11.37.

Average value, $F_{av} = \dfrac{\text{Area of trapezium}}{\pi}$

Now, Area of trapezium = $2 \times$ Area of $\triangle OAE +$ Area of rectangle $ABDE$

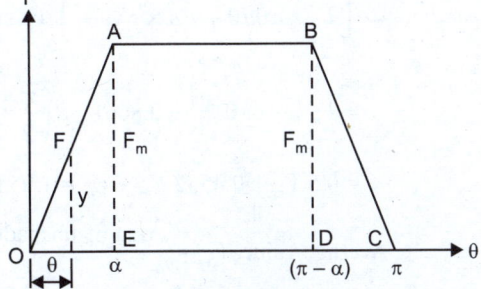

Fig. 11.37

$$= 2 \times \left(\frac{1}{2} \times F_m \times \alpha \right) + (\pi - 2\alpha) F_m$$

$$= F_m \alpha + (\pi - 2\alpha) F_m = (\pi - \alpha) F_m$$

$$\therefore \qquad F_{av} = \frac{(\pi - \alpha) F_m}{\pi}$$

R.M.S. value. From similar triangles in Fig. 11.37, we have,

$$\frac{y}{\theta} = \frac{F_m}{\alpha} \quad \text{or} \quad y^2 = \frac{F_m^2}{\alpha^2} \theta^2 \qquad \qquad ...(i)$$

Exp. (*i*) gives the equation of the signal over the triangles *OAE* and *DBC*. The signal remains constant ($= F_m$) over the angle α to $(\pi - \alpha)$ *i.e.* over the angular displacement of $(\pi - \alpha) - \alpha = (\pi - 2\alpha)$.

Sum of the squares $= 2(\text{Sum of squares from 0 to } \alpha) + \text{Sum of squares over } (\pi - 2\alpha)$

$$= 2 \times \int_0^\alpha y^2 d\theta + F_m^2 (\pi - 2\alpha) = 2 \times \int_0^\alpha \frac{F_m^2}{\alpha^2} \theta^2 d\theta + F_m^2 (\pi - 2\alpha)$$

$$= \frac{2 F_m^2}{\alpha^2} \int_0^\alpha \theta^2 d\theta + F_m^2 (\pi - 2\alpha) = \frac{2 F_m^2}{\alpha^2} \left. \frac{\theta^3}{3} \right|_0^\alpha + F_m^2 (\pi - 2\alpha)$$

$$= \frac{2 F_m^2}{\alpha^2} \left(\frac{\alpha^3}{3} \right) + F_m^2 (\pi - 2\alpha) = \frac{2}{3} F_m^2 \alpha + F_m^2 (\pi - 2\alpha)$$

$$= F_m^2 \left(\frac{2}{3} \alpha + \pi - 2\alpha \right) = F_m^2 \left(\pi - \frac{4\alpha}{3} \right)$$

Mean value of squares $= \dfrac{1}{\pi} \times F_m^2 \left(\pi - \dfrac{4\alpha}{3} \right) = F_m^2 \left(1 - \dfrac{4\alpha}{3\pi} \right)$

$$\therefore \qquad \text{R.M.S. value} = F_m \sqrt{\left(1 - \frac{4\alpha}{3\pi} \right)}$$

Example 11.33. *Determine the form factor and peak factor for the unshaded waveform in Fig. 11.38.*

Solution. $v = V_m \sin\theta$ except for the region between $\theta = 60°$ to $\theta = 90°$.

Area of unshaded portion of the wave in Fig. 11.38

    $=$ Area *OAF* + Area *FABG* + Area *GBCD*

$$= \int_0^{\pi/3} V_m \sin\theta d\theta + FA \times FG + \int_{\pi/2}^{\pi} V_m \sin\theta d\theta$$

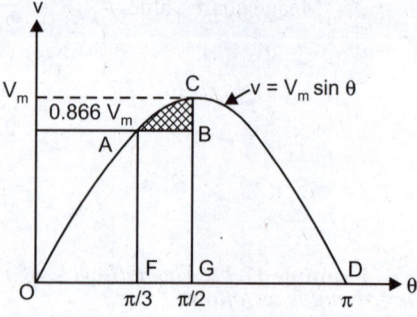

Fig. 11.38

$$= V_m \left| -\cos\theta \right|_0^{\pi/3} + 0.866 V_m \times \left(\frac{\pi}{2} - \frac{\pi}{3} \right) + V_m \left| -\cos\theta \right|_{\pi/2}^{\pi}$$

$$= 0.5\, V_m + 0.4532\, V_m + V_m = 1.9532\, V_m$$

$$\therefore \qquad \text{Average value, } V_{av} = \frac{\text{Area of unshaded portion of wave}}{\pi}$$

$$= \frac{1.9532 V_m}{\pi} = 0.622\, V_m$$

Area of squared unshaded wave

$$= \int_0^{\pi/3} V_m^2 \sin^2\theta \, d\theta + (0.866V_m)^2 \times \left(\frac{\pi}{2} - \frac{\pi}{3}\right) + \int_{\pi/2}^{\pi} V_m^2 \sin^2\theta \, d\theta$$

$$= \int_0^{\pi/3} V_m^2 \frac{1-\cos 2\theta}{2} \, d\theta + 0.75V_m^2 \times \frac{\pi}{6} + \int_{\pi/2}^{\pi} V_m^2 \frac{1-\cos 2\theta}{2} \, d\theta$$

$$= 0.307\, V_m^2 + 0.3925 V_m^2 + 0.785 V_m^2 = 1.4845 V_m^2$$

$$\text{R.M.S. value} = \sqrt{\frac{\text{Area of squared wave}}{\text{Base}}} = \sqrt{\frac{1.4845 V_m^2}{\pi}} = 0.6876\, V_m$$

$\therefore$ $$\text{Form factor} = \frac{\text{R.M.S. value}}{\text{Average value}} = \frac{0.6876\, V_m}{0.622\, V_m} = \textbf{1.105}$$

$$\text{Peak factor} = \frac{\text{Maximum value}}{\text{R.M.S. value}} = \frac{V_m}{0.6876\, V_m} = \textbf{1.454}$$

Example 11.34. *Find the r.m.s. and average values of the saw-tooth waveform shown in Fig. 11.39.*

Solution. The *equation of the straight line from $t = 0$ to $t = 1$ in Fig. 11.39 is

$$f(t) = 4t - 2$$

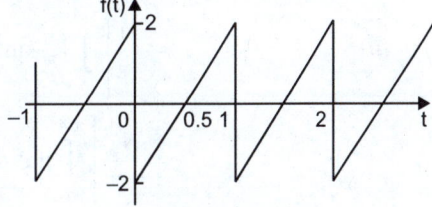

Fig. 11.39

$\therefore$ Average value $= \dfrac{1}{T}\displaystyle\int_0^T (4t-2)\,dt = \dfrac{1}{1}\displaystyle\int_0^1 (4t-2)\,dt$

$$= \frac{1}{1}\left|\frac{4t^2}{2} - 2t\right|_0^1 = \textbf{0}$$

This is expected because from $t = 0$ to $t = 1$, the wave completes one cycle. Now area of negative half-cycle of the wave is exactly equal to the area of the positive half-cycle of the wave so that average value of the wave over complete cycle is zero.

Now, $$f^2(t) = (4t-2)^2 = 16t^2 - 16t + 4$$

$\therefore$ $(\text{R.M.S. value})^2 = \dfrac{1}{T}\left[\displaystyle\int_0^T (16t^2 - 16t + 4)\,dt\right]$

$$= \frac{1}{1}\left[\int_0^1 (16t^2 - 16t + 4)\,dt\right]$$

$$= \frac{1}{1}\left|\frac{16t^3}{3} - \frac{16t^2}{2} + 4t\right|_0^1 = \frac{4}{3}$$

$\therefore$ R.M.S. value $= \sqrt{\dfrac{4}{3}} = \textbf{1.15}$

---

*

$$y = mx + c$$

Here, $y = f(t)$; $m = \dfrac{2}{0.5} = 4$; $x = t$; $c = -2$

$\therefore$ $f(t) = 4t - 2$

Example 11.35. *A full-wave rectified wave is clipped at 70.7% of its maximum value as shown in Fig. 11.40. Find the (i) average and (ii) r.m.s. values.*

Solution. The equation for the wave is

$v = V_m \sin \omega t$ for $0 < \omega t < \dfrac{\pi}{4}$; $v = 0.707\, V_m$

for $\dfrac{\pi}{4} < \omega t < \dfrac{3\pi}{4}$; $v = V_m \sin \omega t$ for $\dfrac{3\pi}{4} < \omega t < \pi$

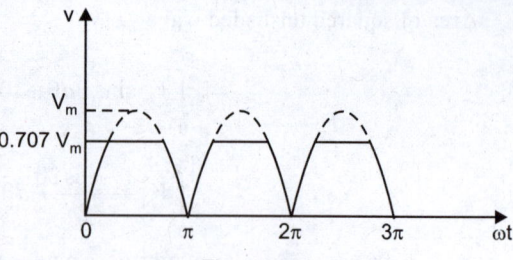

Fig. 11.40

(i)
$$V_{av} = \frac{1}{\pi}\left[\int_0^{\pi/4} V_m \sin \omega t\, d(\omega t) + \int_{\pi/4}^{3\pi/4} 0.707 V_m d(\omega t) + \int_{3\pi/4}^{\pi} V_m \sin \omega t\, d(\omega t) \right]$$

$$= \frac{1}{\pi}\left[V_m[-\cos \omega t]_0^{\pi/4} + 0.707 V_m[\omega t]_{\pi/4}^{3\pi/4} + V_m[-\cos \omega t]_{3\pi/4}^{\pi} \right]$$

$$= \frac{1}{\pi}[V_m(1-0.707) + 0.707 V_m(1.57) + V_m(1-0.707)] = \mathbf{0.54\ V_m}$$

(ii)
$$V^2_{r.m.s.} = \frac{1}{\pi}\left[\int_0^{\pi/4} V_m^2 \sin^2 \omega t\, d(\omega t) + \int_{\pi/4}^{3\pi/4} (0.707 V_m)^2 d(\omega t) + \int_{3\pi/4}^{\pi} V_m^2 \sin^2 \omega t\, d(\omega t) \right]$$

$$= \frac{V_m^2}{\pi}\left[\int_0^{\pi/4} \sin^2 \omega t\, d(\omega t) + \int_{\pi/4}^{3\pi/4} (0.707)^2 d(\omega t) + \int_{3\pi/4}^{\pi} \sin^2 \omega t\, d(\omega t) \right]$$

or $\qquad V^2_{r.m.s.} = 0.341\, V_m^2$ $\qquad \therefore\ V_{r.m.s.} = \sqrt{0.341 V_m^2} = \mathbf{0.584\ V_m}$

Example 11.36. *A moving coil ammeter, a hot-wire ammeter and a resistance of 100Ω are connected in series with a rectifier across 200 V sinusoidal source. The resistance of the rectifier is 100Ω in the forward direction and 500Ω in the backward direction. Find the reading on each ammeter.*

Solution. The rectifier is doing full-wave rectification for which the form factor is 1.11.

R.M.S. current in forward direction = $\dfrac{200}{100+100} = 1\text{A}$

R.M.S. current in backward direction = $\dfrac{-200}{100+500} = -\dfrac{1}{3}\text{A}$

Average current in forward direction = $\dfrac{1}{1.11} = 0.9\text{ A}$

Average current in backward direction = $-\dfrac{1}{3} \times \dfrac{1}{1.11} = -0.3\text{ A}$

Moving coil ammeter measures the average value of current over the full-cycle.

$\therefore$ Reading of moving coil ammeter = $\dfrac{0.9 + (-0.3)}{2} = \mathbf{0.3A}$

Hot-wire ammeter reads the r.m.s. value over full-cycle.

$\therefore$ Reading of hot-wire ammeter = $\sqrt{\dfrac{1}{2}\left[(1)^2 + \left(\dfrac{1}{3}\right)^2 \right]} = \mathbf{0.745A}$

Tutorial Problems

1. An alternating voltage $e = 100 \sin 314\ t$ is applied to a device which offers an ohmic resistance of 50 Ω in one direction while entirely preventing the flow of current in the opposite direction. Calculate the average and r.m.s. value of current. Also find the form factor.   **[0·637 A ; 1 A; 1·57]**

2. An alternating current was measured by a d.c. milliammeter in conjunction with a full-wave rectifier. The reading on the milliammeter was 7 mA. Assuming the waveform of the alternating current to be sinusoidal, calculate (*i*) the r.m.s. value and (*ii*) the maximum value of the alternating current.

    [(*i*) 7·77 mA (*ii*) 11 mA]

3. A hot-wire ammeter, a moving coil ammeter and a rectifier are connected in series. The combination is connected across a sinusoidal source of 50 V r.m.s. The forward resistance of rectifier is 20 Ω and the reverse resistance is infinite. Calculate the reading on each ammeter.   **[1·767 A ; 1·124 A]**

4. A moving coil ammeter and a moving-iron ammeter are connected in series with a rectifier across a 110 V (r.m.s.) supply. The total resistance of the circuit in the conducting direction is 60 Ω and that in the reverse direction may be taken as infinity. Assuming the waveform of the supply voltage to be sinusoidal, calculate the reading on each ammeter.   **[Moving coil 0·825 A ; Moving iron 1·296 A]**

5. An osglim lamp lights when the p.d. applied to it exceeds 170 V and goes out when the p.d. drops to 140 V. If such a lamp is connected to a 240 V a.c. supply, for what fraction of time is the lamp lit?

    [0·7 second]

6. Find (*i*) the r.m.s. value (*ii*) form factor for a symmetrical triangular wave shown in Fig. 11.41.

    [(*i*) 5·77 V (*ii*) 1·155]

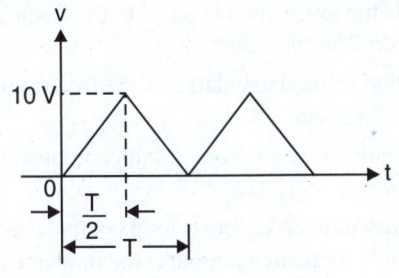

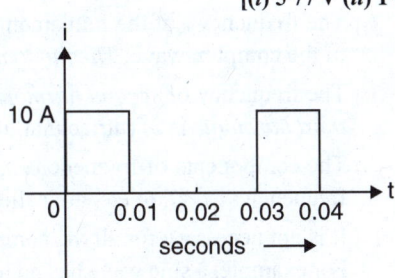

Fig. 11.41   Fig. 11.42

7. Find the form factor for the square wave shown in Fig. 11.42.   **[1·732]**

8. A generator produces a triangular voltage wave as a function of time as shown in Fig. 11.43. This voltage is impressed across a 10 ohm resistor. How much energy is delivered to the resistor in 1 second ? **[333·3 J]**

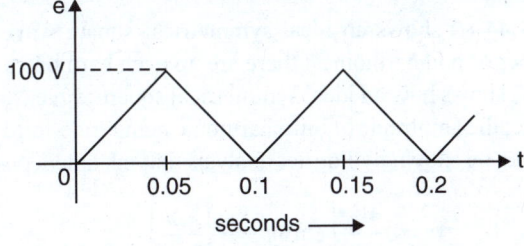

Fig. 11.43

9. The positive half-cycle of a symmetrical alternating current waveform varies as follows :

Time (ms)	0–2	2–6	6–8
Current	Uniform increase from 0 to 1A	Constant at 1A	Uniform decrease from 1A to 0

    Plot the current waveform over one cycle and determine the r.m.s. value of the current.   **[0·79 A]**

10. The voltage waveform produced by a generator is represented by ;

    $$v = 50 + 141 \cdot 4 \sin \theta + 35 \cdot 5 \sin 3\ \theta$$

    Find the effective value of the voltage wave.   **[114·6 V]**

11·17. Complex Waveforms

So far we have considered sinusoidal waves. However, many electrical waveforms encountered in electrical and electronic equipment (*e.g.*, square wave, sawtooth wave etc.) are not sinusoidal. **Complex waves** *are those which depart from sinusoidal form i.e., they are non-sinusoidal.*

A **periodic* **non-sinusoidal wave** *of any form* (*complex waveform*) *can be considered to be made up of sinusoidal waveforms of various frequencies and amplitudes.*

The above statement is the basis of *Fourier series.* Accordingly, a square waveform can be considered to be made up from a combination of sinusoidal waveforms. Thus any non-sinusoidal quantity i can be expressed as the sum of a number of sinusoidal quantities $i_1, i_2, i_3,....$. The frequency of i_1 will be the same as that of i and the frequencies of $i_2, i_3...$ will be integral multiple (2, 3, 4, 5...n) of that of i. In symbols,

$$i = i_1 + i_2 + i_3 + ...$$
$$= I_{m1} \sin(\omega t + \phi_1) + I_{m2} \sin(2\omega t + \phi_2) + I_{m3} \sin(3\omega t + \phi_3) + ... \qquad ...(i)$$

If we add ****d.c. component I_0 to equation (*i*), it is known as *Fourier series.*

The term that has the same frequency as i, namely $I_{m1} \sin(\omega t + \phi_1)$ is called *fundamental* component; $I_{m2} \sin(2\omega t + \phi_2)$, whose frequency is twice that of i (or i_1) is called the *second harmonic* component; $I_{m3} \sin(3\omega t + \phi_3)$ is the *third harmonic*. The following points may be noted :

(*i*) The frequency f of the fundamental component is the lowest and is equal to the frequency of the complex wave. *The fundamental is also called first harmonic.*

(*ii*) The frequency of *second harmonic* is 2f (twice that of fundamental) and the frequency of *third harmonic* is 3f (thrice that of fundamental) and so on.

(*iii*) The components of frequencies f, 3f, 5f etc. are called *odd harmonics* and components of frequencies 2f, 4f, 6f etc. are called *even harmonics.*

(*iv*) It is not necessary for all the components to be present in an equation for a complex wave. For example, a sine wave has no harmonics; the only frequency present is the fundamental.

(*v*) The harmonics are integer multiples of the fundamental frequency. The amplitude of each harmonic decreases as the frequency increases.

(*vi*) Any periodic waveform can be produced by summing sine waves that have proper frequency, amplitude and phase.

Illustration. Fig. 11.44 (*i*) shows an ideal symmetrical square wave. An ideal square wave contains an infinite number of odd harmonics; there are no even harmonics and there is no constant component. Fig. 11.44 (*ii*) shows how an ideal symmetrical square wave can be built up with fundamental and third harmonic; the amplitude of third harmonic being †one-third that of the fundamental. For this square wave, it can be shown by Fourier analysis that instantaneous voltage (*v*) at any time t is given by ;

$$v = \frac{4V_m}{\pi}\left[\sin\omega t + \frac{\sin 3\omega t}{3}\right]$$

where V_m = Peak value of the square wave

           ω = 2π × fundamental frequency = $2\pi f$

---

*   A periodic quantity is one whose waveforms in successive periods are identical.

**   A pulsating quantity is the sum of alternating quantity and a constant quantity.

†   The amplitude of third harmonic component of a square wave is one-third that of fundamental; the amplitude of 5th harmonic is one-fifth that of fundamental and so on

The sin ωt component is the fundamental and sin $3\omega t$ component is the third harmonic.

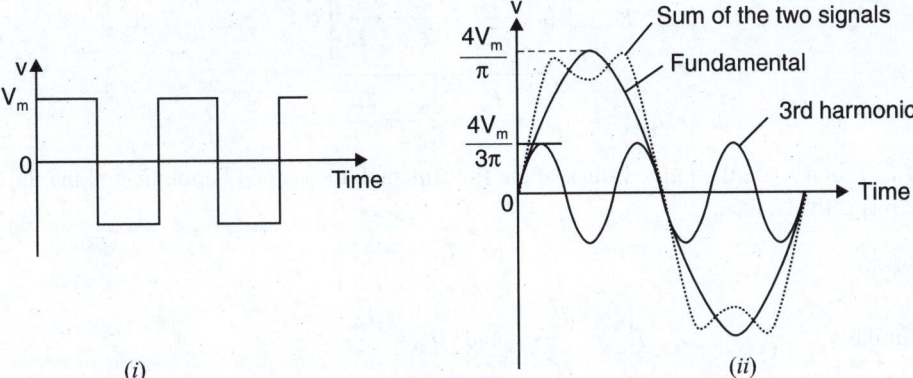

Fig. 11.44

If we also add higher harmonics, the equation becomes :

$$v = \frac{4V_m}{\pi}\left[\sin \omega t + \frac{\sin 3\omega t}{3} + \frac{\sin 5\omega t}{5} + ...\right]$$

Note. *Any variation from a pure sine wave produces harmonics.* A non-sinusoidal wave is composed of the fundamental and the harmonics. Some types of waveforms have only odd harmonics, some have only even harmonics and some contain both. Generally, only the fundamental, and the first few harmonics are of significant importance in determining the waveshape of the wave.

11.18.   R.M.S. Value of a Complex Wave

When a voltage has a complex waveform, the total average power is the arithmetic sum of the average powers developed by the voltages and the currents of like frequencies. Let us assume that the current in a resistor R is given by ;

$$i = I_0 + I_{m1} \sin (\omega t + \phi_1) + I_{m2} \sin (2\omega t + \phi_2) + I_{m3} \sin (3\omega t + \phi_3)$$

The average power produced by the d.c. term is

$$P_0 = I_0^2 R$$

The average power produced by the fundamental is

$$P_1 = \left(\frac{I_{m_1}}{\sqrt{2}}\right)^2 R = \frac{I_{m_1}^2}{2} R$$

The average power produced by the second harmonic is

$$P_2 = \left(\frac{I_{m_2}}{\sqrt{2}}\right)^2 R = \frac{I_{m_2}^2}{2} R$$

The average power produced by the third harmonic is

$$P_3 = \left(\frac{I_{m_3}}{\sqrt{2}}\right)^2 R = \frac{I_{m_3}^2}{2} R$$

The total power P_T is the sum of these average values.

$$\therefore \qquad P_T = I_0^2 R + \frac{I_{m_1}^2}{2} R + \frac{I_{m_2}^2}{2} R + \frac{I_{m_3}^2}{2} R$$

or

$$P_T = \left(I_0^2 + \frac{I_{m_1}^2}{2} + \frac{I_{m_2}^2}{2} + \frac{I_{m_3}^2}{2}\right) R$$

The total power must be the power that is developed by the r.m.s. current $I_{r.m.s.}$.

$$\therefore \qquad P_T = I_{r.m.s.}^2 R$$

or
$$I_{r.m.s.}^2\, R = \left(I_0^2 + \frac{I_{m_1}^2}{2} + \frac{I_{m_2}^2}{2} + \frac{I_{m_3}^2}{2} \right) R$$

$$\therefore \qquad I_{r.m.s.} = \sqrt{ I_0^2 + \frac{I_{m_1}^2}{2} + \frac{I_{m_2}^2}{2} + \frac{I_{m_3}^2}{2} } \qquad\qquad ...(i)$$

If I_1, I_2 and I_3 are the r.m.s. values of the fundamental, the second harmonic and the third harmonic respectively, then,

$$I_1 = \frac{I_{m_1}}{\sqrt{2}} \quad \text{or} \quad I_1^2 = \frac{I_{m_1}^2}{2}$$

Similarly,
$$I_2^2 = \frac{I_{m_2}^2}{2} \quad \text{and} \quad I_3^2 = \frac{I_{m_3}^2}{2}$$

$\therefore$ Equation (i) becomes : $\quad I_{r.m.s.} = \sqrt{ I_0^2 + I_1^2 + I_2^2 + I_3^2 }$

The current $I_{r.m.s.}$ is the root-mean square value of this complex waveform. Similarly, the r.m.s. value $V_{r.m.s.}$ of a complex voltage wave is

$$V_{r.m.s.} = \sqrt{ V_0^2 + V_1^2 + V_2^2 + V_3^2 }$$

Hence, the r.m.s. value of a complex wave (current or voltage) is equal to the square root of the sum of the squares of the r.m.s. values of its individual components.

$$\text{Circuit power factor} = \frac{\text{Active power}}{\text{Apparent power}} = \frac{P_T}{V_{r.m.s.}\, I_{r.m.s.}}$$

Example 11.37. *The following current wave passes through a 20 Ω resistor.*

$$i = 4 + 5\sin\omega t - 3\cos 2\,\omega t \text{ amperes}$$

Calculate (i) the r.m.s. value of current wave (ii) total power produced and (iii) power factor of the circuit.

Solution. *(i)* The r.m.s. values of the various components of this complex wave are 4, $5/\sqrt{2}$ and $3/\sqrt{2}$ amperes.

$$\therefore \qquad I_{r.m.s.} = \sqrt{ (4)^2 + \left(\frac{5}{\sqrt{2}} \right)^2 + \left(\frac{3}{\sqrt{2}} \right)^2 } = \mathbf{5{\cdot}74\ A}$$

(ii)   Total power, $P_T = I_{r.m.s.}^2\, R = (5{\cdot}74)^2 \times 20 = \mathbf{660\ W}$

(iii) $V_{r.m.s.} = I_{r.m.s.}R = 5.74 \times 20 = 114{\cdot}8$ volts

$$\therefore \qquad \text{Power factor} = \frac{P_T}{V_{r.m.s.}I_{r.m.s.}} = \frac{660}{114{\cdot}8 \times 5{\cdot}74} = \mathbf{1} \text{ or } \mathbf{100\%}$$

Example 11.38. *A complex voltage contains fundamental and third harmonic of amplitudes of 100 V and 20V respectively. Find the effective (r.m.s.) value of complex voltage wave.*

Solution. R.M.S. value of fundamental is

$$V_1 = 100/\sqrt{2} = 70.7\ V$$

R.M.S. value of third harmonic is

$$V_3 = 20/\sqrt{2} = 14.14\ V$$

$\therefore$ R.M.S. value of voltage of the complex wave is

$$V = \sqrt{ V_1^2 + V_3^2 } = \sqrt{ (70.7)^2 + (14.14)^2 } = \mathbf{72{\cdot}1\ V}$$

Example 11.39. *A complex voltage wave has an effective value of 172 V. The fundamental has a frequency of 40 Hz and an effecitve value of 160 V. Find the effective value of all the harmonics combined.*

Solution. The effective value of voltage V of the complex wave is given by ;

$$V = \sqrt{V_1^2 + V_2^2 + V_3^2 + ... + V_n^2} = \sqrt{V_1^2 + V_h^2}$$

where, V_h = effective value of all the harmonics

$$\therefore \quad 172 = \sqrt{(160)^2 + V_h^2} \quad \text{or} \quad V_h = \sqrt{(172)^2 - (160)^2} = \textbf{63.1 V}$$

11.19. Phase

Waves of alternating voltage and current are continuous. They do not stop after one cycle is completed but continue to repeat as long as the generator is operating. Consider an alternating voltage wave of time period T second as shown in Fig. 11.45. Note that the time is counted from the instant the voltage is zero. The maximum positive value ($+ V_m$) occurs at $T/4$ second or $\pi/2$ radians. We say that *phase* of maximum positive value is $T/4$ second or $\pi/2$ radians. It means that as the fresh cycle starts, $+ V_m$ will occur at $T/4$ second or $\pi/2$ radians. Similarly, the phase of negative peak ($-V_m$) is $3T/4$ second or $3\pi/2$ radians.

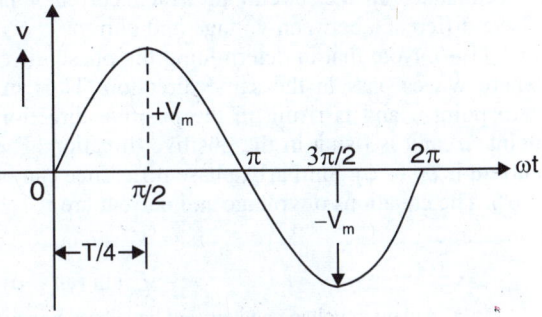

Fig. 11.45

*Hence, **phase** of a particular value of an alternating quantity is the fractional part of time period or cycle through which the quantity has advanced from the selected zero position of reference.*

The following points may be noted carefully :

(*i*) The phase of an alternating quantity (voltage or current) depends upon the instant from which the time is measured. Thus in Fig. 11.45, the time is measured from the instant the voltage is zero. Had the time been measured from the instant the voltage was positive maximum, the phase of maximum positive value would have been zero.

(*ii*) An alternating quantity (voltage or current) is completely known if we know its (*a*) maximum value (*b*) frequency and (*c*) phase.

In electrical engineering, we are more concerned with relative phases or phase difference between different alternating quantities rather than with their absolute values.

11.20. Phase Difference

When an alternating voltage is applied to a circuit, an alternating current of the same frequency flows through the circuit. In most of practical circuits, for reasons we will discuss later, voltage and current have different phases. In other words, they do not pass through a particular point, say *zero point, in the same direction at the same instant. Thus voltage may be passing through its zero point while the current has passed or it is yet to pass through its zero point in the same direction. We say that voltage and current have a phase difference.

*Hence, when two alternating quantities of the same frequency have different zero points, they are said to have a **phase difference.***

---

*   We may select any point but it is more convenient to determine the phase difference by considering zero points or positive maximum values or negative maximum values of the two alternating quantities.

The angle between zero points is the angle of phase difference ϕ. It is generally measured in degrees or radians. The quantity which passes through its zero point earlier is said to be leading while the other is said to be lagging. It should be noted that those zero points of alternating quantities are to be considered where they pass in the same direction. Thus if voltage has passed through its zero point and is rising in the positive direction, then zero point considered for the current should have similar situation. Since both alternating quantities have the same frequency, the phase difference between them remains the same.

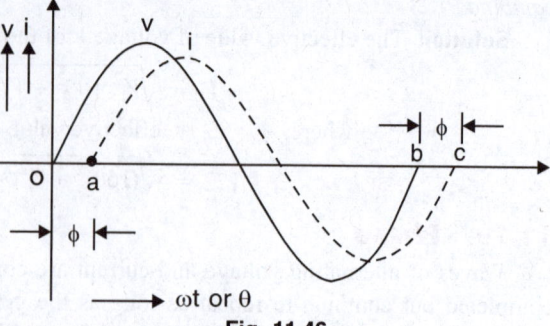

Fig. 11.46

Consider an a.c. circuit in which current i lags behind the voltage v by $\phi°$. We say that phase difference between voltage and current is $\phi°$. This phase relationship is shown by waves in Fig. 11.46. Note that in determining the phase difference, those zero points have been considered where waves pass in the same direction. Thus in Fig. 11.46, voltage v is passing through its zero point o and is rising in the positive direction. Similarly, current i passes through its zero point 'a' and is rising in the positive direction. Therefore, phase difference between voltage and current is oa (= ϕ). Similarly, phase difference can be determined by considering points b and c (bc = $\phi°$). The equations of voltage and current are :

$$v = V_m \sin \omega t \qquad \qquad ...(i)$$

$$i = I_m \sin (\omega t - \phi) \qquad \qquad ...(ii)$$

Note. Although voltage and current have been considered to explain the concept of phase difference, it is equally valid for two or more currents or voltages.

11.21. Representation of Alternating Voltages and Currents

So far we have discussed that an alternating voltage or current may be represented in the form of (*i*) waves and (*ii*) equations. The waveform presents to the eye a very definite picture of what is happening at every instant. But it is difficult to draw the wave accurately. No doubt the current flowing at any instant can be determined from the equation form $i = I_m \sin \omega t$ but this equation presents no picture to the eye of what is happening in the circuit.

The above difficulty has been overcome by representing sinusoidal alternating voltage or current by a line of definite length rotating in *anticlockwise direction at a constant angular velocity (ω). Such a rotating line is called a **phasor**. The length of the phasor is taken equal to the maximum value (on a suitable scale) of the alternating quantity and angular velocity equal to the angular velocity of the alternating quantity. As we shall see presently, this phasor (*i.e.* rotating line) will generate a sine wave.

11.22. Phasor Representation of Sinusoidal Quantities

Consider an alternating current represented by the equation $i = I_m \sin \omega t$. Take a line OP to represent to scale the maximum value I_m. Imagine the line OP (or **phasor, as it is called) to be rotating in anticlockwise direction at an angular velocity ω rad/sec about the point O. Measuring the time from the instant when OP is horizontal, let OP rotate through an angle θ (= ωt) in the anticlockwise direction. The projection of OP on the Y-axis is OM.

---

*   It is a standard convention that the phasor is rotated anticlockwise—a convention that is in harmony with the general use of polar co-ordinates.

**   An arrowhead is drawn at the outer end of phasor OP partly to indicate which end is assumed to move and partly to indicate precise length of the phasor when two or more phasors happen to coincide.

$$OM = OP \sin \theta$$
$$= I_m \sin \omega t$$
$$= i, \text{ the value of current at that instant}$$

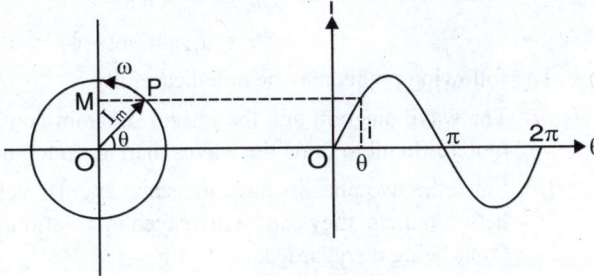

Hence the projection of the phasor OP on the Y-axis at any instant gives the value of current at that instant. Thus when $\theta = 90°$, the projection on Y-axis is OP ($= I_m$) itself. That the value of current at this instant (*i.e.* at θ or $\omega t = 90°$) is I_m can be readily established if we

Fig. 11.47

put $\theta = 90°$ in the current equation. If we plot the projections of the phasor on the Y-axis *versus* its angular position point-by-point, a sinusoidal alternating current wave is generated as shown in Fig. 11.47. Thus the phasor represents the sine wave for every instant of time.

The following points are worth noting :

(*i*) The length of the phasor represents the maximum value and the angle with axis of reference (*i.e.*, X-axis) indicates the phase of the alternating quantity *i.e.* current in this case.

(*ii*) The phasor representation enables us to quickly obtain the numerical values and, at the same time, have a picture before the eye of the events taking place in the circuit. Thus in the position of the phasor OP shown in Fig. 11.47, the instantaneous value is OM, the phase is θ and frequency is $\omega/2\pi$.

(*iii*) A phasor diagram permits addition and subtraction of alternating voltages or currents with a fair degree of ease.

Note. *Alternating voltages and currents are not vector quantities.* Voltage is simply energy or work per coulomb and cannot be classified as a vector. Current is also not a vector quantity because it is merely the flow of electrons through a wire. When we insert an ammeter in a circuit to measure current or connect a voltmeter between two points to measure the potential difference (*i.e.* voltage), direction with reference to any set of axes is of no consequence. Therefore, neither alternating voltage nor current is a vector quantity. Instead, they are *phasors.

11·23. Phasor Diagram of Sine Waves of Same Frequency

Consider a sinusoidal voltage wave v and sinusoidal current wave i of the same frequency. Suppose the current lags behind the voltage by $\phi°$. The two alternating quantities can be represented on the same phasor diagram because the phasors V_m and I_m [See Fig. 11.48 (*i*)] rotate at the same angular velocity ω** and hence phase difference ϕ between them remains the same at all times. When each phasor completes one revolution, it generates the corresponding cycle [See Fig. 11.48 (*ii*)]. The equations of the two waves can be represented as :

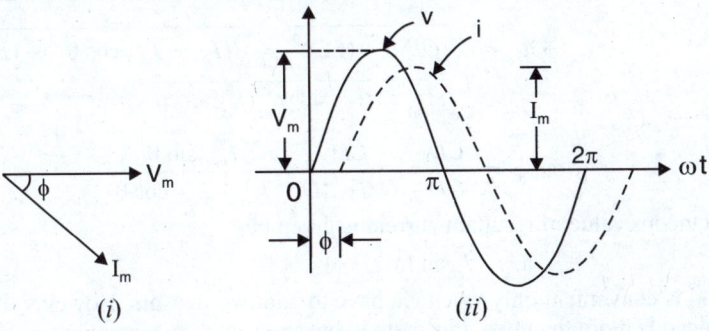

Fig. 11.48

---

* Remember that a vector has space co-ordinates while a phasor is derived from the time varying sinusoid.

** Remember $\omega = 2\pi f$.

$$v = V_m \sin \omega t$$

$$i = I_m \sin (\omega t - \phi)$$

The following points may be noted carefully :

  (*i*) The wave diagram and the phasor diagram convey the same information. However, it is more difficult to draw the waves than to sketch the phasor diagram.

  (*ii*) Since the two phasors have the same angular velocity (ω) and there is no relative motion between them, they can be displayed in a stationary diagram, the common angular rotation (*ωt) being disregarded.

11·24. Addition of Alternating Quantities

Alternating voltages and currents are phasors. They are added in the same manner as forces are added. Only phasors of the same kind may be added. Common sense tells us that we should not try to add volts to amperes. Addition of alternating currents or voltages can be accomplished by one of the following methods :

  1. Parallelogram method   2. Method of components

  1. Parallelogram method. This method is used for the addition of two phasors at a time. The two phasors are represented in magnitude and direction by the adjacent sides of a parallelogram. Then the diagonal of the parallelogram represents the maximum value of the resultant.

Consider two alternating currents i_1 and i_2 flowing in the two branches of a circuit [See Fig. 11.49 (*i*)]. Let they be represented by :

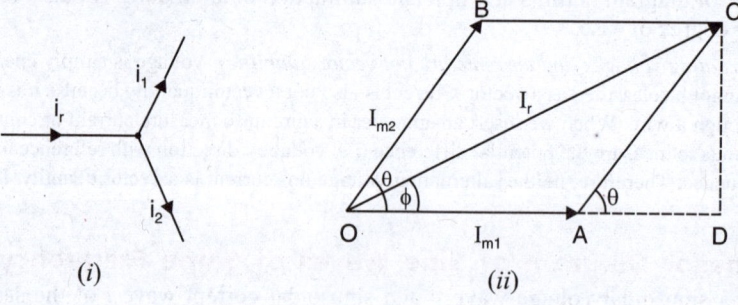

Fig. 11.49

$$i_1 = I_{m1} \sin \omega t \quad ; \quad i_2 = I_{m2} \sin (\omega t + \theta)$$

The two phasors I_{m1} and I_{m2} are represented by the adjacent sides *OA* and *OB* respectively of the parallelogram *OACB* [See Fig. 11.49 (*ii*)]. The phase difference between the phasors is $\theta°$; I_{m2} leading I_{m1}. The maximum value of resultant is I_r. It is represented by the diagonal *OC* and leads the phasor I_{m1} by $\phi°$.

$$OC = \sqrt{(OD)^2 + (CD)^2} = \sqrt{(I_{m1} + I_{m2} \cos \theta)^2 + (I_{m2} \sin \theta)^2}$$

$$\therefore \qquad I_r = \sqrt{I^2{}_{m1} + I^2{}_{m2} + 2 I_{m1} I_{m2} \cos \theta}$$

Also $\qquad\qquad \tan \phi = \dfrac{CD}{OD} = \dfrac{CD}{OA + AD} = \dfrac{I_{m2} \sin \theta}{I_{m1} + I_{m2} \cos \theta}$

The instantaneous value of resultant current is given by ;

$$i_r = I_r \sin (\omega t + \phi)$$

This method is convenient only when we have to add two phasors. However, if the number of phasors to be added is more than two, this method becomes quite inconvenient.

---

*   The term ωt is essential in the equations but not in the phasor diagram. It is because a phasor diagram represents the relative phase difference between two alternating quantities. Thus in the equations of v and i, we have ωt and $\omega t - \phi$. The phase difference between the two will be $= \omega t - (\omega t - \phi) = \phi$.

2. Method of Components. This method provides a very convenient means to add two or more phasors. Each phasor is resolved into horizontal and vertical components. The horizontals are summed up algebraically to give the resultant horizontal component X. The verticals are likewise summed up algebraically to give the resultant vertical component Y.

Then, Resultant $= \sqrt{X^2 + Y^2}$

Phase angle of resultant, $\tan \phi = Y/X$

The previously considered currents are represented as phasors in Fig. 11.50.

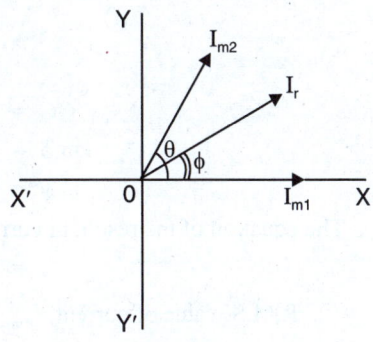

$$X = I_{m1} + I_{m2} \cos \theta$$

$$Y = 0 + I_{m2} \sin \theta$$

$\therefore$ Resultant, $I_r = \sqrt{(I_{m1} + I_{m2} \cos \theta)^2 + (I_{m2} \sin \theta)^2}$

$$= \sqrt{I_{m1}^2 + I_{m2}^2 + 2 I_{m1} I_{m2} \cos \theta}$$

which is the same as derived by parallelogram method.

$$\tan \phi = \frac{Y}{X} = \frac{I_{m2} \sin \theta}{I_{m1} + I_{m2} \cos \theta}$$

$\therefore$ $i_r = I_r \sin (\omega t + \phi)$

Fig. 11.50

11·25. Subtraction of Alternating Quantities

If difference of two phasors is required, then one of the phasors is reversed and this reversed phasor is then compounded with the other phasor using parallelogram method or method of components.

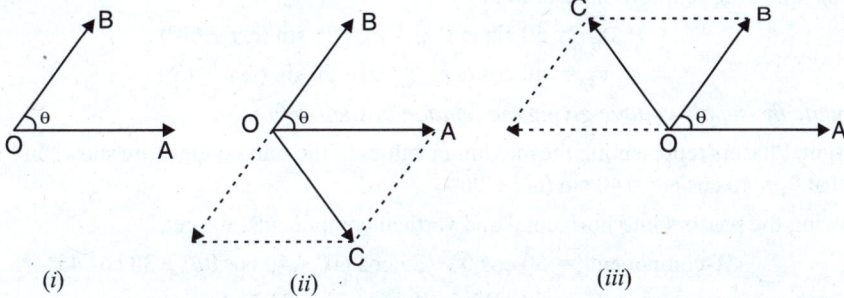

(i)   (ii)   (iii)

Fig. 11.51

Consider two phasors OA and OB representing two alternating quantities of the same kind [See Fig. 11.51 (i)]. The phasor OB leads the phasor OA by θ. If it is required to subtract the phasor OB from OA, then OB is reversed and is compounded with phasor OA as shown in Fig. 11.51 (ii). The phasor difference $OA–OB$ is given by the phasor OC. In Fig. 11.51 (iii), phasor OC represents the phasor difference $OB–OA$.

Example 11.40. *Three circuits in parallel take the following currents :*

$i_1 = 20 \sin 314 t$; $i_2 = 30 \sin (314 t - \pi/4)$; $i_3 = 40 \cos (314 t + \pi/6)$

Find (i) the expression for the resultant current and (ii) its r.m.s. value and frequency. If the circuit has a resistance of 2 Ω, what is the energy loss in 10 hours ?

Solution. Phasors representing the maximum values of the three currents are shown in Fig. 11.52 (i). Resolving these phasors into horizontal and vertical components, we get,

X-component $= 20 + 30 \cos 45° - 40 \cos 60° = 21·2$ A

Y-component $= 20 \cos 90° + 40 \cos 30° - 30 \cos 45° = 13·4$ A *(i)* As shown in Fig. 11.52 *(ii)*, *OC* represents the maximum value of the resultant current.

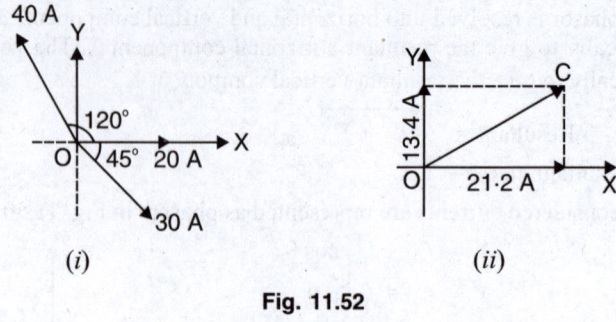

(*i*)   (*ii*)

Fig. 11.52

$$OC = \sqrt{(21.2)^2 + (13.4)^2} = 25.1 \text{ A}$$
$$\tan \phi = 13 \cdot 4/21 \cdot 2 = 0 \cdot 632$$
$$\therefore \qquad \phi = \tan^{-1} 0.632 = 32.3°$$

The equation of the resultant current is

$$i = 25 \cdot 1 \sin (314\, t + 32 \cdot 3°) \quad \textbf{Ans.}$$

(ii) R.M.S. value of current, $I_{r.m.s.} = 25.1/\sqrt{2} = \textbf{17·75 A}$

Frequency $= \omega/2\pi = 314/2\pi = \textbf{50 Hz}$

The frequency of the resultant current is the same as that of the three currents.

Energy loss in 10 hours $= I_{r.m.s.}^2 \times R \times t = (17 \cdot 75)^2 \times 2 \times 10 = \textbf{6301·25 Wh}$

Example 11.41. *A circuit consists of four loads in series ; the voltage across these loads are given by the following relations measured in volts :*

$$v_1 = 50 \sin \omega t \; ; \quad v_2 = 25 \sin (\omega t + 60°)$$
$$v_3 = 40 \cos \omega t \; ; \quad v_4 = 30 \sin (\omega t - 45°)$$

Calculate the supply voltage giving the relation in similar form.

Solution. Phasors representing the maximum values of the four voltages are shown in Fig. 11.53 (*i*). Note that $v_3 = 40 \cos \omega t = 40 \sin (\omega t + 90°)$.

Resolving the phasors into horizontal and vertical components, we get,

$$\text{X-component} = 50 \cos 0° + 25 \cos 60° + 40 \cos 90° + 30 \cos 45°$$
$$= 50 + 12 \cdot 5 + 0 + 21 \cdot 21 = 83 \cdot 71 \text{ V}$$
$$\text{Y-component} = 50 \sin 0° + 25 \sin 60° + 40 \sin 90° - 30 \sin 45°$$
$$= 0 + 21 \cdot 65 + 40 - 21 \cdot 21 = 40 \cdot 44 \text{ V}$$

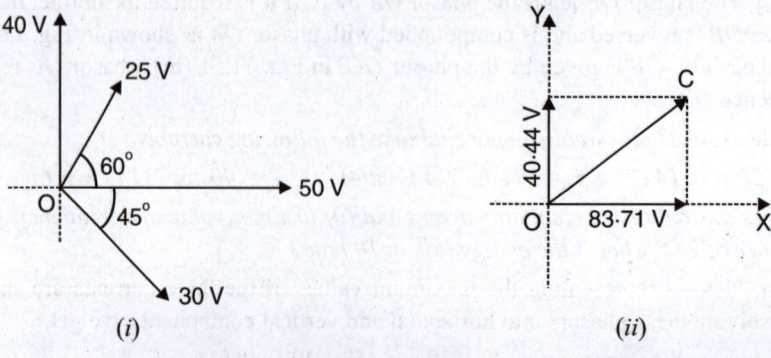

(*i*)   (*ii*)

Fig. 11.53

As shown in Fig. 11.53 (*ii*), *OC* represents the maximum value of supply voltage.

$$OC = \sqrt{(83 \cdot 71)^2 + (40 \cdot 44)^2} = 93 \text{ V}$$
$$\tan \phi = 40 \cdot 44/83 \cdot 71 = 0 \cdot 483$$

∴
$$\phi = \tan^{-1} 0 \cdot 483 = 25 \cdot 8°$$

The equation of the supply voltage is

$$v = 93 \sin (\omega t + 25 \cdot 8°) \quad \textbf{Ans.}$$

Example 11.42. *The e.m.f. of each of the two coils, as read by a voltmeter, is 220 V. When the coils are connected in series at random, the resulting e.m.f. is 381 V. What is the phase angle of the e.m.f.s of the two coils? If the connections of one of the coils are reversed, what will be the phase angle between the e.m.f.s and what will be the resulting voltage?*

Solution. The voltmeter reads the r.m.s. value. Therefore, phasor diagrams shown in Fig. 11.54 represent the r.m.s. values. The phase relationship between the phasors is not changed if we use the r.m.s. values instead of the maximum values.

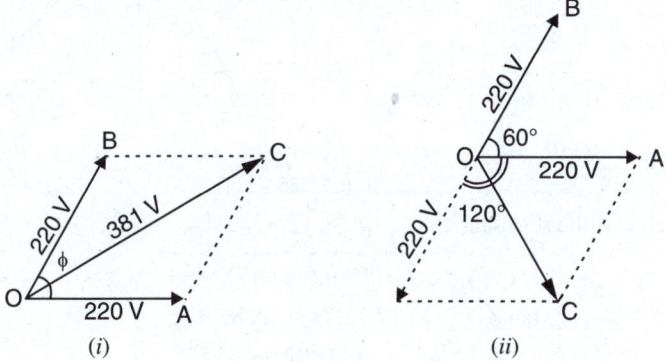

Fig. 11.54

Series connected. Fig. 11.54 (*i*) shows the phasors of the two e.m.f.s. The phase angle ϕ between the e.m.fs. of the coils can be readily obtained by parallelogram method.

$$(381)^2 = (220)^2 + (220)^2 + 2 \times 220 \times 220 \cos \phi$$

∴
$$\cos \phi = \frac{(381)^2 - (220)^2 - (220)^2}{2 \times 220 \times 220} = 0.5$$

or
$$\phi = \cos^{-1} 0 \cdot 5 = \textbf{60°}$$

When connections of one coil are reversed. When we reverse the connection of one coil (say *OB*), then the phasor diagram will be as shown in Fig. 11.54 (*ii*). It is clear that phase angle between the e.m.fs. of the two coils is $180 - 60 = \textbf{120°}$. The resultant e.m.f. is *OC*.

∴
$$OC = \sqrt{(220)^2 + (220)^2 + 2 \times 220 \times 220 \times \cos 120°}$$
$$= \sqrt{(220)^2 + (220)^2 - (220)^2} = \textbf{220 V}$$

Example 11.43. *The following four e.m.fs. act together in a circuit :*

$$e_1 = 10 \sin \omega t \; ; \quad e_2 = 8 \sin (\omega t + \pi/3)$$
$$e_3 = 4 \sin (\omega t - \pi/6) \; ; \quad e_4 = 6 \sin (\omega t + 3\pi/4)$$

Calculate the e.m.f. represented by $e_1 - e_2 + e_3 - e_4$.

Solution. Phasors representing the maximum values of the four e.m.fs. are shown in Fig. 11.55 (*i*). In order to determine $e_1 - e_2 + e_3 - e_4$, reverse the phasors representing e_2 and e_4 as shown in Fig.

11.55 (*ii*). Referring to Fig. 11.55 (*ii*) and resolving the phasors into horizontal and vertical components, we get,

$$X\text{-component} = 10 \cos 0° + 4 \cos 30° - 8 \cos 60° + 6 \cos 45°$$
$$= 10 + 3.464 - 4 + 4.243 = 13.71 \text{ V}$$
$$Y\text{-component} = 10 \cos 90° - 4 \cos 60° - 6 \cos 45° - 8 \cos 30°$$
$$= 0 - 2 - 4.243 - 6.928 = -13.17 \text{ V}$$

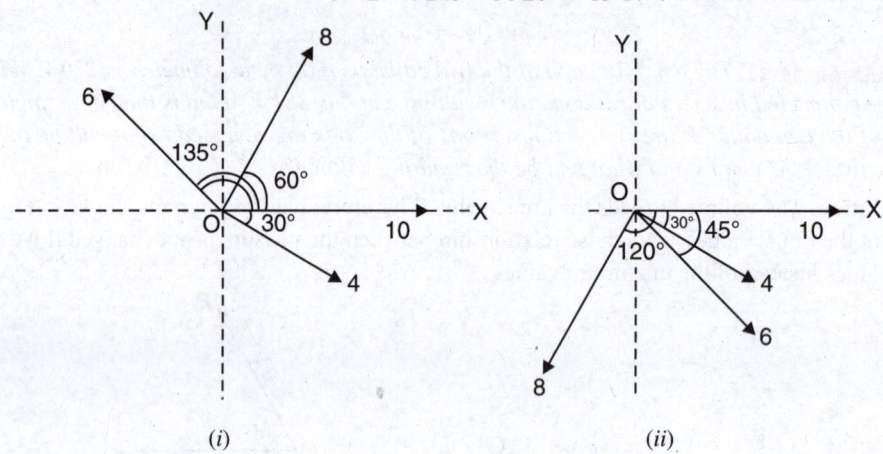

(*i*)   (*ii*)

Fig. 11.55

Maximum value of the resultant e.m.f.

$$= \sqrt{(13.71)^2 + (-13.17)^2} = 19 \text{ V}$$
$$\tan \phi = -13.17/13.71 = -0.9606$$
$$\therefore \qquad \phi = \tan^{-1} - (0.9606) = -43.8°$$

The equation of the resultant e.m.f. is

$$e = 19 \sin (\omega t - 43.8°) \quad \textbf{Ans.}$$

Example 11.44. *The maximum values of alternating voltage and current are 400V and 20A respectively in a circuit connected to a 50 Hz supply. The instantaneous values of voltage and current are 283 V and 10A respectively at time t = 0, both increasing positively.*

(*i*) *Write down the expression for voltage and current at time t.*

(*ii*) *Determine the power consumed in the circuit.*

Assume the voltage and current to be sinusoidal.

Solution. $V_m = 400$ volts ; $I_m = 20$A ; $\omega = 2\pi f = 2\pi \times 50 = 314$ rad s^{-1}

(*i*) The voltage and current can be expressed as :

$$v = V_m \sin (\omega t + \phi_1) = 400 \sin (314\, t + \phi_1)$$
$$i = I_m \sin (\omega t + \phi_2) = 20 \sin (314\, t + \phi_2)$$

At $t = 0$; $283 = 400 \sin (314 \times 0 + \phi_1)$ $\therefore$ $\phi_1 = \sin^{-1} 283/400 = 45°$

At $t = 0$; $10 = 20 \sin (314 \times 0 + \phi_2)$ $\therefore$ $\phi_2 = \sin^{-1} 10/20 = 30°$

Therefore, expressions for voltage and current at time *t* are :

$$v = 400 \sin (314\, t + 45°) \textbf{ Ans.}$$
$$i = 20 \sin (314\, t + 30°) \textbf{ Ans.}$$

(*ii*) R.M.S. value of voltage, $V = 400/\sqrt{2} = 283$ volts

R.M.S. value of current, $I = 20/\sqrt{2} = 14.14$ A

Circuit power factor, $\cos \phi = \cos(45° - 30°) = \cos 15°$

∴   Power consumed, $P = VI \cos \phi = 283 \times 14.14 \times \cos 15° =$ **3865 W**

Example 11.45. *Three sinusoidally alternating currents of r.m.s. values 5A, 7.5A and 10A are having the same frequency with phase angles of 30°, – 60° and 45°.*

(i) Find the average values (ii) Write equations for their instantaneous values (iii) Draw phasor diagram taking first current as the reference. (iv) Find their instantaneous values at 100 ms from the original reference.

Solution. *(i)* For an alternating quantity varying sinusoidally,

$$\text{Average value} = \frac{\text{R.M.S. value}}{\text{Form factor}} = \frac{\text{R.M.S. value}}{1.11}$$

Average value of first current = 5/1.11 = **4.50A**

Average value of second current = 7.5/1.11 = **6.76 A**

Average value of third current = 10/1.11 = **9.01 A**

(ii) $\omega = 2\pi f = 2\pi \times 50 = 314$ rad/sec.

Instantaneous value of sinusoidal current is given by ;

$$i = I_m \sin(\omega t \pm \phi) = I_m \sin(314t \pm \phi)$$

∴
$$i_1(t) = 5\sqrt{2} \sin(314t + 30°) \text{ A}$$
$$i_2(t) = 7.5\sqrt{2} \sin(314t - 60°) \text{ A} \quad \left.\right\} \text{ **Ans**}$$
$$i_3(t) = 10\sqrt{2} \sin(314t + 45°) \text{ A}$$

(iii) From the statement of the problem, the second current lags behind the first current by 90° while the third current leads the first current by 15° (= 45° – 30° = 15°). Fig. 11.56 shows the phasor diagram taking I_1 as the reference phasor.

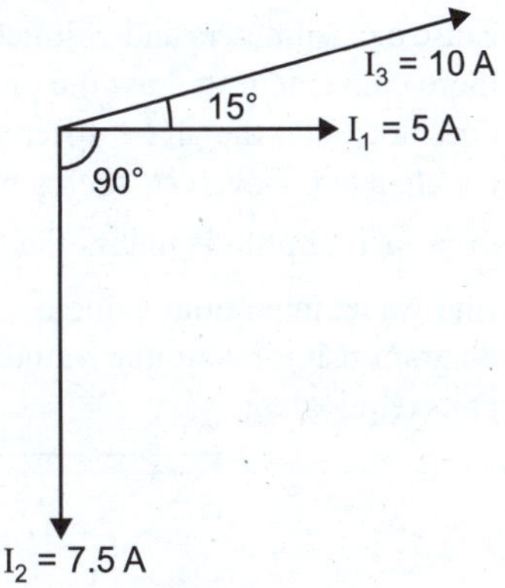

(iv) The time period of 50 Hz alternating current is $T = 1/50$ = 0.02 sec. = 20 ms. It means that 50 Hz a.c. completes one cycle in 20 ms. Therefore, in 100 ms, the a.c. will complete 5 cycles and we reach the conditions of original references. Therefore, the instantaneous values of the three currents after 100 ms are :

$$i_1 = 5\sqrt{2} \sin 30° = \textbf{3.536 A}$$

$$i_2 = 7.5\sqrt{2} \sin(-60°) = \textbf{– 9.816 A}$$

$$i_3 = 10\sqrt{2} \sin 45° = \textbf{10A}$$

Fig. 11.56

Tutorial Problems

1. Two currents represented by $i_1 = 50 \sin 314\,t$ and $i_2 = 30 \sin(314\,t - \pi/6)$ are fed into a common conductor. Find the expression for the resultant current in the form $i = I_m \sin(314\,t \pm \phi)$.

   $[i = \textbf{77.5 sin} (314\,t - 11°10')]$

2. Find the sum of the following two e.m.fs. and express the answer in the similar form :

   $$e_1 = 100 \sin \omega t \quad ; \quad e_2 = 100 \cos \omega t \qquad [e = \textbf{141.4 sin} (\omega t + 45°)]$$

3. The following four currents are fed into a common conductor :

   $$i_1 = 50 \sin \omega t \quad ; \quad i_2 = 25 \sin(\omega t + \pi/3)$$
   $$i_3 = 40 \cos \omega t \quad ; \quad i_4 = 30 \sin(\omega t - \pi/2)$$

   Find the expression for the resultant in the form $i = I_m \sin(\omega t \pm \phi)$.

   $[i = \textbf{94 sin} (\omega t + 25°)]$

4. The instantaneous voltage across each of four series connected coils are given by ;

$$v_1 = 100 \sin \omega t \quad ; \quad v_2 = 250 \cos \omega t$$

$$v_3 = 150 \sin (\omega t + \pi/6) \quad ; \quad v_4 = 200 \sin (\omega t - \pi\sqrt{4})$$

Find the total p.d. and express the answer in the similar form. What will be the resultant p.d. if the polarity of v_2 is reversed ? $[v = 414 \sin (\omega t + 26 \cdot 5^\circ) ; v = 486 \sin (\omega t - 40^\circ)]$

5. Each of the three coils generates an e.m.f. of 230 V. The e.m.f. of second leads that of the first by 120°, and the third lags behind the first by the same angle. What is the resultant e.m.f. across a series combination of the coils ?   **[0 volt]**

6. What is the resultant of a series arrangement of four coils A, B, C and D generating e.m.fs. of 25, 30, 30 and 20 volts respectively ? The e.m.f. of coil B leads that of A by 30°, e.m.f. of coil C leads B by 60° and e.m.f. of D lags 45° behind A. $[e = 72 \cdot 02 \sin (\omega t + 25^\circ 22') V]$

7. Find the value of V_{AB} in Fig. 11.57. $[42 \cdot 4 \angle 0^\circ V]$

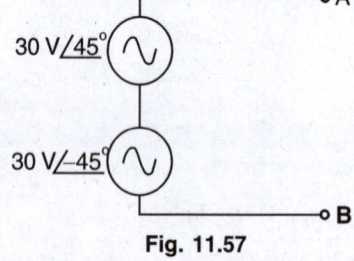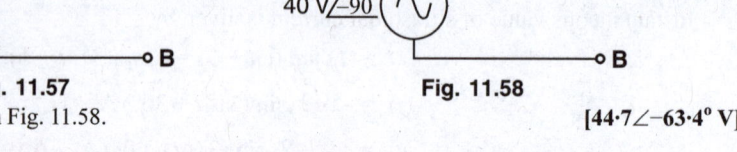

Fig. 11.57   **Fig. 11.58**

8. Find the value of V_{AB} in Fig. 11.58. $[44 \cdot 7 \angle -63 \cdot 4^\circ V]$

11.26. Phasor Diagrams Using R.M.S. Values

The r.m.s. values of voltage and current are much more important than maximum values because a.c. ammeters and voltmeters are calibrated to read the r.m.s. values. Consequently, it is much more convenient to draw the phasor diagram using r.m.s. values rather than maximum values. This does not *alter the phase difference (ϕ) between the phasors because only the lengths of the phasors are changed. However, such a phasor diagram (using r.m.s. values) will not generate the sine wave of proper amplitude unless the lengths of the phasors are first multiplied by $\sqrt{2}$. Remember, for a sine wave, maximum value is equal to $\sqrt{2}$ times the r.m.s. value. Fig. 11.59 (*i*) shows the phasor diagram using maximum values while Fig. 11.59 (*ii*) shows the same phasor diagram in terms of r.m.s. equivalent.

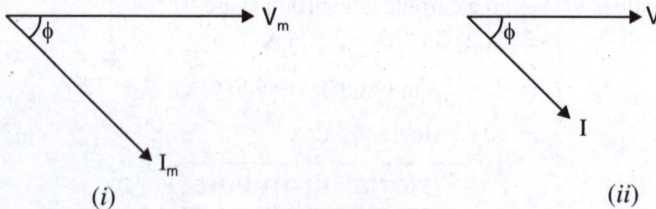

(*i*)   (*ii*)

Fig. 11.59

In all the phasor diagrams from now onwards, we shall use the r.m.s. values.

11.27. Instantaneous Power

The instantaneous power supplied to a circuit is simply the product of the instantaneous voltage and instantaneous current. The instantaneous power is always expressed in watts, irrespective of the type of circuit used. The instantaneous power may be positive or negative. A positive value means

---

*   The phase angle between the phasors is not affected by a conversion factor applied to the lengths of the phasors.

that power flows from the source to the load. Consequently, a negative value means that power flows from the load to the source.

11.28. A.C. Circuit Containing Resistance Only

When an alternating voltage is applied across pure resistance, then free electrons flow (*i.e.* current) in one direction for the first half-cycle of the supply and then flow in the opposite direction during the next half-cycle, thus constituting alternating current in the circuit.

Consider a circuit containing a pure resistance of R Ω connected across an alternating voltage source [See Fig. 11.60]. Let the alternating voltage be given by the equation :

$$v = V_m \sin \omega t \qquad \ldots(i)$$

As a result of this voltage, an alternating current i will flow in the circuit. The applied voltage has to overcome the drop in the resistance only *i.e.*

$$v = i R$$

or $\quad i = \dfrac{v}{R}$

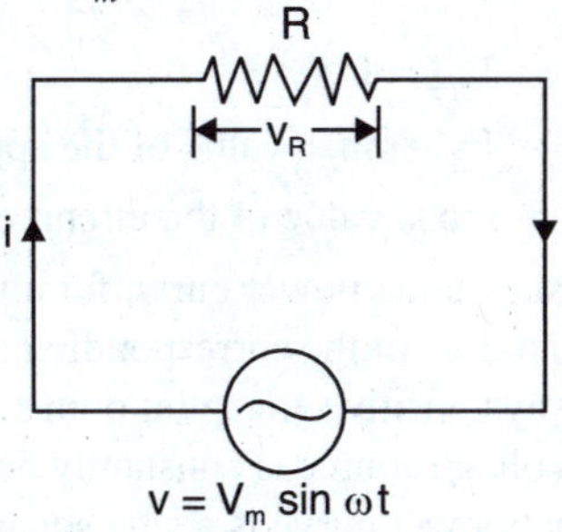

Substituting the value of v, we get,

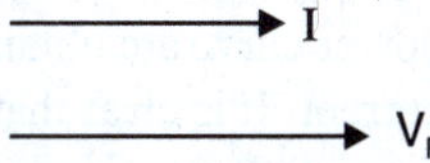

$$i = \dfrac{V_m}{R} \sin \omega t \qquad \ldots(ii)$$

Fig. 11.60

Fig. 11.61

The value of i will be maximum (*i.e.* I_m) when $\sin \omega t = 1$.

$\therefore \qquad\qquad I_m = V_m/R$

$\therefore$ Eq. (*ii*) becomes : $\quad i = I_m \sin \omega t \qquad \ldots(iii)$

In terms of r.m.s. values, $\dfrac{V_m}{\sqrt{2}} = \dfrac{I_m}{\sqrt{2}} \times R$

or $\qquad\qquad\qquad V = V_R = I R$

(*i*) **Phase angle.** It is clear from eqs. (*i*) and (*iii*) that the applied voltage and the circuit current are in phase with each other *i.e.* they pass through their zero values at the same instant and attain their positive and negative peaks at the same instant. This is also indicated by the phasor diagram shown in Fig. 11.61. Note that r.m.s. values have been used in drawing the phasor diagram. The wave diagram shown in Fig. 11.62 also depicts that current is in phase with the applied voltage.

(*ii*) **Power.** In any circuit, electric power consumed at any instant is the product of voltage and current at that instant *i.e.*

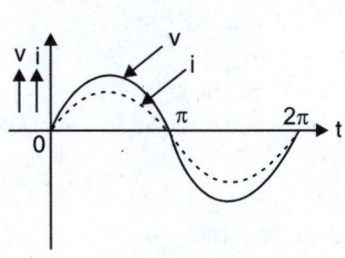

Fig. 11.62

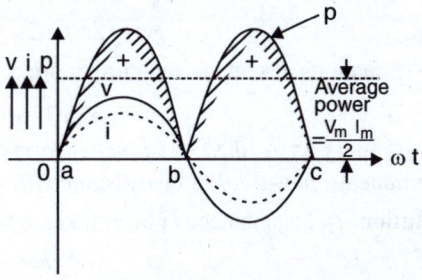

Fig. 11.63

Instantaneous power, $p = vi = (V_m \sin \omega t)(I_m \sin \omega t) = V_m I_m \sin^2 \omega t$

$$= V_m I_m \frac{(1 - \cos 2\omega t)}{2} = \frac{V_m I_m}{2} - \frac{V_m I_m}{2} \cos 2\omega t$$

Thus power consists of two parts *viz.* a constant part $(V_m I_m / 2)$ and a fluctuating part $(V_m I_m / 2) \cos 2\omega t$. Since power is a scalar quantity, average power over a complete cycle is to be considered.

$\therefore$ Power consumed, $P = \dfrac{1}{2\pi} \displaystyle\int_0^{2\pi} \dfrac{V_m I_m}{2} d(\omega t) + \dfrac{1}{2\pi} * \displaystyle\int_0^{2\pi} \dfrac{V_m I_m}{2} \cos 2\omega t \, d(\omega t)$

$$= \frac{V_m I_m}{2} + 0 = \frac{V_m}{\sqrt{2}} \times \frac{I_m}{\sqrt{2}}$$

$\therefore$ $P = V_R I = VI$

where $V = V_R$ = r.m.s. value of the applied voltage

I = r.m.s. value of the circuit current

(iii) Power curve. Fig. 11.63 shows the power curve for a pure resistive circuit. Points on the power curve are obtained from the product of the corresponding instantaneous values of voltage and current. It is clear that power is always positive except at points *a*, *b* and *c* at which it drops to zero for a moment. This means that the voltage source is constantly delivering power to the circuit which is consumed by the circuit. Note that power curve is a sine square wave. The average value of sine squared wave over a complete cycle is one-half of the maximum value. Since the maximum value of instantaneous power is $V_m I_m$,

$\therefore$ Average power, $P = \dfrac{V_m I_m}{2} = \dfrac{V_m}{\sqrt{2}} \times \dfrac{I_m}{\sqrt{2}} = VI$

(iv) Conductance. **Conductance** *is the reciprocal of resistance i.e.*

Conductance, $G = \dfrac{1}{R}$

The SI unit of conductance is siemen (S). The conductance of a purely resistive circuit is its ability to pass current through it. The greater the conductance (*i.e.* the smaller the resistance) of a purely resistive circuit, the greater is its ability to pass current through it and *vice-versa*.

Example 11.46. *An a.c. circuit consists of a pure resistance of 10 Ω and is connected across an a.c. supply of 230 V, 50 Hz. Calculate (i) current (ii) power consumed and (iii) equations for voltage and current.*

Solution. (*i*)   Current, $I = V/R = 230/10 = $ **23 A**

(*ii*)   Power, $P = VI = 230 \times 23 = $ **5290 W**

(*iii*) Now, $V_m = \sqrt{2}V = \sqrt{2} \times 230 = 325.27$ volts

$I_m = \sqrt{2}I = \sqrt{2} \times 23 = 32.52$ A

$\omega = 2\pi f = 2\pi \times 50 = 314$ rad/s

$\therefore$ Equations of voltage and current are :

$v = $ **325·27 sin 314 *t*** ; $i = $ **32·52 sin 314 *t***

Example 11.47. *A 100Ω resistance is carrying a sinusoidal current given by 3cos ωt. Determine (i) instantaneous power taken by resistance (ii) average power.*

Solution. (*i*) Instantaneous power taken by resistance is

$p = vi = iR \times i = i^2 R = (3 \cos \omega t)^2 \times 100$

$= 900 \cos^2 \omega t = $ **450 (1 + cos 2 ωt) watts**

---

*   If you carry out the integration, the result for this part will be zero.

(ii)   Average power, P = Average of $450 (1 + \cos 2\ \omega t)$ over one cycle

$$= 450 + 0 = \textbf{450 W}$$

Example 11.48. *The current in a 2·2 kΩ resistor is*

$$i = 5 \sin (2\pi \times 100t + 45^{\circ})\ mA$$

(i) Write the mathematical expression for the voltage across the resistor.

(ii) What is the r.m.s. value of the resistor voltage ?

(iii) What is the instantaneous value of resistor voltage at t = 0·4 ms ?

Solution. I_m = 5 mA = 5×10^{-3} A ; R = 2.2 kΩ = 2.2×10^{3} Ω

(i) $V_m = I_m R = 5 \times 10^{-3} \times 2.2 \times 10^{3} = 11$ V

Since voltage across the resistor R is in phase with current,

∴ $v = V_m \sin (2\pi \times 100\ t + 45^{\circ})$

or $v = 11 \sin (2\pi \times 100\ t + 45^{\circ})$V **Ans.**

(ii) R.M.S. value of resistor voltage, $V_{r.m.s.}$ $\dfrac{V_m}{\sqrt{2}} = \dfrac{11}{\sqrt{2}} = = \textbf{7.78 V}$

(iii) The instantaneous value of resistor voltage at t = 0·4 ms is

$$v(0·4\ \text{ms}) = 11\sin [2\pi \times 100 \times 0·4 \times 10^{-3} + 45^{\circ}]\text{V}$$

$$= 11\sin [0·2513\ \text{rad} + 45^{\circ}]\text{V}$$

$$= 11 \sin [*14·4^{\circ} + 45^{\circ}]\ \text{V} = 11\sin 59·4^{\circ} = \textbf{9·47 V}$$

Example 11.49. *The a.c. voltage across 150 Ω resistor is $39 \sin (2\pi \times 10^{3}\ t)$ V. At what value of t does the current through the resistor equal – 0.26A?*

Solution. $I_m = \dfrac{V_m}{R} = \dfrac{39}{150}$ A

Since current through the resistor is in phase with the voltage across the resistor,

∴ $i = \dfrac{39}{150} \sin (2\pi \times 1000t)$A

or $-0·26 = \dfrac{39}{150} \sin 6280\ t$

or $\sin 6280\ t = \dfrac{-0·26 \times 150}{39} = -1$

or $6280\ t = \sin^{-1} -1 = \dfrac{270 \times 2\pi}{360}$ rad $= 4·71$ rad

∴ $t = \dfrac{4.71}{6280} = 0·75 \times 10^{-3}$ s $= \textbf{0.75 ms}$

Tutorial Problems

1. A 500 V sine wave appears across a 10 k Ω resistor. What is the instantaneous current in the resistor at a phase angle of 35° ?   **[40·6 mA]**

2. Find the magnitude and the angle of current in R_2 for Fig. 11.64.   **[0·265 ∠– 80·5° A]**

3. A sinusoidal source has a peak-to-peak value of

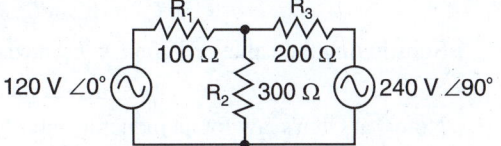

Fig. 11.64

---

*   The radians can be converted into degrees as :

$$0.2513\ \text{rad} = \dfrac{360^{\circ}}{2\pi} \times 0.2513 = 14.4^{\circ}$$

679 V. Calculate the r.m.s. power that it will deliver to a 40 Ω heating element. **[1·44 kW]**

4. The current in a 2·2 k Ω resistor is

$$i = 5 \sin (2\pi \times 100\, t + 45°) \text{ mA}$$

(*i*) Write the mathematical expression for the voltage across the resistor.

(*ii*) What is the effective value of resistor voltage ?

(*iii*) What is the instantaneous value of the resistor voltage at $t = 0·4$ ms ?

[(*i*) $e_R = 11 \sin (2\pi \times 100\, t + 45°)$ V (*ii*) 7·78 V (*iii*) 9·47 V]

5. The a.c. voltage across a 150 Ω resistor is $39 \sin (2\pi \times 10^3 t)$ V. At what value of *t* does the current through the resistor equal − 0·26 A ? **[0·75 ms]**

11·29. A.C. Circuit Containing Pure Inductance Only

When an alternating current flows through a pure *inductive coil, a back e.m.f. $(= L\, di/dt)$ is induced due to the inductance of the coil. This back e.m.f. at every instant opposes the change in current through the coil. Since there is no ohmic drop, the applied voltage has to overcome the back e.m.f. only.

∴ Applied alternating voltage = Back e.m.f.

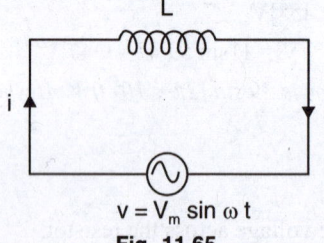

$v = V_m \sin \omega t$

Fig. 11.65

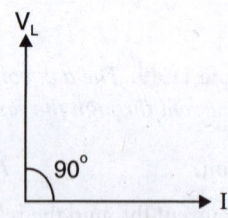

Fig. 11.66

Consider an alternating voltage applied to a pure inductance of *L* henry as shown in Fig. 11.65. Let the equation of the applied alternating voltage be :

$$v = V_m \sin \omega t \qquad \qquad ...(i)$$

Clearly, $\qquad V_m \sin \omega t = L \dfrac{di}{dt}$

or $\qquad di = \dfrac{V_m}{L} \sin \omega t\, dt$

Integrating both sides, we get, $i = \dfrac{V_m}{L} \int \sin \omega t\, dt = \dfrac{V_m}{\omega L}(-\cos \omega t)$

∴ $\qquad i = \dfrac{V_m}{\omega L} \sin(\omega t - \pi/2) \qquad \qquad ....(ii)$

The value of *i* will be maximum (*i.e.* I_m) when $\sin(\omega t - \pi/2)$ is unity.

∴ $\qquad I_m = V_m/\omega L$

Substituting the value of $V_m/\omega L = I_m$ in eq. (*ii*), we get,

$$i = I_m \sin(\omega t - \pi/2) \qquad \qquad ..:(iii)$$

Note that Ohm's law for an inductor states that peak current (I_m) through the inductor equals the peak voltage (V_m) across the inductor divided by the inductive reactance $(X_L = \omega L)$.

---

* Any circuit that is capable of producing flux has inductance as was pointed out in chapter 9. When alternating current flows through such a circuit, there is change in flux linking it and hence back e.m.f. $(= L\, di/dt)$ is induced in the circuit. This e.m.f. opposes the applied voltage at every instant.

(*i*) **Phase angle.** It is clear from eqs. (*i*) and (*iii*) that current lags behind the voltage by $\pi/2$ radians or 90°. *Hence in a pure inductance, current lags the voltage by 90°.* This is also indicated by the phasor diagram shown in Fig. 11.66. Note that r.m.s. values have been used in drawing the phasor diagram. The wave diagram shown in Fig. 11.67 also depicts that current lags the voltage by 90°. There is also physical explanation for the lagging of current behind voltage in an inductive coil. Inductance opposes the change in current and serves to delay the increase or decrease of current in the circuit. This causes the current to lag behind the applied voltage.

(*ii*) **Inductive reactance.** Inductance not only causes the current to lag behind the voltage but it also limits the magnitude of current in the circuit. We have seen above that :

$$I_m = V_m/\omega L$$

or $$\frac{V_m}{I_m} = \omega L$$

Clearly, the opposition offered by inductance to current flow is ωL. This quantity ωL is called the *inductive reactance* X_L of the coil. It has the same *dimensions as resistance and is, therefore, measured in Ω.

$\therefore$ $$I_m = V_m/X_L$$

or $$\frac{I_m}{\sqrt{2}} = \frac{V_m/\sqrt{2}}{X_L}$$

or $$I = \frac{V_L}{X_L} \qquad (V = V_L)$$

where inductive reactance $X_L = \omega L = 2\pi f L$

Note that X_L will be in Ω if L is in henry and f in Hz.

(*iii*) **Inductive susceptance (B_L).** Just as conductance is the reciprocal of resistance (*i.e.* $G = 1/R$), similarly, **inductive susceptance** *is the reciprocal of inductive reactance i.e.*

$$\text{Inductive susceptance, } B_L = \frac{1}{X_L}$$

Like conductance, the unit of inductive susceptance is also siemen (S). The inductive susceptance of a purely inductive circuit is its ability to pass current through it. The greater the inductive susceptance (*i.e.* the smaller the inductive reactance) of a purely inductive circuit, the greater is its ability to pass current through it and *vice-versa*.

(*iv*) **Power**

$$\text{Instantaneous power, } p = v\,i = V_m \sin \omega\, t \times I_m \sin (\omega\, t - \pi/2)$$

$$= -V_m I_m \sin \omega\, t \cos \omega\, t = -\frac{V_m I_m}{2} \sin 2\omega t$$

$\therefore$ $$\text{Average power, } P = \text{Average of } p \text{ over one cycle}$$

$$= \frac{1}{2\pi} \int_0^{2\pi} -\frac{V_m I_m}{2} \sin 2\omega t\, d(\omega t) = 0$$

Hence power absorbed in pure inductance is zero.

(*v*) **Power curve.** Fig. 11.68 shows the power curve for a pure inductive circuit. During the first 90° of the cycle, the voltage is positive and the current is negative. Therefore, the power supplied is negative. This means the power is flowing from the coil to the source. During the next 90° of the cycle, both voltage and current are positive and the power supplied is positive. Therefore, power flows from the source to the coil. Similarly, for the next 90° of the cycle, power flows from the coil to the source and during the last 90° of the cycle, power flows from the source to the coil. An examination of the power curve over one cycle shows that positive power is equal to the negative power. Hence the resultant power over one cycle is zero *i.e.* a pure inductance consumes no power. The electric power merely flows from the source to the coil and back again.

---

* But the two concepts should not be confused. In a resistance, current and p.d. are in phase and power is absorbed. In an inductance, current and p.d. are 90° out of phase and there is no net power consumed.

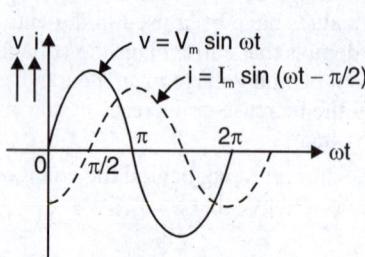

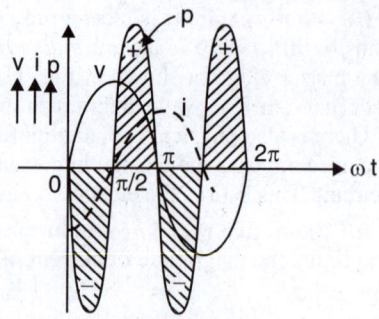

Fig. 11.67   **Fig. 11.68**

Example 11.50. *A pure inductive coil allows a current of 10 A to flow from a 230 V, 50 Hz supply. Find (i) inductive reactance (ii) inductance of the coil (iii) power absorbed. Write down the equations for voltage and current.*

Solution. **(i)** Circuit current, $I = V/X_L$ $(V_L = V)$

∴   Inductive reactance, $X_L = V/I = 230/10 = $ **23 Ω**

(ii) Now, $X_L = 2\pi f L$ ∴ $L = \dfrac{X_L}{2\pi f} = \dfrac{23}{2\pi \times 50} = $ **0.073 H**

(iii)   Power absorbed = **Zero**

$V_m = 230 \times \sqrt{2} = 325.27$ V ; $I_m = 10 \times \sqrt{2} = 14.14$ A ; $\omega = 2\pi \times 50 = 314$ rad/s

Since in a pure inductive circuit, current lags behind the voltage by $\pi/2$ radians, the equations are :

$$v = 325.27 \sin 314\, t \;\; ; \;\; i = 14.14 \sin (314\, t - \pi/2)$$

Example 11.51. *The current through an 80 mH inductor is 0·1 sin (440 t − 25°) A. Write the mathematical expression for the voltage across it.*

Solution. Inductive reactance is

$$X_L = 2\pi f L = 400 \times 80 \times 10^{-3} = 32\ \Omega$$
$$V_m = I_m X_L = 0\cdot1 \times 32 = 3\cdot2\ \text{V}$$

Since the voltage leads the current by 90°, we must add 90° to the phase angle of voltage.

∴ $v = V_m \sin (400\, t - 25° + 90°)$

or $v = \mathbf{3\cdot2 \sin (400\, t + 65°)\ V}$

Example 11.52. *The voltage across and current through a circuit element are :*

v = 100 sin (314t + 45°) volts ; i = 10 sin (314t + 315°) amperes

(i) Identify the circuit element. (ii) Find the value. (iii) Obtain expression for power.

Solution. $v = 100 \sin (314\, t + 45°)$; $i = 10 \sin (314t + 315°)$

The expression for i can be written as : $i = 10 \sin(314t - 45°)$.

(i) From the equations of voltage and current, it is clear that i lags behind v by 90°. Therefore, the circuit element is **pure inductor.**

(ii) Inductive reactance of the element is

$$X_L = \frac{V_m}{I_m} = \frac{100}{10} = 10\ \Omega$$

∴ $L = \dfrac{X_L}{\omega} = \dfrac{10}{314} = 0.0318\ \text{H} = \mathbf{3.18\ mH}$

(iii) Expression for instantaneous power p is

$$p = -\frac{V_m I_m}{2}\sin 2\omega t = -\frac{100 \times 10}{2}\sin(2 \times 314 t)$$

∴ $p = \mathbf{-500 \sin 628 t}$

Excample 11.53. *Two coils having self inductances of 0.1H and 0.2H and a coupling coefficient of 0.35 are connected in series across 125 V, 50 Hz supply. Find the circuit current.*

Solution. $V = 125$ volts ; $f = 50$ Hz ; $k = 0.35$

$$M = k\sqrt{L_1 L_2} = 0.35 \times \sqrt{0.1 \times 0.2} = 0.05 \text{ H}$$

Total inductance in series-aiding connection is

$$L_T = L_1 + L_2 + 2M = 0.1 + 0.2 + 2 \times 0.05 = 0.4 \text{ H}$$

$$\therefore \quad I = \frac{V}{2\pi f L_T} = \frac{125}{2\pi \times 50 \times 0.4} = \mathbf{1A}$$

Total inductance in series-opposing connection is

$$L'_T = L_1 + L_2 - 2M = 0.1 + 0.2 - 2 \times 0.05 = 0.2 \text{H}$$

$$\therefore \quad I' = \frac{V}{2\pi f L'_T} = \frac{125}{2\pi \times 50 \times 0.2} = \mathbf{2A}$$

Tutorial Problems

1. What is the inductive reactance of a 270 mH inductor at 4 kHz ? How much current will flow if this inductor is connected to a source of voltage of 25 V at 4 kHz ? **[6782·4 Ω ; 3·69 mA]**

2. The voltage across a 4 H inductor is 18 sin $(2\pi \times 10^3 \, t - 30°)$ V. Write the mathematical expression for the current through it. **[$i = 0·716$ sin $(2\pi \times 10^3 \, t - 120°)$ mA]**

3. Find the inductive reactance of a 1·5 H inductor when the voltage across it is 65 sin $(320 \, t)$ mV. **[480Ω]**

11·30. A.C. Circuit Containing Capacitance Only

When an alternating voltage is applied across the plates of a capacitor, the capacitor is charged in one direction and then in the other as the voltage reverses. The result is that electrons move to and fro around the circuit, connecting the plates, thus constituting alternating current.

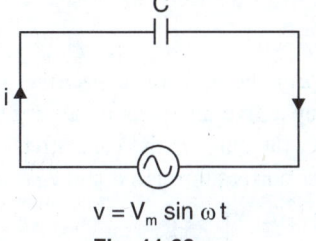

v = V_m sin ωt

Fig. 11.69

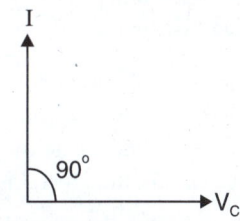

Fig. 11.70

Consider an alternating voltage applied to a capacitor of capacitance C farad as shown in Fig. 11.69. Let the equation of the applied alternating voltage be :

$$v = V_m \sin \omega t \qquad \qquad ...(i)$$

As a result of this alternating voltage, alternating current will flow through the circuit. Let at any instant i be the current and q be the charge on the plates.

Charge on capacitor, $q = C v = C V_m \sin \omega t$

$$\therefore \quad \text{Circuit current, } i = \frac{d}{dt}(q) = \frac{d}{dt}(C V_m \sin \omega t) = \omega C V_m \cos \omega t$$

$$\therefore \quad i = \omega C V_m \sin (\omega t + \pi/2) \qquad \qquad ...(ii)$$

The value of i will be maximum (*i..e.* I_m) when sin $(\omega t + \pi/2)$ is unity.

$$\therefore \quad I_m = \omega C V_m$$

Substituting the value $\omega C V_m = I_m$ in eq. (ii), we get,

$$i = I_m \sin (\omega t + \pi/2) \qquad \qquad ...(iii)$$

(*i*) **Phase angle.** It is clear from eqs. (*i*) and (*iii*) that current leads the voltage by $\pi/2$ radians or 90°. *Hence in a pure capacitance, current leads the voltage by 90°.* This is also indicated in the phasor diagram shown in Fig. 11.70. The wave diagram shown in Fig. 11.71 also reveals the same fact. There is also physical explanation for the lagging of voltage behind the current in a capacitor. Capacitance *opposes the change in voltage and serves to delay the increase or decrease of voltage across the capacitor. This causes the voltage to lag behind the current.

(*ii*) **Capacitive reactance.** Capacitance not only causes the voltage to lag behind current but it also limits the magnitude of current in the circuit. We have seen above that :

$$I_m = \omega C V_m$$

or
$$\frac{V_m}{I_m} = \frac{1}{\omega C}$$

If V_C and I are the r.m.s. values, then,

$$\frac{V_m}{I_m} = \frac{V_C}{I} = \frac{1}{\omega C} \qquad (V = V_C)$$

Clearly, the opposition offered by capacitance to current flow is $1/\omega C$. This quantity $1/\omega C$ is called the *capacitive reactance X_C* of the capacitor. It has the same dimensions as resistance and is, therefore, measured in Ω.

$$\therefore \qquad I = V_C/X_C$$

where capacitive reactance is $X_C = \dfrac{1}{\omega C} = \dfrac{1}{2\pi f C}$

Note that X_C will be in Ω if C is in farad and f in Hz.

(*iii*) **Capacitive susceptance (B_C).** Just as inductive susceptance is the reciprocal of inductive reactance (*i.e.* $B_L = 1/X_L$), similarly, **capacitive susceptance** *is the reciprocal of capacitive reactance* *i.e.*

$$\text{Capacitive susceptance, } B_C = \frac{1}{X_C}$$

Like conductance (G) and inductive susceptance (B_L), the unit of capacitive susceptance is siemen (S). The capacitive susceptance of a purely capacitive circuit is its ability to pass current through it. The greater the capacitive susceptance (*i.e.* the smaller the capacitive reactance) of a purely capacitive circuit, the greater is its ability to pass current through it and *vice-versa*.

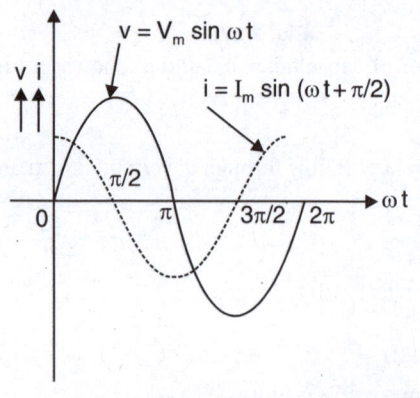

Fig. 11.71

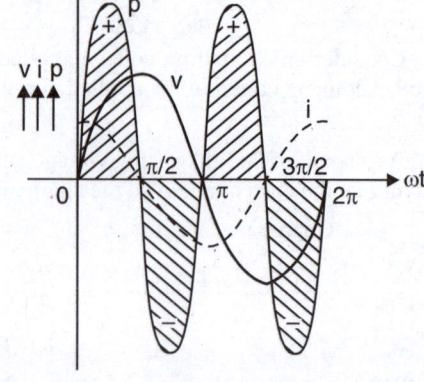

Fig. 11.72

---

*   Note that inductance opposes the change of current in the circuit whereas capacitance opposes the change of voltage in the circuit. In the study of a.c. circuits, these two parameters must be viewed from this aspect.

(*iv*) Power. Instantaneous power is given by ;

$$p = vi = V_m \sin \omega t \times I_m \sin (\omega t + \pi/2) = V_m I_m \sin \omega t \cos \omega t$$

$$\therefore \qquad p = \frac{V_m I_m}{2} \sin 2\omega t$$

$\therefore$ Average power, P = Average of p over one cycle

$$= \frac{1}{2\pi} \int_0^{2\pi} \frac{V_m I_m}{2} \sin 2\omega t \, d(\omega t) = 0$$

Hence power absorbed in a pure capacitance is zero.

(*v*) Power curve. Fig. 11.72 shows the power curve for a pure capacitive circuit. The power curve is similar to that for a pure inductor because now current leads the voltage by 90°. It is clear that positive power is equal to the negative power over one cycle. Hence net power absorbed in a pure capacitor is zero.

Example 11.54. *A 318 µF capacitor is connected across a 230 V, 50 Hz system. Determine (i) the capacitive reactance (ii) r.m.s. value of current and (iii) equations for voltage and current.*

Solution. (*i*) Capacitive reactance, $X_C = \dfrac{1}{2\pi f C} = \dfrac{10^6}{2\pi \times 50 \times 318} = \mathbf{10\Omega}$

(*ii*) R.M.S. value of current, $I = V/X_C = 230/10 = \mathbf{23 \ A}$

(*iii*) $V_m = 230 \times \sqrt{2} = 325 \cdot 27$ volts ; $I_m = \sqrt{2} \times 23 = 32 \cdot 53$ A ; $\omega = 2\pi \times 50 = 314$ rad/s

$\therefore$ Equations for voltage and current are :

$$v = \mathbf{325 \cdot 27 \sin 314 \, t} \quad ; \quad i = \mathbf{32 \cdot 53 \sin (314 \, t + \pi/2)}$$

Example 11.55. *The voltage across a 0.01 µ F capacitor is 240 sin (1·25 × 10^4 t − 30°) V. Write the mathematical expression for the current through it.*

Solution. Capacitive reactance, $X_C = \dfrac{1}{\omega C} = \dfrac{1}{(1 \cdot 25 \times 10^4) \times (0 \cdot 01 \times 10^{-6})} = 8000 \ \Omega$

$$\text{Peak current, } I_m = \frac{V_m}{X_C} = \frac{240}{8000} = 0.03 \text{ A}$$

$\therefore$ Expression for current through the capacitor is

$$i = I_m \sin (1 \cdot 25 \times 10^4 t - 30° + 90°) \text{ A}$$

or $\qquad\qquad i = \mathbf{0 \cdot 03 \sin (1 \cdot 25 \times 10^4 t + 60°) \ A}$

Tutorial Problems

1. How much capacitance is needed to produce 1200 Ω of reactance when the frequency is 2·3 MHz ?

   [**57·7 pF**]

2. A 0·1 µF capacitor and a 0·068 µF capacitor are connected in series to a 12 V, 1 kHz source. Determine the circuit current and the voltage across the 0·1 µF capacitor. [**3·05 mA ; 4·86 V**]

3. The current through a 15 pF capacitor is 0·45 sin (10^8 t + 45°) mA. Write the mathematical expression for the voltage across it. [v_C = **0·3 sin (10^{8}t − 45°) V**]

11·31. Complex Waves and A.C. Circuit

When a complex (*i.e.* non-sinusoidal) alternating voltage is applied to a circuit containing constant resistance, inductance or capacitance or combination of them, the principle of superposition can be applied to find the current drawn from the supply. That is to say we calculate the currents that would be produced by the fundamental and harmonic voltages if they were applied separately. The current in any circuit element is equal to the sum of fundamental and harmonic currents in that element. Consider the following complex voltage wave :

$$v = V_{1m} \sin \omega t + V_{3m} \sin 3\omega t + V_{5m} \sin 5\omega t$$

Note that this complex wave contains the fundamental, third harmonic and fifth harmonic only. Further, the equation tells us that harmonics have no individual phase differences.

(*i*) **Circuit contains resistance only.** Suppose the above complex voltage wave is applied to a circuit that contains resistance R only. Since resistance is independent of frequency,

$$\therefore \qquad I_{1m} = V_{1m}/R \quad ; \quad I_{3m} = V_{3m}/R \quad ; \quad I_{5m} = V_{5m}/R$$

As the circuit has resistance only, each current is in phase with its voltage. Therefore, complex current equation is

$$i = \frac{V_{1m}}{R} \sin \omega t + \frac{V_{3m}}{R} \sin 3\omega t + \frac{V_{5m}}{R} \sin 5\omega t$$

If V_1, V_3 and V_5 are the r.m.s. values of the fundamental, third harmonic and fifth harmonic respectively, then r.m.s. value of voltage of the complex wave is given by ;

$$V = \sqrt{V_1^2 + V_3^2 + V_5^2}$$

Similarly, r.m.s. value of current of complex wave is

$$I = \sqrt{I_1^2 + I_3^2 + I_5^2}$$

$$\therefore \qquad I = \frac{V}{R} = \frac{1}{R}\sqrt{V_1^2 + V_3^2 + V_5^2}$$

(*ii*) **Circuit contains inductance only.** Consider that above complex voltage wave is applied to a circuit that contains inductance only. Since inductive reactance $X (= \omega L)$ depends upon frequency, the reactance offered to the fundamental and harmonics will be different.

For the fundamental, $X_1 = \omega L$; For third harmonic, $X_3 = 3\omega L$ and for fifth harmonic, $X_5 = 5\omega L$. Since in case of pure inductance, current lags behind the voltage by $\pi/2$, the various currents will lag behind their respective voltages by $\pi/2$.

$$\therefore \quad i = \frac{V_{1m}}{\omega L} \sin (\omega t - \pi/2) + \frac{V_{3m}}{3\omega L} \sin (3\omega t - \pi/2) + \frac{V_{5m}}{5\omega L} \sin (5\omega t - \pi/2)$$

Note that the harmonics in the current wave are much smaller in magnitude than in the voltage wave. Thus 3rd harmonic in the current wave is 1/3rd of the harmonic in the voltage wave. This means that current wave is a closer approximation to a sine wave than is the voltage wave.

The r.m.s. values are : $I_1 = V_1/\omega L$; $I_3 = V_3/3\omega L$ and $I_5 = V_5/5\omega L$

$\therefore$ R.M.S. value I of complex current wave is

$$I = \frac{1}{\omega L}\left(V_1^2 + \frac{V_3^2}{9} + \frac{V_5^2}{25} \right)^{1/2}$$

(*iii*) **Circuit contains capacitance only.** It is easy to see that current equation is given by ;

$$i = \omega C V_{1m} \sin \left(\omega t + \frac{\pi}{2} \right) + 3\omega C V_{3m} \sin \left(3\omega t + \frac{\pi}{2} \right) + 5\omega C V_{5m} \sin \left(5\omega t + \frac{\pi}{2} \right)$$

11·32. Fundamental Power and Harmonic Power

When a complex alternating voltage is applied to a circuit, the power associated with each circuit element is equal to the sum of the fundamental power and all the individual harmonic powers. In an a.c. circuit, the fundamental current and the fundamental voltage together produce fundamental power. This fundamental power is the useful power that causes a motor to rotate, for example. On the other hand, the product of harmonic voltage and the corresponding harmonic current produces a harmonic power. The harmonic power is usually dissipated as heat and does no useful work. Therefore, harmonic voltages and currents should be kept as small as possible.

Example 11.56. *The voltage applied to a circuit is*

$$v = 215 \sin 377\,t - 140 \cos 745\,t + 70 \sin 1131\,t$$

and the current in the circuit is

$$i = 15 \cos (377\,t - 30°) + 4 \cos (754\,t - 60°) + 3 \sin (1131\,t + 50°).$$

Find (i) the frequency of the fundamental (ii) the r.m.s. values of the voltage and current.

Solution. (*i*) Frequency of fundamental, $f_1 = \omega_1/2\pi = 377/2\pi = $ **60 Hz**

(*ii*)   R.M.S. value of voltage, $V = \sqrt{\left(\dfrac{215}{\sqrt{2}}\right)^2 + \left(\dfrac{140}{\sqrt{2}}\right)^2 + \left(\dfrac{70}{\sqrt{2}}\right)^2} = $ **188 volts**

R.M.S. value of current, $I = \sqrt{\left(\dfrac{15}{\sqrt{2}}\right)^2 + \left(\dfrac{4}{\sqrt{2}}\right)^2 + \left(\dfrac{3}{\sqrt{2}}\right)^2} = $ **11.17 A**

Example 11.57. *If the effective value of the voltage of a complex waveform is 130 V, determine the peak value of the fundamental if the harmonic content of the waveform is 20% referred to the complex wave.*

Solution. R.M.S. value of complex wave, $V = 130$ volts

R.M.S. values of harmonic content, $V_h = $ 20% of $130 = 26$ volts

Let V_1 volts be the r.m.s. value of the fundamental.

∴ $V^2 = V_1^2 + V_h^2$ or $V_1 = \sqrt{V^2 - V_h^2} = \sqrt{(130)^2 - (26)^2} = 127{\cdot}37$ volts

∴   Peak value of the fundamental $= \sqrt{2}\,V_1 = \sqrt{2} \times 127.37 = $ **180·13V**

Example 11.58. *The current through a purely inductive coil of self-inductance of 15·9 mH is given by ;*

$$i = 20 \sin (314\,t - \pi/2) + 5 \sin (942\,t - \pi/2) + 2 \sin (1570\,t - \pi/2)$$

Find the equation of the applied voltage.

Solution. The current wave contains the fundamental, the third harmonic and the fifth harmonic. The reactances offered to these three components are :

$X_1 = \omega L = 314 \times 15{\cdot}9 \times 10^{-3} = 5\Omega$; $X_3 = 3\omega L = 3 \times 5 = 15\ \Omega$; $X_5 = 5\omega L = 25\ \Omega$

∴ $V_{1m} = 20 \times X_1 = 20 \times 5 = 100$ V   ; $V_{3m} = 5 \times 15 = 75$ V   ; $V_{5m} = 2 \times 25 = 50$ V

∴   The equation of the applied voltage is

$$v = 100 \sin 314\,t + 75 \sin 942\,t + 50 \sin 1570\,t \text{ \textbf{Ans.}}$$

Objective Questions

1. The a.c. system is preferred to d.c. system because ....

    (*i*) a.c. voltages can be easily changed in magnitude

    (*ii*) d.c. motors do not have fine speed control

    (*iii*) high-voltage a.c. transmission is less efficient

    (*iv*) d.c. voltage cannot be used for domestic appliances

2. In a.c. system, we generate sine waveform because ....

    (*i*) it can be easily drawn

    (*ii*) it produces least disturbance in electrical circuits

    (*iii*) it is nature's standard

    (*iv*) other waves cannot be produced easily

3. .... will work only on d.c. supply.

    (*i*) Electric lamp   (*ii*) Refrigerator

    (*iii*) Heater   (*iv*) Electroplating

4. .... will produce a.c. voltage.

    (*i*) Friction

(*ii*) Photoelectric effect

(*iii*) Thermal energy   (*iv*) Crystal

5. A coil is rotating in the uniform field of an 8-pole generator. In one revolution of the coil, the number of cycles generated by the voltage is ....

    (*i*) one   (*ii*) two

    (*iii*) four   (*iv*) eight

6. An alternating voltage is given by $v = 20 \sin 157\,t$. The frequency of the alternating voltage is ....

    (*i*) 50 Hz   (*ii*) 25 Hz

    (*iii*) 100 Hz   (*iv*) 75 Hz

7. An alternating current is given by $i = 10 \sin 314\,t$. The time taken to generate two cycles of current is ....

    (*i*) 0·02 second   (*ii*) 0·01 second

    (*iii*) 0·04 second   (*iv*) 0·05 second

8. An alternating voltage is given by $v = 30 \sin 314\,t$. The time taken by the voltage to reach -30V for the first time is ....

    (*i*) 0·02 second   (*ii*) 0·1 second

    (*iii*) 0·03 second   (*iv*) 0·015 second

9. A sine wave has a maximum value of 20 V. Its value at $135°$ is ....

    (*i*) 10 V   (*ii*) 14·14 V

    (*iii*) 15 V   (*iv*) 5 V

10. A sinusoidal current has a magnitude of 3A at $120°$. Its maximum value will be ....

    (*i*) $\sqrt{3}$ A   (*ii*) $\sqrt{3}/2$ A

    (*iii*) $2\sqrt{3}$ A   (*iv*) 6 A

11. An alternating current is given by $i = 10 \sin 314\,t$. Measuring time from $t = 0$, the time taken by the current to reach $+10$ V for the second time is ....

    (*i*) 0·05 second   (*ii*) 0·1 second

    (*iii*) 0·025 second   (*iv*) 0·02 second

12. An a.c. generator having 10 poles and running at 600 r.p.m. will generate an alternating voltage of frequency ....

    (*i*) 25 Hz   (*ii*) 100 Hz

    (*iii*) 50 Hz   (*iv*) 200 Hz

13. We have assigned a frequency of 50 Hz to power system because it ....

    (*i*) can be easily obtained

    (*ii*) gives best result when used for operating both lights and machinery

(*iii*) leads to easy calculations

(*iv*) none of the above

14. An alternating voltage is given by $v = 100 \sin 314\,t$ volts. Its average value will be.............

    (*i*) 70.7 V   (*ii*) 50 V

    (*iii*) 63·7 V   (*iv*) 100 V

15. An alternating current whose average value is 1 A will produce .... 1 A d.c. under similar conditions.

    (*i*) less heat than   (*ii*) more heat than

    (*iii*) the same heat as (*iv*) none of the above

16. A sinusoidal alternating current has a maximum value of I_m. Its average value will be ....

    (*i*) I_m/π (*ii*) $I_m/2\pi$

    (*iii*) $2\,I_m/\pi$ (*iv*) none of the above

17. The area of a sinusoidal wave over a half-cycle is ....

    (*i*) max. value $\div 2$ (*ii*) $2 \times$ max. value

    (*iii*) max. value $\div \pi$ (*iv*) max. value $\div 2\pi$

18. An alternating voltage is given by $v = 200 \sin 314\,t$. Its r.m.s value will be ....

    (*i*) 100 V   (*ii*) 282·8 V

    (*iii*) 141·4 V   (*iv*) 121·4 V

19. The r.m.s. value of sinusoidally varying current is .... that of its average value.

    (*i*) more than   (*ii*) less than

    (*iii*) same as   (*iv*) none of the above

20. Alternating voltages and currents are expressed in r.m.s. values because ....

    (*i*) they can be easily determined

    (*ii*) calculations become very simple

    (*iii*) they give comparison with d.c.

    (*iv*) none of the above

21. The average value of $\sin^2\theta$ over a complete cycle is ....

    (*i*) $+1$ (*ii*) -1

    (*iii*) 1/2   (*iv*) zero

22. The average value of $\sin \theta$ over a complete cycle is ....

    (*i*) zero   (*ii*) $+1$

    (*iii*) -1 (*iv*) 1/2

23. An alternating current is given by $i = I_m \sin \theta$. The average value of squared wave of this current over a complete cycle is

    (*i*) $I^2_m/2$ (*ii*) I_m/π

    (*iii*) $2I_m/\pi$ (*iv*) $2\,I_m$

24. The form factor of a sinusoidal wave is ....

   (i) 1·414   (ii) 1·11

   (iii) 2   (iv) 1·5

25. The filament of a vacuum tube requires 0·4 A d.c. to heat it. The r.m.s. value of a.c. required is ....

   (i) 0·4 × $\sqrt{2}$ A   (ii) 0·4 ÷ 2 A

   (iii) 0·8 ÷ $\sqrt{2}$ A   (iv) 0·4 A

26. A 100 V peak a.c. is as effective as .... d.c.

   (i) 100 V   (ii) 50 V

   (iii) 70·7 V   (iv) none of the above

27. The form factor of a .... wave is 1.

   (i) sinusoidal   (ii) square

   (iii) triangular   (iv) saw tooth

28. Out of the following .... wave is the peakiest.

   (i) sinusoidal   (ii) square

   (iii) rectangular (iv) triangular

29. The peak factor of a sine waveform is ....

   (i) 1.11   (ii) 1.414

   (iii) 2   (iv) 1.5

30. When a 15-V square wave is connected across a 50-V a.c. voltmeter, it will read ....

   (i) 15 V   (ii) 15 × $\sqrt{2}$ V

   (iii) 15/$\sqrt{2}$ V   (iv) none of the above

Answers

1. (i)	2. (ii)	3. (iv)	4. (iv)	5. (iii)
6. (ii)	7. (iii)	8. (iv)	9. (ii)	10. (iii)
11. (iii)	12. (iii)	13. (ii)	14. (iii)	15. (i)
16. (iii)	17. (ii)	18. (iii)	19. (i)	20. (iii)
21. (iii)	22. (i)	23. (i)	24. (ii)	25. (iv)
26. (iii)	27. (ii)	28. (iv)	29. (ii)	30. (i)

12
Series A.C. Circuits

Introduction

An a.c. circuit differs from a d.c. circuit in many respects. First, in a d.c. circuit we consider resistances only whereas in an a.c. circuit, in addition to resistance (R), inductance (L) and capacitance (C) also play the part. The elements L and C offer opposition (*i.e.* X_L and X_C) to current flow in an a.c. circuit. Secondly, the magnitude of current in an a.c. circuit is affected by the supply frequency because $X_L (= 2\pi f L)$ and $X_C (= 1/2\pi f C)$ are frequency dependent. However, such a situation is not encountered in a d.c. circuit. Thirdly, in a d.c. circuit, voltages or currents can be added or subtracted arithmetically. But in an a.c. circuit, there is a phase difference of 90° between voltage across and current through L or C. This implies that for the addition or subtraction of alternating voltages or currents, phase difference has to be taken into account. All these features make the analysis of an a.c. circuit quite different from that of a d.c. circuit. In this chapter, we shall confine our attention to series a.c. circuits only.

12.1. R-L Series A.C. Circuit

This is the most general case met in practice as nearly all a.c. circuits contain both resistance and inductance. Fig. 12.1 (*i*) shows a pure resistance of R ohms connected in series with a coil of pure inductance L henry.

Let $\qquad\qquad\qquad V =$ r.m.s. value of the applied voltage

$\qquad\qquad\qquad\qquad I =$ r.m.s. value of the circuit current

$\therefore\qquad\qquad\qquad V_R = I R$ .... where V_R is in phase with I

$\qquad\qquad\qquad\qquad V_L = I X_L$ .... where V_L leads I by 90°

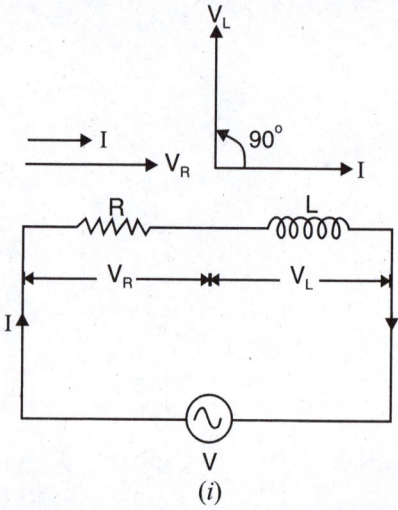

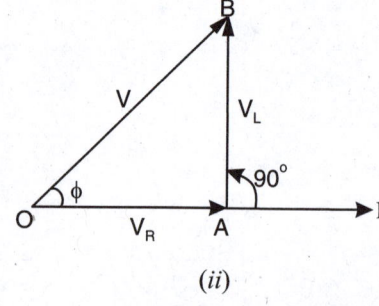

Fig. 12.1

Taking *current as the reference phasor, the phasor diagram of the circuit can be drawn as shown in Fig. 12.1 (*ii*). The voltage drop V_R ($= I R$) is in phase with current and is represented in magnitude and direction by the phasor OA. The voltage drop V_L ($= I X_L$) leads the current by 90° and is represented in magnitude and direction by the phasor AB. The applied voltage V is the phasor sum of these two drops *i.e.*

$$V = \sqrt{V_R^2 + V_L^2} = \sqrt{(IR)^2 + (IX_L)^2} = I\sqrt{R^2 + X_L^2}$$

$$\therefore \qquad I = \frac{V}{\sqrt{R^2 + X_L^2}}$$

The quantity $\sqrt{R^2 + X_L^2}$ offers opposition to current flow and is called **impedance** of the circuit. It is represented by Z and is measured in ohms (Ω).

$$\therefore \qquad I = \frac{V}{Z} \quad \text{where } Z = \sqrt{R^2 + X_L^2}$$

(*i*) Phase angle. It is clear from the phasor diagram that circuit current I lags behind the applied voltage V by ϕ°. This fact is also illustrated in the wave diagram shown in Fig. 12.2. The value of phase angle ϕ can be determined from the phasor diagram.

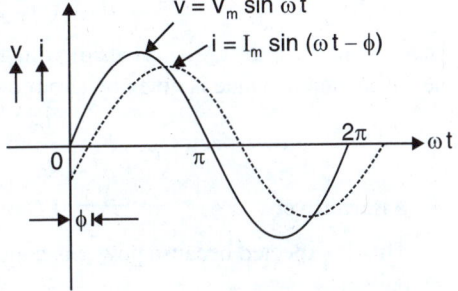

$v = V_m \sin \omega t$

$i = I_m \sin (\omega t - \phi)$

Fig. 12.2

$$\tan \phi = \frac{V_L}{V_R} = \frac{I X_L}{I R} = \frac{X_L}{R}$$

Since X_L and R are known, ϕ can be calculated.

If the applied voltage is $v = V_m \sin \omega t$, then equation for the circuit current will be :

$$i = I_m \sin (\omega t - \phi) \quad \text{where } I_m = V_m / Z$$

We arrive at a very important conclusion that *in an inductive circuit, current lags behind the applied voltage.* The angle of lag (*i.e.* ϕ) is greater than 0° but less than 90°. It is determined by the ratio of inductive reactance to resistance (tan $\phi = X_L/R$) in the circuit. The greater the value of this ratio, the greater will be the phase angle ϕ and *vice-versa.*

(*ii*) Impedance. *The total opposition offered to the flow of alternating current by a circuit is called* **impedance** *Z of the circuit.* In *R-L* series circuit,

Impedance, $Z = \sqrt{R^2 + X_L^2}$ where $X_L = 2\pi f L$

The magnitude of impedance in *R-L* series circuit depends upon the values of R, L and the supply frequency f.

(*iii*) Admittance (*Y*). **Admittance** *of an a.c. circuit is the reciprocal of its impedance i.e.*

$$\text{Admittance, } Y = \frac{1}{Z}$$

The unit of admittance is siemen (*S*). The admittance of an a.c. circuit is its ability to pass current through it. The greater the admittance (*i.e.* the smaller the impedance) of a circuit, the greater is its ability to pass current through it and *vice-versa.*

(*iv*) Power

Instantaneous power, $p = v i = V_m \sin \omega t \times I_m \sin (\omega t - \phi)$

---

*   It is always convenient to take that quantity (voltage or current) as the reference phasor which is common in the circuit. Since current is common in a series circuit, it shall be taken as the reference phasor in drawing the phasor diagrams.

$$= \frac{1}{2} V_m\, I_m\, [2 \sin \omega t \sin (\omega t - \phi)]$$

$$= \frac{1}{2} V_m\, I_m\, [\cos \phi - \cos (2\omega t - \phi)]$$

$$= \frac{1}{2} V_m\, I_m\, \cos \phi - \frac{1}{2} V_m\, I_m\, \cos (2\omega t - \phi)$$

Thus instantaneous power consists of two parts :

(*a*) Constant part $\frac{1}{2} V_m\, I_m \cos \phi$ whose average value over a cycle is the same.

(*b*) A pulsating component $\frac{1}{2} V_m I_m \cos (2\omega t - \phi)$ whose average value over one complete cycle is zero.

$$\therefore \qquad \text{Average power, } P = \frac{V_m\, I_m}{2} \cos \phi = \frac{V_m}{\sqrt{2}} \times \frac{I_m}{\sqrt{2}} \times \cos \phi$$

or $$P = VI \cos \phi$$

where *V* and *I* are the r.m.s. values of voltage and current. The term cos ϕ is called **power factor** of the circuit and its value is given by (from phasor diagram) ;

$$\text{Power factor, } \cos \phi = \frac{IR}{IZ} = \frac{R}{Z}$$

Alternatively. $\qquad P = VI \cos \phi = (IZ)\, I\, (R/Z) = I^2 R \qquad [\because V = IZ \text{ and } \cos \phi = R/Z]$

This is expected because power is consumed in resistance only; inductance does not consume any power.

(*v*) **Power curve.** The power curve for a *R-L* series circuit is shown in Fig. 12.3. To make the treatment more illustrative, it is assumed that current lags behind the voltage by 30° *i.e.* $\phi = 30°$. It is clear that power is negative between 0° and 30° and between 180° and 210°. During the rest of the cycle, the power is positive. Since the area under the positive loops is greater than that under the negative loops, the net power over the cycle is positive. This means that power is being consumed by the circuit. However, it is less than the average power consumed by a comparable circuit where voltage and current are in phase (*i.e.* pure resistive circuit). The negative area simply indicates that inductance of the circuit returns power to the source during that interval.

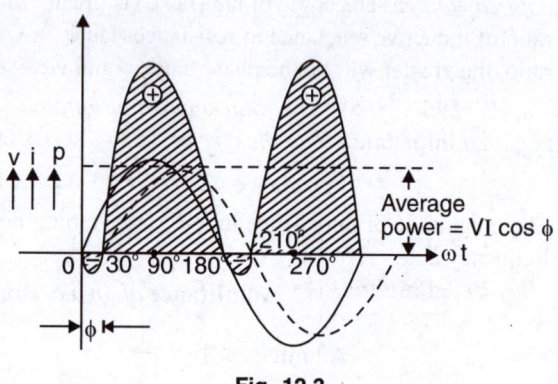

Fig. 12.3

12.2. Impedance Triangle

The phasor diagram of a *R-L* series circuit is shown in Fig. 12.4. Dividing each side of the phasor diagram by the same factor *I*, we get a triangle whose sides represent *R*, X_L and *Z*. Such a triangle is known as *impedance triangle* (See Fig. 12.5). Just as in Fig. 12.4, the impedance triangle is also a right-angled triangle.

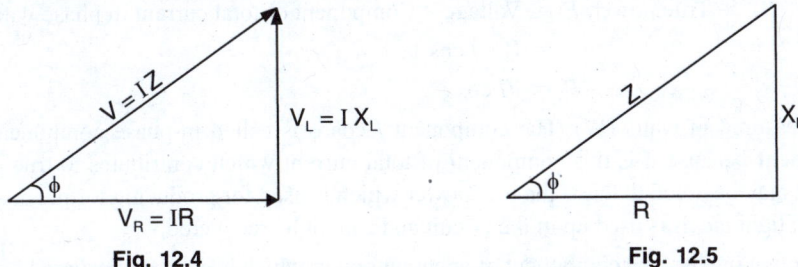

Fig. 12.4   **Fig. 12.5**

Impedance triangle is a useful concept in a.c. circuits as it enables us to calculate :

(*i*) the impedance of the circuit *i.e.,* $Z = \sqrt{R^2 + X_L^2}$

(*ii*) power factor of the circuit *i.e., cos* $\phi = R/Z$

(*iii*) phase angle ϕ *i.e.,* *tan $\phi = X_L/R$

(*iv*) whether current leads or lags the voltage.

Therefore, it is always profitable to draw the impedance triangle while analysing an a.c. circuit.

12.3. Apparent, True and Reactive Powers

Consider an inductive circuit in which circuit current I lags behind the applied voltage V by $\phi°$. The phasor diagram of the circuit is shown in Fig. 12.6. The current I can be resolved into two rectangular components *viz.*

(*i*) $I \cos \phi$ in phase with V.

(*ii*) $I \sin \phi$; 90° out of phase with V.

1. Apparent power. *The total power that appears to be transferred between the source and load is called* **apparent power.** It is equal to the product of applied voltage (V) and circuit current (I) *i.e.*

Apparent power, $S = V \times I = VI$

It is measured in volt-ampers (VA).

Apparent power has two components *viz* true power and reactive power.

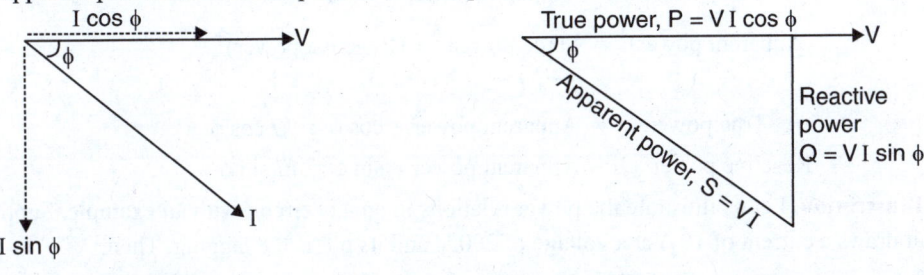

Fig. 12.6   **Fig. 12.7**

2. True power. The power which is actually consumed in the circuit is called **true power** or **active power.** We know that power is consumed in resistance only since neither pure inductor (L) nor pure capacitor (C) consumes any active power. Now, current and voltage are in phase in a resistance. Therefore, current in phase with voltage produces true or active power. It is the useful component of apparent power.

The product of voltage (V) and component of total current in phase with voltage ($I \cos \phi$) is equal to **true power** *i.e.*

---

*   Note that value of ϕ can also be determined from cos ϕ. But for greater accuracy, value of ϕ is found from tan ϕ.

True power, P = Voltage × Component of total current in phase with voltage

$$= V \times I \cos \phi$$

$$\therefore \qquad P = VI \cos \phi$$

It is measured in watts (W). The component $I \cos \phi$ is called in-phase component or **watt-ful component** because it is this component of total current which contributes to true power (*i.e.* $VI \cos \phi$). It may be noted that it is the true power which is used for producing torque in motors and supply heat, light *etc*. It is used up in the circuit and cannot be recovered.

3. Reactive power. The component of apparent power which is neither consumed nor does any useful work in the circuit is called **reactive power.** The power consumed (or true power) in L and C is zero because all the power received from the source in one quarter-cycle is returned to the source in the next quarter-cycle. This circulating power is called *reactive power. Now, current and voltage in L or C are 90° out of phase. Therefore, current 90° out of phase with voltage contributes to reactive power.

The product of voltage (V) and component of total current 90° out of phase with voltage ($I \sin \phi$) *is equal to* **reactive power** *i.e.*

Reactive power, Q = Voltage × Component of total current 90° out of phase with voltage

$$= V \times I \sin \phi$$

$$\therefore \qquad Q = VI \sin \phi$$

It is measured in volt-amperes reactive (VAR). The component $I \sin \phi$ is called the **reactive component** (or **wattless component**) and contributes to reactive power (*i.e.* $VI \sin \phi$). It does no useful work in the circuit and merely flows back and forth in both directions in the circuit. A wattmeter does not measure the reactive power.

Power triangle. If we multiply each of the current phasors in Fig. 12.6 by V, we get the power triangle shown in Fig. 12.7. This is a right-angled triangle and indicates the relation among apparent power, true power and reactive power. It reveals the following facts about the circuit :

(*i*)   Power factor, $\cos \phi = \dfrac{\text{True power}}{\text{Apparent power}} = \dfrac{VI \cos \phi}{VI}$

(*ii*) $(\text{Apparent power})^2 = (\text{True power})^2 + (\text{Reactive power})^2$

or $S^2 = P^2 + Q^2$

(*iii*)   True power, P = Apparent power × $\cos \phi = VI \cos \phi$

Reactive power, Q = Apparent power × $\sin \phi = VI \sin \phi$

Illustration. Let us illustrate the power relations in an a.c. circuit with an example. Suppose a circuit draws a current of 10 A at a voltage of 200 V and its p.f. is 0.8 lagging. Then,

Apparent power, $S = VI = 200 \times 10 = 2000$ VA

True power, $P = VI \cos \phi = 200 \times 10 \times 0.8 = 1600$ W

Reactive power, $Q = VI \sin \phi = 200 \times 10 \times 0{\cdot}6 = 1200$ VAR

The circuit receives an apparent power of 2000 VA and is able to convert only 1600 watts into true power. The reactive power of 1200 VAR does no useful work, it merely flows into and out of the circuit periodically. In fact, reactive power is a liability on the source because the source has to supply the additional current (*i.e.* $I \sin \phi$) to provide for this power.

---

*   This is the power that flows back and forth in both directions in the circuit or reacts upon itself. Hence the name.

12.4. Power Factor

The power factor (*i.e.* cos ϕ) of a circuit can be defined in one of the following ways :

(*i*) Power factor = cos ϕ = cosine of angle between V and I

(*ii*) Power factor = $\dfrac{R}{Z} = \dfrac{\text{Resistance}}{\text{Impedance}}$ [See Fig. 12.5]

(*iii*) Power factor = $\dfrac{VI\cos\phi}{VI} = \dfrac{\text{True power}}{\text{Apparent power}}$

For example, in a resistor, the current and voltage are in phase *i.e.* ϕ = 0°. Therefore, power factor of a pure resistive circuit is cos 0° = 1. Similarly, phase difference between voltage and current in a pure inductance or capacitance is 90°. Hence power factor of pure L or C is zero. This is the reason that power consumed by pure L or C is *zero. For a circuit having R, L and C in varying proportions, the value of power factor will lie between 0 and 1. It may be noted that power factor can never have a value greater than 1.

(*a*) It is a usual practice to attach the word 'lagging' or 'leading' with the numerical value of power factor to signify whether the current lags or leads the voltage. Thus if a circuit has a p.f. of 0.5 and the current lags the voltage, we generally write p.f. as 0.5 lagging.

(*b*) Sometimes power factor is expressed as a percentage. Thus 0.8 lagging power factor may be expressed as 80% lagging.

12.5. Significance of Power Factor

The apparent power drawn by a circuit has two components *viz.* (*i*) true power and (*ii*) reactive power. True power component should be as large as possible because it is this component which does useful work in the circuit. This is possible only if the reactive power component is small. As seen from the power triangle in Fig. 12.7, the smaller the phase angle ϕ (*i.e.* greater the p.f. cos ϕ), the smaller is the reactive power component. Thus when ϕ = 0° (i.e. cos ϕ = 1), the reactive power component is zero and the true power is **equal to the apparent power. That means the whole of apparent power drawn by the circuit is being utilised by it. *Thus power factor of a circuit is a measure of its effectiveness in ***utilising the apparent power drawn by it*. The greater the power factor of a circuit, the greater is its ability to utilise the apparent power. Thus 0·5 p.f. (*i.e.* 50% p.f.) of a circuit means that it will utilise only 50% of the apparent power whereas 0·8 p.f. would mean 80% utilisation of apparent power. For this reason, we wish that the power factor of the circuit to be as near to 1 as possible.

12.6. Q-factor of a Coil

The ratio of the inductive reactance (X_L) of a coil to its resistance (R) at a given frequency is known as †Q-factor of the coil at that frequency *i.e.*,

$$Q\text{-factor} = \frac{X_L}{R} = \frac{\omega L}{R}$$

Also, $Q\text{-factor} = 2\pi \times \dfrac{\text{maximum energy stored}}{\text{energy dissipated per cycle}}$

The Q-factor is used to describe the quality or effectiveness of a coil. A coil is usually designed to have high value of L compared to its resistance R. The greater the value of Q-factor of a coil, the greater is its inductance (L) as compared to its resistance (R). Many of the equations to be developed in a.c. circuit analysis can be simplified by the substitution of Q for the ratio X_L/R.

---

* $P = VI \cos \phi$. For pure L or C, cos ϕ = 0. Hence, $P = 0$.

** $P = VI \cos \phi$. If voltage and current are in phase (*i.e.* ϕ = 0°), then cos ϕ = 1. Therefore, $P = VI$.

*** Power factor is a factor which must be multiplied to apparent power to obtain true power. Hence the name power factor.

† Q-factor = 1/power factor = $\dfrac{Z}{R}$

If R is small compared to reactance, then $Q = \omega L/R$.

Example 12.1. *A coil having a resistance of 7 Ω and an inductance of 31·8 mH is connected to 230 V, 50 Hz supply. Calculate (i) the circuit current (ii) phase angle (iii) power factor (iv) power consumed and (v) voltage drop across resistor and inductor.*

Solution. *(i)* Inductive reactance, $X_L = 2\pi f L = 2\pi \times 50 \times 31.8 \times 10^{-3} = 10\ \Omega$

$$\text{Coil impedance, } Z = \sqrt{R^2 + X_L^2} = \sqrt{7^2 + 10^2} = 12.2\ \Omega$$

∴   Circuit current, $I = V/Z = 230/12.2 = \mathbf{18·85\ A}$

(ii) $\tan\phi = X_L/R = 10/7$

∴   Phase angle, $\phi = \tan^{-1}(10/7) = \mathbf{55°\ lag}$

(iii)   Power factor $= \cos\phi = \cos 55° = \mathbf{0.573\ lag}$

(iv)   Power consumed, $P = VI\cos\phi = 230 \times 18.85 \times 0.573 = \mathbf{2484·24\ W}$

(v)   Voltage drop across $R = IR = 18.85 \times 7 = \mathbf{131.95V}$

        Voltage drop across $L = IX_L = 18.85 \times 10 = \mathbf{188.5\ V}$

Example 12.2. *An inductor coil is connected to a supply of 250 V at 50 Hz and takes a current of 5 A. The coil dissipates 750 W. Calculate (i) power factor (ii) resistance of coil and (iii) inductance of coil.*

Solution. *(i)* Power consumed, $P = VI\cos\phi$

∴   Power factor, $\cos\phi = \dfrac{P}{VI} = \dfrac{750}{250 \times 5} = \mathbf{0.6\ lag}$

(ii)   Impedance of coil, $Z = V/I = 250/5 = 50\ \Omega$

       Resistance of coil, $R = Z\cos\phi = 50 \times 0.6 = \mathbf{30\ \Omega}$

(iii)   Reactance of coil, $X_L = \sqrt{Z^2 - R^2} = \sqrt{(50)^2 - (30)^2} = 40\ \Omega$

∴   Inductance of coil, $L = \dfrac{X_L}{2\pi f} = \dfrac{40}{2\pi \times 50} = \mathbf{0.127\ H}$

Example 12.3. *A pure inductance of 318 mH is connected in series with a pure resistance of 75 Ω. The circuit is supplied from 50 Hz source and the voltage across 75 Ω resistor is found to be 150 V. Calculate the supply voltage and the phase angle.*

Solution. The circuit diagram and the phasor diagram are shown in Fig. 12.8.

       Circuit current, $I = V_R/R = 150/75 = 2\ A$

    Reactance of coil, $X_L = 2\pi f L = 2\pi \times 50 \times 318 \times 10^{-3} = 100\ \Omega$

     Voltage across L, $V_L = IX_L = 2 \times 100 = 200\ V$

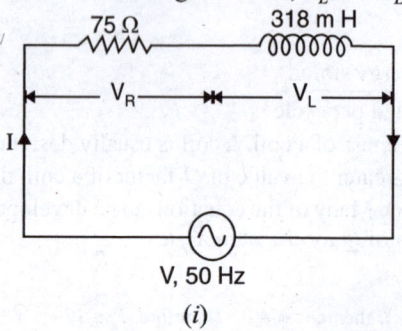

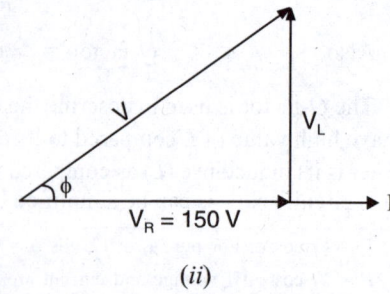

(i)   (ii)

Fig. 12.8

Referring to the phasor diagram of the circuit in Fig. 12.8 *(ii)*,

      Supply voltage, $V = \sqrt{V_R^2 + V_L^2} = \sqrt{150^2 + 200^2} = \mathbf{250\ V}$

Alternatively. $Z = \sqrt{R^2 + X_L^2} = \sqrt{75^2 + 100^2} = 125\ \Omega$

$\therefore$ $V = IZ = 2 \times 125 = \textbf{250 V}$

Now, $\tan \phi = X_L/R = 100/75 = 1.33$

$\therefore$ Phase angle, $\phi = \tan^{-1} 1.33 = \textbf{53.06° lag}$

Example 12.4. *A coil when connected across a 100 V d.c. supply dissipates 500 W of power. When connected across a 100 V a.c. supply of frequency 50 Hz, it dissipates 200 W. Calculate the values of resistance and inductance of the coil.*

Solution. For d.c. supply, the reactance of the coil is zero ($\because f = 0$, hence $X_L = 2\pi f L = 0$). Hence for d.c. supply, we have to take into account the resistance (R) of coil only. However, for a.c. supply, both resistance (R) and inductance (L) offer opposition to current flow.

D.C. supply. Resistance of coil, $R = V^2/P = 100^2/500 = \textbf{20 }\Omega$

A.C. supply. Power consumed, $P = VI \cos \phi = V \times \dfrac{V}{Z} \times \dfrac{R}{Z} = \dfrac{V^2 R}{Z^2}$

$\therefore$ Impedance of coil, $Z = \sqrt{\dfrac{V^2 R}{P}} = \sqrt{\dfrac{100^2 \times 20}{200}} = 31.62\ \Omega$

Reactance of coil, $X_L = \sqrt{Z^2 - R^2} = \sqrt{(31.62)^2 - 20^2} = 24.5\ \Omega$

$\therefore$ Inductance of coil, $L = \dfrac{X_L}{2\pi f} = \dfrac{24\cdot5}{2\pi \times 50} = \textbf{0.078 H}$

Example 12.5. *Two coils A and B are connected in series across a 240 V, 50 Hz supply. The resistance of A is 5 Ω and the inductance of B is 0.015H. If the input from the supply is 3 kW and 2 kVAR, find the inductance of A and the resistance of B. Calculate the voltage across each coil.*

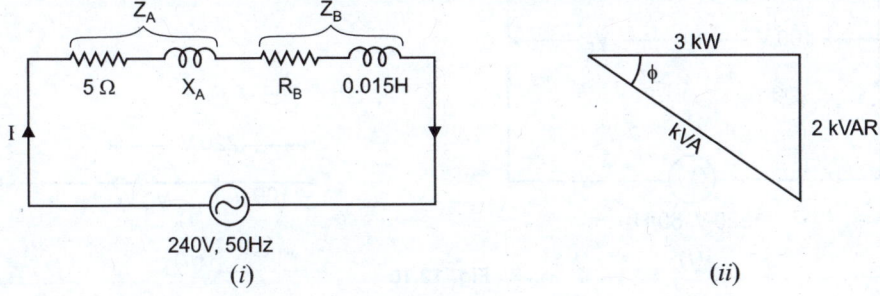

240V, 50Hz

(i)   *(ii)*

Fig. 12.9

Solution. Fig. 12.9 (*i*) shows the circuit diagram while Fig. 12.9 (*ii*) shows the power traingle of the circuit.

kVA drawn from supply $= \sqrt{(kW)^2 + (kVAR)^2} = \sqrt{3^2 + 2^2} = 3.606\ kVA$

Circuit current, $I = \dfrac{kVA \times 10^3}{V} = \dfrac{3.606 \times 10^3}{240} = 15.02\ A$

Circuit impedance, $Z = \dfrac{V}{I} = \dfrac{240}{15.02} = 15.975\ \Omega$

Now, $I^2(R_A + R_B) = $ Active power $= 3 \times 10^3$

or $R_A + R_B = \dfrac{3 \times 10^3}{(15.02)^2} = 13.3\ \Omega$

$\therefore$ $R_B = 13.3 - R_A = 13.3 - 5 = \textbf{8.3 }\Omega$

Total circuit reactance, $X = \sqrt{Z^2 - (R_A + R_B)^2} = \sqrt{(15.975)^2 - (13.3)^2} = 8.86\ \Omega$

Now, $X_B = 2\pi f L_B = 2\ \pi \times 50 \times 0.015 = 4.17\ \Omega$

$\therefore$ $X_A = X - X_B = 8.86 - 4.71 = 4.15\ \Omega$

$\therefore$ Inductance of coil A, $L_A = \dfrac{X_A}{2\pi f} = \dfrac{4.15}{2\pi \times 50} = \mathbf{0.0132\ H}$

Impedance of coil A, $Z_A = \sqrt{R_A^2 + X_A^2} = \sqrt{5^2 + 4.15^2} = 6.5\ \Omega$

$\therefore$ Voltage across coil $A = IZ_A = 15.02 \times 6.5 = \mathbf{97.6\ V}$

Impedance of coil B, $Z_B = \sqrt{R_B^2 + X_B^2} = \sqrt{8.3^2 + 4.71^2} = 9.545\ \Omega$

$\therefore$ Voltage across coil $B = IZ_B = 15.02 \times 9.545 = \mathbf{143.4\ V}$

Example 12.6. *A 100 volt, 60 W lamp is to be operated on 220 V, 50 Hz mains. Find what value of (i) non-inductive resistance (ii) pure inductance would be required in order that the lamp is run on the correct voltage. Which method is preferable and why ?*

Solution. The voltage across lamp (pure resistance) is to remain 100 V in each case. The current flowing in each case is

$$I = W/V = 60/100 = 0.6\ A$$

(*i*) Fig 12.10 (*i*) shows the required resistance $R\ \Omega$ in series with the lamp. The voltage drop across the lamp (i.e. 100 V) and that across R are in phase with current as shown in the phasor diagram in Fig. 12.10 (*ii*).

Referring to the phasor diagram in Fig. 12.10 (*ii*),

Voltage across R, $V_R = 220 - 100 = 120\ V$

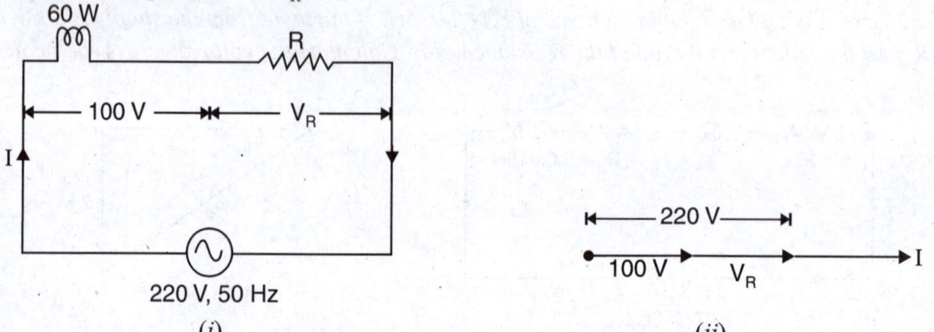

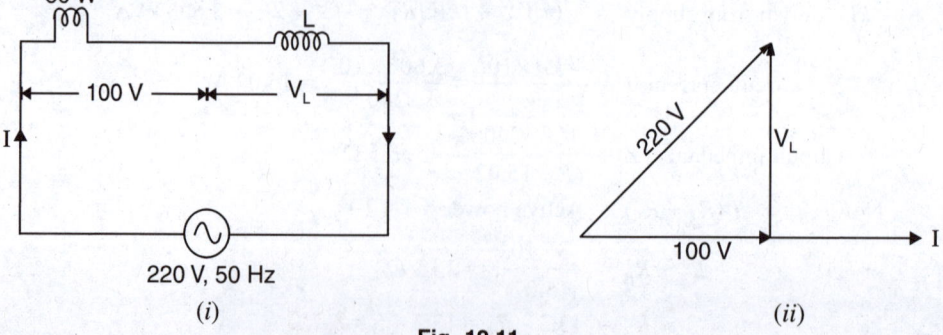

Fig. 12.10

$\therefore$ $R = V_R/I = 120/0.6 = \mathbf{200\ \Omega}$

(*ii*) Fig. 12.11 (*i*) shows the required inductance L henry in series with the lamp. The phasor diagram of the circuit is shown in Fig. 12.11 (*ii*).

Fig. 12.11

From the phasor diagram shown in Fig. 12.11 (*ii*), we have,

$$\text{Voltage across } L, \ V_L = \sqrt{(220)^2 - (100)^2} = 195.9 \text{ V}$$

$$\text{Inductive reactance, } X_L = V_L/I = 195.9/0.6 = 326.5 \ \Omega$$

$$\therefore \qquad \text{Required inductance, } L = \frac{X_L}{2 \pi f} = \frac{326 \cdot 5}{2\pi \times 50} = \textbf{1.039 H}$$

Method (*ii*) is preferable because there is no power loss in an inductance. If the first method is used, there will be a large power loss in R [Loss in $R = I^2 R = (0.6)^2 \times 200 = 72$ W].

Example 12.7. *A coil is connected in series with a non-inductive resistance of 30 Ω across 240 V, 50 Hz supply. The reading of a voltmeter across the coil is 180 V, and across the resistance is 130 V. Calculate (i) power absorbed by the coil (ii) inductance of the coil (iii) resistance of the coil and (iv) power factor of the whole circuit.*

Solution. A coil possesses both resistance (*R*) and inductance (*L*). The conditions of the problem are shown in Fig. 12.12 (*i*). The phasor diagram of the circuit is shown in Fig. 12.12 (*ii*). It is clear that p.f. of the coil is cos θ and that of the whole circuit is cos ϕ.

$$\text{Circuit current, } I = \text{Current through 30 } \Omega \text{ resistor} = 130/30 = 4.33 \text{ A}$$

(*i*) Referring to the phasor diagram and applying cosine formula to Δ *OAC*, we have,

$$OC^2 = OA^2 + AC^2 - 2.OA \cdot AC \cos (180 - \theta)$$

or

$$240^2 = 130^2 + 180^2 + 2 \times 130 \times 180 \cos \theta$$

$$\therefore \qquad \cos \theta = \frac{240^2 - 130^2 - 180^2}{2 \times 130 \times 180} = 0.177 \text{ lag}$$

i.e.   p.f. of the coil $= \cos \theta = 0.177$ lag

$$\therefore \quad \text{Power absorbed by coil} = \text{Voltage across coil} \times I \times \cos \theta$$

$$= 180 \times 4 \cdot 33 \times 0 \cdot 177 = \textbf{137.95 W}$$

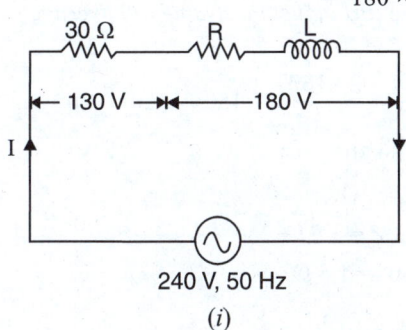

240 V, 50 Hz

(*i*)

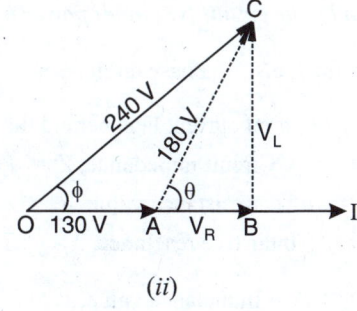

(*ii*)

Fig. 12.12

(*ii*)   Voltage across L, $V_L = BC = AC \sin\theta = 180 \sqrt{1 - (0 \cdot 177)^2} = 177 \cdot 16$ V

$$X_L = V_L/I = 177.16/4.33 = 40.9 \ \Omega$$

$$\therefore \qquad L = \frac{X_L}{2 \pi f} = \frac{40.9}{2\pi \times 50} = \textbf{0.13 H}$$

(*iii*)   Voltage across R, $V_R = AB = AC \cos \theta = 180 \times 0.177 = 31 \cdot 86$ V

$$\therefore \qquad R = V_R/I = 31.86/4.33 = \textbf{7.36 } \Omega$$

(*iv*) Circuit power factor, $\cos \phi = \dfrac{OB}{OC} = \dfrac{OA + AB}{OC} = \dfrac{130 + 31.86}{240} = \textbf{0.674 lag}$

Example 12.8. *When 1A flows through three air-cored coils A, B and C in series, the voltage drops are 6V, 3V and 8V on d.c. and 7V, 5V and 10V on a.c. Find (i) power factor and power dissipated in each coil (ii) power factor of entire circuit.*

Solution. D.C. values give resistance while a.c. values give impedance.

(*i*) **For coil A.** $R = \dfrac{6}{1} = 6\Omega$; $Z = \dfrac{7}{1} = 7\Omega$; Power factor $= \dfrac{R}{Z} = \dfrac{6}{7} = \textbf{0.857}$ *lagging*

Power dissipated $= I^2R = (1)^2 \times 6 = \textbf{6W}$

For coil B. $R = \dfrac{3}{1} = 3\Omega$; $Z = \dfrac{5}{1} = 5\Omega$; Power factor $= \dfrac{R}{Z} = \dfrac{3}{5} = \textbf{0.6}$ *lagging*

Power dissipated $= I^2R = (1)^2 \times 3 = \textbf{3W}$

For coil C. $R = \dfrac{8}{1} = 8\Omega$; $Z = \dfrac{10}{1} = 10\Omega$; Power factor $= \dfrac{R}{Z} = \dfrac{8}{10} = \textbf{0.8}$ *lagging*

Power dissipated $= I^2R = (1)^2 \times 8 = \textbf{8W}$

(*ii*) $X_A = \sqrt{7^2 - 6^2} = 3.6\Omega$; $X_B = \sqrt{5^2 - 3^2} = 4\Omega$; $X_C = \sqrt{10^2 - 8^2} = 6\Omega$

∴　　　　Total reactance, $X_T = X_A + X_B + X_C = 3.6 + 4 + 6 = 13.6\ \Omega$

Also,　　Total resistance, $R_T = 6 + 3 + 8 = 17\Omega$

Total impedance, $Z_T = \sqrt{R_T^2 + X_T^2} = \sqrt{17^2 + 13.6^2} = 21.77\ \Omega$

∴　　Power factor of entire circuit $= \dfrac{R_T}{Z_T} = \dfrac{17}{21.77} = \textbf{0.781}$ *lagging*

Example 12.9. *In a series circuit containing pure resistance and a pure inductance, the current and the voltage are expressed as :*

$$i(t) = 5 \sin (314\ t + 2\pi/3) \text{ and } v(t) = 15 \sin (314\ t + 5\ \pi/6)$$

Find (i) impedance of the circuit (ii) resistance value (iii) inductance value (iv) average power drawn by the circuit (v) circuit power factor.

Solution.　　Phase angle, $\phi = \dfrac{5\pi}{6} - \dfrac{2\pi}{3} = \dfrac{5 \times 180°}{6} - \dfrac{2 \times 180°}{3} = 150° - 120° = 30°$

∴　Circuit current lags behind the applied voltage by 30°.

(*i*)　　Circuit impedance, $Z = V/I = V_m/I_m = 15/5 = \textbf{3}\Omega$

(*ii*)　　Resistance value, $R = Z \cos \phi = 3 \times \cos 30° = \textbf{2.6}\ \Omega$

(*iii*)　Inductive reactance, $X_L = Z \sin \phi = 3 \times \sin 30° = 1.5\ \Omega$

∴　　　　Inductance value, $L = \dfrac{X_L}{2\pi f} = \dfrac{1.5}{314} = 4.78 \times 10^{-3}\ \text{H} = \textbf{4.78 mH}$

(*iv*)　Average power drawn, $P = I^2R = \left(5/\sqrt{2}\right)^2 \times 2.6 = \textbf{32.5 W}$

(*v*)　　Circuit power factor $= \cos \phi = \cos 30° = \textbf{0.866 lag}$

Example 12.10. *A circuit operating at a p.f. 0·8 lagging takes a current of 10 A from 230 V, 50 Hz supply.*

(*i*) *Find the expression for the instantaneous values for voltage and current.*

(*ii*) *Calculate the value of current 5 ms after the voltage has reached its positive maximum value.*

Solution. (*i*)　Phase angle, $\phi = \cos^{-1} 0.8 = 37°$; $V_m = 230 \times \sqrt{2} = 325$ V ;

$$I_m = 10 \times \sqrt{2} = 14.14\ \text{A} ; \omega = 2\pi f = 2\pi \times 50 = 314\ \text{rad/sec.}$$

∴ The equations for voltage and current are :

$$v = 325 \sin 314\, t \; ; \; i = 14.14 \sin (314\, t - 37°) \text{ Ans.}$$

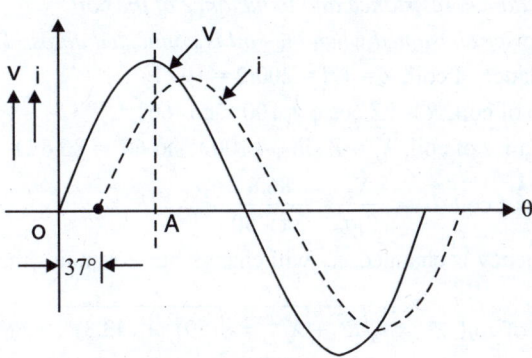

Fig. 12.13

(*ii*) The above equations are applicable if time is counted from the instant the voltage is zero *i.e.*, point *O* in Fig. 12.13. If time is counted from the instant the voltage is maximum positive (*i.e.* point *A* in Fig. 12.13), then voltage and current equations are modified. The new equations with point *A* as reference are :

$$v = 325 \sin (314\, t + \pi/2) = 325 \cos 314\, t$$

$$i = 14.14 \sin (314\, t - 37° + \pi/2) = 14.14 \sin (314\, t + 53°)$$

Value of current 5 ms after the voltage has reached its positive maximum value is

$$
\begin{aligned}
i &= 14.14 \sin (314 \times 5 \times 10^{-3} + 53°) \\
&= 14.14 \sin (314 \times 5 \times 10^{-3} \times 180/\pi + 53°) \\
&= 14.14 \sin (90° + 53°) = \mathbf{8.51\ A}
\end{aligned}
$$

Example 12.11. *A coil, having both resistance and inductance, has a total effective impedance of 50 Ω and the phase angle of the current through it with respect to the voltage across it is 45° lag. The coil is connected in series with a 40 Ω resistor across a sinusoidal supply. The circuit current is 3A. Find (i) supply voltage and (ii) circuit phase angle.*

Solution. Fig. 12.14 shows the conditions of the problem. The coil impedance $Z_C = 50\ \Omega$ and

coil phase angle is $\phi_C = 45°$ lag.

(*i*) Coil resistance, $R_C = Z_C \cos \phi_C$

$$= 50 \cos 45°$$

$$= 35.35\ \Omega$$

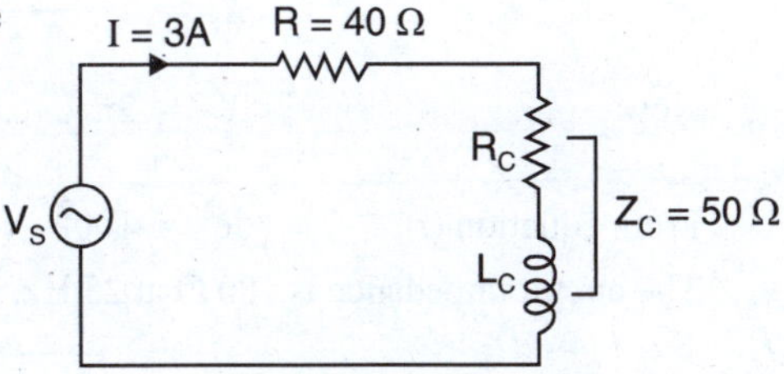

Coil inductive reactance is given by ;

$$X_{LC} = 50 \sin \phi_C$$

$$= 50 \sin 45° = 35.35\ \Omega$$

Fig. 12.14

Total circuit resistance, $R_T = R + R_C = 40 + 35.35 = 75.35\ \Omega$

∴ Circuit impedance, $Z = \sqrt{R_T^2 + X_{LC}^2} = \sqrt{(75.35)^2 + (35.35)^2} = 83.23\ \Omega$

∴ Supply voltage $= IZ = 3 \times 83.23 = \mathbf{250\ V}$

(*ii*) Circuit power factor, $\cos\phi = R_T/Z = 75.35/83.23 = 0.905$

∴ Circuit phase angle, $\phi = \cos^{-1} 0.905 = \mathbf{25°}$

Example 12.12. *A choke coil takes a current of 2A lagging 60° behind the applied voltage of 200 V at 50Hz.*

(i) Calculate impedance, resistance and inductance of the coil.

(ii) Also find the power consumed when the coil is connected across 100 V, 25 Hz supply.

Solution. (*i*) Impedance of coil, $Z = V/I = 200/2 = $ **100 Ω**

Resistance of coil, $R = Z \cos \phi = 100 \times \cos 60° = $ **50 Ω**

Inductive reactance of coil, $X_L = Z \sin \phi = 100 \times \sin 60° = 86.6 \, \Omega$

$\therefore$ Inductance of coil, $L = \dfrac{X_L}{2\pi f} = \dfrac{86.6}{2\pi \times 50} = $ **0.275 H**

(*ii*) Since the frequency is changed, X_L will change but R remains the same. Therefore, $X'_L = X_L/2 = 86.6/2 = 43.3 \, \Omega$.

New impedance of coil, $Z' = \sqrt{R^2 + X'^2_L} = \sqrt{(50)^2 + (43.3)^2} = 66.1 \, \Omega$

Circuit current, $I' = V'/Z' = 100/66.1 = 1.5 \, A$

$\therefore$ Power consumed, $P = VI' \cos \phi' = 100 \times 1.5 \times 50/66.1 = $ **113.5 W**

Example 12.13. *In an inductive series circuit, a voltage of 10V at 25 Hz produces 100 mA while the same voltage at 75 Hz produces 60 mA. Find the values of circuit constants. At what frequency will the value of the impedance be twice that at 25Hz?*

Solution. Suppose the resistance of the circuit is R and inductive reactance at 25 Hz is X_L. Then inductive reactance at *75 Hz is $3X_L$ ($\because X_L = 2\pi f L$).

For the first case, impedance, $Z_1 = 10/0.1 = 100 \, \Omega$

For the second case, impedance, $Z_2 = 10/0.06 = 500/3 \, \Omega$

$$Z_1^2 = R^2 + X^2_L$$

or $(100)^2 = R^2 + X^2_L$ ...(i)

Also, $(500/3)^2 = R^2 + (3X_L)^2$ ...(ii)

Subtracting equation (*i*) from equation (*ii*), we have,

$$8X^2_L = (500/3)^2 - 10,000 = 1.78 \times 10^4$$

$\therefore$ $X_L = \sqrt{\dfrac{1.78 \times 10^4}{8}} = 47.16\Omega$

Now $X_L = 2\pi f L \quad \therefore \quad L = \dfrac{X_L}{2\pi f} = \dfrac{47.16}{2\pi \times 25} = $ **0.3 H**

From equation (*i*), $R^2 = 100^2 - (47 \cdot 16)^2 = 7775 \quad \therefore \quad R = \sqrt{7775} = $ **88.1 Ω**

The circuit impedance is 100 Ω at 25 Hz. Suppose the impedance becomes 200 Ω at frequency f'.

$\therefore$ $X'_L = \sqrt{(200)^2 - (88.1)^2} = 179.5 \, \Omega$

Now $\dfrac{X'_L}{X_L} = \dfrac{179.5}{47.16} \quad$ or $\quad \dfrac{2\pi f' L}{2\pi f L} = 3.81$

$\therefore$ $f' = 3.81 f = 3 \cdot 81 \times 25 = $ **95 Hz**

Example 12.14. *A circuit has a fixed resistance of 2Ω and a reactance of 10 Ω in series with a resistor R across 100V constant frequency mains. For what value of R is the power consumed in the circuit a maximum ?*

Solution. Circuit impedance, $Z = \sqrt{(R+2)^2 + 10^2}$

---

*   Note that R is independent of frequency at low values of f so that its value at 75 Hz is the same as at 25Hz.

$$\therefore \quad \text{Circuit current, } I = \frac{100}{\sqrt{(R+2)^2 + 100}}$$

Power consumed in the circuit is given by ;

$$P = I^2 R = \frac{R \times 10^4}{(R+2)^2 + 100}$$

For P to be maximum, dP/dR should be zero.

Now,
$$\frac{dP}{dR} = \frac{10^4 \times [(R+2)^2 + 100] \times 1 - R[2(R+2)] \times 10^4}{(\text{Deno.})^2}$$

$\therefore \quad (R+2)^2 + 100 - 2R(R+2) = 0$

or $\quad R^2 + 4 + 4R + 100 - 2R^2 - 4R = 0$

or $\qquad\qquad\qquad R^2 = 104 \quad \therefore \quad R = \sqrt{104} = \mathbf{10.2\ \Omega}$

$\therefore \quad$ Power consumed in the circuit is maximum when $R = 10.2\ \Omega$.

Example 12.15. *When a resistor and inductor in series are connected to a 240 V supply, a current of 3A flows, lagging 37° behind the supply voltage while the voltage across the inductor is 171 V. Find the resistance of the resistor and the resistance and reactance of inductor.*

Solution. Supply voltage, $V = 240$ volts ; Circuit current, $I = 3$A ; Phase angle, $\phi = 37°$

Total resistance drop $= V \cos\phi = 240 \times 0.8 = 192$ volts

$\therefore \quad$ Total circuit resistance $= 192/I = 192/3 = 64\ \Omega$

Reactance drop $= V \sin\phi = 240 \times 0.6 = 144$ V

$\therefore \quad$ Reactance of inductor, $X_L = 144/I = 144/3 = \mathbf{48\ \Omega}$

Impedance of inductor, $Z_L = 171/I = 171/3 = 57\ \Omega$

$\therefore \quad$ Resistance of inductor $= \sqrt{Z_L^2 - X_L^2} = \sqrt{(57)^2 - (48)^2} = \mathbf{30.74\ \Omega}$

Resistance of resistor $= 64 - 30.74 = \mathbf{33.26\ \Omega}$

Example 12.16. *When a voltage of 100 V at 50 Hz is applied to coil A, the current and power taken are 8 A and 120 W respectively. When applied to coil B, the current and power taken are 10 A and 500 W respectively. What current and power will be taken when 100 V is applied to the two coils connected in series ?*

Solution. Coil A. $\qquad Z_A = 100/8 = 12.5\ \Omega\ ; R_A = P/I^2 = 120/(8)^2 = 1.875\ \Omega$

$\therefore \qquad\qquad X_A = \sqrt{Z_A^2 - R_A^2} = \sqrt{(12.5)^2 - (1.875)^2} = 12.36\ \Omega$

Coil B. $\qquad\qquad Z_B = 100/10 = 10\ \Omega\ ; R_B = P/I^2 = 500/(10)^2 = 5\ \Omega$

$\therefore \qquad\qquad X_B = \sqrt{Z_B^2 - R_B^2} = \sqrt{(10)^2 - (5)^2} = 8.66\ \Omega$

When the two coils are connected in series

Total resistance, $R_T = R_A + R_B = 1.875 + 5 = 6.875\ \Omega$

Total reactance, $X_T = X_A + X_B = 12.36 + 8.66 = 21.02\ \Omega$

Total impedance, $Z_T = \sqrt{R_T^2 + X_T^2} = \sqrt{(6.875)^2 + (21.02)^2} = 22.12\ \Omega$

$\therefore \qquad$ Circuit current, $I = \dfrac{V}{Z_T} = \dfrac{100}{22.12} = \mathbf{4.52\ A}$

Power taken $= I^2 R_T = (4.52)^2 \times 6.875 = \mathbf{140\ W}$

Example 12.17. *A bulb is rated at 100W and 110V. Calculate the impedance of a choke which should be connected in series with the bulb so that it may be used on 230V a.c. supply. Find total*

active power and overall power factor. The reactance to resistance ratio of the choke is 10. Draw the impedance triangle.

Solution. Rated current of bulb, $I = 100/110 = 0.909$A

Resistance of bulb, $R = 110/0.909 = 121\Omega$

Let R_1 be the resistance of the choke. Then inductive reactance of the choke is $X_1 = 10R_1$. When the bulb is used on 230V a.c. supply, the circuit current must remain equal to the rated current of the bulb ($= 0.909$A).

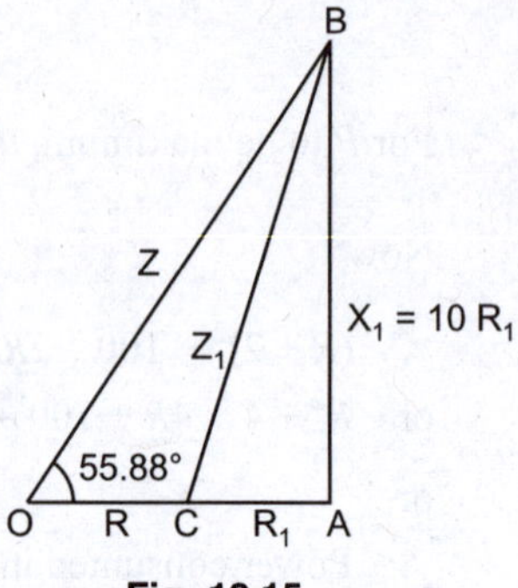

$$\therefore \quad \text{Circuit impedance, } Z = \frac{230}{0.909} = 253.02\ \Omega$$

Also,

$$Z^2 = (121 + R_1)^2 + (10R_1)^2$$

or

$$(253.02)^2 = 101R^2 + 242R_1 + (121)^2$$

On solving, we get, $R_1 = 20.95\ \Omega$

$$\therefore \quad X_1 = 10R_1 = 10 \times 20.95 = 209.5\ \Omega$$

$$\therefore \quad \text{Impedance of choke, } Z_1 = \sqrt{R_1^2 + X_1^2} = \sqrt{(20.95)^2 + (209.5)^2} = \textbf{210.54}\ \Omega$$

Total circuit resistance, $R_T = R + R_1 = 121 + 20.95 = 141.95\ \Omega$

$$\therefore \quad \text{Overall power factor, } \cos\phi = \frac{R_T}{Z} = \frac{141.95}{253.02} = \textbf{0.561 } \textit{lagging}$$

$$\text{Total active power, } P = VI \cos\phi = 230 \times 0.909 \times 0.561 = \textbf{117.29W}$$

Fig. 12.15

Fig. 12.15 shows the impedance triangle for the choke (ΔCAB) as well as for the entire circuit (ΔOAB). Note that phase angle for the entire circuit is $\phi = \cos^{-1} 0.561 = 55.88°$.

Example 12.18. *A single phase motor operating from 400 V, 50 Hz supply is developing 7·46 kW output with an efficiency of 84% and p.f. of 0·7 lagging. Calculate (i) input kVA (ii) active and reactive components of current and (iii) reactive kVAR.*

Solution. Motor input $= 7.46/0.84 = 8.88$ kW

(i) kVA drawn by motor $= 8.88/0.7 = \textbf{12.68 kVA}$

(ii) Current drawn by motor, $I = 12.68 \times 10^3/400 = 31.7$ A

Active component of current $= I \cos\phi = 31.7 \times 0.7 = \textbf{22.2 A}$

Reactive component of current $= I \sin\phi = 31.7 \times 0.7 = \textbf{22.2 A}$

(iii) kVAR drawn by motor $= $ kVA $\times \sin\phi = 12.68 \times 0.7 = \textbf{8.88 kVAR}$

Example 12.19. *An alternating voltage of v = 100 sin 376·8 t volts is applied to a circuit consisting of a coil having a resistance of 6 Ω and an inductance of 21·22 mH.*

(i) Express the current flowing through the circuit in the form :

$$i = I_m \sin (376.8\ t \pm \phi)$$

(ii) If a moving iron voltmeter, a wattmeter and a frequency meter are connected in the circuit what would be the respective readings on the instruments?

Solution. *(i)* $X_L = \omega L = 376.8 \times 21.22 \times 10^{-3} = 8\ \Omega$; $Z = \sqrt{R^2 + X_L^2} = \sqrt{(6)^2 + (8)^2} = 10\ \Omega$

$$\therefore \quad I_m = V_m/Z = 100/10 = 10\text{A} ; \phi = \tan^{-1} X_L/R = \tan^{-1} 8/6 = 53.1° \text{ lag}$$

$$\therefore \quad i = 10 \sin (376.8t - 53.1°)\ \textbf{Ans.}$$

(ii) A moving-iron voltmeter reads r.m.s. voltage.

$$\therefore \quad \text{Voltmeter reading} = V_m / \sqrt{2} = 100/\sqrt{2} = \textbf{70.7 V}$$

$$\text{Wattmeter reading} = I^2_{r.m.s.} R = \frac{1}{2} I^2_m R = \frac{1}{2} \times (10)^2 \times 6 = \textbf{300 W}$$

Frequency meter reading $= 376.8/2\pi = \textbf{60 Hz}$

Example 12.20. *A series circuit consists of a resistance of 6 Ω and an inductive reactance of 8 Ω. A potential difference of 141.4V (r.m.s.) is applied to it. At a certain instant, the applied voltage is +100 V and is increasing. Calculate at this instant (i) the current (ii) the voltage drop across resistance and (iii) voltage drop across the inductive reactance.*

Solution. Circuit impedance, $Z = \sqrt{6^2 + 8^2} = 10\ \Omega$; Circuit phase angle, $\phi = \tan^{-1} 8/6 = 53.1°$

Since the circuit is inductive, circuit current lags behind the applied voltage by 53.1°. Taking voltage as the reference phasor, the voltage and current equations are :

$$v = 141.4 \times \sqrt{2} \sin \omega t = 200 \sin \omega t$$
$$i = V_m/Z \sin(\omega t - \phi) = 20 \sin(\omega t - 53.1°)$$

(i) When the voltage is + 100 V and increasing,

$$100 = 200 \sin \omega t \text{ or } \sin \omega t = 0.5 \therefore \omega t = 30°$$

At this instant, the current is given by ;

$$i = 20 \sin(30° - 53.1°) = \textbf{-7.847 A}$$

(ii) Drop across R, $v_R = i R = -7.847 \times 6 = \textbf{-47 V}$

(iii) We shall first find the equation of the voltage across inductive reactance (X_L). Maximum voltage across reactance $= I_m X_L = 20 \times 8 = 160$ V. As shown in the phasor diagram in Fig. 12.16, current lags the applied voltage (reference phasor) by 53.1° while voltage v_L (= voltage across X_L) leads I by 90°. Therefore, angle between v_L and $V = 90° - 53.1° = 36.9°$ lead.

∴

$$v_L = 160 \sin(\omega t + 36.9°)$$
$$= 160 \sin(30° + 36.9°) = \textbf{147 V}$$

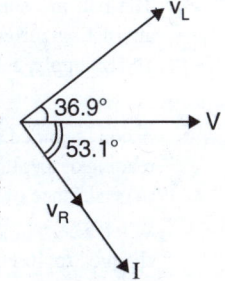

Fig. 12.16

Alternatively. $V = v_R + v_L$ or $100 = -47 + v_L \therefore v_L = \textbf{147 V}$

Example 12.21. *It is desired to operate a 100 W, 120 V lamp at its current rating from a 240 V, 50 Hz supply by using an inductor having resistance of 10 Ω. Find (i) the value of inductance (ii) circuit power factor and (iii) power consumed.*

Solution. Fig. 12.17 shows the conditions of the problem.

(i) Rated current of bulb, $I = 100/120 = (5/6)$ A

Voltage across R, $V_R = IR = (5/6) \times 10 = 25/3$ volts

$$\text{Voltage across } L, V_L = \sqrt{(240)^2 - \left(120 + \frac{25}{3}\right)^2} = 203 \text{ volts}$$

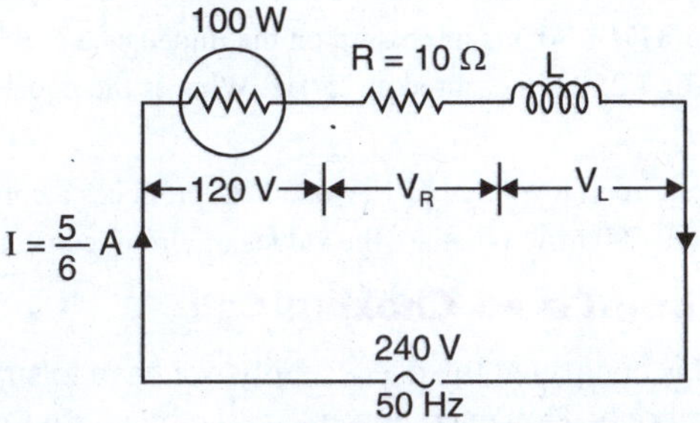

100 W

R = 10 Ω   L

←120 V→←V$_R$→←V$_L$→

$I = \dfrac{5}{6}$ A

240 V
50 Hz

Fig. 12.17

Now $\qquad V_L = IX_L \therefore X_L = V_L/I = 203 \times 6/5 = 243.6\ \Omega$

$\therefore \qquad$ Inductance, $L = \dfrac{X_L}{2\pi f} = \dfrac{243.6}{2\pi \times 50} = \mathbf{0.775\ H}$

(ii)   Total resistive drop $= 120 + (25/3) = 128.3$ volts

$\therefore$ Circuit power factor, $\cos \phi = \dfrac{128.3}{240} = \mathbf{0.535\ lag}$

(iii)   Power consumed, $P = VI \cos \phi = 240 \times (5/6) \times 0.535 = \mathbf{107\ W}$

Tutorial Problems

1. A resistance of 5 Ω is connected in series with a pure inductance of 0.01 H to a 100 V, 50 Hz supply. Calculate (*i*) impedance (*ii*) current and (*iii*) power absorbed.   **[(*i*) 5.9 Ω (*ii*) 16.94 A (*iii*) 1435 W]**

2. A 200-V, 50 Hz inductive circuit takes a current of 10 A lagging the voltage by 30°. Calculate (*i*) resistance (*ii*) reactance and (*iii*) inductance of the circuit.   **[(*i*) 10 Ω (*ii*) 17.3 Ω (*iii*) 31.8 mH]**

3. An inductive coil connected to a 200-V, 50 Hz supply takes a current of 10 A. If the power dissipated in the coil is 1000 W, calculate (*i*) inductance of the coil (*ii*) power factor and (*iii*) angle of lag.   **[(*i*) 0.0552 H (*ii*) 0.5 (*iii*) ϕ = 60°]**

4. The p.d. measured across a coil is 20 V when a direct current of 2 A is passed through it. With an alternating current of 2 A at 40 Hz, the p.d. across the coil is 140 V. If the coil is connected to a 230 V, 50 Hz supply, calculate (*i*) current (*ii*) power and (*iii*) the power factor.   **[(*i*) 2·64 A (*ii*) 69.7 W (*iii*) 0.1147 lag]**

5. A coil is joined in series with a pure resistor of resistance 800 Ω across a 100 V, 50 Hz supply. The reading of a voltmeter across the coil is 45 V and across the pure resistor is 80 V. Find (*i*) inductance and (*ii*) resistance of the coil.   **[(*i*) L = 1.4 H (*ii*) R = 98.5 Ω]**

6. When a certain inductive coil is supplied at 240 V, 50 Hz, the current is 6.45 A. When the frequency is changed to 40 Hz at 240 V, the current taken is 7.48 A. Calculate the inductance and resistance of the coil.   **[L = 0.1 H ; R = 20 Ω]**

7. A coil has a resistance of 75 Ω and an inductance of 1.4 H. When the applied voltage is 240 V a.c., at what frequency is the current 0.3 A ? What is the power factor at this frequency ?   **[90.55 Hz ; 0.0938 lag]**

8. A two-element series circuit is connected across an a.c. source $e = 200\sqrt{2}\ \sin(\omega t + 20°)$V. The current in the circuit then is found to be $i = 10\sqrt{2}\ \cos(314t - 25°)$ A. Determine the parameters of the circuit.   **[R = 14.1 Ω; C = 226 μF]**

9. A resistance and an inductance are connected in series across a voltage $v = 283 \sin 314t$. The current expression is found to be $i = 4\sin(314t - \pi/4)$. Find the values of inductance, resistance and p.f.   **[L = 0.1592 H; R = 50 Ω ; 0.707 (lag)]**

10. A coil has resistance of 10 Ω and draws a current of 5A when connected across 100V, 50 Hz source. Determine the reactive power of the circuit.   **[433 VAR]**

11. An e.m.f. $e_o = 141.4 \sin(377t + 30°)$ is impressed on the impedance coil having a resistance of 4 Ω and an inductive reactance of 1.25 Ω, measured at 25 Hz. What is the equation of the current? Sketch the waves for i, e_R, R_L and e_o.   **[i = 28.28 sin (377t − 6.87°)]**

12. In a particular *R-L* series circuit a voltage of 10V at 50 Hz produces a current of 700mA while the same voltage at 75 Hz produces 500 mA. What are the values of circuit constants?   **[R = 6.9 Ω; L = 40 mH]**

12.7. Power in an Iron-Cored Choking Coil

When an air-cored coil is connected to an a.c. supply, we have to supply power to meet the loss in the resistance R of the coil only. However, power P taken by an iron-cored coil has to supply :

(*i*) power loss in ohmic resistance *i.e.*, I^2R where R is the d.c. resistance or true resistance of the coil.

(*ii*) the iron loss P_i (*i.e.*, hysteresis and eddy current losses)

$$\therefore \qquad\qquad P = I^2R + P_i \qquad\qquad\qquad ...(i)$$

The iron loss can be thought as additional resistance in series with the resistance of the coil. Thus the effect of iron core is to increase the effective resistance of the coil. We can say that input to an iron-cored coil is $I^2 R_{eff}$ *i.e.*,

$$P = I^2R_{eff} \qquad\qquad\qquad ...(ii)$$

From equations (*i*) and (*ii*), we have,

$$I^2R_{eff} = I^2R + P_i$$

or $$R_{eff} = R + \frac{P_i}{I^2}$$

i.e.,   Effective resistance = True (or d.c.) resistance + $\dfrac{P_i}{I^2}$

Example 12.22. *An iron-cored choking coil has a resistance of 4Ω when measured by a d.c. supply. On a 240 V, 50 Hz mains supply, it dissipates 500 W, the current taken being 10A. Calculate (i) impedance (ii) the power factor (iii) the iron loss and (iv) inductance of the coil.*

Solution. (*i*) Impedance of coil, $Z = V/I = 240/10 = $ **24 Ω**

(*ii*)   Power factor, $\cos\phi = \dfrac{P}{VI} = \dfrac{500}{240\times10} = $ **0.208 lag**

(*iii*)   Total loss = Loss in resistance + Iron loss

or $500 = 10^2 \times 4 + P_i$ $\therefore$ $P_i = 500 - 400 = $ **100 W**

(*iv*) Effective resistance, $R_{eff} = R + P_i/I^2 = 4 + 100/10^2 = 5\ \Omega$

Reactance of coil, $X_L = \sqrt{Z^2 - R_{eff}^2} = \sqrt{24^2 - 5^2} = 23.47\ \Omega$

$\therefore$ Inductance of coil, $L = \dfrac{X_L}{2\pi f} = \dfrac{23.47}{2\pi\times50} = $ **74.71 × 10⁻³H**

Example 12.23. *A voltmeter, ammeter and wattmeter are suitably connected to measure the power input to an iron-cored coil. If the readings on the instruments are 110V, 2·5 A and 150 W respectively and the d.c. resistance of the copper windings of coil is 15Ω, calculate the inductance of the coil and the power loss in the core. The supply frequency is 50 Hz.*

Solution. Impedance of coil, $Z = V/I = 110/2.5 = 44\ \Omega$

Effective resistance of coil, $R_{eff} = P/I^2 = 150/(2.5)^2 = 24\ \Omega$

Reactance of coil, $X_L = \sqrt{Z^2 - R_{eff}^2} = \sqrt{44^2 - 24^2} = 36.88\ \Omega$

$\therefore$ Inductance of coil, $L = \dfrac{X_L}{2\pi f} = \dfrac{36.88}{2\pi\times50} = $ **117 × 10⁻³ H**

Total power loss = Loss in resistance + Iron loss

or $150 = (2.5)^2 \times 15 + $ Iron loss

$\therefore$ Iron loss $= 150 - (2.5)^2 \times 15 = $ **56.25 W**

Example 12.24. *An iron-cored choking coil takes 5A at a power factor of 0.6 when supplied at 100V, 50 Hz. When the iron-core is removed and the supply reduced to 15V, the current rises to 6A at a power factor of 0.9. Determine the iron loss in the core, the copper loss at 5A and the inductance of the choking coil with core when carrying a current of 5A.*

Solution. With iron core removed, the impedance of the coil is

$$Z = 15/6 = 1.5\Omega$$

∴ True resistance of coil, $R = Z \cos\phi = 2.5 \times 0.9 = 2.25\Omega$

With iron core. With iron core, $V' = 100$ volts, $I' = 5A$ and $\cos\phi' = 0.6$ lag.

∴   Input power to choke coil $= V'I' \cos\phi' = 100 \times 5 \times 0.6 = 300$ W

Power loss in R (= Cu loss) when current is 5A

$$= 5^2 \times 2.25 = \mathbf{56.2W}$$

∴   Iron loss $= 300 - 56.2 = \mathbf{243.8\ W}$

With iron core, the impedance Z' of the choke coil is

$$Z' = 100/5 = 20\Omega$$

Reactance of choke coil $= Z' \sin\phi' = 20 \times 0.8 = 16\Omega$

∴   Inductance of coil, $L = \dfrac{16}{2\pi f} = \dfrac{16}{2\pi \times 50} = \mathbf{0.051\ H}$

12.8. R-C Series A.C. Circuit

Fig. 12.18 shows a resistance of R ohms connected in series with a capacitor of C farad.

Let V = r.m.s. value of applied voltage

I = r.m.s. value of the circuit current

$V_R = IR$ .... where V_R is in phase with I

$V_C = IX_C$ .... where V_C lags I by 90°

Taking current as the reference phasor, the phasor diagram of the circuit can be drawn as shown in Fig. 12.19. The voltage drop V_R ($= IR$) is in phase with current and is represented in magnitude and direction by the phasor OA. The voltage drop V_C (= IX_C) lags behind the current by 90° and is represented in magnitude and direction by the phasor AB. The applied voltage V is the phasor sum of these two drops *i.e.*

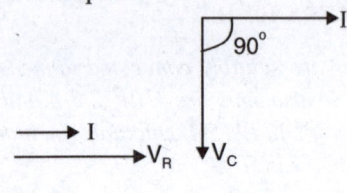

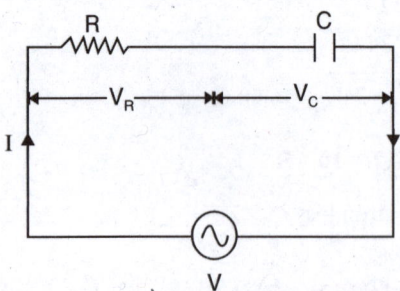

Fig. 12.18

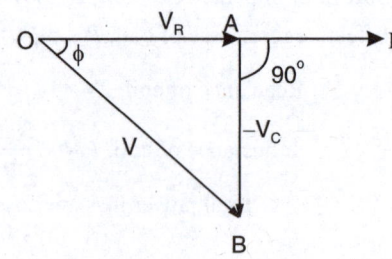

Fig. 12.19

$$V = \sqrt{V_R^2 + (-V_C)^2} = \sqrt{(IR)^2 + (-IX_C)^2} = I\sqrt{R^2 + X_C^2}$$

∴ $I = \dfrac{V}{\sqrt{R^2 + X_C^2}}$

The quantity $\sqrt{R^2 + X_C^2}$ offers opposition to current flow and is called **impedance** of the circuit.

$$\therefore \qquad I = V/Z \text{ where } Z = \sqrt{R^2 + X_C^2}$$

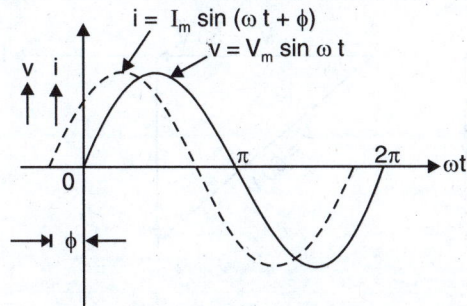

Fig. 12.20

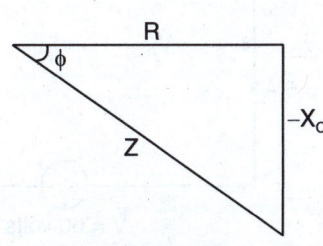

Impedance triangle

Fig. 12.21

(*i*) **Phase angle.** It is clear from the phasor diagram that circuit current I leads the applied voltage V by $\phi°$. This fact is also illustrated in the wave diagram (See Fig. 12.20) and impedance triangle (See Fig. 12.21) of the circuit. The value of the phase angle can be determined as under :

$$\tan \phi = -\frac{V_C}{V_R} = -\frac{IX_C}{IR} = -\frac{X_C}{R}$$

Since current is taken as the reference phasor, negative phase angle implies that voltage lags behind the current. This is the same thing as current leads the voltage.

If the applied voltage is $v = V_m \sin \omega t$, then equation for the circuit current will be :

$$i = I_m \sin (\omega t + \phi) \text{ where } I_m = V_m/Z$$

(*ii*) **Power.** The equations for voltage and current are :

$$v = V_m \sin \omega t \ ; \ i = I_m \sin (\omega t + \phi)$$

$\therefore$ Average power, P = Average of vi

$$= VI \cos \phi \qquad\qquad \text{[By same way as in Art. 12.1]}$$

Alternatively. P = Power in R + Power in C

$$= I^2 R + 0 = IR \times I = IR \times \frac{V}{Z} = VI \times \frac{R}{Z} = VI \cos \phi$$

Example 12.25. *A capacitor of capacitance 79·5 μ F is connected in series with a non-inductive resistance of 30 Ω across 100 V, 50 Hz supply. Find (i) impedance (ii) current (iii) phase angle and (iv) equation for the instantaneous value of current.*

Solution. (*i*) Capacitive reactance, $X_C = \dfrac{1}{2\pi f C} = \dfrac{10^6}{2\pi \times 50 \times 79 \cdot 5} = 40 \ \Omega$

Circuit impedance, $Z = \sqrt{R^2 + X_C^2} = \sqrt{30^2 + 40^2} = 50 \ \Omega$

(*ii*)   Circuit current, $I = V/Z = 100/50 = $ **2 A**

(*iii*) $\tan \phi = X_C/R = 40/30 = 1.33$

$\therefore$ Phase angle, $\phi = \tan^{-1} 1.33 = $ **53° *lead***

(*iv*) $I_m = 2 \times \sqrt{2} = 2.828 \ A$

$$\omega = 2\pi f = 2\pi \times 50 = 314 \text{ rad/sec.}$$

$\therefore$ $i = $ **2.828 sin (314 t + 53°)**

Example 12.26. *A 10 Ω resistor and 400 μF capacitor are connected in series to a 60-V sinusoidal supply. The circuit current is 5 A. Calculate the supply frequency and phase angle between the current and voltage.*

Solution. Fig. 12.22 (*i*) shows the circuit diagram whereas Fig. 12.22 (*ii*) shows phasor diagram.

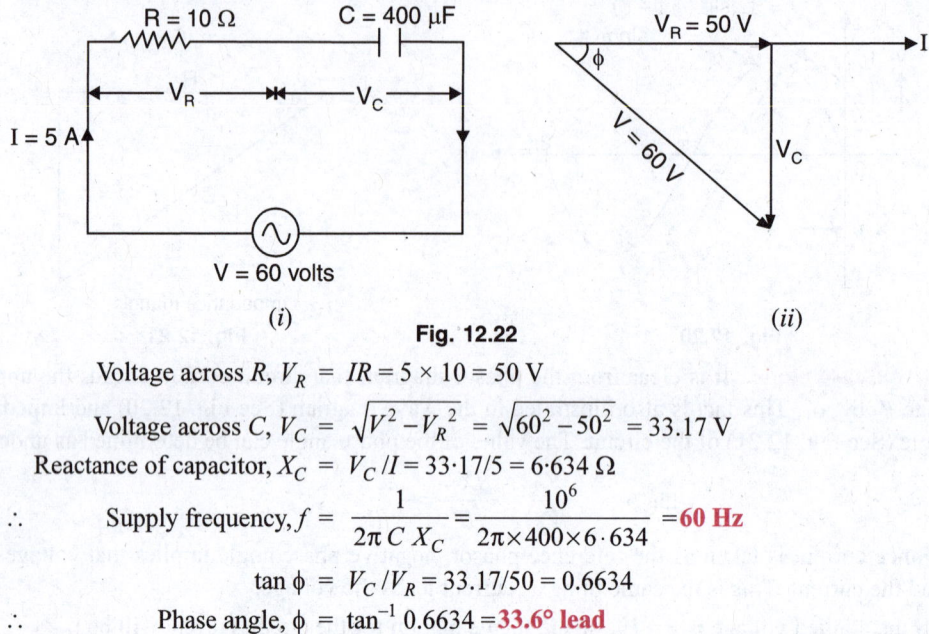

(*i*)

Fig. 12.22

(*ii*)

Voltage across R, $V_R = IR = 5 \times 10 = 50$ V

Voltage across C, $V_C = \sqrt{V^2 - V_R^2} = \sqrt{60^2 - 50^2} = 33.17$ V

Reactance of capacitor, $X_C = V_C/I = 33.17/5 = 6.634$ Ω

∴ Supply frequency, $f = \dfrac{1}{2\pi C X_C} = \dfrac{10^6}{2\pi \times 400 \times 6.634} = \mathbf{60\ Hz}$

$\tan \phi = V_C/V_R = 33.17/50 = 0.6634$

∴ Phase angle, $\phi = \tan^{-1} 0.6634 = \mathbf{33.6°\ lead}$

Example 12.27. *A two-element series circuit consumes 700 W and has a p.f. of 0.707 leading. If the applied voltage is v = 141.1 sin (314 t + 30°), find the circuit constants.*

Solution. Since the circuit p.f. is leading, one circuit element must be a capacitor. Further, power consumed in the circuit is 700 W. This suggests that other circuit element is a resistor. Therefore, it is *RC* series circuit.

R.M.S. value of applied voltage, $V = V_m/\sqrt{2} = 141.1/\sqrt{2} = 100$ volts

Power consumed, $P = VI \cos \phi$ or $700 = 100 \times I \times 0.707$ ∴ $I = 10$A

Also, $P = I^2 R$ ∴ $R = P/I^2 = 700/(10)^2 = 7$ Ω

Now, $Z = V/I = 100/10 = 10$Ω ∴ $X_C = \sqrt{Z^2 - R^2} = \sqrt{(10)^2 - (7)^2} = 7$Ω

∴ Capacitance, $C = \dfrac{1}{2\pi f X_C} = \dfrac{1}{2\pi \times 50 \times 7} = 450 \times 10^{-6}$ F = **450 μF**

Example 12.28. *A circuit when connected to 200 V, 50 Hz mains takes a current of 10 A, leading the voltage by one-twelfth of time period. Calculate (i) resistance (ii) capacitive reactance and (iii) capacitance of the circuit.*

Solution. One time period corresponds to a phase difference of 2π radians or 360°. Hence one-twelfth of time period corresponds to a phase difference of $\phi = 360°/12 = 30°$. This means that current leads the voltage by 30°.

(*i*)   Circuit impedance, $Z = V/I = 200/10 = 20$ Ω

Circuit resistance, $R = Z \cos \phi = 20 \cos 30° = \mathbf{17.32\ Ω}$

(*ii*) Capacitive reactance, $X_C = Z \sin \phi = 20 \sin 30° = \mathbf{10\ Ω}$

(*iii*)   Capacitance, $C = \dfrac{1}{2\pi f\, X_C} = \dfrac{1}{2\pi \times 50 \times 10} = \mathbf{318 \times 10^{-6}\ F}$

Example 12.29. *A resistor R in series with a capacitor C is connected to 50 Hz, 240V source. Find the value of C so that R absorbs 300W and voltage across R is 100V. Also find the maximum charge and the maximum stored energy in C.*

Solution. Supply voltage, $V = 240$ volts ; $V_R = 100$ volts ; Power in $R = 300$W

Now, $V^2 = V_R^2 + V_C^2$ or $(240)^2 = (100)^2 + V_C^2$ $\therefore V_C = 218.17$ volts

$$\text{Value of } R = \frac{V_R^2}{300} = \frac{(100)^2}{300} = 33.33\ \Omega$$

$$\text{Circuit current, } I = \frac{V_R}{R} = \frac{100}{33.33} = 3\text{A}$$

Now, $X_C = \dfrac{V_C}{I} = \dfrac{218.17}{3} = 72.72\ \Omega$

or $\dfrac{1}{2\pi f C} = 72.72$ $\therefore C = \dfrac{1}{2\pi \times 50 \times 72.72} = \mathbf{43.77 \times 10^{-6}\ F}$

$$\text{Max. charge on } C = CV_{C(max)} = (43.77 \times 10^{-6}) \times \left(\sqrt{2} \times 218.17\right) = \mathbf{0.0135\ C}$$

$$\text{Max. energy stored in } C = \frac{1}{2}CV_{C(max)}^2 = \frac{1}{2} \times 43.77 \times 10^{-6} \times \left(\sqrt{2} \times 218.17\right)^2 = \mathbf{2.08\ J}$$

Example 12.30. *A capacitor of 8 µF takes a current of 1A when alternating voltage applied across it is 250 V. Calculate (i) frequency of the applied voltage (ii) the resistance to be connected in series with the capacitor to reduce the current in the circuit to 0.5 A at the same frequency (iii) phase angle of the resulting circuit.*

Solution. (*i*) Capacitive reactance, $X_C = V/I = 250/1 = 250\ \Omega$

$\therefore$ Frequency of applied voltage, $f = \dfrac{1}{2\pi\, C\, X_C} = \dfrac{1}{2\pi \times 8 \times 10^{-6} \times 250} = \mathbf{79.5\ Hz}$

(*ii*)   When a resistance is connected in series, the circuit becomes as shown in Fig. 12.23. Since frequency remains the same, the value of X_C is unchanged.

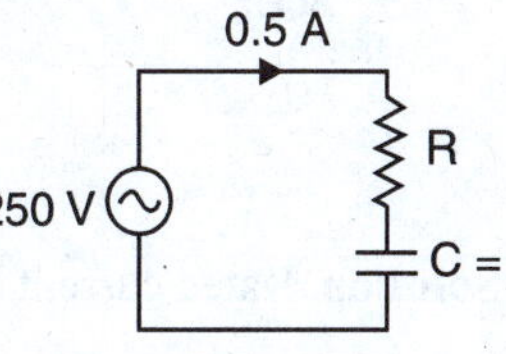

Circuit impedance, $Z = V/I = 250/0.5 = 500\ \Omega$

Now, $Z^2 = R^2 + X_C^2$

$\therefore$ $R = \sqrt{Z^2 - X_C^2} = \sqrt{(500)^2 - (250)^2} = \mathbf{433\ \Omega}$

(*iii*)   Circuit p.f., $\cos\phi = R/Z = 433/500 = 0.866$

$\therefore$ Circuit phase angle, $\phi = \cos^{-1} 0.866 = \mathbf{30°\ lead}$

Fig. 12.23

Example 12.31. *A 240 V, 50 Hz, RC series circuit takes an r.m.s. current of 20A. The maximum value of current occurs 1/900 second before the maximum value of the voltage. Calculate (i) the power factor (ii) average power (iii) parameters of the circuit.*

Solution. The time period of the alternating voltage is 1/50 second. This means that time interval of 1/50 second corresponds to 2π rad. or 360°. Therefore, a time interval of 1/900 second corresponds to a phase difference $= 360 \times (50/900) = 20°$. Therefore, current leads the voltage by 20° *i.e.*, circuit phase angle $\phi = 20°$.

(*i*) Power factor $= \cos\phi = \cos 20° = \mathbf{0.9397\ lead}$

(*ii*) Average power, $P = VI \cos\phi = 240 \times 20 \times 0.9397 = \mathbf{4510\ W}$

(iii) Circuit impedance, $Z = V/I = 240/20 = 12 \, \Omega$

$\therefore \quad R = Z \cos \phi = 12 \times \cos 20° = \mathbf{11.2 \, \Omega}$; $X_C = Z \sin \phi = 12 \sin 20° = 4.1 \, \Omega$

$$\therefore \qquad C = \frac{1}{2\pi f \, X_C} = \frac{1}{2\pi \times 50 \times 4.1} = 775 \times 10^{-6} \, \text{F} = \mathbf{775 \, \mu F}$$

Example 12.32. *A coil of power factor 0·6 is in series with a 100 μF capacitor. When connected to a 50Hz supply, the P.D. across the coil is equal to the P.D. across the capacitor. Find the resistance and inductance of the coil.*

Solution. $\omega = 2\pi f = 2\pi \times 50 = 314$ rad/sec. ; $\cos \phi = 0.6$; $\sin \phi = 0.8$

Reactance of capacitor, $X_C = \dfrac{1}{2\pi f C} = \dfrac{1}{314 \times 100 \times 10^{-6}} = 31.8 \, \Omega$

It is a series circuit so that current through the coil and capacitor is the same. Since P.D. across coil is equal to P.D. across the capacitor, it means that impedance of coil, $Z = X_C = 31.8 \, \Omega$.

$\therefore \qquad$ Resistance of coil, $R = Z \cos \phi = 31.8 \times 0.6 = \mathbf{19.1 \, \Omega}$

Reactance of coil, $X_L = Z \sin \phi = 31.8 \times 0.8 = 25.4 \, \Omega$

$\therefore \qquad$ Inductance of coil, $L = \dfrac{X_L}{2\pi f} = \dfrac{25.4}{2\pi \times 50} = \mathbf{0.081 \, H}$

Example 12.33. *It is desired to operate a 100W, 120V electric lamp at its current rating from a 240V, 50 Hz supply. Give details of the simplest manner in which this could be done using (i) a resistor (ii) a capacitor and (iii) an inductor having resistance of 10 Ω. What power factor would be presented to the supply in each case and which method is the most economical of power?*

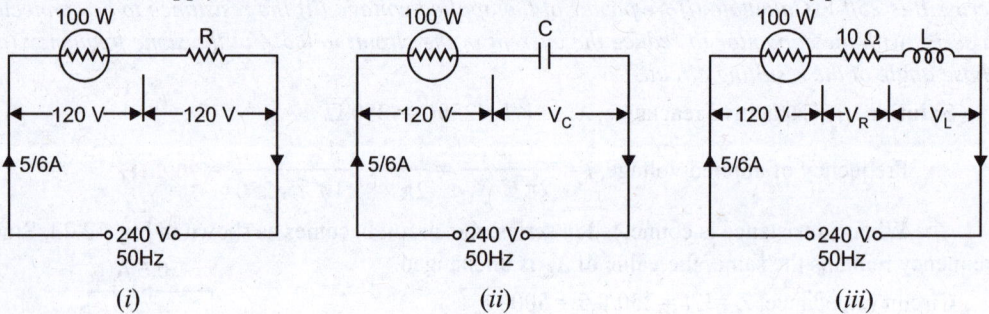

Fig. 12.24

Solution. Rated current of the bulb, $I = \dfrac{100}{120} = \dfrac{5}{6} A$. For proper operation of the bulb, the rated voltage (= 120V) and the rated current (= 5/6 A) of the bulb must be the same in the three cases.

(i) Fig. 12.24 *(i)* shows the circuit conditions for the first case.

Voltage across $R = 240 - 120 = 120$V

$\therefore \qquad$ Value of $R = \dfrac{120}{5/6} = \mathbf{144 \, \Omega}$; Power loss $= 240 \times (5/6) = 200$W

In this case, the circuit power factor is **unity.**

(ii) Fig. 12.24 *(ii)* shows the circuit conditions for the second case.

Voltage across $C = \sqrt{240^2 - 120^2} = 207.5$V

Reactance of capacitor, $X_C = \dfrac{207.5}{5/6} = 249 \, \Omega$

$\therefore \qquad$ Value of $C = \dfrac{1}{2\pi f \, X_C} = \dfrac{1}{2\pi \times 50 \times 249} = 12.8 \times 10^{-6} \, \text{F} = \mathbf{12.8 \, \mu F}$

Circuit power factor $= \dfrac{120}{240} = $ **0.5 lead** ; Power loss $= 240 \times (5/6) \times 0.5 = 100\text{W}$

(iii) Fig. 12.24 (*iii*) shows the circuit conditions for the third case. Here $V_R = IR = (5/6) \times 10 = 25/3\,\text{V}$.

$\therefore$ Voltage across $L = \sqrt{240^2 - (120 + 25/3)^2} = 203\text{V}$

$\therefore$ $X_L = \dfrac{203}{5/6} = 243.6\,\Omega$

$\therefore$ $L = \dfrac{X_L}{2\pi f} = \dfrac{243.6}{2\pi \times 50} = $ **0.775 H**

Power factor $= \dfrac{120 + 25/3}{240} = $ **0.535 lag**

Power loss $= 240 \times (5/6) \times 0.535 = 107\text{ W}$

Method (*ii*) is the most economical because power wasted in this case is minimum.

Example 12.34. *A capacitor and a non-inductive resistance are connected in series to a 200V, single-phase supply. When a voltmeter having a non-inductive resistance of 13,500Ω is connected across the resistor, it reads 132V and the current taken from the supply is 22.35 mA. Indicate on a vector diagram, the voltages across the two components and also the supply current (i) when the voltmeter is connected and (ii) when it is disconnected.*

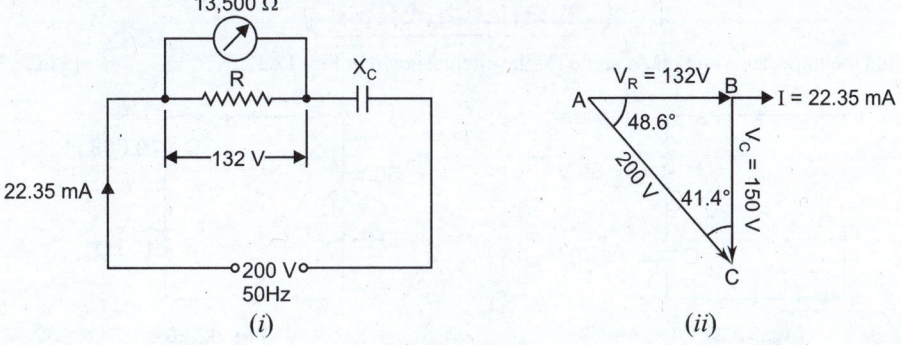

Fig. 12.25

Solution. **(*i*) When voltmeter is connected.** The circuit diagram for this condition is shown in Fig. 12.25 (*i*) while Fig. 12.25 (*ii*) shows the phasor diagram. It is clear from Fig. 12.25 (*ii*) that :

$$V_C = \sqrt{(200)^2 - (132)^2} = 150\text{ V}$$

$$\tan \phi = \dfrac{V_C}{V_R} = \dfrac{150}{132} \text{ or } \phi = \tan^{-1} 150/132 = 48.6°$$

The supply current is 22.35 mA (given). It is clear from the phasor diagram that supply voltage lags behind the circuit current by 48.6°; V_R leads supply voltage by 48.6° and V_C lags behind supply voltage by $90° - 48.6° = 41.4°$.

Current through voltmeter $= \dfrac{132}{13500} = 9.78 \times 10^{-3}\,\text{A} = 9.78\text{ mA}$

Current through $R = 22.35 - 9.78 = 12.57\text{ mA}$

$\therefore$ Value of $R = \dfrac{132}{12.57 \times 10^{-3}} = 10,500\,\Omega$

Also,   Value of $X_C = \dfrac{V_C}{I} = \dfrac{150}{22.35 \times 10^{-3}} = 6711\ \Omega$

(*ii*) **When voltmeter disconnected.** In this case, $R = 10{,}500\ \Omega$ and $X_C = 6711\ \Omega$ ($\because X_C$ depends on f and C which are not changed when voltmeter is disconnected). The phasor diagram for this condition is shown in Fig. 12.26.

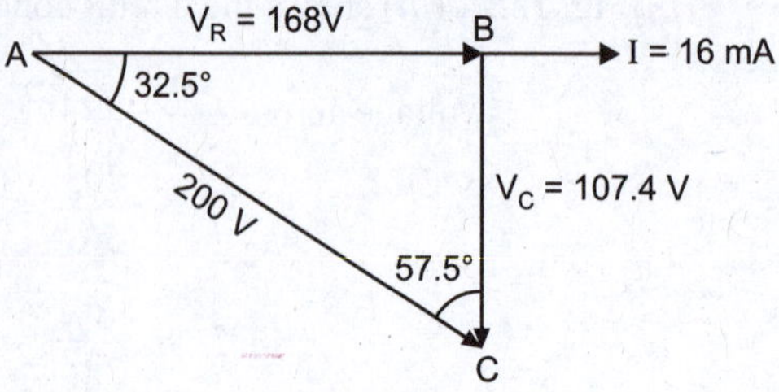

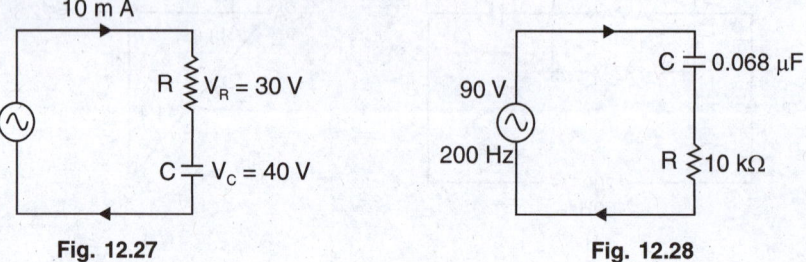

Circuit impedance, $Z = \sqrt{R^2 + X_C^2}$

$$= \sqrt{10500^2 + 6711^2}$$

$$= 12461\ \Omega$$

Fig. 12.26

Supply current $= \dfrac{200}{12461} = 16 \times 10^{-3}\,\text{A} = 16\ \text{mA}$

$\therefore$ $V_R = 16 \times 10^{-3} \times 10500 = 168\text{V}\ ;\ V_C = 16 \times 10^{-3} \times 6711 = 107.4\ \text{V}$

$$\tan \phi = \dfrac{V_C}{V_R} = \dfrac{107.4}{168} \quad \therefore \phi = 32.5^\circ$$

In this case, supply voltage lags behind circuit current by 32.5°; V_R leads supply voltage by 32.5° and V_C lags behind supply voltage by $90^\circ - 32.5^\circ = 57.5^\circ$.

Tutorial Problems

1. Find the impedance and phase angle for the circuit shown in Fig. 12.27.   **[5 kΩ ; 53.1°]**

10 m A

R V_R = 30 V

C V_C = 40 V

Fig. 12.27

C 0.068 µF

90 V

200 Hz

R 10 kΩ

Fig. 12.28

2. For the circuit in Fig. 12.28, determine (*i*) circuit impedance (*ii*) circuit current (*iii*) phase angle and (*iv*) voltage across capacitor.   **[(*i*) 11.7 kΩ (*ii*) 5·84 mA (*iii*) 49·5° (*iv*) 68.3 V]**

3. A capacitor when in series with a 145 ohm resistor has a circuit impedance of 208 Ω. Determine (*i*) the size of capacitor (*ii*) the power and the (*iii*) power factor when the circuit is connected to a 130 V, 60 Hz source.   **[(*i*) 17.8 µF (*ii*) 56.64 W (*iii*) 69.7%]**

4. A 120 Ω resistor is placed in series with a capacitor and the circuit is connected to a 120 V, 60 Hz source. If the power factor is 91%, determine the power, the current and the value of *C*.

   [99.37 W ; 0.91 A ; 48.52 µF]

5. A resistor is placed in series with a 12 µF capacitor. The supply is 120 V at 60 Hz. What two values of *R* yield a circuit power of 10 watts ?   **[1405.23 Ω ; 34.78 Ω]**

12.9. Equivalent Circuit for a Capacitor

No capacitor is ideal or perfect *i.e.,* electrical energy supplied to a capacitor is not entirely stored in the electric field between the capacitor plates. A part of the supplied electrical energy is converted into heat due to the resistances of the various parts of the capacitor. The total effective resistance is made up of the following resistances :

(*i*) Resistance of the lead wires

(*ii*) Resistance of the plates

(*iii*) Insulation resistance of the dielectric

(*iv*) *Resistance due to polarisation of the dielectric.

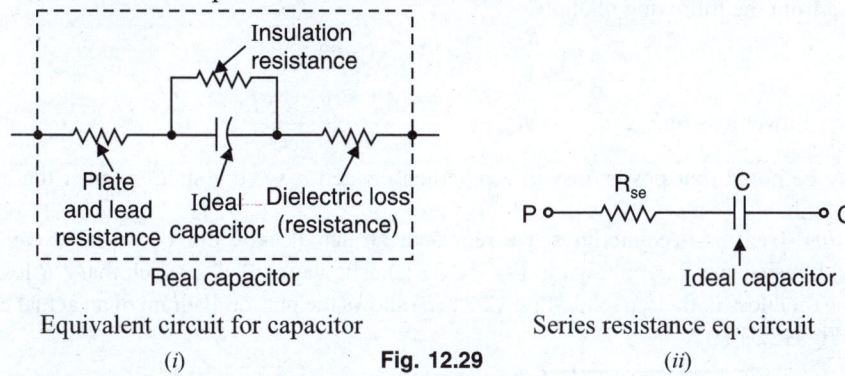

Equivalent circuit for capacitor

(*i*) **Fig. 12.29** (*ii*)

Series resistance eq. circuit

Fig. 12.29 (*i*) shows the equivalent circuit of a capacitor. We can replace all the resistances by a single equivalent series resistance R_{se} as shown in Fig. 12.29 (*ii*). The value of R_{se} is such that I^2R loss in it is equal to the total loss in the capacitor. Note that the equivalent circuit is, in effect, a series *RC* circuit.

Power Factor of a capacitor. *An ideal (or perfect) capacitor is one in which there are no losses* (*i.e.,* $R_{se} = 0$) *and the current through the capacitor leads the applied voltage by* 90°. However, in practice, it is not possible to get such a capacitor so that capacitor power factor angle ϕ_C is less than 90°. The angle by which ϕ_C falls short of 90° is called **loss angle** δ.

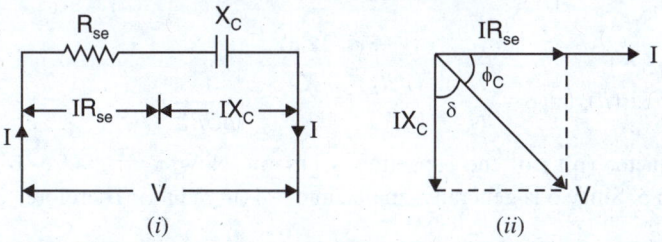

Fig. 12.30

Fig. 12.30 (*i*) shows voltage *V* applied to a capacitor. The resulting circuit current is *I*. Fig. 12.30 (*ii*) shows the phasor diagram for the circuit shown in Fig. 12.30 (*i*). Note that true power factor of the capacitor is cos ϕ_C.

Circuit impedance, $Z = \sqrt{R_{se}^2 + X_C^2}$

∴ True power factor of the capacitor is

$$\cos \phi_C = \frac{R_{se}}{Z} = \frac{R_{se}}{\sqrt{R_{se}^2 + X_C^2}}$$

Now, $\phi_C = 90° - \delta$ so that cos $\phi_C = \cos(90° - \delta) = \sin \delta$

Since δ is generally small, sin δ = δ (in radian). Therefore, tan δ = δ = cos ϕ_C.

Referring to Fig. 12.30 (*ii*), tan δ = $\dfrac{IR_{se}}{IX_C} = \dfrac{R_{se}}{X_C}$

∴ $R_{se} = X_C \tan \delta = \dfrac{1}{\omega C} \cos \phi_C$ (∵ tan δ = cos ϕ_C)

---

* This loss is called dielectric hysteresis loss and can be represented by a small resistance in series with the capacitor.

or $$R_{se} = \frac{\cos \phi_C}{\omega C}$$

When $X_C \gg R_{se}$, the capacitor power factor $\cos \phi_C = R_{se}/X_C$. The loss angle δ can be easily determined from the following relation :

$$\tan \delta = \frac{R_{se}}{X_C} = \omega C R_{se} = 2\pi f C R_{se}$$

Power loss in R_{se}, $$P = I^2 R_{se} = I^2 \frac{\cos \phi_C}{\omega C} = \frac{I^2 \times \text{p.f.}}{\omega C}$$

It may be noted that power loss in a practical capacitor is very small even if the capacitor voltage is high.

Another treatment. Sometimes, we represent an actual capacitor by an ideal capacitor in parallel with resistance R_P as shown in Fig. 12.31 (*i*). The value of R_P is such that I^2R loss in it is equal to the total loss in the capacitor. Fig. 12.31 (*ii*) shows the phasor diagram of an actual capacitor shown in Fig. 12.31 (*i*).

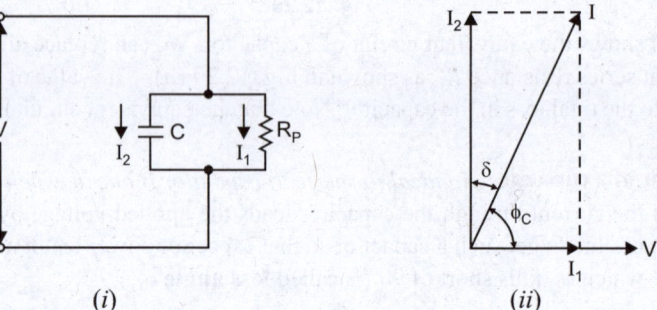

(*i*)   (*ii*)

Fig. 12.31

From Fig. 12.31 (*ii*), $$\tan \delta = \frac{I_1}{I_2} = \frac{V/R_P}{V/X_C} = \frac{X_C}{R_P} = \frac{1}{\omega C R_P}$$

The power factor (p.f.) of the capacitor is $\cos \phi_C$. Now $\phi_C = 90° - \delta$ so that $\cos \phi_C = \cos(90° - \delta) = \sin \delta$. Since δ is generally small, $\sin \delta = \delta$ (in radian). Therefore, $\tan \delta = \delta = \cos \phi_C$.

$\therefore$ $$\cos \phi_C = \frac{1}{\omega C R_P}$$

or $$R_P = \frac{1}{\omega C \times \cos \phi_C} = \frac{1}{\omega C \times \text{p.f.}}$$

Power loss in R_P is $$P = \frac{V^2}{R_P} = \omega C V^2 \times \text{p.f.}$$

Example 12.35. *A 0.068 µF capacitor has an equivalent series resistance R_{se} of 4Ω at 1000 Hz. Find the power factor of the capacitor.*

Solution. $$X_C = \frac{1}{2\pi f C} = \frac{1}{2\pi \times 1000 \times 0.068 \times 10^{-6}} = 2342 \ \Omega$$

Since $X_C \gg R$, the power factor $\cos \phi_C$ of the capacitor is given by ;

$$\cos \phi_C = \frac{R_{se}}{X_C} = \frac{4}{2342} = \mathbf{0.0017}$$

Note that the value of power factor of a capacitor depends upon the values of R_{se} and X_C. It is reminded that power factor of an ideal capacitor = $\cos \phi_C = \cos 90° = 0$.

Example 12.36. *A capacitor of capacitance 10 µF has a loss angle of 10° at 50 Hz. Find (i) power factor angle (ii) power factor and (iii) equivalent series resistance R_{se}.*

Solution. (*i*) Power factor angle, $\phi_C = 90° - \delta = 90° - 10° = \mathbf{80°}$

(*ii*)   Power factor, $\cos \phi_C = \cos 80° = \mathbf{0.174}$

(*iii*) $$X_C = \frac{1}{2\pi f C} = \frac{1}{2\pi \times 50 \times 10 \times 10^{-6}} = 318.3\ \Omega$$

$\therefore$ Equivalent series resistance, $R_{se} = X_C \tan \delta = 318.3 \times \tan 10° = \mathbf{56.1\ \Omega}$

Example 12.37. *Dielectric heating is to be employed to heat a slab of insulating material 2cm thick 150 cm^2 in area. The power required is 200 W and a frequency of 30 MHz is to be used. The material has a relative permittivity of 5 and a power factor of 0.05. Determine the voltage necessary and the current which will flow through the material. If the voltage were to be limited to 600 V, to what would the frequency have to be raised?*

Solution. We shall use parallel equivalent circuit for the capacitor.

The capacitance of the parallel-plate capacitor with insulating slab between the plates is

$$C = \frac{\varepsilon_0 \varepsilon_r A}{d} = \frac{8.854 \times 10^{-12} \times 5 \times 150 \times 10^{-4}}{2 \times 10^{-2}} = 33.2 \times 10^{-12}\ F$$

The value of parallel resistance R_P (See Art. 12.9) is

$$R_P = \frac{1}{\omega C \times \text{p.f.}} = \frac{1}{(2\pi \times 30 \times 10^6) \times 33.2 \times 10^{-12} \times 0.05} = 3196\ \Omega$$

Power loss in R_P, $P = \dfrac{V^2}{R_P}$ $\therefore V = \sqrt{P \times R_P} = \sqrt{200 \times 3196} = \mathbf{800\ V}$

Current I through the capacitor is given by ;

$$I = \frac{V}{X_C} = \omega CV = (2\pi \times 30 \times 10^6) \times 33.2 \times 10^{-12} \times 800 = \mathbf{5A}$$

Now, $P = \dfrac{V^2}{R_P} = V^2\, \omega C \times \text{p.f.}$ so that $P \propto V^2 f$

$\therefore$ $V_1^2 f_1 = V_2^2 f_2$ or $f_2 = \left(\dfrac{V_1}{V_2}\right)^2 \times f_1 = \left(\dfrac{800}{600}\right)^2 \times 30 = \mathbf{53.3\ MHz}$

12.10. R-L-C Series A.C. Circuit

This is a general series a.c. circuit. Fig. 12.32 shows R, L and C connected in series across a supply voltage V (*r.m.s.*). The resulting circuit current is I (*r.m.s.*).

$\therefore$ Voltage across R, $V_R = I R$ ... V_R is in phase with I

Voltage across L, $V_L = I X_L$ ... where V_L leads I by 90°

Voltage across C, $V_C = I X_C$ ... where V_C lags I by 90°

As before, the phasor diagram is drawn taking current as the reference phasor. In the phasor diagram (See Fig. 12.33), OA represents V_R, AB represents V_L and AC represents V_C. It may be seen that V_L is in phase opposition to V_C. It follows that the circuit can either be effectively inductive or capacitive depending upon which voltage drop (V_L or V_C) is predominant. For the case considered, $V_L > V_C$ so that net voltage drop across L-C combination is $V_L - V_C$ and is represented by AD. Therefore, the applied voltage V is the phasor sum of V_R and $V_L - V_C$ and is represented by OD.

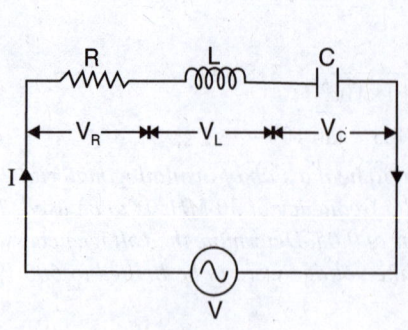

Fig. 12.32   **Fig. 12.33**

$$\therefore \qquad V = \sqrt{V_R^2 + (V_L - V_C)^2} = \sqrt{(IR)^2 + (IX_L - IX_C)^2}$$

$$= I\sqrt{R^2 + (X_L - X_C)^2}$$

$$\therefore \qquad I = \frac{V}{\sqrt{R^2 + (X_L - X_C)^2}}$$

The quantity $\sqrt{R^2 + (X_L - X_C)^2}$ offers opposition to current flow and is called **impedance** of the circuit.

$$\text{Circuit power factor, } \cos\phi = \frac{R}{Z} = \frac{R}{\sqrt{R^2 + (X_L - X_C)^2}} \qquad \ldots(i)$$

Also, $$\tan\phi = \frac{V_L - V_C}{V_R} = \frac{X_L - X_C}{R} \qquad \ldots(ii)$$

Since X_L, X_C and R are known, phase angle ϕ of the circuit can be determined.

$$\text{Power consumed, } P = VI\cos\phi = {}^*I^2 R$$

Three cases of R-L-C series circuit. We have seen that the impedance of a R-L-C series circuit is given by ;

$$Z = \sqrt{R^2 + (X_L - X_C)^2}$$

   (i) **When $X_L - X_C$ is positive** (i.e. $X_L > X_C$), phase angle ϕ is positive and the circuit will be inductive. In other words, in such a case, the circuit current I will lag behind the applied voltage V by ϕ ; the value of ϕ being given by eq. (ii) above.

  (ii) **When $X_L - X_C$ is negative** (i.e. $X_C > X_L$), phase angle ϕ is negative and the circuit is capacitive. That is to say the circuit current I leads the applied voltage V by ϕ ; the value of ϕ being given by eq. (ii) above.

 (iii) **When $X_L - X_C$ is zero** (i.e. $X_L = X_C$), the circuit is purely resistive. In other words, circuit current I and applied voltage V will be in phase i.e. $\phi = 0°$. The circuit will then have unity power factor.

If the equation for the applied voltage is $v = V_m \sin \omega t$, then equation for the circuit current will be :

---

* $P = VI\cos\phi = (IZ)I \times \dfrac{R}{Z} = I^2 R$

This is expected because there is no power loss in L or C.

$$i = I_m \sin (\omega t \pm \phi) \quad \text{where } I_m = V_m / Z$$

The value of ϕ will be positive or negative depending upon which reactance (X_L or X_C) predominates.

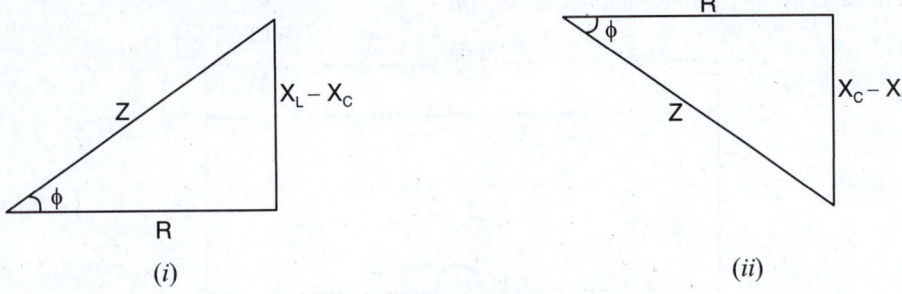

Fig. 12.34

Fig. 12.34 (*i*) shows the impedance triangle of the circuit for the case when $X_L > X_C$ whereas impedance triangle in Fig. 12.34 (*ii*) is for the case when $X_C > X_L$.

Example 12.38. *A 230 V, 50 Hz a.c. supply is applied to a coil of 0.06 H inductance and 2.5Ω resistance connected in series with a 6·8 μF capacitor. Calculate (i) impedance (ii) current (iii) phase angle between current and voltage (iv) power factor and (v) power consumed.*

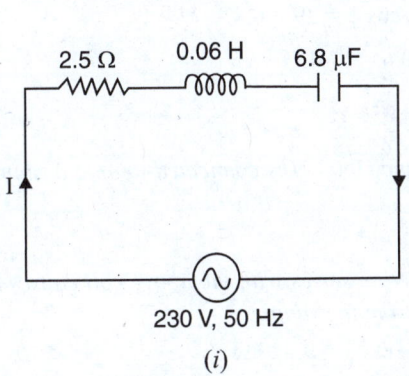
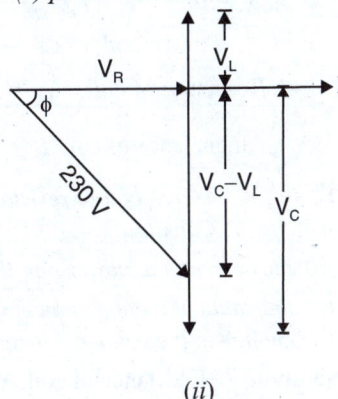

Fig. 12.35

Solution. Fig. 12.35 (*i*) shows the conditions of the problem.

$$X_L = 2\pi f L = 2\pi \times 50 \times 0.06 = 18.85\ \Omega$$

$$X_C = \frac{1}{2\pi f C} = \frac{10^6}{2\pi \times 50 \times 6.8} = 468\ \Omega$$

(*i*) Circuit impedance, $Z = \sqrt{R^2 + (X_L - X_C)^2} = \sqrt{(2.5)^2 + (18.85 - 468)^2} = \mathbf{449.2\ \Omega}$

(*ii*) Circuit current, $I = V/Z = 230/449.2 = \mathbf{0.512\ A}$

(*iii*) $\tan \phi = \dfrac{X_L - X_C}{R} = \dfrac{18.85 - 468}{2.5} = -179.66$

∴ Phase angle, $\phi = \tan^{-1} -179.66 = -89.7° = \mathbf{89.7°\ \textit{lead}}$

The negative sign with ϕ shows that current is leading the voltage [See the phasor diagram in Fig. 12.35 (*ii*)].

(*iv*) Power factor, $\cos \phi = \dfrac{R}{Z} = \dfrac{2.5}{449.2} = \mathbf{0.00557\ \textit{lead}}$

(*v*) Power consumed, $P = VI \cos \phi = 230 \times 0.512 \times 0.00557 = \mathbf{0.656\ W}$

Example 12.39. *A coil of p.f. 0.8 is connected in series with a 110 μF capacitor. The supply frequency is 50 Hz. The p.d. across the coil is found to be equal to the p.d. across the capacitor. Calculate the resistance and inductance of the coil.*

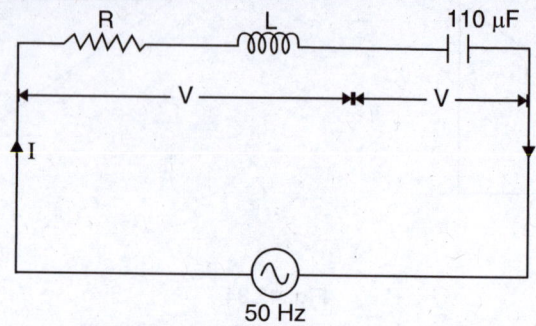

Fig. 12.36

Solution. Fig. 12.36 shows the conditions of the problem.

$$\text{Reactance of capacitor, } X_C = \frac{1}{2\pi f C} = \frac{10^6}{2\pi \times 50 \times 110} = 29\,\Omega$$

Now, $\qquad I\,Z_{coil} = IX_C \therefore Z_{coil} = X_C = 29\,\Omega$

For the coil, $\cos\phi = R/Z_{coil} \therefore R = Z_{coil}\cos\phi = 29 \times 0.8 = \mathbf{23.2\,\Omega}$

Reactance of coil, $*X_L = Z_{coil}\sin\phi = 29 \times 0.6 = 17.4\,\Omega$

$\therefore \qquad$ Inductance of coil, $L = \dfrac{X_L}{2\pi f} = \dfrac{17.4}{2\pi \times 50} = \mathbf{0.055\,H}$

Example 12.40. *A coil of resistance 8Ω and inductance 0·03 H is connected to an a.c. supply at 240 V, 50 Hz. Calculate :*

(i) the current, the power and the power factor

(ii) the value of a capacitance which, when connected in series with the above coil causes no change in the value of current and power taken from the supply.

Solution. *(i)* Reactance of coil, $X_L = 2\pi f L = 2\pi \times 50 \times 0.03 = 9.424\,\Omega$

$$\text{Impedance of coil, } Z = \sqrt{R^2 + X_L^2} = \sqrt{(8)^2 + (9.424)^2} = 12.36\,\Omega$$

$$\text{Circuit current, } I = V/Z = 240/12.36 = \mathbf{19.42\,A}$$

$$\text{Power consumed, } P = I^2 R = (19.42)^2 \times 8 = \mathbf{3017\,W}$$

$$\text{Power factor, } \cos\phi = R/Z = 8/12.36 = \mathbf{0.65}\ \textit{lag}$$

(ii) To maintain the same current and power, the impedance of the circuit should remain unchanged. Thus the value of capacitance in the series circuit should be such so as to cause the current to lead by the same angle as it previously lagged. This can be achieved if the series capacitor has a capacitive reactance equal to twice the inductive reactance [See Fig. 12.37].

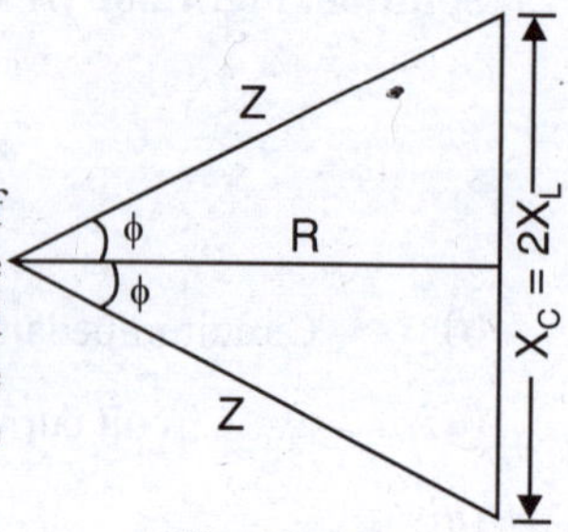

$\therefore \qquad X_C = {**}2X_L$

$\qquad\qquad = 2 \times 9.424 = 18.848\,\Omega$

Fig. 12.37

---

* $\quad$ or $\quad X_L = \sqrt{Z_{coil}^2 - R^2} = \sqrt{29^2 - 23 \cdot 2^2} = 17.4\Omega$

** $\quad$ The reactance of the capacitor should be such that reactance of $L - C$ combination is 9.424 Ω. This is possible only if $X_C = 2\,X_L = 2 \times 9.424 = 18.848\,\Omega$.

$\therefore$ Capacitance, $C = \dfrac{1}{2\pi f\, X_C} = \dfrac{1}{2\pi \times 50 \times 18.848} = 168.9 \times 10^{-6}\,\text{F} = \textbf{168.9 μF}$

Example 12.41. *A resistance R, an inductance L = 0.01 H and a capacitance C are connected in series. When an alternating voltage v = 400 sin (3000 t – 20°) is applied to the series combination, the current flowing is* $10\sqrt{2}$ *sin (3000t – 65°). Find the values of R and C.*

Solution.

$\phi = 65° - 20° = 45°$ lag *i.e.*, circuit is inductive

$X_L = \omega L = 3000 \times 0.01 = 30\,\Omega$

$\tan 45° = X/R \therefore X = R$

$Z = V_m/I_m = 400/10\sqrt{2} = 28.3\,\Omega$

$Z^2 = R^2 + X^2 = R^2 + R^2 = 2R^2$

$\therefore \quad R = Z/\sqrt{2} = 28.3/\sqrt{2} = \textbf{20 Ω}$

Now $\quad X = X_L - X_C \therefore X_C = X_L - X = 30 - 20 = 10\,\Omega$

$\therefore \quad C = \dfrac{1}{\omega X_C} = \dfrac{1}{3000 \times 10} = \textbf{33.3 × 10}^{-6}\,\textbf{F}$

Example 12.42. *In the circuit shown in Fig. 12.38 (i), find the values of (i) the current I (ii)* V_1 *and* V_2 *and (iii) power factor. Draw the phasor diagram.*

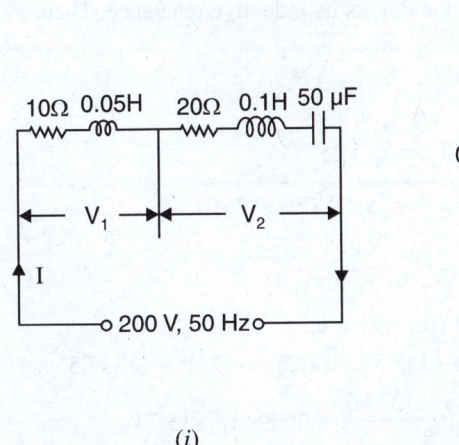

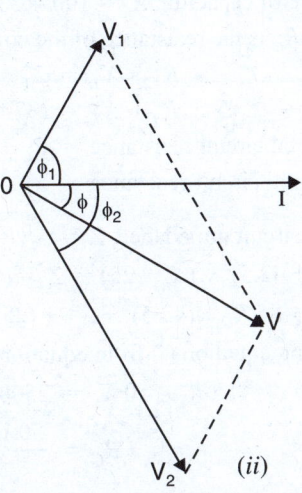

(*i*)   (*ii*)

Fig. 12.38

Solution. (*i*) Total $L = 0.05 + 0.1 = 0.15\,\text{H} \therefore X_L = 2\pi f L = 2\pi \times 50 \times 0.15 = 47.1\,\Omega$

$X_C = \dfrac{1}{2\pi f C} = \dfrac{1}{2\pi \times 50 \times 50 \times 10^{-6}} = 63.7\,\Omega$

Net circuit reactance, $X = X_L - X_C = 47.1 - 63.7 = -16.6\,\Omega$

Circuit impedance, $Z = \sqrt{R^2 + X^2} = \sqrt{(10+20)^2 + (-16.6)^2} = 34.3\,\Omega$

$\therefore \quad$ Circuit current, $I = 200/34.3 = \textbf{5.83 A}$

(*ii*) Considering the branch across which terminal voltage is V_1, we have,

$X_{L1} = 2\pi f_{L1} = 2\pi \times 50 \times 0.05 = 15.7\,\Omega$

$Z_1 = \sqrt{(10)^2 + (15.7)^2} = 18.6\,\Omega$

$\therefore \quad V_1 = IZ_1 = 5.83 \times 18.6 = \textbf{108.4 V}$

Considering the branch across which terminal voltage is V_2, we have,

$$X_{L2} = 2\pi f L_2 = 2\pi \times 50 \times 0.1 = 31.4\ \Omega\ ;\ X_{C2} = 63.7\ \Omega\ \text{(found above)}$$

∴ Net reactance, $X_2 = X_{L2} - X_{C2} = 31.4 - 63.7 = -32.2\ \Omega$

Now

$$Z_2 = \sqrt{R_2^2 + X_2^2} = \sqrt{(20)^2 + (-32.2)^2} = 38\ \Omega$$

∴

$$V_2 = IZ_2 = 5.83 \times 38 = \textbf{221 V}$$

(*iii*) Circuit power factor, $\cos\phi = \dfrac{\text{Total circuit resistance}}{\text{Total circuit impedance}} = \dfrac{30}{34.3} = 0.875\ lead$

Fig. 12.38 (*ii*) shows the phasor diagram of the circuit. Note that V_1 leads I by $\phi_1 = \cos^{-1} R_1/Z_1$ $= \cos^{-1} 10/18.6 = 57.5°$.

Also V_2 lags behind I by $\phi_2 = \cos^{-1} R_2/Z_2 = \cos^{-1} 20/38 = 58.2°$.

Note that circuit current I leads the applied voltage by $\phi = \cos^{-1} 0.875 = 28.95°$.

Example 12.43. *A voltage of 200 V is applied to a series circuit consisting of a resistor, an inductor and a capacitor. The respective voltages across these components are 170 V, 150 V and 100 V and the circuit current is 4A. Find the power factor of the inductor and of the circuit.*

Solution.

Resistance of resistor, $R = 170/4 = 42.5\ \Omega$; Impedance of coil (inductor), $Z_C = 150/4 = 37.5\ \Omega$

Reactance of capacitor, $X_C = 100/4 = 25\ \Omega$; Circuit impedance, $Z = 200/4 = 50\ \Omega$

Suppose R_C is the resistance of the coil and X_L is its inductive reactance. Then,

$$R_C^2 + X_L^2 = Z_C^2$$

or

$$R_C^2 + X_L^2 = (37.5)^2 \qquad\qquad\qquad ...(i)$$

Total circuit resistance $= R_C + R = R_C + 42.5$

Total circuit reactance $= X_L - X_C = X_L - 25$

Circuit impedance, $Z = \sqrt{(R_C + 42.5)^2 + (X_L - 25)^2}$

or $(R_C + 42.5)^2 + (X_L - 25)^2 = Z^2 = (50)^2 = 2500$

or $R_C^2 + 85\,R_C + (42.5)^2 + X_L^2 + (25)^2 - 50\,X_L = 2500 \qquad ...(ii)$

Subtracting equation (*i*) from equation (*ii*), we have,

$$85\,R_C - 50\,X_L = 2500 - (37.5)^2 - (42.5)^2 - (25)^2 = -1337.5$$

∴

$$R_C = \dfrac{50 X_L - 1337.5}{85} = 0.588\,X_L - 15.74 \qquad\qquad ...(iii)$$

Putting this value of R_C in equation (*i*), we have,

$$(0.588\,X_L - 15.74)^2 + X_L^2 = (37.5)^2$$

On solving, $X_L = 37\ \Omega$ (Taking + ve value)

From equation (*iii*), $R_C = 0.588 \times 37 - 15.74 = 6\ \Omega$

∴ Power factor of inductor $= \dfrac{R_C}{Z_C} = \dfrac{6}{37.5} = 0.16\ lag$

Total circuit resistance, $R_T = R + R_C = 42.5 + 6 = 48.5\ \Omega$

∴ Circuit power factor $= R_T/Z = 48.5/50 = 0.97\ lag$

Example 12.44. *A 230 V, 50 Hz voltage is applied to a coil having L = 5H and R = 2Ω. This coil is connected in series with a capacitance C. What value C must have in order that voltage across the coil shall be 250 V ?*

Solution. Inductive reactance of coil, $X_L = 2\pi f L = 2\pi \times 50 \times 5 = 1570\ \Omega$

Impedance of coil, $Z_{coil} = \sqrt{R^2 + X_L^2} = \sqrt{(2)^2 + (1570)^2} \simeq 1570\ \Omega$

In order that p.d. across coil is 250 V, the circuit current I is given by ;

$$I = \frac{250}{Z_{coil}} = \frac{250}{1570} = 0.159 \text{ A}$$

Total circuit impedance, Z_T = $230/I = 230/0.159 = *1444 \ \Omega$

∴ Reactance of capacitor, X_C = $1570 - 1444 = 126 \ \Omega$

∴
$$C = \frac{1}{2\pi f \ X_C} = \frac{1}{2\pi \times 50 \times 126} = 26 \times 10^{-6} \text{ F} = \textbf{26 } \boldsymbol{\mu}\textbf{F}$$

Example 12.45. *A voltage v(t) = 100 sin 314 t is applied to a series circuit consisting of 10Ω resistance, 0·0318 H inductance and a capacitor of 63.6 μF. Find (i) expression for i(t) (ii) phase angle between voltage and current (iii) power factor (iv) active power consumed (v) peak value of pulsating power.*

Solution. (*i*)
$$X_L = \omega L = 314 \times 0.0318 = 10 \ \Omega$$

$$X_C = \frac{1}{2\pi f C} = \frac{1}{314 \times 63.6 \times 10^{-6}} = 50 \ \Omega$$

Net reactance, X = $X_L - X_C = 10 - 50 = -40 \ \Omega$

Circuit impedance, Z = $\sqrt{R^2 + X^2} = \sqrt{(10)^2 + (-40)^2} = 41.2 \ \Omega$

Circuit current, I = $\dfrac{100/\sqrt{2}}{41.2} = 1.716 \text{ A}$

Max. value of current = $\sqrt{2} \times I = \sqrt{2} \times 1.716 = 2.43 \text{ A}$

Circuit phase angle, ϕ = $\tan^{-1} - 40/10 = 76° \ lead$

Since the circuit is capacitive, the equation of current is

$$i = I_m \sin(314 \ t + \phi) = 2.43 \sin(314 \ t + 76°) \textbf{ Ans.}$$

(*ii*) Circuit phase angle = **76°, current leading voltage**

(*iii*) Circuit p.f., $\cos \phi$ = $\cos 76° = \textbf{0.24 } \textit{lead}$

(*iv*) Active power consumed, $P = VI \cos \phi = \left(100/\sqrt{2}\right) \times 1.716 \times 0.24 = \textbf{29.16 W}$

(*v*) Peak value of pulsating power = $\dfrac{V_m I_m}{2} \cos \phi + \dfrac{V_m I_m}{2} = \dfrac{V_m I_m}{2}(1 + \cos \phi)$

$$= \frac{100 \times 2.43}{2}(1 + 0.24) = \textbf{151 W}$$

Example 12.46. *Two impedances Z_1 and Z_2 when connected separately across a 230 V, 50 Hz supply consumed 100 W and 60 W at power factors of 0.5 lagging and 0.6 leading respectively. If these impedances are now connected in series across the same supply, find (i) total power absorbed and overall p.f. (ii) the value of the impedance to be added in series so as to raise the overall p.f. to unity.*

Solution. The statement of the problem suggests that Z_1 is inductive while Z_2 is capacitive.

Z_1. $V_1 = 230$ volts; $P_1 = 100$ W ; $\cos \phi_1 = 0.5$ lag

Now, $P_1 = V_1 I_1 \cos \phi_1 \quad \therefore I_1 = \dfrac{P_1}{V_1 \cos \phi_1} = \dfrac{100}{230 \times 0.5} = 0.87 \text{ A}$

Also $P_1 = I_1^2 R_1 \quad \therefore R_1 = \dfrac{P_1}{I_1^2} = \dfrac{100}{(0.87)^2} = 132 \ \Omega$

---

* This is nearly the reactance of the circuit because R is small.

$$\text{Value of } Z_1 = \frac{V_1}{I_1} = \frac{230}{0.87} = 264 \ \Omega$$

$$\therefore \qquad X_1 = \sqrt{Z_1^2 - R_1^2} = \sqrt{264^2 - 132^2} = 229 \ \Omega \text{ inductive}$$

Z_2.

$$X_1 = \sqrt{Z_1^2 - R_1^2} = \sqrt{264^2 - 132^2} = 229 \ \Omega \text{ inductive}$$

Z_2. $V_2 = 230$ volts ; $P_2 = 60$ W ; $\cos \phi_2 = 0.6$ lead

Now, $P_2 = V_2 I_2 \cos\phi_2 \quad \therefore I_2 = \dfrac{P_2}{V_2 \cos\phi_2} = \dfrac{60}{230 \times 0.6} = 0.434$ A

Also $P_2 = I_2^2 R_2 \quad \therefore R_2 = \dfrac{P_2}{I_2^2} = \dfrac{60}{(0.434)^2} = 318 \ \Omega$

$$\text{Value of } Z_2 = \frac{V_2}{I_2} = \frac{230}{0.434} = 530 \ \Omega$$

$$\therefore \qquad X_2 = \sqrt{Z_2^2 - R_2^2} = \sqrt{530^2 - 318^2} = 424 \ \Omega \text{ capacitive}$$

When Z_1 and Z_2 connected in series

     Total resistance, $R = R_1 + R_2 = 132 + 318 = 450 \ \Omega$

     Total reactance, $X = X_1 + (-X_2) = 229 - 424 = -195 \ \Omega$

     Total impedance, $Z = \sqrt{R^2 + X^2} = \sqrt{(450)^2 + (-195)^2} = 490 \ \Omega$

     Circuit current, $I = \dfrac{V}{Z} = \dfrac{230}{490} = 0.47$ A

(i)   Total power absorbed $= I^2 R = (0.47)^2 \times 450 = $**99 W**

     Overall power factor, $\cos \phi = \dfrac{R}{Z} = \dfrac{450}{490} = $**0.92 *lead***

(ii) Power factor will become unity when net reactance in the circuit is zero. Now reactance of the circuit is $-195 \ \Omega$.

$$\therefore X_L + X_C = 0 \quad \text{or} \quad X_L - 195 = 0 \quad \therefore X_L = 195 \ \Omega$$

Therefore, pure inductive reactance of 195 Ω should be connected in series with the circuit.

Example 12.47. *A 230 V, 50 Hz alternating p.d. supplies a choking coil having an inductance of 0.06 H in series with a capacitance of 6.8 µF, the effective resistance of the circuit being 2.5 Ω. Estimate the current and the angle of the phase difference between it and the applied p.d. If the p.d. has a 10% harmonic of 5 times the fundamental frequency, estimate (i) the circuit current and (ii) the p.d. across the capacitance.*

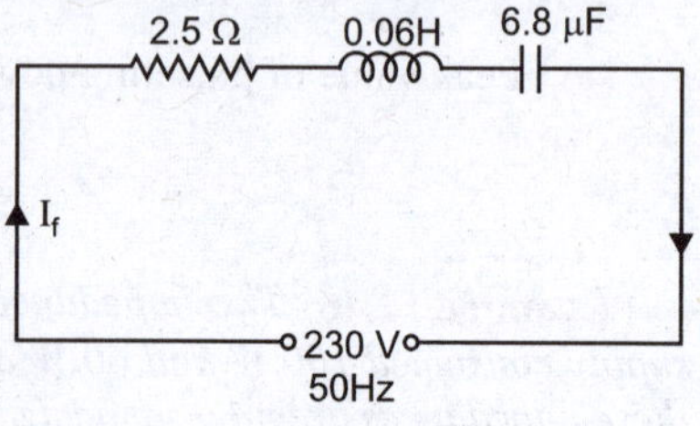

Fig. 12.39

Solution. Fig. 12.39. shows the required circuit diagram.

Fundamental frequency. Frequency, $f = 50$ Hz

$$X_L = 2\pi f L = 2\pi \times 50 \times 0.06 = 18.85 \ \Omega \ ; \quad X_C = \frac{1}{2\pi f C} = \frac{1}{2\pi \times 50 \times 6.8 \times 10^{-6}} = 468 \ \Omega$$

$\therefore$ Net reactance, $X = 18.85 - 468 = -449.15 \ \Omega$

     Circuit impedance, $Z = \sqrt{R^2 + X^2} = \sqrt{(2.5)^2 + (-449.15)^2} = 449.2 \ \Omega$

     Circuit current, $I_f = \dfrac{V}{Z} = \dfrac{230}{449.2} = $**0.512 A**

     Phase angle, $\phi = \tan^{-1} \dfrac{X}{R} = \tan^{-1} - \dfrac{449.15}{2.5} = $**− 89.68°**

∴ Current leads the applied voltage by 89.68°.

Fifth harmonic frequency. The frequency of fifth harmonic is $5f$. Since $X_L \propto f$ and $X_C \propto 1/f$,

$$\therefore \qquad X_{Lh} = 5 \times 18.85 = 94.25 \ \Omega \ ; \ X_{Ch} = 468 \div 5 = 93.6 \ \Omega$$

Net reactance, $X_h = 94.25 - 93.6 = 0.65 \ \Omega$

∴ Impedance offered by the circuit to harmonic current is

$$Z_h = \sqrt{R^2 + X_h^2} = \sqrt{2.5^2 + (0.65)^2} = 2.585 \ \Omega$$

Harmonic voltage, $V_h = 10\%$ of 230 volts = 23 volts

$$\therefore \qquad \text{Harmonic current, } I_h = \frac{V_h}{Z_h} = \frac{23}{2.585} = 8.893 \ A$$

(i) The r.m.s. value I of the circuit current is

$$I = \sqrt{I_f^2 + I_h^2} = \sqrt{(0.512)^2 + (8.893)^2} = \textbf{8.9A}$$

(ii) P.D. across capacitor due to fundamental frequency current is

$$V_f = I_f \times X_C = 0.512 \times 468 = 239.6 \text{ volts}$$

P.D. across capacitor due to harmonic current is

$$V_h = I_h \times X_{Ch} = 8.893 \times 93.6 = 832.6 \text{ volts}$$

Resultant p.d. across capacitor is given by ;

$$V_C = \sqrt{V_f^2 + V_h^2} = \sqrt{(239.6)^2 + (832.6)^2} = \textbf{866.4 volts}$$

Example 12.48. *For the series circuit shown in Fig. 12.40, the current and voltages are indicated. Find (i) the values of R, r, f and L (ii) magnitude of applied voltage (iii) p.f. of the whole circuit and (iv) active power consumed.*

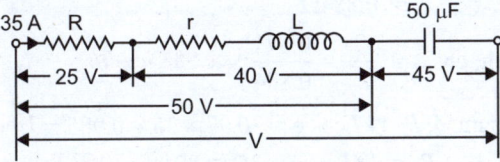

Fig. 12.40

Solution. *(i)* In the circuit diagram in Fig. 12.40, it is given that circuit current is $I = 35$ A.

$$\therefore \qquad \text{Value of } R = \frac{\text{Voltage across } R}{I} = \frac{25}{35} = \textbf{0.714} \ \Omega$$

Current through $C = I = 2\pi f C \times V_C$

$$\therefore \qquad f = \frac{I}{2\pi C V_C} = \frac{35}{2\pi \times 50 \times 10^{-6} \times 45} = \textbf{2475 Hz}$$

Fig. 12.41 shows the phasor diagram of the required part of the circuit. Here circuit current I is taken as the reference phasor. The voltage drops across the various circuit elements are indicated on the phasor diagram. For want of space, the phasor diagram is not drawn to scale. We can find the value of r and hence that of the applied voltage V if we know the value of θ in Fig. 12.41.

From $\Delta \ OAB$ in Fig. 12.41, we have,

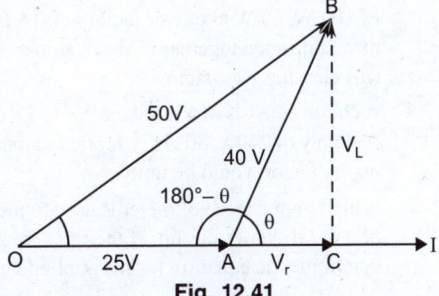

Fig. 12.41

$$OB^2 = OA^2 + AB^2 - 2OA \times AB \times \cos(180° - \theta) \quad \text{[cosine formula]}$$

or

$$50^2 = 25^2 + 40^2 + 2 \times 25 \times 40 \times \cos\theta$$

∴

$$\cos\theta = \frac{50^2 - 25^2 - 40^2}{2 \times 25 \times 40} = 0.1375 \therefore \sin\theta = 0.9905$$

∴ Value of $r = \dfrac{V_r}{I} = \dfrac{40\cos\theta}{I} = \dfrac{40 \times 0.1375}{35} = \mathbf{0.157\ \Omega}$

Also, $X_L = \dfrac{V_L}{I} = \dfrac{40\sin\theta}{I} = \dfrac{40 \times 0.9905}{35} = 1.132\ \Omega$

∴ Inductance, $L = \dfrac{X_L}{2\pi f} = \dfrac{1.132}{2\pi \times 2475} = 72.8 \times 10^{-6}\,\text{H} = \mathbf{72.8\ \mu H}$

(ii) Now, $V_r = 40\cos\theta = 40 \times 0.1375 = 5.5$ volts

$V_L = 40\sin\theta = 40 \times 0.9905 = 39.62$ volts

∴ Magnitude of applied voltage V is given by ;

$$V = \sqrt{(V_R + V_r)^2 + (V_L - V_C)^2}$$

$$= \sqrt{(25 + 5.5)^2 + (39.62 - 45)^2} = \mathbf{30.97\ volts}$$

(iii) $V_C = IX_C \quad \therefore X_C = \dfrac{V_C}{I} = \dfrac{45}{35} = 1.285\ \Omega$

Total circuit resistance, $R_T = R + r = 0.714 + 0.157 = 0.871\ \Omega$

Total circuit impedance, $Z_T = \sqrt{R_T^2 + (X_L - X_C)^2} = \sqrt{(0.871)^2 + (1.132 - 1.285)^2}$

$$= 0.884\,\Omega$$

P.F. of the whole circuit, $\cos\phi = \dfrac{R_T}{Z_T} = \dfrac{0.871}{0.884} = \mathbf{0.985\ leading}$

(iv) Active power consumed, $P = VI\cos\phi = 30.97 \times 35 \times 0.985 = \mathbf{1067W}$

Alternatively. $P = I^2(R + r) = 35^2(0.871) = \mathbf{1067\ W}$

Tutorial Problems

1. A coil of resistance 20 Ω is in series with an inductance of 0·04 H. A supply of 230 V, 50 Hz is applied to the combination. Determine the capacitance which when connected in series with the coil causes no change in the magnitude and power taken from the supply. **[67.5 μF]**

2. When a certain inductive coil is supplied at 240 V, 50 Hz, the current is 6.45 A. When the frequency is changed to 40 Hz at 240 V, the current taken is 7.48 A. Calculate the inductance and resistance of the coil. **[L = 0.1 H; R = 20 Ω]**

3. Two impedances Z_1 and Z_2 when connected separately across a 200 V, 50 Hz supply consume powers of 100 W, 60 W at power factors of 0.5 lagging and 0.6 leading respectively. If the two impedances are now connected together in series across the same supply, calculate *(i)* circuit current *(ii)* power absorbed *(iii)* circuit power factor. **[(i) 0.54 A (ii) 99 W (iii) 0.787 lead]**

4. A circuit consists of a resistance of 12 Ω, capacitance of 320 μF and an inductance of 0.08 H, all in series. A supply of 240 V, 50 Hz is applied to the ends of the circuit. Calculate the frequency at which the circuit power factor would be unity. **[32 Hz]**

5. A high impedance voltmeter is used to measure the voltage drop across each of the ideal circuit elements of a R-L-C series circuit. If the r.m.s. readings are 40 V, 25 V and 60 V for V_L, V_R and V_C respectively, determine the equation for the applied voltage. The frequency of the applied voltage is 50 Hz. **[v = 45·28 sin (314 t – 38.66°)]**

6. A coil is in series with a 20 μF capacitor across a 230 V, 50 Hz supply. The current taken by the circuit is 8 A and the power consumed is 200 W. Calculate the inductance of the coil if the p.f. of the circuit is (*i*) leading (*ii*) lagging. Sketch a vector diagram for each condition and calculate the coil p.f. in each case.

[(*i*) 0.416H; 0.9997 (*ii*) 0.597H; 0.9998]

7. A non-inductive resistor is connected in series with a coil and a capacitor. The circuit is connected to a single phase a.c. supply. If $V_R = 25$ V, $V_L = 40$ V and $V_C = 55$ V when current flowing through the circuit is 0.345 A, find the applied voltage and the power loss in the coil.   **[34.2V ; 1.9 W]**

8. A series *RLC* circuit consists of a 100 Ω resistor, an inductor of 0.318 H and capacitor of unknown value. When the circuit is energised by $230\sqrt{2}$ sin ω*t* volts supply, the current was found to be $i = 2.3\sqrt{2}$ sin ω*t* amperes. Find (*i*) the value of the capacitor (*ii*) the voltage across the inductor (*iii*) the total power consumed. Assume ω = 314.5 rad/sec.   **[(*i*) 31.8 μF (*ii*) 230 V (*iii*) 529 W]**

9. A coil of 0.8 p.f. is connected in series with 110 μF capacitor. Supply frequency is 50 Hz. The potential difference across the coil is found to be equal to that across the capacitor. Calculate the resistance and inductance of the coil. Calculate the net p.f.   **[23.15 Ω; 55.3 mH; 0.894 (leading)]**

10. A resistance of 20 Ω, inductance of 0.2 H and capacitance of 150 μF are connected in series and are fed by a 230 V, 50 Hz supply. Find X_L, X_C, Z, Y, p.f., active power and reactive power.

[62.8 Ω; 21.2 Ω; 46.16 Ω; 0.0217 S; 0.433 (lag); 497 W; 1030 VAR]

12.11. Resonance in A.C. Circuits

An a.c. circuit containing reactive elements (L and C) is said to be in **resonance** *when the circuit power factor is unity.*

Resonance means to be in step with. When applied voltage and circuit current in an a.c. circuit are in step with (*i.e.* phase angle is zero or p.f. is unity), the circuit is said to be in electrical resonance. If this condition exists in a series a.c. circuit, it is called **series resonance.** On the other hand, if this condition exists in a parallel a.c. circuit, it is called **parallel resonance.** The frequency at which resonance occurs is called *resonant frequency* (f_r).

12.12. Resonance in Series A.C. Circuit (Series Resonance)

A series R-L-C a.c. circuit is said to be in **resonance** *when circuit power factor is unity.*

Consider a series *R-L-C* circuit connected to a.c. supply of *V* volts (r.m.s.). The circuit impedance (*Z*) and circuit current (*I*) are given by ;

$$Z = \sqrt{R^2 + (X_L - X_C)^2}$$

$$I = \frac{V}{Z} = \frac{V}{\sqrt{R^2 + (X_L - X_C)^2}}$$

Resonance will occur in this circuit when circuit power factor is unity. This will happen when $X_L = X_C$. Regardless of the values of inductance (*L*) and capacitance (*C*), there is one frequency at which these two reactances are equal because X_L and X_C are frequency dependent. The frequency at which $X_L = X_C$ (*i.e.* circuit power factor becomes unity) is called resonant frequency f_r.

The **resonant frequency** *(f_r) for R – L – C series a.c. circuit is defined as the frequency at which $X_L = X_C$.*

At series resonance, $X_L = X_C$

or $2\pi f_r L = \dfrac{1}{2\pi f_r C}$

∴   Resonant frequency, $f_r = \dfrac{1}{2\pi\sqrt{LC}}$

If *L* and *C* are measured in henry and farad respectively, then f_r will be in Hz.

Effects of series resonance. The key points concerning series resonance are :

(*i*) $X_L = X_C$

(*ii*) $f_r = \dfrac{1}{2\pi\sqrt{LC}}$

(*iii*) $Z_r = \text{Minimum} = R$ $(\because X_L = X_C)$

(*iv*)   Circuit current, $I_r = \dfrac{V}{Z_r} = \dfrac{V}{R} = \text{Maximum}$

(*v*)   Circuit power factor $= 1$ $(\because$ Circuit is purely resistive)

(*vi*) Power dissipated in the circuit is maximum.

(*vii*) Since at series resonance the current flowing in the circuit is very large, the voltage drops across L and C are also very large. In fact, these drops are much greater than the applied voltage. However, voltage drop across L–C combination as a whole will be zero because these drops are equal in magnitude but 180° out of phase with each other.

Series resonance should be avoided in power circuits because the possibility of excessive voltages across the inductive and capacitive elements of the circuit may cause considerable damage. Resonance in power circuits may blow protective fuses, trip circuit breakers or cause damage to other equipment. However, in radio, television and electronic circuits, principles of series resonance are used to increase the signal voltage and current at a desired frequency (*i.e.* at f_r).

Graphical Explanation. Fig. 12.42 shows the graphical explanation of series resonance. We know that $X_L = 2\pi\, fL$ so that $X_L \propto f$. Therefore, graph between X_L and f is a straight line passing through the origin. Also $X_C = 1/2\pi fC$ so that $X_C \propto 1/f$. Therefore, graph between X_C and f is a hyperbola.

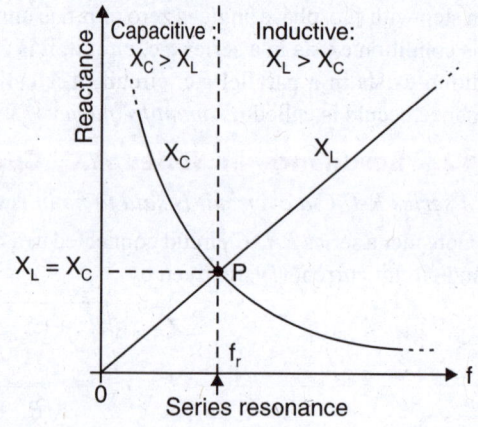

Starting at a very low frequency, X_C is high and X_L is low and the circuit is predominantly capacitive. As the frequency is increased, X_C decreases and X_L increases until a value is reached (point P) where $X_L = X_C$. At this frequency (f_r), the two reactances cancel, making the circuit purely resistive. This condition is **series resonance.** As frequency is increased further (*i.e.* beyond f_r), X_L becomes greater than X_C and the circuit is predominantly inductive. Note that at series resonance, the circuit impedance is minimum and is equal to circuit resistance R.

Fig. 12.42

12.13. Resonance Curve

The curve between current and frequency is known as resonance curve. Fig. 12.43 shows the resonance curve of a typical *R-L-C* series circuit. Note that current reaches its maximum value at the resonant frequency (f_r), falling off rapidly on either side at that point. It is because if the frequency is below f_r, $X_C > X_L$ and the net reactance is no longer zero. If the frequency is above f_r, then $X_L > X_C$ and the net reactance is again not zero. In both cases, the circuit impedance will be more than the impedance Z_r (= R) at resonance. The result is that the magnitude of circuit current decreases rapidly as the

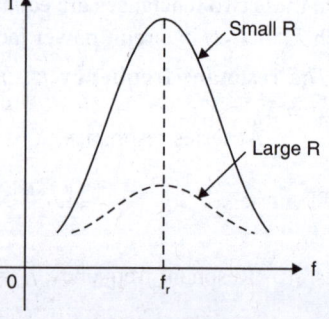

Fig. 12.43

frequency changes from the resonant frequency. Note also the effect of resistance in the circuit. The smaller the resistance, the greater the current at resonance and sharper the curve. On the other hand, the greater the resistance, the lower the resonant peak and flatter the curve (See Fig. 12.43).

12.14. Q-Factor of Series Resonant Circuit

At series resonance, the voltage across L or C (the two drops being equal and opposite) builds up to a value many times greater than the applied voltage $V(= *I_r R)$.

At series resonance, $V_L = I_r X_L$ and $V_C = I_r X_C$

$$\therefore \qquad \frac{V_L}{V} = \frac{I_r X_L}{I_r R} = \frac{X_L}{R}$$

and

$$\frac{V_C}{V} = \frac{I_r X_C}{I_r R} = \frac{X_C}{R}$$

The ratio V_L/V or V_C/V at resonance is a measure of the *quality* of a series resonant circuit. This is called the **Q-factor** (Q stands for quality) of the circuit. It is also known as the **voltage magnification factor.**

$$\therefore \qquad \text{Q-factor} = \frac{X_L}{R} = \frac{\omega_r L}{R} \qquad \qquad ...(i)$$

Also,

$$\text{Q-factor} = \frac{X_C}{R} = \frac{1}{\omega_r CR} \qquad \qquad ...(ii)$$

Since the coil resistance is often the only resistance in a series resonant circuit, the Q is sometimes referred to as the *Q-factor of the coil*. Therefore, we use the expression X_L/R for Q in series resonant circuit.

The Q-factor of a series resonant circuit can also be expressed in terms of L and C.

We know that :

$$f_r = \frac{1}{2\pi\sqrt{LC}} \quad \text{or} \quad 2\pi f_r = \frac{1}{\sqrt{LC}}$$

or

$$\omega_r = \frac{1}{\sqrt{LC}}$$

Substituting the value of ω_r in equation (i), we get,

$$\text{Q-factor} = \frac{1}{R}\sqrt{\frac{L}{C}}$$

It is clear that the Q-factor of a series resonant circuit may be increased either by reducing R or by increasing the L/C ratio.

Notes. (i)

$$Q = \frac{X_L}{R} = \frac{2\pi f L}{R}$$

It is clear that value of Q varies with frequency. *In this chapter, Q means Q at resonant frequency f_r.* At any other frequency f (called off resonance frequency), it will be represented by Q'.

(*ii*) Since Q is a ratio of like units, it has no units *i.e.*, it is a number.

The series resonant circuit is also called an **acceptor circuit** because such a circuit accepts currents at one particular frequency (*i.e.*, f_r) but virtually rejects currents of other frequencies. Such circuits are used in radio and television receivers.

12.15. Bandwidth of a Series Resonant Circuit

Consider the current versus frequency graph of a R-L-C series circuit shown in Fig. 12.44. It is clear from the graph that the current reaches maximum value $(= I_r)$ at resonance. It is also clear

---

* At series resonance, applied voltage $V = I_r R$ = voltage across R.

that at frequencies close to resonance, the current level is only a little below its maximum value. Thus the resonant circuit is said to select a *band* (*i.e.*, range) of frequencies rather than just one frequency f_r. We *arbitrarily* select frequency f_1 below f_r and frequency f_2 above f_r such that at f_1 and f_2, the circuit current = 0.707 I_r where I_r is the circuit current at resonance as shown in Fig. 12.44. Then,

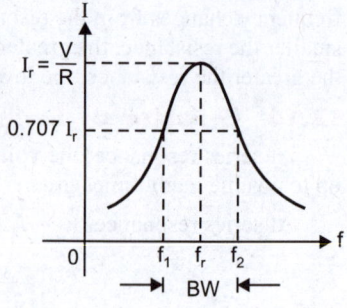

Fig. 12.44

Bandwidth of the series resonant circuit is

$$\text{Bandwidth, } BW = \Delta f = f_2 - f_1$$

*Hence **bandwidth** of a series resonant circuit is the range of frequencies for which the circuit current is equal to or ****greater** than 70.7% of the circuit current at resonance (i.e., I_r).*

Note that f_1 and f_2 are the limiting frequencies at which current is exactly equal to 70.7% of the maximum value. The frequency f_1 (*i.e.*, on the lower side) is called the **lower cut off frequency** and the frequency f_2 (*i.e.*, on the higher side) is called the **upper cut off frequency**. The frequencies f_1 and f_2 are also called **half-power frequencies** (or **half-power points**) or **–3dB frequencies**.

(*i*)　The frequencies f_1 and f_2 are called half-power frequencies as explained hereafter. At series resonance, the circuit current is maximum ($= I_r$) and circuit impedance is R. Also power delivered at resonance is maximum (P_{max}) and is given by ;

$$P_{max} = I_r^2 R$$

At f_1 or f_2, circuit current = 0.707 I_r so that power delivered at f_1 or f_2 is

$$P_{f_1 \text{ or } f_2} = (0.707 \, I_r)^2 R = \frac{I_r^2 R}{2} = \frac{P_{max}}{2}$$

Hence frequencies f_1 and f_2 may also be defined as those frequencies at which the power delivered to the circuit is half the power delivered at resonance.

(*ii*)　The frequencies f_1 and f_2 are also called – 3dB frequencies as explained hereafter.

Power delivered at resonance $= P_{max}$

Power delivered at f_1 or f_2 $= \dfrac{P_{max}}{2}$

∴ Change in power level from resonance to f_1 or f_2

$$= 10 \log_{10} \frac{P_{max}}{P_{max}/2} = 10 \log_{10} 2$$

$$= 3 \text{ dB below resonance} = -3 \text{ dB}$$

Hence frequencies f_1 and f_2 can also be defined as those frequencies at which the power is 3 dB below the power at resonance.

Other names for f_1 and f_2 are *critical frequencies* and *band frequencies*.

Important points about f_1 and f_2. At f_1 and f_2,

(*a*)　Circuit current is $\dfrac{I_r}{\sqrt{2}}$ where I_r = current at resonance.

(*b*)　Circuit impedance is $\sqrt{2}R$ or $\sqrt{2} Z_r$.

(*c*)　$P_1 = P_2 = \dfrac{P_{max}}{2}$

(*d*)　Circuit phase angle is $\phi = \pm 45°$.

12.16. Expressions for Half-power Frequencies

In *R-L-C* series circuit, the current is maximum at resonance and its value is $I_m (= I_r) = V/R$. The magnitude of circuit current at any frequency is given by ;

$$I = \frac{V}{\sqrt{R^2 + (X_L - X_C)^2}}$$

This equation can be rewritten as :

$$I = \frac{V}{R\sqrt{1 + \left(\dfrac{X_L - X_C}{R}\right)^2}}$$

or $\qquad\qquad I = \dfrac{I_m}{\sqrt{1 + \left(\dfrac{X_L - X_C}{R}\right)^2}}$ $\qquad\qquad\qquad(\because V/R = I_m)$

At half-power points, $\quad I = I_m / \sqrt{2}$

$\therefore \qquad\qquad \dfrac{I_m}{\sqrt{2}} = \dfrac{I_m}{\sqrt{1 + \left(\dfrac{X_L - X_C}{R}\right)^2}}$ $\qquad\qquad\qquad ...(i)$

$\therefore$ At half-power points, $\dfrac{*X_L - X_C}{R} = 1$

or $\qquad\qquad X_L - X_C = R$ $\qquad\qquad\qquad\qquad\qquad ...(ii)$

But at f_r, $X_L - X_C = 0$. Therefore, when the frequency increases from f_r to f_2, X_L must increase by $R/2$ and X_C must decrease by $R/2$ to satisfy equation (*ii*).

$\therefore \qquad\qquad 2\pi f_2 L - 2\pi f_r L = \dfrac{R}{2}$

or $\qquad\qquad f_2 - f_r = \dfrac{R}{4\pi L}$

$\therefore \qquad\qquad f_2 = f_r + \dfrac{R}{4\pi L}$ $\qquad\qquad\qquad\qquad ...(iii)$

Similarly, when the frequency decreases from f_r to f_1, X_C must increase by $R/2$ and X_L must decrease by $R/2$.

$\therefore \qquad\qquad 2\pi f_r L - 2\pi f_1 L = \dfrac{R}{2}$

or $\qquad\qquad f_r - f_1 = \dfrac{R}{4\pi L}$

$\therefore \qquad\qquad f_1 = f_r - \dfrac{R}{4\pi L}$ $\qquad\qquad\qquad\qquad ...(iv)$

Eqs. (*iii*) and (*iv*) give expressions for half-power frequencies f_2 and f_1 respectively. It may be noted that *R* is the total resistance of the series resonant circuit, including the resistance of a generator if one is present.

---

*   If we put $\dfrac{X_L - X_C}{R} = 1$ in eq. (*i*), the R.H.S. becomes $I_m / \sqrt{2}$ and equals L.H.S.

12.17. To Prove : $f_r = \sqrt{f_1 f_2}$

At half-power frequencies, the circuit current becomes $1/\sqrt{2}$ times the circuit current at resonance. Now circuit impedance at resonance $= R$ so that circuit impedance at half-power frequencies is $\sqrt{2}R$ ($\because Z = V/I$).

$$\therefore \qquad \sqrt{2}R = \sqrt{R^2 + X^2} \quad \text{or} \quad R = X$$

$\therefore$ Reactance X_1 at lower half-power frequency f_1 is

$$X_1 = \omega_1 L - \frac{1}{\omega_1 C} = -R \qquad \qquad \text{(Here } \omega_1 = 2\pi f_1)$$

The negative sign on the right hand side of the equation is given because below f_r ($f_1 < f_r$), the capacitive reactance exceeds the inductive reactance. The above equation can be written as :

$$\omega_1 L + R - \frac{1}{\omega_1 C} = 0$$

or

$$\omega_1^2 + \frac{R}{L}\omega_1 - \frac{1}{LC} = 0$$

or

$$\omega_1 = -\frac{R}{2L} \pm \sqrt{\left(\frac{R}{2L}\right)^2 + \frac{1}{LC}} = -\alpha \pm \sqrt{\alpha^2 + \omega_r^2}$$

where $\dfrac{R}{2L} = \alpha$. Also we know that $\omega_r = \dfrac{1}{\sqrt{LC}}$.

The negative angular frequency is meaningless and is discarded.

$$\therefore \qquad \omega_1 = -\alpha + \sqrt{\alpha^2 + \omega_r^2} \qquad \qquad ...(i)$$

Similarly, reactance X_2 at upper half-power frequency f_2 is

$$X_2 = \omega_2 L - \frac{1}{\omega_2 C} = +R$$

or

$$\omega_2 L - R - \frac{1}{\omega_2 C} = 0$$

or

$$\omega_2^2 - \frac{R}{L}\omega_2 - \frac{1}{LC} = 0$$

Considering the positive value of angular frequency, we have,

$$\omega_2 = \alpha + \sqrt{\alpha^2 + \omega_r^2} \qquad \qquad ...(ii)$$

Multiplying eqs. (i) and (ii), we have,

$$\omega_1 \omega_2 = \left(-\alpha + \sqrt{\alpha^2 + \omega_r^2}\right)\left(\alpha + \sqrt{\alpha^2 + \omega_r^2}\right) = \alpha^2 + \omega_r^2 - \alpha^2 = \omega_r^2$$

$$\therefore \qquad \omega_1 \omega_2 = \omega_r^2$$

or

$$(2\pi f_1)(2\pi f_2) = (2\pi f_r)^2$$

$$\therefore \qquad f_r = \sqrt{f_1 f_2}$$

Hence the resonant frequency f_r of a series resonant circuit is equal to the geometric mean of half-power frequencies (f_1, f_2).

12.18. Expressions for Bandwidth

We have seen above that (See Art. 12.16) :

$$f_2 = f_r + \frac{R}{4\pi L} \quad \text{and} \quad f_1 = f_r - \frac{R}{4\pi L}$$

$\therefore$ $$f_2 - f_1 = \frac{R}{2\pi L}$$ ...(i)

But $f_2 - f_1$ is the bandwidth (*BW*) of the series resonant circuit.

$\therefore$ Bandwidth, **BW** $= \dfrac{R}{2\pi L}$

It is clear that smaller the ratio R/L, the narrower will be the bandwidth and vice-versa.

We can also express the bandwidth of a series resonant circuit in terms of the *Q*-factor of the series resonant circuit. Thus multiplying and dividing the R.H.S. of eq. (*i*) by f_r, we have,

$$f_2 - f_1 = \frac{Rf_r}{2\pi f_r L}$$

But $Q = \dfrac{2\pi f_r L}{R}$ so that $\dfrac{R}{2\pi f_r L} = \dfrac{1}{Q}$

$\therefore$ $$f_2 - f_1 = \frac{f_r}{Q}$$

or Bandwidth, **BW** $= f_2 - f_1 = \dfrac{f_r}{Q}$

Clearly, the greater the value of Q, the narrower will be the bandwidth. The circuit effectively rejects currents whose frequencies are outside $f_2 - f_1$. The ability of a resonant circuit to select one particular frequency and to discriminate against other frequencies is termed as **selectivity** of the circuit. The circuits with the narrowest bandwidths obviously have greatest selectivity. Note that series resonance is also called **voltage resonance** because at series resonance, very large voltages are developed across *L* and *C* — *Q* times the applied voltage.

Notes. (*i*) We can also express f_2 and f_1 in terms of f_r and *BW*. Now $BW = R/2\pi L$ so that $BW/2 = R/4\pi L$. Therefore, we have,

$$f_2 = f_r + \frac{R}{4\pi L} = f_r + \frac{BW}{2}$$

$$f_1 = f_r - \frac{R}{4\pi L} = f_r - \frac{BW}{2}$$

(*ii*) Now, $$f_2 = f_r + \frac{BW}{2} = f_r + \frac{f_r}{2Q} \qquad \left(\because BW = \frac{f_r}{Q} \right)$$

Multiplying the two sides by 2π, we get,

$$\omega_2 = \omega_r + \frac{\omega_r}{2Q} = \omega_r \left(1 + \frac{1}{2Q} \right)$$

Similarly, $$\omega_1 = \omega_r \left(1 - \frac{1}{2Q} \right)$$

12.19. Important Relations in R-L-C Series Circuit

We now derive important relations between circuit values at resonant frequency f_r and at any other frequency *f* (called *off-resonance frequency*) in *R-L-C* series circuit.

(*i*) **Circuit current at any frequency *f*.** In *R-L-C* series circuit, the magnitude of circuit current *I* at any frequency *f* (*i.e.*, other than f_r) is given by ;

$$I = \frac{V}{\sqrt{R^2 + (X_L - X_C)^2}}$$ ...(i)

At series resonance, the inductive reactance X_{Lr} ($= 2\pi f_r L$) is equal to the capacitive reactance X_{Cr} ($= 1/2\pi f_r C$) i.e., $X_{Lr} = X_{Cr}$. Let $X_{Lr} = X_{Cr} = X_r$ (say).

Let the frequency f be related to the resonant frequency f_r of the circuit as :

$$f = k f_r \quad \text{or} \quad k = \frac{f}{f_r}$$

Now,

$$X_L = 2\pi f L = 2\pi k f_r L = k(2\pi f_r L) = k X_r$$

where

$$X_r = \text{Inductive reactance at series resonance}$$

Also,

$$X_C = \frac{1}{2\pi f C} = \frac{1}{2\pi k f_r C} = \frac{1}{k(2\pi f_r C)} = \frac{X_r}{k}$$

where

$$X_r = \text{Capacitive reactance at series resonance.}$$

Putting the values of X_L and X_C in eq. (i), we have,

$$I = \frac{V}{\sqrt{R^2 + \left(kX_r - \dfrac{X_r}{k} \right)^2}} = \frac{V}{\sqrt{R^2 + X_r^2 \left(k - \dfrac{1}{k} \right)^2}}$$

$$= \frac{V/R}{\sqrt{\dfrac{R^2}{R^2} + \left(\dfrac{X_r}{R} \right)^2 \left(k - \dfrac{1}{k} \right)^2}}$$

Now, $\dfrac{X_r}{R} = Q$ and $\dfrac{V}{R} = I_r = $ Current at series resonance.

$$\therefore \qquad I = \frac{I_r}{\sqrt{1 + Q^2 \left(k - \dfrac{1}{k} \right)^2}}$$

Now,

$$k = \frac{f}{f_r} \quad \text{and} \quad \frac{1}{k} = \frac{f_r}{f}$$

$$\therefore \qquad I = \frac{I_r}{\sqrt{1 + Q^2 \left(\dfrac{f}{f_r} - \dfrac{f_r}{f} \right)^2}} \qquad \qquad ...(ii)$$

Eq. (ii) gives the expression for circuit current I at any frequency f other than f_r in terms of I_r ($=$ current at series resonance) and Q at series resonance. It is clear that circuit current is maximum at resonance ($= I_r$) and is less at any other frequency f. Thus if we put $f = f_r$ in eq. (ii), $I = I_r$.

(ii) Circuit power dissipation at any frequency f. In R-L-C series circuit, the power dissipated by the circuit is maximum at resonance and is given by ;

$$P_r = I_r^2 R = \left(\frac{*V}{R} \right)^2 \times R = \frac{V^2}{R}$$

At any other frequency f, the power dissipated (P) by the circuit is less than P_r and is given by ;

$$P = I^2 R = \left(\frac{V}{Z} \right)^2 \times R = \frac{V^2 R}{Z^2} = \frac{V^2 R}{R^2 + X^2} = \frac{V^2 R}{R^2 + X^2 R^2 / R^2}$$

---

*   At series resonance, $Z_r = R$ so that $I_r = \dfrac{V}{Z_r} = \dfrac{V}{R}$

Now $\dfrac{X}{R} = Q' = Q$-factor of the circuit at frequency f

$$\therefore \qquad P = \frac{V^2 R}{R^2 + R^2 Q'^2} = \frac{V^2 R}{R^2 (1 + Q'^2)} = \frac{V^2}{R(1 + Q'^2)}$$

or $\qquad\qquad P = \dfrac{P_r}{1 + Q'^2} \qquad \left(\because \dfrac{V^2}{R} = P_r \right) \qquad\qquad \ldots(iii)$

Eq. (*iii*) gives the expression for power dissipation (P) by the circuit at any frequency f other than f_r in terms of P_r (= power dissipation at resonance) and Q' at frequency f. Note that power dissipation is maximum at resonance and is reduced by a factor $(1 + Q'^2)$ at any other frequency f.

12.20. Applications of Series Resonant Circuits

We have seen that in a R–L–C series circuit, there is one frequency (called resonant frequency) at which the circuit offers minimum impedance. For all other frequencies, the circuit offers considerable impedance. Thus if alternating currents of various frequencies are supplied to such a circuit, only those close to the resonant frequency would pass through the circuit. This is the key point in series resonance. There are many applications of series resonance in electronics but only two are discussed below by way of illustration :

(*i*) **As a wave trap.** Fig. 12.45 shows a series resonant circuit connected to the antenna circuit of a radio receiver. This series resonant circuit will offer a very low impedance to any signal current whose frequency is equal to f_r (*i.e.* $1/2\pi\sqrt{LC}$). However, for all other frequencies, the circuit will offer a high impedance. Thus the circuit will provide a low impedance path to the ground for any signal current of frequency f_r. In this way, the circuit prevents the signal currents of undesired frequency from entering the radio receiver and thus acts as a wave trap.

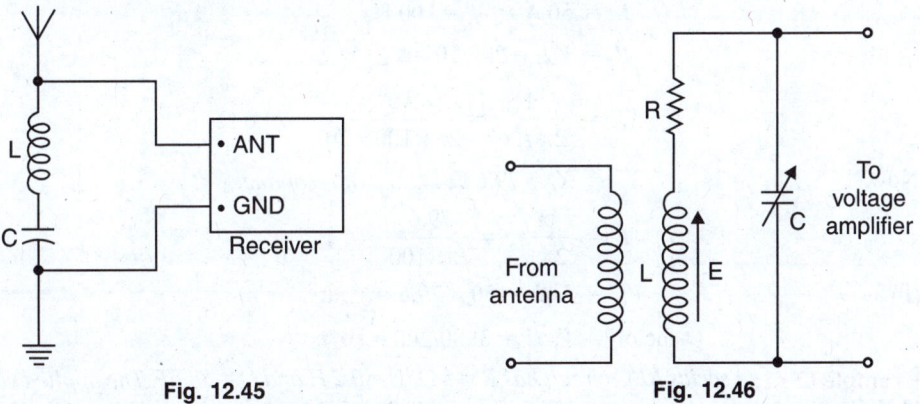

Fig. 12.45   **Fig. 12.46**

(*ii*) **As a tuning circuit.** A series resonant circuit is used for tuning purposes. Fig. 12.46 shows such an application in a radio receiver. The input signal comes from the antenna and induces a voltage E in L of the *series resonant circuit. The voltage across the capacitor becomes **$V_C = QE$

---

*   The circuit has a general appearance of a parallel circuit but actually it is a series circuit. It is because no separate voltage is applied to L, but instead a voltage E is induced in it which is considered as a voltage in series with L and C.

** $\dfrac{V_C}{V_R} = Q$ where V_R is the voltage across R.

But at series resonance, $V_R = E$, the applied voltage (induced voltage E in this case).

$$\therefore \qquad\qquad \frac{V_C}{E} = Q \quad \text{or} \quad V_C = QE.$$

where Q is the quality factor of the circuit. As the value of Q is generally large, the original signal received by the antenna increases many times in value and appears across C. The value of V_C is much more than that could have been obtained by direct transformer ratio. Thus the amplifier receives a greatly increased signal.

Example 12.49. *A coil of resistance 100 Ω and inductance 100 μH is connected in series with a 100 pF capacitor. The circuit is connected to a 10 V variable frequency source. Calculate (i) the resonant frequency (ii) current at resonance (iii) voltage across L and C at resonance and (iv) Q-factor of the circuit.*

Solution. (*i*) Resonant frequency, $f_r = \dfrac{1}{2\pi\sqrt{LC}} = \dfrac{1}{2\pi\sqrt{10^{-4}\times 10^{-10}}} = \mathbf{1.59 \times 10^6\ Hz}$

(*ii*)   Current at resonance, $I_r = V/R = 10/100 = \mathbf{0.1\ A}$

(*iii*)   At resonance, $X_L = 2\pi f_r L = 2\pi \times 1.59 \times 10^6 \times 10^{-4} = 1000\ \Omega$

          At resonance, $V_L = I_r X_L = 0.1 \times 1000 = \mathbf{100\ V}$

          At resonance, $V_C = I_r \times X_C = 0.1 \times 1000 = \mathbf{100\ V}$

(*iv*) $Q\text{-factor} = \dfrac{1}{R}\sqrt{\dfrac{L}{C}} = \dfrac{1}{100}\sqrt{\dfrac{10^{-4}}{10^{-10}}} = \mathbf{10}$

The reader may note that at resonance, voltage across L or C is Q times the applied voltage.

Example 12.50. *A choking coil is connected in series with a 20 μF capacitor. With a constant supply voltage of 200 V, it is found that the circuit takes its maximum current of 50 A when the supply frequency is 100 Hz. Calculate (i) resistance and inductance of the choking coil and (ii) voltage across the capacitor. What is the Q-factor of the coil ?*

Solution. Since current is maximum, the circuit is in resonance *i.e.*

$$I_r = 50\ A\ ;\ f_r = 100\ Hz$$

(*i*)   ∴ $R = V/I_r = 200/50 = \mathbf{4\ \Omega}$

$$X_C = \frac{1}{2\pi f_r C} = \frac{10^6}{2\pi \times 100 \times 20} = 79.6\ \Omega$$

Now, $X_C = X_L = 79.6\ \Omega$ .... *at resonance*

∴ $L = \dfrac{X_L}{2\pi f_r} = \dfrac{79.6}{2\pi \times 100} = \mathbf{0.127\ H}$

(*ii*) $V_C = I_r X_C = 50 \times 79.6 = \mathbf{3980\ V}$

              $Q\text{-factor} = V_C/V = 3980/200 = \mathbf{19.9}$

Example 12.51. *A series RLC circuit has R = 5 Ω, L = 0.2 H and C = 50 μF. The applied voltage is 200 V. Find (i) resonant frequency (ii) Q-factor (iii) bandwidth (iv) upper and lower half-power frequencies (v) current at resonance (vi) current at half-power points (vii) voltage across inductance at resonance.*

Solution. $R = 5\ \Omega\ ;\ L = 0.2\ H\ ;\ C = 50 \times 10^{-6}\ F\ ;\ V = 200$ volts

(*i*)   Resonant frequency, $f_r = \dfrac{1}{2\pi\sqrt{LC}} = \dfrac{1}{2\pi\sqrt{0.2 \times 50 \times 10^{-6}}} = \mathbf{50.33\ Hz}$

(*ii*)   Quality factor, $Q = \dfrac{\omega_r L}{R} = \dfrac{2\pi f_r L}{R} = \dfrac{2\pi \times 50.33 \times 0.2}{5} = \mathbf{12.65}$

(*iii*)   Bandwidth, $BW = \dfrac{f_r}{Q} = \dfrac{50.33}{12.65} = \mathbf{3.98\ Hz}$

(*iv*) Upper half-power frequency, $f_2 = f_r + \dfrac{BW}{2} = 50.33 + \dfrac{3.98}{2} = $ **52.32 Hz**

Lower half-power frequency, $f_1 = f_r - \dfrac{BW}{2} = 50.33 - \dfrac{3.98}{2} = $ **48.34 Hz**

(*v*) Current at resonance, $I_r = \dfrac{V}{R} = \dfrac{200}{5} = $ **40 A**

(*vi*) Current at half-power points $= 0.707\, I_r = 0.707 \times 40 = $ **28.28 A**

(*vii*) Voltage across L at resonance $= I_r X_L = I_r \times 2\pi f_r L$

$$= 40 \times 2\pi \times 50.33 \times 0.2 = \textbf{2529.87 V}$$

Example 12.52. *Determine the parameters of a R-L-C series circuit that will resonate at 10,000 Hz, has a bandwidth of 1000 Hz and draws 15.3 W from a 200 V generator operating at the resonant frequency of the circuit.*

Solution. At resonance, the voltage V_R across resistance is equal to the applied voltage *i.e.,* $V_R = 200$ V.

$$\therefore \qquad R = \frac{V_R^2}{P} = \frac{(200)^2}{15.3} = \textbf{2614}\ \Omega$$

$$Q\text{-factor of the circuit} = \frac{f_r}{BW} = \frac{10,000}{1000} = 10$$

But

$$Q = \frac{X_L}{R} = \frac{2\pi f_r L}{R}$$

$$\therefore \qquad L = \frac{QR}{2\pi f_r} = \frac{10 \times 2614}{2\pi \times 10,000} = \textbf{416} \times 10^{-3}\ \textbf{H}$$

At resonance, $2\pi f_r L = \dfrac{1}{2\pi f_r C}$

$$\therefore \qquad C = \frac{1}{4\pi^2 f_r^2 L} = \frac{1}{4\pi^2 \times (10,000)^2 \times 416 \times 10^{-3}} = \textbf{609} \times 10^{-12}\textbf{F}$$

Example 12.53. *An R-L-C series circuit has R = 2.5 Ω, C = 100 μF and a variable inductance. The applied voltage is 50 V at 800 rad/sec. The inductance is varied till the voltage across resistance is maximum. Under this condition, find (i) value of inductance (ii) Q-factor (iii) current (iv) voltages across resistance, capacitance and inductance.*

Solution. In a series resonant circuit, the voltage across resistance is maximum (*i.e.* equal to supply voltage) at resonance. Supply voltage, $V = 50$ volts.

(*i*)

$$\omega_r = \frac{1}{\sqrt{LC}} \text{ or } L = \frac{1}{\omega_r^2 C} = \frac{1}{(800)^2 \times (100 \times 10^{-6})} = \textbf{0.01563 H}$$

(*ii*)

$$Q = \frac{\omega_r L}{R} = \frac{800 \times 0.1563}{2.5} = \textbf{5}$$

(*iii*)

$$I_r = \frac{V}{R} = \frac{50}{2.5} = \textbf{20 A}$$

(*iv*)

$$\text{Voltage across } R = I_r R = 20 \times 2.5 = \textbf{50 volts}$$

$$\text{Voltage across } L = Q \times V = 5 \times 50 = \textbf{250 volts}$$

$$\text{Voltage across } C = Q \times V = 5 \times 50 = \textbf{250 volts}$$

Alternatively. $\quad \text{Voltage across } L = I_r X_L = \textbf{250 volts}$

$$\text{Voltage across } C = I_r X_C = \textbf{250 volts}$$

Example 12.54. *A resistor and a capacitor are in series with a variable inductor. When the circuit is connected to a 200 V, 50 Hz supply, the maximum current obtainable by varying the inductance is 0·314 A, the voltage across the capacitor is then 300 V. Find the circuit constants.*

Solution. At series resonance, $I_r = V/R$ and $V_L = V_C = 300$ V

Now
$$I_r = \frac{V_C}{X_C} = 2\pi f C V_C$$

$\therefore$
$$C = \frac{I_r}{2\pi f\, V_C} = \frac{0.314}{2\pi \times 50 \times 300} = 3.33 \times 10^{-6}\ \text{F} = \mathbf{3.33\ \mu F}$$

Also
$$I_r = \frac{V_L}{X_L} = \frac{V_L}{2\pi f L}$$

$\therefore$
$$L = \frac{V_L}{2\pi f_r I_r} = \frac{300}{2\pi \times 50 \times 0.314} = \mathbf{3.03\ H}$$

$$R = V/I_r = 200/0.314 = \mathbf{637\ \Omega}$$

Example 12.55. *A coil of resistance 40 Ω and inductance 0.75 H forms part of a series circuit for which the resonant frequency is 55 Hz. If the supply is 250 V, 50 Hz, find (i) the line current (ii) circuit power factor and (iii) voltage across the coil.*

Solution.

(i) At resonance (55 Hz), $X_L = 2\pi f_r L = 2\pi \times 55 \times 0.75 = 259\ \Omega$

At resonance, $X_L = X_C = 259\ \Omega$. Since $X_L \propto f$ and $X_C \propto 1/f$, these values at 50 Hz are :

$$X'_L = \frac{259 \times 50}{55} = 235.45\ \Omega\ ;\ X'_C = \frac{259 \times 55}{50} = 285\ \Omega$$

Net reactance at 50 Hz, $X = X'_C - X'_L = 285 - 235.45 = 49.55\ \Omega$

Circuit impedance at 50 Hz, $Z = \sqrt{R^2 + X^2} = \sqrt{(40)^2 + (49.55)^2} = 63.6\ \Omega$

$\therefore$ Line current, $I = V/Z = 250/63.6 = \mathbf{3.93\ A}$

(ii) Circuit power factor, $\cos \phi = R/Z = 40/63 \cdot 6 = \mathbf{0.63\ leading}$

(iii) Impedance of coil $= \sqrt{(40)^2 + (235.45)^2} = 239\ \Omega$

$\therefore$ Voltage across coil $= 3.93 \times 239 = \mathbf{939\ V}$

Example 12.56. *A series LCR circuit which resonates at f_r = 500 kHz has L = 100 μH, R = 25 Ω and C = 1000 pF. Determine (i) the Q-factor of the circuit (ii) the new value of C required to resonate at 500 kHz when the value of L is doubled and new Q-factor.*

Solution. *(i)*
$$Q_1 = \frac{1}{R}\sqrt{\frac{L}{C}} = \frac{1}{25}\sqrt{\frac{100 \times 10^{-6}}{1000 \times 10^{-12}}} = \mathbf{12.6}$$

(ii) The new value of L is 200 μH.

Now
$$f_r = \frac{1}{2\pi\sqrt{LC}} \quad \text{or} \quad C = \frac{1}{4\pi^2 f_r^2 L}$$

$\therefore$
$$C = \frac{1}{4\pi^2 \times (500 \times 10^3)^2 \times 200 \times 10^{-6}} = 500 \times 10^{-12}\ \text{F} = \mathbf{500\ pF}$$

$$Q_2 = \frac{1}{R}\sqrt{\frac{L}{C}} = \frac{1}{25}\sqrt{\frac{200 \times 10^{-6}}{500 \times 10^{-12}}} = \mathbf{25}$$

Example 12.57. *In the above example, calculate the half-power frequencies and bandwidth when L = 100 μH.*

Solution.

$$f_2 = f_r + \frac{R}{4\pi L} = 500 \times 10^3 + \frac{25}{4\pi \times 100 \times 10^{-6}} = 520 \times 10^3 \text{ Hz} = \textbf{520 kHz}$$

$$f_1 = f_r - \frac{R}{4\pi L} = 500 \times 10^3 - \frac{25}{4\pi \times 100 \times 10^{-6}} = 480 \times 10^3 \text{ Hz} = \textbf{480 kHz}$$

Bandwidth, $BW = f_2 - f_1 = 520 - 480 = \textbf{40 kHz}$

Example 12.58. *Find the expression for the frequency at which the circuit current is half as much as at series resonance.*

Solution. At series resonance, the circuit current is maximum (I_m) and is given by ;

$$I_m = \frac{V}{R} \qquad \qquad \text{...}R \text{ is circuit resistance}$$

At any frequency $f\,(= \omega/2\pi)$ above or below the resonant frequency $f_r\,(= \omega_r/2\pi)$, the circuit impedance is more than that at resonance and is given by ;

$$Z = \sqrt{R^2 + (X_L - X_C)^2}$$

$\therefore$ Circuit current, $I = \dfrac{V}{Z}$

$\therefore$ $\dfrac{I_m}{I} = \dfrac{V}{R} \times \dfrac{Z}{V} = \dfrac{Z}{R}$

But it is given that $I_m/I = 2$ so that $Z/R = 2$.

$\therefore$ $Z^2 = 4R^2$ or $R^2 + (X_L - X_C)^2 = 4R^2$

or $(X_L - X_C)^2 = 3R^2$ or $\left(\omega L - \dfrac{1}{\omega C}\right)^2 = 3R^2$

or $\omega L - \dfrac{1}{\omega C} = \sqrt{3}R$

But $\omega_r^2 = \dfrac{1}{LC}$ so that $C = \dfrac{1}{\omega_r^2 L}$.

$\therefore$ $\omega L - \dfrac{\omega_r^2 L}{\omega} = \sqrt{3}R$

or $(\omega^2 - \omega_r^2) = \dfrac{\sqrt{3}R}{L}\omega$ ...(i)

Hence the desired frequency $f\,(= \omega/2\pi)$ can be found out from exp. (*i*).

Example 12.59. *A series $R - L - C$ circuit consists of $R = 1000\,\Omega$, $L = 100\,mH$ and $C = 10\,\mu F$. The applied voltage across the circuit is 100 V.*

(i) Find the resonant frequency of the circuit.

(ii) Find the quality factor of the circuit at the resonant frequency.

(iii) At what angular frequencies do the half power points occur?

(iv) Calculate the bandwidth of the circuit.

Solution. (*i*) Resonant frequency, $f_r = \dfrac{1}{2\pi\sqrt{LC}} = \dfrac{1}{2\pi\sqrt{100 \times 10^{-3} \times 10 \times 10^{-12}}}$

$$= 159 \times 10^3 \text{ Hz} = \textbf{159 kHz}$$

(*ii*) $Q = \dfrac{1}{R}\sqrt{\dfrac{L}{C}} = \dfrac{1}{1000} \times \sqrt{\dfrac{100 \times 10^{-3}}{10 \times 10^{-12}}} = \textbf{100}$

(*iii*) $\omega_1 = \omega_r \left(1 - \dfrac{1}{2Q}\right) = 2\pi \times 159 \times 10^3 \left(1 - \dfrac{1}{2 \times 100}\right) = $ **994.03 × 10³ rad/sec.**

$$\omega_2 = \omega_r \left(1 + \dfrac{1}{2Q}\right) = 2\pi \times 159 \times 10^3 \left(1 + \dfrac{1}{2 \times 100}\right) = \textbf{1004.02 × 10}^\textbf{3}\textbf{ rad/sec.}$$

(*iv*) Bandwidth, $BW = \omega_2 - \omega_1 = (1004.02 - 994.03)\,10^3$ rad/sec = **9.99 × 10³ rad/sec.**

Example 12.60. *A resistor and a capacitor are connected in series across a 150 V a.c. supply. When the frequency is 40 Hz, the circuit draws 5A. When the frequency is increased to 50 Hz, it draws 6A. Find the values of resistance and capacitance. Also find the power drawn in the second case.*

Solution. We shall use suffix 1 for 40 Hz and suffix 2 for 50 Hz.

$$Z_1 = 150/5 = 30\,\Omega \quad \therefore\ R^2 + X_{C1}^2 = 900 \qquad\qquad \text{...(}i\text{)}$$
$$Z_2 = 150/6 = 25\,\Omega \quad \therefore\ R^2 + X_{C2}^2 = 625 \qquad\qquad \text{...(}ii\text{)}$$

From eqs. (*i*) and (*ii*), $X_{C1}^2 - X_{C2}^2 = 900 - 625 = 275$

Now $\dfrac{X_{C1}}{X_{C2}} = \dfrac{50}{40}$ or $X_{C1} = 1.25\,X_{C2}$ $(\because X_C \propto 1/f)$

∴ $(1.25\,X_{C2})^2 - X_{C2}^2 = 275$ or $X_{C2} = 22.11\,\Omega$

∴ $X_{C1} = 1.25\,X_{C2} = 1.25 \times 22.11 = 27.64\,\Omega$

or $\dfrac{1}{2\pi f_1 C} = 27.64$

∴ $C = \dfrac{1}{2\pi \times 40 \times 27.64} = 144 \times 10^{-6}\ \text{F} = $ **144 μF**

Alternatively. $C = \dfrac{1}{2\pi\,f_2 \times C_2} = \dfrac{1}{2\pi \times 50 \times 22.11} = 144 \times 10^{-6}\ \text{F} = $ **144 μF**

From eq. (*i*), $R^2 + X_{C1}^2 = 900$ or $R^2 + (27.64)^2 = 900$ ∴ $R = $ **11.662 Ω**

Power drawn in second case = $6^2 \times 11.662 = $ **420 W**

Example 12.61. *A coil of inductance 9 H and resistance 50 Ω in series with a capacitor is supplied at constant voltage from a variable frequency source. If the maximum current of 1 A occurs at 75 Hz, find the frequency when the current is 0.5 A.*

Solution. In a series circuit, the maximum current occurs at resonance.

∴ $75 = \dfrac{1}{2\pi\sqrt{9C}}$ or $C = 0.5 \times 10^{-6}\ \text{F}$

Supply voltage, V = Current at resonance × $R = 1 \times 50 = 50$ volts

When circuit current is 0.5A, circuit impedance Z is

$$Z = \dfrac{V}{0.5} = \dfrac{50}{0.5} = 100\,\Omega$$

Now $Z^2 = R^2 + \left(\omega L - \dfrac{1}{\omega C}\right)^2$

or $100^2 = 50^2 + \left(\omega L - \dfrac{1}{\omega C}\right)^2$

or $\omega L - \dfrac{1}{\omega C} = \pm 86.6\,\Omega$

If $\omega L - \dfrac{1}{\omega C} = 86.6$, then,

$$\omega^2 LC - 1 = 86.6 \,\omega C$$

or $\quad \omega^2 (9 \times 0.5 \times 10^{-6}) - 1 = 86.6 \,\omega(0.5 \times 10^{-6})$

On solving and accepting the positive value, we have,

$$\omega = 476.26 \text{ rad/s} \quad \therefore f = \frac{476.26}{2\pi} = \textbf{75.8 Hz}$$

If $\quad \omega L - \dfrac{1}{\omega C} = -86.6$, then,

$$\omega^2 LC - 1 = -86.6 \,\omega C$$

or $\quad \omega^2 (9 \times 0.5 \times 10^{-6}) - 1 = -86.6 \,\omega(0.5 \times 10^{-6})$

On solving and accepting the positive value, we have,

$$\omega = 466.63 \text{ rad/s} \quad \therefore f = \frac{466.63}{2\pi} = \textbf{74.26 Hz}$$

Example 12.62. *A circuit is made up of a 10 Ω resistance, a 1 μF capacitance and 1 H inductance all connected in series. A voltage of 100 V at varying frequencies is applied to the circuit. Find the frequency (or frequencies) at which the circuit would consume only 10% of the power it consumed at resonance.*

Solution. Here, $R = 10 \,\Omega$; $L = 1\text{H}$; $C = 10^{-6}\,\text{F}$; $V = 100$ volts

Circuit current at resonance, $I_r = \dfrac{V}{R} = \dfrac{100}{10} = 10 \text{ A}$

Power consumed at resonance, $P_r = I_r^2 R = 10^2 \times 10 = 1000 \text{ W}$

Let f be the supply frequency at which the circuit current is such that power consumed in the circuit is 10% of that consumed at resonance. Suppose the circuit current at this frequency is I.

$$\therefore \qquad I^2 R = \frac{P_r}{10} = \frac{1000}{10} = 100 \quad \therefore I = \sqrt{\frac{100}{R}} = \sqrt{\frac{100}{10}} = \sqrt{10}\text{A}$$

Circuit impedance when circuit current $= \sqrt{10}$ A is

$$Z = \frac{V}{I} = \frac{100}{\sqrt{10}} = 10\sqrt{10}\,\Omega$$

$\therefore \qquad$ Circuit reactance $= X_L - X_C = \sqrt{Z^2 - R^2} = \sqrt{(10\sqrt{10})^2 - 10^2} = \pm 30 \,\Omega$

or $\qquad \omega L - \dfrac{1}{\omega C} = \pm 30$

or $\qquad 2\pi f L - \dfrac{1}{2\pi f C} = \pm 30$

$\therefore \qquad 2\pi f \times 1 - \dfrac{1}{2\pi f \times 10^{-6}} = \pm 30$

On solving and accepting the positive values, $f = \textbf{161.56 Hz}$ or **156. 78 Hz**

Example 12.63. *Voltages across resistance, inductance and capacitance connected in series are 3V, 4V and 5V respectively. If supply voltage has 50 Hz frequency, what is the magnitude of supply voltage? Find the resonant frequency of this series RLC circuit.*

Solution. $V_R = 3$ volts ; $V_L = 4$ volts ; $V_C = 5$ volts

$\therefore$ Supply voltage, $V = \sqrt{V_R^2 + (V_L - V_C)^2} = \sqrt{3^2 + (4-5)^2} = \textbf{3.162 volts}$

Resonant frequency, $f_r = \dfrac{1}{2\pi\sqrt{LC}}$ $\qquad\qquad\qquad …(i)$

We can easily find the value of LC from the given data.

Now
$$V_L = 2\pi fLI = 4 \qquad \qquad ...(ii)$$

and
$$V_C = \frac{I}{2\pi fC} = 5 \qquad \qquad ...(iii)$$

Dividing eq. (ii) by eq. (iii), we have,

$$4\pi^2 f^2 LC = \frac{4}{5} = 0.8$$

$$\therefore \qquad LC = \frac{0.8}{4\pi^2 \times f^2} = \frac{0.8}{4\pi^2 \times 50^2} = 8.105 \times 10^{-6} \qquad (\because f = 50 \text{ Hz})$$

Putting the value of LC in eq. (i), we have,

$$f_r = \frac{1}{2\pi\sqrt{8.105 \times 10^{-6}}} = \textbf{55.9 Hz}$$

Tutorial Problems

1. What is the BW of a circuit in which the half-power points occur at 150 kHz and 180 kHz ?   **[30 kHz]**

2. Determine the inductance required to series-resonate with a 180 pF capacitor at 320 kHz. Also determine BW and Q if R = 20 Ω.   **[L = 1·373 mH; BW = 2·32 kHz ; Q = 138]**

3. A series LC circuit has an inductance of 200 µH, a capacitance of 300 pF and a Q of 60. Determine f_r, BW and upper and lower cut-off frequencies.

   [f_r = 649.75 kHz ; BW = 10·83 kHz ; 655.16 kHz ; 644.34 kHz]

4. A resistance of 15 Ω and an inductance of 4H and a capacitance of 25 µF are connected in series across 230 V a.c. supply. Calculate (i) the frequency at which the current shall be maximum (ii) current at this frequency and (iii) p.d. across inductance.   **[(i) 15.9 Hz (ii) 15.33 A (iii) 6123 V]**

5. A 1200 Ω resistor, a 0.7 H coil and a 0.001 µF capacitor are in series across a 120 V source. Determine (i) the resonant frequency (ii) the voltage across the capacitor at resonance and (iii) Q-factor of the circuit at resonance.   **[(i) 6015 Hz (ii) 2646 V (iii) 22]**

6. A capacitor C is in series with a 75 Ω resistor and a 12 H coil across a 220 V, 60 Hz supply. Determine the value of C that resonates the circuit.   **[0.587 µF]**

7. A series circuit consists of a 325 Ω resistor, a 0.6 pF capacitor and a coil of 2.93 H. Determine the bandwidth if the circuit resonates at 120×10^6 Hz.   **[17.643×10^6 Hz]**

8. A constant voltage at a frequency of 1 MHz is applied to an inductor in series with a variable capacitor. When the capacitor is set 500 pF, the current has its maximum value while it is reduced to one-half when the capacitance is 600 pF. Find (i) resistance (ii) inductance and (iii) Q-factor of the inductor.

   [(i) 30.62 Ω (ii) 50.66 µH (iii) 10.4]

9. It is desired to design a series resonant circuit with the following specifications $C = 250 \times 10^{-12}$ F; f_r = 600 kHz; BW = 20 kHz.

   Calculate Q, R and L of the circuit. Also calculate the current at 500 kHz as a fraction of current at resonance.   **[30 ; 35.37 Ω ; 0.28 mH; 0.09 I_r]**

10. A series RLC circuit with R = 250 Ω and L = 0.6 H results in a leading phase angle of 60° at a frequency of 40 Hz. At what frequency will the circuit resonate?   **[81.2 Hz]**

11. For a series RLC circuit, the inductor is variable. Source voltage is $200\sqrt{2} \sin 100\pi t$ volts. Maximum current obtainable by varying the inductance is 0.314 A and the voltage across the capacitor then is 300 V. Find the circuit elements values.   **[R = 637 Ω; L = 3.03 H ; C = 3.33 µF]**

12. A series circuit consists of a resistance of 10 Ω, an inductance of 8 mH and a capacitance of 500 pF. A sinusoidal e.m.f. of constant amplitude 5 V is introduced into the circuit and its frequency varied over a range including the resonant frequency. At what frequencies will current be (i) a maximum (ii) one-half the maximum?   **[(i) 79.6 kHz (ii) 79.87 kHz; 79.53 kHz]**

12.21. Decibels

It is sometimes found convenient to express the ratio of two powers P_2 and P_1 in logarithmic units called *bel*. Mathematically, the bel is defined as under :

$$\text{Power ratio in bels} = \log_{10} \frac{P_2}{P_1}$$

where

$$\log_{10} = \text{Common (base 10) log}$$
$$P_2 = \text{Output power}$$
$$P_1 = \text{Input power}$$

It is found that the bel is rather a large unit and the **decibel** (1 bel = 10 decibels) is more common in use.

∴ Power ratio in decibels $(dB) = 10 \log_{10} \dfrac{P_2}{P_1}$ ...(*i*)

If $P_2 > P_1$, then power ratio will be $+dB$ and if $P_2 < P_1$, then power ratio will be $-dB$. If $P_2 = P_1$, then ratio will be 0 dB because $\log_{10}(1) = 0$.

If P_1 and P_2 are the powers dissipated in equal resistances (say R), then,

$$P_1 = \frac{V_1^2}{R} \;\; ; \;\; P_2 = \frac{V_2^2}{R}$$

Also,

$$P_1 = I_1^2 R \;\; ; \;\; P_2 = I_2^2 R$$

∴ Power ratio in $dB = 10 \log_{10} \dfrac{V_2^2/R}{V_1^2/R} = 20 \log_{10} \dfrac{V_2}{V_1}$ ...(*ii*)

Eq. (*ii*) gives the power ratio in terms of voltages.

Also, Power ratio in $dB = 10 \log_{10} \dfrac{I_2^2 R}{I_1^2 R} = 20 \log_{10} \dfrac{I_2}{I_1}$ ...(*iii*)

Eq. (*iii*) gives the power ratio in terms of currents.

Example 12.64. *Two output voltage levels from an amplifier are measured as $V_1 = 1V$ and $V_2 = 2V$. Calculate the output power change. Also determine the new level for V_2 which will give a power output change of $-3dB$ when $V_1 = 1V$.*

Solution. Output power change

$$= 20 \log_{10} \frac{V_2}{V_1} = 20 \log_{10} \left(\frac{2V}{1V} \right) = \textbf{6 dB}$$

When power changes by $-3\ dB$, we have,

$$-3dB = 20 \log_{10} \frac{V_2}{V_1}$$

$$\text{antilog} \left(-\frac{3dB}{20} \right) = \frac{V_2}{V_1}$$

∴

$$V_2 = V_1 \times \text{antilog} \left(-\frac{3dB}{20} \right)$$

$$= 1 \times \text{antilog} \left(-\frac{3dB}{20} \right) = \textbf{0.707 V}$$

Example 12.65. *An amplifier receives 0.1 W of input signal and delivers 15 W of signal power. What is its power gain as a number and power gain in dB?*

Solution. Power gain $= \dfrac{15\ W}{0.1\ W} = \textbf{150}$

Power gain in $dB = 10 \log_{10} \dfrac{15}{0.1} = 10 \times 2.18 = \textbf{21.8 dB}$

Example 12.66. *The output of a microphone is rated at –52 dB. The reference level is 1V under specified sound conditions. What is the output voltage of this microphone under the same sound conditions?*

Solution.
$$-52 \text{ dB} = 20 \log_{10} \frac{V_2}{V_1}$$

Dividing by 20 ;
$$-2.6 \text{ dB} = \log_{10} \frac{V_2}{V_1}$$

∴
$$\frac{V_2}{V_1} = \text{antilog} \, (-2.6) = 0.0025$$

or
$$V_2 = V_1 \times 0.0025 = 1V \times 0.0025 = 0.0025 \text{ V} = \textbf{2.5 mV}$$

Tutorial Problems

1. The input power to a circuit is 5W and the output is 4W. Determine the decibel power loss. **[–0.97 dB]**

2. A circuit provides –5dB of attenuation, the input voltage is 10V. What is the output voltage if $R_{in} = R_{out}$? **[5.6 V]**

3. What is the decibel power gain of an amplifier that requires 0.2 W input to provide 30 W of output? **[21.76 dB]**

Objective Questions

1. In an a.c. circuit shown in Fig. 12.47, the r.m.s. current in the circuit is 5 A. The impedance of the circuit is

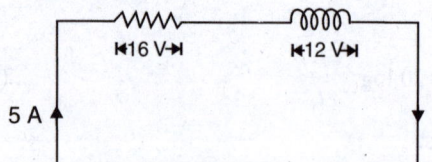

Fig. 12.47

(i) 10 Ω   (ii) 5 Ω

(iii) 25 Ω   (iv) 4 Ω

2. In the circuit shown in Fig. 12.48, the p.d.s. across the various elements are shown. What is the magnitude of the applied voltage ?

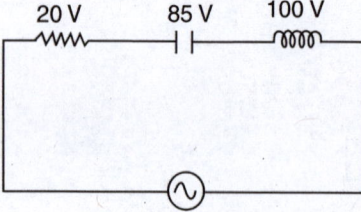

Fig. 12.48

(i) 205 V   (ii) 35 V

(iii) 25 V   (iv) none of above

3. In the circuit shown in Fig. 12.49, the p.d.s across the various elements are shown. What is the power factor of the circuit ?

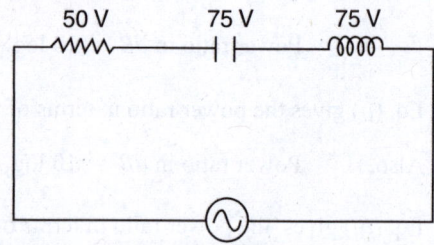

Fig. 12.49

(i) 0   (ii) 1

(iii) 0.5   (iv) cannot be predicted

4. In Fig. 12.50, what will be the reading of the ammeter ?

(i) 2 A   (ii) 1.2 A

(iii) 2·5 A   (iv) 4 A

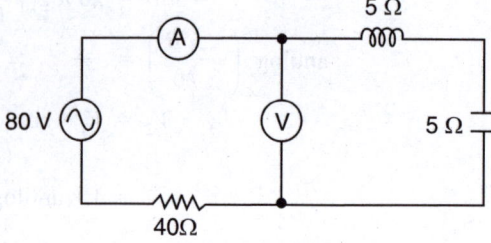

Fig. 12.50

5. In Fig. 12.50, what will be the reading of the voltmeter ?

(i) 80 V   (ii) 20 V

(iii) 0 V   (iv) none of above

6. A lamp consumes only 25% of peak power in an a.c. circuit. What is the phase difference between the applied voltage and the circuit current ?

    (i) $\pi/6$ (ii) $\pi/3$

    (iii) $\pi/4$ (iv) $\pi/2$

7. In an L-C-R series circuit, the capacitance is changed from C to $4\,C$. For the same resonant frequency, the inductance should be changed from L to

    (i) $2\,L$ (ii) $L/2$

    (iii) $4\,L$ (iv) $L/4$

8. In a circuit element, the p.d. is higher than the applied voltage of the source. That will be

    (i) an a.c. circuit   (ii) a d.c. circuit

    (iii) an a.c. or a d.c. circuit

    (iv) neither a.c. nor a d.c. circuit

9. Fig. 12.51 shows an R-L-C series circuit connected across an a.c. source E_v. If the frequency of a.c. source is more than the resonant frequency, the circuit current (I_v)

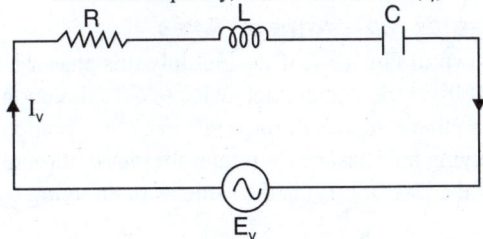

Fig. 12.51

    (i) is in phase with the applied voltage (E_v)

    (ii) leads the applied voltage (E_v)

    (iii) lags the applied voltage (E_v)

    (iv) none of the above

10. In the above question, if the frequency of source is less than the resonant frequency, then circuit current (I_v)

    (i) is in phase with the applied voltage

    (ii) leads the applied voltage

    (iii) lags behind the applied voltage

    (iv) none of the above

11. In a series L-C-R circuit, the voltages across R, L and C are shown in Fig. 12.52. The voltage of the applied source is

    (i) 110 V   (ii) 10 V

    (iii) 50 V   (iv) 70 V

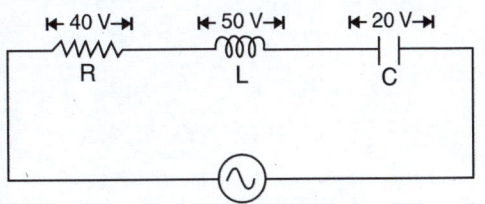

Fig. 12.52

12. In the L-C-R circuit shown in Fig. 12.53, the voltmeter and ammeter readings are :

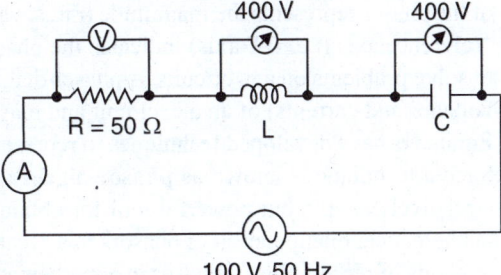

100 V, 50 Hz

Fig. 12.53

    (i) 100 V, 2 A   (ii) 100 V, 5 A

    (iii) 100 V, 3 A   (iv) 100 V, 1 A

13. A 20 V a.c. is applied to a circuit consisting of a resistance and a coil with a negligible resistance. If the voltage across the resistance is 12 V, the voltage across the coil is

    (i) 10 V   (ii) 8 V

    (iii) 6 V   (iv) 16 V

14. In a series LCR circuit, the voltages read at resonance across R, L and C are 40 V, 60 V and 60 V respectively. Then the applied voltage is

    (i) 60 V   (ii) 40 V

    (iii) 160 V   (iv) $\sqrt{40^2 + 120^2}$ V

15. An ideal choke takes a current of 8 A when connected to an a.c. source of 100 V, 50 Hz. A pure resistor under the same conditions takes a current of 10 A. If the two are connected in series to an a.c. supply of 100 V, 40 Hz, then the current in the series combination of above resistor and inductor is

    (i) 10 A   (ii) 8 A

    (iii) $5\sqrt{2}$A   (iv) $10\sqrt{2}$A

Answers

1. (iv)	**2.** (iii)	**3.** (ii)	**4.** (i)	**5.** (iii)
6. (ii)	**7.** (iv)	**8.** (i)	**9.** (iii)	**10.** (ii)
11. (iii)	**12.** (i)	**13.** (iv)	**14.** (ii)	**15.** (iii)

13

Phasor Algebra

Introduction

We have seen that an alternating voltage or current can be represented by a phasor. The length of the phasor represents the magnitude (r.m.s. value) of the alternating quantity and angle θ from the *reference axis (*i.e.* OX-axis) indicates the phase of that quantity. So far we used phasor diagrams to solve problems on a.c. circuits. A phasor diagram is a graphical representation of the phasors (*i.e.* voltages and currents) of an a.c. circuit and may not yield quick results in case of complex circuits. Engineers have developed techniques to represent a phasor in an algebraic (*i.e.* mathematical) form. Such a technique is known as **phasor algebra** or **complex algebra.** Phasor algebra has provided a relatively simple but powerful tool for obtaining quick solution of a.c. circuits. It simplifies the mathematical manipulation of phasors to a great extent. In this chapter, we shall discuss the various methods of representing a phasor in a mathematical form and its application to a.c. circuits.

13.1. Notation of Phasors on Rectangular Co-ordinate Axes

Consider a phasor **V** lying along OX-axis as shown in Fig. 13.1. If we multiply this phasor by –1, the phasor is reversed *i.e.* it is rotated through 180° in the counterclockwise (*CCW*) direction. Let us see with what factor the phasor be multiplied so that it rotates through 90° in *CCW* direction. Suppose this factor is j. As shown in Fig. 13.1, multiplying the phasor by j^2 rotates the phasor through 180° in *CCW* direction. This means that multiplying the phasor by j^2 is the same as multiplying by –1. It follows, therefore, that

$$j^2 = -1 \quad \text{or} \quad **j = \sqrt{-1}$$

We arrive at a very important conclusion that *when a phasor is multiplied by j (= $\sqrt{-1}$), the phasor is rotated through 90° in CCW direction.* Each successive multiplication by j rotates the phasor through an additional 90° in *CCW* direction. It is easy to see that multiplying a phasor by

$$j = \sqrt{-1} \qquad \qquad \text{...90° } CCW \text{ rotation from } OX\text{-axis}$$

$$j^2 = -1 \qquad \qquad \text{...180° } CCW \text{ rotation from } OX\text{-axis}$$

$$j^3 = j^2.j = -j \qquad \qquad \text{...270° } CCW \text{ rotation from } OX\text{-axis}$$

$$j^4 = j^2.j^2 = 1 \qquad \qquad \text{...360° } CCW \text{ rotation from } OX\text{-axis}$$

Fig. 13.1 shows the effect of multiplying a phasor by j. The following points may be noted carefully :

(i) **j is an operator** *which rotates the phasor through 90° in CCW direction without changing the magnitude of the phasor.*

(ii) The operator j occurs with the phasor only when it is along Y-axis (See Fig. 13.1). Thus when the phasor is along OY-axis, it is jV and when along OY'-axis, it is $-jV$. The phasor lying along X-axis is not associated with j.

---

*   We have already seen that in the phasor diagram, OX-axis is taken as the reference axis and phase angles are measured from this reference axis.

**   In mathematics, $\sqrt{-1}$ is denoted by i but electrical engineers represent it by j to avoid confusion with the symbol i for current.

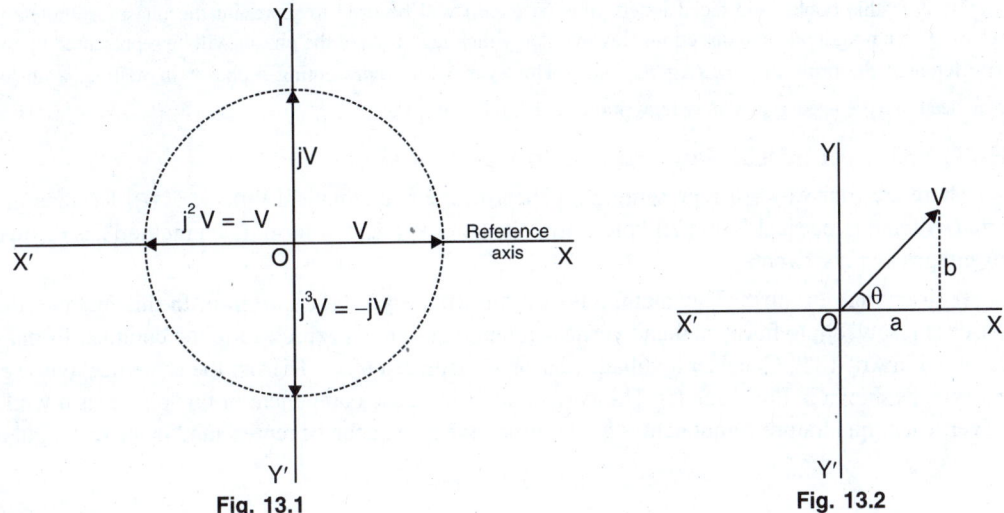

Fig. 13.1   Fig. 13.2

(*iii*) Since $j = \sqrt{-1}$ and its value cannot be determined, it is called an imaginary number. For this reason, any phasor (or its component) associated with j is called the imaginary part. *A phasor (or its component) along X-axis is not associated with j and is called the real part.*

13.2. Significance of Operator j

Just as the symbol + indicates the operation of adding two numbers, similarly j indicates an operation of rotating the phasor through 90° in *CCW* direction. The operator j does not change the magnitude of the phasor. For example, jV means that V is 90° *CCW* from *OX*-axis (*i.e.* it is lying along *OY* axis) while $j^2V (= -V)$ means that V is turned through 180° *CCW* from *OX*-axis *i.e.* it is lying along *OX'*-axis. As seen in Fig. 13.1, j will be associated with a phasor lying along *OY*-axis and $-j$ with that lying along *OY'*-axis.

The utility of j becomes more apparent if we consider a phasor **V** displaced θ° counter-clockwise from *OX*-axis as shown in Fig. 13.2. This phasor is not lying along the rectangular co-ordinate axes. However, it can be resolved into two components *viz.* the horizontal component 'a' along *X*-axis and the vertical component 'b' along *Y*-axis. It can be seen that vertical component is displaced 90° *CCW* from *OX*-axis. Therefore, mathematically, we can express this component as jb, meaning that component b is displaced 90° *CCW* from the component a (*i.e. OX*-axis).

$\therefore$ $$\mathbf{V} = a + j\,b$$

Magnitude of **V**, $V = \sqrt{a^2 + b^2}$

Its angle with *OX*-axis, $\theta = \tan^{-1}(b/a)$

The reader may note that $a + j\,b$ is the mathematical form of the phasor **V**. This form describes the magnitude as well as the phase angle of the phasor. The following points may be noted carefully :

(*i*) The quantity $a + j\,b$ is called a **complex number** or **complex quantity.** It is because it consists of a real component (a) and an imaginary component (jb).

(*ii*) To call the component lying along *Y*-axis (*i.e.* the component associated with j or $-j$) as imaginary merely because the value of $j(= \sqrt{-1})$ cannot be determined is really *unfortunate. The so called imaginary components can be represented graphically on the *Y*-axis just as real components can be represented on the *X*-axis. Electrical engineers rightly call the horizontal and vertical components as *in-phase* and *quadrature* components respectively.

---

* This is an unfortunate choice of name given by the mathematicians. Electrical engineers do not agree with this name for the so called imaginary numbers are no more imaginary than the real numbers.

Note. In this book, bold-faced letters (*e.g.* **V**, **I** *etc.*) will be used to represent the phasor completely, including both magnitude and direction. However, only the magnitude of the phasor will be represented by the same letter in the ordinary type (*e.g. V, I etc.*). However, while representing a phasor in writing, a dot or arrowhead may be used. *e.g.* **V** may be written as $\dot{V}$ or $\vec{V}$ and so on.

13.3. Mathematical Representation of Phasors

There are four ways of representing a phasor in the mathematical form *viz.* (*i*) Rectangular form (*ii*) Trigonometrical form (*iii*) Polar form and (*iv*) Exponential form. Each method has its own advantages and disadvantages.

(*i*) **Rectangular form.** This method is also known as **symbolic notation.** In this method, the phasor is resolved into horizontal and vertical components and is expressed in the complex form as discussed in Art. 13.2. Consider a voltage phasor **V** displaced $\theta°$ *CCW* from the reference axis (*i.e. OX*-axis) as shown in Fig. 13.3. (*i*). The horizontal or in-phase component of this phasor is *a* while the vertical or quadrature component is *b*. Therefore, the phasor can be represented in the rectangular form as :

$$\mathbf{V} = a + j\,b$$

Magnitude of phasor, $V = \sqrt{a^2 + b^2}$

Its angle *w.r.t. OX*-axis, $\theta = \tan^{-1}(b/a)$

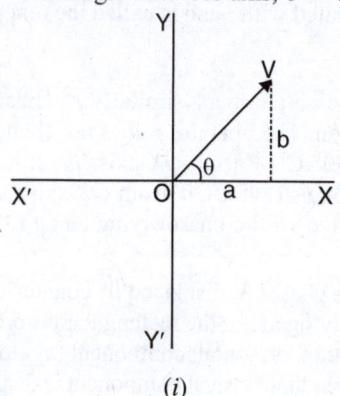

(*i*)

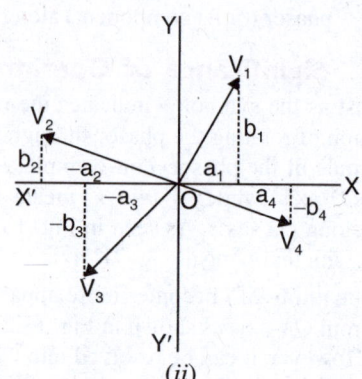

(*ii*)

Fig. 13.3

Fig. 13.3 (*ii*) shows the phasors in the various quadrants. The phasors have been resolved into horizontal and vertical components. The various phasors can be represented in the rectangular form as :

$$\mathbf{V_1} = a_1 + j\,b_1 ; \qquad \mathbf{V_2} = -a_2 + j\,b_2$$
$$\mathbf{V_3} = -a_3 - j\,b_3 ; \qquad \mathbf{V_4} = a_4 - j\,b_4$$

The magnitude and phase angles of these phasors can be found out as explained above. Remember phase angles are to be measured from the reference axis *i.e. OX*-axis. If the angle is measured *CCW* from this reference axis, it is assigned a + ve sign. If the angle is measured clockwise from the reference axis, it is assigned a – ve sign.

(*ii*) **Trigonometrical form.** It is similar to the rectangular form except that in-phase and quadrature components of the phasor are expressed in the trigonometrical form. Thus referring to Fig. 13.3 (*i*) above, $a = V \cos\theta$ and $b = V \sin\theta$ where V is the magnitude of the phasor **V**. Hence phasor **V** can be expressed in the trigonometrical form as :

$$\mathbf{V} = V(\cos\theta + j\sin\theta)$$

In general, the phasor **V** may be represented in trigonometrical form as :

$$\mathbf{V} = V(\cos\theta \pm j\sin\theta)$$

It may be noted that in this form, we express the phasor in terms of its magnitude and phase angle θ.

(*iii*) **Polar form.** It is a usual practice to write the trigonometrical form $\mathbf{V} = V(\cos \theta + j \sin \theta)$ in what is called polar form as :

$$*\mathbf{V} = V \angle \theta$$

where V is the magnitude of the phasor and θ is its phase angle measured *CCW* from the reference axis *i.e. OX*-axis. A negative angle in the polar form indicates clockwise measurement of the angle from the reference axis. Hence polar form can be written in general as :

$$\mathbf{V} = V \angle \pm \theta$$

(*iv*) **Exponential form.** According to Euler's equation :

$$e^{j\theta} = \cos \theta + j \sin \theta$$

Similarly, $e^{-j\theta} = \cos \theta - j \sin \theta$

The trigonometrical form of representing a phasor $\mathbf{V}$ is $\mathbf{V} = V (\cos \theta \pm j \sin \theta)$. Therefore, the phasor $\mathbf{V}$ may be represented in the exponential form as :

$$\mathbf{V} = V e^{\pm j\theta}$$

Here V is the magnitude (or modulus) of the phasor and ± θ (or argument) is the phase angle of the phasor from the reference axis (*i.e. OX*-axis).

13.4. Conversion from One Form to the Other

The reader may see that the above four mathematical forms of representing a phasor convey the same information *i.e.* magnitude of the phasor and its phase angle. Therefore, it is possible to convert one form to the other. Consider a phasor $\mathbf{V}$ having in-phase and quadrature components as 3 and 4 respectively as shown in Fig. 13.4. (*i*).

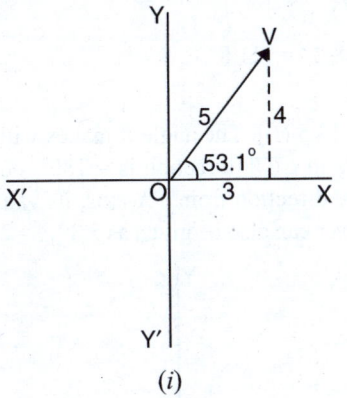
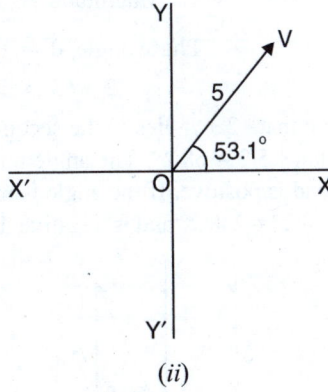

(*i*)   (*ii*)

Fig. 13.4

Magnitude of the phasor, $V = \sqrt{3^2 + 4^2} = 5$

Its angle *w.r.t. OX*-axis, $\theta = \tan^{-1} 4/3 = 53.1°$

In rectangular form, $\mathbf{V} = 3 + j\,4$

In trigonometrical form, $\mathbf{V} = 5 (\cos 53.1° + j \sin 53.1°)$

In polar form, $\mathbf{V} = 5 \angle 53.1°$ [See Fig. 13.4. (*ii*)]

In exponential form, $\mathbf{V} = 5\, e^{j53.1°}$

This numerical illustrates how one form can be converted to the other.

---

*   Meaning $\mathbf{V}$ equals V at angle θ (in *CCW* direction). There is no mathematical explanation for this form. It is a short hand form of trigonometrical form.

Example 13.1. *Express the polar form of voltage* $\mathbf{V} = 50 \angle 36.87°$ *V in trigonometrical and rectangular forms.*

Solution. Magnitude of voltage, $V = 50$ V ; Phase angle, $\theta = 36.87°$

Trigonometrical form. $V = V (\cos \theta + j \sin \theta) = \mathbf{50 \ (cos \ 36.87° + j \ sin \ 36.87°) \ volts}$

Rectangular form. In-phase component $= V \cos \theta = 50 \times \cos 36.87° = 40$

Quadrature component $= V \sin \theta = 50 \times \sin 36.87° = 30$

$\therefore$ $\mathbf{V} = \mathbf{(40 + j \ 30) \ volts}$

Example 13.2. *Express the following in polar form* (i) $3 + j \ 7$ (ii) $-2 + j \ 5$ (iii) $-50 - j \ 75$ *and* (iv) $6 - j \ 8$.

Solution. In the polar form $A \angle \pm \theta$, the phase angle θ is measured from the *OX*-axis. If measurement is in *CCW* direction from *OX*-axis, θ is positive. If angle is measured from *OX*-axis in the clockwise direction, θ is negative. The best way is that the reader should see the quadrant in which the phasor (or complex number) lies and then determine its phase angle *w.r.t. OX*-axis.

(*i*)   Magnitude $= \sqrt{3^2 + 7^2} = 7.62$

Phase angle, $\theta = \tan^{-1} (7/3) = 66.8°$

$\therefore$ $3 + j \ 7 = \mathbf{7.62 \angle 66.8°}$

Note that $3 + j \ 7$ lies in the first quadrant. As seen in Fig. 13.5 (*i*), θ measured in *CCW* direction from *OX*-axis is 66.8° and is therefore positive. If θ is measured in clockwise direction from *OX*-axis, its value is $= 360° - 66.8° = 293.2°$ and this angle is negative. Hence the above answer can also be given as $7.62 \angle - 293.2°$.

(*ii*)   Magnitude $= \sqrt{(-2)^2 + (5)^2} = 5.39$

Phase angle, $\theta = \tan^{-1} - 5/2 = 180° - 68.2° = 111.8°$

$\therefore$ $-2 + j \ 5 = \mathbf{5.39 \angle 111.8°}$

Note that $-2 + j5$ lies in the second quadrant [See Fig. 13.5 (*i*)]. The angle it makes with *OX*-axis is $\tan^{-1} 5/2 = 68.2°$. The angle measured from *OX*-axis in *CCW* direction is $= 180° - 68.2° = 111.8°$ and is positive. If the angle is measured in clockwise direction from *OX*-axis, its value is $= 180° + 68·2° = 248.2°$ and is negative. Hence the above answer can also be given as $5.39\angle - 248.2°$.

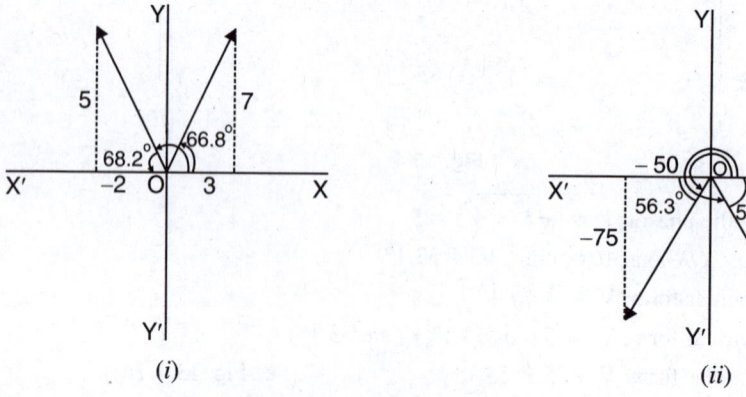

(*i*)   (*ii*)

Fig. 13.5

(*iii*)   Magnitude $= \sqrt{(-50)^2 + (-75)^2} = 90.1$

Phase angle, $\theta = \tan^{-1} - 75/-50 = 180° + 56.3° = 236.3°$

∴ $-50 - j\,75 \ = \ \mathbf{90.1\ \angle - 236.3°}$

If the angle is measured in clockwise direction from OX-axis, its value is $180° - 56.3° = 123.7°$ and is negative. Hence the above answer can also be written as $90.1 \angle - 123.7°$.

(*iv*)   Magnitude $= \ \sqrt{6^2 + (-8)^2} \ = 10$

          Phase angle, $\theta \ = \ \tan^{-1} - 8/6 = 360° - 53.13° = 306.87°$

∴ $6 - j\,8 \ = \ \mathbf{10\ \angle\ 306.87°}$

If the angle is measured in clockwise direction from OX-axis, its value is $53.13°$ and is negative. Hence the above answer can also be written as $10 \angle - 53.13°$.

Example 13.3. *A phasor is represented by* $20\ e^{-j2\pi/3}$. *Write the equivalent trigonometrical, rectangular and polar forms.*

Solution. Magnitude of phasor $= 20$; Its phase angle *w.r.t.* OX-axis is $\theta = -2\pi/3 = -120°$.

∴   Trigonometrical form : $20(\cos\theta + j\sin\theta)$

                    $= \mathbf{20\ [\cos\ (-120°) + j\sin\ (-120°)] = (-10 - j\,17.32)}$

          Polar form : $\mathbf{20\ \angle - 120°}$

13.5. Addition and Subtraction of Phasors

The rectangular form is best suited for addition or subtraction of phasors. If the phasors are given in polar form, they should be first converted into rectangular form. Once the phasors are in rectangular form, the in-phase components and the quadrature components can be algebraically added or subtracted. The answer (sum or difference) can be left in rectangular form or it can be converted back to polar form if desired.

(*i*) **Addition.** For the addition of phasors in the rectangular form, the in-phase components are *added together and the quadrature components are added together. Consider two voltage phasors represented as :

$$\mathbf{V_1} \ = \ a_1 + j\,b_1 \ ; \ \ \mathbf{V_2} = a_2 + j\,b_2$$

Then resultant, $\mathbf{V} \ = \ \mathbf{V_1} + \mathbf{V_2} = (a_1 + j\,b_1) + (a_2 + j\,b_2)$

                    $= \ (a_1 + a_2) + j\,(b_1 + b_2)$

Magnitude of resultant, $V = \ \sqrt{(a_1 + a_2)^2 + (b_1 + b_2)^2}$

Its angle from OX-axis, $\theta \ = \ \tan^{-1}\left(\dfrac{b_1 + b_2}{a_1 + a_2}\right)$

Fig. 13.6 shows graphically the addition of two phasors $\mathbf{V_1}$ and $\mathbf{V_2}$ in rectangular form.

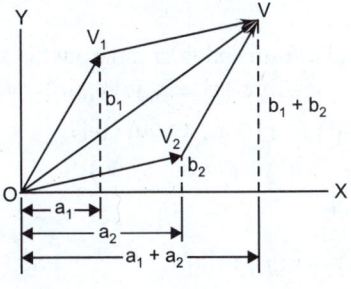

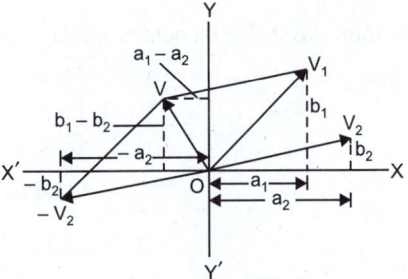

Fig. 13.6   **Fig. 13.7**

---

*   You cannot add or subtract components of different kinds. Thus real components cannot be added to or subtracted from imaginary components.

(*ii*) Subtraction. Like addition, the subtraction of phasors is done by using ordinary rules of phasor algebra.

$$\mathbf{V} = \mathbf{V_1} - \mathbf{V_2} = (a_1 + j\, b_1) - (a_2 + j\, b_2)$$
$$= (a_1 - a_2) + j(b_1 - b_2)$$

∴   Magnitude, $V = \sqrt{(a_1 - a_2)^2 + (b_1 - b_2)^2}$

Phase angle, $\theta = \tan^{-1}\left(\dfrac{b_1 - b_2}{a_1 - a_2}\right)$

Fig. 13.7 shows graphically the phasor difference of phasors $\mathbf{V_1}$ and $\mathbf{V_2}$ in rectangular form.

Example 13.4. *Two current phasors are given in the rectangular form as* $\mathbf{I_1} = (15 + j\,10)$ *A and* $\mathbf{I_2} = (12 + j\,6)$ *A. Perform the operation (i)* $\mathbf{I_1} + \mathbf{I_2}$ *and (ii)* $\mathbf{I_1} - \mathbf{I_2}$.

Solution. (*i*) Resultant current, $\mathbf{I} = \mathbf{I_1} + \mathbf{I_2} = (15 + j\,10) + (12 + j\,6) = \mathbf{(27 + j\,16)\ A}$

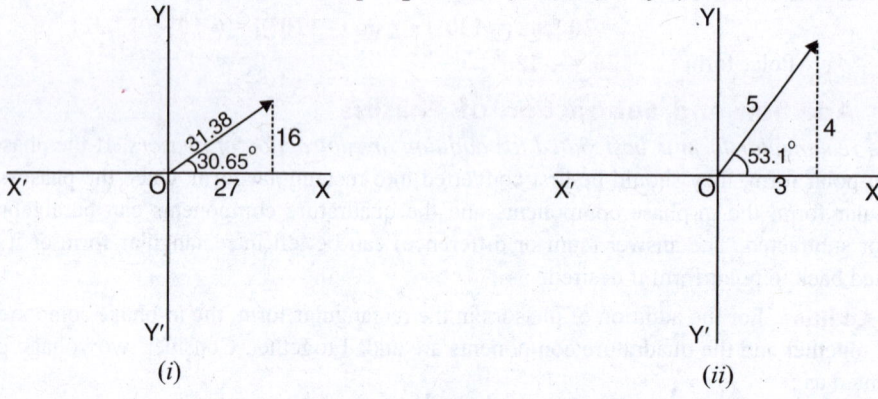

Fig. 13.8

The magnitude of resultant current is $= \sqrt{27^2 + 16^2} = 31.38$ A and it makes an angle $\theta = \tan^{-1} 16/27 = 30.65°$ with *OX*-axis as shown in Fig. 13.8 (*i*).

(*ii*)   Resultant current, $\mathbf{I} = \mathbf{I_1} - \mathbf{I_2} = (15 + j\,10) - (12 + j\,6) = \mathbf{(3 + j\,4)\ A}$

The magnitude of resultant is $= \sqrt{3^2 + 4^2} = 5$A and it makes an angle $\theta = \tan^{-1} 4/3 = 53.1°$ with *OX*-axis as shown in Fig 13.8 (*ii*).

Example 13.5. *Determine the resultant voltage of two sinusoidal generators in series whose voltages are* $\mathbf{V_1} = 25\angle15°$ *V and* $\mathbf{V_2} = 15\angle60°$ *V.*

Solution. We shall first convert polar form to rectangular form and then carry out the addition.

$$\mathbf{V_1} = 25(\cos 15° + j \sin 15°) = (24.15 + j\,6.47)\ \text{volts}$$
$$\mathbf{V_2} = 15(\cos 60° + j \sin 60°) = (7.5 + j\,12.99)\ \text{volts}$$

∴ $\mathbf{V} = \mathbf{V_1} + \mathbf{V_2} = (24.15 + j\,6.47) + (7.5 + j\,12.99)$
$$= (31.65 + j\,19.46)\ \text{volts}$$

∴ $V = \sqrt{(31.65)^2 + (19.46)^2} = 37.15$ volts

Phase angle, $\theta = \tan^{-1}(19.46/31.65) = 31.6°$

∴ $\mathbf{V = 37.15 \angle 31{\cdot}6°\ volts}$

Example 13.6. *Given two currents* $i_1 = 10 \sin(\omega t + \pi/4)$ *and* $i_2 = 5 \cos(\omega t - \pi/2)$, *find the r.m.s. value of* $i_1 + i_2$ *using the complex number representation.*

Solution.
$$i_1 = 10 \sin(\omega t + \pi/4)$$
$$i_2 = 5 \cos(\omega t - \pi/2) = 5 \sin(90° + \omega t - \pi/2) = 5 \sin \omega t$$

The maximum value of first current is 10 A and it leads the reference axis by 45° or $\pi/4$ rad. while the maximum value of second current is 5A and is in phase with the reference axis.

$$\therefore \qquad \boldsymbol{I}_{m1} = 10(\cos 45° + j \sin 45°) = (7.07 + j\, 7.07)A$$
$$\boldsymbol{I}_{m2} = 5(\cos 0° + j \sin 0°) = (5 + j0)\, A$$

$\therefore$ Maximum value of resultant current is

$$\boldsymbol{I}_m = \boldsymbol{I}_{m1} + \boldsymbol{I}_{m2} = (7.07 + j\, 7.07) + (5 + j0)$$
$$= (12.07 + j\, 7.07) = 14 \angle 30.4° \, A$$

$$\therefore \qquad \text{R.M.S value} = 14/\sqrt{2} = \mathbf{10\ A}$$

Tutorial Problems

1. Add $\mathbf{V_1} = 70 \angle 45°$ V to $\mathbf{V_2} = 67 \angle -63°$ V and record the sum in rectangular form.   **[(79.9–j 10.2)V]**
2. Add $\mathbf{V_1} = (-10 + j\, 50)$V to $\mathbf{V_2} = (30 + j\, 20)$V and express the result in polar form.   **[78.2 $\angle$ 74° V]**
3. Subtract $\mathbf{I_1} = 3.6 \angle -56.3°$ A from $\mathbf{I_2} = 5.83 \angle 31°$A and record the difference in polar form.   **[6.7 $\angle$ 63.4° A]**
4. Add $(-4 + j5)$ A to $11 \angle 240°$ A and express the result in polar form.   **[10.5 $\angle$ 205.5° A]**
5. Subtract $9 \angle 45°$ V from $(8 - j7)$ V and express the result in polar form.   **[13.5 $\angle$ –83°V]**

13.6. Conjugate of a Complex Number

Two complex numbers (or phasors) are said to be conjugate if they differ only in the algebraic sign of their quadrature components. Thus the conjugate of $2 + j\, 3$ is $2 - j\, 3$. The conjugate of $-4 + j\, 3$ is $-4 - j\, 3$. In the polar form, conjugate of $5 \angle 30°$ is $5 \angle -30°$ and conjugate of $10 \angle -40°$ is $10 \angle 40°$. It is a usual practice to use an asterik (*) to indicate the conjugate.

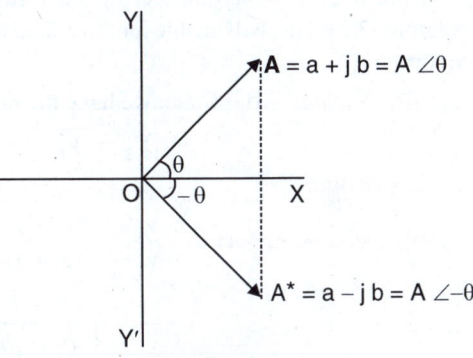

Consider a complex number $\mathbf{A} = a + jb = A\angle\theta$. Then its conjugate will be $\mathbf{A*} = a - jb = A \angle -\theta$. The two conjugate numbers are shown graphically in Fig. 13.9. The conjugate numbers have the following properties :

Fig. 13.9

(*i*) *The sum of two conjugate numbers results in in-phase component only (i.e. no j part).*
$$\mathbf{A} + \mathbf{A*} = (a + j\, b) + (a - j\, b) = 2a$$

(*ii*) *The difference of two conjugate numbers results in quadrature component only.*
$$\mathbf{A} - \mathbf{A*} = (a + j\, b) - (a - j\, b) = j2b$$

(*iii*) *When two conjugates are multiplied, the result has no j-part (i.e. no quadrature component).*
$$\mathbf{A} \times \mathbf{A*} = (a + j\, b)(a - j\, b) = a^2 - j^2 b^2 = a^2 + b^2 \qquad (\because j^2 = -1)$$

The conjugate of a complex number is used in determining the apparent power of a circuit in complex form (See Art. 13.14).

13.7. Multiplication and Division of Phasors

It is easier to multiply and divide the phasors when they are in polar or exponential form than in the rectangular form. This will become apparent from the following discussion. Consider two phasors given by ;

$$V_1 = a_1 + j\,b_1 = V_1 \angle \theta_1 \;; \quad V_2 = a_2 + j\,b_2 = V_2 \angle \theta_2$$

1. Multiplication.

(i) Rectangular form. $V_1 \times V_2 = (a_1 + j\,b_1)(a_2 + j\,b_2) = a_1 a_2 + j\,a_1 b_2 + j\,a_2 b_1 + j^2\,b_1 b_2$

$$= (a_1 a_2 - b_1 b_2) + j\,(a_1 b_2 + a_2 b_1) \qquad (\because j^2 = -1)$$

$$\text{Magnitude of resultant} \;=\; \sqrt{(a_1 a_2 - b_1 b_2)^2 + (a_1 b_2 + a_2 b_1)^2}$$

$$\text{Its angle } w.r.t. \; OX\text{-axis, } \theta \;=\; \tan^{-1}\!\left(\frac{a_1 b_2 + a_2 b_1}{a_1 a_2 - b_1 b_2}\right)$$

(ii) Polar form. To multiply the phasors that are in polar form, just multiply their magnitudes and *algebraically add* the phase angles.

$$V_1 \times V_2 \;=\; V_1 \angle \theta_1 \times V_2 \angle \theta_2 = V_1 V_2 \angle \theta_1 + \theta_2$$

Fig. 13.10 shows the multiplication of above two phasors graphically. The reader may see that multiplication of phasors becomes easier when they are expressed in the polar form. Another advantage is that multiplication in this form provides immediate information regarding the magnitudes and phase angles of the components and the resultant. Such advantages are not claimed when we multiply the phasors in the rectangular form. Therefore, as a rule, multiplication should always be done by expressing the phasors in the polar form. The reader is strongly advised to have a fair practice of converting rectangular form to polar form (See example 13.2). This will enable him to obtain quick solution of a.c. circuits.

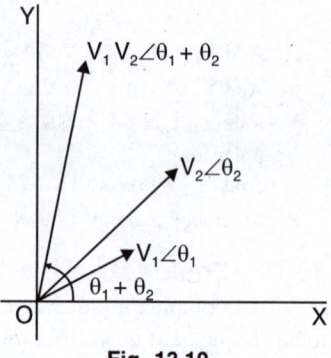

Fig. 13.10

(iii) In exponential form, we have the product as :

$$V_1 V_2 \;=\; V_1 e^{j\theta_1} \times V_2 e^{j\theta_2} = V_1 V_2 \, e^{j(\theta_1 + \theta_2)}$$

2. Division.

(i) Rectangular form.
$$\frac{V_1}{V_2} \;=\; \frac{a_1 + j\,b_1}{a_2 + j\,b_2}$$

$$= \frac{a_1 + j\,b_1}{a_2 + j\,b_2} \times \frac{a_2 - j\,b_2}{a_2 - j\,b_2} \qquad [\text{*Rationalising the denominator}]$$

$$= \frac{(a_1 a_2 + b_1 b_2) + j(b_1 a_2 - a_1 b_2)}{a_2^2 + b_2^2}$$

$$= \frac{(a_1 a_2 + b_1 b_2)}{a_2^{\,2} + b_2^{\,2}} + j\,\frac{(b_1 a_2 - a_1 b_2)}{a_2^{\,2} + b_2^{\,2}}$$

(ii) Polar form. To divide the phasors that are in polar form, divide the magnitudes of phasors and subtract the denominator angle from the numerator angle.

$$\therefore \qquad \frac{V_1}{V_2} \;=\; \frac{V_1 \angle \theta_1}{V_2 \angle \theta_2} = \frac{V_1}{V_2} \angle \theta_1 - \theta_2$$

(iii) In exponential form, we have the division as :

---

*   In order to express the result in the standard $a \pm jb$ form, multiply the numerator and denominator by the conjugate of denominator $(a_2 + jb_2)$ which is $a_2 - jb_2$. This operation ; called *rationalising the denominator*, eliminates the j component in the denominator.

$$\frac{V_1}{V_2} = \frac{V_1 e^{j\theta_1}}{V_2 e^{j\theta_2}} = \frac{V_1}{V_2} e^{j(\theta_1 - \theta_2)}$$

Practically without exception, division is the easiest in the polar or exponential form.

Example 13.7. *Two phasors are given in the following form :*

$$V_1 = (4 + j\,3)\,V\,;\,V_2 = (5 + j\,6)\,V$$

Evaluate $V_1 \times V_2$ and V_1/V_2 in (i) rectangular form (ii) polar form.

Solution. (*i*) **Rectangular form.**

$$V_1 \times V_2 = (4 + j\,3)\,(5 + j\,6) = 20 + j\,24 + j\,15 + j^2 18 = \mathbf{(2 + j\,39)\,V} \qquad (\because j^2 = -1)$$

$$\frac{V_1}{V_2} = \frac{4 + j3}{5 + j6} = \frac{4 + j3}{5 + j6} \times \frac{5 - j6}{5 - j6} \qquad\qquad \text{[Rationalising the denominator]}$$

$$= \frac{20 - j24 + j15 - j^2 18}{5^2 - j^2 6^2} = \frac{38 - j9}{61} = \frac{38}{61} - j\frac{9}{61}$$

$$= \mathbf{(0.623 - j\,0.147)\ volts}$$

(*ii*) **Polar form.** Convert the phasors to polar form.

$$*V_1 = (4 + j\,3)\,V = 5 \angle 36.87°\ V$$

$$V_2 = (5 + j\,6)\,V = 7.81 \angle 50.19°\ V$$

$$\therefore \qquad V_1 \times V_2 = 5 \angle 36.87° \times 7.81 \angle 50.19°$$

$$= 5 \times 7.81 \angle 36.87° + 50.19° = \mathbf{39.05 \angle 87.06°\ volts}$$

$$\frac{V_1}{V_2} = \frac{5 \angle 36.87°}{7.81 \angle 50.19°} = \frac{5}{7.81} \angle 36.87° - 50.19°$$

$$= \mathbf{0.64 \angle -13.32°\ volts}$$

The negative angle means that the phasor is below *OX*-axis. The reader may note the relative ease of multiplying and dividing the phasors in polar form compared with the same operations in the rectangular form.

Example 13.8. *The following three phasors are given :*

$$A = 5 + j\,5\,;\ B = 50 \angle 40°\,;\ C = 4 + j\,0$$

Perform the following indicated operations :

(*i*) $\dfrac{AB}{C}$ (*ii*) $\dfrac{BC}{A}$

Solution. Expressing the three phasors in polar form, we have, $A = 5 + j\,5 = 7.07 \angle 45°$; $B = 50 \angle 40°$; $C = 4 + j\,0 = 4 \angle 0°$

(*i*) $\qquad\qquad \dfrac{AB}{C} = \dfrac{7.07 \angle 45° \times 50 \angle 40°}{4 \angle 0°} = \mathbf{88.37 \angle 85°}$

(*ii*) $\qquad\qquad \dfrac{BC}{A} = \dfrac{50 \angle 40° \times 4 \angle 0°}{7.07 \angle 45°} = \mathbf{28.29 \angle -5°}$

Example 13.9. *Perform the following operations and express the final result in polar form :*

(*i*) $(8 + j\,6) \times (-10 - j\,7.5)$

(*ii*) $5 \angle 30° + 8 \angle -30°$

---

* Magnitude of V_1 is $= \sqrt{4^2 + 3^2} = 5$

Its angle *w.r.t. OX*-axis, $\theta = \tan^{-1}(3/4) = 36.87°$

Solution. (*i*) $(8 + j\,6) \times (-10 - j\,7.5) = (-80 - j\,60 - j\,60 + 45) = (-35 - j\,120)$

Magnitude $= \sqrt{(-35)^2 + (-120)^2} = 125$

The phasor quantity lies in the third quadrant because its in-phase and quadrature components are negative.

∴   Phase angle *w.r.t. OX*-axis is given by ;

$$\theta = \tan^{-1}\frac{-120}{-35} = 180° + 73.74° = 253.74°$$

∴   Final result in polar form is **125 ∠ 253.74°.**

(*ii*) $5 \angle 30° + 8 \angle - 30° = 5(\cos 30° + j \sin 30°) + [8 \cos(-30°) + 8j \sin(-30°)]$
$$= (4.33 + j\,2.5) + (6.928 - j\,4)$$
$$= (11.258 - j\,1.5)$$

$$\text{Magnitude} = \sqrt{(11.258)^2 + (-1.5)^2} = 11.358$$

$$\text{Phase angle, } \theta = \tan^{-1}\frac{-1.5}{11.258} = -7.59°$$

∴ $5 \angle 30° + 8 \angle - 30° = $ **11.358 ∠ – 7.59°**

Example 13.10. *Divide (6 + j 7) by (5 + j 3) and express the result in (i) rectangular form (ii) polar form.*

Solution. (*i*) $\dfrac{6 + j\,7}{5 + j\,3} = \dfrac{6 + j\,7}{5 + j\,3} \times \dfrac{5 - j\,3}{5 - j\,3} = \dfrac{30 - j\,18 + j\,35 + 21}{34}$

$$= \frac{51 + j\,17}{34} = (1.5 + j\,0.5)$$

(*ii*) The result in rectangular form is $(1.5 + j\,0.5)$. The phasor quantity lies in the first quadrant because its in-phase and quadrature components are positive.

∴ $\text{Magnitude} = \sqrt{(1.5)^2 + (0.5)^2} = 1.581$

Phase angle *w.r.t. OX*-axis, $\theta = \tan^{-1}\dfrac{0.5}{1.5} = 18.44°$

∴   Result in polar form is **1.581 ∠ 18.44°.**

Alternatively. $\dfrac{6 + j\,7}{5 + j\,3} = \dfrac{9.22 \angle 49.40°}{5.83 \angle 30.96°} = $ **1.581 ∠ 18.44°**

(Tutorial Problems)

1. Multiply $16 \angle - 30°$ by $7 + j\,2$.   [116.48 ∠ –14.05°]
2. Determine the voltage across an inductive reactance of $5 \angle 90°$ kΩ if the current   [20 ∠ 90° V]
    is $4 \angle 0°$ A.
3. Divide $30 - j\,10$ by $10.54 \angle 60°$.   [3 ∠ –78.43°]
4. Find the impedance of a circuit in which **I** $= (2 + j\,3)$A and **V** $= (10 - j\,6)$V.   [3.24 ∠ – 87.26° Ω]
5. Divide $(3 + j\,4)$V by $10 \angle - 40°$ A.   [0.5 ∠ –93.1° Ω]

13.8. Powers and Roots of Phasors

Powers and roots of complex numbers can be found very conveniently in polar form. If the complex number is not in polar form, it is always advisable to convert the number to this form and then carry out these algebraic operations.

(i) **Powers.** Suppose it is required to find the cube of the phasor $2 \angle 10°$. For this purpose, the phasor has to be multiplied by itself three times.

$$\therefore \qquad (2 \angle 10°)^3 = 2 \times 2 \times 2 \angle 10° + 10° + 10° = 8 \angle 30°$$

In general, $\qquad \mathbf{A^n} = A^n \angle n\,\theta$

(ii) **Roots.** Let us proceed backward with the above given example. Suppose we are to find the cube root of $8 \angle 30°$. It is clear that :

$$\sqrt[3]{8 \angle 30°} = 2 \angle 10°$$

In general, $\qquad \sqrt[n]{A} = \sqrt[n]{A} \angle \dfrac{\theta}{n}$

Thus the square root of $64 \angle 40°$ is $= \sqrt[2]{64} \angle \dfrac{40°}{2} = 8 \angle 20°$.

Example 13.11. *A current phasor is given by* $\mathbf{I} = 64 \angle 30°$ *A.*

Find (i) $\mathbf{I}^2$ *(ii)* $\mathbf{I}^3$ *(iii)* $\sqrt{\mathbf{I}}$ *(iv)* $\sqrt[3]{\mathbf{I}}$

Solution. $\quad$ *(i)* $\qquad \mathbf{I}^2 = (64 \angle 30°)^2 = 64^2 \angle 30° + 30° = \mathbf{4096 \angle 60°}$ **A**

(ii) $\qquad \mathbf{I}^3 = (64 \angle 30°)^3 = 64^3 \angle 30° + 30° + 30° = \mathbf{262144 \angle 90°}$ **A**

(iii) $\qquad \sqrt{\mathbf{I}} = \sqrt[2]{64 \angle 30°} = \sqrt[2]{64} \angle \dfrac{30°}{2} = \mathbf{8 \angle 15°}$ **A**

(iv) $\qquad \sqrt[3]{\mathbf{I}} = \sqrt[3]{64} \angle \dfrac{30°}{3} = \mathbf{4 \angle 10°}$ **A**

Example 13.12. *Find the square root of* $(5 + j8.7)$.

Solution. $\qquad 5 + j\,8.7 = A \angle \theta$

Here $\qquad A = \sqrt{5^2 + 8.7^2} = 10 \; ; \; \theta = \tan^{-1} \dfrac{8.7}{5} = 60°$

$\therefore \qquad 5 + j\,8.7 = 10 \angle 60°$

$\therefore \qquad (5 + j\,8.7)^{1/2} = 10^{\frac{1}{2}} \angle \dfrac{60°}{2} = 3.16 \angle 30°$

$$= 3.16 \,(\cos 30° + j \sin 30°) = \mathbf{2.7 + j\,1.6}$$

13.9. Applications of Phasor Algebra to A.C. Circuits

We know that alternating voltages and currents are phasor quantities. Therefore, they can be expressed in the complex form *e.g.* rectangular, trigonometrical or polar form. The magnitude and phase angle of voltage or current can be readily obtained from its complex form. We shall now see how phasor algebra applied to a.c. circuits yields quick solution. While applying phasor algebra to a.c. circuits, the following points must be kept in mind :

(i) If circuit current is taken as the reference phasor (*i.e.* along *OX*-axis), then,

$$\mathbf{I} = I + j\,0$$

It is because phasor lying along *OX*-axis has no *j* part.

(ii) If voltage is taken as the reference phasor (*i.e.* along *OX*-axis), then,

$$\mathbf{V} = V + j\,0$$

(iii) The angles of voltage or current are measured from *OX*-axis which is taken as the axis of reference. If the angle is measured in *CCW* direction from *OX*-axis, then it is considered a positive angle. If measurement is made in clockwise direction from *OX*-axis, the angle is considered negative.

(iv) Magnitudes of phasors (current or voltage) mean the r.m.s. values unless stated otherwise.

13.10. R-L Series A.C. Circuit

Fig. 13.11 shows an *R-L* series circuit and its phasor diagram. Since the circuit current is taken as the reference phasor (*i.e.* along *OX*-axis), we have,

$$\mathbf{I} = I + j\,0$$

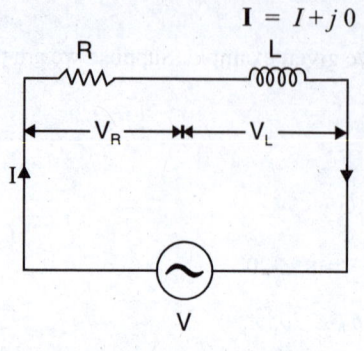
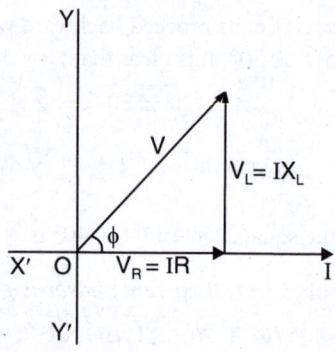

Fig. 13.11

Now, $\mathbf{V} = \mathbf{V_R} + \mathbf{V_L} = (\mathbf{I}\,R + j\,0) + (0 + j\,\mathbf{I}\,X_L)$

or $\mathbf{V} = \mathbf{I}\,(R + j\,X_L)$

But $\mathbf{V} = \mathbf{I}\,\mathbf{Z}$ where $\mathbf{Z}$ is the impedance of the circuit

∴ $*\mathbf{Z} = R + j\,X_L$

Magnitude of impedance, $Z = \sqrt{R^2 + X_L^{\,2}}$

where $R = Z\cos\phi$ and $X_L = Z\sin\phi$

Phase angle, $\phi = \tan^{-1}(X_L/R)$

It is clear that current lags behind the applied voltage by $\phi°$.

Polar form. In the polar form, we have,

$$\mathbf{I} = I + j\,0 = I\angle 0°$$

$$\mathbf{Z} = R + j\,X_L = Z\angle\phi°$$

∴ $\mathbf{V} = \mathbf{I}\,\mathbf{Z} = I\angle 0° \times Z\angle\phi° = IZ\angle\phi°$

This shows that applied voltage leads the current by $\phi°$. This is the same thing as the current lags the voltage by $\phi°$.

In practice, we are given the applied voltage and circuit impedance and it is desired to find the circuit current. It is then profitable to take voltage as the reference phasor (*i.e.,* along *OX*-axis).

$$\mathbf{V} = V + j\,0 = V\angle 0°$$

$$\mathbf{Z} = R + j\,X_L = Z\angle\phi°$$

∴ $\mathbf{I} = \dfrac{\mathbf{V}}{\mathbf{Z}}$

$$= \frac{V + j\,0}{R + j\,X_L} \qquad\qquad ... \textit{Rectangular form}$$

$$= \frac{V}{Z}\angle -\phi° \qquad\qquad ...\textit{Polar form}$$

Note that phase angle of current is conjugate of impedance angle.

---

*   Although impedance is not a phasor quantity, it may be represented in the complex form because it has both magnitude and angle.

Fig. 13.12 shows the phasor diagram of the circuit with voltage as the reference phasor.

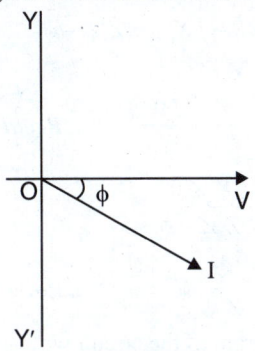

Fig. 13.12

13.11. R-C Series A.C. Circuit

Fig. 13.13 shows *R-C* series circuit and its phasor diagram. Since circuit current is taken as the reference phasor (*i.e.* along *OX*-axis), we have,

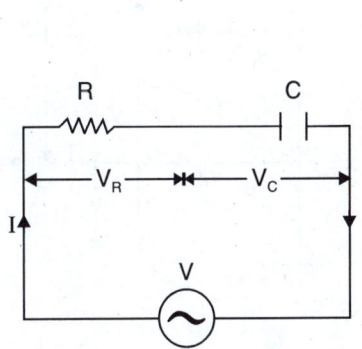 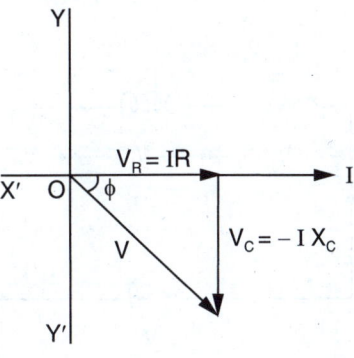

Fig. 13.13

$$\mathbf{I} = I + j\,0$$

Now,
$$\mathbf{V} = \mathbf{V_R} + \mathbf{V_C} = (\mathbf{I}\,R + j\,0) + (0 - j\,\mathbf{I}\,X_C)$$
or
$$\mathbf{V} = \mathbf{I}\,(R - j\,X_C)$$

But $\mathbf{V} = \mathbf{I}\,\mathbf{Z}$ where $\mathbf{Z}$ is the impedance of the circuit.

$\therefore$
$$\mathbf{Z} = R - j\,X_C$$

Its magnitude, $Z = \sqrt{R^2 + X_C^2}$; $R = Z\cos\phi$ and $X_C = Z\sin\phi$

Phase angle, $\phi = \tan^{-1}(-X_C/R)$

It is clear that current leads the applied voltage by $\phi°$.

Polar form. In the polar form, we have,

$$\mathbf{I} = I + j\,0 = I\angle 0°$$
$$\mathbf{Z} = R - j\,X_C = Z\angle -\phi°$$
$\therefore$
$$\mathbf{V} = \mathbf{I}\,\mathbf{Z} = I\angle 0° \times Z\angle -\phi° = IZ\angle -\phi°$$

This shows that the applied voltage lags the current by $\phi°$. This is the same thing as current leads the applied voltage by $\phi°$.

If applied voltage is taken as the reference phasor, then circuit calculations are carried out as under :

$$V = V + j0 = V \angle 0°$$
$$Z = R - jX_C = Z \angle -\phi°$$
$$\therefore \quad I = V/Z$$
$$= \frac{V + j0}{R - jX_C} \qquad ...Rectangular \ form$$
$$= \frac{V \angle 0°}{Z \angle -\phi°}$$
$$= \frac{V}{Z} \angle \phi° \qquad ...Polar \ form$$

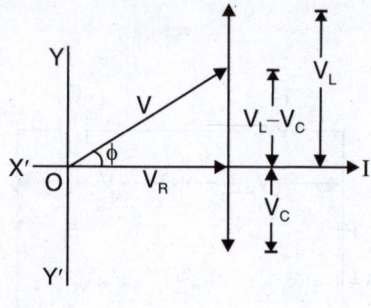
Fig. 13.14

Fig. 13.14 shows the phasor diagram of the circuit with applied voltage as the reference phasor.

13.12. R-L-C Series A.C. Circuit

Fig. 13.15 shows R-L-C series circuit and its phasor diagram. Since the circuit current is taken as the reference phasor (*i.e.* along *OX*-axis), we have,

$$I = I + j0$$

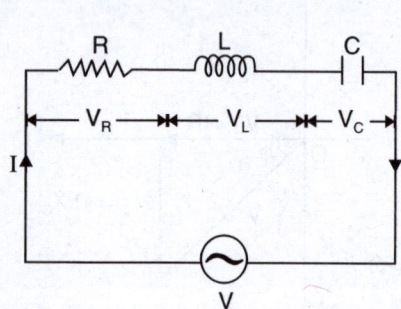

Fig. 13.15

Now,
$$V = V_R + j(V_L - V_C) = IR + j(IX_L - IX_C)$$
$$= I[R + j(X_L - X_C)] = IZ$$
$$\therefore \quad Z = R + j(X_L - X_C)$$

Magnitude of impedance, $Z = \sqrt{R^2 + (X_L - X_C)^2}$

Phase angle, $\phi = \tan^{-1}\left(\dfrac{X_L - X_C}{R}\right)$

If $X_L > X_C$, then ϕ is positive *i.e.* current lags the voltage. If $X_C > X_L$, then ϕ is negative *i.e.* current leads the voltage.

· **Polar form.** In the polar form, we have,

$$I = I + j0 = I \angle 0°$$
$$Z = R + j(X_L - X_C) = Z \angle \pm \phi°$$
$$\therefore \quad V = IZ = I \angle 0° \times Z \angle \pm \phi° = IZ \angle \pm \phi°$$

The applied voltage will lead or lag the circuit current depending upon whether X_L or X_C is greater.

If applied voltage is taken as the reference phasor, the circuit calculations are carried out as under :

$$\mathbf{V} = V + j0 = V \angle 0°$$

$$\mathbf{Z} = R + j(X_L - X_C) = Z \angle \pm \phi°$$

$$\therefore \quad \mathbf{I} = \mathbf{V}/\mathbf{Z}$$

$$= \frac{V + j0}{R + j(X_L - X_C)} \qquad ...Rectangular\ form$$

$$= \frac{V}{Z} \angle \mp \phi° \qquad ...Polar\ form$$

Note that the phase angle of current is the conjugate of its impedance angle. This is an important point to keep in mind.

13.13. Power Determination Using Complex Notation

If the voltage and current in an a.c. circuit are expressed in complex form, we can easily determine the true power or active power consumed in the circuit. Let the circuit voltage and circuit current be expressed as :

$$\mathbf{V} = V_1 + j\,V_2$$

$$\mathbf{I} = I_1 - j\,I_2$$

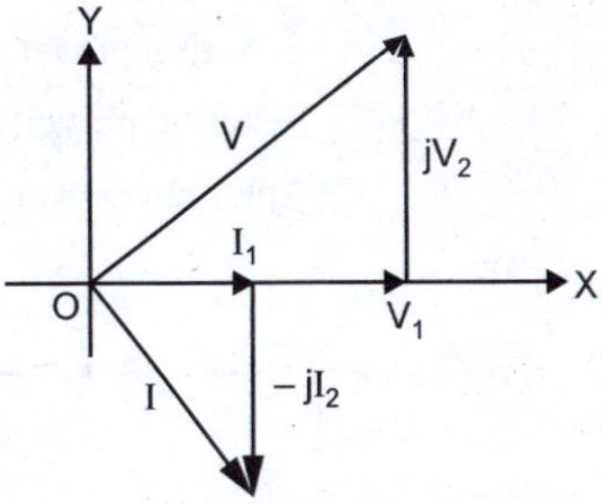

Fig. 13.16 shows the phasors $\mathbf{V}$ and $\mathbf{I}$ resolved into rectangular components. The component V_1 of voltage and component I_1 of current are in phase and contribute to true power given by ;

$$P_1 = V_1 I_1$$

Fig. 13.16

The component V_2 of voltage and component I_2 of current are 180° out of phase and, therefore, contribute to negative true power given by ;

$$P_2 = -V_2 I_2$$

The components V_2 and I_1 as well as V_1 and I_2 are 90° out of phase and contribute to no power. Therefore, total power P consumed in the circuit is the *algebraic sum* of P_1 and P_2 because power is a scalar quantity.

$$\therefore \quad \text{Total power consumed, } P = P_1 + P_2 = (V_1 I_1) + (-V_2 I_2)$$

$$\therefore \quad P = V_1 I_1 - V_2 I_2$$

*Hence the total power consumed in an a.c. circuit is the algebraic sum of product of in-phase components of **V** and **I** and the product of quadrature components of **V** and **I**. Note that operator j is not to be included in the multiplication.*

Example 13.13. *A circuit is connected across (100 + j40) volts supply and the resulting current is (3 − j4) amperes. Find the power consumed in the circuit.*

Solution. $\mathbf{V} = (100 + j\,40)$ volts ; $\mathbf{I} = (3 - j4)$ amperes

$$\therefore \qquad \text{Power consumed, } P = V_1 I_1 + V_2 I_2 = (100 \times 3) + (40 \times -4) = \mathbf{140\ W}$$

13.14. Power Determination By Conjugate Method

We have seen in Art. 12.3 (chapter 12) that the apparent power, active power and reactive power drawn by a circuit can be represented by a right angled triangle called **power triangle.** Fig. 13.17 shows the power triangle for an inductive circuit. Here $S (= VI)$ is the apparent power drawn by the circuit, $P (= VI \cos \phi)$ is the active power (*i.e.* power consumed by the circuit) and $Q (= VI \sin \phi)$ is the reactive power. Using complex algebra, we can express the apparent power in complex form as :

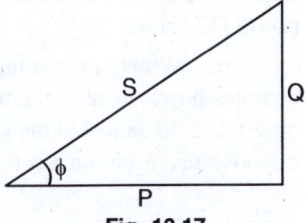

Fig. 13.17

$$\mathbf{S} = P + jQ$$

It can be shown mathematically *that apparent power in complex form is equal to the product of phasor voltage and conjugate of phasor current i.e.*

$$\mathbf{S} = \text{Phasor voltage} \times \text{Conjugate of phasor current}$$

or $$P + jQ = \mathbf{V} \times \mathbf{I}^*$$

where **V** is phasor voltage in complex form and **I*** is the conjugate of phasor current **I** in complex form.

$$\text{Active power, } P = \text{In-phase component}$$

$$\text{Reactive power, } Q = \text{Quadrature component}$$

Proof. Fig. 13.18 shows an inductive circuit and its phasor diagram. It is clear from the phasor diagram that phase angle $\phi° = \alpha° - \beta°$. Note that V and I are the r.m.s. values.

∴ $$\text{Active power, } P = VI \cos (\alpha° - \beta°)$$

$$\text{Reactive power, } Q = VI \sin (\alpha° - \beta°)$$

$$\text{Apparent power, } S = VI$$

or $$\mathbf{S} = P + jQ$$

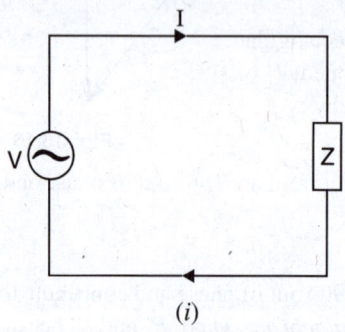

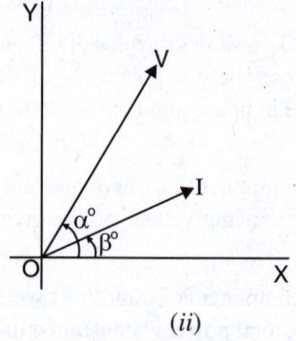

(i)   (ii)

Fig. 13.18

Using phasor algebra, we shall prove that product of phasor voltage and conjugate of phasor current is equal to the apparent power in complex form. Referring to the phasor diagram in Fig. 13.18 (*ii*),

$$\mathbf{V} = V \angle \alpha° ; \mathbf{I} = I \angle \beta°$$

Conjugate of **I** is $$\mathbf{I}^* = I \angle - \beta°$$

∴ $$\mathbf{V} \times \mathbf{I}^* = V \angle \alpha° \times I \angle - \beta° = VI \angle \alpha° - \beta°$$

$$= VI \cos (\alpha° - \beta°) + j \, VI \sin (\alpha° - \beta°)$$

But $\alpha° - \beta° = \phi°$, the phase angle between voltage and current.

∴ $$\mathbf{V} \times \mathbf{I}^* = VI \cos \phi + j \, VI \sin \phi$$

or $$\mathbf{S} = P + jQ$$

This shows that in-phase component of $\mathbf{V} \times \mathbf{I}^*$ (*i.e.* product of phasor voltage and conjugate of phasor current) gives the active power ($VI \cos \phi$) while quadrature component gives the reactive power ($VI \sin \phi$).

Note. We may use conjugate of **I** (*i.e.* **I***) or conjugate of **V** (*i.e.* **V***) to determine expression for complex power in an a.c. circuit. In both cases, the magnitudes of P (active power) and Q (reactive power) are the same but the sign of Q changes. Let us illustrate this point with a numerical example. Suppose an a.c. circuit has the following values :

$$\mathbf{V} = 115 \angle 0° \text{ V} ; \mathbf{I} = 1.37 \angle - 26.9° \text{ A}$$

Clearly, the load is inductive.

Using current conjugate. The conjugate of $\mathbf{I}$ $(= 1.37 \angle -26.9°$ A) is $\mathbf{I^*} = 1.37 \angle 26.9°$A.

∴ Complex power, $\mathbf{S} = \mathbf{V}\,\mathbf{I^*} = 115 \angle 0° \times 1.37 \angle 26.9° = 157.5 \angle 26.9°$ VA

or $\qquad\qquad\qquad\qquad \mathbf{S} = 140.8 + j\,71.3$

∴ $\qquad\qquad\qquad\qquad P = 140.8$ W ; $Q = +71.3$ VAR

Using voltage conjugate. The conjugate of $\mathbf{V}$ $(= 115 \angle 0°$ V) is $\mathbf{V^*} = 115 \angle -0°$ V.

∴ $\qquad$ Complex power, $\mathbf{S} = \mathbf{V^*}\mathbf{I} = 115 \angle -0° \times 1.37 \angle -26.9° = 157.5 \angle -26.9°$ VA

or $\qquad\qquad\qquad\qquad \mathbf{S} = 140.8 - j\,71.3$

∴ $\qquad\qquad\qquad\qquad P = 140.8$ W ; $Q = -71.3$ VAR

Note that magnitudes of P and Q are the same in the two cases but the sign of Q changes. When current conjugate is used, Q is positive for inductive load and negative for capacitive load. However, when voltage conjugate is used, Q is negative for inductive load and positive for capacitive load.

Example 13.14. *The current in a circuit is given by (4.5 + j12) A when the applied voltage is (100 + j150) V. Determine (i) the magnitude of impedance and (ii) phase angle.*

Solution. $\qquad\qquad\qquad\qquad \mathbf{V} = (100 + j\,150)$ V $= 180.28 \angle 56.31°$ V

$\qquad\qquad\qquad\qquad\qquad \mathbf{I} = (4.5 + j\,12)$ A $= 12.82 \angle 69.44°$ A

(i) ∴ $\qquad\qquad\qquad \mathbf{Z} = \dfrac{\mathbf{V}}{\mathbf{I}} = \dfrac{180.28 \angle 56.31°}{12.82 \angle 69.44°} = 14.06 \angle -13.13°\ \Omega$

∴ $\qquad\qquad\qquad\qquad Z = \mathbf{14.06\ \Omega}$

(ii) $\qquad\qquad$ Phase angle, $\phi = \mathbf{13.13°\ lead}$

It is clear that current is leading the voltage by 13.13°. Therefore, the circuit is capacitive.

Note. The problem can also be solved using rectangular form. Since division is involved, it is easier to work out in polar form.

Example 13.15. *In an R-L series circuit, R = 10 Ω and X_L = 8.66 Ω. If current in the circuit is (5 − j 10)A, find (i) the applied voltage (ii) power factor and (iii) active power and reactive power.*

Solution. $\qquad\qquad \mathbf{Z} = R + j\,X_L = (10 + j\,8.66)\ \Omega$

$\qquad\qquad\qquad\qquad\quad = 13.23 \angle 40.9°\ \Omega$

$\qquad\qquad\qquad\qquad \mathbf{I} = (5 - j\,10)$ A $= 11.18 \angle -63.43°$ A

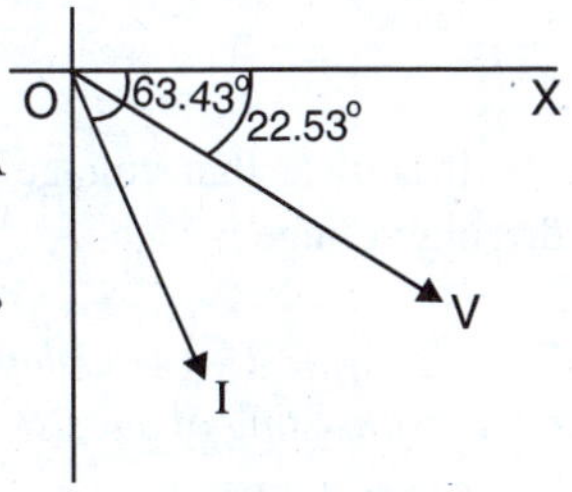

(i) $\qquad$ Applied voltage, $\mathbf{V} = \mathbf{I}\,\mathbf{Z}$

$\qquad\qquad\qquad\qquad\quad = 11.18 \angle -63.43° \times 13.23 \angle 40.9°$

$\qquad\qquad\qquad\qquad\quad = 148 \angle -22.53°$ V

∴ $\qquad\qquad\qquad\qquad V = \mathbf{148\ volts}$

(ii) $\qquad$ Phase angle, $\phi = 63.43° - 22.53° = 40.9°$

∴ $\qquad\qquad$ Power factor $= \cos \phi = \cos 40.9° = \mathbf{0.756\ lag}$

(iii) $\qquad$ Complex VA, $\mathbf{S} = $ Phasor voltage × Conjugate of phasor current

or $\qquad\qquad P + j\,Q = 148 \angle -22.53° \times 11.18 \angle 63.43° = 1654.64 \angle 40.9°$ VA

$\qquad\qquad\qquad\qquad\quad = 1654.64\,(\cos 40.9° + j \sin 40.9°) = (1250.66 + j\,1083.36)$ VA

∴ $\qquad$ Active power, $P = \mathbf{1250.66\ W}$; Reactive power, $Q = \mathbf{1083.36\ VAR}$

Fig. 13.19

Example 13.16. *The complex volt amperes in a series circuit are (4330 − j 2500) and the current is (25 + j 43·3) A. Find the applied voltage.*

Solution. Let the applied voltage be $\mathbf{V} = (a + j\,b)$

Complex VA = Phasor voltage × Conjugate of phasor current

or $\quad 4330 - j\,2500 = (a + jb) \times (25 - j\,43.3) = (25a + 43.3\,b) + j\,(-43.3\,a + 25\,b)$

∴ $\qquad\qquad 25a + 43.3\,b = 4330$ ...(i)

and $\qquad -43.3\,a + 25\,b = -2500$ ...(ii)

Solving eqs. (i) and (ii), we have, $a = 86.6$ and $b = 50$.

∴ $\qquad\qquad\qquad \mathbf{V = (86.6 + j\,50)\ volts}$

Example 13.17. *A coil of resistance 12 Ω and inductive reactance of 25 Ω is connected in series with a capacitive reactance of 41 Ω. The combination is connected to a supply of 230 V, 50Hz. Using phasor algebra, find (i) circuit impedance (ii) current and (iii) power consumed.*

Solution. (i) $\qquad\qquad \mathbf{Z} = R + j\,(X_L - X_C)$

$\qquad\qquad\qquad = 12 + j\,(25 - 41) = (12 - j\,16)\ \Omega = 20\ \angle -53.13°\ \Omega$

∴ $\qquad\qquad\qquad \mathbf{Z = 20\ \Omega}$

(ii) Taking voltage as the reference phasor,

$\qquad\qquad\qquad \mathbf{V} = (230 + j\,0)\ V = 230\ \angle\,0°\ \text{volts}$

∴ $\qquad\qquad \mathbf{I} = \dfrac{\mathbf{V}}{\mathbf{Z}} = \dfrac{230\ \angle 0°}{20\ \angle -53.13°} = 11.5\ \angle\,53.13°\ A$

∴ $\qquad\qquad\qquad I = \mathbf{11.5\ A}$

(iii) It is clear that current leads the voltage by 53.13° i.e. $\phi = 53.13°$.

∴ Power factor = $\cos\phi = \cos 53.13° = 0.6$ lead

$\qquad$ Power consumed, $P = VI\cos\phi = 230 \times 11.5 \times 0.6 = \mathbf{1587\ W}$

Example 13.18. *A high-impedance voltmeter is used to measure voltage drop across each of three series connected ideal circuit elements. If the r.m.s. readings are 25 V, 40 V and 60 V for V_R, V_L and V_C respectively, determine the equation for the voltage wave representing the driving voltage whose frequency is 50 Hz.*

Solution. Applied voltage, $\mathbf{V} = V_R + j\,(V_L - V_C) = 25 + j\,(40 - 60)$

$\qquad\qquad\qquad = (25 - j\,20)\ V = 32\ \angle -38.66°\ \text{volts}$

∴ $\qquad\qquad\qquad V = 32\ \text{volts}$

$\qquad\qquad V_m = 32 \times \sqrt{2} = 45.25\ V\ ;\ \ \omega = 2\pi f = 2\pi \times 50 = 314\ \text{rad s}^{-1}$

It is clear that voltage lags behind the current by 38.66° i.e. $\phi = 38.66°$. The equation of the driving voltage is :

$\qquad\qquad\qquad \mathbf{\mathit{v} = 45.25\ sin\ (314t - 38.66°)\ volts}$

Example 13.19. *Calculate the admittance Y, the conductance G and the susceptance B of a circuit consisting of a resistor of 10 Ω in series with an inductor of 0.1 H when the frequency is 50 Hz.*

Solution. Here, $\qquad R = 10\ \Omega\ ;\ L = 0.1\ H\ ;\ f = 50\ Hz$

$\qquad\qquad \mathbf{Z} = R + jX_L = R + j\,2\pi\,fL = 10 + j\,2\pi \times 50 \times 0.1$

$\qquad\qquad\qquad = 10 + j\,31.4 = 32.95\ \angle\,72.3°\ \Omega$

∴ $\qquad\qquad \mathbf{Y} = \dfrac{1}{\mathbf{Z}} = \dfrac{1}{32.95\ \angle\,72.3°} = 0.0303\ \angle -72.3°\ S$

$\qquad\qquad\qquad = 0.0092 - j\,0.029 = G + j\,B$

∴ $Y = \mathbf{0.0303\ S}\ ;\ G = \mathbf{0.0092\ S}\ ;\ B = \mathbf{-0.029\ S}$

Example 13.20. *A 2 Ω resistance, 0.125 F capacitance and 3 H inductance are connected in series across a voltage v = 12 sin (2t + 30°). (i) Find Z, I, V_L, V_C, power factor and active power. (ii) Write time functions for i, v_L and v_c.*

Solution. Here, $\mathbf{V} = \dfrac{12}{\sqrt{2}} \angle 30° = 8.487 \angle 30°$ volts ; $\omega = 2$ rad/sec.

$$X_L = \omega L = 2 \times 3 = 6\Omega \ ; X_C = \frac{1}{\omega C} = \frac{1}{2 \times 0.125} = 4\,\Omega$$

∴ Circuit impedance, $\mathbf{Z} = R + j(X_L - X_C) = 2 + j(6 - 4)$

$$= 2 + j2 = 2.828 \angle 45°\ \Omega$$

Circuit current, $\mathbf{I} = \dfrac{\mathbf{V}}{\mathbf{Z}} = \dfrac{8.487 \angle 30°}{2.828 \angle 45°} = 3 \angle -15°$ A

Voltage across L, $\mathbf{V_L} = j\,\mathbf{I}\,X_L = j\,(3 \angle -15°) \times 6 = 18 \angle 75°$ volts

Voltage across C, $\mathbf{V_C} = -j\,\mathbf{I}\,X_C = -j\,(3 \angle -15°) \times 4 = 12 \angle -105°$ volts

(*i*) ∴ $Z = \mathbf{2.828\ \Omega}\ ; I = \mathbf{3\ A}\ ; V_L = \mathbf{18\ volts}; V_C = \mathbf{12\ volts}$

Power factor $= \cos \phi = \cos (30° + 15°) = \mathbf{0.707}$ *lagging*

Active power, $P = VI \cos \phi = 8.487 \times 3 \times 0.707 = \mathbf{18\ W}$

(*ii*) The time functions for i, v_L and v_c are :

$i = \mathbf{3\sqrt{2}\ \sin(2t - 15°)}\ ; v_L = \mathbf{18\sqrt{2}\ \sin(2t + 75°)}\ ; v_c = \mathbf{12\sqrt{2}\ \sin(2t - 105°)}$

Example 13.21. *An alternating current of frequency 100 Hz is passed through a non-inductive 10 Ω resistor in series with a coil of resistance 1.3 Ω and inductance 0.018 H. When the terminal voltage is at its maximum value of 100 V, what will be the voltage across the resistor ?*

Solution.

Total circuit resistance, $R_T = 10 + 1.3 = 11.3\ \Omega$

Circuit reactance, $X_L = 2\,\pi f L = 2\pi \times 100 \times 0.018 = 11.3\ \Omega$

∴ Circuit impedance, $Z = (11.3 + j\,11.3)\ \Omega = 15.98 \angle 45°\ \Omega$

Since the circuit is inductive, circuit current lags behind the applied voltage by 45°.

Max. value of current, $I_m = V_m/Z = 100/15.98 = 6.25$ A

If $v = 100 \cos \omega t$, then current equation will be

$$i = I_m \cos(\omega t - 45°) = 6.25 \cos(\omega t - 45°)$$

Now v will be maximum when $t = 0$. The corresponding value of current is

$$i = 6.25 \cos(\omega \times 0 - 45°) = 6.25 \cos 45° = 4.42\ \text{A}$$

∴ Voltage drop across resistor $= iR = 4.42 \times 10 = \mathbf{44.2\ V}$

Example 13.22. *In the circuit shown in Fig. 13.20, calculate (i) current (ii) voltage drops V_1, V_2 and V_3 (iii) power absorbed by each impedance and total power absorbed by the circuit.*

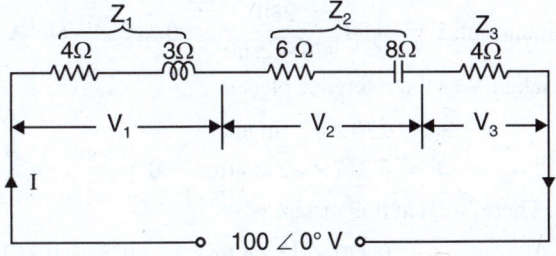

Fig. 13.20

Solution.

$$\mathbf{Z_1} = (4 + j\,3)\,\Omega\ ; \mathbf{Z_2} = (6 - j\,8)\,\Omega\ ; \mathbf{Z_3} = (4 + j\,0)\,\Omega$$

$\therefore$ Total circuit impedance, $\mathbf{Z_T} = \mathbf{Z_1} + \mathbf{Z_2} + \mathbf{Z_3}$

$$= (4 + j\,3) + (6 - j\,8) + (4 + j\,0) = (14 - j\,5)\,\Omega$$

Taking voltage as the reference phasor, $\mathbf{V} = 100\,\angle\,0°$ volts $= (100 + j\,0)$ volts.

(i)   Circuit current, $\mathbf{I} = \dfrac{\mathbf{V}}{\mathbf{Z_T}} = \dfrac{100}{(14 - j5)} = \dfrac{100\,(14 + j5)}{(14 - j5)\,(14 + j5)}$

$$= (6.34 + j\,2.26)\,A$$

$\therefore$ Magnitude of current, $I = \sqrt{(6.34)^2 + (2.26)^2}\ = \textbf{6.73 A}$

(ii) $\mathbf{V_1} = \mathbf{I}\,\mathbf{Z_1} = (6.34 + j\,2.26)\,(4 + j\,3)$

$$= (18.58 + j\,28.06)\ \text{volts} = \textbf{33.65}\ \angle\ \textbf{56.49°}\ \textbf{volts}$$

$$\mathbf{V_2} = \mathbf{I}\,\mathbf{Z_2} = (6.34 + j\,2.26)\,(6 - j\,8)$$

$$= (56.12 - j\,37.16)\ \text{volts} = \textbf{67.3}\ \angle - \textbf{33.51°}\ \textbf{volts}$$

$$\mathbf{V_3} = \mathbf{I}\,\mathbf{Z_3} = (6.34 + j\,2.26)\,(4 + j\,0)$$

$$= (25.36 + j\,9.04)\ \text{volts} = \textbf{26.92}\ \angle\ \textbf{19.62°}\ \textbf{volts}$$

(iii) $P_1 = (6.73)^2 \times 4 = \textbf{181.13 W};\ \ P_2 = (6.73)^2 \times 6 = \textbf{271.74 W};\ \ P_3 = (6.73)^2 \times 4 = \textbf{181.13 W}$

Total power, $P = P_1 + P_2 + P_3 = 181.13 + 271.74 + 181.13 = \textbf{634 W}$

Example 13.23. *A resistor of 5 Ω and an inductive reactance of 10 Ω are connected in series. Find the current and the power dissipated in the 5 Ω resistor if an alternating voltage of 200 V is applied across the circuit.*

Solution. Circuit impedance, $\mathbf{Z} = R + j\,X_L = (5 + j\,10)\,\Omega$

Taking voltage as reference phasor, $\mathbf{V} = 200\,(1 + j\,0) = 200$ volts

$\therefore$ Circuit current, $\mathbf{I} = \dfrac{\mathbf{V}}{\mathbf{Z}} = \dfrac{200}{5 + j10} = \dfrac{200\,(5 - j10)}{(5 + j\,10)\,(5 - j10)}$

$$= \dfrac{200\,(5 - j10)}{25 + 100} = (8 - j\,16)\,A$$

Magnitude of current, $I = \sqrt{8^2 + 16^2}\ = \textbf{17.9 A}$

Power dissipated in 5 Ω, $P = I^2 R = (17.9)^2 \times 5 = \textbf{1600 W}$

Example 13.24. *A coil with L = 2 H and R = 362 Ω is connected to a 230 V, 50 Hz supply. Calculate the energy stored when the instantaneous voltage is (i) zero (ii) a maximum.*

Solution. Reactance of coil, $X_L = 2\,\pi f L = 2\,\pi \times 50 \times 2 = 628\,\Omega$

Circuit impedance, $\mathbf{Z} = R + j\,X_L = (362 + j\,628)\,\Omega = 724.4\,\angle\,60°\,\Omega$

Circuit current, $\mathbf{I} = \dfrac{\mathbf{V}}{\mathbf{Z}} = \dfrac{230}{724.4\,\angle\,60°} = 0.317\,\angle - 60°\,A$

Note that we have taken $\mathbf{V}$ as the reference phasor.

(i) Now, $v = 230 \times \sqrt{2}\ \sin\,\omega t$

$\therefore$ $i = 0.317 \times \sqrt{2}\ \sin\,(\omega t - 60°)$

Now $v = 0$ at $t = 0$. Therefore, i at this instant is

$$i = 0.317 \times \sqrt{2}\ \sin\,(\omega \times 0 - 60°) = -0.317 \times \sqrt{2}\ \sin 60°$$

$$= -0.388\,A$$

$\therefore$ Energy stored $= \dfrac{1}{2} i^2 L = \dfrac{1}{2} \times (0.388)^2 \times 2 = \textbf{0.152 J}$

(ii) v is maximum when $\omega t = 90°$. Therefore, current at this instant is

$$i = 0.317 \times \sqrt{2} \sin(90° - 60°) = 0.317 \times \sqrt{2} \sin 30° = 0.224 \text{ A}$$

$\therefore$ Energy stored $= \dfrac{1}{2}i^2 L = \dfrac{1}{2} \times (0.224)^2 \times 2 = \textbf{0.051 J}$

Example 13.25. *If the potential drop across a circuit be represented by (40 + j 25) V with reference to the circuit current and power absorbed by the circuit is 160 W, find the complex expression for the impedance. Find also (i) the power factor of the circuit and (ii) the magnitude of the impedance.*

Solution. Fig. 13.21 shows the phasor diagram of the circuit. The drop across resistance $= I R$ = 40 volts and that across inductive reactance $= I X_L = 25$ volts.

Now, $I R = 40$ volts and $I^2 R = 160$ W

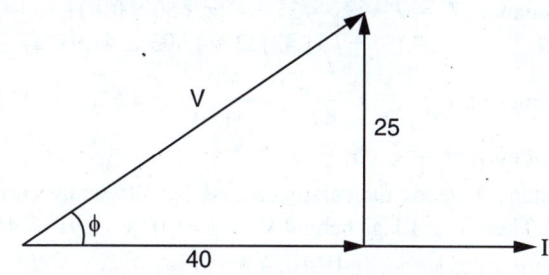

Fig. 13.21

$\therefore$ $I = I^2 R / I R = 160/40 = 4$A

Now $\mathbf{I} = 4 \angle 0° \text{ A} ; \mathbf{V} = (40 + j\,25) \text{ volts} = 47.2 \angle 32° \text{ volts}$

$\therefore$ $\mathbf{Z} = \dfrac{\mathbf{V}}{\mathbf{I}} = \dfrac{47.2 \angle 32°}{4 \angle 0°} = \textbf{11.8} \angle \textbf{32°}\ \Omega$

(ii) $\therefore$ $Z = \textbf{11.8}\ \Omega$

(i)   Power factor $= \cos \phi = \cos 32° = \textbf{0.848}\ lag$

Example 13.26. *For the circuit shown in Fig. 13.22 (i), determine the (i) magnitude of current (ii) power factor of the circuit.*

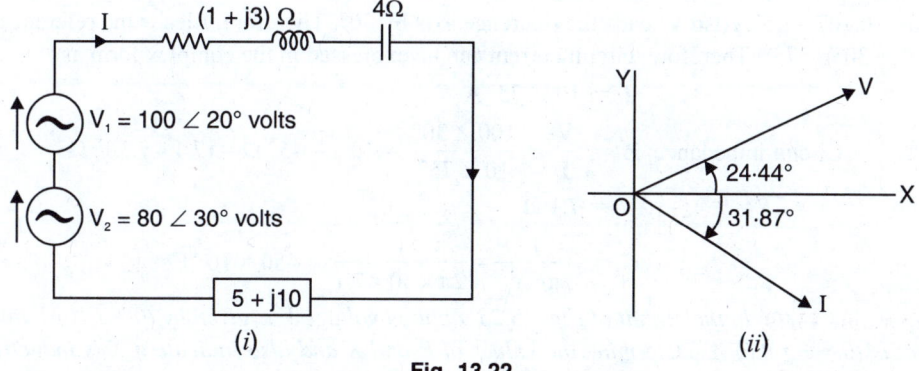

(i)   *(ii)*

Fig. 13.22

Solution. *(i)* Resultant voltage, $\mathbf{V} = \mathbf{V_1} + \mathbf{V_2} = 100 \angle 20° + 80 \angle 30°$

$= (93.97 + j\,34.2) + (69.28 + j\,40)$

$= (163.25 + j\,74.2) \text{volts} = 179.32 \angle 24.44° \text{ volts}$

Total impedance, $\mathbf{Z} = (1 + j\,3) + (-j\,4) + 5 + j\,10$

$= (6 + j\,9)\ \Omega = 10.82 \angle 56.31°\ \Omega$

$\therefore$ Circuit current, $\mathbf{I} = \dfrac{\mathbf{V}}{\mathbf{Z}} = \dfrac{179.32 \angle 24.44°}{10.82 \angle 56.31°} = 16.57 \angle -31.87°$ A

$\therefore$ $I = \mathbf{16.57\ A}$

(ii)   Phase angle, $\phi = 24.44° + 31.87° = 56.31°$

$\therefore$ Power factor $= \cos \phi = \cos 56.31° = \mathbf{0.554}\ lag$

Fig. 13.22 *(ii)* shows the phasor diagram of the circuit.

Example 13.27. *In a given R–L series circuit, R = 35 Ω and L = 0.1 H. Find (i) current through the circuit (ii) power factor if a 50 Hz frequency, voltage V = 220 ∠ 30°V is applied across the circuit.*

Solution. $\mathbf{V} = 220 \angle 30°$ volts ; $R = 35\ \Omega$; $L = 0.1\text{H}$; $f = 50\ \text{Hz}$

Circuit impedance, $\mathbf{Z} = R + jX_L = 35 + j\,2\pi \times 50 \times 0.1$ H

$\qquad\qquad\qquad\qquad = (35 + j\,31.42)\ \Omega = 47.03 \angle 41.91°\ \Omega$

(i)   Circuit current, $\mathbf{I} = \dfrac{\mathbf{V}}{\mathbf{Z}} = \dfrac{220 \angle 30°}{47.03 \angle 41.91°} = 4.678 \angle -11.91°$ A

Magnitude of current $= \mathbf{4.678\ A}$

(ii) The applied voltage **V** leads the reference axis by 30° while current **I** lags behind the reference axis by 11.91°. Therefore, **I** lags behind **V** by $\phi = 30° + 11.91° = 41.91°$.

$\therefore$ Circuit power factor $= \cos \phi = \cos 41.91° = \mathbf{0.744}\ lagging$

Example 13.28. *A two-element series circuit consumes 700 W and has a p.f. = 0.707 leading. If applied voltage is v = 141.1 sin (314t + 30°) V, find the circuit constants.*

Solution. $v = 141.1 \sin (314t + 30°)$ volts

Now, $V = 141.1/\sqrt{2} = 100$ volts. The applied voltage leads the reference axis by 30°.

$\therefore$ $\mathbf{V} = 100 \angle 30°$ volts

Now, $P = 700$ W ; p.f., $\cos \phi = 0.707$ *lead*

$\therefore$ $P = VI \cos \phi$ or $700 = 100 \times I \times 0.707$ $\therefore I = 10$ A

As the p.f. is 0.707 leading, it means that circuit current **I** leads the applied voltage **V** by $\phi = \cos^{-1} 0.707 = 45°$. Also **V** leads the reference axis by 30°. Therefore, **I** leads the reference axis by $(45° + 30°) = 75°$. Therefore, circuit current can be expressed in the complex form as :

$\mathbf{I} = 10 \angle 75°$ A

$\therefore$ Circuit impedance, $\mathbf{Z} = \dfrac{\mathbf{V}}{\mathbf{I}} = \dfrac{100 \angle 30°}{10 \angle 75°} = 10 \angle -45°\ \Omega = (7.1 - j\,7.1)\ \Omega$

$\therefore$ $R = \mathbf{7.1\ \Omega}$; $X_C = 7.1\ \Omega$

$\therefore$ $C = \dfrac{1}{2\pi f\,X_C} = \dfrac{1}{2\pi \times 50 \times 7.1} = 450 \times 10^{-6} \text{F} = \mathbf{450\ \mu F}$

Example 13.29. *In the circuit of Fig. 13.23, applied voltage V is given by (0 + j 10) V and the current is (0.8 + j 0.6) A. Determine the values of R and X and also indicate if X is inductive or capacitive.*

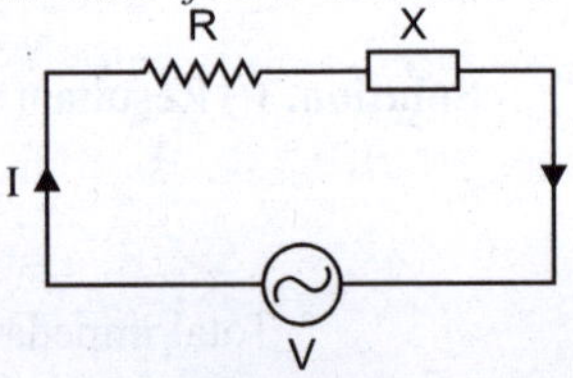

Solution. Applied voltage, $\mathbf{V} = (0 + j\,10) = 10 \angle 90°$ volts ; Circuit current, $\mathbf{I} = (0.8 + j\,0.6)$ A $= 1 \angle 36.9°$ A. It is clear that **V** leads the reference axis by 90° while **I** leads the reference axis by 36.9°. Therefore, **I** lags behind **V** by an angle ϕ given by ;

$\phi = 90° - 36.9° = 53.1°$

Fig. 13.23

$$\text{Circuit impedance, } \mathbf{Z} = \frac{\mathbf{V}}{\mathbf{I}} = \frac{10\angle 90°}{1\angle 36.9°}$$

$$= 10\angle 53.1° \,\Omega = (6 + j\,8)\,\Omega$$

∴ $R = 6\,\Omega$ and $X_L = 8\,\Omega$ (*inductive*)

Example 13.30. *In a circuit, the applied voltage is found to lag the current by 30°.*

(a) Is the power factor lagging or leading? (b) What is the value of the power factor? (c) Is the circuit inductive or capacitive? In the diagram of Fig. 13.24, the voltage drop across Z_1 is (10 + j 0) V. Find out (i) the current in the circuit (ii) the voltage drops across Z_2 and Z_3 (iii) the voltage of the generator.

Solution. (*a*) Since circuit current leads the applied voltage, the power factor is **leading.**

(*b*) Power factor, $\cos\phi = \cos 30° = $ **0.866** *leading*

(*c*) Since circuit current leads the applied voltage, the circuit is **capacitive.**

(*i*) Since $\mathbf{V_1}$ and $\mathbf{Z_1}$ are known, the circuit current **I** can be found from these values *i.e.,*

$$\mathbf{I} = \frac{\mathbf{V_1}}{\mathbf{Z_1}} = \frac{10 + j\,0}{3 + j\,4} = \frac{10\angle 0°}{5\angle 53.1°}$$

$$= 2\angle -53.1°\text{A} = \mathbf{(1.2 - j\,1.6)\,A}$$

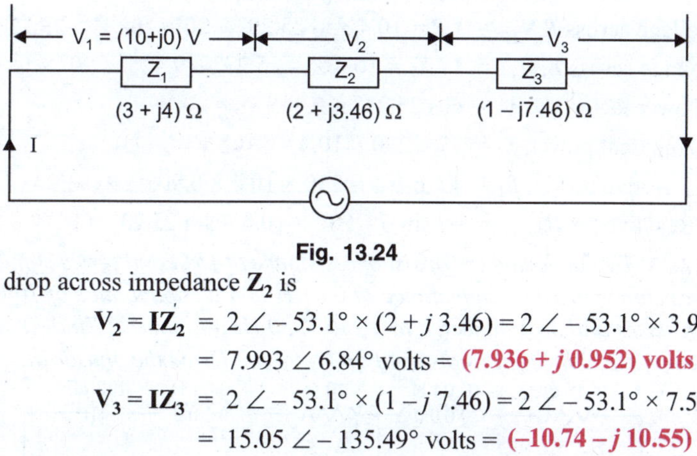

Fig. 13.24

(*ii*) Voltage drop across impedance $\mathbf{Z_2}$ is

$$\mathbf{V_2} = \mathbf{IZ_2} = 2\angle -53.1° \times (2 + j\,3.46) = 2\angle -53.1° \times 3.996\angle 59.97°$$

$$= 7.993\angle 6.84° \text{ volts} = \mathbf{(7.936 + j\,0.952)\ volts}$$

Similarly, $\mathbf{V_3} = \mathbf{IZ_3} = 2\angle -53.1° \times (1 - j\,7.46) = 2\angle -53.1° \times 7.527\angle -82.36°$

$$= 15.05\angle -135.49° \text{ volts} = \mathbf{(-10.74 - j\,10.55)\ volts}$$

(*iii*)   Generator voltage, $\mathbf{V} = \mathbf{V_1} + \mathbf{V_2} + \mathbf{V_3} = (10 + j0) + (7.936 + j0.952) + (-10.74 - j10.55)$

$$= (7.2 - j\,9.6) \text{ volts} = \mathbf{12\angle -53.1°\ volts}$$

Example 13.31. *A load having impedance of (1 + j1) Ω is connected to an a.c. voltage represented as v = $20\sqrt{2}$ cos(ωt + 10°) V.*

(i) Find the current in load, expressed in the form of i = I_m sin (ωt + ϕ)A.

(ii) Find real power consumed by the load.

Solution. (*i*) Load impedance, $\mathbf{Z} = (1 + j\,1)\,\Omega = \sqrt{2}\angle 45°\,\Omega$

Now $v = 20\sqrt{2}\,\cos(\omega t + 10°) = 20\sqrt{2}\,\sin(\omega t + 90° + 10°) = 20\sqrt{2}\,\sin(\omega t + 100°)$

Also $V_m = 20\sqrt{2}$ volts ; $V\text{(r.m.s.)} = \dfrac{20\sqrt{2}}{\sqrt{2}} = 20$ volts

∴   Load current, $\mathbf{I} = \dfrac{\mathbf{V}}{\mathbf{Z}} = \dfrac{20\angle 100°}{\sqrt{2}\angle 45°} = 10\sqrt{2}\angle 55°$

Peak value of load current is given by ;

$$I_m = I \times \sqrt{2} = 10\sqrt{2} \times \sqrt{2} = 20 \text{ A}$$

$\therefore \qquad i = I_m \sin(\omega t + \phi) = \textbf{20 sin }(\boldsymbol{\omega}\textbf{t + 55°})$

(*ii*) Real power consumed, $P = VI \cos\phi = 20 \times 10\sqrt{2} \times \cos 45° = \textbf{200 W}$

Example 13.32. *A 20 Ω resistance and a 30 mH inductance are connected in series across 230V, 50 Hz supply. Find :*

(i) Inductive reactance, impedance and admittance in polar form.

(ii) Current, voltage across resistance and voltage across inductance in polar form.

(iii) Apparent power, active power, reactive power and power factor.

Solution. $X_L = 2\pi fL = 2\pi \times 50 \times 30 \times 10^{-3} = 9.425 \ \Omega$

(*i*) Inductive reactance $= j X_L = j\,9.425 \ \Omega = \textbf{9.425} \angle \textbf{90°} \ \boldsymbol{\Omega}$

Circuit impedance, $\mathbf{Z} = R + jX_L = (20 + j\,9.425)\ \Omega = \textbf{22.11} \angle \textbf{25.23°} \ \boldsymbol{\Omega}$

Circuit admittance, $\mathbf{Y} = \dfrac{1}{\mathbf{Z}} = \dfrac{1}{22.11 \angle 25.23°} = \textbf{0.04523} \angle \textbf{– 25.23°} \ \textbf{S}$

(*ii*) Circuit current, $\mathbf{I} = \dfrac{\mathbf{V}}{\mathbf{Z}} = \dfrac{230 \angle 0°}{22.11 \angle 25.23°} = \textbf{10.4} \angle \textbf{– 25.23°} \ \textbf{A}$

Voltage across $R, \mathbf{V_R} = \mathbf{I}\,R = 10.4 \angle -25.23° \times 20 = \textbf{208} \angle \textbf{– 25.23°} \ \textbf{V}$

Voltage across $L, \mathbf{V_L} = \mathbf{I}\,jX_L = 10.4 \angle -25.23° \times 9.425 \angle 90° = \textbf{98.02} \angle \textbf{64.77°} \ \textbf{V}$

(*iii*) Power factor, $\cos\phi = \cos 25.23° = \textbf{0.905} \textit{ lagging}$

Apparent power, $S = VI = 230 \times 10.4 = \textbf{2392 VA}$

Active power, $P = VI \cos\phi = 230 \times 10.4 \times 0.905 = \textbf{2164.7 W}$

Reactive power, $Q = VI \sin\phi = 230 \times 10.4 \times \sin 25.23° = \textbf{1019.6 VAR}$

Example 13.33. *Two impedances consist of resistance of 15 Ω and series connected inductance of 0.04 H and resistance of 10 Ω, inductance of 0.01 H and a capacitance of 100 μF, all in series are connected in series and are connected to a 230 V, 50 Hz a.c. source. Find (i) current drawn (ii) voltage across each impedance (iii) individual and total p.f. Draw the phasor diagram.*

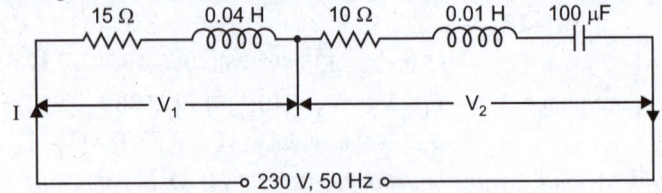

Fig. 13.25

Solution. Fig.13.25 shows the conditions of the problem. Let us assign suffix 1 to coil 1 and suffix 2 to coil 2.

(*i*) $\qquad R_1 = 15\ \Omega \ ; X_{L1} = 2\pi \times 50 \times 0.04 = 12.56\ \Omega$

$\therefore \qquad \mathbf{Z_1} = R_1 + jX_{L1} = (15 + j\,12.56)\ \Omega = 19.56 \angle 40°\ \Omega$

$\therefore \qquad$ Magnitude, $Z_1 = 19.56\ \Omega$; Impedance angle, $\theta_1 = 40°$

$$R_2 = 10\ \Omega\ ; X_{L2} = 2\pi \times 50 \times 0.01 = 3.14\ \Omega\ ; X_{C2} = \dfrac{1}{2\pi \times 50 \times 100 \times 10^{-6}} = 31.85\ \Omega$$

$\therefore \qquad \mathbf{Z_2} = R_2 + j(X_{L2} - X_{C2}) = 10 + j(3.14 - 31.85)$

$$= (10 - j\,28.71)\ \Omega = 30.4 \angle -70.8°\ \Omega$$

∴   Magnitude, Z_2 = 30.4 Ω ; Impedance angle, $\theta_2 = -70.8°$

Total impedance, $\mathbf{Z} = \mathbf{Z_1} + \mathbf{Z_2} = (15 + j\,12.56) + (10 - j\,28.71)$

$$= (25 - j\,16.15)\,\Omega = 29.76 \angle -32.86°\,\Omega$$

Magnitude, Z = 29.76 Ω ; Impedance angle, $\theta = -32.86°$

Circuit current, $\mathbf{I} = \dfrac{\mathbf{V}}{\mathbf{Z}} = \dfrac{230 \angle 0°}{29.76 \angle -32.86°}$ = **7.73 ∠ 32.86° A**

(*ii*)   Voltage across $\mathbf{Z_1}$, $\mathbf{V_1} = \mathbf{I}\,\mathbf{Z_1}$ = 7.73 ∠ 32.86° × 19.56 ∠ 40° = **151.2 ∠ 72.86° V**

Voltage across $\mathbf{Z_2}$, $\mathbf{V_2} = \mathbf{I}\,\mathbf{Z_2}$ = 7.73 ∠ 32.86° × 30.4 ∠ – 70.8° = **235 ∠ – 37.94° V**

(*iii*)   P.F. of first coil $= \dfrac{R_1}{Z_1} = \dfrac{15}{19.56}$ = **0.766** *lagging*

P.F. of second coil $= \dfrac{R_2}{Z_2} = \dfrac{10}{30.4}$ = **0.329** *leading*

Total P.F. $= \dfrac{R_1 + R_2}{Z} = \dfrac{15 + 10}{29.76}$ = **0.84 leading**

The total p.f. is leading because net reactance in the circuit is capacitive.

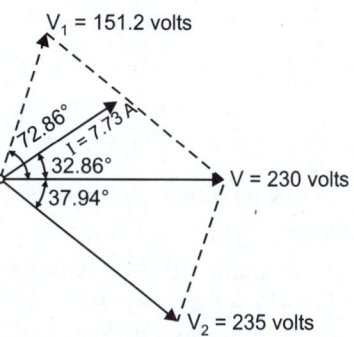

Fig. 13.26

Fig. 13.26 shows the phasor diagram of the circuit. Note that applied voltage V has been taken as the reference phasor. The voltage V_1 across first coil leads the applied voltage V by 72.86° while the voltage V_2 across second coil lags behind the applied voltage V by 37.94°. The circuit current I leads the applied voltage by 32.86° so that the circuit is capacitive.

Example 13.34. *In the circuit of Fig. 13.27 (i), find the values of R and C so that $V_b = 3\,V_a$ and phase angle between V_b and V_a is 90°. Draw phasor diagram of the circuit taking current I as the reference phasor.*

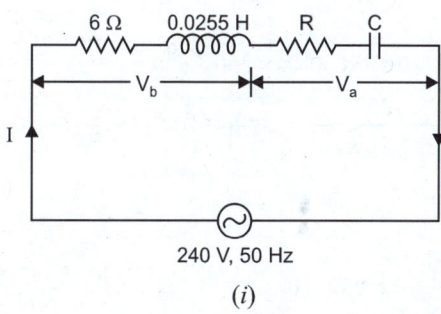

(*i*)

Fig. 13.27

(*ii*)

Solution. $X_L = 2\pi\,fL = 2\pi \times 50 \times 0.0255 = 8\ \Omega$

∴ $\mathbf{Z_b} = (6 + j\,8)\,\Omega = 10 \angle 53.13°\ \Omega$

Let $\mathbf{I} = I \angle \theta$

∴ $\mathbf{V_b} = I \angle \theta \times 10 \angle 53.13° = 10\,I \angle \theta + 53.13°$ volts

and $\mathbf{V_a} = I \angle \theta \times (R - jX_C) = I\sqrt{R^2 + X_C^2}\ \angle\,\theta - \tan^{-1}\dfrac{X_C}{R}$

It is given that : $V_b = 3\,V_a$

$\therefore$ $10\,I = 3 \times I\sqrt{R^2 + X_C^2}$

or $R_2 + X_C^2 = \dfrac{100}{9}$ ...(i)

Moreover, phase angle between V_b and V_a is to be 90° i.e.

$(\theta + 53.13°) - \left(\theta - \tan^{-1}\dfrac{X_C}{R}\right) = 90°$

or $\dfrac{X_C}{R} = 0.75$ ...(ii)

From eqs. (i) and (ii), $R = \textbf{2.67 } \Omega$; $X_C = 2\,\Omega$

Now, $X_C = \dfrac{1}{2\pi\,fC}$ or $2 = \dfrac{1}{2\pi \times 50 \times C}$ $\therefore C = \textbf{1591} \times \textbf{10}^{-6}$ **F**

Total impedance, $\mathbf{Z} = (6 + j\,8) + (2.67 - j\,2) = (8.67 + j\,6)\,\Omega = 10.54\,\angle\,34.68°\,\Omega$

$\therefore$ Circuit current, $\mathbf{I} = \dfrac{\mathbf{V}}{\mathbf{Z}} = \dfrac{240\,\angle\,0°}{10.54\,\angle\,34.68°} = 22.77\,\angle - 34.68°$ A

$\therefore$ $\mathbf{V}_b = \mathbf{I\,Z_b} = 22.77\,\angle - 34.68° \times 10\,\angle\,53.13° = 227.7\,\angle\,18.45°$ V

$\mathbf{V}_a = \mathbf{I\,Z_a} = 22.77\,\angle - 34.68° \times (2.67 - j\,2)$

$= 22.77\,\angle - 34.68° \times 3.34\,\angle - 36.83° = 76.05\,\angle - 71.5°$ V

Taking current (I) as the reference phasor, the phasor diagram of the circuit will be as shown in Fig. 13.27 (ii).

Example 13.35. *An inductor having a resistance of 25 Ω and Q of 10 at a resonant frequency of 10 kHz is fed from a 100 $\angle$ 0° V supply. Calculate the (i) value of series capacitance required to produce resonance with the coil (ii) inductance of the coil (iii) Q using the L/C ratio (iv) voltage across the capacitor (v) voltage across the coil.*

Solution. (i) $X_L = QR = 10 \times 25 = 250\,\Omega$

At resonance : $X_C = X_L = 250\,\Omega$

$\therefore$ $C = \dfrac{1}{2\pi f_r\,X_C} = \dfrac{1}{2\pi \times 10 \times 10^3 \times 250} = \textbf{63.6} \times \textbf{10}^{-9}$ **F**

(ii)   Inductance, $L = \dfrac{X_L}{2\pi\,f_r} = \dfrac{250}{2\pi \times 10 \times 10^3} = 3.98 \times 10^{-3}$ H $= \textbf{3.98 mH}$

(iii)   Quality factor, $Q = \dfrac{1}{R}\sqrt{\dfrac{L}{C}}$

Now, $\dfrac{L}{C} = \dfrac{3.98 \times 10^{-3}}{63.6 \times 10^{-9}} = 6.25 \times 10^4$

$\therefore$ $Q = \dfrac{1}{25} \times \sqrt{6.25 \times 10^4} = 10$

This value agrees with the given value.

(iv)   Voltage across $\mathbf{V_C} = -j\,QV = -j \times 10 \times 100\,\angle\,0° = \textbf{- 1000} \angle \textbf{ 90° volts}$

(v) Since $\mathbf{V_L}$ and $\mathbf{V_C}$ are equal and opposite,

$\therefore$ $\mathbf{V_L} = -\mathbf{V_C} = -(-1000\,\angle\,90°) = +1000\,\angle\,90°$ volts

$V_{coil} = \mathbf{V_R} + \mathbf{V_L} = 100 + 1000\,\angle\,90° = 100 + j\,1000 = \textbf{1005} \angle \textbf{ 84.3° volts}$

Tutorial Problems

1. Write the polar form for the following :

   (*i*) $- 4 + j\,6{\cdot}928$ (*ii*) $-50 - j\,75$ [(*i*) 8 $\angle$ 120° (*ii*) 90.1 $\angle$ –123.7°]

2. Obtain the sum of $\mathbf{E}_1 = 100 \angle 45°$ and $\mathbf{E}_2 = 80 \angle 120°$ in (*i*) rectangular form and (*ii*) polar form.

                                                     [(*i*) 30.7 + *j*140 (*ii*) 143 $\angle$ 77.6°]

3. Perform the indicated operations and express the results in polar form :

   (*i*) $(5 + j12)^2$

   (*ii*) Add $10 \angle 40°$ to $15 \angle 60°$ and divide the result by $(20 \angle 45° - 12 \angle 180°)$

   (*iii*) $3 \angle 30° + 5 \angle - 50° - 7 \angle 20° + (6 + j5)$

                                   [(*i*) 169 $\angle$ 134.8° (*ii*) 1.715 $\angle$ – 29.35° (*iii*) 5.24 $\angle$ 3.06°]

4. Determine the resistance and inductance of series connected elements that will draw a current of $20 \angle - 30°$ A from a 60 Hz sinusoidal generator whose voltage is $100 \angle 0°$ V.   [4.35Ω ; 6.6 mH]

5. A coil, a capacitor and a resistor are connected in series and supplied by 1.5 kHz generator. The capacitive reactance is 4 Ω, the resistor is 4 Ω and the coil has a resistance of 3 Ω and an inductive reactance of 4 Ω. If the current in the circuit is $10 \angle 20°$ A, determine (*i*) complex impedance (*ii*) voltage drop across coil in polar form.   [(*i*) 7 $\angle$ 0° Ω (*ii*) 50 $\angle$ 73.13° V]

6. The p.d. across and the current in a circuit are represented by $(100 + j\,200)$ V and $(10 + j5)$ A respectively. Calculate the power consumed and the reactive power.   **[2000 W ; 1500 VAR]**

7. In an R-L-C series circuit, $L = 25$ mH, $C = 59$ mF. Find the value of R so that current leads the voltage by 63.4° at a supply frequency of $400/2\pi$.   **[20 Ω]**

8. The current in a circuit is given by $(4.5 + j12)$A when the applied voltage is $(100 + j150)$ V. Determine (*i*) the complex expression for the impedance (*ii*) power (*iii*) the phase angle between voltage and current.

                                       [(*i*) (13.7 – *j*3.2) Ω (*ii*) 2250 W (*iii*) 13.9°]

9. A resistance of 20 Ω in series with an inductance of 0.1 H is connected to 230 V, 50 Hz supply. Using complex notation, determine (*i*) the magnitude of impedance (*ii*) circuit current (*iii*) power factor and (*iv*) power consumed.   [(*i*) 37.23 Ω (*ii*) 6.18 A (*iii*) 0.537 *lag* (*iv*) 763.3 W]

10. An impedance of $(3 + j4)$ Ω is connected in series with another impedance of $(5 + j8)$ Ω. If a voltage of $(80 + j60)$ V is applied, find the power dissipated in the circuit.   **[383 W]**

13.15. A.C. Voltage Divider

The voltage-divider rule as applied to series resistors in a d.c. circuit can also be applied to series impedances in an a.c. circuit. As always in an a.c. circuit, care must be taken to ensure that the phase angles are taken into account. Fig. 13.28 shows a series a.c. circuit consisting of a number of impedances. The voltage $\mathbf{v}_x$ across any impedance $\mathbf{Z}_x$ can be found by multiplying the applied circuit voltage by the ratio $\mathbf{Z}_x/\mathbf{Z}_T$ i.e.

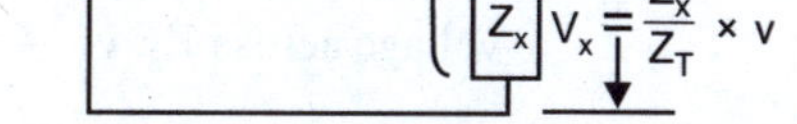

$$\text{Voltage across } \mathbf{Z}_x, \mathbf{v}_x \;=\; \mathbf{v} \times \frac{\mathbf{Z}_x}{\mathbf{Z}_T}$$

where $\mathbf{Z}_T$ = Total circuit impedance = $\mathbf{Z}_1 + \mathbf{Z}_2 + ...$

Fig. 13.28

$$\therefore \text{ Voltage across } \mathbf{Z}_1, \; \mathbf{v}_1 = \mathbf{v} \times \frac{\mathbf{Z}_1}{\mathbf{Z}_T}$$

$$\text{Voltage across } \mathbf{Z}_2, \; \mathbf{v}_2 = \mathbf{v} \times \frac{\mathbf{Z}_2}{\mathbf{Z}_T} \text{ and so on.}$$

Note that voltage-divider rule for a.c. circuits is identical in concept to that for d.c. circuits. In fact, voltage-divider rule is a useful shortcut to determine how series impedances will divide voltage.

Example 13.36. *Use the voltage-divider rule to find the voltage across 0.05 μF capacitor in Fig. 13.29.*

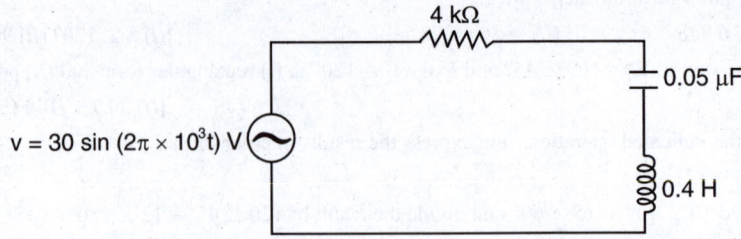

Fig. 13.29

Solution.
$$X_C = \frac{1}{\omega C} = \frac{1}{\left(2\pi \times 10^3\right)\left(5 \times 10^{-8}\right)} = 3183 \; \Omega$$

$$X_L = \omega L = (2\pi \times 10^3)(0.4) = 2513 \; \Omega$$

∴   Total impedance of the circuit is given by ;

$$Z_T = 4000 + j(2513 - 3183) = (4000 - j670) \; \Omega = 4055 \angle -9.51° \; \Omega$$

By voltage-divider rule, voltage across the capacitor is

$$v_C = \frac{X_C}{Z_T} \times v = \frac{3183 \angle -90°}{4055 \angle -9 \cdot 51°} \times 30 \angle 0° = 23.55 \angle -80.49° \; V$$

∴ $$v_C = 23.55 \sin (2\pi \times 10^3 t - 80.49°) \; V \; \textbf{Ans.}$$

Example 13.37. *An a.c. voltage divider consists of three impedances $Z_1 = 70.7 \angle 45° \; \Omega$; $Z_2 = 92.4 \angle 330° \; \Omega$ and $Z_3 = 67 \angle 60° \; \Omega$ in series. Find the voltage across each-impedance if the supply voltage is 100 V.*

Solution. The total circuit impedance Z_T is

$$Z_T = Z_1 + Z_2 + Z_3 = 70.7 \angle 45° + 92.4 \angle 330° + 67 \angle 60°$$

$$= (70.7 \cos 45° + 70.7 \sin 45°) + (92.4 \cos 330° + 92.4 \sin 330°)$$
$$+ (67 \cos 60° + 67 \sin 60°)$$

$$= (50 + j \, 50) + (80 - j \, 46.2) + (33.5 + j \, 58)$$

$$= (163.5 + j \, 61.8) \; \Omega = 174.8 \angle 20.7° \; \Omega$$

Voltage across Z_1, $v_1 = v \times \dfrac{Z_1}{Z_T} = 100 \angle 0° \times \dfrac{70.7 \angle 45°}{174.8 \angle 20.7°} = \mathbf{40.4 \angle 21.3° \; V}$

Voltage across Z_2, $v_2 = v \times \dfrac{Z_2}{Z_T} = 100 \angle 0° \times \dfrac{92.4 \angle 330°}{174.8 \angle 20.7°} = \mathbf{52.9 \angle 309.3° \; V}$

Voltage across Z_3, $v_3 = v \times \dfrac{Z_3}{Z_T} = 100 \angle 0° \times \dfrac{67 \angle 60°}{174.8 \angle 20.7°} = \mathbf{38.3 \angle 39.3° \; V}$

Objective Questions

1. When a phasor is multiplied by $-j$, it is rotated through .... in the counter-clockwise direction.

   (i) 90°   (ii) 180°

   (iii) 270°   (iv) none of the above

2. The value of j^5 is equal to ....

   (i) 1   (ii) $\sqrt{-1}$

   (iii) −1   (iv) $-\sqrt{-1}$

3. When a phasor is multiplied by j^6, it is rotated through .... in counterclockwise direction.

   (i) 540°   (ii) 360°

   (iii) 90°   (iv) 270°

4. If a phasor is multiplied by j, then ....

   (i) only its magnitude changes

   (ii) only its direction changes

   (iii) both magnitude and direction change

   (iv) none of the above

5. In the complex number $4 + j\,7$, 7 is called the .... component.

   (i) real   (ii) imaginary

   (iii) in-phase   (iv) quadrature

6. The reciprocal of a complex number results in a ....

   (i) complex number

   (ii) real component only

   (iii) quadrature component only

   (iv) none of the above

7. Rationalising the denominator of a complex number eliminates ....

   (i) j component in the numerator

   (ii) j component in the denominator

   (iii) j component in both numerator and denominator

   (iv) none of above

8. If two complex numbers are equal, then ....

   (i) only their magnitudes will be equal

   (ii) only their angles will be equal

   (iii) their in-phase and quadrature components will be separately equal

   (iv) none of the above

9. A phasor $2 \angle 180°$ can be expressed as ....

   (i) $j\,2$ (ii) $-j\,2$

   (iii) -2 (iv) 2

10. A current of $(3 + j\,4)$ amperes is flowing through a circuit. The magnitude of current is ....

   (i) 7 A   (ii) 5 A

   (iii) 1 A   (iv) 1.33 A

11. A complex number is given by $a + j\,b$. The magnitude of this complex number will be ....

   (i) less than either a or b

   (ii) more than the sum of a and b

   (iii) less than 1.414 × whichever is the greater a or b

   (iv) none of the above

12. The voltage applied in a circuit is given by $100 \angle 60°$ volts. It can be written as ....

   (i) $100 \angle -60°$ volts

   (ii) $100 \angle 240°$ volts

   (iii) $100 \angle -300°$ volts

   (iv) none of the above

13. The phasor $8 \angle 60°$ is identical to ....

   (i) $8 \angle 240°$ (ii) $8 \angle 780°$

   (iii) $8 \angle 310°$ (iv) none of the above

14. For addition or subtraction of phasors, we use .... form.

   (i) polar   (ii) rectangular

   (iii) trigonometrical (iv) none of the above

15. The quantity $2 + j\,2$ can be expressed in the polar form as ....

   (i) $2\sqrt{2} \angle 45°$ (ii) $4 \angle 90°$

   (iii) $2\sqrt{2} \angle -45°$ (iv) $1 \angle 90°$

16. For multiplication or division of phasors, we use .... form.

   (i) rectangular   (ii) trigonometrical

   (iii) polar   (iv) none of the above

17. If the impedance of an a.c. circuit is $10 \angle 60°$ Ω, then resistance in the circuit is ....

   (i) 5 Ω   (ii) 8.66 Ω

   (iii) 10 Ω   (iv) none of the above

18. The conjugate of $-4 + j\,3$ is ....

   (i) $4 - j\,3$ (ii) $-4 - j\,3$

   (iii) $4 + j\,3$ (iv) none of the above

19. The conjugate of $10 \angle -40°$ is ....

   (i) $0.1 \angle 40°$ (ii) $0.1 \angle -40°$

   (iii) $10 \angle 40°$ (iv) $5 \angle -20°$

20. The sum of two conjugate numbers results in ....

   (i) a complex number

   (ii) in-phase component only

   (iii) quadrature component only

   (iv) none of the above

21. The difference of two conjugate numbers results in ....

   (i) a complex number

   (ii) in-phase component only

   (iii) quadrature component only

   (iv) none of the above

22. When two conjugate numbers are multiplied, the result is ....

   (i) a complex number

    (*ii*) no in-phase component

    (*iii*) no quadrature component

    (*iv*) none of the above

23. The value of $5 \angle 40° \times 3 \angle 20°$ is ....

    (*i*) $15 \angle 60°$ (*ii*) $15 \angle 20°$

    (*iii*) $1.6 \angle 2°$ (*iv*) $15 \angle 800°$

24. The value of $9 \angle 30° \div 3 \angle 10°$ is

    (*i*) $27 \angle 40°$ (*ii*) $3 \angle 20°$

    (*iii*) $3 \angle 40°$ (*iv*) $3 \angle 3°$

25. The square root of $64 \angle 36°$ is ....

    (*i*) $8 \angle 6°$ (*ii*) $8 \angle 18°$

    (*iii*) $8 \angle 36°$ (*iv*) none of the above

26. The cube root of $8 \angle 30°$ is ....

    (*i*) $2 \angle 10°$ (*ii*) $2 \angle 30°$

    (*iii*) $8 \angle 10°$ (*iv*) none of the above

27. The kVA drawn by an a.c. circuit is given by $(3 + j\,4)$ kVA. The active power drawn by the circuit is ....

    (*i*) 3 kW (*ii*) 4 kW

    (*iii*) 5 kW (*iv*) none of the above

28. The impedance of an a.c. circuit is $45 \angle -30°\Omega$. The circuit will be ....

    (*i*) resistive   (*ii*) inductive

    (*iii*) capacitive   (*iv*) *none of the above

29. In an a.c. circuit, the apparent power in complex form is equal to ....

    (*i*) phasor voltage $\times$ conjugate of phasor current

    (*ii*) phasor voltage $\div$ conjugate of phasor current

    (*iii*) phasor voltage $\times$ phasor current

    (*iv*) none of the above

30. A series a.c. circuit has $R = 4 \Omega$ and $X_L = 3 \Omega$. It will be expressed in the rectangular form as ....

    (*i*) $(-4 - j\,3)\,\Omega$ (*ii*) $(-4 + j\,3)\,\Omega$

    (*iii*) $(4 + j\,3)\,\Omega$ (*iv*) $(4 - j\,3)\,\Omega$

31. The two complex numbers are given as :

    $5 (\cos 30° + j \sin 30°)$; $10 (\cos 15° + j \sin 15°)$

    The value of product of these two numbers is equal to ....

    (*i*) $50 (\cos 45° + j \sin 45°)$

    (*ii*) $50 (\cos 15° + j \sin 15°)$

    (*iii*) $50 (\cos 450° + j \sin 450°)$

    (*iv*) none of the above

32. The voltage and current in an a.c. series circuit are $230 \angle 0°$ volts and $100 \angle 30°$ A respectively.

The circuit will be ....

    (*i*) resistive   (*ii*) inductive

    (*iii*) capacitive   (*iv*) in resonance

33. The value of $\dfrac{10 \angle 60° \times 2 \angle 20°}{5 \angle 40°}$ will be

....

    (*i*) $4 \angle 40°$ (*ii*) $4 \angle 120°$

    (*iii*) $4 \angle 300°$ (*iv*) none of the above

34. The current through a 4 H inductor is $i = 0.6 \sin (100\,t - 10°)$ A. Using phasor algebra, voltage across the inductor is

    (*i*) $24 \angle 100°$ V   (*ii*) $340 \angle 70°$ V

    (*iii*) $48 \angle -90°$ V   (*iv*) $240 \angle 80°$ V

35. The polar form of the following waveform is

    $v_1 = 16 \sin (\omega t + 35°)$ V

    $v_2 = 32 \sin (\omega t - 55°)$ V

    (*i*) $v_1 = 16 \angle 70°$ V; $v_2 = 32 \angle 110°$ V

    (*ii*) $v_1 = 16 \angle 35°$ V; $v_2 = 32 \angle -55°$ V

    (*iii*) $v_1 = 8 \angle 35°$ V; $v_2 = 32 \angle 55°$ V

    (*iv*) none of above

36. The current through a 400Ω resistor is $i = 0.06 \sin (\omega t - 30°)$A. The voltage across the resistor is

    (*i*) $12 \angle 30°$ V   (*ii*) $48 \angle 45°$ V

    (*iii*) $24 \angle -30°$ V   (*iv*) none of above

37. The reciprocal of $12 - j\,16$ in polar form is

    (*i*) $0.05 \angle 53.13°$ (*ii*) $5 \angle -53.13°$

    (*iii*) $20 \angle -31.8°$ (*iv*) none of above

38. The reciprocal of j is

    (*i*) j (*ii*) $-j$

    (*iii*) $j\,2$ (*iv*) none of above

39. In Fig. 13.30, $R = 300 \Omega$, $L = 0.2$ H and $v = 17 \sin (2000\,t)$ volts. The total equivalent impedance in rectangular form is

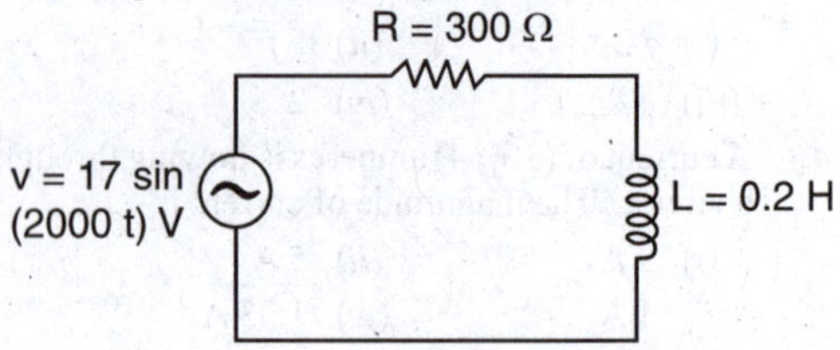

Fig. 13.30

    (*i*) $(150 + j\,200)\,\Omega$ (*ii*) $(300 - j\,400)\,\Omega$

    (*iii*) $(150 - j\,200)\,\Omega$ (*iv*) $(300 + j\,400)\,\Omega$

40. In the above question, what is the equivalent impedance in polar form ?

    (*i*) $250 \angle 53.13° \Omega$ (*ii*) $500 \angle -30° \Omega$

    (*iii*) $500 \angle 53.13° \Omega$ (*iv*) none of above

Answers

1. (*iii*)	**2.** (*ii*)	**3.** (*i*)	**4.** (*ii*)	**5.** (*iv*)
6. (*i*)	**7.** (*ii*)	**8.** (*iii*)	**9.** (*iii*)	**10.** (*ii*)
11. (*iii*)	**12.** (*iii*)	**13.** (*ii*)	**14.** (*ii*)	**15.** (*i*)
16. (*iii*)	**17.** (*i*)	**18.** (*ii*)	**19.** (*iii*)	**20.** (*ii*)
21. (*iii*)	**22.** (*iii*)	**23.** (*i*)	**24.** (*ii*)	**25.** (*ii*)
26. (*i*)	**27.** (*i*)	**28.** (*iii*)	**29.** (*i*)	**30.** (*iii*)
31. (*i*)	**32.** (*iii*)	**33.** (*i*)	**34.** (*iv*)	**35.** (*ii*)
36. (*iii*)	**37.** (*i*)	**38.** (*ii*)	**39.** (*iv*)	**40.** (*iii*)

14

Parallel A.C. Circuits

Introduction

As in parallel d.c. circuits, the voltage is the same across each branch of a parallel a.c. circuit. But current in any branch depends upon the impedance of that branch. The total line current supplied to the circuit is the *phasor sum* of the branch currents. Parallel circuits are used more frequently in electrical systems than are the series circuits. For example, electrical devices and equipment are connected in parallel across a.c. mains. There are two principal reasons for it. First, the operation of each device becomes independent of the other. Therefore, it is possible to turn on or off any device without disturbing the operation of other devices. Secondly, most of the electrical appliances requiring different currents at the same voltage are to be connected to the same power source. This necessitates parallel connections. In this chapter, we shall discuss the various methods of solving parallel a.c. circuits.

14.1. Methods of Solving Parallel A.C. Circuits

While analysing a parallel a.c. circuit, two important points must be kept in mind. First, a parallel circuit, in fact, consists of two or more series circuits connected in parallel. Therefore, each branch of the circuit can be analysed separately as a series circuit and then the effect of the separate branches can be combined. Secondly, alternating voltages and currents are phasor quantities. This implies that both magnitudes and phase angles must be taken into account while carrying out circuit calculations. There are four principal methods of solving parallel a.c. circuits, namely ;

    (*i*) By phasor diagram   (*ii*) By phasor algebra

    (*iii*) Equivalent impedance method   (*iv*) Admittance method

The use of a particular method will depend upon the conditions of the problem. *However, in general, that method should be used which yields quick results.*

14.2. By Phasor Diagram

In this method, we find the magnitude and phase angle of each branch current. We then draw the phasor diagram taking voltage as the *reference phasor. The circuit or line current is the phasor sum of the branch currents and can be determined either (*i*) by parallelogram method or (*ii*) by the method of components. The second method is preferred because it yields quick results.

Consider a parallel circuit consisting of two branches and connected to an alternating voltage of *V* volts (r.m.s.) as shown in Fig. 14.1.

Branch 1. $\qquad\qquad Z_1 = \sqrt{R_1^2 + X_{C1}^2} \; ; I_1 = \dfrac{V}{Z_1} \; ; \phi_1 = \tan^{-1}\dfrac{X_{C1}}{R_1}$

The current I_1 in branch 1 leads the applied voltage V by ϕ_1^0 as shown in the phasor diagram in Fig. 14.2.

Branch 2. $\qquad\qquad Z_2 = \sqrt{R_2^2 + X_{L2}^2} \; ; I_2 = \dfrac{V}{Z_2} \; ; \phi_2 = \tan^{-1}\dfrac{X_{L2}}{R_2}$

---

* Since voltage is common (*i.e.* same) in a parallel circuit, it is taken as the reference phasor in drawing the phasor diagram.

The current I_2 in branch 2 lags behind the applied voltage V by ϕ_2^0 as shown in the phasor diagram in Fig. 14.2.

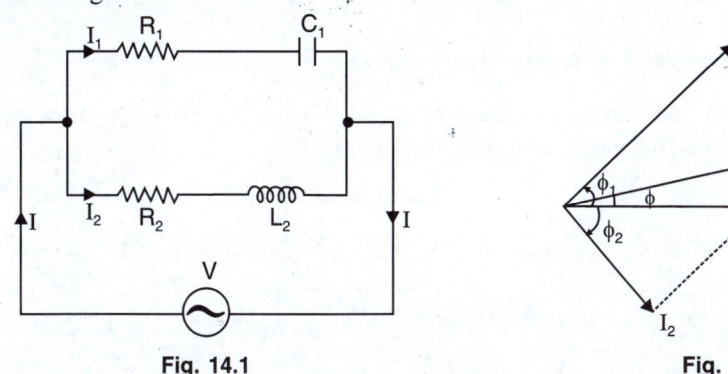

| Fig. 14.1 | Fig. 14.2 |

The line current I is the phasor sum of I_1 and I_2. Suppose its phase angle is ϕ^0 as shown in Fig. 14.2. The values of I and ϕ can be determined by resolving the currents into rectangular components.

$$I \cos \phi = \text{Algebraic sum of components of } I_1 \text{ and } I_2 \text{ along } x\text{-axis}$$
$$= I_1 \cos \phi_1 + I_2 \cos \phi_2$$
$$I \sin \phi = \text{Algebraic sum of components of } I_1 \text{ and } I_2 \text{ along } y\text{-axis}$$
$$= I_1 \sin \phi_1 - I_2 \sin \phi_2$$

But
$$I^2 = (I \cos \phi)^2 + (I \sin \phi)^2$$

$\therefore$
$$I^2 = (I_1 \cos \phi_1 + I_2 \cos \phi_2)^2 + (I_1 \sin \phi_1 - I_2 \sin \phi_2)^2$$

or
$$I = \sqrt{(I_1 \cos\phi_1 + I_2 \cos\phi_2)^2 + (I_1 \sin\phi_1 - I_2 \sin\phi_2)^2}$$

$$\tan \phi = \frac{I \sin \phi}{I \cos \phi} = \frac{I_1 \sin\phi_1 - I_2 \sin\phi_2}{I_1 \cos\phi_1 + I_2 \cos\phi_2} = \frac{Y - \text{comp.}}{X - \text{comp.}}$$

If ϕ is positive, line current I leads the voltage and if ϕ is negative, I lags behind the voltage.

$$\text{Circuit p.f.} = \frac{I \cos \phi}{I} = \frac{I_1 \cos\phi_1 + I_2 \cos\phi_2}{I} = \frac{X - \text{comp.}}{I}$$

The phasor diagram method is suitable only when the parallel circuit is simple and contains two branches. *However, if the parallel circuit is complex having more than two branches, this method becomes very inconvenient.* In such a case, use of phasor algebra is recommended to solve parallel-circuit problems.

14.3. By Phasor Algebra

In this method, voltages, currents and impedances are expressed in the complex form *i.e.* either in the rectangular or polar form. Since complex form includes both magnitude and phase angle, the solution of parallel-circuit problems can be obtained mathematically by using the rules of phasor algebra. This eliminates the need of phasor diagram. Referring back to the parallel circuit shown in Fig. 14.1, we have,

$$\mathbf{V} = V + j\,0 = V \qquad \qquad ...Reference\ phasor$$
$$\mathbf{Z_1} = R_1 - j\,X_{C1}\ ;\ \ \mathbf{Z_2} = R_2 + j\,X_{L2}$$

(i) **Rectangular form.** $\quad \mathbf{I_1} = \dfrac{\mathbf{V}}{\mathbf{Z_1}} = \dfrac{V}{R_1 - j\,X_{C1}}$

$$\mathbf{I_2} = \frac{\mathbf{V}}{\mathbf{Z_2}} = \frac{V}{R_2 + jX_{L2}}$$

∴ Line current, $\mathbf{I} = \mathbf{I_1} + \mathbf{I_2} = \frac{V}{R_1 - jX_{C1}} + \frac{V}{R_2 + jX_{L2}}$

The solution of **I** can be obtained in the standard form $a \pm jb$ by using the rules of phasor algebra. Then it is an easy task to find the magnitude and phase angle of **I**.

(*ii*) **Polar form.** $\mathbf{V} = V \angle 0°$ ...*Reference phasor*

$$\mathbf{Z_1} = Z_1 \angle -\phi_1^0 \text{ where } Z_1 = \sqrt{R_1^2 + X_{C1}^2} \; ; \; \phi_1 = \tan^{-1}\frac{X_{C1}}{R_1}$$

$$\mathbf{Z_2} = Z_2 \angle \phi_2^0 \text{ where } Z_2 = \sqrt{R_2^2 + X_{L2}^2} \; ; \; \phi_2 = \tan^{-1}\frac{X_{L2}}{R_2}$$

∴ $$\mathbf{I_1} = \frac{\mathbf{V}}{\mathbf{Z_1}} = \frac{V\angle 0°}{Z_1 \angle -\phi_1^0} = \frac{V}{Z_1}\angle \phi_1^0$$

$$\mathbf{I_2} = \frac{\mathbf{V}}{\mathbf{Z_2}} = \frac{V\angle 0°}{Z_2 \angle \phi_2^0} = \frac{V}{Z_2}\angle -\phi_2^0$$

∴ Line current, $\mathbf{I} = \mathbf{I_1} + \mathbf{I_2} = \frac{V}{Z_1}\angle \phi_1^0 + \frac{V}{Z_2}\angle -\phi_2^0$

It may be noted that the phase angle of any current is the *conjugate of its impedance angle.* This is an important point. The reader is strongly advised to use polar form for multiplication and division of complex quantities. However, for addition and subtraction, rectangular form should be preferred.

14.4. Equivalent Impedance Method

In this method, we find the equivalent or total impedance of the parallel circuit. The line current is equal to the applied voltage divided by the equivalent impedance. Consider several impedances connected in parallel as shown in Fig. 14.3.

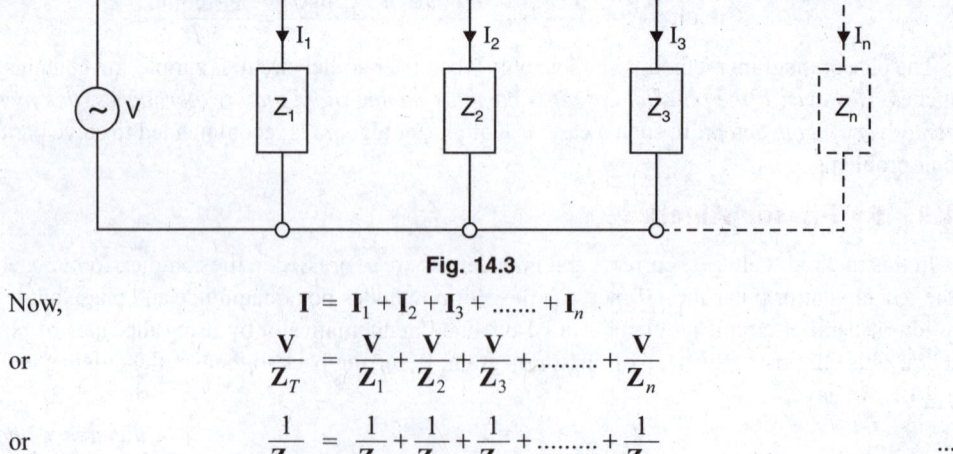

Fig. 14.3

Now, $\mathbf{I} = \mathbf{I_1} + \mathbf{I_2} + \mathbf{I_3} + ... + \mathbf{I_n}$

or $\dfrac{\mathbf{V}}{\mathbf{Z_T}} = \dfrac{\mathbf{V}}{\mathbf{Z_1}} + \dfrac{\mathbf{V}}{\mathbf{Z_2}} + \dfrac{\mathbf{V}}{\mathbf{Z_3}} + ... + \dfrac{\mathbf{V}}{\mathbf{Z_n}}$

or $\dfrac{1}{\mathbf{Z_T}} = \dfrac{1}{\mathbf{Z_1}} + \dfrac{1}{\mathbf{Z_2}} + \dfrac{1}{\mathbf{Z_3}} + ... + \dfrac{1}{\mathbf{Z_n}}$ ...(*i*)

where $\mathbf{Z_T}$ is the total or equivalent impedance of the parallel circuit.

---

*   For example, the first branch has an impedance of $Z_1 \angle -\phi_1^0$. The impedance angle of this branch is $-\phi_1^0$. The phase angle of current in this branch is ϕ_1^0 *i.e.* conjugate of $-\phi_1^0$.

$\therefore$ Line current, $\mathbf{I} = \dfrac{\mathbf{V}}{\mathbf{Z}_T}$

Note that relation (*i*) compares with that for parallel resistors but with one important difference. Here each impedance is in complex form and takes care of magnitude as well as impedance angle. Therefore, all algebraic operations (*e.g.* addition, division, subtraction etc.) must be in complex form. No attempt should be made to carry out these operations airthmetically.

Special case. If only two impedances are in parallel, then the total or equivalent impedance is given by ;

$$\frac{1}{\mathbf{Z_T}} = \frac{1}{\mathbf{Z}_1} + \frac{1}{\mathbf{Z}_2} = \frac{\mathbf{Z}_1 + \mathbf{Z}_2}{\mathbf{Z}_1 \mathbf{Z}_2}$$

or

$$\mathbf{Z_T} = \frac{\mathbf{Z}_1 \mathbf{Z}_2}{\mathbf{Z}_1 + \mathbf{Z}_2}$$

$$\text{Line current, } \mathbf{I} = \frac{\mathbf{V}}{\mathbf{Z}_T} = \mathbf{V}\,\frac{\mathbf{Z}_1 + \mathbf{Z}_2}{\mathbf{Z}_1\,\mathbf{Z}_2}$$

$$\text{Branch current, } \mathbf{I_1} = \frac{\mathbf{V}}{\mathbf{Z}_1} = \mathbf{I}\,\frac{\mathbf{Z}_2}{\mathbf{Z}_1 + \mathbf{Z}_2}$$

$$\text{Branch current, } \mathbf{I_2} = \frac{\mathbf{V}}{\mathbf{Z}_2} = \mathbf{I}\,\frac{\mathbf{Z}_1}{\mathbf{Z}_1 + \mathbf{Z}_2}$$

The reader may note that finding the equivalent impedance in complex form involves lengthy calculations. Such an approach to solve parallel a.c. circuits is not recommended particularly when there are more than two branches in the circuit.

14.5. Admittance (Y)

The **admittance** *of an a.c. circuit is defined as the reciprocal of its impedance i.e.*

$$\text{Admittance, } \mathbf{Y} = \frac{1}{\mathbf{Z}} = \frac{\mathbf{I}}{\mathbf{V}}$$

The unit of admittance is siemen (S). Whereas impedance (**Z**) is the opposition to alternating current flow, admittance (**Y**) is the inducement to alternating current flow. In chapter 11, we defined the following terms :

Conductance, $G = \dfrac{1}{R}$; Inductive susceptance, $B_L = \dfrac{1}{X_L}$; Capacitive susceptance, $B_C = \dfrac{1}{X_C}$

The units of G, B_L and B_C are siemen (S). The admittance approach is quite useful in the solution of parallel a.c. circuits.

Components of admittance. Depending upon the nature of reactance, the impedance of an a.c. circuit can be expressed in the complex form as :

$$\mathbf{Z} = R + j\,X_L \quad \text{or} \quad \mathbf{Z} = R - j\,X_C$$

Here, R is the resistive or in-phase component of **Z** while X_L or X_C is the reactive or quadrature component of **Z**. Let us find the components of **Y**. We shall discuss two cases by way of illustration.

(*i*) **R and L in parallel.** Fig. 14.4 (*i*) shows the circuit. Here **V** is the applied voltage and **Z** is total circuit impedance.

Here, $\mathbf{I} = \dfrac{\mathbf{V}}{\mathbf{Z}}$; $\mathbf{I_1} = \dfrac{\mathbf{V}}{R}$; $\mathbf{I_2} = \dfrac{\mathbf{V}}{j\,X_L} = \dfrac{\mathbf{V}}{j\,X_L} \times \dfrac{j}{j} = -j\dfrac{\mathbf{V}}{X_L}$

Now, $\mathbf{I} = \mathbf{I_1} + \mathbf{I_2}$

or
$$\frac{V}{Z} = \frac{V}{R} + \left(-j\frac{V}{X_L}\right)$$

or
$$\frac{1}{Z} = \frac{1}{R} - j\left(\frac{1}{X_L}\right)$$

Now,
$$\frac{1}{Z} = Y \; ; \; \frac{1}{R} = G \; ; \; \frac{1}{X_L} = B_L$$

∴
$$Y = G - jB_L$$

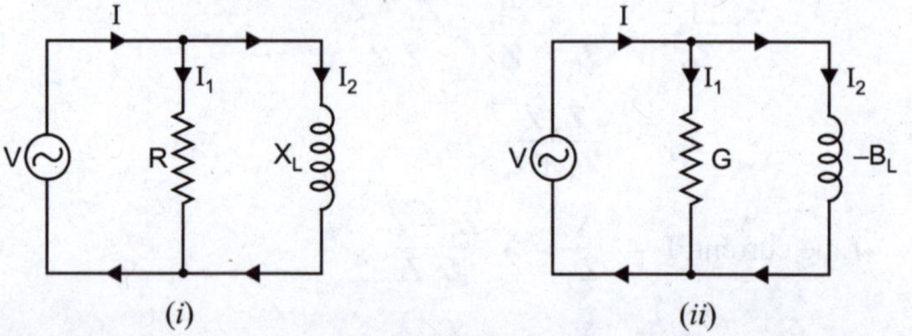

Fig. 14.4

Fig. 14.4 (*ii*) shows the components G and B_L of **Y**. Here, G is the in-phase component of **Y** while B_L is the quadrature component of **Y**. Note that G (conductance) is **positive** and B_L (inductive susceptance) is **negative**.

(*ii*) **R and C in parallel.** Fig. 14.5 (*i*) shows the circuit. Here **V** is the applied voltage and **Z** is total circuit impedance.

Now, $\mathbf{I} = \dfrac{V}{Z} \; ; \; \mathbf{I_1} = \dfrac{V}{R} \; ; \; \mathbf{I_2} = \dfrac{V}{-jX_C} = \dfrac{V}{-jX_C} \times \dfrac{j}{j} = j\dfrac{V}{X_C}$

Now,
$$\mathbf{I} = \mathbf{I_1} + \mathbf{I_2}$$

or
$$\frac{V}{Z} = \frac{V}{R} + j\frac{V}{X_C}$$

or
$$\frac{1}{Z} = \frac{1}{R} + j\left(\frac{1}{X_C}\right)$$

Now,
$$\frac{1}{Z} = Y \; ; \; \frac{1}{R} = G \; ; \; \frac{1}{X_C} = B_C$$

∴
$$Y = G + jB_C$$

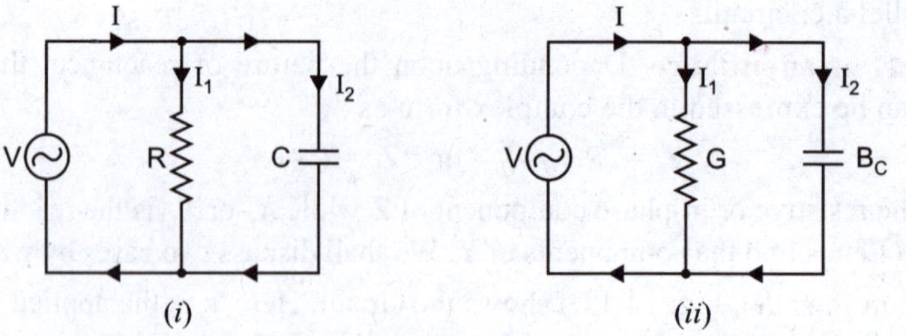

Fig. 14.5

Fig. 14.5 (*ii*) shows the components G and B_C of **Y**. Here, G is the in-phase component of **Y** while B_C is the quadrature component of **Y**. Note that G (conductance) is **positive** and B_C (capacitive susceptance) is also **positive**.

Note. Conductance G is always positive. However, B_L (inductive susceptance) is negative while B_C (capacitive susceptance) is positive. Moreover, it is a usual practice to omit the suffix L or C with susceptance B. *The positive sign with B indicates capacitive susceptance and the negative sign with B implies inductive susceptance.*

14.6. Importance of Admittance in Parallel A.C. Circuit Analysis

In the analysis of parallel circuits, it is quite useful to use admittance values instead of impedance values. Consider several impedances Z_1, Z_2, Z_3 ... connected in parallel across an a.c. supply of **V** volts. We convert these parallel-connected impedances into equivalent parallel-connected admittances Y_1, Y_2, Y_3 ... in rectangular form. The total impedance Z_T of the circuit is given by ;

$$\frac{1}{Z_T} = \frac{1}{Z_1} + \frac{1}{Z_2} + \frac{1}{Z_3} + ...$$

Total circuit admittance, $Y_T = Y_1 + Y_2 + Y_3 + ...$

The admittances in rectangular form ($Y_1 = G_1 \pm jB_1$, $Y_2 = G_2 \pm jB_2$ ...) of parallel branches can be added to give the resultant conductance G and the resultant susceptance B. Note that G is always positive while inductive susceptance is negative and capacitive susceptance is positive. Thus admittance method of parallel circuits makes the approach somewhat similar to a series circuit where impedances (in rectangular form) are added. For this reason, the admittance approach in the analysis of parallel a.c. circuits is for more convenient than impedance approach.

Magnitude of total admittance, $Y = \sqrt{G^2 + B^2}$

Magnitude of total current, $I = VY$

Magnitudes of various branch currents are :

$$I_1 = VY_1 \;\; ; \;\; I_2 = VY_2 \;\; ; \;\; I_3 = VY_3 \text{ and so on.}$$

Phase angle between V and I is

$$\tan \phi = \frac{B}{G}$$

Thus admittance method permits us to find the various circuit values of the parallel a.c. circuit with a fair degree of ease.

14.7. Admittance Triangle

Since admittance has in-phase component (*i.e.* G) as well as quadrature component (*i.e.* B_L or B_C), it can be represented by a right angled triangle, called *admittance triangle.*

(*i*) For an inductive circuit (*i.e.* $R + jX_L$), the impedance and admittance triangles will be as shown in Fig. 14.6. Note that admittance angle is equal to the impedance angle but is *negative. For this reason, B_L will be along OY'-axis and hence negative.

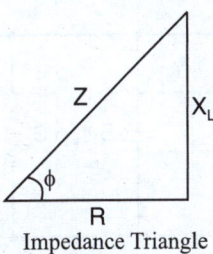

Impedance Triangle

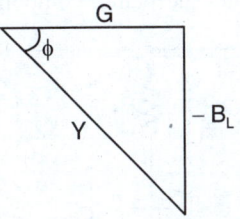

Admittance Triangle

Fig. 14.6

---

* The reciprocal of a complex number does not change the magnitude of the angle but it changes the sign of the angle.

Conductance, $G = Y\cos\phi = \dfrac{1}{Z}\times\dfrac{R}{Z}$ $\therefore G = \dfrac{R}{Z^2} = \dfrac{R}{R^2 + X_L^2}$

Susceptance, $B_L = Y\sin\phi = \dfrac{1}{Z}\times\dfrac{X_L}{Z}$ $\therefore B_L = \dfrac{X_L}{Z^2} = \dfrac{X_L}{R^2 + X_L^2}$ *(negative)*

Phase angle, $\phi = \tan^{-1} - B_L/G$

(*ii*) For a capacitive circuit (*i.e.* $R - jX_C$), the impedance and admittance triangles will be as shown in Fig. 14.7. Note that admittance angle is equal to the impedance angle but of opposite sign. For this reason, B_C will lie along OY-axis and hence positive.

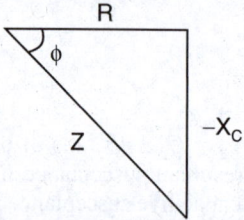 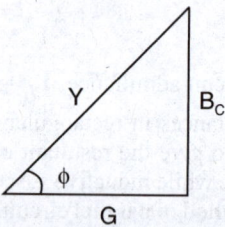

Impedance Triangle   Admittance Triangle

Fig. 14.7

Following the above procedure, we have,

$$G = \dfrac{R}{Z^2} = \dfrac{R}{R^2 + X_C^2} \quad ; \quad B_C = \dfrac{X_C}{Z^2} = \dfrac{X_C}{R^2 + X_C^2} \qquad (positive)$$

Phase angle, $\phi = \tan^{-1} B_C/G$

14.8. Admittance Method for Parallel A.C. Circuit Solution

Fig. 14.8 (*i*) shows two impedances $Z_1 = R_1 - j X_{C1}$ and $Z_2 = R_2 + jX_{L2}$ in parallel across an a.c. supply of **V** volts. We can convert the impedances into equivalent admittances as under :

$$Z_1 = \sqrt{R_1^2 + X_{C1}^2} \; ; G_1 = \dfrac{R_1}{Z_1^2} ; B_1 = \dfrac{X_{C1}}{Z_1^2}$$

$$\therefore \quad \mathbf{Y_1} = G_1 + jB_1$$

$$Z_2 = \sqrt{R_2^2 + X_{L2}^2} \; ; G_2 = \dfrac{R_2}{Z_2^2} ; B_2 = -\dfrac{X_{L2}}{Z_2^2}$$

$$\therefore \quad \mathbf{Y_2} = G_2 - j B_2$$

Fig. 14.8 (*ii*) shows $\mathbf{Y_1}$ and $\mathbf{Y_2}$ resolved into conductances and suceptances. It may be noted that conductance and suceptance components of each admittance are paralleled elements.

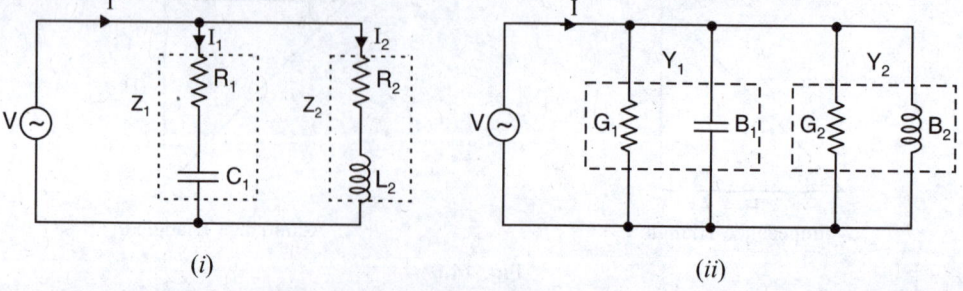

(*i*)   (*ii*)

Fig. 14.8

Total circuit admittance, $\mathbf{Y} = \mathbf{Y_1} + \mathbf{Y_2} = (G_1 + jB_1) + (G_2 - jB_2)$

$$= (G_1 + G_2) + j(B_1 - B_2)$$

$$\therefore \qquad \mathbf{Y} = G + jB$$

where $G = G_1 + G_2$ and $B = B_1 - B_2$

The other circuit values can be determined quite easily as discussed in the next article.

14.9. Application of Admittance Method

Consider a parallel a.c. circuit consisting of three branches as shown in Fig. 14.9. The total conductance of the circuit is the sum of the conductances of the individual branches. However, total susceptance is the *algebraic sum* of susceptances of the individual branches. Remember, inductive susceptance is negative while capacitive susceptance is positive. It is an important point.

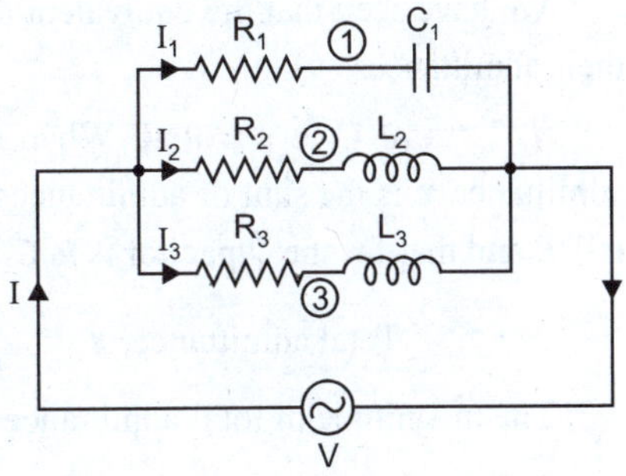

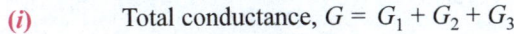

Fig. 14.9

The magnitudes of various circuit values are :

(*i*)   Total conductance, $G = G_1 + G_2 + G_3$

(*ii*)   Total susceptance, $B = B_1 - B_2 - B_3$

(*iii*)   Admittance, $Y = \sqrt{G^2 + B^2}$

(*iv*)   Circuit current, $I = VY$

(*v*)   Power factor, $\cos \phi = G/Y$ [See Fig. 14.6 or 14.7]

(*vi*)   Power consumed, $P = VI \cos \phi = V(VY)(G/Y) = V^2 G$

It can be seen that only conductance of the circuit is responsible for power dissipation.

(*vii*) The currents in the individual branches are :

$$I_1 = VY_1 \text{ where } Y_1 = \sqrt{G_1^2 + B_1^2}$$

$$I_2 = VY_2 \text{ where } Y_2 = \sqrt{G_2^2 + (-B_2)^2}$$

$$I_3 = VY_3 \text{ where } Y_3 = \sqrt{G_3^2 + (-B_3)^2}$$

(*viii*)   Phase angle, $\phi = \tan^{-1} \dfrac{B}{G}$

It ϕ is positive, then current leads the voltage and if ϕ is negative, current lags behind the voltage.

Polar form. $\mathbf{Z_1} = R_1 - jX_{C1} = Z_1 \angle -\phi_1^0$

$\therefore \qquad \mathbf{Y_1} = \dfrac{1}{\mathbf{Z_1}} = \dfrac{1}{Z_1 \angle -\phi_1^0} = Y_1 \angle \phi_1^0$

$\mathbf{Z_2} = R_2 + jX_{L2} = Z_2 \angle \phi_2^0$

$\therefore \qquad \mathbf{Y_2} = \dfrac{1}{\mathbf{Z_2}} = \dfrac{1}{Z_2 \angle \phi_2^0} = Y_2 \angle -\phi_2^0$

$\mathbf{Z_3} = R_3 + jX_{L3} = Z_3 \angle \phi_3^0$

$\therefore \qquad \mathbf{Y_3} = \dfrac{1}{\mathbf{Z_3}} = \dfrac{1}{Z_3 \angle \phi_3^0} = Y_3 \angle -\phi_3^0$

Total admittance, $\mathbf{Y} = \mathbf{Y_1} + \mathbf{Y_2} + \mathbf{Y_3} = Y \angle \pm \phi^0$

Line current, $\mathbf{I} = \mathbf{VY}$

Power in admittance, $P = VI \cos \phi = V^2 Y \cos \phi = V^2 G$ $\qquad$ ($\because I = VY$; $Y \cos \phi = G$)

It can be seen that only conductance of the circuit is responsible for power dissipation.

14.10. Some Cases of Parallel Connected Elements

We have seen that the equivalent admittance of a number of parallel admittances is the sum of their admittances.

(*i*) **R and C in parallel.** When resistance (R) and capacitance (C) are in parallel, the total admittance **Y** is the sum of admittances of the two components. Since the admittance for resistance is $1/R$ and that for the capacitor is $j\omega C$,

$\therefore \qquad$ Total admittance, $\mathbf{Y} = \dfrac{1}{R} + j\omega C$

The magnitude of total admittance is

$$Y = \sqrt{(1/R)^2 + (\omega C)^2}$$

Phase angle, $\phi = \tan^{-1}(\omega CR)$

(*ii*) **R and L in parallel.** When resistance (R) and inductance (L) are in parallel, the total admittance **Y** is the sum of admittances of the two components. Since admittance for the resistance is $1/R$ and that for the inductor is $-j/\omega L$,

$\therefore \qquad$ Total admittance, $\mathbf{Y} = \dfrac{1}{R} - \dfrac{j}{\omega L}$

The magnitude of total admittance is

$$Y = \sqrt{\left(\dfrac{1}{R}\right)^2 + \left(\dfrac{1}{\omega L}\right)^2}$$

Phase angle, $\phi = \tan^{-1}(R/\omega L)$

(*iii*) **R, L and C in parallel.** For resistance (R), inductance (L) and capacitance (C) all in parallel, the total admittance **Y** is

$$\mathbf{Y} = (1/R) - j/\omega L + j\omega C$$

or $\qquad\qquad \mathbf{Y} = (1/R) + j(\omega C - 1/\omega L)$

The magnitude of total admittance is

$$Y = \sqrt{(1/R)^2 + (\omega C - 1/\omega L)^2}$$

Phase angle, $\phi = \tan^{-1}[R(\omega C - 1/\omega L)]$

Example 14.1. *A resistance of 20 Ω and a coil of inductance 31·8 mH and negligible resistance are connected in parallel across 230 V, 50 Hz supply. Find (i) the line current (ii) power factor and power consumed by the circuit.*

Solution. Fig. 14.10 (*i*) shows the circuit diagram while Fig. 14.10 (*ii*) shows the phasor diagram of the circuit.

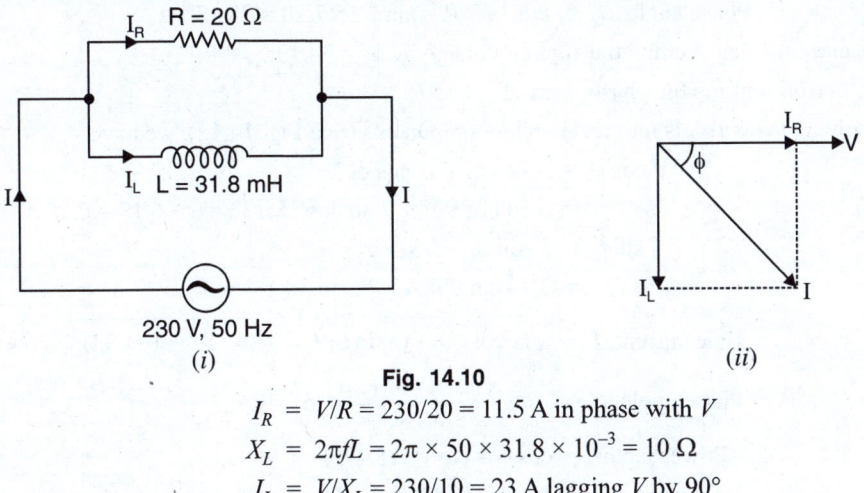

Fig. 14.10

$$I_R = V/R = 230/20 = 11.5 \text{ A in phase with } V$$
$$X_L = 2\pi fL = 2\pi \times 50 \times 31.8 \times 10^{-3} = 10 \ \Omega$$
$$I_L = V/X_L = 230/10 = 23 \text{ A lagging } V \text{ by } 90°$$

The line current I is the phasor sum of I_R and I_L.

(i)   Line current, $I = \sqrt{I_R^2 + I_L^2} = \sqrt{(11.5)^2 + (23)^2} = \textbf{25.71 A}$

(ii)   Power factor, $\cos \phi = I_R/I = 11.5/25.71 = \textbf{0.447 lag}$

            Power consumed, $P = VI \cos \phi = 230 \times 25.71 \times 0.447 = \textbf{2643 watts}$

Example 14.2. *A capacitor of 50 μF is connected in parallel with a coil that has a resistance of 20 Ω and inductance of 0·05 H. If this parallel combination is connected across 200 V, 50 Hz supply, calculate (i) the line current (ii) power factor and (iii) power consumed.*

Solution. Fig. 14.11 shows the conditions of the problem.

Branch 1. $Z_1 = X_C = \dfrac{1}{2\pi fC} = \dfrac{10^6}{2\pi \times 50 \times 50} = 63.7 \ \Omega$

$\therefore$ $I_1 = V/X_C = 200/63.7 = 3.14 \text{ A}$

The current I_1 leads the applied voltage by $\phi_1 = 90°$ as shown in the phasor diagram in Fig. 14.12.

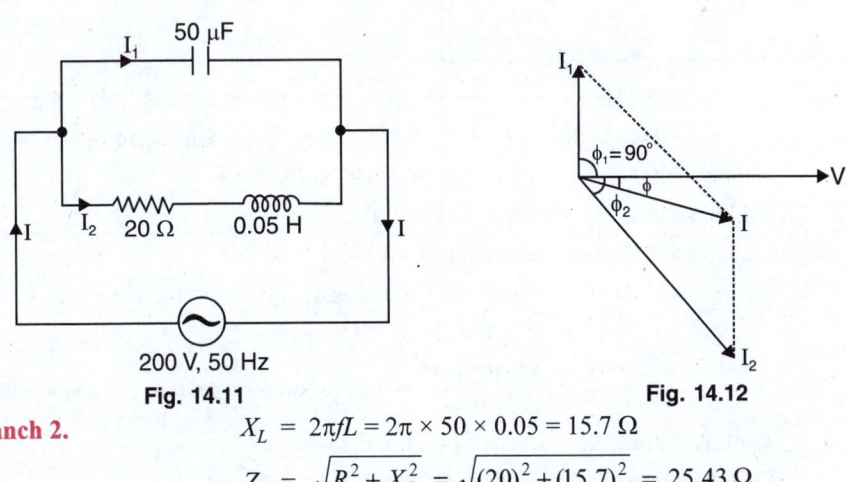

Fig. 14.11   **Fig. 14.12**

Branch 2. $X_L = 2\pi fL = 2\pi \times 50 \times 0.05 = 15.7 \ \Omega$

                   $Z_2 = \sqrt{R^2 + X_L^2} = \sqrt{(20)^2 + (15.7)^2} = 25.43 \ \Omega$

$\therefore$ $I_2 = \dfrac{V}{Z_2} = \dfrac{200}{25.43} = 7.86 \text{ A}$

Phase angle, $\phi_2 = \tan^{-1} X_L/R = \tan^{-1} 15.7/20 = 38.13°$

The current I_2 lags behind the applied voltage by $\phi_2 = 38.13°$.

The line current I is the phasor sum of I_1 and I_2.

Resolving the currents into rectangular components (See Fig. 14.12), we have,

$$I \cos \phi = I_1 \cos \phi_1 + I_2 \cos \phi_2$$
$$= 3.14 \cos 90° + 7.86 \cos 38.13° = 0 + 6.18 = 6.18 \text{ A}$$
$$I \sin \phi = I_1 \sin \phi_1 - I_2 \sin \phi_2$$
$$= 3.14 \sin 90° - 7.86 \sin 38.13° = 3.14 - 4.85 = -1.71 \text{ A}$$

(i) ∴   Line current, $I = \sqrt{(I \cos \phi)^2 + (I \sin \phi)^2} = \sqrt{(6.18)^2 + (-1.71)^2} = \textbf{6.41 A}$

Phase angle, $\phi = \tan^{-1} \left(-\dfrac{1.71}{6.18} \right) = -*15.47°$

(ii)   Power factor $= \cos \phi = \cos(-15.47°) = \textbf{0.964 } \textit{lag}$

(iii)   Power consumed, $P = VI \cos \phi = 200 \times 6.41 \times 0.964 = \textbf{1235.85 W}$

Example 14.3. *An inductive coil is connected in parallel with a pure resistor of 30 Ω and this parallel circuit is connected to a 50 Hz supply. The total current taken from the circuit is 8 A while the current in the resistor is 4 A and that in inductive coil is 6 A. Calculate (i) resistance and inductance of the coil (ii) power factor of the circuit and (iii) power taken by the circuit.*

Solution. The second branch has a pure resistance ($Z_2 = 30 Ω$) so that current $I_2 (= 4 \text{ A})$ will be in phase with the applied voltage. The first branch has an impedance of Z_1 and current $I_1 (= 6 \text{ A})$ through it will lag behind the applied voltage by $\phi_1°$. The line current $I (= 8 \text{ A})$ is the phasor sum of I_1 and I_2 as shown in the phasor diagram in Fig. 14.14.

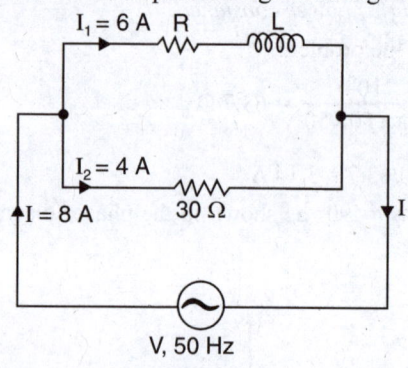

Fig. 14.13

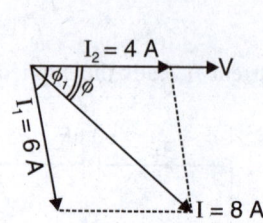

Fig. 14.14

(i)   Supply voltage, $V = I_2 Z_2 = 4 \times 30 = 120$ volts

Coil impedance, $Z_1 = V/I_1 = 120/6 = 20 Ω$

Referring to the phasor diagram shown in Fig. 14.14,

$$I^2 = I_1^2 + I_2^2 + 2I_1 \times I_2 \cos \phi_1$$

or $8^2 = 6^2 + 4^2 + 2 \times 6 \times 4 \times \cos \phi_1$

∴ $\cos \phi_1 = \dfrac{8^2 - 6^2 - 4^2}{2 \times 6 \times 4} = 0.25$; $\sin \phi_1 = \sin[\cos^{-1} 0.25] = 0.968$

∴   Coil resistance, $R = Z_1 \cos \phi_1 = 20 \times 0.25 = \textbf{5 Ω}$

Coil reactance, $X_L = Z_1 \sin \phi_1 = 20 \times 0.968 = 19.36 Ω$

---

*   In a parallel circuit, voltage is taken as the reference phasor. Therefore, negative phase angle means that circuit current lags behind the voltage.

$$\text{Coil inductance, } L = \frac{X_L}{2\pi f} = \frac{19.36}{2\pi \times 50} = \textbf{0.0616 H}$$

(*ii*) Resolving the currents along *X*-axis (See Fig. 14.14),

$$I \cos \phi = I_2 + I_1 \cos \phi_1$$

∴ Circuit p.f., $\cos \phi = \dfrac{I_2 + I_1 \cos \phi_1}{I} = \dfrac{4 + 6 \times 0.25}{8} = \textbf{0.687 } lag$

(*iii*) Power consumed, $P = VI \cos \phi = 120 \times 8 \times 0.687 = \textbf{660 W}$

Example 14.4. *A 10 Ω resistor, a 15.9 mH inductor and 159 μF capacitor are connected in parallel to a 200 V, 50 Hz source. Calculate the supply current and power factor.*

Solution. Fig. 14.15 shows the circuit diagram.

$$X_L = 2\pi f L = 2\pi \times 50 \times 15.9 \times 10^{-3} = 5 \ \Omega$$

$$X_C = \frac{1}{2\pi f C} = \frac{10^6}{2\pi \times 50 \times 159} = 20 \ \Omega$$

$$I_R = V/R = 200/10 = 20 \text{ A} \qquad \text{...in phase with } V$$

$$I_L = V/X_L = 200/5 = 40 \text{ A} \qquad \text{...lags } V \text{ by } 90°$$

$$I_C = V/X_C = 200/20 = 10 \text{ A} \qquad \text{...leads } V \text{ by } 90°$$

Fig. 14.16 shows the phasor diagram of the circuit. Note that I_L and I_C are 180° out of phase with each other. The supply current *I* is the phasor sum of I_R and $(I_L - I_C)$.

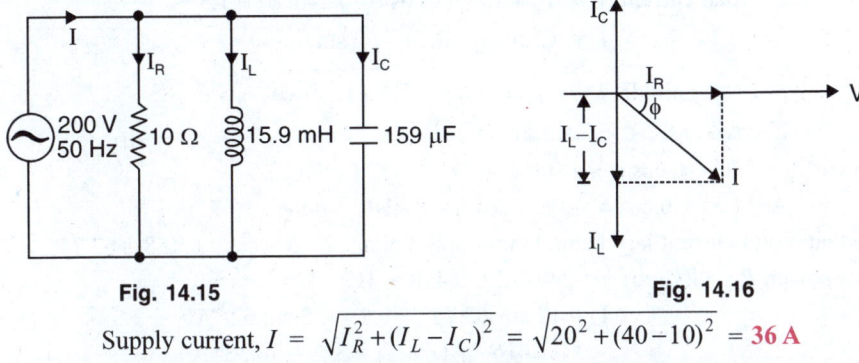

Fig. 14.15	**Fig. 14.16**

$$\text{Supply current, } I = \sqrt{I_R^2 + (I_L - I_C)^2} = \sqrt{20^2 + (40 - 10)^2} = \textbf{36 A}$$

$$\text{Circuit p.f.} = \cos \phi = I_R/I = 20/36 = \textbf{0.56 } lag$$

Example 14.5. *A coil of resistance 50 Ω and inductance 318 mH is connected in parallel with a circuit consisting of a 75 Ω resistor in series with a 159 μF capacitor. The circuit is connected to a 230 V, 50 Hz supply. Determine the supply current and circuit power factor.*

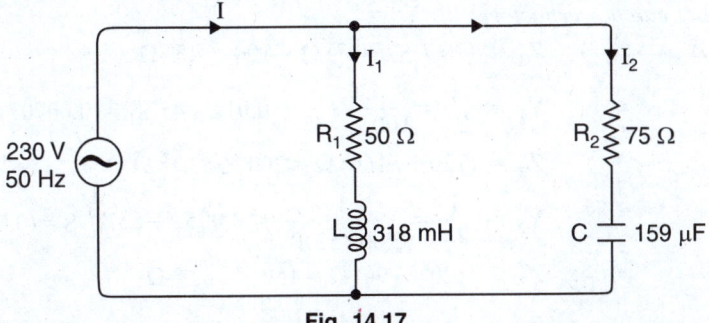

Fig. 14.17

Solution. Fig. 14.17 shows the conditions of the problem. Taking supply voltage as the reference phasor, we have, **V** = 230 ∠ 0° V.

$$X_L = 2\pi f L = 2\pi \times 50 \times 318 \times 10^{-3} = 100\ \Omega$$

$$\mathbf{Z_1} = R_1 + j X_L = (50 + j\,100)\ \Omega = 112\ \angle\,63.5°\ \Omega$$

$$\mathbf{I_1} = \frac{\mathbf{V}}{\mathbf{Z_1}} = \frac{230\ \angle\,0°}{112\ \angle\,63.5°} = 2.05\ \angle -63.5°\ \text{A} = (0.91 - j\,1.83)\ \text{A}$$

$$X_C = \frac{1}{2\pi f C} = \frac{10^6}{2\pi \times 50 \times 159} = 20\ \Omega$$

$$\mathbf{Z_2} = R_2 - j X_C = (75 - j\,20)\ \Omega = 77.6\ \angle -15°\ \Omega$$

$$\mathbf{I_2} = \frac{\mathbf{V}}{\mathbf{Z_2}} = \frac{230\angle 0°}{77.6\angle -15°} = 2.96\ \angle 15°\text{A} = (2.86 + j\,0.766)\ \text{A}$$

Supply current, $\mathbf{I} = \mathbf{I_1} + \mathbf{I_2} = (0.91 - j\,1.83) + (2.86 + j\,0.766)$

$$= (3.77 - j\,1.06)\ \text{A} = \mathbf{3.92\ \angle -15.7°\ A}$$

Power factor $= \cos\phi = \cos 15.7° = \mathbf{0.963\ \textit{lag}}$

Example 14.6. *A small single phase induction motor is tested in parallel with a 160 Ω resistor. The motor takes 2A and the total current is 3A. Find the power and power factor of (i) motor (ii) whole circuit if V = 240 volts.*

Solution. (*i*) Let the power factor of the induction motor be $\cos\phi$ lagging.

$$\therefore\quad \mathbf{I_m} = 2\cos\phi - j2\sin\phi\ ;\quad \mathbf{I_R} = \frac{240\ \angle\,0°}{160} = 1.5\ \text{A}$$

Total current, $\mathbf{I} = \mathbf{I_m} + \mathbf{I_R} = (2\cos\phi - j\,2\sin\phi) + 1.5$

or $\qquad\qquad\qquad\mathbf{I} = (2\cos\phi + 1.5) - j\,2\sin\phi$

Magnitude, $I = \sqrt{(2\cos\phi + 1.5)^2 + (2\sin\phi)^2}$

$\therefore\qquad (2\cos\phi + 1.5)^2 + (2\sin\phi)^2 = I^2$

or $\ 4\cos^2\phi + 2.25 + 6\cos\phi + 4\sin^2\phi = 9\qquad (\because I = 3\text{A})$

or $\qquad 4 + 2.25 + 6\cos\phi = 9\ \therefore \cos\phi = \mathbf{0.458\ \textit{lagging}}$

Note that motor current lags behind the supply voltage by $\phi = \cos^{-1} 0.458 = 62.74°$.

Motor power, $P_m = VI_m \cos\phi = 240 \times 2 \times 0.458 = \mathbf{219.84\ W}$

(*ii*) $\qquad\qquad\qquad \mathbf{I} = (2\cos 62.74° + 1.5) - j\,2\sin 62.74°$

$$= (2.416 - j\,1.78)\ \text{A} = 3\ \angle -36.38°\ \text{A}$$

$\therefore\qquad$ P.F. of whole circuit $= \cos(-36.38°) = \mathbf{0.805\ \textit{lagging}}$

Power drawn by circuit, $P = VI \times 0.805 = 240 \times 3 \times 0.805 = \mathbf{579.6\ W}$

Example 14.7. *Three impedances of (70.7 + j 70.7) Ω, (120 + j 160) Ω and (120 + j 90) Ω are connected in parallel across a 250 V supply. Determine (i) admittance of the circuit (ii) supply current and (iii) circuit power factor.*

Solution. (*i*) $\qquad\qquad \mathbf{Z_1} = (70.7 + j\,70.7)\ \Omega = 100\ \angle\,45°\ \Omega$

$$\therefore\qquad \mathbf{Y_1} = \frac{1}{\mathbf{Z_1}} = \frac{1}{100\ \angle\,45°} = 0.01\angle -45°\ \text{S} = (0.00707 - j\,0.00707)\ \text{S}$$

$$\mathbf{Z_2} = (120 + j\,160)\ \Omega = 200\ \angle\,53.1°\ \Omega$$

$$\therefore\qquad \mathbf{Y_2} = \frac{1}{\mathbf{Z_2}} = \frac{1}{200\angle 53.1°} = 0.005\angle -53.1°\ \text{S} = (0.003 - j\,0.004)\ \text{S}$$

$$\mathbf{Z_3} = (120 + j\,90)\ \Omega = 150\ \angle\,36.9°\ \Omega$$

$$\therefore\qquad \mathbf{Y_3} = \frac{1}{\mathbf{Z_3}} = \frac{1}{150\angle 36.9°} = 0.00667\angle -36.9°\ \text{S}$$

$$= (0.0053 - j\,0.004)\ \text{S}$$

Total admittance, $\mathbf{Y_T} = \mathbf{Y_1} + \mathbf{Y_2} + \mathbf{Y_3} = (0.0154 - j\,0.015)^*\ \text{S} = \mathbf{0.0215 \angle - 44.3°\ S}$

(ii) Taking supply voltage as the reference phasor, we have, $\mathbf{V} = 250 \angle 0°\ \text{V}$.

Supply current, $\mathbf{I} = \mathbf{V\,Y_T} = 250 \angle 0° \times 0.0215 \angle -44.3° = \mathbf{5.37 \angle - 44.3°\ A}$

(iii)   Power factor $= \cos \phi = \cos 44.3° = \mathbf{0.71\ lagging}$

Example 14.8. *Three loads are placed across 230V, 50Hz supply. The loads are $10 \angle -30°\ \Omega$; $20 \angle 60°\ \Omega$ and $40 \angle 0°\ \Omega$. Determine (i) the admittance (ii) equivalent impedance (iii) power consumed and (iv) power factor.*

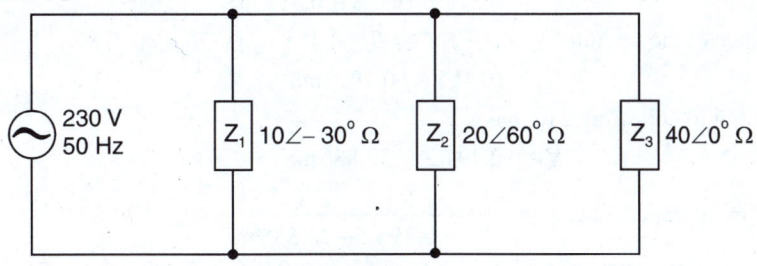

Fig. 14.18

Solution. *(i)* $\mathbf{Y_1} = \dfrac{1}{\mathbf{Z_1}} = \dfrac{1}{10 \angle -30°} = 0.1 \angle 30°\ \text{S} = (0.0866 + j\,0.05)\ \text{S}$

$\mathbf{Y_2} = \dfrac{1}{\mathbf{Z_2}} = \dfrac{1}{20 \angle 60°} = 0.05 \angle -60°\ \text{S} = (0.025 - j\,0.0433)\ \text{S}$

$\mathbf{Y_3} = \dfrac{1}{\mathbf{Z_3}} = \dfrac{1}{40 \angle 0°} = 0.025 \angle -0°\ \text{S} = (0.025 - j\,0)\ \text{S}$

$\mathbf{Y} = \mathbf{Y_1} + \mathbf{Y_2} + \mathbf{Y_3}$

$= (0.0866 + j\,0.05) + (0.025 - j\,0.0433) + (0.025 - j\,0)$

$= (0.1366 + j\,0.0067)\ \text{S} = 0.137 \angle 3°\text{S}$

∴   Circuit admittance, $Y = \mathbf{0.137\ S}$

(ii)   Circuit impedance, $\mathbf{Z} = \dfrac{1}{\mathbf{Y}} = \dfrac{1}{0.137 \angle 3°} = 7.3 \angle -3°\ \Omega$

Its magnitude, $Z = \mathbf{7.3\ \Omega}$

(iii)   Power consumed, $P = V^2\,G = (230)^2 \times 0.1366 = \mathbf{7226\ W}$

(iv)   Power factor $= \cos \phi = \cos 3° = \mathbf{0.998\ lead}$

Example 14.9. *Using admittance method, determine (i) circuit impedance and (ii) circuit current for the circuit shown in Fig. 14.19.*

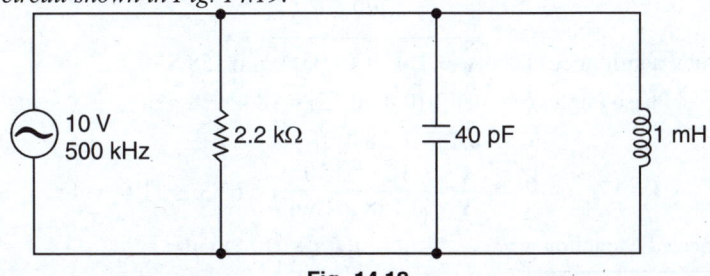

Fig. 14.19

---

*   If you carry out the sum, you will get this result.

Solution. (*i*)
$$G = \frac{1}{R} = \frac{1}{2.2\times10^3} = 0.455\times10^{-3} \text{ S} = 0.455 \text{ mS}$$

$$+jB = B_C = \frac{1}{X_C} = 2\pi f C = 2\pi\times(500\times10^3)\times40\times10^{-12}$$

$$= 0.126 \times 10^{-3} \text{ S} = 0.126 \text{ mS}$$

$$-jB = B_L = \frac{1}{X_L} = \frac{1}{2\pi f L} = \frac{1}{2\pi\times(500\times10^3)\times1\times10^{-3}}$$

$$= 0.318 \times 10^{-3} \text{ S} = 0.318 \text{ mS}$$

∴ Admittance of the circuit, $\mathbf{Y} = G + jB - jB = 0.455 + j\,0.126 - j\,0.318$

$$= (0.455 - j\,0.192) \text{ mS}$$

Converting it to polar form, we have,

$$\mathbf{Y} = 0.494 \angle - 22.88° \text{ mS}$$

∴ Circuit impedance, $\mathbf{Z} = \dfrac{1}{\mathbf{Y}} = \dfrac{1}{0.494\angle-22.88° \text{ mS}} = 2 \angle 22.88° \text{ k}\Omega$

(*ii*) Circuit current, $\mathbf{I} = \mathbf{VY} = 10 \angle 0° \times 0.494 \angle - 22.88° = 5 \angle - 22.88° \text{ mA}$

Example 14.10. *Determine the potential difference across the capacitor in Fig. 14.20 when there is a current input of 10 sin (1000 t + 30°) A. Use admittance method.*

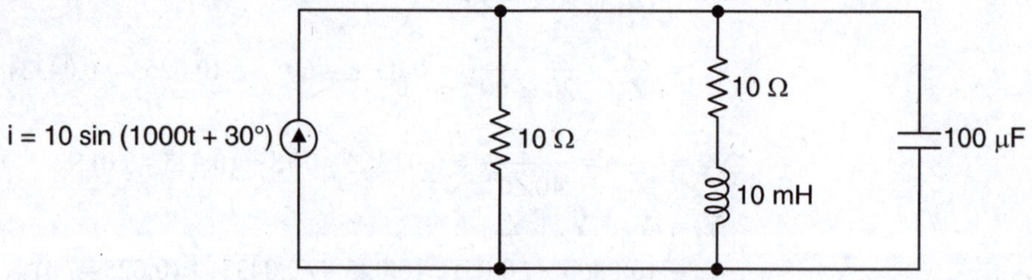

Fig. 14.20

Solution. The admittance of single resistor is $1/10 = 0.1$ S. The impedance of branch containing $10\,\Omega$ and 10 mH is $R + jX_L = (10 + *j\,10)\,\Omega$. Therefore, admittance is $1/(10 + j\,10)$ S. The admittance of the capacitor $= j\,\omega C = j \times 1000 \times 100 \times 10^{-6} = j\,0.1$ S.

∴ Total admittance, $\mathbf{Y} = 0.1 + \dfrac{1}{10 + j10} + j\,0.1$

$$= 0.1 + \frac{10 - j10}{(10 + j10)\,(10 - j10)} + j\,0.1$$

$$= 0.1 + \frac{10 - j10}{200} + j\,0.1 = (0.15 + j\,0.05) \text{ S}$$

Magnitude of total admittance, $Y = \sqrt{(0.15)^2 + (0.05)^2} = 0.158$ S

Phase angle, $\phi = \tan^{-1} (0.05/0.15) = 18.4°$

∴ $\mathbf{Y} = 0.158 \angle 18.4°$ S

Now $\mathbf{I} = \mathbf{VY}$ or $\mathbf{V} = \dfrac{\mathbf{I}}{\mathbf{Y}} = \dfrac{10 **\angle 30°}{0.158 \angle 18.4°} = 63.3 \angle 11.6°$ volts

∴ P.D. across capacitor, $v = \mathbf{63.3 \sin (1000\, t + 11.6°)}$ **volts**

---

* $X_L = \omega L = 1000 \times 10 \times 10^{-3} = 10\,\Omega$

** Note that OX-axis has been taken as the reference.

Example 14.11. *Two circuits having the same ohmic impedance are joined in parallel. The power factor of one circuit is 0.8 lag and that of the other is 0·6 lag. Find the power factor of the whole circuit.*

Solution. The phase angle of current I_1 in the first circuit is $\phi_1 = \cos^{-1} 0.8 = 36.87°$ and that of current I_2 is $\phi_2 = \cos^{-1} 0.6 = 53.13°$. The resultant current I is the phasor sum of I_1 and I_2 as shown in Fig. 14.21. The angle θ between I_1 and I_2 is given by ;

$$\theta = \phi_2 - \phi_1 = 53.13° - 36.87° = 16.26°$$

Since the impedances of the two circuits are equal, I_1 and I_2 will be equal. Obviously, the resultant current I divides the angle θ between I_1 and I_2 equally.

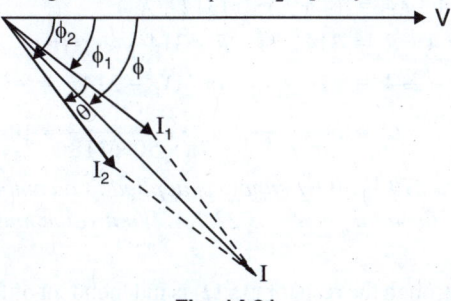

Fig. 14.21

$\therefore$ Phase angle of I, $\phi = \phi_1 + (\theta/2) = 36.87° + (16.26°/2) = 45°$

$\therefore$ P.F. of entire circuit $= \cos \phi = \cos 45° = \mathbf{0.707}$ *lag*

Example 14.12. *Two impedances $Z_1 = (8 + j6)\ \Omega$ and $Z_2 = (3 - j4)\ \Omega$ are in parallel. If the total current of this combination is 25 A, find the power taken by each impedance.*

Solution. $\mathbf{Z_1} = (8 + j\,6)\ \Omega = 10\ \angle\ 36.87°\ \Omega$; $\mathbf{Z_2} = (3 - j\,4)\ \Omega = 5\ \angle -53.13°\ \Omega$

$$\mathbf{Z_1 + Z_2} = (8 + j\,6) + (3 - j\,4) = (11 + j\,2)\ \Omega = 11.18\ \angle\ 10.3°\ \Omega$$

$\therefore$
$$\mathbf{I_1} = \mathbf{I}\frac{\mathbf{Z_2}}{\mathbf{Z_1 + Z_2}} = 25 \times \frac{5\ \angle -53.13°}{11.18\ \angle\ 10.3°} = 11.18\ \angle -63.43°\ \text{A}$$

$$\mathbf{I_2} = \mathbf{I}\frac{\mathbf{Z_1}}{\mathbf{Z_1 + Z_2}} = 25 \times \frac{10\ \angle\ 36.87°}{11.18\ \angle\ 10.3°} = 22.36\ \angle\ 26.57°\ \text{A}$$

Power taken by first branch $= I_1^2 R_1 = (11.18)^2 \times 8 = \mathbf{1000\ W}$

Power taken by second branch $= I_2^2 R_2 = (22.36)^2 \times 3 = \mathbf{1500\ W}$

Example 14.13. *For the circuit shown in Fig. 14.22, find the circuit current when the voltage input is 250 sin 1000 t volts.*

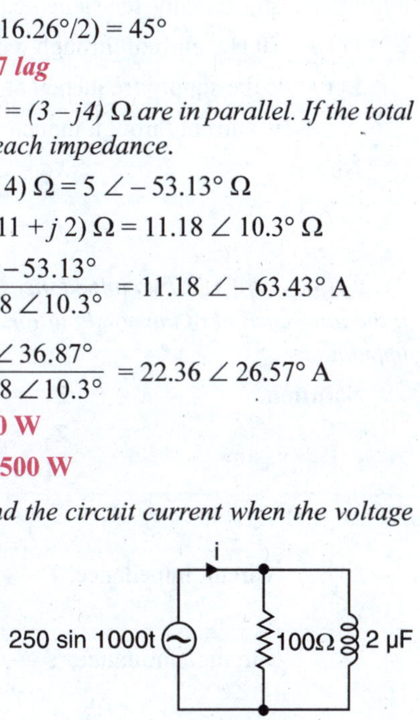

Solution. The admittance of the resistor is $1/100 = 0.01$ S. The admittance of the capacitor is $= j\,\omega C = j \times 1000 \times 2 \times 10^{-6} = j\,0.002$ S. As the two admittances are in parallel,

250 sin 1000t 100Ω 2 μF

$\therefore$ Total admittance, $\mathbf{Y} = (0.01 + j\,0.002)$ S

Taking $\mathbf{V}$ as the reference phasor, $\mathbf{V} = (250 + j0)$ volts.

Fig. 14.22

Circuit current, $\mathbf{I} = \mathbf{VY} = (250 + j\,0)\,(0.01 + j\,0.002)$

$$= (2.5 + j\,0.5)\,\text{A} = 2.55\ \angle\ 11.3°\ \text{A}$$

$\therefore$ $i = \mathbf{2.55\ sin\ (1000\ t + 11.3°)\ A}$

Example 14.14. *A coil having a resistance of 4 Ω and an inductance of 1 H is connected in parallel with a circuit comprising a similar coil in series with a capacitor C and a non-inductive resistor R. Calculate the values of C and R so that the currents in either branch of the arrangement are equal but differ in phase by 90°. The frequency is 50 Hz.*

Solution. The inductive reactance of each branch is the same and is

$$X_L = 2\pi f L = 2\pi \times 50 \times 1 = 314 \, \Omega$$

Impedance of first branch, $\mathbf{Z_1} = (4 + j\, 314) \, \Omega$

Impedance of second branch, $\mathbf{Z_2} = [(4 + R) + j(314 - X_C)] \, \Omega$ ...(i)

Here X_C is the reactance of the capacitor.

Since the currents and hence the impedances are equal in magnitude and differ in phase by 90°,

∴ $*\mathbf{Z_2} = (314 - j\, 4) \, \Omega$ ...(ii)

From eq. (i) and eq. (ii), we have,

$$314 - j\, 4 = (4 + R) + j\,(314 - X_C)$$

∴ $4 + R = 314$ or $R = 314 - 4 = \mathbf{310 \, \Omega}$

and $-4 = 314 - X_C$ or $X_C = 314 + 4 = 318 \, \Omega$

∴ $C = \dfrac{1}{2\pi f X_C} = \dfrac{1}{2\pi \times 50 \times 318} = 10 \times 10^{-6} \, F = \mathbf{10 \, \mu F}$

Example 14.15. *When a 240 V, 50 Hz supply is applied to a combination of a resistor of 15 Ω in parallel with an inductor, the total current is 22.1 A. What value must the frequency have for the total current to be 34 A ?*

Solution. The current through the resistor (15 Ω) is independent of frequency. Therefore, current through resistor remains the same i.e., current through resistor, $I_R = 240/15 = 16$ A.

For $f = 50$ Hz, current through inductor, $I_L = 22.1 - I_R = 22.1 - 16 = 6.1$ A

Let f' be the supply frequency at which the total current drawn from the supply is 34 A.

∴   New current through inductor, $I'_L = 34 - I_R = 34 - 16 = 18$ A

Now $I_L X_L = I'_L X'_L$ or $6.1 \times 2\pi f L = 18 \times 2\pi f' L$

∴ $f' = \dfrac{6.1 \times 2\pi \times 50}{18 \times 2\pi} = \mathbf{16.94 \, Hz}$

Example 14.16. *Two impedances $Z_1 = (6 - j8)$ ohms and $Z_2 = (16 + j12)$ ohms are in parallel. If the total current of the combination is $(20 + j\, 10)$ amperes, find the complex power taken by each impedance.*

Solution. $\mathbf{Z_1} = (6 - j\, 8) \, \Omega = 10 \angle -53.13° \Omega \; ; \; \mathbf{Y_1} = 0.1 \angle 53.13° \, S$

$\mathbf{Z_2} = (16 + j\, 12) \, \Omega = 20 \angle 36.87° \, \Omega \; ; \; \mathbf{Y_2} = 0.05 \angle -36.87° \, S$

$\mathbf{I} = (20 + j\, 10) \, A = 22.36 \angle 26.56° \, A$

$\mathbf{Z_1 + Z_2} = (6 - j\, 8) + (16 + j\, 12) = (22 + j\, 4) \, \Omega = 22.36 \angle 10.3° \, \Omega$

Circuit impedance, $\mathbf{Z} = \dfrac{\mathbf{Z_1 Z_2}}{\mathbf{Z_1 + Z_2}} = \dfrac{10 \angle -53.13° \times 20 \angle 36.87°}{22.36 \angle 10.3°} = 8.94 \angle -26.56° \Omega$

Circuit admittance, $\mathbf{Y} = \dfrac{1}{\mathbf{Z}} = \dfrac{1}{8.94 \angle -26.56°} = 0.1118 \angle 26.56° \, S$

Supply voltage, $\mathbf{V} = \dfrac{\mathbf{I}}{\mathbf{Y}} = \dfrac{22.36 \angle 26.56°}{0.1118 \angle 26.56°} = 200 \angle 0° \, volts$

Branch current, $\mathbf{I_1} = \mathbf{V Y_1} = 200 \angle 0° \times 0.1 \angle 53.13° = 20 \angle 53.13° \, A$

Branch current, $\mathbf{I_2} = \mathbf{V Y_2} = 200 \angle 0° \times 0.05 \angle -36.87° = 10 \angle -36.87° \, A$

Using the method of conjugates and taking current conjugate, the complex power taken by each branch is given by ;

---

* $\mathbf{Z_1} = (4 + j\, 314) \, \Omega = 314 \cdot 025 \angle 89.27° \, \Omega \; ; \; \mathbf{Z_2} = (314 - j\, 4) \, \Omega = 314.025 \angle -0.73° \, \Omega$

$$\mathbf{S_1} = 200 \angle 0° \times 20 \angle -53.13° = 4000 \angle -53.13° \text{ VA}$$
$$= 4000 [\cos(-53.13°) + j\sin(-53.13°)] = \mathbf{(2400 - j\,3200) \text{ VA}}$$
$$\mathbf{S_2} = 200 \angle 0° \times 10 \angle 36.87° = 2000 \angle 36.87° \text{ VA}$$
$$= 2000 [\cos 36.87° + j \sin 36.87°] = \mathbf{(1600 + j\,1200) \text{ VA}}$$

Example 14.17. *The admittance of 0.2 S is connected in parallel with a pure inductive reactance of susceptance 0·15 S. The combined admittance is 0·314 S. What is the value of resistance in the circuit?*

Solution. Suppose the admittance 0.2 S has components x and y as shown in Fig. 14.23. Clearly, $\triangle ABC$ is the admittance triangle for the circuit having 0.2 S admittance. When a pure inductive reactance is connected in parallel with this admittance, only the vertical component increases so that now $\triangle ABD$ is the admittance triangle.

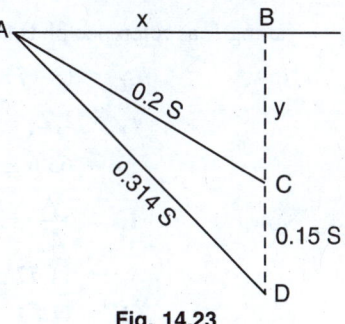

From $\triangle ABC$, $x^2 + y^2 = (0.2)^2$ ...(i)

From $\triangle ABD$, $x^2 + (y + 0.15)^2 = (0.314)^2$ ...(ii)

Solving eqs. (i) and (ii), we get, $x = 0.16$; $y = 0.12$

Fig. 14.23

Let R and X be the resistance and reactance respectively of 0.2 S admittance. Then,

$$x = \frac{R}{R^2 + X^2} = 0.16 \qquad\qquad ...(iii)$$

$$y = \frac{X}{R^2 + X^2} = 0.12 \qquad\qquad ...(iv)$$

Solving eqs. (iii) and (iv), we get, $\dfrac{R}{X} = \dfrac{4}{3}$ or $X = \dfrac{3R}{4}$

$\therefore$ $\dfrac{R}{R^2 + (3R/4)^2} = 0.16$ or $R = 4\,\Omega$

Since the values of R, X and Z and their reciprocals G, B and Y are not time-dependent, they are not true phasors. Yet their values are determined by voltage phasors and current phasors.

Example 14.18. *Three loads are connected in parallel across of $250 \angle 0°$ V r.m.s. supply. Load 1 takes a power of 8 kW at a lagging power factor of 0.8; load 2 a power of 12 kW at a leading power factor of 0.6 and load 3 a power of 5 kW at a power factor of 0.5 lagging. What are the currents through each load? Give the answers in polar form.*

Solution. The complex power S for any load is $S = P + jQ$ where P is the active power and Q the reactive power.

For Load 1. $\mathbf{VI_1^*} = P + jQ$ or $250\,\mathbf{I_1}^* = 8000 + jQ$

Note that $\mathbf{I_1}^*$ is the conjugate of the current in this load. Now $\phi_1 = \cos^{-1} 0.8 = 36.9°$ so that $\sin \phi_1 = \sin 36.9° = 0.6$. The reactive power, $Q = VI_1 \sin \phi_1 = (8000/0.8) \times 0.6 = 6000 \text{ VAR}$.

$\therefore$ $250\,\mathbf{I_1^*} = 8000 + j\,6000$ or $\mathbf{I_1^*} = (32 + j\,24) \text{ A}$

$\therefore$ Current through load, $\mathbf{I_1} = (32 - j\,24) \text{ A} = \mathbf{40 \angle -36.9° \text{ A}}$

For Load 2. $\mathbf{VI_2^*} = P - jQ$ or $250\,\mathbf{I_2^*} = 12000 - jQ$

Now, $\phi_2 = \cos^{-1} 0.6 = 53.1°$ so that $Q = (12000/0.6) \sin 53.1° = 16000 \text{ VAR}$

$\therefore$ $250\,\mathbf{I_2^*} = 12000 - j\,16000$ or $\mathbf{I_2^*} = (48 - j\,64) \text{ A}$

Note that minus sign is because of leading power factor.

$\therefore$ Current through load, $\mathbf{I_2} = (48 + j\,64) \text{ A} = \mathbf{80 \angle 53.1° \text{ A}}$

For Load 3. $\mathbf{VI_3^*} = P + jQ$ or $250\,\mathbf{I_3^*} = 5000 + jQ$

Now, $\phi_3 = \cos^{-1} 0.5 = 60°$ so that $Q = (5000/0.5) \sin 60° = 8660$ VAR

∴ $250\, \mathbf{I}_3^* = 5000 + j\, 8660$ or $\mathbf{I}_3^* = (20 + j\, 35)$ A

∴   Current through load, $\mathbf{I}_3 = (20 - j\, 35)$ A $= \mathbf{40 \angle - 60.3° \text{ A}}$

Example 14.19. *In the circuit shown in Fig. 14.24, the voltmeter across 3Ω resistor reads 45 V. Find the reading of the ammeter.*

Solution. The current $I_1 = 45/3 = 15$ A.

Taking I_1 as reference phasor, $\mathbf{I}_1 = 15 \angle 0°$ A.

Now, $\mathbf{Z}_1 = (3 - j\, 3)\, \Omega = 4.24 \angle - 45°\, \Omega$

∴ $\mathbf{V}_1 = \mathbf{I}_1 \mathbf{Z}_1 = 15 \angle 0° \times 4.24 \angle - 45°$

                      $= 63.6 \angle - 45°$ volts

Now, $\mathbf{I}_2 = \dfrac{\mathbf{V}_1}{\mathbf{Z}_2} = \dfrac{63.6 \angle - 45°}{5 + j2} = \dfrac{63.6 \angle - 45°}{5.4 \angle 21.8°}$

                      $= 11.77 \angle -66.8°$ A $= (4.64 - j\, 10.8)$ A

∴ $\mathbf{I} = \mathbf{I}_1 + \mathbf{I}_2 = (15 + j\, 0) + (4.64 - j\, 10.8) = (19.64 - j\, 10.8)$ A

                      $= \mathbf{22.4 \angle - 28.8° \text{ A}}$

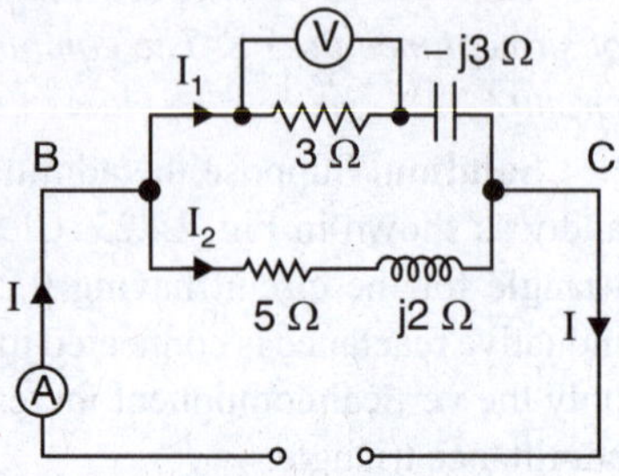

Therefore, the ammeter will read 22.4 A.

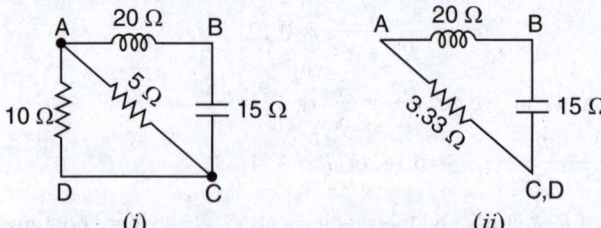

Fig. 14.24

Example 14.20. *In the circuit shown in Fig. 14.25 (i), find the impedance between points B and D.*

Fig. 14.25

Solution. The points C and D are at the same potential so that $\mathbf{Z}_{BD} = \mathbf{Z}_{BC}$. Referring to Fig. 14.25 (i), 10 Ω and 5 Ω are in parallel and their equivalent resistance $= 5 \times 10/(5 + 10) = 3.33\, \Omega$. The circuit then reduces to the one shown in Fig. 14.25 (ii). Referring to Fig. 14.25 (ii),

$$\mathbf{Z}_{BC} = (3.33 + j\, 20)\, \|\, (-j\, 15)$$

$$= \frac{(3.33 + j\, 20)(-j\, 15)}{(3.33 + j\, 20) + (-j\, 15)} = \frac{304.1 \angle -9.5°}{6 \angle 56.3°} = \mathbf{50.7 \angle - 65.8°\, \Omega}$$

Example 14.21. *In the circuit shown in Fig. 14.26, find*

(i) the voltage v_T across the a.c. current source and (ii) the currents i_1 and i_2.

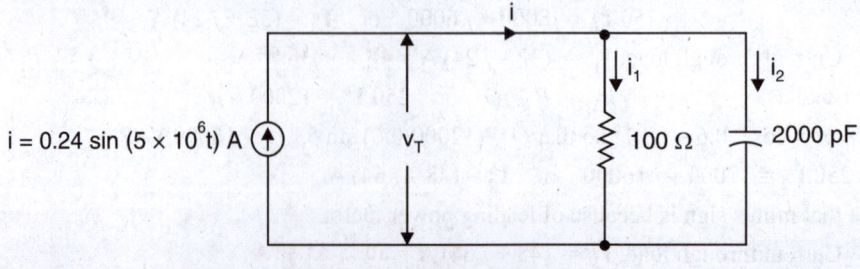

Fig. 14.26

Solution. Like a d.c. current source, an (ideal) *a.c. current source supplies the same constant current to any circuit connected across its terminals. *In the a.c. case, the current is constant in the sense that its peak value does not change.*

(*i*) Magnitude of capacitive reactance is

$$X_C = \frac{1}{\omega C} = \frac{1}{(5 \times 10^6) \times 2000 \times 10^{-12}} = 100\ \Omega \qquad (\because \omega = 5 \times 10^6 \text{rad.s}^{-1})$$

Now, $\qquad \mathbf{Z_1} = 100\ \angle\ 0°\ \Omega = (100 + j\ 0)\ \Omega\ ;\ \mathbf{Z_2} = 100\ \angle -90°\ \Omega = (0 - j\ 100)\ \Omega$

Total circuit impedance, $\mathbf{Z_T} = \dfrac{\mathbf{Z_1}\,\mathbf{Z_2}}{\mathbf{Z_1} + \mathbf{Z_2}} = \dfrac{100\ \angle\ 0° \times 100\ \angle -90°}{(100 + j\ 0) + (0 - j\ 100)}$

$$= \frac{10^4\ \angle -90°}{100 - j\ 100} = \frac{10^4\ \angle -90°}{141.4\ \angle -45°} = 70.7\ \angle -45°\ \Omega$$

∴ Voltage across the a.c. current source is

$$v_T = i\ \mathbf{Z_T} = 0.24\ \angle\ 0° \times 70.7\ \angle -45° = \mathbf{16.97\ \angle -45°\ volts}$$

(*ii*) Since v_T is the voltage across each component,

∴ $\qquad i_1 = \dfrac{v_T}{\mathbf{Z_1}} = \dfrac{16.97\ \angle -45°}{100\ \angle\ 0°} = \mathbf{0.1697\ \angle -45°\ A}$ or $\mathbf{(0.12 - j\ 0.12)\ A}$

$$i_2 = \frac{v_T}{\mathbf{Z_2}} = \frac{16.97\ \angle -45°}{100\ \angle -90°} = \mathbf{0.1697\ \angle\ 45°\ A}\ \text{or}\ \mathbf{(0.12 + j\ 0.12)\ A}$$

Note that $i** = i_1 + i_2$ so that Kirchhoff's current law (as well as voltage law) is also applicable to a.c. circuits.

Example 14.22. *Four identical motors are connected in parallel across 240 V, 60 Hz supply. Each motor is equivalent to 0·1 H inductor. The magnitude of current drawn from the supply was measured to be 27 A peak. Determine if one or more motors are not operating.*

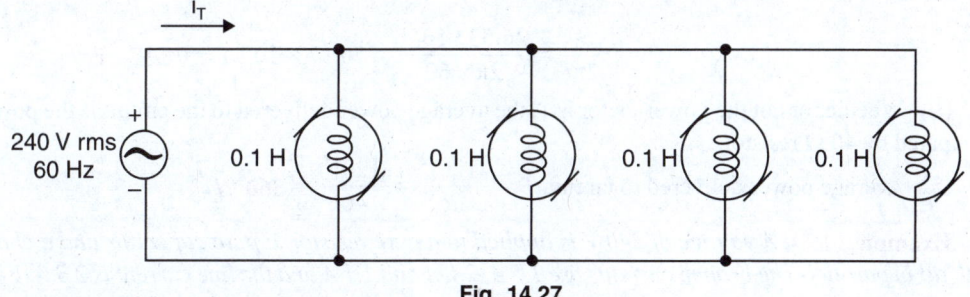

Fig. 14.27

Solution. The equivalent circuit is shown in Fig. 14.27.

Inductive reactance of each motor, $X_L = \omega L = 2\pi f L = 2\pi \times 60 \times 0.1 = 37.7\ \Omega$

Peak value of supply voltage, $V_m = \sqrt{2} \times 240 = 339.4\ V$

∴ Peak current drawn by each motor is given by ;

$$I_m = V_m / X_L = 339.4 / 37.7 = 9A$$

---

* Note that we will use the same symbol for an a.c. current source that we used for a d.c. current source. Since the current reverses direction every half-cycle, the arrow in the symbol is simply a phase reference that proves useful when there is more than one a.c. current source in a circuit. Reversing the arrow is the same as multiplying the current by –1 which is the same as adding or subtracting 180° from its phase angle.

** $i_1 + i_2 = (0.12 - j\ 0.12) + (0.12 + j\ 0.12) = 0.24 + j\ 0 = 0.24\ \angle\ 0°\ A = i$

Since the motors are identical, *the currents through the motors are identical in magnitude and phase angle.* Remember that magnitudes of phasor quantities can be added if all have the same phase angle.

∴   Peak current drawn from supply = $9 \times 4 = 36$ A

Since the actual peak current drawn from the supply is 27 A, it is concluded that **one motor is not operating.**

Example 14.23. *In the circuit shown in Fig. 14.28, find (i) the value of capacitance C to make the circuit power factor equal to 1 (ii) the average power delivered to the circuit.*

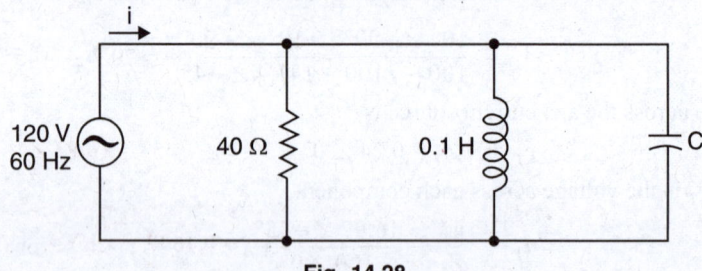

Fig. 14.28

Solution. (*i*) Inductive susceptance, $B_L = \dfrac{1}{\omega L} = \dfrac{1}{2\pi f L} = \dfrac{1}{2\pi \times 60 \times 0.1} = 26.53 \times 10^{-3}$ S

∴ Admittance of inductor, $\mathbf{Y_1} = G - j\, B_L = (0 - j\, 26.53 \times 10^{-3})$ S

Admittance of capacitor, $\mathbf{Y_2} = G + jB_C = (0 + jB_C)$ S

When capacitive susceptance is exactly equal to inductive susceptance, the circuit is then purely resistive and the power factor will be 1. Therefore, the value of $B_C = B_L = 26.53 \times 10^{-3}$ S.

Now, $B_C = 2\pi f C$

∴ $C = \dfrac{B_C}{2\pi f} = \dfrac{26.53 \times 10^{-3}}{2\pi \times 60} = 70.37 \times 10^{-6}$ F = **70.37 μF**

(*ii*)   Whether or not the power factor is 1, the average power delivered to the circuit is the power dissipated by 40 Ω resistor.

∴   Average power delivered to circuit, $P = \dfrac{(120)^2}{R} = \dfrac{(120)^2}{40} = $ **360 W**

Example 14.24. *A voltage of 240 V is applied to a pure resistor, a pure capacitor and a choke coil, all in parallel. The branch currents are 1.5 A, 2.0 A and 1.1 A and the line current is 2.3 A. Find the power factor of inductor, total volt amperes, active and reactive power and the overall p.f.*

Solution. Let the phase angle of the choke coil be $\theta°$ lagging.

Now, $\mathbf{I_R} = 1.5$ A ; $\mathbf{I_C} = + j2$A ; $\mathbf{I_L} = 1.1 \angle - \theta° = 1.1 \cos \theta - j\, 1.1 \sin \theta$

∴   Line current, $\mathbf{I} = \mathbf{I_R} + \mathbf{I_C} + \mathbf{I_L} = 1.5 + j2 + 1.1 \cos \theta - j\, 1.1 \sin \theta$

                    $= (1.5 + 1.1 \cos \theta) + j(2 - 1.1 \sin \theta)$

∴ $I^2 = (1.5 + 1.1 \cos \theta)^2 + (2 - 1.1 \sin \theta)^2$

But it is given that $I = 2.3$ A so that $I^2 = (2.3)^2$.

∴ $(1.5 + 1.1 \cos \theta)^2 + (2 - 1.1 \sin \theta)^2 = (2.3)^2$

On solving, $\cos \theta = 0.5$ and $\sin \theta = 0.866$.

$\therefore$ Power factor of choke coil = cos θ = **0.5** *lagging*

Total volt amperes = VI = 240 × 2.3 = **552 VA**

Overall power factor, cos ϕ = $\dfrac{1.5 + 1.1\cos\theta}{I} = \dfrac{1.5 + 1.1 \times 0.5}{2.3}$ = **0.89** *leading*

Active power, P = $VI\cos\phi$ = 240 × 2.3 × 0.89 = **491.28 W**

Reactive power, Q = $VI\sin\phi$ = 240 × 2.3 × 0.4555 = **251.44 VAR** *capacitive*

Example 14.25. *A lamp requires 3.05 A, 410 W at unity p.f. Find (i) the value of inductance to be placed in series with the lamp so that it can be operated from 230 V, 50 Hz supply (ii) the value of capacitance which should be placed across the supply to raise the p.f. to unity.*

Solution. (*i*) Resistance of lamp, $R = \dfrac{P}{I^2} = \dfrac{410}{(3.05)^2}$ = 44.074 Ω

Let L henry be the value of inductance to be placed in series with the lamp so that the lamp can be operated from 230 V, 50 Hz supply.

$\therefore$ Impedance of the circuit, $Z = \dfrac{V}{I} = \dfrac{230}{3.05}$ = 75.41 Ω

Now, $X_L = \sqrt{Z^2 - R^2} = \sqrt{(75.41)^2 - (44.074)^2}$ = 61.19 Ω

$\therefore$ $L = \dfrac{X_L}{2\pi f} = \dfrac{61.19}{2\pi \times 50}$ = **0.195 H**

(*ii*) The current drawn by the lamp and inductance is

$\mathbf{I} = \dfrac{230\angle 0°}{R + jX_L} = \dfrac{230\angle 0°}{44.074 + j\,61.19}$

$= \dfrac{230\angle 0°}{75.41\angle 54.24°}$ = 3.05 $\angle -54.24°$ A = (1.783 $- j\,2.475$) A

Therefore, the capacitor must draw a current of 2.475 A to raise the power factor to unity. If C is the required capacitance of the capacitor, then,

$X_C = \dfrac{230}{2.475}$ = 92.93 Ω

or $\dfrac{1}{2\pi f C}$ = 92.93 $\therefore$ $C = \dfrac{1}{2\pi f \times 92.93} = \dfrac{1}{2\pi \times 50 \times 92.93}$ = **34.2 × 10⁻⁶F**

It is always desirable that power factor of an a.c. circuit should be as close to unity as possible.

Example 14.26. *The currents in each branch of a two branched parallel circuit are given by the expressions :*

$i_a = 7.07 \sin (314t - \pi/4)$ *and* $i_b = 21.2 \sin (314t + \pi/3)$

The supply voltage is given by the expression v = 354 sin 314t. Derive a similar expression for the supply current and calculate the ohmic values of components assuming two pure components in each branch. State whether the reactive components are inductive or capacitive.

Solution. It is clear from the expressions of supply voltage and branch currents that i_a lags behind the supply voltage by $\pi/4$ radian or 45° while i_b leads the supply voltage by $\pi/3$ radian or 60°. Therefore, the branch a consists of a resistance in series with pure inductive reactance and branch b consists of resistance in series with pure capacitive reactance as shown in Fig. 14.29 (*i*).

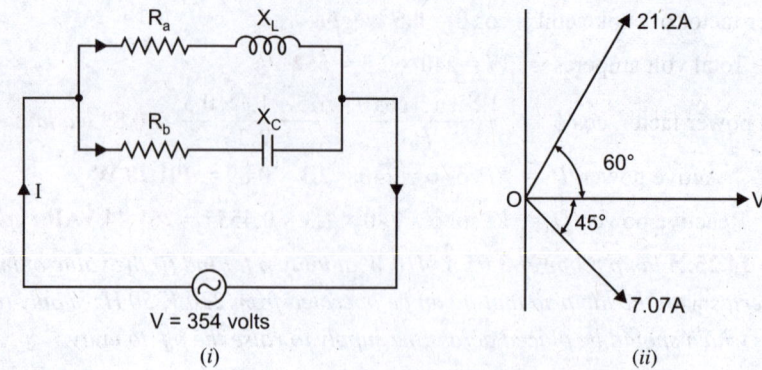

Fig. 14.29

The maximum value of current in branch a is 7.07 A and that in branch b is 21.2 A as shown in the phasor diagram in Fig. 14.29 (ii).

$$X\text{-component} = 21.2 \cos 60° + 7.07 \cos 45° = 15.6 \text{ A}$$

$$Y\text{-component} = 21.2 \sin 60° - 7.07 \sin 45° = 13.36 \text{ A}$$

∴ Maximum value of resultant current

$$= \sqrt{15.6^2 + 13.36^2} = 20.55 \text{ A}$$

Its phase angle is $\phi = \tan^{-1} 13.36/15.6 = 40.5°$.

∴ Expression for supply current is $i = 20.55 \sin (314 t + 40.5°)$

Now, $Z_a = \dfrac{354}{7.07} = 50 \ \Omega$; $\cos \phi_a = \cos 45° = \dfrac{1}{\sqrt{2}}$; $\sin \phi_a = \dfrac{1}{\sqrt{2}}$

∴ $R_a = Z_a \cos \phi_a = 50 \times 1/\sqrt{2} = 35.4 \ \Omega$; $X_a = Z_a \sin \phi_a = 50 \times 1/\sqrt{2} = 35.4 \ \Omega$

Also, $Z_b = \dfrac{354}{21.2} = 16.7 \ \Omega$; $\phi_b = 60°$

∴ $R_b = Z_b \cos 60° = 16.7 \cos 60° = 8.35 \ \Omega$; $X_b = Z_b \sin 60° = 16.7 \sin 60° = 14.46 \ \Omega$

Example 14.27. *An inductive coil of $(6 + j 8) \ \Omega$ impedance is connected to 100 V, 50 Hz a.c. supply. It is desired to improve p.f. of supply current to 0.8 lagging by connecting a capacitor (i) in series with the coil (ii) in parallel with the coil. Determine the value of the capacitance in each case.*

Solution. (*i*) Let a capacitor of capacitance C be connected in series with the coil so as to raise the power factor of supply current to 0.8 lagging as shown in Fig. 14.30.

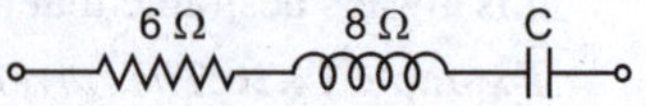

Fig. 14.30

Net inductive reactance of the series circuit is

$$X = X_L - X_C = 8 - X_C$$

The p.f. is to be improved to 0.8 lagging *i.e.* $\cos \phi = 0.8$ so that :

$$\tan \phi = \tan (\cos^{-1} 0.8) = 0.75$$

Now, $\tan \phi = \dfrac{X}{R} = \dfrac{8 - X_C}{R}$

or $0.75 = \dfrac{8 - X_C}{6}$ ∴ $X_C = 3.5 \ \Omega$

∴ $C = \dfrac{1}{2\pi f X_C} = \dfrac{1}{2\pi \times 50 \times 3.5} = 909.46 \times 10^{-6} \text{ F} = 909.46 \ \mu F$

(*ii*) Let a capacitor of capacitance C be connected in parallel with the coil so as to raise the power factor of supply current to 0.8 lagging as shown in Fig. 14.31.

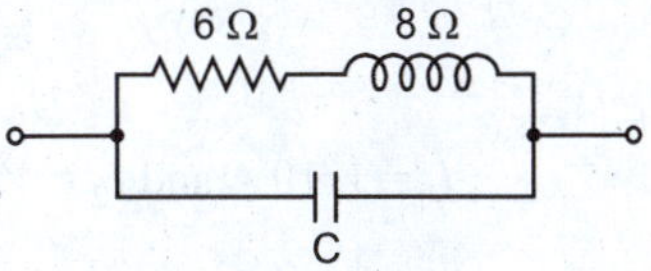

Fig. 14.31

Conductance of coil, $G = \dfrac{R}{R^2 + X_L^2} = \dfrac{6}{6^2 + 8^2} = 0.06$ S

Inductive susceptance of coil, $B_L = \dfrac{X_L}{R^2 + X_L^2} = \dfrac{8}{6^2 + 8^2} = 0.08$ S

Net inductive susceptance of the parallel circuit is

$$B = B_L - B_C$$

Now $\tan \phi = \dfrac{B}{G} = \dfrac{B_L - B_C}{G}$

or $0.75 = \dfrac{0.08 - B_C}{0.06}$ $\therefore B_C = 0.035$ S

Now, $B_C = \dfrac{1}{X_C} = 2\pi f C$ $\therefore C = \dfrac{B_C}{2\pi f} = \dfrac{0.035}{2\pi \times 50} = 111.4 \times 10^{-6}$ F = **111.4 μF**

Example 14.28. *The impedances $Z_1 = (6 + j\,8)\Omega$, $Z_2 = (8 - j\,6)\ \Omega$ and $Z_3 = (10 + j\,0)\ \Omega$ measured at 50 Hz form three branches of a parallel circuit. This circuit is fed from a 100 V, 50 Hz supply. A purely reactive (inductive or capacitive) circuit is added as the fourth parallel branch to the above three-branched parallel circuit so as to draw minimum current from the source. Determine the value of L or C to be used in the fourth branch and also find the minimum current.*

Solution. Total admittance of the 3-branch parallel circuit is

$$Y = \frac{1}{6 + j8} + \frac{1}{8 - j6} + \frac{1}{10 + j0}$$

$$= (0.06 - j\,0.08) + (0.08 + j\,0.06) + (0.1 + j\,0)$$

$$= (0.24 - j\,0.02)\ \text{S}$$

The current drawn from the source will be minimum when the net susceptance of circuit is zero. For this to happen, we should add a capacitor of capacitance C in parallel as the fourth branch. The capacitive susceptance of this capacitor should be $B_C = + 0.02$ S.

Now $B_C = \dfrac{1}{X_C} = 2\pi f C$

$\therefore$ $C = \dfrac{B_C}{2\pi f} = \dfrac{0.02}{2\pi \times 50} = 63.7 \times 10^{-6}$ F = **63.7 μF**

Admittance of four parallel branches is

$$Y' = (0.24 - j\,0.02) + j\,0.02 = 0.24\ \text{S}$$

$\therefore$ Minimum current drawn from the source is

$$I = VY' = 100 \times 0.24 = \textbf{24 A}$$

Example 14.29. *The power consumed by both branches of the circuit shown in Fig. 14.32 is 2200 W. Calculate power of each branch and the reading of the ammeter.*

Solution. Let the applied voltage be $\mathbf{V} = V \angle 0°$ volts. Now, $\mathbf{Z_1} = (6 + j\,8)\ \Omega = 10 \angle 53.1°\ \Omega$ and $\mathbf{Z_2} = 20\ \Omega$.

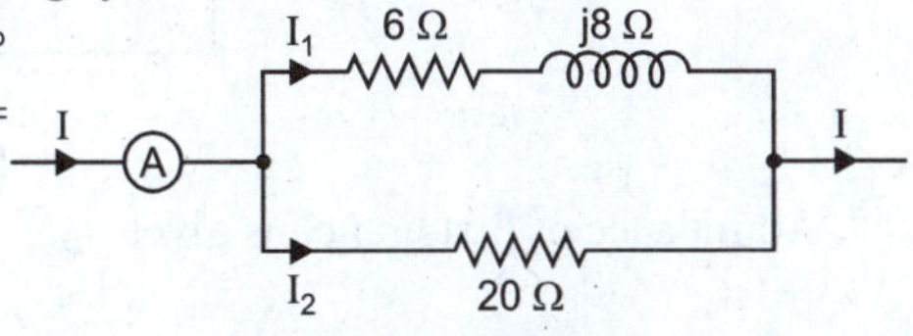

$\therefore$ $\mathbf{I_1} = \dfrac{\mathbf{V}}{\mathbf{Z_1}} = \dfrac{V \angle 0°}{10 \angle 53.1°} = \dfrac{V}{10} \angle -53.1° \text{A}$

Fig. 14.32

$$\mathbf{I_2} = \frac{\mathbf{V}}{\mathbf{Z_2}} = \frac{V \angle 0°}{20} = \frac{V}{20} \text{A}$$

$I_1 = V/10$ A and $I_2 = V/20$ A

$$\therefore \qquad \frac{I_1}{I_2} = \frac{V/10}{V/20} = 2$$

Now,

$$\frac{P_1}{P_2} = \frac{I_1^2 R_1}{I_2^2 R_2} = (2)^2 \times \frac{6}{20} = \frac{6}{5} \qquad \qquad ...(i)$$

Also,

$$P_1 + P_2 = 2200 \qquad \qquad ...(iii)$$

From eqs. (i) and (ii), $P_1 =$ **1200 W** and $P_2 =$ **1000 W**

$$\therefore \qquad I_1^2 = \frac{P_1}{R_1} \text{ or } I_1^2 = \frac{1200}{6} \quad \therefore I_1 = 14.14 \text{ A}$$

and

$$\frac{I_1}{I_2} = 2 \text{ or } \frac{14.14}{I_2} = 2 \quad \therefore I_2 = 7.07 \text{ A}$$

$$\therefore \qquad \mathbf{I_1} = 14.14 \angle -53.1° \text{ A} ; \mathbf{I_2} = 7.07 \angle 0° \text{ A}$$

Line current, $\mathbf{I} = \mathbf{I_1} + \mathbf{I_2} = 14.14 \angle -53.1° + 7.07 \angle 0°$

$$= (8.48 - j11.31) + (7.07 + j0) = (15.55 - j11.31) = 19.3 \angle -36° \text{A}$$

Therefore, the ammeter will read **19.3 A.**

Example 14.30. *The active and lagging reactive components of current taken by an a.c. circuit from a 250 V supply are 50 A and 25 A respectively. Calculate the conductance, susceptance, admittance and p.f. of the circuit.*

Solution. Given: $I \cos \phi = 50$ A ; $I \sin \phi = 25$ A

$$\therefore \qquad \text{Circuit current, } I = \sqrt{(I \cos \phi)^2 + (I \sin \phi)^2} = \sqrt{50^2 + 25^2} = 55.9 \text{ A}$$

$$\text{Admittance of circuit, } Y = \frac{I}{V} = \frac{55.9}{250} = \mathbf{0.224 \text{ S}}$$

$$\text{P.F. of the circuit, } \cos \phi = \frac{I \cos \phi}{I} = \frac{50}{55.9} = \mathbf{0.8944} \text{ *lagging*}$$

$$\text{Circuit conductance, } G = Y \cos \phi = 0.224 \times 0.8944 = \mathbf{0.2 \text{ S}}$$

$$\text{Circuit susceptance, } B = \sqrt{Y^2 - G^2} = \sqrt{(0.224)^2 - (0.2)^2} = \mathbf{0.1 \text{ S}} \text{ *(inductive)*}$$

Example 14.31. *Obtain the power factor of a two branch parallel circuit where the first branch has $Z_1 = (2 + j 4) \Omega$ and second $Z_2 = (6 + j 0) \Omega$. To what value must the 6 Ω resistor be changed to result in the overall p.f. 0.9 lagging?*

Solution. Fig. 14.33 shows the conditions of the problem.

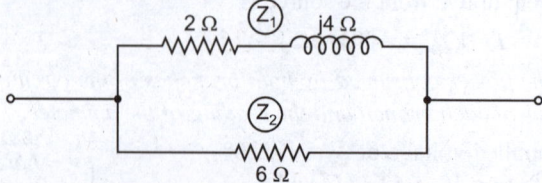

Fig. 14.33

Admittance of first branch is given by ;

$$\mathbf{Y_1} = \frac{1}{\mathbf{Z_1}} = \frac{1}{2 + j4} = (0.1 - j\,0.2) \text{ S}$$

Admittance of the second branch is given by ;

$$\mathbf{Y_2} = \frac{1}{\mathbf{Z_2}} = \frac{1}{6+j0} = (0.167+j\,0)\text{ S}$$

Total admittance, $\mathbf{Y} = \mathbf{Y_1} + \mathbf{Y_2} = (0.1-j\,0.2)+(0.167+j\,0) = (0.267-j\,0.2)\text{S}$

∴

$$Y = \sqrt{(0.267)^2 + (0.2)^2} = 0.333\text{ S}$$

Circuit power factor, $\cos\phi = \dfrac{G}{Y} = \dfrac{0.267}{0.333} = \textbf{0.8 }\textit{lagging}$

Let the resistance of Z_2 ($= 6\ \Omega$ resistor) be changed to R so as to improve the power factor $\cos\phi = 0.8$ lagging to $\cos\phi' = 0.9$ lagging. Now, conductance of second branch is $G'_2 = 1/R$.

∴   Total circuit conductance, $G' = G_1 + G'_2 = \left(0.1 + \dfrac{1}{R}\right)\text{S}$

Total circuit susceptance, $B' = -0.25$ ... *same as before*

Circuit phase angle, $\phi' = \cos^{-1} 0.9 = 25.84°$ lagging

Now,

$$\tan\phi' = \frac{B'}{G'} = \frac{-0.2}{G'}$$

or

$$\tan 25.84° = \frac{-0.2}{0.1+(1/R)} \qquad \therefore R = \textbf{3.2 }\Omega$$

Example 14.32. *A 159.23 µF capacitor in parallel with a resistance R draws a current of 25 A from 300 V, 50 Hz mains. Using phasor diagram, find the frequency f at which this combination draws the same current from 360 V mains.*

Solution. Fig. 14.34 (*i*) shows the circuit diagram.

When supply voltage V = 300 volts and f = 50 Hz. The supply current is $I = 25$ A.

$$X_C = \frac{1}{2\pi f\,C} = \frac{1}{2\pi\times50\times159.23\times10^{-6}} = 20\ \Omega$$

∴

$$I_C = \frac{V}{X_C} = \frac{300}{20} = 15\text{ A leading }V\text{ by }90°$$

The phasor diagram for the circuit is shown in Fig. 14.34 (*ii*). From the phasor diagram in Fig. 14.34 (*ii*), we have,

$$I_R = \sqrt{I^2 - I_C^2} = \sqrt{25^2 - 15^2} = 20\text{ A}$$

∴

$$R = \frac{V}{I_R} = \frac{300}{20} = 15\ \Omega$$

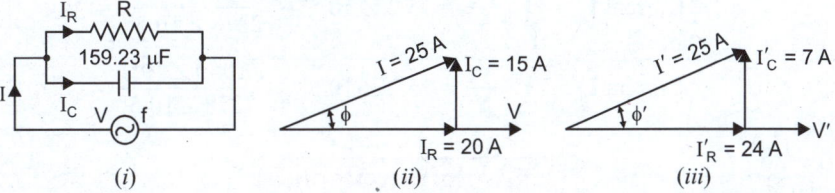

(*i*)   (*ii*)   (*iii*)

Fig. 14.34

When supply voltage V' = 360 volts and frequency is f'. The supply current is the same *i.e.* $I' = I = 25$ A.

$$I'_R = \frac{V'}{R} = \frac{360}{15} = 24\text{ A}$$

The phasor diagram for the circuit is shown in Fig. 14.34 (*iii*). From the phasor diagram in Fig. 14.34 (*iii*), we have,

$$I'_C = \sqrt{(I')^2 - (I'_R)^2} = \sqrt{25^2 - 24^2} = 7 \text{ A}$$

$$\therefore \qquad X'_C = \frac{V'}{I'_C} = \frac{360}{7} = 51.43 \ \Omega$$

$$\therefore \qquad \text{Supply frequency}, f' = \frac{1}{2\pi C \, X'_C} = \frac{1}{2\pi \times 159.23 \times 10^{-6} \times 51.43} = \textbf{19.4 Hz}$$

Example 14.33. *The three impedances in Fig. 14.35 are* $Z_1 = 1606 \angle 51° \ \Omega$, $Z_2 = 977 \angle -33° \ \Omega$ *and* $Z_3 = 953 \angle -19° \ \Omega$. *The supply current is* $I = 75.2 \angle 10.5° \text{ mA}$. *Determine* I_1, I_2 *and* I_3.

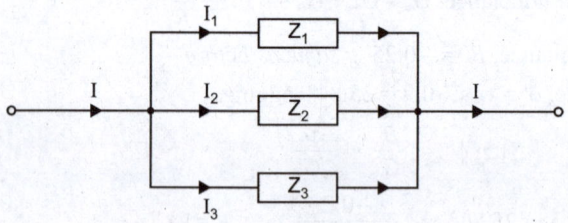

Fig. 14.35

Solution. $\mathbf{Z_1} = 1606 \angle 51° \ \Omega$; $\mathbf{Z_2} = 977 \angle -33° \ \Omega$; $\mathbf{Z_3} = 953 \angle -19° \ \Omega$

$$\therefore \qquad \mathbf{Y_1} = \frac{1}{\mathbf{Z_1}} = \frac{1}{1606 \angle 51°} = 622.7 \angle -51° \ \mu S = (392 - j\,484) \ \mu S$$

$$\mathbf{Y_2} = \frac{1}{\mathbf{Z_2}} = \frac{1}{977 \angle -33°} = 1.02 \angle 33° \text{ mS} = 1020 \angle 33° \ \mu S$$

$$= (855 + j\,556) \ \mu S$$

$$\mathbf{Y_3} = \frac{1}{\mathbf{Z_3}} = \frac{1}{953 \angle -19°} = 1.05 \angle 19° \text{ mS} = 1050 \angle 19° \ \mu S$$

$$= (993 + j\,343) \ \mu S$$

Total admittance, $\mathbf{Y} = \mathbf{Y_1} + \mathbf{Y_2} + \mathbf{Y_3} = (392 - j\,484) + (855 + j\,556) + (993 + j\,343)$

$$= (2240 + j\,415) \ \mu S = 2278 \angle 10.5° \ \mu S$$

$$\text{Current } \mathbf{I_1} = \mathbf{I} \times \frac{\mathbf{Y_1}}{\mathbf{Y}} = 75.2 \angle 10.5° \times \frac{622.7 \angle -51°}{2278 \angle 10.5°} = \textbf{20.5} \angle -\textbf{51° mA}$$

$$\text{Current } \mathbf{I_2} = \mathbf{I} \times \frac{\mathbf{Y_2}}{\mathbf{Y}} = 75.2 \angle 10.5° \times \frac{1020 \angle 33°}{2278 \angle 10.5°} = \textbf{33.6} \angle \textbf{33° mA}$$

$$\text{Current } \mathbf{I_3} = \mathbf{I} \times \frac{\mathbf{Y_3}}{\mathbf{Y}} = 75.2 \angle 10.5° \times \frac{1050 \angle 19°}{2278 \angle 10.5°} = \textbf{34.6} \angle \textbf{19° mA}$$

Tutorial Problems

1. A resistor of 10 Ω is connected in parallel with an inductor of 31·8 mH. A 200 V, 50 Hz supply is connected to the circuit. Determine (*i*) the line current (*ii*) power factor and (*iii*) power consumed by the circuit. **[(*i*) 28.28 A (*ii*) 0.71 lag (*iii*) 4015.76 W]**

2. A coil of resistance 15 Ω and inductance 0.05 H is connected in parallel with a non-inductive resistance of 20 Ω. If the circuit is connected to 200 V, 50 Hz, find (*i*) current in each branch (*ii*) supply current (*iii*) power factor. **[(*i*) 9.22 A ; 10 A (*ii*) 17.7A (*iii*) 0.926 lag]**

3. A 10 Ω resistor, a 31.8 mH inductor and a 318 μF capacitor are connected in parallel and supplied from 200 V, 50 Hz supply. Determine the supply current and power factor. **[20 A ; 1]**

4. The magnitude of an impedance at 15 kHz is 50 Ω. When the frequency is raised to 60 kHz, the magnitude of the impedance is 20 Ω. What is the parallel combination of R and C to yield these results ?

[62.02 Ω ; 0.1255 μF]

5. A parallel circuit consists of two branches A and B. Branch A has a resistance of 10 Ω and an inductance of 0.1 H in series. Branch B has a resistance of 20 Ω and a capacitor of 100 μF in series. The circuit is connected to 250 V, 50 Hz supply. Determine (*i*) the supply current and power factor (*ii*) power consumed by the circuit. **[(*i*) 6.04 A ; 0.965 *lag* (*ii*) 1457 W]**

6. An inductive circuit, in parallel with a non-inductive circuit of 20 Ω is connected across a 50 Hz supply. The inductive current is 4.3 A and the non-inductive current is 2.7 A. The total current is 5.8 A. Find (*i*) the power absorbed by the inductive branch (*ii*) its inductance and (*iii*) power factor of the combined circuit. **[(*i*) 78.6 W (*ii*) 0.0376 H (*iii*) 0.719 *lag*]**

7. Two impedances $Z_1 = (10 + j\,5)$ Ω and $Z_2 = (8 + j\,6)$ Ω are in parallel and connected to a 200 V, 50 Hz supply. Calculate (*i*) the supply current (*ii*) circuit power factor and (*iii*) power consumed by the circuit. **[(*i*) 37.74 A (*ii*) 0·848 *lag* (*iii*) 6400 W]**

8. Two impedances $Z_1 = (10 + j\,5)$ Ω and $Z_2 = (25 - j\,10)$ Ω are connected in parallel to a 100 V, 50 Hz supply. Find (*i*) circuit admittance (*ii*) circuit current (*iii*) the phase angle between circuit current and the applied voltage. **[(*i*) 0.1174 S (*ii*) 11.74 A (*iii*) 12.9° *lagging*]**

9. Three impedances 65 ∠ 45° Ω, 42 ∠ – 20° Ω and 30 ∠ – 35° Ω are joined in parallel and connected to 240 V, 60 Hz supply. Determine (*i*) total power (*ii*) circuit power factor (*iii*) series equivalent circuit.

[(*i*) 3613.5 W (*ii*) 0.965 *lead* (*iii*) R = 15.38 Ω, X_C = 4.16 Ω]

10. Two impedances of 20 ∠ – 45° Ω and 30 ∠ 30° Ω are connected in series across a certain supply and the resulting current is found to be 10 A. If the supply voltage remains unchanged, calculate the supply current when the impedances are connected in parallel. **[26.8 A]**

11. A voltage of 120 ∠ 60° volts is supplied to two impedances in parallel. The impedances are $Z_1 = (2 + j\,3)$ Ω and $Z_2 = (4 - j\,8)$ Ω. Determine (*i*) current in each branch (*ii*) line current (*iii*) circuit power factor. **[(*i*) I_1 = 33.2 A; I_2 = 13.42 A (*ii*) 29 A (*iii*) 0.888 *lag*]**

14.11. Series-Parallel A.C. Circuits

The a.c. circuits containing combinations of series connected components and parallel connected components are called *series-parallel a.c. circuit*. The solution of problems in series-parallel a.c. circuits is obtained in the same way as for series-parallel d.c. circuits. We shall first convert all parallel a.c. circuits into equivalent series a.c. circuits. This will enable us to make the entire circuit a series a.c. circuit. We can then use phasor algebra, admittance method or other techniques to analyse the circuit. We shall solve a few problems on series-parallel a.c. circuits to illustrate the methods for their analysis.

14.12. Series-to-Parallel Conversion and Vice-Versa

In many circuit applications, it is desired to convert series-connected parameters to the equivalent parallel-connected parameters and *vice-versa*. Fig. 14.36 shows equivalent RL series and RL parallel circuits. The series circuit has resistance R_S and reactance X_S while the parallel circuit has resistance R_P and reactance X_P. *Since the two circuits are equivalent, they must have equal impedance (Z) or admittance (Y) and phase angle* ϕ. This general principle can be applied while carrying out the conversion.

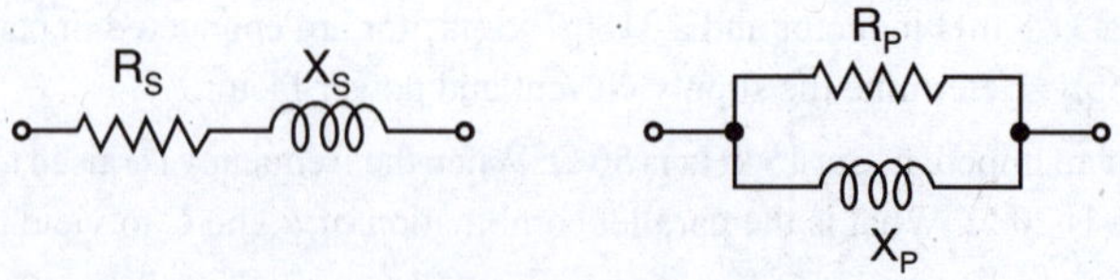

Fig. 14.36

The same principle can be applied to two *RC* circuits. But an *RL* circuit cannot be made equivalent to an *RC* circuit. It is because even if the phase angles were made numerically equal, one of them would be an angle of lag and the other an angle of lead.

Note. **We can easily develop formulas for series-to-parallel conversion and vice-versa. However, there is no need of remembering such formulas in view of the general principle explained above.** The general principle discussed above gives more insight than the derived formulas for above conversions.

Example 14.34. *Fig. 14.37 shows equivalent series and parallel circuits. Show that condition for their equivalence is*

$$R_S R_P = X_S X_P = Z^2$$

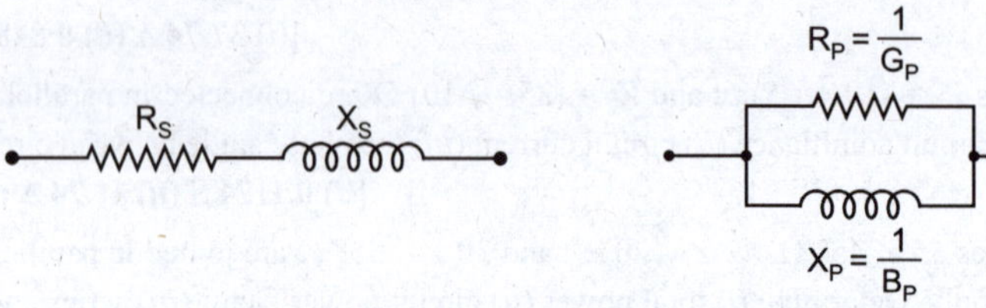

Fig. 14.37

Solution. Since the two circuits are equivalent, they have equal impedance (*Z*) or admittance (*Y*) and phase angle ϕ.

Now, $\qquad\qquad\qquad R_S = Z \cos \phi$

and $\qquad\qquad\qquad G_P = Y \cos \phi \quad$ or $\quad \dfrac{1}{R_P} = \dfrac{1}{Z} \cos \phi \qquad \left(\because G_P = \dfrac{1}{R_P} \text{ and } Y = \dfrac{1}{Z} \right)$

From above eqs. : $\qquad R_S = \dfrac{Z^2}{R_P} = \dfrac{G_P}{Y^2}$ $\qquad\qquad\qquad\qquad\qquad\qquad$ ...(*i*)

Also, $\qquad\qquad\qquad X_S = Z \sin \phi$

and $\qquad\qquad\qquad B_P = Y \sin \phi \quad$ or $\quad \dfrac{1}{X_P} = \dfrac{1}{Z} \sin \phi$

From these eqs. : $\qquad X_S = \dfrac{Z^2}{X_P} = \dfrac{B_P}{Y^2}$ $\qquad\qquad\qquad\qquad\qquad\qquad$ ...(*ii*)

From eqs. (*i*) and (*ii*), $\quad R_S R_P = Z^2$ and $X_S X_P = Z^2$

$\therefore \quad$ Condition for equivalence is

$$R_S R_P = X_S X_P = Z^2$$

Example 14.35. *Calculate the values of resistance and reactance, which when connected in parallel with one another, will be equivalent to a circuit consisting of a resistance of 10 Ω in series with an inductive reactance of 5 Ω.*

Solution. Fig. 14.38 shows the series circuit.

Conductance of series circuit is $G = R/Z^2 = 10/10^2 + 5^2 = 0.08$ S

Susceptance of series circuit is $B = X/Z^2 = 5/10^2 + 5^2 = -0.04$ S

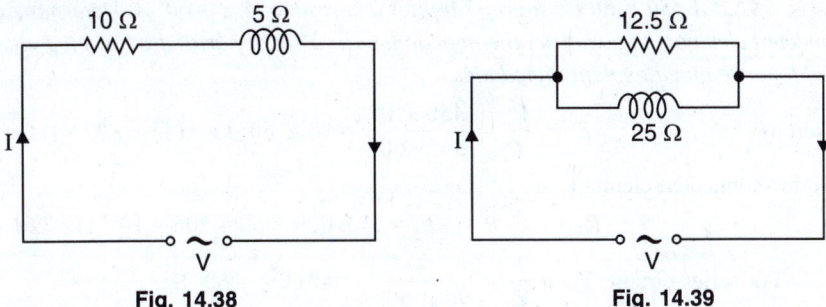

Fig. 14.38													**Fig. 14.39**

Since the parallel circuit is to have the same conductance and susceptance as the series circuit, the values of parameters of parallel circuit are :

$$R = 1/G = 1/0.08 = \textbf{12.5 } \Omega \; ; \; X_L = 1/B = 1/0.04 = \textbf{25 } \Omega$$

Fig. 14.39 shows the equivalent parallel circuit.

Alternatively. $\qquad\qquad R_S R_P = X_S X_P = Z^2$

Now, $\qquad\qquad\qquad\quad Z^2 = 10^2 + 5^2 = 100 + 25 = 125$

∴ $\qquad\qquad\qquad\quad R_P = \dfrac{Z^2}{R_S} = \dfrac{125}{10} = \textbf{12.5 } \Omega$

$$X_P = \dfrac{Z^2}{X_S} = \dfrac{125}{5} = \textbf{25 } \Omega$$

Example 14.36. *A parallel circuit consisting of a 15 Ω resistance in parallel with an inductance of 0·06 H is connected to 230 V, 50 Hz source. Find the equivalent series circuit.*

Solution. Fig. 14.40 shows the parallel circuit.

Conductance of the parallel circuit is

$$G = 1/R = 1/15 = 0.0667 \text{ S}$$

Susceptance of the parallel circuit is

$$B = 1/X_L = 1/2\pi \times 50 \times 0.06 = -0.0531 \text{ S}$$

$$Y = \sqrt{G^2 + B^2} = \sqrt{(0.0667)^2 + (-0.0531)^2} = 0.0852 \text{ S}$$

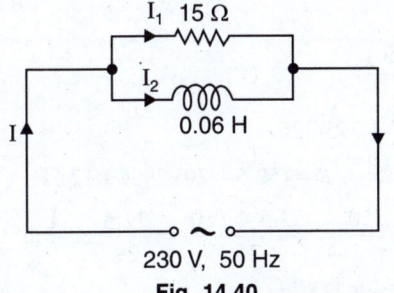

230 V, 50 Hz

Fig. 14.40

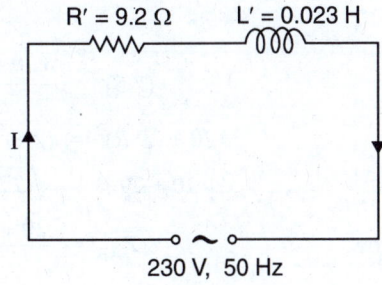

230 V, 50 Hz

Fig. 14.41

Equivalent resistance of the series circuit is

$$R' = G/Y^2 = 0.0667/(0.0852)^2 = \textbf{9.2 } \Omega$$

Equivalent inductive reactance of the series circuit is

$$X'_L = B/Y^2 = 0.053/(0.0852)^2 = 7.3 \ \Omega$$

∴ $\qquad\qquad\qquad\quad L' = X'_L/2\pi f = 7.3/ 2\pi \times 50 = \textbf{0.023 H}$

The equivalent series circuit is shown in Fig. 14.41.

Example 14.37. *A 450 V, 60 Hz source supplies a current of 5 ∠ – 60° A. Determine (i) a set of series-connected elements to construct the impedance (ii) a set of parallel-connected elements that can be used to make an equivalent impedance.*

Solution. *(i)*
$$Z = \frac{V}{I} = \frac{450 \angle 0°}{5 \angle -60°} = 90 \angle 60° \; \Omega = (45 + j\,77.94)\,\Omega$$

∴   Series-connected elements are :
$$R_S = 45\,\Omega \; ; \; L_S = 77.94/2\pi \times 60 = 206 \times 10^{-3}\,\text{H} = 206\,\text{mH}$$

(ii)   For series circuit, $Y = \dfrac{1}{Z} = \dfrac{1}{90 \angle 60°} = 0.011 \angle -60°\,\text{S}$

$$= (0.0055 - j\,0.0096)\,\text{S} \qquad\qquad\qquad ...(i)$$

For equivalent parallel circuit, $Y = \dfrac{1}{R_P} + \dfrac{1}{jX_{LP}}$ $\qquad\qquad\qquad ...(ii)$

From equations *(i)* and *(ii)*, $1/R_P = 0.0055$ ∴ $R_P = 1/0.0055 = $ **181.8 Ω**

Also, $\dfrac{1}{jX_{LP}} = -j\,0.0096$ ∴ $X_{LP} = \dfrac{1}{(-j\,0.0096)j} = 104.2\,\Omega$

or $L_P = 104.2/2\pi \times 60 = 276 \times 10^{-3}\,\text{H} = $ **276 mH**

Example 14.38. *A voltage of 100 V is applied across AB in Fig. 14.42 to produce a circuit current I = 40 A. Find the value of R when $R_P = 5\Omega$. Also find the power factor.*

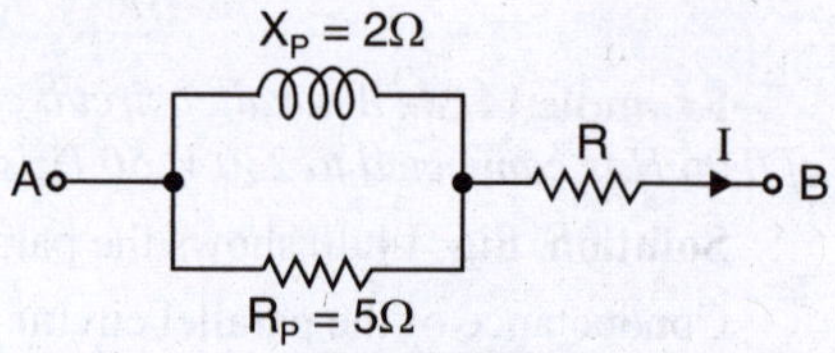

Solution. $Z_{AB} = 100/40 = 2.5\,\Omega$

Also $Z_{AB} = \dfrac{R_P(j\,2)}{R_P + j\,2} + R$

$$= \frac{j\,2\,R_P\,(R_P - j\,2)}{R_P^2 + 4} + R$$

$$= \frac{j\,2 \times (5 - j\,2)}{5^2 + 4} + R = \frac{j\,10 \times (5 - j\,2)}{29} + R = \frac{20 + 29R + j\,50}{29}$$

The magnitude of $Z_{AB} = 2.5\,\Omega$ as found above.

∴ $Z_{AB}^2 = \dfrac{(20 + 29R)^2 + (50)^2}{29^2}$

or $(2.5)^2 = \dfrac{(20 + 29R)^2 + (50)^2}{29^2}$

∴ $(20 + 29\,R)^2 = (2.5)^2 \times 29^2 - 50^2 = 2756.25$

or $20 + 29\,R = \sqrt{(2756.25)} = 52.5$ ∴ $R = (52.5 - 20)/29 = $ **1.12 Ω**

Now, $Z_{AB} = \dfrac{20 + 29 \times 1.12 + j\,50}{29} = \dfrac{52.5 + j\,50}{29} = \dfrac{72.5}{29} \angle 43.6°\,\Omega$

∴   Circuit power factor $= \cos\phi = \cos 43.6° = $ **0.725** *lag*

Example 14.39. *A capacitor of 50 μF shunted across a non-inductive resistance of 100 Ω is connected in series with a resistor of 50 Ω to a 200 V, 50 Hz supply. Find (i) the circuit current (ii) the current in the capacitor (iii) the current through the shunted resistor.*

Solution. Fig. 14.43 shows the conditions of the problem.

Fig. 14.42

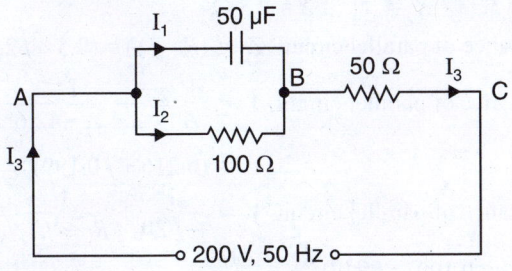

Fig. 14.43

$$Y_{AB} = \frac{1}{R} + j\omega C = \frac{1}{100} + j\,314 \times 50 \times 10^{-6}$$

$$= (0.01 + j\,0.0157)\ S = 0.0186 \angle 57.5°\ S$$

$$\therefore \qquad Z_{AB} = \frac{1}{Y_{AB}} = \frac{1}{0.0186 \angle 57.5°} = 53{\cdot}76 \angle -57.5°\ \Omega = (28.89 - j\,45.36)\ \Omega$$

$$\therefore \qquad Z_{AC} = Z_{AB} + Z_{BC} = (28.89 - j\,45.36) + (50 + j\,0)$$

$$= (78.89 - j\,45.36)\ \Omega = 91 \angle -29.9°\ \Omega$$

(*i*) Circuit current, $I_3 = 200/91 = $ **2.20 A**

(*ii*) Capacitor current $= I_3 \times \dfrac{\text{Capacitive admittance}}{Y_{AB}} = 2.20 \times \dfrac{0.0157}{0.0186} = $ **1.85 A**

(*iii*) Current in $100\ \Omega = I_3 \times \dfrac{\text{Conductance}}{Y_{AB}} = 2.20 \times \dfrac{0.01}{0.0186} = $ **1.18 A**

Example 14.40. *The circuit shown in Fig. 14.44 takes 12 A at a lagging power factor and dissipates 1.8 kW when the voltmeter reading is 200 V. Find the values of R_1, X_1 and X_2.*

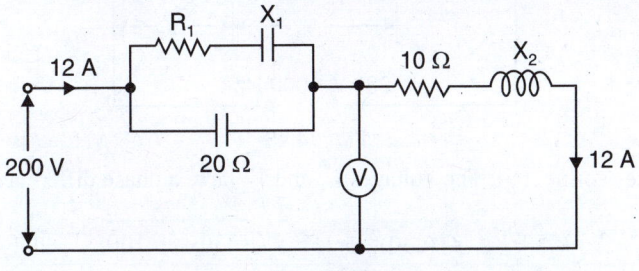

Fig. 14.44

Solution. The voltage across series circuit consisting of $10\ \Omega$ and X_2 is 200 V and current through it is 12 A. Therefore, impedance of this branch $= 200/12 = 50/3\ \Omega$.

$$\therefore \quad (50/3)^2 = (10)^2 + X_2^2 \quad \text{or} \quad X_2 = \sqrt{(50/3)^2 - (10)^2} = \textbf{13.33 }\Omega$$

Let $(R - j\,X)^*\ \Omega$ be the equivalent series impedance of the parallel circuit. Therefore, total circuit resistance $= (R + 10)\ \Omega$.

Now, $\qquad\qquad I^2(R + 10) = 1800 \quad \text{or} \quad 12^2\,(R + 10) = 1800 \qquad \therefore\ R = 2.5\ \Omega$

Total circuit impedance $= (R + 10) + j\,(X_2 - X)$

or $\qquad (R + 10)^2 + (X_2 - X)^2 = (50/3)^2$

or $\quad (2.5 + 10)^2 + (13.33 - X)^2 = 2500/9$

or $\qquad\qquad (13.33 - X)^2 = 2500/9 - 12.5^2 = 121.5$

---

* Since the parallel circuit is capacitive, the series equivalent circuit must be capacitive.

or $\qquad$ $13.33 - X = 11$ $\therefore$ $X = 2.3\,\Omega$

$\therefore$ $\quad$ Series impedance of parallel circuit, $\mathbf{Z} = (R - jX) = (2.5 - j\,2.3)\,\Omega = 3.4\,\angle - 42.6°\,\Omega$

$\quad$ Admittance of parallel circuit, $\mathbf{Y} = \dfrac{1}{\mathbf{Z}} = \dfrac{1}{3.4\,\angle - 42.6°} = 0.294\,\angle\,42.6°\,\text{S}$

$$= (0.216 + j\,0.199)\,\text{S}$$

$\quad$ Also admittance of parallel circuit, $\mathbf{Y} = \dfrac{1}{-j\,20} + \dfrac{1}{R_1 + jX_1} = j\,0.05 + \dfrac{1}{R_1 + jX_1}$

or $\qquad$ $0.216 + j\,0.199 = j\,005 + \dfrac{1}{R_1 + jX_1}$

$\therefore$ $\qquad$ $\dfrac{1}{R_1 + jX_1} = 0.216 + j\,0.199 - j\,0.05 = 0.216 + j\,0.149 = 0.262\,\angle\,34.6°$

or $\qquad$ $R_1 + jX_1 = \dfrac{1}{0.262\angle 34.6°} = (3.82\,\angle - 34.6°)\,\Omega = (3.14 - j\,2.17)\,\Omega$

$\therefore$ $\qquad$ $R_1 = \mathbf{3.14\,\Omega}$ and $X_1 = \mathbf{2.17\,\Omega}$

Example 14.41. *A 230 V, 1000 Hz voltage is applied to a resistor in series with $C = 0.05\ \mu F$. When C is shunted by a voltmeter of capacitance 0.06 μF, the reading is 100 V. Find the circuit current when the voltmeter is disconnected.*

Solution. Fig. 14.45 shows the conditions of the problem.

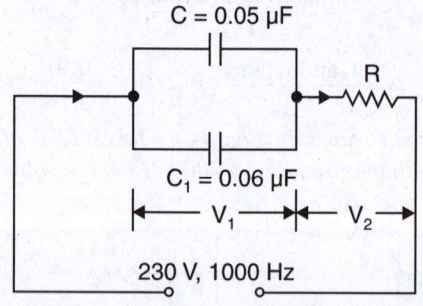

Fig. 14.45

With voltmeter connected. The voltages V_1 and V_2 have a phase difference of $90°$. Now $V_1 = 100$ volts.

$\therefore$ $\qquad$ $V_1^2 + V_2^2 = (230)^2$ or $V_2 = \sqrt{(230)^2 - (100)^2} = 207$ volts

Now, $\qquad$ $C_T = C + C_1 = 0.05 + 0.06 = 0.11\ \mu F = 0.11 \times 10^{-6}\,\text{F}$

$\quad$ Circuit current, $I = V_1 \omega C_T = 100 \times 2\pi \times 1000 \times 0.11 \times 10^{-6} = 0.069\,\text{A}$

$\therefore$ $\qquad$ Value of $R = V_2/I = 207/0.069 = 2996\,\Omega$

When voltmeter is disconnected. When the voltmeter is disconnected, the circuit becomes C-R series circuit.

$\therefore$ $\quad$ Circuit impedance, $\mathbf{Z} = R - jX_C = 2996 - j\left(\dfrac{1}{2\pi \times 1000 \times 0.05 \times 10^{-6}}\right)$

$$= (2996 - j\,3180)\,\Omega = 4369\,\angle - 46.7°\,\Omega$$

$\therefore$ $\qquad$ Circuit current $= 230/4369 = \mathbf{0.0527\,A}$

Example 14.42. *In the circuit shown in Fig. 14.46, determine what 50 Hz voltage applied across AB will cause a current of 10 A to flow in the capacitor.*

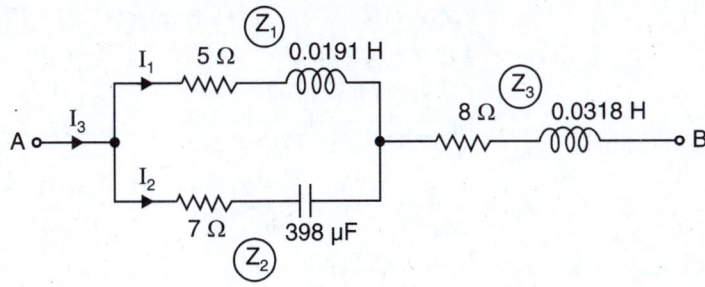

Fig. 14.46

Solution. $X_{L1} = 2\pi f L_1 = 2\pi \times 50 \times 0.0191 = 6\,\Omega$; $X_{C2} = \dfrac{1}{\omega C_2} = \dfrac{1}{2\pi \times 50 \times 398 \times 10^{-6}} = 8\,\Omega$

$$X_{L3} = 2\pi f L_3 = 2\pi \times 50 \times 0.0318 = 10\,\Omega$$

$\therefore$
$$\mathbf{Z_1} = (5 + j\,6)\,\Omega = 7.81 \angle 50.2°\,\Omega$$
$$\mathbf{Z_2} = (7 - j\,8)\,\Omega = 10.63 \angle -48.3°\,\Omega$$
$$\mathbf{Z_3} = (8 + j\,10)\,\Omega = 12.8 \angle 51.3°\,\Omega$$

Taking capacitor current I_2 as reference phasor, $\mathbf{I_2} = 10 \angle 0°$ A.

Voltage across parallel branch $= \mathbf{I_2\,Z_2} = 10 \angle 0° \times 10.63 \angle -48.3° = 106.3 \angle -48.3°$ volts
$$= (70 - j\,80)\text{ volts}$$

$\therefore$ Current $\mathbf{I_1} = \dfrac{106.3 \angle -48.3°}{7.81 \angle 50.2°} = 13.6 \angle -98 \cdot 5°$ A $= (-2.01 - j\,13.45)$A

$\therefore$
$$\mathbf{I_3} = \mathbf{I_1} + \mathbf{I_2} = (-2.01 - j13.45) + (10 + j\,0)$$
$$= (7.99 - j\,13.45)\text{ A} = 15.65 \angle -59.3°\text{ A}$$

Voltage drop across $\mathbf{Z_3} = \mathbf{I_3\,Z_3} = 15.65 \angle -59.3° \times 12.8 \angle 51.3°$
$$= 200.3 \angle -8°\text{ volts} = (198.4 - j\,27)\text{ volts}$$

Voltage across $AB = $ Drop across $\mathbf{Z_2}$ + Drop across $\mathbf{Z_3}$
$$= (70 - j\,80) + (198.4 - j\,27) = (268.4 - j\,107)\text{ volts}$$
$$= 288 \angle -21.7°\text{ volts}$$

$\therefore$ Magnitude of supply voltage $= \mathbf{288\ V}$

Example 14.43. *In the circuit shown in Fig. 14.47, find (i) supply voltage and (ii) supply current.*

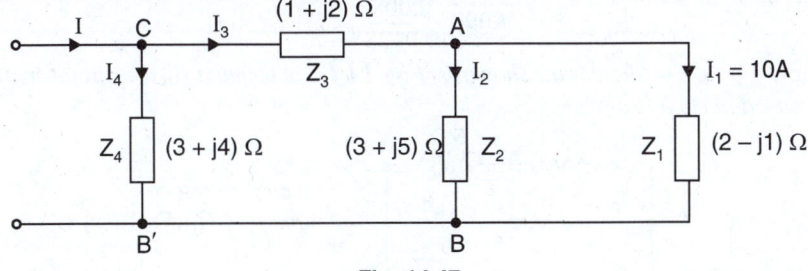

Fig. 14.47

Solution. (*i*) Taking I_1 as reference phasor, $\mathbf{I_1} = 10 \angle 0°$ A $= (10 + j\,0)$ A.

$$\mathbf{V_{AB}} = \mathbf{I_1 Z_1} = 10(2 - j\,1) = (20 - j\,10)\text{ volts}$$

$$\mathbf{I_2} = \frac{\mathbf{V_{AB}}}{\mathbf{Z_2}} = \frac{(20 - j\,10)}{3 + j5} = \frac{(20 - j\,10)\,(3 - j5)}{3^2 + 5^2} = (0.29 - j\,3.82)\text{ A}$$

$$\mathbf{I_3} = \mathbf{I_1} + \mathbf{I_2} = (10 + j\,0) + (0.29 - j\,3.82) = (10.29 - j\,3.82)\text{ A}$$

$$V_{CA} = I_3 Z_3 = (10.29 - j\,3.82)\,(1 + j\,2) = (17.93 + j\,16.77) \text{ volts}$$
$$V_{CB'} = V_{CA} + V_{AB} = (17.93 + j\,16.77) + (20 - j\,10)$$
$$= (37.93 + j\,6.77) \text{ volts}$$

Magnitude, $V_{CB'} = \sqrt{(37.93)^2 + (6.77)^2} = \mathbf{38.5\ V}$

(*ii*)
$$I_4 = \frac{V_{CB'}}{Z_4} = \frac{37.93 + j\,6.77}{3 + j\,4} = \frac{(37.93 + j\,6.77)\,(3 - j\,4)}{3^2 + 4^2}$$
$$= (5.63 - j\,5.28)\ \text{A}$$

$\therefore$
$$I = I_3 + I_4 = (10.29 - j\,3.82) + (5.63 - j\,5.28) = (15.92 - j\,9.1)\ \text{A}$$

$\therefore$ Magnitude, $I = \sqrt{(15.92)^2 + (9.1)^2} = \mathbf{18.3\ A}$

Example 14.44. *A 100 Ω resistor shunted by a 0.4H inductor is in series with a capacitor C. A voltage of 250 V at 50 Hz is applied to the circuit. Find (i) the value of C to give unity power factor (ii) the total current and (iii) current in the inductor.*

Solution. Fig. 14.48 shows the conditions of the problem.

(*i*) $R = 100\ \Omega$; $X_L = 2\pi fL = 2\pi \times 50 \times 0.4 = 125.6\ \Omega$

$$Y_{AB} = \frac{1}{R} + \frac{1}{jX_L} = \frac{1}{100} + \frac{1}{j\,125.6}$$
$$= (0.01 - j\,0.008)\ \text{S} = 0.0128\ \angle -38.6°\ \text{S}$$

$\therefore$
$$Z_{AB} = \frac{1}{Y_{AB}} = \frac{1}{0.0128\ \angle -38.6°}$$
$$= 78.1\ \angle\ 38.6°\ \Omega = (61.06 + j\,48.7)\ \Omega$$

In order that circuit power factor is unity, X_C should be 48.7 Ω.

Now, $X_C = \dfrac{1}{2\pi fC}$ $\therefore$ $C = \dfrac{1}{2\pi f\,X_C} = \dfrac{1}{2\pi \times 50 \times 48.7} = 64.9 \times 10^{-6}\ \text{F} = \mathbf{64.9\ \mu F}$

(*ii*)   Circuit impedance, $Z_{AO} = 61.06 + j\,48.7 - j\,48.7 = 61.06\ \Omega$

$\therefore$ Circuit current $= 250/61.06 = \mathbf{4.09\ A}$

(*iii*)   Current in inductor $= $ Total current $\times \dfrac{\text{Admittance of inductor}}{Y_{AB}}$

$$= 4.09 \times \frac{0.008}{0.0128} = \mathbf{2.56\ A}$$

Example 14.45. *For the circuit shown in Fig. 14.49, determine (i) the circuit impedance (ii) power consumed in each branch.*

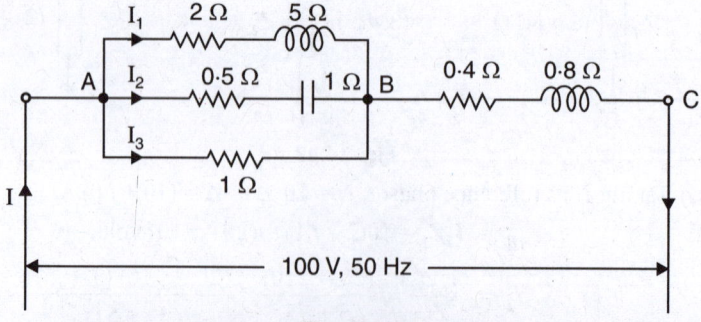

Fig. 14.49

Solution. *(i)*

$$Y_1 = \frac{1}{2+j5} = \frac{2-j5}{29} = (0.069 - j\,0.172)\text{ S}$$

$$Y_2 = \frac{1}{0.5-j1} = \frac{0.5+j1}{1.25} = (0.4 + j\,0.8)\text{ S}$$

$$Y_3 = (1+j\,0)\text{ S}$$

$$Y_{AB} = Y_1 + Y_2 + Y_3 = (0.069 - j\,0.172) + (0.4 + j0.8) + (1 + j\,0)$$

$$= (1.469 + j\,0.628)\text{ S} = 1.598 \angle 23.1°\text{ S}$$

$$\therefore \quad Z_{AB} = \frac{1}{Y_{AB}} = \frac{1}{1.598 \angle 23.1°} = \frac{1}{1.598}\angle -23.1° = (0.575 - j\,0.246)\Omega$$

$$Z_{BC} = (0.4 + j\,0.8)\ \Omega$$

$\therefore$ Total circuit impedance, $Z_{AC} = Z_{AB} + Z_{BC} = (0.575 - j\,0.246) + (0.4 + j\,0.8)$

$$= (0.975 + j\,0.554)\ \Omega = \textbf{1.12} \angle \textbf{29.5°}\ \Omega$$

(ii) Total circuit current, $I = \dfrac{100}{Z_{AC}} = \dfrac{100}{1.12} = 89.3\text{ A}$

Voltage across AB, $V_{AB} = \dfrac{I}{Y_{AB}} = \dfrac{89.3}{1.598} = 55.9\text{ volts}$

$$Z_1 = \sqrt{2^2 + 5^2} = \sqrt{29}\ \Omega\ ;\ Z_2 = \sqrt{(0.5)^2 + (1)^2} = \sqrt{1.25}\ \Omega\ ;\ Z_3 = 1\ \Omega$$

$$I_1 = 55.9/\sqrt{29} = 10.4\text{ A}\ ;\ I_2 = 55.9/\sqrt{1.25} = 50\text{ A}\ ;\ I_3 = 55.9/1 = 55.9\text{ A}$$

Note that power is consumed in resistors only. Therefore, powers consumed in the various branches of the parallel circuit between A and B are :

$$P_1 = I_1^2\,R = (10.4)^2 \times 2 = \textbf{216 W}\ ;\ P_2 = I_2^2\,R = (50)^2 \times 0.5 = \textbf{1250 W}$$

$$P_3 = I_3^2\,R = (55.9)^2 \times 1 = \textbf{3120 W}\ ;\ \text{Power consumed in branch } BC = I^2 R = (89.3)^2 \times 0.4 = \textbf{3190 W}$$

Example 14.46. *Fig. 14.50 shows an a.c. circuit which draws an r.m.s. current of 1 A from the source. The total power dissipated in the circuit is 136 W. Find current through C_2 if $R_1 = R_2 = 100\ \Omega$, $L_2 = 477\ mH$ and $C_2 = 15.9\ \mu F$.*

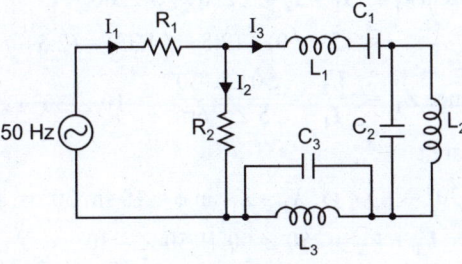

Fig. 14.50

Solution. Power dissipated in the circuit is 136 W. It is given that $I_1 = 1$A.

$$\therefore \quad I_1^2 R_1 + I_2^2 R_2 = 136 \quad \text{or} \quad (1)^2 \times 100 + I_2^2 \times 100 = 136$$

$$\therefore \quad I_2^2 = \frac{136-100}{100} = 0.36 \quad \text{or} \quad I_2 = 0.6\text{ A}$$

It is clear from the circuit that I_3 is a reactive current so that I_2 and I_3 are 90° out of phase.

$$\therefore \quad I_2^2 + I_3^2 = I_1^2 \quad \text{or} \quad I_3^2 = 1 - (0.6)^2 \therefore I_3 = 0.8\text{ A}$$

Impedance of $L_2 = j\,2\pi f L_2 = j \times 2\pi \times 50 \times 477 \times 10^{-3} = j\,149.85\ \Omega$

Impedance of $C_2 = -j\dfrac{1}{2\pi fC} = -j \times \dfrac{1}{2\pi \times 50 \times 15.9 \times 10^{-6}} = -j\,200\;\Omega$

Current in $C_2 = I_3 \times \dfrac{j149.85}{j149.85 - j\,200} = 0.8 \times \dfrac{j149.85}{-j\,50.15} = \mathbf{2.39\ A}$ *(magnitude)*

Example 14.47. *For the circuit shown in Fig. 14.51, given that L = 0.159 H, C = 0.3183 mF, I_2 = 5 ∠ 60° A, V_1 = 250 ∠ 90° V. Find (i) impedance Z_1 with its components (ii) source voltage in the form of V_{max} cos (ωt + φ) (iii) impedance Z_2 with its components so that source p.f. is unity without adding the circuit power loss (iv) power loss in the circuit. Draw the phasor diagram.*

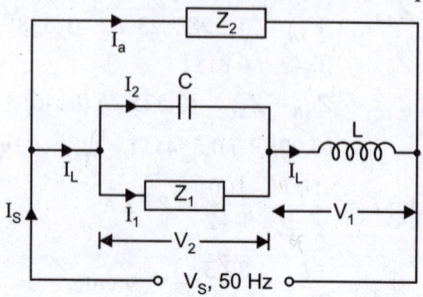

Fig. 14.51

Solution. $X_L = 2\pi fL = 2\pi \times 50 \times 0.159 = 50\ \Omega$

$$X_C = \dfrac{1}{2\pi fC} = \dfrac{1}{2\pi \times 50 \times 0.3183 \times 10^{-3}} = 10\ \Omega$$

Current in L, $\mathbf{I_L} = \dfrac{V_1}{jX_L} = \dfrac{250 \angle 90°}{50 \angle 90°} = 5 \angle 0°\ A$

Voltage V_2 = Voltage drop across capacitor C

$= -j\,\mathbf{I_2}\,X_C = 1 \angle -90° \times 5 \angle 60° \times 10$

$= 50 \angle -30°\ V = (43.3 - j\,25)\ V$

Current $I_1 = \mathbf{I_L} - \mathbf{I_2} = 5 \angle 0° - 5 \angle 60°$

$= (5 + j\,0) - (2.5 + j\,4.33) = (2.5 - j\,4.33)\ A = 5 \angle -60°\ A$

(i)   Impedance $Z_1 = \dfrac{V_2}{I_1} = \dfrac{50 \angle -30°}{5 \angle -60°} = \mathbf{10 \angle 30°}\ \Omega$

The components of Z_1 are :

$R_1 = Z_1 \cos\phi = 10 \cos 30° = \mathbf{8.66\ \Omega}$; $X_1 = Z_1 \sin\phi = 10 \sin 30° = \mathbf{5\Omega}$

(ii) Source voltage, $V_S = V_1 + V_2 = 250 \angle 90° + 50 \angle -30°$

$= (0 + j\,250) + (43.3 - j\,25) = (43.3 + j\,225)\ V = 229.13 \angle 79.1°$ volts

Peak value of supply voltage $= \sqrt{2} \times 229.13 = 324$ volts

Taking I_L as reference phasor, the supply voltage is given by ;

$v_S = 324 \sin(\omega t + 79.1°) = 324 \cos(\omega t + 79.1° - 90°)$

∴ $v_S = \mathbf{324 \cos(\omega t - 10.9°)}$ **volts**

(iii) We want the source p.f. to be unity (*i.e.* source current I_S is in phase with source voltage V_S) and there should be no power loss in Z_2. This is possible only if Z_2 is purely capacitive and the current I_a (capacitive) drawn by it is of such magnitude to just neutralise the lagging reactive component of I_L *i.e.*

$$I_a = \text{Lagging reactive component of } I_L$$

or $\qquad I_a = I_L \sin 79.1° = 5 \sin 79.1° = 4.91 \text{ A}$

$\therefore$ Magnitude of $\mathbf{Z_2}$ is given by ;

$$Z_2 = \frac{V_S}{I_a} = \frac{229.1}{4.91} = 46.66 \ \Omega \ (\text{pure capacitance})$$

$\therefore \qquad C_2 = \dfrac{1}{2\pi f \, X_{C2}} = \dfrac{1}{2\pi \times 50 \times 46.66} = 68.34 \times 10^{-6} \text{ F} = \mathbf{68.34 \ \mu F}$

(iv) $\qquad$ Circuit power loss $= I_1^2 R_1 = 5^2 \times 8.66 = \mathbf{216.5 \ W}$

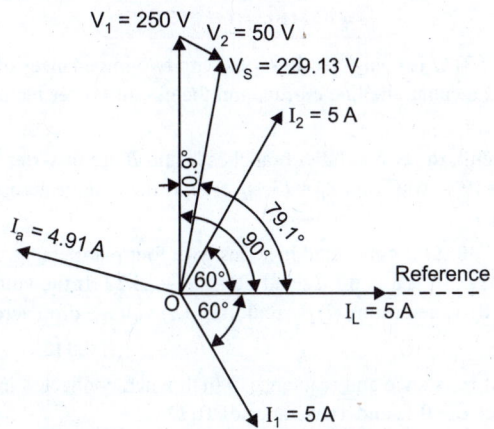

Fig. 14.52

Fig. 14.52 shows the phasor diagram of the circuit.

Example 14.48. *Draw admittance triangles between terminals AB of Fig. 14.53 labelling its sides with appropriate values and units in case of (i) $X_L = 4 \ \Omega$ and $X_C = 8 \ \Omega$ (ii) $X_L = 10 \ \Omega$ and $X_C = 5\Omega$.*

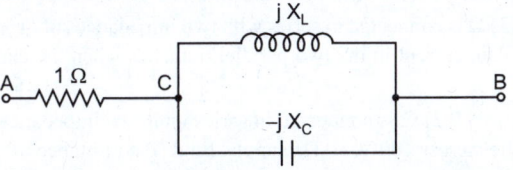

Fig. 14.53

Solution. (i) When $X_L = 4 \ \Omega$ and $X_C = 8 \ \Omega$.

$$\mathbf{Z_{CB}} = j X_L \| (-j X_C) = \frac{j X_L (-j X_C)}{j(X_L - X_C)} = \frac{-j^2 (4 \times 8)}{j(4-8)} = j \, 8 \ \Omega$$

$\therefore \quad$ Circuit impedance, $\mathbf{Z_{AB}} = \mathbf{Z_{AC}} + \mathbf{Z_{CB}} = (1 + j \, 0) + (0 + j \, 8) = (1 + j \, 8) \ \Omega$

$\therefore \quad$ Circuit admittance, $\mathbf{Y_{AB}} = \dfrac{1}{\mathbf{Z_{AB}}} = \dfrac{1}{1 + j8} = \left(\dfrac{1}{65} - j \dfrac{8}{65} \right) \text{S}$

Therefore, the total circuit conductance, $G = 1/65$ S and total circuit susceptance (inductive) is $B = -8/65$ S. Fig. 14.54 (i) shows the admittance triangle for the circuit.

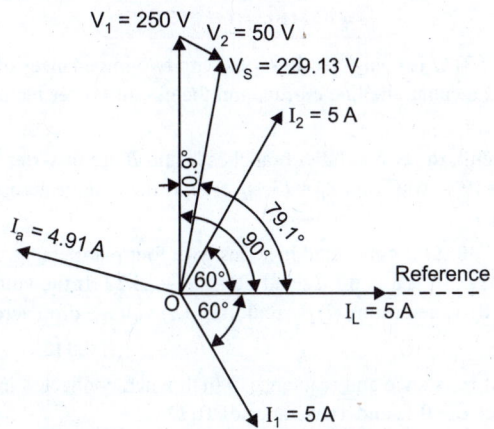

Fig. 14.54

(ii) **When $X_L = 10\ \Omega$ and $X_C = 5\ \Omega$.**

$$Z_{CB} = j X_L \| (-j X_C) = \frac{j X_L (-j X_C)}{j (X_L - X_C)} = \frac{-j^2 10 \times 5}{j (10 - 5)} = -j\,10\ \Omega$$

∴   Circuit impedance, $Z_{AB} = Z_{AC} + Z_{CB} = (1 + j\,0) + (0 - j\,10) = (1 - j\,10)\ \Omega$

∴   Circuit admittance, $Y_{AB} = \dfrac{1}{Z_{AB}} = \dfrac{1}{(1 - j10)} = \left(\dfrac{1}{101} + j\dfrac{10}{101} \right)$ S

Therefore, the total circuit conductance is $G = 1/101$ S and total circuit susceptance (capacitive) is $B = +\,10/101$ S. Fig. 14.54 *(ii)* shows the admittance triangle for the circuit.

Tutorial Problems

1. An impedance of $(2 + j\,3)\ \Omega$ is connected in series with two impedances of $(4 + j\,2)\ \Omega$ and $(1 - j\,5)\ \Omega$, which are in parallel. Calculate the line current and the circuit power factor when the combined circuit is supplied at 10 V. **[1.71 A ; 0.963 *lag*]**

2. In a series-parallel circuit, the two parallel branches A and B are in series with C. The impedances are $Z_A = (10 + j\,8)\ \Omega$; $Z_B = (9 - j\,6)\ \Omega$ and $Z_C = (3 + j\,2)\ \Omega$. If the voltage across C is $100\ \angle\ 0°$ V, determine the values of I_A and I_B. **[I_A = 15.7 A ; I_B = 18.6 A]**

3. An impedance of $40 \angle 30°\ \Omega$ is connected in series with four paralleled group of impedance : $Z_1 = 3 \angle 60°\ \Omega$; $Z_2 = 5 \angle 40°\ \Omega$; $Z_3 = 8 \angle 60°\ \Omega$ and $Z_4 = 4 \angle - 25°\Omega$. If the voltage drop across $40 \angle 30°\ \Omega$ impedance is $(200 + j\,0)$ V, determine *(i)* current in Z_3 *(ii)* voltage drop across the paralleled section. **[*(i)* 0.845 $\angle - 55.4°$ A *(ii)* 6.76 $\angle 4.58°$ V]**

4. Calculate the values of resistance and reactance, which when connected in parallel, are equivalent to a coil having a resistance of $20\ \Omega$ and a reactance of $10\ \Omega$. **[R_P = 25 Ω ; X_P = 50 Ω]**

5. The admittance of a parallel circuit of R and L is $(0.0001 - j\,0.01)$ S at 2 MHz. Determine the values of R and L. **[10,000 Ω; 7.96 μH]**

6. An impedance of $(3 + j\,6)\ \Omega$ is connected in series with two impedances of $(4 + j\,8)\ \Omega$ and $(5 - j\,8)\ \Omega$, which are in parallel. Calculate *(i)* total impedance *(ii)* total current *(iii)* total power absorbed and p.f. Draw the vector diagram. **[*(i)* 14.13 $\angle$ 29.2° Ω *(ii)* 7.08 $\angle - 29.2°$ A *(iii)* 618 W ; 0.873 (*lag*)]**

7. An impedance $(2 + j\,3)\ \Omega$ is connected in series with two impedances of $(4 + j\,2)\ \Omega$ and $(1 - j\,5)\ \Omega$, which are in parallel. Calculate current in the two parallel branches when the combined circuit is supplied at 10 V. **[1.5 $\angle$ 10.3° A ; 1.3 $\angle - 41.8°$ A]**

8. An impedance of $(1.6 + j\,7.2)\ \Omega$ is connected in series with two impedances of $(4 + j\,3)\ \Omega$ and $(6 - j\,8)$ Ω, which are in parallel across 100 V, 50 Hz supply. Find *(i)* admittance of each parallel branch *(ii)* total circuit impedance *(iii)* supply current and p.f. *(iv)* total power supplied by source. **[*(i)* 0.2 $\angle - 36.87°$S ; 0.1 $\angle$ 53.13° S *(ii)*10 $\angle$ 53.13° Ω *(iii)* 10A ; 0.6 (lag)*(iv)* 600 W]**

9. In a series-parallel circuit, the parallel branches A and B are in series with C. The impedances are $Z_A = (4 + j\,3)\ \Omega$, $Z_B = (4 - j\,16/3)\ \Omega$ and $Z_C = (2 + j\,8)\ \Omega$. If the current $I_C = (25 + j\,0)$ A, draw the complete phasor diagram determining the power (active and reactive power) for each branch and the whole circuit. **[S_A = (1600 $- j$ 1200) VA; S_B = (900 $+ j$ 1200) VA ; S_C = (1250 $- j$ 5000) VA; S = (3750 $- j$ 5000) VA]**

10. Impedances Z_2 and Z_3 in parallel are in series with an impedance Z_1 across a 100 V, 50 Hz a.c. supply. $Z_1 = (6.26 + j\,1.25)\ \Omega$; $Z_2 = (5 + j\,0)\ \Omega$ and $Z_3 = (5 - j\,X_C)\ \Omega$. Determine the value of capacitance of X_C such that the total current of the circuit will be in phase with the total voltage. What is then the circuit current and power ? **[318 μF ; 10 A ; 1000 W]**

14.13. Resonance in Parallel A.C. Circuits (Parallel Resonance)

A parallel a.c. circuit containing reactive elements (L and C) is said to be in resonance when the circuit p.f. is unity i.e. reactive component of line current is zero. The frequency at which it occurs is called the resonant frequency f_r. It is called parallel resonance because it concerns a parallel circuit.

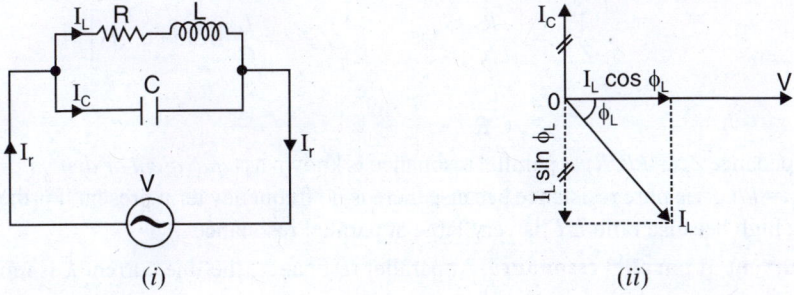

Fig. 14.55

Expression for f_r. Consider the most practical parallel circuit consisting of a coil shunted by a capacitor as shown in Fig. 14.55 (*i*). The phasor diagram of this circuit is shown in Fig. 14.55 (*ii*). The circuit will be in resonance when the reactive component of line current is zero *i.e.* $I_C - I_L \sin \phi_L = 0$. This can be achieved by changing the supply frequency because both I_C and $I_L \sin \phi_L$ are frequency* dependent. At some frequency f_r, called resonant frequency, the reactive component of line current will be zero and resonance takes place.

At parallel resonance, the circuit condition is :

$$I_C - I_L \sin \phi_L = 0 \quad \text{or} \quad I_C = I_L \sin \phi_L$$

$$\therefore \qquad \frac{V}{X_C} = \frac{V}{Z_L} \times \frac{X_L}{Z_L} \quad \text{or} \quad X_L X_C = Z_L^2$$

$$\therefore \qquad \omega L/\omega C = Z_L^2 \qquad \qquad ...(i)$$

$$\text{or} \qquad L/C = R^2 + (2\pi f_r L)^2$$

$$\text{or} \qquad (2\pi f_r L)^2 = \frac{L}{C} - R^2$$

$$\text{or} \qquad 2\pi f_r L = \sqrt{\frac{L}{C} - R^2}$$

$$\text{or} \qquad f_r = \frac{1}{2\pi L}\sqrt{\frac{L}{C} - R^2}$$

$$\therefore \qquad f_r = \frac{1}{2\pi}\sqrt{\frac{1}{LC} - \frac{R^2}{L^2}}$$

If the coil resistance is small (as generally is the case), then,

$$f_r = \frac{1}{2\pi\sqrt{LC}} \qquad .....same \ as \ for \ series \ resonance$$

The resonant frequency will be in Hz if R, L and C are measured in ohms, henry and farad respectively.

Impedance at resonance. Line current, $I_r = I_L \cos \phi_L \quad$ or $\quad \dfrac{V}{Z_r} = \dfrac{V}{Z_L} \times \dfrac{R}{Z_L}$

$$\therefore \qquad \frac{1}{Z_r} = \frac{R}{Z_L^2}$$

---

* $\quad I_C = \dfrac{V}{X_C} = 2\pi f \, CV$

$I_L \cos \phi_L = \dfrac{V}{Z_L}\cos \phi_L$ where $Z_L = \sqrt{R^2 + (2\pi f L)^2}$ and $\phi_L = \tan^{-1}\dfrac{X_L}{R}$

or
$$\frac{1}{Z_r} = \frac{R}{L/C} \qquad \left[\because Z_L^2 = \frac{L}{C} \text{ from eq. } (i)\right]$$

$$\therefore \qquad Z_r = \frac{L}{CR}$$

The impedance $Z_r (= L/CR)$ at parallel resonance is known as *equivalent or dynamic impedance*. Note that $Z_r (= L/CR)$ is pure resistance because there is no frequency term present. Further, the value of Z_r is very high because ratio L/C is very large at parallel resonance.

Line current at parallel resonance. At parallel resonance, the line current I_r is minimum and is given by ;

$$I_r = V/Z_r \quad \text{where } Z_r = L/CR$$

Because Z_r is very high, I_r will be very small. The small current I_r is only the amount needed to meet the resistance losses in the circuit. The parallel resonance is also **current resonance** because the current circulating between the two branches of the circuit is many times greater than the line current taken from the supply. Parallel resonant circuit is also called **rejector circuit** because it rejects (or takes minimum current) that frequency to which it resonates.

14.14.   Graphical Representation of Parallel Resonance

The action of a resonant circuit is best explained by referring to the curves illustrating the variation in the circuit conditions at or near resonance. We shall discuss three such curves *viz* impedance-frequency curve, current-frequency curve and susceptance-frequency curve.

(i) Impedance-frequency curve. If we plot impedance-frequency graph for a parallel circuit shown in Fig. 14.55 (*i*), the shape of the curve will be as shown in Fig. 14.56. Note that the impedance of the circuit is maximum at resonance. As the frequency changes from resonance, the circuit impedance decreases very rapidly. This behaviour can be explained as follows. For frequencies below resonance, the capacitive reactance will be higher and thus *more current will flow through the coil.

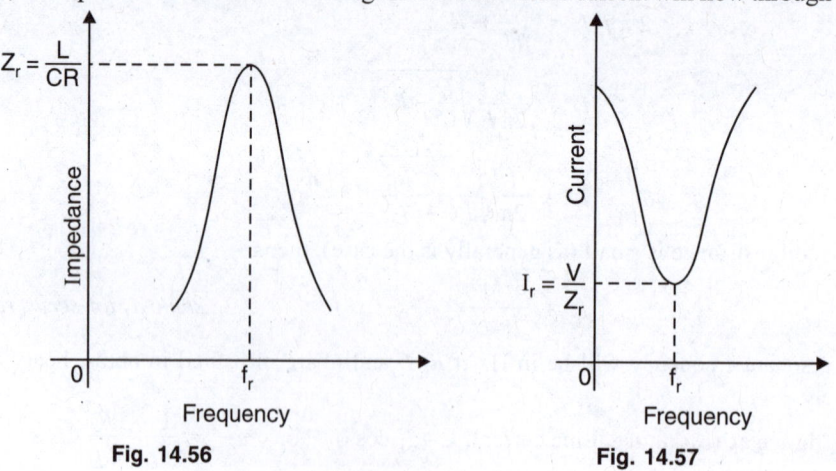

Fig. 14.56   Fig. 14.57

This causes the line current to lag behind the applied voltage and the circuit appears inductive. For frequencies above resonance, inductive reactance is higher and more current will flow through the capacitor. Consequently, line current leads the applied voltage and circuit appears capacitive. In either case, the circuit impedance is far less than its value at resonance.

---

*   In the two parallel branches, more current will flow through that branch which has less impedance.

(ii) Current-frequency curve. Fig. 14.57 shows the current-frequency curve of the parallel circuit shown in Fig. 14.55 (i). Note that value of line current is minimum at resonance $(I_r = V/Z_r)$. As the frequency changes from resonance, the line current increases rapidly. This action can be explained as follows. For frequencies other than the resonance, the reactive currents in the two branches of the circuit are not equal. The resultant reactive current must be supplied by the line. As the difference of the reactive currents in the two branches increases with the amount of deviation from the resonant frequency, the line current will increase in the same manner.

(iii) Susceptance-frequency curve. Fig. 14.58 shows the graph between susceptance (inductive and capacitive) of the two parallel branches and the supply frequency.

$$\text{Inductive susceptance, } B_L = -\frac{1}{X_L} = -\frac{1}{2\pi fL} \text{ or } B_L \propto \frac{1}{f}$$

It is clear that inductive susceptance B_L is inversely proportional to the frequency f. Hence $B_L - f$ graph is a rectangular hyperbola as shown in Fig. 14.58. Note that graph lies in the fourth quadrant because B_L is considered negative.

$$\text{Capacitive susceptance, } B_C = \frac{1}{X_C} = 2\pi fC \text{ or } B_C \propto f$$

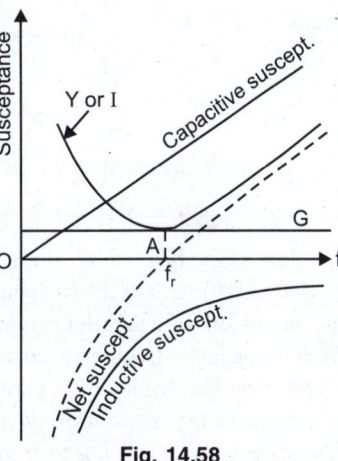

Fig. 14.58

It is clear that capacitive susceptance B_C is directly proportional to the frequency f. Hence $B_C - f$ graph is a straight line passing through the origin as shown in Fig. 14.58. Note that graph lies in the first quadrant because B_C is considered positive. The net circuit susceptance is the difference of the two susceptances and is represented by the dotted hyperbola (not rectangular). At point A in Fig. 14.58, the supply frequency $f = f_r$ and the circuit susceptance is zero. Therefore, parallel resonance occurs. Under this condition :

(a) the circuit admittance is minimum (*i.e.* circuit impedance is maximum) and is equal to the conductance G of the circuit.

(b) the circuit current is minimum.

(c) the circuit current is in phase with the supply voltage *i.e.* circuit p.f. is unity.

At supply frequency $f > f_r$, the capacitive susceptance of the circuit becomes greater than the inductive susceptance of the circuit. Consequently, the circuit is effectively capacitive and the circuit current leads the supply voltage. On the other hand, at $f < f_r$, inductive susceptance predominates and the circuit becomes effectively inductive. Therefore, the circuit current lags behind the supply voltage. Fig. 14.58 also shows the $Y - f$ or $I - f$ graph. ($\because I = VY$ so that $I \propto Y$) as well as $G - f$ graph. Note that conductance G of the circuit is independent of the supply frequency f.

14.15. Q-factor of a Parallel Resonant Circuit

At parallel resonance, the line current drawn from the supply is $I_r (= V/Z_r)$ which is in phase with supply voltage V. Also, I_r is very much less than I_L and I_C. Thus in a parallel resonant circuit, there is a **current amplification** which is analogous to the voltage amplification that occurs in a series resonant circuit.

*The **Q-factor** or **current magnification factor** of a parallel resonant circuit is the ratio of I_C *(not I_L) to the line current I_r, i.e.*

---

* In a practical resonant circuit, $I_L > I_C$ and that part of I_L is resistive current (*i.e.* I_r). Therefore, use $Q = I_C/I_r$ and not I_L/I_r.

$$Q\text{-factor} = \frac{I_C}{I_r}$$

Now,

$$I_C = V/X_C = \omega_r\, CV \text{ and } I_r = V/(L/CR)$$

$$\therefore \qquad Q\text{-factor} = \omega_r\, CV \div V/(L/CR) = \frac{2\pi f_r\, L}{R} \qquad \textit{... same as for series circuit}$$

Note that expression for Q-factor is the same as the Q factor for a series resonant circuit $i.e.$ Q is once again the Q-factor of the coil.

Neglecting R, the value of resonant frequency f_r is

$$f_r = \frac{1}{2\pi\sqrt{LC}}$$

$$\therefore \qquad Q\text{-factor} = \frac{2\pi f_r\, L}{R} = 2\pi \times \frac{1}{2\pi\sqrt{LC}} \times \frac{L}{R}$$

or $\qquad Q\text{-factor} = \dfrac{1}{R}\sqrt{\dfrac{L}{C}} \qquad\qquad\qquad \textit{... same as for series circuit}$

This is the expression for the Q-factor in terms of R, L and C.

14.16. Bandwidth of Parallel Resonant Circuit

Just as the bandwidth of a series resonant circuit is defined from current-frequency curve, similarly the bandwidth of a parallel resonant circuit is defined from impedance-frequency curve as shown in Fig. 14.59. The bandwidth of a parallel resonant circuit is defined *as the range of frequencies over which the circuit impedance is equal to or greater than 70.7% of maximum circuit impedance (i.e. Z_r, the impedance at resonance)*. The circuit impedance is maximum (*i.e.* Z_r) at f_r while half-power points (*i.e.* f_1 and f_2) occur at 0.707 Z_r. Using the same procedure as was used for series resonant circuit, it can be shown that :

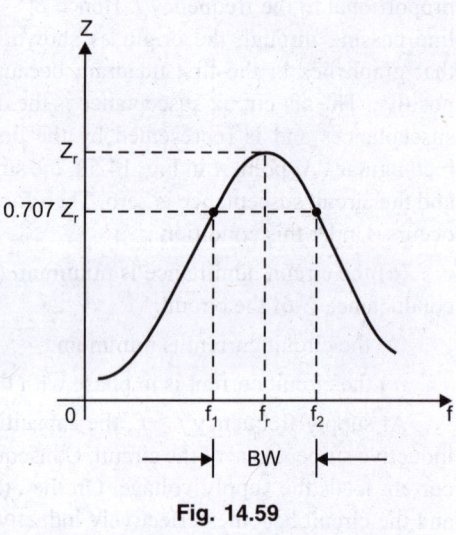

Fig. 14.59

$$\text{Bandwidth, } BW = \frac{f_r}{Q} = f_2 - f_1$$

$$\text{Lower cut-off frequency, } f_1 = f_r - \frac{BW}{2}$$

$$\text{Upper cut-off frequency, } f_2 = f_r + \frac{BW}{2}$$

The points on the curve in Fig. 14.59 where the upper and lower cut-off frequencies intersect the curve (at 70.7% of Z_r) are called the *half-power points*. Thus the bandwidth (BW) of a parallel resonant circuit is often referred to as the band of frequencies between the half-points on the impedance-frequency curve.

At half-power frequencies (*i.e.* f_1 and f_2), the power dissipated in the circuit is one-half of that dissipated at resonant frequency (f_r). A signal at a frequency outside the bandwidth is transmitted with less than 1/2 the output power of a signal of similar strength at the resonant frequency.

Note. The power at cut off frequencies is half that at the resonant frequency and are known as half power frequencies. They are also called – 3 dB frequencies because power decreases by 1/2 and in decibels it is

$$10 \log_{10} \frac{1}{2} = -3dB$$

14.17. Key Points About Parallel Resonance

The key points concerning parallel resonant circuits are :

(*i*) The circuit power factor is unity (*i.e.* 1) so that parallel resonant circuit behaves as a resistance.

(*ii*) Since parallel resonant circuit is resistive, it means that :

(*a*) Circuit admittance = Circuit conductance

(*b*) Circuit susceptance is zero.

(*iii*) The circuit impedance of a parallel resonant circuit is maximum.

(*iv*) At parallel resonance, the circuit current is minimum and is given by ;

$$I_r = \frac{V}{Z_r}$$

where Z_r = Circuit impedance at parallel resonance

(*v*) *The resonant frequency (f_r) of a parallel resonant circuit can be found by equating susceptance of the circuit at f_r to zero.*

14.18. General Case for Parallel Resonance

We have discussed above parallel resonance for a circuit consisting of a coil ($R - L$ series circuit) in parallel with an ideal capacitor. However, no capacitor is perfect and can be considered to consist of an ideal capacitor in series with a resistor. The circuit then becomes as shown in Fig. 14.60. The total admittance $\mathbf{Y_T}$ of this circuit is given by ;

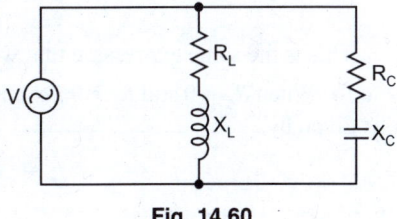

Fig. 14.60

$$\mathbf{Y_T} = \left(\frac{R_C}{R_C^2 + X_C^2} + j\frac{X_C}{R_C^2 + X_C^2} \right) + \left(\frac{R_L}{R_L^2 + X_L^2} - j\frac{X_L}{R_L^2 + X_L^2} \right)$$

or
$$\mathbf{Y_T} = \left(\frac{R_C}{R_C^2 + X_C^2} + \frac{R_L}{R_L^2 + X_L^2} \right) + j\left(\frac{X_C}{R_C^2 + X_C^2} - \frac{X_L}{R_L^2 + X_L^2} \right) \qquad ...(i)$$

Resonance is defined as a circuit condition of unity power factor. Under this condition, the circuit admittance is a pure conductance. In other words, *circuit susceptance is zero.* Therefore, at resonance, the *j* part of eq. (*i*) must be zero *i.e.*

$$\frac{X_C}{R_C^2 + X_C^2} - \frac{X_L}{R_L^2 + X_L^2} = 0$$

or
$$R_L^2 X_C + X_C X_L^2 - R_C^2 X_L - X_C^2 X_L = 0$$

or
$$R_C^2 X_L - R_L^2 X_C - X_C X_L^2 + X_C^2 X_L = 0$$

or $\quad R_C^2 (2\pi f_r L) - R_L^2 \dfrac{1}{(2\pi f_r C)} - \dfrac{1}{(2\pi f_r C)}(2\pi f_r L)^2 + \dfrac{1}{(2\pi f_r C)^2}(2\pi f_r L) = 0$

or
$$R_C^2 2\pi f_r L - \frac{R_L^2}{2\pi f_r C} - \frac{L}{C} 2\pi f_r L + \frac{L}{C} \frac{1}{2\pi f_r C} = 0$$

or
$$R_C^2 4\pi^2 f_r^2 LC - R_L^2 - \frac{L}{C} 4\pi^2 f_r^2 LC + \frac{L}{C} = 0$$

or
$$4\pi^2 f_r^2 LC \left[R_C^2 - \frac{L}{C} \right] = \left[R_L^2 - \frac{L}{C} \right]$$

or $$4\pi^2 f_r^2 LC = \frac{R_L^2 - (L/C)}{R_C^2 - (L/C)}$$

or $$f_r^2 = \frac{1}{4\pi^2 LC} \frac{R_L^2 - (L/C)}{R_C^2 - (L/C)}$$

∴ $$f_r = \frac{1}{2\pi\sqrt{LC}} \sqrt{\frac{R_L^2 - (L/C)}{R_C^2 - (L/C)}}$$

Special cases. $$f_r = \frac{1}{2\pi\sqrt{LC}} \sqrt{\frac{R_L^2 - (L/C)}{R_C^2 - (L/C)}}$$

(*i*) When capacitor is pure or perfect, $R_C = 0$ so that :

$$f_r = \frac{1}{2\pi\sqrt{LC}} \sqrt{\frac{R_L^2 - (L/C)}{0 - (L/C)}}$$

or $$f_r = \frac{1}{2\pi} \sqrt{\frac{1}{LC} - \frac{R_L^2}{L^2}}$$

This is the same expression that was derived in Art. 14.13.

(*ii*) When $R_C = 0$ and $R_L = 0$ *i.e,* pure L and C are connected in parallel, then resonant frequency f_r is given by ;

$$f_r = \frac{1}{2\pi\sqrt{LC}} \sqrt{\frac{0^2 - (L/C)}{0^2 - (L/C)}}$$

or $$f_r = \frac{1}{2\pi\sqrt{LC}} \qquad \textit{... same as for series resonance}$$

14.19. Comparison of Series and Parallel Resonant Circuits

S. No.	Particular	Series circuit	Parallel circuit
1.	Impedance at resonance	Minimum ($Z_r = R$)	Maximum ($Z_r = L/CR$)
2.	Current at resonance	Maximum ($I_r = V/R$)	Minimum ($I_r = V/Z_r$)
3.	p.f. at resonance	Unity	Unity
4.	Resonant frequency	$f_r = \dfrac{1}{2\pi\sqrt{LC}}$	$f_r = \dfrac{1}{2\pi}\sqrt{\dfrac{1}{LC} - \dfrac{R^2}{L^2}}$
5.	When $f < f_r$	Circuit is capacitive	Circuit is inductive
6.	When $f > f_r$	Circuit is inductive	Circuit is capacitive
7.	Q-factor	X_L/R	X_L/R
8.	It magnifies	Voltage	Current

Example 14.49. *The dynamic impedance of a parallel resonant circuit is 500 kΩ. The circuit consists of a 250 pF capacitor in parallel with a coil of resistance 10 Ω. Calculate (i) the coil inductance (ii) the resonant frequency and (iii) the Q-factor of the circuit.*

Solution. (*i*) Dynamic impedance, $Z_r = \dfrac{L}{CR}$

∴ Inductance of coil, $L = Z_r CR = (500 \times 10^3) \times 250 \times 10^{-12} \times 10$

$$= 1.25 \times 10^{-3} \text{ H} = \textbf{1.25 mH}$$

(ii)   Resonant frequency, $f_r = \dfrac{1}{2\pi}\sqrt{\dfrac{1}{LC} - \dfrac{R^2}{L^2}} = \dfrac{1}{2\pi}\sqrt{\dfrac{10^{12}}{1.25\times10^{-3}\times250} - \dfrac{10^2}{(1.25\times10^{-3})^2}}$

$$= 284.7 \times 10^3 \text{ Hz} = \textbf{284.7 kHz}$$

(iii)   Q-factor of the circuit $= \dfrac{2\pi f_r L}{R} = \dfrac{2\pi\times(284.7\times10^3)\times1.25\times10^{-3}}{10} = \textbf{223.6}$

Example 14.50. *A tuned circuit consisting of a coil having an inductance of 200 µH and a resistance of 20 Ω is in parallel with a variable capacitor. This combination is in series with a resistor of 8000 Ω. The entire circuit is connected to a 230 V, 1MHz supply. Calculate (i) the value of C to give resonance (ii) the dynamic impedance and Q-factor of the tuned circuit and (iii) the current in each branch.*

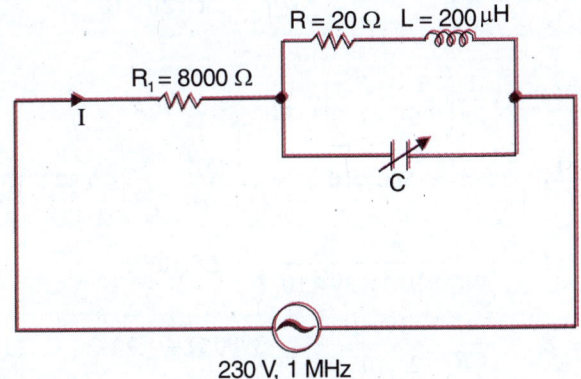

Fig. 14.61

Solution. **(i)** Reactance of coil, $X_L = 2\pi f_r L = 2\pi \times 10^6 \times 200 \times 10^{-6} = 1256\ \Omega$

Since the resistance of the coil is very small as compared to its reactance, the resonant frequency is given by ;

$$f_r = \frac{1}{2\pi\sqrt{LC}} \quad \text{or} \quad f_r^2 = \frac{1}{4\pi^2 LC}$$

$\therefore$ $C = \dfrac{1}{4\pi^2 f_r^2 L} = \dfrac{1}{4\pi^2 \times (10^6)^2 \times 200\times10^{-6}}$

$$= 126.65 \times 10^{-12}\text{ F} = \textbf{126.65 pF}$$

(ii) Dynamic impedance, $Z_r = \dfrac{L}{CR} = \dfrac{200\times10^{-6}}{126.65\times10^{-12}\times20} = \textbf{78957 } \Omega$

Q-factor $= \dfrac{2\pi f_r L}{R} = \dfrac{2\pi\times10^6\times200\times10^{-6}}{20} = \textbf{62.8}$

(iii)   Total circuit resistance $= 8000 + 78957 = 86957\ \Omega$

Line current, $I = \dfrac{230}{86957} = 2.64\times10^{-3}\text{ A}$

P.D. across tuned circuit $= I Z_r = 2.64 \times 10^{-3} \times 78957 = 208.4\text{ V}$

Current in inductive branch $= \dfrac{208.4}{\sqrt{20^2 + (1256)^2}} = \textbf{0.1659 A}$

Reactance of capacitor, $X_C = \dfrac{1}{2\pi f_r C} = \dfrac{10^{12}}{2\pi \times 10^6 \times 126.65} = 1256.6\ \Omega$

Current through capacitor $= \dfrac{208.4}{1256.6} = \textbf{0.1659 A}$

Note that current in each parallel branch is 62.8 times (*i.e.* Q-factor times) the line current.

Example 14.51. *Calculate the impedance of parallel-tuned circuit shown in Fig. 14.62 at a frequency of 500 kHz and for bandwidth of operation equal to 20 kHz.*

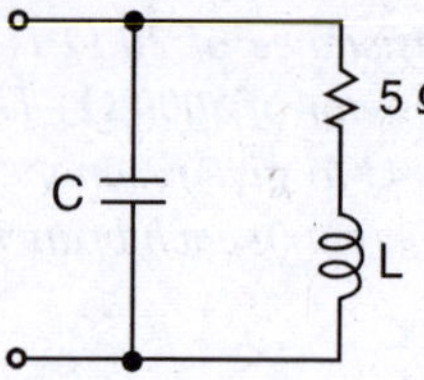

Solution. $f_r = 500\ \text{kHz} = 500 \times 10^3\ \text{Hz}$; $BW = 20\ \text{kHz} = 20 \times 10^3\ \text{Hz}$

Now, $BW = \dfrac{R}{2\pi L} \quad \therefore\ L = \dfrac{R}{2\pi \times BW} = \dfrac{5}{2\pi \times 20 \times 10^3}$

$= 39 \times 10^{-6}\ \text{H}$

$f_r = \dfrac{1}{2\pi}\left(\dfrac{1}{LC} - \dfrac{R^2}{L^2}\right)^{1/2}$ or $\dfrac{1}{LC} = 4\pi^2 f_r^2 + \dfrac{R^2}{L^2}$

Fig. 14.62

$\therefore$ $\dfrac{1}{LC} = 4\pi^2 \times (500 \times 10^3)^2 + \dfrac{(5)^2}{(39 \times 10^{-6})^2} \simeq 9.87 \times 10^{12}$

or $C = \dfrac{1}{987 \times 10^{12} \times 39 \times 10^{-6}} = 2 \times 10^{-9}\ \text{F}$

Circuit impedance, $Z_r = \dfrac{L}{CR} = \dfrac{39 \times 10^{-6}}{2 \times 10^{-9} \times 5} = 3900\ \Omega = \textbf{3.9 k}\boldsymbol{\Omega}$

Example 14.52. *An inductor of resistance 10 Ω and inductance 100 mH is in parallel with a 10 nF capacitor. Find (i) the resonant angular frequency (ii) the Q-factor and (iii) the bandwidth.*

Solution. $R = 10\ \Omega$; $L = 100\ \text{mH} = 100 \times 10^{-3}\ \text{H} = 10^{-1}\ \text{H}$; $C = 10\ \text{nF} = 10 \times 10^{-9}\ \text{F} = 10^{-8}\ \text{F}$

(i) The resonant angular frequency $\omega_r\ (= 2\pi f_r)$ is given by ;

$$\omega_r = \sqrt{\left(\dfrac{1}{LC} - \dfrac{R^2}{L^2}\right)} = \sqrt{\left(\dfrac{1}{10^{-1} \times 10^{-8}} - \dfrac{(10)^2}{(10^{-1})^2}\right)} = \textbf{3.16} \times \textbf{10}^4\ \textbf{rad/s}$$

(ii) $Q\text{-factor} = \dfrac{\omega_r L}{R} = \dfrac{3.16 \times 10^4 \times 10^{-1}}{10} = \textbf{316}$

(iii) $\text{Bandwidth} = \dfrac{\omega_r}{Q} = \dfrac{3.16 \times 10^4}{316} = \textbf{100 rad/s}$

Example 14.53. *An inductor of resistance R and inductance L is in parallel with a capacitor C. Show that resonant angular frequency ω_r is given by ;*

$$\omega_r = \dfrac{1}{\sqrt{LC}}\sqrt{\left(1 - \dfrac{1}{Q^2}\right)}$$

Solution. $\omega_r = \sqrt{\left(\dfrac{1}{LC} - \dfrac{R^2}{L^2}\right)}$ ...(*i*)

This equation can be rearranged as :

$$\omega_r = \dfrac{1}{\sqrt{LC}}\sqrt{\left(1 - \dfrac{R^2 C}{L}\right)}$$

Now,
$$Q = \frac{1}{R}\sqrt{\frac{L}{C}} \text{ or } Q^2 = \frac{L}{CR^2}$$

$\therefore$
$$\omega_r = \frac{1}{\sqrt{LC}}\sqrt{\left(1 - \frac{1}{Q^2}\right)}$$

When Q is large, this equation approximates to $\omega_r = \dfrac{1}{\sqrt{LC}}$

Example 14.54. *A 230 V, 50 Hz source supplies energy to a parallel circuit consisting of a 25 $\angle 30°$ Ω branch and a 12 $\angle -40°$ Ω branch. Determine the capacitance or inductance of a pure circuit element that, if connected in series with the source, will cause the system to be in resonance.*

Solution.
$$\mathbf{Z_1} = 25 \angle 30° \, \Omega = (21.65 + j\,12.5)\,\Omega$$
$$\mathbf{Z_2} = 12 \angle -40° \, \Omega = (9.19 - j\,7.71)\,\Omega$$
$$\mathbf{Z_1 + Z_2} = (21.65 + j\,12.5) + (9.19 - j\,7.71)$$
$$= (30.84 + j\,4.79)\,\Omega = 31.21 \angle 8.83° \, \Omega$$
$$\mathbf{Z_T} = \frac{\mathbf{Z_1 Z_2}}{\mathbf{Z_1 + Z_2}} = \frac{25\angle 30° \times 12 \angle -40°}{31.21 \angle 8.83°} = 9.61 \angle -18.83° \, \Omega$$
$$= (9.1 - j\,3.1)\,\Omega$$

The equivalent series circuit of the parallel circuit is a resistance of 9.1 Ω in series with a capacitive reactance of 3.1 Ω. In order that the system is in resonance, the net reactance should be zero. Therefore, we should connect pure inductance of reactance $X_L = 3.1 \, \Omega$ in series with the circuit.

$\therefore$
$$L = \frac{X_L}{2\pi f} = \frac{3.1}{2\pi \times 50} = 9.87 \times 10^{-3} \, H = \mathbf{9.87\ mH}$$

Example 14.55. *In the circuit shown in Fig. 14.63, $R_L = 400\,\Omega$; $R_C = 700\,\Omega$; $C = 3.6\ \mu F$ and $L = 0.24\ H$. Find the resonant frequency. What will be the frequency if the resistances of the branches were zero ?*

Solution. The resonant frequency of this circuit is given by ;

Fig. 14.63

$$f_r = \frac{1}{2\pi\sqrt{LC}}\sqrt{\frac{R_L^2 - (L/C)}{R_C^2 - (L/C)}} \quad \text{...See Art. 14.18}$$

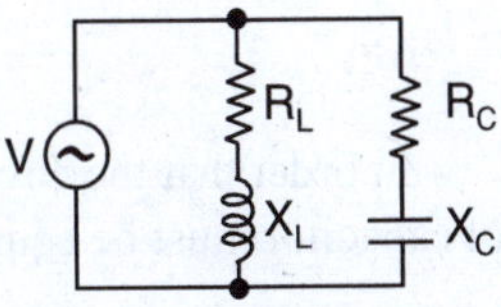

Now,
$$\frac{1}{2\pi\sqrt{LC}} = \frac{1}{2\pi\sqrt{0.24 \times 3.6 \times 10^{-6}}} = 171\ Hz$$

66700

$$R_L^2 - (L/C) = (400)^2 - (0.24/3.6 \times 10^{-6}) = 400^2 -$$
$$R_C^2 - (L/C) = (700)^2 - (0.24/3.6 \times 10^{-6}) = 700^2 - 66700$$

$\therefore$
$$f_r = 171\sqrt{\frac{400^2 - 66700}{700^2 - 66700}} = 171\sqrt{\frac{93300}{423300}} = \mathbf{80\ Hz}$$

If R_C and R_L were zero, $f_r = \mathbf{171\ Hz}$. Note that the effect of branch resistances is to shift the resonant frequency to 80 Hz.

Example 14.56. *A coil of resistance 10 Ω and inductance 0.5 H is connected in series with a capacitor. On applying a sinusoidal voltage, the current is maximum when the frequency is 50 Hz. A second capacitor is connected in parallel with the circuit. What capacitance it should have so that the combination acts as a non-inductive resistor at 100 Hz ? Calculate the total current supplied in each case if the applied voltage is 220 V.*

Solution. Since current is maximum at 50 Hz, series resonance occurs. Therefore, $X_L = X_C$. Now $X_L = 2\pi f L = 2\pi \times 50 \times 0.5 = 157\ \Omega = X_C$.

At 100 Hz. At 100 Hz, the inductive reactance ($= 2\pi f L$) is doubled and capacitive reactance ($= 1/2\pi f C$) is halved.

∴   Impedance of series branch is given by ;

$$\mathbf{Z} = 10 + j(2 \times 157 - 157/2) = (10 + j\ 236)\ \Omega = 236.2\ \angle\ 87.57°\ \Omega$$

$$\mathbf{Y} = \frac{1}{\mathbf{Z}} = \frac{1}{236.2\ \angle\ 87.57°} = 4.2 \times 10^{-3}\ \angle -87.57°\ S$$

$$= (0.179 \times 10^{-3} - j\ 4.2 \times 10^{-3}\ )S$$

When capacitor C is connected in parallel with the series branch, it is desired that resonance should occur at 100 Hz. For this to happen, capacitive susceptance must be 4.2×10^{-3} S *i.e.*, equal to the net inductive susceptance of the series branch.

∴

$$C = \frac{4.2 \times 10^{-3}}{2\pi f} = \frac{4.2 \times 10^{-3}}{2\pi \times 100} = 6.7 \times 10^{-6}\ F = \mathbf{6.7\ \mu F}$$

$$\text{At 50 Hz, } I = 220/10 = \mathbf{22\ A} \qquad\qquad\qquad \textit{...series resonance}$$

$$\text{At 100 Hz, } I = 220 \times \text{Real part of } \mathbf{Y} = 220 \times 0.179 \times 10^{-3} = \mathbf{0.04\ A}$$

Example 14.57. *A circuit has an inductive reactance of 20 Ω at 50 Hz ; its resistance being 15 Ω. For an applied voltage of 200 V at 50 Hz, calculate (i) the angle of phase difference between current and applied voltage (ii) the value of current (iii) the value of shunting capacitance to bring the resulting current in phase with the voltage (iv) the resultant current in case (iii).*

Solution. *(i)* $\mathbf{Z} = R + j\ X_L = (15 + j\ 20)\ \Omega = 25\ \angle\ 53°\ \Omega$

Therefore, phase angle between applied voltage and circuit current is **53°**; current lags behind voltage.

(ii)   Circuit current, $\mathbf{I} = \dfrac{\mathbf{V}}{\mathbf{Z}} = \dfrac{200\ \angle\ 0°}{25\ \angle\ 53°} = 8\ \angle -53°\ A$

Therefore, magnitude of circuit current is **8A.**

(iii) $\mathbf{Y} = \dfrac{1}{\mathbf{Z}} = \dfrac{1}{25\ \angle\ 53°} = 0.04\ \angle -53°\ S = (0.024 - j\ 0.032)\ S$

In order that the circuit current is in phase with the applied voltage, the capacitive susceptance of capacitor must be equal to the inductive susceptance (= 0.032 S).

∴ $C = \dfrac{0.032}{2\pi f} = \dfrac{0.032}{2\pi \times 50} = 102 \times 10^{-6}\ F = \mathbf{102\ \mu F}$

(iv) $\text{Current} = \mathbf{V} \times \text{Real part of } \mathbf{Y} = 200 \times 0.024 = \mathbf{4.8\ A}$

Example 14.58. *Find active and reactive components of the current taken by a series circuit consisting of a coil of inductance 0.1H and resistance 8 Ω and a capacitor of 120 μF, connected to a 240 V, 50 Hz supply mains. Find the value of the capacitor that has to be connected in parallel with the above series circuit so that power factor of the entire circuit is unity.*

Solution. $X_L = 2\pi \times 50 \times 0.1 = 31.4\ \Omega$; $X_C = 1/\omega C = 1/314 \times 120 \times 10^{-6} = 26.5\ \Omega$

Net circuit reactance, $X = X_L - X_C = 31.4 - 26.5 = 4.9\ \Omega$

Magnitude of impedance, $Z = \sqrt{R^2 + X^2} = \sqrt{8^2 + 4.9^2} = 9.43\ \Omega$

Magnitude of circuit current, $I = V/Z = 240/9.43 = 25.45\ A$

Circuit power factor, $\cos\phi = R/Z = 8/9.43 = 0.848\ \text{lag}$

Active component $= I \cos\phi = 25.45 \times 0.848 = \mathbf{21.58\ A}$

Reactive component $= I \sin \phi = 25.45 \times 0.53 = $ **13.49 A**

Admittance of series circuit, $\mathbf{Y} = \dfrac{1}{Z} = \dfrac{1}{R + jX} = \dfrac{1}{8 + j\,4.9}$

$$= \dfrac{8 - j\,4.9}{8^2 + 4.9^2} = (0.09 - j\,0.0557)\ \text{S}$$

In order that the circuit p.f. is unity, capacitive susceptance of the capacitor should be equal to the net inductive susceptance (= 0.0557 S) of the series branch.

$$\therefore \qquad C = \dfrac{0.0557}{2\pi f} = \dfrac{0.0557}{2\pi \times 50} = 177 \times 10^{-6}\ \text{F} = \textbf{177 μF}$$

Example 14.59. *A parallel LC circuit has an inductance L = 100 μH which has a coil resistance of 12 Ω. The capacitor C is adjustable over the range 200 pF to 300 pF. Find (i) the maximum and minimum resonance frequencies for the circuit (ii) the Q-factor and (iii) bandwidth of the circuit at the two resonance frequency extremes.*

Solution. Since $Q > 10$, the following formula can be used for the resonance frequency :

$$f_r = \dfrac{1}{2\pi \sqrt{LC}}$$

(i) **With C = 200 pF ;** $\quad f_r = \dfrac{1}{2\pi \sqrt{100 \times 10^{-6} \times 200 \times 10^{-12}}} = 1.13 \times 10^6\ \text{Hz} = \textbf{1.13 MHz}$

This is the value of maximum resonance frequency *i.e.*, $f_{r(max)} = 1.13$ MHz.

With C = 300 pF ; $\quad f_r = \dfrac{1}{2\pi \sqrt{100 \times 10^{-6} \times 300 \times 10^{-12}}} = 919 \times 10^3\ \text{Hz} = \textbf{919 kHz}$

This is the value of minimum resonance frequency *i.e.*, $f_{r(min)} = 919$ kHz.

(ii) At $f_r = 1.13$ MHz ; $\quad Q = \dfrac{\omega_r L}{R} = \dfrac{2\pi \times 1.13 \times 10^6 \times 100 \times 10^{-6}}{12} = \textbf{59.2}$

At $f_r = 919$ kHz ; $\quad Q = \dfrac{\omega_r L}{R} = \dfrac{2\pi \times 919 \times 10^3 \times 100 \times 10^{-6}}{12} = \textbf{48}$

(iii) Bandwidth, $BW = \dfrac{R}{2\pi L} = \dfrac{12}{2\pi \times 100 \times 10^{-6}} = 19.1 \times 10^3\ \text{Hz} = \textbf{19.1 kHz}$

Example 14.60. *Calculate the value of C which results in resonance for the circuit shown in Fig. 14.64 when frequency is 1000 Hz. Also find Q-factor of each branch.*

Solution. Admittance of inductive branch is

$$\mathbf{Y}_{ind} = \dfrac{1}{4 + j8} = (0.05 - j\,0.1)\ \text{S}$$

Admittance of capacitive branch is

$$\mathbf{Y}_{cap} = \dfrac{1}{5 - jX_C} = \left(\dfrac{5}{25 + X_C^2} + j\,\dfrac{X_C}{25 + X_C^2} \right) \text{S}$$

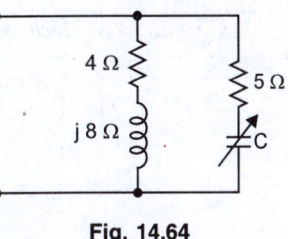

Fig. 14.64

For resonance, the susceptance of the whole circuit should be zero *i.e.*

$$(-0.1) + \left(\dfrac{X_C}{25 + X_C^2} \right) = 0$$

or $\qquad 0.1\,X_C^2 - X_C + 2.5 = 0 \qquad \therefore X_C = 5\ \Omega$

$\therefore$ $$C = \frac{1}{2\pi f\, X_C} = \frac{1}{2\pi \times 1000 \times 5} = 31.83 \times 10^{-6}\,\text{F} = \textbf{31.83 }\boldsymbol{\mu}\textbf{F}$$

$$Q\text{-factor of } R-L \text{ branch} = \frac{\omega L}{R} = \frac{8}{4} = \textbf{2}$$

$$Q\text{-factor of } R-C \text{ branch} = \frac{1}{\omega RC} = \frac{1}{\omega C} \times \frac{1}{R} = 5 \times \frac{1}{5} = \textbf{1}$$

Example 14.61. *In the circuit shown in Fig. 14.65, find the value of R such that the impedance of the whole circuit should be independent of the supply frequency. If voltage V = 200 volts, L = 0.16H and C = 100 μF, calculate the power loss in the circuit.*

Solution. Impedance of the inductive branch is

$$Z_L = R + j\omega L$$

Impedance of the capacitive branch is

$$Z_C = R - \frac{j}{\omega C}$$

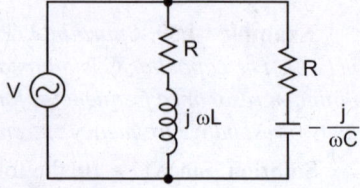

$\therefore$ Impedance of the whole circuit is   **Fig. 14.65**

$$Z = \frac{Z_L\, Z_C}{Z_L + Z_C} = \frac{(R + j\omega L)(R - j/\omega C)}{(R + j\omega L) + (R - j/\omega C)}$$

or $$Z = \frac{R^2 + \dfrac{L}{C} + jR\left(\omega L - \dfrac{1}{\omega C}\right)}{2R + j\left(\omega L - \dfrac{1}{\omega C}\right)}$$

Note that quadrature component of numerator in the above expression is R times the quadrature component of the denominator. Therefore, the impedance Z of the circuit will be independent of ω or frequency ($\omega = 2\pi f$) if the in-phase component of numerator is R times the in-phase component of the denominator *i.e.*

$$R^2 + \frac{L}{C} = R \times 2R \quad \text{or} \quad \boxed{R = \sqrt{\frac{L}{C}}}$$

$\therefore$ $$R = \sqrt{\frac{L}{C}} = \sqrt{\frac{0.16}{100 \times 10^{-6}}} = 40\,\Omega$$

Power loss in the circuit is given by ; $$P = \frac{V^2}{R} = \frac{(200)^2}{40} = \textbf{1000 W}$$

Example 14.62. *Determine the frequency at which the voltage V_{out} is zero in Fig. 14.66 shown below.*

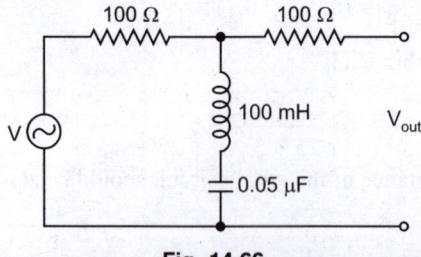

Fig. 14.66

Solution. A little reflection shows that it is $R - L - C$ series a.c. circuit. The output voltage V_{out} will be zero when voltage drop across 100 mH inductor becomes equal to the voltage drop across

0.05 µF capacitor. This happens in $R - L - C$ series a.c. circuit when the supply frequency becomes equal to the resonant frequency (f_r) of the circuit.

∴ Frequency at which V_{out} will be zero is

$$f_r = \frac{1}{2\pi\sqrt{LC}} = \frac{1}{2\pi\sqrt{100\times10^{-3}\times0.05\times10^{-6}}} = \textbf{2250 Hz}$$

Example 14.63. *Find the expression for the resonant frequency of the circuit shown in Fig. 14.67 (i). Hence determine the resonant frequency if R = 10 Ω, C = 3 µ F, L_1 = 40 mH, L_2 = 10 mH and M = 10 mH*

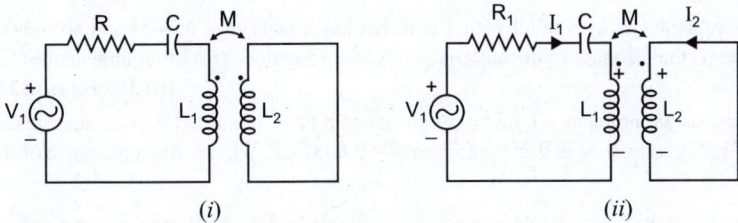

Fig. 14.67

Solution. Fig. 14.67. (*i*) is redrawn in Fig. 14.67 (*ii*) showing the various polarities.

Applying *KVL* to the loop carrying phasor current I_1, we have,

$$V_1 = I_1 R_1 + j\omega L_1 I_1 + \frac{I_1}{j\omega C} + j\omega M I_2 \qquad \qquad ...(i)$$

Applying *KVL* to the loop carrying phasor current I_2, we have,

$$0 = j\omega L_2 I_2 + j\omega M I_1 \quad \text{or} \quad I_2 = -\frac{MI_1}{L_2}$$

Putting the value of $I_2 \left(= -\dfrac{MI_1}{L_2} \right)$ in eq. (*i*), we have,

$$V_1 = I_1 \left(R + j\omega L_1 + \frac{1}{j\omega C} - j\omega\frac{M^2}{L_2} \right)$$

∴ $$I_1 = \frac{V_1}{R_1 + j\omega L_1 + \dfrac{1}{j\omega C} - j\omega\dfrac{M^2}{L_2}}$$

At resonance (*i.e.* at ω_r), the circuit power factor is unity. For this to happen :

$$j\omega_r L_1 + \frac{1}{j\omega_r C} - j\omega_r\frac{M^2}{L_2} = 0$$

or $$j\omega_r \left(L_1 - \frac{M^2}{L_2} \right) = -\frac{1}{j\omega_r C} = \frac{j}{\omega_r C}$$

∴ $$\omega_r = \sqrt{\frac{1}{C\left(L_1 - \dfrac{M^2}{L_2} \right)}} \qquad \qquad ...(ii)$$

Eq. (*ii*) gives the required expression.

It is given that L_1 = 40 mH = 0.04 H ; L_2 = 10 mH = 0.01 H ;

C = 3 µF = 3 × 10^{-6} F ; M = 10 mH = 0.01 H.

$$\therefore \qquad f_r = \frac{\omega_r}{2\pi} = \frac{1}{2\pi} \times \sqrt{\frac{1}{3\times10^{-6}\left(0.04 - \frac{(0.01)^2}{0.01}\right)}} = 530.52 \text{ Hz}$$

Tutorial Problems

1. A coil of 1000 Ω resistance and 0.15 H inductance is connected in parallel with a variable capacitor across a 2 V, 10 kHz a.c. supply. Calculate (i) the capacitance of the capacitor when the supply current is minimum (ii) the dynamic impedance of the network and (iii) the supply current at resonance.

[(i) 1690 µF (ii) 890 kΩ (iii) 2.25 µA]

2. A 0.8 µF capacitor is in parallel with a coil that has a resistance of 4 Ω and an inductance of 0.2 H. Determine (i) the resonant frequency (ii) Q-factor at resonance (iii) input impedance at resonance.

[(i) 398 Hz (ii) 125 (iii) 62.5 k Ω]

3. A coil that has an inductance L and a resistance of 8 Ω is in parallel with a variable capacitor. At $\omega = 1.5 \times 10^6$ rad/s, resonance is achieved when $C = 0.0037$ µF. What is the inductance of the coil ?

[L = 120 µH]

4. An inductor for which $L = 40$ mH is to be used in parallel with a 1 µF capacitor in a circuit that must have Q-factor atleast as large as 25 when the frequency is 1 kHz. What is the minimum amount of resistance the coil may have ? What will the input impedance be when the minimum R is used ?

[R = 12.56 Ω ; Z = 3185 Ω]

5. A coil has a resistance of 400 Ω and inductance of 318 µH. Find the capacitance of a capacitor which, when connected in parallel with the coil, will produce resonance with a supply frequency of 1 MHz. If a second capacitor of capacitance 23.5 pF is connected in parallel with the first capacitor, find the frequency at which resonance will occur.   [76.5 pF ; 870 kHz]

6. A parallel circuit consists of a 2.5 µF capacitor and a coil whose resistance and inductance are 15 Ω and 260 mH respectively. Determine (i) the resonant frequency (ii) Q-factor of the circuit at resonance (iii) dynamic impedance of the circuit.   [(i) 197 Hz (ii) 21.45 (iii) 6933 Ω]

7. A coil of 10 Ω resistance and 0.2 H inductance is connected in parallel with a variable capacitor across a 220 V, 50 Hz supply. Calculate (i) the capacitance of the capacitor for resonance (ii) the effective impedance of the circuit and (iii) the supply current.   [(i) 49.41 µF (ii) 404.78 Ω (iii) 0.543A]

8. A coil of resistance 25 Ω and inductance 0.5 H is in parallel with a 5µF capacitor. Determine (i) resonant frequency (ii) total impedance of the circuit at resonance (iii) bandwidth (iv) Q-factor.

[(i) 100.34 Hz (ii) 4000 Ω (iii) 7.96 Hz (iv) 12.65]

9. A circuit consists of resistance R_1 and 4 mH inductor in parallel with R_2 and 160 µF capacitor. Find the values of R_1 and R_2 which will make the circuit to resonate at all frequencies

[R_1 = R_2 = 5]

10. A parallel R– L – C circuit is fed by a constant current source of variable frequency. The circuit resonates at 100 kHz and the Q-factor measured at this frequency is 5. Find the frequencies at which the amplitude of the voltage across the circuit falls to (i) 70.7% (ii) 50% of the resonant frequency amplitude.

[(i) 90.5 kHz; 110.5 kHz (ii) 84.18 kHz; 118.8 kHz]

Objective Questions

1. Domestic appliances are connected in parallel across a.c. mains because ....

   (i) it is a simple arrangement

   (ii) operation of each appliance becomes independent of the other

   (iii) appliances have same currents ratings

   (iv) this arrangement occupies less space

2. When a parallel a.c. circuit contains a number of branches, then it is convenient to solve the circuit by ....

   (i) phasor diagram

   (ii) phasor algebra

   (iii) equivalent impedance method

   (iv) none of the above

3. The power taken by the circuit shown in Fig. 14.68 is ....

    (*i*) 480 W   (*ii*) 1920 W

    (*iii*) 1200 W   (*iv*) none of the above

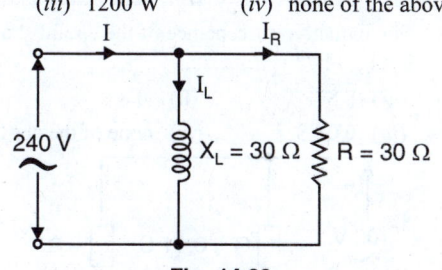

Fig. 14.68

4. The active component of line current in Fig. 14.68 is ....

    (*i*) 8 A   (*ii*) 4 A

    (*iii*) 5.3 A   (*iv*) none of the above

5. The power factor of the circuit shown in Fig. 14.68 is ....

    (*i*) 0.707 lagging   (*ii*) 0.5 lagging

    (*iii*) 0.866 lagging   (*iv*) none of the above

6. The total line current drawn by the circuit shown in Fig. 14.68 is ....

    (*i*) $8/\sqrt{2}$ A   (*ii*) 16 A

    (*iii*) $8\sqrt{2}$ A   (*iv*) none of the above

7. The power consumed in the circuit shown in Fig. 14.69 is ....

    (*i*) 480 W   (*ii*) 960 W

    (*iii*) 1200 W   (*iv*) none of the above

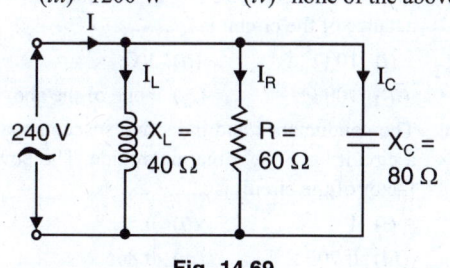

Fig. 14.69

8. The active component of line current in Fig. 14.69 is ....

    (*i*) 6A   (*ii*) 3 A

    (*iii*) 13 A   (*iv*) 4 A

9. The line current drawn by the circuit shown in Fig. 14.69 is ....

    (*i*) 13 A   (*ii*) 6 A

    (*iii*) 5 A   (*iv*) none of the above

10. The power factor of the circuit shown in Fig. 14.69 is ....

    (*i*) 0.8   (*ii*) 0.5

    (*iii*) 0.707   (*iv*) none of the above

11. The impedance of the circuit shown in Fig. 14.69 is ....

    (*i*) 180 ohms   (*ii*) 24 ohms

    (*iii*) 48 ohms   (*iv*) none of the above

12. The circuit shown in Fig. 14.69 is ....

    (*i*) resistive   (*ii*) capacitive

    (*iii*) inductive   (*iv*) in resonance

13. If in Fig. 14.69, X_L is made equal to X_C, then line current will be ....

    (*i*) 10 A   (*ii*) 6 A

    (*iii*) 4 A   (*iv*) none of the above

14. The power consumed in the circuit shown in Fig. 14.70 is ....

    (*i*) 8400 W   (*ii*) 3600 W

    (*iii*) 4000 W   (*iv*) none of the above

15. If the circuit shown in Fig. 14.70 is connected to 120 V d.c., the current drawn by the circuit is ....

    (*i*) 24 A   (*ii*) 70 A

    (*iii*) 48 A   (*iv*) 30 A

16. The circuit shown in Fig. 14.70 is ....

    (*i*) capacitive   (*ii*) inductive

    (*iii*) resistive   (*iv*) in resonance

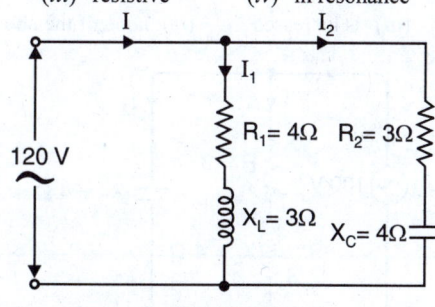

Fig. 14.70

17. If the source frequency in Fig. 14.71 is low, then, ...

    (*i*) coil takes a high lagging current

    (*ii*) coil takes a low lagging current

    (*iii*) capacitor takes a high leading current

    (*iv*) circuit offers high impedance

18. If the source frequency in Fig. 14.71 is high, then, ...

    (*i*) coil takes a high lagging current

    (*ii*) capacitor takes a high leading current

    (*iii*) capacitor takes a low leading current

    (*iv*) circuit offers high impedance

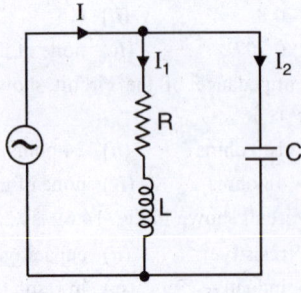

Fig. 14.71

19. The circuit shown in Fig. 14.71 will be in resonance when ....

    (i) $X_L = X_C$ (ii) $I_1 = I_2$

    (iii) V and I are in phase

    (iv) none of the above

20. The circuit shown in Fig. 14.72 is ....

    (i) in resonance   (ii) resistive

    (iii) inductive   (iv) capacitive

21. The circuit shown in Fig. 14.72 will consume a power of ....

    (i) 1200 W   (ii) 2400 W

    (iii) 500 W   (iv) none of the above

22. If the admittance of a parallel a.c. circuit is increased, the circuit current ....

    (i) remains constant (ii) is decreased

    (iii) is increased   (iv) none of the above

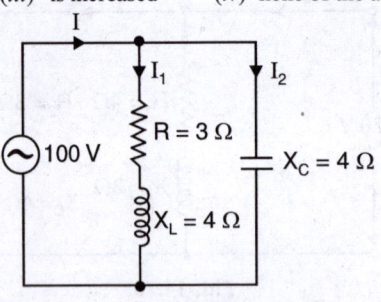

Fig. 14.72

23. The admittance of the circuit shown in Fig. 14.73 is ....

    (i) 10 S   (ii) 14 S

    (iii) 0.1 S   (iv) none of the above

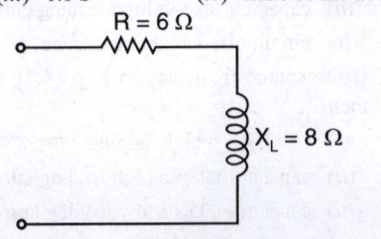

Fig. 14.73

24. The conductance of the circuit shown in Fig. 14.73 is ....

    (i) 14 S   (ii) 0.6 S

    (iii) 0.06 S   (iv) none of the above

25. The inductive susceptance of the circuit shown in Fig. 14.73 is ....

    (i) 8 S   (ii) 0.8 S

    (iii) 0.08 S   (iv) none of the above

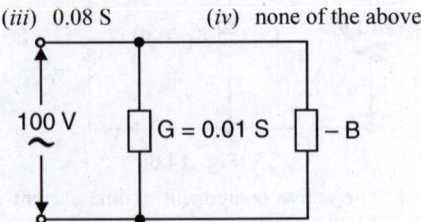

Fig. 14.74

26. The circuit shown in Fig. 14.74 is ....

    (i) resistive   (ii) inductive

    (iii) capacitive   (iv) none of the above

27. The power loss in the circuit shown in Fig. 14.67 is ....

    (i) 100 W   (ii) 10,000 W

    (iii) 10 W   (iv) none of the above

28. The conductance and susceptance components of admittance are ....

    (i) series elements (ii) parallel elements

    (iii) series-parallel elements

    (iv) none of the above

29. The impedance of a circuit is 10 ohms. If the inductive susceptance is 1 S, then inductive reactance of the circuit is ....

    (i) 10 Ω   (ii) 1 Ω

    (iii) 100 Ω   (iv) none of the above

30. The conductance and inductive susceptance of a circuit have the same magnitude. The power factor of the circuit is ....

    (i) 1   (ii) 0.5

    (iii) 0.707   (iv) 0.866

31. The admittance of a circuit is $(0.1 + j\,0.8)$ S. The circuit is ....

    (i) resistive   (ii) capacitive

    (iii) inductive   (iv) none of the above

32. In a parallel a.c. circuit, power loss is due to ....

    (i) conductance alone

    (ii) susceptance alone

    (iii) both conductance and susceptance

    (iv) none of the above

33. The admittance of a parallel circuit is 0.12 ∠ –30° S. The circuit is ....

    (i) inductive   (ii) capacitive

    (iii) resistive   (iv) in resonance

34. A circuit has an impedance of $(1 - j\,2)$ Ω. The susceptance of the circuit is ....

    (i) 0.1 S   (ii) 0.2 S

    (iii) 0.4 S   (iv) none of the above

35. A circuit has admittance of 0.1 S and conductance of 0.08 S . The power factor of the circuit is ....

    (i) 0.1   (ii) 0.8

    (iii) 0.08   (iv) none of the above

Answers

1. (ii)	2. (ii)	3. (ii)	4. (i)	5. (i)
6. (iii)	7. (ii)	8. (iv)	9. (iii)	10. (i)
11. (iii)	12. (iii)	13. (iii)	14. (iv)	15. (iv)
16. (i)	17. (i)	18. (ii)	19. (iii)	20. (iv)
21. (i)	22. (iii)	23. (iii)	24. (iii)	25. (iii)
26. (ii)	27. (i)	28. (ii)	29. (iii)	30. (iii)
31. (ii)	32. (i)	33. (i)	34. (iii)	35. (ii)

15

Polyphase Circuits

Introduction

The a.c. circuits discussed so far in the book are termed as single phase circuits because they contain a single alternating current and voltage wave. The generator producing a single phase supply (called single-phase generator) has only one armature winding. But if the generator is arranged to have two or more separate windings displaced from each other by equal electrical angles, it is called a *polyphase generator and will produce as many independent voltages as the number of windings or phases. The electrical displacement between the windings depends upon the number of windings or phases. For example, a 2-phase generator has two separate but identical windings that are 90° electrical apart and rotate in a common uniform magnetic field. Obviously, such a generator will produce two alternating voltages of the same magnitude and frequency having a phase difference of 90°. Similarly, a 3-phase generator has three separate but identical windings that are 120° electrical apart and rotate in a common uniform magnetic field. A 3-phase generator will, therefore, produce three voltages of the same magnitude and frequency but displaced 120° electrical from one another. Although several polyphase systems are possible, *the 3-phase system is by far the most popular because it is the most efficient of all the supply systems.* In this chapter, we shall confine our attention to 3-phase system only.

15.1. Polyphase System

A polyphase alternator has two or more separate but identical windings (called **phases**) displaced from each other by **equal electrical angle and acted upon by the common uniform magnetic field. Each winding or phase produces a single alternating voltage of the same magnitude and frequency. However, these voltages are displaced form one another by equal electrical angle.

(*i*) Fig. 15.1 (*i*) shows an elementary **single-phase alternator.** It has one winding or coil A rotating in anticlockwise direction with an angular velocity ω in the 2-pole field. The equation of the e.m.f. induced in the coil is given by ;

$$e_{a_1 a_2} = E_m \sin \omega t$$

(*ii*) Fig. 15.1 (*ii*) shows an elementary **two-phase alternator.** It has two identical windings or coils A and B displaced ***90 electrical degrees from each other and rotating in anticlockwise direction with an angular velocity ω in the 2-pole field. Here a_1 and b_1 are the start and a_2 and b_2 are the finish terminals of the two coils respectively. Note that the corresponding terminals a_1 and b_1 are 90 electrical degrees apart. Likewise terminals a_2 and b_2 are 90° apart. Since the two coils are identical and have the same angular velocity, the e.m.f.s induced in them will be of the same magnitude and frequency. However, these

---

*   Poly means many and phases mean windings so that a polyphase generator has many phases or windings.

**   The electrical displacement between the windings is determined by the number of phases or windings. For 2-phase alternator, the two windings are displaced by 90 electrical degrees. For other polyphase systems (*e.g.*, 3-phase, 6-phase), the electrical displacement between different phases or windings is equal to 360/N where N is the number of phases. Thus for a 3-phase alternator, the three windings are 120 electrical degrees apart.

*** For a 2-pole machine, mechanical degrees are equal to electrical degrees.

e.m.f.s will have a phase difference of 90° as shown in the wave diagram in Fig. 15.1 (*ii*). Note that e.m.f. in coil *A* leads that in coil *B* by 90°. The equations of the two e.m.f.s. are :

$$e_{a_1 a_2} = E_m \sin \omega t$$
$$e_{b_1 b_2} = E_m \sin (\omega t - 90°)$$

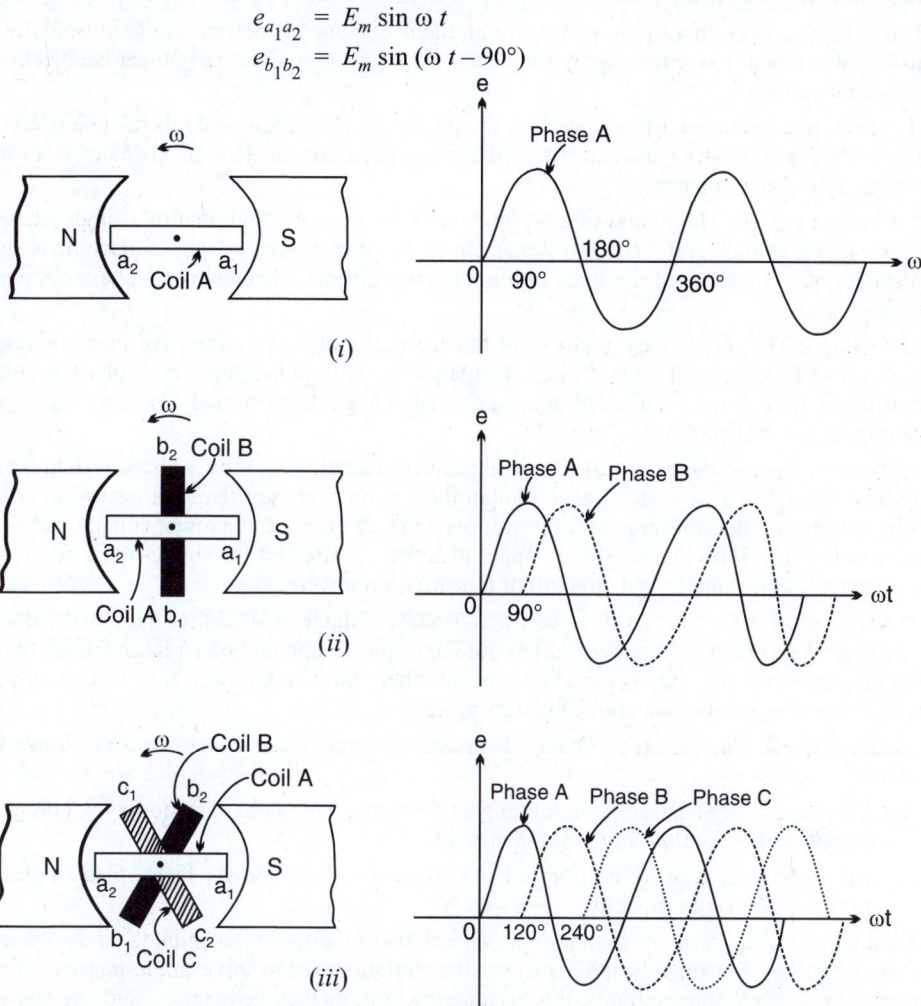

Fig. 15.1

(*iii*) Fig. 15.1 (*iii*) shows an elementary **3-phase alternator.** It has three identical windings or coils *A*, *B* and *C* displaced 120 electrical degrees from each other and rotating in anticlockwise direction with an angular velocity ω in the 2-pole field. Note that the corresponding terminals a_1, b_1 and c_1 are 120° apart. Likewise the terminals a_2, b_2 and c_2 are 120 electrical degrees apart. Since the three coils are identical and have the same angular velocity, the e.m.f.s induced in them will be of the same magnitude and frequency. However, the three e.m.f.s will be displaced from one another by 120°. Note that e.m.f. in coil *B* will be 120° behind that of coil *A* and the e.m.f. in coil *C* will be 240° behind that of coil *A*. This is shown in the wave diagram in Fig. 15.1(*iii*). The equations of the three e.m.f.s can be represented as :

$$e_{a_1 a_2} = E_m \sin \omega t$$
$$e_{b_1 b_2} = E_m \sin (\omega t - 120°)$$
$$e_{c_1 c_2} = E_m \sin (\omega t - 240°)$$

15.2. Reasons for the Use of 3-phase System

The following are the advantages of 3-phase system over the single phase system :

1. Constant power. In a single-phase circuit, the instantaneous power varies sinusoidally from zero to a peak value at twice the supply frequency. This pulsating nature of power is objectionable for many applications.

However, in a balanced 3-phase system, the power supplied at all instants of time is constant. Because of this, the operating characteristics of 3-phase apparatus, in general, are *superior to those of similar single-phase apparatus.

2. Greater output. The output of a 3-phase machine is greater than that of a single-phase machine for a given volume and weight of the machine. In other words, a 3-phase machine is smaller than a single-phase machine of the same rating. This is a distinct advantage of 3-phase system over single-phase system.

3. Cheaper. The three-phase motors are much smaller and less expensive than single-phase motors because less material (copper, iron, insulation) is required. Moreover, 3-phase motors are self-starting *i.e.* they do not require any special provision to get them started. However, single phase motors require internal starting device.

4. Power transmission economics. Transmission of electric power by 3-phase system is cheaper than that of single-phase system, even though three conductors are required instead of two. For example, to transmit the same amount of power over a fixed distance at a given voltage, the 3-phase system requires only 3/4th the weight of copper than that required by the single-phase system. This means a saving in the number and strength of transmission towers.

5. Three-phase rectifier service. Rectified 3-phase voltage is smoother than rectified single-phase voltage. As a result, it is easier to filter out the ripple component of 3-phase voltage than that of a single-phase voltage. This is especially useful where large a.c. power is to be converted into steady d.c. power *e.g.* radio and television transmitters.

6. Miscellaneous advantages. Other advantages of three-phase system over the single-phase system are :

(*i*) A 3-phase system can set-up a rotating uniform magnetic field in stationary windings. This cannot be done with a single-phase current.

(*ii*) The 3-phase motors are more efficient and have a higher power factor than single phase motors of the same capacity.

A knowledge of 3-phase power and 3-phase circuits is, therefore, essential to an understanding of power technology. Fortunately, the basic circuit techniques used to solve single-phase circuits can be directly applied to 3-phase circuits. It is because the three phases are identical and one phase (*i.e.*, single phase) represents the behaviour of all the three.

15.3. Elementary Three-Phase Alternator

In an actual 3-phase alternator, the three windings or coils are stationary and the field **rotates. Fig. 15.2 (*i*) shows an elementary 3-phase alternator. The three identical coils A, B and C are symmetrically placed in such a way that e.m.f.s induced in them are displaced 120 electrical degrees from one another. Since the coils are identical and are subjected to the same uniform rotating field, the e.m.f.s induced in them will be of the same magnitude and frequency. Fig. 15.2 (*ii*) shows the wave diagram of the three e.m.f.s whereas Fig.15.2 (*iii*) shows the phasor diagram. Note that r.m.s. values have been used in drawing the phasor diagram. Thus E_A is the r.m.s. value of the e.m.f. induced in coil A. The equations of the three e.m.f.s are :

$$e_A = E_m \sin \omega t$$

---

* For example, the torque on the rotor of a 3-phase motor is much more constant (because power supplied is constant) than that of a single phase motor. Therefore, 3-phase motors are subjected to less mechanical vibrations.

** This arrangement has many technical and economical advantages.

$$e_B = E_m \sin (\omega t - 120°)$$
$$e_C = E_m \sin (\omega t - 240°) = E_m \sin (\omega t + 120°)$$

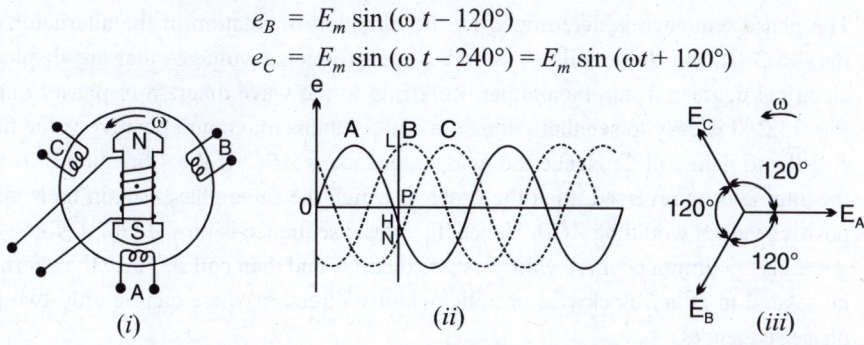

Fig. 15.2

It can be proved in many ways that the sum of the three e.m.f.s at every instant is zero.

(i)   Resultant $= e_A + e_B + e_C$

$$= E_m [\sin \omega t + \sin (\omega t - 120°) + \sin (\omega t - 240°)]$$
$$= E_m [\sin \omega t + 2 \sin (\omega t - 180°) \cos 60°]$$
$$= E_m [\sin \omega t - 2 \sin \omega t \cos 60°]$$
$$= 0$$

(ii) Referring to the wave diagram in Fig. 15.2 (*ii*), the sum of the three e.m.f.s at any instant is zero. For example, at the instant *P*, ordinate *PL* is positive while the ordinates *PN* and *PH* are negative. If you make actual measurements, it will be seen that :

$$PL + (-PN) + (-PH) = 0$$

(iii) Since the three windings or coils are identical, $E_A = E_B = E_C = E$ (in magnitude). As shown in Fig. 15.3, the resultant of E_A and E_B is E_r and its magnitude is $= 2E \cos 60° = E$. This resultant is equal and opposite to E_C.

Hence the resultant of the three e.m.f.s is zero.

(iv) Using complex algebra, we can again prove that the sum of the three e.m.f.s is zero. Thus taking $\mathbf{E_A}$ as the reference phasor, we have,

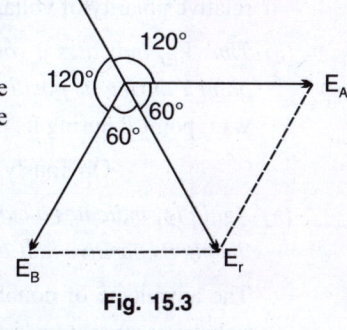

Fig. 15.3

$$\mathbf{E_A} = E \angle 0° \text{ V} = (E + j\,0)\text{V}$$
$$\mathbf{E_B} = E \angle -120° \text{ V}$$
$$= E (-0.5 - j\,0.866) \text{ V}$$
$$\mathbf{E_C} = E \angle -240° \text{ V}$$
$$= E (-0.5 + j\,0.866) \text{ V}$$

∴ $\mathbf{E_A} + \mathbf{E_B} + \mathbf{E_C} = (E + j\,0) + E (-0.5 - j\,0.866) + E (-0.5 + j\,0.866) = 0$

The reader may wonder how we can get power from the terminals of a 3-phase alternator if the sum of the voltages it delivers is zero at every instant ? This will be explained when we come to the connections of the three phases or windings.

15.4. Some Concepts in 3-Phase System

In the analysis of 3-phase system, we often come across the following terms :

    (i) Phase sequence. *The order in which the voltages in the three phases (or coils) of an alternator reach their *maximum positive values is called* **phase sequence** *or* **phase order.**

---

*   Instead of the positive maximum value, any other instantaneous value can be used to determine the phase sequence.

The phase sequence is determined by the direction of rotation of the alternator. Thus in Fig. 15.2 (*i*), the three coils *A*, *B* and *C* are producing voltages that are displaced 120 electrical degrees from one another. Referring to the wave diagram or phasor diagram in Fig. 15.2, it is easy to see that voltage in coil *A* attains maximum positive value first, next coil *B* and then coil *C*. Hence the phase sequence is *ABC*. If the *direction of rotation of the alternator is reversed, then the order in which the three phases attain their maximum positive values would be *ACB*. Hence, the phase sequence is now *ACB i.e.* voltage in coil *A* attains maximum positive value first, next coil *C* and then coil *B*. Since the alternator can be rotated in either clockwise or anticlockwise direction, there can be only two possible phase sequences.

Phase sequence is important in certain applications. For example, in 3-phase induction motors, the phase sequence of the supply determines whether the motor turns clockwise or anticlockwise.

(*ii*) **Naming the phases.** The three phases or windings may be numbered (1, 2, 3) or lettered (*A, B, C*). *However, it is a usual practice to name the three phases or windings after the three natural colours viz. Red (R), yellow (Y) and blue (B).* In that case, the phase sequence is *RYB i.e.* voltage in phase *R* attains maximum positive value first, next phase *Y* and then phase *B*. It may be noted that there are only two possible phase sequences *viz RYB* and *RBY*. *By convention, sequence RYB is taken as positive and RBY as negative.* Throughout this book, the phase sequence considered is *RYB* unless stated otherwise.

(*iii*) **Double-subscript notation.** The double-subscript notation is a very useful concept and may be found advantageous in the analysis of 3-phase system. In this notation, two letters are placed at the foot of the symbol for voltage or current. The two letters indicate the two points between which voltage (or current) exists and the order of the letters indicates the relative polarity of voltage (or current) during its **positive half-cycle.

(*a*) *Thus V_{RY} indicates a voltage V between points R and Y with point R being positive w.r.t. point Y during its positive half-cycle.* On the other hand, V_{YR} means that point *Y* is positive w.r.t. point *R* during its positive half-cycle.

$$\text{Obviously, } V_{RY} = - V_{YR}$$

(*b*) *Again I_{RY} indicates a current I between points R and Y and that its direction is from R to Y during its positive half-cycle.*

The advantage of double-subscript notation lies in the fact that a formal description of voltage or current under consideration is not necessary ; the subscripts and the order of the subscripts describe the quantity completely.

15.5. Interconnection of Three Phases

In a 3-phase alternator, there are three windings or phases. Each phase has two terminals *viz.* start and finish. If a separate load is connected across each winding as shown in Fig. 15.4, 6 conductors are required to transmit power. This will make the whole system complicated and expensive. In practice, the three windings are interconnected to give rise to two methods of connections *viz.*

(*i*) Star or Wye (*Y*) connection   (*ii*) Mesh or Delta (Δ) connection

---

*   Phase sequence can also be reversed by changing any two of the three lines (not a line and neutral wire). The reader will soon study that three phases (or coils) are suitably connected to obtain three lines.

**   One may consider negative half-cycle. This criterion must be applied to all the three phases.

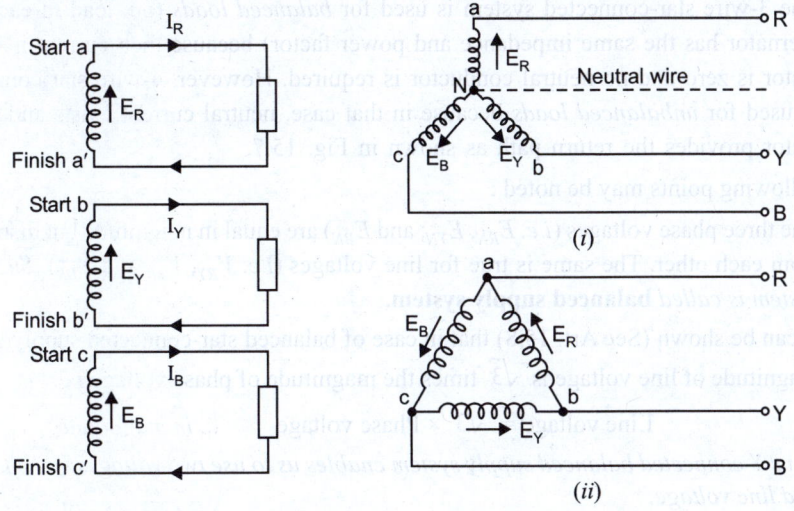

Fig. 15.4   Fig. 15.5

(*i*) **In Y-connection,** similar ends (start or finish) of the three phases are joined together within the alternator and three lines are run from the other free ends as shown in Fig. 15.5 (*i*). The common point N is called neutral point. In *Y*-connection, neutral conductor (shown dotted) may or may not be brought out. If a neutral conductor exists, the system is called *3-phase, 4-wire system.* If there is no neutral conductor, it is called *3-phase, 3-wire system.*

(*ii*) **In Δ-connection,** dissimilar ends (start to finish) of the phases are joined to form a closed mesh and the three lines are run from the junction points as shown in Fig. 15.5 (*ii*). *In a Δ-connection, no neutral point exists and only 3-phase, 3-wire system can be formed.*

The *Y* or delta connection serves substantially all the functions of three separate single phase circuits but with one important advantage that the number of conductors required is reduced. This results in the saving of conductor material and hence leads to economy.

15.6. Star or Wye Connected System

In this method, *similar ends* (start or finish) of the three phases of the alternator are joined together to form a common junction N as shown in Fig. 15.6. The common junction N is called the ***star point** or **neutral point.**

The three line conductors are run from the three ends (finish ends F in this case) and are designated as R, Y and B. This constitutes a **3-phase, 3-wire star-connected system.** The voltage between any line and the neutral point (*i.e.,* voltage across each winding) is called the **phase voltage** while the voltage between any two lines is called the **line voltage.** The currents flowing in the phases are called the **phase currents** while those flowing in the lines are called the **line currents.** Note that the phase sequence is RYB. Sometimes, a 4th wire, called **neutral wire,** is run from the neutral point as shown in Fig. 15.7. This gives **3-phase, 4-wire star-connected system.**

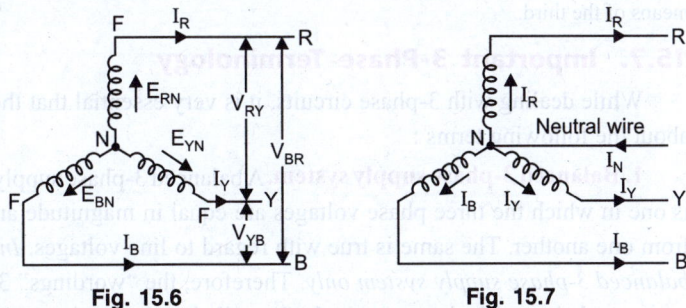

Fig. 15.6   Fig. 15.7

---

*   The reader may note that the figure looks like a star or inverted *Y*. Hence, the name. But it is a misnomer. It is because star-connected windings can be represented diagramatically in a manner which does not look like a star or inverted *Y*.

The 3-wire star-connected system is used for *balanced loads* (*i.e.* load in each phase of the alternator has the same impedance and power factor) because then current in the neutral conductor is zero and no neutral conductor is required. However, 4-wire star-connected system is used for *unbalanced loads* because in that case, neutral current exists and the neutral conductor provides the return path as shown in Fig. 15.7.

The following points may be noted :

(*i*) The three phase voltages (*i.e.* E_{RN}, E_{YN}, and E_{BN}) are equal in magnitude but displaced 120° from each other. The same is true for line voltages (*i.e.* V_{RY}, V_{YB} and V_{BR}). *Such a supply system is called* **balanced supply system.**

(*ii*) It can be shown (See Art. 15.8) that in case of balanced star-connected supply system, the magnitude of line voltage is $\sqrt{3}$ times the magnitude of phase voltage *i.e.*

$$\text{Line voltage} = \sqrt{3} \times \text{Phase voltage} \qquad \text{... in magnitude}$$

Thus Y-connected balanced supply system enables us to use two voltages viz. phase voltage and line voltage.

(*iii*) In star connection, the lines are in series with their respective phases. Therefore, magnitude of line current is equal to the magnitude of phase current *i.e.*

$$\text{Line current} = \text{Phase current} \qquad \text{... in magnitude}$$

(*iv*) For balanced loads, all line currents (or phase currents) are equal in magnitude but displaced 120° from each other.

(*v*) For 3-phase, 4-wire star-connected system, the current I_N in the neutral wire is the phasor sum of the three line currents. For a balanced load, $I_N = 0$. If the load is not balanced, the neutral wire will carry current equal to the phasor sum of the three line currents.

Note. The arrowheads alongside currents (or voltages) indicate their directions when they are assumed to be *positive* and not their actual directions at a particular instant. *At no instant will all the three line currents flow in the same direction either outwards or inwards.* This is expected because the three line currents are displaced 120° from one another. When one is positive, the other two might both be negative or one positive and one negative. Thus, at any one instant, current flows from the alternator through one of the lines to the load and returns through the other two lines. Or else current flows from the alternator through two of the lines and returns by means of the third.

15.7. Important 3-Phase Terminology

While dealing with 3-phase circuits, it is very essential that the reader has clear understanding about the following terms :

1. Balanced 3-phase supply system. A balanced 3-phase supply system (star or delta connected) is one in which the three phase voltages are equal in magnitude and frequency but displaced 120° from one another. The same is true with regard to line voltages. *In this chapter, we shall deal with balanced 3-phase supply system only.* Therefore, the "wordings" 3-phase supply, 3-phase voltages *etc.* mean balanced 3-phase supply. In fact, all electric supply companies make efforts to ensure the availability of 3-phase balanced supply at all times.

2. Types of 3-phase loads. There are two types of 3-phase loads *viz* (*i*) **star-connected load** (*ii*) **delta-connected load** as shown in Fig. 15.8. The 3-phase load (star or delta connected) is said to be **balanced** if load in each phase is the same (*i.e.* load in each phase has the same impedance and power factor). If a 3-phase load (star or delta connected) does not meet any one of the above requirements, it is said to be **unbalanced.**

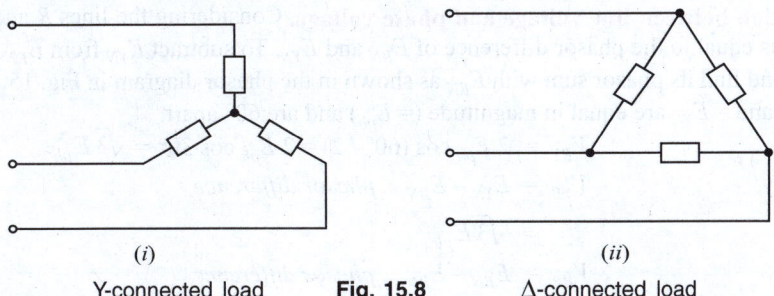

Y-connected load   **Fig. 15.8**   Δ-connected load

A 3-phase star-connected alternator may be connected to a star-connected or delta-connected load. Similarly, a 3-phase delta-connected alternator may be connected to a star-connected or delta-connected load.

3. Balanced 3-phase system. A balanced 3-phase system is one which meets the following requirements :

    (*i*) It should have balanced 3-phase supply.

    (*ii*) It should have balanced 3-phase or single phase loads on the balanced 3-phase supply.

    (*iii*) It should have equal active power and equal reactive power flow in each phase.

Note. The great majority of problems facing the electric supply companies are to ensure balanced 3-phase system. Fortunately, most of the industrial loads are 3-phase loads (*i.e.* 3-phase motors) that are inherently balanced. However, problems arise while suppling single-phase loads from 3-phase supply. Here, definite efforts are made to keep the 3-phase system balanced by putting approximately equal single-phase loads on each of the three phases.

15.8. Voltages and Currents in Balanced Y-Connected Supply System

Fig. 15.9 shows a balanced 3-phase, Y-connected supply system in which the r.m.s. values of the e.m.f.s generated in the three phases are E_{RN}, E_{YN} and E_{BN}. Since the supply system is balanced, these e.m.f.s will be equal in magnitude (say each equal to E_{ph}, the phase voltage) but displaced 120° from one another as shown in the phasor diagram in Fig. 15.10. It is clear from the circuit diagram (See Fig. 15.9) that p.d. between any two line terminals (*i.e.* line voltage) is the phasor difference between the potentials of these two terminals w.r.t. the neutral *i.e.*

P.D. between lines R and Y, $*V_{RY} = E_{RN} - E_{YN}$ ... *phasor difference*

P.D. between lines Y and B, $V_{YB} = E_{YN} - E_{BN}$ ... --do--

P.D. between lines B and R, $V_{BR} = E_{BN} - E_{RN}$ ... --do--

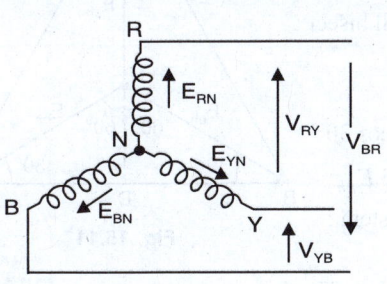

Fig. 15.9

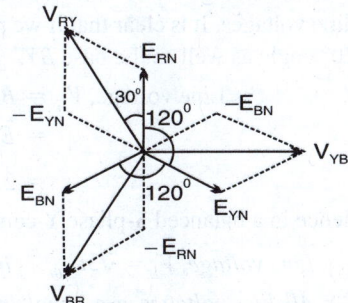

Fig. 15.10

---

   *

$$V_{RY} = E_{RN} + E_{NY} \quad \text{...phasor sum}$$
$$= E_{RN} - E_{YN} \quad \text{...phasor difference}$$

The reader may note the utility of double-subscript notation. The voltage between lines R and Y can be found without reference to the circuit diagram. Thus, between lines R and Y, there is a point N (*i.e.* neutral point) of our interest. The voltage V_{RY} is the phasor sum of voltages R to N and N to Y.

1. Relation between line voltage and phase voltage. Considering the lines R and Y, the line voltage V_{RY} is equal to the phasor difference of E_{RN} and E_{YN}. To subtract E_{YN} from E_{RN}, reverse the phasor E_{YN} and find its phasor sum with E_{RN} as shown in the phasor diagram in Fig. 15.10. The two phasors E_{RN} and $-E_{YN}$ are equal in magnitude ($= E_{ph}$) and are 60° apart.

$$\therefore \qquad V_{RY} = 2\,E_{ph}\cos(60°/2) = 2\,E_{ph}\cos 30° = \sqrt{3}\,E_{ph}$$

Similarly,
$$V_{YB} = E_{YN} - E_{BN} \quad ...phasor\ difference$$
$$= \sqrt{3}\,E_{ph}$$

and
$$V_{BR} = E_{BN} - E_{RN} \quad ...phasor\ difference$$
$$= \sqrt{3}\,E_{ph}$$

By phasor algebra. The above relation can be easily established by phasor algebra. Suppose the magnitude of each phase voltage is E_{ph}. Then,

$$E_{RN} = E_{ph}\angle 0° = E_{ph}(1+j\,0)$$
$$E_{YN} = E_{ph}\angle -120° = E_{ph}(-0.5 - j\,0.866)$$
$$E_{BN} = E_{ph}\angle -240° = E_{ph}(-0.5 + j\,0.866)$$

Now,
$$V_{RY} = E_{RN} - E_{YN} = E_{ph}(1+j\,0) - E_{ph}(-0.5 - j\,0.866)$$
$$= E_{ph}(1.5 + j\,0.866) = \sqrt{3}\,E_{ph}\angle 30°$$

$$\therefore \qquad V_{RY} = \sqrt{3}E_{ph}\angle 30°$$

Therefore, magnitude of V_{RY} is
$$V_{RY} = \sqrt{3}\,E_{ph}$$

Similarly,
$$V_{YB} = \sqrt{3}\,E_{ph} \text{ and } V_{BR} = \sqrt{3}\,E_{ph}$$

Note that line voltages are 30° ahead of their respective phase voltages.

By geometrical method. We have used graphical method and phasor algebra to prove the relation: $V_L = \sqrt{3}\,E_{ph}$. The same result can also be obtained by geometrical method as shown in Fig. 15.11. The distance of the neutral N to any line R, Y or B represents the phase voltage E_{ph}. The triangle RYB will be an equilateral triangle with N at the centre of the triangle. The sides of this triangle represent the three line voltages. It is clear that if we produce RN, it will bisect the 120° angle as well as the base BY.

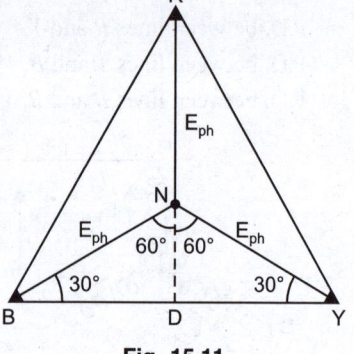

Fig. 15.11

$$\therefore \qquad \text{Line voltage, } V_L = BD + DY$$
$$= E_{ph}\cos 30° + E_{ph}\cos 30°$$
$$= 2\,E_{ph}\cos 30° = \sqrt{3}\,E_{ph}$$

Hence in a balanced 3-phase Y-connected supply system :

(i) *Line voltage, $V_L = \sqrt{3}E_{ph}$ ... in magnitude*

(ii) *All line voltages are equal in magnitude (i.e. $= \sqrt{3}E_{ph}$) but displaced 120° from one another (See phasor diagram in Fig. 15.10).*

(iii) *Line voltages are 30° ahead of their respective phase voltages.*

2. Relation between line current and phase current. In Y-connected supply system, each line conductor is connected in series to a separate phase as shown in Fig. 15.12. Therefore, current in a line conductor is the same as that in the phase to which the line conductor is connected.

Line current, $I_L = I_{ph}$ ...in magnitude

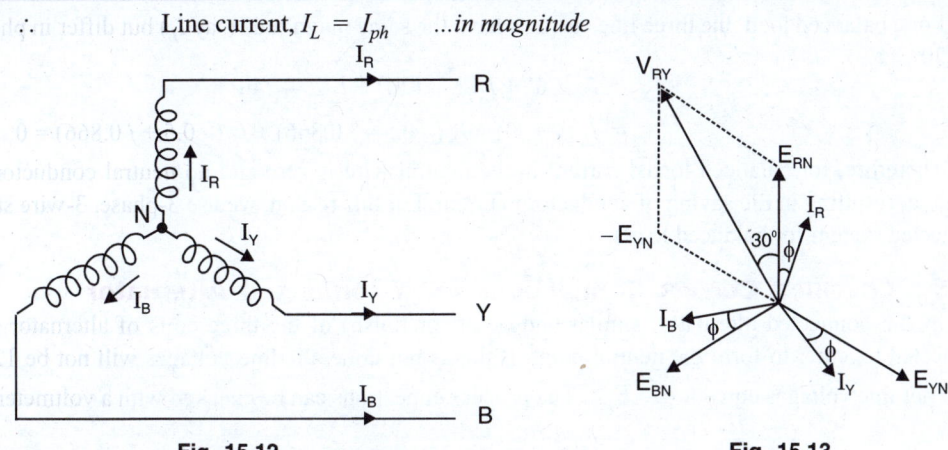

Fig. 15.12   Fig. 15.13

For a balanced load, all the phase currents are equal in magnitude but displaced 120° from one another. Fig. 15.13 shows the phasor diagram for a balanced lagging load ; the phase angle being ϕ. *Note that ϕ is the angle between phase voltage and the corresponding phase current.* However, the angle between the line current and the corresponding line voltage (say I_R and V_{RY}) is 30° + ϕ. If the balanced load has a leading power factor of cos ϕ, then the angle between the line current and the corresponding line voltage will be 30° – ϕ.

Hence in a balanced 3-phase Y-connected supply system :

(*i*) *Line current, $I_L = I_{ph}$*

(*ii*) *All the line currents are equal in magnitude (i.e. $= I_{ph}$) but displaced 120° from one another.*

(*iii*) *The angle between the line currents and the corresponding line voltages is 30° $\pm \phi$; + if p.f. is lagging and – if it is leading.*

3. Power. The total power in the circuit is quite logically, the sum of powers in the three phases. For a balanced load, the power consumed in each load phase is the same.

∴   Total power, $P = 3 \times$ Power in each phase $= 3 \times V_{ph} I_{ph} \cos \phi$ ...(*i*)

For a star connection, $V_{ph} = V_L / \sqrt{3}$; $I_{ph} = I_L$

∴ $P = 3 \times \dfrac{V_L}{\sqrt{3}} \times I_L \cos \phi$

or $P = \sqrt{3}\, V_L I_L \cos \phi$ ...(*ii*)

Also,   Reactive power, $Q = \sqrt{3} V_L I_L \sin \phi$

Either of relations (*i*) and (*ii*) can be used to determine the power. *It may be noted that ϕ is the phase difference between a phase voltage and the corresponding phase current and not between the line current and corresponding line voltage.*

Apparent power, $S = \sqrt{3}\, V_L I_L$

The relationship between active power (P), reactive power (Q) and apparent power (S) is the same for balanced 3-phase circuits as for single-phase circuits.

∴ $S = \sqrt{P^2 + Q^2}$ and power factor cos $\phi = \dfrac{P}{S}$

Note. The current I_N in the neutral wire in 4-wire, star-connected system will be the phasor sum of currents in the three lines *i.e.*

$$I_N = I_R + I_Y + I_B$$

For a balanced load, the three line currents have the same magnitude (say I_L) but differ in phase by 120°.

$$\therefore \qquad I_N = I_L \angle 0° + I_L \angle -120° + I_L \angle -240°$$

$$= I_L(1 + j0) + I_L(-0.5 - j\,0.866) + I_L(-0.5 + j\,0.866) = 0$$

Therefore, for balanced loads, current in the neutral wire is zero and no neutral conductor is required, resulting in the saving of conductor material. For this reason, we use 3-phase, 3-wire star-connected system for balanced loads.

15.9. Checking Correct Connections for Y-connected Alternator

In star-connected alternator, similar ends (start or finish) of the three coils of alternator are connected together to form the neutral point. If this is not done, the line voltages will not be 120° apart nor line voltages equal to $\sqrt{3}\,E_{ph}$. The proper connections can be checked with a voltmeter as under :

 (*i*) Connect two coils in series and measure the voltage across the free ends. This voltage should be $\sqrt{3}\,E_{ph}$ where E_{ph} is the voltage of each coil. If it is less, reverse the connections of one coil.

 (*ii*) Now connect one end of the third coil to the common junction (*i.e.*, neutral point). Again the voltage from the free end of this coil to the free end of each of the other two coils should be $\sqrt{3}\,E_{ph}$. If this value is not obtained, reverse the third coil connections.

Example 15.1. *Phase voltages of a star-connected alternator are $E_R = 231 \angle 0°$ V; $E_Y = 231 \angle -120°$ V and $E_B = 231 \angle +120°$ V. What is the phase sequence of the system? Compute the line voltages E_{RY} and E_{YB}.*

Solution. The phase voltage $E_B = 231 \angle +120°$ V can be written as $E_B = 231 \angle -240°$ V. Therefore, the three phase voltages are :

$$E_R = 231 \angle 0° \text{ V}\,;\, E_Y = 231 \angle -120° \text{ V}\,;\, E_B = 231 \angle -240° \text{ V}$$

It is clear that E_R is the reference voltage. Now E_Y lags behind E_R by 120° while E_B lags behind E_R by 240°. Therefore, the phase sequence is **RYB**. Further, the 3-phase system is balanced so that the magnitudes of line voltages E_{RY} and E_{YB} are :

$$E_{RY} = E_{YB} = \sqrt{3} \times 231 = \textbf{400 V}$$

Example 15.2. *Three-phase star-connected load when supplied from 400 V, 50 Hz source takes a line current of 10 A at 66.86° w.r.t. its line voltage. Calculate (i) impedance parameters (ii) power factor and active power consumed.*

Solution. Fig. 15.14 shows the phasor diagram of the circuit for R-phase ; the conditions in the other two phases are similar. We know that line voltages are 30° ahead of their respective phase voltages. This fact is illustrated in Fig. 15.14. It is clear that phase angle between phase voltage and phase current is $\phi = 66.86° - 30° = 36.86°$.

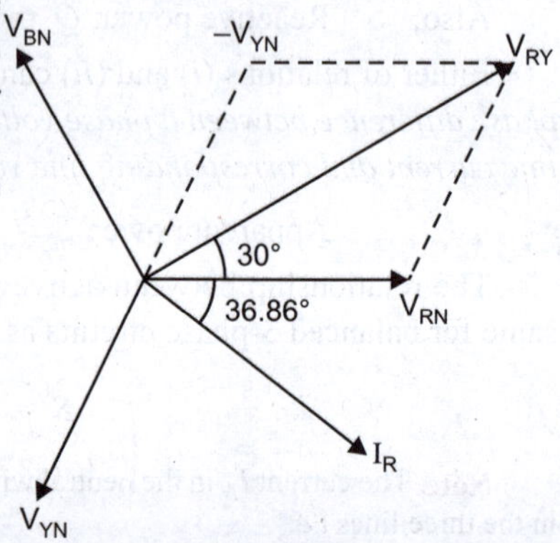

(*i*) $$V_{ph} = \frac{V_L}{\sqrt{3}} = \frac{400}{\sqrt{3}} = 231 \text{ V}$$

$$I_L = I_{ph} = 10 \text{ A}$$

$$Z_{ph} = \frac{V_{ph}}{I_{ph}} = \frac{231}{10} = 23.1 \ \Omega$$

Fig. 15.14

$$\therefore \qquad R_{ph} = Z_{ph}\cos\phi = 23.1 \times \cos 36.86° = \textbf{18.48 } \Omega$$
$$X_{Lph} = Z_{ph}\sin\phi = 23.1 \times \sin 36.86° = \textbf{13.86 } \Omega$$

(ii) Power factor $= \cos\phi = \cos 36.86° = \textbf{0.8 lagging}$

Alternatively. p.f. $= \dfrac{R_{ph}}{Z_{ph}} = \dfrac{18.48}{23.1} = \textbf{0.8 lagging}$

Active power consumed, $P = \sqrt{3}V_L I_L \cos\phi = \sqrt{3} \times 400 \times 10 \times 0.8 = \textbf{5544 W}$

Example 15.3. *Three coils, each having a resistance of 20 Ω and an inductive reactance of 15 Ω, are connected in star to a 400 V, 3-phase, 50 Hz supply. Calculate (i) the line current (ii) power factor and (iii) power supplied.*

Solution. Fig. 15.15 shows the circuit diagram whereas Fig. 15.16 shows its phasor diagram.

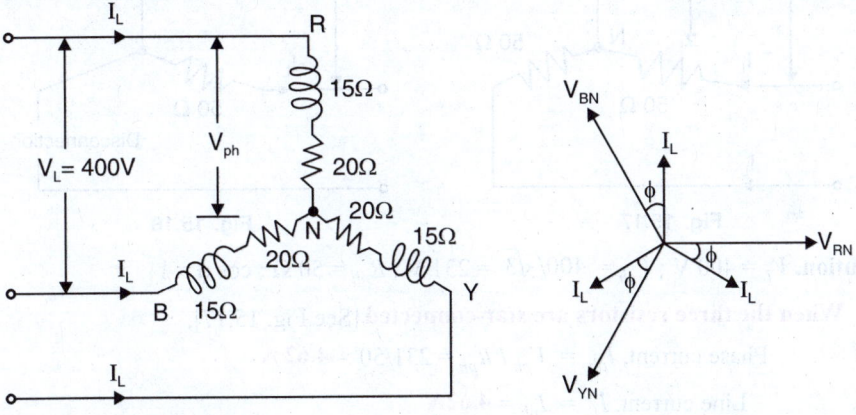

| **Fig. 15.15** | **Fig. 15.16** |

(i) Phase voltage, $*V_{ph} = V_L/\sqrt{3} = 400/\sqrt{3} = 231$ V

Impedance / phase, $Z_{ph} = \sqrt{20^2 + 15^2} = 25\ \Omega$

Phase current, $I_{ph} = V_{ph}/Z_{ph} = 231/25 = 9.24$ A

$\therefore$ Line current, $I_L = I_{ph} = \textbf{9.24 A}$

(ii) Power factor, $\cos\phi = \dfrac{\text{Resistance/phase}}{\text{Impedance/phase}} = 20/25 = \textbf{0.8 } \textit{lag}$

(iii) Power supplied, $P = \sqrt{3}\ V_L I_L \cos\phi = \sqrt{3} \times 400 \times 9.24 \times 0.8 = \textbf{5121 W}$

Alternatively. $P = 3 I^2_{ph} R_{ph} = 3 \times (9.24)^2 \times 20 = \textbf{5121 W}$

Example 15.4. *Calculate the active and reactive components of current in each phase of a star-connected 10,000 volts, 3-phase generator supplying 5,000 kW at a lagging p.f. of 0.8. Find the new output if the current is maintained at the same value but the p.f. is raised to 0.9 lagging.*

Solution. Power supplied, $P = \sqrt{3}\ V_L I_L \cos\phi$

or Line current, $I_L = \dfrac{P}{\sqrt{3}\ V_L \cos\phi} = \dfrac{5{,}000 \times 10^3}{\sqrt{3} \times 10{,}000 \times 0.8} = 360.8$ A

Phase current, $I_{ph} = I_L = 360.8$ A

Active component $= I_{ph}\cos\phi = 360.8 \times 0.8 = \textbf{288.64 A}$

Reactive component $= I_{ph}\sin\phi = 360.8 \times 0.6 = \textbf{216.48 A}$

---

* Note that phase voltage may be represented by V_{ph} or E_{ph}.

Power output when p.f. is raised to 0.9 lagging

$$= \sqrt{3}\ V_L I_L \cos \phi = \sqrt{3} \times 10{,}000 \times 360.8 \times 0.9$$
$$= 5624 \times 10^3 \text{ W} = \textbf{5624 kW}$$

Example 15.5. *Three 50 ohm resistors are connected in star across 400 V, 3-phase supply.*

(i)   Find phase current, line current and power taken from the mains.

(ii) What would be the above values if one of the resistors were disconnected ?

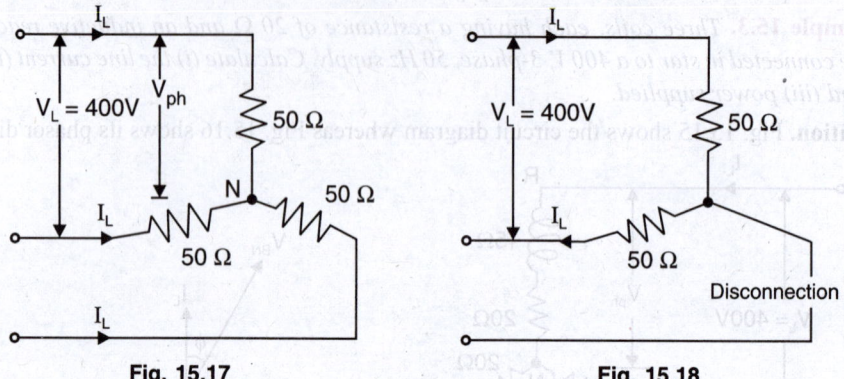

Fig. 15.17   **Fig. 15.18**

Solution. $V_L = 400$ V ; $V_{ph} = 400/\sqrt{3} = 231$ V ; $R_{ph} = 50\ \Omega$; $\cos \phi = 1$

(i)   When the three resistors are star-connected [See Fig. 15.17].

Phase current, $I_{ph} = V_{ph}/R_{ph} = 231/50 = \textbf{4.62 A}$

Line current, $I_L = I_{ph} = \textbf{4.62A}$

Power taken, $P = \sqrt{3}\ V_L I_L \cos \phi = \sqrt{3} \times 400 \times 4.62 \times 1 = \textbf{3200 W}$

(ii) When one of the resistors is disconnected. When one of the resistors is disconnected [See Fig. 15.18], the remaining two resistors behave as if they were connected in series across the line voltage. In fact the circuit behaves as a single phase circuit.

$$\therefore \qquad I_{ph} = I_L = \frac{400}{50 + 50} = \textbf{4A}$$

Power taken, $P = V_L I_L \cos \phi = 400 \times 4 \times 1 = \textbf{1600 W}$

Hence, by disconnecting one of the resistors, power consumption is reduced by half.

Example 15.6. *Three similar coils, connected in star, take a total power of 1.5 kW at a p.f. of 0.2 lagging from a 3-phase 400 V, 50 Hz supply. Calculate (i) the resistance and inductance of each coil and (ii) the line currents if one of the coils is short-circuited.*

Solution. Phase voltage, $V_{ph} = V_L/\sqrt{3} = 400/\sqrt{3} = 231$ V

(i)   When the three coils are star-connected [See Fig. 15.19].

Total power taken, $P = \sqrt{3}\ V_L I_L \cos \phi$

$$\therefore \qquad \text{Line current, } I_L = \frac{P}{\sqrt{3}\ V_L \cos \phi} = \frac{1500}{\sqrt{3} \times 400 \times 0.2} = 10.83 \text{ A}$$

Phase current, $I_{ph} = I_L = 10.83$ A

Impedance/phase, $Z_{ph} = V_{ph}/I_{ph} = 231/10.83 = 21.33\ \Omega$

Resistance/phase, $R_{ph} = Z_{ph} \cos \phi = 21.33 \times 0.2 = \textbf{4.266}\ \Omega$

Reactance/phase, $X_{ph} = \sqrt{(21.33)^2 - (4.266)^2} = 20.9\ \Omega$

Inductance/phase, $L_{ph} = \dfrac{X_{ph}}{2\pi f} = \dfrac{20.9}{2\pi \times 50} = \mathbf{0.0665\ H}$

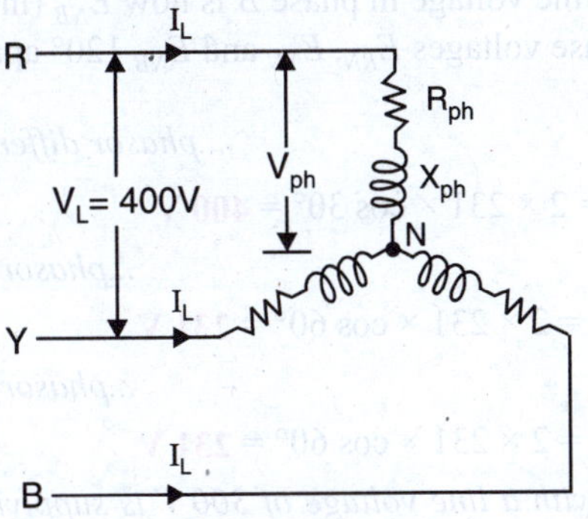

Fig. 15.19

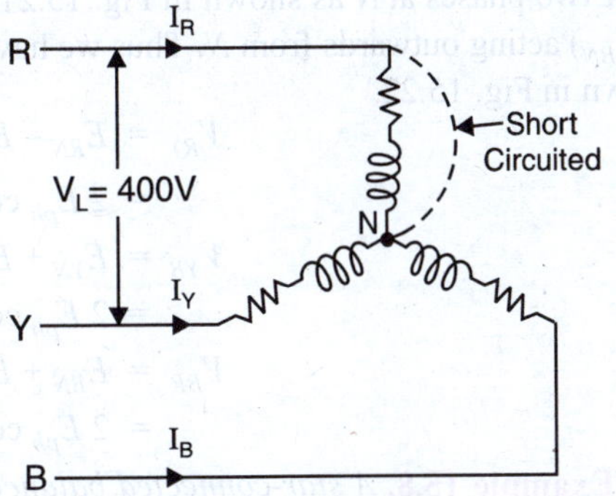

Fig. 15.20

(ii) When one of the coils is short-circuited. Suppose the phase load connected between terminals R and N is short-circuited as shown in Fig. 15.20. Clearly, the terminal N will be at the same potential as the terminal R. Therefore, the phase voltages V_{YN} and V_{BN} become *equal to the line voltages.

Current in phase Y, $I_Y = V_L/Z_{ph} = 400/21.33 = 18.75$ A

Current in phase B, $I_B = V_L/Z_{ph} = 400/21.33 = 18.75$ A

Therefore, line currents in each of the unfaulted sections (*i.e.*, lines Y and B) are **18.75 A.**

The magnitudes of the two phase currents are equal (being 18.75 A) and they are 60° apart. The current in phase R is equal to the phasor sum of the two.

∴　　Current in phase R, $I_R = 2\,I_{ph}\cos(60°/2) = 2 \times 18.75 \times \sqrt{3}/2 = \mathbf{32.47\ A}$

Example 15.7. *If the phase voltage of a three-phase star connected alternator be 231 V, what will be the line voltages (i) when the phases are correctly connected (ii) when the connections of one of the phases are reversed ?*

Solution. (i) When the phases are correctly connected. When the phases are connected correctly (*i.e.*, start or finish ends are joined together), the phasor diagram will be as shown in Fig. 15.10 (refer back). Note that the phase sequence is *RYB*. As proved in Art. 15.8,

$$V_{RY} = V_{YB} = V_{BR} = \sqrt{3}E_{ph}$$

Each line voltage, $V_L = \sqrt{3}E_{ph} = \sqrt{3} \times 231 = \mathbf{400\ V}$

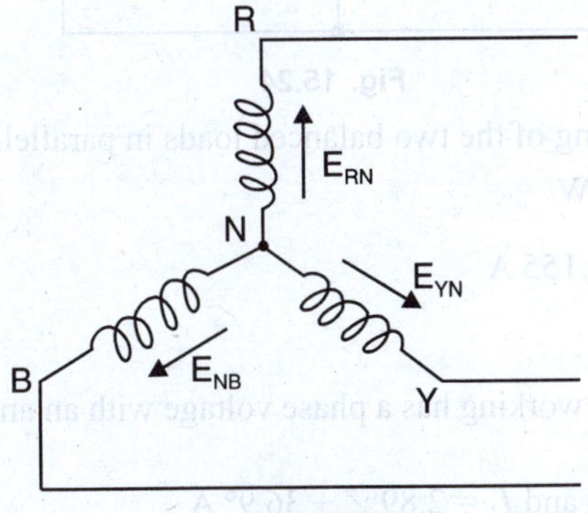

Fig. 15.21

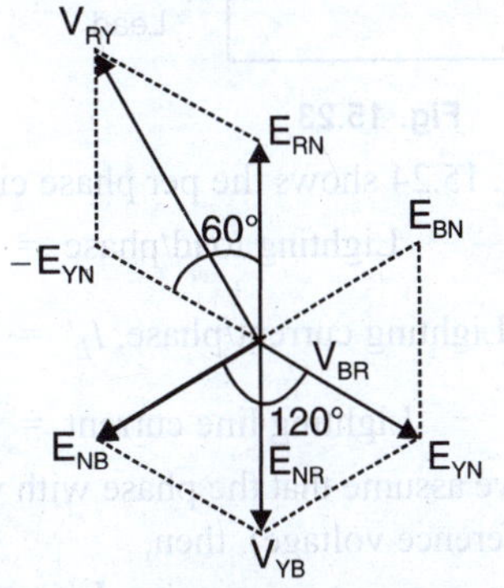

Fig. 15.22

---

*　Since the magnitudes and phase angles of the line voltages are fixed by the source, the voltages of unfaulted phases (*i.e.*, Y and B) of the load must rise until they become equal to the line voltages.

(ii) **When one phase connection is reversed.** Suppose connections to phase B have been reversed. It simply means that the finish end of phase B has been connected to the start ends of the other two phases at N as shown in Fig. 15.21. Note that the voltage in phase B is now E_{NB} (instead of E_{BN}) acting outwards from N. Thus we have three phase voltages E_{RN}, E_{YN} and E_{NB} 120° apart as shown in Fig. 15.22.

$$V_{RY} = E_{RN} - E_{YN} \qquad\qquad\qquad \textit{...phasor difference}$$
$$= 2\,E_{ph}\cos(60°/2) = 2 \times 231 \times \cos 30° = \textbf{400 V}$$
$$V_{YB} = E_{YN} + E_{NB} \qquad\qquad\qquad \textit{...phasor sum}$$
$$= 2\,E_{ph}\cos(120°/2) = 2 \times 231 \times \cos 60° = \textbf{231 V}$$
$$V_{BR} = E_{BN} + E_{NR} \qquad\qquad\qquad \textit{...phasor sum}$$
$$= 2\,E_{ph}\cos(120°/2) = 2 \times 231 \times \cos 60° = \textbf{231 V}$$

Example 15.8. *A star-connected balanced system with a line voltage of 300 V is supplying a balanced Y-connected load of 1200 W at a leading p.f. of 0·8. What is the line current and per phase impedance ? If a balanced 600 W lighting load is added in parallel, find the line current.*

Solution. For balanced load, we need to solve one phase only; the conditions in the other two phases being similar. Fig. 15.23 shows the per phase circuit of the balanced load.

$$\text{Phase voltage, } V_{ph} = 300/\sqrt{3} = 173.2\text{ V}$$
$$\text{Power per phase, } P_{ph} = 1200/3 = 400\text{ W}$$
$$\text{Now,} \qquad\qquad P_{ph} = V_{ph}I_{ph}\cos\phi$$
$$\text{or} \qquad\qquad I_{ph} = \frac{P_{ph}}{V_{ph}\cos\phi} = \frac{400}{173.2 \times 0.8} = 2.89\text{ A}$$
$$\therefore \qquad \text{Line current, } I_L = I_{ph} = \textbf{2.89 A}$$
$$\text{Load impedance/phase, } Z_{ph} = V_{ph}/I_{ph} = 173.2/2.89 = \textbf{59.93 }\Omega$$
$$\text{Impedance phase angle} = \cos^{-1} 0.8 = 36.9°\ (lead)$$

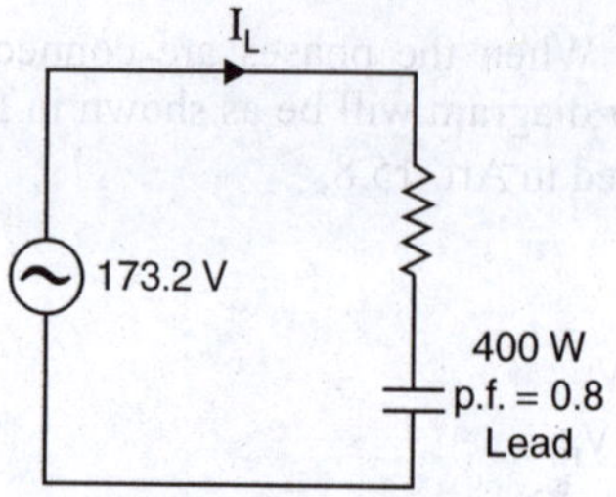

Fig. 15.23 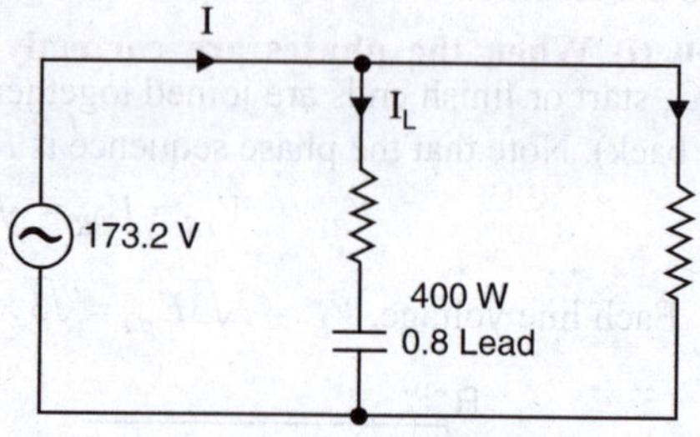   Fig. 15.24

Fig. 15.24 shows the per phase circuit consisting of the two balanced loads in parallel.

$$\text{Lighting load/phase} = 600/3 = 200\text{ W}$$
$$\text{Lighting current/phase, } I_L' = \frac{200}{173.2 \times 1} = 1.155\text{ A}$$
$$\therefore \qquad \text{Lighting line current} = I_L' = 1.155\text{ A}$$

If we assume that the phase with which we are working has a phase voltage with an angle of 0° *(i.e.,* reference voltage), then,

$$I_L' = 1.155 \angle 0°\text{ A and } I_L = 2.89 \angle +36.9°\text{ A}$$
$$\therefore \qquad \text{Total line current, } I = I_L' + I_L = 1.155 \angle 0° + 2.89 \angle 36.9°$$
$$= 1.155 + (2.312 + j\,1.734)$$

$$= (3.467 + j\,1.734)\ \text{A} = 3.87\ \angle\ 26.6°\ \text{A}$$

∴   Total line current, I = **3.87 A**

Check. We can easily check the correctness of the solution.

       Power/phase = $V_{ph} \times I \times \cos 26.6° = 173.2 \times 3.87 \times \cos 26.6° = 600\ \text{W}$

which tallies with our problem.

Example 15.9. *The load to a 3-phase supply consists of three similar coils connected in star. The line currents are 25 A and the kVA and kW inputs are 20 and 11 respectively. Find (i) the phase and line voltages (ii) the kVAR input (iii) resistance and reactance of each coil.*

Solution.

Apparent power, $S = 20\ \text{kVA} = 20000\ \text{VA}$; Active power, $P = 11000\ \text{W}$; $I_L = I_{ph} = 25\ \text{A}$

(*i*) $S = 3V_{ph}I_{ph}$ ∴ $V_{ph} = \dfrac{S}{3I_{ph}} = \dfrac{20{,}000}{3 \times 25} =$ **267 V**

       Line voltage, $V_L = \sqrt{3}\,V_{ph} = \sqrt{3} \times 267 =$ **462 V**

(*ii*)   Input kVAR $= \sqrt{S^2 - P^2} = \sqrt{(20)^2 - (11)^2} =$ **16.7 kVAR**

(*iii*)   Power factor, $\cos\phi = P/S = 11/20$

       Impedance/phase, $Z_{ph} = V_{ph}/I_{ph} = 267/25 = 10.68\ \Omega$

∴   Resistance/phase, $R_{ph} = Z_{ph}\cos\phi = 10.68 \times (11/20) =$ **5.87 Ω**

       Reactance/phase, $X_{ph} = \sqrt{Z_{ph}^2 - R_{ph}^2} = \sqrt{(10.68)^2 - (5.87)^2} =$ **8.92 Ω**

Example 15.10. *A balanced star-connected load of impedance $(6 + j\,8)\ \Omega$ per phase is connected to a 3-phase, 230 V, 50Hz supply. Using phasor algebra, find the line current and power absorbed by each phase.*

Solution. Phase voltage $= 230/\sqrt{3} = 133\ \text{V}$. Taking V_{RN} as the reference phasor, we have,

$$V_{RN} = 133\ \angle\ 0°\ \text{V}\ ;\ V_{YN} = 133\ \angle - 120°\ \text{V}\ ;\ V_{BN} = {}^*133\ \angle\ 120°\ \text{V}$$

Impedance/phase, $Z_{ph} = (6 + j\,8)\ \Omega = 10\ \angle\ 53.13°\ \Omega$

∴ $I_R = \dfrac{V_{RN}^{**}}{Z_{ph}} = \dfrac{133\ \angle 0°}{10\ \angle 53.13°} = 13.3\ \angle - 53.13°\ \text{A}$

This current lags behind V_{RN} by 53.13°.

$$I_Y = \dfrac{V_{YN}}{Z_{ph}} = \dfrac{133\ \angle -120°}{10\ \angle 53.13°} = 13.3\ \angle -173.13°\ \text{A}$$

This current lags behind V_{RN} by 173.13°.

$$I_B = \dfrac{V_{BN}}{Z_{ph}} = \dfrac{133\ \angle 120°}{10\ \angle 53.13°} = 13.3\ \angle\ 66.87°\ \text{A}$$

This current leads V_{RN} by 66.87°.

Note that the three phase currents are equal in magnitude (each being 13.3 A) and displaced 120° from one another. Since the load is star-connected, these currents also represent the line currents.

∴   Line current, I_L = **13.3 A**

---

* $V_{BN} = 133\ \angle - 240°\ \text{V} = 133\ \angle\ 120°\ \text{V}$

**   Note that phase voltage may be represented as E_{RN} or V_{RN}.

To find the power absorbed by each phase, consider the R-phase.

$$V_{RN} = 133 \angle 0° \text{ V} ; I_R = 13.3 \angle -53.13° \text{ A}$$

$\therefore$ Power absorbed/phase $= 133 \times 13.3 \times \cos(-53.13°) = $ **1061 W**

Example 15.11. *When the three identical star-connected coils are supplied with 440 V, 50 Hz, 3-phase supply, the $1-\phi$ wattmeter whose current coil is connected in the line R and the pressure coil across the phase R and neutral reads 6 kW and ammeter connected in R-phase reads 30 A. Assuming RYB phase, find (i) the resistance and reactance of each coil (ii) reactive power of 3–ϕ load (iii) power factor.*

Solution. $V_{ph} = 440/\sqrt{3} = 254$ V ; $I_{ph} = 30$ A

Now, Power/phase $= V_{ph} I_{ph} \cos\phi$ $\therefore \cos\phi = \dfrac{6 \times 10^3}{254 \times 30} = 0.787$

Impedance/phase, $Z_{ph} = V_{ph}/I_{ph} = 254/30 = 8.47 \ \Omega$

(i) Resistance of coil, $R = Z_{ph}\cos\phi = 8.47 \times 0.787 = $ **6.66 Ω**

Reactance of coil, $X_L = Z_{ph}\sin\phi = 8.47 \times 0.616 = $ **5.22 Ω**

(ii) Reactive power, $Q = \sqrt{3} V_L I_L \sin\phi = \sqrt{3} \times 440 \times 30 \times 0.616 = $ **14083 VAR**

(iii) Power factor, $\cos\phi = R/Z_{ph} = 6.66/8.47 = $ **0.787** *lag*

Example 15.12. *Each phase of a star-connected load consists of a non-reactive resistance of 100 Ω in parallel with a capacitance of 31.8 μF. Calculate the line current, the power absorbed, the total kVA and power factor when connected to a 416 V, 3-phase, 50 Hz supply.*

Solution. Fig. 15.25 shows the conditions of the problem. The phase voltage V_{ph} $= 416/\sqrt{3} = 240$ V. Let us take R-phase as the reference. Its phase voltage is given by ;

$$V_{ph} = 240 \angle 0° \text{ V} = (240 + j\,0) \text{ V}$$

Admittance/phase, $Y_{ph} = \dfrac{1}{R} + j\omega C$

Fig. 15.25

$$= \dfrac{1}{100} + j \times 314 \times 31.8 \times 10^{-6}$$

$$= (0.01 + j\,0.01)\text{S}$$

$\therefore$ Phase current, $I_{ph} = V_{ph} Y_{ph} = 240\,(0.01 + j\,0.01)$

$$= (2.4 + j\,2.4) \text{ A} = 3.39 \angle 45° \text{ A}$$

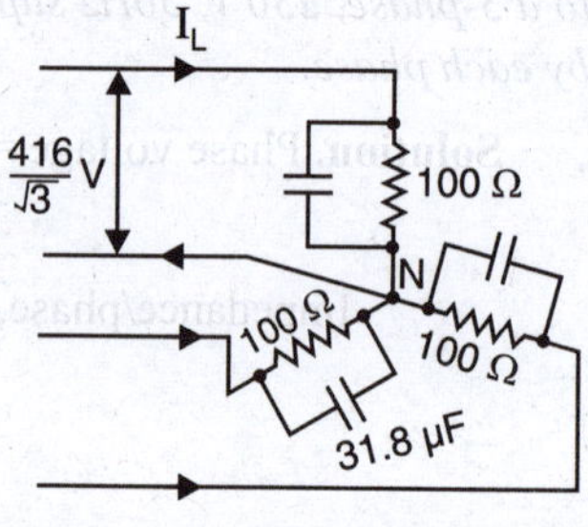

For star connection, $I_{ph} = I_L = $ **3.39 A**

Circuit power factor $= \cos\phi = \cos 45° = $ **0.707** *lead*

Now, $S = V_{ph} I_{ph} = (240 + j\,0)\,(2.4 + j\,2.4)$

$$= (576 + j\,576) \text{ VA} = 814.4 \angle 45° \text{ VA}$$

$\therefore$ Total active power $= 3 \times 576 = 1728$ W $= $ **1.728 kW**

Total apparent power $= 3 \times 814.4 = 2443$ VA $= $ **2.443 kVA**

Example 15.13. *Three similar coils, arranged symmetrically in space, are fed from a 400 V, 3-phase, 50 Hz supply. The coils are connected in star and each coil has a resistance of 100 Ω and inductive reactance of 250 Ω. The mutual reactance (ωM) between each pair of coils is 95 Ω. Find the current taken by each coil and its power factor.*

Solution. Fig.15.26 shows the conditions of the problem. Assuming the phase sequence to be RYB and taking V_{RY} as the reference phasor,

$$V_{RY} = 400 \angle 0° \text{ V} ; V_{YB} = 400 \angle -120° \text{ V} ;$$
$$V_{BR} = 400 \angle 120° \text{ V}$$

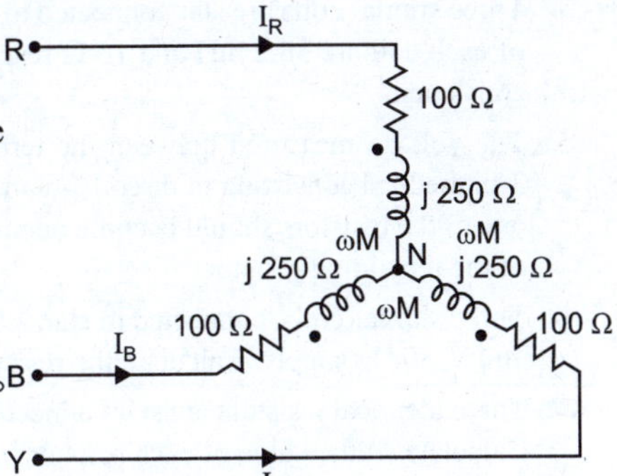

Since the system is balanced, the magnitude of all phase voltages is the same.

$$\therefore \quad V_{RN} = \frac{V_{RY}}{\sqrt{3}} = \frac{400 \angle 0°}{\sqrt{3}} = 231 \angle 0° \text{ V}$$

Voltage drop in R-phase $= V_{RN}$

or $I_R(100 + j\,250) + *I_Y(j\,95) + I_B(j\,95) = 231 \angle 0°$

or $I_R(100 + j\,250) + (I_Y + I_B)(j\,95) = 231 \angle 0°$

Since the system is balanced, we have,

$$I_R + I_Y + I_B = 0 \quad \therefore I_Y + I_B = -I_R$$

$$\therefore I_R(100 + j\,250) + (-I_R)(j\,95) = 231 \angle 0°$$

or $\qquad I_R(100 + j155) = 231 \angle 0°$

$$\therefore \qquad I_R = \frac{231 \angle 0°}{100 + j155} = 1.25 \angle -57.17° \text{A}$$

Fig. 15.26

i.e. magnitude of phase current, $I_R = 1.25$ A

$$\therefore \qquad I_R = I_Y = I_B = \mathbf{1.25\ A}$$

Power factor $= \cos \phi = \cos(-57.17°) = \mathbf{0.542}$ *lagging*

Example 15.14. *A 3-phase voltage is applied to a load consisting of two equal resistors R in series, phase B being connected to the junction. Find the ratio of the currents in the three lines and their relative phase position.*

Solution. Fig. 15.27 shows the conditions of the problem. Since the supply system is balanced, the three line voltages can be represented as :

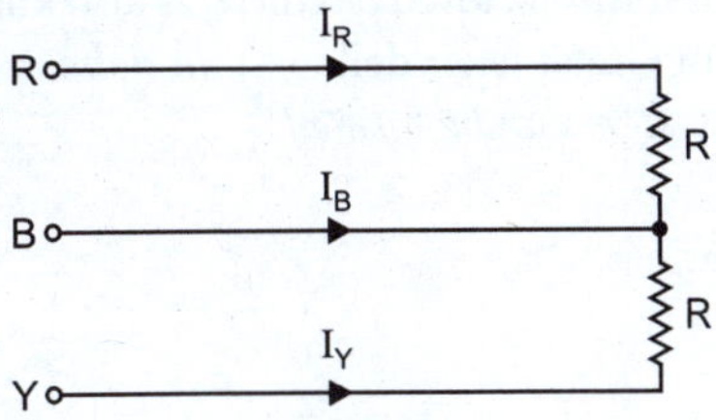

$$V_{RY} = V \angle 0° ; V_{YB} = V \angle -120° ; V_{BR} = V \angle 120°$$

$$I_R = \frac{V_{RB}}{R} = \frac{-V_{BR}}{R} = \frac{-V \angle 120°}{R} = \frac{V}{R} \angle -60°$$

$$I_Y = \frac{V_{YB}}{R} = \frac{V \angle -120°}{R} = \frac{V}{R} \angle -120°$$

Fig. 15.27

Applying *KCL* to the junction point of resistors, we have,

$$I_R + I_Y + I_B = 0$$

or $\qquad I_B = -(I_R + I_Y) = -\frac{V}{R}(\angle -60° + \angle -120°)$

$$= -\frac{V}{R}(0.5 - j\,0.866 - 0.5 - j\,0.866)$$

$$= -\sqrt{3}\frac{V}{R} \angle -90° = \sqrt{3}\frac{V}{R} \angle 90°$$

$\therefore$ Ratio of magnitudes of line currents is : $\mathbf{1 : 1 : \sqrt{3}}$

Also relative phases are $I_R : I_Y : I_B = \mathbf{-60° : -120° : 90°}$

(**Tutorial Problems**)

1. The phase voltage of a 3-phase, 12 MVA star-connected alternator is 6000 V. Calculate (*i*) the line voltage (*ii*) full load line current of the alternator. **[(*i*) 10392 V (*ii*) 666.67 A]**

---

* There is mutual reactance between phase R and phase Y as well as between R phase and B phase and it is equal to 95 Ω *i.e.* $\omega M = 95\ \Omega$.

2. Three similar coils are star connected to a 3-phase, 400 V, 50 Hz supply. If the inductance and resistance of each coil are 38.2 mH and 16 Ω respectively, determine (*i*) line current (*ii*) power factor (*iii*) power consumed. **[(i) 11.55 A (ii) 0.8 lag (iii) 6.4 kW]**

3. The voltage measured between the terminals of a 3-phase, 3-wire alternator is recorded as 208 V. A 3-phase load consisting of three 10-ohm resistors in star is connected to the terminals of the alternator. If one of the resistors should become open-circuited, what would be the current, the voltage and the power of the remaining resistors ? **[10.4 A ; 104 V ; 1081.6 W]**

4. Three similar coils, connected in star, take a total power of 3 kW at a p.f. of 0.8 lagging from a 3-phase, 400 V, 50 Hz supply. Calculate the resistance and reactance of each coil. **[33.92 Ω ; 25.68 Ω]**

5. Three identical resistors are star connected across a 440 V, 3-phase supply and a line current of 4 A flows. Calculate (*i*) the value of each resistor (*ii*) the power consumed. **[(i) 63.5 Ω (ii) 3048.4 W]**

6. The resistance between two terminals of a balanced star-connected load is measured as 12 Ω. What is the resistance of each phase ? **[6 Ω]**

7. Calculate the active and reactive components in each phase of a star-connected 5000 V, 3-phase alternator supplying 3000 kW at power factor 0.8. **[346A ; 260 A]**

8. Calculate the power absorbed by three similar choking coils each of resistance 150 Ω and inductance 1.275 H when star-connected to a 440 V, 3-phase, 25 Hz supply. **[463 W]**

15.10. Delta (Δ) or Mesh Connected System

In this method of interconnection, the *dissimilar ends* of the three phase windings of the alternator are joined together *i.e.*, finishing end of one phase is connected to the starting end of the other phase and so on to obtain mesh or delta as shown in Fig. 15.28. The three line conductors are taken from the three junctions of the mesh or delta and are designated as *R*, *Y* and *B*. This is called *3-phase, 3-wire, delta connected system*. The arrangement is referred to as mesh connection because it forms a closed circuit. It is also known as delta connection because the diagram has the appearance of Greek letter delta (Δ). *In delta connection, no neutral exists and, therefore, only 3-phase, 3-wire system can be formed.*

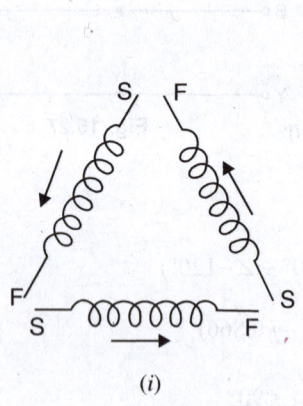

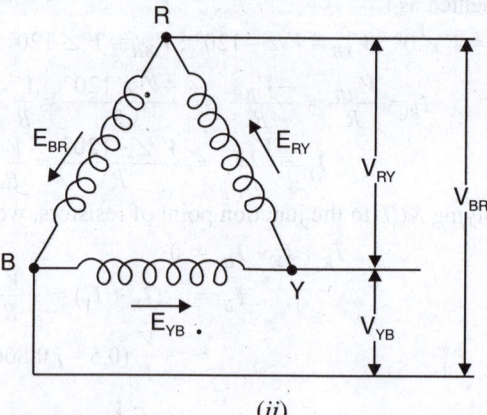

(*i*)　　　　　　　　　　　　　　　　　(*ii*)

Fig. 15.28

The following points may be noted :

(*i*) As can be seen from Fig. 15.28 (*ii*), only one phase is included between any two lines. *Hence magnitude of voltage between any two lines (i.e.* **line voltage***) is equal to the magnitude of* **phase voltage** *i.e.*

Line voltage magnitude, V_L = Phase voltage magnitude (E_{ph})

The three phase voltages (= line voltages) are equal in magnitude but displaced 120° from one another.

(*ii*) When 3-phase load (star or delta connected) is connected to the 3-phase Δ-supply, currents flow through the phases (called **phase currents**) as well as through the lines (called **line currents**). An examination of Fig. 15.28 (*ii*) shows that current in any line is equal to the *phasor difference* of the currents in the two phases connected to that line. Therefore, magnitude of line currents is different from the magnitude of phase currents.

(*iii*) For balanced load, the three phase currents (I_R, I_Y and I_B) are equal in magnitude but displaced 120° from one another. In such a case, it can be proved (See Art. 15.12) that :

$$\text{Line current} = \sqrt{3} \times \text{Phase current} \qquad ... \text{ in magnitude}$$

The three line currents will be equal in magnitude but displaced 120° from one another.

Note. At no instant will all the three line currents flow in the same direction either outwards or inwards. This is expected because the three line currents are displaced 120° from one another. When one is positive, the other two might both be negative or one positive and one negative. Thus at any one instant, current flows from the alternator through one of the lines to the load and returns through the other two lines. Or else current flows from the alternator through two of the lines and returns by means of the third. It may be noted that arrows placed alongside currents (or voltages) in the diagram indicate the directions of currents (or voltages) when they are assumed to be positive and not their actual directions at a particular instant.

15.11. Correct and Incorrect Δ Connections of Alternator

When the three phase windings of an alternator are connected into a delta, they must be "properly phased" *i.e.* finishing end of one phase winding should be connected to the starting end of the other phase winding and so on. Any wrong connection may develop high voltage in the windings. Let us illustrate this point.

(*i*) **Correct Δ connections.** We first consider the case when the phase windings of a Δ-connected alternator are connected correctly. Suppose the magnitude of each phase voltage is E_{ph}. Then,

$$E_{RY} = E_{ph} \angle 0° = E_{ph} (1 + j\,0)$$
$$E_{YB} = E_{ph} \angle -120° = E_{ph}(-0.5 - j\,0.866)$$
$$E_{BR} = E_{ph} \angle -240° = E_{ph}(-0.5 + j\,0.866)$$

The phasor sum of the voltages around the mesh is

$$E_{RY} + E_{YB} + E_{BR} = E_{ph} (1 + j\,0) + E_{ph}(-0.5 - j\,0.866) + E_{ph} (-0.5 + j\,0.866) = 0$$

Thus the phasor sum of three phase voltages is zero and there is no circulating current in the mesh. Therefore, line voltage is equal to the phase voltage.

(*ii*) **Incorrect Δ connections.** If the phase windings of a Δ-connected alternator are not connected properly, there will be circulating current in the mesh. This can be easily proved. Suppose connections to one of the phase windings, say *blue phase winding* are reversed. Taking E_{RY} as reference phasor, we have,

$$E_{RY} = E_{ph} \angle 0° = E_{ph} (1 + j\,0)$$
$$E_{YB} = E_{ph} \angle -120° = E_{ph}(-0.5 - j\,0.866)$$
$$E_{BR} = -E_{ph} \angle -240° = -E_{ph} (-0.5 + j\,0.866)$$
$$\therefore \quad E_{RY} + E_{YB} + E_{BR} = E_{ph} (1 + j\,0 - 0.5 - j\,0.866 + 0.5 - j\,0.866)$$
$$= E_{ph} (1 - j\,1.732) = 2\,E_{ph} \angle -60°$$

Thus the resultant loop voltage has a magnitude twice the value of phase voltage. This high voltage may cause large current to flow through the windings even without external load and may damage the windings. Therefore, when the coils of an alternator or secondary of a transformer bank are connected into a delta connection, a check must be made with a voltmeter before closing the delta.

15.12. Voltages and Currents in Balanced Δ Connected Supply System

Fig. 15.29 shows a balanced 3-phase Δ-connected supply system. It is desired to find the relation between (*i*) line voltage and phase voltage (*ii*) line current and phase current.

(*i*) **Line voltage and phase voltage.** Since the system is balanced, the three phase voltages are equal in magnitude (say each equal to V_{ph}, the phase voltage) but displaced 120° from one another. An examination of Fig. 15.29 shows that only one phase winding is included between any pair of lines. Hence, in Δ connection, the line voltage is equal to the phase voltage *i.e.*

$$V_L = V_{ph} \qquad \text{... in magnitude}$$

Since the phase sequence is *RYB, the line voltage V_{RY} is 120° ahead of V_{YB} and 240° ahead of V_{BR}. Incidentally, these are also the phase voltages.

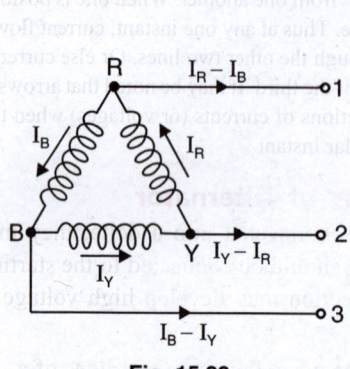

Fig. 15.29

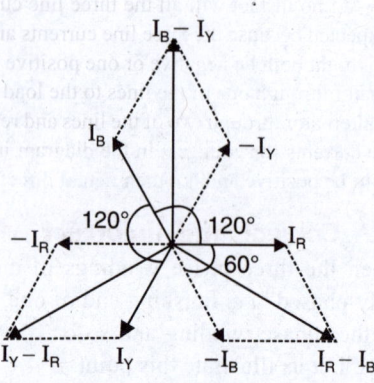

Fig. 15.30

(*ii*) **Line current and phase current.** Since the system is balanced, the three phase currents I_R, I_Y and I_B are equal in magnitude (say each equal to I_{ph}, the phase current) but displaced 120° from one another as shown in the phasor diagram in Fig. 15.30. An examination of the circuit diagram in Fig. 15.29 shows that current in any line is equal to the *phasor difference* of the currents in the two phases attached to that line. Thus :

$$\text{Current in line 1, } **I_1 = I_R - I_B \qquad \text{...phasor difference}$$
$$\text{Current in line 2, } I_2 = I_Y - I_R \qquad \text{...do...}$$
$$\text{Current in line 3, } I_3 = I_B - I_Y \qquad \text{...do...}$$

The current I_1 in line 1 is the phasor difference of I_R and I_B. To subtract I_B form I_R, reverse the phasor I_B and find its phasor sum with I_R as shown in Fig. 15.30. The two phasors I_R and $- I_B$ are equal in magnitude ($= I_{ph}$) and are 60° apart.

∴ $$I_1 = 2 I_{ph} \cos (60°/2) = 2 I_{ph} \cos 30° = \sqrt{3}\ I_{ph}$$

Similarly, $$I_2 = I_Y - I_R \qquad \text{...phasor difference}$$

$$= \sqrt{3}\ I_{ph}$$

and $$I_3 = I_B - I_Y \qquad \text{...phasor difference}$$

$$= \sqrt{3}\ I_{ph}$$

The three line currents I_1, I_2 and I_3 are equal in magnitude ; each being equal to $\sqrt{3}\ I_{ph}$. Therefore, in a Δ connected supply system :

---

*   We always assume the phase sequence to be *RYB* unless stated otherwise.

**   Consider the line current I_1 in line 1 connected to the common point R of red and blue phase windings. It is clear that I_1 is equal to phasor difference of I_R and I_B since 'positive' direction for I_R is towards point R and for I_B it is away from point R.

Line current, $I_L = \sqrt{3}\ I_{ph}$

By phasor algebra. The above relation can be easily established by phasor algebra. Suppose the magnitude of each phase current is I_{ph}. Then,

$$I_R = I_{ph} \angle\ 0° = I_{ph}\ (1 + j\ 0)$$

$$I_Y = I_{ph} \angle - 120° = I_{ph}\ (-0.5 - j\ 0.866)$$

$$I_B = I_{ph} \angle - 240° = I_{ph}\ (-0.5 + j\ 0.866)$$

Now, line current $I_1 = I_R - I_B = I_{ph}(1 + j\ 0) - I_{ph}(-0.5 + j\ 0.866)$

$$= I_{ph}\ (1.5 - j\ 0.866) = \sqrt{3}\ I_{ph} \angle - 30°$$

$$\therefore \qquad I_1 = \sqrt{3}\ I_{ph} \angle - 30°$$

Therefore, magnitude of I_1 is given by ;

$$I_1 = \sqrt{3}\ I_{ph}$$

Similarly, $I_2 = \sqrt{3}\ I_{ph}$ and $I_3 = \sqrt{3}\ I_{ph}$.

Note that line currents are 30° behind the respective phase currents.

Hence in a balanced Δ connected supply system :

(*a*) *Line current, $I_L = \sqrt{3}\ I_{ph}$*

(*b*) *All the line currents are equal in magnitude (= $\sqrt{3}\ I_{ph}$) but displaced 120° from one another as seen from Fig. 15.30.*

(*c*) *Line currents are 30° behind the respective phase currents.*

(*d*) *The angle between the line currents and the corresponding line voltages is 30° ± ϕ; + if p.f. is lagging and – if it is leading.*

(*e*) *Line voltage (V_L) is equal to phase voltage (V_{ph}).*

(*iii*) **Power.** Total power, $P = 3 \times$ Power per phase $= 3 \times V_{ph}\ I_{ph} \cos \phi$

For a delta connection, $V_{ph} = V_L\ ;\ I_{ph} = I_L/\sqrt{3}$

$$\therefore \qquad P = 3 \times V_L \times \frac{I_L}{\sqrt{3}} \times \cos \phi$$

or $$P = \sqrt{3}\ V_L\ I_L \cos \phi$$

where $\cos \phi$ is the power factor of each phase. Note that in either case, star or delta, the expression for the total power is the same provided that the system is balanced.

Reactive power, $Q = \sqrt{3}\ V_L I_L \sin \phi$

Apparent power, $S = \sqrt{3}V_L I_L$

The relationship between active power (*P*), reactive power (*Q*) and apparent power (*S*) is the same for balanced 3-phase circuits as for single-phase circuits.

$$\therefore \qquad S = \sqrt{P^2 + Q^2} \text{ where power factor } \cos \phi = \frac{P}{S}$$

15.13. Advantages of Star and Delta Connected Systems

In 3-phase system, the alternators may be star or delta connected. Likewise, 3-phase loads may be star or delta connected. It is desirable to list the advantages of star and delta connections.

Advantages of Star Connection

(i) In star connection, phase voltage $V_{ph} = V_L/\sqrt{3}$. Since the induced e.m.f. in the phase winding of an alternator is directly proportional to the number of turns, a star-connected alternator will require less number of turns than a Δ-connected alternator for the same line voltage.

(ii) For the same line voltage, a star-connected alternator requires less insulation than a delta-connected alternator.

Due to above two reasons, 3-phase alternators are generally star-connected. One can hardly find *delta-connected alternators.

(iii) In star connection, we can get 3-phase, 4-wire system. This permits to use two voltages viz, phase voltages as well as line voltages. Remember that in star connection, $V_L = \sqrt{3}\, E_{ph}$.

Single phase loads (e.g. lights etc) can be connected between any one line and the neutral wire while the 3-phase loads (e.g. 3-phase motors) can be put across the three lines. Such a flexibility is not available in Δ-connection.

(iv) In star connection, the neutral point can be earthed. Such a measure offers many advantages. For example, in case of line to earth fault, the insulators have to bear $1/\sqrt{3}$ (i.e. 57.7%) times the line voltage. Moreover, earthing of neutral permits to use protective devices (e.g. relays) to protect the system in case of ground faults.

Advantages of Delta Connection

(i) This type of connection is most suitable for rotary convertors.

(ii) Most of the 3-phase loads are Δ-connected rather than Y-connected. One reason for this, atleast for the case of an unbalanced load, is the flexibility with which loads may be added or removed on a single phase. This is difficult (or impossible) to do with a Y-connected 3-wire load.

(iii) Most of 3-phase induction motors are delta-connected.

Example 15.15. *A 3 – ϕ, Δ-connected alternator drives a balanced 3 – ϕ load whose each phase current is 10 A in magnitude. At the time when $I_a = 10 \angle 30°$ A, determine the following for a phase sequence of abc.*

(i) Polar expressions for I_b and I_c

(ii) Polar expressions for the three line currents.

Solution. (i) The phase sequence is abc. Since the 3-phase system is balanced, I_b lags behind I_a by 120° and I_c lags I_a by 240° or leads it by 120°. Therefore, when current in phase a is $10 \angle 30°$ A, the currents in the other phases are :

$$I_b = I_a \angle - 120° = I_a \angle 30° - 120° = 10 \angle 30° - 120° = \mathbf{10 \angle - 90°\ A}$$

$$I_c = I_a \angle 120° = 10 \angle 30° + 120° = \mathbf{10 \angle 150°\ A}$$

---

*   Another drawback of Δ-connected alternator is that some currents always circulate around the three phase windings, even without an external load. It is because, in practice, the voltages in the three phases are neither exactly equal nor are they an exactly sine wave.

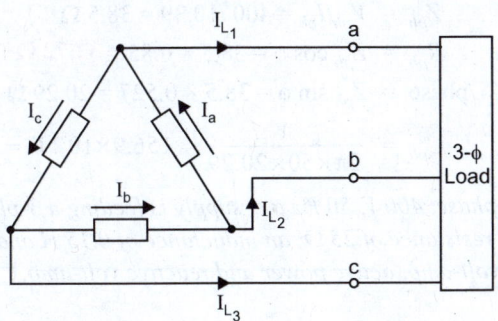

Fig. 15.31

(*ii*) Now the line currents lag behind their respective phase currents by 30°. Therefore, the three line currents are given by ;

$$I_{L1} = \sqrt{3}\, I_a \angle 30° - 30° = \mathbf{17.32 \angle 0° \ A}$$

$$I_{L2} = \sqrt{3} I_b \angle -90° - 30° = \mathbf{17.32 \angle -120° \ A}$$

$$I_{L3} = \sqrt{3}\, I_c \angle 150° - 30° = \mathbf{17.32 \angle 120° \ A}$$

Example 15.16. *Three similar coils each having a resistance of 5Ω and an inductance of 0.02H are connected in delta to a 440V, 3-phase, 50Hz supply. Calculate the line current and total power absorbed.*

Solution. Reactance of coil, $X_L = 2\pi f L = 2\pi \times 50 \times 0.02 = 6.28 \ \Omega$

Impedance / phase, $Z_{ph} = \sqrt{R^2 + X_L^2} = \sqrt{(5)^2 + (6.28)^2} = 8.05 \,\Omega$

Power factor of each coil, $\cos \phi = \dfrac{R_{ph}}{Z_{ph}} = \dfrac{5}{8.05} = 0.622 \ lag$

Phase voltage, $V_{ph} = V_L = 440 \ V$

∴   Phase current, $I_{ph} = \dfrac{V_{ph}}{Z_{ph}} = \dfrac{440}{8.05} = 54.6 \ A$

∴   Line current, $I_L = \sqrt{3}\ I_{ph} = \sqrt{3} \times 54.6 = \mathbf{94.8 \ A}$

Power absorbed, $P = \sqrt{3}\ V_L I_L \cos \phi = \sqrt{3} \times 440 \times 94.8 \times 0.622 = \mathbf{45000 \ W}$

Note that if the coils had been star-connected and connected to the same supply, then,

$$I_{ph} = \dfrac{V_{ph}}{Z_{ph}} = \dfrac{440/\sqrt{3}}{8.05} = 31.6A = I_L \ \ ; \cos\phi = 0.622 \ lag$$

∴ $P = \sqrt{3}\ V_L I_L \cos \phi = \sqrt{3} \times 440 \times 31.6 \times 0.622 = 15000 \ W$

Here is an important conclusion. *The kW delivered to three equal impedances which are delta connected to a 3-phase supply is always three times the kW delivered to them if they are star-connected.*

Example 15.17. *A balanced Δ-connected load takes a line current of 18A at a p.f. of 0.85 leading from a 400 V, 3-phase, 50 Hz supply. Calculate the resistance and capacitance of each leg of the load.*

Solution. $V_{ph} = V_L = 400 \ V \ ; I_L = 18 \ A \ ; \cos\phi = 0.85 \ lead$

$$I_{ph} = I_L / \sqrt{3} = 18/\sqrt{3} = 10.39 \ A$$

$$Z_{ph} = V_{ph}/I_{ph} = 400/10.39 = 38.5 \ \Omega$$

$$\therefore \qquad R_{ph} = Z_{ph} \cos \phi = 38.5 \times 0.85 = \textbf{32.72} \ \boldsymbol{\Omega}$$

$$X_C/\text{phase} = Z_{ph} \sin \phi = 38.5 \times 0.527 = 20.29 \ \Omega$$

$$\therefore \qquad C_{ph} = \frac{1}{2\pi \times 50 \times 20.29} = 156.9 \times 10^{-6} \text{F} = \textbf{156.9} \ \boldsymbol{\mu}\textbf{F}$$

Example 15.18. *A 3-phase, 400 V, 50 Hz a.c. supply is feeding a 3-phase delta-connected load with each phase having a resistance of 25 Ω, an inductance of 0.15 H and a capacitor of 120 μF in series. Find line current, volt-amp, active power and reactive volt-amp.*

Solution.

$$X_L = 2\pi f L = 2\pi \times 50 \times 0.15 = 47.1 \ \Omega \ ; \ X_C = \frac{1}{2\pi f C} = \frac{1}{2\pi \times 50 \times 120 \times 10^{-6}} = 26.54 \ \Omega$$

Net reactance/phase, $X = X_L - X_C = 47.1 - 26.54 = 20.56 \ \Omega$

Impedance/phase, $Z_{ph} = \sqrt{R^2 + X^2} = \sqrt{(25)^2 + (20.56)^2} = 32.37 \ \Omega$

Power factor, $\cos \phi = R/Z_{ph} = 25/32.37 = 0.772 \ lag$

Phase current, $I_{ph} = V_{ph}/Z_{ph} = 400/32.37 = 12.36 \ \text{A}$

$\therefore \qquad$ Line current, $I_L = \sqrt{3} \ I_{ph} = \sqrt{3} \times 12.36 = \textbf{21.4 A}$

Total active power, $P = \sqrt{3} \ V_L I_L \cos \phi = \sqrt{3} \times 400 \times 21.4 \times 0.772 = \textbf{11446 W}$

Total apparent power, $S = P/\cos \phi = 11446/0.772 = \textbf{14830 VA}$

Total reactive power, $Q = \sqrt{S^2 - P^2} = \sqrt{(14830)^2 - (11446)^2} = \textbf{9430 VAR}$

Example 15.19. *Three identical resistances, each of 15 Ω, are connected in delta across 400 V, 3-phase supply. What value of resistance in each leg of balanced star-connected load would take the same line current ?*

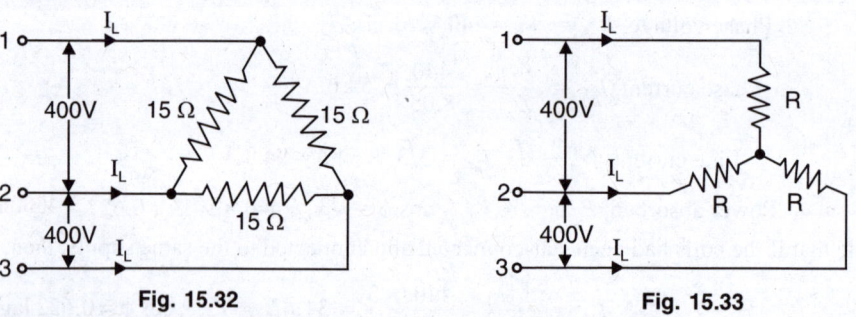

Fig. 15.32   Fig. 15.33

Solution. Let R ohms be the required resistance in each leg of the star-connection (See Fig. 15.33). Since the two circuits have the same line voltage and line current, the resistance between any two corresponding terminals of the two circuits is the same. Considering the terminals 1 and 3,

For delta connection, $R_{13} = 15 \ || \ (15 + 15) = \dfrac{15 \times 30}{15 + 30} = 10 \ \Omega$

For star connection, $R_{13} = R + R = 2R$

$\therefore \qquad 2R = 10 \quad \text{or} \quad R = 10/2 = \textbf{5} \ \boldsymbol{\Omega}$

Example 15.20. *Three 40 Ω non-inductive resistances are connected in delta across 400 V, 3-phase lines. Calculate the power taken from the mains. If one of the resistances is disconnected, what would be the power taken from the mains ?*

Solution. When the three resistances are delta connected.

Fig. 15.34 shows three delta connected resistances.

$$V_{ph} = V_L = 400 \text{ V} \; ; R = 40 \; \Omega$$
$$I_{ph} = V_{ph}/R = 400/40 = 10 \text{ A}$$
$$I_L = \sqrt{3} \, I_{ph} = \sqrt{3} \times 10 = 17.32 \text{ A}$$
$$\therefore \qquad P = \sqrt{3} \, V_L I_L \cos \phi = \sqrt{3} \times 400 \times 17.32 \times 1 = \mathbf{12000 \text{ W}}$$

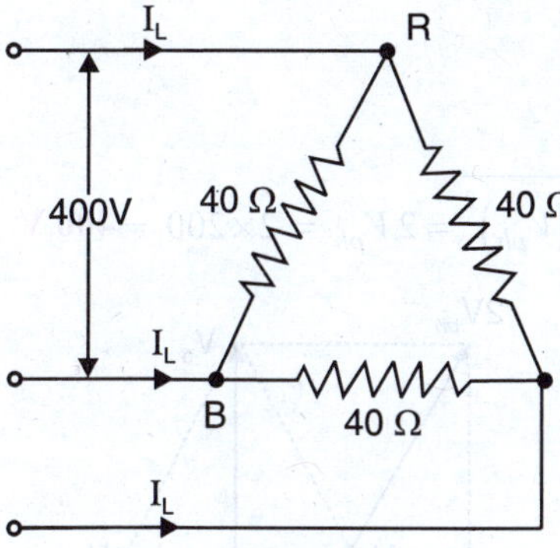

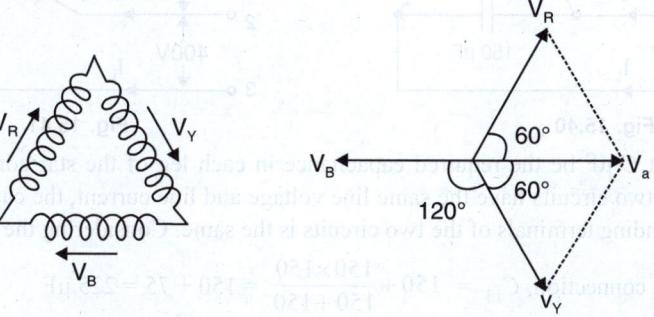

Fig. 15.34 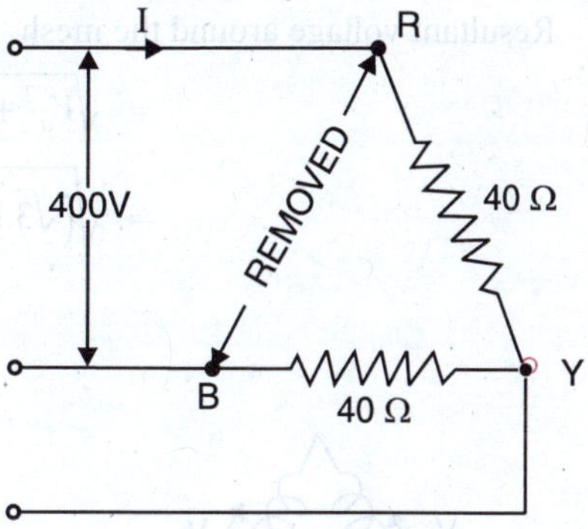   Fig. 15.35

When one resistor is removed. Fig. 15.35 shows the circuit with one resistor removed. In this case, each of the two resistors acts independently as if 400 V were applied to 40 Ω.

Current in each resistor, $I = 400/40 = 10$ A

Power consumed in the two $= 2 \, I^2 R = 2 \times (10)^2 \times 40 = \mathbf{8000 \text{ W}}$

Hence, by disconnecting one resistor, the power consumption is reduced by one-third.

Example 15.21. *A delta connected alternator is on no load. The voltage per phase is 200 V. What will be the resultant voltage around the mesh (i) when the phases are connected correctly (ii) when the connections of one of the phases are reversed ?*

Solution. Each phase voltage is 200 V *i.e.*

$$V_R = V_Y = V_B = V_{ph} = 200 \text{ V} \qquad \text{...in magnitude}$$

(i) When the phases are connected correctly. When the phases are connected correctly (*i.e.* finishing end of one winding is connected to the starting end of the second winding and so on), the direction of each of the phase voltages when positive is in the same direction around the mesh as shown in Fig. 15.36. The phasor diagram for this arrangement is shown in Fig. 15.37. As can be seen, the resultant of V_R and V_Y, shown as V_a, is *equal and opposite to V_B and the resultant voltage around the mesh is zero *i.e.*

Resultant voltage around the mesh = **0 V**

Fig. 15.36   Fig. 15.37

*
$$V_R = V_Y = V_B = V_{ph} = 200 \text{ V and}$$
$$V_a = 2 \, V_{ph} \cos 60° = V_{ph}$$

Note that angle between V_B and V_a is 180°.

   (ii) When one phase connection is reversed. Suppose connections to phase Y have been reversed. Then the directions of the phase voltages when positive will be as shown in Fig. 15.38. Note that the phasor V_Y will now act in the reversed direction as shown in the phasor diagram in Fig. 15.39. The resultant of V_R and V_Y, shown as V_a, is $= 2\,V_{ph} \cos 30° = \sqrt{3}\;V_{ph}$. The resultant voltage around the mesh is the phasor sum of V_a and V_B. These two phasors are 90° apart.

   ∴   Resultant voltage around the mesh

$$= \sqrt{V_a^2 + V_B^2}$$

$$= \sqrt{\left(\sqrt{3}\,V_{ph}\right)^2 + \left(V_{ph}\right)^2} = 2\,V_{ph} = 2 \times 200 = \textbf{400 V}$$

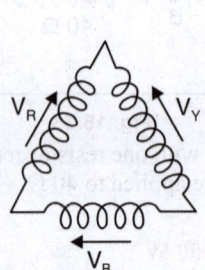

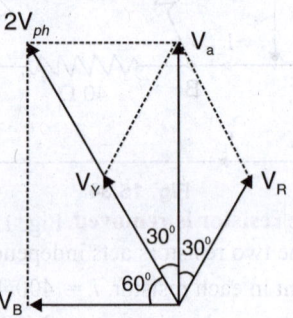

Fig. 15.38   **Fig. 15.39**

   Thus if one phase of delta connection is reversed, the resultant voltage around the mesh is equal to *twice* the phase voltage. This high voltage may cause high current to flow through the windings, even without external load, and may damage the windings.

   Example 15.22. *Three capacitors, each of 150 µF, are connected in delta to a 400 V, 3-phase, 50 Hz supply. What will be the capacitance of each of the three capacitors such that when connected in star across the same supply, the line current remains the same ?*

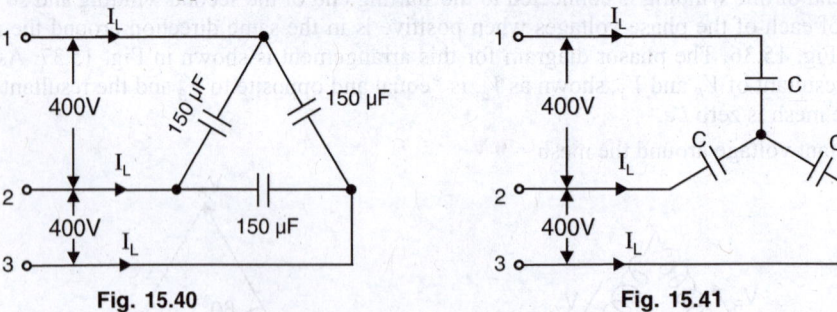

Fig. 15.40   **Fig. 15.41**

   Solution. Let C µF be the required capacitance in each leg of the star connection (See Fig. 15.41). Since the two circuits have the same line voltage and line current, the capacitance between any two corresponding terminals of the two circuits is the same. Considering the terminals 1 and 3,

$$\text{For delta connection, } C_{13} = 150 + \frac{150 \times 150}{150 + 150} = 150 + 75 = 225\ \mu F$$

$$\text{For star connection, } C_{13} = \frac{C \times C}{C + C} = \frac{C}{2}$$

∴ $C/2 = 225$ or $C = 2 \times 225 = \textbf{450 µF}$

Example 15.23. *Three equal impedances are connected in delta to a 440 V, 50 Hz supply. The p.f. is 0.8 lagging and $25\sqrt{3}$ kVA is drawn from the supply. Calculate the line current and total power drawn from the supply when the same impedances are star-connected.*

Solution. Delta connection. $I_L = \dfrac{25\sqrt{3} \times 10^3}{\sqrt{3} \times 440} = 56.82$ A

$$I_{ph} = I_L / \sqrt{3} = 56.82/\sqrt{3} = 32.8 \text{ A}$$
$$Z_{ph} = V_{ph}/I_{ph} = 440/32.8 = 13.4 \ \Omega$$

Star connection.

$$Z_{ph} = 13.4 \ \Omega ; V_L = 440 \text{ V}$$
$$V_{ph} = V_L / \sqrt{3} = 440/\sqrt{3} = 254 \text{ V}$$
$$I_{ph} = V_{ph}/Z_{ph} = 254/13.4 = 18.95 \text{ A}$$
$$\therefore \qquad I_L = I_{ph} = \textbf{18.95 A}$$

Also, $\qquad P = \sqrt{3} \ V_L I_L \cos\phi = \sqrt{3} \times 440 \times 18.95 \times 0.8 = \textbf{11553 W}$

Example 15.24. *Three similar resistors are connected in star across 400 V, 3-phase supply. The line current is 5 A. Calculate the value of each resistance. To what value should the line voltage be changed to obtain the same line current with the resistors connected in delta ?*

Solution. Star connection. $V_{ph} = V_L / \sqrt{3} = 400/\sqrt{3} = 231$ V

$$\therefore \qquad I_{ph} = I_L = 5 \text{ A}$$
$$R_{ph} = V_{ph}/I_{ph} = 231/5 = \textbf{46.2} \ \Omega$$

Delta connection. $\qquad I_L = 5\text{A} \qquad\qquad\qquad ...given$

$$I_{ph} = I_L / \sqrt{3} = 5/\sqrt{3} = 2.88 \text{ A}$$
$$V_{ph} = I_{ph} R_{ph} = 2.88 \times 46.2 = \textbf{133 V}$$

It may be seen that the line voltage required is one-third that of the star value.

Example 15.25. *A 3-phase, 440 V, 50 Hz delta-connected motor on full-load has an efficiency of 90%, a power output of 15 kW and a power factor of 0·82 lagging. Find (i) the line current (ii) the cost of running the motor for 10 hours, price of 1 kWh being 80 paise.*

Solution. $V_L = 440$ V ; $\cos\phi = 0.82$ lag ; $\eta = 0.9$

(i) Input power to motor, $P = 15 \times 10^3/0.9 = 16.67 \times 10^3$ W

Now, $\qquad\qquad P = \sqrt{3} \ V_L I_L \cos\phi$

$\therefore \qquad$ Line current, $I_L = \dfrac{P}{\sqrt{3} \ V_L \cos\phi} = \dfrac{16.67 \times 10^3}{\sqrt{3} \times 440 \times 0.82} = \textbf{26.67 A}$

(ii) Energy consumed in running the motor for 10 hours

$$= 16.67 \times 10 = 166.7 \text{ kWh}$$

Energy charges $= 166.7 \times 0.8 = \textbf{Rs. 133.36}$

Example 15.26. *A 220 V, 3-ϕ voltage is applied to a balanced Δ-connected 3-ϕ load of phase impedance $(15 + j 20)$ Ω.*

(i) Find the phasor current in each line.

(ii) What is the power consumed per phase?

(iii) What is the phasor sum of the three line currents? Why does it have this value?

Solution. Phase voltage, $V_{ph} = V_L = 220$ volts

Impedance/phase, $Z_{ph} = \sqrt{R_{ph}^2 + X_{ph}^2} = \sqrt{15^2 + 20^2} = 25 \ \Omega$

$$\text{Phase current, } I_{ph} = \frac{V_{ph}}{Z_{ph}} = \frac{220}{25} = 8.8 \text{ A}$$

$$\text{Phase angle, } \phi = \cos^{-1}\frac{R_{ph}}{Z_{ph}} = \cos^{-1}\frac{15}{25} = 53.13°$$

$$\text{Line current, } I_L = \sqrt{3} I_{ph} = \sqrt{3} \times 8.8 = 15.24 \text{ A}$$

(*i*) The phase sequence is *RYB*. Taking phase voltage V_{RY} as the reference phasor, $V_{RY} = 220 \angle 0°$ volts. Now line currents are 30° behind their respective phase currents. Therefore, line current I_R lags behind V_{RY} by (53.13° + 30°) = 83.13°.

$$\therefore \qquad\qquad I_R = \mathbf{15.24 \angle -83.13° A}$$

Similarly, $\qquad I_Y = 15.24 \angle -203.13° A = \mathbf{15.24 \angle 156.87° A} \; ; \; I_B = \mathbf{15.24 \angle 36.87° A}$

(*ii*) Power consumed/phase = $V_{ph} I_{ph} \cos \phi = 220 \times 8.8 \times 0.6 = \mathbf{1161.6 \text{ W}}$

(*iii*) The phasor sum of the three line currents will be **zero** *i.e.*

$$\text{Phasor sum} = I_R + I_Y + I_B = 0$$

This is expected because the three line currents are equal in magnitude and have a phase difference of 120° from one another.

Tutorial Problems

1. Calculate the phase and line currents in a balanced delta connected load taking 75 kW at a power factor 0.8 from a 3-phase 440 V supply. **[71.02 A ; 123 A]**

2. Three identical inductive loads of resistance 15 Ω and reactance 40 Ω are connected in star to a 440 V, 3-phase supply. Calculate (*i*) phase current (*ii*) line current and (*iii*) power absorbed.
   [(*i*) 5.94 A (*ii*) 5.94 A (*iii*) 1589.5 W]

3. Three identical resistances, each of 18 Ω, are connected in delta across 400 V, 3-phase supply. What value of resistance in each leg of balanced star-connected load would take the same line current ? **[6 Ω]**

4. Three similar resistors are connected in star across a 415 V, 3-phase supply. The line current is 10 A. Calculate (*i*) the value of each resistance (*ii*) the line voltage required to give the same line current if the resistors were delta-connected. **[(*i*) 23.96 Ω (*ii*) 138.33 V]**

5. A balanced 3-phase load consists of three coils, each of resistance 4 Ω and inductance 0.02 H. Determine the total power when the coils are (*i*) star-connected (*ii*) delta-connected to a 400 V, 3-phase, 50 Hz supply. **[(*i*) 11.56 kW (*ii*) 34.68 kW]**

6. Three capacitors, each 50 μF, are connected in delta to a 400 V, 3-phase 50 Hz supply. What will be the capacitance of the three capacitors such that when connected in star across the same supply, the line current remains the same ? **[150 μF]**

7. Three 6 Ω non-inductive resistances are connected in (*i*) star (*ii*) delta across 400 V, 50 Hz 3-phase supply. Calculate the total power drawn in each case. In the event of one of the three resistances getting open-circuited, what would be the value of the total power taken from the mains in each of the two cases? **[(*i*) Star : 26.66 kW ; 13.33 kW (*ii*) Delta : 80 kW ; 53.33 kW]**

8. A 440 V, 3-phase delta-connected induction motor has an output of 14.92 kW at a p.f. of 0.82 and efficiency 85%. If another star-connected load of 10 kW at 0.85 p.f. is added in parallel, find (*i*) current drawn from the supply (*ii*) total power. **[(*i*) 43.56 A (*ii*) 27.6 kW]**

9. A 3-phase delta connected motor connected to 400 V, 50 Hz supply has an output of 29.84 kW, the efficiency and power factor being 95% and 0.9 respectively. Calculate the phase current in each winding.
   [29.08 A]

10. A 3-phase delta-connected load, each phase of which has an inductive reactance of 40 Ω and a resistance of 25 Ω, is fed from the secondary of a 3-phase, star connected transformer which has a phase voltage of 240 V. Calculate (*i*) the current in each phase of the load (*ii*) voltage across each phase of the load (*iii*) current in transformer secondary (*iv*) total power taken from the supply.

[(*i*) **8.812 A** (*ii*) **415.68 V** (*iii*) **15.26 A** (*iv*) **5800 W**]

15.14. Constancy of Total Power in Balanced 3-phase System

It was shown in chapter 12 that the instantaneous power p supplied by a single-phase source to a load (*lagging p.f.) is given by ;

$$p = \frac{V_m I_m}{2} \cos \phi - \frac{V_m I_m}{2} \cos (2 \omega t - \phi)$$

In terms of effective or r.m.s. values, it becomes :

$$p = VI \cos \phi - VI \cos (2 \omega t - \phi) \qquad \left(\because \frac{V_m I_m}{2} = \frac{V_m}{\sqrt{2}} \frac{I_m}{\sqrt{2}} = VI \right)$$

Therefore, instantaneous power (p) varies sinusoidally from zero to the peak value at twice the supply frequency. This pulsating nature of power is objectionable for many applications.

We now prove that power delivered by 3-phase balanced supply is constant; being 3 $V_{ph} I_{ph}$ cos ϕ or $\sqrt{3} \, V_L I_L$ cos ϕ. The above expression for single phase power can be applied to each of the three phases (*viz R, Y* and *B*) of the 3-phase system. The only modification needed is introducing 120° displacement that exists between the phases. Taking *R*-phase (the phase sequence is *RYB*) as the reference, the instantaneous powers in the three phases can be written as :

$$p_R = V_{ph} I_{ph} \cos \phi - V_{ph} I_{ph} \cos (2 \omega t - \phi)$$
$$p_Y = V_{ph} I_{ph} \cos \phi - V_{ph} I_{ph} \cos (2 \omega t - \phi - 120°)$$
$$p_B = V_{ph} I_{ph} \cos \phi - V_{ph} I_{ph} \cos (2 \omega t - \phi - 240°)$$

$\therefore$ Total instantaneous power is given by ;

$$p = p_R + p_Y + p_B$$
$$= 3 \, V_{ph} I_{ph} \cos \phi - V_{ph} I_{ph} [\cos (2 \omega t - \phi)$$
$$+ \cos (2 \omega t - \phi - 120°) + \cos (2 \omega t - \phi - 240°)]$$

If the second term in the bracket [*i.e.*, cos $(2\omega t - \phi)$ + cos $(2 \omega t - \phi - 120°)$ + cos $(2 \omega t - \phi - 240°)$] is expanded, it will be found that its value is zero.

$$\therefore \quad p = 3 \, V_{ph} I_{ph} \cos \phi$$
$$= \sqrt{3} \, V_L I_L \cos \phi$$

Since phase and line values and power factor for a balanced 3-phase system are fixed, **total 3-phase power shall remain constant at all times**. Therefore, an important advantage of a 3-phase system over the single-phase system is this constant and smooth power flow into a load such as a 3-phase motor. For this reason, 3-phase motors are used when the load is greater than 3 h.p. Moreover, 3-phase system presents overall economic and operating advantages compared with single-phase system or other polyphase systems. *Therefore, the 3-phase system is the most widely used of all the polyphase systems.*

---

* For leading p.f., the expression is $p = \frac{V_m I_m}{2} \cos \phi - \frac{V_m I_m}{2} \cos(2 \omega t + \phi)$

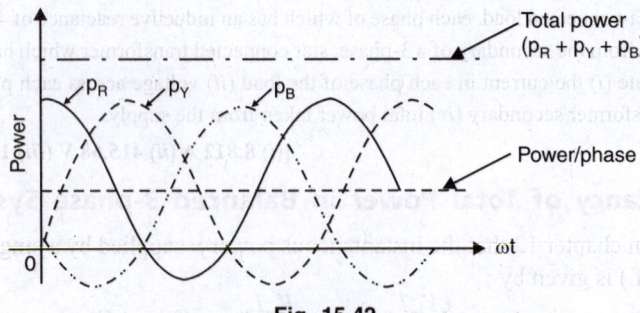

Fig. 15.42

Graphical Explanation. Fig. 15.42 shows the graphical explanation for the constancy of total power in a 3-phase balanced system. Thus p_R represents the power supplied to phase R. The power graph for phase Y lags by 120° behind p_R and is represented by p_Y. Similarly, graph p_B represents the power supplied to phase B. If we sum up the three curves point by point, it will be seen that total power is constant as shown by the dotted horizontal line in Fig. 15.42. Thus, although the three component powers fluctuate, the total power is constant. This is due to the phase difference between the fluctuations of the components. Due to the constant power supplied by a balanced 3-phase system, large loads are always driven by 3-phase motors rather than single phase motors.

15.15. Effects of Phase Sequence

The order in which the voltages in the three phases (R, Y and B) of an alternator reach their maximum positive values is called the *phase sequence*. It is determined by the direction of rotation of the alternator. Suppose an alternator rotating in an anticlockwise direction produces voltages of phase sequence *RYB*. If this alternator rotates in the clockwise direction, the voltages produced will have a phase sequence *RBY*. Since an alternator can be rotated in either anticlockwise or clockwise direction, there can be only two possible phase sequences *viz. RYB* and *RBY*. By convention, phase sequence RYB is taken as positive and RBY as negative. Suppose each phase of an alternator generates 240 V (r.m.s.).

Phase sequence RYB. In phasor notation, the phase voltages can be expressed as :

$E_R = 240 \angle 0°$ V ; $E_Y = 240 \angle -120°$ V ; $E_B = 240 \angle -240°$ V

Phase sequence RBY. In phasor notation, the phase voltages can be expressed as :

$E_R = 240 \angle 0°$ V ; $E_B = 240 \angle -120°$ V ; $E_Y = 240 \angle -240°$ V

At first thought, it might seem that phase sequence is immaterial in 3-phase system. But this is not always correct as is clear from the following discussion :

(*i*) Where balanced resistive loads are involved, the phase sequence is normally not important.

(*ii*) The direction of rotation of a 3-phase induction motor depends upon the phase sequence of the 3-phase supply. If the phase sequence is reversed by interchanging any two lines of the 3-phase supply, the motor would rotate in the *opposite direction.

(*iii*) The currents in the three phases of an unbalanced Y-connected load depend on the phase sequence of the 3-phase supply. If the phase sequence is reversed, we get completely different currents in the three phases of the Y-connected load.

---

*   An induction motor is usually required to run in a particular direction. When initially connected to 3-phase supply, its direction of rotation can be checked. If it is not correct, the direction can be reversed by interchanging any two lines of the 3-phase supply.

(*iv*) Reversing the phase sequence of a 3-phase alternator which is to be paralleled with a similar alternator can cause extensive damage to both the machines.

From the above discussion, it follows that phase sequence is very important in certain applications. While dealing with such applications, the phase sequence must be clearly specified in order to avoid unnecessary confusion.

15.16. Phase Sequence Indicator

The fact that the currents in the three phases of an unbalanced 3-wire *Y*-connected load change with the reversal of phase sequence of the 3-phase supply can be employed to determine the phase sequence of the supply *i.e.* whether the phase sequence is *RYB* or *RBY*. Fig. 15.43 shows the circuit of a simple phase sequence tester. Each lamp in the *R*-phase and *Y*-phase has a filament resistance of 50 Ω while the *B*-phase has known inductive load. Suppose the line voltage is 200 V, 60 Hz.

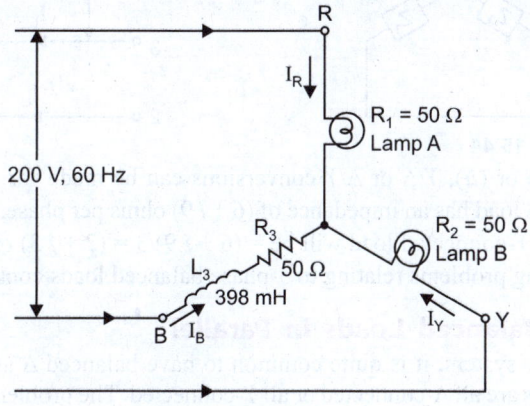

Fig. 15.43

When phase sequence is RYB. When the phase sequence is *RYB* as shown in Fig. 15.43, it can be shown mathematically (See Art. 15.35) that the currents in the two lamps are :

Current in lamp A, I_R = 1.55 A

Current in lamp B, I_Y = 2.46 A

Therefore, lamp *B* will be brighter than lamp *A*.

When phase sequence is RBY. When the phase sequence is *RBY* (*i.e.* phase sequence is reversed), it can be shown mathematically that currents in the two lamps are :

Current in lamp A, I_R = 2.46 A

Current in lamp B, I_B = 1.55 A

Therefore, lamp *A* is brighter than lamp *B*.

Thus, when lamp B is brighter than lamp A, the phase sequence is RYB. However, when lamp A is brighter than lamp B, the phase sequence is RBY.

15.17. Y/Δ or Δ/Y Conversions for Balanced Loads

Two circuits are electrically equivalent if they draw the same line current and power when connected to the same line voltage. Consider a balanced *Y*-connected load having an impedance Z_1 in each phase as shown in Fig. 15.44. Let the equivalent Δ-connected load have an impedance of Z_2 in each phase as shown in Fig. 15.45. Since the two circuits are equivalent, the impedance between any *two corresponding terminals of the two circuits is the same. Considering terminals 1 and 2,

---

*   Since the two circuits take the same line current and have the same line voltage, the impedance between any two corresponding terminals is the same. Not only the total power, line current and line voltage are equal in the two cases but power factor, reactive power and apparent power are also equal. In fact, one cannot distinguish between the two circuits from their terminal quantities.

For star connection, $Z_{12} = Z_1 + Z_1 = 2Z_1$

For delta connection, $Z_{12} = Z_2 \parallel (Z_2 + Z_2) = \dfrac{(Z_2)(2Z_2)}{Z_2 + 2Z_2} = \dfrac{2}{3}Z_2$

$\therefore \qquad 2Z_1 = \dfrac{2}{3}Z_2 \quad \text{or} \quad Z_2 = 3Z_1$ ...(i)

$\therefore \qquad Z_\Delta = 3\,Z_Y \quad \text{or} \quad Z_Y = \dfrac{Z_\Delta}{3}$ ...(ii)

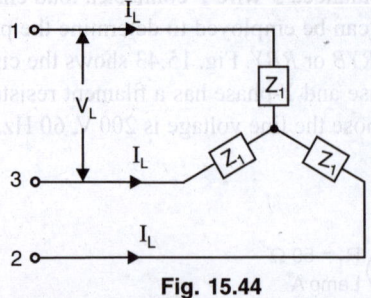

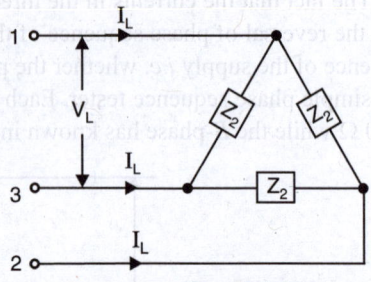

Fig. 15.44   Fig. 15.45

Using equations (i) or (ii), Y/Δ or Δ/Y conversions can be made for the balanced loads. For example, if a balanced Δ load has an impedance of $(6 + j\,9)$ ohms per phase, then impedance of each phase of the equivalent Y-connected load will be $= (6 + j\,9)/3 = (2 + j\,3)$ ohms. These conversions may be helpful in solving problems relating to 3-phase balanced loads connected in parallel.

15.18. 3-phase Balanced Loads in Parallel

In a 3-phase supply system, it is quite common to have balanced Δ and Y loads connected in parallel. Rarely the loads are all Δ-connected or all Y-connected. The problems on such parallel loads can be solved by one of the following three ways :

(*i*) All the Y-loads in the problem may be converted into equivalent Δ loads. Then these loads in parallel (all now being Δ loads) can be treated on a single-phase basis (line-to-neutral) to find the various circuit values *e.g.*, currents, power factor etc.

(*ii*) All the Δ loads in the problem may be converted into equivalent Y loads and treated as in (*i*) above.

Out of the above two methods, usually the latter is prefered because it is more convenient to handle Y-connection.

(*iii*) A still shorter and more commonly used method is to treat each load on a complete 3-phase basis. The active, reactive and apparent powers in each 3-phase load are determined. The active powers (watts) of the parallel loads can be added directly to give the total active power P. The reactive powers (VAR) of the parallel loads can be added algebraically (with due regard to the *signs of inductive VAR and capacitive VAR) to give the total reactive power Q. Then total apparent power S (in VA) is given by ;

$$S = \sqrt{P^2 + Q^2}$$

The various characteristic quantities (*e.g.*, power factor, line current etc.) can be easily determined for the combined loads.

Example 15.27. *Three star-connected impedances $Z_1 = (20 + j\,37.7)\ \Omega$ per phase are connected in parallel with three delta connected impedances $Z_2 = (30 - j\,159.3)\ \Omega$ per phase. The line voltage is 398 V. Find (i) the total line current (ii) circuit power factor (iii) active power and reactive power taken by the combination.*

---

*   Inductive VAR is assigned + ve sign and capacitive VAR the – ve sign.

Solution. (*i*) The phase voltage $= 398/\sqrt{3} = 230$ V

$$Z_1 = (20 + j\,37.7)\,\Omega = 42.7 \angle 62.05°\,\Omega$$

$\therefore$ Line current, $I_1 = \dfrac{230\angle 0°}{Z_1} = \dfrac{230\angle 0°}{42.7\angle 62.05°} = 5.39 \angle -62.05°\,$A $= (2.52 - j\,4.76)$A

Equivalent star impedance per phase of Δ-load

$$= \frac{Z_2}{3} = \frac{30 - j\,159.3}{3} = (10 - j\,53.1)\,\Omega = 54 \angle -79.3°\,\Omega$$

$\therefore$ Line current, $I_2 = \dfrac{230 \angle 0°}{54 \angle -79.3°} = 4.26 \angle 79.3°\,$A $= (0.79 + j\,4.19)$A

Total line current, $I = I_1 + I_2 = (2.52 - j\,4.76) + (0.79 + j\,4.19)$

$$= (3.31 - j\,0.57)\text{A} = \textbf{3.36} \angle \textbf{-- 9.8°A}$$

(*ii*) Power factor $= \cos 9.8° = \textbf{0.985} \; lag$

(*iii*) Total active power, $P = 3 \times$ power/phase $= 3 \times 230 \times 3.36 \times 0.985 = \textbf{2284 W}$

Total reactive VA, $Q = 3\,VI \sin \phi = 3 \times 230 \times 3.36 \times 0.17 = \textbf{401 VAR}$

Example 15.28. *A 3-phase star-connected alternator feeds a 2000 h.p. delta-connected induction motor having a p.f. of 0·85 and an efficiency of 0·93. Calculate the current, active and reactive components in (i) each alternator phase (ii) each motor phase. The line voltage is 2200 V.*

Solution. Input to motor $= \dfrac{h.p \times 746}{\eta} = \dfrac{2000 \times 746}{0.93} = 1.6 \times 10^6$ W

If *I* is the line current, then, we have,

$$I = \frac{1.6 \times 10^6}{\sqrt{3} \times 2200 \times 0.85} = 495 \text{ A}$$

(*i*) **Alternator** (star-connected)

Phase current $=$ Line current $= \textbf{495 A}$

Active component $= I \cos \phi = 495 \times 0.85 = \textbf{421 A}$

Reactive component $= I \sin \phi = 495 \times 0.527 = \textbf{261 A}$

(*ii*) **Induction motor** (delta-connected)

Phase current $= 495/\sqrt{3} = \textbf{286 A}$

Active component $= I \cos \phi = 286 \times 0.85 = \textbf{243 A}$

Reactive component $= I \sin \phi = 286 \times 0.527 = \textbf{151 A}$

Example 15.29. *Three 3-phase balanced loads are connected in parallel across a 400 V, 3-phase, 3-wire balanced supply.*

Load 1 : 12000 W, Δ-connected, p.f. = 1

Load 2 : 9000 VA, Y-connected, p.f. = 0.866 lag

Load 3 : 6000 VAR, Δ-connected, p.f. = 0, capacitive

Find (i) the total power (ii) combined power factor and (iii) current drawn from the line.

Solution. Fig. 15.46 shows the given 3-phase balanced loads connected to 400 V, 3-phase, 3-wire supply.

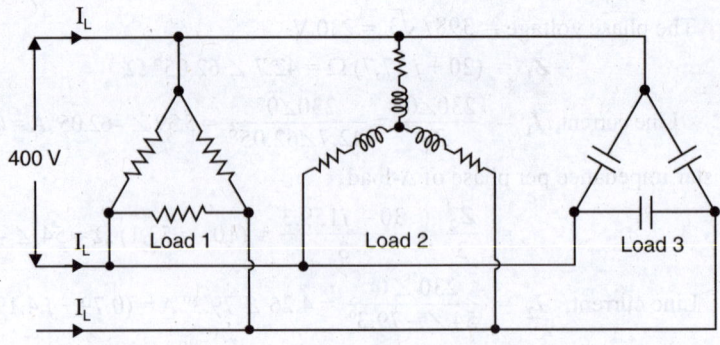

Fig. 15.46

The active power, apparent power and reactive power drawn by each 3-phase load are calculated in the table below :

Loads	Active power P (watts)	Apparent power, S = P/cos φ (VA)	Reactive power, Q = S sin φ (VAR)	Power factor
Load 1	12000	12000	0	1
Load 2	7800	9000	4500	lag
Load 3	0	6000	– 6000	lead
Total	19800	cannot add	– 1500	lead

Since the net reactive power is capacitive, the combined power factor is leading.

Total apparent power, $S = [P^2 + Q^2]^{1/2} = [(19800)^2 + (1500)^2]^{1/2} = 19856$ VA

(i) Total power, $P =$ **19800 W**

(ii) P.F. of combined loads $= \dfrac{P}{S} = \dfrac{19800}{19856} =$ **0.997 lead**

(iii) $P = \sqrt{3}\, V_L I_L \cos \phi$

∴ Line current, $I_L = \dfrac{P}{\sqrt{3}\, V_L \cos \phi} = \dfrac{19800}{\sqrt{3} \times 400 \times 0.997} =$ **28.66 A**

Example 15.30. *Three equal impedances (60 + j 30) ohms are connected in delta at the end of a 3-phase transmission line which has an impedance of (1 + j 2) ohms per conductor. If the line voltage at the generating station is maintained at 2300 V, find (i) line-to-line voltage at the load (ii) efficiency of transmission.*

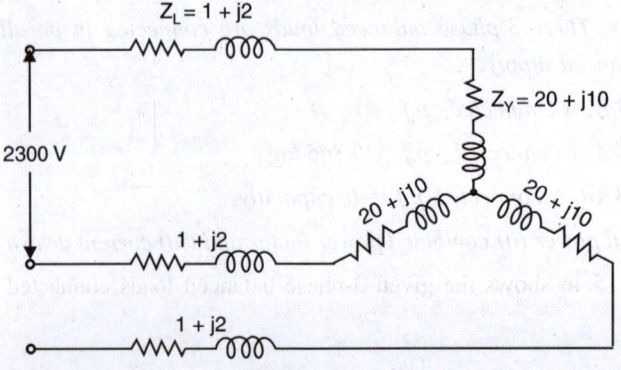

Fig. 15.47

Solution. Fig. 15.47 shows delta-connected load replaced by the equivalent Y-connected load. The impedance per phase of this equivalent Y-load is :

$$Z_Y = \frac{Z_\Delta}{3} = \frac{(60 + j\,30)}{3} = (20 + j\,10)\,\Omega = 22.36\,\angle\,26.56°\,\Omega$$

$$Z_L = (1 + j\,2)\,\Omega$$

Total impedance per conductor is given by ;

$$Z = Z_L + Z_Y = (1 + j\,2) + (20 + j\,10) = (21 + j\,12)\,\Omega = 24.19\,\angle\,29.74°\,\Omega$$

Voltage per phase at the sending end *i.e.*, generator end is

$$(V_{ph})_{sending} = 2300/\sqrt{3} = 1328\,V$$

Line current, $\quad I_L = \dfrac{(V_{ph})_{sending}}{Z} = \dfrac{1328}{24.19} = 54.9\,A$

Voltage per phase at the load end

$$= I_L \times Z_Y = 54.9 \times 22.36 = 1227.56\,V$$

(*i*)   Line-to-line voltage at the load end

$$= \sqrt{3} \times 1227.56 = \textbf{2126.2 V}$$

(*ii*)   Transmission efficiency $= \dfrac{3 I_L^{\,2} \times R_Y}{3 I_L^{\,2}\,(R_Y + R_L)} = \dfrac{R_Y}{R_Y + R_L} = \dfrac{20}{20 + 1} = \textbf{0.952}$

Example 15.31. *Each phase of a star-connected load consists of a coil of resistance 20 Ω and inductance 0.07 H. The load is connected to a 415V, 3-phase, 50Hz supply. Calculate :*

(i)   The line current, the power and the power factor

(ii)   The capacitance per phase of a delta-connected capacitor bank which would improve the overall power factor to unity.

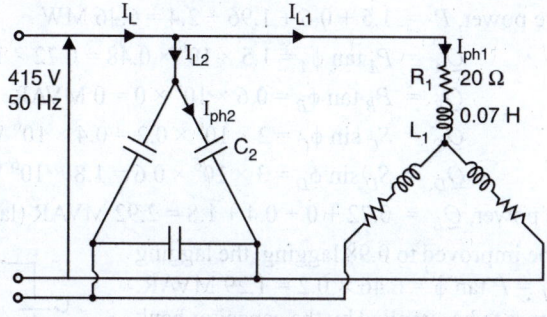

Fig. 15.48

Solution. We shall consider only the magnitudes of the quantities.

(*i*)   **Without capacitor bank**

$$X_{L1} = 2\pi f L_1 = 2\pi \times 50 \times 0.07 = 22\,\Omega$$

$$Z_{ph1} = \sqrt{R_1^2 + X_{L1}^2} = \sqrt{20^2 + 22^2} = 29.8\,\Omega$$

$$I_{L1} = I_{ph1} = \frac{V_{ph1}}{Z_{ph1}} = \frac{415/\sqrt{3}}{29.8} = \textbf{8.05 A}$$

$$P_1 = 3 I_{ph1}^2 R_1 = 3 \times (8.05)^2 \times 20 = \textbf{3890 W}$$

$$\cos\phi_1 = R_1/Z_{ph1} = 20/29.8 = \textbf{0.672} \textit{ lag}$$

We can also find power in another way.

$$P_1 = \sqrt{3}\,V_L I_L \cos\phi = \sqrt{3} \times 415 \times 8.05 \times 0.672 = 3890\,W$$

(ii) With capacitor bank. Fig. 15.48 shows the delta-connected capacitor bank (balanced) in parallel with the star-connected load. In order that the overall power factor becomes unity, the reactive component of line current should be zero *i.e.*, $I_L \sin \phi = 0$. This will be so if $I_{L2} = I_{L1} \sin \phi_1$ *i.e.*,

$$I_{L2} = I_{L1} \sin \phi_1 = I_{L1} \frac{X_{L1}}{Z_{ph1}} = 8.05 \times \frac{22}{29.8} = 5.95 \text{ A}$$

$$\therefore \quad I_{ph2} = I_{L2}/\sqrt{3} = 5.95/\sqrt{3} = 3.44 \text{ A}$$

$$X_{C2} = V_{ph}/I_{ph2} = 415/3.44 = 121 \ \Omega$$

$$\therefore \quad C_2 = \frac{1}{2\pi f X_{C_2}} = \frac{1}{2\pi \times 50 \times 121} = 26.3 \times 10^{-6} \text{ F} = \textbf{26.3 μF}$$

Example 15.32. *A factory is supplied at 11 kV, 3-phase, 50 Hz system and has the following balanced loads :*

        Load A : 1.5 MW at 0·9 power factor lagging

        Load B : 600 kW at unity power factor

        Load C : 2MVA at 0·98 power factor lagging

        Load D : 3MVA at 0·8 power factor lagging

A 3-phase bank of star-connected capacitors is connected at supply terminals to give power factor correction. Find the required capacitance per phase to give an overall power factor of 0·98 lagging when the factory is operating at maximum load.

Solution.
$$P_A = 1.5 \text{ MW} ; P_B = 600 \text{ kW} = 0.6 \text{ MW}$$
$$P_C = S_C \cos \phi_C = 2 \times 10^6 \times 0.98 = 1.96 \times 10^6 \text{ W} = 1.96 \text{ MW}$$
$$P_D = S_D \cos \phi_D = 3 \times 10^6 \times 0.8 = 2.4 \times 10^6 \text{ W} = 2.4 \text{ MW}$$

Total active power, $P = 1.5 + 0.6 + 1.96 + 2.4 = 6.46 \text{ MW}$

$$Q_A = P_A \tan \phi_A = 1.5 \times 10^6 \times 0.48 = 0.72 \times 10^6 \text{ VAR} = 0.72 \text{ MVAR}$$
$$Q_B = P_B \tan \phi_B = 0.6 \times 10^6 \times 0 = 0 \text{ MVAR}$$
$$Q_C = S_C \sin \phi_C = 2 \times 10^6 \times 0.2 = 0.4 \times 10^6 \text{ VAR} = 0.4 \text{ MVAR}$$
$$Q_D = S_D \sin \phi_D = 3 \times 10^6 \times 0.6 = 1.8 \times 10^6 \text{ VAR} = 1.8 \text{ MVAR}$$

Total reactive power, $Q_L = 0.72 + 0 + 0.4 + 1.8 = 2.92 \text{ MVAR (lagging)}$

Since the p.f. is to be improved to 0.98 lagging, the lagging MVAR at this p.f. is $Q_L = P \tan \phi = 6.46 \times 0.2 = 1.29$ MVAR. The leading reactive power to be supplied by the capacitor bank is $Q_C = 2.92 - 1.29 = 1.63$ MVAR.

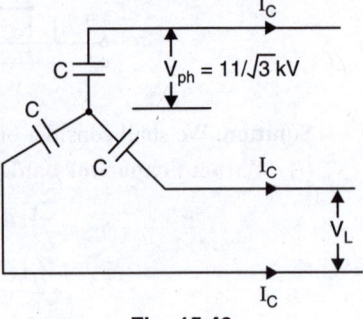

Fig. 15.49

Referring to the star-connected capacitor bank in Fig. 15.49, we have,

$$Q_C = \sqrt{3} V_L I_C$$

or
$$I_C = \frac{Q_C}{\sqrt{3} V_L} = \frac{1.63 \times 10^6}{\sqrt{3} \times 11 \times 10^3} = 85.5 \text{ A}$$

Capacitive reactance/phase is given by ;

$$X_C = \frac{V_{ph}}{I_C} = \frac{(11 \times 10^3)/\sqrt{3}}{85.5} = 74.2 \ \Omega$$

$$\therefore \quad \text{Capacitance/phase, } C = \frac{1}{2\pi f X_C} = \frac{1}{2\pi \times 50 \times 74.2}$$

$$= 42.9 \times 10^{-6} \text{ F} = \textbf{42.9 μF}$$

Example 15.33. *The circuit shown in Fig. 15.50 has L = 530 mH; R = 200 Ω; C = 1 μF, V_L =*
400 V and f = 60 Hz. Find P, I_L and cos ϕ with (i) only the inductors (ii) only the capacitors in the
circuit (iii) all components present.

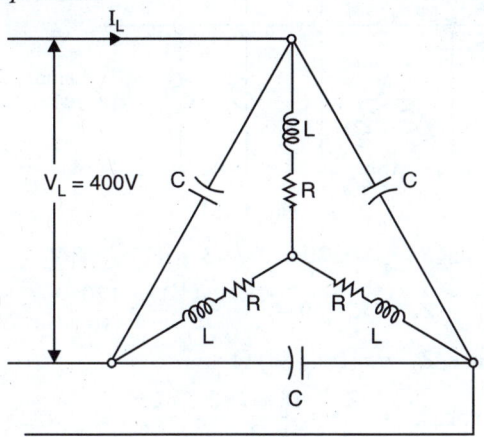

Fig. 15.50

Solution. (*i*) **Inductor and resistor circuit.**

$$X_L = 2\pi f L = 2\pi \times 60 \times 530 \times 10^{-3} = 200 \ \Omega$$

$$Z_{ph} = \sqrt{R^2 + X_L^2} = \sqrt{200^2 + 200^2} = 283 \ \Omega$$

$$\cos \phi = R/Z_{ph} = 200/283 = \textbf{0.707 } lag$$

$$I_L = \frac{V_{ph}}{I_{ph}} = \frac{400/\sqrt{3}}{283} = 816 \times 10^{-3} \text{A} = \textbf{816 mA}$$

$$P_L = \sqrt{3} V_L I_L \cos \phi = \sqrt{3} \times 400 \times 816 \times 10^{-3} \times 0.707 = \textbf{400 W}$$

$$Q_L = \sqrt{3} \ V_L I_L \sin \phi = \sqrt{3} \times 400 \times 816 \times 10^{-3} \times 0.707 = 400 \text{ VAR } (lag)$$

(*ii*) **Capacitor Circuit.**
$$X_C = \frac{1}{2\pi f C} = \frac{1}{2\pi \times 60 \times 1 \times 10^{-6}} = 2650 \ \Omega$$

$$I_{ph} = V_L/X_C = 400/2650 = 151 \times 10^{-3} \text{A} = 151 \text{ mA}$$

$$\therefore \qquad I_L = \sqrt{3} \ I_{ph} = \sqrt{3} \times 151 = \textbf{261 mA}$$

$$\cos \phi = R/Z = 0/2650 = \textbf{0}$$

$$P_C = \sqrt{3} \ V_L I_L \cos \phi = \sqrt{3} \times 400 \times 261 \times 10^{-3} \times 0 = \textbf{0 W}$$

$$Q_C = \sqrt{3} \ V_L I_L \sin \phi = \sqrt{3} \times 400 \times 261 \times 10^{-3} \times 1 = 181 \text{ VAR } (lead)$$

(*iii*) **All components present.**

$$P = P_L + P_C = 400 + 0 = \textbf{400 W}$$

$$Q = Q_L - Q_C = 400 - 181 = 219 \text{ VAR } (lag)$$

$$\phi = \tan^{-1} Q/P = \tan^{-1} 219/400 = 28.7°$$

$$\therefore \qquad \cos \phi = \cos 28.7° = \textbf{0.877 } lag$$

Now
$$P = \sqrt{3} \ V_L I_L \cos \phi$$

$$\therefore \qquad I_L = \frac{P}{\sqrt{3} \ V_L \cos \phi} = \frac{400}{\sqrt{3} \times 400 \times 0.877} = 658 \times 10^{-3} \text{A} = \textbf{658 mA}$$

Note that the capacitors increase the power factor from 0.707 to 0.877 and this produced a
reduction in line current from 816 mA to 658 mA without any reduction in load power dissipation.

Example 15.34. *In the circuit shown in Fig. 15.51, find the line current and total power.*

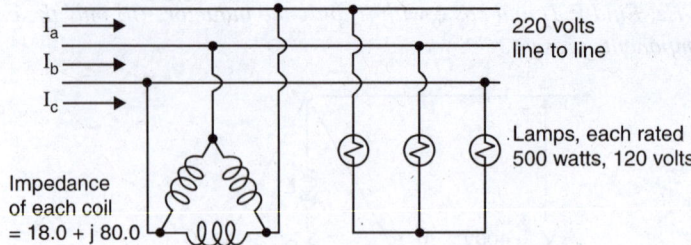

Fig. 15.51

Solution. The phase voltage V_{an} = $220 \angle 0° / \sqrt{3}$ = $127 \angle 0°$ V

$$Z_1 = R + j\,0 = \frac{(\text{Rated } V)^2}{P} = \frac{120^2}{500} = 28.8 \ \Omega$$

Coil impedance, Z_2 = $(18 + j\,80)\ \Omega$

Equivalent star impedance/phase = $\dfrac{Z_2}{3} = \dfrac{18 + j\,80}{3} = \dfrac{82 \angle 77.6°}{3} = 27.3 \angle 77.6°\ \Omega$

Since lamp and coil impedances are in parallel, it is convenient to work considering admittances.

Total admittance/phase, Y_{ph} = $\dfrac{1}{R} + \dfrac{1}{27.3 \angle 77.6°} = \dfrac{1}{28.8} + 0.0366 \angle -77.6°$

$$= (0.0347) + (0.008 - j\,0.0357) = (0.0427 - j\,0.0357)\ \text{S}$$

∴   Line current, I_a = $Y_{ph} V_{an}$ = $(0.0427 - j\,0.0357)\,(127)$

$$= (5.42 - j\,4.54)\ \text{A} = \mathbf{7.06 \angle -39.9°\ A}$$

Power per phase = $127 \times 7.06 \times \cos 39.9° = 688$ W

Total power, P = $3 \times 688 = \mathbf{2064\ W}$

Example 15.35. *A 440 V, 3-phase supply is connected through three single phase transformers to a load consisting of three 10 Ω resistors connected in delta. The transformers are connected in delta on the primary side and in star on the secondary side. The turn ratio is 2 : 1. Neglecting magnetising current and losses, calculate (i) the current in each resistor (ii) the load in kW (iii) the load line current (iv) the transformer primary and secondary phase currents.*

Solution. The primary phase voltage = 440 V. Since the turn ratio is 2 : 1, the secondary phase voltage = 440/2 = 220 V.

(i) ∴ Secondary (star connection) line voltage

$$= 220\sqrt{3}\ \text{V} = \text{Voltage across each } 10\ \Omega \text{ resistor}$$

∴   Current in 10 Ω resistor = $220\sqrt{3}/10$ = **38.1 A**

(ii)   Load = $3 \times I^2 R = 3 \times (38.1)^2 \times 10 = 43.6 \times 10^3$ W = **43.6 kW**

(iii)   Load line current = $38.1 \times \sqrt{3}$ = **66 A**

(iv) Transformer secondary phase current = **66A**

             Primary current = 66/2 = **33 A**

Example 15.36. *A star-connected load having R = 42.6 Ω/phase and X_L = 32 Ω/phase is connected across 400 V, 3-phase supply. Calculate :*

(i) Line current, reactive power and power loss.

(ii) Line current when one of the load becomes open-circuited.

Solution. (i) Fig. 15.52 (i) shows the conditions of the problem.

          Impedance/phase, Z_{ph} = $(42.6 + j\,32)\ \Omega$

$$\therefore \qquad Z_{ph} = \sqrt{(42.6)^2 + (32)^2} = 53.28 \ \Omega \ ; \ \theta = \tan^{-1}\frac{32}{42.6} = 36.9°$$

Therefore, phase angle $\phi = -36.9°$ *i.e.* p.f. is lagging.

$$\text{Phase current, } I_{ph} = \frac{V_{ph}}{Z_{ph}} = \frac{400/\sqrt{3}}{53.28} = 4.336 \ A$$

$\therefore \qquad$ Line current, $I_L = I_{ph} = $ **4.336 A**

$$\text{Total reactive power, } Q = 3 \ V_{ph} I_{ph} \sin \phi = 3 \times \frac{400}{\sqrt{3}} \times 4.336 \times \sin 36.9° = \textbf{1803 VAR}$$

$$\text{Total power loss, } P = 3 \ V_{ph} I_{ph} \cos \phi = 3 \times \frac{400}{\sqrt{3}} \times 4.336 \times \cos 36.9° = \textbf{2404 W}$$

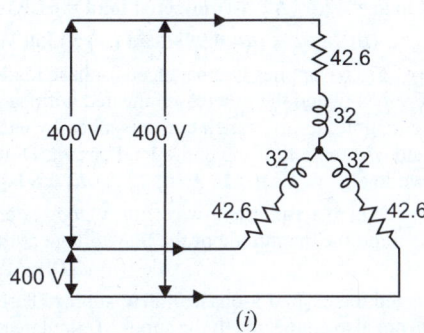 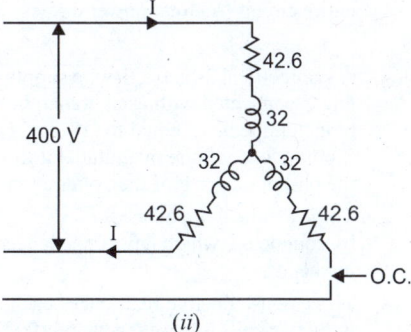

(*i*) (*ii*)

Fig. 15.52

(*ii*) When one of the loads is open-circuited, the circuit becomes as shown in Fig. 15.52 (*ii*). It is clear that now two "phase-impedances" are in series across the line voltage $V_L = 400$ V. Therefore, the circuit impedance is $2Z_{ph}$ and the supply voltage is 400 V.

$\therefore$ Current in the two unaffected lines is

$$I = \frac{V_L}{2Z_{ph}} = \frac{400}{2 \times 53.28} = \textbf{3.754 A}$$

Note that third line, being open, does not carry any current.

Example 15.37. *A 415 V, 3-ϕ generator supplies power to both a delta and a star-connected load connected in parallel. All the phase impedances are identical and specifically equal to (5 + j 8.66) Ω. Compute the total generator current in each line.*

Solution. The phase sequence is *abc*. Line voltage, $V_L = 415$ V.

Impedance/phase, $\mathbf{Z}_{ph} = (5 + j \ 8.66) \ \Omega = 10 \angle 60° \ \Omega$

Line current drawn by delta-connected load in line *a* is

$$\mathbf{I}_\Delta = \frac{\sqrt{3}V_{ph}}{Z_{ph}} = \sqrt{3} \times \frac{V_a}{Z_{ph}} = \sqrt{3} \times \frac{415 \angle 0°}{10 \angle 60°} = 71.88 \angle -60° \ A$$

Line current drawn by star-connected load is line *a* is

$$\mathbf{I}_{star} = \mathbf{I}_{ph} = \frac{V_a/\sqrt{3}}{Z_{ph}} = \frac{415/\sqrt{3} \angle 0°}{10 \angle 60°} = 23.96 \angle -60° \ A$$

$\therefore$ Total line current in line *a* is

$$\mathbf{I}_a = \mathbf{I}_\Delta + \mathbf{I}_{star} = 71.88 \angle -60° + 23.96 \angle -60°$$
$$= (35.94 - j \ 62.26) + (11.98 - j \ 20.75)$$
$$= (47.92 - j \ 83.01) \ A = \textbf{95.84} \ \angle -\textbf{60° A}$$

Since the system is balanced, the three line currents are equal in magnitude but are displaced 120° from one another. Therefore, currents in the other two lines *b* and *c* are :

$$I_b = 95.84 \angle -60° - 120° = \mathbf{95.84 \angle -180° \, A}$$

$$I_c = 95.84 \angle -180° - 120° = 95.84 \angle -300° \, A = \mathbf{95.84 \angle 60° \, A}$$

Tutorial Problems

1. A 440 V, 3-phase, Δ-connected induction motor has an output of 14.96 kW at a p.f. of 0.82 and efficiency 85%. If another balanced Y-connected load of 10 kW at 0.85 p.f. lagging is added in parallel, find (*i*) the total power drawn (*ii*) power factor of the combined load (*iii*) current drawn from the line.

    [(*i*) 27.6 kW (*ii*) 0.83 lag (*iii*) 43.63 A]

2. A 400 V, 3-phase, 3-wire system feeds balanced 3-phase loads in parallel; one Y-connected with each branch impedance equal to $(5 + j\,5)\ \Omega$ and the other Δ-connected with each branch impedance equal to $(10 - j\,5)\ \Omega$. Find (*i*) current in each load phase (*ii*) line current drawn from the supply (*iii*) the p.f. of the entire circuit (*iv*) total power drawn.   **[(*i*) Y-connected load = 32.67 A ; Δ-connected load = 61.93 A**

    (*ii*) 78.62 A (*iii*) 0.998 lead (*iv*) 54360 W]

3. A symmetrical 3-phase, 3-wire supply with a line voltage of 173 V supplies two balanced 3-phase loads; one Y-connected with each branch impedance equal to $(6 + j\,8)\ \Omega$ and the other Δ-connected with each branch impedance equal to $(18 + j\,24)\Omega$. Calculate (*i*) the magnitudes of branch currents taken by each 3-phase load (*ii*) the magnitude of the total line current and (*iii*) the p.f. of the entire load circuit. Draw the phasor diagram of the voltages and currents for the two loads.   **[(*i*) 10 A (*ii*) 20 A (*iii*) 0.6 lag]**

4. Three identical impedances of $30 \angle 30°\ \Omega$ are connected in delta to a 3-phase, 3-wire, 208 V *abc* system by conductors which have impedances of $(0.8 + j\,0.63)\ \Omega$. Find the magnitude of the line voltage at the load end.   **[109.3 A]**

5. Three non-inductive resistances each of 100 Ω are connected in star to a 3-phase, 440 V supply. Three inductive coils each of reactance 100 Ω connected in delta are also connected to the supply. Calculate (*i*) magnitude of line currents and (*ii*) p.f. of the system.   **[(*i*) 8.03 A (*ii*) 0.316 (*lag*)]**

15.19. Use of Single-Phase Wattmeter

Before discussing the special techniques employed to measure power in a 3-phase circuit, it is desirable to consider how a single-phase wattmeter is used to measure power in a single-phase circuit.

A single-phase wattmeter consists of two coils, one fixed and one movable carrying a pointer which moves over a scale. The fixed coil (called **current coil**) is of low resistance and is inserted in series with the line so that it carries the line current. The movable coil (called **potential coil**) is of high resistance and is connected like a voltmeter across the line. The small current in the potential coil is thus equal to the input voltage divided by the resistance of the potential coil. The wattmeter *deflection is, therefore, proportional to the average power

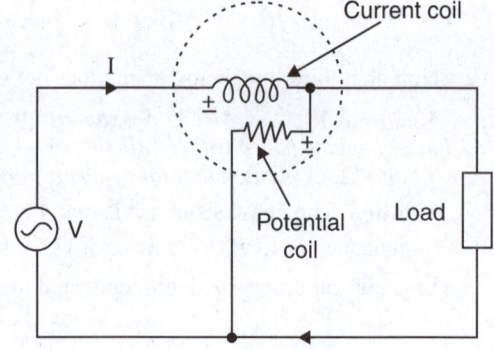

Current coil

Potential coil

Load

Fig. 15.53

---

*   Let the alternating voltage and current be :

$$v = V_m \sin \omega t \quad ; \quad i = I_m \sin (\omega t - \phi)$$

     Instantaneous power, $p = v\,i$

But the wattmeter reads average power *i.e.* power over one cycle.

$$\therefore \qquad \text{Wattmeter reading} = \frac{1}{2\pi} \int_0^{2\pi} v\,i\,d(\omega t) = \frac{1}{2\pi} \int_0^{2\pi} V_m I_m \sin \omega t \, \sin (\omega t - \phi) \, d(\omega t)$$

$$= \frac{V_m I_m}{2} \cos \phi = VI \cos \phi$$

where V = r.m.s. value of voltage across potential coil

                      I = r.m.s. value of current in current coil

                      ϕ = phase angle between V and I

(*i.e. V I* cos φ) delivered to the circuit. Fig. 15.53 shows connections of the wattmeter to measure power in a single phase circuit.

It is clear that a wattmeter has four terminals ; two for current coil and two for potential coil. When connected in the circuit to measure power, sometimes it gives backward deflection *i.e.* it reads downscale. This is due to the improper connections. As shown in Fig. 15.53, one current coil and one potential coil terminals are marked, usually ± on an actual instrument. To obtain upscale reading on the meter, ± terminal of current coil is connected to the *line* side of the circuit and ± terminal of potenital coil is connected to the same line lead as the current coil (See Fig. 15.53). A reversal of either coil connection will result in a backward deflection.

15.20. Power Measurement in 3-phase Circuits

At first glance, the measurement of the power drawn by a 3-phase load seems to be a simple problem. We need place only one wattmeter in each of the three phases and add the three readings algebraically to obtain the total power. This procedure for the measurement of power in a 3-phase circuit is certainly impracticable if not impossible. It is because the neutral of a Y-connected load is usually not accessible and the phases of Δ-connected load are not available for insertion of wattmeter in each phase. *In most of the *practical 3-phase circuits, only three line terminals are available.* It is obvious that we need a method for measuring the total power drawn by a 3-phase load having only three accessible terminals (*i.e.* the three lines). Two such methods available are :

(*i*) Three-wattmeter method   (*ii*) Two-wattmeter method

Both these methods can measure power in a 3-phase load (*Y* or Δ) whether the load is balanced or not. *The two-wattmeter method is widely used for the measurement of 3-phase power due to its many advantages.*

Note. If the load is balanced, the power in any phase can be measured by a single wattmeter. The total circuit power is given by multiplying the wattmeter reading by three. This method can only be used if the load is balanced. Its principal disadvantage is that it is not always possible to make the required connections.

15.21. Three-Wattmeter Method

In this method, the three wattmeters are connected in such a way that each has its current coil in one line and its potential coil between that line and some **common point *x* as shown in Fig. 15.54. It can be shown mathematically that the algebraic sum of the readings of the three wattmeters gives the total power whether the load is balanced or not *i.e.*

$$\text{Total power} = W_1 + W_2 + W_3 \qquad ...algebraic\ sum$$

Although a *Y*-connected load has been considered, the above arguments are equally valid for a Δ-connected load, balanced or unbalanced. A *caution* if neutral wire is available. Then common point *x* should be at the neutral wire.

Proof. We shall now prove mathematically that the algebraic sum of the three wattmeter readings is equal to the total power drawn by the *Y*-load in Fig. 15.54.

---

*   For example, in a 3-phase induction motor, usually three line terminals are available.

**   The point *x* may be some specified point in the three-phase system or it may be merely a point in space at which the three potential coils have a common junction.

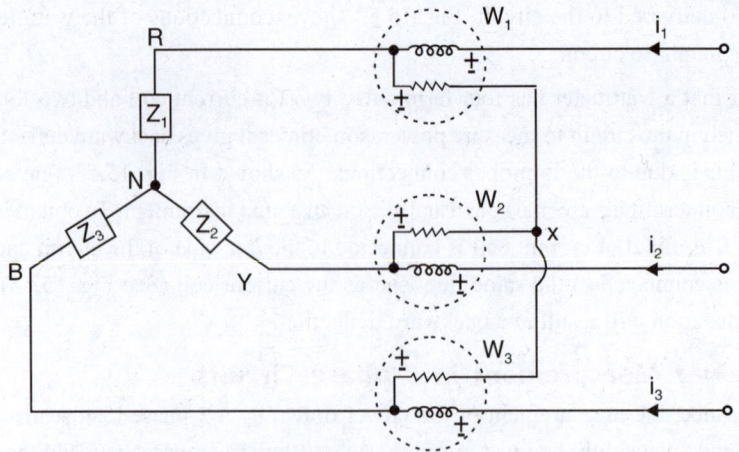

Fig. 15.54

Average power indicated by wattmeter W_1 is

$$W_1 = \frac{1}{T}\int_0^T {}^*v_{Rx}\, i_1\, dt$$

where T is the period for all the voltages.

Similarly,

$$W_2 = \frac{1}{T}\int_0^T v_{Yx}\, i_2\, dt \text{ and } W_3 = \frac{1}{T}\int_0^T v_{Bx}\, i_3\, dt$$

$$\therefore \quad W_1 + W_2 + W_3 = \frac{1}{T}\int_0^T (v_{Rx}\, i_1 + v_{Yx}\, i_2 + v_{Bx}\, i_3)\, dt \qquad \ldots(i)$$

Each of the three voltages in the above expression can be written in terms of phase voltage and voltage between point x and the neutral N as :

$$v_{Rx} = v_{RN} + v_{Nx} \; ; \; v_{Yx} = v_{YN} + v_{Nx} \; ; \; v_{Bx} = v_{BN} + v_{Nx}$$

Substituting the values of v_{Rx}, v_{Yx} and v_{Bx} in exp. (i), we have,

$$W_1 + W_2 + W_3 = \frac{1}{T}\int_0^T (v_{RN}\, i_1 + v_{YN}\, i_2 + v_{BN}\, i_3)\, dt + \frac{1}{T}\int_0^T v_{Nx}\,(i_1 + i_2 + i_3)\, dt$$

Since $i_1 + i_2 + i_3 = 0$,

$$\therefore \quad W_1 + W_2 + W_3 = \frac{1}{T}\int_0^T (v_{RN}i_1 + v_{YN}i_2 + v_{BN}i_3)\,dt \qquad \ldots(ii)$$

The R.H.S. of expression (ii) is the sum of the average powers taken by each phase of the load.

$$\therefore \quad W_1 + W_2 + W_3 = \text{Total power}$$

Note. If neutral wire is available, then point x is to be connected to the neutral. In that case [replacing x by N in exp. (i)],

$$W_1 + W_2 + W_3 = \frac{1}{T}\int_0^T (v_{RN}\, i_1 + v_{YN}\, i_2 + v_{BN}\, i_3)\, dt \; = \text{Total Power}$$

It may be noted that a single-phase wattmeter reads the product of the r.m.s. current in its current coil, the r.m.s. voltage across its potential coil and the cosine of the phase angle between these two quantities.

---

* The current in the current coil of wattmeter W_1 is i_1 and voltage across its potential coil is v_{Rx} (*i.e.* voltage between points R and x).

15.22. Two-Wattmeter Method

In fact, one of the three wattmeters used to measure 3-phase power above is superfluous. We can measure power in a 3-phase load (Y or Δ, balanced or unbalanced) by using two wattmeters only. This can be easily proved. The point x, the common connection of the potential coils (See Fig. 15.54) can be located anywhere without affecting the algebraic sum of the three wattmeters. It is because the point x can be at any potential. Thus, if one end of each potential coil in Fig. 15.54 is connected to Y-line, then voltage across the potential coil of wattmeter W_2 will be zero and this wattmeter reads zero.

Now, $W_1 + W_2 + W_3$ = Total power

or $W_1 + 0 + W_3$ = Total power

or $W_1 + W_3$ = Total power

The wattmeter W_2 may, therefore, be removed and the algebraic sum of the remaining two wattmeters readings is still equal to total power.

As a matter of fact, 2-wattmeter method has become universal method for the measurement of power in a 3-phase circuit. The principal advantage is that the algebraic sum of the readings of the two wattmeters indicates the total power regardless of (*i*) load unbalance (*ii*) source unbalance (*iii*) difference in wattmeters (*iv*) wave-form of the source and (*v*) phase sequence.

15.23. Proof for Two-Wattmeter Method

In two-wattmeter method for the measurement of 3-phase power, *the current coils of the two wattmeters are connected in any two lines and the potential coil of each joined to the third line* as shown in Fig. 15.55. We now prove from first principles that the algebraic sum of the readings of the two wattmeters gives the total power drawn by the 3-phase load (Y or Δ, balanced or unbalanced). Fig. 15.55 shows the Y-connected load.

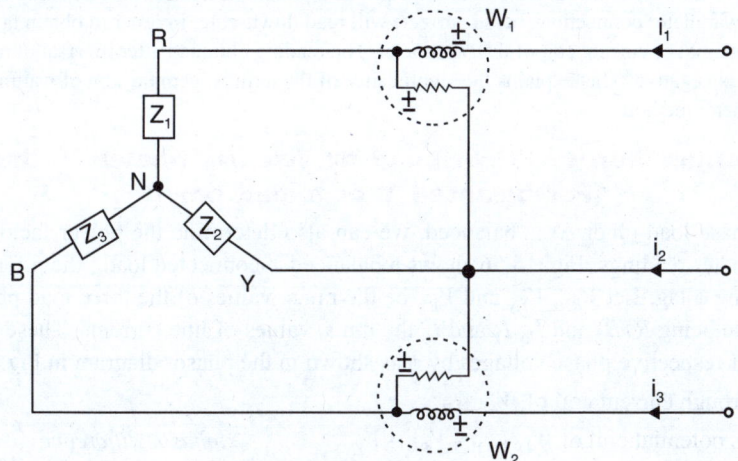

Fig. 15.55

Average power indicated by wattmeter W_1 is

$$W_1 = \frac{1}{T}\int_0^T v_{RY}\, i_1\, dt$$

where T is the period for all the voltage sources.

Similarly, $$W_2 = \frac{1}{T}\int_0^T v_{BY}\, i_3\, dt$$

$$\therefore \quad W_1 + W_2 = \frac{1}{T}\int_0^T (v_{RY}\, i_1 + v_{BY}\, i_3)dt \qquad \ldots(i)$$

Each of the above two voltages can be written as :

$$v_{RY} = v_{RN} + v_{NY}\ ;\quad v_{BY} = v_{BN} + v_{NY}$$

Substituting the values of v_{RY} and v_{BY} in exp. (i), we have,

$$W_1 + W_2 = \frac{1}{T}\int_0^T (v_{RN}\, i_1 + v_{BN}\, i_3)dt + \frac{1}{T}\int_0^T v_{NY}(i_1 + i_3)dt$$

Since $i_1 + i_2 + i_3 = 0$ $\quad \therefore\ i_1 + i_3 = -i_2$

$$\therefore \quad W_1 + W_2 = \frac{1}{T}\int_0^T (v_{RN}\, i_1 + v_{BN}\, i_3)dt + \frac{1}{T}\int_0^T v_{NY}(-i_2)dt$$

$$= \frac{1}{T}\int_0^T (v_{RN}\, i_1 + v_{BN}\, i_3)dt + \frac{1}{T}\int_0^T v_{YN}(i_2)dt$$

$$\therefore \quad W_1 + W_2 = \frac{1}{T}\int_0^T (v_{RN}\, i_1 + v_{YN}\, i_2 + v_{BN}\, i_3)dt \qquad \ldots(ii)$$

The R.H.S. of exp. (ii) is the sum of average powers taken by each phase of the load.

$$\therefore \quad W_1 + W_2 = \text{Total power}$$

Note. With single phase loads, a wattmeter will deflect backward only if it is improperly connected. This is also true when the three-wattmeter method is used to measure power in a 3-phase circuit. *However, in two-wattmeter method, the reverse deflection may be due to improper connections of the wattmeter or low power factor of the load.* As we shall see in Art. 15.25, if the p.f. of 3-phase balanced load is less than 0.5, then one of the wattmeters (wattmeter connections being correct) will read downscale. In order to obtain upsacle reading, reverse either potential or current coil of this wattmeter. The reading obtained after reversal of coil connection should be taken as negative. This explains the significance of the term "algebraic sum of wattmeter readings" in the two-wattmeter method.

15.24. Determination of P.F. of Load by Two-wattmeter Method (For balanced Y or Δ load only)

If the 3-phase load (Y or Δ) is balanced, we can also determine the power factor of the load from the wattmeter readings. Fig. 15.56 shows a balanced Y-connected load ; the p.f. angle of load impedance being ϕ lag. Let V_{RN}, V_{YN} and V_{BN} be the r.m.s. values of the three load phase voltages (phase sequence being RYB) and I_R, I_Y and I_B the r.m.s. values of line currents. These currents will lag behind their respective phase voltages by ϕ as shown in the phasor diagram in Fig. 15.57.

Current through current coil of $W_1 = I_R$

P.D. across potential coil of W_1, $V_{RY} = V_{RN} - V_{YN}$ ...*phasor difference*

To obtain V_{RY}, find the phasor sum of V_{RN} and $- V_{YN}$ as shown in Fig. 15.57. It is clear from the phasor diagram that phase angle between V_{RY} and I_R is $(30° + \phi)$.

$$\therefore \quad W_1 = V_{RY}\, I_R \cos(30° + \phi)$$

Current through current coil of $W_2 = I_B$

P.D. across potential coil of W_2 is

$$V_{BY} = V_{BN} - V_{YN} \qquad \text{...\textit{phasor difference}}$$

To obtain V_{BY}, find the phasor sum of V_{BN} and $- V_{YN}$ as shown in Fig. 15.57. It is clear from the phasor diagram that phase angle between V_{BY} and I_B is $(30° - \phi)$.

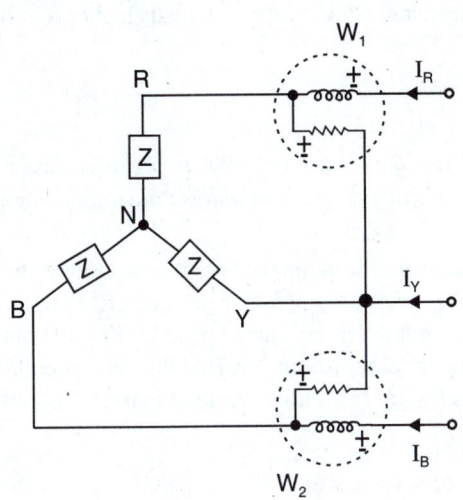

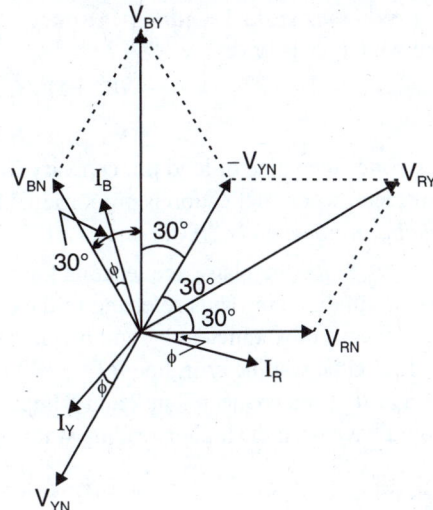

Fig. 15.56   **Fig. 15.57**

$\therefore$ $\qquad\qquad\qquad W_2 = V_{BY} I_B \cos (30° - \phi)$

Since the load is balanced, $V_{RY} = V_{BY} =$ Line voltage, V_L and $I_R = I_B =$ Line current, I_L

$\therefore$ $\qquad\qquad\qquad W_1 = V_L I_L \cos (30° + \phi)$

and $\qquad\qquad\qquad W_2 = V_L I_L \cos (30° - \phi)$

Total power. $\qquad W_1 + W_2 = V_L I_L [\cos (30° + \phi) + \cos (30° - \phi)]$

$\qquad\qquad\qquad\qquad = V_L I_L [(\cos 30° \cos \phi - \sin 30° \sin \phi)$

$\qquad\qquad\qquad\qquad\quad + (\cos 30° \cos \phi + \sin 30° \sin \phi)]$

$\qquad\qquad\qquad\qquad = V_L I_L (2 \cos 30° \cos \phi)$

$\qquad\qquad\qquad\qquad = \sqrt{3} \, V_L I_L \cos \phi$

$\qquad\qquad\qquad\qquad =$ Total power in the 3-phase load

Hence the algebraic sum of the two wattmeter readings gives the total power consumed in the 3-phase load.

Power factor. $\qquad\qquad W_2 = V_L I_L \cos (30° - \phi)$

$\qquad\qquad\qquad\qquad W_1 = V_L I_L \cos (30° + \phi)$

$\therefore$ $\qquad\qquad W_2 + W_1 = \sqrt{3} \, V_L I_L \cos \phi$ $\qquad$ ...as proved above

Now let us take the difference between the two readings, *subtracting the smaller reading from the larger*. Calling W_2 the higher-reading wattmeter, we get,

$\qquad\qquad\qquad *W_2 - W_1 = V_L I_L \sin \phi$

$\therefore$ $\qquad\qquad\qquad \tan \phi = \sqrt{3} \dfrac{W_2 - W_1}{W_2 + W_1}$

Thus from the two wattmeter readings, we can find ϕ and hence the load power factor $\cos \phi$. It is interesting to note that the two wattmeters give equal readings when the load power factor is unity *i.e.* $\cos \phi = 1$ (See Art. 15.25).

---

* $\qquad\qquad W_2 - W_1 = V_L I_L [\cos (30° - \phi) - \cos (30° + \phi)]$

$\qquad\qquad\qquad\qquad = V_L I_L [(\cos 30° \cos \phi + \sin 30° \sin \phi) - (\cos 30° \cos \phi - \sin 30° \sin \phi)]$

$\qquad\qquad\qquad\qquad = V_L I_L (2 \sin 30° \sin \phi) = V_L I_L \sin \phi$

Which is higher-reading wattmeter ? How can we tell which wattmeter reads higher (*i.e.* W_2) and which reads lower (*i.e.* W_1) ?

(i)
$$W_2 = V_L I_L \cos (30° - \phi)$$
$$W_1 = V_L I_L \cos (30° + \phi)$$

Since the value of load p.f. can vary from 0 to 1 (*i.e.* ϕ can vary from 90° to 0°), it is clear that wattmeter whose deflection is proportional to (30° − ϕ) is *always positive and is *always the higher reading wattmeter* (*i.e.* W_2 in this case).

(ii) Since the phase sequence is known, we can even tell the higher-reading wattmeter from the circuit diagram. Thus referring to the circuit in Fig. 15.56, the phase sequence is *RYB* and that current coils of wattmeters W_1 and W_2 are connected in *R* and *B* lines. Since V_{BN} leads V_{RN} (of course by 120°), the wattmeter in line *B* (*i.e.* W_2) is the higher reading wattmeter. Had the current coils of W_1 and W_2 been connected in *R* and *Y* lines, V_{RN} leads V_{YN} by 120°, the wattmeter in line *R* (*i.e.*, W_1) would have been the higher-reading wattmeter.

15.25. Effect of Load p.f. on Wattmeter Readings

We have seen above (See Fig. 15.57) that for *lagging* load (balanced) power factor of cos ϕ, the two wattmeter readings are :
$$W_2 = V_L I_L \cos (30° - \phi)$$
$$W_1 = V_L I_L \cos (30° + \phi)$$

It is clear that readings of the two wattmeters depend upon load p.f. angle ϕ.

(i) **When p.f. is unity** (*i.e.* ϕ = 0°).
$$W_2 = V_L I_L \cos 30°$$
$$W_1 = V_L I_L \cos 30°$$

Both wattmeters indicate equal and positive (*i.e.* upscale) readings.

(ii) **When p.f. is 0.5** (*i.e.* ϕ = 60°).
$$W_2 = V_L I_L \cos 30°$$
$$W_1 = V_L I_L \cos 90° = 0$$

Hence total power is measured by wattmeter W_2 alone.

(iii) **When p.f. is less than 0.5 but greater than 0** (*i.e.* 90° > ϕ > 60°).
$$W_2 = \text{positive reading}$$
$$W_1 = \text{negative reading}$$

The wattmeter W_2 reads positive (*i.e.* upscale) because for the given conditions (*i.e.* 90° > ϕ > 60°), the phase angle between voltage and current will be less than 90°. However, for wattmeter W_1, the phase angle between voltage and current shall be more than 90° and hence this wattmeter gives negative (*i.e.* downscale) reading. In order to obtain upscale reading on wattmeter W_1, reverse its **potential or current coil. The reading obtained after reversal of coil connection should be taken as negative.

$$\text{Total power} = W_2 + (-W_1) \qquad \text{...}algebraic\ sum$$
$$= W_2 - W_1$$

Note that total power is obtained by subtracting the reading of W_1 from W_2. We arrive at a very important conclusion that *if load p.f. is less than 0.5, then lower reading wattmeter (i.e. W_1 in this case) will give negative reading.*

---

*   Except for the case when load p.f. is unity (*i.e.* ϕ = 0°) at which the two wattmeters have equal readings.

**   In practice, potential coil connections are reversed because it may be done more quickly than the current coil.

(*iv*) **When p.f. is zero** (*i.e.* $\phi = 90°$). Such a case will occur when the load consists of pure inductance and/or capacitance.

$$W_2 = V_L I_L \cos (30° - 90°) = V_L I_L \sin 30°$$
$$W_1 = V_L I_L \cos (30° + 90°) = - V_L I_L \sin 30°$$

Thus the two wattmeters will read equal and opposite.

$$\therefore \qquad W_1 + W_2 = 0$$

The above facts are summarised below in the tabular form.

ϕ	0°	60°	More than 60°	90°
$\cos \phi$	1	0.5	< 0.5	0
W_2	positive	positive	positive	positive
W_1	positive	0	negative	negative
Conclusion	$W_1 = W_2$ Total power $= W_1 + W_2$	$W_1 = 0$ Total power $= W_2$	Total power $= W_2 - W_1$	$W_2 = - W_1$ Total power $= 0$

The following points may be noted carefully :

(*i*) *The wattmeter whose deflection is proportional to (30° – ϕ) (i.e. W_2 in this case) is always positive and is the higher-reading wattmeter.*

(*ii*) *The wattmeter whose deflection is proportional to (30° + ϕ) (i.e., W_1 in this case) is the lower-reading wattmeter.*

(*iii*) *The negative reading will only be obtained on the lower-reading wattmeter (i.e. W_1 in this case) and that too when the load p.f. is less than 0.5.*

15.26. Leading Power Factor

In discussing two-wattmeter method for measuring power in a 3-phase balanced load, we have considered lagging load power factor *i.e.* p.f. angle ϕ is considered positive. For leading power factor, angle ϕ becomes negative. Therefore, by putting the value of load p.f. angle as $- \phi$ in the readings of the wattmeters above, we have,

$$W_2 = V_L I_L \cos (30° + \phi)$$
$$W_1 = V_L I_L \cos (30° - \phi)$$

Note that effect of leading power factor is that the readings of the two wattmeters are interchanged. Now wattmeter W_1 has become the higher-reading wattmeter. All other discussions remain the same as for the lagging power factor.

$$W_1 + W_2 = V_L I_L [\cos (30° - \phi) + \cos(30° + \phi)]$$
$$= \sqrt{3} \, V_L I_L \cos \phi \qquad \text{...As proved in Art. 15.24}$$
$$W_1 - W_2 = V_L I_L [\cos (30° - \phi) - \cos (30° + \phi)]$$
$$= V_L I_L \sin \phi \qquad \text{...As proved in Art. 15.24}$$
$$\therefore \qquad \tan \phi = \sqrt{3} \, \frac{W_1 - W_2}{W_1 + W_2}$$

Compare this expression with that for lagging power factor. Here W_1 is the higher-reading wattmeter instead of W_2 in case of lagging power factor.

15.27. How to Apply p.f. Formula ?

As shown above, the value of tan ϕ (and hence the p.f. cos ϕ) can be determined from the two wattmeter readings.

$$\tan \phi = \sqrt{3}\,\frac{W_2 - W_1}{W_2 + W_1} \qquad\qquad \textit{...lagging p.f.}$$

$$\tan \phi = \sqrt{3}\,\frac{W_1 - W_2}{W_1 + W_2} \qquad\qquad \textit{...leading p.f.}$$

It may be noted that for lagging p.f., W_2 is the higher-reading wattmeter whereas it is W_1 for leading p.f. To avoid any confusion, the reader is well advised to remember the following general formula for both lagging and leading p.f.

$$\tan \phi = \sqrt{3}\,\frac{\text{(Higher reading)} - \text{(Lower reading)}}{\text{(Higher reading)} + \text{(Lower reading)}}$$

You may recall that higher-reading wattmeter always reads positive. This means that positive value in the problem is to be substituted here. The lower-reading wattmeter may read positive or negative depending upon the load power factor. This value is substituted in the above formula with its proper sign ; + *ve* or negative.

For example, in the two-wattmeter method if the two wattmeter readings are 12·5 kW and – *4.8 kW, then,

$$\tan \phi = \sqrt{3}\,\frac{(12.5) - (-4.8)}{(12.5) + (-4.8)} = \sqrt{3}\,\frac{17.3}{7.7} = 3.89$$

$$\therefore \qquad\qquad \phi = \tan^{-1} 3.89 = 75.6°$$

$$\therefore \qquad\qquad \text{Power factor} = \cos \phi = \cos 75.6° = 0.2487$$

Note that we can only find the magnitude of p.f. with the above formula. It does not tell about the **nature of the load *i.e.* whether the p.f. is lagging or leading. In many cases, this can be judged from the nature of the load. For example, if the above load is a 3-phase induction motor, the p.f. must be lagging. If the load is an over-excited synchronous motor, the p.f. must be leading.

15.28. One-Wattmeter Method — Balanced Load

If the 3-phase load (Y or Δ) is balanced, then one wattmeter is sufficient to measure the power drawn by the load. There are two methods in which one wattmeter may be applied.

(*i*) **Artificial-star method.** In this method, two impedances each equal in value (magnitude and phase) to the impedance of the potential coil of the wattmeter are so connected that these two impedances and the potential coil form a Y-connected load as shown in Fig. 15.58. The wattmeter will read the phase power and the total power will be three times the wattmeter reading.

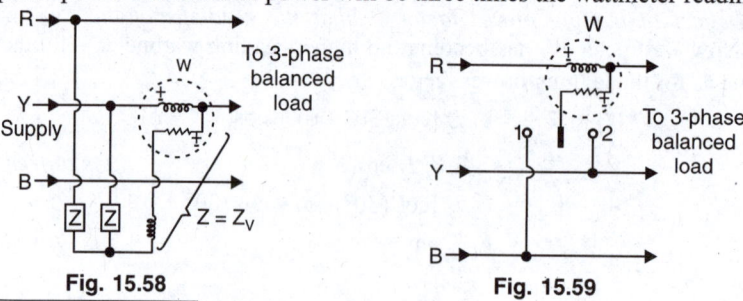

Fig. 15.58   **Fig. 15.59**

---

*   Note that a wattmeter cannot give negative reading. This reading is obtained after reversing the connections of either potential or current coil.

**   If there is no knowledge of the nature of 3-phase load, there is a simple method to resolve it. A high-impedance reactive load (*e.g.* 3-phase capacitor load) is connected across the 3-phase load. The load must become more capacitive. If the magnitude of $\tan \phi$ (or magnitude of ϕ) decreases, then the 3-phase load is inductive and p.f. is lagging. If the magnitude of $\tan \phi$ (or magnitude of ϕ) increases, then the 3-phase load is capacitive and p.f. is leading.

(ii) **Double-reading method.** In this method, two readings of the two-wattmeter method are taken with a single wattmeter as shown in Fig. 15.59. The current coil of the wattmeter is connected in any one line and the pressure coil is connected alternately between this and the other two lines. The algebraic sum of the two readings gives the total power drawn by the balanced 3-phase load.

15.29. Reactive Power with Two-Wattmeter Method

The reactive power (*i.e.* VAR) in a 3-phase balanced load can be determined from the readings of the two-wattmeter method.

$$W_2 - W_1 = V_L I_L \sin \phi \qquad \qquad \text{...See Art. 15.24}$$

Multiplying throughout by $\sqrt{3}$, we get,

$$\sqrt{3}(W_2 - W_1) = \sqrt{3}\, V_L I_L \sin \phi$$

But $\sqrt{3}\, V_L I_L \sin \phi$ is equal to the reactive power taken by the 3-phase balanced load.

$$\therefore \qquad \text{Reactive power} = \sqrt{3}\, (*W_2 - W_1)\, \text{VAR}$$

15.30. Reactive Power with One Wattmeter

If the 3-phase load is balanced, then the reactive power drawn by the load can be determined by using a single wattmeter. For this purpose, the current coil of the wattmeter is connected in one line and the potential coil is connected across the other two lines as shown in Fig. 15.60. It can be proved that the total reactive power drawn by the load is equal to $\sqrt{3}$ times the reading of the wattmeter. Fig. 15.60 shows the connections of a single wattmeter for measuring reactive power in a 3-phase balanced load having lagging p.f. cos ϕ. Here current coil of the wattmeter is connected in the R-line and the potential coil is connected across the lines Y and B.

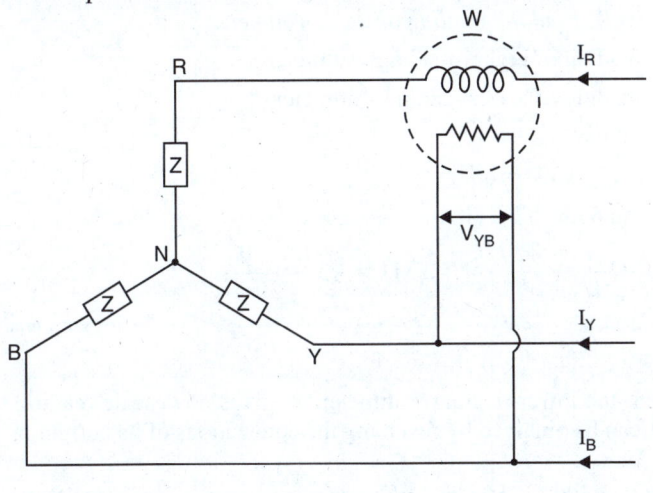

| Fig. 15.60 | Fig. 15.61 |

Current through current coil $= I_R$

P.D. across potential coil, $V_{YB} = V_{YN} - V_{BN}$ *...phasor difference*

To obtain V_{YB}, find the phasor sum of V_{YN} and $- V_{BN}$ as shown in the phasor diagram in Fig. 15.61. It is clear from the phasor diagram that the phase angle between V_{YB} and I_R is $(90° - \phi)$**.

---

*   Note that W_2 is the higher-reading wattmeter. Further $W_2 - W_1$ is the algebraic difference of two wattmeter readings.

**   Clearly angle (See phasor diagram in Fig. 15.61) between V_{RN} and V_{YB} is 90°. Since I_R lags behind V_{RN} by ϕ, angle between V_{YB} and I_R is $(90° - \phi)$.

$\therefore$　　　Wattmeter reading, $W = V_{YB} I_R \cos(90° - \phi) = V_{YB} I_R \sin \phi$

Since the load is balanced, $V_{YB} = V_L$, the line voltage and $I_R = I_L$, the line current.

$\therefore$　　　　　　　　　　$W = V_L I_L \sin \phi$

or　　　　　　$\sqrt{3} V_L I_L \sin \phi = \sqrt{3} \ W$

or　　　　　　Reactive power $= \sqrt{3} \times$ Wattmeter reading

Thus, if we multiply the wattmeter reading by $\sqrt{3}$, we get the total reactive power drawn by the 3-phase balanced load.

Example 15.38. *Two-wattmeter method is used to measure the power taken by a 3-phase induction motor on no load. The wattmeter readings are 375 W and –50 W. Calculate (i) power factor of the motor at no load (ii) phase difference of voltage and current in two wattmeters (iii) reactive power taken by the load.*

Solution. Higher-reading wattmeter, $W_2 = 375$ W ; Lower-reading wattmeter, $W_1 = -50$ W

(i)　　　　　$\tan \phi = \sqrt{3} \dfrac{W_2 - W_1}{W_2 + W_1} = \sqrt{3} \times \dfrac{375 - (-50)}{375 + (-50)} = \sqrt{3} \times \dfrac{425}{325} = 2.265$

$\therefore$　　　　　$\phi = \tan^{-1} 2.265 = 66.18°$

$\therefore$　No load p.f. $= \cos \phi = \cos 66.18° = \textbf{0.404} \textit{ lag}$

(ii) Phase angle in wattmeter $W_2 = 30° - \phi = 30° - 66.18° = \textbf{–36.18°}$

Phase angle in wattmeter $W_1 = 30° + \phi = 30° + 66.18° = \textbf{96.18°}$

(iii)　　　　Reactive power $= \sqrt{3}(W_2 - W_1) = \sqrt{3}(375 + 50) = \textbf{736.12 VAR}$

Example 15.39. *A 3-phase motor load has a p.f. of 0.397 lagging. Two wattmeters connected to measure power show the input as 30 kW. Find the reading on each wattmeter.*

Solution. Let　　　　$W_2 = $ reading of higher-reading wattmeter

　　　　　　　　$W_1 = $ reading of lower-reading wattmeter

$\therefore$　　　　$W_2 + W_1 = 30$ kW　　　　　　　　　　...(i)

Power factor angle, $\phi = \cos^{-1} 0.397 = 66.6°$

$\therefore$　　　　$\tan \phi = \tan 66.6° = 2.311$

Now,　　　　$\tan \phi = \sqrt{3} \dfrac{W_2 - W_1}{W_2 + W_1}$　or　$2.311 = \sqrt{3} \dfrac{W_2 - W_1}{30}$

$\therefore$　　　　$W_2 - W_1 = 40$ kW　　　　　　　　　　...(ii)

Solving eqs. (i) and (ii), we get, $W_2 = \textbf{35 kW}$; $W_1 = \textbf{–5 kW}$

Since the load p.f. is less than 0.5, the lower-reading wattmeter W_1 gives downscale reading. The upscale reading of this wattmeter can be obtained by reversing the connections of its current or potential coil.

Example 15.40. *The power input to a 2000 V, 50 Hz, 3-phase motor running on full-load at an efficiency of 90% is measured by two wattmeters which indicate 300 kW and 100 kW respectively. Find (i) the input (ii) the power factor (iii) the line current and output.*

Solution. Higher-reading wattmeter, $W_2 = 300$ kW ; Lower-reading wattmeter, $W_1 = 100$ kW

Since the motor is running on full-load, its p.f. must be greater than 0.5. Hence reading of W_1 will also be positive.

(i)　　　　Input to motor $= W_2 + W_1 = 300 + 100 = \textbf{400 kW}$

(ii)　　　　$\tan \phi = \sqrt{3} \dfrac{W_2 - W_1}{W_2 + W_1} = \sqrt{3} \dfrac{300 - 100}{300 + 100} = 0.866$

$\therefore$　　　　$\phi = \tan^{-1} 0.866 = 40.9°$

∴   Power factor $= \cos \phi = \cos 40.9° = $ **0.756** *lag*

(iii) $400 \times 10^3 = \sqrt{3}\, V_L I_L \cos \phi$

∴   Line current, $I_L = \dfrac{400 \times 10^3}{\sqrt{3} \times 2000 \times 0.756} = $ **152.74 A**

Motor output $=$ Input $\times 0.9 = 400 \times 0.9 = $ **360 kW**

Example 15.41. *Each of the two wattmeters connected to measure the input to a 3-phase induction motor reads 10 kW. If the power factor of the motor be changed to 0.866 lagging, determine the readings of the two wattmeters, the total input power remaining unchanged. Draw the phasor diagram in each case.*

Solution. In the first case, both wattmeters read equal and positive. Hence, the motor must be running at unity p.f.

$$W_1 = W_2 = 10 \text{ kW}$$

∴   Total power $= W_2 + W_1 = 10 + 10 = 20 \text{ kW}$

For the circuit diagram, refer back to Fig. 15.55. The phasor diagram at unity p.f. is shown in Fig. 15.62.

At 0.866 lagging p.f. As the power factor of the motor is changed to 0.866 lagging, one wattmeter, say W_2, reads higher and the other (*i.e.* W_1) reads lower. Since the load p.f. is greater than 0.5, both readings are positive and the sum of the two is 20 kW. Fig. 15.63 shows the phasor diagram for 0.866 lagging p.f.

$$W_2 + W_1 = 20 \text{ kW} \qquad\qquad\qquad ...(i)$$

$$\text{p.f. angle, } \phi = \cos^{-1} 0.866 = 30°$$

∴ $\tan \phi = \tan 30° = 0.577$

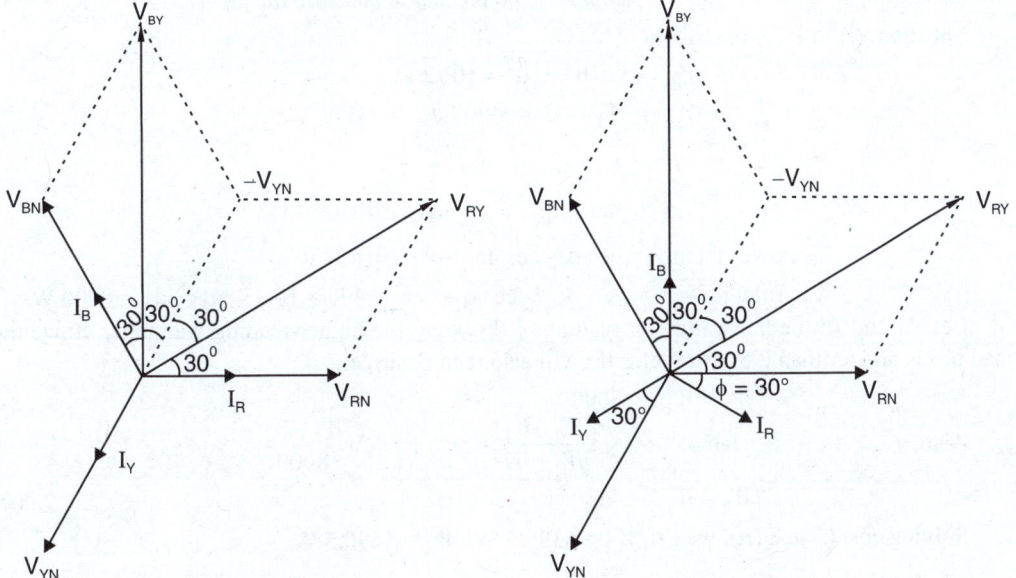

Fig. 15.62   **Fig. 15.63**

Now, $\tan \phi = \sqrt{3}\,\dfrac{W_2 - W_1}{W_2 + W_1}$ or $0.577 = \sqrt{3}\,\dfrac{W_2 - W_1}{20}$

∴ $W_2 - W_1 = 6.66 \text{ kW}$ ...(ii)

Solving eqs. (*i*) and (*ii*), we get, $W_2 = $ **13.33 kW**; $W_1 = $ **6.67 kW**

Example 15.42. *A 3-phase induction motor working on 400 V takes a line current of 30 A at a p.f. of 0.866 lagging. Two wattmeters are connected to measure the input power to the motor. What will be the wattmeter readings ? If the p.f. were changed from lagging to leading, the value remaining the same, find the new readings of the wattmeters, assuming the input power remains the same.*

Solution. When p.f. is 0.866 lagging.

$$\text{p.f. angle, } \phi = \cos^{-1} 0.866 = 30°$$

$$W_2 = V_L I_L \cos (30° - \phi) \qquad \qquad \text{...higher reading}$$

$$= 400 \times 30 \times \cos (30° - 30°)$$

$$= 400 \times 30 \times \cos 0° = \mathbf{12000 \ W}$$

$$W_1 = V_L I_L \cos (30° + \phi) \qquad \qquad \text{...lower-reading}$$

$$= 400 \times 30 \times \cos 60° = \mathbf{6000 \ W}$$

When p.f. is 0·866 leading.

$$W_1 = V_L I_L \cos (30° - \phi) \qquad \qquad \text{...higher reading}$$

$$= 400 \times 30 \times \cos 0° = \mathbf{12000 \ W}$$

$$W_2 = V_L I_L \cos (30° + \phi) \qquad \qquad \text{...lower-reading}$$

$$= 400 \times 30 \times \cos 60° = \mathbf{6000 \ W}$$

Hence W_1 reads 12000 W and W_2 reads 6000 W *i.e.* wattmeter readings are interchanged.

Example 15.43. *Three identical coils, each having a resistance of 10 Ω and a reactance of 10 Ω are connected in (i) star (ii) delta, across 400 V, 3-phase supply. Find in each case the line current and the readings on each of the two wattmeters connected to measure the power.*

Solution. (*i*) Star connection.

$$Z_{ph} = \sqrt{10^2 + 10^2} = 10\sqrt{2} \ \Omega$$

$$V_{ph} = V_L / \sqrt{3} = 400 / \sqrt{3} = 231 \ V$$

$$I_L = I_{ph} = 231 \div 10\sqrt{2} = \mathbf{16.33 \ A}$$

$$\phi = \tan^{-1} X_{ph} / R_{ph} = \tan^{-1} 10/10 = 45°$$

$$\text{Power factor} = \cos \phi = \cos 45° = 1 / \sqrt{2} = 0.707$$

$$\text{Total power} = \sqrt{3} \ V_L I_L \cos \phi = \sqrt{3} \times 400 \times 16.33 \times 1/\sqrt{2} = 8000 \ W$$

Let W_1 and W_2 be the wattmeter readings ; W_2 being the higher-reading wattmeter. Since the load p.f. is greater than 0.5, wattmeter W_1 will also read positive.

$$W_2 + W_1 = 8000 \qquad \qquad \text{...(i)}$$

Also, $$\tan 45° = \sqrt{3} \frac{W_2 - W_1}{W_2 + W_1} \quad \text{or} \quad 1 = \sqrt{3} \frac{W_2 - W_1}{8000}$$

∴ $$W_2 - W_1 = 4619 \qquad \qquad \text{...(ii)}$$

Solving eqs. (*i*) and (*ii*), we get, $W_2 = \mathbf{6309.5 \ W}$; $W_1 = \mathbf{1690.5 \ W}$

(*ii*) Delta connection. $\quad Z_{ph} = 10\sqrt{2} \qquad \qquad \text{...same as for star}$

$$V_{ph} = V_L = 400 \ V$$

∴ $$I_{ph} = V_{ph}/Z_{ph} = 400 \div 10\sqrt{2} = 28.28 \ A$$

$$I_L = \sqrt{3} \ I_{ph} = \sqrt{3} \times 28.28 = \mathbf{48.98 \ A}$$

$$\text{Power factor} = \cos \phi = \cos 45° = 1 / \sqrt{2}$$

Total power $= \sqrt{3}\ V_L\,I_L \cos\phi = \sqrt{3} \times 400 \times 48.98 \times 1/\sqrt{2} = 23995$ W

Now, $\qquad W_2 + W_1 = 23995$ ...(i)

Also, $\qquad \tan 45° = \sqrt{3}\dfrac{W_2 - W_1}{W_2 + W_1}$ or $1 = \sqrt{3}\dfrac{W_2 - W_1}{23995}$

∴ $\qquad W_2 - W_1 = 13854$ ...(ii)

Solving eqs. (i) and (ii), we get, $W_2 = \mathbf{18924.5\ W}$; $W_1 = \mathbf{5070.5\ W}$

Example 15.44. *The ratio of the readings of the two wattmeters connected to measure power in a 3-phase balanced load is 3 : 1. The load is known to be inductive with a lagging power factor. Calculate the power factor of the load.*

Solution. $\qquad W_2/W_1 = 3$; W_2 being the higher-reading.

Now, $\qquad \tan\phi = \sqrt{3}\dfrac{W_2 - W_1}{W_2 + W_1} = \sqrt{3}\dfrac{(W_2/W_1)-1}{(W_2/W_1)+1} = \sqrt{3}\dfrac{3-1}{3+1} = \dfrac{\sqrt{3}}{2}$

∴ $\qquad \phi = \tan^{-1}\left(\sqrt{3}/2\right) = 40.9°$

∴ $\qquad$ Load p.f. $= \cos\phi = \cos 40.9° = \mathbf{0.756}$ *lag*

Example 15.45. *Find the reading on the wattmeter when the network shown in Fig. 15.64 is connected to a symmetrical 440 V, 3-phase supply. The phase sequence is RYB. Neglect electrostatic effects and instrument losses.*

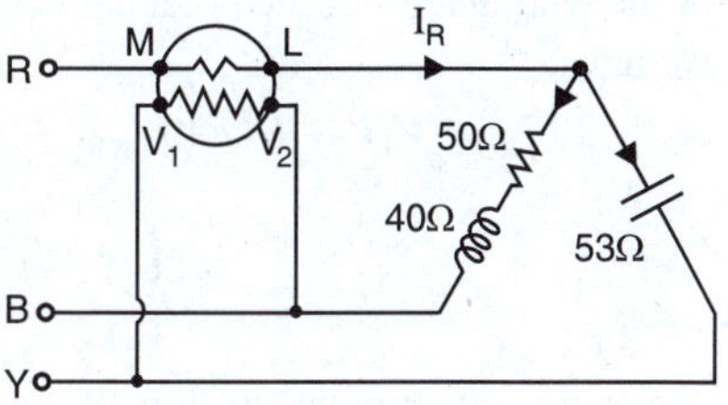

Solution. The phase sequence is *RYB*.

∴ $\quad \mathbf{V_{RY}} = 440 \angle 0°\ \text{V}$; $\mathbf{V_{YB}} = 440 \angle -120°\ \text{V}$;

$\qquad \mathbf{V_{BR}} = 440 \angle 120°\ \text{V}$

Current in current coil,

$$I_{ML} = I_R = \frac{V_{RB}}{50 + j\,40} + \frac{V_{RY}}{53\angle -90°}$$

$$= -\frac{440\angle 120°}{64\angle 38.7°} + \frac{440\angle 0°}{53\angle -90°}$$

$$= j\,8.3 - 6.875 \angle 81.3°$$

$$= j\,8.3 - (1.03 + j\,6.796)$$

$$= (-1.03 + j\,1.5)\ \text{A} = 1.825 \angle 124.7°\ \text{A}$$

Fig. 15.64

Voltage across potential coil, $V_1V_2 = V_{YB} = 440 \angle -120°$ V

∴ $\quad$ Wattmeter reading, $W = I_{ML} \times V_1V_2 = I_{ML}\cdot V_{YB}$

$$= 1.825 \times 440 \times \cos 244.7°$$

$$= -0.343 \times 10^3\ \text{W} = \mathbf{-0.343\ kW}$$

Example 15.46. *In a balanced 3-phase 400 V circuit, the line current is 115.5 A. When power is measured by two wattmeter method, one meter reads 40 kW and the other zero. What is the power factor of the load ? If the power factor were unity and the line current is the same, what would be the reading of each wattmeter ?*

Solution. $\qquad P = \sqrt{3}V_L I_L \cos\phi$

∴ $\qquad \cos\phi = \dfrac{P}{\sqrt{3}\ V_L I_L} = \dfrac{40 \times 10^3}{\sqrt{3} \times 400 \times 115.5} = \mathbf{0.5}$

Also, as proved in Art. 15.25, when p.f. = 0.5, total power is measured by one wattmeter (W_2) alone.

At unity p.f. (*i.e.* $\cos \phi = 1$), $\phi = 0°$.

$$\therefore \qquad \tan \phi = \sqrt{3} \frac{(W_2 - W_1)}{W_2 + W_1} = 0 \qquad \therefore W_1 = W_2$$

Also, $\qquad W_1 + W_2 = \sqrt{3} \times 400 \times 115.5 \times 1 = 80 \text{ kW}$

$\therefore \qquad\qquad W_1 = $ **40 kW**; $W_2 = $ **40 kW**

Example 15.47. *Find the reading on the wattmeter when the network shown in Fig. 15.65 is connected to a symmetrical 400 V, 3-phase supply of phase sequence RYB. Neglect instrument losses.*

Solution. The phase sequence is *RYB*.

$\therefore \qquad V_{RY} = 400 \angle 0° \text{ V}$; $V_{YB} = 400 \angle -120° \text{ V}$; $V_{BR} = 400 \angle 120° \text{ V}$

$$I_{RY} = \frac{V_{RY}}{30 - j\,40} = \frac{400}{50 \angle -53°}$$

$$= 8 \angle 53° \text{ A} = (4.8 + j\,6.4)\,\text{A}$$

$$I_{RB} = \frac{V_{RB}}{40} = \frac{-400 \angle 120°}{40}$$

$$= 10* \angle -60° \text{ A} = I_{ML}$$

As seen in Fig. 15.65, I_{RB} is the current in the current coil of the wattmeter. For the potential coil $V_1 V_2$ of the wattmeter, we have,

$$20\,I_{RB} + V_1 V_2 - (-j\,40\,I_{RY}) = 0$$

or $\qquad\qquad 0 = 200 \angle -60° + V_1 V_2 + 40 \angle 90° \times 8 \angle 53°$

or $\qquad -V_1 V_2 = (100 - j\,173) + 320 \angle 143° = (100 - j\,173) + (-255.6 + j\,192.6)$

or $\qquad\quad V_1 V_2 = (155.6 - j\,20) \text{ volts} = 156.9 \angle -7.3° \text{ volts}$

$\therefore$ Wattmeter reading, $W = I_{ML} \cdot V_1 V_2 = 10 \times 156.9 \times \cos 52.7°$

$$= 0.94 \times 10^3 \text{ W} = \textbf{0.94 kW}$$

Example 15.48. *The total power absorbed by a certain 3-phase balanced load is 60 kW ; the reactive power being 100 kVAR. Determine the power factor of the load and readings on two wattmeters connected to read the power.*

Solution. Let $\qquad\qquad W_2 = $ higher-reading wattmeter

$$W_1 = \text{lower-reading wattmeter}$$

$\therefore \qquad\qquad W_2 + W_1 = 60000 \qquad\qquad\qquad\qquad\qquad\qquad\qquad ...(i)$

And $\qquad\quad \sqrt{3}(W_2 - W_1) = 100000 \qquad\qquad\qquad\qquad\qquad\quad ...\text{See Art. 15.25}$

or $\qquad\qquad\quad W_2 - W_1 = 57735 \qquad\qquad\qquad\qquad\qquad\qquad\qquad\quad ...(ii)$

Solving equations (*i*) and (*ii*), we get, $W_2 = $ **58867.5 W** ; $W_1 = $ **1132.5 W**

$$**\tan \phi = \frac{\text{Reactive power}}{\text{Active power}} = \frac{100}{60} = 1.67$$

$\therefore \qquad\qquad\qquad \phi = \tan^{-1} 1.67 = 59.1°$

$\therefore \qquad$ Load power factor $= \cos \phi = \cos 59.1° = $ **0.514**

Since the load power factor is greater than 0.5, the reading of W_1 is expectedly positive.

---

*   When phasor V_{BR} is reversed, it becomes phasor V_{RB}. Now phasor V_{RB} lags behind the reference phasor V_{RY} by 60°.

** $\tan \phi = \sqrt{3} \dfrac{W_2 - W_1}{W_2 + W_1}$

Example 15.49. *A 3-phase, 3-wire 415 V system supplies a balanced load of 20 A at a power factor of 0.8 lag. The current coil of wattmeter 1 is in phase R and of wattmeter 2 in phase B. Calculate (i) the reading on wattmeter 1 when its voltage coil is across R and Y (ii) reading on wattmeter 2 when its voltage coil is across B and Y and (iii) reading on wattmeter 1 when its voltage coil is across Y and B.*

Solution. Fig. 15.66 (*i*) shows the conditions of the problem. Since $\cos \phi = 0.8$; $\phi = \cos^{-1} 0.8 = 36.87°$.

(*i*) $$W_1 = V_{RY}I_R \cos (30° + \phi) = 415 \times 20 \cos (30° + 36.87°) = \textbf{3260 W}$$

(*ii*) $$W_2 = V_{BY}I_B \cos (30° - \phi) = 415 \times 20 \times \cos (30° - 36.87°) = \textbf{8240 W}$$

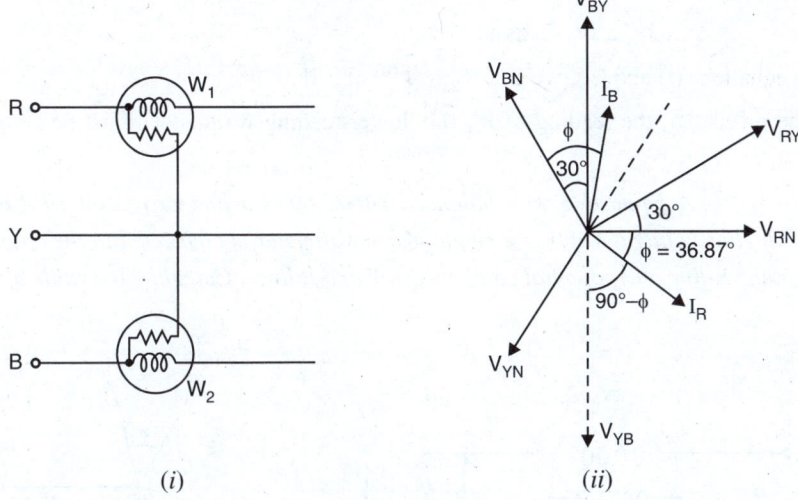

(*i*)   (*ii*)

Fig. 15.66

(*iii*) As shown in phasor diagram in Fig. 15.66 (*ii*), the angle between I_R and V_{YB} is $(90° - \phi)$.

$\therefore$ $$W_1 = V_{YB}I_R \cos (90° - \phi) = 415 \times 20 \times \sin \phi = \textbf{4980 VAR}$$

Note that connected in this way, the wattmeter will indicate *VAR*.

Example 15.50. *Use of two wattmeters to measure the total power taken by a 400 V balanced Δ-connected load leads to readings of 1000 W and 600 W. The installation of equal capacitors across each phase causes the readings to become 1030 W and 570 W respectively. What is the nature of the 3-phase load ; inductive or capacitive ?*

Solution. With capacitors across each phase, the p.f. of the load changes but the total power drawn remains the same. Hence the wattmeter readings are changed.

For the first case, $$\tan \phi = \sqrt{3}\frac{W_2 - W_1}{W_2 + W_1} = \sqrt{3}\frac{1000 - 600}{1000 + 600} = 0.433$$

For the second case, $$\tan \phi = \sqrt{3}\frac{W_2 - W_1}{W_2 + W_1} = \sqrt{3}\frac{1030 - 570}{1030 + 570} = 0.498$$

Since with capacitors across each phase the magnitude of $\tan \phi$ increases, the 3-phase load is ***capacitive**. Had the value of $\tan \phi$ decreased, the load would have been inductive.

Example 15.51. *Each phase of a 3-phase delta load has an impedance of $Z = 50 \angle 60°$ ohms. The line voltage is 400 V. Calculate the power consumed by each phase and the total power. What will be the readings of the two wattmeters connected to measure the power ?*

---

*   See foot note on page 770.

Solution. $Z_{ph} = 50\ \Omega\ ;\ \phi = 60°\ ;\ V_{ph} = V_L = 400\ V$

$\therefore$ $I_{ph} = V_{ph}/Z_{ph} = 400/50 = 8\ A$

Power factor $= \cos\phi = \cos 60° = 0.5$

$R_{ph} = Z_{ph}\cos\phi = 50 \times 0.5 = 25\ \Omega$

$\therefore$ Power/phase $= I^2_{ph}R_{ph} = 8^2 \times 25 = \mathbf{1600\ W}$

Total power $= 3 \times 1600 = \mathbf{4800\ W}$

Now, $W_2 + W_1 = 4800$ ...(i)

Also, $\tan 60° = \sqrt{3}\dfrac{W_2 - W_1}{W_2 + W_1}$ or $\sqrt{3} = \sqrt{3}\dfrac{W_2 - W_1}{4800}$

$\therefore$ $W_2 - W_1 = 4800$ ...(ii)

Solving equations (i) and (ii), we get, $W_2 = \mathbf{4800\ W}$; $W_1 = \mathbf{0\ W}$

Since the p.f. is 0.5, the reading of W_1 (i.e. lower-reading wattmeter) must be zero (See Art. 15.25).

Example 15.52. *A 3-phase, 3-wire balanced Y-load takes a line current of 30 A at a p.f. of 0.7 lagging. The line voltage is 400 V. A single phase wattmeter is connected in the circuit with its current coil in the R-line and potential coil between Y and B lines. Calculate the value of wattmeter reading.*

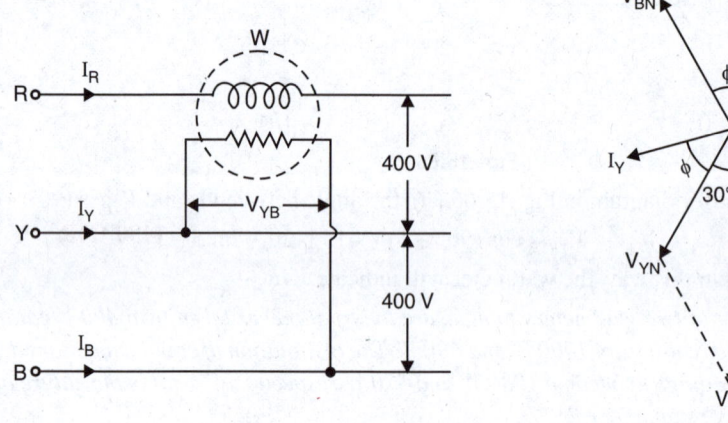

Fig. 15.67

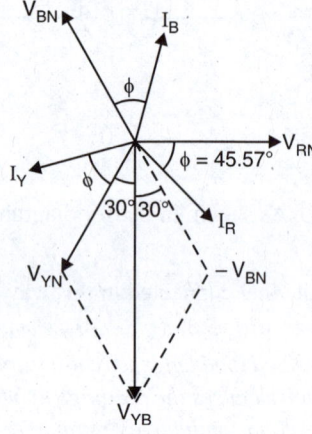

Fig. 15.68

Solution. The current through current coil of the wattmeter is I_R and voltage across its potential coil is V_{YB} as shown in Fig. 15.67. As seen from the phasor diagram in Fig. 15.68, the phase angle between V_{YB} and I_R is $(90° - \phi)$.

$\therefore$ Wattmeter reading, $W = V_{YB}\ I_R \cos(90° - \phi)$

Since $V_{YB} = V_L$, the line voltage

and $I_R = I_L$, the line current

$\therefore$ $W = {}^*V_L I_L \sin\phi = 400 \times 30 \times 0.714 = \mathbf{8568\ VAR}$ ($\because \sin\phi = 0.714$)

It may be noted that if the reading of the wattmeter is multiplied by $\sqrt{3}$, we shall get the total reactive volt-amperes drawn by the load.

---

* Connected in this way, the wattmeter will indicate VAR (See Art. 15.30).

Example 15.53. *A 3-phase, 3 wire balanced load with a lagging p.f. is supplied at 400 V (between lines). A single phase wattmeter when connected with its current coil in the R-line and voltage coil between R and Y lines gives a reading of 6 kW. When the same terminals of the voltage coil are switched over to Y and B lines, the current coil connections remaining the same, the reading of the wattmeter remains unchanged. Calculate the line current and power factor of the load. The phase sequence is RYB.*

Solution. In the first case, current through the current coil of the wattmeter is I_R and p.d. across its potential coil is V_{RY}. As seen from the phasor diagram in Fig. 15.69 (*ii*), the phase angle between V_{RY} and I_R is $(30° + \phi)$.

∴   Wattmeter reading, $W_1 = V_{RY}I_R \cos(30° + \phi) = V_L I_L \cos(30° + \phi)$

In the second case, the current through current coil of wattmeter is I_R and p.d. across its potential coil is V_{YB}. As seen from the phasor diagram in Fig. 15.69. (*ii*), the angle between V_{YB} and I_R is $(90° - \phi)$.

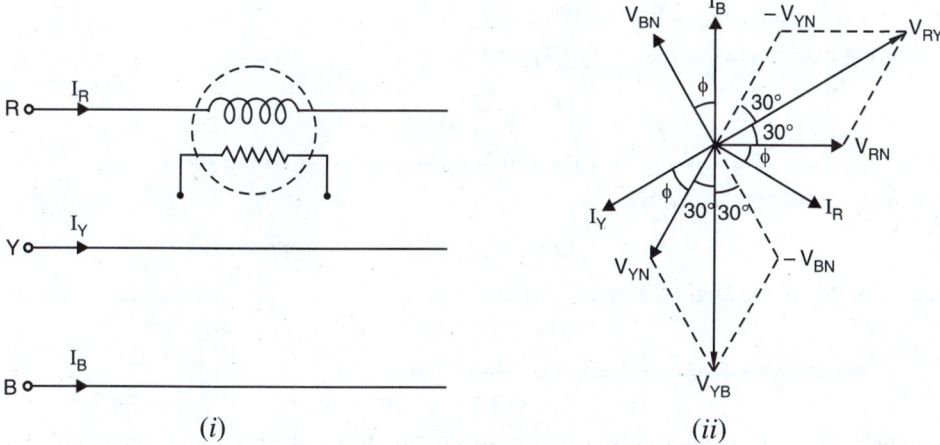

(*i*)   (*ii*)

Fig. 15.69

∴   Wattmeter reading, $W_2 = V_{YB} I_R \cos(90° - \phi) = V_L I_L \cos(90° - \phi)$

Since $W_1 = W_2$ ...*given*

∴ $V_L I_L \cos(30° + \phi) = V_L I_L \cos(90° - \phi)$

or $\cos(30° + \phi) = \cos(90° - \phi)$

or $30° + \phi = 90° - \phi$

∴ $\phi = 60°/2 = 30°$

∴   Load p.f. $= \cos\phi = \cos 30° = $ **0.866** *lag*

Now, $W_1 = V_L I_L \cos(30° + \phi)$

or $6 \times 10^3 = 400 \times I_L \times \cos(30° + 30°)$

∴ $I_L = \dfrac{6 \times 10^3}{400 \times \cos 60°} = $ **30 A**

Example 15.54. *Find the readings on each of the two similar wattmeters when the network shown in Fig. 15.70 is connected to 400 V, 3-phase supply. The phase sequence is RYB.*

Solution. The phase sequence is *RYB*.

∴ $V_{RY} = 400 \angle 0°$ V ; $V_{YB} = 400 \angle -120°$ V ; $V_{BR} = 400 \angle 120°$ V

To find the reading on the wattmeter, we find current through its current coil and potential across its potential coil.

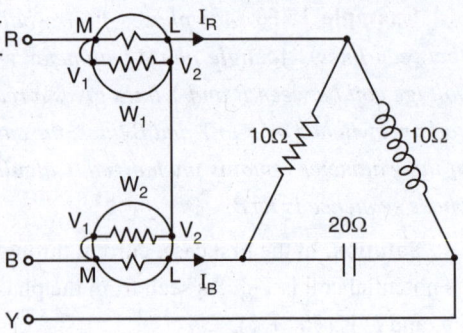

Wattmeter 1. Referring to Fig. 15.70,

Current in current coil, $I_{ML} = I_R$

$$= \frac{V_{RB}}{10} + \frac{V_{RY}}{10\angle 90°} = \frac{-400\angle 120°}{10} + \frac{400\angle 0°}{10\angle 90°}$$

$$= -40\angle 120° + 40\angle -90° = (20 - j\,74.64)\text{A}$$

$$= 77.3\angle -75°\text{A}$$

Now, $V_1 V_2 = \dfrac{1}{2} V_{RB} = -200\angle 120°$ V

$\therefore$ Wattmeter reading, $W_1 = I_{ML} \cdot V_1 V_2 \cos 195°$

$$= 77.3 \times (-200) \times \cos 195°$$

$$= 14.93 \times 10^3 \text{ W} = \textbf{14.93 kW}$$

Fig. 15.70

Wattmeter 2. Referring to Fig. 15.70,

Current in current coil, $I_{ML} = I_B = I_{BR} + I_{BY}$

$$= \frac{V_{BR}}{10} + \frac{V_{BY}}{20\angle 90°} = \frac{400\angle 120°}{10} + \frac{-400\angle -120°}{20\angle -90°}$$

$$= 40\angle 120° - 20\angle -30° = 58.19\angle 130°\text{ A}$$

Potential across potential coil is

$$V_1 V_2 = \frac{1}{2} V_{BR} = \frac{1}{2} \times 400\angle 120° = 200\angle 120° \text{ V}$$

It is clear that phase angle ϕ between I_{ML} and $V_1 V_2$

$$= 130° - 120° = 10°$$

$\therefore$ Wattmeter reading, $W_2 = (V_1 V_2) \times I_{ML} \times \cos \phi$

$$= 200 \times 58.19 \times \cos 10° = 11.46 \times 10^3 \text{ W} = \textbf{11.46 kW}$$

Example 15.55. *Two wattmeters connected to measure the power input to a synchronous motor working on a constant load and with a lagging p.f. indicate 10 kW and 5 kW. Determine the power factor at which the motor is working.*

If the excitation of the motor is adjusted so that the power factor changes to 0·6 leading, what are the respective readings of the wattmeters, assuming the total input to the motor remains constant?

Solution. When p.f. is lagging.

$$W_2 = 10 \text{ kW}; W_1 = 5 \text{ kW}$$

$$\tan \phi = \sqrt{3}\frac{W_2 - W_1}{W_2 + W_1} = \sqrt{3}\frac{10 - 5}{10 + 5} = \frac{1}{\sqrt{3}}$$

$\therefore$ $\phi = \tan^{-1}\left(1/\sqrt{3}\right) = 30°$

$\therefore$ Motor p.f. $= \cos \phi = \cos 30° = \textbf{0.866} \textit{ lagging}$

When p.f. is 0.6 leading. When the p.f. of the motor is changed to 0.6 leading [See Fig. 15.71], W_1 becomes the higher-reading wattmeter while W_2 is the lower-reading wattmeter. The total input power is the same.

$\therefore$ $W_1 + W_2 = 15 \text{ kW}$ ...(*i*)

P.F. angle, $\phi = \cos^{-1} 0.6 = 53.13°$

$\tan \phi = \tan 53.13° = 1.333$

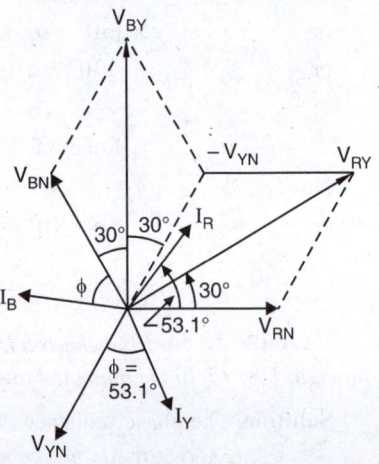

Fig. 15.71

$$\tan\phi \;=\; \sqrt{3}\frac{W_1-W_2}{W_1+W_2} \quad \text{or} \quad 1.333 = \sqrt{3}\frac{W_1-W_2}{15}$$

$$\therefore \qquad\qquad W_1 - W_2 \;=\; 11.54 \text{ kW} \hspace{5cm} ...(ii)$$

Solving equations (*i*) and (*ii*), we get, W_1 = **13.27 kW**; W_2 = **1.73 kW**

Example 15.56. *Three equal impedances, each consisting of R and L in series are connected in star and are supplied from a 400 V, 50 Hz, 3 – phase, 3 – wire balanced supply mains. The input power to the load is measured by two wattmeters method and the two wattmeters read 3 kW and 1 kW. Determine the values of R and L connected in each phase.*

Solution. Total power, $P \;=\; W_2 + W_1 = 3 + 1 = 4 \text{ kW} = 4 \times 10^3 \text{ W}$

Now, $\tan\phi \;=\; \sqrt{3}\dfrac{W_2-W_1}{W_2+W_1} = \sqrt{3}\dfrac{3-1}{3+1} = \dfrac{\sqrt{3}}{2}$

$\therefore$ Phase angle, $\phi \;=\; \tan^{-1}\sqrt{3}/2 = 40.89°$ lagging

$\therefore$ Power factor of load $=\cos\phi = \cos 40.89° = 0.756$ lagging

Line current, $I_L \;=\; \dfrac{P}{\sqrt{3}\,V_L\cos\phi} = \dfrac{4\times10^3}{\sqrt{3}\times400\times0.756} = 7.64 \text{ A}$

Load impedance/phase, $Z_{ph} \;=\; \dfrac{V_{ph}}{I_{ph}} = \dfrac{V_L/\sqrt{3}}{I_L} = \dfrac{400/\sqrt{3}}{7.64} = 30.23 \ \Omega$

$\therefore$ Load resistance/phase, $R_{ph} = Z_{ph}\cos\phi = 30.23 \times 0.756 = $ **22.85 Ω**

Load inductive reactance/phase $X_{Lph} = Z_{ph}\sin\phi = 30.23 \times \sin 40.89° = 19.8 \ \Omega$

$\therefore$ Load inductance/phase, $L_{ph} \;=\; \dfrac{X_{Lph}}{2\pi f} = \dfrac{19.8}{2\pi\times50} = $ **0.063 H**

Example 15.57. *In a 2 wattmeter method, power measured was 30 kW at 0.7 p.f. lagging. Find the reading of each wattmeter.*

Solution. Total power, $P \;=\; 30 \text{ kW} = 30 \times 10^3 \text{ W}$; p.f., $\cos\phi = 0.7$ lagging

$\therefore$ Phase angle, $\phi \;=\; \cos^{-1} 0.7 = 45.57°$ lagging

Now, $P \;=\; \sqrt{3}V_L I_L \cos\phi$

$\therefore$ $V_L I_L \;=\; \dfrac{P}{\sqrt{3}\cos\phi} = \dfrac{30\times10^3}{\sqrt{3}\times0.7} = 24743.6 \text{ VA}$

Let W_2 be the reading of the higher-reading wattmeter and W_1 that of lower-reading wattmeter.

$\therefore$ $W_2 \;=\; V_L I_L \cos(30° - \phi) = 24743.6 \cos(30° - 45.57°) = $ **23835 W**

$W_1 \;=\; V_L I_L \cos(30° + \phi) = 24743.6 \cos(30° + 45.57°) = $ **6165 W**

Example 15.58. *The two wattmeters A and B give readings as 5000 W and 1000 W respectively during the power measurement of 3 – ϕ, 3 wire, balanced load system.*

(a) Calculate the power and p.f. if (i) both meters read direct and (ii) one of them reads in reverse.

(b) If the voltage of the circuit is 400 V, what is the value of capacitance which must be introduced in each phase to cause the whole of the power to appear on A. The frequency of supply is 50 Hz.

Solution. (*a*) (*i*) **When both wattmeters read direct.** This means that both wattmeters read positive *i.e.* W_2 = 5000 W ; W_1 = 1000 W.

Now, $\tan \phi = \sqrt{3}\dfrac{W_2 - W_1}{W_2 + W_1} = \sqrt{3}\dfrac{5000 - 1000}{5000 + 1000} = \sqrt{3} \times \dfrac{4000}{6000} = 1.1547$

∴   Power factor $= \cos \phi = \cos \tan^{-1}(1.1547) = \cos 49.1° = \mathbf{0.655}$ *lag*

Total power, $P = W_2 + W_1 = 5000 + 1000 = \mathbf{6000\ W}$

(*ii*) **When one wattmeter reads in reverse.** This means that higher-reading wattmeter (W_2) reads positive and lower-reading wattmeter (W_1) reads negative *i.e.* $W_2 = 5000$ W ; $W_1 = -1000$ W.

Now, $\tan \phi = \sqrt{3}\dfrac{W_2 - W_1}{W_2 + W_1} = \sqrt{3}\dfrac{5000 - (-1000)}{5000 + (-1000)} = \sqrt{3} \times \dfrac{6000}{4000} = 2.598$

∴   Power factor $= \cos \phi = \cos \tan^{-1}(2.598) = \cos 68.95° = \mathbf{0.36}$ *lag*

Total power, $P = W_2 + W_1 = 5000 + (-1000) = \mathbf{4000\ W}$

(*b*) The reading of wattmeter B (*i.e.* W_1) will be zero and the whole power would appear on wattmeter A (*i.e.* W_2) when p.f. is $\cos \phi = 0.5$ or $\phi = 60°$. This means p.f. is to be improved from 0.36 to 0.5. For this purpose, capacitive reactance will have to be introduced in each phase of the load. When not given, it is understood that the load is star-connected.

At p.f. = 0.36. $I_L = \dfrac{P}{\sqrt{3}V_L \cos \phi} = \dfrac{4000}{\sqrt{3} \times 400 \times 0.36} = 16.04\text{A}$

∴ $I_{ph} = I_L = 16.04$ A

Load impedance/phase, $Z_{ph} = \dfrac{V_{ph}}{I_{ph}} = \dfrac{400/\sqrt{3}}{16.04} = 14.4\ \Omega$

Resistance/phase, $R_{ph} = Z_{ph} \cos \phi = 14.4 \times 0.36 = 5.185\ \Omega$

Reactance/phase, $X_{ph} = Z_{ph} \sin \phi = 14.4 \sin \cos^{-1}(0.36) = 13.4\ \Omega$

At p.f. = 0.5. There is no change in resistance per phase due to the introduction of capacitive reactance in each phase. Only reactance per phase changes. When p.f. is improved to 0.5, $\phi = 60°$.

∴   Reactance/phase, $X'_{ph} = R_{ph} \tan 60° = 8.98\ \Omega$

∴   Capacitive reactance to be introduced in each phase is

X_C/phase $= 13.4 - 8.98 = 4.42\ \Omega$

or $\dfrac{1}{2\pi f C} = 4.42$ or $C = \dfrac{1}{2\pi f \times 4.42} = \dfrac{1}{314 \times 4.42} = 720 \times 10^{-6}\text{F} = \mathbf{720\ \mu F}$

Example 15.59. *A 3-phase, 3-wire, 100 V, ABC system supplies a balanced Δ-connected load with impedance of 20 ∠45° Ω. (i) Determine the phase and line currents and draw the phasor diagram. (ii) Find the wattmeter readings when the two wattmeter method is applied to the system.*

Solution. Fig. 15.72 (*i*) shows the circuit diagram. Taking V_{AB} as the reference phasor, $V_{AB} = 100 \angle 0°$ V. As the phase sequence is *ABC*,

∴ $V_{AB} = 100 \angle 0°$ V ; $V_{BC} = 100 \angle -120°$ V ; $V_{CA} = 100 \angle -240°$ V $= 100 \angle 120°$ V

(*i*)   ∴ $I_{AB} = \dfrac{V_{AB}}{Z_{AB}} = \dfrac{100 \angle 0° \text{V}}{20 \angle 45° \Omega} = \mathbf{5 \angle -45° A}$

$I_{BC} = \dfrac{V_{BC}}{Z_{BC}} = \dfrac{100 \angle -120° \text{V}}{20 \angle 45° \Omega} = \mathbf{5 \angle -165° A}$

$I_{CA} = \dfrac{V_{CA}}{Z_{CA}} = \dfrac{100 \angle 120° \text{V}}{20 \angle 45° \Omega} = \mathbf{5 \angle 75° A}$

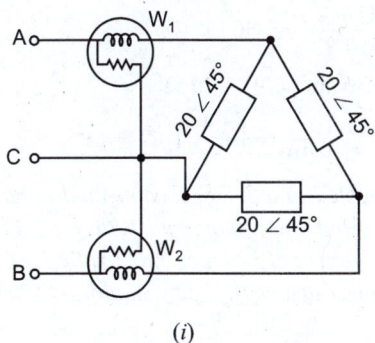 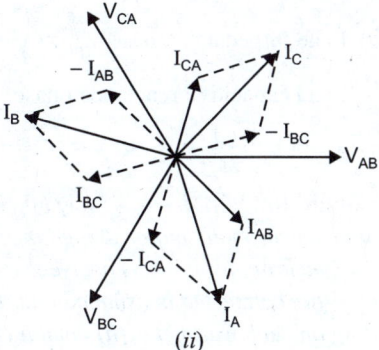

Fig. 15.72

Applying *KCL* to junction *A* in Fig. 15.72 (*i*), we have,

$$I_A + I_{CA} = I_{AB} \qquad \therefore \quad I_A = I_{AB} - I_{CA}$$

∴   Line current, $I_A = 5\angle -45° - 5\angle 75° = \mathbf{8.66 \angle -75° A}$

Since the system is balanced, the three line currents are equal in magnitude but differ in phase by 120°.

∴
$$I_B = 8.66\angle -75° - 120° = \mathbf{8.66\angle -195° A}$$
$$I_C = 8.66\angle -75° - 240° = 8.66\angle -315° A = \mathbf{8.66\angle 45° A}$$

Fig. 15.72 (*ii*) shows the phasor diagram of the circuit.

(*ii*) As seen in Fig. 15.72 (*i*), reading of wattmeter W_1 is

$$W_1 = V_{AC} I_A \cos \phi$$

Here ϕ is the phase angle between V_{AC} and I_A. Now V_{CA} leads the reference phasor (*i.e.* V_{AB}) by 120° so that V_{AC} lags behind the reference phasor by $(180° - 120°) = 60°$. Also I_A lags behind the reference phasor by 75°. Therefore, phase angle between V_{AC} and I_A is $\phi = 75° - 60° = 15°$.

∴ $W_1 = 100 \times 8.66 \times \cos 15° = \mathbf{836.5\ W}$

Also, $W_2 = V_{BC} I_B \cos \phi$

Now, V_{BC} lags behind the reference phasor (*i.e.* V_{AB}) by 120° while I_B lags behind the reference phasor by 195°. Therefore, phase angle between V_{BC} and I_B is $\phi = 195° - 120° = 75°$.

∴ $W_2 = 100 \times 8.66 \cos 75° = \mathbf{224.1\ W}$

Example 15.60. *A Y-connected balanced load is supplied from a 3 – ϕ balanced supply with a line voltage of 416 V at a frequency of 50 Hz. Each phase of the load consists of a resistance and a capacitor joined in series and the readings on two wattmeters connected to measure the total power supplied are 782 W and 1980 W, both positive. Calculate (i) p.f. of the circuit (ii) the line current and (iii) the capacitance of each capacitor.*

Solution. Since the load is capacitive, the load p.f. is leading. Higher-reading wattmeter, $W_1 = 1980$ W; Lower reading wattmeter, $W_2 = 782$ W.

(*i*) Now, $\tan \phi = \sqrt{3}\dfrac{W_1 - W_2}{W_1 + W_2} = \sqrt{3}\dfrac{1980 - 782}{1980 + 782} = 0.75$

or $\phi = \tan^{-1} 0.75 = 36.9°$

∴   Load power factor $= \cos \phi = \cos 36.9° = \mathbf{0.8\ \textit{leading}}$

(*ii*)   Total load power, $P = W_1 + W_2 = 1980 + 782 = 2762$ W

∴   Line current, $I_L = \dfrac{P}{\sqrt{3}\,V_L \cos \phi} = \dfrac{2762}{\sqrt{3}\times 416 \times 0.8} = \mathbf{4.8\ A}$

(iii) Load impedance/phase, $Z_{ph} = \dfrac{V_{ph}}{I_{ph}} = \dfrac{V_L/\sqrt{3}}{I_L} = \dfrac{416/\sqrt{3}}{4.8} = 50\ \Omega$

     Load capacitive reactance/phase, $X_C = Z_{ph} \sin \phi = 50 \sin 36.9° = 30\ \Omega$

or $\dfrac{1}{2\pi fC} = 30$ $\therefore C = \dfrac{1}{2\pi f \times 30} = \dfrac{1}{2\pi \times 50 \times 30} = 106 \times 10^{-6}\ \text{F} = \textbf{106 μF}$

Example 15.61. *A 3 – ϕ, 434 V, 50 Hz supply is connected to a 3 – ϕ, Y-connected induction motor and synchronous motor. Impedance of each phase of induction motor is (1.25 + j 2.17) Ω. The 3 – ϕ synchronous motor is over-excited and it draws a current of 120 A at 0.87 leading p.f. Two wattmeters are connected in usual manner to measure the power drawn by the two motors. Calculate (i) reading on each wattmeter (ii) combined p.f.*

Solution. Line voltage, $V_L = 434$ V ; Phase voltage, $V_{ph} = 434/\sqrt{3} = 250$ V

     Impedance of induction motor per phase is

$$Z_{ph} = (1.25 + j\,2.17)\ \Omega = 2.5 \angle 60°\ \Omega$$

Taking phase voltage V_R in R-phase as the reference phasor, $V_R = 250 \angle 0°$ V.

$\therefore$ Phase current in induction motor is

$$I_{ph} = \frac{V_{ph}}{Z_{ph}} = \frac{V_R}{Z_{ph}} = \frac{250 \angle 0°}{2.5 \angle 60°} = 100 \angle -60°\text{A}$$

$\therefore$ Motor line current, $I_1 = I_{ph} = 100 \angle -60°\text{A} = (50 - j\,86.6)$ A

The synchronous motor draws a line current I_2 leading V_R by an angle $= \cos^{-1} 0.87 = 29.5°$.

$\therefore$ $I_2 = 120 \angle 29.5°\text{A} = (104.6 + j\,59)$ A

$\therefore$ Total line current, $I_L = I_1 + I_2 = (50 - j\,86.6) + (104.6 + j\,59)$

                $= (154.6 - j\,27.6)$ A $= 156.8 \angle -10.1°$ A

Therefore, magnitude of line current in each line is $I_L = 156.8$ A and the phase angle is $\phi = 10.1°$.

(i) $\therefore$ Higher-reading wattmeter will read :

$$W_2 = V_L I_L \cos(30° - \phi) = 434 \times 156.8 \times \cos(30° - 10.1°) = \textbf{63988 W}$$

     Lower-reading wattmeter will read :

$$W_1 = V_L I_L \cos(30° + \phi) = 434 \times 156.8 \times \cos(30° + 10.1°) = \textbf{52054 W}$$

(ii)   Combined power factor $= \cos \phi = \cos 10.1° = \textbf{0.9845} \ lagging$

Example 15.62. *A wattmeter reads 5.54 kW when its current coil is connected in R-phase and its voltage coil is connected between the neutral and the R-phase of a symmetrical 3 – ϕ system supplying a balanced load of 30A at 400V. What will be the reading on the instrument if the connections to the current coil remain unchanged and the voltage coil be connected between B and Y phases? Take phase sequence RYB. Draw the corresponding phasor diagram.*

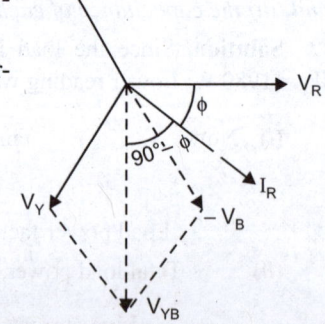

Solution. When not given, it is understood that the load is star-connected.

         Phase current, $I_{ph} = $ Line current, $I_L = 30$ A

         Phase voltage, $V_{ph} = V_L/\sqrt{3} = 400/\sqrt{3} = 231$ V

    Power supplied/phase, $P_{ph} = $ Wattmeter reading

                 $= 5.54$ kW $= 5540$ W

Fig. 15.73

$$\therefore \quad \text{Load power factor, } \cos \phi = \frac{P_{ph}}{V_{ph} I_{ph}} = \frac{5540}{231 \times 30} = 0.8$$

or $$\phi = \cos^{-1} 0.8 = 36.87°$$

When connections to the current coil remain unchanged and the voltage coil is connected between B and Y phases, the phasor diagram will be as shown in Fig. 15.73. Now I_R lags behind V_R by ϕ (= 36.87°) while V_{YB} lags behind V_R by 90°. Therefore, phase angle between V_{YB} and I_R is $(90° - \phi)$.

$$\therefore \quad \text{Wattmeter reading} = I_R V_{YB} \cos (90° - \phi)$$

$$= 30 \times 400 \times \cos (90° - 36.87°) = \textbf{7200 W}$$

Example 15.63. *A 440 V, 3-phase, Δ-connected induction motor has an output of 14.92 kW at a p.f. of 0.82 and efficiency 85%. Calculate the readings on each of the two wattmeters connected to measure the input. If another Y-connected load of 10 kW at 0.85 p.f. lagging is added in parallel to the motor, what will be the current drawn from the line and the power taken from the line?*

Solution. $$\text{Motor input} = \frac{14.92}{\eta} = \frac{14.92}{0.85} = 17.6 \text{ kW}$$

$$\therefore \quad W_2 + W_1 = 17.6 \text{ kW} \qquad\qquad ...(i)$$

$$\text{Motor p.f., } \cos \phi_m = 0.82$$

$$\therefore \quad \phi_m = \cos^{-1} 0.82 = 34.9°$$

Now, $$\tan \phi_m = \sqrt{3} \frac{W_2 - W_1}{W_2 + W_1}$$

or $$\tan 34.9° = \sqrt{3} \frac{W_2 - W_1}{17.6}$$

$$\therefore \quad W_2 - W_1 = 7.09 \text{ kW} \qquad\qquad ...(ii)$$

From eqs. (*i*) and (*ii*), $W_2 = \textbf{12.35 kW}$; $W_1 = \textbf{5.26 kW}$

$$\text{Active power of motor, } P_m = 17.6 \text{ kW}$$

$$\text{Reactive power of motor, } Q_m = \text{Motor kVA} \times \sin \phi_m = \frac{\text{Motor kW}}{\cos \phi_m} \times \sin \phi_m$$

$$= \frac{17.6}{0.82} \times \sin 34.9° = 12.28 \text{ kVAR } (inductive)$$

$$\text{Active power of load, } P_L = 10 \text{ kW} ; \phi_L = \cos^{-1} 0.85 = 31.8°$$

$$\text{Reactive power of load, } Q_L = \frac{10}{\cos \phi_L} \times \sin \phi_L = \frac{10}{0.85} \times \sin 31.8° = 6.20 \text{ kVAR } (inductive)$$

$$\text{Total active power, } P = P_m + P_L = 17.6 + 10 = 27.6 \text{ kW}$$

$$\text{Total reactive power, } Q = Q_m + Q_L = 12.28 + 6.20 = 18.48 \text{ kVAR } (inductive)$$

$$\text{Total kVA, } S = \sqrt{P^2 + Q^2} = \sqrt{(27.6)^2 + (18.48)^2} = 33.21 \text{ kVA}$$

$$\therefore \quad \text{Line current, } I = \frac{S}{\sqrt{3}V} = \frac{33.21 \times 10^3}{\sqrt{3} \times 440} = \textbf{43.57 A}$$

$$\text{Power taken from line} = P = \textbf{27.6 kW}$$

Tutorial Problems

1. Two wattmeters are used to measure power in a 3-phase balanced load. The wattmeter readings are 8.2 kW and 7.5 kW. Calculate (*i*) total power (*ii*) power factor and (*iii*) total reactive power.

[(*i*) **15.7 kW** (*ii*) **0.997** (*iii*) **1.21 kVAR**]

2. A balanced 3-phase load takes 10 kW at a p.f. of 0.9 lagging. Calculate the readings on each of the two wattmeters connected to read the input power. **[6398 W ; 3602 W]**

3. A 440 V, 3-phase induction motor has an output of 20.7 kW at a p.f. of 0.82 and efficiency 85%. Calculate the readings on each of the two wattmeters connected to measure input. **[12.35 kW ; 5.25 kW]**

4. The power input to a 400 V, 3-phase, 50 Hz induction motor is measured by two wattmeter method. The readings of the two wattmeters are 40 kW and − 10 kW. Calculate (*i*) the input power (*ii*) power factor (*iii*) line current. **[(*i*) 30 kW (*ii*) 0.327 (lag) (*iii*) 132.42 A]**

5. Three identical coils, each having a resistance of 20 Ω and a reactance of 20 Ω are connected in (*i*) star (*ii*) delta across 440 V, 3-phase lines. Calculate for each method of connection the line current and readings on each of the two wattmeters connected to measure the power.

[(*i*) 8.98 A ; 3817.5 W ; 1022.5 W (*ii*) 26.95 A ; 12452.5 W ; 3067.5 W]

6. A delta-connected load consists of a 25 Ω resistor in series with 15 mH inductor in each phase. Two-wattmeter method is used to measure the power and power factor of the load. Determine (*i*) load power factor and (*ii*) wattmeter readings. **[(*i*) 0.983 (*lag*) (*ii*) 11.11 kW ; 8.92 kW]**

7. Each of the two wattmeters connected to measure the input to a balanced 3-phase load reads 10 kW. What does each instrument read when the load power factor is changed to (*i*) 0.866 lagging (*ii*) 0.5 lagging (*iii*) 0.866 leading ? **[(*i*) 13.33 kW, 6.67 kW (*ii*) 20 kW, 0 (*iii*) 6.67 kW, 13.33 kW]**

8. The ratio of the readings of two wattmeters connected to measure power in a balanced 3-phase, 3-wire load is 5 : 3. The load is known to be inductive. Calculate the power factor of the load. **[0.917 *lag*]**

9. A single-phase wattmeter is used to measure the total power taken by a 2 kV, 3-phase induction motor having an efficiency of 90% and a p.f. of 0.756 lagging at full-load. The full-load motor output is 360 kW. The current coil of the wattmeter is connected in the yellow line. Calculate the wattmeter reading when the potential coil is connected (*i*) between yellow and red lines (*ii*) between yellow and blue lines. The phase sequence is *RYB*. **[(*i*) 300 kW (*ii*) 100 kW]**

10. A 3-phase, 3-wire balanced load with a lagging power factor is supplied at a line voltage of 400 V. A single-phase wattmeter when connected with its current coil in the *R*-line and voltage coil between *R* and *Y* lines gives a reading of 12 kW. When the same terminals of the voltage coil are switched over to *Y* and *B* lines, the current coil connections remaining the same, the reading of the wattmeter remains unchanged. Calculate the power factor and the line current. Phase sequence is RYB.

[0.866 lag ; 60 A]

15.31. Unbalanced 3-Phase Loads

We know that there are two types of 3-phase loads *viz* Y-connected load and Δ-connected load. So far we have considered **balanced 3-phase loads** *i.e. the loads that have the same impedance and power factor in each phase*. In case of balanced 3-phase loads, the three load phase voltages are equal in magnitude displaced 120° from each other. The same is true with regard to the three load phase currents and the three line currents. Also power delivered to each of the three load impedances is the same. Therefore, the problems on balanced 3-phase loads can be solved by considering one phase only ; the conditions in the other two phases being similar.

In practice, we come across situations where **3-phase loads are unbalanced** *i.e. loads in one or more phases have different impedances and/or power factor*. In such a case, the three load phase voltages are unequal in magnitude and displaced by unequal angles from each other. The same is true with regard to the three load phase currents and the three line currents. Also power delivered to each of the three load impedances is different. The problems on unbalanced 3-phase loads are more difficult to deal with. It is because the conditions in the three phases are different and we have to determine the circuit values (*i.e.* phase current, line current, load phase voltage, power *etc.*) for each phase separately. In practice, we may come across the following unbalanced 3-phase loads :

(*i*) Four-wire star-connected unbalanced load

(ii) Unbalanced Δ-connected load

(iii) Unbalanced 3-wire, *Y*-connected load.

15.32. Four-Wire Star-Connected Unbalanced Load

We can obtain this type of load in two ways. First, we may connect a 3-phase, 4-wire unbalanced load to a 3-phase, 4-wire supply as shown in Fig. 15.74. Note that star point *N* of the supply is connected to the load star point *N'*. Secondly, we may connect single phase loads between any line and the neutral wire as shown in Fig. 15.75. This will also result in a 3-phase, 4-wire *unbalanced load because it is rarely possible that single phase loads on all the three phases have the same magnitude and power factor. Since the load is unbalanced, the line currents will be different in magnitude and displaced from one another by unequal angle. The current in the neutral wire will be the phasor sum of the three line currents *i.e.*

Current in neutral wire, $I_N = I_R + I_Y + I_B$ *...phasor sum*

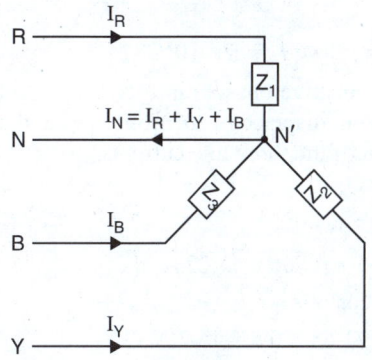

Fig. 15.74

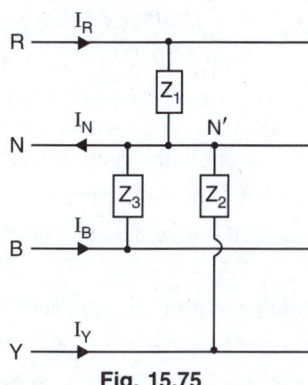

Fig. 15.75

The following points may be noted carefully :

(i) *Since the neutral wire has negligible impedance, supply neutral N and load neutral N' will be at the same potential.* It means that voltage across each impedance is equal to the phase voltage of the supply. However, current in each phase (or line) will be different due to unequal impedances.

(ii) The amount of current flowing in the neutral wire will depend upon the magnitudes of line currents and their phase relations. *In most circuits encountered in practice, the neutral current is equal to or smaller than any one of the line currents.* The exceptions are those circuits having severe unbalance.

Example 15.64. *Non-reactive loads of 10 kW, 8 kW and 5 kW are connected between the neutral and the red, yellow and blue phases respectively of a 3-phase, 4-wire system. The line voltage is 400 V. Calculate (i) the current in each line and (ii) the current in the neutral wire.*

Solution. This is a case of unbalanced load so that the line currents (and hence the phase currents) in the three lines will be different. The current in the **neutral wire will be equal to the phasor sum of the three line currents as shown in Fig. 15.76.

---

*   In actual practice, we never have an unbalanced 3-phase, 4-wire load. Most of the 3-phase loads (*e.g.*, 3-phase motors) are 3-phase, 3-wire and are balanced loads. In fact, these are the single phase loads on the 3-phase, 4-wire supply which constitute unbalanced, 4-wire *Y* load.

**   Had the load been balanced (*i.e.*, each phase having identical load), the current in the neutral wire would have been zero.

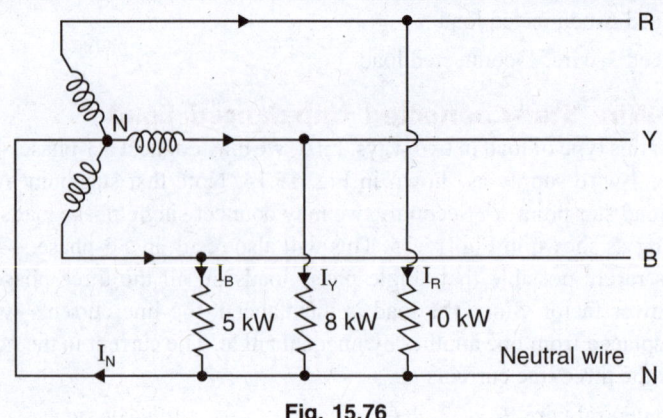

Fig. 15.76

(i)   Phase voltage $= 400/\sqrt{3} = 231$ V

$I_R = 10 \times 10^3/231 = \mathbf{43.3\ A}$; $I_Y = 8 \times 10^3/231 = \mathbf{34.6\ A}$; $I_B = 5 \times 10^3/231 = \mathbf{21.65\ A}$

(ii) The three line currents are represented by the respective phasors in Fig. 15.77. Note that the three line currents are of different magnitude but displaced 120° from one another. The current in the neutral wire will be the phasor sum of the three line currents.

Resolving the three currents along x-axis and y-axis is

Resultant horizontal component $= I_Y \cos 30° - I_B \cos 30°$

$\qquad\qquad\qquad = 34.6 \times 0.866 - 21.65 \times 0.866 = 11.22$ A

Resultant vertical component $= I_R - I_Y \cos 60° - I_B \cos 60°$

$\qquad\qquad\qquad = 43.3 - 34.6 \times 0.5 - 21.65 \times 0.5 = 15.2$ A

As shown in Fig. 15.78, current in neutral wire is

$$I_N = \sqrt{(11.22)^2 + (15.2)^2} = \mathbf{18.9\ A}$$

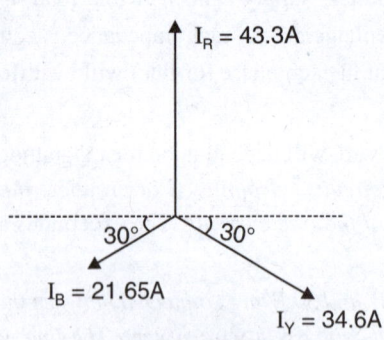

Fig. 15.77

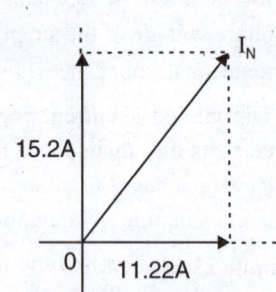

Fig. 15.78

Example 15.65. *A 3-phase, 4-wire system supplies power at 400 V and lighting at 230 V. If the lamps in use require 70, 84 and 33 amperes in each of the three lines, what should be the current in the neutral wire ? If a 3-phase motor is now started, taking 200 A from the lines at a p.f. of 0·2 lagging, what should be the total current in each line and the neutral wire ? Find also the total power supplied to the lamps and the motor.*

Solution. Fig. 15.79 shows the lamp load and motor load on 400 V/230 V, 3-phase, 4-wire supply.

Lamp load alone. If there is lamp load alone, the line currents in phases *R*, *Y* and *B* are 70 A, 84 A and 33 A respectively. These currents will be 120° apart (assuming phase sequence *RYB*) as shown in Fig. 15.80.

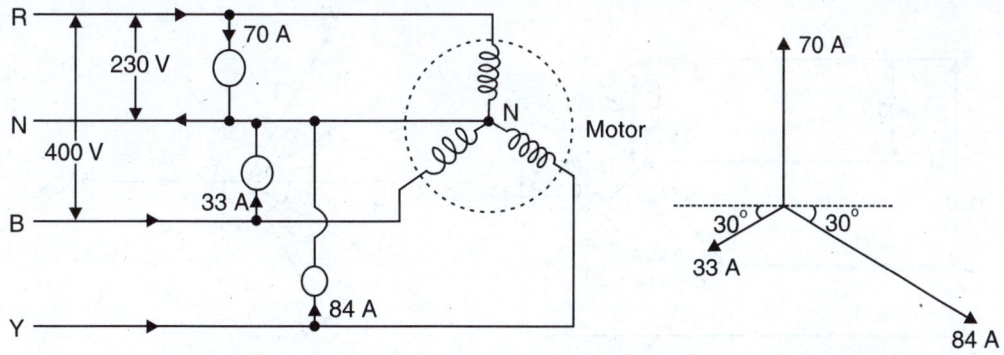

Fig. 15.79   Fig. 15.80

Resultant H-component $= 84 \cos 30° - 33 \cos 30° = 44.17$ A

Resultant V-component $= 70 - 33 \cos 60° - 84 \cos 60° = 11.5$ A

$\therefore$ Neutral current, $I_N = \sqrt{(44.17)^2 + (11.5)^2} = \textbf{45.64 A}$

Both lamp load and motor load.

When motor load is also connected along with lighting load, there will be no change in current in the neutral wire. It is because the motor load is balanced and hence no current will flow in the neutral wire due to this load.

$\therefore$ Neutral current, $I_N = \textbf{45.64 A}$ ...*same as before*

The current in each line is the phasor sum of the line currents due to lamp load and motor load.

Active component of motor current

$$= 200 \times \cos \phi_m = 200 \times 0.2 = 40 \text{ A}$$

Reactive component of motor current

$$= 200 \times \sin \phi_m = 200 \times 0.98 = 196 \text{ A}$$

$\therefore$ $I_R = \sqrt{(\text{sum of active comp.})^2 + (\text{reactive comp.})^2}$

$$= \sqrt{(40+70)^2 + (196)^2} = \textbf{224.8 A}$$

$$I_Y = \sqrt{(40+84)^2 + (196)^2} = \textbf{232 A}$$

$$I_B = \sqrt{(40+33)^2 + (196)^2} = \textbf{209.15 A}$$

Power supplied.

Power supplied to lamps $= 230(70 + 84 + 33) \times 1 = \textbf{43010 W}$ ($\because \cos \phi_L = 1$)

Power supplied to motor $= \sqrt{3} \; V_L I_L \cos \phi_m = \sqrt{3} \times 400 \times 200 \times 0.2$ ($\because \cos \phi_m = 0.2$)

$$= \textbf{27712 W}$$

Example 15.66. *The three line leads of a 400/230 V, 3-phase, 4-wire supply are designated as R, Y and B respectively. The fourth wire or neutral wire is designated as N. The phase sequence is RYB. Compute the currents in the four wires when the following loads are connected to this supply :*

From R to N : 20 kW, unity power factor

From Y to N : 28.75 kVA, 0.866 lag

From B to N : 28.75 kVA, 0.866 lead

If the load from B to N is removed, what will be the value of currents in the four wires ?

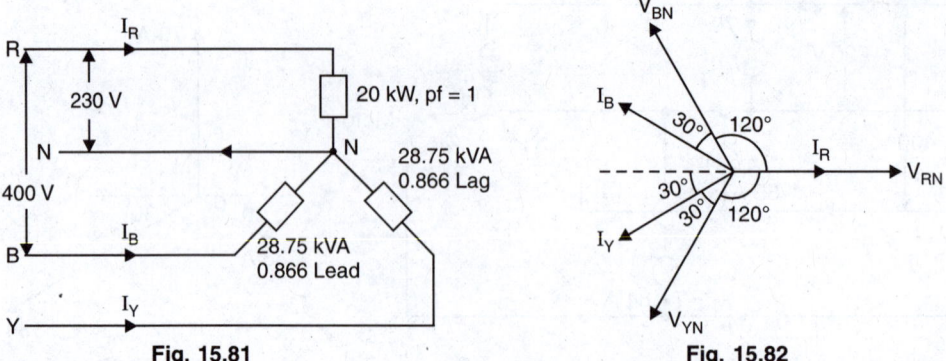

Fig. 15.81   **Fig. 15.82**

Solution. Fig. 15.81 shows the circuit diagram whereas Fig. 15.82 shows its phasor diagram. The current I_R is in phase with V_{RN}, current I_Y lags behind its phase voltage V_{YN} by $\cos^{-1} 0.866 = 30°$ and the current I_B leads its phase voltage V_{BN} by $\cos^{-1} 0.866 = 30°$.

$$I_R = 20 \times 10^3/230 = \textbf{86.96 A} \; ; I_Y = 28.75 \times 10^3/230 = \textbf{125 A} \; ; I_B = 28.75 \times 10^3/230 = \textbf{125 A}$$

The current in the neutral wire will be equal to the phasor sum of the three line currents I_R, I_Y and I_B. Referring to the phasor diagram in Fig. 15.82 and resolving these currents along x-axis and y-axis, we have,

$$\text{Resultant H-component} = 86.96 - 125 \cos 30° - 125 \cos 30°$$
$$= 86.96 - 108.25 - 108.25 = -129.54 \text{ A}$$
$$\text{Resultant V-component} = 0 + 125 \sin 30° - 125 \sin 30° = 0$$

∴   Neutral current, $I_N = \sqrt{(-129.54)^2 + (0)^2} = \textbf{129.54 A}$

When load from B to N removed. When load from N to N is removed, the various line currents are :

$I_R = \textbf{86.96 A}$ in phase with V_{RN} ; $I_Y = \textbf{125 A}$ lagging V_{YN} by 30° ; $I_B = \textbf{0 A}$

The current in the neutral wire is equal to the phasor sum of these three line currents. Resolving the currents along x-axis and y-axis, we have,

$$\text{Resultant H-component} = 86.96 - 125 \cos 30°$$
$$= 86.96 - 108.25 = -21.29 \text{ A}$$
$$\text{Resultant V-component} = 0 - 125 \sin 30° = 0 - 125 \times 0.5 = -62.5 \text{ A}$$

∴   Neutral current, $I_N = \sqrt{(-21.29)^2 + (-62.5)^2} = \textbf{66.03 A}$

Example 15.67. *An unbalanced 4-wire star connected load has balanced voltages of 440 V. The loads are : $Z_R = 10 \ \Omega$; $Z_Y = (5 + j \ 10) \ \Omega$; $Z_B = (15 - j \ 5) \ \Omega$. Calculate the current in the neutral wire and its phase relationship to the voltage across the red phase. The phase sequence is RYB.*

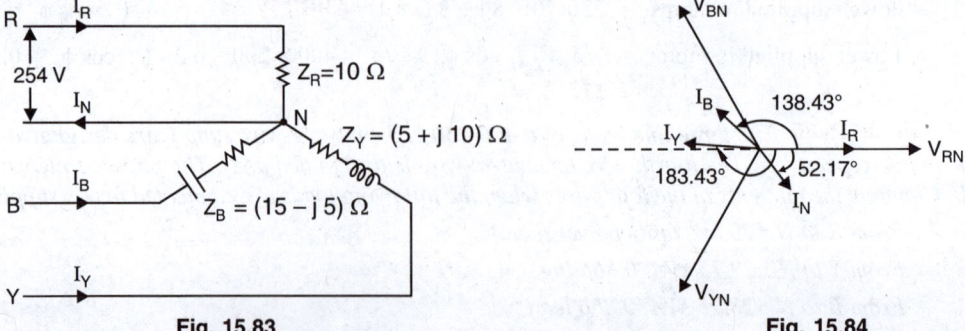

Fig. 15.83   **Fig. 15.84**

Solution. Fig. 15.83 shows the circuit diagram whereas Fig. 15.84 shows its phasor diagram.

$$Z_R = (10 + j\,0)\,\Omega = 10 \angle 0°\,\Omega$$
$$Z_Y = (5 + j\,10)\,\Omega = 11.18 \angle 63.43°\,\Omega$$
$$Z_B = (15 - j\,5)\,\Omega = 15.81 \angle -18·43°\,\Omega$$

Phase voltage $= 440/\sqrt{3} = 254$ V

Taking V_{RN} as the reference phasor, we have,

$$V_{RN} = 254 \angle 0°\text{ V}\,;\;V_{YN} = 254 \angle -120°\text{ V}\,;\;V_{BN} = 254 \angle 120°\text{ V}$$

$$\therefore \qquad I_R = \frac{V_{RN}}{Z_R} = \frac{254 \angle 0°}{10 \angle 0°} = 25.4 \angle 0°\text{ A} = (25.4 + j\,0)\text{ A}$$

$$I_Y = \frac{V_{YN}}{Z_Y} = \frac{254 \angle -120°}{11.18 \angle 63.43°} = 22.72 \angle -183.43°\text{ A} = (-22.68 + j\,1.36)\text{ A}$$

$$I_B = \frac{V_{BN}}{Z_B} = \frac{254\angle120°}{15.81\angle-18.43°} = 16.07 \angle 138.43°\text{ A} = (-12.02 + j\,10.66)\text{ A}$$

$$I_N = -*[I_R + I_Y + I_B]$$
$$= -[(25.4 + j\,0) + (-22.68 + j\,1.36) + (-12.02 + j\,10.66)]$$
$$= -[-9.3 + j\,12.02] = (9.3 - j\,12.02)\text{ A} = 15.2 \angle -52.17°\text{ A}$$

Magnitude, $I_N = $ **15.2 A**

Its phase angle w.r.t. V_{RN} is 52.17° as shown in Fig. 15.84.

Example 15.68. *Determine the current in the neutral wire for the star-connected load shown in Fig. 15.85. Also find the total power consumed. The line voltage is 415 V.*

Solution. The phase voltage $= 415/\sqrt{3} = 240$ V
The phase sequence is *RYB*.

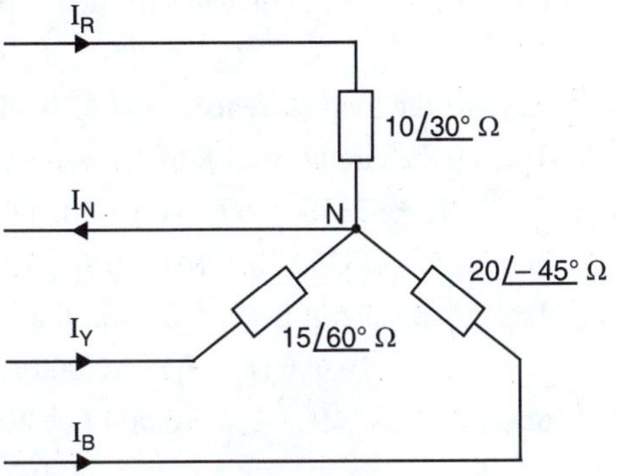

$$\therefore\quad V_{RN} = 240 \angle 0°\text{ V}\,;\;V_{YN} = 240 \angle -120°\text{ V}\,;$$
$$V_{BN} = 240 \angle 120°\text{ V}$$

Fig. 15.85

$$I_R = \frac{V_{RN}}{10\angle30°} = \frac{240\angle0°}{10\angle30°}$$
$$= 24 \angle -30°\text{ A} = (20.8 - j\,12)\text{ A}$$

$$I_Y = \frac{V_{YN}}{15\angle60°} = \frac{240\angle-120°}{15\angle60°}$$
$$= 16 \angle -180°\text{ A} = (-16 + j\,0)\text{ A}$$

$$I_B = \frac{V_{BN}}{20\angle-45°} = \frac{240\angle120°}{20\angle-45°} = 12 \angle 165°\text{ A} = (-11.6 + j\,3.1)\text{ A}$$

$\therefore$ Current in neutral wire, $I_N = I_R + I_Y + I_B$

$$= (20.8 - j\,12) + (-16 + j\,0) + (-11.6 + j\,3.1)$$
$$= (-6.8 - j\,8.9)\text{ A} = \textbf{11.2} \angle \textbf{-127° A}$$

The total power consumed will be the sum of the powers consumed by each phase load.

Power for red phase $= 240 \times 24 \times \cos 30° = 4988$ W
Power for yellow phase $= 240 \times 16 \times \cos 60° = 1920$ W
Power for blue phase $= 240 \times 12 \times \cos 45° = 2036$ W

$\therefore$ Total power consumed $= 4988 + 1920 + 2036 = $ **8944 W**

---

*   To find the neutral current, we must add the three currents. The neutral current must then be equal and *opposite* to this sum, so that total current at the junction (*i.e.*, at star point *N*) is zero.

Example 15.69. *A voltmeter with a resistance of 10,000 Ω is connected with two 10,000 Ω resistors across a 200 V, 3-phase system as shown in Fig. 15.86 (i)*

    (i) Find the reading of voltmeter.

    (ii) If the voltmeter is shunted by a 5000 Ω resistor, find the new reading of voltmeter.

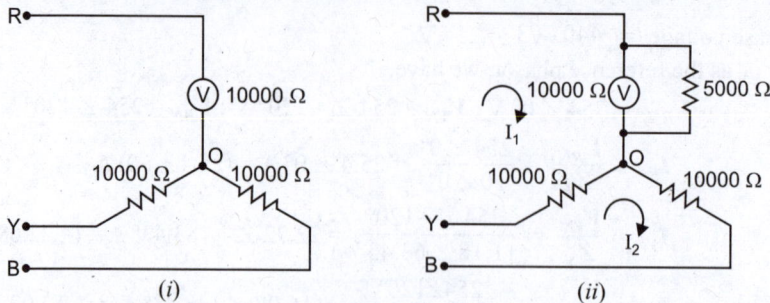

Fig. 15.86

Solution. (*i*) It is a case of balanced Y load with 10,000 Ω resistance in each phase. Since a voltmeter measures voltage between two points across which it is connected,

$\therefore$ Reading of voltmeter = Phase voltage = $200/\sqrt{3}$ = **115.47 V**

(*ii*) When the voltmeter is shunted by 5000 Ω resistor, the circuit becomes as shown in Fig. 15.86 (*ii*).

New R-phase resistance, Z_R = 10,000 Ω || 5,000 Ω = $\dfrac{10,000 \times 5,000}{10,000 + 5,000}$ = 3333.33 Ω

It is now a case of unbalanced Y-load. Taking V_{RY} as the reference phasor,

$$V_{RY} = 200 \angle 0° \text{ V} \; ; \; V_{YB} = 200 \angle -120° \text{ V} \; ; \; V_{BR} = 200 \angle 120 \text{ V}$$

Assume the loop currents I_1 and I_2 as shown in Fig. 15.86 (*ii*).

Tracing the circuit from R to Y *via* the neutral point O and applying *KVL*, we have,

$$-I_1 \times 3333.33 - (I_1 - I_2) \times 10,000 + V_{RY} = 0$$

or $(3333.33 + 10,000) I_1 - 10,000 I_2 = 200 \angle 0°$ ...(*i*)

Tracing the circuit from Y to B *via* the neutral point O and applying *KVL*, we have,

$$-10,000 (I_2 - I_1) - 10,000 I_2 + V_{YB} = 0$$

or $-10,000 I_1 + 20,000 I_2 = 200 \angle -120°$ ...(*ii*)

From eqs. (*i*) and (*ii*), $I_1 = \dfrac{300 - j\,173.2}{16666.66}$

$\therefore$ Reading of voltmeter = $I_1 Z_R = \dfrac{300 - j173.2}{16666.66} \times 3333.33$ = **69.27 $\angle -30°$ V**

Example 15.70. *Two lamps of equal resistance are connected across lines R and B of a symmetrical 3-phase system. The junction of the two lamps is connected to the neutral N of the system through a capacitor of reactance numerically equal to the resistance of either lamp. Show that the lamp connected to B-phase takes 59.7% more current than in R-phase.*

Solution. Fig. 15.87 shows the conditions of the problem. Taking V_{RN} as reference phasor, we have,

$$V_{RN} = V_{ph} \angle 0°$$
$$V_{YN} = V_{ph} \angle -120°$$
$$V_{BN} = V_{ph} \angle 120°$$

Tracing the circuit from R to N and back to R and applying KVL,

$$-I_R R - (I_R + I_B)(-jR) + V_{RN} = 0$$

or $\quad I_R (R - jR) + I_B (-jR) = V_{ph} \angle 0°$

or $\quad I_R (1-j) + I_B (-j) = \dfrac{V_{ph}}{R} \angle 0°$ ... (i)

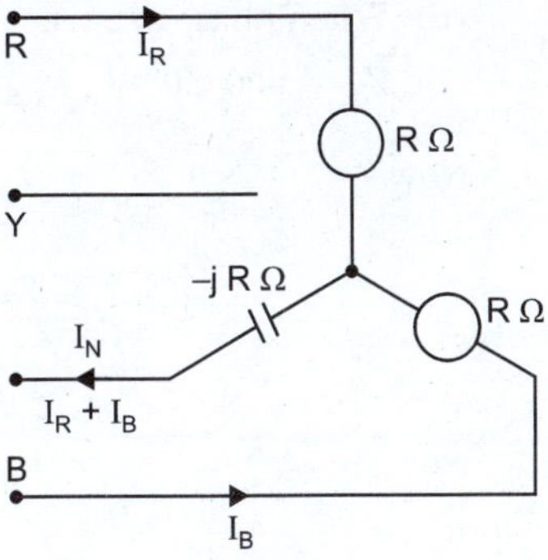

Tracing the circuit from B to N and back to B and applying KVL,

$$-I_B R - (I_R + I_B)(-jR) + V_{BN} = 0$$

or $\quad I_R (-jR) + I_B (R - jR) = V_{ph} \angle 120°$

or $\quad I_R (-j) + I_B (1-j) = \dfrac{V_{ph}}{R} \angle 120°$ ...(ii)

Fig. 15.87

Eqs. (i) and (ii) can be written in the matrix form as :

$$\begin{bmatrix} 1-j1 & -j1 \\ -j1 & 1-j1 \end{bmatrix} \begin{bmatrix} I_R \\ I_B \end{bmatrix} = \begin{bmatrix} \dfrac{V_{ph}}{R} \angle 0° \\ \dfrac{V_{ph}}{R} \angle 120° \end{bmatrix}$$

or $\qquad \dfrac{I_R}{I_B} = \dfrac{\begin{vmatrix} 1\angle 0° & -j1 \\ 1\angle 120° & 1-j1 \end{vmatrix}}{\begin{vmatrix} 1-j1 & 1\angle 0° \\ -j1 & 1\angle 120° \end{vmatrix}}$

On solving, $\dfrac{I_B}{I_R} = \textbf{1.597}$

Thus current in B-phase is 59.7% more than the R-phase.

Example 15.71. *A resistance of 300 Ω and a capacitance of 8 μF are connected in series across lines A and B of a 50 Hz, 400 V, 3-phase system. Find the voltage between the junction of resistance and capacitance and terminal C if phase sequence is (i) ABC (ii) ACB.*

Solution. Fig. 15.88 shows the conditions of the problem. The point D is the junction of resistance and capacitance.

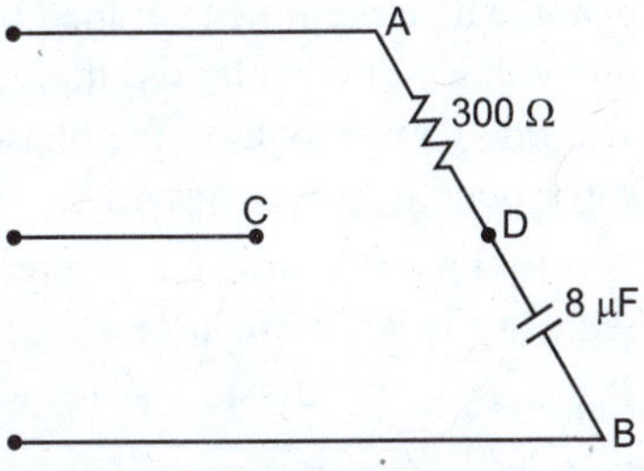

Now, $\quad X_C = \dfrac{1}{2\pi fC} = \dfrac{1}{2\pi \times 50 \times 8 \times 10^{-6}} = 398\ \Omega$

∴ $\quad Z_{AB} = R - jX_C = (300 - j398)\ \Omega = 498.4 \angle -53°\ \Omega$

(i) When phase sequence is ABC. Taking V_{AB} as reference phasor, we have,

$$V_{AB} = 400 \angle 0°\ V\ ;\ V_{BC} = 400 \angle -120°\ V\ ;\ V_{CA} = 400 \angle 120°\ V$$

Fig. 15.88

Now, $\qquad I_{AB} = \dfrac{V_{AB}}{Z_{AB}} = \dfrac{400 \angle 0°}{498.4 \angle -53°}$

∴ $\qquad V_{AD} = I_{AB} \times R = \dfrac{400 \angle 0°}{498.4 \angle -53°} \times 300 = 240.77 \angle 53°\ V$

∴ $\qquad V_{DA} = -V_{AD} = -240.77 \angle 53°\ V$

Now, $\qquad V_{DC} = V_{DA} + V_{AC} = V_{DA} - V_{CA} = -240.77 \angle 53° - 400 \angle 120°$

$$= \textbf{541.5} \angle \textbf{-84.16°\ V}$$

(ii) When phase sequence is ACB. Taking V_{AC} as reference phasor, we have,

$$V_{AC} = 400 \angle 0° \text{ V} \;;\; V_{CB} = 400 \angle -120° \text{ V} \;;\; V_{BA} = 400 \angle +120° \text{ V}$$

Now, $$I_{AB} = \frac{V_{AB}}{Z_{AB}} = \frac{-400 \angle 120°}{498.4 \angle -53°}$$

∴ $$V_{AD} = I_{AB} \times R = \frac{-400 \angle 120°}{498.4 \angle -53°} \times 300 = -240.77 \angle 173° \text{ V}$$

∴ $$V_{DA} = -V_{AD} = 240.77 \angle 173° \text{ V}$$

Now, $$V_{DC} = V_{DA} + V_{AC} = 240.77 \angle 173° + 400 \angle 0° = \mathbf{163.68 \angle 10.3° \text{ V}}$$

Tutorial Problems

1. Non-reactive loads of 10 kW, 6 kW and 4 kW are connected between the neutral and red, yellow and blue phases respectively of a 3-phase, 4-wire supply. Find the current in each line and in the neutral wire.
$$[I_R = 43.3 \text{ A} \;;\; I_Y = 26 \text{ A} \;;\; I_B = 17.3 \text{ A}; \; I_N = 22.9 \text{ A}]$$

2. A factory has the following loads with a power factor of 0.9 lagging in each case. Red phase 40 A, yellow phase 50 A and blue phase 60 A. If the supply is 400 V, 3-phase, 4-wire, calculate the current in the neutral wire and the total power. **[17.3 A, 31.2 kW]**

3. In a 3-phase, 4-wire system, two phases have currents of 10 A and 6 A at lagging power factors of 0.8 and 0.6 respectively, while the third phase is open-circuited. Calculate the current in the neutral wire.
[7 A]

4. A 3-phase, 4-wire system supplies a lighting load of 40 A, 30 A and 20 A respectively in the three phases. If the line voltage is 400 V, determine the current in the neutral wire. **[17.32 A]**

5. An unbalanced 4-wire star-connected load has a phase voltage of 120 V. The loads are $Z_R = (8 + j\,6)\ \Omega$, $Z_Y = 12\ \Omega$ and $Z_B = (12 - j\,16)\ \Omega$. Calculate the value of neutral current and its phase relationship to the voltage across the red phase. The phase sequence is *RYB*. **[8.04 A ; 90.3° *lagging*]**

15.33. Unbalanced Δ-Connected Load

Fig. 15.89 shown an unbalanced Δ-load connected to a balanced 3-phase supply. It is clear that line voltage of the supply appears across each load phase. The current in each load phase is equal to the line voltage divided by the impedance of that phase. The line current will be the phasor difference of the corresponding phase currents.

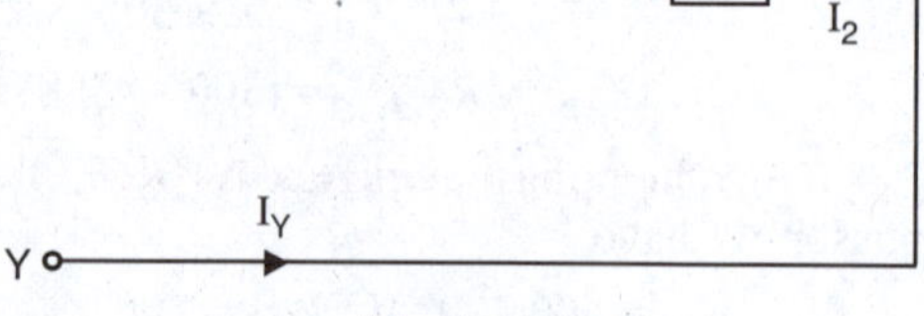

Taking V_{RY} as the reference phasor (Phase sequence is *RYB*), we have,

$$V_{RY} = V_L \angle 0° \text{ V} \;;\; V_{YB} = V_L \angle -120° \text{ V} \;;\; V_{BR} = V_L \angle 120° \text{ V}$$

∴ $$I_1 = \frac{V_{RY}}{Z_1} \;;\; I_2 = \frac{V_{YB}}{Z_2} \;;\; I_3 = \frac{V_{BR}}{Z_3}$$

$$I_R = I_1 - I_3 \;;\; I_Y = I_2 - I_1 \;;\; I_B = I_3 - I_2$$

Fig. 15.89

The reader may note that problems on unbalanced Δ loads can be solved with a fair degree of ease. It is because the problem simply reduces to three independent single phase circuits supplied with equal voltages that are 120° apart in phase.

Example 15.72. *Three impedances Z_1, Z_2 and Z_3 are delta-connected to a symmetrical 3-phase, 400 V supply of phase sequence RYB.*

$$Z_1 = (8 - j\,6)\ \Omega \text{ and is connected between lines R and Y}$$

Z_2 = $(6 + j\,8)$ Ω *and is connected between lines Y and B*

Z_3 = $(10 + j\,0)$ Ω *and is connected between lines B and R*

Calculate (i) line currents (ii) power per phase (iii) total power.

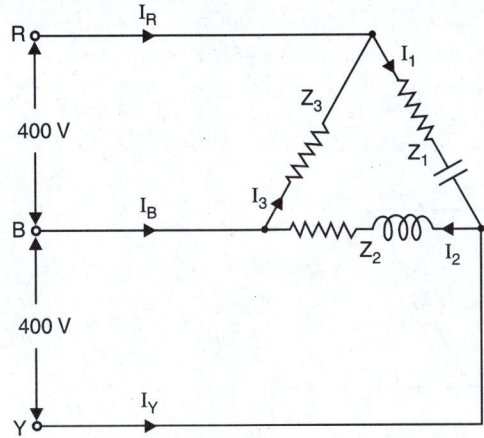

Fig. 15.90

Solution. Fig. 15.90 shows the circuit diagram.

$Z_1 = (8 - j\,6)\,\Omega = 10\,\angle - 36.9°\,\Omega$; $Z_2 = (6 + j\,8)\,\Omega = 10\,\angle 53.1°\,\Omega$; $Z_3 = (10 + j\,0)\,\Omega = 10\,\angle 0°\,\Omega$

(i) Taking V_{RY} as the reference phasor, we have,

$$V_{RY} = 400\,\angle 0°\,V \ ; \ V_{YB} = 400\,\angle - 120°\,V \ ; \ V_{BR} = 400\,\angle 120°\,V$$

$$I_1 = \frac{V_{RY}}{Z_1} = \frac{400\,\angle 0°}{10\,\angle - 36.90°} = 40\,\angle 36.9°\,A = (31.99 + j\,24.02)\,A$$

$$I_2 = \frac{V_{YB}}{Z_2} = \frac{400\,\angle - 120°}{10\,\angle 53.1°} = 40\,\angle - 173.1°\,A = (-39.71 - j\,4.8)\,A$$

$$I_3 = \frac{V_{BR}}{Z_3} = \frac{400\,\angle 120°}{10\,\angle 0°} = 40\,\angle 120°\,A = (-20 + j\,34.64)\,A$$

$\therefore$ $\quad I_R = I_1 - I_3 = (31.99 + j\,24.02) - (-20 + j\,34.64)$

$$= (51.99 - j\,10.62)\,A = \textbf{53.06}\,\angle - \textbf{11.5°\,A}$$

$$I_Y = I_2 - I_1 = (-39.71 - j\,4.81) - (31.99 + j\,24.02)$$

$$= (-71.7 - j\,28.83)\,A = \textbf{77.28}\,\angle - \textbf{158.1°\,A}$$

$$I_B = I_3 - I_2 = (-20 + j\,34.64) - (-39.71 - j\,4.81)$$

$$= (19.71 + j\,39.45)\,A = \textbf{44.1}\,\angle\,\textbf{63.5°\,A}$$

(ii) The current in each phase of Δ load is 40 A.

$\therefore$ $\quad P_{RY} = (40)^2 \times 8 = \textbf{12800 W}$; $P_{YB} = (40)^2 \times 6 = \textbf{9600 W}$; $P_{BR} = (40)^2 \times 10 = \textbf{16000 W}$

(iii) $\quad\quad\quad$ Total power = $P_{RY} + P_{YB} + P_{BR} = 12800 + 9600 + 16000 = \textbf{38400 W}$

Example 15.73. *Three impedances Z_1, Z_2 and Z_3 are Δ-connected to a symmetrical 3-phase, 400 V, 50 Hz supply of phase sequence RYB.*

$$Z_1 = 15\,\angle\,30°\,\Omega \ and \ is \ connected \ between \ lines \ R \ and \ Y$$

$Z_2 = 15 \angle 0° \; \Omega$ *and is connected between lines Y and B*

$Z_3 = 15 \angle -30° \; \Omega$ *and is connected between lines B and R*

Two wattmeters are used to measure the circuit power. The current coils of the wattmeters are connected in the lines R and B and the corresponding pressure coils across R and Y and B and Y respectively. Calculate (i) readings of the two wattmeters (ii) total power.

Solution. Fig. 15.91 shows the connections of two wattmeters.

Taking V_{RY} as the reference phasor, we have,

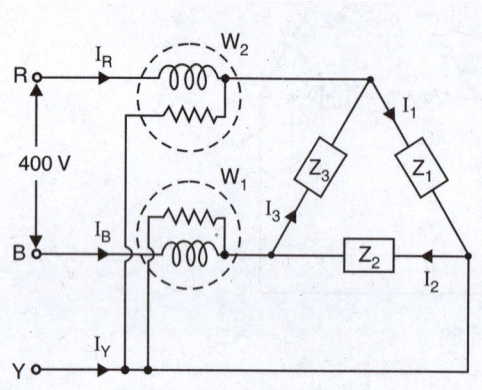

Fig. 15.91	**Fig. 15.92**

$$V_{RY} = 400 \angle 0° \text{ V} \; ; \; V_{YB} = 400 \angle -120° \text{ V} \; ; \; V_{BR} = 400 \angle 120° \text{ V}$$

The various phase currents (See Fig. 15.91) are :

$$I_1 = \frac{V_{RY}}{Z_1} = \frac{400 \angle 0°}{15 \angle 30°} = 26.67 \angle -30° \text{ A} = (23.1 - j\,13.34) \text{ A}$$

$$I_2 = \frac{V_{YB}}{Z_2} = \frac{400 \angle -120°}{15 \angle 0°} = 26.67 \angle -120° \text{ A} = (-13.34 - j\,23.1) \text{ A}$$

$$I_3 = \frac{V_{BR}}{Z_3} = \frac{400 \angle 120°}{15 \angle -30°} = 26.67 \angle 150° \text{ A} = (-23.1 + j\,13.34) \text{ A}$$

Line current $I_R = I_1 - I_3 = (23.1 - j\,13.34) - (-23.1 + j\,13.34)$

$$= (46.2 - j\,26.68) \text{ A} = 53.35 \angle -30° \text{ A}$$

Line current $I_B = I_3 - I_2 = (-23.1 + j\,13.34) - (-13.34 - j\,23.1)$

$$= (-9.76 + j\,36.44) \text{ A} = 37.72 \angle 105° \text{ A}$$

(i) As seen in Fig. 15.91, the current through the current coil of wattmeter W_2 is I_R and p.d. across its pressure coil is V_{RY}. The phase angle between V_{RY} and I_R is $\phi_2 = 30°$ as seen from the phasor diagram in Fig. 15.92.

∴ Reading of $W_2 = V_{RY} I_R \cos \phi_2 = 400 \times 53.35 \times \cos 30° =$ **18481 W**

Now current in the current coil of wattmeter W_1 is I_B and p.d. across its pressure coil is V_{BY}. As seen from the phasor diagram in Fig. 15.92, the phase angle between V_{BY} and I_B is *ϕ_1 = 105° − 60° = 45°.

Reading of $W_1 = V_{BY} I_B \cos \phi_1 = 400 \times 37.72 \times \cos 45° =$ **10669 W**

(ii) Total power $= W_2 + W_1 = 18481 + 10669 =$ **29150 W**

---

* The three line voltages (*i.e.*, V_{RY}, V_{YB} and V_{BR}) of the supply are symmetrical and 120° apart in phase. Since V_{RY} has been taken as the reference phasor, all angles are measured w.r.t. this voltage. Now I_B leads V_{RY} by 105°. The voltage V_{BY} (reverse the phasor V_{YB}) leads V_{RY} by 60°. Hence angle between I_B and V_{BY} is = 105° − 60° = 45°.

Example 15.74. *Two resistors of 100 Ω each are connected in series. The phases a and c of a 3-phase, 400 V supply are connected to the two ends. The phase b is connected to the junction of two resistors. The phase sequence is abc. Find the line currents.*

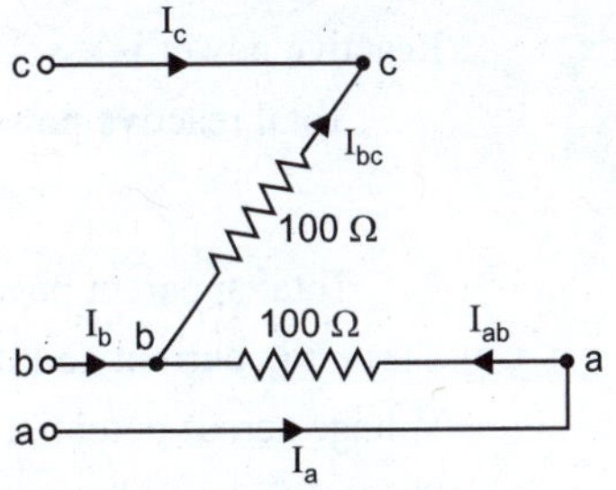

Solution. Fig. 15.93 shows the conditions of the problem.

$$V_{ab} = 400 \angle 0° \text{ V} ; V_{bc} = 400 \angle -120° \text{ V}$$

$$I_{ab} = \frac{V_{ab}}{100} = \frac{400 \angle 0°}{100} = 4 \angle 0° \text{ A}$$

$$I_{bc} = \frac{V_{bc}}{100} = \frac{400 \angle -120°}{100} = 4 \angle -120° \text{ A}$$

$$I_{ca} = 0 \text{ A}$$

Now,

$$I_b = I_{bc} - I_{ab} = 4 \angle -120° - 4 \angle 0° = \mathbf{6.928 \angle -150° A}$$

$$I_c = I_{ca} - I_{bc} = 0 - 4 \angle -120° = \mathbf{4 \angle 60° A}$$

$$I_a = I_{ab} - I_{ca} = 4 \angle 0° - 0 = \mathbf{4 \angle 0° A}$$

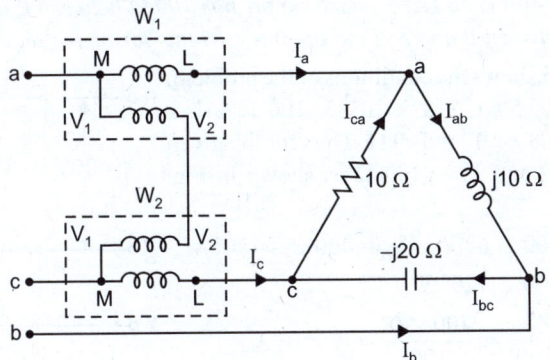

Fig. 15.93

Example 15.75. *Fig. 15.94 shows a 3 phase load fed by a symmetrical 400V, 3-phase supply having phase sequence abc. Find (i) phase currents (ii) line currents (iii) total active power, reactive power and voltamperes (iv) readings of wattmeters W_1 and W_2. The wattmeters are similar.*

Fig. 15.94

Solution. Taking V_{ab} as reference phasor,

$$V_{ab} = 400 \angle 0° \text{ V} ; V_{bc} = 400 \angle -120° \text{ V} ; V_{ca} = 400 \angle 120° \text{ V}$$

(i)

$$I_{ab} = \frac{V_{ab}}{j10} = \frac{400 \angle 0°}{j10} = \mathbf{-j\,40 \text{ A}}$$

$$I_{bc} = \frac{V_{bc}}{-j20} = \frac{400 \angle -120°}{20 \angle -90°} = \mathbf{20 \angle -30° A}$$

$$I_{ca} = \frac{V_{ca}}{10} = \frac{400 \angle 120°}{10} = \mathbf{40 \angle 120° A}$$

(ii) To find line currents, we shall apply *KCL* at the load junctions.

$$I_a = I_{ab} - I_{ca} = -j\,40 - 40 \angle 120°$$

$$= -j\,40 + 20 - j\,34.64 = (20 - j\,74.64) \text{ A} = \mathbf{77.27 \angle -75° A}$$

$$I_b = I_{bc} - I_{ab} = 20 \angle -30° - (-j\,40)$$

$$= 17.32 - j\,10 + j\,40 = (17.32 + j\,30) \text{ A} = \mathbf{34.64 \angle 60° A}$$

$$I_c = I_{ca} - I_{bc} = 40 \angle 120° - 20 \angle -30°$$

$$= -20 + j\,34.64 - 17.32 + j\,10 = (-37.32 + j\,44.64) \text{ A}$$

$$= \mathbf{58.19 \angle 129.9° A}$$

(*iii*) Active power is absorbed in resistance only.

∴   Total active power, $P = I_{ca}^2 \times 10 = (40)^2 \times 10 =$ **16×10^3 W**

Reactive power is associated with reactances (X_L and X_C) only.

∴   Total reactive power, $Q = I_{ab}^2 \times 10 - I_{bc}^2 \times 20 = 40^2 \times 10 - 20^2 \times 20$

$$= 16000 - 8000 = \textbf{8000 VAR} \ (\textit{inductive})$$

Total apparent power, $S = \sqrt{P^2 + Q^2} = \sqrt{(16 \times 10^3)^2 + (8000)^2} =$ **17.89×10^3 VA**

(*iv*) Current in current coil of $W_1 = I_a = 77.27 \angle -75°$ A

Voltage across potential coil of $W_1 = 0.5\ V_{ac} = -0.5\ V_{ca} = -200 \angle 120° = 200 \angle -60°$ V

Phase difference between V_{ac} and I_a, $\phi_1 = 75° - 60° = 15°$

∴   Reading of $W_1 = 200 \times 77.27 \times \cos 15° =$ **14927.4 W**

Current in current coil of $W_2 = I_c = 58.19 \angle 129.9°$A

Voltage across potential coil of $W_2 = 0.5\ V_{ca} = 200 \angle 120°$

Phase difference between V_{ca} and I_c, $\phi_2 = 129.9° - 120° = 9.9°$

∴   Reading of $W_2 = 200 \times 58.19 \times \cos 9.9° =$ **11464.7 W**

Example 15.76. *A 400 V, 50 Hz, 3-phase supply has 100 Ω between R and Y, 318 mH between Y and B and 31·8 μF between B and R. Find the line currents for phase sequence RYB.*

Solution. Fig. 15.95 shows the conditions of the problem. Now $X_L = 2\pi fL = 2\pi \times 50 \times 318 \times 10^{-3} = 100$ Ω. Also, $X_C = 1/\omega C = 1/314 \times 31.8 \times 10^{-6} = 100$ Ω. Thus for the given conditions $X_L = 100$ Ω and $X_C = 100$ Ω as shown in Fig. 15.95.

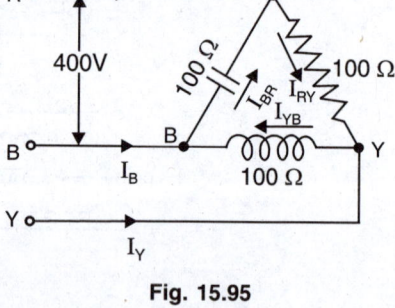

Fig. 15.95

Now, $V_{RY} = 400 \angle 0°$ V ; $V_{YB} = 400 \angle -120°$ V ;
$V_{BR} = 400 \angle 120°$ V

∴ $I_{RY} = \dfrac{V_{RY}}{100} = \dfrac{400 \angle 0°}{100} = 4 \angle 0°$ A

$= (4 + j\,0)$ A

$I_{YB} = \dfrac{V_{YB}}{100 \angle 90°} = \dfrac{400 \angle -120°}{100 \angle 90°} = 4 \angle -210°$ A $= (-3.46 + j\,2)$ A

$I_{BR} = \dfrac{V_{BR}}{100 \angle -90°} = \dfrac{400 \angle 120°}{100 \angle -90°} = 4 \angle 210°$ A $= (-3.46 - j\,2)$ A

∴ $I_R = I_{RY} - I_{BR} = (4 + j\,0) - (-3.46 - j\,2) = (7.46 + j\,2)$ A = **$7.73 \angle 15°$ A**

$I_Y = I_{YB} - I_{RY} = (-3.46 + j\,2) - (4 + j\,0) = (-7.46 + j\,2)$ A = **$7.73 \angle 165°$ A**

$I_B = I_{BR} - I_{YB} = (-3.46 - j\,2) - (-3.46 + j\,2) = 4 \angle -90°$ **A**

Tutorial Problems

1. Three impedances $Z_1 = (10 + j\,10)$ Ω, $Z_2 = (8.66 + j\,5)$ Ω and $Z_3 = (12 + j\,16)$ Ω are Δ-connected to a 380 V, 3-phase system. Determine the line currents and draw the phasor diagram.

    [$I_R = 38.3 \angle -72.5°$A; $I_Y = 51.9 \angle 180°$A; $I_B = 54.4 \angle 42.1°$A]

2. Three loads of $(31 + j\,59)$ Ω, $(30 - j\,40)$ Ω and $(80 + j\,60)$ Ω are connected on delta to a 3-phase, 200 V supply. Find the phase currents, line currents and power when the phase sequence is *RYB*.

    [$I_{RY} = 3 \angle -62.82°$ A; $I_{YB} = 4 \angle -66.87°$ A; $I_{BR} = 2 \angle 83.13°$ A; $I_R = 4.8 \angle -76°$A;

    $I_Y = 1.03 \angle -80.2°$ A; $I_B = 5.8 \angle 103.23°$A; 1079 W]

3. A 3-phase, 3-wire, *CBA* system supplies a Δ-connected load in which $Z_{AB} = 25 \angle 90°$ Ω, $Z_{BC} = 15$ $\angle 30°$ Ω and $Z_{CA} = 20 \angle 0°$ Ω. Line voltage is 230 V. Find the wattmeter readings when the two watt-meter method is used with meters in lines *A* and *B*.   **[812.64 W ; 4886.63 W]**

15.34. Unbalanced 3-Wire Star-Connected Load

Fig. 15.96 shows an unbalanced *3-wire star load connected to a balanced 3-phase supply. Note that supply star point *N* (normally at zero potential) is isolated from the load star point *N'*. Since the *Y* load is unbalanced and *N* is isolated from *N'*, the potential of *N'* will be different from *N*. The result is that (*i*) load phase voltages will not be equal to the supply phase voltages *e.g.*, V_{RN} is not **equal to V_{RN}' (*ii*) the three load phase voltages will be unequal and will differ in phase by unequal

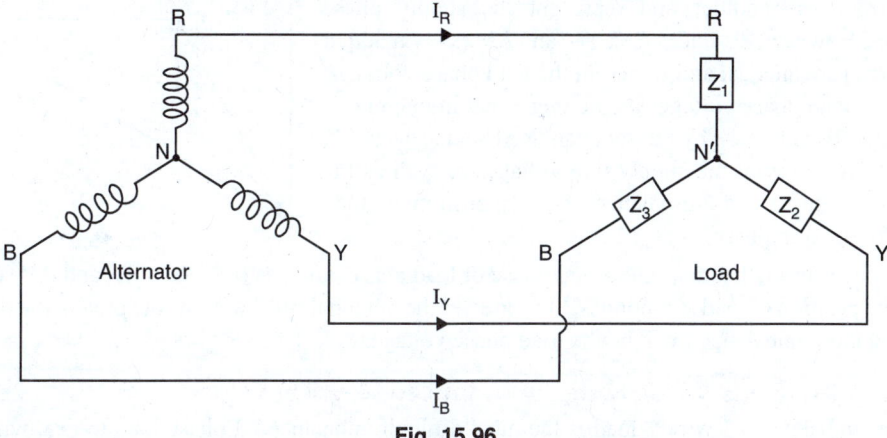

Fig. 15.96

angle. The magnitude of each load phase voltage will depend upon the degree of load unbalance. If the load impedance in one or more phases changes, the potential of *N'* also changes, causing the load phase voltages to change in magnitude and in phase. The load neutral *N'* is sometimes called the **floating neutral** because its potential changes according to the imbalance of the load.

If the load unbalance is severe, the potential of *N'* w.r.t. *N* may vary considerably and the load phase voltages may become highly unbalanced in magnitude as well as phase. Such a condition is definitely undesirable. Some loads will operate inefficiently because of low voltage, while other equipment may be damaged due to overvoltage. Because of the variations on an unbalanced 3-wire *Y* load, individual single phase loads should not be connected in Y unless they form a balanced 3-phase load (or nearly balanced 3-phase load) at all times.

The following points may be noted carefully :

(*i*)   *The phasor sum of the three unbalanced line currents is zero i.e.,*

$$I_R + I_Y + I_B = 0 \qquad\qquad \textit{...phasor sum}$$

This is expected because the unbalanced line currents meet at load star point *N'* and they cannot pile up there. According to Kirchhoff's current law, their phasor sum must be †zero.

---

*   This type of load (*i.e.*, unbalanced 3-wire star load) is rarely found in practice because all the 3-wire star loads (*e.g.*, 3-phase star-connected motor) are balanced loads. When does such a load occur ? It occurs as a result of some fault on the system. For example, if in a 3-phase, 4-wire system, the connection between supply neutral and load neutral is broken, this would result in an unbalanced 3-wire star load. Hence an unbalanced 3-wire star load is usually the result of a fault on the system.

**   Had *N'* and *N* been connected through a wire, both *N* and *N'* would have been at the same potential. In that case, phase voltage of supply would have appeared across each branch of the unbalanced *Y* load. This would have made the whole problem a simple affair—three independent single phase circuits supplied with balanced voltages.

†   A phasor sum of zero can be obtained with unbalanced line currents also.

(ii) *Only load phase voltages change in magnitude and in phase. However, line voltages remain balanced i.e., they are equal in magnitude and 120° apart in phase.*

Triangular phasor diagram. The behaviour of unbalanced 3-wire star load can be best explained from the triangular phasor diagram shown in Fig. 15.97. Here N is the neutral (normally at zero potential) of the supply and is located at the centre of the equilateral triangle RYB. The point N' is the load neutral. Due to load unbalance, N' has some potential *w.r.t.* N and is shifted to some other position. The distances RN, YN and BN are equal but 120° apart from one another and represent the supply phase voltages. However, distances RN', YN' and BN' (shown dotted for clarity) are unequal and represent the load phase voltages. Note that load phase voltages are neither equal in magnitude nor they differ in phase by the same angle. The distances RY, YB and BR represent the supply line voltages as well as the load line voltages and are balanced *i.e.*, equal in magnitude and 120° apart in phase.

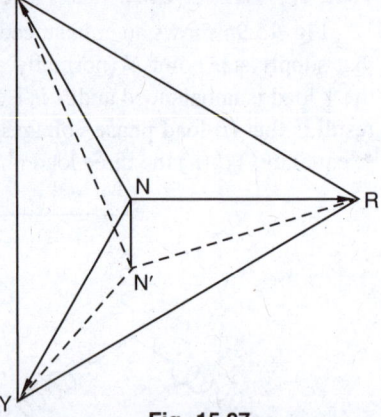

Fig. 15.97

It may be seen that magnitude and phase of load phase voltages (*i.e.*, RN', YN' and BN') depend upon the position of load star point N'. The greater the potential of N' *w.r.t.* N, the greater the distance NN' and more unbalanced will be the load phase voltages.

15.35. Methods of Solving Unbalanced 3-wire Y load

The unbalanced 3-wire Y load is the most difficult unbalanced 3-phase load to deal with. It is because load phase voltages cannot be determined directly from the given supply line voltages. Several methods have been developed to handle such unbalanced loads. Four frequently used methods are :

(i) By Kirchhoff's laws

(ii) By Maxwell's mesh current method

(iii) **By Y/Δ Conversion.** In this method, unbalanced 3-wire Y load is converted into equivalent Δ load. The problem is then solved as a Δ-connected system. The line currents calculated in the equivalent Δ load will be equal in magnitude and phase to those taken by the original unbalanced Y load.

(iv) **By using Millman's theorem.** By using Millman's theorem, we can determine $V_{N'N}$ (*i.e.*, potential of load neutral N' *w.r.t.* supply neutral N) in an unbalanced 3-wire Y load. Once $V_{N'N}$ is known, the load phase voltages and line currents of unbalanced Y load can be easily calculated (See Art. 15.39).

15.36. Solving Unbalanced 3-Wire Y Load by Kirchhoff's Laws

We can solve problems related to unbalanced 3-wire star-connected loads by using Kirchhoff's laws. Consider an unbalanced 3-wire star-connected load fed from 3-phase balanced supply as shown in Fig. 15.98. Using Kirchhoff's voltage law, we have,

$$*V_{RY} = I_R Z_R - I_Y Z_Y \qquad ...(i)$$

$$V_{YB} = I_Y Z_Y - I_B Z_B \qquad ...(ii)$$

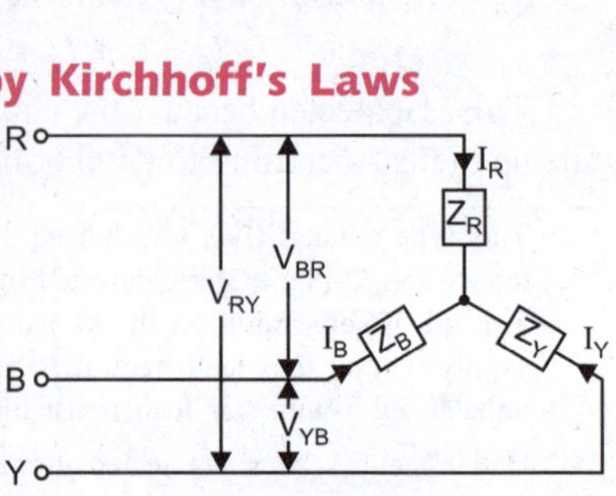

Fig. 15.98

---

* $\quad V_R - I_R Z_R + I_Y Z_Y = V_Y \quad \therefore\ V_R - V_Y = I_R Z_R - I_Y Z_Y$

But $V_R - V_Y = V_{RY} \quad \therefore\ V_{RY} = I_R Z_R - I_Y Z_Y$

$$V_{BR} = I_B Z_B - I_R Z_R \qquad ...(iii)$$

Also, applying Kirchhoff's current law to load star point, we have,

$$I_R + I_Y + I_B = 0 \qquad ...(iv)$$

Adding eqs. (i) and (iii), $V_{RY} + V_{BR} = (I_R Z_R - I_Y Z_Y) + (I_B Z_B - I_R Z_R)$

or $\qquad V_{RY} + V_{BR} = I_B Z_B - I_Y Z_Y$

$\therefore \qquad I_B = \dfrac{V_{RY} + V_{BR} + I_Y Z_Y}{Z_B} \qquad ...(v)$

Now $- I_R = I_Y + I_B$. Putting this value of $- I_R$ in eq. (iii), we have,

$$V_{BR} = I_B Z_B + (I_Y + I_B) Z_R$$

or $\qquad V_{BR} = I_B (Z_B + Z_R) + I_Y Z_R \qquad ...(vi)$

Putting the value of I_B from eq. (v) into eq. (vi), we have,

$$V_{BR} = \dfrac{V_{RY} + V_{BR} + I_Y Z_Y}{Z_B} (Z_B + Z_R) + I_Y Z_R$$

or $\quad V_{BR} Z_B = V_{RY} Z_B + V_{BR} Z_B + I_Y Z_Y Z_B + V_{RY} Z_R + V_{BR} Z_R$
$$+ I_Y Z_Y Z_R + I_Y Z_R Z_B$$

or $\quad V_{BR} Z_B = I_Y (Z_R Z_Y + Z_Y Z_B + Z_B Z_R) + Z_R (V_{RY} + V_{BR})$
$$+ V_{RY} Z_B + V_{BR} Z_B$$

Now, $V_{RY} + V_{YB} + V_{BR} = 0$ so that $V_{RY} + V_{BR} = - V_{YB}$

$\therefore \qquad 0 = I_Y (Z_R Z_Y + Z_Y Z_B + Z_B Z_R) - V_{YB} Z_R + V_{RY} Z_B$

or $I_Y (Z_R Z_Y + Z_Y Z_B + Z_B Z_R) = V_{YB} Z_R - V_{RY} Z_B$

$\therefore \qquad I_Y = \dfrac{V_{YB} Z_R - V_{RY} Z_B}{Z_R Z_Y + Z_Y Z_B + Z_B Z_R}$

From the symmetry of above equation, the other line currents are :

$$I_B = \dfrac{V_{BR} Z_Y - V_{YB} Z_R}{Z_R Z_Y + Z_Y Z_B + Z_B Z_R}$$

$$I_R = \dfrac{V_{RY} Z_B - V_{BR} Z_Y}{Z_R Z_Y + Z_Y Z_B + Z_B Z_R}$$

15.37. Solving Unbalanced 3-wire Y Load By Loop Current Method

We can also solve problems related to unbalanced 3-wire star-connected loads by using Maxwell's loop current method. Consider an unbalanced 3-wire star-connected load fed from 3-phase balanced supply as shown in Fig. 15.99. Let the loop or mesh currents be I_1 and I_2. It is clear that :

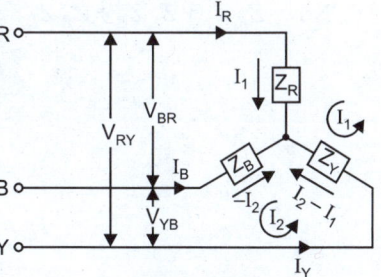

Fig. 15.99

$$I_R = I_1 \; ; \; I_Y = I_2 - I_1 \text{ and } I_B = - I_2$$

Considering the phases R and Y and applying Kirchhoff's voltage law, we have,

$$V_{RY} = I_R Z_R - I_Y Z_Y = I_1 Z_R - (I_2 - I_1) Z_Y$$

or $\qquad V_{RY} = I_1 (Z_R + Z_Y) - I_2 Z_Y \qquad ...(i)$

Considering the phases Y and B and applying Kirchhoff's voltage law, we have,

$$V_{YB} = I_Y Z_Y - I_B Z_B = (I_2 - I_1) Z_Y - (-I_2 Z_B)$$

or $\qquad V_{YB} = I_2 (Z_Y + Z_B) - I_1 Z_Y \qquad ...(ii)$

Solving eqs. (*i*) and (*ii*) for I_1 and I_2, we have,

$$I_1 = \frac{V_{RY}(Z_Y + Z_B) + V_{YB}Z_Y}{(Z_R + Z_Y)(Z_Y + Z_B) - Z_Y^2}$$

$$I_2 = \frac{V_{YB}(Z_R + Z_Y) + V_{RY}Z_Y}{(Z_R + Z_Y)(Z_Y + Z_B) - Z_Y^2}$$

Once we know the mesh currents I_1 and I_2, the line currents I_R, I_Y and I_B can be easily determined.

Example 15.77. *A Y-connected load is supplied from a 400 V, 3-phase, 3-wire symmetrical system RYB. The branch circuit impedances are :*

$$Z_R = \left(10\sqrt{3} + j10\right)\Omega \; ; Z_Y = \left(20 + j20\sqrt{3}\right)\Omega \; ; Z_B = (0 - j\,10)\ \Omega$$

The phase sequence is RYB. Determine the currents in each branch by (i) Kirchhoff's laws (ii) mesh current method.

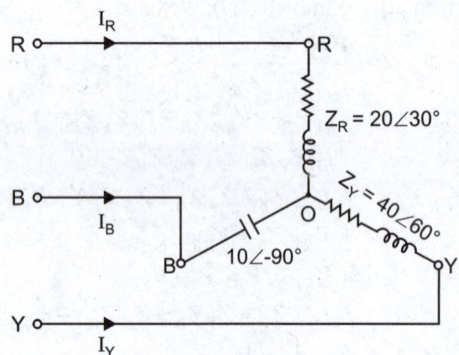

Fig. 15.100

Solution. Fig. 15.100 shows the conditions of the problem.

$$Z_R = \left(10\sqrt{3} + j10\right)\Omega = (17.32 + j\,10)\ \Omega = 20\ \angle\ 30°\ \Omega$$

$$Z_Y = \left(20 + j20\sqrt{3}\right)\Omega = (20 + j\,34.64)\ \Omega = 40\ \angle\ 60°\ \Omega$$

$$Z_B = (0 - j\,10)\ \Omega = 10\ \angle -90°\ \Omega$$

Now, $V_{RY} = 400\ \angle\ 0°$ V ; $V_{YB} = 400\ \angle -120°$ V ; $V_{BR} = 400\ \angle\ 120°$ V

(i) By applying Kirchhoff's laws. As proved in Art. 15.36, the various line currents are :

$$I_R = \frac{V_{RY}Z_B - V_{BR}Z_Y}{Z_R Z_Y + Z_Y Z_B + Z_B Z_R}$$

Now, $Z_R Z_Y + Z_Y Z_B + Z_B Z_R = 20\ \angle\ 30° \times 40\ \angle\ 60° + 40\ \angle\ 60° \times 10\ \angle -90°$

$$+ 10\ \angle -90° \times 20\angle 30°$$

$$= 800\ \angle\ 90° + 400\ \angle -30° + 200\ \angle -60°$$

$$= (0 + j\,800) + (346.41 - j\,200) + (100 - j\,173.2)$$

$$= 446.41 + j\,426.8 = 617\angle\ 43.7°$$

$$\therefore \qquad I_R = \frac{400\ \angle\ 0° \times 10\ \angle -90° - 400\ \angle 120° \times 40\ \angle\ 60°}{617\ \angle\ 43.7°}$$

$$= \mathbf{26.73\ \angle -57.7°A}$$

Also, $\qquad I_Y = \dfrac{V_{YB}Z_R - V_{RY}Z_B}{Z_R Z_Y + Z_Y Z_B + Z_B Z_R}$

$$= \frac{400\ \angle -120° \times 20\ \angle\ 30° - 400\ \angle\ 0° \times 10\ \angle -90°}{617\ \angle\ 43.7°}$$

$$= \mathbf{6.48\ \angle -133.7°A}$$

And
$$I_B = \frac{V_{BR}Z_Y - V_{YB}Z_R}{Z_RZ_Y + Z_YZ_B + Z_BZ_R}$$

$$= \frac{400 \angle 120° \times 40 \angle 60° - 400 \angle -120° \times 20 \angle 30°}{617 \angle 43.7°}$$

$$= \mathbf{29 \angle 109.73° \ A}$$

(*ii*) **By mesh current method.** The unbalanced 3-wire star-connected load in Fig. 15.100 is reshown in Fig. 15.101 with mesh currents I_1 and I_2. As proved in Art. 15.37,

$$I_1 = \frac{V_{RY}(Z_Y + Z_B) + V_{YB}Z_Y}{(Z_R + Z_Y)(Z_Y + Z_B) - Z_Y^2}$$

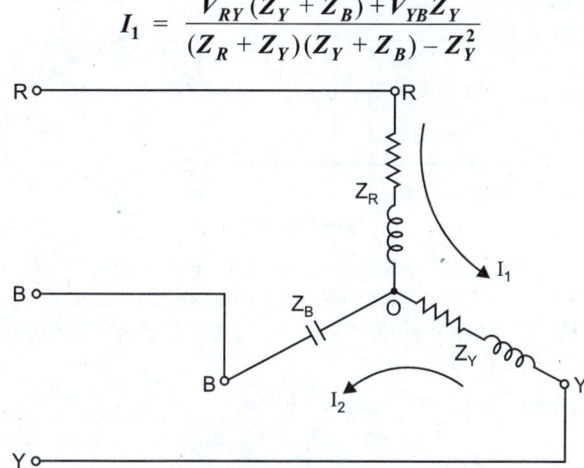

Fig. 15.101

Here, $Z_Y + Z_B = (20 + j\,34.64) + (0 - j\,10)$

$$= (20 + j\,24.64) \ \Omega = 31.74 \angle 50.9° \ \Omega$$

$$Z_R + Z_Y = (17.32 + j\,10) + (20 + j\,34.64) = (37.32 + j\,44.64) \ \Omega = 58.2 \angle 50.1° \ \Omega$$

∴ $$I_1 = \frac{400 \angle 0° \times 31.74 \angle 50.9° + 400 \angle -120° \times 40 \angle 60°}{58.2 \angle 50.1° \times 31.74 \angle 50.9° - 1600 \angle 120°}$$

$$= 26.5 \angle -57.7° \ A$$

$$= (14.2 - j\,22.4) \ A$$

Also, $$I_2 = \frac{V_{YB}(Z_R + Z_Y) + V_{RY}Z_Y}{(Z_R + Z_Y)(Z_Y + Z_B) - Z_Y^2}$$

Now, $Z_R + Z_Y = 58.2 \angle 50.1° \ \Omega \ ; \ Z_Y + Z_B = 31.74 \angle 50.9° \ \Omega$

∴ $$I_2 = \frac{400 \angle -120° \times 58.2 \angle 50.1° + 400 \angle 0° \times 40 \angle 60°}{58.2 \angle 50.1° \times 31.74 \angle 50.9° - 1600 \angle 120°} = 28.94 \angle -70.3° \ A$$

$$= (9.76 - j\,27.2) \ A$$

∴ $I_R = I_1 = \mathbf{26.5 \angle -57.7° A}$

$I_Y = I_2 - I_1 = (9.76 - j\,27.2) - (14.2 - j\,22.4)$

$$= (-4.44 - j\,4.8) \ A = \mathbf{6.5 \angle -133° \ A}$$

$I_B = -I_2 = -(9.76 - j\,27.2) \ A = (-9.76 + j\,27.2) \ A = \mathbf{28.9 \angle 109.74° \ A}$

Example 15.78. *Three resistors 10 Ω, 20 Ω and 20 Ω are connected in star to the terminals A, B and C of a 3 – ϕ, 3-wire supply through two single-phase wattmeters for measurement of total power with current coils in lines A and C and pressure coils between A and B and C and B. Calculate (i) the line currents I_A and I_C (ii) the readings of each wattmeter. The line voltage is 400 V.*

Solution. (i) Fig. 15.102 shows the conditions of the problem. It is understood that phase sequence is *ABC* so that :

$V_{AB} = 400 \angle 0° \text{ V} ; V_{BC} = 400 \angle -120° \text{ V} ; V_{CA} = 400 \angle 120° \text{ V}$

Also, $Z_A = 10 \angle 0° \Omega ; Z_B = 20 \angle 0° \Omega ; Z_C = 20 \angle 0° \Omega$

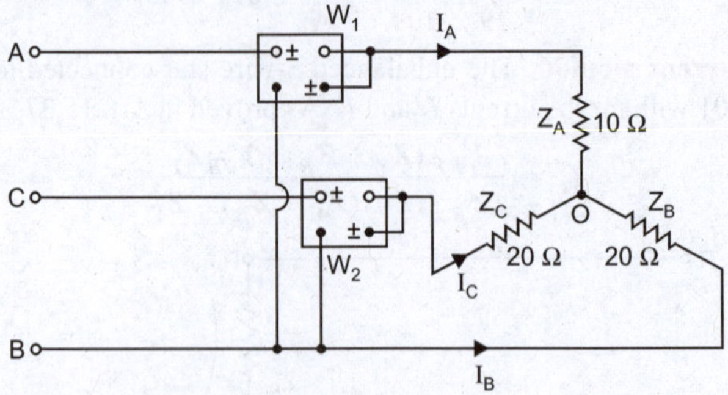

Fig. 15.102

As seen from Fig. 15.102, the current through the current coil of W_1 is I_A and the current through the current coil of W_2 is I_C. Further, voltage across potential coil of W_1 is V_{AB} and that across potential coil of W_2 is V_{CB}. We can find the values of I_A and I_C either by applying Kirchhoff's laws or Maxwell's mesh current method.

Using Kirchhoff's laws (See Art. 15.36), we have,

$$I_A = \frac{V_{AB}Z_C - V_{CA}Z_B}{Z_AZ_B + Z_BZ_C + Z_CZ_A}$$

$$= \frac{400 \angle 0° \times 20 \angle 0° - 400 \angle 120° \times 20 \angle 0°}{(10 \times 20) + (20 \times 20) + (20 \times 10)}$$

$$= (15 - j\,8.65) \text{ A} = \mathbf{17.32 \angle -30° \text{ A}}$$

$$I_C = \frac{V_{CA}Z_B - V_{BC}Z_A}{Z_AZ_B + Z_BZ_C + Z_CZ_A}$$

$$= \frac{400 \angle 120° \times 20 \angle 0° - 400 \angle -120° \times 10 \angle 0°}{(10 \times 20) + (20 \times 20) + (20 \times 10)}$$

$$= (-2.5 + j\,13) \text{ A} = \mathbf{13.24 \angle 100.9° \text{ A}}$$

Using Maxwell's mesh current method (See Art. 15.37), we have,

$$I_1 = \frac{V_{AB}(Z_B + Z_C) + V_{BC}Z_B}{(Z_A + Z_B)(Z_B + Z_C) - Z_B^2}$$

Here, $Z_A + Z_B = 10 + 20 = 30 \Omega ; Z_B + Z_C = 20 + 20 = 40 \Omega$

∴ $I_1 = \dfrac{400 \angle 0° \times (40) + 400 \angle -120° \times 20}{30 \times 40 - (20)^2} = (15 - j\,8.65) \text{ A}$

Also, $I_2 = \dfrac{V_{BC}(Z_A + Z_B) + V_{AB}Z_B}{(Z_A + Z_B)(Z_B + Z_C) - Z_B^2}$

                 $= \dfrac{400 \angle -120° \times 30 + 400 \angle 0° \times 20}{30 \times 40 - (20)^2} = (2.5 - j\,13) \text{ A}$

∴ $I_A = I_1 = \mathbf{(15 - j\,8.65) \text{ A}}$

           $I_C = -I_2 = -(2.5 - j\,13) = \mathbf{(-2.5 + j\,13) \text{ A}}$

(ii) Reading of $W_1 = V_{AB} I_A \cos \phi$ where ϕ = Angle between V_{AB} and I_A.

Here, $\qquad V_{AB} = 400$ V ; $I_A = 17.32$ A ; $\phi = -30°$

$\therefore \qquad W_1 = 400 \times 17.32 \times \cos(-30°) = \mathbf{6000\ W}$

$\qquad$ Reading of $W_2 = V_{CB} I_C \cos \phi'$

Here $\qquad V_{CB} = 400$ V ; $I_C = 13.24$ A

Now, V_{BC} lags behind V_{AB} by 120° so that V_{CB} leads V_{AB} by 60°. Also I_C leads V_{AB} by 100.9°. Therefore, phase difference between V_{CB} and I_C is $\phi' = 100.9° - 60° = 40.9°$.

$\therefore \qquad W_2 = 400 \times 13.24 \times \cos 40.9° = \mathbf{4002\ W}$

Tutorial Problems

1. A 3-phase, 3-wire, Y-connected load of $(0 + j\,20)$ Ω, $(0 + j\,1)$ Ω and $(0 - j\,1)$ Ω is connected to 400 V, $3 - \phi$, 3-wire symmetrical *RYB* system. Calculate the voltage drop across each impedance and the potential of star point above ground. The phase sequence is *RYB*.

   $[V_R = 8000 \angle 60°\text{ V}; V_Y = 7807 \angle 62.5°\text{ V} ; V_B = 8207\angle 62.4°\text{ V}; V_{NN} = 8005 \angle -118.3°\text{ V}]$

2. A Y-connected load comprising two resistors and a pure inductor is connected to a symmetrical 3-phase *RYB* system. If the numerical impedance of all branches is the same, find the voltage across each branch as a percentage of the line voltage. Phase sequence is *RYB*.

   $[V_R = 23.15\% ; \ V_Y = 85.52\% ; V_B = 77.46\%]$

3. An unbalanced Y-connected load is fed from a 3-phase, 3-wire balanced 400 V supply. The load impedances are $Z_R = (4 + j\,8)$ Ω, $Z_Y = (3 + j\,4)$ Ω and $Z_B = (15 + j\,20)$ Ω. Assuming the phase sequence to be *RYB*, calculate the total power absorbed by the load. **[8.45 kW]**

15.38. Solving Unbalanced 3-Wire Y Load by Y/Δ Conversion

Fig. 15.103 shows an unbalanced 3-wire Y-connected load while Fig. 15.104 shows the equivalent Δ-connected load. It was shown in chapter 3 that Y connection of resistances can be replaced by an equivalent Δ and *vice-versa*. With a.c., the same formulas hold good except that resistances are replaced by impedances (in complex form).

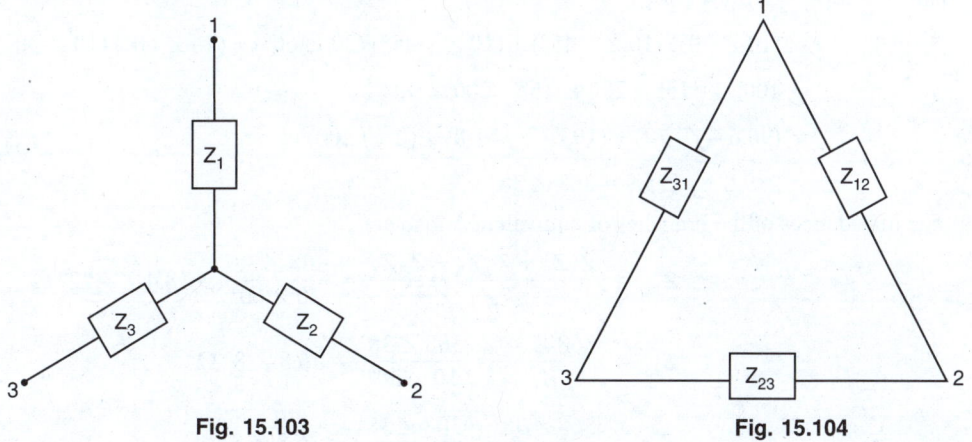

Fig. 15.103 $\qquad\qquad$ **Fig. 15.104**

Thus the Y-load of Fig. 15.103 can be replaced by equivalent Δ-load having branch impedances as :

$$Z_{12} = \frac{Z_1 Z_2 + Z_2 Z_3 + Z_3 Z_1}{Z_3} \ ; \ \ Z_{23} = \frac{Z_1 Z_2 + Z_2 Z_3 + Z_3 Z_1}{Z_1} \ ;$$

$$Z_{31} = \frac{Z_1 Z_2 + Z_2 Z_3 + Z_3 Z_1}{Z_2}$$

The equivalent Δ load can be solved to obtain the line currents. The line currents so calculated will be the line currents taken by the original unbalanced Y-load.

How to remember above relations ? The symmetry of above relations shows that one should not try to memorise them. It may be seen that numerator is the same in the three cases *i.e.*, it is the sum of the cross products of star impedances viz. $Z_1 Z_2 + Z_2 Z_3 + Z_3 Z_1$. However, denominator is that star impedance which is not included in the considered terminals. Thus, for Z_{12}, the terminals are 1 and 2 and the star impedance not included in these terminals is Z_3. Hence in this case Z_3 is the denominator.

Example 15.79. *A symmetrical 3-phase, 400 V, 3-wire supply feeds an unbalanced star-connected load. The branch impedances of the load are $Z_1 = (8 \cdot 66 + j\, 5)$ Ω, $Z_2 = (7 \cdot 07 - j\, 7 \cdot 07)$ Ω and $Z_3 = (10 + j\, 17 \cdot 32)$ Ω. Find the line currents and the voltage across each impedance. The phase sequence is RYB.*

Solution. The unbalanced Y-connected load and its equivalent Δ-connected load are shown in Fig. 15.105 (*i*) and 15.105 (*ii*) respectively.

$$Z_1 = 10 \angle 30° \ \Omega \ ; Z_2 = 10 \angle -45° \ \Omega \ ; Z_3 = 20 \angle 60° \ \Omega$$

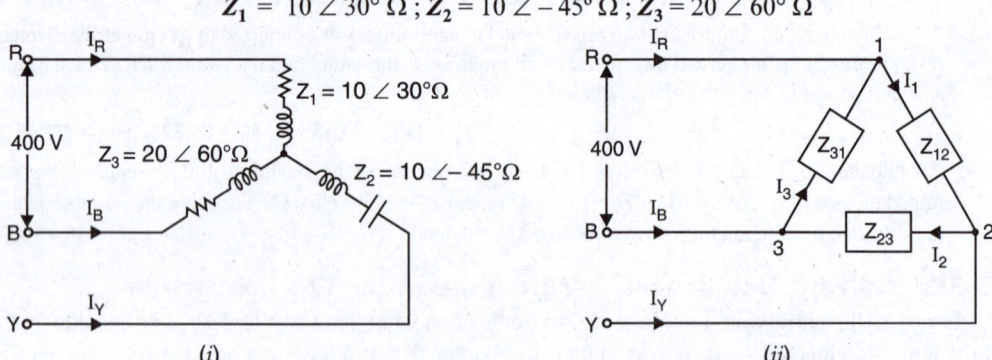

Fig. 15.105

Now, $Z_1 Z_2 + Z_2 Z_3 + Z_3 Z_1$

$$= (10 \angle 30°)(10 \angle -45°) + (10 \angle -45°)(20 \angle 60°) + (20 \angle 60°)(10 \angle 30°)$$

$$= 100 \angle -15° + 200 \angle 15° + 200 \angle 90°$$

$$= (96.6 - j\, 25.9) + (193.2 + j\, 51.8) + (0 + j\, 200)$$

$$= 289.8 + j\, 225.9 = 368 \angle 38°$$

The impedances of the branches of equivalent Δ load are :

$$Z_{12} = \frac{Z_1 Z_2 + Z_2 Z_3 + Z_3 Z_1}{Z_3} = \frac{368 \angle 38°}{20 \angle 60°} = 18.4 \angle -22° \ \Omega$$

$$Z_{23} = \frac{368 \angle 38°}{Z_1} = \frac{368 \angle 38°}{10 \angle 30°} = 36.8 \angle 8° \ \Omega$$

$$Z_{31} = \frac{368 \angle 38°}{Z_2} = \frac{368 \angle 38°}{10 \angle -45°} = 36.8 \angle 83° \ \Omega$$

Taking V_{RY} as the reference phasor, we have,

$$V_{RY} = 400 \angle 0° \text{ V} \ ; V_{YB} = 400 \angle -120° \text{ V} \ ; V_{BR} = 400 \angle 120° \text{ V}$$

The various phase currents in the equivalent Δ load [See Fig. 15.105 (*ii*)] are :

$$I_1 = \frac{V_{RY}}{Z_{12}} = \frac{400 \angle 0°}{18.4 \angle -22°} = 21.74 \angle 22° \text{ A} = (20.16 + j \, 8.14) \text{ A}$$

$$I_2 = \frac{V_{YB}}{Z_{23}} = \frac{400 \angle -120°}{36.8 \angle 8°} = 10.87 \angle -128° \text{ A} = (-6.69 - j \, 8.57) \text{ A}$$

$$I_3 = \frac{V_{BR}}{Z_{31}} = \frac{400 \angle 120°}{36.8 \angle 83°} = 10.87 \angle 37° \text{ A} = (8.68 + j \, 6.54) \text{ A}$$

The various line currents in the equivalent Δ load [See Fig. 15.105 (*ii*)] are :

$$I_R = I_1 - I_3 = (20.16 + j \, 8.14) - (8.68 + j \, 6.54)$$

$$= (11.48 + j \, 1.6) \text{ A} = 11.59 \angle 7.93° \text{ A}$$

$$I_Y = I_2 - I_1 = (-6.69 - j \, 8.57) - (20.16 + j \, 8.14)$$

$$= (-26.85 - j \, 16.71) \text{ A} = 31.62 \angle -148.1° \text{ A}$$

$$I_B = I_3 - I_2 = (8.68 + j \, 6.54) - (-6.69 - j \, 8.57)$$

$$= (15.37 + j \, 15.11) \text{ A} = 21.6 \angle 44.51° \text{ A}$$

The line currents in the equivalent Δ load are also the line currents (and hence phase currents) of the *Y*-load. Hence the magnitudes of line currents in the *Y*-load are :

$$I_R = \textbf{11.59 A} ; I_Y = \textbf{31.62 A} ; I_B = \textbf{21.6 A}$$

Referring to the *Y*-load in Fig. 15.105 (*i*), we have,

Voltage drop across $Z_1 = I_R Z_1 = (11.59 \angle 7.93°)(10 \angle 30°) = 115.9 \angle 37.93° \text{ V}$

Voltage drop across $Z_2 = I_Y Z_2 = (31.62 \angle -148.1°)(10 \angle -45°) = 316.2 \angle -193.1° \text{ V}$

Voltage drop across $Z_3 = I_B Z_3 = (21.6 \angle 44.51°)(20 \angle 60°) = 432 \angle 104.51° \text{ V}$

Hence the magnitudes of voltage drops across Z_1, Z_2 and Z_3 are **115.9 V**, **316.2 V** and **432 V** respectively.

Example 15.80. *A 440 V symmetrical 3-phase supply feeds a star-connected load consisting of three non-inductive resistances 10 Ω, 5 Ω and 12 Ω connected to R, Y and B phases respectively. Calculate the line currents and voltage across each resistor. The phase sequence is RYB.*

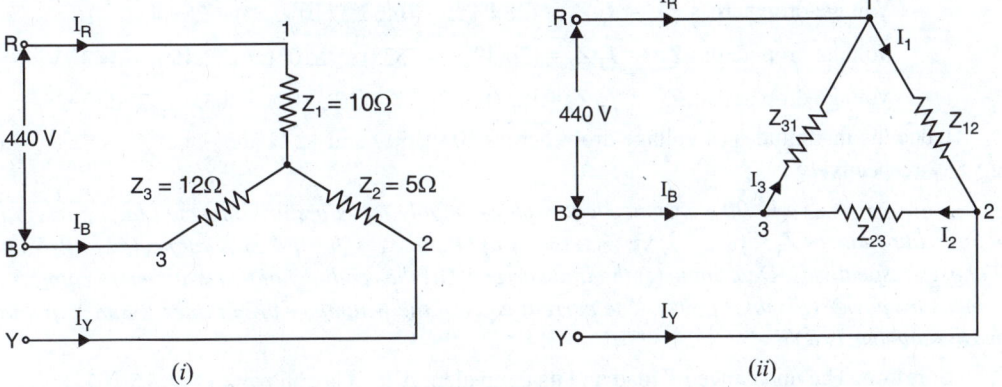

(*i*)   (*ii*)

Fig. 15.106

Solution. The unbalanced *Y*-connected load and its equivalent Δ-connected load are shown in Fig. 15.106. Note that star-connected impedances are pure resistances and quite logically the branches of the equivalent Δ load will also consist of pure resistances.

Now, $Z_1Z_2 + Z_2Z_3 + Z_3Z_1 = 10 \times 5 + 5 \times 12 + 12 \times 10 = 230$

The impedances of the branches of equivalent Δ-load [See Fig. 15.106 (*ii*)] are :

$$Z_{12} = \frac{Z_1Z_2 + Z_2Z_3 + Z_3Z_1}{Z_3} = 230/Z_3 = 230/12 = 19.17\ \Omega$$

$$Z_{23} = 230/Z_1 = 230/10 = 23\ \Omega$$

$$Z_{31} = 230/Z_2 = 230/5 = 46\ \Omega$$

Taking V_{RY} as the reference phasor, we have,

$$V_{RY} = 440\ \angle\ 0°\ V\ ;\ V_{YB} = 440\ \angle -120°\ V\ ;\ V_{BR} = 440\ \angle\ 120°\ V$$

The various phase currents in the equivalent Δ load [See Fig. 15.106 (*ii*)] are :

$$I_1 = \frac{V_{RY}}{Z_{12}} = \frac{440\ \angle\ 0°}{19.17\ \angle\ 0°} = 22.95\ \angle 0°\ A = (22.95 + j\ 0)\ A$$

$$I_2 = \frac{V_{YB}}{Z_{23}} = \frac{440\angle -120°}{23\angle 0°} = 19.13\ \angle -120°\ A = (-9.56 - j\ 16.5)\ A$$

$$I_3 = \frac{V_{BR}}{Z_{31}} = \frac{440\angle 120°}{46\angle 0°} = 9.56\ \angle\ 120°\ A = (-4.78 + j\ 8.28)\ A$$

The various line currents in the equivalent Δ load [See Fig. 15.106 (*ii*)] are :

$$I_R = I_1 - I_2 = (22.95 + j\ 0) - (-4.78 + j\ 8.28)$$
$$= (27.73 - j\ 8.28)\ A = 28.94\ \angle -16.63°\ A$$

$$I_Y = I_2 - I_1 = (-9.56 - j\ 16.57) - (22.95 + j\ 0)$$
$$= (-32.51 - j\ 16.57)\ A = 36.49\ \angle -153°\ A$$

$$I_B = I_3 - I_2 = (-4.78 + j\ 8.28) - (-9.56 - j\ 16.57)$$
$$= (4.78 + j\ 24.85)\ A = 25.3\ \angle\ 79.11°\ A$$

The line currents in the equivalent Δ load are also the line currents (and hence phase currents) of the Y load. Hence the magnitudes of line currents in the Y load are :

$$I_R = \textbf{28.94 A}\ ;\ I_Y = \textbf{36.49 A}\ ;\ I_B = \textbf{25.3 A}$$

Referring to the Y load in Fig. 15.106 (*i*), we have,

Voltage drop across $Z_1 = I_R Z_1 = (28.94\ \angle -16.63°)\ (10\ \angle\ 0°) = 289.4\ \angle -16.63°\ V$

Voltage drop across $Z_2 = I_Y Z_2 = (36.49\ \angle -153°)\ (5\ \angle\ 0°) = 182.45\ \angle -153°\ V$

Voltage drop across $Z_3 = I_B Z_3 = (25.3\ \angle\ 79.11°)\ (12\ \angle\ 0°) = 303.6\ \angle\ 79.11°\ V$

Hence the magnitudes of voltage drops across 10 Ω, 5 Ω and 12 Ω are **289.4 V**, **182.45 V** and **303.6 V** respectively.

Example 15.81. *A 400 V symmetrical 3-phase supply feeds a star-connected load consisting of three impedances $Z_1 = (4 + j\ 3)\ \Omega$, $Z_2 = (8 + j\ 6)\ \Omega$ and $Z_3 = (6 + j\ 4.5)\ \Omega$ connected to R, Y and B phases respectively. Determine (i) the line currents (ii) the readings of two wattmeters connected to measure power (iii) total power. The current coils of the wattmeter are in lines R and Y and the phase sequence is RYB.*

Solution. The unbalanced Y load and its equivalent Δ-load are shown in Fig. 15.107.

$$Z_1 = (4 + j\ 3)\ \Omega = 5\ \angle\ 36.87°\ \Omega\ ;\ Z_2 = (8 + j\ 6)\ \Omega = 10\ \angle\ 36.87°\ \Omega\ ;$$

$$Z_3 = (6 + j\ 4.5)\ \Omega = 7.5\ \angle\ 36.87°\ \Omega$$

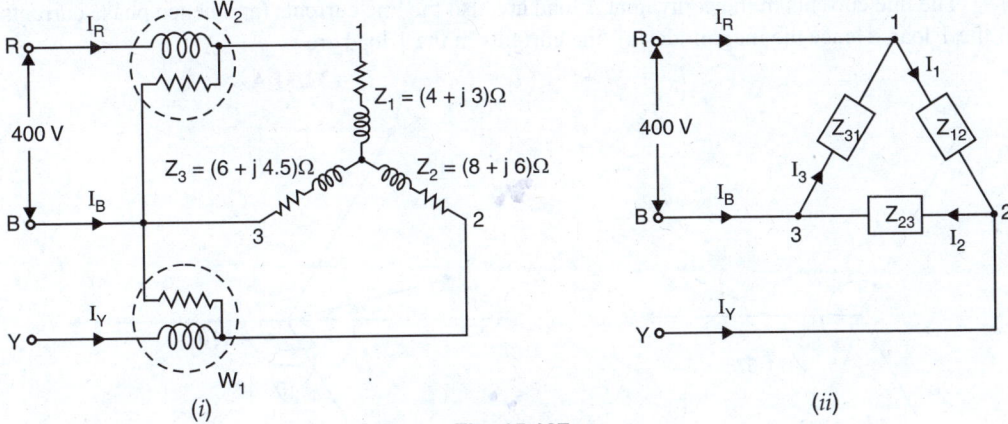

Fig. 15.107

Now, $Z_1Z_2 + Z_2Z_3 + Z_3Z_1$

$= (5 \angle 36.87°)(10 \angle 36.87°) + (10 \angle 36.87°)(7.5 \angle 36.87°) + (7.5 \angle 36.87°)(5 \angle 36.87°)$

$= 50 \angle 73.74° + 75 \angle 73.74° + 37.5 \angle 73.74°$

$= (14 + j48) + (21 + j72) + (10.5 + j36)$

$= 45.5 + j156 = 162.5 \angle 73.74°$

The branch impedances of the equivalent Δ load [See Fig. 15.107 (ii)] are :

$$Z_{12} = \frac{Z_1Z_2 + Z_2Z_3 + Z_3Z_1}{Z_3} = \frac{162.5 \angle 73.74°}{7.5 \angle 36.87°} = 21.67 \angle 36.87° \ \Omega$$

$$Z_{23} = \frac{162.5 \angle 73.74°}{Z_1} = \frac{162.5 \angle 73.74°}{5 \angle 36.87°} = 32.5 \angle 36.87° \ \Omega$$

$$Z_{31} = \frac{162.5 \angle 73.74°}{Z_2} = \frac{162.5 \angle 73.74°}{10 \angle 36.87°} = 16.25 \angle 36.87° \ \Omega$$

(*i*) Taking V_{RY} as the reference phasor, we have,

$$V_{RY} = 400 \angle 0° \text{ V} ; V_{YB} = 400 \angle -120° \text{ V} ; V_{BR} = 400 \angle 120° \text{ V}$$

The various phase currents in the equivalent Δ load [See Fig. 15.107 (ii)] are :

$$I_1 = \frac{V_{RY}}{Z_{12}} = \frac{400 \angle 0°}{21.67 \angle 36.87°} = 18.46 \angle -36.87° \text{ A} = (14.77 - j11.07) \text{A}$$

$$I_2 = \frac{V_{YB}}{Z_{23}} = \frac{400 \angle -120°}{32.5 \angle 36.87°} = 12.31 \angle -156.87° \text{ A} = (-11.32 - j4.84) \text{A}$$

$$I_3 = \frac{V_{BR}}{Z_{31}} = \frac{400 \angle 120°}{16.25 \angle 36.87°} = 24.62 \angle 83.13° \text{ A} = (2.94 + j24.44) \text{ A}$$

The various line currents in the equivalent Δ load [See Fig. 15.107 (ii)] are :

$$I_R = I_1 - I_3 = (14.77 - j11.07) - (2.94 + j24.44)$$

$$= (11.83 - j35.51) \text{ A} = 37.43 \angle -71.57° \text{ A}$$

$$I_Y = I_2 - I_1 = (-11.32 - j4.84) - (14.77 - j11.07)$$

$$= (-26.09 + j6.23) \text{ A} = 26.82 \angle -193.43° \text{ A}$$

$$I_B = I_3 - I_2 = (2.94 + j24.44) - (-11.32 - j4.84)$$

$$= (14.26 + j29.28) \text{ A} = 32.57 \angle 64.03° \text{ A}$$

The line currents in the equivalent Δ load are also the line currents (and hence phase currents) of the Y load. Hence the magnitudes of line currents in the Y load are :

$$I_R = \textbf{37.43 A} \; ; I_Y = \textbf{26.82 A} \; ; I_B = \textbf{32.57 A}$$

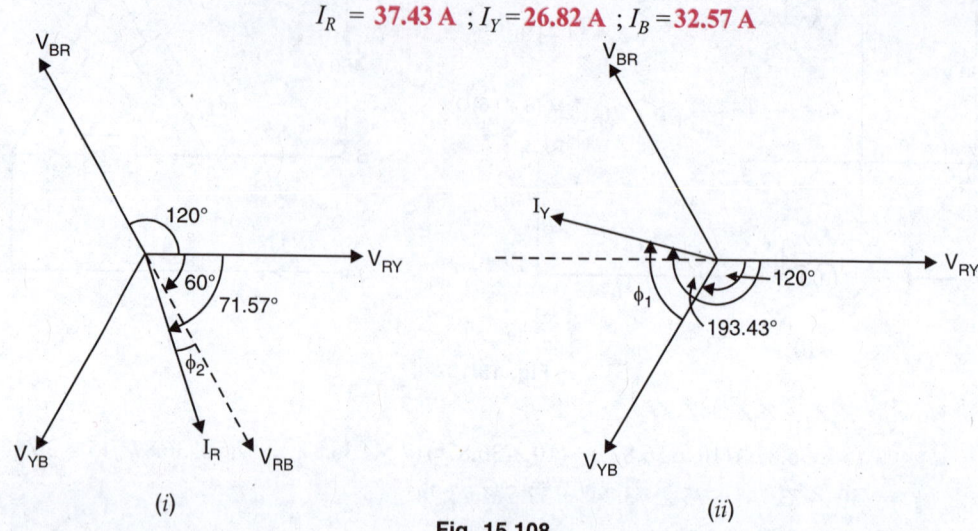

Fig. 15.108

(ii) Referring to Fig. 15.107 (*i*), current through the current coil of wattmeter W_2 is I_R and p.d. across its potential coil is V_{RB}. As seen in the phasor diagram in Fig. 15.108 (*i*), the phase angle between I_Y and V_{RB} is *$\phi_2 = 71\cdot57° - 60° = 11.57°$.

$\therefore$ Reading of $W_2 = V_{RB}I_R \cos \phi_2 = 400 \times 37.43 \times \cos 11.57° = \textbf{14668 W}$

The current in the current coil of wattmeter W_1 is I_Y and p.d. across its potential coil is V_{YB}. As seen in the phasor diagram in Fig. 15.108 (*ii*), the phase angle between I_Y and V_{YB} is $\phi_1 = 193.43° - 120° = 73.43°$.

$\therefore$ Reading of $W_1 = V_{YB} I_Y \cos \phi_1 = 400 \times 26.82 \times \cos 73.43° = \textbf{3059 W}$

(iii)   Total power $= W_2 + W_1 = 14668 + 3059 = \textbf{17727 W}$

Example 15.82. *An unbalanced star load is supplied from a balanced 3-wire system. The current taken by one branch is 20 A at a p.f. 0·8 lagging. The current taken by the second branch is 10 A at a p.f. 0·75 lagging. Find the current and p.f. in the third branch.*

Solution. Let the first, second and third branch of the Y-load be connected to R, Y and B lines of the supply respectively.

Assuming the phase sequence to be *RYB* and taking the phase voltage V_{RN} as the reference phasor, the phasor diagram will be as shown in Fig. 15.109. The current I_R (= 20 A) in the phase R lags V_{RN} by $\phi_1 = \cos^{-1} 0.8 = 36.8°$. The current I_Y (= 10 A) in the phase Y lags V_{YN} by $\phi_2 = \cos^{-1} 0.75 = $ 41.4°.

$\therefore$ $I_R = 20\angle{-36\cdot80°}\,A = (16 - j\,12)\,A$

$I_Y = 10\angle{-}**161\cdot4°\,A = (-9.48 - j\,3.19)\,A$

But $I_R + I_Y + I_B = 0$

or $(16 - j\,12) + (-9.48 - j\,3.19) + I_B = 0$

Fig. 15.109

---

* The three line voltages (*i.e.* V_{RY}, V_{YB} and V_{BR}) of the supply are symmetrical and 120° apart in phase. Since V_{RY} has been taken as the reference phasor, all angles are measured w.r.t. this voltage. Now I_R lags V_{RY} by 71.57°. The voltage V_{RB} (reverse the phasor V_{BR}) lags V_{RY} by 60°. Hence phase angle between I_R and V_{RB} is $= \phi_2 = 71.57° - 60° = 11.57°$.

** Note that $\phi_2 = 41.4°$. Therefore, I_Y makes an angle (in *CW* direction) of 41.4° + 120° = 161.4° with the reference phasor V_{RN}.

∴ $I_B = (-6.52 + j\,15.19)\,A$

            $= 16.53 \angle -246.77°\,A$

∴ Magnitude of current in phase B (*i.e.*, third branch) is **16.53 A.**

Phase angle between V_{BN} and I_B is

$$\phi_3 = 246.77° - 240° = 6.77°$$

∴   Power factor $= \cos \phi_3 = \cos 6.77° = \textbf{0.993}\,lag$

Tutorial Problems

1. A 440 V symmetrical 3-phase supply feeds a star-connected load consisting of impedances $Z_1 = 5 \angle 30°\,\Omega$, $Z_2 = 10 \angle 45°\,\Omega$ and $Z_3 = 10 \angle 60°\,\Omega$ connected to R, Y and B phases respectively. The phase sequence is *RYB*. Determine the line currents.   **[35.7 A, 32.8 A, 27.7 A]**

2. A 230 V symmetrical 3-phase supply feeds a star-connected load consisting of three non-inductive resistances of 5 Ω, 10 Ω and 15 Ω connected to R, Y and B phases respectively. Calculate (*i*) the line currents (*ii*) voltage across each resistor. The phase sequence is *RYB*.

    [(*i*) 18.25 A, 15.08 A, 11.06 A (*ii*) 91.25 V, 150.8 V, 165.9 V]

3. A 400 V symmetrical 3-phase supply feeds a star-connected load consisting of 10 Ω, 20 Ω and 20 Ω connected to R, Y and B phases respectively. Two wattmeters connected to measure the input power have their current coils in lines R and B respectively. Determine the wattmeter readings.   **[6 kW, 4 kW]**

4. A symmetrical 3-phase 400 V source feeds a star-connected load consisting of impedances $Z_1 = 15 \angle 0°\,\Omega$, $Z_2 = 15 \angle 30°\,\Omega$ and $Z_3 = 15 \angle -30°\,\Omega$ connected to R, Y and B phases respectively. Determine the total power taken by the load.   **[10.65 kW]**

15.39. Solving Unbalanced 3-Wire Y Load by Millman's Theorem

We have seen that problems on unbalanced 3-wire, Y load can be solved by converting the given Y load into equivalent Δ-load. However, such a procedure involves lengthy calculations. J.E. Millman proposed a new technique to obtain the solution of unbalanced Y loads. Consider an unbalanced Y load connected to a balanced 3-phase supply as shown in Fig. 15.110. Here O is the star point of the supply (normally at zero potential) and O' is the load star point. Due to load unbalance, the potential of O' is different from that of O. If we know $V_{O'O}$ (*i.e.*, voltage of O' w.r.t. O), we can easily determine the load phase voltages and the line currents of the unbalanced Y load. According to *Millman's theorem, the voltage $V_{O'O}$ is given by ;

$$V_{O'O} = \frac{V_{RO}\,Y_R + V_{YO}\,Y_Y + V_{BO}\,Y_B}{Y_R + Y_Y + Y_B}$$

where V_{RO}, V_{YO} and V_{BO} are the phase voltages of the supply and are equal in magnitude but 120° apart in phase. The quantities Y_R, Y_Y and Y_B are the admittances of the branches of the unbalanced Y load.

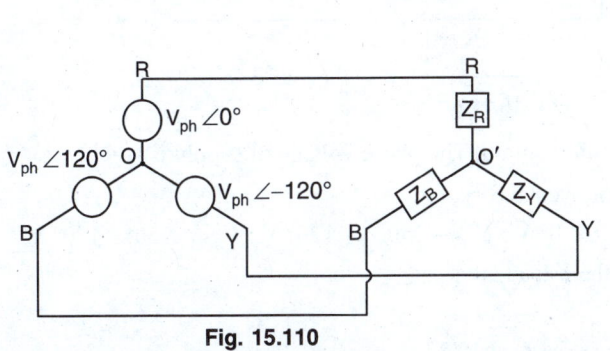

Fig. 15.110

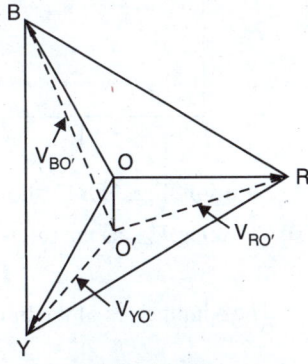

Fig. 15.111

***

*   For proof of this theorem, the reader may refer to any advanced book on the subject.

The voltage across load phase R is $V_{RO'}$. Using double-subscript notation, this is given by ;

$$V_{RO'} = V_{RO} + V_{OO'}$$
$$= V_{RO} - V_{O'O} \qquad ...(i)$$

Thus load phase voltage $V_{RO'}$ is obtained by subtracting $V_{O'O}$ from the supply phase voltage V_{RO} (phasor difference).

Similarly, $V_{YO'} = V_{YO} - V_{O'O}$ ...(ii)

and $V_{BO'} = V_{BO} - V_{O'O}$ ...(iii)

The line currents in the unbalanced Y load are :

$$I_{RO'} = (V_{RO} - V_{O'O})\, Y_R \;\; ; \;\; I_{YO'} = (V_{YO} - V_{O'O})\, Y_Y \;\; ; \;\; I_{BO'} = (V_{BO} - V_{O'O})\, Y_B$$

Fig. 15.111 shows the triangular phasor diagram. Here O is the star point of the supply and is located at the centre of the equilateral triangle RYB. The point O' is the load star point. Due to load unbalance, O' has some potential w.r.t. O and is shifted away from the centre of the triangle. Such a diagram is very useful in analysing an unbalanced 3-wire star load because it gives at a glance the picture of the happenings in the circuit. Thus :

(*i*) distances OR, OY and OB are equal in magnitude but 120° apart from one another and represent the phase voltages of the supply.

(*ii*) distance $O'O$ represents $V_{O'O}$ i.e., potential of O' w.r.t. O.

(*iii*) distances $O'R$, $O'Y$ and $O'B$ represent the load phase voltages. Note that load phase voltages are unequal in magnitude as well as differ in phase by unequal angle.

(*iv*) *distances RY, YB and BR are equal in magnitude but *120° apart in phase and represent the supply line voltages as well as load line voltages.*

Example 15.83. *A 433 V symmetrical 3-phase supply feeds a star-connected load consisting of impedances $(2 + j\,0)\,\Omega$, $(2 + j\,2)\,\Omega$ and $(0 - j\,5)\,\Omega$ connected to the R, Y and B phases respectively. Find (i) the potential of load star point w.r.t. the supply neutral (ii) the load phase voltages (iii) line currents. The phase sequence is RYB.*

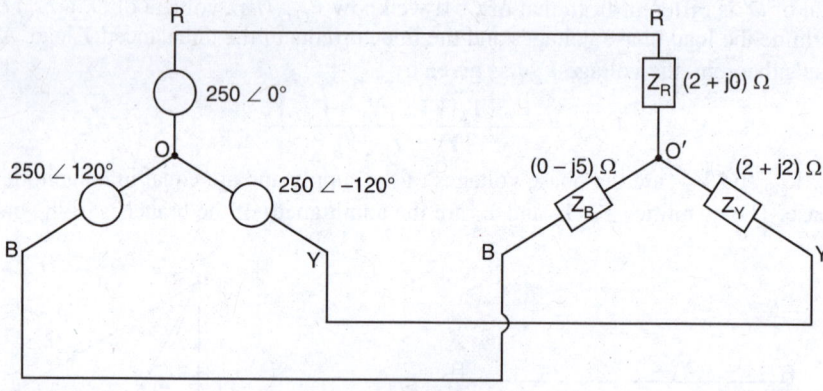

Fig. 15.112

Solution. Fig. 15.112 shows the circuit diagram. The phase voltage of supply is $= 433/\sqrt{3} = $ 250 V. Taking V_{RO} as the reference phasor, we have,

$$V_{RO} = 250 \angle 0°\; \text{V} \; ; \; V_{YO} = 250 \angle -120°\; \text{V} \; ; \; V_{BO} = 250 \angle 120°\; \text{V}$$

The admittances of the branches of the Y-load are :

---

*   Note that angle between RY (*i.e.* RY line extended) and YB is 120°.

$$Y_R = \frac{1}{Z_R} = \frac{1}{2+j\,0} = (0.5 - j\,0)\ S = 0.5 \angle 0°\ S$$

$$Y_Y = \frac{1}{Z_Y} = \frac{1}{2+j\,2} = (0.25 - j\,0.25)\ S = 0.353 \angle -45°\ S$$

$$Y_B = \frac{1}{Z_B} = \frac{1}{0-j\,5} = (0 + j\,0.2)\ S = 0.2 \angle 90°\ S$$

Now, $\quad V_{RO}Y_R + V_{YO}Y_Y + V_{BO}Y_B$

$$= (250 \angle 0°)\,(0.5 \angle 0°) + (250 \angle -120°)\,(0.353 \angle -45°) + (250 \angle 120°)\,(0.2 \angle 90°)$$

$$= 125 \angle 0° + 88.25 \angle -165° + 50 \angle 210°$$

$$= (125 + j\,0) + (-85.24 - j\,22.84) + (-43.3 - j\,25)$$

$$= (-3.54 - j\,47.84)\ A = 47.97 \angle -94.23°\ A$$

Also, $\qquad Y_R + Y_Y + Y_B = (0.5 - j\,0) + (0.25 - j\,0.25) + (0 + j\,0.2)$

$$= (0.75 - j\,0.05)\ S = 0.75 \angle 3.81°\ S$$

(*i*) According to Millman's theorem,

$$V_{O'O} = \frac{V_{RO}\,Y_R + V_{YO}\,Y_Y + V_{BO}\,Y_B}{Y_R + Y_Y + Y_B}$$

$$= \frac{47.97 \angle -94.23°}{0.75 \angle -3.81°} = 63.96 \angle -90.42°\ V$$

$$\simeq {}^* 64 \angle -90°\ V = -j\,64\ V$$

(*ii*) The load phase voltages are :

$$V_{RO'} = V_{RO} - V_{O'O} = (250 + j\,0) - (-j\,64)$$

$$= (250 + j\,64)\ V = \mathbf{258 \angle 14.36°\ V}$$

$$V_{YO'} = V_{YO} - V_{O'O} = 250 \angle -120° - 64 \angle -90°$$

$$= (-125 - j\,216.5) - (-j\,64)$$

$$= (-125 - j\,152.5)\ V = \mathbf{197 \angle -129.34°\ V}$$

$$V_{BO'} = V_{BO} - V_{O'O} = 250 \angle 120° - 64 \angle -90°$$

$$= (-125 + j\,216.5) - (-j\,64)$$

$$= (-125 + j\,280.5)\ V = \mathbf{307 \angle -245.98°\ V}$$

Note that load phase voltages are not only unequal in magnitude but also subtend angles other than 120° with one another.

(*iii*) The line currents are given by ;

$$I_{RO'} = V_{RO'}Y_R = (258 \angle 14.36°)\,(0.5 \angle 0°) = \mathbf{129 \angle 14.36°\ A}$$

$$I_{YO'} = V_{YO'}Y_Y = (197 \angle -129.34°)\,(0.353 \angle -45°) = \mathbf{69.54 \angle -174.34°\ A}$$

$$I_{BO'} = V_{BO'}Y_B = (307 \angle -245.98°)\,(0.2 \angle 90°) = \mathbf{61.4 \angle -155.98°\ A}$$

Note that phasor sum of the three line currents will come out to be zero.

Example 15.84. *A symmetrical 3-phase supply feeds a star-connected load of impedances* 10 Ω, (6 + j 8) Ω *and* (6 − j 8) Ω *connected to the R, Y and B phases respectively. The supply phase voltage is 110 V and the phase sequence is RYB. Calculate* (i) *the potential of load star point w.r.t. the supply neutral and* (ii) *the line currents.*

---

* The error introduced by this approximation is negligible but calculations are very much simplified.

Solution. Taking V_{RO} as the reference phasor, we have,

$$V_{RO} = 110\angle0° \text{ V} ; V_{YO} = 110 \angle-120° \text{ V} ; V_{BO} = 110 \angle120° \text{ V}$$

The admittances of the branches of the Y load are :

$$Y_R = \frac{1}{Z_R} = \frac{1}{10 \angle0°} = 0.1\angle0° \text{ S}$$

$$Y_Y = \frac{1}{Z_Y} = \frac{1}{6+j\,8} = \frac{1}{10 \angle53.1°} = 0.1 \angle-53.1° \text{ S}$$

$$Y_B = \frac{1}{Z_B} = \frac{1}{6-j\,8} = \frac{1}{10 \angle-53.1°} = 0.1 \angle 53.1° \text{ S}$$

Now, $V_{RO}Y_R + V_{YO}Y_Y + V_{BO}Y_B$

$$= (110\angle0°)(0.1 \angle0°) + (110 \angle-120°)(0.1 \angle-53.1°) + (110\angle120°)(0.1 \angle53.1°)$$

$$= 11\angle 0° + 11\angle- 173.1° + 11 \angle 173.1°$$

$$= (11 +j\,0) + (- 10.92 -j\,1.32) + (-10.92 +j\,1.32)$$

$$= (- 10.84 +j\,0) \text{ A} = 10.84 \angle 180° \text{ A}$$

Also, $Y_R + Y_Y + Y_B = 0.1 \angle0° + 0.1 \angle-53.1° + 0.1 \angle53.1°$

$$= (0.1 +j\,0) + (0.06 -j\,0.08) + (0.06 +j\,0.08)$$

$$= (0.22 +j\,0) \text{ S} = 0.22 \angle 0° \text{ S}$$

(*i*) $$V_{O'O} = \frac{V_{RO} Y_R +V_{YO} Y_Y +V_{BO} Y_B}{Y_R +Y_Y +Y_B} = \frac{10.84\angle180°}{0.22\angle0°} = \mathbf{49.27 \angle180° \text{ V}}$$

(*ii*)   The load phase voltages are :

$$V_{RO}' = V_{RO} - V_{O'O} = (110 +j\,0) - 49.27 \angle180°$$

$$= (110 +j\,0) - (-49.27 +j\,0) = (159.27 +j\,0) \text{ V} = 159.27 \angle0° \text{ V}$$

$$V_{YO}' = V_{YO} - V_{O'O} = 110 \angle-120° - 49.27 \angle180°$$

$$= (- 55 -j\,95.26) - (- 49.27 +j\,0)$$

$$= (-5.73 -j\,95.26) \text{ V} = 95.43 \angle - 93.44° \text{ V}$$

$$V_{BO}' = V_{BO} - V_{O'O} = 110 \angle120° - 49.27 \angle180°$$

$$= (- 55 +j\,95.26) - (- 49.27 +j\,0)$$

$$= (- 5.73 +j\,95.26) \text{ V} = 95.43 \angle 93.44° \text{ V}$$

The line currents are given by ;

$$I_{RO}' = V_{RO'} Y_R = (159.27 \angle0°)(0.1 \angle0°) = \mathbf{15·927 \angle 0° \text{ A}}$$

$$I_{YO}' = V_{YO'} Y_Y = (95.43 \angle -93.44°)(0.1 \angle - 53.1°)$$

$$= 9.543 \angle - 146.54° \text{A}$$

$$I_{BO}' = V_{BO'} Y_B = (95.43 \angle93.44°)(0.\dot{1} \angle53.1°) = \mathbf{9.543 \angle146.54° \text{ A}}$$

Example 15.85. *A 400 V symmetrical 3-phase supply feeds a star-connected load consisting of impedances Z_1, Z_2 and Z_3 connected to the R, Y and B phases respectively. If the potential of load star point w.r.t. supply neutral is $100\angle125°V$ and currents in the phases Y and B are $5\angle180°$ A and $5\angle30°$ A respectively, calculate the values of Z_1, Z_2 and Z_3. The phase sequence is RYB.*

Solution. The phase voltage of supply is $= 400/\sqrt{3} = 231$ V. Taking V_{RO} as the reference phasor, we have,

$$V_{RO} = 231\angle0° \text{ V} ; V_{YO} = 231 \angle-120° \text{ V} ; V_{BO} = 231 \angle120° \text{ V}$$

The potential of load star point O' w.r.t. supply neutral O is $V_{O'O} = 100\angle 125°$ V.

The load phase voltages are :

$$\begin{aligned}
V_{RO}' &= V_{RO} - V_{O'O} = 231\angle 0° - 100\angle 125° \\
&= (231 + j\,0) - (-57.36 + j\,81.92) \\
&= (288.36 - j\,81.92)\ \text{V} = 299.77\angle -15.86°\ \text{V} \\
V_{YO}' &= V_{YO} - V_{O'O} = 231\angle -120° - 100\angle 125° \\
&= (-115.5 - j\,200.1) - (-57.36 + j\,81.92) \\
&= (-58.14 - j\,282.02)\ \text{V} = 287.95\angle -101.65°\ \text{V} \\
V_{BO}' &= V_{BO} - V_{O'O} = 231\angle 120° - 100\angle 125° \\
&= (-115.5 + j\,200.1) - (-57.36 + j\,81.92) \\
&= (-58.14 + j\,118.18)\ \text{V} = 131.7\angle 116.2°\ \text{V}
\end{aligned}$$

Now,
$$I_{RO}' + I_{YO}' + I_{BO}' = 0$$

or
$$I_{RO}' + 5\angle 180° + 5\angle 30° = 0$$

or
$$I_{RO}' + (-5 + j\,0) + (4.33 + j\,2.5) = 0$$

or
$$I_{RO}' + (-0.67 + j\,2.5) = 0$$

$\therefore$
$$I_{RO}' = -(-0.67 + j\,2.5) = (0.67 - j\,2.5)\ \text{A} = 2.59\angle -75°\ \text{A}$$

$\therefore$
$$Z_1 = \frac{V_{RO}'}{I_{RO}'} = \frac{299.77\angle -15.86°}{2.59\angle -75°} = \mathbf{115.74\angle 59.14°\ \Omega}$$

$$Z_2 = \frac{V_{YO}'}{I_{YO}'} = \frac{287.95\angle -101.65°}{5\angle 180°} = \mathbf{57.6\angle 78.35°\ \Omega}$$

$$Z_3 = \frac{V_{BO}'}{I_{BO}'} = \frac{131.7\angle 116.2°}{5\angle 30°} = \mathbf{26.34\angle 86.2°\ \Omega}$$

Tutorial Problems

1. Three impedances Z_R, Z_Y and Z_B are connected in star across a 440 V, 3-phase supply. If the voltage of star-point relative to the supply neutral is $200\angle 150°$ V and Y and B line currents are $10\angle -90°$ A and $20\angle 90°$ A respectively, all with respect to the voltage between the supply neutral and R line, calculate the values of Z_R, Z_Y and Z_B. The phase sequence is RYB.

   [$43.85\angle 76.8°\ \Omega$; $32.3\angle 8.4°\ \Omega$; $6.43\angle -21°\ \Omega$]

2. Three non-inductive resistances of 5 Ω, 10 Ω and 15 Ω are connected in star and supplied from a 230 V symmetrical 3-phase system. Calculate the magnitude of line currents. The phase sequence is RYB.

   [$I_R = 18.22$ A ; $I_Y = 15.1$ A ; $I_B = 11.7$ A]

3. A symmetrical 440 V, 3-phase system supplies a Y-connected load. The branch impedances are $Z_R = 10\angle 30°\ \Omega$, $Z_Y = 12\angle 45°\ \Omega$ and $Z_B = 15\angle 40°\ \Omega$. Assuming the neutral of the supply to be earthed, calculate the voltage to earth of the star point. Assume the phase sequence RYB.   [$18.59\angle -11.9°$ V]

15.40. Significance of Power Factor

The apparent power S drawn by a circuit has two components viz (i) true power P and (ii) reactive power Q. True power component should be as large as possible because it is this component which does the useful work in the circuit. This is possible if the reactive power component is small. As seen from the power triangle in Fig. 15.113, the smaller the phase angle ϕ ($i.e.$ greater the power factor cos ϕ), the smaller is the reactive power component. Thus when $\phi = 0°$ ($i.e.$, cos $\phi = 1$),

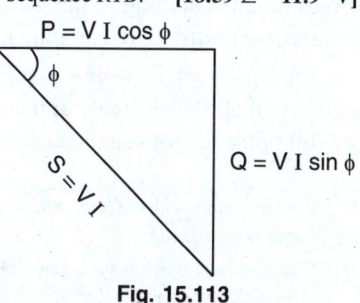

Fig. 15.113

the reactive power component is zero and the true power is equal to the apparent power ($P = VI \cos \phi$ $= VI \cos 0° = VI = S$). This means the whole of apparent power drawn by the circuit is being utilised by it. *Thus power factor of a circuit is a measure of its effectiveness in *utilising the apparent power drawn by it.* The greater the power factor of a circuit, the greater is its ability to utilise the apparent power. Thus 0·5 p.f. (*i.e.* 50 % p.f.) of a circuit means that it will utilise only 50% of the apparent power whereas 0·8 p.f. would mean 80% utilisation of apparent power. For this reason, we wish the power factor of the circuit to be as near to 1 as possible.

$$\text{True power, } P = VI \cos \phi \text{ watts (W)}$$

$$\text{Apparent power, } S = VI \text{ volt-amperes (VA)}$$

$$\text{Reactive power, } Q = VI \sin \phi \text{ volt-amperes reactive (VAR)}$$

15.41. Disadvantages of Low Power Factor

The power factor plays an important role in a.c. circuits since power consumed depends upon this factor. The greater the power factor (*i.e.* closer to 1) of the load, the larger the active power delivered to the load.

$$I_L = \frac{P}{V_L \cos \phi} \qquad \qquad \text{...for single phase supply}$$

$$I_L = \frac{P}{\sqrt{3} \, V_L \cos \phi} \qquad \qquad \text{...for 3-phase supply}$$

It is clear from above that for fixed power and voltage, the load current is inversely proportional to the power factor. Smaller the power factor, higher is the load current and *vice-versa*. The large current due to poor power factor results in the following disadvantages :

(*i*) **Large kVA rating of equipment.** The electrical machinery (*e.g.*, alternators, transformers, switchgear *etc.*) is always rated in **kVA.

$$\text{Now,} \qquad\qquad \text{kVA} = \frac{\text{kW}}{\cos \phi}$$

It is clear that kVA rating of the equipment is inversely proportional to power factor. The smaller the power factor, the larger is the kVA rating. Therefore, at low power factor, the kVA rating of the equipment has to be made more, making the equipment larger and expensive.

(*ii*) **Greater conductor size.** To transmit or distribute a given amount of power at constant voltage, the conductor will have to carry more current at low power factor. This necessitates large conductor size.

(*iii*) **Large copper losses.** The large current at low power factor causes more I^2R losses in all elements of power supply. This results in poor efficiency.

(*iv*) **Poor voltage regulation.** The large current at low lagging power factor causes greater voltage drops in alternators, transformers, transmission lines and distributors. This results in the reduced voltage available to the utilisation devices, impairing their performance *e.g.* lighting becomes dimmer, starting torque of motors is reduced *etc.* In order to keep the supply voltage within permissible limits, extra equipment (*i.e.* voltage regulators) is required.

(*v*) **Reduced handling capacity of system.** The lagging power factor reduces the handling capacity of all the elements of the system. It is because the reactive component of current prevents the full utilisation of installed capacity.

---

*   Power factor is that factor which must be multiplied to apparent power to obtain true power. Hence the name power factor.

**   The electrical machinery is rated in kVA because the power factor of load is not known when the machinery is manufactured in the factory.

The above discussion leads to the conclusion that low power factor is an objectionable feature in the supply system.

15.42. Causes of Low Power Factor

Low power factor is undesirable from economic point of view. Normally, the power factor of the whole load on the supply system in lower than 0·8. The following are the causes of low power factor:

(*i*) Most of the a.c. motors are of induction type (1ϕ and 3ϕ induction motors) which have low lagging power factor. These motors work at a power factor which is extremely small on light load (0·2 to 0·3) and rises to 0·8 or 0·9 at full load.

(*ii*) Arc lamps, electric discharge lamps and industrial heating furnaces operate at low lagging power factor.

(*iii*) The load on the power system is varying ; being high during morning and evening and low at other times. During low load period, supply voltage is increased which increases the magnetisation current. This results in the decreased power factor.

15.43. Power Factor Improvement

The low power factor is mainly due to the fact that most of the power loads (*e.g.*, induction motors *etc.*) are inductive in nature and, therefore, take lagging currents. In order to improve the power factor, some device taking leading current should be connected in parallel with the load. One of such devices can be a capacitor. The capacitor draws a leading current and partly or completely neutralises the lagging reactive component of load current. This raises the power factor of the load.

Illustration. To illustrate the power factor improvement by a capacitor, consider a *single phase load taking lagging current I at a p.f. of cos ϕ_1 as shown in Fig. 15.114 (*i*). The capacitor C is connected in parallel with the load and draws current I_C which leads the supply voltage by 90°. The resulting line current I' is the phasor sum of I and I_C and its angle of lag is ϕ_2 as shown in the phasor diagram in Fig.15.114 (*iii*). It is clear that ϕ_2 is less than ϕ_1, so that cos ϕ_2 is greater than cos ϕ_1. Hence, the power factor of the load is increased.

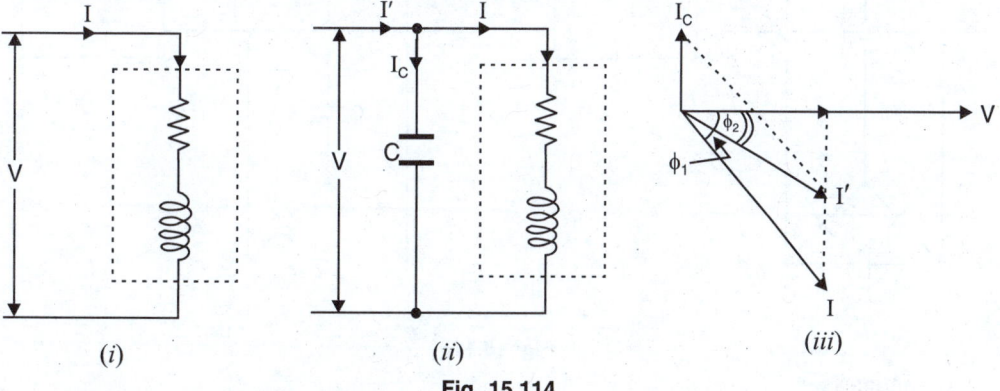

(*i*)　　　　　　　(*ii*)　　　　　　　(*iii*)

Fig. 15.114

The following points are worth nothing :

(*i*) *The circuit current I' after p.f. correction is less than the original circuit current I.*

(*ii*) *The active or wattful component remains the same before and after p.f. correction because only the lagging reactive component of load is reduced by the capacitor i.e.*

$$I \cos \phi_1 = I' \cos \phi_2$$

or　　　　　　$$V I \cos \phi_1 = V I' \cos \phi_2$$　　　　　　(Multiply by V)

---

*　The treatment can be used for 3-phase balanced loads *e.g.* 3-phase induction motor. In a balanced 3-phase load, analysis of only one phase leads to the desired results.

Therefore, active power (kW) remains unchanged during p.f. improvement.

 (*iii*) The original reactive component of load is $I \sin \phi_1$. However, after p.f. correction, it is reduced by I_C and becomes $I' \sin \phi_2$ *i.e.*,

$$I' \sin \phi_2 = I \sin \phi_1 - I_C$$

or $$V I' \sin \phi_2 = V I \sin \phi_1 - V I_C$$

i.e. Net kVAR after p.f. correction = Lagging kVAR before p.f. correction–Leading kVAR of capacitor

 (*iv*) $$I_C = I \sin \phi_1 - I' \sin \phi_2$$

 $\therefore$ Capacitance of the capacitor to improve p.f. from $\cos \phi_1$ to $\cos \phi_2$

$$= \frac{I_C}{\omega V} \qquad\qquad (\because X_C = V/I_C = 1/\omega\, C)$$

Note that improving the power factor means making it closer to 1.

15.44. Power Factor Improvement Equipment

 Normally, the power factor of the whole load on a large generating station is in the region of 0·8 to 0·9. However, sometimes it is lower and in such cases it is generally desirable to take special steps to improve the power factor. This can be achieved by the following equipment :

 1. Static capacitors. **2.** Synchronous condenser. **3.** Phase advancers.

 1. Static capacitors. The power factor can be improved by connecting capacitors in parallel with the equipment operating at lagging power factor. The capacitor (generally known as static* capacitor) draws a leading current and party or completely neutralises the lagging reactive component of load current. This raises the power factor of the load. For three-phase loads, the capacitors can be connected in delta or star as shown in Fig. 15.115. Static capacitors are invariably used for power factor improvement in factories.

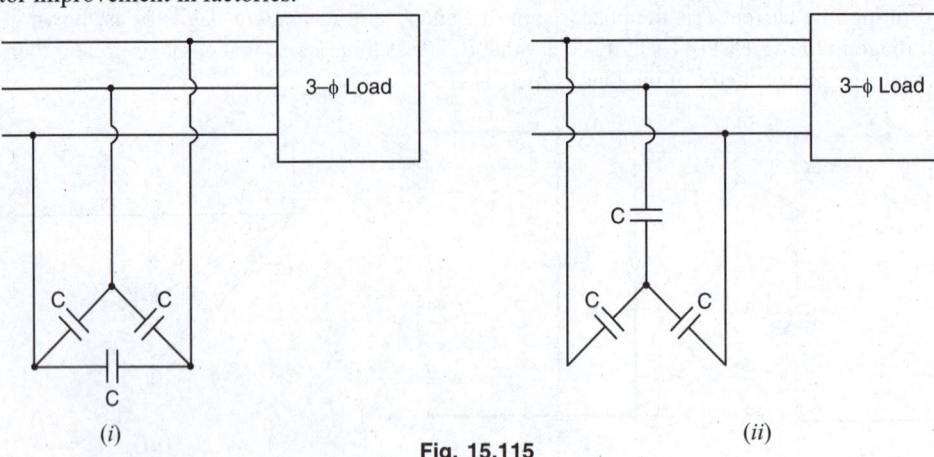

Fig. 15.115

Advantages

 (*i*) They have low losses.

 (*ii*) They require little maintenance as there are no rotating parts.

 (*iii*) They can be easily installed as they are light and require no foundation.

 (*iv*) They can work under ordinary atmospheric conditions.

Disadvantages

 (*i*) They have short service life ranging from 8 to 10 years.

---

 * To distinguish from the so called *synchronous condenser* which is a synchronous motor running at no load and taking leading current.

(*ii*) They are easily damaged if the voltage exceeds the rated value.

(*iii*) Once the capacitors are damaged, their repair is uneconomical.

2. Synchronous condenser. A synchronous motor takes a leading current when over-excited and, therefore, behaves as a capacitor. *An over-excited synchronous motor running on no load is known as* **synchronous condenser.** When such a machine is connected in parallel with the supply, it takes a leading current which partly neutralises the lagging reactive component of the load. Thus the power factor is improved.

Fig. 15.116 shows the power factor improvement by synchronous condenser method. The 3ϕ load takes current I_L at low lagging power factor $\cos \phi_L$. The synchronous condenser takes a current I_m which leads the voltage by an angle ϕ_m*. The resultant current I is the phasor sum of I_m and I_L and lags behind the voltage by an angle ϕ. It is clear that ϕ is less than ϕ_L so that $\cos \phi$ is greater than $\cos \phi_L$. Thus the power factor is increased from $\cos \phi_L$ to $\cos \phi$. Synchronous condensers are generally used at major bulk supply substations for power factor improvement.

Advantages

(*i*) By varying the field excitation, the magnitude of current drawn by the motor can be changed by any amount. This helps in achieving †stepless control of power factor.

(*ii*) The motor windings have high thermal stability to short circuit currents.

(*iii*) The faults can be removed easily.

Disadvantages

(*i*) There are considerable losses in the motor.

(*ii*) The maintenance cost is high.

(*iii*) It produces noise.

(*iv*) Except in sizes above 500 kVA, the cost is greater than that of static capacitors of the same rating.

(*v*) As a synchronous motor has no self-starting torque, therefore, an auxiliary equipment has to be provided for this purpose.

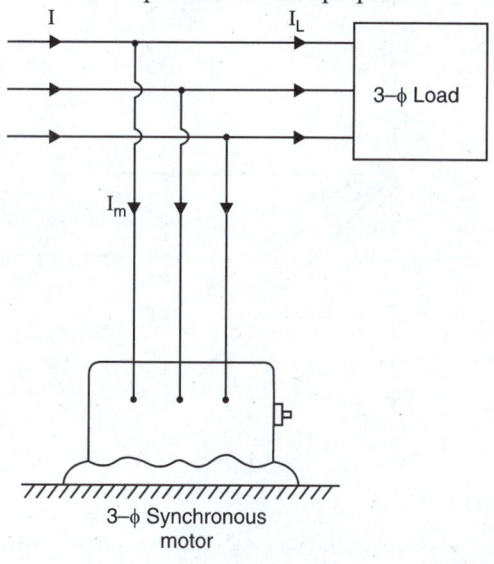

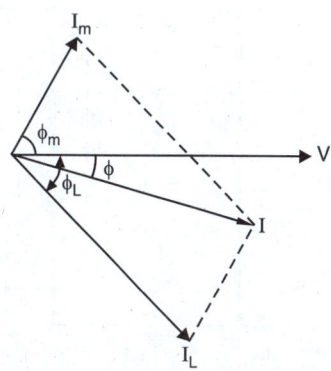

Fig. 15.116

---

*   If the motor is ideal *i.e.*, there are no losses, then $\phi_m = 90°$. However, in actual practice, losses do occur in the motor even at no load. Therefore, the current I_m leads the voltage by an angle less than $90°$.

†   The p.f. improvement with capacitors can only be done in steps by switching on the capacitors in various groupings. However, with synchronous motor, any amount of capacitive reactance can be provided by changing the field excitation.

Note. The reactive power taken by a synchronous motor depends upon two factors, the d.c. field excitation and the mechanical load delivered by the motor. Maximum leading power is taken by a synchronous motor with maximum excitation and zero load.

3. Phase advancers. Phase advancers are used to improve the power factor of induction motors. The low power factor of an induction motor is due to the fact that its stator winding draws exciting current which lags behind the supply voltage by 90°. If the exciting ampere turns can be provided from some other a.c. source, then the stator winding will be relieved of exciting current and the power factor of the motor can be improved. This job is accomplished by the phase advancer which is simply an a.c. exciter. The phase advancer is mounted on the same shaft as the main motor and is connected in the rotor circuit of the motor. It provides exciting ampere turns to the rotor circuit at slip frequency. By providing more ampere turns than required, the induction motor can be made to operate on leading power factor like an over-excited synchronous motor.

Phase advancers have two principal advantages. Firstly, as the exciting ampere turns are supplied at slip frequency, therefore, lagging kVAR drawn by the motor are considerably reduced. Secondly, phase advancers can be conveniently used where the use of synchronous motors is unadmissible. However, the major disadvantage of phase advancers is that they are not economical for motors below 200 h.p.

15.45. Calculations of Power Factor Correction

Consider an inductive load taking a lagging current I at a power factor $\cos \phi$. In order to improve the power factor of this circuit, the remedy is to connect such an equipment in parallel with the load which takes a leading reactive component and partly cancels the lagging reactive component of the load. Fig. 15.117 (*i*) shows a capacitor connected across the load. The capacitor takes a current I_C which leads the supply voltage V by 90°. The current I_C partly cancels the lagging reactive component of the load current as shown in the phasor diagram in Fig. 15.117 (*ii*). The resultant circuit current becomes I' and its angle of lag is ϕ_2. It is clear that ϕ_2 is less than ϕ_1 so that new p.f. $\cos \phi_2$ is more than the previous p.f. $\cos \phi_1$.

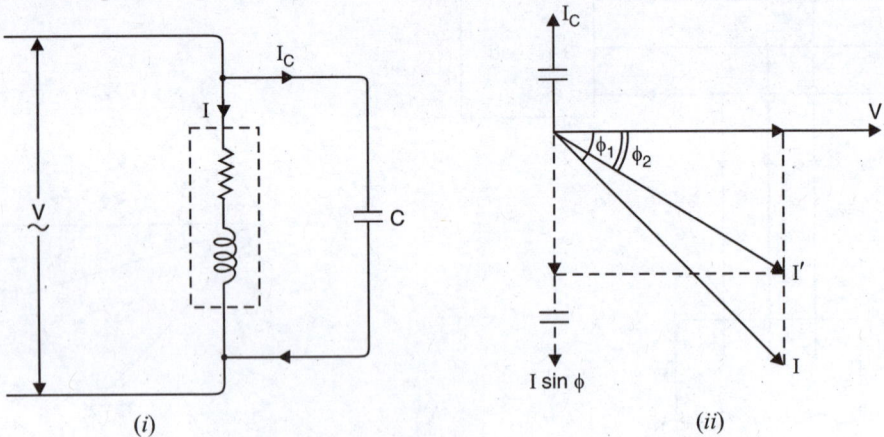

(*i*)   (*ii*)

Fig. 15.117

From the phasor diagram, it is clear that after p.f. correction, the lagging reactive component of the load is reduced to $I' \sin \phi_2$.

Obviously, $I' \sin \phi_2 = I \sin \phi_1 - I_C$

or $I_C = I \sin \phi_1 - I' \sin \phi_2$

∴   Capacitance of capacitor to improve p.f. from $\cos \phi_1$ to $\cos \phi_2$ is

$$C = \frac{I_C}{\omega V} \qquad \left(\because X_C = \frac{V}{I_C} = \frac{1}{\omega C} \right)$$

Power triangle. The power factor correction can also be illustrated from power triangle. Thus referring to Fig. 15.118, the power triangle OAB is for the power factor $\cos \phi_1$, whereas power triangle OAC is for the improved power factor $\cos \phi_2$. It may be seen that active power (OA) does not change with power factor improvement. However, the lagging kVAR of the load is reduced by the p.f. correction equipment, thus improving the p.f. to $\cos \phi_2$.

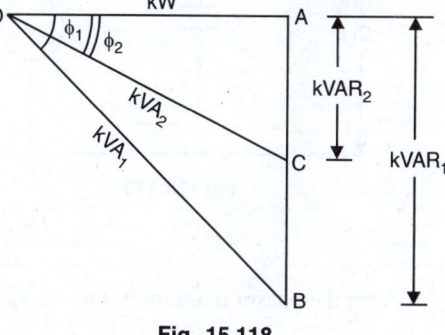

Fig. 15.118

Leading kVAR supplied by p.f. correction equipment

$$
\begin{aligned}
&= BC = AB - AC \\
&= kVAR_1 - kVAR_2 \\
&= OA \,(\tan \phi_1 - \tan \phi_2) \\
&= kW \,(\tan \phi_1 - \tan \phi_2)
\end{aligned}
$$

Knowing the leading kVAR supplied by the p.f. correction equipment, the desired results can be obtained. The p.f. of an electric system is very important to power companies and large industries because p.f. determines how efficiently the power distribution equipment (*e.g.* transformer, power lines etc.) is used.

Example 15.86. *An alternator is supplying a load of 300 kW at a p.f. of 0·6 lagging. If the power factor is raised to unity, how many more kilowatts can alternator supply for the same kVA loading ?*

Solution.

$$kVA = \frac{kW}{\cos \phi} = \frac{300}{0.6} = 500 \, kVA$$

$$kW \text{ at } 0.6 \text{ p.f.} = 300 \, kW$$

$$kW \text{ at } 1 \text{ p.f.} = 500 \times 1 = 500 \, kW$$

$\therefore$ Increased power supplied by the alternator

$$= 500 - 300 = \textbf{200 kW}$$

Note the importance of power factor improvement. When the p.f. of the alternator is unity, the 500 kVA are also 500 kW and the engine driving the alternator has to be capable of developing this power together with the losses in the alternator. But when the power factor of the load is 0·6, the power is only 300 kW. Therefore, the engine is developing only 300 kW, though the alternator is supplying a rated output of 500 kVA.

Example 15.87. *A single phase motor operating at 400V, 50 Hz draws 31.7A at a power factor of 0.7 lagging. Calculate the capacitance required in parallel to raise the power factor to 0.9 lagging.*

Solution. The circuit diagram and the phasor diagram are shown in Figs. 15.119 and 15.120 respectively. The motor takes a current $I_M = 31.7$ A.

The current I_C taken by the capacitor should be such that when combined with I_M, the resultant current I lags the voltage by ϕ when $\cos \phi = 0.9$. Since active component of current before and after power factor correction remains the same,

$\therefore$ $$I_M \cos \phi_M = I \cos \phi$$

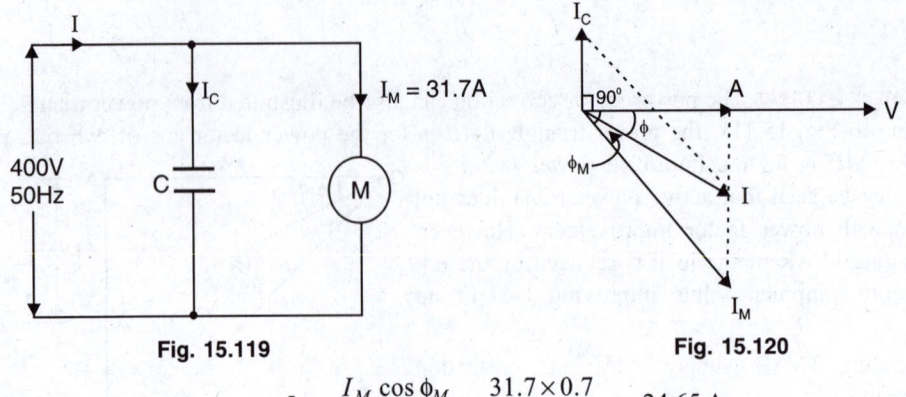

Fig. 15.119   Fig. 15.120

or $$I = \frac{I_M \cos \phi_M}{\cos \phi} = \frac{31.7 \times 0.7}{0.9} = 24.65 \text{ A}$$

From the phasor diagram in Fig. 15.120, we have,

$$I_C = I_M \sin \phi_M - I \sin \phi = 31.7 \times 0.714 - 24.65 \times 0.436 = 11.85 \text{ A}$$

Now, $$I_C = 2\pi f C V$$

or $$11.85 = 2\pi \times 50 \times C \times 400$$

∴ $$C = \frac{11.85}{2\pi \times 50 \times 400} = 94.3 \times 10^{-6} \text{ F} = \textbf{94.3 μF}$$

Note the effect of improving the power factor. The current taken from the supply reduces from 31·7 A to 24·65 A without altering the current or the power taken by the motor. An obvious advantage is the economy in the size of the generating plant and in the cross-sectional area of the conductors.

Example 15.88. *A 3-phase, star-connected motor, connected across a 3-phase, star-connected supply of 400V, 50 Hz takes a current of 20A at 0.8 p.f. lagging. Determine :*

(i) *the capacitance of the capacitors per phase that are to be connected in delta across the terminals of the motor to raise the p.f. to unity.*

(ii) *the new value of supply line current with capacitors connected.*

Solution. Fig. 15.121 (*i*) shows the power triangle before the connection of capacitor bank. The active power drawn by the motor is *OA* while the lagging reactive power is *AB* ; the p.f. being $\cos \phi_1$ (= 0·8). When the capacitor bank is connected across the motor terminals, the p.f. is raised to unity *i.e.* $\cos \phi_2$ = 1. The power triangle then becomes as shown in Fig. 15.121 (*ii*). Note that active power (= *OA*) drawn by the motor remains the same. The capacitor bank reduces the lagging reactive power *AB* to zero.

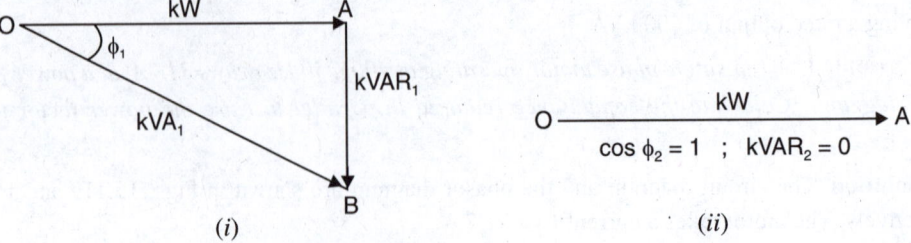

(i)   (ii)

Fig. 15.121

(i)   Power input (*OA*), $P = \sqrt{3} V_L I_L \cos \phi_1 = \sqrt{3} \times 400 \times 20 \times 0.8$

$$= 11.085 \times 10^3 \text{ W} = 11.085 \text{ kW}$$

Leading kVAR supplied by delta-connected capacitor bank

$$= AB = P \tan \phi_1 = 11.085 \tan (\cos^{-1} 0.8) = 8.314 \text{ kVAR}$$

If I_C is the phase current in the delta-connected bank, then,

$$\text{Total VAR} = 3 V_{ph} I_C = 3 \times 400 \times I_C = 1200 I_C$$

$\therefore \qquad\qquad 1200 I_C = 8314 \quad \text{or} \quad I_C = 8314/1200 = 6.93\text{A}$

Capacitive reactance/phase $= X_C = V_{ph}/I_C = 400/6.93 = 57.72 \ \Omega$

Capacitance/phase, $C = \dfrac{1}{2\pi f \, X_C} = \dfrac{1}{2\pi \times 50 \times 57.72} = 55.17 \times 10^{-6} \text{ F} = \mathbf{55.17 \ \mu F}$

(ii) Since the p.f. is raised to unity, the new line current I_L is given by ;

$$P = \sqrt{3} \ V_L I_L \qquad\qquad\qquad (\because \cos \phi_2 = 1)$$

$\therefore \qquad\qquad I_L = \dfrac{P}{\sqrt{3} \ V_L} = \dfrac{11.085 \times 10^3}{\sqrt{3} \times 400} = \mathbf{16A}$

Example 15.89. *A 3-phase, 50 Hz, 400 V motor develops 100 h.p. (74·6 kW), the power factor being 0.75 lagging and efficiency 93%. A bank of capacitors is connected in delta across the supply terminals and power factor raised to 0·95 lagging. Each of the capacitance units is built of 4 similar 100 V capacitors. Determine the capacitance of each capacitor.*

Solution. Original p.f., $\cos \phi_1 = 0.75$ lag ; Final p.f., $\cos \phi_2 = 0.95$ lag

$$\text{Motor input, } P = \text{Output}/\eta = 74.6/0.93 = 80 \text{ kW}$$

$$\phi_1 = \cos^{-1} (0.75) = 41.41°$$

$$\tan \phi_1 = \tan 41.41° = 0.8819$$

$$\phi_2 = \cos^{-1} (0.95) = 18.19°$$

$$\tan \phi_2 = \tan 18.19° = 0.3288$$

Leading kVAR supplied by the condenser bank

$$= P (\tan \phi_1 - \tan \phi_2)$$

$$= 80(0.8819 - 0.3288) = 44.25 \text{ kVAR}$$

Leading kVAR supplied by each of three sets

$$= 44.25/3 = 14.75 \text{ kVAR} \qquad\qquad\qquad ...(i)$$

Fig. 15.122 shows the delta* connected condenser bank. Let C farad be the capacitance of 4 capacitors in each phase.

Phase current of capacitor is

$$I_{CP} = V_{ph}/X_C = 2\pi f C V_{ph}$$

$$= 2\pi \times 50 \times C \times 400$$

$$= 1,25,600 \ C \text{ amperes}$$

$$\text{kVAR/phase} = \dfrac{V_{ph} I_{CP}}{1000}$$

$$= \dfrac{400 \times 1,25,600 \ C}{1000}$$

$$= 50240 \ C \qquad\qquad ...(ii)$$

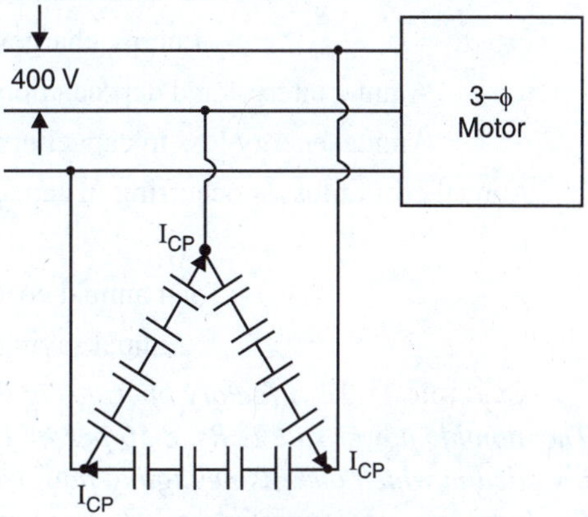

Fig. 15.122

Equating exps. (i) and (ii), we get,

$$50240 \ C = 14.75$$

or $\qquad\qquad C = 14.75/50,240 = 293.4 \times 10^{-6} \text{ F} = 293.4 \ \mu F$

Since it is the combined capacitance of four equal capacitors joined in series,

$\therefore$ Capacitance of each capacitor $= 4 \times 293.4 = \mathbf{1173.6 \ \mu F}$

---

* In practice, capacitors are always connected in delta since the capacitance of the capacitor required is one-third of that required for star connection.

Example 15.90. *A factory takes a load of 200 kW at 0.85 p.f. lagging for 2500 hours per annum. The tariff is Rs. 150 per kVA plus 5 paise per kWh consumed. If the p.f. is improved to 0·9 lagging by means of capacitors costing Rs. 420 per kVAR and having a power loss of 100 W per kVA, calculate the annual saving effected by their use. Allow 10% per annum for interest and depreciation.*

Solution.   Factory load, P_1 = 200 kW

$$\cos \phi_1 = 0.85 \; ; \tan \phi_1 = 0.62$$
$$\cos \phi_2 = 0.9 \; ; \tan \phi_2 = 0.4843$$

Suppose the leading kVAR supplied by the capacitors is x.

∴   Capacitor loss = $\dfrac{100 \times x}{1000} = 0.1x$ kW

          Total power, P_2 = $(200 + 0.1x)$ kW

Leading kVAR supplied by the capacitors is

$$x = P_1 \tan \phi_1 - P_2 \tan \phi_2$$
$$= 200 \times 0.62 - (200 + 0.1x) \times 0.4843$$

or $x = 124 - 96.96 - 0.0483\,x$

∴ $x = 27.14/1.04843 = 25.89$ kVAR

Annual cost before p.f. improvement.

          Max. kVA demand = 200/0.85 = 235.3

     kVA demand charges = Rs. 150 × 235.3 = Rs. 35, 295

     Units consumed/year = 200 × 2500 = 5,00,000 kWh

          Energy charges = Rs. 0.05 × 5,00,000 = Rs. 25,000

      Total annual cost = Rs. (35, 295 + 25,000) = Rs. 60,295

Annual cost after p.f. improvement.

          Max. kVA demand = 200/0.9 = 222.2

     kVA demand charges = Rs. 150 × 222.2 = Rs. 33,330

          Energy charges = same as before, *i.e.*, Rs. 25,000

 Annual interest and depreciation = Rs. 420 × 25·89 × 0.1 = Rs. 1087

 Annual energy loss in capacitors = 0.1 x × 2500 = 0.1 × 25.89 × 2500 = 6472 kWh

Annual cost of losses occurring in capacitors

                    = Rs. 0.05 × 6472 = Rs. 323

∴   Total annual cost = Rs. (33,330 + 25,000 + 1087 + 323) = Rs. 59,740

          Annual saving = Rs. (60,295 – 59,740) = **Rs. 555**

Example 15.91. *A factory operates at 0·8 p.f. lagging and has a monthly demand of 750 kVA. The monthly power rate is Rs. 8.50 per kVA. To improve the power factor, 250 kVA capacitors are installed in which there is negligible power loss. The installed cost of equipment is Rs. 20,000 and fixed charges are estimated at 10% per year. Calculate the annual saving effected by the use of capacitors.*

Solution. Monthly demand is 750 kVA.

          $\cos \phi = 0.8 \; ; \sin \phi = \sin (\cos^{-1} 0.8) = 0.6$

 kW component of demand = kVA × $\cos \phi$ = 750 × 0.8 = 600

 kVAR component of demand = kVA × $\sin \phi$ = 750 × 0.6 = 450

Leading kVAR supplied by the capacitors is 250 kVAR. Therefore, net kVAR after p.f. improvement is $450 - 250 = 200$.

$$\therefore \quad \text{kVA after p.f. improvement} = \sqrt{(600)^2 + (200)^2} = 632.45$$

$$\text{Reduction in kVA} = 750 - 632.45 = 117.5$$

$$\text{Monthly saving on kVA charges} = \text{Rs. } 8.5 \times 117.5 = \text{Rs. } 998.75$$

$$\text{Yearly saving on kVA charges} = \text{Rs. } 998.75 \times 12 = \text{Rs. } 11,985$$

$$\text{Fixed charges/year} = \text{Rs. } 0.1 \times 20,000 = \text{Rs. } 2000$$

$$\text{Net annual saving} = \text{Rs. } (11,985 - 2000) = \textbf{Rs. 9,985}$$

Example 15.92. *A synchronous motor improves the power factor of a load of 200 kW from 0·8 lagging to 0·9 lagging. Simultaneously the motor carries a load of 80 kW. Find (i) the leading kVAR supplied by the motor (ii) kVA rating of the motor and (iii) power factor at which the motor operates.*

Solution. Load, $P_1 = 200$ kW ; Motor load, $P_2 = 80$ kW

p.f. of load, $\cos \phi_1 = 0.8$ lag ; p.f. of combined load, $\cos \phi_2 = 0.9$ lag

Combined load, $P = P_1 + P_2 = 200 + 80 = 280$ kW

In Fig. 15.123, $\triangle OAB$ is the power triangle for load, $\triangle ODC$ for combined load and $\triangle BEC$ for the motor.

(i) Leading kVAR supplied by the motor

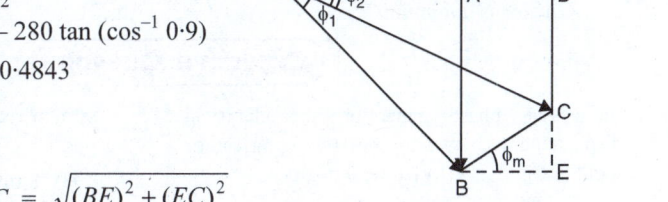

$$= CE = DE - DC = AB - DC \quad [\because AB = DE]$$

$$= P_1 \tan \phi_1 - P^* \tan \phi_2$$

$$= 200 \tan (\cos^{-1} 0·8) - 280 \tan (\cos^{-1} 0·9)$$

$$= 200 \times 0·75 - 280 \times 0·4843$$

$$= \textbf{14.14 kVAR}$$

(ii) kVA rating of the motor

$$BC = \sqrt{(BE)^2 + (EC)^2}$$

$$= \sqrt{(80)^2 + (14.4)^2} = \textbf{81.28 kVA}$$

(iii) p.f. of motor, $\cos \phi_m = \dfrac{\text{Motor kW}}{\text{Motor kVA}} = \dfrac{80}{81.28} = \textbf{0.984} \textit{ leading}$

Fig. 15.123

> ### Tutorial Problems

1. What should be the kVA rating of a capacitor which would raise the power factor of load of 100 kW from 0.5 lagging to 0.9 lagging ? **[125 kVA]**

2. A 3-phase, 50 Hz, 3300 V star connected induction motor develops 250 h.p. (186.5 kW), the power factor being 0.707 lagging and the efficiency 0.86. Three capacitors in delta are connected across the supply terminals and power factor raised to 0.9 lagging. Calculate :

    (*i*) the kVAR rating of the capacitor bank.

    (*ii*) the capacitance of each unit. **[(*i*) 111.8 kVAR (*ii*) 10.9 μF]**

3. A 3-phase, 50 Hz, 3000 V motor develops 600 h.p. (447.6 kW), the power factor being 0.75 lagging and the efficiency 0.93. A bank of capacitors is connected in delta across the supply terminals and power factor raised to 0.95 lagging. Each of the capacitance units is built of five similar 600-V capacitors. Determine the capacitance of each capacitor. **[156 μF]**

---

* In right angled triangle OAB, $AB = P_1 \tan \phi_1$

In right angled triangle ODC, $DC = OD \tan \phi_2 = (P_1 + P_2) \tan \phi_2 = P \tan \phi_2$

4. A factory takes a load of 800 kW at 0.8 p.f. (lagging) for 3000 hours per annum and buys energy on tariff of Rs. 100 per kVA plus 10 paise per kWh. If the power factor is improved to 0.9 lagging by means of capacitors costing Rs. 60 per kVAR and having a power loss of 100 W per kVA, calculate the annual saving effected by their use. Allow 10% per annum for interest and depreciation on the capacitors.

[Rs. 3972]

5. A station supplies 250 kVA at a lagging power factor of 0.8. A synchronous motor is connected in parallel with the load. If the combined load is 250 kW with a lagging p.f. of 0.9, determine :

    (*i*) the leading kVAR supplied by the motor.

    (*ii*) kVA rating of the motor.

    (*iii*) p.f. at which the motor operates.   **[(*i*) 28.9 kVAR (*ii*) 57.75 kVA (*iii*) 0.866 lead]**

6. A generating station supplies power to the following :

    (*i*) a lighting load of 100 kW ;

    (*ii*) an induction motor 800 h.p. (596·8 kW) p.f. 0·8 lagging, efficiency 92% ;

    (*iii*) a rotary converter giving 150 A at 400 V at an efficiency of 0.95.

What must be the power factor of the rotary converter in order that power factor of the supply station may become unity ?   **[0.128 leading]**

7. A 3-phase, 400 V synchronous motor having a power consumption of 50 kW is connected in parallel with an induction motor which takes 200 kW at a power factor of 0.8 lagging.

    (*i*) Calculate the current drawn from the mains when the power factor of the synchronous motor is unity.

    (*ii*) At what power factor should the synchronous motor operate so that the current drawn from the mains is minimum ?   **[(*i*) 421 A (*ii*) 0.316 leading]**

Objective Questions

1. In a two phase generator, the electrical displacement between the two phases or windings is .... electrical degrees.

    (*i*) 120   (*ii*) 90

    (*iii*) 180   (*iv*) none of the above

2. In a six-phase generator, the electrical displacement between different phases or windings is .... electrical degrees.

    (*i*) 60   (*ii*) 90

    (*iii*) 120   (*iv*) 45

3. The torque on the rotor of a 3-phase motor is more constant than that of a single phase motor because ....

    (*i*) single phase motors are not self-starting

    (*ii*) single phase motors are small in size

    (*iii*) 3-phase power is of constant value

    (*iv*) none of the above

4. For the same rating, the size of a 3-phase motor will be .... single phase motor.

    (*i*) less than that of (*ii*) more than that of

    (*iii*) same as that of (*iv*) none of the above

5. To transmit the same amount of power over a fixed distance at a given voltage, the 3-phase system requires .... the weight of copper.

    (*i*) 3 times   (*ii*) 3/4th times

    (*iii*) 1·5 times   (*iv*) 0·5 times

6. The phase sequence of a three-phase system is *RYB*. The other possible phase sequences can be ....

    (*i*) *BRY*   (*ii*) *YRB*

    (*iii*) *RBY*   (*iv*) none of the above

7. If in Fig. 15.124, the phase sequence is RYB, then, ...

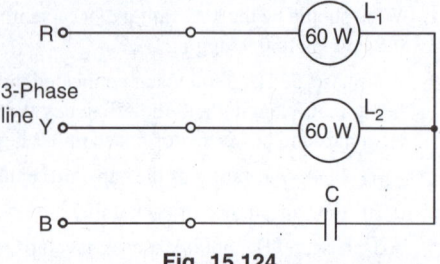

Fig. 15.124

    (*i*) L_1 will burn more brightly than L_2

    (*ii*) L_2 will burn more brightly than L_1

    (*iii*) both lamps will be equally bright

    (*iv*) none of the above

8. If the phase sequence of the 3-phase line in Fig. 15.124 is reversed, then, ...

    (i) L_1 will be brighter than L_2

    (ii) L_2 will be brighter than L_1

    (iii) both lamps will be equally bright

    (iv) none of the above

9. The advantage of star-connected supply system is that ....

    (i) line current is equal to phase current

    (ii) two voltages can be used

    (iii) phase sequence can be easily changed

    (iv) it is a simple arrangement

10. In a balanced star-connected system, line voltages are .... ahead of their respective phase voltages.

    (i) 30°   (ii) 60°

    (iii) 120°   (iv) none of the above

11. In a star-connected system, the relation between the line voltage V_L and phase voltage V_{ph} is ....

    (i) $V_L = V_{ph}$ (ii) $V_L = V_{ph}/\sqrt{3}$

    (iii) $V_L = \sqrt{3}\, V_{ph}$ (iv) none of the above

12. Fig. 15.125 shows a balanced star-connected system. The line voltage V_{RY} is given by ....

    (i) $V_{RY} = E_{RN} - E_{NY}$ .... phasor difference

    (ii) $V_{RY} = E_{RN} - E_{YN}$ ...... phasor difference

    (iii) $V_{RY} = E_{RN} + E_{YN}$ ....phasor sum

    (iv) none of the above

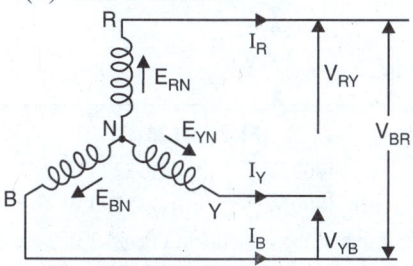

Fig. 15.125

13. If the load connected to the 3-phase generator shown in Fig. 15.125 has a lagging p.f. of $\cos \phi$, then phase angle between V_{RY} and I_R is ....

    (i) $30° + \phi$ (ii) $30° - \phi$

    (iii) $60° + \phi$ (iv) $120° - \phi$

14. Voltage $V_{CD} = 48 \angle 30°$ V. What is V_{DC} ?

    (i) $48 \angle -150°$ V   (ii) $48 \angle -180°$ V

    (iii) $24 \angle 30°$ V   (iv) $48 \angle -210°$ V

15. In a 3-phase system, if the instantaneous values of phases R and Y are +60 V and – 40 V respectively, then instantaneous voltage of phase B is

    (i) – 20 V   (ii) 40 V

    (iii) 120 V   (iv) none of the above

16. In a 3-phase system, $V_{YN} = 100 \angle -120°$ V and $V_{BN} = 100 \angle 120°$ V. What is V_{YB} ?

    (i) $170 \angle 90°$ V   (ii) $173 \angle -90°$ V

    (iii) $200 \angle 60°$ V   (iv) none of the above

17. In the above question, does V_{YB} lead or lag V_{YN}?

    (i) lead   (ii) lag

    (iii) Sometimes leads, sometimes lags

    (iv) none of the above

18. In question 16, what is the phase angle between V_{YB} and V_{YN} ?

    (i) 60°   (ii) 45°

    (iii) 120°   (iv) 30°

19. Voltage $V_{Ee} = 15 \angle 0°$ V and $V_{Ff} = 12 \angle 180°$ V. These voltages are connected in series with e connected to f. What is V_{FE} ?

    (i) $18 \angle -120°$ V (ii) $9 \angle -60°$ V

    (iii) $27 \angle 180°$ V   (iv) $18 \angle -30°$ V

20. The algebraic sum of instantaneous phase voltages in a three-phase circuit is equal to

    (i) zero   (ii) line voltage

    (iii) phase voltage (iv) none of the above

21. If the load connected to the 3-phase generator shown in Fig. 15.126 has a leading p.f. of $\cos \phi$, then angle between V_{RY} and I_R is ....

    (i) $90° - \phi$ (ii) $90° + \phi$

    (iii) $60° + \phi$ (iv) $30° - \phi$

22. Each phase voltage in Fig. 15.126 is 230 V. If connections of phase B are reversed, then, ...

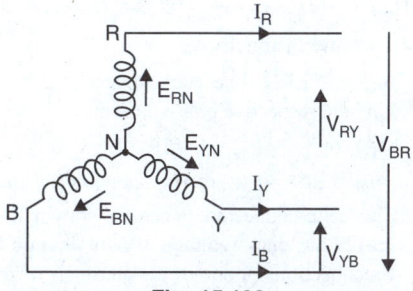

Fig. 15.126

(i) $V_{RY} = 230$ V   (ii) $V_{RY} > 230$ V

(iii) $V_{RY} < 230$ V   (iv) $V_{RY} = 0$

23. The power delivered by the 3-phase system shown in Fig. 15.126 is $\sqrt{3}\ V_L I_L \cos \phi$. Here ϕ is the phase difference between ....

   (i) line voltage and corresponding line current

   (ii) phase voltage and corresponding phase current

   (iii) phase current and line current

   (iv) none of the above

24. A 3-phase load is balanced if all the three phases have the same ....

   (i) impedance   (ii) power factor

   (iii) impedance and power factor

   (iv) none of the above

25. Three 50-ohm resistors are connected in star across 400 V, 3-phase supply. If one of the resistors is disconnected, then, line current will be ....

   (i) 8 A   (ii) 4 A

   (iii) $8\sqrt{3}$ A   (iv) $8/\sqrt{3}$ A

26. Fig. 15.127 shows balanced delta-connected supply system. The current in line 1 is ....

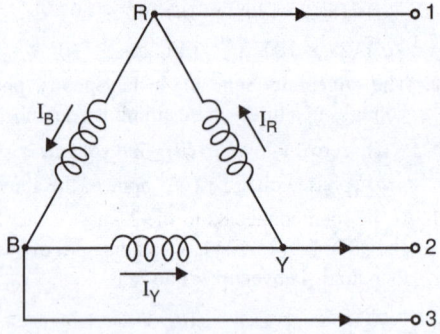

Fig. 15.127

(i) $I_R - I_B$ - - - - phasor difference

(ii) $I_B - I_R$ - - - - phasor difference

(iii) $I_Y - I_R - I_B$ - - - - phasor difference

(iv) none of the above

27. In Fig. 15.127, line currents are .... behind the respective phase currents.

   (i) 60°   (ii) 30°

   (iii) 120°   (iv) none of the above

28. The delta-connected generator shown in Fig. 15.127 has phase voltage of 200 V on no load. If connections of one of the phases is reversed,

then, resultant voltage across the mesh is ....

   (i) 200 V   (ii) $200 \times \sqrt{3}$ V

   (iii) 400 V   (iv) none of the above

29. If one line conductor of a 3-phase line is cut, the load is then supplied by .... voltage.

   (i) single phase   (ii) two phase

   (iii) three phase   (iv) none of the above

30. The resistance between any two terminals of a balanced star-connected load is 12 Ω. The resistance of each phase is ....

   (i) 12 Ω   (ii) 24 Ω

   (iii) 6 Ω   (iv) none of the above

31. The resistance between any two terminals of a balanced delta-connected load is 12 Ω. The resistance of each phase is ....

   (i) 12 Ω   (ii) 18 Ω

   (iii) 6 Ω   (iv) 36 Ω

32. The voltage rating of each resistor in Fig. 15.128 should be ....

   (i) 400 V   (ii) $400 \times \sqrt{2}$ V

   (iii) 230 V   (iv) none of the above

33. The power rating of each resistor in Fig. 15.128 is ....

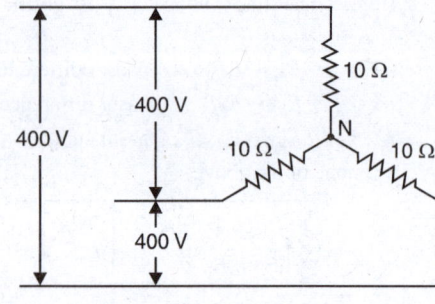

Fig. 15.128

   (i) 4000 W   (ii) 2300 W

   (iii) 4600 W   (iv) 5290 W

34. If one of the resistors in Fig. 15.128 were open-circuited, then, power consumed in the circuit is ....

   (i) 8000 W   (ii) 4000 W

   (iii) 16000 W   (iv) none of the above

35. The power consumed in the star-connected load shown in Fig. 15.129 is 690 W. The line current is ....

   (i) 2.5 A   (ii) 1 A

   (iii) 1.725 A   (iv) none of the above

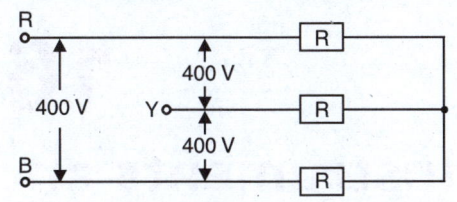

Fig. 15.129

36. If one of the resistors in Fig. 15.129 is open-circuited, power consumption will be ....

    (i) 200 W   (ii) 300 W

    (iii) 345 W   (iv) none of the above

37. The power factor of the star-connected load shown in Fig. 15.130 is ....

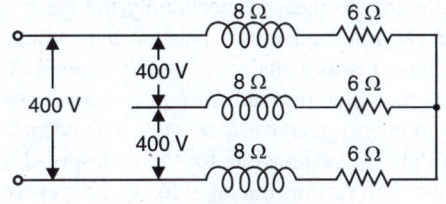

Fig. 15.130

    (i) 0.8 lagging   (ii) 0.6 lagging

    (iii) 0.75 lagging   (iv) none of the above

38. The voltage drop across each inductor in Fig. 15.130 is ....

    (i) 184 V   (ii) 138 V

    (iii) 400 V   (iv) none of the above

39. The power consumed in each phase of the circuit shown in Fig. 15.130 is ....

    (i) 2300 W   (ii) 4000 W

    (iii) 3174 W   (iv) none of the above

40. Three identical resistances connected in star consume 4000 W. If the resistances are connected in delta across the same supply, the power consumed will be ....

    (i) 4000 W   (ii) 6000 W

    (iii) 8000 W   (iv) 12000 W

Answers

1. (ii)	2. (i)	3. (iii)	4. (i)	5. (ii)
6. (iii)	7. (i)	8. (ii)	9. (ii)	10. (i)
11. (iii)	12. (ii)	13. (i)	14. (i)	15. (i)
16. (ii)	17. (i)	18. (iv)	19. (iii)	20. (i)
21. (iv)	22. (ii)	23. (ii)	24. (iii)	25. (ii)
26. (i)	27. (ii)	28. (iii)	29. (i)	30. (iii)
31. (ii)	32. (iii)	33. (iv)	34. (i)	35. (ii)
36. (iii)	37. (ii)	38. (i)	39. (iii)	40. (iv)

Electrical Instruments and Electrical Measurements

Introduction

Electrical energy is being used in the manufacture of many commodities. In order to ensure quality and efficiency, it is important that we should be able to measure accurately the electrical quantities involved. The instruments used to measure electrical quantities (*e.g.* current, voltage, power, energy etc.) are called electrical instruments. These instruments are generally named after the electrical quantity to be measured. Thus the instruments which measure current, voltage, power and energy are called ammeter, voltmeter, wattmeter and energy meter respectively. The accuracy, convenience and reliability of electrical instruments are mainly responsible for the widespread use of electrical methods of measurements. In this chapter, we shall confine our attention to the construction, working and applications of some important electrical instruments.

16.1. Classification of Electrical Measuring Instruments

The various electrical measuring instruments can be broadly divided into two classes *viz.*

(*i*) Absolute instruments (*ii*) Secondary instruments.

(*i*) **Absolute instruments.** Those instruments which give the value of the quantity to be measured in terms of the constants of the instrument and its deflection are called *absolute instruments*. For example, a tangent galvanometer which is used to measure current is an absolute instrument. It is because a tangent galvanometer gives the value of current being measured in terms of the tangent of the angle of deflection, the radius and number of turns of the coil and the horizontal component of earth's magnetic field. No previous calibration or comparison is necessary in case of absolute instruments. They are used only in standard laboratories as standardising instruments.

(*ii*) **Secondary instruments.** Those instruments in which the electrical quantity being measured is given directly by the deflection of the instrument are called *secondary instruments*. These instruments are provided with a calibrated scale. The calibration is done with the help of an absolute instrument or another calibrated instrument. These are the secondary instruments which are most generally used in everyday work. For example, they are very commonly used in laboratories, power stations, substations, industries *etc.*

16.2. Types of Secondary Instruments

Secondary instruments may be classified according to their functions as (*i*) indicating instruments (*ii*) integrating instruments and (*iii*) recording instruments.

(*i*) **Indicating instruments.** Those instruments which directly indicate the value of the electrical quantity *at the time* when it is being measured are called *indicating instruments e.g.* **ammeters**, **voltmeters** and **wattmeters**. In such instruments, a pointer moving over a graduated scale directly gives the value of the electrical quantity being measured. For example, when an ammeter is connected in the circuit, the pointer of the meter directly indicates the value of current flowing in the circuit at that time.

(*ii*) **Integrating instruments.** Those instruments which measure the total quantity of electricity (in ampere-hours) or electrical energy (in watt-hours) in a given time are called *integrating instruments e.g.* **ampere-hour meter** and **watt-hour meter**. In such instruments, there are sets of dials and pointers which register the total quantity of electricity or electrical energy supplied to the load.

(*iii*) **Recording instruments.** Those instruments which give a continuous record of the variations of the electrical quantity to be measured are called *recording instruments.* A recording instrument is merely an indicating instrument with a pen attached to its pointer. The pen rests lightly on a chart wrapped over a drum moving with a slow uniform speed. The motion of the drum is in a direction perpendicular to the direction of the pointer. The path traced out by the pen indicates the manner in which the quantity, being measured, has varied during the time of the record. Recording voltmeters are used in supply stations to record the voltage of the supply mains during the day. Recording ammeters are employed in supply stations for registering the current taken from the batteries.

16.3. Principles of Operation of Electrical Instruments

An electrical instrument essentially consists of a movable element and a scale to indicate or register the electrical quantity being measured. The movable element is supported on jewelled bearings and carries a pointer or sets of dials. The movement of the movable element is caused by utilising one or more of the following effects of current or voltage :

1. *Magnetic effect*	.....	Moving-iron instruments
2. *Electrodynamic effect*	.....	(*i*) Permanent-magnet moving coil
		(*ii*) Dynamometer type
3. *Electromagnetic-induction*	.....	Induction type instruments
4. *Thermal effect*	.....	Hot-wire instruments
5. *Chemical effect*	.....	Electrolytic instruments
6. *Electrostatic effect*	.....	Electrostatic instruments

S. No.	Type	Effect	Suitable for	Instrument
1.	Moving-iron	Magnetic effect	d.c. and a.c.	Ammeter, Voltmeter
2.	Permanent-magnet moving coil	Electrodynamic effect	d.c. only	Ammeter, Voltmeter
3.	Dynamometer type	Electrodynamic effect	d.c. and a.c.	Ammeter, Voltmeter, Wattmeter
4.	Induction type	Electro-magnetic induction effect	a.c. only	Ammeter, Voltmeter, Wattmeter, Energy-meter
5.	Hot-wire	Thermal effect	d.c. and a.c.	Ammeter, Voltmeter
6.	Electrolytic meter	Chemical effect	d.c. only	Ampere-hour meter
7.	Electrostatic type	Electrostatic effect	d.c. and a.c.	Voltmeter only

The principles of operation of electrical instruments are given in the above table for facility of reference.

16.4. Essentials of Indicating Instruments

An indicating instrument essentially consists of moving system pivoted in jewel bearings. A pointer is attached to the moving system which indicates on a graduated scale, the value of the electrical quantity being measured. In order to ensure proper operation of indicating instruments, the following three torques are required :

(*i*) Deflecting (or operating) torque

(*ii*) Controlling (or restoring) torque

(*iii*) Damping torque

The deflecting torque is produced by utilising the various effects of electric current or voltage and causes the moving system (and hence the pointer) to move from zero position. The controlling torque is provided by spring or gravity and opposes the deflecting torque. The pointer comes to rest at a position where these two opposing torques are equal. The damping torque is provided by air friction or eddy currents. It ensures that the pointer comes to the final position without oscillations, thus enabling accurate and quick readings to be taken.

16.5. Deflecting Torque

One important requirement in indicating instruments is the arrangement for producing deflecting or operating torque (T_d) when the instrument is connected in the circuit to measure the given electrical quantity. This is achieved by utilising the various effects of electric current or voltage mentioned in Art. 16.3. The deflecting torque causes the moving system (and hence the pointer attached to it) to move from zero position to indicate on a graduated scale the value of electrical quantity being measured. The actual method of producing the deflecting torque depends upon the type of instrument and shall be discussed while dealing with particular instrument.

16.6. Controlling Torque

If deflecting torque were acting alone, the pointer would continue to move indefinitely and would swing over to the maximum deflected position irrespective of the magnitude of current (or voltage or power) to be measured. This necessitates to provide some form of controlling or opposing torque (T_C). This controlling torque should oppose the deflecting torque and should increase with the deflection of the moving system. The pointer will be brought to rest at a position where the two opposing torques are equal *i.e.* $T_d = T_C$. The controlling torque performs two functions :

(*i*) It increases with the deflection of the moving system so that the final position of the pointer on the scale will be according to the magnitude of current (or voltage or power) to be measured.

(*ii*) It brings the pointer back to zero position when the deflecting torque is removed. If it were not provided, the pointer once deflected would not return to zero position on removing the deflecting torque.

The controlling torque in indicating instruments may be provided by one of the following two methods :

   1. By one or more springs   ... *Spring control*

   2. By weight of moving parts   ... *Gravity control*

1. Spring Control. This is the most common method of providing controlling torque in electrical instruments. A spiral *hairspring made of some non-magnetic material like phosphor bronze is attached to the moving system of the instrument as shown in Fig. 16.1. With the deflection of the pointer, the spring is twisted in the opposite direction. This twist in the spring provides the controlling torque. Since the torsion torque of a spiral spring is proportional to the angle of twist, the controlling torque is directly proportional to the deflection of the pointer *i.e.* $T_C \propto \theta$.

---

*   In some instruments, two spiral hairsprings wound in the opposite directions are used. The two springs serve the additional purpose of leading current to the moving system (*i.e.,* operating coil).

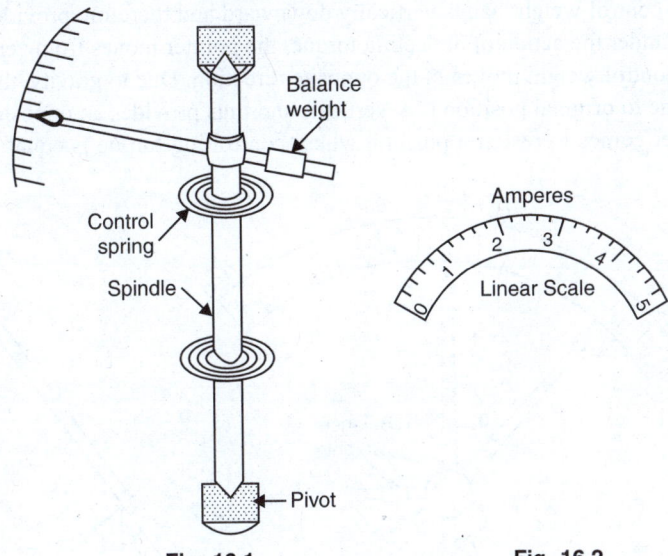

Fig. 16.1   **Fig. 16.2**

The pointer will come to rest at a position where controlling torque T_C is equal to the deflecting torque T_d i.e. $T_d = T_C$.

In an instrument where the deflecting torque is uniform, spring control provides a linear or evenly-spaced scale over the whole range. For example, in a permanent-magnet moving coil instrument, the deflecting torque is directly proportional to the current flowing through the operating coil *i.e.*

$$T_d \propto I$$

With spring control, $T_C \propto \theta$

In the final deflected position, $T_d = T_C$

$$\therefore \qquad \theta \propto I$$

Since the deflection is directly proportional to I, scale of such an instrument will be *linear (uniform) as shown in Fig. 16.2.

Advantages

(*i*) The levelling of the instrument is not required if the moving parts are balanced.

(*ii*) In some instruments (*e.g.* permanent-magnet moving coil and dynamometer type), springs also serve as the current leads to the moving coil.

(*iii*) There is practically no increase in the weight of the moving system.

(*iv*) In instruments where deflecting torque is uniform, spring control provides uniform scale.

Disadvantages

(*i*) Change of temperature affects the spring length and hence the controlling torque.

(*ii*) Controlling torque cannot be adjusted easily.

(*iii*) Accidental stresses in the springs may damage them.

2. Gravity Control. In this method, a small adjustable weight W is attached to the moving system [See **Fig. 16.3 (*i*)] which provides the necessary controlling torque. In the zero position

---

*   Linear scales are not necessarily straight as the name implies, but are usually in the form of an arc. They are called linear (or uniform) to denote that they are evenly graduated over the whole length of the arc, with equal subdivisions as shown in Fig. 16.2.

**   Note that another weight (called balance weight) is attached to counterbalance the weight of the pointer and other parts.

of the pointer, the control weight hangs vertically downward and therefore provides no controlling torque. However, under the action of deflecting torque, the pointer moves from zero position (from left to right) and control weight moves in the opposite direction. Due to gravity, the control weight would tend to come to original position (*i.e.* vertical) and thus provides an opposing or controlling torque. The pointer comes to rest at a position where controlling torque is equal to the deflecting torque.

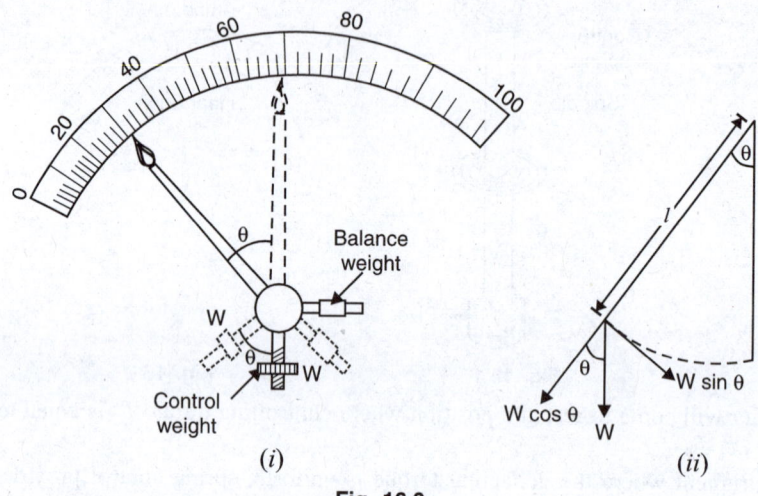

Fig. 16.3

In the deflected position shown in Fig. 16.3 (*ii*), weight W can be resolved into two rectangular components *viz* $W \cos \theta$ and $W \sin \theta$. Only the component $W \sin \theta$ provides the controlling torque T_C.

$$\therefore \qquad T_C = W\,l\,\sin \theta$$

or $\qquad\qquad\qquad T_C \propto \sin \theta \text{ (for fixed } W \text{ and } l)$

Thus, the controlling torque is proportional to the sine of angle of deflection. It may be seen that the value of T_C can be varied by changing l *i.e.* by changing the position of control weight W on the arm.

Taking the case of an instrument in which the deflecting torque is directly proportional to current (*e.g.* permanent-magnet moving coil instrument), we have for gravity control,

$$T_d \propto I \quad \text{and} \quad T_C \propto \sin \theta$$

In the final deflected position, $T_d = T_C$.

$$\therefore \qquad\qquad I \propto \sin \theta$$

Since I is proportional to the sine of angle of deflection, gravity-controlled instruments have non-uniform scales ; being crowded at the beginning as shown in Fig. 16.3 (*i*).

Advantages

    (*i*) It is slightly cheaper in manufacturing costs than spring control.

    (*ii*) It is unaffected by temperature variations.

    (*iii*) It is not subjected to fatigue.

    (*iv*) The controlling torque can be changed easily.

Disadvantages

    (*i*) The instrument has to be kept in vertical position.

    (*ii*) The control weight increases the weight of the moving system.

    (*iii*) Gravity-controlled instruments have non-uniform scale.

Example 16.1. *The deflecting torque of an ammeter is directly proportional to the current passing through it, and the instrument has full scale deflection of 70° for a current of 10 A. What deflection will occur for a current of 5 A when the instrument is (i) spring-controlled (ii) gravity-controlled ?*

Solution. Deflecting torque, $T_d \propto I$

(i) Spring-controlled.

Controlling torque, $T_C \propto \theta$ $\therefore$ $I \propto \theta$

In the first case, $10 \propto 70°$; In the second case, $5 \propto \theta$

$\therefore$ $\theta = (5/10) \times 70° = \mathbf{35°}$

(ii) Gravity-controlled.

Controlling torque, $T_C \propto \sin \theta$ $\therefore I \propto \sin \theta$

In the first case, $10 \propto \sin 70°$; In the second case, $5 \propto \sin \theta$

$\therefore$ $\sin \theta = (5/10) \sin 70° = 0.47$

or $\theta = \sin^{-1} 0.47 = \mathbf{28°}$

Example 16.2. *A moving-coil ammeter has springs giving a control constant of 0.3×10^{-6} Nm per degree. If the deflecting torque on the instrument is 28.8×10^{-6} Nm, find the angular deflection of the pointer.*

Solution. Controlling torque, $T_C = k\theta = 0.3 \times 10^{-6} \times \theta$ Nm

Deflecting torque, $T_d = 28.8 \times 10^{-6}$ Nm

In the final deflected position, $T_d = T_C$.

$\therefore$ $0.3 \times 10^{-6} \times \theta = 28.8 \times 10^{-6}$ or $\theta = \dfrac{28.8 \times 10^{-6}}{0.3 \times 10^{-6}} = \mathbf{96°}$

Example 16.3. *In a gravity-controlled instrument, the controlling weight is 0.005 kg and acts at a distance of 2.4 cm from the axis of the moving system. Determine the deflection in degrees corresponding to deflecting torque of 1.05×10^{-4} kg m.*

Solution. Controlling weight, $W = 0.005$ kg ; Lever arm, $l = 2.4$ cm $= 0.024$ m

$\therefore$ Controlling torque, $T_C = Wl \sin \theta = 0.005 \times 0.024 \times \sin \theta = 0.00012 \sin \theta$

In the final deflected position, $T_d = T_C$.

$\therefore$ $1.05 \times 10^{-4} = 0.00012 \sin \theta$ or $\theta = \mathbf{61°}$

Example 16.4. *The torque of an ammeter varies as the square of current through it. If a current of 10 A produces a deflection of 90°, what deflection will occur for a current of 5 A when the instrument is (i) spring-controlled (ii) gravity-controlled.*

Solution. Deflecting torque, $T_d \propto I^2$

(i) In spring-controlled instrument, controlling torque $T_C \propto \theta$.

In the final deflected position, $T_d = T_C$ so that $\theta \propto I^2$.

For first case, $\theta_1 \propto 10^2$; For second case, $\theta_2 \propto 5^2$

$\therefore$ $\dfrac{\theta_2}{\theta_1} = \dfrac{5^2}{10^2}$ or $\theta_2 = \theta_1 \times \dfrac{5^2}{10^2} = 90° \times \dfrac{25}{100} = \mathbf{22.5°}$

(ii) In gravity-controlled instrument, $T_C \propto \sin \theta$ so that $\sin \theta \propto I^2$.

For first case, $\sin \theta_1 \propto 10^2$; For second case, $\sin \theta_2 \propto 5^2$

$\therefore$ $\dfrac{\sin \theta_2}{\sin \theta_1} = \dfrac{5^2}{10^2}$ or $\sin \theta_2 = \sin \theta_1 \times \dfrac{5^2}{10^2} = \sin 90° \times \dfrac{25}{100} = 0.25$

$\therefore$ $\theta_2 = \sin^{-1} 0.25 = \mathbf{14.5°}$

Tutorial Problems

1. If the deflecting torque of an instrument is directly proportional to the current to be measured and the maximum current produces a deflection of 90°, compare the deflections in a spring-controlled instrument with a similar instrument having gravity control for a current equal to half the maximum value.   **[45° ; 30°]**

2. The torque of an ammeter varies as the square of the current through it. If a current of 5 A produces a deflection of 90°, what deflection will occur for a current of 3 A when the instrument is (i) spring-controlled (ii) gravity-controlled ?   **[(i) 32.4° (ii) 21.1°]**

16.7. Damping Torque

If the moving system is acted upon by deflecting and controlling torques alone, then pointer, due to inertia, will oscillate about its final deflected position for quite some time before coming to rest. This is often undesirable because it makes difficult to obtain quick and accurate readings. In order to avoid these oscillations of the pointer and to bring it quickly to its final deflected position, a damping torque is provided in the indicating instruments. This damping torque acts only when the pointer is in motion and always opposes the motion. The position of the pointer when stationary is, therefore, not *affected by damping.

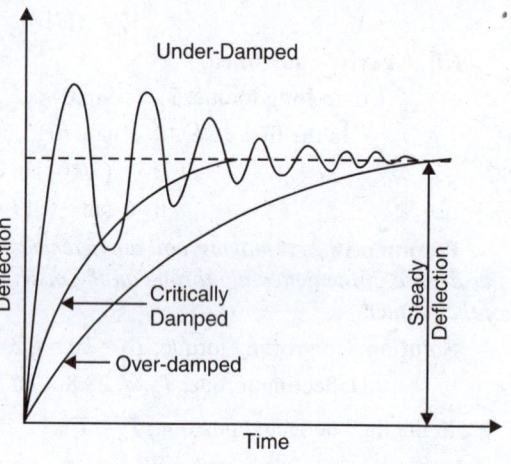

Fig. 16.4

The degree of damping decides the behaviour of the moving system. If the instrument is under-damped, the pointer will oscillate about the final position for some time before coming to rest. On the other hand, if the instrument is over-damped, the pointer will become slow and lethargic. However, if the degree of damping is adjusted to such a value that the pointer comes up to the correct reading quickly without passing beyond it or oscillating about it, the instrument is said to be *dead-beat* or *critically damped*. Fig. 16.4 shows graph for under-damping, over damping and critical damping (dead-beat). The damping torque in indicating instruments can be provided by (i) air-friction (ii) fluid friction and (iii) eddy currents.

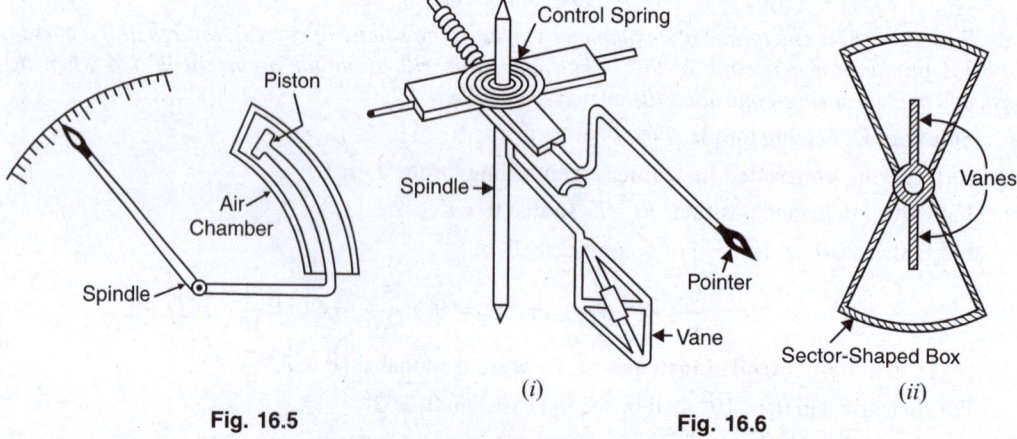

Fig. 16.5   (i)   (ii)

                                                      Fig. 16.6

---

*   The damping torque acts only when the pointer is in *motion*. When the pointer is in a particular deflected position, though deflecting and controlling torques are acting on the moving system but the damping torque is zero. It is because the pointer is steady and there is no movement of the moving system.

(i) Air friction damping. Two arrangements of air friction damping are shown in Fig. 16.5 and Fig. 16.6. In the arrangement shown in Fig. 16.5, a light aluminium piston is attached to the spindle that carries the pointer and moves with a very little clearance in a rectangular or circular air chamber closed at one end. The cushioning action of the air on the piston damps out any tendency of the pointer to oscillate about the final deflected position. This method is not favoured these days and the one shown in Fig. 16.6 (*i*) is preferred.

The arrangement shown in Fig. 16.6 (*ii*) is widely used. In this method, one or two light aluminium vanes are attached to the same spindle that carries the pointer. The vanes are permitted to swing in a sector-shaped closed box that is just large enough to accommodate the vanes. As the pointer moves, the vanes swing in the box, compressing the air in front of them. The pressure of compressed air in the vanes provides the necessary damping force to reduce the tendency of the pointer to oscillate.

(ii) Fluid friction damping. In this method, discs or vanes attached to the spindle of the moving system are kept immersed in a pot containing oil of high viscosity (See Fig. 16.7). As the pointer moves, the friction between the oil and vanes opposes the motion of the pointer and thus necessary damping is provided. This method is not used because of several disadvantages such as objectionable creeping of oil, the necessity of using the instrument always in the vertical position and its unsuitability for portable instruments. In general, fluid friction damping is not employed in indicating instruments, although one can find its use in Kelvin electrostatic voltmeter.

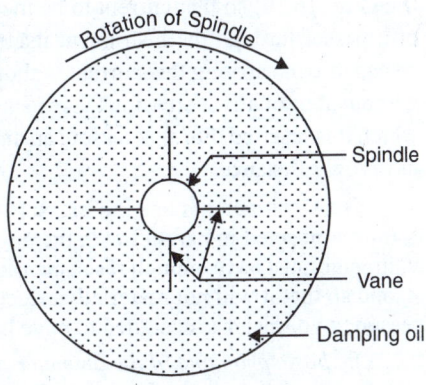

Fig. 16.7

(iii) Eddy current damping. Eddy current damping is the most efficient form of damping. Two methods of eddy current damping are shown in Fig. 16.8 and Fig. 16.9. In Fig. 16.8, a thin *aluminium or copper disc attached to the moving system is allowed to pass between the poles of a permanent magnet. As the pointer moves, the disc cuts across the magnetic field and **eddy currents are induced in the disc. These eddy currents react with the field of the magnet to produce a force which opposes the motion (Lenz's Law). In this way, eddy currents provide the damping torque to reduce the oscillations of the pointer.

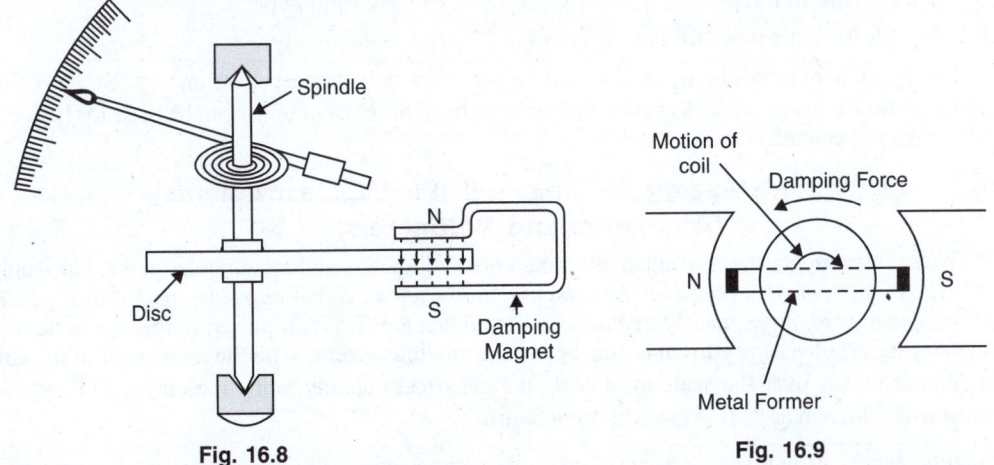

Fig. 16.8   **Fig. 16.9**

---

*   The disc must be a conductor but non-magnetic.

**   The currents induced in the disc assume the form of little whirls or eddies and hence the name eddy currents.

In Fig. 16.9, the operating coil (*i.e.* the coil which produces the deflecting torque) is wound on the aluminium former. As the coil moves in the field of the instrument, eddy currents are induced in the aluminium former to provide the necessary damping torque.

Note. The moving system of an indicating instrument experiences *three* torques while in motion and *two* torques when it has come to rest in the final deflected position.

(*a*) During motion, the *deflecting torque* due to the quantity being measured is opposed by *controlling torque* and *damping torque*.

(*b*) When the pointer comes to rest in the final deflected position, damping torque is zero and the *controlling torque* is equal to the *deflecting torque*.

16.8. Ammeters and Voltmeters

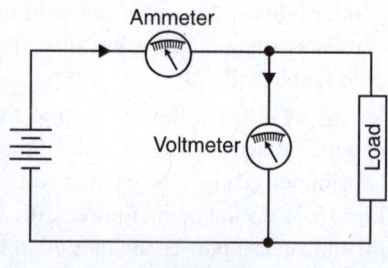

Fig. 16.10

(*i*) An ammeter is used to measure the flow of current in a circuit. It is thus connected in series with the circuit under test (See Fig. 16.10) so that current to be measured or a *fraction of it passes through the instrument itself. The ammeter must be capable of carrying this current without injury to itself and without abnormally increasing the resistance of the circuit into which it is inserted. For this reason, an ammeter is designed to have low resistance.

(*ii*) A voltmeter is used to measure the potential difference between two points of a circuit. It is thus connected in parallel with the circuit or some part of the circuit as shown in Fig. 16.10. The voltmeter must have enough resistance so that it will not be injured by the current that flows through it, and so that it will not materially affect the current in the circuit to which it is connected. For this reason, a voltmeter is designed to have high resistance.

The basic principle of the ammeter and of the voltmeter is the same. Both are current operated devices i.e. deflecting torque is produced when current flows through their operating coils. In the ammeter, the deflecting torque is produced by the current we wish to measure, or a certain fraction of that current. In the voltmeter, the deflecting torque is produced by a current which is proportional to the potential difference we wish to measure. Thus, the same instrument can be used as an ammeter or voltmeter with proper design.

The following types of instruments are used for making voltmeters and ammeters :

(*i*) Permanent-magnet moving coil type   (*ii*) Dynamometer type

(*iii*) Moving-iron type   (*iv*) Hot-wire type

(*v*) Electrostatic type (for voltmeter only)   (*vi*) Induction type

The instrument at Sr. No. (*i*) can be used for d.c. work only whereas instrument at Sr. No. (*vi*) is employed for a.c. work only. However, instruments from Sr. No. (*ii*) to (*v*) can be used for both d.c. and a.c. measurements.

16.9. Permanent-Magnet Moving Coil (PMMC) Instruments
(Ammeters and Voltmeters)

These instruments are used either as ammeters or voltmeters and are suitable for d.c. work only. *This type of instrument is based on the principle that when a current carrying conductor is placed in a magnetic field, mechanical force acts on the conductor.* The coil placed in the magnetic field and carrying the operating current is attached to the moving system. With the movement of the coil, the pointer moves over the scale to indicate the electrical quantity being measured. This type of movement is known as **D'Arsonval movement.**

---

*   If the circuit current is large, a shunt is used to divert a major portion of current so that only a small current flows through the instrument.

**   After the French physicist Arsene D'Arsonval who first conceived such a movement of the instrument.

Construction. Fig. 16.11 shows the various parts of a permanent-magnet moving coil instrument. It consists of a light rectangular coil of many turns of fine wire wound on an aluminium former inside which is an iron core as shown in Fig. 16.11 (*i*). The coil is delicately pivoted upon jewel bearings and is mounted between the poles of a permanent horse-shoe magnet. Attached to these poles are two soft-iron pole pieces which concentrate the magnetic field. The current is led into and out of the coil by means of two control hair-springs, one above and the other below the coil, as shown in Fig. 16.11 (*ii*). These springs also provide the controlling torque. The damping torque is provided by eddy currents induced in the aluminium former as the coil moves from one position to another.

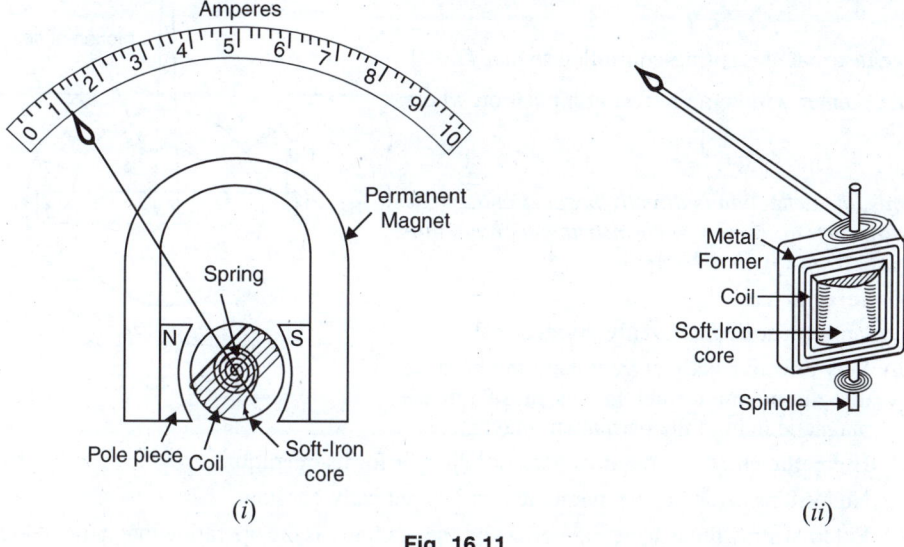

Fig. 16.11

Working. When the instrument is connected in the circuit to measure current or voltage, the operating current flows through the coil. Since the coil is carrying current and is placed in the magnetic field of the permanent magnet, a mechanical torque acts on it. As a result, the pointer attached to the moving system moves in a clockwise direction over the graduated scale to indicate the value of current or voltage being measured. If the current in the coil is reversed, the deflecting torque will also be reversed since the direction of the field of the permanent magnet is the same. Consequently, the pointer will try to deflect below zero. Deflection in this direction (*i.e.* reverse direction) is prevented by a spring "stop". *Since the deflecting torque reverses with the reversal of current in the coil, such instruments can be used to measure direct currents and voltages *only.

Deflecting torque. The magnetic field in the air gap is radial due to the presence of **soft-iron core. This means that conductors of the coil will always move at right angles to the field. When current is passed through the coil, forces act on its both sides which produce the deflecting torque. Referring to Fig. 16.12, let,

$$B = \text{flux density in Wb/m}^2$$
$$l = \text{length or depth of coil in m}$$
$$b = \text{breadth of coil in m}$$
$$N = \text{No. of turns in the coil}$$

---

*   The instrument can be used to measure alternating currents and voltages by using a rectifier. The given alternating quantity is converted to d.c. by using a bridge type rectifier. This d.c. or average value is measured by the permanent-magnet moving coil instrument. Since a.c. meters are calibrated to indicate *effective* or *r.m.s.* values, we must multiply the average value registered on this meter by 1.11 (the form factor of sinusoidal wave) to obtain the r.m.s. values. Some instruments are scaled to read r.m.s. values directly.

** The soft-iron results in the decreased reluctance of the magnetic circuit and in a uniform, radial flux in the air gap.

If a current of I amperes flows in the coil, then force acting on each coil side is given by ;

Force on each coil side, $F = BIlN$ newtons

Deflecting torque, T_d = Force × perpendicular distance

$$= (BIlN) \times b$$

$\therefore$ $\qquad T_d = BINA$ newton-metre

where $A (= b \times l)$ is the area of the coil in m².

Since the values of B, N and A are fixed,

$\therefore$ $\qquad T_d \propto I$

The instrument is spring-controlled so that $T_C \propto \theta$.

The pointer will come to rest at a position where $T_d = T_C$.

$\therefore$ $\qquad \theta \propto I$

Thus, the deflection is directly proportional to the operating current. Hence, such instruments have uniform scale [See. Fig. 16.11 (*i*)].

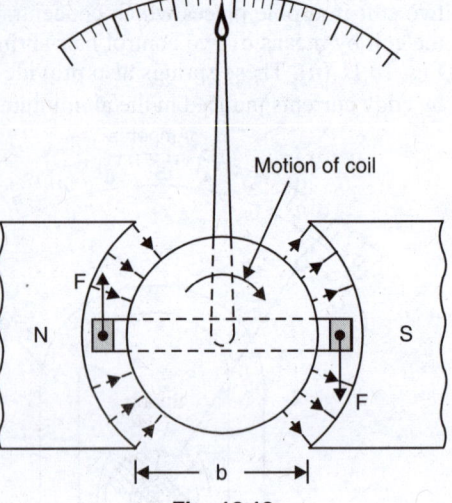

Fig. 16.12

Advantages

(*i*) Uniform scale *i.e.*, evenly divided scale.

(*ii*) Very effective eddy current damping because the aluminium former moves in an intense magnetic field of the permanent magnet.

(*iii*) High efficiency as it requires very little power for its operation.

(*iv*) No hysteresis loss as the magnetic flux is practically constant.

(*v*) External stray fields have little effect on the readings as the operating magnetic field is very strong.

(*vi*) Very accurate and reliable.

Disadvantages

(*i*) Such instruments cannot be used for a.c. measurements.

(*ii*) About 50% more expensive than moving-iron instruments because of their accurate design.

(*iii*) Some errors are caused due to variations (with time or temperature) either in the strength of permanent magnet or in the control springs.

Applications. *Permanent-magnet moving coil instruments are acknowledged to be the best type for all d.c. measurements.* They are very sensitive and maintain a high degree of accuracy over long periods. The chief applications of such instruments are :

(*i*) In the measurement of direct currents and voltages.

(*ii*) In d.c. galvanometers to detect small currents.

(*iii*) In ballistic galvanometers used mainly for measuring changes of magnetic flux linkages.

Example 16.5. *The resistance of a moving coil voltmeter is 12,000 Ω. The moving coil has 100 turns and is 4 cm long and 3 cm wide. The flux density in the air gap is 6×10^{-2} Wb/m². Find the deflection produced by 300 V if the spring control gives a deflection of one degree for a torque of 25×10^{-7} Nm.*

Solution. Applied voltage, $V = 300$ volts ; Resistance of voltmeter, $R_V = 12,000$ Ω

$\therefore$ $\qquad$ Current through coil, $I = \dfrac{V}{R_V} = \dfrac{300}{12,000} = 0.025$ A

$\qquad$ Area of coil, $A = 0.04 \times 0.03 = 12 \times 10^{-4}$ m²

$\qquad$ Deflecting torque, $T_d = BINA$ Nm

Here, $B = 6 \times 10^{-2}$ Wb/m² ; $I = 0.025$ A ; $N = 100$ turns ; $A = 12 \times 10^{-4}$ m²

$$\therefore \qquad T_d = 6 \times 10^{-2} \times 0.025 \times 100 \times 12 \times 10^{-4} = 18 \times 10^{-5} \text{ Nm}$$

If θ is the deflection, then controlling torque T_C is

$$T_C = k\theta = 25 \times 10^{-7} \theta \text{ Nm}$$

In the final deflected position, $T_C = T_d$.

$$\therefore \qquad 25 \times 10^{-7} \theta = 18 \times 10^{-5} \quad \text{or} \quad \theta = \frac{18 \times 10^{-5}}{25 \times 10^{-7}} = \textbf{72°}$$

Example 16.6. *A moving coil millivoltmeter has a resistance of 200 Ω and full-scale deflection is reached when a potential difference of 100 mV is applied across its terminals. The moving coil has effective dimensions of 30 × 25 mm and is wound with 100 turns. The flux density in the gap is 0.2 Wb/m². Determine the control constant of the spring if the final deflection is 100° and suitable diameter of copper wire for the coil winding if 20 % of total instrument resistance is due to coil winding. Resistivity of copper is 1.7 × 10⁻⁸ Ω m.*

Solution. Full-scale deflection (F.S.D.), $\theta = 100°$; Voltage for F.S.D., $V = 100$ mV = 0.1 volt; Resistance of voltmeter, $R_V = 200 \, \Omega$

$$\text{F.S.D. current, } I_g = \frac{V}{R_V} = \frac{0.1}{200} = 0.0005 \text{ A}$$

F.S.D. deflecting torque, $T_d = BI_g NA$

Here, $B = 0.2$ Wb/m² ; $I_g = 0.0005$ A ; $N = 100$ turns ; $A = 0.03 \times 0.025$ m²

$$\therefore \qquad T_d = 0.2 \times 0.0005 \times 100 \times 0.03 \times 0.025 = 7.5 \times 10^{-6} \text{ Nm}$$

In the final deflected position, controlling torque, $T_C = T_d = 7.5 \times 10^{-6}$ Nm.

Now, $T_C = k\theta$ $\therefore$ Control spring constant, $k = \dfrac{T_C}{\theta} = \dfrac{7.5 \times 10^{-6}}{100} = \textbf{7.5} \times \textbf{10}^{-8} \text{ \textbf{Nm/degree}}$

$$\text{Length of coil, } l = 2(30 + 25) \times 10^{-3} \times 100 = 11 \text{ m}$$

$$\text{Resistance of coil, } R_C = 20\% \text{ of } R_V = \frac{20}{100} \times 200 = 40 \, \Omega$$

Now, $\qquad\qquad\qquad R_C = \dfrac{\rho l}{a}$

$$X\text{-sectional area of wire, } a = \frac{\rho l}{R_C} = \frac{1.7 \times 10^{-8} \times 11}{40} = 0.4675 \times 10^{-8} \text{ m}^2$$

Now, $\qquad\qquad\qquad a = \dfrac{\pi}{4} d^2$

$$\therefore \qquad \text{Diameter of wire, } d = \sqrt{\frac{4a}{\pi}} = \sqrt{\frac{4 \times 0.4675 \times 10^{-8}}{\pi}} = 0.077 \times 10^{-3} \text{ m} = \textbf{0.077 mm}$$

The small diameter of the instrument coil shows that it can carry very small current (a few mA) without being burnt due to excessive heat.

Example 16.7. *The coil of a moving coil permanent magnet voltmeter is 40 mm long and 30 mm wide and has 100 turns on it. The control spring exerts a torque of 120 × 10⁻⁶ Nm when the deflection is 100 divisions on full-scale. If the flux density of the magnetic field in the air-gap is 0.5 Wb/m², estimate the resistance that must be put in series with the coil to give one volt per division. The resistance of the voltmeter coil may be neglected.*

Solution. Let I_m be the full-scale meter current.

Full-scale deflecting torque is

$$T_d = BI_m NA = 0.5 \times I_m \times 100 \times 0.04 \times 0.03 = 0.06 \, I_m \text{ Nm}$$

Full-scale controlling torque is

$$T_C = 120 \times 10^{-6} \text{ Nm} \quad \ldots \text{ given}$$

In the final full-scale position, $T_d = T_C$.

$$\therefore \qquad 0.06 \, I_m = 120 \times 10^{-6} \quad \text{or} \quad I_m = 2 \times 10^{-3} \text{ A}$$

Full-scale reading required from the instrument is

$$V = 100 \times 1 = 100 \text{ volts}$$

$\therefore$ External resistance R_s to be put in series with the coil is

$$R_s = \frac{V}{I_m} - R_m = \frac{100}{2 \times 10^{-3}} - 0 = \mathbf{50{,}000 \ \Omega}$$

Tutorial Problems

1. The coil of a permanent-magnet moving coil instrument has 50 turns and its effective length and breadth are 2.8 cm and 2 cm. The control springs have a total torque of 0.8×10^{-6} Nm per degree of deflection. Find the air-gap flux density that will give a full-scale deflection of 100° at 15 mA. **[0.19 Wb/m²]**

2. A permanent-magnet moving coil instrument used as a voltmeter has a coil of 50 turns with a width of 3 cm and active length of 3 cm. The gap flux density is 0.15 Wb/m². If the full-scale reading is 150 V and the total resistance is 100 kΩ, find the torque exerted by the springs at full-scale. **[8.1 × 10⁻⁵ Nm]**

3. A permanent-magnet moving coil instrument has an air-gap flux density of 0.08 Wb/m². There are 60 turns of wire on the moving coil which has an effective length in the gap of 4 cm, a width of 2.5 cm and negligible resistance. The control springs exert a torque of 49.05×10^{-6} Nm at full-scale deflection. Find the resistance to be connected in series with the coil to enable the instrument to be used as a voltmeter reading upto 500 V. **[49,000 Ω]**

16.10. Extension of Range of PMMC Instruments

In a permanent-magnet moving coil (PMMC) instrument, the moving coil and the springs used as coil connections have a very delicate design and can carry maximum current of about 10 mA *i.e.*, full-scale deflection (*f.s.d.*) will occur when about 10 mA current flows through the instrument coil. If current through the coil exceeds this value (*i.e.* 10 mA), the coil may be burnt due to excessive heat. However, in practice, we have to measure large currents and voltages. In such situations, some means are adopted to increase the range of the instruments.

16.11. Extension of Range of PMMC Ammeter

The range of a permanent-magnet moving coil ammeter can be extended by connecting a low resistance, called *shunt in parallel with the moving coil of the instrument as shown in Fig. 16.13. The shunt bypasses most of the line current and allows a small current through the meter which it can handle without burning.

Let R_m = meter resistance

S = shunt resistance

I_m = full-scale deflection (*f.s.d.*) current

I = full range current of the meter

Voltage across shunt = Voltage across the meter

or $(I - I_m)S = I_m R_m$

$\therefore \qquad S = \dfrac{I_m R_m}{I - I_m} \qquad \ldots(i)$

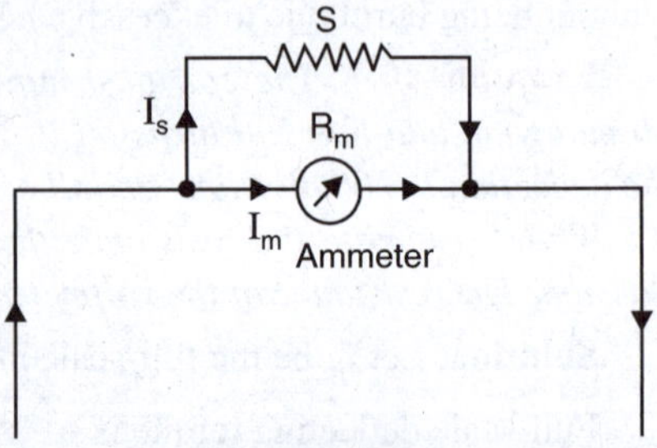

Fig. 16.13

---

* Shunts are made of materials such as manganin having low temperature coefficient of resistance.

$$\text{Multiplying power of shunt} = \frac{I}{I_m} = \frac{R_m + S}{S}$$

Note that *multiplying power of a shunt* is the ratio of circuit current to be measured to the meter current. The multiplying power of a shunt is constant and indicates the factor by which the meter current must be multiplied to obtain the circuit current. Suppose the meter has a resistance of 5 Ω and requires 15 mA for full-scale deflection. In order that the meter may read 1A, the value of shunt is given by ;

$$S = \frac{I_m \, R_m}{I - I_m} = \frac{0.015 \times 5}{1 - 0.015} = 0.0761 \ \Omega$$

$$\text{Multiplying power of shunt} = \frac{I}{I_m} = \frac{1}{0.015} = 66.67$$

Note that multiplying power of a shunt plays an important role in the extension of ammeter range.

Resistance of ammeter. *A shunted PMMC galvanometer is called a PMMC* **ammeter.**

$$\text{Resistance of ammeter} = \frac{R_m S}{R_m + S}$$

Clearly, the value of ammeter resistance will be less than S. Since the value of S is very small, the ammeter resistance will also be very low. Thus shunt has not only extended the current range but it has also lowered the resistance of the ammeter—a desirable requirement for an ammeter.

Multirange ammeter. The single movement may be used with several shunts to make a multirange ammeter as shown in Fig. 16.14. The switch selects which shunt (R_1, R_2 or R_3) is to be used across the meter. With the switch at position 1, only resistance R_1 acts as the shunt. Likewise with switch at position 2, only resistance R_2 acts as the shunt. Note that the shape of the moving contact of the switch is short-circuiting type. This is to ensure that when this moving contact is shifted from one position to another, it makes contact with the new fixed contact before it breaks contact with its previous fixed contact. This will ensure that, at all times, there is a shunt across the meter. Otherwise, when the switch is changed from one position to another, there might be an instant when there would be no shunt across the meter. Under such circumstances, the full circuit current flowing through the meter coil may burn it.

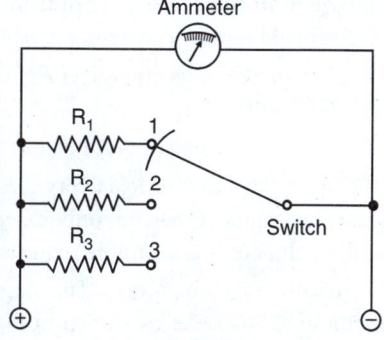

Fig. 16.14

16.12. Extension of Range of PMMC Voltmeter

The range of a permanent-magnet moving coil voltmeter can be increased by connecting a high resistance R_s called **multiplier,** in *series with it as shown in Fig. 16.15.

Let R_m = meter resistance

R_s = series resistance *i.e.*, multiplier

I_m = full-scale deflection current

V = full-range voltage of the meter

Voltage across AB = Voltage across the meter and R_s

or $V = I_m(R_s + R_m)$

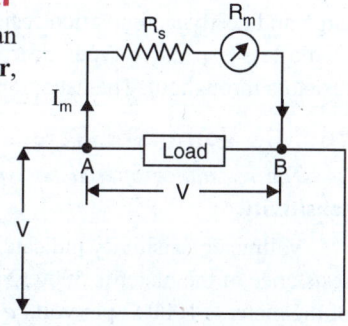

Fig. 16.15

---

*   A voltmeter is connected in parallel with the load. Consequently, current through the meter will be proportional to the voltage across the load *i.e.*, voltage we wish to measure. In order to restrict the current through the meter (about 10 mA), a high resistance R_s is connected in series with it.

$$\therefore \qquad R_s = \frac{V}{I_m} - R_m \qquad\qquad\qquad ...(i)$$

$$\text{Voltage amplification} = \frac{\text{Voltage to be measured}}{\text{Voltage across meter}}$$

$$= \frac{V}{I_m R_m} = 1 + \frac{R_s}{R_m} \qquad\qquad \text{[From equation } (i)\text{]}$$

Suppose the meter has a resistance of 5 Ω and requires 15 mA for full-scale deflection. In order that the meter may read 15 V, the value of series resistance R_s is given by ;

$$R_s = \frac{V}{I_m} - R_m = \frac{15}{0.015} - 5 = 995 \;\Omega$$

$$\text{Voltage amplification} = 1 + \frac{R_s}{R_m} = 1 + \frac{995}{5} = 200$$

Clearly, greater the value of R_s, greater is the voltage amplification. For this reason, R_s is called **voltage multiplier** or simply **multiplier.** The important requirement of a multiplier is that its resistance should remain constant, *i.e.* it should have low temperature coefficient of resistance.

Resistance of voltmeter. *A PMMC galvanometer in series with a high resistance (R_s) is called a PMMC* **voltmeter**.

$$\text{Resistance of voltmeter} = R_m + R_s$$

Since the value of R_s is very large, the resistance of the voltmeter will be very high. Thus the series resistance R_s has not only extended the voltage range but it has also increased the resistance of the voltmeter—a desirable requirement for a voltmeter.

Multirange voltmeter. The single movement may be used with several multipliers to make a multirange voltmeter as shown in Fig. 16.16. The different ranges are determined by a selector switch connecting the proper multiplier into the circuit. With switch at position 1, only multiplier R_1 is in series with the meter. Similarly, with switch at position 2, only multiplier R_2 is in series with the meter. It may be noted that here short-circuiting type switch is not desirable and a *non-shorting type switch* must be used. The multipliers are constructed of high resistance wire, such as manganin wound on wooden spools. Where the resistance required is so high as to make the windings prohibitively large and costly, composition resistors are employed. Some of the less expensive types of voltmeters use composition resistors throughout. The usual ranges of such instruments are from 0–1000 V in two or more steps.

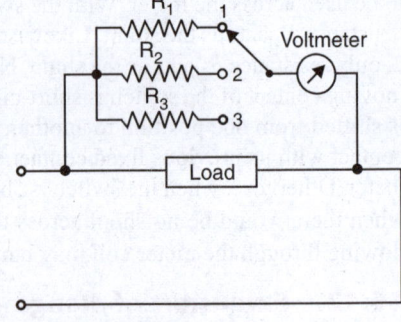

Fig. 16.16

16.13. Voltmeter Sensitivity

The resistance offered per volt of full scale deflection by the voltmeter is called **voltmeter sensitivity.**

Voltmeter sensitivity indicates the internal resistance of the voltmeter. For example, if the total resistance of the meter is 5000 Ω and the meter is to read 5 volts full scale, then internal resistance of the meter is 1000 Ω per volt *i.e.* meter sensitivity is 1000 Ω per volt. Conversely, if the voltmeter sensitivity is 400 Ω per volt which reads from 0 to 100 V, then meter resistance is 40, 000 Ω. If the voltmeter is to read V volts full scale and I_m is the full scale deflection (f.s.d.) current, then,

$$\text{Voltmeter resistance} = \frac{V}{I_m}$$

∴ Voltmeter sensitivity = Resistance per volt full scale deflection

$$= \frac{V}{I_m} \div V = \frac{1}{I_m}$$

Sensitivity is the most important characteristic of a voltmeter. If the sensitivity of a voltmeter is high, it means that it has high internal resistance. When such a meter is connected in the circuit to measure voltage, it will draw a very small current. Consequently, there will be practically no change in the circuit current due to the introduction of the meter. Hence, it will measure the voltage correctly. On the other hand, if the sensitivity of the voltmeter is low, it would cause serious error in voltage measurements. The sensitivity of voltmeters available in the market ranges from 5 kΩ per volt to 20 kΩ per volt.

Example 16.8. *The meter element of a permanent-magnet moving coil instrument has a resistance of 5 ohms and requires 15 mA for full-scale deflection. Calculate the resistance to be connected (i) in parallel to enable the instrument to read upto 1A (ii) in series to enable it to read upto 15 V.*

Solution. Meter resistance, $R_m = 5\ \Omega$

Full-scale meter current, $I_m = 15\ mA = 0.015\ A$

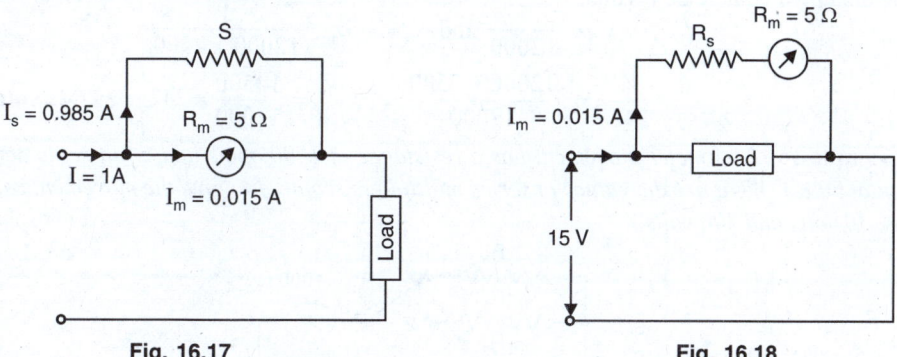

Fig. 16.17   Fig. 16.18

(i) As Ammeter. Full-scale circuit current, $I = 1\ A$

Let S ohms be the required value of the shunt [See Fig. 16.17].

Now, $I_m R_m = (I - I_m) S$

∴ $S = \dfrac{I_m R_m}{I - I_m} = \dfrac{0.015 \times 5}{1 - 0.015} = \textbf{0.0761}\ \boldsymbol{\Omega}$

(ii) As Voltmeter. Desired full-scale reading, $V = 15$ volts

Let R_s ohms be the required series resistance [See Fig. 16.18].

Now, $V = I_m (R_s + R_m)$

or $R_s = \dfrac{V}{I_m} - R_m = \dfrac{15}{0.015} - 5 = \textbf{995}\ \boldsymbol{\Omega}$

Example 16.9. *A moving coil milliammeter with a resistance of 1·6 Ω is connected with a shunt of 0·228 Ω. What will be the current flowing through the instrument if it is connected in a circuit in which a current of 200 mA is flowing ?*

Solution. Multiplying power of shunt $= \dfrac{I}{I_m} = \dfrac{R_m + S}{S} = \dfrac{1.6 + 0.228}{0.228} = 8$

∴ $I_m = \dfrac{I}{8} = \dfrac{200}{8} = \textbf{25 mA}$

The shunt should have a very low temperature coefficient of resistance so that its multiplying power remains practically constant with temperature changes.

Example 16.10. *What should be the resistance of the moving coil of an ammeter which requires 2·5 mA for full-scale deflection so that it may be used with a shunt having a resistance of 0·0025 Ω for a range of 0 – 10 A ?*

Solution. Multiplying power of shunt $= \dfrac{I}{I_m} = \dfrac{10}{2.5 \times 10^{-3}} = 4000$

Now, $\qquad 4000 = \dfrac{R_m + S}{S} = \dfrac{R_m + 0.0025}{0.0025}$

∴ $\qquad\qquad R_m = \mathbf{10\ \Omega}$

As far as possible, the temperature coefficient of resistance of the shunt should be nearly equal to that of the instrument coil.

Example 16.11. *When a 250-volt moving coil voltmeter that has a resistance of 12 kΩ is used to measure an unknown voltage, the pointer just goes off scale. When a resistance of 2500 Ω is placed in series with this voltmeter, the instrument reads 242 volts. What is the unknown voltage ?*

Solution. The current through the meter and hence deflection of the pointer is inversely *proportional to the total meter resistance (i.e., the sum of series resistance and the coil resistance). Let the unknown voltage be V volts.

$$V \propto \frac{1}{12000} \quad \text{and} \quad 242 \propto \frac{1}{12000 + 2500}$$

∴ $\qquad \dfrac{V}{242} = \dfrac{12000 + 2500}{12000} \quad \text{or} \quad V = \dfrac{14500}{12000} \times 242 = \mathbf{292.42\ volts}$

Example 16.12. *A meter movement has a resistance of 1000 ohms and a full-scale deflection current of 50 μA. What are the values of series multipliers required to give the movement ranges of 3 volts, 30 volts and 300 volts ?*

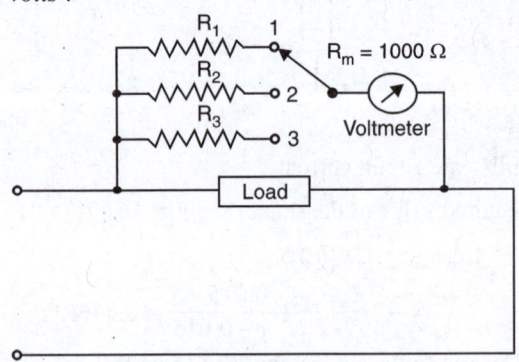

Fig. 16.19

Solution. Let R_1, R_2 and R_3 be the required series multipliers to give the meter movement ranges of 3 volts, 30 volts and 300 volts respectively (See Fig. 16.19).

Meter resistance, $R_m = 1000\ \Omega$

Full-scale meter current, $I_m = 50\ \mu A = 50 \times 10^{-6}\ A$

For 3 volts range, $R_1 = \dfrac{V}{I_m} - R_m = \dfrac{3}{50 \times 10^{-6}} - 1000 = \mathbf{59 \times 10^3\ \Omega}$

For 30 volts range, $R_2 = \dfrac{V}{I_m} - R_m = \dfrac{30}{50 \times 10^{-6}} - 1000 = \mathbf{599 \times 10^3\ \Omega}$

---

* $\qquad$ Meter current $= \dfrac{\text{Voltage}}{\text{Total meter resistance}}$

Note that voltage reading is directly proportional to meter current and meter current is inversely proportional to meter circuit resistance.

For 300 volts range, $R_3 = \dfrac{V}{I_m} - R_m = \dfrac{300}{50 \times 10^{-6}} - 1000 = \mathbf{5.99 \times 10^6 \, \Omega}$

Example 16.13. *The moving coil of a permanent magnet type voltmeter has a resistance of 4000 Ω at 20 °C at which it reads correctly when connected to a d.c. supply of 200 volts. If the coil is wound with a copper wire whose temperature co-efficient is 0·004 at 20 °C, find the percentage error in the reading when the temperature of the moving coil increases to 60 °C.*

Solution. Resistance of the coil at 20°C, $R_{20} = 400 \, \Omega$

$$\text{Resistance of the coil at 60°C, } R_{60} = R_{20} [1 + \alpha_{20} (60 - 20)]$$
$$= 4000[1 + 0.004 \times 40] = 4640 \, \Omega$$

$$\text{Current through the coil at 20°C} = 200/4000 = 0.05 \text{ A}$$
$$\text{Current through the coil at 60°C} = 200/4640 = 0.043 \text{ A}$$

At 60 °C, the current through the coil is reduced and hence the voltmeter will give low reading. With 0.05 A through the coil, the voltmeter indicates 200 V, therefore, 0.043 A will indicate

$$\dfrac{200 \times 0.043}{0.05} = 172 \text{ volts}$$

∴ $\qquad$ % error $= -28 \times \dfrac{100}{200} = \mathbf{-14\%}$ *of correct value*

i.e., the voltmeter indicates 14% low. This example shows that the coil of moving coil instrument should be made from materials having small temperature coefficients of resistance so that the accuracy of the meter does not change dramatically due to temperature variations.

Example 16.14. *An unknown voltage is measured with a 100-volt voltmeter having a rating of 1000 ohms per volt. This meter indicates a reading of 45 volts. When a 100-volts, 5000 ohms-per volt meter is used, the indication is 75 volts. Find (i) the true unknown voltage (ii) resistance of the source.*

Solution.

Total resistance of first voltmeter $= 1000 \times 100 = 10^5 \, \Omega$

Full-scale current of first voltmeter $= 100/10^5 = 10^{-3} \text{ A} = 1 \text{ mA}$

Total resistance of second voltmeter $= 5000 \times 100 = 5 \times 10^5 \, \Omega$

Full-scale current of second voltmeter $= 100/5 \times 10^5 = 0.2 \times 10^{-3} \text{ A} = 0.2 \text{ mA}$

Let V volts be the unknown voltage and R ohms be the resistance of the source.

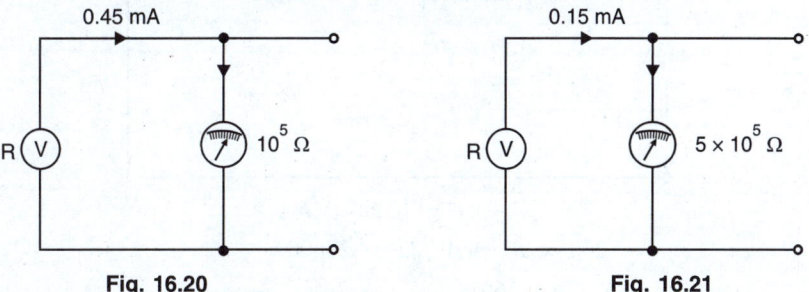

Fig. 16.20 $\qquad\qquad\qquad\qquad$ Fig. 16.21

(*i*) Since the first meter reads 45 V, the current through it will be 0.45* mA as shown in Fig. 16.20. Again the second meter reads 75 V and therefore current through it will be **0.15 mA as shown in Fig. 16.21.

---

* $\quad$ Note that for full-scale deflection (*i.e.,* 100 V), the meter current is 1 mA. For 45 V reading, the meter current will be $= (1/100) \times 45 = 0.45$ mA.

** $\quad$ For full-scale deflection (*i.e.,* 100 V), the meter current is 0.2 mA. For 75 V reading, the meter current will be $= (0.2/100) \times 75 = 0.15$ mA.

From Fig. 16.20, $R = \dfrac{V - 45}{0.45}$; From Fig. 16.21, $R = \dfrac{V - 75}{0.15}$

$\therefore$ $\dfrac{V - 45}{0.45} = \dfrac{V - 75}{0.15}$ or $V = \textbf{90 volts}$

(*ii*)   Source resistance, $R = \dfrac{V - 45}{0.45 \text{ mA}} = \dfrac{(90 - 45) \text{ volts}}{0.45 \text{ mA}} = \textbf{100 k}\Omega$

Example 16.15. *While determining the resistance R of a conductor, a student by mistake connects ammeter in parallel with R and voltmeter in series as shown in Fig. 16.22. What are the readings of ammeter and voltmeter ? The resistance of the voltmeter is 2000 Ω and the resistance of the ammeter is negligible.*

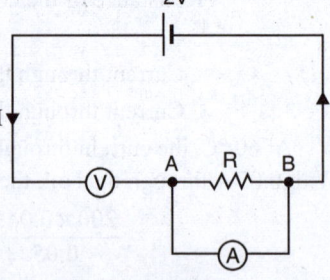

Solution. Refer to Fig. 16.22. The resistance R_{AB} is negligible because ammeter has negligible resistance. This means that circuit resistance is equal to the resistance of the voltmeter (*i.e.,* 2000 Ω).

$\therefore$ Circuit current, $I = \dfrac{2 \text{ V}}{2000 \ \Omega} = \textbf{1 mA}$

Fig. 16.22

Hence, the reading of the ammeter will be 1 mA.

P.D. across voltmeter = $I \times$ Resistance of voltmeter = 1 mA $\times$ 2000 Ω = **2 V**

Hence the voltmeter will read 2 V.

Example 16.16. *A permanent-magnet moving coil voltmeter has a resistance of 28600 Ω. When connected in series with an external resistance across a 480 V d.c. supply, the instrument reads 220 V. What is the value of the external resistance ?*

Solution. Since the voltmeter reads 220 V, the voltage across the external resistance *R* is = 480 – 220 V = 260 V [See Fig. 16.23]. As the current in the circuit is the same, therefore, voltage drops across *R* and the voltmeter are proportional to their resistances *i.e.,*

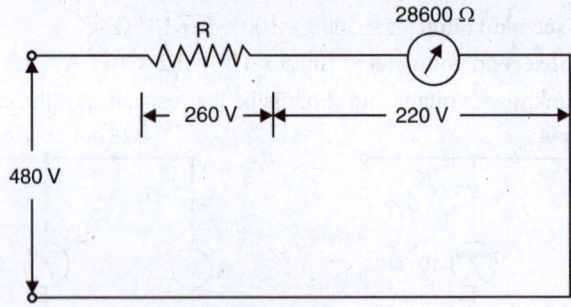

Fig. 16.23

$260 \propto R$ and $220 \propto 28600$

$\therefore$ $\dfrac{R}{28600} = \dfrac{260}{220}$ or $R = \dfrac{260}{220} \times 28{,}600 = \textbf{33800 }\Omega$

Example 16.17. *A 35 V d.c. supply is connected across a resistance of 600 Ω in series with an unknown resistance R. A voltmeter having a resistance of 1200 Ω is connected across 600 Ω and shows a reading of 5 V. Calculate the value of unknown resistance R.*

Solution. Fig. 16.24 shows the conditions of the problem. The reading of the voltmeter is equal to p.d. across points *A* and *B*. The total resistance of parallel resistor is

$$R_P = \frac{1200 \times 600}{1200 + 600}$$

$$= 400 \; \Omega$$

∴   Circuit current, $I = \dfrac{V_{AB}}{R_P} = \dfrac{5}{400}$

$$= 0.125 \; A$$

P.D. across $R = 35 - 5 = 30 \; V$

∴   Value of $R = 30/I = 30/0.0125 = \textbf{2400} \; \boldsymbol{\Omega}$

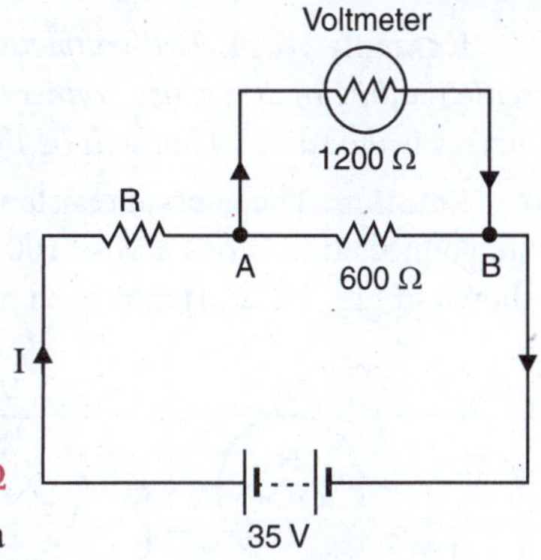

Whenever an ammeter or a voltmeter is connected in a circuit, the internal resistance of the meter changes the circuit resistance and hence the circuit current. This is called *loading effect* of the meter.

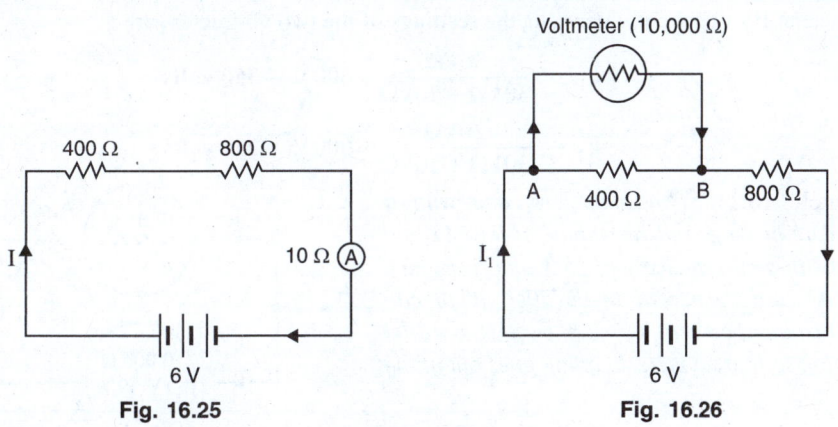

Fig. 16.24

Example 16.18. *Two resistors of 400 Ω and 800 Ω are connected in series with a battery of 6V of negligible resistance. It is desired to measure current in the circuit by means of an ammeter of resistance of 10 Ω. What is the reading of the ammeter ? If a voltmeter of resistance 10,000 Ω is connected across 400 Ω resistor, what will be the reading of the voltmeter ?*

Solution. Reading of ammeter. Fig. 16.25 shows the conditions of the problem. The reading of the ammeter will be equal to the magnitude of circuit current.

Fig. 16.25   **Fig. 16.26**

Total circuit resistance, $R_T = 400 + 800 + 10 = 1210 \; \Omega$

∴   Circuit current, $I = V/R_T = 6/1210 = \textbf{0.00496 A}$

Therefore, the ammeter will read 0.00496 A.

Reading of voltmeter. Fig. 16.26 shows the conditions of the problem. The reading of the voltmeter will be equal to p.d. between points A and B. Total resistance of parallel resistors is

$$R_P = \frac{10,000 \times 400}{10,000 + 400} = 384.6 \; \Omega$$

∴   Total circuit resistance $= R_P + 800 = 384.6 + 800 = 1184.6 \; \Omega$

∴   Circuit current, $I_1 = 6/1184.6 = 0.0051 \; A$

∴   P.D. between A and B, $V_{AB} = I_1 R_P = 0.0051 \times 384.6 = \textbf{1.96 V}$

Hence the voltmeter will read 1.96 V.

Example 16.19. *Two voltmeters, one with a full-scale reading of 100 V and another with a full-scale reading of 200 V, are connected in series across a 100 V supply. The internal resistance of both meters is the same. What will be the readings of the voltmeters ?*

Solution. The internal resistance of both the meters is the same. Therefore, when the two meters are connected in series across 100 V supply, the voltage drop across each meter = 100/2 = 50 V as shown in Fig. 16.27. Hence each meter will read **50 V.**

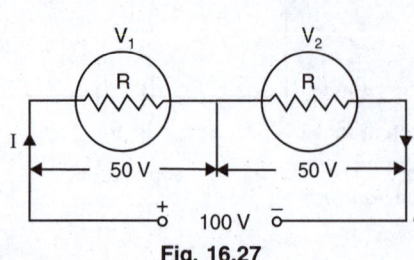

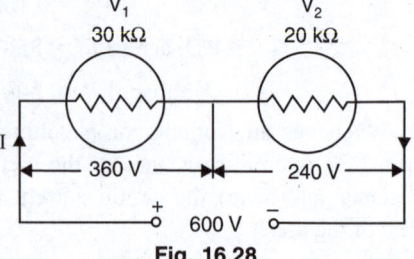

Fig. 16.27   Fig. 16.28

Example 16.20. *Two voltmeters have the same range 0-400 V. The internal impedances are 30 kΩ and 20 kΩ. If they are connected in series and 600 V be applied across them, what will be their readings ?*

Solution. Fig. 16.28 shows the conditions of the problem. It is a series resistor circuit and acts as a *voltage divider circuit*. It is because the voltage across any voltmeter is a fraction of the total voltage applied across the series combination ; the fraction being determined by the resistances of the voltmeters. By voltage divider-rule, the readings of the two voltmeters are :

$$V_1 = \frac{30 k\Omega}{30 k\Omega + 20 k\Omega} \times 600 V = \textbf{360 volts}$$

$$V_2 = \frac{20 k\Omega}{30 k\Omega + 20 k\Omega} \times 600 V = \textbf{240 volts}$$

Example 16.21. *Two ammeters, one with a current scale of 10 A and resistance of 0.01 Ω and the other with a current scale of 15 A and resistance of 0.005 Ω, are connected in parallel. What can be the maximum current carried by this parallel combination so that no meter reading goes out of the scale ?*

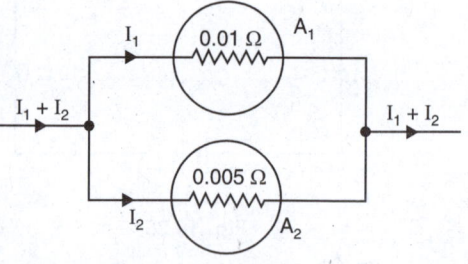

Fig. 16.29

Solution. Fig. 16.29 shows the two ammeters (A_1 and A_2) connected in parallel. Suppose the *applied voltage is such that meter A_2 carries 15 A. Then according to current-division rule, meter A_1 will carry a current of 7.5 A ($\because$ resistance of A_1 is twice that of A_2). Therefore, I_1 = 7.5 A and I_2 = 15 A so that maximum current carried by the combination = $I_1 + I_2$ = 7.5 A + 15 A = **22.5 A.**

Example 16.22. *A moving coil ammeter has a fixed shunt of 0.02 Ω with a coil circuit resistance of R = 1 kΩ and needs p.d. of 0.5 V across it for full-scale deflection.*

(i) To what total current does this correspond?

(ii) Calculate the value of shunt to give full-scale deflection when the total current is 10 A and 75 A.

Solution. (i) Voltage across shunt = voltage across meter coil = 0.5 V

$$\text{Shunt current, } I_s = \frac{0.5}{S} = \frac{0.5}{0.02} = 25 A$$

---

*   We want maximum current to be carried by this parallel combination. Therefore, applied voltage should be such that A_2 carries 15A.

$$\text{Meter current, } I_m = \frac{0.5}{R_m} = \frac{0.5}{1 \times 10^3} = 0.0005 \text{ A}$$

$$\therefore \qquad \text{Line current, } I = I_s + I_m = 25 + 0.0005 = \textbf{25.0005 A}$$

(ii) $\qquad$ Value of shunt, $S = \dfrac{I_m R_m}{I - I_m}$

Here, $I_m = 0.0005$ A ; $R_m = 1$ k$\Omega = 1000 \ \Omega$; $I = $ Full-scale circuit current

$$\text{When } I = 10 \ A, \quad S = \frac{0.0005 \times 1000}{10 - 0.0005} = \textbf{0.05 } \boldsymbol{\Omega}$$

$$\text{When } I = 75 \ A, \quad S = \frac{0.0005 \times 1000}{75 - 0.0005} = \textbf{0.00667 } \boldsymbol{\Omega}$$

Tutorial Problems

1. A permanent-magnet moving-coil instrument gives full-scale deflection with 5 mA and has a resistance of 5 Ω. Calculate the resistance of the necessary components in order that the instrument may be used as *(i)* a 2 A ammeter *(ii)* a 100-V voltmeter. $\qquad$ **[(*i*) 0.0378 Ω in parallel (*ii*) 6662 Ω in series]**

2. The coil of a permanent-magnet moving coil instrument has a resistance of 25 Ω and gives full-scale deflection when a current of 2 mA passes through it. Calculate the value of shunt required to convert the instrument to an ammeter having a full-scale deflection current of 1.5 A. $\qquad$ **[0.0334 Ω]**

3. The coil of a permanent-magnet moving coil voltmeter has a resistance of 5000 Ω at 15°C at which it reads correctly when connected to a supply of 200 V. If the coil is wound with a wire whose temperature co-efficient at 15°C is 0.004, find the percentage error in the reading when the temperature is 40°C.

$\qquad\qquad\qquad\qquad\qquad\qquad\qquad\qquad\qquad\qquad\qquad\qquad$ **[9.8% low]**

4. A moving coil galvanometer of 10 Ω resistance has a 50 division scale and indicates one microampere per division. Show how it can be used as a milliammeter of range 50 mA and as a voltmeter of range of 5 volts. $\qquad$ **[$S = 0.01001 \ \Omega$; $R_s = 99{,}990 \ \Omega$]**

5. A PMMC instrument gives a reading of 25 mA when the p.d. across its terminals is 75 mV. Calculate shunt resistance for full-scale deflection corresponding to 50 A. $\qquad$ **[3/1999 Ω]**

6. A moving coil instrument has a resistance of 5 Ω and gives a full-scale deflection of 100 mV. Show how this instrument may be used to measure *(i)* voltages upto 50 V *(ii)* currents upto 10 A.

$\qquad\qquad\qquad\qquad\qquad\qquad\qquad$ **[(*i*) $R_s = $ 2495 Ω (*ii*) $S = $ 5/499 Ω]**

7. A permanent-magnet moving coil instrument gives full-scale reading of 25 mA when p.d. across its terminals is 75 mV. Show how it can be used *(i)* as an ammeter for a range of 0–100 A *(ii)* as a voltmeter for a range of 0–750 V. Also find the multiplying power of the shunt and voltage amplification.

$\qquad\qquad\qquad\qquad\qquad$ **[(*i*) $S = $ 0.00075 Ω (*ii*) $R_s = $ 29997 Ω; 4000; 10,000]**

8. A basic D'Arsonval movement with internal resistance $R_m = 100 \ \Omega$ and full scale deflection current $I_f = 1$ mA is to be converted into a multirange d.c. voltmeter with voltage ranges of 0–10 V, 0–50 V, 0–250 V and 0–500 V. Find the values of suitable multipliers. $\qquad$ **[9 kΩ; 40 kΩ; 200 kΩ ; 250 kΩ]**

16.14. Dynamometer Type Instruments (Ammeters and Voltmeters)

These instruments are the modified form of permanent-magnet type. Here magnetic field is not produced by a permanent magnet but by two air-cored fixed coils placed on either side of the moving coil. Such instruments can be used as ammeters or as voltmeters but are generally used as wattmeters. They are suitable for d.c. as well as a.c. work.

Principle. *These instruments are based on the principle that mechanical force exists between the current carrying conductors.*

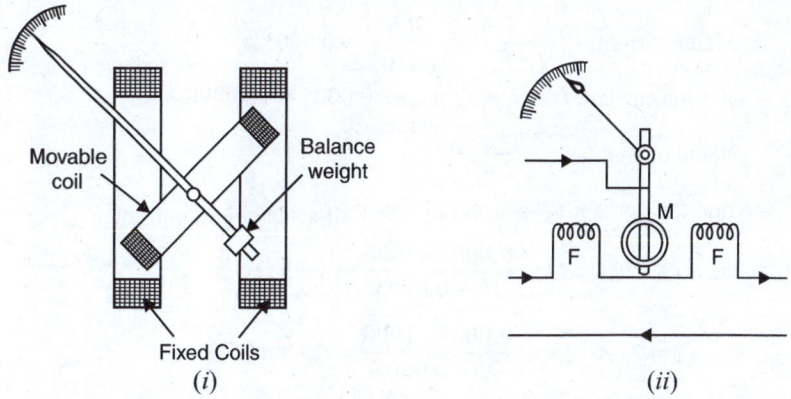

Fig. 16.30

Construction. It essentially consists of a fixed coil and a moving coil. The fixed coil is split into two equal parts (*F, F*) which are placed close together and parallel to each other. The moving coil (*M*) is pivoted inbetween the two fixed coils and carries a pointer as shown in Fig. 16.30. The current is led into and out of the moving coil by means of two spiral hair-springs which also provide the controlling torque. *Air friction damping is provided by means of the aluminium vanes that move in the sector shaped champer at the bottom of the instrument.

Working. For use as an ammeter or voltmeter, the fixed coils *FF* and the moving coil *M* are so connected that the **same current flows through the two coils. Due to these currents, mechanical force exists between the coils. The result is that moving coil *M* moves the pointer over the scale. The pointer comes to rest at a position where deflecting torque is equal to the controlling torque. Since the polarity of the fields produced by both fixed and moving coils is reversed by the reversal of current, the deflection of the moving system is always in the same direction regardless of the direction of current through the coils. For this reason, dynamometer instruments can be used for both d.c. and a.c. measurements.

Deflecting torque. The force of attraction or repulsion between the fixed and moving coils is directly proportional to the product of ampere-turns of fixed coils and the moving coil *i.e.*,

$$\text{Deflecting torque, } T_d \propto N_f I_f \times N_m I_m$$

Since N_m and N_f are constant, $T_d \propto I_f I_m$

Since the instrument is spring-controlled, the controlling torque is proportional to the angular deflection θ *i.e.*,

$$T_C \propto \theta$$

In the steady position of deflection, $T_d = T_C$.

∴ $$\theta \propto I_f I_m$$

Thus deflection (θ) is directly proportional to the product of currents in the fixed coils and the moving coil.

---

*   Since the coils are air-cored, the operating magnetic field is very weak. For this reason, eddy current damping cannot be provided.

**   For measuring large current, the moving coil *M* of the ammeter is shunted so that a portion of the main current flows through the moving coil. However, the fixed coils carry the circuit current.

(*i*) **As ammeter.** When the instrument is used as an ammeter, the fixed coils and the moving coil are connected in series so that the same current flows through the two coils as shown in Fig. 16.31. In that case, $I_f = I_m = I$ so that :

$$\theta \propto I^2$$

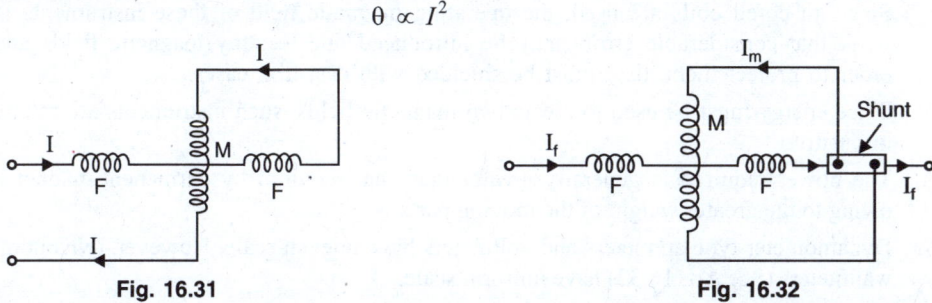

| Fig. 16.31 | Fig. 16.32 |

For measuring large currents, moving coil is shunted; the shunt being in series with the fixed coils as shown in Fig. 16.32. The fixed coils carry the main current while the moving coil carries a current proportional to the main current.

(*ii*) **As voltmeter.** When the instrument is used as a voltmeter, both fixed coils and the moving coil are connected in series together with a high resistance R_s (called multiplier) having a negligible temperature co-efficient as shown in Fig. 16.33. Therefore, current in both the coils is the same and is proportional to the voltage V being measured.

$$\theta \propto V^2$$

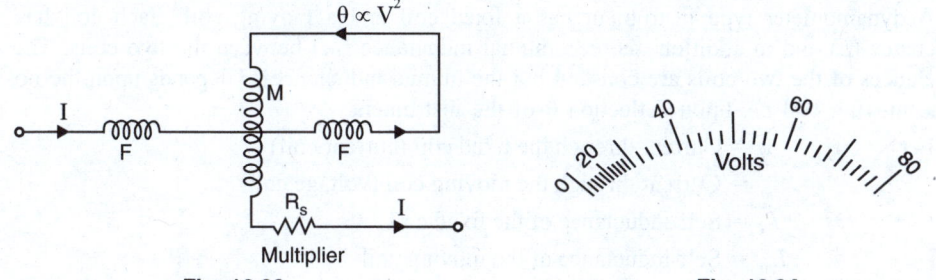

| Fig. 16.33 | Fig. 16.34 |

It may be seen that whether the instrument is used as an ammeter or voltmeter, the deflection is directly proportional to the square of quantity (current or voltage) being measured. Doubling the current will make the deflection four times as large. Hence, the scale of dynamometer type instruments is *not uniform; being crowded at the beginning and open at the upper end of the scale as shown in Fig. 16.34. The obvious disadvantage of such a scale is that the divisions near the start of the scale are small and cannot be read accurately.

Sources of errors. The errors that generally occur in dynamometer type instruments are (*i*) frictional error (*ii*) temperature error (*iii*) error due to stray magnetic fields.

Since the coils are air-cored, the operating magnetic field produced is small. For producing an appreciable deflecting torque, a large number of turns is necessary for the moving coil. This leads to a heavy moving system resulting in considerable frictional losses. Since the operation of dynamometer type instruments requires considerable power, a large amount of heat is produced. This increases the resistance of the coils resulting in error due to temperature. Since the coils are air-cored, the magnetic flux density hardly exceeds 6×10^{-3} Wb/m^2. With such a weak operating magnetic field, the stray magnetic fields may affect the reading of the instrument.

Advantages

(*i*) These instruments can be used for both d.c. and a.c. measurements.

---

* For ammeters and voltmeters only. However, dynamometer wattmeters have uniform scale (See Art. 16.37).

(ii) Since air cored coils are used, they are generally free from hysteresis and eddy current errors when used on a.c. circuits.

Disadvantages

(i) Since air-cored coils are used, the operating magnetic field of these instruments is so weak that considerable errors may be introduced due to stray magnetic fields and in order to protect them, they must be shielded with cast-iron cases.

(ii) Since energy must be used to create two magnetic fields, such instruments are relatively insensitive.

(iii) The power required is generally greater than that required by permanent-magnet type owing to the greater weight of the moving parts.

(iv) Dynamometer type ammeters and voltmeters have uneven scale. However, dynamometer wattmeters (See Art. 16.37) have uniform scale.

(v) They are more expensive than the permanent-magnet type instruments.

Applications. It is clear from above that dynamometer ammeters and voltmeters are inferior to the permanent-magnet moving coil instruments and are seldom used for d.c. measurements. Their chief sphere of application is in a.c. measurements and particularly when the same instrument is required to read both direct and alternating currents as in the a.c. potentiometer.

16.15. Deflecting Torque (T_d) of Dynamometer Type Instruments in Terms of Mutual Inductance

A dynamometer type instrument has a fixed coil and a moving coil. Each coil has self-inductance (L) and in addition, there is mutual inductance (M) between the two coils. The self-inductances of the two coils are constant but the mutual inductance M depends upon the position of the moving coil *i.e.* upon deflection θ of the instrument.

Let $\quad\quad i_f$ = Current through the fixed coil (current coil)

$\quad\quad\quad i_m$ = Current through the moving coil (voltage coil)

$\quad\quad\quad L_f$ = Self-inductance of the fixed coil

$\quad\quad\quad L_m$ = Self-inductance of the moving coil

$\quad\quad\quad M$ = Mutual inductance between the two coils

$\therefore\quad$ Magnetic energy stored at this instant is

$$w = \frac{1}{2}i_f^2 L_f + \frac{1}{2}i_m^2 L_m + i_f i_m M$$

Since the mutual inductance between the two coils depends upon deflection θ, the instantaneous torque T_d' is given by ;

$$T_d' = \frac{dw}{d\theta} = \frac{1}{2}i_f^2 \frac{dL_f}{d\theta} + \frac{1}{2}i_m^2 \frac{dL_m}{d\theta} + i_f i_m \frac{dM}{d\theta}$$

Since L_f and L_m are constant, $\dfrac{dL_f}{d\theta} = 0$ and $\dfrac{dL_m}{d\theta} = 0$.

$$\therefore\quad\quad\quad\quad T_d' = i_f i_m \frac{dM}{d\theta}$$

Average deflecting torque, T_d = Average of T_d' over one cycle

$$= \frac{1}{T}\int_0^T i_f i_m \frac{dM}{d\theta} dt = \frac{dM}{d\theta}\frac{1}{T}\int_0^T i_f i_m \, dt$$

$$\therefore\quad\quad\quad\quad T_d = i_f i_m \frac{dM}{d\theta}$$

For direct current. $T_d = I_f I_m \dfrac{dM}{d\theta}$

where I_f and I_m are direct currents in the fixed coil and the moving coil respectively.

For alternating current. The instantaneous torque T'_d is

$$T'_d = i_f i_m \frac{dM}{d\theta}$$

If $i_f = I_{f\,max} \sin \omega t$ and $i_m = I_{m\,max} \sin(\omega t - \phi)$, then,

$$T'_d = I_{f\,max} \cdot I_{m\,max} \sin \omega t \sin(\omega t - \phi) \frac{dM}{d\theta}$$

∴   Average torque, T_d = Average of T'_d over one cycle

$$= \frac{1}{T}\int_0^T I_{f\,max}\, I_{m\,max} \sin \omega t \sin(\omega t - \phi) \frac{dM}{d\theta}\, dt$$

$$= \frac{dM}{d\theta}\frac{1}{T}\int_0^T I_{f\,max} I_{m\,max} \sin \omega t \sin(\omega t - \phi)\, dt$$

$$= \frac{I_{f\,max} I_{m\,max}}{2} \cos\phi\, \frac{dM}{d\theta}$$

∴ $$T_d = I_f I_m \cos\phi \frac{dM}{d\theta}$$

where I_f and I_m are the r.m.s. values of alternating currents in fixed and moving coils respectively.

Example 16.23. *The mutual inductance of a 25 A electrodynamic ammeter changes uniformly at a rate of 0.2 µH/degree. The torsion constant of the controlling spring is 10^{-6} Nm per degree. Determine the angular deflection for full-scale.*

Solution. Spring torsion constant, $k = 10^{-6}$ Nm/degree

Full-scale meter current, $I = 25$ A

If θ is the full-scale deflection in degrees, then,

Full-scale deflecting torque, $T_d = k\theta = 10^{-6}\,\theta$ Nm   ...(i)

Now, $$T_d = I_C I_V \frac{dM}{d\theta}$$

When the instrument is used as an ammeter, the same current passes through both fixed and moving coils *i.e.* $I_C = I_V = I$.

∴ $$T_d = I^2 \frac{dM}{d\theta}$$

Here $I = 25$ A ; $\dfrac{dM}{d\theta} = 0.2$ µH/degree $= 0.2 \times 10^{-6}$ H/degree

∴ $$T_d = 25^2 \times 0.20 \times 10^{-6} \qquad\qquad\qquad ...(ii)$$

From eqs. (i) and (ii), $10^{-6} \times \theta = 25^2 \times 0.20 \times 10^{-6}$ ∴ $\theta = \mathbf{125°}$

Example 16.24. *The spring constant of a 10 A dynamometer wattmeter is 10.5×10^{-6} Nm/radian. The variation of inductance with angular position of moving system is practically linear over the operating range, the rate of change being 0.078 m H/rad. If the full scale deflection of the instrument is 83°, calculate the current required in the voltage coil at full scale on d.c. circuit.*

Solution. Spring constant, $k = 10.5 \times 10^{-6}$ Nm/rad $= \dfrac{10.5 \times 10^{-6} \times \pi}{180} = 0.18326 \times 10^{-6}$ Nm/degree

$$\text{Controlling torque, } T_C = k \times \text{deflection} = 0.18326 \times 10^{-6} \times 83 = 15.21 \times 10^{-6} \text{ Nm}$$

$$\text{Deflecting torque, } T_d = I_C I_V \frac{dM}{d\theta} = 10 \times I_V \times 0.078 \times 10^{-3}$$

For steady state deflection, $T_d = T_C$.

$$\therefore \qquad 10 \times I_V \times 0.078 \times 10^{-3} = 15.21 \times 10^{-6}$$

$$\therefore \quad \text{Current in voltage coil, } I_V = \frac{15.21 \times 10^{-6}}{10 \times 0.078 \times 10^{-3}} = 0.0195 \text{ A} = \mathbf{19.5 \text{ mA}}$$

16.16. Range Extension of Dynamometer Type Instruments

The chief application of dynamometer type instruments is as wattmeters. However, they can be used as ammeters and voltmeters.

(*i*) **Ammeter.** The usual range of an unshunted dynamometer ammeter is 0–50 mA because the moving coil of the instrument cannot carry current more than 50 mA. However, the range can be increased by connecting a suitable shunt across the moving coil of the instrument. In an a.c. ammeter, shunts are used only for a range upto 0–5 A as the division of *current depends upon the inductance and resistance of the instrument. For the measurement of higher alternating currents, a transformer is used in conjunction with 0–5 A a.c. ammeter. Suppose we wish to measure a very large alternating current. By means of a transformer (known as **current transformer**), we step this current down to say 50 times. Assume that, as the output of the transformer is measured by the a.c. ammeter, a reading of 4A is indicated. The original or line current is then = 4 × 50 = 200 A. By putting current transformer of different ratios, different current ranges of the instrument can be obtained.

(*ii*) **Voltmeter.** The principal use of dynamometer voltmeter is for the measurement of alternating voltages. The range of the a.c. voltmeter can be increased from 0–750 V by the use of multipliers *i.e.,* by connecting a suitable high resistance R_s in series with the instrument. For ranges higher than 0–750 V, where the power wasted in the multiplier would be excessive, a transformer is used in conjunction with a 0–110 V a.c. voltmeter. Suppose we wish to measure a very large alternating voltage. By means of a transformer (known as **potential transformer**), we step this voltage down to say 20 times. Assume that, as the output of the transformer is measured by the 0–110 V a.c. voltmeter, a reading of 100 V is indicated. The original voltage is then = 20 × 100 = 2000 V. By putting voltage transformer of different ratios, different voltage ranges of the instrument can be obtained.

Example 16.25. *In a high-voltage a.c. circuit, current through the load and voltage across the load are measured by means of current and voltage transformers. The current transformer ratio is 10 : 1 and the potential transformer ratio is 20 : 1. If current and voltage indicated on the meters are 3A and 100 V respectively, find the load current and load voltage.*

---

*   For measuring a.c. currents larger than 5 A, shunts are not practicable with a dynamometer ammeter. It is because the division of current between the shunt and the coils varies with frequency (since reactance of coils depends upon frequency). Therefore, the instrument will be accurate only at the frequency at which it is calibrated.

**   The primary of this transformer is in series with the load and carries the current to be measured. The a.c. ammeter is connected across the secondary of the transformer.

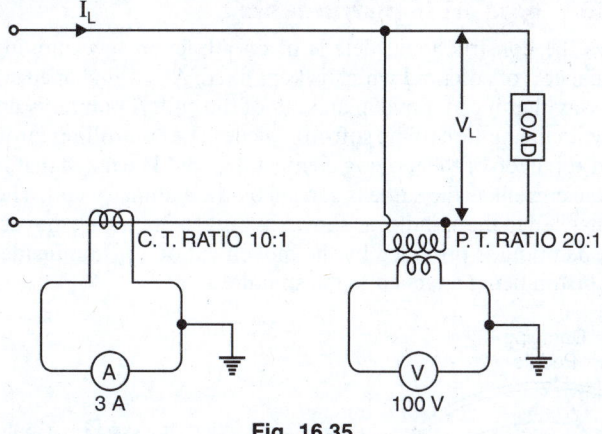

Fig. 16.35

Solution. Fig. 16.35 shows the connections for measuring current and voltage using current and potential transformers in conjunction with the instruments.

Load Current. The purpose of current transformer is to provide a means of reducing the line current to values that may be used to operate standard low-current a.c. ammeter. An a.c. ammeter (usually with a range of 0–5 A) is connected across the secondary winding of the *current transformer whose primary is in series with the load as shown in Fig. 16.35. Since in the given problem the current transformer ratio is 10 : 1, it means the line (or load) current is equal to 10 times the reading on the a.c. ammeter.

$$\therefore \qquad \text{Load current, } I_L \; = \; 3 \times 10 = \textbf{30 A}$$

Load Voltage. The purpose of potential transformer is to reduce the line voltage to values that may be used to operate low range a.c. voltmeter. The a.c. voltmeter (usually with a range of 0–110 V) is connected across the secondary of the transformer whose primary is connected across the load as shown in Fig. 16.35. Since in the given problem the potential transformer ratio is 20 : 1, the load voltage is equal to 20 times the reading on the a.c. voltmeter.

$$\therefore \qquad \text{Load voltage, } V_L \; = \; 100 \times 20 = \textbf{2000 V}$$

Note that both secondaries of the instrument transformers are grounded as a safety measure.

16.17. Moving-Iron (M.I.) Ammeters and Voltmeters

This type of instrument is principally used for the measurement of alternating currents and voltages, though it can also be used for d.c. measurements. There are two types of moving-iron instruments.

(*i*) **Attraction type** in which a single soft-iron vane (or moving iron) is mounted on the spindle and is attracted towards the coil when operating current flows through it.

(*ii*) **Repulsion type** in which two soft-iron vanes are used; one fixed and attached to the stationary coil while the other is movable (*i.e.* moving iron) and mounted on the spindle of the instrument. When operating current flows through the coil, the two vanes are magnetised, developing similar polarity at the same ends. Consequently, repulsion takes place between the vanes and the movable vane causes the pointer to move over the scale.

In case the instrument is to be used as an ammeter, the coil has a fewer turns of thick wire so that the ammeter has low resistance–a desirable requirement. In case it is to be used as a voltmeter, the coil has a large number of turns of fine wire so that the voltmeter has high resistance–a desirable requirement.

---

* Since a current transformer is used to reduce the current, the secondary contains more turns than the primary. Usually, the primary winding of the transformer consists of a single turn or at the most a few turns so that it may be able to carry the line current.

16.18. Attraction Type M.I. Instruments

Fig. 16.36 shows the constructional details of an attraction type moving-iron instrument. It consists of a cylindrical coil or solenoid which is kept fixed. An oval-shaped soft-iron is attached to the spindle in such a way that it can move in and out of the coil. A pointer is attached to the spindle so that it is deflected with the motion of the soft-iron piece. The controlling torque is provided by one spiral spring arranged at the top of the moving element. It should be noted that in this instrument, the springs do not carry the current as the same is carried by the stationary coil. The *damping device is an aluminium vane attached to the spindle, as shown in Fig. 16.36, which moves in a closed chamber. In some instruments, damping is provided by the movement of a piston inside the curved chamber [See Fig. 16.37]; the piston being attached to the spindle.

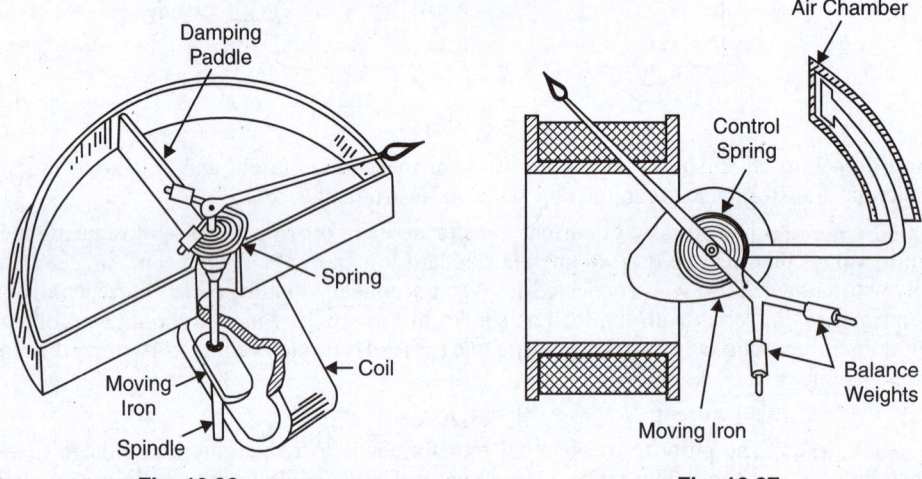

| Fig. 16.36 | Fig. 16.37 |

Working. When the instrument is connected in the circuit to measure current or voltage, the operating current flowing through the coil sets up a magnetic field. In other words, the coil behaves like a magnet and therefore it attracts the soft-iron piece towards it. The result is that the pointer attached to the moving system moves from zero position. The pointer will come to rest at a position where deflecting torque is equal to the controlling torque. If current in the coil is reversed, the direction of magnetic field also reverses and so does the magnetism produced in the soft-iron piece. Hence, the direction of the deflecting torque remains unchanged. For this reason, such instruments can be used for both d.c. and a.c. measurements.

Deflecting torque. The force F pulling the soft-iron piece towards the coil is directly proportional to :

   (*i*) magnetic field strength H produced by the coil

   (*ii*) pole strength m developed in the iron piece

$\therefore$ $F \propto mH$

                            $\propto H^2$ $(\because m \propto H)$

$\therefore$ Instantaneous deflecting torque $\propto H^2$

     If the permeability of iron is assumed constant, then,

                     $H \propto i$ where i is the instantaneous coil current.

$\therefore$ Instantaneous deflecting torque $\propto i^2$

     Average deflecting torque, $T_d \propto$ mean of i^2 over a cycle

   Since the instrument is spring controlled,

---

*   Eddy current damping is not provided in moving-iron instruments because the presence of the permanent magnet required for the purpose would affect the field due to the coil. This will, in turn, affect the reading of the instrument.

$$T_C \propto \theta$$

In the steady position of deflection, $T_d = T_C$.

∴ $\theta \propto$ mean of i^2 over a cycle

$$\propto I^2 \qquad \text{...for } d.c.$$

$$\propto I^2_{r.m.s.} \qquad \text{...for } a.c.$$

Since the deflection is proportional to the *square of coil current, the scale of such instruments is non-uniform; being crowded in the beginning and spread out near the finish end of the scale.

16.19. Repulsion Type M.I. Instruments

Fig. 16.38 (*i*) shows the constructional details of a repulsion type moving iron instrument. It consists of two soft-iron pieces or vanes surrounded by a fixed cylindrical hollow coil which carries the operating current. One of these vanes is fixed and the other is free to move as shown in Fig. 16.38 (*ii*). The movable vane is of cylindrical shape and is mounted axially on a spindle to which a pointer is attached. The fixed vane, which is wedge-shaped and has a larger radius, is attached to the stationary coil. The controlling torque is provided by one spiral spring at the top of the instrument. It may be noted that in this instrument, springs do not provide the electrical connections. Damping is provided by air friction due to the motion of a piston in an air chamber.

Working. When current to be measured or current proportional to the voltage to be measured flows through the coil, a magnetic field is set up by the coil. This magnetic field magnetises the two vanes in the same direction *i.e.*, similar polarities are developed at the same ends of the vanes as shown in Fig. 16.38 (*iii*). Since the adjacent edges of the vanes are of the same polarity, the two vanes repel each other. As the fixed vane cannot move, the movable vane deflects and causes the pointer to move from zero position. The pointer will come to rest at a position where deflecting torque is equal to the controlling torque provided by the spring. If the current in the coil is reversed, the direction of deflecting torque remains unchanged. It is because reversal of the field of the coil reverses the magnetisation of both iron vanes so that they repel each other regardless of which way the current flows through the coil. For this reason, such instruments can be used for both d.c. and a.c. applications.

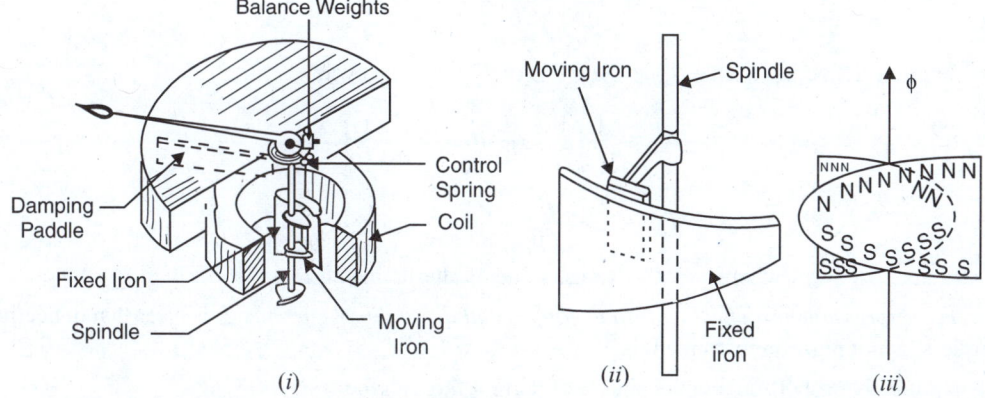

Fig. 16.38

Deflecting torque. The deflecting torque results due to the repulsion between the similarly charged soft-iron pieces or vanes. If the two pieces develope pole strengths of m_1 and m_2 respectively, then,

---

*   Such a scale is called a **squared scale**. The "expanded" or spread out scale is often useful when the values of current and voltage tend to fluctuate about particular values. For example, if a line voltage fluctuates between 210 and 220 V, it can be monitored more easily on a squared scale meter than on a linear scale if both meters have a full-scale deflection of 225 V.

Instantaneous deflecting torque $\propto m_1\, m_2 \propto {}^*H^2$

If the permeability of iron is assumed constant, then,

$$H \propto i \quad \text{where } i \text{ is coil current.}$$

Instantaneous deflecting torque $\propto i^2$

$\therefore$ Average deflecting torque, $T_d \propto$ mean of i^2 over a cycle

Since the instrument is spring-controlled,

$\therefore$ $T_C \propto \theta$

In the steady position of deflection, $T_d = T_C$.

$\therefore$ $\theta \propto$ mean of i^2 over a cycle

$$\propto I^2 \qquad\qquad\qquad \text{...for d.c.}$$

$$\propto I^2_{r.m.s.} \qquad\qquad \text{... for a.c.}$$

Thus, the deflection is proportional to the square of coil current as is the case with attraction type moving-iron instruments. Therefore, the scale of such instruments is also non-uniform; being crowded in the beginning and spread out near the finish end of the scale. However, the non-linearity of the scale can be corrected to some extent by the accurate shaping (*e.g.*, using tongue-shaped vanes) and positioning of iron vanes in relation to the operating coil.

16.20. T_d of M.I. Instruments in Terms of Self-Inductance

In a moving-iron instrument, soft-iron vane moves in the magnetic field set up by the operating current in the instrument coil. Therefore, self-inductance (L) of the instrument coil changes with the position of the moving iron *i.e.* coil inductance changes with deflection θ. Suppose at any instant, the alternating current flowing through the coil is i.

$\therefore$ Energy stored in the magnetic field at this instant is

$$w = \frac{1}{2}Li^2$$

Since the inductance L of the coil depends upon deflection θ, the instantaneous deflecting torque T'_d is given by ;

$$**T'_d = \frac{dw}{d\theta} = \frac{d}{d\theta}\left(\frac{1}{2}Li^2\right) = \frac{1}{2}i^2\frac{dL}{d\theta}$$

Average deflecting torque, $T_d =$ Average of T'_d over one cycle

$$= \frac{1}{T}\int_0^T \frac{1}{2}i^2\frac{dL}{d\theta}\,dt = \frac{1}{2}\frac{dL}{d\theta}\frac{1}{T}\int_0^T i^2 dt$$

$\therefore$ $T_d = \frac{1}{2}I^2\frac{dL}{d\theta}$

where $I =$ r.m.s. value of alternating current

This expression for deflecting torque applies equally to direct currents. It is clear that deflecting torque in a moving-iron instrument is

(*i*) directly proportional to the square of current through the coil.

(*ii*) directly proportional to $dL/d\theta$ *i.e.* change of inductance of coil with deflection θ.

---

*   Since pole strengths developed are proportional to the field strength H produced by the operating current in the coil.

**   Mechanical energy = Torque × Angular displacement

     or $dw = T \times d\theta \quad \therefore T = \dfrac{dw}{d\theta}$

Scale. In a moving-iron instrument, the controlling torque is provided with control springs so that :

Controlling torque, $T_C = k\theta$

where k = Control spring constant

θ = Deflection of the instrument

In the final deflected position, $T_d = T_C$.

$\therefore$ $\qquad \dfrac{1}{2} I^2 \dfrac{dL}{d\theta} = k\theta$

or $\qquad$ Deflection, $\theta = \dfrac{1}{2} \dfrac{I^2}{k} \dfrac{dL}{d\theta}$

If $dL/d\theta$ were constant, $\theta \propto I^2$ so that the instrument would have a squared scale. However, in actual practice, $dL/d\theta$ is not constant and the scale of a moving-iron instrument is not square-law type.

16.21. Sources of Errors in Moving Iron Instruments

The errors which may occur in moving-iron instruments can be divided into two categories *viz..* **(1)** errors with both d.c. and a.c. work **(2)** errors with a.c. work only.

1. Errors with both d.c. and a.c. work

(*i*) **Errors due to hysteresis.** Since the iron parts move in the magnetic field, hysteresis loss occurs in them. The effect of this error is that the readings are higher when current increases than when it decreases. This error can be reduced by employing vanes of *mumetal and by working it over a low range of flux densities.

(*ii*) **Error due to stray fields.** Since the operating field is comparatively weak (say 7×10^{-3} Wb/m^2), such instruments are readily affected by stray fields. This may give rise to wrong readings. This error can be reduced by enclosing the movement in an iron case.

(*iii*) **Error due to temperature.** Change of temperature affects the instrument resistance and stiffness of the control spring.

(*iv*) **Error due to friction.** Due to the friction of moving parts, a slight error may be introduced. This can be avoided by making torque-weight ratio high.

2. Errors with a.c. work only. With the change in frequency, the impedance of the instrument coil changes. This will cause a change of current in the coil. This is particularly important in case of voltmeters since these are connected in parallel with the circuit. The indicated voltage V is given by:

$$V = I_m \sqrt{(R_m + R_s)^2 + \omega^2 L_m^2}$$

where I_m is the meter current, R_m and L_m are the resistance and inductance of the coil of the meter and R_s is multiplier. For high frequencies, the meter gives low readings and *vice-versa*. This error can be eliminated by connecting a capacitor of suitable value in parallel with the series resistance R_s of the voltmeter. The value of capacitor is given by $C = L_m/R_s^2$.

16.22. Characteristics of Moving-Iron Instruments

Advantages

(*i*) These are less expensive, robust and simple in construction.

---

* Mumetal is a magnetic material having high permeability and low hysteresis loss. However, this material saturates at even medium flux density which affects the deflecting torque. For this reason, working flux density should have a low value.

(*ii*) These can be used for both d.c. and a.c. measurements. However, when used with *d.c., they are liable to small errors due to residual magnetism.

(*iii*) These instruments have high operating torque.

(*iv*) These instruments are reasonably accurate.

Disadvantages

(*i*) These instruments have non-linear scales.

(*ii*) These instruments are not as sensitive as the permanent-magnet moving coil instruments.

(*iii*) Errors are introduced due to change in frequency in case of a.c. measurements.

Applications. *The moving-iron instruments are primarily used for a.c. measurements viz.,* alternating currents and voltages. They are not used to measure direct currents and voltages because their characteristics are inferior to permanent-magnet moving coil instruments.

16.23.   Extending Range of Moving-Iron Instruments

As explained above, moving-iron instruments are used mainly on a.c. circuits. Therefore, range extension shall be discussed with reference to a.c. measurements.

1.   Ammeter. Shunts are not used to extend the range of moving-iron a.c. ammeters. It is because the division of current between the operating coil and the shunt varies with frequency (since reactance of the coil depends upon frequency). In practice, the range of moving-iron a.c. ammeter is extended by one of the following two methods :

(*i*) **By changing the number of turns** of the operating coil. For example, suppose that full-scale deflection is obtained with 400 ampere-turns. For full-scale reading with 100 A, the number of turns required would be = 400/100 = 4.

Similarly, for full-scale reading with 50 A, the number of turns required is = 400/50 = 8. Thus the ammeter can be arranged to have different ranges by merely having different number of turns on the coil. Since the coil carries the whole of the current to be measured , it has a few turns of thick wire. The usual ranges obtained by this method are 0–250 A.

(*ii*) For ranges above 0–250 A, a **current transformer** is used in conjunction with 0–5 A a.c. ammeter as shown in Fig. 16.39. The current transformer is a step-up transformer *i.e.*, number of secondary turns is more than the **primary turns. The primary of this transformer is connected in series with the load and carries the load current. The a.c. ammeter is connected across the secondary of the transformer. Since in Fig. 16.39, the current transformer ratio is 10 : 1, it means that the line (or load) current is equal to 10 times the reading on the a.c. meter.

$$\therefore \qquad \text{Load current, } I_L = 3 \times 10 = 30 \text{ A}$$

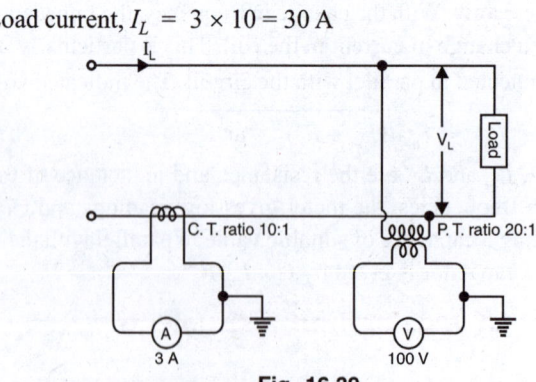

Fig. 16.39

---

*   A moving-iron instrument can be used to measure direct currents and voltages as long as the effects of a possible residual magnetism in the iron are considered. Thus a second reading should be made with the leads interchanged when the meter is used on d.c. circuits. The average of the two readings gives the true value.

**   Usually, the primary winding of the transformer consists of a single turn or at the most a few turns.

2. Voltmeter. The range of a moving-iron a.c. voltmeter is extended by connecting a high resistance (multiplier) in series with it. For ranges higher than 0–750 V, where power wasted in the multiplier would be excessive, a 0–110 V a.c. voltmeter is used in conjunction with a **potential transformer** as shown in Fig. 16.39. The potential transformer is a step-down transformer *i.e.*, number of primary turns is more than the secondary turns. The primary of the transformer is connected across the load across which voltage is to be measured. The a.c. voltmeter is connected across the secondary. Since in Fig. 16.39, the potential transformer ratio is 20 : 1, the load voltage is equal to 20 times the reading on the a.c. voltmeter.

$$\therefore \qquad \text{Load voltage, } V_L = 100 \times 20 = 2000 \text{ V}$$

Note that both secondaries of the instrument transformers are grounded as a safety measure.

Example 16.26. *A 15 V moving-iron voltmeter has a resistance of 300 Ω and an inductance of 0·12 H. Assume that the voltmeter reads correctly on d.c., what will be the percentage error when the instrument is placed on 15V a.c. supply at 100 Hz ?*

Solution.

$$\text{Resistance of meter, } R_m = 300 \text{ Ω} \quad ; \quad \text{Inductance of meter, } L_m = 0.12 \text{ H}$$

On d.c. circuit. Meter current, $I_m = V/R_m = 15/300 = 0.05$ A

Since the meter reads correctly on d.c., it means that a current of 0.05A flowing through the instrument will give a reading of 15 V.

On a.c. circuit.

$$\text{Meter impedance, } Z_m = \sqrt{R^2 + (2\pi f \, L_m)^2} = \sqrt{(300)^2 + (2\pi \times 100 \times 0.12)^2} = 309.33 \, \Omega$$

$$\therefore \qquad \text{Meter current, } I'_m = V/Z_m = 15/309.33 = 0.0485 \text{ A}$$

$$\therefore \qquad \text{Meter reading} = \frac{15}{0.05} \times 0.0485 = 14.55 \text{ V}$$

$$\therefore \qquad \text{\% age error} = \frac{15 - 14.55}{15} \times 100 = \mathbf{3\%}$$

Example 16.27. *A moving-iron instrument gives full-scale deflection with 200 V. It has a coil of 20,000 turns and a resistance of 2000 Ω. If the instrument is used as an ammeter to give full-scale deflection at 10 A, calculate the number of turns required.*

Solution. As voltmeter.

Full-scale deflection voltage, $V = 200$ volts ;   Resistance of instrument coil, $R_m = 2000$ Ω

Full-scale meter current, $I_m = V/R_m = 200/2000 = 0.1$ A

Full-scale AT required $= NI_m = 20,000 \times 0.1 = 2000$

As ammeter. The ampere-turns required to give full-scale deflection should be the same *i.e.* 2000 AT. Since the full-scale deflection current is 10 A,

$$\therefore \qquad \text{No. of turns required} = 2000/10 = \mathbf{200}$$

Example 16.28. *A moving-iron instrument requires 250 ampere turns to give full-scale deflection. Calculate (i) the number of turns required if the instrument is to be used as an ammeter reading upto 50 A and (ii) the number of turns and total resistance if the instrument is to be used as a voltmeter reading upto 300 V with a current of 20 mA.*

Solution. Since the instrument is used both as ammeter and voltmeter, it is essential to provide the same strength of magnetic field (*i.e.*, same ampere-turns) for both when giving full-scale deflection.

Full-scale deflection AT $= 250$

(*i*) As ammeter.

Full-scale deflection current, $I = 50$ A

∴   No. of turns required $= 250/50 = \mathbf{5}$

This means that if the coil is wound with 5 turns, then a current of 50 A through it will give full-scale deflection.

(ii) As voltmeter.

Full-scale deflection voltage, $V = 300$ volts

Full-scale meter current, $I_m = 20$ mA $= 0.02$ A

No. of turns required $= 250/0.02 = \mathbf{12500}$

Total resistance of meter $= V/I_m = 300/0.02 = \mathbf{15000}\ \Omega$

Example 16.29. *The working coil of a 0–400 V moving-iron voltmeter requires 300 ampere-turns to give full-scale deflection. The added resistance is to be three times the coil resistance. Find the diameter of the wire for the coil if the wire be of copper having a resistivity of $1.7 \times 10^{-6}\ \Omega$ cm ; the mean length of 1 turn = 13.5 cm.*

Solution. The p.d. across the working coil will be one-quarter of the applied voltage, since the added resistance is three times the coil resistance.

∴   Voltage across the coil is given by ;

$$V = 400/4 = 100 \text{ volts}$$

Suppose the number of turns on the coil is N and the full-scale deflection current through the coil is I.

∴ $NI = 300$ (given)

Resistance of coil, $R = \dfrac{\rho * l}{a} = \dfrac{\rho \times l_m \times N}{(\pi/4)d^2}$

or $\dfrac{V}{I} = \dfrac{\rho \times l_m \times N}{(\pi/4)d^2}$ $\left[\because R = \dfrac{V}{I}\right]$

∴   Diameter of wire, $d = \sqrt{\dfrac{4 \times \rho \times l_m \times NI}{\pi \times V}} = \sqrt{\dfrac{4 \times 1.7 \times 10^{-6} \times 13.5 \times 300}{\pi \times 100}}$

$$= 0.0094 \text{ cm} = \mathbf{0.094 \text{ mm}}$$

Example 16.30. *The inductance L of a spring-controlled moving-iron ammeter is given by the following expression in which θ is the deflection in radians,*

$$L = 1.0 + 0.0716\,\theta - 0.0114\,\theta^2 \text{ mH}$$

The control torque constant is 0.57×10^{-3} Nm/rad. Calculate (i) the current for full-scale deflection of 120° (ii) current for one-half full-scale deflection and (iii) deflection for one-quarter of full-scale current.

Solution. $L = 1.0 + 0.0716\,\theta - 0.0114\,\theta^2 \text{ mH}$

∴ $\dfrac{dL}{d\theta} = 0 + 0.0716 - 0.0288\,\theta \text{ mH/rad.}$

(i)   For full-scale of deflection of 120° (= **2.1 rad.),

$$\dfrac{dL}{d\theta} = 0.0716 - 0.0228 \times 2.1 = 0.024 \text{ mH/rad.} = 0.024 \times 10^{-3} \text{ H/rad.}$$

Deflecting torque, $T_d = \dfrac{1}{2}\dfrac{dL}{d\theta}I^2 = \dfrac{1}{2} \times 0.024 \times 10^{-3} \times I^2 \text{ Nm}$

---

*   Length of coil = Mean length of 1 turn × No. of turns $= l_m \times N$

** $120° = \dfrac{\pi}{180°} \times 120° = 2.1$ rad.

Controlling torque, $T_C = 0.57 \times 10^{-3} \times 2.1$ Nm

In the final deflected position, $T_d = T_C$.

$$\therefore \qquad \frac{1}{2} \times 0.024 \times 10^{-3} \times I^2 = 0.57 \times 10^{-3} \times 2.1$$

or $\qquad\qquad I = \sqrt{\dfrac{0.57 \times 2.1 \times 2}{0.024}} = \textbf{10 A}$

(*ii*) For half full-scale deflection (*i.e.* $\theta = 60° = 1.05$ rad.),

$$\frac{dL}{d\theta} = 0.716 - 0.0228 \times 1.05$$
$$= 0.048 \text{ mH/rad.} = 0.048 \times 10^{-3} \text{ H/rad.}$$

$$\therefore \qquad T_d = \frac{1}{2}\frac{dL}{d\theta}I^2 = \frac{1}{2} \times 0.48 \times 10^{-3} \times I^2 \text{ Nm}$$

and $\qquad\qquad T_C = 0.57 \times 10^{-3} \times 1.05$ Nm

In the steady position of deflection, $T_d = T_C$.

$$\therefore \qquad \frac{1}{2} \times 0.048 \times 10^{-3} \times I^2 = 0.57 \times 10^{-3} \times 1.05$$

or $\qquad\qquad I = \sqrt{\dfrac{0.57 \times 1.05 \times 2}{0.048}} = \textbf{4.99 A}$

(*iii*) Full-scale current = 10 A ... as calculated in (*i*) above

One-quarter full-scale current = 10/4 = 2.5 A

Let the required deflection be θ radians.

$$\therefore \qquad T_d = \frac{1}{2}\frac{dL}{d\theta}I^2 = \frac{1}{2}(0.0716 - 0.0228\,\theta) \times 10^{-3} \times 2.5^2$$

and $\qquad\qquad T_C = 0.57 \times 10^{-3} \times \theta$

In the steady position of deflection, $T_d = T_C$.

$$\therefore \qquad \frac{1}{2}(0.0716 - 0.0228\,\theta) \times 10^{-3} \times 2.5^2 = 0.57 \times 10^{-3} \times \theta$$

or $\quad (0.0358 - 0.0114\,\theta) \times 6.25 = 0.57\,\theta$

or $\qquad\qquad 0.64\,\theta = 0.224 \qquad \therefore \quad \theta = 0.224/0.64 = 0.35 \text{ rad.} = \textbf{20°}$

Example 16.31. *The change of inductance for a moving iron ammeter is 2 µH/degree. The control spring constant is 5×10^{-7} Nm/degree. The maximum deflection of the pointer is 100°. What is the current corresponding to maximum deflection?*

Solution. Control spring constant, $k = 5 \times 10^{-7}$ Nm/degree

Maximum deflection, $\theta_m = 100°$

$\therefore \qquad$ Controlling torque, $T_C = k\theta_m = 5 \times 10^{-7} \times 100$

Now, $\quad$ Deflecting torque, $T_d = T_C = 5 \times 10^{-5}$ Nm

Also, $\qquad\qquad T_d = \frac{1}{2}\frac{dL}{d\theta}I^2$

or $\qquad\qquad I = \sqrt{2T_d \times \dfrac{d\theta}{dL}}$

Here, $T_d = 5 \times 10^{-5}$ Nm ; $\dfrac{dL}{d\theta} = 2\mu\text{H/degree} = 2 \times 10^{-6}$ H/degree

$$\therefore \qquad I = \sqrt{2 \times 5 \times 10^{-5} \times \frac{1}{2 \times 10^{-6}}} = \textbf{7.07 A}$$

Example 16.32. *A 150 V moving iron voltmeter intended for 50 Hz has an inductance of 0.7 H and a resistance of 3 kΩ. Find the series resistance required to extend the range of the instrument to 300 V. If this 300 V, 50 Hz instrument is used to measure a d.c. voltage, find the d.c. voltage when the scale reading is 200 V.*

Solution. Inductive reactance of voltmeter $= 2\pi \times 50 \times 0.7 = 220\ \Omega$

Impedance of voltmeter $= \sqrt{(3000)^2 + (220)^2} = 3008\ \Omega$

Now, voltmeter range is to be doubled. Therefore, the impedance of the voltmeter must be doubled in order to have the same current for full-scale deflection. To do so, the series resistance R required is given by ;

$$(3000 + R)^2 + (220)^2 = (2 \times 3008)^2 \qquad \therefore R = \textbf{3012}\ \boldsymbol{\Omega}$$

When this 300 V, 50 Hz voltmeter is used on d.c. supply, it reads 200 V.

$\therefore$ Actual applied d.c. voltage $= 200 \times \dfrac{\text{Total a.c. impedance}}{\text{Total d.c. resistance}}$

$$= 200 \times \frac{2 \times 3008}{3000 + 3012} = \textbf{200.134 V}$$

Tutorial Problems

1. A 150 V moving-iron voltmeter is correct at 50 Hz ; it has a resistance of 2000 Ω and an inductance of 0·35 H. If it were used on 400 Hz system in an aircraft, what would be its full-scale error ?   **[8.25%]**

2. A 50 V moving-iron voltmeter has a resistance of 1000 Ω and an inductance of 0·765 H. Assuming the instrument reads correctly on d.c., what will be the percentage error when the instrument is used to measure alternating voltage of 50 V at 25 Hz ?   **[0.72%]**

3. A moving-iron voltmeter gives full-scale deflection with 100 V. It has a coil of 10,000 turns and a resistance of 2000 Ω. If the instrument is to be used as an ammeter to give full-scale deflection at 20 A, calculate the number of turns required in the coil.   **[25]**

4. A moving-iron instrument requires 310 ampere-turns to produce a full-scale deflection. It is proposed to use it as a voltmeter reading upto 260 volts. Find the diameter of the copper wire with which the working coil is wound if in series with it is connected a manganin resistance to absorb 220 volts. Mean length of one turn of coil is 14.5 cm and the resistivity of copper is $1.7 \times 10^{-6}\ \Omega$ cm.   **[0.15 mm]**

5. The working coil of a moving-iron voltmeter has a resistance of 6000 Ω at 20° C. At this temperature, it was calibrated and indicated correctly on 300 V. Find the error as a percentage of the indication when the temperature of the coil increases to 40° C. The coil is wound with copper wire having a temperature co-efficient of 0.004 at 20° C.   **[8%]**

16.24. Comparison of Moving Coil, Dynamometer type and Moving Iron Voltmeters and Ammeters

Table below shows the comparison between permanent magnet moving coil (PMMC), dynamometer type and moving iron voltmeters and ammeters.

S.No.	Particular	Moving coil	Dynamometer type	Moving iron
1.	Construction	Delicate construction	Heavy moving system	Simple
2.	Cost	Very high	High	Low
3.	Power consumption	Very low	High	Less than dynamometer type

4	*Scale*	Uniform	Non-uniform	Non-uniform
5	*Torque-weight ratio*	High	Small	More than dynamometer type
6.	*Suitable for*	d.c. only	d.c. as well as a.c.	d.c. as well as a.c.
7.	*Effect of stray magnetic fields*	Not affected	Affected	Not affected
8.	*Application*	Voltmeters and Ammeters	Generally for wattmeters	Voltmeters and ammeters
9.	*Accuracy*	High	Poor	Reasonable
10	*Sensitivity*	High	Poor	Reasonable

16.25. Hot-Wire Ammeters and Voltmeters

The operation of these instruments depends upon the expansion of a wire which is heated due to the passage of electric current. The expansion of the wire is taken up by the moving system which causes the pointer to move over a graduated scale. They are suitable for d.c. as well as a.c. measurements.

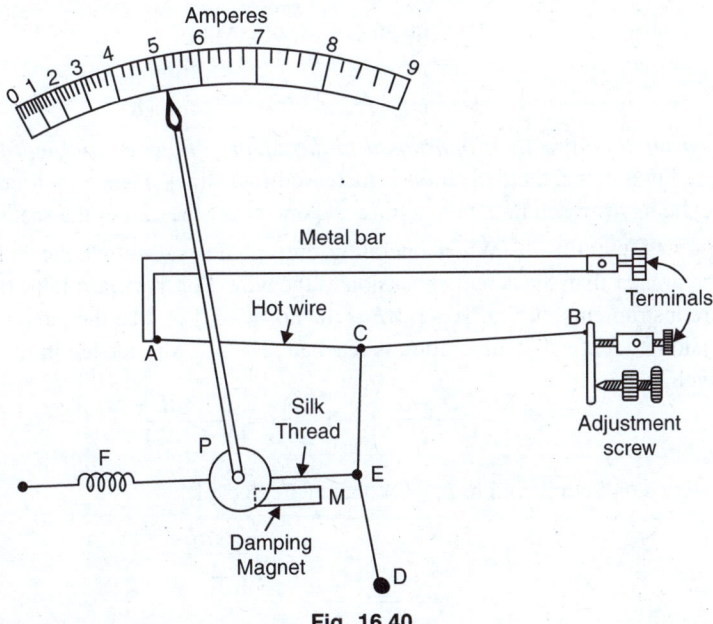

Fig. 16.40

Construction. Fig. 16.40 shows the simplified diagram of a hot-wire instrument. It consists of a very thin platinum-iridium wire *AB* stretched between terminals *A* and *B*. The wire *AB* is generally called the hot-wire and carries the operating current of the instrument. A fine phosphor-bronze wire *CD* (called **tension wire**) is attached to the centre *C* of the hot-wire; the other end *D* being fixed to the base of the instrument. A fine silk thread attached to *CD* at *E* passes around a pulley on the spindle and is connected to a spring *F*. The effect of wire *CD* and the silk thread is to *magnify the expansion of the wire *AB*. A pointer is attached to the spindle which moves over the scale as the hot-

---

*   The co-efficient of linear expansion of most metals is quite small, say 0.000015. Therefore, passage of current through the hot-wire will cause a very little expansion of the wire. With this arrangement, we measure the sag (*i.e.* downward movement of point *C* due to the expansion of the wire *AB*) which would be much larger than the actual expansion of the wire *AB*.

wire *AB* expands. A thin light aluminium disc is attached to the pulley and moves between the poles of a permanent magnet *M*. Eddy currents produced in the disc provide the necessary damping. The instrument is generally spring controlled.

Working. When the instrument is connected in the circuit to measure current or voltage, the operating current flows through the hot-wire *AB*. This causes the hot-wire to expand. The slack in the hot-wire is taken up by the spring *F*, causing the pulley *P* to turn and move the pointer over the scale. When the operating current ceases to flow, the wire *AB* cools and contracts, and the silk thread is pulled to the right causing the pointer to return to zero. Since the tension of wire *AB* changes with temperature, the pointer is set to zero (with adjustment screw) practically each time the instrument is to be used. As the heating effect of electric current is independent of the direction of current flow, therefore, such instruments can be used for both d.c. and a.c. work.

Deflecting torque. The expansion of the hot-wire depends upon i^2R loss where i is the instantaneous current through the hot wire *AB*.

∴   Average deflecting torque, $T_d \propto$ mean of i^2 over a cycle

Since the instrument is spring-controlled,

∴ $T_C \propto \theta$

In the steady position of deflection, $T_d = T_C$.

∴ $\theta \propto$ mean of i^2 over a cycle

$$\propto I^2 \qquad \qquad \text{...for d.c.}$$

$$\propto I^2_{r.m.s.} \qquad \qquad \text{...for a.c.}$$

Note that deflection is directly proportional to the square of current through the hot-wire. If current is increased two times, the deflection is increased four times. Hence such instruments have square-law scale; being crowded in the beginning and open near the end of the scale.

Magnification of expansion. When operating current flows through the hot-wire, the sag produced is much greater than the actual expansion of the wire. For this reason, we utilise the effect of sag in hot-wire instruments. In Fig. 16.41, *AB* is the hot-wire. Let *L* be the length of the wire *AB* and *dL* its expansion after steady temperature is reached. The sag *S* produced in the wire (Refer to Fig. 16.41) is given by ;

$$S^2 = \left(\frac{L+dL}{2}\right)^2 - \left(\frac{L}{2}\right)^2 = \frac{2L \times dL + (dL)^2}{4}$$

Since *dL* is very small compared to *L*, $(dL)^2$ can be neglected.

∴ $$S^2 = \frac{2L \times dL}{4} \quad \text{or} \quad S = \sqrt{\frac{L \times dL}{2}}$$

$$\text{Magnification} = \frac{S}{dL} = \frac{\sqrt{\dfrac{L \times dL}{2}}}{dL} = \sqrt{\frac{L}{2dL}}$$

It may be seen that the value of *S*/*dL* is much greater than 1.

Further magnification is obtained by connecting a phosphor bronze wire of length L_1 to the centre of wire *AB* as shown in Fig. 16.42. Let S_1 be the sag of phosphor bronze wire. Referring to Fig. 16.42,

$$S_1 = \sqrt{\left(\frac{L_1}{2}\right)^2 - \left(\frac{L_1 - S}{2}\right)^2} = \sqrt{\frac{L_1^2}{4} - \frac{L_1^2}{4} + \frac{2L_1 S - S^2}{4}}$$

or $\qquad\qquad S_1 = \sqrt{\dfrac{L_1 S}{2}}$ $\qquad$ ...Neglecting S^2 being very small

Putting the value of S in the above expression, we get,

$$S_1 = \sqrt{\frac{L_1}{2}\sqrt{\frac{LdL}{2}}} = \sqrt{\frac{L_1}{2\sqrt{2}}} \times (LdL)^{1/4}$$

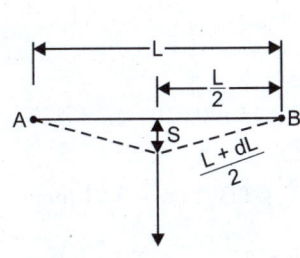

Fig. 16.41

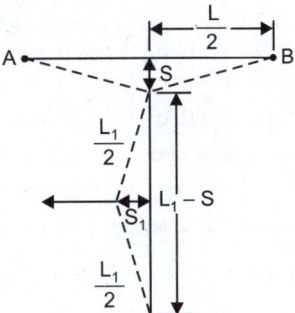

Fig. 16.42

Advantages

(*i*) They can be used for both d.c. and a.c. measurements.

(*ii*) Their readings are independent of waveform and frequency.

(*iii*) They are unaffected by stray magnetic fields.

Disadvantages

(*i*) They have non-uniform scale.

(*ii*) They have a sluggish action since the hot-wire takes some time to reach its final temperature.

(*iii*) Their zero position needs frequent adjustment.

(*iv*) They are liable to burn out or the hot-wire melts, if over-loaded.

(*v*) They have a high power consumption.

Range extension. The range of a hot-wire voltmeter can be extended by using a high resistance (multiplier) in series with the hot-wire. The ammeter range can be extended by using a shunt. However, for use at radio frequency, special precautions must be observed in the design of the instrument and the shunt; in particular inductance must be avoided as far as possible.

Applications

(*i*) Hot-wire instruments are particularly useful for a.c. measurements and are as a rule cheaper than instruments based on dynamometer principle.

(*ii*) Hot-wire instruments are used for high-frequency alternating currents (*e.g.*, in wireless work) because the inductance of the hot-wire is very small. Dynamometer or moving-iron instruments are unsuitable at such high frequencies.

(*iii*) Hot-wire ammeters are not, however, as accurate as dynamometer ammeters, and must be calibrated by comparison with a d.c. ammeter, say of the moving-coil type.

Example 16.33. *The working wire of a single-sag hot-wire instrument is 20 cm long and has co-efficient of linear expansion of $15 \times 10^{-6}/°C$. When connected in the circuit to measure current, the temperature of the wire rises by $50°C$. Assuming the sag is taken up at the centre, find the magnification if the working wire has no initial sag.*

Solution. Length of wire at room temp., $L = 20$ cm

Length when heated through $50°C = 20 (1 + 15 \times 10^{-6} \times 50) = 20.015$ cm

Increase in length, $dL = 20.015 - 20 = 0.015$ cm

$\therefore$ Magnification $= \sqrt{\dfrac{L}{2 \times dL}} = \sqrt{\dfrac{20}{2 \times 0.015}} = \mathbf{25.82}$

Example 16.34. *The working wire of a single-sag hot wire instrument is 15 cm long and is made up of platinum-silver with a coefficient of linear expansion of 16×10^{-6}. The temperature rise of the wire is $85°C$ and the sag is taken up at the centre. Find the magnification (i) with no initial sag and (ii) with an initial sag of 1 mm.*

Solution. Length of wire at room temp., $L = 15$ cm

Increase in length of wire due to a temperature rise of $85°C$ is

$$dL = \alpha \, \theta \, L = 16 \times 10^{-6} \times 85 \times 15 = 0.0204 \text{ cm} = 0.204 \text{ mm}$$

(*i*) With no initial sag, we have,

$$S = \sqrt{\dfrac{LdL}{2}} = \sqrt{\dfrac{15 \times 0.0204}{2}} = 0.391 \text{ cm} = 3.91 \text{ mm}$$

$\therefore$ Magnification, $M = \dfrac{S}{dL} = \dfrac{3.91 \text{ mm}}{0.204 \text{ mm}} = \mathbf{19.2}$

(*ii*) With initial sag of 1 mm, we have,

Change in sag, $S' = $ Final sag–Initial sag $= 3.91 - 1.0 = 2.91$ mm

$\therefore$ Magnification, $M' = \dfrac{S'}{dL} = \dfrac{2.91 \text{ mm}}{0.204} = \mathbf{14.25}$

16.26. Thermocouple Instruments

The hot-wire instruments are obsolete and have been superseded by thermocouple instruments. A thermocouple instrument consists of (*i*) a permanent-magnet moving coil instrument and (*ii*) a thermoelement *i.e.*, an evacuated glass bulb containing a heater wire and a *thermocouple whose active junction *J* is in contact with the heater as shown in Fig. 16.43. When operating current flows through the heater, the heat produced is applied to the active junction *J* of the thermocouple. Due to thermo-electric effect, a direct voltage (directly proportional to the heat) appears across the cool ends 1 and 2 of the thermocouple. The permanent-magnet moving coil instrument connected across ends 1 and 2 will give the indication of the current flowing in the line. Since the heating effect in a resistance (*i.e.*, heater in this case) is independent of current direction, thermocouple instruments can be used for both d.c. and a.c. measurements.

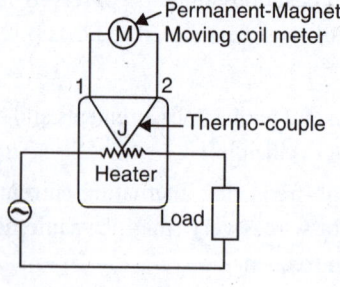

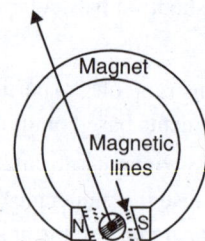

Fig. 16.43   **Fig. 16.44**

---

* **Thermocouple.** If two dissimilar metal wires or strips are joined at one end and this junction is heated, a small direct voltage will appear between the cool, open ends. The voltage between the open ends is directly proportional to the temperature difference between the hot and cold ends. This phenomenon is known as *thermo-electric effect* and the combination of metal wires or strips is known as a *thermocouple.* Any two dissimilar metals may be used for the thermocouple. However, the two different alloys, *constantan* and *manganin* are frequently employed.

As the amount of direct voltage (called thermo-e.m.f.) appearing across the cold ends of the thermocouple is directly proportional to the heating effect (I^2R) at the active junction J, the deflection is directly proportional to the square of current *i.e.,*

$$\theta \propto I^2 \qquad \qquad \text{...for d.c.}$$
$$\propto I^2_{r.m.s.} \qquad \qquad \text{...for a.c.}$$

Scale. *It may be seen that scale of such an instrument is of square-law type i.e., crowded in the beginning and open near the end of the scale.* The scale can be modified to a uniform type by changing the shape of the pole pieces of the moving-coil meter as shown is Fig. 16.44. When the moving coil is in its low-scale position (*i.e.,* pointer is to the left), it is cutting across stronger portion of the magnetic field as indicated by the concentration of the magnetic lines. Consequently, the deflecting torque *increases which makes the meter more sensitive during the initial portion of the scale. When the coil is in its high-scale position (*i.e.,* pointer is to the right), it is in the weaker portion of the field. This decreases the deflecting torque and hence the sensitivity of the meter for that portion of the scale. The effect of this arrangement is that the scale of the meter tends to change from square-law relation to the linear one.

Range extension. A thermo-couple meter may be used to measure direct as well as alternating currents and voltages. Their ranges may be extended through the use of shunts or voltage multipliers.

Applications. The most useful application is in high-frequency circuits (*e.g.,* radio work) because the forces operating the pointer are practically independent of frequency. The moving-iron or dynamometer instruments will be simply useless in such situations. The thermocouple meter is normally employed for accurate measurement of radio-frequency aerial and feeder currents.

Note. An interesting variation of the thermocouple meter is the **pyrometer,** an instrument used to measure high temperatures. Here, only the thermocouple and permanent-magnet moving coil millivoltmeter are used. The junction of the thermocouple is applied to the object whose temperature is to be measured. The meter scale is marked in degrees to indicate the temperature of the object.

16.27. Electrostatic Voltmeters

These instruments are based on the fact that an electric force (attraction or repulsion) exists between charged plates or objects. An electrostatic voltmeter is essentially an air capacitor; one plate is fixed while the other, which is coupled to the pointer, is free to rotate on jewelled bearings. When p.d. to be measured is applied across the plates, the electric force between the plates gives rise to a deflecting torque. Under the action of deflecting torque, the movable plate moves and causes the deflection of the pointer to indicate the voltage being measured. Such instruments can be used to measure direct as well as alternating voltages.

There are three types of electrostatic voltmeters *viz* :

(*i*) Attracted disc type	—usual range from 500 V to 500 kV
(*ii*) Quadrant type	—usual range from 250 V to 10 kV
(*iii*) Multicellular type	—usual range from 30 V to 300 V

Two things are worth noting about electrostatic voltmeters. First, the deflecting torque is very small for low voltages. For this reason, they are not very sensitive to measure small voltages. Secondly, the instrument is only available for the measurement of p.d., that is to say as voltmeter. *It cannot be used as an **ammeter.*

---

*   It may be recalled that the deflecting toque in a permanent-magnet moving coil meter is given by ;

$$T_d = BINA$$

**   When used as an ammeter, there will be a few millivolts voltage across the instrument. This extremely small p.d. is insufficient to produce any deflecting torque.

16.28. Attracted Disc Type Voltmeter

Fig. 16.45 shows the simplified diagram of an attracted disc electrostatic voltmeter. It consists of two mushroom-shaped plates *A* and *B*, each mounted on insulated pedestal. The plate *B* is fixed while the plate *A* (negative, for direct voltage) has a movable central portion—the attracted disc. The movable plate *A* is attached to a horizontal rod which is suspended by two phosphor bronze strips. When p.d. to be measured is applied across the plates, the plate *A* moves towards the fixed plate *B* and actuates the pointer *via* a pulley or link mechanism. The control force is provided by gravity and damping force by air dash pot. If the plates are too close together or if the applied voltage is too high, a spark discharge may occur. In order to prevent such a possibility, a *ballast resistor is included in the circuit. The function of this resistor is to limit the current if any sparkingover occurs. If the applied voltage reverses in polarity, there is a simultaneous change in the sign of charge on the plates so that the direction of deflecting force remains unchanged. Hence such instruments can be used for both d.c. and a.c measurements.

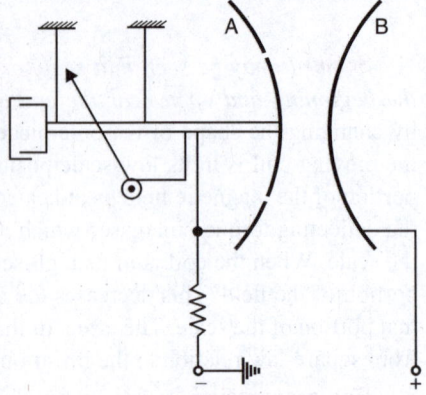

Fig. 16.45

Theory. The force of attraction *F* between the charged plates is given by ;

$$F = \frac{**1}{2}\frac{dC}{dx}V^2$$

where x = distance between the plates

                             C = capacitance between the plates

                             V = applied voltage

Since *x* is always small, *dC/dx* is practically constant.

∴ $F \propto V^2$

Obviously, the scale of the instrument will be non-uniform.

16.29. Quadrant Type Voltmeter

Fig. 16.46 shows the simplified diagram of a quadrant electrostatic voltmeter. It consists of a light aluminium vane *A* suspended by a phosphor bronze string mid-way between two inter-connected quadrant shaped brass plates *BB*. One terminal is joined to fixed plates *BB* (positive for direct voltage) and the other to the movable plate *A* (negative for direct voltage). The controlling torque is provided by the torsion of the suspension string. Damping is provided by air friction due to the motion of another vane in a partially closed box.

Working. When the instrument is connected in the circuit to measure the p.d., an electric force exists between the plates. Consequently, the movable vane *A* moves inbetween the fixed plates and causes the deflection of the pointer. The pointer comes to rest at a position where deflecting torque is

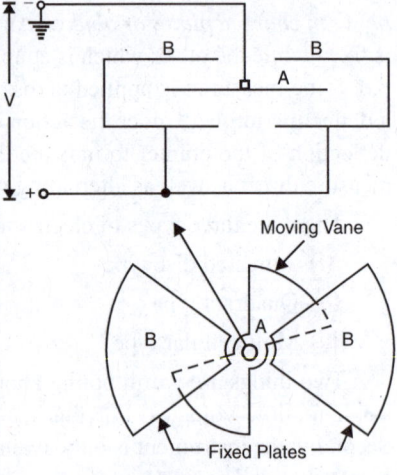

Fig. 16.46

---

*   When direct voltage is being measured, there will be no voltage drop across this resistor. For alternating voltages, the drop will be negligible at low frequencies. Hence, this resistor will not cause any error in the measurement of voltage.

**   For this expression, refer to Art. 6.24.

equal to the controlling torque. Since the force of attraction between the movable plate A and the fixed plates BB is directly proportional (p.d.)^2, the instrument can be used to measure either direct or alternating voltages. When used in an a.c. circuit, it reads the r.m.s. values. More robust but *less accurate voltmeters are made by pivoting the moving system. In pivoted voltmeters, the controlling torque is provided by a spiral spring.

Theory. The capacitance C between the plates depends upon deflection θ *i.e.*, upon the position of the movable plate (or vane) A. Suppose that at any instant, the applied alternating voltage is v.

Electrostatic energy at this instant $= \dfrac{1}{2}Cv^2$

Since the capacitance between the plates depends upon deflection θ, the instantaneous deflecting torque T'_d is given by ;

$$T'_d = \frac{1}{2}\frac{dC}{d\theta}v^2$$

Average deflecting torque, $T_d =$ Average of T'_d over a cycle

$$= \frac{1}{T}\int_0^T \frac{1}{2}\frac{dC}{d\theta}v^2\,dt = \frac{1}{2}\frac{dC}{d\theta}\cdot\frac{1}{T}\int_0^T v^2\,dt$$

$\therefore$
$$T_d = \frac{1}{2}\frac{dC}{d\theta}V^2$$

where $V =$ r.m.s. value of alternating voltage

This equation equally applies to direct voltages. If $dC/d\theta$ were constant, then,

$$T_d \propto V^2$$

Hence the instrument has non-uniform scale. The non-linearity in the scale can be corrected by shaping the movable vane A in such a way as to increase $dC/d\theta$ for small deflections and to make the scale nearly uniform for larger ones.

16.30. Multicellular Electrostatic Voltmeter

The major drawback of quadrant type voltmeter is that deflecting torque is very **small for low voltages. Therefore, such an instrument cannot measure accurately voltages below 250 V. This difficulty has been overcome in a multicellular electrostatic voltmeter which can read as low as 30 volts.

Fig. 16.47. shows the constructional details of a multicellular voltmeter. It is essentially a quadrant type voltmeter with the difference that it has ten moving vanes instead of one and eleven fixed plates forming "cells" in and out of which the vanes move. The moving vanes are fixed to a vertical spindle and suspended by a phosphor-bronze wire so that the vanes are free to move, each between a pair of fixed plates. At the lower end of the spindle, an aluminium disc hangs horizontally in an oil bath and provides damping torque

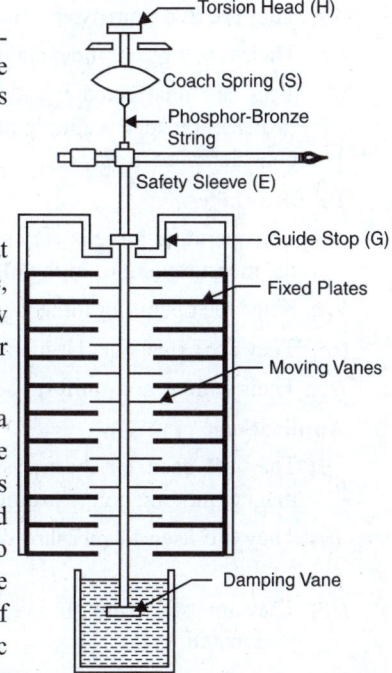

Torsion Head (H)

Coach Spring (S)

Phosphor-Bronze String

Safety Sleeve (E)

Guide Stop (G)

Fixed Plates

Moving Vanes

Damping Vane

Fig. 16.47

---

*   Due to pivot friction, the pivoted voltmeters are less accurate than the suspension type. For this reason, low voltage electrostatic voltmeters are always of suspension type.

**   Deflecting torque also depends upon the capacitance between plates [*i.e.*, $T_d = \dfrac{1}{2}(dC/d\theta)V^2$]. In a quadrant voltmeter, the capacitance cannot be increased since the number of vanes is limited by space consideration.

due to fluid friction. The controlling torque is provided by the torsion of the suspension wire as the moving system rotates. The upper end of the suspension wire is attached through a coach spring S to a torsion head H. The torsion head is provided with a tangent screw for zero adjustment. The function of the coach spring is to prevent the suspension wire from breaking when accidentally jerked. Should the moving vanes be jerked downward, then the coach spring yields sufficiently to allow the safety sleeve E to come into contact with the guide stop G before the suspension wire is over strained. The scale is horizontal if the pointer is straight but the indications can be given on a vertical scale by bending the pointer at right angles.

The working principle of multicellular voltmeter is exactly similar to the quadrant type. By using a number of inter-leaved stationary and moving plates, we are able to increase the capacitance and hence the deflecting torque. Consequently, the multicellular voltmeter is much more sensitive than the quadrant type and can accurately measure low voltages.

16.31. Characteristics of Electrostatic Voltmeters

It is worthwhile to mention the advantages, disadvantages and applications of electrostatic voltmeters.

Advantages

(*i*) They can be used for both d.c. and a.c. measurements.

(*ii*) They draw *negligible power from the mains. Hence such voltmeters do not alter the condition of the circuit to which they are connected.

(*iii*) They are free from hysteresis and eddy current losses as no iron is used in their construction.

(*iv*) Their readings are independent of waveform and frequency.

(*v*) They are unaffected by stray magnetic fields, although electrostatic fields (set up for instance, by such a simple process as rubbing the glass of the case to clean it) may cause considerable errors.

Disadvantages

(*i*) The operating force is very small for low voltages so that they are particularly suitable for the measurement of high voltages.

(*ii*) Since the operating force is generally small, errors due to friction are difficult to avoid.

(*iii*) They are expensive, large in size and are not robust in construction.

(*iv*) Their scale is non-uniform ; being crowded in the beginning of the scale.

Applications

(*i*) They are used for the measurement of very high direct voltages at which a permanent magnet moving coil instrument and the multiplier would be unsuitable.

(*ii*) They are used to measure direct low voltages when it is necessary to preserve an open circuit.

(*iii*) They are used to measure very high alternating voltages when the use of a transformer must be avoided.

16.32. Range Extension of Electrostatic Voltmeters

The range of electrostatic voltmeters can be increased by the use of multipliers. Two types of multipliers are employed for this purpose *viz*.

---

* With direct voltage, the instrument draws only the initial charging current. With alternating voltages, the alternating current drawn is extremely small.

   (*i*) Resistance potential divider —for ranges upto 40 kV

   (*ii*) Capacitance potential divider — for ranges upto 1000 kV

The first method can be used for both direct and alternating voltages whereas the second method is suitable only for alternating voltages.

   (*i*)   **Resistance potential divider.** This divider consists of a high resistance with tappings taken off at intermediate points. The voltage *V* to be measured is applied across the whole of the potential divider and the electrostatic voltmeter connected across part of it (resistance *r* in this case) as shown in Fig. 16.48. Since voltmeter *practically carries no current, the p.d. *v* across it is the same fraction of the applied voltage *V* as the resistance across it (*i.e.*, *r*) is of the whole resistance (*i.e.*, *R*) *i.e.*,

$$\text{Multiplying factor, } \frac{V}{v} = \frac{R}{r}$$

Thus if the voltmeter is connected across 1/5 of the whole resistance (*i.e.*, $R/r = 5$), then voltage *V* to be measured is 5 times the reading of the voltmeter. The advantage of this method is that there is no shunting effect of the voltmeter. The drawback is that there is power loss in the resistance divider.

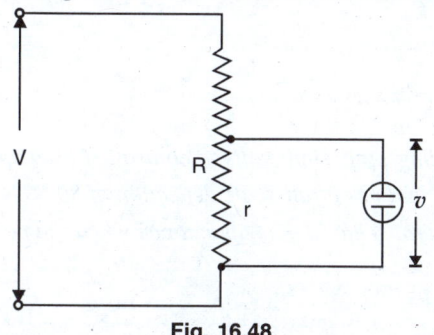

Fig. 16.48

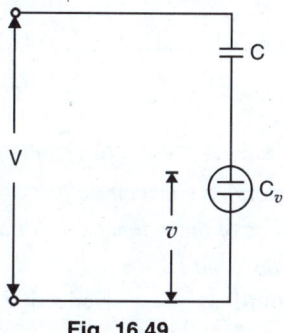

Fig. 16.49

   (*ii*)   **Capacitance potential divider.** In this method, a single capacitor of capacitance *C* is connected in series with the voltmeter and the whole circuit is connected across the voltage *V* to be measured as shown in Fig. 16.49. Let *v* volts be the reading of the voltmeter. Since the voltage across a capacitor is inversely proportional to its capacitance,

∴
$$V \propto \frac{{}^{**}C + C_v}{C \times C_v}$$

and
$$v \propto \frac{1}{C_v}$$

∴   Multiplying factor, $\dfrac{V}{v} = \dfrac{C + C_v}{C} = 1 + \dfrac{C_v}{C}$

By using capacitors of different capacitances, different voltage ranges can be obtained. This method has the advantage that the circuit consumes no power. However, the drawback is that capacitance current taken is greatly increased.

   Example 16.35. *A non-inductive coil AB is connected across 440 V mains and an electrostatic voltmeter of negligible capacitance is connected across a portion CD of this coil. The resistance of the coil AB is 8000 Ω and that of portion CD is 2000 Ω. Find (i) the reading of the voltmeter and (ii) multiplying factor.*

---

*   An electrostatic voltmeter is essentially an air capacitor. For direct voltages, no current can flow through it. For alternating voltages, the current through the voltmeter is extremely small.

**   Total circuit capacitance = $\dfrac{C \times C_v}{C + C_v}$ because *C* and C_v are in series.

Solution. (*i*) $\dfrac{V}{v} = \dfrac{R}{r}$

Here $V = 440$ volts ; $R = 8000\ \Omega$; $r = 2000\ \Omega$ $\therefore$ $\dfrac{440}{v} = \dfrac{8000}{2000}$

or   Voltmeter reading, $v = \dfrac{2000}{8000} \times 440 = \textbf{110 volts}$

(*ii*)   Multiplying factor $= V/v = 440/110 = \textbf{4}$

Example 16.36. *An electrostatic voltmeter has a capacitance of 0.1 µF and full-scale deflection of 10 kV. Determine the capacitance of the capacitor to be connected in series which will make the full-scale deflection represent 60 kV.*

Solution. Let C µF be the desired capacitance of the capacitor connected in series with the voltmeter.

$$\dfrac{V}{v} = 1 + \dfrac{C_v}{C}$$

Here $V = 60$ kV ; $v = 10$ kV ; $C_v = 0.1$ µF

$\therefore$ $\dfrac{60}{10} = 1 + \dfrac{0.1}{C}$ or $C = \textbf{0.02 µF}$

Example 16.37. *An electrostatic voltmeter reading upto 1000 volts is controlled by a spring with a torsion constant of $9{\cdot}81 \times 10^{-8}$ Nm per degree and has a full-scale deflection of 80°. The capacitance of the voltmeter is 10 µµF when it reads zero. What is the capacitance when the pointer indicates 1000 V ?*

Solution.   Deflecting torque, $T_d = \dfrac{1}{2} \dfrac{dC}{d\theta} V^2$

Controlling torque, $T_C = k\theta$

In the steady position of deflection, $T_d = T_C$.

$\therefore$ $k\theta = \dfrac{1}{2} \dfrac{dC}{d\theta} V^2$

or $\dfrac{dC}{d\theta} = \dfrac{2k\theta}{V^2} = \dfrac{2 \times (9.81 \times 10^{-8}) \times 80}{(1000)^2} = 15.7 \times 10^{-12}$ F per radian

$= 15.7$ µµF per radian $= 0.274$ µµF per degree

Change in capacitance when voltmeter reads 1000 V

$$= \dfrac{dC}{d\theta} \times \theta = 0.274 \times 80 = 21.92\ \text{µµF}$$

Total capacitance of the voltmeter when it reads 1000 V

$$= 10 + 21.92 = \textbf{31.92 µµF}$$

Example 16.38. *An electrostatic voltmeter is constructed with six parallel, semi-circular fixed plates equi-spaced at 4 mm intervals and five interleaved semi-circular movable plates that move in planes mid-way between the fixed plates in air. The instrument is spring controlled. If the radius of the movable plates is 4 cm, calculate the spring constant if 10 kV corresponds to full-scale deflection of 100°. Neglect fringing, edge effects and plate thickness.*

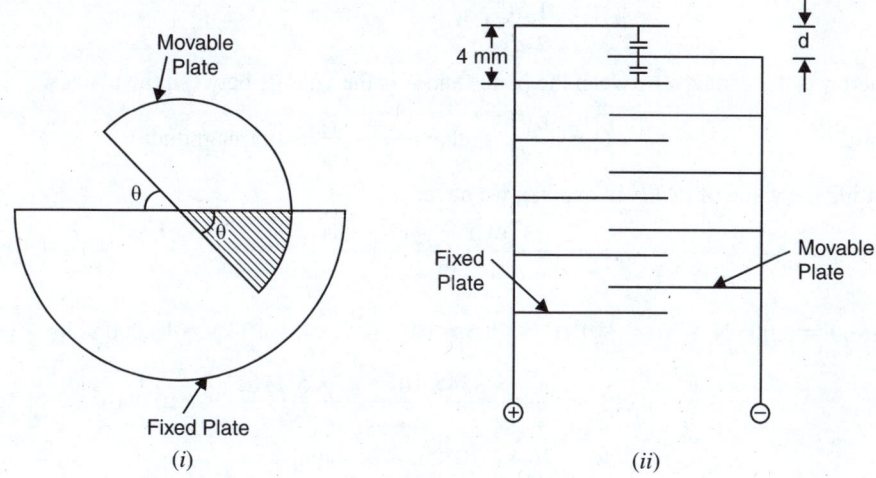

Fig. 16.50

Solution. Suppose the deflection is θ radian. The fixed and movable plates overlap each other (shaded portion) over an angle θ as shown in Fig. 16.50 (*i*). The total number of plates (both fixed and movable) is 11. Hence there are 10 parallel plate capacitors as shown in Fig. 16.50 (*ii*).

Overlap area between one fixed plate and one movable plate is

$$*A = \frac{1}{2}r^2\,\theta = \frac{1}{2}(0.04)^2 \times \theta = 8 \times 10^{-4}\,\theta\ \text{m}^2$$

$$d = 4/2 = 2\ \text{mm} = 2 \times 10^{-3}\ \text{m}$$

Capacitance of each of 10 parallel-plate capacitors

$$= \frac{\varepsilon_0\,A}{d} = \frac{(8.854 \times 10^{-12}) \times 8 \times 10^{-4}\,\theta}{2 \times 10^{-3}} = 3.54 \times 10^{-12}\,\theta\ \text{F}$$

Total capacitance, $C = 10 \times 3.54 \times 10^{-12}\,\theta = 35.4 \times 10^{-12}\,\theta\ \text{F}$

$$\therefore \qquad \frac{dC}{d\theta} = 35.4 \times 10^{-12}\ \text{F/radian}$$

Deflecting torque, $T_d = \dfrac{1}{2}\dfrac{dC}{d\theta}V^2\ \text{Nm} = \dfrac{1}{2} \times (35.4 \times 10^{-12}) \times (10{,}000)^2\ \text{Nm}$

$$= 17.7 \times 10^{-4}\ \text{Nm}$$

If k is the spring constant in Nm/radian, then, $T_C = k\theta$. In the steady position of deflection, $T_d = T_C$.

$$\therefore \qquad\qquad k\theta = 17.7 \times 10^{-4}$$

or $\qquad\qquad\qquad k = \dfrac{17.7 \times 10^{-4}}{1.74} \qquad [\theta = 100° = 100 \times \pi/180 = 1.74\ \text{rad.}]$

$$= \textbf{\textcolor{red}{10.17} } \times 10^{-4}\ \textbf{Nm/rad.}$$

Example 16.39. *An electrostatic voltmeter has two parallel plates. The movable plate is 10 cm in diameter. With 10 kV between the plates, the pull is 0.005 N. Find the change in capacitance for a movement of 1 mm of the movable plates.*

Solution. Force F between the charged plates is

---

* For an angle of 2π, area $= \pi r^2$. Therefore, for angle θ, $A = \dfrac{\pi r^2}{2\pi} \times \theta = \dfrac{1}{2}r^2\theta.$

$$F = \frac{1}{2}\frac{dC}{dx}V^2 \qquad \qquad ...(i)$$

where x is the distance between the plates and V is the voltage between the plates.

Now, $\qquad \qquad C = \frac{\varepsilon_0 A}{x}$ so that $\frac{dC}{dx} = \frac{\varepsilon_0 A}{x^2}$ (in magnitude)

Putting the value of dC/dx in exp. (i), we have,

$$F = \frac{1}{2}\times\frac{\varepsilon_0 A}{x^2}\times V^2.$$

Here, $F = 0.005$ N ; $A = \frac{\pi}{4}(0.1)^2 = 78.5\times10^{-4}$ m^2 ; $V = 10$ kV $= 10,000$ volts

$\therefore \qquad \qquad 0.005 = \frac{1}{2}\times\dfrac{8.854\times10^{-12}\times78.5\times10^{-4}}{x^2}\times(10,000)^2$

or $\qquad \qquad x = 26.4\times10^{-3}$ m $= 26.4$ mm

$\therefore$ Change in capacitance due to change in distance between plates from 26.4 mm to 25.4 mm

$$= \varepsilon_0 A\left(\frac{1}{x_2}-\frac{1}{x_1}\right)$$

$$= 8.854\times10^{-12}\times78.5\times10^{-4}\left(\frac{1}{25.4\times10^{-3}}-\frac{1}{26.4\times10^{-3}}\right)\text{F}$$

$$= 0.103\times10^{-12}\text{ F} = \textbf{0.103 }\boldsymbol{\mu\mu}\textbf{F}$$

Tutorial Problems

1. An electrostatic voltmeter has two flat parallel plates, each 12 cm^2 in area. If these plates are 5 mm apart, estimate the force of attraction when there is a p.d. of 2000 V between them.   **[8 × 10^{-4} N]**

2. An absolute electrostatic voltmeter has a movable circular plate 80 mm in diameter. If the distance between the plates during a measurement is 4 mm, find the p.d. when the force of attraction is 2×10^{-3} N. The dielectric is air having a permittivity of 8.85×10^{-12} F/m.   **[1203 V]**

3. The reading '100' of a 120 V electrostatic voltmeter is to represent 10,000 V when its range is extended by the use of a capacitor in series. If the capacitance of the voltmeter at the above reading is 70 pF, find the capacitance of the capacitor multiplier required.   **[0.707 pF]**

16.33. Induction Type Instruments

This class of instruments is suitable only for a.c. measurements. These instruments may be used either as ammeter or voltmeter or wattmeter or energy meter. Perhaps the widest application of induction principle is in watt-hour or energy meter.

Principle. Fig. 16.51 illustrates the principle of induction type instruments. Two alternating fluxes ϕ_1 and ϕ_2 (whose magnitudes depend upon the current or voltage to be measured) having a phase difference θ pass through a metallic disc, usually of copper or aluminium. These alternating fluxes induce currents in the disc. The current produced by one flux reacts with the other flux, and *vice-versa*, to produce the deflecting torque that acts on the disc. It can be proved that net deflecting torque on the disc is given by ;

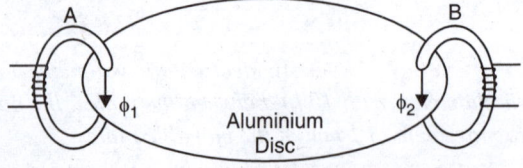

Fig. 16.51

$$T_d \propto \phi_{1m} \, \phi_{2m} \sin \theta$$

where $\quad \phi_{1m}$ = maximum value of alternating flux ϕ_1

ϕ_{2m} = maximum value of alternating flux ϕ_2

Obviously, to obtain maximum deflecting torque, the angle θ (*i.e.,* phase angle between ϕ_1 and ϕ_2) should be 90°.

The induction type instruments are worked on single phase. The question arises how to obtain two fluxes having a phase difference θ (= 90° as far as possible) from a single phase supply. This can be achieved in two ways *viz.*

1.   Splitting the phase (*i.e.,* Ferrari's Principle)

2.   Shaded-pole arrangement

1. Splitting the phase. In this method, two flux-producing windings are connected in parallel across a single phase supply; an inductive coil L in series with one and a resistance R in series with the other. The values of R and L are so selected that currents through the two windings [*i.e.,* I_R and I_L in Fig. 16.52 (*i*)] have a phase difference of nearly 90°. The result is that we have two alternating fluxes with a relative phase shift of 90°. These fluxes pass through the aluminium disc and induce currents in it to produce the necessary driving torque.

The fluxes produced by the two currents may be represented as :

$$\phi_1 = \phi_{1m} \sin \omega t$$
$$\phi_2 = \phi_{2m} \sin (\omega t + \theta)$$

where θ is the phase angle by which ϕ_2 leads ϕ_1.

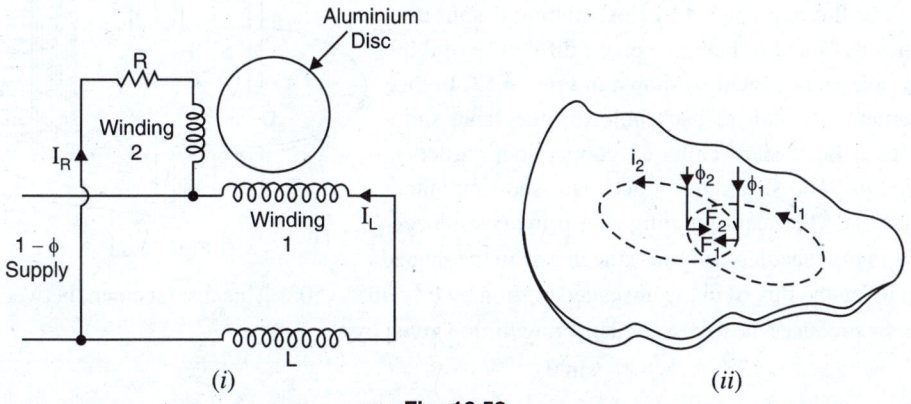

Fig. 16.52

The two fluxes ϕ_1 and ϕ_2 will induce e.m.f.s e_1 and e_2 respectively in the disc. Assuming r to be the resistance offered by the disc to each induced e.m.f., the induced currents are given by ;

$$i_1 = \frac{e_1}{r} = \frac{d\phi_1/dt}{r} = \frac{1}{r} \frac{d}{dt} (\phi_{1m} \sin \omega t)$$

$$= \frac{\omega \, \phi_{1m} \cos \omega t}{r}$$

or $\quad\quad\quad\quad\quad i_1 \propto \phi_{1m} \cos \omega t \quad\quad .... \text{ as } r \text{ and } \omega \text{ are constant}$

Similarly, $\quad\quad\quad\quad i_2 \propto \phi_{2m} \cos (\omega t + \theta)$

The portion of the disc which is traversed by flux ϕ_1 and carries current i_2 experiences a force F_1 along the direction indicated in Fig. 16.52 (*ii*). The magnitude of this force is given by ;

$$F_1 \propto \phi_1 \, i_2$$

Similarly, $\qquad\qquad\qquad F_2 \propto \phi_2 i_1$

Since the direction of both fluxes and both currents are the same, these forces will be in the opposite directions. This can be easily ascertained by applying left hand rule.

$\therefore \qquad$ Resultant force, $F \propto F_2 - F_1$

$$\propto (\phi_2 i_1 - \phi_1 i_2)$$
$$\propto \phi_{1m} \phi_{2m} [\sin (\omega t + \theta) \cos \omega t - \sin \omega t \cos (\omega t + \theta)]$$
$$\propto \phi_{1m} \phi_{2m} \sin \theta$$

This resultant force will produce the deflecting torque T_d which is directly proportional to it.

$\therefore \qquad\qquad\qquad T_d \propto \phi_{1m} \phi_{2m} \sin \theta$

  (*i*) If $\theta = 0°$ (*i.e.,* the two fluxes are in phase), then deflecting torque is zero. The deflecting torque will be maximum when $\theta = 90°$ *i.e.,* when the alternating fluxes have a phase difference of 90°.

 (*ii*) The deflecting torque is the same at every instant since ϕ_{1m}, ϕ_{2m} and θ are fixed for a given condition.

(*iii*) The direction of deflecting torque depends upon which flux is leading the other. The deflecting torque acts in such a direction so as to rotate the disc from the point where the leading flux passes the disc towards the point where the lagging flux passes the disc.

    2. Shaded-pole arrangement. The shaded-pole structure differs from the split-phase type in that there is only one flux-producing winding connected to a.c. supply. The flux ϕ produced by this winding is split into two portions ϕ_1 and ϕ_2 having a phase difference of θ by shaded-pole arrangement as shown in Fig. 16.53. In this arrangement, one-half of each pole (on the same side) embraces a thick short-circuited copper loop called a *shading coil*. The shading coil acts as a short-circuited secondary and the main winding as a primary. Induced currents in the *shading coil cause the flux ϕ_1 in the shaded

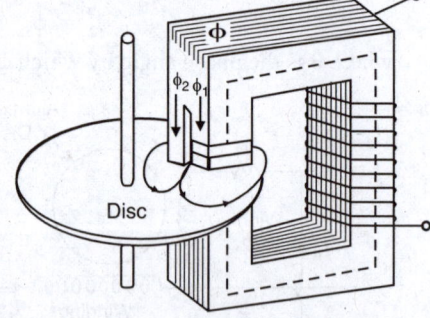

Fig. 16.53

portion to lag the flux ϕ_2 in the unshaded portion by θ (= 40° to 50°). This displacement between the two fluxes produces the necessary deflecting torque given by ;

$$T_d \propto \phi_{1m} \phi_{2m} \sin \theta$$

    The principle of operation of this type is the same as that of split-phase type.

16.34. Induction Ammeters and Voltmeters

    Induction ammeters and voltmeters can be used for a.c. measurement only and can be of shaded-pole type or split-phase type.

    1. Shaded-pole type. Fig. 16.54 shows the principal parts of a shaded-pole type instrument. It consists of a specially shaped aluminium disc coupled to a pointer and suspended in jewelled bearings. The disc passes through two air-gaps ; the

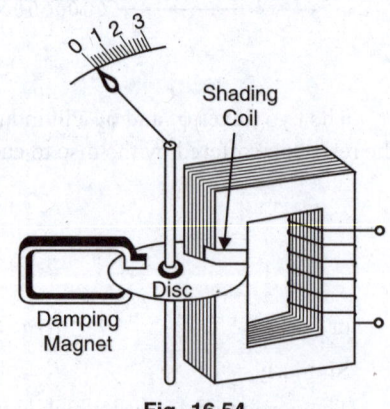

Fig. 16.54

---

*   Note that shading coil serves the same purpose as connecting resistance and inductance in split-phase arrangement.

first located in an electromagnet having a shading coil and the second in a permanent magnet. The permanent magnet provides the necessary damping torque. The controlling torque is provided by a spiral spring attached to the moving system. As shown above, the deflecting torque is given by ;

$$T_d \propto \phi_{1m}\, \phi_{2m} \sin \theta$$

(*i*) When used as an **ammeter**, the current to be measured or a part of it is passed through the operating coil of the instrument. Since both the fluxes are produced by the same alternating current I (r.m.s. value),

$\therefore$
$$T_d \propto I^2$$

As the instrument is spring controlled, $T_C \propto \theta$.

$\therefore$
$$\theta \propto I^2$$

(*ii*) When used as a **voltmeter**, current proportional to the voltage to be measured is passed through the operating coil.

$\therefore$
$$T_d \propto V^2$$

Again the instrument is spring controlled so that $T_C \propto \theta$.

$\therefore$
$$\theta \propto V^2$$

It is clear that shaded-pole instruments have uneven scale, being crowded in the beginning and open near the end of the scale. However, the non-linearity in the scale can be corrected to a considerable extent by modifying the shape of the disc *i.e.*, by using cam-shaped disc.

2. Split-phase type. In this method, the windings of the two electromagnets A and B are connected in parallel across a single phase supply ; an inductive coil L in series with one and a resistance R in series with the other. The values of R and L are so selected that the currents through the two windings [See Fig. 16.55] have a phase difference of nearly 90°. This produces the deflecting torque on the aluminium disc. The permanent magnet provides the necessary damping torque. The controlling torque is provided by a spiral spring attached to the moving system. As shown above, the deflecting torque is given by ;

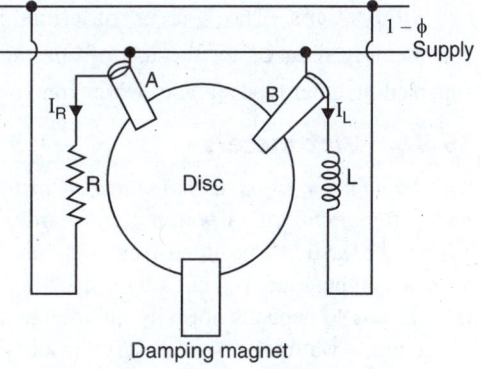

Fig. 16.55

$$T_d \propto \phi_{1m}\, \phi_{2m} \sin \theta$$

Both fluxes are proportional to current or voltage to be measured.

$\therefore$
$$T_d \propto I^2 \qquad ...for\ ammeter$$
$$\propto V^2 \qquad ...for\ voltmeter$$

As the instrument is spring controlled, $T_C \propto \theta$.

$\therefore$
$$\theta \propto I^2 \qquad ...for\ ammeter$$
$$\propto V^2 \qquad ...for\ voltmeter$$

Obviously, the scale of this type of instrument is also non-uniform.

16.35. Characteristics of Induction Ammeters and Voltmeters

The characteristics of induction ammeters and voltmeters are given below :

Advantages

(*i*) There is no moving iron in the instrument.

(*ii*) The moving element (*i.e.,* disc) is not electrically connected to the circuit.

    (*iii*) A full-scale deflection of over 250° can be obtained.

    (*iv*) They provide very effective damping.

    (*v*) They are not easily affected by stray magnetic fields owing to the intense concentration of instrument's own magnetic field.

Disadvantages

    (*i*) They have high cost.

    (*ii*) They have non-linear scales.

    (*iii*) They can be used for a.c. measurements only.

    (*iv*) They introduce fairly large errors due to temperature, frequency and waveform variations.

    (*v*) They consume fairly large power because of relatively large power losses in the shading coil.

Range extension. In voltmeters, the operating coil is of fine wire, a non-inductive resistance R being connected in series with it. In ammeters, the coil is comparatively of thick wire. A voltmeter is connected in the circuit either directly or through a potential transformer depending upon the voltage to be measured. An ammeter is connected directly in the circuit for currents upto 10 A provided the circuit voltage does not exceed 650 V. Beyond that it is usual to employ a current transformer.

Applications. The sources of errors in induction ammeters and voltmeters have been considerably reduced by the use of special alloys and modern design. Consequently, they are superseding other types (*e.g.* moving-iron voltmeters and ammeters) for switchboards and panels.

16.36. Wattmeters

A wattmeter, as its name implies, measures electric power given to or developed by an electric apparatus or circuit. A wattmeter is hardly ever required in a d.c. circuit because power ($P = VI$) can be easily determined from voltmeter and ammeter readings. However, in an a.c. circuit, such a computation is generally *speaking impossible. It is because in an a.c. circuit, power ($P = VI \cos \phi$) depends not only on voltage and current but also on the phase shift between them. Therefore, a wattmeter is necessary for a.c. power measurement. There are two principal types of wattmeters *viz.*,

    (*i*) Dynamometer wattmeter — for both d.c. and a.c. power

    (*ii*) Induction wattmeter — for a.c. power only

16.37. Dynamometer Wattmeter

A dynamometer wattmeter is almost universally used for the measurement of d.c. as well as a.c. power. *It works on the dynamometer prinicple i.e., mechanical force exists between two current carrying conductors or coils.*

Construction. When a dynamometer instrument is used as a wattmeter, the fixed coils are connected in series with the load and carry the load current (I_1) while the moving coil is connected across the load through a series multiplier R and carries a current (I_2) proportional to the load voltage as shown in Fig. 16.56. The fixed coil (or coils) is called the **current coil** and the movable coil is known as **potential coil**. The controlling torque is provided by two spiral springs which also serve the additional purpose of leading current into and out of the moving coil. Air friction damping is provided in such instruments. A pointer is attached to the movable coil.

---

*   Except for the case of pure resistance when $P = VI$ ($\because \cos \phi$ is 1 for pure resistance).

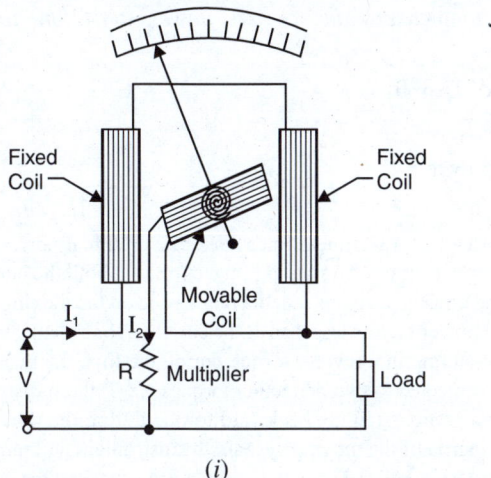

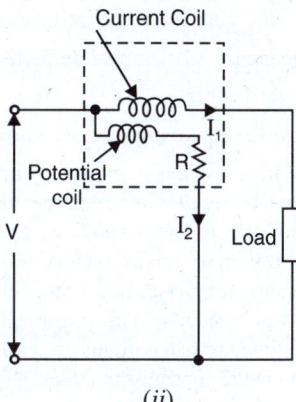

(i)   (ii)

Fig. 16.56

Working. When the wattmeter is connected in the circuit to measure power (See Fig. 16.56), the current coil carries the load current and potential coil carries current proportional to the load voltage. Due to currents in the coils, mechanical force exists between them. The result is that movable coil moves the pointer over the scale. The pointer comes to rest at a position where deflecting torque is equal to the controlling torque. Reversal of current reverses currents in both the fixed coils and the movable coil so that the direction of deflecting torque remains unchanged. Hence, such instruments can be used for the measurement of d.c. as well as a.c. power.

Deflecting torque. We shall now prove that deflecting torque is proportional to load power in a d.c. as well as a.c. circuit.

(i) Consider that the wattmeter is connected in a d.c. circuit to measure power as shown in Fig. 16.56 (ii). The power taken by the load is VI_1.

Deflecting torque, $T_d \propto I_1 I_2$

Since I_2 is directly proportional to V,

∴   Deflecting torque, $T_d \propto VI_1 \propto$ Load power

(ii) Consider that the wattmeter is connected in an a.c. circuit to measure power. Suppose at any instant, current through the load is i and voltage across the load is v. Let the load power factor be $\cos \phi$ lagging. Then,

$$v = V_m \sin \theta \quad ; \quad i = I_m \sin (\theta - \phi)$$

Instantaneous deflecting torque $\propto vi$

The pointer cannot follow the rapid changes in the instantaneous power owing to the large inertia of the moving system. Hence the instrument indicates the mean or average power.

∴   Average deflecting torque, $T_d \propto$ Average of vi over a cycle

$$\propto \frac{1}{2\pi} \int_0^{2\pi} V_m I_m \sin \theta \sin (\theta - \phi) \, d\theta$$

$$\propto \frac{V_m I_m}{2} \cos \phi$$

$$\propto VI \cos \phi$$

∴ $T_d \propto$ Load power

Thus whether the instrument is used to measure d.c. or a.c. power, deflecting torque is proportional to load power.

Since the instrument is spring-controlled, $T_C \propto \theta$

In the steady position of deflection, $T_d = T_C$.

$\therefore$ $\theta \propto$ Load power

Hence such instruments have uniform scale.

Note. It has been mathematically proved above that a wattmeter measures true or actual power (*i.e.*, $VI \cos\phi$) in an a.c. circuit. The physical explanation is as follows. At unity p.f., the current in both the current and potential coils of the wattmeter reverses at the same time, resulting in a deflecting torque on the moving element which is always in the same direction. However, at power factors less than 1, the current in one element reverses before the current reverses in the other element, resulting in a reverse torque during the time the two currents are in opposite directions. The instrument is now subjected to two deflecting torques *viz* (*i*) the forward torque during the time the two currents are in the same directions (*ii*) the backward torque during the time the two currents are in the opposite directions. Due to the inertia of the moving system, the instrument will indicate the resultant of the two torques. For a phase angle of 90° (*i.e.*, p.f. = 0), the two torques are equal and the wattmeter indicates zero power.

16.38. Characteristics of Dynamometer Wattmeters

It is worthwhile to give the advantages, disadvantages, errors, ratings and range extension c. dynamometer wattmeters.

Advantages

(*i*) They can be used for the measurement of a.c. as well as d.c. power.

(*ii*) They have uniform scale unlike dynamometer ammeters and voltmeters which have square-law scales.

(*iii*) By careful design, high accuracy can be obtained.

Disadvantages

(*i*) At low power factors, the inductance of the potential coil causes serious errors.

(*ii*) The readings of the instrument may be affected by stray magnetic fields. In order to prevent it, the instrument is shielded from the external magnetic fields by enclosing it in a soft-iron case.

Errors. A wattmeter may not give true reading due to several sources of errors such as (*i*) error due to connection of potential coil circuit (*ii*) error due to inductance of potential coil (*iii*) error due to capacitance in potential coil circuit (*iv*) error due to stray fields and (*v*) error due to eddy currents. These are fully discussed in Art. 16.39.

Wattmeter ratings. A wattmeter has two circuits *viz.* current coil circuit and potential coil circuit. These circuits (fixed coils and the movable coil) may burn out on carrying too much current. For this reason, current and voltage ratings are marked on the instrument and care should be taken that neither of this limits is exceeded.

It has been observed that coils of wattmeters are burnt up while their readings stood much less than full-scale value. This particularly happens when the load power factor is *low. At low power factors, the current through the current coil may exceed the safe value, thus burning the current coils of the instrument. Because a wattmeter gives no indication of current and voltage values, an ammeter and a voltmeter should always be used in conjuction with wattmeter when the load p.f. is suspected to be low or whenever the load power factor is not known.

---

*   Current through the current coil, $I = \dfrac{P}{V\cos\phi}$

At low p.f., current I is more for the same P and V.

Range extension. Since a dynamometer wattmeter is principally used for a.c. power measurement, range extension shall be discussed with reference to a.c. power measurement. The usual ranges of dynamometer wattmeters are from 0–25 A to 0–50 A for current coil circuit and from 0–15 V to 0–750 V for potential coil circuit. The current coil (*i.e.*, fixed coils) carries the whole of load current and for this reason it is made of thick copper wire of comparatively few turns. The current through the potential coil (*i.e.*, movable coil) is limited to about 30 mA by means of a high series resistance (*i.e.*, multiplier).

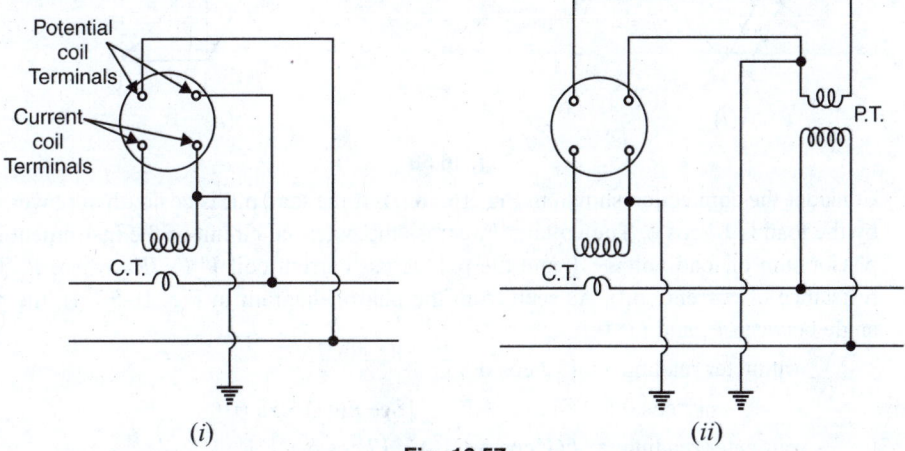

Fig. 16.57

For higher current ranges, a 0–5 A instrument is used with a current transformer (C.T.) as shown in Fig. 16.57 (*i*). On high voltage a.c. circuits, besides the current transformer, a potential transformer is used with a 0–110 V instrument as shown in Fig. 16.57 (*ii*). When instrument transformers (*i.e.*, C.T. and P.T.) are used in this way, indications of the pointer must be multiplied by the "transformation ratios" (N_S/N_P) to get the true power, just as the reading of a shunted ammeter must be multiplied by the multiplying power of the shunt to get the equivalent unshunted deflection.

16.39. Wattmeter Errors

When a dynamometer wattmeter is connected in a circuit to measure power, it may not indicate the true power due to several types of errors. A few of them are discussed below :

1. Error due to potential circuit connections. There are two methods of connecting a wattmeter in a single phase a.c. circuit as shown in Fig. 16.58. In Fig. 16.58 (*i*), the potential coil circuit of the wattmeter is connected on the supply side. In this case, the current through the current coil is the same as through the load but voltage applied across the potential coil circuit is larger than the load voltage by an amount equal to the voltage drop across the current coil. In Fig. 16.58 (*ii*), the potential coil circuit is connected on the load side. In this case, the same voltage acts on the potential coil circuit as on the load but current in the current coil is greater than the load current by an amount equal to the current taken by the potential coil circuit. For either connection, the wattmeter indicates a power greater than that actually taken by the load due to losses in the instrument circuits.

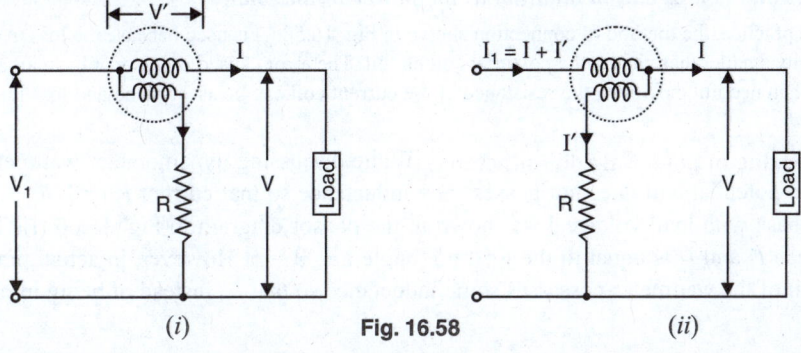

Fig. 16.58

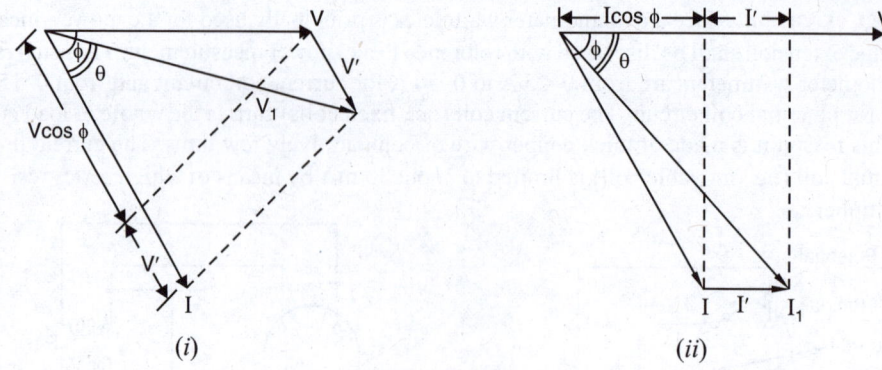

Fig. 16.58

(*i*) Consider the connection shown in Fig. 16.58 (*i*). If the load p.f. is cos ϕ, then power taken by the load is $VI \cos \phi$. The voltage V_1 across the potential circuit of the instrument is the phasor sum of load voltage V and the p.d. across current coil V' ($= IR_C$ where R_C is the resistance of current coil). As seen from the phasor diagram in Fig. 16.58 (*i*), the phase angle between V_1 and I is θ.

∴   Wattmeter reading $= V_1 I \cos \theta$

Now, $V_1 \cos \theta = V \cos \phi + V'$ [See Fig. 16.58 (*i*)]

∴   Wattmeter reading $= I(V \cos \phi + V') = VI \cos \phi + V'I$

                       $= VI \cos \phi + I^2 R_C$ $(\because V' = IR_C)$

                       $=$ Load power + Power in current coil

Thus with this type of connection, the wattmeter measures the power consumed by its current coil in addition to the power of the load.

(*ii*) Now consider the connection shown in Fig. 16.58 (*ii*). The power taken by the load is again $VI \cos \phi$. The current I_1 through the current coil of the instrument is the phasor sum of load current I and current I' ($= V/R_T$ where R_T is the total resistance of potential coil circuit) taken by the potential coil circuit. As seen from the phasor diagram in Fig. 16.58 (*ii*), the phase angle between V and I_1 is θ.

∴   Wattmeter reading $= VI_1 \cos \theta$

Now, $I_1 \cos \theta = I \cos \phi + I'$ [See Fig. 16.58 (*ii*)]

∴   Wattmeter reading $= V(I \cos \phi + I') = VI \cos \phi + VI'$

                       $= VI \cos \phi + V^2/R_T$ $(\because I' = V/R_T)$

                       $=$ Load power + Power in potential coil

Thus with this type of connection, the wattmeter measures power consumed by its potential coil circuit in addition to the power of the load.

Note. In practice, the method of connection shown in Fig. 16.58 (*i*) is used because the loss in the current coil is generally smaller than that in the potential coil circuit. The error caused is reasonably small and may be neglected. When needful, however, the resistance of the current coil can be ascertained and loss in the current coil calculated.

2. Error due to potential-coil inductance. While discussing dynamometer wattmeter, it was assumed that potential coil does not possess any inductance so that current I_2 ($= V/R_p + R$) drawn by it is in phase with load voltage V as shown in the phasor diagram in Fig. 16.60 (*i*). The phase angle between I_1 and I_2 is equal to the load p.f. angle *i.e.*, $\theta = \phi$. However, in actual practice, the potential coil of the wattmeter possesses some inductance so that I_2, instead of being in phase with load voltage V, will lag behind it by an angle α as shown in the phasor diagram in Fig. 16.60 (*ii*). Consequently, the phase angle between I_1 and I_2 is decreased and becomes $\theta = \phi - \alpha$. The result of this phase error is that wattmeter indicates power higher than actually taken by the load.

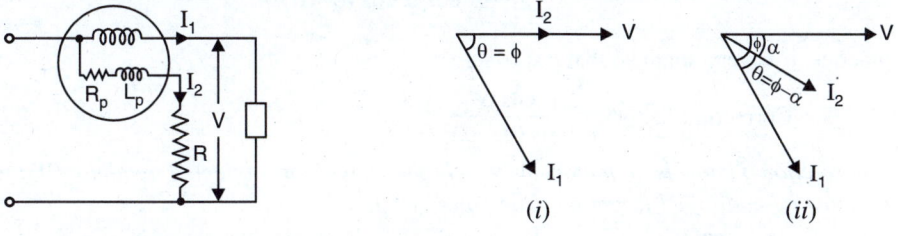

Fig. 16.59   **Fig. 16.60**

(*i*) Consider first that potential coil of the wattmeter does not possess any inductance.

$$\therefore \qquad I_2 = \frac{V}{R + R_p} = \frac{V}{R} \text{ app.}^*$$

$$\theta = \phi = \text{load p.f. angle}$$

The phasor diagram for such a case is shown in Fig. 16.60 (*i*).

Note that I_2 is in phase with the load voltage V.

$$\text{Wattmeter reading} \propto I_1 I_2 \cos \theta$$

$$\propto I_1 \frac{V}{R} \cos \phi \qquad \qquad \text{...(i)}$$

(*ii*) Next consider that potential coil of the wattmeter has inductance L_p.

$$\therefore \qquad I_2 = \frac{V}{\sqrt{(R + R_p)^2 + X_p^2}} = \frac{V}{\sqrt{R^2 + X_p^2}} = \frac{V}{Z_p}$$

The current I_2 now lags behind the load voltage V [See phasor diagram in Fig. 16.60 (*ii*)] by an angle α given by ;

$$\tan \alpha = \frac{X_p}{R + R_p} \simeq \frac{X_p}{R} = \frac{\omega L_p}{R}$$

Also, $\qquad \qquad \theta = \phi - \alpha$

$$\therefore \qquad \text{Wattmeter reading} \propto I_1 I_2 \cos(\phi - \alpha)$$

$$\propto I_1 \frac{V}{Z_p} \cos(\phi - \alpha)$$

Now, $\qquad \qquad \cos \alpha = \dfrac{R_p + R}{Z_p} \simeq \dfrac{R}{Z_p}$

$$\therefore \qquad Z_p = \frac{R}{\cos \alpha}$$

$$\therefore \qquad \text{Wattmeter reading} \propto I_1 \frac{V}{R} \cos \alpha \cos(\phi - \alpha) \qquad \qquad \text{...(ii)}$$

Exp. (*i*) gives the reading of the wattmeter when inductance of the potential coil is neglected and exp. (*ii*) gives the reading when it is taken into account.

Correction factor. The true reading (W_t) is given by exp. (*i*) whereas the actual or indicated reading (W_a) is given by exp. (*ii*). The ratio W_t/W_a is known as correction factor.

---

*   In practice, resistance R_p of potential coil is very small as compared to R (multiplier) so that R_p can be neglected in comparison to R.

$$\text{Correction factor} = \frac{W_t}{W_a} = \frac{\dfrac{VI_1}{R}\cos\phi}{\dfrac{VI_1}{R}\cos\alpha\cos(\phi-\alpha)} = \frac{\cos\phi}{\cos\alpha\cos(\phi-\alpha)}$$

In practice, α is very small so that $\cos\alpha = 1$.

$$\therefore \qquad \text{Correction factor} = \frac{\cos\phi}{\cos(\phi-\alpha)}$$

The correction factor is a factor by which the actual or indicated reading (W_a) of the wattmeter must be multiplied to get the true power (W_t).

Percentage error. The percentage error in terms of actual or indicated wattmeter reading (W_a) is given by ;

$$\% \text{ age error} = \frac{W_a - W_t}{W_a}\times 100 = \left[1-\frac{W_t}{W_a}\right]\times 100$$

$$= \left[1-\frac{\cos\phi}{\cos(\phi-\alpha)}\right]\times 100$$

$$= \left[1-\frac{\cos\phi}{\cos\phi\cos\alpha+\sin\phi\sin\alpha}\right]\times 100$$

$$= \left[1-\frac{\cos\phi}{\cos\phi+\sin\phi\sin\alpha}\right]\times 100 \qquad (\because \cos\alpha = 1)$$

$$= \frac{\sin\phi\sin\alpha}{\cos\phi+\sin\phi\sin\alpha}\times 100$$

$$\therefore \qquad \% \text{ age error} = \frac{\sin\alpha}{\cot\phi+\sin\alpha}\times 100$$

Note. The above expressions are for lagging p.f. of the load as is the case in actual practice. However, if the load p.f. happens to be leading, then change load p.f. angle ϕ to $-\phi$ in the above expressions. The reader may note that for leading load p.f., the wattmeter will indicate power less than actually taken by the load.

3. Error due to capacitance in potential coil circuit. The potential coil circuit of the wattmeter may possess capacitance in addition to inductance. This capacitance is mainly due to the inter-turn capacitance of the series resistance (*i.e.,* multiplier). The effect of this capacitance is to reduce angle α and thus reduce error due to inductance of the potential coil circuit. Efforts are made to design the multiplier in such a way that errors due to inductance and capacitance of the potential circuit *neutralise each other.

4. Error due to eddy currents. Another important source of error in wattmeters is that due to the eddy currents induced by the alternating flux in the metal parts of the instrument or in the thick current coils. These eddy currents will create a flux that lags the main flux produced by the current in the current coils (*i.e.,* load current). Since this flux lags the load current, it will produce an error opposite in sign to the inductance-error of potential coil. The wattmeter will read low for lagging p.f. and high for leading p.f. Great care should therefore be taken to avoid metal case or supports for the coils unless they are of specially chosen metals of high resistivity or are laminated or otherwise arranged so that no appreciable eddy currents are induced in them.

---

*   Often inductance of the potential coil circuit predominates capacitance. For this reason, in some wattmeters, a small capacitor is purposely connected in parallel with a portion of the series resistance. This helps in eliminating the error due to inductance of potential coil circuit.

5. Error due to stray magnetic fields. Since the dynamometer wattmeter has a relatively weak operating field (as no iron is used in the magnetic circuit), it is easily affected by stray magnetic fields and must be shielded. Laminated iron shields are used in portable instruments and steel cases sometimes provide shielding in switchboard instruments.

Example 16.40. *A dynamometer type wattmeter with its voltage coil connected across the load side reads 192 W. The load voltage is 208 V and the resistance of the potential coil circuit is 3825 Ω. Calculate (i) true load power and (ii) percentage error due to wattmeter connection.*

Solution. Since the voltage coil of the wattmeter is connected on the load side (See Fig. 16.61), the power consumed by it is also included in the reading of the wattmeter.

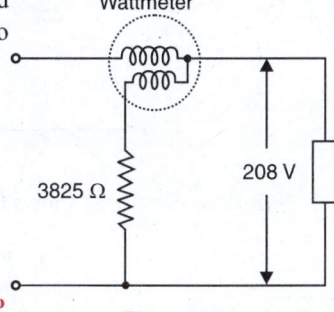

Fig. 16.61

Wattmeter reading = 192 W

Power taken by the potential coil circuit

$$= \frac{(208)^2}{3825} = 11.3 \text{ W}$$

(i)   True load power $= 192 - 11.3 = \textbf{180.7 W}$

(ii)   % age error $= \dfrac{192 - 180.7}{180.7} \times 100 = \textbf{6.25\%}$

Example 16.41. *The resistances of the two coils of a wattmeter are 0·01 Ω and 1000 Ω respectively and both are non-inductive. The load is taking a current of 20 A at 200 V and 0·8 p.f. lagging. Show the two ways in which the voltage coil can be connected and find the error in the reading of the meter in each case.*

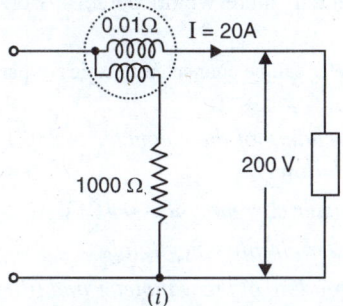
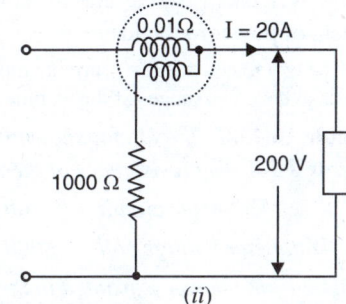

(i)   *(ii)*

Fig. 16.62

Solution. Fig. 16.62 shows the two possible ways of connecting the voltage coil of the wattmeter.

Load power $= VI \cos \phi = 200 \times 20 \times 0.8 = 3200$ W

(i) Consider the connections shown in Fig. 16.62 *(i)*.

Power loss in current coil $= I^2 R_C = (20)^2 \times 0.01 = 4$ W

∴   Wattmeter reading $= 3200 + 4 = 3204$ W

      %age error $= (4/3200) \times 100 = \textbf{0.125\%}$

(ii) Consider the connections shown in Fig. 16.62 *(ii)*.

Power loss in voltage coil $= V^2/R_p = (200)^2/1000 = 40$ W

∴   Wattmeter reading $= 3200 + 40 = 3240$ W

∴   %age error $= (40/3200) \times 100 = \textbf{1.25\%}$

Example 16.42. *A wattmeter has two current coils connected in parallel, each having a resistance of 0·7 Ω. The wattmeter is connected in a circuit to measure power with its potential coil on the supply side. The reading on the wattmeter is 150 W and the reading on the ammeter connected*

in series with the current coils is 3 A. Calculate (i) power loss in the wattmeter (ii) true load power and (iii) percentage error due to wattmeter connections.

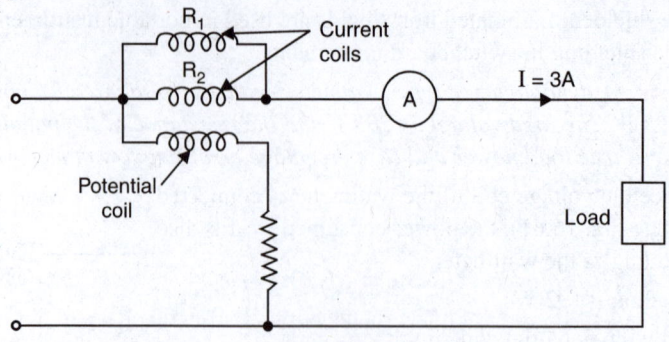

Fig. 16.63

Solution. Fig. 16.63 shows the wattmeter connections. Effective resistance of the current coils,

$$R_C = \frac{R_1 R_2}{R_1 + R_2} = \frac{0.7 \times 0.7}{0.7 + 0.7} = 0.35 \ \Omega$$

(*i*)   Wattmeter loss $= I^2 R_C = (3)^2 \times 0.35 =$ **3.15 W**

(*ii*)   True load power $= 150 - 3.15 =$ **146.85 W**

(*iii*)   %age error $= \dfrac{3.15}{146.85} \times 100 =$ **2.14%**

If the power taken by the current coils were neglected, there would be 2.14% error in the measurement of power to the load.

Note. The two fixed coils (*i.e.,* current coils) of the wattmeter can be connected in series or parallel. This will enable us to have two ranges of the wattmeter.

Example 16.43. *In the circuit shown in Fig. 16.64, reading of the voltmeter is 230 V and that of ammeter is 2·5 A. The resistances of the meters are as follows :*

Voltmeter circuit = 2000 Ω   ;   Wattmeter current coil = 0·46 Ω

Wattmeter voltage coil = 8000 Ω   ;   Ammeter circuit = negligible

Calculate (i) the power expended in the load (ii) the reading of the wattmeter and (iii) the reading of the wattmeter if its voltage coil is connected to point A instead of point B.

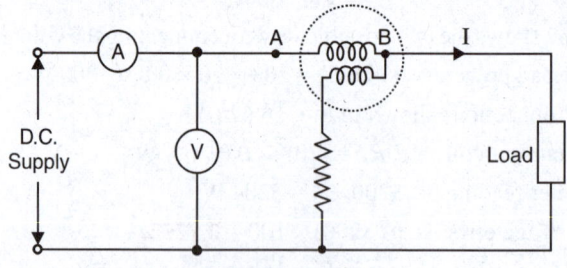

Fig. 16.64

Solution. (*i*)   Current through voltmeter $= 230/2000 = 0.115$ A

Current through current coil $= 2.5 - 0.115 = 2.385$ A

Voltage drop in current coil $= 2.385 \times 0.46 = 1.1$ V

P.D. across load and voltage coil $= 230 - 1.1 = 228.9$ V

Current taken by voltage coil $= 228.9/8000 = 0.029$ A

Load current, I = 2.385 – 0.029 = 2.356 A

Power expended in load = 228.9 × 2.356 = **539.29 W**

(*ii*)   Wattmeter reading = 228.9 × 2.385 = **545.93 W**

(*iii*) When the voltage coil is connected to point A instead of point B, voltage across the potential coil is 230 V and current through the current coil is 2.356 A.

∴   Wattmeter reading = 230 × 2.356 = **541.9 W**

Example 16.44. *A wattmeter is connected in a single phase circuit through a current transformer ; the potential coil being connected on the load side. The C.T. ratio is 5/1 and the resistance of the potential coil of the wattmeter is 10150 Ω. If the wattmeter indicates 45 W and load voltage indication is 230 V, find (i) wattmeter losses and (ii) power taken by the load.*

Solution. Fig. 16.65 shows the wattmeter connections in a single phase circuit through a current transformer.

(*i*) **Wattmeter losses.** Since the potential coil is connected across the load, the power consumed by it is also included in the wattmeter reading.

$$\text{Power loss in potential coil} = \frac{V^2}{R_{PC}} = \frac{(230)^2}{10150} = \textbf{5.21 W}$$

(*ii*) **Load power.** Since current coil of the wattmeter is connected through a current transformer of ratio 5/1, the actual reading of the wattmeter is given by ;

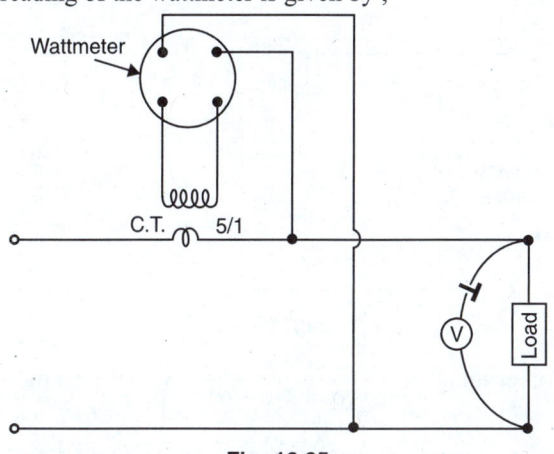

Fig. 16.65

Actual reading of wattmeter = Indicated reading × C.T. ratio

= 45 × 5 = 225 W

Power drawn by load = 225 – 5.21 = **219.79 W**

Note. The push button in series with the voltmeter enables the wattmeter to be read without adding the voltmeter load. The button is pushed to obtain the voltmeter reading.

Example 16.45. *The potential coil of an electrodynamic wattmeter has an inductance of 8 mH and a resistance of 2000 Ω. What is the percentage error of the instrument when measuring power in an inductive load having a p.f. of 0·707 lagging at 50 Hz ? Neglect the impedance of current coil and assume the current drawn by potential circuit to be negligible.*

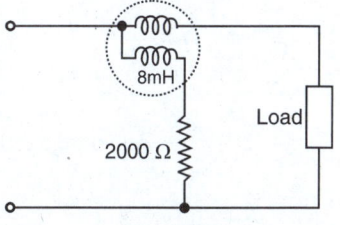

Fig. 16.66

Solution. Fig. 16.66 shows the circuit connections. When not mentioned, it is understood that potential coil is connected on the supply side.

Resistance of P.C., $R_p = 2000 \ \Omega$

Reactance of P.C., $X_p = 2\pi f L_p = 2\pi \times 50 \times 8 \times 10^{-3} = 2.5 \ \Omega$

Phase angle of P.C., $\alpha = \tan^{-1} X_p/R_p = \tan^{-1} 2.5/2000 = 0.072°$

Load p.f. angle, $\phi = \cos^{-1} 0.707 = 45°$

$\therefore$ Precentage error $= \dfrac{\sin \alpha}{\cot \phi + \sin \alpha} \times 100$

$$= \dfrac{\sin 0.072°}{\cot 45° + \sin 0.072°} \times 100 = \textbf{0.125\%}$$

Example 16.46. *The current coil of a wattmeter is connected in series with an ammeter and an inductive load. A voltmeter and voltage coil are connected across a 100 Hz supply. The ammeter reading is 4·5 A and voltmeter and wattmeter readings are 240 V and 23 W respectively. The inductance of the voltage coil circuit is 10 mH and its resistance is 2000 Ω. If the voltage drops across the ammeter and the current coil are negligible, what is the percentage error in the wattmeter reading ?*

Solution. The conditions of the problem are represented in Fig. 16.67.

Reactance of P.C., $X_p = 2\pi f L_p = 2\pi \times 100 \times 10 \times 10^{-3} = 6.28 \ \Omega$

$\therefore$ Phase angle of P.C., $\alpha = \tan^{-1} \dfrac{X_p}{R_p} = \tan^{-1} \dfrac{6.28}{2000} = 0.18°$

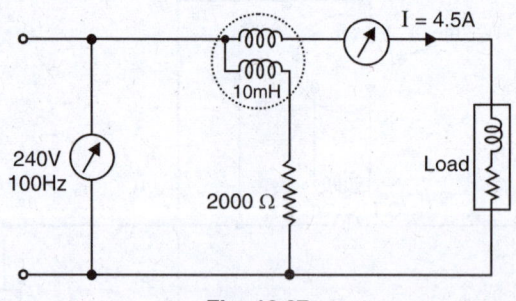

Fig. 16.67

Now,   True reading $= \dfrac{\cos \phi}{\cos \alpha \cos (\phi - \alpha)} \times$ Actual reading   *...See Art. 16.39*

or $VI \cos \phi = \dfrac{\cos \phi}{\cos \alpha \cos (\phi - \alpha)} \times$ Actual reading

or $VI = \dfrac{\text{Actual reading}}{\cos (\phi - \alpha)}$ [Taking *$\cos \alpha = 1$]

$\therefore$ $\cos (\phi - \alpha) = \dfrac{\text{Actual reading}}{VI} = \dfrac{23}{240 \times 4.5} = 0.0213$

or $\phi - \alpha = \cos^{-1} 0.0213 = 88.78°$

$\therefore$ $\phi = 88.78° + \alpha = 88.78° + 0.18° = 88.96°$

Percentage error $= \dfrac{\sin \phi}{\cot \phi + \sin \alpha} \times 100$

$$= \dfrac{\sin 0.18°}{\cot 88.96° + \sin 0.18°} \times 100 = \textbf{14.75\%}$$

---

*   Since $\alpha \ (= 0.18°)$ is very small, $\cos \alpha \simeq 1$.

Example 16.47. *The inductive reactance of the pressure coil circuit of a dynamometer wattmeter is 0.4% of its resistance at normal frequency and the capacitance is negligible. Calculate the percentage error and correction factor due to reactance for loads at (i) 0.707 p.f. lagging and (ii) 0.5 p.f. lagging.*

Solution. It is given that $X_p/R_p = 0.4\% = 0.004$.

Now, $\qquad\qquad \tan \alpha = \dfrac{X_p}{R_p} = 0.004 \quad \therefore \alpha = 0.23° \text{ so that } \sin \alpha = 0.004$

(i) When p.f. = 0.707 (i.e., $\phi = 45°$).

$$\text{Correction factor} = \frac{\cos \phi}{\cos(\phi - \alpha)} = \frac{\cos 45°}{\cos(45° - 0.23°)} = \mathbf{0.996}$$

$$\text{Percentage error} = \frac{\sin \alpha}{\cot \phi + \sin \alpha} \times 100 = \frac{\sin 0.23°}{\cot 45° + \sin 0.23°} \times 100$$

$$= \frac{0.004}{1 + 0.004} \times 100 = \mathbf{0.4\ \%}$$

(ii) When p.f. = 0.5 (i.e., $\phi = 60°$).

$$\text{Correction factor} = \frac{\cos \phi}{\cos(\phi - \alpha)} = \frac{\cos 60°}{\cos(60° - 0.23°)} = \mathbf{0.993}$$

$$\text{Percentage error} = \frac{\sin \alpha}{\cot \phi + \sin \alpha} \times 100 = \frac{\sin 0.23°}{\cot 60° + \sin 0.23°} \times 100$$

$$= \frac{0.004}{0.577 + 0.004} \times 100 = \mathbf{0.7\%}$$

Tutorial Problems

1. An electrodynamic wattmeter has a voltage circuit of resistance of 8 kΩ and inductance of 63.6 mH which is connected directly across a load carrying 8 A at a 50 Hz voltage of 240 V and p.f. of 0.1 lagging. Estimate the percentage error in the wattmeter reading caused by the loading and inductance of the voltage circuit. **[6.46% high]**

2. A 250 V, 10 A dynamometer type wattmeter has resistance of current and potential coils of 0.5 Ω and 12,500 Ω respectively. Find the percentage error due to each of the two methods of connections when unity p.f. loads at 250 V are of (i) 4 A (ii) 12 A. **[(i) 0.8 % ; 0.5% (ii) 2.4% ; 0.167%]**

3. A dynamometer type wattmeter with its voltage coil connected across the load side of the instrument reads 250 W. If the load voltage be 200 V, what power is taken by load? The voltage coil branch has a resistance of 2 kΩ. **[230 W]**

16.40. Induction Wattmeters

The induction type wattmeter can be used to measure a.c. power only in contrast to dynamometer wattmeter which can be used to measure d.c. as well as a.c. power. The principle of operation of an induction wattmeter is the same as that of induction ammeter and voltmeter *i.e.,* induction principle. However, it differs from induction ammeter or voltmeter in so far that two separate coils are used to produce the rotating magnetic field in place of one coil with phase split arrangement. Fig. 16.69 shows the physical arrangement of the various parts of an induction wattmeter.

Construction. Fig 16.68 shows the principal parts of an induction wattmeter.

(i) It consists of two laminated electromagnets. One electromagnet, called *shunt magnet* is connected across the supply and carries current proportional to the supply voltage. The coil of this magnet is made highly inductive so that the current (and hence the flux produced) in it lags behind the supply voltage by 90°. The other electromagnet, called *series magnet* is connected in series with the supply and carries the load current. The coil of this magnet is made highly non-inductive so that angle of lag or lead is wholly determined by the load.

(*ii*) A thin aluminium disc mounted on the spindle is placed between the two magnets so that it cuts the flux of both the magnets. The controlling torque is provided by spiral springs. The damping is electro-magnetic and is usually provided by a permanent magnet embracing the aluminium disc (See Fig. 16.69). Two or more closed copper rings (called *shading rings*) are provided on the central limb of the shunt magnet. By adjusting the position of these rings, the shunt magnet flux can be made to lag behind the supply voltage by exactly 90°.

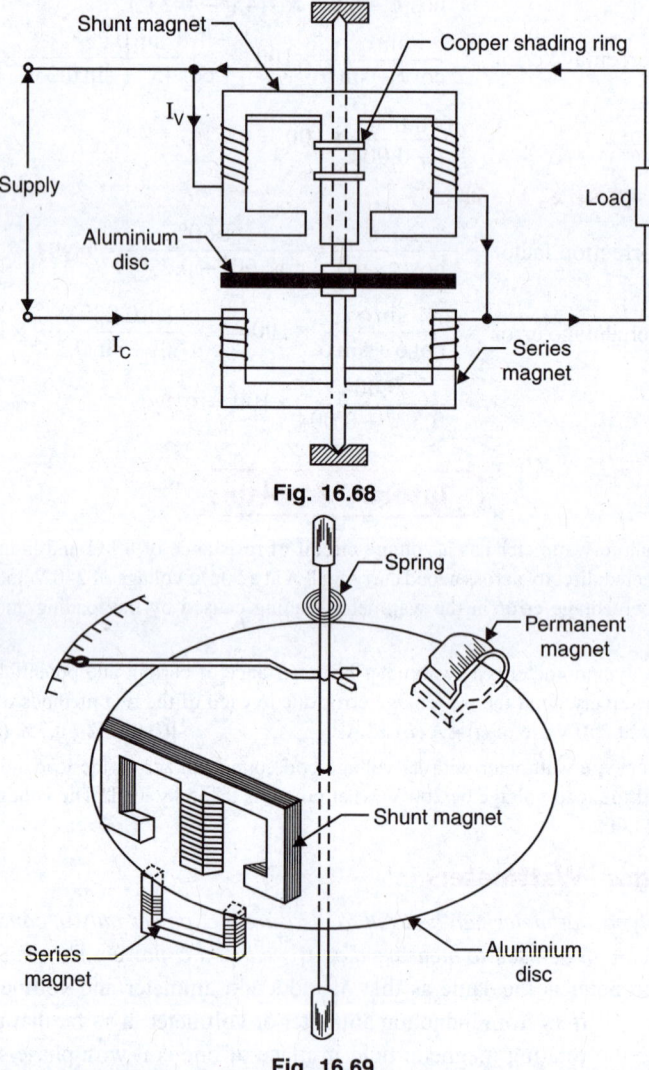

Fig. 16.68

Fig. 16.69

Working. When the wattmeter is connected in the circuit (See Fig. 16.68) to measure a.c. power, the shunt magnet carries current proportional to the supply voltage and the series magnet carries the load current. The two fluxes produced by the magnets induce eddy currents in the aluminium disc. The interaction between the fluxes and eddy currents produces the deflecting torque on the disc, causing the pointer connected to the moving system to move over the scale. The pointer comes to rest at a position where deflecting torque is equal to the controlling torque.

Let $\quad\quad V$ = supply voltage

$\quad\quad\quad I_V$ = current carried by shunt magnet

$\quad\quad\quad I_C$ = current carried by series magnet (= load curent I)

$\quad\quad\cos\phi$ = lagging power factor of the load

The phasor diagram is shown in Fig. 16.70. The current I_V in the shunt magnet lags the supply voltage V by 90° and so does the flux ϕ_V produced by it. The current I_C in the series magnet is the load current and hence lags behind the supply voltage V by ϕ. The flux ϕ_C produced by this current (*i.e.*, I_C) is in phase with it. It is clear that phase angle θ between the two fluxes is $90° - \phi$ *i.e.*,

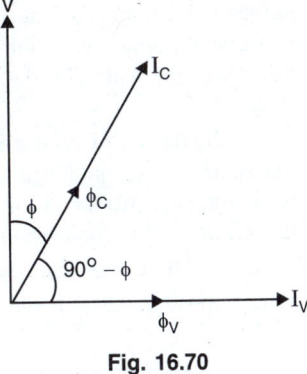

Fig. 16.70

$$\theta = 90° - \phi$$

∴ Mean deflecting torque, $T_d \propto \phi_V \phi_C \sin\theta$ ...*See Art. 16.33*

$$\propto VI \sin(90° - \phi)$$

$$[\because \phi_V \propto V \text{ and } \phi_C \propto I]$$

$$\propto VI \cos\phi$$

$$\propto \text{ a.c. power}$$

Since the instrument is spring controlled, $T_C \propto \theta$

For steady deflected position, $T_d = T_C$.

∴ $$\theta \propto \text{ a.c. power}$$

Hence such instruments have uniform scale.

Range. Circuit currents upto 100 A are handled by such wattmeters directly but for currents greater than this value, they are used in conjunction with current transformers (C.T.). Voltages upto 750 V are handled directly by such wattmeters but for voltages greater than this value, they are used in conjunction with potential transformers (P.T.).

Advantages

(*i*) They have a uniform scale.

(*ii*) They have a long scale (extending over 300°), making possible to take accurate readings.

(*iii*) They are free from the effects of stray fields.

(*iv*) They provide very good damping.

Disadvantages

(*i*) They can be used to measure a.c. power only.

(*ii*) They have low accuracy due to the heavy moving system.

(*iii*) They cause serious errors due to temperature variations.

(*iv*) They have high power consumption.

(*v*) Variation of frequency affects the reactance of the windings.

Applications. *Due to low accuracy and high power consumption, the characteristics of induction wattmeters are inferior to those of dynamometer wattmeters.* For this reason, dynamometer wattmeters are almost universally used for the measurement of a.c. as well as d.c. power. However, induction wattmeters have their chief application as panel instruments where the variations in frequency are not too much.

16.41. Three-phase Wattmeter

Two single phase wattmeters are required to measure the total power taken by a 3-phase circuit (See Art. 15.22). Two such wattmeters may be combined into one so that total 3-phase power is read from a single scale. Such an arrangement gives rise to a 3-phase wattmeter. Fig. 16.71 shows a 3-phase dynamometer type wattmeter which is most commonly used in polyphase circuits. The arrangement consists of two similar dynamometer wattmeters; the two voltage coils being mounted on the same shaft and rotate in their respective current coils. The wattmeter elements are connected according to two-wattmeter method and the resulting deflection of the instrument pointer is a function of the algebraic summation of the torques produced by the individual element. The deflection of the instrument pointer is therefore a function of the total power.

In the design of such a 3-phase wattmeter, care must be taken that the two elements have no mutual action *i.e.*, field from one element must not produce torque in the other. This may be checked by exciting the current circuit of one element and the voltage circuit of the other. There should be no deflection for this connection. Also the two elements must be matched in characteristics. This may be checked by connecting the elements with voltage circuits in parallel and the current circuits in series opposing. Again there should be no deflection.

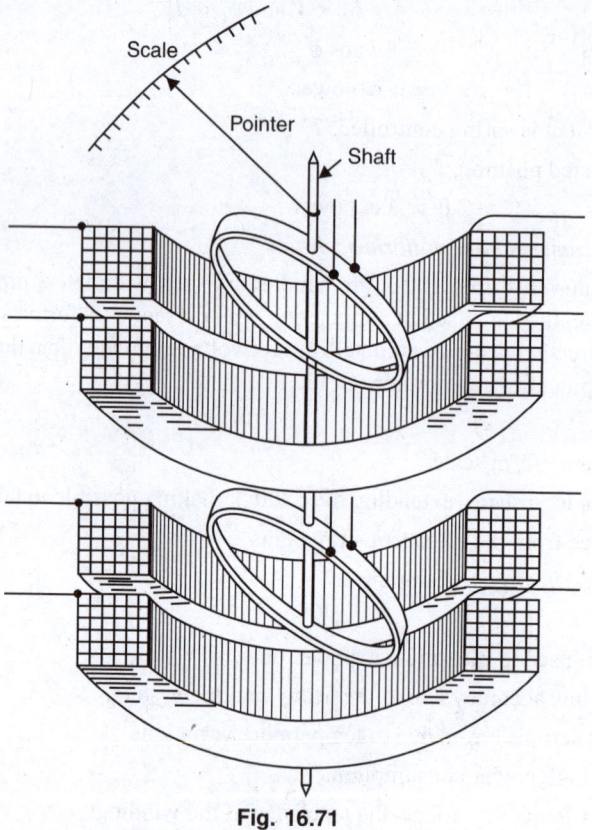

Fig. 16.71

Fig. 16.72 shows the terminals of a 3-phase wattmeter. It has two current coils (L_1S_1 and L_3S_3) and two voltage coils (V_1V_2 and V_2V_3). The letters L and S respectively stand for load and supply. Note that current coils and voltage coils are connected in the two supply lines (S_1 and S_3 in the present case) of the 3-phase circuit. Observing *polarity is essential in a 3-phase wattmeter because if the

---

*   Polarity is not important in a single phase wattmeter because if the meter tends to indicate in the reverse, this condition is corrected by merely reversing connections of current coil or potential coil.

instrument is not connected correctly, it may indicate upscale but incorrectly. A convenient rule to follow is to connect the instrument so that, as the current flows from the supply, it enters both the current and voltage circuits of the wattmeter at the marked (0 or ±) terminals. Note the arrowheads in the diagram. The arrowhead between S_1 and L_1 indicates that positive direction of current is from S_1 to L_1. Similarly positive direction of voltage is from V_1 to V_2 and from V_2 to V_3.

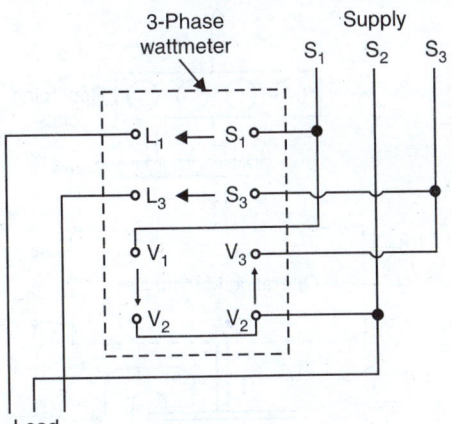

Fig. 16.72

16.42. Watthour Meters or Energy Meters

The measurement of electrical energy is, from the viewpoint of economics, the most important of all electrical measurements. An instrument which measures electrical energy is called an energy meter or a watthour meter. Since electrical energy consumed by a load adds up as the time goes on (watt-hours = watts × hours), *it is evident that watthour meter is an integrating instrument.* The following three types of energy meters are most commonly used :

(*i*) Commutator motor meter .... for d.c. and a.c. circuits.

(*ii*) Mercury motor meter ...... for d.c. circuits.

(*iii*) Induction motor meter .... for a.c. circuits only.

In general, energy meters designed for d.c. circuits can be used on a.c. circuits but the reverse is not true. *All energy meters are essentially * wattmeters with control spring and pointer removed but braking torque and counting mechanism provided.* Since there is **no controlling torque, the moving system will rotate continuously like an electric motor and hence the name motor meters. The driving torque on the rotor will obviously be proportional to the power being supplied. A braking or retarding torque proportional to the disc speed is provided by a brake magnet. Consequently, the disc speed will be proportional to the power and the number of revolutions made by the disc in a given time which is a measure of the electrical energy taken by the load in that time. The counting mechanism is so designed that it counts the revolutions of the disc in terms of kilowatthours (kWh).

General Theory. The general operating principle of all watthour meters is the same. Fig. 16.73 (*i*) shows the general construction of a watthour meter. As indicated, the operating mechanism has a shunt circuit and a series circuit connected in the same manner as in a wattmeter. The shunt circuit carries current proportional to the supply voltage and the series circuit carries the load current. Consequently, the driving torque on the ***rotor is proportional to the power being supplied *i.e.,*

---

* As in a wattmeter, the driving torque in an energy meter is provided by a shunt circuit (carrying current proportional to supply voltage) and a series circuit (carrying the load current).

** Unlike an indicating instrument in which the controlling torque is provided which allows the moving system to rotate through a fraction of a revolution.

*** The rotor may be an aluminium disc or an armature mounted on the shaft ; depending upon the type of energy meter.

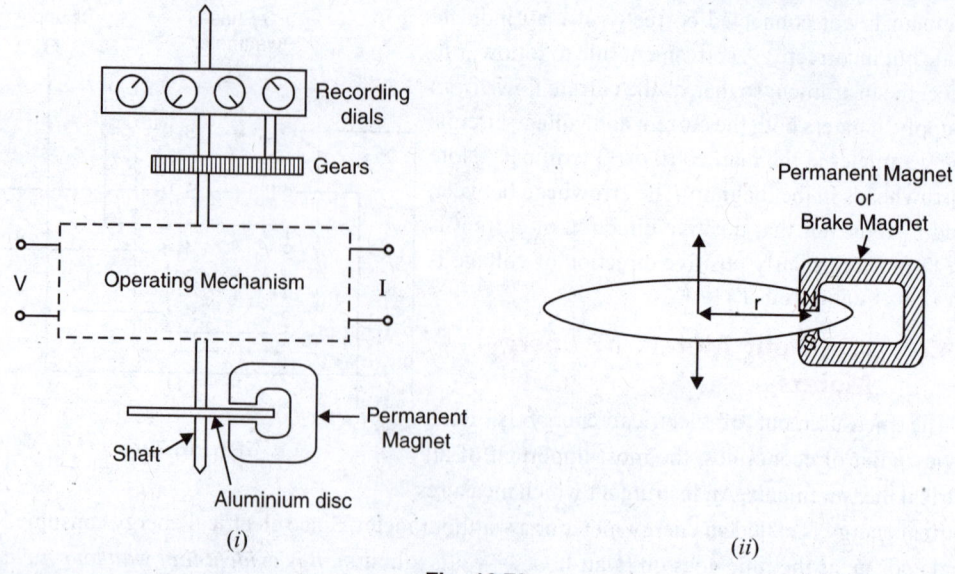

Fig. 16.73

Driving torque, $T_d \propto$ Power

Since there is no controlling torque, the disc will rotate continuously like an electric motor. If the speed of rotation of the rotor is made proportional to the driving torque (*i.e.*, power), then its rate of rotation can be used as a measure of power and the total number of revolutions in a given time is a measure of the electrical energy taken by the load in that time. This is achieved by providing a braking torque proportional to the rotor speed. Such a braking torque is produced by rotating an aluminium disc mounted on the same spindle between the poles of a permanent magnet (called brake magnet).

Braking torque. Fig. 16.73 (*ii*) shows how braking torque is provided in a watthour meter. An aluminium disc is attached to the spindle and intercepts the flux of a permanent magnet (*i.e.*, brake magnet). As the disc rotates, eddy currents are induced in the disc. These eddy currents react with the flux of the permanent magnet to produce a braking (or retarding) torque proportional to the disc speed.

Let ϕ = flux of permanent magnet

n = disc speed

∴ E.M.F. induced in disc, $e \propto \phi n$

If R is the resistance of eddy current path, then,

Induced eddy current, $i = \dfrac{e}{R}$ or $i \propto \dfrac{\phi n}{R}$ ($\because e \propto \phi n$)

Braking torque, $T_B \propto \phi i$ or $T_B \propto \dfrac{\phi^2 n}{R}$ ...(*i*)

The braking mechanism is so designed that the values of ϕ and R remain constant.

∴   Braking torque, $T_B \propto n$

For a *steady speed of rotation, $T_d = T_B$.

∴   Power $\propto n$ (*i.e.*, disc speed)

---

*   The strength of the brake magnet is so designed that braking torque will balance the driving torque at a certain given speed which is usually dictated by the past practice on the speed of meters in service.

Multiplying both sides by t, the time for which the power is supplied, we have,

$$\text{Power} \times t \propto n\,t$$

or

$$\text{Energy} \propto N$$

where $N (= n\,t)$ is the total number of revolutions in time t.

The number of revolutions of the rotor or disc are recorded on dials which are geared to the shaft [See Fig. 16.73 (i)]. The counting mechanism is so arranged that it indicates kilowatthours (kWh) directly and not revolutions.

Note. Eq. (i) above reveals that braking torque will be proportional to disc speed only if the values of ϕ and R are constant. This implies that strength of the brake magnet should remain constant throughout the life of the meter. Also the resistance of the disc should be substantially constant.

16.43. Commutator Motor Meter

The first electrical energy meter was the commutator meter developed by Elihu Thomson in 1889. This meter can be used to measure d.c. as well as a.c. energy, although it is rarely used to measure a.c. energy. It is because induction watthour meter (See Art. 16.46) is far superior to the commutator meter for the measurement of a.c. energy.

Construction. Fig. 16.74 shows the various parts of a commutator motor meter. It is essentially a small motor (no iron is used in the fields or armature) with a magnetic brake. The field coils FF consist of a few turns of heavy copper wire and are connected in series with the load so that they carry the load current. In the field of these coils rotates an armature pivoted in jewelled bearings. The armature has a number of coils connected to the segments of a small commutator. The commutator is made of silver and the brushes are silver tipped in order to reduce friction. The armature A and a high resistance R (multiplier) are in series and connected across the supply. Therefore, armature carries current proportional to the supply voltage. An aluminium disc mounted on the spindle (or shaft) rotates between the poles of two permanent magnets and provides the necessary braking torque. The braking torque prevents the armature from rotating too fast and from continuing to rotate after the load is disconnected.

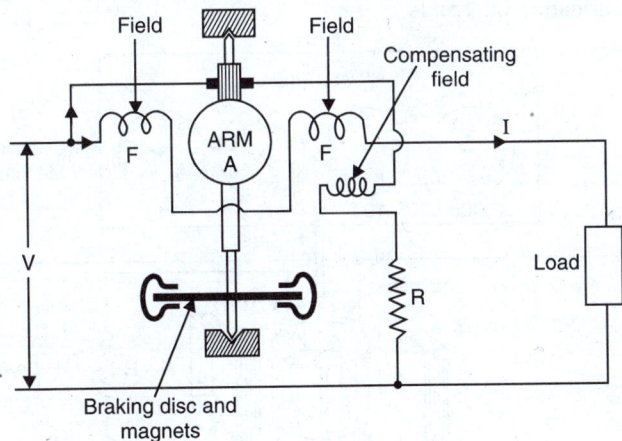

Fig. 16.74

Theory. When the meter is connected in the circuit to measure energy, the field coils carry the load current and the armature (or rotor) carries current proportional to the supply voltage. In this respect, it resembles dynamometer wattmeter so that driving torque is proportional to the power being supplied *i.e.,*

$$\text{Driving torque, } T_d \propto \text{Power}$$

The braking torque is due to the eddy currents induced in the aluminium disc. Since the magnitude of eddy currents is proportional to disc speed, the braking torque will also be proportional to the disc speed n i.e.,

$$\text{Braking torque, } T_B \propto n \qquad \qquad ...See Art. 16.42$$

For steady speed of rotation, $T_d = T_B$.

$$\therefore \qquad \text{Power} \propto n$$

Multiplying both sides by t, the time for which the power is supplied,

$$\text{Power} \times t \propto n\,t$$

or

$$\text{Energy} \propto N$$

where $N (= n\,t)$ is the total number of revolutions in time t.

The counting mechanism is so arranged that the meter indicates kilowatthours (kWh) directly and not the revolutions as shown in Fig. 16.75. The armature is connected through a set of gears to a series of dials that indicate the electrical energy (in kWh) consumed. The right-hand dial indicates kWh in units of 1, the next in units of 10, the next in units of 100 and the left-hand dial in units of 1000.

Friction Compensation. There are a number of sources of mechanical friction in the Thomson watt-hour meter *viz* (*i*) bearing (*ii*) gear train and meter register (*iii*) brush pressure on the commutator and (*iv*) windage. The frictional effects may cause considerable error unless compensation for friction is provided. This is accomplished by means of a *compensating coil* (See Fig. 16.74) placed co-axially with the two current coils and connected in series with the armature. Its flux adds to that of the field coils which carry the load current. Since the compensating coil carries current proportional to the supply voltage, its reaction with the armature current (constant at constant voltage) will contribute a substantially constant driving torque on the motor. Its torque contribution may be adjusted by changing its position relative to the armature to alter the portion of its flux which the armature intercepts.

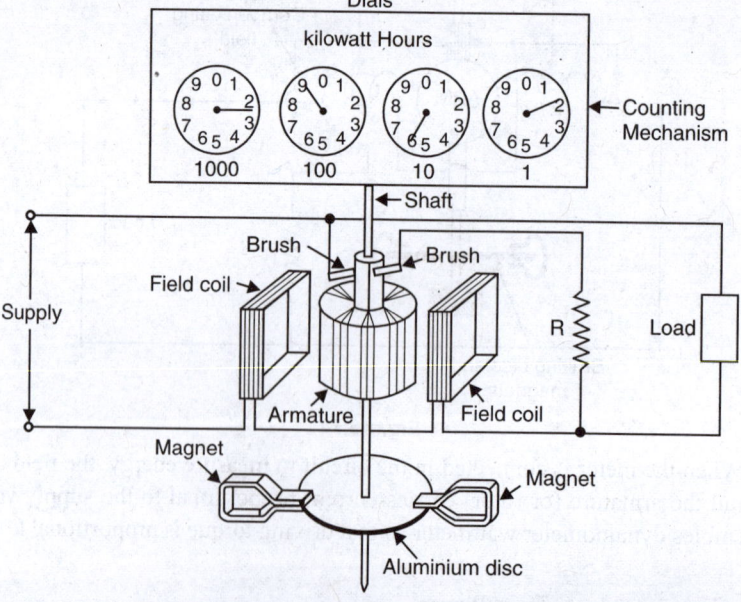

Fig. 16.75

16.44. **Mercury Motor Watthour Meter**

The weight of the moving system of a commutator is quite large and the torque-weight ratio is low. Also the friction is large because of the brush pressure required for good commutation. Both these disadvantages are largely overcome in the mercury motor meter. This type of instrument is used for direct currents only.

Principle. The working principle of this meter is the same as that of commutator motor meter. Here the rotor is a copper disc (instead of armature) floating in a mercury chamber through which current is passed. The mercury serves the same purpose as the brushes and commutator in the commutator meter.

Construction. Fig. 16.76 shows the essential parts of a mercury motor watt-hour meter. It consists of a thin copper disc mounted on the spindle and floating in a mercury chamber. Below this disc is placed an electromagnet whose coils are in series with a high resistance and connected across the supply. Therefore, current in the coils of this magnet is proportional to the supply voltage. The circuit current is passed through the disc at its circumference through mercury. This current leaves the disc at a point on the circumference diametrically opposite to the entrance point. In order to ensure radial flow of current through the disc, radial slots are cut in the disc. As the disc is slotted, it cannot be used for braking purposes. Another aluminium disc is used to provide the necessary braking torque. This disc is mounted on the same spindle and rotates between the poles of two permanent magnets.

Theory. When the meter is connected in the circuit to measure energy, the coils of the electromagnet carry current proportional to the supply voltage. The flux produced by the electromagnet is, therefore, proportional to the supply voltage. The reaction between the current in the disc and magnetic field due to the electromagnet produces the necessary driving torque.

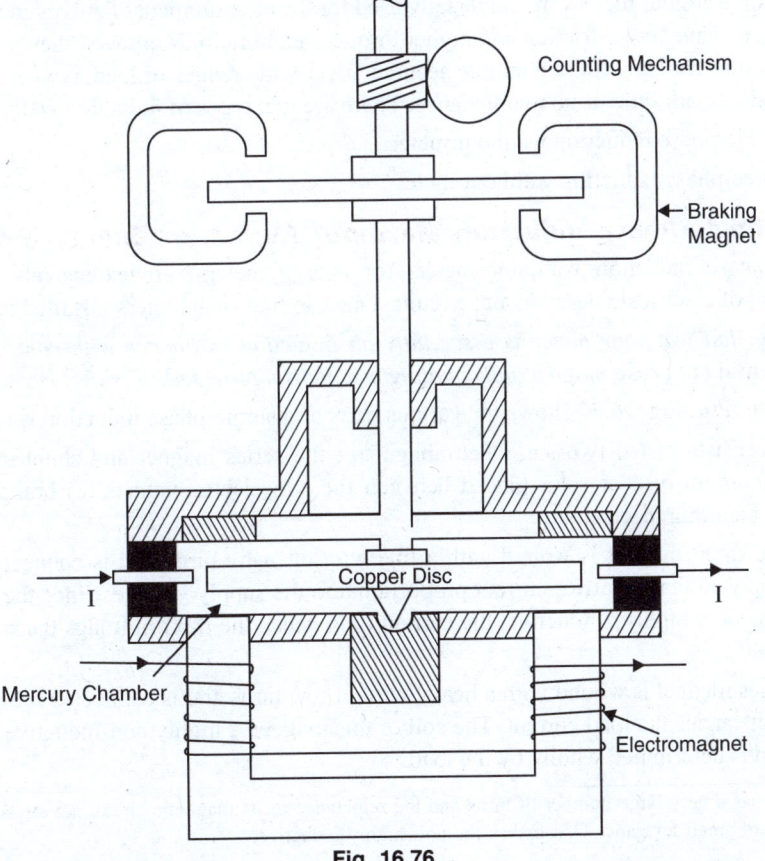

Fig. 16.76 .

Driving torque, $T_d \propto$ Flux of electromagnet × Load current

$$\propto VI$$

$$\propto \text{Power}$$

The braking torque is due to the eddy currents induced in the aluminium disc. Since the magnitude of eddy currents is proportional to the disc speed, the braking torque will also be proportional to the disc speed n *i.e.,*

Braking torque, $T_B \propto n$

For steady speed of rotation, $T_d = T_B$.

$\therefore$ Power $\propto n$

Multiplying both sides by t, the time for which the power is supplied,

Power × $t \propto nt$

or Energy $\propto N$

where $N (= n\,t)$ is the total number of revolutions in time t.

The counting mechanism is so arranged that the meter indicates kilowatthours (kWh) directly and not the revolutions.

Note. The meter can be used as an ampere-hour meter with slight modifications. The construction of mercury ampere-hour meter is the same except that a permanent magnet is used with disc in place of the electromagnet to provide the driving torque.

16.45. Induction Watthour Meters or Energy Meters

Induction watthour meters are universally used for the measurement of energy in a.c. circuits. It is because they have lower friction and higher torque-weight ratio. Moreover, they are inexpensive and can measure energy with remarkable accuracy over wide ranges of load, power factor, voltage and temperature with little or no maintenance. There are two types of induction watthour meter *viz*:

(*i*) Single-phase induction watthour meter

(*ii*) Three-phase induction watthour meter

16.46. Single-Phase Induction Watthour Meters or Energy Meters

Single-phase induction watthour meters (or energy meters) are extensively used for the measurement of electrical energy in a.c. circuits. One can find such meters installed in homes.

An induction watthour meter is essentially an induction wattmeter with control spring and pointer removed but brake magnet and counting mechanism provided.

Construction. Fig. 16.77 shows the various parts of a single-phase induction watthour meter.

(*i*) It consists of (*a*) two a.c. electromagnets ; the series magnet and shunt magnet (*b*) an aluminium disc or rotor placed between the two electromagnets (*c*) brake magnet and (*d*) counting mechanism.

(*ii*) The shunt magnet is wound with a fine wire of many turns and is connected across the supply so that it carries current proportional to the supply voltage. Since the coil of shunt magnet is highly *inductive, the current (and hence the flux) in it lags the supply voltage by 90°.

The series magnet is wound with a heavy wire of few turns and is connected in series with the load so that it carries the load current. The coil of this magnet is highly non-inductive so that angle of lag or lead is determined wholly by the load.

---

* The coil has a very large number of turns and the reluctance of its magnetic circuit is very small due to the presence of small air gaps. This makes the coil highly inductive.

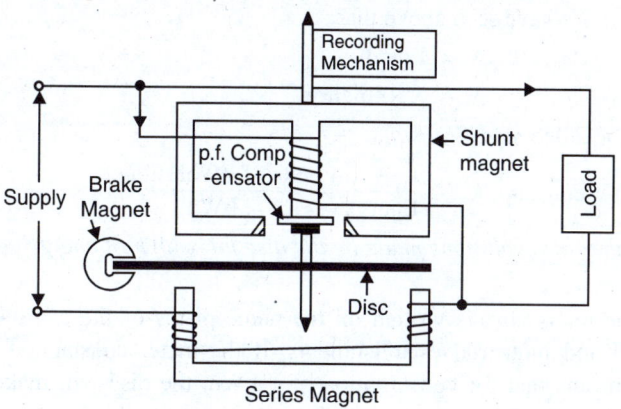

Fig. 16.77

(iii) A thin aluminium disc mounted on the spindle is placed between the shunt and series magnets so that it cuts the fluxes of both the magnets.

(iv) The braking torque is obtained by placing a permanent magnet near the rotating disc so that the disc rotates in the field established by the permanent magnet. Eddy currents induced in disc produce a braking or retarding torque that is proportional to the disc speed.

(v) A short-circuited copper loop (also known as *power factor compensator*) is provided on the central limb of the shunt magnet. By adjusting the position of this loop, the shunt magnet flux can be made to lag behind the supply voltage exactly by 90°.

Frictional compensation is obtained by means of two adjustable short-circuited loops placed in the leakage gaps of the shunt magnet. Geared to the rotating element is counting mechanism which indicates the energy consumed directly in kilowatthours (kWh).

Theory. When induction watthour meter is connected in the circuit to measure energy, the shunt magnet carries current proportional to the supply voltage and the series magnet carries the load current. Therefore, expression for the driving torque is the same as for induction wattmeter. Referring back to the phasor diagram in Fig. 16.70,

$$\text{Driving torque, } T_d \propto \phi_V \phi_C \sin \theta$$
$$\propto VI \sin (90° - \phi)$$
$$\propto VI \cos \phi$$
$$\propto \text{Power}$$

The braking torque is due to the eddy currents induced in the aluminium disc. Since the magnitude of eddy currents is proportional to the disc speed, the braking torque will also be proportional to the disc speed n i.e.,

$$\text{Braking torque, } T_B \propto n$$

For steady speed of rotation, $T_d = T_B$.

∴ $$\text{Power} \propto n$$

Multiplying both sides by t, the time for which power is supplied,

$$\text{Power} \times t \propto n t$$

or $$\text{Energy} \propto N$$

where $N (= n t)$ is the total number of revolutions in time t.

The counting mechanism is so arranged that the meter indicates kilowatthours (kWh) directly and not the revolutions.

Meter constant. We have seen above that :

$$N \propto \text{Energy}$$

or
$$N = K \times \text{Energy}$$

where K is a constant called *meter constant*.

$$\therefore \qquad \text{Meter constant, } K = \frac{N}{\text{Energy}} = \frac{\text{No. of revolutions}}{\text{kWh}}$$

Hence the number of revolutions made by the disc for 1 kWh of energy consumption is called **meter constant.**

The meter constant is always written on the name plates of the energy meters installed in homes, commercial and industrial establishments. If the meter constant of an energy meter is 1500 rev./kWh, it means that for consumption of 1 kWh, the disc will make 1500 revolutions.

16.47. Errors in Induction Watthour Meters

The users of electrical energy are charged according to the readings of the energy meters installed in their premises. It is, therefore, very important that construction and design of energy meters should be such as to ensure long-time accuracy *i.e.,* they should give correct readings over a period of several years under normal use conditions. Some of the common errors in energy meters and their remedial measures are discussed below :

(i) **Phase error.** The meter will read correctly only if the shunt magnet flux lags behind the supply voltage by exactly 90°. Since the shunt magnet coil has some resistance and is not completely reactive, the shunt magnet flux does not lag the supply voltage by exactly 90°. The result is that the meter will not read correctly at all power factors.

Adjustment. The flux in the shunt magnet can be made to lag behind the supply voltage by exactly 90° by adjusting the position of the shading coil placed around the lower part of the central limb of the shunt magnet. A current is induced in the shading coil by the shunt magnet flux and causes a further displacement of the flux. By moving the shading coil up or down the limb, the displacement between shunt magnet flux and the supply voltage can be adjusted to 90°. This adjustment is known as *lag adjustment or power factor adjustment.*

(ii) **Speed error.** Sometimes the speed of the disc of the meter is either fast or slow, resulting in the wrong recording of energy consumption.

Adjustment. The speed of the disc of the energy meter can be adjusted to the desired value by changing the position of the brake magnet. If the brake magnet is moved towards the centre of the spindle, the braking torque is reduced and the disc speed is *increased. Reverse would happen should the brake magnet be moved away from the centre of the spindle.

(iii) **Frictional error.** Frictional forces at the rotor bearings and in the counting mechanism cause noticeable error especially at light loads. At light loads, the torque due to friction adds considerably to the braking torque. Since friction torque is not proportional to the speed but is roughly constant, it can cause considerable error in meter reading.

Adjustment. In order to compensate for this error, it is necessary to provide a constant addition to the driving torque that is equal and opposite to the friction torque. This is produced by means of two adjustable short-circuited loops placed in the leakage gaps of the shunt magnet. These loops upset the symmetry of the leakage flux and produce a small torque to oppose the friction torque. This adjustment is known as *light-load adjustment.* The loops are adjusted so that when no current is passing through the current coil (*i.e.,* exciting coil of the series magnet), the torque produced is just sufficient to overcome the friction in the system, without actually rotating the disc.

---

*   Moving the brake magnet inwards means that the speed of the part of disc under the pole face of brake magnet will be less. This results in the lesser induced voltage in the disc and hence reduced braking torque.

(iv) **Creeping.** Sometimes the disc of the meter makes slow but continuous rotation at no load *i.e.*, when potential coil is excited but with no current flowing in the load. This is called *creeping*. This error may be caused due to overcompensation for friction, excessive supply voltage, vibrations, stray magnetic fields etc.

Adjustment. In order to prevent this creeping, two diametrically opposite holes are drilled in the disc. This causes sufficient distortion of the field. The result is that the disc tends to remain stationary when one of the holes comes under one of the poles of the shunt magnet.

(v) **Temperature error.** Since watthour meters are frequently required to operate in outdoor installations and are subject to extreme temperatures, the effects of temperature and their compensation are very important. The resistance of the disc, of the potential coil and characteristics of magnetic circuit and the strength of brake magnet are affected by the changes in temperature. Therefore, great care is exercised in the design of the meter to eliminate the errors due to temperature variations.

(vi) **Frequency variations.** The meter is designed to give minimum error at a particular frequency (generally 50 Hz). If the supply frequency changes, the reactance of the coils also changes, resulting in a small error. Fortunately, this is not of much significance because commercial frequencies are held within close limits.

(vii) **Voltage variations.** The shunt magnet flux will increase with an increase in voltage. The driving torque is proportional to the first power of flux whereas braking torque is proportional to the square of the flux. Therefore, if the supply voltage is higher than the normal value, the braking torque will increase much more than the driving torque and *vice-versa*. The result is that the meter has the tendency to run slow at higher than normal voltages and fast at reduced voltages. However, the effect is small for most of the meters and is not more than 0·2 % to 0·3 % for a voltage change of 10 % from the rated value. The small error due to voltage variations can be eliminated by the proper design of the magnetic circuit of the shunt magnet.

16.48. Three-Phase Watthour Meter

In a 3-phase system, energy, like power, can be measured by means of two single-phase watthour meters. The total energy supplied will be equal to the algebraic sum of the two readings (a negative sign is used for the reading of the meter which runs backward). However, this is never done commercially as it would be more expensive and more troublesome than the use of a 3-phase meter. A 3-phase meter is merely a combination of two single-phase meters (See Fig. 16.78), with their moving elements mounted on the same spindle. The total driving torque is equal to the sum of the torques exerted by both the moving elements. Thus only one counting mechanism is required which will directly indicate the energy being supplied to the 3-phase circuit. Fig. 16.78 shows how a 3-phase watthour meter is connected in a 3-phase circuit to measure energy. The current coils are connected in any two lines and each potential coil is joined to the third line. In fact, the connections are similar to 2-wattmeter method used to measure power in a 3-phase circuit.

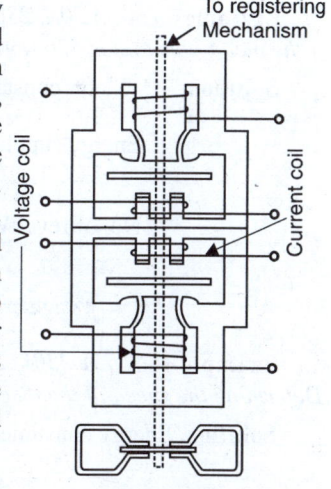

Fig. 16.78

It is very important that the two elements are "balanced" *i.e.*, the driving torque of the two elements be exactly equal for equal amounts of power flowing through each. If this is not done, the meter will not indicate correct reading on unbalanced load. The balancing adjustment is most conveniently made with the potential coils connected in parallel and the current coils in series opposition. If the elements are *balanced, there will be no rotation of the disc for this condition. The usual lag, load and power factor adjustments are made independently for each element.

---

*   If the disc rotates for this condition, it means that the elements are not balanced. Balancing can be easily done by adjusting the disc air gap.

Note. In some 3-phase meters, a single disc is used and the two elements drive this disc. The disc is slotted radially to prevent interaction between eddy currents produced by one element with the flux produced by other element.

Example 16.48. *An energy meter whose constant is 1500 revolutions per kWh makes 20 revolutions in 30 seconds. Calculate the load in kW.*

Solution. Energy consumed when disc makes 20 revolutions

$$= (1/1500) \times 20 = 1/75 \text{ kWh}$$

Now energy consumed is equal to load in kW multiplied by time in hours *i.e.*,

$$\text{Load} \times \frac{30}{3600} = \frac{1}{75} \quad \therefore \quad \text{Load} = \frac{1}{75} \times \frac{3600}{30} = \mathbf{1.6 \text{ kW}}$$

Example 16.49. *A 230V single phase energy meter has a constant load current of 4A passing through it for 5 hours at unity power factor. If the meter makes 1104 revolutions during this period, what is the meter constant in revolutions per kWh ? If the load power factor is 0·8, what number of revolutions the disc will make in the above time ?*

Solution.   Energy supplied $= \dfrac{VI \cos \phi}{1000} \times t = \dfrac{230 \times 4 \times 1}{1000} \times 5 = 4.6 \text{ kWh}$

No. of revolutions $= 1104$

$\therefore$ Meter constant $= 1104/4.6 = \mathbf{240 \text{ rev./kWh}}$

Energy supplied when the load p.f. is 0.8

$$= \frac{230 \times 4 \times 0.8}{1000} \times 5 = 3.68 \text{ kWh}$$

$\therefore$ No. of revolutions $= 240 \times 3.68 = \mathbf{883.2}$

Example 16.50. *A 50A, 230V energy meter on full load test makes 61 revolutions in 37 seconds. If the meter constant is 520 rev./kWh, what is the percentage error ?*

Solution.   Meter constant $= 520 \text{ rev./kWh}$

Energy supplied $= \dfrac{230 \times 50}{1000} \times \dfrac{37}{60 \times 60} = 0.1182 \text{ kWh}$

No. of rev./kWh $= \dfrac{61}{0.1182} = 516.07$

$\therefore$ % age error $= \dfrac{520 - 516.07}{520} \times 100 = \mathbf{0.76\% \text{ slow}}$

Example 16.51. *A 230V, 50Hz single phase energy meter has a constant of 1200 rev./kWh. Determine the speed of the disc in r.p.m. for current of 10A at a p.f. of 0·8 lagging.*

Solution. Energy consumed by the load in 1 minute

$$= \frac{VI \cos \phi}{1000} \times t = \frac{230 \times 10 \times 0.8}{1000} \times (1/60) = 0.0307 \text{ kWh}$$

Revolutions made by the disc in 1 minute

$$= 0.0307 \times 1200 = 36.84$$

$\therefore$ Disc speed $= \mathbf{36.84 \text{ r.p.m.}}$

Example 16.52. *A single phase energy meter has a constant of 1200 rev./kWh. When a load of 200 watts is connected, the disc rotates at 4·2 r.p.m. If the load is on for 10 hours, how many units are recorded as error ? Also find the percentage error.*

Solution. Actual energy consumed by the load in 10 hours

$$= (200 \times 10^{-3}) \times 10 = 2 \text{ kWh}$$

No. of revolutions made by the disc in 10 hours

$$= \text{speed in r.p.m.} \times \text{minutes in 10 hours}$$

$$= 4.2 \times 10 \times 60 = 2520 \text{ revolutions}$$

Energy recorded by the energy meter = 2520/1200 = 2.1 kWh

The meter records 2.1 kWh whereas the actual energy consumed by load during the given period (*i.e.*, 10 hours) is 2 kWh. Therefore, the meter records **0.1 kWh** or **0.1 unit** more.

$$\% \text{ age error } = \frac{0.1}{2} \times 100 = \textbf{5\%}$$

Example 16.53. *A 230V, 10A single phase energy meter has a meter constant of 600 rev./kWh when correctly adjusted. If the lag adjustment is disturbed so that the phase angle between shunt magnet flux and the applied voltage is 86°, calculate the error introduced at (i) unity p.f. (ii) 0·5 p.f. lagging.*

Solution. As proved in Art. 16.40, the driving torque is given by ;

$$T_d \propto \phi_V \phi_C \sin \theta$$

where θ is the phase angle between the two fluxes.

The meter will read correctly if *θ = 90° – φ. In other words, the meter will register correctly if phase angle between shunt magnet flux and supply voltage is 90°. If this is not so, the meter will give a wrong reading.

(*i*) **At unity p.f.** Power factor angle, $\phi = \cos^{-1} 1 = 0°$

∴   θ = 86° – 0° = 86° whereas it should be 90° – 0° = 90°

∴ $\% \text{ age error } = \dfrac{\sin 90° - \sin 86°}{\sin 90°} \times 100 = \textbf{0.24\%}$

(*ii*) **At 0.5 p.f. lagging.** Power factor angle, $\phi = \cos^{-1} 0.5 = 60°$

∴   θ = 86° – 60° = 26° whereas it should be 90° – 60° = 30°

∴ $\% \text{age error } = \dfrac{\sin 30° - \sin 26°}{\sin 30°} \times 100 = \textbf{12.32\%}$

The reader may note that at low power factors, the error due to shunt magnet flux not being in quadrature with supply voltage is considerable.

Example 16.54. *A 3-phase energy meter having a meter constant of 0·12 rev./kWh is used with a potential transformer of ratio 22000/110 volts and a current transformer of ratio 500/5 amperes. When connected to a load of unity p.f., the disc makes 40 revolutions in 61 seconds. If the instrument readings are 110 V and 5.25 A, find the percentage error.*

Solution. Actual load voltage,*V* = Indicated reading × P.T. ratio

$$= 110 \times (22000/110) = 22,000 \text{ volts}$$

Actual load current, *I* = Indicated reading × C.T. ratio

$$= 5.25 \times (500/5) = 525 \text{ A}$$

Actual energy consumption $= \sqrt{3} \; VI \cos \phi \; t$

$$= \sqrt{3} \times (22000) \times 525 \times 1 \times (61/3600)$$

$$= 338.98 \times 10^3 \text{ Wh} = 338.98 \text{ kWh}$$

*   In that case, $T_d \propto \phi_V \phi_C \sin (90° - \phi) \propto VI \cos \phi \propto$ True power.

Energy recorded by the meter $= 40/0.12 = 333.33$ kWh

$$\therefore \qquad \text{\%age error} = \frac{338.98 - 333.33}{338.98} \times 100 = \textbf{1.67\% slow}$$

Example 16.55. *An ampere-hour meter, calibrated at 210 V, is used on 230 V circuit and indicates a consumption of 730 units in a certain period. What is the actual energy supplied ? If this period is reckoned as 200 hours, what is the average value of the current?*

Solution. The ampere-hour meters are calibrated to read directly in kWh at the *declared voltage*. However, their readings will be incorrect when used on any other voltage.

$$\text{Reading on 210 V} = 730 \text{ kWh}$$

$$\text{Reading on 230 V} = \frac{730 \times 230}{210} \simeq \textbf{800 kWh}$$

Now, $\qquad\qquad V I t = 800 \times 10^3$

$$\therefore \qquad\qquad I = \frac{800 \times 10^3}{Vt} = \frac{800 \times 10^3}{230 \times 200} = \textbf{17.4 A}$$

Example 16.56. *The testing constant of a supply meter of the ampere hour type is given as 60 coloumbs/revolution. It is found that with a steady current of 50 A, the spindle makes 153 revolutions in 3 minutes. Calculate the factor by which dial indications of the meter must be multiplied to give the consumption.*

Solution. Charge supplied in 3 min $= It = 50 \times (3 \times 60) = 9000$ C

Correct number of revolutions for the supply of 9000 C

$$= \frac{9000}{60} = 150$$

But the meter records 153 revolutions. This means that meter is fast. Therefore, for correct consumption, the registered readings should be multiplied by $150/153 = \textbf{0.9804.}$

Example 16.57. *A d.c. ampere hour meter is rated at 5A, 250 V. The declared constant is 5 As/rev. Express this constant in rev/kWh. Also calculate the full-load speed of the meter.*

Solution.   Meter constant $= 5$ A-s/revolution

Now, $\qquad\qquad 1$ kWh $= 10^3$ Wh $= 10^3 \times 3600$ volt $\times$ amp. $\times$ second

$\therefore \qquad\qquad 1$ kWh $= 36 \times 10^5$ volt $\times$ amp. $\times$ second

On a 250 V circuit this corresponds to $36 \times 10^5/250 = 14400$ A-s.

Now, for every 5 A-s, there is one revolution.

$\therefore$ No. of revolutions in 1 kWh $= 14400/5 = 2880$ revolutions

or $\qquad\qquad$ Meter constant $= \textbf{2880 rev./kWh}$

The full-load current is 5 A and meter constant is 5A-s/rev. Therefore, the meter will make one revolution in 1 second.

$$\therefore \qquad\qquad \text{Full-load speed} = \textbf{60 r.p.m.}$$

Example 16.58. *A 230 V, single-phase domestic energy meter has a constant load of 4 A passing through it for 6 hours at unity p.f. If the meter disc makes 2208 revolutions during this period, what is the constant in rev/kWh? Calculate the p.f. of the load if the number of revolutions made by the meter are 1472 when operating at 230 V and 5 A for 4 hours.*

Solution. Energy consumption in 6 hours is

$$E = VI \cos \phi \times 6 = (230 \times 4 \times 1) \times 6 = 5.52 \text{ kWh}$$

$\therefore$ Meter constant $= \dfrac{2208}{5.52} = \textbf{400 rev/kWh}$

Energy consumed when the meter makes 1472 revolutions is

$$E' = 1472/400 = 3.68 \text{ kWh} = 3680 \text{ Wh}$$

Now, $230 \times 5 \times \cos\phi \times 4 = 3680$ $\therefore \cos\phi = \textbf{0.8}$

Example 16.59. *The constant of a 25 A, 220 V meter is 500 rev/kWh. During a test at full-load of 4400 W, the disc makes 50 revolutions in 83 seconds. Calculate the meter error.*

Solution. In 1 hour at full load, the energy consumption $= 4400 \times 1 = 4400$ Wh $= 4.4$ kWh. Therefore, 1 hour ($= 3600$ s), the meter disc should make $4.4 \times 500 = 2200$ revolutions.

$$\therefore \text{Correct time for 1 revolution} = \dfrac{3600}{2200} \text{ s}$$

$$\text{Correct time for 50 revolutions} = \dfrac{3600}{2200} \times 50 = 81.8 \text{ s}$$

But the meter disc makes 50 revolutions in 83 s. This means that the meter is slow by $83 - 81.8 = 1.2$ s.

$$\therefore \qquad \text{Percentage error} = \dfrac{1.2}{81.8} \times 100 = \textbf{1.47 \%}$$

Example 16.60. *An a.c. energy meter is tested for half hour run at a supply voltage of 230 V with a current of 12 A at 0.8 p.f. lag. The dial reading at the beginning of the test was 58.5 and at the end it was 59.5. The meter constant is 1200 rev/kWh. The meter revolutions registered during the test were 1150. Find out (i) error in registration and (ii) error in r.p.m. of energy meter. How error in r.p.m. can be rectified?*

Solution. Actual energy consumed in half an hour ($= 0.5$ hr)

$$= VI \cos\phi \times t = 230 \times 12 \times 0.8 \times 0.5 = 1104 \text{ Wh} = 1.104 \text{ kWh}$$

Energy consumption recorded by meter in half an hour

$$= 59.5 - 58.5 = 1 \text{ kWh}$$

(i) Error in registration $= 1.104 - 1 = \textbf{0.104 kWh less}$

(ii) Error in r.p.m. $= \dfrac{1150 - 1200}{30} = \textbf{-1.667 r.p.m.}$

The meter is slow by $1200 - 1150 = 50$ revolutions per kWh. The error can be rectified by bringing the brake magnet nearer to the centre of the disc.

Tutorial Problems

1. A meter whose constant is 600 revolutions/kWh makes 5 revolutions in 20 seconds. Calculate the load in kW. **[1.5 kW]**

2. The number of revolutions per kWh for a 230 V, 10 A watthour meter is 900. On test at half-load, the time for 20 revolutions is found to be 69 seconds. Determine the meter error at half-load. **[0.82% *fast*]**

3. A correctly adjusted single phase 240 V energy meter has a meter constant of 600 rev./kWh. Determine the speed of the disc for a current of 10 A at a p.f. of 0.8 lagging. **[19.2 r.p.m.]**

4. A single phase energy meter has a meter constant of 200 rev./kWh. When supplying a non-inductive load of 4.4A at normal voltage, the meter takes 3 minutes for 10 revolutions. Find the percentage error.

    [1.186%]

5. An energy meter has a meter constant of 600 rev./kWh when correctly adjusted. If the lag adjustment is altered so that the phase angle between shunt magnet flux and the supply voltage is 85°, calculate the percentage error at 0.8 p.f. lagging. **[7%]**

9. D.C. Potentiometer

A **potentiometer** *is a device used to measure the e.m.f. of a voltage source or p.d. between two points in an electrical circuit.* It basically measures the potential difference but on this basis, it can be used to make several other measurements. A potentiometer uses null point method (described below) and has the advantage of being able to measure the potential difference of a source without drawing any current from it. Hence it can be used to measure the e.m.f. of a source directly.

Principle. If a wire of uniform area of cross-section is carrying a steady current, then fall in potential across it is uniform and voltage drop between any two points on the wire is directly proportional to the distance between these points.

$$\text{Voltage drop, } V = Irl$$

where I = steady current through the wire

$\quad r$ = resistance per unit length of wire

$\quad l$ = length of wire over which drop occurs

Now, I and r are constant so that $V \propto l$.

Theory. Fig. 16.79 shows a simple type of d.c. potentiometer. It consists of german silver or manganin wire AC (called slide wire) of uniform area of cross-section. A battery B (called driver battery) sends a steady current through the slide wire. The slide-wire current can be changed with the help of rheostat Rh. The cell whose e.m.f. E is to be measured is connected in series with a galvanometer G and a key K. *Note that positive terminals of battery B and the cell must be connected together.* The jockey J can slide along the slide wire AC and can make contact at any

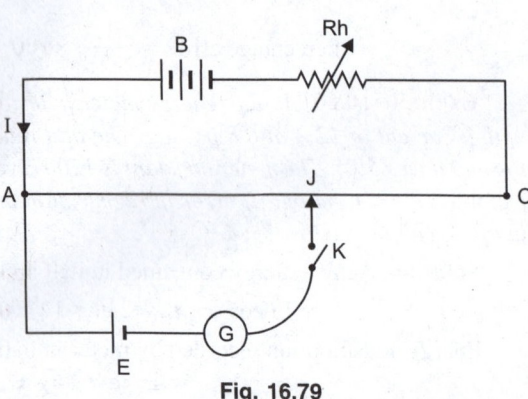

Fig. 16.79

point on the wire. Suppose that r is the resistance per unit length of the slide wire AC.

With key K open, the rheostat Rh is set at the desired position to obtain suitable steady current I in the slide wire AC. Now key K is closed so that the cell is put in the circuit. The jockey is moved over the slide wire till we get a point J' on the wire at which the galvanometer reads zero. Under this condition (*i.e.*, zero galvanometer current), we say that **null-point** or **balance point** is reached. At null point, the current in the slide wire is again I. It is clear that at null point, the e.m.f. E of the cell just balances the potential difference across the length AJ' (= l) of the slide wire *i.e.*,

E = Voltage drop across length $AJ' = Irl$

Thus e.m.f. E of the cell can be determined. Note that at null point, there is no current in the cell so that voltage drop across its internal resistance is zero. Therefore, the potentiometer measures the e.m.f. of the cell. Further, at null point, there is no change in circuit current (*i.e.*, current in the slide wire in this case). This is the main advantage of null point method.

Comparison of e.m.f.s of two cells. Fig. 16.80 shows the arrangement for comparing the e.m.f.s E_1 and E_2 of the two cells with a potentiometer. *Note that positive terminals of the cells and the positive terminal of the driver battery B must be connected together.* The cells are joined to the galvanometer G through a two-way key.

With both cells out of the circuit, the rheostat Rh is set at the desired position to obtain suitable steady current I in the slide wire AC. *The rheostat setting is not to be changed throughout the experiment otherwise wrong results will be obtained.* First, *only* cell of e.m.f. E_1 is put in the circuit. The jockey is moved on the slide wire till galvanometer reads zero (null point). Suppose null point is obtained when jockey is at point J_1 on the slide wire. If $AJ_1 = l_1$, then,

$$E_1 = Irl_1$$

where r = resistance per unit length of slide wire. Now, *only* cell E_2 is put in the circuit and null point is obtained when jockey is at point J_2 on the slide wire. If $AJ_2 = l_2$, then,

$$E_2 = Irl_2$$

$$\therefore \quad \frac{E_2}{E_1} = \frac{Irl_2}{Irl_1} = \frac{l_2}{l_1}$$

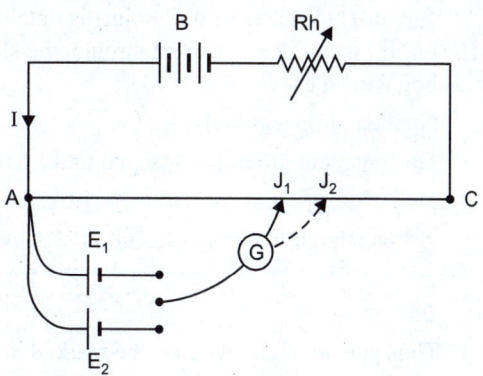

Fig. 16.80

Since the lengths l_1 and l_2 are known, the ratio E_2/E_1 can be determined. Further, if the e.m.f. of one cell is known, the e.m.f. of the other cell can be found out.

Note. One important disadvantage of potentiometer is that it cannot measure a potential difference greater than the e.m.f. of the driver battery B (about 2 V). For measuring higher voltages, *volt ratio box* is used in conjunction with the potentiometer (See Art. 16.53).

16.50. Direct Reading Potentiometers

The simple potentiometer described above for determining or comparing e.m.f.s involves some computations. Therefore, it is used for educational purposes only. For convenience, modern potentiometers are always made *direct reading i.e.,* the potentiometer wire is marked in terms of voltages. Therefore, the voltages being measured can be directly read off from the calibrated potentiometer. *The calibration of a potentiometer (i.e., making it direct reading) is called standardisation of potentiometer.*

Standardising the potentiometer. The calibration (*i.e.,* standardisation) of a simple potentiometer is done with the help of a standard cell as shown in Fig. 16.81. The standard cell employed is Weston cadmium cell whose e.m.f. is 1.0183 volts. A high variable resistance is put in the standard cell circuit to keep the current through it to permissible limit. This high resistance is gradually cut off as the null point condition is approached. Ultimately, this high resistance is entirely cut off by short-circuiting it with key K when final null point is obtained.

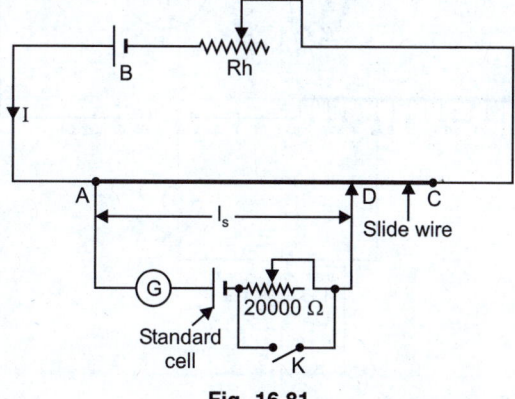

Fig. 16.81

Suppose balance (*i.e.*, null point) is obtained when jockey is at point D on the slide wire. Let $AD = l_s$. If I is the steady current through the slide wire AC and r is the resistance per unit length of the slide wire, then,

Potential drop across $AD = Il_s r$

This potential difference is equal to the e.m.f. of the standard cell.

$$\therefore \qquad\qquad Il_s r = 1.0183$$

$\therefore$ Potential drop per unit length of slide wire is

$$e_s = Ir = \frac{1.0183}{l_s}$$

Thus potentiometer wire can be marked in terms of voltages and the instrument becomes direct reading. Note that standardisation of the potentiometer remains the same as long as the current I through the slide wire is constant. *Therefore, once the potentiometer is standardised, the setting of rheostat should not be changed in any case.* After standardising, the potentiometer can be used for various purposes.

16.51. Modern D.C. Potentiometers

The accuracy of a potentiometer depends on the length of the slide wire used. The greater the length of the slide wire, the greater is the accuracy and *vice-versa*. For precision work, the total length of the slide wire required would be inconveniently long. In modern potentiometers designed for precision work, the effect of a very long slide wire is obtained by connecting a number of resistance coils in series with a comparatively short slide wire. Each resistance coil has resistance equal to the resistance of the whole slide wire. One such potentiometer is Crompton d.c. potentiometer.

16.52. Crompton D.C. Potentiometer

Fig. 16.82 shows the general arrangement of Crompton d.c. potentiometer. A graduated slide-wire AC is connected in series with 14 (or more) resistance coils, each of which has a resistance exactly equal to the resistance of the whole slide wire. There are two moving contacts P_1 and P_2. The contact P_1 can slide over the slide wire and contact P_2 can slide over the studs connected to the resistance coils. B is the supply battery (2 volts) and R_1 and R_2 are two variable resistors. The resistor R_1 consists of a number of coils in series and is meant for coarse adjustment of potentiometer current. The resistor R_2 is in the form of a slide wire and is meant for fine adjustment of potentiometer current. The connections C_1, C_2 are taken to selector switch S (double-pole, three position switch) for connecting either the standard cell (S.C.) or cell of unknown e.m.f. The terminals for cell connections are marked positive (+) and negative (–) for correct polarity connection.

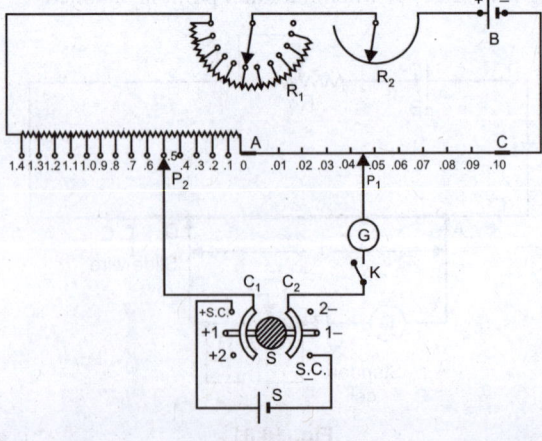

Fig. 16.82

Working. The potentiometer is first "standardised" (*i.e.,* made direct reading) by adjustment of the current from the supply battery B. For this purpose, we connect Weston type standard cell whose e.m.f. is 1.0183 volts to the terminals marked S.C. (Be sure to connect correct polarity). The moving contacts P_1 and P_2 are set to a reading equal to the e.m.f. of the standard cell *i.e.,* contact P_2 is placed on stud 1.0 and contact P_1 on 0.0183 on the slide wire. The regulating resistances R_1 and R_2 are varied until the galvanometer reads zero. Now the potentiometer is standardised *i.e.,* reading of the potentiometer (= 1.0 + 0.0183 = 1.0183) is directly read as the voltage being measured. Therefore, the potentiometer becomes direct reading. *Note that after standardisation of the potentiometer, the setting of R_1 and R_2 is not to be changed.*

The source of e.m.f. whose e.m.f. is to be measured is connected between the terminals + 1 and – 1 (or + 2 and – 2) and the selector switch is switched over to the position + 1 and – 1. The moving contacts P_1 and P_2 are adjusted until the galvanometer reads zero. The reading of the potentiometer will give the e.m.f. or potential difference to be measured directly.

16.53. Volt Ratio Box

We have already seen that a potentiometer cannot measure a potential difference or e.m.f. greater than the e.m.f. of the supply battery B (about 2 volts). Therefore, the range of potentiometer is less than 2 volts. For measuring voltages above 2 volts, we use a volt ratio box (or volt-box) in conjunction with the potentiometer. The volt-box consists of a high resistance (30 Ω to 50 Ω per volt) having a number of tappings. Its function is that of a potential divider.

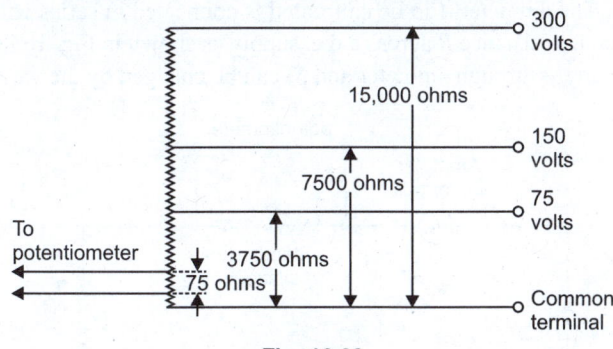

Fig. 16.83

Fig. 16.83 shows the connections of a volt-box having resistance of 50 Ω per volt. The connections to the potentiometer are taken from two tapping points, which include between them (say) 75 Ω. If a voltage greater than 75 V but less than 150 V is to be measured, then this voltage is connected between the two terminals marked 150 V and "common". Thus if the measured value of voltage on the potentiometer is 1.25 V, then the value of unknown voltage connected across the terminals of volt box

$$= 1.25 \times \frac{7500}{75} = 125 \text{ V}$$

16.54. Applications of D.C. Potentiometers

A potentiometer basically measures the potential difference between two points in an electrical circuit. On this basis, it can make several other measurements. A few applications of a potentiometer are discussed below by way of illustration.

(*i*) Measurement of current. A potentiometer can be used to measure the current accurately. For this purpose, the unknown current I is passed through a standard resistance S of known value as shown in Fig. 16.84. The potential difference (P.D.) across the standard resistance S is measured with a potentiometer. The value of I is given by ;

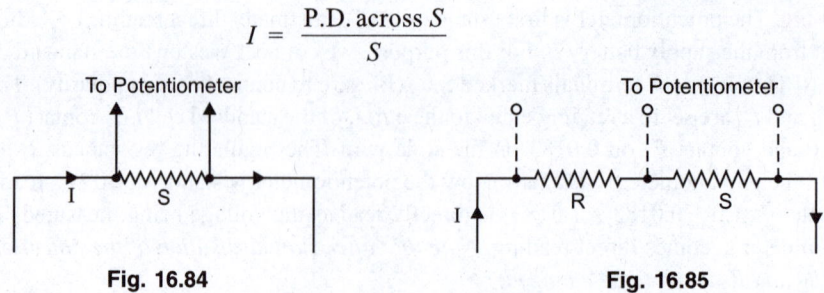

$$I = \frac{\text{P.D. across } S}{S}$$

Fig. 16.84　　　　　　　　　　　　　　　　**Fig. 16.85**

(*ii*) **Measurement of resistance.** A potentiometer can also be employed to measure the unknown resistance. For this purpose, the unknown resistance R is connected in series with a known standard resistance S and a suitable current is passed through this series combination as shown in Fig. 16.85. The potential differences across R and S are measured with a potentiometer. Then,

$$\frac{R}{S} = \frac{\text{P.D. across } R}{\text{P.D. across } S}$$

Since the p.d.s across R and S are measured and the value of S is known, the value of unknown resistance R can be determined.

(*iii*) **Calibration of ammeter.** A potentiometer can be usefully employed to calibrate an ammeter. By calibration of ammeter means to determine the extent of error in the reading of the ammeter throughout its range. The ammeter A to be calibrated is connected in series with a standard known resistance S and variable resistance R across a d.c. supply as shown in Fig. 16.86. The magnitude of circuit current (*i.e.*, current through ammeter and S) can be changed by the variable resistance R.

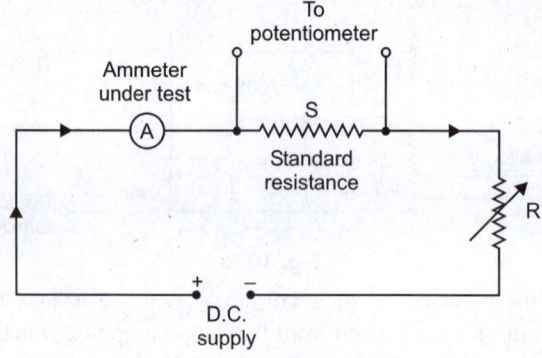

Fig. 16.86

The circuit current is adjusted with the help of variable resistance R to give a particular value of current on the ammeter. This current is now measured by the potentiometer and is given by p.d. across S indicated by the potentiometer divided by S. The current measured by the potentiometer is accurate and is compared with that indicated by the ammeter to find the error in the ammeter reading. The test is repeated for various values of current over the entire range of the ammeter.

Note. The value of standard resistance S should be such that when full scale current of ammeter is flowing in the circuit, the potential difference across S does not exceed the range of potentiometer.

(*iv*) **Calibration of voltmeter.** A voltmeter can be calibrated with a potentiometer in conjunction with a volt-box as shown in Fig. 16.87. A high variable voltage to the volt-box is provided by the potential divider arrangement. The voltmeter V to be calibrated is connected across the output of the potential divider and indicates the voltage fed to the volt-box. We can vary the voltage across the voltmeter through the adjustment of the potential divider.

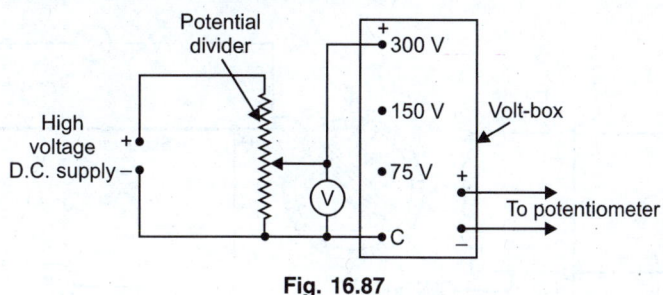

Fig. 16.87

The potential divider is adjusted to give a particular value of voltage on the voltmeter. This voltage is now measured with a *potentiometer. The voltage measured by the potentiometer is accurate and is compared with that indicated by the voltmeter to find the error in the voltmeter reading. The test is repeated for various values of voltages over the entire range of the voltmeter.

16.55. A.C. Potentiometer

An a.c. potentiometer basically works on the same principle as a d.c. potentiometer *i.e.*, unknown e.m.f. and slide-wire voltage drop are made equal to obtain the balance. However, there is one important difference in the operation of the two. In a d.c. potentiometer, only the *magnitudes* of the unknown e.m.f. and slide-wire voltage drop are made equal to obtain the balance. But in an a.c. potentiometer, the *magnitudes as well as phases* of the two voltages are made equal for balance. Further, in an a.c. potentiometer, the frequency and waveform of slide-wire voltage must be the same as that of the voltage to be measured. This necessitates that the a.c. supply for the slide-wire must be taken from the same source as the voltage or current to be measured. These considerations suggest that for a.c. measurements, there should be some modifications in the potentiometer used for d.c. work. The practical field of use of a.c. potentiometer is in engineering measurements where an accuracy of 0.5% to 1% is acceptable.

16.56. Drysdale A.C. Potentiometer

This potentiometer was developed by C.V. Drysdale for a.c. measurements. In this potentiometer, the unknown a.c. voltage is measured in the polar form $V\angle\theta$ *i.e.*, in magnitude and relative phase. For this reason, it is called *polar potentiometer*.

Construction. Fig. 16.88 shows the basic construction of Drysdale a.c. potentiometer. The slide-wire PQ is supplied from a **phase-shifting circuit** so arranged that the magnitude of a.c. voltage supplied by it remains constant but its phase can be changed from 0° to 360°. The phase-shifting circuit consists of (*i*) two-phase stator winding and (*ii*) a movable single-phase rotor winding. The stator is supplied from a single-phase supply which is converted into 2-phase supply by using a phase-splitting device consisting of capacitor C and resistance r as shown. The two-phase winding produces a rotating magnetic field which induces a secondary e.m.f. in the rotor winding. The e.m.f. induced in the rotor winding is of constant magnitude but its phase can be changed by rotating the rotor to any desired position. The rotor moves over a graduated circular scale marked in degrees to indicate the phase of the rotor e.m.f. relative to the stator.

---

*   A volt-box is simply a voltage reducer which reduces the voltage applied to it to a value within the range of the potentiometer. The voltage measured on the potentiometer is multiplied by the ratio of volt-box and gives the actual value of voltage being measured.

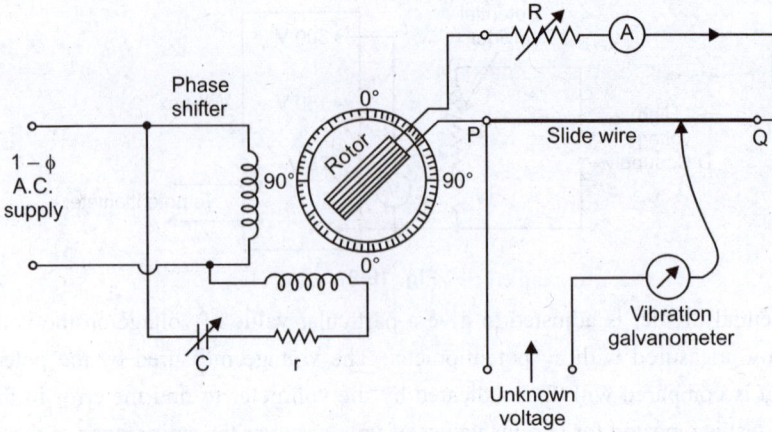

Fig. 16.88

Working. The working of Drysdale a.c. potentiometer is as under :

(*i*) First, the a.c. potentiometer is standardised *i.e.*, it is made direct reading. For this purpose, the slide-wire circuit is connected to d.c. supply and the standard current is obtained as usual by using a standard cell. This standard current makes the potentiometer direct reading and is measured by a dynamometer ammeter which is included in the battery supply circuit of the potentiometer. This ammeter remains connected for a.c. operation also because the r.m.s. value of current in the slide-wire must be maintained at the same value (*i.e.*, standard current) as was required on direct current. Since dynamometer ammeter reads correctly on both direct and alternating current, the potentiometer remains direct reading with an a.c. supply.

(*ii*) Once the a.c. potentiometer is standardised, the d.c. supply is removed and the slide-wire is connected to the rotor winding of the phase-shifting circuit. The r.m.s. value of alternating current in the slide-wire is made (through the adjustment of rheostat R) the same as on the d.c. supply. Now, the unknown a.c. voltage is applied to the slide-wire through the vibration galvanometer (detector). Balance is obtained by varying the position of slide-wire contact and position of the phase-shifting rotor. When the vibration galvanometer reads zero, it means that balance is achieved. Now, slide-wire reading gives the magnitude V of the unknown a.c. voltage and the rotor position its phase angle θ. Therefore, the unknown a.c. voltage is represented as $V\angle\theta$.

16.57. Ballistic Galvanometer

The basic construction of the ballistic galvanometer is the same as that of the moving-coil galvanometer. *The ballistic galvanometer is used to measure the quantity of electricity (i.e., coulombs) which passes through it.* This quantity of electricity is the result of e.m.f. induced in a search coil when the magnetic flux linking the coil changes.

The basic difference between the conventional galvanometer and the ballistic galvanometer is that while the conventional galvanometer carries a certain steady current which produces deflection, the passage of current through the ballistic galvanometer is momentary and the moving system will be at rest during the flow of current. After the quantity of electricity has passed through the moving coil of the ballistic galvanometer, it gives first deflection or 'throw' which is recorded by lamp and scale arrangement. After the first deflection or throw has been observed, the moving coil is rapidly brought to rest by eddy current damping. The ballistic galvanometer is so designed that the first deflection or throw is directly proportional to the quantity of electricity Q that has passed through it

i.e. $$Q = k\,\theta$$

where k = Ballistic constant of the galvanometer

θ = First deflection or throw

Construction. Fig. 16.89 shows the simplified diagram of the ballistic galvanometer. It consists of a moving coil suspended between the poles of a permanent magnet. The coil is wound on a non-magnetic former so that there is very little damping. A fine wire is used to lead the current into and out from the coil. The instrument is provided with eddy current damping. The first deflection or throw is recorded by lamp and scale arrangement. The farther is the scale from the ballistic galvanometer, the more is the amplification of the deflected angle which is indicated.

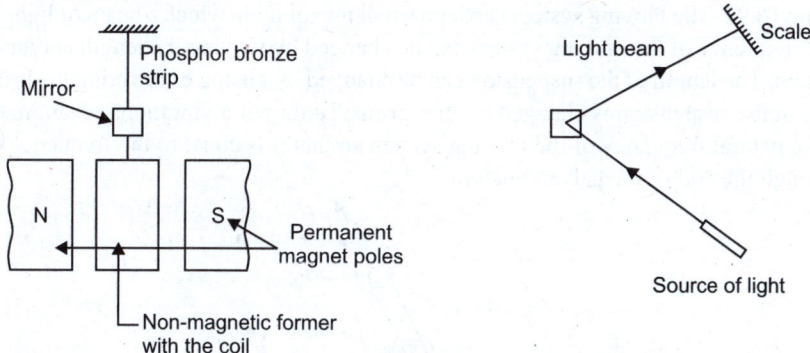

Fig. 16.89

Theory. The theory involved in the working of the ballistic galvanometer is the same as that of the conventional moving coil galvanometer with the following constructional modifications :

(i) The moving system of a ballistic galvanometer is designed to have a large moment of intertia. This is achieved by attaching small weights to the moving system.

(ii) The eddy current damping provided in a ballistic galvanometer is very small. This is achieved by winding the moving coil on a non-magnetic former.

The large moment of inertia of the moving system makes its period of oscillation quite long, usually 10 to 15 seconds in practice. The result is that whole of the charge passes through the galvanometer before its coil has had time to move from rest *i.e.,* the moving system will be almost at rest during the passage of whole of the charge. This quick flow of current through the coil of the galvanometer produces an *impulsive torque.* This torque is applied while the movement is still effectively at rest. The impulse causes the coil to move from its initial position of rest but there is no longer a driving torque because the current has already ceased to flow. Therefore, the first deflection or 'throw' of the ballistic galvanometer will be directly proportional to the impulse of current or the quantity of charge passed through the coil.

It may be noted that this mode of operation is valid only if the charge passes through the coil of the ballistic galvanometer in a very short period of time. After the first deflection or 'throw', the movement will continue to oscillate for some time because of small damping torque. However, these oscillations are quickly brought to rest by the shunt switch provided in the instrument. Only the first deflection may be used for measurement purposes.

16.58. Vibration Galvanometer

Vibration galvanometers are widely used as null-point detectors in a.c. bridges and a.c. potentiometers. These galvanometers are very sensitive because they utilise the principle of mechanical resonance.

The most commonly used vibration galvanometer is of moving coil type in which a coil is suspended between the poles of a permanent magnet. When alternating current is passed through the coil, alternating deflecting torque is produced which causes the coil to vibrate with a frequency equal to the frequency of current passing through the coil. Due to inertia of the moving system, the amplitude of vibrations is small. However, if the natural frequency of the moving system is made equal to the frequency of the current passing through the coil, mechanical resonance occurs and the moving system vibrates with a large amplitude.

Construction. Fig. 16.90 shows the basic construction of a typical moving coil vibration galvanometer. The moving coil consists of a single loop of a fine bronze or platinum silver wire. This wire passes over a small pulley at the top and is pulled tight by a spring attached to the pulley as shown in Fig. 16.90. The moving system carries a small mirror upon which a beam of light is thrown. The natural frequency of the moving system can be changed by varying the length and/or tension of the suspension. The length of the suspension can be changed by raising or lowering the bridge piece. The tension in the suspension is changed by the spring. Tuning of a vibration galvanometer means adjusting the natural frequency of the moving system so that it is equal to the frequency of current passing through the coil of the galvanometer.

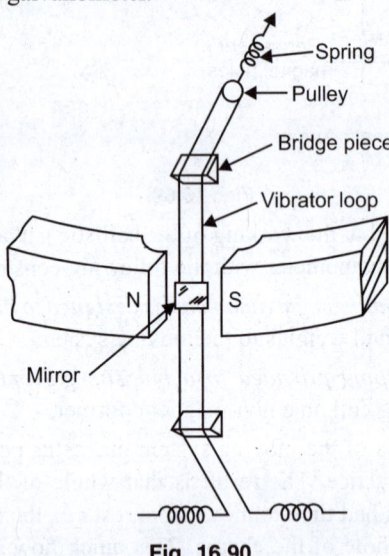

Fig. 16.90

Working. The vibration galvanometer is *tuned and a beam of light is thrown on the mirror. When current through the coil of the galvanometer is zero, a single spot of light is observed on the scale. However, when a small alternating current flows through the coil, the amplitude of vibration of the moving system is very large due to mechanical resonance. As a result, a large band of light is produced on the scale. The size of this band of light decreases as the current through the coil decreases and *vice-versa*. The bridge or potentiometer circuit is adjusted until a single spot of light is observed on the scale. Under this condition, the current in the coil of the galvanometer is zero and null-point is obtained. Some practice is necessary in observing when this condition has been obtained. It is usually best to switch the galvanometer in and out of the circuit and to note if there is any observable difference in the size of the spot in the two cases.

Note. The galvanometer should be well shunted to protect it from excessive current when the circuit is out of balance. When the balance is almost obtained, the shunt should be removed in order to achieve maximum sensitivity.

---

*   **Tuning.** To tune the vibration galvanometer, a small current of supply frequency is passed through it and the tuning adjustments (variation of the length and tension of the suspension) are continued until the reflected band of light reaches its maximum length.

16.59. Frequency Meters

Now-a-days electrical energy is exclusively generated, transmitted and distributed in the form of alternating current. It is very important that frequency in power networks is maintained strictly constant as any change in frequency may affect the operation of a number of automatic control devices. Many *different systems are in common use for the measurement of frequency. However, in power system, frequency is generally measured by an instrument called *frequency meter*. A frequency meter is a direct reading instrument and is used where frequency variations are in close limits *e.g.* in a power system. There are a number of different types of frequency meters but the most commonly used ones are :

   (*i*) Vibrating-reed frequency meter

  (*ii*) Electrodynamic frequency meter

 (*iii*) Moving-iron frequency meter

16.60. Vibrating-Reed Frequency meter

It is the simplest type of frequency meter and uses the principle of mechanical resonance for the measurement of frequency. In this type, a number of thin, flat steel strips (called *reeds*) of different **natural frequency are excited electromagnetically from the source whose frequency is to be determined. The amplitude of vibration will be greatest for the reed whose natural frequency (or resonant frequency) is equal to the frequency of the source. Thus a row of such reeds with a calibrated scale is a means of measuring frequency.

Construction. Fig. 16.91 (*i*) shows the principal parts of a vibrating-reed frequency meter. It consists of a number of thin steel strips (called *reeds*) attached to a bar which has flexible supports. An electromagnet with a laminated core is mounted close to the reeds as shown. The coil of this electromagnet, in series with a resistance R, is connected across the supply [like a voltmeter as shown in Fig. 16.91 (*ii*)] whose frequency is to be determined. The lengths of the reeds are such that every reed has a natural frequency 0.5 Hz greater than the preceding one. The free end of each reed faces the observer. White painted flags are attached to the free ends of the reeds so that their vibrations may be seen more easily.

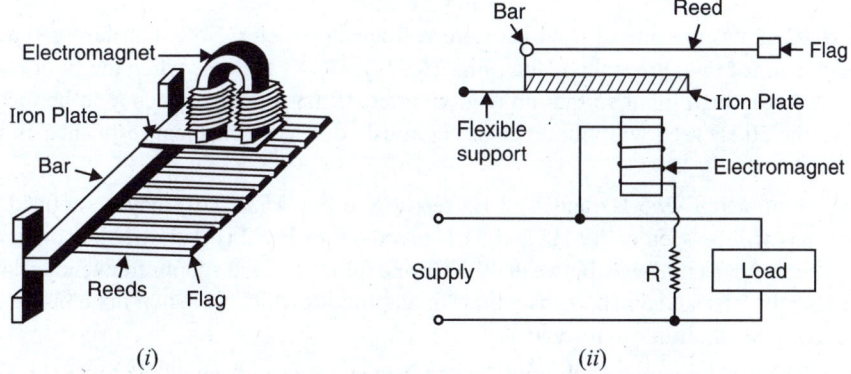

(*i*)   (*ii*)

Fig. 16.91

Working. When the electromagnet is connected across the supply whose frequency is to be determined, the bar is attracted twice per cycle as the current reaches its positive and negative

---

*   Frequency of a.c. can also be measured by frequency bridges.

**   If a reed, or any other object, is free to vibrate, it will normally do so at a frequency that is determined by its physical characteristics such as nature of material, length, thickness etc. This frequency is known as natural frequency or resonant frequency of the object.

peaks. Thus the bar makes two vibrations for each cycle of the alternating current through the electromagnet. Since the reeds are attached to the bar, they also tend to vibrate accordingly. But the reed whose natural frequency is exactly *double the supply frequency will vibrate with maximum amplitude due to mechanical resonance. Thus, by noting which reed vibrates most vigorously, we can read the frequency of the supply from the scale.

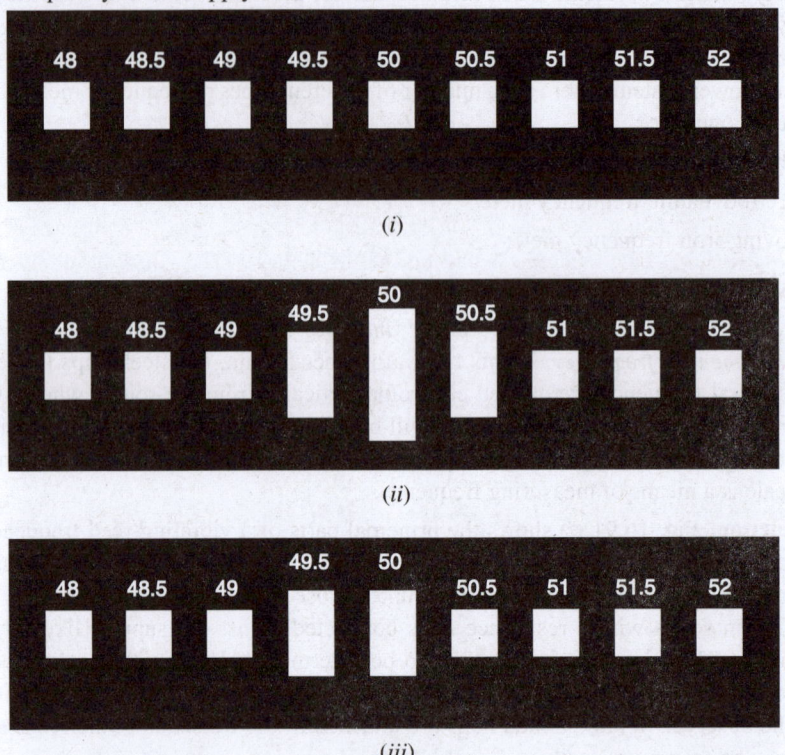

Fig. 16.92 shows the dial of a vibrating-reed frequency meter. Note that the free ends of the reeds (white painted flags) face the reader. Fig. 16.92 (*i*) shows the case when the frequency meter is not connected in the circuit so that no reed vibrates. If the supply frequency to be measured is 50 Hz, then the 50 Hz reed will vibrate most vigorously due to mechanical resonance as shown in Fig. 16.92 (*ii*).

Some vibrations of 49·5 Hz and 50·5 Hz reeds [See Fig. 16.92 (*ii*)] may be noticed but very little vibrations will be seen on 49 Hz and 51 Hz reeds. Fig. 16.92 (*iii*) shows the condition when the supply frequency is mid-way between 49·5 Hz and 50 Hz *i.e.,* the supply frequency is 49.75 Hz. Note that both 49·5 Hz and 50 Hz reeds vibrate at amplitudes which are equal but considerably less than the maximum amplitude at resonance.

Range. The usual range of such frequency meters is 6 Hz say from 47 Hz to 53 Hz. They may be used in electric power system where the maximum variation of frequency is only a few cycles.

The frequency range of this meter may be doubled by polarising the electromagnet *i.e.,* by using a d.c. winding on the electromagnet in addition to the a.c. winding. The magnitude of direct current through the d.c. winding is so adjusted that steady flux produced by it is equal in magnitude to the alternating flux of the a.c. winding.

---

*   In fact, the reed marked 50 Hz has a natural frequency of 100 Hz so that on 50 Hz supply, this reed will vibrate with maximum amplitude.

(i) If only a.c. winding is excited, the alternating flux acting alone will cause two vibrations of each reed for one cycle of supply as explained above. The reed whose natural frequency is twice that of the alternating current in the electromagnet will vibrate most vigorously due to mechanical resonance.

(ii) If both a.c. and d.c. windings are excited, the fields (fluxes being equal in magnitude) will cancel during one half-cycle and reinforce during the opposite half-cycle. In other words, the resultant flux will be zero in one half-cycle and double in the other half-cycle. The result is that each reed will receive one impulse for each cycle of the supply. Consequently, the reed whose natural frequency is equal to that of the alternating current in the electromagnet will respond the most. It is clear that range of the frequency meter will be doubled with polarisation.

Let us illustrate polarisation of the electromagnet with a numerical example. Suppose the frequency to be measured has a value of 50 Hz. If the electromagnet is unpolarised (*i.e.,* only a.c. winding is excited), then the reed whose natural frequency is 100 Hz will respond the most. On the other hand, if the electromagnet is polarised (*i.e.,* both a.c. and d.c. windings are excited), then the reed whose natural frequency is 50 Hz will respond the most.

Advantages

    (i) It has a very simple construction.

    (ii) The indications are independent of the wave-form of supply voltage.

Disadvantages

    (i) It cannot read closer than half the frequency difference between the adjacent reeds. Consequently, precise measurement of frequency cannot be made with such a meter.

    (ii) Any change in the physical characteristics of reeds (*e.g.,* length, weight etc.) may change their natural frequency to a considerable extent.

16.61. Electrodynamic Frequency Meter

This is also known as a moving-coil frequency meter and is a ratiometer type of instrument.

Principle. The working principle of a ratiometer type frequency meter is shown in Fig. 16.93. It consists of two coils X and Y rigidly fixed together with their planes at right angles to each other and mounted on the same spindle. The coils are placed in the field of a permanent magnet. There is no mechanical control torque and the current leads to the two coils are fine ligaments designed to have a negligible effect on the position of the coils.

If G_1 and G_2 are the displacement constants of the two coils and I_1 and I_2 are the two currents, then their respective torques are $T_1 = G_1 I_1 \cos \theta$ and $T_2 = G_2 I_2 \sin \theta$. The currents in the two coils are in such directions (See Fig. 16.93) that the two torques oppose each other. It is clear that if angular deflection θ increases, torque $T_1 (= G_1 I_1 \cos \theta)$ decreases while the torque $T_2 (= G_2 I_2 \sin \theta)$ increases. However, an equilibrium position is possible for some angle θ for which :

$$G_1 I_1 \cos \theta = G_2 I_2 \sin \theta$$

or
$$\tan \theta = \frac{G_1}{G_2} \cdot \frac{I_1}{I_2}$$

or
$$\tan \theta \propto \frac{I_1}{I_2}$$

By modifying the shape of the pole faces and the angle between the coils, the ratio I_1/I_2 is made proportional to θ instead of $\tan \theta$.

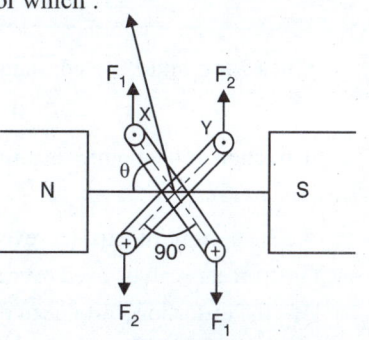

Fig. 16.93

$$\therefore \qquad \theta \propto \frac{I_1}{I_2}$$

i.e., deflection is proportional to the *ratio* of the two currents I_1 and I_2 and hence the name ratiometer type instrument.

Construction. Fig. 16.94 shows the various parts of an electrodynamic frequency meter. It consists of two coils X and Y rigidly fixed at right angles to each other and placed in the magnetic field of a permanent magnet. The two coils are connected through respective bridge rectifiers across the supply whose frequency is to be measured. The direct current I_1 flowing through coil X is the rectified current I_C drawn by the capacitor C. Similarly, direct current I_2 flowing through coil Y is the rectified current I_R drawn by the resistor R. A pointer is attached to the coils which moves over a calibrated scale.

Working. The meter is connected across the supply whose frequency f is to be measured. The direct currents I_1 and I_2 flowing through coils X and Y respectively produce two torques which oppose each other. The pointer will come to rest at angle θ for which the two torques are equal.

In the equilibrium position, $\theta \propto I_1/I_2$ *...as proved above*

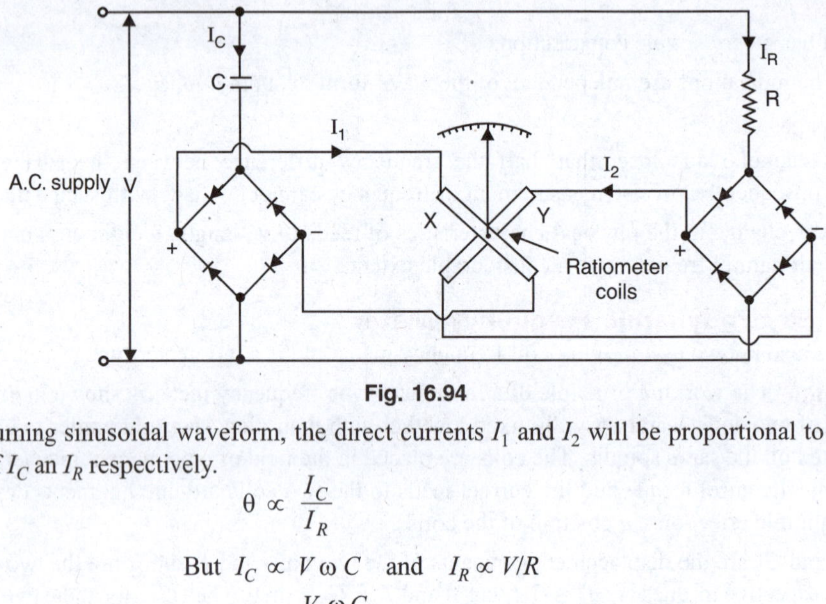

Fig. 16.94

Assuming sinusoidal waveform, the direct currents I_1 and I_2 will be proportional to the r.m.s. values of I_C an I_R respectively.

$$\therefore \qquad \theta \propto \frac{I_C}{I_R}$$

But $I_C \propto V\omega C$ and $I_R \propto V/R$

$$\therefore \qquad \theta \propto \frac{V\omega C}{V/R}$$

$$\propto \omega CR$$

$$\propto 2\pi fCR \qquad (\because \omega = 2\pi f)$$

Since 2π, C and R are constants,

$$\therefore \qquad \theta \propto f \qquad\qquad\qquad\qquad\qquad ...(i)$$

i.e., deflection of the pointer is directly proportional to the supply frequency f. *Clearly, such meters have linear scale.*

An inspection of exp. (*i*) reveals that deflection is independent of supply voltage. Therefore, such instruments can be used over a fairly wide range of voltages. However, if the voltage becomes too low, the distortion introduced by the rectifiers prevents an accurate indication of frequency.

16.62. Moving-Iron Frequency Meter

This meter is also known as Weston frequency meter and operates on the principle of variation of impedance of a circuit with the change in supply frequency.

Principle. The action of this meter depends upon variation of current distribution in two parallel circuits — one inductive and the other non-inductive — when supply frequency changes.

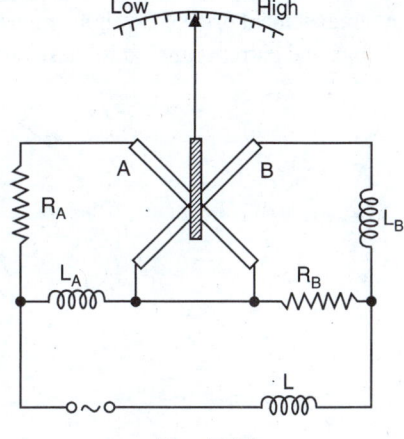

Fig. 16.95

Construction. Fig. 16.95 shows the constructional details of a moving-iron frequency meter. It consists of two coils A and B mounted perpendicular to each other. The coil A has a resistor R_A in series while coil B has an inductance L_B in series. Circuit A is in parallel with inductance L_A while B is in parallel with resistance R_B. At the centres of the coils is pivoted a long and thin soft-iron needle which carries the pointer. No controlling torque is required; damping being provided by air friction. The series inductance L helps to suppress higher harmonics in the current waveform and therefore tends to minimise the waveform errors in the indications of the instrument.

The currents flowing through coils A and B produce opposing torques. The circuit parameters (*i.e.*, R_A, L_A, R_B and L_B) are so designed that at normal frequency of supply, the currents in the two coils are equal and the pointer is at the centre of the scale.

Working. When the meter is connected to the supply whose frequency is to be determined, the currents flowing through coils A and B set up magnetic fields. The magnitude of magnetic fields will depend upon the amount of current flowing through the coils. The moving system would tend to align itself along the resultant magnetic field of two coils :

(*i*) If the supply frequency increases above the normal value, the current through coil A *increases whereas that through coil B decreases. Thus the coil A exerts more force on the moving system so that the pointer is turned more nearly parallel with the magnetic axis of coil A.

(*ii*) If the supply frequency is below the normal value, the current through coil A decreases whereas that through coil B increases. Thus the coil B exerts more force on the moving system so that the pointer is pulled towards the magnetic axis of coil B.

16.63. Power Factor Meters

In an a. c. system, it is important to know the power factor because it affects the operation of the entire system. For instance, a low lagging p. f. leads to poor voltage regulation and low efficiency. The power factor in a **single phase circuit can be calculated ($\cos \phi = P/VI$) from the readings of ammeter, voltmeter and wattmeter. However, such a method of calculations is neither convenient nor practical because the system power factor is changing continuously. The power factor of a circuit can be determined by a single instrument called power factor meter. The following two important power factor meters will be discussed :

---

*   At frequency above the normal value, the impedance of circuit B becomes more than (due to increased voltage drop across L_B) that of circuit A. Hence current in coil A increases whereas that in coil B decreases.

**   In a 3-phase balanced circuit, the p.f. can be determined from the readings of two wattmeters (See Art. 15.24). In case of an unbalanced 3-phase system, the p.f. may be different in the different phases and determination of mean p.f. becomes a matter of difficulty. For most practical purposes, it is sufficient to take the p.f. of an unbalanced 3-phase system as the value indicated by a 3-phase power factor meter or the value calculated from the ratio: True power/Apparent power.

(*i*) Electrodynamic power factor meter

(*ii*) Moving-iron power factor meter

16.64. Single-Phase Electrodynamic Power Factor Meter

This instrument is based on the dynamometer principle. The mechanism is somewhat similar to that used in dynamometer wattmeter ; the principal differences being (*i*) there are two coils on the moving element oriented at right angles to each other as shown in Fig. 16.96, (*ii*) there are no control springs and currents are led to these coils through fine ligaments which exert no controlling torque.

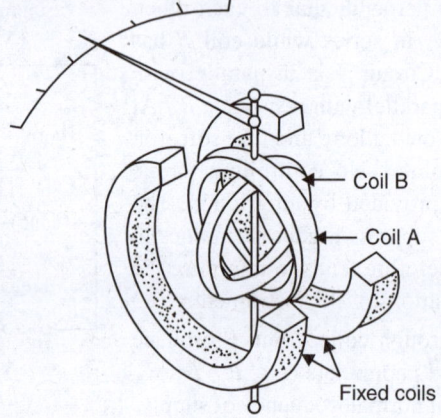

Fig. 16.96

Construction. Fig. 16.97 shows the essential parts of a single phase electrodynamic power factor meter. Like a dynamometer wattmeter, the fixed coil is split into two equal parts and carries the load current I. Pivoted between the fixed coils are two coils A and B rigidly fixed at angle of 90° apart. These coils move together and carry the pointer which indicates the power factor of the circuit directly on the scale. The coil A is connected through a resistor R across the line so that its current I_A is proportional to and in phase with the supply voltage V. The coil B is connected through an inductance L across the line so that its current I_B (proportional to supply voltage V) lags the supply voltage by 90°. The values of R and L are so adjusted that the two coils carry the same current at normal frequency. Note that there is no controlling torque acting on the moving system ; the current being led into coils A and B by fine ligaments which exert no control.

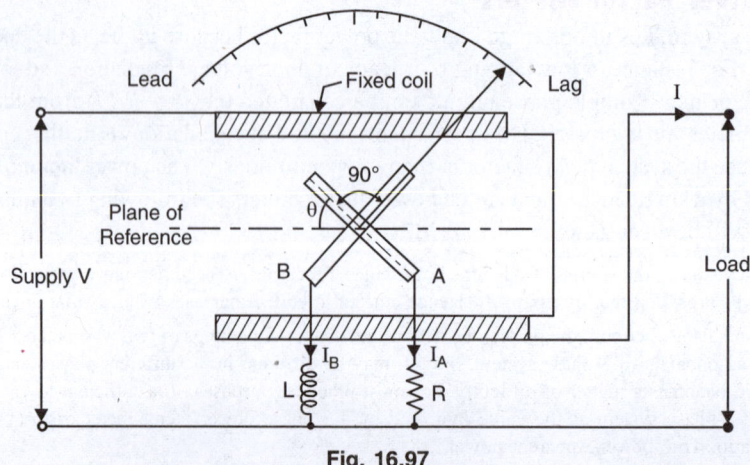

Fig. 16.97

Theory. The currents in each of the moving coils (*i.e.*, coil A and coil B) react with the current in the fixed coils to produce two torques. Connections to the coils are such that the two torques oppose each other. The moving system will come to rest at a position where the two torques are equal.

Suppose the instrument is connected in a circuit whose p.f. is cos ϕ lagging as shown in Fig. 16.97. As the currents are flowing through coils A and B as well as through the fixed coils, there will be two torques acting on the moving system.

$$\text{Torque on coil } A, \; T_A \propto I_A * I \cos \phi \sin \theta$$

where $\qquad\qquad \theta$ = angular deflection from the plane of reference

or $\qquad\qquad\qquad T_A \propto VI \cos \phi \sin \theta$ $\qquad\qquad\qquad\qquad$ $[\because I_A \propto V]$

This torque will act in one direction, say clockwise.

$$\text{Torque on coil } B, \; T_B \propto I_B ** I \cos (90° - \phi) \sin (90° + \theta)$$

$$\propto I_B I \sin \phi \cos \theta$$

$$\propto VI \sin \phi \cos \theta \qquad\qquad [\because I_B \propto V]$$

This torque will act in the opposite direction *i.e.*, anticlockwise.

The moving system will come to rest where $T_A = T_B$.

$\therefore \qquad\qquad VI \cos \phi \sin \theta = VI \sin \phi \cos \theta$

or $\qquad\qquad \cos \phi \sin \theta = \sin \phi \cos \theta$

or $\qquad\qquad \sin \theta / \cos \theta = \sin \phi / \cos \phi$

or $\qquad\qquad \tan \theta = \tan \phi \qquad \therefore \quad \theta = \phi$

Thus the angular deflection of the moving system in degrees is numerically equal to the power factor angle ϕ (*i.e.*, phase angle between supply voltage V and line current I). The scale of the instrument is calibrated in terms of cos ϕ to directly indicate the p.f. of the circuit.

Note. The meter will indicate correctly only at one frequency *i.e.*, at the frequency at which the meter is calibrated. At the frequency of calibration, the values of R and L are so adjusted that $I_A = I_B$. At any other frequency, the two currents will be ***different. However, this is not a serious disadvantage because in most power distribution systems, the line frequency is held within narrow limits.

16.65. 3-Phase Electrodynamic Power Factor Meter

For the measurement of power factor of a balanced 3-phase load, a 3-phase power factor meter may be used. The general principle of operation is the same as for the single-phase instrument.

Construction. The construction of a 3-phase electrodynamic power factor meter (See Fig. 16.98) is almost the same as for a single phase instrument. The two fixed coils (*i.e.*, current coils) are connected in series with one phase (phase R in the present case) and carry the line current I. The two moving coils A and B are fixed at 120° to each other and are connected across two different

---

* It was shown in Art. 16.15 that in an electrodynamic mechanism,

$$T \propto I_1 I_2 \cos \phi \; \frac{dM}{d\theta}$$

where $dM/d\theta$ = rate of change of mutual inductance.

The mutual inductance between the two coils (*i.e.*, moving and fixed) is maximum when they are parallel to each other *i.e.*, when $\theta = 0°$.

$\therefore \qquad\qquad\qquad M \propto \cos \phi \quad$ or $\quad dM/d\theta \propto \sin \theta$

$\therefore \qquad\qquad\qquad T \propto I_1 I_2 \cos \phi \sin \theta$

** The current I_B in coil B lags 90° behind the supply voltage. Therefore, angle between I_B and line current I is $(90° - \phi)$. The coil B is 90° ahead of coil A. Hence coil B is $(90° + \theta)$ ahead of plane of reference.

*** The coil B has inductance L in its circuit so that current through it depends upon the supply frequency. But current in coil A (being purely resistive) is independent of supply frequency.

phases through high non-inductive resistances. In the figure shown, the coil A is connected across R and Y and coil B across R and B. Since the moving coils A and B are connected to the two phases of the 3-phase supply, *splitting device is not required.

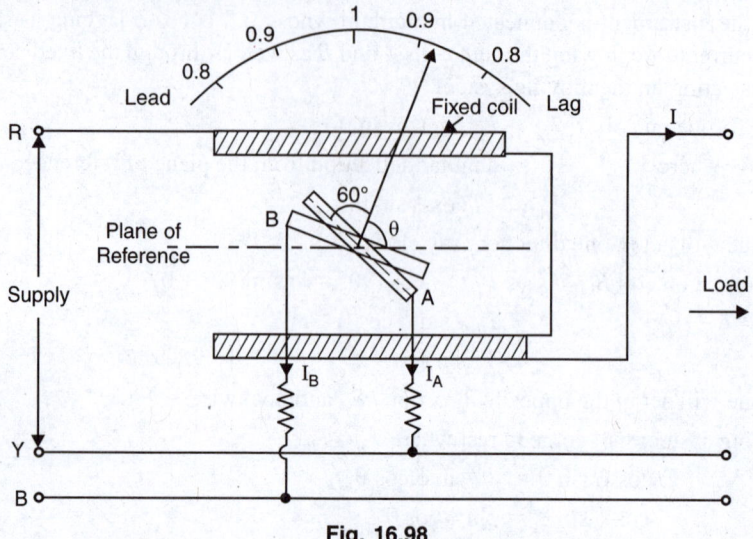

Fig. 16.98

Theory. Fig. 16.99 shows the **phasor diagram of the circuit. Here V_{RN}, V_{YN} and V_{BN} are the phase voltages of the supply that are equal in magnitude but 120° apart. The voltage across coil A is the line voltage $V_{RY} (= V_{RN} - V_{YN} ...$ phasor difference) and that across coil B is the line voltage V_{RB} $(= V_{RN} - V_{BN} ...$ phasor difference).

Let ϕ = load p.f. angle ; θ = angular deflection from the plane of reference

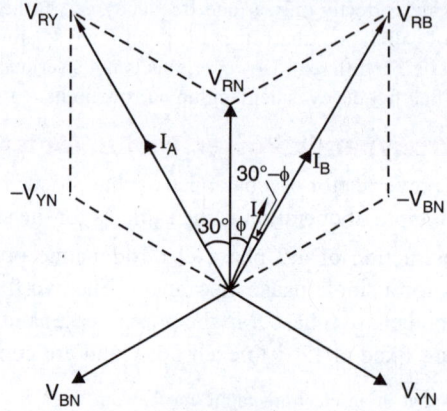

Fig. 16.99

Torque on coil A, $\quad T_A \propto I_A I \cos (30° + \phi) \sin (60° + \theta)$

$$\propto V_{RY} I \cos (30° + \phi) \sin (60° + \theta) \quad [\because I_A \propto V_{RY}]$$

$$\propto \sqrt{3} \, V_{ph} I \cos (30° + \phi) \sin (60° + \theta)$$

---

*   In the single phase instrument (Refer back to Fig. 16.97), an inductor is used in the circuit of one moving coil to cause the phase difference in the currents drawn by the two moving coils.

** Note that currents I_A and I_B are in phase with the voltages (*i.e.*, V_{RY} and V_{RB}) across them. It is because the circuits of the two coils are essentially resistive.

Torque on coil B, $T_B \propto I_B I \cos (30° - \phi) \sin (60° - \theta)$

$$\propto V_{RB} I \cos (30° - \phi) \sin (60° - \theta) \qquad [\because I_B \propto V_{RB}]$$

$$\propto \sqrt{3} \, V_{ph} I \cos (30° - \phi) \sin (60° - \theta)$$

The two torques T_A and T_B act in the opposite directions. For steady deflection of the pointer, $T_A = T_B$.

$\therefore \quad \sqrt{3} \, V_{ph} I \cos (30° + \phi) \sin (60° + \theta) = \sqrt{3} \, V_{ph} I \cos (30° - \phi) \sin (60° - \theta)$

or $\quad \cos(30° + \phi) \sin (60° + \theta) = \cos (30° - \phi) \sin (60° - \theta)$

Solving the above expression, we have, $\theta = \phi$.

Thus the angular deflection of the pointer is equal to the phase angle (ϕ) of the circuit to which the meter is connected. The scale is graduated in terms of power factor (*i.e.*, $\cos \phi$) instead of ϕ.

16.66. Moving-Iron Power Factor Meter

Several types of moving-iron power factor meters are available. The arrangement shown in Fig. 16.100 is of rotating field type and is suitable for the measurement of p.f. of balanced 3-phase circuits. The mechanism consists of :

(*i*) *a stationary field-coil assembly* consisting of coils A, B and C (similar to the stator of a 3-phase motor) with axes mutually at 120°. These coils are connected respectively in R, B and Y lines of the 3-phase supply through current transformers. When energised from a 3-phase supply, these coils produce a rotating field.

(*ii*) *a moving mechanism* consisting of spindle carrying an iron cylinder to which are fixed sector-shaped vanes VV. The moving mechanism is surrounded by a fixed coil D placed at the centre of the rotating field system and connected between any two phases (phase R and Y in the present case) through a high resistance. The spindle also carries damping vanes and pointer but there are no control springs.

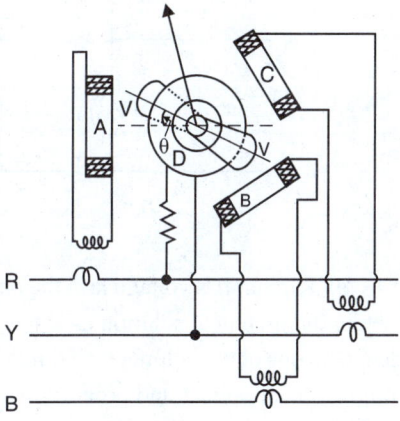

Fig. 16.100

Working. The moving iron system is placed inside the fixed coil D so that the moving iron carries an alternating flux. This alternating flux interacts with the fluxes produced by the three coils. The total torque on the moving system is equal to the sum of the torques due to currents in the coils A, B and C. The pointer will come to rest at a position where total torque is equal to zero. By constructing the phasor diagram and using the rules of trigonometry, it can be proved that for steady deflection of the pointer,

$$\theta = \phi, \text{ the load p.f. angle}$$

Thus deflection (θ) of the pointer from the reference axis is equal to phase angle (ϕ) between each current and the corresponding phase voltage. The meter is scaled to read power factor ($\cos \phi$) directly.

Advantages

(*i*) They are robust and simple in construction.

(*ii*) The deflecting torque produced is greater than that of electrodynamic type.

(*iii*) Since all the coils in the instrument are stationary, there are no electrical connections to the moving parts.

(*iv*) A wider scale extending upto 360° is available.

Disadvantages

(*i*) Errors are introduced due to eddy current and hysteresis loss in the iron parts.

(*ii*) The calibration of these instruments is affected by supply frequency, voltage and waveform.

(*iii*) They are less accurate than the electrodynamic type.

16.67. 3-Voltmeter Method of Determining Phase Angle

It is sometimes very convenient to determine the phase angle between two voltages V_{12} and V_{23} with no equipment other than 3 a.c. voltmeters connected as shown in Fig. 16.101 (*i*). The method may be used to determine the phase angles between line voltages of 3-phase circuits by connecting the lines directly to the junctions of the voltmeter leads as shown in Fig. 16.101 (*ii*).

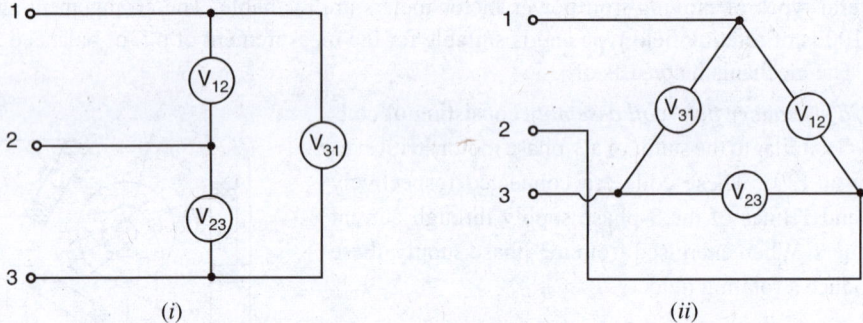

Fig. 16.101

By Kirchhoff's voltage law, the phasor sum of the three voltmeter readings is zero. We can thus obtain the graphical solution as shown in Fig. 16.102. First, the base line is drawn proportional in length to one of the voltmeter readings, in this case V_{31}. Next two arcs are drawn ; one having a radius originating at point 1 and proportional in length to V_{12}, and the other having a radius originating at point 3 and proportional in length to V_{23}. The intersection of the two arcs becomes point 2 and locates the ends of phasors V_{12} and V_{23}. From familiar principles of trigonometry, we have,

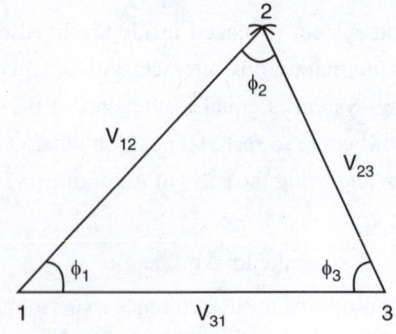

Fig. 16.102

$$\phi_1 = \cos^{-1} (V_{12}^2 + V_{31}^2 - V_{23}^2)/(2V_{12} V_{31})$$
$$\phi_2 = \cos^{-1} (V_{12}^2 + V_{23}^2 - V_{31}^2)/(2V_{12} V_{23})$$
$$\phi_3 = \cos^{-1} (V_{31}^2 + V_{23}^2 - V_{12}^2)/(2V_{31} V_{23})$$

16.68. Ohmmeter

A device that measures the resistance directly is called an **ohmmeter.** The simplest direct reading ohmmeter is the basic *series ohmmeter* circuit shown in Fig. 16.103 (*i*). It consists of a permanent magnet moving coil (PMMC) instrument in series with a battery and a rheostat R. Note that A and B are the terminals of ohmmeter.

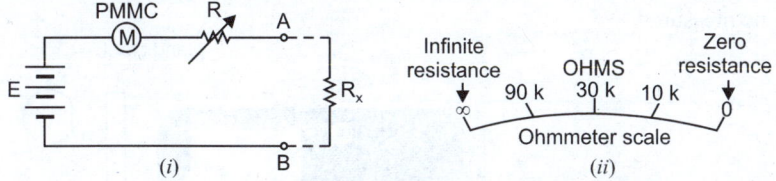

Fig. 16.103

(*i*) With terminals A and B shorted together, R is adjusted for full-scale deflection (f.s.d.). Since terminals A and B are shorted, the ohmmeter should read zero resistance. Thus as shown in Fig. 16.103 (*ii*), the full-scale deflection should read zero resistance.

Full-scale deflection current, $I_g = \dfrac{E}{R}$ (neglecting meter resistance)

(*ii*) When terminals A and B are open-cricuited, the pointer should indicate infinity. Therefore, zero deflection point on the scale is marked as infinite resistance. [See Fig. 16.103 (*ii*)].

(*iii*) When an unknown resistance R_x is connected to terminals A and B, the meter current I_m is

$$I_m = \frac{E}{R + R_x}$$

Since the value of I_m is more than zero and less than I_g (f.s.d.), the meter will give a reading between zero and infinity. Thus the value of unknown resistance R_x can be determined.

Note. The circuit shown in Fig. 16.103 (*i*) relies upon battery voltage remaining absolutely constant. When the battery terminal voltage falls (as they all do with use), the instrument scale is no longer accurate. Thus some means of adjusting for battery voltage variations must be built in the circuit.

Example 16.61. *The series ohmmeter shown in Fig. 16.103 is made up of a 3 V battery, a 100 μA meter and a resistance R which has a fixed value of 30 kΩ. Determine the value of the unknown resistance R_x when the pointer indicates (i) 1/2 F.S.D. (ii) 3/4 F.S.D.*

Solution. The meter will indicate F.S.D. (full-scale deflection) when $R_x = 0$ *i.e.*,

Full-scale deflection current is

$$I_g = \frac{E}{R + R_x} = \frac{3V}{30k\Omega + 0} = 100 \ \mu A \quad (given)$$

(*i*) **At 1/2 F.S.D.** Meter current, $I_m = 100/2 = 50 \ \mu A$

Now, $I_m = \dfrac{E}{R + R_x}$

or $R + R_x = \dfrac{E}{I_m} = \dfrac{3\,V}{50\mu A} = 60 \ k\Omega$

∴ $R_x = 60 - R = 60 - 30 = \mathbf{30 \ k\Omega}$

(*ii*) **At 3/4 F.S.D.** Meter current, $I_m = (3/4) \times 100 = 75 \ \mu A$

Now, $I_m = \dfrac{E}{R + R_x}$

or $R + R_x = \dfrac{E}{I_m} = \dfrac{3\,V}{75\mu A} = 40 \ k\Omega$

∴ $R_x = 40 - 30 = \mathbf{10 \ k\Omega}$

16.69. Megger

The megaohmmeter (or megger) is an instrument for measuring very high resistance such as the insulation resistance of electrical cables. A megger is essentially an ohmmeter with its own hand-cranked high voltage generator (See Fig. 16.104). The generated voltage may be anything from 100 V to 2.5 kV. The high voltage source is required to pass a measurable current through the high resistance to be measured.

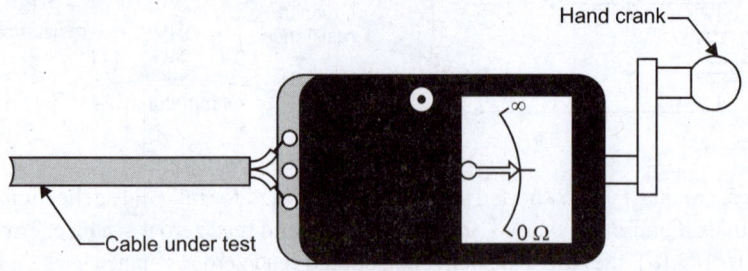

Fig. 16.104

As in the case of ohmmeter, the scale of the megger indicates infinity (∞) when its terminals are open-circuited and zero when the megger terminals are shorted. When the unknown resistance is connected to megger terminals, it gives the value of resistance between zero and infinity. The range of the instrument can be changed by switching different values of standard resistor in the circuit.

16.70. Instrument Transformers

For measuring a large current in a d.c. circuit, we use low-range ammeter with a suitable shunt. The measurement of high d.c. voltage is made using a low-range voltmeter with a multiplier. However, this method is not used for the measurement of high alternating currents and voltages for many good reasons. In order to measure high alternating currents and voltages, we employ specially designed transformers, called *instrument transformers*. These transformers facilitate the a.c. measurements with low-range a.c. instruments. There are two types of instrument transformers *viz*.

(*i*) Current transformers (*ii*) Potential transformers.

16.71. Current Transformer (C.T.)

A current transformer (C.T.) is used to measure high alternating current in a power system. The primary of this transformer has a few turns of thick wire whereas the secondary has many turns of very fine wire as shown in Fig. 16.105. It is clear from the figure that a current transformer is simply a well-designed step-up transformer. Since voltage is stepped up, the current is stepped down which can be measured with a low-range a.c. ammeter.

The primary of the current transformer is connected in *series* with the line whose current is to be measured as shown in Fig. 16.105. The secondary of the transformer is connected across a low-range (0 –5A) a.c. ammeter. The line current (I_P) and a.c. ammeter current (I_S) are *related as :

$$N_P I_P = N_S I_S$$

or

$$\frac{I_P}{I_S} = \frac{N_S}{N_P}$$

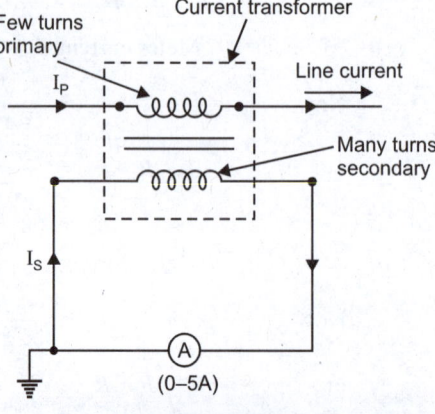

Fig. 16.105

_____
*   The magnetising component of primary current is very small and can be neglected with reasonable accuracy.

The primary to secondary current ratio (*i.e.*, I_P/I_S) is called C.T. ratio (current transformation ratio).

$$\therefore \qquad \frac{I_P}{I_S} = \text{C.T. ratio}$$

or $\qquad\qquad I_P = I_S \times \text{C.T. ratio}$

i.e., Line current $(I_P) = \text{A.C. ammeter reading} \times \text{C.T. ratio}$

Thus if the reading of a.c. ammeter is 1 A and the C.T. ratio is 100 : 1 (or 100/1), then line current $= 1 \times 100 = 100$ A.

Similarly, if the a.c. ammeter reads 2 A, the line current $= 2 \times 100 = 200$ A. One most commonly used current transformer is the clamp-on or clip-on type shown in Fig. 16.106. It consists of a ring-shaped laminated core which carries the secondary winding. The current carrying conductor itself acts as a one-turn primary that simply passes through the centre of the ring. The position of the primary is unimportant as long as it is more or less centred. This current transformer has the arrangement to open and close the ring-shaped core so that current can be measured without opening the line. The clamp-on current

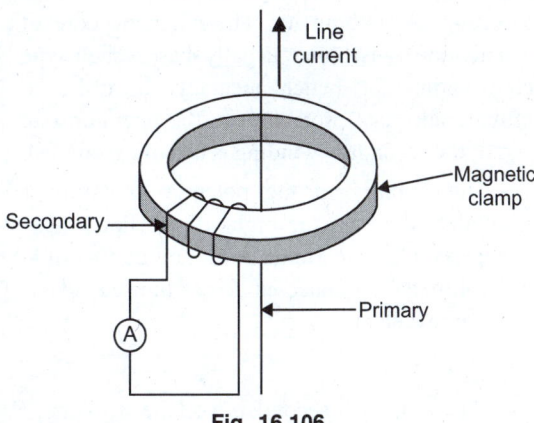

Fig. 16.106

transformers are simple and inexpensive and are widely used in low-voltage (LV) and medium-voltage (MV) lines in the power system.

C.T. secondary should never be left open. Every precaution must be taken to *never* open the secondary circuit of a current transformer (C.T.) while current is flowing in the primary circuit. Opening of the secondary under these conditions removes the *opposing m.m.f. produced in the secondary and allows the flux in the core to reach a value determined only by the primary m.m.f. The result is that the flux in the core is greatly increased. This situation leads to the following harmful effects :

(*i*) The increased flux in the core may saturate it. The effect of saturation is to leave the transformer with a large value of residual flux. This will seriously ** impair the accuracy of the current transformer.

(*ii*) The flux in the core changes at a very high rate inducing a large voltage in the open secondary winding. The effect of excessive voltage may be to puncture the insulation of the transformer or to provide dangerous shock to the operator opening the circuit.

In view of the above, if we have to remove a meter in the secondary circuit of a C.T., we must first short-circuit the secondary *and then* remove the meter. The short circuit across the winding may be removed after the secondary circuit is again closed.

Note. Unlike an ordinary transformer, the primary current of C.T. is not controlled by secondary but remains constant at a value determined by the current in the main circuit. Therefore, short-circuiting a current transformer does not harm because the primary current remains unchanged and the secondary current depends only upon the turns ratio.

---

*   With secondary circuit closed, the secondary current sets up m.m.f. which opposes the primary m.m.f. As a result, the value of flux in the core is small.

**   Due to change in the magnetic characteristics of the core, the ratio of transformation of the transformer may change.

16.72. Potential Transformer (P.T.)

A potential transformer (P.T.) is used to measure high alternating potential difference (voltage) in a power system. The primary of this transformer has many turns while the secondary has few turns as shown in Fig. 16.107. It is clear from the figure that a potential transformer is simply a well-designed step-down transformer. The stepped down voltage is measured with a low-range a.c. voltmeter. The magnetic core of a potential transformer usually has a shell-type construction for better accuracy. In order to provide adequate protection to the operator, one end of the secondary winding is usually grounded.

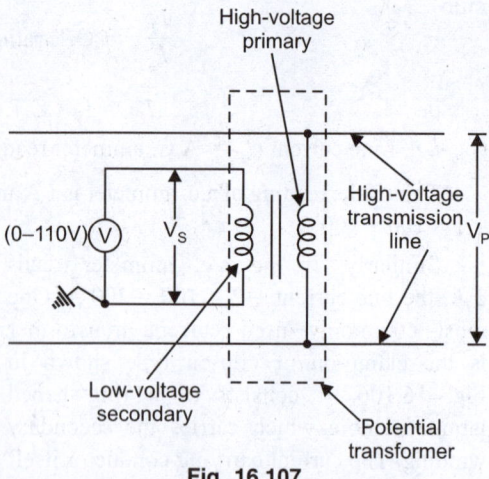

Fig. 16.107

The primary of the potential transformer is connected across the high-voltage line whose voltage is to be measured. A low-range (0-110 V) a.c. voltmeter is connected across the secondary. The line voltage (V_P) and a.c. voltmeter voltage (V_S) are related as :

$$\frac{V_P}{V_S} = \frac{N_P}{N_S}$$

The primary to secondary voltage ratio (*i.e.*, V_P/V_S) is called P.T. ratio (potential transformation ratio).

$$\therefore \qquad \frac{V_P}{V_S} = \text{P.T. ratio}$$

or $$\qquad V_P = V_S \times \text{P.T. ratio}$$

i.e.,   Line voltage (V_P) = A.C. voltmeter reading × P.T. ratio

Thus if the reading of a.c. voltmeter is 50 V and the P.T. ratio is 100 : 1 (or 100/1), then line voltage = 50 × 100 = 5,000 V. Similarly, if the a.c. voltmeter reads 100 V, then line voltage = 100 × 100 = 10,000 V.

16.73. Advantages of Instrument Transformers

In order to measure high alternating currents and voltages in a power system, we prefer instrument transformers to shunts and multipliers for the following reasons :

(*i*) The errors due to stray inductance and capacitance in shunts, multipliers and their leads are eliminated.

(*ii*) The measuring circuit is isolated from the mains by the transformer.

(*iii*) We can use low-range and accurate a.c. instruments.

(*iv*) The length of the connecting leads from the transformer to the instrument is of lesser importance and leads may be of small cross-sectional area.

(*v*) By using a clip-on type of transformer core, the current in a heavy-current conductor can be measured without breaking the circuit.

Example 16.62. *Fig. 16.108 shows the connections of instrument transformers in a high-voltage circuit. The C.T. ratio is 80 : 5 while P.T. ratio is 100 : 1. If the ammeter, voltmeter and wattmeter read 4 A, 110 V and 352 W respectively, determine (i) current (ii) voltage and (iii) the power supplied to the load.*

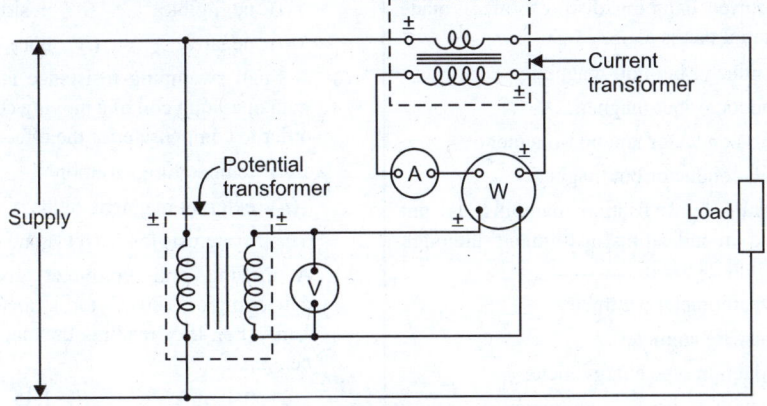

Fig. 16.108

Solution. *(i)*   Current, I = Ammeter reading × C.T. ratio

$$= 4 × (80/5) = \textbf{64 A}$$

(ii)   Voltage, V = Voltmeter reading × P.T. ratio

$$= 110 × (100/1) = \textbf{11,000 V}$$

(iii)   Power supplied, P = Wattmeter reading × C.T. ratio × P.T. ratio

$$= 352 × (80/5) × (100/1) = \textbf{563.2 × 10}^3 \textbf{ W}$$

Example 16.63. *A potential transformer of 14400/115 V and a current transformer of 75/5 A are used to measure voltage and current in a transmission line. If the voltmeter indicates 111 V and ammeter reads 3 A, what is the voltage and current in the line ?*

Solution.   Line voltage, V = Voltmeter reading × P.T. ratio

$$= 111 × (14400/115) = \textbf{13900 V}$$

Line current, I = Ammeter reading × C.T. ratio

$$= 3 × (75/5) = \textbf{45 A}$$

Objective Questions

1. An ammeter is .... instrument.

    (i) an indicating   *(ii)* an integrating

    (iii) a recording   *(iv)* none of the above

2. The controlling torque of an indicating instrument .... as the deflection of the moving system increases.

    (i) remains unchanged

    (ii) decreases

    (iii) increases   *(iv)* none of the above

3. When the pointer of an indicating instrument comes to rest in the final deflected position, ...

    (i) only controlling torque acts

    (ii) only deflecting torque acts

    (iii) both deflecting and controlling torques act

    (iv) none of the above

4. When the pointer of an indicating instrument is in motion, then deflecting torque is opposed by ....

    (i) controlling torque only

    (ii) damping torque only

    (iii) both damping & controlling torques

    (iv) none of the above

5. The pointer of an indicating instrument is generally made of ....

    (i) copper   *(ii)* aluminium

    (iii) silver   *(iv)* soft steel

6. When the pointer of an indicating instrument is in the final deflected position ,..............

    (i) deflecting torque is zero

    (ii) controlling torque is zero

    (iii) damping torque is zero

    (iv) both deflecting & controlling torques are zero

7. In eddy current damping, disc or former is made of a material that is a ....
   (i) conductor but non-magnetic
   (ii) conductor but magnetic
   (iii) non-conductor and non-magnetic
   (iv) non-conductor but magnetic

8. In general, fluid friction damping is not employed in indicating instruments although one can find its use in ....
   (i) dynamometer wattmeter
   (ii) hot-wire ammeter
   (iii) induction type energy meter
   (iv) Kelvin electrostatic voltmeter

9. Permanent-magnet moving coil instrument can be used for ....
   (i) a.c. work only   (ii) d.c. work only
   (iii) both d.c. and a.c. work
   (iv) none of the above

10. The scale of a permanent-magnet moving coil instrument is uniform because ....
    (i) of effective eddy current damping
    (ii) external magnetic fields have no effect
    (iii) it is spring controlled
    (iv) it has no hysteresis loss

11. Shunts are generally made of ....
    (i) copper   (ii) aluminium
    (iii) silver   (iv) manganin

12. The range of a permanent-magnet moving coil instrument is 0-10 A. If the full-scale deflection current of the meter is 2 mA, then multiplying power of the shunt is ....
    (i) 2500   (ii) 10000
    (iii) 5000   (iv) none of the above

13. A moving coil instrument having meter resistance of 5Ω is to be used as a voltmeter of range 0-100 V. If the full-scale deflection current is 10 mA, then required series resistance is ....
    (i) $20\,\Omega$ (ii) $1000\,\Omega$
    (iii) $9995\,\Omega$ (iv) none of the above

14. The mutliplying power of the shunt of a milliammeter is 8. If the circuit current is 200 mA, then current through the meter is ....
    (i) 200 mA   (ii) 25 mA
    (iii) 1600 mA   (iv) none of the above

15. The material of the shunt should have .... temperature co-efficient of resistance.

    (i) negligible   (ii) positive
    (iii) negative   (iv) none of the above

16. A small swamping resistance is put in series with operating coil of a moving coil ammeter in order to compensate for the effects of ....
    (i) temperature variation
    (ii) external magnetic fields
    (iii) hysteresis loss   (iv) none of the above

17. A moving coil voltmeter gives full-scale deflection of 100 V for a meter current of 1 mA. For 45 V reading, the meter current will be ....
    (i) 0.45 mA   (ii) 1.45 mA
    (iii) 2.22 mA   (iv) none of the above

18. Dynamometer type instruments can be used for ....
    (i) a.c. work only   (ii) d.c. work only
    (iii) for both d.c. and a.c. work
    (iv) none of the above

19. A dynamometer instrument is chiefly used as a ....
    (i) d.c. ammeter   (ii) d.c. voltmeter
    (iii) wattmeter   (iv) none of the above

20. In a dynamometer type instrument, damping is provided by ....
    (i) air friction   (ii) eddy currents
    (iii) fluid friction (iv) none of the above

21. Dynamometer type .... has uniform scale.
    (i) ammeter   (ii) wattmeter
    (iii) voltmeter   (iv) none of the above

22. The instrument in which springs provide the controlling torque as well as serve to lead current into and out of the operating coil is .... instrument.
    (i) moving-iron   (ii) hot-wire
    (iii) permanent-magnet moving coil
    (iv) none of the above

23. If current through the operating coil of a moving-iron instrument is doubled, the operating force becomes ....
    (i) two times   (ii) four times
    (iii) one-half time (iv) three times

24. The full-scale deflection current of a moving coil instrument is about ....
    (i) 10 mA   (ii) 1 A
    (iii) 3 A   (iv) 2 A

25. For measuring high values of alternating current with a dynamometer ammeter, we use a ....

   (*i*) shunt   (*ii*) multiplier

   (*iii*) potential transformer

   (*iv*) current transformer

26. Hot-wire instruments have .... scale.

   (*i*) uniform   (*ii*) log

   (*iii*) squared   (*iv*) none of the above

27. The full-scale voltage across a moving coil voltmeter is about ....

   (*i*) 10 V   (*ii*) 5 V

   (*iii*) 100 V   (*iv*) 10 mV

28. Moving-iron instruments have .... scale.

   (*i*) uniform   (*ii*) squared

   (*iii*) log   (*iv*) none of the above

29. The range of a moving-iron a.c. ammeter is extended by ....

   (*i*) a shunt   (*ii*) a multiplier

   (*iii*) changing number of turns of operating coil

   (*iv*) none of the above

30. To measure high-frequency currents, we mostly use .... ammeter.

   (*i*) hot-wire   (*ii*) dynamometer

   (*iii*) moving-iron   (*iv*) thermocouple

31. For the measurement of high direct voltage (say 10 kV), one would use .... voltmeter.

   (*i*) permanent-magnet moving coil

   (*ii*) electrostatic

   (*iii*) hot-wire   (*iv*) moving iron

32. .... movement is most expensive.

   (*i*) D'Arsonval

   (*ii*) Moving-iron

   (*iii*) Dynamometer   (*iv*) none of the above

33. Electrostatic instruments are used as ....

   (*i*) voltmeters only   (*ii*) ammeters only

  S(*iii*) both ammeters and voltmeters

   (*iv*) wattmeters only

34. An electric pyrometer is an instrument used to measure ....

   (*i*) phase   (*ii*) frequency

   (*iii*) high temperatures

   (*iv*) none of the above

35. The best type of meter movement is .... movement.

   (*i*) iron-vane   (*ii*) D'Arsonval

   (*iii*) dynamometer   (*iv*) none of the above

36. ....instruments are most sensitive.

   (*i*) Moving-iron   (*ii*) Hot-wire

   (*iii*) Dynamometer

   (*iv*) Permanent-magnet moving coil

37. In induction type ammeter, ... damping is provided.

   (*i*) air friction   (*ii*) eddy current

   (*iii*) fluid friction   (*iv*) none of the above

38. The most commonly used induction type instrument is ....

   (*i*) induction voltmeter

   (*ii*) induction wattmeter

   (*iii*) induction watt-hour meter

   (*iv*) induction ammeter

39. All voltmeters except .... voltmeters are operated by the passage of current.

   (*i*) moving-iron   (*ii*) electrostatic

   (*iii*) dynamometer

   (*iv*) permanent-magnet moving coil

40. The watt-hour meter is .... instrument.

   (*i*) an integrating   (*ii*) an indicating

   (*iii*) a recording   (*iv*) a transfer

Answers

1. (*i*)	**2.** (*iii*)	**3.** (*iii*)	**4.** (*iii*)	**5.** (*ii*)
6. (*iii*)	**7.** (*i*)	**8.** (*iv*)	**9.** (*ii*)	**10.** (*iii*)
11. (*iv*)	**12.** (*iii*)	**13.** (*iii*)	**14.** (*ii*)	**15.** (*i*)
16. (*i*)	**17.** (*i*)	**18.** (*iii*)	**19.** (*iii*)	**20.** (*i*)
21. (*ii*)	**22.** (*iii*)	**23.** (*ii*)	**24.** (*i*)	**25.** (*iv*)
26. (*iii*)	**27.** (*iv*)	**28.** (*ii*)	**29.** (*iii*)	**30.** (*iv*)
31. (*ii*)	**32.** (*iii*)	**33.** (*i*)	**34.** (*iii*)	**35.** (*ii*)
36. (*iv*)	**37.** (*ii*)	**38.** (*iii*)	**39.** (*ii*)	**40.** (*i*)

17

A.C. Network Analysis

Introduction

An a.c. circuit differs from a d.c. circuit in two important respects. First, in a d.c. circuit, we have *resistances only whereas in an a.c. circuit, in addition to resistance (R), inductance (L) and capacitance (C) also play the part. Therefore, in an a.c. circuit, we have to deal with impedances instead of resistances as in a d.c. circuit. Secondly, in a d.c. circuit, voltages or currents can be added or subtracted arithmetically. However, **alternating voltages and currents are phasors so that the addition or subtraction of alternating voltages or currents requires the use of phasor algebra. The techniques and theorems used to solve d.c. network problems in chapter 3 of this book can also be applied to a.c. networks provided if we keep in mind that in an a.c. circuit, we have impedances instead of resistances and that alternating voltages or currents can be added or subtracted by the use of phasor algebra. In this chapter, we shall discuss the various techniques and network theorems to solve a.c. network problems.

17.1. A.C. Network Analysis

The techniques and theorems employed to solve d.c. network problems can also be applied to a.c. network problems keeping in view the following points :

(*i*) In an a.c. network, we have impedances (R, L and C in varying proportions) instead of resistances only as in a d.c. circuit.

(*ii*) *The alternating voltages and currents are phasors. Therefore, for the addition or subtraction of alternating voltages or currents, we have to use phasor algebra.*

(*iii*) *In solving a.c. networks, the phase angles of all impedances, voltages and currents must be carefully considered.* Where two quantities are to be multiplied or divided, they should be stated in polar form. Where they are to be added or subtracted, they must be converted to rectangular form. Because of the necessity of converting polar to rectangular form and vice-versa, there is much more work involved in analysing an a.c. network than for a similar d.c. network.

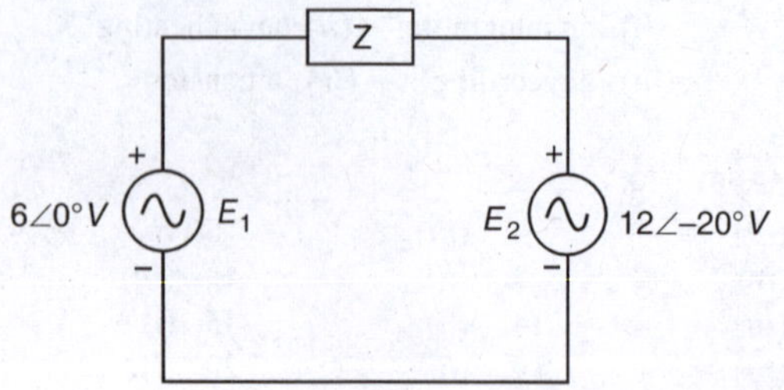

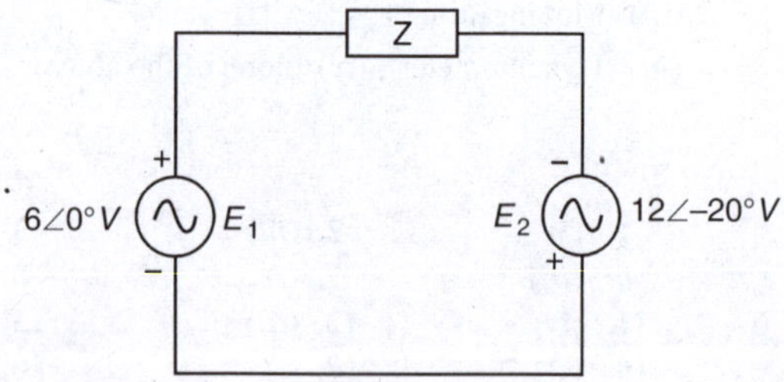

Fig. 17.1

---

* Other passive elements *viz.* inductance (L) and capacitance (C) are relevant only in a.c circuits.

** Note that alternating voltage or current means sinusoidal voltage or current unless stated otherwise. Remember that a sinusoidal voltage or current can be represented by a phasor.

(iv) Although alternating voltages are continuously reversing polarity, + and – terminals must be identified at each voltage source. This is necessary because if a voltage source is connected in reverse, the phase angle of the source output is changed by 180°. This point is illustrated in Fig. 17.1.

Fig. 17.1*(i)* shows a circuit that has two voltage sources ; *$E_1 = 6\angle 0°$V and $E_2 = 12 \angle -20°$ V. Since both E_1 and E_2 have **identical polarities (+ and –), the phase angle of E_2 is – 20° w.r.t. E_1. In Fig. 17.1.*(ii)*, the source E_2 is now connected in reverse. The reversal of the output of E_2 has added a further 180° to the phase of E_2 w.r.t. E_1 *i.e.,* the phase angle of E_2 is now – 200° w.r.t. E_1. It is for this reason that the terminals of all a.c. sources in a circuit diagram must have a polarity identification.

17.2. Kirchhoff's Laws for A.C. Circuits

Kirchhoff's laws for d.c. circuits can also be applied to a.c. circuits with a slight modification. In d.c. circuits, we take *algebraic sum* of voltages and currents whereas in a.c. circuits, we take *phasor sum* of voltages and currents.

1. Kirchhoff's Current Law (KCL). Kirchhoff's current law for a.c. circuits may be stated as under :

The phasor sum of the currents entering a point in an a.c. circuit is equal to the phasor sum of the currents leaving that point.

The term 'phasor sum' is used because we are dealing with alternating currents. This law is based on the principle of conservation of charge.

2. Kirchhoff's Voltage Law (KVL). Kirchhoff's voltage law for a.c.circuits may be stated as under :

In any closed electric circuit, the phasor sum of voltage drops plus the phasor sum of voltage rises is zero.

By convention, a voltage rise is given a *positive sign* and a voltage drop is assigned a *negative sign.* Remember that passing through a source or impedance from + to – indicates a voltage drop, while passing through it from – to + indicates a voltage rise.

Example 17.1. *Using Kirchhoff's voltage law, determine the circuit current in Fig. 17.2.* .

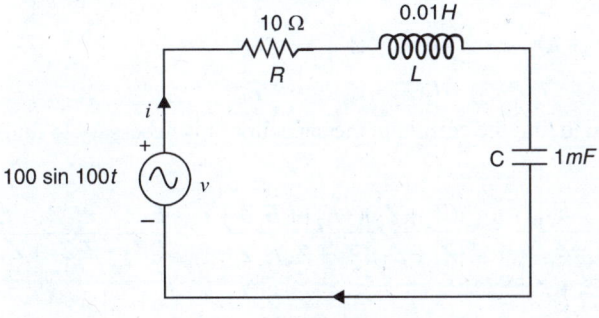

Fig. 17.2

Solution. The phasor voltage $V = 100\angle 0°$ volts. Let I be the phasor current in the circuit. Applying Kirchhoff's voltage law to the circuit in Fig. 17.2, we have,

$$V - IR - Ij\,X_L + {}^{\dagger}Ij\,X_C = 0$$

---

*　Note that instantaneous alternating voltages and currents are represented by e (or v) and i respectively. However, a phasor voltage or a phasor current is represented by E (or V) and I respectively.

**　Both E_1 and E_2 are increasing in the positive direction from zero.

†　Drop in capacitor $= -I\,Z_C = -I(-jX_C) = +I\,jX_C$

or
$$V = I(R + jX_L - jX_C)$$

or
$$V = I\left[R + j\omega L - \frac{j}{\omega C}\right]$$

Here $\omega L = (100) \times (0.01) = 1$; $\dfrac{1}{\omega C} = \dfrac{1}{(100) \times 1 \times 10^{-3}} = 10$

$$\therefore \qquad V = I[10 + j1 - j10] = I[10 - j9]$$

or
$$I = \frac{V}{10 - j9} = \frac{100\angle 0°}{13.5\angle -42°} = 7.4\angle 42°\,\text{A}$$

Therefore, the circuit current i is $i = 7.4 \sin(100\,t + 42°)$ A **Ans.**

Example 17.2. *Using Kirchhoff's laws, find current in the capacitor in Fig. 17.3.*

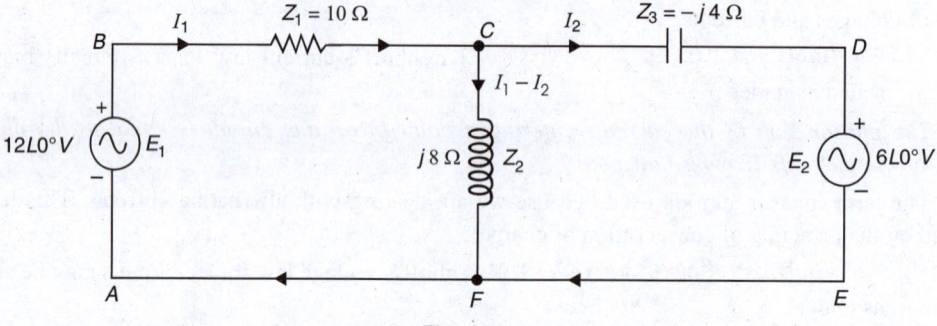

Fig. 17.3

Solution. We shall apply Kirchhoff's voltage law to the loops *ABCFA* and *CDEFC*. The loop equations are obtained by setting the voltage drops equal to the voltage rises.

Loop ABCFA. $\qquad I_1 Z_1 + (I_1 - I_2) Z_2 = E_1$

Loop CDEFC. $\qquad (I_2 - I_1) Z_2 + I_2 Z_3 + E_2 = 0$

Rearranging these equations and collecting terms so that they can be solved by Cramer's rule, we have,

$$I_1(Z_1 + Z_2) + I_2(-Z_2) = E_1$$

and
$$I_1(-Z_2) + I_2(Z_2 + Z_3) = -E_2$$

Since we are required to find the current in the capacitor, it is necessary to find I_2. By Cramer's rule,

$$I_2 = \frac{\begin{vmatrix} Z_1 + Z_2 & E_1 \\ -Z_2 & -E_2 \end{vmatrix}}{\begin{vmatrix} Z_1 + Z_2 & -Z_2 \\ -Z_2 & Z_2 + Z_3 \end{vmatrix}} = \frac{(Z_1 + Z_2)(-E_2) + E_1 Z_2}{(Z_1 + Z_2)(Z_2 + Z_3) - Z_2^2} = \frac{-E_2 Z_1 - E_2 Z_2 + E_1 Z_2}{Z_1 Z_2 + Z_1 Z_3 + Z_2 Z_3}$$

Now $E_1 = 12\angle 0°$ V ; $E_2 = 6\angle 0°$V ; $Z_1 = 10\angle 0°\ \Omega$; $Z_2 = 8\angle 90°\ \Omega$; $Z_3 = 4\angle -90°\ \Omega$

$$\therefore \qquad E_2 Z_1 = 6\angle 0° \times 10\angle 0° = 60\angle 0°$$

$$E_2 Z_2 = 6\angle 0° \times 8\angle 90° = 48\angle 90°$$

$$E_1 Z_2 = 12\angle 0° \times 8\angle 90° = 96\angle 90°$$

$$Z_1 Z_2 = 80\angle 90° \; ; \; Z_1 Z_3 = 40\angle -90° \; ; Z_2 Z_3 = 32\angle 0°$$

$$\therefore \quad I_2 = \frac{-60\angle 0° - 48\angle 90° + 96\angle 90°}{80\angle 90° + 40\angle -90° + 32\angle 0°} = \frac{-60 + j48}{32 + j40} = \frac{76.84\angle 141.34°}{51.22\angle 51.34°} = \textbf{1.5}\angle \textbf{90°A}$$

Example 17.3. *Determine the impedance seen by the source $V_s = 24 \angle 0°$ volts in the network shown in Fig. 17.4.*

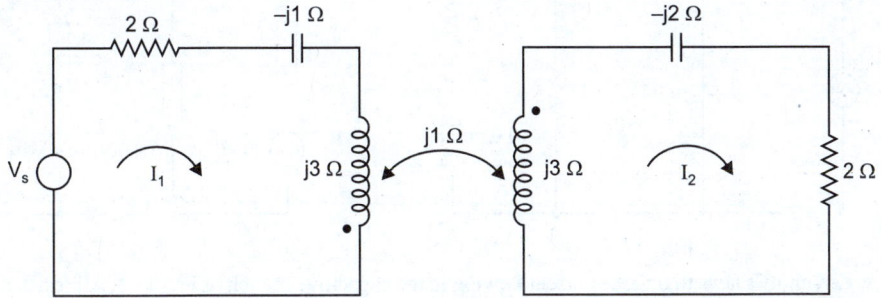

Fig. 17.4

Solution. Let I_1 and I_2 be the phasor currents in the two loops as shown in Fig. 17.4.

Loop 1. Applying *KVL*, we have,

$$\text{Voltage rise} = \text{Voltage drops}$$

or $$V_s = (2 - j1 + j3) I_1 + jI_2 \qquad \qquad ...(i)$$

Loop 2. Applying *KVL*, we have,

$$\text{Voltage drops} = \text{Voltage rise}$$

or $$(j3 - j2 + 2) I_2 + jI_1 = 0 \qquad \qquad ...(ii)$$

From eq. (*ii*), $$I_2 = \frac{-jI_1}{2 + j1} = \frac{-jI_1}{2 + j1} \times \frac{2 - j1}{2 - j1} = \frac{-(1 + j2)I_1}{5}$$

Putting this value of I_2 in eq. (*i*), we have,

$$V_s = (2 - j1 + j3) I_1 + \frac{-j(1 + j2)}{5} I_1 = (2.4 + j1.8) I_1$$

$\therefore$ Impedance seen by source is given by ;

$$Z_s = \frac{V_s}{I_1} = (2.4 + j1.8)\ \Omega = \mathbf{3 \angle 36.87°\ \Omega}$$

Tutorial Problems

1. Using Kirchhoff's voltage law, determine the current in Fig. 17.5. [$i = 2 \sin (1000\, t + 76°)$ A]

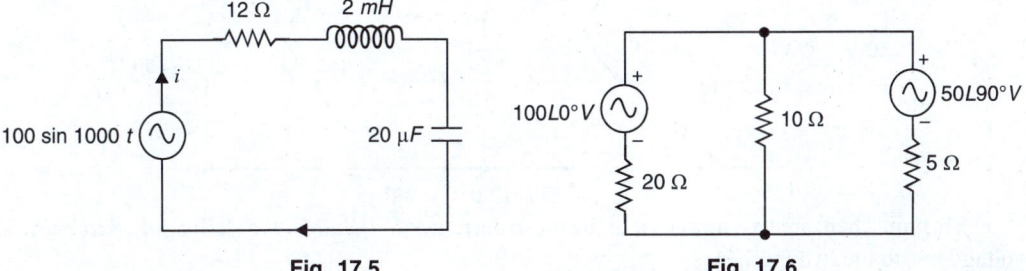

Fig. 17.5 **Fig. 17.6**

2. Using Kirchhoff's laws, determine the current in 10 Ω resistor in Fig. 17.6. [$(1.4 + j3.8)$A]

3. Using Kirchhoff's laws, determine the current flowing through the branch containing 4 Ω resistance in Fig. 17.7. [$3.17 \angle -33.9°$ A]

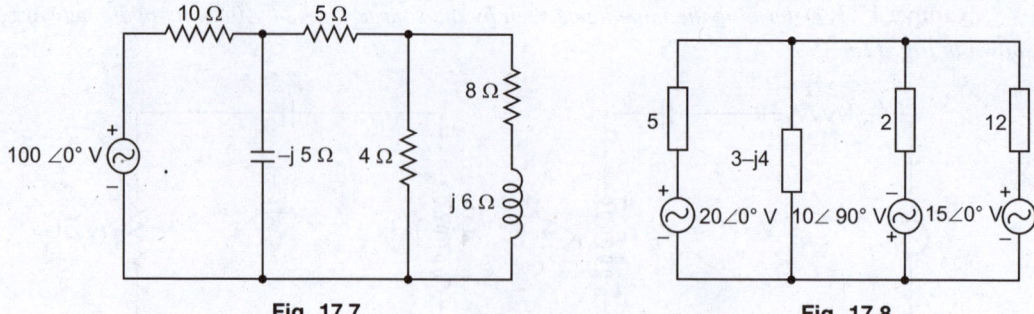

Fig. 17.7   **Fig. 17.8**

4. Use Kirchhoff's laws to find the current flowing in the capacitive branch of Fig. 17.8. All resistances and impedances are in ohms.   **[5.87 A]**

17.3. A.C. Mesh Current Analysis

The procedure for the analysis of a.c. networks is exactly the same as that for d.c. networks except that impedances, voltages and currents are expressed as complex numbers. Remember that mesh current method consists of the following steps :

(*i*) Each mesh is assigned a separate mesh current. The same direction must be chosen for all the mesh currents in a circuit. A usual convention is to make all the mesh currents clockwise currents.

(*ii*) When two mesh currents are flowing through the same circuit element, the net current in that circuit element is found in a manner identical to that used for d.c. mesh analysis.

(*iii*) Kirchhoff's voltage law for a.c. circuits is applied to write equation for each mesh in terms of mesh currents. Remember, while writing mesh equations, voltage rise is assigned positive sign and voltage drop negative sign.

(*iv*) When the analysis is complete, those branch currents that come out as positive quantities are (instantaneously) in the same direction as that selected for the mesh currents. The branch currents that have negative signs have an additional 180° phase shift in relation to the mesh currents.

Example 17.4. *Use mesh analysis to find the currents through the source and capacitor in the circuit in Fig. 17.9. Also find the power delivered by the source.*

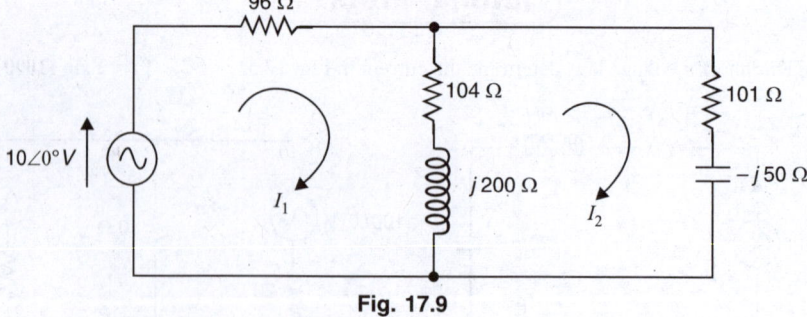

Fig. 17.9

Solution. There are two meshes and the mesh currents are I_1 and I_2. We shall apply Kirchhoff's voltage law to the two meshes.

Mesh 1. $10 - 96 I_1 - 104 (I_1 - I_2) - j\,200(I_1 - I_2) = 0$

Mesh 2. $- 101 I_2 - (- j\,50) I_2 - j\,200 (I_2 - I_1) - 104(I_2 - I_1) = 0$

Rearranging these equations and collecting terms so that they can be solved by Cramer's rule, we have,

$$I_1(200 + j\,200) - I_2(104 + j\,200) = 10$$

and $\quad -I_1(104 + j\,200) + I_2(205 + j\,150) = 0$

$$\therefore \quad I_1 = \dfrac{\begin{vmatrix} 10 & -(104 + j200) \\ 0 & (205 + j150) \end{vmatrix}}{\begin{vmatrix} (200 + j200) & -(104 + j200) \\ -(104 + j200) & (205 + j150) \end{vmatrix}} = \dfrac{10(205 + j150) - 0}{(200 + j200)(205 + j150) - (104 + j200)^2}$$

$$= \dfrac{2050 + j1500}{40184 + j29400} = \dfrac{2540.18\angle 36.19°}{49790\angle 36.19°} = 5.1 \times 10^{-2}\angle 0° \text{ A}$$

Also, $\quad I_2 = \dfrac{\begin{vmatrix} (200 + j200) & 10 \\ -(104 + j200) & 0 \end{vmatrix}}{40184 + j29400} = \dfrac{0 + 10(104 + j200)}{40184 + j29400}$

$$= \dfrac{1040 + j2000}{40184 + j29400} = \dfrac{2254.24\angle 62.53°}{49790\angle 36.19°} = 4.526 \times 10^{-2}\angle 26.34° \text{A}$$

$\therefore \quad$ Source current $= I_1 = \mathbf{5.1 \times 10^{-2} \angle\, 0°\ A}$

Current through capacitor $= I_2 = \mathbf{4.526 \times 10^{-2} \angle\, 26.34°A}$

Power supplied by source, $P = VI_1 \cos\phi = 10 \times 5.1 \times 10^{-2} \times \cos 0° = \mathbf{0.51\ W}$

Example 17.5. *Use mesh analysis to find the voltage across the inductor in Fig. 17.10.*

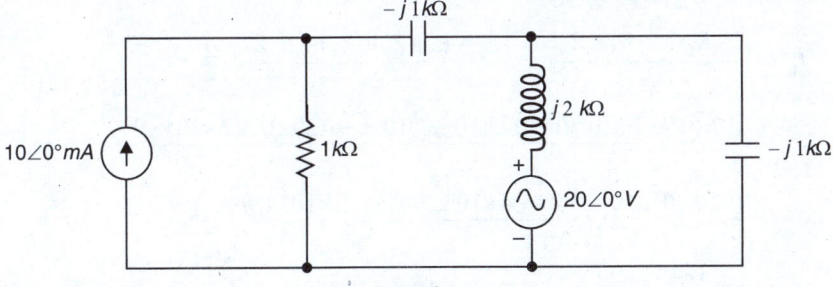

Fig. 17.10

Solution. The current source will first be converted to an equivalent voltage source. Since the current source is in parallel with a 1 kΩ resistor, it is equivalent to a voltage source having value

$$E_1 = (10\angle 0° \text{ mA})(1\angle 0° \text{ k}\Omega) = 10\angle 0° \text{ V}$$

Fig. 17.11 shows the equivalent voltage source (E_1), drawn in place of current source and in series with 1 kΩ resistor.

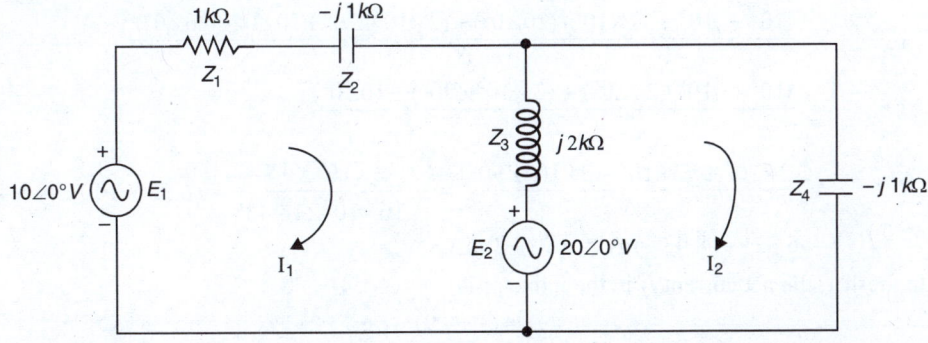

Fig. 17.11

Referring to Fig. 17.11, there are two meshes and the mesh currents are I_1 and I_2. We now apply Kirchhoff's voltage law to the two meshes.

Mesh 1. $I_1(Z_1 + Z_2) + (I_1 - I_2)Z_3 + E_2 = E_1$

Mesh 2. $(I_2 - I_1)Z_3 + I_2 Z_4 = E_2$

Rearranging and collecting terms, we have,

$$I_1(Z_1 + Z_2 + Z_3) + I_2(-Z_3) = E_1 - E_2$$

and $I_1(-Z_3) + I_2(Z_3 + Z_4) = E_2$

The denominator matrix Δ is

$$\Delta = \begin{vmatrix} Z_1 + Z_2 + Z_3 & -Z_3 \\ -Z_3 & Z_3 + Z_4 \end{vmatrix} = (Z_1 + Z_2 + Z_3)(Z_3 + Z_4) - Z_3^2$$

Putting $Z_1 = 10^3 \,\Omega$; $Z_2 = -j10^3 \,\Omega$; $Z_3 = j\,2 \times 10^3 \,\Omega$ and $Z_4 = -j10^3 \,\Omega$, we have,

$$\begin{aligned} \Delta &= (10^3 - j10^3 + j2 \times 10^3)(j\,2 \times 10^3 - j10^3) - (j2 \times 10^3)^2 \\ &= (10^3 + j10^3)(10^3 \angle 90°) + (4 \times 10^6) \\ &= (1.414 \times 10^3 \angle 45°)(10^3 \angle 90°) + (4 \times 10^6) \\ &= 1.414 \times 10^6 \angle 135° + (4 \times 10^6) \\ &= -10^6 + j10^6 + 4 \times 10^6 = 3 \times 10^6 + j10^6 = 3.16 \times 10^6 \angle 18.43° \end{aligned}$$

By Cramer's rule, we have,

$$\begin{aligned} I_1 &= \frac{\begin{vmatrix} E_1 - E_2 & -Z_3 \\ E_2 & Z_3 + Z_4 \end{vmatrix}}{\Delta} = \frac{(E_1 - E_2)(Z_3 + Z_4) + E_2 Z_3}{\Delta} \\ &= \frac{(10\angle 0° - 20\angle 0°)(0 + j2 \times 10^3 - j10^3) + (20\angle 0°)(2 \times 10^3 \angle 90°)}{\Delta} \\ &= \frac{(-10\angle 0°)(10^3 \angle 90°) + 4 \times 10^4 \angle 90°}{\Delta} = \frac{3 \times 10^4 \angle 90°}{\Delta} \\ &= \frac{3 \times 10^4 \angle 90°}{3.16 \times 10^6 \angle 18.43°} = 9.49 \times 10^{-3} \angle 71.57° \text{A} \\ &= 9.49 \angle 71.57° \text{mA} = (3 + j\,9) \text{ mA} \end{aligned}$$

Also, $\begin{aligned} I_2 &= \frac{\begin{vmatrix} Z_1 + Z_2 + Z_3 & E_1 - E_2 \\ -Z_3 & E_2 \end{vmatrix}}{\Delta} = \frac{(Z_1 + Z_2 + Z_3)E_2 + Z_3(E_1 - E_2)}{\Delta} \\ &= \frac{(10^3 - j10^3 + j2 \times 10^3)(20\angle 0°) + (2 \times 10^3 \angle 90°)(10\angle 0° - 20\angle 0°)}{\Delta} \\ &= \frac{(10^3 + j10^3)(20\angle 0°) + (2 \times 10^3 \angle 90°)(-10\angle 0°)}{\Delta} \\ &= \frac{2 \times 10^4 + j2 \times 10^4 - 2 \times 10^4 \angle 90°}{\Delta} = \frac{2 \times 10^4 \angle 0°}{3.16 \times 10^6 \angle 18.43°} = 6.32 \times 10^{-3} \angle -18.43° \text{A} \\ &= 6.32\angle -18.43° \text{ mA} = (6 - j\,2) \text{mA} \end{aligned}$

In mesh 1, the net current I_L in the inductor is

$$\begin{aligned} I_L &= I_1 - I_2 = (3 + j\,9) - (6 - j\,2) \\ &= -(3 - j\,11)\text{mA} = -11.4 \angle -74.74° \text{ mA} \end{aligned}$$

∴   Voltage V_L across inductor is given by ;

$$V_L = I_L X_L = (-11.4 \angle -74.74° \text{ mA}) (2 \times 10^3 \angle 90° \ \Omega)$$
$$= -22.8 \angle 15.26° \text{ V} = *22.8 \angle -164.74° \text{ V}$$

Example 17.6. *Use mesh analysis to find currents in the various branches of the circuit shown in Fig. 17.12.*

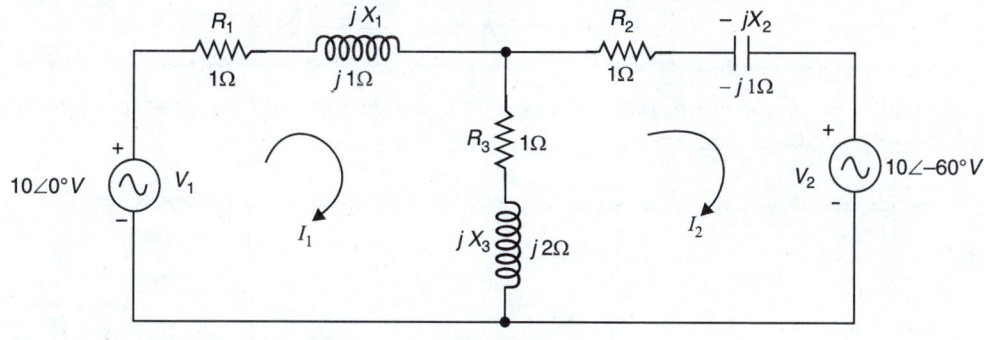

Fig. 17.12

Solution. There are two meshes and the mesh currents are I_1 and I_2.

$$**Z_{11} = (R_1 + R_3) + j (X_1 + X_3) = (2 + j\ 3)\Omega$$
$$Z_{12} = Z_{21} = Z_3 = (1 + j2)\Omega$$
$$Z_{22} = (R_2 + R_3) + j(X_3 - X_2) = (2 + j\ 1)\Omega$$

We now apply Kirchhoff's voltage law to the two meshes.

Mesh 1. $\quad I_1 Z_1 + (I_1 - I_2)Z_3 = V_1$

or $\quad\quad I_1 (Z_1 + Z_3) - I_2 Z_3 = V_1$ .....(i)

Mesh 2. $\quad I_2 Z_2 + (I_2 - I_1)Z_3 = -V_2$

or $\quad\quad -I_1 Z_3 + I_2 (Z_2 + Z_3) = -V_2$ ......(ii)

Now $Z_1 + Z_3 = Z_{11}$, $Z_3 = Z_{12} = Z_{21}$ and $Z_{22} = Z_2 + Z_3$ so that eqs. (i) and (ii) can be written as :

$$I_1 Z_{11} - I_2 Z_{12} = V_1$$

and $\quad\quad -I_1 Z_{21} + I_2 Z_{22} = -V_2$

By Cramer's rule, we have,

$$I_1 = \frac{\begin{vmatrix} V_1 & -Z_{12} \\ -V_2 & Z_{22} \end{vmatrix}}{\begin{vmatrix} Z_{11} & -Z_{12} \\ -Z_{21} & Z_{22} \end{vmatrix}} = \frac{V_1 Z_{22} - V_2 Z_{12}}{Z_{11} Z_{22} - Z_{12}^2}$$

$$= \frac{10(2 + j1) - (10\angle -60°)(1 + j2)}{(2 + j3)(2 + j1) - (1 + j2)^2} = \frac{(20 + j10) - (5 - j5\sqrt{3})(1 + j2)}{(1 + j8) + (3 - j4)}$$

$$= \frac{(20 + j10) - (22.32 + j1.34)}{4 + j4} = \frac{8.97\angle 105°}{5.66\angle 45°} = 1.58\angle 60°\text{A} = (\ 0.79 + j1.37)\text{A}$$

---

*   If we consider mesh 2, then,

$V_L = -22.8 \angle -164.74° \text{ V} = 22.8 \angle 15.26°$ V

**   Recall that Z_{11} is the sum of impedances in mesh 1 and Z_{12} (= Z_{21}) is the common impedance between meshes 1 and 2.

Also,
$$I_2 = \frac{\begin{vmatrix} Z_{11} & V_1 \\ -Z_{21} & -V_2 \end{vmatrix}}{Z_{11}Z_{22} - Z_{12}^2} = \frac{-V_2 Z_{11} + Z_{21}V_1}{5.66\angle 45°}$$

$$= \frac{-10\angle -60°(2 + j3) + (1 + j2)10}{5.66\angle 45°} = \frac{34.1\angle 139.3°}{5.66\angle 45°}$$

$$= 6.03\angle 94.3°\,\text{A} = \mathbf{(-0.46 + j\,6.01)\,A}$$

$$\therefore \quad \text{Current thro' } Z_3 = I_1 - I_2 = (0.79 + j1.37) - (-0.46 + j6.01) = \mathbf{(1.25 - j\,4.64)A}$$

Example 17.7. *In the circuit shown in Fig. 17.13, find the value of V_2 so that current through* $(2 + j3)$ *ohms impedance is zero.*

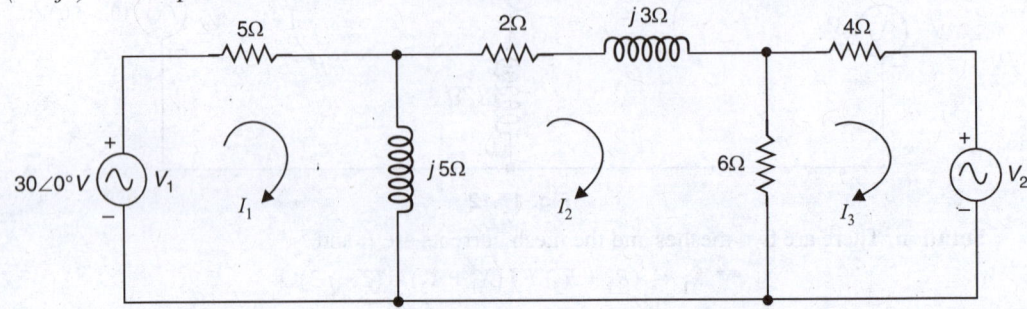

Fig. 17.13

Solution. There are three meshes and the mesh currents are I_1, I_2 and I_3. Applying Kirchhoff's voltage law to the three meshes, we have,

*Mesh 1. $(5 + j5)\, I_1 - j5 I_2 = 30\angle 0°$

Mesh 2. $-j5 I_1 + (8 + j8)I_2 - 6 I_3 = 0$

Mesh 3. $-6 I_2 + 10\, I_3 = -V_2$

By Cramer's rule, we have,

$$I_2 = \frac{\begin{vmatrix} 5 + j5 & 30 & 0 \\ -j5 & 0 & -6 \\ 0 & -V_2 & 10 \end{vmatrix}}{\begin{vmatrix} 5 + j5 & -j5 & 0 \\ -j5 & 8 + j8 & -6 \\ 0 & -6 & 10 \end{vmatrix}} = \frac{(5 + j5)(-6V_2) - 30(-j50)}{\text{Denominator}}$$

$$\therefore \qquad I_2 = \frac{-V_2(30 + j30) + j1500}{\text{Denominator}}$$

It is given that current through $(2 + j3)\,\Omega$ impedance is zero *i.e.*, $I_2 = 0$.

$$\therefore \quad -V_2(30 + j30) + j1500 = 0 \quad \text{or} \quad V_2 = \frac{j1500}{30 + j30} = \mathbf{35.36\,\angle 45°\,V}$$

Example 17.8. *Use mesh analysis to find voltage across 10 ohm resistor in the network shown in Fig. 17.14. The coupling co-efficient is 0.8.*

---

* $5 I_1 + (I_1 - I_2)\, j\, 5 = 30 \angle 0°$

    or $(5 + j\, 5)\, I_1 - j\, 5\, I_2 = 30 \angle 0°$

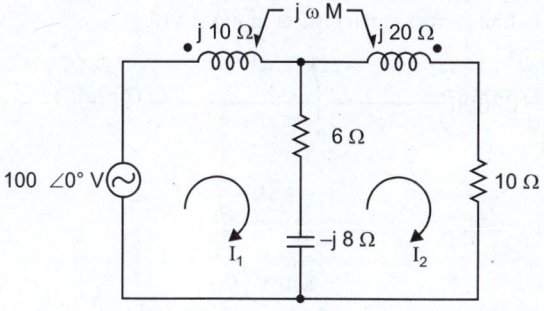

Fig. 17.14

Solution. $\omega M = \omega k \sqrt{L_1 L_2} = k\sqrt{\omega L_1 \times \omega L_2} = 0.8\sqrt{10 \times 20} = 11.32\ \Omega$

Let I_1 and I_2 be the phasor currents in the two meshes as shown in Fig. 17.14. Applying *KVL*, we obtain mesh equations by setting voltage drops equal to voltage rises.

Mesh 1. Applying *KVL*, we have,

$$I_1(j\,10 + 6 - j\,8) - I_2(6 - j\,8) - j\,11.32\,I_2 = 100\ \angle\ 0°$$

or $\quad I_1\,(6 + j\,2) + I_2\,(-6 - j3.32) = 100\ \angle\ 0°$ $\qquad\qquad$ ...(*i*)

Mesh 2. Applying *KVL*, we have,

$$-I_1(6 - j\,8) - j\,11.32\,I_1 + I_2\,(6 + 10 - j\,8 + j\,20) = 0$$

or $\quad I_1\,(-6 - j\,3.32) + I_2\,(16 + j\,12) = 0$ $\qquad\qquad$ ...(*ii*)

Applying Cramer's rule to eqs. (*i*) and (*ii*), we have,

$$I_2 = \frac{\begin{vmatrix} 6+j2 & 100\angle 0° \\ -6-j3.32 & 0 \end{vmatrix}}{\begin{vmatrix} 6+j2 & -6-j3.32 \\ -6-j3.32 & 16+j12 \end{vmatrix}}$$

$$= \frac{100\,(6 + j\,3.32)}{47.04 + j\,64.16} = \frac{685.73\ \angle\ 28.96°}{79.56\ \angle\ 53.75°}$$

$\therefore \qquad\qquad I_2 = 8.619\ \angle\ -24.79°\ \text{A}$

Voltage across 10 Ω resistor = $10\,I_2 = 10 \times 8.619\ \angle\ -24.79° =$ **86.19 $\angle$ –24.79° V**

Tutorial Problems

1. Use mesh analysis to find current in the inductor in Fig. 17.15. **[2.67∠–71.36°A →]**

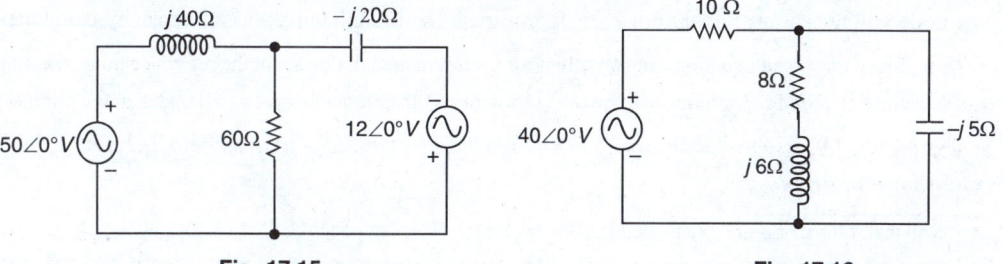

Fig. 17.15 $\qquad\qquad\qquad\qquad\qquad\qquad$ **Fig. 17.16**

2. Use mesh analysis to find current in the capacitor in Fig. 17.16. **[3.5∠52.1°A]**

3. Use mesh analysis to find current in the resistor in Fig. 17.17.   **[8.82∠8.74° mA↓]**

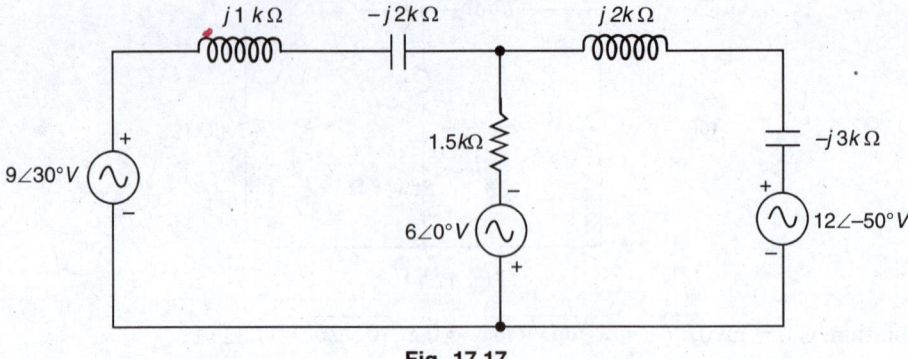

Fig. 17.17

4. Using mesh analysis, determine the values of I, I_1 and I_2 in Fig. 17.18.

[2.7 ∠ – 58.8° A ; 0.1 ∠ 97° A ; 2.8 ∠ – 59.6° A]

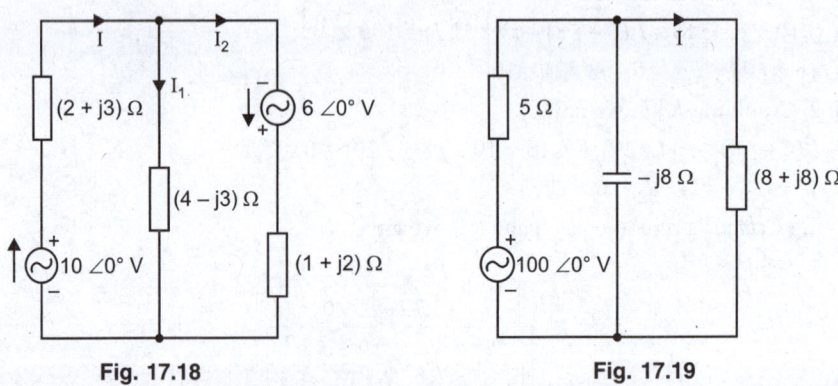

Fig. 17.18   **Fig. 17.19**

5. Using mesh analysis, find the value of current I and active power output of the voltage source in Fig. 17.19.   **[7 ∠– 50° A ; 645 W]**

17.4. A.C. Nodal Analysis

The procedure for nodal analysis of an a.c. network is the same as that for a d.c. network except that we have to deal with impedances instead of resistances and that the phase angle of all quantities (*e.g.* impedances, voltages and currents) must be taken into consideration. Remember that nodal analysis requires the selection of one of the * principal nodes as the **reference node. We can apply Kirchhoff's current law to any principal node (except the reference node) to find its node voltage. Once node voltages at principal nodes are determined, the branch currents can be easily calculated.

Note. Since one nodal equation can be written for each principal node except the reference node, the number of equations required for solution will be $(n-1)$ where n is the number of principal nodes in the circuit.

Example 17.9. *Use nodal analysis to find the voltage at node 'a' and currents I_1, I_2 and I_3 in the circuit shown in Fig. 17.20.*

---

*   Recall that if three or more circuit elements join at a node, then that node is called a principal node.

**   It does not matter which principal node is selected as the reference node but it is easiest to take that node with the greatest number of branches connected to it.

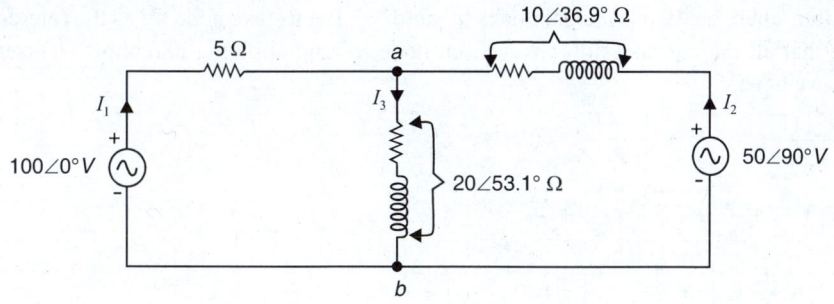

Fig. 17.20

Solution. There are two principal nodes '*a*' and '*b*' in the circuit. Taking node '*b*' as the reference node and applying Kirchhoff's current law to node '*a*', we have,

$$I_1 + I_2 = I_3$$

or $\dfrac{100 - V_a}{5} + \dfrac{50\angle 90° - V_a}{10\angle 36.9°} = \dfrac{V_a}{20\angle 53.1°}$

or $(20 - 0.2\,V_a) + (5\angle 53.1° - 0.1 V_a \angle - 36.9°) = 0.05 V_a \angle - 53.1°$

or $V_a(0.2 + 0.1 \angle - 36.9° + 0.05\angle - 53.1°) = 20 + 5 \angle 53.1°$

or $V_a[0.2 + (0.08 - j0.06) + (0.03 - j0.04)] = 20 + (3 + j4)$

or $V_a[0.31 - j0.1] = 23 + j4$

∴ $V_a = \dfrac{23 + j4}{0.31 - j0.1} = \dfrac{23.35\angle 9.87°}{0.326\angle -17.88°} = $ **71.62 ∠ 27.75° V**

$I_1 = \dfrac{100 - 71.62\angle 27.75°}{5} = \dfrac{100 - (63.38 + j33.35)}{5}$

$= \dfrac{36.62 - j33.35}{5} = \dfrac{49.53\angle -42.32°}{5} = $ **9.9∠ –42.32°A**

$I_2 = \dfrac{50\angle 90° - 71.62\angle 27.75°}{10\angle 36.9°} = \dfrac{(j50) - (63.38 + j33.35)}{10\angle 36.9°}$

$= \dfrac{-63.38 + j16.65}{10\angle 36.9°} = \dfrac{65.53\angle 165.28°}{10\angle 36.9°} = $ **6.55 ∠ 128.38°A**

$I_3 = \dfrac{71.62\angle 27.75°}{20\angle 53.1°} = $ **3.58 ∠ – 25.35° A**

Example 17.10. *Using nodal analysis, find current in Z_3 (= $R_3 + jX_3$) branch in the circuit shown in Fig. 17.21.*

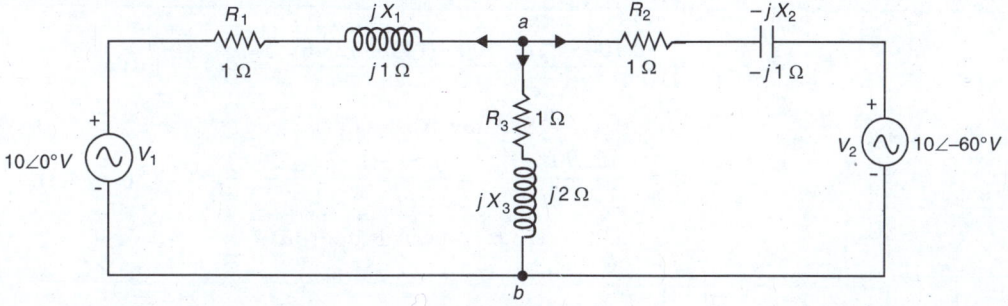

Fig. 17.21

Solution. There are two principal nodes '*a*' antd '*b*'. Let us take node '*b*' as the reference node. Assuming that all the currents flow away from node '*a*' and applying Kirchhoff's current law to anode '*a*', we have,

$$\frac{V_a - V_1}{Z_1} + \frac{V_a}{Z_3} + \frac{V_a - V_2}{Z_2} = 0$$

or $\quad V_a\left(\frac{1}{Z_1} + \frac{1}{Z_2} + \frac{1}{Z_3}\right) = \frac{V_1}{Z_1} + \frac{V_2}{Z_2}$ ....(*i*)

Now, $Y = \dfrac{1}{Z_1} + \dfrac{1}{Z_2} + \dfrac{1}{Z_3} = {}^*\left(\dfrac{1}{2} - j\dfrac{1}{2}\right) + \left(\dfrac{1}{2} + j\dfrac{1}{2}\right) + \left(\dfrac{1}{5} - j\dfrac{2}{5}\right)$

$$= (1.2 - j\,0.4)\,\text{S} = 1.263 \angle -18.4°\,\text{S}$$

Also, $\quad \dfrac{V_1}{Z_1} + \dfrac{V_2}{Z_2} = \dfrac{10\angle 0°}{\sqrt{2}\angle 45°} + \dfrac{10\angle -60°}{\sqrt{2}\angle -45°} = (5 - j5) + (6.81 - j1.83)$

$$= (11.81 - j6.83)\,\text{A} = 13.65\angle -30°\text{A}$$

Putting the proper values in eq. (*i*), we have,

$$V_a(1.263\angle -18.4°) = 13.65\angle -30°$$

∴ $\qquad\qquad V_a = \dfrac{13.65\angle -30°}{1.263\angle -18.4°} = 10.8 \angle -11.6°\,\text{V}$

∴ $\qquad$ Current in $Z_3 = \dfrac{V_a}{Z_3} = \dfrac{10.8\angle -11.6°}{\sqrt{5}\angle 63.5°} = $ **4.82∠−75°A**

Example 17.11. *Use nodal analysis to find the current in each branch of the network shown in Fig. 17.22.*

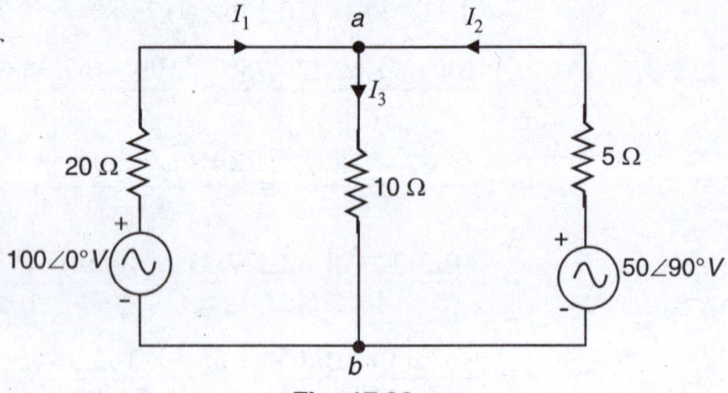

Solution. There are two principal nodes '*a*' and '*b*' in the circuit. Taking node '*b*' as the reference node and applying Kirchhoff's current law to node '*a*', we have,

Fig. 17.22

$$I_1 + I_2 = I_3$$

or $\quad \dfrac{100 - V_a}{20} + \dfrac{50\angle 90° - V_a}{5} = \dfrac{V_a}{10}$

or $\quad (5 - 0.05\,V_a) + (10 \angle 90° - 0.2V_a) = 0.1\,V_a$

or $\qquad\qquad 0.35V_a = 5 + 10\angle 90°$

∴ $\qquad\qquad V_a = \dfrac{5 + 10\angle 90°}{0.35} = \dfrac{5 + j10}{0.35} = (14.3 + j\,28.6)\,\text{V} = 32 \angle 63.4°\,\text{V}$

$$I_1 = \dfrac{100 - V_a}{20} = \dfrac{100 - (14.3 + j28.6)}{20} = (4.3 - j\,1.4)\,\text{A}$$

$$= \textbf{4.5} \angle \textbf{−18° A towards node '}a\textbf{'}$$

$$I_2 = \dfrac{50\angle 90° - V_a}{5} = \dfrac{j\,50 - (14.3 + j\,28.6)}{5} = (-2.86 + j\,4.3)\,\text{A}$$

$$= \textbf{5.16} \angle \textbf{123.6°A towards node '}a\textbf{'}$$

${}^* \quad \dfrac{1}{Z_1} = \dfrac{1}{1 + j1} = \dfrac{1}{1 + j1} \times \dfrac{1 - j1}{1 - j1} = \dfrac{1 - j1}{2} = \dfrac{1}{2} - j\dfrac{1}{2}$

$$I_3 = \frac{V_a}{10} = \frac{32\angle 63.4°}{10} = \textbf{3.2} \angle \textbf{63.4°A from node 'a' to node 'b'}$$

Example 17.12. *Determine the voltages at nodes 'a' and 'b' with respect to the reference node 'c' in the circuit of Fig. 17.23. Also find current through j2Ω and −j2Ω.*

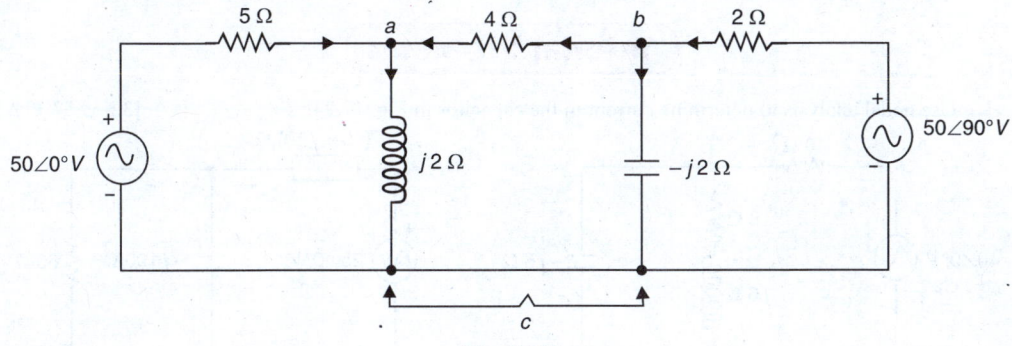

Fig. 17.23

Solution. There are two principal nodes 'a' and 'b' other than the reference node 'c'. Therefore, we require two nodal equations for the solution of this circuit.

At node 'a'. $\dfrac{50-V_a}{5}+\dfrac{V_b-V_a}{4}=\dfrac{V_a}{j2}$

or $(10 - 0.2V_a) + (0.25V_b - 0.25V_a) = -j0.5*V_a$

or $10 - 0.45V_a + 0.25V_b = -j\,0.5V_a$

∴ $V_a(0.45 - j\,0.5) - 0.25V_b = 10$ ...(i)

At node 'b'. $\dfrac{V_b-V_a}{4}+\dfrac{V_b}{-j2}=\dfrac{50\angle 90°-V_b}{2}$

or $(0.25\,V_b - 0.25V_a) + j\,0.5V_b = j\,25 - 0.5V_b$

∴ $-0.25V_a + V_b(0.75 + j\,0.5) = j25$ ...(ii)

Applying Cramer's rule to eqs. (i) and (ii), we have,

$$V_a = \frac{\begin{vmatrix} 10 & -0.25 \\ j25 & 0.75+j0.5 \end{vmatrix}}{\begin{vmatrix} 0.45-j0.5 & -0.25 \\ -0.25 & 0.75+j0.5 \end{vmatrix}} = \frac{(7.5+j5)+j6.25}{(0.45-j0.5)(0.75+j0.5)-(0.25)^2}$$

$$= \frac{7.5+j11.25}{0.525-j0.15} = \frac{13.5\angle 56.3°}{0.546\angle -15.95°} = \textbf{24.7}\angle\,\textbf{72.25°V}$$

Also, $V_b = \dfrac{\begin{vmatrix} 0.45-j0.5 & 10 \\ -0.25 & j25 \end{vmatrix}}{0.546\angle -15.95°} = \dfrac{(0.45-j0.5)(j25)+2.5}{0.546\angle -15.95°}$

$$= \frac{15+j11.23}{0.546\angle -15.95°} = \frac{18.75\angle 36.87°}{0.546\angle -15.95°} = \textbf{34.34}\angle\,\textbf{52.82° V}$$

---

* $\dfrac{V_a}{j2} = \dfrac{V_a}{j2}\times\dfrac{-j2}{-j2} = -j\,0.5\,V_a$

Current through $j2\ \Omega = \dfrac{V_a}{j2} = \dfrac{24.7\angle 72.25°}{2\angle 90°} = \mathbf{12.35\ \angle -17.75°\ A}$

Current through $-j2\Omega = \dfrac{V_b}{-j2} = \dfrac{34.34\angle 52.82°}{2\angle -90°} = \mathbf{17.17\ \angle\ 142.82°A}$

Tutorial Problems

1. Use nodal analysis to determine current in the capacitor in Fig. 17.24. $[3.5\ \angle\ 52.1°\ A]$

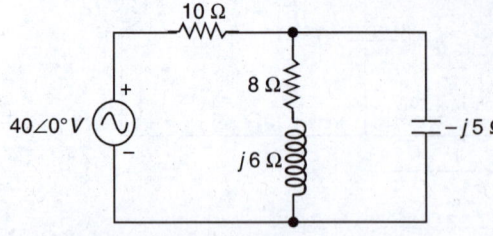

Fig. 17.24

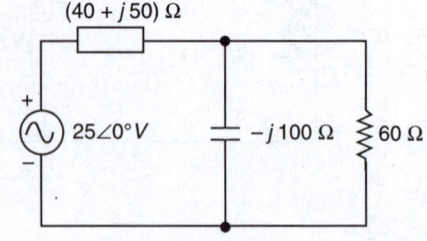

Fig. 17.25

2. Use nodal analysis to determine current in the capacitor in Fig. 17.25. $[0.1473\ \angle\ 43.43°\ A]$

3. Use nodal analysis to determine current in 5Ω resistor in Fig. 17.26. $[1.1\ \angle -109.5°\ A]$

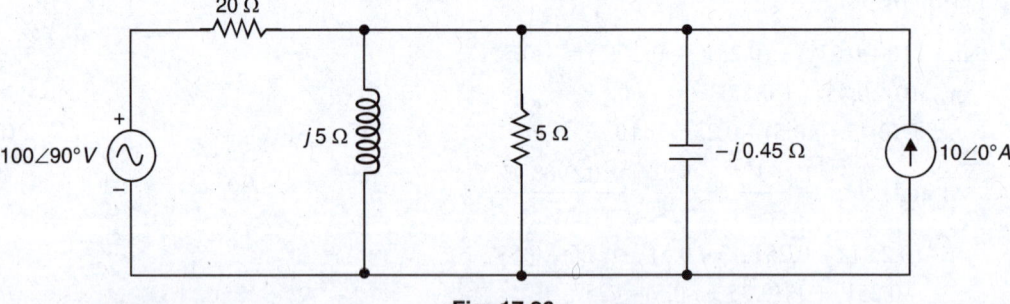

Fig. 17.26

4. Use nodal analysis to determine the value of voltages at nodes 1 and 2 in Fig. 17.27.

$[V_1 = 88.1\ \angle\ 33.88°\ V\ ;\ V_2 = 58.7\ \angle\ 72.34°\ V]$

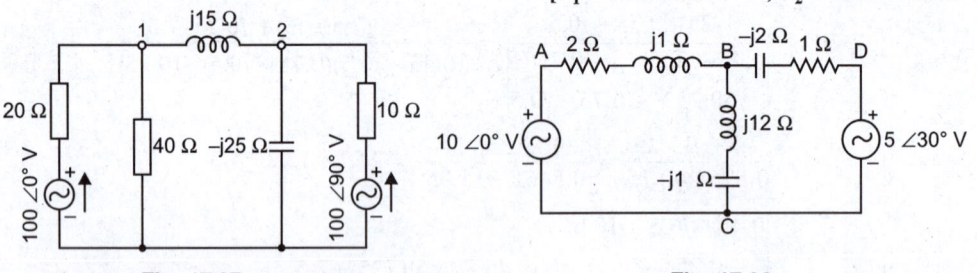

Fig. 17.27   **Fig. 17.28**

5. Determine the current flowing through branch *BC* of the network shown in Fig. 17.28 by using nodal analysis. $[3.9\ \angle -32.75°\ A]$

17.5. Superposition Theorem for A.C. Circuits

The superposition theorem for an a.c. circuit is the same as that for a d.c. circuit except that phase angle of all quantities (impedances, voltages and currents) must be taken into consideration. Therefore, superposition theorem for a.c. circuits can be stated as under :

In an a.c. network containing more than one source of voltage or current, the total current or voltage in any branch of the network is the phasor sum of currents or voltages produced in that branch by each source acting independently.

The procedure for using this theorem is as under :

(*i*) Select one source and replace all other sources with their *internal impedances.

(*ii*) Determine current through or voltage across the desired branch as a result of the single source acting alone.

(*iii*) Repeat steps (*i*) and (*ii*) using each source in turn until the branch current/voltage components have been calculated for all sources.

(*iv*) Determine the phasor sum of current/voltage components to obtain the actual current through or voltage across that branch.

Example 17.13. *Use superposition theorem to find current through 120Ω resistor in Fig. 17.29.*

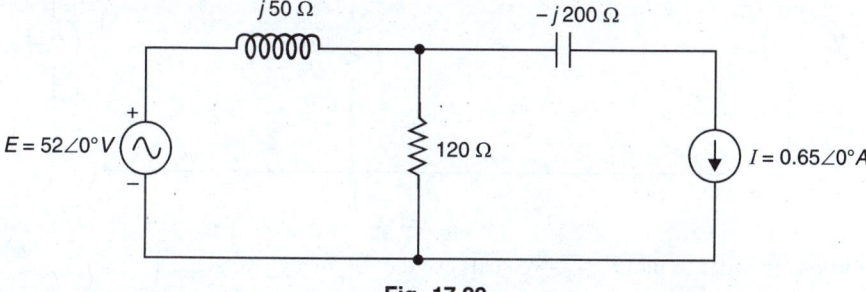

Fig. 17.29

Solution. The use of superposition theorem requires the following two steps :

(*i*) **E = 52 ∠ 0° V source acting alone.** To find the current through 120Ω resistor due to $E = 52\angle 0°$V source acting alone, we **open-circuit the current source as shown in Fig. 17.30. Since no current can flow through the capacitor, the total impedance Z_T across voltage source is

$$Z_T = (120 + j\,50)\ \Omega = 130 \angle 22.62°\ \Omega$$

Therefore, current I_1 in 120Ω resistor due to voltage source alone is

$$I_1 = \frac{E}{Z_T} = \frac{52\angle 0°}{130\angle 22.62°} = 0.4 \angle -22.62°\text{A}\!\downarrow \text{ (downward)}$$

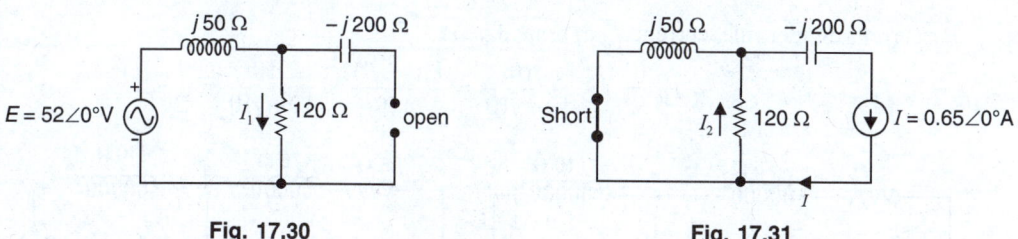

Fig. 17.30   **Fig. 17.31**

(*ii*) **I = 0.65 ∠ 0° A source acting alone.** In order to find current through 120Ω resistor due to $I = 0.65\angle 0°$A source acting alone, we replace the voltage source by a short-circuit as shown in Fig. 17.31. Referring to Fig. 17.31 and applying current-divider rule, current I_2 through 120Ω resistor is

---

*   Recall that an ideal voltage source has zero internal impedance so that an ideal voltage source is replaced by a short circuit. An ideal current source has infinite internal impedance so that an ideal current source is replaced by an open circuit.

**   When the internal impedance of the source is not given, it is understood that the source is ideal one. Since the current source is ideal, it is replaced by an open-circuit.

$$I_2 = \frac{j50}{120 + j50} \times I = \frac{50 \angle 90°}{130 \angle 22.62°} \times 0.65 \angle 0°$$

$$= 0.25 \angle 67.38° \, A \uparrow \text{ (upward)}$$

$\therefore$ Current thro' 120Ω resistor $= I_1 - I_2 = 0.4 \angle -22.62° - 0.25 \angle 67.38°$

$$= (0.369 - j\,0.154) - (0.096 + j0.231)$$

$$= (0.273 - j0.385) \, A = \mathbf{0.472 \angle -54.66° \, A\downarrow \textbf{ (downward)}}$$

Example 17.14. *In the network shown in Fig. 17.32, the two sources acting separately produce equal currents in the branch ab. Find the ratio V_1/V_2.*

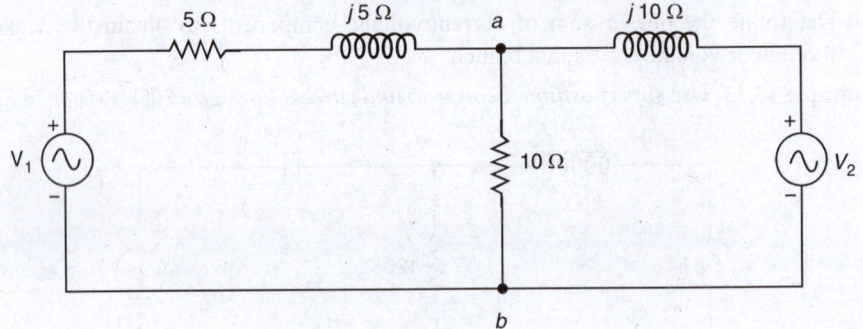

Fig. 17.32

Solution. In order to find V_1/V_2, we shall use the following two steps :

(i) V_1 acting alone. In order to find current through branch *ab* due to source V_1 acting alone, we replace V_2 by a short-circuit. Then circuit becomes as shown in Fig. 17.33. The total impedance Z_T across the voltage source V_1 is given by ;

$$Z_T = (5 + j5) + (10 \, \| \, j10) = (5 + j5) + \frac{10 \times j10}{10 + j10}$$

$$= (5 + j5) + \frac{j100}{10 + j10} = \frac{j20}{1 + j1}$$

Current *I* delivered by source V_1 is given by ;

$$I = \frac{V_1}{Z_T} = \frac{V_1(1 + j1)}{j20}$$

By current-divider rule, current I_{ab} in branch *ab* is

$$I_{ab} = I \times \frac{j10}{10 + j10} = \frac{V_1(1 + j1)}{j\,20} \times \frac{j10}{10 + j10} = \frac{V_1}{20}$$

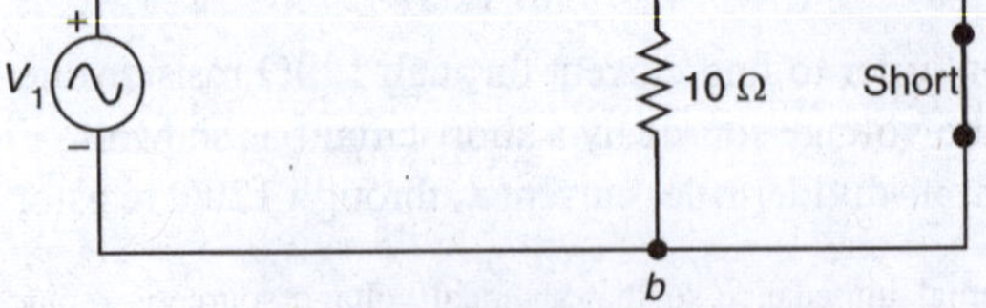

Fig. 17.33 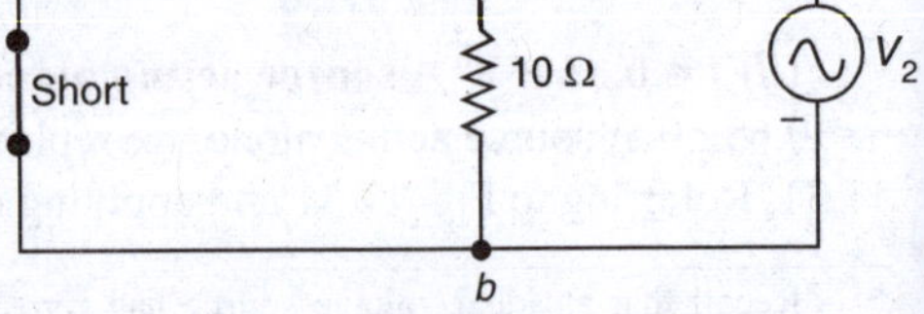   **Fig. 17.34**

(ii) V_2 acting alone. In order to find current through branch *ab* due to source V_2 acting alone, we replace V_1 by a short-circuit as shown in Fig. 17.34. The total impedance Z'_T across the voltage source V_2 is

$$Z'_T = (j\,10) + (10\,\|\,5 + j\,5) = (j\,10) + \frac{10(5 + j5)}{10 + 5 + j5}$$

$$= j\,10 + \frac{50 + j50}{15 + j5} = \frac{j200}{15 + j5}$$

Current I' delivered by source V_2 is

$$I' = \frac{V_2}{Z'_T} = \frac{V_2(15 + j5)}{j200}$$

By current-divider rule, current I'_{ab} in branch ab is

$$I'_{ab} = I' \times \frac{5 + j5}{10 + 5 + j5} = \frac{V_2(15 + j5)}{j200} \times \frac{5 + j5}{15 + j5} = \frac{V_2(5 + j5)}{j200}$$

It is given that $I_{ab} = I'_{ab}$.

$$\therefore \qquad \frac{V_1}{20} = \frac{V_2(5 + j5)}{j200}$$

or

$$\frac{V_1}{V_2} = \frac{20(5 + j5)}{j200} = \frac{20 \times 7.07\angle 45°}{200\angle 90°} = \mathbf{0.707\angle -45°}$$

Example 17.15. *Use superposition theorem to find the voltage across the capacitor in Fig.* 17.35.

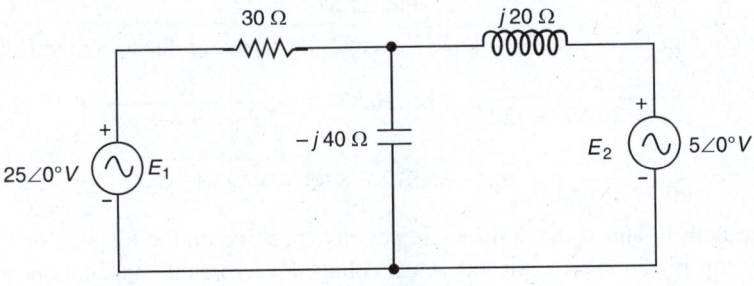

Fig. 17.35

Solution. The use of superposition theorem requires the following two steps :

(i) $E_1 = 25 \angle 0°$ **V source acting alone.** To find voltage across capacitor due to $E_1 = 25 \angle 0°$ V source acting alone, we replace source E_2 (= $5\angle 0°$ V) by a short-circuit as shown in Fig. 17.36. (*i*). Referring to Fig. 17.36(*i*), the inductor (= $j20$ Ω) is in parallel with the capacitor (= $-j40$ Ω) and the equivalent impedance of this parallel combination

$$= \frac{j20 \times (-j40)}{j20 - j40} = \frac{20\angle 90° \times 40\angle -90°}{20\angle -90°} = 40 \angle 90° \text{ Ω} = j\,40 \text{ Ω}$$

The circuit shown in Fig.17.36 (*i*) then reduces to the one shown in Fig.17.36 (*ii*).

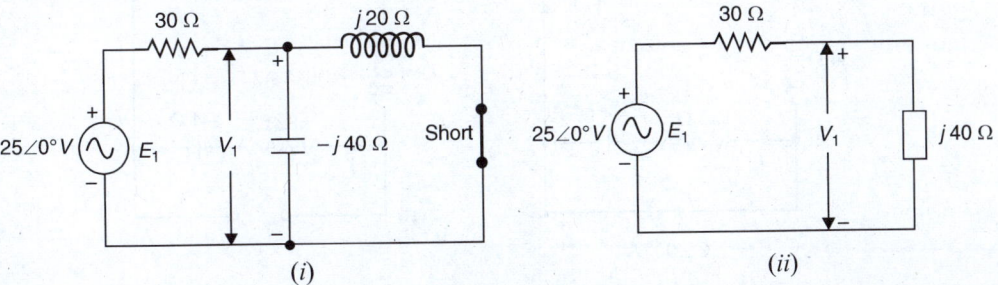

Fig. 17.36

Referring to Fig. 17.36 (*ii*), the voltage V_1 across the capacitor due to source E_1 acting alone is

$$V_1 = \frac{E_1}{30 + j40} \times j40 = \frac{25\angle 0°}{50\angle 53.13°} \times 40 \angle 90° = 20\angle 36.87°\text{V} = (16 + j\,12)\text{V}$$

(ii) $E_2 = 5 \angle 0°\text{V}$ source acting alone. To find voltage across capacitor due to $E_2 = 5\angle 0°$ V source acting alone, we replace source $E_1(= 25\angle 0°\text{V})$ by a short-circuit as shown in Fig. 17.37 (*i*). Referring to Fig. 17.37 (*i*), the resistor (= 30Ω) and the capacitor (= $-j$ 40Ω) are in parallel and the equivalent impedance of this parallel combination

$$= \frac{30\angle 0° \times (-j40)}{30\angle 0° - j40} = \frac{30\angle 0° \times 40\angle -90°}{30 - j40} = \frac{1200\angle -90°}{50\angle -53.13°} = 24 \angle -36.87° \ \Omega$$

The circuit shown in Fig. 17.37 (*i*) then reduces to the one shown in Fig. 17.37 (*ii*).

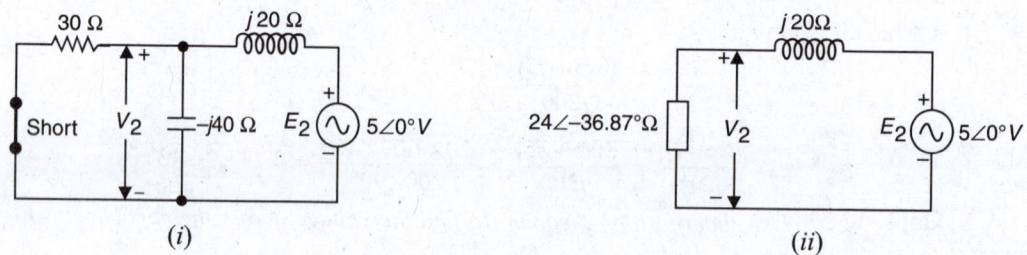

(*i*)   (*ii*)

Fig. 17.37

Referring to Fig. 17.37 (*ii*), the voltage V_2 across the capacitor due to source E_2 acting alone is

$$V_2 = \frac{E_2}{24\angle -36.87° + j20} \times 24\angle -36.87° = \frac{5\angle 0°}{20 - j14.4 + j20} \times 24 \angle -36.87°$$

$$= \frac{5\angle 0°}{20.77\angle 15.64°} \times 24\angle -36.87° = 5.78 \angle -52.51° \text{ V} = (3.52 - j\,4.59)\text{V}$$

Note that both V_1 and V_2 have the same polarity (positive on the top side of the capacitor). Therefore, by superposition principle, the actual voltage V_C across the capacitor due to both sources E_1 and E_2 is the sum of V_1 and V_2.

$$\therefore \qquad V_C = V_1 + V_2 = (16 + j\,12) + (3.52 - j4.59)$$
$$= (19.52 + j\,7.41) \text{ V} = \mathbf{20.88 \angle 20.79° \ V}$$

Example 17.16. *Determine the current in the capacitor branch by superposition theorem in the circuit of Fig. 17.38 (i).*

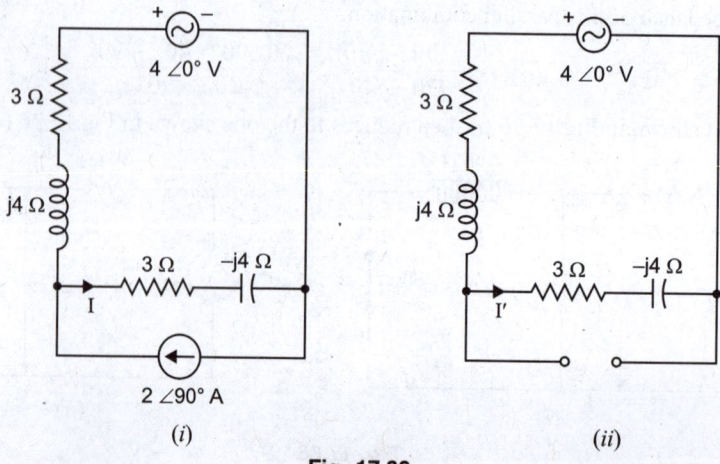

(*i*)   (*ii*)

Fig. 17.38

Solution. First open-circuit the current source so that voltage source is acting alone. The circuit then becomes as shown in Fig. 17.38 (*ii*). Referring to Fig. 17.38 (*ii*), current *I′* in the capacitor branch is

$$I' = \frac{4\angle 0°}{(3+j4)+(3-j4)}A$$

$$= \frac{2}{3}\angle 0°A = \left(\frac{2}{3}+j0\right)A$$

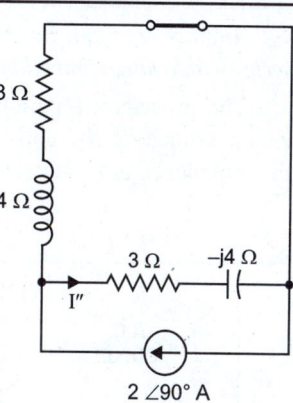

Now short circuit the voltage source so that current source is acting alone. The circuit then becomes as shown in Fig. 17.38 (*iii*). Referring to Fig. 17.38 (*iii*), the current *I″* in the capacitor branch by current-divider rule is

$$I'' = 2\angle 90°\times\frac{3+j4}{(3+j4)+(3-j4)}A$$

Fig. 17.38 (*iii*)

$$= \frac{j2(3+j4)}{6}A = \left(-\frac{4}{3}+j1\right)A$$

By superposition theorem, total current *I* in the capacitor branch is

$$I = I' + I'' = \left(\frac{2}{3}+j0\right)+\left(-\frac{4}{3}+j1\right) = \left(-\frac{2}{3}+j1\right)A = \textbf{1.20}\angle \textbf{123.69° A}$$

Tutorial Problems

1. Use superposition theorem to determine current in $(4+j3)\Omega$ impedance in Fig. 17.39. **[0.7∠ – 34.1°A]**

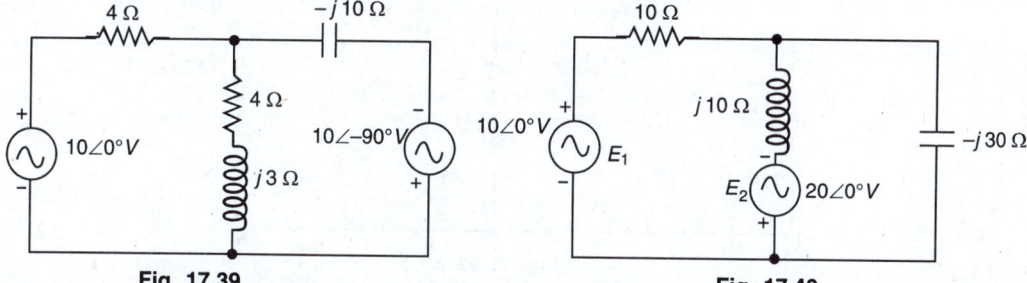

Fig. 17.39   **Fig. 17.40**

2. Use superposition theorem to find current in 10Ω resistor in Fig. 17.40. **[2.21∠ – 56.24°A]**

3. Use superposition theorem to find the total average power dissipated in the circuit shown in Fig. 17.41. **[37.29 W]**

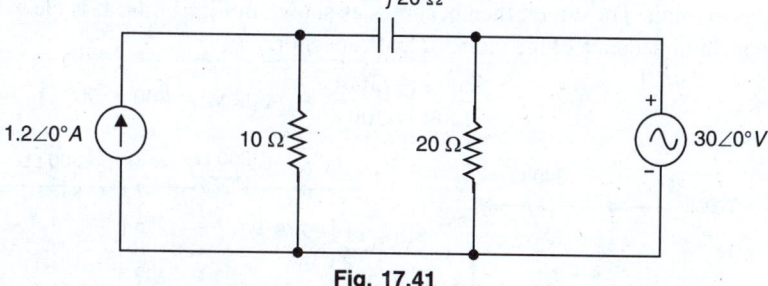

Fig. 17.41

17.6. Thevenin's Theorem for A.C. Circuits

Thevenin's theorem for an a.c. circuit is the same as that for a d.c. circuit except that phase angle of all quantities (impedances, voltages and currents) must be taken into consideration. Therefore, Thevenin's theorem for a.c. circuits can be stated as under :

Any two-terminal linear a.c. circuit can be replaced by a single a.c. voltage source (V_{Th}) in series with a single impedance (Z_{Th}).

The procedure for finding V_{Th} and Z_{Th} is the same as for d.c. circuits *i.e.* V_{Th} is the open-circuit voltage at the considered terminals (say terminals *AB*) and the series impedance Z_{Th} is the impedance at terminals *AB* with all sources replaced by their internal impedances.

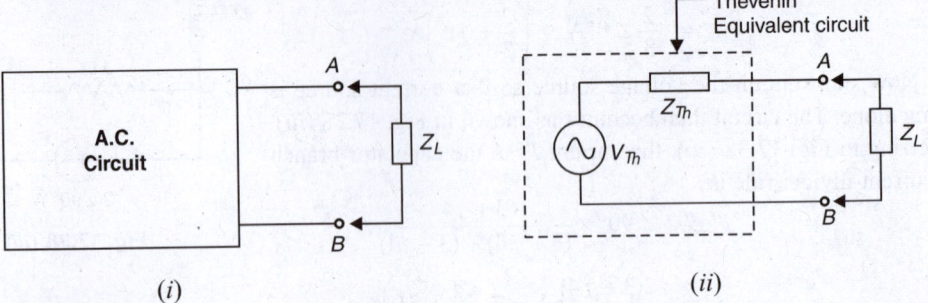

Fig. 17.42

Fig. 17.42 (*ii*) shows the Thevenin equivalent circuit of the a.c. circuit shown in Fig. 17.42 (*i*). Note that behind terminals *AB*, the a.c. circuit is replaced by a single a.c. voltage source V_{Th} in series with a single impedance Z_{Th}.

Example 17.17. *Using Thevenin's theorem, find current in the resistor R in Fig. 17.43.*

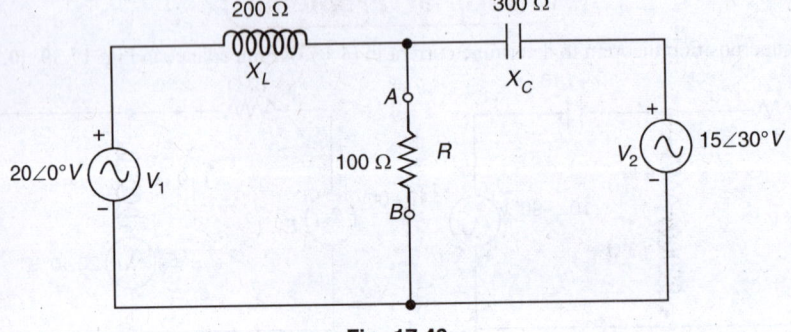

Fig. 17.43

Solution. Consider the resistor *R* to be the load.

$$Z_1 = j\,200\ \Omega = 200 \angle 90°\ \Omega\ ;\ Z_2 = -j\,300\ \Omega = 300 \angle -90°\ \Omega$$

Finding Z_{Th}. To find Z_{Th} at the terminals *AB*, we remove the resistor *R* and replace the voltage sources by short-circuit. The circuit then becomes as shown in Fig. 17.44. It is clear that Z_{Th} is the parallel equivalent impedance of the inductor and capacitor.

$$\therefore \qquad Z_{Th} = \frac{j200 \times (-j300)}{j200 - j300} = j\,600\ \Omega = 600 \angle 90°\ \Omega$$

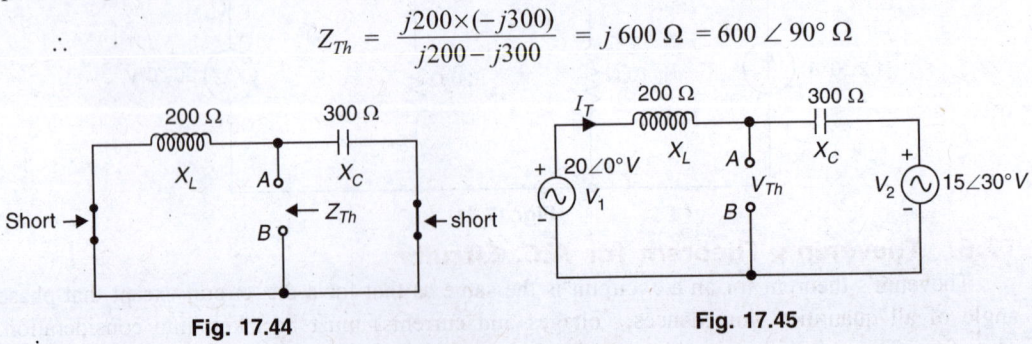

Fig. 17.44   **Fig. 17.45**

Finding V_{Th}. To find V_{Th}, remove the load R from the original circuit. The circuit then becomes as shown in Fig. 17.45. The voltage across the open terminals AB is V_{Th}. It can be found as under :

Net circuit voltage $= V_1 - V_2 = 20\angle 0° - 15 \angle 30° = 10.3 \angle -46.9°$ V

Total circuit impedance, $Z_T = 200 \angle 90° + 300 \angle -90° = 100 \angle -90°$ Ω

$$\text{Circuit current, } I_T = \frac{V_1 - V_2}{Z_T} = \frac{10.3\angle -46.9°}{100\angle -90°}$$

Voltage across X_L, $V_L = I_T X_L = \dfrac{10.3\angle -46.9°}{100\angle -90°} \times 200 \angle 90° = 20.6 \angle 133.1°$ V

∴ $\quad V_{Th} = V_1 - V_L = 20 \angle 0° - 20.6 \angle 133.1° = 37.2 \angle -23.8°$ V

Without load R, Thevenin equivalent circuit at terminals AB is an a.c. source of $37.2\angle -23.8°$V in series with an impedance of $600 \angle 90°$ Ω. When load R is connected to terminals AB, the circuit becomes as shown in Fig. 17.46. The current I through R is given by ;

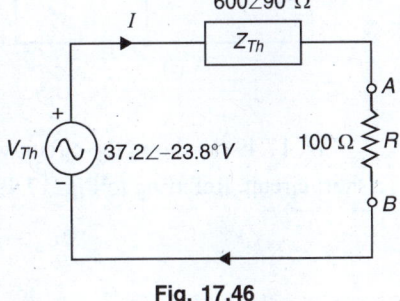

Fig. 17.46

$$I = \frac{V_{Th}}{Z_{Th} + R} = \frac{37.2\angle -23.8°}{600\angle 90° + 100\angle 0°}$$

$$= \frac{37.2\angle -23.8°}{608.3\angle 80.5°}$$

$$= 61 \times 10^{-3}\angle -104.3° \text{ A} = \textbf{61} \angle \textbf{-- 104.3° mA}$$

Remember that for many a.c. circuits, the internal impedance of the source is very small so that internal impedance of the source can be neglected.

Example 17.18. *Using Thevenin's theorem, find the voltage across the inductor in Fig. 17.47.*

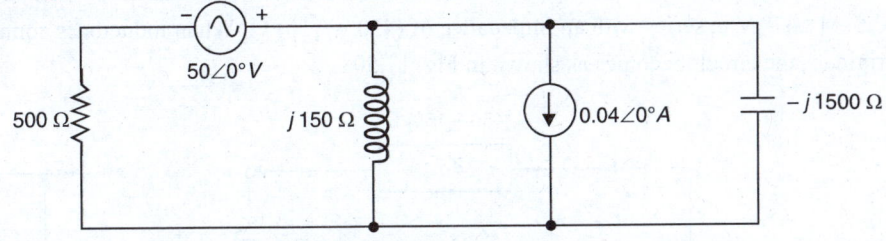

Fig. 17.47

Solution. Consider the inductor to be the load. We shall replace the circuit external to the inductor with its Thevenin equivalent circuit. For this purpose, we are to find Z_{Th} and V_{Th}.

Finding Z_{Th}. To find Z_{Th} at inductor terminals, remove the inductor, short-circuit the voltage source and open-circuit the current source. The circuit then becomes as shown in Fig.17.48. It is clear that Z_{Th} is equal to the equivalent impedance of the parallel combination of the resistor and the capacitor.

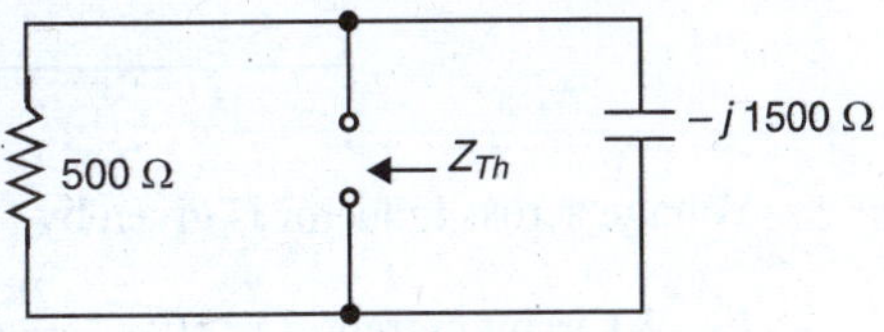

Fig. 17.48

$$Z_{Th} = \frac{500\angle 0° \times 1500\angle -90°}{500 - j1500}$$

$$= \frac{7.5 \times 10^5 \angle -90°}{1581\angle -71.57°} = 474.38 \angle -18.43° \text{ Ω} = (450 - j\,150)\text{Ω}$$

Finding V_{Th}. To find V_{Th}, remove the inductor (load) from the original circuit. Then voltage across open terminals is V_{Th}. We shall use superposition principle to find V_{Th}. Fig. 17.49 (*i*) shows the circuit with current source replaced by an open circuit so that now voltage source is acting alone. Referring to Fig. 17.49 (*i*),

$$E_1 = \frac{50\angle 0°}{500 - j1500} \times 1500 \angle -90° = \frac{50\angle 0°}{1581 \angle -71.57°} \times 1500 \angle -90°$$

$$= 47.44 \angle -18.43° \text{ V} = (45 - j15) \text{ V}$$

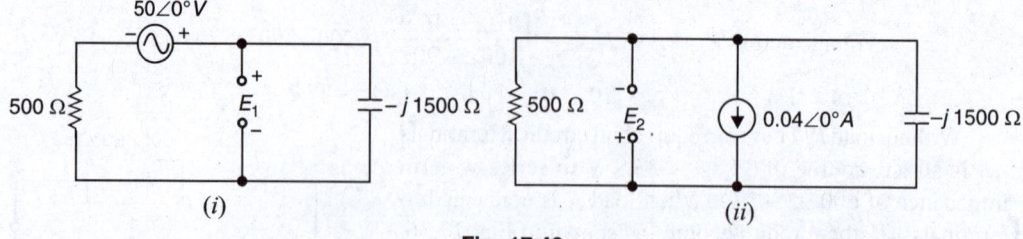

Fig. 17.49

Fig. 17.49 (*ii*) shows the circuit with current source restored and the voltage source replaced by a short circuit. Referring to Fig. 17.49 (*ii*),

$$*E_2 = 0.04 \angle 0° \times Z_{Th} = 0.04 \angle 0° \times 474.38 \angle -18.43°$$

$$= 18.98 \angle -18.43° \text{ V} = (18 - j6) \text{ V}$$

The net voltage at open terminals is $\dot{E}_1 - E_2$ and it is equal to V_{Th}.

$$\therefore \quad V_{Th} = E_1 - E_2 = (45 - j15) - (18 - j6) = (27 - j9) \text{ V} = 28.46 \angle -18.43° \text{ V}$$

Without inductor, Thevenin equivalent circuit at inductor terminals is an a.c. source of $28.46 \angle -18.43°$ V in series with an impedance of $(450 - j150)$ Ω. When inductor is connected to the terminals, the circuit becomes as shown in Fig. 17.50.

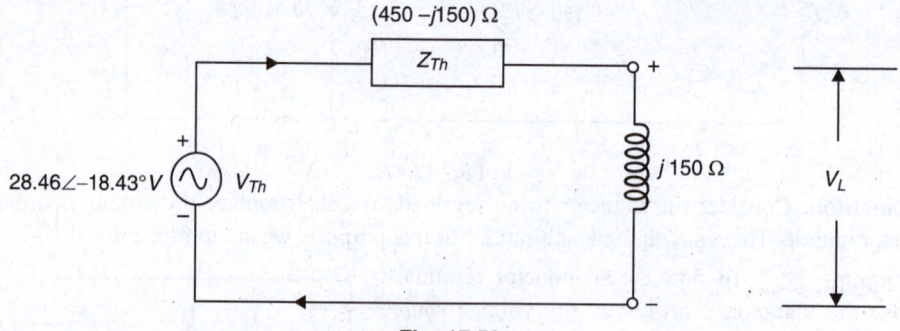

Fig. 17.50

Voltage across inductor is given by ;

$$V_L = \text{Circuit current} \times j\,150 = \frac{28.46 \angle -18.43°}{(450 - j150) + j150} \times 150 \angle 90°$$

$$= \frac{28.46 \angle -18.43°}{450 \angle 0°} \times 150 \angle 90° = \mathbf{9.49 \angle 71.57° \text{ V}}$$

---

* E_2 is equal to current of current source multiplied by equivalent impedance of the parallel combination of the resistor and capacitor. Now equivalent impedance of parallel combination of resistor and capacitor is equal to Z_{Th}.

Example 17.19. *Find the Thevenin equivalent circuit at terminals AB of the circuit shown in Fig. 17.51.*

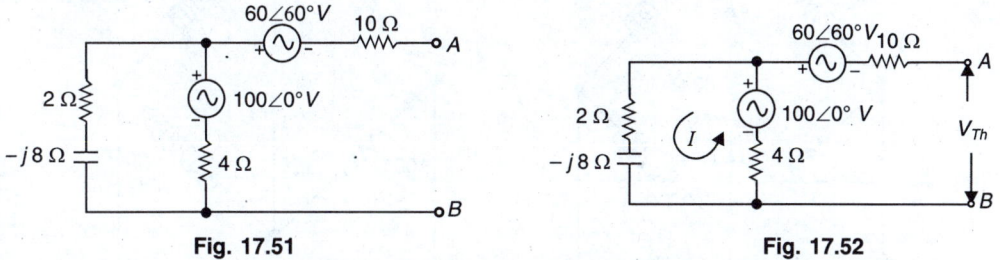

Fig. 17.51   Fig. 17.52

Solution. In order to find Thevenin equivalent circuit, we require V_{Th} and Z_{Th}.

Finding V_{Th}. The open-circuit voltage at terminals AB is the V_{Th} as shown in Fig. 17.52. Referring to Fig. 17.52, the circuit current I is given by ;

$$I = \frac{100 \angle 0°}{2 + 4 - j8} = \frac{100}{6 - j8} = (6 + j8)\text{A}$$

Drop across 4Ω resistor $= 4 \times I = 4 \times (6 + j8) = (24 + j\,32)$ V

∴ $V_{Th} = {}^*V_{AB} = -(24 + j32) + (100 + j0) - 60\,(0.5 + j\,0.866)$

$$= (46 - j84) \text{ V} = 96 \angle -61.3° \text{ V}$$

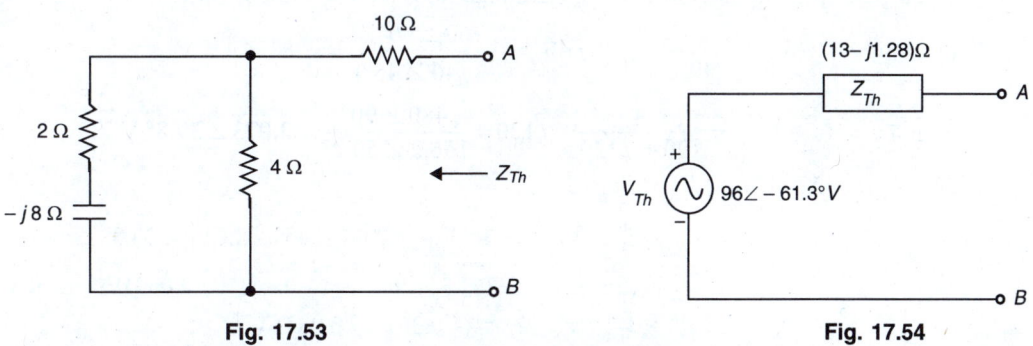

Fig. 17.53   Fig. 17.54

Finding Z_{Th}. To find Z_{Th}, replace the voltage sources in the original circuit by short circuits. The circuit then becomes as shown in Fig. 17.53.

∴ $$Z_{Th} = 10 + [4 \,||\, (2 - j8)] = 10 + \frac{4\,(2 - j8)}{4 + 2 - j8}$$

$$= 10 + \frac{8 - j\,32}{6 - j\,8} = (13 - j\,1.28)\Omega$$

Fig. 17.54 shows the Thevenin equivalent circuit at terminals AB of the given circuit.

Example 17.20. *Find current through 60Ω resistor in the circuit of Fig. 17.55 using Thevenin's theorem. The applied voltage is 4V at a frequency of $2000/2\pi$ Hz and $L_1 = 20mH$; $L_2 = 60mH$.*

---

* $V_{AB} = V_A - V_B$

Now, $V_B - (24 + j\,32) + (100 + j\,0) - 60\,(0.5 + j\,0.866) = V_A$

∴ $V_A - V_B = V_{AB} = -(24 + j\,32) + (100 + j\,0) - 60\,(0.5 + j\,0.866)$

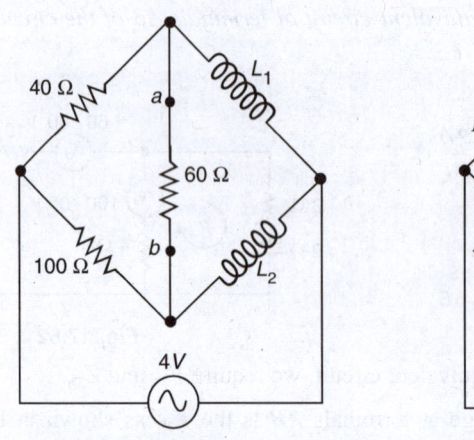

Fig. 17.55   **Fig. 17.56**

Solution. Consider the 60 Ω resistor to be the load.

$$X_{L_1} = 2\pi f L_1 = 2\pi \times \frac{2000}{2\pi} \times 20 \times 10^{-3} = 40 \ \Omega$$

$$X_{L_2} = 2\pi f L_2 = 2\pi \times \frac{2000}{2\pi} \times 60 \times 10^{-3} = 120 \ \Omega$$

Finding V_{Th}. To find V_{Th}, remove the load (60 Ω resistor). The circuit then becomes as shown in Fig. 17.56. The open circuit voltage across terminals *ab* is the V_{Th}. Now $V_{Th} = V_{ab} = V_a - V_b$.

Drop in $L_1 = I_1 \times X_{L_1} = \dfrac{4}{40 + j\,40} \times j\,40 = \dfrac{160 \angle 90°}{\sqrt{2}\,(40 \angle 45°)} = 2.83 \angle 45° \ V$

Drop in $L_2 = I_2 \times X_{L_2} = \dfrac{4}{100 + j\,120} \times j\,120 = \dfrac{480 \angle 90°}{156.2 \angle 50.2°} = 3.073 \angle 39.8° \ V$

Now, $V_a - I_1 \times X_{L_1} + I_2 \times X_{L_2} = V_b$

∴ $V_a - V_b = I_1 \times X_{L_1} - I_2 \times X_{L_2} = 2.83 \angle 45° - 3.073 \angle 39.8°$

$$= (2 + j2) - (2.361 + j1.967) = (-0.361 + j\,0.033)V$$

∴ $V_{Th} = V_{ab} = V_a - V_b = (-0.361 + j\,0.033) \ V$

Finding Z_{Th}. To find Z_{Th}, remove the load (60Ω resistor) and replace the voltage source by a short circuit in the original circuit. The circuit then becomes as shown in Fig. 17.57. The impedance between terminals *a* and *b* is equal to Z_{Th}.

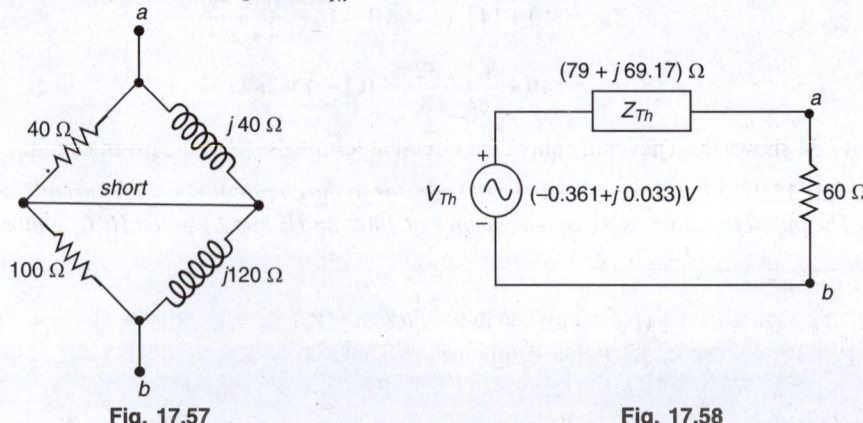

Fig. 17.57   **Fig. 17.58**

$$\therefore \qquad Z_{Th} = Z_{ab} = (40\Omega \,||\, j\,40\Omega) + (100\Omega \,||\, j\,120\Omega)$$

$$= \frac{40 \times j40}{40 + j40} + \frac{100 \times j120}{100 + j120} = 28.29 \angle 45° + 76.82 \angle 39.8°$$

$$= (20 + j\,20) + (59 + j\,49.17) = (79 + j\,69.17)\Omega$$

The Thevenin equivalent circuit along with the load is shown in Fig. 17.58.

$$\therefore \qquad \text{Current thro' } 60\Omega = \frac{V_{Th}}{Z_{Th} + 60} = \frac{-0.361 + j\,0.033}{79 + j\,69.17 + 60}$$

$$= \frac{0.3625 \angle 174.8°}{155.26 \angle 26.46°} = \mathbf{2.33 \times 10^{-3} \angle 148.34° A}$$

Example 17.21. *Find the magnitude and phase angle of current through a short circuit placed across terminals ab in the network of Fig. 17.59 (i).*

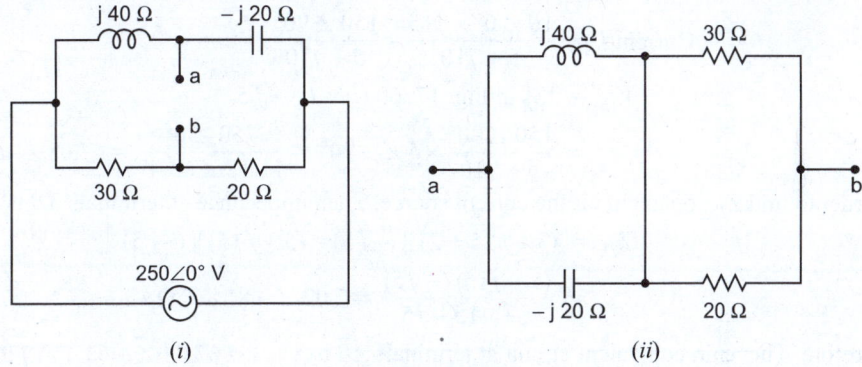

Fig. 17.59

Solution. To find the desired answer, we should calculate V_{Th} and Z_{Th} at terminals *ab*.

$$\text{Now,} \qquad V_a = \frac{250 \angle 0°}{j\,40 - j\,20} \times (-j\,20) = -250 \angle 0° \text{ V}$$

$$V_b = \frac{250 \angle 0°}{30 + 20} \times 20 = 100 \angle 0° \text{ V}$$

$$\therefore \qquad V_{Th} = V_a - V_b = -250 \angle 0° - 100 \angle 0° = -350 \angle 0° \text{ V}$$

In order to find Z_{Th}, replace the voltage source by a short in the original circuit. The circuit then becomes as shown in Fig. 17.59 (*ii*). Then impedance at open terminals *ab* is Z_{Th}.

$$\therefore \qquad Z_{Th} = (j\,40 \,||\, -j\,20) + (30 \,||\, 20)$$

$$= \frac{(j\,40)(-j\,20)}{j\,40 - j\,20} + \frac{30 \times 20}{30 + 20} = (12 - j\,40)\,\Omega$$

Therefore, Thevenin equivalent circuit at terminals *ab* is V_{Th} (= – 350 V) in series with Z_{Th} [= (12 – *j* 40) Ω]. When short circuit is placed across terminals *ab*, the short circuit current I_{sc} is

$$I_{sc} = \frac{V_{Th}}{Z_{Th}} = \frac{-350}{12 - j\,40} = \frac{-350}{41.76 \angle -73.3°}$$

$$= \mathbf{8.38 \angle 73.3° \text{ A } from \ b \ to \ a}$$

Example 17.22. *Obtain Thevenin equivalent circuit with respect to terminals AB in the network shown in Fig. 17.60 (i).*

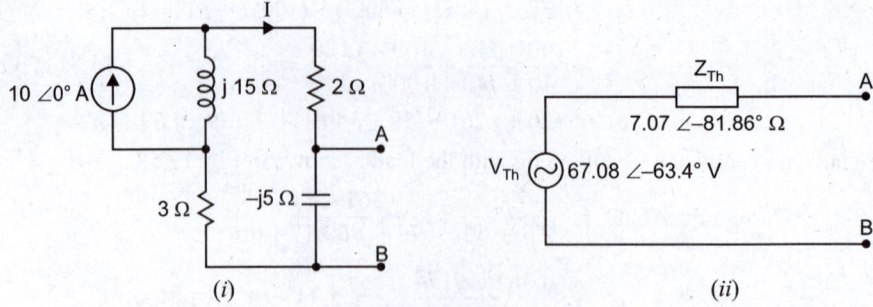

Fig. 17.60

Solution. With terminals AB open, the circuit is a series a.c. circuit of voltage $= 10 \angle 0°$ A $\times j\,15\,\Omega$ volts and impedance $Z = 3 + j\,15 + 2 - j\,5 = (5 + j\,10)\,\Omega$.

$\therefore$ Current $I = \dfrac{10 \angle 0° \times j\,15}{5 + j\,10} = \dfrac{150 \angle 90°}{5 + j\,10}$

$\therefore$ $V_{Th} = V_{AB}$ in Fig. 17. 60 $(i) = I \times -j\,5$

$$= \frac{150 \angle 90°}{5 + j\,10} \times 5 \angle -90° = \frac{750 \angle 0°}{11.18 \angle 63.4°} = 67.08 \angle -63.4° \text{ V}$$

In order to find Z_{Th}, open circuit the current source. Then impedance at terminals AB is Z_{Th}.

$\therefore$ $Z_{Th} = (3 + j\,15 + 2) \parallel (-j\,5) = (5 + j\,15) \parallel (-j\,5)$

$$= \frac{(5 + j\,15)\,(-j\,5)}{5 + j\,15 - j\,5} = 7.07 \angle -81.86°\,\Omega$$

Therefore, Thevenin equivalent circuit at terminals AB is **V_{Th} (= 67.08 $\angle$ – 63.4° V) in series with Z_{Th} (= 7.07 $\angle$ – 81.86° Ω)** as shown in Fig. 17.60 (ii).

Example 17.23. *Fig. 17.61 (i) shows an a.c. circuit. Find (i) V_{Th} and Z_{Th} with respect to terminals AB (ii) active and reactive power absorbed by an impedance of $(3 + j\,4)$ Ω connected across AB.*

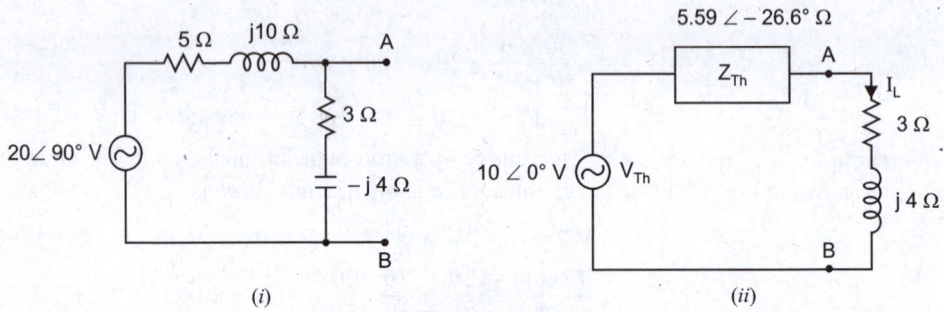

Fig. 17.61

Solution. (i) V_{Th} and Z_{Th} at terminals AB are found as under :

$$V_{Th} = V_{AB} \text{ in Fig. 17.61 } (i) = \text{Current in } (3 - j\,4) \times (3 - j\,4)$$

$$= \frac{20 \angle 90°}{5 + j\,10 + 3 - j\,4} \times (3 - j\,4) = \frac{20 \angle 90°}{8 + j\,6} \times 5 \angle -53.1° = \mathbf{10 \angle 0° \text{ V}}$$

In order to find Z_{Th}, replace the voltage source by a short. Then impedance at terminals AB is Z_{Th}.

$\therefore$ $Z_{Th} = (5 + j\,10) \parallel (3 - j\,4) = \dfrac{(5 + j\,10)\,(3 - j\,4)}{5 + j\,10 + 3 - j\,4}$

$$= \frac{(5 + j\,10)\,(3 - j\,4)}{8 + j\,6} = \mathbf{5.59 \angle -26.6°\,\Omega}$$

(*ii*) Therefore, Thevenin equivalent circuit at *AB* is V_{Th} (= 10 ∠ 0° V) in series with Z_{Th} (= 5.59 ∠ – 26.6° Ω). When impedance $Z = (3 + j4)\,\Omega$ is connected across terminals *AB*, the circuit becomes as shown in Fig. 17.61 (*ii*).

$$\text{Load current, } I_L = \frac{V_{Th}}{Z_{Th} + Z} = \frac{10\angle 0°}{5.59\angle -26.6° + (3 + j4)} = 1.23\angle -10.6° \text{ A}$$

Active power absorbed $= I_L^2 \times 3 = (1.23)^2 \times 3 = \textbf{4.54 W}$

Reactive power absorbed $= I_L^2 \times 4 = (1.23)^2 \times 4 = \textbf{6.05 VAR}$

Tutorial Problems

1. Using Thevenin's theorem, find current in the 50Ω resistor for the circuit shown in Fig. 17.62.

   [0.14 ∠ 52.13° A]

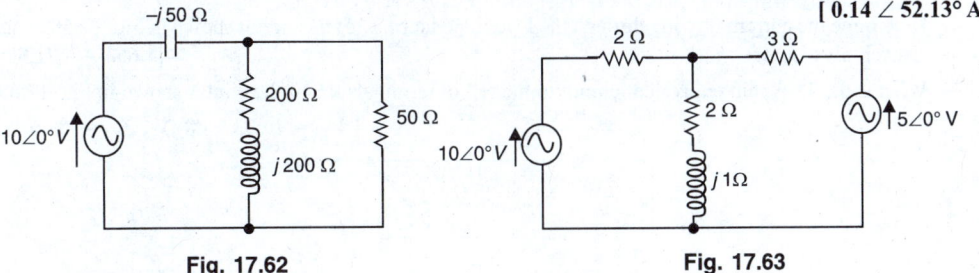

Fig. 17.62   **Fig. 17.63**

2. Using Thevenin's theorem, find current in $(2 + j1)\,\Omega$ impedance for the circuit shown in Fig. 17.63.

   [2.39 ∠ – 17.4° A]

3. Find the Thevenin equivalent circuit at terminals *AB* for the circuit shown in Fig. 17.64.

   [V_{Th} = 8.2 ∠ – 14° V; Z_{Th} = (7 + j 1) Ω]

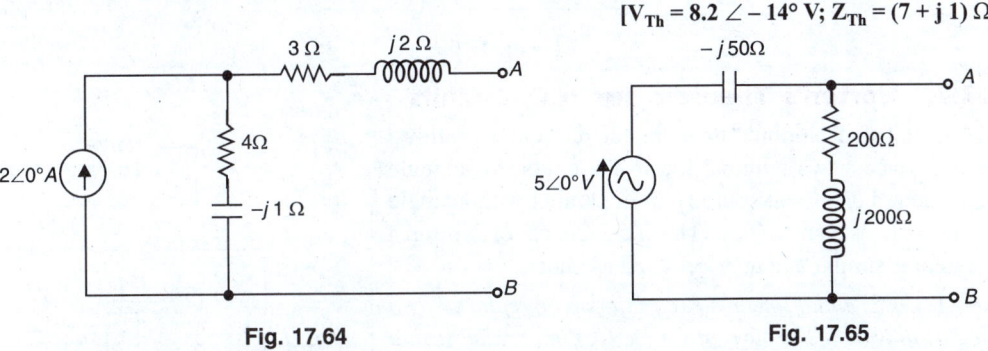

Fig. 17.64   **Fig. 17.65**

4. Find the Thevenin equivalent circuit at terminals *AB* for the circuit shown in Fig. 17.65.

   [V_{Th} = 5.7 ∠ 8.1° V; Z_{Th} = (7.98 – j 56) Ω]

5. Use Thevenin's theorem to find current through the capacitor in the circuit shown in Fig. 17.66.

   [0.2 ∠ 78.7° A]

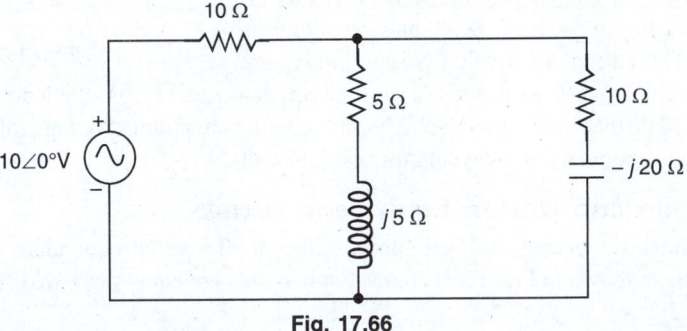

Fig. 17.66

6. Determine the current flowing through 2 Ω resistor of the circuit shown in Fig. 17.67 by Thevenin's theorem.   **[1.387 ∠ 56.31° A]**

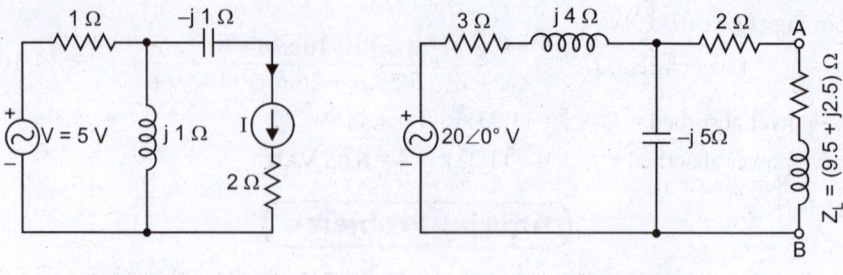

              Fig. 17.67   **Fig. 17.68**

7. Determine the current flowing through the load impedance Z_L of the circuit shown in Fig. 17.68 by using Thevenin's theorem.   **[1.663 ∠ – 71.56° A]**

8. What is the Thevenin equivalent circuit to the left of terminals *AB* of the circuit shown in Fig. 17.69 ?

                                [V_{Th} = (9.33 + *j* 8) V ; Z_{Th} = (8 – *j* 11)Ω]

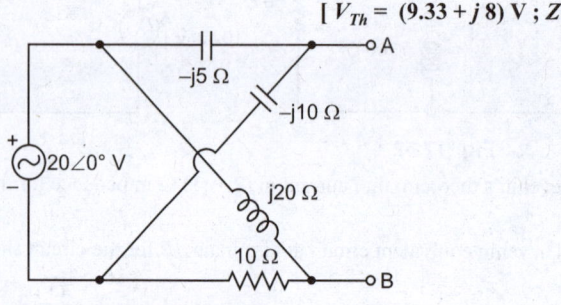

                                Fig. 17.69

17.7. Norton's Theorem for A.C. Circuits

Recall that Norton's theorem for d.c. circuits allows us to replace a two-terminal linear d.c. circuit by a single equivalent d.c. current source (I_N) in parallel with a single equivalent resistance (R_N). The a.c. version of Norton's theorem is similar and may be stated as under :

A two-terminal linear a.c. circuit can be replaced by a single equivalent a.c. current source (I_N) in parallel with a single equivalent impedance (Z_N).

Fig. 17.70 shows the Norton equivalent circuit of a two-terminal a.c. circuit. The impedance Z_N (called Norton equivalent impedance) has exactly the same value as the Thevenin equivalent impedance (Z_{Th}) and is found in the same way. The current I_N (called Norton equivalent

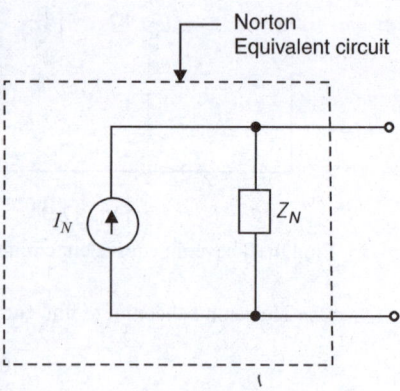

                              Fig. 17.70

current) is the current that flows through a short circuit connected across the Norton terminals (*i.e.,* load terminals). Note that the Thevenin and Norton circuits are alternative equivalents for a circuit. Norton's theorem is popular for analysing transistor circuits.

17.8. Thevenin and Norton Equivalent Circuits

A two-terminal a.c. circuit can be replaced either by Thevenin equivalent circuit or Norton equivalent circuit as shown in Fig. 17.71. Since both circuits are equivalent, we can show that V_{Th} = $I_N Z_N$ and Z_{Th} = Z_N.

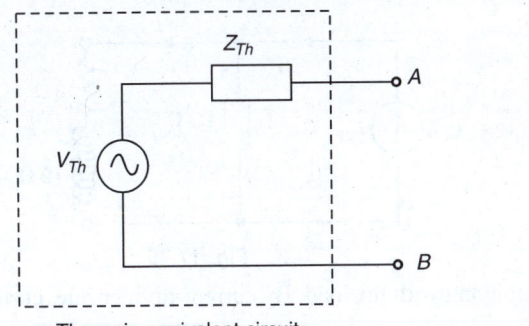

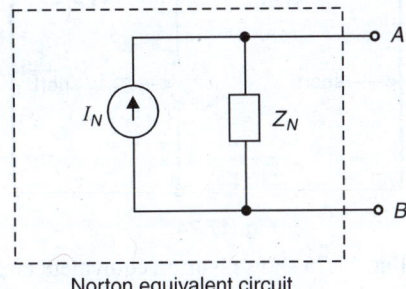

Thevenin equivalent circuit   Norton equivalent circuit

Fig. 17.71

(*i*) Since the two circuits are equivalent, their open-circuit voltages must be the same. For the Thevenin circuit, the open-circuit voltage is V_{Th} while for Norton circuit, it is $I_N Z_N$.

$$\therefore \qquad V_{Th} = I_N Z_N$$

(*ii*) If both the equivalent circuits are short-circuited, then same short-circuit current must flow through each. For the Norton circuit, the short-circuit current is I_N while for Thevenin circuit, it is V_{Th}/Z_{Th}.

$$\therefore \qquad I_N = \frac{V_{Th}}{Z_{Th}}$$

It is clear that $Z_N = Z_{Th}$. For the solution of a circuit, you can use either equivalent circuit. The choice is purely a matter of convenience.

Example 17.24. *Use Norton's theorem to find current in* $(3 + j\,2)\ \Omega$ *impedance in the circuit shown in Fig. 17.72.*

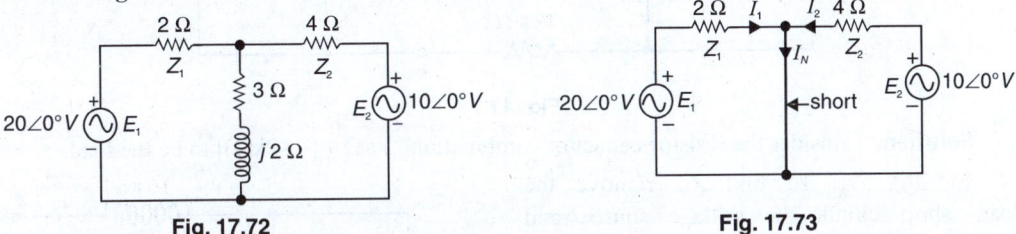

Fig. 17.72   **Fig. 17.73**

Solution. Consider the $(3 + j2)\Omega$ impedance to be the load.

Finding I_N. To find I_N, short-circuit the load terminals. The circuit then becomes as shown in Fig. 17.73. The current that flows in the short-circuited terminals is the Norton current I_N. It is clear that :

$$I_N = I_1 + I_2 = \frac{E_1}{Z_1} + \frac{E_2}{Z_2} = \frac{20\angle 0°}{2\angle 0°} + \frac{10\angle 0°}{4\angle 0°} = 10 + 2.5 = 12.5\text{A}$$

Finding Z_N. In order to find $Z_N (= Z_{Th})$, remove the load [*i.e.*, $(3 + j\,2)\Omega$ impedance] and replace each voltage source by a short circuit. The circuit then becomes as shown in Fig. 17.74. It is clear that Z_N is equal to parallel combination of $2\ \Omega$ and $4\ \Omega$ *i.e.*

$$Z_N = 2\Omega\,||\,4\Omega = \frac{2\times 4}{2 + 4} = 1.3\Omega$$

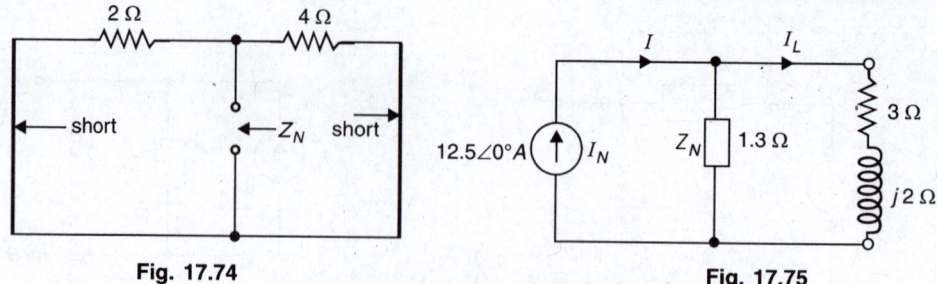

Fig. 17.74   **Fig. 17.75**

Fig. 17.75 shows Norton equivalent circuit along with the load. By current-divider rule, current I_L in the load is given by ;

$$I_L = I \times \frac{Z_N}{Z_N + Z_L} = 12.5 \times \frac{1.3}{1.3 + 3 + j\,2} = 12.5 \times \frac{1.3}{4.3 + j\,2}$$

$$= 12.5 \times \frac{1.3}{4.74 \angle 24.9°} = \mathbf{3.43 \angle -24.9° \, A}$$

Example 17.25. *Find the Norton equivalent circuit lying to the left of terminals a – b in Fig. 17.76. Also find Thevenin equivalent of the circuit.*

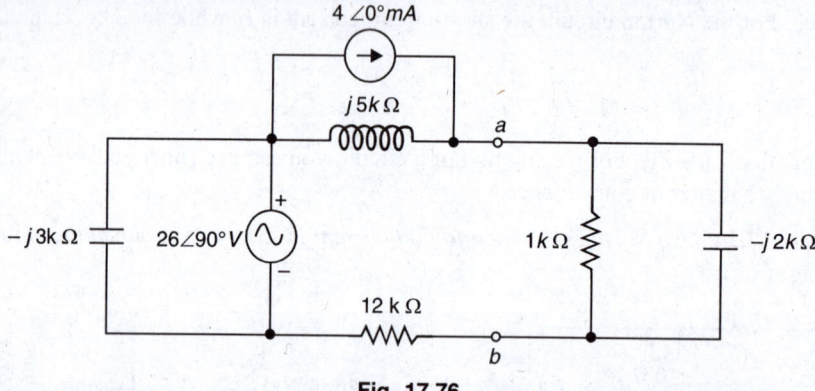

Fig. 17.76

Solution. Consider the resistor-capacitor combination [1 kΩ || (– j 2kΩ)] to be the load.

Finding Z_N. To find Z_N, remove the load, short circuit the voltage source and open circuit the current source. The circuit then becomes as shown in Fig. 17.77. Since the capacitive reactance (– j 3kΩ) in parallel with the voltage source is shorted out, Z_N is simply the series combination of resistance (12 kΩ) and inductive reactance (j 5kΩ).

∴ $Z_N = 12 \text{ k}\Omega + j\,5 \text{ k}\Omega = 13 \angle 22.62° \text{ k}\Omega$

Fig. 17.77

Finding I_N. We shall use the superposition principle to find the Norton current I_N that flows in the shorted terminals *ab*. To find current I_1 that flows in the shorted terminals *ab* due to voltage source alone, remove the load and replace the current source by an open circuit. The circuit then becomes as shown in Fig. 17.78 (*i*). It is clear that short-circuit current I_1 is given by ;

$$I_1 = \frac{26 \angle 90° \text{ V}}{12\text{k}\Omega + j\,5\text{k}\Omega} = \frac{26 \angle 90°\text{V}}{13 \angle 22.62°\text{k}\Omega} = 2 \angle 67.38° \text{ mA}$$

Note that the capacitive reactance $(-j3\ \text{k}\Omega)$ in the circuit has no effect in the value of the short-circuit current.

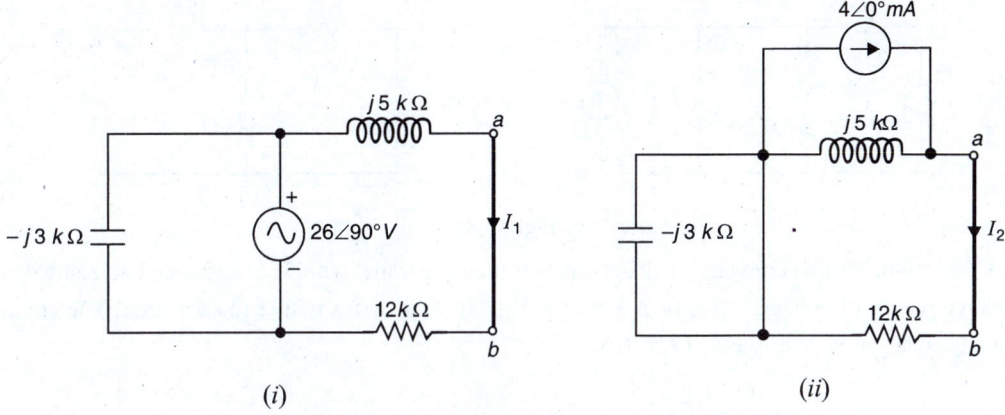

Fig. 17.78

To find current I_2 that flows in the shorted terminals ab due to current source alone, remove the load and replace the voltage source by a short circuit. The circuit then becomes as shown in Fig. 17.78 (*ii*). By current-divider rule,

$$I_2 = 4 \angle 0° \text{ mA} \times \frac{5 \angle 90° \text{ k}\Omega}{12 \text{ k}\Omega + j\ 5 \text{ k}\Omega} = 4 \angle 0° \text{ mA} \times \frac{5 \angle 90° \text{ k}\Omega}{13 \angle 22.62° \text{ k}\Omega}$$

$$= 1.54 \angle 67.38° \text{ mA}$$

By superposition principle, the Norton current I_N is the sum of short circuit currents I_1 and I_2.

$$\therefore \qquad I_N = I_1 + I_2 = 2 \angle 67.38° \text{ mA} + 1.54 \angle 67.38° \text{ mA}$$

$$= 3.54 \angle 67.38° \text{ mA}$$

The Norton equivalent circuit is shown in Fig. 17.79. From Norton equivalent circuit, we can find Thevenin equivalent circuit as under :

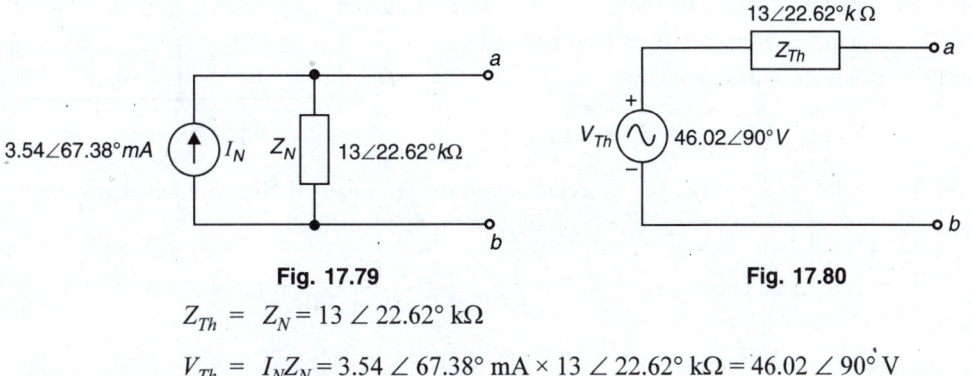

Fig. 17.79   Fig. 17.80

$$Z_{Th} = Z_N = 13 \angle 22.62° \text{ k}\Omega$$

$$V_{Th} = I_N Z_N = 3.54 \angle 67.38° \text{ mA} \times 13 \angle 22.62° \text{ k}\Omega = 46.02 \angle 90° \text{ V}$$

Therefore, Thevenin equivalent circuit is **an a.c. voltage source of 46.02 $\angle$ 90° V in series with 13 $\angle$ 22.62° kΩ impedance** as shown in Fig. 17.80.

Example 17.26. *Using Norton's theorem, find the active and reactive powers supplied to Z in the circuit of Fig. 17.81 (i).*

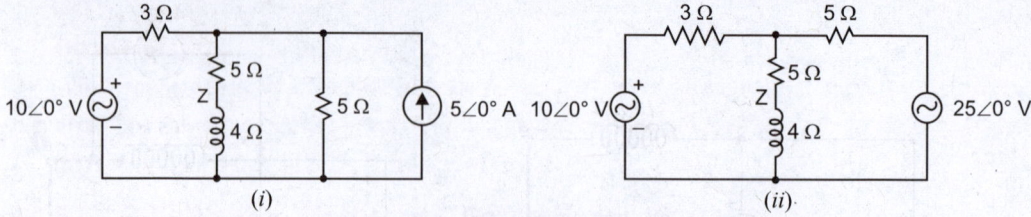

Fig. 17.81

Solution. We first convert $5 \angle 0°$ A current source in parallel with 5Ω resistance into equivalent voltage source of voltage $= 5 \angle 0°$ A $\times 5 \Omega = 25 \angle 0°$ V in series with 5Ω resistance. The circuit then becomes as shown in Fig. 17.81 (*ii*).

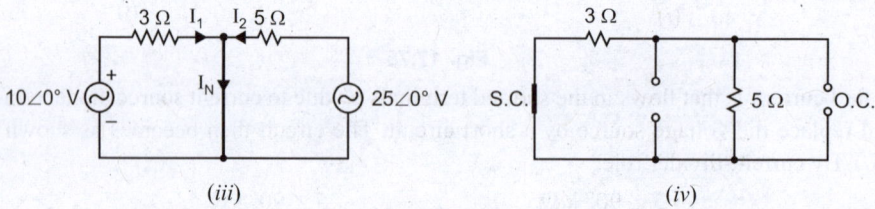

Fig. 17.81

In order to find Norton current I_N, we short circuit Z(*i.e.,* load) in Fig. 17.81 (*ii*). The circuit then becomes as shown in Fig. 17.81 (*iii*). The current in the short circuit is I_N.

$$\therefore \qquad I_N = I_1 + I_2 = \frac{10 \angle 0°}{3} + \frac{25 \angle 0°}{5} = 8.33 \text{ A}$$

In order to find Norton impedance Z_N, open circuit the current source, short circuit the voltage source and remove the load Z in the original circuit. The circuit then becomes as shown in Fig. 17.81 (*iv*). The impedance at open-circuit load terminals in Fig. 17.81 (*iv*) is Z_N and is given by ;

$$\therefore \quad Z_N = 3 \ \Omega \| \ 5 \ \Omega = \frac{3 \times 5}{3 + 5} = 1.875 \ \Omega$$

Fig. 17.81 (*v*)

When load $Z = (5 + j \ 4)\Omega$ is connected across the terminals of Norton equivalent circuit, the circuit becomes as shown in Fig. 17.81 (*v*). By current-divider rule,

$$\text{Load current, } I_L = 8.33 \times \frac{1.875 \angle 0°}{1.875 \angle 0° + (5 + j4)} = 1.964 \angle -30.2° \text{ A}$$

$$\therefore \qquad \text{Active power} = I_L^2 \times 5 = (1.964)^2 \times 5 = \textbf{19.286 W}$$

$$\text{Reactive power} = I_L^2 \times 4 = (1.964)^2 \times 4 = \textbf{15.43 VAR}$$

Tutorial Problems

1. Use Norton's theorem to determine current through $(5 + j\,3)\Omega$ impedance in Fig. 17.82.

[0.85 ∠ 25.6° A]

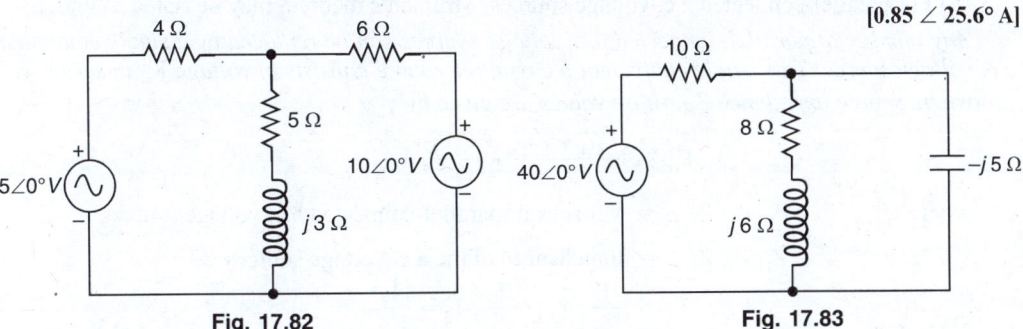

Fig. 17.82   **Fig. 17.83**

2. Use Norton's theorem to determine current through the capacitor in Fig. 17.83.   **[3.5 ∠ 52.1° A]**

3. Convert the Thevenin circuit in Fig. 17.84 to an equivalent Norton circuit.

[I_N = 1.8 ∠ − 26.6° A in parallel with 11.2 ∠ 26.6° Ω impedance]

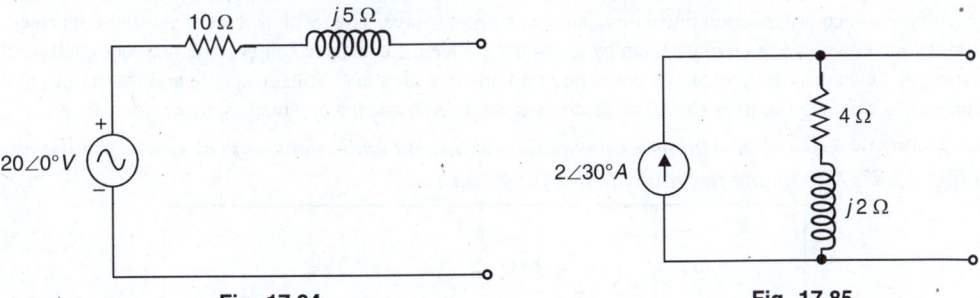

Fig. 17.84   **Fig. 17.85**

4. Convert the Norton circuit in Fig. 17.85 to an equivalent Thevenin circuit.

[V_{Th} = 9 ∠ 56.6° V in series with 4.5 ∠ 26.6° Ω impedance]

5. Use Norton's theorem to find current in the capacitor in Fig. 17.86.   **[0.35 ∠ − 81.6° A]**

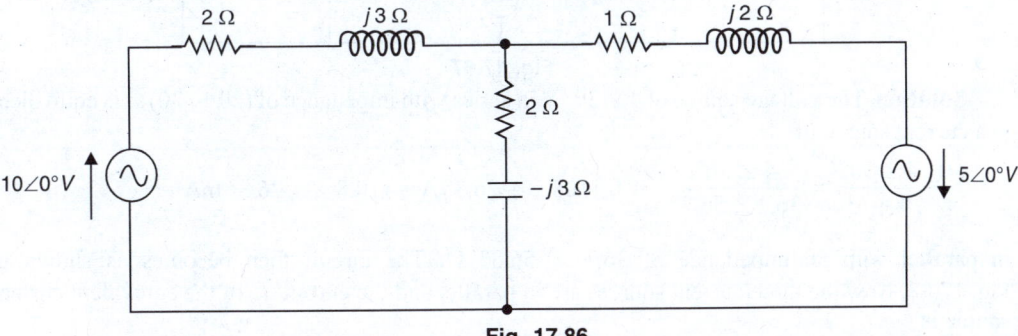

Fig. 17.86

17.9. Millman's Theorem for A.C. Circuits

Millman's theorem is used to replace a number of parallel-connected a.c. voltage/current sources by a single equivalent a.c. voltage/current source.

(i) For parallel-connected a.c. current sources, Millman's theorem may be stated as under :

Any number of parallel-connected a.c. current sources can be replaced by a single equivalent a.c. current source. This single equivalent a.c. current source consists of an ideal a.c. current source and a parallel equivalent source impedance.

The current of the equivalent a.c. current source is equal to the phasor sum of individual source currents. The parallel equivalent source impedance is the parallel combination of individual source impedances.

(*ii*) For parallel-connected a.c. voltage sources, Millman's theorem may be stated as under :

Any number of parallel-connected a.c. voltage sources can be replaced by a single equivalent a.c. voltage source. This single equivalent a.c. voltage source consists of voltage V_m in series with equivalent source impedance Z_m whose values are given by ;

$$V_m = \frac{V_1Y_1 + V_2Y_2 + V_3Y_3 + ...}{Y_1 + Y_2 + Y_3 ...}$$

where $V_1, V_2, V_3 ...$ = Voltages of parallel-connected a.c. voltage sources

$Z_1, Z_2, Z_3 ...$ = Impedances of the a.c. voltage sources

$$Y_1 = \frac{1}{Z_1}; Y_2 = \frac{1}{Z_2}; Y_3 = \frac{1}{Z_3} ...$$

$$Z_m = \frac{1}{Y_1 + Y_2 + Y_3 + ...}$$

Note. If the circuit has a combination of parallel a.c. voltage and current sources, each parallel-connected a.c. voltage source is converted into equivalent a.c. current source. The result is a set of parallel-connected a.c. current sources and we can replace them by a single equivalent a.c. current source. Alternatively, each parallel-connected a.c. current source can be converted into an equivalent a.c. voltage source and the set of parallel-connected a.c. voltage sources can be replaced by a single equivalent a.c. voltage source.

Example 17.27. *Use Millman's theorem to obtain an equivalent current source for the circuit of Fig. 17.87. Also obtain the equivalent voltage source.*

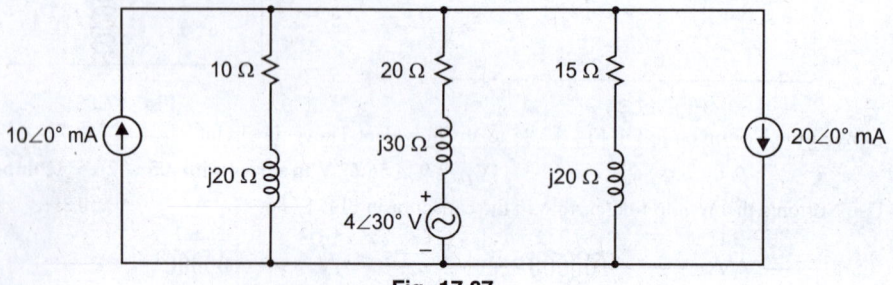

Fig. 17.87

Solution. The voltage source of $4 \angle 30°$ V in series with impedance of $(20 + j\,30)\,\Omega$ is equivalent to a current source of

$$\frac{4 \angle 30°}{20 + j\,30} = \frac{4 \angle 30°}{36.1 \angle 56.3°} = 0.1108 \angle -26.3° \text{ A} = 110.8 \angle -26.3° \text{ mA}$$

in parallel with an impedance of $36.1 \angle 56.3°\,\Omega$. The circuit then becomes as shown in fig. 17.88. Now the three current sources are in parallel and the current I_N of the equivalent current source is

$$I_N = (10 \angle 0° + 110.8 \angle -26.3° - 20 \angle 0°) \text{ mA}$$
$$= (10 + j\,0) + (99.33 - j\,49.1) - (20 + j\,0)$$
$$= (89.33 - j\,49.1) \text{ mA} = 101.93 \angle -28.8° \text{ mA}$$

The Norton's impedance Z_N is given by ;

$$\frac{1}{Z_N} = \frac{1}{10 + j\,20} + \frac{1}{36.1 \angle 56.3°} + \frac{1}{15 + j\,20}$$
$$= (0.02 - j\,0.04) + (0.0154 - j\,0.023) + (0.024 - j\,0.032)$$
$$= (0.0594 - j\,0.095) = 0.112 \angle -57.98°$$

$$\therefore \qquad Z_N = \frac{1}{0.112 \angle -57.98°} = 8.93 \angle 57.98° \ \Omega = (4.73 + j\,7.57)\ \Omega$$

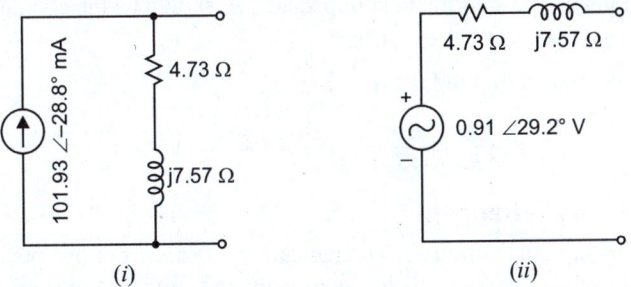

Fig. 17.88

Therefore, the equivalent current source for the circuit in Fig. 17.87 is I_N [= **101.93** $\angle$ **– 28.8° mA] in parallel with** Z_N [= **(4.73 + j 7.57) Ω**] as shown in Fig. 17.89 (*i*).

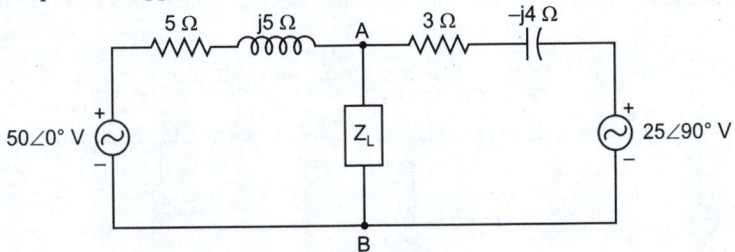

(*i*)   (*ii*)

Fig. 17.89

The equivalent voltage source is V_{Th} in series with Z_{Th} where

$$V_{Th} = I_N Z_N = (101.93 \angle -28.8°) \times (4.73 + j\,7.57) = \mathbf{0.91 \angle 29.2° \ V}$$
$$Z_{Th} = Z_N = \mathbf{(4.73 + j\,7.57)\ \Omega}$$

Fig. 17.89 (*ii*) shows the equivalent voltage source of the circuit.

Example 17.28. *In the circuit shown in Fig. 17.90, two voltage sources act on the load impedance connected to terminals AB. If the load is variable in both reactance and resistance, what load Z_L will receive maximum power ? Apply Millman's theorem and also calculate maximum power.*

Fig. 17.90

Solution. $\qquad\qquad V_1 = 50 \angle 0° \ V \ ; \ V_2 = 25 \angle 90° \ V$

$$Y_1 = \frac{1}{Z_1} = \frac{1}{(5 + j5)} = (0.1 - j\,0.1)\ S = 0.1414 \angle -45° \ S$$

$$Y_2 = \frac{1}{Z_2} = \frac{1}{(3 - j4)} = (0.12 + j\,0.16)\ S = 0.2 \angle 53.13° \ S$$

According to Millman's theorem, the two parallel-connected voltage sources can be replaced by a single equivalent voltage source of voltage V_m in series with equivalent source impedance Z_m where

$$V_m = \frac{V_1 Y_1 + V_2 Y_2}{Y_1 + Y_2} \; ; \; Z_m = \frac{1}{Y_1 + Y_2}$$

$$\therefore \qquad V_m = \frac{50 \angle 0° \times 0.1414 \angle -45° + 25 \angle 90° \times 0.2 \angle 53.13°}{(0.1 - j0.1) + (0.12 + j0.16)}$$

$$= \frac{7.07 \angle -45° + 5 \angle 143.13°}{(0.22 + j0.06)} = \frac{5 - j5 - 4 + j3}{(0.22 + j0.06)}$$

$$= \frac{1 - j2}{(0.22 + j0.06)} = 9.8 \angle -78.66° \text{ V}$$

$$Z_m = \frac{1}{(0.1 - j0.1) + (0.12 + j0.16)} = \frac{1}{0.22 + j0.06} = 4.386 \angle -15.26° \, \Omega$$

$$= (4.23 - j1.15) \, \Omega$$

For maximum power transfer, the load impedance Z_L should be the conjugate of Z_m i.e.,

$$Z_L = \mathbf{(4.23 + j1.15) \, \Omega}$$

Maximum power transferred to load is

$$P_{max} = \frac{V_m^2}{4 R_L} = \frac{(9.8)^2}{4 \times 4.23} = \mathbf{5.68 \text{ W}}$$

17.10. Reciprocity Theorem

This theorem permits us to transfer voltage/current source from one position in the circuit to another and is applicable to circuits which contain only one voltage source or one current source. It may be stated as under :

In any linear bilateral circuit, if a voltage source acting in branch X causes a current I in branch Y, then shifting the voltage source (but not its impedance) to branch Y will cause the same current I in branch X. However, currents in the other parts of the circuit will not remain the same after this interchange.

The dual is also true : If current applied at node A of the circuit produces voltage V at another node B, then the same current applied at node B will produce the same voltage V at node A. The reciprocity theorem seems quite strange but it is true. We shall verify this theorem in the following example.

Example 17.29. *Verify reciprocity theorem for the network shown in Fig. 17.91 (i).*

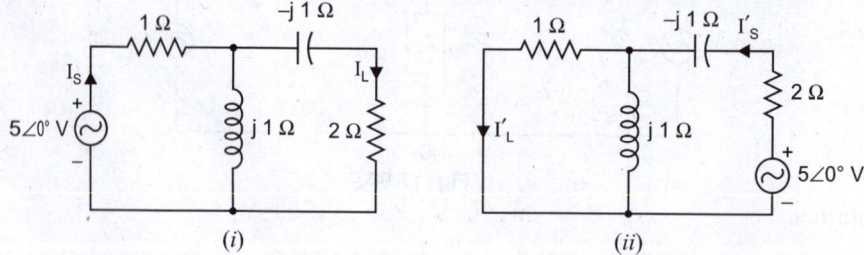

Fig. 17.91

Solution. Referring to Fig. 17.91 (*i*), the impedance offered to the source is

$$Z = 1 + [(2 - j1) \| j1] = 1 + \frac{(2 - j1) j1}{(2 - j1) + j1} = 1 + \frac{1 + j2}{2}$$

$$= (1.5 + j1) \, \Omega = 1.8 \angle 33.7° \, \Omega$$

$\therefore$ Current supplied by the source in Fig. 17.91 (*i*) is

$$I = \frac{V_S}{\sqrt{(R_S + R_L)^2 + (X_S + X_L)^2}}$$

Power delivered to load, $P_L = I^2 R_L = \dfrac{V_S^2 R_L}{(R_S + R_L)^2 + (X_S + X_L)^2}$

For maximum power transfer, $\dfrac{dP_L}{dR_L} = 0$.

or $\dfrac{[(R_S + R_L)^2 + (X_S + X_L)^2]V_S^2 - V_S^2 R_L \times 2(R_S + R_L)}{[(R_S + R_L)^2 + (X_S + X_L)^2]^2} = 0$

or $(R_S + R_L)^2 + (X_S + X_L)^2 = 2R_S R_L + 2R_L^2$

or $R_S^2 + R_L^2 + 2R_S R_L + (X_S + X_L)^2 = 2R_S R_L + 2R_L^2$

or $R_L^2 = R_S^2 + (X_S + X_L)^2$

Hence when load has variable R_L but fixed X_L, power transferred to the load is maximum when

$$R_L^2 = R_S^2 + (X_S + X_L)^2$$

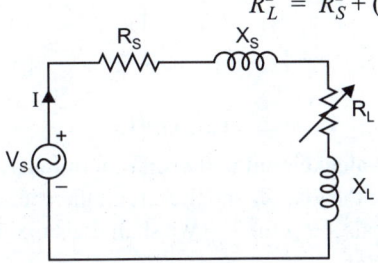

Fig. 17.94

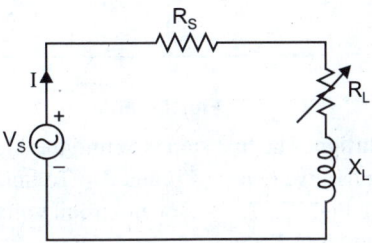

Fig. 17.95

4. When source impedance is resistive and load has variable R_L but fixed X_L. This case is shown in Fig. 17.95. Here the source impedance is purely resistive (*i.e.*, $Z_S = R_S$) and the resistive component of load (R_L) is variable while reactive component of load (X_L) is fixed. The circuit current is given by ;

$$I = \frac{V_S}{\sqrt{(R_S + R_L)^2 + X_L^2}}$$

Power delivered to load, $P_L = I^2 R_L = \dfrac{V_S^2 R_L}{(R_S + R_L)^2 + X_L^2}$

For maximum power transfer, $\dfrac{dP_L}{dR_L} = 0$.

or $\dfrac{[(R_S + R_L)^2 + X_L^2]V_S^2 - V_S^2 R_L \times 2(R_S + R_L)}{[(R_S + R_L)^2 + X_L^2]^2} = 0$

or $(R_S + R_L)^2 + X_L^2 - R_L \times 2\,(R_S + R_L) = 0$

or $R_L^2 = R_S^2 + X_L^2$

Hence when the source impedance is purely resistive and the load has variable R_L but fixed X_L, power transferred to the load is maximum when

$$R_L^2 = R_S^2 + X_L^2$$

5. When magnitude of load impedance is variable but not its phase angle. This case has practical importance because it is the kind of change in impedance that can be produced by an ordinary transformer. In such a situation, it can be proved that maximum power is transferred to the load when :

Magnitude of load impedance (Z_L) = Magnitude of source impedance (Z_S)

Hence if only the magnitude of load impedance can be changed but not its phase angle, maximum power is transferred to the load when the magnitude of load impedance is made equal to the magnitude of the source impedance.

However, maximum power transferred to the load in this case is not as great as when conjugate matching is used. Nevertheless, it is the best that can be done under the restriction that load phase angle is fixed.

Example 17.30. *For the circuit shown in Fig. 17.96, determine the load impedance for maximum power, the current through the load and the power at maximum power.*

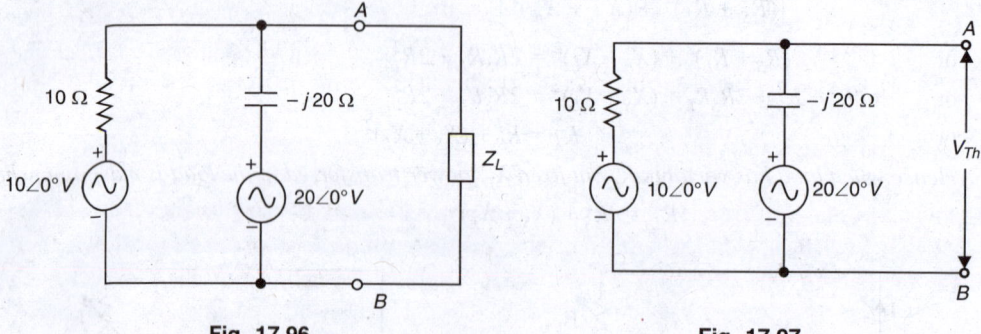

Fig. 17.96   **Fig. 17.97**

Solution. The first step is to find the Thevenin equivalent circuit to the left of terminals *AB*. For this purpose, we require V_{Th} and Z_{Th}. To find V_{Th}, remove the load Z_L and the circuit then becomes as shown in Fig. 17.97. Then open-circuit voltage at terminals *AB* is the V_{Th}. We shall use superposition theorem to find V_{Th}.

With the $20\angle0°$V source replaced by a short circuit in Fig. 17.96, the circuit current due to $10\angle0°$V source acting alone is $10/(10 - j20)$ and voltage across terminals *AB* is

$$V_1 = \frac{10}{10 - j20} \times (-j20) = \frac{-j20}{1 - j2}$$

With the $10\angle0°$V source replaced by a short circuit in Fig. 17.96, the circuit current due to $20\angle0°$V source acting alone is $20/(10 - j20)$ and voltage across terminals *AB* is

$$V_2 = \frac{20}{10 - j20} \times 10 = \frac{20}{1 - j2}$$

Therefore, open-circuit voltage due to the two sources acting simultaneously is

$$V_{Th} = V_1 + V_2 = \frac{-j20}{1 - j2} + \frac{20}{1 - j2} = \frac{20 - j20}{1 - j2}$$

$$= \frac{20 - j20}{1 - j2} \times \frac{1 + j2}{1 + j2} = \frac{60 + j20}{5} = (12 + j4)\text{V}$$

To find Z_{Th}, remove the load and replace each voltage source by a short circuit in the original circuit. The circuit then becomes as shown in Fig. 17.98. It is clear that :

$$\frac{1}{Z_{Th}} = \frac{1}{10} + \frac{1}{-j20} = \frac{1 - j2}{-j20}$$

$$\therefore \qquad Z_{Th} = \frac{-j20}{1 - j2} = \frac{-j20(1 + j2)}{(1 - j2)(1 + j2)} = (8 - j4)\ \Omega$$

For maximum power transfer, the load impedance should be the conjugate of source impedance (Z_{Th}).

$$\therefore \qquad Z_L = \mathbf{(8 + j4)\ \Omega}$$

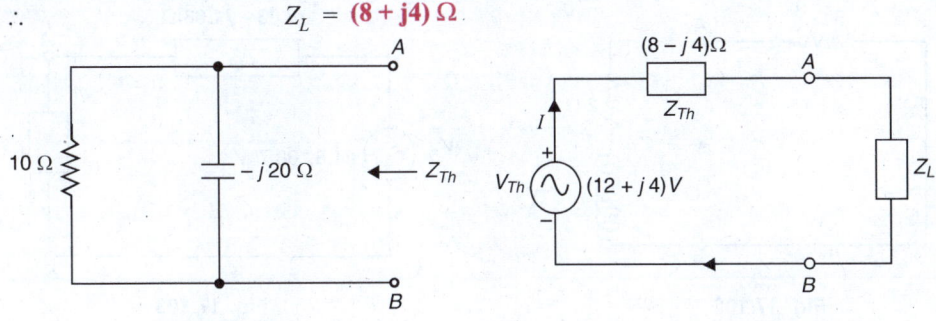

Fig. 17.98   **Fig. 17.99**

Fig.17.99 shows the Thevenin equivalent circuit at terminals *AB* along with the load Z_L. The current *I* through the load at maximum power is given by ;

$$I = \frac{V_{Th}}{Z_{Th} + Z_L} = \frac{12 + j4}{(8 - j4) + (8 + j4)} = \frac{12 + j4}{16}$$

$$= (0.75 + j0.25)\ \text{A} = \mathbf{0.79 \angle 18.43°\ A}$$

$$P_{max} = I^2 R_L = (0.79)^2 \times 8 = \mathbf{5W}$$

Example 17.31. *In the circuit shown in Fig. 17.100, find the value of load to be connected across terminals AB consisting of variable resistance R_L and variable capacitive reactance X_C which would result in maximum power transfer.*

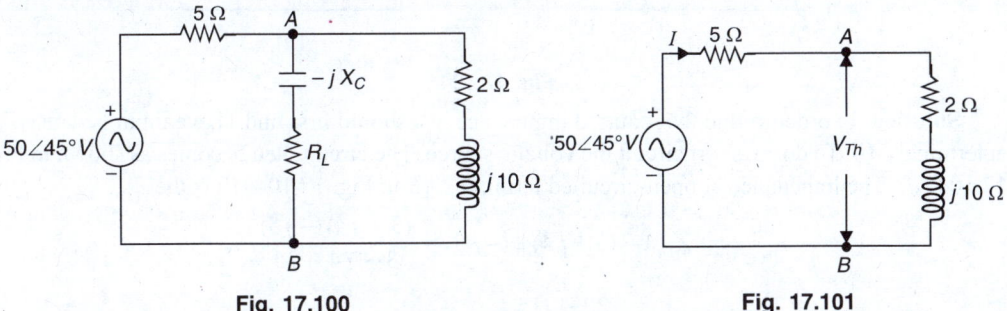

Fig. 17.100   **Fig. 17.101**

Solution. Let us find the Thevenin equivalent circuit at the terminals *AB*. For this purpose, we require V_{Th} and Z_{Th}. To find V_{Th}, remove the load. The circuit then becomes as shown in Fig. 17.101. The open-circuit voltage between terminals *A* and *B* is the V_{Th}.

$$\text{Circuit current, } I = \frac{50 \angle 45°}{5 + 2 + j10} = \frac{50 \angle 45°}{(7 + j10)}$$

$$\therefore \qquad V_{Th} = I(2 + j10) = \frac{50 \angle 45°}{(7 + j10)} \times (2 + j10)$$

$$= \frac{50 \angle 45°}{12.2 \angle 55°} \times 10.2 \angle 78.7° = 41.8 \angle 68.7°\ \text{V}$$

To find Z_{Th}, remove the load and replace the voltage source by a short circuit in the original circuit. The circuit then becomes as shown in Fig. 17.102. It is clear that :

$$Z_{Th} = 5 \| (2 + j\,10) = \frac{5(2 + j10)}{5 + 2 + j10} = \frac{10 + j50}{7 + j10}$$

$$= \frac{51 \angle 78.7°}{12.2 \angle 55°} = 4.18 \angle 23.7°\ \Omega = (3.83 + j1.68)\ \Omega$$

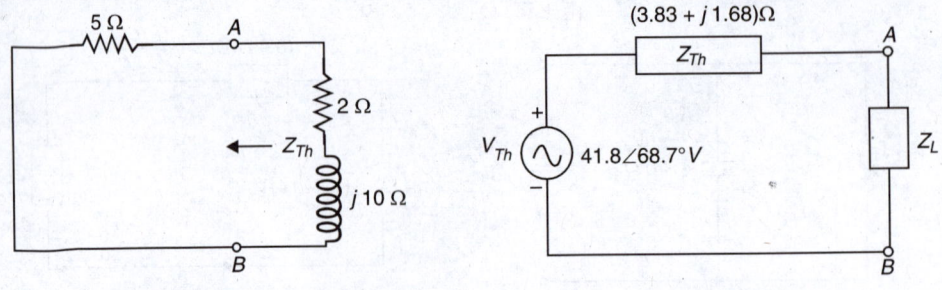

Fig. 17.102   **Fig. 17.103**

Fig. 17.103 shows the Thevenin equivalent circuit at terminals *AB* along with the load Z_L. For maximum power transfer, load impedance should be the conjugate of source impedance (Z_{Th}).

∴ $Z_L = (3.83 - j\,1.68)\ \Omega$

Example 17.32. *A loudspeaker is connected across terminals A and B of the network shown in Fig. 17.104 (i). What should its impedance be to obtain maximum power dissipation in it ?*

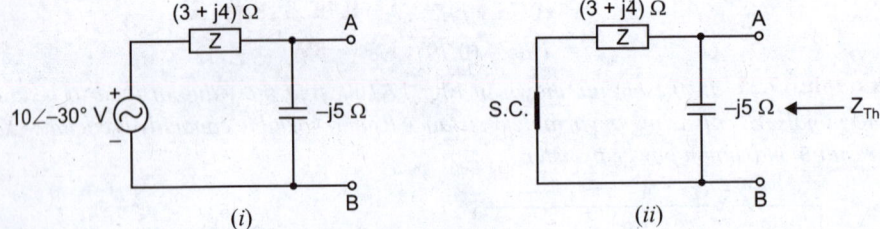

Fig. 17.104

Solution. In order to find the required impedance, we should first find Thevenin impedance Z_{Th} at terminals *AB*. To do so, short circuit the voltage source. The circuit then becomes as shown in Fig. 17.104 (*ii*). The impedance at open-circuited terminals *AB* in Fig. 17.104 (*ii*) is the Z_{Th}.

∴ $Z_{Th} = (3 + j\,4) \,\|\, (-j\,5) = \dfrac{(3 + j\,4)(-j\,5)}{3 + j\,4 - j\,5}$

$$= \frac{20 - j\,15}{3 - j\,1} = (7.5 - j\,2.5)\ \Omega$$

For maximum power dissipation in the loudspeaker, its impedance should be conjugate of $(7.5 - j\,2.5)\Omega$ i.e., impedance of loudspeaker $= (7.5 + j\,2.5)\ \Omega$.

Example 17.33. *For the circuit shown in Fig. 17.105, which load impedance of p.f. = 0.8 lag when connected across terminals A and B will draw the maximum power from the source. Also find the power developed in the load and the power loss in the source.*

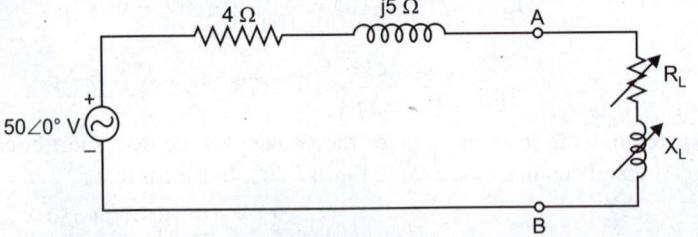

Fig. 17.105

Solution. Here the magnitude of load impedance can be varied but not its phase angle because load p.f. is fixed. In this case, the condition for maximum power transfer [See Art. 17.11] is

Magnitude of load impedance = Magnitude of source impedance

or $\qquad Z_L = \sqrt{4^2 + 5^2} = 6.40\ \Omega$

The load p.f. = 0.8 lag so that $\cos\phi = 0.8$ and $\sin\phi = 0.6$.

$\therefore \qquad\qquad R_L = Z_L \cos\phi = 6.40 \times 0.8 = 5.12\ \Omega$

$\qquad\qquad\qquad X_L = Z_L \sin\phi = 6.40 \times 0.6 = 3.84\ \Omega$

$\therefore$ Load impedance for maximum power transfer is

$$Z_L = \mathbf{(5.12 + j\ 3.84)}\ \Omega$$

Total circuit impedance, $Z_T = \sqrt{(R_S + R_L)^2 + (X_S + X_L)^2}$

$$= \sqrt{(4 + 5.12)^2 + (5 + 3.84)^2} = 12.70\ \Omega$$

$\therefore \qquad\qquad$ Circuit current, $I = \dfrac{V}{Z_T} = \dfrac{50}{12.70} = 3.94\ \text{A}$

Power developed in load, $P = I^2 R_L = (3.94)^2 \times 5.12 = \mathbf{79.48\ W}$

Power loss in the source $= (3.94)^2 \times 4 = \mathbf{62.09\ W}$

(Tutorial Problems)

1. A source represented by a voltage of $100\ \angle\ 0°$ V and internal impedance $(10 + j\ 10)\ \Omega$ is connected to a load by means of a circuit with an impedance of $(2 + j\ 0.5)\Omega$. What is the load impedance which would give maximum power transfer? $\qquad$ **[(12 – j 10.5)Ω]**

2. For the circuit shown in Fig. 17.106, determine the value of the load impedance for maximum power, the load current at maximum power and the value of maximum power.

$$\mathbf{[(10 - j5)\Omega\ ;\ 0.25\ \angle\ 0°\ A\ ;\ 0.63\ W]}$$

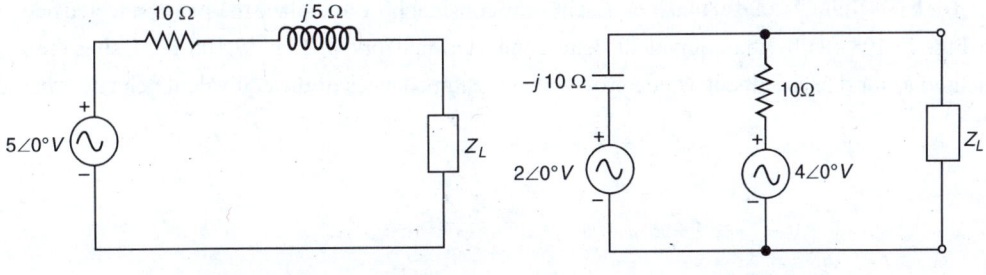

| Fig. 17.106 | Fig. 17.107 |

3. For the circuit shown in Fig. 17.107, determine the value of the load impedance for maximum power, the load current at maximum power and the value of maximum power.

$$\mathbf{[(5 + j\ 5)\Omega\ ;\ 0.32\ \angle - 18.4°A\ ;\ 0.51W]}$$

17.12. A.C. Network Transformations

There are some a.c. networks in which the impedances are neither in series nor in parallel. A familiar case is a three-terminal network such as a delta network or a star network. In such situations, it is not possible to simplify the network by series and parallel circuit rules. However, converting delta network into star and vice versa often simplifies the network and makes it possible to apply series-parallel circuit rules.

(*i*) **Delta / Star Transformation.** Consider three impedances Z_{AB}, Z_{BC} and Z_{CA} connected in delta to three terminals A, B and C as shown in Fig. 17.108 (*i*). Let the equivalent star-connected network has impedances Z_A, Z_B and Z_C as shown in Fig. 17.108. (*ii*). Using the same method as for d.c. circuits, it can be proved that the impedances of the equivalent star network are :

$$Z_A = \frac{Z_{AB}Z_{CA}}{Z_{AB}+Z_{BC}+Z_{CA}}$$

$$Z_B = \frac{Z_{BC}Z_{AB}}{Z_{AB}+Z_{BC}+Z_{CA}}$$

$$Z_C = \frac{Z_{CA}Z_{BC}}{Z_{AB}+Z_{BC}+Z_{CA}}$$

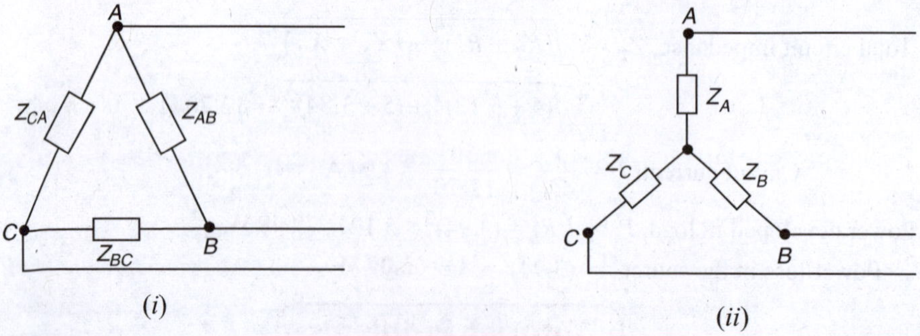

Fig. 17.108

Note that these equations are identical to their d.c. counterparts with impedances substituted for resistances. As for d.c. circuits, the equivalent star impedances are found as under :

Each star impedance is equal to the product of the two delta impedances connected to it, divided by the sum of the delta impedances.

(ii) Star/Delta Transformation. Let us now consider how to replace the star-connected network in Fig. 17.108 (*ii*) by the equivalent delta-connected network of Fig. 17.108 (*i*). Using the same method as for d.c. circuits, it can be proved that the impedances of the equivalent delta network are:

$$Z_{AB} = Z_A + Z_B + \frac{Z_A Z_B}{Z_C}$$

$$Z_{BC} = Z_B + Z_C + \frac{Z_B Z_C}{Z_A}$$

$$Z_{CA} = Z_C + Z_A + \frac{Z_C Z_A}{Z_B}$$

Note that these equations are identical to their d.c. counterparts with impedances substituted for resistances. As for d.c. circuits, the equivalent delta impedances are found as under :

Each delta impedance is equal to the sum of two star impedances connected to it plus the product of the same two impedances divided by the third star impedance.

Example 17.34. *Simplify the circuit shown in Fig. 17.109 by delta -star transformation.*

Solution. Recall that in order to convert a delta (Δ) network into equivalent star network (Y), the following rule is used :

Each Y impedance is the product of two delta impedances connected to it, divided by the sum of delta impedances. Therefore, the impedances of the equivalent star network (See Fig. 17.110) are :

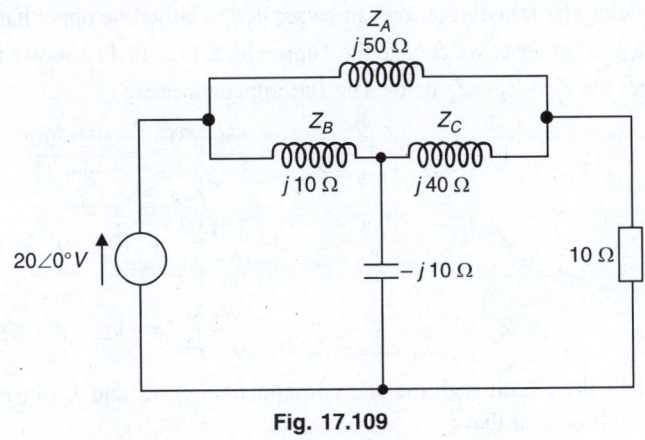

Fig. 17.109

$$Z_1 = \frac{Z_A Z_B}{Z_A + Z_B + Z_C} = \frac{(j50)(j10)}{j50 + j10 + j40}$$

$$= -\frac{5}{j} = j5\,\Omega$$

$$Z_2 = \frac{Z_B Z_C}{Z_A + Z_B + Z_C} = \frac{(j10)(j40)}{j50 + j10 + j40}$$

$$= -\frac{4}{j} = j4\,\Omega$$

$$Z_3 = \frac{Z_A Z_C}{Z_A + Z_B + Z_C} = \frac{(j50)(j40)}{j50 + j10 + j40}$$

$$= -\frac{20}{j} = j\,20\,\Omega$$

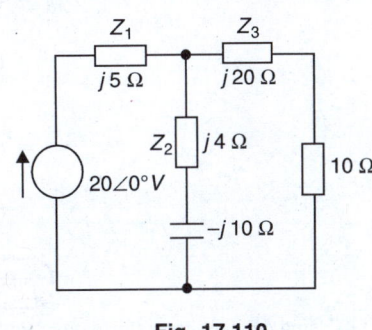

Fig. 17.110

Fig. 17.110 shows the circuit with delta network of the original circuit replaced by the equivalent star network. Note the utility of this transformation. The circuit is now simplified and we can use series -parallel circuit rules to find the solution.

Example 17.35. *Find the impedance of the bridge circuit shown in Fig. 17.111. Use delta /star transformation.*

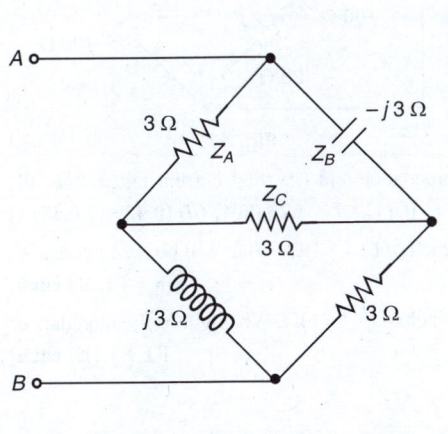

Fig. 17.111

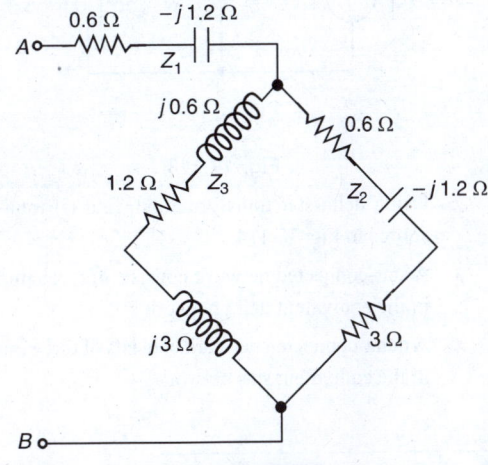

Fig. 17.112

Solution. The delta-star transformation can be applied to either the upper half or the lower half of the circuit. In the present case, we choose the *upper half. Fig. 17.112 shows the equivalent star network that replaces the $Z_A - Z_B - Z_C$ delta. The star impedances are :

$$Z_1 = \frac{Z_A Z_B}{Z_A + Z_B + Z_C} = \frac{3(-j3)}{3 - j3 + 3} = \frac{9\angle -90°}{6 - j3} = (0.6 - j1.2)\Omega$$

$$Z_2 = \frac{Z_B Z_C}{Z_A + Z_B + Z_C} = \frac{(-j3)3}{6 - j3} = (0.6 - j1.2)\Omega$$

$$Z_3 = \frac{Z_A Z_C}{Z_A + Z_B + Z_C} = \frac{3 \times 3}{6 - j3} = (1.2 + j0.6)\Omega$$

Fig. 17.112 shows the circuit with the star components (Z_1, Z_2 and Z_3) inserted and the delta components removed. It is clear that :

$$Z_{AB} = (0.6 - j1.2) + (0.6 - j1.2 + 3) \| (1.2 + j 0.6 + j 3)$$

$$= (0.6 - j1.2) + \frac{(3.6 - j1.2)(1.2 + j3.6)}{(3.6 - j1.2) + (1.2 + j3.6)}$$

$$= (0.6 - j1.2) + \frac{(3.6 - j1.2)(1.2 + j3.6)}{(4.8 + j2.4)}$$

$$= (0.6 - j1.2) + (2.4 + j1.2) = \mathbf{3 \angle 0° \, \Omega}$$

Tutorial Problems

1. Using delta-star transformation, find the current taken from the supply in Fig.17.113. **[5.65 $\angle -45°$ A]**

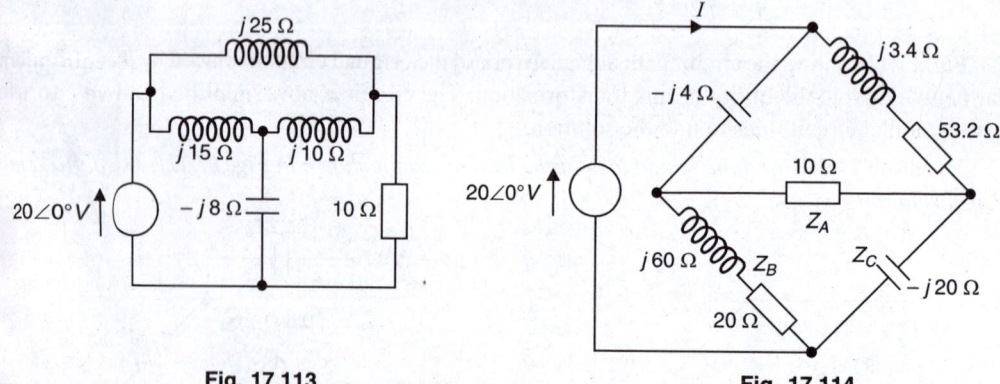

Fig. 17.113   **Fig. 17.114**

2. Using delta-star transformation, find (*i*) total circuit impedance and (*ii*) total current taken from the source in Fig. 17.114.   **[(*i*) (17.7 − j 23.06)Ω (*ii*) (0.42 + j 0.55)A]**

3. A star-connected network consists of three impedances each of $(2 + j 4)\Omega$. What will be the impedances in the equivalent delta network ?   **[(6 + j 12)Ω each]**

4. A delta-connected network consists of three impedances each of $(3 + j 6)\Omega$. What will be the impedances in the equivalent star network ?   **[(1 + j 2)Ω each]**

---

* Since we are not required to find the voltage or current in any particular component, it does not matter which of the two deltas is transformed.

Objective Questions

1. Kirchhoff's current law for a.c. circuits is based on the principle of conservation of
   - (i) charge
   - (ii) energy
   - (iii) momentum
   - (iv) none of above

2. Kirchhoff's voltage law for a.c.circuits is based on the principle of conservation of
   - (i) charge
   - (ii) energy
   - (iii) momentum
   - (iv) none of above

3. Alternating voltages and currents are
   - (i) vectors
   - (ii) scalars
   - (iii) phasors
   - (iv) none of above

4. In Fig. 17.115, there are
   - (i) two loops
   - (ii) three loops
   - (iii) four loops
   - (iv) none of above

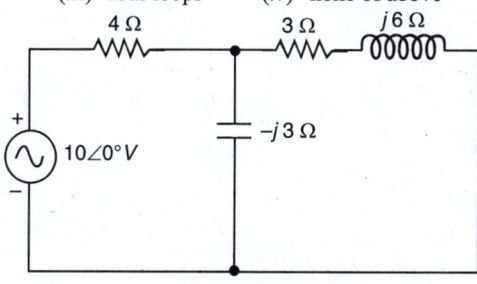

Fig. 17.115

5. In Fig. 17.115, there are
   - (i) two meshes
   - (ii) three meshes
   - (iii) four meshes
   - (iv) none of above

6. The number of principal nodes in Fig. 17.115 is
   - (i) 1
   - (ii) 3
   - (iii) 4
   - (iv) 2

7. In Fig. 17.115, the solution by nodal analysis will require
   - (i) one equation
   - (ii) two equations
   - (iii) three equations
   - (iv) none of above

8. There are 4 principal nodes in a network. For the solution of this network, the number of equations required is
   - (i) 2
   - (ii) 4
   - (iii) 3
   - (iv) 5

9. In Fig. 17.116, the Norton current I_N is
   - (i) $0.75 \angle 25.7°$ A
   - (ii) $0.89 \angle 63.4°$ A
   - (iii) $3.5 \angle 52.1°$ A
   - (iv) none of above

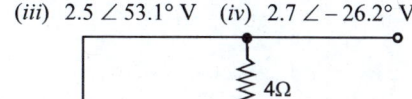

Fig. 17.116

10. In Fig. 17.116, the Norton impedance Z_N is
    - (i) $(4 + j\,2)\,\Omega$
    - (ii) $(1.4 + j\,0.35)\,\Omega$
    - (iii) $(1 - j\,4.2)\,\Omega$
    - (iv) $(10 - j\,20)\,\Omega$

11. In Fig. 17.117, the Thevenin voltage V_{Th} is
    - (i) $8.9 \angle 26.6°$V
    - (ii) $5.7 \angle 8.1°$ V
    - (iii) $2.5 \angle 53.1°$ V
    - (iv) $2.7 \angle -26.2°$ V

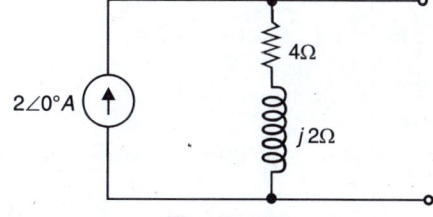

Fig. 17.117

12. In Fig. 17.117, the Thevenin impedance Z_{Th} is
    - (i) $(4 - j\,2)\,\Omega$
    - (ii) $(2 + j\,1)\,\Omega$
    - (iii) $(4 + j\,2)\Omega$
    - (iv) $(8 + j\,4)\,\Omega$

13. A star-connected network consists of three impedances each of $(1 + j\,2)\,\Omega$. The impedance of each equivalent delta network is
    - (i) $(1 + j\,2)\,\Omega$
    - (ii) $(3 + j\,6)\,\Omega$
    - (iii) $(0.5 + j\,1)\,\Omega$
    - (iv) $(6 + j\,12)\,\Omega$

14. The value of load impedance for maximum power transfer in Fig. 17.118 is
    - (i) $(10 - j\,5)\,\Omega$
    - (ii) $(20 + j\,10)\,\Omega$
    - (iii) $(2 - j\,2.5)\,\Omega$
    - (iv) none of above

Fig. 17.118

15. In Fig. 17.118, the load current at maximum power transfer is
    - (i) 0.25 A
    - (ii) 0.5A
    - (iii) 0.75 A
    - (iv) 1 A

Answers

1. (i)	2. (ii)	3. (iii)
4. (ii)	5. (i)	6. (iv)
7. (i)	8. (iii)	9. (ii)
10. (iv)	11. (i)	12. (iii)
13. (ii)	14. (i)	15. (i)

Index